OXFORD MEDICAL PUBLICATIONS

CUNNINGHAM'S
TEXTBOOK OF
ANATOMY

CUNNINGHAM'S TEXTBOOK OF ANATOMY

ELEVENTH EDITION

EDITED BY

G. J. ROMANES, C.B.E.

B.A., Ph.D., M.B., Ch.B., F.R.C.S.Ed., F.R.S.E.

Professor of Anatomy in the University of Edinburgh

LONDON
OXFORD UNIVERSITY PRESS
NEW YORK · TORONTO
1972

Oxford University Press, Ely House, London W. 1

GLASGOW NEW YORK TORONTO MELBOURNE WELLINGTON
CAPE TOWN IBADAN NAIROBI DAR ES SALAAM LUSAKA ADDIS ABABA
DELHI BOMBAY CALCUTTA MADRAS KARACHI LAHORE DACCA
KUALA LUMPUR SINGAPORE HONG KONG TOKYO

ISBN 0 19 263123 3
New material in this edition

© Oxford University Press 1964, 1972

First Edition 1902
Eleventh Edition 1972

*Printed in Great Britain
by R. & R. Clark, Ltd., Edinburgh*

CONTENTS

PLATES

ix

CONTRIBUTORS

R. G. HARRISON, M.A., D.M.
Derby Professor of Anatomy in the University of Liverpool.

R. J. HARRISON, M.A., M.D., D.Sc.
Professor of Anatomy in the University of Cambridge.

J. B. KING, M.D., F.R.C.P.ED., D.R., F.F.R.
Lecturer in Radiographic Anatomy in the University of Edinburgh.

G. J. ROMANES, *C.B.E.*, B.A., PH.D., M.B., CH.B., F.R.C.S.ED., F.R.S.E.
Professor of Anatomy in the University of Edinburgh.

ERIC SAMUEL, B.Sc., M.D., F.R.C.S.ENG., F.R.C.S.ED., F.R.C.P.ED., F.F.R., D.M.R.E.
Professor of Medical Radiology in the University of Edinburgh; Radiologist in charge,
X-ray Diagnostic Department, Royal Infirmary, Edinburgh.

D. C. SINCLAIR, M.A., M.D., D.Sc., F.R.C.S.Ed.
Regius Professor of Anatomy in the University of Aberdeen.

E. W. WALLS, M.D., B.Sc., F.R.S.E.
Courtauld Professor of Anatomy, University of London at the Middlesex Hospital Medical School.

ROBERT WALMSLEY, M.D., F.R.C.S.ED., F.R.S.E.
Bute Professor of Anatomy in the University of St. Andrews.

G. M. WYBURN, M.B., CH.B., D.Sc., F.R.C.S.(GLASG.), F.R.S.E.
Regius Professor of Anatomy in the University of Glasgow.

PREFACE TO THE ELEVENTH EDITION

THE 11th Edition of *Cunningham's Textbook of Anatomy* follows the same general plan as the 10th Edition, but some of the sections have been entirely rewritten and others extensively modified so as to decrease the length of purely descriptive passages while adding to the amount of functional information available. For example, the muscle section now gives concise accounts of the manner in which the various muscles co-operate in most everyday actions (together with tables showing the functions and nerve supplies of associated muscles) so adding to the more limited information usually given for each muscle and helping to clarify the effects of paralysis of individual muscles or groups of muscles even where such paralyses produce effects remote from the affected muscle. Taken together with the effects of injury to each of the peripheral nerves or their branches, which are given in the appropriate section, this should help to highlight the significance of the distribution of these nerves even for the junior undergraduate and so help to prepare him for accurate diagnosis and prognosis in the increasing number of cases of injury which he will subsequently meet, while forming a sound basis for his understanding of the appropriate rehabilitation procedures.

A considerable number of the older illustrations have been replaced with new drawings, and some entirely new illustrations have been added where it seemed likely that they would help to facilitate the student's understanding. The Nomina Anatomica (1968) has been used throughout, but as this nomenclature has been available with only minor modifications since 1955 it was not felt desirable to continue the practice of including the older terms in brackets in the text as this led to confusion for the student and was no longer necessary for his teachers. For the same reason and in order to avoid, so far as possible, unnecessarily increasing the cost of the *Textbook*, the glossary which was given in the 10th Edition to assist with the introduction of the new nomenclature has not been included.

The index has been entirely revised and every effort made to increase the usefulness of the book by double entries of many of the terms so that some of the advantages of an adjectival index can be obtained in the more conventional format now used.

It is a pleasure to record the debt which I owe to many colleagues for the valuable suggestions and corrections which they have made from time to time ; to the Contributors, new and old, for their patient helpfulness ; to the Staff of the Medical Department of the Oxford University Press and to Mrs. C. Clarke, artist, without whom much of this would not have been possible ; and to Dr. H. S. Barrett for reading and correcting my own manuscripts.

Edinburgh
December, 1971
G. J. ROMANES

EXTRACT FROM THE PREFACE TO THE TENTH EDITION

IN the period since the publication of the 9th Edition of *Cunningham's Textbook of Anatomy* the medical curriculum has been under special scrutiny in Great Britain as in other parts of the world, and as a result the time devoted to Anatomy has been restricted in many universities to make way for the growth of other subjects in an already overcrowded curriculum. This situation has led to the demand for shorter textbooks to fit the new courses; yet as the time available to the undergraduate student for study in departments of anatomy is curtailed, there is an increasing need for a book of reference which is sufficiently comprehensive to allow the student to expand his knowledge of the structural details of the human body and their functional implications, without the need for a special textbook on each subsidiary aspect of the subject. Such a textbook should be adequate not only for the undergraduate student but also for postgraduate studies, and should contain sufficient references to the literature to allow those interested in particular aspects of the subject a route of entry to the relevant original publications. With these points in view a considerable part of the *Textbook* has been entirely rewritten and the remainder considerably modified so as to produce some reduction in overall length without the sacrifice of any necessary material.

The reduction in length has been achieved partly by the removal of the section on 'Surface and Surgical Anatomy' and the incorporation of the relevant parts of surface anatomy in the appropriate sections. It is hoped that this will prove to be a more satisfactory arrangement and should not detract from the value of the *Textbook* for postgraduate students in surgery, since an adequate surgical anatomy section would require repetition of most of the material already presented in other sections of the book. There is also no separate chapter on Radiographic Anatomy but, where appropriate, each section has a short terminal note on the use of X-rays in demonstrating the parts of the system concerned, and there is a general account of the principles governing the application of radiology to the study of normal and abnormal anatomy at the end of the Introduction. The number of radiographs has been increased considerably, all are new and designed to form an atlas of normal radiographic anatomy to demonstrate those features which are best shown by this technique.

The illustrations have been extensively revised, new blocks having been prepared of all the old illustrations which are retained and 217 new half-tone and line blocks added. These include a number of histological preparations and electronmicrographs of those tissues in which such techniques help to clarify the relation between structure and function. The total number of figures appears to be less than in the 9th Edition, but the considerable increase is hidden partly because similar illustrations have been grouped together in one figure, and partly because there is less repetition of illustrations in different parts of the text, the extensive use of figure references rendering this repetition unnecessary. The colours used throughout the textbook have been rationalized as far as possible, so that, in addition to differentiating one tissue from another, each colour is used to indicate a limited variety of structures and in any one type of illustration may show either structures belonging to the same functional group, or those which have a common embryological origin. Care has been taken to choose colours which are as visible in artificial light as in daylight. The illustrations also include a brief atlas of the human brain stem to show the surface features and internal structure of this region, and a series of Plates showing horizontal and coronal sections of the brain placed in their correct position within that organ—a feature designed to give a clearer picture of the relation between the surface configuration and the deep structures of the brain, and to form an introduction to detailed stereotaxic atlases.

In order to avoid the tedious repetition of certain terms of position, alternatives have been used throughout the *Textbook*, thus 'above' has been used as a synonym for 'superior' or 'cephalic'; 'below' for 'inferior' or 'caudal'; 'in front' for 'anterior' or 'ventral', and 'behind' for 'posterior' or 'dorsal'. It should be noted that these terms have not been used in any other sense, and though they are not recognized as anatomical terms, they have the same precise meaning in the text.

Edinburgh
February, 1964

G. J. ROMANES

INTRODUCTION

ANATOMY deals with all those branches of knowledge which are concerned with the study of bodily structure. Originally it meant the cutting up of the body for the purpose determining the character and arrangement of its parts. It has come to mean knowledge of the form and relations of the parts of the body which may be obtained by this fundamental method of study and includes not only those aspects of information which may be obtained by dissection with scalpel and forceps, but also the more minute structural details which are uncovered with the aid of the hand lens and microscope.

It is natural that in the process of gross dissection the body is studied region by region, and when the knowledge so acquired is formulated in descriptions a sort of geography of the body is obtained; this is **regional** or **topographical anatomy**; but each region tends to consist of the same kinds of tissues (bones, muscles, nerves, etc.) arranged in different ways. Thus the whole body is seen to be composed of a limited number of different tissues, each of which comprises a system. The study of these groups of similar structures throughout the body constitutes **systematic anatomy**. Within each system uniformity of structure throughout its parts indicates similarity of function; whereas structural differences between the systems, visible even to the naked eye, indicate that each performs a different function. Much can be inferred about the functional significance of a system from a knowledge of its structural details and such inferences are susceptible to experimental investigation. This is **functional anatomy**, which deals not only with the structural basis of functions in the various systems but also with the interrelationships of the systems with each other, thus merging insensibly into Physiology, which is the study of the normal functions of the body.

The structural details elucidated with the naked eye, **gross** or **macroscopic anatomy**, can be extended with the hand lens, the microscope, and the electron microscope, **microscopic anatomy**. This differs from gross anatomy only in the size of stucture which may be studied and deals with the architecture of tissues, **histology**, and of the basic elements or cells of which they are constructed—**cytology**. The great increase in magnification which is obtained with the electron microscope has uncovered structural details within individual cells which are at once of great functional importance and structural beauty, and has made it possible even to visualize some of the larger molecules within cells, thus bringing anatomy within the realms of biochemistry and allowing of the elucidation of structural details invisible through the light microscope. It should be stressed that such magnifications enlarge not only the true structures but also the artefacts which arise from the processes necessary in the preparation of tissues, but it should also be clear that at this level, as in histology and in gross anatomy, no functional theory can be accepted until it has been shown to be compatible with the structure of the organ or system concerned.

No living body is a static structure, but is subject to change from the moment of conception till the time of death, and these changes, which are common knowledge, represent the alterations in structure which differentiate the fertilized egg from the embryo, the child, the adult, and the aged. These processes, which constitute the development and growth of the individual (**ontogeny**), proceed at very different rates and involve different tissues at the various stages. They can be divided into two major periods, prenatal or intra-uterine development or **embryology**, and the postnatal changes. Of the latter, some are immediate and allow the foetus to adapt itself to extra-uterine life, while others are continuing, such as the growth processes which involve every tissue for the purpose of increasing the size of the individual, for replacing those tissues which are subject to wear and tear, and for altering the strength or volume of tissues in relation to the demands placed upon them. The latter two continue throughout life, and are exemplified by the continuous replacement of the skin and the increased size of the muscles and strength of the bones in manual labourers.

In the growing individual the rate of growth is not uniform in all tissues at the same time; thus the postnatal growth of the limbs is proportionately much greater than that of the head, and the rate of growth is not equal in all parts of the limbs at the same time, either in the same or different individuals. Rates of growth are different in males and females, the latter reaching their final height and puberty earlier than the former, and showing periods of rapid growth at different ages from those in the male. Some structures such as the sex organs lie dormant only to make a rapid spurt of growth at puberty when the secondary sex characters appear, while the placenta, which is derived from the fertilized egg and is useful only for the period of intra-uterine life, shows many age changes by the end of that period (10 lunar months) which the tissues of the body, developed from the same egg, will show only scores of years later.

The structural changes which involve the individual (ontogeny) are superimposed on another developmental process which involves every member of the animal group collectively, and through which gradual changes in the structure of animals are believed to take place. This evolution constitutes the ancestral history or **phylogeny** of the individual, and just as Man is part of the Animal Kingdom, so Human Anatomy is part of a larger subject, the study of which is known as **comparative anatomy** and **comparative embryology**. These subjects have produced much of the evidence on which the theory of evolution is based and, drawing information from a wide spectrum of animals, have tended to formulate laws governing the relation of form and structure which, for the vertebrates, is known as Vertebrate Morphology.

The student of medicine is not in a position to become conversant with every aspect of anatomy, but he should understand that a knowledge of anatomy is essential to him in the practice of medicine, since he will be unable properly to examine a patient and recognize abnormality without a sound

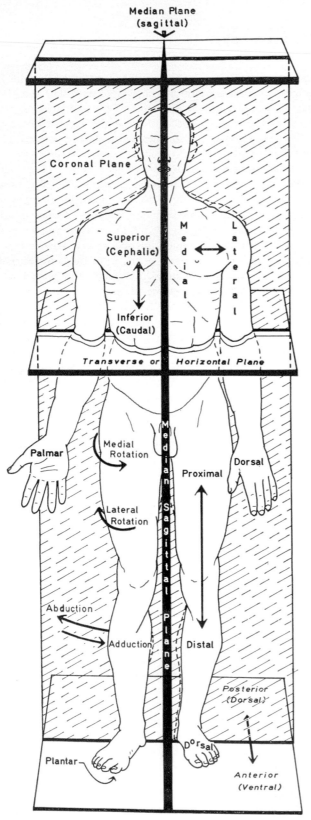

FIG. 1 A diagram illustrating the use of some anatomical terms referring to position and movement.

knowledge of the normal. He must be able to transfer information obtained on the cadaver to the living body so that he can visualize the arrangement of structures within the body from the surface—**surface anatomy**. In this he can derive considerable help from the use of X-rays and sections of the body, but he should lose no opportunity of confirming on himself and his friends the various structures which can be felt or seen and of trying to analyse the movements which the body can perform. The student will inevitably fail to gain from his studies any real interest and enlightenment if he allows the learning of detailed facts from a book to obscure the primary purpose of attempting to discover the function of the body by the examination of its structure.

TOPOGRAPHICAL ANATOMY

Anatomy is a descriptive science which uses a series of clearly defined and unambiguous terms to indicate the positions of structures relative to each other and to the body as a whole. In human anatomy the body is always described erect with the palms of the hands facing forwards, that is, in the standard or anatomical position, shown on the left of FIGURE 1. Three artificial sets of planes are described which are the co-ordinates forming the basis of any description; of these, the sagittal and coronal planes run parallel to the long axis of the body but at right angles to each other, while the transverse or horizontal plane cuts across the body at right angles to the other two.

1. **Sagittal** planes run from the front to the back of the body, and one of these divides the body into two apparently equal halves, the middle or **median plane** [FIG. 1]. Any structure which lies in a sagittal plane nearer to the median plane than another is said to be **medial** to it, while the other is said to be **lateral**; structures lying in the median plane are said to be **median**.

2. **Coronal** planes also run parallel to the long axis of the body and pass through it from side to side. Those nearer the front of the body are **anterior** or **ventral**, while those which lie nearer the back are **posterior** or **dorsal**. The terms anterior and posterior are used in human anatomy while dorsal and ventral are applicable also to the quadrupedal position and are used particularly in comparative anatomy. Anterior and posterior are used for the most part as relative terms but are also absolute when they refer to the surfaces of organs or of the body. The surfaces of the latter are cut by the median plane at the anterior and posterior median lines respectively.

3. **Transverse** or horizontal planes need no description. Of any two, that lying nearer the head is said to be **superior** or **cephalic**, while the other is **inferior** or **caudal**.

Thus the relative position of any structure in the body is described as medial, lateral, anterior, posterior, superior or inferior to another, or any combination may be used, e.g., superomedial, posterolateral, etc. It should be appreciated that in quadrupeds the terms anterior and posterior are synonymous with cephalic or headward and caudal or tailward. Certain advantages are therefore to be gained in using the latter terms together with dorsal and ventral since they are applicable to all circumstances.

Certain other terms are also used:

1. In any two adjacent structures, that nearer to the surface

of the body is **superficial** or external while the other is **deep** or internal, irrespective of the plane in which they may lie; thus the teeth are deep to the lips and the scalp is superficial to the skull.

2. The relation of a third structure lying between two others is described as intermediate (middle) in whatever plane they may be arranged; thus they may be medial, intermediate and lateral, anterior, intermediate and posterior, etc.

3. In the limbs certain additional terms are commonly used to overcome the difficulties of description arising out of the mobility of these structures, and though all the above terms are applicable in the anatomical position, they become difficult to apply when, for example, the arm is raised above the head.

i. The parts of the limbs further from the trunk are **distal** while those nearer are **proximal**. The hand is distal to the forearm and the leg is proximal to the foot.

ii. In the hand and foot the surfaces corresponding to the palm and sole are known as the **palmar** and **plantar** surfaces, while the opposite surfaces, the dorsum of the hand and the dorsum of the foot, are the **dorsal** surfaces, even though the dorsal surface of the foot is superior in the anatomical position; a situation which arises from the rotation of the lower limbs which brings the plantar surfaces into contact with the ground.

iii. The lateral or thumb side of the forearm and hand may be designated the radial side from the radius, a bone which occupies this position in the forearm, while the medial or little finger side similarly may be called the ulnar side.

Movements

The terms used to describe movements are also related to the above planes. Thus *movements of the trunk* or neck in the sagittal plane are known as **flexion**, forward bending, and **extension**, backward bending, and in a coronal plane as **lateral flexion**. Turning movements around the long axis are known as **rotation**.

The *movements of the limbs* are more complicated, but again those in the sagittal plane are known as flexion and extension, the former carrying the limbs anteriorly and folding them, the latter the reverse movement. On the other hand movement of the limbs in a coronal plane is almost limited to the proximal joint and carries the limb either further away from the median plane, **abduction**, or closer to it, **adduction**. Similar movements occur at the wrist when the hand is slewed on the forearm in the plane of the palm, and these are known as abduction or **radial deviation** and adduction or **ulnar deviation**. Abduction and adduction in the fingers and toes are not named with reference to the median plane of the body but to two other sagittal planes—the median plane of the hand, which passes through the middle finger; and the median plane of the foot, which passes through the second toe. Rotation in the limbs, as in the trunk, is represented by a movement of the whole limb or a part of it around its long axis, and when this rotation brings the lateral side of the limb anteriorly it is termed **medial rotation**; the opposite is **lateral rotation**.

A special example of rotation occurs in the thumb and to a lesser extent in the little finger. Medial rotation of the thumb brings its palmar surface into opposition with the palmar surfaces of the fingers. This movement of **opposition** is characteristic of the hand, is an essential part of grasping and holding, and the basis of many skilled movements.

A very special form of rotation is the movement of one forearm bone (the radius) around the other (the ulna). This is not simple rotation of the forearm on its long axis since the ulna remains stationary during the movement, but at the limits of this movement the palm of the hand, which is carried with the radius, faces either anteriorly, the **supine position**, or posteriorly, the **prone position**. Thus movement to the prone position is known as **pronation** while **supination** describes the reverse movement. At the ankle the main movements are in the sagittal plane and in either direction are known as flexion; **plantar flexion** if the movement is towards the sole, **dorsiflexion** if it is towards the dorsum. In the region of the ankle movements exist which allow the sole of the foot to be turned to face inwards and to a lesser degree outwards; these movements are named **inversion** and **eversion** respectively.

SYSTEMATIC ANATOMY

The description of the several systems of organs separately and in logical order, constitutes systematic anatomy, and is the plan upon which this treatise is based. The several parts of each system not only present a certain similarity of structure but are also associated in specialized functions. As already pointed out, functional anatomy merges insensibly into physiology. It begins with simple ideas, such as that the skeleton has the primary function of a supporting framework of the body, and the muscles have the primary function of moving the parts of the framework: it advances by deductions about the function of parts from their anatomical arrangement (such as Harvey's famous discovery of the circulation of the blood (1628), from observations and simple experiments on the valves of the veins and of the heart): it is also concerned with the wider field of the interrelations of parts belonging to different systems—for example, the anatomical localization in the nervous system of the origin of nerve fibres concerned with the regulation and control of the functions of different organs. Anatomy and physiology are indeed but two different aspects of one subject, separated as a matter of convenience for investigation and study. Structure and function are in reality indissolubly associated; and that is the basis of systematic anatomy. Thus there are:

1. THE LOCOMOTOR SYSTEM, which includes:

i. The bones and certain cartilaginous and membranous parts associated with them.

ii. The joints or articulations.

iii. The muscular system. With the muscles are usually included fasciae, synovial sheaths of tendons, and bursae.

2. THE DIGESTIVE SYSTEM, which consists of the alimentary canal and its associated glands from the mouth of the anus.

3. THE RESPIRATORY SYSTEM, the nasal passages, larynx, windpipe, and lungs.

4. THE UROGENITAL SYSTEM, composed of the urinary organs and the genital organs—the latter differing in the two sexes.

5. THE DUCTLESS GLANDS, which, though heterogeneous in their origin, structure, and particular functions, are conveniently grouped together as a 'system', as they share the common feature of releasing into the blood stream secretions which are distributed throughout the body and have a profound

influence on its functions. They include the thyroid and parathyroid glands, the thymus, the hypophysis cerebri and pineal body (these two are attached to the brain), and the suprarenal glands.

The term splanchnology denotes the knowledge of the organs included in the digestive, respiratory, and urogenital systems, and the ductless glands.

6. THE NERVOUS SYSTEM, which is divided into:
   i. The central nervous system—the brain and the spinal medulla (*cord*).
   ii. The peripheral nervous system—the nerves and their ganglia.
   iii. The autonomic nervous system, comprising the sympathetic and parasympathetic systems of nerves and ganglia.
   With the nervous system may be included:
   iv. The organs of the special senses (sight, hearing, smell, taste).
   v. The common integument (skin, nails, hair, etc.), which is also the important sense organ.

7. THE BLOOD VASCULAR SYSTEM, including the heart and blood vessels (arteries, veins, and capillaries).

8. THE LYMPHATIC SYSTEM of lymph vessels, lymph nodes, and spleen.

MICROSCOPIC ANATOMY (HISTOLOGY)

The organs which constitute the various systems of the body are constructed of tissues which can only be studied satisfactorily under the microscope. This reveals them to be formed of two distinct but closely interrelated elements: the cells, and the intercellular materials which are the products of the cells themselves. Each **cell** consists of a minute mass of colourless, watery, living substance or protoplasm whose external surface is a delicate, lipoid, cell-membrane about 1/1000,000 mm. thick which separates the cell from its neighbours and from the intercellular substances. The cell contains one (occasionally several) smaller, often centrally placed, body, the **nucleus**, which is surrounded by a membrane and embedded in the remainder of the protoplasm, cytoplasm. Cells vary in size from 8 to 200μm. in diameter (a μm. = a micron, which is 1/1,000 of a millimetre); the majority lie between 10 and 30μm. and are invisible to the naked eye. They also show many different shapes and some have processes which extend over considerable distances from the cell body, up to 1 m. in the case of some nerve cells, while striated muscle cells may be several centimetres long and have many nuclei. It is a mistake to assume that cells have a static shape just because they always appear similar after the processes of preparing them for microscopy have been undertaken; in fact, all cells undergo changes in shape and constitution and many of them are capable of movement. The nucleus differs from the cytoplasm in containing all the deoxyribonucleic acid of the cell, and has within it one or more small bodies (the nucleoli) which contain ribose nucleic acid. The nuclear membrane has numerous pores through which nuclear material is believed to enter the cytoplasm [FIG. 2], but these are not simple apertures for they do not allow free ionic exchange between nucleus and cytoplasm. The cytoplasm on the other hand

contains a number of structures within it which are similar in all cells but of different degrees of development in each.

1. **Mitochondria** [FIG. 2.9] are minute rod-like structures which appear at high magnification to consist of a sac of membranes the inner layer of which is folded into the interior (cristae) to a greater or lesser degree. These structures contain the energy-producing oxidative enzymes present in the cytoplasm.

2. **Golgi complex** [FIG. 2.2] has the appearance of a collection of flattened vesicles within the cell, usually aggregated near the nucleus. It is intimately associated with the formation of secretory granules in many cells though the exact part which it plays in cellular secretion is not at all clear.

3. **Endoplasmic reticulum** [FIG. 2.8] consists of membrane-like structures arranged in tubes or sheets carrying minute granules of ribonucleic acid (ribosomes). It seems to be concerned with protein synthesis in the cells and is present in large amounts in secretory cells such as those of the pancreas, liver, and nervous system. Such a high content of nucleic acid makes the cytoplasm take up basic dyes and thus appear blue or purple when stained with haematoxylin. Endoplasmic reticulum without ribosomes is present in many cells, but its function is unknown. Ribosomes are also found free in the cytoplasm, as are granules of glycogen.

4. **Fibrils** [FIG. 330] of many kinds are found in several different types of cells, but particularly in muscle cells where each myofibril is formed of a number of even finer filaments, myofilaments (in this case composed of actin and myosin). In nerve cells there are fine neurofibrillae and also neurotubules. Tubules are also found arranged in pairs in cilia (the motile processes of some cells) and in sperm tails, and in triplets in the basal bodies of these structures as in the centrioles from which they appear to be formed.

5. The **centrosome** is a minute structure which contains two small, dense, cylindrical bodies, the **centrioles** [FIG. 2.10], which lie at right angles to each other. It lies close to the nucleus and forms the achromatic spindle in cell division. It is absent from those cells which have become highly differentiated and have lost the power of division, e.g., most nerve cells.

6. **Vesicles** in scavenging or phagocytic cells may be ingestion vacuoles which are produced where a part of the cell membrane is engulfed by the surrounding cell surface and nipped off, thus containing material which lay outside the cell. This process is very active in some cells and is a method whereby particulate and other matter may enter the cell and even be transferred across it. In the latter case the vacuole is discharged at another surface of the cell; a process which is believed to occur in many situations where transport of large molecules or particles through a cellular membrane is involved, notably in the endothelium of capillaries and in the lining cells of the intestine. Vesicles may also contain substances produced by the cells, and their contents can likewise be discharged at the cell surface, the lining of the vesicle being incorporated in the cell membrane. One special type of membrane-enclosed vesicle, the lysosome, is very variable in appearance under the electron microscope. It contains numerous enzymes which destroy proteins (acid hydrolases) and are known to discharge their contents into phagocytic vesicles, thus digesting their contents. Such action is important in the

FIG. 2 Some of the surface and cytoplasmic modifications of cells. Not all the structures shown are present in every cell, and the diagrammatic cells shown are unlikely to be found adjacent to each other.

1. Cilia. These long, motile processes are found in many epithelia, (e.g. respiratory epithelium). Each contains nine pairs of tubules around its periphery (5 of the 9 are shown in the longitudinally sectioned cilium) and a central pair. The basal body from which each arises appears to be a centriole. Thus in ciliated epithelium the usual two centrioles which make up the centrosome (10) seem to have multiplied considerably.

2. The Golgi complex lies close to the nucleus, towards the free surface in secretory cells, but may surround the nucleus, e.g., in nerve cells. It consists of layers of flattened, membranous sacs, and gives rise to the secretion granules of the cell. It is believed that materials produced by the rough endoplasmic reticulum (8) are further elaborated in the Golgi complex to produce the secretory material.

3. Secretory granules are enclosed in a membrane derived from the Golgi complex.

4. Where absorption is taking place the surface of the cells frequently shows an array of fine, non-motile, finger-like processes, **microvilli**, which greatly increase the surface area of the cell. These are found in the epithelial cells of the intestine and the proximal convoluted tubules of the kidney.

5. Three types of adhesions between cells. a. A tight junction where the outer layer of the trilaminar cell membranes fuse. b. A zone of adherence forms a continuous strip where intestinal epithelial cells come together. The cell cytoplasm is granular close to the membrane, and fibrils in the cytoplasm pass into the granular region. c. A desmosome or point of adherence between cells. The inner layer of the cell membrane is thickened and fibrils in the cytoplasm pass into it. These are widely distributed in the tissues of the body and may appear as half desmosomes where an epithelial cell is attached to the underlying connective tissue, e.g., in skin.

6. Lipid droplets in the cytoplasm.

7. Part of the nuclear membrane to show the nuclear pores.

8. Rough endoplasmic reticulum. These membrane sacs are studded with ribosomes (particles of ribonucleoprotein). They are mainly responsible for the basophilia of the cytoplasm, e.g. in secretory and nerve cells, and are concerned with protein synthesis, including enzyme production. Scattered ribosomes are also found free in the cytoplasm, and there is endoplasmic reticulum which does not carry ribosomes; smooth endoplasmic reticulum.

9. Mitochondria. These structures, found in all cells, are principally concerned with oxidative mechanisms, but they may also play a part in protein synthesis.

10. Centrioles. All cells capable of division contain a pair of cylindrical centrioles which form the centrosome and usually lie close to the nucleus at right angles to each other. Early in mitosis they are duplicated, one pair passing to each end of the cell and forming the basis of the spindle. Each consists of nine triplets of tubules, and where it forms the basal body of a cilium, each triplet appears to be continuous with a twin tubule of the cilium. There is no central structure in the centriole corresponding to the central pair of tubules in the cilium.

The cytoplasm also contains fibrils, glycogen granules, lysosomes, etc. The latter contain proteolytic enzymes in a membrane, but may cause autolysis of the cell if they escape into the cytoplasm. They are probably concerned, amongst other activities, with the digestion of phagocytosed foreign material.

destruction of phagocytosed bacteria, etc. Normally the enzymes are isolated from the cell cytoplasm, but may escape when the cell is damaged, thus leading to its destruction.

7. The **cell-membrane** is a lipoid-containing structure which undergoes many alterations in shape and may be thrown into complicated folds or even into numerous thin finger-like processes or microvilli [FIG. 2.4] which markedly increase the surface area of the cell. These are therefore found in situations where cells are concerned with absorption, as in the gastro-intestinal tract and in parts of the urinary apparatus. Here they constitute what is known as a brush border because of the appearance of the microvilli under the light microscope. Other thinner, longer, and motile processes of the cell called cilia [FIG. 2.1] are present in many situations, including the surface cells of the respiratory tract and parts of the genital tract. In the former they are concerned with the movement of a surface film of moisture so that particles of dust which are breathed in may be carried away, and in the latter with the transport of ova along the oviduct. The cell membranes of adjacent cells may be separated by variable amounts of intercellular substance or they may be applied and even adherent to each other, e.g., in epithelia and endothelia [FIG. 2], and where the former are attached to connective tissue. Apart from well-defined structures such as the cilia, the cell membrane during life is in a state of constant movement.

Growth of the body as a whole is achieved by the division of cells and their consequent increase in number, and in some cases, e.g., muscle, by the growth of the individual cells, as well as by the increase in the amount of intercellular substances of all kinds. The normal division of cells is known as **mitosis** and is a complicated process whereby the material in a cell is equally shared between the two daughter cells. The stages of this are shown diagrammatically in FIGURE 3A. The first sign of division, the prophase, is seen in the duplication and movement of the centrosomes to opposite ends of the cell. Condensation of the nuclear material into a series of long, nodular, intertwined, double filaments occurs, and these pairs of identical filaments, the **chromosomes**, steadily shrink in length and increase in thickness and density. The nuclear membrane disappears, a series of fibrils radiates from each centrosome, and apparently passing through the condensed chromosomes, forms the **achromatic spindle** on the equator of which the chromosomes are arranged in a disc.

The two parts of each chromosome, formed by the longitudinal splitting of a single chromosome, now begin to separate, apparently drawn apart (the metaphase) by the fibrils of the achromatic spindle, and thus the single disc of chromosomes splits into two discs, one moving towards each centrosome; this is the anaphase. Towards the end of the anaphase the cell becomes constricted in the region of the original equatorial disc of chromosomes and this constriction gradually deepens until the cell finally divides into two daughter cells, each containing one disc of chromosomes. These gradually elongate to form a tangled skein of fine filaments as in the prophase, except that each filament is now single. The nuclear membrane re-forms and all evidence of the chromosomes as separate bodies disappears, leaving two daughter cells which are small replicas of the original cell. These grow to the parent size and synthesize the materials necessary for a further division, including duplication of the chromosomal threads.

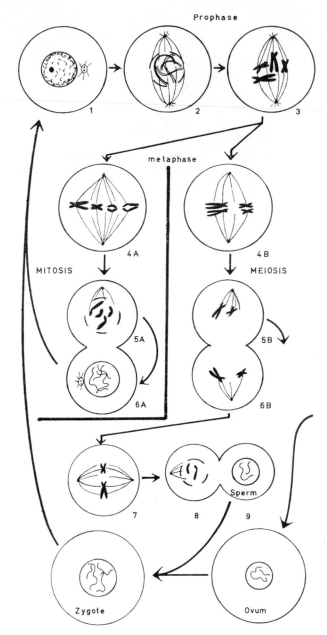

FIG. 3A A diagram to indicate the processes involved in the division of a cell. On the left, the common division or mitosis which gives rise to two daughter cells with the same number of chromosomes as the parent cell. On the right, reduction division or meiosis, the process used in the production of ova and spermatozoa whereby four daughter cells are formed, each with half the number of chromosomes present in the parent cell. The figure shows the formation of spermatozoa, four cells of equal size, the only difference in the production of ova being the formation of one large ovum and three small polar bodies. To avoid confusion, a hypothetical cell containing only four chromosomes is shown.

Mitotic division is found in all growing cells with one exception, the **germ cells** of the ovary and testis, which undergo a very similar process [FIG. 3A] except that in the prophase the double chromosome filaments do not arrange themselves in a single disc but in similar pairs, each pair representing four filaments of the mitotic prophase. In the subsequent separation of metaphase and anaphase these double pairs separate from each other; as a result half of the original chromosomes passes into each daughter cell. This process constitutes that sex. In mitotic division each of the forty-six chromosomes is split longitudinally into two equal parts so that a replica passes to each daughter cell; thus their chromosome content and therefore the material which they inherit from the parent cell are identical and consist of an equal share of the chromosome material derived originally from the ovum and sperm at fertilization. In meiosis, on the other hand, the chromosomes are not split but arrange themselves in pairs which are similar in appearance (with the exception of the X and Y chromosomes in the male cell) but different in origin, one of each pair having been derived from the ovum, the other from the sperm. When these pairs of chromosomes separate as the cell divides, each daughter cell receives half the number that the parent cell had, and since some of the chromosomes which each daughter cell receives were originally derived from the ovum and some from the sperm, the chromosomal content of the two cells is not the same and the inheritance which each carries is different.

In addition to this relatively simple process the chromosomes lying in pairs prior to the metaphase of meiosis may fuse at a point along their length and separate again in such a fashion that they exchange a corresponding part. Thus some of the chromosomes passing into each daughter cell may be a mixture of the chromosomes derived originally from the ovum and sperm and therefore different from both. The separation of homologous chromosomes in meiosis is brought out most clearly by the pairing of the X and Y chromosomes and their passage into opposite halves of the dividing testicular germ cells from which the spermatozoa are derived, so that half the sperms contain X and half Y chromosomes, and the chromosomal determination of sex depends on the type of spermatozoon, X- or Y-containing, which fertilizes the X containing ovum.

It will be appreciated that all cells derived from the fertilized ovum by mitotic division must of necessity have an equal complement of chromosomes and yet a wide variety of different tissues are formed from these apparently uniform cells. This is an interesting but unexplained phenomenon since all the daughter cells of the developing ovum are capable of forming any tissue up to a certain phase in development, but after this stage is reached they are capable of forming only the tissue which is appropriate to their position in the developing embryo. Thus separation of the first two daughter cells derived from the amphibian ovum leads to the formation of two complete animals. At a later stage, transplantation of cells, to a new situation, is followed by their development into a tissue appropriate to their new position, provided that the transplantation is made at a sufficiently early phase. Each cell is, however, a complex of structures, and the differences between cells represent a difference in degree of development of these different characters (e.g., mitochondria, endoplasmic reticulum, fibrils, etc.) rather than a basic difference in the constituents of the cells. Once a cell has developed the characteristics of a particular tissue, it and its daughter cells retain these unless some profound change occurs in its basic constitution, as can arise in cancer.

The apparent similarity of the chromosomal pattern in a single species no doubt determines the similarity of its members, but does not prevent the cells which make up any individual member from developing the marked variations which occur in different tissues. It should be appreciated, however, that just as individual members of a species are different, so all the

FIG. 3B Above. A photograph of human male chromosomes arrested in mitosis. Below. The same chromosomes arranged in corresponding pairs. By courtesy of Dr. W. M. Court Brown.

1   2   3   4 & 5

6 – 12 & X

13 – 15   16   17 & 18

19 & 20   21 & 22   Y

meiosis or reduction division and is a preparation of germ cells with half the normal complement of chromosomes ready to have this number made up by a similar reduced complement from the germ cell of the opposite sex at fertilization.

In each species of animal the number, shape, and arrangement of the chromosomes is constant and in the normal cell is always an even number. In human cells there are forty-six chromosomes, composed in the male of twenty-two similar pairs and one dissimilar pair known as the X (large) and Y (small) chromosomes [FIG. 3B], and in the female of twenty-three similar pairs, there being a pair of X chromosomes in cells of each individual have certain common chemical characteristics which are unique to that individual and which represent, it is believed, minor differences in the molecular arrangements of the complex molecules which make up the chromosomes and which are faithfully reproduced from cell to cell. Only two individuals derived from a single fertilized ovum, identical twins, have exactly the same chromosomal pattern and it is only in these cases that tissues can be transferred from one to another and survive. In all other cases, including non-identical twins, brothers and sisters, the chemical specificity of the individual tissue is recognized by the host to which it is transplanted and it is unable to survive because of the host reaction to this foreign, though similar, tissue.

TISSUES OF THE BODY

The animal body consists essentially of a thick-walled cylinder lined internally and covered externally with a continuous layer of cells, or epithelium, called entoderm and ectoderm respectively. The tissue which they enclose is the mesoderm and the two epithelia are continuous at the mouth and anus, the entoderm forming the lining of the alimentary and respiratory passages and the ectoderm the skin. Both epithelia send prolongations into the mesoderm which are thereby divorced from the protective or absorptive functions of the surface layer and undertake different activities, including the formation of glands which discharge their secretions to the surface of the epithelium from which they are derived along the epithelial stalk, or duct, which connects them. Examples of these are the liver, pancreas, sweat and mammary glands, and hair follicles. Some downgrowths, however, become separated from the surface and discharge their secretions into the blood stream, endocrine glands, or take on quite other functions, e.g., the nervous system, which is derived in this fashion from the ectoderm.

The mesoderm forms all the cells which originate neither in ectoderm nor in entoderm, thus including all the epithelia lining the spaces which develop in mesoderm, the blood and lymphatic vessels, the serous cavities which surround the heart, lungs, alimentary canal and central nervous system, bursae, tendon sheaths and joint cavities. It also forms muscle, blood, and connective tissue cells, the gonads, and the lining epithelia of the urogenital tract.

Connective tissue cells are the source of the intercellular substances of the body including connective tissue fibres, the numerous inelastic, white, collagen fibres and their slender counterparts, the reticular fibres, as well as the yellow elastic fibres.

It should be appreciated that the intercellular substance fills the entire extracellular space, permeates every organ and is continuous with the surfaces of all the cells which are exposed to it. It is responsible for holding the various organs in place and yet for allowing the requisite amount of movement between them. Thus in some situations where free movement is essential the extracellular material is in the form of a fluid, as in the blood and lymph vessels, in the cavities of joints, tendon sheaths, and bursae and in the spaces surrounding the heart, lungs, abdominal viscera, and central nervous system. Even these fluids are variable in consistence, ranging from the thin watery lymph to the thick glutinous material in some joints. Elsewhere the fluid contains a fine meshwork of connective tissue fibres (loose **areolar tissue**) which is sufficiently delicate to allow a considerable amount of movement between adjacent tissues. This is found between the bundles of fibres in a muscle and around the muscle and its tendon. It also surrounds many other tissues such as nerves and ligaments which have to slide on adjacent structures during movement, and organs which have to be capable of considerable distension such as the gullet and the urinary bladder. This loose areolar tissue forms most of the planes of cleavage in the body and allows the rapid spread of infection through it.

In other situations the extracellular material is stronger. This can be either by an increase in the number of fibres which it contains, or by the deposition of different substances produced by the connective tissue cells in the interstices of the meshwork it forms. Thus in many situations the connective tissue cells become loaded with fat or the fibres are surrounded with mucopolysaccharides to form the gelatinous basement membranes which surround some cells, or with the more solid sulphated mucopolysaccharides to form cartilage, or with calcium salts to form bone.

Where both strength and flexibility are required, only the fibrous elements are increased to form :

1. Compact layers of interlacing collagen fibres. This is the deep **fascia** which invests the entire body and sends sheets inwards between the various organs to surround them with a sheath of greater or lesser density which, because it is felted, is equally strong in all directions. It therefore produces structures such as the epimysium and peritendineum (within which each muscle and tendon respectively slides) the epineurium which confers on peripheral nerves their great strength, and the tough fibrous capsules of such organs as the kidneys and testes.

2. Muscles may be attached to some of the sheets described above, and where they are thus subjected to a pull in one direction, large numbers of collagenous fibres are laid down in this direction, conferring on the sheet a glistening appearance and forming a flat tendon or **aponeurosis**. In an exactly similar fashion the fibrous capsule which surrounds each type of movable joint is subjected to forces unequally in its several parts ; where these are greatest there is a similar addition of fibres to form the **ligaments** which give stability to the joints and which are usually therefore a part of the capsule. Most of the ligaments so formed are inelastic or collagenous but a few contain a preponderance of elastic fibres and are yellowish in colour ; to this group belong some of the ligaments of the vertebral column. The thickness of such ligaments and aponeuroses is directly related to the forces which they are called upon to resist. The thick rounded **tendons** which arise from muscles of fusiform shape are formed in the same way and have the same structure as ligaments and aponeuroses.

Since the ligaments, aponeuroses, and tendons are but part of the general extracellular material, they are continuous with the same tissue which forms the fibrous basis of cartilage and bone and are thereby directly attached to these tissues by fibres which penetrate them as part of their structure. In a similar fashion, extracellular materials are continuous with the surfaces of the cells which they surround but do not penetrate. Thus the collagen fibres of a tendon are attached to the outer membrane of muscle cells (sarcolemma), and the collagen fibres of the dermis are attached to the deepest cells of the epithelium (epidermis) which covers them, binding the outer protective tissue to the underlying structures.

Reticular tissue is a special kind of areolar tissue formed of very fine collagen fibres in a loose network supporting the proper tissue of various organs. It is especially seen in lymphoid tissue, lymph nodes, the tonsils, the spleen and in the liver and bone marrow. The connective tissue cells associated with reticular fibres are frequently phagocytic (macrophages), that is, they are capable of ingesting particles and taking up dyes injected into the living animal. These cells and others of similar properties in the blood and elsewhere are collectively known as the reticulo-endothelial system.

The cells of the connective tissue in some situations may take up and store large quantities of fat, thus forming adipose tissue. The individual cells become swollen with the accumulation of fat which forms a single globule within the cytoplasm. Groups of them form small nests enclosed in the collagen network which may be either very delicate, as in perirenal fat, or dense as in the palm of the hand and the back of the neck and scalp. Another type of fat storage, brown fat, is found especially in hibernating animals in which the fat forms a large number of fine globules in the cytoplasm giving the cells a foamy appearance.

In the embryo the connective tissue forms a loose cellular network or mesenchyme and from these cells are developed, in the adult, the connective tissues, cartilage, bone, fat, smooth muscle, lymph nodes, endothelium, blood cells, and macrophages. All these cells appear very different from each other under the microscope though the fibrous tissue cells differ least from the mesenchyme cells of the embryo. As a general rule the greater the degree of differentiation in the development of a cell, the more it loses its primitive potentiality to form cells of various kinds and in a few cases it may even lose its ability to undergo mitosis (e.g., nerve cells). The similarity of fibrous tissue cells in the adult to the mesenchyme cells of the embryo, has led to the assumption that some of them at least retain the ability to form other kinds of cells, but it is doubtful if in fact they have more than a very limited ability to do so. Certainly if the blood-forming or lymphoid cells are destroyed they are not replaced, and injuries involving smooth muscle and other specialized tissues are repaired with fibrous tissue. However, it does seem that the connective tissue cells which are most closely allied to each other can, in some circumstances, fulfil different roles. Thus fibrous tissue cells, cartilage cells, and bone cells, which share the ability to form collagen and probably also elastic fibres, seem to be partly interchangeable, since cartilage masses can be formed in a broken bone, even when this is developed originally in membrane and no cartilage cells would normally be present (Girgis and Pritchard, 1955), and ossification can occur in connective tissue in abnormal situations. It should be appreciated that ossification is difficult to produce experimentally in abnormal situations, and cannot be produced in many sites where there is an adequate supply of connective tissue cells no different in appearance from those in areas, such as the kidney, where abnormal ossification is easily induced.

Whether they are interchangeable or not, the cells derived from mesenchyme retain to a very full measure the ability to react to changing circumstances. Thus, in the rabbit's ear,

Clark and Clark (1934, 1940) have shown the remarkable changes which can occur in the vascular pattern; growth of new capillaries, the development in their wall of muscle fibres, and a complete change in the pattern of blood vessels, even to the temporary formation of arteriovenous anastomoses, as a result of changes in the flow of blood through the ear. Again the thickness of tendons and bones can be modified in relation to the stresses placed upon them, the blood-forming bone marrow can react to blood loss by the rapid production of new cells, and under an adequate stimulus the fixed tissue macrophages appear in large numbers, and becoming motile, move towards and ingest foreign material.

These changes and many others are going on constantly in the body, and the study of the cadaver and of histological preparations alone gives very little idea of the vital processes which no student of anatomy should ignore if he is to understand the significance of the structures he studies.

One of the most formidable tasks which face the medical student at the beginning of his studies is the need to acquire the vocabulary of medicine and to use it with precision. For this purpose it is essential to obtain a clear understanding of anatomical terms and this the student can obtain only by repeated reference to a good medical dictionary. The following short glossary defines some of the terms in general use but cannot replace the need for a medical dictionary.

A SHORT GLOSSARY OF SOME GENERAL ANATOMICAL TERMS

In some cases the translation of the term from Latin or Greek only is given, in others this is followed by its modern meaning where this is different, and in a third group where there is no clear derivation or the original meaning has been distorted, only the modern usage is given.

ACINUS. A berry growing in clusters. Hence the smallest unit of a compound gland; more commonly called an alveolus.

ADITUS. The entrance into a cavity.

AFFERENT. Used to indicate that a structure leads towards the organ it supplies. Of nerves, corresponds to sensory nerves, cf. efferent.

ALA. A wing-like projection.

ALVEOLUS. A small pail. Hence applied to a tooth socket, the smallest air spaces of the lungs; synonymous with acinus.

AMPULLA. A flask. Hence the dilated portion of a tube.

ANSA. A loop.

ANTRUM. A cave. Hence a cavity or hollow filled with air and lined with mucous membrane in the interior of a bone.

ANULUS. A ring.

APONEUROSIS. A tendinous sheet covering a muscle or extending from it to the attachment of the muscle.

—BLAST. A builder. Hence an ending indicating an immature or stem cell.

BRACHIUM. An arm.

BURSA. A collapsed sac of fluid found especially where a tendon or skin slides over bone. Such structures enclose the tendons where these lie in osteofascial tunnels and are then known as synovial sheaths. In general arrangement they are similar to the cavities of synovial joints.

CANALICULUS. Diminutive of canalis, a canal. Also used to indicate a tunnel.

COMMISSURE. A joining together. Used in the nervous system to indicate bundles of nerve fibres crossing the midline from side to side, but also for a ridge, etc., joining the two halves of an organ across the median plane.

CONDYLE. A knuckle. A smooth rounded eminence covered with articular cartilage.

CORONA. A crown.

CORTEX. A rind or outer covering, cf. medulla.

CRISTA. A sharp upstanding ridge.

CRUS. A leg. Used (as is 'brachium') to indicate any projecting process.

CYSTIC. Appertaining to a bladder (cyst).

DENS. A tooth. Odontoid—tooth-like.

EFFERENT. The reverse of afferent. Of nerves, corresponds to 'motor' nerves.

EPICONDYLE. The prominence or projection situated above a smooth articular eminence, though that eminence may not be called a condyle.

FACIES. A surface.

FALCIFORM. Sickle-shaped. FALX, a sickle.

FASCIA. A bandage or swaddling-cloth. Hence the membrane of fibrous tissue which sheaths the tissues.

FILUM. A thread.

FOLLICULUS. A small bag.

FORAMEN. A hole.

FORNIX. An arch.

FOSSA. A shallow depression.

FOVEA. A pit. Also used in place of facet.

FRENULUM. A small bridle or ligament. Usually applied to a fold of skin limiting the separation of the structures to which it is attached.

FUNDUS. The base of a hollow organ, usually opposite to its outlet.

GANGLION. A swelling, usually small and round. In the nervous system any peripheral collection of nerve cells.

GENU. A knee. Hence any bend on a structure.

GLOMUS. A ball or tight meshwork of vessels. Diminutive, glomerulus.

HAMULUS. A hook.

HIATUS. A slit or gap.

HILUS. A depression where blood vessels and nerves enter an organ.

INCISURA. A notch.

INFUNDIBULUM. A funnel.

ISTHMUS. A narrow strip of tissue joining two larger pieces. Hence a narrowing of a canal.

LABIUM OR LIP. The raised margin of an orifice.

LABRUM. A brim.

LABYRINTH. A maze of communicating spaces or canals.

LACUNA. A pit or hole.

LAMINA OR LAMELLA. A thin plate or sheet.

LIGAMENT. A band or tie joining two structures. Most commonly a fibrous band but may be composed of any tissue.

LIMBUS. A border or margin.

LINGULA. A tongue-like projection.

MACULA. A spot or stain.

MEATUS. A passage. A short canal.

MEDULLA. Bone marrow or the pith of plants. Hence the central portion of any organ where its structure is different from the outer layer or cortex.

MESENTERY. The fold of tissue which supports the bowel (enteron) in the belly. Has been expanded to include any such supporting structure, e.g., the mesentery of the ovary, more correctly called the mesovarium.

NODUS OR NODE. Diminutive, nodulus or nodule. A knot. A swelling or protuberance, a spherical aggregation of cells. A discontinuity in the myelin sheath of a nerve fibre.

NUCLEUS. The kernel of a nut. In the central nervous system refers to a collection of nerve cells.

OSTIUM. An entrance. Hence the opening into a tube or space.

PAPILLA. The nipple. Any nipple-shaped elevation.

PARENCHYMA. The proper tissue of an organ as distinct from accessory structures such as its fibrous capsule.

PARIES. A wall. The walls of the abdomen. The parietal bone.

PEDICLE. Diminutive of pes, a foot. PEDICULUS. The stalk of a fruit. A narrow rod or tube joining two structures.

PELVIS. A basin.

PLEXUS. A plaited or braided structure of vessels or nerves.

PLICA. A fold.

PORUS. A pore or opening, e.g., of a meatus.

RADIX. A root or origin.

RAMUS. A branch.

RAPHE. A seam. The line of union of two soft tissues.

RETE. A net. Often a labyrinth of communicating channels.

RETINACULUM. A stay or tie. Usually a band of connective tissue (ligament) which holds a tendon in place.

RIMA. A chink or cleft, e.g., the slit between the eyelids.

SCAPHOID. Scapha, a boat. Any hollowed-out structure.

SEPTUM. A hedge, fence or dividing wall. A partition, usually thin.

SINUS. A hollow or creek are two of its many meanings. In anatomy it is applied to the air-filled cavities of the cranial bones, to large venous spaces within the skull and elsewhere, and to dilatations of blood vessels.

SPINA OR SPINE. A sharp-pointed projection.

SOMATIC. Belonging to the body wall.

SPLANCHNIC. Belonging to the gut tube.

SQUAMA. A fish scale or armour plate. SQUAMOUS, scaly.

STRIA. A stripe or line.

STYLOID. Any structure resembling a stylus or stake.

SULCUS. A rut or furrow.

SYNOVIA. The fluid in joint cavities, bursae, and tendon sheaths.

TAENIA. A band or ribbon. Hence any ribbon-like structure.

TEGMEN. A cover.

TELA. A web. Hence any thin mesh, usually of fibrous tissue.

TORUS. A heap or protuberance.

TRABECULA. A small beam. Used especially for the pieces which make up the lattice of cancellous bone.

TROCHLEA. A pulley or pulley-shaped surface.

TUBER. A bump or swelling. Also tubercle and tuberosity. Three terms used, without much distinction, for any kind of rounded swelling or eminence.

VAGINA. A sheath.

VELUM. A covering or curtain.

VESICA. A bladder, especially the urinary bladder, cf. cystic.

VILLUS. Shaggy hair. Hence applicable to fine processes visible to the naked eye, projecting from a surface.

VISCERAL. Belonging to the gut tube.

ZONA. A girdle or belt. A circular or ring-like structure.

RADIOGRAPHIC ANATOMY

J. B. KING

GENERAL

In the radiographic section on each system only those aspects are described which are better demonstrated by radiography than by classical anatomical methods. For example, the dates of appearance of ossification centres and epiphysial union were accurately known before the use of radiography and are therefore left to the general text of the section on osteology, though radiographs are used to illustrate such features.

Radiography is the process of making a photographic image by the use of X-rays. These are of the same nature as visible light and have the same effect on photographic emulsion, but by their short wavelength they are able to penetrate substances and yet cast shadows representing the varying densities of the materials through which they have passed. Certain examinations can be made by direct observation of the shadow cast by X-rays on a fluorescent screen, the process being known as fluoroscopy, but it is more usual to record the image on film even though this precludes the visualization of movement unless cineradiography is carried out.

There is therefore an apparent simplicity about the formation of the radiographic image which is a complex process. The rays from an X-ray tube, no matter how produced, are not all of the same wavelength, those which are longer being by convention called soft and the shorter hard. In traversing an object the soft rays are progressively absorbed more rapidly than the hard, and the beam therefore is quickly stripped of much of its intensity, but what remains is more penetrating as it reaches the deeper layers. It is for this reason that a radiograph of, say, half the thickness of a bone may actually have a density little different from that of the whole bone even with a standard exposure.

Fundamentally, and without the use of special substances, there are only three significant densities involved in radiography of the living body. Bone has an intrinsic density due to the high atomic weight of its calcium content, and air has no significant absorption. Otherwise the tissues of the body, including the blood, lymph, and cerebrospinal fluid, have virtually the specific gravity of, and therefore the same radiographic absorption as, water, with the notable exception of fat which has a specific gravity of 0·9.

Radiographic demonstration without the use of contrast material is possible only when two adjacent tissues absorb X-rays to a different degree. Thus, the heart is readily and accurately demonstrable by virtue of the surrounding air in the lungs [PLATES 34-36], but the liver, spleen, and pancreas, although of the same relative density as the heart, cannot be outlined in simple ('straight') radiographs [PLATE 5C], whereas the kidney can be visualized in proportion to the amount of surrounding fat. An understanding of this is necessary to appreciate why even very fine blood vessels surrounded by air in the lungs are visible in straight radiographs of the chest, but even the largest of vessels in the abdomen, skull or limbs cannot be seen. It must be emphasized that in the taking of any radiograph the technique requires to be related with precision to the anatomical structure to be examined. The simplest example is an anteroposterior view of the chest in which, because of their different densities, either the heart or the thoracic spine may be shown clearly, but not both on the same film, and in most regions the need for accuracy of technique is equally great.

A radiograph is always a representation in one plane of a three-dimensional object and the consequent distortion may be sufficient to make analysis difficult, this being particularly true at the roots of the lungs where the overlapping vascular structures form a complex pattern when reduced to one plane [PLATES 34-36]. In theory two radiographs at right angles to one another, the traditional anteroposterior and lateral views, should overcome the difficulty of three-dimensional study, but in actual practice these are frequently either not feasible or inadequate. In radiography, as in photography, the detail that can be shown and the technique to be used must be related to the presence or absence of movement. Where complete immobility is obtainable as with the skeleton, radiography is the method of choice for accurate examination. Where movement is beyond control, as in the heart, or gastrointestinal tract, fluoroscopy plays an important part. Many organs have an intermediate position in regard to the possibility of immobilization, as with the lungs where respiration may be suspended, and in the examination of organs such as the kidney and gall-bladder where transmitted respiratory movement may amount to an inch or more. Because of the particle size in fluorescent screens the anatomical detail obtainable on screening can never equal that attainable on X-ray film. There is, therefore, no place for the X-ray examination of bones or joints by fluoroscopy.

Since adjacent 'soft tissues' cannot be distinguished in straight radiographs, certain procedures have been developed to permit of such differentiation.

1. Hollow organs may be filled with substances having a greater or lesser density, the most familiar example being the use of a meal of barium sulphate for the examination of the oesophagus, stomach, and small intestine [PLATES 28-30]. A barium enema is used for the large intestine [PLATES 31 and 32].

2. A hollow organ to which there is no ready access may be visualized, as can the gall-bladder or kidney, by using their functional activity to secrete and concentrate an iodine-containing contrast medium which may be given by mouth or intravenous injection as in cholecystography and intravenous pyelography respectively [PLATES 33, 37 and 38].

3. Solid organs such as the liver, spleen, and pancreas, for which no functional contrast media have been found, can be studied, at least in their gross external anatomy, by the radiographic demonstration of structures which abut on them; an instance being the examination of the head of the pancreas by outlining with an opaque meal the loop of the duodenum which surrounds it. Air also may be used to fill potential spaces around organs or to replace fluid in certain body cavities. Such a method is used to outline the ventricles of the brain or to make the surface of the brain visible by partially

filling the subarachnoid space which surrounds it [PLATE 45]. It may also be used to make the kidney more readily visible by allowing air to infiltrate into the surrounding retroperitoneal tissues [PLATE 37A].

4. The structure of certain organs, both hollow and solid, may be examined indirectly by arteriography. By this is meant the injection of an X-ray opaque medium into the arterial system of the organ, immediately following which a series of radiographs is taken which can cover the whole period in which the material is passing through arteries, capillaries, and veins [PLATE 49].

Frequently, more than one of these methods is applicable to any particular organ, and indeed in the case of the kidney all four methods as well as straight radiography may be used [PLATES 37 and 38].

In many radiographs the most striking shadows are often those produced by what are essentially artefacts, some of which are avoidable while others are not, but many are so constantly present in certain areas that they must be regarded as features of normal radiographic anatomy. The swallowed air which is always present in the gastro-intestinal tract is an example of such an artefact which cannot be avoided entirely and which may produce relative densities obscuring other features [PLATE 5C]. Again, there may be the avoidable artefact of increased radiographic density due to inspissated bowel content which may include opaque dietetic or medicinal

material. Other formations are so constantly found that, while neither strictly anatomical structures nor truly artefacts, must yet be recognized on the radiograph. Such are the small calcified deposits within veins known as phleboliths.

Loss of definition due to movement is used in a technique generally known as body section radiography or tomography, by which improved radiographic definition is obtained of parts normally obscured by the complexity of the shadows of overlapping structures. During the exposure of the film the tube is moved in one direction and the film in the other. There is therefore between these a plane at which no relative movement takes place and adjustment can be made such that this plane is at the level of the structure to be examined [PLATES 34A and 38A]. The technique is of value but interpretation requires knowledge of the fallacies involved and of the dangers of multiple exposures involving excessive irradiation. Such dangers are of two distinct kinds. There is firstly the risk to the individual being examined, which is related to the area of irradiation and the total dose. Hazards of this kind are fairly well understood. Secondly, and quite distinctly, there is the theoretical and probably genuine risk of genetic damage by irradiation of the gonads either parental or foetal, and this is obviously greater the nearer the field of exposure is to the pelvis. Until more is known with certainty it remains desirable to limit all exposure to X-rays to a minimum and no X-ray examination should be made without good cause.

REFERENCES AND FURTHER READING

ABBIE, A. A. (1946) *The Principles of Anatomy*, 2nd ed., Sydney.

BLOOM, W., and FAWCETT, D. W. (1968) *A Textbook of Histology*, 9th ed., Philadelphia.

CLARK, E. R., and CLARK, E. L. (1934) The new formation of arteriovenous anastomoses in the rabbit's ear, *Amer. J. Anat.*, **55,** 407.

CLARK, E. R., and CLARK, E. L. (1940) Microscopic observations on the extra-endothelial cells of living mammalian blood vessels, *Amer. J. Anat.*, **66,** 1.

CLARK, W. E. LE GROS (1971) *The Tissues of the Body*, 6th ed., Oxford.

ELLIS, R. W. B. (1956) *Child Health and Development*, 2nd ed., London.

FIELD, E. J., and HARRISON, R. J. (1968) *Anatomical Terms*, 3rd ed., Cambridge.

GARRISON, F. H. (1929) *An Introduction to the History of Medicine*, 4th ed., Philadelphia.

GIRGIS, F. G., and PRITCHARD, J. J. (1955) Cartilage in repair of the skull vault, *J. Anat.* (*Lond.*), **90,** 573.

GRANT, J. C. B., and BASMAJIAN, J. V. (1965) *An Atlas of Anatomy*, 7th ed., Baltimore.

GRANT, J. C. B. (1958) *A Method of Anatomy, Descriptive and Deductive*, 6th ed., Baltimore.

GUTHRIE, D. (1958) *A History of Medicine*, 2nd ed., Edinburgh.

HAM, A. W. (1969) *Histology*, 6th ed., Philadelphia.

HARVEY, W. (1628) *Exercitatio Anatomica de Motu Cordis et Sanguinis in Animalibus*, Frankfurt.

HILTON, J. (1863) *Lectures on Rest and Pain*, 6th ed. (1950) ed. Walls, E. W., and Philipp, E. E., London.

JONES, F. W. (1941) *The Principles of Anatomy as seen in the Hand*, 4th ed., London.

JONES, F. W. (1949) *Structure and Function as seen in the Foot*, 2nd ed., London.

KEITH, A. (1948) *Human Embryology and Morphology*, 6th ed., London.

MAXIMOW, A. A., and BLOOM, W. (1957) *A Textbook of Histology*, 7th ed., Philadelphia.

SCAMMON, R. E. (1923) A Summary of the Anatomy of the Infant and Child, in *Pediatrics*, ed. Abt, J. A., Vol 1, Chapter 3, Philadelphia.

SHELDON, W. H. (1940) *The Varieties of Human Physique*, New York.

SINCLAIR, D. C. (1961) *A Student's Guide to Anatomy*, Oxford.

SINCLAIR, D. C. (1969) *Human Growth After Birth*, London.

SINGER, C. (1925) *The Evolution of Anatomy. A Short History of Anatomical and Physiological Discovery to Harvey*, London

SINGER, C., and UNDERWOOD E. A. (1962) *A Short History of Medicine*, 2nd ed., Oxford.

SYMINGTON, J. (1887) *The Topographical Anatomy of the Child*, Edinburgh.

SYMINGTON, J. (1917) *An Atlas illustrating the Topographical Anatomy of the Neck, Thorax, Abdomen and Pelvis* (with supplement: *Head*), Belfast. Reprinted (1956) for the *Anat. Soc. Gt Brit. and Ireland*, Edinburgh.

TONER, P. G., and CARR, K. E. (1968) *Cell Structure*, Edinburgh.

WHITNALL, S. E. (1939) *The Study of Anatomy*, 4th ed., London.

INTRODUCTION TO
HUMAN EMBRYOLOGY

R. G. HARRISON

Human embryology—the study of the human embryo and its development—is concerned with the phenomena of the intra-uterine period of life, from the fertilization of the ovum to the birth of the child. But there is a cycle of events in succeeding generations; and so it is necessary first to consider the life-history of the germ cells which link one generation to the next.

All the cells of the body are the descendants of a single cell which is the starting-point of each individual. The innumerable cells derived from the fertilized ovum are separable into two main groups, a larger group—the somatic cells—from which the body of the individual is developed, and a much smaller group—the germ cells—upon which the continuation of the species depends. The germ cells are lodged in the body of the individual in the sex glands (**gonads**)—ovaries in the female, testes in the male.

The germ cells undergo a process of ripening, known as **maturation;** and when mature they are called **gametes.** A new individual is initiated by the union of two gametes—a male gamete, or spermatozoon, and a female gamete, or ovum. The new cell thus formed is a fertilized ovum and is known as a **zygote**—a term conveniently indicating that it is formed by the union of two individuals and that it contains therefore the potentialities of both.

The newly formed human zygote is about 150μm. in diameter and just visible to the naked eye. From it are produced not only a new individual and the germ cells of the next generation, but also a series of membranes and appendages necessary for growth and development during intra-uterine life.

The ontogenetic or developmental history of the individual is usually considered to extend from the zygote to the adult form. The period of development thus defined is divided into an intra-uterine or prenatal period and an extra-uterine or postnatal period.

The intra-uterine life of Man lasts for about 9 months, and is itself divided into three secondary periods:

1. The pre-embryonic period, during which the growing zygote is **implanted** in the mucous membrane of the uterus (**endometrium),** and is differentiated into embryonic and non-embryonic portions; this period lasts rather less than 2 weeks.

2. The embryonic period, in which the rudiments of all the main organs of the adult are developed, although the embryo has not yet assumed a definitely human form. This period extends to the end of the second month.

3. The foetal period, from the end of the second month, when the embryo begins to assume a definitely human appearance and is called, thenceforth, a **foetus.** The foetal period ends at birth, when the foetus becomes a child and passes into the stage of **postnatal development.**

During the foetal period growth proceeds rapidly. It is especially rapid during the fourth month; and at birth the child usually weighs about 7 pounds.

During the first stages of the postnatal period—infancy, childhood, adolescence—growth and development still proceed until the adult condition is attained; then follows a period of maturity, which passes insensibly into the last stage of all—senescence—which ends in natural death. In the following pages it is the intra-uterine period of life which will be considered and more especially the phenomena of the first 2 months—the pre-embryonic and embryonic periods.

The aim of this Section is to provide the student with a general account of the formation of the embryo—**embryogenesis**—so that he may be in a position to understand the more detailed paragraphs on the development of organs—**organogenesis**—which he will find throughout the textbook. Some account will be given also of those important extra-embryonic organs peculiar to intra-uterine life—the **foetal membranes**—which are responsible for the protection of the embryo and foetus and provide for its physiological needs by the connexion established between one of them and the uterus (Barcroft, 1946, 1952; Huggett and Hammond, 1952).

For further information on human development the student should consult the special works on the subject included in the list of References at the end of this Section.

GROWTH AND DIFFERENTIATION

In the development of the embryo two main processes go hand in hand, but growth—cell division leading to augmentation of size—must be carefully distinguished from differentiation—the specialization of cells in the formation of tissues and the rudiments of organs. Growth depends upon a proper supply of nutriment, and the growth of some parts of the body depends further upon special chemical factors, either derived from food-material or secreted into the blood by organs of the body itself. The differentiation of groups of cells to form the rudiments of organs is a progressive diversification, and may be due to potentialities located in each particular group—such rudiments of organs are said to be 'self-differentiating'; or it may depend on the situation of the group of cells and the influence of neighbouring cells and rudiments upon them. Experiments have been made by the simple excision, or the transplantation into other regions, of portions of embryos, and by the cultivation of excised portions, either as grafts on other growing tissues or inserted in suitable nutritive media (tissue-culture). It has thus been shown that the type of differentiation which embryonic cells undergo in a particular situation depends on the influence of adjacent structures which have already differentiated or are differentiating.

The formation of the axial structures in the embryo depends upon the influence exerted by chemical compounds (**organizers**) residing in the tissues of the primitive streak : these are believed to reside in the dorsal lip of the blastopore, the corresponding structure in lower forms. The formation of the lens vesicle of the eye from the ectoderm depends, in some species, on the presence of the underlying optic cup—a self-differentiating structure which grows out from the forebrain ; if the optic cup is removed the lens does not appear, but a lens vesicle may be induced to form from another part of the ectoderm by transplanting the cup (Lewis, 1904).

Progressive differentiation of the cells of a region is illustrated by experiments on limb buds in Amphibia. Part of a limb bud which, if left *in situ*, would form only a portion of the limb, may form a complete limb if transplanted early enough to another site. After a time, however, the cells of the limb bud differentiate, chemically if not morphologically, so that any part has its destiny fixed and is incapable of forming a whole limb ; yet, for a time, it may still be capable of producing the whole of the segment of the limb of which it would normally form only a part. Thus, a portion of the thigh region of a chick embryo, transplanted (Murray and Huxley, 1925) or grown by tissue-culture methods (Fell and Robison, 1929), will produce not a complete limb but only a femur.

For further information on the control of development see the works of Spemann (1938), Weiss (1939), and Waddington (1956).

There are also many general chemical problems in the development of the embryo, and for information on this aspect of the subject and on the history of embryology the student should consult Needham (1931 ; 1934 ; 1950).

GERM CELLS

The majority of the cells derived from a zygote are somatic cells, which form the tissues of the body ; but a minority, which retain all the inherited potentialities of their parents, remain as primordial germ cells or stem cells. These primordial germ cells segregate from the somatic cells at a quite early period, and migrate from the wall of the yolk sac to the genital ridge of the embryo (Witschi, 1948). Though the functions and the life-histories of germ cells are quite different from those of somatic cells, the structural characters of the two groups of cells are very similar ; both contain cytoplasmic inclusions, the Golgi complex and mitochondria, and both contain the same number (forty-six) of chromosomes in their nuclei. The forty-six chromosomes in human cells consist of twenty-two pairs of chromosomes, the **autosomes,** which are common to both males and females, and one pair of sex chromosomes.

Each primordial germ cell produces many descendants which are lodged in the gonads—female sex cells or oogonia in the ovaries, male sex cells or spermatogonia in the testes—where they undergo further multiplication to form **primary oocytes** or **primary spermatocytes** respectively. Each of these cells is capable of producing four descendants by two final divisions—the maturation divisions. The four descendants of an oocyte are a mature ovum and three small structures called **polar bodies ;** the four descendants of a spermatocyte are four equal **spermatids** which undergo metamorphosis into spermatozoa.

Meiosis

The number of chromosomes in the dividing cells varies greatly in different species, but is characteristically constant in each. The chromosomes differ in size and shape, but are arranged in pairs of **homologous chromosomes,** one member of each pair being derived from the male parent and the other from the female parent. Human chromosomes have been carefully classified according to their length and other characteristics (Böök *et al.*, 1960 ; Lennox, 1960).

The first maturation division is heterotypical because each chromosome does not split in the process, as in normal cell division or mitosis, but each pair of chromosomes separates, one passing to each resultant cell. This is termed meiotic division because the outcome is a reduction of the number of chromosomes in each resultant cell to half that in the dividing cell. In Man and other mammals meiosis occurs in the first of the two maturation divisions. The result of this division is two secondary spermatocytes or one secondary oocyte and a polar body.

During the early prophase of meiosis which in the primary spermatocyte is short, but in the primary oocyte may be very prolonged, the chromosomes appear and lie side by side—a stage known as **conjugation of the chromosomes.** At the end of the prophase of meiotic division the paired chromosomes are assembled at the equator of the spindle, where they are attached to the achromatic fibrils.

There is no true metaphase, since the chromosomes do not split longitudinally as in ordinary mitosis, and the anaphase begins when the homologous chromosomes separate from each other, a process of **disjunction.** They thus pass undivided to the opposite ends of the achromatic spindle ; therefore, in the telophase, when the whole cell divides, one resultant cell contains a group of chromosomes which are the homologues of the chromosomes in the other resultant cell, and each of the two resultant cells has only half (the haploid number) of the chromosomes which were present in the parent cell (the diploid number).

The result of this reduction division is that the dividing cell's complement of chromosomes is equally divided between the two resultant cells (secondary spermatocytes, or secondary oocyte and polar body). The second maturation division is not meiotic : each haploid secondary spermatocyte divides mitotically to form two haploid spermatids, and each haploid secondary oocyte undergoes mitosis to form one haploid mature ovum and a second polar body. The first polar body also divides, so that three polar bodies are present at the end of the second maturation division [FIG. 3A].

Mechanism of Inheritance and Determination of Sex

In human primary oocytes and spermatocytes there are twenty-two pairs of autosomes, in addition to a special pair in each (oocyte, XX : spermatocyte, XY) which has a specific relation to the inheritance of sex. The statistical study of the mode of inheritance of pairs of characters, of which one may be dominant and the other recessive, is in accord with the behaviour of the chromosomes. As it appears to be a matter of chance which resultant cell will receive which of any pair of chromosomes, it may be calculated that in the maturation and union in fertilization of any two human germ cells there are over 281 billions of different combinations of the chromosomes alone.

Since interchange of genes or groups of genes—crossing-over—is known to occur between pairs of chromosomes during their conjugation in the prophase of the reduction division, it is abundantly clear that the chromosome theory of inheritance is consistent with the infinite variety of individual characteristics.

Sex is a heritable character, and is determined at the time of the union of the gametes. It is dependent upon special chromosomes, which are known as **sex chromosomes,** although they bear other genes than those that determine sex. In Man, the male sex is characterized by the presence of sex chromosomes of unequal size, termed X (large) and Y (small), whereas each female cell contains two X chromosomes.

The human oocyte has twenty-two pairs of autosomes and one pair of X chromosomes—forty-six in all; its chromosome-constitution may thus be written 44XX. The secondary oocyte and mature ovum therefore each possess an X chromosome; the chromosome formula of the mature ovum is 22X. The formula of the primary spermatocyte, on the other hand, is 44XY; when the reduction division occurs the X chromosome passes into one secondary spermatocyte, the Y chromosome to the other. Therefore, of the four spermatids two will have an X chromosome and two a Y chromosome. There are thus two kinds of spermatozoa, of 22X and 22Y constitution respectively. If an X-bearing spermatozoon unites with an ovum a female results, but if a Y-bearing spermatozoon fertilizes an ovum a male results.

All these changes depend on normal disjunction of the sex chromosomes during the first maturation division. If this fails to occur, abnormal spermatozoa with the constitution 22XY and 22 may occur, and non-disjunction in the meiosis of primary oocytes can result, theoretically, in ova having the constitution 22XX or 22. Such non-disjunction is said to be responsible for the production of certain clinical syndromes. Thus, if a 22Y sperm fertilizes an ovum with a 22XX chromosome constitution, or a normal 22X ovum is fertilized by a non-disjunctive 22XY sperm, the resultant zygote will have 44XXY chromosomes, a condition found in Klinefelter's syndrome. This syndrome consists of an apparent male having gynaecomastia (enlarged mammary glands), small testes, poor facial hair growth and a high-pitched voice. In like manner, Turner's syndrome, with a constitution 44X (22X sperm fertilizing a 22 ovum or a 22 sperm fertilizing a 22X ovum), and the 'super female' with 44XXX chromosomes, may result, the last condition being caused by fertilization of a 22XX ovum with a 22X sperm. Chromosomal anomalies may also occur in autosomes, and mongolism is a condition in which the body cells each contain forty-seven chromosomes; the sex chromosomes are normal but forty-five autosomes are present. It is also possible for an individual to be a mosaic—his body cells being of two or more stem lines with different chromosome numbers.

The cytological and other evidence upon which the chromosome theory of sex is based is confirmed by the observation that uniovular twins, i.e., twins developed from a single ovum, are always of the same sex, whilst binovular twins, resulting from fertilization of two ova, may be of the same or of opposite sex. It is also strikingly confirmed by the phenomena of 'sex linkage' in inheritance. The sex chromosomes are not solely concerned with sex; they bear genes which control other heritable characters, and these characters are linked to sex in their transmission from generation to generation. Sex-linked characters are of some importance in human inheritance, the best-known example being the condition called haemophilia, in which there is an abnormal tendency to spontaneous or excessive bleeding. It is believed that the determining factor of this condition is recessive and is carried on the X chromosome, which explains its characteristic mode of inheritance; it is not transmitted from an affected father to his son, but only to a grandson through a 'carrier' daughter who remains herself unaffected. The gene for red-green colour-blindness is a sex-linked recessive on the X chromosome, and many cases of Turner's syndrome show such colour-blindness.

Gametogenesis

Gametes are the final descendants of a line of germ cells which pass through successive stages of specialization and are finally transformed into mature ova and spermatozoa by means of the two maturation divisions; the whole process is known as gametogenesis.

Both ovum and spermatozoon are specialized cells, inasmuch as each possesses only half the number of chromosomes present in the oogonium or spermatogonium; moreover, each is incapable, under ordinary circumstances, of undergoing cell division, and must either die or unite with the other to produce a new, rejuvenated cell—the zygote.

The ovum and the spermatozoon have their own special characteristics. The phenomena of oogenesis and spermatogenesis must, therefore, be considered separately.

OOGENESIS AND THE OVUM

The structure of the ovary and the relation of the primary oocytes to the 'germinal epithelium' are described with the Urogenital System.

In the cortical part of the ovary, each primary oocyte lies surrounded by a single layer of follicular cells. Together they constitute a primary ovarian follicle [FIG. 4].

Gradually the oocyte increases in size until it attains a

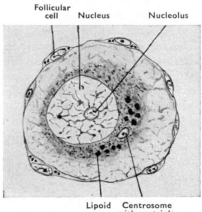

FIG. 4 Human oocyte at beginning of growth period. Diameters, 38μm. × 33μm.

Stratum granulosum   Nucleus of primary oocyte   Nucleolus

Cytoplasm   Theca interna   Theca externa

FIG. 5   Human ovarian follicle from the ovary of a woman aged 38. The follicular cells of the stratum granulosum have multiplied to form a layer of granulosa cells 5–6 cells thick (× 390).

Theca externa   Theca interna   Oocyte

Zona pellucida   Cumulus oophorus   Liquor folliculi

FIG. 6   Human vesicular ovarian follicle (× 180).

diameter of 100μ to 200μ, and so becomes a relatively large cell. During the growth period changes occur both in the surroundings and in the contents of the oocyte.

At first the follicular cells are flattened. Then they become cubical and increase in number by mitosis until they form several layers [**the stratum granulosum**; FIG. 5]. Thereafter, a cavity filled with a fluid called the **liquor folliculi** appears amidst the cells. It increases rapidly in size and separates the cells into two multicellular parts. One part forms the boundary of the cavity and is still called the **stratum granulosum**. The other, with the oocyte embedded in it, projects into the cavity and is the **cumulus oophorus**. This results in a **vesicular ovarian (or Graafian follicle)** [FIG. 6].

When the cells of the stratum granulosum have formed two layers, a clear envelope, the **zona pellucida,** begins to appear around the oocyte, separating it from the cells of the cumulus. It increases in thickness until the maturation of the oocyte begins, when it is an elastic, pellucid membrane varying from 7μm. to 10μm. in thickness [FIGS. 8 and 9].

The changes which take place in the oocyte itself during the growth period concern the contents of the cytoplasm, and the position and relative size of the nucleus.

At the beginning of the growth period the oocyte (called a **primary oocyte** to avoid confusion with its two immediate descendants) is an almost spherical cell; but it soon becomes

ovoid and retains this form in all subsequent periods. The cell body contains a relatively large, eccentrically placed nucleus and a centrosome. The centrosome lies near the central pole of the nucleus and contains one or two centrioles [FIG. 4].

The nucleus possesses one or two nucleoli. The Golgi complex, mitochondria, and globules of lipoid are to be found in the cytoplasm around the nucleus, and are most abundant around the centrosome [FIG. 4]. As growth proceeds they become more diffused throughout the cytoplasm.

When growth is completed the full-grown primary oocyte, enclosed in the zona pellucida, lies in the cumulus of a vesicular follicle in the cortical part of the ovary [FIG. 9]. It consists of a cell body which contains a nucleus, mitochondria, and numerous globules of yolk; but during maturation a centrosome may not be obvious.

The diameter of the cell body, measured along its major axis, varies from 100μm. to 200μm., for full-grown oocytes are not all of the same size; the average diameter is 130–140μm. (Hartman, 1929). The diameter of the nucleus varies from 25μm. to 50μm.

The amount of **yolk** in human and most mammalian ova is relatively small compared with those of fishes, amphibians, reptiles, and birds. The amount of yolk determines whether the segmentation of the fertilized ovum is complete or incom-

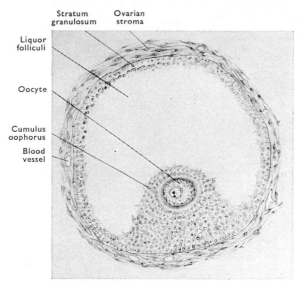

FIG. 7   Human vesicular ovarian follicle.   Diameters of follicle, 620μm. × 465μm. × 465μm.

FIG. 8   Human primary oocyte from the ovary of a woman aged 25 (× 450).

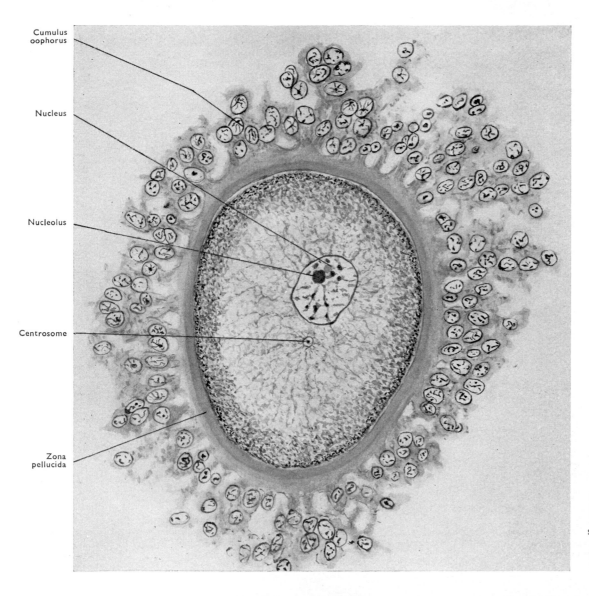

FIG. 9   Human oocyte near the end of the growth period. Diameters of oocyte exclusive of zona pellucida, 140μm. × 110μm. × 84μm. Average thickness of zona pellucida, 5μm.

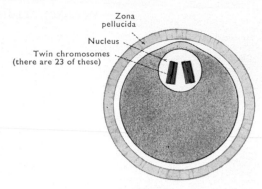

FIG. 10 Schema of maturation of ovum. Early part of prophase of first division. Only four chromosomes are detailed, whereas the actual number is 46.

FIG. 11 Later prophase of first division. The chromatic thread has divided into twin chromosomes.

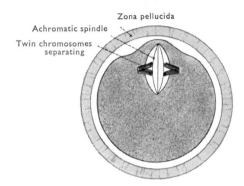

FIG. 12 End of prophase of first division. The twin chromosomes lie at the equator of the achromatic spindle.

FIG. 13 Metaphase of first division. One pole of the spindle lies in the first polar projection, and the paired chromosomes are separating from one another.

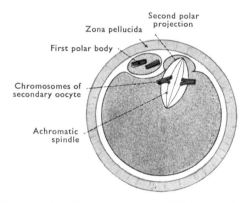

FIG. 14 End of anaphase of first division. One of each pair of chromosomes now lies in the polar projection. The other remains in the larger part of the cell body which becomes the secondary oocyte.

FIG. 15 Beginning of metaphase of second division. The chromosomes of the secondary oocyte are dividing.

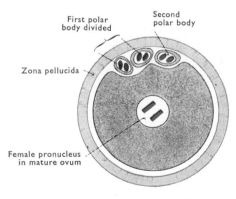

FIG. 16 End of anaphase of second division. The chromosomes of the secondary oocyte have separated into equal parts which have passed to the opposite poles of the spindle.

FIG. 17 Schema of maturation of ovum. End of telophase of second division. The four descendants of the primary oocyte are the mature ovum, with half the original number of chromosomes, and three polar bodies.

plete, and special terms are employed in comparative embryology to denote differences in the amount and distribution of the yolk. It is sufficient here to note that the amount of yolk in the human ovum is so small that it does not in the least interfere with its complete division in segmentation.

The yolk serves as a store of nutritive material which is utilized during the early stages of the growth of the zygote in mammals, and until the time of hatching in oviparous vertebrates.

Apart from the question whether any new ova are produced in the human ovary after birth—as is apparently the case in some animals—there is evidence that a large number of developing follicles atrophy and disappear so that 'radical selective elimination of ova occurs in human ovaries' (Allen *et al.*, 1930).

Maturation of the Ovum

Maturation is the term applied to the phenomena of the two cell divisions which take place after the primary oocyte has attained its full growth.

The first of the two divisions is heterotypical and meiotic. During it the nuclear membrane and the nucleolus or nucleoli disappear, and an achromatic spindle appears at one pole of the oocyte, in the situation previously occupied by the nucleus.

The chromosomes aggregate together in pairs as twin chromosomes [FIGS. 10 and 11]. There appear therefore to be only half the number of the chromosomes originally present in the oocyte; at the end of the prophase they lie at the periphery of the equator of the spindle. When this condition is attained the spindle rotates on its transverse axis [FIGS. 12 and 13].

During the metaphase the two halves of each twin chromosome separate from each other, the two chromosomes thus formed being each equivalent to a whole chromosome of an ordinary cell.

In the anaphase the halves of each twin chromosome travel to the opposite poles of the achromatic spindle [FIG. 14].

Following the telophase the first maturation division is completed, the primary oocyte being divided into a large segment, the **secondary oocyte,** and a small segment called the **first polar body,** both of which lie inside the zona pellucida and each of which contains half the number of chromosomes originally present in the primary oocyte [FIGS. 15 and 18]. Human ova which have attained this stage of maturation have been seen (Rock and Hertig, 1942).

After the first maturation division is completed a new achromatic spindle is formed in the secondary oocyte; this lies at the periphery of the secondary oocyte near the first polar body, the chromosomes become grouped at its equator [FIG. 15], and the second maturation division is commenced.

The ovarian follicle then ruptures, and the liquor folliculi carrying with it the secondary oocyte undergoing its second maturation division and surrounded by its **corona radiata,** composed of attached cumulus cells, is forced through the breach in the surface of the ovary and is swept into the uterine tube. This process, known as **ovulation,** has a definite relation to the menstrual cycle.

If the oocyte is penetrated by a spermatozoon it then completes the second maturation division. The second maturation division is a homotypical mitosis. The achromatic spindle,

already present in the secondary oocyte, carries chromosomes (which, it must be remembered, are only half as numerous as those originally present in the primary oocyte), each of which splits longitudinally into equal parts. The opposite halves of the divided chromosomes then travel to the ends of the achromatic spindle; consequently, the secondary oocyte is divided into a mature ovum and the **second polar body** [FIG. 16], each of which contains half the number of chromosomes (the haploid number) present in the primary oocyte, which is diploid.

As soon as the second polar body is separated off, the maturation is completed [FIG. 17], and a mature ovum is formed.

The Ovum

The mature ovum, like the full-grown oocyte, differs from a typical animal cell on account of its large size, and also because it is surrounded by a special protective envelope—the zona pellucida. An additional and important difference is the fact that the mature ovum under ordinary circumstances is incapable of further cell division because it possesses no centrosome—a deficiency which is supplied by the spermatozoon if fertilization occurs.

The fine surface layer of the cytoplasm of the ovum is a membrane which becomes apparent when the polar bodies are given off [FIG. 23] and has been termed the **vitelline membrane.** The space between the body of the oocyte and the zona pellucida, into which the polar bodies are discharged, is called the **perivitelline space;** in fixed specimens it may appear greatly increased by retraction of the cell body.

As the second polar body is separated off, the first polar body not uncommonly divides into two parts. When that

Cells of cumulus oophorus

First polar body

Second maturation spindle

Perivitelline space

Zona pellucida

FIG. 18 Section of a human ovum after the first maturation division. (Dixon, 1927.)

occurs, the result of the two maturation divisions of the oocyte is the formation of one large cell (the mature ovum) and three polar bodies, all of which are enclosed within the zona pellucida [FIG. 17].

If the secondary oocyte does not meet with a spermatozoon it degenerates or passes through the genital passages and is cast off and lost; but if it unites with a spermatozoon a zygote is formed, from which a new individual may arise, and in that case the polar bodies persist until the zygote has undergone one or two divisions; but sooner or later they disappear, probably breaking down into fragments which are absorbed by the cells of the zygote.

The polar bodies were originally so named in the belief that they indicate the line along which the first segmentation of the fertilized ovum took place; but their exact position does not appear to be of any significance in the mammalian ovum. The production of a polar body ensures conservation of the cytoplasm in a single ovum which has ejected half its chromosomes.

SPERMATOGENESIS AND THE SPERMATOZOON

The **spermatogonia** lie on the membrana propria of the seminiferous tubules of the testes. Their descendants become converted into spermatozoa [FIG. 19]. Primary spermatocytes are first formed and by two maturation divisions these form secondary spermatocytes and then spermatids (Bishop and Walton, 1960).

Spermatocytes differ from oocytes in four important respects : (1) they have no protective membrane corresponding to the zona pellucida of the oocyte; (2) they are not enclosed in follicles; (3) they are not surrounded by definite encircling layers of cells similar to the cells of the stratum granulosum; (4) the spermatozoa to which they give origin are smaller and motile and the centrosome is concerned in the formation of the tail of the spermatozoon.

In the tubules of the testes the spermatocytes are inter-

FIG. 19   Drawing of a section through parts of three human seminiferous tubules, showing the stages of spermatogenesis.

mingled with **sustentacular** (supporting) **cells (cells of Sertoli),** amidst which they undergo their maturation divisions. The spermatids become embedded in the supporting cells, where they mature into spermatozoa, and probably receive nutrition from material in the cytoplasm of these cells.

After it has reached its full growth each primary spermatocyte, like each primary oocyte, can produce only four descendants; the descendants, as in the case of the primary oocyte, are formed by two successive divisions, of which the first is meiotic and produces reduction of the chromosomes, and the second is mitotic.

These divisions differ from the corresponding divisions of the oocyte in the important respect that the four spermatids are of equal size and value, so that they are all eventually capable of uniting with an ovum to form a zygote.

In the prophase of the first or heterotype division the nucleus and nucleolus disappear in the ordinary way. The centrosome divides, and an achromatic spindle appears, which has the daughter centrosomes at its poles and apparently half the typical number of chromosomes at its equator. But the chromosomes are arranged in pairs, and during the metaphase the two segments of each twin chromosome separate from each other. In the anaphase they travel to the opposite poles of the achromatic spindle, and consequently, when the cell divides in the telophase, each daughter cell or secondary spermatocyte contains a centrosome and half the typical number of chromosomes. It is therefore a haploid cell.

The second maturation division, which takes place without the intervention of a resting stage, is homotypical. The centrosome divides, a new achromatic spindle appears, and the chromosomes gather at its equator. In the metaphase the chromosomes divide longitudinally into equal parts, which travel to the opposite poles of the spindle during the anaphase, and when the telophase is completed each granddaughter cell, now called a spermatid, possesses a centrosome and half the typical number of chromosomes.

The Spermatozoon

Ova become ready for conjugation with spermatoza directly after the first maturation division is completed, but spermatids which result from the second maturation division have still to undergo a complicated process of transformation before they become converted to spermatozoa. The process of transformation takes place in association with the supporting cells in which the developing spermatozoa become embedded. Before these details are considered it is necessary that the student should be acquainted with the anatomy of an adult spermatozoon [FIG. 20].

A spermatozoon is a minute organism possessing a head, a middle piece, and a flagellum or tail. Its total length is about 50μm., that is, its length is about the same as the diameter of the nucleus of the ovum.

The head has the form of a compressed ovoid. It contains the nucleus of the spermatid and is completely covered by a head-cap or galea capitis (Schultz-Larsen, 1958). The mean length of the head is 4·4μm. (Duijn, 1958).

The middle piece is about 5μm. in length, and its constituent parts are: (1) the centriole, (2) a portion of the axial fibre, and (3) the mitochondrial sheath.

The axial fibre, which extends from the centriole through the middle piece into the flagellum, consists of one pair of

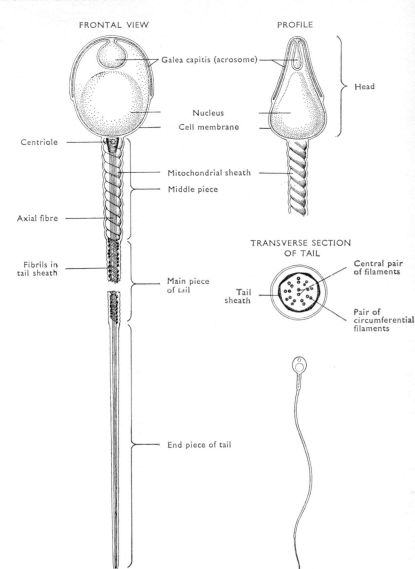

FIG. 20 Diagram of the structure of a human spermatozoon as shown by electron microscopy. Modified from Schultz-Larsen (1958). The appearance as seen by light microscopy is shown in the lower right corner.

central filaments surrounded by nine pairs of peripheral filaments. The mitochondrial sheath surrounds the axial fibre and is formed by numerous fused mitochondria; it is composed of two bands twisted spirally five to seven times round the axial fibre.

The flagellum or tail is long. It consists of prolongations of the axial fibre surrounded by a tail sheath, and terminates in a short thin end-piece, in which the axial fibre appears free from the sheath.

Spermiogenesis

The phase of spermatogenesis involving successive cell divisions up to the production of spermatids is termed **spermatocytogenesis.** The maturation of a spermatid into **a** spermatozoon involves no further cell division, but a modification of the spermatid cell into a highly organized cell, lacking cytoplasm and modified for motility and eventually fertilization. This process of maturation is termed **spermiogenesis** or **spermateliosis.** The head of the spermatozoon

1 SPERMATID IN THE
GOLGI PHASE

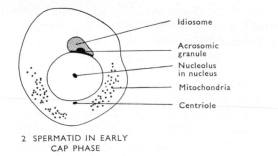

2 SPERMATID IN EARLY
CAP PHASE

Fig. 21   Diagram of the process of spermiogenesis.

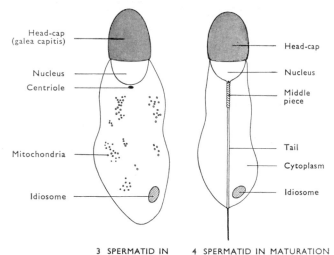

3 SPERMATID IN
ACROSOME PHASE

4 SPERMATID IN MATURATION
PHASE. FORMATION OF
MIDDLE PIECE AND TAIL.
THE CYTOPLASM IS NOW
DEVOID OF MITOCHONDRIA
WHICH HAVE CONTRIBUTED
TO THE FORMATION OF THE
MIDDLE PIECE.

is formed from the nucleus of the spermatid, which becomes encased in a cap at the future anterior end of the spermatozoon. The Golgi complex assembles on the surface of the nucleus as a bead which is deposited on the nucleus to form the **acrosome** from which the head-cap is derived [Fig. 21].

After the second maturation division the centriole of the spermatid passes towards the side of the cell opposite the attachment of the acrosome and from the centriole the axial fibre grows out through the surface of the cell.

The middle piece has still to be formed and the flagellum completed. The mitochondria crowd around the anterior 5μm. of the axial fibre to form the mitochondrial sheath. The remains of the cytoplasm of the spermatid, carrying other mitochondria and a remnant of the Golgi body, are then stripped off, and the spermatozoon is complete (Clermont and Leblond, 1955).

FERTILIZATION

The process of union of male and female gametes to form a zygote is known as fertilization of the ovum. It begins [Fig. 22] when a spermatozoon enters an ovum (either mature or in the last stage of maturation), and it is completed when the nuclear elements of the two have combined.

It is believed that fertilization takes place in the ampulla of

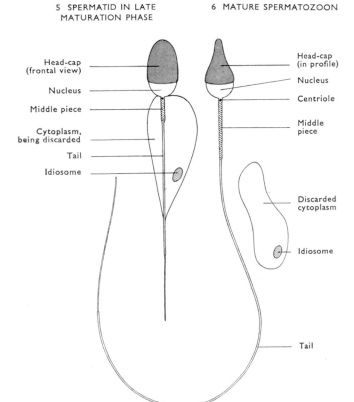

5 SPERMATID IN LATE
MATURATION PHASE

6 MATURE SPERMATOZOON

Fig. 22   Oocyte of a rat at the moment of fertilization. The oocyte is surrounded by cells of the corona radiata, and the head of the spermatozoon has penetrated into the perivitelline space; most of its tail still lies outside the oocyte ($\times$ 700). By permission of Dr. C. R. Austin.

Polar bodies   Zona pellucida   Oocyte   Spermatozoon head

FIG. 23 Guinea-pig ovum, showing polar bodies, and spermatozoon head lacking acrosome, embedded in zona pellucida (×900). By permission of Dr. C. R. Austin.

FIG. 25 Segmentation of intact human ovum at two-cell stage. The smaller of the two polar bodies is visible between the two blastomeres. ×500. (Carnegie No. 8698; by special permission of Dr. Arthur T. Hertig.)

the uterine tube, and that the entry of one spermatozoon causes some change in the surface of the ovum that prevents the entry of others [FIG. 23].

It has already been pointed out that the maturation of the oocyte is usually not completed until after the spermatozoon has entered; thereafter, the second polar body is extruded [FIG. 23] and the nucleus of the mature ovum, now called the **female pronucleus,** takes up a central position. It contains half the number of chromosomes present in the nucleus of the oocyte before the maturation began. As the female pronucleus forms, the tail of the spermatozoon disappears, the head is transformed into the **male pronucleus,** and two centrosomes

arise from the middle piece. The male pronucleus is smaller than the female pronucleus, and it also contains only half the number of chromosomes present in the primary spermatocyte from which it is derived.

As soon as the female and male pronuclei are established they approach each other, meet, and lose their nuclear membranes. The chromosomes of each pronucleus appear, become arranged on the equator of a spindle, and split along their length; the half chromosomes then pass towards their corresponding centrosomes, and the first cleavage division of the zygote ensues.

When fertilization is completed, therefore, a new structure

Polar body   Zona pellucida   Spermatozoa   Polar body

Polar body   Polar body

Cells about to divide

FIG. 24 Two-cell, four-cell, and six-cell stages of segmentation of living ovum of macaque monkey. (After Lewis and Hartman, 1933.) The two larger cells in the six-cell stage were about to divide to produce the eight-cell stage. For the two-cell human stage, see Fig. 25.

Zona pellucida   Future cell of inner cell mass   Future trophoblast cell

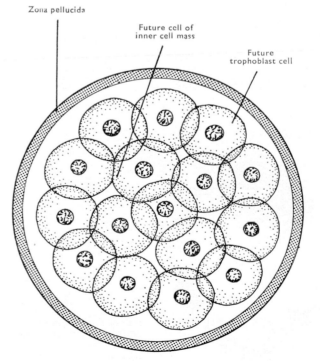

FIG. 26 Diagram of a morula.

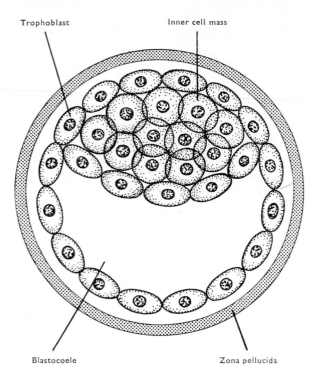

Trophoblast   Inner cell mass

Blastocoele   Zona pellucida

FIG. 27   Diagram of a blastocyst.

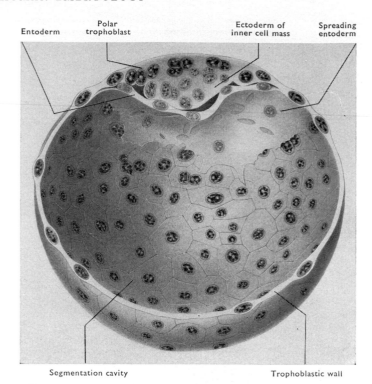

Entoderm   Polar trophoblast   Ectoderm of inner cell mass   Spreading entoderm

Segmentation cavity   Trophoblastic wall

FIG. 29   Reconstruction of 9-day-old blastula of macaque monkey. The entodermal part of the inner cell mass is flattened and its cells are spreading over the inner surface of the trophoblast. ×280. (After Heuser and Streeter, 1941.)

—the **zygote**—is formed. It lies, together with the polar bodies, inside the zona pellucida. There are several phenomena occurring at fertilization. The first is the reconstitution of the diploid number of chromosomes that is typical for the cells of the species to which the zygote belongs, half of them derived from the female parent and half from the male. The second factor is the stimulus to division (**segmentation** or **cleavage**) which the spermatozoon provides. Thirdly, the sex of the resultant embryo is determined by the chromosome constitution of the sperm effecting fertilization—whether it contains an X or a Y chromosome. Finally, various changes occur in the structure and chemistry of the fertilized ovum; these have been collectively termed **activation** (see Rothschild, 1956; Austin and Bishop, 1957; Austin and Walton, 1960).

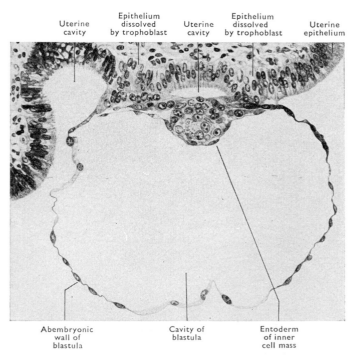

Uterine cavity   Epithelium dissolved by trophoblast   Uterine cavity   Epithelium dissolved by trophoblast   Uterine epithelium

Abembryonic wall of blastula   Cavity of blastula   Entoderm of inner cell mass

FIG. 28   Section of free human blastula, 5–6 days old, from uterine cavity. ×600. Note the arrangement of the inner cell mass in relation to the expanding trophoblastic wall and the segmentation cavity. (Carnegie No. 8663; by special permission of Dr. Arthur T. Hertig.)

FIG. 30   Early stage of attachment of macaque blastula (about 9 days) to the uterine epithelium. Note the dissolution of the endometrial cells in contact with the trophoblast and the spread of the entoderm from the inner cell mass. ×185. (After Heuser and Streeter, 1941.)

Dilated uterine gland   Implanting blastula   Uterine epithelium   Uterine gland

× 100

Dilated uterine gland   Syncytiotrophoblast

Edge of uterine   Ectoderm of inner   Wall of blastula   Amniotic   Cavity of
epithelium   cell mass   resting on entoderm   cavity   blastula

× 300

FIG. 31   Section of human zygote, $7\frac{1}{2}$ days old, in early superficial stage of implantation. The trophoblast is proliferating wherever it is in contact with the endometrium, and the thin abembryonic wall of the blastula, still exposed to the uterine cavity, is collapsed on the inner cell mass.   (Hertig and Rock, 1945.)

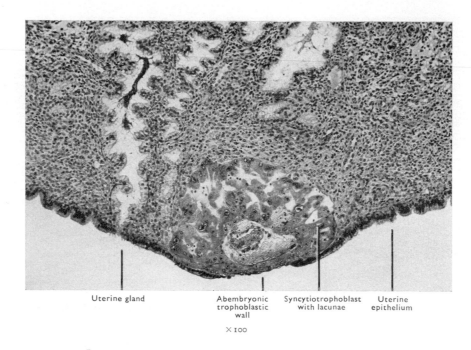

Uterine gland   Abembryonic   Syncytiotrophoblast   Uterine
                     trophoblastic   with lacunae   epithelium
                     wall

× 100

Ectoderm of   Lacuna in
germ disc   Amnion   syncytiotrophoblast

Edge of uterine epithelium   Abembryonic trophoblast   Entoderm of germ disc   Coagulum in cavity

× 300

FIG. 32 Section of human zygote, 9½ days old, more deeply implanted than the 7½-day specimen shown in Fig. 31. Only a small portion of the trophoblastic wall is still exposed and the uterine epithelium is beginning to regenerate over it. There are numerous anastomosing lacunae in the syncytiotrophoblast. (Hertig and Rock, 1945.)

THE PRE-EMBRYONIC PERIOD

Segmentation

At the time of fertilization, segmentation of the zygote is initiated. There follows a series of consecutive mitotic divisions to result in the formation of twelve to sixteen cells called **blastomeres**; these are grouped together in the form of a spherical mass, called a morula [FIGS. 24 and 26] on account of the mulberry-like appearance of its surface. The segmentation of the human ovum has not yet been followed in detail, but a perfect two-cell stage [FIG. 25], an eight-cell stage (Rock and Hertig, 1948), and a morula containing twelve cells (Hertig et al., 1954) have been described.

The **morula** is formed inside the zona pellucida so that its total size does not exceed that of the mature ovum; consequently, the size of the individual cells is progressively reduced by each division and the proportion of cytoplasm to nucleus becomes steadily smaller during segmentation. The proportion is very high in the ovum, and its reduction within the normal range of the somatic cells appears to be an essential preliminary to the next stage of development. This proportionate change involves a transfer of cell proteins, cytoplasmic ribonucleic acid being transferred into the nucleus. At the end of segmentation the zona pellucida disintegrates.

Trophoblast and Inner Cell Mass

The first structural change that occurs in the morula is the differentiation of its cells into an outer layer, the trophoblast, and an inner cell mass [FIGS. 26 and 27] by the appearance of a cavity filled with fluid (the segmentation cavity or **blastocoele**) in between the cells of the morula. As soon as this segmentation cavity appears, the zygote becomes a **blastula** or **blastocyst.** The fluid which fills the blastocoele also acts as a source of nutrition, and is derived from the uterine secretions. By this time the zona pellucida has disintegrated and the blastula is free to expand with the further intake of fluid so that the trophoblastic cells are flattened out into a thin membrane, except where the inner cell mass is attached [FIG. 29]. At this stage—the zygote having entered the uterus as a morula, probably on the third or fourth day from ovulation (see Hertig, Rock, and Adams, 1956) —the blastula becomes attached to the uterine epithelium [FIG. 30]. It is probable that this occurs on the sixth or seventh day, as the earliest implanting human zygote yet observed [FIG. 31] was recovered seven and a half days after the presumed date of fertilization; it is still in the blastula stage.

The trophoblast or trophoblastic ectoderm, by its active proliferation [FIGS. 31–36], plays a most important part by enabling the zygote to burrow into and embed itself in the **decidua** (the term applied to the endometrium after it has become hypertrophied in preparation for implantation of the zygote), and in the nutrition of the embryo and foetus. It enters into the formation of the chorion or outermost envelope of the growing zygote. The chorion serves, in the first instance, as both a protective and a nutritive covering [p. 33], but later a part of it forms the **placenta,** the organ through which gaseous and nutritional exchange between mother and foetus takes place.

The inner cell mass contains the cells from which the embryo itself will be developed; but it is concerned also with the formation of the other extra-embryonic organs—amnion, yolk sac, and allantois—which, with the placenta and chorion, constitute the **foetal membranes.**

Differentiation of Inner Cell Mass

As the blastula begins to attach itself to the decidua and to be implanted, the fate of the cells of the inner cell mass is already indicated by the relative positions they occupy [FIGS. 31 and 32]. The group that adheres for a time to the trophoblast, itself of ectodermal origin, produces the **ectoderm,** destined to take part in the formation of both the embryo and its amnion; the remainder, projecting at first into the blastocoele but rapidly flattening out into a plate of cells one layer thick, constitutes the **entoderm,** which takes part in the formation of the embryo, and of its yolk sac.

The delimitation of the ectodermal and entodermal elements of the inner cell mass is followed by the formation of two vesicles, one from each group of cells. The cavity in the ectodermal group is formed by a loosening of the cells in the middle of the group and by the partial separation of a layer of 'amniogenic' cells from the overlying trophoblast; the cells are reoriented around a space in which fluid accumulates. This **amniotic cavity** is already present in the $7\frac{1}{2}$-day human blastula [FIG. 31].

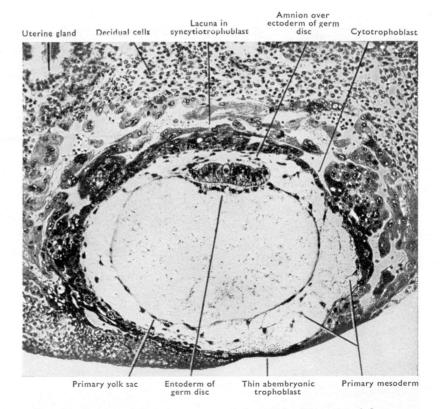

Uterine gland   Decidual cells   Lacuna in syncytiotrophoblast   Amnion over ectoderm of germ disc   Cytotrophoblast

Primary yolk sac   Entoderm of germ disc   Thin abembryonic trophoblast   Primary mesoderm

FIG. 33 Section of human zygote, $12\frac{1}{2}$ days old. The general features are the same as in the $11\frac{1}{2}$-day specimen, but the primary yolk sac is much larger and the 'germ disc' more advanced. The implantation, however, is more superficial. × 125. (Hertig and Rock, 1941.)

The entodermal vesicle or yolk sac is at first very much larger than the ectodermal vesicle and the question of its origin is complicated by the simultaneous appearance of the loose, semifluid tissue known as the **primary (or extra-embryonic) mesoderm.**

The entodermal portion of the inner cell mass appears in the $7\frac{1}{2}$-day human blastula as a simple layer of cells [FIG. 31]. In a $9\frac{1}{2}$-day specimen [FIG. 32] there is little change in this layer beyond a slight extension at its edges, although the ectodermal plate has grown thicker, the amniotic cavity is better defined, and the implantation of the whole zygote has advanced considerably. Yet in quite a number of slightly older specimens a complete 'entodermal' vesicle is present, larger than the ectodermal vesicle with its amniotic cavity, and occupying a varying, but considerable, part of the original cavity of the blastula, which is now filled with 'primary mesoderm' [FIGS. 4 and 39]. In mammals, in general, the entoderm grows round the inside of the trophoblastic shell before any mesoderm—which arises later from the primitive streak of the embryonic disc—is present, but in Man the development of primary mesoderm precedes the formation of the yolk sac.

There is, however, an obvious difference between the cuboidal cells that form the roof of the vesicle (primary yolk sac) in contact with the ectodermal plate and the extremely tenuous remainder of the wall of the vesicle. The thin, drawn-out cells that form it are of mesothelial character and are indeed indistinguishable from the cells of the primary mesoderm that now line the trophoblastic shell and occupy the rest of the cavity in wisps of mesenchymatous tissue.

FOETAL MEMBRANES AND PLACENTA

NUTRITION AND PROTECTION OF EMBRYO

While the zygote is passing along the uterine tube, before the zona pellucida disintegrates and the blastula stage is reached, and for a brief period after it enters the uterus, it depends for its nutrition upon the store of lipoid and other food material in the cytoplasm of the ovum.

As the human ovum is small, and as it contains but little food store, there is urgent necessity for an external source of nutritive supply, of oxygen for tissue respiration, and a provision for the removal of waste products of its metabolism. The morula obtains some food material from the uterine secretions as well as the fluid essential for its transformation into a blastula; but development cannot proceed beyond that stage until an intimate connexion between the zygote and the uterus is established by the critical operation of **implantation** or **embedding** in the decidua. This connexion is provided by the placenta, which is one of the **foetal membranes.**

EVOLUTION OF FOETAL MEMBRANES

The foetal membranes are parts of the original body wall (amnion and chorion) and parts of the original gut wall (yolk sac and allantois) which are set aside to provide organs that adapt the embryo and foetus to its temporary environment. These organs are extra-embryonic structures concerned with protection, nutrition, respiration, and excretion; and the

Dilated gland　　Lacuna in syncytiotrophoblast　　Sinusoid　　Amnion　　Sinusoid　　Lacuna in syncytiotrophoblast

Mouth of dilated gland　　Coagulum　　Entoderm　　Primary yolk sac　　Ectoderm　　Primary mesoderm　　Uterine epithelium

FIG. 34 Sectional reconstruction of the $12\frac{1}{2}$-day human zygote in Fig. 33. The reconstruction shows the main features of the embryonic rudiment and of the trophoblast, in particular its relation to the endometrium. × 125. (After Hertig and Rock, 1941, from a drawing by James F. Didusch.)

Labels on Fig. 35 (clockwise from top): Amnion, Uterine gland, Villi, Germ disc, Intervillous space, Haemorrhagic coagulum, Blood-filled gland, Coagulum in coelom, Remnant of primary yolk sac, Uterine blood sinus, Mesoderm lining cytotrophoblast, Secondary yolk sac, Syncytiotrophoblast.

FIG. 35 Section of human zygote 13½ days old. The embryonic rudiment is clearly defined with amnion and secondary yolk sac, and a remnant of the primary yolk sac is seen. There is an unusually large 'closing coagulum', and chorionic villi have begun to form. × 30. (Heuser, Rock, and Hertig, 1945.)

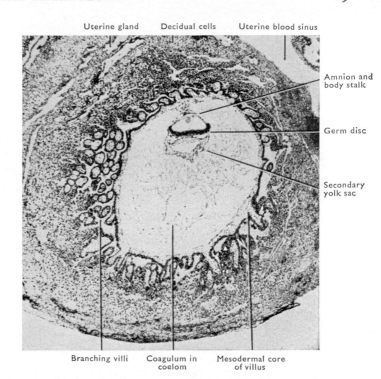

Labels on Fig. 36: Uterine gland, Decidual cells, Uterine blood sinus, Amnion and body stalk, Germ disc, Secondary yolk sac, Branching villi, Coagulum in coelom, Mesodermal core of villus.

FIG. 36 Section of human zygote, 16½ days old. The secondary yolk sac is fully formed with an outer covering of primary mesoderm. The 'germ disc' is more advanced than in the 13½-day specimen (Fig. 35) with a primitive streak, early formation of intra-embryonic mesoderm, and a head-process. There is an indication of an 'amnion stalk' in the condensation of mesoderm that attaches amnion to chorion, and an allantoic diverticulum was present. The chorionic villi with their mesodermal cores are branching freely. × 30. (Heuser, Rock, and Hertig, 1945.)

manner in which they perform these functions, as well as the details of their arrangement, differ in different groups of animals. The four membranes are found only in the higher classes of vertebrates—the reptiles, the birds, and the mammals—whose embryos develop either within an egg-shell or within the body of the mother. Since they all possess the characteristic protective membrane called the amnion, the reptiles, birds, and mammals are grouped together as the Amniota.

The **amnion** provides a cavity full of fluid in which the embryo may develop under conditions which resemble the watery environment of the embryos of lower vertebrates. The embryos of fishes and amphibia do not require an amnion, since for the most part they hatch from the egg and develop in water; and the only 'foetal membrane' they possess is a yolk sac which contains enough food material for the initial stages of development and, becoming highly vascularized, also forms a 'respiratory foetal membrane'. All the Amniota also possess a **yolk sac,** and in the embryos of all those that are oviparous—including the lowest order of mammals (the monotremes)—it is filled with food material. Although the yolk sac ceases to have any importance as a food store in the viviparous mammals, including Man, it is retained in them because of its intimate relation to the development of the gut and also because the earliest blood vessels and blood cells are formed in its wall. In some mammals it is retained, at least temporarily, as a yolk sac placenta.

The third membrane—the **allantois**—is as characteristic of the Amniota as the amnion itself; and it has a very remarkable history. It grows out as a diverticulum from the ventral wall of the hindgut [FIGS. 38 and 41], and in reptiles and birds it serves primarily as a urinary bladder for the embryo; its relation to the gut is exactly the same as that of the permanent

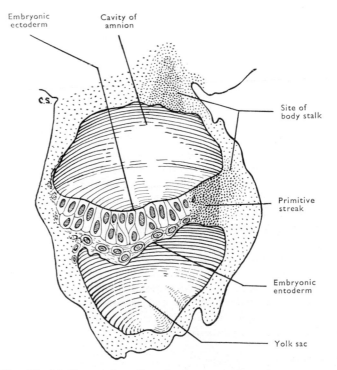

Labels on Fig. 37: Embryonic ectoderm, Cavity of amnion, C.S., Site of body stalk, Primitive streak, Embryonic entoderm, Yolk sac.

FIG. 37 Median section of 15-day human embryo (reconstruction) to show the earliest stage of the primitive streak at the caudal end of the embryonic area. (After Brewer, 1938.)

FIG. 38 Diagrams of stages in the evolution of the foetal membranes. For detailed explanation, see text. (Adapted from Graham Kerr's *Text-Book of Embryology: Vertebrata.*) Trophoblast, amnion and embryonic ectoderm, blue; entoderm, red; mesoderm, yellow.

urinary bladder of the amphibia, with which it is probably homologous. But the allantois in these animals, as it increases in size, comes into contact with the fourth foetal membrane, the **chorion** [FIG. 38], and fuses with it to form a combined membrane which lines the inside of the egg-shell and surrounds the embryo. This membrane, by virtue of the blood vessels it contains, provides for the respiration of the embryo by gaseous exchange through the porous shell between the air and the blood. Thus the allantois of oviparous animals comes secondarily, through fusion of its wall with the chorion, to act functionally as an embryonic 'lung'. In most viviparous mammals the chorion is vascularized in the same manner by blood vessels carried to it by the allantois and thus is enabled to act likewise as a respiratory membrane; but in this case the exchange of oxygen and carbon dioxide is not with the air but with the blood of the mother; and, in addition, food materials also can be obtained by the embryo from the same source by the formation of a **placenta.**

Amnion

The human amnion is formed from that portion of the wall of the ectodermal or amnio-embryonic vesicle which does not take part in the formation of the embryo. It consists of ectoderm cells covered externally by a layer of extra-embryonic mesoderm, and it is continuous with the margin of the embryonic area or germ disc [FIGS. 39–41].

The cavity of the ectodermal vesicle, enclosed by the amnion and the embryonic area, is the amniotic cavity; it is filled with fluid which raises the amnion in the form of a cupola over the embryonic area [FIG. 38]. The amnion progressively enlarges as the amount of fluid increases (Harrison and Malpas, 1953) until it fuses with the chorion; by that time its attachment to the embryo is around the umbilical orifice and it forms a sheath for the umbilical cord [FIG. 45].

Yolk Sac

The wall of the primary yolk sac is not entirely of entodermal origin, since its greater part—the so-called exocoelomic membrane—is composed of cells which appear to be derived from those of the primary mesoderm. But the whole of this vesicle does not participate in the formation of the 'yolk sac' of later stages. The primary yolk sac reaches its final state of development about the twelfth day, when it is much larger than the

FIG. 39 Diagram of early
implanted human zygote.

Lacunae in
syncytiotrophoblast

Syncytiotrophoblast

Decidua

Amnion

Ectoderm of
germ disc

Entoderm of
germ disc

Extra-embryonic
mesoderm

Trophoblast

Primary
yolk sac

ectodermal vesicle [FIGS. 33 and 34]; thereafter, it is quite
rapidly reduced in size so that in a 13½-day embryo [FIG. 35]
it is not much larger than the ectodermal vesicle, and soon,
by the growth of that vesicle, the yolk sac appears about the
same size or even, for a time, relatively smaller [FIGS. 36 and
37]. It may now be called the secondary yolk sac. The term
definitive yolk sac is retained for the yolk sac after separation
from the alimentary canal [p. 48].

The reduction in size of the primary yolk sac to a smaller
secondary yolk sac is occasioned by a considerable portion of
the primary yolk sac being cut off and subsequently degenerating. In several young embryos small detached vesicles, sometimes connected by strands, can be recognized as degenerating
portions of the primary yolk sac [FIG. 35]. The earliest record
of the stage of constriction of the primary yolk sac, with a
suggestion of its evolutionary significance, is in the description

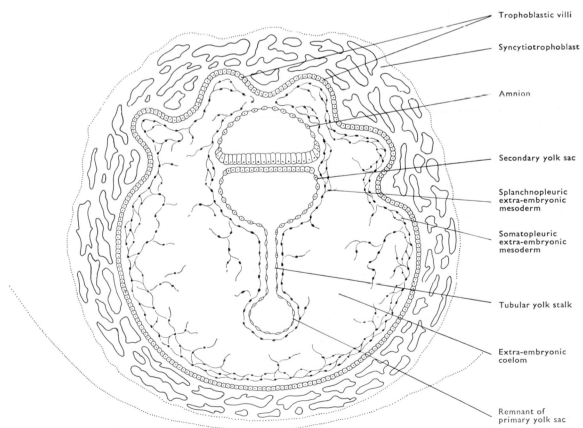

Trophoblastic villi

Syncytiotrophoblast

Amnion

Secondary yolk sac

Splanchnopleuric
extra-embryonic
mesoderm

Somatopleuric
extra-embryonic
mesoderm

Tubular yolk stalk

Extra-embryonic
coelom

Remnant of
primary yolk sac

FIG. 40 Diagram of im-
planted human zygote
showing formation of
extra-embryonic coelom.

of the 'Teacher-Bryce II' embryo of Bryce (1924). Stieve (1931) figured hypothetical stages in the reduction in size of the yolk sac; and his interpretation is borne out by the 13½–14-day specimens described by Lewis and Harrison (1966) and Morton (1949). The primary yolk sac appears to correspond to the yolk sac of the lower mammals, and its replacement by the secondary yolk sac is a specialized developmental process in Man and other Primates.

When the embryonic area is folded into the form of the embryo [p. 48 : see FIG. 66] the entodermal vesicle (secondary yolk sac) is differentiated into three parts : (1) a part enclosed in the embryo, where it forms the primitive entodermal alimentary canal; (2) a part which lies external to the embryo in the extra-embryonic coelom [p. 33]—this is the yolk sac proper or definitive yolk sac; (3) the third portion is the vitello-intestinal duct, which connects the yolk sac with the primitive alimentary canal [FIGS. 62 and 66].

The cavity of the yolk sac is therefore in free communication with that of the alimentary canal, and their walls are continuous with each other and identical in structure, each consisting of an internal layer of entodermal cells and an external layer of mesoderm.

Free communication between the yolk sac and the primitive alimentary canal appears to exist in the human embryo until it is 4 weeks old and about 2·5 mm. long [FIG. 66]. During the fifth week the vitello-intestinal duct is elongated into a relatively long, narrow tube, lodged in the umbilical cord, and the yolk sac, which has become a relatively small vesicle, is placed between the outer surface of the amnion and the inner surface of the chorion, in the region of the placenta. By the end of the fifth week, when the embryo has attained a length of about 5 mm., the vitello-intestinal duct begins to undergo atrophy, and it separates from the intestine when the embryo is about 11 mm. long; but remnants of it may be found in the umbilical cord up to the third month.

The definitive yolk sac itself persists until birth, when it may

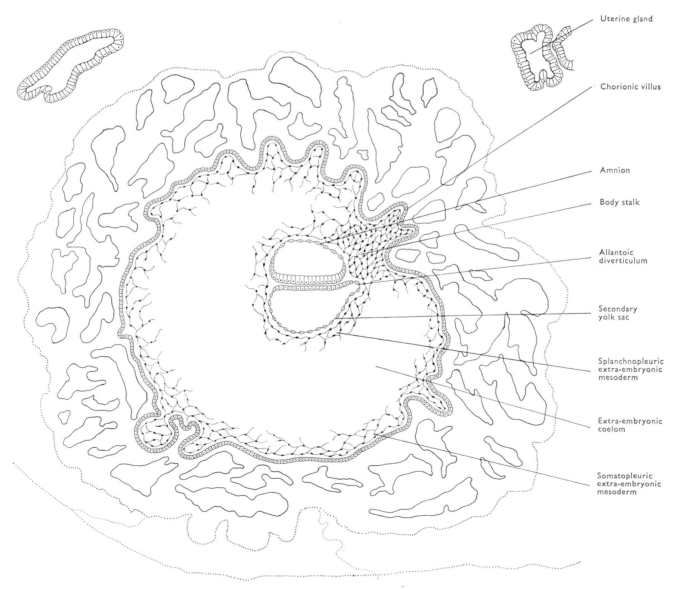

FIG. 41   Diagram of implanted human zygote with bilaminar germ disc.

be found as a minute object lying between the amnion and the placenta near the attachment of the umbilical cord.

The human yolk sac, since it contains no yolk, is in a sense a vestige from the evolutionary point of view; yet, in the early stages of human development, it is relatively quite large [FIG. 92], and it retains its special importance in the development of the blood and blood vessels [p. 62]. The vessels which develop in the wall of the yolk sac are soon connected with others in the body of the embryo to form a **vitelline circulation.** This circulation is mainly responsible for the transfer of nutriment from the yolk to the embryos of birds and reptiles; in mammals the vessels are formed in the same way since they are the basis of the vascular arrangements of the gut. At a very early period, a number of primitive vitelline arteries are distributed to the yolk sac from the primitive aortae, and the blood is returned from the yolk sac to the embryo by a pair of vitelline veins [FIG. 78].

After a time the arteries are reduced to a single pair, which then become converted into a single trunk; this passes through the umbilical orifice along the vitello-intestinal duct to the definitive yolk sac [FIG. 80]. After the umbilical cord is formed the extra-embryonic parts of the vitelline veins disappear and can no longer be traced in the cord. The same fate overtakes the vitelline artery—all except a portion of its intra-embryonic part, which persists as the **superior mesenteric artery.**

Allantois

The human allantois, like the yolk sac, is a greatly reduced structure, and may therefore be termed the **allantoic diverticulum.** In its relation to the chorion and the formation of the placenta, it too may be considered an evolutionary vestige; but it does not disappear entirely, for it is related to the development of the urinary bladder [p. 62]. The allantoic diverticulum grows into the body stalk but does not reach the chorion [FIG. 52]; it arises from the primary entodermal vesicle and so appears at first to come from the yolk sac; but its site of origin is included in the tail fold of the embryo and so is transferred to the ventral wall of the cloacal part of the hindgut [FIGS. 61 and 62]. The allantoic diverticulum is included in the umbilical cord [FIG. 61], where it shrivels and disappears; its intra-abdominal portion becomes the **urachus,** which atrophies to form the **median umbilical ligament.**

Chorion

When the human blastocyst is first implanted, its wall consists solely of trophoblast [FIGS. 31 and 32]; but the primary mesoderm soon appears and, after the formation of the secondary yolk sac [p. 31], the primary mesoderm splits into two layers—one lining the inner surface of the trophoblast and the other covering the outer surfaces of the inner cell mass [FIGS. 40, 41, and 52]; the space between them is the **extra-embryonic coelom.** The extra-embryonic coelom, however, never completely separates the inner cell mass from the inner surface of the trophoblast; continuity is maintained at one point to provide a pathway for blood vessels between the chorion and the embryo. The basis of that pathway is provided by a part of the primary mesoderm which is not split by the extra-embryonic coelom and continues to connect the inner surface of the chorion with the inner cell mass in the

region that will become the caudal end of the embryonic area. This mesodermal connecting link is called the **body stalk** [FIGS. 41, 52, and 61].

The trophoblast and its lining of mesoderm together constitute the chorion, and the essential function of that membrane, in addition to providing an outer protective sac within which the embryo may develop in its amnion, is the formation of the placenta. The development and vascularization of the chorion are described with the account of the development of the placenta itself.

Body Stalk

In the body stalk lies the allantoic diverticulum of the entodermal vesicle [FIG. 61], on either side of which pass the umbilical arteries and veins (one of which atrophies very early in development), by means of which blood is conveyed between the embryo and the chorion. It is important to note that the blood vessels which thus pass through the body stalk enter or leave the body of the embryo at the margin of the embryonic disc. As the embryo enlarges into the amniotic cavity, this margin becomes the edge of the umbilical orifice which is, at first, a relatively large aperture [FIG. 59] through which these vessels therefore pass.

As the embryonic area is folded into the form of the embryo, the amnion increases in extent, filling more and more of the extra-embryonic coelom, and the embryo rises into its cavity Thus the walls of the amnion bulge ventrally round the margin of the umbilical orifice [FIGS. 45, 61, and 62], tending to cover the body stalk and its enclosed structures and force the yolk sac further and further away from the embryo. The vitello-intestinal duct is elongated and surrounded by the external surface of the amnion and its covering layer of extra-embryonic mesoderm, which form a hollow tube around the duct. The cavity of this tube—the proximal part of it becomes the **umbilical sac** (see below)—is an elongated part of the extra-embryonic coelom, surrounded by the amnion [FIGS. 75–77]. The caudal wall of the tube necessarily consists of the elongated body stalk.

Umbilical Cord

As the distension of the amnion still continues, the walls of the tube are forced against the vitello-intestinal duct, and the amniotic mesoderm fuses with the mesoderm surrounding that duct and with the body stalk. When the fusion is complete, a solid cord—the **umbilical cord**—is formed [FIG. 45]; it consists of an external covering of amniotic ectoderm, and a core of mesoderm in which lie two umbilical arteries, a single umbilical vein, and the remains of the vitello-intestinal duct and the vitelline vessels. One of the arteries is absent in about 1 per cent. of umbilical cords (Armitage et al., 1967), and there is considerable evidence to suggest that this is associated with congenital malformations of the foetus. One end of the umbilical cord is connected with the embryo; the other end is attached to the chorion. Near the latter attachment, in the mesoderm of the chorion, lies the yolk sac, now a relatively small vesicle; and the allantoic diverticulum, continuous with the urachus of the urogenital chamber, projects into the umbilical cord from the embryo [FIGS. 45, 76, and 77]. For some time the part of the extra-embryonic coelom in the cord where it springs from the belly of the embryo is distended to form the **umbilical sac.** This is continuous with the intra-embryonic coelom and the intestines develop in it for a time [p. 61].

As the amnion grows still larger, all that part of its outer surface which does not take part in the formation of the umbilical cord is pressed into contact with the inner surface of the chorion, with which it fuses, and the cavity of the extra-embryonic coelom is thus obliterated [FIG. 45].

The outer wall of the complete vesicle now consists of the fused chorion and amnion, and contains in its interior the amniotic cavity and the embryo, which is attached to the chorion by the umbilical cord.

When the umbilical cord is first formed, it is comparatively short, but as the amniotic cavity increases the cord elongates, until it attains a length of from 30 to 129 cm. (average 61 cm. (Malpas, 1964)); it then allows the embryo to float freely in the fluid in the amniotic cavity, and to obtain nutrition from the placenta by the flow and return of blood, through the vessels within the umbilical cord.

THE PLACENTA

The placenta is an organ developed for the purpose of providing the foetus with food and oxygen, and removing the waste products of metabolism. It is formed partly from the chorion and partly from the mucous membrane (endometrium) of the uterus, altered as explained below to form the uterine 'decidua'.

In the development of the placenta the blood vessels of the foetus and the blood of the mother are brought into intimate relation, so that free interchanges may readily take place between the two blood streams; but *the blood of the foetus and the blood of the mother are always separated from each other by two or more layers of foetal cells.*

Endometrial Preparation

The endometrium undergoes changes in preparation for the reception and retention of the zygote, and if implantation occurs the modified endometrium is known as the **uterine decidua,** a term based on the belief that it is to a great extent 'cast off' when the placenta is ultimately detached.

The changes which take place are, for the most part, hypertrophic in character. The vascularity of the endometrium is increased, mainly by the dilatation of its capillaries. The tubular glands are elongated; they become tortuous and dilated, and the interglandular tissue increases in amount. As a result of these changes the decidua is thicker, softer, more spongy, and more vascular than secretory endometrium.

Partly on account of the dilatation of the deep parts of the glands and partly on account of differences in texture, the decidua is usually described as consisting of three layers: (1) an internal layer, next the cavity, containing the necks of the uterine glands and called the **compact layer;** (2) a middle layer—the **spongy layer**—formed largely by the dilated parts of the glands; (3) an external unchanged layer, in which lie the comparatively unaltered bases of the glands— the **basal layer.**

These changes are an intensification of the periodic hypertrophy and secretory changes in the endometrium which precede menstruation or implantation. During menstruation the superficial part of the mucous membrane disintegrates, and the whole of the compact layer with much of the spongy layer is cast off together with some blood, constituting the menstrual flow which lasts a variable period, but usually about 3 to 4 days. Repair ensues and the endometrium is built up again to the premenstrual state, being regenerated from the basal layer under the influence first of the oestrogenic hormones and later of progesterone—produced respectively by the growing ovarian follicle and by the **corpus luteum,** the structure formed in the ovary from the ruptured follicle after ovulation. The evidence that ovulation occurs about 14 days before the expected onset of the next menstruation (Rock and Hertig, 1944, 1948) is reviewed by Davies (1948). This relation is of great importance in estimating the age of young embryos. If the ovum is not fertilized and implantation does not occur, the corpus luteum begins to degenerate about twelve days after ovulation, menstruation occurs, and thereafter the corpus luteum becomes infiltrated with fibrous tissue to form a **corpus albicans.** If fertilization and implantation take place, the corpus luteum continues to grow and maintains its secretion of progesterone; this induces the final changes in the endometrium, constituting the **decidual reaction,** which characteristically accompany the implantation of the blastula.

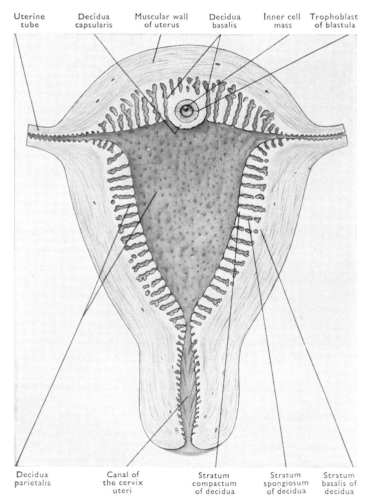

| Uterine tube | Decidua capsularis | Muscular wall of uterus | Decidua basalis | Inner cell mass | Trophoblast of blastula |

| Decidua parietalis | Canal of the cervix uteri | Stratum compactum of decidua | Stratum spongiosum of decidua | Stratum basalis of decidua |

FIG. 42 Diagrammatic coronal section of uterus, showing the various parts of the decidua and a blastula embedded in it.

Implantation and Formation of Placenta

When the blastula is formed, its trophoblast first sticks to the surface of the decidua, proliferates and produces enzymes which erode the epithelium of the decidua, so as to implant itself in the compact layer [Figs. 31 and 32]. It may penetrate the decidua at any point on the wall of the uterine cavity; but it usually enters at some point on the wall of the fundus or of the intestinal or vesical surface. The aperture through which it passes is closed by proliferation of a portion of the tropho-blast which acts as a sealing plug—the **operculum** (Teacher, 1924); and there is sometimes a **closing coagulum** [Fig. 34], which may be partly blood clot, superficial to the operculum and projecting from the aperture. Abnormally, implantation may occur lower down on the uterine wall, so that the placenta overlies the cervical canal: this is termed **placenta praevia.** Implantation may also occur in a site outside the uterus, such as the uterine tube; this is known as ectopic gestation.

The portion of the decidua in which the blastula is embedded soon becomes thicker than the other parts of the decidua;

it is separated by the blastula into an internal part, between the zygote and the uterine lumen, called the **decidua capsularis,** and an external part between the blastula and the myometrium—the **decidua basalis.** The remainder of the decidua, lining the uterus—by far the larger portion—is the **decidua parietalis** [Fig. 42].

As soon as the blastula becomes embedded in the decidua its trophoblast undergoes rapid proliferation and becomes several cells thick wherever it is in direct contact with maternal tissue. Presently the trophoblast differentiates into two layers—an inner cellular layer or **cytotrophoblast,** and an outer syncytical layer or **syncytiotrophoblast** (sometimes incorrectly called plasmoditrophoblast). In the syncytial layer, remnants of cell boundaries are visible by electron microscopy—an indication of the origin of the multinucleated syncytioplasm by fusion of cells which are derived from the division of cytotrophoblast cells (Boyd and Hamilton, 1970).

As development proceeds, the trophoblast, now the **chorionic epithelium,** increases in thickness and continues to invade and destroy the decidua, probably by enzyme action.

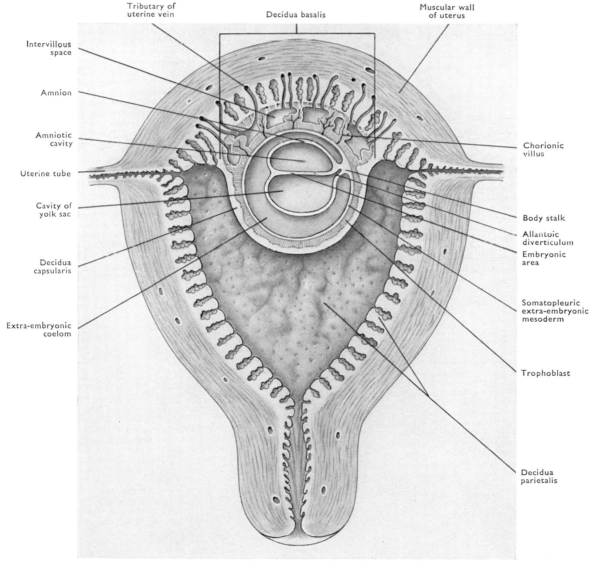

Fig. 43   Diagrammatic coronal section of a pregnant uterus after the formation of the intervillous spaces.

At this stage the decidual reaction becomes apparent. Its characteristic feature is the multiplication of the connective tissue cells of the endometrium which become swollen and have the appearance of epithelial rather than connective tissue cells. These **decidual cells** [FIGS. 33 and 36] contain glycogen and lipoids, and they congregate in close proximity to the advancing trophoblast. If the balance between trophoblast and decidua is upset, then invasion may proceed without check—a rare occurence that may result in a chorionepithelioma, one of the most malignant tumours known—or, on the other hand, the zygote may fail to implant properly and be cast off as an early abortion.

As the invasion of the decidua extends, the syncytiotrophoblast erodes the walls of the maternal blood vessels and itself becomes permeated with **lacunae** or spaces into which the maternal blood flows. The maternal blood does not clot in the trophoblastic lacunae and thus begins to circulate slowly through them, so forming a **haemochorial** type of placenta. The type of placenta depends on the extent to which the maternal tissues are destroyed by the action of the trophoblast: when the uterine blood vessels are broken into, so that chorion comes into direct contact with the maternal blood, the placenta is of the haemochorial type (Mossman, 1937; Boyd and Hamilton, 1952; Amoroso, 1952).

The spaces in the syncytium enlarge rapidly after the maternal blood begins to circulate within them, and the trophoblast becomes divided into three parts : (1) the **primary chorionic villi,** which lie between adjacent blood spaces and are formed by the proliferation and extension of cytotrophoblastic cells into the syncytiotrophoblastic trabeculae separating the lacunae; (2) the parts which lie in contact with the mesoderm of the chorion and form, with the mesoderm, the

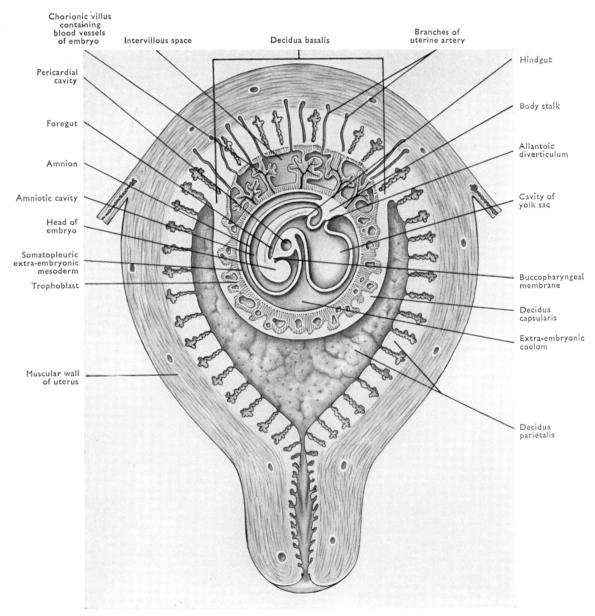

FIG. 44 Diagrammatic coronal section of a pregnant uterus at period of formation of embryo. The amnion has begun to enlarge.

chorionic plate; (3) the **cytotrophoblastic shell** (Hamilton and Boyd, 1960), which covers the maternal tissues and forms the outer boundaries of the blood spaces. The cytotrophoblastic shell is produced by further proliferation of the cytotrophoblastic cells within the primary villi towards the junctional zone between the syncytiotrophoblast of these villi and the decidua basalis. Here, the cytotrophoblastic cells mushroom outwards, and fusion between these lateral proliferations from adjacent villi constitutes the cytotrophoblastic shell. The lacunar blood spaces in the trophoblastic syncytium then become confluent to form the **intervillous space** of the placenta [FIGS. 45 and 47].

After a time each primary villus differentiates into a cellular core and syncytial periphery, and thereafter the villi are invaded by the mesoderm of the chorion and are thus con-

verted into **secondary chorionic villi** [FIGS. 43–45]. The secondary villi consist, therefore, of a mesodermal core continuous with the mesoderm of the chorion, and covered by a layer of cytotrophoblast and a layer of syncytiotrophoblast.

The umbilical arteries extend through the body stalk into the mesoderm of the chorion, and branches from them enter the mesodermal cores of the villi, which thus become vascularized [FIG. 46].

The proximal end of each villus is continuous with the chorionic plate of the intervillous space, formed by the chorion; the distal end is attached to the cytotrophoblastic shell, which forms the outer boundary of the intervillous space and is fused with the maternal decidual tissue.

In the region of the decidua basalis, the villi become longer and more complex, each villus then constituting a **truncus**

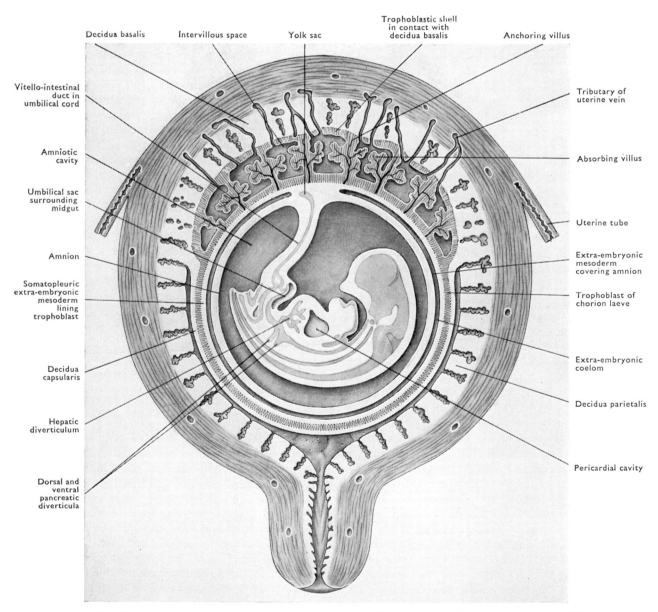

FIG. 45 Diagrammatic coronal section of a pregnant uterus after formation of the umbilical cord. The expanding amnion has almost obliterated the part of the extra-embryonic coelom which lies between it and the chorion. The expansion of the decidua capsularis has brought it in contact with the decidua parietalis, so that the uterine cavity is almost obliterated.

Syncytiotrophoblast   Cytotrophoblast   Extra-embryonic mesoderm

Branches of umbilical blood vessels

FIG. 46   Chorionic villus of a 5-mm. human embryo (× 760).

chorii or main stem villus of the later, definitive placenta. With increasing length of the trunci there is concomitant longitudinal division of their maternal extremities to produce **rami chorii** of second or third order, which in turn subdivide to form **ramuli chorii** or minor villous stems. After a time the ramuli send out numerous branches into the intervillous space, and thus increase greatly in complexity [FIGS. 43–45]. In this way two sets of secondary villi are differentiated: (1) the **anchoring villi** which cross from the chorion to the trophoblastic shell and are attached to it by cell columns which are the outer parts of the primary villi not yet invaded by the primary mesoderm; (2) **free** or **absorbing villi** which extend from the sides of the original ramuli into the blood in the intervillous space [FIGS. 45 and 47].

The development of the chorionic villi does not proceed equally in all parts of the membrane. The trophoblast thickens only where it is in direct contact with the uterine tissues. The part of the blastocyst containing the inner cell mass (the embryonic pole of the blastocyst) advances first into the decidua. Thus this trophoblast which has the body stalk of the embryo—through which vascular connexions are established—attached to it, is the first part to be affected, and maintains the lead thus acquired [FIGS. 31–34]. As implantation proceeds and the decidua closes over the abembryonic surface of the zygote, the trophoblast proliferates on that surface also. For a time, therefore, the whole surface of the chorion is covered with villi. But the capsular portion of the decidua has a relatively poor maternal blood supply which is progressively diminished as the chorionic vesicle continues to

expand; thus the villi in that region undergo atrophy and disappear. When these degenerative changes have occurred, the portion of the chorion in association with the thinned decidua capsularis presents a relatively smooth surface and is known as the **chorion laeve**. In the meantime the decidua basalis increases in thickness, and the villi associated with it increase in size and in the complexity of their branches. The portion of the chorion from which these large villi spring is termed the **chorion frondosum**. It is this portion of the chorion which takes part in the formation of the **foetal portion of the placenta**, the **maternal (uterine) part** of that organ being formed by the decidua basalis.

The placenta, therefore, is formed mainly by the chorion frondosum and to a small extent only by maternal tissues; the interchanges between the foetal and the maternal blood take place in the substance of the foetal part of the placenta through the trophoblast which covers the surfaces of the villi. Food and oxygen pass from the maternal to the foetal blood, whilst the waste products of foetal metabolism pass from the foetal to the maternal blood.

As the growth of the embryo and the distension of the amnion continue, the amnion is gradually forced against the chorion and fuses with it [FIG. 97]. When fusion is complete the extra-embryonic coelom is obliterated and the zygote contains only one extra-embryonic cavity—the amniotic cavity—which contains the foetus surrounded and supported by the amniotic fluid [FIG. 45].

At this period the cavity of the amnion is bounded by a wall composed of the fused amnion, chorion, and decidua. As the distension of the amnion proceeds still further, the decidua capsularis bulges more and more into the cavity of the uterus, until it is forced against the surrounding wall of the uterine cavity, and fuses with the decidua parietalis, obliterating the cavity of the uterus. This fusion takes place towards the end of the second month, and as soon as it has occurred the discoid mass of placental tissue is continuous at its margin with the fused amnion, chorion, and decidua parietalis [FIG. 45].

After the second month the foetus lies in the **amniotic cavity**, which is bounded by the fused amnion, chorion, and uterine wall except where the cavity of the body of the uterus communicates with the canal of the cervix; there the amniotic cavity is bounded by a membrane formed by the fused amnion, chorion laeve, and decidua capsularis only.

Completion of Placenta

Each absorbing villus consists of a vascular, mesodermal core covered by cytotrophoblast and syncytiotrophoblast, the latter lying next to the maternal blood in the intervillous space. As development proceeds and the intervillous space becomes larger, the villi become longer and more complicated, and at the same time the cellular layer of the trophoblast largely disappears, until in the majority of the villi the syncytiotrophoblast alone covers the mesodermal core.

In still later stages, degeneration occurs not only in the villi, but also in the chorionic plate and in the cytotrophoblastic shell. One of the results of the degenerative process is the deposition of fibrinoid material in the place originally occupied by the trophoblast; the fibrinous layers on the surfaces of adjacent villi adhere together, and the villi thus connected fuse into masses of intermingled fibrinous and vascular tissue.

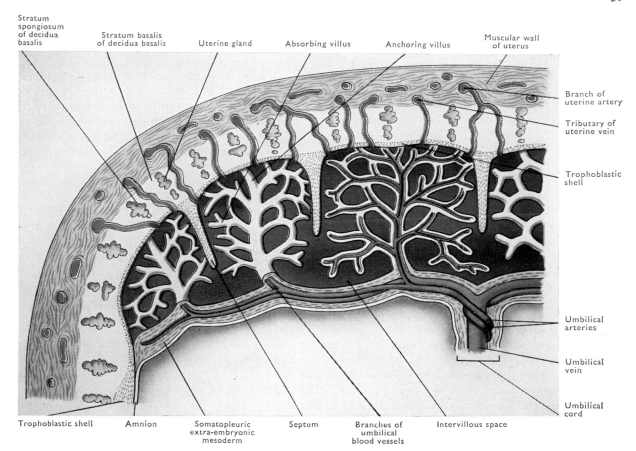

Stratum spongiosum of decidua basalis

Stratum basalis of decidua basalis

Uterine gland

Absorbing villus

Anchoring villus

Muscular wall of uterus

Branch of uterine artery

Tributary of uterine vein

Trophoblastic shell

Umbilical arteries

Umbilical vein

Umbilical cord

FIG. 47 Diagram of the structure of the complete placenta.

Trophoblastic shell   Amnion   Somatopleuric extra-embryonic mesoderm   Septum   Branches of umbilical blood vessels   Intervillous space

When the **foetal (chorionic) part** of the placenta is completed it consists of: (1) the chorionic plate closing the intervillous space internally; (2) the villi; (3) the intervillous space; and (4) the cytotrophoblastic shell, which closes the intervillous space externally and is perforated by the maternal vessels passing to and from the space. The cytotrophoblastic shell, together with the decidua basalis, constitutes the **basal plate** of the placenta, the site of junction between maternal and foetal tissues. Since the embryo and trophoblast constitute a graft with a genetic endowment which is different from that of the mother, some immunological reaction might be expected. Such a reaction would damage the embryo or placenta, and perhaps even cause their death and expulsion from the uterus. There are many possible reasons why this does not normally occur, but the most likely explanation is the absence of antigenicity in the trophoblast.

The **maternal (uterine) part** of the complete placenta consists of the decidua basalis of earlier stages. It is fused internally with the cytotrophoblastic shell to constitute the basal plate.

Projections of the basal plate into the intervillous space form incomplete septa which serve to separate the placenta into lobes or cotyledons—from ten to thirty-eight in number. The lobes are spoken of as placental units, and each contains several trunci chorii from which the free villi branch in the manner of a tree [FIG. 47]. Most of the free villi take a recurved course back towards the chorion (Spanner, 1935).

The maternal blood vessels pass from the muscular wall of the uterus directly into the placenta, where they traverse the basal plate and open into the intervillous space. The arteries usually open on or near the septa and the veins begin in the areas between them. The veins of the pregnant uterus are greatly enlarged.

At the end of pregnancy the membranes, under the increased pressure of the amniotic fluid produced by muscular contraction of the uterus, dilate the cervix, and when they rupture the fluid is expelled through the vagina—the 'breaking of the waters'. Next, the child is born, but it remains attached to the placenta by the umbilical cord, which is usually ligatured in two places and then divided between the ligatures. Later the ruptured membranes and placenta (the after-birth) are expelled from the uterus.

Detachment of the placenta is caused by contraction of the myometrium and rupture of the stratum spongiosum of the decidua. As the detached placenta is expelled, the decidua parietalis is torn through along the plane of the stratum spongiosum. The fused amnion and chorion and the inner part of the decidua parietalis, which are attached to the margin of the placenta, constitute 'the membranes', and are expelled with it.

Placenta at Birth

At birth the placenta weighs about 500 g., has a diameter of 10 to 24 cm., and is about 3 cm. thick at its centre. It may be fairly uniform in thickness throughout, but usually thins off rapidly at the periphery where it is continuous with the 'membranes'. Its inner surface is covered with the amnion, which fused with the chorion towards the end of the second month of pregnancy. Its rough, outer surface is formed by the remains of the stratum spongiosum of the decidua, and is divided into areas by fissures which correspond in position with the septa by which the organ is divided into lobes. One or more accessory (succenturiate) lobes of the placenta may occur occasionally.

EMBRYONIC AREA

When the extra-embryonic coelom is developed, the zygote then consists of three vesicles, one large and two small. The large **chorionic vesicle** is bounded by the chorion, composed of the trophoblast lined with primary mesoderm; it contains the two small vesicles projecting into the coelom and connected by the body stalk to the part of the chorion furthest from the uterine cavity. The **ectodermal** or **amniotic vesicle** and the **entodermal** or **yolk sac vesicle** remain in contact where the primary mesoderm has not penetrated between the original ectodermal and entodermal elements of the inner cell mass; here they form a bilaminar plate known as the **embryonic area** or **germ disc**—the rudiment of the embryo itself. The ectodermal lamina is **embryonic ectoderm**—it will take part in the formation of the embryo; the **amniotic ectoderm** completes the ectodermal vesicle. Similarly, the entodermal vesicle consists of a lamina of **embryonic entoderm** and extra-embryonic or **yolk sac entoderm** [FIGS. 36 and 41]. The embryonic area is indicated as soon as the amniotic cavity is formed; it is well defined in the classical Peters embryo of 13 days, and is clearly demarcated in all young embryos in which the secondary yolk sac is formed, for example, the $13\frac{1}{2}$-day embryo [FIG. 35] and the embryos of 15 days [FIG. 37] and $16\frac{1}{2}$ days [FIG. 36] in which the primitive streak [see below] is already in evidence.

With the formation and spread of the intra-embryonic mesoderm from the primitive streak [p. 43], the embryonic area becomes trilaminar.

DIFFERENTIATION OF THE EMBRYONIC AREA AND FORMATION OF PRIMARY AXIAL STRUCTURES

As the embryonic area is the area of contact between the ectodermal and the entodermal vesicles it is, at first, circular in outline. As growth continues the area becomes oval, and a linear, median thickening of the ectoderm, called the **primitive streak,** appears in that part of the oval which becomes the caudal end of the area [FIGS. 48, 51, 54, and 86]. Bilateral symmetry is thus impressed upon the embryonic area and the line of the axial structures of the developing embryo is indicated. From this point in development we can speak of cephalic and caudal ends of the embryonic area, which elongates and for a time assumes a pear-shaped outline [FIGS. 48 and 86].

The deeper cells of the primitive streak adhere to the entoderm and migrate laterally into the interval between the ectoderm and entoderm of the embryonic area to form the **intra-embryonic** or **secondary mesoderm;** they also give rise, after a series of remarkable developmental events, to the **notochord**—an axial structure characteristic of all vertebrates

—the formation of which is presaged by a proliferation of the cephalic end of the primitive streak, the **primitive node** (Hensen's node). A small dimple, the **primitive pit,** soon appears in the dorsal aspect of this node.

Immediately after the formation of the primitive streak an ectodermal groove, called the **neural groove,** appears in the cephalic part of the embryonic area [FIGS. 53, 60, and 88]. It is formed by the longitudinal folding of a thickened plate of ectoderm—the **neural plate**—which is the rudiment of almost the whole nervous system, the only exceptions being the olfactory nerves, some parts of the ganglia of the cranial nerves, and the end-organs of the sensory nerves. From it also are derived the neurolemmal cells of peripheral nerve fibres, the medullary cells of the suprarenal glands, and other structures [p. 51]. The lateral parts of the neural plate form the **neural folds.** Almost from the first, the cephalic extremities of the neural folds are united a short distance behind the cephalic end of the embryonic area. Their caudal ends, which remain separate for a time, embrace the cephalic part of the primitive streak, and later converge and fuse together.

Formation of Notochord and Intra-embryonic Mesoderm

The notochord and the intra-embryonic mesoderm are formed from the primitive streak—the notochord from its

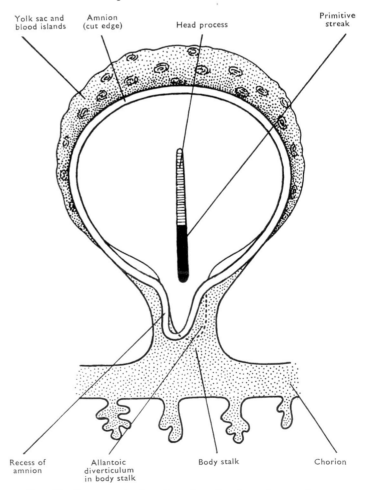

Yolk sac and blood islands   Amnion (cut edge)   Head process   Primitive streak

Recess of amnion   Allantoic diverticulum in body stalk   Body stalk   Chorion

FIG. 48 Reconstructed surface view of embryonic area of young human embryo of 18 days to show position of head process and its relation to primitive streak.

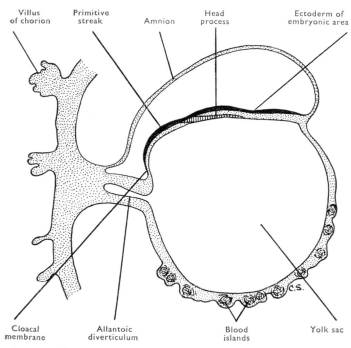

Villus of chorion — Primitive streak — Amnion — Head process — Ectoderm of embryonic area

Cloacal membrane — Allantoic diverticulum — Blood islands — Yolk sac

C.S.

FIG. 49 Reconstructed median section of same young human embryo as in Fig. 48. ×50. (Thompson and Brash, 1923.)

cephalic end, and the mesoderm from its lateral margins and later its caudal end. The primitive streak probably represents the fused and elongated lips of the blastopore in lower vertebrates, and as a continued line of contact between ectoderm and endoderm remains, in part, as the cloacal membrane [p. 49].

As soon as the primitive streak is established its cephalic end becomes a node or centre of growth—the **primitive (Hensen's) node**—by means of which mesoderm is produced. The portion of the mesoderm formed by the activity of the primitive node in the midline is the dorsomedian portion, from the posterior part of the roof of the nose to the caudal end of the trunk. The mesoderm of the perineum and the ventral body wall from the perineum to the umbilicus are formed from the caudal part of the primitive streak. Nevertheless, the primitive streak undergoes little or no increase in length; indeed, as growth continues, it becomes relatively shorter compared with the total length of the embryonic region; for the new material, formed by its lateral and cephalic borders, is transformed into the tissues of the embryo as rapidly as it is created.

Notochord

The notochord, or primitive skeletal axis, is formed indirectly from the primitive node. It appears first as a narrow, median rod of cells, called the head process, which projects headwards from the primitive node between the ectoderm and the entoderm [FIGS. 48, 49, and 53], and continued proliferation from the primitive node [FIG. 53] adds to its caudal end.

The head process is at first solid, but presently a cavity appears in the rod of cells and tunnels it from the ectodermal surface of the primitive pit headwards, though not quite to its cephalic end [FIG. 50], to form the **notochordal canal**. The floor of the tunnel, i.e., the cells next the cavity of the entodermal vesicle, together with the roof of the entodermal vesicle in the midline, then break down, the roof flattens out, and the head process is then represented by a plate of cells

intercalated in the dorsal wall of the entodermal vesicle—the **notochordal plate** [FIGS. 50 and 55]; for a time, therefore, there is a passage from the cavity of the amniotic vesicle through the primitive pit to the cavity of the entodermal vesicle, the **amnio-enteric canal,** later to become the transitory **neurenteric canal** when the neural folds [p. 51] close over it.

At a later period the notochordal cells become separated from the entoderm, and they again form a cylindrical rod of cells in the median plane which has been more or less moulded off from the dorsal part of the entodermal sac [FIG. 57 B], and lies between the floor of the ectodermal neural groove and the reformed entodermal roof of the primitive alimentary canal. This is the true notochord. The cephalic end of the notochord is continuous with the ectoderm and entoderm of a small portion of the embryonic area which lies immediately beyond the cephalic end of the neural groove. This part of the embryonic area, indicated at an earlier date by enlargement of its cells, has been named the **prochordal plate** [FIG. 50]. The ectoderm and entoderm of the prochordal plate are in direct contact without mesoderm between its two layers. This bilaminar region, because it afterwards forms the boundary membrane between the cephalic end of the primitive entodermal canal and the primitive oro-nasal cavity or **stomodaeum,** is called the **buccopharyngeal membrane** [FIG. 62]. It disappears in the fourth week of embryonic life. The caudal end of the notochord remains continuous with the primitive streak until the formation of the neural tube is completed.

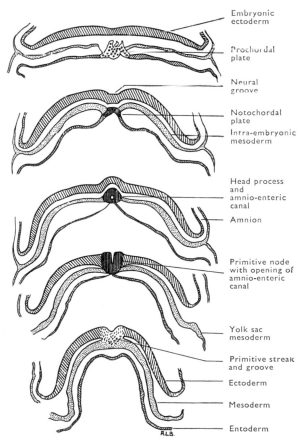

Embryonic ectoderm

Prochordal plate

Neural groove

Notochordal plate

Intra-embryonic mesoderm

Head process and amnio-enteric canal

Amnion

Primitive node with opening of amnio-enteric canal

Yolk sac mesoderm

Primitive streak and groove

Ectoderm

Mesoderm

Entoderm

FIG. 50 Transverse sections of 18-day presomite embryo to show relation of head process to primitive streak and prochordal plate. ×70. (After Heuser, 1932a.)

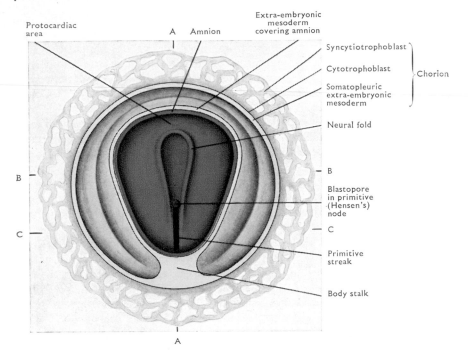

FIG. 51 Diagram of dorsal surface of embryonic area of early zygote after removal of part of the chorion and amnion.

of the intervertebral discs. The extension of the notochord into the region of the head is of interest from a morphological and possibly also from a practical point of view. It extends through the base of the cranium from the anterior border of the foramen magnum into the posterior part of the body of the sphenoid bone. Its presence in the posterior part of the skull-base suggests that that region was, primitively, of vertebral nature. As the notochord lies partly within the occipital bone and partly between it and the dorsal pharyngeal wall, it pierces the basilar portion of the occipital region first from within outwards and then in the reverse direction. It lies, therefore, for a short distance, on the ventral surface of the rudiment of the occipital bone, in the dorsal wall of the pharynx; proliferation of remnants of its cranial portion may give rise to tumours known as chordomata.

After a time the cylindrical notochordal rod is surrounded by secondary mesoderm which becomes converted into the vertebral column. As the vertebral column is formed the notochord is enlarged in the regions of the intervertebral discs and for a time assumes a nodulated appearance.

Ultimately the notochord disappears as a distinct structure, but remnants of it are believed to persist as the pulpy centres

DIFFERENTIATION OF THE INTRA-EMBRYONIC MESODERM

It has already been noted that the **primary mesoderm**—derived either from the inner cell mass or directly from the trophoblast—is formed outside the embryonic area, and is therefore **extra-embryonic mesoderm.**

After the formation of the embryonic area, the ectodermal thickening which forms the primitive streak becomes the main source of the mesoderm; and as this mesoderm appears after the primary mesoderm, and is formed in the embryonic area,

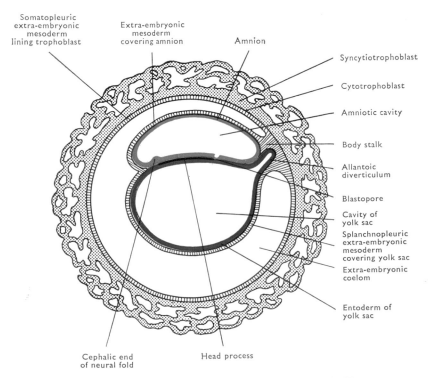

FIG. 52 Longitudinal section of zygote along line A–A in Fig. 51.

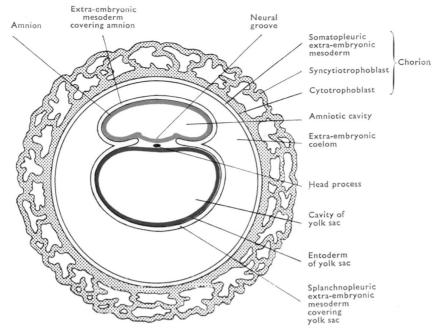

FIG. 53 Transverse section of zygote along line B–B in Fig. 51.

it may be termed **secondary mesoderm** or **intra-embryonic mesoderm.**

The deeper cells of the primitive streak [FIGS. 51 and 54], which are in contact with the entoderm, produce the intra-embryonic mesoderm, which passes from the primitive streak into the interval between the ectoderm and the entoderm [FIG. 54]. The superficial cells of the primitive streak remain as part of the surface ectoderm of the embryo.

The intra-embryonic mesoderm soon forms a continuous sheet of cells which spreads laterally and headwards in the embryonic area on each side of the median plane. Each of these lateral sheets is thickest where it abuts against the notochord and the wall of the neural groove, and thinnest at its peripheral margin, where it is continuous with the extra-embryonic mesoderm [FIG. 55].

At the cephalic end of the embryonic area the medial margins of the mesodermal sheets pass around the prochordal plate and fuse together across the median plane cephalic to it, forming a transverse bar called the **pericardial (cardiogenic) mesoderm** [FIG. 61 A] because the pericardium is

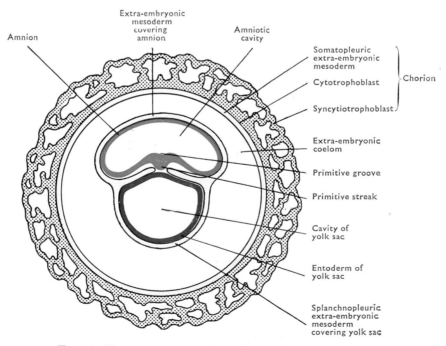

FIG. 54 Transverse section of zygote along line C–C in Fig. 51.

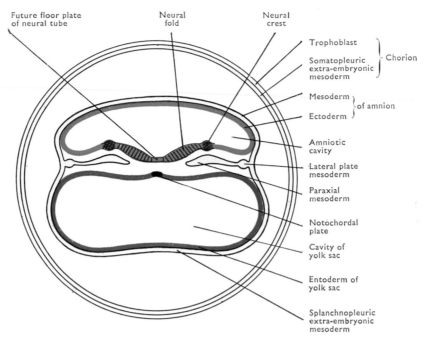

FIG. 55 Diagrammatic transverse section of zygote to show formation of neural folds and position of intra-embryonic mesoderm.

afterwards developed in it. The area in which this mesoderm lies is named the **pericardial (protocardiac; cardiogenic) region** of the embryonic area [FIG. 51].

Between the prochordal plate and the cephalic end of the primitive streak the medial margins of the mesodermal plates are separated from each other by the notochord and the neural groove [FIG. 56], and still more caudally they are continuous with the sides of the streak [FIG. 54].

Intra-embryonic Coelom

After the intra-embryonic mesodermal sheets are definitely established, a series of cleft-like cavities appears in their peripheral parts. These cavities, on each side, soon run together, splitting the mesoderm to form the rudiments of the intra-embryonic coelom [FIG. 56].

At the lateral border of the embryonic area on each side the mesoderm remains unsplit for a time, and separates the intra-embryonic from the extra-embryonic coelom. This septum soon disappears, and the two coelomic cavities become continuous [FIG. 57B].

The intra-embryonic coelom extends medially also, but the medial extension ceases at some distance from the median plane, except at the cephalic end of the embryonic area, where the two halves of the coelom become continuous with each

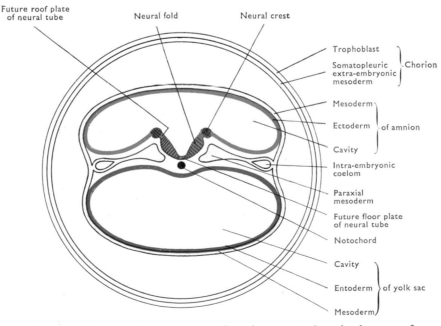

FIG. 56 Diagrammatic transverse section of zygote to show development of neural groove and intra-embryonic coelom.

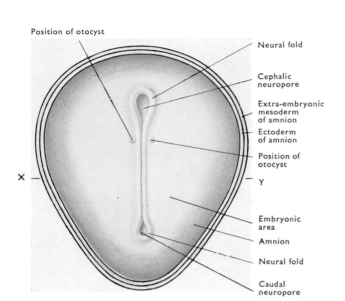

FIG. 57A Diagram of embryonic area showing closure of neural groove with cephalic and caudal neuropores.

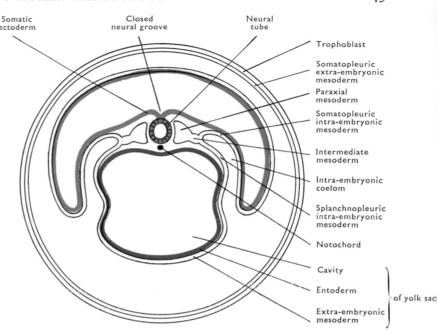

FIG. 57B Corresponding transverse section through the plane X–Y in Fig. 57A. The intra-embryonic coelom has appeared and communicates with the extra-embryonic coelom.

other through the interior of the pericardial mesodermal bar [FIGS. 61B and 62]. In this region the septum of mesoderm separating the intra- and extra-embryonic parts of the coelom persists at the anterior margin of the embryonic disc to become the **septum transversum.**

As the intra-embryonic coelom is forming and extending, a longitudinal constriction appears in each half of the mesoderm a short distance from its medial border. This constriction separates each plate into three parts : (1) a medial bar called the **paraxial mesoderm,** which lies at the side of the neural groove and the notochord ; (2) the constricted portion, called the **intermediate cell mass** (intermediate mesoderm) ; and (3) the part lateral to the constriction called the **lateral plate** [FIG. 57B].

The intra-embryonic coelom is confined, in the human embryo, to the lateral plate, which it divides into a superficial layer next the ectoderm, called the **somatic (somatopleuric) mesoderm,** and a deeper layer next the entoderm, called the **splanchnic (splanchnopleuric) mesoderm.** Ectoderm and somatic mesoderm together constitute the **somatopleure ;** entoderm and splanchnic mesoderm together are known as the **splanchnopleure.**

The medial borders of the somatic and splanchnic mesoderm are continuous with each other round the medial border of the coelom. The lateral border of the somatic mesoderm is continuous, at the margin of the embryonic area, with the extra-embryonic mesoderm which covers the outer surface of the amnion ; the lateral border of the splanchnic layer is continuous with the extra-embryonic mesoderm on the wall of the extra-embryonic portion of the entodermal vesicle or yolk sac.

Paraxial Mesoderm

Each paraxial mesodermal bar soon assumes the shape of a prism lying lateral to the neural tube. The cephalic portion of each paraxial bar, as far as the middle of the hindbrain [p. 51], remains unsegmented, but the more caudal part is divided by

a series of transverse clefts into a number of segments, called the **mesodermal somites** [FIGS. 58 and 64]. The first cleft appears in the region of the hindbrain, and the others are formed successively, each caudal to its predecessor. Only three or four somites lie in the region of the head. The segmentation of the paraxial bars begins before they have reached their full length, and somites continue to be separated off as the paraxial bars are extended by addition to their caudal extremities from the nodal point at the cephalic end of the primitive streak. The total number of somites in the human embryo is thirty-eight or thirty-nine, not counting a few that develop in the tail and soon disappear.

When the somites are first defined they are solid masses of

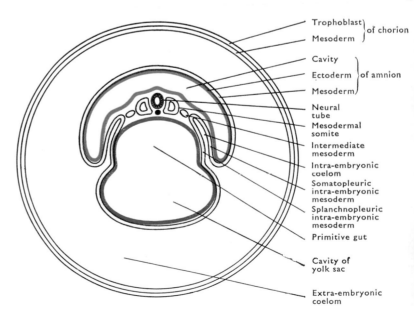

FIG. 58 Diagrammatic transverse section of zygote shown in Fig. 64.

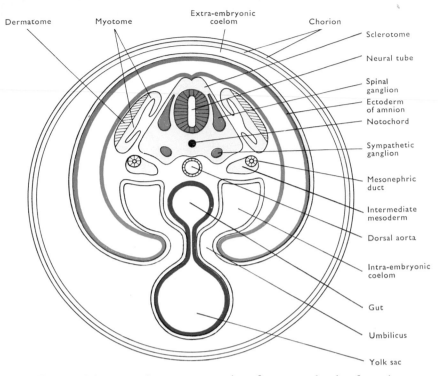

FIG. 59 Diagrammatic transverse section of a zygote showing formation
of umbilicus and ventral extension of amnion.

cells, but in a short time a cavity appears in each mass. This is called the **myocoele** because, in addition to other structures, muscles are derived from the somites.

The ventromedial portion of the hollow mesodermal somite is known as the **sclerotome** (Goodsir, 1857), since it is responsible for the formation of 'hard' skeletal structures. The cells of the scleratogenous section of the somite undergo rapid proliferation and assume the character of embryonic connective tissue to which the special name of **mesenchyme** has been given. Some of these cells invade the myocoele while others migrate towards the notochord. Finally, the scleratogenous cells separate from the remainder of the somite, and, as they increase in number, they migrate around the sides of the notochord and neural tube (which has been formed in the meantime from the neural groove) and mingle with those of

the opposite side and with those derived from adjoining cephalic and caudal somites [FIG. 59]. In this way a continuous sheath of mesoderm is formed around the neural tube and the notochord; it is called the **membranous vertebral column,** and from it differentiate the vertebral column and its ligaments, and the dura mater of the brain and spinal medulla.

The part of a mesodermal somite left after the separation of the sclerotome is called a **myotome.** Each myotome gives rise to a flat plate with incurved dorsal and ventral margins, known as a muscle plate because from these plates voluntary muscle fibres are derived [FIG. 59]. The outer portion of each of the myotomes develops into the cells of subcutaneous connective tissue and dermis; consequently it is spoken of as a cutis plate, or **dermatome.** The dermis of the skin is

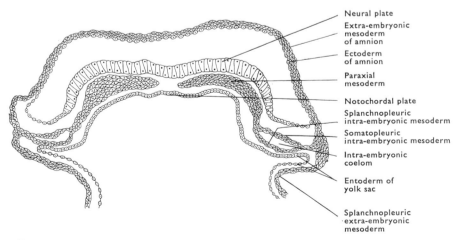

FIG. 60 Diagram of a transverse section through the M'Intyre embryo (Bryce, 1924).

thought thus to arise segmentally, each segmental contribution derived from a dermatome corresponding to the area of skin supplied by one spinal nerve—also called a dermatome.

Intermediate Cell Mass

The continuous tract of cells, lateral to the paraxial mesoderm on each side, to which this name is given, remains unsegmented; but as it gives rise to a series of excretory structures it corresponds to the segmented portions of mesoderm in lower forms known as **nephrotomes.** The intermediate cell mass gives rise to the urogenital system [FIG. 670], with the exception of the germinal cells of the gonads, the lining of the urinary bladder and the urethra, and the prostate.

Lateral Plates

The cells of the lateral plates give origin to : (1) the lining mesothelial cells of the great serous cavities of the body—the pleurae, the pericardium, and the peritoneum; (2) the majority of the connective tissues (with the exception of those of the vertebral column and the head); (3) the greater part or the whole of the mesoderm of the limbs; and (4) the plain muscle fibres of the walls of the alimentary canal and the blood vessels. Most of these tissues are derived from mesenchymatous cells budded off from the lateral plates.

Cephalic Mesoderm. It has already been noted that the mesoderm of the head becomes segmented only in the region of the caudal part of the hindbrain, where four cephalic mesodermal somites are formed on each side. From the scleratogenous portions of these somites the occipital part of the skull and the corresponding part of the dura mater of the brain are developed; and their muscle plates give rise to the intrinsic muscles of the tongue. Three myotomes develop in front of the otic capsule on each side and give rise to the orbital muscles (musculi bulbi).

The unsegmented part of the cephalic mesoderm gives rise to the remaining muscles and connective tissues of the head region. It has been suggested that this mesoderm, which is concerned in the development of the pharyngeal arches, arises from the neural crest [p. 51].

Mesenchyme

In the early stages of development many cells, of irregular shape and wandering habits, appear as a loose mesh between the more definite layer of mesoderm and the adjacent ectoderm or entoderm. They are called mesenchyme cells. The scleratogenous cells that wander out from the somites are of this nature; the remainder of the mesenchyme is derived largely from the lateral plates of mesoderm, though it may have other sources, possibly even from ectoderm and entoderm.

The complete role of the mesenchyme in development has not yet been elucidated, but in addition to connective tissues in general, the tissues of the vascular and lymphatic systems, the reticulo-endothelial system in general, and plain muscle fibres in the walls of the alimentary canal and elsewhere, it appears to be concerned with the development of striated muscles in the limbs, which have not been proved to be directly derived from myotomes.

Germ Layers and their Derivatives

The formation and development of the embryonic area have now been traced to a point at which the general relation of the three main cellular sheets or layers to each other can be understood [FIGS. 53–59]. The **ectoderm,** the **entoderm,** and the **mesoderm** are distinct sheets of cells in the early stage of development, each with its own part in the formation of the embryo and its organs; they are therefore known as the primary **germ layers.** It must be emphasized, however, that the germ layers are neither independent elements in the structural organization of the embryo, nor even completely specific; on the contrary, their interaction and co-operation are essential for normal development; and, since the destiny of any particular group of cells is partly due to its position in the embryo, one group may even, under abnormal or experimental conditions, deputize for another of the same or a different germ layer.

Before the description of the formation of the embryo and of the initial stages in the development of the principal organs, it is useful to summarize the parts played by the three germ layers. In general, it may be said that there is a functional distinction between them : thus the **ectoderm,** the original external covering of the body, has a protective function, and, since it is in contact with the external environment, it is also the source of the essential elements of the sense organs and of the whole of the nervous system. The **entoderm,** the lining of the alimentary and respiratory tracts, has digestive and absorptive functions; and the **mesoderm,** between the other two, gives origin to connective and muscular tissues, is the basis of the circulatory system, and provides for excretion and reproduction by means of the urogenital organs. The table on page 48 gives some details of the derivatives of the germ layers.

THE FORMATION OF THE EMBRYO

The transformation of the relatively flat embryonic area into the form of the embryo is due, in the first instance, to the rapid extension of the area compared with the slow growth of the immediately adjacent parts; and the later modelling of the various parts of the embryo is due to different rates of growth in different regions.

By the rapid proliferation of cells from the cephalic end of the primitive streak, the surface length of the embryonic area is increased, whilst its cephalic and caudal ends remain relatively fixed; consequently the area bulges dorsally into the amnion. At the same time, the cephalic end of the neural groove becomes convex dorsally, so that it appears to rise out of the embryonic area and projects beyond its cephalic border. As a result of this movement the buccopharyngeal and the pericardial areas become reversed in position, and a head fold is formed. This fold is bounded dorsally by what is now the head of the embryo, and ventrally by the reversed pericardial region [FIG. 61B].

The same rapid growth process not only produces a **head fold,** but at the same time it forces the rest of the primitive streak over the caudal end of the embryonic area, thus forming a **tail fold.**

As the head fold and the tail fold are being produced by the longitudinal increase of the embryonic area, so flexing it

DERIVATIVES OF GERM LAYERS

ECTODERM	MESODERM	ENTODERM
Epidermis of skin; hair follicles and hairs; nails.	Connective tissues, including bone and cartilage.	Epithelium of alimentary tract (except ectodermal parts of mouth and anal canal); tongue and taste-buds.
Epithelium and myoepithelial cells of sebaceous, sweat, and mammary glands.	Synovial membranes; bursae and tendon sheaths.	
Epithelium of sense organs. Eye: retina (including ciliary and iridial parts); lens; conjunctiva; lacrimal gland and ducts. Ear: epithelium of membranous labyrinth and external acoustic meatus. Nose: epithelial lining and glands of nasal cavity and paranasal sinuses, including olfactory area and vomeronasal organ.	Serous membranes; pleura; pericardium; peritoneum; tunica vaginalis testis. Muscle—plain, striated, and cardiac (except plain muscle of iris and sweat glands).	Epithelium of glands of alimentary canal (except salivary): liver; gall-bladder; bile-passages; pancreas. Epithelium of pharyngeal organs: thyroid; parathyroids; thymus.
Nervous system (central and peripheral): nerve cells; nerve fibres; neuroglia (except microglia); ependyma; neurolemmal cells; leptomeninges (pia and arachnoid); pineal body; both lobes of hypophysis cerebri; medulla of suprarenal glands; chromaffin tissues.	Endothelium of heart, blood vessels, and lymph vessels. Blood and bone marrow. Lymphoid tissue: lymph; tonsils; lymphatic nodules.	Epithelium of respiratory tract (except nasal cavity); nasal part of pharynx; auditory tube; tympanum; mastoid antrum; mastoid air cells.
Epithelium of lips, cheeks, gums, and hard palate (including glands); salivary glands.	Spleen. Coats and contents of eye (except ectodermal parts specified).	Epithelium of urinary bladder (except trigone); urethra (except terminal part in male); prostate; bulbo-urethral glands; vagina (part).
Enamel of teeth.	Dura mater and microglia.	
Epithelium of lower one-third of anal canal; terminal part of male urethra; vestibule of vagina and vestibular glands.	Teeth (except enamel). Cortex of suprarenal glands.	
Plain muscle of iris.	Urogenital organs (except epithelium of most of bladder, of prostate, urethra, and part of vagina).	

ventrally, transverse growth of the area results in the formation of right and left **lateral folds** [FIGS. 57B and 58], the embryo rising like a mushroom into the cavity of the amnion.

The portion of the secondary yolk sac which becomes enclosed within the embryo by the folding of the embryonic area in this way is the **primitive alimentary canal.** The part which remains outside the embryo is the definitive yolk sac; and the connecting passage between the two is the vitello-intestinal duct [FIG. 66].

That portion of the primitive alimentary canal which lies in the head fold is termed the **foregut**, the part in the tail fold is the **hindgut,** and the middle portion, which at first is in open communication with the yolk sac, is the **midgut.**

As the expansion of the embryonic area and its folding proceed, the margin of the area, which remains relatively stationary, becomes the edge of an orifice on the ventral aspect of the embryo (the **umbilical orifice**) through which the primitive alimentary canal of the embryo communicates with the yolk sac, the intra-embryonic coelom is continuous with the extra-embryonic coelom, and the component structures of the body stalk enter or leave the embryo.

At first the umbilical orifice is relatively large, but as the

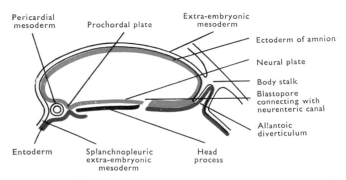

FIG. 61A Diagram of longitudinal median section of embryonic area before folding.

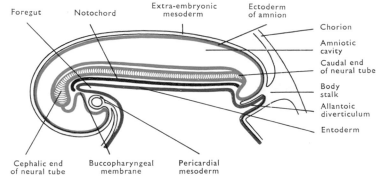

FIG. 61B Diagram of longitudinal median section of embryo after folding has commenced.

embryo rapidly extends in all directions from its margins the orifice soon becomes relatively small. Ultimately the margins of the orifice fuse together, closing the opening and forming the area of attachment of the umbilical cord which, after birth, becomes a scar on the ventral abdominal wall known as the **umbilicus** or navel.

THE EMBRYO

While the embryonic area is folding into the form of the embryo, the neural groove on its ectodermal surface is being converted into the neural tube. After the neural tube is completely closed and separated from the surface, during the fourth week, the embryo elongates, has a larger cephalic end and a smaller caudal end, and is attached by the body stalk to the chorion. Its dorsal surface is continuous and unbroken, but its ventral surface is separated into cephalic and caudal portions by the umbilical orifice. The embryo contains three cavities : (1) the cavity of the neural tube ; (2) the primitive alimentary canal, which is the embryonic portion of the entodermal vesicle included in the head, body, and tail of the embryo during the folding of the embryonic area to form the foregut, midgut, and hindgut [Figs. 61 B and 62] ; (3) the intra-embryonic coelom.

The intra-embryonic coelom consists of a median pericardial portion, which lies ventral to the foregut between the buccopharyngeal membrane and the cranial margin of the umbilical orifice, and is continued dorsolaterally into right and left portions, which extend caudally at the sides of the foregut, the midgut, and the hindgut.

The right and left portions, in addition to communicating with each other through the pericardial coelom, ventral to the foregut, also communicate with the extra-embryonic coelom through an aperture between the midgut and the lateral and caudal margins of the umbilical orifice.

By this time the embryo has become easily distinguishable from the remainder of the zygote, and indications of its general plan of organization are discernible. As yet, it has no limbs, but the general contour of the head and body are defined. It possesses a primitive skeletal axis—the notochord. On the dorsal aspect of the notochord lies the neural tube, which is the rudiment of the brain and the spinal medulla.

At the sides of the neural tube and the notochord are the mesodermal somites and the ganglia of the cranial and spinal nerves [Figs. 58 and 59].

Ventral to the notochord is the primitive alimentary canal closed at its cephalic and caudal ends by the **buccopharyngeal** and **cloacal membranes,** which separate the amniotic cavity from the foregut and the hindgut respectively [Figs. 62 and 66]. The caudal part of the hindgut becomes the **entodermal cloaca** into which open the primary excretory (pronephric) ducts [p. 61].

At the sides of the primitive alimentary canal are the right and left parts of the intra-embryonic coelom, and between the dorsal angle of each half of the coelom and the mesodermal somites of the same side lies the intermediate cell mass mesoderm, which is the rudiment of the greater part of the urogenital system [Figs. 58 and 59].

Ventral to the foregut is the mesoderm surrounding the pericardial cavity, the caudal part of which thickens to form the **septum transversum** [p. 45] which, as an unsplit portion of mesoderm, attaches the caudal portion of the foregut to the ventral abdominal wall. Ventral to the hindgut is the cloacal membrane. Between the septum transversum cranially and the cloacal membrane caudally is the umbilical orifice [Figs. 59 and 62].

THE LIMBS

For some time after its general form is well defined the embryo is entirely devoid of limbs [Figs. 90–92]. During the fifth week a slight ridge appears along each side of the embryo. On this ridge the rudiments of the limbs—the **limb buds—**

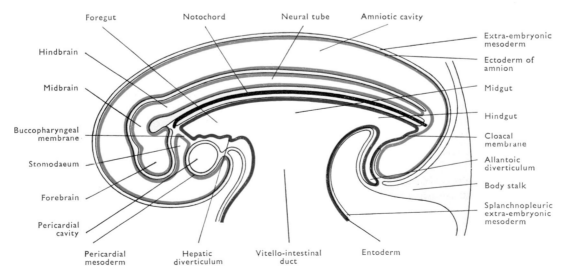

FIG. 62 Diagrammatic longitudinal section of embryo during folding

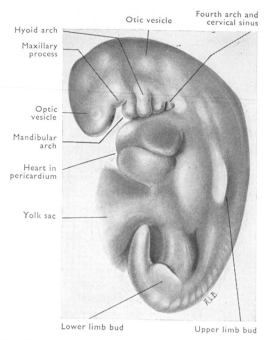

FIG. 63   Human embryo, 5·3 mm. long, to show the development of the limb buds. The rudiment of the upper limb, more advanced than that of the lower limb, appears as a longitudinal ridge on the side of the embryo. (After G. L. Streeter, *Contr. Embryol. Carneg. Instn*, 1945; Embryo no. 8066.)

are formed as secondary elevations [FIG. 63]; the upper limb precedes the lower limb in time of appearance and in the development of each of its corresponding features [FIGS. 93–95].

Each limb bud early assumes a semilunar outline, and projecting at right angles from the surface of the body, possesses dorsal and ventral surfaces, and cephalic or preaxial, and caudal or postaxial borders.

As the limb rudiment increases in length the segments of the limb are differentiated; at the same time the limbs are folded ventrally, so that their original ventral surfaces become medial, and their original dorsal surfaces lateral, with the convexities of the elbow and knee directed laterally. At a later period, on account of a rotation which takes place in opposite directions in the two limbs, the convexity of the elbow is turned towards the caudal end of the body and that of the knee towards the head [FIG. 96].

The distal segment of each limb is, at first, a flat plate with a rounded margin, but it soon differentiates into a proximal or basal part and a more flattened marginal portion [FIG. 94]. It is along the line where these two parts are continuous that the rudiments of the **digits** appear as small elevations on the dorsal surface of the limb bud about the sixth week; the ridges extend peripherally, and by the seventh week the fingers project beyond the margins of the hand segment [FIG. 95], but the toes do not attain a corresponding stage of development until the early part of the eighth week [FIG. 96]. The nails are later developments.

Each limb bud is essentially an extension from a definite number of segments of the body. It consists, at first, of a core of mesenchyme covered with ectoderm. As it grows, the ventral rami of the spinal nerves of the corresponding segments are prolonged into it, together with branches of a number of

intersegmental blood vessels. The nerves remain as the nerves of the fully developed limb, but the blood vessels are reduced in number and are modified to form the permanent main vascular trunks.

The mesenchymatous core of the primitive limb rudiment is derived mainly from somatic mesoderm of the lateral plate. As development proceeds it is differentiated into the cartilaginous and other connective tissue elements which are the rudiments of the skeletal framework and the fasciae of the fully formed limb.

The rudiments of the muscles of the limbs appear in the mesenchyme as pre-muscle masses which have no direct connexion with the myotomes of the segments to which the limb buds are related. Nevertheless, the limb muscles become innervated by the nerves of the segments to which they belong.

OUTLINE OF DEVELOPMENT OF NERVOUS SYSTEM

Early Stages

No definite trace of the nervous system is present until the primitive streak has appeared and the embryonic area has passed from a circular to a pear-shaped form. An area of thickened ectoderm, called the **neural plate,** then appears in the longitudinal axis of the embryonic area dorsal to the notochord. It begins to differentiate just caudal to the bucco-

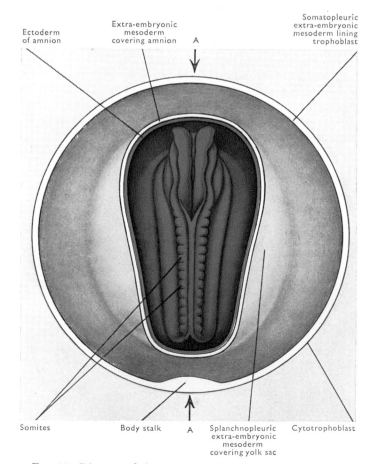

FIG. 64   Diagram of dorsal aspect of embryo, showing partial closure of the neural groove.

pharyngeal membrane, and its caudal extremity embraces the nodal end of the primitive streak. Its margins fade into the surrounding ectoderm; but, while the plate lengthens with the elongation of the embryonic area, its margins are elevated by the thickening paraxial mesoderm beneath them, and so they become distinct.

As its margins are raised the plate is necessarily folded longitudinally, and the sulcus so formed is called the **neural groove** [FIGS. 53 and 57A]. Each half of the neural plate, as it is raised to form a lateral wall for the groove, is then named a **neural fold.** After the neural groove is thus defined, the neural folds approach each other until they meet and their dorsal margins fuse in the median plane, so converting the neural plate into the **neural tube.** The ventral wall of the tube, formed by the central part of the original neural plate, is called the **floor plate:** the dorsal wall, formed where the neural folds unite, is called the **roof plate,** and it is soon separated from the surface ectoderm, which is closed over and completed dorsal to the neural tube by this fusion.

The fusion of the margins of the neural folds to form the roof plate begins in the cervical region, and from there it extends headwards and tailwards. The last parts of the roof plate to be formed are, therefore, its cephalic and its caudal extremities; consequently, for a time, the **neural canal,** which is the cavity of the tube, opens on the surface at the two ends; the openings are called the **cephalic** and the **caudal neuropores** [FIG. 57A]. Eventually, however, about the third week of embryonic life, both apertures are closed and, for a time, the neural canal becomes a completely closed cavity.

Failure of union of the neural folds to form a closed neural canal (myeloschisis or rachischisis) may occur at any point but is commonest at the two ends, and it is often associated with malformations of the skull (anencephaly) or vertebral column (spina bifida). The union of the neural folds is an example of the process of 'embryonic healing' by which growing edges come together and unite, either to form a continuous structure—as in the case of the face and palate [p. 53]—or to enclose a cavity—as in the case of the neural canal, or the lens of the eye [FIG. 998], or the otic vesicle [FIG. 1027].

As the neural folds rise and approach each other their dorsal margins give origin on each side to a column of cells which comes to lie in the angle between the neural tube and the surface ectoderm. This column is called the **neural crest** [FIGS. 55 and 56].

The neural crest is the rudiment of the cranial and spinal nerve ganglia, the sympathetic ganglia, the cells of chromaffin tissue, the medulla of the suprarenal glands and the sheath (Schwann) cells of the peripheral nerves. Melanoblasts and even some of the mesoderm of pharyngeal arches (Horstadius, 1950) are said to develop from the neural crest.

The neural tube becomes transformed into the brain and spinal medulla, and the brain portion gives origin also on each side to the essential part of the eye—the retina—and the optic nerve.

Differentiation of the Neural Tube

Before the neural plate is converted into a closed tube, an expansion of its cephalic part indicates the distinction between brain and spinal medulla.

While the cerebral portion is still unclosed, three secondary dilatations of its walls indicate its separation into three vesicles —the primitive **forebrain,** the **midbrain,** and the **hindbrain** [FIGS. 66 and 74], which are continuous with each other.

Shortly after the three sections of the brain are defined, and before it becomes a closed tube, a vesicular evagination forms at the cephalic end of each side-wall of the primitive forebrain. These evaginations are the **primary optic vesicles,** and they are the rudiments of the **optic nerves,** the **retinae,** and the **posterior epithelium of the ciliary body** and of the **iris.** As the optic vesicle approaches the surface of the head, the lens of the eye develops from the overlying ectoderm [FIG. 998], which thickens and forms a depression which is finally cut off from the surface by the union of its edges. The lens vesicle, thus produced, is received into a depression in the optic vesicle which deepens until the vesicle is transformed into the **optic cup.**

After the three brain vesicles are formed, diverticula grow out from the cephalic end of the primitive forebrain; these are the right and left **cerebral vesicles,** which are the rudiments of the **cerebral hemispheres** of the adult brain.

After their formation the cerebral vesicles expand rapidly in all directions. They soon overlap the primitive forebrain and the midbrain, and eventually the hindbrain also, and each gives off from the cephalic end of its ventral wall an **olfactory diverticulum,** which becomes converted, later, into the olfactory bulb and olfactory tract.

Flexures of Brain and Head. [FIGS. 63, 65, and 75.]

In conformity with the general curvature of the embryo, but due mainly to the rapid, unequal growth of the three brain vesicles, the head region of the embryo bends ventrally as the brain develops. At first the curvature is continuous, but very soon it is accentuated opposite the midbrain to form the **cephalic flexure;** the midbrain, which remains relatively small, becomes the most prominent part of the head, and as the

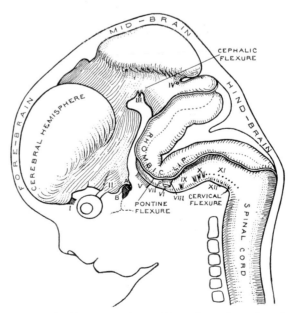

FIG. 65 Head of human embryo of 8 weeks to show the primary parts of the brain and the cephalic, cervical, and pontine flexures. (His.)
A and B, Cerebral and buccal rudiments of hypophysis cerebri
The cranial nerves are indicated by numerals.

flexure increases until its angle becomes quite acute, the fore-brain and the hindbrain approach each other until their ventral walls lie almost parallel.

At the same time a **cervical flexure** develops at the junction of the hindbrain with the spinal portion of the neural tube; it affects the whole head and keeps the facial region in close relation with the pericardium.

At a later stage the exuberant growth of the hindbrain causes that vesicle to bend in a direction opposite to the other flexures, so that its ventral wall forms a prominent convexity. This bend affects the brain only, and it is called the **pontine flexure** as it occurs in the region where the pons is formed.

The cervical flexure eventually disappears as the neck is formed; but the cephalic and pontine flexures have a permanent influence on the form of the brain.

Cavities and Meninges of Neural Tube

The cavity of the spinal portion of the neural tube becomes the central canal of the spinal medulla. The cavities of the primitive brain vesicles are transformed into the ventricles, the interventricular foramina, and the **aqueduct of the cerebrum** in the midbrain. All these cavities are filled with **cerebrospinal fluid,** secreted by the **choroid plexuses** of blood vessels which invaginate the persistently thin parts of the walls of the ventricles at certain places.

The cavities of the developing cerebral hemispheres become the right and left **lateral ventricles** of the brain. The cavity of the central portion of the forebrain becomes the **third ventricle,** and the region where the third and lateral ventricles communicate is the **interventricular foramen.**

The cavity of the hindbrain vesicle becomes the **fourth ventricle,** and is connected to the third ventricle by the aqueduct of the cerebrum which traverses the midbrain.

After the neuropores [p. 51] are closed, the cavity of the neural tube is, for a time, a completely enclosed space. Subsequently three membranous sheaths or **meninges** are developed around the tube; they are the **pia mater,** the **arachnoid mater,** and the **dura mater.** The dura mater, which is the outermost, is derived from the sclerotomes; but the pia mater, which is the innermost, and the intervening arachnoid mater are possibly formed in part by ectodermal cells derived from the neural crest.

As the meninges are differentiated, narrow **subdural** and more extensive **subarachnoid spaces** are formed between them. After a time, a **median** and two **lateral apertures of the fourth ventricle** appear in the dorsal wall of the fourth ventricle and in the pia mater which covers it, and thus the fourth ventricle becomes connected with the subarachnoid space. By these apertures, cerebrospinal fluid passes into the subarachnoid space.

FIG. 66 Diagrammatic reconstruction of 2·5 mm. human embryo (Thompson, 1907) with twenty-three pairs of somites. 1-7: neuromeres.

EARLY DEVELOPMENT OF ALIMENTARY CANAL AND FORMATION OF STOMODAEUM AND PROCTODAEUM

The greater part of the permanent alimentary canal is derived from the entodermal vesicle and is therefore lined with entodermal cells. The foregut and hindgut parts are entirely enclosed in the embryo, while the midgut remains in continuity with the extra-embryonic part of the yolk sac [Figs. 62 and 70].

The cephalic extremity of the alimentary canal is formed from a depression of the ectodermal surface, the **stomodaeum,** which lies, at first, between the ventrally bent head and the bulging pericardial region [Figs. 62 and 67]. When the stomodaeum first appears it is separated from the entodermal alimentary canal by the buccopharyngeal membrane, but when this disappears, during the fourth week, the stomodaeum communicates with the foregut. Later, it is separated into nasal and oral portions by the formation of the palate, and the oral portion forms that part of the mouth in which the gums and the teeth are developed.

The terminal portion of the permanent canal is formed by a pit-like, ectodermal hollow, the **proctodaeum** or **anal pit** [Fig. 77], which is separated from the blind end of the entodermal alimentary canal until about the eighth week by the **anal membrane,** which is the dorsal portion of a more extensive cloacal membrane [p. 49].

Derivatives of the Stomodaeum

When the stomodaeum is definitely established, it is bounded cranially by the frontal end of the head, caudally by the conjoined ends of the **mandibular arches** [p. 56], and laterally by the **maxillary processes,** which grow ventrally from the dorsal parts of the mandibular arches. The space is closed dorsally by the buccopharyngeal membrane until that membrane disappears [Fig. 66].

In the roof of the stomodaeum, anterior to the attachment of the buccopharyngeal membrane, there is a depression which is deepened by the growth of mesoderm around it so that it appears as a diverticulum of the roof. It is lined with ectoderm and is known as Rathke's pouch.

Rathke's Pouch

The blind end of the pouch is in contact with the floor of the third ventricle of the brain where the hypophysial diverticulum arises and forms the posterior lobe of the hypophysis. Rathke's pouch forms the anterior lobe of the **hypophysis** or **pituitary gland** [Figs. 75–77].

DEVELOPMENT OF FACE AND SEPARATION OF NOSE AND MOUTH

The frontal end of the head lies in the cephalic boundary of the stomodaeum and is called the **frontonasal process.** The upper part of the frontonasal process remains undivided and represents the region of the future forehead; but a pair of shallow depressions—the **olfactory pits**—divide the lower part into two **lateral nasal processes** and a **median nasal process** which is divided by a median groove into right and left **globular processes** [Fig. 67A]. As the margins of these processes increase in height the olfactory pits deepen.

The lateral boundary of the stomodaeum is at first formed by the maxillary process which springs from the dorsal part of the mandibular arch and grows forward to approach the lateral nasal process. Between these two processes lie the projecting eye and a groove leading downwards and forwards from it to the olfactory pit—the **nasolacrimal sulcus** [Fig. 67A].

A

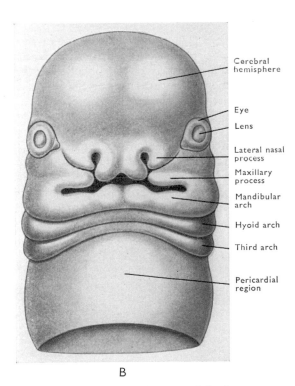

B

FIG. 67 Two stages in development of the face.

A, Boundaries of stomodaeum before completion of primitive upper lip.

B, Completion of upper lip. Note the union of maxillary process with lateral and median nasal processes to form cheek and upper lip.

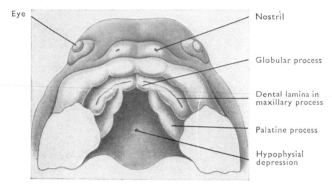

FIG. 68   Portion of the head of a human embryo about 8 weeks old (His).   The lips are separated from the gums, and the line of the dental lamina is visible in the gums.   The palatine processes are growing inwards from the maxillary processes.

Each maxillary process grows ventrally to fuse with the lateral nasal process along the line of the nasolacrimal sulcus, burying the surface epithelium of the sulcus which thus forms a cord of cells between the eye and the olfactory pit; this later canalizes to become the **nasolacrimal duct.**   Continuing its growth medially, and carrying the lateral nasal process with it, the maxillary process fuses with and overgrows the globular process, reaching the opposite maxillary process in the midline.   The olfactory pits are thus completely separated, for a time, from the stomodaeum, and they lie above the newly constituted ledge which now forms its superior boundary.   This ledge consists on each side of the maxillary process fused with the globular process near the midline, and excludes the lateral nasal process from the margin of the stomodaeum [FIG. 67B].

Thus, the upper parts of the **cheeks** are formed by the maxillary processes and the **upper lip** by their fusion with the lower part of the median nasal process.   At a later stage the **maxillae**—the bones of the upper jaw—are formed in the mesodermal core of these united processes; the **premaxillae** ossify separately, they unite with the maxillae and do not form separate bones on the facial surface [FIG. 132].   The lower parts of the cheeks and the lower lip are derived from the mandibular arch; the **aperture of the mouth,** which at first is wide, is gradually reduced by the progressive fusion of the maxillary processes with the mandibular arch.

The **external nose** is developed from the nasal processes; the lateral nasal process forms the lateral part of the external nose, including the ala, and the prominence of the nose rises gradually from the root of the median nasal process where it blends with the forehead.   The external openings of the olfactory pits become the **nostrils.**

The olfactory pits at first are blind and are separated from the cavity of the stomodaeum by thin **bucconasal membranes;** but these soon disappear, and the olfactory pits then communicate directly with the cavity of the stomodaeum through openings which are called the **primitive posterior apertures (choanae) of the nose.**   The olfactory pits now form the primitive nasal cavities; they extend upwards and backwards towards the roof of the main cavity of the stomodaeum on each side.   Mesoderm growing dorsally in the dorsolateral walls of the stomodaeum (future nasal cavities) from each maxillary process forms the roof of this region after meeting its fellow; this roof mesoderm then grows ventrally in the midline to form the posterior part of the **nasal septum.**

After the formation of the primitive choanae of the nose, a horizontal ledge grows from the medial surface of each maxillary process towards the median plane.   These ledges—the **palatine processes**—are separated for a while by the developing tongue, which bulges superiorly between them, but they meet and fuse with each other and with the nasal septum during the third month of intra-uterine life, the fusion beginning anteriorly and being completed posteriorly in the region of the uvula.   In this way the palate is formed, and bone develops in the anterior portion of it to form the **hard palate;** the posterior portion becomes the **soft palate.**   The original stomodaeum is thus separated into definitive nasal and oral cavities.   The **nasal cavity** is divided into lateral halves by the nasal septum [FIG. 69].   The **oral cavity** is continuous with the ventral part of the primitive pharynx.   It forms the cavity and the vestibule of the mouth and from its walls the gums and teeth are developed [FIG. 68].   These serial changes have been illustrated by Kraus, Kitamura, and Latham (1966).

Failure of union of the various elements from which the face and palate are formed causes a variety of malformations.   The commonest are **hare lip** and **cleft palate,** which often occur together.   Hare lip is due to failure of union of the

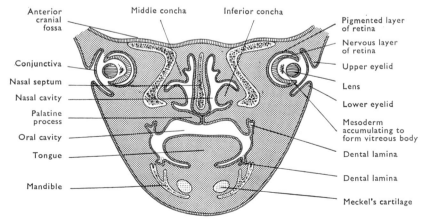

FIG. 69   Diagrammatic coronal section through face of human embryo about 9 weeks old.

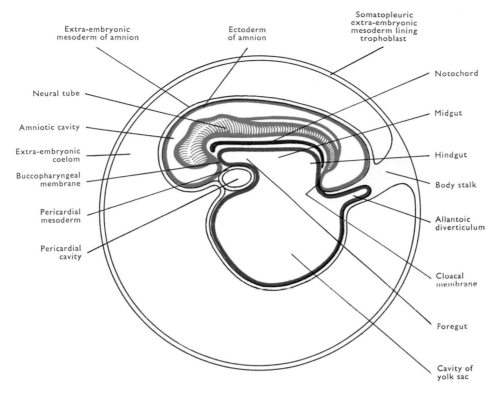

Extra-embryonic mesoderm of amnion

Ectoderm of amnion

Somatopleuric extra-embryonic mesoderm lining trophoblast

Notochord

Neural tube

Midgut

Amniotic cavity

Hindgut

Extra-embryonic coelom

Buccopharyngeal membrane

Body stalk

Pericardial mesoderm

Allantoic diverticulum

Pericardial cavity

Cloacal membrane

Foregut

Cavity of yolk sac

FIG. 70 Median longitudinal section through plane A–A of the embryo shown in Fig. 64.

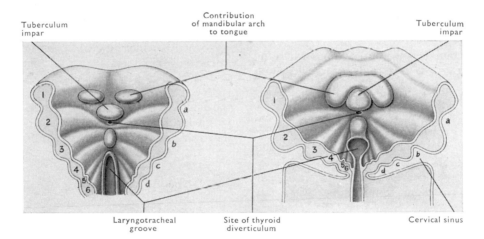

Tuberculum impar

Contribution of mandibular arch to tongue

Tuberculum impar

Laryngotracheal groove

Site of thyroid diverticulum

Cervical sinus

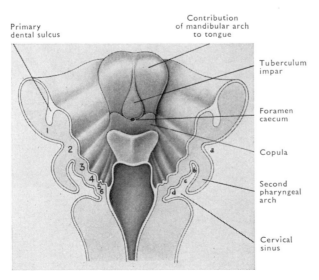

Primary dental sulcus

Contribution of mandibular arch to tongue

Tuberculum impar

Foramen caecum

Copula

Second pharyngeal arch

Cervical sinus

FIG. 71 Diagrams showing formation and closure of cervical sinus, and the development of the tongue. Numbers 1–6: first to sixth pharyngeal arches; a–d: first to fourth pharyngeal grooves.

maxillary and median nasal processes; it may occur on one or both sides and rarely in the midline due to failure of the median nasal process to play its part in the formation of the upper lip. When it is bilateral, the premaxillary part of the upper jaw is isolated.

DIFFERENTIATION OF FOREGUT

Derivatives of the Side-wall

While still separated from the stomodaeum by the bucco-pharyngeal membrane the cephalic part of the foregut, situated dorsal to the pericardium [FIG. 66], dilates to form the primitive pharynx, the walls of which develop alternating ridges and depressions by proliferation of bars of mesoderm. In each lateral wall there are five such depressions, the **pharyngeal pouches.** In the ventral wall there are two: the first is the rudiment of the thyroid gland, and the other, situated more caudally, is the origin of the respiratory system and provides the epithelial lining of larynx, trachea, bronchi, and lungs.

Simultaneously with the formation of the pharyngeal pouches internally a series of grooves appears externally. They correspond in position with the first four pharyngeal pouches, and they are called the **external pharyngeal grooves** [FIGS. 63 and 92]. There is no external groove corresponding to the fifth pouch, which has a common opening into the pharynx with the fourth [FIGS. 71 and 73].

The pharyngeal pouches and the external grooves divide each side-wall of the primitive pharynx into a series of mesodermal bars called the **pharyngeal** (branchial) **arches.** The bars are six in number, but the fifth and sixth can only be seen on the inner aspect of the pharynx, and are less well defined than the others.

The first pharyngeal bar is the rudiment of the maxillary and mandibular processes; it is called the **mandibular arch.** The second is the **hyoid arch,** and between them lie the first pharyngeal groove and pouch. The others are numbered as the third to sixth arches. Each pharyngeal arch consists at first of simple mesoderm—covered externally by ectoderm, internally by entoderm—but a series of similar structures soon develop in them. Each arch then contains a cartilaginous skeleton, a muscle rudiment, two nerves (main and subsidiary), and an artery. The cartilage of the first arch (Meckel's cartilage) forms a basis for the **mandible** which ossifies intramembranously on its external surface [FIG. 143]; those of the second and third give rise to the **hyoid bone** [FIG. 145]; the fourth to sixth arches belong to the **larynx.** The arteries are the **aortic arches** which connect the ventral and dorsal aortae [FIG. 81].

When the arches first appear, they extend from the level of the dorsal wall of the foregut to the pericardium; but as growth proceeds, and the neck develops, the mandible is separated from the pericardium, and the ventral ends of the other arches are lifted away from it and exposed on the surface, though the mandibular arch is the only one that meets its fellow of the opposite side externally. The growth of the mandibular and the hyoid arches soon greatly exceeds that of the other arches, which gradually recede from the surface until, on each side, they lie at the bottom of a depression—the **cervical sinus**—overlapped by the caudal border of the hyoid arch [FIGS. 71 and 93]. The increasing growth of the

hyoid arch reduces the opening of the cervical sinus to a narrow channel called the **cervical duct,** but this is soon obliterated, and then the sinus becomes the **cervical vesicle.** The cervical vesicle, lying at the side of the third pharyngeal groove, is associated with the second and fourth grooves by narrow canals called the 'branchial' ducts, which are the remains of the external pharyngeal grooves or 'branchial clefts'. Ultimately this submerged space (lined with ectoderm) is obliterated; but it may on occasion give rise to a **branchial cyst,** or to a **branchial sinus** (if the communication with the exterior remains) or to a **branchial fistula** (if a separating membrane [see below] also breaks down).

The portion of the wall of the primitive pharynx which lies between any two adjacent arches and separates the external groove from the internal pouch is called the **separating membrane.** In the earliest stages it consists of ectoderm and entoderm separated by mesenchyme; then the mesenchyme disappears so that, for a time, the membranes consist of ectoderm and entoderm only. At a later period the ectoderm and entoderm are again separated by ingrowing mesoderm.

In vertebrates with gills the membranes break down so that **gill-clefts** are formed; but, except as a malformation, the separating membranes are not usually perforated in mammalian development, so that a complete 'cleft' rarely exists.

The first external pharyngeal groove is the only one that is not submerged in the cervical sinus [FIG. 71]; it is the site of the formation of the **external acoustic meatus,** and a series of tubercles which appear at its margins develop into the **auricle** of the external ear. In Man, as in other mammals, the organ of hearing consists of: the **internal ear** or labyrinth; the **middle ear** or tympanum, which is connected to the pharynx by the **auditory tube;** and the external ear, which includes the external acoustic meatus and the auricle. The **tympanic cavity** and the **auditory tube** are developed

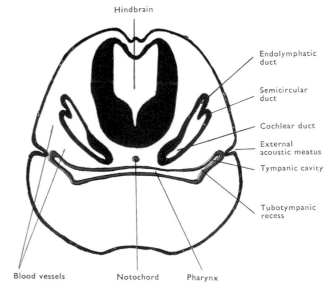

FIG. 72 Diagram of transverse section through the head of an embryo. Showing the rudiments of the three parts of the ear and their relation to the tubotympanic recess and the first pharyngeal groove.

as a lateral extension of the upper part of the cavity of the primitive pharynx, between the first and third arches, which contains in its posterior wall and floor the second arch and in its medial wall the third arch with the dorsal parts of the first and second pouches intervening; it is called the **tubotympanic recess** (Frazer, 1914). A part of the cavity of the second pharyngeal pouch may be represented in the adult by the **intratonsillar cleft,** which passes into the upper part of the tonsil in the side-wall of the pharynx [FIG. 73]. The **tympanic membrane,** which separates the tympanic cavity from the external acoustic meatus, is formed in the position of a 'separating membrane' between the tubotympanic recess and the first external groove [FIGS. 72 and 73]. But it is a secondary formation, since the deeper part of the meatus is first formed by a solid plug of ectodermal cells which later breaks down in its centre to form the meatus.

The development of the **auditory ossicles** is intimately associated with the development of the primitive pharynx, since they develop from the cartilages of the first and second pharyngeal arches. The organ of hearing and balance, the internal ear, has already developed dorsomedial to the tubotympanic recess from the ectoderm of the side of the head in close relation to the hindbrain and the vestibulocochlear (eighth cranial) nerve [FIGS. 66 and 74]. This **otic vesicle,** formed by thickening and invagination of the ectoderm, which produces a small epithelial sac at the side of the hindbrain, becomes converted into the membranous labyrinth which

First pharyngeal pouch
First separating membrane

First pharyngeal groove
Site of thyroid diverticulum
Tonsil
Second pharyngeal groove

Second pharyngeal pouch
Hypobranchial eminence

Inferior parathyroid

Third pharyngeal pouch
Laryngotracheal groove
Superior parathyroid
Thymus
Fourth pharyngeal pouch

Duct of fourth and fifth pouches

Fifth pharyngeal pouch

Thyroid gland

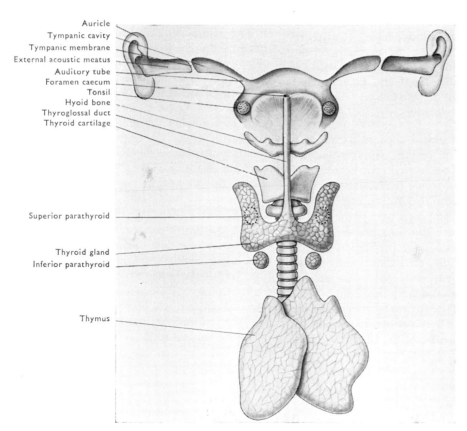

Auricle
Tympanic cavity
Tympanic membrane
External acoustic meatus
Auditory tube
Foramen caecum
Tonsil
Hyoid bone
Thyroglossal duct
Thyroid cartilage

Superior parathyroid

Thyroid gland
Inferior parathyroid

Thymus

FIG. 73 Diagram showing the floor and lateral walls of the embryonic pharynx and their derivatives.

includes the **utricle,** the **saccule,** and the three **semicircular ducts,** all concerned with balance, and the **duct of the cochlea** which contains the essential organ of hearing.

The third pharyngeal pouch opens, like the first and second, directly into the cavity of the foregut, but as the pharyngeal wall thickens the communication is drawn out into a duct-like passage ; the fourth and fifth pouches lie in the side-wall of a common recess which opens by a single aperture or 'duct' into the cavity of the primitive pharynx [FIG. 73].

The cavities of the third, fourth, and fifth pouches ultimately disappear. Before they disappear a diverticulum, at first hollow but afterwards solid, grows from the ventrolateral wall of each, and a solid **epithelial body** buds from the dorso-lateral wall of the third and fourth pouches [FIG. 73].

Organs derived from Pharyngeal Pouches

The diverticula and the epithelial bodies of the third and fourth pouches lose their attachment to the pharyngeal wall and give rise to a series of important organs [FIG. 73]. The diverticulum from the third pouch is concerned with the development of the **thymus ;** and the epithelial bodies of the third and fourth pouches become the **parathyroid glands.** Changes in the situation of these organs occur as the head grows away from the pericardium and the neck is formed ; thus the thymus, which maintains its relation to the pericardium, extends into the thorax, and the parathyroids alter their relative position, that derived from the third pouch being drawn down with the thymus so that it comes to lie inferior to the other and may occasionally become intrathoracic.

The diverticulum from the fifth pharyngeal pouch is trans-formed into the **ultimobranchial body,** which receives its name because it is the last of the series of organs derived from the branchial region of the pharynx. It is associated with the corresponding lobe of the thyroid gland and contributes a specific type of cell to it [see thyroid gland].

Derivatives of Ventral Wall

The rudiment of the **thyroid gland** appears at first as a diverticulum from the ventral wall of the primitive foregut. It begins in the median plane and passes ventrally between the mandibular and hyoid arches. It grows caudally in the sub-stance of the neck, passing ventral to the arch cartilages which develop into the hyoid bone and the cartilages of the larynx. When the caudal end of the diverticulum reaches the level of the origin of the trachea its epithelium proliferates and it becomes bilobed, thus expanding into the isthmus and the two lateral lobes of the permanent gland [FIG. 73]. The stalk of the diverticulum is the **thyroglossal duct ;** the position of its original upper end is indicated by the **foramen caecum** of the tongue. The caudal end of the stalk sometimes persists and is transformed into the **pyramidal lobe** of the **thyroid gland,** and persistence of other parts of the duct may give rise, at any point between the gland and the tongue, to swellings known as **thyroglossal cysts.**

Respiratory System

A more caudally situated diverticulum from the ventral wall of the foregut is the rudiment of the respiratory system [FIGS. 71 and 75]. It first appears behind the developing tongue as a longitudinal groove bounded on each side by the sixth pharyngeal arches. The caudal end of this **laryngotracheal groove** soon dilates into a pouch, the **tracheobronchial diverticulum.** As the foregut elongates with the growth of the neck, the pouch and groove are separated from the more dorsal part of the foregut, which becomes the **oesophagus.** The tracheobronchial diverticulum elongates, but remains in communication with the pharynx at the permanent **inlet of the larynx.** As the groove elongates and is separated from the primitive foregut it is converted into a tube which is gradually differentiated into the **larynx** and the **trachea.** The pouch at the caudal end of the tube soon divides into right and left lobes, each of which is the rudiment of the epi-thelium lining the bronchi, bronchioles and alveoli of the lung of the corresponding side.

At the cranial margin of the laryngotracheal groove is a swelling, the **hypobranchial eminence** [FIG. 73], which develops into the epiglottis.

The Tongue

The mucous membrane of the tongue is formed by four separate rudiments which appear in the ventral wall of the primitive pharynx. Two of these are elevations formed on the ventral ends of the mandibular arches, the **lateral lingual swellings.** The third is a median elevation, called the **tuberculum impar,** which is situated on the caudal surface of the conjoined ventral ends of the mandibular arches. The fourth rudiment, the **copula** of His, formed by the conjoined ventral ends of the third pair of arches which overgrow the second arches, is separated from the tuberculum impar by the orifice of the thyroid rudiment [FIG. 71].

The lateral lingual swellings submerge the tuberculum impar and unite to form the greater part of the anterior two-thirds of the tongue—the part on which all the papillae are developed. The posterior third of the tongue, which lies in the ventral or anterior wall of the permanent pharynx, is formed by a V-shaped swelling which rises from the copula and, overgrowing the second arch, blends with the anterior two-thirds along the line of the **sulcus terminalis.** The foramen caecum is a depression left in the median plane at the junction of the two parts of the tongue, and it therefore in-dicates the site of origin of the thyroid gland.

Abdominal Portion of the Foregut

The portion of the foregut that has now been considered gives rise to the floor of the mouth (with the exception of the lips, teeth, and gums), the pharynx, the thyroid gland, the thymus, the parathyroid glands, the respiratory organs, and the oesophagus. The oesophagus extends through the part of the embryo that will become the thoracic region of the body, and the remainder of the foregut is situated in the upper part of the abdominal region. The abdominal portion of the fore-gut is differentiated into the **stomach,** the superior part and upper portion of the descending part of the **duodenum,** and from it the liver and pancreas take origin as outgrowths.

The gut tube is attached to the posterior wall by its dorsal mesentery (splanchnopleure), the part attached to the stomach being known as the **dorsal mesogastrium** and that support-ing the duodenum as the **mesoduodenum.** It lies on the caudal surface of the **septum transversum,** the mesodermal partition which separates the abdominal foregut from the pericardium [FIG. 75] and from which a **ventral mesentery**

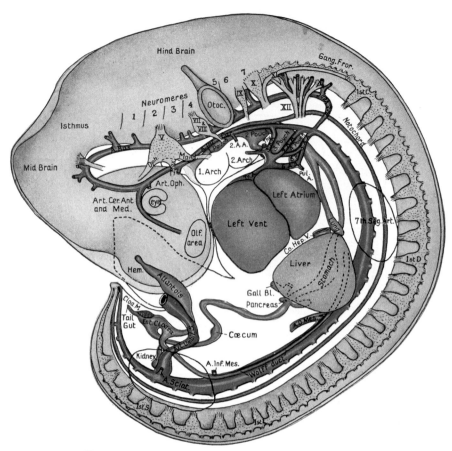

FIG. 74   Reconstruction of human embryo, 7 mm. long.

1st D, First thoracic spinal ganglion; *Co. Hep. V.*, common hepatic vein; *Gang. Fror.*, Froriep's ganglion (XII); *Hem.*, cerebral hemisphere; *Otoc.*, otic vesicle; *Wolff. duct*, mesonephric duct.

(Peter Thompson, *Studies in Anatomy*, University of Birmingham, 1915.)

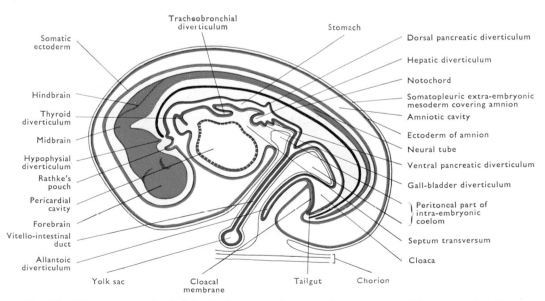

FIG. 75   Diagram of median longitudinal section of a 5-mm. human embryo (about 5 weeks old). (After Mall.)

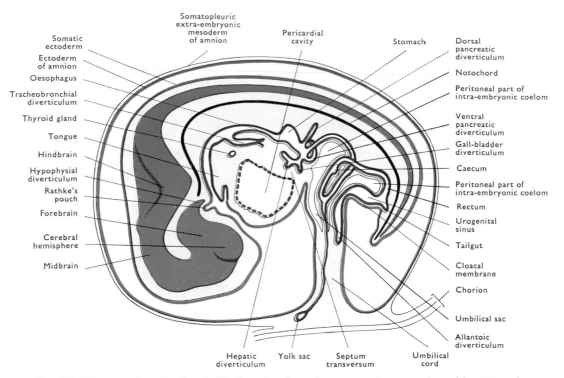

FIG. 76   Diagram of median longitudinal section through a 10-mm. human embryo (about 6 weeks old), showing rotation of the midgut loop lying in umbilical sac.   (After Mall.)

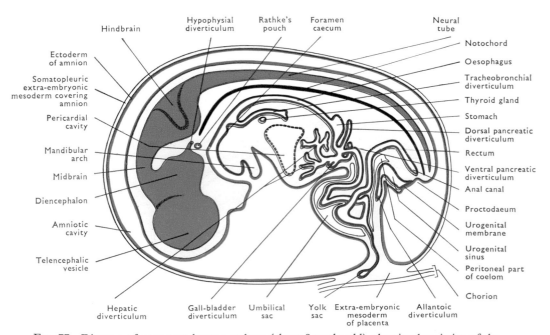

FIG. 77   Diagram of a 25-mm. human embryo (about 8 weeks old), showing herniation of the midgut loop into the umbilical sac.

for the abdominal foregut is produced as that structure shrinks away from the septum.

Liver and Pancreas

When the embryo is about 4 weeks old and has attained a length of 2·5 mm. a ventral diverticulum, which forms the **liver,** the **gall-bladder,** the system of **bile-ducts** and part of the **pancreas,** appears in the angle between the widely open vitello-intestinal duct and the ventral wall of the foregut [FIG. 66] and grows into the septum transversum between the pericardium and the foregut. The septum transversum is thus divided into three parts: a midzone containing this liver diverticulum; a cranial part separating the liver from the pericardium, which forms part of the diaphragm; and a caudal portion attaching the liver to the foregut, the ventral mesentery [FIG. 76]. When the length of the embryo has increased to about 4 mm. another diverticulum is formed from the dorsal wall of the foregut a little nearer the stomach. This is the **dorsal pancreas** which forms the greater part of the adult pancreas [FIGS. 76 and 77].

DIFFERENTIATION OF MIDGUT AND HINDGUT

Derivatives of Midgut

The midgut is that part of the primitive alimentary tract which is in free communication with the yolk sac by the vitello-intestinal duct. It is transformed into the greater part of the **small intestine** (from the entrance of the liver diverticulum, the bile-duct, into the duodenum to the end of the ileum) and part of the large intestine (caecum and appendix, ascending colon, and most of the transverse colon).

Derivatives of Hindgut

The part of the alimentary tract found in the tail of the embryo is the hindgut. Like the foregut, it is divided into dorsal and ventral parts. The dorsal part forms the remainder of the large intestine, except a small portion of the anal canal. The ventral part forms: (1) the urachus and part of the urinary bladder; (2) the urethra in the female, and part of the urethra in the male.

As development proceeds the midgut elongates and separates from the posterior abdominal wall to which it remains attached by the dorsal mesentery. It then forms a U-shaped tube with cephalic and caudal limbs, and a ventral knuckle which points towards the umbilical orifice, through which it remains connected with the yolk sac by the narrowed and elongated **vitello-intestinal duct** [FIG. 75].

In embryos about 8 or 9 mm. long (between 5 and 6 weeks old) there appears about the middle of the caudal limb of the loop an enlargement which is the rudiment of the **caecum** and **vermiform appendix,** thus demarcating the large intestine from the small.

But before this rudiment appears rotation of the loop has begun. The growth of other structures in the abdomen, especially of the liver, is relatively so great that there is insufficient room for the developing intestines. When the embryo is between 5 and 8 mm. long, the loop begins to pass out through the umbilical orifice into an **umbilical sac** which has been formed in the umbilical cord by the enclosure of part of the extra-embryonic coelom [FIGS. 75–77].

As the loop enters this sac its limbs already lie side by side—the cephalic limb on the right, the caudal on the left—and there it remains for a considerable time while a further stage of its rotation and further development take place.

The chief change that now occurs is the rapid elongation of the part of the loop (mainly the cephalic limb) that becomes small intestine, which is thrown into the numerous coils of the jejunum and ileum. The vitello-intestinal duct has meanwhile separated from the loop and disappeared together with the vitelline artery accompanying it, thus leaving the coiling gut free in the umbilical sac. The part of that artery remaining in the mesentery of the loop is the superior mesenteric artery, which supplies the loop and therefore the greater parts of both the small and the large intestine.

As the intestines return into the abdomen the last stage of the rotation of the loop occurs. The small intestine, returning first, falls back into the abdominal cavity inferior and to the right of the hindgut, which is thereby pushed to the left and superiorly. Thus when the large intestine part of the midgut loop returns to the abdominal cavity it lies transversely, above and in front of the coils of small intestine, with the caecum immediately below the right lobe of the liver. The rotation as a whole is usually described as taking place round the superior mesenteric artery as an axis. It is completed by the passage of the caudal limb of the loop to the right, in front of the trunk of that artery and in front of the most cephalic part of the midgut loop which, because of the rotation around the artery, loops behind it to form the horizontal part of the duodenum.

Subsequently, certain parts of the large intestine are fixed in their permanent positions by peritoneal adhesions. The last 60 cm. of the **ileum,** the **ascending colon** and the **transverse colon** are derived from the caudal limb of the loop, whilst the proximal part of the hindgut is transformed within the abdomen into the **descending colon,** and the **sigmoid colon.** The developing intestines remain in the umbilical sac until the embryo is about 40 mm. long and about 10 weeks old, when they return rather suddenly to the abdomen, and the coelomic space in the umbilical cord is obliterated.

Occasionally the site of attachment of the vitello-intestinal duct is indicated by the persistence of a portion of it as a **diverticulum of the ileum** (Meckel's diverticulum) some 60 cm. from the caecum. This diverticulum may rarely be attached to the back of the umbilicus by a fibrous thread; such a thread, from the distal small intestine to the umbilicus and continuous with the end of the superior mesenteric artery, is normally present in many new-born animals (e.g., cats).

The umbilical sac and its communication with the peritoneal cavity may fail to close and the midgut may remain in the umbilical coelom. Thus the child is born with the malformation of the abdominal wall known as congenital umbilical hernia.

Development of Caudal Part of Hindgut

When the hindgut is first enclosed, its blind end and its ventral wall are bounded by the caudal portion of the primitive streak—the **cloacal membrane**—which is bent ventrally during the folding-off of the embryo [FIG. 75].

The terminal part of this portion of the gut becomes expanded to form a chamber called the **entodermal cloaca,** because the **pronephric ducts** from the primitive kidneys open into it, one on each side, as in the permanent **cloaca** of the lower vertebrates.

The ventral part of the cephalic end of the cloaca is continuous with the allantoic diverticulum, and the dorsal part with that portion of the hindgut which forms the descending colon and the sigmoid colon.

As the temporary tail is formed and turned ventrally, a diverticulum of the caudal end of the dorsal part of the cloaca is prolonged into it, forming the **tailgut.** This soon becomes shut off from the cloaca and entirely disappears before the temporary tail is absorbed into the caudal end of the body Figs. 75–77].

At a later period the cloaca itself is divided by a septum (the urorectal septum) into a dorsal part which becomes the **rectum,** and part of the anal canal, and a ventral part called the **urogenital chamber** [Fig. 77]. The septum begins in the angle between the allantoic diverticulum and the ventral wall of the hindgut, and grows towards the surface till it reaches and fuses with the cloacal membrane ; the membrane is thus separated into urogenital and anal portions, both of which disappear about the eighth week.

In both sexes the urogenital chamber is separable into three portions : (1) a cephalic portion called the **urachus,** which is continuous with the remnant of the allantoic diverticulum and becomes the **median umbilical ligament ;** (2) a middle portion, the **vesico-urethral canal,** which becomes the greater part of the urinary bladder, the whole of the female urethra and the upper part of the prostatic urethra in the male ; (3) a caudal portion, known as the **urogenital sinus,** which becomes the vestibule of the vagina in the female and the lower part of the prostatic urethra and a large part of the penile urethra in the male.

Urogenital System

The greater part of the urogenital system is derived from the intermediate cell mass of the mesoderm [p. 47]. That includes the series of three excretory organs or kidneys known as the **pronephros,** the **mesonephros,** and the **metanephros**—the last forming the permanent kidney—and the two ducts, the mesonephric and paramesonephric ducts, which have a different fate in the two sexes [Figs. 675–677].

Derivative of Proctodaeum

The proctodaeum is a surface depression which owes its origin to the elevation of the surface around the margin of the anal portion of the cloacal membrane [Fig. 77]. It forms the lowest portion of the **anal canal** of the adult.

BLOOD VASCULAR SYSTEM

In the early stages of implantation, the trophoblast utilizes the dissolving decidual tissues as a food supply, and fluid is absorbed and transmitted into the interior of the zygote to fill its expanding cavities.

Nutritive materials in the fluid suffice for the requirements of the embryonic and non-embryonic parts of the zygote so long as both consist of comparatively thin layers of cells. But when the embryonic area increases in thickness and begins to be moulded into the embryo, and as the development of its various parts progresses, the supply of food and oxygen required is greater than can be provided by diffusion from the adjacent fluid media. A method of food supply adequate for the increasing requirements of continued development and growth becomes necessary and the blood vascular system is formed. In its earliest stages it consists of a series of simple vessels but, after a short time, parts of the vessels undergo changes which result in the formation of the heart, contractions of the myocardium then producing circulation of blood.

The tissues of the vascular system are derived from **mesenchyme cells** [p. 46]. The development of the blood vessels takes place first in the wall of the yolk sac and in the body stalk, but it occurs later in the mesenchyme of the embryo also.

On the wall of the yolk sac the mesenchyme cells which are the rudiments of the blood vascular system become spherical and their nuclei enlarged ; at the same time they are aggregated together into rounded patches called **blood islands** [Fig. 49]. As soon as the cells have attained their distinctive appearance they are known as **angioblasts,** and from them are derived both the endothelial walls of blood vessels and blood corpuscles.

When the blood islands first become distinct in the walls of the yolk sac they are isolated from one another, but the endothelium which develops around the different islands soon joins together to form a network of endothelial tubes in which fluid appears ; there is thus established a plexus of anastomosing tubes filled with blood plasma in which the primitive nucleated blood corpuscles are suspended.

Most of the corpuscles of the blood, however, eventually become non-nucleated and form the red corpuscles or **erythrocytes** which in the early stages are being formed very rapidly and retain their nuclei while circulating. A small proportion (about 0·2 per cent.) remain nucleated and form the white corpuscles or **leucocytes,** which are separable into several groups having distinctive characteristics. The blood vessels also change as development proceeds, for the original plexuses are transformed into distinct stems and branches of varying size and importance. Therefore two groups of events in the development of the vascular system have to be considered— the evolution of the different kinds of blood corpuscles and the development of the main embryonic blood vessels.

Development of Blood Corpuscles

The view that the angioblast is the parent cell not only of the endothelium of blood vessels but also of all the blood corpuscles both red and white was elaborated by Maximow (1924). It implies that blood formation in the growing embryo and foetus takes place by the development of angioblastic cells from the mesenchyme in the same way as on the yolk sac, and that the stem cells that give rise to all the varieties of blood corpuscles (haemocytoblasts) in the adult are carried from the yolk sac into the embryo to multiply there.

The red blood corpuscles pass through several stages to maturity ; the original nucleated cells (**erythroblasts**) become laden with haemoglobin, pass through a **normoblast** stage (in which the nuclei are contracted), and after the loss of their nuclei become **reticulocytes,** which are anucleate cells containing a fine network stainable with vital dyes. With the loss of this reticulum these cells become the mature red blood corpuscles or erythrocytes.

This process goes on throughout life ; it begins in the wall of the yolk sac, is very active in the liver from the third month onwards, and occurs in the spleen in the later stages of intra-uterine life. When ossification begins, formation of blood cells is actively carried on in the developing bone marrow; and after birth the continuous formation of red blood corpuscles is entirely confined to that special haemopoietic (blood-forming) tissue.

The three main kinds of white blood corpuscles are **granular leucocytes, monocytes** and **lymphocytes.** The bone marrow is the great factory of the first kind (and its varieties) but the monocytes and lymphocytes are normally formed in lymphoid tissue wherever it may be situated—in the spleen, in the mucous membrane of the alimentary canal, and in the lymph nodes. All these cells reach the blood stream by passing through the endothelial walls of blood capillaries, but lymphocytes also reach it in great numbers via the lymph vessels when the lymph nodes have developed, mainly after birth.

In very young embryos both immature and mature blood cells are present in the blood stream. After birth only mature erythrocytes pass into the blood stream; but under pathological conditions, in which there may be excessive formation of red corpuscles in the bone marrow, normoblasts may again appear in the blood.

Formation of Primitive Blood Vessels

It has been pointed out that a plexiform system of blood vessels is formed from angioblasts in the yolk sac area of the entodermal vesicle.

Similar but less easily demonstrated plexuses are formed in the mesenchyme of the body stalk, and of the chorion, and also later in the embryonic area.

It is probable that the various plexuses become connected with one another very soon after they appear, so that a continuous capillary network is formed. By enlargement in some places and diminution in others the general vascular network becomes more open, and from this **retiform** arrangement, by further enlargement of definite channels, there are evolved what may be called the primitive stem vessels of the embryo, through which the blood circulates from one area to another.

In the embryonic area two sets of vessels are differentiated—the **vitelline veins,** which emerge from the vascular plexus in the wall of the yolk sac, and the **primitive aortic vessels,** which extend from the pericardial area along the dorsal aspect of the entodermal vesicle as the **primitive dorsal aortae.** In the pericardial area the primitive aortic vessels form the paired primitive **endocardial heart tubes.**

As the aortae pass tailwards in the embryonic area, ventral to the paraxial bars of mesoderm, they give off a ventral series of paired branches, called **vitelline arteries,** which terminate in the vascular plexus in the yolk sac wall [p. 33]; and from the caudal and dorsal part of that plexus a pair of vessels—the **umbilical arteries**—pass along the body stalk, one on each side of the allantoic diverticulum, to the vascular plexus in the chorion [FIG. 78]. Another stem vessel, called the **unpaired umbilical vein,** begins in the vascular plexus in the chorion and passes along the body stalk to the caudal margin of the embryonic area, where it divides into **right** and **left umbilical veins.** These two umbilical veins run along the corresponding margins of the embryonic area in the somatopleure and fuse with the corresponding vitelline veins in the septum transversum, cranial to the pericardial area, but caudal to it when reversal of the embryo occurs. Together they form the **vitello-umbilical trunks,** each of which, right and left, enters the caudal aspect of the developing heart. Therefore, when rhythmic contraction begins in the cardiogenic region, the blood is driven through the primitive aortic vessels and passes through their vitelline and umbilical branches to the yolk sac and the chorion, whence it is returned to the endocardial heart tubes by the vitelline and umbilical veins.

In the meantime lateral branches have been given off from the dorsal aortae, and in later stages, as the mesodermal somites are defined, they become the intersegmental arteries [FIG. 79]; but for a time no purely intra-embryonic veins are distinguishable.

As the head fold is formed and the pericardial region is reversed into the ventral wall of the foregut [p. 48], the cephalic parts of the primitive aortic vessels are carried ventrally in the fold, and so are bent into the shape of hooks. That condition is well seen in an embryo of 1·35 mm. length which has six mesodermal somites [FIG. 78].

Several parts of the bilaterally symmetrical embryonic vascular system are now defined: the short ventral limb of the hook is the **primitive ventral aorta;** the long or dorsal

FIG. 78 Diagram of the vascular system of a human embryo, 1·35 mm. long. Although the umbilical arteries and veins are shown red and blue, they are carrying deoxygenated and oxygenated blood respectively. (After Felix, 1910.)

Common trunk of right vitelline veins

Dorsal aortae

Dorsolateral intersegmental arteries

Umbilical veins

Umbilical arteries

Unpaired umbilical vein

Vitelline veins

Vitelline arteries

First aortic arch

Common trunk formed by umbilical and vitelline veins

Yolk sac

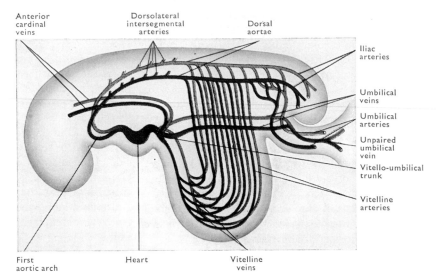

FIG. 79 Diagram of the vascular system of a human embryo, 2·6 mm. long. (After Felix, 1910.)

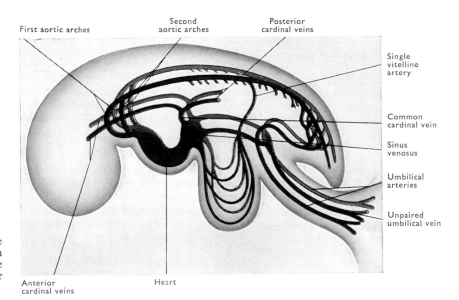

FIG. 80 Diagram of the vascular system of a human embryo showing a single vitelline artery. (After Felix, 1910.)

limb is the **primitive dorsal aorta.** The bend, which connects the two limbs, is the **first aortic arch**; it runs along the side of the buccopharyngeal membrane in the substance of the mandibular arch. The primitive ventral aorta emerges from the cephalic end of the endocardial heart tube.

Heart. Three further important changes take place in the vascular system during the time in which the embryo grows little more than another millimetre in length, the tail fold and lateral folds are formed and the number of its mesodermal somites increases to fourteen pairs.

1. The paired endocardial heart tubes fuse together to form a single, median tube, which comes to be surrounded by a loose reticulum (the **subendocardial tissue**) and by developing muscle (the future myocardium). The single heart tube now formed is divided by dilatations and constrictions into six parts, named, from the caudal towards the cephalic end, the **sinus venosus** in the septum transversum, the **atrium,** the **atrioventricular canal,** the **ventricle, and the bulbus cordis** in the pericardial cavity, and the **truncus arteriosus**

in the mesoderm cephalic to the pericardium. At the same time the heart tube increases in length more rapidly than the pericardial cavity; it therefore becomes bent both in the longitudinal and the transverse direction, and its caudal and cephalic ends come close together [FIGS. 79 and 80].

2. The origins of the umbilical arteries are transferred from the vitelline plexus to the primitive dorsal aortae from which they arise by three roots for each [FIG. 79].

3. The **anterior cardinal veins** are defined [FIG. 79], and are the first pair of purely intra-embryonic veins to appear. They begin in the head and end in the sinus venosus, which they enter, in the septum transversum, close to the common vitello-umbilical veins. They bring back to the heart the blood which has been distributed to the head, neck, and headward part of the trunk of the embryo.

Thereafter, numerous changes take place in the cardiac, the arterial, and the venous parts of the vascular system, and the rudiments of all the main blood vessels of the adult are defined [for details see The Blood Vascular System].

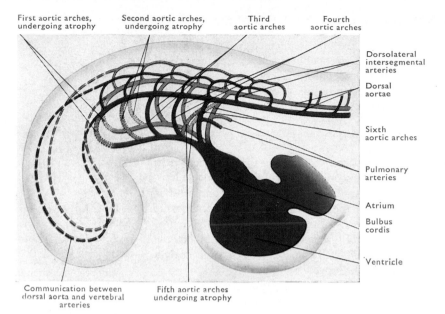

First aortic arches, undergoing atrophy

Second aortic arches, undergoing atrophy

Third aortic arches

Fourth aortic arches

Dorsolateral intersegmental arteries

Dorsal aortae

Sixth aortic arches

Pulmonary arteries

Atrium

Bulbus cordis

Ventricle

Communication between dorsal aorta and vertebral arteries

Fifth aortic arches undergoing atrophy

FIG. 81 Diagram showing the aortic arches of a human embryo.

Arteries

The primitive dorsal aortae fuse, from the region of the tenth (seventh cervical) somite tailwards, to form the permanent descending aorta [FIG. 80]. In this region also, as the gut tube becomes relatively smaller, the paired ventral branches of the aortae fuse to form single, median trunks, and become reduced in number. In the abdominal region they are reduced to three main vessels—the coeliac artery, the superior mesenteric artery (which for a time passes to the yolk sac as the vitelline artery) [FIG. 80], and the inferior mesenteric artery.

Branches are given off from the lateral aspects of the descending aorta to the rudiments of the urogenital system, developing from and in association with the intermediate cell mass.

Four additional pairs of **aortic arches** connecting the unfused portions of the primitive ventral and dorsal aortae appear in the following sequence, the second, the third, the fourth, and the sixth—the temporary fifth pair appearing later [FIG. 81]. These aortic arterial arches run in the **pharyngeal arches** of the embryo [FIG. 74], and they undergo a series of transformations as the arterial system of the thorax and neck develops. The first and second arches disappear, pulmonary arteries arise from the sixth arches [FIG. 81], and the truncus arteriosus and bulbus cordis divide so that a main pulmonary trunk communicates with the right side of the divided heart and the permanent aorta with the left side.

Veins

Two additional pairs of intra-embryonic venous trunks are formed for the drainage of the hinder parts of the body.

The **posterior cardinal veins** [FIG. 80] appear in the thoracic and abdominal regions of the embryo dorsolateral to the intermediate cell mass.

They drain blood from the body walls and from the intermediate mesoderm, and their cephalic ends join the anterior cardinal veins in the thoracic region at the point where the latter turn ventrally to enter the septum transversum [FIG. 80].

As soon as the union of the anterior and posterior cardinal veins is completed, the parts of the anterior cardinals ventral to the points of union are called the **common cardinal veins**. The right common cardinal vein later becomes the terminal part of the **superior vena cava**. In the meantime the common vitello-umbilical venous trunks have been absorbed into the sinus venosus, and the vitelline and umbilical veins open independently into it; so that three pairs of veins now open into the sinus venosus of the heart.

The **subcardinal veins** appear in the abdominal region ventromedial to the intermediate cell mass. They communicate both at their cephalic and caudal ends with the posterior cardinal veins, and drain the intermediate mesoderm.

In addition to the posterior cardinal and subcardinal veins, two further pairs of intra-embryonic veins are established at a later period in the abdominal and thoracic regions. They are called the **supracardinal** and **azygos veins** and they appear dorsolateral to the descending aorta. These veins are concerned in the complex manner of development of the **inferior vena cava** and azygos veins [FIG. 1142].

COELOM AND DIAPHRAGM

Now that an outline of the formation of the body of the embryo and of the principal organs of the trunk—alimentary canal, respiratory system, excretory organs, heart and great vessels—has been given, the division of the coelomic cavity into peritoneal, pleural, and pericardial cavities, and the manner in which these are separated from each other with the formation of the diaphragm may be considered.

The extra-embryonic coelom and the intra-embryonic coelom are derived independently as clefts in the mesoderm and are at first separate from each other [FIGS. 56 and 60]; they become continuous, for a time, through the umbilical orifice [FIG. 57B], but are separated from each other again when the umbilical orifice closes and the extra-embryonic coelom disappears; the intra-embryonic coelom remains as the cavities of the pericardium, pleura, and peritoneum.

EXTRA-EMBRYONIC COELOM

The extra-embryonic coelom is formed in the primary mesoderm [p. 33], but is entirely obliterated, except for a small rudiment in the root of the umbilical cord, when the outer surface of the expanding amnion fuses with the body stalk and the inner surface of the chorion [FIGS. 44 and 45].

INTRA-EMBRYONIC COELOM

When the cleft-like spaces in the lateral plate mesoderm become confluent at their cephalic extremities, the resultant ∩-shaped cavity is the intra-embryonic coelom. The bend of the ∩-shaped cavity is in the cephalic part of the embryonic region, and has no direct communication with the extra-embryonic coelom, but the greater part of each limb of the cavity soon opens laterally into the extra-embryonic coelom [FIGS. 57B and 82].

The transverse portion of the ∩ is the **pericardial cavity.** The adjacent parts of the limbs are named the **pericardio-peritoneal canals;** they are invaded later by the growing lungs and are transformed into the **pleural cavities.** The

FIG. 82 Diagram of intra-embryonic coelom seen from dorsal surface before folding of embryonic area. Portions of the somatopleure are cut away to show the continuity of the intra-embryonic coelom of the two sides through the pericardial canal.

remaining portions of the two limbs unite ventrally, before the umbilical orifice closes, to form the single **peritoneal cavity.**

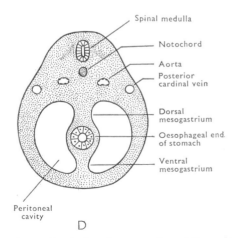

FIG. 83 Diagrammatic representation of embryonic coelom after folding of embryonic area but before the separation of the various parts A. from dorsal surface; B, C, and D (after Patten, 1953), at levels indicated in A.

As the head fold forms, the pericardial part of the cavity is carried ventrally and caudally into the ventral wall of the foregut [FIGS. 62 and 83]. The mesoderm which originally formed its peripheral boundary then lies in the cephalic margin of the umbilical orifice, and is thickened to form an important mass, the **septum transversum** [FIGS. 66 and 75].

On each side, the pericardial cavity is still continuous with the pericardioperitoneal canals which lie dorsal to the septum transversum, between the foregut and the body wall [FIG. 83 C].

Separation of Pericardial, Pleural, and Peritoneal Parts of the Coelom

Each pericardioperitoneal canal lies lateral to the oesophagus and trachea in the mesentery, ventral to the intermediate mesoderm, and dorsomedial to the curved dorsal edge of the septum transversum which contains the common cardinal vein and the phrenic nerve [FIG. 83 C]. On each side a lung bud grows laterally from the caudal end of the trachea into the corresponding canal which becomes the primitive **pleural cavity.** The cavity and lung bud then extend laterally into the dorsomedial edge of the septum transversum, caudal to the common cardinal vein and the phrenic nerve. Thus they split this part of the septum transversum into two ridges, a cranial **pleuropericardial membrane** and a caudal **pleuroperitoneal membrane,** which meet ventrally in the unsplit part of the septum transversum. Having split this part of the septum transversum to form the definitive pleural cavity within it, that cavity and the contained lung bud expand into the body wall, passing cranially (lateral to the common cardinal vein and the phrenic nerve), caudally, and ventrally round the pericardium, till the two pleural cavities abut on each other ventral to the pericardium in the midline, separated only by mesoderm which becomes the anterior mediastinum and anterior part of the superior mediastinum [FIGS. 84 and 85].

Thus the definitive pleural cavity is formed by an extension of the pericardioperitoneal canal into the body wall, the canal remaining as a narrow space around the lung root. As the growing lung passes upwards it lies to the lateral side of the fold containing the common cardinal vein and phrenic nerve, which is thus pressed against the pleuropericardial opening, compressing it towards the median plane until it is obliterated by fusion of the fold with the mesoderm surrounding the trachea and lung bud. When this occurs the pericardial cavity is entirely shut off from the remainder of the coelom and it becomes a completely closed space. Very rarely, as an abnormal condition, the pericardial cavity remains in communication on one or other side, more often the left, with the pleural cavity.

The pleuroperitoneal folds, which increase in size as the pleural cavities and lung buds expand caudally lateral to them, are rudiments of the lateral parts of the diaphragm; they are connected ventrally with the septum transversum and grow medially until they fuse with the mesoderm of the side wall of the foregut and with the dorsal mesentery [FIGS. 84 and 85]. When this fusion is completed and the **pleuroperitoneal opening** closed, the cavity of the portion of the coelom surrounding the lung root—the original pericardioperitoneal canal—is separated from the more caudal part of the coelom, which now becomes the peritoneal cavity.

Formation of Diaphragm

There are four main developmental parts of the diaphragm— ventral, dorsal, and right and left lateral.

The ventral part is formed from the septum transversum, which is gradually differentiated into caudal, middle, and cephalic parts. The caudal part is transformed into: (1) the mesodermal tissue of the liver; and (2) the ventral mesentery of the abdominal part of the foregut. The cephalic part becomes the caudal or diaphragmatic wall of the pericardium.

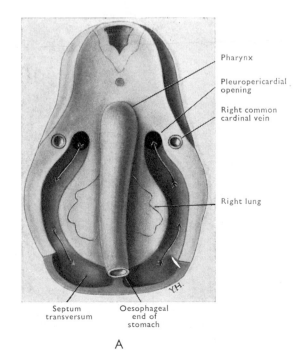

Pharynx

Pleuropericardial opening

Right common cardinal vein

Right lung

Septum transversum

Oesophageal end of stomach

A

Right vagus nerve

Spinal medulla

Notochord

Left dorsal aorta

Oesophagus

Posterior cardinal vein

Lung

Phrenic nerve

Inferior vena cava

Heart in pericardium

Phrenic nerve

B

FIG. 84 Diagram of later stage of differentiation of coelom. A, from dorsal surface; B (after Patten, 1953), transverse section cut at the level of lung bud in A, showing commencing ventral extension of pleural cavities.

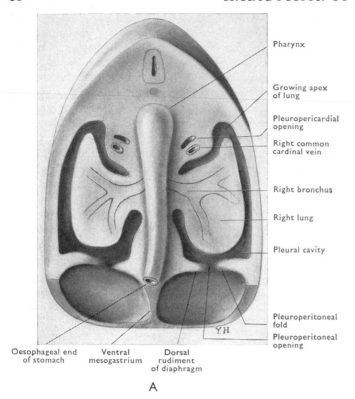

Labels for left figure (A): Pharynx; Growing apex of lung; Pleuropericardial opening; Right common cardinal vein; Right bronchus; Right lung; Pleural cavity; Pleuroperitoneal fold; Pleuroperitoneal opening; Oesophageal end of stomach; Ventral mesogastrium; Dorsal rudiment of diaphragm

A

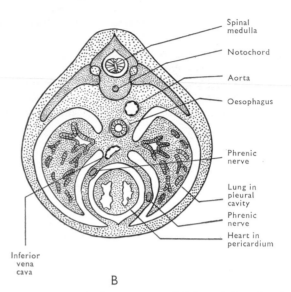

Labels for right figure (B): Spinal medulla; Notochord; Aorta; Oesophagus; Phrenic nerve; Lung in pleural cavity; Phrenic nerve; Heart in pericardium; Inferior vena cava

B

FIG. 85 Diagram of still later stage of differentiation of coelom. The pleurae are separated from the pericardium, but still communicate with the peritoneum.

A, from dorsal surface; B (after Patten, 1953), transverse section at the level of lung roots in A, showing ventral extension of the pleural cavities.

The middle part is transformed into the ventral portion of the diaphragm.

The dorsal part of the diaphragm, including the part surrounding the oesophagus, is developed from the mesoderm of the dorsal mesentery of the foregut. The left gastric artery, which originally lay in the dorsal mesentery, thus comes to lie on the oesophageal part of the diaphragm. Each lateral part is derived from the pleuroperitoneal folds mentioned above. The two lateral portions grow towards the median plane till they fuse with the dorsal portion; but sometimes, especially on the left side, the fusion is not completed. The pleuroperitoneal opening then remains unclosed (in the region of the vertebrocostal trigone [FIG. 416] close to the mesoderm forming the cranial pole of the kidney and suprarenal), and a portion of the abdominal contents may pass through it into the pleural sac, constituting a congenital diaphragmatic hernia.

SUMMARY OF EXTERNAL FEATURES OF HUMAN EMBRYO AND FOETUS AT DIFFERENT PERIODS OF DEVELOPMENT

First Month

During the first 14 days after fertilization the human ovum descends through the uterine tube, enters the uterus as a morula, and is transformed into a blastula which attaches itself to the uterine epithelium [FIG. 30] and penetrates the decidua compacta [FIG. 31]. About the **seventeenth day** the primitive node has appeared; secondary mesoderm is spreading in the embryonic area from the primitive streak (now well formed); and the allantoic diverticulum has grown into the body stalk [FIG. 86].

By the **eighteenth** or **nineteenth day** the area, though variable in both size and shape, may be from 1 to 1·5 mm. long and 1 mm. or less in breadth [FIGS. 48 and 87]. The primitive streak is now fully developed with a very distinct primitive groove on its surface. The head process is growing headwards from the primitive node and is progressively

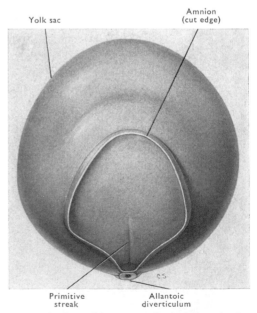

Labels: Yolk sac; Amnion (cut edge); Primitive streak; Allantoic diverticulum

FIG. 86 Dorsal view of human embryo (Mateer) of 17 days. ×35. The amnion is cut away to show the embryonic area from above. The primitive streak is seen in the caudal part of the area and beyond that is a section of the body stalk with the allantoic diverticulum in it. (After G. L. Streeter, *Contr. Embryol. Carneg. Instn*, 1920.)

tunnelled by the notochordal canal. The tail fold is carrying the primitive streak over the caudal margin of the embryonic area, and the body stalk, containing the allantois, appears to be bent dorsally at right angles to the region of the cloacal membrane [FIG. 49]. There may be an indication of the formation of neural folds [FIG. 87].

By the **nineteenth** or **twentieth day** the embryonic area is still variable in size, but the length (1·5–2 mm.) is then definitely greater than the breadth. The tail fold and neural folds are very evident [FIG. 88].

By the end of the **third week** the head fold and tail fold are distinctly formed, the neural folds are well developed especially in the head region, the neural groove is still completely open or just beginning to close, and the formation of mesodermal somites has begun [FIG. 89]. The length of the embryo may now be about 2 mm., but there is some variation in the relation of length to state of development at this stage.

In the next few days the primary parts of the brain begin to be evident, the neural groove closes except in the cephalic and caudal regions where the neuropores [p. 51] are seen. The mesodermal somites increase in number, and the cephalic region begins to bend ventrally as the cephalic flexure forms [FIGS. 90 and 91].

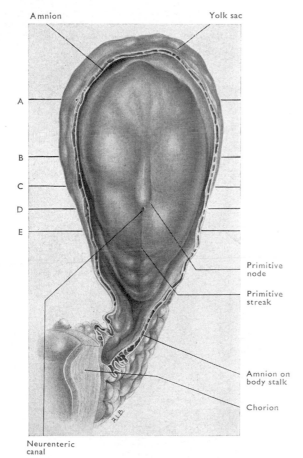

Amnion Yolk sac

A
B
C
D
E

Primitive node

Primitive streak

Amnion on body stalk

Chorion

Neurenteric canal

FIG. 87 Dorsal view of human embryo of 18 days. × 50.
(After C. H. Heuser, *Contr. Embryol. Carneg. Instn*, 1932a.)

The amnion is cut away to show the embryonic area now elongated (1·53 × 0·75 mm.) and with the caudal part slightly bent ventrally. The primitive node, with opening of neurenteric canal, and primitive streak are seen in the caudal half of the area and there is an indication of commencing neural folds in the cranial half. Cf. Fig. 48; and for sections of this embryo, see Fig. 50.

By the end of the **fourth week** the length of the embryo is about 2·5–3 mm., the head is bent at right angles to the body, and shows the projection of the forebrain. The number of somites increases to over twenty. The rudiments of the otic vesicles have appeared as slight ectodermal depressions in the region of the hindbrain, and the cephalic and caudal neuropores are closed or closing. The yolk sac is still relatively large and in free communication with the midgut. The pharyngeal arches and grooves are appearing in the wall of the foregut dorsal to the bulging pericardium [FIGS. 66 and 92].

Second Month

During the **fifth week** the embryo attains a length of 5 to 6 mm. The somites increase to thirty-eight or thirty-nine; the rudiments of the limbs appear and become quite distinct, the fore limb in advance of the hind limb; the otic vesicles sink into the interior of the head. The tail becomes a very definite appendage, and the bulgings caused by the optic vesicles are quite obvious on the surface of the head. The cervical flexure remains acute, and the head bends at right angles upon itself in the region of the midbrain, forming the cephalic flexure, with the result that the frontal region is turned towards the tail [FIG. 93].

By the end of the **sixth week** the length of the embryo has increased to 11 or 12 mm. CR (CR indicates the crown-rump measurement which corresponds with the sitting-height [Mall, 1910]). During this week the lens of the eye appears as a thickening of the surface ectoderm which becomes a vesicle and separates from the surface, sinking into the interior of the eyeball. The three segments of the upper limb become visible, and the rudiments of the fingers appear. The lower limb is less advanced. The third and fourth pharyngeal arches lie in the depths of the cervical sinus, overlapped by the caudal margin of the second arch [FIG. 71].

During this week also the olfactory pits appear between the median and the lateral nasal processes, and grow dorsally into the roof of the stomodaeum; the maxillary processes of the mandibular arches, growing towards the median plane, fuse with the lateral nasal and the globular processes, so completing the lateral parts of the primitive upper lip [FIGS. 67 and 68].

The nodular outgrowths which form the rudiments of the auricle appear on the margins of the first pharyngeal groove and begin to fuse together [FIG. 94].

By the **seventh week** the embryo has attained a length of 17–18 mm. (CR). The cervical flexure has begun to straighten. The rudiments of the eyelids have appeared. The margins of the auricles are now well defined; the thighs and the toes have appeared [FIGS. 95 and 97].

At the end of the **eighth week,** when the embryo becomes a foetus, it has attained a length of about 25 to 30 mm. (CR). The external genital organs are developing and a **genital tubercle** is evident [FIG. 678].

Third Month [FIG. 96].

The head grows less rapidly, and is relatively smaller in proportion to the whole body. The eyelids close, and their margins fuse together. The neck increases in length. Nails appear on the fingers and toes. The proctodaeum is formed and the external genital organs are sufficiently differentiated for the sex to be distinguished. By the end of the third month

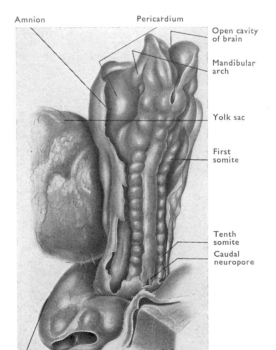

FIG. 88   Dorsal view of human embryo of 19 days.   × 50.
(After W. C. George, *Contr. Embryol. Carneg. Instn*, 1942.)

The amnion is cut away to show the embryonic area (1·16 mm. long) with primitive node and streak, neurenteric canal, and developing neural groove.

FIG. 90   Dorsal view of 2·2 mm. human embryo of 22 days.
× 40.   (After F. Payne, *Contr. Embryol. Carneg. Instn*, 1925.)

The neural groove is closed in the future cervical region, the brain vesicles though still open are beginning to take shape, the site of origin of the optic vesicle is indicated, and there are seven pairs of somites present.

FIG. 89   Dorsal view of human embryo of 21 days. × 50.
(After N. W. Ingalls, *Contr. Embryol. Carneg. Instn*, 1920.)

The broad cephalic end of the neural plate is elevated with commencing formation of the brain.   The neural groove is still open and its caudal end is enclosing the cephalic end of the primitive streak.   The length of the embryonic area has increased to 1·4 mm. and somite formation has begun.

FIG. 91   Dorsal view of human embryo of 23 days.   × 30.
(After G. W. Corner, *Contr. Embryol. Carneg. Instn*, 1929.)

The closure of the neural groove has advanced almost to the stage of cephalic and caudal neuropores.   The three brain vesicles are forming, the pericardium and the mandibular arch are evident, and there are ten pairs of somites present.

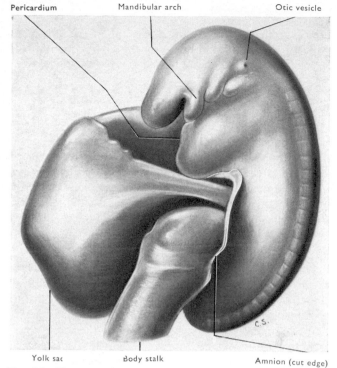

Pericardium   Mandibular arch   Otic vesicle

Yolk sac   Body stalk   Amnion (cut edge)

FIG. 92 Human embryo, 2·5 mm. long, with twenty-three pairs of somites. ×34. (After Thompson, 1907.)

The head is small and flattened; the optic vesicles project slightly from the forebrain, and the position of the otic vesicle is indicated. The yolk sac, still relatively large, is connected to the alimentary canal through the vitelline duct which, with the body stalk, emerges from the umbilical orifice. The position of the somites is indicated by slight elevations of the surface. There is no trace of the limbs.

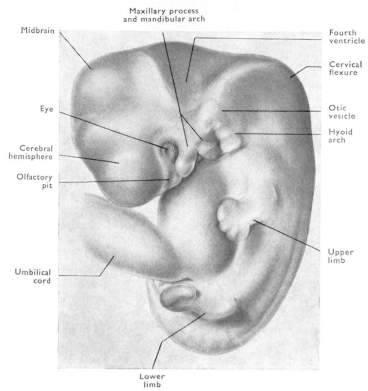

Maxillary process and mandibular arch   Fourth ventricle

Midbrain   Cervical flexure

Eye   Otic vesicle

Hyoid arch

Cerebral hemisphere

Olfactory pit

Upper limb

Umbilical cord

Lower limb

FIG. 94 Human embryo, 13 mm. CR. length. (After G. L. Streeter, *Contr. Embryol. Carneg. Instn*, 1948; Embryo no. 8101.)

The limbs are bent ventrally and show differentiation into segments. The cervical sinus is closing and rudiments of the auricle of the external ear are present on the first and second arches. The pontine flexure of the hindbrain, the fourth ventricle, and the olfactory pit are evident.

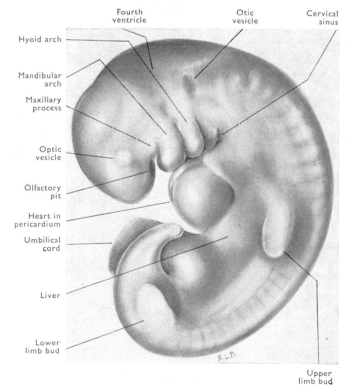

Fourth ventricle   Otic vesicle   Cervical sinus

Hyoid arch

Mandibular arch

Maxillary process

Optic vesicle

Olfactory pit

Heart in pericardium

Umbilical cord

Liver

Lower limb bud

Upper limb bud

FIG. 93 Human embryo, 5·5 mm. CR length. (After G. L. Streeter, *Contr. Embryol. Carneg. Instn*, 1945; Embryo no. 6830.)

The limb buds are now quite distinct [cf. Fig. 63] and the upper limb is beginning to elongate and curve forwards. The maxillary process has grown forwards below the eye, and the cervical sinus is evident.

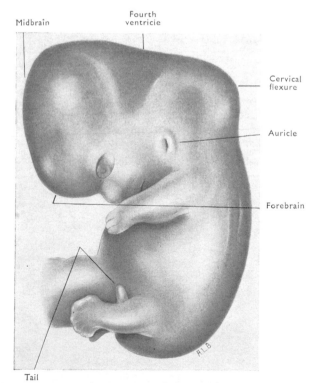

Midbrain   Fourth ventricle

Cervical flexure

Auricle

Forebrain

Tail

FIG. 95 Human embryo, 18·5 mm. greatest length, about 7 weeks old. The abdomen is very prominent on account of the rapid increase of the liver. The digits of the hand and foot are distinct but not separated from one another. The margin of the auricle is completed. The eyelids have begun to form. [Cf. Fig. 97.]

Umbilical   Genital   Tail
cord   tubercle

FIG. 96 Human foetus at the beginning of the third month. The cervical flexure is now slight. The upper limbs have assumed a characteristic position with the hands in front of the face, and the toes are free. The tail has almost disappeared, and the genital tubercle is evident.

the CR length of the foetus is about 100 mm. and it weighs about 50 g.

Fourth Month

Fine, downy hairs (lanugo) are developed. The CR length of the foetus is about 145 mm. and it weighs 200 g.

Between the **fifth** and **eighth months** the hairs become more developed, there is less disproportion between the upper and lower limbs and a sebaceous deposit (**vernix caseosa**) increases in amount to cover the skin. The CR length increases from 190 to 200 mm. at the end of the fifth month, to 300 to 320 mm. at the end of the eighth month. The weight of the foetus increases from a mean of 460 g. to 2 to 2·5 kg. in the same period.

Ninth Month

The hair begins to disappear from the trunk, but it often remains long and abundant on the head. The skin becomes paler, the plumpness increases, and the umbilicus reaches the centre of the trunk. During this month the nails reach the ends of the fingers. At the end of the month, when the foetus is born, its CR length is from 340 to 360 mm. and it weighs from 3 to 3·5 kg.

Age, Length, and Weight of Embryo and Foetus

The lengths and weights of embryos and foetuses at increasing ages given in the foregoing summary are compiled from a number of sources. It must be understood, however, that they are only approximate, since embryos and foetuses of the same length, or of the same weight, or even at the same stage of development, are not necessarily of the same age (Streeter, 1921, 1942). Moreover, it is usually impossible to calculate the **actual age** of an embryo (dating from the time of fertilization) and the age assigned (except in the case of some young embryos for which the data necessary for the calculation of fertilization age are available) is then the **menstrual age,** dating from the first day of the last menstruation. Owing to the average time of ovulation in relation to the menstrual cycle, the menstrual age of the majority of embryos is probably at least 14 days more than the actual age. The usual method of calculating the duration of pregnancy by counting 280 days from the first day of the last menstruation gives in any case only the approximate date of parturition.

It should also be noted that the 'greatest length' of an embryo (excluding the lower limbs) depends upon the cervical flexure. While the head of the embryo is fully flexed the crown-rump measurement (CR) is less than the greatest length, but later the two measurements are the same.

Formulae for the calculation of age from length have been devised, but it is more satisfactory to use tables and graphs (constructed from extensive data) from which the probable range of age can be read at a glance (Streeter, 1921).

FIG. 97 Photograph of 21 mm. human embryo in amnion and chorion. ×2·5. For the form of the embryo, cf. Fig. 95; and note the position of the definitive yolk sac between the expanding amnion and the chorion. (Carnegie no. 8537A; by special permission of Dr. George W. Corner.)

REFERENCES

ALLEN, E., PRATT, J. P., NEWELL, Q. U., and BLAND, L. J. (1930) Human ova from large follicles : including a search for maturation divisions and observations on atresia, *Amer. J. Anat.*, **46**, 1.

AMOROSO, E. C. (1952) Placentation, in *Marshall's Physiology of Reproduction*, Vol. II, Chapter 15, ed. Parkes, A. S., London.

ARMITAGE, P., BOYD, J. D., HAMILTON, W. J., and ROWE, B. C. (1967) A statistical analysis of birth-weights and placental weights, *Hum. Biol.*, **39**, 430.

AUSTIN, C. R., and BISHOP, M. W. H. (1957) Fertilization in mammals, *Biol. Rev.*, **32**, 296.

AUSTIN, C. R., and WALTON, A. (1960) Fertilisation, in *Marshall's Physiology of Reproduction*, Vol. I, Part 2, Chapter 10, ed. Parkes, A. S., London.

BARCROFT, J. (1946) *Researches on Pre-Natal Life*, Oxford.

BARCROFT, J. (1952) Foetal respiration and circulation, in *Marshall's Physiology of Reproduction*, Vol. II, Chapter 17, ed. Parkes, A. S., London.

BISHOP, M. W. H., and WALTON, A. (1960) Spermatogenesis and the structure of mammalian spermatozoa, in *Marshall's Physiology of Reproduction*, Vol. I, Part 2, Chapter 7, ed. Parkes, A. S., London.

BÖÖK, J. A., LEJEUNE, J., LEVAN, A., CHU, E. H. Y., FORD, C. E., FRACCARO, M., HARNDEN, D. G., HSU, T. C., HUNGERFORD, D. A., JACOBS, P. A., MAKINO, S., PUCK, T. T., ROBINSON, A., TJIO, J. H., CATCHESIDE, D. G., MULLER, H. J., and STERN, CURT (1960) A proposed standard system of nomenclature of human mitotic chromosomes, *Lancet*, i, 1063.

BOYD, J. D., and HAMILTON, W. J. (1952) Cleavage, early development and implantation of the egg, in *Marshall's Physiology of Reproduction*, Vol. II, Chapter 14, ed. Parkes, A. S., London.

BOYD, J. D., and HAMILTON, W. J. (1970) *The Human Placenta*, Cambridge.

BREWER, J. I. (1938) A human embryo in the bilaminar blastodisc stage. (The Edwards-James-Brewer ovum), (*Contr. Embryol. Carneg. Instn.* (No. 162), **27**, 85.

BRYCE, T. H. (1924) Observations on the early development of the human embryo, *Trans. roy. Soc. Edinb.*, **53**, 533.

CLERMONT, Y., and LEBLOND, C. P. (1955) Spermiogenesis of man, monkey, ram and other mammals as shown by the ' Periodic Acid-Schiff' technique, *Amer. J. Anat.*, **96**, 229.

DAVIES, F. (1948) Ovulation and the menstrual cycle, *Lancet*, ii. 720.

DIXON, A. F. (1927) Normal oocyte showing first polar body and metaphase stage in formation of second polar body, *Irish J. med. Sci.*, 6th Ser., p. 149.

DUIJN, C. VAN, JR. (1958) *Fertilization*, London.

FELIX, W. (1910) Zur Entwicklungsgeschichte der Rumpfarterien des menschlichen Embryo, *Morph. Jb.*, **41**, 577.

FELL, H. B., and ROBINSON, R. (1929) The growth, development and phosphatase activity of embryonic avian femora and limb-buds cultivated *in vitro*, *Biochem. J.*, **23**, 767.

FRAZER, J. E. (1914) The second visceral arch and groove in the tubotympanic region, *J. Anat. Physiol.*, **48**, 391.

GOODSIR, J. (1857) On the morphological constitution of the skeleton of the vertebrate head, *Anatomical Memoirs*, ed. Turner, W., (1868), Vol. 2, 88, Edinburgh.

HAMILTON, W. J., and BOYD, J. D. (1960) Development of the human placenta in the first three months of gestation, *J. Anat. (Lond.)*, **94**, 297.

HAMILTON, W. J., BOYD, J. D., and MOSSMAN, H. W. (1962) *Human Embryology*, Cambridge.

HARRISON, R. G. (1959) *A Textbook of Human Embryology*, Oxford.

HARRISON, R. G., and MALPAS, P. (1953) The volume of human amniotic fluid, *J. Obstet. Gynaec. Brit. Emp.*, **60**, 632.

HARTMAN, C. G. (1929) How large is the mammalian egg?, *Quart. Rev. Biol.*, **4**, 373.

HERTIG, A. T., and ROCK, J. (1941) Two human ova of the pre-villous stage, having an ovulation age of about eleven and twelve days respectively, *Contr. Embryol. Carneg. Instn*, (No. 184), **29**, 127.

HERTIG, A. T., and ROCK, J. (1945) Two human ova of the pre-villous stage, having a developmental age of about seven and nine days respectively, *Contr. Embryol. Carneg. Instn*, (No. 200) **31**, 65.

HERTIG, A. T., ROCK, J., and ADAMS, E. C. (1956) A description of 34 human ova within the first 17 days of development, *Amer. J. Anat.*, **98**, 435.

HERTIG, A. T., ROCK, J., ADAMS, E. C., and MULLIGAN, W. J. (1954) On the preimplantation stages of the human ovum, *Contr. Embryol. Carneg. Instn*, (No. 240) **35**, 199.

HEUSER, C. H. (1932a) A presomite human embryo with a definite chorda canal, *Contr. Embryol. Carneg. Instn*, (No. 138) **23**, 251.

HEUSER, C. H. (1932b) An intrachorionic mesothelial membrane in young stage of the monkey (macacus rhesus), *Anat. Rec.*, **52**, Suppl. 15.

HEUSER, C. H., ROCK, J., and HERTIG, A. T. (1945) Two human embryos showing early stages of the definitive yolk-sac, *Contr. Embryol. Carneg. Instn*, (No. 201) **31**, 85.

HEUSER, C. H., and STREETER, G. L. (1941) Development of the macaque embryo, *Contr. Embryol. Carneg. Instn*, (No. 181) **29**, 15.

HORSTADIUS, S. (1950) *The Neural Crest*, London.

HUGGETT, A. ST. G., and HAMMOND, J. (1952) Physiology of the placenta, in *Marshall's Physiology of Reproduction*, Vol. II, Chapter 16, ed. Parkes, A. S., London.

KRAUS, B. S., KITAMURA, H., and LATHAM, R. A. (1966) *Atlas of Developmental Anatomy of the Face*, New York.

LENNOX, B. (1960) Chromosomes for beginners, *Lancet*, i, 1046.

LEWIS, B. V., and HARRISON, R. G. (1966) A presomite human embryo showing a yolk-sac duct, *J. Anat. (Lond.)*, **100**, 389.

LEWIS, W. H. (1904) Experimental studies on the development of the eye in Amphibia. I. On the origin of the lens. Rana palustris, *Amer. J. Anat.*, **3**, 505.

LEWIS, W. H., and HARTMAN, C. G. (1933) Early cleavage stages of the egg of the monkey (macacus rhesus), *Contr. Embryol. Carneg. Instn*, (No. 143) **24**, 187.

MALL, F. P. (1910) Determination of the age of human embryos and fetuses, in *Manual of Human Embryology*, ed. Keibel, F., and Mall, F. P., Vol. I, 180.

MALPAS, P. (1964) Length of the human umbilical cord at term, *Brit. med. J.*, **1**, 673.

MAXIMOW, A. A. (1924) Relation of blood cells to connective tissues and endothelium, *Physiol. Rev.*, **4**, 533.

MORTON, W. R. M. (1949) Two early human embryos, *J. Anat. (Lond.)*, **83**, 308.

MOSSMAN, H. W. (1937) Comparative morphogenesis of the fetal membranes and accessory uterine structures, *Contr. Embryol. Carneg. Instn*, (No. 158) **26**, 129.

MURRAY, P. D. F., and HUXLEY, J. S. (1925) Self-differentiation in the grafted limb-bud of the chick, *J. Anat. (Lond.)*, **59**, 379.

NEEDHAM, J. (1931) *Chemical Embryology*, Cambridge.

NEEDHAM, J. (1934) *A History of Embryology*, Cambridge.

NEEDHAM, J. (1950) *Biochemistry and Morphogenesis*, Cambridge.

PATTEN, B. M. (1953) *Human Embryology*, Philadelphia.

ROCK, J., and HERTIG, A. T. (1942) Some aspects of early human development, *Amer. J. Obstet. Gynec.*, **44**, 973.

ROCK, J., and HERTIG, A. T. (1944) Information regarding the time of human ovulation derived from a study of 3 unfertilized and 11 fertilized ova, *Amer. J. Obstet. Gynec.*, **47**, 343.

ROCK, J., and HERTIG, A. T. (1948) The human conceptus during the first two weeks of gestation, *Amer. J. Obstet. Gynec.*, **55**, 6.

ROTHSCHILD, LORD (1956) *Fertilization*, London.

SCHULTZ-LARSEN, J. (1958) The morphology of the human sperm, *Acta path. microbiol. scand.*, Suppl. **128**.

SPANNER, R. (1935) Mütterlicher und kindlicher Kreislauf der menschlichen Placenta und seine Strombahnen, *Z. Anat. EntwGesch.*, **105**, 163.

SPEMANN, H. (1938) *Embryonic Development and Induction*, New Haven.

STIEVE, H. (1931) Die Dottersackbildung beim Ei des Menschen, *Anat. Anz.*, **72**, Ergänzhft. 44.

STREETER, G. L. (1920) A human embryo (Mateer) of the presomite period, *Contr. Embryol. Carneg. Instn*, (No. 43) **9**, 389.

STREETER, G. L. (1921) Weight, sitting height, head size, foot length, and menstrual age of the human embryo, *Contr. Embryol. Carneg. Instn*, (No. 55) **11**, 143.

STREETER, G. L. (1942) Developmental horizons in human embryos. Description of age group XI, 13 to 20 somites, and age group XII, 21 to 29 somites, *Contr. Embryol. Carneg. Instn*, (No. 197) **30**, 211.

TEACHER, J. H. (1924) On the implantation of the human ovum and the early development of the trophoblast, *J. Obstet. Gynaec. Brit. Emp.*, **31**, 166.

THOMPSON, P. (1907) Description of a human embryo of twenty-three paired somites, *J. Anat. Physiol.*, **41**, 159.

THOMPSON, P., and BRASH, J. C. (1923) A human embryo with head-process and commencing archenteric canal, *J. Anat. (Lond.)*, **58**, 1.

WADDINGTON, C. H. (1956) *Principles of Embryology*, London.

WEISS, P. A. (1939) *Principles of Development*, New York.

WITSCHI, E. (1948) Migration of the germ cells of human embryos from the yolk sac to the primitive gonadal folds, *Contr. Embryol. Carneg. Instn*, (No. 209) **32**, 67.

BONES

R. J. HARRISON

INTRODUCTION

OSTEOLOGY is the study of bone and bones, the hard supporting tissues of the body. All the bones present in an animal are together called the **skeleton.** The endoskeleton is that part of the supporting structure within the animal; the exoskeleton is the outer superficial component, well developed in many invertebrates but represented in Man only by hair, nails, and teeth.

The word skeleton is derived from *skeletos* (Gk.) = dried and regrettably it implies that all that is worth studying is the dead, dry bones left after crude preparation. In fact, living bones are plastic tissues, with organic and inorganic components, that change dramatically with age, that exhibit remarkable variations in shape and size, that show sexual dimorphism and, although mainly supporting, have other important functions. Bones can be affected by genetic, external, and internal environmental factors, and are thus subject not only to direct effects of disease but also indirectly to physiological and pathological processes involving other organs. When studying any bone, or a whole skeleton, a student must realize that much information can be derived about the original owner of the bones on matters pertaining to age, sex, stature, body habitus, health, diet, race, genetic and endocrinological conditions, and even personal idiosyncracies. Moreover, techniques have been developed to assess the antiquity of skeletal remains, thus providing much information of anthropological interest.

The bony skeleton cannot be considered solely as a collection of variously shaped components. Each bone is linked to another, or to several others, by joints or **articulations;** such an arrangement enables the skeleton, apart from providing support for soft tissues, to act as a system of levers to which muscles and other structures are attached. Where bones surround structures such as the brain, heart, lungs, or other organs they provide considerable protection. The interior of bones is used to lodge and protect the bone **marrow** in which blood cells develop and the bone substance itself provides a store of calcium and other salts that can be drawn upon when necessary.

Living bone consists essentially of connective tissue fibres impregnated with mineral substances, making it almost equally resistant to compression and to tension (Thompson, 1942), while giving it considerable elasticity. If a bone is decalcified by soaking it in dilute acid, it retains its shape completely but becomes flexible owing to the removal of mineral substances from the fibrous framework. If, on the other hand, the fibrous tissue is destroyed by burning, it still retains its shape but becomes brittle, inelastic and may crumble.

The connective tissue element consists of collagen fibrils in a ground substance, containing polysaccharides, which serves to bind the fibrils together. The mineral element resembles a phosphoric limestone or apatite, the principal ions being calcium and phosphate, though others are present in smaller amounts. The mineral salts are present as minute flattened crystallites which present an enormous surface area. This, along with the vascularity of bone, facilitates the rapid ion exchange which is known to occur in bone. Radioactive isotopes of calcium and phosphorus and certain other elements (e.g. lead and strontium) may be included in this process.

Articulations

To provide movement in an otherwise rigid framework, and to allow for growth, the bony skeleton is divided into separate parts joined, as a rule, by tissues which are sufficiently flexible to allow movement to take place. The union between two or more adjacent bones is called a joint or **articulation.** Bones are said to articulate with each other at any joint. The surface of a bone which meets another at a joint is its **articular surface** and, where the bones can move freely on each other, a joint space is present. The articular surfaces are then covered by an adherent layer of smooth, well-lubricated, **articular cartilage.** The underlying bone is also smooth so that the articular surface of a dried bone is easily identified although the cartilage has gone. Where less movement occurs, the joint is made by some form of connective tissue such as fibrous tissue or fibrocartilage, and the articular surface is usually rough. The union may, however, be made complete by fusion of the bones so that movement or further growth between them is prevented.

Movements at joints are controlled and produced mainly by muscles which are attached to the bones, but the bones are held together by all the tissues which surround them. Each bone is enclosed in a dense layer of fibrous tissue (**periosteum**), and this sleeve runs from bone to bone, constituting the fibrous membrane of the capsules of joints. Here it may be thickened to form connecting bands or ligaments. These ligaments, together with the muscles and other soft tissues, maintain the continuity of the skeleton and transmit tensile stresses. Compression stresses demand something more solid and the ends of bones are usually modified in shape and enlarged so as to provide a good bearing surface which distributes the pressure in any normal position.

Architecture of Bone

The form and structure of any bone are adapted to its functions of support and of resisting mechanical stresses. During its growth, and during adult life, each bone is continually being modified to maintain these functions as the stresses to which it is subjected change. Some bones, such as the mandible, alter dramatically in shape with advancing age, others exhibit little change once established. Alteration in shape and structure is brought about by changes in the rates of deposition and absorption of bone substance. The factors controlling these

FIG. 98   Diagram of the architecture of the bones of the lower
limb showing the direction of trabeculae.   (After Meyer, 1867).

processes are complex and not only involve local activity by
various cell types but are also affected by more general hormonal and metabolic changes.

The structural arrangements in any bone must not be considered without reference to its articulations with adjacent
bones, and the attachment of muscles and ligaments to it.
These related structures can cause or transmit either compression or tension forces which vary when a joint is moved to
a different position (Carey, 1929).   The internal structure of
bones is arranged so as to resist all such forces.

Two fundamental factors lead to strength and economy of
material.   The first of these is the intimate combination of
mineral salts and fibrous tissue.   The other is the nature of the
basic constructional units (osteones) of concentric microscopic
tubular lamellae [FIG. 100] which resist bending equally in
any direction and, being hollow, allow access of blood vessels
to living tissue.   Pritchard (1956) makes a further point that
fibres in each lamella change in direction as it is traced longitudinally, the fibre bundles being plaited or woven, and that
some regularly leave one lamella to pass into the next, binding
them together so that bone is a continuum.

The sections of femur and calcaneus shown in PLATE 3 and
FIGURE 98 demonstrate the essential features of the architecture of bone.   The bone substance forms **trabeculae** running
in directions suited to their functions.   In the **body** (*shaft*) of

the femur the trabeculae are concentrated at the periphery and
merge to form thick **compact bone** capable of resisting bending and other stresses, while the interior is free of bone substance and constitutes the **marrow cavity.**   At the ends of
long bones the thick tubular body is continued into an expanded meshwork of thin trabeculae, **cancellous** or spongy
bone, within a thinned-out cortex.   The trabeculae are strongly
developed in regions subjected to compression or tensile
stresses, as is well shown in the upper end of the femur in
PLATE 3.   The trabeculae spring from the compact bone,
diverge, and arch over to reach the head of the bone and
greater trochanter.

Lining the compact body there is a transitional zone of
coarse cancellous bone, and when the bone is fresh the cavity
is filled with soft, yellow marrow.   The cylindrical structure of
the body of a long bone provides almost as much strength as a
heavier solid rod of the same diameter and material.   Where
a body is curved, as it usually is to a slight extent, there is a
compensatory thickening of the compact bone in the concavity.
This thickening may show as a ridge or pilaster on the surface
of the bone.

At the expanded ends of a long bone, the wide surface area
distributes pressure over the articular cartilage and transmits
shocks through a large number of suitably arranged trabeculae.
The spaces between trabeculae accommodate red bone
marrow.

Bones without a body, exemplified by the calcaneus or heel
bone (PLATE 3), have a thin shell of compact bone, thickened
where necessary, and are filled with cancellous bone and red
marrow.   Trabeculae can be seen which transmit the weight
of the body from its upper articulations back towards the heel
and forwards to the ball of the foot.   Others curve forwards
from the heel to the front of the bone, intersecting with other
systems, transmitting the pull of the tendon of the calf muscles,
and knitting together the other two columns.   In the space
between these systems there is less stress and consequently a
more open texture of trabeculae.

Classification of Bones

For descriptive purposes bones are classified according to
their shape.

Long bones are found in the limbs and where they are
small, as in the fingers and toes, are **miniature long bones.**
Between the ends of each long bone is the **body,** a thick-
walled tube of compact bone with a continuous, central
medullary (marrow) cavity.

The remaining bones are classed as short, flat, or irregular,
according to their shape, and consist of cancellous bone within
a shell of compact bone.   They have no body and no medullary
cavity, but the interstices of the cancellous bone are filled with
very vascular red marrow.

Short bones include the carpal and tarsal bones of the
wrist and foot respectively, and the **sesamoid bones.**   The
sesamoids are found in the substance of tendons and are
named from their resemblance to sesame seeds.   The patella
or kneecap is usually regarded as a sesamoid bone.

Flat bones are thin rather than flat, for all are curved to
some extent.   They include the ribs, the shoulder blade, and
the bones of the skull vault.   In the latter the spongy substance
is sandwiched between plates of compact bone, contains many

venous channels, and is referred to as **diploë**, the veins being called diploic veins.

Irregular bones include the vertebrae and many of the skull bones.

Pneumatic bones. In birds, in particular, many bones are invaded by air-containing sacs from the respiratory system to provide lightness with strength. In Man, a number of the skull bones are so invaded from the region of the nose. The spongy part of the bone is absorbed and replaced by air-filled cavities with walls of compact bone lined by mucous membrane. This membrane is continuous through small openings with the lining of the nose and the cavities thus communicate permanently with the nose. These cavities are the **paranasal sinuses.** In some instances all the cancellous bone is not replaced, with the result that the air spaces form a series of smaller pockets or 'cells' as in the ethmoid bone. Of a similar nature are the **mastoid cells** which extend from the middle ear cavity into the mastoid part of the temporal bone.

Fresh Bone

Bone which has recently been removed from the body differs from the usual dried bones available for study.

The articular parts of most bones are covered by articular cartilage and, where the bones have been freely movable, the cartilage is smooth and slippery. But many bones have no articular cartilage and in the skull cap, for example, the rough articulating margins are united by fibrous tissue.

The whole of the bone, except the articular part, is invested by a membranous covering of **periosteum.** This consists of a layer of fibrous tissue containing connective tissue cells, with a second, deeper, osteogenic layer containing bone-forming cells, the **osteoblasts.** These cells are evident only when bone is actively being formed, during which process many of them are enclosed in the bone substance they produce, thus becoming **osteocytes.**

The periosteum is adherent to the bone because many of its collagenous bundles penetrate the bone and become incorporated in it. Vessels (and some nerves) pass from the periosteum into the bone to supply both the bone substance and the bone marrow. If periosteum is partially stripped off a living bone by accident or disease the loss of blood supply may be severe enough to cause death of the underlying bone.

Structures which are attached to bone (muscles, tendons, intermuscular septa and ligaments) are in continuity with it. The constituent fibres blend with the periosteum and many pass into the bone to join its constituent collagen fibre bundles. Muscles with 'fleshy' attachments have many minute fibrous continuations into the periosteum and bone. Vessels which enter bone tend to follow ligaments or muscles to their point of entry so that tearing of these structures is likely to be associated with damage to the blood supply.

The medullary cavity is filled with yellow marrow and the spaces in the cancellous bone with red marrow. The cavities in cancellous bone and the medullary cavity are lined with a layer of osteogenic cells continuous with the lining of the canals which traverse compact bone [p. 78] and so indirectly with the periosteum. This lining layer of cells is spoken of as **endosteum.**

Bone Marrow

Red and white blood cells are formed in the red bone marrow and after birth the marrow is the only normal source of red corpuscles and the chief source of those white blood cells which contain distinctly stainable granules (granular leucocytes).

In infants, red marrow pervades all the bones including the medullary cavities and even the larger canals. It is gradually replaced by yellow marrow, and at puberty red marrow is found only in the cancellous bone. As age advances even this is replaced by yellow marrow in the peripheral parts, such as the distal ends of long bones, and in the lower end of the vertebral column. Red marrow persists throughout life in most vertebrae, in the hip bones, proximal ends of the femur and humerus, ribs, and sternum. Samples of this marrow may be obtained by sternal puncture.

Yellow marrow is simple in structure and is mainly composed of fatty tissue. A few blood-forming cells remain, however, and may increase in numbers as a result of prolonged blood loss or continued blood destruction.

Red marrow is an active blood-forming organ and its colour is due to red cells. It consists of large numbers of cells contained in a spongework of reticular tissue in which numerous blood vessels ramify and anastomose.

In addition to reticular cells and normal blood cells, the main groups of cells found in red marrow are haemocytoblasts which give rise to white and red blood cells by division, nucleated red blood cells at an intermediate stage of development not normally found in the blood stream, and megakaryocytes credited with the production of blood platelets. Large multinucleated **osteoclasts** are found on or near the bone surfaces and are believed to absorb bone.

Gelatinous marrow is the degenerated marrow found in the skull bones of aged people.

Nutrient blood vessels piercing the shafts of bones are few and relatively small. They supply the yellow marrow and the inner two-thirds of the bone substance, the remainder of the compact bone being supplied by periosteal vessels. In adult long bones, these nutrient vessels extend to the cancellous ends, but during growth they are separated by a plate of cartilage (**epiphysial plate**) from the separate, cancellous extremities, i.e., epiphyses [p. 83], which have an independent blood supply.

Vessels supplying cancellous bone pierce the thin shell of compact bone and ramify in the marrow. Where the marrow is red, these vessels are numerous, relatively large, and form a terminal mesh of wide, thin-walled **sinusoids** which slow the blood flow and discharge into large veins. Many of the sinusoids appear to be partially or completely collapsed, and are referred to as intersinusoidal capillaries. Small nodules of lymphoid tissue are present throughout red marrow but lymph vessels have not been found in either yellow or red marrow. Bones containing red marrow often have large foramina, but this is not invariably true e.g. ribs, cranium, and sternum.

Dried Bones

Bones are prepared for anatomical study by soaking in alkali solutions, degreasing and drying (Edwards and Edwards, 1959). Attached structures, including periosteum and cartilage, are removed in the processing; the endosteum and marrow shrivel, and fat is removed.

Where articular cartilage of a synovial joint [FIG. 276] has been present the underlying bone is smooth with a polished appearance so that articular areas can usually be distinguished. Where the cartilage on the end of a bone is adherent to a disc of fibrocartilage, as in the case of the upper and lower surfaces of a vertebra, the underlying bone is flat and has the appearance of unglazed porcelain. When bones have been joined together by fibrous tissue without intervening cartilage the articular surfaces are rough and frequently serrated.

Where fibrous tissue has been attached to bone, a rough line or ridge is likely to be raised, but a solid mass such as a tendon or ligament may leave a smooth circumscribed area, depressed or elevated, which can resemble an articular facet. Where fleshy muscle is directly attached, the bone is smooth unless the fibrous sheath forms a raised outline or a fibrous intramuscular septum raises a ridge.

For any one bone the markings and proportions vary in different individuals, but they are sufficiently alike for a general description to be applicable to all.

CARTILAGE

Cartilage (gristle) is supplementary to bone. In many places it is a temporary formation which is later replaced by bone, but in others it is permanent, persisting throughout life.

Cartilage is not hard like bone; nor is it as strong. It consists of living cells, **chondrocytes,** contained in spaces or lacunae and surrounded by intercellular substance containing collagen. The amorphous component stains with basic dyes due to the chondroitin sulphates present. These mucopolysaccharides and the collagen give firmness and some elasticity to cartilage but it can easily be cut or damaged by pressure over a small area.

Cartilage is relatively non-vascular and is nourished normally by tissue fluids. The chondrocytes divide and grow within the cartilage but this interstitial growth is limited to a certain size by the need for diffusion of nutrients through the matrix. Increase beyond this limiting size can only be achieved by penetration of blood vessels into the cartilage. Such a vascular invasion, however, is often associated with the deposition of calcium salts in the ground substance, and this **calcification of cartilage** usually results in death of its cells followed by removal of the calcified tissue and its replacement with bone, **ossification of cartilage.** Growth at the surface is provided for by the presence of chondrogenic cells underlying a more fibrous layer of its covering, the **perichondrium.** The articular cartilage of synovial joints is an exception in that it has no perichondrium on its articular surface, which is in effect a raw surface of cartilage.

Classification of Cartilage

There are three main varieties of cartilage, namely, hyaline cartilage, and white and yellow fibrocartilage.

Hyaline cartilage appears homogenous, is bluish-white and translucent owing to the quality of its intercellular substance. It forms the temporary cartilage models from which many bones develop, and the remains of these are: (1) the articular cartilages of synovial joints; (2) the plates of cartilage

between separately ossifying parts of a bone during growth [p. 83]; and (3) the xiphoid process of the sternum and the costal cartilages, all of which ossify late or not at all.

Hyaline cartilage is also found in the respiratory system; in the nasal septum, in most of the cartilages of the larynx and in the supporting rings of the trachea and bronchi. Calcification occurs frequently in old age in the costal cartilages and in those of the larynx, and may be followed by ossification. Such deposition of calcium salts can be detected radiologically.

White fibrocartilage contains strong bundles of white fibrous tissue with more or less widely distributed groups of cartilage cells, but little matrix, among the bundles. This structure gives the tissue greater tensile strength than found in hyaline cartilage and also sufficient elasticity to resist considerable pressure. It is found: (1) as sesamoid cartilages in a few tendons; (2) as articular discs in the wrist joint and the joints at each end of the clavicle; (3) as a labrum or rim deepening the sockets of the shoulder and hip joints; (4) as two semilunar cartilages in each knee joint; (5) as intervertebral discs joining the adjacent surfaces of the vertebral bodies; (6) in a plate or disc which unites the two hip bones at the pubic symphysis; and (7) as a covering to the **articular** surfaces of certain bones that ossify in membrane such as the clavicle and mandible. Almost any grade of cartilage between hyaline and virtually pure collagenous tissue can be encountered at these sites.

Yellow fibrocartilage contains bundles of yellow elastic fibres but little or no white fibrous tissue. It is found in the external ear, in the auditory tube associated with the middle ear, and in the epiglottis.

BONE AS A TISSUE

Compact bone is hard but possesses some resilience. It consists of numerous cylindrical units, called **Haversian systems** after Clopton Havers who described them in 1691. Each system, also known as an osteone, has a **central canal,** about 0·05 mm. in diameter, which is surrounded by up to ten, or more, concentric **lamellae** of bony tissue [FIG. 100]. The central canal contains a small artery and vein, areolar tissue, bone forming cells or **osteoblasts,** and fine nerve filaments. The larger canals contain lymph vessels. The Haversian systems are orientated in line with the long axis of a bone [FIG. 101] and may branch and communicate at irregular intervals.

Between the lamellae of a Haversian system are numerous minute spaces or **lacunae,** connected with each other and with the central canal by fine radiating **canaliculi** [FIG. 100]. Lying in the intervals between the Haversian systems are varying numbers of irregularly arranged **interstitial lamellae** [FIG. 100], which also possess lacunae and canaliculi and often have the appearance of partially eroded Haversian systems.

Bone has two principal components, blended together physically but each with its own functional importance. There is an organic matrix, largely of collagen, but with some mucopolysaccharide, which accounts for about 25 per cent of the weight of fully formed bone. There is also a mineral matrix consisting of calcium, phosphate and variable amounts of magnesium, sodium, carbonate, citrate and fluoride. The mineral matrix has a crystalline structure, and X-ray and electron microscope studies have shown it to be characteristic of crystals of hydroxyapatite. The mineral component is influenced by various factors, such as parathyroid hormone, and is able to exchange its ions with those of the body fluids; thus forming a ready store of calcium.

DEVELOPMENT OF BONE

Bone develops in mesoderm [p. 47] as a sequel to the deposition of mineral salts in connective tissue or in a previously formed cartilage model.

When this process of calcification, followed by ossification, takes place initially in cartilage, the bone so formed is referred to as **cartilage bone.** When it takes place in fibrous tissue without the intervention of cartilage formation, the bone so formed is referred to as **membrane bone.**

Formation of Cartilage Bone

The commonest method of bone formation is for a preliminary foundation or model of cartilage to be laid down. This becomes calcified and then ossified to give rise to true bone.

The Cartilage Model. The first visible step is an aggregation of mesodermal cells where the bone will develop. The cells become visibly modified and rounded, **chondroblasts,** and lay down an amorphous intercellular substance between them.

Each cell enlarges and the surrounding amorphous substance forms a capsule round it. The cells divide and a capsule is laid down round each daughter cell, two new capsules being formed inside each older one. As the cells multiply and the formation of intercellular substance further separates them, the cartilage enlarges by interstitial growth, that is, growth of the cells and of the intercellular substance of the cartilage.

The surrounding mesenchyme forms a sheath of **perichondrium** with a superficial, more fibrous layer and a deeper cellular layer which forms chondroblasts. These provide for growth on the surface of the cartilage by apposition.

Bone Formation. The cartilage model of a **long bone** grows largely at its ends. The oldest part is near the middle and it is here that the oldest cartilage cells lie. After a time the amount of intercellular substance has increased considerably. The cartilage cells swell, alkaline phosphatase appears in them (Greep, 1948) and in the extracellular substance in which lime salts are then deposited as fine crystals less than 50 nm. in length (Robinson and Watson, 1952). The cartilage matrix thus becomes calcified and the cartilage cells are cut off from their nutriment and degenerate. In the meantime the perichondrium becomes vascularized, and a clear ground substance is deposited between it and the cartilage. This substance differs from cartilage and is uncalcified bone or **osteoid tissue** which later becomes impregnated with calcium salts. The perichondrial chondroblasts survive as **osteoblasts** and some of them are buried in the osteoid tissue which they now produce and are then called **osteocytes.** In this way the body of the model

comes to be encased in bone derived from the erstwhile perichondrium, now known as **periosteum.**

The calcified cartilage is short-lived. Cells from the periosteal region, including multinucleated **osteoclasts,** which can destroy bone, eat their way into the calcified cartilage. Blood vessels with bone-forming osteoblasts and blood-forming cells follow into the channels and spaces so formed, and the calcified cartilage is excavated and largely removed. The osteoblasts begin forming bone on the surface of the remaining spicules of calcified cartilage to form trabeculae. Some of the osteoblasts are trapped in the bone substance to become osteocytes, each of which lies in a **lacuna** and sends many fine processes into the bone forming **canaliculi.** Other cells multiply to fill the spaces between bone trabeculae with marrow.

Ultimately, the continued process of excavation by osteoclasts and rebuilding by osteoblasts leads to complete removal of all the calcified cartilage and, if the bone is long enough and wide enough, the excavated spaces will run together to form a **marrow cavity** supported externally by a sheath of periosteal bone.

The cartilage at the ends of the bone continues to grow by multiplication of its cells, particularly where it is continuous with the newly formed bone of the body. Cartilage thus formed calcifies and is, in its turn, excavated by the cells advancing from the new bone, invaded by blood vessels, and converted into cancellous bone. The multiplication of the cartilage cells abutting on the calcified part keeps up a supply of cartilage, the cells of which are arranged in the characteristic columns seen in FIGURE 99. The calcified matrix lying between the columns forms the basis of the bony trabeculae which are later laid down around them.

The bone grows in length as a result of the invasion of

Epiphysial cartilage

Osteoblasts

Periosteum

Endochondral bone

Calcified cartilage

Subperiosteal bone

Osteogenic layer of periosteum

J.L. BOLDERO.

FIG. 99 Ossification in cartilage (Le Gros Clark, 1971). Longitudinal section of metatarsal bone of foetal kitten. Note the arrangement of the zones described in the text.

Haversian canal

Interstitial lamellae

Concentric lamellae

Lacunae and canaliculi

FIG. 100 A   Photograph of ground transverse section of compact bone, showing Haversian systems (Haversian canal, concentric lamellae, lacunae, and canaliculi).   ×c. 75.

Periosteal surface

Periosteal lamellae

Oblique Haversian system

Lacunae and canaliculi

FIG. 101   Photograph of ground longitudinal section of compact bone, showing periosteal lamellae parallel to surface above and Haversian systems cut obliquely below.   ×c. 75.

Haversian canal

Canaliculi

Lacunae

FIG. 100 B   Photograph of a ground section of bone showing a single Haversian system.   The lacunae, canaliculi, and the central tubular cavity all appear black owing to their content of air.   Note how the canaliculi stream towards the central cavity from which the blood supply of the osteocytes is derived.

FIG. 102   Ossification in membrane (Leriche and Policard, 1926).

Schematic representation of the stages of formation of bone without the pre-formation of cartilage.

1, Oedema of connective tissue.
2, Multiplication of fibrils.
3, Deposit of pre-osseous ground substance.
4, Deposit of lime salts.

PLATE 1

Radiograph of full-term (nine-month) foetus.

PLATE 2

Composite radiograph of adult skeleton. (*Courtesy of Messrs. Ilford Ltd.*)

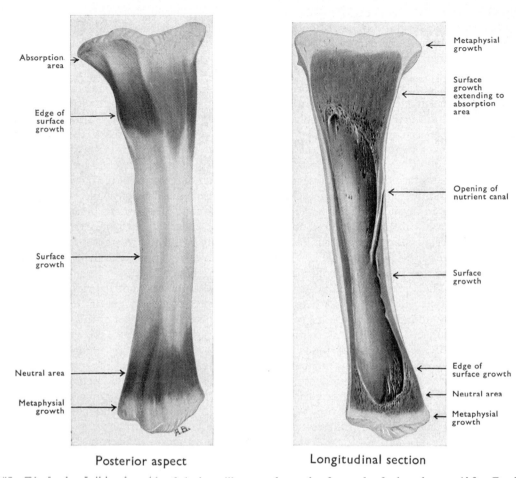

Posterior aspect   Longitudinal section

FIG. 103   Diaphysis of tibia of madder-fed pig to illustrate the mode of growth of a long bone.   (After Brash, 1934).

The animal was 40⅖ weeks old, and the madder had been omitted from the food for 30 days.   The new bone added during that period is white, in contrast to the old, red bone.

calcified cartilage produced by the active cartilage at its ends, and in width by the deposition of periosteal bone on its surface.   The cartilaginous ends increase in size partly by interstitial growth and partly by addition of new cells from the surrounding perichondrium.

Bone when first laid down is cancellous in texture consisting of a framework of delicate trabeculae surrounding a labyrinth of vascular spaces.   It is known as woven, or woven fibred bone, and is seen in foetal bones, in certain situations in adult bones where cartilage is ossifying, and in the early stages of repair in fractured bones.

New bone is laid down in two ways :

1. As successive thin concentric lamellae round the blood vessels to form Haversian systems.   These are associated with blood vessels which grow into the relatively wide channels between the bony trabeculae.   The osteoblasts line these channels and lay down successive layers of bone substance so as to fill the space available.   Some of them become trapped between the lamellae so formed and, with their processes, produce **lacunae** and **canaliculi** in the bone substance.   The innermost lamella is a tube surrounding the blood vessel and is lined with cells which through their processes link with others further out in the system.   Completion of the Haversian systems leads to the formation of compact bone from cancellous bone.   Small vessels which pass vertically in from the periosteal surface into the bone, are surrounded by the growing bone and produce spaces known as Volkmann's canals.   Longitudinal periosteal vessels on the surface become submerged and give rise to Haversian canals near the surface of the bone.

2. As successive thin layers or lamellae under the periosteum.   The periosteal lamellae are also referred to as circumferential lamellae because they lie parallel to the surface and surround the bone [FIG. 101].   Bundles of collagen fibres from the surrounding tissue which become calcified and are incorporated in the bone at right angles to the periosteal lamellae are called **perforating fibres** (Sharpey, 1856).

Bone is a vascular tissue richly supplied with capillaries, and canals containing capillaries in bone must be close enough to adjacent cells to keep them alive.   About 0·2 mm. is the greatest distance that can be tolerated and in practice the thickness of bone substance between an osteocyte and its supply is not more than 0·1 mm.

Osteoblasts and osteoclasts are the cells which lay down and destroy bone respectively.   The osteoblast was noted as long ago as 1845 by Goodsir and the osteoclasts in 1849 by Robin. Many details about the origin, differentiation, and activities of these cells remain obscure, and it may well be that one type can act like the other under certain conditions (Bourne, 1956; Ham, 1969).

Osteoblasts remain near the surface of bone and form a layer of cuboidal cells with large nuclei and basiphilic cytoplasm. They increase by mitotic activity and possibly by recruitment from neighbouring mesenchymal cells. They lay down the organic collagenous matrix and their ultrastructural characteristics are similar to those of other cells that produce protein. When active, alkaline phosphatase is present in osteoblasts but its function in mineralization is not clear. When inactive, osteoblasts take on an inconspicuous spindle-shaped appearance but are believed capable of renewed osteoblastic activity in response to certain stimuli, e.g. fractures.

Osteoclasts are usually large and multinucleated. The current view is that they arise by fusion of smaller cells (Tonna and Cronkite, 1961). Their presence is indicated by eroded areas in nearby bone, especially at places where remodelling occurs, and they are found in small pits called resorption or Howship's lacunae. Their characteristics include a conspicuous brush border, numerous mitochondria and vacuolated cytoplasm. They increase in number in response to parathyroid hormone. After their activity has ceased, they seem to disappear or to revert to their original cell type. Recent reviews by Blackwood (1964), Frost (1964), Richelle and Dallemagne (1965) are concerned with these cells and with various aspects of mineralization.

Intramembranous Ossification

When bone is laid down in connective tissue other than cartilage the site is indicated by a condensation of cells and their associated collagen fibres; organic bone matrix is laid down, and later impregnated with bone salts. In this manner spicules or trabeculae are formed in association with cells which differentiate into osteoblasts. Some of these cells become incorporated in the ground substance laid down and so become osteocytes. Others continue to lay down what can now be called bone so as to thicken and increase the number of the trabeculae. Vascular connective tissue is incorporated in the spaces between the trabeculae, and as more and more bone is laid down concentrically in the spaces, Haversian systems are produced and the bone becomes compact where so induced; elsewhere it remains cancellous.

After development in membrane, the formation of new bone continues much as it does in bone developed in cartilage. The surrounding connective tissue condenses to form a limiting layer of periosteum. In the deeper part of this are osteogenic cells and blood vessels which permit the formation of Haversian systems and continued growth or reconstruction as required. Intramembranous ossification occurs in certain skull bones, including the vault and the mandible, and in part of the clavicle.

Growth of Bone

The absorption of bony substance in some places, and its deposition in others, are processes which continue after the initial formation of the bone to provide for growth. These processes may be observed in bones taken from animals fed for suitable periods on substances, such as madder, which colour the bone, so that new bone laid down is white by contrast [FIGS. 103 and 144]. This continuation of controlled growth and modelling absorption, as it is termed, permits alterations in the shape of bones, alterations which are necessary if proper proportions and the correct relative position of individual parts of the bone are to be preserved.

The principles involved have been stated by Brash (1934), Lacroix (1949), and Frost (1963). One essential feature is that bone is laid down on or absorbed from a surface, whether this is an outer surface, the surface of a cavity, or the surface of a lamella in cancellous bone. Thus, if bone is laid down on one side of a given part of a bone and absorbed on the opposite side, the effect will be to move that part bodily in one direction. This will hold good whether it is a projection or a hollow, the wall of a cavity such as the skull, or the whole thickness of part of a long bone.

There is no interstitial growth of bone, as distinguished from interstitial changes in density which are constantly taking place. The proof of this is threefold:

1. The insertion of two metallic marks in a long bone of an animal, as performed by John Hunter (1772) and others, shows that these remain the same distance apart as the bone grows in length. Metallic rings placed round the bone under the periosteum become embedded in the bone by surface accretion and may be found later in the marrow cavity which has enlarged by absorption.

2. Sometimes the long bones of a child show 'lines of arrested growth'. These are transverse planes of greater density caused by heavier calcification at the ends of a bone where it has grown at a slower rate during an illness. Repeated radiographic observations at increasing ages show that these markings remain the same distance apart as the bone grows in length (Harris, 1933).

3. The indirect madder method of colouring the bones completely and then omitting the dye for a period, leads to the same conclusion. The increase in length is accounted for by the new white bone which appears at the ends, and the increase in girth by the subperiosteal deposit [FIG. 103.]

All three methods, as well as modern techniques using radioactive isotopes and tetracycline as markers, show that growth in length of a long bone occurs at the ends only.

Measurements can be used to calculate the rates of growth (Payton, 1932) and show that growth is greater at one end than the other. This is termed the 'growing end' of the bone. Both ends may be growing but the 'growing end', if damaged during growth, will lead to greater deformity than would the other if similarly damaged.

Modelling of Bone

Long bones grow in girth by subperiosteal deposition of bone, and the thickness of the cortex increases in relation to the stresses placed on the bone. This is achieved by the slower absorption of bone from the medullary aspect, so that the cavity enlarges less rapidly. Growth in length is by the continued ossification of the growing cartilage at the ends of the bone. As growth takes place at the junction between bone and cartilage, the bone which is laid down and not absorbed by the extension of the marrow cavity is left behind to become part of the body, as the process of ossification moves further away from the centre of the body. This bone has, however, been formed at a wide part of the bone, and modelling absorption must take place on the surface if the proper shape of the bone is to be maintained. This process is particularly important at the necks of long bones and in the ramus of the mandible and in similar situations where it is essential that the overall *shape* of the bone remains the same while the bone grows. Failure of such absorption may occur occasionally, even in bones that appear to be normal otherwise, and the result is a characteristic club-shaped appearance at the ends of the bones.

In bones such as those of the skull which are developed in membrane, growth takes place under the periosteum, mainly by deposition of new bone on the outer surface and edges and absorption from the inner surface. Either process occurs, however, wherever it is required to maintain the proper proportions of the bone.

CENTRES OF OSSIFICATION

Places where bone begins to be laid down are referred to as centres of ossification and the process spreads from such centres. The earliest, and usually principal centre in the body or shaft of a bone is referred to as a **primary centre.** This may result from fusion of several closely spaced smaller centres and, if they are further apart, there may be more than one primary centre appearing about the same time in one bone. Ossification centres can be detected by dissection, by staining with alizarin, and, when large enough, on X-ray films.

Primary centres of ossification appear at different dates in different developing bones but most of them before the end of the fourth month (135 mm. stage) of intra-uterine life; the majority appear between the seventh and twelfth weeks. Virtually all are present before birth. The date is fairly constant in different subjects for any one bone, and centres appear in a very orderly sequence (Noback and Robertson, 1951). In the following descriptions all dates of ossification stated in weeks or months refer to the antenatal period.

Secondary centres occur at a much later date than primary centres and are formed in parts of the cartilage model into which ossification from the primary centre has not spread. All long bones and many others acquire secondary centres. Nearly all of these secondary centres appear after birth, and the ossification which extends from them is the same as from primary centres, except that in this case the whole of the ossification takes place within the cartilage, which continues to enlarge by multiplication of its peripheral cells until growth ceases. The bone so formed is almost entirely cancellous.

The part of a bone ossified from a primary centre is called the **diaphysis.** The part ossified from a secondary centre is called an **epiphysis,** and a secondary centre is therefore often referred to as an epiphysial centre.

Ossification in the cartilage proceeds from the primary and secondary centres until only certain parts of the cartilage model are left.

1. A thin plate persists for a time between the diaphysis and each epiphysis. It is called the epiphysial cartilage, growth cartilage, or epiphysial plate [FIG. 276], and its edge at the surface of the bone forms the **epiphysial line.** So long as this plate of cartilage persists the diaphysis can continue to grow in length [PLATE 16, C and D], for it is the growth of this cartilage which produces the material into which the diaphysial ossification extends.

2. Where each epiphysis articulates with another bone, the articular layer of cartilage survives on that surface.

Epiphysial Growth. Since the epiphysis develops within cartilage it is initially surrounded by this substance which can be said to have three parts, all of them proliferating but in different ways.

1. A plate of cartilage lying, as mentioned above, between the epiphysis and the diaphysis, the margin of which forms the epiphysial line at the surface of the bone. This cartilage produces new cells which allow for longitudinal growth of the diaphysis only.

2. If an epiphysis is articular, the cartilage from the margin of the epiphysial plate to the edge of the articular area is covered by perichondrium from which new cartilage cells may be formed to allow lateral expansion of the epiphysis.

3. The articular surface is devoid of perichondrium and here the cartilage cells grow in two directions:

i. Towards the epiphysis. These cells calcify and ossify, allowing for expansion of the epiphysis towards the articular surface.

ii. Towards the articular surface. These cells form the articular surface; they neither calcify nor ossify but act as replacements for the worn cartilage.

With the exception of the latter group, which persist throughout life, but do not proliferate after growth of the epiphysis as a whole ceases, all the other parts of the epiphysial cartilage cease to grow, are invaded by the ossification process and are turned into bone. Extension of diaphysial ossification into the epiphysial plate finally obliterates it, leading to fusion of the epiphysis and the diaphysis, and thus preventing any further growth of the diaphysis in length. Extension of the epiphysial ossification to the perichondrium converts it to periosteum and prevents further endochondral growth of the epiphysis in this direction. With the cessation of growth, no further cartilage cells proliferate towards the epiphysis from its articular surface and the epiphysis therefore ceases to grow towards this surface. The cartilage which remains is entirely concerned with the continued production of new material for the bearing surface of the bone, though no new cells are formed by mitosis.

If an epiphysis is non-articular, as in the iliac crest, the cartilage is usually all replaced by bone.

Diaphysial Growth. The diaphysis continues to grow by virtue of cartilage growth at each epiphysial plate, and this part of the diaphysis where ossification is spreading rapidly is sometimes called the **metaphysis.** Growth in width and thickness takes place under the periosteum and endosteum. Healthy periosteum and endosteum retain their ability to produce new bone when required, and absorption continues to take place where necessary. Bones are remodelled to meet increased exercise, or may atrophy from lack of it, and broken bones can be repaired by the renewed activity of osteoblasts beneath the periosteum and in the medullary cavity of the bone. During growth, the longitudinal trabeculae are consolidated before the transverse trabeculae are laid down, and conversely, in cases of atrophy, the transverse trabeculae are absorbed before changes occur in the longitudinal ones (Harris, 1933). This keeps the length of the formed bone constant and so prevents interference with the function of joints and muscles.

Ossification and Fusion of Epiphyses

Every long bone has an epiphysis at one end, most have an epiphysis at each end, and some have more than one epiphysis at one or both ends, e.g., femur and humerus.

Most epiphyses are ossified from a single centre but if there is more than one centre they usually fuse before union with the diaphysis.

Secondary centres of ossification have been classified into: pressure, traction, and atavistic epiphyses (Parsons, 1904, 1905, 1908). These are supposed to be associated respectively with pressure on articular surfaces, with the pull of muscles, and with bone elements which have lost their independence during evolution, but the evidence is lacking. Section of a tendon or muscle, before the appearance of its 'traction epiphysis', does not arrest the subsequent development of that epiphysis (Barnett and Lewis, 1958). Members of the third group may on occasion remain as separate ossicles and cause doubt as to whether a fracture has occurred.

Dates of Ossification and Fusion of Epiphyses

The dates at which certain epiphysial centres appear and those at which they fuse with each other or with the diaphysis, particularly the latter, may be of considerable use clinically, but no student should ever attempt to memorize lists of dates either of appearance of centres or of their fusion.

The apposed surfaces of diaphyses and epiphyses have pronounced elevations and depressions which fit into each other so that the bones are to some extent locked together, and can withstand knocks and twists during the period of growth. An epiphysis may, however, be damaged or broken off as a result of violence. When this occurs the epiphysial cartilage may be injured so that the bone stops growing prematurely at that part or grows unevenly. It is therefore important, when a bone is broken near one end in a young person, to recognize at the outset whether the condition is an ordinary fracture or a separation of an epiphysis. It is consequently also important to remember the approximate times at which epiphyses usually join.

Postnatal changes in epiphyses of long bones may be placed in three main periods.

1. Secondary centres of ossification appear from the time of birth to 5 years of age (Francis et al., 1939).

2. Ossification spreads from these centres until the age of about 12 years in girls or 15 years in boys (Francis, 1940).

3. From 12 or 14 years to 25 years epiphyses fuse with the diaphyses, and growth ceases as fusion occurs.

The process is speeded up in girls at 5 years and again at 10 years (Todd, 1933, 1937), so that after these ages the female is in advance of the male and epiphysial fusions are earlier.

Exceptions of some practical use include the appearance of a centre in the distal end of the femur in the last month of foetal life. The proximal end of the tibia often develops a centre late in that month, as may the upper end of the humerus. A child is assumed to be 'full-term' if one of these centres is present.

At the other end of the scale, the clavicle has an epiphysis at its medial end the centre of which appears late (18–20 years). Fusion is delayed until between 20 and 30 years. A clavicle with an epiphysis present but unfused would be from a person between 18 and 30 years old.

As a general rule an epiphysis which appears early fuses late and vice versa. In the humerus, for example, a centre is present in the head shortly after birth but fusion does not take place until the age of 18 to 21 years. The centres at the lower end appear later (2–14 years) and fuse comparatively early (14–18 years).

The end of a bone with an epiphysis which fuses late grows for a longer time and also grows more rapidly than the other end. It is the 'growing end' already referred to [p. 82].

Most of the epiphysial centres in the vertebrae and ribs, and in the shoulder and pelvic girdles do not appear until puberty, and fuse about the age of 25. The dates of ossification of the parts of the scapula and hip bone entering into the formation of the shoulder and hip joints conform more with the ends of the adjacent long bones than with the rest of the shoulder and hip girdles.

The ring-like epiphyses on the upper and lower surfaces of the vertebral bodies fuse at the fully adult stage, which may not be reached till the twenty-fifth year or later. The back, therefore, continues to grow after the limb bones have ceased, for these usually fuse before the twenty-first year.

Variations. For any one epiphysis, the date of appearance of its centre varies in different children by weeks or months, and sometimes by 1 or 2 years. The date of union varies even more, especially in the vertebral column. It is common knowledge that some people reach their full height in their teens, while others do not do so until after the age of 25.

The sequence of dates of union is remarkably constant and the intervals between them remain proportionately the same in different people (Stevenson, 1924). If the first union takes place earlier than usual, ossification and growth will be completed at an early age. Dates given for individual bones in this text are approximations based largely on Schinz et al. (1971), but also on Paterson (1929), Todd (1930), Flecker (1932), Greulich and Pyle (1959), and Krogman (1962).

Factors Affecting Bone Growth

As a result of the growth processes described above, the bones of an individual skeleton normally reach a particular length, thickness, and shape by the time epiphysial fusion marks the end of bone growth. These features are usually maintained until affected by age or pathological changes. There are many factors, however, that influence bone growth while the skeleton is attaining its adult characteristics. They exert their actions chiefly on the regions of growth, and essentially affect the rate and duration of the proliferation of cartilage at these regions.

Genetic, hormonal, nutritional, mechanical, and neural factors influence bone growth. The general shape and over-all dimensions of a bone are genetically determined and genetic factors are primarily responsible for the final adult height or stature. Experiments have shown that the development of the main anatomical features of a bone depends on intrinsic factors residing in the tissue itself (Murray, 1936). The pattern and multiplicity of ossification centres in the skeleton reveal much about its phylogenetic history in that the centres are surprisingly persistent and may indicate where fusion has occurred between elements which are separate in other animals.

The somatotrophic or growth hormone from the anterior pituitary has an effect on proliferation of cartilage and thus on growth of long bones. Hypersecretion of the hormone early in the growth period, while the epiphysial cartilages are yet active, leads to gigantism, and if later, to acromegaly. In the latter condition, growth continues in bones which grow by apposition, such as the mandible and the carpal and tarsal bones. Reduced secretion can produce a form of dwarfism associated with premature fusion of epiphyses. Reduction of testicular androgen secretion during the growing period, as after castration in the young, results in prolonged growth at the epiphysial cartilages of long bones, and thus accounts for the disproportionately long limb bones of some eunuchs.

Thyroid hormones affect cartilage cell proliferation, thus thyroidectomy in immature animals results in defective growth of long bones. Parathyroid hormone causes proliferation of osteoclasts, mobilizes bone calcium and phosphate (causing an increase of their concentrations in the blood) and later resorbs the collagen framework. Calcitonin, from the parafollicular or light cells of the thyroid, appears to inhibit the breakdown of bone, perhaps by an effect on the protein matrix, and to cause a fall in blood calcium.

Normal bone growth will obviously depend on an adequate availability of calcium, phosphorus, and vitamin D in appropriate form. Deficiency of these interrelated substances leads to failure of proper ossification, osteoporosis, and loss of mechanical strength; the cartilage, continuing to grow or even

hypertrophy, causes enlarged and weakened ends to the soft bones (rickets). The region of the calcifying cartilage displays irregularity in the arrangement of cartilage cells, the bone formed there is imperfectly ossified and an increased amount of vascular tissue is often present. This poor quality bone is unable to withstand normal stresses and may become grossly deformed. Vitamin C is required to stimulate normal production of the collagenous matrix. Healing of fractures is delayed even in mild scurvy. Lack of vitamin A has been shown experimentally to lead to abnormal skeletal growth with thickening of bones, and to cause interference in remodelling. High concentrations of vitamin A lead to rarefaction of explanted skeletal rudiments.

The influence of mechanical factors in certain situations indicates how plastic growing bone can be. The primary shape and main features of a bone are genetically determined but less conspicuous markings and final modelling are in the main conditioned by extrinsic factors (Le Gros Clark 1971, Ham 1969). Removal of the temporal muscles and some neck muscles in newborn rats results in bony markings such as the temporal line and nuchal crests failing to appear. Certain postural activities, such as squatting, may be the cause of extension of articular facets and changes in shape of a bone. Abnormal pressure, exerted by an enlarging tumour or an aneurysm, can cause extensive bone erosion. A completely denervated bone does not grow as much in thickness as it should, nor show such pronounced markings, though there is only slight loss in length: muscular paralysis plays a major part in this type of atrophy. Severe illness may leave its mark in lines of arrested growth discernible in radiographs as a thin region of dense bone deposition.

THE SKELETON

The main subdivision of the skeleton is into axial and appendicular parts. The **axial skeleton** includes vertebral column, the sternum, twelve pairs of ribs, and the skull. With the skull are associated the lower jaw or mandible, the hyoid bone, and three pairs of small ear bones or tympanic ossicles. The vertebral column is common to all vertebrate animals—fishes, amphibians, reptiles, birds, and mammals. The **appendicular skeleton** includes the bones of the upper and lower limbs.

The terms used for the bones of the skeleton have been derived from Greek and Latin words (*skeletos*, Gk., means

dried). Some have been taken directly and are used in their original sense; scapula, for example, is a shoulder-blade, the sternum is the breast-bone. Others have an interesting, and sometimes obscure history. The sacrum (L. sacred, holy) was at one time supposed to be the bone from which the body would be resurrected. The coccyx became so called because of its resemblance to a cuckoo's bill; the fibula is the brooch-bone; the clavicle was likened to a little key on account of its shape and because it locked the shoulder girdle to the body (Field and Harrison, 1968).

THE VERTEBRAL COLUMN

The vertebral column extends from the base of the skull through the whole length of the neck and trunk. It consists of a number of vertebrae, placed in series and connected together by ligaments and discs of fibrocartilage to form a flexible curved support for the trunk. In a child, the normal number of separate vertebrae is 33. In an adult, 5 have fused to form the

sacrum and 4 to form the coccyx [FIGS. 112 and 114]. Of the 24 vertebrae remaining free, 7 are in the cervical region, 12 articulate with ribs and are thoracic, and 5 are in the lumbar region [FIG. 115]. They are numbered from the head end of each region.

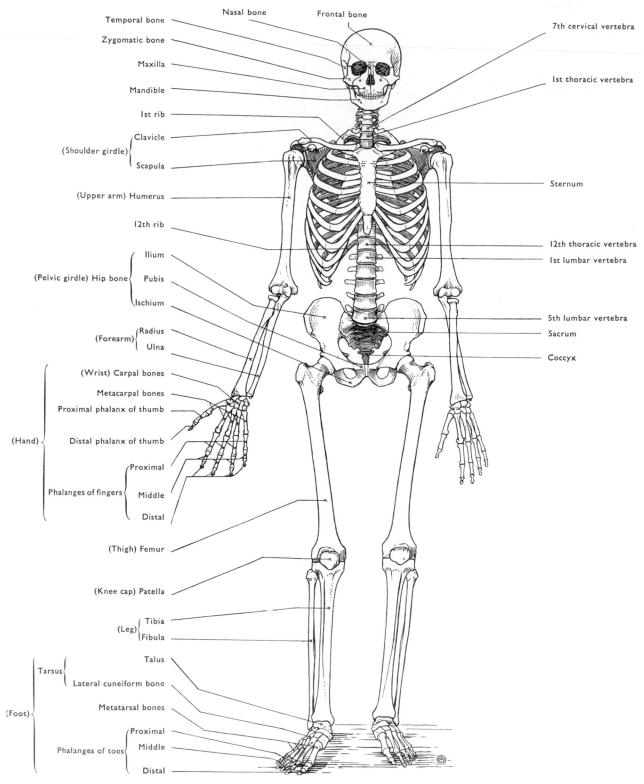

Temporal bone
Zygomatic bone
Maxilla
Mandible
1st rib
Clavicle
(Shoulder girdle)
Scapula
(Upper arm) Humerus
12th rib
Ilium
(Pelvic girdle) Hip bone { Pubis
Ischium
Radius
(Forearm) { Ulna
(Wrist) Carpal bones
Metacarpal bones
Proximal phalanx of thumb
(Hand) { Distal phalanx of thumb
Proximal
Phalanges of fingers { Middle
Distal
(Thigh) Femur
(Knee cap) Patella
Tibia
(Leg) { Fibula
Talus
Tarsus { Lateral cuneiform bone
(Foot) { Metatarsal bones
Proximal
Phalanges of toes { Middle
Distal

Nasal bone
Frontal bone

7th cervical vertebra
1st thoracic vertebra
Sternum
12th thoracic vertebra
1st lumbar vertebra
5th lumbar vertebra
Sacrum
Coccyx

Fig. 104 Anterior view of the skeleton of a man. The bones of
the left forearm and hand are in the position of pronation.

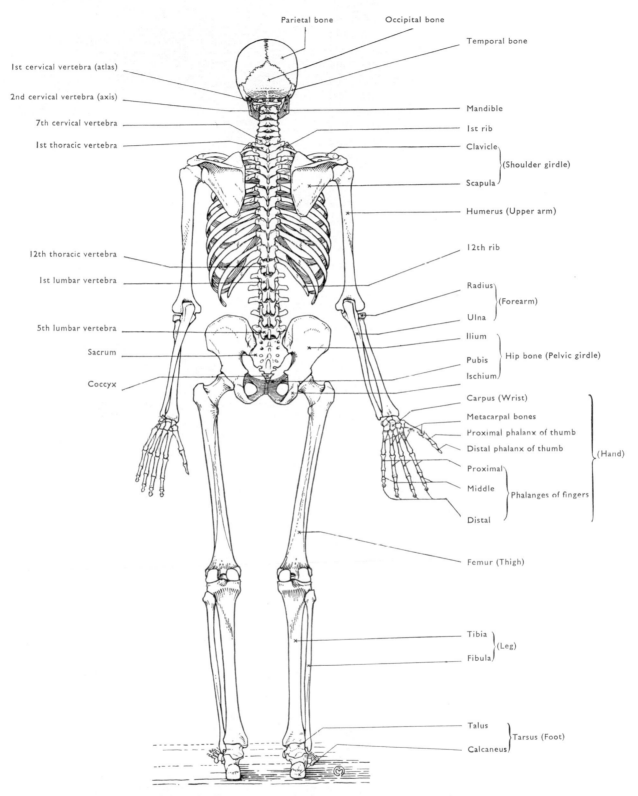

Parietal bone
Occipital bone
Temporal bone
1st cervical vertebra (atlas)
2nd cervical vertebra (axis)
7th cervical vertebra
1st thoracic vertebra
Mandible
1st rib
Clavicle
(Shoulder girdle)
Scapula
Humerus (Upper arm)
12th thoracic vertebra
1st lumbar vertebra
12th rib
Radius
(Forearm)
Ulna
5th lumbar vertebra
Ilium
Sacrum
Hip bone (Pelvic girdle)
Pubis
Coccyx
Ischium
Carpus (Wrist)
Metacarpal bones
Proximal phalanx of thumb
Distal phalanx of thumb
(Hand)
Proximal
Middle
Phalanges of fingers
Distal
Femur (Thigh)
Tibia
(Leg)
Fibula
Talus
Tarsus (Foot)
Calcaneus

FIG. 105 Posterior view of the skeleton of a man. The bones
of the left forearm and hand are in the position of pronation.

FEATURES COMMON TO MOST VERTEBRAE

With the exception of the first two cervical vertebrae, all vertebrae consist of a large, anterior, weight-bearing **body,** and a posteriorly placed **vertebral arch.** The arch springs from the posterolateral aspects of the body, and with it surrounds a large hole, the **vertebral foramen.** When the vertebrae are placed in series, these foramina, together with the ligamenta flava that unite the adjacent laminae (q.v.), form the **vertebral canal** which lodges the spinal medulla and its coverings.

The **vertebral arch** consists of two pedicles and two laminae [FIG. 106]. The sloping **laminae** meet in the midline posteriorly where they are continuous with the posteriorly directed spinous process, many of which can be felt in the back of the neck and trunk in the midline.

Each **pedicle** joins the anterolateral extremity of the corresponding lamina to the upper part of the posterolateral aspect of the vertebral body. Where each pedicle and lamina meet is a mass of bone from which three processes arise—the **transverse process,** directed laterally, and two **articular processes,** one directed superiorly, the other inferiorly. When adjacent vertebrae are fitted together [FIG. 110] the superior articular processes are in contact with the anterior surfaces of the inferior articular processes of the vertebra above, forming synovial joints with them. The apposed surfaces are, therefore, smooth articular **foveae** (facets) which are covered with articular cartilage during life.

Since the movements between vertebrae are about axes passing through the intervertebral discs, these movements are controlled to a great extent by the arrangement of the joints between the articular processes.

The vertical extent of the pedicles is considerably less than that of the vertebral body to which they are attached. Thus there is a deep notch (**incisura vertebralis inferior**) on each side, between the inferior articular process and the vertebral body [FIG. 106A] and this forms the intervertebral foramen [FIG. 110] with the shallow notch (**incisura vertebralis superior**) on the upper aspect of the pedicle below when the vertebrae are articulated. The intervertebral foramen is closed antero-inferiorly by the **intervertebral disc** which unites the vertebral bodies. Thus the disc, the joint between the articular processes, and (in the thoracic region) the costo-vertebral joint all abut on the foramen, which transmits the spinal nerve and various blood vessels entering and leaving the vertebral canal.

The Body

This is approximately cylindrical, but its flat superior and inferior surfaces are slightly flared so that the anterior and lateral surfaces are concave from above downwards. These surfaces are perforated by many small vascular foramina, while the posterior surface is perforated by foramina for the large veins from the red marrow in the body, and is flat or concave from side to side.

Each superior and inferior surface is rough for the attachment of the fibrocartilaginous intervertebral disc, and is not usually marked by the nucleus pulposus [FIG. 285]. At the periphery there is a smooth, slightly raised ring produced by the fused **epiphysis.**

A

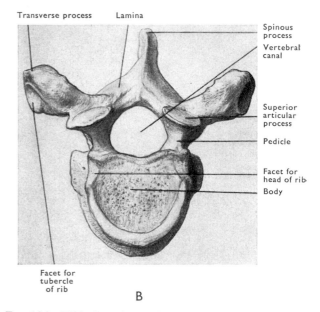

B

FIG. 106   Fifth thoracic vertebra.   A, As viewed from the right side ; B, as viewed from above.

The vertebral bodies increase progressively in size from above downwards, commensurate with the increasing weight transmitted, but decrease rapidly to the coccyx below the level of the transmission of this weight to the hip bones through the sacro-iliac joints [FIG. 104].

The Spines

These posterior projections act as attachments for the interspinous and supraspinous ligaments (which strengthen the vertebral column posteriorly) and give attachments to many muscles. They are thickest and strongest where these attachments are maximal, particularly in the lumbar region.

Transverse Processes

These act as attachments for muscles, helping to increase their leverage over the vertebral column. They form one articulation for most of the ribs, thus supporting them and controlling their movements. In regions other than the thoracic, the rib elements are fused with the transverse processes [FIG. 119].

Structure

The vertebrae are irregular bones consisting of cancellous bone enclosed in a thin shell of compact bone which is thickest at the margins of the intervertebral foramen, the inner surfaces of the laminae, and the superior edge of the spine. The blood supply is by numerous small arterial foramina, and large basivertebral veins emerge from the back of the body.

CERVICAL VERTEBRAE

[PLATE 4]

The cervical vertebrae form a bony axis for the neck and are in a flexible part of the column. Their distinctive character is the foramen transversarium in each transverse process. They are small and do not normally carry much weight. The free ends of the transverse processes are little more than 2·5 cm. from the median plane and can be felt as an uneven bony resistance at the side of the neck. The large transverse process of the first is easily felt below the mastoid process of the skull [FIG. 129]. The second and seventh cervical spines can be felt easily in the midline at the back of the neck. The seventh is sufficiently prominent for its vertebra to be called the **vertebra prominens.** The lower border of the third cervical vertebra lies at the level of the upper border of the thyroid cartilage. The sixth is at the level of the cricoid cartilage (felt below the thyroid cartilage).

Third to Sixth Cervical Vertebrae

The small body of these cervical vertebrae is about one and a half times wider from side to side than from front to back. The upper surface is lipped up at the sides and the lower surface is correspondingly bevelled off. Between the lip and the bevel on adjacent vertebrae there may be a small synovial joint at each side of the intervertebral disc.

The anterior surface is marked by the anterior longitudinal ligament near the midline and projects down to overlap the disc below. Laterally it is slightly hollowed by the attachment of longus colli muscle.

The posterior surface is almost flat and the short rounded pedicles project laterally as well as backwards so that the **vertebral foramen** is nearly triangular and is relatively large and roomy.

The **spine** is short and *except* in the first, sixth, and seventh is bifid at its free end where it gives attachment to the ligamentum nuchae and various muscles. The first has no spine.

The **transverse processes** are composite and project from the lateral side of the body and pedicle. The anterior portion is a rib element which has fused with the body and with the true transverse process. It passes laterally and slightly forwards to end in a prominent **anterior tubercle,** in a typical cervical vertebra. From this, a curved lamella of bone, representing the tubercle of the rib, passes back to fuse with the true transverse process. The true transverse process element passes forwards and laterally from the end of the pedicle to join the rib element. It then terminates as a prominent **posterior tubercle** which projects laterally to give attachment to muscles. The bone between the two

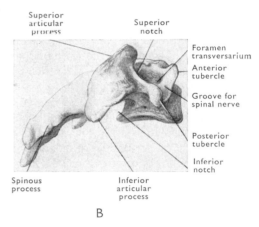

FIG. 107 Fourth cervical vertebra. A, From above; B, from the right side.

tubercles is the costotransverse lamella. It forms the floor of a groove which accommodates the ventral ramus of a spinal nerve as it emerges from behind the vertebral artery to pass between the tubercles.

The **foramen transversarium** is bounded by the pedicle posteriorly and by the two roots of the transverse process laterally and anteriorly. It transmits the vertebral artery and plexuses of veins and of sympathetic nerves from the inferior cervical ganglion, except in the seventh vertebra which transmits the accessory vertebral vein.

The **articular processes** are large, and on each side they form the ends of a short pillar of bone, the articular mass, which carries the flat, oval facets lying in an oblique coronal plane on its extremities.

The **superior and inferior incisurae** are shallow and nearly equal in depth, the emergent nerves lie very close to the intervertebral articulations, especially their dorsal rami; those of the third and fourth nerves grooving the corresponding articular masses.

The Atlas

The atlas or first cervical vertebra is a ring of bone consisting of slender anterior and posterior arches united on each side by a lateral mass on which are situated the articular facets and the transverse process. Most of the body (centrum) of the atlas joins the axis to form the dens, and the anterior arch and lateral masses represent that part of the body normally formed by the vertebral arch ossification [FIGS. 118, 119]. The facets articulate with the skull above and the axis below, and correspond to the small synovial joints at the lateral edges of the bodies of most of these vertebrae. Hence the spinal nerves emerge posterior to them. The normal articular facets are missing.

The **posterior arch** represents the pedicles and laminae of a typical vertebra. It is a curved bar of bone which joins the lateral mass and root of the transverse process anteriorly. Posteriorly, a **posterior tubercle** represents the spine and gives attachment to the nuchal ligament and a rectus capitis posterior minor muscle on each side.

Where it joins the lateral mass, the superior surface of the arch is grooved obliquely by the vertebral artery curving medially below the posterior edge of the articular facet to enter the skull. Here the first cervical nerve lies between the bone and the artery. The upper border gives attachment to the posterior atlanto-occipital membrane. The fibres bordering the gap which transmits the vertebral artery may be ossified from the arch to the edge of the articular facet, so as to convert the groove into a foramen.

The **anterior arch** is shorter, less curved, and flattened anteroposteriorly. A facet on the middle of its posterior surface is for articulation with the dens of the axis. An **anterior tubercle** on the middle of its anterior surface gives attachment to the longus cervicis muscles [FIG. 108].

Its upper and lower margins give attachment to the anterior atlanto-occipital membrane and anterior longitudinal ligament respectively.

The **lateral masses** project medially into the vertebral foramen, and have a rounded tubercle on the anterior part of the projection to which the transverse ligament of the atlas is attached. This ligament holds the dens against the facet on the anterior arch and separates it from the vertebral foramen.

The superior surface of each mass carries an elongated, kidney-shaped **articular facet** which is concave to receive the corresponding occipital condyle. The articular surfaces have their anterior ends closer together than the posterior ends, and and face slightly medially as well as upwards. They may be notched or even divided into two facets. The lateral mass is

slightly larger than the facets, leaving narrow rough areas laterally and medially and a tubercle posteriorly which overhangs the vertebral artery. The medial area usually has a vascular pit. The capsular ligament of the joint is attached immediately beyond the margins of the facet.

The anterior surface of the lateral mass merges with the anterior arch and with the root of the transverse process. Near that root is a small swelling which corresponds to the anterior tubercle of a typical cervical transverse process, and has rectus capitis anterior attached to its anterior surface.

The **lateral aspect** is grooved below the superior articular facet by the vertebral artery, and immediately inferior to that the transverse process takes origin.

The strong **transverse process** gives attachment to muscles which stabilize and help to rotate the head and atlas. Its free end may be tuberculated, or even bifid, but represents only the posterior tubercle of a typical cervical transverse process. It can be felt in the neck a little below and in front of the mastoid process. The **foramen transversarium** lies close to the lateral mass and transmits the vertebral artery, its veins and sympathetic nerves.

The bar of bone limiting the foramen anteriorly represents the costotransverse lamella; the ventral ramus of the first cervical nerve crosses over it after having passed between the vertebral artery and the bone.

The **inferior surface** of the lateral mass carries a nearly circular facet for articulation with the axis. It faces downwards and medially and is slightly concave.

The Axis

The second is the strongest of the cervical vertebrae. The separated centrum [p. 96] of the atlas is fused with the upper surface of the axis as a toothlike process, or **dens** [FIG. 109].

The **laminae** and **spine** are thick and strong, and the inferior surface of the body and the inferior articular process and facet resemble those of a typical cervical vertebra.

The **transverse process** is the smallest of the series, but resembles that of the atlas in that its rounded end represents the posterior tubercle, while the anterior boundary of the foramen transversarium is a bar of bone representing the costotransverse lamella. There is no anterior tubercle.

The **superior articular facets** are large, slightly convex, and face upwards and laterally. They lie at the junction of body and pedicle, with the second cervical nerves posterior to them, and they transmit the weight of the head to the body of the axis, leaving the dens free to act as a pivot. The vertebral arteries deviate laterally to pass round these large superior facets.

The **dens** projects up to lie within the 'ring' of the atlas. It is widened a little at its base where it blends with the upper surface of the body of the axis. Above this, it is constricted where the transverse ligament crosses its posterior surface, and then widens slightly to form a head which tapers sharply to an apex.

The anterior aspect of the head carries

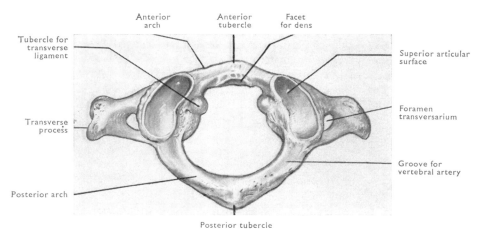

Anterior arch   Anterior tubercle   Facet for dens

Tubercle for transverse ligament

Superior articular surface

Foramen transversarium

Transverse process

Groove for vertebral artery

Posterior arch

Posterior tubercle

FIG. 108   The atlas from above.

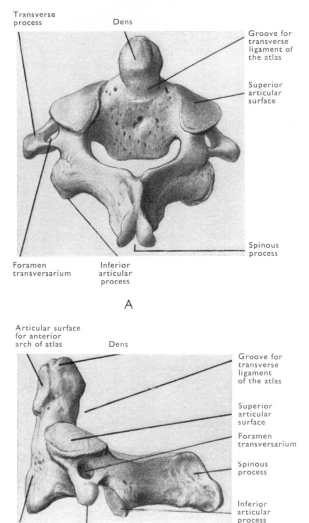

A

B

FIG. 109   Axis vertebra. A, From behind and above ; B, from the left side.

a curved facet which is held against the facet on the anterior arch of the atlas by the transverse ligament, thus leaving the dens free to rotate. An apical ligament continues the line of the bodies from the apex of the dens to the anterior margin of the foramen magnum. A pair of strong alar ligaments run from the sloping sides of the apex to tubercles on the medial sides of the occipital condyles.

The Seventh Cervical Vertebra

This has a long non-bifid spine which forms the upper of two, or the middle of three, prominences at the back of the root of the neck (vertebra prominens). The anterior root of the transverse process is slender and the anterior tubercle is small or absent. Occasionally it may be large (cervical rib, PLATE 7 A) extending as a fibrous or bony strip which usually ends on the upper surface of the first rib, but may pass to the sternum inferior to the subclavian artery and the first thoracic ventral ramus passing to the brachial plexus. The posterior root is large and triangular. Its posterior tubercle extends as far laterally as the first thoracic transverse process.

The foramen transversarium is usually small and only transmits the accessory vertebral vein which is occasionally large.

THORACIC VERTEBRAE

[PLATE 5A and B]

The twelve thoracic vertebrae lie in the posterior wall of the thorax, each with a pair of ribs attached to it.

Their distinctive feature is the presence of articular facets on the sides of the body for the heads of these ribs. Most of the ribs (2 to 9) articulate with the corresponding body, with an intervertebral disc *and* with the body superior to it. The first rib articulates with the body of the first thoracic vertebra, and rarely just reaches the lower border of the seventh cervical. The last three ribs articulate only with the bodies to which they correspond numerically.

The **bodies** of the thoracic vertebrae increase in size from the third downwards and bear increasing weight. Typically heart-shaped, they vary in size and shape from that of a cervical vertebra superiorly, to that of a lumbar vertebra inferiorly, the first having a lipped upper surface like a cervical vertebra. The body of the first and ninth thoracic vertebrae each have an articular facet near the upper border and a small part of one at the lower border, but the upper facet is on the body in the case of the first and at the base of the pedicle in the ninth. The single facet is practically on the pedicle in the twelfth [FIG. 110].

The **spines** are long, three-sided, and sloping, especially in the middle of the series where they are almost vertical. The upper and lower ones are shorter and much less sloping, the upper ones resembling lower cervical spines, and the lower ones approaching the lumbar type. The twelfth, shaped like the first lumbar spine, lies at a level half-way between the inferior angle of the scapula and the iliac crest at the back.

The **pedicles** share in the general increase in size and strength from above downwards. They pass almost directly back and are attached to the upper part of the body so that the inferior notch is deep.

The **laminae,** though narrow from side to side, are deep and overlap protectively like roof tiles.

The **vertebral foramen** is nearly circular and smaller than in the other regions.

The **transverse process** is long, thick, and rounded, and the free end is roughened for ligaments and muscle attachments. It is inclined backwards and a little upwards, and has the articular facet for the tubercle of the corresponding rib facing anterolaterally near its tip. The lower flatter, facets also face slightly upwards, but the eleventh and twelfth transverse processes have no facets. A typical twelfth transverse process is very short and has a trituberculated appearance owing to the presence of lumbar features, the mamillary and accessory processes.

The facets on the **superior articular processes** face mainly backwards but slightly upwards and laterally, with the surface on the circumference of a circle with its centre on the body of the vertebra. This permits rotation round the long axis of the vertebral column, as well as flexion and extension. These are overlapped by similar facets, facing the opposite direction, situated on the short **inferior articular processes** and extending on to the laminae. In the case of the twelfth

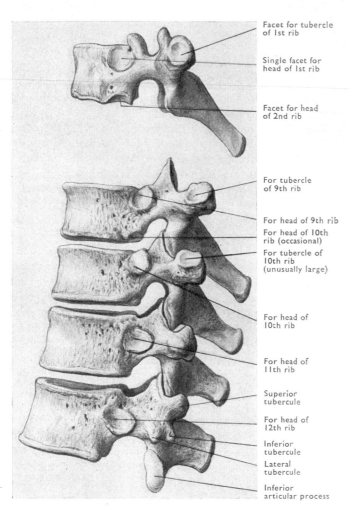

Facet for tubercle
of 1st rib

Single facet for
head of 1st rib

Facet for head
of 2nd rib

For tubercle
of 9th rib

For head of 9th rib

For head of 10th
rib (occasional)

For tubercle of
10th rib
(unusually large)

For head of
10th rib

For head of
11th rib

Superior
tubercle

For head of
12th rib

Inferior
tubercle

Lateral
tubercle

Inferior
articular process

FIG. 110   Left side of the first, ninth, tenth, eleventh and twelfth thoracic vertebrae.

Superior ⎫
Inferior ⎬ tubercles correspond to ⎰ mamillary ⎱ processes of
Lateral tubercle = posterior part of root of ⎰ accessory ⎱ lumbar
　　transverse.　　　　　　　　　　　　　　　vertebrae.

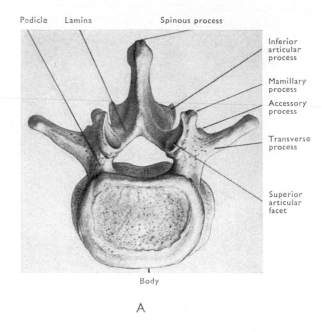

Pedicle　　Lamina　　　　　Spinous process

Inferior
articular
process

Mamillary
process

Accessory
process

Transverse
process

Superior
articular
facet

Body

A

Superior articular process

Mamillary
process

Transverse
process

Spinous
process

Body　　　　Inferior articular process

B

FIG. 111   Third lumbar vertebra. A, From above ; B, from the left side.

thoracic vertebra the inferior articular processes are lumbar in type and fit between the first pair of lumbar processes, thus limiting rotation.

LUMBAR VERTEBRAE
[PLATES 5 and 6]

The lumbar vertebrae lie in the 'small of the back' or loins, carry considerable weight, and are large. The rib element is completely incorporated in the transverse process, thus there are no articular facets for ribs and no foramina transversaria.

The fourth lumbar vertebral body is at the level of the highest part of the iliac crest [FIG. 104], whilst the umbilicus normally lies at the level of the disc above it.

The **bodies** are about 3 cm. in depth and nearly 5 cm. in width. The upper and lower surfaces are kidney-shaped in outline. They are flat and almost parallel to each other, except in the case of the fifth, which is deeper in front than behind. Consequently, the normal lumbar curvature forwards is accounted for mainly by slight wedging of the intervertebral discs.

The **pedicles** are short and stout and project almost straight backwards. The laminae are relatively narrow. The vertebral foramen is triangular, but larger than in the thoracic region, because of the wide lumbar body, and smaller than in the cervical region.

The **superior vertebral incisurae** are distinct but shallow ; the inferior ones are deep and the structures emerging through the intervertebral foramen are related to two joints, the intervertebral disc anteriorly and the joint between inferior and superior articular processes posteriorly.

The **articular processes** are reciprocally curved in a horizontal plane but are almost straight in a vertical plane. The inferior facets face laterally and anteriorly, and fit between the superior ones, which are more widely set and face medially and posteriorly, thus allowing flexion and some extension, but limiting rotation severely. The inferior facets on the fifth are more widely set, flatter, and face in an anterolateral direction to fit the superior articular facets of the sacrum.

The **transverse processes** are flat and spatula-shaped. They project laterally and slightly backwards ; the lower two appearing more oblique because their thick bases extend forwards towards the body. The third is the longest and the fourth and fifth are inclined upwards. The fifth is stout, short, and pyramidal, mainly for the attachment of the iliolumbar ligament and it may be fused with the lateral part of the sacrum.

Occasionally the rib element in a first lumbar vertebra may be separate as a **lumbar rib.** Posteriorly, at the root of each transverse process is a small tubercle called the accessory process (true transverse process) and a more rounded mamillary process lies on the posterior edge of the superior articular process. They give attachment to muscles.

The **spine** of a lumbar vertebra projects almost horizontally back, but owing to the slope of the laminae is level with the lower half of the body. It is wide from above downwards with a rounded posterior edge to give it a hatchet-shaped appearance. The spine of the fifth is frequently rounded off.

The **laminae** in the lumbar region are widely separated, leaving diamond-shaped spaces between them filled in by the ligamenta flava, through which a lumbar puncture may be made.

In the lumbar region, psoas muscle lies in the angle between the bodies and the transverse processes; erector spinae fills the angle between the spines and transverse processes.

SACRUM AND COCCYX
[PLATE 7]

Sacrum

The sacrum consists of five fused sacral vertebrae [FIG. 112]. These provide strength and stability to the pelvis and transmit the weight of the body to the pelvic girdle through the sacro-iliac joints. The joint involves the upper three vertebrae and the bone tapers below that level to a blunt apex formed by the body of the fifth sacral vertebra about 2 cm. above the level of the upper border of the symphysis pubis.

The sacrum is roughly wedge-shaped from base to apex and from pelvic to dorsal surface, and the base is directed more forwards than upwards. It is bent so that the pelvic surface of the anterior part of the bone makes an angle of about 10 degrees with the horizontal when the body is erect, while the posterior part lies at right angles to it.

The **pelvic surface** faces downwards and forwards in the upright position and the **dorsal surface** faces in the reverse direction [FIG. 634].

The **base of the sacrum** is the superior surface of the first sacral vertebra. The body of the vertebra forms an oval elevation occupying about one-third of the width or less in the female, more than one-third in the male. Its anterior border projects into the superior aperture of the pelvis to form the **promontory** of the sacrum [FIG. 222]. On each side the costal elements are fused with the body, pedicle, and transverse processes, and are enlarged to form the **lateral part** of the sacrum which is wing-shaped on its abdominal aspect. Its rounded inferior edge forms part of the linea terminalis of the pelvis. Posteriorly, the pedicle and transverse process are incorporated in the lateral part but a pit or, occasionally, a foramen may be present between them and the enlarged rib element. From the region of the pedicles large flattish articular processes project to articulate with the fifth lumbar inferior

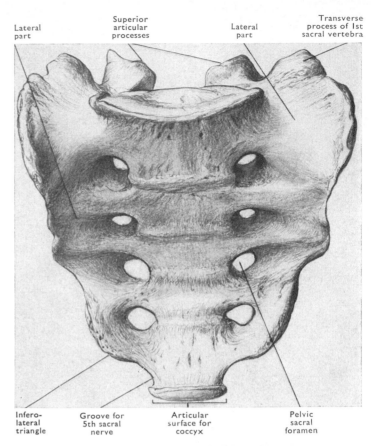

FIG. 112   The sacrum (pelvic surface).

processes, and beyond these the laminae converge to fuse in a short spine. The vertebral foramen so formed is triangular and is the upper end of the **sacral canal.**

The dorsal surface is rough, and has a series of longitudinal ridges formed from the fused posterior parts of the vertebrae [FIG. 113]. In the midline, the **median sacral crest** carries tubercles representing the spines, but is missing to a variable extent inferiorly where the laminae do not fuse, leaving the **sacral hiatus.** During life this is filled with fibrous tissue through which local anaesthetic may be injected into the sacral canal (epidural anaesthesia). The second spinous tubercle is level with the posterior superior iliac spine and with the middle of the sacro-iliac joint.

Lateral to the median sacral crest is a rough plate of bone formed by the fused laminae and ossified ligamenta flava. Immediately lateral to this is a row of small tubercles **(intermediate sacral crest)** representing the articular masses, and beyond this a row of four dorsal sacral foramina. These transmit the dorsal rami of the sacral nerves and small blood vessels, and are separated by bars of bone representing parts of the transverse processes. Lateral to this is a double row of small tubercles representing the tips of the transverse processes **(lateral sacral crest).** The most lateral part, representing the ribs, is well developed in the upper three sacral vertebrae. It is this part which articulates with the ilium, but is roughened by the powerful interosseous sacro-iliac ligaments immediately behind the articular, **auricular surface** [FIG. 113]. Below this region the sacrum narrows, and ends in a small, oval apex which is joined to the coccyx by a narrow

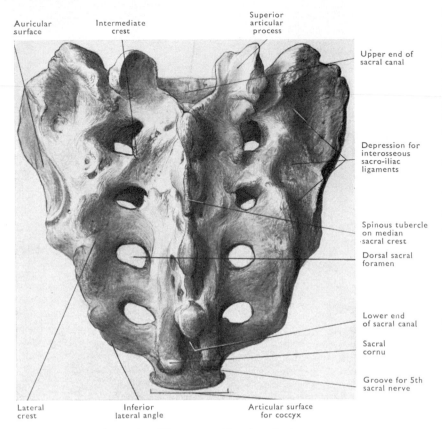

FIG. 113   The sacrum (dorsal surface).

Labels for Fig. 113: Auricular surface; Intermediate crest; Superior articular process; Upper end of sacral canal; Depression for interosseous sacro-iliac ligaments; Spinous tubercle on median sacral crest; Dorsal sacral foramen; Lower end of sacral canal; Sacral cornu; Groove for 5th sacral nerve; Lateral crest; Inferior lateral angle; Articular surface for coccyx.

four **pelvic sacral foramina** which transmit the ventral rami of the sacral nerves, branches of the lateral sacral arteries, and communications between the pelvic veins and the internal vertebral venous plexus. These foramina are larger than the dorsal ones and lead from the sacral canal towards the greater sciatic notch, grooving the lateral part of the sacrum. The fifth sacral nerve sends its ventral ramus round the projecting apex of the bone.

Coccyx

The coccyx can be felt as the lowest part of the backbone in the groove between the buttocks. It is roughly triangular in outline, tapering to an apex below, and consists of four more or less fused coccygeal vertebrae [FIG. 114]. The **base** is formed by the first vertebra and shows an oval body for articulation with the apex of the sacrum, two **cornua,** or superior articular processes, articulating with the sacral cornua and two transverse processes. These parts become progressively more indefinite in the succeeding vertebrae until the terminal one is only a nodule of bone. The number may be reduced to three or increased to five.

Sexual Differences

In the female the sacrum is often wider in proportion to its length than in the male and the lateral parts form nearly two-thirds of the width of the base. The body of the first sacral vertebra is large in the male. In the female, the auricular surface tends to be limited to the first two vertebrae, and the joints, including the sacrococcygeal, are more mobile and less likely to be fused. The curvature of the sacrum in the male is fairly uniform; in the female it is flatter anteriorly and its angulation is more pronounced.

intervertebral disc. At the sides of the lower end of the sacral hiatus the **sacral cornua** project down to articulate with the corresponding processes of the coccyx [FIG. 113].

In lateral view the angulation is obvious between the massive, horizontal, anterior part and the thinner, postero-inferior, vertical part. The latter descends so that the tip of the coccyx is level with the upper margin of the pubic symphysis [FIG. 223]. The L-shaped, auricular surface, covered by cartilage, forms a tight-fitting synovial joint between the first three (or two) sacral vertebrae and the corresponding surface on the ilium. It has ventral sacro-iliac ligaments attached to its anterior and inferior borders, and the interosseous sacro-iliac ligament attached to its posterosuperior border [FIG. 309]. The dorsal part of the sacrum is slightly narrower than the ventral part and the auricular surfaces are sloped accordingly. The interosseous ligaments hold it wedged between the ilia.

The remainder of the lateral part of the bone is an edge which gives attachment to ligaments. The sacrospinous and sacrotuberous ligaments, in particular, are attached to this edge and to the dorsal surface. They hold down the posterior part of the sacrum and resist the tendency of the body weight to rotate the base into the pelvis [FIG. 309]. The pelvic surface is concave and relatively smooth. It is limited anteriorly by the promontory and rounded margin of the lateral part of the sacrum on each side. In the midline is a column of bone formed by the fused bodies of the five sacral vertebrae and ossified intervertebral discs which form ridges between the bodies.

On each side, lateral to the intervertebral discs, lies a row of

THE VERTEBRAL COLUMN AS A WHOLE

The vertebral column varies in length but is about 70 cm. in a man and 60 cm. in a woman.

FIG. 114   The coccyx.

A, Dorsal surface.   B, Pelvic surface.
1, Transverse process.   2, Transverse process.
3, For sacrum.   4, Cornu.

FIG. 115   Left side of the vertebral column.

FIG. 116   Vertebral column from behind.

It is said to diminish by nearly 2 cm. during the day and measurements of height on a supine cadaver require a deduction of 2 cm. to obtain the living height owing to straightening of the curves after death. Length diminishes in old age owing to reduction in thickness of the discs and to exaggeration of the curvatures, particularly in the thoracic region. The discs account for nearly a quarter of the total length.

The column can bear a weight of nearly 355 kg. without crushing and a tearing strain of nearly 152 kg. Its weakest part is in the neck, which normally carries least weight, but the mode of application of a force is most important.

Damage is particularly apt to occur at the junction of relatively fixed and freely movable parts such as in the thoracolumbar region or where a localized force can be applied by leverage (e.g., the dens or an intervertebral disc) or by direct contact (e.g., the coccyx).

Taken as a whole the vertebral column appears as in FIGURES 115 and 116. The vertebral bodies increase in size from the skull down to the point where the weight of the body is passed from the sacrum to the pelvic girdle, and then diminish rapidly.

The transverse processes form a line of projections on each

side which vary considerably in form, size, and direction. The first cervical is large and easily palpable below the mastoid process; the remaining cervical transverse processes are felt as a deep bony resistance at the side of the neck. The thoracic transverse processes decrease in size from the first to the twelfth, and are buried between erector spinae and the ribs. The lumbar ones are again more prominent but are mainly equivalent to small fixed ribs and are completely obscured by muscles.

Posteriorly, the spines project to a varying degree and are palpable in the midline from the seventh or sixth cervical down to the sacrum and coccyx. The large second cervical spine is palpable through the ligamentum nuchae and muscles. Below this, the spines become increasingly oblique to the mid-thoracic region, and thus, like the laminae, overlap progressively, though both are widely separated in the lumbar region. In the cervical region gaps are opened up when the neck is flexed. The widest interlaminar gaps are between the atlas and axis, the fourth and fifth lumbar, and fifth lumbar and sacrum.

The seventh cervical spine is identifiable as the **vertebra prominens** at the base of the neck. The third thoracic is level with the root of the spine of the scapula. The twelfth is midway between the levels of the lower angle of the scapula and the top of the iliac crest. The second sacral spine is level with the posterior superior iliac spines and the sacral promontory. The sacrum forms a bony resistance between the two iliac bones but its details are masked by strong connective tissue and ligaments.

The **vertebral groove** is the wide furrow alongside the spines. Its floor is formed by the laminae, the articular processes and the transverse processes.

Below the atlanto-axial joints the articular masses and the superior and inferior processes form a series of short rods articulated together and joined to the vertebral arch, but from the back they are scarcely distinguishable from the laminae. They lie deeply under the erector spinae muscle.

Curvatures

The vertebral column has four anteroposterior curvatures of which two are regarded as primary.

The fundamental curvature is concave ventrally from an early stage of development. At about the fifth month of intra-uterine life an additional curve, also concave ventrally, appears in the sacral and coccygeal regions and accommodates the pelvic viscera. As the head is raised from the thorax in the first few months after birth, a secondary curve, convex anteriorly, is formed in the neck and extends as far as the second thoracic vertebra. Later, towards the end of the first year, when a sitting posture, followed by an upright walking posture, is adopted, the cervical curvature is increased and the lumbar part of the column also becomes convex forwards to produce another secondary curve. This curve extends from the twelfth thoracic vertebra to the sacral promontory.

The cervical curve is the least marked and is undone as the neck is flexed. The lumbar curve is pronounced in the erect position and brings the bodies of the vertebrae very close to the anterior abdominal wall, its most anterior point being opposite the umbilicus, between the third and fourth lumbar vertebrae, or a little lower.

In an upright attitude the alternating curves of the vertebral column provide for the absorption of the vertically directed shocks. These are not transmitted straight through the column but are absorbed by the intervertebral discs and by slight bending of the curvatures which is resisted by the muscles and ligaments.

OSSIFICATION OF VERTEBRAE

A typical vertebra is ossified in cartilage from three primary centres and five epiphyses.

The **primary centres** appear in the body and in each half of the vertebral arch. That for the **centrum,** the larger, median part of the body, appears dorsal to the notochord first in the lower thoracic region (about 10th intra-uterine week), then spreads up and down the column so that virtually all centra have begun to ossify by the twentieth week. The last two pieces of the sacrum are delayed as late as the thirtieth week and the coccygeal centres appear after birth. The centra are usually formed from two ossific centres; they may fail to unite so that the centrum is formed in two separate halves, or one half only may ossify, producing a hemivertebra.

A primary centre appears in each half of the **vertebral arch** (at the junction of pedicle and lamina) in the upper cervical region (approximately seventh week) and sequentially in the lower vertebrae to reach the sacrum (approximately twentieth week). Ossification from each centre spreads into the lamina and pedicle, the latter extending on to the centrum to complete the body and articulate with the rib (thoracic region).

At birth the centrum and the two halves of the vertebral arch are ossified to form three pieces which are united by cartilage [FIG. 117]. The cartilage between the centrum and each vertebral arch forms a **neurocentral joint** (vertebral arch = neural arch) ossification of which begins in the cervical region (third year) and extends to the other regions by the seventh year. The cartilage between the laminae is continuous with the cartilaginous spine. The laminae begin to unite soon after birth in the lumbar region, the process spreading to the cervical region by the second year except, perhaps, for the arch of the atlas, while it is not completed in the sacrum till the seventh or tenth year. The degree of fusion in the sacrum is variable. After fusion of the laminae, ossification spreads into the spine.

Secondary centres, the vertebral epiphyses, appear about the age of puberty; those for the bodies at about 9 years, those for the tips of the spines and transverse processes at about 18 years. Those for the bodies are multiple and fuse to form flat rings on the circumferential parts of the superior and inferior surfaces.

FIG. 117 Ossification of vertebrae.

PLATE 3

A

B

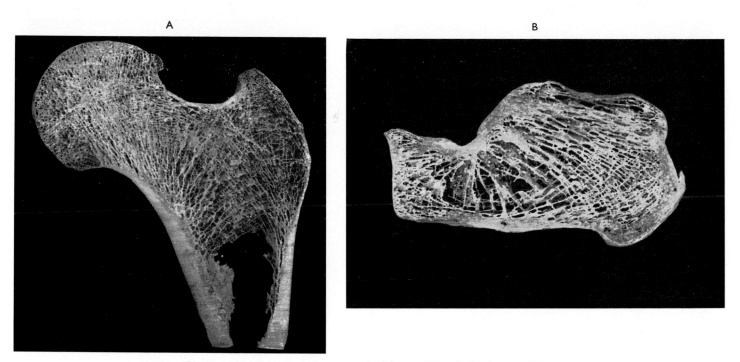

Photographs of sections of upper end of femur (A) and of calcaneus (B).

C

D

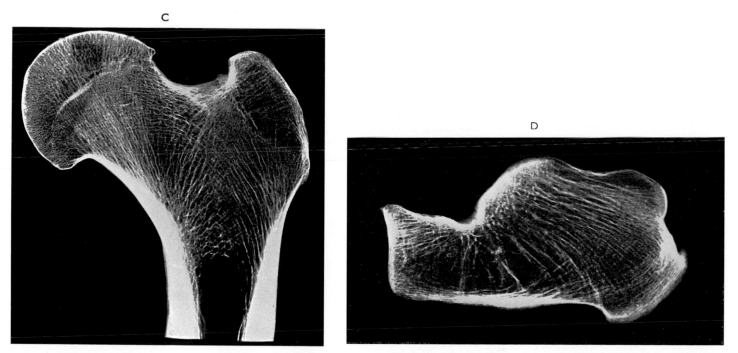

C and D. Radiographs of the same specimens as A and B.

Note the thick compact bone of the cortex of the femur tapering to the thin but intact cortex covering the head, upper part of the neck, and greater trochanter. Note also how the trabeculae in both bones follow the lines of maximum stress and are widely spaced in the central region of the calcaneus. The epiphysial 'scar' of the head of the femur is well shown in C.

PLATE 4

A

B

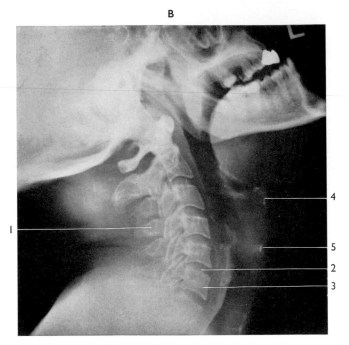

A and B.   Lateral radiographs of the cervical vertebral column in flexion and extension.

1. Spinous process of the 4th cervical vertebra.
2. Intervertebral disc between the 5th and 6th cervical vertebrae.

3. Body of the 6th cervical vertebra.
4. Hyoid bone.
5. Calcification in the thyroid cartilage.

C

D

C.   Oblique radiograph of the cervical vertebral column.

1. Hyoid bone.
2. Calcification in the thyroid cartilage.
3. Transverse process of the 3rd cervical vertebra.
4. Intervertebral foramina.

D.   Anteroposterior view of the cervical vertebral column.

Note that the shadow of the mandible has been obliterated by making it move during the taking of the radiograph.

1. Lateral mass of the 1st cervical vertebra.
2. Dens of the 2nd cervical vertebra.
3. 7th cervical vertebra.

PLATE 5

A

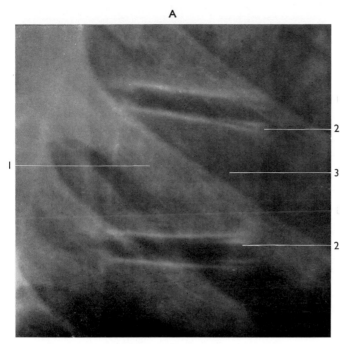

A. Lateral radiograph of a single mid-thoracic vertebra. Age 16.

    1. 6th rib.
    2. Epiphysial rings seen edge on.
    3. Vertebral body.

B

(i)   (ii)

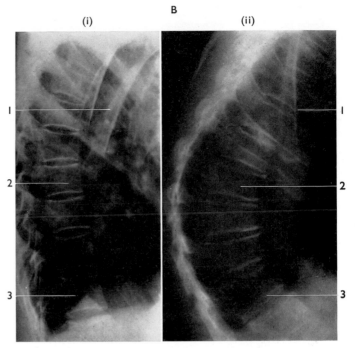

B. Lateral radiographs of thoracic vertebral column:
    (i) in mid-adult life,
    (ii) in old age.

Note the increase in the curvature (kyphosis), the 'wedging' of the vertebral bodies and anterior narrowing of the intervertebral discs in (ii).

    1. Scapula.
    2. 7th thoracic vertebra.
    3. 10th thoracic vertebra.

C

C. Anteroposterior radiograph of the abdomen and lumbar vertebral column.

    1. 12th rib.
    2. 1st lumbar vertebra.
    3. Lateral border of the psoas major muscle.
    4. Air in the stomach.

D

(i)   (ii)

D. Lateral radiographs of lumbar vertebral column:
    (i) in mid-adult life,
    (ii) in old age.

Note the increase in the curvature (lordosis), 'wedging' of the vertebral bodies and posterior narrowing of the intervertebral discs in (ii).

    1. 1st lumbar vertebra.   2. 5th lumbar vertebra.

PLATE 6

A

B

A. Oblique radiograph of the lower lumbar vertebral column.

1. Inferior articular process of the 3rd lumbar vertebra.
2. Superior articular process of the 4th lumbar vertebra.
3. Body of the 4th lumbar vertebra.
4. Disc between the 4th and 5th lumbar vertebrae.

B. Anteroposterior radiograph of the upper lumbar vertebral column.

1. Inferior articular facet of 2nd lumbar vertebra.
2. Spinous process of the 3rd lumbar vertebra.
3. Disc between the 3rd and 4th lumbar vertebrae and inter-laminar space.
4. Superior articular facet of 3rd lumbar vertebra.
5. Transverse process of 3rd lumbar vertebra.

C

D

C. Lateral radiograph of the mid-lumbar vertebral column. Note shape of the intervertebral disc.

1. Disc between the 2nd and 3rd lumbar vertebrae.
2. Pedicle of the 3rd lumbar vertebra.
3. Body of the 3rd lumbar vertebra.

D. Lateral radiograph of the lumbosacral region. Note shape of lumbosacral intervertebral disc.

1. Disc between the 4th and 5th lumbar vertebrae.
2. Margin of iliac crest.
3. Body of the 5th lumbar vertebra.
4. Disc between the 5th lumbar vertebra and sacrum.
5. Body of the 1st sacral vertebra.

The various epiphyses so formed begin to fuse with the rest of the bone at 18 years and fusion should be complete at 25 years of age. Primary centres for the coccygeal bodies appear between birth and puberty and ossification spreads without formation of secondary centres. Epiphyses on the bodies and cornua have been described.

In the **atlas** a primary centre for each half of the vertebral arch and the corresponding lateral mass appears about the ninth week, and the two halves unite in the third or fourth year, but union may be delayed. A secondary centre may appear in the cartilage between them. The anterior arch is cartilaginous at birth and a centre appears during the first or second year, to fuse with the lateral masses soon after the seventh year. It includes part of the superior articular surface[FIG. 118]. An epiphysis for the transverse process unites after puberty.

In the **axis** a primary centre for the vertebral arch appears about the seventh week, and one or two for the lower part of the body early in the fifth month. Two more appear side by side for the dens and upper part of the body later in that

month, and fuse together by the seventh month.

At birth the axis is therefore in four pieces which unite between 3 and 6 years of age. Secondary centres appear in the tip of the dens between 2 and 6 years and for the lower surface of the body about puberty. These epiphyses fuse before 12 years and between 18 and 25 years respectively. A number of additional centres of ossification may be present.

The tip of a bifid cervical spine has a pair of epiphyses and the sixth and seventh cervical vertebrae may have separate primary centres for the costal elements. These normally join the rest of the bone early at the fifth year but the seventh in particular may remain separate to form a cervical rib.

Lumbar vertebrae have centres for the mamillary processes and the first lumbar may have separate primary centres for its transverse processes. These may remain separate to form a lumbar rib on either or both sides.

The fifth may have two primary centres in each half of the vertebral arch, the two parts of each half then being united by cartilage set obliquely between the superior and inferior

FIG. 118   Ossification of vertebrae.

Cervical vertebra

1. Centre for body.
2. Superior epiphysial ring.
3. Anterior bar of transverse process ossified by lateral extension from pedicle.
4. Neurocentral joint.
5. Inferior epiphysial ring.

Lumbar vertebra

6. Centre for body.
7. Superior epiphysial ring.
8. Centre for mamillary process.
9. Centre for transverse process.
10. Centre for spine.
11. Neurocentral joint.
12. Inferior epiphysial ring.

Thoracic vertebra

13. Centre for body.
14. Superior epiphysial ring.
15. Neurocentral joint.
16. Centre for transverse process.
17. Centre for spine.
18. Inferior epiphysial ring.

Axis

19. Centre for transverse process and vertebral arch.
20. Joints close about 3rd year.
21. Centre for summit of dens.
22. Centre for lower part of dens.
23. Neurocentral joint.
24. Inferior epiphysial ring.
25. Single or double centre for body.

Atlas

26. Posterior arch and lateral masses ossified from a single centre on each side. In this figure the posterior arch is represented complete by the union posteriorly of its posterior elements.
27. Anterior arch and portion of superior articular surface ossified from single or double centre.

Thoracic vertebra

28. Centre for transverse process.
29. Centre for spine.
30. Centre for vertebral arch on each side. The arch is here shown complete posteriorly.
31. Centre for body.

articular processes. There is a risk of separation through this joint which is usually temporary. The fifth transverse process may have two epiphyses.

The number of bones incorporated in the **sacrum** results in the presence of numerous centres but they are of little importance, except for the varying degrees of fusion or consolidation which may occur. Primary centres appear between the third and eighth months, one for each of the centra and half vertebral arches and for each costal element in the upper three or four vertebrae. The costal parts fuse with the arches at about 5 years, the arches with the centra a little later, and the halves of the arches join between the seventh and tenth years.

The segments of the lateral mass fuse together at puberty and epiphysial centres for the bodies appear about the same time. The epiphyses and the bodies all fuse together between 18 years and 25 years. Numerous epiphysial centres appear on the ends of the costal and transverse processes and from them an epiphysis is formed which covers the auricular surface, while another completes the margin below that.

VARIATIONS IN VERTEBRAE

Variation in the total number of vertebrae is usually due to a reduction or, more rarely, an increase in number of the coccygeal vertebrae. The number of cervical vertebrae is exceedingly constant although a seventh cervical vertebra may carry a **cervical rib** which is usually incompletely formed.

The number of thoracic vertebrae may be increased by the formation of ribs on the first lumbar vertebra, and the number of the lumbar vertebrae may be reduced in this manner or by fusion of a fifth lumbar vertebra to the sacrum. This **sacralization** may be partial or complete and quite frequently is incomplete on one side. The **sacrum** may have an additional piece, becoming longer than usual, or may lose a vertebra at one end while gaining one at the other, so retaining five pieces. It is not possible to define the numerical condition of a vertebral column unless all the vertebrae are available.

In addition, it is not uncommon to find suppression of half a vertebra or intercalation of an additional half, making a wedge-shaped body.

The vertebral column is produced from so many parts that the range of variation is large. The laminae may fail to meet and fuse at any or all parts (spina bifida), but most commonly in the lumbosacral region. Hemivertebrae (wedge-shaped vertebral bodies) cause an abnormal lateral curvature (scoliosis) of the column. The dens of the axis may be a separate bone, an os odontoideum, connected to the body of the atlas by

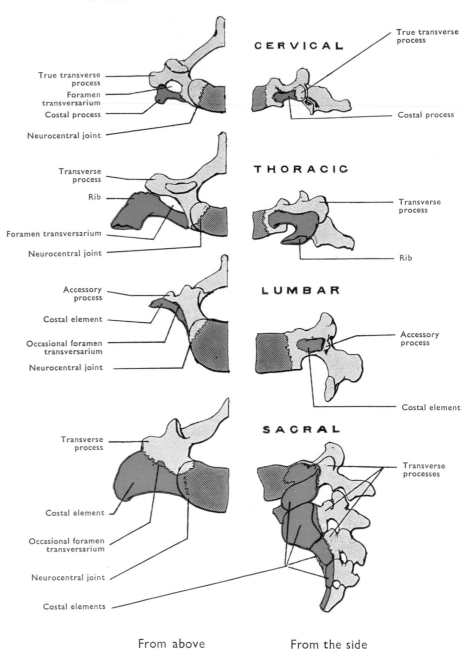

FIG. 119   Diagram to illustrate homologous parts of vertebrae.

ligaments. The spine, laminae, and inferior articular processes of the fourth and fifth lumbar vertebrae may only be united to the rest of the vertebra by cartilage, thus predisposing to a separation which leaves the vertebral body (particularly the fifth) free to slide forwards (spondylolisthesis).

The standard work on vertebral variations is by Le Double (1912); see also Brash (1915) and Brailsford (1948).

SERIAL HOMOLOGIES OF VERTEBRAE

Since each part of a vertebra is potentially represented in every other vertebra it is possible to work out the homologies in detail for each region. The essential features are shown in FIGURE 119.

The body is the only part represented throughout the series, and even there the centrum of the atlas has been detached and fused with that of the axis. The upper cervical vertebrae require special mention in that the articular processes proper are not represented above the inferior process of the axis. The missing joints are replaced by an enlargement of those found in the neurocentral region lower down. The anterior arch of the atlas is ossified from a persistent **hypochordal bow** [FIG. 127] which disappears in all the other vertebrae. The ribs are represented down to the lower pieces of the sacrum but are normally free only in the thoracic region. There is very considerable variation in the size of the different projections from the vertebral arch.

THE STERNUM

The sternum is an elongated, nearly flat bone, 15–20 cm. long. It lies in the midline of the anterior wall of the thorax and extends from the root of the neck into the abdominal wall. It can be felt through the skin in its whole length and articulates on each side with the clavicles at its upper end and with the upper seven costal cartilages at its lateral borders. Its three main parts are the manubrium, body and xiphoid process. These are united by cartilage and in youth the body is in four pieces, also united by cartilage.

The Manubrium

This is the widest piece of the sternum, is roughly quadrilateral, lies inferior to the root of the neck, and forms the anterior wall of the superior mediastinum. It is usually a separate bone, for it only fuses with the body in advanced age.

The middle part of the superior border forms the rounded **jugular notch.** On each side of this is the **clavicular notch,** the sternal part of the sternoclavicular joint. Immediately inferior to each clavicular notch is a rough projection which receives the first costal cartilage, and below this the manubrium narrows to its lower border, the lateral ends of which each carry a half facet for the second costal cartilage. The posterior surface, smooth and slightly concave, has the sternohyoid muscle attached at the level of the clavicular notch and the sternothyroid at the level of the first costal cartilage.

The anterior surface is rougher and has a slightly raised area below the medial part of the clavicular notch for the tendon of sternocleidomastoid muscle. Below that, a concave area on each side gives rise to the upper sternal fibres of the pectoralis major muscle.

The manubrium is joined to the superior border of the body by the fibrocartilage of the manubriosternal joint. The two parts are angulated to each other. This obtuse **sternal angle** is easily palpated even in the obese. It forms a useful landmark, for it indicates the level of the articulation of the second costal cartilage with the manubrium and body of the sternum, and the junction of the superior and inferior mediastina in the thorax.

The Body

This consists of four segments fused together and is rather more than twice as long as the manubrium in most men, rather less than twice as long in many women. These relationships are, however, too unreliable for medico-legal or anthropological purposes (Ashley, 1956b). Its shape is variable but it usually widens from about 2·5 cm. at its upper end to about 4 cm. at its fourth segment, which tapers rapidly to its lower end. The posterior surface is slightly concave in its long axis and is smooth, the transversus thoracis muscle [FIG. 415] and the sternopericardial ligaments are attached to it. The rougher anterior surface is marked by three transverse ridges indicating the lines of union between the segments or sternebrae. The attachment of pectoralis major is continued down from the manubrium as far as the sixth costal cartilage just short of the xiphoid process.

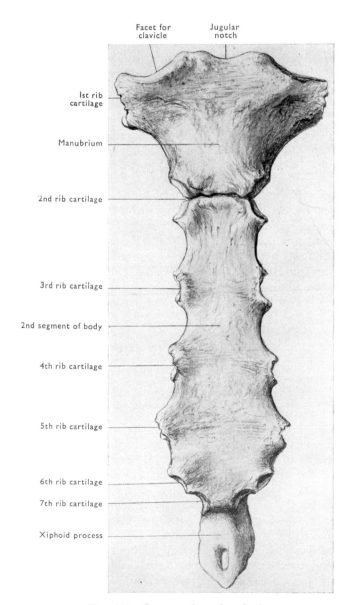

FIG. 120   Sternum (anterior view).

Articular facets for the synovial joints with the second to seventh costal cartilages are present on the lateral margins : at the manubriosternal joint (half of the second), between the segments of the body (third to fifth), on the fourth segment (sixth), and at the xiphisternal joint (seventh).

The lower end of the body is continuous with the **xiphoid process** at the xiphisternal joint. The process is a flat, pointed piece of cartilage which ossifies slowly from a central core until this joins the body of the sternum after middle age. Its shape is very variable and it is often perforated in the middle. It is thinner than the body of the sternum and is set flush with its posterior surface. The depression so formed is the **epigastric fossa,** or pit of the stomach, and the site of the joint can be felt as a ridge in the upper margin of the fossa. The xiphoid process lies in the anterior abdominal wall in front of the liver, and has the linea alba attached to its inferior end. It therefore has not only transversus thoracis but also all the adjacent abdominal muscles attached to its edge and anterior surface and slips of the diaphragm to its deep surface.

The sternum as a whole slopes antero-inferiorly, the manubrium slanting about 20 degrees more than the body so as to be almost in the same plane as the first ribs. The jugular notch is level with the lower border of the body of the second thoracic vertebra [FIG. 125], the sternal angle with the upper border of the fifth and with the interval between the third and fourth spines. The xiphisternal joint is level with the ninth thoracic vertebra and the tip of the eighth spine.

Ossification

Ossification of the sternum is variable (Ashley, 1956a, 1956b) and any centre of ossification may be single or double and can affect the width of the bone. Centres appear in the manubrium in the fifth intra-uterine month and then in the sternebrae in order from above downwards in the sixth, seventh, eighth, and ninth months [FIG. 121]. The xiphoid process does not ossify until the third year and does not join the body until middle age.

The remainder join in sequence from below in childhood, at puberty, and at 21 years. The manubrium should not fuse until old age is reached, leaving some flexibility during respiration.

At birth

At 3 years

FIG. 121   Ossification of sternum.
1, Appears at 5th intra-uterine month. 2, At 6th and 7th month. 3, At 8th and 9th month. 4, At 3rd year.

Structure and Variations

The sternum is composed of cancellous bone enclosed in thin compact bone and is used to obtain samples of red marrow by sternal puncture. The vascular foramina are mostly on the deep surface of the manubrium and body where the bone is supplied by the internal thoracic artery.

The sternum arises from two separate mesodermal masses. Failure of fusion leads to centres of ossification being double, with a hole in the body, or more rarely a cleft sternum, the gap being wider above.

Some degree of asymmetry is common and may be associated with development of the limbs. The sternal angle may be situated one rib lower down than usual. The seventh costal cartilage may fail to reach the sternum or the eighth may reach it.

Small suprasternal ossicles are sometimes found in the ligaments of the sternoclavicular joint.

THE RIBS

The ribs (costae) are curved and slightly twisted strips of bone which lie between the segments of the body. There are twelve of them on each side in both sexes and, though elongated, are classified as 'flat' bones. They curve round from the thoracic vertebrae posteriorly, slant downwards and forwards at the sides, and anteriorly each becomes continuous with a costal cartilage. With the vertebral column and sternum they thus form a thoracic cage which supports the thoracic wall and encloses part of the upper abdomen.

Since the thorax widens from the root of the neck downwards, the upper seven ribs increase similarly in length and are referred to as **true ribs** because their cartilages reach the sternum. The lower five ribs fail to reach the sternum, become progressively shorter from above downwards, and are referred to as **false ribs.** The cartilages of the eighth, ninth, and tenth articulate, each with the cartilage next above, but the cartilages of the eleventh and twelfth lie free and the ribs are called floating ribs [FIG. 126].

The first, second, tenth, eleventh, and twelfth ribs have special features requiring a separate description but the third to the ninth have sufficient in common to be described together.

A typical rib has its posterior end slightly enlarged to form a **head** which articulates with the upper border of the body of its own vertebra and the lower border of the vertebral body next in series above it. It has therefore two sloping articular facets with a ridge between [FIG. 123]. The articular surfaces are surrounded by a joint capsule and the ridge gives attachment to an intra-articular ligament which connects it to the intervertebral disc.

A **neck** about 2·5 cm. long immediately succeeds the head and is continuous with a body. The rough posterior surface is connected to the transverse process directly behind by the **costotransverse ligament;** and the **crest** on the upper border gives attachment to a **superior costotransverse ligament** passing to the vertebra above.

The smooth part of the anterior surface, covered by pleura, is limited above by a faint ridge produced by the internal intercostal membrane [FIG. 414].

The **tubercle,** on the posterior surface at the junction of neck and body, articulates with the corresponding transverse process, and has a rough, lateral part for the attachment of the **lateral costotransverse ligament** [FIG. 123].

The **body** continues laterally to reach a roughened vertical line on its external surface made by the thoracolumbar fascia covering the erector spinae muscle, and situated where the rib curvature is accentuated to form the **angle of the rib.** The angle lies where the anterior part of the body is bent downwards. It coincides with the tubercle of the first rib but is placed increasingly further laterally, by as much as 7·5 cm., as the ribs are followed down to the eighth or ninth. Below this, both the line and the angle become indistinct. This lateral displacement corresponds with the increasing bulk of erector spinae (attached between the tubercle and the angle) as the lumbar region is approached.

Beyond the angle of the rib the body continues in an even curve round the thorax, sloping down, until it reaches the costal cartilage anteriorly. It has a convex external or superficial surface and an internal or deep surface which is concave lengthwise. Anteriorly the rib is somewhat twisted on its long axis so as to evert the lower border slightly.

The rounded superior border of the rib has the intercostal muscles and internal intercostal membrane attached to it [FIG. 414]. The sharp inferior border forms the external margin of the **costal groove** (which fades out anteriorly), the internal margin is higher on the deep surface of the rib. Both margins and the groove give attachment to the intercostal muscles. The rest of the deep surface of a rib is smooth and lined with pleura, except posteriorly, where it may be covered by subcostal muscles and is often marked by their attachments.

The body of a rib widens at its anterior end and terminates as a rough oval pit where it becomes continuous with its costal cartilage. The external surface of the shaft is covered with muscles and is smooth except near its anterior end where there may be an oblique ridge between the interdigitating origins of serratus anterior and obliquus externus abdominis muscles [FIG. 375], and a sudden increase in curvature may form the **anterior angle.**

The internal surface of the thorax is concave superoinferiorly, hence the pleural surfaces of the various ribs face in different directions: the first inferiorly, the twelfth anterosuperiorly, and the intermediate ribs with varying degrees of obliquity [FIG. 125].

The **first rib** is short, sharply curved, and flattened so as to present an antero-superior surface and inner and outer borders

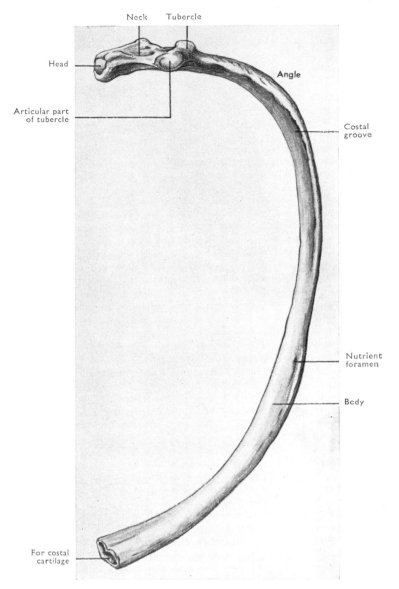

FIG. 122   Fifth right rib from below.

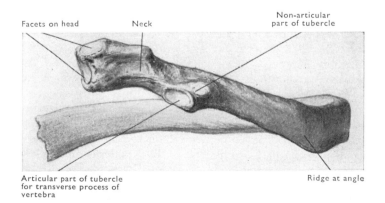

FIG. 123   Fifth right rib from behind.

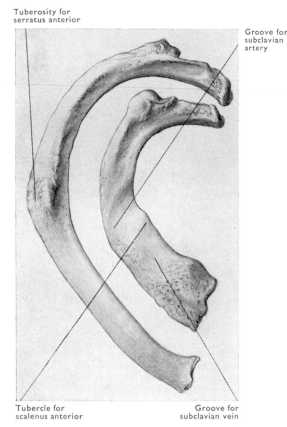

Tuberosity for
serratus anterior

Groove for
subclavian
artery

Tubercle for
scalenus anterior

Groove for
subclavian vein

FIG. 124   First and second right ribs as seen from above.

on its body.   It is felt as a bony resistance in the neck behind the middle of the clavicle.   Its cartilage lies deep to the pectoralis major muscle in the angle between the sternum and the clavicle.   Its **head** is small and normally articulates only with the first thoracic vertebra by means of a single facet.   The relatively long, narrow and horizontal **neck** has the sympathetic trunk, the highest intercostal artery, and the first thoracic ventral ramus separating it from the pleura and the apex of the lung.   The large tubercle lies on the outer border, and from this level the body slopes downwards and forwards forming the lateral margin of the superior aperture of the thorax.

The body has anterosuperior and postero-inferior surfaces, and passes beneath the medial part of the clavicle to reach the sternum.   The postero-inferior (pleural) surface is smooth, but may be roughened externally by the internal intercostal muscle.   The anterosuperior surface is grooved obliquely by the subclavian artery and vein [FIG. 124].   Between these grooves, the **tubercle for scalenus anterior** projects from the inner border, and a triangular, rough area, also for that muscle, extends laterally separating the artery and vein.   The arterial groove also contains most of the ventral ramus of the first thoracic nerve which curves up over the first rib and separates the artery from the scalenus medius muscle which is attached to the first rib between the nerve and the tubercle of the rib.   The first digitation of serratus anterior muscle also arises from this surface close to its outer border, level with and posterior to the grooves.   The costoclavicular ligament binds the clavicle down to the rib at the costochondral junction with subclavius muscle close to it.

The **second rib** is about twice as long as the first rib but is much narrower, though still sharply curved.   It is not twisted and the outer surface of its body faces distinctly upwards and outwards.   The outer surface has a rough tubercle for serratus anterior [FIG. 124].

The second rib is easily recognized in the living subject because its cartilage is level with the sternal angle.   Both the angle and the cartilage may be visible as well as palpable 5 cm. below the jugular notch.

The **tenth, eleventh,** and **twelfth ribs** articulate only with their own vertebrae and so each has only one facet on its head.   The tenth may, however, resemble a typical rib.   The eleventh has only a small tubercle without an articular facet and its angle is poorly defined.   The twelfth rib is small and usually slender.   It has no definite tubercle, angle, or groove, and the body tapers off at its free end in contrast to the widening of the others.

Costal Cartilages

The costal cartilages are of hyaline cartilage and the perichondrium is continuous with the periosteum of the ribs.   Rupture of the cartilage may occur within the perichondrium without displacement of the cartilage.

The cartilages continue the line of the ribs towards the sternum and the upper seven pairs reach the side of the sternum.   The first is continuous with the sternum, the perichondrium being again continuous with the periosteum.   The remaining six are joined by means of synovial joints to the edge of the sternum.   The fifth to the tenth continue their direction for 2–3 cm. and then turn up.   The eighth to the tenth taper to a point and articulate edge to edge with the cartilage next above by a synovial joint in the case of the eighth and ninth, by fibrous tissue in the case of the tenth.

The cartilages of the eleventh and twelfth ribs do not articulate and form tips to the ends of the ribs.

THORACIC CAGE

The wall of the thorax surrounds and protects many structures and its chief function is to provide support against the elastic pull of the lungs, which would otherwise collapse.   It is also part of a suction apparatus which can counteract that elasticity actively and so allow air to enter the lungs.

The necessary rigidity for this purpose is provided by the thoracic cage, which is formed by the vertebral column behind, the sternum in front and, between them, the curving ribs and cartilages.   The slope of the ribs from their evenly spaced posterior ends means that, when the anterior ends of the ribs are raised, the ribs are separated and the intercostal spaces become wider.   At the same time the sternum is pushed forwards and so increases the anteroposterior diameter of the chest cavity.   The cage also gives attachment, particularly along the rib margin, to the diaphragm and muscles of the abdominal wall, which are active in respiration.   For movements of the ribs see page 222.

The thorax has the shape of a cone with the apex and base cut off obliquely : so the superior aperture slopes downwards and forwards, and the inferior aperture (base) slopes downwards and backwards.   It is, therefore, much shorter in front than at the back and sides, and is also flattened antero-

FIG. 125   Front of portion of skeleton showing thorax.

FIG. 126   Back of portion of skeleton showing thorax.

posteriorly, being wider from side to side than antero-posteriorly. In children it is more circular in transverse section.

The **superior aperture of the thorax** lies at the root of the neck and measures about 5 cm. anteroposteriorly and some 10 cm. transversely. It is bounded by the manubrium sterni, the first thoracic vertebra, and the first ribs and costal cartilages. The projection of the vertebral body gives it a kidney-shaped outline and the slope of the ribs leaves the first two thoracic vertebral bodies above the level of the sternum [FIG. 125]. In the middle lie the trachea, oesophagus, and the vessels and nerves entering and leaving the thorax. The more circular, lateral parts are occupied by the apex of the lung and pleura, supported by the suprapleural membrane [FIG. 598].

The **inferior aperture** is large and is bounded by the xiphisternal junction and the lower six pairs of costal cartilages which form the **infrasternal angle.** The shafts of the twelfth ribs slope up to the twelfth thoracic vertebra to complete the circumference which gives attachment to the edge of the diaphragm (except for extensions down to the xiphoid process in front and three lumbar vertebrae behind) and to the muscles of the anterior abdominal wall.

The **posterior wall** is formed by the vertebral column and by the posterior parts of the ribs which sweep back to their angles before turning medially and then forwards, thus forming a deep, paravertebral, **pulmonary sulcus.** Posteriorly, the curve of the ribs is nearly flush with the not very prominent spines, and this allows Man to lie easily on his back.

The **side-walls** are formed by bodies of the ribs which are not quite equally spaced. The upper spaces are wider than the lower ones and the obliquity of the ribs increases from the first to the tenth, which makes the spaces slightly wider towards the front. The degree of obliquity varies with muscu-

larity, age, and sex. The side walls of the thorax slope outwards to the level of the eighth or ninth rib and, as a rule, slightly inwards below that level.

The **anterior wall** is covered like the rest by muscles except near the midline. It is formed by the anterior ends of the ribs, the costal cartilages, and the sternum, and is shorter than the other walls. The third cartilage is almost horizontal; those above slope slightly down, those below are angulated and slope up towards the sternum, the spaces between them becoming increasingly narrow towards the lower end of the sternum [FIG. 125].

Apart from counting down from the second rib, other guides may be used to estimate the position of some of the ribs. Except in the adult female, the nipple usually lies opposite the fourth intercostal space, between the ends of the fourth and fifth ribs. The lower border of pectoralis major muscle slopes medially to the sixth costal cartilage and the seventh is normally the lowest to reach the sternum. The tip of the ninth cartilage usually lies in the midclavicular line, where the lateral border of rectus abdominis crosses the costal margin. The tenth cartilage is the lowest seen from the front and lies 2–3 cm. from the iliac crest. The costochondral junctions lie deep to a line between the medial end of the clavicle and a point 2·5 cm. or so behind the bend of the tenth costal cartilage.

At the back [FIG. 126] the scapula overlies the second to the seventh ribs, or even the eighth, and the triangular area at the medial end of the root of the scapular spine overlies the fourth rib, level with the tip of the third thoracic spine. The twelfth rib is seldom palpable because of the mass of the overlying erector spinae muscle. It lies about half-way between the inferior angle of the scapula and the crest of the ilium, level with the twelfth thoracic or first lumbar spine.

OSSIFICATION OF RIBS

In its ossification, a rib behaves like a long bone, the first ossification being subperichondrial and appearing almost throughout its length. At the same time, an endochondral centre appears near the angle, first in the sixth rib about the sixth week of intra-uterine life. Secondary centres appear about puberty and fuse with the rest of the bone by the twenty-fifth year. The first rib and the seventh to the tenth have one for the head and one for the tubercle. The second to the sixth have one for the head and one each for the articular and non-articular parts of the tubercle. The eleventh and twelfth have one for the head.

STRUCTURE AND VARIATIONS

A rib is composed of highly vascular cancellous bone in a flattened tube of compact bone which is thicker on its two surfaces, thinner at its borders. The compact bone is thickened at the angle where indirect bending stresses tend to cause fracture. The vascular foramina are numerous at the back of the neck, the vessels following the ligaments, and the nutrient canals are in the costal groove, directed towards the angle.

The most important costal variation is the presence of a cervical rib, which may reach the sternum or fuse with the first costal cartilage or with the first rib at its vertebral end to form a bicipital rib. Part of it may be represented by a fibrous band and it may be present only at its sternal end.

Occasionally an extra rib may be present in the lumbar region but a reduction in numbers is rare. The first rib may be rudimentary or fuse with the second to form a bicipital rib.

Ribs may be wide and partly fused together, particularly near their anterior ends, and a widened rib may have a hole in it, as may a costal cartilage. The posterior parts of two ribs may be united by projections which articulate with an intervening nodule of bone.

There may be some variation in the number of cartilages reaching the sternum. In old age, bone tends to be laid down around the cartilages especially the first where this normally happens. Some degree of asymmetry is normal.

DEVELOPMENT OF VERTEBRAE, RIBS, AND STERNUM

Vertebrae

During development the paraxial mesoderm becomes divided by transverse clefts into segments or mesodermal somites and the ventromedial, or scleratogenous, parts of these somites proliferate to form a sheath round the previously formed notochord and neural tube [p. 40]. This sheath is the **membranous stage** of the vertebral column and is soon followed by **procartilaginous** and **cartilaginous stages** which ultimately lead to ossification.

In the procartilaginous stage the vertebrae form as a series of horseshoe-shaped vertebral bows [FIG. 127] which spread round the notochord but are at first open dorsally. The tissue which will form a vertebral centrum surrounds the notochord opposite an intersegmental septum. The **hypochordal bow** is ventral to the cephalic part of the centrum to which it

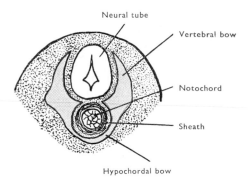

FIG. 127 Transverse section showing development of the membranous basis of a vertebra. (After Keith, 1948).

belongs. The limbs of the **vertebral bow** slant back to enclose the neural tube dorsal to the caudal part of the vertebral centrum and ultimately meet to enclose the neural tube and form a **vertebral arch.**

About the fourth week of intra-uterine life the scleratogenous tissue between the vertebral centra undergoes fibrous and chondrifying changes to form **intervertebral discs.** This tissue is opposite the middle of a somite in each case and the main part of each vertebral body is derived from halves of two adjacent somites. The beginning of chondrification in the centrum and vertebral bow is followed rapidly by fusion of centrum and bow together; chondrification of the bow is completed during the fourth month of intra-uterine life. The limbs of the bow form the vertebral arch and its processes and also the adjacent part of the vertebral body. The hypochordal portion of the bow disappears except in the case of the atlas where it forms the anterior arch. The atlas is therefore a vertebral ring and its centrum becomes the dens. The transverse processes and spine project into the intersegmental septum between somites.

The parts of the notochord included in the bodies of the vertebrae become constricted and disappear after ossification begins, but the part enclosed in each disc takes part in the formation of a **nucleus pulposus.** For the evolutionary history of the notochord and vertebral column see Gadow (1933).

The coccygeal vertebrae develop from small rudiments in which the vertebral bows are deficient. The same deficiency causes the sacral hiatus, and also **spina bifida** at higher levels.

Ribs

The ribs develop in the intersegmental septa and pass through the usual stages of condensation, chondrification, and ossification. Each rib element is first formed opposite the corresponding vertebra, later the vertebral ends of most are shifted slightly headwards to reach their final articulations.

Sternum

The sternum is formed from mesodermal tissue in front of the pericardium, the manubrium probably in association with the shoulder girdle, the remainder from a pair of sternal bars which grow from the manubrial rudiment and fuse in the median plane. After chondrification it becomes connected with the costal cartilages by synovial joints. See also Paterson (1904) and Ashley (1956a).

THE SKULL

The skull is the skeleton of the head and face. Its main bulk is made up of twenty-one bones which are closely fitted together. In addition, there are other bones, developmentally associated with the pharynge alarches [p. 56], which remain separate but are grouped with the skull bones. The **mandible** forms the skeleton of the lower jaw and so forms part of the face. The **hyoid bone** lies in the root of the tongue where it gives attachment to lingual muscles and to those of the floor of the mouth. Three small ossicles are found in each middle ear in the temporal bone but, since they serve the function of hearing, they are described in the section on the Sensory Organs.

There can be confusion with the terms skull, cranium, and calvaria. **Skull** denotes the skeleton of the head, with or without the mandible. The word **cranium** has been used in the same general sense, but it is better to restrict its meaning to skull without mandible. The word **calvaria** usually means the top part of the skull or skull-cap.

The skull consists mainly of a large cranial cavity, the bony walls of which enclose the brain, plus the bones of the face which complete the walls of the orbits and nasal cavity and form the roof of the mouth, and a separate lower jaw or mandible [FIG. 129].

Phylogenetically the skull evolved from cartilage enclosing the brain and the organs of special sense. This origin is reflected by the cartilage bone retained in the base of the skull. As a more recent development, membrane bone is added at the vault of the skull to cover the enlarged space occupied by the brain. The cartilage capsules of the ear and nose are incorporated in the base. In a young person the skull bones are separable though with difficulty. Except in the case of the mandible and the ear ossicles, the joints are either **sutures** of fibrous tissue or joined by cartilage (synchondroses) [FIG. 276]. With advancing years the sutures and synchondroses become ossified.

Most skulls used for study have had the upper part of the skull-cap removed by a circumferential saw cut and the compact outer and inner tables can be seen with the **diploë** (cancellous bone and marrow spaces) sandwiched in between. The thickness of the skull varies in different places in any one skull and the sites of thickening vary in different skulls. Areas covered with muscles are thinner than more exposed parts, e.g., the temporal fossa and the floor of the posterior cranial fossa. The thickness of the skull also varies with age, sex, and race, being thin in children and in the aged. In the former the cortex is thin, in the latter the diploë is reduced. In the young a greater elasticity reduces the risk of fracture. The skull is usually thinner in the female than in the male.

THE SKULL AS A WHOLE

When first examining a skull the following should be identified as they are used as landmarks for reference purposes.

1. The top of the skull is rounded or oval and is relatively smooth. The inferior aspect is rough and irregular.

2. The orbits are the pair of pyramidal cavities in the front of the skull [FIG. 132], so named because the eyes rotate in them.

3. The anterior bony aperture of the nose or **piriform aperture** lies between and below the orbits.

4. The **external acoustic meatus** [FIG. 129] is the small round or oval passage of the ear seen in the lower part of the side of the skull about midway between front and back.

5. The **zygomatic arch** spans the interval between the acoustic meatus and the orbit.

6. The foramen magnum is the large round hole in the base of the skull a little behind its centre [FIG. 133].

The exterior of the skull is described as it is seen from five aspects, and each is spoken of as a **norma.** From the front it is norma frontalis; from the back, norma occipitalis; from above, norma verticalis; and from below, norma basalis; from the side, norma lateralis. There is considerable overlap of the areas visible from the different points of view.

For purposes of description or measurement the skull is orientated in a standard position with the lower border of the orbit and the upper border of the external acoustic meatus in a horizontal plane. This is known as the Frankfurt plane, and the coronal and sagittal planes at right angles to it are used to orientate the skull in obtaining the normae.

NORMA VERTICALIS

The top of the skull [FIG. 128] is rounded or ovoid in most Europeans, the widest part being towards the back, but there is considerable variation in the proportions and often some asymmetry. Its convexity serves to distribute and so minimize the effects of a blow, but hard contact with a flat surface can lead to splits in the scalp resembling cuts.

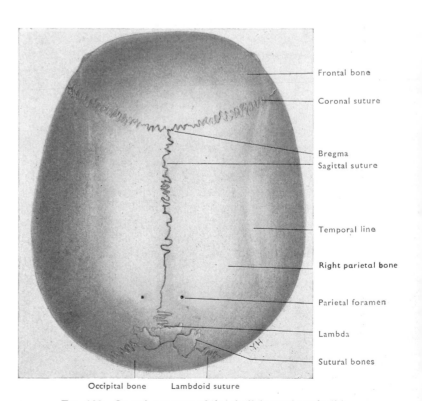

FIG. 128 Superior aspect of the skull (norma verticalis).

The frontal bone forms the forehead and part of the top of the skull. Behind this, separated from it by the **coronal suture,** the two parietal bones extend up from the sides to meet in the medial **sagittal suture** [FIG. 128]. The latter suture extends backwards from the point where it meets the coronal suture **(bregma)** to the apex of a protuberant, triangular part of the occipital bone. Though the occipital bone forms the back of the skull, this triangular part can be seen from above where it intrudes between the parietal bones posteriorly, being separated from them by the **lambdoid suture;** the meeting point of the lambdoid and sagittal sutures is termed **lambda.** The coronal suture extends anteroinferiorly from bregma to **pterion,** where the parietal, frontal, sphenoid and temporal bones almost meet at one point [FIG. 129]. Near the coronal suture, the parietal bone is flattened or slightly concave, whilst it may form a low ridge along the sagittal suture. A parietal foramen, which transmits an emissary vein, may be found in each parietal bone close to the sagittal suture.

At bregma and lambda are two parts of the skull into which ossification from the surrounding bones spreads relatively late. They remain membranous even after birth, constituting respectively the **anterior and posterior fonticuli** which ossify in the second and first years. Bregma lies just anterior to a vertical plane passing through the external acoustic meatus; lambda lies a little behind the site of the parietooccipital sulcus of the brain [FIG. 881].

The most convex parts of the frontal and parietal bones, **frontal and parietal tubera** [FIGS. 150, 152], are easily identified by feeling with the palm of the hand and are useful as landmarks.

The **temporal lines** [FIG. 129] arch up over the side of the frontal bone and across the parietal to run down towards its postero-inferior angle. They indicate the attachment of the temporal fascia (superior line); and the outer limit of the origin of the temporalis muscle (inferior line) [FIG. 355].

The **vertex,** the highest point of the skull placed in the Frankfurt plane, lies near the middle of the sagittal suture.

NORMA OCCIPITALIS
[FIG. 105]

The back of the skull is markedly convex and is usually widest at the parietal tubera. The bones seen are the posterior parts of the two parietal and two temporal bones with the occipital bone between them. The mastoid portion of the temporal bone and its mastoid process are visible at the lower corner on each side.

The parietal tubera and foramina, the posterior part of the sagittal suture, lambda, and the entire lambdoid suture are visible in this view.

The **lambdoid suture** passes inferolaterally to the **mastoid angle** of the parietal bone, and is then continued as the **occipitomastoid suture** between the occipital and temporal bones [FIG. 133]. The **mastoid foramen,** which varies in size, lies in the latter suture or in the mastoid temporal near it, 5–6 cm. behind the external acoustic meatus. It transmits an emissary vein from the sigmoid sinus and a branch of the occipital artery.

Half-way between lambda and the foramen magnum is the median **external occipital protuberance,** from which a curved **superior nuchal line** passes laterally to the mastoid process on each side. The lines, which give origin to trapezius and sternocleidomastoid muscles, mark the junction between the scalp region and that for attachment of the muscles of the back of the neck. Above each of these lines is a poorly marked **highest nuchal line** for the attachment of the epicranial aponeurosis of the scalp and the occipital belly of occipitofrontalis muscle. The centre of the external occipital protuberance is called **inion.**

NORMA LATERALIS
[FIG. 129, PLATE 8 C]

The side view of the skull shows clearly the division into a large, ovoid brain-case and a smaller, uneven but almost triangular facial skeleton attached to it antero-inferiorly by the **zygomatic bone.** This triradiate bone extends: (1) *upwards* (forming the lateral margin of the orbit and the anterior margin of the temporal fossa) to unite with the **zygomatic process** at the antero-inferior corner of the frontal bone; (2) *backwards* (forming the anterior part of the zygomatic arch and the lower margin of the temporal fossa) to unite by an oblique suture running downwards and backwards with the **zygomatic process of the temporal bone,** which completes the zygomatic arch; (3) *antero-inferiorly*, where it is buttressed to the maxilla (the main part of the facial skeleton) and forms a large part of the inferior margin of the orbit. It thus forms the main lateral strut attaching the facial skeleton to the braincase and it is not uncommonly fractured in severe blows to the face.

The superior temporal line begins at the posterior edge of the zygomatic process of the frontal bone (there is a palpable notch at the suture with the zygomatic bone), arches backwards over the frontal and parietal bones, and curves forwards above the external acoustic meatus where it forms the supramastoid crest before becoming continuous with the upper margin of the zygomatic arch. At the anterior end, this margin bends sharply upwards (jugal point) and continues as the posterior margin of the frontal process of the zygomatic bone. The temporal fascia is attached to this entire margin, and covers the temporalis muscle which is attached below the inferior temporal line.

The region enclosed by the superior temporal line is the **temporal fossa.** The anterior wall of the fossa is formed by the zygomatic bone, by the zygomatic process of the frontal bone and by the greater wing of the sphenoid. A small foramen in the zygomatic bone here transmits the zygomaticotemporal nerve.

The medial wall of the fossa is formed by the parietal and frontal bones, and by the upturned parts of two bones in the base of the skull. One is the **squamous part of the temporal bone,** a scale-like plate of bone which lies above the acoustic meatus and overlaps the lower edge of the parietal bone to form the **squamosal suture.** The other is the **greater wing of the sphenoid** which turns upwards from the base of the skull, between the squamous temporal and frontal bones to meet the antero-inferior corner of the parietal bone **(sphenoidal angle)** at the **sphenoparietal suture.**

Four bones thus articulate within an area less than 3 cm. in diameter. Usually the group of sutures is H-shaped as shown in FIGURE 129, but variations occur and epipteric (sutural) bones may be present. The area is named **pterion.** Its centre lies approximately 4 cm. above the zygomatic arch and 3 cm. behind the zygomatic process of the frontal bone. In this region the frontal branch of the middle meningeal artery lies in a groove or tunnel on the inside of the bone and it is liable to be torn in fractures.

The temporal surface of the greater wing of the sphenoid is limited below by an **infratemporal crest,** a horizontal ridge which separates the temporal surface from the pterygoid region on the inferior surface of the skull, and forms a boundary for the temporal fossa.

The **zygomatic process of the temporal bone** arises from the squamous part anterosuperior to the external acoustic meatus. Posteriorly continuous with the supramastoid crest, its root forms part of the articular fossa for the head of the mandible inferiorly, together with a small postglenoid tubercle which projects downwards behind the fossa. Further anteriorly, the inferior surface of the root of the zygomatic process forms a prominent **articular tubercle** which provides the anterior part of the articular area for the head of the mandible. The tubercle is palpable laterally, and medially it continues as a ridge to the squamous temporal bone.

The external acoustic meatus lies in the angle between the squamous part of the temporal and its mastoid process. The bony meatus is 1·5 cm. long, but cartilage attached to its edge during life increases its length to 2·5 cm. The external opening is oval, 5–6 mm. across its narrow anteroposterior diameter. The canal passes medially and slightly upwards, with a gentle superior convexity, to end at the tympanic membrane which separates it from the middle ear cavity during life [FIG. 999]. In a dried skull, the medial wall of the latter can be seen directly through the meatus.

The bony meatus is formed superiorly and to a small extent posteriorly (postauditory process) by the squamous part of the temporal bone. The remaining walls are formed by the tympanic part of the temporal bone, a curved, triangular plate which fits between the mastoid and squamous parts of the same bone. Between it and the mastoid process is the **tympanomastoid fissure.** Anteriorly, the edge of the plate fits in behind the postglenoid tubercle and, more medially, is separated from the mandibular fossa of the squama by the **squamotympanic fissure** [p. 113]. In the infant the tympanic plate is a mere ring, so that the meatus is short. The **suprameatal triangle** is a small depression lying at the meeting of tangents to the superior and posterior walls of the meatus. It lies inferior to the suprameatal crest and marks

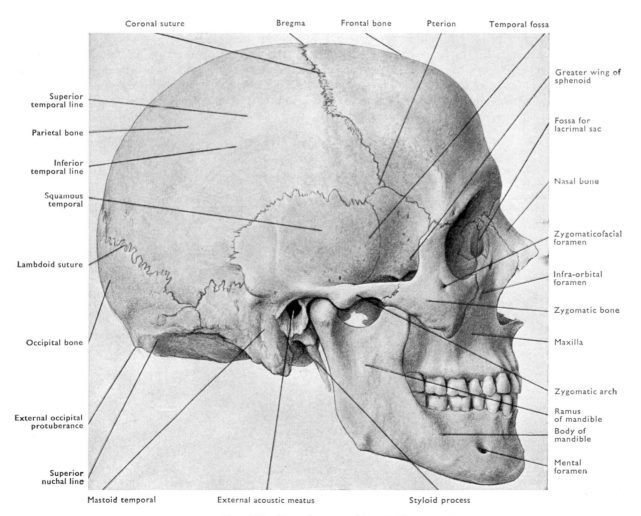

FIG. 129 Lateral aspect of the skull.

the position of the mastoid antrum (an extension of the middle ear cavity to the air cells in the mastoid process) which is 12 mm. deep to it [FIG. 130].

The **styloid process** is a tapering spicule of variable length projecting from behind the tympanic plate which partially surrounds its base. It is continuous antero-inferiorly with the stylohyoid ligament and it lies deeply in the interval between the mastoid process and the mandible [FIG. 129].

The **mastoid region of the temporal bone** is the lateral part of the petrous temporal bone where it appears on the surface of the skull between the external acoustic meatus in front, the squamous part of the temporal bone and the mastoid angle of the parietal bone above, and the occipital bone behind. It is fused with the squamous part along a line behind and below the supramastoid crest [FIG. 157] but is separated from the others by sutures. It carries the prominent **mastoid process** which projects antero-inferiorly behind the lower part of the ear and is readily palpable. The process varies in size with age and muscularity, is absent at birth and small in a child. It is associated with an upright posture and the resultant balance of the head. After puberty the mastoid process becomes permeated by small **mastoid cells** which are air-filled spaces extending from the mastoid antrum, a process which has started at birth but does not extend into the mastoid until this time. They then increase rapidly and reach their full size in a year or two. The first appearance and extent of these air cells is very variable [PLATE 9 D].

The auricular branch of the vagus nerve pierces the **tympanomastoid fissure,** anterior to the mastoid process, and the mastoid emissary foramen lies in or near the occipito-mastoid suture.

The mastoid process has sternocleidomastoid, splenius capitis, and longissimus capitis muscles attached to it [FIG. 344]. Behind the mastoid, the superior nuchal line forms an almost horizontal line running to the external occipital protuberance.

The **infratemporal fossa** lies below the horizontal part of the greater wing of the sphenoid and posterior to the maxilla. The anterior wall of the fossa is the rounded posterior aspect of the maxilla with the **tuberosity of the maxilla** projecting from its lowest part above and behind the last molar tooth. About the middle of this surface are small **alveolar foramina** for the posterior superior alveolar vessels and nerves entering to supply the molar teeth.

Projecting antero-inferiorly from the sphenoid is the **lateral lamina** of its **pterygoid process** which forms the medial wall of the infratemporal fossa. It is an almost flat plate of bone 3 cm. in length and over 1 cm. wide. Anteriorly and medially, it is continuous with a **medial lamina,** concealed from the present view-point. The pterygoid process and maxilla are separated by a narrow V-shaped gap, the **pterygomaxillary fissure.** It leads into the narrow **pterygopalatine fossa,** so called because it lies between the pterygoid process behind and the palatine bone in front, where the vertical plate of the palatine extends laterally over the back of the maxilla to form the anterior wall of the fossa.

At the lower end of the pterygopalatine fossa, the **pyramidal process of the palatine bone** intervenes between the maxilla and the pterygoid process of the sphenoid, and all three are fused. At the upper end of the fossa, the greater wing of the sphenoid passes laterally, separated from the maxilla by a horizontal cleft which leads into the orbit, the **inferior orbital fissure.**

The vertical plate of the palatine bone has a notch at its upper end, converted into a foramen by contact with the sphenoid and therefore called the **sphenopalatine foramen.** A probe passed through the pterygopalatine fossa and through this foramen enters the nasal cavity.

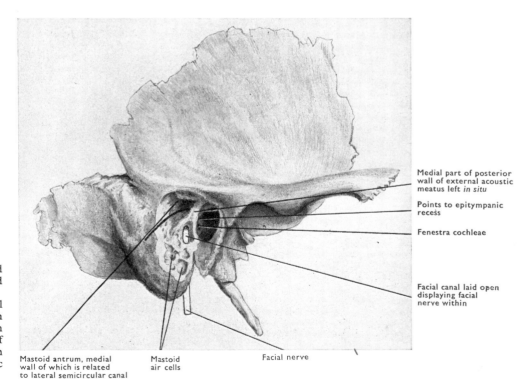

FIG. 130 Right temporal bone prepared to show the position of the mastoid antrum.

The greater part of the posterior wall of the external acoustic meatus has been removed, but a bridge of bone has been left at its medial end; under cover of this a bristle (thick black) passes through the mastoid antrum to the tympanic cavity.

Medial part of posterior wall of external acoustic meatus left *in situ*

Points to epitympanic recess

Fenestra cochleae

Facial canal laid open displaying facial nerve within

Mastoid antrum, medial wall of which is related to lateral semicircular canal

Mastoid air cells

Facial nerve

Sphenopalatine foramen

FIG. 131 Lateral view of the pterygoid region to show the pterygomaxillary fissure and sphenopalatine foramen.

NORMA FRONTALIS

[FIG. 132]

The front of the skull comprises the anterior part of the brain-case above, and the skeleton of the face below. It varies very considerably in its shape and proportions with age, race, and sex.

The adult **mandible** carries a set of sixteen teeth and closes these against the same number in the two **maxillae** which together form the upper jaw. Anteriorly the upper parts of the maxillae diverge to enclose the **piriform aperture** of the nose, and then pass up to reach the frontal bone on each side of the nasal bones which form the bridge of the nose. The maxilla forms the greater part of the floor of an orbit, but only a small part of its inferior margin for it is overlapped laterally by the zygomatic bone.

The Maxillary Region

Below the piriform aperture the two maxillae meet in the median plane and a sharp median spur, the **anterior nasal spine,** projects forwards at the lower margin of the nasal

aperture. It may be bifid, and it is embedded in the lower, mobile part of the septum of the nose.

The external margin of each maxilla projects down to form, with its counterpart, a U-shaped **alveolar process** which carries a set of eight upper teeth on each side [FIG. 476] embedded in a row of sockets on its inferior surface. If the walls of the sockets have not been absorbed, vertical ridges caused by the roots of the teeth extend upwards from the alveolar margin. The canine tooth causes the largest of these, the **canine eminence,** and above and lateral to it is a wide, shallow depression—the **canine fossa**—which reaches almost to the orbital margin. Varying degrees of fullness here, or variations in the shape of the alveolar process, can markedly affect the facial appearance.

The **infra-orbital foramen** lies 6 mm. below the orbital margin and 1·5 cm. from the side of the nose. It transmits the infra-orbital vessels and nerve.

The margin of the **piriform aperture** is formed by the two maxillae and the two nasal bones. It is oval or pear-shaped and lies in a plane which slopes downwards and backwards before curving forwards to the region of the anterior nasal spine. The upper part of the margin is irregular but the edges of the wider, lower part are smooth. They give attachment to the soft parts of the external nose [FIG. 568], including cartilage.

Above the piriform aperture, the two **nasal bones** form the bridge of the nose which varies considerably in size and prominence. They articulate: (1) with the frontal bone (**nasion**) superiorly; (2) with the frontal processes of the maxillae laterally; and (3) with each other in the midline. Their inferior ends are wider and irregular and form the upper margin of the piriform aperture. The external nasal nerve passes to the skin between this margin and the nasal cartilage [FIG. 901]. A minute foramen may transmit a vein from the skin through the nasal bone to reach the superior sagittal sinus inside the skull, thus providing a possible pathway for infection.

To the lateral side of the face the maxilla has a buttress-like projection, **zygomatic process,** which supports the zygomatic bone.

The **zygomatic bone** forms the hard, prominent part of the cheek and is often called the cheek bone [FIG. 129]. It has three surfaces, lateral, orbital, and temporal. The lateral surface is rounded and subcutaneous. It presents a small zygomaticofacial foramen, for the nerve of that name. Anteriorly, this surface blends with that of the maxilla; antero-superiorly, it extends to the lower margin of the orbit forming its lateral half; posteriorly, it is continued by means of the **temporal process** into the zygomatic arch; while superiorly, the **frontal process** extends to articulate with the frontal bone and, separating the temporal fossa from the orbit, forms most of the lateral orbital margin. A flange passes deeply from the frontal process to articulate with the greater wing of the sphenoid and, with it, to complete the separation of the **orbit** (orbital surface) from the temporal fossa (temporal surface). The flange also articulates inferiorly with the maxilla and, closing off the lateral end of the inferior orbital fissure, extends medially into the floor of the orbit.

Thus the orbital surface of the zygomatic bone forms part of the floor and lateral walls of the orbit, and it transmits the zygomatic nerve through a small foramen of that name. The temporal surface is best seen from the side of the skull and is described on page 106.

The **margins of the orbit** are formed by the zygomatic bone below and laterally, by the maxilla below and medially. Above, the frontal bone forms both margin and roof of the orbit and sends down maxillary and zygomatic processes to articulate on the medial and lateral sides with the corresponding process from below.

The whole of the **frontal region** is formed by the frontal bone. Superiorly it merges with the top of the skull and the frontal tubera are visible on each side [FIG. 132]. The narrowest part of the brain-case normally lies immediately posterior to the zygomatic process. The upper margin of the orbital opening (supra-orbital margin) forms a curved border. Its outer two-thirds are sharp and definite. Its medial third is much less so and, at the junction of these two parts there is

usually a **supra-orbital notch** which can be felt from below. The supra-orbital vessels and nerves lie in the notch as they turn up into the scalp. A small foramen in the floor of the notch transmits an emissary vein to connect the frontal diploic vein with the supra-orbital vein. The notch may be divided or converted into a foramen.

Glabella is the smooth rounded eminence between the two supra-orbital margins, above nasion [FIG. 149] and, in the living subject, between the two eyebrows. From glabella a smooth elevation on each side arches laterally above the orbital margin to form the **superciliary arch.** This overlies the **frontal sinus** [p. 122]. The surface of glabella frequently shows the remains of the lower end of a suture between the two halves of the frontal bone. Fusion normally occurs at this frontal suture between 6 and 10 years. When persistent, it is referred to as a **metopic suture.**

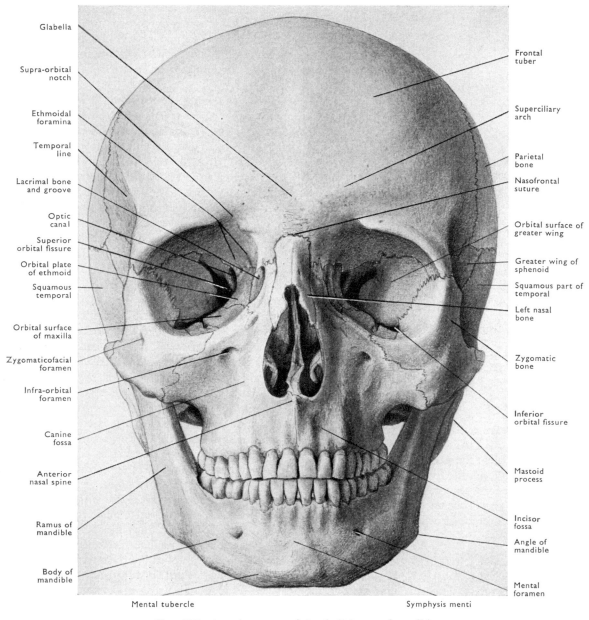

FIG. 132   Anterior aspect of the skull (norma frontalis).

NORMA BASALIS

[FIG. 133]

After removal of the mandible, the inferior surface of the skull shows the under surface of the upper jaw in front, with the palate and teeth, and the supporting zygomatic arches curving back on each side. Behind is the wide nuchal plane of the occipital bone, marked by the muscles of the neck. In between, the surface is exceedingly irregular because of numerous foramina, articulations, and processes.

The posterior part of the base of the skull is formed mainly by the occipital bone, but the mastoid processes of the temporal bone is easily identified [FIGS. 129 and 133]. Level with the mastoid processes the **foramen magnum** pierces the base

of the skull. It measures about 3 cm. wide by 3·5 cm. long and on each side has its anterior half encroached on by an **occipital condyle** which articulates with the atlas. The condyles are oval, and are set obliquely on the anterolateral aspect of the foramen magnum. They are convex longitudinally, but are relatively flat in the transverse plane which slopes inferomedially.

The anterior edge of the foramen is slightly thickened and lies between the anterior ends of the condyles. Its midpoint is used in taking measurements and is named **basion** since it is on the posterior edge of the basi-occipital.

The posterior half of the edge is thin and semicircular. The **atlanto-occipital membranes** are attached to the anterior and posterior margins of the foramen and joint capsules are

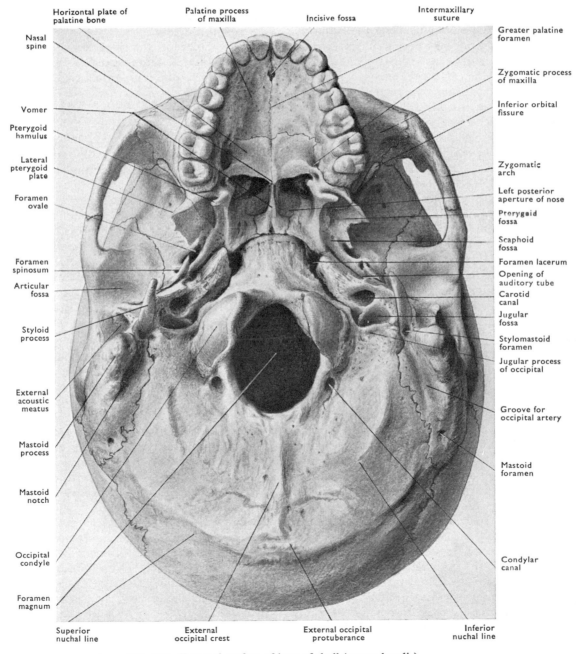

FIG. 133   External surface of base of skull (norma basalis).

attached to the edges of the condyles. The **apical ligament** of the dens is also attached to the anterior margin, but the upper fasciculus of the **cruciate ligament** and the **membrana tectoria** pass through the foramen before gaining attachment [Fig. 290.]

Superior to each condyle, in its anterior half, a **hypoglossal canal** transmits the hypoglossal nerve from the cranial cavity. Behind the posterior end of the condyle is a depression, the **condylar fossa,** usually with a **condylar canal** which carries a venous communication backwards from the sigmoid sinus in the skull to the vertebral veins in the neck. Below this region the vertebral artery, with its veins and sympathetic nerves, curves round the lateral mass of the atlas to reach the foramen magnum and enter the skull. The spinal medulla becomes continuous with the medulla oblongata at this level and the covering meninges also pass through. Spinal vessels run down, and the spinal roots of the accessory nerve pass up through the foramen.

The **nuchal plane** of the occipital bone lies behind the foramen magnum and is divided into right and left halves by an **external occipital crest** which runs from the external occipital protuberance to the edge of the foramen in the midline. Each half is separated into two areas by an **inferior nuchal line** which curves laterally from the crest, midway between the superior nuchal line and the foramen magnum. Each of the four areas so demarcated gives attachment to two muscles and the external protuberance and the nuchal crest give attachment to the elastic nuchal ligament [FIGS. 133, 154, 340]. The muscles and the nuchal ligament support and balance the skull on the vertebral column.

The anterior part of this nuchal area is continued forwards lateral to the condyle on to the condylar part of the bone and ends abruptly at a transverse edge as the jugular foramen is reached. The quadrilateral area behind this edge and lateral to the condyle is the **jugular process** [FIG. 133]. It gives attachment to rectus capitis lateralis which connects it to the transverse process of the atlas.

The part anterior to the foramen magnum, including the anterior ends of the articular facets, is the **basilar part** or basi-occiput, as distinguished from the **lateral** (condylar) and **squamous** (posterior) **parts** of the occipital bone. Anteriorly, the basilar part tapers to a flattened bar of bone which is joined to the body of the sphenoid by a growth cartilage until that cartilage ossifies at puberty. A small **pharyngeal tubercle** projects downwards from the basilar part in the midline, about 1 cm. in front of the edge of the foramen magnum, and gives attachment to the pharyngeal raphe which lies in the posterior wall of the pharynx. Longus capitis muscle is attached anterolateral to it with rectus capitis anterior on each side, immediately anterior to the condyle [FIG. 340].

Thus the occipital and sphenoid bones form the solid central part of the base of the skull, and the **petrous temporal bones** are wedged anteromedially between their lateral parts. The line of junction between the occipital bone and the petrous part of the temporal (mastoid portion) runs forward, a short distance medial to the mastoid process, as far as the anterior edge of the jugular process. Here the large **jugular foramen** faces downwards and forwards between the two bones; the edge of the occipital bone forms a **jugular notch,** and the petrous temporal is excavated to form a **jugular fossa,** which accommodates the superior bulb of the jugular vein. The fossa and the vein are close to the middle ear, as can be judged by looking through the external acoustic meatus, and infection may spread through the bone from the middle ear to the vein.

The jugular foramen lies medial to the external acoustic meatus, and almost level with its lower border. The foramen may be partly or completely divided into three compartments by small spicules of bone. The anteromedial compartment transmits the inferior petrosal sinus and a meningeal branch of the ascending pharyngeal artery. The middle compartment contains the ninth, tenth, and eleventh cranial nerves, from before backwards. The large, posterolateral compartment transmits the sigmoid sinus on its way to become the jugular vein, meningeal branches of the occipital artery and of the vagus nerve, and some lymphatics.

On the lateral wall of the jugular foramen, a small pinpoint hole forms the medial end of the **mastoid canaliculus** for the auricular branch of the vagus which emerges through the **tympanomastoid fissure.**

On the edge of the ridge separating the jugular foramen from the carotid canal lie small foramina for vessels. One is the entrance to a **tympanic canaliculus** for the tympanic branch of the glossopharyngeal nerve passing to the tympanic cavity.

Anteromedial to the jugular foramen, the articulation between the petrous temporal and the occipital bones runs forwards and medially and leads to an irregular space enclosed by the apex of the petrous temporal, the basi-occipital, and the body of the sphenoid bone. This **foramen lacerum** is closed with cartilage, which is pierced vertically by small blood and lymph vessels only.

All the parts of the temporal bone are visible from below. The petrous part extends anteromedially from the mastoid process to the foramen lacerum. Anterior to the jugular foramen its inferior surface is pierced by the round opening of the **carotid canal** which turns at right angles within the bone to reach the side of the foramen lacerum. It transmits the internal carotid artery with its plexus of veins and sympathetic nerves.

Immediately above the lower aperture of the carotid canal, on its posterolateral wall, are the openings of two small **caroticotympanic canaliculi** which convey tympanic branches of the carotid artery and sympathetic nerves.

The tympanic plate lies under the petrous part of the bone and joins the anterior surface of the mastoid process at the **tympanomastoid fissure.** This may be reduced to a small canaliculus for the auricular branch of the vagus nerve.

More medially, behind and partly sheathed in the tympanic plate, the tapering **styloid process** projects downwards and forwards lateral to the jugular vein [FIGS. 133 and 347] to be continued as the stylohyoid ligament. Between the styloid and the mastoid processes lies the **stylomastoid foramen.** This transmits the facial nerve, emerging from the **facial canal** in the petrous temporal. With the nerve are small branches of the posterior auricular vessels. Swelling and consequent pressure in the narrow rigid canal can produce a facial paralysis.

Behind the stylomastoid foramen and medial to the mastoid process, a distinct groove, the **mastoid notch,** gives attachment to the posterior belly of the digastric muscle. Medial to the notch a slight groove may be produced by the occipital artery which ascends to the skull deep to the posterior belly of the digastric.

Anterior to the tympanic plate lies the **squamotympanic fissure** and the fossa for the head of the mandible on the squamous part of the bone [FIGS. 133 and 160]. Between the two edges of the fissure a down-turned edge of the roof of the tympanic cavity (tegmen tympani) appears in the fissure [FIG. 157] and divides it into **petrotympanic** and **petrosquamous** parts. The chorda tympani branch of the facial nerve emerges from the middle ear cavity, through the medial end of the petrotympanic fissure, and passes medial to the spine of the sphenoid.

The **mandibular fossa** is wide and smooth but short from front to back. The articular surface extends over the posterior and inferior surfaces of the **articular tubercle** [p. 107] which limits the fossa anteriorly. Laterally, the fossa extends on to the root of the zygomatic process.

In front of the articular tubercle, the bone slopes upwards into the temporal fossa, whilst anteromedial to it is the roof of the infratemporal fossa. The latter extends anteriorly to the inferior orbital and pterygomaxillary fissures, laterally to the infratemporal crest on the greater wing of the sphenoid, and medially to the lateral pterygoid lamina.

The posterior part of the greater wing of the sphenoid projects between the squamous and petrous parts of the temporal bone, articulating with the edge of the petrous temporal posteromedially. The **auditory tube** lies in a groove along this line of articulation and enters the bone in the apex of the angle between the squamous and petrous parts of the temporal bone, immediately behind the posterior tip of the greater wing of the sphenoid which projects downwards as the **spine of the sphenoid.** The bony semicanal for the auditory tube is partially separated by a thin bony shelf from the semicanal for the tensor tympani muscle superior to it [FIG. 1011].

Immediately anteromedial to the spine of the sphenoid is the **foramen spinosum** which transmits the middle meningeal artery and the meningeal branch of the mandibular nerve. Further anteromedially, in the greater wing of the sphenoid immediately lateral to the groove for the auditory tube, lies the **foramen ovale** which transmits the mandibular nerve, an accessory meningeal artery, and venous communications between the cavernous sinus and the pterygoid plexus. The foramen ovale lies 4 cm. deep to the tubercle of the root of the zygoma.

On each side, the **pterygoid processes** hang down from the root of the greater wing of the sphenoid, lateral to the posterior aperture of the nose [choana; FIG. 133]. Each consists of a medial and a lateral **pterygoid lamina** which are joined anteriorly in their upper half but separated by the **pyramidal process of the palatine bone** inferiorly. The hollow between the two plates is the **pterygoid fossa.**

The **lateral pterygoid lamina** gives attachment to the pterygoid muscles; the lateral muscle arising from its lateral surface, the medial muscle from its medial surface. It slopes laterally as well as backwards to a thin edge which often bears a prominent spine. A layer of fascia extends from this posterior edge to the spine of the sphenoid and may be ossified.

The **medial pterygoid lamina** is narrower and lies in an anteroposterior plane. Where its posterior edge meets the base of the skull, it divides to enclose a **scaphoid fossa** which extends along the posterior edge of the greater wing of the sphenoid, medial to the foramen ovale and lateral to the groove for the auditory tube. The lower end of the posterior border of the medial lamina is continued as a slender, curved **pterygoid hamulus.** The hamulus and the edge of the lamina immediately above it give rise to fibres of the superior constrictor muscle of the pharynx; below the level of the hamulus, these fibres interdigitate with those of the buccinator muscle to form the pterygomandibular raphe. This raphe connects the hamulus to the mandible posterior to the last molar tooth. The medial lamina forms the lateral wall of the nasal cavity where the cavity becomes continuous with the nasal part of the pharynx, and has the opening of the auditory tube against the upper part of its posterior margin. The tensor palati muscle arises from the scaphoid fossa and the spine of the sphenoid, as well as from the auditory tube. Its tendon hooks round the hamulus to reach the palate.

The base of the pterygoid process has a **pterygoid tubercle** which projects posteriorly beneath the foramen lacerum and the pterygoid canal, which runs through the base of the process from the foramen lacerum to the pterygopalatine fossa. Medially, the process extends across the inferior surface of the body of the sphenoid and forms a **vaginal process** which articulates with the vomer. A vomerovaginal sulcus with a small artery may lie between them.

In the midline the **vomer** [p. 142] articulates with the inferior surface of the sphenoid by expanding laterally (**alae of the vomer**). From this articulation the vomer projects antero-inferiorly to form the posterior part of the nasal septum. Thus the vomer in the midline, the medial pterygoid plates laterally, the body of the sphenoid superiorly with the vaginal and alar processes applied to it, and the palate inferiorly surround the posterior apertures of the nose or **choanae.** These oval openings (2·5 cm. high and 1·2 cm. wide) connect the nasal cavities with the nasal part of the pharynx.

The spheno-occipital synchondrosis lies on a line joining the two pterygoid tubercles.

A prominent feature in the anterior part of the norma basalis is the inferior surface of the upper jaw [FIG. 133]. The maxillae bear the alveolar processes which together project down as a thick, arched ridge. Filling in the concavity of the arch is the bony palate which forms the floor of the nose and the roof of the mouth. Each maxilla in a normal adult carries a set of eight teeth: three molars, two premolars, one canine, and two incisors [FIG. 476].

If teeth have been removed some time before death the walls of their sockets are absorbed so that the corresponding part of the alveolar process becomes nearly flush with the palate.

Behind the last molar tooth on the free end of the alveolar margin, is the **tuberosity of the maxilla,** well developed only if the third molar, or wisdom tooth, has erupted. Posterior to the tuberosity, the **pyramidal process of the palatine bone** is wedged between the medial and lateral pterygoid laminae and the maxilla. It is usually fused with the laminae and with the maxilla and, medially, is continuous with the horizontal plate of the palatine bone. Its inferior surface has, as a rule, two **lesser palatine foramina** which transmit lesser palatine nerves and vessels to the soft palate and adjacent structures from the lesser palatine canals, branches of the greater palatine canal.

The **bony palate** is formed from three sources; a premaxilla, two maxillary palatine processes, and, posteriorly, a horizontal strip about 6 mm. in anteroposterior width, is formed from the two **horizontal plates of the palatine bones** joined together in the midline. The common posterior margin is sharp and concave on each side with a **nasal spine** projecting back from it in the midline.

The soft palate is attached to the posterior margin, and the **palatine crest** on the inferior surface just in front of it takes part of the insertion of tensor palati muscle. Laterally, opposite the last molar tooth, in front of the end of the crest, lies the **greater palatine foramen,** the lower end of a **greater palatine canal.** The canal leads from the pterygopalatine fossa, down between the maxilla and a groove on the perpendicular plate of the palatine bone. It conveys the greater palatine vessels and nerves which then run forwards on the bony palate medial to the alveolar process.

The **maxillary part of the palate** lies in front of the palatine bones, and its two halves join in the midline. It is slightly concave and the hollow appearance is increased by the prominence of the alveolar borders. The surface is rough and pitted by the many mucous glands in the mucoperiosteum. Scattered small foramina are more numerous in front.

Anteriorly, the median suture divides and in a young bone may be seen to demarcate a triangular area of bone carrying the incisor teeth. This area is the inferior surface of the **premaxilla** which is a separate bone in some animals.

Behind the incisor teeth in the midline is a pit of varying depth and width, called the **incisive fossa,** into which four foramina open. Two in the midline transmit nasopalatine nerves, the left being in front. The lateral foramina transmit terminal branches of the greater palatine arteries to anastomose with sphenopalatine vessels in the nose.

The median suture and its two continuations forward, to the spaces between the incisor teeth and the canines, indicate the lines along which a developmental failure to fuse may lead to a cleft palate.

THE INTERIOR OF THE SKULL

The walls of the cranial cavity are formed by the frontal, parietal, occipital, temporal, and sphenoid bones, already seen on the exterior of the skull, with the ethmoid which also enters into the formation of orbit and nose.

The capacity of the cranial cavity has been much used in anthropological work. Over 1,450 cm³. is a large capacity; below 1,350 cm³. is small.

Measurements of the thickness of 448 male white American crania (Todd, 1924) ranged from 3 mm. to 6 mm. at the posterior border of the foramen magnum and at the vertex. At the position of greatest thickness the range was from nearly 5 mm. to 11 mm. There is no known relation between thickness of skull bones and intelligence, but a thick skull obviously provides better protection.

The cranial cavity is lined by two closely adherent layers of fibrous membrane sometimes referred to as outer and inner layers of the dura mater. The outer layer, in fact periosteum (or **endocranium**), lines the interior of the cranium and is continuous with the periosteum on the outside (or pericranium) through the foramina and fissures.

The inner layer is the **dura mater** proper and forms a continuous covering for the brain. It is folded between the major parts of the brain, but elsewhere is attached to the endocranium.

THE ROOF OF THE CRANIAL CAVITY

(Skull cap, or calvaria)

It is formed by the frontal, parietal, and occipital bones, and a small part of the squamous temporal bone on each side.

The internal or cerebral surface is fairly smooth and highly concave, particularly from side to side. Except at the sutures, the endocranium can be stripped from it quite easily, particularly in a young skull.

The **sutures** may be indistinct or obliterated if the skull is from an old person, for fusion begins on the inside between the ages of 20 years and 30 years; 10 years earlier than on the outside. The sutures are the same as on the outside, but the squamous suture is at a lower level because of the considerable overlap of the squamous temporal on the outside of the parietal bone.

The **frontal crest** [FIG. 134] is a median ridge on the lower part of the internal surface of the frontal bone. It extends upwards to become continuous with the **sulcus for the superior sagittal sinus;** a shallow, median furrow extending upwards and backwards on the frontal bone, then along the internal surface of the sagittal suture, and down on the occipital bone as far as the internal occipital protuberance, where it turns to one side [FIG. 155]. The groove becomes progressively wider as the sinus enlarges in its course towards the occiput [FIG. 1117]. The dura mater leaves the endocranium at each edge of the furrow to form a fold (falx cerebri) which extends inferiorly between the cerebral hemispheres to a free edge below. The sinus lies between its layers and the endocranium [FIG. 873].

Arachnoid granulations [FIG. 876] project into the cerebral sinuses and particularly into venous spaces at the side of the superior sagittal sinus (lateral lacunae). They project sufficiently to indent the bone [FIG. 153] and form **granular pits,** small irregular depressions placed here and there alongside the groove for the sagittal sinus. They increase in size and number with age.

The **parietal foramen,** if present, lies close to the groove for the sinus on one or both sides 2·5–4 cm. above lambda. In some bones a groove for meningeal vessels leads up to the foramen.

Vascular grooves for the meningeal vessels are narrow and branching and extend from the lower edge of the vault up to the top. The largest are for the middle meningeal vessels and their direction is upwards and backwards [FIG. 137]. Small vascular foramina are numerous especially in or near the grooves. Shallow, ill-defined depressions found all over the interior of the vault correspond more or less to the convolutions of the brain.

THE FLOOR OF THE CRANIAL CAVITY

[FIG. 134]

This is formed by the base of the skull and is divided into three main **fossae,** separated by prominent ridges and set at different levels. The ridges are arranged as a flattened X, and the posterior fossa which lies behind and below its posterior limbs is easily identified by the presence of the foramen magnum in its floor.

The middle fossa lies at a slightly higher level. It has a median part hollowed out in the middle of the X and two lateral parts between its limbs, one on each side. The anterior fossa lies between the flattened anterior limbs of the X and at a still higher level, above the nose and the two orbits.

The Posterior Cranial Fossa

This is the largest and deepest of the three. The floor and posterior wall are formed chiefly by the concave surface of the occipital bone, whilst the posterior surface of the petrous part

of the temporal bone forms the anterolateral wall, and the body of the sphenoid, continuous with the anterior surface of the basi-occipital, forms the anterior wall with it. The mastoid angle of the parietal bone enters the fossa at the side. Posteriorly, an **internal occipital protuberance** corresponds to the external protuberance and is joined to the edge of the foramen magnum by the **internal occipital crest** which divides the posterior part of the fossa into two. A small fold of dura mater, the falx cerebelli, projects from it between the two cerebellar hemispheres, each of which occupies a cerebellar fossa, lateral to the crest and to the posterior half of the foramen magnum. The floor of the fossa is thin-walled, and is situated above the attachment of the muscles of the back of the neck. The **sulcus for the superior sagittal sinus** passes down to the internal occipital protuberance and then turns abruptly to one side, usually the right, as a transverse sinus. This forms a wide groove on the bone and a similar groove for the opposite transverse sinus is formed by the continuation of the straight sinus. The **groove for the transverse sinus** [FIG. 155].

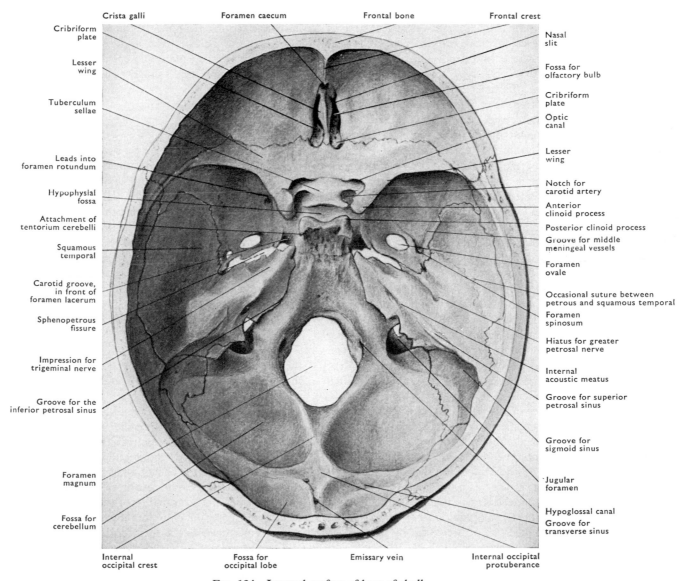

FIG. 134   Internal surface of base of skull.

passes laterally above the cerebellar fossa, from the external occipital protuberance to the superior margin at the base of the petrous temporal bone, and then turns down to continue as the S-shaped **groove for the sigmoid sinus** which leads to the jugular foramen.

The transverse sinus crosses and grooves the mastoid angle of the parietal bone. The sigmoid sinus grooves the petrous temporal (mastoid part) close to the mastoid air cells and to the mastoid antrum [FIG. 130]. It then grooves the occipital bone as it curves towards the jugular foramen. A foramen is usually present in the floor of the groove for the sigmoid sinus and transmits a vein which communicates with the mastoid veins.

The superior margin of the **petrous temporal bone** runs anteromedially from the upper margin of the transverse sinus to the apex of the petrous part of the temporal bone where it articulates with the body of the sphenoid. This margin has a **groove for the superior petrosal sinus** which runs between the cavernous sinus [FIG. 1119] and the transverse sinus. A fold of dura mater forming the **tentorium cerebelli** extends inwards in a continuous line (attached margin) from the bone at the edges of the groove for the transverse and superior petrosal sinuses and from the posterior clinoid process.

The groove for the superior petrosal sinus is absent near its medial end where the trigeminal nerve runs inferior to the edge of the tentorium and the sinus and grooves the bone inferior to them.

The inferior edge of the posterior surface **(posterior margin of the petrous part)** of the petrous temporal bone is irregular but mainly convex downwards. Medially it articulates with the sphenoid and basi-occipital and along the suture both bones are grooved by the **inferior petrosal sinus.** The groove leads into the anterior part of the jugular foramen, whilst the groove for the sigmoid sinus enters its posterior part. Just lateral to the end of the sulcus for the inferior sinus the margin is notched for the glossopharyngeal nerve, and from the bottom of a pit in the notch, a canaliculus leads up to the cochlea **(canaliculus cochleae).**

The jugular foramen lies between the edge of the petrous temporal bone and, on this aspect, the **jugular notch** in the occipital bone [FIG. 155]. Medial to the foramen the occipital bone forms a ridge, with a rounded **jugular tubercle** and a shallow groove behind it for the ninth, tenth, and eleventh cranial nerves passing to the intermediate part of the jugular foramen. Between this raised edge of bone and the lower margin of the foramen magnum is the internal opening of the **hypoglossal canal** for transmission of the twelfth cranial nerve.

In the posterior surface of the petrous part of the temporal bone, is the circular opening of the **internal acoustic meatus,** a cylindrical canal about 10 mm. long and 3–5 mm. wide. Its opening is oblique, but the canal runs directly laterally in the same line as its fellow of the opposite side. It is also opposite the external meatus at a depth of 50 mm. The **fundus** or bottom of the meatus is formed by the thin bony wall of the internal ear and this is pierced by the facial nerve, by branches of the vestibulocochlear nerve and by small blood vessels [FIG. 1016]. The facial nerve passes laterally in the bone, then turns backwards (at its genu) to a point above the stylomastoid foramen where it turns vertically downwards

through the bone to emerge at that foramen [p. 112]. The notch for the glossopharyngeal nerve is inferior to the internal acoustic meatus.

Above and posterolateral to the opening of the meatus is an irregular depression with one or two small foramina for vessels opening into it. It is best seen in young bones [FIG. 162] and is the **subarcuate fossa,** which indicates the concavity of the anterior semicircular canal. Further laterally and at a lower level is a narrow fissure which conceals the external aperture of the **aqueduct of the vestibule.** This carries small vessels and the endolymphatic duct which ends in a dilated endolymphatic sac.

Anterior to the foramen magnum, the basilar part of the occipital bone slopes up to join the body of the sphenoid, and together form the **clivus,** or slope. This passes upwards and forwards from the anterior margin of the foramen magnum to the **dorsum sellae.** This upstanding, square plate of bone, forms the posterior wall of the **sella turcica,** the 'Turkish saddle', the median part of the middle fossa. The upper corners of the dorsum sellae project as **posterior clinoid processes** which give attachment to the anterior ends of the fixed margin of the tentorium cerebelli.

The clivus is separated by meninges from the pons and the basilar artery above, and from the medulla oblongata and vertebral arteries below. The abducent nerve runs up on the clivus for more than 1 cm., pierces the dura mater, crosses behind the inferior petrosal sinus to the apex of the petrous temporal bone, and turns forwards under a small petrosphenoid ligament which may be ossified.

The Middle Cranial Fossa

This lies anterior and superior to the posterior fossa and consists of a median and two lateral parts and is formed by the temporal and sphenoid bones.

The lateral parts contain the temporal lobes of the brain, and extend from the superior margin of the petrous part of the temporal bone to the lateral wall of the orbit formed by the greater wing of the sphenoid bone. Here its anterior part is overlapped by the sharp posterior edge of the anterior cranial fossa, the lesser wing of the sphenoid bone [FIG. 134].

Laterally, the greater wing of the sphenoid and the squamous part of the temporal bone turn up to form the lateral wall of the skull in the temporal region. Medially is the raised body of the sphenoid bone.

The floor of the lateral part of the fossa sinks to the level of the upper border of the zygomatic arch. Posteriorly it is formed by the sloping anterior surface of the **petrous part of the temporal bone.** Laterally it articulates with the squamous part at the **petrosquamous suture,** medially with the greater wing of the sphenoid to reach the **foramen lacerum.** Medial to the petrosquamous suture, the petrous part forms the roof of the middle ear cavity, the **tegmen tympani.** Further medially and near the superior border, a distinct prominence, the **eminentia arcuata,** overlies the anterior semicircular canal of the internal ear [FIG. 158]. Medial to this again, and close to the petrosphenoid suture, a groove runs towards the foramen lacerum. At the posterior end of the groove, midway between the foramen lacerum and the side of the skull, lies a **hiatus for the greater petrosal nerve.** The nerve comes from the geniculate ganglion of the

Cochlea   Geniculate ganglion   Malleus   Incus

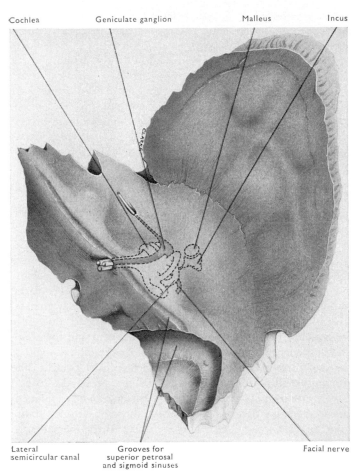

Lateral   Grooves for   Facial nerve
semicircular canal   superior petrosal
                         and sigmoid sinuses

FIG. 135   Internal aspect of a right temporal bone to show the
position of structures within the bone.

facial nerve and runs in the groove to enter the cartilage filling
in the foramen lacerum and join the deep petrosal nerve there.

Anterolateral to the hiatus is a second, smaller **hiatus for
the lesser petrosal nerve.** The nerve runs to the petrosphenoid suture and passes through it, through the foramen
ovale, or through a small foramen posterior to the foramen
ovale, to reach the otic ganglion. The anterior surface of the
petrous temporal bone is slightly hollowed near the foramen
lacerum where the trigeminal ganglion lies against it on the
trigeminal impression.

The lateral part of the middle fossa in front of the petrous
part of the temporal bone is formed by the squamous temporal
and the greater wing of the sphenoid bone and is marked by
the under surface of the brain which produces the digital
impressions. Anteriorly the **greater wing of the sphenoid**
passes laterally and sweeps posterosuperiorly in the lateral wall
of the skull. Above it the lesser wing of the sphenoid extends
horizontally from the body to join the greater wing [FIG. 163].
The comma-shaped gap between them is the **superior orbital
fissure.** It is wide below and medially, and tapers to a point
laterally where the wings meet. The fissure opens directly
into the back of the orbit and transmits numerous structures to
and from the orbital cavity [FIG. 895].

Immediately below the wide medial end of the superior
orbital fissure, the **foramen rotundum** passes forwards
through the greater wing to open into the upper part of the

pterygopalatine fossa. It transmits the maxillary nerve which
crosses the fossa to reach the orbit. The larger **foramen
ovale** lies further back, lateral to the foramen lacerum. The
mandibular nerve, including its motor branch, passes vertically down through the foramen ovale to the pterygoid region,
and the accessory meningeal artery ascends through it. Veins
pass through the foramen ovale and a small foramen for a vein
may be present between it and the body of the sphenoid.
Posterolateral to the foramen ovale, passing through the corner
of the greater wing of the sphenoid, is the **foramen spinosum**
for the middle meningeal vessels and a meningeal branch of the
mandibular nerve.

The **middle meningeal vessels,** the largest nutrient
vessels of the skull, lie in the periosteum and groove the
interior of the skull deeply. They pass forwards and upwards
and then divide. A parietal branch passes almost horizontally
back across the squamous temporal and parietal towards the
occipital region. A frontal branch curves upwards and
usually becomes enclosed in a bony tunnel in the vicinity of
the end of the lesser wing of the sphenoid and of the pterion
[p. 107]. Thereafter, the frontal branch runs posterosuperiorly (separated from the motor area of the brain by the
meninges) towards the top of the skull midway between
nasion and inion. Fractures of the skull may tear the vessels,
particularly where the grooves are deep.

At the medial side of the foramen lacerum the internal
carotid artery, with its plexus of veins and sympathetic nerves,
emerges from the cartilage and turns forwards and then
upwards and backwards in an S-bend. It grooves the body of
the sphenoid and the edge of the groove projects back over the
foramen lacerum as the **lingula.**

The median part of the middle cranial fossa is formed
largely by the sella turcica of the sphenoid bone. Posteriorly
it is limited by the dorsum sellae [p. 116] with its posterior
clinoid processes. The wide smooth depression in the body
of the sphenoid is the **hypophysial fossa** which lodges the
hypophysis [FIG. 137]. Anteriorly an oval elevation, the
tuberculum sellae forms the 'pommel of the saddle'
[PLATE 8 D].

In front of the tuberculum a transverse sulcus runs between
two round openings (the **optic canals**) which transmit the
optic nerves and the ophthalmic arteries. The groove between
is the **sulcus chiasmatis,** but does not lodge the optic
chiasma.

Each optic canal runs forwards and laterally between the
two roots of the lesser wing of the sphenoid, to the corresponding orbit. The roots run to the body of the bone, and the
posterior root and edge of the wing continue back as a projection with a rounded point, the **anterior clinoid process.**
This gives attachment to the free border of the tentorium
cerebelli, and is grooved on its medial aspect by the internal
carotid artery.

The term clinoid, bed-shaped, refers to the old four-
poster bed, but a third, or **middle clinoid process,** may be
present on each side. This is a tubercle found posterior to the
lateral part of the tuberculum sellae. It may be large enough
to reach the anterior clinoid process and convert the carotid
groove at that point into a caroticoclinoid foramen.

A **craniopharyngeal canal** is very rarely found running
from the bottom of the hypophysial fossa through the body of

the sphenoid. It is a survival of the canal through which the pharyngeal part of the hypophysis passed during development.

The cavernous sinus, which is a lateral relation of the body of the sphenoid, is illustrated in FIGURE 1119.

The Anterior Cranial Fossa

This lodges the lower part of the frontal lobes of the brain and lies above and in front of the middle fossa, separated from it on each side by the concave, posterior edge of the lesser wing of the sphenoid, which is wedged into the lateral sulcus of the brain, between the frontal and temporal lobes [FIG. 793]. The walls of the fossa and most of the floor are formed by the frontal bone. In the floor, the convex **orbital parts** of the frontal, roughened in a manner corresponding to the overlying convolutions of the brain, are separated in the midline by a narrow **ethmoidal notch** in which lie the **cribriform plates of the ethmoid bone** with its **crista galli** projecting between them. Posteriorly, the orbital parts of the frontal articulate with the lesser wings of the sphenoid, while the smooth **jugum sphenoidale** which unites these wings across the midline, articulates with the posterior margin of the ethmoid, thus completing the floor of the anterior cranial fossa [FIG. 134]. The posterior edge of the jugum, **limbus sphenoidalis,** is the anterior margin of the sulcus chiasmatis.

The **ethmoid bone** is recessed in the ethmoidal notch but its lateral parts (ethmoidal cells) are overlapped by the orbital parts of the frontal bone [FIG. 151]. The crista galli is the upward continuation of the perpendicular plate of the ethmoid, which forms the upper part of the nasal septum inferior to the cribriform plates [FIG. 136]. The crista galli gives attachment to the falx cerebri [FIG. 873].

Immediately in front of the crista galli is the **foramen caecum,** a small pit which occasionally transmits a vein through an aperture leading to the nose.

Anterior to the foramen caecum, the **frontal crest,** the median ridge on the frontal bone [p. 114], extends upwards on to the vault. The edges of the foramen caecum and the crest also give attachment to the anterior end of the falx cerebri.

Each narrow cribriform plate of the ethmoid forms part of the roof of the corresponding nasal cavity and transmits the olfactory nerve fibres through its numerous foramina from olfactory mucous membrane to the olfactory bulb on the superior surface of the plate. On each side of the crista galli is a nasal slit [FIG. 167] which transmits the anterior ethmoidal vessels and nerve to the nasal mucous membrane. These reach the cribriform plate by passing from the orbit between the ethmoid bone and the overlapping orbital part of the frontal bone, in the **anterior ethmoidal foramen** [FIG. 132]. The **posterior ethmoidal foramina** also transmit vessels and nerves from the orbit but these are distributed mainly to the ethmoidal cells. Either of the ethmoidal vessels may become of importance in nasal haemorrhage. Meningeal branches of the anterior ethmoidal vessels may groove the anterior wall of the anterior cranial fossa.

THE ORBIT

The orbit is an irregular pyramidal cavity in which the eyeball rotates [FIG. 132]. The pyramid is laid on its side; the apex is placed posteriorly and to the medial side and the base opens forwards on to the face. It has, in consequence, a roof, a floor, a medial wall, and a lateral wall, all of which are approximately triangular. The medial walls of the two orbits are nearly parallel.

The concave **roof** is formed mainly by the orbital part of the frontal bone. Posteriorly, it is also formed by the lesser wing of the sphenoid which articulates with the orbital part of the frontal bone. At its lateral side the roof turns down to articulate with the greater wing of the sphenoid bone behind and with the zygomatic bone in front. These three bones form the **lateral wall** and separate the orbit from the temporal fossa. At the back the lesser wing of the sphenoid is separated from the greater wing by the **superior orbital fissure,** which leads into the middle cranial fossa. The orbital surface of the zygomatic bone is pierced by a foramen for the zygomatic nerve, which divides in the bone to emerge as zygomaticotemporal and zygomaticofacial branches.

The zygomatic process of the frontal bone [FIG. 150] forms the lateral margin of the orbit, with the frontal process of the zygomatic bone. Within the margin of the orbit, about the middle of the lateral wall (10 mm. below the frontozygomatic suture), a slight eminence may be felt with a finger-tip. It is for attachment of the lateral palpebral ligament and the check and suspensory ligaments of the eye. At the lateral angle of the roof there is a wide **fossa for the lacrimal gland.** The lateral two-thirds of the supra-orbital margin of the frontal bone are sharp; its medial one-third is smooth and ill defined, and at their junction the supra-orbital notch or foramen transmits the nerves and vessels of that name. At the medial angle of the roof, on the frontal bone, is a small **trochlear fossa** or a **trochlear spine,** where the trochlea or pulley for the superior oblique muscle of the eye-ball is attached.

The indefinite medial margin of the orbit is formed by the maxillary process of the frontal bone and the frontal process of the maxilla which meet at the frontomaxillary suture.

If the **medial wall** of the orbit is traced back from the margin, an edge (**anterior lacrimal crest**) is reached which gives attachment to orbicularis oculi and the medial palpebral ligament. It forms the anterior limit of the lacrimal groove which passes from the orbital surface of the maxilla on to its medial surface and so reaches the nose. In the orbit, the groove forms the anterior part of the **fossa for the lacrimal sac.** The fossa is completed by a groove in the **lacrimal bone** which lies posterior to the frontal process of the maxilla. The lacrimal bone is a thin flake of bone which fills in a rectangular area between the maxilla in front and below, the ethmoid behind and the frontal bone above. A vertical **posterior lacrimal crest** separates its anterior grooved part, which forms part of the lacrimal fossa, from a smooth posterior orbital surface. The lower part of the posterior lacrimal crest turns forwards to meet the anterior crest and so closes the lower border of the fossa for the lacrimal sac laterally [FIG. 170], leaving room for the **nasolacrimal canal** to pass down to the nose.

Posterior to the lacrimal bone the **ethmoid** forms the medial wall of the orbit with the adjacent parts of the frontal and maxillary bones with which it articulates; while posteriorly it meets the sphenoid where the two roots of the lesser wing enclose the optic canal.

The orbital surface, or **lamina orbitalis of the ethmoid,** separates the orbit from the ethmoidal cells and is so thin that it has been called the lamina papyracea. The suture between it and the frontal bone is interrupted by two foramina, the anterior and posterior **ethmoidal foramina,** which transmit vessels and nerves of the same name [FIG. 895].

The **optic canal,** about 4 mm. in diameter, lies at a slightly lower level than the ethmoidal foramina as it enters the orbit through the most posterior part of the medial wall. It transmits the optic nerve and the ophthalmic artery.

Between the lesser and greater wings of the sphenoid, at the back of the orbit, the **superior orbital fissure** separates the roof from the posterior end of the lateral wall. The fissure is narrow above and laterally and curves downwards and medially to become wide and rounded at the apex of the orbit. It passes between the middle cranial fossa and the orbit. Branches of the ophthalmic nerve, the cranial nerves supplying the muscles of the eye-ball, and veins pass through it.

The **inferior orbital fissure** passes forwards and laterally, from the upper part of the pterygopalatine fossa between the lateral wall and floor of the orbit. The maxillary nerve crosses the fossa, passes laterally along the medial half of the fissure, and turns anteriorly into the floor of the orbit as the infra-orbital nerve. Through the fissure, veins pass from the orbit to the pterygoid plexus, the infra-orbital artery enters the infra-orbital sulcus, and small branches pass up to the orbit from the pterygopalatine ganglion.

The **floor of the orbit** is formed by the orbital surface of the maxilla, with the zygomatic bone anterolaterally, and the orbital process of the palatine bone at the apex. The floor also forms the roof of the maxillary sinus [FIG. 138]. The anterior parts of the zygomatic bone and maxilla are thick and together form the inferior margin of the orbit.

Posteriorly, the **infra-orbital sulcus** runs forwards and medially from the middle of the inferior orbital fissure and becomes continuous with the **infra-orbital canal** which emerges on the face 6 mm. below the medial one-third of the lower margin of the orbit. The sulcus and canal transmit the infra-orbital nerve and vessels. In the canal an anterior superior alveolar branch of the nerve enters a canaliculus a few millimetres from the infra-orbital foramen and takes a sinuous course through the bone to the lower part of the nasal septum. It gives off branches which run through the bone to the teeth as far back as the first molar.

A shallow depression on the medial part of the floor, just lateral to the lacrimal fossa, gives attachment to the inferior oblique muscle. The remaining extrinsic ocular muscles take origin far back in the orbit round the optic canal and across the adjacent part of the superior orbital fissure.

THE CAVITY OF THE NOSE

In the dried skull the cavity of the nose extends backwards from the piriform aperture to the choanae or posterior nares with a nearly horizontal floor [FIG. 136].

The cavity is divided by a median septum into right and left halves and the term nasal cavity is used to mean either the whole cavity or one or other of the halves. The interior of the nose has little space in it between the septum and the lateral

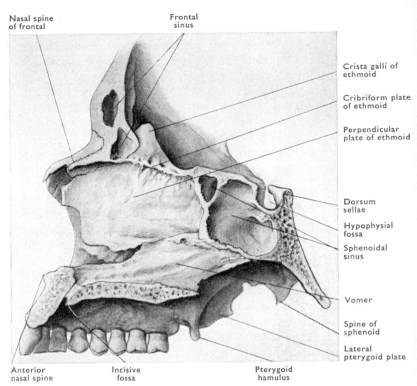

FIG. 136   Roof, floor, and septum of nose.

wall because of projections from the wall and because, on each side, the roof of the cavity is narrow while the floor is only 10 mm. wide.

The **external nose** consists largely of soft parts, supported by cartilage and attached to the edges of the piriform aperture and septum [FIG. 568]. The elongated **nasal bones** support the bridge of the nose and represent the external nose in the skull. They articulate with the frontal bone above, with the frontal process of the maxilla at each side, and with each other in the midline [FIG. 172]. Here their edges are turned down to increase their area of contact, to form a minute part of the nasal septum, and to articulate with the nasal spine of the frontal bone, the perpendicular plate of the ethmoid, and the septal cartilage [FIG. 571]. Laterally, the internal surface of the bone is grooved by the anterior ethmoidal nerve [p. 118].

The **septum of the nose** is vertical and separates the cavity into approximately equal halves, but it is seldom quite in the midline and is often considerably distorted. It is cartilaginous in front [FIG. 571]. The bony part is formed to a small extent by the crests of the nasal bones and the nasal spine of the frontal bone above and in front. Inferior to this is the **perpendicular plate of the ethmoid.** This is attached between the cribriform plates of the ethmoid above, and reaches the cartilaginous septum below and in front, the body of the sphenoid posteriorly, and the vomer below and behind.

The **vomer** slopes downwards and forwards from the inferior surface of the body of the sphenoid to the superior surface of the bony palate in the midline [FIG. 136]. Its superior border is widened to form everted **alae** which articulate with the sphenoid, enclose its rostrum, and reach over the roof of the nose to be overlapped by the expanded bases of the pterygoid processes. The posterior border slopes downwards and forwards and forms a posterior edge for the nasal septum. Its inferior edge fits between the two parts of the **nasal crest** on the superior surface of the bony palate. The anterior edge slopes downwards and forwards to the anterior part of the palate and articulates with the perpendicular plate of the ethmoid and the septal cartilage. The surfaces of the vomer are grooved by the sphenopalatine vessels and by the nasopalatine nerve [FIG. 898].

The **roof of the nasal cavity** on each side of the septum is only 1 or 2 mm. wide except posteriorly where it measures about 10 mm. between a pterygoid process and the septum. It consists of a sloping, anterior part formed by the frontal and nasal bones, a middle, horizontal part formed by the cribriform plate of the ethmoid, and a steeply sloping, angulated, posterior part formed by the anterior and inferior surfaces of the body of the sphenoid [FIG. 136].

The posterior part of the roof forms the anterior and inferior walls of a **sphenoidal sinus** on each side of the septum. The anterior wall of the body of the sphenoid is perforated by an opening, 4 or 5 mm. in diameter, through which the sinus opens into the **spheno-ethmoidal recess** [FIG. 137]. The original size of the opening is much reduced by a thin piece of bone, the sphenoidal concha [p. 139], which is originally separate but joins with the bones adjacent to it in early youth. The inferior surface of the body of the sphenoid is partly separated from the mucosa by the alae of the vomer, the vaginal processes of the pterygoid process, and part of the sphenoidal process of the palatine bone.

The middle, horizontal part of the roof is formed by the cribriform plate through which olfactory nerves from the adjacent epithelium pass via fifteen or twenty foramina. The close proximity of the nasal mucosa to the membranes of the brain constitutes a danger and the plate is a weak spot mechanically.

The anterior part of the roof lies inferior to the frontal sinus, the frontal bone forming only a narrow strip leading down to the nasal bones.

The **floor of the nasal cavity** is wider than the roof, measuring 10–12 mm. across, and is smooth. Its posterior third is formed by the horizontal plates of the palatine bones [FIG. 177] which meet in the midline [FIG. 133] and form the posterior edge of the hard palate. Anterior to the palatine bone, the palatine processes of the two maxillae likewise meet in the midline. Where the pairs of processes meet they project up a little to form a continuous **nasal crest** which lodges the vomer in a longitudinal groove. The anterior end of the crest is prominent and leads to the **anterior nasal spine.**

Near the anterior end of the septum an **incisive canal** passes down through the palate on each side and divides into a median and a lateral part so that four foramina open into the **incisive fossa** on the inferior surface of the palate. The median foramina transmit the terminal parts of the right and left nasopalatine nerves to the oral surface of the palate; the lateral ones convey the termination of the greater palatine arteries up into the nose.

The **lateral wall of the nose** is a fairly flat surface with a number of openings which lead to the paranasal air sinuses. It is usually described as uneven because of three folds which hang down from it. These are supported by thin, curved laminae or **conchae** which project from the bony wall [FIG. 137].

Normally there are three well formed conchae. Each projects inwards from the lateral surface and turns down, thus partially enclosing a space below it which forms a **meatus,** or passage, through the nose [FIG. 137].

The **inferior concha,** with the **inferior meatus** beneath it, passes back, sloping slightly downwards, from near the piriform aperture almost to the posterior edge of the bony palate. The **middle concha** lies above, but separated from the inferior one by the **middle meatus.** It is shorter and the uncovered space in front of it is called the **atrium** of the middle meatus. The small **superior concha** is situated far back in the nose, above and behind the middle concha, and separated from it by the **superior meatus.** The space above the superior concha is the **spheno-ethmoidal recess.**

The **lower half of the lateral wall** of the nose is formed in front by the nasal surface of the maxilla which separates the nose from the **maxillary sinus** [FIG. 138]. The large opening of this sinus is in the upper and posterior part of this surface, and is much reduced in size by the lacrimal bone, ethmoid bone, inferior concha, and palatine bone, all of which overlap it.

Behind and below the opening of the sinus is a rough area on the maxilla for articulation with the **perpendicular plate of the palatine bone.** This area is grooved reciprocally with the palatine bone by a **greater palatine sulcus,** in which the greater palatine nerve and vessels pass down into the palate [FIG. 174]. The posterior nasal branches passing forwards, pierce and supply the lateral wall of the nose. The perpendicular plate encroaches on the opening of the sinus from behind and forms part of the lateral wall of the nose.

Posteriorly, the palatine bone articulates with the medial pterygoid lamina which in turn forms the most posterior part of the lateral wall of the nose [FIG. 137].

Two horizontal crests project from the nasal surface of the palatine bone. The lower or **conchal crest** supports the inferior concha. The upper or **ethmoidal crest** is reached by the middle concha and it is just above and behind this that posterior nasal and nasopalatine vessels and nerves reach the nose through the **sphenopalatine foramen** [FIG. 176].

The **nasal surface of the maxilla** is smooth where it forms part of the inferior meatus below and in front of the large opening into the sinus. Above this area the root of the frontal process carries a **conchal crest** for the inferior concha. Posterior to the frontal process, and between it and the opening of the sinus, is the lacrimal groove leading down to the inferior meatus.

The **inferior concha** is a separate bone which articulates with the two conchal crests, maxillary and palatine, and bridges over the lacrimal groove so that the **nasolacrimal canal** opens under cover of the concha into the inferior meatus 30–35 mm. from the edge of the nostril. The base of the intermediate part of the concha articulates below with the inferior margin of the opening in the maxilla, above with the ethmoid, in front with the lacrimal, and behind with the palatine. Thus it closes the greater part of the opening in the medial surface of the maxilla, and thus forms a considerable part of the medial wall of the maxillary sinus [FIG. 138].

The *upper half of the lateral wall of the nose* contains the openings of most of the paranasal sinuses. It consists largely of ethmoid bone, the labyrinth of which lies between the orbit and the nose on each side [FIG. 138]. The labyrinth consists of small air sinuses (cells) enclosed by exceptionally thin bone so that access from the nose to the orbit can be perilously easy if the bone is damaged by injury or disease.

The *anterior part* of this region is formed by the nasal bone and the frontal process of the maxilla and has the frontal bone and frontal sinus immediately above it. The frontal sinus may open directly into this part of the nose, but it usually passes down via one of the ethmoidal cells.

The *posterior part* is formed by the ethmoid bone which fills in the space between sphenoid bone behind, maxilla below and in front, and frontal bone above. Like other adjacent bones, including the lacrimal which it covers from the nasal aspect, the ethmoid sends down a process to assist in the reduction of the upper part of the opening through the maxilla into its sinus. The residual opening therefore lies inferior to the ethmoid labyrinth and superior to the inferior concha.

The **middle concha** and the **superior concha** project from the ethmoid bone. If the middle concha is removed, the lateral wall of the **middle meatus** shows a swelling **(bulla ethmoidalis)** caused by the lower middle ethmoidal air cells which open on its surface. Immediately below and in front of the bulla is a curved groove, the **hiatus semilunaris.** The maxillary sinus opens into the posterior part of the hiatus by one or sometimes two small openings.

Under cover of the superior concha a posterior group of ethmoidal cells open into the **superior meatus** through one or more small openings.

The frontal sinus opens into, or anterior to, the upper, anterior end of the hiatus via the **infundibulum.** This is formed from an ethmoid air cell, or cells, modified to form a funnel which connects the frontal sinus with the hiatus. Unmodified anterior and some middle ethmoidal cells open into the wall of the hiatus. The lower border of the hiatus semilunaris is formed by a thin curved piece of bone, the **uncinate process,** which extends postero-inferiorly from the ethmoid to reach the inferior concha [FIG. 168].

THE PARANASAL SINUSES

[PLATE 9]

The paranasal sinuses are air-filled cavities produced by evagination of mucous membrane from the nasal cavity into adjacent skull bones. They all communicate, therefore, with the nasal cavity and their mucous linings are continuous with the nasal mucosa. Infection may spread from the nose to the sinuses with the production of localized pain or headache. The relation of nerves to the walls is thus of importance.

The walls of the sinuses are of compact bone, the cancellous bone being displaced by the sinuses. At birth, the sinuses are small or absent and they grow only slowly until puberty, after which they grow rapidly to their adult size. In old age,

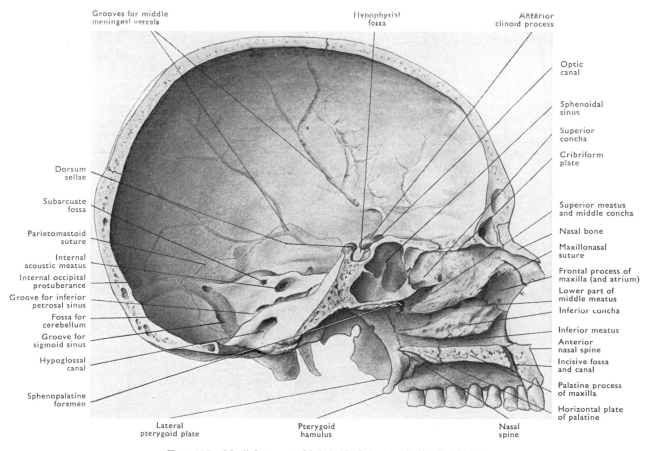

FIG. 137   Medial aspect of left half of skull sagitally divided.

absorption of diploë leads to further enlargement. They are normally larger in men, and serve to lighten what would otherwise be rather heavy bone. They also act as resonating chambers which affect the quality of the voice.

The Maxillary Sinus [FIGS. 138 and 174]

This is a large, pyramidal cavity in the maxilla. It is set with its apex in the zygomatic process of the maxilla and its base or medial wall is the lower part of the lateral wall of the nose. Its roof is the floor of the orbit. Its narrow floor lies over the alveolar process of the maxilla and the roots of the molar and premolar teeth, with its deepest part overlying the second premolar and first molar teeth. Variations in size may limit its extent to the three molars or increase it to include all the molar, premolar, and canine teeth. The roots of the teeth, particularly the first two molars, may produce eminences in the floor or even penetrate the bone.

The lowest part of the floor lies as much as 10 mm. below the level of the floor of the nose. The opening of the sinus passes through the upper part of its medial wall so that, in the absence of effective ciliary action, drainage cannot occur in the erect position.

The very large aperture seen in FIGURE 174 is much reduced by the overlap of adjacent bones and opens into the hiatus semilunaris or into the middle meatus posterior to it. The opening in the mucosa is normally single, though it may be double in the bone.

At birth the maxillary sinus is still no more than a groove between the conchae although it is detectable at the fourth month of intra-uterine life. Growth is slow until puberty. After puberty it grows rapidly to reach an approximate size of 35 mm. in height, 30 mm. anteroposteriorly and 25 mm. in width. It usually extends anterior to the nasolacrimal canal as a **lacrimal recess.** Vertical ridges on the interior may subdivide the cavity partially.

The greater palatine, posterior superior alveolar, infra-orbital, and anterior superior alveolar nerves with accompanying vessels lie in the posterior wall, roof, and anterior wall of the sinus respectively [FIGS. 901 and 902].

The Frontal Sinuses

These are situated in the frontal bone near the midline, above the supra-orbital margin and the root of the nose. They are separated by a bony septum which usually deviates to one or other side. They vary considerably in size and are frequently lobulated. Their presence may be detectable at the second year but they are, as a rule, not clearly recognizable until the seventh year. They grow slowly until shortly before

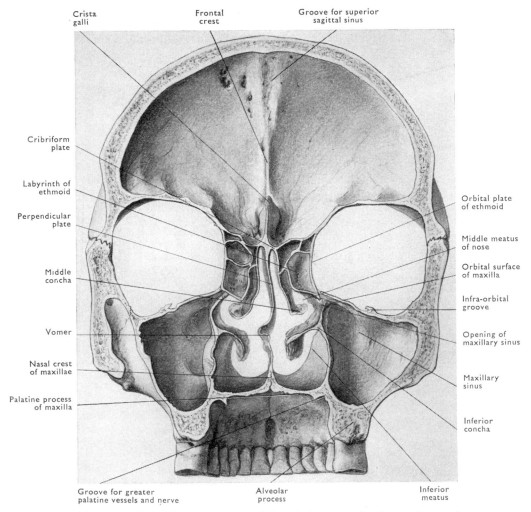

FIG. 138 Coronal section passing inferiorly between the first and second molar teeth.

puberty and more rapidly to adult size thereafter. In the adult, however, they vary from the size of a pea to large spaces extending, though rarely, close to the zygomatic process and up into the diploë of the frontal bone or in the roof of the orbit to the lesser wing of the sphenoid. An average size would be about 25 mm. by 25 mm. as seen from the front, but the shape of the supra-orbital ridges and the forehead are no indication of their size.

The frontal sinuses normally drain down through the ethmoid to the middle meatus, but may open directly at the side of the nasal septum [p. 121].

The Ethmoidal Sinuses

They constitute the ethmoidal labyrinth, between the nasal cavity and the orbit. They lie near the midline, inferior to the anterior cranial fossa, but are separated from it by the orbital plate of the frontal bone. This roofs in some of the superior cells which extend into the frontal bone, and the same extension occurs where the ethmoid articulates with other bones [p. 140].

For descriptive purposes the ethmoidal cells are divided into **posterior, middle,** and **anterior** groups which drain, respectively, on to the surface of the superior meatus, the bulla ethmoidalis and the hiatus semilunaris, and the infundibulum [p. 121].

The ethmoidal cells begin to develop after the fifth month of intra-uterine life and are formed in the cartilaginous nasal capsule. Ossification spreads into the cell walls before birth. The cells are, therefore, already formed at birth.

The Sphenoidal Sinuses

These are two cavities which lie side by side in the body of the sphenoid bone with a bony septum between them which is almost always bent to one side [FIG. 136]. The surrounding bony walls separate them from the nasopharynx and posterior part of the nose below, from the upper part of the nose and posterior ethmoidal cells in front, from the frontal lobes of the brain, olfactory tracts, and hypophysis above, and from the pons and basilar artery behind. Laterally the bony wall separates it from the optic nerve in the optic canal, the cavernous sinus and the structures which pass through it. The paired apertures [FIG. 164] are about 4 mm. in diameter and open on each side of the nasal septum directly into the **spheno-ethmoidal recesses** [p. 120].

Like most other sinuses the rudiments of the sphenoidal sinuses appear early, in the fifth month of intra-uterine life, in this case as recesses enclosed in the sphenoidal conchae [p. 139]. These extend into the body of the sphenoid in the

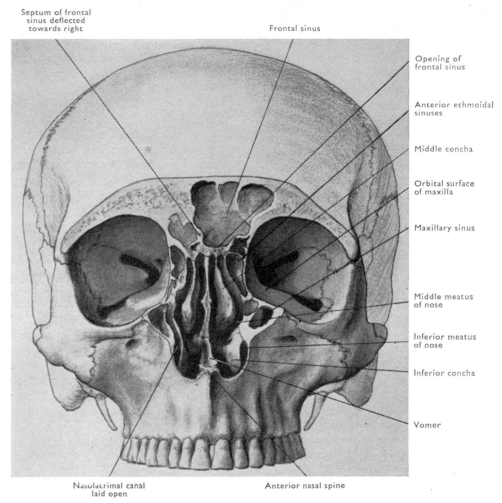

FIG. 139 Part of the frontal, nasal, and maxillary bones removed in order to display the relation of the various cavities exposed.

seventh or eighth year, and undergo their main growth after puberty.

The Palatine Sinuses

When present these are air spaces, one on each side, in the orbital processes of the palatine bones. They open into the sphenoidal sinus or into a posterior ethmoidal cell, and can be regarded as extensions of these into the palatine bone [FIG. 176].

THE MANDIBLE

[PLATE 10]

The mandible or lower jaw bone is the skeleton of the lower part of the face and reaches from the chin to the mandibular fossa [FIG. 129]. It consists of a horseshoe-shaped body joined to a pair of flat rami which project up at the posterior ends of the body. The upper end of each ramus carries an anterior, triangular **coronoid process** and a posterior **condylar process,** separated by the **mandibular notch.** The condylar process consists of a **neck** surmounted by an oval **head** with an articular surface.

The right and left halves of the mandible are joined at their anterior ends by fibrous tissue at birth, but bony fusion occurs during the second year at the **symphysis menti.** The posterior parts of the two halves are widely separated [FIG. 480].

The upper part of the body is a curved bar of bone, the **alveolar arch,** which carries the lower teeth. It normally contains sockets, or **alveoli,** for the roots of the teeth, unless the teeth have been lost during life [p. 113]. The sockets for the canine teeth produce a vertical ridge on the surface on each side, and the incisor teeth may do the same. In old age, when all the teeth have been lost, the alveolar arch becomes absorbed almost completely [FIG. 142].

The lower part of the body is strong and rounded. It extends from the ramus to the symphysis and forms a wider curve than the alveolar arch, thus projecting beyond it and producing the prominence of the chin. This forward projection of the lower part of the mandible is a human characteristic and is absent in other mammals including apes.

The Body of the Mandible

On the outer surface of the bone, along the line of the symphysis, a faint ridge runs down into a triangular **mental protuberance.** The lower angles of this form a **mental tubercle** on each side, the forward curvatures of which produce the chin. From the tubercle on each side an **oblique line** runs back and slightly upwards to become continuous with the anterior border of the ramus. Between the tubercle and the incisor teeth, is an **incisor fossa.** Further back, above the oblique line and below the second premolar tooth, or the space between the first and second, the end of a **mandibular canal** opens on the surface as a **mental foramen** which transmits a terminal branch of the inferior alveolar nerve, the mental nerve, and the corresponding vessels. The opening of the foramen is oblique and faces upwards and slightly backwards. It lies midway between the upper and lower borders of the bone, provided the alveolar process has not been absorbed. If it has been absorbed the foramen will be close to the upper border. The anterior part of the body, the oblique line and the lower margin give attachment to facial muscles [FIG. 349].

The lower border of the body is the **base** of the mandible. It is smooth and rounded except close to the symphysis where a small but distinct depression, the **digastric fossa,** lies on or immediately posterior to it. Posteriorly the border passes back to the angle of the mandible.

The Ramus of the Mandible

On each side, this flat plate of bone forms the mandible behind the last molar tooth. It extends from the base to the coronoid and condylar processes, and its anterior limit at the base is marked by the slight groove produced by the facial artery crossing it.

Its sharp **anterior border** is continued down from the coronoid process to become continuous laterally with the oblique line on the body [FIG. 140], whilst a rounded ridge leads to the posterior end of the mylohyoid line on the medial surface [FIG. 141]. The triangular area so formed behind the last molar tooth is often ridged where it gives attachment to the buccinator and superior constrictor muscles and the inferior end of the pterygomandibular raphe. The lingual nerve slips under the lower border of the superior constrictor muscle close to the medial side of the bone.

The **posterior border** of the ramus is rounded. It descends from the back of the neck, and turning through approximately a right angle, **angle of the mandible,** becomes continuous with the lower border of the mandible [FIG. 140]. The middle of the convexity of the angle of the mandible is used as a landmark.

The **superior border** of the ramus exhibits the coronoid and condylar processes. Between them is the sharp concave edge of the **mandibular notch** which runs from the coronoid process in front, to the lateral aspect of the condylar process behind.

The **coronoid process** is a flattened, triangular plate of bone which gives attachment to the temporalis muscle. Its base is continuous with the anterosuperior corner of the ramus, and the muscle is inserted into its medial surface and margins as well as to the anterior margin of the ramus and the blunt ridge on its medial side [FIG. 345].

The **condylar process** consists of an articular **head** supported on a neck which joins the posterosuperior corner of the ramus. The **neck** is flattened anteroposteriorly and widens as it passes up to the head. The posterior surface is smooth and is continuous with the posterior surface of the head as well as the posterior border of the ramus. The anterior surface is rough, concave, and overhung by the anterior margin of the head. This **pterygoid fovea** is the insertion of part of the tendon of the lateral pterygoid muscle. Lateral to the fovea a strengthening flange of bone sweeps downwards and forwards to become continuous with the ramus. The lateral ligament of the joint is attached to the lateral and posterior aspects of the neck [FIG. 291].

The **head** of the mandible is convex but considerably elongated from side to side and narrow from front to back. Its long axis is slightly oblique and passes medially and a little backwards. It is separated from the mandibular fossa by an articular disc [FIG. 292].

The flat lateral surface of the ramus [FIG. 140] is roughened by the insertion of the masseter muscle which is attached to the whole surface, except the condylar process which is not

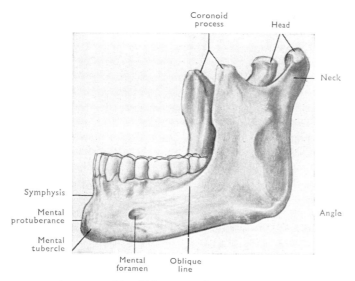

FIG. 140   Mandible as seen from the left side.

Coronoid process   Head   Neck   Angle   Symphysis   Mental protuberance   Mental tubercle   Mental foramen   Oblique line covered by the muscle [FIG. 354]. The facial artery may groove the bone at the antero-inferior corner of the area for the muscle.

The Inner Surface of the Mandible

This includes the medial surface of the ramus and body and the posterior aspect of the region near the symphysis [FIG. 141].

The *medial surface of the ramus* is strongly marked near the angle of the jaw by the insertion of the medial pterygoid muscle. The **stylomandibular ligament** reaches the posterior border behind this area [FIG. 291] while anterior to it a **mylohyoid groove** slopes down towards the lower part of the ramus and is occupied by the mylohyoid vessels and nerve.

The groove runs down from an obliquely set **mandibular foramen** which leads into the **mandibular canal** for the inferior alveolar vessels and nerve. The **lingula** is a spur of bone which overhangs the foramen and projects upwards and backwards to give attachment to the **sphenomandibular ligament** which also spreads out to the adjacent bone [FIG. 293] and bridges the mylohyoid groove. The smooth bone behind this, and at the posterior border, is in contact with the parotid gland.

On the *inner surface of the body* the **mylohyoid line** extends as an oblique ridge which runs downwards and forwards from a point adjacent to the socket of the third molar tooth to the anterior part of the lower border. It divides this area into two triangular zones, one above the line and the other below it. The two zones are separated by the mylohyoid muscle which is attached to the whole length of the line. Above the mylohyoid line, near the symphysis, is a smooth depression, the **sublingual fossa,** which accommodates part of the sublingual gland [FIG. 141]. The remainder of this area is covered by mucous membrane except at its posterior extremity where the lingual nerve intervenes. Below the line and extending back on to the ramus and down to the lower borders of the mandible is a shallow depression, the **submandibular fossa** for the submandibular gland.

On the inner surface of the symphysial region, between the two sublingual fossae, there is a rough prominence which is sometimes divided into two pairs of **mental spines** (genial tubercles). Genioglossus and geniohyoid muscles are attached to the upper and lower pairs, respectively. These are above the level of the mylohyoid line, while the small **digastric fossae** lie inferior to the mylohyoid line.

The Structure of the Mandible

The mandible consists of cancellous bone with an outer layer of compact bone which is exceptionally thick, especially at the lower border. The labial walls of the tooth sockets are considerably thinner than their lingual walls except in the case of the third molar, a point to be remembered in the extraction of teeth. The mandibular canal is in the spongy substance about the level of the mylohyoid line but does not have a definite wall. Branches of the canal carry vessels and nerves to the teeth and gums and to the mental foramen.

Age Changes in the Mandible

At birth the two halves of the mandible are separate and the body is a shell of bone enclosing the developing teeth, which

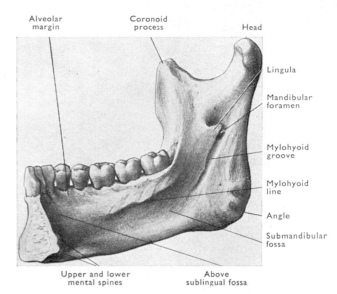

FIG. 141   Medial surface of right half of mandible.

are not completely separated from each other. The mandibular canal and mental foramen are near the lower border and the foramen is opposite the cavity for the first deciduous molar tooth. The ramus is short and the angle is very obtuse so that the coronoid process is almost in line with the body [FIG. 142].

The halves begin to unite in the first year and union is completed in the second year. As the teeth erupt, the body becomes stronger and deeper, the rami enlarge and the angle becomes reduced, to about 140 degrees by the fourth year.

Progressive increase in depth and elongation, especially behind the mental foramen, provides room for the permanent molars, the mental foramen assumes its adult position, and the angle becomes reduced to 110 degrees or less.

With the loss of teeth in old age, absorption of the sockets makes the chin appear prominent and the mental foramen is approached by the upper border. The angle opens out again by remodelling to 140 degrees and the condylar process is bent back so that the mandibular notch is widened.

Development of the Mandible

The mandible is laid down for the most part as membrane bone which lies lateral to the cartilage of the first pharyngeal arch (Meckel's cartilage). Some of the cartilage becomes incorporated in the bone, and small accessory cartilages may appear [FIG. 143]. Ossification starts during the sixth week of intra-uterine life, spreads through the membrane and envelops the nerves and vessels. Small **mental ossicles** may appear at the symphysis shortly before birth, but soon fuse with the mandible.

The upper and posterior end of Meckel's cartilage reaches the skull and forms the malleus, one of the ear ossicles. The original joint of the lower jaw lies between malleus and incus (Scott, 1951). The **sphenomandibular ligament** between the spine of the sphenoid and the lingula of the mandible is usually regarded as representing a part of Meckel's cartilage.

Growth of the Mandible

The mandible grows by surface accretion and absorption. It grows in width between its angles as well as in length,

FIG. 142 Form of mandible at different ages. By the 6th year the first permanent molar has erupted behind the milk teeth; by the 12th year the milk teeth have been replaced by the corresponding permanent teeth, and the second permanent molar has erupted.

height, and thickness. Increase of thickness is brought about chiefly by addition to the outer surfaces. Increase in height of the body is due mainly to growth at the alveolar border. There is an associated continuous upward and forward movement of the teeth in the bone before, during, and after their eruption. The lengthening process takes place partly by

FIG. 143 Development of mandible.

A, As seen from the medial side; B, from the lateral side; C, showing accessory (metaplastic) cartilage (blue).

(In A and B Meckel's cartilage is coloured blue and mandibular nerve is shown.)

addition to the posterior border, and, owing to the lateral slope of the body from before backwards, that accounts for increase in width also. The main increase depends upon the obliquity of the ramus and its processes in a young jaw. The growth of the condylar process is not only upwards but also backwards and sideways, and it thus contributes to the total length as well as to the height and width. As the coronoid process grows, its sloping, anterior margin is encroached upon by the rising alveolar border, which thus becomes longer. This, along with absorption from the anterior edge of the ramus, is the chief means of providing more room for the teeth as they rise with the alveolar border. Modelling maintains the shape of the condyle and the undulating curves of the anterior margin of the ramus and coronoid process. During growth the mandibular foramen maintains its relative position by a corresponding extension of its anterior lip, and the mandibular canal is thus lengthened. The mental foramen also changes its position during growth by moving backwards relative to the teeth [FIG. 142]; its sharp edge is below and behind in the young jaw, but in the adult it is below and in front, an indication of its direction of movement. (For further details and historical references see Brash, 1934; Brash et al., 1956.)

THE TEETH

[PLATE 10]

There are normally two sets of teeth. The deciduous or milk teeth, of which there are twenty, appear between 6 months of age and the end of the second year. These are replaced between the sixth year and the twelfth year by most of the permanent set of thirty-two teeth. The four third molars appear later, usually between 17 and 21 years. The

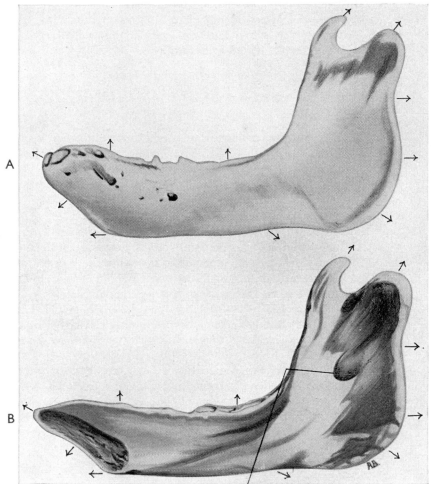

Growing edge of mandibular foramen

FIG. 144 Mandible of madder-fed pig to show the sites of growth. (See Brash, 1934.)

The animal was $20\frac{1}{7}$ weeks old and the madder had been omitted from the food for 19 days. The new bone added during that period is white in contrast to the old, red bone; and the main directions of growth are indicated by arrows.

A. Lateral surface of left half to show the distribution of growth, neutral and absorption areas.

B. Medial surface of right half. Note particularly the indication of growth at the alveolar and posterior borders of the condyle and coronoid process, and of the upward and backward growth-movement of the mandibular foramen.

teeth, their arrangement, and their dates of eruption are described under Digestive System.

THE HYOID BONE

The hyoid bone is a U-shaped piece of bone situated below the tongue and above the larynx [FIG. 579]. It can be felt in the front of the neck 1·5–2 cm. above the laryngeal prominence when the chin is held up. It is attached to the tongue, moves up and down with it during swallowing and is held in position by muscles and by ligaments [FIG. 346].

It consists of a **body,** the curved piece of bone felt in the front of the neck, a pair of greater horns, and a pair of lesser horns [FIG. 145]. The **greater horns** are continued backwards and upwards from the ends of the body and can be felt through the skin below the mandible. They lie just external to the mucous membrane and can be palpated by a finger on the internal pharyngeal wall. The tips of the greater horns are

rounded and slightly swollen, but are not always easily felt because they are overlapped by the anterior edges of the sternocleidomastoid muscles. The **lesser horns** are small nodules of bone or cartilage which project upwards and backwards from the junction of body and greater horn on each side.

The infrahyoid muscles, attached at its lower border, pull the hyoid down. The thyrohyoid membrane passes up, deep to the muscles and the bone, to be attached near its upper border. The stylohyoid ligament runs from the styloid process to the lesser horn and the middle constrictor of the pharynx sweeps back from the angle between the greater horn and the lesser horn and that ligament. The remaining muscles connect the hyoid to the tongue, the mandible, and the base of the skull and so tend to depress the tongue or raise the hyoid bone. The epiglottis is connected to the deep surface of the body by a hyo-epiglottic ligament in the midline [FIG. 588].

Development and Ossification. The greater horn and most of the body of the hyoid bone are derived from the cartilage of the third arch. The lesser horn and probably the upper part of the body are from the cartilage of the second, or hyoid, arch which continues up as the stylohyoid ligament.

Ossification of the hyoid bone begins shortly before birth as a pair of centres for the body

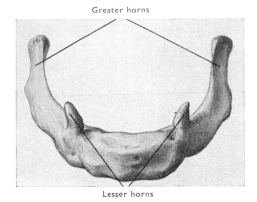

Greater horns

Lesser horns

FIG. 145 Hyoid bone.

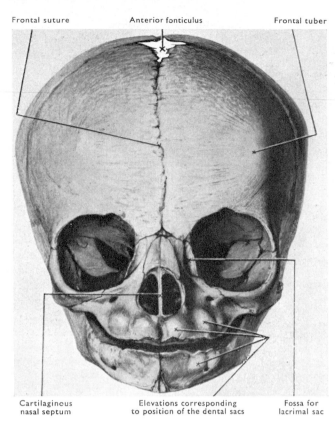

Frontal suture   Anterior fonticulus   Frontal tuber

Cartilaginous   Elevations corresponding   Fossa for
nasal septum   to position of the dental sacs   lacrimal sac

FIG. 146   Frontal aspect of skull at birth.

Sphenoidal fonticulus   Position of anterior fonticulus

Parietal tuber

Position of
posterior
fonticulus

Suture between
interparietal and
supra-occipital parts
of occipital bone

Mastoid fonticulus

Cartilaginous nasal septum

FIG. 147   Lateral aspect of skull at birth.

(which soon unite) and one for each greater horn. During the first year, or considerably later, centres appear for the lesser horn. The greater horn and body unite in middle age; the lesser horn is united with them by a synovial joint which only disappears in old age.

THE SKULL AT BIRTH
[PLATE 8]

At birth the cranium is relatively large but the face is small, about one-eighth of the whole cranium as compared with one third in the adult. The teeth are not fully formed and the paranasal sinuses are rudimentary so that both the jaws and the nasal cavities are small.

The edges of the bones in the vault are not yet serrated and are separated by strips of fibrous tissue continuous with the periosteum covering the bones (pericranium and endocranium). This provides some mobility for the bones and allows overlap during birth of the child. The bones are thin but are not liable to fracture because of their mobility and flexibility.

In certain places, principally at the angles of the parietal bones, ossification does not spread into the connective tissue as rapidly as elsewhere, thus leaving large, unossified areas named **fonticuli** at birth.

The **anterior fonticulus** is large, diamond-shaped, and easily felt in the newborn child. It lies at bregma between the separate halves of the frontal bone and the two parietal bones. It is about 4 cm. long by 2·5 cm. wide. Ossification spreads from the edges of the bones into the membrane of the fonticulus until the gap is closed some time during the second year.

The **posterior fonticulus** is at the apex of the occipital, between the two parietal bones, and is closed 2 months after birth. A **sphenoidal fonticulus** lies at the sphenoid angle of each parietal bone and is closed 3 months after birth. Similarly a **mastoid fonticulus** is found at the mastoid angle of each parietal and does not close until the end of the first year.

The parietal foramina are derived from, and indicate the position of, an unossified space which extends laterally into both parietal bones. It is often closed at birth but, if present, is called a **sagittal fonticulus.**

Sutural bones may result from separate ossification in various sutures and fontanelles. Those formed in the sphenoidal fontanelle, in the region of pterion, are called **epipteric bones.**

At birth the individual bones are separable, being joined by cartilage or membrane, and in many cases they are in several pieces in accor-

PLATE 7

A

B

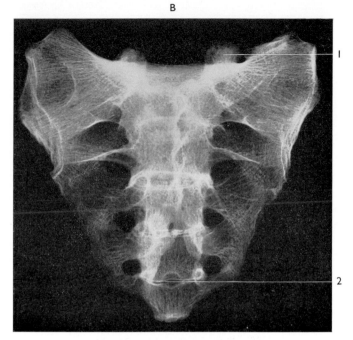

A. Cervical ribs. There is a short one on the right and a long one on the left.

Note that these are both views of the same specimen, the radiographs having been taken from slightly different angles. Though clearly visible in the upper picture, the cervical ribs are hardly apparent in the lower because their shadows overlap those of the normal first ribs.

1. Long cervical rib.

B. Normal sacrum.

1. Superior articular facet.
2. Remnant of disc between 4th and 5th sacral vertebrae.

C

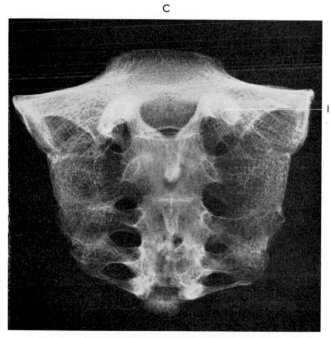

D

C and D. Common anomalies of the sacrum. These are important in that they upset the balance of the entire vertebral column.

C. High 1st sacral body.

1. Superior articular facet.

D. Asymmetrical 1st sacral body.

PLATE 8

A

B

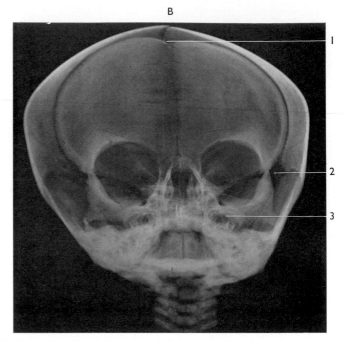

A and B. Lateral (A) and anteroposterior (B) radiographs of the skull at birth. Note the relatively large calvarium compared with the facial skeleton.

1. Anterior fonticulus.   2. Sphenoidal fonticulus.   3. Tooth germs in maxilla.

C

D

C. Lateral radiograph of the skull in late adult life.

1. Anterior temporal diploic vein.
2. Shadow of the auricle.
3. Sphenoidal air sinus.
4. Posterior temporal diploic vein.
5. Calcification in the pineal body.

D. Lateral radiograph (natural size) of the sphenoid bone and adjacent bony structures.

1. Anterior wall of middle cranial fossa.
2. Indentations on the inner table of the calvarium by the middle meningeal vessels.
3. Lesser wing of sphenoid.
4. Posterior clinoid process.
5. Hypophysial fossa.
6. Dorsum sellae.
7. Body of the sphenoid bone.

PLATE 9

A

B

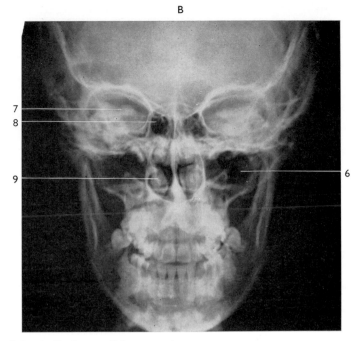

A and B.   Anteroposterior radiographs of the skull taken at different angles.

1. Zygomaticofrontal suture.
2. Nasal cavity.
3. Left frontal air sinus.

4. Orbit.
5. Zygomatic bone.
6. Maxillary air sinus.

7. Superior margin of petrous temporal bone.
8. Ethmoidal cells.
9. Inferior nasal concha.

C

D

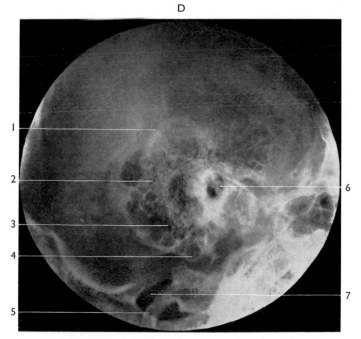

C.   Lateral 'soft tissue' radiograph of nose.

1 and 2. Superimposed shadows of nasal bones.

D.   Radiograph of mastoid process of temporal bone.

1. Shadow of the auricle.
2. Small air cell in mastoid process.
3. Large air cell in mastoid process.
4. Tip of mastoid process.
5. Posterior arch of atlas.
6. External and internal acoustic meatus overlapping.
7. Foramen magnum.

PLATE 10

A

B

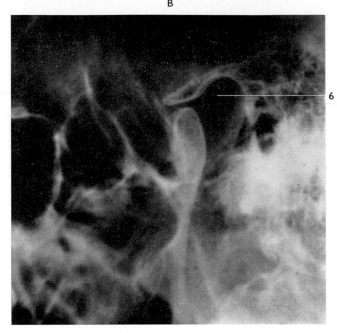

A and B.   Radiographs of temporomandibular joint with the mouth closed (A) and open (B).

1. Temporal bone.
2. Articular tubercle.
3. Articular disc in mandibular fossa.

4. Condyle of mandible.
5. External acoustic meatus.
6. Region occupied by condyle when mouth is closed.

C
(i)

(ii)

D

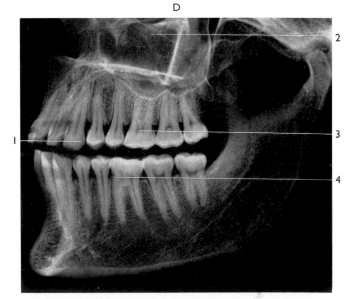

C.   Radiographs of half mandible : (i) middle age, (ii) old age.

1. Coronoid process.
2. An interalveolar septum.
3. Mental foramen.   Compare its position in (i) and (ii).
4. Condyle.
5. Mandibular foramen.
6. Mandibular canal.

D.   Radiograph of maxilla and half mandible (middle age).

1. Maxillary canine.
2. Maxillary sinus.
3. Maxillary first molar. Note proximity of its roots to the maxillary sinus.
4. Space (interdental) between mandibular second premolar and first molar. Note mental foramen between apices of roots of first and second premolars.

dance with their ossification. On the outer surface of the vault, the parietal and frontal tubera are prominent, the frontal bone is in two parts and the occipital squama is partly separated into an upper (interparietal) part and a lower (supraoccipital) part by a fissure on each side.

The tympanic part of the temporal bone is still only an incomplete **tympanic ring** and the acoustic meatus is therefore short. Also, the tympanic membrane is set obliquely facing more downwards than laterally, and is almost on the interior surface of the skull. There is no **mastoid process** at this time. Thus the styloid process and stylomastoid foramen are near the side of the head, but gradually recede after the second year as the mastoid process and the tympanic plate begin to grow.

The maxilla is shallow because of the absence of its sinus. It consists largely of the alveolar process with its developing teeth. Of the other air sinuses only the ethmoidal are well formed.

The nasal bones are flat so that the child has little bridge to its nose and this, with the absence of superciliary arches, adds to the prominent appearance of the forehead. The orbits are relatively large and the nasal cavity lies almost entirely between them.

GROWTH AND AGE CHANGES IN THE SKULL

The skull grows rapidly from birth until the seventh year, but the greatest increase in the size of the brain case takes place during the first year. At the seventh year the orbits are almost as large as in the adult; the cribriform plates of the ethmoid, the body of the sphenoid, the petrous parts of the temporal bones, and the foramen magnum have reached their full size. The jaws have enlarged in preparation for the eruption of the permanent teeth and enlarge during their eruption, chiefly in the alveolar processes. For some years after the seventh, growth is slower, but at puberty a rapid increase in the rate of growth takes place in all directions, especially in the frontal and facial regions, related to the great increase in size of the **paranasal sinuses.**

In the base of the skull, growth takes place in the cartilage between centres of ossification and so, for a time, continues between bones such as the basisphenoid and basi-occipital [FIG. 148]. After the cartilage has ossified, growth can only take place on a surface of the bone covered with membrane—periosteum, endosteum, or endocranium, according to its position.

In the vault of the skull growth takes place between the edges of the bones, but the essential mode of growth in the vault is by deposition of bone on the external surface and absorption from the internal surface. Consequently, sutures which are perpendicular to the surface, such as the sagittal suture or, to a less extent, the coronal, maintain their position during growth. Sutures which are set obliquely to the surface, with a bevel to the edges of the bones, migrate as the skull grows. This is because growth of the overlapping superficial edge of one bone, such as the squamous temporal, carries it over the surface of its neighbour. The deep edge of the neighbouring bone, being part of its internal surface, is absorbed according to the rate of absorption on the inner

surface of the skull. If the two processes keep pace the suture will retain its width.

Accretion outside and absorption inside produce enlargement of the cranial cavity with a simultaneous reduction in curvature. Differential growth and modelling produce and maintain the proportions and shape of the skull and the correct thickness of the bones.

In the bones of the face, growth and modelling are also accomplished by surface accretion and absorption wherever the surface contours change. It has been known since the time of John Hunter that modelling absorption [p. 82] can occur in a marrow cavity or on the surface of a bone. Brash (1934) used bones from madder-fed animals to demonstrate the essential principles of bone growth in the skull and the similarity of the growth processes in the skull to those in long bones. Enlow (1968) and Duterloo and Enlow (1970) should be consulted for further details and references on growth in the human skull.

Up to maturity, age can be ascertained approximately from the skull and the teeth. For the dates of eruption of the teeth see under Digestive System.

During the *first year* the separate parts unite in the temporal and sphenoid bones; the jugum sphenoidale is ossified by extension medially from the lesser wings, and the perpendicular plate of the ethmoid begins to ossify. During the *second year*, the halves of the mandible unite more firmly, the anterior fonticulus closes, the cribriform plate ossifies and unites with the other parts of the ethmoid, and the mastoid process appears. The lateral parts of the occipital bone unite with the occipital squama during the *third year*, and with the basi-occiput during the *fourth or fifth year*. The halves of the frontal bone have almost completely united at the *sixth year*, and during that year the first of the permanent teeth, a first molar, erupts.

If the third molar is present, and also the plate of cartilage between the body of the sphenoid and the occipital bone, the age is between 17 and 25. If the sphenoid and occipital are completely fused the age is likely to be over 25, for the cartilages that unite the body of the sphenoid to the ethmoid and to the occipital bone normally disappear before that.

After maturity, the wear of the teeth and the degree of obliteration of the sutures may give a rough indication of the age. The condition of the sutures in the interior of the skull vault is a more reliable indication of age than the condition on the exterior. Fusion of the outer tables is less regular, slower, and often incomplete. It is commonly stated that the inner table starts to fuse about 10 years earlier than the outer table (Todd and Lyon, 1924-5).

Fusion of Sutures

In the interior of the skull, fusion of the sagittal suture normally begins at about 22 years of age and should be completed by 31 years. The coronal suture follows at 24 years and is for the most part fused by about 30 years, though not completely so till about 40 years. Its lower, pteric part is delayed about 2 years. The lambdoid suture does not start to fuse until about 26 years, but joins rapidly at first and then lags so that fusion is not complete until after 40 years. Other sutures close considerably later: there is often considerable variation in the times of suture closure, even on the two sides of one skull.

Failure of one suture to conform to the usual pattern is likely to be associated with similar anomalies in the remaining sutures, i.e., there is likely to be a slowing up or an acceleration of closure in other sutures (Todd and Lyon, 1924–5; Bolk, 1915). Bolk found that 19 per cent. of 1,820 Dutch skulls aged 3–20 years showed some premature closure and there were seventy-one cases of premature closure of the sagittal suture. However, the normal mode of growth by accretion and absorption can produce a normally shaped skull without the assistance of sutures. Complete obliteration of the suture does not normally take place till an advanced age.

In old age the skull vault may be thicker owing to deposit of bone on the inner surface; but in nearly all cases the skull bones are thinner, and the skull is lighter owing to absorption of the diploë and the associated extension and enlargement of the paranasal sinuses. Consequent on loss of teeth there is diminution of the size of the jaws owing to absorption of the walls of the sockets; the chin protrudes and the angle of the mandible becomes more obtuse. For details on age changes in the skull see Montagu (1938) and Krogman (1962).

SEX DIFFERENCES IN THE SKULL

There is little difference between the skulls of boys and girls till the age of puberty; but the skull of a woman is, on the whole, smaller than that of a man, and the paranasal sinuses are small relative to the size of the skull. The capacity of the cranial cavity averages 200 cm³. less than in a man of the same race, and that is more or less in conformity with the relative size of the whole body of women and men of average build.

Very often it is not possible to say with certainty whether a given adult skull is that of a man or of a woman. Several of the criteria appear or become marked at puberty and many are affected as the individual ages. The sex differences given below are best seen, if present at all, in skulls of individuals 30–50 years of age.

The skull of a woman is lighter than that of a man and the architecture is smoother; the muscular ridges are less pronounced; the mastoid processes are relatively small. The glabella and superciliary arches are less prominent; the forehead is therefore more vertical, and the frontal eminences appear to bulge more; the upper margin of the more rounded orbital opening is sharper (a fact appreciated better by touch than by sight); the parietal eminences are more convex. The facial region is rounder, the zygomatic region is lighter and more compressed. The palate tends to be smaller and to be parabolic in shape. The mandible is smaller and the height at the symphysis is less.

The smoother the external surface of the squama of the occipital bone, the more likely the skull is female. The occipital condyles are smaller. The vertex is said to be more flattened and the relative height of the skull to be less.

Attempts to sex skulls by comparing selected dimensions have revealed considerable overlap in skulls of known sex. No single dimension is absolutely reliable but if all dimensions are either large or small then confidence increases in the assessment of sex of an unknown skull (Krogman, 1962). Cranial traits and dimensions have been assessed by Keen (1950) and discriminant analysis has been used to sex Japanese skulls by Hanihara (1959).

DEVELOPMENT OF THE SKULL

At an early stage of mammalian development the cerebral vesicles are enclosed in a membranous envelope, the **primordial membranous cranium,** derived from the mesoderm. In the elasmobranch fishes a complete cartilaginous capsule, the **primordial cartilaginous cranium,** is developed. In mammals, the cartilaginous part is confined to the base and the rapidly growing brain is covered by membrane in which bone is laid down. The roof, most of the sides, and even part of the base of the skull are formed from this **membrane bone.**

The notochord [p. 40] is embedded in the basal part of the mesenchyme from which the skull develops, reaching cranially to the developing hypophysis [FIG. 734]. In the mammalian cartilaginous base it is related to the part extending from the foramen magnum to the root of the dorsum sellae. This divides the base into a chordal and a prechordal part.

In the human embryo chondrification begins early in the second month and is well developed by the end of the third

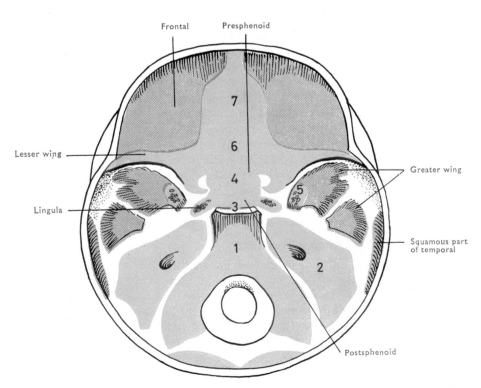

FIG. 148   Ossification on base and side-walls of skull of a 4½ months foetus (Schultze's method).

Cartilage, blue; membrane bone, red. Cartilage bone not shown. For explanation of numbers, see text.

month. In the following description the numbered paragraphs correspond to the numbered areas in FIGURE 148.

1. A basal or parachordal plate of cartilage forms behind the hypophysis cerebri and round the notochord. The notochord grooves the cerebral surface of the plate, pierces it to reach its pharyngeal surface and then turns up to terminate in the anterior end of the plate. Extensions from the sides of the basal plate form the condylar parts of the occipital bone and pass back to form the plate of cartilage in which ossification of the occipital squama takes place. This ossification in cartilage is all below the superior nuchal lines. The halves of this plate are slow to unite posteriorly and so the region behind the foramen magnum remains membranous till shortly before birth. This is the commonest site for a cerebral meningocele.

2. Cartilage appears in the mesoderm round the internal ear to produce the auditory capsule in which the petrous (including mastoid) part of the temporal bone ossifies.

3. A small strip of cartilage in front of the parachordal plate forms the dorsum sellae.

4. Cartilage around and in front of the end of the notochord forms the body of the sphenoid. It sends an extension forwards on each side to surround the craniopharyngeal canal (by which the pharyngeal part of the hypophysis enters the cranial cavity) and these extensions join to form the anterior part of the body of the sphenoid. They usually occlude the canal during the third month, but failure to do so may result in a patent craniopharyngeal canal [p. 117].

5. Chondrification occurs in the region of the root of the pterygoid process and the greater wing of the sphenoid including the lingula and the foramen rotundum.

6. Chondrification occurs in the whole of the region of the lesser wings.

7. A cartilaginous nasal capsule forms during the third month. A paranasal cartilage appears in the lateral wall of the nasal cavity and gives rise to the ethmoidal labyrinth, all the conchae, and part of the maxilla. Part of it persists as the cartilage in the side of the nose but the septum is formed from the cartilaginous sphenoid and from independent paraseptal cartilages on each side of the anterior part of the septum. A pair of subvomerine cartilages are persistent parts of the paraseptal cartilages.

The cartilaginous roof of the nose is formed by fusion of the side-wall with the septum and by chondrification round the olfactory nerve filaments. Chondrification of the floor is very incomplete.

All the chondrified areas fuse to form a platform in the base of the skull from which the pharyngeal arches are suspended.

OSSIFICATION

In FIGURE 148 all the parts coloured red are ossified in membrane as are all the parts above this level. They include the occipital squama above the nuchal lines, which is designated interparietal, and the zygomatic bone.

Most of the cartilage ossifies from centres which are indicated in the description of individual bones but some in the nose are ossified from the membrane overlying the cartilage, namely, the vomer, lacrimal, and nasal bones and the upper part of the maxilla.

The lower part of the maxilla, the palatine bones, and the greater part of the pterygoid processes are ossified from the membrane external to the mucous lining of the oropharyngeal region.

The tympanic plate ossifies in membrane overlying the auditory capsule.

Segmentation

Although the skull never shows conclusive evidence of segmentation, probably owing to the need for stability even in its early forms, it is assumed that the chordal part of the base has arisen by fusion of segments equivalent to vertebrae. The reasons for this assumption are the presence of myotomes in the head region, the connexion with the series of pharyngeal arches, and the segmental arrangement of nerves. The points of exit of nerves through the dura mater, which represents the primitive brain envelope, are the best indication of the disposition of segments. Evolution has altered some of the apertures in the bone, e.g., in the case of the oculomotor and abducent nerves.

The mammalian occipital bone is usually regarded as the equivalent of four fused vertebrae and some vertebrates may have acquired extra occipital elements from the vertebral column. The chordal part of the primordial cranium is related to the primitive cranial nerves, excluding the olfactory and optic nerves which are parts of the brain. The latter are enclosed in capsules developed in the prechordal part of the skull and protect the enlarging brain and the organs of smell and sight.

The outstanding features that distinguish the human skull are the large size of the brain-case, the small size of the face, and the way the skull is balanced on the vertebral column. This poising of the skull on a vertebral column is associated with alteration in its shape and, in particular, its muscular attachments. The occipitovertebral joints are so placed that the fore and hind parts of the head nearly balance each other.

The small size of the face is associated with a reduction in the size of the teeth and a corresponding reduction in the bulk and length of the jaws. Diminution of the muscles of mastication also reduces the fossae, surfaces, and crests which provide attachment for them in animals with stronger jaws.

The human cranium also exhibits a number of minor morphological variations: some have already been mentioned in the text and others are noted in the description of individual skull bones which follows (highest nuchal line, metopic suture, sutural bones, parietal foramen, incomplete or accessory foramina). Although some variants are the result of pathological changes, most result from normal development processes and are genetically determined (Berry and Berry, 1967).

CRANIOMETRY

The external size of the skull and cranial capacity vary considerably in different individuals. Varying thicknesses of bone, and size of the paranasal sinuses, reduce the degree of correlation between the two. Skull capacity is best measured in practice by filling the skull with such things as beads or lead shot, after closing up the larger foramina. The volume required to fill the cavity is then measured. For classification and comparison skulls are grouped according to their cranial capacity as follows.

Microcephalic skulls have a capacity below 1,350 cm³.
Mesocephalic skulls range from 1,350 to 1,450 cm³.
Megacephalic skulls have a capacity over 1,450 cm³.
Other systems of grouping have been suggested which take

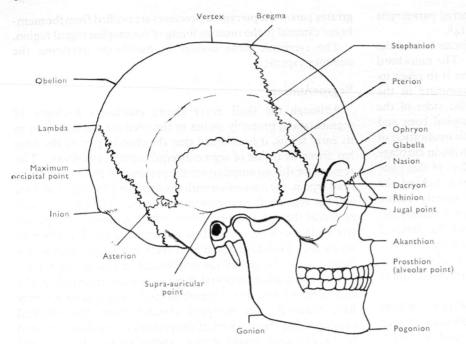

FIG. 149 Points on the skull from which the most important measurements are taken.

means of classifying skulls. These differences are expressed by comparing measurements between given points and the more important of these points are indicated in FIGURE 149.

The relation of the maximum breadth of a skull to its maximum length is a commonly used index and is expressed as the cephalic index.

$$\frac{\text{Maximum breadth}}{\text{Maximum length}} \times 100 = \text{cephalic index.}$$

The results are classified as follows:

Dolichocephalic skulls have an index below 75.

Mesaticephalic skulls range from 75 to 80.

Brachycephalic skulls have an index above 80.

The skulls of most Europeans fall within the mesaticephalic range. Many other measurements and indices from the skull have been established and tabulated for the various groups of mankind (Duckworth, 1915; Hrdlicka, 1952; Trevor, 1955; Comas, 1960). The height/breadth index of the orbit and the breadth/height index of the nasal aperture are of use together with the cephalic index. Discriminant analysis has been applied to Negro and White skulls with 90 per cent. success in one sample (see Krogman, 1962).

account of the difference in average cranial capacity between the sexes.

The shape, as well as the size, of the cranium is used as a

THE INDIVIDUAL BONES OF THE CRANIUM

It is more important that a student should understand the construction of the skull as a whole than that he should acquire detailed knowledge of individual bones disarticulated from the cranium. It is assumed that the reader is already acquainted with the skull as a whole and in the following brief descriptions of each bone reference should be made to the earlier section and especially to the illustrations.

THE FRONTAL BONE

The main part of the frontal bone is the convex shell which forms the front of the skull above the orbital openings. It is deeply concave to accommodate the frontal lobes of the brain.

The **external surface** [FIG. 150] is rounded with an added convexity forming a frontal tuber on each side. Above and posteriorly, it is limited by the serrated edge which forms the coronal suture with both parietal bones. Inferolaterally the surface narrows to reach a prominent zygomatic process which projects downwards and laterally to articulate with the zygomatic bone.

Inferiorly, the surface is limited by the concave supra-orbital margins, each with a supra-orbital notch, or a foramen, two-thirds of the way from its lateral end. The medial end of each orbital margin leads down to a nasal part from which a **nasal spine** projects downwards. The anterior edge of this

area and the nasal spine articulate with a nasal bone on each side of the midline and with the maxilla posterolaterally. Above this the smooth bulge of glabella is marked by the remains of a frontal suture and is continued laterally as the **superciliary arch.**

The **temporal surface** is nearly flat and passes down from the temporal line to the lower edge of the bone where it meets the sphenoid.

The **inferior aspect** [FIG. 151] shows the articular area for the zygomatic bone continued posteriorly to articulate with the greater wing of the sphenoid. The apex of this articulation reaches the end of the coronal suture; its base reaches the roof of the orbit. The **orbital surface** is on the inferior aspect of a horizontal orbital plate which projects back from the supra-orbital margin. It is smooth, concave, and triangular in outline and its blunt apex reaches the posterior edge of the plate where it articulates with the lesser wing of the sphenoid. Near the lateral end is the **fossa for the lacrimal gland,** and near the medial end is the small **trochlear fossa** or spine [p. 118].

Medially each orbital surface has a series of depressions where it articulates with the ethmoid bone and roofs in the ethmoidal cells. Anteriorly, the **nasal spine** projects downwards and forwards between two small, smooth, nasal areas which help to make up the roof of the nose. Posterior to this a deep **ethmoid notch** separates the two ethmoidal areas and is occupied by the cribriform plate of the ethmoid bone. Running across from the notch to the orbit, two grooves between the ethmoidal cells on each side form the **anterior** and **posterior ethmoidal foramina**.

The **nasal margin** is a small rough area, anterior to the ethmoidal cells and the nasal spine, for articulation with the nasal bones.

The **internal surface** of the frontal bone, above the orbital parts, shows structures already described [pp. 114 and 118], impressions for cerebral gyri on the convex upper surface of the orbital plate, the frontal crest, and sulcus for the superior sagittal sinus, markings for meningeal vessels and pits for arachnoid granulations. The **foramen caecum** may be wholly in the frontal bone. The frontal sinuses are described on page 122.

Ossification

Ossification in membrane spreads from centres appearing above the zygomatic processes in the sixth or seventh week of foetal life and from additional centres for the zygomatic processes, the nasal spine, and the trochlear fossae. Fusion in each half takes place in the sixth or seventh foetal month. The **frontal suture** is present at birth but is normally closed by the fifth or sixth year, only traces of it being left above and below.

Variations

The zygomatic process is dense throughout but varies in strength with that of the jaws. The frontal suture may persist as a **metopic suture** (up to 8 per cent. in Europeans) and sutural bones may alter the line of the sutures.

THE PARIETAL BONES

The parietal bones are the curved plates of bone, convex on their outer surface, which fill in the vault of the skull between the frontal, occipital, and temporal bones [FIG. 129]. Below and in front they also reach the sphenoid. They have, then, four borders and four corners spoken of as angles.

The **squamosal border** is inferior and has a sharp, concave, bevelled edge where it is overlapped by the squamous part of the temporal bone and, in its anterior part, by the greater wing of the sphenoid. The outer surface is radially grooved where it is covered by these bones. Where it is covered by the sphenoid the border runs down to join the frontal border at the **sphenoidal angle**. The frontal border is bevelled in this region on its deep surface and overlaps the frontal bone. This corner of the bone projects downwards and forwards a little and is deeply grooved on its interior aspect by meningeal vessels.

The **frontal border** runs up from the sphenoidal angle to meet the sagittal border at a right angle. It articulates with the frontal bone to form half of the coronal suture and is slightly overlapped by the frontal bone in its upper part, and vice versa in its lower part. The

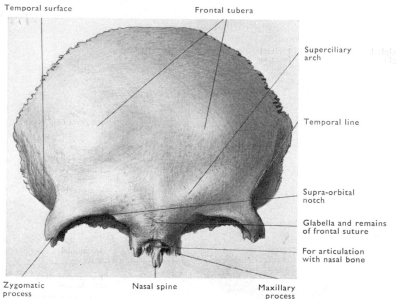

Temporal surface   Frontal tubera
Superciliary arch
Temporal line
Supra-orbital notch
Glabella and remains of frontal suture
For articulation with nasal bone
Zygomatic process   Nasal spine   Maxillary process

FIG. 150 Frontal bone (anterior view).

right-angled anterosuperior angle is named the **frontal angle**.

The **sagittal border** articulates with its neighbour by a series of small pegs and sockets to form the sagittal suture [p. 106]. The **occipital angle** of the parietal lies between the sagittal and occipital borders.

The **occipital border** articulates with the occipital bone to form half of the lambdoid suture [p. 106]. The **mastoid angle** is at the posterior end of the squamosal border where the parietal articulates with the mastoid region of the temporal bone, and the transverse sinus grooves the deep surface of the bone close to the parietomastoid suture.

The **outer surface** is convex with an increased convexity in its posterior half, the **parietal tuber** or eminence. A **parietal foramen** for an emissary vein is usually found near

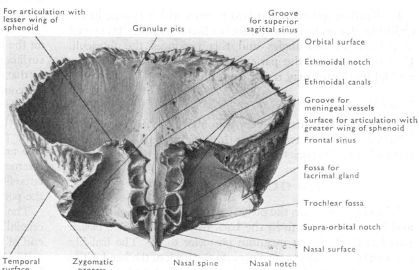

For articulation with lesser wing of sphenoid   Granular pits   Groove for superior sagittal sinus
Orbital surface
Ethmoidal notch
Ethmoidal canals
Groove for meningeal vessels
Surface for articulation with greater wing of sphenoid
Frontal sinus
Fossa for lacrimal gland
Trochlear fossa
Supra-orbital notch
Nasal surface
Temporal surface   Zygomatic process   Nasal spine   Nasal notch

FIG. 151 Frontal bone (inferior aspect).

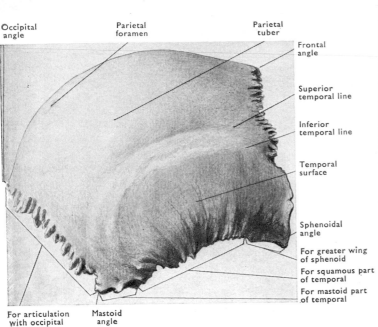

FIG. 152   Right parietal bone (external surface).

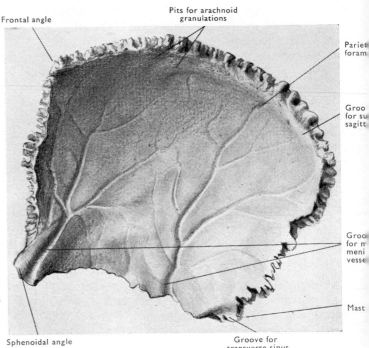

FIG. 153   Right parietal bone (internal surface).

the sagittal suture, about 3·5 cm. from its posterior end. At a lower level two **temporal lines**, superior and inferior, arch upwards from the mastoid angle and then forwards almost parallel to the squamosal border to the zygomatic process. The superior one is the attachment of the temporal fascia; the inferior one the upper limit of temporalis muscle.

The **internal surface** of the bone is concave and is strongly marked by meningeal vessels. The frontal branch of the middle meningeal artery and its vein groove the bone near the frontal border, the parietal branches are further back and the transverse sinus grooves it near the mastoid angle. The superior sagittal sinus and arachnoid granulations mark the inner surface near the sagittal border.

Ossification

Ossification spreads from two centres which appear in the eighth week and unite in the fourth foetal month. Its spread is delayed at the angles and this produces the fonticuli. Ossification round the parietal foramina is also slow and a sagittal fonticulus may result from this.

Variations

The outer table may be very thin above the temporal lines. The bone may be in two or even three pieces and ossification in the region of the parietal foramina may be poor or absent.

THE OCCIPITAL BONE

The occipital forms the posterior part of the base of the skull and is turned up to form the back of the skull. The basilar part has the foramen magnum posterior to it. The leaf-like squama is bent so as to turn up into the back of the head. The two condylar parts lie between the squama and the basilar part, one on each side of the foramen magnum.

The superior part of the **squama**, or interparietal part, is wedged between the parietal bones and curves up to reach the posterior end of the sagittal suture at lambda. Inferiorly, the surface curves down to reach the **superior nuchal lines** where it bends almost through a right angle to reach the base of the skull [FIG. 129]. A **highest nuchal line** for occipitofrontalis may be seen above the superior nuchal line. The main features of the inferior aspect are shown in FIGURE 154. It is almost horizontal and the squama narrows so that each lateral border is concave where the mastoid region of the petrous temporal articulates with the **mastoid border**. The surface posterior to the centre of the foramen magnum is marked by the curved superior and **inferior nuchal lines** [FIG. 340], the two halves being separated by an **external occipital crest** which terminates posteriorly as the **external occipital protuberance**.

In the **lateral parts** the two **occipital condyles** encroach on the anterior half of the foramen magnum and the articular surfaces, elongated and convex in shape, converge anteriorly so that rotation is severely limited. Posterior to each condyle is a condylar fossa, to receive the lateral mass of the atlas during extension at the joint. Opening into the fossa there is usually a **condylar canal** for transmission of a vein. The bone lateral to the posterior half of the condyle is the **jugular process**. Its free anterior margin is the posterior margin of the jugular foramen. Its inferior surface gives attachment to rectus capitis lateralis. Laterally it articulates, by means of cartilage, with the petrous part of the temporal bone.

The **hypoglossal canal**, often double, for the twelfth cranial nerve comes through the bone superior to the anterior end of the condyle to open medial to the jugular notch.

The lateral parts of the occipital bone converge to join the **basilar part** which continues as a flattened bar of bone to the body of the sphenoid. The inferior surface of the basilar part presents, at its centre, a **pharyngeal tubercle**, which receives the upper fibres of the superior pharyngeal constrictor, and the bone may be marked by rectus capitis anterior and by longus capitis muscles [FIG. 340].

The **internal surface** of the occipital shows the features already described for the posterior cranial fossa [p. 115]. The squama is divided into four fossae by ridges and grooves. The groove for the superior sagittal sinus runs down to the **internal occipital protuberance** and swings to one or other side as a groove for a transverse sinus. A groove for the other transverse sinus runs laterally from the protuberance to the other side. These mark off two cerebral fossae in the upper part of the squama and two cerebellar fossae lie inferior to the sinuses [p. 115]. The falx cerebri, tentorium cerebelli, and a small falx cerebelli are attached to the edges of the grooves. The superior surface of the jugular process is grooved by the sigmoid sinus. This groove passes behind a spur of bone into the **jugular notch** which forms part of the jugular fossa. Medial to this a raised ridge carries the **jugular tubercle**. Its posterior part is often grooved by the ninth, tenth, and eleventh cranial nerves. In front of the tubercle the ridge is marked by a groove for the inferior petrosal sinus. The basilar part of the bone forms the floor of a wide groove which lodges the medulla with its coverings and the vertebral arteries. At the side, between the foramen magnum and the ridge, the hypoglossal canal enters the bone. A small bony tubercle on the anterior margin of the foramen magnum may be present to indicate the position of the apical ligament of the dens.

Ossification

The method of ossification is indicated in FIGURE 156. The basilar part begins to ossify in cartilage in the sixth week of intra-uterine life, and includes a portion of each condyle. Centres for the lateral part appear in the eighth week and the squama below the highest nuchal line (supra-occipital) ossifies from centres near the protuberance in the seventh week. The interparietal part is ossified in membrane from centres appearing in the eighth week. Union of the two parts of the squama may be incomplete at birth. The four main components are united by cartilage until after birth. The lateral parts fuse with the squama during the third year and with the basilar part during the fourth or fifth year. The basilar part begins to fuse with the sphenoid between 18 and 20 years, or earlier in the female, and fusion should be complete by the twenty-fifth year.

Variations

The superior nuchal line and external protuberance may form a torus occipitalis transversus and an emissary vein may pierce the protuberance. The hypoglossal canal may be double or multiple. The upper part of the

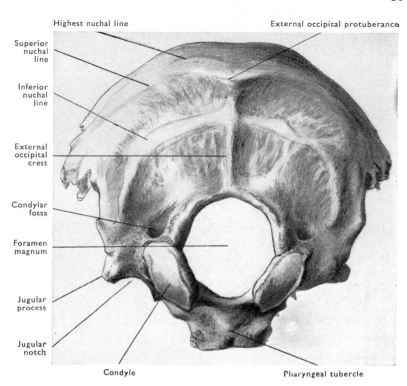

FIG. 154 Occipital bone (inferior aspect).

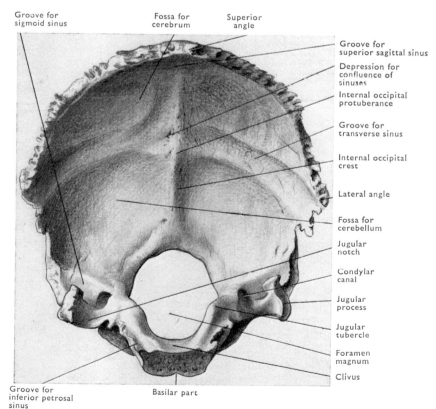

FIG. 155 Occipital bone (internal surface).

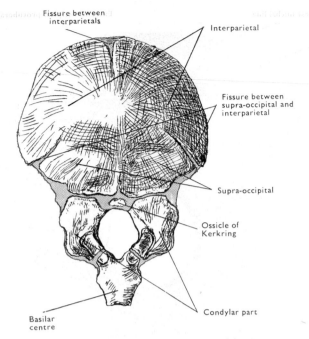

FIG. 156   Ossification of occipital bone.

The occasional ossicle of Kerkring fuses with the squama before birth. The basilar and the squamous centres may be bilateral or multiple.

squama may remain incompletely united laterally, may be separated as an interparietal bone, or it may be in right and left halves or multiple. An epiphysis may occur between the basi-occipital and the sphenoid.

A third occipital condyle may project from the anterior border of the foramen magnum to articulate with the dens and a **paramastoid process** may project from the jugular process and even be long enough to articulate with the transverse process of the atlas.

THE TEMPORAL BONES

Each temporal bone articulates with the lower lateral margin of the squama of the occipital bone, and with the side of its basilar part. It is wedged between these behind, and the sphenoid bone in front, and helps in the formation of the lateral wall of the skull by passing up to join the parietal bone.

It consists of squamous, tympanic, and petrous parts, the last including the mastoid region and mastoid process which are continuous with it.

The **squamous part** is a thin, vertical plate of bone applied to the lateral aspect of the petrous part. Its lateral aspect is smooth above with a thin semicircular **parietal border**. This merges anteriorly with a thicker **sphenoidal border** which articulates with the greater wing of the sphenoid. Posteriorly the **squamoso-mastoid suture** can usually be seen passing downwards and forwards [FIG. 157]. The external acoustic meatus forms a landmark

below the squama which projects down behind it as a postauditory process and in front of it as a postglenoid process. Above this, the root of the zygomatic process is continued forward from the supramastoid crest, to pass above the meatus and the mandibular fossa. The suprameatal triangle, above and behind the meatus, the **articular tubercle** in front of the fossa, and the root of the zygoma are described on page 107.

The **zygomatic process** curves forwards to reach the zygomatic bone [FIG. 129]. The squama seen above the level of the zygomatic process is the temporal surface. In front and below, it runs medially to form a small infratemporal surface.

The **cerebral surface** of the squama is marked by gyri of the cerebrum and by grooves for middle meningeal vessels. Above is the bevelled articulation for the parietal bone and below a thin line indicates the **petrosquamous suture** [FIG. 158].

The **tympanic part,** referred to as the tympanic plate, [FIGS. 157, 160] is a bent, triangular plate of bone which has the petrous part behind and medial to it and the squamous part above and anteriorly. A lamina (the **tegmen tympani**) projecting from the petrous part intervenes between squama and tympanic part at the petrotympanic fissure [FIG. 157].

The upper surface of the tympanic plate is concave and forms most of three walls of the external acoustic meatus, the squama forming its roof and a little of the posterior wall. The curved lateral edge gives attachment to cartilage of the meatus [FIG. 1003]. At the medial end of the meatus a **tympanic sulcus** grooves the plate and gives attachment to the tympanic membrane. Posteriorly the plate reaches the mastoid process at the **tympanomastoid fissure** which transmits the auricular branch of the vagus. More medially it is moulded over the styloid process as a sheath. The upper border of the

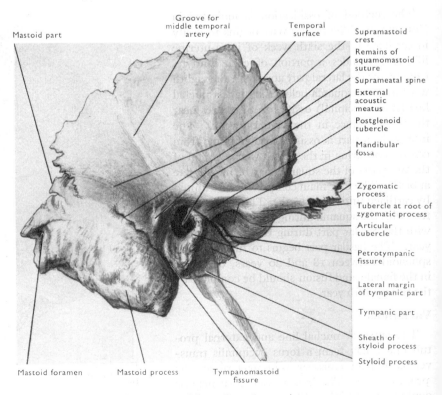

FIG. 157   Right temporal bone (lateral aspect).

anterior surface joins the postglenoid tubercle laterally but is separated from the squama more medially by the **squamo-tympanic fissure** [p. 113].

The **petrous part** of the temporal bone consists of a mastoid portion from which the mastoid process projects downwards and forwards [FIG. 157], and a three-sided pyramid of hard bone which contains the internal ear. It tapers a little as it passes forwards and medially towards the foramen lacerum at the side of the body of the sphenoid [FIG. 133]. The lateral part, anterior to the mastoid region, is excavated to form the **cavity of the middle ear** which is closed in laterally by the tegmen tympani which turns down between the cavity and the squamous part and by the tympanic membrane attached to the tympanic plate [FIGS. 1003–1007]. The cavity is continued forwards as the auditory tube, and backwards through the antrum to the air cells in the mastoid process [FIG. 159].

The mastoid region of the petrous part forms the rounded bony resistance behind the ear. A **mastoid foramen** is usually present in its upper part near the occipitomastoid suture. Sternocleidomastoid muscle is its principal attachment with splenius and longissimus capitis deep to it.

Inferior Surface [FIG. 160]

This is irregular and rough and most of the structures have already been noted [p. 112], including the grooves for the digastric muscle and the occipital artery. The **styloid process**, with the **stylomastoid foramen** behind it, emerges between the tympanic plate and digastric groove on the petrous part. Medial to the foramen a deep notch receives the corner of the jugular process of the occipital bone and must not be confused with the **jugular fossa** which lies still further medially and helps to form the jugular foramen. In the lateral part of the wall of the fossa is the medial end of the **mastoid canaliculus** for the auricular branch of the vagus which runs to the tympanomastoid fissure. The lower end of the **carotid canal** is immediately anterior to the jugular fossa, and between them, a **tympanic canaliculus** transmits a tympanic branch of the glossopharyngeal nerve to the tympanic cavity [FIG. 160]. Two **caroticotympanic canaliculi** for sympathetic nerves from the carotid canal also lead to the tympanic cavity. Towards the apex of the petrous temporal the floor of the carotid canal is formed by a rough quadrilateral area, and at the tip the opening of the carotid canal can be seen; the canal having turned at right angles to run anteromedially in the bone. In the angle between the petrous and squamous parts, anterior to the carotid canal, lie the openings of the pair of **semi-canals** for the auditory tube and tensor tympani muscle.

Between the inferior surface and the posterior surface the articular surface for the occipital bone is divided into two by the jugular fossa.

Anterior and Posterior Surfaces [FIG. 158; pp. 116, 117]

On the medial part of the **anterior surface** in the middle cranial fossa there is a distinct impression for the trigeminal nerve and ganglion. About halfway between this and the base of the pyramid, the **arcuate eminence** indicates the position of the anterior semicircular canal. The bone between this and the squama

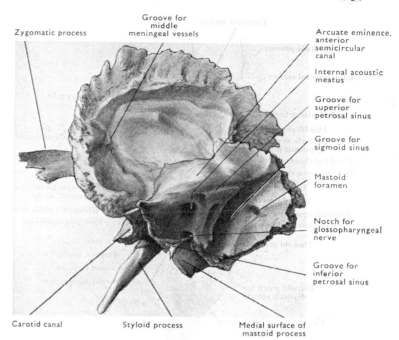

FIG. 158   Right temporal bone (medial aspect).

is the tegmen tympani forming a roof for the middle ear [FIG. 159]. The bone anterior to the eminence covers the cochlea and the facial nerve. Passing anteromedially from this region, two small grooves carry the greater and lesser petrosal nerves from two small openings towards the foramen lacerum.

The anterior surface is separated from the posterior surface by a **superior border** grooved by the superior petrosal sinus, except where the trigeminal nerve intervenes near the medial end, and from the squama by a faint irregular line, the **petro-squamous suture.**

The **posterior surface** has as its most prominent feature the **internal acoustic meatus** which passes laterally so that its opening is oblique. It is about 10 mm. deep and its lateral end or fundus has openings for the facial nerve and for vessels and nerves of the internal ear [FIG. 1016]. Above and lateral to

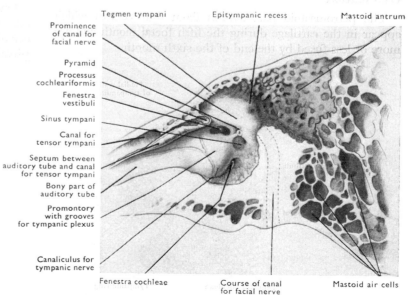

FIG. 159   Section through left temporal bone, showing medial wall of tympanic cavity.

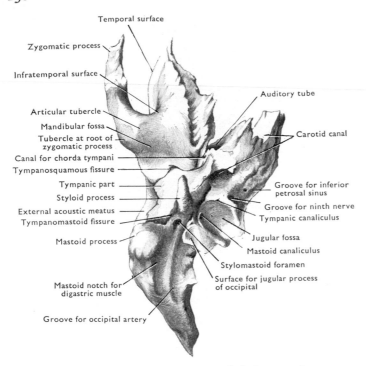

FIG. 160   Right temporal bone (inferior aspect).

The **styloid process** ossifies from two centres which appear in the upper part of the second pharyngeal arch cartilage. The upper, or tympanohyal, centre appears before birth and fuses with the petrous part during the first year. The lower or stylohyal one appears shortly after birth but ossifies slowly. It fuses with the upper part after puberty but may fail to fuse at all.

The **tympanic part** ossifies from a centre in the lateral wall of the tympanum in the third month and forms a curved tympanic ring which is, however, incomplete superiorly [FIG. 161]. After birth ossification spreads medially, laterally, and downwards to form the tympanic plate.

The **squamous part** is ossified in membrane from a centre at the root of the zygoma which appears about the end of the second month.

Variations

The markings vary considerably in different bones. A venous sinus may lie along the petrosquamous suture internally, leading through the bone to the transverse sinus posteriorly. Anteriorly, it may pass through a canal which traverses the root of the zygoma to appear above the squamotympanic fissure. Variations in the ossification of the squamous part includes absence of the zygoma. Abnormalities of the petrous part are likely to be associated with deaf-mutism or idiocy. The squama may articulate with the frontal bone, as in apes.

THE SPHENOID BONE

The sphenoid bone is wedged in among other bones of the skull and consists of a central body, two wings extending out on each side of the body, and a pterygoid or wing-like pair of processes extending down from the body, one on each side. Most of the bone has been described as part of the skull as a whole. The main features which become evident in a separate bone are sutures and openings which were difficult to see in the intact skull.

On the *posterior aspect* [FIG. 163] the body presents a sawn-off surface or an articular area for the basilar part of the occipital. Its upper surface is deeply excavated by the **hypophysial fossa**. The **dorsum sellae** slopes up to overhang the

the meatus lies a **subarcuate fossa** [p. 116]. More laterally, between the meatus and the sigmoid sinus, is another fossa with a ridge above it which indicates the position of the upper part of the posterior semi-circular canal. A small slit overhung by a scale of bone is the opening of the aqueductus vestibuli. Inferior to the meatus a notch in the lower border of the surface accommodates the glossopharyngeal nerve and has at its apex a canaliculus cochleae which leads up to the cochlea of the internal ear. Medial to this the edge of the bone is grooved by the inferior petrosal sinus. The **groove for the sigmoid sinus** lies deep to the posterior half of the mastoid process, and frequently has a foramen for a vein opening into it.

Ossification

The petromastoid part ossifies from four centres which appear in the cartilage during the fifth foetal month and are more or less fused by the end of the sixth month.

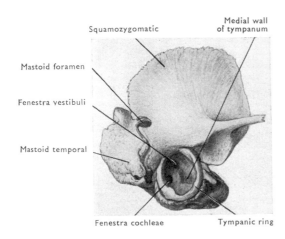

FIG. 161   Temporal surface of right temporal bone at birth.

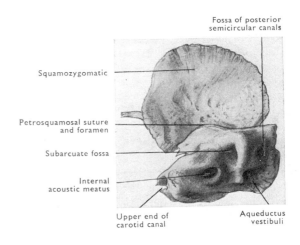

FIG. 162   Cerebral surface of right temporal bone at birth.

fossa and extend laterally as the **posterior clinoid processes.**

The **lesser wings** form tapering processes, each attached to the body by two roots with the optic canal between them. The anterior roots are joined by the jugum sphenoidale. The anterior edge articulates with the frontal bone and the ethmoid. The posterior roots join the body, one on each side, and the tuberculum sellae lies between them and behind the sulcus chiasmatis. The posterior edge of the wing curves back to the prominent **anterior clinoid process.** The sharp tip of the wing reaches towards the greater wing and the region of pterion. The gap between the lesser and greater wings is the **superior orbital fissure.**

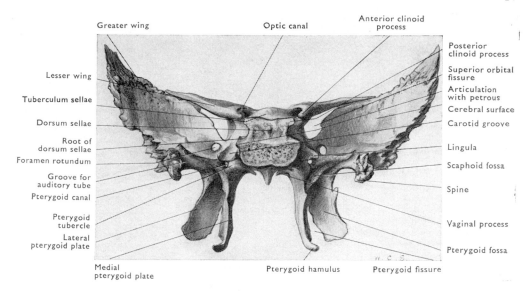

FIG. 163   Sphenoid bone (posterior aspect).

The **greater wing** springs from the lower part of the body and, expanding laterally, passes forwards and then upwards to reach the frontal bone and the corner of a parietal bone. It has a concave **cerebral surface,** a concave **temporal surface,** and a nearly flat **orbital surface.** The edge between the orbital and temporal surfaces articulates with the zygomatic bone to separate the orbit from the temporal fossa. The posterior margin of the greater wing terminates as a sharp angle which fits in between the squamous and petrous parts of the temporal bone. Its end projects down as the **spine of the sphenoid.** The foramen spinosum and **foramen ovale** pierce the bone near the posterior edge of the wing. The most medial part of the edge is prolonged backwards as a **lingula** over the carotid groove on the side of the body. This almost horizontal part of the wing helps to form the floor of the middle cranial fossa and the roof for the infratemporal fossa. Where it joins the body and below the superior orbital fissure, the **foramen rotundum** passes forwards through the wing to reach the pterygopalatine fossa. Below the lingula and the carotid groove a **pterygoid canal** passes horizontally through the root of the pterygoid process from the foramen lacerum to the pterygopalatine fossa.

The squamosal margin of the greater wing which articulates with the squamous temporal is concave and is crossed near its upper end by the frontal branch of the middle meningeal artery which then grooves or penetrates the deep surface of the adjacent corner of the wing.

The *anterior aspect* [FIG. 164] shows the posterior parts of the nose and orbit and the posterior wall of the pterygopalatine fossa.

The thin anterior wall of the body is perforated on each side by the opening of a **sphenoidal sinus** [p. 120] into the spheno-ethmoidal recess of the nose. Each opening is normally covered by a thin, inverted triangle of bone. This **sphenoidal concha** is perforated by a smaller opening 4 or 5 mm. in diameter. Laterally this wall articulates with the ethmoidal labyrinth and the orbital process of the palatine bone, and its inferolateral edge forms the upper margin of the sphenopalatine foramen. Between the conchae, the median **crest of the sphenoid**

continues down to become the **sphenoidal rostrum.** The crest articulates with the perpendicular plate of the ethmoid. The rostrum fits into a recess between the alae of the vomer.

The roof and medial wall of the orbit articulate with the lesser wing and with the sphenoidal concha respectively. The smooth orbital surface of the greater wing passes postero-medially from the zygomatic articulation, and forms the curved inferior margin of the superior orbital fissure. Superiorly, it reaches the frontal articulation. The foramen rotundum opens into the upper part of the posterior wall of the pterygopalatine fossa, and the pterygoid canal opens a little inferomedial to it. This wall of the fossa is continued down on the **pterygoid process** [FIG. 163] the medial and lateral laminae of which are united above, but separated below by a V-shaped pterygoid fissure which is filled by the pyramidal process of the palatine bone. This is fused anteriorly to the posterior surface of the maxilla and thus the pterygopalatine fossa is closed off below.

The **lateral pterygoid lamina** is turned outwards and gives attachment to medial and lateral pterygoid muscles, from its corresponding surfaces. The **medial lamina** forms the most posterior part of the lateral wall of the nose and gives attachment, in the lower half of its posterior margin, to the superior constrictor of the pharynx. Anterior to it the wall of the nose is formed by the palatine bone. The other features of the medial lamina are described on page 113.

Ossification

The sphenoid ossifies in cartilage except for the lateral part of the greater wing and part of the pterygoid processes. Centres appear from the eighth to the tenth foetal week for most of the bone. Those for the conchae are delayed until the fifth month or later. The hamulus chondrifies in the third month.

At birth the bone is in three pieces, the greater wing and pterygoid process on each side, the body and lesser wings between them. These have fused by the end of the first year.

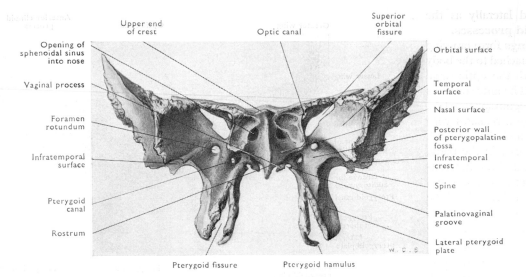

FIG. 164   Sphenoid bone (anterior aspect).

Variations

Foramen spinosum and foramen ovale are frequently incomplete posteriorly and the superior orbital fissure may extend down into foramen rotundum. The optic canal may be divided into compartments for the optic nerve and ophthalmic artery. Accessory ossification in relation to the anterior clinoid process may convert the groove for the internal carotid artery on its medial side into a caroticoclinoid foramen.

Rarely, a craniopharyngeal canal [p. 131] may persist between the pharynx and the hypophysial fossa.

THE ETHMOID BONE

The ethmoid bone consists mainly of two lateral masses, one on each side of the nasal cavity, excavated by air cells and called **ethmoidal labyrinths**. Each is joined by a cribriform plate to a median perpendicular plate [FIG. 166].

The **perpendicular plate** forms part of the nasal septum [FIG. 136] and has irregular edges for articulation with the crest of the sphenoid and the vomer behind and below, the cartilaginous septum in front and below, and the nasal spine of the frontal bone and the median crest of the nasal bones, above and in front. The plate is continued up as the crista galli in the anterior cranial fossa [p. 118].

On each side a **cribriform plate** forms part of the nasal roof through which the olfactory and anterior ethmoidal nerves pass by way of the perforations.

Each **labyrinth** consists of three groups of thin-walled air cells, anterior, middle, and posterior, between nasal cavity and orbit. Many of the cells are incomplete when the bones are disarticulated. Posteriorly it articulates with the body of the sphenoid and the orbital process of the palatine bone; anteriorly, with the lacrimal bone and the frontal process of the maxilla; superiorly, with the orbital plate of the frontal bone; and inferiorly, with the maxilla. On the upper aspect, two narrow grooves correspond to those seen on the frontal bone for the anterior and posterior ethmoidal vessels and nerves. The smooth, thin **orbital plate** is on the lateral surface and forms part of the medial wall of the orbit [FIG. 138]. The inferior surface of the ethmoidal labyrinth reaches the maxilla and the **uncinate process** passes down on its medial side to help in the closure of the opening of the maxillary sinus, and to form the inferior margin of the hiatus semilunaris.

The labyrinth forms part of the lateral wall of the nasal cavity and from it the **superior and middle conchae** project towards the septum and then turn down over the superior and middle meatuses [FIG. 137]. The middle concha extends far enough back to articulate with the palatine bone

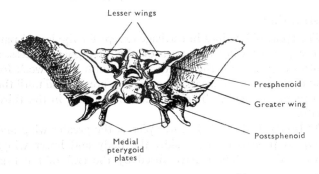

FIG. 165   Ossification of sphenoid.

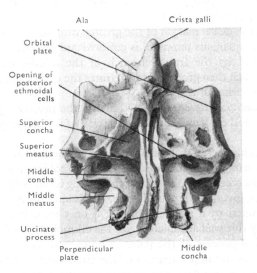

FIG. 166   Ethmoid (posterior aspect).

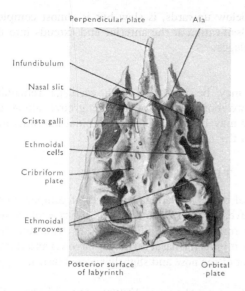

FIG. 167   Ethmoid (superior aspect).

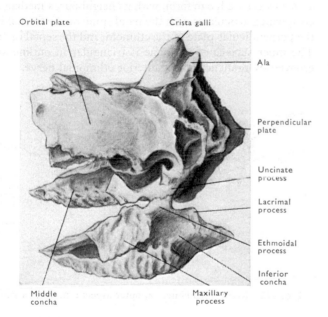

FIG. 168   Lateral surface of ethmoid and inferior concha.

and forwards to reach the maxilla. The middle cells bulge into the middle meatus as the **ethmoidal bulla** and open into the meatus through it. The groove below the bulla and above the uncinate process is the **hiatus semilunaris** which receives the openings of the anterior cells, sometimes including the one which forms the ethmoidal infundibulum [p. 121]. The posterior cells open into the superior meatus.

Ossification

Centres of ossification appear in the cartilage of the nasal capsule in the fourth and fifth months for each labyrinth and the labyrinth is completely ossified at birth, including the walls of its cells which have already appeared. Bone appears in the perpendicular plate at the end of the first year and has spread through it by 5 or 6 years. Bone extends from the labyrinth

to the cribriform plate during the second year but does not fuse with the sphenoid until after 25 years.

Variations

The number of conchae may vary from one to four. The orbital plate varies in size and shape : it may be narrowed from above downwards or may fail to articulate with the lacrimal, as in anthropoid apes. A vertical suture may separate it into two parts.

THE INFERIOR NASAL CONCHAE

Each inferior concha is a thin, curled leaf of bone, pointed at both ends. It has a rough, convex, free, medial surface [FIG. 169] and a concave lateral surface attached by its upper edge to the lateral wall of the nasal cavity [FIG. 168]. Here it articu-

FIG. 169   Medial surface of right inferior concha.

lates with the conchal crests on the maxilla anteriorly, on the perpendicular plate of the palatine bone posteriorly. Between these the attachment expands into a flat plate extending upwards to the lacrimal and ethmoid as **lacrimal and ethmoidal processes** and down to the maxilla (**maxillary process**) where it becomes part of the medial wall of the maxillary sinus [FIG. 168]. The lacrimal process articulates with the lacrimal bone and the maxilla to complete the wall of the **nasolacrimal canal.** The ethmoidal process reaches the uncinate process of the ethmoid bone. The lateral surface is smooth below the lacrimal process where the nasolacrimal canal enters the inferior meatus.

Ossification

Ossification is in the cartilage of the nasal capsule from a centre which appears about the fifth foetal month.

THE LACRIMAL BONES

Each lacrimal bone is a thin scale of bone found in the medial wall of the orbit behind the frontal process of the maxilla. It is roughly quadrangular and its two surfaces are medial and lateral. It articulates with the frontal bone, the maxilla, the inferior concha, and the ethmoid. The medial surface is uneven where it articulates with and helps to close some ethmoidal sinuses. Anterior to this it is smooth where it enters into the lateral wall of the nose above the inferior concha. The lateral surface is shown in FIGURE 170. Its posterior part is the smooth orbital surface. Its anterior part, or **fossa for the lacrimal sac,** is deeply concave to form the posterior half of the fossa and is limited behind by the prominent **posterior lacrimal crest.** At the lower end, the orbital surface and the

FIG. 170   Right lacrimal bone (lateral surface).

crest curve forwards as a **lacrimal hamulus** which reaches the maxilla and forms the posterolateral boundary of the beginning of the nasolacrimal canal. A small descending process forms the medial boundary of the opening.

Ossification and Variations

One centre, appearing in the third foetal month, spreads in the membrane covering the nasal capsule.

The lacrimal bone may be absent, small, or divided. The hamulus may reach the orbital margin, as in lemurs.

THE VOMER

The vomer is shaped, as its name indicates, like a plough-share [FIG. 171]. It lies in the nasal septum and has one smooth, free edge which forms the posterior margin of the septum. The remaining edges articulate as shown in FIGURES 136 and 571. The superior border is split into a pair of everted **alae** which fit over the rostrum of the sphenoid and a small sphenovomerine canal begins at the posterior end of the groove between the alae. It runs horizontally through the bone and transmits nutrient vessels. The inferior border fits into the nasal crest on the upper surface of the palate and the truncated anterior end reaches the region of the incisive canals where it sends a pointed process between them.

Grooves for vessels mark the surfaces of the bone and one, usually more distinct, lodges the nasopalatine nerve.

Ossification

The vomer begins to ossify in the eighth week from centres in the membrane on each side of the septal cartilage. They form plates of bone in each side of the cartilage and these begin to unite along their lower edges in the third foetal month.

FIG. 171   Vomer, from the right side.

Fusion, from below upwards, is slow but is almost complete by puberty. Ossification at the anterior end extends into the paraseptal cartilages.

Variations

Ossification may be imperfect with cartilage remaining between the two bony laminae and the groove along the anterior border may be deep. The sphenoidal air sinus may extend between the laminae.

THE NASAL BONES

The two nasal bones are elongated pieces of compact bone, smooth and convex on their outer surfaces, concave and a little rough on their inner surfaces. Each bone is wide and thin inferiorly where it joins the nasal cartilage, narrower and thicker above at the root of the nose and slightly waisted near its upper end.

The articulations are shown in FIGURES 132 and 172. The medial border which articulates with its fellow is deepened above to increase its area. This widened part projects a little at its posterior edge to form, with its neighbour, a median nasal crest which articulates with the nasal spine of the frontal bone, the perpendicular plate of the ethmoid, and the septal cartilage. The inner surface of the bone is triangular in outline and is grooved by branches of the anterior ethmoidal nerve.

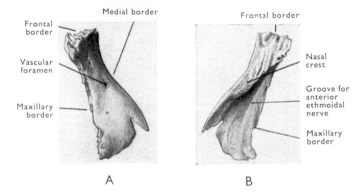

FIG. 172   Right nasal bone.   A, outer aspect;   B, inner aspect.

Ossification

Each nasal bone ossifies from a single centre, appearing about the end of the second foetal month, in the membrane covering the nasal capsule.

Variations

Both racially and individually the nasal bones vary considerably. They tend to be prominent in white races, small and flat in Mongolians and Negroes. They may be absent or divided. Obliteration of the nasal suture is rare though it is the usual condition in apes.

THE MAXILLAE

The two maxillae join in the midline to form the upper jaw and each has a hollow pyramidal body with four processes projecting from it.

The surfaces of the body correspond to the walls of the

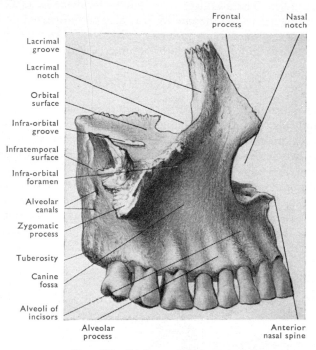

FIG. 173 Right maxilla (lateral view).

Labels (Fig. 173): Lacrimal groove; Lacrimal notch; Orbital surface; Infra-orbital groove; Infratemporal surface; Infra-orbital foramen; Alveolar canals; Zygomatic process; Tuberosity; Canine fossa; Alveoli of incisors; Alveolar process; Frontal process; Nasal notch; Anterior nasal spine.

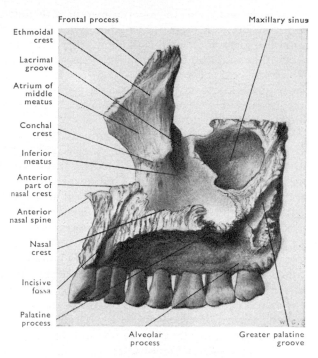

FIG. 174 Right maxilla (medial aspect).

Labels (Fig. 174): Ethmoidal crest; Lacrimal groove; Atrium of middle meatus; Conchal crest; Inferior meatus; Anterior part of nasal crest; Anterior nasal spine; Nasal crest; Incisive fossa; Palatine process; Alveolar process; Frontal process; Maxillary sinus; Greater palatine groove.

pyramidal cavity of the maxillary sinus [p. 122]. The **nasal surface** is the base of the pyramid and has the wide opening of the sinus in its posterosuperior part [FIG. 174]. Its smooth antero-inferior part forms most of the lateral wall of the inferior meatus of the nose. Behind these two parts the surface articulates with the perpendicular plate of the palatine, which thereby forms the medial wall of the greater palatine canal [FIG. 175].

The anterosuperior corner of this surface has a projecting frontal process and immediately behind that is a short, deep, vertical **lacrimal groove**. The angle between the frontal process and the body of the bone is filled in by the lacrimal bone so that the upper end of the groove is converted into a canal. The medial surface of the frontal process has on its lower part a **conchal crest** for the inferior concha and, in its upper part, an **ethmoidal crest** for the middle concha. Between them is the atrium of the middle meatus. The nasolacrimal canal thus lies lateral to the lacrimal, ethmoid, the middle meatus, and the inferior concha, and opens into the inferior meatus. The opening of the maxillary sinus is reduced by encroaching bones.

Each **palatine process** extends medially to articulate with its neighbour in the midline, and with it forms a **nasal crest**, which projects forwards as the **anterior nasal spine**. The rough posterior edge of the process articulates with the horizontal plate of the palatine bone. The anterior part of its upper surface curves up a little in front to form a lower edge for the piriform aperture, but its main bulk turns down at the anterior end to become continuous with the alveolar process. Behind the alveolar process the two articulating surfaces of the maxillae are grooved to form an inverted, funnel-shaped, **incisive fossa** with a small **incisive canal** leading down into it on each side [p. 114; FIG. 899]. The inferior surface of the palatine process is pitted by palatine glands and curves down

to merge with the alveolar process. Near its lateral margin a groove may be present for the greater palatine vessels and nerve.

The **alveolar process** projects down from the margin of the inferior surface of the pyramidal body [FIG. 138]. It meets its fellow in front to form a semi-elliptical **alveolar arch** with sockets for eight teeth on each side. Posteriorly, the process ends as the **tuber** (*tuberosity*) of the maxilla. Laterally, the alveolar process is ridged by the roots of the teeth, particularly the canine.

The **anterior surface** [FIG. 173] extends from the midline to the zygomatic process. The body has two other surfaces: a horizontal, **orbital surface,** and a posterior or **infratemporal surface.** These three surfaces meet at the hollow **zygomatic process,** and this has a rough articular surface for the zygomatic bone. This articular area extends back on to the edge between the orbital and infratemporal surfaces so that the anterior part of the orbit is separated from the infratemporal region by the junction of the two bones.

The **infratemporal surface,** on the posterior aspect, is convex and rather rough but becomes continuous laterally with the smooth, concave, posterior wall of the zygomatic process and forms the anterior wall of the infratemporal fossa and pterygomaxillary fissure. The orbital and pyramidal processes of the palatine bone articulate with the superomedial and inferomedial parts of this surface respectively; the latter separating it from the pterygoid laminae. Above the tuberosity of the maxilla one or two foramina transmit the posterior superior alveolar vessels and nerves to the molar teeth [FIG. 902].

The posterolateral margin of the smooth, slightly concave, **orbital surface** articulates with the zygomatic bone anterolaterally and the orbital process of the palatine bone posteromedially, and forms the anterior margin of the inferior orbital fissure between these. The latter is notched by the **infra-orbital groove,** which is continuous anteriorly with the **infra-orbital canal and foramen.** They transmit the infra-orbital vessels and nerve, which send anterior superior alveolar branches into a small opening in the floor of the canal. The medial edge of the orbital surface articulates with the ethmoid and the lacrimal bones.

The sloping **anterior surface** is continued up from the alveolar process to the infra-orbital margin and merges laterally with the zygomatic process. Medially it is limited by the piriform aperture with the **anterior nasal spine.** The incisor fossa is below the edge of the piriform aperture and above the incisor teeth. The wider fossa between the ridge produced by the canine and the zygomatic process is called the **canine fossa.** The infra-orbital foramen [p. 109] lies above this fossa. Above and medially, the surface is continued on to the frontal process, which extends up to articulate with the frontal bone. The lower part of the anterior edge of the frontal process bounds part of the piriform aperture of the nose; its upper, posterior margin articulates with a lacrimal bone down the middle of the lacrimal fossa. The anterior edge of the fossa is formed by the **anterior lacrimal crest,** a ridge continuing the orbital margin upwards. The groove behind the ridge passes down medial to the body of the bone as the **lacrimal sulcus.** The medial surface of the process is smooth between the ethmoidal and conchal crests already noted and forms the side-wall of the atrium of the middle meatus. The anterior end of the ethmoidal crest produces the agger nasi, a slight elevation seen in the living nose.

The **premaxillae** are normally fused with the maxillae in Man. They lie in front of the incisive fossa and carry the incisor teeth on each side. Up to middle age their independent origin may be indicated by a faint suture from the hinder part of the incisive fossa to the interval between the canine and incisor sockets.

Ossification

The maxilla ossifies in membrane from a centre which appears in the sixth week in the wall of the oral cavity above the germ of the canine tooth. The infra-orbital nerve is above the bone until the second foetal month when the groove and canal begin to form. The maxillary sinus first appears as a shallow fossa on the medial side of the bone in the fourth month of intra-uterine life.

The premaxilla ossifies from at least two centres, the first appearing above the incisor tooth germs in the sixth week. It forms most of the premaxilla and a small part of the frontal process and fusion with the maxilla proper begins almost at once. The second centre appears in the twelfth week, to form the wall of the incisive canal, and soon joins the rest.

Variations

The infra-orbital canal may remain open as a groove or a vertical suture may be present in the infra-orbital margin above it. A ridge on the oral surface of the median suture of the palate, or torus palatinus, is not uncommon. Cleft palate may occur along the suture joining the two halves of the maxilla or on one or both sides of the premaxilla, i.e., between canine and incisor teeth [p. 54]. The lacrimal groove may be constricted

in its middle. The maxillary sinus may be loculated by septa to a varying extent and the anterior part may be constricted off by the lacrimal canal to produce a lacrimal recess. Absence of the premaxilla and incisor teeth has been recorded.

THE PALATINE BONES

Each palatine bone is L-shaped with a horizontal and a perpendicular plate. At their junction, posteriorly, there is a pyramidal process and at its upper end the perpendicular plate carries two processes, sphenoidal and orbital, separated by the sphenopalatine notch.

The **horizontal plate** projects medially to form the posterior third of the hard palate [FIG. 177]. It has a smooth upper **nasal surface,** a rougher lower **palatal surface,** and an anterior articular border for the palatine process of the maxilla. Medially, where it meets its neighbour, the articular part is

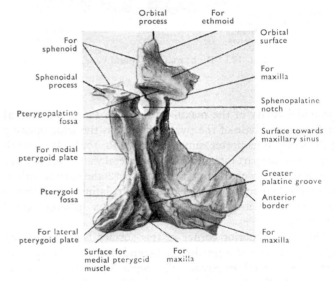

FIG. 175   Right palatine bone (lateral surface).

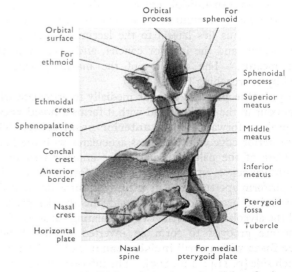

FIG. 176   Right palatine bone (medial surface).

raised up to form the posterior third of the **nasal crest.** The posterior border is a smooth concave edge which ends medially on the prominent **nasal spine** of the palate. Seen from below, a sharp palatine crest runs laterally near the posterior border, and curves towards the pyramidal process, passing the inferior end of the greater palatine groove on the way [FIG. 133].

The thin **perpendicular plate** is continued up from the horizontal plate to form part of the lateral wall of the nose. It is applied to the surface of the maxilla and articulates with the pterygoid process of the sphenoid posteriorly [FIG. 137].

The medial or nasal surface has a smooth area which is part of the inferior meatus, a **conchal crest** for the inferior concha, another smooth area forming part of the middle meatus and above that, a smaller, **ethmoidal crest** for the middle concha [FIG. 176].

At the upper anterior corner of the plate, an irregular, more or less triangular, **orbital process** (sometimes pneumatic) is applied to the posterior end of the orbital surface of the maxilla and so enters the floor of the orbit. Medially it reaches the sphenoid. Its concave, posterior edge forms the anterior limit of the deep sphenopalatine notch.

The **sphenoidal process** projects from the posterosuperior corner of the perpendicular plate. It curves upwards and medially and its surface is applied to the sphenoid at the root of the medial pterygoid plate. Its upper surface is grooved by the palatovaginal canal. The canal carries pharyngeal vessels and nerves from the pterygopalatine fossa to the roof of the pharynx. The medial edge of the process reaches the ala of the vomer.

The **sphenopalatine notch,** converted into a foramen by the edge of a sphenoidal concha, is the main communication between the pterygopalatine fossa and the nose.

The lateral or **maxillary surface** is applied to the maxilla [FIG. 175] and helps to reduce the opening of the maxillary sinus, but posterosuperiorly it forms the medial wall of the pterygopalatine fossa. From this free surface, which is smooth, a distinct **greater palatine groove** runs downwards and forwards. With the corresponding groove on the maxilla it forms the **greater palatine canal.**

Posterior to the palatine canal the **pyramidal process** projects backwards from the lower end of the perpendicular plate, and articulating with the lower ends of the pterygoid plates, fills the gap between them and separates them from the maxilla. From the greater palatine groove two **lesser palatine canals** descend through the process to open on to its inferior surface.

Ossification

Ossification spreads from a centre which appears in the eighth intra-uterine week in the membrane of the side-wall of the nasal cavity, but the orbital process may ossify separately. Until the maxilla enlarges in the third year the vertical height is small.

Variations

The maxillary sinus may extend between plates of cortical bone forming the horizontal plate. The orbital process varies in size, and may join the sphenoidal process, and bridge the sphenopalatine notch.

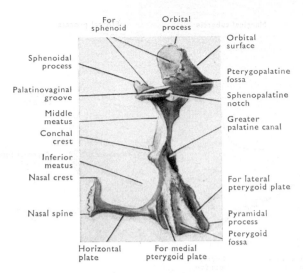

FIG. 177   Right palatine bone (posterior aspect).

THE ZYGOMATIC BONES

Each zygomatic or cheek bone lies below and lateral to the orbit in the most prominent part of the cheek. It connects the maxilla with the temporal bone behind and the frontal bone above by means of a horizontal temporal process and an ascending frontal process.

Its **lateral surface** is slightly convex. Its anterior border is the edge of its rough triangular articulation with the maxilla, and it meets the evenly curved orbital border which forms most of the lateral and inferior margins of the orbit at an acute angle. The smooth border joining the two processes is more sharply curved than the orbital margin. The free inferior border continues back on to the temporal process.

The **temporal process** extends back to a very oblique, serrated articulation with the zygomatic process of the temporal bone. The **frontal process** narrows sharply as it approaches its articulation with the frontal bone and this usually produces a **marginal tubercle** on its posterior border [FIG. 178].

The deep or **medial aspect** of the bone is separated into two concave surfaces by the rough, triangular, articular area for the maxilla and its extension upwards as a crest which articulates with the sphenoid. The more anterior smooth **orbital surface** can be identified by the orbital margin. The more irregular posterior area is the **temporal surface.** The named foramina are shown in FIGURES 178-9.

Ossification

The zygomatic bone ossifies in membrane from a centre which appears between the eighth and tenth weeks of intra-uterine life. The bone may be divided horizontally or vertically by a suture.

SUTURAL BONES

Sutural bones are isolated pieces of bone sometimes found in the sutures or in the region of the fontanelles. They arise from independent centres, may occur in any suture, and usually include the whole thickness of the bone. They are most

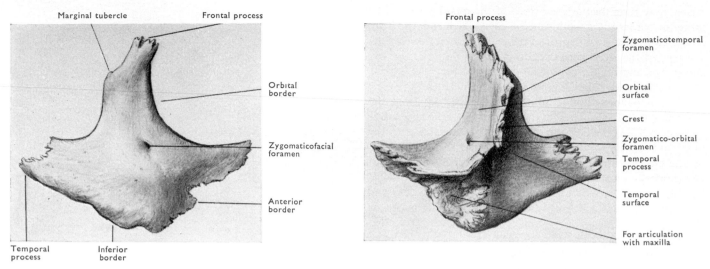

FIG. 178   Right zygomatic bone (lateral surface).

FIG. 179   Right zygomatic bone (medial surface).

frequent at lambda and in the lambdoid suture, but are also common in the region of pterion as **epipteric bones.** They are infrequent in the face but have been noted around the lacrimal bone and at the lateral end of the inferior orbital fissure where a nodule may be wedged in between the greater wing of the sphenoid, the zygomatic bone, and the maxilla.

THE BONES OF THE UPPER LIMB

The shoulder region is supported by the two bones of the shoulder girdle, which are the shoulder blade or **scapula** and the collar bone or **clavicle.** The clavicle acts as a prop for the scapula which is otherwise connected to the trunk only by muscular tissues and is freely movable. The **humerus** is the bone of the upper arm or brachium and the **radius** and **ulna** are the two bones in the forearm or antebrachium.

The hand or **manus** is divided into three parts. The first or proximal part is the wrist or **carpus** made up of eight small carpal bones arranged in two rows. It differs from the colloquial 'wrist' which refers to the distal part of the forearm bones. The second is the hand proper, including the palm and the back of the hand, which is supported by five **metacarpal bones,** largely concealed by soft tissues but clearly seen when they form the knuckles of the clenched fist. They are numbered, one to five, beginning on the lateral side with the thumb. The third or distal part is composed of five **digits** numbered, like the metacarpals, from the lateral side. The first digit is the thumb or **pollex;** the second is the forefinger or index; the third or middle finger is digitus medius; the fourth is the ring finger or anularis; the fifth or 'little finger' is digitus minimus. The bones of the digits are the fourteen **phalanges** of which **two** are in the thumb and three are in each 'finger'.

All the upper limb bones exhibit variations in size and degree of marking depending on sex and race but the differences are seldom reliable criteria for accurate diagnosis. They are often shorter, smoother, and more slender in women but the degree of muscular development in both sexes influences the bony characteristics. Rarely, the clavicle may be incompletely

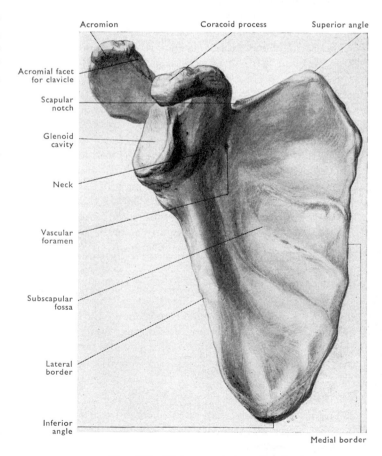

FIG. 180   Right scapula (costal surface).

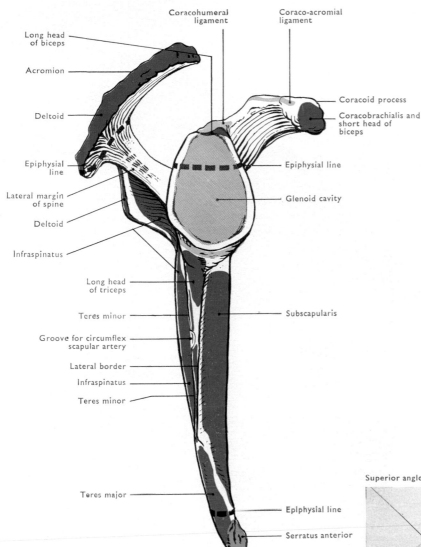

Labels for Fig. 181 (clockwise): Long head of biceps, Acromion, Deltoid, Epiphysial line, Lateral margin of spine, Deltoid, Infraspinatus, Long head of triceps, Teres minor, Groove for circumflex scapular artery, Lateral border, Infraspinatus, Teres minor, Teres major, Coracohumeral ligament, Coraco-acromial ligament, Coracoid process, Coracobrachialis and short head of biceps, Epiphysial line, Glenoid cavity, Subscapularis, Epiphysial line, Serratus anterior

FIG. 181 Right scapula (lateral aspect).

medial border at the **inferior angle** which is the lower end of the scapula.

The **medial border,** a thin edge, ascends to the superior angle, and is slightly angulated at the medial end of the spine. The levator scapulae muscle is attached above the root of the spine rhomboideus minor opposite the root of the spine, and rhomboideus major from there to the inferior angle. The **superior angle,** between the medial and superior borders, is nearly a right angle and the thin **superior border** passes almost horizontally towards the lateral angle. From the upper part of the head and neck a thick bent **coracoid process** projects forwards, and medial to the root of this process, a distinct **scapular notch** interrupts the superior border and is converted into a foramen by the superior transverse scapular ligament. The suprascapular nerve lies below the ligament, the artery lies above it, and the omohyoid muscle is attached to the ligament and the adjoining border.

The **costal surface** [FIGS. 180 and 183] has a smooth narrow strip along the entire medial border to which the serratus anterior muscle is attached. The rest of the surface is marked

developed or absent, as may the ulna or radius (sometimes together with the thumb). The number of carpal bones can vary and fusion of two or more is common in some races.

THE SCAPULA

The scapula is a triangular bone which lies obliquely over the second to the seventh ribs on the back and in the posterior wall of the axilla. Its lower end is easily felt and may be used as a landmark. The main flattened part is the body and is concave on its **costal surface,** convex and divided obliquely into two parts by a triangular **spine** on its **dorsal surface.** The spine is continuous laterally with a shelf of bone, the **acromion,** which turns forwards and overhangs the shoulder joint. The **lateral angle** can be identified because it is truncated and thickened to form a head. This has a concave articular surface, the **glenoid fossa** for the humerus and is continuous with the body through an ill-defined **neck** [FIG. 180]. A thickened **lateral border** runs down to meet the

Labels for Fig. 182: Superior angle, Scapular notch, Facet for clavicle, Acromion, Coracoid process, Supraspinous fossa, Acromial angle, Head and glenoid cavity, Neck, Spinoglenoid notch, Vascular foramen, Groove for circumflex scapular artery, Lateral border, Spine, Infraspinous fossa, Medial border, Inferior angle

FIG. 182 Right scapula (dorsal surface).

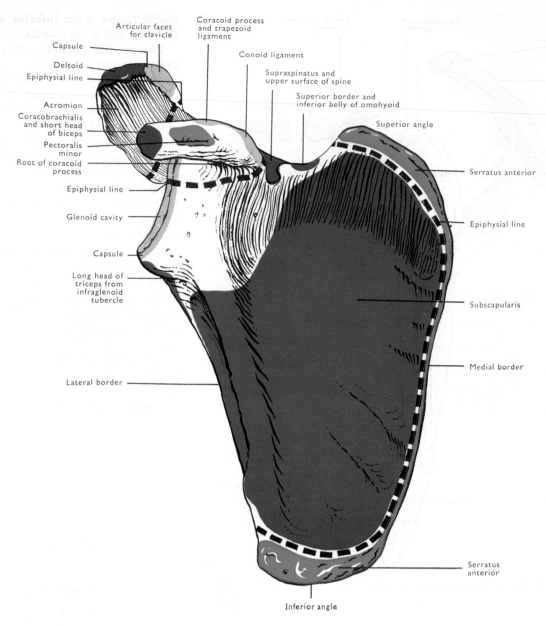

Articular facet
for clavicle

Capsule

Deltoid

Epiphysial line

Acromion

Coracobrachialis
and short head
of biceps

Pectoralis
minor

Root of coracoid
process

Epiphysial line

Glenoid cavity

Capsule

Long head of
triceps from
infraglenoid
tubercle

Lateral border

Coracoid process
and trapezoid
ligament

Conoid ligament

Supraspinatus and
upper surface of spine

Superior border and
inferior belly of omohyoid

Superior angle

Serratus anterior

Epiphysial line

Subscapularis

Medial border

Serratus
anterior

Inferior angle

FIG. 183   Right scapula (costal surface).

by the fibrous septa in the subscapularis muscle, and gives attachment to it except for a smooth area on and near the neck. A bursa from the joint lies between the muscle and the bone in this region.

The **dorsal surface** [FIGS. 182 and 184] is divided by the spine into a **supraspinous fossa** and an **infraspinous fossa** for the attachment of supraspinatus and infraspinatus muscles respectively. These muscles extend on to the sides of the spine, but leave a smooth area on and near the neck. A bursa usually lies under infraspinatus at the neck.

Between infraspinatus and the lower end of the lateral border is a smooth, oval area for the attachment of the teres major muscle. Inferior to this, latissimus dorsi gets a small attachment. Between teres major and the neck of the bone, are two areas for teres minor separated by a groove for the circumflex scapular vessels.

The **spine of the scapula** has two free borders, lateral and

posterior and where they come together laterally, it is continuous with the acromion. The lateral border is the rounded edge [FIG. 181] which curves towards the neck of the bone and makes with it a **spinoglenoid notch** through which vessels and nerves pass to the infraspinous fossa.

The posterior border (or crest) is flattened, readily palpable, and makes edges for the supraspinous and infraspinous fossae. At its medial end, covered by fibres of trapezius, it forms a smooth triangular area which is easily felt and is usually at the level of the third thoracic vertebral spine. The upper edge of the posterior border becomes continuous with the medial border of the acromion and both give attachment to part of trapezius which marks the surface as well as the edge of the bone.

Near its medial end, the inferior edge projects as a tubercle and the lower fibres of trapezius run down from this area [FIG. 184]. The inferior edge becomes continuous with the

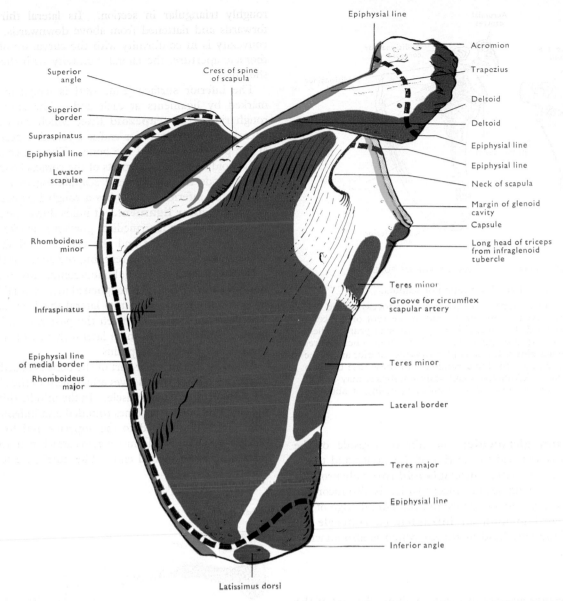

Epiphysial line
Acromion
Trapezius
Deltoid
Deltoid
Epiphysial line
Epiphysial line
Neck of scapula
Margin of glenoid cavity
Capsule
Long head of triceps from infraglenoid tubercle
Teres minor
Groove for circumflex scapular artery
Teres minor
Lateral border
Teres major
Epiphysial line
Inferior angle

Superior angle
Crest of spine of scapula
Superior border
Supraspinatus
Epiphysial line
Levator scapulae
Rhomboideus minor
Infraspinatus
Epiphysial line of medial border
Rhomboideus major
Latissimus dorsi

FIG. 184   Right scapula (dorsal surface).

lateral border of the acromion, and (lateral to the tubercle) both give attachment to deltoid muscle. Where the edge of the spine becomes continuous with the lateral border of the acromion is an easily felt **acromial angle.**

The **acromion** slopes up to its tip from the acromial angle [FIG. 182]. From the tip, the coraco-acromial ligament fans out to reach the lateral border of the coracoid process and complete the coraco-acromial arch. A small, oval, articular facet for the clavicle lies on the anterior part of the medial border of the acromion. The facet faces slightly upwards and medially, and the clavicle projecting above it makes the joint easily felt through the skin. The acromion and the posterior border of the spine can be felt in their whole length between the attached muscles.

The **coracoid process** has a broad base which is directed upwards and forwards, and a narrow horizontal part which passes forwards, laterally, and slightly downwards from the upper edge of the base. Its tip can be felt inferior to the junc-

tion of middle and lateral thirds of the clavicle, deep to the edge of the deltoid muscle.

The concave inferior surface of the coracoid process is smooth where a bursa intervenes between it and the subscapularis tendon, and the lateral margin of the upper edge of the base provides attachment for the coracohumeral ligament. Coracobrachialis and the short head of biceps come from its tip. The horizontal part is slightly flattened and its lateral margin receives the wide end of the coraco-acromial ligament. The medial margin in its anterior half gives attachment to pectoralis minor muscle, and the clavipectoral fascia reaches the rest of that border. The posterior part of the upper surface of the coracoid (**conoid tubercle** and **trapezoid line**) is marked by strong coracoclavicular ligaments.

The shallow articular **glenoid cavity** is pear-shaped with the narrow end uppermost. Its anterior border is slightly notched where developmentally separate parts of the bone join. The glenoid fossa has its depth slightly increased by a fibrous

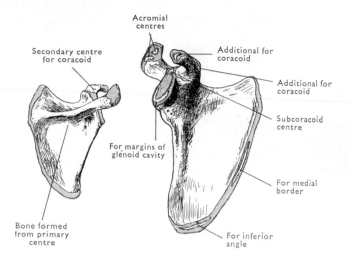

FIG. 185   Ossification of scapula.

Secondary centres appear in the coracoid process in the first year (sometimes even before birth), and in the root of the coracoid and upper third of the glenoid cavity at 10 or 11 years. These fuse during or just after puberty, at which time other centres appear for the margins of the glenoid cavity, the inferior angle, the medial border, and two for the acromion; all these fuse between 20 and 25 years. Additional small scaly epiphyses may appear on the surface of the coracoid after puberty, fusing at about 20 years.

lip, the **labrum glenoidale;** the fibrous capsule of the shoulder joint is attached to the rim of the cavity and to the labrum. Above the cavity, a **supraglenoid tubercle** provides attachment for the long head of biceps which is also incorporated in the labrum. Below the cavity, the lateral edge of the bone is thickened to form an **infraglenoid tubercle** for attachment of the long head of triceps which is also attached to the labrum.

Ossification

A primary centre appears in cartilage about the end of the eighth week and at birth the acromion, coracoid process, glenoid cavity, inferior angle, and medial border are still cartilaginous. The times of appearance and fusion of the numerous secondary centres are given in FIGURE 185.

Variations

The acromion may remain separate throughout life, united to the spine by cartilage or by a synovial joint. The coracoid process may remain separate but this is much less frequent. In old people the bone may be absorbed in parts, though the periosteum remains. In infants the scapula is relatively wider than in the adult.

THE CLAVICLE

The clavicle is the curved rod of bone which lies almost horizontally on each side of the root of the neck [FIG. 104]. It extends from the sternum to the acromion and serves as a strut to prevent the shoulder falling forwards and downwards. The part forming its medial two-thirds is convex forwards and

roughly triangular in section. Its lateral third is concave forwards and flattened from above downwards. The medial convexity is in conformity with the curvature of the superior thoracic aperture, the lateral concavity with the shape of the shoulder.

The inferior surface [FIG. 187] is irregular and strongly marked by ligaments at each end. The lateral part has a rough ridge, the **trapezoid line,** which runs forwards and laterally. At its posterior end is a rounded **conoid tubercle** at the posterior edge of the bone. These give attachment to the conoid and trapezoid parts of the coracoclavicular ligament which binds the clavicle and scapula together. The medial part of the inferior surface has a rough **impression for the costoclavicular ligament.** It holds down the medial end of the clavicle. The intermediate portion is marked by a groove for subclavius muscle and the clavipectoral fascia splits to enclose subclavius before reaching the edges of the groove.

Looked at from above, the bone is comparatively smooth. The sinuous anterior border is roughened in its medial half, or more, by pectoralis major and in its lateral third by deltoid muscle. The posterior border, between the posterior and inferior surfaces, is smooth except in its lateral third which is marked by the anterior fibres of trapezius.

Medially, the superior part of the bone is raised so as to form a superior border which gives attachment to the clavicular head of sternocleidomastoid muscle. In the middle third of the bone the superior border becomes rounded and indefinite. Cervical deep fascia is attached to the superior and to the posterior borders. Sternohyoid muscle gains attachment to the posterior surface near the medial end. The nutrient artery enters the

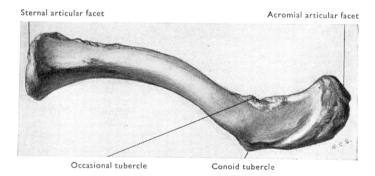

FIG. 186   Right clavicle (superior surface).

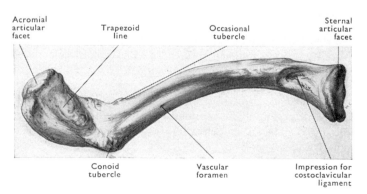

FIG. 187   Right clavicle (inferior surface).

posterior surface at the junction of middle and lateral thirds, and the curvature of the clavicle in this region allows the brachial plexus and subclavian vessels to pass freely between it and the first rib.

The lateral or **acromial end** has an oval articular facet for the acromion process and is set a little obliquely, facing downwards and laterally to fit the articular disc between it and the acromial facet. The capsule of the joint is attached to the margin of the facet and is partly responsible for the ridge which marks the position of the joint.

The medial or **sternal end** is enlarged. The inferior three quarters are bevelled off where it moves on an articular disc, separating it from the clavicular notch on the sternum and on the first costal cartilage. The superior quarter is rough for attachment of the disc, which acts as a ligament, and for the interclavicular ligaments. The articular capsule is attached to the edge of the facet. Since the notch on the sternum is shallow, the medial end of the clavicle stands up from it and is easily felt.

The double curvature of the body of the bone is a source of weakness when it is compressed lengthwise as may occur in a fall on the shoulder or outstretched arm.

Ossification

The clavicle is the first bone to begin ossification. Two primary centres appear during the fifth embryonic week in mesoderm which is still in the procartilage stage. When procartilage appears in the intervening tissue the two centres unite and ossification then spreads into the formed cartilage nearer the ends. A secondary centre appears in the cartilage at the medial end [FIG. 190], and is often the last to unite.

Variations

Its outer end may be raised or depressed according to muscular development or position of the body. Occasionally one of the supraclavicular nerves pierces the clavicle. Macerated bones often have the medial epiphyses missing and have a flat, roughened appearance.

THE HUMERUS
[PLATES 11 and 12]

The humerus extends from the scapula to the elbow joint and has a **body** and two ends. It is almost covered with muscles but can be felt through them in all its length. The upper end includes an almost hemispherical **head** and two distinct **tubercles** with the deep **intertubercular groove** between them [FIG. 191].

The smooth, rounded surface of the head articulates with the shallow and smaller glenoid cavity on the scapula and may be felt indistinctly if the fingers are pushed well up into a relaxed axilla. With the arm by the side, the articular surface faces medially and upwards and also slightly backwards to meet the glenoid surface which is turned slightly forwards [FIG. 192].

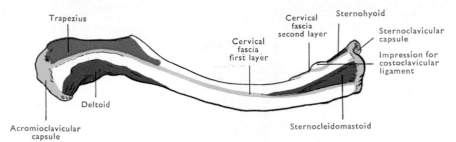

FIG. 188   Right clavicle (superior surface).

FIG. 189   Right clavicle (inferior surface).

FIG. 190   Ossification of clavicle.

A secondary centre appears in cartilage at the medial end from about 14 to 18 years : it fuses as early as 18 to 20 years in females and about 23 to 25 years in males. A small epiphysis may appear at the acromial end about puberty but it soon fuses.

The head is joined to the upper end by the **anatomical neck,** a slightly constricted region which encircles the bone at the edge of the articular surface and separates it from the tubercles. The **surgical neck** is the region below the head and tubercles where they join the shaft, and where fractures often occur. The capsule of the shoulder joint is attached to the anatomical neck above but extends down 1·5 cm. on to the surgical neck inferiorly [FIG. 195].

The **greater tubercle** is the prominence on the lateral side of the upper end of the bone. It merges with the body below, and can be felt through deltoid muscle a little below the edge of the acromion process. The tendons of the supraspinatus and infraspinatus are inserted into impressions on its superior aspect, and teres minor posteriorly [FIG. 198]. Teres minor also extends down on to the surgical neck by fleshy fibres which leave little impression.

The **lesser tubercle** is the distinct prominence on the front of the upper end of the bone. It can be felt through the deltoid muscle just lateral to the tip of the coracoid process [p. 149], moving further away from it when the arm is rotated laterally. The medial side of the tubercle has a well-marked impression

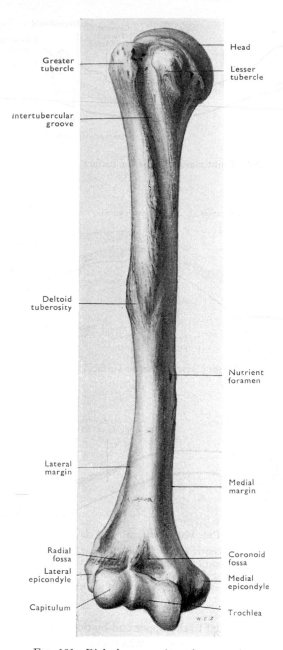

FIG. 191   Right humerus (anterior aspect).

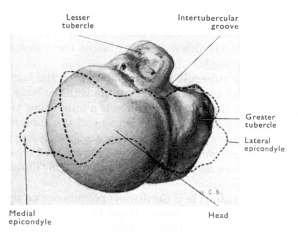

FIG. 192   Proximal aspect of right humerus (with the outline of lower end in relation to it in dotted line).

on it for the tendon of subscapularis muscle. The muscle insertion is continued down on to the surgical neck.

The **intertubercular groove** passes on to the body between the **crests of the greater and lesser tubercles** which continue down from the anterior borders of the tubercles to form lateral and medial lips to the groove. The tendon of the long head of biceps brachii, surrounded by its synovial sheath, lies in the groove with small vessels passing up to the joint. It is held down by a transverse ligament running between the two tubercles.

Pectoralis major crosses over to reach the crest of the greater tubercle, teres major is inserted into the crest of the lesser tubercle, and latissimus dorsi curves round the latter to reach the floor of the groove [FIG. 195].

The **body** is triangular with distinct **medial and lateral borders** in its lower part. From the front, two **surfaces** are visible, anteromedial and anterolateral, separated by a smooth, rounded, median ridge which divides below and is continuous with the crest of the greater tubercle above. The intertubercular groove is thus in continuity with the anteromedial surface and the **medial border** begins as the crest of the lesser tubercle. It ends below by curving medially to the prominent medial epicondyle, and its lower third is referred to as the medial supracondylar ridge or line. Half-way down, the medial border is roughened by the insertion of coracobrachialis muscle [FIG. 195].

The **lateral border** is prominent in its lower third where it runs down to the lateral epicondyle as the lateral supracondylar ridge or line. Above this it is interrupted by the smooth **groove for the radial nerve** which passes downwards and forwards. The groove is limited above and in front by a part of the **deltoid tuberosity**. The remainder of the lateral border continues up to the posterior border of the greater tubercle and gives attachment to the lateral head of the triceps muscle [FIG. 197].

The borders of the body give attachment to the corresponding intermuscular septa which separate the anterior and posterior compartments of the arm and give an increased area of attachment to adjacent muscles.

The anteromedial and anterolateral surfaces of the body give origin to fleshy fibres of brachialis muscle in the lower half. The smooth anterolateral surface has a rough **deltoid tuberosity** about its middle. [FIG. 191]. This insertion of deltoid muscle can be felt half-way up the lateral aspect of the arm. The anterior surface of the lateral supracondylar line gives attachment to brachioradialis muscle in its upper two-thirds and to extensor carpi radialis longus in its lower third.

The **posterior surface** of the body is marked by a spiral **groove for the radial nerve** which runs obliquely across the upper half of the body to reach the lateral border below the deltoid tuberosity, but it is often poorly marked. The radial nerve and profunda brachii vessels lie in the groove. The body between the groove and the inferior end of the bone widens out below and is smooth because it gives attachment to the fleshy fibres of the medial head of triceps brachii [FIG. 197].

The **nutrient foramina** are usually situated anterior to the middle of the medial border and in the upper part of the groove for the radial nerve.

The **lower end** of the humerus is wide, flattened anteroposteriorly, and bent slightly forwards. Approximately the middle third of its distal edge articulates with the ulna and is occupied by a pulley-shaped **trochlea** with the bone above its centre thin or even defective so that the trochlea is nearly circular in sagittal section. The fossae which cause the thinning are the **coronoid** in front and the **olecranon** behind, into which the corresponding parts of the ulna fit during extreme flexion or extension. The medial lip of the trochlea stands out as a prominent rim and its medial surface is non-articular.

Lateral to the trochlea, a rounded **capitulum** for articulation with the radius is continuous with the anterior and distal aspects of the trochlea but is not continued on to the posterior surface of the bone. A small **radial fossa** lies above it anteriorly. Behind the capitulum, the bone is roughened and projects laterally as the **lateral epicondyle.** Its posterior surface is smooth, subcutaneous, and easily felt; its distal aspect is marked by the radial collateral ligament with anconeus attached posterior to it [FIG. 198]. The anterior surface is marked by the common extensor tendon of the forearm muscles and above this the extensor carpi radialis longus continues the line of attachments up on to the supracondylar part of the lateral border.

The **medial epicondyle** is very prominent, partly because the supracondylar part of the medial border is less prominent than the lateral one. Its posterior surface is smooth and has on it a shallow **sulcus for the ulnar nerve** and collateral vessels. Its distal margin gives attachment to the ulnar collateral ligament and, in front of this, the anterior surface has an impression for the common flexor tendon, while pronator teres muscle comes from the flatter upper part and from the adjacent lower end of the medial border [FIG. 195].

The anterior part of the fibrous capsule is attached to the upper margins of the coronoid and radial fossae. The posterior part crosses the olecranon fossa. Between these it is attached to the epicondyles close to the trochlea and capitulum in front, and close to each side of the trochlea behind.

Ossification

A primary centre appears in the body in the eighth week and spreads until, at birth, only the ends are cartilaginous [FIG. 199]. Secondary centres for the head, greater tubercle, and lesser tubercle may ossify by spreading from the centre in the head of the bone (Paterson, 1929) to form a single epiphysis. Details of appearance and fusion of secondary centres are given in FIGURE 199. Note that the centre for the medial epicondyle is entirely outside the joint capsule, and does not fuse with the adjacent epiphysis directly. Much variation can occur in times of fusion: most of the growth in length of the humerus occurs at its upper end.

Variations

The only distinct variation is the presence of a **supracondylar process.** This is found occasionally 3–4 cm. above the medial epicondyle. It is a hooked process connected to the epicondyle by a fibrous band under which the median nerve and brachial vessels pass. The ligament may be ossified, a condition normally found in Carnivora.

FIG. 193   Right humerus (distal aspect).

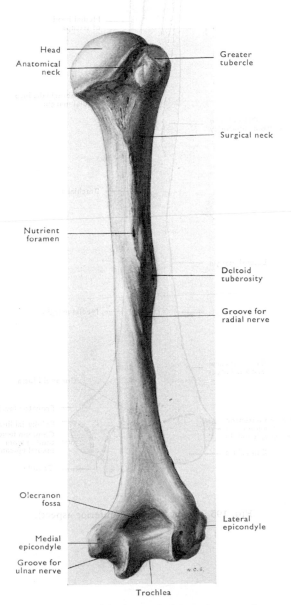

FIG. 194   Right humerus (posterior aspect).

FIG. 195   Right humerus (anterior aspect).

FIG. 196   Right humerus (medial aspect).

Capsule

Head

Synovial
area

Infraspinatus

Capsule
Greater tubercle

Epiphysial line

Teres minor

Lateral head of
triceps from
lateral border

Deltoid

Groove for
radial nerve

Brachialis

Medial head
of triceps

Medial
margin

Lateral
margin

Epiphysial
line

Medial
epicondyle

Epiphysial
line

Olecranon fossa

Epiphysial line

Lateral epicondyle

Anconeus

Trochlea

FIG. 197   Right humerus (posterior aspect).

Supraspinatus

Infraspinatus

Capsule
Greater tubercle

Teres minor

Head

Subscapularis

Lesser tubercle

Epiphysial line

Crest of
greater tubercle

Crest of
lesser tubercle

Pectoralis
major

Lateral head
of triceps

Deltoid into
deltoid tuberosity

Groove for
radial nerve

Brachialis

Medial head
of triceps

Brachioradialis

Lateral margin

Extensor carpi
radialis longus

Epiphysial line

Common extensor tendon

Anconeus

Synovial area

Trochlea

Capitulum

Capsule

FIG. 198   Right humerus (lateral view).

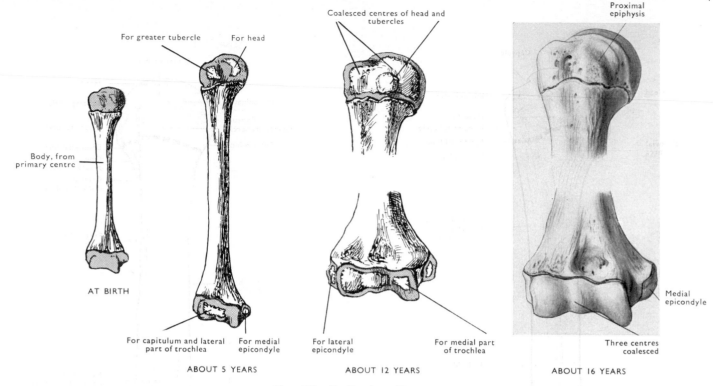

FIG. 199 Ossification of humerus.

Secondary centres appear for the head (early in first year), the greater tubercle, (about 3 years), and lesser tubercle (about 5 years): these join into a single cap of bone at 6 to 8 years and fuse at 18 to 20 years (female) and 20 to 22 years (male).

In the lower end secondary centres appear for the capitulum (2nd year), trochlea (9 to 10 years) and lateral epicondyle (12 to 14 years): these join together at about 14 years and fuse with the shaft about 15 years (female) and 18 years (male). A separate centre forms for the medial epicondyle at 6 to 8 years and fuses at 15 to 18 years with a spicule of the shaft projecting down medial to the trochlea (see third figure above).

THE ULNA
[PLATES 12 and 13]

The ulna is the medial bone of the two in the forearm. It is a long bone with a body, an upper, and a lower end. It can be felt in its whole length down the back of the forearm.

The **upper end** is large with two projecting processes enclosing a concavity [the **trochlear notch**; FIG. 200] which fits on to the trochlea of the humerus.

The **olecranon**, the larger of the two processes, forms the proximal end of the bone. It is continuous with the body and makes the point of the elbow which is felt at the back of the elbow joint. Posteriorly it is smooth and subcutaneous and this area extends down on to the body. The proximal surface extends forwards and is pointed in front. Posteriorly it is marked by the strong tendon of triceps muscle; anteriorly is the attachment of the capsule of the elbow joint, and between these is a smooth area for a bursa.

The anterior surface of the olecranon forms part of the articular surface of the trochlear notch and is divided by a rounded ridge which fits into the groove on the trochlea of the humerus. The borders are thick and rough, the lateral for the attachment of anconeus muscle and the capsule of the elbow joint [FIG. 205]; the medial, for flexor carpi ulnaris, flexor digitorum profundus and the ulnar collateral ligament.

The **coronoid process** projects from the front of the body

and has an articular upper surface which completes the trochlear notch and is divided by a ridge into a larger medial part and a smaller lateral part. The lateral part is continuous over its lateral edge with the articular surface of the radial notch on the lateral surface of the process. The posterior part of the articular surface in the trochlear notch is usually separated from the articular surface on the olecranon by a rough strip.

Where the medial and anterior edges of the articular surface on the coronoid process meet, is a small tubercle to which the anterior part of the ulnar collateral ligament is attached, and from which the medial margin of the process extends down, giving attachment to the flexor digitorum superficialis, pronator teres, and flexor pollicis longus muscles [FIG. 203].

The irregular, anterior surface of the coronoid process ends inferiorly in a rough **tuberosity of the ulna.** Both the surface and the tuberosity provide attachment for brachialis muscle.

On the lateral side of the process the **radial notch** lodges the head of the radius and has the anular ligament attached to its anterior and posterior edges, while the quadrate ligament is inferior to it. Below this and extending on to the body, is a triangular fossa which gives origin to supinator muscle and is bounded posteriorly by a distinct **supinator crest.**

The body of the ulna is triangular in section and tapers down to a slender rounded part which widens again a little to form the **head.**

Olecranon

Trochlear notch

Coronoid process

Radial notch

Tuberosity

Supinator fossa

Posterior border

Interosseous border

Head

Articular circumference
for radius

Groove for extensor
carpi ulnaris

Styloid process

FIG. 200   Right ulna (lateral aspect).

profundus which arises from their upper two-thirds and also from the medial surfaces of both olecranon and coronoid processes [FIG. 203]. An oblique ridge extending downwards and medially marks the lower quarter of the anterior surface where pronator quadratus arises [FIG. 203].

The **posterior surface,** between the interosseous and posterior borders, is divided into areas for muscle attachment. Anconeus is attached above an oblique ridge extending downwards and posteriorly from the back of the radial notch to the posterior border, and this area of attachment extends up on to the olecranon [FIG. 205]. The remainder of the surface is divided into a smooth medial area, covered by extensor carpi ulnaris muscle, and a lateral area which has three faint ridges marking off the origins of abductor pollicis longus, extensor pollicis longus, and extensor indicis muscles [FIG. 205].

The **lower end** of the ulna consists of a small, rounded head and a conical styloid process which projects downwards.

The **head** is the widened lower end of the body, and is the knob felt on the back of the wrist when exposed by the movement of the radius into full pronation. It has an articular surface for the radius round its anterior and lateral aspects, but is smooth medially. Posteriorly, it is separated from the styloid process by a groove which lodges the tendon of extensor carpi ulnaris muscle. The distal surface of the head is smooth and nearly flat, and articulates with an articular disc which intervenes between the head and the os triquetrum of the carpus. The disc is approximately triangular in shape and is attached by its apex to a pit between the head of the ulna and the base of the styloid process.

The **styloid process** projects downwards from the posteromedial part of the distal end of the bone, and its tip is fully 1 cm. proximal to the corresponding part of the radius. It gives attachment to the ulnar collateral ligament of the carpus.

Ossification

The body, coronoid process, and nearly all of the olecranon are ossified from a primary centre, and there are secondary

Its prominent lateral edge is the **interosseous border** where the interosseous membrane is attached. It runs down from the apex of the fossa for supinator muscle and is prominent in its upper part. The **anterior border** runs down from the medial margin of the coronoid process but is indefinite. The **posterior border** is sinuous, subcutaneous, and prominent in its upper part. It is continuous above with the apex of the subcutaneous triangle on the olecranon and upper part of the shaft [FIG. 204], and the aponeurosis of flexor carpi ulnaris is attached to its upper three-quarters and to the medial margin of the triangle. Extensor carpi ulnaris is attached to the posterior border on the lateral side.

The anterior and medial surfaces are continuous at the rounded anterior border, and are covered by flexor digitorum

Fuses 16th-19th year

Appears in 11th year

Appears in 5th year

Appears 8th week
of intra-uterine life

Fuses 20th-22nd year

AT BIRTH   ABOUT 12 YEARS   ABOUT 16 YEARS

FIG. 201   Ossification of ulna.

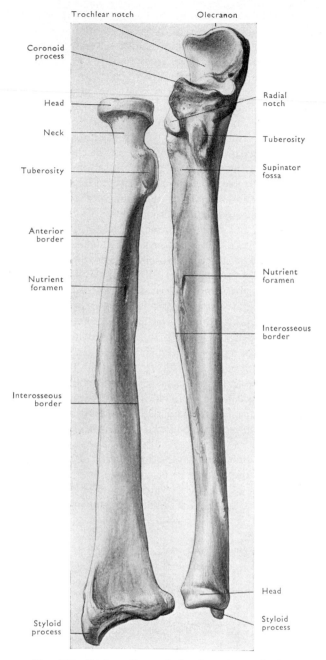

FIG. 202   Right radius and ulna (anterior surface).

centres for the head of the ulna and for the top of the olecranon. The head usually fuses last and is at the 'growing' end. [See FIG. 201 for details.]

Variations

The olecranon may have additional centres; the epiphysis varies considerably in size, and the epiphysial line may pass through the articular part of the olecranon.

THE RADIUS

[PLATES 12 and 13]

The radius lies lateral to the ulna. It extends from the humerus to the carpus but, since it does not overlap the humerus, it is shorter than the ulna. It has a body and two

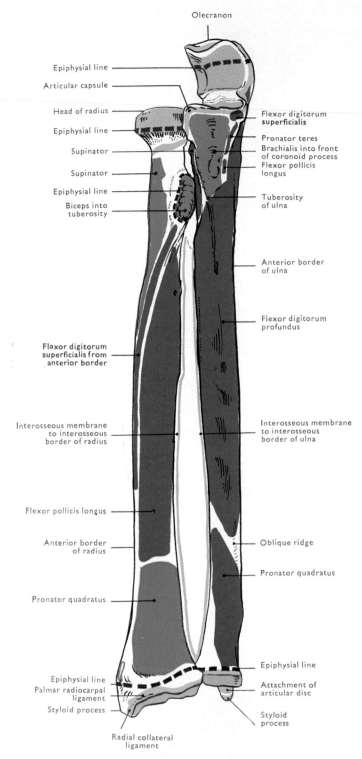

FIG. 203   Right radius and ulna (anterior aspect).

ends of which the inferior one is the larger. The superior end consists of a head which narrows abruptly to join the body by a cylindrical neck.

The **head** is a thick disc with a concave superior surface for articulation with the capitulum of the humerus. Its **articular circumference** is encircled by the radial notch of the ulna with the anular ligament which holds it against the notch while still leaving it free to rotate. The anterior, medial, and posterior parts of the edge bear on the ulna and are distinctly articular.

The remainder is narrower and smooth where it bears on the ligament.

The **neck** is smooth and slopes a little medially as it approaches the body. The anular and quadrate ligaments are attached loosely to the neck so as to permit free rotation.

The **body** is round where it joins the neck but becomes triangular in section lower down. Along with the neck it is curved with a slight medial convexity in its upper fifth and has a lateral convexity in its remaining lower part. The **radial tuberosity** lies on the medial aspect of the uppermost part of the body, at the maximum convexity of the medial curve. This provides leverage for the biceps tendon which is inserted into its posterior part and produces lateral rotation of the radius. The anterior part of the tuberosity is smooth where a bursa intervenes between the tendon and the bone [FIG. 301]. Pronator teres, which has the opposite action, is inserted into a rough mark on the maximum convexity of the lateral curvature, half-way down the lateral surface of the body [FIG. 205].

The **interosseous border** is the prominent edge which extends from just below the tuberosity to the medial side of the inferior extremity of the bone. In its lower part it splits into two diverging ridges which limit the anterior and posterior surfaces of the end of the bone and become continuous with anterior and posterior margins of the ulnar notch. The interosseous membrane is attached to the latter and to the border. The oblique cord passes superomedially from the bone anterior to the upper end of the membrane [FIG. 302].

The **anterior** and **posterior borders** pass in front of and behind the rough mark for the pronator teres on the lateral side of the body. They become more distinct above this level as two oblique lines which ascend superomedially to the anterior and posterior parts of the tuberosity [FIGS. 202 and 204]. The rounded lateral surface thus encroaches on the anterior and posterior aspects of the body above the impression for pronator teres muscle. The supinator muscle is attached here after passing behind the radius from the ulna [FIG. 203]: the insertion scarcely marks the bone.

The **lateral surface** of the body has no attachments below the pronator teres and the posterior border is very indistinct here. The anterior border becomes quite distinct at the inferior end of the bone. It is accentuated in its upper half by the attachment of the radial head of flexor digitorum superficialis.

The **anterior surface** of the body lies between the anterior and interosseous borders and is continous inferiorly with the anterior end of the bone. Its upper three-quarters provide attachment for flexor pollicis longus muscle. Its lower part receives the insertion of pronator quadratus which also reaches the anterior surface of the end of the bone [FIG. 203]. Neither of these muscles marks the bone appreciably.

The **posterior surface** is flatter and merges with the lateral surface in its inferior half where the overlying tendons give the bone a smooth, rounded appearance. In its middle third it gives part origin to abductor pollicis longus muscle and extensor pollicis brevis [FIG. 205].

The **inferior end** has an **anterior surface** which curves forwards to a distinct anterior margin which can be felt about 2·5 cm. above the ball of the thumb. The margin gives attachment to the palmar radiocarpal ligament and is a guide to the position of the wrist joint. The **lateral surface**, about 1 cm.

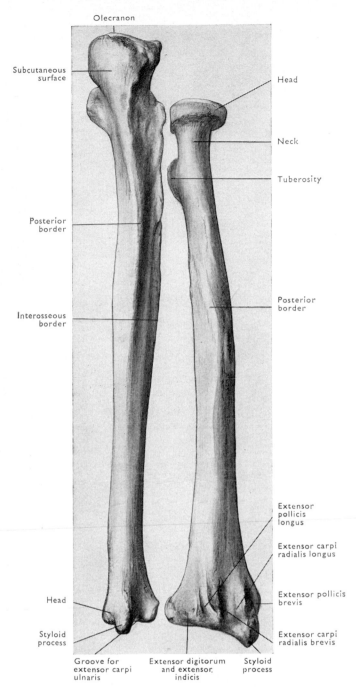

FIG. 204   Right radius and ulna (posterior surface).

wide, is a shallow groove for the tendons of abductor pollicis longus and extensor pollicis brevis, with the tendon of brachioradialis inserted deep to them. The edge between this groove and the anterior surface is sharp and easily felt. It gives attachment to the end of the extensor retinaculum which binds down the tendons.

The lateral part of the inferior end of the radius is prolonged downwards as a thick **styloid process** which is palpable at the lateral side of the wrist in the **anatomical snuff box.** The apex of this styloid process is 1 cm. lower than that of the ulna. Absence of this feature indicates shortening of the radius, e.g. in fractures. The styloid process gives attachment to the radial collateral ligament of the carpus.

Olecranon

Epiphysial line

Subcutaneous surface
of olecranon

Anconeus

Supinator from
supinator crest

Abductor pollicis
longus

Posterior border
of ulna

Extensor pollicis
longus

Flexor digitorum profundus
from medial surface

Posterior surface
of ulna

Extensor indicis

Anterior border

Interosseous membrane
to interosseous border

Pronator quadratus
Groove for extensor
carpi ulnaris tendon
Epiphysial line

Epiphysial line and groove for extensor digiti minimi
Styloid process and ulnar collateral ligament
Groove for extensor digitorum
and extensor indicis

Capsule   Groove for
extensor
pollicis longus

Triceps

Anular ligament

Head of radius
Epiphysial line

Supinator

Epiphysial line
of tuberosity

Interosseous membrane
to interosseous border

Abductor pollicis longus

Posterior border of radius

Pronator teres into
lateral surface

Extensor pollicis brevis

Posterior surface
of radius

Lateral surface

Brachioradialis

Groove for abductor pollicis longus
and extensor pollicis brevis
Groove for extensor
carpi radialis longus
Styloid process

Groove for extensor
carpi radialis brevis

FIG. 205   Right radius and ulna (posterior surface).   The attachment of flexor and extensor carpi
ulnaris to the posterior border of ulna is not shown.

PLATE 11

A

B

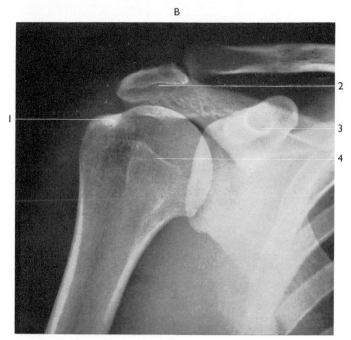

A. Anteroposterior radiograph of the shoulder. Age 16.

1. Epiphysis for the acromion.
2. Epiphysial cartilage of the head of the humerus.
3. Clavicle.
4. Epiphysis for the coracoid process.
5. Margin of the glenoid cavity.

B. Anteroposterior radiograph of the shoulder. Adult.

1. Site of the insertion of the supraspinatus tendon.
2. Acromion.
3. Coracoid process.
4. Line of epiphysial fusion.

C

D

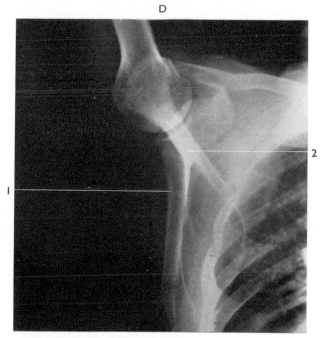

C. Oblique radiograph of the shoulder joint.

1. Crest of lesser tubercle.
2. 'Joint space', representing the total thickness of the articular cartilages of humeral head and glenoid cavity.

D. Anteroposterior radiograph of the shoulder in full abduction.

1. Lateral border of the scapula which has rotated laterally.
2. Spine of the scapula.

PLATE 12

A

A. Anteroposterior radiograph of elbow joint.   Age 10.

1. Epiphysis for medial epicondyle of humerus.
2. Epiphysis for lateral epicondyle of humerus.
3. Epiphysis for head of radius.

B

B. Anteroposterior radiograph of adult elbow joint.

1. Medial epicondyle of humerus.
2. Proximal radio-ulnar joint.
3. Olecranon and coronoid fossae.
4. Lateral epicondyle of humerus.
5. Head of radius.

C

C. Lateral radiograph of elbow joint.   Age 10.

1. Epiphysis of head of radius.
2. Epiphysis of capitulum.
3. Epiphysis of olecranon.

D

D. Lateral radiograph of adult elbow joint.

1. Coronoid process overlapped by head of radius.
2. Coronoid fossa.
3. Medial epicondyle.
4. Olecranon.

PLATE 13

A

B

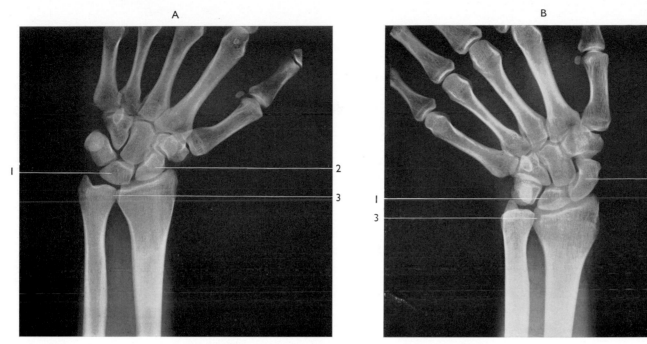

A and B.   Anteroposterior radiographs of the wrist in full abduction (A) and adduction (B).

1. Lunate.   2. Scaphoid.   3. Distal radio-ulnar joint.

C

(i)   (ii)

D

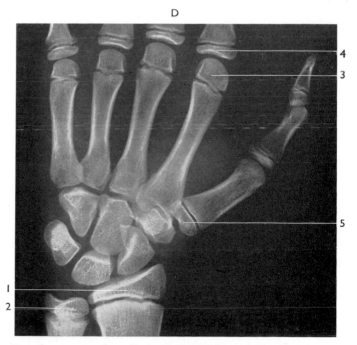

C.   Lateral radiographs of the wrist joint. (i) Straight and (ii) in full flexion.   Note the range of movement of the lunate on the radius.

1. Lunate.   2. Radius.

D.   Anteroposterior radiograph of the wrist and carpus.   Age 12.

1. Epiphysis of radius.
2. Epiphysis of ulna.
3. Epiphysis of second metacarpal.
4. Epiphysis of proximal phalanx of index finger.
5. Epiphysis at proximal end of first metacarpal.

PLATE 14

A

A.   Anteroposterior radiograph of pelvis.   Age 17.

    1. 3rd lumbar vertebra.
    2. Epiphysis of head of femur.
    3. Epiphysis of greater trochanter.

B

B.   Anteroposterior radiograph of adult pelvis.

    1. Gas in caecum overlying ilium.
    2. Sacro-iliac joint.
    3. Symphysis pubis.
    4. Obturator foramen.

C

C.   Oblique radiograph of hip joint.   Age 17.

    1. Margins of epiphysis of head of femur.
    2. Triradiate strip of cartilage (consolidating).
    3. Spine of ischium.

D

D.   Anteroposterior radiograph of hip joint.   Age 4.

    1. Epiphysis of head of femur.
    2. Epiphysial cartilage of head of femur.
    3. Triradiate cartilage of acetabulum.

Note that the angle between the neck and the shaft of the femur
is greater than in the adult.

The **medial surface** forms the concave **ulnar notch** for articulation with the head of the ulna. It is shaped to allow movement of the radius round the ulna and its edges give attachment to the capsule of the joint except inferiorly. The inferior edge has the base of the articular disc attached to it, between the radio-ulnar and radiocarpal surfaces of the radius.

The posterior surface of the lower end of the radius is convex but is much grooved by tendons. Its lateral half continues down on to the styloid process and has two shallow grooves for the tendons of extensor carpi radialis longus, laterally and brevis, medially. The prominent ridge in the middle of the posterior surface projects as a landmark, the **dorsal radial tubercle,** which can be felt above the back of the wrist nearly in line with the forefinger. The ridge and tubercle have the extensor retinaculum attached to them, and lie lateral to the tendon of extensor pollicis longus which changes direction here to reach the thumb, and forms a clear-cut, slightly oblique groove on the tubercle. Medial to this is a wide shallow groove for the extensor digitorum and indicis tendons. The extensor digiti minimi lies over the joint between the radius and ulna and extensor carpi ulnaris lies in the groove between the head and styloid process of the ulna [FIG. 205].

The distal or **carpal articular surface** is concave to fit against the carpal bones. It extends on to the styloid process and is divided by a ridge into two areas; a triangular lateral area for the scaphoid bone, and a more quadrilateral medial one for the lunate bone. In life the articular surface is continuous medially with that of the articular disc separating the os triquetrum from the head of the ulna. The forward curve of the inferior end of the bone brings its posterior border to a more distal level than the anterior border.

Ossification

The radius is ossified from a primary centre and at birth only the head, the lower end, and the tuberosity are cartilaginous. Secondary centres appear first in the lower 'growing' end and later in the head. A centre may appear for the tuberosity but soon fuses with the body. [See FIG. 206 for dates.]

SUPINATION AND PRONATION

[p. 232]

The normal anatomical position of the forearm is that of supination in which the radius and ulna are parallel. The back of the hand is posterior, the palm is anterior and the thumb is directed laterally. In full pronation the hand is turned so as to reverse its position by rotating the upper end of the radius and swinging its lower end round the ulna. The radius and ulna thus become crossed. The slight supination present when the limb hangs naturally is masked by a medial rotation of the humerus.

THE CARPUS

[PLATE 13]

The eight small bones of the wrist are referred to collectively as the **carpus** and individually as the carpal bones. They articulate with each other and are bound together by ligaments so as to form a compact mass which is curved with a posterior convexity and a pronounced anterior concavity, the **carpal sulcus,** which is converted into a **carpal canal** or tunnel by the flexor retinaculum. The proximal surface of the carpus is rounded for articulation with the radius, and with the articular disc which extends from radius to ulna, to form the wrist joint proper; its distal surface is wide and irregular for articulation with the five metacarpal bones [FIG. 307]. The tunnel is completely filled during life by tendons and the median nerve. Because of the mobility of the hand frequent use is made of the terms dorsal and palmar in place of posterior and anterior.

THE CARPUS AS A WHOLE

The student should know the carpus as a whole rather than have a precise knowledge of every carpal bone, but some knowledge of each of them is a necessity in learning the whole. Their clinical importance lies partly in the frequency of injury to some of them, particularly the lunate and scaphoid bones, and partly in the need for recognizable landmarks in the region of the wrist and hand.

The bones of the carpus are arranged round the capitate but they are commonly described as forming two rows. The proximal row contains the scaphoid, lunate, and triquetral bones which together form the distal surface of the wrist joint. The pisiform is also included in this row although it is often regarded as a sesamoid bone (in the tendon of flexor carpi ulnaris) which articulates loosely with the triquetrum. The distal row contains the trapezium, the trapezoid, the capitate and the hamate bones. The trapezium has a saddle-shaped articular surface for the first metacarpal and reaches the edge of the second. The trapezoid bone fits into the deep notch in the second metacarpal. The capitate articulates with the third and reaches the edges of the second and fourth metacarpals.

Fuses 15-17 years

Appears about 14-15 years, fuses soon after

Appears about 6 years

Appears in 8th week of intra-uterine life

Appears in first year

Fuses 20-22 years

AT BIRTH   ABOUT 12 YEARS   ABOUT 16 YEARS

FIG. 206 Ossification of radius.

C.T.A.—G

The hamate bone articulates with both the fourth and fifth metacarpals.

The lateral edges of the carpus are narrow and receive the collateral ligaments of the wrist [p. 233]. For muscle attachments, see FIGURE 209.

The concavity of the **carpal sulcus** is concealed by its contents enclosed by the **flexor retinaculum** which stretches between the pisiform bone and the hook of the hamate bone, medially, and the tubercles of the scaphoid and trapezium, laterally. The hook of the hamate is felt by deep pressure 15–25 mm. along a line directed from the pisiform towards the centre of the palm. The superficial division of the ulnar nerve can be felt slipping from side to side on the hook. The deep division is against its base and cannot be felt. The ulnar vessels and nerve lie on the lateral side of the pisiform bone. The tuberosity of the scaphoid can be found by following the tendon of flexor carpi radialis which passes medial to it. Distal to this, the tubercle on the trapezium can be felt as a deeper bony resistance through the attachments of the thenar muscles.

The dorsal surface of the carpus is convex and uneven but is covered by ligaments and tendons so that individual parts are difficult to identify [FIG. 395]. The styloid process at the lateral side of the base of the third metacarpal can be identified if the body of the bone is traced up to the wrist. It makes a small prominence where it projects between the capitate and trapezoid bones and receives the tendon of extensor carpi radialis brevis. The radial styloid, the trapezium, and a part of the scaphoid bone lie in the floor of the anatomical snuff box.

The blood supply to the carpus reaches the bones along the ligaments, and foramina can be seen on the non-articular surfaces between areas for their attachment and between such areas and the articular facets. Fracture or dislocation, particularly with displacement and tearing of ligaments, may interfere with the blood supply of a bone.

Some details of individual carpal bones are given below [FIGS. 207–209].

THE CAPITATE

The capitate bone is the largest and is centrally placed with a rounded head set into the cavities on the lunate and scaphoid bones. Flatter articular surfaces for the hamate medially and the trapezoid laterally are present at the sides. The dorsal surface is nearly flat but the palmar aspect is rough and protuberant for the attachment of ligaments [FIG. 306]. The distal surface, or base, is articular and is mainly for the base of the middle metacarpal but the capitate reaches and articulates with the second and fourth metacarpals by narrow surfaces at the edges of its distal surface. Looked at from the back, the distal edge of the capitate is set obliquely so that the dorsal surface projects distally and medially. The corresponding slight recession on the lateral side is filled by the styloid process of the third metacarpal which can be felt on the back of the wrist.

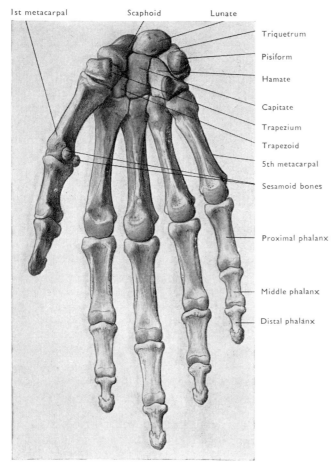

FIG. 207   Palmar aspect of bones of right hand.

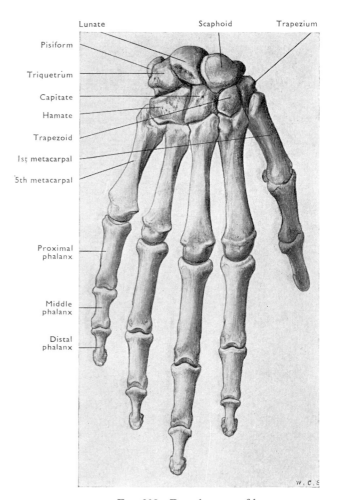

FIG. 208   Dorsal aspect of bones

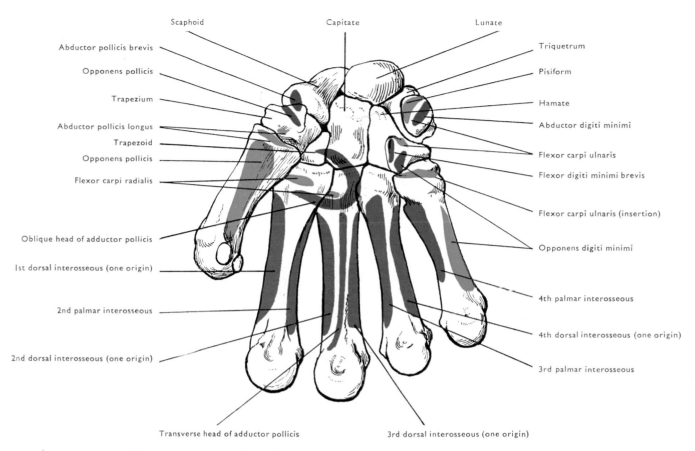

FIG. 209 Palmar surface of right carpus and metacarpus with attachments marked.

THE HAMATE

The hamate bone is distinguished by its wedge shape and by the curved **hamulus** (hook) which projects from its palmar surface near the base of the fifth metacarpal. The hamulus is concave on its lateral side and forms part of the carpal canal. Apart from the muscles, including the pisohamate ligament [FIG. 209], its chief attachment is part of the flexor retinaculum to its free end. The base of the wedge is articular and carries two fused articular surfaces for the base of the fifth metacarpal and most of the base of the fourth. The wedge is thrust between the capitate and the triquetrum and its edge reaches the lunate [PLATE 13 B]. The articulation with the capitate is flat, that with the triquetrum is sinuous. The dorsal surface is flat; the palmar surface curves on to the lateral aspect of the hamulus.

THE TRIQUETRUM

The triquetrum lies in the angle between the hamate and the lunate bones and has, therefore, a sinuous articular surface for the side of the hamate. The articular surface for the lunate is almost square and is set at right angles to that for the hamate. The triquetrum is distinguished by the circular surface for the pisiform which is set on the distal part of the palmar surface close to the hook of the hamate. The remaining surfaces form a convex area which is marked by the ulnar collateral ligament. The proximal part enters the wrist joint and articulates, during ulnar deviation of the hand, with the articular disc which separates it from the ulna [PLATE 13 B].

THE PISIFORM

The pisiform is a small bone which can be felt as a slightly mobile prominence at the medial side of the front of the wrist. It articulates with the palmar surface of the triquetrum by a circular or oval facet with an articular capsule attached to its margin, often in a narrow groove. The remainder of the bone is rounded, but its anterior surface, having the flexor retinaculum attached to it, projects distally and laterally, forming a lateral concavity which is part of the carpal canal.

The flexor carpi ulnaris tendon surrounds the non-articular part of the bone, and is continued to the hook of the hamate and the base of the fifth metacarpal as the **pisohamate** and **pisometacarpal ligaments.**

THE LUNATE

The lunate bone has a square articular surface for the triquetrum on its medial side, and a crescent-shaped surface for the scaphoid on its lateral side. Distally it has a deep concavity for the head of the capitate and proximally it is convex where it articulates with the radius and the articular disc. The dorsal and palmar surfaces have ligaments and nutrient vessels and are small compared with the articular area. The palmar surface is convex, smoother and, unlike the other carpal bones, larger than the dorsal surface. An articular surface for the edge of the hamate may be found as a narrow strip between the areas for triquetrum and the capitate.

THE SCAPHOID

The scaphoid bone has a concave articular surface for the head of the capitate and at the edge of this is a crescentic surface for the corresponding area on the lunate. Proximally, it has a wide convex articular surface for the radius. Distal to these surfaces, the bone has a prominent **tubercle** which projects forwards. It can be felt on the anterior aspect of the wrist immediately proximal to the thenar eminence (especially with the wrist extended) at the junction of middle and lateral thirds of the most distal transverse crease in the wrist. The remaining articular surface is to the lateral side of the tubercle. It faces laterally and carries the trapezium and trapezoid.

The narrow strip between the radial and trapezial surfaces and the tubercle gives attachment to the radial collateral carpal ligament. The tubercle receives part of the flexor retinaculum. This small, non-articular surface is all that is available for the entry of blood vessels. It is a common site of fracture, the result of which may be to deprive the bone, or part of it, of a blood supply.

THE TRAPEZOID

The trapezoid bone is small and irregular and articulates with the second metacarpal. It lies in the space bounded by the metacarpal, the capitate, scaphoid, and trapezium [FIG. 208].

It has a relatively large, slightly convex, non-articular dorsal surface. The small, palmar surface is also non-articular. This surface is rough and extends on to the lateral aspect of the bone as a V-shaped area between the articular surfaces for the second metacarpal and the trapezium.

The distal articular surface is approximately kidney-shaped

and fits into the notch on the base of the metacarpal. The remaining three sides are formed by articular surfaces for the adjacent carpal bones.

THE TRAPEZIUM

The trapezium is the most irregular of the carpal bones. Its main feature is a saddle-shaped articular surface for metacarpal of the thumb. This faces distally, laterally, and slightly forwards. On the opposite side of the bone, are two articular surfaces set at an angle to each other—a larger one for the trapezoid, a smaller one for the scaphoid. Proximal to the surface for the metacarpal of the thumb, the anterior aspect of the bone carries a distinct ridge (**tubercle of the trapezium**) which runs longitudinally and has a groove on its medial side for the flexor carpi radialis tendon.

The tubercle can be felt indistinctly, distal to the tuberosity of the scaphoid, through the muscles attached to it [FIG. 209]. The flexor retinaculum splits, to be attached to the tubercle and the opposite edge of the groove, thus enclosing the tendon of flexor carpi radialis.

OSSIFICATION AND VARIATIONS

The bones of the carpus are ossified from one centre each. These all appear after birth, so that the carpus is cartilaginous at birth. Centres appear in the first 6 months for the capitate and hamate and thereafter for the triquetrum (2–4 years), lunate (3–5 years), scaphoid, trapezium, and trapezoid (all 4–6 years). The pisiform is late in appearance (between 9 and 14 years). Ossification is not complete until between 20 and 25 years.

The hook of the hamate may be separate. An **os centrale,** normally incorporated in the scaphoid, may be present between the scaphoid, capitate, and trapezoid. Other small additional nodules may be present, and the styloid process of the third

FIG. 210 Dorsal surface of the right metacarpus with attachments marked.

metacarpal may be separate or fused with the capitate or trapezium.

THE METACARPUS
[PLATE 13]

The skeleton of the hand, or metacarpus, consists of five miniature long bones, the metacarpal bones. One corresponds to each digit and they are numbered in sequence from the lateral side, that of the thumb being the first. They extend from the carpus to the fingers and are concealed from the palmar aspect but, though covered by tendons [FIG. 402], they can be felt in their whole length on the dorsum of the hand. Their distal ends form the proximal row of the knuckles which become evident when the fist is clenched, and can also be felt in the palm as a series of bony resistances 2 cm. proximal to the crease at the root of each finger.

The metacarpal of the thumb differs considerably from the rest (see below). The main features of the remaining metacarpals are that each has a base, approximately cubical in shape with various articular surfaces, a body with a smooth triangular area on the greater part of the posterior surface and a distal end with a large rounded head for articulation with a phalanx. This is in contrast to the metatarsal bones found in the foot which, again with the exception of the first, have long tapering shafts and narrow heads.

The **head** of the metacarpal is capped by a smooth, rounded articular surface which extends further on the palmar than the dorsal aspect to allow for the pronounced flexion occurring at the joint. The palmar articular margin is notched in the midline. The base of the flat, triangular area on the dorsal surface of the body extends on to the head, and its corners form **tubercles** which project at each side. The collateral ligaments are attached to these tubercles and to the hollow antero-inferior to each of them.

The **body** is slightly curved with a longitudinal palmar concavity. The sides extend from the edges of the smooth posterior triangle forwards to a palmar border and provide areas for attachment of the interosseous muscles [FIGS. 403 and 405]. The lateral surface of the first and the medial surface of the fifth receive the opponens muscles, and the palmar border of the third gives origin to the transverse head of the adductor pollicis muscle.

The **bases** of the metacarpals differ from each other considerably. The metacarpals are arranged fanwise round the distal row of the carpal bones, and their bases articulate with the carpals and with each other [FIGS. 209 and 210]. The consequent variations in shape provide an easy means of distinguishing them and of associating them with the proper carpal bones. Examine FIGURES 211–215 for these details.

The metacarpal bases are bound to the carpals by palmar and dorsal carpometacarpal ligaments and, except in the case of the thumb, by palmar, dorsal, and interosseous ligaments. For the attachment of muscles and tendons see FIGURES 209 and 210.

The Metacarpal of the Thumb

The first metacarpal is shorter than the others and its body is broad and flat [FIG. 211]. It is more freely movable than the others, diverges from them and is set with its dorsal surface looking more in a lateral than a posterior direction. Its position facilitates the characteristically human movement of opposing the thumb to other digits.

The **head** is similar to the heads of the other metacarpals except that it is wider. It has the usual tubercles and markings for ligaments at the sides of the head. Two sesamoid bones, normally found in the short tendons which run over the flexor surface of the joint, articulate with the palmar part of the joint surface and may groove it.

The **body** is nearly as wide as the base and has a slightly rounded dorsal surface. Its palmar surface is divided by a blunt ridge into a larger lateral part, which gives attachment to opponens pollicis, and a smaller medial part for one head of the first dorsal interosseous muscle.

The **base** has a saddle-shaped articular surface which fits the corresponding surface on the trapezium and is surrounded by a loose articular capsule. Its dorsal surface is flush with the dorsal surface of the body but its palmar aspect projects considerably. Abductor pollicis longus is inserted into the lateral side of the base.

Ossification

The metacarpal bones ossify from primary centres which appear in the ninth week and are well ossified at birth. The secondary centres for the second to fifth metacarpals appear in the heads at the age of 2–3 years. The secondary centre for the first appears in the base only slightly later, and all of them fuse between 17 and 19 years.

In rare cases the first metacarpal may have a secondary centre in its head. Even more rarely the second or one of the others may have a secondary centre for its base.

The **styloid process** of the third metacarpal may develop as a separate ossicle.

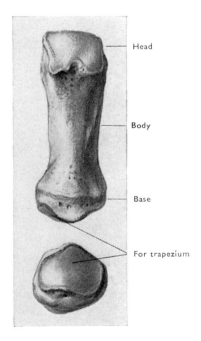

FIG. 211 First right metacarpal bone: its base has a *saddle-shaped* facet for articulation with the trapezium.

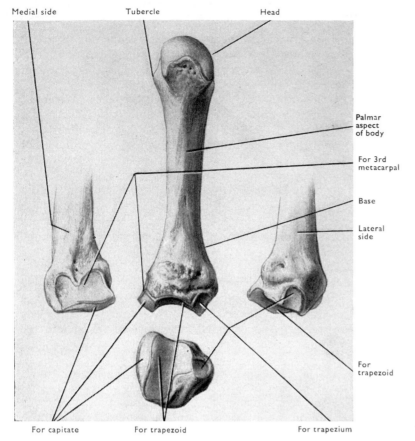

Medial side   Tubercle   Head

Palmar aspect of body

For 3rd metacarpal

Base

Lateral side

For trapezoid

For capitate   For trapezoid   For trapezium

FIG. 212 Second right metacarpal bone: its base is *notched* to receive the trapezoid.

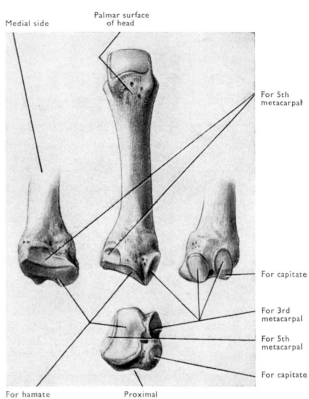

Medial side   Palmar surface of head

For 5th metacarpal

For capitate

For 3rd metacarpal

For 5th metacarpal

For capitate

For hamate   Proximal

FIG. 214 Fourth right metacarpal bone: its base has a *flat facet* for the hamate, sometimes shared with the capitate, and facets on each side for the third and fifth metacarpals.

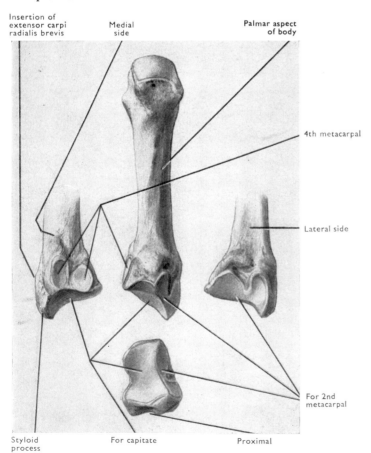

Insertion of extensor carpi radialis brevis   Medial side   Palmar aspect of body

4th metacarpal

Lateral side

For 2nd metacarpal

Styloid process   For capitate   Proximal

FIG. 213 Third right metacarpal bone: its base has a *styloid process* projecting from its posterolateral corner.

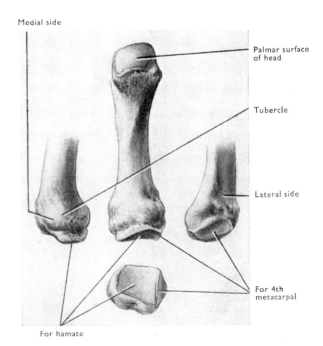

Medial side

Palmar surface of head

Tubercle

Lateral side

For 4th metacarpal

For hamate

FIG. 215 Fifth right metacarpal bone: its base has a proximal facet *concave* in its medial part for articulation with the hamate, a facet on its lateral side for the fourth metacarpal and a *tuberosity* on its medial side for attachment of the extensor carpi ulnaris and ligaments.

THE PHALANGES OF THE FINGERS

Each phalanx is a miniature long bone with a **body,** a larger proximal end or **base,** and a smaller distal end or **head** [FIG. 216]. Each finger has a skeleton of three phalanges, **proximal, middle,** and **distal,** jointed to each other and to the appropriate metacarpal head. The thumb has only two phalanges, proximal and distal.

The sizes of individual bones are so variable that it is not possible to estimate which finger they come from unless a complete set is available for comparisons to be made. Even then there may be some doubt.

The phalanges of the thumb are shorter and broader than those of the fingers. The middle finger is normally the longest; the little finger is the shortest; the ring and index fingers are nearly equal, but the ring finger is usually a little longer than the index. The relative lengths of the phalanges is such that the tips of the fingers fit across the palm when all the interphalangeal and metacarpophalangeal joints are flexed.

Proximal Phalanx

This is the largest (35–45 mm. in length) and has a concave, oval, articular surface on its base for the head of a metacarpal, thus allowing abduction and adduction as well as flexion and extension. Except posteriorly, the **base** has an articular capsule attached round the articular surface and collateral ligaments to its sides. The palmar ligament is attached firmly to the edge of the base, only loosely to the metacarpal bone. The bases receive insertions from some of the interosseous muscles (see Muscle Section) with abductor digiti minimi to the medial side of the fifth in place of an interosseous. Flexor digiti minimi joins the abductor.

The **body** is slightly curved along its length and is convex dorsally. It is markedly convex from side to side on its dorsal surface, flat or concave on its palmar surface, and often has its edges raised by the attachment of the fibrous flexor sheath.

The **head** has a pulley-shaped articular surface, wide from side to side with a groove in the middle and markings for the collateral ligaments on each side.

Middle Phalanx

This is similar to a proximal one except that it is shorter and has a ridge on the articular surface of the base. The ridge fits the groove on the head of a proximal phalanx. The edges of its palmar surface are raised and roughened near the middle by the insertion of flexor digitorum superficialis and the fibrous flexor sheath. The dorsal region of the base receives part of the insertion of an extensor expansion.

Distal Phalanx

This is small and may be referred to as a terminal or ungual phalanx because it is at the end of the finger and carries a finger nail.

Its **base** is relatively large and has a tendon of flexor digitorum profundus inserted into its palmar aspect, collateral ligaments to its sides, and part of an extensor tendon to its dorsal aspect.

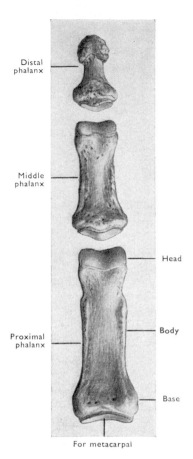

FIG. 216 Phalanges of a finger (palmar view).

Comparison with the phalanges of a foot will show that in the foot the bases of the proximal phalanges are relatively large and the bodies are much thinner. The middle and distal phalanges are so short that they should not be mistaken. The bones of the big toe have special features.

The **body** tapers rapidly to reach a rough spatulate **tuberosity** at the end of the bone.

Phalanges of the Thumb

These are similar to the proximal and distal phalanges of a finger but they are shorter and broader. They are connected to each other and to the metacarpal by similar ligaments.

The muscles attached differ from those going to a finger. The base of the proximal phalanx receives the short muscles of the thumb except the opponens [FIG. 403], but including a palmar interosseous muscle and also the tendon of extensor pollicis brevis. The terminal phalanx has the tendons of flexor and extensor pollicis longus attached to it. The fibrous flexor sheath is attached to the edges of the body of the proximal phalanx and to the base of the distal phalanx.

Ossification

Primary centres for all the phalanges appear between the eighth and twelfth weeks and the phalanges are well formed at birth. The distal phalanges ossify first. Secondary centres appear in the bases in the second or third year and fuse between 17 and 19 years. In rare cases a centre may appear for the head as well as the base.

Occasionally a distal phalanx is bifurcated or a phalanx may be small or absent, and the thumb may have three phalanges.

THE SESAMOID BONES

As a rule the only two sesamoids found in the hand are small nodules in the tendons of insertion of flexor pollicis brevis and adductor pollicis muscles. They blend with the palmar liga-ment of the metacarpophalangeal joint and move on the head of the metacarpal bone. Occasionally small sesamoids are found in the other metacarpophalangeal joints or in the inter-phalangeal joint of the thumb. The sesamoids are cartilag-inous in childhood and begin to ossify after 13 years.

THE BONES OF THE LOWER LIMB

The skeleton of the lower limb is connected to the vertebral column by the pelvic girdle which is formed by an **os coxae** or hip bone on each side. They articulate posteriorly with the **sacrum** and meet below and in front at the symphysis pubis. With the sacrum, the two hip bones form the skeleton of the pelvis which surrounds the lowest part of the abdominal cavity.

The hip joint is on the lateral aspect of the hip bone and from this the **femur** or thigh bone extends to the knee. The bulge of the gluteal muscles running between the hip bone and the upper part of the femur form, with their covering tissues, a **buttock** or natis on each side, but the region is usually named **gluteal,** from the muscles.

The free lower limb is divided into **femur** or thigh between the hip and the knee ; **crus** or leg, between knee and ankle, and **pes** or foot distal to the ankle joint. The term leg refers only to the part between knee and ankle.

The bones of the leg are the medial, **tibia** and the slender, lateral, **fibula.** They articulate with each other at both ends, but, of the two, only the tibia articulates with the lower end of the femur. Both tibia and fibula take part in the ankle joint. The term peroneal (to be distinguished from perineal) is synonomous with the adjective fibular.

The foot is distal to the ankle joint and is divided into the foot proper and the toes. It is normally set almost at right angles to the leg and so has a superior surface or **dorsum** and an inferior, **plantar surface** which is the sole of the foot. The bones of the foot include the seven **tarsal bones** or **tarsus,** corresponding to the carpus in the hand, and five **metatarsals** corresponding to the metacarpals.

The big toe or **hallux** is on the medial side and has two phalanges, like the thumb. The remaining toes each have three phalanges.

The toes and the metatarsals are numbered in sequence from the medial side, starting with the hallux.

VARIATIONS

The skeleton of the lower limb is adapted for bipedal gait and weight bearing. The bones vary in their characteristics in relation to muscular development and build of the body. The proportions of each bone, its robustness, degree of development of muscular markings, and protuberances vary accordingly. Many bones, but especially the hip bone and to a lesser extent the femur, display sexual differences [p. 176]. Variations in the dimensions of the female pelvis [p. 175] may be important to detect in the management of labour.

The length of the femur is an indication of stature : roughly, stature is ×4 the length of a femur. More accurate estimates use regression formulae derived from particular measurements of long bones from individuals whose stature was known before death. Trotter and Gleser (1952, 1958) and Krogman (1962) critically discuss the use of measurements of bones to assess stature, age, sex, race, and identity.

The tibia of prehistoric and primitive peoples may be com-pressed from side to side (platycnemia). The fibula may be incomplete or absent.

Additional ossicles, or divided navicular or cuneiform bones, may be found in the tarsus. An **os trigonum** may be present at the back of the talus.

THE HIP BONE
[PLATE 14]

The hip bone [Figs. 217 and 219] consists of three parts, the **ilium, ischium,** and **pubis,** fused together. The ilium is the broad, sinuously curved portion which joins the other two at the **acetabulum,** a cup-shaped cavity situated on the lateral aspect of the bone. The ilium is superior and is set at right angles to the other two parts.

The ischium and pubis are fused with each other and with the ilium in the acetabulum [Fig. 217]. The ischium is the heavier of the two parts ; the pubis is lighter with a flattened body anteriorly. The two parts, pubis and ischium, surround the oval or triangular obturator foramen by fusing with each other inferiorly.

THE ILIUM

The ilium is the bone felt in the flank 4–6 cm. inferior to the ribs. It is somewhat fan-shaped with the hinge of the fan in the acetabulum. The margins of the fan form anterior and posterior borders with a curved superior border or iliac crest between them.

The sinuous **iliac crest** is palpable in the flank and separates the abdominal wall from the gluteal region. Its anterior end is the **anterior superior iliac spine** and can be felt as a land-mark at the lateral end of the groove of the groin [Fig. 418]. It gives attachment to the inguinal ligament, which lies deep to that groove. The iliac crest has an external and an internal lip, with an intermediate line between them. The **external lip** is prominent 5–7 cm. above and behind the anterior superior spine and this forms another palpable landmark, the **tubercle of the crest.** The **internal lip** becomes very indefinite pos-teriorly. The crest ends posteriorly as a **posterior superior iliac spine** which is at the bottom of a dimple level with the second sacral spine, 4 cm. from the median plane. The muscles of the lateral and posterior abdominal walls and the fascia lata of the thigh are attached to the iliac crest [FIGS. 218 and 220].

The anterior border of the ilium extends downwards and backwards from the anterior superior spine to the anterior margin of the acetabulum. Just above the acetabulum it has a considerable prominence called the **anterior inferior iliac spine** [FIG. 217]. Below and medial to this, a low **iliopubic eminence** on the outer wall of the acetabulum marks the junction of the ilium and pubis. Sartorius is attached to the anterior superior spine and the border below it, rectus femoris to the upper half of the anterior inferior spine, and the strong iliofemoral ligament to the lower half [FIG. 218]

The posterior border of the ilium runs down from the posterior superior spine for a short distance to the **posterior inferior iliac spine.** Thence it arches forwards, and then downwards and backwards (the **greater sciatic notch**) to meet the posterior border of the ischium [FIG. 217]. Where the two bones join, the outer surface of the acetabulum again shows a rounded eminence. The principal attachments are ligamentous to the posterior spines, particularly the sacro-tuberous ligament to the inferior spine, but piriformis muscle, as it comes out of the pelvis, gains some attachment to the edge of the notch.

The **gluteal surface** of the ilium is the undulating, external surface between the acetabulum and the crest. Three distinct **gluteal lines,** formed by the fascia between the gluteal muscles curve upwards and forwards over it from the deepest part of the greater sciatic notch. The **inferior gluteal line** arches towards the anterior inferior iliac spine. The **anterior gluteal line** curves up and then forwards towards the anterior superior iliac spine. The **posterior gluteal line** sweeps backwards and then upwards and forwards close to the external lip of the crest for 5–7 cm. [FIG. 217].

There is a rough mark above the acetabulum for the reflected head of rectus femoris. The areas giving attachment to the glutei and tensor fasciae latae are shown on FIGURE 218.

The medial aspect of the ilium [FIG. 219] consists of: (1) a smooth, concave **iliac fossa** occupied by iliacus muscle, which arises from the upper; larger part of its surface; and (2) a **sacro-pelvic surface,** which lies postero-inferior to the iliac fossa and at an angle to it. The sacral portion has a rough L-shaped articular area **(auricular surface)** with well-defined edges, which articulates with the sacrum at the sacro-iliac joint. Posteriorly the joint reaches the posterior inferior iliac spine, and antero-inferior to the joint surface there may be a pre-auricular sulcus, especially in female bones.

Between the auricular surface and the posterior part of the crest, the bone is rough and raised to form the **iliac tuberosity,** to which the sacro-iliac ligaments and the erector spinae muscle are attached. The iliolumbar ligament reaches the ilium at the triangular area between the iliac fossa, the crest, and the tuberosity [FIG. 220].

The remaining pelvic portion is smooth and forms the part of the bony wall of the lesser pelvis [p. 174] between the greater sciatic notch and the curved **arcuate line** [FIG. 220] which separates iliac fossa from lesser pelvis. The arcuate line is continuous with the pecten pubis to form the iliopubic part of the linea terminalis [p. 174].

THE ISCHIUM

The ischium is the L-shaped part of the hip bone which passes down from the acetabulum and then turns forwards to join the pubis. The thick vertical part which enters into the acetabulum is the **body.** The lighter arm which joins the corresponding part of the pubis is the **ischial ramus.** Posteriorly, on the angle between these parts and extending up on the body, is a rough area, the **ischial tuber** (or tuberosity), which gives attachment to the hamstring tendons. A lateral impression is for semimembranosus, and a vertical line separates it from a more medial impression for semitendinosus and biceps (Martin, 1968). An inferior impression for the hamstring part of adductor magnus is placed laterally on the inferior part of the tuberosity. The medial part is covered by fibrous tissue and a bursa. The medial border of the tuberosity gives attachment to the strong sacrotuberous ligament which is continued (falciform process) along the medial margin of the ischial ramus, raising a small ridge on the bone. Quadratus femoris arises from the lateral border of the tuberosity, which is deep to the edge of gluteus maximus muscle when the thigh is extended.

Superiorly, the **body of the ischium** forms the postero-inferior part of the acetabulum and joins the ilium at the rounded dorsum acetabuli. Between the acetabulum and the ischial tuberosity, the medial margin of the bone projects backwards and medially as a triangular **ischial spine** [FIG. 217]. This separates the **greater sciatic notch** above from a **lesser sciatic notch** below (sciatic is derived from ischiadic). The lesser notch, between the ischial spine and the ischial tuberosity, is smooth where the obturator internus tendon turns over it.

The sacrospinous ligament runs to the tip of the ischial spine and, with the sacrotuberous ligament and sacrum, converts the sciatic notches into foramina.

The **pelvic surface** of the ischium is smooth where it forms the wall of the pelvis and is partly covered by the origin of the obturator internus muscle [FIG. 220]. Coccygeus and levator ani muscles take origin from the pelvic surface of the spine and the two gemelli from the margins of the lesser notch. The ramus of the ischium passes forwards and upwards to meet the inferior ramus of the pubis and form the **conjoined ramus.**

THE PUBIS

Each pubis has a flattened **body,** and together with the disc between them, they form the antero-inferior wall of the lesser pelvis at the lowest part of the front of the abdomen. The **superior ramus** is a bar of bone passing upwards and laterally to form the antero-inferior part of the acetabulum. The **inferior ramus** passes backwards, downwards, and laterally to join the ramus of the ischium.

The body of the pubis has a smooth pelvic surface, slightly convex anteroposteriorly, which faces upwards and only slightly backwards. It has the anterior end of the levator ani muscle and fascial ligaments for the bladder or prostate attached to it, but these do not mark it appreciably. Medially, it has a rough oval **symphysial surface** for the attachment of a fibrocartilaginous **interpubic disc** This connects it to its neighbour and so forms the pubic symphysis. The inferior or femoral surface faces into the thigh and is rough for muscle attachments (see below). Postero-inferiorly, the body becomes continuous with the inferior ramus and the arcuate ligament arches between the two sides. The lateral edge of the body is thin and helps to bound the obturator foramen.

The anterior border of the body of the pubis is thickened to form a **pubic crest** which receives tendons and aponeuroses of the abdominal muscles [FIG. 419]. Its lateral end projects as the **pubic tubercle** 2·5–3 cm. from the median plane. It provides the main pubic attachment for the inguinal ligament and is a useful landmark.

The **superior ramus** widens as it passes laterally to reach the iliopubic eminence and unite with the ilium above and the ischium behind at the acetabulum. The part between the body

and the acetabulum is three-sided. Its pelvic surface is smooth and continuous with the rest of the pelvic surface of the bone. It is limited at the superior aperture of the pelvis by the **pecten pubis,** a sharp edge which runs laterally from the pubic tubercle to the iliopubic eminence. It provides a continuation of the line of attachment of the structures going to the pubic crest, including the conjoint tendon, the lacunar ligament, the reflected ligament [FIG. 422], the pectineal ligament [q.v.], and the fascia transversalis.

The triangular area between the pubic tubercle and the iliopubic eminence **(pectineal surface)** gives attachment to the fleshy fibres of pectineus muscle from its medial part [FIG. 218]. Anteriorly, it is limited by the **obturator crest,** a ridge which runs from the pubic tubercle to the lower part of

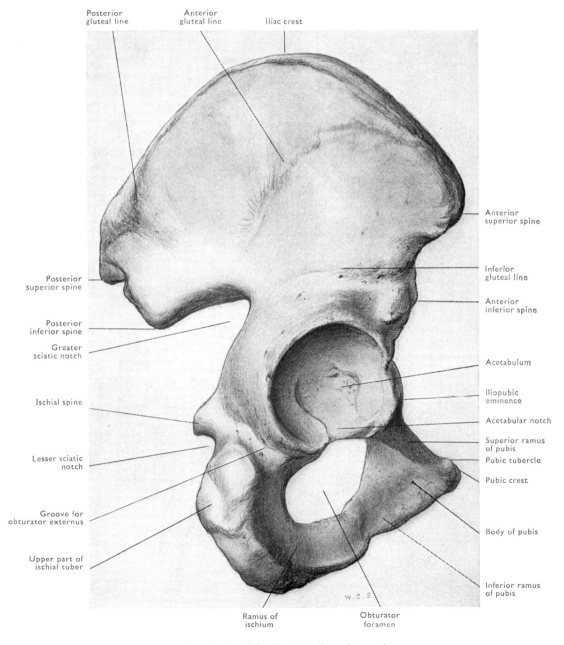

FIG. 217 Right hip bone (lateral aspect).

the acetabulum and provides attachment for the pubofemoral ligament. Below this, the femoral surface of the body becomes continuous with the inferior aspect of the ramus at the **obturator groove** (see below).

The **conjoined ramus** of the pubis and the ischium forms the lower boundary of the obturator foramen, and its superior edge is thin where the obturator membrane is attached. Its *femoral surface* is continuous with that of the body of the pubis in front and the body of the ischium behind. The muscle attachments are continued on to these surfaces. The obturator externus muscle is attached near the edge of the foramen and to the membrane. The adductor group of muscles have their origin inferior and anterior to obturator externus [FIG. 218] and the fascia lata is attached to the inferior edge of the bone.

The **medial surface** gives attachment to obturator internus superiorly [FIG. 220]. Inferiorly, the bone is everted and grooved by the attachment of the crus of the penis or clitoris. The superior edge of the groove has the **inferior fascia of the urogenital diaphragm** (perineal membrane) attached to it, and is continuous posteriorly with the ridge for the falciform part of the sacrotuberous ligament. Ischiocavernosus and transversus perinei superficialis and the fascia covering them are attached to the posterior end of the inferior edge, whilst the muscles of the deep perineal space are attached above the inferior fascia of the urogenital diaphragm.

The **obturator foramen** is the oval or triangular space surrounded by the bodies and rami of the pubis and ischium. It lies inferior to the acetabulum and is nearly closed by the

FIG. 218 Right hip bone (lateral aspect).

obturator membrane. The membrane gives origin to the muscles attached to the margins of the foramen, obturator externus, and obturator internus. The obturator groove under the superior ramus of the pubis leads to a gap between the anterior part of the membrane and the bone, through which the obturator nerve and vessels emerge. This free edge of the membrane may produce a small **anterior obturator tubercle** where it reaches the superior ramus of the pubis and, more frequently, a rough **posterior obturator tubercle** where it reaches the junction of pubis and ischium.

The **acetabulum** is the rounded pit situated on the lateral side of the hip bone over the Y-shaped junction of ilium, ischi-um, and pubis. Its upper part forms a projecting surface on the lower end of the ilium for transmission of weight to the femur. A fibrocartilaginous **labrum acetabulare** is attached to the bony rim within the capsule and deepens the cavity. The inferior wall of the pit is notched **(acetabular notch)** and this is bridged by a transverse acetabular ligament which completes the rim but leaves a gap between the ligament and the bone.

The floor of the acetabulum above the notch is rough and depressed to form the **acetabular fossa.** The remainder, between rim and fossa, is a smooth horseshoe-shaped, articular **lunate surface** for the head of the femur. The synovial

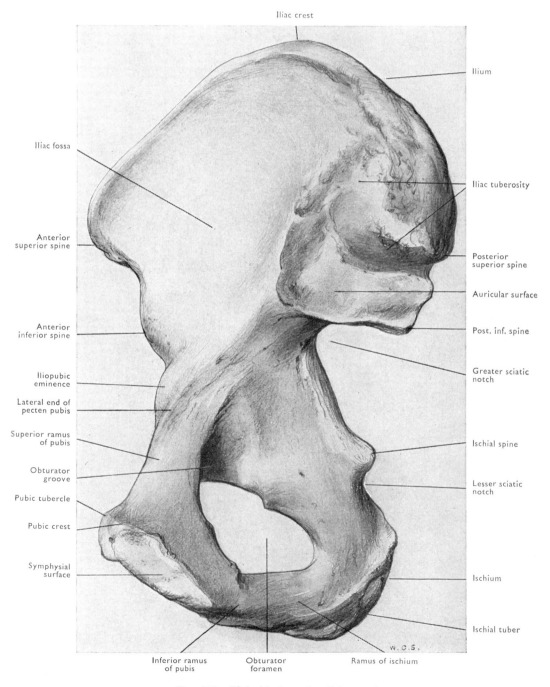

FIG. 219 Right hip bone (medial aspect).

ligament of the head of the femur spreads out and is attached to the transverse ligament and the margins of the notch and fossa.

OSSIFICATION

The hip bone is ossified before birth from three primary centres: (1) above the greater sciatic notch of the ilium early in the third month; (2) in the ischium, below the acetabulum, in the fourth or fifth month; (3) in the superior ramus of the pubis, in the fifth or sixth month.

At birth ossification has spread to include part of the acetabulum [FIG. 221]. By the tenth year the cartilage in the acetabulum is reduced to a triradiate strip, growth of which separates the three components and increases the size of the acetabulum. Secondary centres appear in the acetabular cartilage, and the triradiate strip is usually consolidated by the age of puberty or shortly after. The bones of the ischial and pubic rami normally unite by the tenth or eleventh year, but may be delayed until about 16 years, or may fail to unite.

The remaining secondary centres are shown in FIGURE 221: all appear by the age of puberty and join between 20 and 25 years of age.

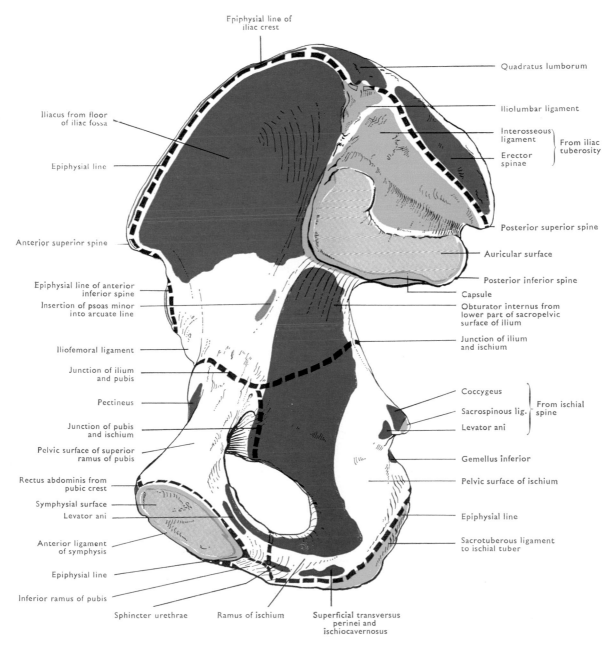

FIG. 220   Right hip bone (medial aspect).

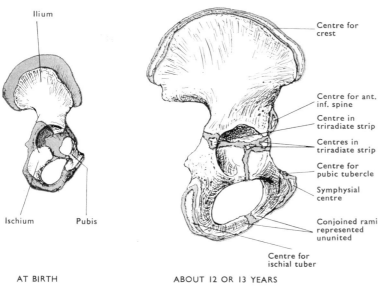

FIG. 221   Ossification of hip bone.
The more anterior of the secondary centres in the acetabular cartilage may remain separate as an os acetabuli.

THE PELVIS
[PLATE 14]

The pelvis is the lower division of the abdomen [FIG. 508] and its skeleton is referred to as the bony pelvis or simply the pelvis. The bony pelvis is made up of the two hip bones, the sacrum, and the coccyx, bound together by ligaments. The walls are padded with muscles and other structures so that the living pelvis is very different from a bony specimen. The term pelvis means a basin, but the pelvis, in the anatomical position, is tilted so that: (1) the anterior superior iliac spines and the pubic tubercles are in the same coronal plane; (2) the coccyx is level with the upper margin of the symphysis pubis; and (3) the anterior part of the pelvic surface of the sacrum is set at an angle of only 10 or 12 degrees from the horizontal.

The primary functions of the bony pelvis are to provide a base for attachment of the lower limb and trunk muscles; to transmit the weight of the body from the vertebral column to the femora; and to contain, support, and protect the pelvic viscera. Childbirth in a female demands a cavity of the pelvis large enough to allow passage of a full term foetus.

The pelvis consists of two parts, the **greater pelvis,** and the **lesser pelvis.** The plane separating these two parts passes through the arcuate line of the ilium and the pecten pubis at each side, the lateral parts and promontory of the sacrum behind, and the upper part of the pubic bodies in front. All these are grouped together as the **linea terminalis,** and surround the **superior aperture** of the pelvis.

The greater pelvis is the upper part, formed mainly by the iliac fossae, and is the skeleton of the lower part of the abdomen proper. It is partly filled by the iliopsoas muscles.

The lesser pelvis is the cavity behind and below the linea terminalis, and its contents are usually referred to as the contents of the pelvis. Its posterosuperior wall is formed by the sacrum and coccyx and so measures 12–15 cm. long and is concave. Its lateral wall is formed by the obturator membrane, the smooth surfaces of all three bones below the linea terminalis, and the gap between them and the sacrum and coccyx, partly filled by the sacrospinous and sacrotuberous ligaments [FIGS. 222 and 309], which complete the sciatic foramina.

The greater sciatic foramen is partly filled by the piriformis, the lesser sciatic foramen by obturator internus. In the lower part of the lateral wall, the obturator membrane is covered, as is most of the bony lateral wall, by obturator internus muscle.

The antero-inferior wall is formed by the pubic bodies and the interpubic disc between them. It is convex and only 4–5 cm. long and faces more upwards than backwards into the cavity. The conjoined rami pass backwards, laterally, and a little inferiorly to form the pubic arch, and where they diverge from the disc is the subpubic angle. The rami are united here by the arcuate ligament of the pubis, and after a slight gap, by the urogenital diaphragm [FIG. 432] which stretches between them.

The **cavity of the lesser pelvis** is a short, wide canal, curved in conformity with the sacrum and coccyx and is continuous with the abdominal cavity at the **superior aperture of the**

FIG. 222   Male pelvis seen from in front.

FIG. 223   Female pelvis seen from in front.

pelvis. This is bounded by the terminal line and looks forwards and slightly upwards. The plane of the superior aperture makes an angle of about 60 degrees with the horizontal.

The **inferior aperture of the pelvis** is bounded by the coccyx behind; by the pubic symphysis in front; and on each side by the sacrotuberous ligament and the conjoined ramus. These are also the boundaries of the perineum. The plane of the inferior aperture of the pelvis makes an angle of only 10–15 degress with the horizontal.

Because the cavity of the lesser pelvis is curved, numerous planes of the cavity can be described in addition to the planes of the apertures, including one through the middle of the third sacral vertebra and the middle of the pubic symphysis.

The **spines of the ischia** project a little into the pelvis and, when this is pronounced, the plane passing through the spines and the posterior margin of the pubis may be of importance as it lies in the narrowest part of the canal.

The **axis of the superior aperture** is at right angles to the centre of its plane and, if produced, would pass approximately through the umbilicus and the tip of the coccyx. The **axis of the inferior aperture** drawn similarly is directed towards the promontory of the sacrum. The **axis of the pelvis** is a line drawn through the centres of all the planes and is therefore curved and nearly parallel to the surface of the sacrum and coccyx.

MEASUREMENTS OF THE PELVIS

The **intercristal diameter** is the furthest distance between the outer lips of the iliac crests. The **interspinous diameter** is the distance between the anterior superior iliac spines. The **external anteroposterior diameter** is measured between the first sacral spine and the anterior (upper) margin of the pubic symphysis. These three measurements are usually slightly greater in a man than in a woman but the difference between intercristal and interspinous measurements is more important than their absolute size and should be more than 3–4 cm. in a normal female.

The **diameters of the lesser pelvis** are of more importance, though not easily measured in the living subject. They are measured anatomically as follows.

Diameters of the Superior Aperture

Anteroposterior or conjugate: from the middle of the sacral promontory to the anterior (upper) margin of the symphysis pubis. Transverse: across the aperture where it is widest. Oblique: from one sacro-iliac joint to the iliopubic eminence of the opposite side.

In the living subject the anteroposterior or conjugate measurement may be increased by extension at the sacro-iliac joint.

Diameters of the Cavity

Anteroposterior: from the centre of the pelvic surface of the middle piece of the sacrum to the middle of the pelvic surface of the symphysis pubis. Transverse: across the cavity where it is widest. Oblique: from the posterior end of the sacro-iliac joint to the centre of the obturator membrane of the other side.

Diameters of the Inferior Aperture

Anteroposterior: from the tip of the coccyx to the posterior (lower) margin of the symphysis pubis (variable owing to mobility of the coccyx). Transverse: between the ischial tuberosities. Oblique: from the crossing of the sacrotuberous and sacrospinous ligaments to the junction of pubic and ischial rami on the other side.

The diameters of the lesser pelvis should be greater in a woman than in a man, but measurements vary considerably within one sex and there may be some overlap. The term **conjugate** applied to the anteroposterior diameters should only be applied to the superior aperture of the lesser pelvis.

In obstetrical work two other conjugates are used in addition to the anatomical or **true conjugate** measured to the anterior margin of the symphysis. The obstetric or **available conjugate** is measured from the promontory of the sacrum to the nearest part of the posterosuperior surface of the symphysis, and is almost 0·5 cm. shorter. The **diagonal conjugate** is taken to the postero-inferior margin of the arcuate ligament of the pubis. It provides a rough estimate of the available conjugate on deduction of 0–3 cm., depending on the depth and angle of the pubic symphysis and the height of the sacral promontory (Chassar Moir, 1964).

The figures in the following table are *average* measurements on skeletal material in millimetres as given by Bryce (1915) and serve only to indicate the differences in proportion.

	Female			Male		
Intercristal	279			286		
Interspinous	250			241		
External anteroposterior	180			184		
True Pelvis	*Superior Aperture*	*Cavity*	*Inferior Aperture*	*Superior Aperture*	*Cavity*	*Inferior Aperture*
Anteroposterior	114	127	108	102	108	85
Oblique	127	140	114	121	114	102
Transverse	133	127	121	127	121	88

The coccyx is normally movable so that it may increase (or reduce) the anteroposterior measurement of the inferior aperture by its movement by as much as 2·5 cm. Ligaments are not rigid structures, thus the oblique measurement of the inferior aperture is not a constant one.

The figures given in clinical texts vary considerably. The following, taken to the nearest centimetre, are the result of a survey of different textbooks.

Diameter	*Superior Aperture*	*Cavity*	*Inferior Aperture*
Anteroposterior (available conjugate)	11	12	13
Oblique	12	12	12
Transverse	13	12	10–11

Nicholson (1943, 1945) measured 640 living subjects, all primigravidae, by means of stereometric radiography. He measured the obstetrical conjugate and the transverse diameter of the pelvis, and found a decrease in the obstetrical conjugate during the First World War (when nutrition was poor) compared with the subsequent 7 years, though the transverse diameter was not similarly affected.

Available figures indicate that not only do pelvic measurements vary between selected groups, but (see Nicholson) may vary within homogeneous groups as a result of various conditions.

SEXUAL DIFFERENCES

The female pelvis differs from the male in many particulars and distinctive sex characters are already present in the foetal pelvis by the third or fourth month of intra-uterine life (Boucher, 1957). They are poorly marked in childhood but become fully developed after puberty (Krogman, 1962).

The essential differences are in the lesser pelvis. In the female it is absolutely wider in all its diameters, particularly those of the inferior aperture. The cavity in the female is roomier and more rounded, shorter and less funnel-shaped.

It is difficult to find samples which are female in all respects. Pelves are often intermediate in many characters, or are female in type in one part (e.g. anteriorly) but show male characters elsewhere. It is advisable to know the types of pelvis to expect and the significance of particular features in obstetrics.

In the normal adult female pelvis the pubic arch is wide with a rounded apex and can be represented fairly well by the angle between outstretched thumb and forefinger. The depth from pectineal line to ischial tuberosity averages 90 mm. In the male the angle is narrow and can be represented by the angle between outstretched index and middle finger. The depth is greater, averaging 101 mm. A wide space between the ischial tuberosities and a shallow pelvis increases the **subpubic angle.** A slender ramus set at the side of a pubic body widened by growth at the pubic symphysis causes a wide, rounded apex to the subpubic angle. The subpubic angle, then, indicates the form of the ischiopubic part of the pelvis and, in particular, a narrow angle indicates that the side-walls converge towards the inferior aperture. A wide angle indicates a female type of pelvis anteriorly but is not a reliable indication of the size of the pelvic cavity.

In the female the **greater sciatic notch** is wide (about 90 degrees); in a massive male pelvis its outline is U-shaped. Massive bone and an extensive sacro-iliac articulation tend to depress the posterior part of the upper margin of the notch and the posterior inferior iliac spine, and a sacrum with its lower end tilted in has a similar effect on the margin of the notch. A wide angle then indicates a roomy posterior part of the pelvis, a narrow angle indicates a long, narrow pelvis.

In the female the space between the spine of the ischium and the sacrum should accommodate three fingers with ease; in the male the distance is about two finger-breadths, indicating a short sacrospinous ligament and a spine which projects into the cavity. For further information on this subject and its clinical and other applications see Caldwell and Moloy (1933), Caldwell, Moloy, and D'Esopo (1934, 1935) and Krogman (1962).

Abnormal softening of bone (e.g., in rickets) can produce a flat pelvis due to compression by the bodyweight, with a reduced anteroposterior diameter while, at the same time, widening the measurement between the anterior superior iliac spines and reducing the difference between it and the intertubercular measurement.

There are numerous differences in detail but the bones *in the female* are lighter and thinner, and the muscular markings are less evident. The ilium is less everted and the iliac fossa shallower. The superior aperture is larger and kidney-shaped rather than heart-shaped. The plane of the superior aperture forms a greater angle with the horizontal and the tilt of the pelvis is greater so that the anterior superior iliac spine may be further forward than the pubic tubercles. The sacral promontory projects less and the pubic tubercles are wider apart. The cavity is less funnel-shaped, shorter, and much roomier. The sacrum is shorter and wider and its curve is less uniform, being flatter anteriorly; each lateral part nearly equals the first sacral body in its transverse measurement. The inferior aperture is much larger, the coccyx is more movable, and the ischial spine is less inturned. The greater sciatic notch is wider. The obturator foramen is smaller, and is often triangular rather than oval. The acetabula are further apart and are smaller, especially relatively, for the femoral head is small and the hip bone is large. The width of the acetabulum is 2–3 cm. less than the distance between its anterior border and the symphysis. A pre-auricular sulcus is more commonly present than in the male.

GROWTH OF THE PELVIS

At birth the pelvis is relatively small in both sexes. The narrow ilium is more upright, and the shallow iliac fossa looks more forwards than in the adult. The subpubic angle is more acute, the ischia are closer together and the cavity is consequently more funnel-shaped. The sacrum is narrow, more upright and less curved and its promontory projects less. The lower limbs are only a quarter of the length of the body but after birth they grow rapidly, especially after the first year when the child learns to walk. The whole pelvis keeps pace with the lower limbs and, since it gives origin to muscles moving the limbs, grows concurrently to provide adequate attachment. At birth, the lesser pelvis is small, and the urinary bladder and the uterus and ovaries are in the abdomen proper. By the sixth year the pelvis contains the greater part of them, but the pelvic organs are not wholly pelvic until puberty.

Radiographic studies of the pelvis in early infancy (Reynolds, 1945) have shown that the pelvis of boys appears to be larger in measurements relating to the outer structures, whereas that of girls is large in measurements relating to the inner structures of the pelvis, including a relatively larger inlet. At puberty growth is modified to produce the distinctive sex characters of the adult pelvis. Such growth takes place at the cartilaginous surfaces of the bones in the acetabulum, at the sacro-iliac joint and the symphysis pubis and between the pubic and ischial rami, and commences in the prepubertal period (Reynolds, 1947). Absorption and accretion on the surfaces produces modelling such as is required to maintain proper strength and shape (Payton, 1935).

Changes occur when the child, learning to walk, straightens out the lower limbs and adopts the erect posture. The upper part of the pelvis tilts forwards and the anterior sacrovertebral angle becomes more obtuse. The base of the sacrum, taking the weight of the trunk obliquely, sinks more deeply between

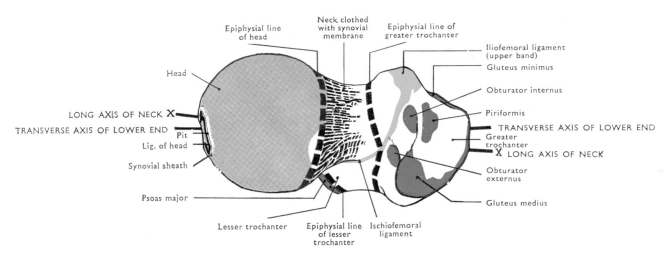

FIG. 226 Right femur (superior aspect).

well as to the lower part of the intertrochanteric line. Part of iliacus is inserted below the lesser trochanter, and pectineus goes to a vertical, **pectineal line** leading down to the linea aspera [FIG. 227]. The **lateral lip of the linea aspera** is continued up as the **gluteal tuberosity**, which reaches the greater trochanter. It is prominent, rounded, or rough, and may be sufficiently raised to form a **third trochanter.** The gluteal tuberosity is for part of gluteus maximus muscle. Vastus lateralis arises from its lateral edge as well as from the lateral lip of the linea aspera, the root of greater trochanter and the upper part of the intertrochanteric line. The attachment of adductor magnus is from the lower border of the insertion of quadratus femoris down to the adductor tubercle [FIG. 228], lying medial to the gluteal tuberosity, on the middle of the linea aspera, and on the medial supracondylar line. The adductor brevis and longus are attached to the medial lip of the linea aspera medial to magnus [FIG. 228] and, in the case of brevis, behind the pectineal line. The short head of the biceps is attached to the lateral lip, lateral to adductor magnus, and medial to vastus lateralis and to vastus intermedius inferiorly. The bone may be grooved by the four perforating arteries which pass horizontally through the muscles near their attachments.

Inferiorly, the linea aspera expands into the flat, triangular, **popliteal surface,** bounded at the sides by the supracondylar lines. The **lateral supracondylar line** is well defined and extends down to the lateral condyle. The **medial supracondylar line** is less well defined because it is smoothed in its upper part by the femoral vessels passing between the bone and adductor magnus to enter the popliteal space as the popliteal vessels. Inferiorly, this line reaches the **adductor tubercle,** a sharp prominence on the upper part of the medial condyle for the insertion of the tendinous, hamstring part of adductor magnus muscle. The tubercle can be found by following down the tendon with the knee flexed.

Inferiorly, the popliteal surface curves back to a condyle on each side and to a prominent **intercondylar line** between them. This lower part of the surface is pitted by foramina and has at its medial side a rough mark for the medial head of gastrocnemius muscle. Plantaris arises from the corresponding area at the lateral side.

The structures attached to the lower part of the linea aspera are continued down on the supracondylar lines ; vastus medialis and adductor magnus on the medial line, biceps and vastus intermedius on the lateral line.

The medial and lateral intermuscular septa pass to the bone ; the former between the anterior, extensor compartment and the medial adductor group of muscles, and the latter between the same extensor compartment and the hamstring group of muscles posteriorly. The septa reach both the linea aspera and the supracondylar lines and increase the area of muscular attachment.

The **inferior end of the femur** consists of two large condyles which blend with each other anteriorly and with the body superiorly. They project backwards and are separated posteriorly by a deep **intercondylar fossa,** and can be felt as large prominences at the sides of the knee. When the femur is in its normal position the distal surfaces of the two condyles should touch the same horizontal plane so that the body is directed obliquely outwards and upwards from the knee to the hip.

The **surfaces of the condyles** which articulate with the tibia and its menisci extend over the distal and posterior aspects so as to provide for full flexion. From below and behind, the surface of the lateral condyle is rounded although the parallel medial and lateral edges may give its outline an oblong appearance. The medial condyle is elongated anteroposteriorly, and turning laterally at its anterior end, is curved, in a horizontal plane, round a centre situated on the middle of the lateral condylar surface. The anteroposterior curvature of the articular surfaces is spiral with the inferior part flatter, the posterior part more convex.

Anteriorly, the two articular surfaces are confluent, but each is separated from this **patellar surface** by a slight groove. In flexion of the knee joint, the patella articulates with the distal aspect immediately anterior to the grooves, but in extension, is in contact with the anterior aspect. The midline of the patellar surface is deeply grooved and its lateral part is much more prominent and extensive than the medial part. The lateral and medial margins of the patellar surface can be felt when the knee is bent.

The **medial surface** of the lower end of the femur is rough and has a prominent **medial epicondyle** which gives attachment to the tibial collateral ligament. Above this is the adductor tubercle.

The **lateral surface** has a prominent **lateral epicondyle** on its posterior half often showing a smooth hollow where the fibular collateral ligament is attached. Above this, an impression marks the attachment of the lateral head of gastrocnemius muscle. Below and behind these markings, running parallel to the margin of the articular surface, is a **groove for popliteus muscle**. The tendon of popliteus is attached to the antero-inferior end of the groove and lies in it only in flexion. A notch in the edge of the articular surface below the attachment of the tendon may show where it crosses the edge during extension of the joint.

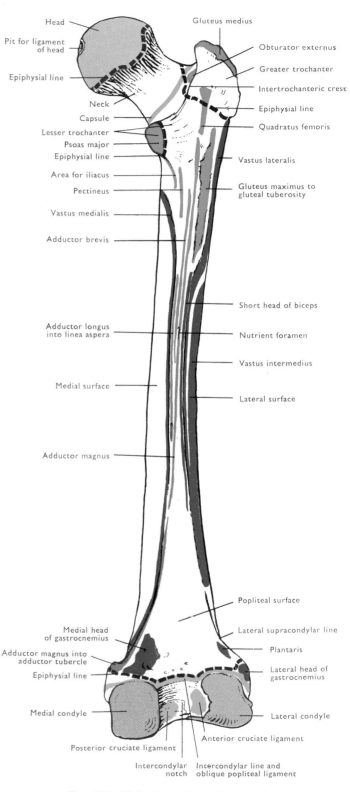

FIG. 228   Right femur (posterior aspect).

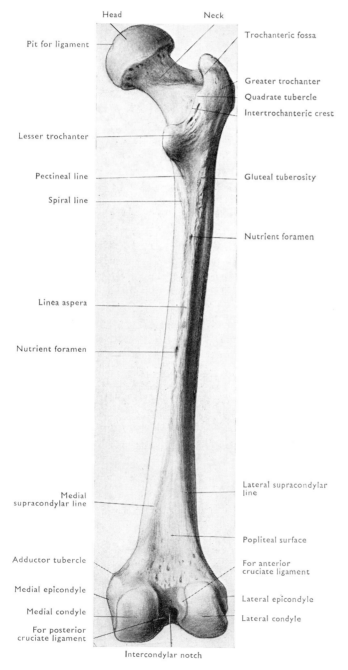

FIG. 227   Right femur (posterior aspect).

Posteriorly, the anterior wall of the intercondylar fossa is rough and marked by vascular foramina. It is limited above by the **intercondylar line**; below, by the articular edge. At the sides its walls are the condyles. The posterior part of the lateral wall of the fossa (lateral condyle) is smooth for the upper attachment of the anterior cruciate ligament. The anterior part of the medial wall (medial condyle) similarly is smooth for the posterior cruciate ligament. The oblique popliteal ligament reaches the intercondylar line and is attached to it.

The Structure of the Femur

The ends are composed of cancellous bone enclosed in a thin shell of compact bone. The body is a thick tube of compact bone, thickest above the middle of the bone and along the linea aspera. It encloses a medullary cavity which extends from the level of the lesser trochanter to approximately 8 cm. above the distal articular surface.

The upper end of the femur is particularly important in walking and weight-bearing and the structure, architecture, and vascularization of the bone have been studied in great detail. Dixon (1910) used stereoscopic radiographs as well as thin sections and dissections of the bone. Koch (1917), Murray (1936), Thompson (1942) and Crock (1967) have all investigated and discussed the anatomy of the region.

The compact bone of the body is continuous with trabeculae which branch off, diverge, and intersect to form the cancellous bone of the neck, the trochanters, and the head. Many of these trabeculae are concentrated to form the walls of a curved tube which is a continuation of the tubular compact bone of the shaft through the neck to the head. If the trochanters and intertrochanteric crest are removed carefully it is possible to demonstrate the continuity of this tube from the body up into the neck. The part which passes deep to the lesser trochanter to reach the posterior aspect of the neck is thickened to form a buttress or spur, the **calcar femorale** [FIG. 232], and is of considerable importance in surgical procedures on the neck. Where the tube lies deep to the greater trochanter it has been referred to as the internal femoral lamina. Trabeculae branch off from the deep and superficial surfaces of the tube to form the adjacent cancellous bone the arrangement of which has been likened to ossified lines of force [PLATES 3 and 15]. The strong trabecular systems are:

1. Trabeculae which pass superomedially from the compact bone of the medial aspect of the body and the inferior aspect of the neck mainly to the upper, weight-bearing part of the articular surface of the head. These are obviously subjected to compression from the weight of the body.

2. Trabeculae which arch across from the compact bone of the lateral aspect of the body, and passing deep to the greater trochanter and through the upper part of the neck, cross the first series approximately at right angles, and reach the articular surface of the head, mainly the lower part. These must be subject to tension when the body weight is applied to the head of the bone.

3. A series of trabeculae arch across from the medial wall of the body to the greater trochanter.

4. Trabeculae also sweep up from the lateral side into the greater trochanter and upper part of the neck of the bone.

The first two of these systems form the inferomedial and

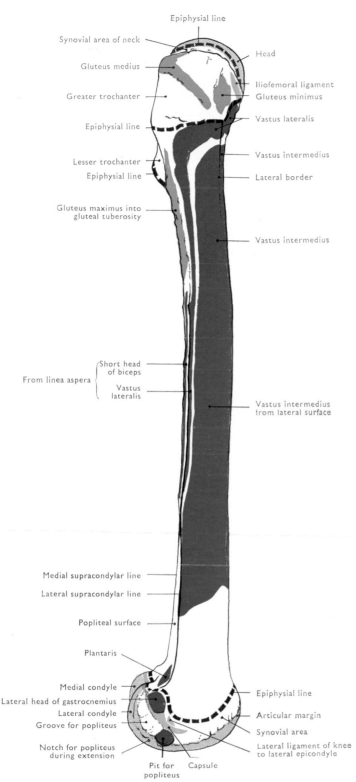

FIG. 229   Right femur (lateral aspect).

Epiphysial line
Synovial area of neck
Gluteus medius
Head
Greater trochanter
Iliofemoral ligament
Gluteus minimus
Epiphysial line
Vastus lateralis
Vastus intermedius
Lesser trochanter
Epiphysial line
Lateral border
Gluteus maximus into gluteal tuberosity
Vastus intermedius
Short head of biceps
From linea aspera
Vastus lateralis
Vastus intermedius from lateral surface
Medial supracondylar line
Lateral supracondylar line
Popliteal surface
Plantaris
Medial condyle
Epiphysial line
Lateral head of gastrocnemius
Lateral condyle
Articular margin
Groove for popliteus
Synovial area
Notch for popliteus during extension
Lateral ligament of knee to lateral epicondyle
Pit for popliteus   Capsule

superolateral walls of the bent tubular continuation of the body.   The area enclosed by the first three systems is relatively free of trabeculae and forms the internal triangle of the femur illustrated by Ward as early as 1838.   Many of the trabeculae are arranged spirally as intersecting right- and left-handed series, an exceedingly efficient arrangement.   Dixon (1910) found that adult femora could stand compression (applied to

the lower end and head of the bone) amounting to between 1,800 and 2,500 lb. Weights of this order produced shearing of the neck before fracture from bending stresses took place.

Ossification

A primary centre for the body and neck appears during the seventh week. A secondary centre for the lower end appears about birth and its presence suggests that a foetus is mature. Details of secondary centres are given in FIGURE 233.   The order in which the epiphyses fuse is lesser trochanter, greater trochanter, head, lower end.   The epiphysial line at the lower end passes through the adductor tubercle, immediately above the patellar surface, and along the intercondylar line [FIG. 228].

THE PATELLA

The patella, or knee-cap, is a sesamoid bone about 5 cm. in diameter, set in the tendon of quadriceps femoris muscle. It is oval in outline but has a pointed **apex,** projecting downwards and so is commonly described as triangular.   The **ligamentum patellae,** which is the continuation of the quadriceps tendon, is attached to the apex and the inferior part of the adjacent posterior surface. *The apex indicates the level of the knee joint when the ligament is taut.* The quadriceps tendon is attached to the upper border or base and some of its fibres are continued over the slightly convex **anterior surface** which is roughened and vertically striated by them as they pass to join the ligamentum patellae.

The lateral and medial edges are rounded and receive fibres of vastus lateralis and medialis respectively.   These, with associated fibres of the fascia lata, form the **retinacula of the patella** [see Knee Joint].

The posterior surface of the patella is largely occupied by an oval **articular surface** for the femur.   This is divided by a vertical ridge into a larger lateral part and a smaller medial part for the corresponding areas on the patellar surface of the femur. The medial and lateral parts are sometimes further subdivided by a slight, transverse ridge into middle and upper facets which are in contact with the condyles during extension and flexion respectively.   A medial facet for the medial condyle during extreme flexion, and an inferior facet for the patellar surface of the femur in extension, may also be seen on the dried bone.

The attachment of the ligamentum patellae below and of the quadriceps tendon above are each separated from the articular surface by

FIG. 230 Right femur (medial aspect).

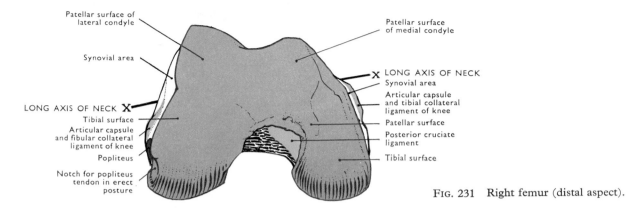

FIG. 231   Right femur (distal aspect).

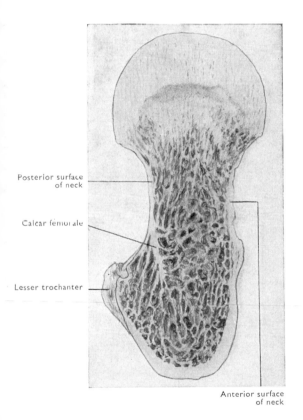

FIG. 232   Section through head and neck of femur to show calcar femorale.

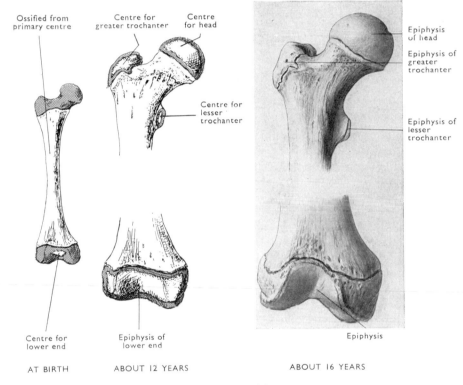

FIG. 233   Ossification of femur.

Secondary centres appear for the head at 6–10 months, for the greater trochanter at 3–5 years, and for the lesser trochanter at 9–12 years: all these centres fuse with the neck or body from 16–20 years, usually in the reverse order of appearance. A centre for the lower end appears just before or just after birth and fuses at 16–19 years.

a non-articular area related to synovial membrane and fat. These areas are frequently perforated by blood vessels.

Ossification

The patella is cartilaginous at birth but ossifies from one or more centres between the third year and puberty (Walmsley, 1940).

Variations

The patella may be absent and this is said to be hereditary. It has also been found with a large upper part and a smaller lower part embedded in the ligamentum patellae. The articular facets can be modified by habitual postures such as squatting.

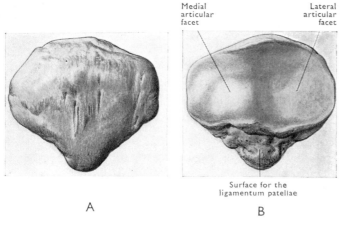

Medial articular facet — Lateral articular facet

Surface for the ligamentum patellae

A   B

FIG. 234   Right patella.

A, Anterior surface.
B, Posterior surface.

THE TIBIA

[PLATES 16 and 17]

The tibia is the larger and medial of the two bones of the leg. It extends from the knee joint to the ankle joint and almost all of its anteromedial surface is subcutaneous.

The proximal end is expanded to form the medial and lateral condyles, which articulate with the corresponding condyles of the femur and the intervening menisci of the knee joint. Both condyles project backwards and the upper part of the anterior surface slopes upwards and backwards, thus giving the upper end a curved appearance. The lateral condyle projects further laterally, and carries a round articular surface which faces downwards and laterally on its posterolateral part, for the head of the fibula. The medial condyle is grooved posteromedially where semimembranosus tendon is attached.

The proximal end has a concave **superior articular surface** on each half with an **intercondylar eminence** between them Uneven anterior and posterior **intercondylar areas** lie in front of and behind the eminence [FIG. 235].

The lateral, superior articular surface is rounded and slightly smaller than the ovoid medial one. Each is concave centrally

to receive a condyle of the femur. The peripheral parts are flatter and crescentic, and each articulates with a meniscus [FIG. 317]. The two menisci are attached to the margins of the condylar articular surfaces via the articular capsule, the medial attachment being reinforced by the tibial collateral ligament.

The **intercondylar eminence** projects up on each side as medial and lateral **intercondylar tubercles** with a ridge passing backwards and laterally from the medial one to give attachment to the posterior horn of the lateral meniscus. Behind this a smooth area gives attachment to the posterior cruciate ligament. A further smooth area in front of the eminence is for the attachment of the anterior cruciate ligament. The remaining attachments between the two articular surfaces [FIG. 317] are: the anterior horn of the medial meniscus is in front; the two horns of the lateral meniscus and the posterior horn of the medial meniscus between the cruciate ligaments; while the posterior cruciate ligament passes well down the sloping hinder part of the posterior intercondylar area [FIG. 316].

Anteriorly, the upper end slopes downwards and forwards to reach the **tuberosity of the tibia** [FIG. 236] which leads down to the anterior border of the body. The smooth upper part of the tuberosity gives attachment to the patellar ligament. The rough lower part is subcutaneous and takes most of the weight of the body in the kneeling position. Its superior edge is raised where it receives the superficial fibres of the patellar ligament.

The anterolateral aspect of the **lateral condyle** is marked by the attachment of the iliotibial tract of the fascia lata: posteriorly its inferior surface carries the **articular surface for the fibula**. Below this, the interosseous border runs down

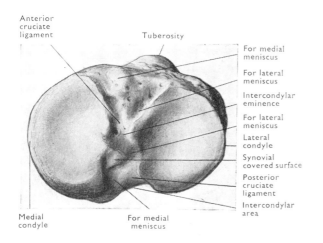

Anterior cruciate ligament
Tuberosity
For medial meniscus
For lateral meniscus
Intercondylar eminence
For lateral meniscus
Lateral condyle
Synovial covered surface
Posterior cruciate ligament
Intercondylar area
Medial condyle
For medial meniscus

FIG. 235   Right tibia (proximal surface).

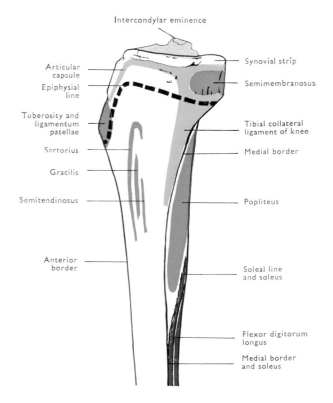

Intercondylar eminence

Articular capsule
Epiphysial line
Tuberosity and ligamentum patellae
Sartorius
Gracilis
Semitendinosus
Anterior border

Synovial strip
Semimembranosus
Tibial collateral ligament of knee
Medial border
Popliteus
Soleal line and soleus
Flexor digitorum longus
Medial border and soleus

FIG. 236   Right tibia (medial aspect of upper half).

the body and a ridge for the fascia covering the tibialis anterior muscle connects its upper end to the area for the iliotibial tract. Biceps femoris gains an attachment to the condyle between the fibular articular surface and the above-mentioned ridge.

The **medial condyle** is roughened medially for the attachment of the articular capsule, fascia lata, and fascia of the leg. It is continuous below with the medial surface of the body and this part is limited posteriorly by the **medial margin of the tibia** where the tibial collateral ligament roughens the bone [FIG. 236]. Behind this, the posterior surface of the medial condyle is grooved horizontally for the tendon of semimembranosus which also inserts into the rough area below the groove [FIG. 241].

Posteriorly, the space between the condyles is smooth and

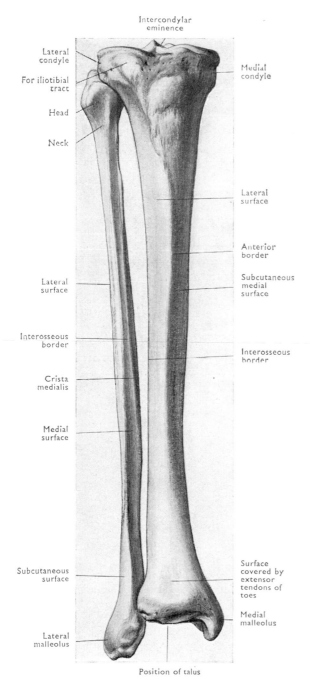

Lateral
condyle

For iliotibial
tract

Head

Neck

Lateral
surface

Interosseous
border

Crista
medialis

Medial
surface

Subcutaneous
surface

Lateral
malleolus

Intercondylar
eminence

Medial
condyle

Lateral
surface

Anterior
border

Subcutaneous
medial
surface

Interosseous
border

Surface
covered by
extensor
tendons of
toes

Medial
malleolus

Position of talus

FIG. 237   Right tibia and fibula (anterior aspect).

passes down as the popliteal surface of the tibia, limited below by the oblique **soleal line** on the body.

The **body of the tibia** tapers slightly from the condyles downwards for two-thirds of its length and then widens again near the lower end.

It has a prominent subcutaneous **anterior border** which begins below the tuberosity and runs down to end by passing medially towards the medial malleolus [FIG. 237].

The **medial surface** slopes posteromedially from the anterior border. It is smooth and almost wholly subcutaneous. At its upper end, the rough mark for the tibial collateral ligament lies at the posterior edge of the surface, and sartorius, gracilis, and semitendinosus muscles are attached between this and the tibial tuberosity without, however, marking the bone appreciably [FIG. 236].

The **lateral surface** slopes back from the anterior border. Its upper two-thirds are slightly concave and give attachment to tibialis anterior muscle. Inferiorly, it is rounded and becomes continuous with the anterior surface of the lower end of the bone. This lateral surface is limited posteriorly by the sharp **interosseous border,** to which the interosseous membrane is attached. The interosseous border begins inferior to the fibular articular surface, and finishes below as a triangular area for the interosseous ligament.

The **posterior surface** of the body lies between the interosseous border and a less defined **medial border** which separates if from the medial surface and runs from the rough mark for the tibial collateral ligament to the posterior edge of the medial malleolus. An oblique **line for the soleus muscle** runs from the fibular articular surface on the lateral condyle to the medial margin a third of the way down. Soleus muscle is attached to this line and to the middle third of the medial margin, whilst the fleshy fibres of popliteus arise from the triangular popliteal part of the posterior surface above the line, and the fascia covering popliteus is attached to the line.

A groove for the tendon of popliteus may be found above the fibular articular surface. Below the oblique line, the posterior surface is divided into a lateral area for tibialis posterior muscle and a medial area for flexor digitorum longus by a vertical line to which the fascia covering tibialis posterior is attached. The nutrient foramen is usually found in the vicinity of the vertical or oblique lines.

The **lower end of the tibia** has a prominent **medial malleolus** projecting down from its medial side. The malleolus has a rounded medial surface which is continuous with the medial surface of the shaft and which forms the prominence at the medial side of the ankle. Its distal border has a pit or notch in its posterior part for attachment of the strong medial (deltoid) ligament of the ankle joint. The distal extremity of the malleolus lies at a level 1–2 cm. higher than that of the lateral malleolus of the fibula.

Posteriorly, the **malleolar sulcus** lodges the tendon of tibialis posterior muscle as it descends on to the medial ligament. The lateral part of the posterior surface of the lower end may also be grooved by the tendon of flexor hallucis longus [FIG. 238]. The inferior edge of the posterior surface and the adjacent margin of the malleolus provide attachment for the articular capsule of the ankle joint and the transverse fibres of the posterior tibiofibular ligament.

Anteriorly, the inferior end of the bone and the malleolus

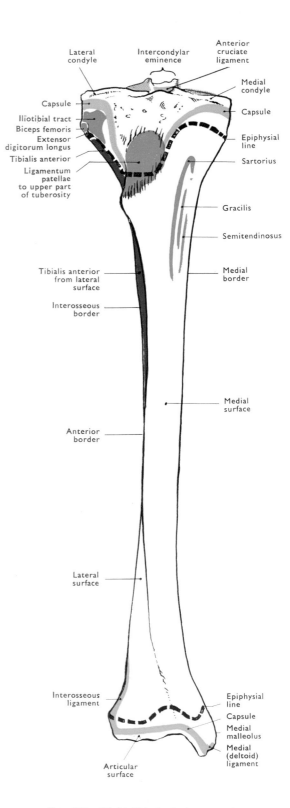

FIG. 238   Right tibia and fibula (posterior aspect).

FIG. 239   Right tibia (anterior aspect).

Intercondylar eminence

Synovial area

Capsule

Iliotibial tract

Facet for fibula

Biceps femoris

Popliteus

Extensor digitorum longus and peroneus longus

Ligamentum patellae into tuberosity

Epiphysial line

Soleus from soleal line

Flexor digitorum longus

Tibialis posterior

Nutrient foramen

Tibialis anterior

Interosseous membrane to interosseous border

Anterior border

Lateral surface

Tibiofibular ligaments in fibular notch

Epiphysial line

Articular area for fibula

Anterior tibiotalar ligament

Medial malleolus

FIG. 240 Right tibia (lateral aspect).

Intercondylar eminence and tubercles

Medial condyle

Posterior cruciate ligament

Synovial area

Capsule

Semimembranosus

Synovial area for popliteus

Articular surface

Capsule of knee joint

Capsule around facet for fibula

Epiphysial line

Tibial collateral ligament of knee

Popliteus

Soleus from soleal line

Nutrient foramen

Soleus from medial border

Flexor digitorum longus

Tibialis posterior

Anterior border

Interosseous border

Tibialis anterior

Interosseous membrane to interosseous border

Posterior surface

Epiphysial line

Groove for tibialis posterior

Medial malleolus

Tibiofibular ligament to floor of fibular notch

Articular area for fibula

Posterior tibiotalar ligament

Medial ligament

Groove for flexor hallucis longus

FIG. 241 Right tibia (posterior aspect).

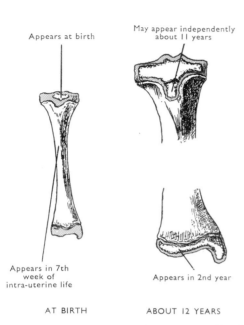

Appears at birth

May appear independently about 11 years

Appears in 7th week of intra-uterine life

Appears in 2nd year

AT BIRTH   ABOUT 12 YEARS

Fuses with shaft between 19-21 years

Fuses 17-19 years

ABOUT 16 YEARS

FIG. 242   Ossification of tibia.

The upper epiphysis includes the tuberosity of the tibia and ossification spreads down to the tuberosity after the tenth year. Fusion with the shaft occurs at 16 to 19 years. A separate centre may appear for the tuberosity.

are smooth and slightly convex. The tendons of the muscles passing down to the foot cover the bone but do not mark it. The inferior edge frequently presents a transverse groove but may be rounded. It may also have a pressure facet near its lateral end caused by full dorsiflexion of the ankle joint.

The **inferior articular surface** has an anteroposterior ridge which fits the groove on the trochlea of the talus. Medially, the articular surface is continuous with the **malleolar articular surface** and laterally it usually turns up on to the lateral surface where the fibula articulates with it. The **lateral surface** is concave anteroposteriorly (**fibular notch**) where it receives the fibula and is continued up as a rough triangular area for the attachment of the interosseous ligament which is continuous above with the interosseous membrane. A synovial fold usually extends up between the lower part of the notch and the fibula.

Ossification

A primary centre for the tibia appears during the seventh week and spreads so that only the ends are cartilaginous at birth. Times of appearance and fusion of the upper and lower epiphyses are given in FIGURE 242.

THE FIBULA

The fibula (brooch bone) is slender but has two slightly expanded ends and is twisted in appearance. The upper end, or head, can be felt as a knob of bone behind and below the lateral side of the knee where the tendon of biceps femoris leads down to it. The lower end is the lateral malleolus, a subcutaneous prominence felt on the lateral side of the ankle joint.

The **head** is roughly cuboidal and tapers below to form a neck and so becomes continuous with the body. It has a conical **apex of the head,** which projects up from the posterolateral part. Medial to this is a round **articular surface of the head** which faces upwards and slightly medially to articulate with the lateral condyle of the tibia. Anterior to the apex, the bone is hollowed where it receives the fibular collateral ligament and part of the insertion of biceps femoris tendon. The main part of the tendon goes to the lateral aspect of the head below the apex [FIG. 245]. Muscles arising from the upper part of the body also arise from the adjacent surface of the neck and head. *The common peroneal nerve lies close to the posterior surface of the head and then runs round the lateral side of the neck of the bone.*

The **body of the fibula** is almost entirely surrounded by muscles and its surfaces correspond to the muscular groups. They are separated by borders which give attachment to the fibrous septa between the muscles.

The **anterior border** is identifiable at its inferior end where is widens into a smooth, triangular, subcutaneous area continuous with the convex lateral surface of the malleolus. This triangle therefore indicates the inferior end of the body, the anterior border, and the lateral aspect of the bone. The border is continued up towards the neck where it becomes indefinite.

The concave **lateral surface** is posterolateral to the anterior border. It extends from the neck down to the lower end of the bone and, becoming convex, winds round posterior to the triangular subcutaneous area to the posterior surface of the malleolus. The whole surface is occupied by the peroneal muscles which arise from its upper two-thirds (peroneus longus) and its lower two-thirds (peroneus brevis) [FIG. 245]. The tendons pass behind the subcutaneous area to reach the back of the malleolus.

The posterior border is roughened above by the attachment of soleus [FIG. 244], lies posterior to the lateral surface, and extends from the lateral aspect of the head and neck down to the medial margin of the posterior surface of the malleolus.

The **interosseous border** is often poorly marked and lies medial to the anterior border. It extends from the neck down to a rough triangle superior to the articular surface on the malleolus. The border gives attachment to the interosseous membrane. The rough triangle is for the interosseous ligament [FIG. 246].

The **medial** (anterior) **surface** is the narrow strip between the anterior and interosseous borders. It gives origin to the extensor group of muscles including extensor digitorum longus, peroneus tertius, and extensor hallucis longus [FIG. 243].

The **posterior surface** occupies the remaining part of the body, between the interosseous and posterior borders. A prominent **medial crest** runs down from the neck of the fibula for two-thirds of the length of the body and then curves forwards to join the interosseous border. The area between

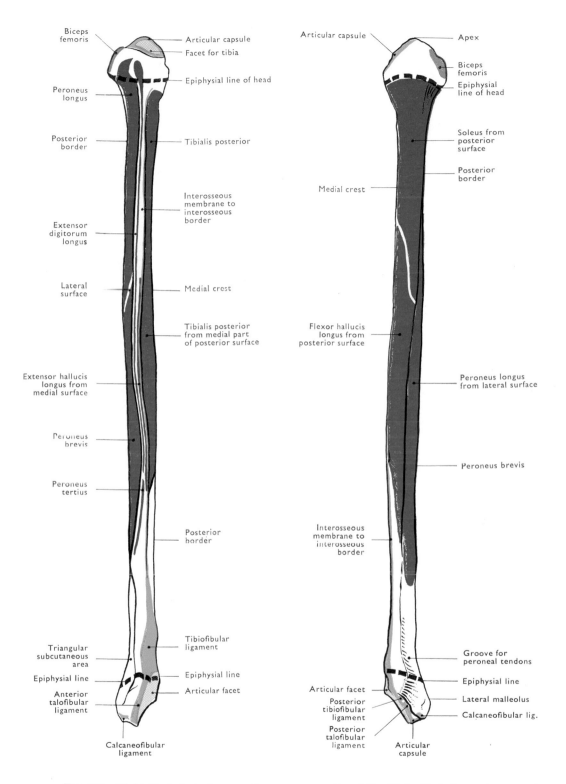

Biceps femoris
Articular capsule
Facet for tibia
Epiphysial line of head
Peroneus longus
Posterior border
Tibialis posterior
Interosseous membrane to interosseous border
Extensor digitorum longus
Lateral surface
Medial crest
Tibialis posterior from medial part of posterior surface
Extensor hallucis longus from medial surface
Peroneus brevis
Peroneus tertius
Posterior border
Triangular subcutaneous area
Tibiofibular ligament
Epiphysial line
Epiphysial line
Anterior talofibular ligament
Articular facet
Calcaneofibular ligament

Articular capsule
Apex
Biceps femoris
Epiphysial line of head
Soleus from posterior surface
Medial crest
Posterior border
Flexor hallucis longus from posterior surface
Peroneus longus from lateral surface
Peroneus brevis
Interosseous membrane to interosseous border
Groove for peroneal tendons
Epiphysial line
Articular facet
Lateral malleolus
Posterior tibiofibular ligament
Calcaneofibular lig.
Posterior talofibular ligament
Articular capsule

FIG. 243 Right fibula (anterior aspect).   FIG. 244 Right fibula (posterior aspect).

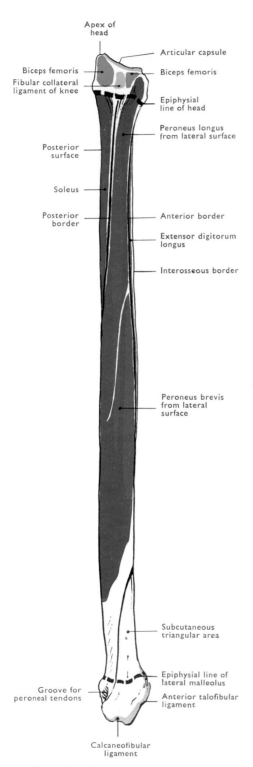

Apex of
head

Articular capsule

Biceps femoris

Biceps femoris

Fibular collateral
ligament of knee

Epiphysial
line of head

Peroneus longus
from lateral surface

Posterior
surface

Soleus

Posterior
border

Anterior border

Extensor digitorum
longus

Interosseous border

Peroneus brevis
from lateral
surface

Subcutaneous
triangular area

Epiphysial line of
lateral malleolus

Groove for
peroneal tendons

Anterior talofibular
ligament

Calcaneofibular
ligament

FIG. 245   Right fibula (lateral aspect).

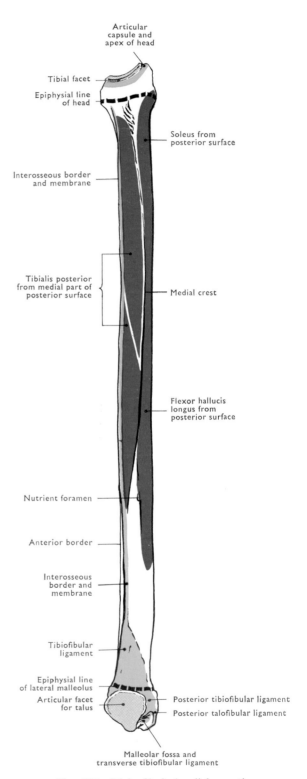

Articular
capsule and
apex of head

Tibial facet

Epiphysial line
of head

Soleus from
posterior surface

Interosseous border
and membrane

Tibialis posterior
from medial part of
posterior surface

Medial crest

Flexor hallucis
longus from
posterior surface

Nutrient foramen

Anterior border

Interosseous
border and
membrane

Tibiofibular
ligament

Epiphysial line
of lateral malleolus

Articular facet
for talus

Posterior tibiofibular ligament

Posterior talofibular ligament

Malleolar fossa and
transverse tibiofibular ligament

FIG. 246   Right fibula (medial aspect).

the crest and the interosseous border is concave and gives origin to tibialis posterior muscle. It is usually subdivided by an oblique line, caused by a fascial septum in the muscle substance. The flat area between the medial crest and the posterior border is roughened in its upper third by soleus muscle. Below that it gives origin to flexor hallucis longus muscle, and turns on to the medial aspect of the body below the lower end of the medial crest [FIG. 246] where it is smooth and has no attachments.

The **lower end** of the fibula forms the **lateral malleolus.** It is more flattened than the upper end and has a rounded subcutaneous lateral surface. The medial aspect has a **malleolar articular surface** which fits the lateral malleolar surface on the talus. It is approximately triangular and the inferior apex is slightly convex. The base of the triangle extends up as far as the tibia and the joint cavity usually extends between the two bones. Here the bone is marked by the anterior and posterior tibiofibular ligaments, anterior and posterior to the articular surface.

Posterior to the articular facet, on the medial surface, is a very definite **fossa of the lateral malleolus.** Here the transverse tibiofibular ligament is attached with the posterior talofibular ligament. The malleolus projects down from the region of the fossa as a conical process which gives attachment to the calcaneofibular ligament and its anterior margin gives attachment to the anterior talofibular ligament.

The tip of the lateral malleolus projects down 1–2 cm. lower and is further back than that of the medial malleolus. The anterior aspect presents a rounded margin which is roughened

for the ligaments mentioned above. The posterior aspect is wide above but narrows inferiorly. It is grooved by the tendon of peroneus brevis which is held to the bone with that of peroneus longus by the superior peroneal retinaculum which is attached to the edges of the groove.

Ossification

A primary centre appears in the body in the seventh week and spreads so that only the ends are cartilaginous at birth. Secondary centres appear and fuse as shown in FIGURE 247. At birth the fibula is relatively thick, half as thick as the tibia in the third prenatal month. Also, the lower end does not reach the adult level, below the medial malleolus, until its ossification has begun.

THE BONES OF THE FOOT

The bones of the foot consist proximally of seven short **tarsal bones** (the talus, calcaneus, cuboid, navicular, and three cuneiforms) collectively known as the **tarsus.** They correspond to the carpal bones of the wrist, and articulate with each other and distally with the five **metatarsal bones** (the metatarsus) each of which carries the **phalanges** of the corresponding toe.

The foot is a lever which is used to raise the body and transmit thrust when walking and running. To give sufficient strength and resilience for the considerable forces involved, and to protect the structures in the sole from compression, the foot is an arched structure. The stability in standing is increased by the bones of the foot being so arranged that each foot is a half dome, the inferior edge of which extends from the heel along the lateral border of the foot, and then medially across the ball of the foot (heads of the lateral four metatarsals) to the ball of the great toe (head of the first metatarsal). The medial side of the foot is highly arched, the lateral side almost flat.

The foot may be divided into two parts. (1) **The talus,** which lies near the highest part of the dome, forms the hinge joint of the ankle with the tibia and fibula above [FIG. 248], and rests on the other tarsal bones, articulating with the **calcaneus** below and the **navicular** in front, and transmitting the weight directly to them. (2) **The tarsal bones, other than the talus,** are bound to each other and to the metatarsals (a) by **plantar ligaments** which maintain their arched form and prevent the talus from being forced down between the pillars of the arch (plantar calcaneonavicular ligament), and (b) by more superficial structures which, passing between the ends of the arch, maintain it much as a bowstring maintains the curvature of a bow. These tarsal and metatarsal bones confer elasticity on the foot by a small degree of movement on each other, but move as a unit on the talus in inversion and eversion (q.v.). Since the movement of inversion produces an effect similar to the presence of a high arch, and eversion does the opposite, the muscles which produce these movements also play a part in the shape and arrangement of the arches.

Fuses 19-21 years

Appears about 3-4 years

Appears at 7th week of intra-uterine life

Appears in 2nd year

Fuses 17-19 years

AT BIRTH ABOUT 12 YEARS ABOUT 16 YEARS

FIG. 247 Ossification of fibula.

Note that the centre for the distal end fuses *before* that for the proximal end.

Distal phalanx

Middle phalanx

Proximal phalanx

Sesamoid bone

1st metatarsal

5th metatarsal

Cuneiforms {

Intermediate

Medial

Lateral

Navicular

Cuboid

Superior surface of
talus for articulation
with tibia

Part of upper surface
of calcaneus covered
with fatty tissue

Lateral process of talus,
at lower part of surface
for articulation with
lateral malleolus

Distal phalanx

Middle phalanx

Proximal phalanx

Sesamoid bones

5th metatarsal

1st metatarsal

Lateral cuneiform

Medial cuneiform

Intermediate
cuneiform

Cuboid

Navicular

Surface of talus
rests on plantar
calcaneonavicular
ligament

Groove for flexor
hallucis longus on
sustentaculum tali

Calcaneus

FIG. 248   Bones of right foot (superior or dorsal surface).   FIG. 249   Bones of right foot (inferior or plantar surface).

PLATE 15

A

(i)   (ii)

(iii)   (iv)

A.   Radiographs of the proximal ends of four dried femora.

(i) Normal angle of neck.
(ii) Normal femoral head in old age.   Note relative demineraliza-
    tion and new bone formation around the margin of the head.
(iii) The angle between the neck and the shaft is increased:
    coxa valga.
(iv) The angle between the neck and the shaft is diminished:
    coxa vara.

B

B.   Anteroposterior radiograph of the hip joint.

1. Greater trochanter of the femur.
2. Lesser trochanter of the femur.
3. Acetabulum.
4. Ischial tuberosity.

C

C.   Anteroposterior radiograph of the knee.

1. Medial condyle of the femur.
2. Medial condyle of the tibia.
3. Lateral condyle of the femur.
4. Intercondylar tubercles of the tibia.
5. Lateral condyle of the tibia.
6. Head of fibula.

D

D.   Lateral radiograph of the knee.

1. Sesamoid (fabella) in the lateral head of gastrocnemius.
2. Apex of head of fibula.
3. Margin of lateral condyle of femur.
4. Margin of medial condyle of femur.

Note different curvatures of femoral condyles.

PLATE 16

A

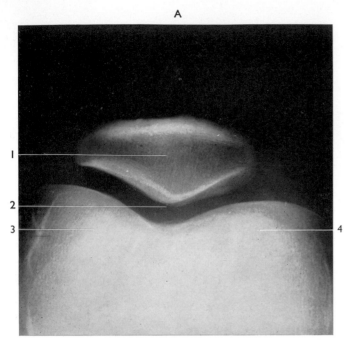

A. Axial radiograph of the joint between the patella and the femur to show the thickness of the cartilage.

1. Patella.
2. Articular surface of the patella.
3. Lateral condyle of the femur.
4. Medial condyle of the femur.

B

B. Anteroposterior radiograph of the knee joint after injection of air.

1. Shadow of the patella superimposed upon that of the femur.
2. Suprapatellar pouch filled with air.
3. 'Joint space' (cartilage) between medial femoral and tibial condyles.

C

C. Anteroposterior radiograph of the knee. Age 12.

1. Epiphysis of the distal end of the femur.
2. Epiphysis of the proximal end of the fibula.
3. Epiphysial cartilage at the proximal end of the tibia.

D

D. Lateral radiograph of the knee. Age 12.

1. Patella.
2. Epiphysial cartilage at the distal end of the femur.
3. Epiphysis of the distal end of the femur.
4. Epiphysis of the proximal end of the tibia.
5. Anterior and distal extension of tibial epiphysis (tibial tuberosity).

PLATE 17

A

B

A. Anteroposterior radiograph of ankle. Age 12.

1. Epiphysial cartilage of distal end of fibula.
2. Lateral malleolus.
3. Epiphysial cartilage of distal end of tibia.
4. Ankle joint ' space '.
5. Medial malleolus.
6. Talus.

B. Lateral radiograph of ankle. Age 12.

1. Epiphysis of distal end of tibia.
2. Calcaneus.
3. Epiphysis of calcaneus.
4. Ankle joint.
5. Talus.

C

D

C. Anteroposterior radiograph of adult ankle.

1. Tibiofibular syndesmosis.
2. Talofibular articulation.
3. Line of arrested bone growth in tibia.
4. Ankle joint.
5. Talus.

D. Lateral radiograph of adult ankle.

1. Ankle joint.
2. Sinus tarsi.
3. Talonavicular joint.

PLATE 18

A

(i)   (ii)

B

A.   Radiograph of hand (i) and foot (ii) at birth.

Note the absence of carpal ossification, but well developed ossification in the calcaneus and talus.

B.   Lateral radiograph of the ankle at the age of 8 years.

1. Distal fibular epiphysis.
2. Epiphysis of calcaneus.
3. Tibia.
4. Distal tibial epiphysis.

C

(i)

D

(ii)

C.   (i) Oblique radiograph of the foot at the age of 12 years.

1. Epiphysis of 1st metatarsal.
2. Epiphysis of tuberosity of 5th metatarsal.
3. Sesamoid bones.

(ii) Oblique radiograph of the foot of an adult.

D.   Plantar view of the metatarsus and phalanges of an adult.

Note the angulation of the phalanges on the metatarsal of the hallux.

Because the foot is a half dome, arches may be traced over it in any direction. However, the architecture of the bones is such that obvious **longitudinal** and **transverse arches** are present. The foot is narrow posteriorly (the heel) and broad anteriorly (the ball of the foot and of the great toe). Hence it is possible to describe lateral and medial parts of the longitudinal arch which have a common posterior end (the calcaneus) but separate anteriorly. The **lateral part of the longitudinal arch** extends from the calcaneus through the cuboid and the lateral two metatarsals, and is sufficiently flat to allow the soft tissues of the lateral border of the foot to touch the ground. The **medial,** highly arched part extends from the calcaneus through the body, neck, and head of the talus, the navicular bone, the three cuneiforms, and the medial three metatarsals. The most medial element of this part, calcaneus, talus, navicular, medial cuneiform, and the massive first metatarsal, is the highest part of the arch. It receives support from the tendons of tibialis posterior and anterior, and from flexor hallucis, all of which hold the bones together and help to maintain this part of the arch, the anterior pillar of which carries the thrust of the kick-off in walking and running.

The anterior part of the **transverse arch** is formed by the metatarsals and is low. It is maintained largely by the subjacent transverse structures. More posteriorly, the transverse arch is formed by the sloping cuboid bone and the wedge-shaped cuneiforms and this part derives considerable support from all the tendons on the medial side of the sole of the foot, particularly tibialis posterior because of its extensions to most of the tarsal bones. The curvature is also maintained by the bow-string effect of peroneus longus tendon.

The weight of the body is transmitted downwards and backwards through the body of the talus to the calcaneus and the heel, downwards and forwards through the head and neck of the talus towards the ball of the foot and the ball of the great toe. Laterally, the weight passes through the cuboid to the metatarsals [see also FIG. 468]. In these bones as elsewhere, the supporting trabeculae are so arranged as to resist the forces to which they are subjected, and are particularly developed in the calcaneus where the load is applied to a relatively small area compared with the fore part of the foot.

Knowledge of the details of each bone is necessary to understand the part each plays in the functioning of the foot.

THE TARSUS

THE TALUS

The talus or ankle bone is characterized by a squarish body, a short neck, and a head with an oval articular surface [FIGS. 250 – 253].

The **body** lies between the malleoli of the tibia and fibula and its upper surface is grooved longitudinally to form a pulley-shaped **trochlea** which articulates with the inferior surface of the tibia and is continuous with the medial and lateral **malleolar surfaces** at the sides. The body is thus held between the two malleoli and its upper surface takes the weight transmitted from the tibia [FIGS. 250 and 252 for details]. The articular surface may extend posteriorly or

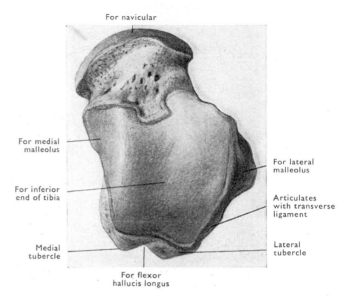

FIG. 250   Right talus (superior surface.)

The superior articular surface is bevelled at its posterior part by contact with the transverse tibiofibular ligament so that the trochlea is considerably narrower at its posterior end.

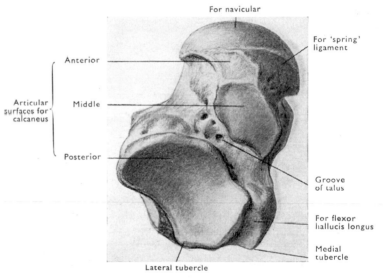

FIG. 251   Right talus (inferior surface).

The groove of talus is the sulcus tali.

anteriorly beyond the normal limit, thus indicating a wide range of movement at the ankle joint.

The **medial malleolar surface** [FIG. 253] is comma-shaped, with the head of the comma extending towards or on to the neck anteriorly, where it may be distinctly concave to receive the medial malleolus in full dorsiflexion.

Inferior to the concave edge of the comma, the medial surface of the bone is uneven and is continuous posteriorly with the prominent **medial tubercle of the posterior process** [FIGS. 250 and 253]. This gives attachment to the deep part of the medial ligament of the ankle joint and the attachment is continued forwards on to the neck.

The **lateral malleolar surface** [FIG. 252] is triangular and is continued from the superior surface down on to a projecting **lateral process.** The narrow area anterior to the surface

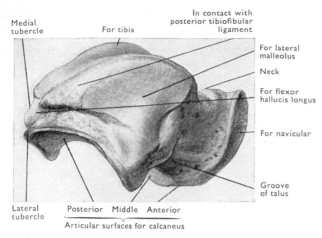

FIG. 252   Right talus (lateral surface).

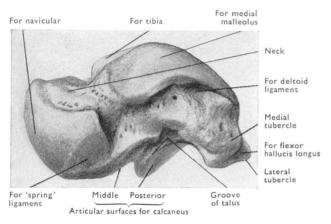

FIG. 253   Right talus (medial surface).

provides attachment for the anterior talofibular ligament. The similar area posterior to the surface is for the posterior talofibular ligament and in each case the cartilage of the surface may be deeply notched by an extension into it. The posterior talofibular ligament is also continued on to the posterior aspect of the body, and the bone in this region projects back as a very variable **lateral tubercle of the posterior process** [see p. 198].

The **posterior process** extends inferiorly from the trochlea, and is grooved obliquely by the tendon of the flexor hallucis longus muscle descending between the lateral and medial tubercles, and covered posteriorly by ligamentous fibres [FIG. 250].

The **inferior surface** of the body is mainly occupied by an oval, deeply concave **posterior articular surface for the calcaneus.** This is set obliquely with its long, concave axis extending from the inferior aspect of the posterior process to that of the lateral process. The surface is frequently modified in shape by habitual movements of the foot.

Anterior to the posterior articular surface is a deep **sulcus tali.** The sulcus froms the **sinus tarsi** with the corresponding sulcus on the calcaneus, and the anterior and posterior walls of the sinus are completed by the capsules of the adjacent joints.

The **middle articular surface for the calcaneus** is medially placed in front of the sulcus tali, and articulates with the sustentaculum tali of the calcaneus. It is flat, nearly oval,

and may meet the posterior articular surface. Anteriorly, it is separated from the anterior articular surface for the calcaneus by a ridge or, occasionally, by a non-articular gap. Laterally, the sulcus tali widens inferolateral to the neck of the talus, and contains the cervical ligament of the talus [p. 251] and the inferior part of the inferior extensor retinaculum [FIG. 325].

The **anterior articular surface for the calcaneus** lies on the inferior surface of the head and articulates with the anterior part on the calcaneus. It is continuous with the main articular surface of the head and is usually continuous with the middle calcanean surface. Medial to the anterior calcanean surface, the articular surface of the head lies against the plantar calcaneonavicular (spring) ligament supported by the tendon of tibialis posterior. This area is more fragile than the adjacent articular surfaces and may be indistinctly outlined.

The **neck of the talus** projects forwards, downwards, and medially, and the medial border is shorter and lower than the lateral border. The angle between the axes of the neck and body, as seen from above, is normally about 18 degrees. It varies, however, between 0 degrees in old age and 30 degrees in the new-born and may be as much as 50 degrees in cases of clubfoot (varus deformity). The downward direction of the neck is best seen from the medial side and is maximal when the arch is high.

The inferior aspect of the neck forms part of the sinus tarsi laterally; medially it is covered by the middle articular surface for the calcaneus.

The prominent lateral border gives attachment to ligaments, including the anterior talofibular and cervical ligaments. The short medial border is frequently encroached on by the medial malleolar surface. It carries a ridge which passes from the medial surface of the body to the superior surface of the neck where it provides attachment for the capsules of the ankle and talonavicular joints. The upper surface frequently shows medial or lateral extensions of the trochlear surface which indicate habitual strong dorsiflexion of the ankle joint [FIG. 250]. There may also be a lateral facet or smoothing on the neck, or, much more rarely, a pressure facet on the medial part of the ridge for the capsule of the ankle joint.

The **head of the talus** is oval in outline and evenly convex, except for the anterior calcanean surface. The articular surface for the navicular faces forwards and slightly medially and downwards, and its long axis slopes downwards and medially. It forms the navicular part of the talocalcaneonavicular joint.

THE CALCANEUS

The calcaneus or heel bone lies inferior to the talus and projects back to form the prominence of the heel. Anteriorly, it articulates with the cuboid bone and is strongly bound to it and to the remaining tarsal bones by ligaments.

The internal, supporting trabeculae of bone pass downwards and backwards to the heel, downwards and forwards to the anterior end, and horizontally from the heel to the anterior end, thus leaving between these an area comparatively transparent to X-rays which may suggest bone resorption by disease [FIG. 98 and PLATE 3].

The anterior end of the calcaneus is raised a little in its normal position to make an angle of a few degrees with the horizontal and so contributes to the longitudinal arch of the foot. The intermediate part of the upper surface projects medially as a thick shelf which supports the neck of the talus and so is called the **sustentaculum tali** [FIGS. 249 and 257].

The **posterior surface** is enlarged and presents three areas. The uppermost is smooth where a bursa lies between it and the tendo calcaneus. The middle, also smooth and convex, receives that tendon and usually has an irregular or jagged inferior edge. The lowest, **tuber calcanei,** is rough and covered by the strong fibrous tissue and fat of the heel pad. Its inferior part transmits the weight of the body from the heel to the ground, and curves forwards as a raised area on the inferior surface. This part of the tuber has a larger **medial process** and a smaller but distinct **lateral process** extending anteriorly from it. These give attachment to the strong fascia and short muscles of the sole of the foot. To the medial process are attached parts of abductor hallucis and flexor digitorum brevis, abductor digiti minimi, the flexor retinaculum, and the

plantar aponeurosis. The abductor digiti minimi and the aponeurosis also arise from the lateral process [FIG. 269].

The **inferior surface** is continued forwards as a rough area which is 2–3 cm. wide and terminates as an **anterior tubercle.** The strong, **long plantar ligament** is attached to this rough area. The **plantar calcaneocuboid ligament** passes forwards from the anterior aspect of the tubercle and from the depression between it and the anterior end of the bone.

The **anterior surface** of the calcaneus is occupied by a concavoconvex articular surface for the cuboid bone, which is overhung superomedially by the anterior surface for the talus. The medial part of the articular surface extends on to the medial surface to accommodate a posterior projection on the cuboid bone The shape prevents any simple sliding movement.

The **superior surface** [FIG. 254] is marked by a large, oval **posterior articular surface for the talus,** which is convex anteroposteriorly. Between this and the anterior

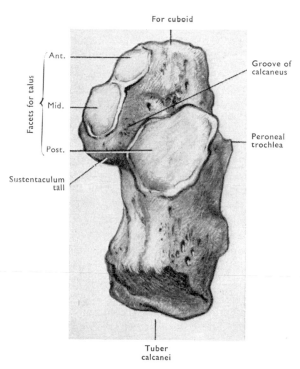

FIG. 254   Right calcaneus (superior surface).

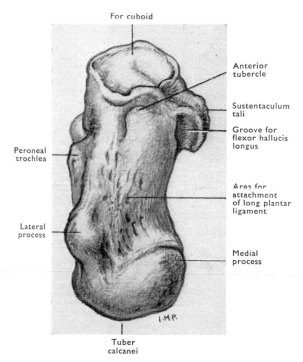

FIG. 255   Right calcaneus (inferior surface).

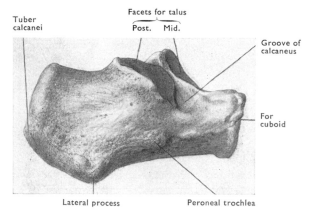

FIG. 256   Right calcaneus (lateral surface).

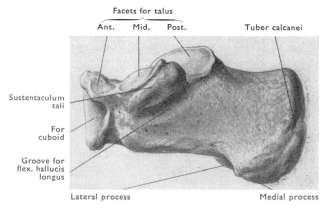

FIG. 257   Right calcaneus (medial surface).

surface is a rough quadrilateral area which gives attachment to the extensor digitorum brevis muscle and to ligaments, particularly the **inferior extensor retinaculum,** the **cervical ligament of the talus** and, near the anterior edge, the **ligamentum bifurcatum.** The medial part of the area has the anterior articular surface for the talus anteriorly, and forms the floor of the **sinus tarsi** posteriorly.

Medial to the main surface for the talus, the **sustentaculum tali** projects with a thick rough medial border which gives attachment to the medial ligament of the ankle joint and, further forwards, to the **plantar calcaneonavicular (spring) ligament.** It also receives a slip from tibialis posterior tendon.

The medial border of the sustentaculum is covered by the flexor digitorum longus tendon but can be felt 2 cm. below the medial malleolus. The upper surface of the sustentaculum carries a **middle articular surface for the talus.** This may be separate or it may be fused with the smaller anterior articular surface for the head of the talus just lateral to the plantar calcaneonavicular ligament. The groove between the posterior and middle articular surfaces is the **sulcus calcanei** which forms the floor of the sinus tarsi. It receives fibres of the inferior extensor retinaculum [FIG. 325] but no single interosseous ligament as such.

The root of the inferior surface of the sustentaculum is grooved by the flexor hallucis longus tendon, and the posterior tibial vessels and nerve lie between this and the edge of the sustentaculum. The **medial surface** of the calcaneus is smooth and concave between the groove and the medial process of the tuber calcanei. It gives origin to part of flexor accessorius muscle [FIG. 463] and is continued upwards and backwards to a saddle-shaped area between the insertion of the tendo calcaneus and the posterior surface for the talus.

The **lateral surface** of the calcaneus is nearly flat but has a tubercle on its lower part half-way along its length. This is the **peroneal trochlea** which intervenes between the two peroneal tendons (brevis and longus) and provides attachment for the inferior peroneal retinaculum which holds them in place.

THE CUBOID

The cuboid bone, approximately cubical in shape [FIG. 248], is placed at the lateral side of the foot between the calcaneus behind and the fourth and fifth metatarsal bones in front.

The **posterior surface** is articular and forms, with the calcaneus, the calcaneocuboid joint. The surface extends on to a pointed **posterior process** at its medial corner and so becomes concavoconvex. The process projects towards the medial surface of the calcaneus under the plantar calcaneonavicular ligament.

The **superior surface** is roughened by the attachments of ligaments passing to adjacent bones but is nearly flat and faces laterally as well as upwards.

The **lateral surface** is short and narrow and lies in the lateral border of the foot between the base of the fifth metatarsal and the calcaneus. It is encroached on by a **groove for peroneus longus tendon** on the inferior surface. The **tuberosity** on the lateral end of the ridge forming the posterior wall of the

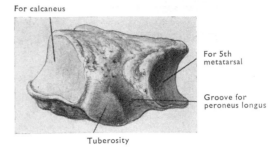

FIG. 258   Right cuboid bone (lateral surface).

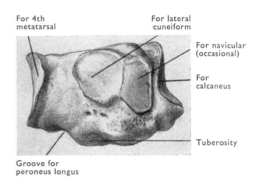

FIG. 259   Right cuboid bone (medial surface).

groove carries an articular surface for a sesamoid bone in the tendon where it turns into the groove.

The **anterior surface** is nearly flat and articulates with the bases of the fourth and fifth metatarsals, the surfaces for these being separated by a slight ridge. The tuberosity on the base of the fifth metatarsal projects back over the lateral surface and the surface of the cuboid slopes back slightly to the lateral margin.

The **inferior surface** tapers to a point on the posterior process and has a pronounced **ridge** which passes forwards and medially across it from the articular surface for the peroneal sesamoid bone. The **groove** between the ridge and the edge of the anterior surface accommodates the peroneus longus tendon. The triangular area behind the ridge gives origin to the flexor hallucis brevis muscle medially and to the plantar calcaneocuboid ligament. Superficial to this, the long plantar ligament passes to the ridge and to the anterior edge of the groove, converting the groove to a tunnel and then passing distally to reach the metatarsals [FIG. 326]. A slip from tibialis posterior tendon reaches the groove for peroneus longus.

The **medial surface** is rough for ligaments and faces upwards as well as medially to meet the corresponding surface of the lateral cuneiform bone and support the transverse arch of the foot. A round articular surface in the posterosuperior angle of the surface articulates with the cuneiform, and a smaller surface is usually present behind it for articulation with the navicular bone.

THE NAVICULAR

The navicular bone is characterized by a concave posterior surface, a convex anterior surface, and a narrow rounded inferomedial end which forms a prominent **tuberosity** [FIGS. 260 and 261].

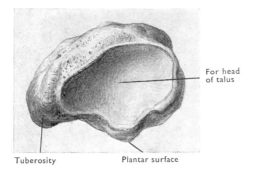

FIG. 260   Right navicular bone (posterior surface).

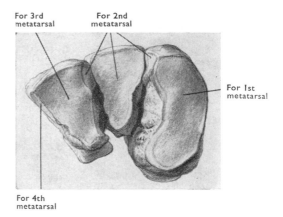

FIG. 262   Anterior aspect of cuneiform bones of right foot.

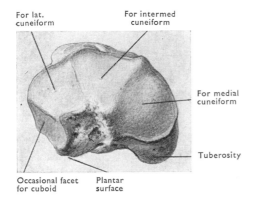

FIG. 261   Right navicular bone (anterior surface).

The dorsal and lateral aspects are rough near their edges for ligaments but together they form a curved surface. The **tuberosity** of the navicular bone can be felt, and is usually visible, on the medial side of the foot 3 cm. in front of the sustentaculum tali, 3–4 cm. below and in front of the medial malleolus. Tibialis posterior tendon has its main insertion into the tuberosity.

The inferior surface is rough and is usually prominent in its middle part for the attachment of ligaments, particularly the plantar calcaneonavicular ligament. Anteriorly it slopes up to the anterior articular surface. The posterior surface is articular and is deeply concave to fit the convexity of the head of the talus.

The anterior surface is also articular and is separated by ridges into three surfaces to articulate with the three cuneiform bones, and like them, arranged in a curve convex inferiorly. At the lateral side there may be an additional articular surface for the cuboid bone.

THE CUNEIFORM BONES

The three cuneiform bones are wedge-shaped. The **medial** one is the largest with a convex medial surface and a large kidney-shaped articular surface on its anterior end. The **intermediate** one is distinctly wedge-shaped and has a square dorsal surface as a base to the wedge. The **lateral** one is also distinctly wedge-shaped but the dorsal surface is longer than it is broad and has an angulated lateral border.

The three cuneiforms articulate with the navicular bone posteriorly and with the bases of the first three metatarsals anteriorly. The line of the tarsometatarsal joints runs obliquely backwards and laterally, the anterior ends of the cuneiforms and the cuboid bone being nearly flush. However, the short intermediate cuneiform allows the base of the second metatarsal to extend proximally between the medial and lateral cuneiforms and the fourth metatarsal overlaps the lateral surface of the lateral cuneiform for a short distance [FIG. 248].

The **medial cuneiform** is thicker and more rounded on its plantar than on its dorsal aspect. Its superior, medial, and inferior aspects form a continuous curved surface which forms part of the medial side of the foot. It is roughened by ligaments and has, at the antero-inferior corner of its medial aspect, a smooth impression over which the tendon of tibialis anterior runs to its insertion on the plantar aspect of the cuneiform and the base of the first metatarsal. A bursa intervenes between the bone and the tendon as it approaches its insertion. The posterior part of the inferior aspect receives a considerable portion of the tibialis posterior tendon [FIG. 454].

The **anterior surface** has a large kidney-shaped articular surface, often divided by a groove, for the base of the first metatarsal. The posterior surface has a smaller pear-shaped concave facet for the medial impression on the navicular bone.

The **lateral surface** is rough for the interosseous ligaments and has an L-shaped articular surface for the intermediate cuneiform bone. This surface is continuous with the surface for the navicular bone, and is continued along the upper border of the lateral surface [FIG. 264] to a small articular surface for the base of the second metatarsal at the anterior end. The peroneus longus tendon is attached to the antero-inferior corner of this surface.

The **intermediate cuneiform** is wedge-shaped with its base uppermost. Its superior surface is roughened and almost square in outline.

Posteriorly, its triangular articular surface is concave to fit the middle impression on the navicular bone. Anteriorly, a flat triangular surface articulates with the base of the second metatarsal bone.

The medial surface is rough for ligaments, but has a large L-shaped articular surface for the medial cuneiform. The lateral surface has a slightly concave, B-shaped articular surface at its posterior border for the lateral cuneiform. Both

FIG. 263 Right medial cuneiform (medial side).

FIG. 266 Right intermediate cuneiform (lateral side).

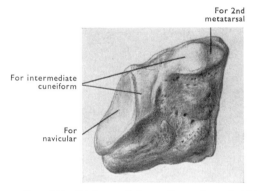

FIG. 264 Right medial cuneiform (lateral side).

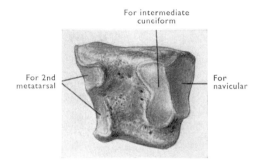

FIG. 267 Right lateral cuneiform (medial side).

FIG. 265 Right intermediate cuneiform (medial side).

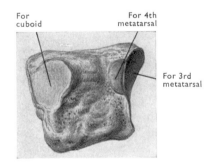

FIG. 268 Right lateral cuneiform (lateral side).

these articular surfaces are continuous with the articular surface for the navicular.

The inferior edge receives a slip from tibialis posterior tendon.

The **lateral cuneiform,** also wedge-shaped, has a rectangular base uppermost, which is roughened by ligaments and has the posterolateral corner cut off by its articulation with the cuboid.

Posteriorly, the surface is oval and slightly concave where it articulates with the navicular bone. Anteriorly the articular surface is triangular where it meets the base of the third metatarsal. This articular surface is prolonged a little on to the medial, and sometimes the lateral, surface where it is overlapped by the second and fourth metatarsals. The medial and lateral surfaces are both rough for ligaments but in addition to the above small articular extensions, they have articular surfaces for the adjacent bones. The posterior part of the medial surface articulates (B-shaped surface) with the intermediate cuneiform, while the posterior part of the lateral surface carries a large round articular surface for the cuboid bone.

The inferior edge receives a slip from tibialis posterior tendon.

OSSIFICATION

The bones of the tarsus are each ossified from one primary endochondral centre, with a secondary centre in the case of the calcaneus.

The **primary centres** for the calcaneus and talus appear before birth, in the sixth and eighth months or a little earlier. The centre for the cuboid may be present at birth or soon after. These can therefore be used to estimate the degree of maturity of a new-born child. The centre for the lateral cuneiform does not appear until the end of the first year, to be followed by the navicular and the other two cuneiforms in the third and fourth years. Ossification is completed after puberty.

The **secondary centre** appears in the posterior end of the calcaneus about the ninth year, and extends to include the medial and lateral processes. Fusion occurs between 15 and 20 years. The lateral tubercle of the posterior process of the talus may ossify separately as an **os trigonum.**

THE METATARSUS

The metatarsus consists of five metatarsal bones. Each metatarsal corresponds to a toe and they are numbered from the medial side. They are miniature long bones, each with a base proximally, a body, and a head distally. The flat triangle seen on the dorsum of a metacarpal body is absent [FIG. 248, cf. FIG. 208].

The **first metatarsal** is shorter than the rest but is much stronger. Its **base** has a kidney-shaped articular surface for the medial cuneiform, the convex border being medial. The bone close to the circumference of the articular surface is rough for ligaments which bind it to the cuneiform and to the second metatarsal. An articular surface may be present on the lateral side if the base touches the second metatarsal.

The inferior part of the base forms a downwards projection which finishes as a **tuberosity** on the lateral side. This projection provides attachment for part of tibialis anterior on the medial side, and the greater part of peroneus longus on the lateral side [FIG. 454], both tendons going also to the adjacent part of the first cuneiform.

The **body** is nearly straight on its dorsal and medial aspects and the smooth, rounded surface has no attachments. Its lateral surface is flat and triangular, owing to the downward

projection of the base, and provides attachment for one head of the first dorsal interosseous muscle. The smooth plantar surface is without attachments, and a strong ridge which curves down to the tuberosity on the base separates it from the lateral surface.

The **head** of the first metatarsal is large, wide, and helps to form the ball of the great toe with the base of the first phalanx and two sesamoid bones. Laterally and medially it is hollowed and tuberculated for the attachment of strong collateral ligaments. It is covered on its dorsal, distal, and plantar aspects by articular cartilage which extends furthest proximally on the plantar surface. This surface is grooved, on each side of a prominent, central ridge, by the sesamoid bones in the tendons of the short muscles which pass inferior to it.

The **second, third, fourth, and fifth metatarsals** are all similar in that they have long, tapering, slightly curved bodies and narrow heads. The second is the longest and provides a long axis for this part of the foot. The concavity of the curvature is plantar.

The bases articulate with each other as well as with the intermediate and lateral cuneiform bones and the cuboid bone. They are quite distinctive and can be identified by their shape and articular surfaces [FIGS. 271 and 272]. The proximal surfaces slope backwards and laterally, in keeping with the sloping tarsometatarsal line of joints. This indicates the medial and lateral sides of each bone, and so the foot from which it has been taken. The two double articular surfaces on the lateral aspect of the second are for the lateral cuneiform and the third metatarsal.

The sides and plantar aspects of the bases are strongly marked by interosseous and plantar ligaments, and their shapes are in conformity with the curve of the transverse arch of the foot. The second and third are triangular in outline. The fourth is cuboidal but is set a little obliquely to articulate with the sloping cuboid bone. The **fifth** is even more obliquely set so as to fit the shape of the lateral border of the foot. It has a large **tuberosity** which projects back over the edge of the cuboid and provides attachment for the peroneus brevis tendon. It is also marked on the medial part of its dorsal surface by the smaller tendon of peroneus tertius which extends on to the body as well. The plantar aspect gives attachment laterally to the strong fascia (with some muscle fibres) which passes from the calcaneus and represents an abductor of the metatarsal, although normally fused with the abductor of the little toe [FIG. 460]. The flexor digiti minimi brevis and a slip from tibialis posterior are attached to the plantar surface medial to this.

The oblique head of adductor hallucis and slips from tibialis posterior are attached to the plantar surface of the second, third, and fourth bases [FIG. 269] and the long plantar ligament reaches them also.

The **bodies** of the metatarsals are sufficiently separated to accommodate the dorsal interosseous muscles, each of which arises from the adjacent surfaces of two metatarsals. The plantar interossei come from the medial side of the third, fourth, and fifth bodies [FIG. 464].

The **heads,** although narrow, are similar to those of the metacarpals with a convex articular surface covering their dorsal, distal, and plantar surfaces, the plantar surface being more extensive than the dorsal surfaces to facilitate flexion.

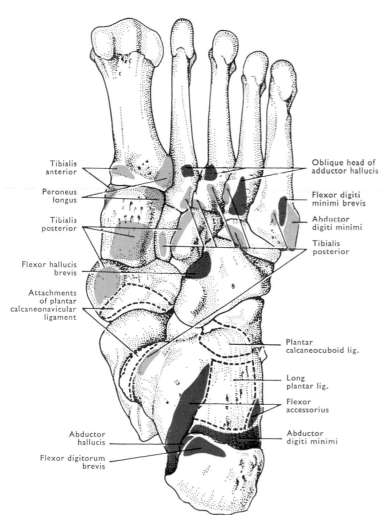

Tibialis anterior

Peroneus longus

Tibialis posterior

Flexor hallucis brevis

Attachments of plantar calcaneonavicular ligament

Abductor hallucis

Flexor digitorum brevis

Oblique head of adductor hallucis

Flexor digiti minimi brevis

Abductor digiti minimi

Tibialis posterior

Plantar calcaneocuboid lig.

Long plantar lig.

Flexor accessorius

Abductor digiti minimi

FIG. 269 Muscle attachments to left tarsus and metatarsus (plantar aspect).

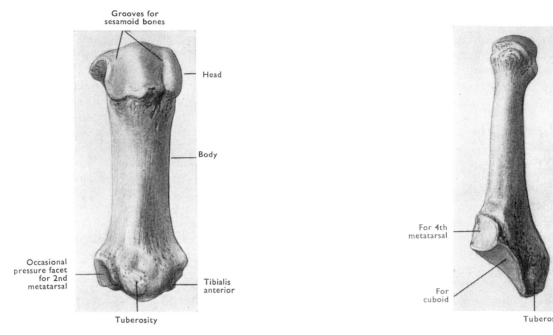

Grooves for
sesamoid bones

Head

Body

Occasional
pressure facet
for 2nd
metatarsal

Tibialis
anterior

Tuberosity

For 4th
metatarsal

For
cuboid

Tuberosity

FIG. 270   Right first metatarsal bone (plantar aspect).   FIG. 272   Right fifth metatarsal bone (dorsal aspect).

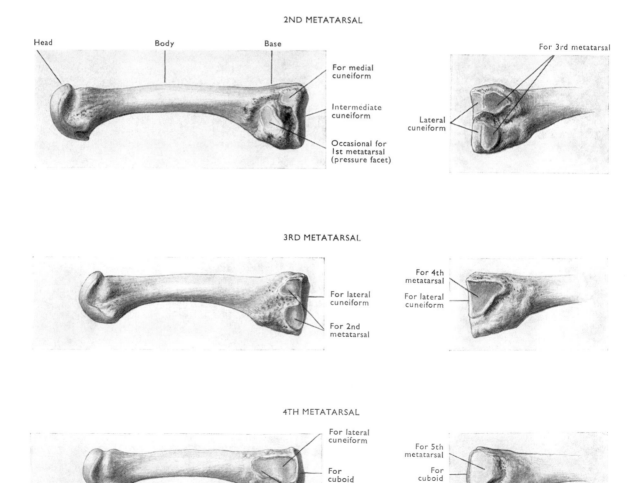

2ND METATARSAL

Head   Body   Base

For medial
cuneiform

Intermediate
cuneiform

Occasional for
1st metatarsal
(pressure facet)

For 3rd metatarsal

Lateral
cuneiform

3RD METATARSAL

For lateral
cuneiform

For 2nd
metatarsal

For 4th
metatarsal

For lateral
cuneiform

4TH METATARSAL

For lateral
cuneiform

For
cuboid

For 3rd
metatarsal

For 5th
metatarsal

For
cuboid

FIG. 271   View of bases and bodies of second, third, and fourth metatarsal bones of right foot.   Note that the second is the longest.

The sides of the heads have, like the metacarpals, a depression for ligamentous attachment and tubercles placed posteriorly [FIG. 271]. They rest on plantar ligaments which have only a loose attachment to the metatarsal, proximal to the head.

Ossification

A primary centre appears for the body of each metatarsal in the ninth week and they are well ossified at birth. Secondary centres appear for the base of the first and for the heads of the others during the second and third years, and they all fuse with the bodies between 15 and 18 years. Epiphyses may, however, be found in the bases of lateral metatarsals instead of in the heads.

THE PHALANGES OF THE TOES

The phalanges are similar to those of the fingers but their proportions vary considerably. They are miniature long bones with a proximal base, a body, and a distal head. The great toe has two strong phalanges, proximal and distal. The remaining toes have three, a proximal, a middle, and a distal, or ungual, carrying the toe nail : they diminish from medial to lateral side of the foot. A proximal phalanx is about 2·5 cm. long ; the others are much shorter and, in the little toe, may be no more than nodules of bone.

The **proximal phalanx** has a relatively large base with a concave articular surface for the metatarsal. It provides attachment for interosseous muscles at the sides and for a plantar ligament inferiorly. Its dorsal aspect is covered by an extensor expansion. The body, except in the great toe, is slender, rounded, and convex dorsally with an ill-defined plantar surface which has the fibrous flexor sheaths attached to its edges. The head is relatively wide and has a pulley-shaped articular surface with rough areas at the sides for collateral ligaments.

The **middle phalanx** is ridged on the articular surface of its base and pulley-shaped on its head. The body is short and frequently ill-formed, especially at the lateral side of the foot, but gives attachment to the fibrous flexor sheath and part of the flexor digitorum brevis to each edge.

The **distal phalanx** only has a proximal articular surface which is slightly ridged to fit the pulley-shaped head of the middle phalanx. Its distal end is expanded and flattened with its plantar surface roughened by the attached fibrous tissue of the toe. The base with its articular surface is slightly expanded and gives attachment to flexor digitorum longus on its plantar aspect, the extensor expansion on its dorsal aspect, and a collateral ligament at each side.

The **phalanges of the great toe** are much larger than the others because of the forces to which the great toe is subjected in walking and the consequent increase in muscle attachments. Because of the stresses there is commonly some degree of deformity, particularly hallux valgus (lateral deviation of the phalanges) even in persons who have never worn shoes (Lake, 1956 ; Barnicott and Hardy, 1955).

FIG. 273   Phalanges of a toe (plantar aspect).

The **first phalanx** of the great toe is short and strong with a large base which assists in producing the ball of the great toe. The base has a large, oval, articular surface for the head of the metatarsal and is marked on each side of its plantar surface by the tendon of flexor hallucis brevis combined with abductor hallucis medially and adductor hallucis laterally. Dorsally it is marked by the insertion of extensor hallucis brevis. The plantar ligament is largely replaced by the tendons and sesamoid bones mentioned above.

The wide **body** is slightly concave on its plantar surface and provides attachment at its edges for a fibrous sheath for flexor hallucis longus tendon. It tapers slightly to a head with a pulley-shaped articular surface and markings for collateral ligaments on each side.

The **terminal phalanx** of the great toe is similar to those of the other toes except for its larger size. The plantar surface of its base is roughened by the attachment of flexor hallucis longus tendon and extensor hallucis longus is attached to the dorsal part of the base.

The **phalanges of the little toe** are similar to those of the other toes except that they are apt to be poorly developed and distorted. Flexor digiti minimi brevis and abductor digiti minimi are attached to the lateral side of the base of the proximal phalanx and the remaining attachments are similar to the other toes.

Ossification

Primary centres for the distal and proximal phalanges appear between the end of the third month and the end of the fourth month, the distal ones appearing first. The centre for the middle phalanx appears later, between the sixth month and the time of birth. Secondary centres for the bases appear during the second and third years and fuse with the bodies between 15 and 21 years.

RADIOGRAPHIC ANATOMY

J. B. KING

Although ossification is present in the clavicle by the sixth week of foetal life, it is not until the sixteenth week that the skeleton of the foetus *in utero* can be demonstrated with certainty, and then it is usually either the vertebrae or the bones of the cranial vault that can be identified. Similarly, although data are available by which the age of the isolated foetus can be estimated accurately by radiography, it is seldom possible in pregnancy to obtain an accurate estimate of maturity to within 4 weeks until the ninth month. During this last month the pelvis, scapulae, and bodies of long bones become clearly identifiable as well as the skull and vertebrae. The epiphysis for the lower end of the femur can usually be seen (though it may not appear till shortly after birth) as may also the ossification centres for the talus, calcaneus, and cuboid. Many of the grosser abnormalities of foetal development are detectable by X-ray examination during pregnancy, and the demonstration of the skeletal distortions which follow foetal death are of value especially in the presence of an excess of liquor amnii.

The symmetry of the right and left halves of the skeleton is commonly used to establish the normal radiographic bone structure for any one individual. A bone on one side of the body can be expected to be similar to that on the other, not only in its general shape, but also in its fine architecture. This is true, of course, only if care is taken to ensure that the parts are in identical positions when the radiographs are being taken, and that, evidence of unequal wear and tear is recognized, particularly in later life.

A radiograph of a bone shows the compact cortical bone as sharply defined white or almost white lines without detail of structure. Within these lines, and overlapping what is actually medullary cavity, is a network of fine lines representing the trabecular structure of the cancellous bone the exact configuration of which is almost constant for any one bone [PLATE 3]. It must be appreciated that bone structure as shown on a radiograph does not represent the Haversian systems of compact bone, these being of an order of magnitude too small for demonstration by routine methods. The periosteum also is not visible in radiographs as it has a density no different from that of the surrounding soft tissues. The actual radiographic density of any one detail of bone structure is dependent on the concentration of calcium and phosphorus, dental enamel casting the densest shadow of all. The metaphysial zones of growing bone have a higher concentration of bone salts than the adjacent organized bone and therefore appear as dense lines with very clear-cut differentiation from the adjacent cartilage. The majority of epiphysial cartilages appear on radiographs as clear-cut transverse zones of soft tissue density which are unmistakable [PLATE 13D], but at certain points, notably at the proximal end of the humerus or base of the first metatarsal, the plane of the epiphysial cartilage is so oblique that bone always overlaps it on the radiograph and the margins of such epiphyses appear as two lines of diminished density [PLATE 11A].

That bone by its hardness should be demonstrable in fine detail by X-ray examination appears so reasonable that an erroneous impression is given of the extent to which abnormalities must necessarily be evident. In the skeleton as a whole a deficit of more than 30 per cent. in calcium content has to take place before unequivocal evidence of decalcification is found, and circumscribed areas of even greater loss of calcium may not be demonstrable. More significant is the fact that extensive infection or even death of bone may take place, and a period of time of great medical importance may lapse, before any radiographic evidence of change can be seen. In the long bones the contents of the medullary cavity may be entirely replaced by abnormal tissue with little or no radiographic evidence of the change. During the last two decades of life a progressive diminution in bone density is very common, especially in women, but certainly cannot be considered normal, inasmuch as many elderly people are not so affected. The causes are not known but among them are involutional changes in the gonads.

THE SKULL

At birth the cranial vault is large relative to the base and face, and the bones appear widely separated, the fonticuli being evident but with poorly defined margins [PLATE 8]. The metopic suture between the frontal bones usually disappears in early childhood but may persist throughout life. The principal sutures of the vault remain clearly visible until middle age when complete fusion takes place. Certain sutures such as the parietotemporal and also the zygomaticotemporal rarely fuse and may closely simulate fractures. Until adolescence a proportion of cranial vaults show oval areas of diminished density about 1·5–2·5 cm. in diameter. These so-called digital or hammer markings are of importance because, if very pronounced or if present in later life, they may be indicative of chronically raised intracranial pressure. Very large emissary foramina in the parietal bone about 4 cm. lateral to the midline may persist throughout life and have no pathological significance. The paranasal sinuses (except of the ethmoid) and mastoid processes do not become significantly pneumatized until about 4 years of age, and therefore no useful radiographic examination can be made before that time.

As the vault assumes its adult form the inner and outer tables become defined and the intervening diploic space is clearly seen particularly in lateral radiographs. It is important to note that on any one such film the bone detail which predominates is that of the structure on the side which has been nearest to the film when the radiograph was being taken.

The variation in thickness in the bones of the vault is much more evident on radiographs than by visual examination. The squama of the temporal bone may be so thin as to appear almost without structure. Apart from the fine granular structure of the diploic bone, and the coarser sieve-like dots due to the arachnoid granulations, the markings most constantly present are due to the grooves for the frontal and parietal branches of the middle meningeal veins and arteries. Very frequently, even more prominent linear markings are present which have a stellate or spider-like appearance and are due to the diploic venous channels [PLATE 8]. These are very variable and either the frontal, temporal, parietal, or occipital may be much larger than any of the others. These variations are only of importance as similar abnormal markings are frequently indicative of changes in adjacent brain tissue. Great variability exists in the extent to which the grooves carrying the venous sinuses of the dura mater appear as wide, clearly defined lines of diminished bone density.

Detailed X-ray examination of the paranasal sinuses [PLATE 9] and temporal bone is of importance in medicine. The sinuses are demonstrable by virtue of their air content, any displacement of which will produce a shadow of either bone or soft tissue density. It must be emphasized again that all body fluids have not only the same radiographic density as solid organs, but are indistinguishable radiographically, there being therefore no possibility of differentiating, either here or elsewhere in the body, between blood and any other fluid. The sinuses show great variability in the extent of their development. The frontal sinuses are not infrequently completely absent and when present are seldom symmetrical. Both the sphenoidal sinuses and ethmoid air cells vary greatly in their size and extent of development. The maxillary sinuses (antra) as a rule are approximately equal in size. Since no one view is adequate for the demonstration of all the paranasal sinuses, radiographs taken in four different positions are used as routine and many more are available for the examination of selected areas.

The X-ray examination of the sella turcica [PLATE 8 D] is of importance since not only may it reflect local pathological changes but because it may, by decalcification, provide the sole evidence of increased intracranial pressure due to remote causes. Here, unfortunately, as elsewhere in the skull, the normal range of variation is very great. While the average sella turcica as shown on a radiograph has a normal antero-posterior diameter of 15 mm. and a vertical one of 12 mm., these may be from 5 to 20 mm. without any abnormality being present. The development of the clinoid processes is equally variable and apparent bridging of the sella by bone is due to a normal ossification of the interclinoid ligaments, and not of the diaphragma sellae. The petroclinoid ligament is also sometimes ossified. The normal skeletal decalcification of ageing may produce very marked thinning of the clinoid processes, particularly the posterior ones. Recognition of the extent to which this can take place is important since similar changes may be produced by any condition which causes a persistently raised intracranial pressure, and also by pathological conditions arising within the sella turcica itself.

In true lateral views the mastoid processes overlap and all radiographic detail is lost, but moderate angulation overcomes this and the extent of the air cells in the peripheral portion of the temporal bone can be shown with accuracy [PLATE 9 D]. The distribution of the pneumatization is extremely variable, the cells sometimes extending into the zygomatic process and sometimes medially along the entire petrous ridge. The cochlea and semicircular canals can also be demonstrated by views in which the shadows of overlapping bone are avoided

as much as possible. Lateral films of the base of the skull produce such a complex of superimposed shadows that no anatomical detail can be visualized, and structures such as the sphenoidal sinus and internal acoustic meatus are usually examined by a view in which the central ray is directed through the base of the skull when the head is tilted backwards. Certain serious conditions such as fractures involving the base of the skull may not be demonstrable at all or only with difficulty.

In common with the rest of the skeleton the skull shows changes in old age. The sutures become obliterated by union during the second half of life but the rate of this change is very variable. The vault shares in the general decalcification of old age, and there is approximation of the inner and outer tables due to loss of the diploë.

The radiographic anatomy of the intracranial contents is dealt with in the section on the Central Nervous System.

THE VERTEBRAL COLUMN
[PLATES 4-7]

References are given which deal in greater detail with the radiographic anatomy of the vertebral column in general. The synovial vertebral joints and intervertebral discs are discussed in the section on Joints.

The radiographic anatomy of the vertebral column is closely related to considerations of technique, only an outline of which is possible as different factors are involved at every level. The dens, for instance, is best examined through the widely open mouth [PLATE 4D], and neither anteroposterior nor lateral radiographs are adequate for the cervicothoracic region where oblique views or tomography are essential, owing to the overlap of ribs, sternum, scapula, and clavicle. The vertebral bodies can be examined in routine antero-posterior and lateral views, but the articular facets only by oblique radiographs, each side being examined independently. Because of the overlapping densities of the bones of the pelvis the fifth lumbar vertebra and its extremely important articulations with the first sacral vertebra require X-ray examination separately from the rest of the lumbar spine. At all levels the accurate demonstration of the intervertebral foramina is difficult because of the variability of their obliquity [PLATE 4C].

At birth the main ossification centres for the vertebrae are present, but the cartilaginous zones are still so extensive that any deviation from normal must be very gross to be demonstrable by X-ray examination. In lateral views the bodies have rounded margins and show a line of diminished bone density which represents the venous drainage of the body and is of no importance except that it disappears in adult life due to overlapping bone but may reappear due to decalcification in old age.

In infancy the superior and inferior surfaces of the vertebral bodies are covered by cartilage which is relatively thicker around the margin and so causes pronounced notching and even right-angled indentation at the edges of the ossifying bodies. At puberty, secondary ossification centres appear for the vertebral bodies. These may not unite until the age of 25 and are of some importance as they are frequently irregular in shape and density and differentiation from significant abnormalities may be difficult.

Developmental anomalies are more common in the vertebral column than anywhere else in the skeleton. Indeed, radiography has shown that only some 40 per cent. of adult vertebral columns conform to the accepted normal pattern of development. The individual importance of any one such abnormality bears no relationship to the frequency of its occurrence. For instance, failure of fusion of the dens with the second cervical vertebra is rare [p. 98] but its recognition may have great significance, whereas some degree of spina bifida [p. 98] is present in 20 per cent. of adult spines and is rarely of any importance. Again, in 90 per cent. of cases of cervical rib its presence is no more than an incidental finding. In the vertebral column it is particularly noticeable that where an anomalous formation is present on one side only it is more likely to produce postural deformity than where it is bilateral or symmetrical. Thus, when two vertebral bodies have fused in the process of development a block vertebra is formed and will be found only incidentally, but a hemivertebra is an important source of lateral curvature of the whole column (scoliosis), and sacralization of the fifth lumbar vertebra [p. 98] is likely to have greater medical importance if present on one side only because of the asymmetrical stresses set up.

THE PELVIS
[PLATE 14]

At birth the three components of the hip bone and the sacrum are well ossified but appear widely separated. During the first 2 years growth is rapid, the acetabulum consolidates and approximates to the normal proportion of full development. At about 10 years of age, ossification of the ischiopubic synchondrosis produces a bulging of cortical new bone in a manner unusual at sites of union. Two epiphyses which may present unusual features are those at the lower border of the inferior ramus of the pubis, and that along the crest of the ilium, both of which appear at about 17 years of age and fuse about 20. The period at which they are evident and the fact that they may be fragmented and dense with a serrated adjacent metaphysis are features which may simulate abnormal growth or even fracture.

The radiographic appearance of the adult pelvis as a whole, and particularly that of the superior aperture, is largely dependent on relatively trivial postural variations which affect the degree of pelvic tilting, and the sexual differences cannot therefore be distinguished accurately in routine anteroposterior radiographs. However, special techniques are available by which accurate measurements of the pelvic diameters and subpubic angle can be made for obstetrical purposes, but all involve some irradiation of the ovaries and foetus which are among the most radiosensitive tissues of the body.

The bony pelvis, more than any other skeletal part, is frequently overlapped by adventitious shadows, such as those due to bowel gas, inspissated faecal content, the scrotum, and gluteal folds, the last particularly tending to cast deceptive lines across the radiograph in the elderly and thin.

THE UPPER LIMB
[PLATES 11-13]

At birth the scapula, clavicle, and humeral body are extensively ossified, but there is no visible secondary centre for the head of the humerus, and those for the greater and lesser tubercles do not appear until the third and fifth year respectively. At this stage, therefore, the gap between the areas of ossification representing the site of the shoulder joint is so wide that no degree of accuracy of examination is possible. By the age of 8 the three centres for the head have fused, and at this stage there are flake-like epiphyses for the tips of the coracoid and acromion processes which deserve attention because of the extent to which they may simulate injuries. Adjacent to the head of the humerus there is frequently present throughout life a zone of dense bone representing the line of epiphysial fusion [PLATE 11 C]. The cortex of the humeral body shows very marked thinning at the ends where it expands in total width. Such thinning is indeed common to most long bones and in old age may be so pronounced as to suggest the presence of some local abnormality when in fact there is no more than generalized decalcification. Local areas of irregular cortical thickening are frequent at the site of attachment of powerful muscles and one such is common at the point of insertion of the deltoid.

Accurate estimation of the normal radiographic anatomy of the lower end of the humerus during the period of bone growth is of unusual importance because of the frequency of injury to this region and because considerable deformity may result from apparently trivial epiphysial displacement [PLATE 12 A].

The tip of the coronoid process of the ulna is one of those sites where bone injury so slight as not to be detectable on X-ray examination may yet subsequently produce prolonged disability from regional new bone formation.

The cortical bone of both radius and ulna is thickest in the middle one-third, tapering off rapidly towards both ends with a relatively indefinite margin along the line of attachment of the interosseous membrane. The foramina for the nutrient vessels may be clearly defined.

The phalanges and metacarpals are well ossified at birth, and the subsequent ossifications of the carpal bones appear in a manner sufficiently orderly for this to be the most favoured area for determining the age of the growing skeleton by X-ray examination. The epiphyses of second, third, fourth, and fifth metacarpals appear at the distal ends early in the second year, and at the proximal end of the first metacarpal at the end of the second year. In all, some twenty-three supernumerary bones have been described in the hand.

THE LOWER LIMB
[PLATES 14-18]

Ossification appears in the epiphysis of the head of the femur a few months after birth, in the greater trochanter at 2 years, and in the lesser trochanter at about the age of 12. In view of the importance and relative frequency of displacements of the hip joint at or immediately after birth, much attention has been paid to the precise normal angulation of the upper acetabular surface at this time, but it is now known that variations within normal limits are sufficient to invalidate the significance of minor abnormalities, and Shenton's line (described in the section on Joints) has little value at birth, the accuracy of correct clinical examination exceeding that of radiography. The epiphysis at the lower end of the femur does not usually fuse until 20 and is one of the last of all the epiphyses of the long bones to do so, as it is the first to appear.

The patella frequently ossifies from more than one centre,

and may remain in two or even three portions throughout life. It is to be noted that such a bipartite or tripartite condition may not be present on both sides, and a similar asymmetry may apply to the sesamoid bone commonly found in the lateral head of the gastrocnemius (fabella).

The epiphysis at the proximal end of the tibia is not visible at birth but ossification commences soon afterwards. This epiphysis has an inferior extension anteriorly (tibial tuberosity) in which the ossification does not take place till the age of 10 and is frequently from multiple, small, irregular centres.

Ossification is present in the calcaneus during the sixth month of foetal life and is usually well enough developed to be clearly seen by radiography during the ninth month of pregnancy. The talus ossifies during the eighth month and if seen by antenatal radiography is therefore a good index of foetal maturity. Occasionally the cuboid is ossified at birth, but the notable feature is the extent to which tarsal ossification is present at an earlier stage than that in the carpus. The secondary centre for the ossification of the calcaneus forms as an arc along its posterior border and is another area where apparent fragmentation and increased density of the opposed bones are normal findings [PLATE 18 B].

A minimum of three radiographic views is necessary for the examination of the tarsus and no standard procedure can be satisfactory because of the differences in normal bones and joints. Radiographic demonstration of the bone structure of the calcaneus, metatarsals, and phalanges is relatively straightforward [PLATE 18 C and D].

In all, some thirteen supernumerary bones have been described in the foot. Because of weight-bearing and liability to injury some of these have a greater importance than the equivalent formations in the hand. Among these are the os trigonum at the posterior border of the talus, the accessory navicular and the os vesalianum at the base of the fifth metatarsal. Sesamoids are invariably present in the tendons of the flexor hallucis brevis and as they are not uncommonly a source of symptoms it is to be noted that the lateral is frequently bifid and that this may be so on one side only.

Bearing as they do the maximum weight of the body throughout life, the bones of the foot and ankle show radiographic changes from infancy to old age. The influence of gravity is indeed such that the bone and joint structure in the foot in the later years of life is seldom, if ever, identical with that in youth and recognition of this is essential in accepting any normal standard for radiographic anatomy. A degree of narrowing of the joint spaces representing degeneration of cartilage of all the tarsal joints is normal in advanced age.

For further details of the radiographic anatomy of bones see: Caffey, 1967; Etter, 1964; Pendergrass, Schaeffer, and Hodes, 1956.

REFERENCES

ASHLEY, G. T. (1956a) The relationship between the pattern of ossification and the definitive shape of the mesosternum in Man, *J. Anat. (Lond.)*, **90**, 87.

ASHLEY, G. T. (1956b) The human sternum. The influence of sex and age on its measurements, *J. forens. Med.*, **3**, 27.

BARNETT, C. H., and LEWIS, O. J. (1958) The evolution of some traction epiphyses in birds and mammals, *J. Anat. (Lond.)*, **92**, 593.

BARNICOTT, N. A., and HARDY, R. H. (1955) The position of the hallux in West Africans, *J. Anat. (Lond.)*, **89**, 355.

BERRY, A. C., and BERRY, R. J. (1967) Epigenetic variation in the human cranium, *J. Anat. (Lond.)*, **101**, 361.

BLACKWOOD, H. J. J. (1964) *Bone and Tooth*, Oxford.

BOLK, L. (1915) On the premature obliteration of the sutures in the human skull, *Amer. J. Anat.*, **17**, 495.

BOUCHER, B. J. (1957) Sex differences in the foetal pelvis, *Amer. J. phys. Anthrop.*, **15**, 581.

BOURNE, G. H. (1956) *The Biochemistry and Physiology of Bone*, New York.

BRAILSFORD, J. F. (1948) *The Radiology of Bones and Joints*, 4th ed., London.

BRASH, J. C. (1915) Vertebral column with six and a half cervical and thirteen true thoracic vertebræ, with associated abnormalities of the cervical spinal cord and nerves, *J. Anat. Physiol.*, **49**, 243.

BRASH, J. C. (1934) Some problems in the growth and developmental mechanics of bone, *Edinb. med. J.*, **41**, 305, 363.

BRASH, J. C., MCKEAG, H. T. H., and SCOTT, J. H. (1956) *The Aetiology of Irregularity and Malocclusion of the Teeth*, The Dental Board of the United Kingdom.

BRYCE, T. H. (1915) Osteology and arthrology, in *Quain's Elements of Anatomy*, 11th ed., Vol. 4, Pt. I, p. 177, London.

CAFFEY, J. (1967) *Pediatric X-ray Diagnosis*, 5th ed., Chicago.

CALDWELL, W. E., and MOLOY, H. C. (1933) Anatomical variations in the female pelvis and their effect on labour, with a suggested classification, *Amer. J. Obstet. Gynec.*, **26**, 479.

CALDWELL, W. E., MOLOY, H. C., and D'ESOPO, D. A. (1934) Further studies on the pelvic architecture, *Amer. J. Obstet. Gynec.*, **28**, 482.

CALDWELL, W. E., MOLOY, H. C., and D'ESOPO, D. A. (1935) A roentgenologic study of the mechanism of engagement of the fetal head, *Amer. J. Obstet. Gynec.*, **28**, 824.

CAREY, E. J. (1929) Studies in the dynamics of histogenesis XIV, *Radiology*, **13**, 1.

CHASSAR MOIR, J. (1964) *Munro Kerr's Operative Obstetrics*, London.

CLARK, W. E. le GROS (1971) *The Tissues of the Body*, 6th ed., Oxford.

COMAS, J. (1960) *Manual of Physical Anthropology*, Springfield, Ill.

CROCK, H. V. (1967) *The Blood Supply of the Lower Limb Bones in Man*, Edinburgh.

DIXON, A. F. (1910) The architecture of the cancellous tissue forming the upper end of the femur, *J. Anat. Physiol.*, **44**, 223.

DUCKWORTH, W. H. L. (1915) *Morphology and Anthropology*, Cambridge.

DUTERLOO, H. S., and ENLOW, D. H. (1970) A comparative study of cranial growth in *Homo* and *Macaca*, *Amer. J. Anat.*, **127**, 357.

EDWARDS, J. J., and EDWARDS, M. J. (1959) *Medical Museum Technology*, London.

ENLOW, D. H. (1968) *The Human Face: An Account of the Postnatal Growth and Development of the Craniofacial Skeleton*, New York.

ETTER, L. E. (1964) *Atlas of Roentgen Anatomy of the Skull*, Springfield, Ill.

FIELD, E. J., and HARRISON, R. J. (1968) *Anatomical Terms: Their Origin and Derivation*, 3rd ed., Cambridge.

FLECKER, H. (1932) Roentgenographic observations of the times of appearance of the epiphyses and their fusion with the diaphyses, *J. Anat. (Lond.)*, **67**, 118.

FRANCIS, C. C. (1940) The appearance of centres of ossification from 6–15 years, *Amer. J. phys. Anthrop.*, **27**, 127.

FRANCIS, C. C., WERLE, P. P., and BEHM, A. (1939) The appearance of centres of ossification from birth to five years, *Amer. J. phys. Anthrop.*, **24**, 273.

FROST, H. M. (1963) *Bone Remodelling Dynamics*, Springfield, Ill.

FROST, H. M. (1964) *Bone Biodynamics*, London.

GADOW, H. F. (1933) In *The Evolution of the Vertebral Column*, ed. Gaskell, J. F., and Green, H. L. H. H., Cambridge.

GOODSIR, J., and GOODSIR, H. D. S. (1845) *Anatomical and Pathological Observations*, Edinburgh.

GREEP, R. O. (1948) Alkaline phosphatase in bone formation, *Anat. Rec.*, **100**, 667.

GREULICH, W. W., and PYLE, S. I. (1959) *Radiographic Atlas of Skeletal Development of the Hand and Wrist*, 2nd ed., Stanford.

HAM, A. W. (1969) *Histology*, 6th ed., Philadelphia.

HANCOX, N. M. (1956) The osteoclast, in *The Biochemistry and Physiology of Bone*, ed. Bourne, G. H., New York.

HANIHARA, K. (1959) Sex diagnosis of Japanese skulls and scapulae by means of discriminant functions, *J. anthrop. Soc. Nippon*, **67**, 21.

HARRIS, H. A. (1933) *Bone Growth in Health and Disease*, London.

HAVERS, C. (1691) *Osteologia Nova*, London.

HRDLICKA, A. (1952) *Anthropometry*, Philadelphia.

HUNTER, J. (1772) Experiments and observations on the growth of bones, in *Palmer's edition of 'Works of John Hunter'*, Vol. IV (1837), p. 315, London.

KEEN, J. A. (1950) Sex difference in skulls. *Amer. J. phys. Anthrop.*, **8**, 65.

KOCH, J. C. (1917) The laws of bone architecture, *Amer. J. Anat.*, **21**, 177.

KOHLER, A., and ZIMMER, E. A. (1956) *Borderlands of the Normal and Early Pathologic in Skeletal Roentgenology*, English trans. Case, J. T., Vol. I, London.

KROGMAN, W. H. (1962) *The Human Skeleton in Forensic Medicine*. Springfield, Ill.

LACROIX, P. (1949) *L'Organisation des os*, Liège.

LAKE, N. C. (1956) The problem of hallux valgus, *Ann. roy. Coll. Surg. Engl.*, **19**, 23.

LE DOUBLE, A. F. (1912) *Traité des variations des os de la colonne vertébrale de l'homme et de leur signification au point de vue de l'anthropologie zoologique*, Paris.

LERICHE, R., and POLICARD, A. (1926) *Les Problèmes de la Physiologie Normale et Pathologique de L'os*, Paris.

MARTIN, B. F. (1968) The origins of the hamstring muscles, *J. Anat. (Lond)*, **102**, 345.

MESCHAN, I. (1959) *An Atlas of Normal Radiographic Anatomy*, 2nd ed., Philadelphia.

MEYER, G. H. (1867) Die Architectur der Spongiosa, *Arch Anat. Physiol. (Lpz.)*, 615.

MONTAGU, M. F. A. (1938) Ageing of the skull, *Amer. J. phys. Anthrop.*, **23**, 355.

MONTAGU, M. F. A. (1960) *An Introduction to Physical Anthropology*, Springfield, Ill.

MURRAY, P. D. F. (1936) *Bones. A study of the development and structure of the vertebrate skeleton*, Cambridge.

NICHOLSON, C. (1943) Accurate pelvimetry, *J. Obstet. Gynaec. Brit. Emp.*, **50**, 37.

NICHOLSON, C. (1945) The two main diameters at the brim of the female pelvis, *J. Anat. (Lond.)*, **79**, 131.

NOBACK, C. R., and ROBERTSON, G. G. (1951) Sequences of appearance of ossification centers in the human skeleton during the first five prenatal months, *Amer. J. Anat.*, **89**, 1.

PARSONS, F. G. (1904) Observations on traction epiphyses, *J. Anat. Physiol.*, **38**, 248.

PARSONS, F. G. (1905) On pressure epiphyses, *J. Anat. Physiol.*, **39**, 402.

PARSONS, F. G. (1908) Further remarks on traction epiphyses, *J. Anat. Physiol.*, **42**, 388.

PATERSON, A. M. (1904) *The Human Sternum*, London.

PATERSON, R. S. (1929) A radiological investigation of the epiphyses of the long bones, *J. Anat. (Lond.)*, **64**, 28.

PAYTON, C. G. (1932) The growth in length of the long bones in the madder-fed pig, *J. Anat. (Lond.)*, **66**, 414.

PAYTON, C. G. (1933) The growth of the epiphyses of the long bones in the madder-fed pig, *J. Anat. (Lond.)*, **67**, 371.

PAYTON, C. G. (1935) The growth of the pelvis in the madder-fed pig, *J. Anat. (Lond.)*, **69**, 326.

PENDERGRASS, E. P., SCHAEFFER, J. P., and HODES, P. J. (1956) *The Head and Neck in Roentgen Diagnosis*, 2nd ed., Springfield, Ill.

PRITCHARD, J. J. (1956) General anatomy and histology of bone, in *The Biochemistry and Physiology of Bone*, ed. BOURNE, G. H., New York.

PRITCHARD, J. J. (1956a) The osteoblast, in *The Biochemistry and Physiology of Bone*, ed. BOURNE, G. H., New York.

REYNOLDS, E. L. (1945) The bony pelvic girdle in early infancy. A roentgenometric study, *Amer. J. phys. Anthrop.*, **3**, 321.

REYNOLDS, E. L. (1947) The bony pelvis in prepuberal childhood, *Amer. J. phys. Anthrop.*, **5**, 165.

RICHELLE, L. J., and DALLEMAGNE, M. J. (1965) *Calcified Tissues*, University of Liège.

ROBIN, C. (1849) Sur l'existence de deux espèces nouvelles d'éléments anatomiques qui se trouve dans le canal médullaire des os, *C.R. Soc. Biol. (Paris)*, **1**, 149.

ROBINSON, R. A., and WATSON, M. L. (1952) Collagen-crystal relationships in bone as seen in the electron microscope, *Anat. Rec.*, **114**, 383.

SCHINZ, H. R., BAENSCH, W. E., FRIEDL, E., and UEHLINGER, E. (1971) *Roentgen Diagnostics*, 2nd ed., Vol. 2, London.

SCOTT, J. H. (1951) The development of joints concerned with early jaw movements in sheep, *J. Anat. (Lond.)*, **85**, 36.

SHANKS, S. C., and KERLEY, P. (1971) *A Textbook of X-ray Diagnosis*, 4th ed,. Vol. 6, London.

SHARPEY, W. (1856) *Quain's Elements of Anatomy*, 6th ed., Vol. 1., London.

STEVENSON, P. H. (1924) Age order of epiphyseal union in Man, *Amer. J. phys. Anthrop.*, **7**, 53.

THOMPSON, D'ARCY W. (1942) *Growth and Form*, 2nd ed., Cambridge.

TODD, T. W. (1924) Thickness of the white male cranium, *Anat. Rec.*, **27**, 245.

TODD, T. W. (1930) The anatomical features of epiphysial union, *Child Develp.*, **1**, 186.

TODD, T. W. (1933) Growth and development of the skeleton, in *Growth and Development of the Child*, Pt. II, Anatomy and Physiology, New York.

TODD, T. W. (1937) *Atlas of Skeletal Maturation*, Introduction: Pt 1, Hand, St. Louis.

TODD, T. W., and LYON, D. W. (1924–5) Endocranial suture closure, its progress and age relationship, *Amer. J. phys. Anthrop.*, **7**, 325; **8**, 23, 149.

TONNA, E. A., and CRONKITE, E. P. (1961) Use of tritiated thymidine for study of the origin of the osteoclast, *Nature (Lond.)*, **190**, 459.

TREVOR, J. C. (1955) Anthropometry, in *Chambers's Encyclopaedia*, **1**, 458–62, London.

TROTTER, M., and GLESER, G. C. (1952) Estimation of stature from long bones of American Whites and Negroes, *Amer. J. phys. Anthrop.*, **10**, 463.

TROTTER, M., and GLESER, G. C. (1958) A re-evaluation of estimation of stature based on measurements of stature taken during life and of long bones after death, *Amer. J. phys. Anthrop.*, **16**, 79.

WALMSLEY, R. (1940) The development of the patella, *J. Anat. (Lond.)*, **74**, 360.

WARD, F. O. (1838) *Outlines of Human Osteology*, p. 370, London.

JOINTS

ROBERT WALMSLEY

THE bones of the body are joined to one another at **joints** or **articulations** and the study of these is called syndesmology or arthrology. Although the term syndesmology is now the recognized anatomical term it is not so correct etymologically as arthrology because 'syndesmos' is the Greek for 'ligament' whereas arthrology specifically indicates a discussion of 'joints'. Furthermore, the Greek word 'arthron' has been widely adopted in standard and accepted clinical terminology as in 'arthritis' and 'arthrodesis' and for that reason it is necessary for medical students to be acquainted with the term 'arthrology'.

Joints vary considerably in structure and function. The form of the articular surfaces and the structure and arrangement of the ligaments that unite the bones at a joint are the principal factors that determine the amount of movement which may occur at it. At many joints in the adult skull the articulating bones are so interlocked and so united by a continuous band of fibrous tissue that little or no movement is permitted at them; this example is given in this introductory statement merely to emphasize that the anatomical designation 'joint' does not always imply movement. Other joints, and especially those in the limbs, are much more complex in their structure, and permit considerable movement. The structure of a joint is an expression of the function that it subserves, and the function and structure of joints should always be considered concurrently.

The variation that exists in the form and function of the various joints of the body allows them, for descriptive purposes, to be grouped into well-defined classes. This classification is, moreover, dependent on the manner in which bones are ossified and some knowledge of the principal features of bone formation is a necessary prelude to a consideration of the classes of joints [pp. 79–85].

The bones and joints of the body are developed from mesenchyme which may be regarded as the most talented of the primary embryonic tissues. Bones which are formed in membrane [p. 80], e.g., the bones of the vault of the skull, are joined by mesenchyme which eventually differentiates into fibrous tissue and such bones are therefore said to be united by **fibrous joints**. Bones that are formed in cartilage are connected by cartilage if two or more ossific centres appear in a single cartilaginous precursor of the bones; these are regarded, therefore, as a type of **cartilaginous joint** (*primary*). If the bones that participate in a joint are represented by separate cartilaginous precursors, then in early development they are separated from each other by a cellular disc of mesenchyme that is called the **primitive articular disc**. The developmental changes that occur in the primitive articular disc vary from joint to joint. The discs between the bodies of the vertebrae, for example, partly differentiate into fibrocartilage and these are regarded as another type of cartilaginous joint (*secondary*). In the freely movable joints of the limbs an articular cavity appears within the central part of the articular disc whereas its peripheral part and the surrounding mesenchyme form a tubular capsule which surrounds the articular ends of the bones and unites them.

CLASSIFICATION OF JOINTS

Three distinct classes of joints are recognized and each class may have two or more types. The three classes are:

1. **Fibrous joints,** where the bones have ossified in membrane or where they are preceded by discrete cartilaginous models and the primitive mesenchymatous connexion differentiates into fibrous tissue.

2. **Cartilaginous joints,** where the bones are united by hyaline cartilage or fibrocartilage.

3. **Synovial joints,** in which the articular surfaces of the bones are covered with articular cartilage and the bones are united by a tubular **articular capsule.** The inner layer of the capsule is formed by **synovial membrane** which secretes **synovial fluid** into the enclosed articular cavity. The synovial membrane and fluid are the specific features which are responsible for this class of joint being designated 'synovial'.

FIBROUS JOINTS

Fibrous joints are of three types, namely, **sutures, gomphoses,** and **syndesmoses.**

Sutures. A suture is a type of fibrous joint which is found only in the skull and at which no active movement occurs. The layers of periosteum [p. 77] on the outer and inner surfaces of the articulating bones bridge the gap between the bones and constitute the main bond between them [FIG. 274]. Between the articular surfaces of the bones there is an intermediate layer of vascular fibrous tissue which also unites the bones and together with the two layers of periosteum constitutes the **sutural ligament.** In the vault of the skull, where the inner periosteal layer is fused with the dura mater, the veins of the sutural ligaments communicate with the dural venous sinuses, the diploic veins of the adjacent bones and the

FIG. 274   Vertical section through a suture (sagittal).

extracranial veins. The histological structure of the sutures has been described by Pritchard, Scott, and Girgis (1956) in some detail and they have stressed that the sutures not only constitute a firm bond of union between neighbouring bones but that they are also one of the sites at which active bone growth occurs. The sutures are not usually permanent structures as obliteration of the simplest of them commences before the thirtieth year. The obliteration is effected by the growth of the articulating bones towards each other and, at a joint which is obliterating, this process of **synostosis** first occurs on the deep aspect of the suture and slowly extends to its superficial part; in a middle-aged skull some sutures may therefore be obliterated on their endocranial aspect but distinctly visible on their external (pericranial) surface. Complete obliteration of the sutures does not occur until an advanced age.

Different varieties of suture are named from the form of the articulating surfaces and the ways in which the opposed surfaces are fitted to each other. The opposed surfaces are seldom smooth and when the roughness is not too exaggerated and does not have a characteristic form it is termed a **plane** suture; an example of this is the suture between the horizontal plates of the two palatine bones. Sutures are commonly strengthened by fairly regular projections of one bone fitting between the projections of the opposing bone and these are termed **serrate** sutures; an example of this is the sagittal suture between the two parietal bones. In a **squamous** suture the margin of one bone overlaps that of the opposing bone to a greater or lesser extent; the joints of the squamous temporal bone represent examples of this variety of suture.

FIG. 275   Surface view of serrate suture (sagittal).

Gomphosis. This is a type of fibrous joint in which a peg fits into a socket; the roots of the teeth in their articulations with the sockets in the maxilla and mandible are the examples of these joints.

Syndesmosis. A syndesmosis is a fibrous joint in which the uniting fibrous tissue, much greater in amount than in a suture, forms an interosseous membrane or ligament. The inferior tibiofibular joint is a syndesmosis where the bones are joined by an interosseous ligament; the attachment of the shaft of the radius to that of the ulna may likewise be regarded as an example of syndesmosis in which an interosseous membrane unites the bones. Movement may be permitted at a syndesmosis by the flexibility of an interosseous membrane or the stretching or twisting of an interosseous ligament. This last occurs to an extremely small extent unless the ligament is composed largely of elastic fibrous tissue, as in

the case of the ligamenta flava between the laminae of adjacent vertebrae.

CARTILAGINOUS JOINTS

In a cartilaginous joint the bones are united by a continuous plate of hyaline cartilage or disc containing fibrocartilage. Cartilaginous joints are of two types, namely **synchondrosis** and **symphysis,** which differ in development, structure, and function.

Synchondrosis. The cartilage of the synchondrosis (*primary cartilaginous*) type of joint is a part of the continuous cartilaginous mass in which the bones are formed and, as long as the joint persists, it retains its original hyaline nature. The commonest synchondroses are the epiphysial (or growth)

A

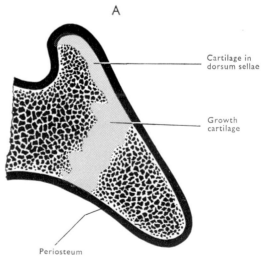

B

FIG. 276   Synchondroses.

A, diagrammatic section of joint between diaphysis and epiphysis of growing long bone.

B, median section of joint between occipital and sphenoid bones in child aged 1 year.

cartilages between the diaphysis and epiphyses of a long bone and the neurocentral joints in a vertebra; and the joint between the occipital and sphenoid bones (spheno-occipital synchondrosis) has the same pattern and fate [FIG. 276]. These joints are concerned with the growth of the bones and are obliterated by the fusion of the related bones certainly before the age of 25 and often considerably earlier. While they persist these joints are virtually immobile because hyaline cartilage is relatively rigid and the plates which they form are not very thick. Two synchondroses, however, are not obliterated but persist throughout life; these are the first sternocostal and the petrobasilar. The first sternocostal joint furthermore differs from a typical synchondrosis in so far as movement may occur at it; the first costal cartilage is sufficiently long (about 2·5 cm.) to allow both bending and twisting.

Symphysis. In a symphysis (*secondary cartilaginous*) the joint is more specialized, occurs only in the median plane and persists throughout life. In its early stages of development it is represented by a plate of cellular mesenchyme between separate cartilaginous rudiments [FIG. 277], but later this plate differentiates into fibrocartilage. When the cartilaginous rudiment ossifies, a thin lamina of the cartilage adjoining the fibrocartilage remains unossified and persists as a lamina of hyaline cartilage between the bone and the fibrocartilage. The joints between the bodies of the vertebrae and the joint between the bodies of the two pubic bones are the only typical examples of symphyses. Fibrocartilage, if sufficient in amount, allows a slight movement and it is for that reason that movement in the vertebral column is possible. Both the symphysis pubis and the intervertebral discs between the bodies of the vertebrae show further modifications and these are considered in the discussion of these joints.

There is one joint in the body which, although belonging to the class of cartilaginous joints, is not typical of either a synchondrosis or a symphysis. It is the manubriosternal joint. The sternum at the fourth month of intra-uterine life is formed of hyaline cartilage and thereafter ossific centres appear in the manubrium and the body. The manubriosternal joint is therefore a synchondrosis in the first instance, but as its hyaline cartilage becomes fibrocartilage at a later date, its definitive form is that of a symphysis.

SYNOVIAL JOINTS

A synovial joint [FIG. 278] is a highly evolved class of joint which is specialized to permit free movement. The great majority of the permanent joints of the limbs are therefore synovial, whereas in the rest of the body many of the joints are either fibrous or cartilaginous. The articular surfaces of the bones are covered by **articular cartilage** which is of such smoothness that it allows the bones to move on each other with the minimum of friction. The bones are connected to each other by a tubular **articular capsule** which surrounds their articular surfaces and encloses the **joint cavity.** The outer layer of the articular capsule is one of the principal structures that holds bone to bone at a joint; it is formed of fibrous tissue and is called the **fibrous capsule.** The inner layer of the articular capsule is formed by **synovial membrane** which secretes the **synovia** or **synovial fluid** into the joint cavity; the conventional term of 'joint cavity' is erroneous in so far as no real 'cavity' exists during life but only a potential cavity.

Articular Cartilage. The articular end of a long bone is usually enlarged, and has a thin but dense outer layer of compact bone within which there is an organized meshwork of cancellous bone. In any joint all of the surface of a bone that comes into contact with another during movement is covered with articular cartilage. In some joints there is a great difference in size between the two contiguous articular surfaces; an example of this is the shoulder joint where the head of the humerus has a much larger articular area than the glenoid cavity; such discrepancies in size of adjacent articular surfaces usually indicate that a wide range of movement can occur at these joints. In most joints articular cartilage is hyaline in type and contains many fine collagen fibres despite its apparently homogeneous, glassy matrix. In a few joints, such as the temporomandibular and the joints of the clavicle, the articular surface is formed of typical fibrocartilage. Articular

FIG. 277 Photomicrograph of symphysis pubis of 35-mm. human foetus to illustrate development of a symphysis. × 60.

Cartilage of pubis

Cellular mesenchyme

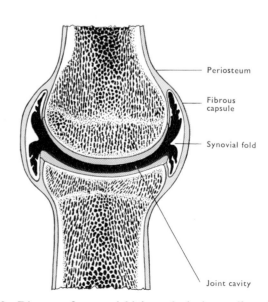

Periosteum

Fibrous capsule

Synovial fold

Joint cavity

FIG. 278 Diagram of a synovial joint. Articular cartilage, blue; synovial membrane, white.

FIG. 279 Section of glenoid cavity of rabbit, showing hyaline articular cartilage and vascular canal in underlying bone. × 170.

Labels on figure: Articular cartilage; Calcified lamina of cartilage; Vascular canal in bone

cartilage has a relatively smooth, free surface which is devoid of perichondrium. It is compressible and regains its original form after the compression stress is removed and for this reason it lessens the jarring shocks at joints (Smith, 1956). Although on naked-eye examination articular cartilage appears perfectly smooth, microscopic examination shows that the free surface has very slight irregularities that are associated with the normal wear and tear that occur at it. Cartilage is a living substance, and its cells continuously replace the matrix lost by surface attrition however slow this may be. It has been postulated that synovial fluid is released into the joint cavity from the articular cartilage during movement, but is reabsorbed by the cartilage when movement ceases (Barnett and Cobbold, 1962). The deepest part of the cartilage is calcified and the strength of its connexion to the underlying bone is enhanced by the irregularity of the junction between cartilage and bone; in many microscopic sections the calcified cartilage is seen as a dark wavy line [FIG. 279]. Articular cartilage has no nerves and is, therefore, insensitive. Likewise, it is avascular and is nourished by diffusion from three sources, namely, the synovia, the vascular ring in the synovial membrane at the periphery of the cartilage [p. 212], and also from blood vessels in the bone adjacent to the deep surface of the cartilage (Inglemark, 1950). It was believed formerly that the cartilage received no nourishment from the underlying bone but the presence of vascular channels in the bone close to the cartilage [FIG. 279] suggests nutrition from this source. The peripheral zone of articular cartilage is particularly well nourished from the adjacent synovial vessels, and is liable to an overgrowth in osteoarthritis, which produces 'lipping' of the articular margin. The central part, on the other hand, is less well favoured and is prone to atrophic degenerative changes so that it is partially or even completely eroded from the articular surface.

Fibrous Capsule (Membrana Fibrosa). The fibrous capsule is the outer layer of the articular capsule. It is a tube of dense, white fibrous (collagenous) tissue which by its flexibility permits movement at a joint but by its great tensile strength resists dislocation. The size and shape of the tube varies with that of the articulating bones. At each end the tube is firmly

attached to the periosteum of the articulating bones at a variable distance from the edge of the articular cartilage. This line of attachment varies in its relation to the epiphysial (growth) cartilage [if present, p. 83], which may be intracapsular, or extracapsular, or partly the one and partly the other. Small nerves and vessels pierce the fibrous capsule, and there may be larger openings through which synovial pouches protrude. The fibrous capsule may be strengthened or even replaced by adjacent muscle tendons or their expansions. The fibres of the capsule may form a dense feltwork; but where any recurring strain must be resisted, the fibres are increased and arranged in parallel bundles in the line of the tensile stress; such specialized parts are usually called ligaments and are given special names. In addition to these, separate ligaments may unite the bones. These **accessory ligaments** are usually **extracapsular** as at the costoclavicular and sternoclavicular joints, but may be **intracapsular,** like the cruciate ligaments of the knee joint. In some joints, such as the acromioclavicular, accessory ligaments constitute the principal ligamentous connexions between the articulating bones. Although ligaments are usually regarded as inelastic they may extend by a small proportion of their length when subjected to a load, and return to their original length when the load is removed (Annovazzi, 1928; Smith, 1954); ligaments are, therefore, elastic in the true sense of the word.

Inside some synovial joints, pads of fibrocartilage lie between the articular surfaces and blend peripherally with the fibrous capsule. Such a fibrocartilage may form a complete **articular disc** and so divide the joint cavity into two separate compartments [FIG. 280]; e.g., the mandibular and sternoclavicular joints. The intra-articular fibrocartilage may be wedge-shaped as in the acromioclavicular joint, or crescentic as in the knee joint, with a free edge projecting into the joint cavity. Though articular discs may add their resilience to that of the articular cartilages in cushioning the bone-ends, their principal function may be to assist in maintaining a layer of synovial fluid between articular surfaces (MacConaill, 1932, see p. 247).

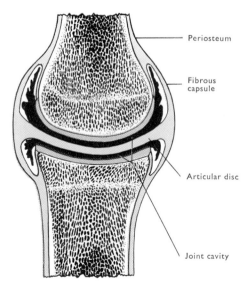

Labels on figure: Periosteum; Fibrous capsule; Articular disc; Joint cavity

FIG. 280 Diagram of a synovial joint and articular disc dividing the joint cavity into two compartments. Articular cartilage, blue; synovial membrane, white.

In the sternoclavicular joint the articular disc is so attached to the bones and to the capsule that it also plays the part of an intracapsular ligament.

Synovial Membrane. Synovial membrane is one of the characteristic features of synovial joints. It is a highly vascular connective tissue which lines the inner surface of the fibrous capsule and with it constitutes the articular capsule. At the attachments of the fibrous capsule the synovial membrane is reflected on to the bones to clothe them as far as the edge of the articular cartilage. Thus, all intracapsular bony surfaces are either covered with articular cartilage on the articular areas, or with synovial membrane on the non-articular areas. The transition between the two coverings is fairly abrupt but the synovial membrane slightly overlaps the edge of the cartilage.

A varying amount of fatty tissue may separate the synovial membrane from the structures which it lines. When plentiful, such extrasynovial fatty tissue bulges the synovial membrane towards the joint cavity and thus may form soft and highly compressible fatty pads or folds. At some joints the fatty pads occupy an intra-articular depression in one bone, but may be displaced during movement to allow a projection on the moving bone to occupy the depression. Good examples of such movable fatty pads are seen at the elbow joint. The synovial membrane is also folded over intracapsular tendons and ligaments, which are thus excluded from the joint cavity. Where fibrocartilages or ligaments within a joint are exposed to fairly continuous pressure, the synovial membrane does not cover them. Thus, the menisci of the knee joint, under constant pressure between the articulating bones, have no synovial covering; nor has that part of the deep surface of the fibrous capsule of the hip joint which is exposed to constant pressure by the head of the femur.

Microscopically, synovial membrane is composed of a vascular collagenous groundwork, showing delicate fibres and containing fibroblasts and mast cells. Its thickness and cell-content vary in different situations: it is thickest where separated from underlying ligaments by a layer of areolar tissue, but where it covers some fatty pads and tendons it may be extremely thin. The more mobile parts of the synovial membrane have numerous elastic fibres which prevent nipping of the membrane by the articular surfaces during movement of the joint (Davies, 1945). Towards its free surface the cells become rather flattened and it is difficult to distinguish the surface cells from the fibroblasts in the underlying stroma of the synovial membrane. The surface cells do not, moreover, constitute a continuous layer, but are absent in places where the synovial cavity communicates with crypt-like spaces in the substance of the synovial membrane. The surface area of the synovial membrane is increased by small **synovial villi-**like projections and larger **synovial folds** [FIG. 281]. The large surface area of the membrane and the richness of its vascular [p. 212] and lymphatic supply accounts for the rapidity with which toxic absorption occurs at acutely infected joints (Ghadially and Roy, 1969).

The **synovia** or **synovial fluid** is a clear, slightly yellow, glairy fluid deriving its names from its likeness in appearance and consistence to white of egg. It is sufficient only to form a thin film over all surfaces within the joint cavity; even in a large joint such as the knee joint there is less than 0·5 ml.

The synovia is a dialysate of blood plasma with mucin which is composed mainly of hyaluronic acid. The hyaluronic acid is secreted by the Golgi complexes of the surface cells of the synovial membrane and gives synovia its viscous character. The function of the synovia is three-fold: it serves to lubricate the joint surfaces; it is a source of nourishment for the articular cartilage; and the cells it contains—constantly migrating between the synovial membrane and the fluid—remove micro-organisms and the debris of wear and tear in the joint by phagocytic action (Coggeshall, Warren, and Bauer, 1940).

It will be clear that the joint or synovial cavity, bounded by synovial membrane and articular cartilage, is a closed cavity. Where an opening in the fibrous capsule allows the synovial layer to protrude, the imperforate continuity of the synovial membrane is always maintained. But the synovial cavity is not completely closed in a histological sense: the synovial membrane, unlike the pleura or the peritoneum of the coelomic cavities, has no complete epithelial lining as a barrier between the cavity of the joint and the tissue spaces around. The cavity is, in effect, an enormously enlarged tissue space from which rapid absorption can take place—an important feature in acutely infected joints.

Bursae. Closely akin to synovial membrane are the walls of certain closed sacs, known as **synovial bursae,** that are found in many situations in the body. The bursae contain fluid resembling synovia, and they serve to facilitate the play of one structure upon another. **Subcutaneous bursae** are found between the skin and underlying bony prominences such as the olecranon and the patella. **Subfascial bursae** are similarly placed beneath the deep fascia. **Subtendinous bursae** are found where one tendon overlies another tendon or a projection of bone. The walls of such bursae in the neighbourhood of a joint may be continuous with the synovial membrane of the joint through an aperture in the fibrous capsule.

Of the same nature are the **synovial sheaths** that surround tendons running in osteofascial tunnels, e.g., the tendons passing

FIG. 281 Synovial folds from human knee joint.
The fibrous capsule (just appearing at lower edge of photograph) is separated from the synovial membrane by loose areolar tissue. Note numerous blood vessels in the synovial membrane. × 100.

Transverse
process

Rib

Primitive articular disc

A

Medial
malleolus

Intermediate
stratum

Talus

B

Tibia

Talus

C

Fig. 282 Photomicrographs to illustrate the development of synovial joints.

A, costotransverse joint of 20-mm. human foetus. × 100.

At this stage the primitive articular disc is very cellular and not sharply demarcated from the adjoining cartilages.

B, ankle joint of 45-mm. human foetus. × 75.

There is an intermediate stratum of loose tissue between the medial malleolus and the talus, but between upper surface of talus and tibia this tissue has been absorbed.

C, ankle joint of 45-mm. human foetus. × 25.

The same specimen as in Fig. 282 B, to show the extent of the joint cavity in relation to upper surface of talus.

behind the flexor retinaculum at the wrist, and the flexor tendons lying on the front of the fingers. The synovial layer lines the tunnel and is also reflected round the tendon that lies within it.

Development of a Synovial Joint. After the cartilaginous precursors of the bones have been formed, their ends are separated from each other by a disc of very cellular mesenchyme that is called the **primitive articular disc** [Fig. 282]. There are no sharp lines of demarcation between the ends of the cartilages and the articular disc, and indeed the cells of a disc that are adjacent to cartilage form a chondrogenetic zone that is associated with the growth in length of the cartilage (Haines, 1947). As development proceeds, a process of differentiation occurs within the disc and the two chondrogenetic layers become separated from each other by an intermediate stratum of loose tissue in which the cells are relatively scanty [Fig. 282]. The intermediate stratum of loose tissue later undergoes liquefaction so that a joint cavity appears between the ends of the cartilages [Fig. 282]. The cartilages in turn lose the chondrogenetic layers over the articular surfaces though a thin fibrillar layer may persist over them until birth. The mesenchymatous tissue that surrounds the primitive articular disc differentiates into a dense outer layer that forms the **fibrous capsule,** and an inner loose layer that forms the **synovial membrane.** The synovial layer is invaded by extensions of the joint cavity so that the typical relationship of cavity to synovial membrane is established. **Intra-articular ligaments** and **fibrocartilages** arise from thickenings that project inwards from the wall of the primitive articular cavity. A complete **articular disc** arises from a transverse mesenchymatous septum between double cavities which appear in the primitive articular disc.

Blood and Lymph Vessels. A synovial joint is freely supplied by branches of the main arteries that are adjacent to it. These branches perforate the fibrous capsule and break up within the synovial membrane into capillaries which form a rich and intricate network. Many of the capillaries are extremely close to the free surface of the synovial membrane [Fig. 283] and this accounts for the frequency with which haemorrhage occurs into the articular cavity after even minor trauma (Davies, 1945). Adjacent to the peripheral margin of the articular cartilage the larger vessels of the synovial membrane branch and anastomose to form a vascular circle, the **circulus articuli vasculosus** of William Hunter (1743), and it is the terminal branches of this circle that overlie the margin of the cartilage and provide it with nourishment [p. 210]. Arteriovenous anastomoses occur in the articular vessels as they do in so many other places in the body.

The **lymph vessels** form a plexus within the synovial membrane, and from it efferent vessels pass towards the flexor aspect of the joint.

Nerve Supply. The fibrous capsule and, to a lesser extent, the synovial membrane are both supplied with nerves. The correlation that exists between the nerves of a joint and the nerves to the overlying tissues is expressed in Hilton's Law (John Hilton, 1863): 'The same trunks of nerves, whose branches supply the groups of muscles moving a joint, furnish also a distribution of nerves to the skin over the insertions of the same muscles; and the interior of the joint receives its nerves from the same source.' Some of the nerves in the fibrous capsule have encapsulated nerve endings and others

FIG. 283 Section of synovial membrane of rabbit (100μ thick) after the vessels had been injected with Indian ink. The close relationship of the vessels to the free (upper) surface of the synovial membrane is apparent. × 100.

have free nerve endings. It is the encapsulated nerve endings that are considered to be proprioceptive, and are concerned not only with the reflex control of posture but also with the constant awareness of the precise position of a joint, as, for example, its degree of flexion or extension. Injury to joint ligaments is notoriously painful and the pain receptors are generally regarded as the free nerve endings. The synovial membrane is much less sensitive and most of the nerves in it are vasomotor.

Joint Approximation Forces. There are several factors which keep the articular surfaces in contact with each other in synovial joints. Two of these are minor factors and are common to all synovial joints, namely, the **force of cohesion** and **atmospheric pressure**—the cohesion because of the smooth joint surfaces in contact with each other except for an intervening capillary layer of fluid, and the pressure because the atmosphere is excluded from the closed joint cavity. Another factor of very different value in different joints is the **interlocking of the co-apted bones;** this is well marked in the hip joint, but is completely absent in plane joints.

All tissues around a joint play a minor part in maintaining the apposition of the joint surfaces, but the most important factors are the **strength of the joint ligaments** and the **tension of the muscles**. These two act together in varying proportions; in the shoulder joint, with its relatively lax capsule, the ligamentous factor is small, whereas in the hip joint it is of greater importance.

At the knee joint the fibrous capsule is largely formed of tendinous expansions from muscles acting on the joint, and this joint complex represents an excellent example of the close structural and functional co-operation of muscles and ligaments at a synovial joint. As the muscles increase in size in graduated exercise, so too, more and more collagen is simultaneously laid down in the ligaments, thereby increasing their strength. Even where the ligaments are not directly connected with muscles, they may increase in thickness if they are subject to the stress of graduated exercise. It must be emphasized, however, that the fibrous tissue of ligaments cannot normally resist continuous stress without the harmonious support of muscles—a feature of many flat-foot deformities.

Limitation of Movement. Four factors are concerned in the limitation of movement at synovial joints : the apposition of soft parts, the locking of the articulating bones, the tightness of ligaments, and the tension of fully stretched muscles. **Apposition of soft parts** obviously limits movements in such joints as the elbow when, in flexion, the front of the forearm is pressed against the front of the upper arm. **Locking of the bones** probably occurs only exceptionally; thus, in the elbow joint, the occasional presence of pressure facets at the sides of the olecranon fossa of the humerus indicates that extension must have been limited by actual contact of the olecranon with the humerus, but in the majority of people the muscles and ligaments in front of the joint are the limiting factors. The **tension of ligaments** is an important factor in the limitation of movement. The various parts of a fibrous capsule are tense only in certain positions of a joint and they are concerned not only in the final limitation of a movement but also in directing the movement of the articular surfaces on each other. In the hip and knee joints the major ligaments are lax in flexion and tense in extension. Furthermore, it is in the position of extension at these joints that the articular surfaces are in fullest contact with each other. Tense ligaments and maximal contact of the articular surfaces are the essential features of a 'locked' joint, and it is when a joint is 'locked' that it is most stable and best adapted for weight-bearing. At many joints **muscular tension** reinforces the restraint on movement imposed by the ligaments, if it does not actually check movement before the ligaments are fully stretched. This effect of the muscles is called their **passive insufficiency** or **ligamentous action**. A good example of limitation of movement by muscular tension is the check imposed on flexion of the hip by the tension of the hamstrings while the knee is fully extended; flexion of the knee lessens the tension of the hamstrings, and further flexion of the hip can then take place [FIG. 335]. The effect of muscles in limiting movement in the ordinary person can be appreciated when one considers the greater range of movement possessed by acrobats, who by constant practice conserve or even increase the full extensibility of muscle possessed by all in early life. In such people, however, the ligaments and articular surfaces also are modified to allow the extreme movements. The restraint that muscles normally exert on movement, apart from that exercised by ligaments, is shown by the extreme movements sometimes produced in the convulsions of tetanus or epileptic fits, when the safeguarding muscular restraint operating in ordinary voluntary action is withdrawn.

CLASSIFICATION OF SYNOVIAL JOINTS

Synovial joints may be subdivided for further consideration in terms of the shape of the articulating surfaces and the movements performed at the joint. These two factors are intimately related and may be used together in a system of classification. From the point of view of form only, a distinction can be made between **homomorphic** and **heteromorphic** joints. The articular surfaces in a homomorphic joint are similar in conformation; those in a heteromorphic joint are dissimilar. Combining further consideration of shape with an analysis of movement, we distinguish **plane** and **saddle** joints of homomorphic pattern, and **hinge, pivot, ball-and-socket, condyloid,** and **ellipsoid** joints in the heteromorphic division.

Articulatio Plana or Plane Joint. In a typical plane joint the opposed surfaces are flat, or relatively flat, and are approximately equal in extent. Movement is of a simple gliding type or the twisting of the one bone on the other within the narrow limits permitted by a slight laxity of the fibrous capsule. Such joints are found between some of the carpal and tarsal bones, and between the articular processes of adjacent vertebrae.

Articulatio Sellaris or Saddle Joint. Here the concavo-convex articular surface of the one bone is reciprocally homomorphic with the convexoconcave surface of the other. The only good example is the carpometacarpal joint of the thumb, and the multiple movements that occur at it are considered in the description of this joint [p. 234].

Ginglymus or Hinge Joint. In this, the articulating surfaces are arranged to permit movement around only one axis—and that a transverse one—as in the hinge of the lid of a box. The elbow and interphalangeal joints are good examples. Usually, the articulating bones are in line with each other in the normal anatomical position; movement around the transverse axis causing the bones to be increasingly bent at the joint is termed **flexion,** while movement in the reverse direction is called **extension.** In cases where the normal position is angular, as at the ankle joint, suitably modified terms may describe the movements with less ambiguity, as **dorsiflexion** and **plantar flexion.** Two features in the disposition of the ligaments are common to all hinge joints. In the first place, the fibrous capsule must be of sufficient length in front and behind to permit the full range of movement; in either full flexion or full extension part of the fibrous capsule will be taut and part will be lax. In the second place, strong ligaments are placed at the sides of the joint and these are called **collateral ligaments.** One end of each ligament is attached in the axis of the hinge and some parts of these ligaments are therefore taut in all positions of the joint without restricting its movement.

Articulatio Trochoidea or Pivot Joint. Here, also, move-ment can take place around only one axis, but it is a vertical one, as in the hinge of a gate. A more or less cylindrical articular surface **rotates** within a ring formed partly by bone, partly by ligament, as in the proximal radio-ulnar joint and in the joint between the dens and the anterior arch and transverse ligament of the atlas.

Articulatio Spheroidea or Ball-and-Socket Joint. Here, as for example in the hip and shoulder joints, a 'ball' formed by a spheroidal surface on the one bone rotates in a 'socket' provided by a concavity in the other. This variety of joint allows a greater range of movement than any other as not only are there an almost infinite number of axes but the amount of movement in any one of them is considerable. It is this freedom of movement at a ball-and-socket joint that distinguishes it functionally from other types of joint. For descriptive convenience three principal axes at right angles to one another are specially considered. Around a transverse axis, movements of **flexion** and **extension** take place as in a hinge joint. Movements sideways around an anteroposterior axis are named **abduction** in the case of a limb raised laterally, away from the median plane of the body, and **adduction** when the limb is brought towards or across the median plane. Around a vertical axis through the centre of the ball, the limb **rotates, medially** or **laterally. Circumduction** is a composite movement around a number of axes whereby the limb describes the side of a cone, the apex of which corresponds to the centre of the 'ball'. The stability of such a joint depends on the depth of the 'socket', but the greater the depth of the socket, the more the movement is limited.

Articulatio Condylaris or Condyloid Joint. In this variety the articular surfaces are of the ball-and-socket type; but, owing to the disposition of the ligaments and muscles, the movement of active rotation around a vertical axis is absent. **Flexion** and **extension, abduction** and **adduction,** and **circumduction** take place. The metacarpophalangeal joints of the fingers (but not of the thumb) are condyloid joints.

Articulatio Ellipsoidea or Ellipsoid Joint. This is a further modification of the ball-and-socket joint; the articular surfaces, instead of being spheroidal, are ellipsoid, as in the radiocarpal joint. The curvature in the long axis of the joint surfaces has a greater radius than in the short axis. Movement takes place around two principal axes at right angles, corresponding to the greatest and least diameters of the joint, and to a slight extent around intermediate axes. Circumduction also is allowed, but the form of the ellipsoid surfaces prevents rotation.

JOINTS OF THE VERTEBRAL COLUMN AND HEAD

THE INTERVERTEBRAL JOINTS

The joints between the free vertebrae are arranged in a common plan with the exception of the specialized joints between the atlas and axis. The bodies of adjacent vertebrae are bound together mainly by the strong and extremely important intervertebral discs. The more posterior parts of the vertebral arches are also united, partly by the synovial joints between the articular processes and partly by ligaments. The intervertebral discs are separated from the joints between the vertebral arches by the intervertebral foramina through which the spinal nerves pass, and the upper and lower boundaries of these foramina are formed by the pedicles of the arches.

The joints between the bodies of the vertebrae and the connexions between the vertebral arches are considered separately, but functionally they must be considered together, as they are both concerned in the structure of the flexible part of the vertebral column which is not only strong but also resilient.

JOINTS BETWEEN VERTEBRAL BODIES

These are cartilaginous joints of the symphysis type as the intervertebral discs are composed of fibrocartilage; anterior and posterior longitudinal ligaments also assist in binding the vertebral bodies to one another.

Intervertebral Discs

The adjacent surfaces of the vertebral bodies are united by a fibrocartilaginous intervertebral disc [FIG. 284]. The peripheral part of this disc is called the **anulus fibrosus,** and is composed of fibrous tissue externally and an increasing amount of fibrocartilage internally, which is so densely fibrous that the cartilaginous element is appreciated only with difficulty under the microscope. The term anulus fibrosus is therefore extremely appropriate. The fibres of the anulus run obliquely between the two vertebrae and are arranged in concentric rings; the fibres in successive rings have different obliquities. The central part of the disc is soft and gelatinous in young people and for that reason is called the **nucleus pulposus;** with increasing age it is gradually replaced by fibrocartilage. On each surface of the disc, above and below, there is a thin plate of hyaline cartilage, in the margin of which the anular (ring) epiphysis of the vertebral body develops. The surface of the vertebra with which the cartilage plate is in contact is formed of modified cancellous bone perforated by numerous small foramina that are most apparent in the zone adjacent to the anular epiphysis. Except at their periphery the discs are avascular and are nourished from the interior of the vertebrae

FIG. 285 Transverse section of the lumbosacral intervertebral disc of a young adult. The laminated form of the anulus fibrosus is apparent but the nucleus pulposus shows no such organization.

(Reproduced by courtesy of the Editor, *Edinb. med. J.*)

by diffusion through the cartilage plates. The discs between the cervical vertebrae do not extend quite to the lateral edges of the vertebral bodies, where there is a small synovial joint between the lipped edge of the body below and the bevelled edge of the one above. The discs account for nearly a quarter of the length of the vertebral column but show considerable difference in their thickness in different regions; they are thinnest in the upper thoracic region and thickest in the lumbar region. The disc between the fifth lumbar vertebra and the sacrum is the largest avascular structure in the body and the degenerative changes which are so liable to occur in it may partly be due to its nutrition being dependent on a process of diffusion. The discs are also slightly wedge-shaped, in conformity with the curvature of the vertebral column in their neighbourhood; this is particularly marked in the cervical and lumbar regions, where the anterior convexities are due chiefly to the greater thickness of the discs in front than behind. It is of considerable practical importance to note that the intervertebral disc forms one of the anterior boundaries of the intervertebral foramen [FIG. 284], and as the spinal nerves pass through the foramina, they lie directly behind the corresponding discs.

The nucleus pulposus is customarily described as a white, glistening body that is composed of fibrocartilaginous tissue embedded in a gelatinous matrix, but from the time of its first appearance in the foetus until late adult life, it undergoes a continuous structural change. It is commonly stated to be a remnant of the notochord but this, as shown by Keyes and Compere (1932) and others, is not strictly correct. Early in foetal life the notochordal cells in each intervertebral region proliferate rapidly but later are surrounded by a plentiful mucoid matrix which results in the formation of a gelatinous nucleus pulposus. At birth and during the first year of life, the nucleus pulposus forms a greater proportion of the intervertebral disc [FIG. 286] than subsequently. Thereafter, the anulus fibrosus grows more rapidly than the nucleus which therefore becomes relatively smaller. As the notochordal cells degenerate the nucleus is gradually invaded by the surrounding fibrocartilaginous tissue. In young subjects the

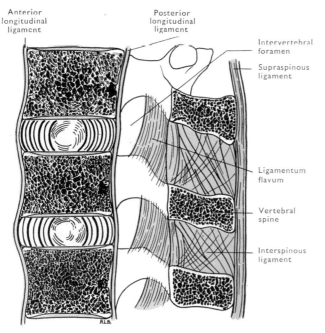

FIG. 284 Median section through part of lumbar region of vertebral column.

Articular cartilage, blue.

FIG. 286 Median microscopic section of lumbar intervertebral disc of full-term foetus. The nucleus pulposus forms the greater part of the disc, but will become relatively smaller after birth as the anulus fibrosus increases in thickness. ×15.

(Reproduced by courtesy of the Editor, *Edinb. med. J.*)

nucleus is still largely formed of gelatinous tissue and hence has a translucent appearance that contrasts sharply with the fibrocartilaginous anulus fibrosus. With increasing age the gelatinous tissue is gradually replaced by fibrocartilage which in old age blends with the anulus fibrosus in such a way that there is no longer a sharp line of demarcation between the two parts of the disc. Coincident with these changes in the structure of the nucleus pulposus there is a progressive decrease in its water content, which at birth is 88 per cent., but at 70 years is only 70 per cent. (Püschel, 1930)—a feature associated with reduced flexibility of the disc.

The nucleus pulposus forms a larger proportion of the intervertebral disc in the more mobile cervical and lumbar regions of the vertebral column. It is normally under considerable pressure and, when the disc of a young or middle-aged subject is cut across, it protrudes above the level of the anulus fibrosus. The nucleus pulposus may be regarded as a water cushion associated with equalizing the distribution of pressure between vertebrae, and the absorption of mechanical shocks that are constantly being transmitted along the vertebral column (Beadle, 1931 ; Bradford and Spurling, 1945 ; Walmsley, 1953).

Longitudinal Ligaments

The **anterior** and **posterior longitudinal ligaments** also help to unite the bodies of the vertebrae. The **anterior ligament** [FIG. 295] is a broad band of some thickness that runs up and down the front of the column over the anterior surfaces of the vertebral bodies. It is narrowest at its upper end, where it has a pointed attachment to the anterior tubercle of the atlas ; it becomes wider as it descends, and ends below by spreading on to the pelvic surface of the sacrum. It is firmly attached to the intervertebral discs and adjacent parts of the vertebral bodies ; the superficial fibres pass over several vertebrae, the deepest join adjacent bones. The **posterior longitudinal ligament** [FIG. 287] is constructed on a similar plan. It is placed on the back of the vertebral bodies, in the

anterior wall of the vertebral canal. Unlike the anterior ligament, it is broader above than below. Its upper end is continued into the membrana tectoria, which covers the dens and its ligaments and is attached to the occipital bone. Opposite the bodies of the thoracic and lumbar vertebrae the posterior ligament narrows markedly, but remains broader opposite the intervertebral discs, so that each edge presents a series of pointed dentations. Its narrow, lower end passes on to the anterior wall of the sacral canal. The posterior ligament is attached only to the discs and adjoining edges of the vertebral bodies ; opposite the middle of each vertebra it is separated from the bone by an interval through which pass the thin-walled basivertebral veins [FIG. 287]. The free surface of the ligament is separated from the spinal dura mater by loose areolar tissue in which there is the internal vertebral plexus of veins ; in the lower part of the canal the ligament is rather more intimately connected to the dura by fibrous bands. The posterior ligament, like the anterior one, is composed of longitudinal fibres of varying length : the most superficial run over several vertebrae ; the deepest pass only from one bone to the next.

JOINTS BETWEEN VERTEBRAL ARCHES

The vertebral arches are united by **synovial joints** between the articular processes, and by **accessory ligaments** which pass between the laminae, between the transverse processes, and over and between the vertebral spines.

The **synovial joints** are of the plane variety ; the shape and orientation of the joint surfaces on the articular processes vary in the different regions of the column, as already described in the section on Bones. Each joint is surrounded by a thin **fibrous capsule** lined with **synovial membrane.** The joint capsules are lax, particularly in the cervical region, so that slight gliding movements may occur between the bones.

The **accessory ligaments** are the supraspinous and interspinous, the intertransverse, and the ligamenta flava. The

Pedicle divided

Posterior longitudinal ligament

Intervertebral disc

FIG. 287 Posterior longitudinal ligament of vertebral column.

interspinous ligaments are relatively weak bands that run between adjacent vertebral spines; they are longest and strongest in the lumbar region [FIG. 284]. The **supraspinous ligament** runs over the tips of the spines and is fused with the posterior edges of the interspinous ligaments; short fibres in the ligament join adjacent spines; longer ones connect spines several vertebrae apart.

In the neck, the supraspinous ligament merges into the triangular **ligamentum nuchae.** In Man, this is merely a thin fibro-elastic septum between the muscles of the two sides of the back of the neck; it is attached above to the external occipital crest and deeply to the cervical spines, while its posterior border runs between the external occipital protuberance and the spine of the last cervical vertebra. In quadrupeds the ligamentum nuchae is a well-developed elastic ligament which relieves the muscles in the support of the head.

The intervals between the laminae of adjacent vertebrae are filled by the **ligamenta flava,** so called because of their yellowish colour due to the presence of much elastic tissue. Each is attached above to the front of the lower border of a lamina [FIG. 288] and below to the superior margin of the next lamina. There are two—a right and a left—in each vertebral interval; their medial borders are separated by a narrow, median chink through which pass veins that connect the venous plexus within the vertebral canal to the posterior external vertebral plexus; laterally, they extend to the capsules of the intervertebral synovial joints, but do not blend with them. The ligamenta flava are the only markedly elastic ligaments in Man. They permit separation of the laminae in flexion of the vertebral column and on account of their elasticity do not form folds when the column returns to the erect position; such folds might press upon the dura mater or be caught between the laminae they connect. The elastic tension in the ligamenta flava also assists the posterior vertebral muscles in maintaining the erect attitude.

The **intertransverse ligaments** are insignificant bands between the transverse processes. They are generally absent in the cervical region and are readily recognizable only in the lumbar region. In the upper part of the column they tend to be replaced by the intertransverse muscles.

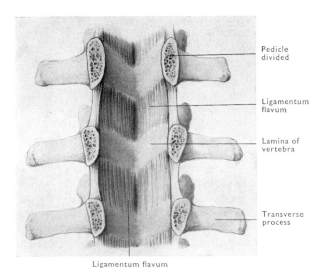

Pedicle divided

Ligamentum flavum

Lamina of vertebra

Transverse process

Ligamentum flavum

FIG. 288 Ligamenta flava seen from the front after removal of the bodies of the vertebrae by saw cuts through the pedicles.

MOVEMENTS OF VERTEBRAL COLUMN

The adult vertebral column in the erect posture has four anteroposterior curves—cervical, thoracic, lumbar, and sacro-coccygeal—and these have already been considered in the section on Bones on page 96. The **compensatory** or **secondary** curves (cervical and lumbar) are due mainly to the shape of the discs, whereas the thoracic curvature, which is a primary one, is caused mainly by the shape of the vertebrae. The sinuous form of the vertebral column, made up of alternate bony vertebrae and resilient intervertebral discs, is a structure that is well adapted to its supporting and shock-absorbing functions. In old age the intervertebral discs shrink and the vertebral column tends to revert to its original primary curve; thus it is that some people in advanced years have a 'bowed back' and appear to become smaller.

The movement between any two successive vertebrae is very slight; but the sum total of movement in the whole column is considerable [PLATE 4A and B]. The possibility of movement is due to the slight flexibility of the intervertebral discs coupled with the laxity of the capsules of the synovial joints between the articular processes. Where the discs are thick, the range of movement is increased, though the arrangement of the fibres in the discs strictly limits movement between adjacent vertebrae in any direction; the semifluid nature of the nucleus pulposus tends to equalize pressure within each disc during movements between the vertebrae.

The movements of the vertebral column are customarily described as forward **flexion,** backward **extension, lateral flexion,** and **rotation.** Lateral flexion and rotation are, however, always associated movements and neither can take place independently of the other. All the above types of movement can take place in all regions of the column, but both the thickness of the intervertebral discs and the 'set' and shape of the articular processes are variable, and largely determine the direction and range of movements in the different regions. During flexion the anterior borders of the vertebrae are approximated, whereas the distance between the posterior borders is increased; as full flexion is reached the anterior parts of the intervertebral discs are tightly compressed, but the posterior parts of the discs, the posterior longitudinal ligament and the ligaments of the vertebral arches, all become taut, though the principal limiting factor to the movement is the tension of the posterior vertebral muscles. In extension, the vertebrae rock backwards on one another so that the distance between the anterior borders of the vertebrae is increased and the posterior borders are approximated [PLATE 4A and B]. During this movement the anterior longitudinal ligament becomes increasingly taut whereas the other ligaments of the column are relaxed. The movement of extension of the column has a wider range than the movement of flexion, and is greatest in the cervical and lumbar regions. Much of the apparent movement of flexion of the vertebral column during the bending forwards of the body is in reality due to flexion at the hip joints and also movement of the head at the atlanto-occipital joints.

In the **cervical region,** where the discs are relatively thick for the length of that part of the column, all movements are fairly extensive. The articular surfaces slope anterosuperiorly, and on the two sides lie parallel to each other in the same transverse plane. They facilitate flexion and extension, but cause lateral flexion and rotation each to be accompanied by some degree of the other movement. Thus, when the neck is bent to one side, each lower articular process on that side glides downwards and backwards on the upper process of the vertebra below, while on the other side each lower process mounts upwards and forwards upon the superior process below it; hence bending to one side is accompanied by a slight rotation to that side.

In the **thoracic region** the intervertebral discs are relatively thin and the presence of the ribs and sternum and the imbrication of the vertebral spines all tend to limit movement. Rotation is possible because the articular processes lie on an arc of a circle with its centre in the vertebral body; but flexion is extremely limited by the presence of the ribs and sternum. Rotation and lateral flexion are, however, always associated, and in complete extension of the vertebral column these movements are greatly diminished, if not altogether lost.

In the **lumbar region** the discs are thick but the 'set' of the articular processes tends to restrict the movement of rotation in particular. Fairly free flexion and extension occur, especially at the lumbosacral joint where the intervertebral disc is thickest. When the vertebral column is flexed the lumbar curve is lost and lateral flexion, which is possible in the erect posture, is greatly diminished.

THE ATLANTO-OCCIPITAL JOINTS

The skull is joined to the atlas by a pair of **atlanto-occipital synovial joints** between the occipital condyles and the superior articular facets of the atlas, and by the **anterior** and **posterior atlanto-occipital membranes,** which connect the arches of the atlas with the margins of the foramen magnum [FIG. 289].

Atlanto-Occipital Membranes. The **anterior** atlanto-occipital membrane is attached above anterior to the margin of

the foramen magnum, and below to the anterior arch of the atlas; its central portion is thickened by fibres prolonged upwards from the anterior longitudinal ligament; the right and left margins blend with the fibrous capsule of the synovial joints. The **posterior** atlanto-occipital membrane likewise reaches the fibrous capsule at each side. Its upper border is attached to the posterior margin of the foramen magnum. The median part of its lower border is attached to the posterior arch of the atlas; but the lateral part, on each side, arches over the vertebral artery and the first cervical nerve as they cross the posterior arch of the atlas; this part of the lower edge of the membrane is thickened and is sometimes ossified.

Atlanto-Occipital Joint. Each of the pair of synovial joints is enclosed by a fibrous capsule which is attached at the margins of the articular surfaces and is lined with **synovial membrane.** The atlantal articular surface is concave; the occipital surface is convex. But the surfaces of the joints of the two sides are really segments of one ellipsoid whose transverse diameter is the longer, and functionally the joints act together as a single ellipsoid.

Movement at the atlanto-occipital joints takes place in two principal directions—nodding or bending the head forwards and backwards and lateral flexion of the head on the neck. No rotation takes place between the skull and the atlas beyond a little slipping of the occipital condyles, the one forwards the other backwards, on the atlantal facets, giving the face a slight oblique tilt; the joint surfaces are then in better apposition, and this is the natural pose of greatest ease and stability.

THE ATLANTO-AXIAL JOINTS

The atlas and the axis have **bilateral synovial joints** between their opposed articular processes, and two **median synovial joints** formed by the articulation of the dens with the anterior arch of the atlas and with the transverse ligament; in addition, various accessory ligaments connect the axis to the atlas, and the dens to the occipital bone.

Lateral Atlanto-Axial Joint. Each of the paired joints is of the plane variety, enclosed by a loose fibrous capsule with a synovial lining. The articular surfaces have slight reciprocal curvatures, and are inclined obliquely laterally and downwards. Applied to the back of the capsule is an **accessory atlanto-axial ligament** which runs obliquely downwards and medially from the back of the lateral mass of the atlas to the back of the body of the axis [FIG. 290].

The anterior arch of the atlas is connected to the front of the body of the axis by a membranous expansion from each side of the pointed upper part of the anterior longitudinal ligament as it descends from the anterior tubercle of the atlas. The gap between the posterior arch of the atlas and the arch of the axis is bridged by a membrane, in series with the ligamenta flava, which is pierced at each side by the second cervical nerve. The position of this nerve behind the atlanto-axial joint and of the first cervical nerve posterior to the atlanto-occipital joint bespeaks the different morphological value of these joints compared with other intervertebral synovial joints; they are held to be homologous with the small synovial joint at the side of each cervical intervertebral disc.

Median Atlanto-Axial Joint. The articulation of the dens with the atlas is a pivot joint provided with two little synovial

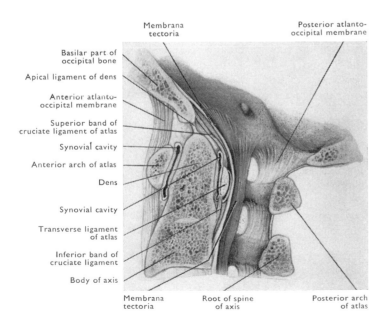

Membrana tectoria
Posterior atlanto-occipital membrane
Basilar part of occipital bone
Apical ligament of dens
Anterior atlanto-occipital membrane
Superior band of cruciate ligament of atlas
Synovial cavity
Anterior arch of atlas
Dens
Synovial cavity
Transverse ligament of atlas
Inferior band of cruciate ligament
Body of axis
Membrana tectoria
Root of spine of axis
Posterior arch of atlas

FIG. 289 Median section through atlanto-occipital and atlanto-axial region.

Dorsum sellae

Superior longitudinal
bundle of
cruciate ligament

Groove for inferior
petrosal sinus

Internal
acoustic
meatus

Apical
ligament
of dens

Jugular
foramen

Hypoglossal
canal

Alar ligament

Transverse
ligament of atlas

Lateral
mass of atlas

Accessory
atlanto-axial
ligament

Inferior
longitudinal bundle

Pedicle of axis (cut)

Membrana tectoria

FIG. 290   Dissection from behind to show the main ligaments that connect the occipital bone, the atlas, and the axis.

cavities [FIG. 289]. The smaller of these cavities is between the facet on the front of the dens and the facet on the back of the anterior arch of the atlas; the larger cavity is between the posterior facet on the dens and the transverse ligament of the atlas, each cavity is enclosed by a thin fibrous capsule lined with synovial membrane.

The **transverse ligament of the atlas** is a stout band that passes behind the dens between the tubercles on the medial sides of the lateral masses of the atlas [FIG. 290]. From the middle of the transverse ligament a small band of longitudinal fibres passes upwards to the anterior edge of the foramen magnum, and another passes downwards to the back of the body of the axis [FIG. 289]. These bands, together with the transverse ligament, constitute the **cruciform ligament of the atlas.**

THE OCCIPITO-AXIAL LIGAMENTS

The axis is connected with the occipital bone indirectly by the longitudinal bands of the cruciform ligament and, more directly, by the apical and alar ligaments of the dens and by the membrana tectoria.

Ligaments of the Dens. Lying immediately in front of the upper band of the cruciform ligament, there is the slender **apical ligament of the dens** which stretches from the apex of the dens to the anterior edge of the foramen magnum. The **alar ligaments** of the dens are short, stout bands that run, one from each side of the apex of the dens, laterally and slightly upwards to the medial side of each occipital condyle [FIG. 290].

Membrana Tectoria. The dens and the cruciform and alar ligaments are covered over posteriorly by the membrana tectoria [FIG. 289]. It is a broad sheet which is continuous below with the upper end of the posterior longitudinal ligament and is attached above to the occipital bone medial to

the hypoglossal canals. A small synovial bursa sometimes lies between the membrana and the median part of the transverse ligament of the atlas. The lateral margin of the membrana tectoria overlies and blends with the accessory atlanto-axial ligament. The upper end of the spinal dura is closely attached in front to the posterior surface of the membrana tectoria, and, behind, to the anterior surface of the posterior atlanto-occipital membrane.

MOVEMENTS OF THE HEAD

The movements of the head at the **atlanto-occipital joints** have been referred to in the description of those joints.

The combined movements at the three **atlanto-axial joints** serve to effect rotation of the head around a vertical axis, the skull and the atlas moving as one. While the ring formed by the anterior arch and transverse ligament of the atlas pivots round the dens, the lateral masses of the atlas glide, the one forwards and the other backwards, on the upper articular facets of the axis. Owing to the oblique disposition of these lateral joint surfaces, rotation of the head to one side is accompanied by a slight vertical descent of the head, so that the median pivot joint around the dens resembles in action a rising-butt hinge. Excessive rotation is checked by the alar ligaments and to a lesser extent by the accessory atlanto-axial ligaments. But the descent of the head, by approximating the attachments of the alar ligaments, serves to delay their check action, and this action can be obviated during a further degree of extreme lateral rotation by tilting the head backwards and to the opposite side. When the face looks straight forward the atlanto-occipital joint surfaces are not fully in apposition; slight rotation to one side secures better apposition and a pose more easily maintained.

THE TEMPOROMANDIBULAR JOINT

[PLATE 10A and B]

The temporomandibular joint is the indirect articulation of the **head of the mandible** with the **mandibular fossa** and the **articular tubercle** of the temporal bone. It is a synovial joint which possesses a complete **articular disc** of fibrous tissue which divides the articular cavity into upper and lower compartments. The articular cartilage on both bones has a large fibrous element and it is usually only its deeper parts that contain cartilage, i.e., the parts away from the free surfaces.

The **temporal articular surface** is concavoconvex from behind forwards, extending from the tympanosquamosal fissure over the articular fossa and eminence to the anterior border of the eminence [FIG. 292]. The **head of the mandible** is a narrow ellipsoid of bone directed medially and slightly backwards, markedly convex from before backwards but only slightly convex from side to side.

The **articular disc** separates the incongruent articular surfaces of the bones and moulds itself upon them during movements at the joint. It is an oval plate of avascular

FIG. 291 Temporomandibular joint.

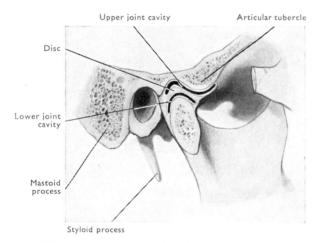

FIG. 292 Anteroposterior section through the temporomandibular joint.

the condyle and the under surface of the articular disc, and at the same time assure the forward and backward movement of the disc and the condyle as one unit on the articular surface of the temporal bone (Sarnat, 1951). A part of the tendon of the lateral pterygoid muscle is inserted into the anterior aspect of the fibrous capsule and thus gains indirect attachment to the articular disc. In the embryo the lateral pterygoid tendon passes backwards between the head of the mandible and the squamous temporal bone to be attached to the malleus, and it is the part between the bones that becomes compressed and forms the disc (Harpman and Woollard, 1938). The fibrous capsule is strengthened laterally by a well-defined **lateral ligament,** which passes downwards and backwards from the lower border and tubercle of the zygoma to the lateral and posterior aspect of the neck of the mandible [FIG. 291].

The **synovial membrane** lines the non-articular structures in each compartment of the joint; it fuses with the periphery of the disc but is absent from its upper and lower surfaces.

Accessory Ligaments. There are two accessory ligaments associated with the temporomandibular joint. The **spheno-mandibular ligament** is a thin band that runs from the spine of the sphenoid bone downwards and forwards to the lingula and the adjoining part of the deep surface of the ramus of the mandible [FIG. 293]. It lies a little medial to the capsule of the joint, and is separated from it by the maxillary artery and the first parts of its middle meningeal and inferior alveolar branches, the auriculotemporal and inferior alveolar nerves, and a process of the parotid gland. The ligament is of interest because it is developed from the sheath of that part of the cartilage of the mandibular arch (Meckel's cartilage) that lies between the base of the skull and the mandibular foramen.

The **stylomandibular ligament** is merely a thickened band of deep cervical fascia between the styloid process and the lower part of the posterior border of the ramus of the mandible [FIG. 293].

Nerve Supply. The mandibular joint is supplied from the mandibular nerve by twigs from its auriculotemporal and masseteric branches.

Movements. An understanding of movements at the temporomandibular joint is dependent, in some measure at least, on an appreciation of the relationship that the teeth of the upper and lower jaws have to each other when the jaws are

fibrous tissue. Thin in its central part, it thickens towards its peripheral attachment to the fibrous capsule and more especially so posteriorly where it is represented by a thick layer of vascularized connective tissue which connects it to the capsule.

The **fibrous capsule** is attached above to the limits of the temporal articular surface and below to the neck of the mandible. The articular disc is attached to the inner aspect of the capsule and above this attachment the capsule is lax to permit free movement of the disc on the temporal bone. Below the attachment of the articular disc, the fibrous capsule is much more taut and the fibres of the medial and lateral parts of the capsule are so short that the disc appears to be attached to the medial and lateral ends of the condyle. These attachments permit the rotatory movement that occurs between

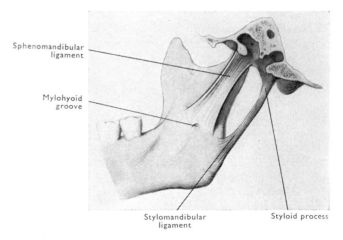

FIG. 293 Accessory ligaments of temporomandibular joint.

closed. The teeth are then in contact with one another and this is called the **occlusal position.** In this position the incisor teeth of the upper jaw lie in front of the lower incisors and there is usually some contact between their opposing surfaces. In opening the mouth from the occlusal position, the edges of the lower incisors pass downwards and forwards over the lingual surfaces of the upper incisors until the edges of the incisors are directed towards each other. This downward and forward movement of the mandible, in initiating opening of the mouth, is indicative of the two distinct movements that then occur at the temporomandibular joint, namely, a forward gliding and a hinge-like rotation. In the gliding movement, the head of the mandible and the articular disc move as a unit downwards and forwards together. The oblique **grinding** movements of chewing are brought about by alternate or opposite movements in the upper compartments

of the two joints, combined with associated hinge movements in the lower compartments.

The upper attachment of the anterior band of the lateral ligament is on the axis of the curved, downward-and-forward gliding movement in the upper compartment, and therefore, to some extent, the ligaments of the two sides resemble the collateral ligaments of a hinge joint. Their tension will not vary much throughout this upper compartment movement, and hence they do not completely prevent the possibility of dislocation of the mandibular head on to the front of the articular eminence. As this danger, however, arises only in excessively wide opening of the mouth, the associated rotation that then occurs in the inferior compartment causes some tension in the ligament which has a certain safeguarding action. The gross, microscopic and functional anatomy of the temporomandibular joint is fully described by Sarnat (1951).

JOINTS OF THE RIBS AND STERNUM

The ribs articulate posteriorly with the thoracic vertebrae. Each typical rib articulates after a common plan with the bodies of two adjacent vertebrae by its head, and with the transverse process of the lower of these vertebrae by its tubercle [FIG. 294]. The first and the last two or three ribs, which themselves deviate in their form from a typical rib, do not conform in their articulations with the vertebrae to the typical rib. Anteriorly, the ribs end in costal cartilages, which, with the exception of the last two, make further connexions with the sternum or with one another.

THE COSTOVERTEBRAL JOINTS

Joints of the Heads of Ribs. The head of each typical rib has two articular facets, namely a superior and inferior, which

are separated by the crest of the head. Between the bodies of each two typical thoracic vertebrae there is situated, posterolaterally, a little cupped depression in the intervertebral disc which is completed above and below by slightly concave facets on the sides of the adjacent vertebrae. Into this depression the head of the rib is fitted, its crest being bound to the intervertebral disc by a short, thick **intra-articular ligament.** The facets on the head articulate with the facets on the vertebral bodies, of which the lower is in numerical correspondence with the rib.

The whole articulation is surrounded by a fibrous capsule within which the intra-articular ligament forms a complete horizontal septum between upper and lower articular cavities, both provided with synovial membranes. The anterior part of the fibrous capsule is thickened to form the **radiate ligament of the head of the rib,** which passes medially from the front of the head to the adjacent parts of the disc and vertebral bodies, spreading fanwise upon them, under cover of the edge of the anterior longitudinal ligament [FIG. 295]. Three radiating bands are often to be seen, especially in the middle joints of the series, a central one blending with the intervertebral disc, and upper and lower bands attached to the vertebral bodies. The back of the capsule is connected to the adjacent denticulation of the posterior longitudinal ligament.

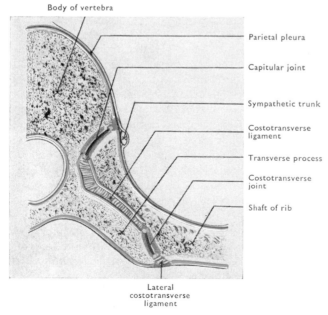

FIG. 294   Transverse section through a typical costovertebral joint.

FIG. 295   The anterior londitudinal ligament and costovertebral joints from in front. One joint has been opened by an oblique slice through the head of the rib.

Costotransverse Joints. Each typical rib also articulates by its **tubercle** with the transverse process of the lower of the two vertebrae to which its head is joined; and it is attached to the transverse processes of both vertebrae by certain accessory ligaments. A small costotransverse synovial joint surrounded by a thin fibrous capsule is formed between the medial part of the costal tubercle and the circular facet on the anterior surface of the corresponding transverse process near the tip. Strengthening this joint, on its posterolateral aspect, there is the **lateral costotransverse ligament,** a stout band running between the rough lateral part of the costal tubercle and the tip of the transverse process. It is in contact with the fibrous capsule of the costotransverse joint but is not fused with it.

The **costotransverse ligament** also binds the rib to the same transverse process; it is composed of very short fibres that bridge the narrow interval between the back of the neck of the rib and the front of the transverse process.

Finally, the rib is attached to the transverse process of the vertebra above by a **superior costotransverse ligament** [FIG. 295]. This ligament consists of an anterior and a posterior band that spring from the upper border (crest) of the neck of the rib and are attached to the lower border of the transverse process above. Laterally, the two bands are separated by the external intercostal muscle and the anterior band blends laterally with the internal intercostal membrane.

The vertebral connexions of the **first rib** and of the **last two** or **three ribs** are atypical. The head of each of these ribs has a single facet for the side of the corresponding vertebra only. There is a **single synovial cavity,** the radiate ligament is poorly developed, and there is no intra-articular ligament. The superior costotransverse ligament of the first rib is represented by feeble bands attached to the seventh cervical transverse process. The tubercles of the lowest ribs do not form synovial joints with the transverse processes; their costotransverse ligaments are progressively less well defined and those of the last rib may be absent altogether.

THE STERNOCOSTAL AND INTERCOSTAL JOINTS

The anterior end of each rib is continuous with a bar of hyaline cartilage, called the **costal cartilage,** which is firmly united to a conical cavity in the end of the rib. The first costal cartilage is directly united to the sternum; the second to the seventh costal cartilages articulate with facets on the side of the sternum; and the lower cartilages, except for the eleventh and twelfth, articulate with one another.

Sternocostal Joints. Little synovial cavities usually develop in the joints between the sternum and the costal cartilages, except in the case of the first, which remains a continuous cartilaginous joint [FIG. 296]. The cavity of each of the others is divided into two by an **intra-articular ligament,** until the corresponding sternal segments become fused together. The second sternocostal joint cavity, however, usually remains double throughout life but the cavities in the remaining joints tend to become obliterated in old age. The **fibrous capsules** of these little joints are strengthened in front and behind by fibres that spread fanwise over the corresponding surface of

the sternum and these constitute the **anterior** and **posterior radiate ligaments.** The fibres of the anterior ligaments interlace with those of the opposite side to form a felted membrane which is fused with tendinous fibres of the pectoralis major muscles.

Interchondral Joints. The tips of the eighth and ninth costal cartilages form little synovial joints, each with the lower border of the cartilage above. In addition, synovial joints are formed between slight bosses developed on the adjacent margins of the fifth to the eighth or ninth cartilages. All these joints between cartilages are enclosed by short fibrous capsules strengthened in front and behind by oblique ligamentous bands. The terminal part of the tenth cartilage is united to the ninth by a syndesmosis.

The last two costal cartilages are short conical structures ending freely amongst the muscles of the flank.

THE STERNAL JOINTS

The **manubriosternal** joint lies between the manubrium and the body of the sternum and is of a singular type. Prior to its ossification the sternum is a continuous bar of hyaline cartilage, but the manubriosternal joint is early demarcated by the transformation of the hyaline cartilage opposite the second costal cartilage into fibrocartilage. When ossification in the sternum is complete a thin layer of the hyaline cartilage above and below the fibrocartilage remains unossified so that the joint has an appearance similar to that of a symphysis. Sometimes the central part of the fibrocartilage is absorbed and a cavity is formed within it, and less frequently the joint is obliterated by synostosis in old age. Special longitudinal fibres and the neighbouring sternocostal ligaments strengthen this joint in front and behind. The cartilaginous **xiphisternal** joint between the body of the sternum and the xiphoid process is usually ossified in early adult life.

MOVEMENTS OF THE RIBS AND STERNUM

In respiration there are two factors concerned in increasing and diminishing the capacity of the thorax: these are, first, the movement of the diaphragm, and secondly the movements of the ribs and sternum. It is the movements of the ribs and sternum that are considered in this section but the respiratory action of the diaphragm should be studied concurrently [FIG. 417]. A rib and its costal cartilage together form a **costal arch** and the upper seven of these articulate with the sternum anteriorly; the eighth, ninth, and tenth cartilages each articulate with the costal cartilage that lies immediately above; the eleventh and twelfth costal cartilages are free so that the corresponding ribs are called 'floating' ribs. The ribs that articulate with the sternum increase in length from above downwards and their anterior ends lie at a lower level than the posterior ends. Owing to the length of the eighth and ninth ribs and the manner in which each of their costal cartilages articulates with the one lying immediately above, the thoracic skeleton continues to increase in width certainly down to the level of the ninth rib.

The capitular and costotransverse joints of each rib together form a hinge with an axis passing through the two joints and, owing to the backward inclination of the transverse process of the typical thoracic vertebra, this axis is directed laterally and backwards. Thus hinged posteriorly, the anterior part of the rib is raised during inspiration and lowered during expiration. As the anterior end of the rib is below the level of the posterior end, raising the rib in inspiration causes the anterior end to travel forwards as well as upwards and, owing to the obliquity of the axis of the hinge, a little laterally as well. The sternum, attached to the anterior ends of the upper costal cartilages, therefore moves upwards and forwards in inspiration, and the lateral deviation of the anterior ends of the ribs causes an opening out of the angles between the sternum and the costal cartilages, thus increasing the circumference of the thorax. The obliquity of the axis of movement posteriorly is most marked in the case of the middle ribs, which possess long, inclined cartilages, and therefore the greater lateral movement of the anterior ends of these ribs can readily take place. It is this lateral movement of the anterior ends of these ribs and their costal cartilages which causes the demonstrable widening of the infrasternal angle in deep inspiration. But in the upper ribs, where the axis is less oblique, the lateral movement is slight or absent and their shorter cartilages meet the sternum approximately at a right angle. As the first rib is so much shorter than those which succeed it, the upward and forward excursion of its anterior end is less in extent, and the movement of the upper border of the sternum is correspondingly reduced (Haines, 1946). The greater travel—especially in the forward direction—imposed upon the rest of the sternum by the longer ribs causes a bending of the breast bone at the manubriosternal joint. The result of all this on the capacity of the thorax is to increase its anteroposterior diameter.

The transverse diameter of the chest is also appreciably increased on inspiration and this is due to the position of the ribs midway between full expiration and inspiration. Each costal arch then has its intermediate part lying below a line joining its anterior and posterior ends. As the sternal and interchondral attachments of the costal arches hamper the upward and forward movement of their anterior ends, the raising of the ribs is in part translated into a twisting of the costal cartilages and a rotation at the sternocostal joints. It is this longitudinal twisting of the costal arch that causes the intermediate part of the rib to be raised upwards and outwards; this again is likened to the raising up of the low-lying bucket handle to a horizontal position. In inspiration, therefore, the increase in anteroposterior diameter of the thorax is accompanied by an increase in its transverse diameter.

Expiration is accompanied by reverse movements whereby the anteroposterior and transverse diameters of the thorax are diminished. Although it is convenient thus to describe the respiratory movements of the thorax as taking place in alternate opposite directions between the extreme inspiratory and expiratory positions, it should be explained that the 'resting' position of the thorax is midway between these extremes. This is the position of balance between the elastic tension of the costal arches tending upwards and the muscular and mechanical pull of the abdominal wall and contents tending downwards. The first half of the extreme range of movement, either between full expiration and full inspiration or vice versa, represents a recoil to this mean position (the expiratory movement assisted by the elastic recoil of the lung); beyond this point, the second half of the movement is the result of positive muscular effort.

Apart from respiratory movements, the ribs move passively with changes in the thoracic part of the vertebral column. Flexion of the column causes a crowding together of the ribs; extension causes a spreading apart; bending of the column to one side brings about a crowding of the ribs on that side of the trunk, and spreading on the opposite side.

JOINTS OF THE UPPER LIMB

THE SHOULDER GIRDLE

The shoulder girdle is formed by the clavicle and the scapula which articulate with each other at the **acromioclavicular joint.** The girdle in turn articulates with the skeleton of the trunk at the **sternoclavicular joint,** but the scapula has no direct articulation with the axial skeleton; the sternoclavicular joint has a degree of mobility that greatly increases the range of free movement of the upper limb.

THE STERNOCLAVICULAR JOINT

At the sternoclavicular joint the medial end of the clavicle articulates with the clavicular notch of the sternum and with the adjacent upper part of the first costal cartilage; an articular disc intervenes, however, between clavicle and sternum. The clavicular articular surface is larger than the sternochondral and for that reason the medial end of the clavicle projects above the upper margin of the manubrium sterni. The articular surfaces are reciprocally concavoconvex with that on the clavicle being convex vertically and slightly concave horizontally. On functional grounds and despite the form of the articular surfaces, the joint should be regarded as a ball and socket around which the clavicle can be moved in many directions. Movement of the clavicle in a vertical plane around the joint is free and the range is commonly about 60 degrees; anteroposterior movement is more restricted, approximately 30 degrees; and the rotation of the clavicle around its long axis which occurs during movement of the shoulder has a range of about 20 degrees. The articular cartilage, unlike that of most articular surfaces, is fibrocartilage and not hyaline cartilage.

A **fibrous capsule** surrounds the whole joint and is attached around the clavicular and the sternochondral articular surfaces. The epiphysis at the sternal end of the clavicle is intracapsular. The inferior part of the capsule passes between the clavicle and the superior surface of the first costal cartilage and is weak; the other parts of the capsule are strong, and are reinforced in front, behind, and above, by thickened bands called the **anterior** and **posterior sternoclavicular ligaments** and the **interclavicular ligament.**

The **anterior** and **posterior sternoclavicular ligaments,** of which the anterior is the stronger, run downwards and medially from clavicle to sternum [FIG. 296]. The origin of the sternohyoid muscle extends across the posterior ligament; this muscle and the sternothyroid separate the back of the left joint from the left brachiocephalic vein and the left common carotid artery, and the right joint from the brachiocephalic trunk. The **interclavicular ligament** is formed by fibres from the strengthening band on the upper aspect of the fibrous capsule that pass across the floor of the jugular notch, to which some of them are attached, to join similar fibres from the joint of the opposite side.

Within the joint there is a complete fibrocartilaginous **articular disc,** which blends with the fibrous capsule in front and behind, but has a firm attachment to the clavicle above and to the first costal cartilage below, near the sternal articular surface. Hence the disc, in addition to cushioning the articular surfaces from shocks transmitted along the clavicle from the shoulder, has an important ligamentous action in preventing such forces from driving the sternal end of the clavicle upwards and medially along the sloping chondrosternal surface. Also, when the clavicle is depressed, as by a heavy weight carried in the hand, the tendency for the medial end of the bone to be forced out of its sternal socket is resisted by the disc and the costoclavicular ligament.

There are two separate **synovial cavities** within the sternoclavicular joint, save when, exceptionally, the thinner central part of the articular disc is perforated.

The **costoclavicular ligament** is an extracapsular accessory ligament placed on the lateral side of the joint. It ascends from the upper surface of the first costal cartilage near its lateral end to a rough tubercle on the under aspect of the medial part of the clavicle. The anterior fibres of the ligament pass upwards and laterally whereas the more posterior fibres pass upwards and medially; the fibres of the ligament have, there-fore, a cruciate arrangement. It limits elevation of the clavicle and compensates for the weakness of the adjacent inferior part of the sternoclavicular fibrous capsule.

The stability of the sternoclavicular joint is dependent on the strength and integrity of its ligaments. The form of the articular surfaces and the surrounding muscles gives little additional security to the joint, and when dislocation does take place it is liable to recur.

Nerve Supply. The sternoclavicular joint is supplied by twigs from the medial supraclavicular nerve and the nerve to the subclavius muscle.

THE ACROMIOCLAVICULAR JOINT

The acromioclavicular joint is a small synovial joint, of the plane type, between oval facets on the lateral end of the clavicle and on the medial border of the acromion [FIG. 298]. The articular surfaces are covered with fibrocartilage and they both slope downwards and medially so that the clavicle tends to override the acromion.

A **fibrous capsule** of no great strength surrounds the joint; it is stoutest above, where it is reinforced by fibres from the trapezius. A wedge-shaped fibrocartilaginous **articular disc,** attached to the upper part of the capsule, partially divides the articular cavity in most joints. Only rarely does the disc form a complete partition and sometimes it is absent altogether.

The strength of the joint is dependent on the extracapsular, accessory, coracoclavicular ligament which connects the superior surface of the coracoid process of the scapula to the inferior surface of the clavicle.

Coracoclavicular Ligament. The powerful coracoclavicular ligament [FIG. 298] anchors the lateral end of the clavicle to the coracoid process and is the principal bond of union between clavicle and scapula. It has two parts which on account of their shape are called the conoid and trapezoid ligaments; they are continuous with each other posteriorly but anteriorly they are separated by an interval which is occupied by a synovial bursa.

The **conoid ligament** passes upwards and slightly backwards from an apical attachment at the angle between the two parts of the angulated coracoid process to a wider insertion on the conoid tubercle on the inferior surface of the clavicle. The **trapezoid ligament** lies anterolaterally from the conoid ligament. Its lower attachment is to a short ridge on the posterior part of the superior surface of the coracoid process; its clavicular end is wider, and is attached to the trapezoid ridge on the under side of the acromial end of the clavicle. The attached borders of the trapezoid ligament are slightly askew, and its surfaces so oblique as to be more nearly horizontal than vertical.

The conoid ligament is so set as to restrain forward movement of the scapula without similar movement of the clavicle, the trapezoid ligament to restrain backward movement. Both ligaments, but more parti-

FIG. 296   Sternoclavicular and sternocostal joints.

Section of clavicle

Interclavicular ligament

Joint cavity

Disc

Joint cavity

Joint cavity

Intra-articular ligament

Costoclavicular ligament

Anterior sternoclavicular ligament

Radiate sternocostal ligament

cularly the trapezoid, tend to prevent the acromion from being carried medially below the lateral end of the clavicle when blows fall upon the lateral surface of the shoulder.

Nerve Supply. The acromioclavicular joint is supplied by the lateral pectoral, suprascapular, and axillary nerves.

MOVEMENTS OF THE SHOULDER GIRDLE

Movements of the shoulder girdle serve to increase the range of movement at the shoulder joint by suitable alteration in the position of the scapular socket for the head of the humerus, brought about by movement of the scapula upon the thoracic wall. Movement of the shoulder girdle accompanies most movements at the shoulder joint and it is almost academic to consider the two separately. Reference is made to shoulder girdle movement at this stage merely as a prelude to shoulder joint movement. Forward movement of the scapula round the chest causes the glenoid cavity to face more directly forwards; backward movement causes the cavity to look more directly laterally; upward movement of the scapula, combined with rotation so that the inferior angle passes forwards and upwards, causes the glenoid cavity to be turned increasingly upwards; while the reverse movement of the scapula returns the cavity to its resting position. But in all these changes of position the shoulder is kept boomed out from the trunk by the thrust of the clavicle upon the acromion, thus securing greater freedom of shoulder movement. The lateral angle of the scapula, carrying the glenoid cavity, must therefore travel in an arc of a circle whose radius is the clavicle : but the medial part of the scapular blade, in its ambit, is held close to the chest wall by the muscles, and must travel in a curve of shorter radius—the radius of the thoracic curvature. Hence, the set of the scapula relative to the clavicle must be capable of modification, and this occurs at the acromioclavicular joint by the gliding of the acromial articular facet on the clavicle. One particular result of the different arc of travel of the acromion and the lateral angle of the scapula from that of the medial part of the blade is to obviate, in forward movement of the shoulder, any marked change in direction of the glenoid cavity which would diminish the power of forward-thrusting movement of the arm.

The clavicle is pivoted at its sternal end, but its acromial end moves, in association with the scapula, forwards, backwards, upwards, downwards, and, of course, in intermediate directions. During elevation of the shoulder a rotation of the clavicle occurs around its long axis so that the anterior surface of the bone is increasingly directed upwards. The range of this rotation has been variously assessed and the figure of 20 degrees which has been given already [p. 223] must be regarded as minimal. Any impairment of the clavicle to rotate due to a lesion at either the sternoclavicular or acromioclavicular joints will interfere with the free movement of the scapula (Inman, Saunders, and Abbott, 1944).

THE SCAPULAR LIGAMENTS

The following ligaments are attached wholly to the scapula and therefore are not connected with any joint. The **superior transverse scapular ligament** bridges the scapular notch, and so continues the line of the superior border of the scapula laterally to the root of the coracoid process. It may be partly

or completely ossified. Fibres of the inferior belly of the omohyoid are attached to its medial portion. The suprascapular nerve enters the supraspinous fossa through the foramen completed by the ligament, while the suprascapular vessels pass over the ligament to reach the fossa.

The **inferior transverse scapular ligament** is a weak band that stretches between the lateral border of the scapular spine and the back of the neck of the bone, over the suprascapular nerve and vessels. It may be regarded as formed from a fusion of the fasciae over the contiguous supraspinatus and infraspinatus muscles.

The **coraco-acromial ligament** is functionally related to the shoulder joint, and will be described with it.

THE SHOULDER JOINT

[PLATE 11]

The shoulder joint is a synovial joint of ball-and-socket variety in which freedom of movement is developed at some expense to stability. The spheroidal surface of the head of the humerus articulates with the shallow glenoid cavity of the scapula, each surface being covered with articular cartilage. The area of the scapular 'socket' is little more than a third of that of the humeral 'ball', so that the co-aptation of these surfaces contributes very little to the security of the joint; neither does the fibrous capsule play a great part, for it is of necessity lax enough to permit the required freedom of movement. The real strength of the joint lies in the muscles which surround it and it is upon these muscles that the stability of the joint depends. They are attached close to the articular areas and are intimately related to the fibrous capsule; they are, therefore, admirably disposed for the purpose of keeping the articular surfaces in firm contact in all positions of the joint. They will receive special consideration when movement at the joint is described.

Labrum Glenoidale [FIG. 298]. The labrum glenoidale is a fibrocartilaginous rim, triangular in cross-section, attached to the edge of the glenoid cavity, which is slightly deepened by it. The tendon of the long head of biceps arises within the joint from the supraglenoid tubercle, and is there fused to the labrum glenoidale, as is the long head of triceps at its attachment to the infraglenoid tubercle.

Fibrous Capsule. The fibrous capsule forms a cylindrical sleeve attached close to the articular area of each bone. On the scapular side it is attached outside to the labrum glenoidale and partly to the labrum itself, especially above and behind. On the humeral side the capsule is attached above to the anatomical neck immediately medial to the tubercles, but below, it is attached to the medial surface of the surgical neck of the humerus at least 1 cm. from the head of the bone. The upper epiphysial line of the humerus is therefore chiefly extracapsular, but comes within the joint on the medial side. When the arm is by the side the lower part of the capsule is lax and forms a redundant fold [FIG. 299], but when the arm is abducted, this part of the capsule becomes increasingly taut. The anterior part of the fibrous capsule is immediately related to the tendon of the subscapularis muscle, the superior part to the supraspinatus tendon, and the posterior part to the tendons of the infraspinatus and teres minor muscles. All these tendons blend with the capsule towards their insertion, and as they are intimately concerned in the maintenance of the head of the humerus in its correct relationship to the glenoid cavity during movements at the shoulder joint, they have been variously termed 'articular' muscles, 'rotator cuff' and 'musculotendinous cuff', but are essentially ligaments of variable length and tension. The inferior part of the fibrous capsule is the weakest and it is also relatively unsupported by muscles; as the arm is raised from the side, however, the long head of the triceps and the teres major are increasingly applied to this surface of the joint.

There are two **openings** in the fibrous capsule [FIG. 297]. One is an interruption in its attachment to the humerus at the upper end of the intertubercular sulcus; the tendon of the long head of biceps escapes through the aperture so formed. The other opening is in front of the fibrous capsule under cover of the subscapularis tendon, and through it the **subscapular bursa** communicates with the synovial cavity of the joint.

The fibres of the fibrous capsule run for the most part longitudinally from bone to bone, though a few fibres run transversely round the fibrous capsule. The **transverse humeral ligament** is a special bundle of these transverse fibres that is attached to the greater and lesser tubercles and bridging over the upper end of the intertubercular sulcus, holds the long head of the biceps brachii in the sulcus as it leaves the joint [FIG. 297].

Glenohumeral Ligaments. Three slightly thickened bands of longitudinal fibres (the **glenohumeral ligaments**) may be seen on the internal surface of the anterior part of the capsule [FIG. 298]. They are seldom prominent, and may be absent. The **superior** glenohumeral ligament is attached to the upper part of the labrum glenoidale immediately anterior to the long tendon of the biceps, and passes laterally, alongside the tendon, to reach the humerus near the upper surface of the lesser tubercle.

The **middle** glenohumeral ligament is attached to the scapula close to the superior band and reaches the humerus at the front of the lesser tubercle. The opening in the fibrous capsule under the subscapularis tendon is between the superior and middle bands. The **inferior** glenohumeral ligament, although usually considered to be the best developed of the three ligaments, is frequently indistinct or even absent. It passes downwards and laterally from the middle of the anterior border of the labrum glenoidale to be attached to the lowest part of the front of the anatomical neck.

Accessory Ligaments. The **coracohumeral ligament** is an accessory ligament on the upper aspect of the joint [FIG. 297]. It passes from the lateral side of the root of the coracoid process to the upper part of the anatomical neck of the humerus. It is most distinct in its medial part and there its anterior border is free, but as it is traced laterally it loses its identity and fuses with the tendon of the supraspinatus as it, too, blends with the capsule. This band greatly strengthens the upper part of the fibrous capsule which is under tension when the arm hangs by the side.

The **coraco-acromial ligament** is in no way connected with the fibrous capsule of the shoulder joint, but as it forms a horizontal shelf above the joint, it may be considered an accessory ligament of the joint [FIGS. 297 and 298]. It is triangular and its base is attached to the lateral border of the horizontal part of the coracoid process and its apex to the tip of the acromion in front of the acromioclavicular joint. The anterior and posterior borders of the ligament are stronger than the middle portion. Sometimes a prolongation of the pectoralis minor tendon pierces the base of the ligament and blends, below it, with the coracohumeral ligament. Superiorly, the coraco-acromial ligament is covered by the deltoid muscle; inferiorly, it is separated by the **subacromial bursa** from the supraspinatus tendon, which lies above the shoulder joint. The acromion and the coraco-acromial ligament form a protective arch against which the head of the humerus is pressed (the articular capsule and overlying muscle intervening) when force is transmitted upwards along the humerus, as when standing at a table, the weight of the body is partly supported on the table by the down-stretched arms.

Synovial Membrane. The synovial membrane of the shoulder joint lines the fibrous capsule and thus extends downwards as a pouch when the arm is by the side. From the fibrous capsule the membrane is reflected upwards on the humerus to the edge of the articular cartilage. The membrane protrudes through the opening in the front of the capsule to form the bursa posterior to subscapularis tendon. The size of this bursa varies; when large it may be wrapped round the upper border of the subscapularis tendon underneath the coracoid process, but this upper, **subcoracoid** extension may be replaced by an independent bursa of that name. Occasionally the synovial membrane protrudes through an opening in the back of the fibrous capsule to form a small bursa beneath the infraspinatus tendon.

The intracapsular part of the **tendon of the long head of biceps** [FIG. 299] is covered with a sheath of synovial membrane which is continued along the

FIG. 297 Articular capsule of left shoulder joint from the front.

tendon in the upper part of the intertubercular sulcus. The synovial sheath is then reflected upwards as a lining to the osteofascial tunnel in which the tendon runs, and so passes into continuity with the synovial lining of the fibrous capsule. As the tendon runs over the upper part of the articular head of the humerus it has a steadying influence upon movements of the shoulder joint.

There is always a large **subacromial bursa** which separates the coraco-acromial arch and the deltoid muscle from the upper and lateral aspect of the shoulder joint and the tendons lying upon it; this bursa does not communicate with the interior of the joint.

Nerve Supply. The shoulder joint is supplied by branches from the suprascapular, axillary, and lateral pectoral nerves (Gardner, 1948c).

MOVEMENTS AT THE SHOULDER JOINT

The form of the articular surfaces and the laxity of the fibrous capsule allow a wider range of movement at the shoulder joint than at any other joint in the body. Being a ball-and-socket joint, movement can take place around an infinite number of axes intersecting in the centre of the globular head of the humerus. The principal axes of descriptive convenience which are sometimes used are : (1) a transverse axis around which the arm moves forward in **flexion** and backwards in **extension**; (2) an anteroposterior axis

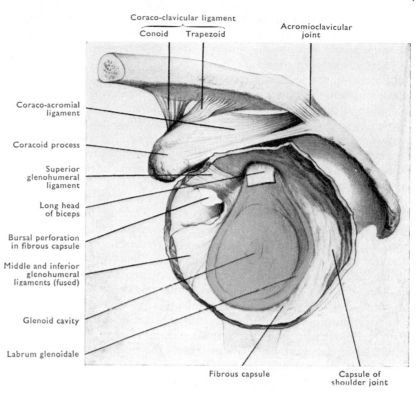

FIG. 298 Articular capsule of shoulder joint cut across and humerus removed.

around which occur the movements of **abduction** (away from the side of the trunk) and **adduction** (towards the side of the trunk); and (3) a vertical axis around which the arm **rotates medially** and **laterally**. **Circumduction** is a combination of anteroposterior and lateral movements around successive axes, when the arm swings round the side of a cone whose apex is the point of intersection of the various axes in the head of the humerus. These axes are not, however (with the exception of axis of rotation), adopted in this account of shoulder joint movement because when the arm is in the normal position by the side, the glenoid cavity looks as much forwards as it does laterally and the humeral head is directed backwards and medially. The true 'plane of the joint' is therefore inclined backwards and laterally at an angle of approximately 45 degrees to the median plane. An appreciation of this angulation of the 'plane' of the shoulder joint is most important for an understanding of the movements that take place at the joint. In the movement of flexion around this axis the arm swings forwards and medially at an angle of 45 degrees to the median plane and in extension it is carried backwards and laterally. Similarly, in true abduction the arm is carried forwards and laterally in the 'plane of the scapula', a plane at right angles to the glenoid fossa through its vertical diameter. Abducting in this plane does not involve torsion of the fibrous capsule and therefore provides the

FIG. 299 Coronal section through the right shoulder joint. The parietal and visceral layers of the synovial sheath of the biceps tendon have been partially left in place.

most easeful position in the treatment of certain injuries at the shoulder (Johnston, 1937). It is in this plane that the subscapularis, supraspinatus, infraspinatus, and teres minor muscles normally function.

Rotation of the humerus takes place around an axis that passes through the centre of the head and the centre of the capitulum at the lower end of the bone. The range of rotation varies with the position of the humerus. When the arm is by the side it can be rotated about 170 degrees but when the arm is vertically upright the range of rotation is greatly reduced ; the elbow must be flexed when the range of these movements is assessed [p. 232].

The amount of movement that occurs at the shoulder joint during movement of the upper limb is usually difficult to estimate because almost all free movements at the shoulder joint are accompanied by associated movements of the shoulder girdle. It is, however, usually considered that, in raising the arm upwards into a vertically upright position, the shoulder joint movement contributes about 120 degrees and the rotation of the scapula about 60 degrees [p. 225]. The total apparent movement of the humerus is therefore made up of its real movement upon the scapula plus movement of the scapula upon the chest wall, with associated movements at the clavicular joints : the effective mobility of this ball-and-socket joint is thus increased by the mobility of the socket itself. The commencement of movement of the scapula is usually co-incident with the commencement of movement at the shoulder joint (Cleland, 1884 ; Lockhart, 1930) and in raising the arm to an upright position the two movements are associated throughout. Shoulder joint movement is, however, more concerned in the raising of the arm to the horizontal position, and scapular movement more concerned in bringing the arm into a vertically upright position from the horizontal [PLATES 11 D and 20]. It must be stressed, however, that in bringing the arm from the horizontal to the vertically upright position there is considerable shoulder joint movement, and it is during this action that the humerus is brought approximately into line with the spine of the scapula, which is to be regarded as the fully abducted position of the shoulder joint proper (Lockhart, 1930).

The association of shoulder girdle movement with shoulder joint movement not only increases the range of movement of the free limb, but also increases the power of this movement. The small muscles of the 'rotator cuff' are attached close to the axis of movement, and have a poor mechanical advantage compared with the muscles acting on the scapula which are powerful and have considerable leverage. Thus the scapular movements add greatly to the power of movements at the shoulder joint. The adjuvant effect of scapular movements is strikingly illustrated in those in whom the shoulder joint is completely fixed as the result of disease or of surgical treatment to combat disease. Provided that care has been exercised to secure fixation in such a position as will allow full advantage to be taken of scapular movement, these patients may retain a considerable degree of mobility of the arm upon the trunk. Nevertheless it must be realized that movements at the shoulder joint and move-ments of the scapula are not dissociated but take place together.

It is to be observed that when the arm is raised to the vertically upright position, the medial epicondyle of the humerus is always directed forwards and slightly medially, and this direction of the epicondyle is constant irrespective of the plane (sagittal, coronal, or any intermediate) through which the limb has been raised. When the limb is raised in the coronal plane the humerus undergoes a lateral rotation of more than 90 degrees, whereas if it is raised to the upright in the sagittal plane it undergoes a slight medial rotation if the initial dependent position of the limb is the anatomical one with the palm of the hand directed forwards. The causal factor of the lateral rotation is most probably the action of the scapulo-humeral muscles, though it has been suggested that a contributory cause is the contact of the greater tubercle of the humerus with the sloping coraco-acromial arch when the humerus nears its limit of full abduction (Lockhart, 1933 ; Johnston, 1937 ; Martin, 1940 ; MacConaill, 1946). In any case it is impossible to raise the arm to the vertical position if the humerus is held in full medial rotation, in which position the greater tubercle abuts on the acromion.

THE ELBOW JOINT
[PLATE 12]

The elbow joint is a large synovial joint, of the ginglymus or hinge variety, in which the bones of the forearm articulate with the lower end of the humerus. The upper ends of the radius and ulna are bound together by the **anular ligament of the radius** in such a way as to permit free movement between these two bones at what is described separately as the proximal radio-ulnar joint. The elbow and the proximal radio-ulnar joints have a common fibrous capsule and synovial cavity, and though the anular ligament plays a part in the structure of both joints, it will be described with the proximal radio-ulnar joint.

Articular Surfaces. The humeral articular surface at the

Medial epicondyle
Epiphysial lines
Distal epiphysis of humerus
Ulnar collateral ligament

Radial collateral ligament
Anular ligament
Head of radius
Epiphysial line
Protrusion of synovial membrane

FIG. 300 Coronal section through the elbow and proximal radio-ulnar joints.

elbow comprises the grooved **trochlea** and the spheroidal **capitulum,** together with the sulcus between them. This composite surface is covered with a continuous layer of articular cartilage. The capitulum is confined to the anterior and distal aspects of the bone, but the trochlea extends from the lower edge of the coronoid fossa on the front of the humerus round the distal end of the bone to the lower edge of the olecranon fossa on the back [FIG. 303]. The ulnar surface of the elbow joint [FIG. 304] is the **trochlear notch,** covered with articular cartilage which is interrupted along a transverse line across the deepest part of the notch; the trochlear notch articulates with the trochlea of the humerus. The radial surface is the slightly cupped surface of the **head** of the radius, covered with articular cartilage continuous with that round the sides of the head in the radio-ulnar joint; the concave part of the upper surface of the head articulates with the capitulum, and the raised margin with the capitulo-trochlear groove. The radial and ulnar surfaces are not strictly congruent with the corresponding humeral surfaces, and the degree of accurate apposition varies slightly in different positions of the joint. The articular surfaces are most fully in contact when the forearm is in a position midway between full pronation and full supination and is flexed to a right angle.

Fibrous Capsule. A fibrous capsule completely invests the joint. Following the usual arrangement in a hinge joint, this ligament is relatively weak in front and behind, but is specially strengthened at the sides to form the radial and ulnar collateral ligaments.

The anterior part of the fibrous capsule [FIG. 301] is composed mostly of fibres that run longitudinally, but among these are some fibres that are arranged both transversely and obliquely : it is thicker in the middle than at the sides. This part of the capsule is attached to the front of the humerus immediately above the radial and coronoid fossae, to the anterior border of the coronoid process of the ulna, and to the anterior part of the anular ligament of the radius ; at the sides it blends with the radial and ulnar collateral ligaments. The brachialis muscle covers the greater part of the front of the capsule, and some of its deep fibres, inserted into the capsule, draw it upwards when the muscle contracts to flex the joint.

The posterior part of the capsule is very weak in its median part, but it is closely attached to the overlying tendon of the triceps, which affords ample support to it and draws it upwards in extension. The fibres in the posterior part of the capsule run mainly from the sides of the olecranon fossa to the margin of the olecranon ; some fibres stretch across the fossa above the olecranon as a transverse band with a free upper edge which falls short of the upper margin of the fossa. Beneath these transverse fibres a few longitudinal strands pass to the upper part of the fossa and afford slight support to the synovial pouch within it. The lateral fibres of the posterior part of the capsule pass from the back of the lateral epicondyle to the ulna at the posterior border of the radial notch and also to the anular ligament. At the lateral and medial sides the posterior part of the capsule blends with the corresponding collateral ligaments.

The **radial collateral ligament** [FIG. 301] is a strong, triangular band whose apex is attached above to the antero-inferior aspect of the lateral epicondyle of the humerus in close relation to the overlying common origin of the extensor muscles·

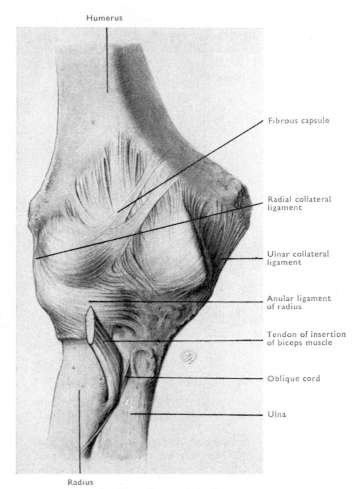

FIG. 301   Right elbow joint (anterior surface).

Labels (top to bottom): Humerus · Fibrous capsule · Radial collateral ligament · Ulnar collateral ligament · Anular ligament of radius · Tendon of insertion of biceps muscle · Oblique cord · Ulna · Radius

Distally, the middle part of the base of the ligament blends with the anular ligament of the radius, and the slightly thicker portions in front and behind sweep forwards and backwards to be attached to the margins of the radial notch on the ulna.

The **ulnar collateral ligament** [FIG. 302] is composed of three fairly distinct bands, continuous with one another. An **anterior band** passes from the front of the medial epicondyle of the humerus to the medial edge of the coronoid process of the ulna ; it is closely associated with the common origin of the superficial flexor muscles and gives rise to fibres of the flexor digitorum superficialis. A **posterior band** is attached above to the back of the medial epicondyle and below to the medial edge of the olecranon. Between these bands there is a middle, thinner part of the ligament, triangular in shape, and presenting a grooved external surface which is crossed by the ulnar nerve as it passes from upper arm to forearm. The apex of the middle part is attached to the under surface of the medial epicondyle, and its base is fixed distally to the upper border of a **transverse band** which stretches between the attachments of the anterior and posterior bands on the coronoid process and the olecranon. The lower edge of this transverse ligament is free, and through the narrow gap between this edge and the bone the synovial membrane protrudes slightly during movement at the joint.

Synovial Membrane. The synovial membrane of the

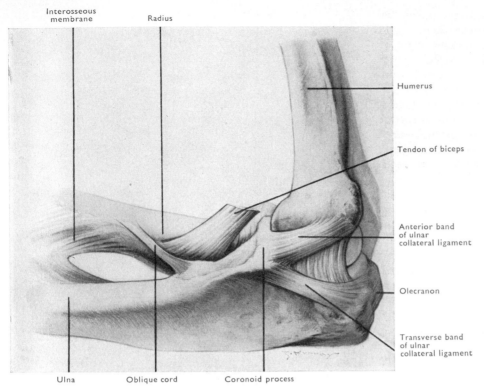

Interosseous membrane   Radius

Humerus

Tendon of biceps

Anterior band of ulnar collateral ligament

Olecranon

Transverse band of ulnar collateral ligament

Ulna   Oblique cord   Coronoid process

FIG. 302   Right elbow joint (medial aspect).   The anterior and posterior parts of the articular capsule were removed.

from the musculocutaneous, median, and radial nerves, and posteriorly from the ulnar nerve and the radial nerve through its branch to the anconeus muscle.

MOVEMENTS AT THE ELBOW JOINT

This being a hinge joint, movement occurs around a transverse axis—a movement of **flexion** when the forearm makes anteriorly a diminishing angle with the upper arm, and of **extension** when this angle is opened out again.   The axis of movement passes through the humeral epicondyles and is not at right angles with either the humerus or the bones of the forearm.   In full extension, with the forearm supinated, the arm and forearm form an angle of more than 180 degrees medially.   The angle by which this is more than 180 degrees is called the 'carrying angle'; although the 'carrying angle' has been stated to be more pronounced in women than in men, this difference is of no real significance as a secondary sexual characteristic.   The axis of movement bisects this angle; hence as the forearm

elbow joint lines the fibrous capsule and is reflected on to the humerus to line the radial and coronoid fossae in front and the olecranon fossa behind [FIG. 303].   Distally, it is prolonged to the upper part of the deep surface of the anular ligament; it covers the lower part of the deep surface of the anular ligament and is then reflected on to the neck of the radius. This reflection is supported by a few loose fibres which pass from the lower border of the anular ligament to the neck of the radius.   The synovial membrane passing from the medial side of the radial neck to the lower border of the radial notch on the ulna is supported by a lax band of fibres called the **quadrate ligament,** which stretches between these two points. Various synovial folds project a little way into the interior of the joint, filling small angular recesses between the edges of the articulating surfaces.   Well-marked extrasynovial fatty pads fill the radial and coronoid fossae in extension of the joint and the olecranon fossa in flexion.   These pads are displaced when the upper ends of the forearm bones occupy the fossae [FIG. 303].   A synovial fold, which forms the greater part of a ring, also overlies the peripheral part of the head of the radius and is interposed between it and the corresponding part of the capitulum of the humerus. There are no openings in the fibrous capsule of the elbow joint, but slight pouching of the synovial membrane may occur beneath the edge of the transverse band of the ulnar collateral ligament, beneath the lower border of the anular ligament, and above the transverse fibres that bridge the upper part of the olecranon fossa.

Nerve Supply.   The elbow joint derives its supply

is flexed it comes into line with the upper arm, the distal end of the ulna moves medially and the 'carrying angle' disappears.   If the humerus is rotated slightly medially, as it is with the arm hanging naturally by the side, the hand approaches the mouth when the forearm is flexed; this movement of bringing the hand to the mouth is also usually

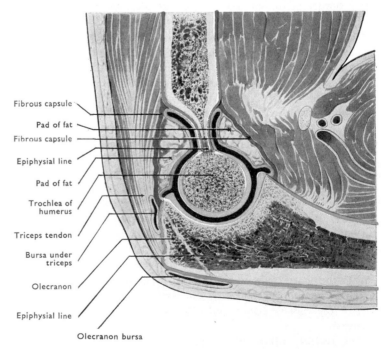

Fibrous capsule

Pad of fat

Fibrous capsule

Epiphysial line

Pad of fat

Trochlea of humerus

Triceps tendon

Bursa under triceps

Olecranon

Epiphysial line

Olecranon bursa

FIG. 303   Sagittal section through the elbow joint.   The synovial membrane is shown in grey, the articular cartilage in blue, and the periosteum in green.   Note that the pads of fat, though intracapsular, are extrasynovial.

associated with some abduction of the upper arm at the shoulder joint. The 'carrying angle' is masked when the forearm is pronated [p. 232], and in that position the wrist, elbow, and shoulder joints are all in line with one another when the elbow joint is extended—this is the usual position of the upper limb in which pushing and pulling movements are performed.

Flexion can proceed until checked by apposition of upper arm and forearm and by the tension of the posterior muscles and collateral ligaments. Extension cannot occur beyond the straight position of the limb; it is then limited by the tension of the anterior muscles and collateral ligaments. Collateral ligaments are fairly tense in all positions but the anterior part of each being specially tight in extension and the posterior part in flexion.

Limitation of movement by locking of the bones seldom occurs. That it does take place in some persons is shown by the occasional presence of small cartilage-covered facets at the bottom of the coronoid fossa and at the sides of the olecranon fossa which obviously must have made contact during life with the coronoid and olecranon processes.

As already mentioned, the joint surfaces are in closest contact in a position of right-angled flexion, with the forearm midway between pronation and supination. This is therefore the position of greatest stability and is the position which is most naturally assumed when the hands are engaged in fine manipulations.

THE RADIO-ULNAR JOINTS

The bones of the forearm are united at their proximal and distal ends by synovial joints which act together to allow movement of the radius on the ulna around a vertical axis; these joints are therefore of the pivot type. In addition, the shafts of the two bones are connected by fibrous interosseous bands in the manner of a syndesmosis.

THE PROXIMAL RADIO-ULNAR JOINT

At the proximal radio-ulnar joint the discoidal head of the radius rotates within the ring formed by the radial notch on the ulna and the anular ligament of the radius. The notch on

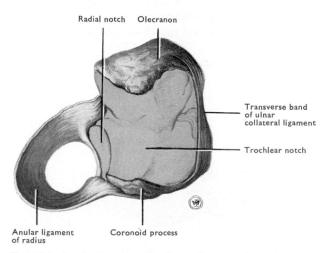

FIG. 304 Anular ligament of radius and proximal articular surfaces of ulna.

the ulna is lined with articular cartilage which is continuous with that on the lower part of the trochlear notch in the elbow joint; the surface of the radial notch is concave anteroposteriorly and is almost flat vertically [FIG. 304]. The head of the radius is covered with a continuous layer of cartilage above and around the sides. The upper surface is slightly cupped and rotates under the capitulum of the humerus, while the sides of the head bear on the anular ligament and radial notch.

Anular Ligament of Radius. The anular ligament of the radius is a strong, well-defined, curved band attached by its ends to the anterior and posterior margins of the radial notch on the ulna so as to form nearly four-fifths of a ring which is completed by the notch itself. This ring is slightly narrower at its lower border than above, forming a segment of a hollow cone; and the radial head, enclosed by the ring, is reciprocally bevelled all round. This arrangement tends to prevent the radius from being pulled downwards through the ring. The anular ligament is supported above by its firm fusion with the radial collateral ligament and the lateral parts of the fibrous capsule of the elbow joint both in front and behind [FIG. 301]. Below, it is only feebly attached to the neck of the radius, distal to the epiphysial line, by fibres which are too loose to interfere with movement. The upper part of the anular ligament is lined with fibrocartilage whose upper margin is continuous with the synovial membrane of the elbow joint. The lower part of the anular ligament is lined with **synovial membrane** which continues downwards on the upper part of the neck of the radius; the membrane lies on the upper surface of the quadrate ligament [p. 230] as it passes medially from the radius to the lower border of the radial notch on the ulna. The synovial cavities of the elbow and proximal radio-ulnar joints are freely continuous with each other.

THE DISTAL RADIO-ULNAR JOINT

The distal radio-ulnar joint is formed between the head of the ulna and the ulnar notch on the radius. The bony surfaces are covered with articular cartilage; the lateral and distal ulnar surfaces are continuous with each other over a rounded border.

Articular Disc. The chief uniting structure is the fibro-cartilaginous, triangular articular disc, which is attached by its base to the sharp edge on the radius between the ulnar and carpal surfaces and by its apex to the lateral side of the root of the ulnar styloid process [FIG. 305]. This disc provides an additional joint surface, for its upper surface articulates with the distal aspect of the head of the ulna. Thus, the joint cavity is L-shaped in vertical section, with a horizontal limb between the ulna and the articular disc [FIG. 307]. The distal surface of the disc forms part of the proximal articular surface of the wrist joint; but the cavities of the two joints do not communicate, except in some cases when the disc is perforated.

Fibrous Capsule. The fibrous capsule is represented only by transverse bands of no great strength which stretch from radius to ulna across the front [FIG. 306] and back of the joint. The distal edges of these bands blend with the margins of the articular disc, but proximally they are separated from each other by a pouch of the synovial lining of the joint, called the **recessus sacciformis,** which extends upwards a little way between the radius and ulna.

Nerve Supply. The distal radio-ulnar joint is supplied by twigs from the anterior and posterior interosseous nerves.

CONNEXION BETWEEN THE SHAFTS OF THE RADIUS AND ULNA

The shafts of the radius and ulna are connected by the oblique cord and the interosseous membrane of the forearm.
Oblique Cord. The oblique cord is a slender fibrous band which passes from the lateral border of the tuberosity of the ulna downwards and laterally to the radius just distal to its tuberosity [FIG. 302]; it is sometimes considered to be a degenerated portion of the flexor pollicis longus but is more probably a degenerate part of the supinator muscle (Martin, 1958).
Interosseous Membrane. The interosseous membrane is a strong fibrous sheet which stretches between the interosseous borders of the radius and ulna. The fibres of the membrane mostly run medially and downwards from radius to ulna and as the upper part of the membrane is deficient it has a free oblique upper border. Proximally this border is attached to the radius about 2·5 cm. below its tuberosity and it passes to a slightly more distal part of the interosseous border of the ulna. Through the gap between the upper border of the interosseous membrane and the oblique cord the posterior interosseous vessels pass to the back of the forearm. Distally, the interosseous membrane is continued into the fascia on the dorsal surface of the pronator quadratus muscle and there it is attached to the posterior of the two lines into which the lower part of the interosseous border of the radius divides. The distal part of the membrane is pierced by the anterior interosseous vessels.

The interosseous membrane serves to increase the surfaces of origin of the deep muscles of the forearm, to brace together the radius and ulna, and, because of the obliquity of most of its fibres, to transmit to the ulna part of any force passing upwards from the hand along the radius. The tension of the membrane varies during radio-ulnar movement but is greatest in the mid position.

MOVEMENTS AT THE RADIO-ULNAR JOINTS

In the **supine** position the bones of the forearm are parallel and the palm is directed forwards if the forearm is extended at the elbow joint; in the **prone** position the radius crosses in front of the ulna and the palm is directed backwards. Movement at the radio-ulnar joints takes place around an axis that passes through the centre of the head of the radius above and the apical attachment of the articular disc below. The movement is chiefly on the part of the radius; its upper end rotates within the ring formed by the anular ligament and the radial notch on the ulna, while its lower end, bearing the hand, travels round the lower end of the ulna, to which it is bound by the articular disc. In the movement of **pronation,** starting from the supine position, the lower end of the radius is carried forwards and medially round the lower end of the ulna until the palm is turned backwards and the shafts of the radius and ulna cross each other. Movement in the reverse direction is termed **supination.** These movements are normally associated with an accompanying rotation of the humerus at the shoulder joint—medially with pronation and laterally with supination—unless the elbow joint is flexed. However, the ulna also, changes position during pronation and supination. Its lower end precedes the lower end of the

radius around the opposite half of the circle of excursion, moving backwards and laterally as the radius travels forwards and medially during pronation, and in the reverse direction during supination. This excursion of the lower end of the ulna is the magnified result of a very slight movement of its upper end at the elbow joint. Pronation is said to be stronger than supination (Darcus, 1951); but screws and screwing instruments are made to be used by supination of the right forearm.

When the upper limb is straight, the axis of humeral rotation is in the same line as the axis of radio-ulnar movement, and therefore pronation and supination can be supplemented by the full extent of rotation at the shoulder joint so that it is then possible to turn the hand through almost 360 degrees.

THE RADIOCARPAL OR WRIST JOINT

[PLATE 13]

The radiocarpal or wrist joint proper is formed between the distal surface of the radius and the articular disc, and the proximal row of carpal bones, the scaphoid, lunate, and triquetrum.
Articular Surfaces. The surface of the radius and the disc together form a concave ellipsoid of shallower curvature in its long axis from side to side than in its short axis from before backwards [FIG. 305]. A similar surface, but convex, is provided distally by the proximal articular areas of the scaphoid, lunate, and triquetrum, closely united by interosseous ligaments which are flush with the articular cartilage on the proximal surface of the bones [FIG. 307]. The joint is therefore ellipsoidal in type. The **articular cartilage on the radius** is divided by a low ridge into a triangular lateral area and a quadrangular medial area. In the normal, straight position of the hand, the scaphoid bone is opposite the lateral radial area, the lunate is opposite the medial area and the disc, and the triquetrum is in contact with the medial portion of the articular capsule [FIG. 307]. When the hand is bent to the ulnar side, the carpus rotates so that the triquetrum lies opposite the disc [PLATE 13B].
Fibrous Capsule. A fibrous capsule surrounds the joint. It is attached close to the articular areas; its proximal attachment is distal to the inferior epiphysial lines of the radius and ulna.

The anterior part of the fibrous capsule has two parts which

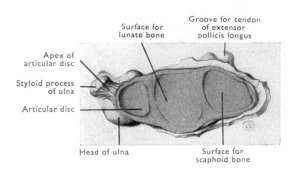

FIG. 305 Proximal articular surface of wrist joint.

are called the palmar radiocarpal ligament and the palmar ulnocarpal ligament.

The **palmar radiocarpal ligament** is formed chiefly of a broad band of fibres that spreads downwards, with a medial inclination, from the anterior edge of the distal end of the radius to the first carpal row—some longer fibres extending to the capitate bone in the second row [FIG. 306].

The **palmar ulnocarpal ligament** is a thickening of the fibrous capsule which extends downwards and laterally from

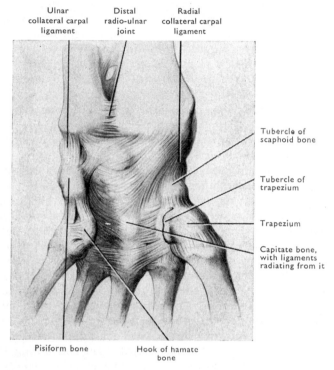

FIG. 306 Ligaments on the front of radiocarpal, carpal, and carpoetamcarpal joints.

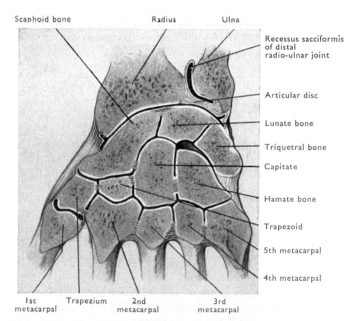

FIG. 307 Coronal section through the radiocarpal, carpal, carpometacarpal, and intermetacarpal joints, to show joint cavities and interosseous ligaments (diagrammatic).

the anterior edge of the articular disc and the base of the ulnar styloid process to the carpus.

The **dorsal radiocarpal ligament** lies on the back of the joint; it is composed of fibres that run mainly from the posterior edge of the lower end of the radius downwards and medially to the bones of the first carpal row and particularly to the triquetrum.

At the sides, the fibrous capsule is specially strengthened to form the radial and ulnar collateral carpal ligaments.

The **radial collateral carpal ligament** runs from the radial styloid process to the scaphoid immediately lateral to its proximal articular surface; the radial artery crosses this ligament deep to the long abductor and short extensor tendons of the thumb.

The **ulnar collateral carpal ligament** passes downwards from the ulnar styloid process to the medial, non-articular border of the triquetrum and to the pisiform bone.

Synovial Membrane. Synovial membrane lines the fibrous capsule. In exceptional cases the synovial cavity communicates with the distal radio-ulnar joint through a perforated articular disc, or with the intercarpal joint when one of the interosseous ligaments of the proximal carpal row forms an incomplete barrier.

Nerve Supply. The radiocarpal joint (in common with the intercarpal and carpometacarpal joints—described later) receives its supply from all the three main nerves of the upper limb: from the median nerve, through its anterior interosseous branch; from the radial nerve, through its posterior interosscous branch; and from the ulnar nerve, through its dorsal and deep branches.

MOVEMENTS AT THE RADIO-CARPAL JOINT

In free movements of the hand at the wrist much of the apparent movement at the radiocarpal joint is the result of movement between the proximal and distal rows of carpal bones (midcarpal or transverse intercarpal joint). Movements at the radiocarpal and midcarpal joints are usually closely associated and for that reason movements at these joints are considered together [p. 234].

THE INTERCARPAL JOINTS

The carpal bones are articulated in two rows—proximal and distal—that form with each other an important joint called the **midcarpal joint.** The joints between the individual bones of the carpus are mostly of the plane or gliding type, but the capitate rotates considerably during movements of the hand.

Joints of the Proximal Row. In the proximal row of carpal bones, the scaphoid, lunate, and triquetrum are bound together by palmar, dorsal, and interosseous bands.

The **palmar** and **dorsal intercarpal ligaments** pass between neighbouring parts on those aspects of the bones. The **interosseous intercarpal ligaments** are short bands that pass between contiguous surfaces, uniting them in their whole palmar-dorsal depth near their proximal margins, and leaving joint clefts between their distal parts, which are therefore coated with articular cartilage [FIG. 307].

Pisiform Joint. The pisiform bone rests on the palmar surface of the triquetrum, and forms with that bone a separate little synovial joint surrounded by a thin but strong fibrous capsule. It is further anchored by strong bands—the **piso-metacarpal** and **pisohamate ligaments**—to the base of the fifth metacarpal and the hook of the hamate bone; these ligaments resist the pull of the flexor carpi ulnaris muscle upon the pisiform bone, and, in effect, provide additional insertions for that muscle.

Joints of the Distal Row. All four bones of the distal carpal row are connected by **palmar, dorsal,** and **interosseous intercarpal ligaments.** The interosseous ligaments [Fig. 307] are, in general, not so extensive as those of the proximal row; they leave joint clefts between the bones both proximally, running into the midcarpal joint, and distally, running into the carpometacarpal joint. These two larger joints sometimes communicate with each other between the bones of the distal row when an interosseous ligament is absent (most often that between the trapezium and trapezoid) or when it does not extend the whole way between the dorsal and palmar surfaces of the adjoining bones.

Midcarpal Joint [Fig. 307]. This is the joint between the two carpal rows. The line of the joint is convex distally on the lateral side, where the trapezium and trapezoid are opposed to the rounded distal surface of the scaphoid. In the medial and larger part of the joint there is a deep concavity facing distally, bounded by the triquetrum and lunate and the medial surface the scaphoid; this concavity is occupied by the rounded head of the capitate and the proximal angle of the hamate.

The joint is surrounded by a fibrous capsule made up, in front and behind, of irregular bands which run between the two rows of bones and constitute the **palmar** and **dorsal intercarpal ligaments;** the bands in the palmar ligament radiate chiefly from the head of the capitate bone [Fig. 306]. At the sides of the midcarpal joint the fibrous capsule is strengthened as it passes between the scaphoid and trapezium and between the triquetrum and hamate.

Intercarpal Joint Cavity. The intercarpal joint synovial cavity [Fig. 307] is large and complicated. The main part extends from side to side between the two rows of bones in the midcarpal joint; this may be partially interrupted by an interosseous ligament connecting contiguous parts of the scaphoid and capitate. Offshoots of the cavity pass proximally between the three main bones of the proximal row as far as the interosseous ligaments. Sometimes, a deficiency in one interosseous ligament may permit the intercarpal joint to communicate with the radiocarpal joint. Other offshoots of the synovial cavity pass distally between the four bones of the distal row, and not infrequently communicate with the carpometacarpal joint cavity, especially on one or other side of the trapezoid.

Nerve Supply. The intercarpal joints are supplied by twigs from the anterior and posterior interosseous nerves and the dorsal and deep branches of the ulnar nerve.

Flexor Retinaculum. The carpus, in each of its rows, forms a transverse arch with a palmar concavity. The concavity is bridged over by the flexor retinaculum, which is attached medially to the pisiform and the hook of the hamate, and laterally to the scaphoid tubercle and the crest of the trapezium. The retinaculum is a major factor in maintaining the carpal bones in their normal position and for that reason it is considered as an accessory ligament of the intercarpal joints.

Through the tunnel deep to the retinaculum, the median nerve and the flexor tendons enter the palm, while the ulnar vessels and nerve pass superficial to it.

MOVEMENTS AT THE RADIOCARPAL AND INTERCARPAL JOINTS

The principal movements between the carpal bones are those that occur at the midcarpal joint and, as movement at this joint is usually associated with movement at the radiocarpal joint, the movements at both joints are considered together.

Movements at these joints occur around transverse and anteroposterior axes. Around transverse axes the hand is bent towards the front of the forearm in **flexion** or towards the back of the forearm in **extension.** Around anteroposterior axes the hand is deflected towards the ulnar border of the forearm in **adduction** (ulnar deviation) or towards the radial border in **abduction** (radial deviation). Oblique movements are also possible around intermediate axes, as well as a combined movement around successive axes bringing about circumduction of the hand on the forearm.

In **flexion** of the hand, movement occurs at both the radiocarpal and midcarpal joints, but in full flexion, as can only occur when the fingers are extended, the greater part of the movement occurs at the midcarpal joint. **Extension** of the hand is also a compound movement, but the range of movement is greater at the radiocarpal than at the midcarpal joint. In both flexion and extension of the hand the distal row of carpal bones moves around a transverse axis that passes through the middle of the head of the capitate, and there is also, in both movements, a compensatory swing of the scaphoid on the lunate for the accommodation of the head of the capitate. In adduction of the hand the greater part of the movement occurs at the radiocarpal joint, and during this movement the proximal row of carpal bones glides laterally so that the entire proximal surface of the lunate lies distal to the radius [Plate 13 B]. Abduction of the hand is not such an extensive movement as adduction and occurs almost entirely at the midcarpal joint. In both adduction and abduction the movement of the distal row of carpal bones upon the proximal row is round an anteroposterior axis that passes through the centre of the head of the capitate. (For a full consideration of these movements see Wright, 1935; MacConaill, 1941.)

It is important to note that slight extension is the position naturally assumed at the wrist when the hand is used for grasping purposes, and therefore, if the wrist joint is likely to become fixed through disease, fixation should be secured in a position of slight extension in order to conserve sufficient grasping power.

THE CARPOMETACARPAL AND INTERMETACARPAL JOINTS

Carpometacarpal Joint of the Thumb [Fig. 307]. This, the only highly mobile carpometacarpal joint, is a separate, self-contained joint between the trapezium and the base of the first metacarpal bone. The surfaces are reciprocally saddle-shaped and, as stated earlier, it is the only good example of a saddle joint. The joint is completely enclosed by a strong but rather loose fibrous capsule [FIG. 306], which is lined with synovial membrane. Deep to the fibrous capsule lie three discrete ligaments that are termed the **anterior** and **posterior oblique** and the **radial carpometacarpal ligaments** (Haines, 1944). The radial ligament is attached to the adjacent surfaces of the trapezium and the first metacarpal on their radial sides. The oblique ligaments are attached respectively to the corresponding surfaces of the trapezium and converge as they pass distally to be attached to the ulnar side of the base of the first metacarpal.

Movements. This is a saddle joint and movement can therefore occur around two principal axes at right angles to each other; oblique movements are also possible around intermediate axes as well as a combined movement around successive axes that allows the movement of circumduction of the thumb. In addition, a limited amount of rotation can take place, and though this movement does not have a wide range it is extremely important as it allows the opposition of the thumb to the other fingers. For the understanding of the movements of the thumb it is necessary to stress the inclination of its palmar and dorsal surfaces. When the thumb is in a position of rest, its dorsal surface is directed laterally and its palmar surface medially. In **flexion** the thumb is moved to the ulnar side in the plane of the palm, and in **extension** it is carried to the radial side in the same plane. In **adduction** the thumb is carried directly backwards and in **abduction** it is carried forwards. In the movement of full flexion the thumb is carried directly in front of the palm of the hand and it is during the latter part of this movement that the metacarpal bone undergoes a **medial rotation** of about 30 degrees so that the palmar surface of the thumb becomes opposed to the corresponding surfaces of the fingers. Conversely, when the thumb is moved into full extension the first metacarpal undergoes a slight **lateral rotation** as the movement nears its completion. The movement of medial rotation is caused by the posterior oblique carpometacarpal ligament becoming increasingly taut during flexion of the metacarpal, and similarly lateral rotation is dependent on the anterior oblique carpometacarpal ligament becoming taut during extension (Haines, 1944). During the movement of opposition a certain amount of medial rotation of the first phalanx occurs at the metacarpophalangeal joint and this supplements the rotation of the metacarpal (Bunnell, 1938).

Common Carpometacarpal Joint [FIG. 307]. The bases of the medial four metacarpal bones form, with the distal row of carpal bones, a common carpometacarpal joint. The line of the joint is highly irregular. Laterally, the base of the second metacarpal fits into a recess formed by the trapezium, trapezoid, and the capitate. The third metacarpal articulates with the capitate bone at a small transverse segment of the joint line. The fourth metacarpal causes the joint line to become angular again as it articulates to a small extent with the capitate and to a greater extent with the hamate bone. Most medially the fifth metacarpal articulates only with the medial, bevelled facet on the hamate bone.

The joint is surrounded by a fibrous capsule in which various **palmar** and **dorsal carpometacarpal** thickenings can be distinguished passing from the distal carpal bones to the metacarpal bases. An **interosseous ligament** is usually present,

stretching from contiguous parts of the capitate and hamate bones to the third or the fourth metacarpal base or to both; this ligament may divide the joint into separate medial and lateral compartments.

The **synovial cavity** of the common carpometacarpal joint extends proximally between the four distal carpal bones, and there usually communicates with the midcarpal joint. Distally, the cavity is continuous with the little joints between the bases of the medial four metacarpal bones [FIG. 307].

Intermetacarpal Joints [FIGS. 306 and 307]. These joints are formed between small articular facets on the contiguous sides of the bases of the medial four metacarpal bones. The three little joints are closed in front, behind, and distally by **palmar, dorsal,** and **interosseous ligaments** that pass transversely between adjacent bones. Proximally, their cavities open into the synovial cavity of the common carpometacarpal joint.

Nerve Supply. The carpometacarpal and intermetacarpal joints are supplied by twigs from the anterior and posterior interosseous nerves and the dorsal and deep branches of the ulnar nerve.

Movements. Very little movement is possible between the carpus and the medial four metacarpal bones. It is more appreciable in the case of the fifth metacarpal, whose articulation with the hamate bone is of a flattened saddle type; around an oblique axis, this metacarpal bone can be flexed a little way across the palm in a fashion similar to the more extensive opposition movement of the metacarpal bone of the thumb.

THE METACARPOPHALANGEAL JOINTS

Each metacarpophalangeal joint is formed between the slightly cupped base of the proximal phalanx and the rounded metacarpal head, which is covered with articular cartilage distally and on the front but not on the back.

The **fibrous capsule** is strengthened on each side by a **collateral ligament** which radiates fanwise from the tubercle and adjacent depression on the side of the metacarpal head to the side of the base of the proximal phalanx and to the front of the capsule [FIG. 308].

On the front there is a dense fibrous pad, called the **palmar ligament,** which blends with the collateral ligaments at the sides and is firmly fixed distally to the base of the phalanx but is only weakly attached proximally to the neck of the metacarpal bone. This pad is grooved in front by the long flexor tendons. In the thumb there are two sesamoid bones

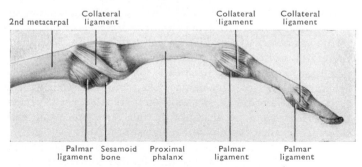

| 2nd metacarpal | Collateral ligament | | Collateral ligament | Collateral ligament |

| Palmar ligament | Sesamoid bone | Proximal phalanx | Palmar ligament | Palmar ligament |

FIG. 308 Metacarpophalangeal and interphalangeal joints.

embedded in it; there is usually one such sesamoid in the radial side of the pad on the index finger and occasionally in the ulnar side in the little finger.

The fibrous capsule is deficient **dorsally,** where it is replaced by the expansion of the long extensor tendon, which blends at the sides with the collateral ligaments.

Deep Transverse Metacarpal Ligaments. Three deep transverse metacarpal ligaments unite the palmar ligaments of the medial four metacarpophalangeal joints. Indirectly they bind the heads of the medial four metacarpals together. The interossei tendons descend behind these bands; the lumbrical tendons pass in front of them. It is to be noted particularly that there is no such binding ligament between the metacarpophalangeal joints of the thumb and the index finger. The thumb is therefore free to move independently of the other fingers.

Movements. The metacarpophalangeal joints are of condyloid type, having all the movements of a ball-and-socket joint except rotation. The proximal phalanges can be **flexed** to at least a right angle with the metacarpal bones but can be **extended** little beyond the line of the metacarpals. In the extended position the fingers can be **adducted** or **abducted** towards or away from the centre line of the middle finger. When the fingers are flexed, lateral movements become impossible because of increased tension of the collateral ligaments, which are fixed nearer to the dorsal and distal surfaces

than to the palmar surface of the metacarpal heads, and are also more stretched in flexion owing to the greater width of the palmar aspect of the metacarpal articular surface. The metacarpophalangeal joint of the thumb has much less extensive movement than the others—hardly any adduction and abduction being possible—though a slight medial rotation occurs at it during the movement of opposition of the thumb to the fingers [p. 235].

INTERPHALANGEAL JOINTS

[FIG 308]

The interphalangeal joints are constructed, as regards ligaments, in exactly the same fashion as the metacarpophalangeal joints. But owing to the bicondyloid shape of the articular surfaces these are pure hinge joints, capable only of **flexion** and **extension.** Although the fingers are of unequal length, the phalanges are so proportioned that in flexion the finger tips come into line and meet the palm simultaneously.

Nerve Supply. The metacarpophalangeal and interphalangeal joints of the thumb, index, and middle fingers are supplied mainly by the median nerve, but where the radial nerve supplies the skin over these joints, it too may send twigs to them. The corresponding joints of the ring and little fingers are supplied mainly by the ulnar nerve.

JOINTS OF THE LOWER LIMB

THE PELVIC GIRDLE

The skeleton of the pelvic or lower limb girdle is joined to the skeleton of the trunk in such a way as to provide great strength in the region of weight transference from trunk to girdle at the sacrifice of almost all mobility. The pelvic girdle is formed by the two hip bones which are firmly, but not quite immovably, joined to the sacrum behind and to each other in front. The pelvic joints are the sacro-iliac, interpubic, sacrococcygeal, and lumbosacral articulations.

THE LUMBOSACRAL AND SACRO-COCCYGEAL JOINTS

Lumbosacral Joint [PLATE 6]. At the lumbosacral joint the fifth lumbar vertebra articulates with the first piece of the sacrum in the same way as any two typical vertebrae do. They are, therefore, united, by an intervertebral disc, anterior and posterior longitudinal ligaments, synovial joints between articular processes, ligamenta flava, and interspinous and supraspinous ligaments. In addition there is a special ligament that springs from the fifth lumbar transverse process on each side and is called the iliolumbar ligament.

The **iliolumbar ligament** [FIG. 310] is a strong ligament that spreads laterally from the fifth lumbar transverse process to the posterior part of the inner lip of the iliac crest; it is

really the thickened lower border of the anterior and middle layers of the thoracolumbar fascia which enclose the quadratus lumborum muscle. Continuous with the lower border of the iliolumbar ligament there is the less distinct fibrous band which is commonly called the **lateral lumbosacral ligament.** It consists of bundles of variable strength that pass downwards and laterally from the lower border of the fifth lumbar transverse process to the ala of the sacrum, where they intermingle with the ventral sacro-iliac ligament.

Sacrococcygeal Joint. At the sacrococcygeal joint there is an intervertebral disc between the last sacral and first coccygeal segments, reinforced all round by longitudinal strands called the **sacrococcygeal ligaments** [FIG. 310]. These have quite unnecessarily been designated as distinct ventral, dorsal, and lateral sacrococcygeal ligaments, but it may be noted that the lateral ligament forms the lateral boundary of the foramen for the fifth sacral ventral ramus. The sacrococcygeal joint frequently becomes obliterated or partially obliterated in old age.

THE SACRO-ILIAC JOINT

The sacro-iliac joint is formed between the **auricular surfaces** on the sacrum and the iliac bone. Reciprocal sinuosities on these two articular surfaces, or even in some cases more pronounced irregularities in the form of tubercles with depressions opposite to them, cause a certain amount of interlocking between the facets. The shape and the irregularity of the auricular surfaces not only vary considerably in different individuals, but also may show marked differences on the two sides in the same individual. The auricular surface of the sacrum is covered with hyaline cartilage, but the cartilage on the corresponding facet of the ilium is usually of a fibrous type (Schunke, 1938). The joint is a synovial one, with a complete fibrous capsule lined with synovial membrane, but movement is greatly restricted by the irregularity of the articular surfaces and the thickness and disposition of the dorsal sacro-iliac ligaments. In later life, particularly in males, it is usual to find fibrous or fibrocartilaginous adhesions between the articular surfaces, with partial or complete obliteration of the cavity of the joint (Brooke, 1924).

The **ventral sacro-iliac ligament** is a broad band, of no great thickness, which closes the joint in front, both above and below the linea terminalis. It stretches from the ala and pelvic surface of the sacrum to the adjoining surfaces of the iliac bone.

The **sacro-iliac ligaments** that lie **behind** and **above** the joint constitute a very much stronger and thicker formation. Two distinct strata may be recognized. The deeper, which is called the **interosseous sacro-iliac ligament,** is short, thick, and very strong. It fills the narrow cleft between the rough areas on the bones immediately above and behind the articular surfaces [FIG. 309]. Occasionally one or two small accessory joint cavities are found in the substance of this ligament between facets near the posterior superior iliac spine and bosses on the sacrum in the position of transverse tubercles (Jazuta, 1929).

Superficial to the interosseous fibres, longer bands run obliquely medially and downwards between more divergent parts of the bones and these constitute the **dorsal sacro-iliac ligament** [FIGS. 309 and 310]. The longest and most superficial fibres of this ligament pass almost vertically downwards from the posterior superior iliac spine to the third and fourth segments of the sacrum. The lateral part of this ligament is indistinguishable from the upper part of the sacrotuberous ligament.

Nerve Supply. The sacro-iliac joint is supplied : (1) by twigs directly from the sacral plexus and the dorsal rami of the first two sacral nerves ; and (2) by branches from the superior gluteal and obturator nerves.

Accessory Ligaments

The gap left in the bony pelvis between the sacrum and the ischial part of the hip bone is bridged by two important accessory ligaments of the sacro-iliac joint called the sacrotuberous and sacrospinous ligaments [FIGS. 309 and 310].

The **sacrotuberous ligament** has an extensive attachment to the posterior superior and posterior inferior iliac spines, the back and side of the lower part of the sacrum and the side of the upper part of the coccyx. From this wide attachment the fibres converge as they pass downwards and laterally towards the ischial tuberosity, but, twisting upon themselves, they diverge again to be attached to the medial margin of the tuberosity and the lower margin of the ramus of the ischium. The fibres to the ramus form a sickle-shaped extension of the ligament called the **falciform process.**

The most superficial of the fibres which are attached to the

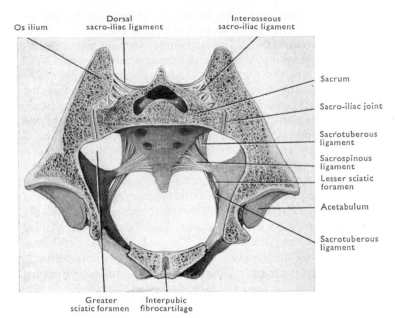

FIG. 309   Coronal section of pelvis.

FIG. 310   Posterior view of pelvic ligaments and hip joint.

ischial tuberosity are intimately associated with the long head of the biceps femoris muscle ; the whole ligament, indeed, is regarded as an isolated proximal portion of this muscle.

The **sacrospinous ligament** lies on the pelvic surface of the sacrotuberous ligament. It is triangular in outline. Its base is attached to the edge of the lower sacral and upper coccygeal segments, in front of the sacrotuberous ligament ; its apex is attached to the ischial spine. This ligament is closely blended with the coccygeus muscle, and may be regarded as a fibrous dorsal part of that muscle.

Sciatic Foramina. The sacrotuberous and sacrospinous ligaments convert the greater and lesser sciatic notches of the hip bone into the greater and lesser sciatic foramina. The **greater sciatic foramen** is bounded above and in front by the greater sciatic notch, behind and medially by the sacrotuberous ligament, and below by the sacrospinous ligament and the ischial spine. The triangular **lesser sciatic foramen** is bounded by the lesser sciatic notch, the ischial spine and sacrospinous ligament above, and the sacrotuberous ligament below [FIG. 310]. The piriformis muscle runs out through the **greater sciatic foramen** with the superior gluteal vessels and nerve above the muscle, and below it the sciatic nerve, the posterior cutaneous nerve of the thigh, the inferior gluteal vessels and nerve, the internal pudendal vessels and the pudendal nerve, and the nerves to the quadratus femoris and obturator internus muscles. The tendon of the obturator internus emerges from the **lesser sciatic foramen**; the nerve to that muscle and the internal pudendal vessels and pudendal nerve enter the perineum above the tendon, by hooking round the ischial spine at its junction with the sacrospinous ligament.

THE PUBIC SYMPHYSIS

The pubic symphysis is a median joint between the bodies of the pubic bones [FIG. 309]. Each pubic articular surface is covered with a thin layer of hyaline cartilage united to the cartilage of the opposite side by a thick fibrocartilaginous **interpubic disc.** In the posterosuperior part of this disc a slit-like cavity appears during early life, and, in women especially, may extend later through the greater part of the fibrocartilage; this cavity has no synovial lining. The joint is therefore of a symphysial type, modified by the appearance of an imperfect joint cavity.

Ligaments of Pubic Symphysis. The fibrocartilaginous disc of the joint is strengthened by the superior and arcuate ligaments.

The **superior ligament** is attached to the pubic crests and tubercles on each side and strengthens the anterior aspect of the symphysis. The **arcuate ligament** arches across between the inferior pubic rami, and so rounds off the subpubic angle. It is separated from the urogenital diaphragm by an interval through which the dorsal vein of the penis (or clitoris) enters the pelvis.

THE OBTURATOR MEMBRANE

The obturator membrane [FIG. 311] is unconnected with any pelvic joint; it is an interlacement of ligamentous bundles that close the obturator foramen and are attached to its margin except anterosuperiorly where the **obturator canal** passes through the foramen. The canal is bounded anterosuperiorly by the obturator sulcus and completed below by the short, free edge of the membrane; it transmits the obturator vessels and nerve. The obturator muscles arise from the surfaces of the membrane and the adjacent bone.

PELVIC MECHANICS

The bony pelvis protects the pelvic viscera and gives attachment to the muscles of the trunk and lower limb, but its primary skeletal function is to provide for stable transference of body weight from the vertebral column to the thigh bones. From the last lumbar vertebra the weight is transmitted through the upper part of the sacrum and the adjoining part of each iliac bone to the head of the femur; these parts constitute a pelvic arch of thickened bone, the concave border of which corresponds to the posterior half of the linea terminalis. The ventral pelvic bar is formed by the superior rami and bodies of the pubic bones and acts as a horizontal tie-beam connecting the bases of the pillars of the arch. The weight of the body tends to displace the anterosuperior part of the sacrum downwards into the pelvis and so tilt the postero-inferior part upwards. This tendency is resisted by a number of factors. (1) As the sacrum is depressed, the interosseous and dorsal sacro-iliac ligaments tighten and draw the ilia together, thus compressing the irregular surfaces of the sacro-iliac joint against each other. (2) The sacrum is wedge-shaped where it articulates with the ilia, the base of the wedge being superior in the anterosuperior part [Fig. 309] and inferior in the postero-inferior part. Thus the tendency to depress the upper part and elevate the lower is resisted by the shape of the articulation alone, particularly when the joint surfaces are tightened together. (3) The strong sacrotuberous ligaments prevent any upward displacement of the postero-inferior part of the sacrum relative to the hip bones.

The weight transmitted downwards to the femora is resisted by a force which is applied to the acetabula by these bones, and this tends to compress the pelvis inwards. Looked at in this way, the ventral pelvic bar and the bone of the linea terminalis are struts which prevent this compression, but which may deform if softened by conditions such as rickets. The slight amount of gliding movement and the cartilaginous plates present in the sacro-iliac joints have a mild cushioning action against jarring shocks, and the iliolumbar ligaments help to prevent the fifth lumbar vertebra from slipping forwards on the sloping upper surface of the first sacral vertebra. Thus the various joints of the pelvis give it a degree of resilience, but tend to make it progressively more rigid as greater loads are applied to it.

The **female pelvis** is modified in association with the requirements of child-bearing. Differences are seen in the joints as in the bones. At the sacro-iliac joint there is less interlocking by reciprocal irregularity of the bones, and more movement is permitted than in the male pelvis; the limitation of movement by fibrous ankylosis in later life is not nearly so marked in women as in men. The larger cavity in the interpubic disc is characteristic of the female joint and allows greater separation of the bones. The coccyx, which in men is usually fused to the sacrum by synostosis in later life, is more mobile and preserves its mobility longer in women.

During pregnancy, softening and relaxation of the pelvic ligaments occur which allow the separation of the pelvic bones and an increase in the movements at the pelvic joints. These joint changes were formerly regarded as of importance during child-birth, but radiographic observations indicate that they do not materially affect the diameters of the pelvis (Young, 1940).

THE HIP JOINT

[PLATE 14]

At the hip joint the globular head of the femur articulates with the cup-like acetabulum of the hip bone, and provides the most striking example in the body of a ball-and-socket joint.

The mechanical requirements at this joint are severe. It must be capable not merely of supporting the entire weight of the body—as in standing on one leg—but of stable transference of the weight, even during movement of the trunk upon the femur, such as occurs with rapid alternation at the two hip joints during walking and running. The joint must therefore possess great strength and stability even at the expense of limitation of range of movement. Accordingly, the deep socket securely holding the femoral head, the strong, tense fibrous capsule, the insertion of the controlling muscles at some distance from the centre of movement, are all in marked contrast to the conditions at the shoulder joint.

Articular Surfaces. The **femoral** articular surface forms nearly two-thirds of a spheroid. The covering cartilage is interrupted at the pit on the head of the femur, and it ends at the commencement of the neck of the femur along a sinuous line which sometimes makes a scalloped excursion upon the front of the neck. The **acetabular** articular surface does not occupy the entire acetabulum but is limited to a broad horseshoe-shaped belt which is covered with cartilage (the **lunate surface**); the cartilage is absent below and in front opposite the acetabular notch. The depressed non-articular part of the acetabulum within the horseshoe is called the **acetabular fossa** and lodges a pad of fat covered with synovial membrane.

Transverse Ligament of the Acetabulum and Labrum Acetabulare. The acetabular notch is bridged by the transverse ligament of the acetabulum. The superficial edge of the ligament is flush with the acetabular rim; its deep edge does not reach the bottom of the notch but helps to bound an aperture which admits articular vessels and nerves. The acetabulum is deepened all round by a fibrocartilaginous lip called the labrum acetabulare, which is firmly attached to the bony rim and the transverse ligament. The thin, free edge of the labrum forms a slightly smaller circle than its attached base, and so is able to grasp the femoral head [FIG. 312].

Fibrous Capsule [FIGS. 310 and 311]. The fibrous capsule encloses the joint and the greater part of the femoral neck as with a cylindrical sleeve. It is very strong and tense in full extension—in contrast to the thin and lax capsule of the shoulder joint. The tendons of surrounding muscles are much less intimately connected with it, but it does receive expansions from the rectus femoris, the gluteus minimus, and the piriformis muscles.

Proximally, the fibrous capsule surrounds the acetabulum and is attached above and behind directly to the hip bone just wide of the labrum, while below and in front it is fixed to the bone and the outer surface of the labrum and the transverse ligament of the acetabulum. Distally, the capsule is attached in front to the inter-

trochanteric line at the junction of the femoral neck and shaft; above and below it is attached to the neck close to its junction with the trochanters; but, behind, it extends over only the medial two-thirds of the neck. Thus, the whole of the neck of the femur is intracapsular in front, but the lateral third is extracapsular behind. The epiphysial line of the head is entirely intracapsular; the trochanteric epiphysial lines are extracapsular.

The fibres of the fibrous capsule run for the most part longitudinally from pelvis to femur. Some deeper fibres, however, pass circularly round the joint, constituting the **zona orbicularis;** this is marked only on the back of the capsule, where it appears on the surface, winding round behind the femoral neck [FIG. 310].

Some of the deepest longitudinal fibres, on reaching the neck of the femur, turn upwards upon the neck towards the articular margin; these reflected fibres form bundles, which are called **retinacula,** and are best marked on the upper and lower surfaces of the neck [FIG. 312].

The main longitudinal fibres of the capsule form certain thickened bands which resist the tensile stresses to which the capsule is subjected. These bands are specially named after the regions of the acetabulum to which they are attached, and are therefore called the iliofemoral, pubofemoral, and ischiofemoral ligaments. It must be clearly understood that these ligaments are thickened parts of the fibrous capsule and are not always readily identifiable.

The **iliofemoral ligament** is of great strength and considerable thickness; it is a triangular band attached proximally by its apex to the lower part of the anterior inferior iliac spine

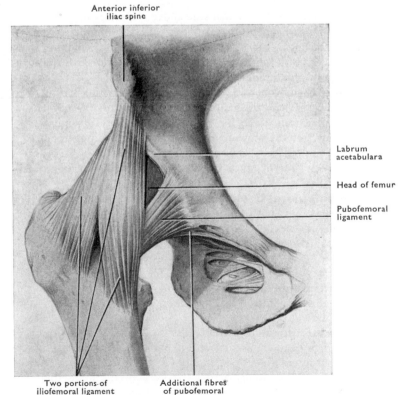

Anterior inferior
iliac spine

Labrum
acetabulara

Head of femur

Pubofemoral
ligament

Two portions of
iliofemoral ligament

Additional fibres
of pubofemoral
ligament

FIG. 311 Dissection of the right hip joint from the front.

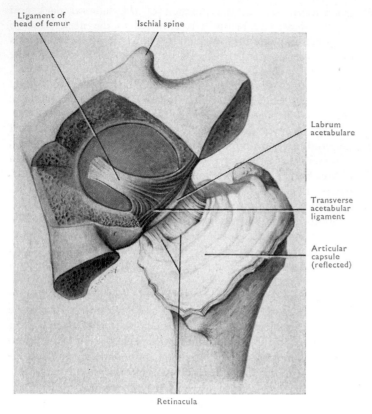

Ligament of
head of femur   Ischial spine

Labrum
acetabulare

Transverse
acetabular
ligament

Articular
capsule
(reflected)

Retinacula

FIG. 312   Dissection of right hip joint from pelvic side. The
floor of the acetabulum has been removed, and the articular
capsule of the joint thrown laterally towards the trochanters.

behind the upper band of the iliofemoral ligament. The
general spiral course of the ischiofemoral ligament upwards
and laterally across the back of the joint causes the ligament,
like the other principal ligaments of the joint, to be most tense
in the position of extension.

Ligament of the Head of the Femur. Within the hip joint
there is a weak, flattened ligament (the ligament of the head of
the femur [FIG. 312]) consisting of delicate connective tissue
surrounded by synovial membrane, which is attached to the
margin of the acetabular fossa and the adjacent edge of the
transverse ligament of the acetabulum. It narrows as it
passes to its apical attachment in the pit on the head of the
femur. Between its attachments the ligament is surrounded
by a synovial sheath so that although it is within the fibrous
capsule of the joint it is extrasynovial. It lies in the acetabular
fossa below the lower part of the femoral head.

The ligament of the head of the femur varies in size, and in
rare cases it is absent. Its function is uncertain. It is stretched
when the flexed thigh is adducted or rotated laterally, but in
many cases it is too weak to have any definite ligamentous
action. No appreciable disability results from its rupture or
absence.

A small artery is always found within its substance; this
vessel takes little, if any, part in the blood supply of the
femoral head in children, though in adults it may supplement
the supply derived from the vessels within the retinacula
(Tucker, 1949). After fusion of the femoral head with the
femoral diaphysis, there is an anastomosis of the vessels of the
ligament of the head of the femur with the metaphysial vessels
(Trueta, 1957).

Synovial Membrane. The synovial membrane of the hip
joint lines the fibrous capsule and covers the labrum acetabulare. At the acetabular notch it is attached to the medial
margin of the transverse ligament, and covering the fatty tissue
in the acetabular fossa, extends as a funnel-shaped sheath upon
the ligament of the head of the femur to the edge of the pit on
the femoral head. Distally, at the femoral attachment of the
fibrous capsule, the synovial membrane is reflected upwards on
the neck of the femur as far as the edge of the articular cartilage
on the head. The retinacula raise this reflected part into prominent folds [FIG. 312], within which blood vessels run upwards
to the head of the bone. A part of the epiphysis for the head of
the femur is non-articular and it is through this part that the
blood vessels in the retinacula enter the head [FIG. 226]. These
vessels are of clinical importance as they constitute the
principal arterial supply to the head; they have been extensively investigated by many workers (Trueta, 1957). At the
back of the joint a fold of synovial membrane may protrude
beneath the lateral border of the fibrous capsule, which is here
composed mainly of fibres passing circularly round the femoral
neck; the few longitudinal fibres that reach the back of the
neck afford some slight support to this synovial protrusion.

Where the ligaments are closely related to the femoral head
and liable to be subjected to pressure, as on the front of the
capsule, the synovial membrane is not usually recognizable as
a definite layer. The presence of articular cartilage on the
front of the neck of the femur is to be regarded as a reaction
to the pressure of the iliopsoas tendon lying on the iliofemoral
ligament.

There may be one extra-articular extension of the synovial

and adjoining part of the acetabular rim, and distally by its
base to the intertrochanteric line. It occupies all the front of
the capsule except at the medial side above. Usually the sides
of this triangular ligament are stronger than the middle part,
so that the stronger bands, diverging below from a common
stem above, give the ligament the appearance of an inverted
Y. The upper or lateral of the two diverging bands is attached
distally to a special tubercle on the front of the greater trochanter at the upper end of the intertrochanteric line [FIG. 311].

In full extension of the hip joint the iliofemoral ligament
becomes taut. It therefore resists the tensile strain put upon
the anterior part of the fibrous capsule in the standing position,
when the body weight tends to roll the pelvis backwards on
the femoral heads.

The **pubofemoral ligament** arises from the pubic part of
the acetabular rim, with additional fibres from the superior
pubic ramus [FIG. 311]. As it passes distally it blends with
the underlying part of the fibrous capsule, and although it
tends to lose its identity, some of its fibres may be traced to
the under part of the femoral neck. This ligament, like the
iliofemoral, becomes taut in extension of the joint and it also
assists the adductor muscles in checking excessive abduction
of the thigh.

The **ischiofemoral ligament,** less well-defined than the
others, springs from the ischial wall of the acetabulum on the
posterior and lower aspect of the joint [FIG. 310]. The upper
fibres pass out horizontally across the back of the joint, the
lower ones ascend spirally; and both sets of fibres, converging
on the upper and lateral part of the femoral neck, are attached
to the neck, medial to the root of the greater trochanter, close

membrane : the psoas tendon passes over the thin portion of the capsule between the upper ends of the iliofemoral and pubofemoral ligaments, with a small bursa intervening ; sometimes there is an opening in the fibrous capsule under the tendon, and the bursa is then continuous with the synovial membrane of the joint [FIG. 311].

Nerve Supply. The hip joint is supplied : (1) from the lumbar plexus by twigs from the femoral and obturator nerves ; and (2) from the sacral plexus by twigs from the superior gluteal nerve and the nerve to the quadratus femoris muscle (Gardner, 1948b).

MOVEMENTS AT THE HIP JOINT

The movements are those typical of a ball-and-socket joint. They can be analysed as : (1) flexion and extension around a transverse axis ; (2) abduction and adduction around an antero-posterior axis ; (3) medial and lateral rotation around a vertical axis ; with movements compounded of these around intermediate axes. In the movement of circumduction, the limb swings round the side of a cone the apex of which is in the hip joint. All axes intersect at the centre of the head of the femur.

Flexion is limited by the tension of the hamstring muscles when the knee joint is extended ; but these muscles are relaxed when the knee is bent, and the thigh can then be brought against the anterior abdominal wall. All parts of the fibrous capsule are relaxed in flexion, and dislocation is therefore most liable to occur then. Extension takes place very little beyond the vertical position with the limb in line with the trunk. The limitation of extension is in part due to the tension of the ligaments and in part to the form of the articular surfaces. The head of the femur is not truly spherical but is an ellipsoid that is only fully congruent with the acetabulum in the position of complete extension. In bringing the limb into full extension from the position of flexion, the principal ligaments become increasingly tense and draw the head of the femur more and more tightly against the acetabulum. When the femur has passed about 15 degrees behind the vertical, the articular surfaces of femur and acetabulum are fully congruent, the major ligaments of the joint are taut and no further extension of the joint normally takes place—the joint is then said to be locked (Walmsley, 1928). Because ligaments are slightly extensible and articular cartilage may be distorted when subjected to a compression stress, the precise position of locking will vary, to a slight extent at least, with the extending force (Smith, 1956). During the screwing home of the head of the femur, part of the pad of fat in the acetabular fossa is displaced from the fossa and passes outwards between the transverse ligament and the acetabular notch, but when the limb is again flexed the fat is drawn into the acetabulum to help to occupy the potential space that is created by the lateral movement of the head of the femur. In the erect position a perpendicular line through the centre of gravity of the trunk falls behind a line joining the centres of the femoral heads ; hence, the pelvis, bearing the trunk, tends to roll backwards on the femora, but is prevented from so doing by the locking that occurs at the hip joints and by the iliofemoral ligaments. Abduction is limited by tension of the adductor muscles and the pubofemoral ligament ; adduction, by tension of the abductors and the lateral part of the iliofemoral ligament.

Rotation is termed medial or lateral according to the direction in which the toes are turned ; it is rather more free when combined with flexion than in the extended position of the hip. Medial rotation is limited by tension of the lateral rotator muscles and the ischiofemoral ligament ; lateral rotation, by the medial rotators and the iliofemoral ligament. A small amount of rotation at the hip joint occurs automatically in association with the termination of extension and the commencement of flexion at the knee joint [p. 246].

THE KNEE JOINT
[PLATES 15 and 16]

The knee joint is the largest joint in the body, and is structurally the most complicated. Although it is generally considered to be a modified hinge joint which permits very free flexion and extension, the movements at it are complex and involve gliding and rotation of the articular surfaces on each other ; rotation of the free leg, moreover, may be demonstrated convincingly when the knee joint is flexed. The human knee joint possesses great stability, especially in extension, and at the same time it has a wide range of movement. In the mechanism of joints in general, stability and mobility are to some extent incompatible qualities, and most joints secure one of these at the expense of the other. At the knee joint both are secured to a remarkable degree by such an alliance of muscle in controlling the joint as prevails nowhere else. Here, to a large extent, the **muscles are the ligaments.**

Articular Surfaces. Functionally, as well as phylogenetically, the knee joint is compounded of three articulations—an intermediate one between the patella and the patellar surface of the femur, and lateral and medial articulations between the femoral and tibial condyles.

The articular surface of the femur comprises : the **condylar areas,** which are opposed to the tibia and are separated behind by the intercondylar fossa ; and the **patellar surface,** which unites the condyles in front and is opposed to the patella. Each condylar surface is delimited from the patellar surface by a shallow groove on the articular cartilage. On the lateral condyle this groove is almost transverse and is emphasized at each end. On the medial condyle the medial end of the groove begins further forward and, passing obliquely backwards across the condyle, disappears before reaching the lateral edge, where a narrow crescentic facet is marked off which engages with the patella in acute flexion [FIG. 313]. The medial condylar surface has two parts, namely a posterior part which lies parallel to the lateral condylar surface and is equal to it in extent, and an anterior, oblique extension which swings laterally ; this area, which is approximately triangular, has no corresponding part on the lateral condyle which is thereby more nearly spherical in shape. Each condylar surface is convex from side to side and from before backwards and when viewed from either lateral or medial side is seen to have a profile which is spiral in form. The manner in which the condyles become increasingly flatter as they are followed from back to front is a factor of particular significance in the mechanism of the joint. The patellar surface is divided by a well-marked groove into a smaller medial part and a larger

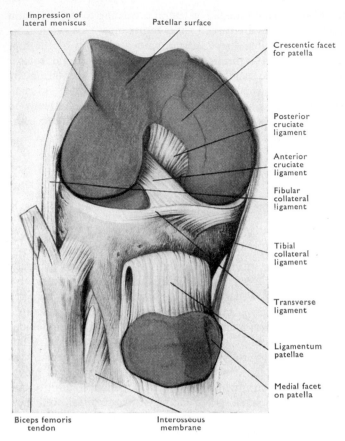

Impression of lateral meniscus

Patellar surface

Crescentic facet for patella

Posterior cruciate ligament

Anterior cruciate ligament

Fibular collateral ligament

Tibial collateral ligament

Transverse ligament

Ligamentum patellae

Medial facet on patella

Biceps femoris tendon

Interosseous membrane

FIG. 313 Dissection of right knee from the front: patella turned down.

and more prominent lateral part [FIG. 313], the importance of which is considered in the next paragraph.

The articular surface on the **patella** is broadly oval, and it is divided into a larger lateral area and a smaller medial area by a rounded vertical ridge apparent even on the macerated bone. The cartilaginous covering reveals a further subdivision of the surface; on each side of the vertical ridge there are two faint transverse ridges which separate three facets on each side; another faint vertical ridge delimits a medial perpendicular facet adjoining the medial border of the articular surface. In acute flexion the medial facet rests on the crescentic facet on the medial femoral condyle; the other facets engage in succession from above downwards with the patellar surface of the femur as the joint moves towards full extension. The manner in which a single facet articulates with a condyle of the femur may be seen in the photograph in FIGURE 315. In the erect posture when the heels are together, each femur is directed downwards and medially at an angle of approximately 10 degrees to the median plane. As the tibia is almost vertical the two bones meet at a definite angle at the knee joint. The quadriceps femoris muscle assumes the alignment of the femur, but the ligamentum patellae, like the tibia, is vertical. It is for this reason that the patella tends to be displaced laterally during forced extension of the knee joint, as in rising from a chair. This is prevented partly by the prominent lateral patellar surface of the femur, and partly by the action of the lower part of the vastus medialis muscle which is inserted into the medial border of the patella.

The articular surfaces of the **tibia** are the cartilage-covered areas on the upper surface of each tibial condyle. These surfaces are separated by the triangular anterior and posterior intercondylar areas which lie in front of, and behind, the intercondylar eminence. The medial articular surface is oval and concave. The lateral surface, smaller and more circular, is concave from side to side and concavoconvex from before backwards; posteriorly it is prolonged downwards over the back of the condyle in relation to the popliteus tendon [FIG. 316]. A peripheral flattened strip on each condyle underlies a meniscus [p. 244]. The tibial articular surfaces are far from congruent with the femoral condyles, which rest upon them; but the effect of this incongruence is lessened by the interposition of the menisci.

Fibrous Capsule. The fibrous capsule of the joint is a complicated structure. There is no complete, independent fibrous capsule uniting the bones. Instead, the joint is surrounded by a thick ligamentous sheath constructed largely of muscle tendons or expansions from them; only here and there is it composed of true capsular fibres running between the articulating bones. Inside the fibrous capsule the synovial lining of the joint is in places separated from the ligamentous stratum by wide intervals occupied by fatty pads or special intra-articular structures.

At the **front of the joint** this ligamentous sheath is composed mainly of the fused tendons of insertion of the rectus femoris and vasti muscles. These descend to the patella from above and from each side to a marginal insertion around the upper half of the bone. Superficial tendinous fibres are continued downwards over the front of the patella into the **ligamentum patellae.** This strong ligament is in reality the distal part of the quadriceps insertion; above it is attached around the apex of the patella, and below it is inserted mainly into the upper part of the tuberosity of the tibia. Other, thinner bands pass down from the sides of the patella and, diverging slightly, are attached to the front of each tibial condyle; these constitute the **medial** and **lateral patellar retinacula.** Tendinous fibres of the vasti muscles pass obliquely downwards across the patella into the retinaculum of the other side. Other, deeper fibres from each side of the patella pass across to the anterior part of each femoral epicondyle. The manner in which the quadriceps muscle, the expansions from it, the patella and the ligamentum patellae participate in replacing the fibrous capsule is a most important feature of the structure of the knee joint.

Superficial to the fibrous bands associated with the quadriceps complex are strong expansions of the fascia lata. The iliotibial tract descends across the anterolateral aspect of the joint to be inserted into the lateral tibial condyle. A strong band passes forwards from the tract to be attached to the upper part of the lateral edge of the patella; this is sometimes called the superior patellar retinaculum.

The whole of the anterior covering of the knee joint, with the patella inset on its deep aspect, is kept tense by the tone of the extensor muscles, and is tightly braced up when these muscles contract in extension. In a typical hinge joint the extensor portion of the capsule must be loose in the extended position to permit the movement of flexion. At the knee, however, this part of the capsule is, as it were, continued into the muscle above, instead of being attached directly to the proximal bone; thus it may be taut, or relatively so, in all positions.

FIG. 314 Right knee joint from behind.

In the specimen the lateral part of the arcuate popliteal ligament was present as a separate band : and the continuity between the fibular collateral ligament and the peroneus longus muscle was specially marked.

The **posterior** part of the fibrous capsule is, like the anterior part, a composite structure. The true capsular fibres are attached to the femur immediately above the condyles and to the intercondylar line, and passing vertically downwards are attached to the posterior border of the upper end of the tibia. The central region of the posterior part of the capsule is strengthened by an expansion of the semimembranosus tendon which passes upwards and laterally and constitutes the **oblique popliteal ligament**. The lower lateral part of the back of the joint is strengthened by the **arcuate popliteal ligament** which springs from the back of the head of the fibula, arches upwards and medially over the popliteus tendon as it emerges from the joint, and spreads out on the back of the capsule. Sometimes the most lateral fibres of the arcuate ligament are specially developed as a separate band that runs from the head of the fibula to the back of the lateral femoral condyle [FIG. 314].

At the **sides of the joint** true capsular fibres descend from the sides of the femoral condyles to the sides of the tibial condyles. They blend posteriorly with the ligamentous felt-work, and anteriorly with the various tendinous expansions already described. Here are found the special **collateral ligaments** characteristic of hinge joints, but the ligaments of the medial and lateral sides differ greatly in form.

The **tibial collateral ligament** is a broad, flat band in the true capsular plane. Its upper end has an extensive attachment to the medial epicondyle of the femur ; some fibres may be traced upwards into the adductor magnus tendon, and the ligament has been regarded as having been formed, in part at least, from an original tibial insertion of this muscle. The liga-

ment passes downwards and slightly forwards and is attached at its lower end to the medial surface of the tibia : the superficial fibres descend below the level of the tibial tubercle ; the deeper fibres have a shorter course from femur to tibia, the deepest of all having an intermediate attachment to the medial meniscus. A downward expansion from the semimembranosus tendon reaches the shaft of the tibia and is partly under cover of the posterior border of the main band of the tibial collateral ligament ; the inferior medial genicular vessels and nerve run forwards between them. The expansions of the semimembranosus tendon to the medial side and back of the fibrous capsule add considerable strength to these aspects of it.

The **fibular collateral ligament** [FIGS. 313 and 314], round and cord-like, stands altogether clear of the thin, lateral part of the fibrous capsule ; it is wrapped round by an expansion of the fascia lata. It is attached above to the lateral epicondyle of the femur ; below, it reaches the head of the fibula in front of its highest point, splitting the tendon of the biceps femoris as it does so. The ligament is sometimes partly continued into the upper end of the peroneus longus muscle [FIG. 314] and may be regarded as a stranded femoral origin of this muscle.

On the lateral side of the joint the true capsular fibres are short and weak ; they bridge the interval between the femoral and tibial condyles, and some of them are separated from the lateral meniscus

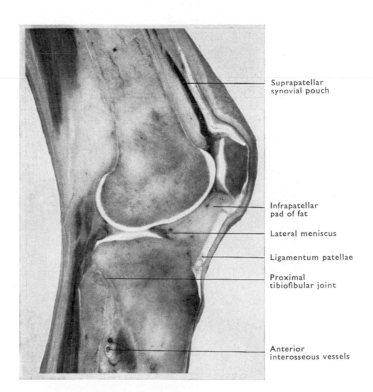

FIG. 315 Sagittal section through the lateral condyle of femur, the tibia and fibula. The section was cut slightly obliquely so as to show the proximal tibiofibular joint and also the extent of the suprapatellar synovial pouch.

by the popliteus tendon. There is a distinct interval between these capsular fibres and the fibular collateral ligament; through the interval the inferior lateral genicular vessels and nerve run forwards.

The tibial and fibular collateral ligaments prevent disruption of the joint at the sides. They are most tightly stretched in extension, and then their direction —the fibular ligament downwards and backwards, the tibial downwards and forwards—is such as to prevent rotation of the tibia laterally or of the femur medially. Rotation may, however, be demonstrated easily in the flexed knee.

Intra-articular Ligaments [FIGS. 313, 316, and 317]. In addition to the capsular ligamentous formation, there are two intra-articular ligaments that stretch between the tibia and femur. These are the cruciate ligaments—strong, rounded bands named from the way in which they cross each other between their attachments, and called anterior and posterior from their relative positions on the tibia, from which they ascend to the sides of the intercondylar fossa of the femur. The **anterior cruciate ligament** extends obliquely upwards and backwards from the rough, anterior intercondylar area of the tibia to the medial side of the lateral femoral condyle. The **posterior cruciate ligament** passes upwards and forwards on the medial side of the anterior cruciate ligament; it extends from the posterior intercondylar area of the tibia to the lateral side of the medial condyle of the femur. The attachments of the cruciate ligaments are such that the anterior cruciate prevents posterior displacement of the femur on the tibia and the posterior ligament prevents anterior displacement.

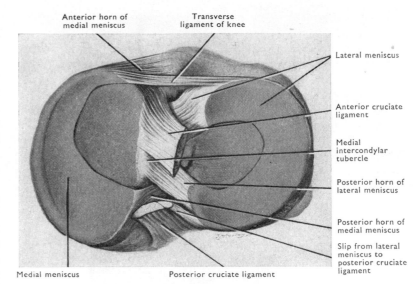

FIG. 317   Upper end of tibia with menisci and attached portions of cruciate ligaments.

In flexion the ligaments are less taut than in extension and they then allow a small amount of passive anteroposterior movement of the tibia on the femur if the thigh is fixed and the leg is forcibly moved. Furthermore, the ligaments become taut when the knee joint is flexed and the tibia is rotated medially, and they become less taut in any position of the tibia, when it is rotated laterally and the ligaments tend to be uncrossed (Haines, 1941). Thus as the cruciate ligaments tighten in full extension, they tend to rotate the tibia laterally.

Menisci [FIGS. 316 and 317]. There are two menisci within the joint, interposed between the femoral and tibial condyles, and they help to compensate for the incongruence of the bones. They are formed almost entirely of fibrous tissue with relatively few cartilage cells interspersed. Each is a flattened pad of crescentic form, hence the name 'meniscus', and is wedge-shaped in cross-section, with the thin edge of the wedge at the concave border.

The thick, convex border of each meniscus is connected to the deep surface of the fibrous capsule, and the capsular fibres which attach the menisci to the condyles of the tibia constitute the **medial** and **lateral coronary ligaments.** The medial meniscus is much more firmly anchored than the lateral, for it is attached to the capsule and hence to the tibial collateral ligament, whereas the lateral meniscus is attached only to the weak fibres which represent the fibrous capsule on the lateral side of the joint and not at all to the fibular collateral ligament. Moreover, where the lateral meniscus is crossed by the popliteus tendon the peripheral edge of the meniscus is not attached to the fibrous capsule. The popliteus muscle arises not only from the femur but also has an attachment to the posterior edge of the lateral meniscus; this attachment shows great variation in size and may be small or relatively large. The extremities of each meniscus are prolonged into fibrous horns which are firmly attached to the intercondylar areas on the upper surface of the tibia [FIG. 317]. The **anterior horn** of the medial meniscus is fixed to the forepart of the anterior intercondylar area in front of the anterior

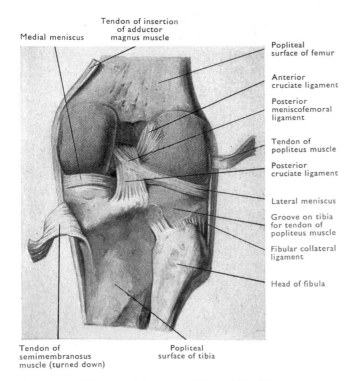

FIG. 316   Right knee joint opened from behind by removal of the posterior part of the articular capsule.

cruciate ligament; its **posterior horn** is attached to the posterior intercondylar area, between the posterior cruciate ligament and the posterior horn of the lateral meniscus. The **two horns** of the lateral meniscus are attached close together, in front of and behind the intercondylar eminence, so that this meniscus comes much nearer to completing a circle than the medial one.

The anterior parts of the two menisci are joined by a fibrous band of variable thickness which is called the **transverse ligament of the knee** [FIG. 317]. The posterior part of the lateral meniscus frequently contributes a ligamentous slip to the posterior cruciate ligament [FIG. 316]; this slip splits on the lateral side of the cruciate ligament and the part which gains the front of it is termed the **anterior meniscofemoral ligament** and the part that lies behind is the **posterior meniscofemoral ligament.** The presence of these ligaments is not constant but the posterior is the more common.

The connexion between the lateral meniscus and the posterior cruciate ligament is in keeping with the typical lower mammalian condition, where the lateral meniscus is attached posteriorly to the medial femoral condyle behind the posterior cruciate ligament and not to the tibia.

Synovial Membrane. The synovial or joint cavity of the knee joint is larger than any other synovial cavity in the body. Its irregular shape makes it desirable, for descriptive purposes, to divide it into four parts which all communicate freely with one another. These parts are a central part and three extensions which are in the form of recesses.

The **central** part lies between the patella in front and the patellar surface of the femur and the cruciate ligaments behind.

Laterally and medially it passes between the femoral and tibial condyles and also both above and below the menisci. In the anterior wall of this central part the synovial membrane passes from the edges of the articular surface of the patella in all directions on the deep aspect of the anterior wall of the joint. Below the patella, the membrane is thrust backwards by a mass of fatty, fibrous tissue, called the **infrapatellar pad of fat,** which lies on the deep surface of the ligamentum patellae and, inferiorly, extends towards the intercondylar notch. From the synovial surface of this pad a vertical, crescentic fold of synovial membrane, called the **infrapatellar fold,** runs backwards and is attached to the intercondylar fossa. On each side of the infrapatellar pad of fat there is a horizontal synovial fold, called the **alar fold,** which is usually readily identifiable [FIG. 319].

The **three recesses** of the synovial cavity open off its central part. One is median, and is frequently called the **suprapatellar synovial pouch** [FIG. 315]. It extends about one hand's breadth above the level of the patella between the quadriceps muscle and the lower part of the shaft of the femur; it develops as a bursa separate from the joint, but it eventually communicates with it. The articularis genus muscle, composed of fleshy slips that arise from the femur above the pouch, is inserted into its wall and draws it upwards during extension of the joint. Of the other two recesses one lies behind the posterior part of each femoral condyle. In these situations the fibrous capsule is overlain by the lateral and medial heads of the gastrocnemius muscle, and there is usually a bursa between each muscle head and the capsule; this is more common on the medial side [FIG. 314], where the bursa sometimes communicates with the joint cavity through

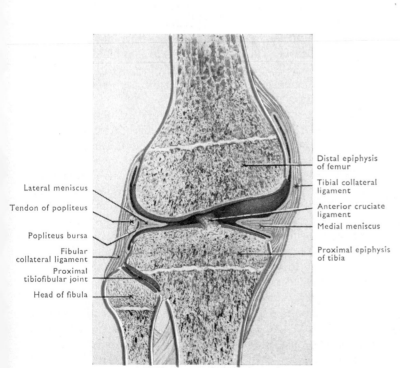

Lateral meniscus

Tendon of popliteus

Popliteus bursa

Fibular collateral ligament

Proximal tibiofibular joint

Head of fibula

Distal epiphysis of femur

Tibial collateral ligament

Anterior cruciate ligament

Medial meniscus

Proximal epiphysis of tibia

FIG. 318   Oblique section through the right knee joint.

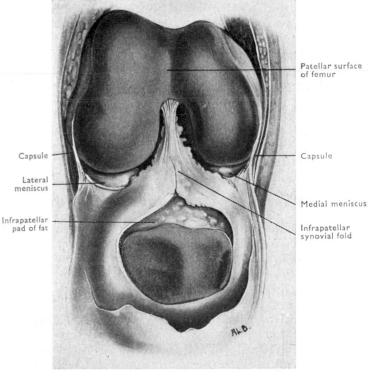

Patellar surface of femur

Capsule

Lateral meniscus

Infrapatellar pad of fat

Capsule

Medial meniscus

Infrapatellar synovial fold

FIG. 319   Right knee joint opened from the front to show the infrapatellar and alar synovial folds. The joint is flexed to a right angle and the patella turned down.
Articular cartilage, blue.

an opening in the capsule. The posterior recesses are separated from each other by a thick, median septum which projects into the joint cavity from behind; this septum is formed by the cruciate ligaments which carry forward a broad, vertical fold of synovial membrane continuous antero-inferiorly with the infrapatellar fold. The posteromedian part of the fibrous capsule is therefore devoid of synovial membrane but the cruciate ligaments, on the other hand, are covered by it on their fronts and sides.

The **synovial relations** of the **popliteus tendon** are important. As the tendon passes backwards from its femoral attachment deep to the lateral part of the fibrous capsule, the synovial membrane is wrapped round its medial side, separating it from the adjacent part of the lateral meniscus except where it is attached to the meniscus. When the tendon leaves the joint through the opening in the posterior part of the capsule, a **synovial pouch** is carried down on its deep surface across the back of the proximal tibiofibular joint. In some cases the cavity of this joint communicates with the bursa-like pouch which accompanies the tendon, and therefore also with the cavity of the knee joint.

Additional Bursae. In addition to the bursae mentioned, numerous others are found in the vicinity of the knee joint. The more constant of these are: the **subcutaneous prepatellar bursa,** under the skin over the lower part of the patella; the **subfascial prepatellar bursa,** between the fascial and tendinous expansions in front of the patella; the **subtendinous prepatellar bursa,** between the superficial and deep layers of the tendinous fibres that cross the patella; the **deep infrapatellar bursa,** between the upper part of the tibia and the lower part of the ligamentum patellae, separated from the knee joint by the infrapatellar pad of fat. A bursa between the tibial collateral ligament and the overlying tendons of the sartorius, gracilis, and semitendinosus forms part of the **anserine bursa;** and there are bursae between the semimembranosus tendon and the tibial collateral ligament, between that tendon and the medial head of the gastrocnemius, and between the biceps tendon and the fibular collateral ligament.

Nerve Supply. The knee joint is supplied by several branches from the femoral nerve and from both divisions—common peroneal and tibial—of the sciatic nerve, and by a filament from the obturator nerve [see Gardner, 1948a].

MOVEMENTS AT THE KNEE JOINT

The movements that occur at the knee joint are flexion and extension and, also, a limited amount of active rotation; the joint is justifiably regarded as a modified hinge joint. It differs from a typical hinge joint not only in that some rotation may occur, but also because the axis of movement and the zone of contact between the articular surfaces move forward during extension and backwards during flexion, whereas in a typical hinge joint, such as the elbow joint, the axis of movement is constant. The axis changes its position during movement owing to a difference in curvature of different parts of the femoral condyles. Towards the end of extension there is also an accompanying rotatory movement of the femur upon the tibia when the foot is on the ground.

In voluntary flexion the leg may be moved through 140

degrees or more from the position of full extension, and this may be increased by forced movement or by the adoption of the squatting position. In full flexion the posterior surfaces of the femoral condyles articulate with the posterior parts of the articular surface of the tibia and the corresponding parts of the menisci. As the joint passes from flexion to extension, if the tibia is fixed as in rising from a chair with the feet planted on the floor, the femoral condyles roll forwards upon the tibia.

The curvatures of the femoral condyles become progressively less as they are traced forwards and they therefore have an increasingly greater contact with the tibia as the femur passes into full extension, and for the same reason the curvatures of the menisci tend to be opened by the displacement forwards of their anterior parts. Just before full extension is reached, the tibial surface of the lateral femoral condyle is almost exhausted, although the anterior, oblique part of the surface on the medial condyle has still to be brought into play. While the lateral condyle completes its forward roll, the medial condyle, continuing the movement, glides backwards until the oblique part in front is in contact with the tibia. The gliding backwards of the medial condyle, assisted by the tightening of the posterior cruciate ligament, occurs in a horizontal arc centred in the lateral condyle (held forwards by the anterior cruciate ligament) so that there is a medial rotation of the femur. This terminal rotatory movement tightens the collateral ligaments of the knee and the joint is then, as it were, screwed home. In the fully extended position the other ligaments of the knee joint, most notably the oblique popliteal, are also taut, and as the articular surfaces of the femur and tibia are then in maximal contact, the joint is said to be locked. On account of the deformation that occurs in articular cartilage and the extensibility of the ligaments the precise position in which locking occurs varies slightly with the extending force. Before flexion can take place the rotatory movement must be undone, and this is accomplished by the action of the popliteus muscle. Rotation of the femur on the tibia occurs when the tibia is fixed during the movement at the knee. But both bones may move if free to do so (e.g., in kicking), the tibia rotating laterally at the end of extension and medially at the beginning of flexion (Goodsir, 1858).

When the body is erect, the axis of gravity of the body falls in front of a line between the knee joints; gravity therefore tends to produce a locking at the knee joint. When standing in the natural upright position, however, the knee joint is not usually locked but is maintained in a position a little short of full extension. In such a position the quadriceps muscle is relaxed and the patella can be moved freely from side to side on the femur. The hamstring and gastrocnemius muscles, on the other hand, are in such a degree of contraction as just to maintain the joint in this position against the action of gravity. In reaching forwards, however, the knee joint often becomes fully extended and locked, and in this position also the quadriceps is relaxed.

In addition to the hinge movements of flexion and extension, some rotation of the leg around a vertical axis can take place at the knee except when the joint is fully extended. In full extension, apart from the special, small rotatory movements already described, all rotation is prevented by the tension of the collateral and cruciate ligaments. Rotation is most free, about

50 degrees (Fick, 1911), when the knee is flexed to about a right angle; for neither cruciate ligament is then very tight and the tension of the collateral ligaments, slightly reduced by the special rotatory movement at the beginning of flexion, is further lessened during flexion by descent of their femoral attachments.

The menisci, as already explained, compensate for the incongruity of the opposed femoral and tibial surfaces and, by slight movement, accommodate themselves to the differently shaped parts of the femoral condyles that rest upon them in different positions of the joint. They shift slightly forwards in extension and open out their curve under the flatter, anterior parts of the condyles then opposed to them. The lateral meniscus, in particular, moves forward, pivoting about its closely set attachments to the tibia, so that its anterior part glides down on the cambered surface of the lateral tibial condyle (Barnett, 1953). In flexion the menisci shift backwards and curl in upon the more rounded posterior parts of the femoral condyles. The attachment of the medial meniscus to the tibial collateral ligament causes this meniscus to be drawn medially when the ligament tightens in extension. In a sudden, unguarded movement, the meniscus may at that moment be pinned between femur and tibia before its normal opening out has occurred. The pull of the tibial collateral ligament may then split the meniscus in its long axis, producing one form of the commonest derangement of the knee joint—torn medial meniscus.

According to MacConaill (1932) the menisci play an important part in relation to efficient lubrication in the joint. No easy movement can take place between the congruent surfaces of a bearing unless they are separated by a layer of viscous fluid, which must maintain a wedge-shaped space between them; as the surfaces roll over one another they squeeze out in advance of their approximation a diminishing wedge of lubricating fluid. The necessity for this 'wedge' of lubricant is greater where, as in the knee joint, there is a considerable thrust transmitted from one joint surface to the other, and here the interposed menisci separate the intervening synovial fluid into two wedges, thereby increasing the efficiency of the mechanism. The slight mobility of the menisci permits them to assume the proper angulation to the joint surfaces necessary for the maintenance of the wedge-shaped synovial films.

THE TIBIOFIBULAR JOINTS

The fibula is closely bound to the tibia by a joint at each end; and the shafts of the two bones are connected by an interosseous membrane.

THE PROXIMAL TIBIOFIBULAR JOINT
[FIGS. 316 and 318]

This is a small synovial joint, of plane type, between a flat, circular or oval facet on the head of the fibula and a similar facet on the tibia placed posterolaterally on the under aspect of the overhanging lateral condyle. The fibrous capsule which surrounds the joint is strengthened in front and behind by the **anterior** and **posterior ligaments of the head of the fibula**; the fibres of these ligaments are directed upwards and medially from the fibula to the adjoining part of the tibia.

The tendon of the popliteus muscle is intimately related to the posterosuperior aspect of the joint; and the pouch of synovial membrane prolonged under the tendon from the knee joint sometimes communicates with the synovial cavity of the tibiofibular joint through an opening in the upper part of the capsule.

Nerve Supply. The superior tibiofibular joint is supplied by twigs from the common peroneal nerve and from the branch of the tibial nerve to the popliteus muscle.

THE DISTAL TIBIOFIBULAR JOINT
[FIGS. 320–324]

This is a fibrous joint. The rough, triangular, opposed surfaces of the bones are united by a strong **interosseous ligament.** In addition, the joint is strengthened in front and behind by longer bands called the **anterior** and **posterior tibiofibular ligaments.** These bands stretch from the lower border of the tibia, in front and behind, to the front and back of the distal end of the fibula, each inclined laterally and downwards. Under cover of the posterior ligament there is a longer band, called the **transverse tibiofibular ligament,** which is attached to the whole length of the posterior edge of the inferior surface of the tibia and to the upper end of the malleolar fossa on the fibula. This band closes the posterior angle between the tibia and fibula and articulates with the posterior part of the lateral border of the trochlea tali [FIG. 320].

A recess of the cavity of the ankle joint usually extends upwards between the tibia and fibula for about 0·5 cm. It is blocked above by the base of the interosseous ligament and is occupied by a fold of synovial membrane of the ankle joint. Sometimes the articular cartilage on the lower ends of the tibia and fibula extends upwards for a little way on the walls of this recess.

Nerve Supply. The inferior tibiofibular joint is supplied by twigs from the deep peroneal and tibial nerves.

THE CRURAL INTEROSSEOUS MEMBRANE

The interosseous membrane of the leg is tightly stretched between the interosseous borders of the two bones; the general direction of its fibres is downwards and laterally from tibia to fibula. The membrane reaches up to the under aspect of the proximal joint and downwards to blend with the upper edge of the interosseous ligament of the distal joint. There is an oval aperture near the upper end of the membrane through which pass the anterior tibial vessels [FIG. 313], and a smaller one at the lower end for the perforating branch of the peroneal artery [FIG. 320].

Movement at the tibiofibular joints is slight and entirely passive. It is occasioned by movement at the ankle joint, with which it will be described [p. 249].

THE TALOCRURAL
OR ANKLE JOINT
[PLATE 17]

The ankle joint is a synovial joint of typical hinge pattern. The lower ends of the tibia and fibula provide a socket in

FIG. 320 Ankle joint dissected from behind and part of articular capsule removed.

which the upper part of the talus rocks around a transverse axis. The only special feature of the hinge is an arrangement whereby the sides of the socket come to exercise a positive grip upon the moving member of the joint as an aid to increased stability.

Articular Surfaces. The **proximal articular surface** provides the socket [FIG. 321]. It is formed by the cartilage-covered areas on the lower ends of the leg bones. The roof of the socket is formed by the **distal surface** of the **tibia**; it is wider in front than behind, is slightly convex from side to side, and is concave from before backwards. The posterior part of this surface which projects slightly downwards is sometimes termed the posterior malleolus. The medial wall is the **lateral surface** of the **medial malleolus** and its articular cartilage is continuous with that on the roof at a rounded angle. The lateral wall is the triangular facet on the fore part of the **medial side of the lateral malleolus.** In the angle between this wall and the roof is the narrow cleft between the tibia and fibula distal to the interosseous ligament; but, posteriorly, this angle is filled up by the transverse tibiofibular ligament.

The **distal articular surface** is formed entirely by the upper part of the body of the talus—the **trochlea tali**—which is shaped like the upper part of a short cylinder placed transversely. The trochlea tali is, however, slightly broader in front than behind. This surface articulates with the roof of the socket (tibia), which also is broader in front. The upper trochlear surface is slightly concave transversely to fit the convexity on the tibial surface. The cartilage on top of the trochlea is continued on to its sides—the ends of the cylinder. The medial articular surface is vertically set, and is shaped like a comma laid on its side, tail backwards; it is opposed to the medial malleolar wall of the proximal articular surface. The lateral articular area of the trochlea is triangular

and very much larger than the medial articular area [FIG. 322]. Inferiorly, it curves laterally to a projecting apex, while posteriorly it slopes medially. It articulates with the facet on the lateral malleolus and, postero-superiorly, with the transverse tibiofibular ligament where the angle between the upper and lateral surfaces of the trochlea is bevelled off [FIG. 320].

Fibrous Capsule. The fibrous capsule surrounds the joint. Above, it is attached to the margins of the articular surfaces on the tibia and fibula and below it passes to the non-articular part of the talus close to the articular areas on the upper, medial, and lateral surfaces of the body of the talus. As is customary in a hinge joint, the capsule is weak in front and behind, but on both the medial and lateral sides it is strengthened by collateral ligaments which are so definite that they are given specific names. The attachments of the collateral ligaments are important.

The **medial (deltoid) ligament** [FIG. 323] is the capsular thickening on the medial side of the joint. This ligament is roughly triangular in shape, and is made up of several bands which are fused in a strong ligamentous formation and differentiated only by their distal attachments. The apical part of the ligament is attached to the pit on the lower border of the medial malleolus. Its thick base has a continuous attachment

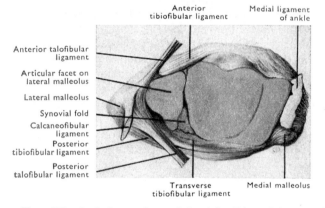

FIG. 321 Articular surfaces of the right tibia and fibula which are opposed to the talus at the ankle joint.

FIG. 322 Coronal section of ankle joint in which the joint has been distended. The lateral malleolus is seen to be much larger than the medial.

which extends from the navicular bone in front to the posterior part of the medial surface of the body of the talus behind. From front to back the distal part of the ligament is attached successively to the rounded upper and medial part of the navicular bone **(tibionavicular part),** to the medial part of the neck of the talus **(anterior tibiotalar part),** to the plantar calcaneonavicular ligament and the medial border of the sustentaculum tali **(tibiocalcaneal part),** and to the medial side of the talus, under the 'tail' of the comma-shaped facet **(posterior tibiotalar part).** The tibiotalar parts are on a deeper plane than the others; the tibionavicular and tibiocalcaneal portions are continuous over the superficial aspect of the anterior tibiotalar parts.

In the *lateral ligament* [FIGS. 320 and 324] there are three bands which are quite distinct from one another. The middle one—the **calcaneofibular ligament**—runs from in front of the tip of the lateral malleolus downwards and slightly backwards to the middle of the lateral surface of the calcaneus a little above and behind the peroneal trochlea. This band is quite free of the fibrous capsule of the joint but the two talofibular ligaments are fused with it. The **anterior talofibular ligament** runs from the anterior border of the lateral malleolus forwards and medially to the neck of the talus. The **posterior talofibular ligament** is attached to the fibula at the bottom of the malleolar fossa and passes medially and slightly backwards to the lateral tubercle of the posterior process of the talus.

Synovial Membrane. The synovial membrane of the joint lines the fibrous capsule and covers well-marked fatty pads that lie in relation to its anterior and posterior parts. A synovial fold [FIG. 321] occupies the cleft between the tibia and

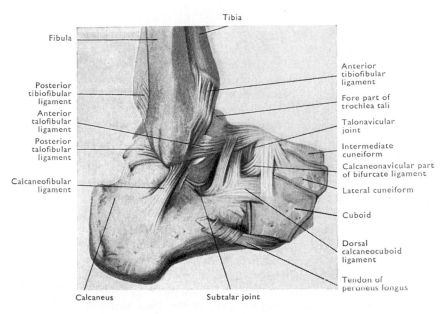

FIG. 324   Ligaments of lateral side of ankle joint and dorsum of tarsus.

fibula below the base of the interosseous tibiofibular ligament; the sides of this cleft may be covered in part by an extension of the articular cartilage on the tibia and fibula.

Nerve Supply. The ankle joint is supplied by twigs from the deep peroneal and tibial nerves.

MOVEMENTS AT THE ANKLE JOINT

Movement takes place around a transverse axis on a level with the tip of the lateral malleolus and slightly below the level of the medial malleolus. In the normal, standing position the foot makes a right angle with the leg. Considerable confusion has resulted in the past from the use of the terms flexion and

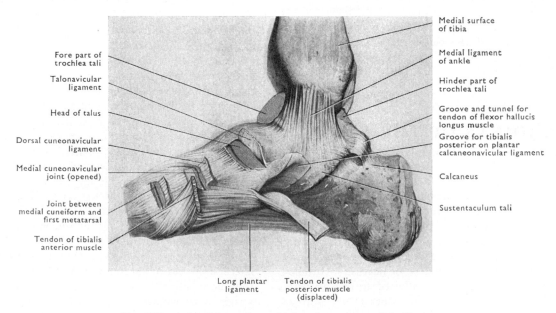

FIG. 323   Ankle joint and tarsal joints from the medial side.

extension in describing the movements that occur at the ankle joint, and for that reason the terms dorsiflexion and plantar flexion are to be preferred. In **dorsiflexion** the foot is drawn upwards, and the trochlea tali turns backwards in its socket; movement in the opposite direction is **plantar flexion.** The range of plantar flexion (about 55 degrees) is greater than that of dorsiflexion (about 35 degrees) but there is considerable individual variation in the extent of the movements.

In **dorsiflexion,** the broader, anterior part of the trochlea tali is forced back into the narrower, posterior part of the tibiofibular socket; this causes slight separation of the tibia and fibula, with increased tension of the interosseous and transverse tibiofibular ligaments. The talus is then most securely held between the malleoli. In **plantar flexion,** as the narrow part of the trochlea turns forward into the broader part of the socket, the malleoli come together again, so retaining, up to a point, their grasp upon the sides of the talus; but in full plantar flexion a little side-play can be demonstrated in the joint. The wedge-shaped form of the joint surfaces helps to prevent backward displacement of the foot on the leg when coming to a sudden stop in running or jumping. The maintenance of this positive grasp upon the talus is the function of the **inferior tibiofibular joint,** the strong ligaments of which provide the fulcrum through which the resilience of the fibula superior to them gives the necessary spring to the lateral malleolus. The **superior tibiofibular joint** permits slight gliding movement in association with the displacement of the lower ends of the bones of the leg.

When the body is erect, muscular effort, mainly by soleus and gastrocnemius, is needed to prevent falling forwards, owing to the incidence of body weight in front of the ankle joints. But this effect of gravity is minimized in so far as the feet are usually turned a little laterally, and the axes of the two ankle joints are inclined to each other so as to meet posteriorly.

The ankle joint is strongly supported by the tendons of the muscles of the leg which descend in close relation to it. In front are the tendons of tibialis anterior, extensor hallucis longus, extensor digitorum longus, and peroneus tertius, in that order mediolaterally; posteriorly and medially are the tibialis posterior and flexor digitorum longus tendons; posteriorly is the tendon of the flexor hallucis longus; and posteriorly and laterally are the tendons of the peroneus longus and peroneus brevis.

THE JOINTS OF THE FOOT
[PLATE 18]

The joints between the tarsal bones, between the tarsus and metatarsus, and between the metatarsal bases are intricate in their arrangement and several of them are complicated in their movements. The significance of the structure of these joints and of the movements that occur at them can only be understood in the light of the architecture of the foot as a whole. The account of the foot in this text is, of necessity, curtailed, and for further study the reader is advised to consult works which have been devoted entirely to it (Morton, 1935; Jones, 1949).

THE ARCHES OF THE FOOT

The human foot is a structure that is highly specialized for the support of the body in the erect posture and for the propulsion of it during movement. To subserve these functions efficiently it must be able to adapt itself to the ground in many different positions and must be capable of absorbing mechanical shocks. This is achieved by the presence of a series of resilient arches, convex above, which become slightly flattened by the incidence of the body weight in standing or during progression, and resume their original curvature when the pressure on them is relieved. The arches are formed by the tarsal and metatarsal bones, and are disposed both longitudinally and transversely.

The **longitudinal arch** is higher on the medial side than on the lateral, and on account of the arrangement of the tarsal and metatarsal bones in two columns [p. 193] it is sometimes, for descriptive purposes, regarded as having two parts called the **lateral** and **medial arches.** The posterior part of the calcaneus forms a common posterior pillar for both arches [PLATE 18 C]. The **lateral part of the arch** proceeds from this through the anterior part of the calcaneus, the cuboid bone, and the lateral two metatarsal bones, which articulate with the cuboid. This part of the longitudinal arch is very low and receives the weight of the body through the talus at the talocalcanean joint. The **medial part of the arch** is higher and more important. The posterior pillar is continued into the sustentaculum tali, and the summit of the arch is then formed by the head of the talus placed above and between the sustentaculum tali and the navicular bone. The anterior pillar is formed by the navicular, the three cuneiforms, and the medial three metatarsal bones.

The **transverse arch** is placed across the anterior part of the tarsus and the posterior part of the metatarsus, and is best marked along the line of the tarsometatarsal joints. The arch is gradually lost as the metatarsals are traced forwards towards their heads which are all in functional contact with the ground.

For the preservation of these arches the integrity of the tarsal, tarsometatarsal, and intermetatarsal joints is important, for at them the bones are held in their proper relationship as segments of the arches. For this purpose certain ligaments of the joints are much more important than others. The important ligaments are especially strong so as to resist undue yielding of the joints and consequent collapse of the arches. They are therefore found on the plantar aspect of the joints, that is, on the concave side of the arches, and are further supported by the more distant plantar aponeurosis. But it must be emphasized once more that unsupported ligaments will, in time, stretch under continuous strain, and the arches of the foot are in every case supported by muscles as well.

The body weight is transmitted from the bones of the leg to the longitudinal arches through the talus. The talus forms two joints with the rest of the tarsus [FIG. 325], each being more particularly concerned in the transference of weight to one of these arches. Together, however, they constitute a single functional unit. The more posterior joint—the **subtalar** —is specially associated with transmission of weight to the lateral arch and to the common posterior pillar—the heel. At the more anterior joint—the **talocalcaneonavicular**—the weight is transmitted to the medial arch. At both joints the talus rests on the calcaneus and between the joints there is the sinus tarsi [see below] to the walls of which parts of the fibrous capsules of the two joints are attached.

THE INTERTARSAL JOINTS

Subtalar Joint [FIGS. 324 and 325]

This joint is formed between the facet on the under surface of the body of the talus and the posterior facet on the upper surface of the calcaneus; each facet is approximately cylindrical in curvature—that on the talus concave anteroposteriorly, the other convex.

The fibrous capsule surrounding the joint is made up of short fibres attached close to the articular surfaces all round. Slightly stronger portions, specially named the **medial** and **lateral talocalcanean ligaments,** are situated in corresponding parts of the capsule.

The anterior part of the fibrous capsule is thin and is attached to the roof and floor of the sinus tarsi. The **sinus tarsi** is a narrow tunnel which is bounded by the grooves on the talus and the calcaneus and runs obliquely forwards and laterally in front of the articular facets of this joint; at its anterolateral end the sinus opens out on to the dorsum of the foot. The anterior part of the fibrous capsule of the subtalar joint and the posterior part of the capsule of the talocalcaneonavicular joint are both attached to the sinus tarsi. The two capsules are thickened where they are adjacent to each other and they together constitute the **interosseous talocalcanean ligament** which forms a connexion between the two bones. Between the two parts of the interosseous ligament lies the deep extension of the lateral limb of the inferior extensor retinaculum which is attached to the floor of the sinus tarsi

(Smith, 1896) [FIGS. 325 and 327]. At the lateral end of the sinus tarsi is a strong, discrete band that is attached above to the neck of the talus and below to the calcaneus; this ligament has been termed by Jones (1944) the **ligamentum cervicis,** and it is of importance because it forms a strong ligamentous bond between the two bones and it becomes taut in the movement of inversion [FIG. 325].

The calcanean parts of the medial and lateral ligaments of the ankle joint act as accessory ligaments of the subtalar joint.

The fibrous capsule is lined with **synovial membrane;** the synovial cavity does not communicate with any other joint [FIG. 327].

Talocalcaneonavicular Joint [FIG. 325]

At this joint a large facet on the head and lower surface of the neck of the talus articulates with a deep and extensive socket that is formed partly by bone and partly by ligaments. The anterior part of the socket is formed by the concave articular surface of the navicular bone, and the posterior part is formed by the upper surface of the sustentaculum tali and a facet that is usually present on the upper surface of the calcaneus anterolateral to the sustentaculum. Between the navicular and the calcaneal articular surfaces there is an appreciable interval that is greatest in its medial part. The medial part of this interval is occupied by the **plantar calcaneonavicular ligament** and the much smaller lateral part by the **calcaneonavicular** fibres of the **bifurcate ligament.**

The **plantar calcaneonavicular ligament** articulates with the medial and inferior surfaces of the head of the talus and is a fibrocartilaginous band of great strength [FIGS. 323, 325, and 326   It is attached behind to the anterior end and medial border of the sustentaculum tali; in front it spreads out on to the plantar, the medial, and the adjacent part of the upper surface of the navicular bone; its medial part is joined and supported by fibres of the medial ligament of the ankle joint. The tendon of the tibialis posterior muscle turns down into the sole under the ligament and acts as a sling for it and for the head of the talus [FIGS. 323 and 326]. The tension of the plantar calcaneonavicular ligament, supported by the tibialis posterior muscle, resists the tendency of the weight of the body to drive the head of the talus downwards between the bones which it braces together, and the resilience it imparts to the medial arch is recognized in its common appellation—the **spring ligament.**

The **calcaneonavicular** part of the **bifurcate ligament** completes the socket on the lateral side. It is composed of short fibres that pass from the upper surface of the anterior end of the calcaneus to the adjacent lateral surface of the navicular bone [FIG. 324].

The rounded **head of the talus** has a continuous covering of articular cartilage that is continued on to the under surface of the neck and the body. This articular surface conforms in shape to the socket and is marked off by faint ridges into a facet for the navicular bone in front, one for the sustentaculum tali postero-inferiorly, with

Talonavicular ligament

Plantar calcaneonavicular ligament

Medial ligament of ankle

Medial talocalcanean ligament

Posterior talocalcanean ligament

Two parts of bifurcate ligament

Ligamentum cervicis

Inferior extensor retinaculum

Lateral talocalcanean ligament

Calcaneofibular ligament

FIG. 325 Ligaments and inferior articular surfaces of the right subtalar and talocalcaneonavicular joints seen from above after removal of the talus.

a facet for the anterior part of the calcaneus anterolateral to it, and a fourth on the infero-medial surface for the plantar calcaneonavicular ligament.

Fibrous Capsule. On account of the extensive form of the bony and ligamentous socket a true fibrous capsule is present only on the posterior and the dorsal aspects of the joint.

The posterior part of the fibrous capsule is slightly thickened and constitutes the anterior component of the interosseous talocalcanean ligament [p. 251].

The fibrous capsule is completed round the dorsal aspect of the joint, between the bifurcate and plantar calcaneonavicular ligaments, by a thin sheet, called the **talonavicular ligament** [FIG. 323], which extends from the neck of the talus to the dorsal surface of the navicular bone. It blends medially with the anterior fibres of the deltoid ligament that go to the navicular bone.

Synovial Membrane. The synovial membrane of this joint lines all the non-articular surface and the enclosed synovial cavity does not communicate with the neighbouring articulations [FIG. 327].

Calcaneocuboid Joint [FIGS. 324 and 327]

The calcaneocuboid joint is the highest point in the lateral part of the longitudinal arch. The body weight falls on this part of the arch behind

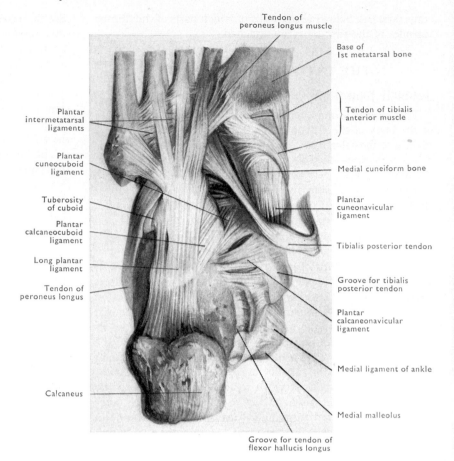

FIG. 326 Plantar aspect of the right tarsal and tarsometatarsal joints.

its summit, at the subtalar joint, and the strain is most felt at the joint immediately in front of the calcaneus—the calcaneocuboid joint. This joint, like the talocalcaneonavicular joint at the summit of the medial arch, must therefore be specially reinforced on its under aspect, and here are placed the strong bands called the long plantar and the plantar calcaneocuboid ligaments. In addition, the tendon of the peroneus longus, passing under the cuboid bone in front of the articulation, acts as a sling after the manner of the tibialis posterior tendon under the medial part of the arch.

The **articular facets** on the opposed surfaces of the calcaneus and cuboid bone are quadrilateral and reciprocally concavoconvex.

Fibrous Capsule. A fibrous capsule completely surrounds the joint, and certain special ligaments are present in connexion with it.

Applied to the dorsomedial aspect of the joint is the **calcaneocuboid** part of the **bifurcate ligament,** which springs from the upper and fore part of the calcaneus, in common with the calcaneonavicular part of the ligament,

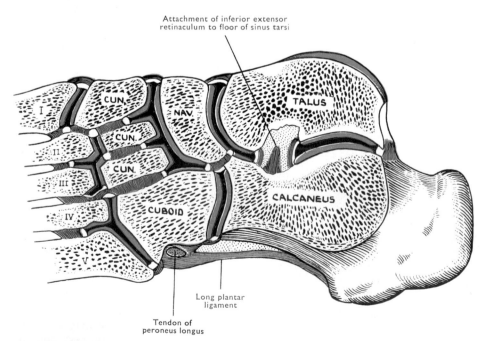

FIG. 327 Oblique section of left foot to show the synovial cavities of the tarsal and tarsometatarsal joints.

The foot was in strong inversion when sectioned and does not reveal the communication that exists between the cuneonavicular and intermediate tarsometatarsal joints. An unusually large cuboideonavicular joint cavity is present.

Articular cartilage, blue ; synovial membrane, white ; ligaments and tendon, green.

and is attached to the adjacent dorsomedial angle of the cuboid bone [FIG. 324].

On the plantar aspect of the joint are two special ligaments, superficial and deep, separated by areolar tissue [FIG. 326]. The deep one is the **plantar calcaneocuboid ligament**. It is a broad band of short fibres immediately applied to the joint; it runs from the anterior part of the inferior surface of the calcaneus to the cuboid bone behind the ridge that bounds the peroneal groove. The superficial and much longer band is the **long plantar ligament**. It arises from the whole length of the rounded, keel-like ridge on the plantar surface of the calcaneus and stretches to the ridge on the cuboid bone, to which its deep fibres are attached. The more superficial fibres pass on over the peroneus longus tendon to the bases of at least the lateral three metatarsal bones. This ligament therefore stretches under nearly the whole length of the lateral part of the arch and strengthens the plantar aspect of all the joints in the arch. As this arch has little height, a long band of this nature stretching, like a tie-beam, from pillar to pillar is more effective in preserving the arch than short ties between adjacent segments of the arch.

Synovial membrane covers the inside of the fibrous capsule. The joint cavity is self-contained, making no communication with any other [FIG. 327].

Transverse Tarsal Joint

The talocalcaneonavicular joint and the calcaneocuboid joint, although they do not communicate with each other, together extend right across the tarsus in an irregular transverse plane, between the talus and calcaneus behind and the navicular and cuboid bones in front [FIG. 327]. This articular plane is termed the transverse tarsal joint.

The anterior two bones involved—navicular and cuboid —are bound together by **dorsal** [FIG. 324] and **plantar cuboideonavicular ligaments** between adjacent parts of the corresponding surfaces, and by an interosseous ligament between their contiguous sides [FIG. 327]. Sometimes there is a synovial cavity amidst these ligaments, between cartilage-covered facets on the bones, and this, when present, opens anteriorly into the cuneonavicular joint.

Cuneonavicular Joint

This is the synovial joint between the convex anterior surface of the navicular bone and the concave articular surface provided by the posterior ends of the three cuneiform bones.

The joint is surrounded by a **fibrous capsule** which is distinct on all sides except laterally, towards the cuboid bone, where the joint may communicate with the cuneocuboid joint and always with a cuboideonavicular joint when present. Distally, the joint cavity forms recesses between the cuneiform bones [FIG. 327]. In the upper and medial parts of the capsule there are short bands, relatively weak, passing from the navicular to each cuneiform bone; these are the **dorsal cuneonavicular ligaments** [FIGS. 323 and 324]. Similar, stronger bands on the under aspect of the joint, inseparable from slips of insertion of the tibialis posterior tendon, constitute the **plantar cuneonavicular ligaments**.

Intercuneiform Joints

At the intercuneiform joints the three cuneiform bones are bound together by weak, transverse **dorsal ligaments** and much stronger **interosseous** and **plantar ligaments**. The last two ligaments form the anterior boundaries of the small joint cavities that lie between the articular surfaces on the posterior parts of the cuneiform bones.

Cuneocuboid Joint

This is a joint between the round facets on the lateral cuneiform bone and the cuboid bone. It is surrounded by a weak **dorsal cuneocuboid ligament** between the dorsal surfaces of the bones, by a **plantar cuneocuboid ligament** [FIG. 326] between their adjacent plantar borders, and by a strong **interosseous cuneocuboid ligament** [FIG. 327] which closes the cavity anteriorly. Posteriorly, the joint may or may not open into the cuneonavicular joint.

The cuboid and cuneiform bones, placed side by side across the distal part of the tarsus, form the tarsal part of the **transverse arch of the foot**. The strong interosseous and plantar intercuneiform and cuneocuboid ligaments maintain the segments of the arch in position, and additional strength is obtained from the peroneus longus tendon as it passes transversely across the sole of the foot.

Nerve Supply. The tarsal joints are supplied on the dorsal aspect by the deep peroneal nerve, and on the plantar aspect by the medial and lateral plantar nerves.

MOVEMENTS IN THE TARSUS

The movements of dorsiflexion and plantar flexion occur at the ankle joint and have been described on page 249. In these movements the talus rotates in the tibiofibular socket and the tarsus and metatarsus move as a unit. In the other movements of the foot the calcaneus and navicular, carrying with them the more distal tarsal bones (between which movements also occur), move on the talus.

The movements that occur at the tarsal joints produce, in summation, the characteristic changes of inversion and eversion.

In **inversion,** the foot is adducted and twisted so that the medial border is raised and the lateral border depressed until the sole is turned slightly medially. Inversion is usually accompanied by plantar flexion. In **eversion,** the foot is abducted, the lateral border is raised and the medial border lowered, so that the sole is turned slightly laterally. Eversion is usually accompanied by dorsiflexion. Inversion and eversion occur principally at the subtalar and talocalcaneonavicular joints, and Inkster (1938) has shown that fixation of the calcaneus greatly limits these movements. The subtalar and talocalcaneonavicular joints have a common axis of movement which runs obliquely forwards, upwards, and medially through the sinus tarsi, in the region of the ligamentum cervicis [p. 251], and the complicated movements that occur at these joints have been described in detail by Manter (1941). Movement occurs also at the other tarsal and at the tarsometatarsal joints during inversion and eversion, but this diminishes in amount in a proximodistal direction (MacConaill, 1945).

THE TARSOMETATARSAL AND INTERMETATARSAL JOINTS

Tarsometatarsal Joints [FIG. 327]

The four anterior tarsal bones—the cuboid and three cuneiform bones—articulate with the bases of the metatarsal bones along an irregular line which 'presents the outline of an indented parapet both on its tarsal and its metatarsal aspects'. The indented form of the tarsometatarsal articulations strengthens the transverse arch of the foot which is most pronounced in this region.

The **first metatarsal** bone articulates only with the anterior surface of the medial cuneiform bone. The **second metatarsal** articulates in a socket with all three cuneiform bones—the lateral side of the medial bone, the anterior end of the intermediate, and the medial side of the lateral bone. The **third metatarsal** articulates with the lateral cuneiform bone. The **fourth metatarsal** articulates with the cuboid and to a small extent with the lateral cuneiform bone. The **fifth metatarsal articulates** with the cuboid bone only.

The articulations are all synovial joints of a gliding type, closed over by dorsal and plantar ligaments. In addition there are certain interosseous ligaments which delimit three separate synovial cavities.

The **dorsal tarsometatarsal ligaments** are short slips that pass between the adjoining dorsal surfaces of the tarsal and metatarsal bones, each metatarsal bone receiving a slip from each tarsal bone with which it articulates.

The **plantar tarsometatarsal ligaments** are similar bands on the under aspect of the joints, but are less regularly disposed. Those for the medial two metatarsal bones are the strongest, but the lateral tarsometatarsal joints are strengthened underneath by fibres from the long plantar ligament. The tarsometatarsal joint of the big toe is strengthened below by the insertion into the two articulating bones of the tibialis anterior medially and the peroneus longus laterally. Slips from the tibialis posterior tendon are applied to the under aspect of the joints of the middle three metatarsal bones [FIG. 326].

Two strong **interosseous tarsometatarsal ligaments** are always present [FIG. 327]. One passes from the anterior and lateral part of the medial cuneiform bone to the contiguous medial side of the second metatarsal base. The other passes from the anterolateral angle of the lateral cuneiform to the adjacent medial surface of the fourth metatarsal base. These two ligaments separate **three tarsometatarsal synovial cavities.**

The **medial synovial** cavity is entirely confined to the joint between the first metatarsal and medial cuneiform bones. The **intermediate cavity** separates the second and third metatarsals from the intermediate and lateral cuneiform bones. This cavity is prolonged forwards for a little way between the second and third, and third and fourth metatarsal bases; posteriorly, it usually communicates, between the medial and intermediate cuneiform bones, with the cuneonavicular joint. The **lateral** joint cavity separates the cuboid from the fourth and fifth metatarsal bones, and is prolonged forwards between these two metatarsal bases.

Intermetatarsal Joints

The intermetatarsal joints are small synovial joints between cartilage-covered facets on the contiguous sides of the bases of all lateral four metatarsal bones. The joint cavities [FIG. 327] are prolonged from the tarsometatarsal cavities—that between the fourth and fifth bones from the lateral tarsometatarsal joint, those on each side of the third bone from the intermediate joint.

Above, below, and in front, these little joints are closed by **dorsal, plantar,** and **interosseous intermetatarsal ligaments,** composed of short, transverse fibres between corresponding surfaces of the bases of the metatarsal bones. The interosseous ligaments are very strong, and play an important part in holding together the metatarsal bases as segments of the metatarsal span of the transverse arch of the foot.

The first metatarsal bone is connected to the second by interosseous fibres only. Sometimes a bursal sac is formed amidst these fibres, between indistinct facets on the bones; its cavity may communicate with the medial tarsometatarsal joint.

Nerve Supply. The tarsometatarsal and intermetatarsal joints are supplied by twigs from the **deep peroneal** and the **lateral** and **medial plantar nerves.**

METATARSAL MOVEMENTS

The interlocking of the bones at the tarsometatarsal joints and the intermetatarsal interosseous ligaments preclude, except at the first tarsometatarsal joint, anything beyond small gliding movements of the metatarsal bones by which they contribute very slightly to the inversion and eversion movements of the foot. At the first tarsometatarsal joint a slight rotatory movement may occur in addition to a limited amount of up and down gliding. Such movements of the first metatarsal take place during the adaptation of the foot to the ground in varying degrees of inversion and eversion.

THE METATARSOPHALANGEAL JOINTS

These joints closely resemble the metacarpophalangeal joints in the shape of the articular surfaces and the disposition of the ligaments. The cupped base of each proximal phalanx is applied to the rounded metatarsal head.

The **fibrous capsule,** lined with synovial membrane, is formed at the sides by strong, fan-shaped **collateral ligaments** attached to neighbouring parts of the bones, and, on the plantar aspect, by a thick, pad-like plantar ligament; dorsally, the capsule is replaced by an expansion of the extensor tendon.

The **plantar ligament** is grooved by the flexor tendons. In the big toe this ligament has two fairly large sesamoid bones developed in it; they are covered with cartilage on their upper surfaces which articulate with grooves on the under surface of the head of the first metatarsal bone. The plantar ligaments of all the joints are connected by the **deep transverse metatarsal ligaments** which are arranged like the deep transverse metacarpal ligaments of the palm, except that in the hand this connexion does not involve the joint of the thumb, whereas in the foot the joint of the big toe is linked up with the others. This difference between hand and foot is of great functional significance.

Movements. The movements of these condyloid joints are similar to the movements at the corresponding joints in the hand but are less extensive. In the foot, however, extension of the proximal phalanges can be carried beyond the line of the metatarsal bones; and abduction and adduction are centred on the second toe because of the slightly different arrangement of the interosseous muscles. In flexion the toes are at the same time drawn together; in extension they tend to spread apart and are inclined slightly in the lateral direction.

THE INTERPHALANGEAL JOINTS

The interphalangeal joints also are similar to the corresponding joints in the hand. At each articulation the proximal bone presents a double convexity within the joint and the distal bone a double concavity.

Collateral ligaments at the sides, a thick **plantar ligament** and the expansion of the extensor tendon dorsally complete the **fibrous capsule,** lined with synovial membrane. The second interphalangeal joint of the little toe is often obliterated by synostosis.

Nerve Supply. The metatarsophalangeal and interphalangeal joints are supplied by the dorsal and plantar digital nerves.

Movements. These are hinge joints. Movement is limited to **flexion** towards the sole of the foot and **extension** in the opposite direction, around a transverse axis.

MECHANISM OF THE FOOT

The foot is specialized to support the body in the erect position, alike in standing and in progression. Its skeletal parts are arranged in arches, longitudinal and transverse, in the manner already described. The presence of the joints in the arches introduces a spring mechanism, so that the arches resemble the half-elliptical springs of a carriage. They yield slightly when weight is put upon them, and recoil when it is removed [FIG. 328]. The medial part of the longitudinal arch is higher than the lateral and has more joints, which give to the corresponding part of the foot a greater resilience. This is in accordance with the propulsive function of the medial part of the foot in walking and running, whereas the lateral part of the foot is more in the nature of a stabilizing flap. The arched form of the foot is established in early foetal life, but the height of the arches shows considerable individual variation.

Attention has been drawn to the ligaments which specially support the arches of the foot [pp. 251–3] and these are supplemented by the plantar aponeurosis [FIG. 472]; the importance of muscles in safeguarding the ligaments from being stretched under continuous strain is considerable. The tendon of the tibialis posterior muscle passes as a sling below the summit of

FIG. 328 Tracing of radiographs superimposed to show the position of the longitudinal arch of the foot during relaxation (red) and contraction (black) of the muscles. [See also Fig. 468, footprints 2 and 3.]

the medial part of the longitudinal arch, and, because of the insertion of that tendon on the plantar aspect of nearly all the tarsal and metatarsal bones, the contraction of the muscle must tend to increase the general concavity of the sole. The peroneus longus acts similarly as a sling beneath the lateral part of the longitudinal arch and also as a contractile tie-beam between the extremities of the transverse arch. The tibialis anterior and peroneus tertius muscles pull upon the medial and lateral sides of the longitudinal arch from above. The long flexors of the toes act as contractile ties to the longitudinal arch. The muscles of the sole help to brace the arches, according to their disposition.

In walking, the weight is first applied to the common posterior pillar of the two longitudinal arches as the heel strikes the ground. As the body moves forward the weight falls on the lateral border of the foot, is rapidly transferred across the metatarsal heads to the medial part of the longitudinal arch as the heel leaves the ground, and a final propulsive thrust by the big toe is obtained by extension of the hip and knee joints, plantar flexion of the ankle, and flexion of the joints of the big toe. In running, the weight reaches the ground only at the distal end of the longitudinal arch, which then acts as a quarter-elliptical spring, and, by its recoil, reinforces the final thrust derived from flexion of the medial toes.

RADIOGRAPHIC ANATOMY

J. B. KING

The radiographic examination of a synovial joint involves the demonstration of the articular cartilage, the adjacent bone structure, the surrounding soft tissues, and the range of movement. To be successful, an X-ray examination must be directed specifically to one or other of these features.

In **infancy** incomplete ossification and extreme mobility make the interpretation of X-rays of most joints a matter of great difficulty, and it is of considerable importance to appreciate that in these years X-ray examination shows much less than

in later life. The greater mass of cartilage, creating as it does wide gaps between the shadows of adjacent bones [PLATES 12 and 18 A], makes it possible for serious joint pathology to be present without any radiographic evidence, provided the cartilage is intact.

Throughout **childhood** the X-ray examination of joints is largely concerned with the structure and relationship of the secondary ossification centres to one another and to the metaphyses. Over this period an accurate assessment of normal radiographic anatomy is even more important than in later life, because deformities, initially slight, may produce gross deformation during later bone growth. The detection of the earliest changes may make it possible to anticipate serious distortions in adult life. During the later stages of ossification and epiphysial union, certain joints do not have the normal architecture of adult life, and the rapidly growing edge of new bone frequently does not have the sharply defined margin to be expected at a metaphysis. An important example of such transient features is the hip joint in early adolescence when it is usual for the acetabulum to protrude into the lesser pelvis, and at the same time for the uniting triradiate cartilage and acetabular margins to have an indefinite, irregular, and fluffy outline [PLATE 14].

The frequency of developmental anomalies at the hip joint and the extreme importance of their early diagnosis has attracted attention to a standard of normality called Shenton's Line. In anteroposterior radiographs the relation between femur and acetabulum is normal when the line of the inferior border of the femur forms part of the same arc as the upper border of the obturator foramen [PLATE 14]. Standards of normality such as this have varying degrees of validity in adolescence and adult life, but because of the incomplete ossification and extreme mobility in infancy they lack accuracy just when their importance is greatest.

In the **adult** the normal, thick, compact cortex of long bones always shows pronounced thinning as the shafts expand to form the articulating ends. Over the actual joint surfaces the cortex may be represented on the radiograph by a very thin line which, however fine, is always intact except at such points as the insertion of the ligament of the head of the femur. The gross and fine trabecular pattern of endochondral cancellous bone adjacent to any one joint adheres closely to a constant design, but even slight alterations in the functional stresses produce changes in the regional bone architecture. The limited way in which such changes can be demonstrated radiographically is mentioned on page 202 in the section on Bones.

Cartilage has a radiographic density neither greater nor less than any other of the soft tissues of the body, and cannot be demonstrated in joints except as an apparent space between the articulating bones. The accuracy therefore with which joint surfaces can be examined by X-ray varies greatly and depends on the extent to which the cartilage, as represented by the so-called **joint space,** can be seen on the radiograph free from overlying bone. If any accurate estimate of the thickness of the articular cartilage is to be made, then the principal plane of the joint must lie exactly parallel to the beam of X-rays in the taking of the radiograph [PLATE 16 A].

It is relatively seldom that the traditional anteroposterior and lateral radiographs are wholly adequate for the X-ray examination of a joint. At the knee, for instance, it is true that two such views will show the space between the femur and tibia and give an accurate indication of the thickness of the weight-bearing cartilage [PLATE 15 C], but the joint between the patella and femur is almost wholly obscured by the femoral condyles. An axial view, however, is possible in which the true relationship of the patella to the intercondylar groove is shown [PLATE 16 A]. In a joint such as the hip, even though the joint space is overlapped by bone, it is still possible to estimate the total thickness of cartilage with some accuracy. In complicated articulations such as those of the vertebral column, especially in the cervical and lower lumbar regions, the articular facets are so oblique and also so variable from one region to another that only very limited visualization of their cartilages is possible even with oblique views, while anteroposterior or lateral radiographs are almost valueless. Similar factors are involved in many other joints, such as the sacro-iliac and subtalar, where X-ray examination shows only the adjacent bone and not the articular cartilage. Many procedures have been described by which joint cartilages, particularly those of the knee, can be demonstrated in detail by the introduction either of air or of opaque media, but these have found little practical application.

The intervertebral discs, being fibrocartilaginous, are represented radiographically by the spaces between adjacent vertebral bodies and in lateral views, being relatively clear of overlapping bone, their dimensions can be demonstrated with accuracy [PLATE 6]. As they participate in the formation of the normal spinal curves the discs in the thoracic region have a greater vertical diameter posteriorly than anteriorly, the converse being the case in the lumbar region. In the normal individual the vertebral column shows changes with increasing age more commonly than any other skeletal structure, and the intervertebral discs are almost invariably involved. The stooping posture of the **aged** represents an increase in the normal anteroposterior curve of the dorsal spine, and since the head must be carried vertically above the feet there is compensatory increase in the curves of the cervical and lumbar spines. The resulting changes in the intervertebral discs are so common that they almost represent the normal appearance of old age and consist of posterior narrowing of the discs in the lower cervical and lumbar spine with anterior narrowing in the mid-thoracic region [PLATE 5 B].

It is not only in the spine that progressive changes in joints with age are an almost constant anatomical feature. In most of the weight-bearing articulations there is some diminution in joint space as a result of decreased thickness of cartilage, and at the same time there is an increase in the radiographic density of the adjacent subchondral bone. Joints which are not subject to the lifelong strain of gravity and the compression forces of powerful muscles show such changes only to a minimal extent.

Tendons have no specific radiographic density but may be visualized on the radiograph either by their actual thickness relative to surrounding tissue, e.g., at the attachment of the tendo calcaneus, or by the presence of adjacent tissue of diminished density, e.g., the pad of fat posterior to the patellar ligament [PLATE 16 D]. The site of attachment of many tendons can be demonstrated radiographically by areas of increased density or spur formation in the cortical bone. Such changes are naturally more marked in muscular individuals and the limits of normality may be difficult to define.

Examination of joints can be extended beyond the purely anatomical demonstration of structure by radiographs taken at the limits of movement. This is in fact the most useful method of examining joints such as the shoulder, where radiographs in adduction and abduction, and sometimes also in medial and lateral rotation, not only give bone detail but also some index of the true joint mobility. Other parts, such as the cervical spine, have such complex articulations that two radiographs at right angles are inadequate, but lateral views in full flexion and extension reveal structure and function in a way not otherwise available. If examination of the intervertebral foramina is required, right and left oblique radiographs are required.

It should be emphasized that immobilization of a joint may be present without any radiographic evidence of abnormality, and that abnormal mobility may not be demonstrable by X-ray examination.

For further details of the radiographic anatomy of joints see: Meschan (1959); Schinz *et al.* (1971); Shanks and Kerley (1971).

REFERENCES

ANNOVAZZI, G. (1928) Osservazioni sulla elasticita dei legamenti, *Arch. Sci. biol. (Bologna)*, **11**, 467.

BARNETT, C. H. (1953) Locking at the knee joint, *J. Anat. (Lond.)*, **87**, 91.

BARNETT, C. H., and COBBOLD, A. F. (1962) Lubrication within living joints, *J. Bone Jt Surg.*, **44B**, 662.

BARNETT, C. H., DAVIES, D. V., and MACCONAILL, M. A. (1961) *Synovial Joints*, London.

BEADLE, O. A. (1931) The intervertebral discs, *Spec. Rep. Ser. med. Res. Coun. (Lond.)*, No. 161, London, H.M.S.O.

BRADFORD, F. K., and SPURLING, R. G. (1945) *The Intervertebral Disc*, 2nd ed., Springfield.

BROOKE, R. (1924) The sacro-iliac joint, *J. Anat. (Lond.)*, **58**, 299.

BRYCE, T. H. (1915) *Quain's Elements of Anatomy*, 11th ed., Vol. IV, Pt I, London.

BUNNELL, S. (1938) Opposition of thumb, *J. Bone Jt Surg.*, **20**, 1072.

CLELAND, J. (1884) Notes on raising the arm, *J. Anat. Physiol.*, **18**, 275.

COGGESHALL, H. C., WARREN, C. F., and BAUER, W. (1940) The cytology of normal human synovial fluid, *Anat. Rec.*, **77**, 129.

DARCUS, H. D. (1951) The maximum torques developed in pronation and supination of the right hand, *J. Anat. (Lond.)*, **85**, 55.

DAVIES, D. V. (1945) Anatomy and physiology of diarthrodial joints, *Ann. rheum. Dis.*, **5**, 29.

FICK, R. (1904–11) *Handbuch der Anatomie und Mechanik der Gelenke. Bardeleben's Handbuch der Anatomie des Menschen*, Bd. II, Jena.

GARDNER, E. (1948a) Innervation of knee joint, *Anat. Rec.*, **101**, 109.

GARDNER, E. (1948b) Innervation of hip joint, *Anat. Rec.*, **101**, 353.

GARDNER, E. (1948c) Innervation of shoulder joint, *Anat. Rec.*, **102**, 1.

GHADIALLY, F. N., and ROY, S. (1969) *Ultrastructure of Synovial Joints in Health and Disease*, London.

GOODSIR, J. (1858) On the mechanism of the knee joint, *Proc. roy. Soc. Edinb.*; also in *Anatomical Memoirs*, ed. Turner, W. (1868), Vol. II, p. 231, Edinburgh.

HAINES, R. W. (1941) A note on the actions of the cruciate ligaments of the knee joint, *J. Anat (Lond.)*, **75**, 373.

HAINES, R. W. (1944) The mechanism of rotation at the first carpometacarpal joint, *J. Anat. (Lond.)*, **78**, 44.

HAINES, R. W. (1946) Movements of the first rib, *J. Anat. (Lond.)*, **80**, 94.

HAINES, R. W. (1947) The development of joints, *J. Anat. (Lond.)*, **81**, 33.

HARPMAN, J. A., and WOOLLARD, H. H. (1938) The tendon of the lateral pterygoid muscle, *J. Anat. (Lond.)*, **73**, 112.

HILTON, J. (1863) *Lectures on Rest and Pain*, 6th ed. (1950), ed. Philipp, E. E., and Walls, E. W., p. 166, London.

HUNTER, W. (1743) On the structure and diseases of articular cartilage, *Phil. Trans. B*, **42**, 514.

INGLEMARK, B. E. (1950) The nutritional supply and nutritional value of synovial fluid, *Acta orthop. scand.*, **20**, 144.

INKSTER, R. G. (1938) Inversion and eversion of the foot and the transverse tarsal joint, *J. Anat. (Lond.)*, **72**, 612.

INMAN, V. T., SAUNDERS, J. B. DE C. M., and ABBOTT, L. C. (1944) Observations on the function of the shoulder joint, *J. Bone Jt Surg.*, **26**, 1.

JAZUTA, K. (1929) Die Nebengelenkflächen am Kreuz- und Hüftbein, *Anat. Anz.*, **68**, 137.

JOHNSTON, T. B. (1937) The movements of the shoulder joint, *Brit. J. Surg.*, **25**, 252.

JONES, F. W. (1944) The talocalcanean articulation, *Lancet*, ii, 241.

JONES, F. W. (1949) *Structure and Function as seen in the Foot*, 2nd ed., London.

KEYES, D. C., and COMPERE, E. L. (1932) The normal and pathological physiology of the nucleus pulposus, *J. Bone Jt Surg.*, **14**, 897.

LOCKHART, R. D. (1930) Movements of the normal shoulder joint, *J. Anat. (Lond.)*, **64**, 288.

LOCKHART, R. D. (1933) A further note on movements of the shoulder joint, *J. Anat. (Lond.)*, **68**, 135.

MACCONAILL, M. A. (1932) The function of intra-articular fibrocartilages, *J. Anat. (Lond.)*, **66**, 210.

MACCONAILL, M. A. (1941) The mechanical anatomy of the carpus, *J. Anat. (Lond.)*, **75**, 166.

MACCONAILL, M. A. (1945) The postural mechanism of the foot, *Proc. roy. Irish Acad. B*, **50**, 14, 265.

MACCONAILL, M. A. (1946) Studies in the mechanics of synovial joints, *Irish J. med. Sci.*, **246**, 190; **247**, 223; **249**, 620.

MANTER, J. T. (1941) Movements of the subtalar and transverse tarsal joints, *Anat. Rec.*, **80**, 397.

MARTIN, B. F. (1958) The oblique cord of the forearm, *J. Anat. (Lond.)*, **92**, 609.

MARTIN, C. P. (1940) The movements of the shoulder joint, *Amer. J. Anat.*, **66**, 213.

MESCHAN, I. (1959) *An Atlas of Normal Radiographic Anatomy*, 2nd ed., Philadelphia.

MORTON, D. J. (1935) *The Human Foot*, New York.

PRITCHARD, J. J., SCOTT, J. H., and GIRGIS, F. G. (1956) The structure and development of cranial and facial sutures, *J. Anat. (Lond.)*, **90**, 73.

PÜSCHEL, J. (1930) Der Wassergehalt normaler und degenerierter Zwischenwirbelscheiben, *Beitr. path. Anat.*, **84**, 123.

SARNAT, B. G. (1951) *The Temporomandibular Joint*, Springfield.

SCHINZ, H. R., BAENSCH, W. E., FRIEDL, E., and UEHLINGER, E. (1971) *Roentgen Diagnostics*, 2nd ed., Vol. 2, London.

SCHUNKE, G. B. (1938) The anatomy and development of the sacro-iliac joint in Man, *Anat. Rec.*, **72**, 313.

SHANKS, S. C., and KERLEY, P. (1971) *A Textbook of X-Ray Diagnosis*, 4th ed., Vol. 6, London.

SMITH, E. BARCLAY (1896) The astragalo-calcaneo-navicular joint, *J. Anat. Physiol.*, **30**, 390.

SMITH, J. W. (1954) The elastic properties of the anterior cruciate ligament of the rabbit, *J. Anat. (Lond.)*, **88**, 369.

SMITH, J. W. (1956) Observations on the postural mechanism of the human knee joint, *J. Anat. (Lond.)*, **90**, 236.

STEINDLER, A. (1955) *Kinesiology of the Human Body*, Springfield.

TRUETA, J. (1957) The normal vascular anatomy of the human femoral head during growth, *J. Bone Jt Surg.*, **39 B**, 358.

TUCKER, F. R. (1949) Arterial supply to the femoral head and its clinical importance, *J. Bone Jt Surg.*, **31 B**, 82.

WALMSLEY, R. (1953) The development and growth of the intervertebral disc, *Edinb. med. J.*, **60**, 341.

WALMSLEY, T. (1928) The articular mechanism of the diarthroses, *J. Bone Jt Surg.*, **10**, 40.

WRIGHT, R. D. (1935) A detailed study of movement of the wrist joint, *J. Anat. (Lond.)*, **70**, 137.

YOUNG, J. (1940) Relaxation of the pelvic joints in pregnancy, *J. Obstet. Gynaec. Brit. Emp.*, **47**, 493.

MUSCLES AND FASCIAE

D. C. SINCLAIR

GENERAL INTRODUCTION

SOMATIC MUSCLE

THE somatic muscles are derived from the mesoderm of the embryo, and make up about 40 per cent. of the total weight of the human body. Somatic muscle is sometimes called skeletal, voluntary, or striated muscle, but these alternative names are not always appropriate. In the first place, some somatic muscles, such as the orbicularis oris [p. 282] which surrounds the mouth, have no connexion with the skeleton. Others, such as the muscles of the pharynx and the upper part of the oesophagus, are not wholly under the control of the will, and normally operate involuntarily. Finally, the word striated, or striped, is unspecific because there are also longitudinal striations in visceral muscle, and cardiac muscle, like somatic muscle, shows both longitudinal and transverse striations [FIG. 329].

Structure

Somatic muscle, like the other types of muscle, is made up of cells in which the ability of protoplasm to contract is both highly developed and organized directionally. When these cells are stimulated they contract along their long axis, and, since their volume remains unaltered, they expand and thicken at the same time.

The functional unit of somatic muscle is the **muscle fibre,** an elongated, cylindrical, multinucleated cell, varying in length from a few millimetres to more than 30 centimetres (Lockhart and Brandt, 1938; Barrett, 1962), and in width from about 10 microns to 100 microns. The fibres do not usually branch.

When examined by the light microscope, a muscle fibre exhibits both a longitudinal striation and a very characteristic transverse striping by alternate dark and light bands. A long fibre may contain several hundred nuclei, which lie towards the side of the fibre rather than centrally; there are no cell boundaries between the nuclei. The cytoplasm of the fibre is called the **sarcoplasm,** and is enclosed within a cell membrane called the **sarcolemma.** This has the typical three-layered structure characteristic of boundary membranes.

Under the electron microscope, the contractile elements of the muscle fibres, which are called **myofibrils,** are revealed. They are long unbranched threads varying from 1 to 3 microns in diameter, and they run the whole length of the fibre. Each myofibril exhibits alternate dark and light banding, and the bands are 'in step' with each other, so producing the typical cross-striation of the muscle fibre. The light bands are called isotropic (I) bands, and the dark ones anisotropic (A) bands. In the middle of the isotropic band is a thin dark line known as the Z disc, and the segment of the fibril lying between two adjacent Z discs is called a **sarcomere.** In the middle of the anisotropic band there is a relatively clear H disc, and in the centre of this lies a dark line known as the M line [FIG. 330].

Higher magnification shows that each myofibril is composed of a number of **myofilaments** [FIG. 330], which are of two kinds. The thicker filaments, made of the protein myosin, are about 10 nm. in diameter, and 1·5 microns long, and lie in the A band. The thinner filaments, about 5 nm. in diameter, are composed of the protein actin, and constitute the I band, extending into the adjacent A bands for some distance and lying in the spaces between the myosin filaments.

Outside the sarcolemma of each muscle fibre is a very delicate connective tissue called **endomysium,** which serves to separate the fibre from its neighbours and at the same time to connect it to them. The fibres are arranged parallel to each other in bundles called **fasciculi,** and each fasciculus is bound together by a rather more dense connective tissue called the the **perimysium.** Finally, the assembly of fasciculi is enclosed in a fibrous wrapping called the **epimysium,** which forms a sheath for the whole muscle. In some muscles this is thick and strong, and in others thin and more or less transparent.

There are two kinds of somatic muscle fibres, **red and white.** The red fibres owe their colour to their content of myoglobin, a substance similar to haemoglobin, and their nuclei are sometimes centrally placed. In man, red and white

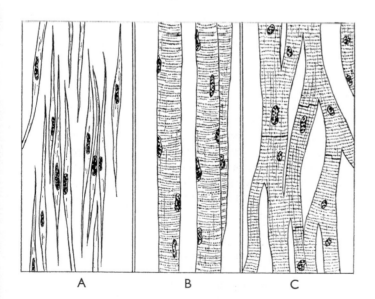

FIG. 329 Three varieties of muscle (human; ×250).

A, Visceral

B, Somatic

C, Cardiac

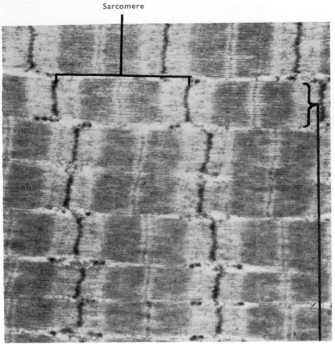

Sarcomere

Myofibril

A

I-Band　　　　　H disc　　　　Mitochondrion

Z disc　　　　　A-Band

B

FIG. 330　Electron micrographs of part of a single muscle fibre in longitudinal section.

A, Parts of seven myofibrils in somatic muscle.

B, A single sarcomere of cardiac muscle. This is the unit of a myofibril between two Z discs. Centrally it contains thick myosin filaments (A band) interspersed with thin actin filaments which reach the Z discs and form the I bands with them. (By courtesy of Professor A. R. Muir.)

fibres are always mixed together in any given muscle, but in some animals the muscles which perform rapid movements are much whiter than the muscles which are concerned with slow and prolonged contractions such as are needed in maintaining posture. The advantage red fibres have over white fibres when contracting in this way is that the myoglobin provides them with a store of oxygen. In general, therefore, red fibres are more numerous in postural muscles and white fibres in muscles which perform rapid movement.

Quite apart from the difference between red and white muscle, some fibres are found to contract more rapidly than others. The structural differences underlying this variation are not yet established (Leeson and Leeson, 1969), but fast contracting fibres within a given muscle tend to be grouped together, and the more slowly contracting ones are found more deeply.

The attachment of a muscle to bone or to other tissues is always indirect, through the medium of the connective tissue elements in the muscle. Sometimes the perimysium and epimysium unite directly with such elements as the periosteum of bone, the connective tissue underlying the skin, or the capsule of a joint. If this intermediate tissue is inconspicuous, the muscle is said to have a fleshy attachment. In other instances the connective tissue components of the muscle fuse together to form an intervening **tendon**, composed of bundles of collagen fibres. Each collagen fibre unites at one end with the fibres of the structure to which the muscle gains attachment, and at the other with the thickened and corrugated end of the sarcolemma of a muscle fibre; there is no direct continuity between an individual muscle fibre and an individual tendon fibre, and the pull is transmitted from one to the other through the sarcolemma.

Tendons, which in general are immensely strong, may be round cords, flat bands, or thin sheets of dense connective tissue; in the latter case the tendon is called an **aponeurosis.** The transition between muscle and tendon does not necessarily take place uniformly across the width of the muscle, so that there may be prolongations of the tendon among the muscle fibres. Sometimes flat sheets of dense connective tissue penetrate the muscle, forming **septa** to which the muscle fibres are attached. In other cases patches of tendon may develop in situations where the muscle is exposed to friction; for example, the deep surface of the trapezius [p. 303] is tendinous where it rubs against the spine of the scapula. If a tendon is subjected to friction it may develop in its substance a **sesamoid bone** [p. 76], as, for example, in the peroneus longus [p. 378] tendon in the sole of the foot. Other tendons, however, although equally exposed to friction, do not develop sesamoid bones, and such bones may appear where friction does not seem to be a primary exciting agent.

Blood Supply

In between the muscle fibres, and largely parallel to them, there runs a very fine plexus of capillaries. These are said to be more numerous in red muscle than in white, and in red muscle they show dilatations which may form a reservoir of blood when the capillaries are compressed by sustained contraction of the muscle. During exercise the capillaries dilate, and the amount of blood they can contain may be increased up to about 800 times.

The vessels which supply the capillary plexus may enter the muscle either towards one end or in the middle: because muscles have to slide on the surrounding connective tissue, it is not possible to have blood vessels entering them at several different points. Muscles supplied or drained by only a single vessel are liable to be wholly or partly destroyed if this vessel is damaged. Other muscles are supplied by a series of anastomosing arteries, and within the muscle itself there are numerous communications between the smaller branches (Blomfield, 1945). However, such collateral circulations are not particularly good, and parts of a muscle can be functionally damaged by tying off one of its arteries of supply.

The tendon of a muscle, being composed of a relatively inactive tissue, has a much less profuse blood supply, but vessels enter it at intervals from the surrounding connective tissue and run longitudinally between its fibres (Edwards, 1946; Brockis, 1953).

Nerve Supply

The nerve supply to a muscle usually enters along with the arterial supply at a neurovascular **hilus** (Brash, 1955), which is usually situated on the deep surface of the muscle, and so protected to some extent from injury. There may be several nerves supplying different parts of the one muscle, as in the segmentally innervated muscles of the abdominal wall [p. 345].

The 'motor' nerve entering the muscle usually contains both sensory and motor nerve fibres in approximately equal proportions, though some muscles may receive separate sensory branches—for example, the brachialis [p. 314]. In the case of muscles supplied by the spinal nerves, both the motor and the sensory fibres are usually derived from more than one segment of the spinal cord, but some muscle groups, such as the small intrinsic muscles of the hand, are supplied almost entirely by a single segment.

The motor nerve fibres form a plexus in the muscle from which individual myelinated branches emerge to supply groups of muscle fibres. One nerve cell through its branching fibres supplies several muscle fibres, but each muscle fibre receives only one terminal branch of a nerve fibre. As this approaches the muscle fibre, it loses its myelin sheath, and divides into a number of terminal arborizations which together form what is called a **motor end plate.** Each separate part of the nerve ending fits into a gutter on the surface of the sarcolemma, which is thrown into folds at right angles to the gutter, and between each part of the nerve ending there are many nuclei which lie in an accumulation of sarcoplasm called the **sole plate.** The whole mechanism is covered over by extensions from the last Schwann cell of the sheath of the nerve fibre.

The apparatus consisting of a single motor nerve cell and the muscle fibres which it innervates is called a **motor unit.** The size of the motor unit varies in different muscles (Feinstein *et al.*, 1955), and determines the possibility of grading the contraction of the muscle coarsely or finely. Since individual muscle fibres probably obey the all-or-none law, and either contract fully or not at all, the only possibility of graduating the contraction of a muscle is by throwing varying numbers of fibres into action, and this can only be done by varying the number of nerve cells stimulated. The strength of contraction can therefore only be increased in 'steps' dependent on the size of the motor units concerned. Where fine gradations of

movement are essential, as in the muscles of the eyeball, the larynx, and the tongue, the number of muscle fibres supplied by a single nerve cell is small. Where, on the other hand, only relatively gross movements are needed, as in the muscles of the lower limb, every nerve cell may supply several hundred muscle fibres.

When a maximum effort is required, nearly all the motor units may be stimulated at the same time, but under normal conditions the motor units probably work in relays, so that in prolonged contractions some are always resting while others are contracting. The muscle fibres comprising a single unit do not necessarily lie side by side, and may be scattered in such a way that adjacent units interlock with each other. In the upper limb muscles each unit is confined to a cylinder of muscle 5-7 mm. in diameter; in the lower limb the corresponding cylinders are 7–10 mm. in diameter. Territories of this size can accommodate the fibres of 25 interlocking units each comprising 500–2000 fibres (Buchthal *et al.*, 1959).

The sensory fibres which enter a muscle also form a dense plexus, and there are simple branching terminals between the muscle fibres (Stacey, 1969). In addition to these there are more complicated endings such as the **neuromuscular spindles** and the **tendon endorgans.** The number of nerve fibres destined for these specialized terminals indicates the importance of a feed-back system informing the nervous system of the state of contraction of the muscle. The sensory organs in a muscle respond either to an increase in tension of the muscle or its tendon, or to stretching of the muscle. The spindles consist basically of several specialized muscle fibres enclosed in a complex connective tissue sheath and profusely innervated [FIG. 976]. Raising the tension in these fibres increases the sensitivity of the apparatus and decreasing the tension lowers it.

Tendons receive their sensory supply from the nerve fibres in the muscle, and also from nearby sensory nerves (Stilwell, 1957).

Muscle Tone. A healthy muscle at rest is firm, with a characteristic elastic resistance to pressure. This is called muscle tone, and was formerly attributed to a resting activity of the muscle fibres. However, no electrical activity can be detected in a resting normal muscle, and what causes tone is now a matter for discussion (Joseph, 1960; Basmajian, 1961, 1967); it may be due to elastic fibres in the connective tissue framework of the muscle. If the nerve supply to the muscle is cut, the normal tone disappears and the muscle becomes flaccid and unresisting. Conversely, if there is destruction of the neurons leading from the brain to the motor neurons in the spinal medulla that innervate the muscle, there is often an increase in tone and the muscle is said to become spastic. In either case postural deformities may be produced; in **flaccid paralysis** the innervated antagonists pull the part out of its normal resting posture, and in **spastic paralysis** the paralysed muscles themselves do so.

Shape and Construction

The fasciculi in a muscle can be arranged in various ways. The simplest arrangement, that of parallel fasciculi extending throughout the length of the muscle, is found in the so-called **strap muscles,** such as the sternohyoid [p. 277]. The rectus abdominis [p. 350] is a strap muscle divided up into shorter segments by three or more fibrous intersections, and a similar arrangement exists in other strap muscles on a microscopic scale. Some of the fasciculi in the human gracilis and sartorius muscles contain units of from two to six short fibres arranged end to end in series and joined by bridges of fibrous tissue (Schwarzacher, 1959, Barrett, 1962).

In a **fusiform muscle** the fleshy belly of the muscle is spindle-shaped, and there is a tendon at one or both ends. In such a muscle not all the fibres extend throughout the whole length of the muscle belly. A modification of this type of muscle is found when a fleshy belly develops at either end, with a tendon in the middle. Such muscles are said to be **digastric.**

The fasciculi of some muscles are attached at an angle to the tendon, an arrangement which, because of its similarity to the structure of a feather, leads to their being called **pennate muscles.** If the tendon develops on one side of the muscle, the pattern is said to be unipennate; if the tendon is in the middle and fibres reach it at an angle from both sides, the muscle is bipennate; if there are several tendinous intrusions into the muscle with fibres reaching them from several directions, the muscle is multipennate [FIG. 331].

When a muscle fibre contracts, it shortens to about 55 per cent. of its original length (Haines, 1934). It follows that the longer the fibres in a given muscle, the greater the range of movement it can produce. The strength of a muscle, on the other hand, depends on the total number of fibres it contains, for a short fibre can pull as hard as a long one.

Now a given volume of muscle can contain either a small number of long fibres or a greater number of short fibres. The arrangement of the fasciculi within a muscle is therefore a compromise between power and range of movement. In the pennate muscles the increased power provided by the greater number of short fibres accommodated is to some extent offset by the fact that these fibres do not exert a straight pull on the tendon. Nevertheless, pennate muscles are generally strong, whereas strap muscles have the greatest possible range of movement for their size.

A long tendon permits a muscle to act at a distance without the movement being impeded by the swelling of the fleshy belly. For example, the long muscles which operate on the fingers have their bellies safely removed from the scene of action and located in the upper part of the forearm. However, for a given length of muscle, the longer the inextensible tendon, the shorter the contractile belly has to be, and this reduces the range of movement considerably. A tendon also permits an alteration of the direction of pull of a muscle, since it can be made to turn round corners. The tendon of the tensor veli palatini [p. 293] converts the vertically delivered pull of the muscle belly into a horizontal force by swinging at right angles round the pterygoid hamulus, and the posteriorly directed pull of the superior oblique muscle of the eye [p. 285] is converted into an anteriorly and medially directed force because the tendon doubles back on itself at the trochlea.

Attachments

Many muscles have only two attachments, one at each end, but more complex muscles may be attached to several different structures. If these structures are separated from each other by an interval, the muscle is said to have two or more **heads.** The biceps brachii [FIG. 332] has two heads at its upper end, one from the coracoid process of the scapula, and the other from the supraglenoid tubercle. The triceps brachii [p. 315] has three heads, and the quadriceps femoris [p. 372] has four. The internal architecture of such muscles is obviously somewhat

Tendinous long head of origin

Tendinous short head of origin

Fusiform belly

Tendon of insertion

Aponeurosis joining investing fascia of forearm

FIG. 332　The construction of the biceps brachii.

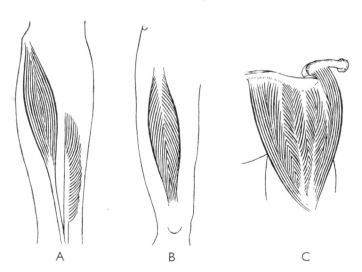

FIG. 331　Types of muscle structure.

A, Fusiform (flexor carpi radialis) and unipennate (flexor pollicis longus).

B, Bipennate (rectus femoris).

C, Multipennate (deltoid).

more complicated than that of simple strap muscles, and patches of tendon may develop between their components to facilitate independent movement.

Most somatic muscles are attached to bone and pass across one or more joints; in this way they can produce or prevent movements of the bones. Some paired muscles may be attached to each other. An example is the levator ani [p. 345], which acts as a diaphragm supporting the floor of the pelvis. Others again, such as the muscles of facial expression [p. 280], may have an attachment to skin. The lumbrical muscles of the fingers and toes [pp. 329, 385] run from the tendon of one muscle to the tendon of another.

Growth and Repair

There is some argument as to whether the number of fibres in a given muscle increases after birth. The fibres certainly grow longer as the muscle increases in length, but it is thought that in most mammals muscles grow thicker simply by increasing the diameter of their individual fibres without increasing their numbers (Rowe and Goldspink, 1969); this may not hold for marsupials (Bridge and Allbrook, 1970). Regular exercise of a muscle readily increases the average diameter of its fibres, and in this way the whole muscle belly can be made to grow in thickness; what seems to happen is that the smaller fibres are brought to about the same diameter as the larger ones, which do not grow appreciably.

In contrast, if a muscle is deprived of its nerve supply the average diameter of its fibres decreases, and if the innervation is not restored they may degenerate and be replaced by a mass of fat or fibrous tissue (Sunderland 1968). This is not wholly explicable on the basis of disuse and consequent vascular deprivation. It is possible that the nerve fibres reaching the muscle may in some way influence its nutrition, perhaps by liberating chemicals within the muscle.

A muscle which has been injured, with destruction of some of its fibres, makes considerable efforts to repair itself, and new fibres are produced by 'satellite cells' lying between the fibres (Church, 1969). Unfortunately, scar formation occurs more easily than regeneration of muscle fibres, and damaged muscle tissue thus tends to be replaced by scar tissue. If this happens, the muscle may exhibit **contractures,** which pull its attachments closer together, so producing deformities. Similar deformities can be produced if the muscle is immobilized for a long time in a position which allows its fibres to shorten, or if continued abnormal stimulation of the muscle occurs, as in cases of spastic paralysis.

Actions and Functions

When a muscle is stimulated, naturally or artificially, its fibres attempt to contract, so as to bring its attachments closer to each other. If such a movement actually takes place, the length of the muscle alters but the tension generated inside it remains approximately constant; such contractions are called **isotonic.** If the muscle is, for one reason or another, unable to approximate its attachments, the tension generated in it increases, but the length remains constant; such contractions are called **isometric.** For example, in the upright position, the location of the centre of gravity of the body produces a tendency for the body to fall forward at the ankle. The muscles of the calf are in constant activity to prevent this (Joseph, 1960),

but they do not shorten appreciably. Again, the centre of gravity of the skull falls in front of the atlanto-occipital joint, and the muscles at the back of the neck contract isometrically to keep the gaze of the eyes level. Isometric contractions are considerably more powerful than isotonic contractions.

Isotonic contractions are of two kinds, **concentric** and **excentric.** In a concentric contraction the muscle shortens, but in an excentric one it actually lengthens. A good example of excentric contraction is the gradual relaxation of the flexors of the elbow during the movement of setting down a glass of water on the table. In such movements the muscle gradually 'pays out', and this active relaxation is of great importance in controlling the movement.

The bones, joints, and muscles form a system of levers in which the muscles apply the force, the joints act as the fulcra, and the bones bear the weight of the part of the body which is to be moved. The **mechanical advantage** of a muscle therefore depends on the distance of its attachment from the fulcrum relative to the distance of the weight which the muscle has to move. A muscle attached close to the fulcrum will be relatively weaker than an equivalent muscle which is attached further away [FIG. 333]. On the other hand, it will be able to produce a faster movement of greater range, since the distance travelled by its moving attachment will be magnified by the length of

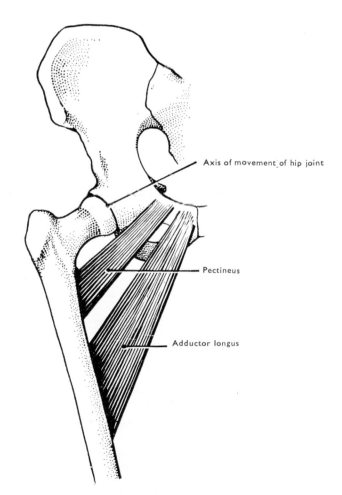

Axis of movement of hip joint

Pectineus

Adductor longus

FIG. 333 Mechanical advantage. The pectineus, attached closer to the axis of movement, is a weaker adductor of the hip joint than the adductor longus, but produces a greater movement of the lower limb for every centimetre of contraction.

the lever. Every muscle, therefore, exhibits a mechanical compromise between the power and the speed of the movement it produces.

Again, the power exerted by a muscle necessarily varies according to the position of the moving bone. In the initial stages of flexion of the elbow, the pull of the brachialis [FIG. 334] is more or less in line with the long axis of the bones of the forearm. Later in the movement, when the elbow has been

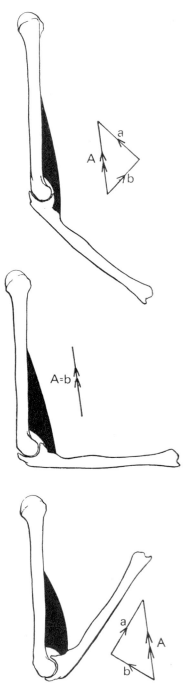

FIG. 334 Action of the brachialis. The force exerted by the muscle (A) can be resolved into two components, one in the line of the moving bone (a), and the other (b) at right angles to it. The latter is the force actually flexing the elbow. When the forearm is more or less at right angles to the arm A = b, and the whole effort of the muscle goes into flexing the elbow. When the forearm is extended or flexed beyond this point, b becomes less than A, and the muscle becomes a less efficient flexor.

bent to about a right angle, the power of the pull becomes maximal. Finally, when the elbow has been flexed past the right angle, the flexing force exerted by the brachialis diminishes again as its direction comes to lie parallel to the forearm bones once more.

A knowledge of the positions of the limbs in which maximum thrust with the foot or pull with the hand is attainable is clearly of importance in industrial design. For a seated subject, using an isometric foot pedal, maximum push is obtained with the hip flexed to 75 degrees and the knee extended to about 160 degrees. Using an isometric hand lever the maximum pull or push is possible when the elbow is extended to approximately 135 degrees, the lever grip is at elbow height, and the lever moves in a vertical plane passing through the shoulder joint (Hugh-Jones, 1947).

In rapid movements of a joint the muscles which exert their pull more or less parallel to the moving bone help to prevent the bony surfaces being pulled apart by centrifugal force. Such muscles are sometimes called '**shunt**' muscles, and must be contrasted with the '**spurt**' muscles, which act at right angles to the moving bone. For example, during rapid flexion of the elbow, the brachioradialis [p. 321] is a shunt muscle, while the biceps brachii [p. 313] and the brachialis [p. 314] are spurt muscles (MacConaill, 1946; Basmajian, 1959).

In many cases the capacity of a muscle to contract is of less functional importance than its ability to maintain its tension throughout the movement of a joint, so acting as what is termed an **extensile ligament.** In contrast to ligaments composed of inextensible collagen, these muscular 'ligaments' can protect the joint during the whole movement and not merely at the end of its range. Perhaps the best example of this function is afforded by the short muscles which surround the very mobile shoulder joint and protect it during movement by holding the head of the humerus into the glenoid socket of the scapula [p. 309]. Indeed, the stability of any joint depends very largely on the tension in the muscles which surround it. If the muscles are paralysed, the ligaments of the joint are unable to resist the forces applied to it, and the joint becomes 'flail', allowing a much greater range of movement than normal, and becoming much more liable to dislocation.

As a joint approaches the extremes of its range of movement, the muscles which oppose the particular movement concerned come into action to slow the movement down to a halt, so preventing damage to the ligaments, which might otherwise give way under the sudden strain. In some instances the contraction of these muscles is a hindrance to the movement. For example, when the knee is straight, the untrained hamstring muscles [p. 367] do not permit full flexion of the hip joint; it is for this reason that it is difficult to perform the 'high kick' and to touch the toes without bending the knees [FIG. 335]. This situation, in which muscles prevent the full movement of the joint, is sometimes called **passive insufficiency.**

The opposite condition, **active insufficiency,** results from the inability of a muscle to contract by more than a fixed percentage of its length. For example, the length of the hamstrings is such that they can fully flex the knee when the hip is flexed, but cannot do so completely when the hip is extended. In the first case the muscles are stretched at the outset of the movement, and the whole of the contraction is used to produce

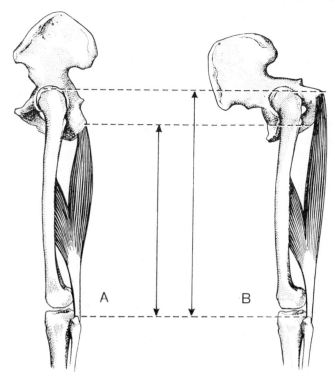

FIG. 335 Passive insufficiency.

A, Normal standing.

B, Bending forward, as in touching the toes.

The origin of the biceps femoris is pulled further away from its insertion by the movement of the hip bone, so stretching the muscle. A continuation of the movement, if the muscle has not been trained to relax, causes the knee to flex.

movement. In the second, the attachments of the muscles are passively approximated at the outset, and some of the contraction has to be used to take up the slack.

In most movements one attachment of a muscle remains more or less stationary while the other attachment moves. The relatively stationary attachment is called the **origin** of the muscle, and the other is known as its **insertion.** A spring which closes a door could be said to have its origin on the door frame and its insertion into the door itself. In the body the arrangement is never so clear cut as this, and the terms origin and insertion are only relative. For example, the muscles which pass between the chest wall and the upper limb are customarily used to move the limb about relative to the trunk; they are therefore said to have their origins on the trunk and their insertions on the limb. But when the upper limb is used in climbing, these same muscles are used to pull the trunk up on the fixed limb. They are thus acting from their now relatively fixed 'insertions' on their relatively movable 'origins'. It is for this reason that many people prefer simply to speak of the attachments of a muscle, not specifying either origin or insertion.

Nevertheless, the attachments of a muscle which lie more proximally, towards the base of a limb, or on the trunk, are usually referred to as the origin, and those which lie more distally as the insertion. An action in which the insertion remains fixed and the origin moves is sometimes called a **reversed action,** and in the lower limb it is often the reversed actions which are the more important. Thus, the ability to

invert the foot while it is off the ground by the use of the tibialis anterior [p. 376] acting from its origin on its insertion is in normal circumstances of little functional value. But when the foot is fixed to the ground by the body weight, the tibialis anterior, acting from its insertion on its origin, plays a most important part in preventing overbalancing during walking.

The action of a muscle necessarily depends on its relationship to the joint which it crosses, and also on the types of movement which that joint permits. At a hinge joint, for example, all the muscles which cross on one aspect are flexors, and all those which cross on the other are extensors; there is no other possibility. But at a ball and socket joint, a muscle may be capable of exerting several different actions, depending on its position relative to the axis of movement. For example, the anterior fibres of the deltoid muscle, which cross the front of the shoulder joint obliquely [p. 308], act as flexors and medial rotators; the posterior fibres, which lie behind the joint, act as extensors and lateral rotators, while the central portion of the muscle, which lies superior and lateral to the joint, acts as a pure abductor. Furthermore, the action of a muscle may depend on the position of the joint at the time. Thus, the obturator internus [p. 356] is a lateral rotator of the thigh when the hip joint is extended, but an abductor of the thigh when the hip joint is flexed.

Very seldom is a movement wholly entrusted to a single muscle; among the few exceptions is the movement of flexion of the distal interphalangeal joints of the fingers and toes. Conversely, it is unusual to find a muscle with but a single action. A common arrangement is that a muscle may be of major importance in relation to one movement, being helped by several other muscles, and at the same time contribute in a minor capacity towards other movements. It follows that if one muscle is paralysed, no single movement will be completely abolished, though several may be weakened. The remaining muscles may satisfactorily compensate for the missing one, so that the resulting disability is small or even unnoticeable.

The fact that several muscles normally collaborate in the production of a movement means that any investigation of weakness of movement must be based on a knowledge of the group concerned, and it is essential to think of muscles as members of functional associations rather than isolated individuals. It also means that if a tendon is to be surgically transplanted to compensate for a paralysed muscle the transplant must be selected from the group of muscles which normally act in association with the paralysed one (Dunn, 1920).

If a muscle has several actions, it normally produces them all when it is activated; it cannot usually contract in such a way as to single out one of them. But the unwanted movements can be neutralized by other members of the group set in action. Thus, the flexor carpi ulnaris [p. 318] both adducts and flexes the wrist, and when stimulated produces both results together. If pure adduction of the wrist is required, the flexion can be cancelled out by setting in action at the same time an extensor of the wrist. The obvious choice in this instance is of course the extensor carpi ulnaris [p. 323], which not only extends the wrist but also contributes to the desired movement of adduction. Similarly, if pure flexion is required, the adduction can be neutralized by the collaboration of an abductor.

A muscle is said to be a **prime mover** when it contracts for the primary purpose of producing the specified movement. A muscle which opposes the specified movement is called an **antagonist,** and often the antagonists play a leading part in the control of the movement [p. 263]. Muscles which help to prevent unwanted movements inherent in the attachments of the prime movers are called **synergists,** and the term is also applied to muscles which contract in order to provide a stable base for the action of the prime movers. A good example is afforded by the muscles which stabilize the scapula during movements of the upper limb [p. 336]. This kind of action is also spoken of as 'fixation' and the muscles performing it are termed **fixators.**

In most movements the role of the prime movers, antagonists, synergists, and fixators does not vary throughout the movement. For example, when the fist is clenched, the extensors of the wrist contract to prevent the flexors of the fingers from also flexing the wrist. This is a vital part of the action, for flexion of the wrist added to flexion of the fingers stretches the extensors of the fingers until they can stretch no more; continued flexion at the wrist then causes the fingers to open out and the grip to relax. This is how an opponent can be made to drop a weapon by forcibly flexing his wrist. The flexors and extensors of the elbow must also contract in order to fix the origins of the flexors of the fingers and the extensors of the wrist, and the scapula must be fixed in order to stabilize in turn the origins of the muscles stabilizing the elbow. In such a manner even an apparently simple movement can have repercussions throughout the musculature of almost the whole body.

In other movements the functional significance of a given muscle may change in successive phases of the movement. For example, in the initial stages of abduction of the humerus the supraspinatus acts as a prime mover, but later it has a more important function as an extensile ligament.

In many instances gravity acts as a prime mover. Thus, in lowering the arm to the side from an abducted position, there is no contraction of the adductor muscles unless the movement is conducted against resistance. Instead, the antagonist to the movement, the deltoid muscle [p. 308] gradually pays out and the movement is effected as a result of the combination of this and the action of gravity. When the arm is adducted against resistance, the deltoid become flaccid, and the adductors are thrown into contraction [PLATE 21 B].

The muscles in one part of the body often act in co-ordination with muscles in more distant parts, a phenomenon known as **associated action.** In walking, the balance of the trunk is maintained by the erector spinae muscles [p. 269] and the upper limb of one side swings in concert with the lower limb of the opposite side. Again, when we look round, the displacement of the head and eyes is accompanied by movements of the feet and trunk. Paralysis of a single muscle may thus upset movements which are not necessarily located in the vicinity of the muscle itself.

There is very little direct voluntary control over individual muscles without training; it is only the movement desired that is voluntary, and the pattern of activity of the executive muscles is automatically organized by the brain. A simple example emphasizes this. Stretch out the upper limb horizontally and then bend the forearm up towards the ceiling. In the phase before the forearm becomes vertical, the biceps can be felt contracting, and the triceps is relaxed. But as soon as the forearm reaches the vertical and starts descending towards the shoulder, the triceps is felt to harden while the biceps relaxes. We are completely unaware of the change-over.

Finally, it is important to distinguish between the actions of which a muscle is capable and the functions which it usually performs (MacConaill and Basmajian, 1969). The interossei of the foot [p. 386] can abduct and adduct the toes, but their chief function is probably to support the anterior arch of the foot and prevent pressure on the vessels and nerves running to the toes. The mere presence of some muscles is of functional importance. For example, in patients who are lying in bed, the gluteus maximus [p. 364] forms a soft pad intervening between the skin and the bony prominence of the buttock. If this muscle is thinned and atrophied by paralysis, the skin is exposed to pressure from the underlying bones, and this may lead to the production of bedsores.

Types of Movement. In most movements, the antagonists are in action throughout, relaxing in a graduated manner to control the activity of the prime movers. But in certain circumstances, as when a sprinter starts a race, a **maximum effort,** unrestrained by the antagonists, is suddenly required. In the initial stages of such movements the antagonists may be almost completely relaxed, so as to reduce the opposing force to a minimum.

Ballistic movements are those in which the prime movers relax once the effort has been made. This type of movement is well seen in serving at tennis, or in kicking a football. The 'follow-through' is due to the momentum produced in the limb at the time of impact.

Contrasting with these movements are the very **fine controlled movements** required by activities such as writing, sewing, painting, etc. Here the prime movers and the antagonists are set to oppose each other equally, and the actual movements are due to temporary slight alterations of the balance of power between the two groups. Movements can be very finely graduated by this means, and the net force exerted can be made very small.

Finally, the occurrence of **involuntary movements** and their occasional dissociation from voluntary movements must be noted. In certain cases of brain injury, muscles which are paralysed for voluntary movement may be found to contract during involuntary or emotional acts. For example, an upper limb which has become paralysed can be moved with its fellow on the other side during the stretching which accompanies a yawn, and facial muscles which cannot be voluntarily activated are found to contract when the patient is asked to smile. The medial rectus muscle of the eye [p. 284] may be unable to contract during convergence of the eyes, but yet be active when the eyes are turned to the opposite side (Beevor, 1904). The explanation of such phenomena is not yet established.

Methods of Investigating Muscle Action. It is often very difficult to determine the part played by a muscle in different movements, and there is still argument about the precise actions of many muscles.

The actions of a muscle may be deduced from its attachments, and it is sometimes helpful to imagine a sheet of elastic passing between them. The bellies and tendons of superficial muscles can be palpated during a given movement to find out whether they are contracting. The standard method of demonstrating a muscle is for the examiner to resist the movement concerned; then the muscles performing it will stand out as they exert their maximum effort. If the resistance is increased additional muscles may be recruited. Electromyography, in which the electrical activity taking place in a muscle during contraction is recorded, is only a more refined example of this method, and has the advantage that it can be applied to muscles lying deeper in the body.

In both cases it is difficult to be sure what the muscle is actually doing when it contracts; is it a prime mover, or a synergist? For example, the tensor fasciae latae [p. 365] contracts during abduction of the thigh, but does not help to abduct the thigh when it is stimulated electrically.

Direct stimulation, whether electrical or mechanical—as when a muscle is pulled upon during a surgical operation—is open to criticism because, in naturally produced movements, muscles are never singled out in this way. However, the method does at least reveal what the muscle is capable of doing, though it does not necessarily show what its functions are in the intact body. A similar objection applies to observations made on the results of paralysis of single muscles or groups of muscles, for the loss of one member of a group at once creates an artificial situation, in which other muscles may take over some or all of its functions.

Testing for Paralysis. When a muscle is paralysed, the movements in which it normally takes part are weakened, some being more affected than others, according to the relative importance of its participation in them. The paralysed muscle may never contract again, or it may gradually recover its power, depending on the nature and extent of the damage inflicted upon it or upon its nerve supply. It is naturally important to be able to detect which muscles are paralysed following a disease or injury, and also to be able to follow the progress of recovery in them.

The only infallible guide to the integrity of a muscle is to see and feel it contract (Sunderland, 1944b), but this is not always possible. True contraction of a muscle must not be confused with displacement or distortion of a paralysed muscle caused by contraction of neighbouring muscles or by a pull imparted by fascial attachments. It is also very necessary to be on guard against '**trick' movements** (Jones, 1919; Sunderland, 1944b). These may be produced in several ways.

An intact muscle may be able to pull on the tendon of a paralysed muscle because it has an accessory slip of attachment to this tendon. Thus, the abductor pollicis brevis can be used to extend the terminal phalanx of the thumb when all the true extensors are paralysed because some of its fibres are inserted into the tendon of the extensor pollicis longus [p. 332].

Again, an intact muscle which does not normally take part in the affected movement may be recruited to help. The abductor pollicis longus, which does not normally flex the wrist, may do so when the flexors are paralysed because its position allows such an action.

A paralysed prime mover may be brought passively into action by its antagonist, providing the muscles concerned are 'two-joint' muscles. Perhaps the best example is seen when active extension of the metacarpophalangeal joints of the hand is paralysed [p. 334]. By flexing the wrist, the extensor tendons can be put on the stretch and the joints can be made to extend passively. Conversely, if the wrist is fully extended, the tendons of the flexors of the digits can be made to pull the fingers into mild flexion at the interphalangeal joints, even

though their muscle bellies may be paralysed. Another important example is afforded in paralysis of the flexor pollicis longus (Sunderland, 1944b); by hyperextending the wrist and fully abducting the thumb the tendon of the flexor pollicis longus is put on the stretch and the terminal joint of the thumb flexes passively.

If the patient is allowed any leverage he may be able to deceive the examiner. For example, when the muscles of the lower limb are being investigated, the heel must not be allowed frictional contact with the couch since movements which are in reality completely paralysed may yet be produced by any muscle capable of taking a leverage from this fixed point (Jones, 1919).

Finally, a sudden relaxation of strongly contracted antagonists may allow a 'rebound' movement in the opposite direction which may simulate an active contraction. Thus, a degree of dorsiflexion of the toes may follow relaxation of the plantar flexors even though the dorsiflexors are paralysed.

When assessing the recovery of a muscle from paralysis, or when attempting to detect a degree of contraction in a grossly weakened muscle, it is obviously inappropriate to resist the movement concerned, as when trying to demonstrate muscle actions. Instead, the weakened muscle is put in a position where gravity does not act against it and the patient attempts to carry out the movement without any resistance. A weak muscle may be incapable of executing a voluntary movement but yet be capable of maintaining the part in a position into which it has been moved passively.

The recovery of muscles is often assessed by a grading system. Thus 0 represents no contraction; 1 represents a flicker or trace of contraction; 2 is active movement with gravity eliminated; 3 is active movement against gravity; 4 is active movement against both gravity and resistance, and 5 is normal power. Unfortunately such grading systems are by no means uniformly accepted (Salter 1955a).

It is a commonplace that muscle strength and size can be increased by use, but there is still controversy about the best method of increasing the power of normal muscles and assisting the recovery of paralysed ones. Isotonic and isometric training both have their proponents, but the results are often equivocal or conflicting (Salter, 1955b; Petersen, 1960).

Nomenclature

The name given to a muscle usually conveys valuable information regarding its position, structure, functions, etc. A muscle may receive its name because of its attachments (sternothyroid), its position (subclavius, supraspinatus), its shape (deltoid, trapezius), its construction (semimembranosus), its action (extensor digitorum), or its size (gluteus maximus). Many names involve a combination of features (extensor carpi radialis longus). Before the eighteenth century few muscles had names, and Galen and Vesalius used numbers, and Leonardo da Vinci letters, to designate the muscles in their illustrations.

Variations

All muscles are subject to a certain amount of variation, but some are more often affected than others. For example, one such muscle, the palmaris longus [p. 318] is not present in every individual, and is thought to be disappearing in the course of evolution. Others are possibly becoming converted into ligaments; with the disappearance of the tail the coccygeus muscle [p. 355] has been partly replaced by fibrous tissue to form the sacrospinous ligament.

Conversely, certain evolutionarily discarded elements may appear in a given individual, either in their primitive form or in an apparently intermediate stage. The muscle belly of the coracobrachialis[p. 314], which normally ends by being attached to the humerus half-way down its length, may be prolonged downwards to the elbow as it is in lower animals. Again, the muscles of the fingers and toes may exhibit anomalous attachments which are characteristic of lower animals.

In a brief account it is impossible to mention more than a very few such variations, and for more details the works of Testut (1884) and Le Double (1897) should be consulted.

SUPERFICIAL FASCIA

All over the body the dermis of the skin merges into a layer of areolar tissue which in most regions allows the skin to move more or less freely over the underlying structures. Except in the eyelids, the nose and external ear, the penis, and the scrotum, this superficial fascia contains a variable quantity of fat, and, from the second year of life onwards, this is more plentiful in females than in males. At puberty the female body stores additional fat in the secondary sexual distribution, chiefly in the breasts and buttocks, but also in the lower abdominal wall, the outer sides of the thighs, and the backs of the arms, as well as in less clearly marked deposits elsewhere. The difference between male and female superficial fascia is thus exaggerated, and in the adult female the subcutaneous fat is almost twice as heavy as in a male of the same total weight. As age advances, more fat may be deposited in the superficial fascia of both sexes.

The deeper part of the superficial fascia is often more fibrous than the rest, and in the lower region of the abdominal wall this stratum condenses into a membranous layer containing elastic fibres, which may play a part in supporting the abdominal viscera against the pull of gravity [p. 360]. In places such as the sole of the foot and the palm of the hand, the deeper layers of the superficial fascia are tightly fused with the deep fascia, but elsewhere the connexion is loose.

The superficial fascia conducts the blood vessels and nerves which supply the skin and is an important factor in preventing the loss of body heat. In the face and neck it contains numerous somatic muscles [p. 280], and in the scrotum the visceral dartos muscle [p. 360], which corrugates the scrotal skin and helps to support the testes, as well as altering the surface area of the scrotum to control heat loss from it.

DEEP FASCIA

Deep to the superficial fascia there is a stratum of much denser fibrous tissue known as the **investing layer** of the deep fascia. The thickness of this relatively inelastic material varies considerably from place to place; in the palm and sole it is very thick and strong, and has a protective function. In con-

trast is the thin and delicate deep fascia on the medial side of the thigh. As the investing layer passes over superficial bony prominences it often becomes attached to them, and from its deep surface there arise sheets of similar material which penetrate between the muscles, forming **intermuscular septa** to which the muscles are often attached [FIG. 336]. These septa are themselves directly or indirectly attached to the underlying bone, and the whole arrangement serves to maintain the shape of the body and to exert a slight compression force on the contents of the **osteofascial compartments** so formed.

The existence of such compartments has an important effect on the return of blood to the heart. As the muscles contract their bellies widen, and because the walls of the compartments are inelastic the veins within them are compressed. The venous valves ensure that the blood can only be squeezed in one direction, towards the heart.

Similar sheets of deep fascia may enclose glands and other viscera, and they serve to conduct to them, and to the muscles, their innervation and vascular supply. Sometimes the fascia covering a muscle is virtually indistinguishable from the epimysium, but often there is a clear separation, so that the muscle is able to move freely against the partition formed by the fascia. In the living body the septa of fascia are very smooth and slippery, so facilitating movement of this kind.

Where the deep fascia gives attachment to muscles it becomes thickened and aponeurotic, and the distinction between such thickened tracts of fascia and true aponeuroses is often impossible to make. A good example is the iliotibial tract of the fascia lata [p. 392].

The deep fascia is not easily penetrated by fluid, and for this reason infection tends to spread from one part of the body to another along 'fascial planes'. Some of these planes are rather artificial products of the process of dissection in dead bodies, and the detailed description which was formerly given of the topography and attachments of the deep fascia can now be abbreviated without serious loss. Connective tissue pervades the whole body and there is no agreed definition of how

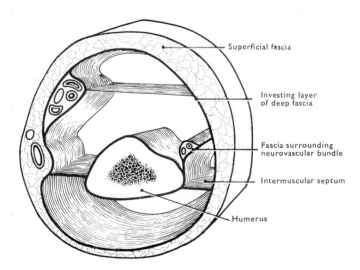

FIG. 336 Schematic cross-section of an arm from which the muscles have been removed, showing the arrangement of superficial and deep fascia.

dense it must be in a given site before it can be regarded as forming a fascial plane. For this reason descriptions of the arrangement of the fascial planes differ from book to book. Similarly the '**fascial spaces**' in which fluid may accumulate, are merely regions of relatively loose connective tissue. Nevertheless, a knowledge of the main fascial 'planes and spaces' is of clinical importance (Gallaudet, 1931; Singer, 1935).

In certain places the deep fascia becomes strengthened to form retention bands or **retinacula,** which serve to strap down the underlying tendons to the bone and so to prevent them from 'bow-stringing' away from the joint on which they operate when their muscle bellies contract. Such retinacula are found in their most characteristic form at the wrist and the ankle [pp. 338, 394]. In the fingers and toes the flexor tendons are confined within a tunnel formed by the attachment to the bones of a dense **fibrous sheath** derived from the deep fascia [p. 339]. Elsewhere specialized **pulleys** may be formed, as for example the trochlea which changes the direction of pull of the superior oblique muscle of the eye [p. 285] and the tunnel

of fascia which holds down the digastric muscle [p. 276] to the hyoid bone.

When a tendon crosses a bone it may be separated from it by a closed 'bag' of connective tissue containing a few drops of lubricant fluid and known as a **bursa.** The fluid which a bursa contains is similar to synovial fluid, and indeed many bursae communicate with the cavity of an adjacent joint. The walls of the bursa are extremely smooth, and ride on each other with a minimum of friction. Similar bursae may separate tendons which are attached very close to each other, and others may separate skin from underlying bone. Where a tendon rubs against the walls of a tight tunnel, as happens at the wrist and ankle, it may be surrounded by an elongated form of bursa called a **synovial sheath.** Such a sheath is wrapped round the tendon, and has a visceral layer adherent to it; the parietal layer of the sheath adheres to the walls of the tunnel. When the tendon moves, the visceral and parietal walls of the sheath slip over each other with a minimum of friction, so facilitating the movement.

MUSCLES OF THE VERTEBRAL COLUMN, NECK, AND HEAD

POSTVERTEBRAL MUSCLES

The muscles of the back can be divided into three groups. Most superficial are those which connect the upper limb with the trunk, comprising the trapezius, the latissimus dorsi, the rhomboids, and the levator scapulae: they belong to and are dealt with in the section on the upper limb [p. 302]. Deep to this group are two small muscles called the serratus posterior superior and the serratus posterior inferior, which are described along with the muscles of the thorax.

The third, and deepest, group consists of muscles which run longitudinally on the vertebral column, and vary considerably in thickness and importance in its different regions. They play a vital part in the maintenance of posture and in the movements of the vertebral column, but so complicated are their attachments that for most purposes only the broad outlines of their origins and insertions are necessary.

The superficial members of this group are arranged in more or less separate longitudinal columns, the fibres of which travel for a considerable distance between origin and insertion. In the deeper layers the fibres are shorter, and the deepest muscles stretch only between one vertebra and its immediate neighbours.

Erector Spinae

This large, complicated, and very powerful muscle takes origin by a strong tendon which is attached along a U-shaped line circumscribing the origin of the multifidus muscle [Figs. 337, 339]. The medial limb of the U springs from the lower two thoracic spines, all the lumbar spines, and the median sacral crest. The lateral limb extends upwards along the lateral sacral crest to the posterior superior iliac spine and the posterior part of the iliac crest. Deep to this lateral limb, the erector spinae has a fleshy attachment to the iliac tuberosity

and the inner lip of the iliac crest. The tendinous fibres blend with the dorsal sacro-iliac, the sacrotuberous, and the sacro-coccygeal ligaments, as well as with the origin of the gluteus maximus.

The muscle has a relatively narrow sacral attachment, but the muscle fibres arising from the tendon swell out to form a broad belly which is prominent in the lumbar region and has a well marked lateral margin which can be detected in the living subject. As this vertically running mass approaches the twelfth rib, it divides into three parallel columns. The most lateral of these is the **iliocostalis,** the intermediate one is the **longissimus,** and the most medial column is the **spinalis.** All three columns are in turn divided into three relays of fibres, each relay arising just as the lower fibres are inserted [Fig. 338].

The **iliocostalis lumborum** inserts by six slips into the lower six ribs near their angles. Medial to the insertion of each of these slips arises the **iliocostalis thoracis,** which is inserted by similar slips into the upper six ribs. Medial to these in turn arises the **iliocostalis cervicis,** a narrow band which inserts into the posterior tubercles of the transverse processes of the lower cervical vertebrae.

The intermediate column, as its name suggests, is the longest (and also the thickest) element of the erector spinae. The **longissimus thoracis** is inserted by two sets of slips, lateral and medial, to the ribs, and the transverse processes of all the thoracic vertebrae and the accessory processes of the upper lumbar vertebrae. The **longissimus cervicis** originates from the transverse processes of the upper six thoracic vertebrae, medial to the insertions of the longissimus thoracis, and inserts into the posterior tubercles of the transverse processes of all the cervical vertebrae except the first and the seventh. In the neck it lies deep to the iliocostalis cervicis. The **longissimus capitis,** the only portion of the erector spinae to reach

Rectus capitis posterior minor

Ligamentum nuchae

Rectus capitis posterior major

Semispinalis capitis

Obliquus capitis superior

Longissimus capitis

Obliquus capitis inferior

Splenius capitis et cervicis

Splenius capitis

Levator scapulae

Sternocleidomastoid

Iliocostalis cervicis

Semispinalis cervicis

Longissimus cervicis

Longissimus cervicis

Iliocostalis thoracis

Semispinalis thoracis

Spinalis thoracis

Levatores costarum

Longissimus thoracis

Iliocostalis lumborum

Quadratus lumborum

Erector spinae

Multifidus

FIG. 337   Deep muscles of the back.

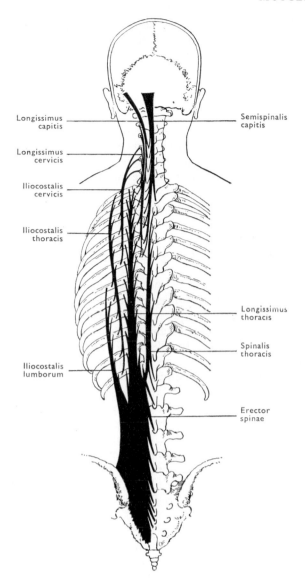

Longissimus capitis

Longissimus cervicis

Iliocostalis cervicis

Iliocostalis thoracis

Iliocostalis lumborum

Semispinalis capitis

Longissimus thoracis

Spinalis thoracis

Erector spinae

FIG. 338   Schematic representation of the parts of the left erector spinae muscle.

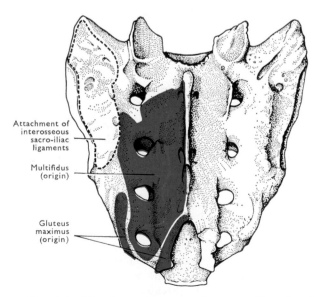

Attachment of interosseous sacro-iliac ligaments

Multifidus (origin)

Gluteus maximus (origin)

FIG. 339   Muscle attachments to the sacrum (dorsal surface).

the skull, takes origin from the transverse processes of the upper thoracic vertebrae, in common with longissimus cervicis, and also from the articular processes of the lower four cervical vertebrae. Its narrow tendon runs between the splenius capitis and the semispinalis capitis muscles, and is inserted into the posterior aspect of the mastoid process, deep to splenius capitis.

The spinalis column is relatively insignificant. The **spinalis thoracis**, which is its only well demarcated component, runs from the vertebral spines in the lower thoracic and upper lumbar regions to the upper thoracic spines. The **spinalis cervicis** is not often present as a distinct entity, and the **spinalis capitis** is usually fused with the most medial part of the semispinalis capitis.

Nerve Supply. The components of the erector spinae are all supplied by the dorsal rami of the spinal nerves according to their situation.

Actions. The main bulk of the erector spinae crosses the secondary lumbar curvature of the spine, and is responsible (particularly the longissimus thoracis) for maintaining this curvature in the erect and sitting positions. It is concerned in extension and rotation of the vertebral column (Morris, Benner, and Lucas, 1962), and comes into play in lateral flexion, but only if forward bending is allowed (Pauly, 1966). It is a most important controlling agent during flexion of the column, which is largely produced by active relaxation of the erector spinae and the other postvertebral muscles. During walking, the erector spinae on both sides contracts the vertebral column to steady on the pelvis. This action is readily seen and felt.

The longissimus capitis is able to extend, and laterally flex the skull on the neck, and rotate the face to the same side.

Splenius Capitis

The splenius capitis arises from the lower part of the ligamentum nuchae, and from the spines of the seventh cervical and the upper three or four thoracic vertebrae. It appears in the floor of the posterior triangle [p. 278] before inserting into the posterior aspect of the mastoid process and the lateral part of the superior nuchal line [FIG. 340], deep to the attachment of the sternocleidomastoid. The splenius capitis is covered posteriorly by the trapezius, the rhomboideus major and minor, and the serratus posterior superior. In turn, it lies superficial to the cervical portions of the erector spinae, and posterior to the semispinalis capitis.

Nerve Supply. Lateral branches of the dorsal rami of the cervical nerves.

Actions. When both muscles act together they extend the head and the neck; when one acts individually, it extends the head and neck, laterally flexes the neck, and turns the face to the same side.

Splenius Cervicis

This muscle arises in common with the splenius capitis, but rather lower down, from the spines of the third to the sixth thoracic vertebrae. It passes upwards to be inserted into the posterior tubercles of the transverse processes of the upper three or four cervical vertebrae.

Nerve Supply. Lateral branches of the dorsal rami of the lower cervical nerves.

Actions. With its fellow it helps to extend the neck; acting on its own it laterally flexes and slightly rotates the cervical portion of the vertebral column to the same side.

Transversospinalis

Deep to the erector spinae complex lies a group of muscles loosely classed together as the transversospinalis system. This name is given them because their fibres in general pass upwards and medially from the transverse processes towards the vertebral spines (contrast the fibres of the erector spinae). Like the erector spinae, the transversospinalis consists of three components. These, however, do not lie side by side, but successively deeper from the surface. The most superficial is the **semispinalis system,** which extends from the loin to the skull, and is described as three separate muscles.

The **semispinalis thoracis** arises from the transverse processes of the lower thoracic vertebrae, and is inserted into the spines of the upper thoracic and the lower cervical vertebrae. The **semispinalis cervicis,** a larger muscle, arises from the transverse processes of the upper thoracic and the articular processes of the lower cervical vertebrae, and is inserted into the spines of the cervical vertebrae. The largest of the three is the **semispinalis capitis,** which lies under cover of the splenius capitis, taking origin from the tips of the transverse processes of the upper six thoracic and the articular processes of the lower four cervical vertebrae. The most medial part of the muscle is often separated from the rest and is sometimes referred to as a separate muscle under the name of **spinalis capitis** [p. 271]; it may be attached to some of the lower cervical spines. The semispinalis capitis broadens as it passes upwards towards the skull, and is inserted into the medial impression between the superior and inferior nuchal lines of the occipital bone [FIG. 340].

The semispinalis capitis forms the bulge on either side of the nuchal furrow in the midline of the neck, and it covers over the semispinalis cervicis and most of the suboccipital triangle.

Usually its vertically running fibres may be seen at the upper angle of the posterior triangle of the neck, appearing from under cover of the splenius capitis [FIG. 344].

The second component of the transversospinalis group is the **multifidus,** which lies in the furrow between the spines of the vertebrae and their transverse processes. Its deeper fasciculi extend only from one vertebra to the next, though the more superficial ones may cross several vertebrae. The muscle arises from the dorsal surface of the sacrum [FIG. 339], under cover of the tendon of the erector spinae, from the posterior sacro-iliac ligaments, and from the mamillary processes of the lumbar vertebrae, the transverse processes of the thoracic vertebrae, and the articular processes of the lower cervical vertebrae. It is inserted into the spines of all the vertebrae from the fifth lumbar up to the axis. Throughout its extent the muscle lies deep to the semispinalis system and the erector spinae.

The third component of the transversospinalis group, the **rotatores,** may be said to represent the deepest layer of the multifidus muscle, with which their fibres are often continuous. They are only developed to any extent in the thoracic region, but some small variable bundles with similar attachments may be found in the cervical and lumbar regions. Each rotator muscle arises from the transverse process of one vertebra and is inserted into the lamina of the vertebra directly above it.

Nerve Supply. The transversospinalis system is supplied by the dorsal rami of the spinal nerves.

Actions. The thoracic and cervical portions of the semispinalis muscle extend the thoracic and cervical parts of the vertebral column. The semispinalis capitis is the most powerful extensor of the head on the neck, and is a strong postural muscle preventing flexion at the atlanto-occipital joint [p. 280].

The multifidus is concerned in extension, lateral flexion, and rotation of the vertebral column, but its main function is probably to act as a series of extensile ligaments guarding the movements of the vertebral column produced by the more powerful and superficial prime movers. As their name implies, the rotatores may be concerned in rotatory movements of the vertebral column, but functionally they are probably extensile ligaments.

Interspinales and Intertransversarii

The **interspinales** are short and insignificant bands extending from one spinous process to the next, lying on either side of the interspinous ligament. They are best developed in the cervical and lumbar regions, and may be absent in the thoracic part of the column. The **intertransverse** muscles are equally small slips extending between the transverse processes in the cervical and lumbar regions. In the cervical portion of the column

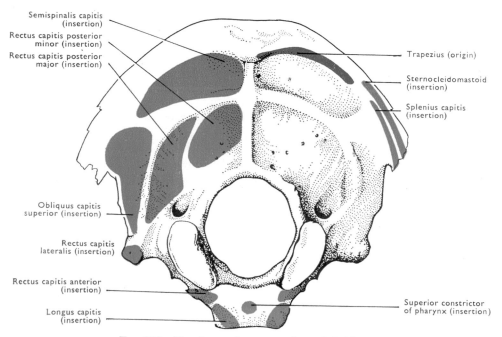

Semispinalis capitis (insertion)
Rectus capitis posterior minor (insertion)
Rectus capitis posterior major (insertion)
Obliquus capitis superior (insertion)
Rectus capitis lateralis (insertion)
Rectus capitis anterior (insertion)
Longus capitis (insertion)
Trapezius (origin)
Sternocleidomastoid (insertion)
Splenius capitis (insertion)
Superior constrictor of pharynx (insertion)

FIG. 340   Muscle attachments to the occipital bone.

they lie both anterior and posterior to the emerging ventral rami of the spinal nerves, but in the lumbar region they are wholly posterior to the ventral rami, divided into a medial and a lateral component.

Nerve Supply. The interspinales are supplied by the dorsal rami of the spinal nerves, as are the medial portions of the lumbar intertransversarii. The lateral portions of the lumbar and all the cervical intertransversarii are supplied by the ventral rami of the spinal nerves.

Actions. They are probably extensile ligaments.

Suboccipital Group

The last group of postvertebral muscles lies deep in the neck [FIGS. 340, 341], anterior to the semispinalis capitis, longissimus capitis, and splenius capitis. They enclose a triangular space known as the **suboccipital triangle,** in which lie the posterior arch of the atlas, the posterior atlanto-occipital membrane, the third part of the vertebral artery, the dorsal ramus of the first cervical nerve, and the suboccipital plexus of veins.

The **rectus capitis posterior major** arises from the spine of the axis and is inserted into the occipital bone deep to the obliquus capitis superior and semispinalis capitis, below the inferior nuchal line. The **rectus capitis posterior minor** arises from the posterior tubercle of the atlas and is inserted into the occipital bone medial and deep to the insertion of the rectus capitis posterior major.

The **obliquus capitis inferior,** the largest muscle of the group, originates from the spine of the axis and passes upwards, laterally and forwards to be inserted into the posterior aspect of the transverse process of the atlas. Despite its name, none of its fibres reaches the skull. The **obliquus capitis superior**

arises from the transverse process of the atlas at the site of insertion of the inferior oblique muscle, and passes backwards, upwards, and somewhat medially, to be inserted into the occipital bone between the superior and inferior nuchal lines, lateral to the semispinalis capitis.

Nerve Supply. All the suboccipital muscles are supplied by the dorsal ramus of the first cervical nerve.

Actions. Both rectus muscles can extend the head on the neck, and the rectus capitis posterior major can rotate the face towards the same side. The obliquus capitis inferior also rotates the atlas on the axis, turning the face to the same side; because of the length of the lever afforded by the atlas this is quite a strong movement. The obliquus capitis superior can extend the head. Nevertheless, all the suboccipital muscles are probably of more importance as extensile ligaments and in the maintenance of posture.

PREVERTEBRAL MUSCLES

The prevertebral muscles form a small group attached to the bodies and transverse processes of the cervical and upper thoracic regions of the vertebral column.

Longus Colli

This is the largest member of the group. Vertically running fibres pass from the fronts of the bodies of the upper thoracic and lower cervical vertebrae to the fronts of the bodies of the upper cervical vertebrae. Some oblique fibres run from the upper three thoracic bodies to the transverse processes of the fifth and sixth cervical vertebrae, and others from the transverse processes of the third, fourth, and fifth cervical vertebrae to the anterior tubercle of the atlas [FIG. 342].

Nerve Supply. From the ventral rami of the cervical nerves.

Actions. The longus colli bends the neck forwards; its inferior oblique fibres may help to produce lateral flexion.

Longus Capitis

The longus capitis travels from the transverse processes of the middle three or four cervical vertebrae to the basilar part of the occipital bone lateral to the pharyngeal tubercle [FIGS. 340, 342]. It lies in front of the upper oblique fibres of the longus colli.

Nerve Supply. From the ventral rami of the first three or four cervical nerves.

Actions. The longus capitis flexes the head and the upper part of the cervical spine. This movement is commonly produced merely by relaxation of the extensor muscles [p. 280], and active flexion is usually needed only against resistance.

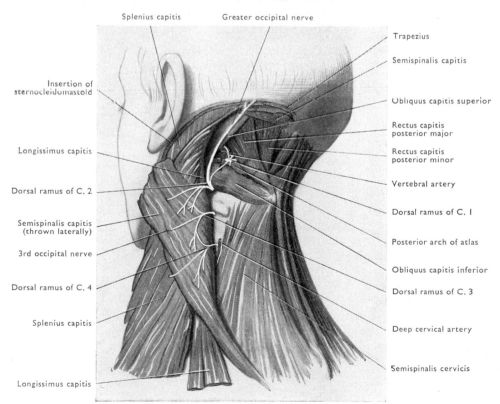

FIG. 341 Suboccipital triangle of the left side.

Splenius capitis

Greater occipital nerve

Insertion of sternocleidomastoid

Longissimus capitis

Dorsal ramus of C. 2

Semispinalis capitis (thrown laterally)

3rd occipital nerve

Dorsal ramus of C. 4

Splenius capitis

Longissimus capitis

Trapezius

Semispinalis capitis

Obliquus capitis superior

Rectus capitis posterior major

Rectus capitis posterior minor

Vertebral artery

Dorsal ramus of C. 1

Posterior arch of atlas

Obliquus capitis inferior

Dorsal ramus of C. 3

Deep cervical artery

Semispinalis cervicis

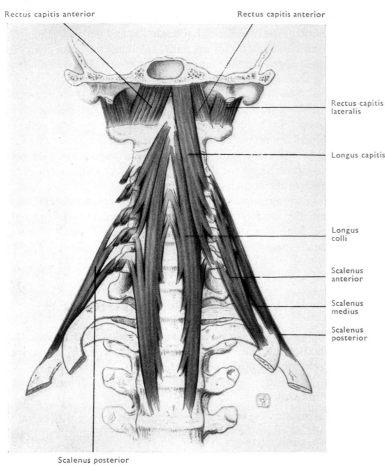

Fig. 342   Prevertebral and scalene muscles.

Rectus Capitis Anterior

This short strap muscle [Figs. 340, 342] passes from the anterior surface of the lateral mass of the atlas to the basilar part of the occipital bone between the longus capitis and the occipital condyle. It corresponds to an anterior intertransverse muscle, but is considerably larger.

Nerve Supply. From the loop between the ventral rami of the first and second cervical nerves.

Actions. It can flex the head on the neck, but probably serves mainly to hold the articular surfaces of the atlanto-occipital joint in close apposition during movements.

Rectus Capitis Lateralis

Another strap muscle, in series with the posterior intertransverse muscles, the rectus capitis lateralis passes from the transverse process of the atlas to the inferior surface of the jugular process of the occipital bone [Figs. 340, 342].

Nerve Supply. From the loop between the ventral rami of the first and second cervical nerves.

Actions. It can bend the head to the same side, but its stabilizing action on the atlanto-occipital joint is probably more important.

LATERAL MUSCLES OF NECK

Scalene Group

Three muscles arise from the transverse processes of the cervical vertebrae and pass downwards to the ribs.

The **scalenus anterior,** an important landmark in the neck, arises from the anterior tubercles of the transverse processes of the third to the sixth cervical vertebrae and runs downwards and laterally behind the prevertebral fascia to be inserted into the scalene tubercle on the first rib [Figs. 342, 343, 412].

The **scalenus medius** [Figs. 342, 343] is the largest of the scalenes, and usually arises from the posterior tubercles of the transverse processes of all the cervical vertebrae. Its fibres descend in the floor of the posterior triangle [p. 278], behind the brachial plexus, to insert into a rough impression on the first rib behind the groove for the subclavian artery.

The **scalenus posterior,** a relatively small muscle, arises behind the scalenus medius from the posterior tubercles of the fourth, fifth, and sixth cervical transverse processes, and is inserted into the outer side of the second rib posterior to the origin of serratus anterior [Fig. 343]. It is often difficult to separate the scalenus posterior from the scalenus medius, which completely hides it, except in the lowest part of the floor of the posterior triangle.

Nerve Supply. By branches from the ventral rami of the cervical nerves : scalenus anterior from the fourth to the sixth, scalenus medius from the lower five or six, and scalenus posterior from the last three.

Actions. Acting from their origins on their insertions, the scalene muscles steady the first two ribs during respiration [p. 357], and may assist in elevating them. Acting from their insertions, they help to produce lateral flexion of the cervical part of the vertebral column.

Sternocleidomastoid

The sternocleidomastoid [Fig. 344] is a long strap muscle with two heads of origin. The narrow tendinous sternal head

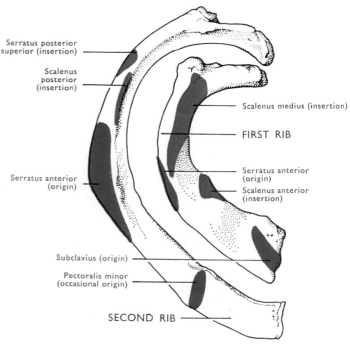

Fig. 343   Muscle attachments to upper surface of first rib and outer surface of second rib (right side).

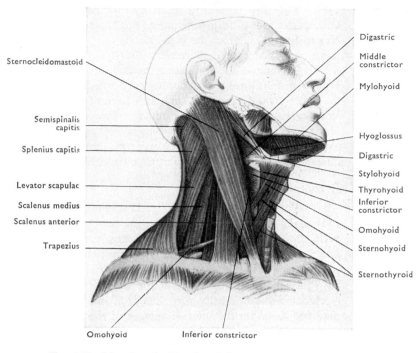

FIG. 344 Muscles of side of neck (anterior and posterior triangles).

HYOID MUSCLES

The hyoid bone is tethered by muscles [FIGS. 145, 344] to the mandible and skull above and to the sternum, thyroid cartilage, and scapula below. These muscles are primarily concerned with steadying or moving the hyoid bone, and hence the larynx and tongue, both of which are attached to it. They are usually described in two groups. The **suprahyoid group** includes the mylohyoid, geniohyoid, stylohyoid, and digastric muscles, and the **infrahyoid group** comprises the sternohyoid, sternothyroid, omohyoid, and thyrohyoid. Members of both groups of hyoid muscles co-operate to depress the mandible when the mouth has to be opened against resistance.

Mylohyoid

The mylohyoid arises from the mylohyoid line on the inner surface of the body of the mandible [FIG. 345], and its fibres run downwards and medially, the posterior ones to be inserted into the body of the hyoid bone and the remainder into a median raphe which extends from the hyoid bone to the chin. The two mylohyoid muscles therefore form a sling or diaphragm which supports the floor of the mouth [FIG. 358]. The geniohyoid lies superior to this sling, and the anterior belly of the digastric muscle lies inferior to it.

Nerve Supply. From the mylohyoid branch of the inferior alveolar nerve.

Actions. The mylohyoids raise the floor of the mouth in swallowing; they elevate and fix the hyoid bone, and help to press the tongue upwards and backwards against the roof of the mouth (Whillis, 1946). They can depress the mandible if the hyoid bone is fixed by the infrahyoid group of muscles,

arises from the anterior surface of the manubrium sterni, and the broad clavicular head from the upper surface of the clavicle in its medial third: there may be an interval between the two. The clavicular head passes deep to the sternal head, and the united muscle is inserted by a short tendon into the outer surface of the mastoid process and into the lateral third of the superior nuchal line of the occipital bone [FIG. 340].

The muscle is the major landmark in the neck. The spinal part of the accessory nerve enters the deep aspect of its upper third, and runs within its substance to emerge about half way down its lateral border.

Nerve Supply. By motor fibres from the spinal part of the accessory nerve, and by sensory fibres from the ventral rami of the second and third cervical nerves through the cervical plexus: these fibres may join the accessory.

Actions. It tilts the head towards the same side and rotates it so that the face is turned towards the opposite side and also upwards. The anterior fibres flex the head on the neck at the atlanto-occipital joint, but the posterior fibres may extend this joint (Last, 1954). The two sternocleidomastoids acting together flex the neck, as when the head is raised from the pillow. If the head is fixed on the neck by other muscles the sternocleidomastoids can act as accessory muscles of inspiration, since they raise the manubrium sterni and consequently the ribs [p. 357].

The muscle is sometimes injured at birth, and may be wholly or partly replaced by fibrous scar tissue, which contracts to produce the condition of torticollis (wry neck). Temporary wry neck may result from irritation of the muscle or its motor nerve, and spasms may occur which cause jerking movements of the head; these are often associated with a similar spasm of the clavicular portion of the trapezius muscle [p. 302].

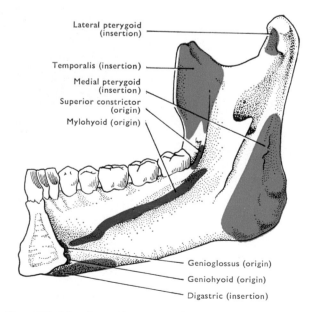

FIG. 345 Muscle attachments to deep surface of mandible.

but this action is not called into play unless there is some resistance to opening the mouth.

Geniohyoid

The geniohyoid muscles arise from the lower part of the mental spine of the mandible [FIG. 345] and run backwards between the mylohyoid and the genioglossus muscles to be inserted into the body of the hyoid bone [FIG. 145]. The muscles of the two sides are often fused.

Nerve Supply. From fibres of the first cervical nerve conveyed by the hypoglossal nerve.

Actions. They pull the hyoid bone forwards and slightly upwards, so shortening the floor of the mouth, and widening the pharynx for the reception of food. If the hyoid bone is fixed, they can help to retract the mandible and possibly to depress it.

Stylohyoid

A small muscle, which arises from the posterior border of the styloid process near its root, the stylohyoid runs down and forwards along the posterior belly of the digastric and splits into two slips enclosing the tendon of this muscle [FIGS. 344,

346]. The slips reunite to insert into the body of the hyoid bone at the root of its greater horn [FIG. 145].

Nerve Supply. From the facial nerve.

Actions. It elevates and retracts the hyoid bone, carrying with it the tongue, and elongating the floor of the mouth. The **stylohyoid ligament** is a fibrous cord which runs from the tip of the styloid process to the lesser horn of the hyoid bone, closely related to the stylohyoid muscle. It represents a portion of the cartilage of the second pharyngeal arch, and is sometimes partly ossified. It usually gives origin to some of the highest fibres of the middle constrictor of thep harynx [FIGS. 346, 360].

Digastric

The posterior belly of the digastric [FIGS. 344, 346, 347] is attached to the mastoid notch of the temporal bone. This belly, closely related to the stylohyoid muscle, runs downwards and forwards to an intermediate tendon which passes through the insertion of the stylohyoid and at this point is strapped down by a fascial sling to the body of the hyoid bone. From this tendon the anterior belly runs forwards and upwards to the digastric fossa, on the inferior surface of the mandible, close to the symphysis.

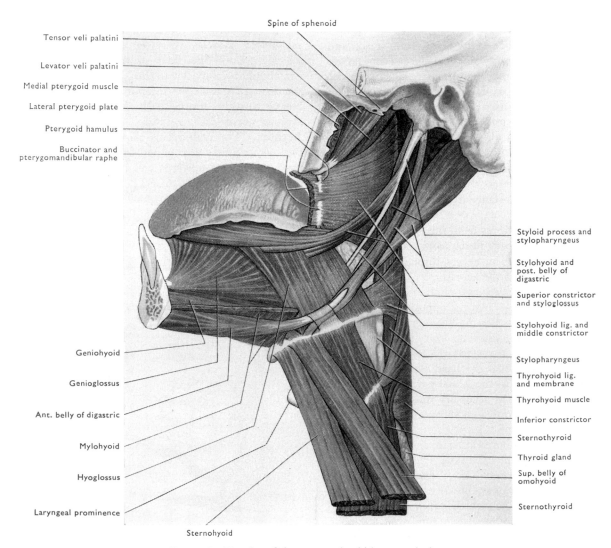

FIG. 346   Muscles of the tongue, hyoid bone, and pharynx.

Nerve Supply. The posterior belly is supplied by the facial nerve, and the anterior belly by the mylohyoid branch of the inferior alveolar nerve.

Actions. It raises the hyoid bone, or, if the hyoid bone is fixed, can depress and retract the mandible (Last, 1954).

Sternohyoid

The sternohyoid [FIGS. 344, 346] takes origin from the posterior surface of the manubrium sterni, the back of the sternoclavicular joint, and the medial end of the clavicle. It is a strap muscle, which runs upwards, inclining towards its fellow of the opposite side, to be inserted into the medial part of the lower border of the body of the hyoid bone [FIG. 145].

Most of the muscle is superficial, but inferiorly it is covered by the sternocleidomastoid and elsewhere by the investing layer of the deep fascia.

Nerve Supply. From the ventral rami of the first three cervical nerves, by fibres conveyed through the ansa cervicalis.

Actions. The sternohyoid depresses the hyoid bone and helps to fix it when other muscles are acting from it. If the hyoid bone is held by the suprahyoid muscles, the sternohyoid may act as an accessory muscle of inspiration [p. 357].

Sternothyroid and Thyrohyoid

The sternothyroid [FIGS. 344, 346], which lies deep to the sternohyoid, arises from the back of the manubrium sterni

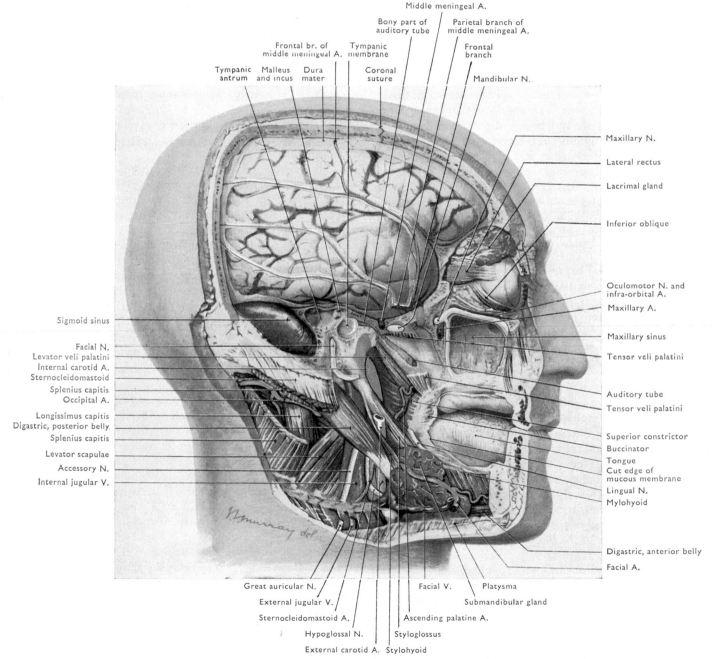

FIG. 347 Dissection of head showing the relations of muscles at its junction with the neck. For enlargement of part of this illustration, see Fig. 361.

below the origin of the sternohyoid, and also from the first costal cartilage. It is broader than the sternohyoid, and instead of converging on its fellow as it rises, it swings laterally in front of the trachea and the thyroid gland to be inserted into the oblique line on the outer surface of the thyroid cartilage. It is deep to the sternocleidomastoid and the sternohyoid, and overlapped in its upper part by the omohyoid.

The thyrohyoid is a short strap muscle which arises from the oblique line of the thyroid cartilage [FIG. 346], continuing the line of the fibres of the sternothyroid. It runs upwards to be inserted into the lower border of the body and greater horn of the hyoid bone [FIG. 145].

Nerve Supply. The sternothyroid is supplied from the ansa cervicalis : the thyrohyoid is supplied from the ventral ramus of the first cervical nerve by fibres brought to the muscle through the hypoglossal nerve.

Actions. The sternothyroid and thyrohyoid muscles can act as a unit which has actions very similar to those of the sternohyoid muscle, except that the thyroid cartilage is pulled down as well as the hyoid bone. This unit can, in conjunction with the mylohyoid and digastric muscles, depress the mandible.

Acting by itself, the thyrohyoid approximates the thyroid cartilage to the hyoid bone. This pulls the larynx up under the base of the tongue, and brings the arytenoid cartilages under cover of the epiglottis ; it is therefore largely responsible for closing the laryngeal orifice and is of great importance in preventing food from entering the larynx in swallowing.

Acting on its own, the sternothyroid pulls the thyroid cartilage away from the hyoid bone, so opening the laryngeal orifice. In forced inspiration the muscle contracts, perhaps to help in raising the sternum, but mainly to keep the laryngeal orifice fully open.

Levator Glandulae Thyroideae

An occasional unpaired slip of muscle passes between the hyoid bone and the isthmus or the pyramidal lobe of the thyroid gland. It presumably helps to elevate the gland, but its frequent absence produces no disability, and the real elevators of the gland, because of its attachment by fascia to the thyroid cartilage, are the suprahyoid and the thyrohyoid muscles.

Omohyoid

This muscle has two bellies ; the inferior one arises from the upper border of the scapula and the superior transverse scapular ligament [FIG. 380] and passes forwards and slightly upwards across the scalene muscles to end in an intermediate tendon under cover of the sternocleidomastoid [FIG. 344]. The tendon is tied down to the clavicle and first rib by a sling of the cervical fascia [p. 301], and from its anterior end the superior belly of the muscle runs upwards, crossing the carotid sheath at the level of the arch of the cricoid cartilage, to be inserted into the lower border of the body of the hyoid bone lateral to the sternohyoid [FIG. 145].

Nerve Supply. Both bellies are supplied from the ansa cervicalis.

Actions. Like the other infrahyoid muscles, the omohyoid pulls down the hyoid bone.

REGIONS OF THE NECK

The sternocleidomastoid divides the side of the neck, for descriptive purposes, into two triangles. The **posterior** triangle [FIG. 344] is bounded by the middle third of the clavicle, the posterior border of the sternocleidomastoid, and the anterior border of the trapezius. Its apex is formed by the occipital bone, and in its floor lie the scalene muscles, the levator scapulae, the splenius capitis, and, usually, a portion of the semispinalis capitis. Across the lower part of the triangle runs the inferior belly of the omohyoid muscle, which divides the triangle into two, a larger **occipital triangle** and a smaller **supraclavicular triangle.**

The **anterior triangle,** bounded by the midline of the neck, the lower border of the mandible, and the anterior border of the sternocleidomastoid [FIG. 344], has its apex at the manubrium sterni, and is subdivided into four smaller triangles. The small **submental triangle** lies between the anterior bellies of the digastric muscles and the body of the hyoid bone, and the **digastric triangle** of each side lies between the mandible and the anterior and posterior bellies of the digastric muscle. The **muscular triangle** is the part of the anterior triangle anterior to the superior belly of the omohyoid muscle, and the remainder of the anterior triangle, between the posterior belly of the digastric muscle above and the superior belly of the omohyoid muscle below, constitutes the **carotid triangle.**

MOVEMENTS OF THE VERTEBRAL COLUMN

The muscles of the vertebral column are involved in the maintenance of the erect posture [p. 388]. The most important parts of the musculature in this respect are those which bridge the secondary curvatures of the vertebral column, for these have less intrinsic stability. In the erect 'stand-at-ease' position no single component of the postvertebral group is continuously active, but the constant small disturbances of posture produced by the normal swaying of the body result in the appropriate muscles contracting to restore balance. Movement of a limb, because it alters the centre of gravity of the body, results in greater activity in the vertebral muscles, and during walking the muscles of the back serve to steady the body above the supporting limb [p. 391]. They are also in action as fixators and synergists in every movement of the upper limb.

Because of the relatively recent acquisition of the upright posture in the course of evolution, the vertebral muscles have a task which imposes on them a considerable strain, and this in part explains the common occurrence of aches and pains referred to the musculature of the lower part of the back and to the neck—the two most mobile parts of the column and the ones in which muscular control is most necessary. Following injury, the muscles of the back readily pass into reflex contraction to prevent painful movements, and the difficulty experienced in the performance of a large variety of movements under such circumstances is indicative of the degree to which these muscles enter into all everyday actions.

In flexion and extension of the vertebral column, each vertebra may be thought of as rocking backwards and forwards on the nucleus pulposus of the intervertebral disc. The spines which protrude backwards from the vertebrae therefore afford a considerable mechanical advantage to the muscles attached to them, and this factor, together with the great bulk of the postvertebral muscles as compared to the prevertebral muscles, indicates the importance of the maintenance of, and the ability to regain, the extended upright position. In active extension of the column, all the muscles posterior to the axis of movement can be involved, but once the normal erect position has been attained, further extension is largely controlled by the active relaxation of the abdominal muscles, the prevertebral muscles, and the sternocleidomastoid. Similarly, the most important factor in flexion of the column from the upright position is active relaxation of the postvertebral group. If, however, flexion is performed against resistance, or against gravity, as in getting up from the supine position, the abdominal, prevertebral, and sternocleidomastoid muscles are called into action as prime movers. The abdominal muscles have a mechanical advantage even greater than that of the postvertebral muscles, since they operate on the ribs, at some considerable distance from the axis of movement. Their action in flexion of the spine involves fixation of the thorax by the muscles which elevate the ribs [p. 357].

During flexion of the trunk, the erector spinae pays out excentrically until the position of full flexion has been reached. In this position, no muscular activity in the back can be detected, and the strain is taken entirely by the ligaments of the vertebral column (Floyd and Silver, 1952; Pauly, 1966). Much of the movement of bending forwards takes place at the hip joint, and a similar excentric contraction of the gluteus maximus and hamstring muscles occurs [p 365].

Extending the trunk from the position of full flexion to the anatomical position is more complicated. Up to about the half-way stage of the movement, the lumbar spine is held compressed on to the sacrum by contraction of the abdominal and back muscles. It does not move very much during this phase, and the main straightening takes place at the hip joints, where the pelvis is rotated backwards on the heads of the femora. After this phase is completed, the abdominal muscles relax, and the erector spinae now pulls the lumbar spine into its normal curvature, so straightening the back. Faults in this mechanism of regaining the erect posture often become apparent at the stage of transition from hip to spinal movement, when the maximum power of the erector spinae is required.

Lateral bending and rotation of the vertebral column are intimately connected [p. 217], and are produced largely by muscles which lie lateral to the vertebral column and obliquely in relation to it, such as the sternocleidomastoid in the neck, the psoas major and the quadratus lumborum in the lumbar region, and the multifidus and rotatores throughout the length of the column. The following table indicates the main muscles which are anatomically in a position to be concerned in the various movements of the vertebral column, but it will be appreciated that there is an extremely complex interaction between them all, and it is not always possible to distinguish prime movers from synergists, or to pick out those whose primary function is that of an extensile ligament.

The table is not intended to be complete, and some muscles have been omitted for the sake of simplicity. Thus, many of the suprahyoid muscles (mylohyoid, stylohyoid, and digastric) may combine with the elevators of the mandible (to fix the jaw) and the infrahyoid group to form a unit capable of flexing the cervical part of the vertebral column.

All the muscles in the table have segmental innervation from the spinal nerves except for sternocleidomastoid and trapezius, whose motor innervation comes from the spinal part of the accessory nerve.

MUSCLES CONCERNED IN MOVEMENTS OF THE INTERVERTEBRAL JOINTS

MOVEMENT	CERVICAL REGION	THORACIC AND/OR LUMBAR REGIONS
Flexion	Sternocleidomastoid Longus colli Longus capitis	Muscles of anterior abdominal wall [p. 345]
Extension	Splenius capitis Splenius cervicis Semispinalis capitis Semispinalis cervicis Iliocostalis cervicis Longissimus capitis Longissimus cervicis Trapezius [p. 302] Interspinales	Erector spinae as a whole Quadratus lumborum Trapezius [p. 302]
Rotation and Lateral Flexion	Sternocleidomastoid Scalene group Splenius capitis Splenius cervicis Longissimus capitis Longissimus cervicis Levator scapulae [p. 305] Longus colli Iliocostalis cervicis Multifidus Intertransversarii	Psoas major [p. 353] Quadratus lumborum [p. 353] Muscles of anterior abdominal wall [p. 345] Multifidus Iliocostalis lumborum Iliocostalis thoracis Intertransversarii Rotatores

MOVEMENTS OF THE HEAD ON THE NECK

The muscles which lie behind the atlanto-occipital joint have a considerable postural responsibility, since the centre of gravity of the skull falls in front of this joint [p. 263]. It also follows that in the upright position flexion of the head on the neck is usually produced by the action of gravity, controlled by the relaxation of the extensor muscles, while extension of the skull is an active movement against gravity. It is therefore not surprising to find that the extensor muscles are much better developed than the flexors.

As in the movements of the vertebral column, the precise role of the muscles acting at the atlanto-occipital and atlanto-axial joints is a matter for conjecture, and the table given is not to be interpreted too rigidly.

Again, some omissions have been made deliberately. The hyoid muscles may combine with the elevators of the mandible to form a unit capable of flexing the head on the neck, and some of the extensors are also capable of a degree of rotation and lateral flexion at the atlanto-axial joint (e.g. rectus capitis posterior major, semispinalis capitis, trapezius).

All the muscles in the table have a segmental innervation from the spinal nerves except for sternocleidomastoid and trapezius, whose motor innervation comes from the spinal part of the accessory nerve.

MUSCLES CONCERNED IN MOVEMENTS OF THE ATLANTO-OCCIPITAL AND ATLANTO-AXIAL JOINTS

MOVEMENT	
Flexion	Longus capitis
	Rectus capitis anterior
	Sternocleidomastoid
	(anterior fibres)
Extension	Semispinalis capitis
	Splenius capitis
	Rectus capitis posterior major
	Rectus capitis posterior minor
	Obliquus capitis superior
	Longissimus capitis
	Trapezius [p. 302]
	Sternocleidomastoid
	(posterior fibres)
Rotation and	Sternocleidomastoid
Lateral Flexion	Obliquus capitis inferior
	Obliquus capitis superior
	Rectus capitis lateralis
	Longissimus capitis
	Splenius capitis

MUSCLES OF THE SCALP AND FACE

These muscles [FIGS. 348, 349] are derived from the mesoderm of the second pharyngeal arch, which in the course of evolution spread in the superficial fascia over the underlying structures of the head and neck. They are all supplied by the facial nerve, the nerve of this arch.

Epicranius (Occipitofrontalis)

The epicranius has two pairs of bellies, frontal and occipital [FIG. 348], which are united by a strong aponeurosis, the **galea aponeurotica,** so named because it forms a sort of helmet for the skull. Each occipital belly arises from the outer part of the superior nuchal line of the occipital bone and the mastoid part of the temporal bone. The larger frontal bellies have no bony attachment; they extend forwards from the galea to the region of the eyebrow, where they become attached to the skin, interlacing with the orbicularis oculi. Unlike the two occipital bellies, the two frontal bellies blend with each other in the midline, and some of the fibres pass down over the nose to become continuous with the procerus muscle [p. 281].

The galea aponeurotica is attached posteriorly, between the two occipital bellies, to the superior nuchal line. Anteriorly it sends a slip for some distance between the two frontal bellies, and laterally it becomes considerably thinner and fuses with the temporal fascia just above the zygomatic arch. It is very firmly bound to the skin by dense superficial fascia containing the blood vessels and nerves, but can move freely over the underlying periosteum of the skull since the areolar tissue between the two is loose.

Nerve Supply. The frontal bellies are supplied by temporal branches, and the occipital bellies by posterior auricular branches of the facial nerve.

Actions. The occipital bellies pull the scalp backwards, and the frontal bellies can move it forwards. On the other hand, the frontal bellies can act from the aponeurosis on the eyebrows and the skin over the root of the nose, which they pull upwards in the expression of surprise, fright, and horror, so producing transverse wrinkling of the forehead.

Extrinsic Muscles of the Auricle

Three small muscles are inserted into the cartilages of the external ear. The **auricularis posterior** arises from the mastoid process; the **auricularis superior,** the largest of the three, comes from the galea aponeurotica and the temporal fascia, and the **auricularis anterior,** the smallest, comes from the temporal fascia [FIG. 348].

Nerve Supply. From the temporal and posterior auricular branches of the facial nerve.

Actions. The muscles are not normally used, but can be trained to move the external ear slightly; this power is greater in some people than in others.

Orbicularis Oculi

The orbicularis oculi is one of the two most complex members of the group of facial muscles and is functionally extremely important. It consists of three parts. The **orbital part** surrounds the orbit and spreads on to the forehead, temple, and cheek [FIG. 348]. Its fibres are reddish in colour, and its fasciculi are thicker than those of the other parts. They take origin from the medial margin of the orbit between the supra-orbital notch and the infra-orbital foramen [FIG. 349], and from the medial palpebral ligament. The fibres arising above this ligament pass round the margin of the orbit in an ellipse and return to the bony attachment below the ligament; there is no lateral attachment to bone.

The **palpebral part** of the muscle lies in the eyelids; it arises from the medial palpebral ligament and the adjacent bone, and runs laterally in the lids, where the fibres eventually interlace to form the **lateral palpebral raphe.** A slip of fine fibres called the **ciliary bundle** runs along the free margin of each lid behind the eyelashes [FIG. 994].

The **lacrimal part** of the orbicularis oculi [FIG. 370] comes from the fascia posterior to the lacrimal sac and from the crest of the lacrimal bone, and passes laterally to be inserted into the tarsal plates of both eyelids. These fibres are closely related to the lacrimal canaliculi.

Nerve Supply. From the temporal and zygomatic branches of the facial nerve.

Actions. The muscle is an important protective mechanism surrounding the eye, and plays a vital part in the disposal of the tears. The orbital part, which can operate either together with the other parts or independently, draws the skin of the forehead and the cheek towards the medial angle of the orbit. By so doing it helps to close the eyelids firmly and 'screws up' the eye. It thus helps to prevent excess light entering the eye, and guards against injury to the eyeball. The contraction of this part of the muscle wrinkles the skin at the lateral margin of the orbit and these wrinkles become permanent and noticeable in old age, particularly in those whose occupation has constantly exposed them to the weather.

The palpebral part of orbicularis oculi closes the lids gently, and comes into action involuntarily in blinking. By drawing the eyelids medially the tears are wiped across the eyeball to the lacrimal puncta, which are held firmly to the conjunctival sac by the contraction of the muscle. Paralysis of the orbicularis oculi causes the punctum of the lower lid to fall away from the eyeball, and tears overflow from the conjunctival sac on to the cheek.

Finally, the lacrimal part of the muscle dilates the lacrimal sac.

Corrugator Supercilii

This small muscle arises from the medial end of the superciliary arch, blending with the orbicularis oculi, and passes upwards and laterally to be inserted into the skin of the eyebrow.

Nerve Supply. From the facial nerve.

Actions. It draws the eyebrow medially and downwards, so producing vertical wrinkles in the forehead; this action is well seen in frowning.

Procerus

The two procerus muscles form a continuous unit, which arises from the lower part of the nasal bones and the upper parts of the lateral nasal cartilages. The fibres insert into the

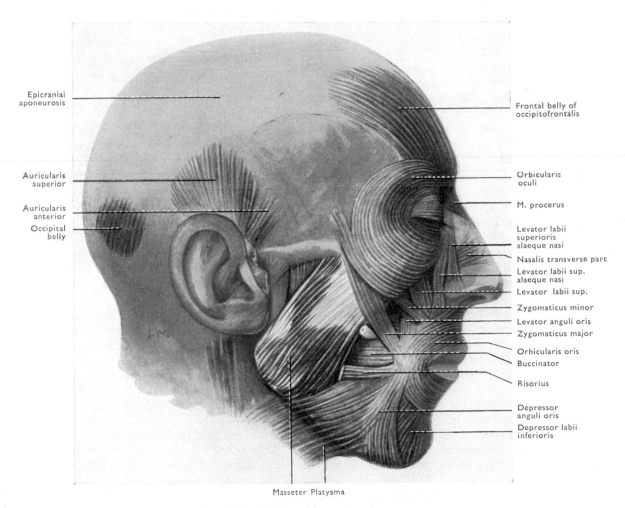

FIG. 348 Muscles of face and scalp.

skin over the lower part of the forehead, interlacing with the frontal belly of the epicranius.

Nerve Supply.   From the facial nerve.

Actions.   The procerus pulls down the medial part of the eyebrow and so produces transverse wrinkles at the root of the nose.

Nasalis

The nasalis consists of a transverse and an alar portion. The transverse portion [FIGS. 348, 349] runs from the upper end of the canine eminence of the maxilla to an aponeurosis on the cartilaginous part of the nose. The alar portion of the muscle [FIG. 349] comes from the maxilla above the lateral incisor tooth, and is inserted into the lateral part of the lower margin of the ala of the nose.

Nerve Supply.   From the facial nerve.

Actions.   The alar part of the muscle draws the ala downwards and laterally, so widening the aperture of the nose : the transverse part has been thought to act as a compressor of the nose, but it is possible that both parts of the muscle are really dilators.

Depressor Septi

This muscle arises with the medial fibres of the alar part of the nasalis, and is inserted into the mobile part of the nasal septum.

Nerve Supply.   From the facial nerve.

Actions.   It draws the septum downwards, so narrowing the nostril.

Orbicularis Oris

The orbicularis oris is a composite sphincter muscle which receives fasciculi from many other muscles. It encircles the mouth, its vertical extent being from the septum of the nose to a point midway between the chin and the free margin of the lower lip [FIG. 348]. The deepest fibres of the muscle are continuous with buccinator [p. 283], the middle fibres of which decussate posterolateral to the angle of the mouth so that the lower fibres of the buccinator run into the upper lip and the upper fibres into the lower lip. Some of these deep fibres are attached to the maxilla and mandible near the lateral incisor teeth. The most superficial fibres of the orbicularis oris are derived from other muscles of facial expression, and some of them decussate like those of the buccinator. The interlacement of the deep and superficial fibres at the angle of the mouth forms a nodule which is easily felt in the living subject and which is known as the **modiolus**. In between the superficial and deep layers the intrinsic fibres of the orbicularis oris sweep in an ellipse round the mouth without being attached to bone or to any other rigid structure.

Nerve Supply.   From the facial nerve.

Actions.   Such a complicated muscle must inevitably have complicated actions. Primarily, the orbicularis oris closes the mouth ; its deep fibres compress the lips against the teeth, and its superficial fibres compress the lips and protrude them. Contraction of the various components which contribute to the orbicularis oris alters the shape of the mouth, and the muscle plays an important part in speech. It is also used during mastication, to prevent food from being forced out of the mouth anteriorly.

Muscles of the Upper Lip

The **levator labii superioris alaeque nasi** arises from the frontal process of the maxilla and runs down to be inserted into the ala of the nose and into the skin and muscle of the upper lip [FIGS. 348, 349]. The **levator labii superioris** takes origin from the maxilla immediately superior to the infraorbital foramen, and is inserted into the orbicularis oris in the upper lip.

The **zygomaticus minor** is a small bundle which comes from the anterior surface of the zygomatic bone to be inserted into the skin and muscle of the upper lip. The **zygomaticus major,** somewhat larger, comes from the zygomatic portion of the zygomatic arch and passes to the angle of the mouth, where it is inserted partly into the skin and partly into the orbicularis oris. The **levator anguli oris** comes from the maxilla inferior to the infraorbital foramen. It passes downwards and laterally to be inserted into the orbicularis oris and the skin at the angle of the mouth ; some of its fibres pass into the orbicularis oris in the lower lip.

Nerve Supply.   All the muscles of the upper lip are supplied by branches of the facial nerve.

Actions.   The two elevators of the upper lip raise and evert the lip ; the levator labii superioris alaeque nasi also acts as a dilator of the nostril by means of its attachment to the ala of the nose. The zygomaticus minor assists in raising the upper lip, and, together with the levator labii superioris and the levator anguli oris, is responsible for producing the nasolabial furrow, which passes from the side of the nose to the upper lip and, like all facial creases, becomes more marked in old age. The zygomaticus major raises the angle of the mouth and pulls it laterally, as in laughing, and in this it is assisted by the levator anguli oris.

Muscles of the Lower Lip

The **depressor labii inferioris** [FIGS. 348, 349] arises from an oblique line on the outer surface of the mandible between the symphysis menti and the mental foramen. Its fibres pass up and medially to the skin of the lower lip and the orbicularis oris, blending with those of its fellow of the opposite side. The **depressor anguli oris** takes origin just inferolateral to the depressor of the lower lip. Its fibres converge on the angle of the mouth to be inserted into the skin and the orbicularis oris ; some of them pass into the upper lip. The muscle is continuous with the platysma and overlaps the depressor labii inferioris.

The **mentalis** is a small muscle which arises from the mandible just below the incisor teeth and passes downwards to be inserted into the skin of the chin. It is the only muscle of the lips which usually has no connexion with the orbicularis oris. The **risorius** is a thin muscle which passes from the fascia covering the parotid gland to the skin at the angle of the mouth : it is often completely fused with the platysma.

Nerve Supply.   All the muscles of the lower lip are supplied by the marginal mandibular branch of the facial nerve.

Actions.   The depressor labii inferioris and the depressor anguli oris pull the mouth downwards and somewhat laterally. The mentalis pulls the skin of the chin upwards, and so causes a protrusion of the lower lip as in pouting. The risorius pulls the angle of the mouth laterally.

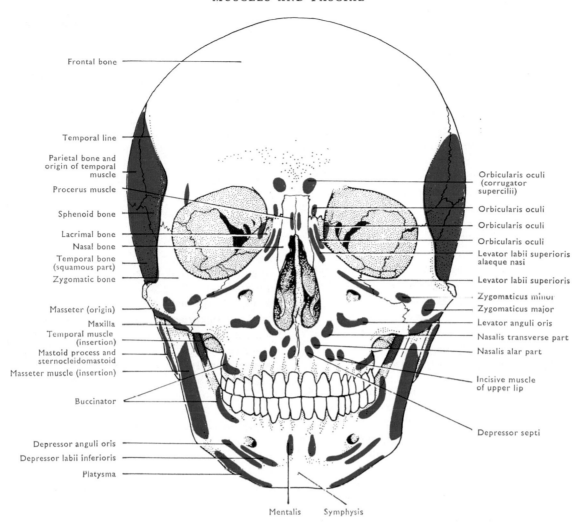

Frontal bone

Temporal line

Parietal bone and
origin of temporal
muscle

Procerus muscle

Sphenoid bone

Lacrimal bone

Nasal bone

Temporal bone
(squamous part)

Zygomatic bone

Masseter (origin)

Maxilla

Temporal muscle
(insertion)

Mastoid process and
sternocleidomastoid

Masseter muscle (insertion)

Buccinator

Depressor anguli oris

Depressor labii inferioris

Platysma

Orbicularis oculi
(corrugator
supercilii)

Orbicularis oculi

Orbicularis oculi

Orbicularis oculi

Levator labii superioris
alaeque nasi

Levator labii superioris

Zygomaticus minor

Zygomaticus major

Levator anguli oris

Nasalis transverse part

Nasalis alar part

Incisive muscle
of upper lip

Depressor septi

Mentalis   Symphysis

FIG. 349   Muscle attachments to the front of the skull.

Platysma

The platysma [PLATE 19A], like the muscles of the lips, lies in the superficial fascia, and it is very variably developed in different people. It arises from the skin and superficial fascia in the upper pectoral and anterior deltoid regions, crosses the clavicle and runs upwards over the front and side of the neck. The anterior fibres of the muscle decussate with those of the opposite side below the chin, while the intermediate fibres are inserted into the lower border of the body of the mandible. The posterior fibres cross the mandible superficially and sweep forwards to blend with the depressor anguli oris and depressor labii inferioris, so becoming inserted into the modiolus and the upper lip [FIGS. 348, 349]. The uppermost fibres are reinforced by the risorius muscle, with which they are continuous.

Nerve Supply. From the cervical branch of the facial nerve.
Actions. The platysma may be seen in action in a runner finishing a race. In such circumstances it is conceivably being used as an accessory muscle of inspiration, helping to draw up the chest wall to the fixed mandible, though its utility in this respect can only be very slight. An alternative theory is that it prevents the soft tissues at the root of the neck from being sucked inwards by the violent inspiratory efforts; in this way

it guards against compression of the great veins returning blood to the heart. The muscle wrinkles the skin of the neck obliquely, and can serve to depress the mandible against resistance; it also depresses the lower lip and the angle of the mouth. If the cervical branch of the facial nerve is cut, the resulting paralysis of the platysma causes the skin to fall away from the neck in slack folds.

Buccinator

The buccinator [FIG. 356] forms the substance of the cheek, and lies in the same plane as the superior constrictor of the pharynx, being covered by an extension forwards of the buccopharyngeal fascia [p. 301]. It takes origin from the outer surfaces of both the maxilla and the mandible opposite the molar teeth [FIGS. 349, 354], and from a fibrous line of union with the superior constrictor of the pharynx, the **pterygomandibular raphe,** which stretches from the pterygoid hamulus to the posterior end of the mylohyoid line.

The fibres of the buccinator proceed forwards to blend with those of the orbicularis oris almost at right angles. The intermediate fibres decussate at the modiolus, but the upper and lower ones pass into the upper and lower lips respectively.

Posteriorly, the buccinator is covered superficially by the masseter muscle [FIG. 348], from which it is separated by the buccal pad of fat. It is pierced by the duct of the parotid gland, and on its inner surface it is covered by the mucous membrane and buccal glands of the vestibule of the mouth.

Nerve Supply. From the facial nerve.

Actions. The buccinator is named from the Latin *buccina* (a trumpet) because if the cheeks are filled with air it is the buccinator which compresses the air through the lips as in whistling or blowing a trumpet. More important, however, is its use during mastication [p. 297], when it forces food escaping into the vestibule of the mouth back into the sphere of action of the teeth and the tongue.

If the lips are protruded by orbicularis oris, the buccinators cause the cheeks to 'cave in', producing the action of sucking. They can be used to pull backwards the orbicularis oris against the teeth, or to retract the angles of the mouth to show the premolar and molar teeth.

FACIAL EXPRESSION AND FACIAL PARALYSIS

The individual actions of the muscles described are combined together to produce emotional expressions. Joy results from the contraction of one set of muscles and grief from the contraction of an opposing set. Determination is accompanied by a 'rigid' expression, and despair is expressed by a general relaxation of the facial musculature. The details of the muscular expression of individual emotions are too complex to discuss here, but a full account may be found in the works of Bell (1847), Darwin (1872), and Huber (1931).

The part of the facial musculature surrounding the eye has the additional important task of protecting the eyeball and disposing of the tears [p. 281]. The muscles connected to the nose take part in contraction and dilation of the nostrils. The alae nasi do not move during quiet respiration, but when there is any obstruction to inspiration there is a visible movement of elevation and lateral expansion, due to the action of the dilators of the nose. If these muscles are weakened, as in terminal illnesses, the alae nasi are drawn inwards with each inspiration, giving to the nose a characteristic pointed appearance.

The functions of the facial muscles are perhaps best illustrated by what happens if they are paralysed. In the common form of peripheral paralysis of the facial nerve (**Bell's palsy**), the mouth is pulled to the opposite side by the unopposed action of the intact muscles. On the paralysed side, the lines of the face are smoothed out, due to the lack of tone in the muscles, and the face remains expressionless. The affected eye cannot be closed firmly, and the lower lid falls away from the eyeball under the pull of gravity. This results in the tears overflowing down the cheek because the lower lacrimal punctum is not functioning. Food and saliva cannot be retained by the affected cheek and angle of the mouth, and they escape through the lips on that side. Speech is usually only slightly affected, but the patient cannot whistle or show his teeth properly—the attempt to do so results in the mouth

being pulled further to the healthy side. Nor can he wrinkle the affected side of his forehead on demand.

MUSCLES OF THE ORBIT

Levator Palpebrae Superioris

This muscle arises from the roof of the orbit just in front of the optic canal. As it passes forwards it expands to form a triangular aponeurosis which covers over the superior rectus muscle [FIGS. 351, 352]. The aponeurosis is loosely attached to the midpoints of the medial and lateral orbital margins, and between these two points passes into the upper lid. Here it divides into three; the most anterior and superficial part runs down into the eyelid to become attached to the front of the tarsus and also into the skin of the eyelid, blending with the orbicularis oculi. The middle part of the aponeurosis, which contains a layer of visceral muscle fibres, is attached to the upper margin of the tarsus, and the deepest of the three layers is attached to the superior fornix of the conjunctiva. The levator palpebrae superioris is unusual in containing both somatic and visceral muscle fibres; a similar condition is found in the middle third of the oesophagus [p. 420].

Nerve Supply. The somatic portion of the muscle is supplied by the superior division of the oculomotor nerve. The visceral component is supplied by the sympathetic system.

Actions. The muscle elevates the upper eyelid, and its attachments ensure that the skin folds up and the conjunctiva is pulled out of the way as the lid is raised. The movement of blinking is normally reflex and involuntary, but the muscle is used under the control of the will when the eyes are raised to look at an object above the head. The levator palpebrae superioris is the antagonist of the palpebral part of the orbicularis oculi, and paralysis of the levator results in the upper lid drooping down over the eyeball, a condition known as **ptosis.**

Rectus Muscles

The four rectus muscles—**superior, inferior, medial,** and **lateral**—all take origin from a common tendinous ring [FIG. 896] which surrounds the opening of the optic canal and bridges across the superior orbital fissure. Inside this ring lie the optic nerve and ophthalmic artery, emerging from the canal,

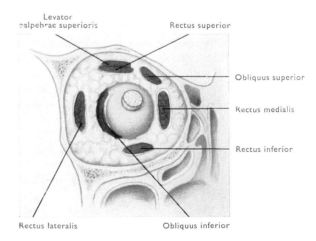

FIG. 350 Coronal section through left orbit behind the eyeball to show arrangement of muscles.

and, passing through the medial part of the superior orbital fissure, the two divisions of the oculomotor nerve, the naso-ciliary nerve, and the abducent nerve.

The inferior rectus comes from the lowest part of the ring, the medial rectus from its medial aspect, and the superior rectus from its upper part. The lateral rectus is sometimes said to have two heads, but there is no separation between them, and the muscle arises from the lateral and inferior parts of the tendinous ring, as it crosses the fissure, and from the adjacent orbital surface of the greater and lesser wings of the sphenoid bone. All four muscles form flattened bands in the fat round the optic nerve and the eyeball [FIGS. 350–3] and end in short tendons which pierce the fascial sheath of the eyeball [p. 302] to be inserted into the sclera some 6 to 8 mm. behind the margin of the cornea. The superior and inferior rectus muscles are inserted slightly medial to the vertical axis of the eyeball, but the lateral and medial rectus muscles are inserted exactly in its horizontal axis; all are attached in front of the equator of the eye.

Nerve Supply. The superior rectus is supplied by the superior division of the oculomotor nerve, and the medial and inferior rectus muscle by its inferior division. The lateral rectus is supplied by the abducent nerve.

Actions. The medial and lateral rectus muscles turn the eyeball so that the cornea looks medially or laterally respectively. The movement is entirely in the horizontal plane. The superior and inferior recti turn the eyeball so as to make the cornea look upwards and downwards respectively, but because the pull of both muscles is on the medial side of the vertical axis of the eyeball, the cornea is also deviated medially [p. 286].

Superior Oblique

The superior oblique arises close to the margin of the optic canal above the origin of the medial rectus. It is a fusiform muscle which passes forwards along the medial wall of the orbit superior to the medial rectus. The narrow tendon of the muscle passes through a special fibrocartilaginous pulley called the **trochlea,** which is attached to the trochlear fossa [FIG. 151]. As it does so, it is enclosed in a synovial sheath. The trochlea changes the direction of pull of the superior

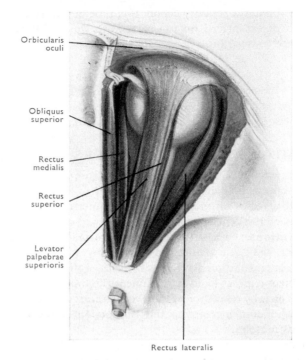

FIG. 352   Muscles of right orbit (from above).

oblique, the tendon of which now passes backwards and laterally between the tendon of the superior rectus and the eyeball, and becomes inserted into the sclera between the superior rectus and the lateral rectus, behind the equator of the eye [FIG. 353].

Nerve Supply. From the trochlear nerve.

Actions. When the eye is looking straight forwards the superior oblique turns it so that the cornea looks downwards and laterally. But when the cornea is already deviated medially, the line of pull of the superior oblique is straightened out, so that in this position it acts as a pure depressor of the cornea. Conversely, when the cornea is already deviated laterally, the line of pull of the superior oblique comes into a more transverse position in relation to the axis of the eye, and in this position it acts as a inward rotator; that is, it turns the eyeball so that an object placed on top of it would fall off towards the nose.

Inferior Oblique

Unlike all the other orbital muscles, the inferior oblique takes origin anteriorly from the floor of the orbit immediately lateral to the nasolacrimal canal. The muscle passes laterally and backwards inferior to the inferior rectus muscle [FIG. 351], and curves up, deep to the lateral rectus, to be inserted between the superior and lateral rectus muscles behind the equator of the eyeball [FIG. 353].

Nerve Supply. From the inferior division of the oculomotor nerve.

Actions. The inferior oblique turns the eye so that the cornea deviates upwards and laterally. But when the cornea is already deviated medially the line of pull of the muscle is such that it acts as a pure elevator, and when the cornea is already deviated laterally it has, like

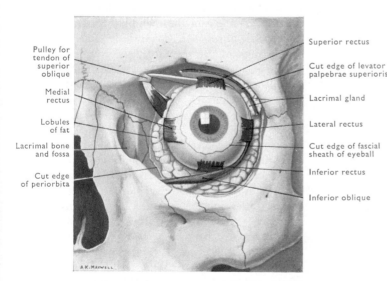

FIG. 351   Dissection of left orbit from the front.

the superior oblique, a rotatory action, this time in a direction such that an object placed on top of the eye would fall off towards the ear (outward rotation).

Orbitalis

A scanty layer of visceral muscle bridges over the inferior orbital fissure and the infraorbital groove; it is supplied by the sympathetic system but its actions are obscure.

MOVEMENTS OF THE EYES

The movements of both eyes are intimately co-ordinated by their nerve supply. When the gaze is directed to the right or left, the medial rectus of one eye acts in conjunction with the lateral rectus of the other; such movements are called **conjugate deviations**. When the eyes are directed towards a distant object, they are maintained parallel by a balance of power in the lateral and medial rectus muscles of the two eyes. When an object close at hand is examined, the eyes are made to **converge** on it by relaxation of the two lateral rectus muscles and a corresponding increase in the contraction of the two medial rectus muscles.

Movements in the vertical plane are more complicated. The cornea of each eye is turned straight downwards by contraction of the inferior rectus and the superior oblique, the superior oblique being able to cancel out the unwanted medial deviation produced by the inferior rectus. Movement of the cornea straight upwards is effected by simultaneous contraction of the superior rectus and the inferior oblique. In addition, the upper eyelid is raised by the levator palpebrae superioris. If this is insufficient to clear the line of gaze, the skin of the eyebrows is raised by the epicranius muscle; still further elevation of the gaze is brought about by extension of the head on the neck.

When the cornea is deviated medially, its elevation by the superior rectus becomes less efficient, since the line of action of the muscle now lies at a considerable angle to the optic axis, and it tends to produce inward rotation of the eyeball on this axis while increasing the medial deviation of the cornea. Accordingly, as the cornea moves medially, so the inferior oblique muscle, whose direction of pull now comes to

lie more nearly in the optic axis, is found to be progressively more important in the movement of elevation of the cornea.

Exactly the converse happens if the cornea is deviated laterally. The superior rectus is now pulling more directly in the line of the axis of the eye and can effect a straight elevation of the cornea. The inferior oblique, on the other hand, is now pulling in a line which has a component at right angles to the axis of the eye, and its rotatory effect becomes more pronounced as its capacity to elevate the cornea weakens.

Similar considerations apply to the movement which depresses the cornea. When the cornea is medially deviated, depression in the vertical plane is largely due to the contraction of the superior oblique, and the inferior rectus becomes less important. Again, when the cornea is deviated laterally, it is the inferior rectus which takes the major part in the movement.

Further complications are introduced by the contraction of two adjacent rectus muscles simultaneously. For example, the lateral rectus can help to cancel out some of the unwanted medial deviation produced by the superior and inferior rectus muscles.

These factors are important when considering the origin of a squint. For example, if the superior oblique muscle is paralysed, the patient will still be able to turn the cornea downwards and laterally, by using his lateral and inferior rectus muscles in conjunction. But if the cornea is deviated medially and the patient is then asked to depress it, double vision at once appears, since in this position the superior oblique is the only muscle which can satisfactorily depress the cornea.

MUSCLES OF MASTICATION

Masseter

The masseter [FIG. 348], the most superficial of the muscles of mastication, is easily felt when the jaw is clenched. It is a flat muscle which stretches from its origin on the lower border and deep surface of the zygomatic arch downwards to the lateral surfaces of the coronoid process, ramus, and angle of the mandible [FIG. 354]. The superficial fibres run downwards and backwards to the angle of the mandible, and the deeper fibres run more vertically, being separated from the superficial ones by a cleft posteriorly, though they blend with them anteriorly. The deepest fasciculi of the muscle often partly unite with the subjacent fasciculi of the temporalis.

On its lateral surface the duct of the parotid gland runs forwards, and when the muscle is contracted the duct can be rolled under the finger as it turns inwards around the anterior border of the masseter to pierce the buccinator. The lateral surface of the masseter is covered posteriorly by the parotid gland, further forward by the accessory parotid gland and the zygomaticus major, and inferiorly by the risorius and platysma muscles. Deep to the masseter lie the ramus and notch of the mandible and anteriorly the

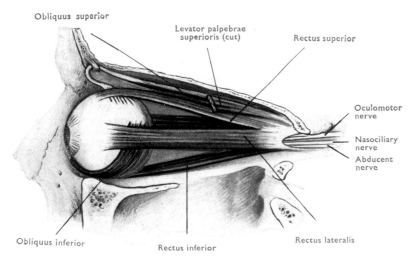

Obliquus superior

Levator palpebrae superioris (cut)

Rectus superior

Oculomotor nerve

Nasociliary nerve

Abducent nerve

Obliquus inferior

Rectus inferior

Rectus lateralis

FIG. 353   Muscles of left orbit (from lateral aspect).

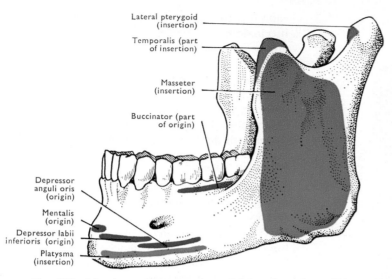

Lateral pterygoid (insertion)

Temporalis (part of insertion)

Masseter (insertion)

Buccinator (part of origin)

Depressor anguli oris (origin)

Mentalis (origin)

Depressor labii inferioris (origin)

Platysma (insertion)

FIG. 354 Muscle attachments to superficial surface of mandible.

buccinator muscle, which is separated from it by the buccal pad of fat.

Nerve Supply. From the anterior division of the mandibular nerve.

Actions. The masseter elevates the mandible, and its superficial fibres help to pull it forwards.

Temporalis

This strong fan-shaped muscle arises from the temporal fossa and from the temporal fascia [p. 302] which covers it.

The fasciculi in the anterior part of the muscle run more or less vertically, while those in the more posterior part are almost horizontal [FIG. 355]. They all converge on a thick tendon which passes medial to the zygomatic arch and is inserted into the apex and the deep surface of the coronoid process and into the anterior border of the ramus of the mandible [FIGS. 345, 354]. The fibres reach down almost as far as the last molar tooth, and some of them become continuous with the buccinator muscle. Others, in the more superficial fasciculi, fuse with the masseter.

Nerve Supply. By the deep temporal nerves from the anterior division of the mandibular nerve.

Actions. The vertical fibres elevate the mandible and close the mouth: they are in constant postural contraction to counteract the effect of gravity. The posterior fibres pull the mandible backwards after it has been protruded (MacDougall and Andrew, 1953). This action is important since the head of the mandible passes forwards on to the articular eminence when the mouth is opened or the mandible protruded, and has to be retracted again when the mouth is closed [p. 221].

Lateral Pterygoid

The lateral pterygoid has two heads; the upper one arises from the inferior surface of the greater wing of the sphenoid bone, and the lower one from the lateral surface of the lateral pterygoid plate. From this double origin the fibres pass backwards and slightly laterally to be inserted into the front of the neck of the mandible, and into the capsule and the disc of the temporomandibular joint [FIGS. 345, 356]. Between the two heads of the muscle pass the buccal nerve and the maxillary artery, and laterally it is covered by the ramus of the mandible, the temporalis, and the masseter.

Nerve Supply. From the anterior division of the mandibular nerve.

Actions. The muscle pulls the head of the mandible, the disc, and the capsule of the temporomandibular joint forwards on to the articular eminence [p. 113]. This movement occurs during: (1) protrusion of the jaw, and (2) opening of the mouth: in the latter case the head of the mandible also rotates on the articular disc. Protrusion occurs when the elevators of the jaw are in action at the same time as the lateral pterygoids, the lower incisors being carried forwards in front of the upper ones [p. 221]. When the elevators are not in action, the lateral pterygoids carry the head of the mandible forwards and rotate it on the disc so that the chin moves downwards and the mouth is opened. Simultaneous contraction of the lateral and medial pterygoid muscles of one side draws the head of the mandible forwards on that side, the depressing action of the lateral pterygoid being cancelled out by

Temporal muscle

Buccinator

FIG. 355 Right temporal muscle. (The zygomatic arch and the masseter muscle have been removed.)

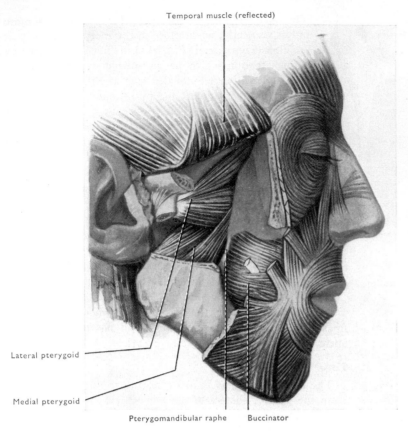

Temporal muscle (reflected)

Lateral pterygoid

Medial pterygoid

Pterygomandibular raphe   Buccinator

FIG. 356   Pterygoid muscles of the right side.

the elevating action of the medial pterygoid. The result is to carry the chin towards the opposite side, so producing a grinding movement between the teeth. Alternate contraction of these muscles, first on one side and then on the other, is an important element of chewing.

The lateral pterygoid is the chief antagonist to the movement of retraction of the jaw (carried out by temporalis), and pays out to control it.

Medial Pterygoid

The main head of the medial pterygoid arises from the medial surface of the lateral pterygoid plate and from the pyramidal process of the palatine bone. A much smaller head comes from the tuber of the maxilla. From these two origins, which embrace the inferior fibres of the lateral pterygoid [FIG. 356], the fasciculi of the muscle pass downwards, backwards, and laterally to be inserted into a triangular rough impression on the medial surface of the mandible between the mylohyoid groove and the angle [FIG. 345].

Nerve Supply. From the main trunk of the mandibular nerve.
Actions. The medial pterygoid elevates and protrudes the mandible. Acting along with the lateral pterygoid of the same side, the medial pterygoid protrudes the mandible and swings the chin to the opposite side. Alternation of such movements is important in chewing.

MUSCLES OF THE TONGUE

The substance of the tongue consists mainly of somatic muscle. Some of the muscles which compose the tongue lie entirely within the organ, and are therefore known as **intrinsic** muscles; others pass into the tongue from origins outside it, and are called **extrinsic** muscles. The tongue is actively concerned in mastication, swallowing, and speaking. The posterior part of the tongue is attached to, and moves with, the hyoid bone, on which it rests. The anterior part lies on the mylohyoid and geniohyoid muscles, and therefore moves with the floor of the mouth.

Intrinsic Muscles

There are four pairs of intrinsic muscles [FIG. 357], arranged on either side of a midline, vertical, septum of fibrous tissue, to which many of their fibres are attached (Abd-el-Malik, 1939). The **superior longitudinal** muscle lies under the mucous membrane of the dorsum of the tongue, and extends from the back of the tongue, close to the epiglottis, to the tip and edges of the tongue, some of its fibres passing into the mucosa. It takes origin in part from the median fibrous septum. The **inferior longitudinal** muscle lies in the lower part of the tongue between the extrinsic muscles, genioglossus and hyglossus, and blends with the fibres of another extrinsic muscle, the styloglossus.

The **transverse** muscle of the tongue arises from the median septum and radiates upwards to the dorsum and laterally to the sides of the tongue, where it inserts into the submucous tissue and mucous membrane. The **vertical** intrinsic muscle arises from the mucous membrane of the dorsum and passes downwards and laterally to the sides of the tongue, mingling with the fibres of the other extrinsic and intrinsic muscles.
Nerve Supply. All the intrinsic muscles are supplied by the hypoglossal nerve.
Actions. The intrinsic muscles alter the shape of the tongue, and are of great importance in speech and in the movements of mastication and swallowing [p. 297]. The longitudinal muscles pull the tip backwards towards the base, and, because the volume of the tongue does not change, this causes it to broaden from side to side and increases its vertical dimensions. Similarly, the transverse muscle narrows the tongue from side to side, so causing it to lengthen and become vertically thicker at the same time. The vertical muscle flattens, lengthens, and broadens the tongue.

Contraction of the superior longitudinal muscles in isolation will turn the tip and sides of the tongue upwards to render the dorsum concave. Contraction of the inferior longitudinal muscles pulls the tip of the tongue downwards and makes the dorsum convex.

Genioglossus

The genioglossus is an extrinsic muscle which arises from the upper part of the mental spine of the mandible [FIG. 345]. From this small origin, its fasciculi radiate forwards, upwards, and backwards into the tongue [FIGS. 346, 357], forming a paramedian sheet. The lowest fibres are inserted into the body of the hyoid bone, and the highest fibres pass forwards into the tip of the tongue. The intermediate fibres permeate the substance of the tongue in its whole length. The two

FIG. 357   Low-power photomicrographs of sections through the mouth of two human foetuses.
A and B, horizontal sections of a 7-months-old foetus; C and D, sagittal sections of a 14-weeks-old foetus.

muscles are separated from each other by the median septum.

Nerve Supply.   From the hypoglossal nerve.

Actions.   Like all the extrinsic muscles, the genioglossus is concerned mainly in moving the tongue bodily about rather than in altering its shape.   The anterior fibres of the muscle retract the tip of the tongue; the posterior fibres protrude it. When the whole muscle contracts, the central part of the tongue is depressed.   If the genioglossus of one side acts alone, the tip of the tongue is protruded and turned to the opposite side.   The genioglossus is the only extrinsic muscle which can protrude the tongue, and its attachment to the mandible prevents the tongue from falling back and obstructing

respiration.   In an unconscious patient the airway can be kept clear by pulling the mandible forwards.

Hyoglossus

The second extrinsic muscle of the tongue arises from the body and the greater horn of the hyoid bone, from which it passes upwards and forwards to be inserted into the side of the tongue, intermingling with the fibres of the inferior longitudinal muscle and the styloglossus [FIGS. 346, 357, 358].   Sometimes a slip of the hyoglossus arises from the lesser horn of the hyoid bone and is called the **chondroglossus.**

Nerve Supply.   From the hypoglossal nerve.

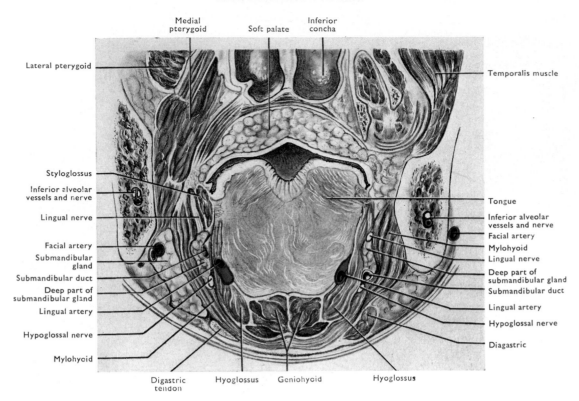

Medial pterygoid · Soft palate · Inferior concha · Lateral pterygoid · Temporalis muscle · Styloglossus · Inferior alveolar vessels and nerve · Lingual nerve · Facial artery · Submandibular gland · Submandibular duct · Deep part of submandibular gland · Lingual artery · Hypoglossal nerve · Mylohyoid · Tongue · Inferior alveolar vessels and nerve · Facial artery · Mylohyoid · Lingual nerve · Deep part of submandibular gland · Submandibular duct · Lingual artery · Hypoglossal nerve · Diagastric · Digastric tendon · Hyoglossus · Geniohyoid · Hyoglossus

FIG. 358   Coronal section through the tongue and submandibular region behind the molar teeth.

Actions. The hyoglossus pulls the side of the tongue downwards and retracts it. Acting with the genioglossus, the muscle depresses the tongue; acting with the transverse intrinsic muscle it arches the tongue transversely.

Styloglossus

This muscle arises mainly from the anterior border of the styloid process near its tip; some of the fibres come from the stylohyoid ligament. It passes forwards, medially and downwards to be inserted into the side of the tongue, mingling with the palatoglossus, the hyoglossus, and the inferior longitudinal muscle [FIG. 346]. It lies superficial to the stylopharyngeus muscle, and also to the lower and anterior part of the superior constrictor.

Nerve Supply. From the hypoglossal nerve.

Actions. The styloglossus retracts and elevates the tongue; acting with the vertical intrinsic muscle and the genioglossus, it raises the sides of the tongue while leaving the centre depressed, so producing a transverse concavity. This movement is important in drinking [p. 298].

Palatoglossus

The palatoglossus, which also contributes fibres to the substance of the tongue, is considered along with the muscles of the soft palate [p. 294].

MUSCLES OF THE PHARYNX AND SOFT PALATE

The wall of the pharynx is a sandwich of muscle covered internally and externally with fascia. The outer layer of fascia is the **buccopharyngeal fascia** [p. 301], which extends forwards on to the buccinator muscle, and the inner layer is the **pharyngobasilar fascia,** which is thicker near the skull than elsewhere. The main bulk of the meat in the sandwich is provided by three constrictor muscles, and these are reinforced internally by a longitudinal layer consisting of the fibres of the stylopharyngeus, salpingopharyngeus, and palatopharyngeus muscles.

In contrast, the soft palate is a sandwich with the fibrous tissue in the middle, surrounded by several muscles. The fibrous **palatine aponeurosis** is attached in front to the posterior border of the bony palate, and on each side to the pharyngobasilar fascia. At first horizontal, the soft palate hangs downwards as it is traced backwards, separating the pharynx from the mouth.

Constrictor Muscles

The **superior constrictor** takes origin from the lower half of the posterior border of the medial pterygoid plate, the pterygoid hamulus, the pterygomandibular raphe, the mylohyoid line of the mandible, and the musculature of the posterior part of the side of the tongue [FIG. 346]. The pterygomandibular raphe lies between it and the buccinator, and is a fibrous line of junction between the two muscles [p. 283].

The fibres of the superior constrictor curve backwards around the pharynx, to unite with their fellows in a median raphe in the posterior wall of the pharynx. The upper fibres curve upwards to reach the pharyngeal tubercle of the occipital bone [FIGS. 340, 359], while the lowest fibres curve slightly downwards deep to the middle constrictor.

On the inner surface of the superior constrictor near its

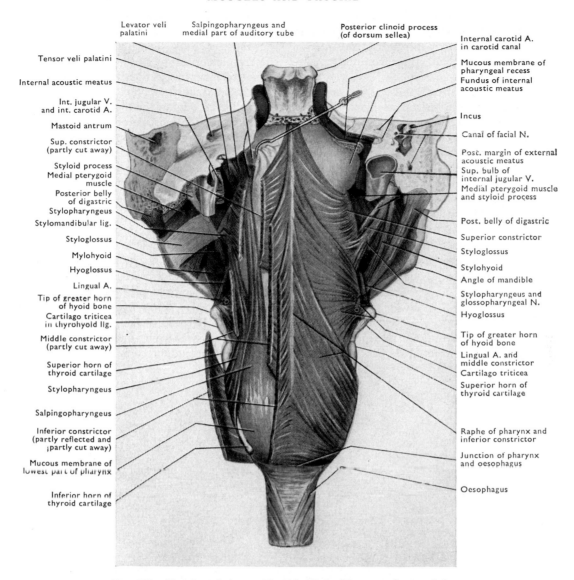

FIG. 359 Muscles of pharynx from behind. The constrictors of the left side have been cut to show the deeper muscles.

upper border there is a horizontal band of muscle fibres which elevates the mucous membrane to form a ridge. These fibres arise from the palatine aponeurosis and sweep backwards to surround the pharynx at the junction of its nasal part and its oral part: they are sometimes referred to as the **palato-pharyngeal sphincter** (Whillis, 1930), which is considerably enlarged in cases of complete cleft palate.

The curved upper border of the superior constrictor is separated from the base of the skull by an interval occupied by the pharyngobasilar fascia, and through this interval pass the auditory tube and the levator veli palatini muscle [FIGS. 361, 362]. Posteriorly and laterally the superior constrictor is overlapped inferiorly by the middle constrictor. Between the lower border of the superior constrictor, the upper border of the middle constrictor, and the posterior border of the hyoglossus is another muscular gap filled with fascia, through which pass the stylopharyngeus, the glossopharyngeal nerve, and the stylohyoid ligament [FIG. 362].

The **middle constrictor** [FIGS. 359, 362] arises from the lower part of the stylohyoid ligament and from the greater and

lesser horns of the hyoid bone. It curves backwards around the pharynx to be inserted into the median raphe of the pharynx, the upper fibres overlapping the superior contrictor and the lower fibres being overlapped posteriorly and laterally by the inferior constrictor. Between the anterior part of the lower border of the middle constrictor and the upper border of the inferior constrictor there is yet another muscular gap, triangular in shape, bounded by the thyrohyoid muscle in front and occupied by the posterior part of the thyrohyoid membrane and the lower part of the stylopharyngeus muscle [FIGS. 346, 362]. Through the gap pass the superior laryngeal vessels and the internal branch of the superior laryngeal nerve.

The **inferior constrictor** [FIGS. 359, 362] comes from the oblique line of the thyroid cartilage, from the side of the cricoid cartilage, and from the fascia on the lateral aspect of the cricothyroid muscle. Like those of the other constrictors, its fasciculi pass upwards and backwards to the median raphe of the pharynx, overlapping those of the middle constrictor. The lowest fasciculi, which arise from the cricoid cartilage, run more horizontally, and blend inferiorly with the beginning of the circular muscle of the oesophagus. These fibres are sometimes classified as forming a separate **cricopharyngeus** muscle, and it has been claimed that this part of the inferior constrictor has somewhat different functions from the remainder of the muscle (Batson, 1955).

Nerve Supply. All three constrictor muscles are supplied by the pharyngeal plexus : the cricopharyngeus receives additional innervation from the recurrent laryngeal nerve, and sometimes also from the external laryngeal.

Actions. The constrictor muscles contract involuntarily during the process of swallowing. This contraction takes place sequentially from above downwards, and propels the bolus of food onwards into the oesophagus [see also p. 298]. Contraction of the palatopharyngeal sphincter helps to shut off the nasal part of the pharynx from the oral part during swallowing. The cricopharyngeus fibres of the inferior constrictor are said to act as a sphincter preventing air from entering the oesophagus, and they relax suddenly during swallowing.

Stylopharyngeus

The stylopharyngeus takes origin from the medial side of the root of the styloid process, and its fibres pass downwards, between the external and internal carotid arteries, to enter the wall of the pharynx between the superior and middle constrictor muscles [FIGS. 360, 362]. Here the muscle is accompanied by the glossopharyngeal nerve winding round its lateral aspect. The stylopharyngeus then spreads out on the inner surface of the middle constrictor to be inserted into the superior and posterior borders of the thyroid cartilage, the side of the epiglottis, and the musculature of the pharynx itself, becoming continuous with the palatopharyngeus.

Nerve Supply. From the glossopharyngeal nerve.

Actions. It elevates the larynx and helps to pull the wall of the pharynx up over the descending bolus of food in the process of swallowing [see also p. 297].

Salpingopharyngeus

This small muscle [FIG. 359] comes from the lower part of the cartilage at the pharyngeal opening of the auditory tube. It passes downwards inside the constrictor muscles and blends with the palatopharyngeus.

Nerve Supply. From the pharyngeal plexus.

Actions. It raises the upper part of the lateral wall of the pharynx in swallowing, and at the same time pulls open the mouth of the auditory tube.

FIG. 360 Dissection of the constrictors of the pharynx and the associated structures which lie adjacent to the mucous membrane lining the mouth, pharynx, and larynx, from the medial side. The tongue has been removed to expose the structures which lie between it and the mylohyoid muscle.

Palatopharyngeus

The palatopharyngeus [FIGS. 360, 363] originates in the soft palate by two layers, between which lie the levator veli palatini and the muscle of the uvula. The stronger of the two layers lies antero-inferiorly, between the levator and tensor veli palatini muscles, and it is attached to the palatine aponeurosis and the posterior border of the hard palate. The thin posterosuperior layer lies under the mucous membrane of the nasopharyngeal surface of the soft palate, joining with its fellow across the midline. The two layers blend at the lateral side of the soft palate and run downwards on the side of the pharynx, forming the substance of the palatopharyngeal arch. Later, the muscle is joined by the posterior part of the salpingopharyngeus, and inserts partly on to the posterior border of the thyroid cartilage. Most of the fibres end by blending with the other muscles of the pharyngeal wall, and gaining insertion to the pharyngobasilar fascia; they decussate across the midline.

Nerve Supply. From the pharyngeal plexus.

Actions. The palatopharyngeus pulls up the walls of the pharynx over the descending bolus during swallowing and also produces a constriction of the palatopharyngeal arch. This movement, assisted by the backward and upward movement of the tongue [p. 297] helps to shut off the pharynx from the mouth during swallowing. The pull of the vertical fibres of the muscle depresses the soft palate to separate the oral part of the pharynx from the mouth, so allowing breathing when the mouth is full: their insertion on the thyroid cartilage assists in raising the larynx.

Tensor Veli Palatini

This thin triangular muscle [FIGS. 360, 362, 363] has a linear origin on the posteromedial edge of the greater wing of the sphenoid bone from the scaphoid fossa to the spine of the sphenoid. It also arises from the adjacent cartilage of the auditory tube. Its fibres converge downwards on a tendon which changes direction abruptly by hooking medially round the pterygoid hamulus; there is a small bursa here to facilitate its movement on the bone. The tendon then expands horizontally to form, with its fellow, the **palatine aponeurosis,** to which most of the other muscles of the soft palate are attached. The palatine aponeurosis is attached to the posterior border of the hard palate and to its under surface behind the palatine crest.

Nerve Supply. From the mandibular nerve by the branch which supplies the medial pterygoid muscle.

Actions. The tensors tighten the soft palate and flatten its arch; this prevents the passage of a bolus of food from forcing the soft palate upwards into the nasal part of the pharynx. If one tensor is paralysed, the other pulls the palate over to its own side. The fibres attached to the auditory tubes may help to pull them open during swallowing.

Levator Veli Palatini

The elevator of the soft palate [Figs. 360, 361, 362] arises from the rough area on the inferior surface of the temporal bone in front of the opening of the carotid canal. Other fibres come from the upper portion of the carotid sheath, and from the inferior aspect of the cartilaginous part of the auditory

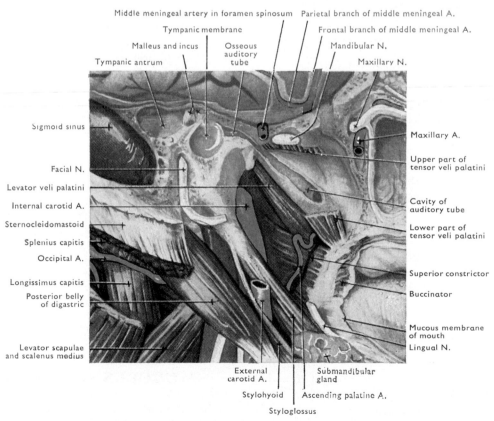

FIG. 361   Region of auditory tube showing the levator and tensor veli palatini muscles. Enlargement of part of Fig. 347.

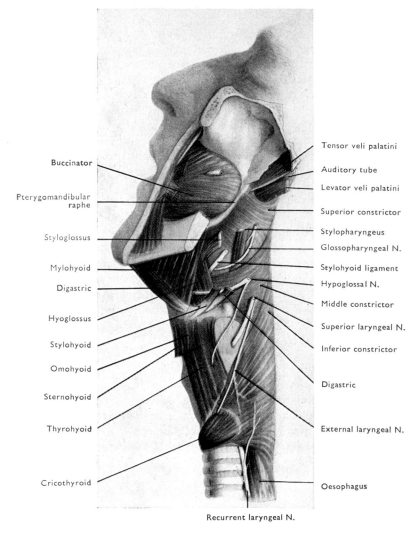

Buccinator

Pterygomandibular
raphe

Styloglossus

Mylohyoid

Digastric

Hyoglossus

Stylohyoid

Omohyoid

Sternohyoid

Thyrohyoid

Cricothyroid

Tensor veli palatini

Auditory tube

Levator veli palatini

Superior constrictor

Stylopharyngeus

Glossopharyngeal N.

Stylohyoid ligament

Hypoglossal N.

Middle constrictor

Superior laryngeal N.

Inferior constrictor

Digastric

External laryngeal N.

Oesophagus

Recurrent laryngeal N.

FIG. 362   Lateral view of wall of pharynx.

tube. The muscle descends medially above the upper border of the superior constrictor to enter the soft palate between the two layers of the palatopharyngeus muscle. Its fibres are inserted into the palatine aponeurosis and blend with those of its fellow. It is separated from the tensor veli palatini at its origin by the pharyngobasilar fascia.
Nerve Supply. From the pharyngeal plexus.
Actions. The levators raise the soft palate and draw it backwards, and the fibres attached to the auditory tubes tend to open them during swallowing. The muscles co-operate with the palatopharyngeal sphincter [p. 291] to separate the nasal and oral parts of the pharynx.

Palatoglossus

This thin muscle [FIG. 363] arises from the under surface of the palatine aponeurosis, being continuous with its fellow of the opposite side, and then turns downwards, forming the substance of the palatoglossal arch, to be inserted into the dorsum and side of the tongue.
Nerve Supply. From the pharyngeal plexus.
Actions. The palatoglossus muscles pull the palatoglossal arches towards the midline, and so narrow the isthmus

of the fauces. In the movement of swallowing, this helps to close off the mouth cavity from the pharynx. They also assist in pulling upwards the posterior part of the tongue.

Musculus Uvulae

The small muscle of the uvula [FIG. 363] is attached anteriorly to the nasal spine, and is enveloped by the palatine aponeurosis, which splits to surround it. The muscle fibres travel backwards to end in the mucous membrane of the uvula.
Nerve Supply. From the pharyngeal plexus.
Actions. The uvular muscle elevates the uvula in the movement of swallowing.

MUSCLES OF THE MIDDLE EAR

Two muscles operate on the auditory ossicles which cross the middle ear between the tympanic membrane and the fenestra vestibuli.

Tensor Tympani

Originating from the upper surface of the cartilage of the auditory tube and the adjacent part of the greater wing of the sphenoid, the tensor tympani runs backwards and laterally in a bony semicanal superior to the bony part of the auditory tube; from this canal additional fibres arise. The tendon, which emerges in the middle ear above the auditory tube, turns laterally round the processus cochleariformis and runs to be inserted into the handle of the malleus, near its root.
Nerve Supply. Like the tensor veli palatini, the tensor tympani is supplied by the branch of the mandibular nerve to the medial pterygoid muscle.
Actions. The muscle draws the handle of the malleus towards the tympanic cavity, so tensing the tympanic membrane.

Stapedius

The stapedius is a tiny muscle which arises from the walls of a cavity in the interior of the pyramidal eminence. The tendon emerges from the apex of the eminence and is inserted directly into the posterior surface of the neck of the stapes [FIG. 1011].
Nerve Supply. From the facial nerve.
Actions. The stapedius draws the stapes backwards, and tilts the anterior part of the base of the bone towards the tympanic cavity, thus tightening the anular ligament and reducing the amount of movement of the footplate.

MUSCLES OF THE LARYNX

The larynx has a set of **extrinsic** muscles which move the apparatus bodily about; these muscles have already been described [pp. 275, 278]. The **intrinsic** muscles of the larynx

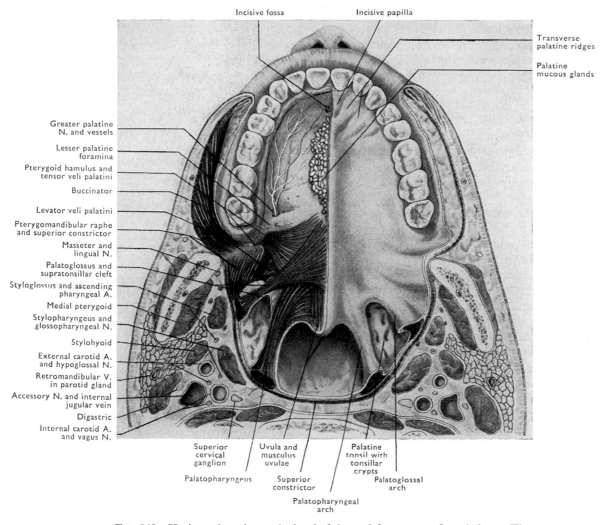

Incisive fossa   Incisive papilla

Transverse
palatine ridges

Palatine
mucous glands

Greater palatine
N. and vessels

Lesser palatine
foramina

Pterygoid hamulus and
tensor veli palatini

Buccinator

Levator veli palatini

Pterygomandibular raphe
and superior constrictor

Masseter and
lingual N.

Palatoglossus and
supratonsillar cleft

Styloglossus and ascending
pharyngeal A.

Medial pterygoid

Stylopharyngeus and
glossopharyngeal N.

Stylohyoid

External carotid A.
and hypoglossal N.

Retromandibular V.
in parotid gland

Accessory N. and internal
jugular vein

Digastric

Internal carotid A.
and vagus N.

Superior
cervical
ganglion

Palatopharyngeus

Uvula and
musculus
uvulae

Superior
constrictor

Palatopharyngeal
arch

Palatine
tonsil with
tonsillar
crypts

Palatoglossal
arch

FIG. 363   Horizontal section at the level of the oral fissure seen from below.   The
right side has been dissected to show the muscles of the soft palate.

are those which alter the length, tension, and distance apart of
the vocal cords ; in this way they alter the size and shape of the
space between the cords (**rima glottidis**).   Some of the
intrinsic muscles also alter the size of the inlet of the larynx.
The intrinsic muscles are concerned in the production of
speech and in the regulation and protection of the air inlet to
the lungs.

With the exception of the cricothyroid muscle, the intrinsic
muscles of the larynx all lie between the cartilaginous framework and the mucous membrane, and with the exception of
the transverse arytenoid muscle all the intrinsic muscles are
paired.   These muscles are also mentioned in the section on
the larynx [p. 480], where their actions are further discussed,
and the cartilages are described.

Cricothyroid

The cricothyroid [FIGS. 362, 364] arises from the anterior
and lateral aspects of the outer surface of the cricoid cartilage.
The fibres radiate upwards and backwards to the lower border
of the lamina and the inferior horn of the thyroid cartilage.
Nerve Supply.   From the external laryngeal branch of the
superior laryngeal nerve.
Actions.   The cricothyroid is the chief tensor of the vocal

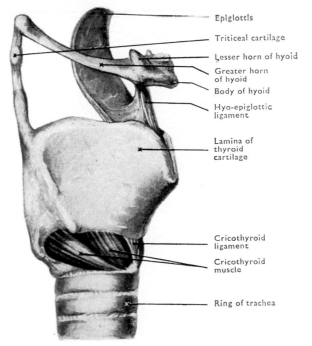

Epiglottis

Triticeal cartilage

Lesser horn of hyoid

Greater horn
of hyoid

Body of hyoid

Hyo-epiglottic
ligament

Lamina of
thyroid
cartilage

Cricothyroid
ligament

Cricothyroid
muscle

Ring of trachea

FIG. 364   Right cricothyroid muscle.

cords. Contraction of the muscle pivots the thyroid and cricoid cartilages on each other, thus tilting back the upper border of the cricoid lamina. These movements separate the two attachments of the vocal ligaments (the vocal processes of the arytenoid cartilages and the angle of the thyroid cartilage), so elongating and tensing the ligaments.

Posterior Crico-Arytenoid

This muscle arises from the back of the lamina of the cricoid cartilage [Figs. 365, 366] and its fibres run upwards and laterally to be inserted into the muscular process of the arytenoid cartilage. The highest fibres are virtually horizontal while the lowest ones are almost vertical.

Nerve Supply. From the recurrent laryngeal branch of the vagus nerve.

Actions. It is the only abductor of the vocal folds, and it is interesting that the adductor muscles of the larynx are four times as heavy as the abductors (Bowden and Scheuer, 1960). The more horizontal fibres rotate the arytenoid cartilages on their vertical axes, swinging the vocal processes outwards, and so separate the posterior ends of the vocal folds. The more vertical fibres draw the arytenoids away from each other by pulling them downwards and laterally on the sloping upper border of the cricoid lamina. When the whole muscle is in action, abduction of the arytenoids by the vertical fibres is largely cancelled out by the tendency of the horizontal fibres to cause adduction: the net result is a pure rotation of the arytenoids around a vertical axis.

Lateral Crico-Arytenoid

Arising from the upper border of the arch of the cricoid cartilage, the fibres of this muscle are inserted into the anterior aspect of the muscular process of the arytenoid cartilage [Fig. 366].

Nerve Supply. From the recurrent laryngeal nerve.

Actions. It is one of the chief adductors of the vocal ligaments, since it rotates the arytenoid cartilage medially and approximates the vocal processes.

Transverse Arytenoid

The only unpaired muscle of the larynx [Fig. 365] stretches between the posterior surfaces of the bodies and muscular processes of the two arytenoid cartilages.

Nerve Supply. From the recurrent laryngeal nerve.

Actions. It pulls the arytenoid cartilages together and thus helps to close the posterior part of the opening of the glottis.

Oblique Arytenoid

Lying posterior to the transverse arytenoid, the oblique arytenoid crosses its colleague of the opposite side in a St. Andrew's cross [Fig. 365], passing from the posterior aspect of the muscular process of one arytenoid cartilage to the apex of the cartilage of the opposite side. Some of the fibres run into the aryepiglottic fold, where they form the **aryepiglotticus** [Figs. 365, 366].

Nerve Supply. From the recurrent laryngeal nerve.

Actions. The continuous fibres of the oblique arytenoid and the aryepiglottic muscle help to close the inlet of the larynx. They approximate the arytenoid cartilages, and pull them upwards towards the epiglottis, while at the same time they help to bend the epiglottis backwards over the laryngeal inlet.

Thyro-Arytenoid

This flat muscle [Fig. 366] takes origin from the lower half of the internal surface of the thyroid cartilage at the angle, and

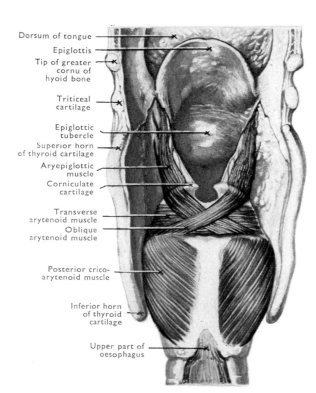

FIG. 365 Muscles on back of larynx.

FIG. 366 Muscles of side-wall of larynx.

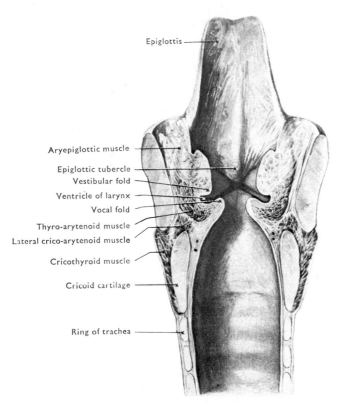

Epiglottis —

Aryepiglottic muscle —

Epiglottic tubercle —
Vestibular fold —
Ventricle of larynx —
Vocal fold —
Thyro-arytenoid muscle —
Lateral crico-arytenoid muscle —

Cricothyroid muscle —

Cricoid cartilage —

Ring of trachea —

FIG. 367   Coronal section of larynx to show situation of muscles.

also from the upper end of the cricothyroid ligament. Its fibres radiate upwards, backwards and also horizontally. The lowest, horizontal fibres are attached to the lateral surface of the vocal process of the arytenoid cartilage, and the deepest members of this group, which are attached near the tip of the process, are called the **vocalis** muscle [FIG. 367]. This lies parallel with and just lateral to the vocal ligament, to which many of its fibres are attached.

The more superior fibres of the thyro-arytenoid run into the aryepiglottic fold, and some of them reach the epiglottis, so forming the **thyro-epiglotticus** [FIG. 366]. The rest merge with the entering fibres of the oblique arytenoid muscle to form the aryepiglotticus.

Nerve Supply.   The whole composite mass is supplied by the recurrent laryngeal nerve.

Actions.   The thyro-arytenoids pull the arytenoid cartilages forwards to the thyroid cartilage, so relaxing the vocal ligaments. They also rotate the arytenoid cartilages medially, approximating the vocal folds. The vocalis portion of the muscle is a fine adjustment on the tension of the vocal folds and is intimately concerned in vocalization.

MOVEMENTS OF MASTICATION AND SWALLOWING

Food is bitten off by the incisor teeth, by means of the contraction of the masseter, temporalis, and medial pterygoid muscles. Once in the mouth, the food is then ground up by the premolar and molar teeth under the control of the muscles of mastication. Chewing movements are produced by the alternate contraction of the pterygoid muscles of first one and

then the other side, and if the food escapes from between the teeth into the vestibule of the mouth it is returned by the contraction of the buccinator muscle. If it escapes medially into the cavity of the mouth it is triturated by the action of the tongue on the hard palate and pushed back between the teeth by the muscles of the tongue, both intrinsic and extrinsic. The mylohyoid helps to keep the tongue braced up towards the hard palate, and the orbicularis oris contracts to prevent food escaping through the lips.

Normally the downward movement of the mandible during mastication results from relaxation of the elevators, and contraction of the lateral pterygoid, but more active depression of the mandible may be required if sticky food is being eaten, and this movement results from contraction of the mylohyoid, digastric, and geniohyoid muscles acting from a hyoid bone fixed by the infrahyoid muscles (sternohyoid, sternothyroid, omohyoid).

Once the food is judged by the sensory endorgans in the mouth and tongue to be of suitable consistency for swallowing, it is collected from the anterior portion of the mouth by the tip of the tongue, which is then squeezed up against the hard palate by the contraction of the superior longitudinal muscle, assisted by the mylohyoid and styloglossus, the elevators of the mandible coming into play at the same time. This forces the food backwards towards the palatoglossal and palatopharyngeal arches, much as toothpaste is squeezed from a tube.

In this phase the hyoid bone, carrying with it the larynx, is raised by means of the mylohyoid, stylohyoid, digastric, palatopharyngeus, and stylopharyngeus muscles. Fractionally later, the geniohyoid pulls the hyoid bone forwards. As the food passes through the palatoglossal and palatopharyngeal arches, the palatoglossus and palatopharyngeus contract reflexly, bringing the pillars of the arches closer together. The dorsum of the tongue bulges backwards, under the influence of the styloglossus, to occupy the remaining space between the arches, and in this way the oral part of the pharynx is shut off from the mouth cavity, so that food cannot be forced back into the mouth during subsequent stages of swallowing. The backward movement of the tongue pulls the hyoid backwards, so that the epiglottis may come into contact with the posterior wall of the pharynx.

Now the soft palate, tensed by the tensor veli palatini, is raised by the levator veli palatini to block off the communication between the oral and the nasal parts of the pharynx, so preventing the subsequent contraction of the constrictor muscles from forcing food up into the nose. In this the palatal muscles are assisted by the muscular contraction of the palatopharyngeal sphincter, which pulls the pharynx forwards to meet the rising soft palate.

The elevation of the tongue and the hyoid bone brings the inlet of the larynx up under the cover of the epiglottis, and at the same time this is bent backwards and downwards by the backward pressure of the tongue and the contraction of the oblique arytenoid and the aryepiglottic muscles. The thyroid cartilage, under the influence of the thyrohyoid muscle, is pulled up to the level of the hyoid bone, so approximating the arytenoid cartilages to the bent epiglottis. The contraction of the aryepiglotticus helps to narrow the inlet, and the whole arrangement is designed to prevent the entry of food into the larynx, for the food now slips over the surface of the epiglottis and the closed inlet of the larynx into the laryngeal part of the pharynx.

Respiration is now reflexly inhibited, and the constrictor muscles of the pharynx contract successively from above downwards, the constriction wave being preceded by a wave of relaxation. At the same time the pharynx is pulled upwards over the descending bolus of food by the palatopharyngeus, stylopharyngeus, and salpingopharyngeus. The action of the constrictors is normally automatic, but they can be brought under voluntary control, so that a bolus of food can voluntarily be recovered from the lower part of the pharynx. Once it has come into the grip of the visceral muscle of the oesophagus, however, it is beyond recall except by vomiting.

Following the descent of the bolus past the aperture of the larynx, the hyoid bone and larynx are pulled down by the infrahyoid muscles, breathing can be resumed, and another mouthful can be dealt with.

The swallowing of fluid is essentially similar, except that in the initial stages the tongue is made to form a tube with the palate by the contraction of the styloglossus and the transverse intrinsic muscle of the tongue. The fluid is then forced backwards in the same way as solid food, and cascades down the sides of the epiglottis, avoiding the entry to the larynx.

Gravity is of considerable assistance in causing the bolus of food to descend through the pharynx. However, the pharyngeal constrictors can still operate successfully even though the individual is standing on his head, and a glass of water can be drunk in this position. During the movement of the bolus through the pharynx, a considerable pressure is developed by its muscular wall, and the mucous membrane may be temporarily protruded through weak points in the pharyngeal wall where the fascial basis of the pharynx is inadequately guarded by musculature. In particular, it is possible for such bulging to take place above, below, or through the inferior constrictor, and a '**pulsion diverticulum**' may develop, in which food lodges during, and remains after, the descent of the bolus. Such diverticula may require surgical attention.

If the innervation of the soft palate or pharynx is damaged, the whole process of swallowing becomes difficult or impossible, and food may be returned through the nose after it is taken in by the mouth. In such cases feeding by means of a tube may be necessary. The movements of the soft palate are important not only in swallowing, but also in speech and in blowing or whistling. All these activities require an airtight seal between the nasal and the oral parts of the pharynx, since otherwise air would escape through the nose. For this reason paralysis of the soft palate results not only in difficulty in swallowing, but in a nasal tone of voice and an inability to develop any pressure on blowing through the mouth.

The complex mechanisms necessitated by the crossing over of the stream of air and the flow of food are quite liable to disturbance, and if food should succeed in penetrating the opening into the larynx, explosive coughing will be set up, and the whole delicate process may be temporarily disorganized, with food being forced up into the nose. The diversion of food or liquid to one or other side of the opening of the larynx forces it into the piriform recess [FIG. 583], and it is here that such objects as fish bones tend to lodge.

During the action of swallowing, the salpingopharyngeus helps to open the orifice of the auditory tube, and in this it is probably assisted by the levator and tensor veli palatini. This opening of the tube equalizes the pressure in the nasal part of the pharynx and the middle ear, and passengers in an aircraft can relieve the differential pressure on their eardrums during ascent or descent by swallowing.

The process of swallowing, like most common functions of the body, is extremely complicated, and the exact pattern of muscular activity varies with the individual and with the material being swallowed (Cunningham and Basmajian, 1969); a useful review of the literature is that of Bosma (1957).

MUSCLES CONCERNED IN MOVEMENTS OF THE MANDIBLE

MOVEMENT	MUSCLE	PRINCIPAL NERVE SUPPLY
Elevation	Temporalis (anterior fibres) Masseter Medial pterygoid	Anterior division of mandibular Anterior division of mandibular Main trunk of mandibular
Depression	Lateral pterygoid Digastric Mylohyoid ⎫ acting in Geniohyoid ⎬ concert Infrahyoid group ⎭	Anterior division of mandibular Facial: mylohyoid branch of inferior alveolar Mylohyoid branch of inferior alveolar C. 1 through hypoglossal Ansa cervicalis: sternothyroid by C.1 through hypoglossal
Protraction	Lateral pterygoid Medial pterygoid Masseter (superficial fibres)	Anterior division of mandibular Main trunk of mandibular Anterior division of mandibular
Retraction	Temporalis (horizontal fibres) Digastric	Anterior division of mandibular Facial: mylohyoid branch of inferior alveolar
Chewing	Medial pterygoid Lateral pterygoid Masseter Temporalis	Main trunk of mandibular Anterior division of mandibular Anterior division of mandibular Anterior division of mandibular

MUSCLES CONCERNED IN MOVEMENTS OF THE LARYNX

MOVEMENT	MUSCLE	PRINCIPAL NERVE SUPPLY
Elevation	Digastric	Facial: mylohyoid branch of inferior alveolar
	Stylohyoid	Facial
	Mylohyoid	Mylohyoid branch of inferior alveolar
	Geniohyoid	C.1 through hypoglossal
	Thyrohyoid	C.1 through hypoglossal
	Palatopharyngeus	Pharyngeal plexus
	Stylopharyngeus	Glossopharyngeal
	Inferior constrictor (possibly)	Pharyngeal plexus: recurrent laryngeal
Depression	Sternohyoid	Ansa cervicalis
	Sternothyroid	Ansa cervicalis
	Omohyoid	Ansa cervicalis
Protraction	Geniohyoid	C.1 through hypoglossal
Retraction	Stylohyoid	Facial
	Inferior constrictor (possibly)	Pharyngeal plexus: recurrent laryngeal
	Middle constrictor (possibly)	Pharyngeal plexus

FASCIAE OF THE VERTEBRAL COLUMN, NECK, AND HEAD

SUPERFICIAL FASCIA

The superficial fascia of the back is thick and contains a considerable amount of fat held within a meshwork of fibres. Laterally, it is loosely connected to the skin, but in the midline, particularly in the upper part of the neck, it tacks the skin more tightly to the deep fascia.

The front and sides of the neck are covered by loose superficial fascia which allows free movement of the skin over the underlying structures, and which contains the fibres of the platysma. This muscle is attached to the skin, and if it is paralysed, the skin of the neck sags downwards [p. 283].

On the scalp, the superficial fascia forms a dense fibro-fatty layer, which is adherent both to the skin and to the galea aponeurotica, which replaces the deep fascia in the scalp [p. 280]. This arrangement means that the three superficial layers of the scalp (skin, superficial fascia, and galea aponeurotica) move as a unit over the periosteum of the skull, which is separated from the galea by a space filled by looser connective tissue. It may therefore be pulled away from the skull in a flap of skin following injuries or during surgical operations. The texture of the superficial fascia becomes somewhat more open as it is traced laterally, and it contains the superficial vessels and nerves of the scalp.

The superficial fascia of the face contains the muscles of facial expression as well as the regional nerves and blood vessels. Superficial to the buccinator and between it and the masseter lies the **buccal pad** of fat (Gaughran, 1957), an accumulation which is large and prominent in the infant. Over most of the face the superficial fascia is fairly tightly attached to the skin, especially in the region of the nose and the external ear where it is very scanty and ties the skin down to the underlying cartilages. But in the eyelids it is lax and open, and a fascial 'space' exists in which fluid can easily accumulate following injuries to the region of the orbit.

DEEP FASCIA

The whole of the back is covered by a layer of deep fascia of variable thickness and strength. In the neck it is strong and dense, and it forms part of the investing layer of the cervical fascia. Lower down it is relatively thin. It covers over and ensheaths the superficial muscles which connect the upper limb to the trunk, and is attached to the spines of the thoracic and lumbar vertebrae, to the spine and the acromion of the scapula and to the iliac crests and the back of the sacrum. More laterally still, it becomes continuous with the deep fascia of the axilla, the thorax, and the abdomen, and blends with the investing deep fascia of the arm.

Deep to the superficial muscles of the back lies a very much stronger development of the deep fascia. This is well developed only in the sacral, lumbar, and lower thoracic regions, and is therefore termed the **thoracolumbar fascia.** It consists of three layers. The **posterior layer,** which is superficial to the erector spinae muscle, is attached medially to the thoracic, lumbar, and sacral spines and their supraspinous ligaments. It extends from the sacrum and the iliac crest upwards to be attached to the angles of the ribs lateral to the iliocostalis part of the erector spinae. In the lower part of the back this layer forms a very strong membranous aponeurosis which gives attachment to the latissimus dorsi muscle.

The **middle layer** of the thoracolumbar fascia extends from the lower border of the twelfth rib and the lumbocostal ligament down to the iliac crest and iliolumbar ligament. It is attached medially to the tips of the transverse processes of the lumbar vertebrae and to the intertransverse ligaments, and laterally it joins the posterior layer at the lateral border of the erector spinae [FIG. 368].

The **anterior layer** of the thoracolumbar fascia is the thinnest of the three. It lies in front of the quadratus lumborum and is attached medially to the anterior aspect of the

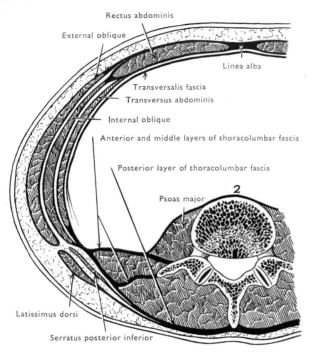

FIG. 368   Transverse section through the abdomen at level of second lumbar vertebra.

cervical vertebra. In front of the trapezius it forms the roof of the posterior triangle [p. 278], and at the anterior margin of the triangle it splits again to invest the sternocleidomastoid [FIG. 369]. Anterior to this muscle the investing layer passes across the structures in the anterior triangle of the neck [p. 278], becoming bound down to the body and the greater horn of the hyoid bone as it does so.

Superiorly, the investing layer is attached in a continuous line to the external occipital protuberance, the superior nuchal line of the occipital bone, and the mastoid process. Further forwards, it divides to form a sheath for the parotid gland, the superficial part of this sheath being attached to the zygomatic arch and disappearing anteriorly on the face. A portion of the deep part of the sheath forms the **stylomandibular ligament,** which ascends to the styloid process as a partition between the parotid and submandibular glands. Anterior to the parotid gland, the investing layer is thin and capable of being stretched ; it is attached along the lower border of the mandible, but splits again to enclose the submandibular gland, the deep layer ascending to the mylohyoid line on the inner surface of the mandible.

Inferiorly, the investing layer is attached, along with the trapezius, to the acromion process of the scapula, and in front

lumbar transverse processes, while laterally it fuses with the middle layer at the lateral border of the quadratus lumborum; here it becomes continuous with the posterior aponeurotic attachments of the internal oblique and the transversus abdominis [FIG. 368]. Superiorly the anterior layer is attached to the twelfth rib and the first lumbar transverse process, between which it is thickened to form the lateral arcuate ligament. Infieriorly, it gains attachment to the iliolumbar ligament and the iliac crest.

Cervical Fascia

In the neck, as elsewhere in the body, the muscles and viscera are surrounded by connective tissue packing material. In some places this is more dense than in others, and on this basis many fascial planes have been described which have at best a somewhat doubtful importance, and at worst a tenuous claim to reality. However, three main layers of cervical fascia can be recognized without much difficulty.

The **investing layer** surrounds the neck like a sleeve ; it is less strong anteriorly, where it is covered by the platysma, and posteriorly it splits to enclose the trapezius muscle. In the midline, the investing layer is attached to the external occipital protuberance, the ligamentum nuchae, and the spine of the seventh

FIG. 369   Transverse section of neck between fourth and fifth cervical vertebrae.

of this, in the roof of the posterior triangle, it splits inferiorly into two layers. The superficial one is attached to the upper surface of the clavicle, and the deeper one surrounds the inferior belly of the omohyoid, forming for this muscle a fascial sling which holds it approximately parallel to the clavicle. This deeper layer passes behind the clavicle, fusing with its periosteum, and is attached to the sheath of the subclavius muscle and the anterior end of the first rib.

In front of the sternocleidomastoid, in the anterior triangle, the inferior part of the investing layer splits again. The superficial layer so formed is attached to the anterior border of the manubrium sterni and the deeper layer to its posterior border, so enclosing between them a small compartment called the **suprasternal space,** which contains the interclavicular ligament, the sternal heads of the sternocleidomastoid muscles and the jugular venous arch.

The deepest of the three layers of deep fascia in the neck is the **prevertebral fascia,** which extends from the basilar part of the occipital bone into the thorax, where it loses its identity by becoming continuous with the anterior longitudinal ligament. It lies in front of the prevertebral muscles [FIG. 369], and is connected to the buccopharyngeal fascia [q.v.] by a loose layer of areolar tissue; in this **retropharyngeal space** lie the retropharyngeal lymph nodes. Laterally, the fascia extends in front of the scalene muscles on to the levator scapulae and the splenius muscles. Here it forms a floor for the posterior triangle of the neck, and lies in front of the cervical plexus. As the subclavian artery and the trunks of the brachial plexus emerge from behind the scalenus anterior, they take with them a prolongation of this fascia which forms the **axillary sheath.** Further laterally, deep to the trapezius, the prevertebral fascia becomes much thinner and loses its identity.

The third, and intermediate, layer of fascia is the **pretracheal fascia,** which is less definite than the other two, and forms a sheath for the thyroid gland. Anteriorly it is attached to the arch of the cricoid cartilage and posterolaterally to the carotid sheath. Superiorly the pretracheal fascia is fixed to the oblique line of the thyroid cartilage, and fuses with the fascia on the inferior constrictor muscle. Inferiorly, the fascia passes downwards from the isthmus of the thyroid gland in front of the trachea into the thorax, where it blends with the fascial sheaths of the great vessels.

The **carotid sheath** is merely a condensation of rather more dense connective tissue around the common carotid artery, the internal jugular vein, and the vagus nerve. The sheath is thick in the region of the artery, and relatively thin elsewhere. Anteromedially it blends loosely with the pretracheal fascia, and anterolaterally with the portion of the investing layer on the deep surface of the sternocleidomastoid [FIG. 369]. Posteriorly, the carotid sheath is connected to the prevertebral fascia by relatively loose tissue in which lies the sympathetic trunk.

These layers and sheaths of fascia to a certain extent determine the way infection will spread in the neck. Perhaps the most important of the compartments which they form is the **visceral compartment,** which contains the larynx and the trachea, the pharynx, and the oesophagus. This is bounded anteriorly by the pretracheal fascia, posteriorly by the prevertebral fascia, and laterally by the carotid sheath. It is open below, so that infection entering the space from the larynx or oesophagus may travel downwards into the superior mediastinum of the thorax. On the other hand, fluid forming posterior to the prevertebral fascia will be directed laterally towards the posterior triangle, where it may burst the fascia and appear on the surface, or may be conducted down the axillary sheath to the axilla.

The **buccopharyngeal fascia** clothes the outer surface of the pharynx; it is continuous above with the fascia on the buccinator muscle and below with that on the surface of the oesophagus. Laterally it fuses with the carotid sheath. Between the mucosa of the pharynx and its muscular wall there is another fibrous layer which is given the name of the **pharyngobasilar fascia.** This is well developed superiorly, where it forms the lateral wall of the pharynx above the superior constrictor [p. 291], and gradually fades out as it is traced downwards, though it is still distinguishable at the level of the tonsil, for which it helps to form a fibrous bed. It is attached to the base of the skull, to the pterygomandibular raphe, and to the posterior end of the mylohyoid line of the mandible, and some of its fibres are attached to the cartilaginous part of the auditory tube. Anteriorly and inferiorly extensions of the fascia pass forwards on to the hyoid bone and the thyroid cartilage.

Fascia of Head

In the central portion of the **scalp,** the deep fascia is replaced by the galea aponeurotica [p. 280], but in the temple a strong layer of fascia extends from the superior temporal line to the zygomatic arch, covering over the temporalis muscle. This

Tarsus of upper eyelid
Back of upper eyelid
Palpebral fissure
Orbicularis oculi
Back of lower eyelid
Lateral palpebral ligament
Lateral check ligament
Zygomatic bone
Sheath of lateral rectus
Sheath of inferior oblique
Tendinous fibres of temporalis
Temporal fascia
Fatty fibrous tissue

Nasal bone
Floor of frontal sinus
Upper end of lacrimal sac
Lacrimal part of orbicularis oculi
Medial palpebral ligament
Cavity of nose
Ethmoidal sinuses
Medial check ligament
Sheath of medial rectus
Retina and choroid
Sclera
Sheath of optic nerve
Orbital fat

Optic nerve and canal   Anterior clinoid process

FIG. 370 Horizontal section through left orbit to show arrangement of fascial sheath of eyeball and check ligaments.

temporal fascia is attached anteriorly to the posterior border of the frontal process of the zygomatic bone, and does not pass into the face. The deep fascia of the face is indeed almost non-existent, though a recognizable layer occurs over the masseter muscle.

In the orbit the deep fascia forms a sort of cup for the eyeball, the **vagina bulbi,** but is separated from it by the **episcleral space,** which allows the eyeball to move slightly within the cup. The vagina bulbi and the eyeball move together in the surrounding orbital fat.

The vagina bulbi adheres anteriorly to the conjunctiva and the sclera close to the corneal margin. It is pierced by the extra-ocular muscles, which carry backward from it sleeves of fascia which are adherent to the walls of the orbit, so anchoring the fascial cup. The strongest of these attachments are the medial and lateral **check ligaments,** derived from the sheaths of the medial and lateral rectus muscles [FIG. 370]. The medial check ligament is attached to the lacrimal bone, to the fascia of the eyelids behind the lacrimal part of the orbicularis oculi, and to the medial fornix of the conjunctiva. The lateral check ligament is attached to the middle of the lateral margin of the orbit and to the lateral palpebral ligament and conjunctival fornix [FIG. 370]. The portion of the vagina bulbi which surrounds the inferior rectus muscle and is connected to the medial and lateral check ligaments is somewhst thickened to form a sling (the **suspensory ligament**) which helps to support the eyeball in its surroundings of soft and yielding fat.

If the eye has to be removed, the vagina bulbi forms a socket for an artificial eye, and it also prevents the severed muscles from retracting very far. It is pierced by the vessels and nerves of the eyeball, and posteriorly it is continuous with the dural sheath of the optic nerve.

MUSCLES OF THE UPPER LIMB

In the course of evolution the hand has become a manipulating instrument of great precision, power, and delicacy, and the upper limb is a system of joints and levers designed to bring this instrument to bear at any desired point in space and to hold it there steadily and securely while it carries out its task. The muscles of the upper limb therefore exert their primary functions from proximal origins on distal insertions. However, the upper limb is still on occasion an instrument of locomotion, and by grasping some convenient immovable object, the body can be pulled towards the hand, as in climbing. In this case the muscles operate from a distal fixed point on a proximal movable one.

Since the upper limb is also used for carrying loads, and occasionally for supporting the weight of the body, it is necessary to think of the way in which force is transmitted—usually by the tension of muscles—from the hand to the trunk and vice versa. Lastly, because the upper limb is itself heavy, it must be borne in mind that every movement of it has to be accompanied by postural contractions of the muscles of the trunk and lower limb to compensate for the shift in the centre of gravity of the body.

Many upper limb muscles cross two or more joints, and can therefore operate on all of them; a complex system of synergists and fixators is thus provided to cancel out the unwanted movements, and the procedure of testing for suspected paralysis of a given muscle is often quite complicated.

MUSCLES ATTACHING THE UPPER LIMB TO THE TRUNK

Movements of the limb relative to the trunk take place at the sternoclavicular, acromioclavicular, and shoulder joints, and this system of three joints is largely responsible for the great mobility of the upper limb. Some of the muscles in this region, those which run from the trunk to the scapula, operate exclusively on the shoulder girdle and have no action on the shoulder joint; these are the trapezius, levator scapulae, the rhomboids, the serratus anterior, pectoralis minor, and subclavius. Two, the latissimus dorsi and the pectoralis major, pass from the trunk to the humerus, so their primary action is on the shoulder joint, but they also produce movements of the shoulder girdle. Others—the deltoid, supraspinatus, infraspinatus, teres minor, subscapularis, teres major, and coracobrachialis—pass from the scapula to the humerus, and thus act exclusively on the shoulder joint.

The muscles passing between the trunk, the scapula and the humerus form the boundaries of a pyramidal space, the **axilla,** which lies between the arm and the upper part of the thorax [FIG. 371]. The apex of the axilla is at the medial side of the root of the coracoid process and the base looks downwards. The axilla is bounded anteriorly by the pectoralis major and minor muscles, posteriorly by the subscapularis, teres major, and latissimus dorsi, medially by the serratus anterior and the upper five ribs, and laterally by the humerus, the coracobrachialis, and the short head of the biceps brachii. Lining this space is the axillary fascia [p. 337].

Trapezius

The trapezius is a roughly triangular muscle, which with its fellow forms a trapezium, so meriting its name. It takes origin from the medial third of the superior nuchal line of the occipital bone, the external occipital protuberance, the ligamentum nuchae, the spines of the seventh cervical and all the thoracic vertebrae, and the intervening supraspinous ligaments. It is inserted into the posterior border of the lateral third of the clavicle [FIGS. 372, 376], the medial border of the acromion, the upper border of the crest of the spine of the scapula, and the tubercle on this crest.

The uppermost fibres, which descend to the clavicle, form the posterior border of the posterior triangle of the neck. The lowest fibres, which run up from the lower thoracic spines to the medial end of the spine of the scapula and the tubercle,

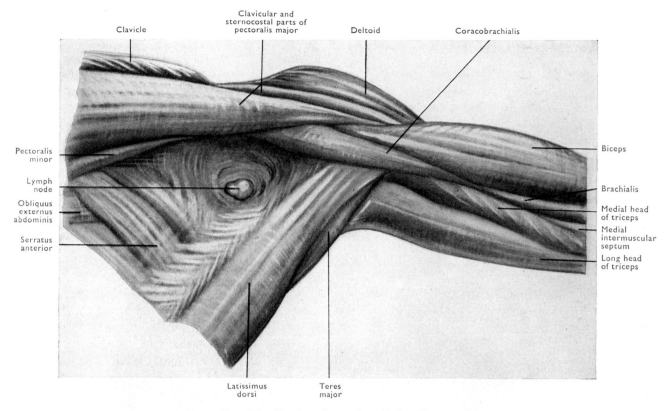

FIG. 371 Fascia and muscles of left axilla.

form the medial boundary of the **triangle of auscultation,** a portion of the chest wall relatively thinly covered by muscle and free of the bony obstruction of the scapula. It is completed below by the upper border of the latissimus dorsi, and laterally by the medial border of the scapula [FIG. 373].

The intermediate fibres of the trapezius run almost horizontally, and are separated by a small bursa from the smooth area at the most medial part of the scapular spine. There is a small diamond-shaped tendinous area over the lower part of the neck and the upper part of the thorax [FIG. 373], and this

corresponds to a hollow in this vicinity seen in the living subject [PLATE 19B].

Nerve Supply. The trapezius receives its motor supply from the spinal portion of the accessory nerve, which enters it from the posterior triangle. In addition it receives sensory fibres from the ventral rami of the third and fourth cervical nerves through the cervical plexus.

Actions. The upper fibres pull the shoulder girdle up towards the skull (elevation) and help to prevent depression of the shoulder girdle when a weight is being carried on the shoulder or in the hand; the importance of this action varies in different subjects (Bearn, 1961), and when the shoulder is lowered to the fully depressed position, activity in the trapezius decreases dramatically and may stop, the strain presumably being taken by the capsule of the sterno-clavicular joint (Bearn, 1967). The lower fibres, on the other hand, depress the medial part of the scapula and lower the shoulder, particularly against resistance, as when using the hands to get up from a chair, or when the weight of the body is taken on the hands by leaning on a table. The intermediate fibres pull the scapula backwards towards the midline (retraction).

Since the upper fibres are attached to the lateral end and the lower fibres to the medial end of the spine of the scapula, the the result of contracting both groups together is to rotate the scapula round a point approximately half way along the spine, the glenoid socket turning to face upwards and forwards. This rotatory action is of importance in the movement of raising the arm above the head [p. 313]. The spine of the scapula affords a good purchase for the muscle, and turns like a winged nut in

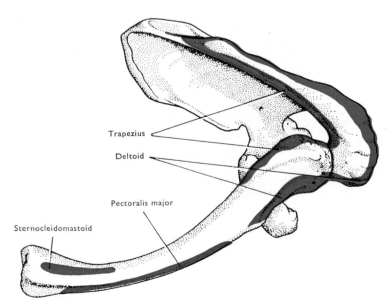

FIG. 372 Shoulder girdle from above showing muscle attachments.

FIG. 373　Superficial muscles of the back, and vertebroscapular muscles.

response to the couple produced by the trapezius [PLATE 19B].

Paralysis of the upper part of the trapezius allows the scapula to drift forwards on the chest wall, and to rotate so that its inferior angle swings medially (Lockhart, 1930); the smooth curve of the upper border of the muscle from occiput to acromion process eventually becomes converted into an unsightly angulation. The usual test for integrity of the trapezius is to ask the patient to shrug the shoulders against resistance, when the upper fibres can be seen and felt in action. Even though the trapezius is paralysed, a full range of abduction of the upper limb is often possible, for the serratus anterior is an even more powerful rotator of the scapula than the trapezius.

Levator Scapulae

The upper third of the levator scapulae is concealed by the sternocleidomastoid; the middle third forms part of the floor of the posterior triangle of the neck [FIG. 344] and the lower third is hidden by the trapezius. It arises from the posterior tubercles of the transverse processes of the first three or four cervical vertebrae posterior to scalenus medius, and descends to the medial margin of the scapula between the superior angle and the spine [FIGS. 373, 374].

Nerve Supply. From the ventral rami of C.3, 4, and 5, the latter via the dorsal scapular nerve.

Actions. The levator scapulae elevates the scapula and helps to retract it. It also helps to support the weight of the limb and to transmit to the axial skeleton the forces produced by weights carried in the hand.

Rhomboideus Minor

This small muscle comes from the lower part of the ligamentum nuchae, the spines of the seventh cervical and first thoracic

vertebrae, and the supraspinous ligament between them. Its fibres pass downwards and laterally, parallel to those of levator scapulae, to the medial border of the scapula at the level of the spine [FIGS. 373, 374].

Nerve Supply. From the ventral ramus of C.5 through the dorsal scapular nerve.

Actions. As for levator scapulae, though it is more of a retractor than an elevator.

Rhomboideus Major

The rhomboideus major is often continuous with the rhomboideus minor, and arises from the spines and supraspinous ligaments of the thoracic vertebrae from the second to the fifth inclusive. It runs parallel to the rhomboideus minor to be inserted into the medial border of the scapula between the root of the spine and the inferior angle. Its lower border appears from under cover of the trapezius to form the floor of the triangle of auscultation.

Nerve Supply. From the ventral ramus of C.5, via the dorsal scapular nerve.

Actions. As for levator scapulae, but principally a retractor. The rhomboid muscles are tested by asking the patient to brace his shoulders back against resistance, when the muscles may be felt and sometimes seen to contract. It is difficult to distinguish their activity from that of the overlying part of the trapezius.

Serratus Anterior

The serratus anterior is a large muscle which arises from the outer surfaces of the upper eight ribs and the fascia covering the spaces between them by a series of slips [FIG. 375]; the lower ones interdigitate with the origin of the external oblique muscle of the abdomen [FIG. 376]. The first slip runs almost

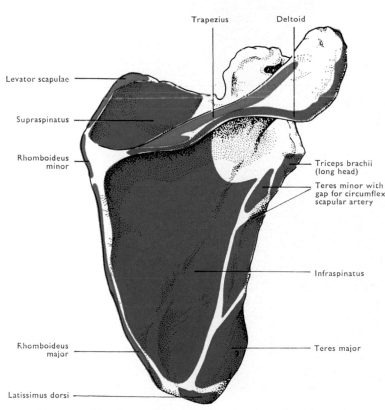

FIG. 374 Muscle attachments to dorsal surface of right scapula.

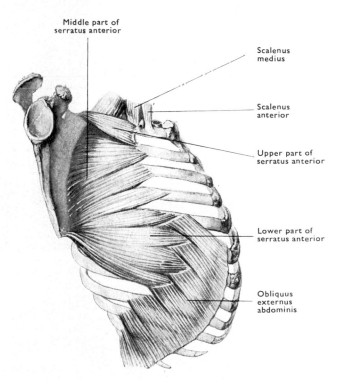

FIG. 375 Serratus anterior and origin of external oblique. The scapula is drawn away from the side of the chest.

straight backwards to the costal surface of the scapula at the superior angle. The next three go to the costal surface of the medial border of the scapula, and the last four pass obliquely upwards and backwards to the inferior angle of the scapula on its costal surface.

The serratus anterior forms the medial wall of the axilla, and is partly covered by the inferolateral part of the mammary gland. Its upper digitation lies behind the clavicle, and its lower border is crossed by the latissimus dorsi.

Nerve Supply. From the ventral rami of C. 5, 6, and 7 via

the long thoracic nerve, which enters the muscle on its superficial aspect. The highest digitations are supplied by the fifth cervical nerve and the lowest by the seventh.

Actions. In quadrupeds the serratus anterior, acting from its insertion on its origin, transfers the weight of the trunk to the scapula and thence to the forelimb. This action is also of some importance in man when the weight is taken on the hands, as in crawling on all fours. In the erect posture, the muscle thrusts the shoulder girdle forwards, and can push a stalled car, or deliver a punch in boxing. In all such movements of

FIG. 376   Muscles of anterior wall of the trunk.

protraction the serratus anterior pulls the scapula forwards on the curve of the chest and holds it closely in to the chest wall. Because the concentration of protracting power brought to bear at the lower angle of the scapula is greater than that achieved at the superior angle, the muscle operates with the trapezius in rotating the scapula so that the glenoid socket looks upwards and forwards [PLATE 20B].

With the rhomboids and levator scapulae, the serratus anterior steadies the scapula to form a stable platform from which other muscles can act. If the serratus anterior is paralysed, the medial border of the scapula stands further away from the chest wall, especially when the body is supported by the arms or when pushing movements are carried out. This is called 'winging' of the scapula, and is an important test for detecting paralysis of the muscle. In the absence of the rotating action of the serratus anterior the upper limb cannot be abducted from the side through more than about 90 degrees. Contrary to former belief, the serratus anterior does not appear to play any part in forced respiration (Catton and Gray, 1951).

Pectoralis Minor

This flat triangular muscle lies in the anterior wall of the axilla posterior to the pectoralis major, and concealed by it [FIGS. 371, 376]. It arises from the outer surfaces of the third, fourth and fifth ribs near their anterior ends and from the fascia of the corresponding intercostal spaces. There may be an additional origin from the second rib or the sixth rib; less commonly from both. The short, flat tendon passes into the medial border and upper surface of the coracoid process of the scapula. Rarely, the tendon continues over the coracoid process, separated from it by a bursa, to pierce the coraco-acromial ligament and become continuous with the coraco-humeral ligament [FIG. 297].

Nerve Supply. The medial pectoral nerve, a branch of the medial cord of the brachial plexus, pierces the muscle and supplies it. The lateral pectoral nerve, a branch of the lateral cord of the plexus, communicates with the medial pectoral nerve in the axilla, and this communication ensures that the pectoralis minor, like the pectoralis major, is supplied by both pectoral nerves. The segmental supply is probably from C. 6, 7, and 8.

Actions. The muscle draws the shoulder girdle forwards and downwards, and helps to transfer the weight of the trunk to the upper limb when leaning on the hands. It could also be used to pull the ribs up as an accessory muscle of inspiration. It is virtually impossible to test for the integrity of the pectoralis minor, since it is wholly under cover of the pectoralis major, and is never paralysed alone.

Subclavius

The tendinous origin of the subclavius is from the junction of the first rib and the first costal cartilage in front of the costoclavicular ligament [FIG. 376]. The small fleshy belly is inserted into the floor of a groove on the inferior surface of the clavicle [FIG. 377]. The muscle lies wholly under cover of the pectoralis major and the clavicle.

Nerve Supply. The nerve to the subclavius comes from the upper trunk of the brachial plexus (C. 5,

6), and descends in front of the subclavian artery to reach the muscle.

Actions. The subclavius depresses the lateral end of the clavicle and draws the bone towards the sternum, thereby steadying the clavicle in movements of the shoulder girdle. Paralysis of the muscle produces no demonstrable effect.

Latissimus Dorsi

The latissimus dorsi passes between the trunk and the humerus, and therefore acts on both the shoulder joint and the shoulder girdle. It arises from the posterior layer of the thoracolumbar fascia, which is attached to the spines of the lower thoracic and all the lumbar and sacral vertebrae, as well as to the intervening supraspinous ligaments. In addition, it takes origin from the posterior part of the outer lip of the iliac crest, from the outer surfaces of the lower three or four ribs [FIG. 373] and the inferior angle of the scapula, for which the muscle forms a kind of pocket.

From this widespread origin the latissimus dorsi narrows down to a thin flattened tendon 3–4 cm. in width, which winds round and adheres to the lower border of the teres major, and inserts into the floor of the intertubercular sulcus of the humerus [FIG. 382] anterior to the tendon of the teres major and separated from it by a bursa. Because of this spiral turn, the anterior surface of the tendon of insertion is continuous with the posterior surface of the rest of the muscle. The superior border of the latissimus dorsi forms the inferior border of the triangle of auscultation, and the lateral border of the muscle forms the medial border of the lumbar triangle [p. 345 and FIG. 373].

Nerve Supply. From the thoracodorsal nerve (C. 6, 7, 8), which supplies it on its deep surface. The nerve comes from the posterior cord of the brachial plexus.

Actions. The muscle retracts the shoulder girdle (if the humerus is fixed to the scapula by the scapulohumeral muscles) and is a strong extensor of the flexed arm. It is also a powerful adductor of the humerus, and helps to rotate it medially. Acting from its insertion, the latissimus dorsi is one of the chief climbing muscles, since it pulls the trunk up on the arms [PLATE 20A], and it is also a powerful factor in rowing and in the downstroke in swimming. This action can also be used in forced inspiration, for the costal attachment serves to raise the lower ribs. But the muscle also contracts during violent expiration, such as coughing or sneezing, apparently to

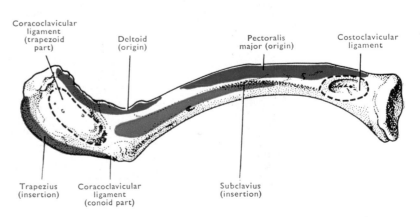

FIG. 377 Muscle attachments to inferior surface of right clavicle.

compress the thorax and abdomen. The best test of the muscle is to ask the patient to cough while holding the muscle between finger and thumb as it reaches the posterior axillary fold. Alternatively, the muscle can be felt when the patient adducts the horizontally abducted arm against resistance.

Pectoralis minor and latissimus dorsi transfer the weight to the upper limb when the body is pushed upwards by the hands. In patients with paralysis of the lower limbs resulting from spinal injury the nerve supply of the muscle, coming from the cervical region, may remain intact, and the latissimus dorsi can be used from its insertion to move the pelvis. Accordingly, efforts are made to strengthen the muscle, through exercise such as archery, in order to help these patients to achieve some independence of movement.

Pectoralis Major

Like the latissimus dorsi, the pectoralis major extends from the thoracic wall to the humerus, and therefore acts on both the shoulder girdle and shoulder joint; its major importance relates to the shoulder joint. The smaller clavicular head of the muscle arises from the front of the clavicle in its medial half or two thirds [FIGS. 372, 376, 377]; the larger sternocostal portion arises from the front of the manubrium and the body of the sternum, and from the anterior aspects of the first six costal cartilages as well as from the anterior wall of the rectus sheath [FIGS. 376, 378]. The two parts of the muscle are usually continuous with each other, but may be separated by a groove.

The pectoralis major is inserted into the crest of the greater tubercle of the humerus. The clavicular portion of the muscle

FIG. 378   Muscle attachments to front of sternum.

is attached in front of the sternocostal portion, the fibres of which twist on themselves, forming the rounded **anterior fold of the axilla**, so that the fibres which arise lowest on the chest wall pass behind the upper fibres to be inserted posterior to them. In this way the tendon comes to resemble a U in cross section. The posterior limb of the U is blended with the capsule of the shoulder joint, and the clavicular fibres blend with the insertion of the deltoid.

The pectoralis major is separated from the deltoid muscle by a depression known as the **infraclavicular fossa** [PLATE 21 C], in which lie the cephalic vein and the deltoid branches of the thoraco-acromial artery.

Nerve Supply. The lateral pectoral nerve (C. 5, 6, 7), from the lateral cord of the brachial plexus, supplies the clavicular and the upper part of the sternocostal portion; the medial pectoral nerve (C. 8, T. 1), from the medial cord, supplies the lower part of the sternocostal portion.

Actions. The pectoralis major is a powerful adductor and medial rotator of the humerus. The clavicular portion helps to flex the humerus to the horizontal, while the sternocostal portion, acting against resistance, extends the flexed humerus [PLATE 22 A]. In such actions the two parts of the muscle act quite independently, but they combine in movements of pushing or throwing. The pectoralis major is one of the main climbing muscles, and pulls the body up to the fixed arm. This action is also used in respiratory difficulties; the upper limb is fixed by the hand grasping the bed rail, and the pectoralis major becomes an accessory muscle of inspiration, pulling up the upper ribs.

Despite its numerous and important actions, absence of the pectoralis major, which may rarely occur as a congenital anomaly, causes virtually no disability, though the deformity caused by the absence of the anterior axillary fold is striking. The integrity of the muscle is tested by adduction of the arm against resistance [PLATE 21 C].

MUSCLES OF THE SHOULDER JOINT

Deltoid

The deltoid [FIGS. 373, 376] is a coarse, thick, multipennate muscle with an extensive origin from the anterior border of the lateral third of the clavicle [FIG. 372], the lateral border of the acromion, the lower lip of the crest of the spine of the scapula, and the fascia covering the infraspinatus. This attachment is similar to that of the trapezius, though the deltoid is on the outside and the trapezius on the inside of the bony arch to which they are attached.

As its name implies, the muscle is triangular, and its apex is inserted into the deltoid tuberosity half way down the lateral surface of the body of the humerus [FIG. 382]. Its fibres blend with the anterior aspect of the tendon of the pectoralis major.

The deltoid is composed of three functional parts, anterior, posterior, and intermediate. Only the intermediate portion, which abducts the shoulder, is multipennate, and this is probably because the mechanical disadvantage under which it must act requires considerable strength.

Nerve Supply. From the axillary nerve, a branch of the

posterior cord of the brachial plexus containing fibres from the ventral rami of C. 5 and 6.

Actions. The intermediate portion is the chief abductor of the humerus, elevating the arm away from the side in the plane of the scapula, but it can only do so efficiently after the movement has been initiated by the supraspinatus, since when the arm is by the side the deltoid of necessity directs its pull upwards rather than outwards. Its tendency to produce upward displacement of the humerus is resisted by the muscles of the 'rotator cuff'. The deltoid is also active in controlling gravitational descent of the upper limb [PLATE 21 A and B].

The anterior portion of the muscle is a strong flexor and medial rotator of the humerus; the posterior portion is a strong extensor and lateral rotator, and can help to transfer to the shoulder girdle the strain imposed by heavy weights carried in the hand (Bearn, 1961). In adduction, the posterior fibres are very active, probably to resist the medial rotation which the latissimus dorsi and pectoralis major would otherwise produce (Scheving and Pauly, 1959). To test the deltoid, the patient's arm is abducted in the plane of the scapula to about 45 degrees, and he is asked to hold it there against resistance. The muscle is then easily seen and felt. It is interesting that when the deltoid is paralysed it is sometimes possible for the upper limb to be abducted through a full normal range (Sunderland, 1944b).

Supraspinatus

The origin of the supraspinatus is from the medial two-thirds of the supraspinous fossa of the scapula [FIG. 374], and from the fascia which covers the muscle. It is inserted into a facet on the superior aspect of the greater tubercle of the humerus and also into the capsule of the shoulder joint [FIG. 382]. The supraspinatus lies deep to the trapezius, the acromion, and the coraco-acromial ligament.

Nerve Supply. From the supra-scapular nerve (C. 5 and 6), a branch of the upper trunk of the brachial plexus.

Actions. The supraspinatus initiates the process of abduction at the shoulder joint, and is more important in the early stages of this movement than it is later, when the deltoid takes over. If it is paralysed, the patient may produce this initial abduction by leaning slightly to the

side, so using gravity to replace the supraspinatus, or he may 'kick' the elbow out by a jerk of his hip. In all movements of the joint the supraspinatus acts as an extensile ligament [p. 264] holding the head of the humerus in contact with the shallow glenoid socket. In this it is assisted by the infraspinatus, teres minor, and subscapularis, and the whole muscle group is sometimes known as the **'rotator cuff'** of the shoulder joint (Moseley, 1951). Damage to the supraspinatus tendon is a common form of derangement of the shoulder joint, and naturally causes pain specially on abduction of the humerus: the tendon occasionally ruptures. To test the muscle, the patient attempts to abduct the arm from the side

FIG. 379   Left scapular muscles and triceps.

against resistance; the muscle belly can be felt above the spine of the scapula.

Infraspinatus

The infraspinatus comes from the infraspinous fossa of the scapula, except for the portion next the neck of the bone [FIG. 374], and from the thick fascia covering the muscle. The tendon inserts into the middle facet on the greater tubercle of the humerus and into the capsule of the shoulder joint. Between the infraspinatus and the neck of the scapula there is a bursa which occasionally communicates with the shoulder joint.

The upper part of the infraspinatus lies deep to the acromion, the deltoid muscle, and the trapezius, but the lower part is superficial [FIG. 373].

Nerve Supply. From the suprascapular nerve (C. 5 and 6), a branch of the upper trunk of the brachial plexus.

Actions. The infraspinatus is a lateral rotator of the humerus, but its main action is as a member of the '**rotator cuff**', in which it co-operates with the supraspinatus, teres minor, and subscapularis. These muscles hold the head of the humerus in contact with the glenoid socket of the scapula during movements of the shoulder and are a most important factor in preventing dislocation of the joint. To test the infraspinatus, the patient is asked to rotate the arm laterally against resistance, while keeping the elbow close to the side. This is conveniently done if the forearm is flexed to a right angle and used to push backwards against the examiner's hand. The muscle belly can be felt contracting during this movement.

Teres Minor

The teres minor comes by fleshy fibres from the upper two-thirds of the lateral border of the dorsal surface of the scapula, alongside the lateral border of the infraspinatus [FIG. 374]. It is inserted into the lowest of the three facets on the greater tubercle of the humerus and into the capsule of the shoulder joint. Some fibres extend downwards on to the body for a short distance. The long head of the triceps separates the tendon of the teres minor from the teres major [FIG. 379], so producing a very small **triangular space,** and a larger **quadrangular space** which is bounded laterally by the surgical neck of the humerus.

Nerve Supply. From the axillary nerve (C. 5 and 6), a branch of the posterior cord of the brachial plexus.

Actions. The teres minor is a lateral rotator and an adductor of the humerus, but, like the other members of the rotator cuff [p. 309], its main importance is in holding the head of the humerus in contact with the glenoid socket during movements of the shoulder joint. It helps to prevent upward displacement of the head of the humerus by such muscles as the deltoid, biceps brachii, and triceps (long head.) It is impossible to test the teres minor satisfactorily.

Subscapularis

The fleshy fibres of the subscapularis arise from the groove along the lateral border of the costal surface of the scapula and from the subscapular fossa, except in the region of the angles of the bone [FIG. 380]. The muscle is reinforced by fibrous septa springing from ridges in the fossa. A broad thick tendon inserts into the lesser tubercle of the humerus, the capsule of

the shoulder joint, and the front of the shaft of the humerus for 2–3 cm. below the tubercle [FIG. 382].

The subscapularis forms the greater part of the posterior wall of the axilla, and it is separated from the neck of the scapula by the subscapular bursa, which nearly always communicates with the shoulder joint. It constitutes the upper boundary of the triangular and the quadrangular spaces as they are seen from the front [FIG. 381].

Nerve Supply. From the upper and lower subscapular nerves (C. 5, 6, and 7), branches of the posterior cord of the brachial plexus.

Actions. The muscle is a medial rotator of the humerus but its chief function is as a member of the 'rotator cuff' [p. 309] of extensile ligaments stabilizing the shoulder joint and holding the head of the humerus into the glenoid socket. It also prevents the head from being pulled upwards by the deltoid, biceps, and long head of triceps. If the humerus is abducted through 90 degrees, the subscapularis, together with the pectoralis major and the anterior part of the deltoid, pulls the arm horizontally forwards, a movement which has been termed horizontal flexion [p. 311].

Teres Major

This muscle comes from a raised oval area on the dorsum of the scapula near the inferior angle [FIG. 374], and also from the adjacent fascia. Its fibres are adherent to those of the latissimus dorsi [FIG. 381] and it is inserted by a broad flat tendon into the crest of the lesser tubercle of the humerus, separated from the tendon of latissimus dorsi by a bursa.

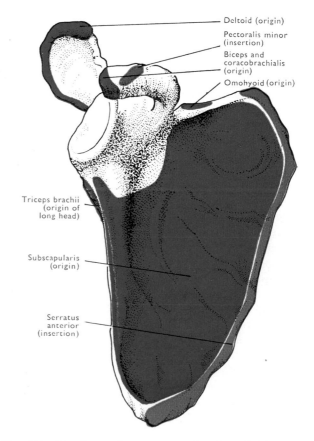

FIG. 380 Muscle attachments to costal surface of right scapula.

Serratus anterior

Subscapularis

Supraspinatus

Pectoralis minor

Coracoid process

Triceps (long head)

Serratus anterior

Latissimus dorsi

Teres major

Coracobrachialis

Biceps (short head)

Teres major

Biceps (long head)

Pectoralis major

Deltoid

Triceps

Biceps

Medial intermuscular septum

Brachialis

Biceps tendon

Supinator muscle

Brachioradialis

Flexor carpi radialis

Pronator teres

FIG. 381   Muscles of posterior wall of left axilla and front of arm.

The teres major, and the tendon of latissimus dorsi which sweeps round it, form the **posterior fold of the axilla** [p. 302]. The muscle also constitutes the lower boundary of the triangular and quadrangular spaces in the posterior wall of the axilla [FIG. 379]. Through the quadrangular space pass the axillary nerve and the posterior humeral circumflex vessels, and through the triangular space pass the circumflex scapular vessels [FIG. 386].

Nerve Supply. From the lower subscapular nerve (C. 6 and 7), a branch of the posterior cord of the brachial plexus which also supplies the lower portion of the subscapularis.

Actions. It adducts and medially rotates the humerus, and can help to extend it from the flexed position. If the arm is adducted against resistance, the teres major can be felt contracting in the posterior axillary fold.

MOVEMENTS OF THE SHOULDER REGION

Movements occurring at the shoulder girdle and movements occurring at the shoulder joint normally complement each other, and it is almost impossible to dissociate them under normal conditions. From the anatomical position, with the upper limb by the side, the scapula can be elevated or depressed, protracted or retracted round the chest wall, or rotated so that the inferior angle travels laterally or medially. All such movements involve the sternoclavicular and acromioclavicular joints, and the muscles concerned in them are shown in the first table below. The second table shows the muscles concerned in movements of the shoulder joint.

From the point of view of muscle actions, the movements of the shoulder joint are best defined in relation to the plane of the scapula rather than in relation to the three basic anatomical planes. Thus abduction is considered to take place away from the body in the plane of the scapula, the arm travelling upwards, outwards, and forwards. Adduction is the reverse of this, and flexion and extension take place at right angles to this plane. Thus, in flexion the arm passes not only forwards but medially, and in extension it travels laterally as well as backwards.

Two additional useful movements require special definition. The swing of the arm from a position of 90 degrees of abduction forwards to a position of 90 degrees flexion is called **horizontal flexion,** and the reverse movement is **horizontal extension.** These movements are used in such activities as drawing horizontal lines on a blackboard, the breast stroke in swimming, and dusting a shelf.

The way in which movement of the shoulder girdle combines with movement of the shoulder joint may be illustrated by considering abduction of the upper limb above the head. The muscles which rotate the scapula are active throughout this movement, and so are the muscles which abduct the humerus at the shoulder joint. About half the total range of

MUSCLES CONCERNED IN MOVEMENTS OF THE SHOULDER GIRDLE

MOVEMENT	MUSCLE	PRINCIPAL NERVE SUPPLY
Elevation	Trapezius (upper fibres)	Spinal part of accessory
	Levator scapulae	Ventral rami of C. 3, 4, 5
	Rhomboideus major	Ventral ramus of C. 5
	Rhomboideus minor	Ventral ramus of C. 5
	Sternocleidomastoid	Spinal part of accessory
Depression	Trapezius (lower fibres)	Spinal part of accessory
	Pectoralis minor	Medial and lateral pectoral
	Pectoralis major (sternocostal portion)	Medial and lateral pectoral
	Latissimus dorsi	Thoracodorsal
Protraction	Serratus anterior	Long thoracic
	Pectoralis minor	Medial and lateral pectoral
	Pectoralis major	Medial and lateral pectoral
Retraction	Trapezius (middle fibres)	Spinal part of accessory
	Rhomboideus major	Ventral ramus of C. 5
	Rhomboideus minor	Ventral ramus of C. 5
	Latissimus dorsi	Thoracodorsal
Lateral displacement of inferior angle of scapula	Serratus anterior	Long thoracic
	Trapezius (upper and lower fibres)	Spinal part of accessory
Medial displacement of inferior angle of scapula	Pectoralis minor	Medial and lateral pectoral
	Rhomboideus major	Ventral ramus of C. 5
	Rhomboideus minor	Ventral ramus of C. 5
	Latissimus dorsi	Thoracodorsal

MUSCLES CONCERNED IN MOVEMENTS OF THE SHOULDER JOINT

MOVEMENT	MUSCLE	PRINCIPAL NERVE SUPPLY
Flexion	Deltoid (anterior portion)	Axillary
	Pectoralis major (clavicular portion: sternocostal portion flexes the extended humerus as far as the position of rest)	Medial and lateral pectoral
	Coracobrachialis	Musculocutaneous
	Biceps brachii	Musculocutaneous
Extension	Deltoid (posterior portion)	Axillary
	Teres major (of flexed humerus)	Lower subscapular
	Latissimus dorsi (of flexed humerus)	Thoracodorsal
	Sternocostal portion of pectoralis major (of flexed humerus)	Medial and lateral pectoral
	Long head of triceps (to position of rest)	Radial
Abduction	Deltoid (middle portion)	Axillary
	Supraspinatus	Suprascapular
	Biceps brachii (long head)	Musculocutaneous
Adduction	Pectoralis major	Medial and lateral pectoral
	Latissimus dorsi	Thoracodorsal
	Teres major	Lower subscapular
	Triceps (long head)	Radial
	Coracobrachialis	Musculocutaneous
Lateral rotation	Deltoid (posterior portion)	Axillary
	Infraspinatus	Suprascapular
	Teres minor	Axillary
Medial rotation	Pectoralis major	Medial and lateral pectoral
	Teres major	Lower subscapular
	Latissimus dorsi	Thoracodorsal
	Deltoid (anterior portion)	Axillary
	Subscapularis	Upper and lower subscapular
Horizontal flexion	Deltoid (anterior portion)	Axillary
	Pectoralis major	Medial and lateral pectoral
	Subscapularis	Upper and lower subscapular
Horizontal extension	Deltoid (posterior portion)	Axillary
	Infraspinatus	Suprascapular

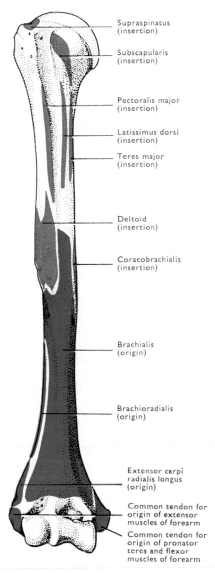

Supraspinatus
(insertion)

Subscapularis
(insertion)

Pectoralis major
(insertion)

Latissimus dorsi
(insertion)

Teres major
(insertion)

Deltoid
(insertion)

Coracobrachialis
(insertion)

Brachialis
(origin)

Brachioradialis
(origin)

Extensor carpi
radialis longus
(origin)

Common tendon for
origin of extensor
muscles of forearm

Common tendon for
origin of pronator
teres and flexor
muscles of forearm

FIG. 382   Muscle attachments to front of right humerus.

lie inferior to the capsule of the joint, and acts as a support preventing downward dislocation of the head of the humerus. If abduction is undertaken in the coronal plane, the summated action of the muscles operating from the scapula and trunk on the humerus causes an automatic lateral rotation of the humerus (Martin, 1940). If this rotation is deliberately prevented, it is impossible to abduct the arm to an angle of more than about 90 degrees because of the obstruction of the greater tubercle of the humerus. If the movement takes place in the plane of the scapula no such rotation takes place.

The deltoid, pectoralis major, latissimus dorsi, and teres major are the chief prime movers of the shoulder joint, while the components of the 'rotator cuff'—supraspinatus, infraspinatus, teres minor, and subscapularis—have the primary role of extensile ligaments. The deltoid and supraspinatus help to transfer the strain produced by weights carried in the hand. All the muscles passing from the trunk or scapula to the humerus can act in the reverse direction, and this is well seen when a gymnast performs exercises on the parallel bars or on the rings. The two chief climbing muscles are pectoralis major and latissimus dorsi, and both are used in the downstroke in swimming.

The shoulder joint is involved in many movements occurring primarily at other joints. For example, in supination and pronation it is brought into play to extend the range of these movements. Paralysis of the lateral rotators of the shoulder means that when writing the patient has to pull the paper along to the left, since without this shoulder movement he cannot write a continuous line. It is perhaps strange that the medial rotators are apparently so much more powerful than the lateral rotators. The infraspinatus, posterior portion of deltoid, and teres minor (lateral rotators) correspond roughly to the subscapularis, anterior portion of deltoid, and teres major (medial rotators). This leaves a surplus of two extremely strong muscles (pectoralis major and latissimus dorsi) favouring medial rotation. Yet this movement does not appear to be, functionally, any more important than lateral rotation.

movement occurs as a result of the glenoid socket being turned to face upwards, the other half is due to abduction of the humerus [PLATE 11, A and D].

When abduction begins, the weight of the upper limb may cause a slight medial movement of the inferior angle of the scapula, but it then begins to move laterally under the influence of the trapezius and the serratus anterior. The levator scapulae steadies the upper angle of the scapula and prevents it moving laterally. The rotation of the scapula is necessarily accompanied by some rotation and elevation of the clavicle, under the influence of the upper fibres of the trapezius, and the subclavius is probably also in action to hold the clavicle into the socket of the sternoclavicular joint [p. 307].

At the same time as this is going on, the supraspinatus and deltoid abduct the humerus at the shoulder joint, while the rotator cuff of muscles, particularly the supraspinatus, holds the head of the humerus into the glenoid socket of the scapula. The downward pull of subscapularis, infraspinatus, and teres minor prevents upward displacement of the head of the humerus due to the action of the strong deltoid muscle. As the movement proceeds, the long head of the triceps comes to

MUSCLES OF THE ARM

Biceps Brachii

This complicated muscle operates on three joints. It has two tendinous heads at its upper end, and two tendinous insertions at its lower end. The short head is attached, in common with the coracobrachialis, to the tip of the coracoid process of the scapula [FIGS. 380, 381]. The long head arises from the supraglenoid tubercle of the scapula and from the labrum glenoidale of the shoulder joint. Its tendon runs through the shoulder joint, invested by a covering of the synovial membrane, and leaves the joint deep to the transverse humeral ligament. It then lies in the intertubercular sulcus of the humerus, covered over anteriorly by the tendon of the pectoralis major [FIG. 381].

The two fleshy bellies derived from the separate heads unite just below the middle of the arm. At the elbow, the resultant fusiform mass gives rise to a strong tendon which is inserted into the posterior part of the tuberosity of the radius [FIGS. 388, 389], from the remainder of which it is separated by a small bursa. The second insertion is the **bicipital apo-neurosis,** a strong membranous band which runs downwards across the cubital fossa [FIG. 387], to join the deep fascia covering the origins of the flexor muscles of the forearm. The upper portion of this band can easily be felt in the living body as a crescentic border crossing in front of the brachial artery and median nerve.

Occasionally there may be a third head of the biceps; this arises at the insertion of the coracobrachialis and passes into the bicipital aponeurosis on the medial side of the main belly of the muscle.

Nerve Supply. From the musculocutaneous nerve (C. 5, 6). **Actions.** The biceps is an important flexor of the elbow joint, and a powerful supinator of the forearm. It has been described as the muscle which puts in the corkscrew and pulls out the cork. Under normal circumstances both actions occur together, and the unwanted one is cancelled out by synergists. The biceps exerts its maximum power in both supination and flexion when the elbow is flexed at right angles, and the usual test of its power is to ask the patient to flex, against resistance, the supinated forearm, bent at right angles to the abducted arm. The muscle is then easily seen and felt.

The biceps can also flex the shoulder joint, and the long head is part of the important stabilizing mechanism for the joint. If the deltoid is paralysed, the long head of the biceps can occasionally effect some useful abduction of the humerus.

Coracobrachialis

The coracobrachialis [FIGS. 371, 381] arises from the tip of the coracoid process of the scapula in company with the short head of the biceps, and ends in a flat tendon inserted into the middle of the medial aspect of the body of the humerus [FIG. 382]. Some of the fibres may continue into the medial intermuscular septum of the arm.

Nerve Supply. The nerve to the coracobrachialis (C. 6, 7) may arise directly from the lateral cord of the brachial plexus, though it is usually a branch of the musculocutaneous nerve.

Actions. It is a flexor and adductor of the shoulder joint.

Brachialis

The main flexor of the elbow lies posterior to the biceps brachii, and comes from the distal two thirds of the front of the body of the humerus and from the anterior aspects of both intermuscular septa [FIGS. 382, 384]; its proximal end clasps the insertion of the deltoid. The muscle inserts into the inferior part of the coronoid process and the tuberosity of the ulna [FIGS. 383, 389] by a tendon which forms the floor of the cubital fossa [p. 316]. Some fibres of the brachialis are inserted into the articular capsule of the elbow joint, and the muscle may be partly fused with the brachioradialis [p. 321].

Nerve Supply. From the musculocutaneous nerve (C. 5, 6). Another branch (C. 5, 6) is received from the radial nerve as it runs along the lateral border of the brachialis. This branch is almost certainly sensory.

Actions. It flexes the elbow. The fibres attached to the capsule of the elbow joint pull the redundant folds of the articular capsule

FIG. 383   Dissection of muscles of front of upper arm. The biceps has been removed to show the brachialis.

Short head of biceps
Deltoid
Long head of biceps
Coracobrachialis
Lateral head of triceps
Brachialis
Long head of triceps
Radial nerve
Ulnar nerve
Medial head of triceps (surface cut from medial intermuscular septum)
Brachialis (surface cut from medial intermuscular septum and triceps)
Brachioradialis
Tendon of biceps
Humeral head
Ulnar head
} of pronator teres
Supinator
Flexor digitorum superficialis

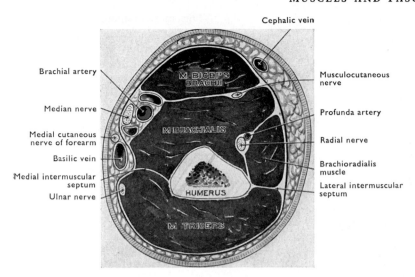

FIG. 384 Section through distal third of right upper arm.

upwards during flexion of the joint, and prevent them from becoming trapped between the moving bones. During flexion of the elbow against resistance, the brachialis can be felt contracting alongside the tendon of the biceps and on either side of its belly.

Triceps Brachii

The triceps is the only muscle on the back of the arm, and arises by three heads. The tendinous long head comes from the infraglenoid tubercle of the scapula and the labrum glenoidale, and its fleshy belly occupies the middle of the back of the arm [FIGS. 379, 386]. The fleshy lateral and medial heads are attached to the posterior aspect of the humerus between the insertion of teres minor and the olecranon fossa [FIG. 385]. The lateral head arises above and lateral to the groove for the radial nerve, and the larger medial head arises inferior and medial to it. The lateral head therefore covers over the groove posteriorly as it joins the medial head. The medial head, which lies deep to the other two, has additional attachments to the posterior aspects of the lateral and medial intermuscular septa.

All three heads are inserted together by a broad flattened tendon into the posterior part of the proximal surface of the olecranon of the ulna and into the deep fascia of the forearm on each side of it. Some of the fibres of the medial head are attached to the posterior part of the capsule of the elbow joint, and a small bursa separates the tendon of the triceps from the remainder of the capsule.

The triceps is superficial in almost its whole length [PLATES 20 C, 22 B], and the long head passes between teres major and teres minor in the posterior wall of the axilla.

Nerve Supply. All three heads are supplied separately by branches of the radial nerve. The lateral head derives its supply from C. 6, 7, 8, and the long and medial heads from C. 7, 8. The medial head receives two branches, one of which has a long extramuscular course in company with the ulnar nerve before it enters the distal part of the muscle; the other enters its proximal part and continues through the substance of the muscle to end in the anconeus [FIG. 386].

Actions. It extends the elbow joint and pulls the posterior part of its capsule out of danger as it does so. The long head can adduct the humer~ tend it from the flexed position; it has a action on the shoulder joint, and is an extensile ligament on the under surface o' during abduction. To test the tricep, abducted to a right angle to bring the fo to the ground and so eliminate the ef on extension of the elbow. The flexed extended against resistance.

MOVEMENTS OF THE ELBOW

There is little difficulty in sorting out the muscles concerned in the movements of flexion and extension of the elbow joint, and they are indicated in the table below. Nevertheless, some points of interest must be mentioned. In the first place the movements are considerably influenced by gravity. From the anatomical position the elbow can be flexed and

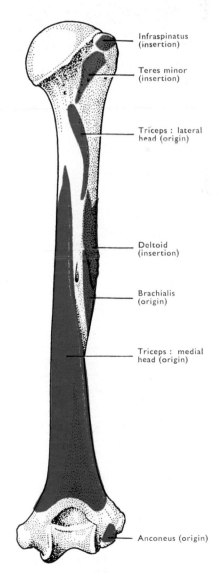

FIG. 385 Muscle attachments to back of right humerus.

FIG. 386   Dissection of back of shoulder and upper arm.

The more important of the two movements is flexion, which depends on several groups of muscles. The muscles arising from the common flexor origin on the front of the medial epicondyle are not normally used to flex the elbow, but can assist in the movement in case of difficulty, particularly in the case of the pronator teres and the flexor carpi radialis. Flexion has thus the safeguard of being mediated by three nerves—musculocutaneous, radial, and median, whereas extension depends on one nerve alone, the radial.

Flexion at the elbow is usually limited by the collision of the soft parts of the forearm with those of the arm : the greater the development of the flexor muscles, the smaller the range of flexion permitted. A similar situation occurs in relation to the knee joint.

MUSCLES OF THE FOREARM

MUSCLES OF THE FRONT OF THE FOREARM

There are three functional groups of muscles on the front of the forearm —the pronators of the forearm, the flexors of the wrist, and the long flexors of the fingers and thumb. They are arranged in three layers. The **superficial layer** consists of four muscles which radiate from a common tendon attached to the front of the medial epicondyle of the humerus : these are the pronator teres, the flexor carpi radialis, the palmaris longus, and the flexor carpi ulnaris [FIG. 387; PLATE 22 B]. The pronator teres forms the medial boundary of a depression in front of the elbow known as the **cubital fossa**; the lateral boundary of this fossa is the medial edge of the brachioradialis

then extended by the flexors alone, even though the triceps and anconeus are paralysed. Again, a patient with paralysed flexors can both flex and extend the elbow by the triceps alone if the upper limb is raised above the head.

MUSCLES CONCERNED IN MOVEMENTS AT THE ELBOW JOINT

MOVEMENT	MUSCLE	PRINCIPAL NERVE SUPPLY
Flexion	Brachialis	Musculocutaneous
	Biceps brachii	Musculocutaneous
	Brachioradialis	Radial
	Extensor carpi radialis longus	Radial
	Pronator teres	Median
	Flexor carpi radialis	Median
Extension	Triceps brachii	Radial
	Anconeus	Radial

muscle [FIG. 387], and its superior boundary is an imaginary line drawn between the medial and lateral epicondyles of the humerus. The floor is formed by the tendon of the brachialis and the supinator, and the fossa contains the tendon of insertion of the biceps brachii, in addition to parts of the brachial, radial, and ulnar arteries and the radial and median nerves.

Deep to the four superficial muscles lies a **middle layer** consisting of one muscle only, the flexor digitorum superficialis. Deep to this again is the **third layer,** comprising the flexor digitorum profundus, the flexor pollicis longus and the pronator quadratus [FIG. 394].

Pronator Teres

The humeral head of the pronator teres comes from the lower third of the medial supracondylar ridge of the humerus, the corresponding part of the medial intermuscular septum, and the common origin of the flexor muscles of the forearm from the front of the medial epicondyle of the humerus and the fascia covering them. The ulnar head of the muscle is much smaller, and runs from the coronoid process of the ulna [FIG. 389] to join the humeral head on its deep surface [FIG. 388]. Between the two heads the median nerve passes into the forearm.

FIG. 387 Superficial muscles of front of left forearm.

FIG. 388 Deeper muscles of front of left forearm.

The fasciculi of the pronator teres pass downwards and laterally, deep to the brachioradialis, to a roughened oval impression on the middle of the lateral surface of the body of the radius [FIGS. 388, 389].

Nerve Supply. From the median nerve by a branch (C. 6, 7) which enters it above the elbow.

Actions. Pronation of the forearm is the chief action, but it also assists in flexing the elbow. To test the muscle the patient resists an attempt to supinate the forearm, the limb lying vertically by the side. The muscle belly can be felt and occasionally seen contracting.

Flexor Carpi Radialis

The radial flexor of the wrist arises from the common flexor origin on the anterior aspect of the medial epicondyle of the humerus, and from the fascia which covers it and separates it from its neighbours. The muscle runs downwards and obliquely across the forearm from the medial to the lateral side. About the middle of the forearm the fleshy belly is replaced by a flattened tendon which becomes cord-like as it approaches the wrist [FIG. 387]. The tendon, surrounded by a synovial sheath [p. 326], runs in a special compartment in the lateral border of the flexor retinaculum [p. 338], and plays in the groove on the trapezium. It inserts into the front of the bases of the second and third metacarpal bones.

At the wrist, the tendon lies between the radial vessels laterally and the median nerve medially, and is an important guide to the position of both.

Nerve Supply. By a branch (C. 6, 7) from the median nerve, which usually enters it just below the elbow joint.

Actions. It flexes and abducts the carpus and also helps to flex the elbow and to pronate the forearm. It is an important synergist during extension of the fingers [pp. 323, 327]. The muscle is tested with the dorsum of the forearm flat on a table : the patient then flexes the wrist against resistance while the examiner feels the tendon of the muscle.

Palmaris Longus

This small elongated muscle is frequently absent. It arises from the common flexor origin on the front of the medial epicondyle, medial to the flexor carpi radialis, and from the septa between it and its neighbours. Half way down the forearm the thin flat tendon appears, and at the wrist it adheres to the superficial surface of the flexor retinaculum and inserts into the apex of the palmar aponeurosis [FIG. 387]. A prolongation from this insertion may cover over the short muscles of the thumb. If the muscle is present, the median nerve lies posterior to its tendon at the wrist.

Nerve Supply. By a branch (C. 8) from the median nerve, which enters it just distal to the elbow joint.

Actions. It is a flexor of the wrist, and, by its attachment to the palmar aponeurosis and thus to the digital fibrous sheaths [p. 339], it may have some slight action in flexing the meta-carpophalangeal joints of the fingers. The muscle is tested like the flexor carpi radialis.

Flexor Carpi Ulnaris

Like the pronator teres [p. 317] the flexor carpi ulnaris has two heads, and between them the ulnar nerve enters the forearm. The humeral head of the muscle is small, and arises from the common flexor origin and the adjacent fascia. The ulnar head takes aponeurotic origin from the medial border of the olecranon and the posterior border of the ulna in its upper two-thirds. The tendon, which appears approximately half way down the forearm, is attached to the pisiform bone, and so, through the **pisohamate** and **pisometacarpal ligaments**, to the hook of the hamate and the base of the fifth metacarpal [FIG. 404]. Some fibres may also be prolonged into the abductor digiti minimi [p. 331]. The tendon is a guide to the ulnar nerve and vessels, which lie on its lateral side.

Nerve Supply. From the ulnar nerve by several branches (C. 7, 8) which enter it at intervals in the forearm.

Actions. It flexes and adducts the carpus, and fixes the pisiform during abduction of the little finger in order to stabilize the origin of the abductor digiti minimi [p. 331]. It is an important synergist during extension of the fingers and extension of the thumb. The muscle is tested like the flexor carpi radialis.

Flexor Digitorum Superficialis

The flexor digitorum superficialis [FIG. 388] has a long linear origin from the common flexor tendon, the adjacent fascia and

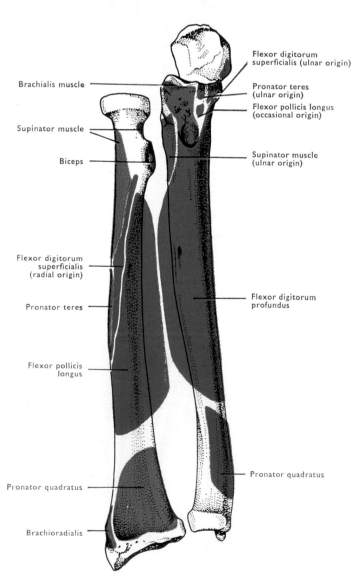

FIG. 389 Muscle attachments to front of right radius and ulna.

septa, the ulnar collateral ligament of the elbow joint, the medial border of the coronoid process of the ulna, and from the upper two-thirds of the anterior border of the radius [FIG. 389]. It also arises from a thin fibrous arch, which crosses the median nerve and ulnar vessels, between the latter two attachments.

Four tendons arise from the muscle half way down the forearm : two, destined for the middle and ring fingers, lie side by side superficial to those for the index and little fingers. They maintain this relationship as they pass under the flexor retinaculum in the carpal tunnel, lying superficial to the tendons of the flexor digitorum profundus. In the palm, the tendons diverge, and each one, in company with the corresponding tendon of the flexor digitorum profundus, enters the fibrous flexor sheath [p. 339] of its own digit [FIG. 390]. At the level of the metacarpophalangeal joint, each tendon of the flexor digitorum superficialis splits into two. The two halves pass posteriorly, with a spiral twist, round the sides of the tendon of the flexor profundus, to be inserted into the margins of the palmar surface of the middle phalanx. Some of the tendinous fibres decussate posterior to the tendon of the profundus, forming for it a tunnel which cannot be obliterated by tension.

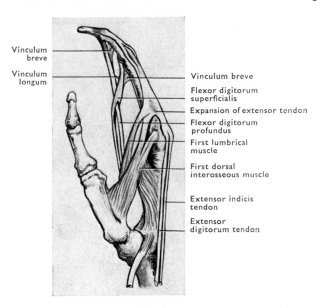

FIG. 391   The tendons attached to the index finger

The tendons of the flexor superficialis are also attached to the bones of the fingers by **vincula tendinum** which convey to them the small blood vessels necessary for their nutrition (Brockis, 1953). The **vincula brevia** are small triangular bands of connective tissue containing some collagen and elastic fibres, and occupy the interval between the tendon and the digit close to the insertion [FIGS. 391, 392]. Each vinculum breve is attached to the front of the interphalangeal joint and the distal part of the proximal phalanx. The **vincula longa** are variable, narrow bands extending from the back of the tendon to the proximal part of the palmar surface of the proximal phalanx.

In the most distal part of the forearm, the carpal tunnel, and the palm, the tendons of the flexor digitorum superficialis and profundus are enclosed in a common synovial sheath [p. 326].

Nerve Supply. By branches (C. 7, 8 ; T. 1) from the median nerve in the forearm.

Actions. The muscle is a flexor of the metacarpophalangeal and proximal interphalangeal joints of the fingers. Because it crosses the radiocarpal and midcarpal joints it can flex the wrist, and it can also help to flex the elbow. It is always difficult to be sure that this muscle is working if the flexor digitorum profundus is active, since the profundus tendon can produce all the actions of the superficialis. The

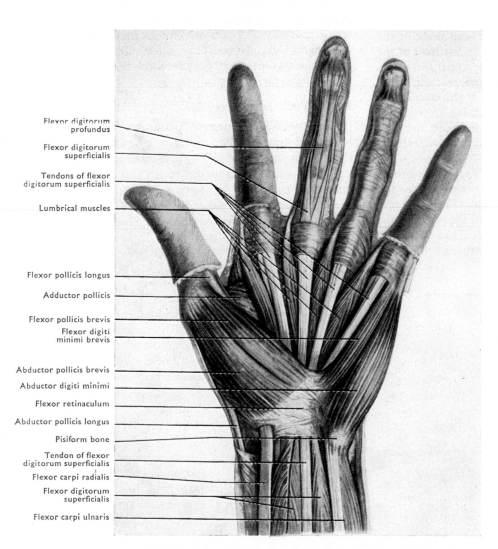

FIG. 390   Superficial muscles and tendons in palm of left hand.

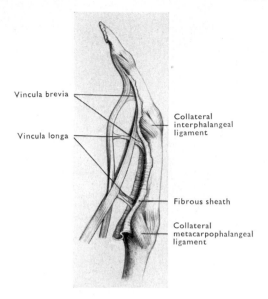

FIG. 392   Flexor tendons of finger with vincula tendinum.

flexor digitorum superficialis. Each tendon enters the fibrous sheath of its digit behind the tendon of the flexor superficialis, and pierces this tendon opposite the proximal phalanx. Eventually it inserts into the anterior surface of the base of the distal phalanx. Like the tendons of the flexor digitorum superficialis, those of the profundus are provided with vincula; the **vincula brevia** are attached to the capsule of the distal interphalangeal joint, and the **vincula longa** to the tendons of the flexor superficialis and their vincula brevia [FIGS. 391, 392].

In the palm, the tendons of the flexor digitorum profundus afford origins for the lumbrical muscles. Proximal to this, they are enveloped with the superficialis tendons in a common synovial sheath [p. 326].

Nerve Supply. The medial part of the muscle, destined for the little and ring fingers, is supplied by branches (C. 8; T. 1) from the ulnar nerve, and the lateral part, destined for the index and middle fingers, is supplied by branches (C. 7, 8; T. 1) from the anterior interosseous nerve. Sometimes the ulnar nerve supplies the whole muscle, and more rarely the anterior interosseous nerve does so.

Actions. The flexor digitorum profundus is the only muscle which can flex the distal interphalangeal joints of the fingers. Although this is its primary action, it helps to flex all the joints across which it passes—the proximal interphalangeal, the metacarpophalangeal, the carpometacarpal, midcarpal, and radiocarpal joints. To test the muscle, the dorsum of the hand and fingers is laid flat on a table, the middle phalanx is pressed firmly against the table by the examiner's finger, and the patient tries to bend the distal phalanx.

best method is to ask the patient to bend the proximal interphalangeal joint without bending the distal one.

Flexor Digitorum Profundus

The deep flexor of the fingers, with the long flexor of the thumb, forms the deepest layer of muscles on the front of the forearm [FIGS. 393, 394]. It arises from the upper two-thirds of the medial and anterior surfaces of the ulna, reaching up on to the medial side of the olecranon process. It also comes from the medial half of the anterior surface of the interosseous membrane in its middle third [FIG. 389], and from the aponeurosis attaching the flexor carpi ulnaris to the posterior border of the ulna.

Half way down the forearm, the portion of the muscle coming from the interosseous membrane forms a separate tendon destined for the index finger [FIG. 394]. Those for the remaining fingers do not usually separate until they arrive at the flexor retinaculum, behind which they pass into the palm through the carpal tunnel, posterior to the tendons of the

Flexor Pollicis Longus

Like the flexor digitorum profundus, the flexor pollicis longus lies deeply in the forearm, arising by fleshy fibres from the anterior surface of the body of the radius between the radial tuberosity superiorly and the upper border of the pronator quadratus inferiorly. It also comes from the corresponding lateral portion of the front of the interosseous membrane, and often also by a small slip from the medial border of the coronoid process of the ulna [FIG. 389].

The muscle descends lateral to the flexor digitorum profundus [FIG. 394] and a tendon leaves it just above the wrist and passes into the hand posterior to the flexor retinaculum, enveloped in a special synovial sheath [p. 326]. The tendon enters the fibrous digital sheath of the thumb and inserts into the anterior aspect of the base of the distal phalanx of the thumb [FIG. 399].

Nerve Supply. By a branch (C. 8; T. 1) from the anterior interosseous nerve.

Actions. The flexor pollicis longus is the only flexor of the interphalangeal joint of the thumb. It also flexes the metacarpophalangeal and carpometacarpal joints of the thumb, and can assist in flexion of

FIG. 393   Distal surface of section through middle third of left forearm.

Bicipital
aponeurosis

Biceps

Brachialis

Pronator teres

Common
flexor tendon

Biceps tendon

Tuberosity
of radius

Supinator muscle

Brachioradialis

Pronator teres

Flexor digitorum
profundus

Flexor carpi
ulnaris

Flexor digitorum
profundus (for
index finger)

Flexor pollicis
longus

Brachioradialis

Flexor digitorum
superficialis

Pronator quadratus

Flexor digitorum
superficialis

Pisiform bone

Flexor carpi
radialis

Abductor pollicis
longus

FIG. 394 Deepest muscles of front of left forearm.

surface of the body of the ulna to the distal quarter of the anterior surface of the body of the radius [FIG. 389].

Nerve Supply. From the anterior interosseous nerve (C. 8 ; T. 1).

Actions. It is said to be a more powerful pronator of the forearm than pronator teres (Basmajian and Travill, 1961), and it also holds the radius and ulna together, thus protecting the inferior radio-ulnar joint. The muscle is tested as for pronator teres, but cannot be seen or felt.

MUSCLES OF THE BACK OF THE FOREARM

The muscles of the back of the forearm are arranged in two groups, superficial and deep. In the **superficial group,** from the radial to the ulnar side, lie the brachioradialis, the extensor carpi radialis longus, the extensor carpi radialis brevis, the extensor digitorum, the extensor digiti minimi, the extensor carpi ulnaris, and the anconeus [FIG. 395]. The **deep group** consists of five muscles, the supinator, the abductor pollicis longus, the extensor pollicis brevis, the extensor pollicis longus, and the extensor indicis. This group is largely concealed by the superficial group until just above the wrist, where the abductor pollicis longus and the extensor pollicis brevis emerge to wind round the lateral side of the radius to the thumb [FIG. 395].

Brachioradialis

One of the flexors of the elbow, the brachioradialis [FIG. 387] arises from the proximal two-thirds of the front of the lateral supracondylar ridge of the humerus [FIG. 382] and the adjoining part of the lateral intermuscular septum. The brachioradialis forms the lateral boundary of the cubital fossa, and its fasciculi converge in the middle of the forearm into a narrow flat tendon which is inserted into the lateral side of the lower end of the radius [FIG. 389], just above the styloid process.

Nerve Supply. From the radial nerve by a branch (C. 5, 6) which enters its medial aspect above the elbow.

Actions. The brachioradialis is a flexor of the elbow, and is regarded as a shunt muscle, in contrast to the biceps and the brachialis, which are spurt muscles [p. 264]. In normal circumstances it plays no part in either pronation or supination, but cases have been reported in which it has been able to effect some pronation from the fully supinated position when the other pronators were paralysed (Sunderland, 1944b). When testing the muscle, the forearm is bent to a right angle in a position midway between pronation and supination, and then flexed against resistance. The muscle belly is readily seen and felt.

Extensor Carpi Radialis Longus

This muscle, which is often blended with the brachioradialis, arises by fleshy fibres from the anterior aspect of the distal third of the lateral supracondylar ridge of the humerus and the adjacent part of the lateral intermuscular septum [FIGS. 382, 395]. The muscle ends, about the middle of the forearm, in a flat tendon closely related to the lateral side of the tendon of the extensor carpi radialis brevis (in lower animals the two muscles are fused). The two tendons descend side by side into the wrist, and are enveloped in a single synovial sheath

the wrist. (Because the thumb is placed at right angles to the fingers, flexion is a movement which takes place in the plane of the palm.) The muscle can be tested by holding the proximal phalanx of the thumb and asking the patient to flex the terminal phalanx.

Pronator Quadratus

This fleshy quadrangular muscle lies deeply in the distal part of the forearm, passing from the distal quarter of the anterior

C.T.A.—M

under the extensor retinaculum [p. 338]. They groove the distal end of the radius and are crossed obliquely by the tendon of the extensor pollicis longus [FIG. 402]. Eventually the extensor carpi radialis longus inserts into the dorsal surface of the base of the second metacarpal bone on its radial side [FIG. 406] ; there may be a small bursa under the tendon close to its insertion.

Nerve Supply. From the radial nerve, by a branch (C. 6, 7) which enters the muscle above the elbow.

Actions. It extends and abducts the radiocarpal and midcarpal joints, and is an important synergist during flexion of the fingers [p. 327]. Because of its origin from the lateral supracondylar ridge, the muscle helps to flex the elbow joint. To test the muscle the patient attempts, with the fingers extended, to extend the wrist to the radial side against resistance. The muscle belly can be felt and sometimes seen.

Extensor Carpi Radialis Brevis

The short radial extensor [FIG. 395], which is often fused with its longer colleague at its origin, comes from the anterior aspect of the lateral epicondyle of the humerus by a tendon which it shares with the extensor digitorum, the extensor digiti minimi, and the extensor carpi ulnaris [FIG. 382]. It also arises from the fascia covering this **common extensor origin,** and from the intermuscular septa which penetrate between it and its neighbours.

The tendon of insertion, which appears approximately half way down the forearm, runs with that of the extensor carpi radialis longus, and is inserted into the base of the dorsal surface of the third metacarpal bone with a small bursa between it and the styloid process of the metacarpal.

Nerve Supply. By the deep branch of the radial nerve (C. 6, 7).

Actions. Extension and abduction of the radiocarpal and midcarpal joints : it helps the long extensor to prevent flexion of the wrist during flexion of the fingers. The muscle is tested in the same way as extensor carpi radialis longus.

Extensor Digitorum

The extensor digitorum arises from the common extensor origin on the anterior surface of the lateral epicondyle of the humerus [FIGS. 382, 395] from the fascia covering it, and from the intermuscular septa at its sides. Above the wrist the muscle ends in four tendons which pass deep to the extensor retinaculum [p. 338] in a compartment along with the extensor indicis and surrounded by a common synovial sheath which ends about the middle of the dorsum of the hand [FIG. 402]. On the back of the hand the tendons of extensor digitorum are interconnected by obliquely placed fibrous bands. The arrangement of these bands is very variable, but the tendons for the little and ring fingers are usually quite closely attached until just proximal to the metacarpophalangeal joints, and that for the index finger is frequently not attached to the tendon for the middle finger [FIGS. 395, 402].

As each individual tendon approaches the metacarpophalangeal joint it forms a triangular membranous expansion (the **extensor hood**) over the metacarpophalangeal joint and the dorsum of the proximal phalanx. This hood forms the dorsal part of the fibrous capsule of the joint, and its base, which lies proximally, extends forwards on each side of the

head of the metacarpal to fuse with the deep transverse metacarpal ligament. Distally it narrows towards the dorsal surface of the proximal interphalangeal joint, and is reinforced at each side by receiving the insertions of the tendons of the interosseous and lumbrical muscles of that particular finger [FIGS. 391, 396].

At the distal end of the proximal phalanx the hood splits into somewhat ill-defined central and collateral slips, which replace the dorsal ligament of the proximal interphalangeal joint. The central slip, directly continuous with the fibres of the extensor tendon itself, inserts into the dorsum of the base of the middle phalanx. The two collateral slips, which are continuous with the fibres of the lumbrical and interossei tendons, cross the dorsal surface of the proximal interphalangeal joint and unite

FIG. 395 Superficial muscles on dorsum of left forearm.

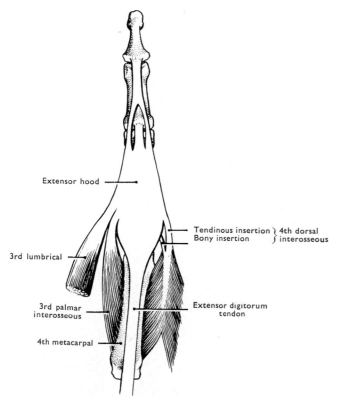

Extensor hood

3rd lumbrical

3rd palmar interosseous

4th metacarpal

Tendinous insertion } 4th dorsal
Bony insertion } interosseous

Extensor digitorum tendon

FIG. 396 The extensor hood of the right ring finger. The complicated system of local strengthening bands in the hood revealed by detailed dissection has been omitted for the sake of clarity.

on the dorsum of the middle phalanx. They are inserted into the base of the distal phalanx (Haines, 1951).

Nerve Supply. By branches (C. 7, 8) from the posterior interosseous nerve.

Actions. The muscle extends the metacarpophalangeal joints of the fingers. By hyperextending these joints, it can abduct (diverge) the fingers away from an imaginary line drawn through the middle finger.

The extensor digitorum can also help to extend the proximal and distal interphalangeal joints. However, the main extensors of these joints, the tendons of which form the collateral slips of the extensor hood, are the interossei and the lumbricals, which also help to prevent the extensor digitorum from hyperextending the metacarpophalangeal joint [p. 331]. If these muscles are paralysed, the extensor digitorum will hyperextend this joint, and, having done so, cannot extend the distal or proximal interphalangeal joints until the hyperextension is undone.

Since the extensor digitorum crosses the radiocarpal and midcarpal joints it will also extend these joints, and if it is desired to extend the fingers without extending the wrist, it is necessary to call into play the flexors of the radiocarpal and midcarpal joints. Although the muscle crosses the elbow joint anteriorly, its action as an elbow flexor is negligible. The tendon running to the index finger can deviate the finger radially when the first dorsal interosseous is paralysed (Duchenne, 1867).

If an attempt is made by the examiner to flex the extended fingers at the metacarpophalangeal joints, the belly of the muscle can be felt contracting as the subject resists the movement.

Extensor Digiti Minimi

The extensor of the little finger arises from the common extensor origin on the anterior surface of the lateral epicondyle of the humerus, and from the fascia surrounding it. It runs distally between the extensor digitorum and the extensor carpi ulnaris [FIG. 395] to end in a tendon which traverses a special compartment of the extensor retinaculum in a groove between the radius and the ulna, surrounded by a synovial sheath [FIG. 402]. On the dorsum of the hand the tendon splits into two, and both parts are inserted into the extensor hood on the back of the proximal phalanx of the little finger [FIG. 397], the lateral one joining the tendon of extensor digitorum to the little finger.

Nerve Supply. By branches (C. 7, 8) from the posterior interosseous nerve.

Actions. The extensor digiti minimi helps to extend the metacarpophalangeal joint of the little finger, and acts similarly to the corresponding tendon of the extensor digitorum, except that it is said to be capable of deviating the little finger in an ulnar direction (Duchenne, 1867; Jones, 1919).

The presence of a separate extensor for the little finger does confer on it some independence of movement, but the tendons of the extensor digitorum passing to the ring and little fingers are usually firmly adherent to each other, and pulling on the one always involves pulling on the other. Like the extensor digitorum, the extensor digiti minimi helps to extend the radiocarpal and midcarpal joints.

Extensor Carpi Ulnaris

This muscle arises from the common extensor origin on the front of the lateral epicondyle of the humerus, and from the adjacent fascia. It has a second aponeurotic head, partly shared with the flexor carpi ulnaris, from the middle part of the posterior border of the ulna. The tendon emerges from the muscle belly nearer to the wrist than the other extensor tendons [FIG. 395] and runs in a groove on the back of the lower end of the ulna in a special compartment of the extensor retinaculum, surrounded by a short synovial sheath [FIG. 402]. It is inserted into the medial side of the base of the fifth metacarpal bone [FIG. 406].

Nerve Supply. By branches (C. 7, 8) from the posterior interosseous nerve.

Actions. Extension and adduction of the carpus. It is an important synergist during flexion of the fingers [p. 327]. To test the muscle, the patient is asked to extend the wrist to the ulnar side against resistance: the tendon stands out and can be palpated by the examiner.

Anconeus

This small muscle arises from the back of the lateral epicondyle of the humerus [FIG. 395] and from the adjacent part of the capsule of the elbow joint. The insertion is to a triangular surface on the lateral side of the olecranon and of the posterior surface of the ulna as far down as an oblique ridge [FIG. 398], and also into the fascia which covers it.

Nerve Supply. From the branch (C. 7, 8) of the radial nerve to the medial head of the triceps.

Actions. It helps to extend the elbow joint.

Supinator

The supinator lies deeply, almost wholly concealed by the superficial muscles. It has two heads. The humeral one arises from the inferior aspect of the lateral epicondyle of the humerus, from the radial collateral ligament of the elbow joint, and from the anular ligament of the radius. The ulnar head comes from the supinator crest of the ulna [Fig. 389]. Between the two layers of muscle derived from these heads the deep branch of the radial nerve runs into the back of the forearm.

From its composite origin the muscle fibres run distally and laterally posterior to the upper third of the radius [Fig. 397]. Wrapping round this almost completely, they become inserted into the posterior, lateral, and anterior aspects of the radius as far forwards as the anterior margin, as far proximally as the neck, and as far distally as the insertion of pronator teres [Figs. 389, 398].

Nerve Supply. From the deep branch of the radial nerve (C. 5, 6).

Actions. Supination of the forearm. The supinator is probably the main prime mover in this movement, the biceps brachii being an auxiliary (Travill and Basmajian, 1961). It is difficult to test the supinator apart from the biceps brachii, but the patient is asked to resist pronation of the forearm while the elbow is held straight.

Abductor Pollicis Longus

The long abductor of the thumb arises from the dorsal surfaces of the ulna and radius distal to the origin and insertion of the supinator [Fig. 398], and also from the intervening portion of the interosseous membrane. Although deeply situated under the other extensor muscles at its origin, the muscle emerges as it passes towards the wrist to become superficial in the distal part of the forearm [Figs. 395, 397]. Its tendon, together with that of the extensor pollicis brevis, crosses those of the extensor carpi radialis brevis and extensor carpi radialis longus and the insertion of the brachioradialis. It then passes through the most lateral compartment of the extensor retinaculum in a common synovial sheath with the tendon of the extensor pollicis brevis [Fig. 401], and after crossing the radial artery, is inserted, sometimes by several slips, into the lateral side of the base of the first metacarpal [Fig. 404]. It usually gives off a slip which passes to the abductor pollicis brevis and to the fascia over the thenar eminence, and sometimes another which goes to the trapezium (Baba, 1954).

Nerve Supply. From the posterior interosseous nerve (C. 7, 8).

Actions. It pulls the metacarpal bone of the thumb into a position midway between extension and abduction [p. 336], and the tendon stands out during this movement. It is also quite a strong assistant in resisted flexion and abduction of the wrist, and this action is important in 'trick' flexion of the wrist when the normal prime movers are paralysed.

Extensor Pollicis Brevis

The short extensor of the thumb comes from an impression on the posterior surface of the radius, and from the adjacent part of the interosseous membrane [Fig. 398]. It lies distal to the abductor pollicis longus, to which it is closely adherent, and

the tendons of the two muscles pass down to the wrist and run deep to the extensor retinaculum in a common synovial sheath [Fig. 401]. The tendon of the extensor pollicis brevis helps to replace the dorsal part of the capsule of the metacarpophalangeal joint of the thumb, and is inserted into the base of the dorsal surface of the proximal phalanx. Occasionally it proceeds further and may be inserted into the distal phalanx along with the extensor pollicis longus.

Nerve Supply. By a branch from the posterior interosseous nerve (C. 7, 8).

Actions. It extends the carpometacarpal and metacarpophalangeal joints of the thumb. Because of its position on the

FIG. 397 Deep muscles on dorsum of left forearm.

anterior aspect of the lateral side of the wrist, the muscle can be called into play to help in flexing and abducting the wrist against resistance. To test the muscle, the patient tries to resist flexion of the thumb at the metacarpophalangeal joint; the tendon of the muscle will then stand out.

Extensor Pollicis Longus

The extensor pollicis longus arises from the lateral part of the posterior surface of the ulna in its middle third, and also from the interosseous membrane somewhat more distally [FIG. 398]. Above the wrist a tendon arises and is enclosed in a synovial sheath as it grooves the posterior surface of the radius in a special compartment deep to the extensor retinaculum [FIG. 402]. At this point it changes direction by turning laterally round the medial side of the **dorsal tubercle of the radius** [FIG. 204], and passes obliquely across the back of the hand over the tendons of the extensor carpi radialis brevis and longus and over the radial artery. The extensor pollicis longus tendon replaces the dorsal part of the capsule of the metacarpophalangeal joint of the thumb, and is here joined by slips from the tendons of the abductor pollicis brevis on the lateral side and adductor pollicis on the medial side, forming a triangular expansion somewhat similar to the extensor hoods of the fingers (Napier, 1952). This expansion may be joined by the tendon of extensor pollicis brevis. It is finally inserted into the dorsal surface of the base of the distal phalanx of the thumb.

As the tendon passes laterally across the back of the hand, it forms the posterior boundary of a triangular hollow known as the **anatomical snuff box,** which is best seen when the muscle is in action. The other boundaries of this hollow are the distal end of the radius proximally, and the tendons of abductor pollicis longus and extensor pollicis brevis anteriorly. In the hollow lies the radial artery, which can be felt pulsating against the scaphoid bone.

Nerve Supply. By a branch (C. 7, 8) from the posterior interosseous nerve.

Actions. The extensor pollicis longus extends all the joints of the thumb and can assist in extension and abduction of the wrist. In full abduction or extension of the thumb it can act as a 'trick' adductor of the thumb. Because of its oblique course the muscle is capable of contributing to supination of the forearm. The tendon stands out when an attempt is made to resist flexion at the interphalangeal joint of the thumb.

Extensor Indicis

Unlike the extensor digiti minimi, the extensor indicis lies deeply in the dorsum of the forearm. It arises from the lowest impression on the posterior surface of the ulna, and sometimes also from the interosseous membrane [FIG. 398]; the origin is just distal to that of the extensor pollicis longus. The tendon of the extensor indicis passes through the extensor retinaculum in common with the extensor digitorum and surrounded by the same synovial sheath [FIG. 402]. On the back of the hand the tendon lies on the ulnar side of the index finger tendon of the extensor digitorum, and is inserted into the extensor hood on the back of the proximal phalanx of the index.

Nerve Supply. By branches (C. 7, 8) from the posterior interosseous nerve.

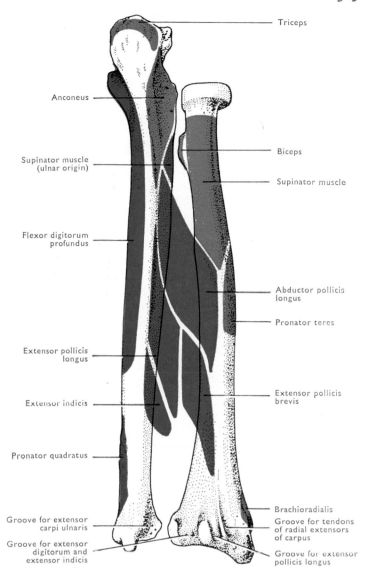

FIG. 398 Muscle attachments to back of right radius and ulna.

Actions. The extensor indicis produces independent extension of the index finger. In all other respects the actions of extensor indicis are similar to those of extensor digitorum, except that it has been claimed as an ulnar deviator of the index.

MOVEMENTS OF THE RADIO-ULNAR JOINTS

The two strong supinators are biceps brachii and supinator, but the extensor pollicis longus can assist them.

The chief pronators are pronator quadratus and pronator teres, but flexor carpi radialis, which passes obliquely from medial to lateral side on the front of the forearm, is capable of assisting. Both supination and pronation are most powerful when the elbow is flexed to a right angle, and in the majority of shoulder-elbow positions isometric pronation is a stronger movement than isometric supination (Darcus, 1951).

MUSCLES CONCERNED IN MOVEMENTS AT THE RADIO-ULNAR JOINTS

MOVEMENT	MUSCLE	PRINCIPAL NERVE SUPPLY
Supination	Supinator	Radial
	Biceps brachii	Musculocutaneous
	Extensor pollicis longus	Posterior interosseous
Pronation	Pronator quadratus	Anterior interosseous
	Pronator teres	Median
	Flexor carpi radialis	Median

SYNOVIAL SHEATHS IN THE WRIST AND HAND

At the wrist and in the fingers, the flexor tendons are surrounded by synovial sheaths [FIGS. 399, 401, and 402]. Each sheath has a visceral layer which surrounds and is attached to the tendon, and a parietal layer, which lines the surrounding fibrous or osteofibrous compartment through which the tendon runs [p. 269]. In front of the wrist there are three such sheaths, all of which begin approximately 2 or 3 cm. proximal to the flexor retinaculum. Two of them, the separate sheaths which surround the flexor pollicis longus and the flexor carpi radialis, terminate at the insertions of these muscles. The third, which is by far the largest, surrounds the tendons of both the flexor digitorum profundus and the flexor digitorum superficialis [FIG. 399]. These tendons may be thought of as being pushed into a closed bursa from the lateral side, so producing a somewhat complicated-looking arrangement [FIG. 400]. In about 50 per cent. of people there is a communication between this common flexor synovial sheath and the synovial sheath of the flexor pollicis longus. Distally, the common flexor sheath terminates about the middle of the palm.

Each fibrous digital sheath [p. 339] is lined by a synovial sheath which envelops the tendons of the flexor digitorum superficialis and the flexor digitorum profundus of that digit. These synovial sheaths extend from the insertion of the profundus tendon distally to the neck of the metacarpal proximally, except that in about 70 per cent. of cases the digital synovial sheath of the little finger communicates with the common flexor synovial sheath [FIG. 399]. The flexor tendons give the appearance of having invaginated their synovial sheath from behind, so that the vincula longa and brevia [FIG. 392], representing the remains of their original mesotendon and conveying their blood supply, reach the tendons posteriorly, and attach them to the bone.

On the dorsum of the wrist and hand, the synovial sheaths of the extensor tendons begin just superior to the extensor retinaculum. Those of the tendons going to the metacarpal bones usually extend to their insertions, and the others reach half way down the hand [FIGS. 401, 402].

The digital synovial sheaths are often infected. Such an infection may, because of the confined space within which the swelling develops, obliterate the blood supply of the tendons and consequently kill them, so disorganizing the finger. An infection of the synovial sheath of the little finger is potentially more dangerous in this respect than those of the other fingers, since the infection may spread to involve the common flexor synovial sheath, so endangering the viability of all the flexor tendons to the fingers. Similarly, an infection of the synovial sheath of the flexor pollicis longus may, in about 50 per cent. of cases, affect not only the thumb, but also all the flexor tendons of the fingers. An infection developing in a digital

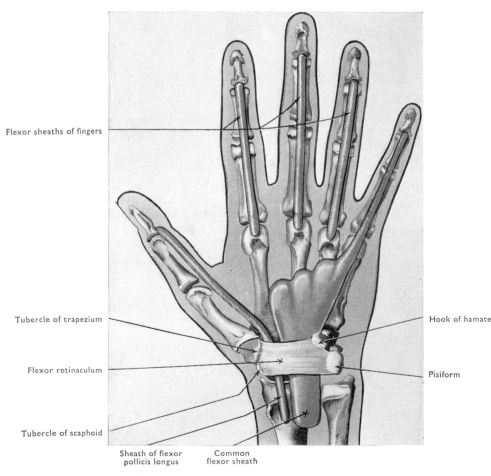

Flexor sheaths of fingers

Tubercle of trapezium

Flexor retinaculum

Tubercle of scaphoid

Hook of hamate

Pisiform

Sheath of flexor pollicis longus   Common flexor sheath

FIG. 399   Synovial sheaths of flexor tendons of digits.

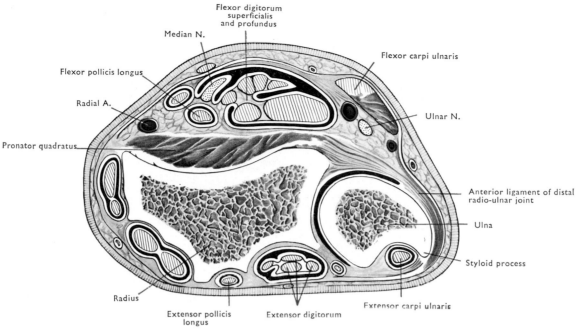

FIG. 400 Transverse section through forearm above flexor retinaculum showing the relation of the synovial sheaths to the tendons.

synovial sheath may burst into one of the potential spaces of the palm [p. 340], whence it may spread up into the forearm posterior to the flexor retinaculum, again endangering by pressure the viability of the tendons and the other structures, such as the median nerve, which lie in the carpal tunnel [p. 338].

A chronic infection in the common flexor synovial sheath may produce a swelling in the palm and another extending up in front of the wrist for a short distance into the forearm, with a constriction between the two at the site of the flexor retinaculum. The fluid may be pressed from one to the other behind the retinaculum.

MOVEMENTS AT THE WRIST

Movements in the region of the wrist take place at the radiocarpal and midcarpal joints. The radiocarpal joint is more concerned in extension and adduction at the wrist than the midcarpal, which is responsible for the greater part of flexion and abduction. The flexors of the carpus fix the wrist during extension of the fingers, and the extensors of the wrist fix it during flexion of the fingers. The object in both cases is to prevent the muscles operating on the fingers from losing power and efficiency through also moving the radiocarpal and midcarpal joints. In powerful movements of the fingers, the wrist is rigidly fixed by simultaneous contraction of the flexors and extensors of the carpus.

The importance of these synergic

FIG. 401 Dissection of lateral side of left wrist and hand, showing synovial sheaths of tendons.

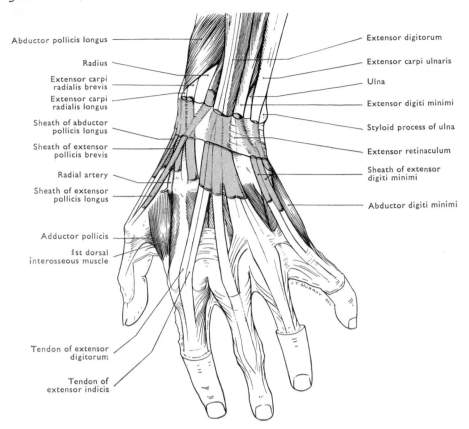

Abductor pollicis longus

Radius

Extensor carpi radialis brevis

Extensor carpi radialis longus

Sheath of abductor pollicis longus

Sheath of extensor pollicis brevis

Radial artery

Sheath of extensor pollicis longus

Adductor pollicis

1st dorsal interosseous muscle

Tendon of extensor digitorum

Tendon of extensor indicis

Extensor digitorum

Extensor carpi ulnaris

Ulna

Extensor digiti minimi

Styloid process of ulna

Extensor retinaculum

Sheath of extensor digiti minimi

Abductor digiti minimi

FIG. 402   Dissection of back of forearm, wrist, and hand, showing synovial sheaths of tendons.

shorten sufficiently, and the power they can exert is greatly diminished. This is the basis of the fact that forcible flexion of the wrist can cause an assailant to drop a weapon. Conversely, extension of the wrist puts the flexors of the fingers on the stretch and the power they can exert is then considerable.

A patient with paralysis of the radial nerve can often extend his wrist to some extent by flexing the fingers fully at the metacarpophalangeal joints ; the tendons of the extensor digitorum are tightened and pull the wrist into extension. This trick movement is inadequate to permit a satisfactory grip.

MUSCLES OF THE HAND

The small 'intrinsic' muscles of the hand include the interossei, which occupy the intermetacarpal spaces and act on all five digits, and the lumbricals, which spring from the flexor digitorum profundus tendons in the middle of the palm and act on the four fingers. The muscles which operate solely on the thumb or on the little finger are grouped together to form the rounded swellings, on either side of the palm, known as the **thenar** and **hypothenar eminences.**

actions of the wrist muscles is easily seen when an attempt is made to grip an object with the long flexors of the fingers when the wrist is fully flexed. Because of the slackening of the flexor tendons caused by the wrist flexion, they cannot

The muscles of the hypothenar eminence—the abductor digiti minimi, flexor digiti minimi brevis, and the opponens digiti minimi—have their thenar counterparts in the abductor pollicis brevis, the flexor pollicis brevis, and the opponens pollicis, but

MUSCLES CONCERNED IN MOVEMENTS AT THE RADIOCARPAL AND MIDCARPAL JOINTS

MOVEMENT	MUSCLE	PRINCIPAL NERVE SUPPLY
Flexion	Flexor carpi radialis	Median
	Flexor carpi ulnaris	Ulnar
	Palmaris longus	Median
	Flexor digitorum profundus	Anterior interosseous : ulnar
	Flexor digitorum superficialis	Median
	Flexor pollicis longus	Anterior interosseous
	Abductor pollicis longus	Posterior interosseous
	Extensor pollicis brevis	Posterior interosseous
Extension	Extensor carpi radialis longus	Radial
	Extensor carpi radialis brevis	Radial
	Extensor carpi ulnaris	Posterior interosseous
	Extensor digitorum	Posterior interosseous
	Extensor pollicis longus	Posterior interosseous
	Extensor indicis	Posterior interosseous
	Extensor digiti minimi	Posterior interosseous
Abduction	Extensor carpi radialis longus	Radial
	Extensor carpi radialis brevis	Radial
	Flexor carpi radialis	Median
	Abductor pollicis longus	Posterior interosseous
	Extensor pollicis brevis	Posterior interosseous
	Extensor pollicis longus	Posterior interosseous
Adduction	Flexor carpi ulnaris	Ulnar
	Extensor carpi ulnaris	Posterior interosseous

the thumb has an additional muscle, the adductor pollicis, which lies deep in the palm. All the small muscles of the hand are supplied by the first thoracic segment of the spinal cord (Highet, 1943).

Abduction of the fingers is defined as movement, in the plane of the nails, away from an imaginary line drawn through the middle finger; adduction is movement towards this line. The middle finger can thus be abducted in two directions, either towards the little finger, or towards the index. Further, adduction of the index finger is a movement in the opposite direction to adduction of the ring finger and little finger. This apparently difficult convention in fact makes the actions and attachments of the interossei easier to remember.

Lumbricals

The four lumbricals, one for each finger, are small cylindrical muscles resembling the earthworm from which they are named. The lateral two lumbricals each arise, just distal to the flexor retinaculum, by a single head from the lateral side of the tendons of the flexor digitorum profundus running to the index and middle fingers. The medial two muscles arise by two heads from the adjacent sides of the profundus tendons between which they lie—that is, from those going to the middle, ring, and little fingers [FIG. 390]. Each lumbrical then passes in a plane anterior to the deep transverse metacarpal ligament and sweeps obliquely dorsally on the lateral side of the metacarpophalangeal joint of the finger concerned, to join the lateral edge of the extensor hood at the side of the proximal phalanx [FIG. 391]. Some of its fibres can be traced to the central attachment of the hood to the dorsum of the base of the middle phalanx, but most of them run in the lateral part of the hood, over the dorsal surface of the proximal interphalangeal joint, and ultimately reach the dorsum of the base of the terminal phalanx [p. 322].

Nerve Supply. This varies. The most usual arrangement is that the lateral two lumbricals are supplied by twigs (T. 1) from the median nerve; the medial two being supplied by twigs (T. 1) from the ulnar nerve. However, the number of lumbricals supplied by the ulnar nerve may be increased to four or decreased to one.

Actions. Each lumbrical is a flexor of the metacarpophalangeal joint of the corresponding finger and an extensor of both interphalangeal joints. Acting from its insertion on its origin, the lumbrical slackens off the tension in the flexor digitorum profundus tendon.

Theoretically, all the lumbricals could rotate the fingers on which they act at the metacarpophalangeal joint (Sunderland, 1953). This action, however, is only well marked in the index finger. It brings the pad of the index finger to face medially, at an angle of about 30 degrees to the pad of the middle finger, and this movement is used in precision gripping [p. 336]. The first dorsal interosseous muscle is associated with the first lumbrical in producing this rotation, and is probably the major factor. Although they run on the lateral side of each digit, it is still not certain whether the lumbricals have any action in radially deviating the fingers (Braithwaite et al., 1948; Backhouse and Catton, 1954; Matheson, Sinclair, and Skene, 1970).

Interossei

Four palmar and four dorsal interossei occupy the spaces between the metacarpal bones. Each palmar interosseous comes from the metacarpal of the digit on which it acts; each dorsal interosseous comes from the adjacent surfaces of the metacarpals between which it lies: the dorsal muscles are about twice the size of the palmar ones (Eyler and Markee, 1954), and are composite muscles evolved from the fusion of primitive dorsal abductors and short flexors of the fingers (Lewis, 1965). The tendons of all the interossei, except the first palmar, wind round the appropriate digit to one or other side of the metacarpophalangeal joint and (except for the first pair) posterior to the deep transverse metacarpal ligament. They become inserted partly into the base of the proximal phalanx and partly into the extensor hood. The attachment to the hood is largely responsible for prolonging the hood to the base of the distal phalanx [p. 322].

The **first palmar interosseous** is a small slip of muscle

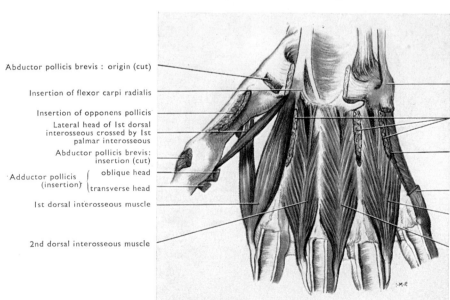

Abductor pollicis brevis : origin (cut)

Insertion of flexor carpi radialis

Insertion of opponens pollicis

Lateral head of 1st dorsal interosseous crossed by 1st palmar interosseous

Abductor pollicis brevis: insertion (cut)

Adductor pollicis (insertion) { oblique head / transverse head

1st dorsal interosseous muscle

2nd dorsal interosseous muscle

Insertion of flexor carpi ulnaris

Origins of 2nd, 3rd, and 4th palmar interosseous muscles

Insertion of opponens digiti minimi

Insertion of abductor digiti minimi

4th dorsal interosseous muscle

3rd dorsal interosseous muscle

FIG. 403   Dorsal interosseous muscles of right hand (seen from the palmar aspect).

which arises from the ulnar side of the base of the first meta-carpal and passes between the first dorsal interosseous and the oblique head of the adductor pollicis to insert into the ulnar side of the base of the proximal phalanx of the thumb in com-mon with the tendon of the adductor pollicis [FIG. 403].

The **second palmar interosseous** arises from the ulnar side of the shaft of the second metacarpal and is inserted into the ulnar side of the extensor hood of the index; the **third and fourth palmar interossei** arise from the radial sides of the shafts of the fourth and fifth metacarpals respectively, and insert into the radial sides of the extensor hoods of the ring and little fingers [FIGS. 404, 405]. (The middle finger has no palmar interosseous, and is deviated to either side by the action of the two dorsal interossei which are attached to it.)

The arrangement of the dorsal interossei is shown in FIGURES 391 and 403, and their attachments in FIGURES 404 and 406. The **first dorsal interosseous,** occupying the space between the thumb and the index finger, is larger than the others, and has an interval between its two heads through which the radial artery passes into the palm: its fibres are mostly inserted into the radial side of the proximal phalanx of the index finger, and though some join the extensor hood, they are not prolonged to the distal phalanx (Braithwaite *et al.,* 1948). The tendon of the **second dorsal interosseous** inserts into the radial side of the extensor hood and proximal phalanx of the middle finger; the **third and fourth** are inserted into the ulnar sides of the extensor hoods of the middle and ring fingers, and in a minority of instances into the proxi-mal phalanges (Eyler and Markee, 1954).

FIG. 405 The second, third, and fourth palmar inter-osseous muscles (right side). See Fig. 403 for first palmar interosseous muscle.

Nerve Supply. All the interossei are supplied by branches (T. 1) from the deep branch of the ulnar nerve.

Actions. The dorsal interossei abduct (diverge) the fingers, in the plane of the nails, away from an imaginary line drawn through the middle finger. The first dorsal interosseous can rotate the index finger at the metacarpophalangeal joint [p. 329].

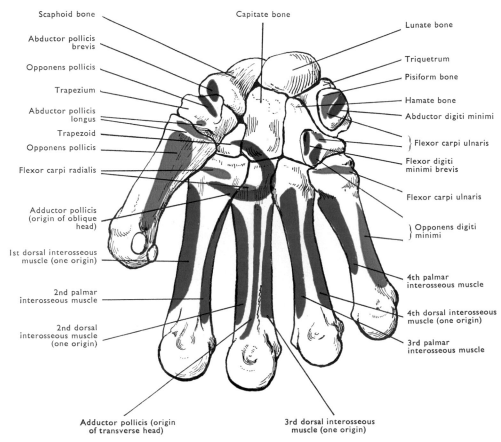

FIG. 404   Muscle attachments to front of carpus and metacarpus.

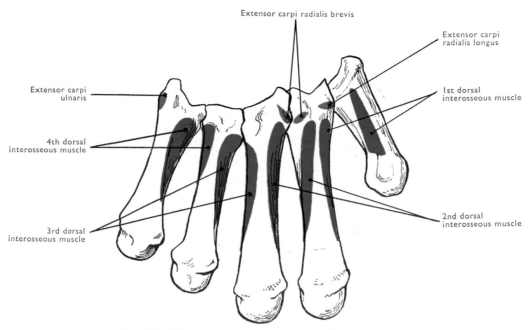

FIG. 406   Muscle attachments to back of right metacarpus.

This muscle may also assist the adductor pollicis in the movement of adduction of the thumb.

The first palmar interosseous assists the flexor pollicis brevis in flexion of the metacarpophalangeal joint of the thumb and the adductor pollicis in adduction of the thumb; the other three palmar interossei adduct (converge) the index, ring, and little fingers towards the line of the middle finger. Such movements play an important part in all kinds of manipulation and grasping.

Along with the lumbricals, the interossei flex the metacarpophalangeal joints and extend the proximal and distal interphalangeal joints. They are responsible for the fine control over these movements, and this action is well seen in the upward stroke in writing; it is essential in all such activities as typing, sewing, and playing musical instruments. When the interossei and lumbricals are paralysed, the extensor digitorum pulls the metacarpophalangeal joints into a position of hyperextension [p. 323], and the interphalangeal joints become flexed, so producing one variety of 'claw hand'. In such a hand the individual, independent movements of the affected fingers, whether abduction or adduction, flexion or extension, are largely lost. For this reason, it becomes very difficult or impossible to do up buttons, to pick up small objects, or to perform any delicate controlled movement with the hand [p. 336].

The interossei and the lumbrical muscle acting on a given finger are tested by hyperextending and fixing the metacarpophalangeal joint and then asking the patient to extend the proximal interphalangeal joint against resistance. Individual interossei are also tested by the patient's ability to make the appropriate movement of abduction or adduction of the finger concerned, the hand and finger being placed palm down on the table.

Abductor Digiti Minimi

This is the most superficial of the muscles which form the hypothenar eminence, and is morphologically in series with the dorsal interossei (Lewis, 1965). It arises from the pisiform bone, from the tendon of the flexor carpi ulnaris, and from the pisohamate and pisometacarpal ligaments [FIGS. 390, 404]; the tendon is inserted into the medial side of the base of the proximal phalanx of the little finger and into the extensor hood of this finger.

Nerve Supply. From the deep branch of the ulnar nerve (T I)

Actions. It abducts the little finger from the ring finger, and helps to flex its metacarpophalangeal joint (Forrest and Basmajian, 1965). The attachment to the extensor hood of the little finger probably helps in extending the interphalangeal joints of that digit. The abductor digiti minimi is a surprisingly powerful muscle, and comes into play when the fingers are spread to grasp a large object [p. 336]. To test it, the back of the hand and fingers is laid flat on the table and the patient abducts the little finger against resistance: the muscle belly can be felt and usually seen. (It is important to remember that a similar but less extensive movement can often be produced by transmitted tension from the flexor carpi ulnaris even though the abductor digiti minimi is paralysed.)

Opponens Digiti Minimi

The opponens [FIG. 407] lies deep to the abductor, and arises from the flexor retinaculum and the hook of the hamate bone. It is inserted into the medial half of the palmar surface of the fifth metacarpal in its distal two-thirds [FIG. 404].

Nerve Supply. From the deep branch of the ulnar nerve (T. I).

Actions. It pulls the metacarpal of the little finger forwards and rotates it laterally, so deepening the hollow of the hand. This is the basis of the test for the muscle, in which the patient is asked to make a cup of the palm while keeping the fingers extended. Contact between the pads of the thumb and the little finger during opposition [p. 335] depends on this muscle as well as upon the thenar musculature (Sunderland, 1944a).

Flexor Digiti Minimi Brevis

This rather variable muscle [FIG 390] which may be absent, or fused with one or other of its neighbours, arises by tendinous fibres from the front of the flexor retinaculum and the hook of the hamate bone [FIG. 404]. It is inserted, along with the abductor digiti minimi, to the medial side of the base of the proximal phalanx of the little finger; the tendon usually contains a sesamoid bone.

Nerve Supply. From the deep branch of the ulnar nerve (T. 1).

Actions. It flexes the little finger at the carpometacarpal and metacarpophalangeal joints. During abduction of the little finger all three hypothenar muscles are active (Forrest and Basmajian, 1965), but what the opponens and flexor are doing to assist this movement is not understood.

Palmaris Brevis

The palmaris brevis is a small subcutaneous muscle [FIG. 409] overlying the hypothenar eminence. It arises from the medial border of the palmar aponeurosis and the front of the flexor retinaculum, and is inserted into the skin of the medial border of the hand, covering the ulnar artery and nerve.

Nerve Supply. From the superficial branch of the ulnar nerve (T. 1).

Actions. It wrinkles up the skin on the medial border of the hand, and deepens the hollow of the hand, assisting the opponens digiti minimi to obtain a firm palmar grip (Kirk, 1924).

Abductor Pollicis Brevis

This muscle [FIGS. 390, 404] is the most superficial of those which form the thenar eminence, and takes origin by fleshy fibres mainly from the front of the flexor retinaculum. Other fibres come from the tubercles of the scaphoid and trapezium, and there may be a contribution from the tendon of the abductor pollicis longus, or from the insertion of palmaris longus. The muscle is inserted by a short tendon into the radial side of the proximal phalanx of the thumb, at its base; some fibres pass dorsally to reach the expansion of the tendon of the extensor pollicis longus.

Nerve Supply. The short abductor is usually supplied by a recurrent branch (T. 1) of the median nerve in the palm. In a minority of cases it may be supplied by the deep branch of the ulnar nerve.

Actions. It abducts the thumb at the carpometacarpal and metacarpophalangeal joints, causing it to travel anteriorly at right angles to the plane of the palm. This action is used in typing or playing the piano. When the thumb is fully abducted, there is an angulation of some 30 degrees between the proximal phalanx and the metacarpal. Because of the set of the articular surfaces of the carpometacarpal joint, abduction involves medial rotation of the metacarpal, and the abductor is used along with the opponens pollicis in the initial stages of the movement of opposition of the thumb.

Like that of an interosseous muscle, the tendon of the abductor pollicis brevis lies in front of the metacarpophalangeal joint but inserts partly into the long extensor, and so the muscle can help in flexion of the metacarpophalangeal joint and in extension of the interphalangeal joint. This extension is often of some importance as a 'trick' movement when the extensors are paralysed.

To test the muscle, the patient's hand is placed with its dorsum flat on the table, the thumb nail being at right angles to the other nails. The patient is then asked to raise the thumb vertically at right angles to the table.

Opponens Pollicis

Deep to the abductor pollicis brevis lies a fleshy mass imperfectly separated into two components. The proximal and lateral of these [FIG. 407] is the opponens pollicis, and the distal and medial part of the mass is the flexor pollicis brevis. The opposer of the thumb arises from the front of the flexor retinaculum and from the tubercle of the trapezium, and is inserted into the whole length of the radial half of the anterior surface of the first metacarpal [FIG. 404].

Nerve Supply. Like the abductor pollicis brevis, the opponens is usually supplied by the recurrent branch of the median nerve (T. 1): in a minority of cases it may be supplied by the ulnar nerve.

Actions. It pulls the metacarpal bone of the thumb into the position of opposition; that is, it rotates it at the carpometacarpal joint so that the pad of the thumb can be brought into contact with the pads of the fingers. This movement involves, first abduction away from the plane of the palm, secondly medial rotation at the carpometacarpal joint, and finally flexion and adduction of the abducted and medially rotated thumb. In the first two phases the opponens is active, helped in abduction by the abductor pollicis brevis, and in medial rotation by the abductor pollicis brevis and the flexor pollicis brevis. In the last phase the thumb, held in position by these muscles, is flexed to the little finger, or adducted towards the index finger by the adductor pollicis [p. 333].

Opposition is tested by asking the patient to touch the palmar surface of the thumb to the palmar surface of the little finger, without bending the tip of the thumb.

Flexor Pollicis Brevis

This small muscle is usually partly fused with the opponens pollicis [FIG. 407], and arises by a superficial head from the distal border of the flexor retinaculum and the tubercle of the trapezium, and a deep head from the trapezoid and capitate bones (Day and Napier, 1961). It is inserted into the radial side of the base of the proximal phalanx of the thumb just medial to the insertion of abductor pollicis brevis. The tendon of insertion contains a scsamoid bone.

Nerve Supply. The flexor pollicis brevis is more variable in its nerve supply than the abductor and the opponens. Usually the recurrent branch of the median nerve (T. 1) is responsible but this is often replaced by the deep branch of the ulnar nerve (T. 1), and occasionally both nerves may supply the muscle.

Actions. It flexes the thumb at the carpometacarpal and metacarpophalangeal joints. It is also involved in the movement of opposition of the thumb, since flexion of the carpometacarpal joint automatically involves medial rotation. It is difficult to test this muscle apart from the flexor pollicis longus and the opponens pollicis.

Adductor Pollicis

The oblique head of the adductor pollicis comes from the sheath of the tendon of the flexor carpi radialis, the bases of the second, third, and fourth metacarpal bones, and the anterior

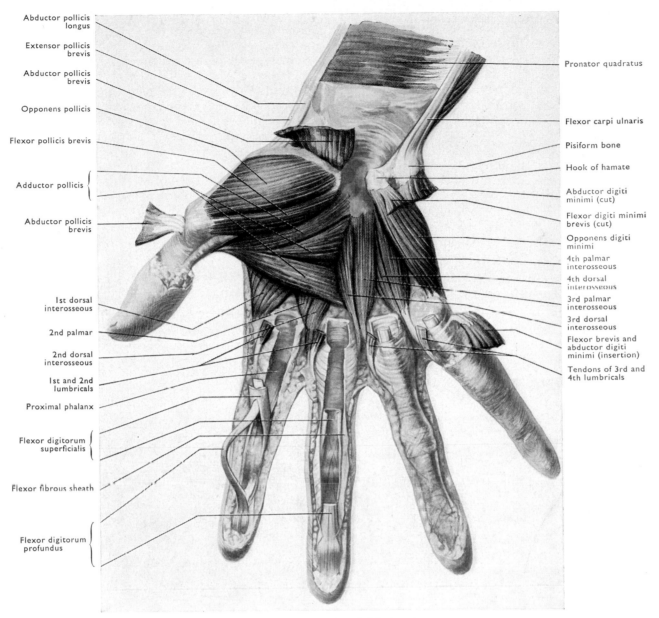

Abductor pollicis longus

Extensor pollicis brevis

Abductor pollicis brevis

Opponens pollicis

Flexor pollicis brevis

Adductor pollicis

Abductor pollicis brevis

1st dorsal interosseous

2nd palmar

2nd dorsal interosseous

1st and 2nd lumbricals

Proximal phalanx

Flexor digitorum superficialis

Flexor fibrous sheath

Flexor digitorum profundus

Pronator quadratus

Flexor carpi ulnaris

Pisiform bone

Hook of hamate

Abductor digiti minimi (cut)

Flexor digiti minimi brevis (cut)

Opponens digiti minimi

4th palmar interosseous

4th dorsal interosseous

3rd palmar interosseous

3rd dorsal interosseous

Flexor brevis and abductor digiti minimi (insertion)

Tendons of 3rd and 4th lumbricals

FIG. 407   Palmar muscles of right hand.

aspects of the trapezoid and capitate bones [FIG. 404]. The transverse head of the muscle comes from the longitudinal ridge on the front of the shaft of the third metacarpal.

The fibres of the two heads, which close to their origin are separated by a gap which transmits the radial artery, converge on a tendon, containing a sesamoid bone, which inserts into the ulnar side of the base of the proximal phalanx of the thumb. Some fibres often pass deep to the tendon of the flexor pollicis longus to be inserted along with the flexor pollicis brevis into the radial side of the proximal phalanx, and others may run into the ulnar side of the extensor tendon.

Nerve Supply. From the deep branch of the ulnar nerve (C. 8; T. I).

Actions. It adducts the thumb into the plane of the palm, and is active in the later stages of the movement of opposition. If the muscle is paralysed the patient will be unable to prevent the examiner withdrawing a piece of paper gripped between the thumb and the index finger, always provided the thumb nail is kept at right angles to the other nails and the tip of the thumb is not allowed to bend (flexor pollicis longus). This test is not reliable unless care is taken to exclude the trick action of the extensor pollicis longus in adducting the thumb [p. 325].

MOVEMENTS OF THE FINGERS

The muscles primarily involved in movements of the fingers are shown in the table below. Because there is so little movement at the carpometacarpal joints of the three radial fingers, no special heading has been included for these joints. But the carpometacarpal joint of the little finger is more mobile, and can be flexed by the flexor digiti minimi brevis and the long flexors, and rotated by the opponens digiti minimi.

Because the knuckle is in each case formed by the proximal bone, the length of the dorsal surface of the finger increases when the finger is bent, and the extensor hoods are pulled distally. This in turn alters the line of pull of the lumbricals and interossei which are inserted into them, and at the same time drags these insertions distally. Contraction of the lumbricals in this situation pulls the flexor profundus tendons distally, so reducing their power and range of movement. The interossei are maximally efficient in abduction and adduction of the fingers when the metacarpophalangeal joints are flexed to an angle of 135 degrees, but they are unable to adduct or abduct when flexion is continued to 90 degrees (Matheson, Sinclair and Skene, 1970).

Movement of the distal part of the finger is further complicated by the existence of an **oblique band** of fibrous tissue which runs from the front of the proximal phalanx and the fibrous digital sheath to the side of the terminal portion of the extensor hood. When the distal interphalangeal joint is flexed, this band is pulled upon, and causes the proximal joint to become flexed also, quite apart from any muscular action. When the proximal joint is extended, the oblique bands cause the distal joint to be extended also.

Clinical evidence has made it clear that extension of the interphalangeal joints depends very largely on the lumbricals and interossei, and that the extensor digitorum, which is the undisputed extensor of the metacarpophalangeal joints, has much less effect on the interphalangeal joints. Exactly why this should be is not easy to explain.

The movements of the fingers in fact present one of the most difficult mechanical problems in anatomy, and the exact actions of the muscles involved are still open to question. For further information see Landsmeer (1949, 1955); Stack (1962).

A patient with paralysis of the radial nerve can weakly extend his metacarpophalangeal joints by the trick action of fully flexing the wrist, so tightening the tendons of the extensor digitorum. Similarly, a patient with a complete paralysis

MUSCLES CONCERNED IN MOVEMENTS OF FINGERS

A. METACARPOPHALANGEAL JOINTS

MOVEMENT	MUSCLE	PRINCIPAL NERVE SUPPLY
Flexion	Flexor digitorum profundus	Anterior interosseous : ulnar
	Flexor digitorum superficialis	Median
	Lumbricals	Median : ulnar
	Interossei	Ulnar
	Flexor digiti minimi	Ulnar
	Abductor digiti minimi	Ulnar
	Palmaris longus (through palmar aponeurosis)	Median
Extension	Extensor digitorum	Posterior interosseous
	Extensor indicis	Posterior interosseous
	Extensor digiti minimi	Posterior interosseous
Abduction and Adduction	Interossei	Ulnar
	Abductor digiti minimi	Ulnar
	Lumbricals may assist in radial deviation	Median : ulnar
	Extensor digitorum (abducts by hyperextending; tendon to index radially deviates)	Posterior interosseous
	Flexor digitorum profundus } Adducts	Anterior interosseous : ulnar
	Flexor digitorum superficialis } by flexing	Median
Rotation	Lumbricals (specially first)	Median : ulnar
	Interossei (movement slight except index : only effective when phalanx is flexed)	Ulnar
	Opponens digiti minimi (rotates little finger at carpometacarpal joint)	Ulnar

B. INTERPHALANGEAL JOINTS

MOVEMENT	MUSCLE	PRINCIPAL NERVE SUPPLY
Flexion	Flexor digitorum profundus (both joints)	Anterior interosseous : ulnar
	Flexor digitorum superficialis (proximal joint only)	Median
Extension	Extensor digitorum	Posterior interosseous
	Extensor digiti minimi	Posterior interosseous
	Extensor indicis	Posterior interosseous
	Lumbricals	Median : ulnar
	Interossei	Ulnar

of the median and ulnar nerves can flex his fingers by hyper-extending the wrist. A patient with an isolated paralysis of the median nerve may sometimes be able to flex the distal interphalangeal joint of the middle or middle and index fingers. This may be due either to the ulnar nerve supplying the whole of the flexor digitorum profundus or to incomplete separation of its tendons from each other. In this case the trick action can only be produced if the distal interphalangeal joints of the ring and little fingers are free to move.

MOVEMENTS OF THE THUMB

Because the metacarpal of the thumb is set at right angles to those of the fingers, all the movements of the thumb take place in planes at right angles to the corresponding movements of the fingers. Thus, **flexion** and **extension** are in the plane of the palm: flexion brings the thumb medially across the palm and extension moves it laterally. In **abduction** the thumb stands away from the palm at right angles; **adduction** closes the thumb on the index finger. **Opposition**, the most important movement, is a composite movement involving first abduction of the thumb, then flexion and medial rotation, usually followed by adduction, resulting in the palmar surface of the thumb being brought into contact with the palmar surface of one of the other digits.

Because of the set of the articular surfaces of the carpometacarpal joint, flexion and abduction of the joint are associated with medial rotation, and extension and adduction with lateral rotation.

MUSCLES CONCERNED IN MOVEMENTS OF THE THUMB

A. CARPOMETACARPAL JOINT

MOVEMENT	MUSCLE	PRINCIPAL NERVE SUPPLY
Flexion	Flexor pollicis brevis Flexor pollicis longus Opponens pollicis	Median or ulnar Anterior interosseous Median
Extension	Extensor pollicis brevis Extensor pollicis longus Abductor pollicis longus	Posterior interosseous Posterior interosseous Posterior interosseous
Abduction	Abductor pollicis longus Abductor pollicis brevis	Posterior interosseous Median
Adduction	Adductor pollicis First dorsal interosseous Extensor pollicis longus } in full extension or abduction Flexor pollicis longus	Ulnar Ulnar Posterior interosseous Anterior interosseous
Opposition	Opponens pollicis Abductor pollicis brevis Flexor pollicis brevis Flexor pollicis longus Adductor pollicis	Median Median Median or ulnar Anterior interosseous Ulnar

B. METACARPOPHALANGEAL JOINT

MOVEMENT	MUSCLE	PRINCIPAL NERVE SUPPLY
Flexion	Flexor pollicis longus Flexor pollicis brevis First palmar interosseous Abductor pollicis brevis	Anterior interosseous Median or ulnar Ulnar Median
Extension	Extensor pollicis longus Extensor pollicis brevis	Posterior interosseous Posterior interosseous
Abduction	Abductor pollicis brevis	Median
Adduction	Adductor pollicis First palmar interosseous	Ulnar Ulnar

C. INTERPHALANGEAL JOINT

MOVEMENT	MUSCLE	PRINCIPAL NERVE SUPPLY
Flexion	Flexor pollicis longus	Anterior interosseous
Extension	Extensor pollicis longus Abductor pollicis brevis Adductor pollicis Extensor pollicis brevis (occasional insertion)	Posterior interosseous Median Ulnar Posterior interosseous

The abductor pollicis brevis produces abduction at both the carpometacarpal and metacarpophalangeal joints. The abductor pollicis longus, which operates only on the carpometacarpal joint, pulls the thumb into a position midway between abduction and extension (i.e. at an angle of about 45 degrees to the plane of the palm). It is accordingly involved in both abduction and extension of the thumb. In full extension the extensor pollicis longus can, as a trick action, be used to adduct the thumb after the palmar ligament of the carpometacarpal joint has become tight (Sunderland, 1944b). If this muscle is paralysed, the interphalangeal joint of the thumb can often be extended, as a trick movement, by the abductor pollicis brevis through its insertion into the extensor tendon. Such extension is, of course, usually accompanied by abduction of the thumb.

In extension of the thumb the flexor and extensor carpi ulnaris contract to cancel the abduction produced at the wrist by the extensor pollicis longus and abductor pollicis longus.

Flexion of the distal phalanx of the thumb is often used as a test for the integrity of the median nerve, which supplies the flexor pollicis longus. It is therefore important to bear in mind a trick movement of flexion produced by strongly extending the wrist and thumb. This puts the flexor pollicis longus on the stretch and so pulls the terminal phalanx into a position of flexion (Sunderland, 1944b).

THE USE OF THE HAND

However fine the movements produced in the hand, they must be controlled from a stable base, and, in order to fix the origins of the intrinsic musculature of the hand, the forearm muscles are brought into play. These in turn require fixation of the elbow by the muscles of the arm, and in turn this requires stabilization of the scapula and the shoulder girdle. Thus even the movements of writing and sewing involve shoulder muscles as well as those of the hand.

The **position of rest** of the hand is one in which the palm of the hand is hollowed, the fingers are flexed, and the thumb slightly opposed; the wrist is adducted, so that when the elbow is extended the axis of the middle finger is parallel to that of the humerus and not to that of the bones of the forearm. Flexion of the fingers increases progressively from the index to the little finger.

From this position the hand can be used in three different ways. In the first place the digits may be used individually. The thumb is fully separated functionally from the fingers, and its movements are not necessarily associated with movements of the other digits. The index finger is less completely autonomous, but it has its own extensor tendon, and the tendon of the extensor digitorum which runs to the index is often not firmly fixed to that of the middle finger. The tendon of the flexor digitorum profundus to the index is also usually separated higher in the forearm. These arrangements allow the finger to be extended separately, and this capacity is used in pointing movements. Again, it has a much greater range of abduction and active rotation (about 30 degrees) at the metacarpophalangeal joint than any other finger except the little finger; this relative freedom is of use in grasping.

When the interphalangeal joints are straight, individual movements of flexion at the metacarpophalangeal joints of the fingers are carried out largely by the small muscles—the lumbricals and interossei (Long and Brown, 1964). Such movements are of very considerable importance in typing, playing musical instruments, etc., though of course the power exerted is relatively small, and when much force is required the long flexors must be called upon. The little finger has a wider range of abduction at the metacarpophalangeal joint than any other finger; its metacarpal is more mobile, and can be opposed towards the thumb.

In the second type of hand movement, the fingers and thumb combine in a '**precision grip**'. This kind of grip is used to handle an object which is small or breakable. The object is seized with the pads of the fingers, which spread around it, the pads adapting themselves to its surface. This entails a considerable amount of rotation, some active, some passive, at the metacarpophalangeal joints. This rotation is sufficient to bring the pad of the index finger at right angles to the pad of the little finger should it be necessary, with the pads of the other fingers occupying an intermediate position. The thumb lies in a position of opposition. The motive power for the precision grip comes largely from the small muscles of the hand—those of the thenar and hypothenar eminences, the lumbricals and the interossei, though the flexor digitorum profundus is probably also in action, since the distal interphalangeal joints are usually slightly flexed. In consequence, the grip depends very largely on the integrity of the ulnar nerve, and is grossly impaired if that nerve is paralysed.

The third type of activity of the hand comes into play when force is required, as in manipulating tools, climbing, and so on. The long flexor muscles, in association with the extensors of the wrist, contract to produce a '**power grip**' in which the fingers are brought towards the palm round the object gripped and the thumb is flexed at the carpometacarpal joint and extended at the interphalangeal joint to grip the shaft or handle of the object concerned. When maximum power is exerted, all the muscles round the wrist, including the wrist flexors, are set in contraction to stabilize it. If the object gripped is small enough, the tips of the fingers come into contact with the palm and in such a clenched fist they all reach the palm together, even though the lengths of the individual fingers are very different. The power grip, involving the long flexor muscles and their synergists, depends very greatly on the integrity of the median nerve, and suffers if this nerve is paralysed. The two types of grip are thus largely independent of each other, and can be separately abolished (Napier, 1956; Bowden and Napier, 1961).

A fourth type of grip, the '**hook grip**' (Napier, 1956), closely resembles the power grip, but the thumb plays little or no part. It is used for carrying suitcases, and in rock climbing, and depends on the long flexor muscles and their synergists.

The versatility of the human hand (Capener, 1956) has in the past been considered evidence of its unspecialized character, in contrast to the foot, which has altered greatly in the course of evolution in response to weight bearing. The contrary view is well expressed by Bishop (1964), who considers that the human hand, 'far from being a primitive organ which has consistently avoided the dangers of specialization, on the contrary, has first adapted to the branch, then to the ground, and lastly specialized beyond all other Primates to the tool'.

PLATE 19

A

A. The platysma in action.

1. Platysma.
2. Sternal head of sternocleidomastoid.
3. Platysma.

B

B. Action of trapezius and deltoid muscles in elevation of the upper limb.

1. Upper fibres of trapezius.
2. Deltoid.
3. Diamond-shaped tendinous area of trapezius muscles.
4. Lower fibres of trapezius.
5. Erector spinae.

C

C. The rectus abdominis in action.

1. Sternocleidomastoid {a sternal head.
                        {b clavicular head.
2. Clavicular fibres of trapezius.
3. Tendinous intersection of rectus abdominis.
4. Linea semilunaris.
5. Tendinous intersections.
6. Linea semilunaris.

D

D. Subject attempting inspiration with closed glottis. Note the effect of atmospheric pressure on the abdomen, the intercostal spaces, the suprasternal and the supraclavicular and infraclavicular fossae.

1. Clavicular fibres of trapezius.
2. Scalene muscles in greater supraclavicular fossa.
3. Heads of sternocleidomastoid.
4. Infraclavicular fossa.

(By courtesy of Professor R. D. Lockhart.)

PLATE 20

A

B

A. Action of latissimus dorsi as depressor and adductor of upper limb, subject pulling on a rope.

1. Posterior part of deltoid.
2. Long head of triceps.
3. Teres major and latissimus dorsi.
4. Rhomboideus major.
5. Serratus anterior.
6. Latissimus dorsi.

B. Action of serratus anterior in elevation of upper limb.

1. Serratus anterior.
2. External oblique.
3. Latissimus dorsi.

C

C. Folds of the axilla and the medial aspect of the upper arm, showing the position of the long and medial heads of the triceps.

Note the contraction of the right sternocleidomastoid and compare with PLATE 19, C and D.

1. Deltoid.
2. Infraclavicular fossa.
3. Biceps.
4. Medial head of triceps.
5. Long head of triceps.
6. Anterior and posterior folds of axilla.
7. Serratus anterior.

(By courtesy of Professor R. D. Lockhart.)

PLATE 21

A

A. Activity of deltoid in controlling gravitational descent of upper limb, shown by weight riding upon the muscle. Cf. B.

1. Deltoid.
2. Sternal head of sternocleidomastoid.
3. Pectoralis major.

B

C

B. Flaccidity of deltoid when antagonists pull on the limb against resistance, shown by weight sinking into the muscle. Note the activity of the pectoralis major in contrast to its passive state in A.

1. Clavicular part of pectoralis major.
2. Sternocostal part of pectoralis major.

C. Activity of the pectoral muscles as adductors against reciprocal resistance of hands.

1. Infraclavicular fossa.
2. Deltoid ⎰a anterior fibres.
⎱b middle fibres.
3. Pectoralis major ⎰a clavicular part.
⎱b sternocostal part.

(By courtesy of Professor R. D. Lockhart.)

PLATE 22

A

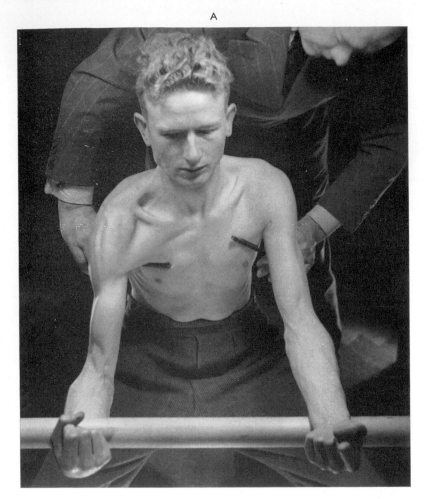

A. Actions of the pectoralis major. The subject's right arm is being raised and the left arm depressed against resistance. The clavicular part of the muscle is active on the right, inactive on the left; the reverse is the case with the sternocostal part.

B

B. Flexors of wrist and elbow in action. In the top left figure the median nerve, escaping from under flexor carpi radialis to lie upon flexor digitorum superficialis tendons, is seen well because palmaris longus is absent.

1. Flexor carpi ulnaris.
2. Median nerve.
3. Flexor digitorum superficialis.
4. Flexor carpi ulnaris.
5. Flexor digitorum superficialis, ring finger tendon.
6. Palmaris longus.
7. Flexor carpi radialis.
8. Biceps, short head.
9. Coracobrachialis.
10. Neurovascular bundle.
11. Triceps, long head.
12. Triceps, medial head.
13. Intermuscular septum.
14. Brachialis.
15. Bicipital aponeurosis.
16. Pronator teres.
17. Brachioradialis.
18. Flexor digitorum superficialis.
19. Flexor carpi radialis.
20. Flexor carpi ulnaris.

From *Living Anatomy* by R. D. Lockhart, 6th ed., 1963, London, Faber and Faber Ltd.

FASCIAE OF THE UPPER LIMB

SUPERFICIAL FASCIA

The superficial fascia of the shoulder region and the arm contains a variable amount of fat; secondary sexual fat is deposited in the back of the female arm, and after middle age large quantities may be found here. At the back of the elbow a subcutaneous bursa separates the skin from the olecranon of the ulna. An enlargement of this bursa in people who lean on their elbow a great deal is called 'student's elbow'.

In the forearm the superficial fascia is in no way remarkable, but in the hand it undergoes several specializations. On the back of the hand it is loose and thin and contains little fat. In the palm it embodies tough strong strands of connective tissue which enclose the fat in loculi and connect the skin firmly to the thickened deep fascia, the palmar aponeurosis [p. 338], especially at the flexure lines. The fixation of the skin by this means is very marked in the centre of the palm, but much less so in the thenar and hypothenar eminences. Here, however, the superficial fascia forms thicker and less fibrous pads of tissue which play an important part in the gripping mechanism of the hand, for they are readily adaptable to the contours of objects which are grasped in it. Other such pads are found in the palm opposite the metacarpophalangeal joints, and here a band of transverse fibres, the **superficial transverse metacarpal ligament,** crosses the distal part of the palm in the superficial fascia of the webs of the fingers, being connected to the palmar surfaces of the fibrous flexor sheaths of the digits [p. 339].

A similar, and very important, pad occurs on the palmar surfaces of the distal phalanges of the digits; here the skin is tacked down to the distal two-thirds of the distal phalanges by strong fibrous strands between which are loculi containing fat. In this way the chief tactile exploratory instruments of the body are provided with a resilient and strong backing. Through the specialized tissue of the digital pads runs the blood supply to the shaft of the distal phalanx, and should the pad become infected, the pressure generated within it by the confined pus is likely to compress the artery and cause death of this part of the bone. On the dorsum of the distal phalanx strength is provided by the nail, and there is no superficial fascia deep to this.

In the superficial fascia of the hypothenar eminence lie the fibres of the palmaris brevis muscle, which help to wrinkle the skin of the medial side of the hand and give it a better grip [p. 332].

DEEP FASCIA

In front of the pectoralis major muscle is a layer of deep fascia continuous with the fascia of the anterior abdominal wall; superiorly, it is attached to the front of the clavicle. Lateral to the pectoralis major it thickens to form the **axillary fascia** [FIG. 371], which constitutes the floor of the axilla. As it is traced laterally it becomes continuous with the deep fascia of the arm.

Posterior to the pectoralis major in the anterior wall of the axilla lies the dense **clavipectoral fascia.** This membrane extends from the first costal cartilage medially to the coracoid process and the coracoclavicular ligament laterally. Superiorly, it splits around the subclavius muscle to become attached to the margins of its insertion to the under surface of the clavicle. Inferiorly it splits to form an investment for the pectoralis minor muscle. The deep surface of the clavipectoral fascia is connected with the sheath of the axillary vessels [FIG. 408].

At the inferolateral border of the pectoralis minor there is a further extension of this stratum of deep fascia downwards to join the fascia of the axillary floor. This extension is continued laterally into the fascia covering the biceps and coracobrachialis muscles.

In the shoulder region, the deep fascia is very strong over the infraspinatus and teres minor muscles, being firmly attached to the medial and lateral borders of the scapula. Superiorly, it forms a sheath for the deltoid muscle, and is attached along the line of the clavicle, the acromion and the spine of the scapula.

In the arm, the deep fascia forms a less dense **investing layer** which encloses the muscles of the arm. The strength of this layer is enhanced by the fact that it receives fibres from the pectoralis major, the latissimus dorsi, and the deltoid. At the elbow the fascia is attached to the medial and lateral epicondyles of the humerus and to the olecranon process, and becomes continuous with the deep fascia of the forearm. From the investing layer two intermuscular septa run to be attached to the supracondylar ridges of the humerus [FIG. 384]. The **medial intermuscular septum** is the stronger of the two; it lies between the brachialis and the medial head of the triceps, and gives origin to both [FIG. 383]. It is usually reinforced by some fibres from the coracobrachialis, and does not extend

FIG. 408   Diagram of clavipectoral fascia.

superiorly beyond the insertion of this muscle into the humerus.

The **lateral intermuscular septum** lies between the brachialis and the brachioradialis in front, and the medial and lateral heads of the triceps behind; it gives origin to fibres of all three muscles. This septum extends up to the insertion of the deltoid and receives some fibres from it.

The deep fascia of the forearm is of varying density and strength in different parts of the region. Superiorly, it is strengthened by the attachment of the bicipital aponeurosis in front and the insertion of the triceps behind. The flexor and extensor muscles of the forearm are covered by a strong layer of fascia from which they take part of their origin. For these reasons, the deep fascia in the upper part of the forearm next the elbow is very strong and dense. Lower down the forearm, as the muscle bellies give way to tendons and are no longer attached to it, the deep fascia becomes thinner, although the attachment to the posterior border of the ulna is thickened by the attachments of the flexor and extensor carpi ulnaris and the flexor digitorum profundus. At the wrist, it develops strong transverse retinacula within its substance; these bands prevent the flexor and extensor muscles of the wrist and digits from 'bowstringing' away from the bones during flexion or extension.

The **extensor retinaculum** [FIGS. 397, 401, 402] is a thickening of the deep fascia on the back of the forearm and wrist; it runs more obliquely than transversely, and is continuous with the deep fascia of the dorsum of the forearm and hand. Laterally it is attached to the distal part of the anterior border of the radius, and medially to the distal end of the ulna, the carpus, and the ulnar collateral ligament of the wrist. From its deep surface septa run to the distal ends of the radius and ulna, so producing six separate compartments, each lined by a synovial sheath. From the lateral to the medial side these compartments transmit: (1) abductor pollicis longus and extensor pollicis brevis; (2) extensors carpi radialis longus and brevis; (3) extensor pollicis longus; (4) extensor digitorum and extensor indicis; (5) extensor digiti minimi; (6) extensor carpi ulnaris. Superficial to the retinaculum lie the dorsal veins of the hand and forearm, the superficial branch of the radial nerve, and the dorsal branch of the ulnar nerve.

On the dorsum of the hand, the deep fascia is arranged in two layers. The superficial layer covers the extensor tendons and is continuous with the distal border of the extensor retinaculum. Over the fingers, this layer fuses with the extensor hoods [p. 322], which replace the deep fascia on the dorsum of the digits. The deep layer of deep fascia is somewhat stronger, and lies over the interossei, blending with the superficial layer at the interdigital clefts.

The **flexor retinaculum** [FIGS. 399, 410] is about 2·5 cm. in length and in breadth. Laterally it is attached to the tubercle of the scaphoid and to both lips of the groove on the anterior surface of the trapezium, so completing an osteofascial compartment through which the flexor carpi radialis tendon and its synovial sheath pass into the hand. Medially, the retinaculum is attached to the hook of the hamate and to the pisiform bone.

The flexor retinaculum converts the hollow on the anterior aspect of the carpus into the **carpal tunnel,** through which run the flexor pollicis longus tendon, together with its synovial sheath, the flexor tendons of the fingers and their synovial sheath, and the median nerve. The anterior surface of the retinaculum gives origin to the muscles of the thenar and hypothenar eminences, and is crossed by the palmar branches of the median and ulnar nerves. The tendon of the palmaris longus muscle is partly inserted into this surface of the retinaculum. The ulnar nerve and artery cross it on its medial side, often covered superficially by a thin layer of fascia which passes from the front of the pisiform bone to the fascia over the palmaris longus. Distally the retinaculum is loosely attached to the deep surface of the palmar aponeurosis.

In the palm of the hand, as on the dorsum, there are two layers of deep fascia. The deep layer covers thè interosseous muscles and encloses the adductor muscle of the thumb [FIG. 410]. The superficial layer has a strong central portion known as the **palmar aponeurosis** [FIG. 409] and on each

Flexor digitorum profundus

Flexor fibrous sheath

Lumbrical muscles

Palmar aponeurosis

Thenar eminence

Hypothenar eminence

Palmaris brevis

Flexor retinaculum

Abductor pollicis longus

Flexor carpi radialis

Palmaris longus

Flexor digitorum superficialis

Flexor carpi ulnaris

FIG. 409 Palmar aponeurosis.

FIG. 410   Oblique cross section through hand, showing fascial compartments.

side it thins out to allow the maximum freedom of movement to the muscles of the thenar and hypothenar eminences. The palmar aponeurosis, which is continuous with the insertion of the palmaris longus [p. 318], is part of the apparatus designed to strengthen the hand for gripping, and to prevent the tendons of the flexor muscles from 'bowstringing' away from the joints they control. It is therefore dense and thick, consisting of longitudinal and transverse fibres which are firmly connected through the superficial fascia to the skin.

The palmar aponeurosis is triangular in shape, with its apex proximal and its base at the webs of the fingers. Four slips are prolonged from it into the fingers to become continuous with the fibrous digital sheaths of the flexor tendons. At their origin, these slips are connected together on their deep surfaces by transverse fibres which cross in front of the lumbrical muscles and the digital nerves and vessels. Each slip is also attached, by dorsally running fibres, to the capsule of the metacarpophalangeal joint and to the deep transverse metacarpal ligament.

The **fibrous digital sheaths** [FIG. 409] are attached to the raised lateral and medial edges of the palmar surfaces of the proximal and middle phalanges and to the palmar surface of the distal phalanx. Between these attachments the fibrous tissue composing each sheath arches from one side of the finger to the other, so completing an osteofibrous tunnel in which the flexor tendons run, surrounded by their synovial sheath. This tunnel is closed distally by the attachment of the

sheath to the distal phalanx, but is open proximally, deep to the palmar aponeurosis. At the interphalangeal joints, the sheath is attached to the palmar ligament, and is relatively thin and loose. On the phalanges, it is thick and strong and holds the tendons closely applied to the bones during flexion of the digit.

In the condition known as **Dupuytren's contracture,** the medial portion of the palmar aponeurosis and the fibrous digital sheaths of the ring and little fingers may become shortened, so pulling these fingers down towards the palm. In severe cases their tips may be pressed firmly into the palm and they become useless. The lateral two fibrous sheaths are less commonly affected.

From the medial and lateral borders of the aponeurosis septa pass backwards to fuse with the fascia on the interossei, so

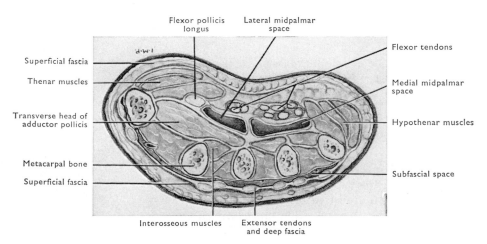

FIG. 411   Transverse section through hand to show the position of the (potential) midpalmar spaces.

dividing the palm into three compartments. The most lateral of these, the **thenar compartment,** contains the short muscles of the thumb; the most medial, the **hypothenar compartment,** contains the short muscles of the little finger, and the **intermediate compartment** contains the long flexor tendons and the lumbrical muscles, surrounded by thin and relatively loose connective tissue. From the deep aspect of this connective tissue sheath a septum runs to the front of the shaft of the third metacarpal [FIG. 411]. Although this septum is relatively fragile, and often incomplete, it serves to divide the intermediate compartment into two potential spaces in which infected fluid may accumulate following injury to the hand. These are the **lateral mid-palmar** and the **medial mid-**palmar spaces, and they communicate proximally, behind the deep flexor tendons, with a similar potential space in front of the pronator quadratus. The spaces are bounded laterally and medially by the septa connecting the palmar aponeurosis with the deep layer of deep fascia; posteriorly, by the interossei and the adductor pollicis [FIG. 411]. Distally, prolongations of loose connective tissue round the tendons of the lumbrical muscles allow a spread of fluid into the web of the finger; this **web space** is freely continuous with the **dorsal subaponeurotic space** on the back of the hand, deep to the extensor tendons, and with the **dorsal subcutaneous space,** in which the fluid may then travel up the dorsum of the forearm in the loose superficial fascia.

MUSCLES OF THE TRUNK

MUSCLES OF THE THORAX

The wall of the thorax is composed of ribs, between which lie the intercostal muscles, and to which are attached many other muscles. Some of these are considered with the upper limb, on which they act; others with the abdominal wall, in which their main bulk lies. The muscles dealt with in this section are all small and are all primarily concerned with movements of the ribs.

Intercostal Muscles

Each **external intercostal** muscle is attached to the lower border of one rib and passes downwards and forwards to gain attachment to the upper border of the rib next below. Posteriorly it extends to the tubercles of the ribs, and anteriorly it thins out gradually to become continuous with the external intercostal membrane, a transition which takes place roughly where the rib becomes continuous with the costal cartilage. The membrane is a thin fibrous sheet which fills the remainder of the intercostal space, bounded by the adjacent costal cartilages and the sternal margin [FIG. 413].

In the lower intercostal spaces the external intercostal muscles may blend with the fibres of the external oblique muscle of the abdomen, which overlap them, but run in an exactly similar direction [FIG. 412]. The external intercostals and the external oblique may indeed be looked on as one continuous sheet of muscle, which in the region of the thorax has become stranded in between the ribs.

The **internal intercostal** muscle of each space is attached to the lower border of the costal cartilage and the costal groove of one rib and to the upper border of the rib below. The fibres lie deep to those of the external intercostal, and run downwards and backwards, crossing those of the external intercostal obliquely [FIGS. 412, 414]. They run in the same direction as those of the internal oblique muscle of the abdomen, and in the lower two spaces the fibres of the two muscles are continuous [FIG. 421].

The internal intercostal is thicker anteriorly than posteriorly,

Scalenus posterior   Scalenus medius

Scalenus anterior

External intercostal muscle

Internal intercostal muscle exposed by removal of external intercostal membrane

Internal intercostal muscles

Rectus abdominis (insertion)

Sheath of rectus abdominis

Obliquus externus abdominis (reflected)

FIG. 412 Muscles of the right side of thoracic wall.

and runs from the side of the sternum to the angle of the rib, where it is replaced by the internal intercostal membrane [FIG. 413]. This extends back to the tubercle of the rib and becomes continuous with the superior costotransverse ligament.

The **innermost intercostals** [FIGS. 413, 414] are variable layers separated off from the internal intercostals by the intercostal nerve and vessels. They lie deep to the internal intercostals, and their fibres run in the same direction. The **subcostal** muscles [FIG. 414] are irregular slips, varying considerably in number and size, which lie on the inner surface of the lower ribs near their angles, and pass over one or two intercostal spaces. They run in the same direction as the innermost intercostal muscles of these spaces, and may be continuous with them.

Nerve Supply. All the intercostals and subcostals are supplied by the corresponding intercostal nerves.

Actions. The anatomical arrangement of the intercostals is simple, but their actions and functions are disputed. It is generally accepted that at least some of the external intercostals are active during inspiration, when they are believed to pull the rib below up towards the rib above [p. 357]. There is also electrophysiological evidence to suggest that they are active during the first phase of expiration, but the reason for this is not established.

The actions of the deeper groups of intercostals are even more uncertain. It has been claimed that they act with the external intercostals to elevate the ribs in inspiration, but others hold that they act from the rib below on the rib above to depress the thoracic cage during forced expiration. The reason for this discrepancy probably lies in the difficulty with which activity in one set of muscles can be distinguished from activity in the other. It is also fairly certain that the type and the amount of such activity may vary with depth of respiration (Campbell, 1955b; Campbell et al., 1970).

Both the external and the internal groups are apparently active during speech, and both groups are inactive in quiet expiration. It is also undisputed that both groups have a function in preventing the bulging or caving-in of the intercostal spaces during the variations in the intrathoracic pressure produced by respiratory movements. Contraction of the intercostals occurs during various movements of the trunk, and it has been suggested that this activity is perhaps more important than their actions in respiration (Jones and Pauly, 1957).

Transversus Thoracis

This muscle originates from the posterior aspect of the xiphoid process and the body of the sternum [FIG. 415] and is inserted into the inner aspect of the costal cartilages, from the second to the sixth inclusive. The lower fibres are more or less horizontal, and are parallel with the uppermost part of the transversus abdominis; the higher fibres pass obliquely upwards and laterally.

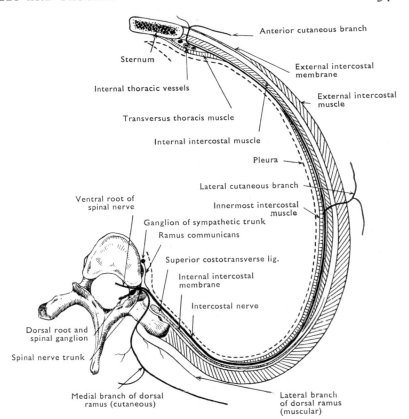

FIG. 413 Diagram of intercostal space to show relations of intercostal muscles, membranes, and nerve.

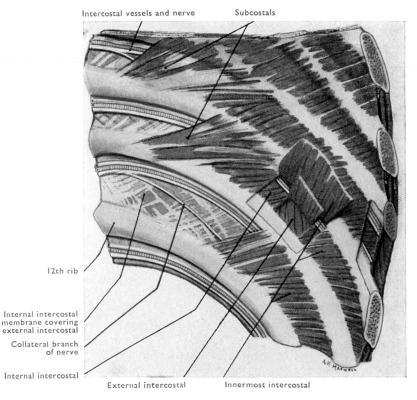

FIG. 414 Dissection of inner surface of chest wall to show intercostal muscles, vessels, and nerves.

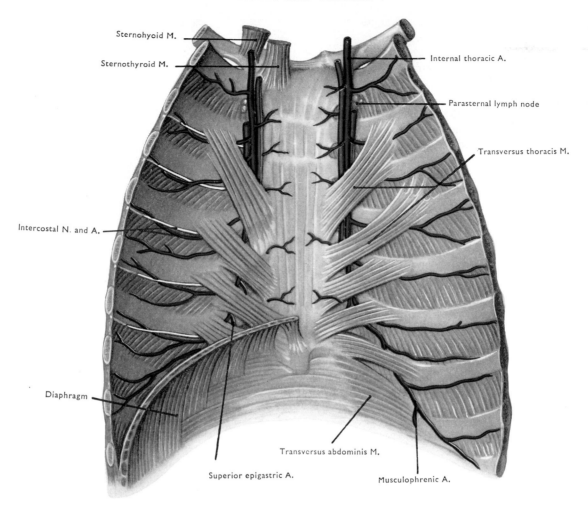

Sternohyoid M.

Sternothyroid M.

Internal thoracic A.

Parasternal lymph node

Transversus thoracis M.

Intercostal N. and A.

Diaphragm

Transversus abdominis M.

Superior epigastric A.

Musculophrenic A.

FIG. 415   A dissection of the posterior surface of the anterior thoracic wall.

Nerve Supply. From the corresponding intercostal nerves.
Actions. The transversus thoracis is thought to depress the costal cartilages articulating with the sternum, and so to contribute to expiration (Taylor, 1960).

Levatores Costarum

Each levator costae [FIG. 337] is a small triangular muscle arising from the transverse processes of a vertebra from the 7th cervical to the 11th thoracic inclusive. It fans out as it descends to be inserted into the external surface of the rib below, between the tubercle and the angle.
Nerve Supply. From the ventral rami of the thoracic nerves.
Actions. Elevation of the ribs [p. 223].

Serratus Posterior Superior

This thin, wide, unimportant muscle lies deep to the vertebroscapular muscles and superficial to the splenius, erector spinae, and semispinalis groups. It is attached to the lower part of the ligamentum nuchae and the spines of the seventh cervical and the upper three or four thoracic vertebrae, and runs downwards and laterally to be inserted into the second, third, fourth, and fifth ribs lateral to the angles.

Nerve Supply. From the second to the fourth intercostal nerves.
Actions. Elevation of the upper ribs.

Serratus Posterior Inferior

Another thin and unimportant muscle, lying deep to the latissimus dorsi, and taking origin from the thoracolumbar fascia [FIG. 373]. The fibres pass almost horizontally to be inserted into the last four ribs.
Nerve Supply. From the ninth to the eleventh intercostal nerves.
Actions. It may help to draw the lower ribs downwards and backwards.

THE DIAPHRAGM

The diaphragm is a dome-shaped partition separating the thoracic and abdominal cavities. It consists of a **central tendon,** shaped not unlike a boomerang, into which muscle fibres are inserted from all parts of the circumference of the inner aspect of the body wall. The fibres attached to the posterior part of the body wall extend much further inferiorly than do the anterior fibres, so that the concavity of the diaphragm looks forwards as well as downwards [FIGS. 416, 417]. The posterior, concave border of the central tendon receives the fibres from the vertebral column, and the convex anterior border receives those from the thoracic margin. The tendon is dense and fibrous, particularly towards the midline, where its fibres interlace in a cruciform arrangement, and the pericardium is firmly attached to its upper surface.

The **sternal portion** of the diaphragm consists of two small slips which arise from the back of the xiphoid process and pass backwards to the central tendon. The **costal portion** arises from the deep surfaces of the lower six costal cartilages on each side, interdigitating with the slips of origin of the transversus abdominis, and is inserted into the anterolateral border of the central tendon.

The **lumbar portion** of the diaphragm arises in part by two **crura** from the anterolateral surfaces of the bodies of the first three lumbar vertebrae on the right side and the first two lumbar vertebrae on the left. The crura embrace the aorta as it enters the abdominal cavity [FIGS. 416, 417], and are attached firmly to the margins of the vertebrae and the intervertebral discs. In between these attachments the upper lumbar arteries separate the fibres from the bodies of the vertebrae. The two crura are connected by a tendinous band, the **median arcuate ligament**, which arches in front of the aorta and gives origin to fibres of the right crus. This, which is usually the larger of the two, spreads out to form a thick triangular sheet of muscle directed upwards and somewhat to the left to be inserted into the middle part of the concave border of the central tendon on both sides of the median plane. Through this part of the right crus passes the oesophagus [FIGS. 416, 417], and the muscle fibres of the crus decussate and surround it (Bowden and El-Ramli, 1967). This arrangement may have a mild sphincteric action, though the importance of this is disputed. Close to the oesophageal opening the **suspensory muscle of the duodenum** is attached to the right crus (Low, 1907). This somewhat unusual mixture of somatic and plain muscle, elastic

and fibrous tissue descends to be attached to the ascending part of the duodenum, which it tethers in place.

The bulk of the fibres of the left crus pass to the left of the oesophageal opening, separated from it by the left margin of the right crus. Some fibres may pass to the right, behind the right crus [FIG. 416] and between the oesophageal and aortic openings.

The remainder of the lumbar portion of the diaphragm arises from the **medial** and **lateral arcuate ligaments,** which are thickenings of the anterior layer of the thoracolumbar fascia covering the psoas major and the quadratus lumborum respectively [p. 362].

The sternocostal and lumbar portions of the diaphragm are distinct developmentally, and in 80 per cent. of bodies they are separated by a hiatus in the muscular sheet called the **vertebrocostal trigone.** This gap [FIG. 416] lies above the twelfth rib, and allows the kidney to come into contact with the pleura, separated from it only by loose areolar tissue. It is the site of the pleuroperitoneal canal of the embryo.

Several structures pass from the thorax to the abdomen or vice versa either through the diaphragm or between it and the body wall. The **aortic opening** lies behind the diaphragm at the level of the lower border of the twelfth thoracic vertebra slightly to the left of the midline. It transmits the aorta and the thoracic duct. Posterolateral to the thoracic duct the azygos vein passes upwards behind the right crus, and each crus is pierced by a greater and a lesser splanchnic nerve. The sympathetic trunk runs from the thorax to the abdomen behind the medial end of the medial arcuate ligament, and the subcostal nerve runs posterior to the lateral arcuate ligament. Between the sternal and costal origins of the diaphragm, where there is a small hiatus, the superior epigastric artery passes in front of the diaphragm to enter the sheath of the rectus abdominis. Similarly, the musculophrenic artery runs between the attachments of the diaphragm to the seventh and eighth costal cartilages.

The **oesophageal opening** in the diaphragm lies to the left

FIG. 416   Dissection showing posterior origin of the diaphragm.

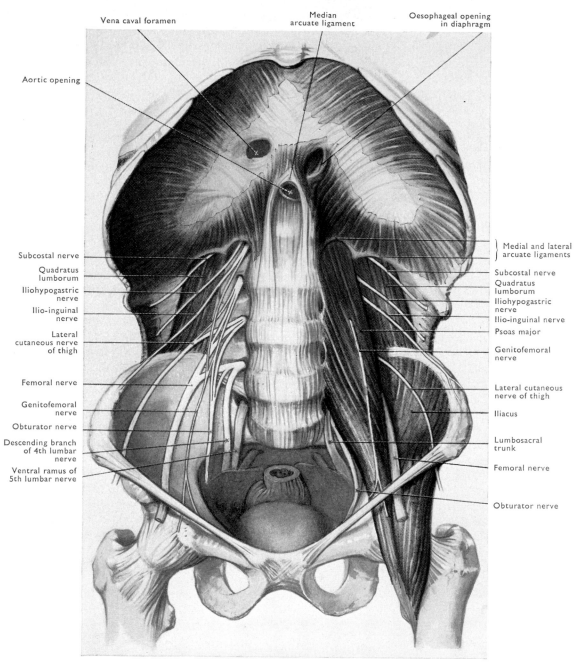

FIG. 417 Diaphragm and posterior wall of abdomen.

of the midline in the muscular part of the right crus at the level of the tenth thoracic vertebra; it is elliptical in shape, and transmits the oesophagus, the trunks of the vagus nerves, and oesophageal branches of the left gastric vessels. The **opening for the inferior vena cava** lies in the central tendon at the level of the disc between the eighth and ninth thoracic vertebrae; its wall is adherent to the margins of the opening. It transmits branches of the right phrenic nerve as well as the inferior vena cava. The left phrenic nerve runs off the pericardium to pierce the muscular part of the diaphragm in front of the left limb of the central tendon.

The three major structures passing between chest and abdomen all have different relationships to the diaphragm and are thus liable to be affected differently by its contraction. The inferior vena cava, passing through the central tendon, is stretched and dilated, so helping the return of blood to the heart. The oesophagus is, if anything, constricted by its muscular surroundings, and the aorta is completely unaffected, since it lies behind the diaphragm altogether.

Nerve Supply. The phrenic nerves (ventral rami of C. 3, 4, 5) supply the diaphragm with motor and sensory innervation. Additional sensory fibres are supplied by branches from the lower six or seven intercostal nerves which run to the peripheral part of the muscle.

Actions. The diaphragm is the chief muscle of inspiration [p. 357]. Acting from the ribs and from its lumbar origins the radiating fibres of the diaphragm pull the central tendon downwards and tend to flatten the dome, thereby increasing the vertical diameter of the chest. The abdominal viscera are compressed against the anterior abdominal wall and pelvic diaphragm, and this eventually brings the tendinous dome to a halt. After this stage, the costal fibres act from the central tendon to pull upwards the lower ribs, thereby raising and everting the sternum and the costal margin. The thoracic capacity is thus increased transversely and anteroposteriorly as well as vertically [p. 223].

Normally the dome of the diaphragm bulges into the thorax higher on the right side than on the left, and on both sides the level of the diaphragm in relation to the ribs and vertebrae varies considerably according to the phase of respiration, the posture adopted, and the size and degree of distension of the abdominal viscera. It is highest when the body is supine, for in this position gravity causes the viscera to press on the under surface of the diaphragm, forcing it up into the chest. In the standing position the diaphragm falls, and it may fall even lower in the sitting position, presumably because of relaxation of the abdominal muscles. When the patient lies on his side the half of the diaphragm which lies uppermost sinks to a lower level still. In contrast the half of the diaphragm nearest the bed rises higher in the thorax because of the gravitational push of the abdominal viscera. As might be imagined, therefore, the excursion of the diaphragm varies very considerably according to circumstances [p. 357].

MUSCLES OF THE ANTERIOR ABDOMINAL WALL

The anterior abdominal wall, unsupported by bone, is required to form a strong but expansible support for the viscera. It must be able to stretch to accommodate a heavy meal, a distended urinary bladder, or a pregnant uterus, and yet to recoil in the expiratory phase of respiration, and to contract to expel the contents of the abdomen in defaecation, micturition, childbirth, or vomiting. In addition, it has to be capable of producing bending and twisting movements of the trunk. These demands are met by a three-layered arrangement of muscle corresponding to the three layers of muscle in the wall of the thorax, and having fibres running in the same directions. The outermost layer, formed by the external oblique muscle, has fibres which cross the middle layer, formed by the internal oblique, like a St. Andrew's cross, and the innermost layer, formed by the transversus abdominis, runs more or less horizontally, like the transversus thoracis. The mechanical strength afforded by this construction is increased by a thickening of the deepest layer of the superficial fascia [p. 360] over the lower part of the abdominal wall, which is subject to the greatest pressure from the pull of gravity on the abdominal contents.

On each side of the midline lies a muscle, the rectus abdominis, which has no counterpart in the thorax. The fibres of each of these muscles run vertically from the pubis to the rib cage and xiphisternum, and are enclosed in a strong fibrous sheath formed by contributions from the other muscles.

Obliquus Externus Abdominis

The external oblique [FIG. 418] is the outermost of the three layers of muscle on the anterior and lateral aspects of the abdominal wall. It arises from the outer surfaces of the lower eight ribs by slips which interdigitate with the serratus anterior and the latissimus dorsi, and which can easily be seen in a muscular subject [PLATE 20 B]. The fibres may be continuous with those of the external intercostal muscles, and they radiate downwards and forwards towards the midline, the most posterior ones being almost vertical.

The posterior vertical fibres are commonly overlapped by the latissimus dorsi, but occasionally there may be an interval between the two, known as the **lumbar triangle**. This triangle is bounded medially by the latissimus dorsi, laterally

FIG. 418 Left obliquus externus abdominis.

Pectoralis major

Serratus anterior

Latissimus dorsi

Obliquus externus abdominis

Sheath of rectus abdominis

Iliac crest

Anterior superior iliac spine

Inguinal ligament

Superficial inguinal ring

Suspensory ligament of penis

Spermatic cord

Anterior superior iliac spine

Pectineal ligament

Rectus abdominis

Pubic tubercle

Transversus abdominis

Internal oblique

Inguinal ligament

Lacunar ligament

Conjoint tendon insertion

External oblique

Pyramidalis

FIG. 419   Muscle attachments to inguinal ligament and pubis.

by the external oblique, and inferiorly by the iliac crest; it is a weak point in the abdominal wall [FIG. 373].

The lower and posterior part of the external oblique is inserted by fleshy fibres into the anterior two thirds of the outer lip of the iliac crest [FIG. 418], and the remainder gives rise to an aponeurosis which is broader below than above. Towards the midline the aponeurosis fuses with the underlying aponeurosis of the internal oblique; above the umbilicus the fusion extends for almost the whole width of the rectus abdominis muscle, and forms the anterior wall of its sheath. Below the umbilicus the area of fusion gradually narrows, and the two layers remain separate until they reach the midline. From the xiphoid process to the pubis the aponeurosis is inserted into the linea alba [p. 351].

The upper part of the aponeurosis of the external oblique covers the insertion of the rectus abdominis, and affords an origin for fibres of the pectoralis major. The lower, free border of the aponeurosis, between the anterior superior iliac spine and the pubic tubercle, forms the **inguinal ligament.** This ligament lies anterior to the structures passing into the lower limb in front of the hip bone—the iliacus, the psoas major, and the pectineus muscles, the femoral nerve and vessels, and the femoral canal [p. 393]. The inguinal ligament is convex downwards because of the pull exerted on it by the

fascia lata (the deep fascia of the thigh), which is attached along its length. It gives attachment in its whole length to the fascia transversalis and in its lateral part to the fascia iliaca [p. 362] and the fibres of the internal oblique and transversus abdominis muscles. Its medial part, which is rolled backwards on itself like a gutter, forms a floor for the inguinal canal [p. 351].

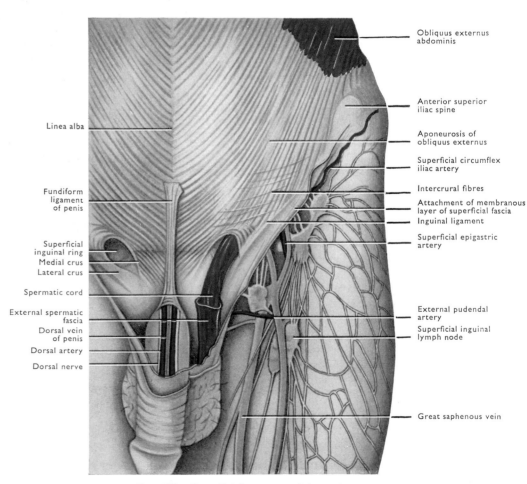

Linea alba

Fundiform ligament of penis

Superficial inguinal ring
Medial crus
Lateral crus

Spermatic cord

External spermatic fascia
Dorsal vein of penis
Dorsal artery
Dorsal nerve

Obliquus externus abdominis

Anterior superior iliac spine

Aponeurosis of obliquus externus

Superficial circumflex iliac artery

Intercrural fibres

Attachment of membranous layer of superficial fascia

Inguinal ligament

Superficial epigastric artery

External pudendal artery

Superficial inguinal lymph node

Great saphenous vein

FIG. 420   Superficial anatomy of the groin.

The medial end of the inguinal ligament has a triangular expansion backwards to the pecten pubis [FIG. 419]; this is the **lacunar ligament.** The free crescentic lateral margin of the lacunar ligament lies close to the medial boundary of the femoral ring, and it sends an extension laterally along the pecten which is termed the **pectineal ligament.** Still other fibres pass obliquely from the pecten across the linea alba [p. 351] to mingle with their opposite equivalents; this is the **reflected** part of the inguinal ligament, and is of comparatively little importance.

Just above the pubic tubercle there is a triangular cleft in the aponeurosis of the external oblique; its base is at the pubic crest, and its apex is directed upwards and laterally. This **superficial inguinal ring,** which transmits the spermatic cord in the male or the round ligament of the uterus in the female, is bounded by lateral and medial **crura.** The lateral crus is formed by the fibres of the external oblique aponeurosis passing to the pubic tubercle. The medial crus is the part of the external oblique aponeurosis attached to the pubic crest and the symphysis pubis. Arching across between the two crura and strengthening the opening are a number of intercrural fibres [FIG. 420], and the margins of the superficial ring are continuous with the **external spermatic fascia.** This thin evagination of the external oblique aponeurosis, which surrounds the spermatic cord (or round ligament) after it has passed through the abdominal wall, is the outermost of the three coverings of the cord. If a finger is placed so as to invaginate the skin of the scrotum upwards and laterally, the spermatic cord can be followed up to the superficial inguinal ring. By pressing backwards, the finger can then detect the crura of the ring, and the size of the aperture can be estimated; normally it just admits the tip of the little finger. **Nerve Supply.** The external oblique is supplied by the ventral rami of the lower six thoracic nerves.

Actions. Like the other muscles of the antero-lateral part of the abdominal wall, the external oblique helps to support the abdominal viscera against the pull of gravity. If the pelvis is fixed, the external oblique muscles forcibly depress the ribs and help to compress the chest, as for instance in the sudden violent expiratory efforts of coughing and sneezing. If both the ribs and pelvis are fixed, the external oblique acting with its fellow of the opposite side can bend the trunk forwards and flex the lumbar vertebral column. If the muscle of one side acts on its own, the trunk is bent laterally to that side and rotated to the opposite side. If these movements are prevented by their antagonists, the external obliques help to compress the abdominal viscera, an action which is brought into play in the movement of expiration against resistance, since it forces the diaphragm upwards, so decreasing the volume of the thoracic cavity. The compressing action of the abdo-

minal muscles is also used in defaecation, parturition, micturition, and vomiting, and the rise in intra-abdominal pressure forces together the walls of the inguinal canal. When heavy weights are lifted, compression of the abdominal contents by the abdominal muscles and the diaphragm helps to make the trunk rigid and protects the vertebral column. Finally, with the other abdominal muscles, the external oblique steadies the pelvis during walking, so providing a stable origin for the muscles operating from the pelvis on the thigh. It can also flex the pelvis on the vertebral column during jumping or climbing.

Obliquus Internus Abdominis

The internal oblique, the middle of the three muscular layers of the abdominal wall, arises by fleshy fibres from the thoracolumbar fascia [FIG. 368], the anterior two thirds of the intermediate line of the iliac crest, and the lateral two-thirds of the inguinal ligament, and the related fascia on the iliopsoas muscle. The fibres fan out from this origin [FIG. 421], the

Serratus anterior

Latissimus dorsi

Obliquus externus abdominis (reflected)

Obliquus internus abdominis

Aponeurosis of obliquus externus (reflected)

Aponeurosis of obliquus internus

Aponeurosis of obliquus externus (reflected)

Cremaster muscle

Suspensory ligament of penis

Spermatic cord

FIG. 421   Right obliquus internus abdominis.

highest and most posterior fibres being inserted into the cartilages of the last three ribs in line with the internal intercostals. The more anterior and lower fibres give way to an aponeurosis along a line which extends downwards and medially from the tenth costal cartilage to the body of the pubis. The lower quarter of this aponeurosis, which represents the fibres arising from the inguinal ligament, passes in front of the rectus abdominis, and unites with the lower part of the aponeurosis of the transversus abdominis to form the **conjoint tendon,** which arches downwards to be attached to the pubic crest and the pecten pubis [FIG. 419]. Its fibres are continuous with those of the anterior wall of the sheath of the rectus abdominis [p. 350].

The upper three-quarters of the internal oblique aponeurosis splits almost immediately into anterior and posterior layers which enclose the rectus abdominis and at its medial margin are attached to the linea alba and the xiphoid process. The anterior layer blends with the aponeurosis of the external oblique and extends up in front of the rectus abdominis on to the chest wall. The posterior layer reaches up as far as the margins of the seventh, eighth, and ninth costal cartilages; above this level the posterior surface of the rectus abdominis is in direct contact with the chest wall.

The lower fibres of the internal oblique, as they arch across the spermatic cord to join the conjoint tendon, lie first in the anterior wall of the inguinal canal [FIG. 422], and later form its roof; medially, the conjoint tendon is the main component of the posterior wall of the canal. Some fibres pass off the inferomedial part of the internal oblique and run along the spermatic cord to form the cremaster muscle.

Nerve Supply. From the ventral rami of the lower six thoracic and the first lumbar nerves.

Actions. Similar to those of the external oblique [p. 347], except that in rotation of the body the internal oblique of one side acts with the external oblique of the opposite side.

Cremaster

This is a loose arrangement of muscle fibres forming festoons connected by the thin cremasteric fascia [p. 361]. The fibres are continuous with the inferior edge of the internal oblique muscle and the adjacent part of the inguinal ligament [FIG. 423], and they loop round the spermatic cord and the testicle, returning to the pubic tubercle, which is the other fixed point of the muscle. The male cremaster is usually reasonably well developed, but in the female it is scanty or absent.

Nerve Supply. From the genital branch of the genitofemoral nerve (L. 1 and 2).

Actions. The cremaster pulls the testis up from the scrotum towards the superficial inguinal ring. Voluntary control over the muscle is not usually possible, but in the **cremasteric reflex** the testis is raised when the medial side of the thigh is stroked. This reflex is very active in the infant

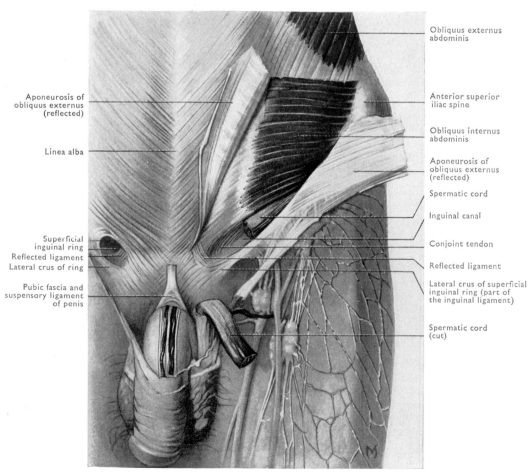

FIG. 422   Left inguinal canal. Structures seen on reflecting obliquus externus.

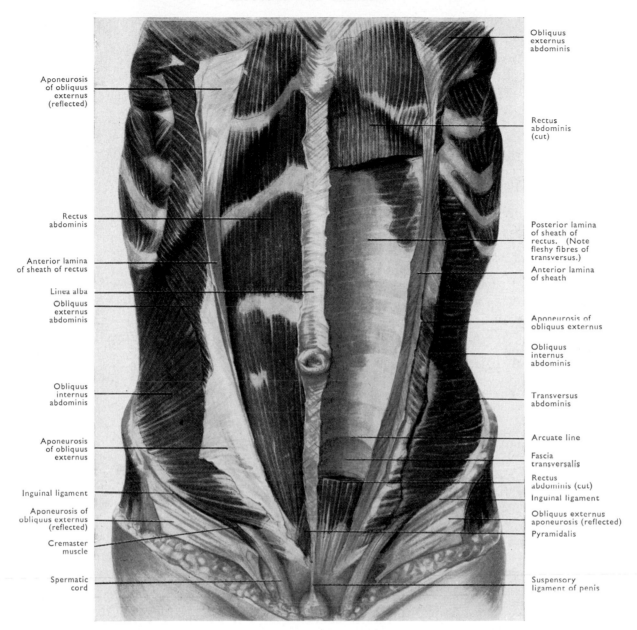

Aponeurosis of obliquus externus (reflected)

Rectus abdominis

Anterior lamina of sheath of rectus

Linea alba

Obliquus externus abdominis

Obliquus internus abdominis

Aponeurosis of obliquus externus

Inguinal ligament

Aponeurosis of obliquus externus (reflected)

Cremaster muscle

Spermatic cord

Obliquus externus abdominis

Rectus abdominis (cut)

Posterior lamina of sheath of rectus. (Note fleshy fibres of transversus.)

Anterior lamina of sheath

Aponeurosis of obliquus externus

Obliquus internus abdominis

Transversus abdominis

Arcuate line

Fascia transversalis

Rectus abdominis (cut)

Inguinal ligament

Obliquus externus aponeurosis (reflected)

Pyramidalis

Suspensory ligament of penis

FIG. 423   Deep dissection of abdominal wall.   Rectus muscle and its sheath.

and young child, and it is important to remember its existence when examining an infant in whom it is suspected that the testis has not descended properly, since the mere performance of the examination may cause the testis to disappear into the superficial inguinal ring or up on to the anterior abdominal wall, leaving that side of the scrotum empty.   By the age of puberty this mobility is much reduced.

Together with the dartos muscle [p. 360], the cremaster forms a mechanism for controlling the temperature of the testis.   When the testis is allowed to hang down at a distance from the superficial inguinal ring, its temperature falls; when it is pulled closer to the ring, it becomes warmer.   This is important, since the proper formation of spermatozoa requires a fairly constant temperature below that of the remainder of the body.

Transversus Abdominis

The transversus abdominis [FIGS. 423, 424, 426], corresponding to the transversus thoracis, arises from the deep surfaces of the costal cartilages of the lower six ribs, interdigitating with the origin of the diaphragm, from the thoracolumbar fascia [FIG. 368], from the anterior two-thirds of the inner lip of the iliac crest, from the lateral third of the inguinal ligament, and from the fascia covering the iliopsoas.   As its name suggests, its fibres run horizontally round the abdominal wall, and they end in an aponeurosis which fuses with the posterior layer of the aponeurosis of the internal oblique.   The upper three-quarters of the resultant fibrous sheet passes behind the rectus abdominis to reach the linea alba [p. 351] and the xiphoid process, but the lower quarter is fused with the entire thickness of the internal oblique

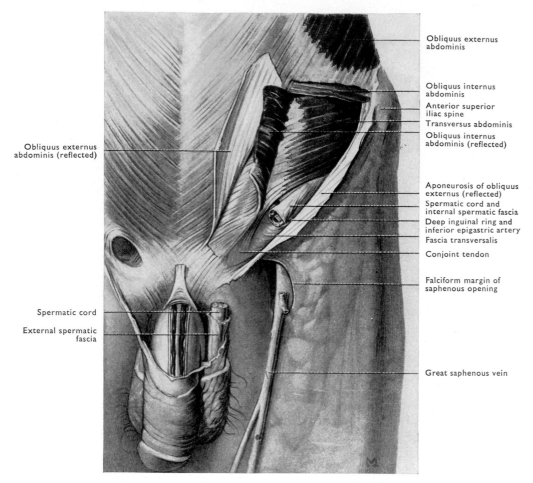

Obliquus externus abdominis

Obliquus internus abdominis

Anterior superior iliac spine

Transversus abdominis

Obliquus internus abdominis (reflected)

Aponeurosis of obliquus externus (reflected)

Spermatic cord and internal spermatic fascia

Deep inguinal ring and inferior epigastric artery

Fascia transversalis

Conjoint tendon

Falciform margin of saphenous opening

Great saphenous vein

Obliquus externus abdominis (reflected)

Spermatic cord

External spermatic fascia

FIG. 424   Dissection of inguinal canal.

aponeurosis and passes with it anterior to the rectus abdominis, to the linea alba. This changeover takes place approximately half way between the umbilicus and the pubis, at the **arcuate line** [p. 351]. Inferior to this there is no aponeurotic posterior layer to the rectus sheath, and the posterior surface of the rectus is in contact with the fascia lining transversus abdominis **(transversalis fascia).** The lowest fibres of the aponeurosis, together with those of the internal oblique, turn downwards to form the **conjoint tendon** [FIGS. 424, 426], and are inserted into the pubic crest, in front of the rectus abdominis, and into the pecten pubis. As they do so, they form the medial boundary of the femoral ring [p. 393].

The lower, free border of the transversus abdominis is concave downwards between its attachment to the inguinal ligament and to the pecten. Between this edge and the inguinal ligament is an elliptical interval in which the transversalis fascia comes into contact with the inferomedial part of the internal oblique, the aponeurosis of the external oblique, and the inguinal ligament. The spermatic cord evaginates the transversalis fascia in the lateral part of this interval at the **deep inguinal ring,** where it enters the inguinal canal, the inferior margin of transversus abdominis arching medially over the cord as it does so.

Nerve Supply. From the ventral rami of the lower six thoracic and the first lumbar nerves.

Actions. These are similar to those of the external and

internal oblique muscles [pp. 347, 348], except that it has little flexing action on the vertebral column, does not depress the ribs, and has virtually no stabilizing effect on the pelvis. The transversus can pull the umbilicus to its own side.

Rectus Abdominis

This long strap muscle arises from the front of the pubic symphysis and the pubic crest. Widening as it passes upwards, it is inserted into the anterior surfaces of the xiphoid process and of the fifth, sixth, and seventh costal cartilages. In the anterior part of its substance, but not extending through the entire depth of the muscle, are three or sometimes more transverse tendinous intersections, which are firmly adherent to the anterior wall of the rectus sheath. The lowest is at the level of the umbilicus, the highest at the xiphoid process, and the third about half way between the other two. The convex lateral border of the muscle shows as a groove on the skin of the lean living subject; this is the **linea semilunaris** [PLATE 19 C].

The muscle is enclosed in a **sheath** formed by the aponeuroses of the other abdominal muscles. At its lateral margin, the aponeurosis of the internal oblique splits into two layers, one of which passes in front of the rectus and the other behind it. The anterior layer is joined by the aponeurosis of the external oblique, and the posterior layer by the aponeurosis of the transversus abdominis, as well as by some fibres of the muscle itself in the upper lateral part of the sheath [FIG. 423]. Above the costal margin, the posterior layer of the sheath is deficient, and the rectus lies directly on the thoracic wall. Below a point about midway between the umbilicus and the symphysis pubis, the internal oblique aponeurosis does not split, and the posterior wall of the sheath is again deficient. Its lower limit is marked by a crescentic border, the **arcuate line** [FIG. 426]. Inferior to this, the rectus is directly in contact with the fascia transversalis, and the anterior wall of the sheath in this region is formed by the aponeuroses of the external oblique, internal oblique, and transversus abdominis, the inferior parts of the last two fusing to form the conjoint tendon [FIG. 426].

Within the sheath are found the pyramidalis muscle, the superior and inferior epigastric vessels, and the terminal portions of the lower five intercostal and the subcostal vessel and nerves.

Nerve Supply. From the ventral rami of the lower six or seven thoracic nerves.

Actions. The rectus abdominis plays comparatively little part in compressing the abdominal contents (Floyd and Silver, 1950), but it is a powerful flexor of the lumbar spine and depressor of the rib cage. During walking, it stabilizes the pelvis, so fixing the origins of the large muscles of the thigh which are concerned in propulsion. A similar fixation of the pelvis occurs when the body is supine and the feet are lifted off the bed; the rectus abdominis contracts to prevent the pelvis being tilted by the weight of the lower limbs.

Pyramidalis

This small muscle, which is not always present, lies on the front of the lower part of the rectus abdominis, arising from the pubic crest and inserting into the linea alba [FIGS. 419, 423].
Nerve Supply. From the ventral ramus of the subcostal nerve.
Actions. It tenses the linea alba, but the purpose this serves is not known.

Linea Alba

The linea alba is a midline raphe extending from the xiphoid process to the symphysis pubis, and formed by the fusion of the aponeuroses of the external oblique, internal oblique, and transversus abdominis. The raphe is relatively narrow below the umbilicus, but widens out to separate the rectus abdominis muscles as they diverge from each other above the umbilicus; here it forms a visible groove in the lean living subject [PLATE 19C]. Approximately half way down the linea alba lies the **umbilicus,** a puckered scar through which, in the foetus, the umbilical vessels, the urachus, and the vitello-intestinal duct were transmitted. On its deep surface remnants of these structures are still attached. The ligamentum teres of the liver is derived from the left umbilical vein, the lateral umbilical ligaments from the umbilical arteries, and the median umbilical ligament from the urachus. In addition, a fibrous band representing the remains of the vitello-intestinal duct may connect the umbilicus with the ileum or with the diverticulum ilei when this is present. If the urachus or the vitello-intestinal duct remains patent, urine or faeces may be discharged through the umbilicus of the newborn infant.

The linea alba permits no transverse stretching under normal circumstances, but stretches longitudinally sufficiently to allow considerable extension of the lumbar vertebral column.

Inguinal Canal

The inguinal canal [FIGS. 422, 424, 426] is the passage in the abdominal wall through which the spermatic cord in the male or the round ligament of the uterus in the female tunnels obliquely from within outwards. In both sexes it transmits in addition the ilio-inguinal nerve. The canal begins at the **deep inguinal ring,** which is an area of evagination in the fascia transversalis [p. 362] about 1·5 cm. above the inguinal ligament midway between the anterior superior iliac spine and the pubic symphysis. From this point the canal travels for about 4 cm. medially and slightly downwards to the **superficial inguinal ring,** which is an area of evagination of the external oblique aponeurosis immediately above the pubic tubercle and the medial end of the inguinal ligament.

The anterior wall of the canal is formed throughout its extent by the aponeurosis of the external oblique, which in its lateral

Rectus muscle

5th costal cartilage

Obliquus externus

Rectus muscle

Anterior lamina of rectus sheath

Obliquus externus

Obliquus internus

Transversus abdominis

Obliquus externus

Obliquus internus

Transversus abdominis

Anterior lamina of rectus sheath — 6th costal cartilage

Xiphoid process   I

Diaphragm

Fascia transversalis

Linea alba   II

Peritoneum

Posterior lamina of rectus sheath

Rectus muscle   III

Anterior lamina of rectus sheath

Inferior epigastric artery

Peritoneum   Fascia transversalis

FIG. 425   Sheath of rectus abdominis muscle.
I. In the thoracic wall. II. In the upper three-quarters of the abdominal wall [cf. Fig. 423]. III. In the lower fourth of the abdominal wall. (Below the umbilicus, the fusion of the aponeuroses of external and internal oblique muscles in front of the rectus is incomplete.)

Transversus abdominis

Lateral cutaneous nerve of thigh

Deep circumflex iliac artery

Transversalis fascia

Deep inguinal ring

Iliacus

External iliac artery and vein

Femoral nerve under fascia iliaca

Psoas muscle

Obturator canal

Arcuate line

Inferior epigastric artery

Internal oblique

Aponeurosis of external oblique

Spermatic cord

Conjoint tendon

Inguinal ligament

Pubic branch of inferior epigastric artery

Lacunar ligament

Fat in femoral canal

Pectineal ligament

Pubic branch of obturator artery

Obturator fascia

FIG. 426 Deep inguinal ring, femoral ring, and obturator canal seen from above and behind inside the abdomen. The anterior abdominal wall has been pulled forwards.

one third is reinforced by the muscular fibres of the internal oblique. The posterior wall of the canal is formed throughout its extent by the fascia transversalis, and in its medial one-third or so this is reinforced by the conjoint tendon, which therefore lies deep to the superficial ring. It follows that the anterior wall is strongest opposite the deep inguinal ring, and the posterior wall strongest opposite the superficial inguinal ring. The weakest portion of the canal is the intermediate one-third. The floor of the canal is formed by the incurved inguinal ligament and the lacunar ligament. The roof is formed by the fibres of the internal oblique and transversus arching medially to cross the spermatic cord obliquely from front to back.

The spermatic cord (or the round ligament of the uterus) enters the canal at the deep inguinal ring, carrying with it a covering of internal spermatic fascia from the fascia transversalis [p. 361]. The cord runs along the canal and acquires from the lower border of the internal oblique muscle a second covering, the cremasteric fascia, which contains the cremaster muscle. A third covering is added as the cord escapes through the superficial inguinal ring, for from the edges of this the external spermatic fascia is carried off as a continuation of the aponeurosis of the external oblique.

The **inguinal triangle** is bounded inferiorly by the inguinal ligament, medially by the lateral border of the rectus abdominis,

and laterally by the inferior epigastric artery, which runs upwards and medially behind the fascia transversalis from the external iliac artery to the rectus sheath, passing immediately medial to the deep inguinal ring. The triangle is divided into medial and lateral parts by the **lateral umbilical ligament.**

The inguinal canal is a weak point in the abdominal wall, and contraction of the abdominal muscles tends to force mobile abdominal contents along it. However, the same muscles, by their contraction, narrow the canal and tend to close the rings. Contraction of the transversus causes the sling of transversalis fascia surrounding the medial margin of the internal ring to move laterally and upwards (Blunt, 1951), while the arched fibres of the internal oblique which form its lateral margin move medially, so closing the ring like the shutter of a camera (Patey, 1949). The pull of the external oblique causes the inguinal ligament to tighten, so supporting the lower edge of the external ring and somewhat narrowing its calibre. The fact that the internal oblique is in more or less continuous contraction during the maintenance of the upright posture (Floyd and Silver, 1950) may thus have a protective as well as a supportive significance. However, it has been suggested that the movement of rotation of the trunk, in which the external oblique of one side co-operates with the internal oblique of the other, may actually pull open the canal. Thus, when the trunk rotates to the left, as the patient pulls sideways on a rope, the linea alba becomes convex to the left (Parry, 1966) and the transversus and internal oblique fibres on the right side are presumably pulled away from the inguinal ligament.

MUSCLES OF THE POSTERIOR ABDOMINAL WALL

Behind the posterior margin of the anterior abdominal muscles, between the iliac crest and the last rib, the abdominal wall is formed by the quadratus lumborum [FIG. 417]. Medial to this, clothing the fronts of the lumbar transverse processes and the sides of the vertebral bodies, lies the origin of the psoas major, a long muscle which runs downwards to be joined by the iliacus, a fleshy mass which lines the iliac fossa. The result is a composite unit, the iliopsoas, which leaves the abdomen behind the inguinal ligament to become the chief flexor of the hip joint. A fourth muscle, the psoas minor, is sometimes present in front of the psoas major.

Together, these muscles form a soft padded bed for many important abdominal viscera, and their position lateral to the lumbar vertebrae allows them to act as lateral flexors of the vertebral column.

Quadratus Lumborum

The quadratus lumborum runs between the posterior part of the iliac crest, the iliolumbar ligament, and the transverse processes of the lower lumbar vertebrae to the medial part of the lower border of the last rib and the transverse processes of the upper lumbar vertebrae. It lies between the anterior and middle layers of the thoracolumbar fascia [FIG. 368], and in front of it lie the ascending or descending colon, the kidney, the psoas major and minor, and the diaphragm. Behind, it is overlapped by the erector spinae.

Nerve Supply. From the ventral rami of the subcostal nerve and the upper three or four lumbar nerves.

Actions. The quadratus lumborum is a lateral flexor of the vertebral column, and fixes the twelfth rib during deep respiration. When both muscles act together they help to extend the lumbar part of the vertebral column, as well as giving it lateral stability.

Psoas Minor

A weak muscle, which is absent in about 40 per cent. of bodies, the psoas minor arises from the sides of the bodies of the twelfth thoracic and first lumbar vertebrae, and from the disc between them. It has a short fleshy belly [FIG. 416] and a long tendon, which runs down anterior to the psoas major to be inserted into the iliopubic eminence and the iliac fascia.

Nerve Supply. From the ventral ramus of the first lumbar nerve.

Actions. It is a weak flexor of the lumbar spine.

Psoas Major

In contrast to the psoas minor, the psoas major is a long thick powerful muscle which comes by fleshy fibres from the adjacent margins of the bodies of the vertebrae from the lower border of the twelfth thoracic to the upper border of the

fifth lumbar, and from the intervertebral discs between them. Some fibres take origin from tendinous arches which pass over the sides of the bodies of the upper four lumbar vertebrae, protecting the lumbar vessels from pressure, and others come from the roots of the transverse processes of all the lumbar vertebrae [FIGS. 368, 417].

The muscle lies in the angle between the transverse processes and the sides of the bodies of the vertebrae, and narrows in a fusiform fashion to pass along the brim of the lesser pelvis. As it does so it is joined by the fibres of the iliacus. It enters the thigh posterior to the inguinal ligament, and ends in a tendon inserted into the lesser trochanter of the femur [FIGS. 437, 448]. The tendon is separated from the pubis and the capsule of the hip joint by a large bursa, which may be continuous with the cavity of the hip joint.

Nerve Supply. From the ventral rami of the upper three or four lumbar nerves.

Actions. In conjunction with the iliacus, it is the chief flexor of the hip joint. Acting from its insertion, the psoas major bends the trunk and pelvis forwards, as in sitting up from the supine position, and it is also a powerful lateral flexor of the lumbar spine. This action is used to maintain the balance of the trunk while sitting. The psoas major may come into play as a postural muscle preventing hyperextension of the hip joint (Basmajian, 1958); but usually it is relaxed in the standing position (Joseph and Williams, 1957; Keagy et al., 1966).

Iliacus

This large fleshy muscle, which lies along the lateral side of the psoas major in the greater pelvis, arises mainly from the floor of the iliac fossa, but fibres also take origin from the ala of the sacrum and the anterior ligaments of the lumbosacral and sacroiliac joints. The fasciculi run downwards and medially behind the inguinal ligament and the majority of them are attached to the lateral side of the tendon of the psoas major, so forming a composite muscle commonly known as the **iliopsoas.** Others run to the lesser trochanter of the femur and to a short linear insertion below this [FIG. 437]. A few of the most lateral fibres are inserted into the capsule of the hip joint.

Nerve Supply. From branches of the femoral nerve (L. 2 and 3).

Actions. The iliacus acts with the psoas major as the chief flexor of the hip joint. It has no action on the vertebral column.

ANATOMY OF HERNIA

The musculature of the wall of the abdominal cavity presents several areas of weakness, through which the more mobile abdominal viscera may occasionally be protruded, so forming a hernia. The most obvious region for such herniation is the inguinal canal, and a contributory factor to the frequency of hernia in this region is the fact that in the erect posture the pressure inside the lower part of the abdomen becomes considerably greater than that inside the upper part. Out of every 100 hernias through the anterior abdominal wall seen in practice, about 90 are in the inguinal region. Anything which raises the intra-abdominal pressure, such as heavy muscular effort, or severe and prolonged coughing, may precipitate the protrusion of some of the abdominal contents through the inguinal canal or other weak spots in the abdominal wall. The most important of these other weak spots are the femoral canal and the umbilicus.

Hernias are about six times as common in males as in females, and in males about 97 per cent. of hernias occur in the inguinal region, 2 per cent. in the femoral canal, and 1 per cent. in the umbilical region. In females, on the other hand, only about 50 per cent. of hernias occur in the inguinal region, 34 per cent. in the femoral canal, and 16 per cent. at the umbilicus.

Inguinal Hernia. The reason for the preponderance of inguinal hernia in males is that in the male the testis and the spermatic cord traverse the inguinal canal; in the female, only the round ligament of the uterus, a much smaller structure, passes through the abdominal wall in this region. The testis in the course of its descent into the scrotum, which it reaches about the time of birth, is preceded by a process of peritoneum, the **processus vaginalis,** which is normally closed off after the descent by adherence of its walls, leaving only a slight dimple in the peritoneum at the site of the deep inguinal ring. But if the funnel-shaped processus vaginalis persists unobliterated, some of the gut may descend within it along the spermatic cord, so acquiring the same coverings as the cord. Because of the course taken by the protrusion, such a hernia is called an **oblique,** or indirect inguinal hernia, and it always emerges from the abdominal cavity lateral to the inferior epigastric artery. Oblique hernias may also occur in the absence of a patent processus vaginalis.

A **direct** inguinal hernia, as its name suggests, bulges outwards in the region of the weak point in the abdominal wall at the inguinal triangle. Here, in the middle of the canal, both its anterior and its posterior walls are relatively weak, and the protrusion occurs medial to the inferior epigastric artery and usually lateral to the edge of the conjoint tendon. There is never a pre-formed tunnel of peritoneum along which the gut may protrude, and it pushes bodily in front of it the peritoneum, the transversalis fascia, and the external oblique aponeurosis.

Umbilical Hernia. The umbilicus is a weak point in the linea alba, through which a hernia of the midgut into the umbilical cord always occurs in the early stages of embryonic development. Usually this hernia is 'reduced' (i.e., the gut returns within the abdominal cavity) by the eleventh week of foetal development, but occasionally it may still remain outside the abdominal wall at birth, forming a **congenital** umbilical hernia. Gross herniation is unusual, but a small knuckle of gut may be protruded in this way, and it normally retreats within the abdomen within a week or so after birth, though in some races it may persist until adolescence.

In the adult, umbilical hernia is most common in females after middle age, and this may be associated with the distension of the abdomen during pregnancy and with the violent efforts of the abdominal wall during childbirth, for the raised abdominal pressure so produced may further weaken the linea alba in this region. The strain is then increased by the deposition of fat inside the abdomen.

Epigastric Hernia. It is much less common for a hernia to occur elsewhere in the linea alba. Such protrusions are usually above the umbilicus, and are known as epigastric hernias.

Femoral Hernia. This is discussed in relation to the anatomy of the femoral canal [p. 393].

Diaphragmatic Hernia. This is most common in middle age, when the intra-abdominal pressure may rise because of the accumulation of fat in the abdominal cavity. This is associated with a weakening of the musculature, which may allow part of the stomach to be pushed from the abdomen into the thoracic cavity. Such a protrusion may rarely occur through a persistent pleuroperitoneal canal, between the costal and vertebral origins of the diaphragm, above the last rib [FIG. 416], but this weak point in the diaphragmatic partition is normally closed by the kidney on the left side and by the kidney and the liver on the right.

Much more usual is herniation through or alongside the oesophageal opening in the diaphragm. Several varieties of this type of hernia are described. As might be expected, the protrusion occurs and symptoms tend to develop when the pressure in the upper part of the abdomen is raised, as for example when the patient lies down in bed or bends down as in gardening.

MUSCLES OF THE PELVIS

The adoption of the erect posture in the course of evolution necessitated the provision of a strong and resilient pelvic floor to maintain the pelvic viscera in position (Thompson, 1899). The levator ani and coccygeus muscles, which are concerned in this function, together make a supporting hammock called the **pelvic diaphragm.**

Levator Ani

The two levator ani muscles [FIGS. 427, 428, 429] form a thin, gutter-like sheet between the pelvic cavity and the perineum. For most of their extent, the two muscles are separated by a narrow interval occupied in the male by the lower part of the prostate gland, and in the female by the urethra and vagina.

On each side the levator ani takes origin anteriorly from the pelvic surface of the body of the pubis, and posteriorly from the pelvic surface of the ischial spine. Between these two bony attachments it arises from the fascia covering the inner surface of the obturator internus muscle [p. 363]. From this linear origin the fibres run backwards, medially, and downwards to be inserted into the central tendon of the perineum, the sidewall of the anal canal, and the fibromuscular **anococcygeal ligament** stretching from the anal canal to the coccyx. The muscle consists of two distinct portions, the iliococcygeus and the pubococcygeus; the latter is somewhat artificially divided for descriptive purposes into three imperfectly separated parts. These are the pubococcygeus, proper, the puborectalis, and the levator prostatae (pubovaginalis in the female).

The **pubococcygeus** is that part of the levator ani arising in front of the obturator canal [FIG. 428]. Its most medial fibres pass back inferior to the capsule of the prostate gland (**levator prostatae**) or to be attached to the side of the vagina (**pubovaginalis**), and are inserted into the central tendon of the perineum [p. 358]. More laterally lie the fibres forming the **puborectalis,** which runs backwards past the prostate or

Piriformis muscle   Arcus tendineus
Coccygeus muscle   Ischial spine   Obturator internus muscle

Anococcygeal ligament
Iliococcygeus muscle
Puborectalis muscle
External sphincter muscle   Pubococcygeus proper
Apex of prostate
Levator prostatae
Central tendon of perineum

FIG. 427   Left levator ani muscle.

the ventral rami of the third and fourth sacral nerves by branches which enter its pelvic surface, and by a branch from the perineal branch of the pudendal nerve, which enters the muscle on its perineal surface.

Actions. With the coccygeus, the levator ani plays an extremely important part in supporting the pelvic viscera and retaining them in position, especially in the female. It is constantly active, even during sleep (Porter, 1962). Contraction of the pelvic diaphragm helps to increase the intra-abdominal pressure, and, conversely, any increase in this pressure results in reflex contraction of the pelvic diaphragm to prevent the viscera being forced downwards. The puborectalis and the pubovaginalis reinforce the sphincters of the anal canal and the vagina respectively. The levator prostatae is responsible for holding the prostate forwards, and can raise it as the pelvic diaphragm is contracted. The insertion of the levator ani into the central tendon of the perineum draws it upwards and forwards in the later stages of defaecation and parturition, and the puborectalis portion of the muscle pulls the anal canal up over the descending faeces. When the rectum is full the puborectalis maintains the angle between the rectum and the anal canal, and may be thought of as an accessory sphincter preventing faeces from entering the anal canal.

Coccygeus. The coccygeus [FIGS. 427, 428, 429] lies posterior to and edge to edge with the levator ani. It stretches as a triangular sheet of muscle and fibrous tissue from the inner surface of the ischial spine to the margin of the lower two pieces of the sacrum and the side of the coccyx. Its fibrous gluteal surface forms the **sacrospinous ligament.**

Nerve Supply. From the ventral ramus of the fourth sacral nerve.

Actions. It assists the levator ani in supporting the pelvic viscera and in maintaining the intra-abdominal pressure. It also pulls forwards the coccyx after it has been pressed backwards during defaecation or parturition.

Obturator Internus

This muscle [FIGS. 427, 428, 435] arises from the inner surface of the obturator membrane and from the surrounding bony margin of the obturator foramen except in the region of the obturator groove. An extension of this origin runs backwards and upwards on the pelvic surface of the ilium. The fasciculi converge on a tendon which develops as the muscle plays on the lesser sciatic notch, round which it turns at an angle, deep to the sacrotuberous ligament. As it does so, the surface of the bone is covered with cartilage and grooved to

vagina and the upper part of the anal canal to form with its fellow a U-shaped loop which holds the anorectal junction forwards towards the symphysis pubis [FIGS. 428, 429]. The lowest fibres of insertion of the puborectalis fuse with the involuntary longitudinal muscle coat of the rectum, which then gives way to a series of fibro-elastic slips which thread their way through the external sphincter of the anus and become attached to the skin round the anus.

More laterally still are the fibres of the **pubococcygeus** proper, which run to be inserted into the anococcygeal ligament and into the sides and front of the coccyx.

The **iliococcygeus** is the most posterior part of the levator ani, and is firmly attached to the strong fascia on the internal surface of the obturator internus [FIG. 435]. This attachment represents the primitive origin of the iliococcygeus from the terminal line of the pelvis. Its fibres run medially to be inserted into the side of the coccyx and the anococcygeal ligament; they are covered on their pelvic aspect by the backward-running fibres of the pubococcygeus [FIGS. 427, 428].

Nerve Supply. The levator ani has a double supply, from

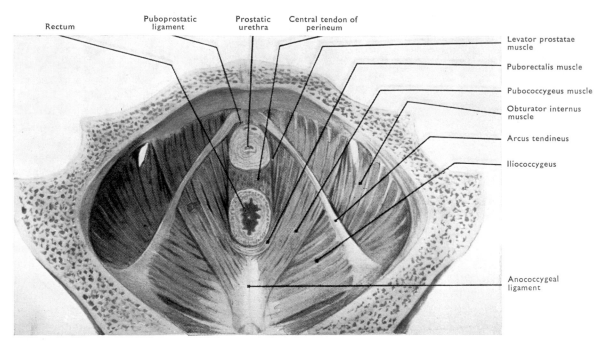

FIG. 428   Levator ani muscles viewed from above.

accommodate the tendon, which is separated from the cartilage by a bursa.   The tendon then passes over the back of the hip joint and is inserted into the medial surface of the greater trochanter of the femur above the trochanteric fossa [FIGS. 438, 440].   It is closely associated with the two gemelli [p. 367], which virtually form another belly for the muscle.

The inner surface of the muscle is covered by the obturator fascia [p. 363] from which arises the levator ani.   The part of the obturator internus above this origin lies in the lateral wall of the lesser pelvis, and the part below lies in the lateral wall of the ischiorectal fossa [p. 357].

Nerve Supply.   From the nerve to the obturator internus,

FIG. 429   Dissection of female pelvic diaphragm from below.

a branch from the anterior aspect of the ventral rami of the fifth lumbar and the first two sacral nerves.

Actions.   It is a lateral rotator of the thigh, but when the thigh is flexed, the alteration in its line of pull allows it to assist in abduction.   Its main function is probably to hold the head of the femur in the acetabulum.

Piriformis

The piriformis [FIGS. 427, 438, 441] comes from the anterior aspect of the second to the fourth pieces of the sacrum inclusive between and lateral to the pelvic sacral foramina.   It leaves the pelvis through the greater sciatic foramen, where it has a further origin from the margin of the greater sciatic notch and the pelvic surface of the sacrotuberous ligament.   The tendon is inserted into the upper border and medial side of the greater trochanter of the femur [FIG. 437]; this insertion is close to the insertion of the obturator internus and the gemelli, with which it may be merged. If the common peroneal nerve leaves the pelvis separately it often pierces the piriformis, the anterior surface of which is directly related to the sacral plexus.

Nerve Supply.   Branches from the ventral rami of the fifth lumbar and the first two sacral nerves.

Actions.   It is a lateral rotator of the thigh when the hip joint is extended; when it is flexed, the piriformis abducts the thigh. Its main function is probably to hold the head of the femur in the acetabulum.

MOVEMENTS OF RESPIRATION

The capacity of the thorax can be altered in a transverse, an anteroposterior, and a vertical direction. Alterations in the vertical diameter are due mainly to movements of the diaphragm, though a limited decrease or increase can be achieved by bending or straightening the thoracic part of the vertebral column; alterations in the other two dimensions are due to movements of the ribs.

A distinction must be made between quiet and forced respiration. In **quiet inspiration** the diaphragm is the main muscle concerned. The central tendon is pulled downwards, and the pressure in the thoracic cavity is thereby lowered at the same time as the pressure in the abdominal cavity is increased. When the resistance of the raised abdominal pressure rises to a certain level the descent of the diaphragm slows and stops, and the contraction of its costal fibres can then raise and evert the costal margin, thus increasing the transverse diameter of the lower part of the chest [p. 223]. The scalene muscles hold the first rib in position during these movements, and the upper external intercostals are in action, probably to prevent the intercostal spaces being sucked in by the changes in pressure within the chest. There is very little actual movement of the upper ribs.

In **quiet expiration** the main factor is the elastic recoil of the lungs and the chest wall. Activity in the diaphragm continues well into the expiratory phase, and the muscle gradually relaxes to control the expiratory movements.

If there is any obstruction to breathing, other muscles are brought into play. In **deep** or **forced inspiration,** the first rib is raised by the scalenus anterior and medius, though the movement is often slight (Jones *et al.*, 1953). The intercostal muscles (certainly the external ones and possibly the internal ones also) pull the second rib up to the first, the third up to the second, and so on, in a cascade effect which reaches its greatest magnitude round about the seventh or eighth rib and then fades away as it continues downwards to the twelfth rib, which is now steadied by the quadratus lumborum to provide a stable origin for increased efforts of the diaphragm. The levatores costarum help to raise the ribs, and it has indeed been claimed that these are the chief muscles of costal inspiration, but their mechanical advantage is poor.

In severe inspiratory difficulties the components of the erector spinae straighten out the thoracic curvature of the vertebral column, and so extend the depth of the thorax. At the same time accessory muscles of inspiration come into action. The sternum is pulled upwards by the sternocleidomastoid acting from its insertion and by the sternohyoid and the sternothyrohyoid complex acting from a fixed hyoid bone; the clavicle is pulled up by the trapezius and sternocleidomastoid, carrying with it the manubrium and the first rib. The scapula may be fixed by the trapezius, the rhomboids and the levator scapulae, so allowing the pectoralis minor to act from its insertion to elevate the ribs; the humerus may be fixed by the patient grasping the head of the bed, so allowing the pectoralis major to raise the sternum and ribs. Towards the end of maximal voluntary inspiration the abdominal muscles contract to limit the movement (Mills, 1950); this contraction does not occur when breathing is increased by asphyxia (Campbell, 1952).

In **forced expiration,** the normal reflex recoil of the anterior abdominal wall and the pelvic diaphragm becomes converted into stronger contraction, particularly of the oblique and transverse abdominal muscles; the rectus abdominis is less concerned. The increased intra-abdominal pressure so produced forces the diaphragm up and decreases the vertical diameter of the chest cavity. At the same time, the abdominal muscles pull down the lower ribs, so decreasing the transverse and anteroposterior diameters of the chest [p. 223]. In sudden sharp expiratory movements, such as coughing, the abdominal muscles contract against the resistance of a closed glottis [p. 480], so building up a considerable expiratory pressure which is then suddenly released by relaxation of the laryngeal muscles. In such movements the latissimus dorsi [p. 307] comes into sudden contraction, probably to help to compress the abdominal and thoracic contents, and the scalene muscles may contract, possibly to prevent downward displacement of the upper ribs (Campbell, 1955a; Jones and Pauly, 1957).

During quiet breathing the central tendon of the diaphragm moves up and down about 1 to 2 cm.; in deep respiration it may achieve as much as 10 cm. Because of the varying level of the diaphragm in different postures [p. 345], and because of the pressure of the abdominal viscera on the under surface of the diaphragm when the body is supine, patients with inspiratory difficulties breathe more easily when propped up in bed.

In summary, the pattern of activity in quiet breathing is a balance between active muscular inspiration and passive elastic expiration; the expiratory muscles are involved only in forced expiration or in expiration against some resistance. In old age, as the elastic recoil of the lungs and body wall becomes progressively impaired, expiration is also impeded and the lungs tend to be incompletely emptied.

The muscles of respiration, although somatic, are not wholly under voluntary control. It is possible to vary the rate and depth of the breathing at will, and also to stop breathing for a short time, but there is very little voluntary control of the degree of contraction of the muscles engaged (Wade, 1954).

MUSCLES OF THE PERINEUM

The perineum is a diamond-shaped area between the pubic arch in front and the coccyx behind. Laterally it is bounded by the inferior ramus of the pubis, the ramus of the ischium, the ischial tuberosity, and the sacrotuberous ligament. It is conveniently divided, for descriptive purposes, into a posterior **anal region** and an anterior **urogenital region,** by a transverse line just in front of the ischial tuberosity [FIGS. 430, 431]. The urogenital part of the perineum is blocked in by the inferior fascia of the urogenital diaphragm [p. 363], and is divided by it into **superficial** and **deep perineal spaces,** each of which contains a group of muscles. The inferior fascia of the urogenital diaphragm is pierced in both sexes by the urethra, and in the female by the vagina. The arrangement of the muscles is therefore somewhat different in the two sexes.

Posteriorly, in the anal compartment, the anus is surrounded by the sphincter ani externus, and on each side of this lies an **ischiorectal fossa** [FIGS. 430, 431] bounded medially by the levator ani, and laterally by the obturator internus, the ischial tuberosity, the sacrotuberous ligament, and the gluteus maximus [p. 364]. Inferiorly, the fossa is bounded by the skin of the perineum, and superiorly by the origin of the levator ani

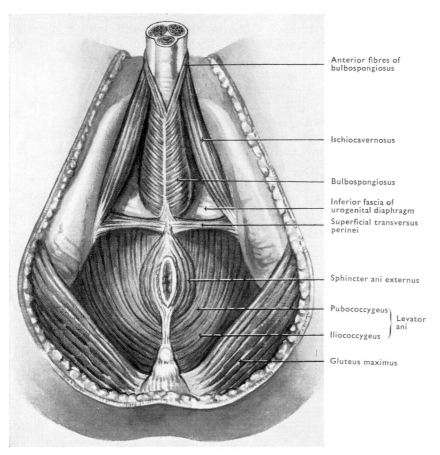

FIG. 430   Muscles of the male perineum.

Anterior fibres of bulbospongiosus
Ischiocavernosus
Bulbospongiosus
Inferior fascia of urogenital diaphragm
Superficial transversus perinei
Sphincter ani externus
Pubococcygeus } Levator ani
Iliococcygeus }
Gluteus maximus

from the fascia covering the obturator internus. Posteriorly, the ischiorectal fossa extends backwards for a little distance above the sacrotuberous ligament and the gluteus maximus, and anteriorly there is a variable extension of the fossa above the deep perineal space [FIG. 435], sometimes as far as the pubis. These extensions, like the rest of the fossa, are occupied by loose fatty tissue [p. 361], and this is sometimes infected from the lower part of the rectum or the anal canal.

Between the anal canal and the urogenital diaphragm lies the **central tendon of the perineum.** This fibromuscular node blends with the urogenital diaphragm and it contains both somatic and visceral muscle fibres. Most of the perineal muscles meet at this central point, which in the female is of great importance during childbirth, since it may be torn and damaged by the descending head of the child [p. 360].

Sphincter Ani Externus

This elliptical muscle surrounds and clasps the conical lower end of the puborectalis part of the levator ani, and consists of three imperfectly separated layers. The subcutaneous lamina contains fibres which surround the anus, a few being attached to the central tendon of the perineum and to the anococcygeal ligament. The main bulk of the muscle is in the superficial lamina which has the only bony attachment. It runs from the coccyx and anococcygeal ligament forwards round the internal sphincter of the anus (a specialization of the circular layer of visceral muscle in the gut wall) to the central tendon of the perineum [FIGS. 430, 431]. The deep layer of the sphincter

ani externus encircles the upper part of the internal sphincter. It blends anteriorly with the central tendon of the perineum; its deeper fibres fuse with those of the puborectalis. The superficial and the subcutaneous parts of the muscle are penetrated by fibres derived from the longitudinal muscle of the rectum [p. 355], on their way to be attached to the skin.

Nerve Supply. From the perineal branch of the fourth sacral nerve, and from the inferior rectal branch of the pudendal nerve (S. 2, 3).

Actions. The muscle is normally contracted (Floyd and Walls, 1953), even in sleep, and with the internal sphincter it holds the anal canal and its orifice closed. It is under voluntary control, and can be made to shut the orifice firmly. It also tethers the central tendon of the perineum to the coccyx.

Transversus Perinei Superficialis

This thin strand of muscle is often difficult to detect in older people. It arises from the medial side of the ischial tuberosity and passes medially to be inserted into the central tendon of the perineum [FIGS. 430, 431]. Like the other muscles of the superficial perineal space it lies inferior to the inferior fascia of the urogenital diaphragm, to which it is attached. Some of its fibres are continuous with decussating fibres of the sphincter ani externus.

Nerve Supply. From the perineal branch of the pudendal nerve (S. 2, 3, 4).

Actions. It helps to steady the central tendon of the perineum.

Bulbospongiosus

The male bulbospongiosus muscles unite with each other in the midline by a fibrous raphe which covers inferiorly the bulb of the penis and the adjoining part of the corpus spongiosum [FIGS. 430, 435]. The muscles arise from this raphe and from the central tendon of the perineum, and curve upwards and forwards round the side of the bulb. The more posterior fibres are inserted into the lower surface of the inferior fascia of the urogenital diaphragm, the intermediate ones to the dorsal surface of the corpus spongiosum of the penis, and the most anterior ones sweep round the corpora cavernosa of the penis and are inserted into the deep fascia on the dorsum of the penis.

In the female, the bulbospongiosus muscles are separated by the openings of the vagina and the urethra [FIG. 431], and each muscle is a thin layer covering the bulb of the vestibule at the side of the lower part of the vagina. They arise from the central tendon of the perineum, and are inserted into the sides of the pubic arch and into the root and dorsum of the clitoris.

Nerve Supply. From the perineal branch of the pudendal nerve (S. 2, 3, 4).

Actions. The male muscles contract at the end of the act of micturition to expel the last drops of urine. They also contract rhythmically during ejaculation, to help to propel the semen along the urethra. The middle fibres of the muscle assist in erection of the penis by compressing the erectile tissue of the bulb, and the anterior fibres compress the corpora cavernosa and the dorsal vein of the penis.

The female bulbospongiosus muscles provide a sphincter for the orifice of the vagina. The most anterior fibres can compress the dorsal vein of the clitoris and are concerned in the erection of the organ.

Ischiocavernosus

The ischiocavernosus comes from the medial side of the ischial tuberosity and passes forwards inferior to the crus of the penis or the clitoris to be inserted into the margin of the pubic arch on each side of the crus, and also into the corpus cavernosum of the penis or clitoris [FIGS. 430, 431, 435].

Nerve Supply. From the perineal branch of the pudendal nerve (S. 2, 3, 4).

Actions. By compressing the crus of the penis or clitoris the muscle contributes to the process of erection.

Sphincter Urethrae

This muscle arises from the inferior pubic ramus, and from the walls of the deep perineal space [FIGS. 432, 435] in which it lies. Its fibres run medially towards the urethra, and in the male are inserted into a median raphe, mostly posterior to the urethra. The fibres which surround the urethra form a muscular collar for the canal and have no bony attachments. In the female the sphincter urethrae encircles the lower end of the urethra as in the male, and the outer fibres decussate with each other between the urethra and the opening of the vagina; some of them are attached to the vaginal wall. Together with the deep transverse muscle of the perineum, with which it is often fused, the sphincter urethrae forms the muscular basis of the **urogenital diaphragm.**

Nerve Supply. From the perineal branch of the pudendal nerve (S. 2, 3, 4).

Actions. The muscle, which is under voluntary control, compresses the membranous portion of the urethra and prevents the onset of micturition when the bladder wall contracts. It is relaxed during micturition but can be brought into action to shut off the stream.

Deep Transversus Perinei

This is another thin muscle which blends with the posterior fibres of the sphincter urethrae and is inserted into the central tendon of the perineum, some of its fibres being inserted into the vaginal wall in the female. Its origin is from the medial surface of the ramus of the ischium [FIG. 432] and it runs parallel with, and superior to, the superficial transverse muscle.

Nerve Supply. From the perineal branch of the pudendal nerve (S. 2, 3, 4).

Actions. It helps to steady the central tendon of the perineum.

MECHANISM OF EXPULSIVE MOVEMENTS

Defaecation

Like most commonplace actions, defaecation is in reality an extremely complicated process. Our know-

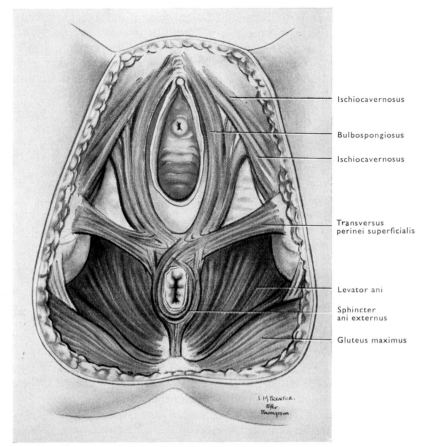

FIG. 431 Muscles of the female perineum.

ledge of the mechanism is derived largely from radiographic examination, and detailed information regarding the function of the various muscles is not available. Nevertheless, it can be said that when the rectum has filled from the sigmoid colon—an entirely involuntary process—the abdominal pressure

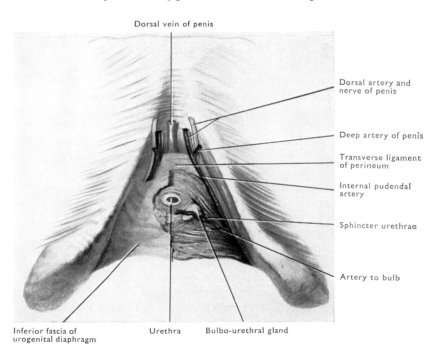

FIG. 432 Inferior fascia of the urogenital diaphragm (perineal membrane) and sphincter urethrae.

is increased by the contraction of the muscles of the abdominal wall. At the same time the puborectalis muscle relaxes, so decreasing the angle between the ampulla of the rectum and the upper portion of the anal canal. The general relaxation of the levator ani allows the colon and the rectum to pass downwards and to become narrow and elongated. The involuntary muscle of the wall of the rectum contracts, the involuntary internal sphincter relaxes, and the sphincter ani externus, which represents the voluntary control, allows the faeces to pass. At the same time fibres of levator ani draw up the walls of the anal canal over the descending faeces. Subsequent to evacuation, the puborectalis and the sphincter ani externus close the anal canal, and the internal sphincter contracts after the passage of the faeces.

Micturition

Even less is known about the movements concerned in micturition. Radiographic studies suggest that urine is held up at the level of the neck of the bladder, not at the sphincter urethrae, and certainly the sphincter urethrae can be destroyed without necessarily producing incontinence. The sphincter is under voluntary control, and can be used to cut off the stream either during or at the end of micturition. In the female the sphincter is much less well defined than in the male, yet there is no diminution of the ability to control the stream of urine.

One theory of the process of micturition suggests that relaxation of the levator ani allows the neck of the bladder to descend, and this initiates reflexes resulting in contraction of the bladder wall. This is supported by the finding that the constant resting activity in the muscles of the pelvic floor falls to a very low level during micturition (Porter, 1962). Voluntary cessation of micturition can be produced by contraction of the levator ani, the stream being cut off at the neck of the bladder.

It is possible that in the male the bulbospongiosus helps to propel the urine along the urethra, and this muscle is certainly concerned in the expulsion of the last few drops of urine from the urethra.

The contraction of the bladder musculature is assisted by a rise in the intra-abdominal pressure due to the contraction of the muscles of the abdominal wall.

Parturition

The main propulsive force in childbirth is the massive involuntary muscle of the wall of the uterus, but this is assisted by strong contractions of the abdominal muscles. As the head of the baby descends through the birth canal, the muscles forming its walls are stretched and may become torn. This is particularly so in the case of the perineal muscles, which have little time to stretch gradually, as once the head has reached them the final stages of expulsion are usually rapid. As a result of this, the central tendon of the perineum is commonly torn in uncontrolled childbirth. It is one of the main objectives of the midwife to prevent this happening, since should it be torn, control of the perineal musculature is lost, and such catastrophes as prolapse of the pelvic organs may result. It is for this reason that before the actual birth an incision is often made through the posterolateral part of the wall of the vaginal orifice and the adjacent part of the perineum lateral to the central tendon. This operation of **episiotomy** enlarges the outlet in a planned direction, and prevents ragged tearing; the incision is sewn up after the birth and the perineum can be well reconstituted.

FASCIA OF THE TRUNK

SUPERFICIAL FASCIA

The fascia of the front and sides of the trunk contains a very variable amount of fat, an accumulation of which surrounds and protects the mammary gland. This mammary fat is laid down at puberty as a secondary sexual characteristic of the adult female, in whom it extends from the side of the sternum to the mid-axillary line, and from the second to the sixth ribs in the vertical plane. The male breast is surrounded by a much smaller amount of fat.

In the superficial fascia of the anterior abdominal wall fat is commonly deposited in middle age, the favoured site in males being the upper part of the abdomen and in females the lower part. In really obese people this layer may reach several centimetres in thickness.

As the superficial fascia is traced down towards the thigh, it separates into two layers, between which lie the superficial vessels and nerves and the superficial inguinal lymph nodes. The superficial layer is continuous with the superficial fascia of the thigh without interruption.

The deeper layer, which is a thin membrane containing elastic fibres, is an elastic tunic which may well have a function in supporting the abdominal contents. In the lower part of the abdominal wall this **membranous layer** is loosely attached to the aponeurosis of the external oblique muscle, and more closely to the linea alba and to the symphysis pubis; there is very little deep fascia in this region, and the membranous layer of the superficial fascia may be looked on as a substitute. Traced downwards, it passes superficial to the inguinal ligament to become attached to the deep fascia [fascia lata, p. 392] of the thigh just distal to, and roughly parallel with, the inguinal ligament.

In the male, the membranous layer is prolonged on to the dorsum of the penis, where it forms a relatively dense band called the **fundiform ligament**, and both layers of the superficial fascia become thinner as they pass downwards over the spermatic cord to become continuous with the superficial fascia of the scrotum. This, like that of the penis, contains very little fat. It is reddish in colour and in its substance lie the fibres of the involuntary **dartos** muscle, which assists in suspending the testis and corrugates the skin of the scrotum. This muscle is supplied by the genital branch of the genito-femoral nerve. The superficial fascia also helps to form the septum of the scrotum, which extends upwards between the two testes and incompletely separates them and their coverings.

In the female, the superficial fascia of this region contains a considerable quantity of fat, which is responsible for much of the bulk of the mons pubis and the labia majora.

In the anterior part of the perineum of both sexes the superficial fascia is again composed of two recognizable components, a superficial fatty layer, and a deep membranous layer. The fatty layer is continuous with the superficial fascia in the thigh, and also posteriorly with the fat of the ischiorectal fossae and buttocks. The membranous layer, on the other hand, is firmly attached laterally to the inferior ramus of the pubis and the ischial ramus [FIG. 433]. Posteriorly it fuses with the posterior border of the inferior fascia of the urogenital diaphragm (perineal membrane) and with the central tendon of the perineum. In the midline it is attached to the septum of the scrotum and to the median raphe of the bulbospongiosus muscles [FIG. 435].

The attachments of the membranous layer of the superficial fascia are of some importance in cases of rupture of the male urethra in the perineum, as may occur when a small boy falls astride a railing. The fascial attachments prevent urine which leaks out into the surrounding tissues from passing backwards into the ischiorectal fossae or sideways into the thigh. The escaping fluid is therefore directed forwards into the subcutaneous tissue of the scrotum. It then fills up the 'space' within the fascial sheath of the penis and mounts upwards round the spermatic cord to the anterior abdominal wall. Here it may pass freely upwards, but it cannot enter the thigh because of the attachment of the membranous layer of the

superficial fascia to the fascia lata and to the body and inferior ramus of the pubis [FIGS. 433, 434, 435].

In each ischiorectal fossa the superficial layer of the superficial fascia forms a pad of fat, which separates the levator ani from the perineal part of the wall of the pelvis, lined by the obturator internus [p. 355]. These ischiorectal pads form a semifluid packing material which allows the anal canal to distend as the faeces pass along it during defaecation. Over the ischial tuberosities the fat is infiltrated by bands of fibrous tissue which adhere to the skin and to the bone.

The membranous layer of the superficial fascia in each ischiorectal fossa is thin and for the most part difficult to define; it lines the walls of the fossa, covering over the fatty pad superiorly. On the lateral wall of the fossa it is somewhat thickened to form the medial wall of the **pudendal canal**, which extends from the lesser sciatic foramen to the posterior margin of the deep perineal space. The canal conveys the internal pudendal vessels, the perineal nerve, and the dorsal nerve of the penis (or clitoris), and its lateral wall is formed by the deep fascia covering the inferior part of the obturator internus [p. 363].

DEEP FASCIA

The deep fascia covering the anterior and lateral surfaces of the walls of the chest and abdomen presents no special features; it is thin and relatively elastic to meet the need for the thorax and abdomen to expand, and in places is virtually replaced by the aponeurosis of the external oblique muscle and by the membranous layer of the superficial fascia. Superiorly it is attached to the clavicle and to the side of the sternum; inferiorly, it is attached to the iliac crest. At the superficial inguinal ring [p. 347] the external oblique aponeurosis and the deep fascia covering it are prolonged downwards and medially as the **external spermatic fascia**, the outermost of the three fascial coats which in the male invest the spermatic cord and the testis, and in the female surround the round ligament of the uterus. Deep to this outer layer lies the **cremasteric fascia**, a layer pulled away from the internal oblique, and containing the fibres of the cremaster muscle; deep to this again is the **internal spermatic fascia**, a coating derived from the transversalis fascia [p. 362].

The cavities of the chest, abdomen, and pelvis all have a fascial lining, though the strength and importance of this fascia vary from place to place. Lining the chest wall, outside the pleura, lies the **endothoracic fascia**. This is a thin layer of loose areolar tissue, which is often only recognizable posteriorly. However, superior to the first rib, it becomes a more imposing fibrous layer which is attached

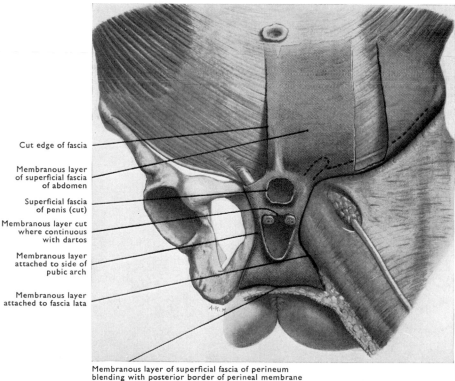

Cut edge of fascia

Membranous layer of superficial fascia of abdomen

Superficial fascia of penis (cut)

Membranous layer cut where continuous with dartos

Membranous layer attached to side of pubic arch

Membranous layer attached to fascia lata

Membranous layer of superficial fascia of perineum blending with posterior border of perineal membrane

FIG. 433 Diagram showing continuity of membranous layer of the superficial fascia of abdominal wall and perineum.

to the inner border of the first rib, and runs from this to the anterior border of the transverse process of the seventh cervical vertebra. This layer is known as the **suprapleural membrane,** and is reinforced and made tense by some fibres from the scalene muscles under cover of which it lies (scalenus minimus).

The fascial lining of the abdominal and pelvic cavities is a continuous membranous bag lying outside the peritoneum; the different regions of the bag are, perhaps confusingly, given different names. It encloses the abdominal and pelvic cavities, the viscera, and the great vessels, but the nerves which emerge from the vertebral canal lie posterior to the bag, as do the lumbar and lumbosacral plexuses which they form. The **transversalis fascia** is the name given to the portion of this fibrous bag which lines the deep surface of the transversus abdominis muscle; it is thick and strong inferiorly, but becomes thinner as it ascends to become continuous with the fascia on the under surface of the diaphragm. Posteriorly the transversalis fascia is continuous with the anterior layer of the thoracolumbar fascia on the quadratus lumborum, and hence

with the fascia on the anterior surface of the psoas muscles [FIG. 368]. Inferiorly it becomes continuous with the pelvic fascia, but along the line of the inguinal ligament and the iliac crest, to both of which it is firmly attached, it changes its name to become the **iliac fascia.** As the femoral vessels pass from their position inside the fascial bag of the abdomen into the upper part of the thigh, they drag out with them an investment of this fascia, the **femoral sheath** [FIGS. 470, 471]. The anterior wall of this sheath is derived from the transversalis part of the abdominal fascia, and the posterior wall from the iliac part. Above the inguinal ligament, at the deep inguinal ring, the transversalis fascia is evaginated by the spermatic cord in the male and the round ligament of the uterus in the female. Its prolongation into the inguinal canal around these structures forms the **internal spermatic fascia.**

The fascia on the posterior wall of the abdomen is strengthened in two places. The **medial arcuate ligament** is a thickening of the fascia covering the psoas major and minor muscles, and stretches from the side of the body of the first lumbar vertebra to its transverse process. The **lateral arcuate**

FIG. 434   Sagittal section of male pelvis showing fasciae in blue (semidiagrammatic).

ligament is a similar band in the anterior layer of the thoraco-lumbar fascia covering the quadratus lumborum, and runs from the transverse process of the first lumbar vertebra to the twelfth rib. Both arcuate ligaments serve to give origin to fibres of the posterior part of the diaphragm [FIG. 417].

Between the lining fascia of the abdominal wall and the peritoneum lies a layer of areolar **extraperitoneal tissue** which is relatively scanty anteriorly, but abundant posteriorly, where it contains a good deal of fat. This tissue surrounds and forms sheaths for the kidneys, the ureters, the suprarenal glands, the abdominal aorta, inferior vena cava, and their branches and tributaries. It is continuous with similar tissue in the posterior mediastinum of the thorax through the aortic opening in the diaphragm, and is also prolonged into the pelvis. In both abdomen and pelvis it is found between the layers of peritoneum forming the mesenteries which tether the viscera to the body wall. The extraperitoneal tissue on the anterior abdominal wall below the level of the umbilicus contains the inferior epigastric vessels, the lateral and median umbilical ligaments, and the upper part of the full bladder as it rises into the abdominal cavity. There is no extraperitoneal tissue on the under surface of the diaphragm.

The ramifications and detailed topography of the fascia in the pelvis are exceedingly complicated. However, in broad terms its arrangement is simple. A **parietal layer** covers the muscles which line the pelvis and form its floor. Some of the pelvic viscera are surrounded by a **visceral layer** which forms a sheath for them. Thirdly, specialized condensations of the fascia or extraperitoneal tissue serve to form 'ligaments' tethering the viscera to the pelvic wall.

The **parietal pelvic fascia** covers the pelvic surfaces of the obturator internus, the piriformis, the coccygeus, the sphincter urethrae, the deep transversus perinei, and the levator ani muscles; the part which lines the obturator internus is thicker than the rest. This **obturator fascia** is continuous with the iliac fascia above, and is attached anteriorly to the back of the body of the pubis. The levator ani takes origin from a thickened part of it, the **arcus tendineus**, and the part of the obturator fascia above this linear origin is tough and strong, being reinforced by the aponeurotic origin of the levator ani. The part below the origin, which forms the lateral wall of the ischiorectal fossa, is thin and relatively weak.

The fascia lining the superior surface of the levator ani is called the **superior fascia of the pelvic diaphragm.** The portion of this fascia which stretches between the symphysis pubis and the neck of the bladder and the urethra in the female or the prostate gland in the male is thickened to form the **pubovesical ligaments** or **puboprostatic ligaments** respectively. These bands lie on each side of the median plane, and between them the superior fascia of the pelvic diaphragm blends with the **superior fascia of the urogenital diaphragm**, which lines the superior surfaces of the deep transversus perinei and the sphincter urethrae muscles [FIGS. 434, 435].

The **inferior fascia of the pelvic diaphragm** forms the medial wall of the ischiorectal fossa, lying on the under surface of the levator ani [FIG. 435]. It is continuous with the fascia on the medial surface of the obturator internus, and with the fascia on the inferior surface of the sphincter ani externus. The fascia on the under surface of the sphincter urethrae and the deep transversus perinei muscles is very dense and strong, and is known as the **inferior fascia of the urogenital diaphragm** (perineal membrane). This membrane stretches almost horizontally across the pubic arch, attached to the inferior ramus of the pubis and the ramus of the ischium: the crura of the penis or clitoris are attached to its inferior surface.

Orifice of right ureter — Fatty areolar tissue — Obturator nerve and vessels — Tendinous arch of levator ani — Prostatic venous plexus embedded in fascia — Inferior fascia of the pelvic diaphragm — Levator ani — Anterior diverticulum from ischiorectal fossa — Sphincter urethrae — Side of pubic arch — Posterior scrotal nerves and artery — Ischiocavernosus — Fascia lata

Ductus deferens — Obturator internus — Obturator membrane — Pectineus — Capsule of prostate — Obturator externus — Obturator fascia — Superior fascia of urogenital diaphragm — Dorsal nerve of penis — Adductor magnus — Adductor brevis — Artery to bulb — Inferior fascia of urogenital diaphragm — Gracilis — Membranous layer of superficial fascia

Bulbo-urethral gland   Bulbospongiosus

FIG. 435 Section of male pelvis in a coronal plane to show the fasciae of pelvis and perineum. The fascia of the prostate has been removed in front to show the prostatic capsule.

It is roughly triangular in shape, and the short anterior border at its apex is thickened to form the **transverse perineal ligament** [FIG. 432]. Between this and the symphysis pubis the dorsal vein of the penis or clitoris enters the pelvis. Posteriorly the base of the triangle is connected to the central tendon of the perineum and becomes continuous with both the superior fascia of the urogenital diaphragm and the membranous layer of the superficial fascia [FIG. 434], thus closing the deep and superficial perineal spaces posteriorly.

Around the prostate gland, the deep fascia becomes thickened and condensed to form the **fascial sheath of the prostate**, which surrounds the veins forming the prostatic venous plexus. This sheath is continuous below with the superior fascia of the urogenital diaphragm, and on each side with the superior and inferior fasciae of the pelvic diaphragm round the free medial border of the levator ani muscle [FIG. 435]. Anteriorly it is attached to the pelvic surface of the symphysis pubis by the puboprostatic ligaments, and superiorly it becomes continuous with the connective tissue surrounding the plexus of veins on the inferolateral surfaces of the bladder. Posteriorly the sheath of the prostate is continuous with the **rectovesical septum** (rectovaginal septum in the female) which extends from the central tendon of the perineum to the floor of the rectovesical pouch of peritoneum, separating the prostate, the seminal vesicles, and the ductus deferentia anteriorly from the ampulla of the rectum posteriorly [FIG. 434]. The fascial

sheath of the prostate is of importance when the prostate has to be removed surgically, since if the sheath and the capsule of the gland are left intact much haemorrhage can be avoided.

In the female, the cervix of the uterus is attached to the front of the sacrum by a thickening of extraperitoneal tissue which forms the **uterosacral ligaments**. These contain a small amount of visceral muscle, and are enclosed in the recto-uterine folds of the peritoneum [FIG. 634].

Elsewhere in the pelvis the extraperitoneal tissue contains the internal iliac vessels and their branches and tributaries, the visceral nerves and plexuses, the ureters, and the ductus deferentia. Around the lower part of the rectum it forms a thick sheath of fat which allows expansion of the rectum during defaecation. The same tissue forms a packing for the parts of the bladder that are not covered with peritoneum. In the female the broad ligament, a double fold of peritoneum which connects the uterus to the side wall of the pelvis [FIG. 659] contains a packing of extraperitoneal tissue called the **parametrium** [p. 539], which surrounds the various ligaments, vessels, and embryonic remnants which the broad ligament contains. Passing from each side of the cervix of the uterus to the wall of the pelvis within the base of the broad ligament is a thickening of the extraperitoneal tissue, the **lateral cervical ligament** [p. 540], around the uterine arteries passing to the cervix uteri. This strong band is probably of importance in holding the cervix of the uterus in place.

MUSCLES OF THE LOWER LIMB

Unlike the upper limb, the lower limb is concerned with support and locomotion, and, although the 'origins' of the muscles are, by convention, located proximally, the fixed points from which they act are as often as not situated distally. In standing the feet are anchored to the ground by gravity and friction, and the tendency of the body to overbalance is corrected by muscles acting from the foot on the lower limb and from the limb on the trunk. In walking, the propulsive thrust derives from the limb in contact with the ground and the actions of a given muscle on its insertion are often less important than the actions on its origin.

Another difference between the lower and the upper limb is that in the lower limb the components of the pelvic girdle are fused together; there is virtually no movement at the sacro-iliac joint, so that all movements between the limb and the trunk take place at a single joint—the hip joint.

Thirdly, the rotation of the lower limb in the course of development results in its extensor and flexor surfaces coming to lie anteriorly and posteriorly respectively. This rotation, which is contrary to that of the upper limb, is reflected in the arrangement of the innervation. Thus the muscles on the front of the thigh and leg are supplied by nerves coming from the posterior aspect of the lumbosacral plexus, and those at the back of the thigh and leg and in the sole of the foot from the anterior aspect of the plexus.

As in the upper limb, many muscles cross over several joints and exert their action on each. It is most exceptional for one joint of the lower limb to be moved in isolation, and in standing

and walking the joints and muscles form a co-ordinated mechanism which cannot be rigidly subdivided into individual functional groups.

MUSCLES OF THE BUTTOCK

The buttock is largely formed by the gluteus maximus muscle, which lies posterior to a number of smaller muscles. Some of these, like the rotator cuff of the shoulder joint, probably have as their main function the protection of the hip joint by their action as extensile ligaments, but others, such as the gluteus minimus and medius, are important prime movers. Two muscles, the obturator internus and the piriformis, whose tendons enter the buttock from the pelvis, have been dealt with elsewhere [pp. 355, 356].

Gluteus Maximus

The gluteus maximus is the heaviest and most coarsely fibred muscle in the body, and forms a soft pad covering the ischial tuberosity [p. 266]. It originates from the area behind the posterior gluteal line of the ilium, from the posterior layer of the thoracolumbar fascia, from the dorsal surfaces of the scarum and coccyx, from the posterior surface of the sacro-tuberous ligament, and from the fascia which covers the muscle [FIGS. 436, 439].

The deep fibres of the lower part of the muscle, comprising about a quarter of its bulk, are inserted into the gluteal tuberosity of the femur [FIG. 437]. The remaining fibres are inserted into the iliotibial tract of the fascia lata [p. 392]. This tract runs down the lateral side of the thigh, and blending with the capsule of the knee joint, becomes fixed to the lateral condyle of the tibia and to the head of the fibula [FIGS. 240, 447, 452; PLATE 25A and B].

The lower border of the gluteus maximus runs obliquely across the fold of the buttock [FIG. 436]. Three bursae separate the muscle from the ischial tuberosity, the lateral side of the greater trochanter of the femur, and the upper part of the origin of the vastus lateralis.

Nerve Supply. From the inferior gluteal nerve (L.5; S.1, 2).

Actions. Acting from its origin it extends and rotates the thigh laterally at the hip joint; its lowest fibres can adduct the femur, and the upper fibres may help in abduction. Its action as an extensor is not much used in ordinary walking, but comes into play in running. The force which the muscle applies to the iliotibial tract is a powerful support on the lateral side of the knee joint (see tensor fasciae latae).

Acting from its insertion, the muscle extends the trunk on the lower limb. This action is of the greatest importance in rising from a seated or stooped position, or in climbing stairs. The gradual relaxation of the gluteus maximus and the hamstring muscles [p. 367] under tension is the chief controlling factor when the body is bent forwards, a movement which takes place largely at the hip joint. The muscle continues to act, this time concentrically, as the stooping posture is replaced by the erect one. The gluteus maximus may be tested by making the patient lie prone and then asking him to lift the thigh off the couch; the gluteus maximus can be felt contracting.

Tensor Fasciae Latae

In front of the gluteus maximus, on the lateral side of the hip, lies the tensor fasciae latae [FIG. 447], which also inserts into the iliotibial tract. The muscle takes origin from the anterior part of the outer lip of the iliac crest [FIG. 439], and joins the iliotibial tract just below the level of the greater trochanter [PLATE 24B and C]. The layer of fascia on its deep surface extends medially to fuse with the capsule of the hip joint.

Nerve Supply. By a branch (L. 4, 5) from the superior gluteal nerve.

Actions. It helps to flex, abduct, and medially rotate the femur; it also straightens out the somewhat backward pull of the gluteus maximus on the iliotibial tract. With the gluteus maximus it affords support to the lateral side of the knee joint: the balanced pull of the gluteus maximus on the posterior end of the iliac crest and the tensor fasciae latae on the anterior end also stabilizes the pelvis on the femur in the anteroposterior plane.

Gluteus Medius

The gluteus medius shows on the surface between the gluteus maximus and the tensor fasciae latae [FIG. 436], but

FIG. 436 Right gluteus maximus muscle.
The broken line indicates the site of the ischial tuberosity.
Note that the curved line indicating the site of the fold of the buttock does not correspond to the lower border of gluteus maximus.

FIG. 437 Muscle attachments to back of upper part of left femur.

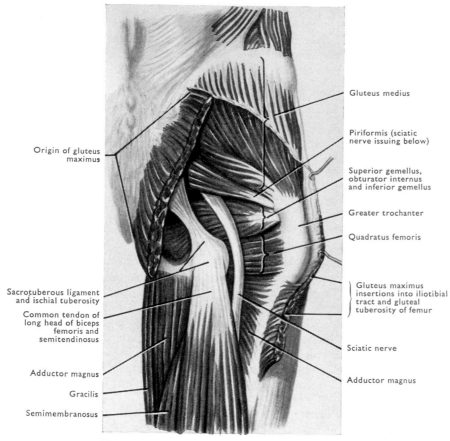

Origin of gluteus maximus

Sacrotuberous ligament and ischial tuberosity

Common tendon of long head of biceps femoris and semitendinosus

Adductor magnus

Gracilis

Semimembranosus

Gluteus medius

Piriformis (sciatic nerve issuing below)

Superior gemellus, obturator internus and inferior gemellus

Greater trochanter

Quadratus femoris

Gluteus maximus insertions into iliotibial tract and gluteal tuberosity of femur

Sciatic nerve

Adductor magnus

FIG. 438   Muscles of right gluteal region deep to gluteus maximus.

most of its substance lies under cover of the gluteus maximus [FIG. 438]. It is shaped like a fan, and arises from the outer surface of the ilium between the posterior gluteal line superiorly, and the anterior gluteal line inferiorly [FIG. 439]. Many of its fibres are attached to the strong fascia covering its anterior portion. The muscle is inserted by tendon into an area running diagonally downwards and forwards across the lateral surface of the greater trochanter of the femur [FIG. 440]; it is separated from the trochanter by a bursa.

Nerve Supply. By a branch (L. 4, 5; S. 1) from the superior gluteal nerve.

Actions. The muscle abducts the thigh in all positions of the lower limb, and, by its anterior fibres, medially rotates it. In walking it abducts the pelvis on the supporting thigh, and rotates it so that the contralateral side of the pelvis is advanced. This movement pulls the centre of gravity of the body forwards and at the same time brings it over the supporting limb [p. 391].

Gluteus Minimus

This muscle lies antero-inferior to and under cover of the gluteus medius [FIG. 441]. Its origin is from the outer surface of the ilium between the anterior and inferior gluteal lines [FIG. 439]; its insertion is into

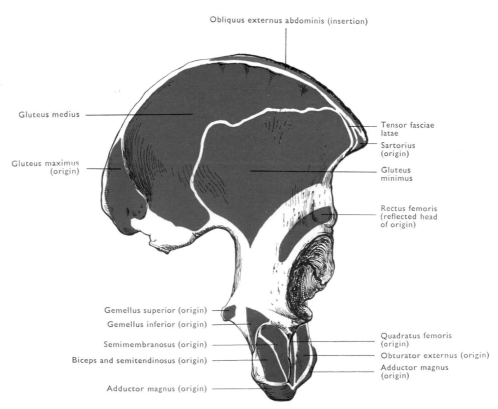

Obliquus externus abdominis (insertion)

Gluteus medius

Gluteus maximus (origin)

Gemellus superior (origin)

Gemellus inferior (origin)

Semimembranosus (origin)

Biceps and semitendinosus (origin)

Adductor magnus (origin)

Tensor fasciae latae

Sartorius (origin)

Gluteus minimus

Rectus femoris (reflected head of origin)

Quadratus femoris (origin)

Obturator externus (origin)

Adductor magnus (origin)

FIG. 439   Muscle attachments to outer surface of right hip bone.

the anterosuperior angle of the greater trochanter, extending on to its anterior surface [FIGS. 440, 449], from which it is separated by a bursa.

Nerve Supply. By a branch (L. 4, 5; S. 1) from the superior gluteal nerve.

Actions. The gluteus minimus is an abductor of the thigh on the pelvis or of the pelvis on the thigh. Its anterior fibres produce medial rotation of the thigh. The gluteus minimus acts with the gluteus medius in walking, to pull the centre of gravity forwards and over the supporting limb; paralysis of the two muscles (by section of the superior gluteal nerve) makes it impossible for the patient to support the pelvis when the opposite foot is raised from the ground. This he overcomes by flexing the trunk strongly to the paralysed side to bring the centre of gravity over the supporting limb—a movement which results in a characteristic waddling gait. The same effect is produced by bilateral congenital dislocation of the hip joints, a condition which prevents the gluteus medius and minimus from functioning properly.

The gluteus medius and minimus and the tensor fasciae latae are tested together. The patient lies on his back with the knee straight, and is asked to abduct the thigh against resistance. The bellies of all three muscles can then usually be felt. Alternatively, the patient lies face down with the knee bent to a right angle and is asked to push the foot laterally against resistance. This is a test of the capacity of the muscles to rotate the femur medially.

Gemelli

These small 'twin' muscles are accessory to the obturator internus [p. 356] and give it additional origins from the margins of the lesser sciatic notch [FIGS. 438, 439]. The superior gemellus arises from the gluteal surface of the ischial spine, and the inferior gemellus from the upper part of the ischial tuberosity. They insert into the tendon of the obturator internus, one above and the other below, and so obtain an attachment to the medial surface of the greater trochanter above the trochanteric fossa [FIG. 440].

Nerve Supply. The superior gemellus is supplied by the nerve to obturator internus (L. 5; S. 1); the inferior gemellus by the nerve to quadratus femoris (L. 5; S. 1).

Actions. They help the obturator internus to rotate the thigh laterally when it is extended, and to abduct it when it is flexed. They most probably function as extensile ligaments [p. 264].

Quadratus Femoris

This is a flat quadrilateral muscle [FIGS. 438, 445] passing from the lateral margin of the ischial tuberosity to be inserted into the intertrochanteric crest and quadrate tubercle of the femur [FIGS. 437, 443]. It lies between the inferior gemellus above and the upper fibres of the adductor magnus below, and is often fused with either or both.

Nerve Supply. By a special branch of the lumbosacral plexus (L. 4, 5; S. 1), which supplies both it and the inferior gemellus.

Actions. It laterally rotates the thigh on the trunk, or

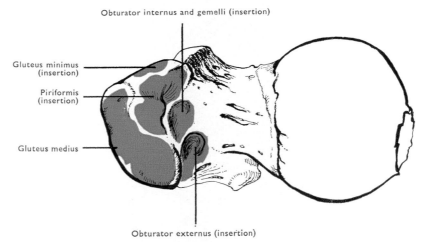

FIG. 440 Femur (looked at from above), showing muscle attachments to greater trochanter.

medially rotates the trunk on the thigh. Like most of the short muscles in relation to the hip joint, it probably has a more important function as an extensile ligament [p. 264].

MUSCLES OF THE THIGH

The muscles of the thigh fall into three main groups by virtue of their situation, actions, and nerve supply. On the back of the thigh (the flexor surface) lies the **hamstring group**, corresponding to the flexors of the elbow in the upper limb, and including the semitendinosus, semimembranosus, and biceps femoris. The second group of muscles in the thigh corresponds to the coracobrachialis in the upper limb, but is much more extensive, and because the main action of its components is adduction of the thigh, it is known as the **adductor group**. To this group belong the adductors magnus, brevis and longus, the gracilis, obturator externus and the pectineus. Finally, on the front of the thigh lies the **anterior group**, corresponding to the triceps brachii of the upper limb and consisting of the sartorius and the four components of the quadriceps femoris. The front of the thigh also contains the terminations of the main flexors of the hip, the iliacus and psoas major, which have been described elsewhere [p. 353].

The boundaries of these groups are not rigidly exclusive; for example, part of the adductor magnus [p. 369] is to be considered as a hamstring muscle in view both of its action and of its nerve supply, and the pectineus [p. 372], though it is functionally an adductor, usually receives its innervation from the femoral nerve, which supplies the flexors of the hip joint.

HAMSTRING GROUP

Semitendinosus

The semitendinosus arises by a short tendon in common with the long head of the biceps femoris from the medial impression (Martin, 1968) on the ischial tuberosity [FIGS. 439, 443]. The long round tendon of insertion begins approximately two thirds of the way down the thigh [FIG. 441] and runs as a cord in a furrow made for it by the semimembranosus.

FIG. 441   Muscles of the back of right thigh.

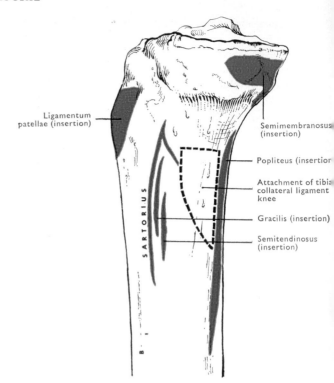

FIG. 442   Muscle attachments to medial surface of upper part of right tibia.

Actions. The semitendinosus helps the gluteus maximus to extend the hip joint, and it is important in controlling flexion of the hip when the body is bent forwards from the erect position. It is a flexor of the knee joint and a medial rotator of the tibia on the femur (particularly when the knee is flexed), or, if the foot is on the ground, a lateral rotator of the pelvis and femur on the tibia. The muscle cannot exert its full power on both hip and knee joints simultaneously; if the knee is fully flexed, it is so shortened that it cannot contract further and act on the hip. Conversely, if it is initially stretched by fully flexing the hip it will not allow enough further stretching to permit full simultaneous extension of the knee unless it has been specially trained to do so [p. 264].

Semimembranosus

The strong membranous tendon which gives the muscle its name arises from the lateral facet on the ischial tuberosity [FIG. 439], and continues downwards along the lateral margin of the muscle (Martin, 1968). The fleshy belly medial to the tendon is deep to the semitendinosus and the long head of the biceps femoris. It runs downwards and medially with the semitendinosus grooving its superficial surface [FIGS. 441, 450], and ends at the back of the knee in a tendon which is inserted into a horizontal groove on the posteromedial surface of the medial condyle of the tibia. A bursa separates it from the medial head of the gastrocnemius, and another from the tibia near its insertion [FIGS. 316, 442].

From the tendon a fascial band runs to join the posterior border of the tibial collateral ligament of the knee joint;

Over the medial side of the knee it spreads out to form a flattened aponeurosis which inserts into the upper part of the medial surface of the body of the tibia, behind the gracilis and sartorius [FIG. 442], and into the deep fascia of the leg. A bursa (the **bursa anserina**) lies deep to the insertion of the semitendinosus and the gracilis, and there is another between them and the sartorius.

Nerve Supply. By two branches (L. 5; S. 1, 2) from the tibial part of the sciatic nerve.

PLATE 23

The subject is 'setting' the muscles of the leg. Note the concavity of the popliteal fossa behind the flexed knee and the fullness of that region behind the straight knee.

1. Tensor fasciae latae.
2. Gluteus medius.
3. Vastus lateralis.
4. Iliotibial tract.
5. Tendon of biceps femoris.
6. Head of fibula.
7. Prominence of side of condyle of tibia.
8. Peroneus longus.
9. Gastrocnemius.
10. Soleus.
11. Extensor digitorum longus.
12. Peroneus brevis.
13. Extensor hallucis longus.
14. Extensor digitorum longus.
15. Lateral malleolus.
16. Peroneus longus superficial to peroneus brevis.
17. Tendo calcaneus.

From *Living Anatomy* by R. D. Lockhart, 6th ed., 1963, London, Faber and Faber Ltd.

PLATE 24

A

B

A. Activity of the calf muscles in rising on the toes.

1. Prominence in popliteal region during extension.
2. Medial and lateral bellies of gastrocnemius.
3. Soleus.
4. Tendo calcaneus.

B. Activity of muscles in extension and resisted flexion of leg.

1. Tensor fasciae latae.
2. Vastus medialis.
3. Vastus lateralis.
4. Semimembranosus.
5. Semitendinosus.
6. Gastrocnemius in action as flexor of knee.
7. Biceps femoris.
8. Expansion from vastus lateralis.
9. Iliotibial tract.

C

C. The sartorius in action.

1. Sartorius.
2. Semitendinosus applied to semimembranosus covered by sartorius and gracilis.
3. Tensor fasciae latae.
4. Vastus lateralis.
5. Vastus medialis.

(By courtesy of Professor R. D. Lockhart.)

PLATE 25

A

B

B. Lateral rotation of leg upon thigh. Note the increased interval between the tendon of biceps femoris and the iliotibial tract covering vastus lateralis.

1. Iliotibial tract.   2. Biceps femoris.

A. Medial rotation of leg upon thigh with inversion of the foot. Contrast the position of the tendon of biceps femoris and iliotibial tract in B.

C

D

C. Eversion with plantar flexion of foot. The peronei and the muscles of the calf are active. Cf. D.

D. Eversion with dorsiflexion of foot. The tendons of the peronei and extensor digitorum longus stand out actively. Cf. C.

1. Tendo calcaneus.   4. Extensor digitorum brevis.
2. Peroneus longus.   5. Extensor digitorum longus.
3. Peroneus brevis.

From *Living Anatomy* by R. D. Lockhart, 6th ed., 1963, London, Faber and Faber Ltd.

PLATE 26

A

B

A. Lateral radiograph of the foot in the tip-toe position. Note extreme plantar flexion at ankle joint with flexion of the calcaneo-cuboid joint and dorsiflexion of the phalanges. The latter tightens the plantar fascia which plantar flexes the calcaneus.

B. Lateral radiograph of the same foot as in A showing the position of 'push off' before the foot leaves the ground in running. Note further dorsiflexion of phalanges than in A and maintenance of the plantar flexion of the calcaneus in spite of some dorsiflexion at the ankle.

C
(i)

(ii)

D
(i)

(ii)

C. Two lateral views of the same foot.
  (i) Non-weight bearing.
 (ii) The foot is carrying the whole weight of the body.

D. Two lateral views of the foot in different young adults. Note that the epiphyses for the calcaneus are not united.
  (i) Is non-weight bearing but is 'flat'.
 (ii) Has a greatly exaggerated longitudinal arch (pes cavus).

another, running downwards and laterally, forms the fascia covering the popliteus muscle [p. 394], and is attached to the soleal line of the tibia. Finally, a strong band extends upwards and laterally to the back of the lateral condyle of the femur, forming the **oblique popliteal ligament** of the knee joint [FIG. 314].

Nerve Supply. By a branch (L.5; S.1, 2) from the tibial part of the sciatic nerve.

Actions. As for semitendinosus, except that its rotatory action on the knee joint only operates when the joint is flexed.

Biceps Femoris

The long head of the biceps arises with the semitendinosus from the medial impression on the ischial tuberosity [FIGS. 438, 439] and from the sacrotuberous ligament. The fleshy belly of the long head becomes tendinous in the distal third of the thigh and is then joined by the short head of the muscle [FIGS. 441, 445]. The short head arises separately from the lateral lip of the linea aspera and the upper half of the lateral supracondylar line of the femur; some of its fibres come from the lateral intermuscular septum.

The rounded tendon of insertion can easily be felt [PLATE 25 A and B] as it passes to its attachment to the head of the fibula; it is split in two by the fibular collateral ligament of the knee joint, which is separated from it by a bursa. Some of its fibres gain insertion to the lateral condyle of the tibia and to the deep fascia on the lateral side of the leg [FIG. 452].

Nerve Supply. The long head of the muscle, like the other hamstring muscles, receives a branch from the tibial portion of the sciatic nerve (L.5; S.1, 2). The short head is separately supplied by a branch (L.5; S.1, 2) from the common peroneal part of the sciatic nerve.

Actions. The long head helps the semitendinosus, semimembranosus and gluteus maximus to extend the hip joint. Both heads of the muscle flex the knee, and, when this has been done, rotate the leg laterally on the thigh or, if the foot is fixed, the thigh and pelvis medially on the leg.

The three hamstrings are tested together. The patient lies on his face and attempts to bend the knee against resistance, when the tendons of the biceps and semitendinosus stand out and can be palpated.

ADDUCTOR GROUP

Adductor Magnus

The adductor magnus [FIG. 444] is the largest of the adductor group of muscles; and is really a composite muscle, part hamstring, part adductor. It arises from the femoral aspect of the conjoint rami of the ischium and pubis, and from the lateral part of the inferior surface of the ischial tuberosity [FIG. 443]. The upper fibres, which arise most anteriorly, run horizontally, parallel to and often fused with the fibres of quadratus femoris; the lowest fibres, which arise from the ischial tuberosity, pass vertically downwards, and the intervening ones radiate out obliquely [FIG. 445]. It inserts linearly along the whole length of the femur from the insertion of quadratus femoris to the adductor tubercle, including the linea aspera and the medial supracondylar line. The posterior fibres of the muscle reach the adductor tubercle, and some

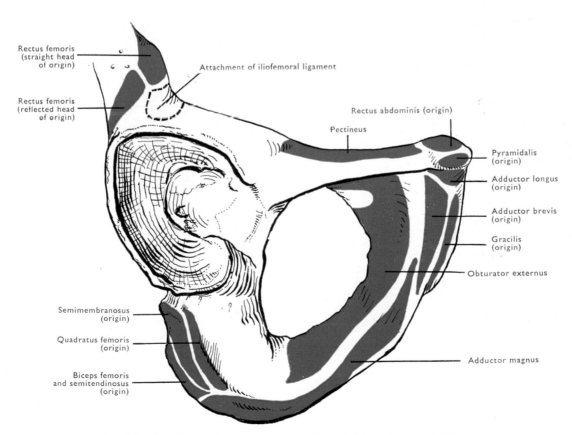

FIG. 443 Muscle attachments to outer surface of right pubis and ischium.

FIG 444 Adductor muscles of the right side
viewed from the front.

Adductor longus, interrupted outline ; adductor
brevis, white shadow ; adductor magnus, red.

Sacrotuberous
ligament

Gluteus maximus
(reflected)

Obturator
internus

Biceps femoris and
semitendinosus

Semimembranosus

Quadratus femoris

Adductor
magnus

Fascia lata

Gracilis

Biceps femoris
(short head)

Biceps femoris
(long head)

Semimembranosus

FIG. 445 Deep muscles of back of right thigh.

continue downwards to fuse with the tibial collateral ligament
of the knee joint [FIG. 442].

The perforating branches of the profunda femoris artery run
between the insertion of the muscle and the femur, and the
attachment of the muscle to the linea aspera is interrupted by a
series of tendinous arches which protect these vessels. A
much larger opening, the **hiatus tendineus**, interrupts the
attachment of the muscle to the supracondylar line ; it allows
the passage of the femoral vessels into the popliteal fossa.

Nerve Supply. The vertical fibres belong morphologically
to the hamstring group, and are supplied by a branch (L. 4)
from the tibial portion of the sciatic nerve ; like the rest of the
adductors, the remainder of the muscle is supplied from the

obturator nerve, in this case by a branch (L. 2, 3, 4) from its posterior division [FIG. 446].

Actions. The proximal portion of the adductor magnus acts with the adductors longus and brevis in adduction and lateral rotation of the thigh, and in preventing lateral overbalancing at the hip during walking. The vertical fibres of the ischial part of the muscle are also weak extensors of the hip joint.

The adductor magnus is tested together with the whole group of adductor muscles. The patient lies on his back with the knee extended, and then adducts the thigh against resistance while the examiner feels the bellies of the adductor muscles.

Adductor Brevis

The adductor brevis, lying in front of the adductor magnus, has a linear origin from the femoral surfaces of the body and inferior ramus of the pubis [FIGS. 443, 444]. Its fibres pass downwards and laterally, and are inserted, behind the pectineus and adductor longus, into the upper half of the linea aspera and the lower two-thirds of the line leading from it to the lesser trochanter of the femur [FIG. 437].

Nerve Supply. From the anterior division (L. 2, 3, 4) of the obturator nerve [FIG. 446]: sometimes the posterior division also supplies a branch to it.

Actions. It adducts the femur and laterally rotates it. It can also flex the extended femur and extend the flexed femur. It is, like the other adductors, an important factor in preventing lateral overbalancing on the supporting leg during walking [p. 390].

Adductor Longus

The triangular adductor longus [FIGS. 444, 447] is the most anterior of the three adductor muscles, arising by a cylindrical tendon from the femoral surface of the body of the pubis in the angle between the crest and the symphysis [FIG. 443]. The fibres, like those of the adductors brevis, pass downwards and laterally, and are inserted into the medial lip of the linea aspera between the adductors brevis and magnus posteriorly and the vastus medialis anteriorly. It is often difficult to separate its fibres from those of the vastus medialis and the adductor magnus [FIG. 450].

Nerve Supply. From the anterior division (L. 2, 3, 4) of the obturator nerve [FIG. 446].

Actions. As for adductor brevis.

Gracilis

The gracilis [FIGS. 450, 455] the last large member of the adductor group, is a long flat muscle which tapers somewhat from above downwards. It takes origin from the femoral surface of the body of the pubis, and from its inferior ramus

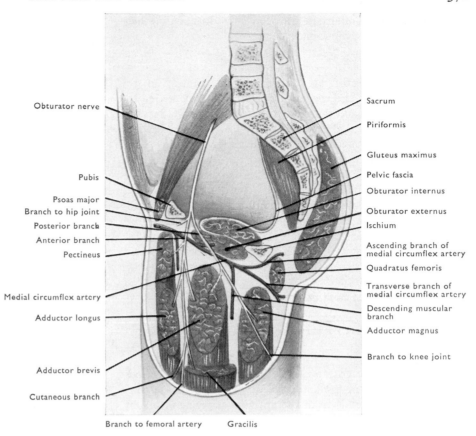

FIG. 446 Scheme of adductor group of muscles and obturator nerve.

[FIG. 443]. It descends, immediately deep to the deep fascia, along the medial side of the thigh between sartorius anteriorly and semimembranosus posteriorly, ending in a tendon which expands to be inserted into the upper part of the medial surface of the body of the tibia between the sartorius and the semitendinosus [FIG. 442]. There is a bursa deep to the tendon, between it and the semitendinosus, and another between it and the sartorius. If they communicate, they are given the name of **bursa anserina**.

Nerve Supply. By a branch (L. 2, 3) from the anterior division of the obturator nerve [FIG. 446].

Actions. It adducts the thigh, flexes the knee joint, and rotates the tibia medially on the femur or, if the foot is on the ground, rotates the pelvis and the femur laterally on the tibia. Like the other adductors, its chief importance is as a balancing muscle during walking [p. 390].

Obturator Externus

The obturator externus lies deeply in the front of the thigh, taking origin from the margins of the pubis and ischium surrounding the obturator membrane and from the inferior surface of the membrane itself [FIG. 443]. The fasciculi of the muscle converge on the groove inferior to the acetabulum, ending in a tendon which passes obliquely across the back of the neck of the femur to be inserted into the trochanteric fossa [FIG. 437].

Nerve Supply. From the posterior division (L. 3, 4) of the obturator nerve.

Actions. It is a lateral rotator and adductor of the thigh.

Probably its most important function is as an extensile ligament [p. 264] of the hip joint.

Pectineus

The last member of the adductor group comes from the pecten of the pubis and the surface in front of it, between the iliopubic eminence and the pubic tubercle. It passes between the psoas major and the adductor longus [FIG. 447], to be inserted into the upper half of a line leading from the back of the lesser trochanter of the femur to the linea aspera [FIG. 437], anterior to the upper part of adductor brevis.

Nerve Supply. This is variable. It is normally supplied by a branch (L. 2, 3) from the femoral nerve, but occasionally receives an additional branch from the obturator nerve or from the accessory obturator nerve (L. 3) when this is present. Sometimes the pectineus is divided into a medial and a lateral part which are innervated respectively by the obturator and the femoral nerves.

Actions. It adducts and flexes the femur.

ANTERIOR GROUP

Sartorius

This is the longest strap muscle in the body, and the most superficial muscle in the anterior compartment of the thigh [FIG. 447]. It extends from the anterior superior iliac spine to the upper part of the medial surface of the body of the tibia, in front of the insertions of the gracilis and semitendinosus [FIG. 442]. A bursa separates it from the insertion of the gracilis, and some fibres run from the tendon of insertion to the tibial collateral ligament of the knee joint and to the fascia of the leg.

The medial border of the upper third of the sartorius forms the lateral boundary of the **femoral triangle**, a region bounded medially by the medial border of the adductor longus and superiorly by the inguinal ligament. The floor of the triangle is formed laterally by the iliopsoas and medially by the pectineus, adductor longus, and, occasionally, a small part of adductor brevis [FIG. 447].

The middle third of the sartorius, and the strong fascia underlying it, form the roof of the **adductor canal**, in which lie the femoral vessels, the saphenous nerve, and the nerve to vastus medialis. The canal is bounded anterolaterally by vastus medialis, and posteromedially by adductor longus and adductor magnus; it is triangular in cross-section [FIG. 450].

Nerve Supply. It receives two branches (L. 2, 3) from the femoral nerve, conveyed to the muscle by its anterior cutaneous branches.

Actions. It is a flexor of the hip and of the knee; it abducts the thigh slightly and rotates it laterally [PLATE 24 C] and it assists in medial rotation of the tibia on the femur. These actions may be summarized by saying that it places the heel on the knee of the opposite limb.

Quadriceps Femoris

The quadriceps femoris is a composite muscle which includes four distinct parts usually described as separate muscles—the rectus femoris, vastus lateralis, vastus medialis, and vastus intermedius [FIGS. 447, 448; PLATE 24 B and C].

The **rectus femoris** is a spindle-shaped bipennate muscle which is described as having two heads of origin, although the two heads are continuous with each other. The straight head of the muscle comes from the anterior inferior iliac spine, and the reflected head from a rough groove on the ilium immediately above the acetabulum. The reflected head is the one which is in the line of pull of the muscle in four-footed animals, and the straight head is a human acquisition, developed with the upright posture. From this continuous origin, a single tendon descends to a fleshy belly on the front of the vastus intermedius [FIG. 447]. From the lower end of this belly a tendon runs into the upper border of the patella; on its lateral and medial sides it receives fibres of insertion of the vastus lateralis and medialis respectively. Because the rectus femoris is capable of movement independent of the vastus intermedius (it acts on the hip joint as well as on the knee) it has a smooth tendinous deep surface which can slip over the similar tendinous superficial surface of the vastus intermedius [FIG. 448].

The patella is really a sesamoid bone in the tendons of rectus femoris and vastus intermedius, and the true insertion of the muscle is via the patellar ligament, a strong band which attaches the inferior angle and the margins of the patella to the tuberosity of the tibia [FIG. 442]. The patella is also anchored to the anterior borders of the tibial condyles by the lateral and medial patellar retinacula [p. 393].

The **vastus lateralis** has a linear origin from the capsule of the hip joint, the upper part of the intertrochanteric line, the lower border of the greater trochanter, the lateral margin of the gluteal tuberosity, and the upper half of the linea aspera; it also comes from the fascia lata, and from the lateral intermuscular septum [FIGS. 225, 228, 449]. The fibres run downwards and forwards, and give way to a broad tendon which is inserted into the tendon of the rectus femoris, the lateral border of the patella, and the front of the lateral condyle of the tibia, blending with the iliotibial tract, and to a large extent replacing the capsule of the knee joint in this region [FIGS. 447, 448]. These aponeurotic fibres blend with the deep fascia to form the **lateral patellar retinaculum** [p. 242]. The muscle exerts its main pull on the tibia through the patellar ligament.

The **vastus medialis** is larger and heavier than the vastus lateralis, and has a long origin from the lower half of the intertrochanteric line, the linea aspera, the upper two-thirds of the medial supracondylar line, the medial intermuscular septum, and the tendon of the adductor magnus [FIGS. 225, 228]. Corresponding to the vastus lateralis on the lateral side, it is inserted into the tendon of the rectus femoris, the upper and medial borders of the patella and the front of the medial condyle of the tibia, replacing the anteromedial part of the capsule of the knee joint, and becoming fused with the deep fascia to form the **medial patellar retinaculum** [p. 242]. Unlike those of the vastus lateralis, its muscular fibres extend to the patella, and the lowest ones are almost horizontal [FIGS. 447, 448].

The **vastus intermedius**, the deepest portion of the quadriceps, arises by fleshy fibres from the upper two-thirds of the body of the femur on its anterior and lateral surfaces. It also comes from the lower half of the lateral lip of the linea aspera and the upper part of the lateral supracondylar line, as well as from the lateral intermuscular septum [FIGS. 225, 228, 449]. The muscle has a membranous tendon on its anterior

Iliacus

Psoas major

Tensor fasciae latae

Pectineus

Adductor longus

Sartorius

Iliotibial tract

Gracilis

Adductor magnus

Vastus lateralis

Rectus femoris

Vastus medialis

Tendon of rectus femoris

Ligamentum patellae

Rectus femoris

Sartorius

Tensor fasciae latae

Gluteus minimus

Gluteus medius

Iliofemoral ligament

Iliotibial

Vastus lateralis turned aside (pointer on edge cut from rectus femoris)

Vastus intermedius

Straight head of rectus femoris

Reflected head

Iliopsoas

Pectineus

Adductor longus

Adductor brevis

Gracilis

Vastus medialis (edge cut from rectus femoris)

FIG. 447 Muscles of front of right thigh.

FIG. 448 Dissection of muscles of front of right thigh. The rectus femoris has been removed and the vastus lateralis and vastus medialis pulled apart to show the vastus intermedius.

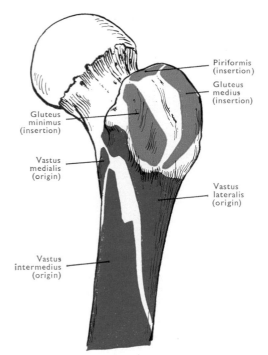

FIG. 449 Muscle attachments to front of upper part of left femur.

surface, which separates it from the deep surface of the rectus femoris, and allows movement between the two [FIG. 448]. It is inserted into the deep surface of the tendons of the rectus and the other vasti muscles. The vastus intermedius is difficult to separate from the vastus lateralis in the middle of the thigh, and impossible to separate from the vastus medialis lower down [FIG. 450]. Some of its deepest fibres in the lower third of the thigh form a muscle called the **articularis genus**; they are inserted into the walls of the **suprapatellar bursa** (synovial pouch) which lies deep to the vastus intermedius in the lowest part of the thigh.

Nerve Supply. All the components of the quadriceps femoris are supplied by branches (L. 2, 3, 4) of the femoral nerve. The branch to vastus medialis is specially large, and reaches it by way of the adductor canal.

Actions. The rectus femoris is a strong flexor of the hip joint, and it assists the iliopsoas to flex the trunk on the thigh. Together with all the other components of the quadriceps, it extends the limb at the knee joint. The vasti, acting without the rectus femoris, pay out to control the movement of sitting down. The quadriceps is used to straighten the knee when rising from the sitting position, during walking, and in climbing. Without it, the knee becomes grossly unstable. Because the pull of the

rectus femoris and vastus intermedius is directed in the line of the femur, at an angle to the line of the patellar ligament and to the shaft of the tibia, there is a tendency for the patella to dislocate laterally during extension of the knee. This tendency is resisted by the pull of the horizontal fibres of the vastus medialis and the raised patellar surface of the lateral condyle. The quadriceps femoris is particularly vulnerable to disuse atrophy, and if the lower limb is enclosed in plaster, perhaps for a fracture, the muscle rapidly loses its bulk and power; for this reason it has to be kept exercised in such circumstances.

The quadriceps is tested with the patient lying on his back; he tries to extend the knee against resistance, and the muscle bellies can be seen and felt.

MOVEMENTS AT THE HIP JOINT

At the hip, as at all joints of the lower limb, the movements must always be looked at from two points of view. The muscles operating on the joint act from one end when they are moving the limb upon the trunk, and from the other when they are moving the trunk upon the limb during walking or running. It is also somewhat unrealistic to consider the hip joint in isolation, since (for example) flexion of the hip very seldom occurs without a corresponding flexion of the knee; indeed, if the knee is voluntarily kept extended, flexion of the hip will be limited because of the tension of the hamstrings [p. 264].

Secondly, the action of certain muscles alters as the hip is moved from its anatomical position of extension. For example, the obturator internus is a lateral rotator of the thigh when the joint is extended, and an abductor when the joint is flexed.

At the hip the extensors are stronger than the flexors, and the adductors and lateral rotators are much stronger than the abductors and medial rotators. As in any ball and socket joint, most muscles acting on the hip joint are active in several of the

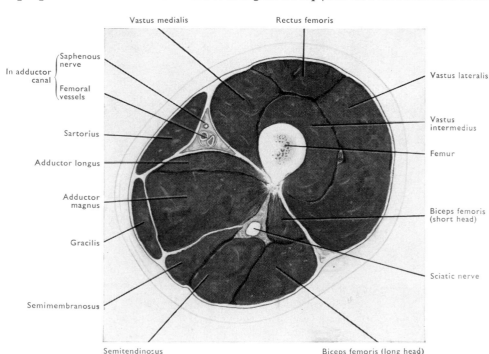

FIG. 450 Transverse section of thigh through the adductor canal.

MUSCLES CONCERNED IN MOVEMENTS OF THE HIP JOINT

MOVEMENT	MUSCLE	PRINCIPAL NERVE SUPPLY
Flexion	Iliopsoas	Ventral rami of L. 1, 2, 3
	Rectus femoris	Femoral
	Tensor fasciae latae	Superior gluteal
	Sartorius	Femoral
	Adductor longus	Obturator
	Adductor brevis	Obturator
	Pectineus	Femoral
Extension	Gluteus maximus	Inferior gluteal
	Semimembranosus	Tibial part of sciatic
	Semitendinosus	Tibial part of sciatic
	Biceps femoris (long head)	Tibial part of sciatic
	Adductor magnus (ischial fibres)	Tibial part of sciatic
Abduction	Gluteus medius	Superior gluteal
	Gluteus minimus	Superior gluteal
	Tensor fasciae latae	Superior gluteal
	Obturator internus (in flexion)	Ventral rami of L. 5 ; S. 1, 2
	Piriformis (in flexion)	Ventral rami of L. 5 ; S. 1, 2
Adduction	Adductor magnus	Obturator
	Adductor longus	Obturator
	Adductor brevis	Obturator
	Pectineus	Femoral
	Gracilis	Obturator
	Gluteus maximus (lower fibres)	Inferior gluteal
	Quadratus femoris	Ventral rami of L. 4, 5 ; S. 1
Lateral rotation	Gluteus maximus	Inferior gluteal
	Obturator internus and gemelli	Ventral rami of L. 5 ; S. 1, 2
	Obturator externus	Obturator
	Quadratus femoris	Ventral rami of L. 4, 5 ; S. 1
	Piriformis	Ventral rami of L. 5 ; S. 1, 2
	Sartorius	Femoral
	Adductor magnus	Obturator
	Adductor longus	Obturator
	Adductor brevis	Obturator
Medial rotation	Iliopsoas (in initial stage of flexion)	Ventral rami of L. 1, 2, 3
	Tensor fasciae latae	Superior gluteal
	Gluteus medius (anterior fibres)	Superior gluteal
	Gluteus minimus (anterior fibres)	Superior gluteal

MUSCLES CONCERNED IN MOVEMENTS OF THE KNEE JOINT

MOVEMENT	MUSCLE	PRINCIPAL NERVE SUPPLY
Flexion	Biceps femoris	Sciatic
	Semitendinosus	Tibial part of sciatic
	Semimembranosus	Tibial part of sciatic
	Gastrocnemius	Tibial
	Plantaris	Tibial
	Sartorius	Femoral
	Gracilis	Obturator
	Popliteus	Tibial
Extension	Quadriceps femoris	Femoral
Medial rotation of tibia on femur (equivalent to lateral rotation of femur on tibia)	Popliteus	Tibial
	Semitendinosus	Tibial part of sciatic
	Semimembranosus	Tibial part of sciatic
	Sartorius	Femoral
	Gracilis	Obturator
Lateral rotation of tibia on femur (equivalent to medial rotation of femur on tibia)	Biceps femoris	Sciatic

artificially defined movements which such joints can undertake. One of the main functions of the muscles around the hip
joint is to stabilize the trunk on the limb when the other foot
is raised from the ground [p. 391].

The table shows the muscles chiefly involved in these movements : the iliopsoas, the piriformis, and the obturator internus
are described elsewhere [pp. 353–356].

MOVEMENTS AT THE KNEE JOINT

The table shows the muscles operating on the knee joint.
The quadriceps femoris is the only extensor of the joint,
whereas there are several flexors, only two of which do not
operate on other joints—the short head of the biceps, and the
popliteus [p. 381]. The hamstrings, which flex the leg at the
knee, are also extensors of the hip. Because of this, many
people are unable to touch their toes while the knee joints are
kept extended. Tight hamstrings may also be a source of dis-
comfort to a patient sitting up in bed, unless the knees are
flexed by a pillow under them. It is possible to train the hamstrings to relax, as any contortionist can demonstrate.

Rotation of the tibia at the knee joint is a much stronger
movement when the knee is flexed to a right angle, since in this
position the tendons of the muscles concerned are placed at
right angles to the tibia. The iliotibial tract, which represents
the main insertion of gluteus maximus and the only insertion
of the tensor fasciae latae, has an important supportive role as
an extensile ligament on the lateral side of the knee joint, and
can also maintain the joint braced back in full extension, so
that standing, and even clumsy walking, may be possible even
though the quadriceps femoris is paralysed.

MUSCLES OF THE LEG

The leg contains three groups of muscles : the extensor
(dorsiflexor) muscles anteriorly, the peroneal muscles on the
lateral side, and the flexor (plantar flexor) muscles posteriorly.
These groups are all contained within osteofascial compartments [FIG. 453].

The muscles in the **anterior compartment** are the
tibialis anterior, the extensor hallucis longus, the extensor
digitorum longus, and the peroneus tertius. In the **peroneal
compartment** are the peroneus longus and brevis, and in the
posterior compartment are the gastrocnemius, soleus, and
plantaris, the popliteus, flexor digitorum longus, flexor hallucis
longus, and the tibialis posterior.

As in the upper limb, the muscles acting on the digits fall
into two categories ; long muscles with their bellies in the leg,
and short muscles fully contained within the foot. The big
and little toes, like the thumb and little fingers, have separate
muscles of their own in the sole, and the other toes share a
common muscular apparatus.

MUSCLES OF THE ANTERIOR COMPARTMENT

Tibialis Anterior

The tibialis anterior [FIGS. 451, 452, 453, 455] arises from
the lateral condyle and the upper two-thirds of the lateral
surface of the body of the tibia, from the interosseous mem-

brane, and from the deep fascia surrounding the muscle. The
tendon, which becomes free in the lower third of the leg, passes
downwards and medially over the front of the distal end of the
tibia, posterior to both extensor retinacula [p. 394], and enclosed
in a separate synovial sheath [FIGS. 465, 467]. It inserts into
the medial sides of the medial cuneiform and the base of the
first metatarsal bone [FIG. 461], reaching the ventral surface
to blend with the insertion of peroneus longus [p. 378].

Nerve Supply. By a branch (L. 4, 5) from the deep peroneal
nerve.

Actions. It dorsiflexes the ankle joint, and is used to pull
the body forwards on the foot of the supporting limb during
walking, as well as to raise the fore part of the foot of the free
limb as it swings forwards. If the tibialis anterior is paralysed,
the remaining dorsiflexors are unable to raise the toes suffi-

Soleus

Extensor digitorum longus

Peroneus longus

Peroneus brevis

Inferior extensor retinaculum

Tendon of peroneus tertius

Tibialis anterior

Extensor hallucis longus

Superior extensor retinaculum

Most medial slip of extensor digitorum brevis (extensor hallucis brevis)

FIG. 451   Muscles of front of leg and dorsum of foot of right
side.

ciently to prevent them dragging along the ground; this is the condition of 'foot-drop', and can be remedied either by flexing the knee more than is normally needed during walking, or by fitting a 'toe-raising' spring to the toe of the shoe.

The tibialis anterior is also a powerful invertor of the foot: that is, it turns the sole to face medially. This action, which occurs at the joints between the talus and the other tarsal bones, allows the foot to adapt itself to uneven ground. Acting from the foot, the tibialis anterior pulls the leg medially, and so prevents lateral overbalancing during walking [p. 391]. With the tendon of the peroneus longus, the tendon of the tibialis anterior forms a sling under the foot which is an important support for the transverse arch of the foot.

The integrity of the muscle is tested by asking the patient to dorsiflex the ankle; the muscle belly and the tendon can be seen and felt if the muscle is working.

Extensor Digitorum Longus

Lying on the lateral side of the tibialis anterior [FIGS. 451, 452], the extensor digitorum longus is a pennate muscle with a tendon which appears on its medial side as it courses down the leg. Its origin is from the lateral side of the lateral condyle of the tibia, the upper two-thirds of the anterior surface of the body of the fibula, the deep fascia, and the upper part of the interosseous membrane.

The tendon passes deep to the superior extensor retinaculum without dividing, but under the inferior extensor retinaculum [FIG. 451] it gives rise to four tendons which run to the lateral four toes. In exactly the same manner as the corresponding tendons in the hand [p. 322] they form extensor hoods on the dorsum of the proximal phalanges, and these hoods are joined by the tendons of the lumbricals and the extensor digitorum brevis (2nd–4th toes), though not by the interossei (Jones, 1944; Manter, 1945). Each hood is then inserted by its central portion into the base of the middle phalanx, and by two collateral portions into the base of the distal phalanx, just as in the hand [p. 322].

A common synovial sheath surrounds the four tendons as they lie deep to the inferior extensor retinaculum [FIGS. 465, 466].

Nerve Supply. By a branch (L. 5; S. 1) from the deep peroneal nerve.

Actions. It extends the lateral four toes at the metatarsophalangeal joints, and assists with the extension of the interphalangeal joints, which are principally extended by the lumbricals. The arrangement is very similar to that in the fingers, but the interossei play no part [p. 386]. Like the extensor digitorum of the upper limb [p. 334] it is unable on its own to extend the interphalangeal joints. If the lumbricals are paralysed, the action of the muscle produces hyperextension of the metatarsophalangeal joint and the interphalangeal joints become passively flexed through tension in the long flexors. Because of its position at the ankle, the

Semitendinosus

Semimembranosus

Common peroneal nerve

Gastrocnemius

Soleus

Peroneus brevis

Tendo calcaneus

Superior peroneal retinaculum

Peroneus longus

Inferior peroneal retinaculum

Peroneus brevis

Extensor digitorum brevis

Iliotibial tract

Vastus lateralis

Tendon of quadriceps femoris

Biceps femoris

Tibialis anterior

Peroneus longus

Extensor digitorum longus

Superior extensor retinaculum

Peroneus tertius

Extensor digitorum longus

Extensor hallucis longus

Inferior extensor retinaculum

Extensor hallucis longus

Extensor digitorum longus

Extensor hallucis brevis

FIG. 452 Muscles of knee, leg, and dorsum of foot seen from lateral side.

extensor digitorum longus is a dorsiflexor of the ankle joint, and helps to pull the body forwards on the foot during walking. To test the muscle, the patient tries to dorsiflex the toes against resistance; the tendons may be seen and felt.

Peroneus Tertius

This small muscle [FIG. 452] is a partially separated portion of the extensor digitorum longus, and arises in common with it from the lower part of the anterior surface of the fibula and from the adjoining fascia. Its tendon runs with those of the extensor digitorum longus but is inserted into the dorsum of the fifth metatarsal near its base, or into the deep fascia nearby [FIG. 466]. The muscle is often absent.

Nerve Supply. By a branch (L. 5; S. 1) from the deep peroneal nerve.

Actions. It assists in dorsiflexion of the ankle joint and in everting the foot. It may be that the peroneus tertius is a muscle which evolution is rendering more important, for eversion (turning the sole so that it faces laterally) is a characteristically human movement.

Extensor Hallucis Longus

Deep to, and between, the tibialis anterior and the extensor digitorum longus lies the extensor hallucis longus, which arises from the anterior surface of the fibula in its middle half, and from the corresponding portion of the interosseous membrane [FIG. 453]. The tendon arises on the medial side of the muscle belly, and appears superficially in the lower part of the leg [FIG. 451]. It occupies the same compartment under the superior extensor retinaculum as the extensor digitorum longus and peroneus tertius, but under the inferior extensor retinaculum it has a special compartment lined with a separate synovial sheath [FIG. 465]. The tendon does not form a fully developed extensor hood on the great toe, but runs to be inserted into the dorsal surface of the base of the distal phalanx, occasionally sending slips to the proximal phalanx and to the metatarsal bone of the great toe.

Nerve Supply. By a branch (L. 5; S. 1) from the deep peroneal nerve.

Actions. It is a strong dorsiflexor of the ankle joint, and it extends all the joints of the big toe. Acting from its insertion, it is used to pull the body weight forwards on the supporting foot during walking.

The great toe has neither a lumbrical nor an interosseous muscle, and extension of the interphalangeal joint thus depends entirely on the extensor hallucis longus. If the extensor hallucis longus is paralysed, the contraction of the flexor hallucis longus in the last phase of the propulsive movement in

walking will therefore flex the joint and buckle up the toe. In testing the muscle the patient tries to dorsiflex the great toe against resistance while the examiner observes the tendon.

MUSCLES OF THE PERONEAL COMPARTMENT

Peroneus Longus

The peroneus longus [FIGS. 452, 453, 454, 463], arises from the lateral condyle of the tibia, from the head and the upper two-thirds of the lateral surface of the body of the fibula, and from the fascia which surrounds the muscle [p. 394]. The tendon lies superficial to that of the peroneus brevis, and is enclosed with it in a common synovial sheath which runs posterior to the lateral malleolus, deep to the superior peroneal retinaculum. A separate sheath prolonged from this envelops the peroneus longus tendon as it runs inferior to the peroneal trochlea on the calcaneus and enters the groove on the inferior surface of the cuboid bone [FIG. 466]. This groove is converted into a tunnel by a floor consisting of fibres derived from the long plantar ligament and the tibialis posterior tendon. The tunnel conveys the peroneus longus tendon obliquely across the sole of the foot to be inserted into the lateral side and plantar surface of the medial cuneiform bone and of the base of the first metatarsal [FIG. 461]. Here it is virtually continuous with the insertion of tibialis anterior [p. 376]. As it enters the sole of the foot the tendon develops a sesamoid

FIG. 454 Insertions of tibialis posterior and peroneus longus in sole of left foot.

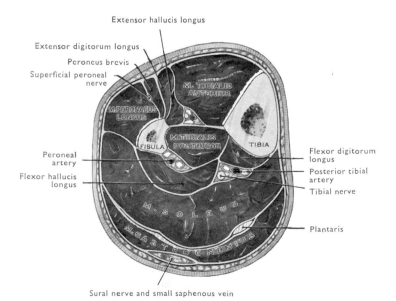

FIG. 453 Transverse section through middle of leg.

fibrocartilage (occasionally a bone) where it plays over the smooth surface of the cuboid bone.

Nerve Supply. By a branch (L. 5; S. 1) from the superficial peroneal nerve.

Actions. The peroneus longus everts the foot; that is, it turns the sole to face laterally [PLATE 25C and D]. During walking, the muscle helps to prevent overbalancing medially on the supporting foot. The course of the tendon obliquely forwards and medially across the sole of the foot means that it helps to maintain the transverse and lateral longitudinal arches of the foot [p. 389], and its position posterior to the fibula allows the peroneus longus to assist in plantar flexing the ankle.

Peroneus Brevis

Enclosed in the same osteofascial compartment as the peroneus longus [FIG. 453], the peroneus brevis arises from the lower two-thirds of the lateral surface of the fibula and from the intermuscular septa at its sides: its upper fibres lie anterior to the lower fibres of peroneus longus [FIG. 452]. In a synovial sheath shared with the tendon of the peroneus longus, its tendon grooves the back of the lateral malleolus [FIG. 458]. Inferior to the superior peroneal retinaculum, it passes above the peroneal trochlea to be inserted into the dorsal surface of the base of the fifth metatarsal bone [FIG. 466]. A slip from the muscle often joins the long extensor tendon of the little toe and is then known as the **peroneus digiti minimi.** Other slips may run to the calcaneus, the cuboid, or the peroneus longus tendon.

Nerve Supply. By a branch (L. 5; S. 1) from the superficial peroneal nerve.

Actions. It plantar flexes the ankle joint, everts the foot, and helps to prevent medial overbalancing of the body on the supporting foot during walking. Except that it has no action on the arches of the foot, its actions are identical with those of peroneus longus.

The peroneal muscles are tested by asking the patient to evert the foot against resistance: the tendons can be felt working behind the lower end of the fibula.

MUSCLES OF THE POSTERIOR COMPARTMENT

Gastrocnemius

The gastrocnemius [FIGS. 452, 453, 455], with the soleus and the plantaris, makes up a composite muscle, sometimes called the **triceps surae**, which forms the prominence of the calf, and in the upper part of the leg conceals the deeper muscles.

The two bellies of origin of the gastrocnemius, together with the plantaris, form the inferior boundaries of a diamond-shaped region at the back of the knee known as the **popliteal fossa**. Superiorly the fossa is bounded by the tendon of the biceps laterally and by semimembranosus and semitendinosus on the medial side [FIG. 441]. The fossa only exists as such when the knee is flexed; when it is extended, the fleshy semimembranosus pushes laterally and bulges posteriorly to help to form, not a depression, but a prominence at the back of the knee [PLATE 24A].

FIG. 455   Muscles of knee, leg, and foot seen from medial side.

The lateral head of the gastrocnemius comes from the upper and posterior part of the lateral surface of the lateral condyle of the femur; the medial head comes from the popliteal surface of the femur above its medial condyle. Either head may contain a sesamoid bone (sometimes known as the **fabella**), and each head has an additional origin from the capsule of the knee joint; inferior to this it is separated from the capsule by a bursa. The bursa related to the medial head frequently communicates with the cavity of the knee joint, but the bursa under the lateral head seldom does so.

The two fleshy bellies arising from the tendons of origin remain separate [PLATE 24A], and insert into the posterior surface of a broad membranous tendon which fuses with the tendon of the soleus to make up the **tendo calcaneus** [FIGS. 455, 456]. This, the strongest tendon in the body, runs downwards for about 15 cm. to be inserted into the middle part of the posterior surface of the calcaneus. A bursa lies between the tendon and the bone above the insertion, and inferior to the insertion lies the fibro-fatty pad of the heel.

Nerve Supply. Each head receives a branch (S. 1, 2) from the tibial nerve.

Actions. The gastrocnemius and the soleus are the chief plantar flexors of the foot, and probably the gastrocnemius plays more part as a propelling force in walking and running; the soleus is more concerned in maintaining the upright posture [p. 389].

The gastrocnemius is also a flexor of the knee joint [PLATE 24B] but, being a 'two-joint' muscle, is incapable of exerting its full power on both joints simultaneously [p. 264]. Thus, if the knee is fully flexed, the gastrocnemius cannot completely plantar flex the foot, and vice versa. The fibres of the gastrocnemius may become considerably shortened if high heels are constantly worn, for this brings the insertion of the muscle closer to its origin. Once such shortening has taken place, difficulties may occur when an attempt is made to walk without a high heel, for the tightness of the muscle may hamper dorsiflexion at the ankle joint. Conversely, full dorsiflexion at the ankle may stretch even a normal gastrocnemius enough to pull the knee into flexion.

Plantaris

The plantaris is a small and variable muscle with a fleshy belly only about 10 cm. in length. It arises from the lowest part of the lateral supracondylar ridge, the adjacent part of the popliteal surface of the femur, and the capsule of the knee joint. The long slender tendon runs obliquely between the gastrocnemius and the soleus [FIG. 456] and later along the medial side of the tendo calcaneus to be inserted into it or into the medial side of the posterior surface of the calcaneus.

Nerve Supply. By a branch (S. 1, 2) from the tibial nerve.

Actions. It is a feeble flexor of the knee and plantar flexor of the ankle joint. In digitigrade animals the plantaris passes into the plantar aponeurosis in the same way as palmaris longus passes into the palmar aponeurosis [p. 318], but the functional significance of the retention of this muscle in Man is unknown.

Soleus

This muscle receives its name from its shape, which resembles that of the fish [FIGS. 455, 456]. It takes origin from the posterior surfaces of the head and the upper third of the body of the fibula, from the soleal line of the tibia and the middle third of its medial border, and from a tendinous arch between these bony attachments [FIG. 457]. The fibres run downwards to join the anterior surface of a membranous tendon which faces posteriorly, and glides on a similar tendon on the anterior surface of the gastrocnemius. This arrangement facilitates independent movement of the two major components of the triceps surae. Inferiorly, the two tendons of soleus and

FIG. 456   Right soleus muscle.

gastrocnemius fuse to form the tendo calcaneus, which is inserted into the middle part of the posterior surface of the calcaneus.

Nerve Supply. By two branches (S.1, 2) from the tibial nerve. The first arises in the popliteal fossa and runs to its superficial surface; the other arises in the back of the leg and enters its deep surface.

Actions. The soleus is a plantar flexor of the ankle joint. Because the centre of gravity of the body is so placed [p. 389] as to cause the body to tend to fall forwards at the ankle during standing, the soleus is frequently in contraction to prevent this happening. It is probably mainly devoted to this postural duty, in contrast to the gastrocnemius, which is a propulsive muscle during walking and running. The soleus stands out very clearly in the living body when the weight is supported on tiptoe, and the gastrocnemius is also recruited to help.

The triceps surae is tested with the patient lying prone: he attempts to plantar flex the foot against resistance, and the gastrocnemius belly and tendon can be seen and felt.

Popliteus

The popliteus takes origin by tendon from a pit on the lateral surface of the lateral condyle of the femur just below the attachment of the fibular collateral ligament of the knee. Other fibres are attached to the back of the lateral meniscus of the knee joint. The tendon, which lies inside the capsule of the joint at its origin, receives an investment from the synovial membrane, which runs along it for some distance. It passes downwards and medially, and emerges from under the arcuate popliteal ligament of the knee, from which the muscle takes an additional fleshy origin [FIG. 314]. The insertion is by fleshy fibres into a triangular surface on the back of the tibia superior to the soleal line [FIGS. 457, 458] and into the strong fascia covering the muscle. In full flexion of the knee, the popliteus tendon lies in a groove on the lateral femoral condyle [FIG. 229].

Nerve Supply. By a branch (L. 5) from the tibial nerve winding round the inferolateral border of the muscle to enter its anterior surface.

Actions. It is a lateral rotator of the femur on the tibia when the foot is fixed on the ground, or a medial rotator of the tibia on the femur in the free limb. When the knee is fully extended and bearing weight the ligaments of the joint are 'wound up' tight by a movement of medial rotation of the femur on the tibia. In the first stage of flexion of the joint this rotation has to be undone, and the popliteus has been said to have a special responsibility for this task. Its attachment to the lateral meniscus of the knee draws this meniscus out of the way during lateral rotation of the femur, so protecting it against being nipped between the moving bones; this is said to be a factor responsible for its relative freedom from injury.

The popliteus can assist in flexion of the knee joint, and is active during crouching (Barnett and Richardson, 1953), though the explanation of this activity is not yet certain.

Flexor Digitorum Longus

The flexor digitorum longus lies in the intermediate layer of muscles on the back of the leg, concealed by the soleus, and only emerging from under cover of it near the ankle. It has a fleshy origin from the medial part of the posterior surface of the body of the tibia inferior to the soleal line, and from the fascia surrounding it [FIGS. 453, 457, 458]. The tendon of the muscle is crossed obliquely and anteriorly by that of the tibialis posterior, which comes to lie medial to it behind the lower end of the tibia. The tendon passes deep to the flexor retinaculum in an individual synovial sheath [FIG. 467], runs over the medial margin of the sustentaculum tali [FIG. 459] and enters the sole superior to the abductor hallucis [FIG. 455].

In the sole it travels forwards superior to the flexor digitorum brevis, and crosses inferior to the tendon of the flexor hallucis longus [FIG. 462] which separates it from the plantar calcaneo-navicular ligament. As the tendons cross each other, the flexor hallucis longus gives to the flexor digitorum longus a slip which passes into the medial two of the four tendons into which the long flexor now divides. In the middle of the sole these tendons receive the insertion of the flexor accessorius [p. 385] which converts their oblique pull towards the medial side of the ankle into a straight one towards the heel. A little further forward the lumbricals [p. 385] have their origin.

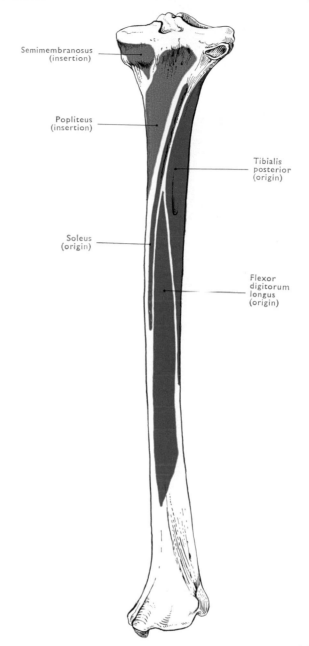

FIG. 457 Muscle attachments to posterior surface of right tibia.

Semimembranosus (insertion)

Popliteus (insertion)

Tibialis posterior (origin)

Soleus (origin)

Flexor digitorum longus (origin)

The insertion of the tendons of the flexor digitorum longus into the lateral four toes exactly parallels the insertion of the flexor digitorum profundus in the hand [p. 320]. Each tendon enters the fibrous flexor sheath of the toe, perforates the tendon of the flexor digitorum brevis, and is inserted into the inferior surface of the base of the distal phalanx. As in the hand, vincula longa and brevia pass to it from the sheath, conveying blood vessels to the tendon.

Nerve Supply. By a branch (S. 1, 2) from the tibial nerve.

Actions. Its primary action is to flex all the joints of the lateral four toes. This is an important action during walking, for it grips the supporting foot firmly to the ground [p. 391]. The muscle also has a plantar flexing effect on the ankle joint, and helps to maintain the longitudinal arch of the foot [p. 389]. To test the muscle, the patient tries to flex the terminal joints of the lateral four toes against resistance: the muscle belly can be felt.

FIG. 458 Deep muscles on back of right leg.

Flexor Hallucis Longus

This powerful bipennate muscle [FIGS. 458, 462] arises from the lower two-thirds of the posterior surface of the body of the fibula and from the adjacent fascia [FIG. 453]. Its tendon passes deep to the flexor retinaculum enclosed in an individual synovial sheath [p. 388], and grooves (1) the posterior surface of the lower end of the tibia, (2) the talus, and (3) the plantar surface of the sustentaculum tali [FIG. 459], to which it is strapped by a fibrous sheath lined by a synovial sheath. It runs forwards in the sole of the foot inferior to the plantar

calcaneonavicular ligament and crosses the superior surface of the tendon of the flexor digitorum longus, sending a slip to its medial two tendons. Eventually it runs into the fibrous digital sheath of the great toe [FIGS. 460, 467] and is inserted into the plantar surface of the base of its distal phalanx.

Nerve Supply. By a branch (S. 1, 2) from the tibial nerve.

Actions. The muscle flexes all the joints of the great toe, and is of great importance in the final propulsive thrust of the foot during walking [p. 391]. It is also a factor in maintaining the medial longitudinal arch of the foot [p. 389], and helps to plantar flex the ankle joint.

The test for integrity of the flexor hallucis longus is similar to that for integrity of the flexor digitorum longus.

Tibialis Posterior

The tibialis posterior, the deepest muscle on the back of the leg, takes origin from the upper and lateral part of the posterior surface of the tibia lateral to the soleal line and the origin of flexor digitorum longus, and from the posterior surface of the shaft of the fibula between the medial crest and the interosseous border [FIGS. 453, 457, 458]. It also arises from the interosseous membrane, and from the fascial septum which covers it posteriorly [FIG. 453].

The tendon of the muscle, enclosed in a synovial sheath, grooves the back of the medial malleolus [FIGS. 459, 467] and turns forwards between the flexor retinaculum and the deltoid ligament of the ankle joint. As it lies inferior to the plantar calcaneonavicular ligament, it usually contains a sesamoid fibrocartilage or bone, and it then spreads out to be inserted chiefly into the tuberosity of the navicular and the plantar surface of the medial cuneiform bone. A tendinous band passes backwards to the tip of the sustentaculum tali, and others pass anterolaterally to all the tarsal bones (except the talus) and to the bases of the middle three metatarsals [FIG. 461].

Nerve Supply. By a branch (L. 4, 5) from the tibial nerve.

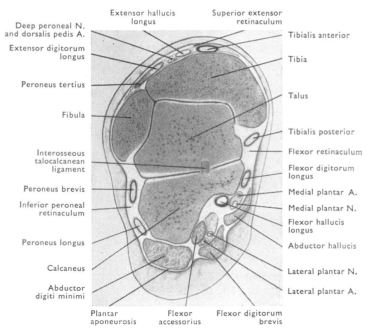

FIG. 459 Frontal section through left ankle joint, talus, and calcaneus.

Actions. It is a plantar flexor of the ankle joint, and because of its attachments to the lateral tarsal bones, is also an invertor of the foot; that is, it turns the sole of the foot to face medially. Its many extensions help to maintain the arches of the foot [p. 389]. Acting from its insertion, the tibialis posterior acts with the tibialis anterior in preventing lateral overbalancing on the supporting foot. The peroneal muscles and the two tibial muscles are of considerable use when walking on rough or irregular ground, since they allow the foot to maintain a wide contact with the ground while the body remains upright.

To investigate the tibialis posterior the patient, lying supine, tries to invert the plantar flexed foot against resistance. The tendon can then be seen and felt behind the medial malleolus.

MOVEMENTS OF THE ANKLE JOINT

The muscles operating on the ankle joint, which permits only dorsiflexion and plantar flexion, are shown in the table. The plantar flexors are much more powerful than the dorsiflexors, partly because in the erect posture the body tends to fall forwards at the ankle, and partly because they have to provide the main propulsive thrust during walking. The main dorsiflexor of the joint is the tibialis anterior, and the main plantar flexor is the composite triceps surae. However, several other muscles which cross the joint can also assist in these movements.

MOVEMENTS AT THE INTERTARSAL JOINTS

The main tarsal movements occur between the talus and the other tarsal bones at the talocalcaneonavicular and subtalar joints, but a limited amount of movement occurs at the other intertarsal joints. The muscles crossing these joints, whether they be ultimately inserted into the digits or into the tarsal bones, are therefore capable of producing intertarsal movements, the most important of which are those of **inversion** and **eversion**. The table shows the muscles chiefly concerned. The flexors and extensors of the digits play a minor part, and the chief prime movers are the peronei and the tibialis anterior and posterior.

The movements of inversion and eversion allow the foot to adapt itself to uneven surfaces during walking [p. 391], and the reversed action of the invertors and evertors is of great importance in preventing overbalancing during standing or walking [p. 390].

MUSCLES OF THE FOOT

Like the hand, the foot contains a large number of important **'intrinsic'** muscles on its plantar surface, but, unlike the hand, it also contains one such muscle on its dorsal surface, the extensor digitorum brevis. Whereas the chief importance of the intrinsic muscles in the hand lies in their control of the

MUSCLES CONCERNED IN MOVEMENTS OF THE ANKLE JOINT

MOVEMENT	MUSCLE	PRINCIPAL NERVE SUPPLY
Dorsiflexion	Tibialis anterior	Deep peroneal
	Extensor digitorum longus	Deep peroneal
	Extensor hallucis longus	Deep peroneal
	Peroneus tertius	Deep peroneal
Plantar flexion	Gastrocnemius	Tibial
	Soleus	Tibial
	Plantaris	Tibial
	Tibialis posterior	Tibial
	Flexor hallucis longus	Tibial
	Flexor digitorum longus	Tibial
	Peroneus longus	Superficial peroneal
	Peroneus brevis	Superficial peroneal

MUSCLES CONCERNED IN MOVEMENTS OF THE INTERTARSAL JOINTS

MOVEMENT	MUSCLE	PRINCIPAL NERVE SUPPLY
Inversion	Tibialis anterior	Deep peroneal
	Tibialis posterior	Tibial
Eversion	Peroneus longus	Superficial peroneal
	Peroneus brevis	Superficial peroneal
	Peroneus tertius	Deep peroneal
Other movements	Sliding movements summating to allow some dorsiflexion, plantar flexion, abduction and adduction, are produced by the muscles acting on the toes. Tibialis anterior, tibialis posterior, and peroneus longus are also concerned.	

movements of individual digits, the intrinsic muscles of the sole are important mainly because of their massed action in the movements of walking; fine control of individual toes is of much less value than fine control of individual fingers. Indeed individuals usually have little ability in this direction, and testing the small muscles of the foot satisfactorily is virtually impossible. In contrast to the corresponding movements in the hand, abduction and adduction of the toes are considered to be carried out relative to an imaginary line drawn through the second (not the third) toe.

The muscles of the sole of the foot are arranged in four layers. The **first**, and most inferior, contains the abductor hallucis, the flexor digitorum brevis, and the abductor digiti minimi. Above this, in the **second layer**, lie the lumbricals and the flexor accessorius, together with the tendons of the flexor hallucis longus and flexor digitorum longus. Superior

to this again, in the **third layer**, lie the flexor hallucis brevis, the adductor hallucis, and the flexor digiti minimi brevis. Finally, in the **uppermost layer**, lie the interossei and the tendons of the tibialis posterior and peroneus longus.

Abductor Hallucis

The abductor of the great toe is a member of the most inferior layer of muscles in the sole, and arises from the medial side of the tuber of the calcaneus, the flexor retinaculum, the plantar aponeurosis, and the intermuscular septum between it and flexor digitorum brevis [FIG. 460]. The muscle lies superficially along the medial border of the sole, and its tendon inserts into the medial side of the base of the proximal phalanx of the great toe, along with the medial head of flexor hallucis brevis.

Nerve Supply. By a branch (S. 1, 2) from the medial plantar nerve.

Actions. It abducts the great toe at the metatarsophalangeal joint and helps to flex it at this joint. Electromyography indicates that it plays no part in maintaining the medial side of the longitudinal arch of the foot, although it is anatomically extremely well placed to do so. However, it comes into action in the final phase of the propulsive thrust in walking [p. 391].

Flexor Digitorum Brevis

The short flexor of the toes corresponds to the flexor digitorum superficialis in the upper limb. It lies between abductor hallucis and abductor digiti minimi, superior to the tough central portion of the plantar aponeurosis, and inferior to the flexor accessorius and the tendons of the flexor digitorum longus [FIG. 460]. It arises from the medial process of the tuber of the calcaneus, the plantar aponeurosis, and the muscular septa on either side of it. The fleshy belly runs directly forwards in the middle of the sole, and separates into four tendons which are inserted into the middle phalanges of the lateral four toes after being perforated by the long flexor tendons, in exactly the same way as the flexor digitorum superficialis inserts in the hand [p. 319].

Nerve Supply. By a branch (S. 2, 3) from the medial plantar nerve.

Actions. It flexes all the joints of each of the lateral four toes except the distal interphalangeal joint. This action supplements that of the flexor digitorum longus in walking [p. 391].

Abductor Digiti Minimi

The abductor of the little toe is the most lateral member of the three muscles in the inferior layer of the sole. It arises by fleshy and tendinous fibres [FIG. 460] from both processes of the tuber of the calcaneus, from the fascia covering it, from the lateral part of the plantar aponeurosis, and from the intermuscular septum between it and flexor digitorum brevis. The tendon runs along the fifth metatarsal bone to be inserted with the flexor digiti minimi brevis into the lateral side of the base of the proximal phalanx of the little toe.

Nerve Supply. By a branch (S. 2, 3) from the lateral plantar nerve.

Actions. It abducts the little toe at the metatarsophalangeal joint, and helps to flex it at this joint.

Flexor hallucis longus

1st lumbrical

2nd lumbrical

3rd lumbrical

4th lumbrical

Flexor hallucis brevis

Flexor digiti minimi brevis

Abductor hallucis

Flexor digitorum brevis

Flexor digitorum accessorius

Abductor digiti minimi

Plantar aponeurosis

FIG. 460   Superficial muscles of sole of foot.

Flexor Accessorius

This muscle, which has no counterpart in the hand, converts the oblique pull of the flexor digitorum longus into a straight one. It lies immediately superior to the flexor digitorum brevis, and arises by two heads, one on each side of the long plantar ligament, and attached both to it and to the medial and lateral surfaces of the calcaneus [FIG. 461]. The two heads converge to form a flattened band inserted into the tendons of the flexor digitorum longus in the middle of the sole [FIG. 462].

Nerve Supply. By a branch (S. 2, 3) from the lateral plantar nerve.

Actions. It helps the long flexor to flex all the joints of the lateral four toes, and alters its direction of pull so that the toes are flexed towards the heel and not towards the medial side of the ankle.

Lumbricals

Like the flexor digitorum profundus in the hand, the flexor digitorum longus has four lumbrical muscles arising from its component tendons [FIG. 462]. The first of these arises from the medial side of the tendon for the second toe, but all the others spring by two heads from the adjacent sides of two tendons. Each then passes forwards on to the tibial side of the corresponding toe, inferior to the deep transverse metatarsal ligament, and winds obliquely upwards and forwards to be inserted by tendinous fibres into the medial side of the base of the proximal phalanx and into the corresponding extensor hood.

Nerve Supply. The lateral three lumbricals are supplied by branches (S. 2, 3) from the lateral plantar nerve, and the first lumbrical is supplied by the medial plantar nerve (S. 1, 2).

Actions. The lumbricals of the foot, like those in the hand, flex the metatarsophalangeal joints and extend the interphalangeal joints of the lateral four toes. This action prevents the long flexors of the toes from 'clawing' the toes underneath themselves during the propulsive phase in walking [p. 391]. If they are paralysed, the extensors of the toes pull the metatarsophalangeal joints into hyperextension and the toes become 'clawed' at rest.

Flexor Hallucis Brevis

The flexor hallucis brevis appears between the flexor digitorum brevis and the abductor hallucis on the sole of the foot [FIG. 460], but it belongs to a deeper layer of muscles and lies superior to the tendon of the flexor hallucis longus. It arises from the medial part of the plantar surface of the cuboid bone [FIG. 461] posterior to the groove for the peroneus longus, from the adjacent part of the lateral cuneiform bone, and from the tendon of tibialis posterior. The fleshy belly separates into two parts, one on each side of the flexor hallucis longus tendon. Each part gives rise to a tendon inserted into the corresponding side of the base of the proximal phalanx of the great toe; the medial one unites with the insertion of the abductor hallucis, and the lateral one with the insertion of the adductor hallucis. In each tendon a sesamoid bone is developed.

C.T.A.—O

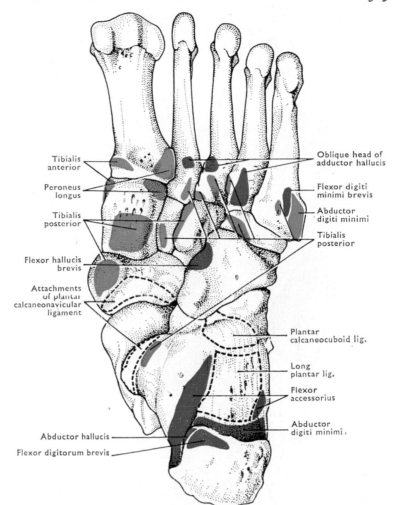

FIG. 461 Muscle attachments to left tarsus and metatarsus (plantar aspect).

Nerve Supply. By a branch (S. 1, 2) from the medial plantar nerve.

Actions. It flexes the metatarsophalangeal joint of the toe. During walking the great toe, which is the most important part of the forefoot, pivots on the sesamoid bones in the tendons of the muscle [p. 391].

Adductor Hallucis

Like the adductor of the thumb, the adductor of the great toe has two heads. The oblique head arises from the sheath of the peroneus longus tendon, and from the plantar surfaces of the bases of the second, third, and fourth metatarsal bones [FIGS. 461, 463]. The transverse head arises from the plantar surface of the lateral four metatarsophalangeal joints and from the deep transverse metatarsal ligament. The two heads unite with the lateral part of flexor hallucis brevis at their insertion into the lateral side of the base of the proximal phalanx of the great toe. The muscle lies inferior to the interosseous muscles, on the same plane as the flexor hallucis brevis, and superior to the long flexor tendons and the lumbricals of the toes.

Nerve Supply. By a branch (S. 2, 3) from the lateral plantar nerve.

Actions. It adducts the great toe towards the second toe (not a powerful movement), flexes the metatarsophalangeal

FIG. 462   Muscles of right foot (second layer).

Lumbricals

Flexor hallucis brevis

Flexor digitorum longus

Flexor digiti minimi brevis

Flexor accessorius

Abductor hallucis

Abductor digiti minimi

Peroneus longus

Flexor digitorum longus

Flexor hallucis longus

Long plantar ligament

FIG. 463   Deep muscles of sole of foot.

Adductor hallucis (transverse head)

Adductor hallucis (oblique head)

Interosseous muscles

Flexor hallucis brevis

Flexor digiti minimi brevis

Peroneus longus

Flexor accessorius (origins)

Flexor digitorum longus

Flexor hallucis longus

Long plantar ligament

joint of the great toe, and assists in maintaining the transverse arch of the foot.

Flexor Digiti Minimi Brevis

The short flexor of the little toe [FIG. 463] arises from the sheath of the peroneus longus tendon and from the base of the fifth metatarsal. It runs along the fifth metatarsal bone to be inserted, in common with the abductor digiti minimi, to the lateral side of the base of the proximal phalanx of the little toe.

Nerve Supply. By a branch (S. 2, 3) from the lateral plantar nerve.

Actions. It flexes the little toe at the metatarsophalangeal joint.

Interossei

In the most superior layer of muscles in the sole lie the interossei, which are essentially similar to those of the hand. There are four dorsal interossei but only three plantar ones. The dorsal ones diverge the toes away from an imaginary line drawn through the second (not the third) toe, and the plantar ones converge the toes on this line.

The four **dorsal** muscles, as in the hand, are larger than their plantar colleagues, and arise by two heads from the adjacent bones. Each runs forwards superior to the deep transverse metatarsal ligament to be inserted into the side of the proximal phalanx and into the capsule of the metatarsophalangeal joint, but not, as in the hand, into the extensor hood on the dorsum of the toe concerned (Jones, 1944; Manter, 1945). The first dorsal interosseus inserts into the medial side, and the second dorsal interosseus into the lateral side of the second toe. The third and fourth muscles are inserted into the lateral sides of the third and fourth toes.

The three **plantar** muscles lie in the lateral three interosseous spaces. Each arises by a single head from the medial sides of the third, fourth, and fifth metatarsal bones respectively. Their tendons pass forwards superior to the deep transverse metatarsal ligament and are inserted in the same manner as those of the dorsal muscles into the medial sides of the third, fourth, and fifth toes [FIG. 464].

Nerve Supply. By branches (S. 2, 3) from the lateral plantar nerve.

Actions. Apart from the abduction and adduction on the line of the second toe, which is of small importance in the foot, the interossei have the function of flexing the metatarso-phalangeal joints. They may also help to hold the metatarsals together to maintain the transverse arch of the foot. By opposing the action of the extensor muscles on the proximal phalanx, they allow the extensors to pull the leg forward on the fixed toes during walking.

Extensor Digitorum Brevis

This, the only muscle springing from the dorsum of the foot [FIGS. 451, 452; PLATE 25 D] has its origin on the anterior part of the upper surface of the calcaneus and the deep surface of the fascia which covers the muscle. Four thin tendons run forwards and medially to be inserted into the medial four toes. The lateral three tendons join the hoods of the extensor digitorum longus, and the most medial tendon, sometimes known as the **extensor hallucis brevis,** is inserted separately into the base of the proximal phalanx of the great toe.

Nerve Supply. By a branch (S. 1, 2) from the deep peroneal nerve.

Actions. It helps the extensor hallucis longus and extensor digitorum longus to extend the metatarsophalangeal joints of the medial four toes. Like the extensor digitorum longus, it is unable to extend the interphalangeal joints without the assistance of the lumbricals.

MOVEMENTS OF THE TOES

The table shows the muscles involved in movements of the toes. The situation is essentially similar to that obtaining in the hand, although in the hand the actions of the muscles are directed towards forming a grasping mechanism, and in the foot they are directed towards providing a stable support for the body in locomotion. This support has to be flexible and able to contribute towards the thrust of the foot while at the same time capable of taking the enormous strains involved in landing on the foot during running, jumping, and other athletic exercises.

SYNOVIAL SHEATHS OF THE ANKLE AND FOOT

The synovial sheaths of the tendons entering the foot [FIGS. 465, 466, 467] are not so important as those in the region of the wrist and hand. They are not so liable to injury or infection, and they do not communicate with the synovial sheaths in the fibrous sheaths of the flexor tendons in the toes.

FIG. 464 Interosseous muscles of right foot.

In front of the ankle and on the dorsum of the foot there are three separate sheaths. One is for the tendon of tibialis anterior, beginning near the upper border of the superior extensor retinaculum and reaching almost to the insertion of the tendon. The second is for the extensor hallucis longus; it begins between the two extensor retinacula and extends to the proximal phalanx of the great toe. The third is a sheath common to the extensor digitorum longus and the peroneus tertius. This begins about the lower border of the superior extensor retinaculum and ends about the middle of the dorsum of the foot.

On the lateral side of the ankle and foot, the peroneus longus and brevis share a common sheath which begins at the back of the lateral malleolus, and bifurcates as it approaches the peroneal trochlea, so that each tendon receives a separate prolongation from it. Both these continue round their respective tendons to their insertions.

On the medial side of the ankle and foot there are three synovial sheaths. That of the tibialis posterior begins about 5 cm. above the tip of the medial malleolus and runs to its insertion into the navicular bone. The sheath of the flexor digitorum longus begins at about the same level, and extends rather further forwards, reaching the attachment of flexor

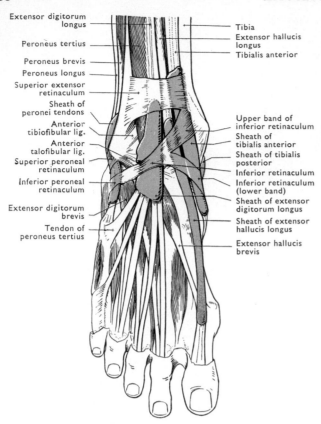

Extensor digitorum longus
Peroneus tertius
Peroneus brevis
Peroneus longus
Superior extensor retinaculum
Sheath of peronei tendons
Anterior tibiofibular lig.
Anterior talofibular lig.
Superior peroneal retinaculum
Inferior peroneal retinaculum
Extensor digitorum brevis
Tendon of peroneus tertius

Tibia
Extensor hallucis longus
Tibialis anterior

Upper band of inferior retinaculum
Sheath of tibialis anterior
Sheath of tibialis posterior
Inferior retinaculum
Inferior retinaculum (lower band)
Sheath of extensor digitorum longus
Sheath of extensor hallucis longus
Extensor hallucis brevis

Fig. 465   Synovial sheaths of dorsum of foot.

accessorius. The third is the sheath of the flexor hallucis longus, which begins rather lower down than the other two, and extends to the point at which the slip is given to the flexor digitorum longus [Fig. 467].

Within the fibrous flexor sheaths of the toes, the long flexor tendons, as in the fingers, are surrounded by short synovial sheaths extending proximally as far as the heads of the metatarsals.

MECHANISM OF STANDING

In the anatomical position the centre of gravity of the skull lies in front of the atlanto-occipital joint, so that the skull tends to fall forwards, and has to be restrained from doing so by postural contraction of the muscles at the back of the neck, chiefly the semispinalis capitis [p. 272]. Similarly the lumbar curvature of the vertebral column is maintained by postural contraction in the various portions of the erector spinae [p. 269]: detectable activity is very variable, but is most marked in the lower thoracic region (Joseph and McColl, 1961). Equilibrium in the erect posture is not very stable, and there are constant slight swaying movements of the trunk. These movements are immediately checked by contraction of the antagonistic muscles of the vertebral column [p. 279].

The centre of gravity of the whole body lies near the body of the second sacral vertebra. A perpendicular dropped from this point to the ground falls behind the axis of movement of the hip joint, and the weight of the body thus tends to cause

MUSCLES CONCERNED IN MOVEMENTS OF THE TOES

Metatarsophalangeal Joints

MOVEMENT	MUSCLE	PRINCIPAL NERVE SUPPLY
Flexion	Flexor hallucis longus	Tibial
	Flexor hallucis brevis	Medial plantar
	Flexor digitorum longus	Tibial
	Flexor accessorius	Lateral plantar
	Flexor digitorum brevis	Medial plantar
	Flexor digiti minimi brevis	Lateral plantar
	Lumbricals	Lateral and medial plantar
	Interossei	Lateral plantar
Extension	Extensor hallucis longus	Deep peroneal
	Extensor digitorum longus	Deep peroneal
	Extensor digitorum brevis	Deep peroneal
Abduction and adduction	Abductor hallucis	Medial plantar
	Adductor hallucis	Lateral plantar
	Interossei	Lateral plantar
	Abductor digiti minimi	Lateral plantar

Interphalangeal Joints

MOVEMENT	MUSCLE	PRINCIPAL NERVE SUPPLY
Flexion	Flexor hallucis longus	Tibial
	Flexor digitorum longus	Tibial
	Flexor accessorius	Lateral plantar
	Flexor digitorum brevis (proximal joint only)	Medial plantar
Extension	Extensor hallucis longus	Deep peroneal
	Extensor digitorum longus	Deep peroneal
	Extensor digitorum brevis (not in great toe)	Deep peroneal
	Lumbricals	Lateral and medial plantar

the pelvis to roll backwards at the hip. This tendency is resisted by the disposition of the ligaments, which, especially the iliofemoral ligament [p. 239], are taut during standing and take most of the strain. The iliopsoas, the chief flexor of the hip joint, is apparently not called upon when standing at ease (Joseph and Williams, 1957; Keagy *et al.*, 1966).

The perpendicular line of gravity passes through the front of the knee joint, and this factor tends to produce full extension of the joint. But in the natural standing position the joint is not fully extended [p. 246], and, because of the apparent instability of the joint surfaces and the fact that the ligaments are not 'wound up' tight unless the joint is fully extended, one would expect the muscles of the thigh to be in continuous postural contraction to prevent overbalancing at this joint. However, no constant activity can be detected in the muscles of the thigh during standing (Joseph and Nightingale, 1954).

At the ankle joint, the line of gravity passes well forward of the axis of movement, and there is thus a tendency to fall forwards. This tendency is resisted by muscular contraction of the triceps surae, in particular of the soleus muscle, which comes into play as a postural muscle more often than any other during standing (Joseph and Nightingale, 1952). Wearing high-heeled shoes, which tip the body still further forwards, increases the activity in the soleus (Joseph and Nightingale, 1956) and may produce a measurable increase in the circumference of the calf.

In the standing position the weight of the body is shared approximately equally by the heel pad and the pads underlying the heads of the metatarsals. About one third of the total weight taken by the metatarsal pads falls on the pad under the ball of the great toe, and the remainder is shared between the other toes (Morton, 1935). Between the posterior and the anterior weight-bearing areas lies the longitudinal arch of the foot [PLATE 26]. The lateral side of the sole is usually in contact with the ground in the normal standing position [FIG. 468], but this contact may be incomplete or absent, and the region does not normally bear weight. Nevertheless, this lateral side of the longitudinal arch is much less elevated than the medial side, under which the vessels and nerves pass into the sole. The longitudinal arch is maintained particularly by the dense intermediate part of the plantar aponeurosis, by the ligaments underlying the intertarsal joints—especially by the plantar calcaneonavicular ligament—and by the long tendons going to the toes, which stretch like tie-beams across the sole of the foot. The action of the flexor hallucis longus and flexor digitorum longus in supporting the medial side of the longi-

FIG. 466 Dissection showing synovial sheaths of tendons of foot.

tudinal arch is reinforced by the tendon of the tibialis posterior as it hooks under the medial side of the ankle [FIG. 467]. The short muscles of the sole appear to be of little importance as supports of the longitudinal arch (Smith, 1954; Basmajian, 1961) though they become very active when standing on tip-toes.

The anterior part of the foot shows a transverse arch, which is particularly well marked between the shafts of the metatarsals. This arch is maintained by the tension of the plantar aponeurosis and by the ligaments, but also by the activity of the tibialis anterior, the adductor hallucis, the interossei, and the tendons of tibialis posterior and peroneus longus, which sling across the sole transversely (Hicks, 1961). Activity in these muscles is minimal with the foot off the ground, but as

FIG. 467 Dissection of leg and foot showing synovial sheaths.

FIG. 468   Comparative series of footprints.

1-4 Footprints of a man. (1) Walking print; (2) stationary, foot highly arched; (3) with the muscles relaxed as much as possible; (4) walking on the toes. Note the prominence of the bar between the ball and phalangeal pad of the big toe.

5. Footprint of male athlete.

6. Athletic girl.

7. Ballet dancer in attention attitude.

8. The same showing the difference between the arched and relaxed state of the foot, the increased area of contact indicated by the dotted area.

9. Negro male.

10. Australian aboriginal woman.

11. Chimpanzee, female, 3½ years of age, walking print.

12. Hamadryas baboon, female.

13. Rhesus macacus monkey. (Observe the narrow heel, 11, 12, and 13.)

14. Infant before walking, at 3 months old.

15. The same child after walking at 2 years 9 months old.

16. The same child at 3 years 7 months old.

17. The same child at 7½ years old.

18. This infant, after beginning to walk, was confined to bed through illness and lost the art. The footprint is taken from a series while the art of walking was being learned anew. The eminence in front of the heel on the medial side is caused by the impression of the area under the head of the navicular bone.

soon as any movement takes place or pressure is applied they come into play to protect the nerves and vessels which run forwards in the interdigital clefts to supply the toes.

The standing posture consists of a cycle of practically static phases, each lasting about half a minute, alternating with brief phases of movement (Smith, 1954; Thomas and Whitney, 1959). During the static phases, all the muscles directly associated with the foot, except for the calf muscle, are inactive, and the support for the arches of the foot is presumably provided mainly by the ligaments.

In the phase of movement, alterations in the distribution of muscle contraction occur to prevent the tendency of the body to overbalance. The only permanently static part of the body is the skin of the sole of the foot in contact with the ground. As postural sway occurs the centre of gravity is pulled back into its original position relative to the feet by contraction of muscles which first of all balance the leg on the foot, then the thigh on the leg, and lastly the trunk on the thigh. For example, should the body begin to topple to the left side, the tibialis anterior and tibialis posterior of the left lower limb, acting together, would pull the left leg medially on the foot, while the adductor group in the thigh, acting from their insertions, would pull the pelvis over again to the medial side. These movements might be assisted by contraction of the corresponding antagonist muscles in the right lower limb; the peroneus longus, the peroneus brevis, and peroneus tertius would pull the right leg laterally on the foot, and the gluteus medius and minimus might come into play to tilt the pelvis over towards the right side again. At the same time the right erector spinae muscle would pull the centre of gravity towards the right.

Anteroposterior sway is dealt with in a similar fashion by the dorsiflexors and plantar flexors of the ankle, the extensors and flexors of the knee, and the extensors and flexors of the hip joint; the balancing action of gluteus maximus and tensor fasciae latae acting on the posterior and anterior ends of the iliac crest is an important factor.

Standing is therefore not merely a matter of attaining a fixed position, and the erect posture necessarily involves a rotation of activity in the various muscles of the whole of the lower limb and trunk. Standing on one foot is a much more compli-

FIG. 469 Tracings of radiographs superimposed to show the position of the longitudinal arch of the foot during relaxation (red) and contraction (black) of the muscles. (See also Fig. 468, footprints 2 and 3.)

cated achievement than standing with the weight equally distributed between the two, and standing on the toes of one foot is more difficult still; the principles, however, are exactly the same. It is worth noting that when standing on one limb with the pelvis level, the force which prevents the pelvis rotating about the supporting hip is not entirely due to the pull of the abductors of the hip; a good deal is contributed by passive tension on the iliotibial tract of the fascia lata (Inman, 1947).

MECHANISM OF WALKING AND RUNNING

In **walking,** the centre of gravity of the body is thrust upwards and forwards by the propulsion of the lower limb muscles. As one foot advances, the upper limb of the opposite side swings forwards, thus helping to balance the trunk, and the foot of the opposite side remains on the ground until the advancing foot re-establishes contact. The trunk is thus always supported by at least one lower limb.

The rise in the centre of gravity with each step amounts to about 5 cm. (Saunders et al., 1953), and is brought about by a combination of extension at the hip and knee with plantar flexion at the ankle, the heel being raised from the ground by the contraction of the triceps surae. In this way an upward and forward impetus is achieved, which is now assisted by gravity, so that the body tends to fall forwards. The thrusting lower limb now leaves the ground, and as it does so the limb is brought forward in front of the body by contraction of the flexors of the hip, so increasing the tendency of the body to fall forward. This flexion, coupled with elevation of the pelvis by the opposite hip abductors, flexion at the knee, and dorsiflexion at the ankle, lifts the advancing foot clear of the ground. After the limb passes the line of the centre of gravity, the hip and knee once more extend, so that the lower limb returns to the ground in the fully extended position, with the ankle partly dorsiflexed. As the foot strikes the ground, the knee flexes, and the dorsiflexors of the ankle again contract strongly, preventing the forepart of the foot from slapping down. The heel therefore comes down first, and the weight of the body is then transferred, first along the lateral side of the longitudinal arch of the foot to the lateral side of the metatarsal pad, and then across the transverse arch to the ball of the great toe, which takes most of the strain as the cycle begins again with another propulsive thrust of the extensors of hip and knee and the plantar flexors of the ankle. The great toe rocks on the two sesamoid bones in the substance of the flexor hallucis brevis, which act as roller bearings. As the heel is raised, the metatarsophalangeal joint of the great toe dorsiflexes, so that the last parts of the lower limb to retain contact with the ground are the ball and phalangeal pad of the great toe. This dorsiflexion 'winds up' the plantar aponeurosis by pulling it forwards round the heads of the metatarsals, like a cable being wound on a windlass (Hicks, 1954). This helps to support the medial side of the longitudinal arch as the main strain is taken by it.

The long flexors of the toes have an important part to play in maintaining the frictional grip between the foot and the ground. Their pull tends to flex all the joints of the toes, but this action is resisted by the extensors of the interphalangeal joints, which hold them in a position of extension. If the extensors of these joints are paralysed, the resulting curling up of the toes causes a considerable loss in the efficiency of the propulsive thrust, as well as corns, sores, and ulcers caused by unspecialized areas of skin being called upon to bear weight and take friction. At the beginning of the thrust, the flexor muscles of the toes are fully extended, and, though exerting themselves powerfully with the toes applied to the ground, they do not in fact shorten [PLATE 26]. The inner toes exert the greatest pressure and the importance of the big toe in walking may be estimated from the fact that the flexor hallucis longus exerts three times more force upon it than is sustained by any other toe [FIG. 468].

The abductor hallucis is active during the final phase of the thrust, and in some adults and in all children it is strong enough to cause abduction of the great toe at this stage. In those who wear unphysiological tight pointed shoes, and who walk 'toeing out', the prevention of this movement of abduction leads to a drift of the great toe laterally, a condition known as hallux valgus (Barnett, 1962). This condition tends to be aggravated by the oblique pull of extensor hallucis brevis.

The short intrinsic muscles of the foot, considered as a whole, are most active during the final propulsive thrust (Basmajian, 1961).

As the limb leaves the ground, its support for the trunk is suddenly withdrawn, so that the trunk tends to fall over away from the supporting limb. Actual falling is prevented by the contraction of the gluteus medius and minimus of the supporting limb, which abduct the trunk so that the centre of gravity is brought back over its supporting pillar. At the same time the erector spinae and the abdominal muscles on the side of the free limb contract, apparently to keep the pelvis from tilting. They are not entirely successful in this, so that there is a sideways lurch of about 5 cm. with each step. If the gluteus medius and minimus are paralysed, this lurch becomes greatly exaggerated, and the patient has a characteristic dipping gait.

As the trunk moves forward, it not only becomes flexed and abducted at the supporting hip, but also rotates to the side of the supporting limb (this is equivalent to a medial rotation of the femur on the pelvis). This movement, which allows the free limb to pass directly forwards in the line of progression, is largely carried out by the gluteus medius and minimus. At the same time, and for the same reason, the thigh of the free limb is rotated laterally on the pelvis. Walking movements at the hips are thus sequences or combinations of abduction, flexion, extension, and rotation.

Rotation movements also occur at the knee. For example, in the final 30 degrees of extension of the knee of the free limb, there is an automatic lateral rotation of the tibia on the femur, and in the similar phase of extension when the foot is on the ground the femur rotates medially on the tibia. As the knee flexes again, these rotations are rapidly undone (Barnett, 1953; Langa, 1963).

In such a complicated situation it is not surprising that there should be a tendency to overbalance, and this is rectified by the lower limb muscles acting from their insertions on their origins, just as they do to prevent overbalancing during standing [p. 390]. The peronei and the tibialis posterior and anterior have an additional value because of their ability to allow the greatest possible surface of the foot to remain in contact with the ground in spite of local irregularities in the surface. For example, a firm grip can be maintained and the body can be held upright even though the feet are traversing a steep incline.

In **running,** the movements are all exaggerated, the time of each movement is diminished, and their force and distance are increased. Both feet are off the ground for a time during each stride, and the whole foot does not strike the ground, but only the metatarsal pads and the toes. The trunk is inclined forward much more than in walking, so that the centre of gravity is well forward of the point of support. The basic principles are, however, similar to those obtaining during walking.

It is a curious and so far unexplained fact that in walking women take shorter steps in relation to the length of their legs than do men, and that this is associated with a lower expenditure of energy (Booyens and Keatinge, 1957). Women increase their walking speed mainly by increasing the frequency of their stride.

Walking on high heels involves greater activity in the soleus than walking on low heels. There is less powerful but more continuous activity in the tibialis anterior, intermittent or continuous contraction of the quadriceps femoris during the supporting phase, and contraction of the gluteus medius in the free limb (Joseph, 1968). These changes are less marked than might be expected.

Walking is one of the most complicated actions the general musculature of the body is called upon to perform, and the mechanisms concerned are readily disturbed by disease or injury, often producing an altered gait so characteristic as to lead to immediate diagnosis. The disturbances of normal gait are discussed by Steindler (1935, 1955), and the effects of flat foot on muscle action by Gray and Basmajian (1968).

FASCIAE OF THE LOWER LIMB

SUPERFICIAL FASCIA

The superficial fascia of the lower limb is continuous with that of the perineum, abdominal wall, and back. The two layers of superficial fascia found in the lower part of the abdominal wall [p. 360] and in the perineum are also represented in the uppermost part of the anterior surface of the thigh, in the groin. The external **fatty layer** presents no special features in this region; the deep **membranous layer** crosses superficial to the inguinal ligament and fuses with the deep fascia of the thigh along a line which runs laterally from the region of the pubic tubercle, roughly parallel to the inguinal ligament, and about a finger's breadth inferior to it. This attachment prevents the passage into the thigh of fluid collected in the perineum or deep to the superficial fascia of the abdominal wall [FIG. 433].

In the gluteal region the superficial fascia is very thick and fatty. In the vicinity of the ischial tuberosity, where it takes the weight of the sitting body, it has the characteristic structure found where pressure is borne, with dense strands of fibrous tissue enclosing between them loculi containing fat. The fat of the buttock contributes to its contour, and is responsible for the formation of the gluteal fold. On the outer side of the female thigh there is a deposit of secondary sexual fat [p. 268], and the whole of the thigh may be the site of considerable fat storage.

The superficial fascia of the leg and foot is not remarkable except in the sole. Here it is greatly thickened by characteristic loculated pads of fat, particularly on the weight-bearing areas of the heel, the balls and pads of the toes. These digital pads are exactly similar to those which occur in the palm and fingers [p. 337], and serve to protect the deeper structures against the pressure of the weight of the body. The heel pad underlying the calcaneus may be as much as 2 cm. in thickness. In the webs of the toes, the superficial fascia contains some weak transversely running fibres which form the superficial transverse ligament of the sole.

On the dorsum of the foot, as on the dorsum of the hand, the superficial fascia is thin and composed of very loose connective tissue containing little fat.

DEEP FASCIA

The investing layer of the deep fascia of the lower limb is attached superiorly to the inguinal ligament, the crest of the ilium, the sacrotuberous ligament, the ischium, the pubic arch, and the anterior surface of the pubis and the pubic tubercle. *In the thigh*, where it is known as the **fascia lata,** there are considerable differences in its thickness. On the medial side of the thigh it is thin and almost transparent, but on the lateral side it is immensely tough and strong, particularly in a band extending from the tubercle of the iliac crest to the lateral condyle of the tibia. This is the **iliotibial tract of the fascia lata,** and into it are inserted three-quarters of the gluteus maximus muscle and the whole of the tensor fasciae latae; it may be looked upon as a common aponeurosis of these muscles.

In the lower part of the thigh the investing layer sends two intermuscular septa to be attached to the femur. The **lateral intermuscular septum** extends from the iliotibial tract to the lateral supracondylar line and the linea aspera of the femur; it gives attachment to the vastus lateralis and vastus intermedius muscles anteriorly, and to the short head of the biceps posteriorly. The **medial intermuscular septum,** which runs to the medial supracondylar line and the linea aspera, is greatly strengthened by the tendon of insertion of the adductor magnus muscle, and separates the hamstrings from the adductors.

Higher up in the thigh the fascial septa become more indistinct, but there is a well-marked thickening of a layer of fascia deep to the sartorius muscle. This layer binds the vastus medialis to the adductor longus and adductor magnus, and forms a roof for the adductor canal [p. 372], which transmits the femoral vessels.

At the knee, the deep fascia is continuous with that of the leg, and is attached to the condyles of the tibia, the head of the fibula, and the patella. The patella is held to the tibial condyles by the **medial and lateral patellar retinacula** [p. 372], which are thickened bands of deep fascia, reinforced by the tendons of the vastus medialis and vastus lateralis respectively. At the back of the knee, covering over the popliteal fossa [p. 379], the deep fascia is strengthened by transverse fibres.

In the upper part of the front of the thigh there is an oval gap in the fascia lata, through which the great saphenous vein and some of its tributaries pass on their way to join the femoral vein. The centre of this **saphenous opening** [FIG. 424] lies approximately three fingerbreadths inferior and lateral to the pubic tubercle. The opening itself is about 4 cm. long and 1–2 cm. wide, and it has a smooth sloping medial margin formed by the fascia covering the anterior surface of the pectineus muscle. In contrast, the upper, lateral, and lower boundaries of the opening form a sharp crescentic edge. This **falciform margin** is joined to the medial boundary of the opening by the **cribriform fascia,** a thin layer of fibrous and fatty tissue which is perforated by the great saphenous vein, small arteries, and lymph vessels [FIG. 470].

Deep to the investing layer in the groin, an oval funnel-shaped tube, the **femoral sheath,** is prolonged downwards from the abdomen into the thigh around the femoral vessels [FIG. 471]. This sheath is formed anteriorly by a prolongation from the transversalis fascia and posteriorly from a prolongation of the fascia iliaca [p. 362]. The femoral sheath is divided into three compartments by septa which pass from the anterior to the posterior wall. In the lateral compartment is the femoral artery; in the intermediate one is the femoral vein, and in the medial one, which is called the **femoral canal,** there is little except some fat, a few lymph vessels, and possibly a lymph node. This canal, which is necessary to allow for the expansion of the femoral vein when the body is upright, is a weak spot in the wall of the abdomen, through which a **femoral hernia** may protrude. Laterally, the femoral sheath is closely applied to the walls of the femoral vessels, and terminates about 4 cm. inferior to the inguinal ligament, deep to the saphenous opening, by becoming fused with their adventitial coats. The upper end of the femoral canal admits the tip of the little finger and is known as the **femoral ring;** its wall is firmly attached to the inguinal ligament, the lacunar ligament [FIG. 471], the pectineal ligament, and the fascia in front of the pectineus muscle. Inferiorly, the canal, which is about 2 cm. long, narrows down to nothing because its medial wall slopes laterally to become fused with the wall of the femoral vein.

A hernia passing down the femoral canal will push in front of it a covering of peritoneum, and the contents of the femoral canal. It will then distend the wall of the canal and bulge outwards and forwards, pushing in front of it the relatively weak cribriform fascia. Finally it turns upwards over the inguinal ligament on to the anterior abdominal wall. As the hernia passes down through the narrow femoral ring it is in a position of considerable danger, since the sharp crescentic edge of the lacunar ligament may interfere with its blood supply.

Lateral to the femoral sheath, a thickening of the fascia passes from the inguinal ligament to the superior ramus of the

FIG. 470 Diagram of fasciae and muscles of inguinal and subinguinal regions in the line of the saphenous opening.

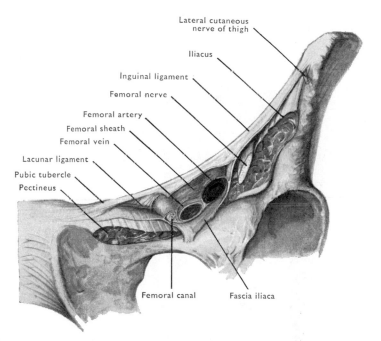

FIG. 471 Dissection to show femoral sheath and structures which pass between inguinal ligament and hip bone.

pubis. This thickening, sometimes known as the **iliopectineal ligament,** divides the space between the inguinal ligament and the pelvis into two. Medial to it lies the femoral sheath and its contents, and lateral to it the psoas major and iliacus pass into the thigh along with the femoral nerve and the lateral cutaneous nerve of the thigh [FIG. 471]. Along the pathway afforded by the fascial sheath of the psoas major muscle, pus produced in the vertebrae to which the muscle is attached may track down into the thigh behind the fascia lata.

The deep fascia of the leg is continuous with the fascia lata and is strengthened in the region of the knee by expansions from the tendons of the sartorius, gracilis, semitendinosus, and biceps femoris. Anteriorly it is reinforced by the fibres of the vastus medialis and the vastus lateralis as they form the patellar retinacula [p. 372].

The investing layer extends round the leg from the anterior border of the tibia to reach its medial border, binding together the muscles and affording them origins. It blends with the periosteum on the subcutaneous surfaces of the tibia and fibula, and superiorly it is attached to the tuberosity of the tibia and the head of the fibula. Two intermuscular septa pass from its deep surface, separating the peroneal group of muscles from the extensor group in front and the flexor group behind. The anterior septum is attached to the anterior border of the fibula, and the posterior septum to its posterior border [FIG. 453].

Thick portion opposite phalanx

Fibrous (flexor) sheath

Thick portion opposite phalanx

Plantar aponeurosis

Calcaneometatarsal portion of deep fascia

FIG. 472   Left plantar fascia.

Another septum extends across the back of the leg, forming a partition between the superficial and deep muscles; superiorly it is attached to the soleal line and to the head of the fibula. Deep to this again is a septum separating the tibialis posterior from the flexors of the toes [FIG. 453]. Above the soleal line a strong fascial covering is prolonged down over the popliteus muscle from the insertion of the semimembranosus [FIG. 457].

At the ankle the deep fascia is strengthened by transverse fibres which help to form the thickened bands of the flexor, extensor, and peroneal retinacula. The **flexor retinaculum** [FIG. 467] runs between the medial malleolus and the medial process of the tuber of the calcaneus. From its deep surface fibrous septa pass to the back of the lower end of the tibia and to the capsule of the ankle joint, dividing the space into four tunnels. These tunnels transmit, from medial to lateral side, the tendon of the tibialis posterior, the tendon of the flexor digitorum longus, the posterior tibial vessels and tibial nerve, and the tendon of the flexor hallucis longus; each of the tendons is enclosed in a separate synovial sheath. The upper border of the flexor retinaculum is continuous with the investing deep fascia of the leg, and also with the septum covering the deep muscles of the calf. Its lower border is continuous with the deep fascia of the sole, and from its outer surface springs the abductor hallucis muscle.

There are two **peroneal retinacula** [FIG. 466]; the superior one holds down the tendons of the peroneus longus and peroneus brevis as they pass together in a single synovial sheath posterior to the lateral malleolus. The inferior peroneal retinaculum, which is attached by a septum to the peroneal trochlea, holds the same two tendons separately to the surface of the calcaneus, each having its own prolongation from the common synovial sheath.

There are also two **extensor retinacula** [FIGS. 465, 466]. The superior one runs between the body of the tibia and the body of the fibula just above the ankle; it binds down the tendons of the tibialis anterior and the extensor muscles. The inferior extensor retinaculum is a Y-shaped thickening of the deep fascia on the dorsum of the foot. The stem of the Y is attached to the anterior part of the upper surface of the calcaneus and to the floor of the sinus of the tarsus [FIG. 325]. The upper of the two limbs of the Y is attached to the medial malleolus, and the lower of the two blends with the deep fascia on the medial side of the foot [FIGS. 465, 467]. Deep to the stem of the Y the peroneus tertius and extensor digitorum longus run distally in a common synovial sheath. The tendons of the extensor hallucis longus and tibialis anterior usually run deep to the limbs of the Y, though they may pierce them; they are enclosed in separate synovial sheaths [p. 387]. Also running deep to the limbs of the Y are the deep peroneal nerve, the anterior tibial artery, and its continuation, the dorsalis pedis artery.

The deep fascia of the sole, like that of the palm, has a greatly thickened central portion, known as the **plantar aponeurosis** [FIG. 472]. There is also a dense thickening of the plantar fascia on the lateral side of the sole, attaching the calcaneus to the base of the fifth metatarsal and covering as well as giving origin to the abductor digiti minimi.

The plantar aponeurosis is attached to the medial process of the tuber of the calcaneus, and spreads out anteriorly into five slips which run forwards to become continuous with the digital fibrous flexor sheaths of the toes, just as in the hand. The slip for each toe is attached on each side to the deep transverse metatarsal ligament, and to the base of the proximal phalanx; other fibres pass to the skin of the web of the toes.

From the margins of the plantar aponeurosis strong septa pass upwards to separate the flexor digitorum brevis from the abductor digiti minimi laterally and the abductor hallucis medially. These septa give origin to the muscles between which they lie and divide the sole of the foot into osteofascial compartments.

The plantar aponeurosis is a most important factor in the maintenance of the longitudinal arch of the foot. When the toes are dorsiflexed by the extensors, or when standing on the toes, the proximal phalanx winds its slip of the aponeurosis round the head of the metatarsal, a windlass effect which tightens the plantar aponeurosis and raises the longitudinal arch (Hicks, 1961).

REFERENCES AND FURTHER READING

General references and review articles are marked with an asterisk

ABD-EL-MALIK, S. (1939) Observations on the morphology of the human tongue, *J. Anat. (Lond.)*, **73**, 201.

BABA, M. A. (1954) The accessory tendon of the abductor pollicis longus, *Anat. Rec.*, **119**, 541.

BACKHOUSE, K. M., and CATTON, W. T. (1954) An experimental study of the functions of the lumbrical muscles in the human hand, *J. Anat. (Lond.)*, **88**, 133.

BARNETT, C. H. (1953) Locking at the knee joint, *J. Anat. (Lond.)*, **87**, 91.

BARNETT, C. H. (1962) The normal orientation of the human hallux and the effect of footwear, *J. Anat. (Lond.)*, **96**, 489.

BARNETT, C. H., and RICHARDSON, A. T. (1953) The postural function of the popliteus muscle, *Ann. phys. Med.*, **1**, 177.

BARRETT, B. (1962) The length and mode of termination of individual muscle fibres in the human sartorius and posterior femoral muscles, *Acta anat. (Basel)*, **48**, 242.

BASMAJIAN, J. V. (1958) Electromyography of iliopsoas, *Anat. Rec.*, **132**, 127.

BASMAJIAN, J. V. (1959) 'Spurt' and 'shunt' muscles: an electromyographic confirmation, *J. Anat. (Lond.)*, **93**, 551.

*BASMAJIAN, J. V. (1961) Electromyography of postural muscles, in *Biomechanical Studies of the Musculoskeletal System*, ed. Evans, F. G., Springfield, Illinois.

*BASMAJIAN, J. V. (1967) *Muscles Alive*, 2nd ed., Baltimore.

BASMAJIAN, J. V., and TRAVILL, A. (1961) Electromyography of the pronator muscles in the forearm, *Anat. Rec.*, **139**, 45.

BATSON, O. V. (1955) The cricopharyngeus muscle, *Ann. Otol. Rhinol. Laryngol.*, **64**, 47.

BEARN, J. G. (1961) An electromyographic study of the trapezius, deltoid, pectoralis major, biceps and triceps muscles, during static loading of the upper limb, *Anat. Rec.*, **140**, 103.

BEARN, J. G. (1967) Direct observations on the function of the capsule of the sternoclavicular joint in clavicular support, *J. Anat. (Lond.)*, **101**, 159.

*BEEVOR, C. E. (1904) *The Croonian Lectures on Muscular Movements*, London; (1903) *Adlard. Abstracts, Brit. med. J.*, **1**, 1357, 1417, 1480: **2**, 12; edited and printed (1951) for the guarantors of *Brain*, London.

*BELL, C. (1847) *The Anatomy and Philosophy of Expression as connected with the Fine Arts*, 4th ed., London.

BISHOP, A. (1964) Use of the hand in lower Primates, in *Evolutionary and Genetic Biology of Primates*, ed. Buettner-Janusch, J. (2 vols.), New York.

BLOMFIELD, L. B. (1945) Intramuscular vascular patterns in man, *Proc. roy. Soc. Med.*, **38**, 617.

BLUNT, M. J. (1951) The posterior wall of the inguinal canal, *Brit. J. Surg.*, **39**, 230.

BOOYENS, J., and KEATINGE, W. R. (1957) The expenditure of energy by men and women walking, *J. Physiol. (Lond.)*, **138**, 165.

*BOSMA, J. (1957) Deglutition: pharyngeal stage, *Physiol. Rev.*, **37**, 275.

*BOURNE, G. H. (1960) *The Structure and Function of Muscle*, vols. 1, 2, 3, New York.

BOWDEN, R. E. M., and EL-RAMLI, H. A. (1967) The anatomy of the oesophageal hiatus, *Brit. J. Surg.*, **54**, 983.

BOWDEN, R. E. M., and NAPIER, J. R. (1961) The assessment of hand function after peripheral nerve injuries, *J. Bone Jt Surg.*, **43B**, 481.

BOWDEN, R. E. M., and SCHEUER, J. L. (1960) Weights of abductor and adductor muscles of the human larynx, *J. Laryng. Otol.*, **74**, 971.

BRAITHWAITE, F., CHANNELL, G. D., MOORE, F. T., and WHILLIS, J. (1948) The applied anatomy of the lumbrical and interosseous muscles of the hand, *Guy's Hosp. Rep.*, **97**, 185.

BRASH, J. C. (1955) *Neurovascular Hila of Limb Muscles*, Edinburgh.

BRIDGE, D. T., and ALLBROOK, D. (1970) Growth of striated muscle in an Australian marsupial (*Setonyx brachyurus*), *J. Anat. (Lond.)*, **106**, 285.

BROCKIS, J. G. (1953) The blood supply of the flexor and extensor tendons of the fingers in man, *J. Bone Jt Surg.*, **35B**, 131.

BUCHTHAL, F., ERMINIO, F., and ROSENFALCK, P. (1959) Motor unit territory in different human muscles, *Acta physiol. scand.*, **45**, 72.

CAMPBELL, E. J. M. (1952) An electromyographic study of the role of the abdominal muscles in breathing, *J. Physiol. (Lond.)*, **117**, 222.

CAMPBELL, E. J. M. (1955a) The role of the scalene and sternomastoid muscles in breathing in normal subjects: an electromyographic study, *J. Anat. (Lond.)*, **89**, 378.

CAMPBELL, E. J. M. (1955b) An electromyographic examination of the role of the intercostal muscles in breathing in man, *J. Physiol. (Lond.)*, **129**, 12.

CAMPBELL, E. J. M., AGOSTINI, E., and DAVIS, J. N. (1970) *The Respiratory Muscles; Mechanisms and Neural Control*, 2nd ed., London.

*CAPENER, N. (1956) The hand in surgery, *J. Bone Jt Surg.*, **38B**, 128.

CATTON, W. T., and GRAY, J. E. (1951) *J. Anat. (Lond.)*, **85**, 412P.

CHURCH, J. C. T. (1969) Satellite cells and myogenesis: a study in the fruit-bat web, *J. Anat. (Lond.)*, **105**, 419.

CUNNINGHAM, D. P., and BASMAJIAN, J. V. (1969) Electromyography of genioglossus and geniohyoid muscles during deglutition, *Anat. Rec.*, **165**, 401.

DARCUS, H. D. (1951) The maximum torques developed in pronation and supination of the right hand, *J. Anat. (Lond.)*, **85**, 55.

*DARWIN, C. (1872) *The Expression of the Emotions in Man and Animals*, 2nd ed. (1889) ed. Darwin, F., London.

DAY, M. H., and NAPIER, J. R. (1961) The two heads of flexor pollicis brevis, *J. Anat. (Lond.)*, **95**, 123.

*DUCHENNE, G. B. A. (1867) *Physiologie des mouvements démontrée à l'aide de l'expérimentation électrique et de l'observation clinique*, Paris. Translated and edited by Kaplan, E. B. as *Physiology of Motion*, Philadelphia, 1949.

DUNN, N. (1920) The causes of success and failure in tendon transplantation, *J. orthop. Surg.*, **2**, 554.

EDWARDS, D. A. W. (1946) The blood supply and lymphatic drainage of tendons, *J. Anat. (Lond.)*, **80**, 147.

EYLER, D. L., and MARKEE, J. E. (1954) The anatomy and function of the intrinsic muscles of the fingers, *J. Bone Jt Surg.*, **36A**, 1.

FEINSTEIN, B., LINDEGÅRD, B., NYMAN, E., and WOHLFART, G. (1955) Morphologic studies of motor units in normal human muscles, *Acta anat. (Basel)*, **23**, 217.

FLOYD, W. F., and SILVER, P. H. S. (1950) Electromyographic study of patterns of activity of the anterior abdominal wall muscles, *J. Anat. (Lond.)*, **84**, 132.

FLOYD, W. F., and SILVER, P. H. S. (1952) Electromyography of the erectores spinae muscles in flexion of the lumbar vertebrae, *J. Anat. (Lond.)*, **86**, 484.

FLOYD, W. F., and SILVER, P. H. S. (1955) The function of the erectores spinae muscles in certain movements and postures in man, *J. Physiol. (Lond.)*, **129**, 184.

FLOYD, W. F., and WALLS, E. W. (1953) Electromyography of the sphincter ani externus in man, *J. Physiol. (Lond.)*, **122**, 599.

FORREST, W. J., and BASMAJIAN, J. V. (1965) Functions of human thenar and hypothenar muscles, *J. Bone Jt Surg.*, **47A**, 1585.

GALLAUDET, B. B. (1931) *A Description of the Planes of Fascia of the Human Body*, New York.

GAUGHRAN, G. R. L. (1957) Fasciae of the masticator space, *Anat. Rec.*, **129**, 383.

GRAY, E. G., and BASMAJIAN, J. V. (1968) Electromyography and cinematography of leg and foot ('normal' and flat) during walking, *Anat. Rec.*, **161**, 1.

HAINES, R. W. (1934) On muscles of full and of short action, *J. Anat. (Lond.)*, **69**, 20.

HAINES, R. W. (1951) The extensor apparatus of the fingers, *J. Anat. (Lond.)*, **85**, 251.

HICKS, J. H. (1954) The mechanics of the foot. II. The plantar aponeurosis and the arch, *J. Anat. (Lond.)*, **88**, 25.

*HICKS, J. H. (1961) The three weight-bearing mechanisms of the foot, in *Biomechanical Studies of the Musculo-skeletal System*, ed. Evans, F. G., Springfield, Illinois.

HIGHET, W. B. (1943) Innervation and function of the thenar muscles, *Lancet*, i, 227.

HILL, A. V. (1956) The design of muscles, *Brit. med. Bull.*, **12**, 165.

*HUBER, E. (1931) *Evolution of Facial Musculature and Facial Expression*, Baltimore.

HUGH-JONES, P. (1947) The effect of limb position in seated subjects on their ability to utilize the maximum contractile force of the limb muscles, *J. Physiol. (Lond.)*, **105**, 332.

INMAN, V. T. (1947) Functional aspects of the abductor muscles of the hip, *J. Bone Jt Surg.*, **29**, 607.

JONES, D. S., BEARGIE, R. J., and PAULY, J. E. (1953) An electromyographic study of some muscles of costal respiration in man, *Anat. Rec.*, **117**, 17.

JONES, D. S., and PAULY, J. E. (1957) Further electromyographic studies on muscles of costal respiration in man, *Anat. Rec.*, **128**, 733.

JONES, F. W. (1919) Voluntary muscular movements in cases of nerve lesions, *J. Anat. (Lond.)*, **54**, 41.

*JONES, F. W. (1941) *The Principles of Anatomy as seen in the Hand*, 2nd ed., London.

*JONES, F. W. (1944) *Structure and Function as seen in the Foot*, London.

*JOSEPH, J. (1960) *Man's Posture*, Springfield, Illinois.

JOSEPH, J. (1968) The pattern of activity of some muscles in women walking on high heels, *Ann. phys. Med.*, **9**, 295.

JOSEPH, J., and McCOLL, I. (1961) Electromyography of muscles of posture: posterior vertebral muscles in males, *J. Physiol. (Lond.)*, **157**, 33.

JOSEPH, J., and NIGHTINGALE, A. (1952) Electromyography of muscles of posture: leg muscles in males, *J. Physiol. (Lond.)*, **117**, 484.

JOSEPH, J., and NIGHTINGALE, A. (1954) Electromyography of muscles of posture: thigh muscles in males, *J. Physiol. (Lond.)*, **126**, 81.

JOSEPH, J., and NIGHTINGALE, A. (1956) Electromyography of muscles of posture: leg and thigh muscles in women, including the effects of high heels, *J. Physiol. (Lond.)*, **132**, 465.

JOSEPH, J., and WILLIAMS, P. L. (1957) Electromyography of certain hip muscles, *J. Anat. (Lond.)*, **91**, 286.

KEAGY, R. D., BRUMLIK, J., and BERGAN, J. J. (1966) Direct electromyography of the psoas major muscle in man, *J. Bone Jt Surg.*, **48A**, 1377.

KIRK, T. S. (1924) Some points in the mechanism of the human hand, *J. Anat. (Lond.)*, **58**, 228.

LANDSMEER, J. M. F. (1949) The anatomy of the dorsal aponeurosis of the human finger and its functional significance, *Anat. Rec.*, **104**, 31.

*LANDSMEER, J. M. F. (1955) Anatomical and functional investigations on the articulations of the human fingers, *Acta anat. (Basel)*, Suppl. **24**, 1.

LANGA, G. S. (1963) Experimental observations and interpretations on the relationship between the morphology and function of the human knee joint, *Acta anat. (Basel)*, **55**, 16.

*LAST, R. J. (1954) The muscles of the mandible, *Proc. roy. Soc. Med.*, **47**, 571.

*LE DOUBLE, A. F. (1897) *Traité des variations du système musculaire de l'homme*, Paris.

LEESON, C. R., and LEESON, T. S. (1969) Mitochondrial organization in skeletal muscle of the rat soft palate, *J. Anat. (Lond.)*, **105**, 363.

LEWIS, O. J. (1965) The evolution of the mm. interossei in the Primate hand, *Anat. Rec.*, **153**, 275.

LOCKHART, R. D. (1930) Movements of the normal shoulder joint and of a case with trapezius paralysis studied by radiogram and experiment in the living, *J. Anat. (Lond.)*, **64**, 288.

*LOCKHART, R. D. (1963) *Living Anatomy*, 6th ed. (paperback, 1970), London.

LOCKHART, R. D., and BRANDT, W. (1938) The length of striated muscle fibres, *J. Anat. (Lond.)*, **72**, 470.

LONG, C., and BROWN, M. E. (1964) Electromyographic kinesiology of the hand: muscles moving the long finger, *J. Bone Jt Surg.*, **46A**, 1683.

LOW, A. (1907) A note on the crura of the diaphragm and the muscle of Treitz, *J. Anat. Physiol.*, **42**, 93.

MACCONAILL, M. A. (1946) Some anatomical factors affecting stabilising functions of muscles, *Irish J. med. Sci.*, March, 160.

MACCONAILL, M. A., and BASMAJIAN, J. V. (1969) *Muscles and Movements*, Edinburgh.

MACDOUGALL, J. D. B., and ANDREW, B. L. (1953) An electromyographic study of the temporalis and masseter muscles, *J. Anat. (Lond.)*, **87**, 37.

MANTER, J. T. (1945) Variations of the interosseous muscles of the human foot, *Anat. Rec.*, **93**, 117.

MARTIN, B. F. (1968) The origins of the hamstring muscles, *J. Anat. (Lond.)*, **102**, 345.

MARTIN, C. P. (1940) The movements of the shoulder joint, with special reference to rupture of the supraspinatus tendon, *Amer. J. Anat.*, **66**, 213.

MATHESON, A. B., SINCLAIR, D. C., and SKENE, W. G. (1970) The range and power of ulnar and radial deviation of the fingers, *J. Anat. (Lond.)*, **107**, 439.

*MEDICAL RESEARCH COUNCIL (1951) *War Memorandum No. 7: Aids to the Investigation of Peripheral Nerve Injuries*, 2nd ed., London.

MILLS, J. N. (1950) The nature of the limitation of maximal inspiratory and expiratory efforts, *J. Physiol. (Lond.)*, **III**, 376.

MORRIS, J. M., BENNER, G., and LUCAS, D. B. (1962) An electromyographic study of the intrinsic muscles of the back in man, *J. Anat. (Lond.)*, **96**, 509.

*MORTON, D. J. (1935) *The Human Foot*, New York.

MOSELEY, H. F. (1951) Ruptures of the rotator cuff, *Brit. J. Surg.*, **38**, 340.

NAPIER, J. R. (1952) The attachments and function of the abductor pollicis brevis, *J. Anat. (Lond.)*, **86**, 335.

*NAPIER, J. R. (1956) The prehensile movements of the human hand, *J. Bone Jt Surg.*, **38B**, 902.

PARRY, E. (1966) The influence of rotation of the trunk on the anatomy of the inguinal canal, *Brit. J. Surg.*, **53**, 205.

PASSAVANT, G. (1869) Ueber die Verschliessung des Schlundes beim Sprechen, *Virchows Arch. path. Anat.*, **46**, 1.

PATEY, D. H. (1949) Some observations on the functional anatomy of inguinal hernia and its bearing on the operative treatment, *Brit. J. Surg.*, **36**, 264.

PAULY, J. E. (1966) An electromyographic analysis of certain movements and exercises. 1. Some deep muscles of the back, *Anat. Rec.*, **155**, 223.

PETERSEN, F. B. (1960) Muscle training by static, concentric, and eccentric contractions, *Acta physiol. scand.*, **48**, 406.

PORTER, N. H. (1962) A physiological study of the pelvic floor in rectal prolapse, *Ann. roy. Coll. Surg. Engl.*, **31**, 379.

ROWE, R. W. D., and GOLDSPINK, G. (1969) Muscle fibre growth in five different muscles in both sexes of mice, *J. Anat. (Lond.)*, **104**, 519.

*SALTER, N. (1955a) Methods of measurement of muscle and joint function, *J. Bone Jt Surg.*, **37B**, 474.

SALTER, N. (1955b) The effect on muscle strength of maximum isometric and isotonic contractions at different repetition rates, *J. Physiol. (Lond.)*, **130**, 109.

*SAUNDERS, J. B. de C. M., INMAN, V. T., and EBERHART, H. D. (1953) The major determinants in normal and pathological gait, *J. Bone Jt Surg.*, **35A**, 543.

SCHEVING, L. E., and PAULY, J. E. (1959) An electromyographic study of some muscles acting on the upper extremity of man, *Anat. Rec.*, **135**, 239.

SCHWARZACHER, H. G. (1959) Über die Länge und Anordnung der Muskelfasern in menschlichen Skeletmuskeln, *Acta anat. (Basel)*, **37**, 217.

*SINGER, E. (1935) *Fasciae of the Human Body, and their Relations to the Organs they Envelop*, Baltimore.

SMITH, J. W. (1954) Muscular control of the arches of the foot in standing: an electromyographical assessment, *J. Anat. (Lond.)*, **88**, 152.

STACEY, M. J. (1969) Free nerve endings in skeletal muscle of the cat, *J. Anat. (Lond.)*, **105**, 231.

*STACK, A. J. (1962) Muscle function in the fingers, *J. Bone Jt Surg.*, **44B**, 899.

*STEINDLER, A. (1935) *Mechanics of Normal and Pathological Locomotion in Man*, London.

*STEINDLER, A. (1955) *Kinesiology of the Human Body*, Springfield, Illinois.

STILWELL, D. L. (1957) The innervation of tendons and aponeuroses, *Amer. J. Anat.*, **100**, 289.

SUNDERLAND, S. (1944a) The significance of hypothenar elevation in movements of opposition of the thumb, *Aust. N.Z. J. Surg.*, **13**, 155.

*SUNDERLAND, S. (1944b) Voluntary movements and the deceptive action of muscles in peripheral nerve lesions, *Aust. N.Z. J. Surg.*, **13**, 160.

SUNDERLAND, S. (1953) Rotation of the fingers by the lumbrical muscles, *Anat. Rec.*, **116**, 167.

*SUNDERLAND, S. (1968) *Nerves and Nerve Injuries*, Edinburgh.

TAYLOR, A. (1960) The contribution of the intercostal muscles to the effort of respiration in man, *J. Physiol. (Lond.)*, **151**, 390.

*TESTUT, L. (1884) *Les anomalies musculaires chez l'homme expliquées par l'anatomie comparée*, Paris.

THOMAS, D. P., and WHITNEY, R. J. (1959) Postural movements during normal standing in man, *J. Anat. (Lond.)*, **93**, 524.

*THOMPSON, P. (1899) *The Myology of the Pelvic Floor*, London.

TRAVILL, A., and BASMAJIAN, J. V. (1961) Electromyography of the supinators of the forearm, *Anat. Rec.*, **139**, 557.

WADE, O. L. (1954) Movements of the thoracic cage and diaphragm in respiration, *J. Physiol. (Lond.)*, **124**, 193.

WHILLIS, J. (1930) A note on the muscles of the palate and the superior constrictor, *J. Anat. (Lond.)*, **65**, 92.

WHILLIS, T. (1946) Movements of the tongue in swallowing, *J. Anat. (Lond.)*, **80**, 115.

*WHITNALL, S. E. (1932) *The Anatomy of the Human Orbit and Accessory Organs of Vision*, 2nd ed., London.

*WRIGHT, W. G. (1928) *Muscle Function*, New York.

*ZIMMERMAN, L. M., and ANSON, B. J. (1967) *Anatomy and Surgery of Hernia*, 2nd ed., Baltimore.

THE DIGESTIVE SYSTEM

G. M. WYBURN

THE term digestive system includes the organs concerned with the reception, digestion, and absorption of food, and the excretion of the unabsorbed residue. The basis of the digestive system is the digestive tube or **alimentary canal,** some 9 metres in length from the mouth opening to the anal opening. The successive parts of this tube are the mouth, the **pharynx, oesophagus, stomach, small intestine,** and **large intestine.** The mouth and pharynx are used for respiration as well as for digestion, and the alimentary canal proper begins at the upper end of the oesophagus. In addition, there are several glandular organs whose ducts open into the mouth—the **salivary glands**—and into the small

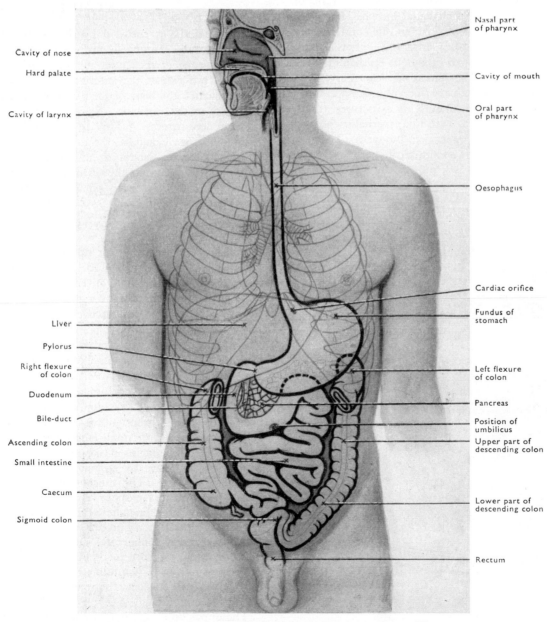

Cavity of nose

Hard palate

Cavity of larynx

Liver

Pylorus

Right flexure of colon

Duodenum

Bile-duct

Ascending colon

Small intestine

Caecum

Sigmoid colon

Nasal part of pharynx

Cavity of mouth

Oral part of pharynx

Oesophagus

Cardiac orifice

Fundus of stomach

Left flexure of colon

Pancreas

Position of umbilicus

Upper part of descending colon

Lower part of descending colon

Rectum

FIG. 473   Diagram of general arrangement of digestive system. The transverse colon is not represented in order that the duodenum and pancreas, which lie behind it, may be seen.

intestine—**liver** and **pancreas.** These also are included in the digestive system. The general arrangement, position, and size of these various parts are shown in FIGURE 473.

The earliest evidence of a special organ to deal with the digestion of food is in the Coelenterata, where there is a cavity or enteron opening to the exterior. A further stage is the evolution in the worms of an epithelial-lined, muscular tube capable of independent movement, with two openings, mouth and anus, and separated by a coelom or body cavity from the body wall. Despite major modifications in detail, this remains the basic plan of the alimentary canal, though the respiratory system and parts of the urogenital appartus are developed from it as separate entities, and a number of accessory glands are also formed from its walls.

THE MOUTH, PHARYNX, AND OESOPHAGUS

THE MOUTH

The characteristic feature of the mammalian mouth is the movable muscular lips and cheeks, correlated with the functions of mastication and, in Man, of articulated speech. The mouth is the first part of the alimentary canal, and its contribution to digestion is the secretion of saliva [FIG. 474]. It is lined by a mucous membrane of stratified squamous epithelium with a fibrous lamina propria. The blood vessels, nerves, and lymphatics run in the connective tissue submucosa attaching the mucous membrane to the underlying structures. There are two parts of the mouth—the vestibule external to the teeth, and the mouth cavity proper internal to the teeth. When the teeth are closed these two parts communicate with one another by a narrow passage on each side behind the third molar tooth.

VESTIBULE

The vestibule is enclosed by the cheeks and by the lips which surround the oral fissure with its angle at the level of the outer edge of the upper incisor teeth. A finger in the vestibule can palpate the anterior and infratemporal surfaces of the maxilla, the body of the mandible, and at the posterior extremity of the vestibule, the ramus of the mandible up to the coronoid process. This is covered by the tendon of temporalis muscle, and has the anterior border of the clenched masseter palpable lateral to it. The duct of the parotid salivary gland opens into the vestibule at the level of the second upper molar tooth on each side, and its opening can be felt with the tip of the tongue.

LIPS

The junction of the upper lip and cheek is indicated externally by the nasolabial sulcus directed laterally from the ala of the nose to the angle of the mouth, and the junction of the lower lip and chin by the mentolabial sulcus. The upper lip has a median vertical groove—the **philtrum**—widening below to end opposite the labial tubercle, a projection on the free edge of the lip which is prominent in children and responsible for the characteristic curve of the oral fissure. The lower lip is usually longer and more movable than the upper lip. From without inwards the structure of the lip consists of: (1) skin with underlying fascia; (2) voluntary muscle (mainly the orbicularis oris) some of whose fibres are attached to the skin; (3) a submucous layer with small labial glands which can be felt with the tip of the tongue on the oral surface; and (4) a mucous membrane connected by strands of connective tissue to the orbicularis muscle. The mucous membrane has a stratified but non-cornified squamous epithelium. An arterial ring formed by the labial branches of the facial arteries is situated in the submucous layer. The red transitional zone between the outer skin and the inner oral surface is found only in Man, and here large capillary loops lie close to the surface of the mucous membrane. A midline fold of mucous membrane—the **frenulum**—connects the inner surface of each lip to the corresponding gum.

The upper lip is developed by the fusion of the median part of the frontonasal process—the globular or premaxillary process—with the maxillary processes of each side, and the lower

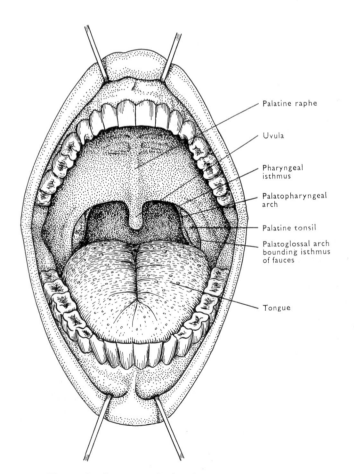

Palatine raphe

Uvula

Pharyngeal isthmus

Palatopharyngeal arch

Palatine tonsil

Palatoglossal arch bounding isthmus of fauces

Tongue

FIG. 474 Open mouth showing palate and tonsils.

lip from the mandibular arch [p. 53]. The ectoderm on the apposed surfaces of the mandibular arch below, and of the fused maxillary and premaxillary processes above, invades the underlying mesoderm, thickens, and then splits following the degeneration of the central cells. The split is known as the **alveololabial groove** and frees the tissues of the lips from those of the gums. The failure of fusion of the maxillary and premaxillary processes of the upper lip on one or both sides is responsible for the congenital abnormality of unilateral, or bilateral, hare-lip.

CHEEKS

The cheeks have essentially the same structure as the lips. The muscle is the buccinator covered by the buccopharyngeal fascia, superficial to which is the buccal pad of fat giving the cheeks the rounded contour apparent in children. The **buccal glands** of the cheek, like the labial glands, are in the submucosa and are both serous and mucous. Four or five of the larger ones, called **molar glands,** are situated in the vicinity of the entry of the **parotid duct.** Movements of the lips and cheeks are a necessary part of mastication, forcing the food from the vestibule into the mouth cavity proper. Thus in paralysis of the facial nerve which supplies the muscles of the lips and cheeks, food tends to accumulate in the vestibule. While the mucous membrane of both lips and cheeks is firmly attached to the underlying muscles, it is loosely fixed at the reflection on to the alveolar surface of the jaws, thus permitting freedom of movement.

The **sensory nerves** of the upper lip, the cheeks, and the lower lip are the infra-orbital, the buccal, and the mental branches of the trigeminal nerve respectively. Lymph vessels from both lips drain into the submandibular lymph nodes [FIG. 1172].

GUMS

The gums, which consist of dense fibrous tissue covered with mucous membrane, have a partially segmented appearance on their labial (external) aspect. They are firmly bound to the periosteum of the underlying bone by their fibrous tissue, especially at the edges of the alveolar processes and at the necks of the teeth by the fibres of the periodontal membrane [FIG. 482]. The epithelium of the margins of the gums, which may be cornified in part, continues into the epithelial lining of the sulcus surrounding each tooth, and terminates on the surface of the enamel of the tooth near the amelocemental junction, as an epithelial attachment. In later life, when the gums tend to recede, the epithelial attachment may be to the dental cement rather than to the enamel.

In addition to their supply from the nerves which innervate the teeth, the gums receive nerve fibres from adjacent sensory nerves. The labial aspect of the upper gums as far forward as the canines is supplied by branches of the **buccal nerve.** More anteriorly, the nerve supply is the labial branches of the **infra-orbital nerve** supplying the mucous membrane of the upper lip. The lingual aspect of the gums of the upper teeth is supplied by the **greater palatine nerve,** with branches of the nasopalatine nerve in the region of the incisor teeth. The labial aspect of the lower gums is supplied by twigs from the buccal nerve, a small area in front by the mental nerve, and the lingual aspect by branches of the **lingual nerve.**

MOUTH CAVITY PROPER

The mouth cavity proper is roofed by the **palate,** which separates it from the nasal cavities. Below is the tongue, and beneath the tongue the floor of the mouth is formed by the mucous membrane covering the mylohyoid and genioglossus muscles. A fold of mucous membrane—**the frenulum**—connects the under surface of the tongue to the floor of the mouth in the midline. At the base of the frenulum the ducts of the **submandibular salivary glands** open on the **sublingual papillae,** and the **sublingual folds,** raised by the underlying **sublingual salivary gland,** run posterolaterally [FIG. 475]. Posteriorly, the narrower isthmus of the fauces connects the mouth with the pharynx, and has each lateral wall formed by a palatoglossal and a palatopharyngeal fold with a tonsillar fossa between them [FIG. 477].

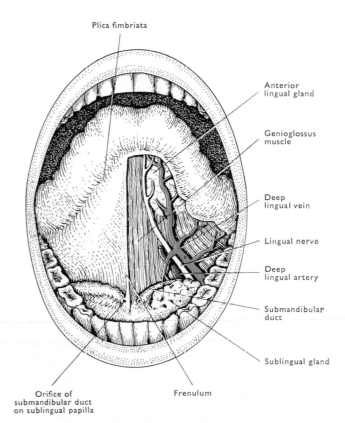

Plica fimbriata

Anterior lingual gland

Genioglossus muscle

Deep lingual vein

Lingual nerve

Deep lingual artery

Submandibular duct

Sublingual gland

Orifice of submandibular duct on sublingual papilla

Frenulum

FIG. 475 Open mouth with tongue raised and mucous membrane removed from right side to show deep lingual vessels.

PALATE

The complete palate is characteristic of mammals, and its development is associated with the ability to suck. The palate forms the roof of the mouth. It separates the mouth from the nasal cavities and the isthmus of the fauces from the nasal part of the pharynx, and it extends backwards into the cavity of the pharynx, forming a partial division between the oral and the nasal parts of the pharynx. It consists of two portions—the anterior two-thirds with a bony foundation, termed the **hard palate,** and the posterior third, with a fibromuscular basis, termed the **soft palate.** The palate is arched anteroposteriorly and also transversely. The transverse curvature is the more pronounced in the hard palate.

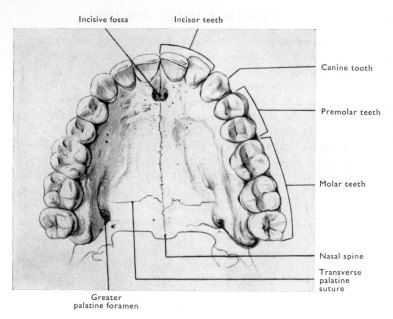

FIG. 476   Hard palate and upper permanent teeth.

Hard Palate

The hard palate is formed by the palatine processes of the maxilla, including the premaxilla in front and the horizontal plates of the palatine bones behind [FIG. 476]. Anteriorly and laterally, it is continuous with the maxillary alveolar arch, and posteriorly with the soft palate. There is a median ridge (raphe) ending in front in the incisive papilla with rudimentary transverse ridges fanning out from the anterior part of the raphe. These are more prominent in lower mammals and are well marked in the human foetus. Sometimes a small pit which will admit the point of a pin is seen on each side immediately behind the central incisor teeth and about 2 mm. from the median plane. These pits correspond to the inferior openings of the incisive canals, with which they are occasionally continuous. The mucous membrane of the oral surface of the palate is firmly fixed to the underlying bone and periosteum, has a lamina propria of dense connective tissue, and the stratified squamous epithelium is cornified. The submucous layer

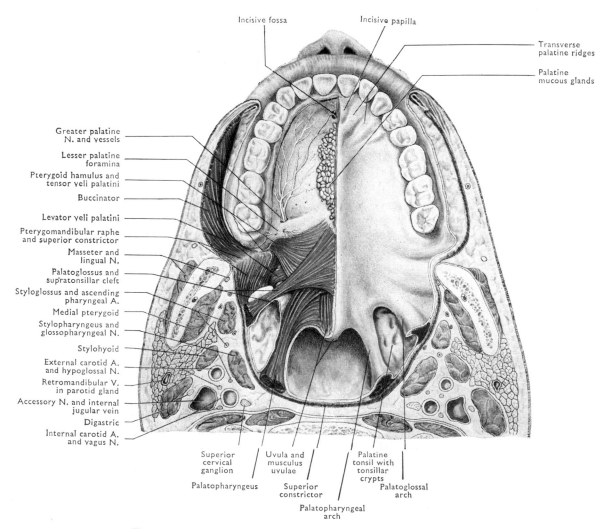

FIG. 477   Horizontal section at level of oral fissure showing palate from below.

The right half of the palate has been dissected to show the muscles of the soft palate by the removal of the majority of the palatal glands.

contains fat in front, and is packed with mucous and serous glands continuous with the glandular layer in the soft palate. On its nasal surface most of the epithelium of the mucous membrane is ciliated.

Soft Palate

The soft palate, a muscular flap, is attached anteriorly to the posterior edge of the hard palate and slopes postero-inferiorly to a free, curved margin on which is a median conical projection—**the uvula**—which varies in length from 5 to 25 mm. Laterally, the soft palate is continuous with the pharyngeal wall and has extensions to the sides of the tongue and the pharynx, the **palatoglossal** and **palatopharyngeal folds** respectively.

The anterior part of the soft palate has an aponeurotic framework, formed by the tendons of the tensor veli palatini muscles, and attached to the posterior border of the bony palate [FIG. 477]. The posterior part is muscular. The anterior part remains horizontal in position, while the more freely movable posterior part, when at rest, is curved downwards. Of the five pairs of palatal muscles [p. 293] the **tensor veli palatini** and the **palatoglossus** act on the anterior part of the soft palate, and the **palatopharyngeus**, the **levator veli palatini,** and the **musculus uvulae** on the posterior part of the palate.

The covering mucous membrane is stratified squamous epithelium on the oral surface and on most of the nasal surface. The palatal glands lie deep to the oral mucous membrane of the anterior part of the soft palate. The uvula has some fatty tissue and longitudinal muscle fibres.

Movements of the Soft Palate. The posterior part of the soft palate moves freely like a door on its hinges. When raised by the action of the levator muscles it becomes a horizontal continuation of the plane of the hard palate, projecting backwards across the cavity of the pharynx. Movements of the soft palate take place during speech and swallowing. For proper speech the nasopharynx has to be sealed off by the apposition of the soft palate to the posterior wall of the pharynx (except in the production of nasal sounds, e.g., *m* and *n*, when the soft palate is relaxed). This closure is accomplished by the palatal muscles pulling the soft palate upwards and backwards until it is almost in contact with the posterior wall of the pharynx. Meanwhile, the posterior pharyngeal wall is elevated by the action of the palatopharyngeus muscle and this, together with an active contraction of its horizontal fibres and those of the superior constrictor muscle, produces a ridge— the so-called ridge of Passavant—which completes the separation of the nasopharynx. The soft palate can also be drawn down by the palatoglossus and the palatopharyngeus muscles so that its under surface is brought into contact with the posterior part of the upper surface of the tongue. During quiet breathing the soft palate is curved backwards and downwards. When making the sound 'ah', the isthmus of the fauces widens, bringing the posterior pharyngeal wall into view.

Vessels, Nerves and Lymphatics of the Palate

The **arteries** of the palate are branches of the maxillary artery. The **descending palatine artery** gives its largest branch, the **greater palatine artery**, through the greater palatine foramen. It runs along the lateral margin of the hard palate near the alveolus to the incisive fossa where it anastomoses with a terminal branch of the sphenopalatine artery. **Lesser palatine arteries** from the descending palatine artery emerge from the lesser palatine foramina to supply the soft palate and anastomose with the ascending palatine branch of the facial artery. The soft palate receives an additional blood supply from the facial artery via its ascending palatine and tonsillar branches, and from the ascending pharyngeal artery.

The sensory **nerves** are branches of the maxillary nerves. The largest—the **greater palatine nerve**—accompanies the artery through the greater palatine foramen, divides in the roof of the mouth into branches which run in grooves in the hard palate, and extend forwards nearly to the incisor teeth where they may communicate with branches of the naso-palatine nerve. The **lesser palatine nerves** emerge from the lesser palatine foramina and are distributed to the hard and the soft palate. The palatine nerves may contain sensory fibres from the facial as well as from the maxillary nerve, and are accompanied by secretory and vasomotor fibres.

The **lymph vessels** of the palate run lateral to the tonsil and the palatoglossal arch, to drain into the upper deep cervical lymph nodes.

Development of the Palate [see also p. 54]

At an early stage the primitive mouth is directly continuous with the nasal pits. The **palatine processes,** horizontal folds projecting into the primitive mouth from the medial aspect of the maxillary processes and at first in contact with the sides of the tongue, later meet in the midline above the tongue and fuse with one another and with the lower end of the nasal septum. Anteriorly, the palatine processes fuse with the **frontonasal process** (premaxilla) which is described as forming the primitive palate, separating the nasal pits from the oral orifice, and carrying the incisor teeth. Posteriorly, the fused palatine processes extend beyond the nasal septum. In the congenital defect of cleft palate there is a failure of fusion of the palatine processes, giving rise to a defect ranging from a bifid uvula or cleft soft palate, to a wide separation along the entire length of the palatine processes, leaving a large gap in the roof of the mouth, with a fissure on each side separating the premaxilla and palatine processes. Such a complete cleft palate is generally associated with a bilateral hare-lip.

Development of the Mouth

The mammalian mouth is developed from the **ectodermal stomodaeum** and the cephalic end of the **entodermal fore-gut.** The stomodaeum [FIG. 62] is a depression on the ventral surface of the embryo between the bulging forebrain and the developing heart. The cephalic end of the foregut extends to the stomodaeum, but the two cavities are at first separated by a double-layered membrane composed of the foregut entoderm and the stomodaeal ectoderm, i.e., the **buccopharyngeal membrane,** which breaks down as early as the fourth week. It is not, however, possible to indicate any plane of division between ectoderm and entoderm in the adult mouth. In general, the roof and sides of the mouth, the vestibule, the tooth enamel, and probably the salivary glands, are considered to be ectodermal in origin.

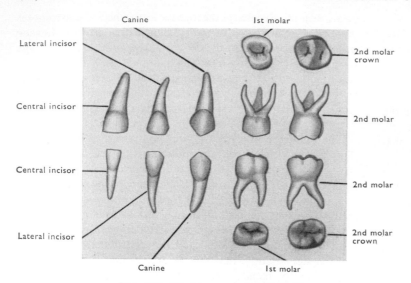

FIG. 478　Deciduous teeth of left side.

The masticating surfaces of the two upper molars are shown above. In the second row the upper teeth are viewed from the outer or labial side. In the third row the lower teeth are shown from the same side; and below are the masticating surfaces of the two lower molars. In the specimen from which the first upper molar was drawn the two labial tubercles were not distinctly separated, as is often the case.

TEETH

[PLATE 27]

The teeth are developed from the oral mucous membrane, and like the placoid scales of Selachians can be regarded as modified skin papillae with calcified elements derived from the embryonic ectoderm and mesoderm. In general, the evolution of the teeth has been from an indefinite number embedded loosely in part of the mucous lining of the mouth and all alike, to a definite number, securely set into bony sockets in the jaws and grouped as **incisors** (cutting), **canines** (piercing), and **molars** (grinding). To adjust the arrangement of the teeth to the growth of the jaws, mammals have in general two sets

of teeth—the milk or deciduous teeth, and the second or permanent teeth. In Man there are ten deciduous teeth in each jaw, the first of which erupts about the sixth month after birth and the last at the end of the second year. They are shed from the sixth to the twelfth year as they are replaced by the sixteen permanent teeth in each jaw, which begin to appear at about the sixth year and are completed with the eruption of the third molar at the age of 18 or later.

Each tooth has a **crown,** the part visible above the gum, connected by the **neck,** flush with the gum, to the **root** fixed in a socket in the alveolar process of the jaw by the fibrous periodontal membrane. The deciduous teeth are distinguished from the permanent teeth by their smaller size, their whiter colour, their constricted neck and, in the case of the molar teeth, by their widely divergent roots. The deciduous teeth on each side of the upper and lower jaws are indicated by the formula :

M_2	M_1	C	I_2	I_1	I_1	I_2	C	M_1	M_2
M_2	M_1	C	I_2	I_1	I_1	I_2	C	M_1	M_2

Distal　　　　　　　Mesial　　　　　　　Distal

where I_1 = the central incisors, I_2 = the lateral incisors, C = the canines, M_1 = the first molars, and M_2 = the second molars [FIG. 478]; and the permanent teeth by the formula :

M_3	M_2	M_1	PM_2	PM_1	C	I_2	I_1	I_1	I_2	C	PM_1	PM_2	M_1	M_2	M
M_3	M_2	M_1	PM_2	PM_1	C	I_2	I_1	I_1	I_2	C	PM_1	PM_2	M_1	M_2	M

Distal　　　　　　　Mesial　　　　　　　Distal

where I_1, I_2 = the central and lateral incisors, C = the canines, PM_1, PM_2 = the premolars, and M_1, M_2, and M_3 = the molars.

The surfaces of the incisor and canine teeth are : labial (lip side), lingual (tongue side), incisal (cutting edge), mesial (medial), and distal (lateral). The molar and bicuspid (premolar) teeth have the following surfaces—buccal (cheek side), lingual, occlusal (chewing surface), mesial (anterior), and distal (posterior).

Crowns

The crowns of human teeth are believed to have evolved from a tricuspid tooth, and basically each crown has three cusps, two labial and one lingual. In the incisors and the canines the labial cusps unite as a cutting edge, forming in the upper incisors and canines a distinct V-shaped ridge on the lingual surfaces, known as the cingulum, with a small lingual tubercle at the apex of the V near the neck [FIG. 479].

The occlusal surfaces of the premolars have two cusps, representing the fused labial and lingual cusps, with a dividing fissure in the upper premolars and an intervening ridge in the

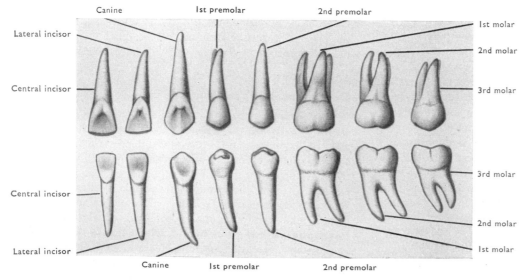

FIG. 479　Permanent teeth of right side, lingual surfaces.

The cingulum is distinct on the upper incisors and both canines, the lingual tubercle on the upper lateral incisor and the upper canine.

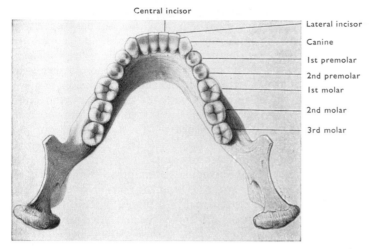

Central incisor
Lateral incisor
Canine
1st premolar
2nd premolar
1st molar
2nd molar
3rd molar

FIG. 480   Mandible and lower permanent teeth.

lower premolars. The upper molars have three or four cusps, invariably four on the first, and frequently three on the third. The lower molars have four or five cusps, frequently five in the first molar, and usually four on the second and third [FIG. 480].

Roots

The roots of the incisors, canines, and premolars are single ; commonly the first upper premolar root is bifid. The first and second lower molars have two roots, a mesial and a distal, while the third usually has its roots fused into one. The upper molars have three roots, two on the buccal, and one on the lingual side, all divergent ; the third molar has fused roots [FIG. 481].

The central upper incisors are the largest of the incisor teeth and the large upper canines, known as the eye teeth, have the longest roots, producing an eminence on the anterior surface of the maxilla. The roots of the upper first and second molar teeth form prominences on the floor of the maxillary air sinus, and exceptionally so may the roots of the premolars. Occasionally, the roots of the first molars are separated from the floor of the sinus by a thin lamina of bone, and when such teeth are extracted a direct opening from mouth to sinus may be produced.

Sockets

The external walls of the sockets of the upper teeth, especially those of the more mesial teeth, are quite thin, as evidenced by the prominent ridges which the teeth make on the labial surface of the alveolar process, but the lingual walls are strengthened by the attachment of the hard palate to the alveolar process. The third upper molar, closely related to the posterior surface of the maxilla, may have particularly thin socket walls, and efforts to remove it may result in fracture of the maxillary tuberosity.

Occlusion

The **upper dental arch,** the curve formed by the upper teeth, is elliptical, and the **lower dental arch,** the curve formed by the lower teeth, is parabolic.

When the teeth are in occlusion the upper incisors and canines overlap the corresponding lower teeth, and only the buccal cusps of the lower premolars and molars are in actual contact with the occlusal surface of the corresponding upper teeth. The two dental arches end together behind, but due to the large upper central incisors each tooth in occlusion is in contact with two teeth, except the lower central incisors.

Structure of the Teeth

The bulk of a tooth is **dentine,** a hard, calcified substance enclosing the pulp cavity. The dentine of the crown of the teeth is coated with **enamel,** and **cement** covers the dentine of the roots [FIG. 482]. The teeth are held in separate sockets in the alveolar processes of the maxilla and mandible by the periodontal membrane.

Dentine. Dentine consists of organic substance in the form of fibrils, and hard inorganic calcium and phosphorus as apatite. In microscopic sections dentine has a striated appearance due to a system of minute canals—the **dentinal tubules** —running from the pulp cavity to its outer surface. Each tubule contains a **Tomes's fibre,** a protoplasmic process of the cells lining the pulp cavity—the **odontoblasts**—which form the dentine. Although hard, dentine is exquisitely sensitive to touch and to thermal change. Dentine formation continues slowly throughout life, and the secondary dentine formed in later years may be demarcated by an abrupt change in the direction of its less numerous tubules. In old teeth

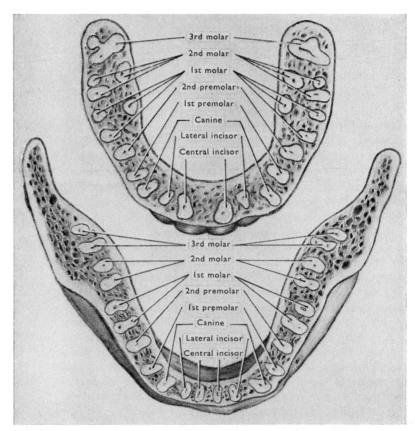

3rd molar
2nd molar
1st molar
2nd premolar
1st premolar
Canine
Lateral incisor
Central incisor

3rd molar
2nd molar
1st molar
2nd premolar
1st premolar
Canine
Lateral incisor
Central incisor

FIG. 481   Horizontal sections through maxilla and mandible to show the roots of the teeth.

The sections were carried through the bones a short distance from the edge of their alveolar borders. Note the flattened roots of the lower incisors, the two root canals in the anterior root of each lower molar, and the confluence of the three roots of the upper third molars.

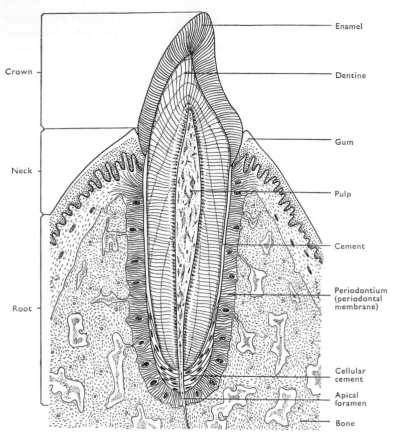

Crown

Neck

Root

— Enamel

— Dentine

— Gum

— Pulp

— Cement

Periodontium
(periodontal
membrane)

Cellular
cement

Apical
foramen

— Bone

FIG. 482   Vertical section through a tooth.

type of bone (the inorganic matrix consists mainly of calcium salts) but it is not so hard as either enamel or dentine. Some parts of the cement are cellular with cementocytes, like the osteocytes, embedded in its matrix. Cement is more resistant to resorption than bone, and thus it is possible to move the teeth in the jaw with the help of an orthodontic appliance, when bone is resorbed on the pressure side and new bone formed on the tension side of the appliance.

Periodontal Membrane. The periodontal membrane is the connective tissue surrounding the root of a tooth and attaching it to the bony alveolus. Fibres from the periodontal membrane, like the Sharpey's fibres of bone, are embedded in the cement. They attach the tooth to the tissues of the gum and to the bony wall of the socket. There is continuous wear and tear and replacement of the fibres with formation of new cement. Hyperplasia of the cement, often secondary to inflammation of the root of the tooth, can make tooth extraction difficult, requiring sometimes the removal of bone.

Pulp Cavity. The pulp cavity, enclosed by dentine, extends into the cusps, and is prolonged into the root or roots as the **root canals,** narrowing to the **apical foramen** which is usually at the apex, but may have accessory openings to one or other side of the root canal. The cavity is lined by odontoblasts, contains connective tissue elements, fibres and cells, together with blood vessels, lymphatics, and nerves entering through the apical foramen. The nerve of a tooth is composed of thick bundles of myelinated fibres with fine non-myelinated terminals extending into the predentine, some of which may run in the dentinal tubules to the dentine. The pulp cavity tends to become narrower in later years through continued formation of dentine, and denticles, or small, calcified stones, may be found in the pulp cavity of old teeth.

there are areas which are more translucent in sections of the dentine where the Tomes's fibres have degenerated and the dentinal tubules are obliterated. Whether nerve fibres enter the dentinal tubules from the pulp is uncertain, but the evidence for their presence is poor.

Enamel. The enamel covering the crown of the tooth is the hardest substance in the body, and over 90 per cent. of it by weight is inorganic salts in the form of apatite crystals. Enamel is laid down by epithelial cells—**ameloblasts**—in minute rods or prisms from 2 to 5 μm. in diameter, enclosed in a sheath and cemented together. They are fixed to the dentine at the **amelodentinal junction,** terminate at the free enamel surface, and vary in direction from the radial arrangement of the enamel rods of the cusps to the horizontal rods at the amelocemental junction. The thickest part of the enamel is at the cusps, which therefore have the longest rods. Each rod is made up of submicroscopic units arranged with their long axis approximately parallel with the long axis of the rods. Sections of the enamel show contour growth lines illustrating the rhythmic formation of the enamel matrix. There are clefts and fissures on the occlusal surfaces of adjoining cusps where the enamel is very thin and mastication wears away the enamel on the edges and contact areas of the teeth. Developmental and nutritional disturbances cause defective formation of the enamel of the teeth.

Cement. The cement is the hard covering of the anatomical roots of the teeth. It begins where the enamel finishes, continues to the apex of the root, and serves as a medium for the attachment of the fibres of the periodontal membrane that bind the teeth to the surrounding bone. It is a modified

Development of the Teeth

The enamel is derived from the ectoderm, and all other dental elements, i.e., dentine, cement, and the tissues of the pulp cavity, from the mesoderm. In the human embryo, about the sixth week, a thickening of the ectoderm of the maxillary and mandibular processes projects into the underlying mesoderm. This forms the **labiogingival groove** of the upper and lower jaws, with a continuous ridge, the **dental lamina,** on its lingual side. Ten thickenings of the ectoderm spaced along the deep edge of the dental laminae represent the early tooth germs of the deciduous teeth in each jaw, and are invaginated by proliferating mesoderm to form the **enamel organs.** These consist of an outer layer of epithelium, separated by a cavity filled with stellate reticulum from an inner layer of epithelium capping the invaginating mesoderm, the **dental papilla** [FIG. 483]. The dental lamina disintegrates, except for epithelial cords connecting the enamel organs to the surface. About the fifth month of foetal life the enamel organs of the permanent teeth start to develop from this epithelial cord and lie on the lingual aspect of the deciduous tooth germs, while those for the three permanent molars arise from a backward extension of the dental lamina, the last of them representing the third permanent molar, appearing at the fifth postnatal year.

FIG 483 Diagram to illustrate development of a tooth.

I. Shows the dental lamina (2), the surface epithelium (1), and the enamel germ (4).

II. Shows the further growth of the enamel germ and its invagination to form the enamel organ.

III. The enamel organ is more invaginated, and its inner layer of cells (8) becomes columnar. The dental lamina is thinner, but near its posterior or lingual edge there is an enlargement (7) which is the germ of a permanent tooth. The superficial cells of the dental papilla (9) are becoming columnar.

IV. The inner columnar cells of the enamel organ (ameloblasts, 8) have formed a cap of enamel (11) inside which the superficial cells of the papilla (odontoblasts, 13) have formed a layer of dentine (12).

V. Shows a still more advanced stage. The deposit of dentine is extending downwards and enclosing the papilla to form the future pulp, in which a vessel (16) is seen.

1. Epithelium.	9. Dental papilla.
2. Dental lamina.	10. Outer layer of enamel
3. Mesoderm.	organ.
4. Enamel germ or organ.	11. Enamel.
5. Primary dental lamina or	12. Dentine.
furrow.	13. Odontoblasts.
6. Labiogingival lamina.	14. Stellate reticulum.
7. Germ of permanent tooth.	15. Periodontal membrane.
8. Ameloblasts.	16. Blood vessel.

The enamel of the crown of the tooth is laid down by the **ameloblasts** of the inner epithelial layer of the enamel organ, and the dentine by the mesodermal cells capping the dental papilla—the **odontoblasts.** The roots of the teeth begin to form from an extension of the dental papilla shortly before the crowns erupt through the gums, and an important part of the process is a downgrowth of the epithelium of the enamel organ to form a **root sheath**—Hertwig's sheath. This later breaks up, and a covering layer of cement is formed by the mesoderm of the dental sac. The enamel organ and the dental papilla become enclosed in a fibrous capsule—the **dental sac**—part of which forms the periodontal membrane. Mesodermal septa, which subsequently ossify, divide the alveolus into separate compartments for each tooth.

While the teeth are developing, the upper and lower jaws are growing, and accordingly there is some shifting of the position of the tooth germs. The deciduous tooth germs grow in length at about the same rate as the jaws grow in height, and there are changes in the relative positions of the tooth germs of the deciduous and permanent teeth. Just before eruption the teeth undergo intricate movements of tilting and rotation to rectify their position.

Eruption. Eruption of teeth [FIG. 484], i.e., the extrusion of the crown through the tissues of the gum, is accompanied by an active absorption of these tissues, and also by growth of the roots of the teeth. When the eruption of the deciduous teeth is about to take place, the labial wall of the crypt is absorbed; the tooth passes through the neck of the dental sac and appears above the gum, and the socket is reformed closely around the tooth. The root, which is only partly formed at the time of the eruption, continues to elongate. The crown of the newly erupted tooth is covered by a membrane—Nasmyth's membrane—representing the remnants of the epithelial enamel organs. Shedding of the teeth is due to the resorption of their roots by osteoclasts differentiated from the connective tissue cells in response to the stimulus of pressure exerted by growing and erupting permanent tooth germs, first against the separating bony septa, and then against the root itself. When the permanent tooth is about to erupt it makes its way from its own bony compartment through the lingual wall of the socket of its deciduous predecessor, and meanwhile, the root of the deciduous tooth undergoes absorption.

The socket, now occupied by both teeth, is further enlarged by absorption, particularly on the labial side, and what remains of the temporary tooth is shed. The permanent tooth then passes onwards through the enlarged socket and, making its way to the surface, appears above the gum. After some time, when the tooth has taken up its final position, the socket is again re-formed, first around its neck and later round the fully formed root, and the tooth becomes permanently fixed. All these processes occur while the alveolar parts of the jaws are themselves growing in height.

There is a considerable range of variation in the time of shedding and eruption of the teeth, influenced by genetic and environmental factors. Girls attain dental maturity earlier than boys. The following table gives average times for the eruption and shedding of deciduous teeth and the eruption of the permanent teeth.

The mouth of the infant at birth usually contains no visible teeth, although tooth buds of both deciduous and permanent teeth are embedded in the jaws beneath the gum. At 6 months

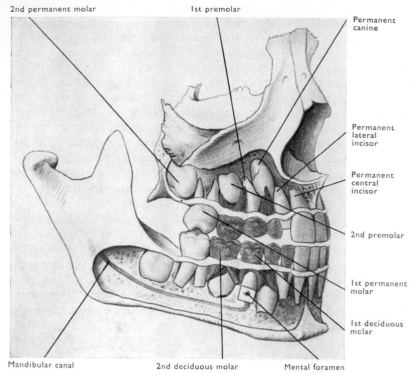

2nd permanent molar　　　　1st premolar

Permanent canine

Permanent lateral incisor

Permanent central incisor

2nd premolar

1st permanent molar

1st deciduous molar

Mandibular canal　　　2nd deciduous molar　　　Mental foramen

FIG. 484　Teeth of child over 7 years old. (Modified from Testut.)

By the removal of the bony outer wall of the alveoli, the roots of the teeth which have been erupted, and the permanent teeth which are still embedded in the mandible and maxilla, have been exposed. The deciduous teeth are shown dark, the permanent teeth light. It will be seen that the first permanent molars have appeared, the central and lateral deciduous incisors have been replaced by the corresponding permanent teeth in the maxilla, but the deciduous canine and molars have not yet been shed. In the mandible the central deciduous incisor has been replaced by the permanent central; the lateral has not yet been shed, but its permanent successor is making its way up to the surface on its lingual side. In addition, the canine and two molars of the deciduous set persist. The position of the crowns of the permanent teeth between the roots of the deciduous molars, and the deep situation occupied by the permanent canines, should be noted. Observe also the absorption of the root of the lower lateral incisor.

the teeth begin to appear, and by the end of the second year all the deciduous or milk teeth have been 'cut'. After a pause of

DECIDUOUS TEETH

	I_1	I_2	C	M_1	M_2
Eruption	6± months	8± months	18± months	12± months	24± months
Shedding	6–7 years	7–8 years	10–12 years	9–11 years	10–12 years

PERMANENT TEETH

	I	I_2	C	PM_1	PM_2	M_1	M_2	M_3
Eruption	7–8 years	8–9 years	10–12 years	10–11 years	11–12 years	6–7 years	12 years	13–25 years

4 years, i.e., about the sixth year, the next stage of dentition begins, and consists in the eruption of four new teeth—the first permanent molars—one on each side, above and below, behind those of the deciduous set, and this is followed by the gradual shedding of the twenty deciduous teeth and their replacement by twenty new teeth which take up the vacancies one by one. The teeth of the lower jaw erupt before those of the upper jaw, and it will be observed that the eruption of the permanent canine is delayed until the two premolars, which succeed it in the row, are cut, so that it breaks the otherwise regular order of eruption. All the permanent teeth have appeared by the end of the twelfth or thirteenth year, except the four 'wisdom teeth' which are usually cut between the eighteenth and the twenty-fifth year, but are often delayed until a very much later period, and occasionally never appear. The first molar is popularly known as the 6-year-old tooth and the second molar as the 12-year-old tooth.

The common dental abnormalities are mostly the result of delayed dentition and faulty position of the teeth. The state of the eruption of the teeth is an index of dental maturity, and defective nutrition and abnormal conditions of various glands such as the thyroid may be responsible for a general delay in dental maturity. Retarded eruption is much more frequent than accelerated eruption, and may be due to local or general causes. Deciduous teeth may be retained for a long time if the germ of the permanent tooth is congenitally missing. Premature loss of a deciduous tooth and closure of the gap by a shift of neighbouring teeth may retard the eruption of the permanent tooth. Local injury may lead to fusion of alveolar bone and tooth root so that the movement of the tooth is arrested, resulting in a shortened tooth. A common abnormality is the failure of the third molar tooth to erupt, generally due to some lack of correlation between the growth of the jaws and tooth development. Such impacted molars, by pressing on the roots of the second molar, may cause their resorption.

Nerve Supply of the Teeth

The **upper molars** are supplied by the **posterior superior alveolar nerves** which enter canals on the posterior and lateral wall of the maxilla. They can be anaesthetized before entering their canals by injecting the mucosa behind the last molar tooth, and upwards along the tuberosity of the maxilla. The **middle superior alveolar nerve,** when present, supplies the bicuspid teeth, but in its absence the incisors, canines, premolars, and part of the first molar receive branches from the **anterior superior alveolar nerves,** which run in canals in the anterior wall of the maxillary sinus. As these nerves in no part of their course lie on the outer surface of the maxilla they cannot be injected directly, except at their origin in the infra-orbital canal, and more usually anaesthesia is obtained by infiltration of the gum tissues. The nerve supply of individual upper teeth is not, however, accurately known, as the small branches of the nerves concerned form a **superior dental plexus** through which it is not possible to trace twigs to particular teeth. However, local anaesthesia of the posterior superior alveolar nerves rarely causes complete anaesthesia of the first permanent molar.

The lower teeth are supplied by branches given off from the **inferior alveolar nerve** in the mandibular canal. Soon after

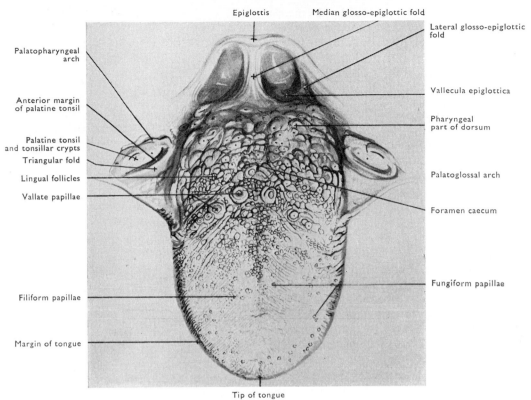

Epiglottis
Median glosso-epiglottic fold
Lateral glosso-epiglottic fold
Palatopharyngeal arch
Vallecula epiglottica
Anterior margin of palatine tonsil
Pharyngeal part of dorsum
Palatine tonsil and tonsillar crypts
Triangular fold
Palatoglossal arch
Lingual follicles
Vallate papillae
Foramen caecum
Fungiform papillae
Filiform papillae
Margin of tongue
Tip of tongue

FIG. 485 Dorsum of tongue and the palatine tonsils.

it enters the mandibular foramen this nerve gives off a branch to supply teeth posterior to the incisors. At the level of the canine tooth other fibres are said to leave the nerve and form an intricate plexus external to the roots of the incisor teeth. The incisor teeth, and sometimes the canines, may be supplied from branches of this **inferior dental plexus.** Good anaesthesia of the lower teeth can be obtained by injection of the inferior alveolar nerve before it enters the mandible, and this is usually done from the mouth by inserting a needle behind the oblique ridge which can be felt running upwards from the back of the last molar tooth on the inner border of the mandible.

TONGUE

The tongue is concerned with mastication, articulation, swallowing, and taste. The substance of the tongue is mainly muscle, with a covering mucous membrane of stratified squamous epithelium, a lamina propria of connective tissue, and lingual glands and lymph nodules. There are two **parts** of the tongue—oral and pharyngeal—separated by a V-shaped furrow, the **sulcus terminalis,** on the upper surface of the dorsum, with the apex directed backwards, and indicated by the **foramen caecum** marking the site of the embryonic thyroglossal duct. The oral part, which is freely movable, forms the anterior two-thirds of the tongue and is loosely attached to the floor of the mouth by the frenulum, a fold of mucous membrane seen on the under aspect when the tip of the tongue is raised. On each side of the frenulum, the deep lingual veins are visible through the thin mucous membrane, and fringed mucous folds—the **plica fimbriata**—converge towards the apex of the tongue [FIG. 475].

On the margins of the tongue, immediately in front of the attachment of the palatoglossal arches, there are five or six

vertical furrows termed the **foliate papillae** of the tongue, studded with taste buds. They represent the prominent foliate papillae found on the sides of the tongue in rabbits, hares, and other animals.

The more fixed pharyngeal part of the tongue is developed in the floor of the pharynx, largely from the elements of the third pharyngeal arch. It lies behind the hyoid bone and in

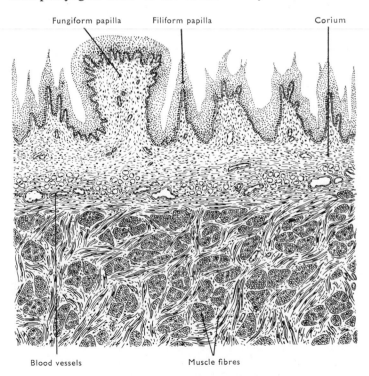

Fungiform papilla
Filiform papilla
Corium
Blood vessels
Muscle fibres

FIG. 486 Fungiform and filiform papillae of tongue.

front of the epiglottis, to which it is connected by the median glosso-epiglottic and the two lateral glosso-epiglottic folds [FIG. 485]. It includes the root of the tongue, formed by its extrinsic muscles by means of which it is attached to the hyoid, the mandible, and the styloid processes. Laterally, there are the attachments of the palatoglossal and palatopharyngeal arches with their respective muscles.

The **dorsum** of the oral part of the tongue has a velvety appearance due to numerous fine **papillae,** each with a core of connective tissue and a covering epithelium. There are three types of papillae—**filiform, fungiform,** and **vallate.** The numerous filiform papillae [FIG. 486], 2–3 mm. long and conical in shape, with hard scales of squamous cells at the apex, are arranged on the dorsum of the tongue in rows, parallel to the sulcus terminalis. The fungiform papillae are inverted cones with flat red surfaces and are found chiefly at the sides, the apex of the tongue, and scattered throughout the dorsum. There are from eight to ten vallate papillae, each 1–2 mm. in diameter, immediately in front of the sulcus terminalis and each surrounded by a sulcus which gives them a punched-out appearance [FIG. 487]. The mucous membrane on the pharyngeal part of the tongue has no papillae, but shows a number of bulges due to underlying nodules of lymphoid tissue. These are the **lingual follicles,** which together form the **lingual tonsil.** On the surface of each follicle there is a small pit which is the entrance to an epithelial-lined crypt surrounded by lymphoid tissue. Lymphocytes migrate into and fill the crypts. The inferior surface of the tongue is smooth, and its thin epithelium is bound tightly by connective tissue to the underlying muscle.

There are **taste buds** on the dorsum, the apex, and the sides of the tongue, associated in particular with the walls of the fungiform and vallate papillae. There are also taste buds in the palatoglossal folds, the posterior wall of the pharynx, the posterior surface of the epiglottis, the palate, and the mucosa covering the arytenoid cartilages.

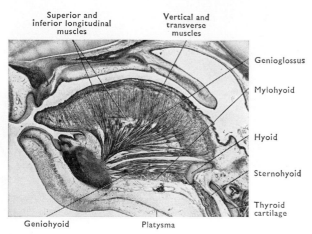

FIG. 488 Sagittal section through tongue of a 14-week-old foetus.

Throughout the substance of the tongue there are serous and mucous glands in the submucosa and extending between the muscle fibres, especially in the posterior part, and an apical group of glands—the **anterior lingual glands**—mainly mucous, on each side of the frenulum, with ducts opening on the plica fimbriata. Small serous glands open into the sulcus surrounding the vallate papillae, and are believed to wash out the sulcus so that new material may come into contact with the taste buds in it.

The muscular substance of the tongue is formed by the extrinsic muscles and the intrinsic muscles, arranged both parallel to and at right angles to the long axis of the tongue [FIGS. 357 and 488]. The tongue is divided into two halves by a median fibrous septum which does not, however, reach the dorsum. Posteriorly, the septum fans out and is attached to the hyoid bone. When the muscles of the tongue are paralysed it may fall back into the pharynx and cause respiratory obstruction. If the muscles on one side only are paralysed the tongue is drawn over to the opposite side, and the protruded part deviates to the paralysed side.

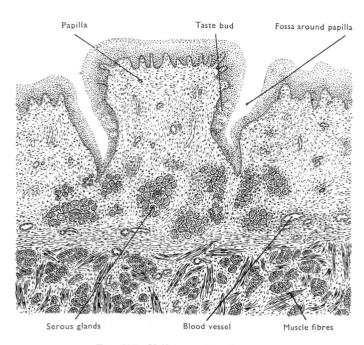

FIG. 487 Vallate papilla of tongue.

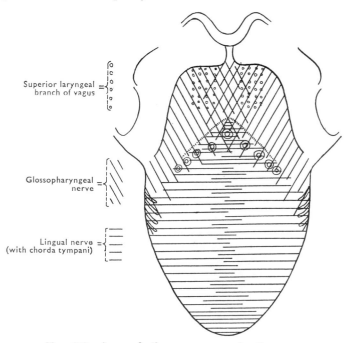

FIG. 489 Areas of afferent nerve supply of tongue.

Blood Vessels, Lymphatics, and Nerves of the Tongue

The main artery, arising from the **external carotid**, is the **lingual**. It gives off dorsal branches supplying the posterior part of the tongue, and continues as the **deep artery** along the inferior surface of the tongue near the frenulum to the apex, where it anastomoses with the corresponding branch of the lingual artery of the opposite side. The posterior part of the tongue also receives arteries from the tonsillar and ascending palatine branches of the **facial artery**. The venous drainage is via the dorsal lingual veins which join the lingual veins accompanying the lingual artery, and via the deep lingual veins which commence at the tip of the tongue and run backwards close to the mucous membrane on its inferior surface to join the sublingual vein.

The **lymph vessels** of the tongue drain into the submandibular and submental lymph nodes, and into the deep cervical lymph nodes.

The sensory **nerves** of the tongue are : (1) the lingual branch of the mandibular division of the **trigeminal nerve** (the nerve of the first pharyngeal arch) which is the nerve of common sensation to the anterior two-thirds of the tongue ; (2) the **chorda tympani** (a branch to the first pharyngeal arch from the **facial nerve,** the nerve of the second pharyngeal arch) which is the nerve of taste of the anterior two-thirds ; (3) the lingual branch of the **glossopharyngeal** (the nerve of the third pharyngeal arch) which is the nerve of common sensation and taste to the pharyngeal part of the tongue, including the vallate papillae. The superior laryngeal branch of the **vagus nerve** supplies the taste buds immediately anterior to the epiglottis [Fig. 489]. The nerve to the muscles of the tongue is the **hypoglossal.**

Development of the Tongue

The tongue is developed from the structures in the floor of the pharynx, the epithelium and connective tissue in the anterior two-thirds from paired **lingual swellings** of the first arch, together with the **tuberculum impar** [p. 419], and in the posterior third from the anterior part of the **hypobranchial eminence** which receives the ventral ends of the third and fourth pharyngeal arches. The third arch elements (with their nerve, the glossopharyngeal) grow forwards, excluding the second arch from the pharyngeal part of the tongue. The line of union of the separate developmental parts of the tongue is at the sulcus terminalis.

Anomalies. A true hypertrophy of the lingual muscles may be an unusual cause of macroglossia or large tongue, and is generally associated with some form of mental deficiency, cretinism, or mongolism. A thyroglossal sinus, running from the region of the foramen caecum through the muscles of the tongue to end blindly somewhere in the midline of the neck, represents a non-obliterated portion of the embryonic thyroglossal duct. A greater or lesser amount of thyroid tissue may be present in the posterior part of the tongue in addition to a normally situated thyroid gland, but these rare cases more commonly have a lingual thyroid with absence of thyroid tissue elsewhere in the neck.

SALIVARY GLANDS

All the glands of the digestive system are developed as buds or outgrowths of the lining epithelium of the embryonic gut.

They have an excreting part—the **ducts**—and a secreting part, which may be tubular as in the gastric and intestinal glands, or enlarged into **alveoli** or **acini**. They are classified as shown in FIGURE 490.

Salivary glands secrete saliva which is a fluid containing protein, salts, mucin, the enzyme ptyalin, and some desquamated epithelial cells and leucocytes. Saliva aids mastication, lubricates, and has a protective action on the oral mucous membrane and teeth. It acts as a pH buffer, and its enzyme ptyalin initiates the preliminary breakdown of the carbohydrates of the food.

All salivary glands are compound **racemose** glands with serous, mucous, or mixed alveoli [Figs. 493, 494]. There are three pairs of large glands—the **partoid,** the **submandibular** and the **sublingual**—and numerous small glands in the palate, the tongue, the lips, and the cheeks, which open directly on to the surface of the overlying oral mucous membrane. The **labial** and **buccal** glands have mostly mixed alveoli, and those in the vicinity of the opening of the partoid duct—the **molar**

Simple tubular gland   Simple alveolar glands

Compound tubular gland   Compound alveolar gland

Tubulo-alveolar gland

FIG. 490 Various types of glands.

glands—represent a group of orbital glands, highly developed in some of the carnivores. The **palatine glands** are independent aggregates, some in the hard palate, but most in the soft palate, and others in the uvula. They are mostly mucous glands. The **lingual glands** are also mainly mucous, but the anterior lingual glands have mixed alveoli. The glands related to the vallate and foliate papillae, sometimes known as the glands of von Ebner, are serous glands.

Parotid Glands [PLATE 28A]

The parotids, the largest of the salivary glands, are serous glands. They are situated on the side of the face in front of the ear and extend from the zygomatic arch above to just below the angle of the jaw. A large part of the gland is superficial, lying on the masseter under the skin and fascia, but its substance penetrates deeply into the space between the ramus of the mandible and the external acoustic meatus [FIG. 491]. It is enclosed in a fascial capsule attached above to the zygomatic arch, and a deep layer thickened to form the **stylomandibular ligament**. The superficial part of the gland is triangular, extending forwards on the masseter and back to the mastoid process and the anterior border of the sternocleidomastoid. There are two or three lymph nodes on the surface of the gland or embedded in its substance, draining the anterior part of the scalp and the face above the level of the mouth. The **parotid duct** emerges from the anterior border, runs transversely across the masseter at the level of a line drawn from the lobule of the ear to a point midway between the ala of the nose and the red margin of the lip, and turning medially, pierces the substance of the cheek at the anterior border of the masseter to open into the mouth opposite the second upper molar tooth. A separate aggregate of lobules, known as the accessory part of the parotid gland, is associated with the upper border of the duct. If the teeth are clenched the duct can be palpated in the living subject against the tense masseter.

The deep part of the gland has an anteromedial and a posteromedial aspect. The anteromedial aspect is related to the posterior border of the mandible, with a deep part extending forwards between the medial pterygoid muscle and the deep surface of the ramus of the mandible. The posteromedial surface is related to the anterior part of the mastoid process and, below, to the anterior border of the sternocleidomastoid muscle. Deep to the parotid is the posterior belly of the digastric and the styloid process and its muscles separating the gland substance from the internal carotid artery and internal jugular vein. Above, the gland is moulded round the temporomandibular joint, and lobules of the gland occupy the non-articular part of the glenoid fossa posterior to the head of the mandible.

Structures within the Gland. The most superficial is the facial nerve, dividing into its terminal branches which radiate out from the anterior border of the gland, along with the duct and the transverse facial artery. On a deeper plane, the retromandibular vein divides to join the facial and posterior auricular veins respectively, and deep to the vein is the external carotid artery and its terminal branches—the maxillary and superficial temporal arteries—the latter emerging from the superior border of the gland along with its vein, the temporal branch of the facial nerve, and the auriculotemporal nerve.

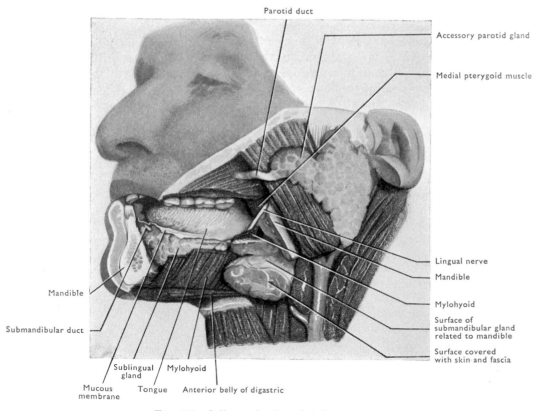

FIG. 491 Salivary glands and their ducts.

Blood Vessels, Lymphatics, and Nerves. The **arteries** are from the external carotid, and its branches : the **veins** join the retromandibular vein and its tributaries. The **lymph vessels** pass to the small parotid lymph nodes and to the superficial and deep cervical lymph nodes. Secretory **nerves** leave the brain stem with the glossopharyngeal nerve to reach the otic ganglion via the tympanic branch and the lesser petrosal nerve, and the postganglionic fibres are distributed to the gland via the auriculotemporal nerve. The postganglionic sympathetic fibres run in the external carotid plexus.

Submandibular Glands

The submandibular gland is smaller than the parotid gland and has more serous than mucous alveoli. It is situated in the submandibular triangle and fills the interval between the mandible and the medial pterygoid muscle laterally, and the mylohyoid and the hyoglossus muscles medially, reaching as far back as the parotid gland, from which it is separated by the **stylomandibular ligament** [FIG. 293]. There is a loose capsule of cervical deep fascia, one layer passing superficially over the gland to the lower border of the mandible, and a deeper layer behind the gland, attached to the mylohyoid line of the mandible.

Three surfaces of the gland are described—lateral, medial, and inferior. The **inferior surface,** covered by the platysma and fascia, is crossed by the facial vein and the cervical branch of the facial nerve. The submandibular lymph nodes are within the capsule or embedded in the substance of the gland. The **lateral surface** is related to the submandibular fossa of the mandible below the mylohyoid line, and more posteriorly to the medial pterygoid muscle. The **medial surface** lies on the

mylohyoid muscle with the mylohyoid artery and nerve between them, and posteriorly is in contact, from above downwards, with the lingual nerve and the submandibular ganglion, the hypoglossal nerve, and its accompanying vein, all lying on the hyoglossus muscle.

The facial artery grooves the posterior part of the gland before turning up across the lower border of the mandible. A small, **deep part** of the submandibular gland hooks round the posterior border of the mylohyoid, and extends forwards superior to that muscle, beneath the mucous membrane of the mouth, and on the hyoglossus with the lingual nerve above and the hypoglossal nerve below.

The **submandibular duct,** about 5 cm. in length, leaves the deep surface of the gland to run forward lateral to the hyoglossus and genioglossus muscles, deep to the mucous membrane of the mouth. It passes medial to the deep part of the gland and the sublingual gland, and opens on the summit of the sublingual papilla on the floor of the mouth at the side of the frenulum of the tongue [FIG. 475]. The lingual nerve loops round the duct, crossing it first superficially, then on its deep aspect. In the terminal part of its course the submandibular duct may receive a sublingual duct before it opens into the mouth. Like the parotid duct, the submandibular duct is narrowest at its origin and may to some extent act as a reservoir for saliva. Obviously, it is most accessible to the surgeon through the mucous membrane of the floor of the mouth.

Blood Vessels, Lymphatics, and Nerves. The supply is from the facial artery and its submental branch, and the veins accompany the arteries. The lymph vessels pass to the submandibular lymph nodes. The secretory nerves leave the

Sublingual gland   Lingual N.

Palatine tonsil

Lingual follicles

Submandibular gland (deep part)

Palatoglossus

Posterior margin of mylohyoid

Styloglossus

Hypoglossal N.

Lingual A.

Root of tongue

Gum

Sublingual papilla

Genioglossus

Geniohyoid

Mylohyoid

Sublingual gland

Submandibular duct

Mylohyoid

Hyoglossus   Hyoid bone

FIG. 492   Side-wall and floor of mouth after removal of tongue.

FIG. 493   Part of a microscopic section
through the submandibular gland.

FIG. 494   Part of a microscopic section
through the parotid gland.

brain stem with the facial nerve, run in the chorda tympani,
and relay in the submandibular ganglion from which the postganglionic fibres reach the gland.   Postganglionic sympathetic
fibres accompany the arteries.

Sublingual Glands

The sublingual glands are the smallest of the three paired
glands and are mainly mucous, but have some serous acini.
The glands are almond-shaped but are much larger (from
above downwards) anteriorly than posteriorly, and are 3–4 cm.
in length.   They are situated immediately below the mucous
membrane of the mouth, the upper border indicated by the
sublingual fold [FIG. 475].

Laterally, the gland lies in the sublingual fossa on the inner
surface of the mandible, and medially is in contact with the
genioglossus and hyoglossus muscles, with the submandibular
duct and the lingual nerve intervening.   The gland has 10–12
small ducts which open in linear series along the summit of
the sublingual fold, one of which may open into the submandibular duct.

The arteries are branches of the sublingual and submental
arteries, and the secretory nerves accompany those of the
submandibular gland.

Structure of the Salivary Glands

The three principal salivary glands are made up of lobules
within a framework of areolar tissue.   Each lobule consists
[FIG. 490] of a branching system of ducts terminating in
saccular or tubular secreting alveoli.   The ducts have a lining
of columnar epithelium.   The cells of a serous alveolus are
small, with a rounded nucleus, and may be filled with secretory
granules [FIG. 493, 494].   In the mucous alveolus, the cells
are larger, irregular in shape, with the flattened nucleus near
the base of the cell, due to the accumulation of mucinogen in
the cell cytoplasm.   In the fixed and stained sections the
droplets of mucinogen are usually destroyed and the cell
body appears clear.   In the submandibular gland there are
mixed alveoli, with both serous and mucous cells.   Here the
large, clear mucous cells line the cavity of the alveolus, and the
smaller serous cells are arranged on the basement membrane,
external to the mucous cells in the form of a crescent, termed
the **crescents** or **demilunes of Gianuzzi** [FIG. 493].   They
communicate with the cavity of the alveolus by small, intercellular channels.

Anomalies.   Occasionally either the submandibular or sublingual glands may be absent, and the sublingual glands in
particular vary considerably in size.   If the mylohyoid muscle

is deficient, as sometimes occurs, the submandibular gland may bulge through this opening into the cavity of the mouth. Cases have been described in which a submandibular duct opens into the pharynx near the isthmus of the fauces and, as has already been mentioned, it is not uncommon to find a duct of the sublingual gland as a tributary of the terminal portion of the submandibular duct.

THE PHARYNX

[PLATE 28B and C]

The pharynx, serving both respiratory and digestive systems, is situated in front of the basilar part of the occipital bone and the bodies of the upper six cervical vertebrae, from which it is separated by the prevertebral muscles and fascia and the loose areolar tissue of the retropharyngeal space. It is 13 cm. long, funnel-shaped, narrowing from a width of 3·5 cm. above to 1·5 cm. below, where it joins the oesophagus at the lower border of the cricoid cartilage.

The carotid sheath structures lie lateral to the posterior part of the pharynx and, superiorly, the ninth, tenth, eleventh, and twelfth cranial nerves, lying with the sheath, are separated from the medial pterygoid muscle on the ramus of the mandible by the styloid process with its attached muscles, and the posterior belly of the digastric muscle [FIG. 500]. Anteriorly, the pharynx communicates with the nasal cavities, the mouth, and the larynx.

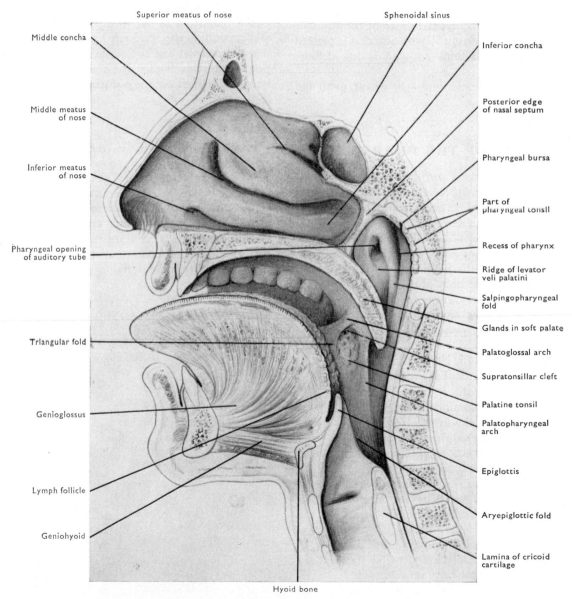

FIG. 495 Sagittal section through mouth, tongue, larynx, pharynx, and nasal cavity.

The section was slightly oblique, and the posterior edge of the nasal septum has been preserved. The specimen is viewed slightly from below; hence the apparently low position of the inferior concha.

The wall of the pharynx consists of an aponeurotic layer—the **pharyngobasilar fascia**—covered externally by the overlapping constrictor muscles (in the posterior and lateral parts of the wall) and their sheath of **buccopharyngeal fascia,** and internally by a mucous membrane which has a columnar ciliated epithelium in the nasal part, and stratified squamous epithelium elsewhere. The pharyngobasilar fascia is fixed above to the basilar part of the occipital bone and the petrous part of the temporal bone. Anteriorly, it is attached to the posterior border of the medial pterygoid plate of the sphenoid, the pterygomandibular raphe, the mylohyoid line of the mandible, the side of the tongue, the hyoid bone, and the thyroid and cricoid cartilages. It is strengthened behind by a fibrous posterior raphe which gives attachment to the constrictor muscles in the midline posteriorly, and is inserted on to the pharyngeal tubercle on the basi-occiput. The pharyngobasilar fascia is a thick, well-defined structure above, particularly between the upper border of the superior constrictor muscle and the base of the skull where it is pierced by the auditory tube, but it thins out and is less definite below.

There are three parts of the pharynx—the **nasal, oral,** and **laryngeal** [FIG. 495].

NASAL PART OF THE PHARYNX

The nasal part of the pharynx, with its rigid bony boundaries, is an open, air-filled cavity continuous anteriorly with the nasal cavities. The interval between the soft palate and the posterior pharyngeal wall can be obliterated by raising the soft palate, and in this way shutting off the nasal from the oral pharynx. The roof, below the base of the skull, slopes downwards and backwards to join the posterior wall lying in front of the arch of the atlas and the body of the axis. In front are the posterior apertures of the nose (choanae), separated by the edge of the nasal septum and bounded laterally by the medial pterygoid plates of the sphenoid.

The auditory tubes open on the lateral walls about 1·5 cm. behind the posterior end of the inferior conchae and immediately above the level of the hard palate. The opening has a prominent superior and posterior margin—the **tubal elevation**—formed by the cartilage of the tube. The **salpingopharyngeal fold** of mucous membrane with its corresponding muscle runs from the lower end of this elevation to the side wall of the pharynx. The slight bulge below the tubal opening is formed by the levator veli palatini muscle and is more prominent when that muscle contracts [FIGS. 495 and 497].

Behind the opening of the auditory tube the **pharyngeal recess** runs posterolaterally, below the petrous portion of the temporal bone on each side. The mucous membrane of the roof and the upper part of the posterior wall has an accumulation of lymphoid tissue—the **pharyngeal tonsil** or **adenoids**—arranged as a number of folds radiating forwards from a blind median recess—the **pharyngeal bursa**—as far as the opening of the auditory tube,

where it forms the **tubal tonsil.** The pharyngeal bursa is associated with the attachment of the embryonic notochord. Enlarged adenoids, common in children, may fill up practically the whole of the nasopharynx, blocking both the nasal cavities and the auditory tubes. The pharyngeal, palatal, tubal, and lingual tonsils together form a ring of lymphoid tissue in the upper pharynx. Pituitary gland tissue is often present in the mucous membrane of the nasopharynx in the infant, a reminder that part of the pituitary gland is developed from a diverticulum—Rathke's pouch—on the roof of the stomodaeum, just anterior to the pharyngeal bursa.

ORAL PART OF THE PHARYNX

The oral part of the pharynx lies behind the mouth, between the soft palate above and the inlet of the larynx below. In front is the isthmus of the fauces and the pharyngeal surface of the tongue, with the median and lateral glosso-epiglottic folds bounding the epiglottic valleculae [FIG. 498]. Laterally, the palatoglossal and palatopharyngeal arches with their respective muscles border the tonsillar fossa where masses of lymphoid tissue form the **palatine tonsil,** or simply the **tonsil** [FIGS. 496 and 500].

Palatine Tonsil

Like all lymphoid tissue, the tonsils tend to be large in earlier years and diminish at or after puberty. The free medial surface of the tonsils, which in children often projects prominently into the pharynx, is lined with stratified squamous epithelium and has a number of crypts encircled with lymph nodules underneath the lining epithelium [FIG. 498]. Superiorly, an intratonsillar or a **supratonsillar fossa,** representing part of the second pharyngeal pouch, penetrates deeply into the lymphoid tissue and may pass into the palate.

Pharyngeal opening of auditory tube

Tubal elevation

Palatoglossal arch

Palatopharyngeal arch

Salpingopharyngeal fold

Palatine tonsil

Triangular fold

FIG. 496 Median section of pharynx to show the right palatine tonsil. The palatine tonsil is large and divided below by an oblique cleft. The orifice of the supratonsillar cleft is near its upper pole, and the extent of the cleft in this specimen is indicated by a dotted line.

PLATE 27

A
(i)

(ii)

A. (i) Radiograph of maxilla at 7 years.

1. Crown of 1st premolar in its bony crypt.
2. 2nd deciduous molar.
3. Crown of 2nd permanent molar in its bony crypt.
4. 1st permanent molar.

(ii) Radiograph of half of mandible at 7 years.

1. Deciduous 1st molar.
2. Crown of permanent canine in its bony crypt.
3. 1st permanent molar.
4. Mental foramen.

B

B. Radiograph (intra-oral) of mandibular molars at 7 years (mesial to left).

1. Deciduous 2nd molar, its roots showing evidence of absorption.
2. 2nd premolar in its bony crypt, only the crown is completely calcified.
3. Permanent 1st molar, its roots have not yet been completely formed.

C

C. Radiograph (intra-oral) of mandibular premolar and molars at 14 years (mesial to left).

1. 2nd premolar, its root has not yet been completely formed.
2. Lamina dura (compact bone) of socket.
3. Periodontal space, for periodontal membrane.
4. Radiopaque filling in crown of 1st molar.
5. Bone trabeculae of interalveolar septum.

D

D. Radiograph of mandibular first permanent molar.

1. Enamel.
2. Dentine.
3. Pulp chamber.
4. Root canal.

PLATE 28

A

B

A. Parotid sialogram produced by injection of X-ray opaque material into the oral opening of the duct.

    1. The temporomandibular joint.
    2. Accessory ducts.
    3. Main duct branching into the substance of the gland.
    4. The third cervical vertebra.
    5. The hard palate.
    6. The opening of the main duct into the mouth.

B. Lateral X-ray of pharynx.

    1. Retropharyngeal tissue (nasal part of pharynx).
    2. Posterior arch of atlas.
    3. Posterior pharyngeal wall (oral part).
    4. Posterior pharyngeal wall (laryngeal part).
    5. Hard palate.
    6. Soft palate.
    7. Dorsum of tongue.

C

(i)   (ii)

D

C. Lateral (i) and anteroposterior (ii) X-rays of the pharynx after the mucosal lining has been coated with barium by a barium swallow.

    1. Dorsum of tongue.   4. Piriform fossa.
    2. Vallecula.   5. Oesophagus.
    3. Epiglottis.

D. Right oblique radiograph of the thorax with the oesophagus filled with barium.

    1. Impression by arch of aorta.
    2. Impression by left principal bronchus.
    3. Impression by left atrium of the heart, below which the diaphragm can be seen.

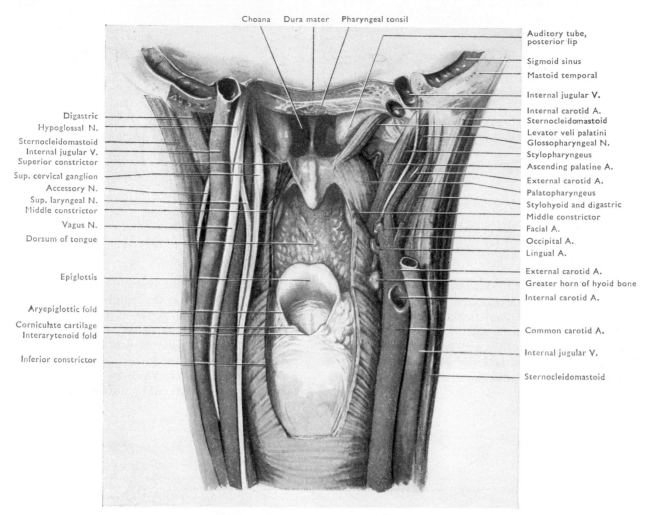

Choana   Dura mater   Pharyngeal tonsil

Auditory tube,
posterior lip
Sigmoid sinus
Mastoid temporal
Internal jugular V.
Internal carotid A.
Sternocleidomastoid
Levator veli palatini
Glossopharyngeal N.
Stylopharyngeus
Ascending palatine A.
External carotid A.
Palatopharyngeus
Stylohyoid and digastric
Middle constrictor
Facial A.
Occipital A.
Lingual A.
External carotid A.
Greater horn of hyoid bone
Internal carotid A.
Common carotid A.
Internal jugular V.
Sternocleidomastoid

Digastric
Hypoglossal N.
Sternocleidomastoid
Internal jugular V.
Superior constrictor
Sup. cervical ganglion
Accessory N.
Sup. laryngeal N.
Middle constrictor
Vagus N.
Dorsum of tongue

Epiglottis

Aryepiglottic fold
Corniculate cartilage
Interarytenoid fold

Inferior constrictor

Fig. 497   Interior of pharynx and the structures in relation to its side-walls, viewed
from behind.

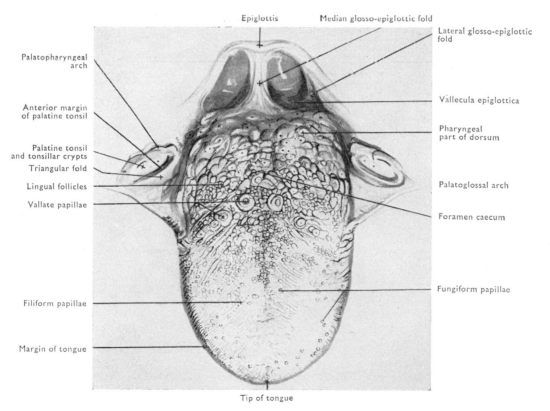

Epiglottis   Median glosso-epiglottic fold

Lateral glosso-epiglottic
fold

Palatopharyngeal
arch

Anterior margin
of palatine tonsil

Palatine tonsil
and tonsillar crypts
Triangular fold
Lingual follicles
Vallate papillae

Vallecula epiglottica

Pharyngeal
part of dorsum

Palatoglossal arch

Foramen caecum

Fungiform papillae

Filiform papillae

Margin of tongue

Tip of tongue

Fig. 498   Dorsum of tongue and the palatine tonsils.

FIG. 499   Diagrams of surroundings of palatine tonsil (Browne, 1928).
A, tonsil *in situ*.   B, after its removal.

In children, a **triangular fold** of mucous membrane extending across from the palatoglossal arch may cover the inferior part of the tonsil.   The lateral surface of the tonsil has a fibrous capsule attached to the side of the tongue below and separating it from the wall of the pharynx, formed here by the superior constrictor muscle and the pharyngobasilar fascia, and is held in place by an attachment of fibres of the palatopharyngeus muscle.

The appearance of the visible tonsil is not necessarily a true indication of its size, as the lymphoid tissue may extend into the palatoglossal arch as far as the dorsum of the tongue below and the soft palate above.   The main **artery** supplying the tonsil is the tonsillar branch of the facial artery, which pierces the superior constrictor muscle and enters the lower pole of the tonsil accompanied by two veins.   A large, paratonsillar vein—the **external palatine vein**—descending from

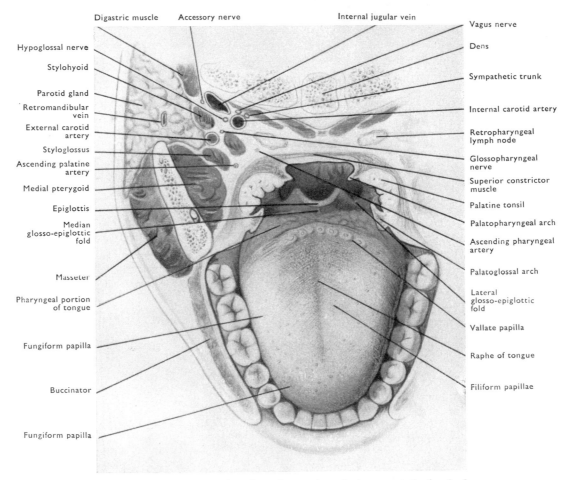

FIG. 500   Horizontal section through mouth and pharynx at the level of the palatine tonsils.

The prevertebral muscles and the stylopharyngeus (which is shown immediately to the medial side of the external carotid artery) are not indicated by reference lines.

the soft palate, is immediately related to the lateral surface of the tonsil before it pierces the superior constrictor to join the **pharyngeal venous plexus,** and is often the cause of haemorrhage in tonsillectomy [FIG. 499]. The ascending palatine artery, the facial artery, and the internal carotid artery are behind and separated from the tonsil by the muscular wall of the pharynx.

There is a tonsillar plexus of **nerves** formed by a branch of the glossopharyngeal nerve which lies against the pharyngeal wall at the level of the tonsil, and other branches from the pharyngeal plexus. The **lymph vessels** pass into the lateral surface of the tonsil and drain into the deep cervical nodes, in particular the jugulodigastric node just below the angle of the mandible [FIG. 1172].

LARYNGEAL PART OF THE PHARYNX

[PLATE 28]

The laryngeal part of the pharynx lies behind the entire length of the larynx. It is the longest part of the pharynx and narrows rapidly to the oesophagus with its slit-like lumen. The inlet of the larynx has the epiglottis anterosuperiorly, the arytenoid cartilages postero-inferiorly, and the aryepiglottic folds laterally [FIG. 497].

On each side of the orifice of the larynx is the mucosa-lined **piriform recess,** bounded medially by the lateral surfaces of the aryepiglottic folds, and laterally by the medial surfaces of the thyroid cartilage and the thyrohyoid membrane. Branches of the internal laryngeal nerve lie deep to the mucous membrane of the recess. Foreign bodies may lodge in the piriform recess, which represents the remains of the third and fourth pharyngeal clefts.

Below the laryngeal inlet, the mucous membrane covering the posterior surfaces of the arytenoid and cricoid cartilages forms the anterior wall of the pharynx, which at this level is a closed cavity opening only during the passage of food. Posteriorly, the laryngeal part of the pharynx is related to the bodies of the third, fourth, fifth, and upper part of the sixth cervical vertebrae.

Blood Vessels, Lymphatics, and Nerves of the Pharynx

The main **arteries** supplying the pharynx are the **ascending pharyngeal,** the **ascending palatine,** the **descending palatine** branch of the maxillary artery, and the **laryngeal branches** of the superior and inferior thyroid arteries. The **veins** form a plexus which is extensive in the posterior pharyngeal wall, and drains into the pterygoid plexus of veins above, and into the internal jugular or facial vein below. The **lymph vessels** of the pharynx pass directly to the deep cervical nodes, the posterior upper deep cervical being especially involved from the nasal part, or indirectly via retropharyngeal and pretracheal nodes. Both sensory and motor nerves are branches of the **pharyngeal plexus** of nerves formed by the vagus, the glossopharyngeal, and the sympathetic nerves.

SWALLOWING

There are two main phases of the act of swallowing. The first is a voluntary phase when the bolus of food is pushed into the pharynx, mainly by the movements of the tongue. The second phase is involuntary. It consists of the separation of the oral from the nasal part of the pharynx by the elevation of the soft palate, followed by the raising of the pharynx (this can be verified by the visible upward movement of the laryngeal prominence of the thyroid cartilage), and during this movement the muscles round the laryngeal inlet approximate the aryepiglottic folds, while at the same time the epiglottis is bent backwards over the inlet. The vocal cords are also closed during swallowing. Successive contractions of the three constrictor muscles move the food down the pharynx. At the lower end of the pharynx the cricopharyngeus muscle [p. 292], which acts as a sphincter, relaxes as the food passes into the oesophagus, where it is then propelled towards the stomach by peristaltic contraction of the oesophageal musculature [see also p. 420].

DEVELOPMENT OF THE PHARYNX

The pharynx is the cranial end of the primitive foregut, and for a short time is separated from the embryonic stomodaeum by the buccopharyngeal membrane. It is at first funnel-shaped, with a broad cranial end and a tapering caudal end. There is no splitting of the mesoderm surrounding the pharynx, which therefore remains in direct contact with the body wall so that the pharyngeal arches on its lateral and ventral aspects bulge into the floor and sides of the pharynx, forming elevations separated by grooves or clefts, which deepen laterally to form the pharyngeal pouches.

A lateral extension of the broad cranial end of the pharynx, involving the dorsal parts of the first and second pharyngeal pouches, forms the **tubotympanic recess,** the adult tympanic cavity and auditory tube. The prominent ventral ends of the first arch meet in the midline of the floor of the oronasal cavity, and are here separated from the first pharyngeal cleft by a central swelling, the **tuberculum impar.** The ventral ends of the third and fourth pharyngeal arches enter a central swelling, the hypobranchial eminence. The remaining pharyngeal arches are less prominent and end at the side of a median, ventral groove, the laryngotracheal groove. They contribute to the formation of the larynx and the lower pharyngeal wall.

The palatine tonsils are developed in the ventral parts of the second pharyngeal pouches, where part of the epithelium grows into the surrounding mesoderm as solid buds which are later excavated to remain as the **tonsillar crypts.** Lymphoid cells accumulate round these crypts to form the lymphoid follicles. The supratonsillar fossa is a remnant of the ventral part of the second pharyngeal pouch.

Outgrowths from the entoderm of the pharynx form the thyroid, parathyroid, and thymus glands. The thyroid outgrowth is in the form of a duct—the **thyroglossal duct**—just behind the tuberculum impar at the site of the adult foramen caecum. The thymus and parathyroids are developed as downgrowths from the third and fourth pharyngeal pouches [FIGS. 73 and 693], while the thyroid arises from the thyroglossal duct.

Anomalies

Congenital sinuses and fistulae of the pharynx generally arise in connexion with the development of the pharyngeal pouches and grooves. Most commonly a congenital fistula extends from the supratonsillar fossa to the skin of the anterior triangle of the neck. Median diverticula of the pharynx arise most commonly in the cricopharyngeal part of the inferior constrictor muscle, but may also occur further cranially.

THE OESOPHAGUS

[PLATE 28 D]

The oesophagus, or gullet, starts as the continuation of the pharynx at the cricoid cartilage opposite the sixth cervical vertebra, and is from 23 to 25 cm. in length (10 cm. in the new-born). It descends in front of the sixth and seventh cervical vertebrae and the upper ten thoracic vertebrae, then passes through the oesophageal opening in the diaphragm. It ends at the cardiac orifice of the stomach 1·5 cm. further on, behind the seventh left costal cartilage 2·5 cm. from the median plane.

The oesophagus follows the longitudinal curve of the vertebral column. In the midline at its origin, it curves slightly to the left to return to the midline at the level of the fifth thoracic vertebra. At the seventh thoracic vertebra it again curves to the left and forwards across the descending thoracic aorta to the diaphragm. The structures immediately in front of the oesophagus are the whole length of the trachea, to which it is connected by some loose tissue, the aortic arch to the left, the left bronchus, the right pulmonary artery, the pericardium posterior to the left atrium, and the diaphragm [FIG. 501]. The details of the relations of the oesophagus as it lies in the lower part of the neck, in the superior and posterior mediastinum, are shown in the series of transverse sections in FIGURE 502. Relations which should be noted in particular are the **recurrent laryngeal nerves** in the angles between the oesophagus and the trachea, the arches of the **vena zygos** and **aorta** in the superior mediastinum, the crossing of the thoracic duct from right to left posteriorly at the lower boundary of the superior mediastinum, and the **pleurae** and descending aorta in the posterior mediastinum.

A small infracardiac bursa is sometimes present to the right of the oesophagus just above the diaphragm. There is considerable variation in the detailed muscular arrangements of the oesophageal opening or hiatus in the diaphragm. Up to eight different types of hiatus have been described, most of them wholly within the right crus, but others with muscular slips from the left crus.

The abdominal part of the oesophagus, about 1·5 cm. long, is conical in shape, and its right border is directly continuous with the lesser curvature of the stomach, but the left border is separated from the fundus of the stomach by the cardiac notch. It is covered with peritoneum in front and on the left side, and anteriorly it lies in the oesophageal groove on the posterior surface of the left lobe of the liver. Behind is the left crus of the diaphragm and the left inferior phrenic artery.

The oesophagus is said to have four constrictions: (1) at its beginning 15 cm. from the incisor teeth; (2) where it is crossed by the arch of the aorta in the superior mediastinum 22·5 cm. from the incisor teeth; (3) where it is crossed by the left principal bronchus a little lower down; and (4) where it pierces the diaphragm 37–40 cm. from the incisor teeth (18 cm. in the new-born).

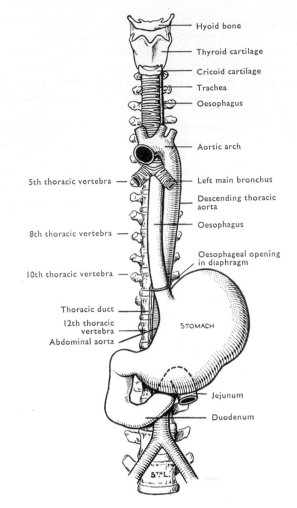

FIG. 501　Diagram to show course of oesophagus in relation to vertebral column, trachea, and aorta.

STRUCTURE OF THE OESOPHAGUS

The oesophagus [FIG. 503] has thick walls made up of an outer fibrous layer with some elastic fibres, a muscular layer, a submucous layer, and a mucous membrane. The outer fibrous layer is loosely arranged and permits considerable freedom of movement and variation in diameter.

The **muscular coat** consists of outer longitudinal and inner circular muscle fibres. In the human oesophagus, the upper third or so has striated muscle only, the lower third non-striated muscle only, and the middle portion a varying proportion of striated and non-striated muscle. The longitudinal muscle is thicker than the circular muscle and, above, its fibres diverge behind to form two longitudinal bands which are attached to the posterior surface of the cricoid cartilage, leaving a potentially weak area posteriorly where pouching or diverticula of the oesophagus may occur [FIG. 504]. Below, the longitudinal muscle of the oesophagus is continuous with that of the stomach. There may be muscular slips connecting the oesophagus to the left pleura, the bronchi, the trachea, the pericardium, and the wall of the aorta. The circular muscular layer is not so well developed as the longitudinal layer, and is continuous with the cricopharyngeal part of the inferior constrictor muscle above, and below with the circular and oblique

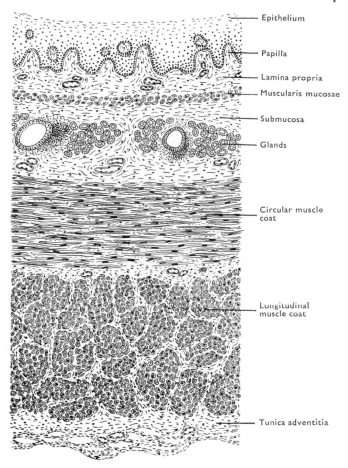

Epithelium

Papilla

Lamina propria

Muscularis mucosae

Submucosa

Glands

Circular muscle coat

Longitudinal muscle coat

Tunica adventitia

FIG. 503   Transverse section through the wall of the oesophagus.

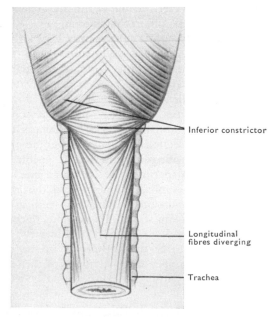

Inferior constrictor

Longitudinal fibres diverging

Trachea

FIG. 504   Arrangement of muscular fibres on back of oesophagus and pharynx.

1st thoracic vertebra

3rd thoracic vertebra

5th thoracic vertebra

8th thoracic vertebra

9th thoracic vertebra

FIG. 502   Tracings from frozen sections to show relations of oesophagus at the levels of the first, third, fifth, eighth, and ninth thoracic vertebrae.

1. At level of the superior part of the first thoracic vertebra, showing the chief relations of the oesophagus in the neck and also its divergence to the left.

2. At the third thoracic vertebra, showing the thoracic duct lying on the left side of the oesophagus. V, right vagus nerve.

3. At level of the fifth thoracic vertebra.   The left principal bronchus is seen in relation to the anterior surface of the oesophagus.

4. At level of the eighth thoracic vertebra, showing the oesophagus passing behind the pericardium.

5. At the ninth thoracic vertebra, showing the oesophagus inclining to the left just before piercing the diaphragm.

A, aorta ; C, common carotid artery ; D, diaphragm ; L.B., left principal bronchus ; L.C., left subclavian artery ; L.R., left recurrent laryngeal nerve ; L.V., left vagus ; OE, oesophagus ; P, pleura ; Pc, pericardium ; R.B., right principal bronchus ; R.R. right recurrent laryngeal nerve ; R.V., right vagus ; T, trachea ; T.D., thoracic duct ; V.A., vena azygos.

Traced upwards, the longitudinal muscle fibres of the oesophagus are seen to separate ; passing round to the sides, they form two longitudinal bands which meet in front and are united to the cricoid cartilage, as shown in Fig. 505.

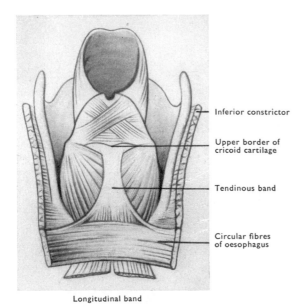

Inferior constrictor

Upper border of
cricoid cartilage

Tendinous band

Circular fibres
of oesophagus

Longitudinal band

FIG. 505 Relation of muscular coat of oesophagus to cricoid
cartilage and pharynx.

The pharynx and oesophagus have been slit up from behind
and the mucous membranes removed to show the muscular
fibres. The two longitudinal bands are seen passing by a common
tendon to the upper border of the cricoid cartilage. The
circular fibres are continuous with the inferior constrictor.

muscular fibres of the stomach [FIG. 506]. While the
cricopharyngeus acts as a sphincter at the commencement of the oesophagus there is no anatomical sphincter
at the gastric end, although radiological examination
shows that there is a functional sphincter with some
hold-up of the food before it enters the stomach, which
may be due to a local contraction of the circular muscle,
or possibly a sphincteric action of the muscle encircling
the oesophageal opening in the diaphragm.

There is a well-marked **submucous layer** of areolar
tissue with the blood vessels, nerves, and mucous
glands. The **mucous coat** is arranged in longitudinal
folds, hence the stellate lumen often seen in transverse
sections of the gullet. It consists of a lining of stratified
squamous epithelium, a lamina propria of connective
tissue, and a longitudinally arranged muscularis mucosae
more marked in sections from the lower oesophagus.
The **gastro-oesophageal junction** is visible to the
naked eye in fresh specimens as a crenated line separating the smooth oesophageal lining from the mammillated gastric mucous membrane, and indicating the
abrupt transition from stratified squamous to columnar
epithelium [FIG. 506]. Glands resembling the cardiac
glands of the stomach may be found in the mucous
membrane of the upper and lower parts of the oesophagus.

Blood Vessels, Lymphatics, and Nerves of the Oesophagus

The cervical oesophagus is supplied mainly by
branches of the **inferior thyroid artery.** The chief
blood supply of the thoracic oesophagus is either by

branches of the bronchial arteries, or direct from the aorta,
and there are branches of the left gastric and inferior phrenic
arteries running along the surface of the lower oesophagus
for 2–3 cm. above the diaphragm. This arterial supply is
variable, and the nature of the distribution of the blood vessels
has some bearing on the mobilization of the oesophagus in
surgical procedures. It is safe to mobilize the cervical oesophagus or the thoracic oesophagus from the arch of the aorta
through the diaphragm, but not the whole length of both
cervical and thoracic oesophagus. There is a submucous plexus
of a dozen or more longitudinal venous channels, with branches
penetrating the oesophageal wall at intervals to join the inferior thyroid and the vertebral veins above, the azygos and the
hemiazygos veins lower down, and the left gastric vein in the
abdomen. There are large submucous venous trunks forming
anastomotic channels between the left gastric vein and the
azygos vein at the lower end of the oesophagus, thus forming a
connexion between the portal and systemic venous systems.

Lymphatic vessels from the cervical part of the oesophagus
drain into the cervical nodes, those in the thoracic part into
the posterior mediastinal nodes, and those from the abdominal
part of the oesophagus into the left gastric lymph nodes. It
has been suggested that some lymph vessels may pass directly
into the thoracic duct.

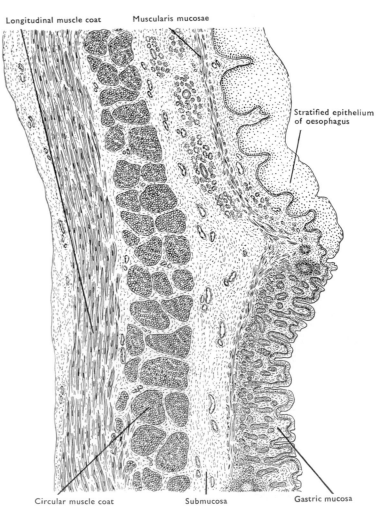

Longitudinal muscle coat   Muscularis mucosae

Stratified epithelium
of oesophagus

Circular muscle coat   Submucosa   Gastric mucosa

FIG. 506 Gastro-oesophageal junction.

The oesophagus receives its nerve supply from branches of the **vagus** and the **sympathetic.** The recurrent laryngeal branches of the vagus supply the striated muscle through a ganglionated plexus, and parasympathetic postganglionic and sympathetic postganglionic branches supply the smooth muscle. The vagus nerves, joined by branches from the thoracic sympathetic chain, the greater splanchnic nerves, and at the lower oesophagus by branches from the plexuses around the left gastric and inferior phrenic arteries, form an oesophageal plexus starting below the level of the tracheal bifurcation. Somewhere in the lower third of the oesophagus the plexus is reconstituted into two definite nerve trunks—the anterior and posterior vagal trunks—each with fibres from the right and left vagus nerves. Within the walls of the oesophagus as elsewhere in the gut, there is a myenteric and submucosal plexus with ganglionic cells. Stimulation of the vagus produces relaxation of the circular muscle at the cardiac orifice, i.e., the **physiological sphincter.** The operation of vagotomy, or section of the reformed vagal trunks, has been performed as one method of the operative treatment of gastric ulcer.

Gastro-oesophageal Junction

When empty, the oesophagus is collapsed. Fluid passes directly, with no hold-up, into the stomach, but with solid food there is a hold-up of about 2 seconds at the gastric end of the oesophagus. Moreover, there is normally no reflux of gastric contents into the oesophagus. There would therefore seem to be some controlling mechanism at the gastro-oesophageal junction. There are three possibilities.

1. The intrinsic circular muscle acts as a functional sphincter at this point.

2. That the muscular fibres of the oesophageal hiatus act as an external sphincter.

3. It has been suggested that a funnel-shaped process of gastric mucous membrane is pulled up by contraction of the muscularis mucosae and acts as a one-way valve.

Anomalies

A short oesophagus is usually accompanied by the type of hiatal hernia where the gastro-oesophageal junction and part of the stomach herniates through the diaphragmatic oesophageal hiatus into the thorax.

Congenital Atresia. There is a narrowing and complete occlusion of a segment of the gullet. A contributory factor here is the coalescence of the mucous membrane of the foetal oesophagus, which for a period has no lumen.

Tracheo-oesophageal Fistula. This is said to be the most common congenital abnormality of the oesophagus. The trachea develops as an outgrowth from the ventral part of the primitive foregut, indicated by a groove—the **laryngotracheal groove**—and the condition is due to interference with the normal processes of separation. Of several varieties of tracheo-oesophageal fistula the most common is that in which the upper segment of the oesophagus ends blindly while a lower segment arises from the trachea.

THE ABDOMINAL CAVITY

The abdominal cavity consists of an upper, larger part, the **abdomen** proper, continuous below with the cavity of the lesser pelvis. The long axis of the abdominal cavity is vertical, while the cavity of the lesser pelvis is directed downwards and backwards. Owing to the obliquity of the bony pelvis in the upright position, most of the pelvic cavity lies above the level of the symphysis pubis and is thus bounded in front by the lower part of the anterior abdominal wall. The walls of the abdominal cavity, anterior and posterior, and those of the pelvis, including the muscular pelvic diaphragm separating it from the perineum, are described elsewhere.

As a way of describing the position of the organs it is customary to subdivide the abdominal cavity into nine separate regions, mapped out on the surface by two vertical and two horizontal planes. The two vertical planes—right and left—are indicated by a perpendicular projected up from a point on the inguinal ligament midway between the symphysis pubis and the anterior superior iliac spines—the **midinguinal point.** The horizontal planes are the transtubercular between the tubercles of the iliac crests at the level of the fifth lumbar vertebra, and the transpyloric plane through a point midway between the suprasternal notch and the symphysis pubis, which more or less corresponds with the midpoint between the xiphoid process and the umbilicus. The regions are named as shown in FIGURE 507. It should be noted that the right and left hypochondriac regions are under the shelter of the lower ribs, extending during expiration as far up as the fifth rib in the right lateral plane, and a little lower on the left lateral plane.

While the position of the **umbilicus** may vary, it is relatively constant in a reasonably firm abdominal wall, and provides a useful landmark indicating the level of the disc between the third and fourth lumbar vertebrae. Posterolaterally, there is an average of two to three finger-breadths between the iliac crests and the eleventh and twelfth ribs, so that the abdominal cavity has considerable protection from the skeleton except in front. The vertebral column projects well forward in the midline posteriorly, its anterior surface being nearer to the anterior than to the posterior surface of the body in the lower abdomen of spare individuals, and a transverse section of the abdominal cavity is kidney-shaped with deep posterolateral recesses.

THE CONTENTS OF THE ABDOMEN

The serous lining of the abdominal cavity is the **peritoneum,** which covers the abdominal organs in whole or in part, and lubricates their surfaces by the secretion of a little fluid, so preventing friction during the constant movement and shifting of the viscera which occurs during respiration and the activity of the organs themselves. The abdominal organs develop external to the peritoneum but some of them, particularly those of the digestive system, grow in the abdominal cavity

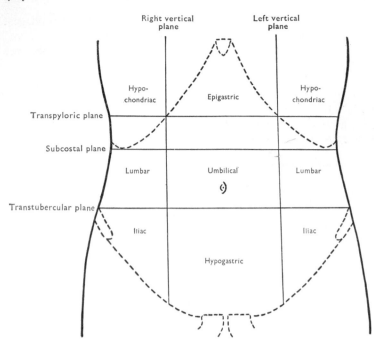

FIG. 507   Planes of subdivision of the abdomen proper, with the names of the nine abdominal regions.

and are surrounded by the peritoneum. Thus, they lie in the **peritoneal cavity,** have a serous covering of visceral peritoneum, and are fixed to some part of the abdominal wall or adjacent organs by sheets of connective tissue between two layers of peritoneum and sometimes referred to as 'ligaments'. The serous lining of the abdominal wall is the **parietal peritoneum.** A peritoneal ligament of the intestines is known as a **mesentery,** and of the stomach as an **omentum.** The vessels and nerves run to the gut wall in these peritoneal ligaments. Fatty tissue tends to accumulate in the mesenteries and omenta in later life.

The abdominal cavity contains the digestive and genitourinary systems, the suprarenal gland, and the spleen. In a general way the digestive organs lie in front of the genitourinary organs within the abdominal cavity proper, but the position is reversed in the pelvis. The parts of the digestive system within the abdominal cavity are the **liver,** the **pancreas,** and the **gastro-intestinal tract** which has the following parts —the stomach; 6 m. or so of small intestine made up of 25 cm. of duodenum, the next two-fifths of jejunum and the remaining three-fifths of ileum, and about 1·8 m. of large intestine, comprising the vermiform appendix, the caecum, the ascending colon, the right colic flexure, the transverse colon, the left colic flexure, the descending colon, the sigmoid colon, the rectum, and the anal canal [PLATE 31].

The **liver,** with a large right lobe and smaller left lobe, fills most of the upper part of the abdominal cavity—the hypochondriac and epigastric regions. Its lower border extends across the epigastric region from the ninth right costal cartilage to the eighth left costal cartilage. The gall-bladder [p. 459] is attached to the inferior surface of the liver, and its fundus lies at the level of the ninth right costal cartilage, where the lateral edge of the rectus abdominis crosses it.

A triangular peritoneal fold—the **falciform ligament**—passes from the deep surface of the anterior abdominal wall in the midline, above the umbilicus, to the anterior surface of the liver, a little to the right of the midline. In its lower free edge there is a fibrous cord—the **ligamentum teres**—passing to the inferior surface of the liver, and representing the remains of the blood vessel—the **umbilical vein**—which brought the blood back from the placenta to the foetus. The liver is in contact with and moves with the diaphragm, to which it is attached by short but broad peritoneal ligaments.

The **stomach** lies below and behind the left lobe of the liver. Its size and position are related to the particular body build, but also vary markedly in the same individual in different postures [PLATE 29]. From the inferior, convex border of the stomach (the **greater curvature**)—an apron-like fold of peritoneum—the **greater omentum**—hangs down in front of the coils of small intestine. The greater omentum has high absorptive powers and a tendency to form adhesions in any inflammatory areas. From its superior concave border—the **lesser curvature**—the stomach is attached to the liver by the **lesser omentum,** which has a right free border extending from the first 2–3 cm. of the duodenum to the inferior surface of the liver. Except for its first 2–3 cm., the **duodenum** is retroperitoneal, lying against the posterior abdominal wall, and can only be fully exposed after removal of the stomach and intestines. The coils of small intestine, covered in part by the greater omentum, lie principally in the umbilical and hypogastric regions against the deep surface of the anterior abdominal wall. Some coils of ileum are in the pelvic cavity.

The **jejunum** commences at the duodenojejunal junction above and to the left of the umbilicus, and the **ileum** terminates by joining the caecum in the right iliac fossa [FIG. 473].

The 1·5–1·8 m. of **large intestine** form three and a half sides of a square enclosing the coils of small intestine. The first part—the **caecum**—to which the vermiform appendix is attached, lies in the right iliac fossa and is continuous with the **ascending colon** which passes upwards through the right lumbar region to the under surface of the liver, where it turns to the left forming the right colic flexure, to become the **transverse colon.** The transverse colon extends across the abdomen towards the spleen in the left hypochondriac region, where it bends downwards at the left colic flexure as the **descending colon.** The descending colon runs vertically downwards to the iliac crest, thence to the superior aperture of the lesser pelvis, and curves into the lesser pelvis as the **sigmoid colon,** becoming continuous in front of the third sacral vertebra with the **rectum,** which terminates in the short **anal canal.**

High up in the left hypochondriac region, sheltered by the lower ribs, is the **spleen,** attached by peritoneal ligaments to the stomach and the left kidney. The **pancreas** lies against the bodies of the first and second lumbar vertebrae, extending across to the spleen. The **kidneys** lie on each side of the vertebral column on the posterior abdominal wall behind the peritoneum, and have the small **suprarenal glands** on their upper poles. In the lesser pelvis, the **rectum** lies against the sacrum, the **urinary bladder** is in front, and in the female the **uterus** with the broad ligaments containing the ovaries and uterine tubes, occupies a mid position, as do the seminal vesicles in the male.

The majority of the abdominal viscera are not fixed, and shift their position during respiration, when changing from the erect to the supine position or vice versa, and according to the state of distension of the different parts of the gastro-intestinal tract. The build and age of the individual also influence the position of the viscera, and a considerable range of variation is compatible with normal health and function and is not regarded today as likely to be the cause of ill health.

THE PERITONEUM

The peritoneum, like all serous membranes, has a basis of fibro-elastic tissue and a surface layer of flattened endothelial cells. The extraperitoneal tissue, which binds the peritoneum to underlying parietal structures, forms a thin layer in the anterior abdominal wall, but a thick mass

FIG. 509 Transverse section of abdomen at level of the opening into the omental bursa.

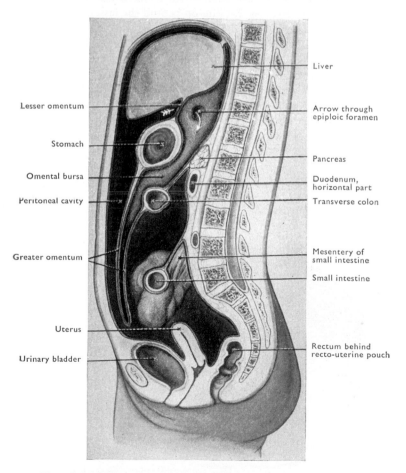

FIG. 508 Diagrammatic median section of female body to show abdominopelvic cavity and peritoneum on vertical tracing. The greater sac of peritoneum is blue and the omental bursa is red; both are represented as much larger spaces than in nature.

of tissue elsewhere, e.g., round the kidneys. The sac enclosed by the peritoneum—the **peritoneal cavity**—is closed, except for the openings of the uterine tubes in the female, and is normally completely filled by the intraperitoneal viscera, everywhere in contact with one another and with the abdominal walls, so that the 'fixed' organs in the embalmed subject have impressions made by neighbouring structures. The peritoneal cavity has a smaller diverticulum, or **omental bursa,** behind the stomach in the upper part of the abdominal cavity. The two parts communicate with one another through the **epiploic foramen,** or aditus, to the omental bursa, which can be found by passing a finger backwards and to the left behind the right free margin of the lesser omentum at the neck of the gall-bladder [FIG. 509].

The peritoneum, as a closed sac, forms a continuous membrane in both the horizontal and the vertical planes, but its arrangement is complicated, not only by the omental bursa, but by the reflexions of the visceral peritoneum as the various ligaments and folds. FIGURES 508 to 511 give some idea of the disposition of the peritoneum in the vertical plane and at different horizontal levels. It will be seen from FIGURE 508 that the greater omentum is made up of two anterior and two posterior layers of peritoneum enclosing a variable extent of the omental bursa, with the stomach in its anterior and the transverse colon in its posterior wall. The two posterior layers of the greater omentum above the transverse colon form a mesentery for this part of the gut—the **transverse mesocolon**—which is attached across the posterior abdominal wall at the level of the pancreas [FIG. 523]. The various reflexions of the peritoneum to the abdominal wall partially divide the peritoneal cavity into a number of compartments which are of some practical importance.

FIG. 510 Transverse section of abdomen immediately below the opening into the omental bursa.

rectus abdominis muscle. Lateral to the inferior part of the inferior epigastric artery is the **lateral inguinal fossa** with the deep inguinal ring in its floor, where a finger-like extension of the peritoneal cavity, the **processus vaginalis,** extended through the anterior abdominal wall during the descent of the testis [FIG. 646].

The **omental bursa** lies behind the stomach and lesser omentum, but in the adult does not usually extend lower than the transverse colon. The presence of a cavity of this kind behind the stomach enables it to undergo changes in size and position, and allows it some freedom of movement on the posterior abdominal wall. The omental bursa is described as hour-glass in shape, with a small superior and a larger inferior recess. The **epiploic foramen** is situated below the caudate process of the liver [FIG. 556], above the superior part of the duodenum, in front of the inferior vena cava, and behind the right margin of the lesser omentum. It opens into a **vestibule** which communicates directly with the superior recess, passing up behind the lesser omentum and the caudate lobe of the liver, and in front of the diaphragm. The vestibule communicates with the larger inferior recess through an aperture limited by superior and inferior **gastropancreatic folds** raised by the left gastric and common hepatic arteries respectively. These run behind the peritoneum and arch forwards to the cardiac and pyloric ends of the stomach. The inferior recess has the stomach and the anterior two layers of the great omentum in front, and the transverse mesocolon behind. It extends over to the left in front of the

The shelf-like projection of the transverse mesocolon incompletely divides the abdominal cavity into a supracolic and an infracolic part. The **supracolic space** lies between the diaphragm above and the transverse colon and its mesocolon below. The upper part of the supracolic space is partly divided into right and left **subphrenic recesses** by the peritoneal attachments of the liver to the diaphragm (falciform, coronary, and triangular ligaments). The posterior part of the right subphrenic recess forms a deep pouch between the liver in front and the right kidney and suprarenal gland behind, known as the **hepatorenal recess.** The **infracolic space** is divided into right superior, and left inferior parts by the attachment of the mesentery of the small intestine. The left infracolic space is in free communication with the pelvis. The appendix vermiformis [FIG. 539] normally lies in this compartment but may extend towards the right infracolic space behind the caecum.

The reflexion of the peritoneum on to the abdominal wall from the ascending and the descending parts of the colon forms the **paracolic sulci** [FIG. 511] which communicate with the pelvic cavity below. The right paracolic sulcus leads superiorly into the hepatorenal recess, which is the most dependent part of the peritoneal cavity in the recumbent position.

The parietal peritoneum of the infra-umbilical portion of the anterior abdominal wall is raised into five folds converging on the umbilicus. In the centre is the **median umbilical fold,** stretching from the apex of the bladder and containing the fibrous remnant of the urachus. Flanking this on each side is the **medial umbilical fold** formed by the peritoneum covering the obliterated umbilical artery (lateral umbilical ligament) running from the superior vesical artery to the umbilicus, and more laterally the **lateral umbilical fold** formed by the inferior epigastric artery before it enters the sheath of the

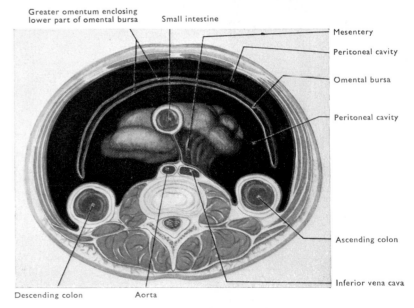

FIG. 511 Transverse section of abdomen through fourth lumbar vertebra.

pancreas and the left kidney as far as the hilus of the spleen (splenic recess), and is closed off by the gastro-splenic and lienorenal ligaments—a fold of peritoneum passing from the stomach to the left kidney with the spleen enclosed in its left wall—which form the left portion of the greater omentum (dorsal mesogastrium).

In the **male pelvis** the peritoneum sweeps off the front of the lower part of the rectum, passing anteriorly on to the back of the bladder, over the seminal vesicles, to form the **recto-vesical pouch,** and continuing over its superior surface to the anterior abdominal wall. In the **female,** since the uterus intervenes between rectum and bladder, there is the **recto-uterine peritoneal pouch** between the rectum posteriorly and the uterus anteriorly, and the **uterovesical pouch** between the uterus superiorly and the bladder inferiorly. Further details about the peritoneum will be given with the description of the individual organs.

THE STOMACH

[PLATE 29]

The stomach secretes digestive juices which have a chemical action on the food, and acts as a reservoir feeding the products of digestion at regular intervals into the next part of the gastro-intestinal tract—the **duodenum.** The stomach lies in the epigastric, umbilical, and left hypochondriac regions [FIG. 507]. Its shape and position not only vary in different individuals, but also in the same individual, depending on the nature and amount of the stomach contents, the condition of the adjacent intestines, and the posture, standing or lying down. The parts of the stomach are : the cardiac orifice where the oesophagus opens into the stomach ; a larger part to the left ; and a smaller pyloric part to the right, separated by an imaginary line running down and to the left from an **angular notch** at the lowest point of the lesser curvature of the stomach [FIG. 513]. In the empty stomach when in the recumbent position there is an uninterrupted smooth curve of the lesser curvature with no angular notch. The left portion is divided further into: (1)

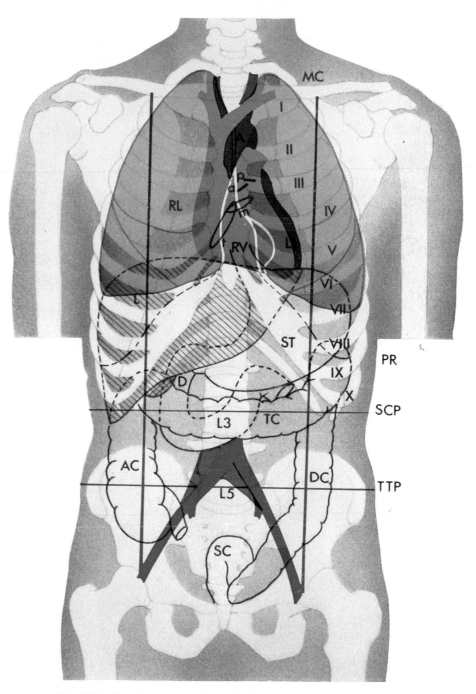

FIG. 512 Anterior aspect of trunk, showing surface topography of viscera.

A.	Aorta.	p.	Pulmonary orifice.
a.	Aortic orifice.	PR.	Opposite line of pleural reflexion.
AC.	Ascending colon.	RL.	Right lung.
D.	Duodenum.	RV.	Right ventricle.
DC.	Descending colon.	SC.	Sigmoid colon.
L.	Liver.	SCP.	Subcostal plane.
LL.	Left lung.	ST.	Stomach.
m.	Mitral valve.	t.	Tricuspid orifice.
MC.	Midclavicular line.	TC.	Tranverse colon.

TTP. Transtubercular plane.

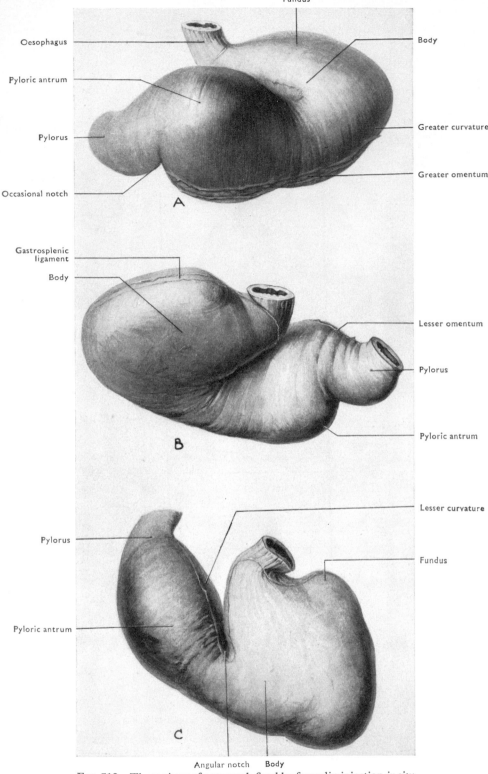

Fundus

Oesophagus

Pyloric antrum

Pylorus

Occasional notch

Body

Greater curvature

Greater omentum

A

Gastrosplenic ligament

Body

Lesser omentum

Pylorus

Pyloric antrum

B

Lesser curvature

Pylorus

Fundus

Pyloric antrum

C

Angular notch   Body

FIG. 513   Three views of a stomach fixed by formalin injection *in situ*.
A, from the front.   B, from the back.   C, from above.

The orientation of the stomach was determined by the insertion of long pins into it in the sagittal, coronal, and transverse planes. These views show the comparatively transverse position assumed by the stomach when the body lies supine. They also show the partial division into chambers produced by temporary constriction of the stomach wall fixed by the action of formalin.

The **pyloric part** has a wider pyloric **antrum** narrowing to the 2·5 cm. long pyloric **canal** which opens into the duodenum at the pyloric **orifice,** where there is a ring of thick muscle—the **pyloric sphincter.** The term pylorus applies to the area with the sphincter and is characteristically thicker to touch. The pyloric sphincter remains in tonic contraction, i.e., it is closed except for periodic emission of stomach contents.

The **cardiac orifice,** the most fixed part of the stomach, is 11 cm. behind a point 2–3 cm. to the left of the median plane on the seventh costal cartilage and corresponding to the level of the eleventh thoracic vertebra. The pyloric orifice, when the subject is lying down, is typically on the transpyloric plane about 1 cm. to the right of the lower border of the first lumbar vertebra, but it can have a 'normal' range of position from the twelfth thoracic to the third lumbar vertebra where it is situated, for example, in the common J-shaped type of stomach, in the standing position. The highest point of the fundus is at the fifth left intercostal space just below the nipple, and the lowest point of the greater curvature can be well below the level of the umbilicus. As the stomach fills up, the fundus expands vertically and also forwards, displacing the pylorus to the right so that the whole organ assumes a more oblique position. In the 'steer horn' type of stomach it lies almost transversely, even in the erect position [PLATE 29 A].

The average capacity of the stomach is 1 litre, but it is capable of considerable dilatation. In the new-born child the stomach is about as large as a hen's egg with a capacity of 30 ml. The full stomach not only becomes J-shaped but also enlarges to the right, and the pylorus may be displaced as much as 5 cm. to the right of the midline. The anterior surface of the body is partly behind the sixth to the ninth costal cartilages. In the living, the diphragm and the fundus move together while the pyloric end moves very slightly with the liver in respiration, but can vary in position from the first to the third lumbar vertebra on changing from the supine to the erect position.

the **cardiac part,** a rather indefinite area round the cardiac orifice; (2) the **fundus,** the part above the level of the cardiac orifice; and the remainder, known as (3) the **body** of the stomach.

The stomach is described as having two surfaces and two borders, but in the living subject, especially when full, it is more or less tubular in shape. Its anterior surface faces into the peritoneal cavity and is directed upwards as well as forwards. It is largely under cover of the left lobe of the liver, but is in contact with the diaphragm above and to the left, and with the anterior abdominal wall below the costal margin. Posteriorly it is directed downwards as well as backwards, facing into the omental bursa and resting on what is known as the stomach bed [FIG. 515]. This is formed by the posterior wall of the omental bursa with the transverse colon, the mesocolon, the gastric surface of the spleen, and on the posterior abdominal wall, the anterior surface of the body and tail of the pancreas, and the left kidney and suprarenal gland. The pylorus is related anterosuperiorly to the liver and postero-inferiorly to the pancreas [FIG. 514].

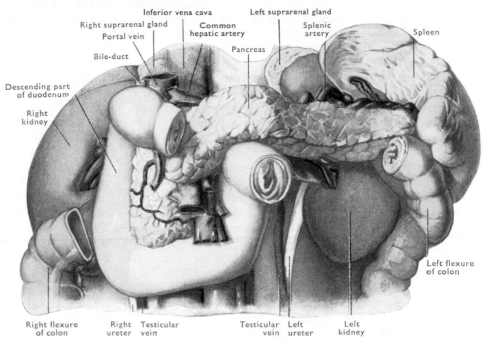

FIG. 514 Dissection to show posterior relations of stomach. The stomach has been cut close to the pylorus and removed. The transverse colon has been taken away, and the small intestine has been cut across close to the duodenojejunal flexure. (From a model by Birmingham.)

Peritoneal Relations of the Stomach

The stomach lies between the two layers of peritoneum forming the anterior wall of the omental bursa. To the left of the oesophageal orifice, a small area of the cardiac part of the stomach is directly in contact with the diaphragm, between the two layers of peritoneum which form the gastrophrenic ligament. The lesser curvature is attached to the liver by the lesser omentum which is very short near the oesophagus, but

FIG. 515 Stomach bed viewed from the front and from below.

becomes longer in the pyloric region and is directed vertically upwards. When the inferior recess of the omental bursa has a reduced vertical extent, the greater omentum is fused to the anterior surface of the transverse mesocolon and is known as the **gastrocolic ligament.** To the left, the greater omentum runs from the stomach to the spleen as the **gastrosplenic ligament.** It should be noted that the stomach is separated from the first part of the jejunum by the transverse mesocolon [FIG. 508].

Structure of the Stomach

The wall of the gastro-intestinal tract consists of four coats. From without inwards these are serous (peritoneum), muscular, submucous, and mucous. The mucous membrane is separated from the submucosa by a thin layer of muscularis mucosae.
Serous Coat. The serous coat of the stomach is formed by the peritoneum, which is closely attached to the subjacent muscular layer except near the curvatures where the connexion is looser.
Muscular Coat. The muscular coat is composed of plain muscle, thinnest in the fundus and body and thickest in the pyloric portion, and is responsible for the production of gastric movements in digestion. It is made up of three complete, or almost complete, layers—an external longitudinal, a middle circular, and an internal oblique layer.

The **longitudinal layer,** continuous with the longitudinal fibres of the oesophagus, sweeps from the cardiac orifice along and on each side of the curvatures, particularly the lesser curvature, but is almost or entirely absent on the central parts of the anterior and posterior surfaces. On each side of the lesser curvature it forms two strong bands, many of whose marginal fibres turn aside and terminate, at intervals, in the deeper layers. Towards the pylorus the longitudinal fibres become much thicker, and also much tougher and more closely united, and they take part in the formation of the pyloric sphincter.

The **circular layer,** continuous with the more superficial of the circular fibres of the oesophagus, is a more complete layer. Its fibres encircle the body from the lesser to the greater curvatures and they are absent only from the fundus. In the wall of the pyloric canal they increase in thickness and pass into the **pyloric sphincter.** The thickened circular muscle of the pyloric sphincter is, at any rate in part, separated from the circular muscle of the duodenum by a fibrous septum. A marked thickening or hypertrophy of the circular muscle in this region is found in the condition known as congenital hypertrophic pyloric stenosis, which occurs in infants.

The oblique fibres form the **inner layer** and are continuous above with the deeper circular fibres of the oesophagus. They are arranged in U-shaped bundles which loop over the stomach immediately to the left of the cardiac orifice and run very obliquely downwards on both surfaces of the organ as far as the angular notch [FIG. 517]. The looped fibres, as they pass to the left, gradually become less oblique and they finally form circular bundles around the superior part of the stomach as far as the fundus. These latter fibres are sometimes ascribed to the circular layer. There is some radiographic evidence that the contraction of the oblique musculature of the stomach forms

FIG. 516 Transverse section of trunk at level of second lumbar vertebra.

a **gastric canal** or gutter along the lesser curvature.

Submucous Coat. The submucous coat is a layer of strong but loose areolar tissue which lies between the muscular and mucous coats and unites them. It is more loosely attached to the muscular, and more closely to the mucous coat, and it forms a bed in which the vessels and nerves break up before they enter the mucous membrane.

Mucous Coat. In the fresh state the mucous coat is of a reddish-grey colour and of moderate consistency. After death the colour turns to a darker grey and the whole membrane becomes softer and more pulpy. It is thicker (over 2 mm.) and firmer in the pyloric part than elsewhere, and it is thinnest at the fundus. When the stomach is empty the mucous coat is thrown into numerous prominent folds or rugae which project into the interior. They are, in general, longitudinal in direction with numerous cross branches, and disappear when the stomach is distended. Along the lesser curvature there are three or four long, narrow, very

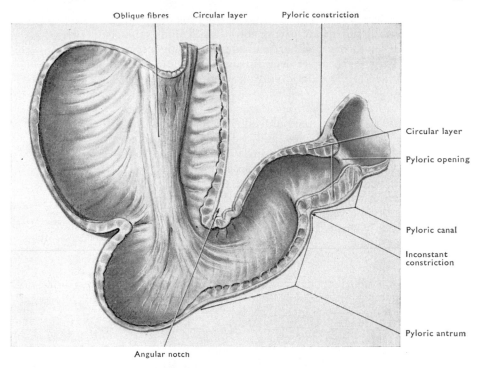

FIG. 517 Muscular coat of stomach, seen from within after removal of mucous and submucous coats (Cunningham, 1906). The anterior half of the stomach is shown, viewed from behind.

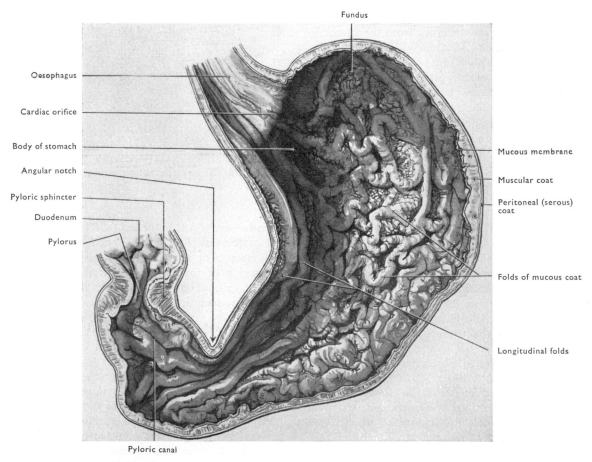

FIG. 518 Interior of stomach, showing its mucous coat and its parts when moderately filled.

constant ridges which pass without a break from the oesopha-geal orifice towards the pyloric part of the stomach [FIG. 518]. They are parallel to one another, and are apparently never obliterated. Between them are deep furrows which lead from the cardiac orifice to the pyloric portion of the stomach, the gastric canal.

When the surface of the mucous coat is examined in a fresh stomach it is seen to be marked out into a number of small, slightlyelevated , polygonal areas, and is consequently said to be mammillated [FIG. 519]. These little areas measure from 1 to 6 mm. in diameter and show numerous small **pits** about 0·2 mm. wide which are the mouths of the closely packed gastric glands. Although the gland mouths cannot be seen with the naked eye, a very slight magnification is sufficient to make them clearly visible. The mucous membrane of the stomach is continuous with that of the duodenum, although

Mouths of gastric pits, with gland tubes at bottom

Depression between two mamillae

Mouth of gastric gland

FIG. 519 Mucous membrane of stomach. The mouths of the pits and the ridges separating them are seen. ×25.

there is an abrupt change in its structure from the gastric to the duodenal type of mucosa. The mucous membrane also thickens in the region of the sphincter.

The mucous coat has a surface epithelial covering of tall columnar cells, and its thickness is occupied by numerous tubular **gastric glands** extending as far as the **muscularis mucosae,** a thin layer of longitudinal and circular muscle fibres separating the mucosa and submucosa.

Gastric Glands. The gastric glands are simple, branched, secreting tubules terminating in ducts lined with columnar epithelium which open in small groups into the gastric pits. The glands in the different regions of the stomach vary in structure.

The **cardiac glands** are found in the region of the oeso-phageal opening. Each gland is composed of a few branching tubules opening into a short duct, and the cells lining the tubules secrete a form of mucus.

Fundic glands in the fundus and body of the stomach have long secreting tubules with short ducts opening into shallow gastric pits [FIG. 520]. The main lining cells are the **mucous neck cells** in the upper part of the gland, and the **chief** or **zymogenic cells** in the remainder of the tubule. These are the cells which secrete pepsin, a proteolytic enzyme of the gastric juice.

At intervals throughout the length of the tubule there are large eosinophilic cells between the zymogenic cells, or, more typically, lying against the basement membrane and separated from the gland lumen by part of the zymogenic cells. These are the **parietal** or **oxyntic cells** responsible for the secretion of the hydrochloric acid of the gastric juice.

The glands of the pyloric part of the stomach—the **pyloric**

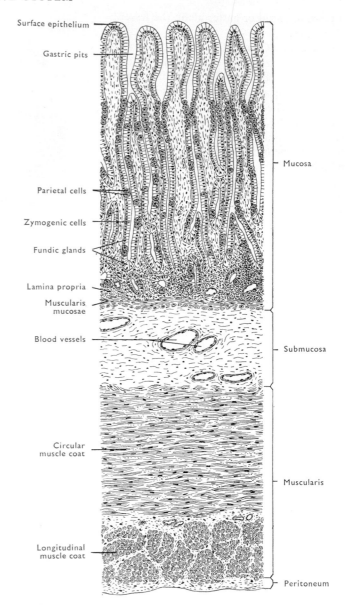

Surface epithelium

Gastric pits

Parietal cells

Zymogenic cells

Fundic glands

Lamina propria

Muscularis mucosae

Blood vessels

Circular muscle coat

Longitudinal muscle coat

Peritoneum

Mucosa

Submucosa

Muscularis

FIG. 520 Transverse section through the wall of the stomach (body).

glands—are short, branched tubules opening into long duct-like gastric pits [FIG. 521]. They have only one type of lining cell, resembling the mucous neck cell, although the glands in the region of the pyloric sphincter may have an occasional oxyntic cell. The exact contribution of the pyloric glands to the gastric secretion is not known.

Blood Vessels, Lymphatics, and Nerves of the Stomach

The **arteries** of the stomach are all derived from the **coeliac trunk.** The left gastric artery arises from that trunk direct, and, reaching the lesser curvature and giving off an oesophageal branch, it divides into two branches which run along the front and the back of the lesser curvature and join with two similarly disposed arteries derived from the right gastric branch of the common hepatic artery. From the two arches thus formed, four or five branches pass to each surface of the stomach and soon pierce the muscular coat. Along the greater curvature the **right** and **left gastro-epiploic arteries,**

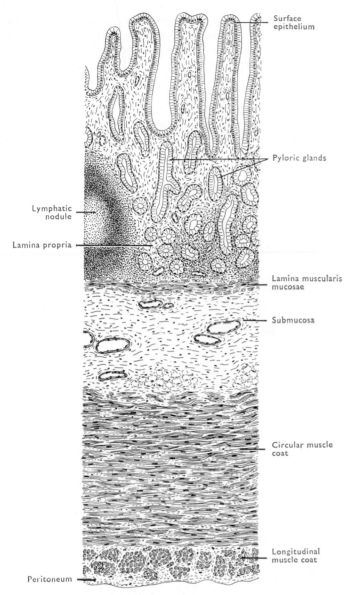

Surface
epithelium

Pyloric glands

Lymphatic
nodule

Lamina propria

Lamina muscularis
mucosae

Submucosa

Circular muscle
coat

Longitudinal
muscle coat

Peritoneum

FIG. 521 Transverse section through the wall of the stomach (pylorus).

which are branches of the gastroduodenal and the splenic arteries respectively, run in the greater omentum a short distance from its attachment to the stomach, and give branches to the surface of the stomach. Four or five **short gastric arteries,** branches of the splenic, are distributed to the fundus of the stomach, which they reach by passing forwards between the layers of the gastrosplenic ligament. At first the arteries lie beneath the peritoneum, but soon pierce the muscular coats and break up into a network of vessels in the submucosa from which small branches enter the mucous membrane and form capillary plexuses around the glands as far as the surface.

The **veins** begin in the capillary plexuses around the glands, uniting to form a network in the submucosa from which branches arise that pierce the muscular coat and end in the following veins : the **right gastro-epiploic** which joins the superior mesenteric vein ; the **left gastro-epiploic,** and four or five short gastric veins which join the splenic vein ; the

left gastric vein which runs along the lesser curvature from right to left and ultimately joins the portal vein. A small, right gastric vein accompanies the corresponding artery, and enters the portal vein. These veins contain numerous valves which, though competent to prevent the return of blood in the child, are rarely so in the adult.

There is a dense **lymphatic** plexus in the gastric submucosa. This communicates with intermuscular lymph vessels which in turn drain into a subserosal network. This network drains by numerous small channels into larger vessels accompanying the four chief arteries along the lesser and greater curvatures, and there are four areas of lymph drainage of the stomach, each with its own regional lymph nodes. The largest area, taking in the whole of the lesser curvature and a large part of the body of the stomach, drains with the left gastric artery to the **left gastric nodes** along the artery.

The pyloric portion of the stomach sends its lymph vessels with the right gastro-epiploic artery to **right gastric nodes** lying in relation to that artery. Of special importance are the **pyloric nodes** which send their efferents to the nodes about the coeliac axis. The fundus and upper portion of the body drain to **pancreaticosplenic nodes** situated along the splenic artery from the hilus of the spleen. A small part of the distal end of the pylorus drains into nodes on the upper border of the head of the pancreas, or into nodes along the bile-duct.

The **nerves** to the stomach are branches of the **anterior** and **posterior vagal trunks.** These are mainly parasympathetic preganglionic fibres. The postganglionic sympathetic fibres come from the coeliac plexus and are distributed along the gastric arteries. Most of the nerves of the stomach tend to be concentrated on the left gastric artery and follow it along the lesser curvature. The pylorus, however, is also innervated by both vagal and sympathetic fibres in the plexus on the common hepatic and right gastric arteries. It is generally stated that stimulation of the vagus increases and accelerates both the movement of the stomach and the secretion of acid, while stimulation of the sympathetic produces inhibition of contraction. There is no evidence that there is reciprocal innervation of the pyloric sphincter. Because of its action, section of the vagus, or vagotomy, may be carried out as part of the treatment of gastric and duodenal ulcers, and is said to result in a lowering of the secretion of acid. The sensory fibres in the stomach run in the splanchnic nerves. As elsewhere in the gastro-intestinal tract there are intramuscular (myenteric) and submucous plexuses, and those of the stomach consist of both sympathetic and vagal fibres. The associated ganglia are parasympathetic. The pyloric sphincter is concerned with the passage of the products of digestion from the stomach into the duodenum, and its relaxation and contraction depend on the presence or absence of free hydrochloric acid in the pyloric part of the stomach or duodenum.

Anomalies

These are relatively rare in the stomach. The stomach may be congenitally herniated through the diaphragm as a thoracic stomach and may be involved in oesophageal hiatal hernias. The most common anomaly is congenital hypertrophic pyloric stenosis due to the hypertrophy of the circular muscle fibres of the pyloric sphincter which may be from 5 to 10 mm. thick in such an infant.

THE SMALL INTESTINE

DUODENUM

[PLATES 29-30]

The duodenum is the first 25 cm. and the widest part of the small intestine, with a diameter of 4–5 cm. Apart from the first 2·5 cm. it lies on the posterior abdominal wall behind the peritoneum. The ducts of the liver and pancreas open into the duodenum which has glands situated in its submucosa. It forms a curve on the upper three lumbar vertebrae, beginning at the pylorus to end at the duodenojejunal junction on the left side of the second lumbar vertebra and enclosing the head of the pancreas [FIG. 522].

Four parts of the duodenum are described. **Superior Part.** The superior part is about 5 cm. long. At its commencement, indicated externally by a groove often occupied by a vein, there is an abrupt transition from the thick walls of the pylorus to the thin walls of the duodenum. The first 2·5 cm. of the duodenum has the same peritoneal relations as the stomach and is suspended from the liver by the lesser omentum with the greater omentum attached to its lower border. The omental bursa separates the posterior surface from the pancreas. The remainder passes backwards behind the peritoneum and forms the lower boundary of the epiploic foramen. The anterior surface is related to the quadrate lobe of the liver and the neck of the gall-bladder. The omental bursa separates the posterior surface from the pancreas. At its upper border the hepatic artery, the bile-duct, and the portal vein lie in the right free margin of the lesser omentum. As the hepatic artery enters the lesser omentum, the right gastric artery runs to the left. The gastro-duodenal artery descending behind this part of the duodenum, with the bile-duct on its right and the portal vein posteriorly, divides at its lower border into the right gastro-epiploic and the superior pancreaticoduodenal branches [FIG. 1092]. The duodenum immediately adjacent to the pylorus is usually prominent in radiographs following a barium meal, and is known as the **duodenal cap** [PLATE 29 D]. **Descending Part.** This part, 7·5 cm. long, descends behind the peritoneum to the right of the first, second, and third lumbar vertebrae, lying on the right suprarenal gland, the right kidney with its vessels and ureter, and the right psoas

major muscle. In front it is crossed by the transverse mesocolon or the transverse colon, so that there is an upper supramesocolic part related to the right lobe of the liver, and a lower inframesocolic part related to the coils of jejunum [FIG. 523]. The medial surface is overlapped by the head of the pancreas with the anterior and posterior pancreaticoduodenal arterial arcades. Inferior to the superior part of the duodenum, the bile-duct is embedded in the back of the head of the pancreas. Here, joined by the pancreatic duct, it pierces obliquely the postero-

FIG. 522   Anterior aspect of trunk, showing surface topography of viscera.

A.	Arch of aorta.	O.	Oesophagus.	RK.	Right kidney.
D.	Duodenum.	P.	Pancreas.	S.	Spleen.
IVC.	Inferior vena cava.	PR.	Line of pleural	SR.	Suprarenal.
LL.	Left lung.		reflexion.	U.	Ureter.

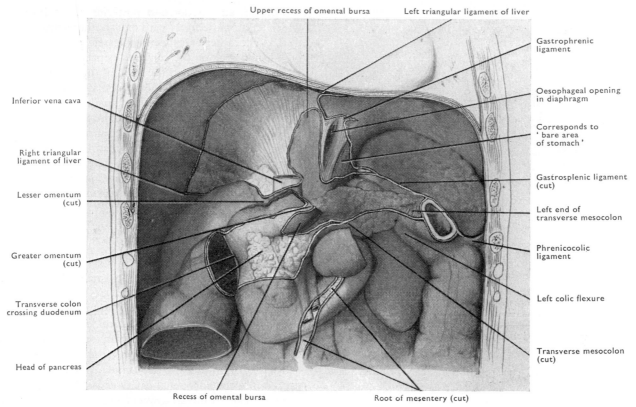

Upper recess of omental bursa · Left triangular ligament of liver

Gastrophrenic ligament

Oesophageal opening in diaphragm

Corresponds to 'bare area of stomach'

Gastrosplenic ligament (cut)

Left end of transverse mesocolon

Phrenicocolic ligament

Left colic flexure

Transverse mesocolon (cut)

Inferior vena cava

Right triangular ligament of liver

Lesser omentum (cut)

Greater omentum (cut)

Transverse colon crossing duodenum

Head of pancreas

Recess of omental bursa · Root of mesentery (cut)

FIG. 523 Peritoneal relations of duodenum, pancreas, spleen, kidneys, etc. From a body hardened by injection of formalin. When the liver, stomach, and intestines were removed the lines of the peritoneal reflexions were carefully preserved. The peritoneum is coloured blue.

medial wall of the duodenum about the middle of its descending part, to open into it through the **greater duodenal papilla** 8–10 cm. from the pylorus, and 2–3 cm. inferior to the opening of the accessory pancreatic duct on the **lesser duodenal papilla.**

Horizontal Part. The horizontal part, 7·5 cm. long, crosses the third lumbar vertebra behind the peritoneum and in front of the right ureter, the right psoas major muscle, the right gonadal vessels, the inferior vena cava, and the aorta. Its anterior and inferior surfaces are separated from the coils of intestine by the peritoneum, which passes superiorly on to the pancreas to be reflected as the posterior layer of the transverse mesocolon. The superior mesenteric vessels within the root of the mesentery cross in front of this part of the duodenum, and its superior surface is applied to the head of the pancreas [FIG. 514].

Ascending Part. The ascending part, about 5 cm. long, turns up the left side of the aorta to the level of the second lumbar vertebra where it terminates by bending abruptly forward as the duodenojejunal flexure. Behind it lie the left sympathetic trunk, the left psoas major muscle, and the left gonadal vessels. The root of the mesentery is attached in front of this part of the duodenum, while the left kidney lies laterally and the uncinate process of the pancreas medially. In about 40 per cent. of individuals the horizontal and ascending parts lie in a straight line passing upwards and to the left.

Duodenojejunal Flexure. This flexure is fixed in position by the suspensory muscle of the duodenum, blending below with the muscular coat of the duodenum, and above passing behind the pancreas to the right crus of the diaphragm. The position of the duodenum varies according to the posture, erect or supine, and the relative amounts of stomach and intestinal contents. The superior part is the most mobile, with a range of movement from the first to the third lumbar vertebra. The duodenojejunal junction is the least mobile and tends to move with the diaphragm.

Peritoneal Recesses. A number of peritoneal recesses [FIG. 524] are found in the region of the duodenojejunal junction, and their formation is related to the rotation of the developing gut. The **inferior duodenal recess,** found in over 50 per cent. of individuals, is formed by a fold of peritoneum extending from the left surface of the terminal portion of the duodenum to the peritoneum of the body wall—the **inferior duodenal fold.** It is the largest of the recesses, with a depth of a centimetre or more, and opens upwards. The **superior duodenal recess** is present in 30 per cent. of individuals, situated above the inferior recess and formed by the superior duodenal fold passing from the duodenum to the peritoneum of the posterior abdominal wall. This recess opens downwards. The **paraduodenal recess** is most commonly found in infants. It is on the left side of the duodenojejunal junction and is formed by a fold of peritoneum connecting the left margins of the superior and inferior duodenal folds, and often enclosing a blood vessel, usually the inferior mesenteric vein with a branch of the left colic artery; it opens to the right. The importance of these recesses is in connexion with internal hernia, now generally regarded as a form of congenital malformation

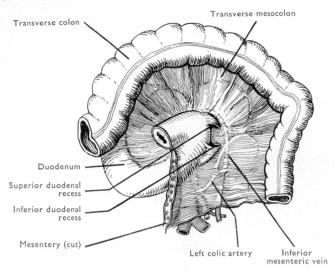

FIG. 524　Duodenal folds and recesses.

The transverse colon and mesocolon have been thrown up, and the mesentery has been turned to the right and cut. The paraduodenal recess is situated to the medial side of the inferior mesenteric vein, between it and the terminal part of the duodenum. It is not shown in the illustration.

occurring at the time of the rotation of the primitive gut [p. 452].

Blood Vessels, Lymphatics, and Nerves of the Duodenum

The main blood supply to the duodenum is from the **superior** and **inferior pancreaticoduodenal** branches of the gastroduodenal and superior mesenteric arteries respectively. These form anterior and posterior arterial arcades which lie in the angle between the duodenum and the pancreas anteriorly and posteriorly, and supply both the duodenum and the head of the pancreas. Surgical mobilization of the duodenum without interference with its blood supply is therefore carried out by incising the peritoneum to the right of the descending part of the duodenum. The superior part of the duodenum may have in addition the rather inconstant supraduodenal artery arising from the hepatic and running over the front of the gut. The veins for the most part follow the arteries, draining into the superior mesenteric, and a few of them running upwards directly into the portal vein. The superior part of the duodenum has numerous small veins on its anterior and posterior surfaces, some of them draining into the superior pancreaticoduodenal veins. One on the anterior surface is known as the **prepyloric vein** and runs upwards to drain into the portal vein. It is said to demarcate the position of the pyloric orifice.

There is an anterior and posterior set of **lymph vessels** which anastomose freely with one another within the duodenal wall. The anterior collecting vessels drain upwards to the lymph nodes along the common hepatic artery, and the posterior ones drain behind the head of the pancreas down to the nodes along the superior mesenteric artery.

The sympathetic and parasympathetic **nerves** from the coeliac and mesenteric plexuses accompany the arteries. There are one or two duodenal nerves coming off directly from the anterior hepatic plexus, and the first part of the duodenum is said to obtain a branch from the anterior vagal trunk. The nerve to the sphincter of the **hepatopancreatic ampulla** is quite distinct from the duodenal plexuses.

JEJUNUM AND ILEUM

[PLATE 30]

The length of the rest of the small intestine, the upper two-fifths jejunum and the lower three-fifths ileum, is on average 6 m., but can vary from 4·5-9 m. in length. Typical ileum is distinct from typical jejunum (see below), but there is no abrupt transition from one to the other. The calibre of the small intestine tapers from 3 to 2 cm. at the lower end where the walls also are thinner. The small intestine is packed into the abdominal cavity as a series of coils and loops lying inferior to the attachment of the transverse mesocolon, and within the enclosing ring of large intestine with some of the coils in the pelvic cavity. The jejunum for the most part occupies the umbilical region, the first coil being in front of the

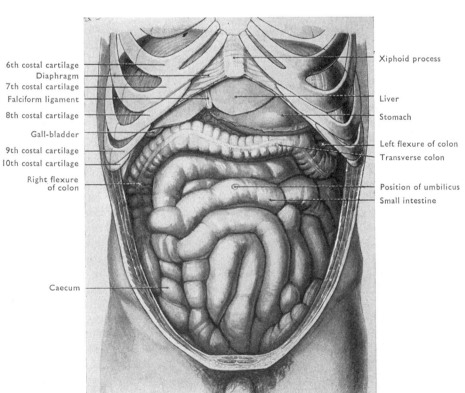

FIG. 525　Abdominal viscera, after removal of greater omentum.

The oblique position of the stomach and the high position of the transverse colon are largely due to the fact that the subject was fixed in the horizontal position.

left kidney, while the ileum is situated in the hypogastric and pelvic regions, the terminal coil usually lying in the pelvis from which it ascends over the right iliac vessels and right psoas muscle, to end in the right iliac fossa by opening into the medial side of the junction of caecum and the ascending colon [FIG. 537] after adhering to this aspect of the caecum for 2–3 cm. The small intestine is completely covered by the peritoneum except for a narrow strip at the attachment of the mesentery where the nerves and blood and lymph vessels reach the wall of the gut.

The root of the mesentery, about 15 cm. long, is attached to the posterior abdominal wall along a line which runs obliquely from the duodenojejunal flexure to the right sacro-iliac joint, crossing the horizontal part of the duodenum, the aorta, the inferior vena cava, the right gonadal vessels, the right ureter and the right psoas muscle. Its average breadth from the vertebral

column to its attachment to the gut is 20 cm., tending to narrow towards the lower end of the ileum. The mesentery contains the blood vessels, lymphatics, lymph nodes, and nerves of the small intestine, and a variable amount of fat which tends to accumulate at its posterior attachment.

STRUCTURE OF THE SMALL INTESTINE

The wall of the small intestine consists of the usual four layers—mucous, submucous, inner circular and outer longitudinal muscular layers, and serous covering. There is an abrupt change at the gastroduodenal junction from the typical gastric mucous membrane to that of the small intestine, with villi, unbranched tubular glands, and goblet cells in the surface epithelium. There is also, of course, the thickening of the circular muscle layer to form the pyloric sphincter [FIG. 526].

Mucous Membrane

The mucous membrane has the basic structure characteristic of the gastro-intestinal tract, i.e., a columnar epithelium, lamina propria, and muscularis mucosae. The small intestine is concerned with the absorption of the products of digestion, and for this purpose the surface area of its mucous membrane is enormously increased by the formation of large and small duplications—the **circular folds** and **villi**—of the mucosa.
Circular Folds. The circular folds give the characteristic pleated appearance to the inner aspect of the small intestine. They are formed by the whole thickness of the mucous membrane, including the muscularis mucosae, have a core of submucosa, and are permanent structures which do not disappear when the gut is distended [FIG. 527]. Commencing 2·5–5 cm. from the pylorus, the circular folds reach their maximum development in the distal half of the duodenum and proximal part of the jejunum where they reach a height of 8 mm., extend two-thirds round the circumference, and are branched. Thereafter they gradually become smaller and less numerous, and are absent from the lower ileum.

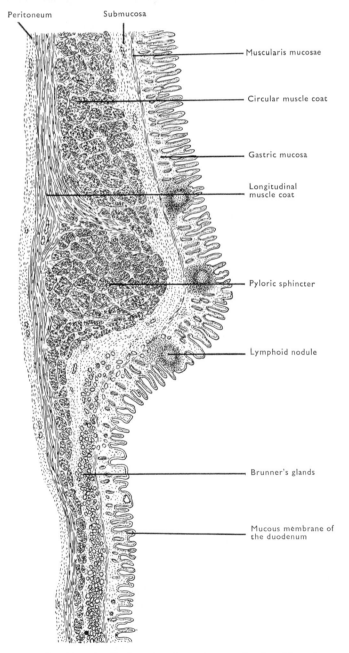

Peritoneum Submucosa
Muscularis mucosae
Circular muscle coat
Gastric mucosa
Longitudinal muscle coat
Pyloric sphincter
Lymphoid nodule
Brunner's glands
Mucous membrane of the duodenum

FIG. 526   Longitudinal section through gastroduodenal junction.

FIG. 527   Circular folds (natural size). Portion of jejunum which has been filled with alcohol and hardened.

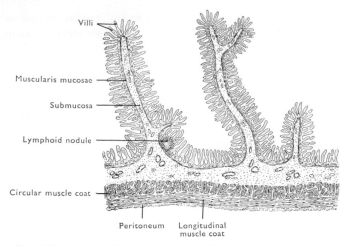

FIG. 528   Longitudinal section of small intestine passing transversely through the circular folds.

papilla is often protected by a hood-like circular fold, and from its lower margin a ridge of mucous membrane, called the longitudinal fold of the duodenum, descends for a considerable distance and acts as a frenulum which fixes the papilla and directs its apex slightly downwards [FIG. 531]. About 2·5 cm. higher up and more ventrally there is a second and smaller **papilla (lesser duodenal)** on whose summit the accessory pancreatic duct opens by a very small orifice. This second papilla is normally present.

Lymphatic Tissue. Small and large aggregates of lymphatic tissue are present in the mucous membrane throughout both the small and the large intestines. The small intestine contains great numbers of solitary lymph follicles, particularly numerous in the lower jejunum where they occupy only the deeper layers of the mucous membrane inside the muscularis mucosae. Solitary nodules may be grouped together to form **aggregated follicles** (Peyer's patches). From thirty to forty aggregated

Intestinal Villi. The intestinal villi are microscopic finger-like projections of the mucous membrane less than half a millimetre in height, and so not normally visible to the naked eye. They begin at the pyloric orifice where they are broad and short, and becoming longer and narrower, finally cease at the ileocaecal valve [FIG. 528]. The villi are present on and between the circular folds but not over the **solitary lymph follicles.** Each villus has a core of lamina propria covered with columnar epithelium. These cells have a finely striated border which the electron microscope shows to consist of microvilli which are longer and more numerous in the upper parts of the small intestine. The epithelium includes numerous goblet cells. In the centre of each villus there is a lymph channel—the **central lacteal**—which joins the submucous lymph plexuses, and a vascular capillary network draining into a vein at the base of the villus, also a strand of muscularis mucosae.

Intestinal Glands. The intestinal glands, or crypts of Lieberkühn, which secrete digestive enzymes, are simple tubular glands penetrating the thickness of the mucous membrane as far as the muscularis mucosae and opening between the bases of the villi [FIG. 530]. The epithelium at the bottom of the crypts is low, columnar, and includes the specialized **argentaffin** and **Paneth cells** with their specific staining granules. The argentaffin cells are particularly common in the duodenum and in the appendix, but their function is still unknown, although currently they are associated with the production of serotonin. The Paneth cells are probably concerned with the secretion of proteolytic enzymes. In the wall of the duodenum there are glands in the submucosa whose ducts pierce the muscularis mucosae to open into the crypts of Lieberkühn. These are the **duodenal glands** (of Brunner) which commence in the region of the pylorus and gradually disappear in the lower horizontal and ascending parts of the duodenum. The glands of Brunner secrete a viscid mucous material containing a proteolytic enzyme which is activated by hydrochloric acid and closely resembles pepsin [FIG. 530].

Duodenal Papillae. The bile-duct and pancreatic duct open by a common orifice at the greater duodenal papilla on the posteromedial wall of the internal surface of the descending part of the duodenum, about 10 cm. from the pylorus. The

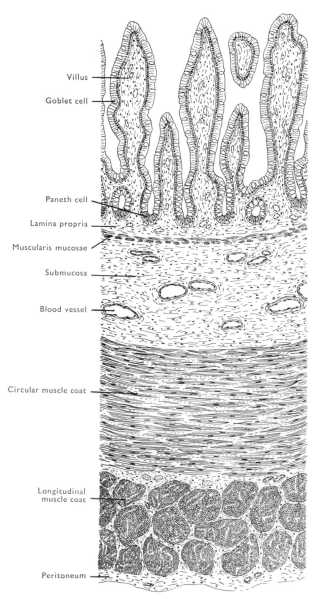

FIG. 529   Transverse section through the wall of the small intestine.

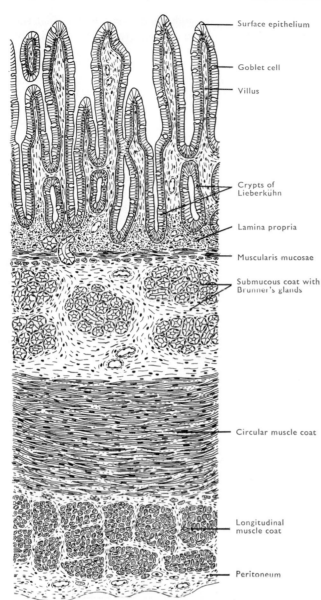

FIG. 530   Transverse section through the wall of the duodenum.

FIG. 532   Internal surface of part of the distal ileum.

follicles up to 2·5 cm. in length are found along the anti-mesenteric border of the ileum [FIG. 532].

Blood Vessels, Lymphatics, and Nerves of the Small Intestine

The small intestine is supplied by the intestinal branches of the **superior mesenteric artery,** ten to sixteen in number. Within the mesentery each intestinal artery divides into an ascending and a descending branch, which anastomose respectively with branches of the vessels above and below, thus forming a series of **arcades.**   The arches or arcades are simple in the upper part of the jejunum and generally single, but become more complex as the vessel and mesentery are traced downwards.   In the lower part of the mesentery the primary arcades give off branches which in turn form secondary arcades, and these in turn form tertiary arcades, and there may be more [FIG. 533 A and B].   From the last of the arcades straight arteries run to the wall of the intestine, and these are much longer in the jejunum than in the ileum.   The straight arteries alternately pass to opposite sides of the gut, and while they are sometimes regarded as end arteries, anastomoses between them are present within the wall of the gut.   It has been shown, for instance, that as much as 15 cm. of intestine can be injected through the intramural anastomosis after ligature of the straight arteries.

The **veins** for the most part follow the course of the arteries and join the superior mesenteric vein.   The arteries of the intestinal wall form a plexus in the submucous layer, from which branches are given to the mucous surface, one

FIG. 531   Interior of second part of duodenum (viewed from the front) showing duodenal papilla and opening of accessory pancreatic duct (lesser duodenal papilla).

FIG. 533   Portion of A, jejunum, and B, ileum.

branch to each villus. There is a corresponding arrangement for the veins.

Lymph vessels begin in the wall of the gut, and connecting vessels drain through a series of nodes lying along the arteries of the mesentery. At the base of the mesentery, they drain into nodes said to number from one to two hundred, near the origin of the superior mesenteric artery (**mesenteric nodes**). The products of fat digestion are absorbed through the lymphatic vessels, and during digestion lymphatic vessels of the mesentery are filled with a fine white emulsion of neutral fats known as chyle. The central lacteal of each villus drains into a sub-mucosal plexus, and large lymph vessels from this plexus run through the muscular coats of the gut into the mesentery.

The **nerves** reach the wall of the gut in plexuses along the arteries and consist of sympathetic and parasympathetic nerve fibres. The pattern of the peripheral nerve distribution is basically the same throughout the gastro-intestinal tract and consists of intrinsic plexuses of nerve cells and their fibres fed by preganglionic parasympathetic (vagus or sacral autonomics) and the postganglionic sympathetic fibres. There are two such plexuses, the **myenteric plexus** (Auerbach) between the circular and longitudinal muscle layers, and the **submucous plexus** (Meissner).

THE LARGE INTESTINE

[PLATES 31 and 32]

The parts of the large intestine and their general position have already been described. The large intestine, so named because it is on average of larger calibre than the small intestine, is distinguished from the small intestine by: (1) the presence in the longitudinal muscle of three thick bands or **taeniae**; (2) its characteristic sacculation, or **haustration**, due to the shorter length of the taeniae compared with that of the gut; and (3) the **appendices epiploicae**, small masses of fat enclosed within the peritoneum and protruding from its surface. The taeniae, which commence at the root of the appendix, are about 6 mm. broad, and on the caecum, ascending and descending colons are situated on the anterior, posterolateral, and posteromedial aspects, and named respectively **t. libera, t. omentalis,** and **t. mesocolica**. On the transverse colon the situation of the three taeniae is inferior, superior corresponding to the attachment of the mesocolon, and posterior [FIG. 534]. They are less prominent on the descending colon, tend to fade out on the sigmoid colon, and spread out as broad anterior and posterior bands on the rectum. The total length of the large intestine varies within the limits of from 1·5 to 1·8 m., the most variable segments being the transverse and sigmoid parts.

FIG. 534   Large intestine.

A piece of transverse colon from a child 2 years old. The three chief characteristics of the large intestine—appendices epiploicae, taenia, and sacculations (haustra)—are shown.

At the junction of the ileum with the large intestine the caecum projects downwards as a blind pouch, while the ascending colon passes upwards. The line of union between the caecum and ascending colon is frequently marked by a groove extending round the large intestine at the level of the upper border of the terminal ileum, and corresponding in position to the frenulum of the ileocaecal valve.

CAECUM

The caecum, which is sacculated, has an average length of 6 cm., is conical in the foetus with the appendix vermiformis at the apex of the cone. It becomes asymmetrical in the infant by a greater growth of the lateral wall which relegates the original apex with the attachment of the appendix vermiformis to the posteromedial aspect of the caecum [FIG. 536].

The caecum lies on the iliopsoas muscle in the right iliac fossa, its midpoint approximately on the level of the anterior superior iliac spine, although in the erect position in most living subjects it lies in the lesser pelvis. Anteriorly it is in contact either with the greater omentum or directly with the anterior abdominal wall. On the medial side it is related to the terminal part of the ileum which, typically, ascends from the lesser pelvis to the ileo-

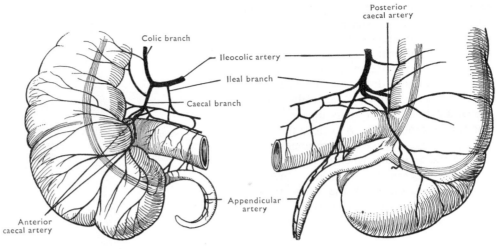

FIG. 536 Blood supply of caecum and vermiform appendix. (Modified from Jonnesco, 1895.)

The illustration on the left is from the front, the right from behind. In the latter the appendicular artery and the three taeniae coli springing from the root of the appendix should be specially noted.

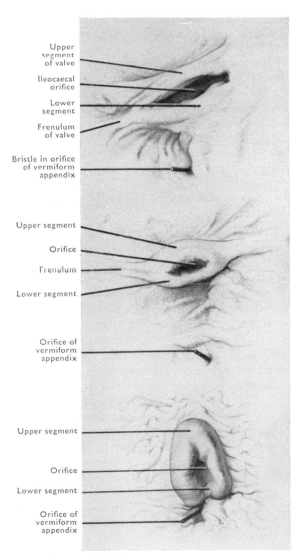

FIG. 535 Three forms of ileocaecal valve from bodies hardened by formalin.

The hardening was not so complete in the uppermost of the three valves represented.

caecal junction. The ileum appears to penetrate the postero-medial wall of the large intestine and projects into its interior as the **ileocaecal valve,** whose position is indicated on the surface of the body by the point of intersection of the transtubercular and right vertical planes. The valve is formed by two folds made up of the mucous and submucous coats enclosing muscular fibres from the circular layer [FIG. 535]. The upper fold is generally the more prominent, and at each end of the orifice the two folds join to form a frenulum (lateral and medial) of the valve which continues for a variable distance round the circumference of the gut. How the valve functions is not yet fully understood; intracaecal tension tightens the frenula and tends to close the orifice. In addition to this passive action, there is said to be some muscular action which presumably can be influenced by its sympathetic and parasympathetic nerve supply, and is involved in the gastrocolic reflex.

APPENDIX VERMIFORMIS

[PLATE 32C]

The appendix vermiformis, which is the apex of the embryonic caecum, arises in the adult from its posteromedial aspect, about 1·5–2 cm. below the ileocaecal junction. It can vary in length from 2 to 20 cm., on average 9 cm., and is about 6 mm. in diameter at its base. The appendix, which is longer, relative to the large intestine, in the child at birth (the proportion being about 1 to 16 compared with 1 to 20 in the adult) attains its greatest absolute length and diameter during adult and middle age, and thereafter atrophies slowly. It is said to be slightly longer in the male than in the female. The frequency of occlusion of its narrow lumen, the physiological atrophy which takes place after middle life, the great variation in length, and other signs of instability, have been considered to point to the retrogressive character of the appendix, but is in keeping with the fact that it is composed largely of lymphoid tissue.

The narrow lumen of the appendix (variable in size) tends to become obliterated in later life, and is found to be totally or partially occluded in at least 25 per cent. of all adult bodies examined. It opens on the posteromedial wall of the cavity of the caecum [FIG. 538], at a point 2·5–4 cm below and slightly posterior to the ileocaecal orifice. The apparent difference in the level of the internal orifice and external root is due to the ileum adhering to the medial side of the caecum for a distance of nearly 2·5 cm. before it opens into it. The orifice has a crescentic fold placed at its upper border, but this is probably of very little functional importance. While the commonest position of the appendix, whether free or fixed, is upwards behind the caecum, it may be directed downwards on the psoas muscle hanging into the lesser pelvis [FIG. 537], or up to the left in front of or behind the ileum.

The position of the appendix and caecum is variable; commonly in children they lie immediately inferior to the liver [p. 452]; rarely in the left iliac fossa due to reversed rotation of the gut.

The surface marking of the base of the appendix—the so-called McBurney's point—is the junction of the lower and middle thirds of a line joining the anterior superior iliac spine and the umbilicus. The exact position of this point, however, with reference to the abdominal cavity, varies with the condition of the anterior abdominal wall, and moreover the degree of mobility of the caecum will also affect the position of the base of the appendix. Where there is difficulty in locating the appendix a reliable guide is the anterior taenia of the caecum, which can be followed back to the junction of the caecum and appendix where the taeniae fuse.

The appendix, as part of the caecum, has no true mesentery, but is usually surrounded by peritoneum and connected to the posterior surface of the ileal mesentery, close to the ileocaecal

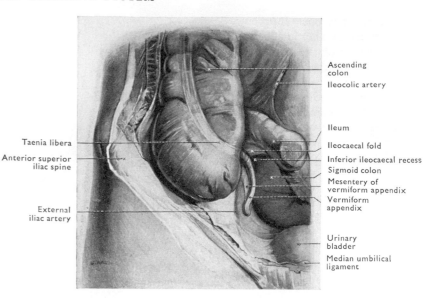

FIG. 537   Caecum and vermiform appendix from the front.

junction, by a triangular fold of peritoneum which contains the artery to the appendix [FIG. 536]. This mesentery of the appendix may not reach its apex.

Because of its posterior origin, some part of the appendix is almost always retrocaecal, but an intraperitoneal retrocaecal appendix is freely movable behind the caecum. It may, however, be fused to the posterior surface of a retroperitoneal caecum and extend up behind the ascending colon almost as far as the lower pole of the right kidney. This is often what is meant clinically by the term 'retrocaecal appendix'.

Peritoneum of the Caecum and Appendix Vermiformis

These vary considerably, but in over 90 per cent. of cases the caecum is completely covered with peritoneum, although as it develops from the antimesenteric border of the embryonic intestinal loop it never has a true mesentery. In a small proportion of cases some part of the posterior surface of the caecum is not covered by peritoneum. Commonly, the first 2.5 cm. or so of the ascending colon is also almost surrounded by peritoneum, and the line of the reflection of visceral on to parietal peritoneum is about the level of the crest of the ilium, indicated sometimes by folds bounding a retrocaecal recess. The following peritoneal recesses may be present in the region of the appendix [FIG. 538].

The **superior ileocaecal recess,** best developed in children and frequently present, opens downwards just above the terminal part of the ileum. It is bounded in front by a vascular fold of peritoneum containing the anterior caecal artery, and behind by the mesentery of the ileum.

The **inferior ileocaecal recess,** also best developed in children, opens downwards and is produced by a peritoneal fold—the ileocaecal fold

FIG. 538   Folds and recesses in the ileocaecal region.

A. The caecum viewed from the front; the mesentery of the vermiform appendix is distinct. B. The caecum turned upwards to show a retrocaecal recess which lies behind it and behind the beginning of the ascending colon.

(bloodless fold of Treves) extending from the front and lower part of the ileum to the mesentery of the appendix.

The **retrocaecal recess** varies considerably in size and extent, and may occasionally be big enough to admit one or even two fingers. The caecum and sometimes the lower part of the ascending colon are anterior and the parietal peritoneum posterior, while folds of peritoneum reflected from the caecum on to the posterior abdominal wall lie on each side.

ASCENDING COLON

[Plate 31]

The ascending colon is continuous with the caecum at the ileocaecal junction and runs upwards and slightly backwards in the paravertebral gutter on the posterior abdominal wall to the inferior surface of the liver, where it bends forwards as the right colic flexure. It is from 12 to 20 cm. long, may have a number of minor curves or flexures, and is wider and more prominent than the descending colon. Anteriorly, it is usually in direct contact with the anterior abdominal wall, but the small intestine may intervene, particularly above. On the medial side are the coils of small intestine and the psoas major muscle.

Lateral to the caecum and the ascending colon is the **right paracolic sulcus,** often divided by peritoneal folds into small recesses. Except for the lower 2·5 cm. or so, the ascending colon is covered by peritoneum on its anterior surface and sides only, and its posterior surface is connected by areolar tissue and sometimes by bands of smooth muscle to the iliacus muscle, the quadratus lumborum, and finally the lower part of the right kidney. Behind it are the vessels and nerves lying on the right side of the posterior abdominal wall. In a small proportion of cases where the ascending colon is intra-peritoneal, it has a short mesentery that permits only a slight amount of movement. An ascending meso-colon is usually associated with what is clinically termed 'a mobile caecum'. Sometimes the ascending colon is firmly bound down to the abdominal wall by pericolic membranes which can cause obstruction. Such a membrane—Jackson's membrane—may stretch across and obliterate the right paracolic recess. The ascending colon is frequently distended with gas or faeces after death, while the descending colon is generally empty.

On reaching the inferior surface of the liver the ascending colon bends, usually acutely but sometimes obtusely, forwards and to the left on the antero-inferior surface of the right kidney to form the **right colic flexure,** which thereafter passes into the transverse colon. The flexure is placed between the duodenum medially and the inferior margin of the liver or the side-wall of the abdomen laterally. Above, it is related to the colic impression on the liver, and posteriorly it rests on the kidney [Fig. 514]. Its peritoneal relations are similar to those of the ascending colon.

TRANSVERSE COLON

[Plate 31]

The transverse colon is curved downwards and has an average length of about 50 cm. which is more than twice the distance between its two extremities in the right and left hypochondriac regions. Initially, its course is transverse and it is relatively fixed, being united to the front of the descending part of the duodenum and the head of the pancreas by a very short mesocolon. Thereafter, the mesocolon lengthens and the colon hangs down in front of the small intestine in the umbilical or hypogastric region, the position varying in the same individual from time to time according to posture, degree of distension of the bowel, etc. To the left, the mesocolon again shortens and the transverse colon extends under cover of the stomach as far as the colic surface of the spleen. Here it bends downwards to form the left colic flexure. Except for the first few inches, the transverse colon is completely covered with peritoneum, and its mesentery, the transverse mesocolon, is fused to the posterior surface of the greater omentum which must therefore be turned up in order to expose it [Fig. 508]. The transverse colon is in

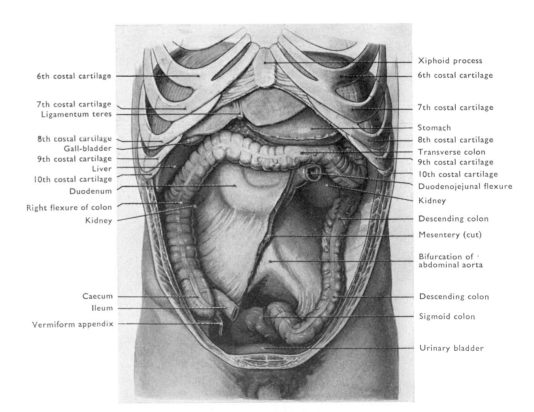

FIG. 539 Abdominal viscera after removal of jejunum and ileum. (From a photograph of the same body as depicted in Fig. 525.) The transverse colon is much less dependent than usual.

contact from right to left, superiorly, with the liver, the gall-bladder, the stomach, and the spleen; anteriorly, with the anterior abdominal wall, the greater omentum, and the greater curvature of the stomach; posteriorly, with the right kidney, the descending part of the duodenum, the head of the pancreas, the small intestine, and the left kidney.

The terminal part of the transverse colon turns backwards and to the left, bending sharply to form the **left colic flexure** which lies deep behind the stomach at a higher level than the right colic flexure, and in contact with the lower part of the spleen [FIG. 514]. The left colic flexure is connected to the posterior abdominal wall by the **phrenicocolic ligament,** attached laterally to the diaphragm.

DESCENDING COLON
[PLATES 31 and 32]

The descending colon, about 25 cm. long, follows the lateral border of the left kidney, through the left hypochondrium and left lumbar regions, as far as the crest of the ilium. It then curves downwards and medially on iliacus to the medial part of the inguinal ligament, and crossing psoas and the external iliac vessels, becomes the **sigmoid colon** at the superior aperture of the lesser pelvis. It is covered by peritoneum in front and at the sides, and its posterior surface is connected by areolar tissue to the fascia on the left kidney, quadratus lumborum, iliacus, and psoas. The descending colon is more deeply placed than the ascending colon, with the coils of jejunum in front, except at its lower part which can be felt through the anterior abdominal wall when the muscles are relaxed. As it crosses the iliopsoas muscle, the femoral and genitofemoral nerves, the left testicular or ovarian vessels, and the external iliac vessels are posterior. Its position, as indicated on the surface of the anterior abdominal wall, is lateral to the vertical plane from a point just above the transpyloric plane down to the groin, where its lower part becomes medial to the vertical plane.

SIGMOID COLON
[PLATE 31]

The sigmoid colon begins at the superior aperture of the lesser pelvis, and ends ventral to the third sacral vertebra in the midline, where it becomes continuous with the rectum. With an average length of 40 cm., it forms a loop within the lesser pelvis, finally bending downwards in the midline [FIG. 540].

The sigmoid colon is covered by the peritoneum and suspended by a mesentery—the **sigmoid mesocolon**—which has a V-shaped attachment with the apex near the point of division of the left common iliac artery, the left limb ascending on the medial side of the left psoas and the right limb passing into the lesser pelvis to end at the third sacral vertebra. The longest part of the sigmoid mesocolon is the centre, and it shortens towards the ends of the loop. The left ureter runs into the lesser

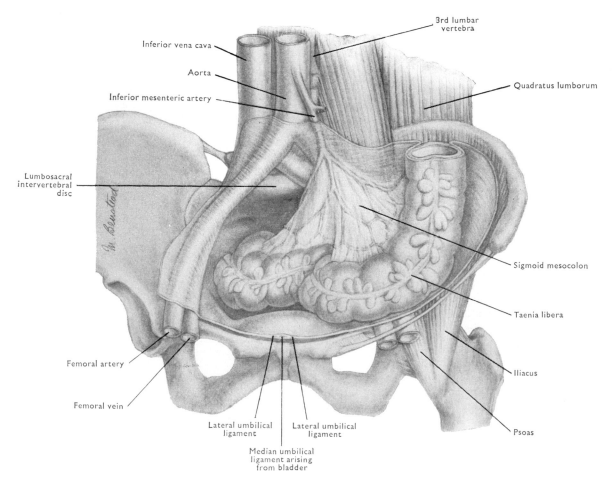

FIG. 540 Sigmoid colon *in situ*.

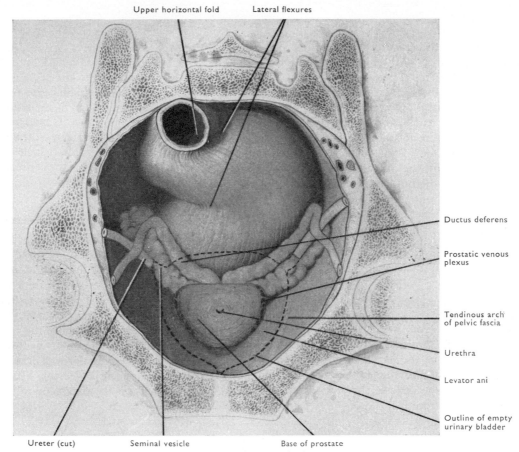

Upper horizontal fold   Lateral flexures

Ductus deferens

Prostatic venous plexus

Tendinous arch of pelvic fascia

Urethra

Levator ani

Outline of empty urinary bladder

Ureter (cut)   Seminal vesicle   Base of prostate

FIG. 541   Distended rectum *in situ.*

From a formalin-hardened male body, aged 58. The peritoneum and extraperitoneal tissue were removed, after the pelvis had been sawn along a plane passing through the upper part of the pubic symphysis in front and the lower part of the second sacral vertebra behind. The bladder, which was empty and contracted, has also been removed, but its form is shown by a dotted line. The rectum was very much distended and almost completely occupied the pararectal fossae.

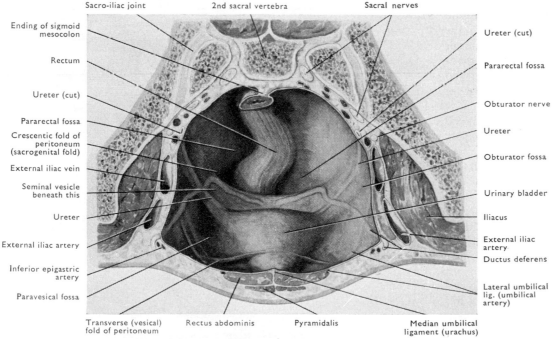

Sacro-iliac joint   2nd sacral vertebra   Sacral nerves

Ending of sigmoid mesocolon

Rectum

Ureter (cut)

Pararectal fossa

Crescentic fold of peritoneum (sacrogenital fold)

External iliac vein

Seminal vesicle beneath this

Ureter

External iliac artery

Inferior epigastric artery

Paravesical fossa

Ureter (cut)

Pararectal fossa

Obturator nerve

Ureter

Obturator fossa

Urinary bladder

Iliacus

External iliac artery

Ductus deferens

Lateral umbilical lig. (umbilical artery)

Transverse (vesical) fold of peritoneum   Rectus abdominis   Pyramidalis   Median umbilical ligament (urachus)

FIG. 542   Peritoneum of pelvic cavity seen from above.

The pelvis of a thin male subject, aged 60, was sawn across obliquely. Owing to the absence of fat the various pelvic organs were visible through the peritoneum, though not quite so distinctly as represented here. The urinary bladder and rectum are both empty and contracted; the paravesical and pararectal fossae, as a result, are very well marked.

pelvis behind its apical attachment, and the sigmoid and superior rectal vessels run respectively in the left and right limbs of its mesentery. An intersigmoid peritoneal recess is commonly present in infancy and tends to disappear later. It lies behind the apical attachment of the mesocolon with its orifice looking downwards, and the peritoneum on its posterior wall covers the ureter at the point where it is crossing the bifurcation of the left common iliac artery. To the left, the sigmoid colon lies on the structures in the lateral wall of the lesser pelvis— the external iliac vessels, the obturator artery and nerve, the ovary in the female, and the ductus deferens in the male. Posteriorly are the left internal iliac vessels, the ureter, and the left piriformis muscle with the sacral plexus, while inferiorly it rests on the bladder in the male and the uterus and bladder in the female, with the terminal coils of the ileum above and to the right.

RECTUM

The rectum begins in contact with the pelvic surface of the third sacral vertebra as a continuation of the sigmoid colon, and descends in the curve of the sacrum and coccyx to end at a point 3·5 cm. in front of and slightly below the tip of the coccyx, corresponding in the male to the level of the apex of the prostate gland. Its terminal part is intimately related to the pelvic diaphragm formed by the levator ani muscle.

The level of the **rectosigmoid junction** is arbitrary, but is usually indicated by one or all of several landmarks, e.g.:

1. The absence of a posterior covering of peritoneum.

2. The appendices epiploicae and sacculations disappear and the taeniae blend to form broad anterior and posterior bands which gradually spread laterally to meet and form a complete, thick layer of longitudinal muscle.

3. Division of the superior rectal artery into a right and a left branch.

4. An external constriction, more apparent in the living, which has been called the sphincter of O'Beirne.

5. There is an abrupt change from the rugose mucous membrane of the sigmoid to the relatively smooth rectal mucous membrane.

The rectum is 12–15 cm. in length and the upper part has the same diameter as the sigmoid colon, but the lower part below the reflexion of the peritoneum widens out to form the **rectal ampulla** which narrows abruptly to the anal canal. The rectum follows the sacral curve, its final part bending sharply backwards to join the anal canal.

There are three lateral flexures, an upper and a lower one convex to the right, and a middle one convex to the left [FIG. 544]. Corresponding to the flexures there are, typically, permanent sickle-shaped, **transverse folds** of mucous membrane projecting into the lumen of the rectum and consisting of the mucous and submucous coats, and some circular muscle fibres. The lowest fold is 4–6 cm. above the anus, and the highest 12 cm. from the anus. The folds, which can vary in number and apparently help to support the contents of the rectum, are more prominent in the distended rectum, and can be seen on rectal examination.

Peritoneal Relations of the Rectum

The upper third of the rectum is covered by the peritoneum in front and at the sides. The middle part has the peritoneum in front only, and the lower third of the rectum has no peritoneal covering. The level of the reflexion of the peritoneum from the front of the rectum to the posterior wall of the bladder in the male, where it forms the floor of the rectovesical pouch, is about 2·5 cm. above the base of the prostate, or 7·5 cm. from the anus, and relatively lower in the female where it is reflected to the posterior wall of the vagina as the floor of the recto-uterine pouch [FIG. 634]. The lateral reflexions of the peritoneum form the pararectal fossae providing ample room

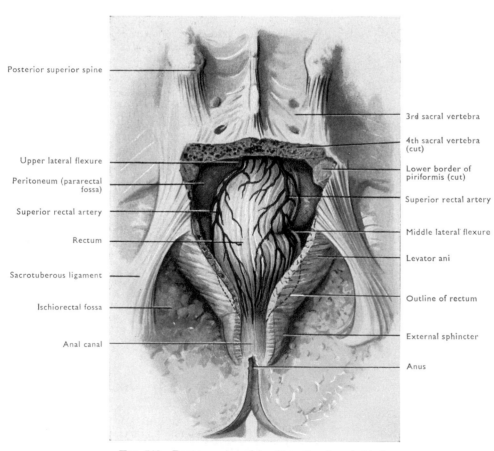

FIG. 543   Rectum exposed by dissection from behind.

The sacrum has been sawn across the fourth sacral vertebra, and its inferior part removed with the coccyx. The posterior portions of the coccygei, levatores ani, and of the external sphincter have been cut away. The 'pinching in' of the lower end of the rectum by the medial edges of the levatores ani, resulting in the formation of the flattened anal canal, is suggested in the illustration, which has been made from a formalin-hardened male body, aged 30. The lateral flexures are also seen.

to accommodate the distended rectum. The level of the peritoneal reflexion is not a fixed point, but is raised when the rectum is distended, and tends to be lower where there is poor muscular development of the pelvic diaphragm. In a male infant, where the bladder is relatively higher in the pelvis, the peritoneum extends down as far as the base of the prostate.

Position of the Rectum [Figs. 541-544]

Posteriorly the rectum lies on the lower three sacral vertebrae, the coccyx, the anococcygeal ligament, the median sacral vessels, the branches of the superior rectal artery, and the lower end of the sympathetic trunks. The rectum is surrounded by a fascial sheath which at the front and the sides is loosely connected to the underlying muscular coat. The fascial sheath surrounding the paired branches of the superior rectal artery is also attached to the front of the sacrum, and the connective tissue which accompanies the middle rectal arteries forms the lateral ligaments of the rectum. In addition, there may be a band of connective tissue from the pelvic surface of the sacrum to the posterior wall of the rectum in the midline, part of the so-called fascia of Waldeyer, but otherwise there is a natural cleavage plane between the rectum and the sacrum. Laterally, the distended rectum obliterates the pararectal fossae and becomes related to the piriformis muscles with the sacral nerves, the lateral sacral vessels, and lower down, the coccygeus and the levator ani muscles [Fig. 543]. In front, the peritoneum is bound firmly to the upper part of the rectum, which in the male is here separated from the upper part of the base of

the bladder and the seminal vesicles by the rectovesical pouch which may contain some of the terminal coils of the ileum or the sigmoid colon. Inferior to the peritoneal reflexion, the rectum is separated, in the male, from the lower part of the base of the bladder with the seminal vesicles, the deferent ducts, and inferiorly the prostate gland, by a fibrous **recto-vesical septum** which is more closely associated with the seminal vesicles and the prostate than with the rectum. This represents a potential cleavage plane in front between the rectum and the prostate. In the female, the rectum is separated from the cervix of the uterus and the posterior fornix of the vagina by the recto-uterine pouch, and below the pouch the **rectovaginal septum** intervenes between it and the vagina.

ANAL CANAL

Anatomically, the anal canal begins where the rectal ampulla narrows abruptly, ends at the anus, and is from 2·5 to 3·5 cm. in length. The **anorectal junction,** therefore, coincides with the sharp change in direction of the bowel and lies on the upper surface of the levator ani muscle through which the anal canal passes and is then surrounded by the external sphincter muscle of the anus [Fig. 544]. Clinically, the anorectal junction is often identified by the wavy line formed by the anal valves —the **pectinate line**—representing the upper limit of the lining of stratified squamous epithelium. This gives a much shorter anal canal, little more than 1 cm.

When empty, the canal is an anteroposterior slit, in contact

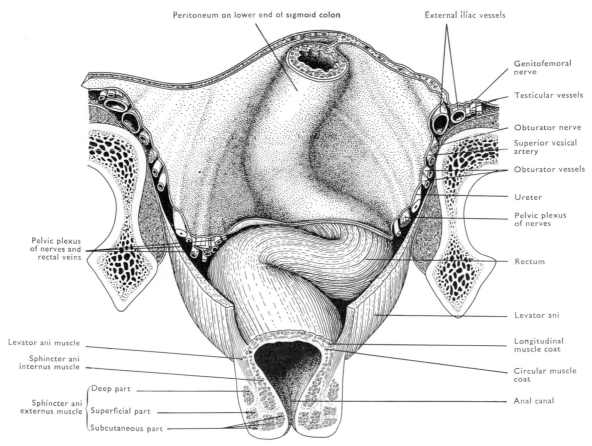

FIG. 544 Drawing of rectum and anal canal seen from the front to show peritoneum and relations to the pelvic wall.

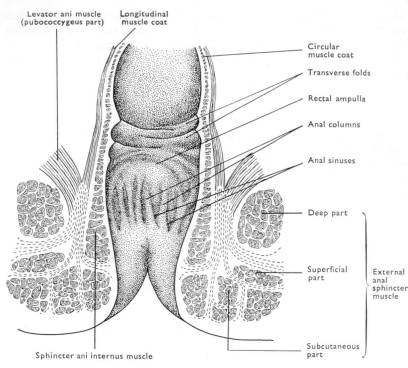

FIG. 545 Rectum and anal canal with the sphincter muscles. See text.

behind with the mass of fibromuscular tissue forming the **anococcygeal ligament,** in front with the **central perineal tendon** to whose anterior surface is attached the penis in the male and the vagina in the female, and on each side with the fat-filled ischiorectal fossae. The lower limit of the anal canal —the **anus**—is indicated by a change in the nature of the skin from the dry, hairy, perianal skin, to the moist, hairless, lining skin. The lining of the anal canal has from six to ten permanent vertical folds, known as the **anal columns,** each containing a terminal branch of the superior rectal artery and vein. The lower ends of the columns are joined together by crescentic folds—the **anal valves**—each forming a small recess or **anal sinus** [FIG. 545]. The line of the valves is the pectinate line, commonly supposed to represent the junction of the endodermal and ectodermal parts of the anal canal. Below the valves is a transitional zone, known as the **pecten,** where septa from the longitudinal muscle of the rectum are inserted into the mucous lining, and the epithelium, although stratified, is thin. The pecten ends at a narrow, wavy white line (of Hilton) representing the level of the lower border of the internal sphincter which can be felt on examination, and the last few millimetres of the anal canal are lined by a true stratified squamous epithelium. The change from stratified to columnar epithelium occurs in the young child at the base of the anal columns, but in the adult a modified stratified epithelium is commonly continued on to the anal columns, and appears to extend upwards with advancing age [FIG. 546]. In a proportion of anal canals, epithelial ducts, sometimes associated with glandular acini, open into the anal sinuses.

Musculature of the Rectum and Anal Canal

The muscle of the rectum is arranged as an outer longitudinal layer and an inner circular layer which in the lower part of the rectum and anal canal thickens to form the **internal**

anal sphincter, about 2.5 cm. in length. As the anal canal passes through the pelvic floor it is closely invested by muscular fibres of the **levator ani muscle** which pass down the wall of the canal to blend with its longitudinal muscle. From the anterior wall of the rectum, smooth muscle fibres turn forward along the upper surface of the levator ani as the **recto-urethralis muscle,** and fibres from the lateral wall of the rectum blend with the superior layer of the levator ani, while posteriorly similar muscle fibres form the **rectococcygeus muscle.** Thus, the anal canal is firmly fixed to the levator ani muscle as it passes through the pelvic diaphragm. Where the rectum bends downwards to become the anal canal, it is supported by a muscle sling—the **puborectalis** —arising from the pubococcygeus part of the levator ani and passing on either side of the rectum to unite posteriorly.

The longitudinal muscle, including the fibres of the levator ani, separate the involuntary muscle fibres of the **internal sphincter** from the voluntary fibres of the **external sphincter** which is arranged as a deep part surrounding the longitudinal muscle of the anal canal, and a subcutaneous part which lies

FIG. 546 Longitudinal section through the anorectal junction.

PLATE 29

A

B

A, B, and C are three anteroposterior views of the stomach filled with barium sulphate suspension.
A and B are taken in the erect posture and show the different shapes of the stomach found in thick-set and lean individuals respectively.

C

D

C is taken in the supine position with air lying in the body of the stomach over the vertebral column and outlining the mucosal pattern.

1. Superior part of the duodenum.
2. Pylorus.
3. Pyloric antrum.
4. Air bubble in the fundus.
5. Mixture of barium and gastric juice, the former outlining the mucosal pattern of the stomach.
6. Body of the stomach.
7. Descending part of duodenum.

D. An oblique view of the open pylorus.

1. Pylorus.   2. Pyloric antrum.   3. Wall of stomach.

PLATE 30

A

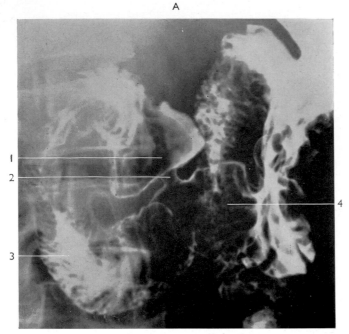

A. Supine radiograph outlining the pyloric antrum of the stomach, the pylorus, and the duodenum during an emptying phase of a barium meal.

1. Superior part of duodenum.
2. Pylorus.
3. Descending part of duodenum.
4. Pyloric antrum of stomach.

B

B. Mucous membrane pattern of jejunum.

1. Circular folds outlined by slight covering of barium.

C

C. Loops of ileum filled with barium.

1. Contracted segments of ileum.

D

D. Loops of ileum in ileocaecal region.

1. Terminal ileum.
2. 4th lumbar vertebra.
3. Transverse process of 5th lumbar vertebra.
4. Loop of ileum, showing longitudinal folds of mucous membrane.

PLATE 31

A

A. Colon and terminal ileum filled by barium enema.

1. Right colic flexure.   4. Left colic flexure.
2. Caecum.   5. Lower end of descending colon.
3. Terminal ileum.   6. Rectum.

B

B. Colon and terminal ileum distended by barium enema.
There is extensive overlapping of loops.

1. Right colic flexure.
2. Left colic flexure.
3. Transverse and sigmoid colon overlapping.
4. Descending colon and sigmoid colon overlapping.

C

C. Small intestine, caecum and proximal colon containing a
small quantity of barium and showing the mucosal pattern in its
different parts.

1. Transverse colon.   4. Jejunum.
2. Terminal ileum.   5. Ileum.
3. Caecum.

D

D. Outline of colon after it has contracted following a barium
enema.

1. Transverse colon.
2. Lower end of descending colon.
3. Sigmoid colon.

PLATE 32

A

A. Segment of transverse colon filled by a barium enema showing the extent of the sacculations when the colon is contracting.

1. Sacculations.

B

B. Segment of transverse colon when contracted after a barium enema. The relative size of the contracted colon can be seen by comparison with a vertebral body.

1. Longitudinal folds of mucous membrane.
2. Transverse folds of mucous membrane.
3. Lumbar vertebra.

C

(i)   (ii)

C. Two views of the caecum and ascending colon :

(i) contracted.   (ii) distended.

1. Right colic flexure.
2. Ascending colon.
3. Ileocaecal junction.
4. Appendix vermiformis.
5. Terminaj ileum.

6. Caecum.
7. Sacro-iliac joint.
8. Brim of lesser pelvis.
9. Hip joint.

D

D. Common appearance of partially contracted large intestine.

1. Terminal ileum.
2. Caecum.
3. Descending colon.
4. Transverse colon.
5. Gas in lower part of descending colon.

below the internal sphincter with a superficial part between it and the deep part [FIG. 545]. Together with the puborectalis, the external anal sphincter forms a muscular anorectal ring which closes the anal canal.

STRUCTURE OF THE LARGE INTESTINE

The wall of the large intestine is made up of the usual four coats, though the peritoneum is absent where the intestine is applied to the body wall. When the large intestine is empty and contracted, its mucous membrane is thrown into numerous folds, longitudinal and transverse, but apart from the transverse folds of the rectum and the anal columns they are not permanent, and disappear when the gut is distended. There are no villi in the mucous membrane of the large intestine, but numerous tubular glands —the crypts of Lieberkühn—open on the surface which is covered with a simple columnar epithelium. They are longer than the corresponding glands of the small intestine, particularly in the rectum, and have a greater proportion of goblet

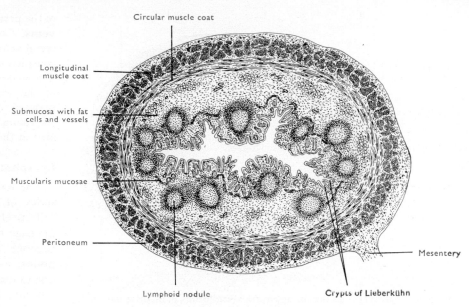

FIG. 548   Transverse section of adult appendix vermiformis.

cells in their lining of columnar epithelium [FIG. 547]. Scattered lymph follicles are always present in varying numbers as far as the rectum.

The appendix is characterized by the presence of large masses of lymphoid tissue surrounded by sinus-like spaces in the submucous coat. These numerous lymphatic follicles correspond to the solitary nodules of the intestine but lie in the submucosa. The mucous lining is like that of the large intestine, but the glands are fewer and more irregular, and the muscularis mucosae is thin and ill defined [FIGS. 548 and 549].

BLOOD VESSELS, LYMPHATICS, AND NERVES OF THE LARGE INTESTINE

Blood Supply of the Colon

The right side of the colon as far as the left colic flexure is supplied from the colic branches of the **superior mesenteric artery,** and the remainder from the colic and sigmoid branches of the **inferior mesenteric artery.** These colic arteries branch to form a continuous arcade a short distance from the inner side of the colon, known as the **marginal artery,** which gives off a series of straight arteries running to the wall of the intestine. The colic arteries are behind the peritoneum, except the ileocolic, the artery to the appendix, the middle colic branch of the superior mesenteric, and the lower sigmoid arteries, which run within the two layers of their respective mesenteries. In surgical resection of the large intestine the marginal artery may maintain the blood supply even when some of the major arteries are ligatured. Special care, however, is required in dealing with the region of the rectosigmoid junction following ligature of the branches of the inferior mesenteric artery in resection of the sigmoid colon or rectum. Here the critical factor is the adequacy of the circulation through the marginal artery, which is very small or missing, between the sigmoid and superior rectal arteries. The veins from the ascending and transverse parts of the colon drain into the **superior mesenteric vein,** and those from the

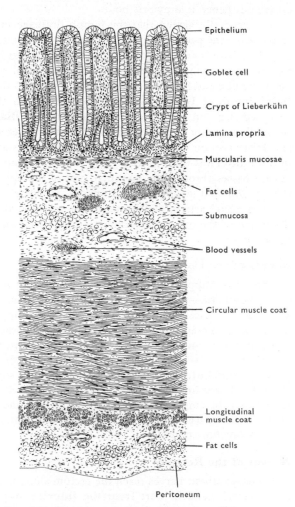

FIG. 547   Transverse section of large intestine.

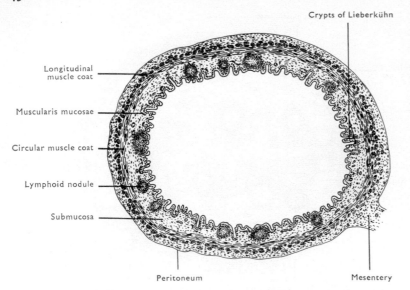

Crypts of Lieberkühn

Longitudinal muscle coat

Muscularis mucosae

Circular muscle coat

Lymphoid nodule

Submucosa

Peritoneum   Mesentery

FIG. 549   Transverse section of appendix vermiformis of infant 4 days old.

descending and sigmoid colons, corresponding to the distribution of the inferior mesenteric artery, drain into the **inferior mesenteric vein,** both tributaries of the portal system of veins.

Blood Supply of the Rectum and Anal Canal

The main **artery** of the rectum is the **superior rectal.** It lies posteriorly where it divides, and each branch then runs down in the fascia of the lateral rectal wall. Branches of the superior rectal arteries supply the mucous membrane of the rectum and the anal canal. The **middle rectal** branches of the internal iliac artery, which with their fascial sheath form the lateral ligaments of the rectum, are said to be distributed to the muscular layers of the rectum and anal canal where they anastomose with branches of the superior rectal arteries. The extent and significance of their contribution to the blood supply of the rectum varies considerably. The anal canal itself, the perianal skin, and probably part of the external anal sphincter muscle, are supplied by the **inferior rectal** branches of the **internal pudendal arteries.** The rectal arteries above and below the pectinate line anastomose with one another by branches running in the substance of the rectal columns, and it is now well established that the inferior rectal arteries alone are sufficient to supply a considerable stump of rectum.

The **veins** of the rectal wall form two plexuses, an internal submucous plexus between the muscle layer and the mucous membrane, and an external plexus situated outside the muscular layer below the level of the peritoneal reflexion. Above the pectinate line the veins of the submucous plexus run up for about 10 cm. in the submucosa before passing through the muscular layer to form the **superior rectal veins.** Dilatation of the relatively unsupported veins of this plexus causes them to bulge into the lumen of this part of the rectum and anal canal (zona haemorrhoidalis) forming internal haemorrhoids. The veins of the inferior part of the submucous plexus become the **inferior rectal veins,** draining into the **internal pudendal veins,** and their dilatation is responsible for external haemorrhoids. The external or perimuscular plexus drains the muscular wall of the rectum below the level

of the peritoneal reflexion and forms the **middle rectal veins.** Communication between the middle and superior rectal veins becomes more marked with age, and more conspicuous when the individual has haemorrhoids. The majority of such connexions are in the submucosa of the rectum and represent one site of anastomosis between the portal and caval venous systems. They correspond to the anastomosis which is similarly situated in the lower oesophagus.

Lymphatics [FIGS. 1161 and 1162]

The lymph vessels of the large intestine drain into a series of lymph nodes arranged along the arteries, collectively known as the **colic nodes.** The lymph drainage of that part of the colon supplied from the superior mesenteric artery is into the **mesenteric nodes,** joining the lymph vessels from the small intestine to form the **intestinal trunk.** The lymph vessels following the branches of the inferior mesenteric artery drain into the **left lumbar nodes** at the level of the origin of this artery, and therefore join the thoracic duct independent of the intestinal lymph trunk. The lymph nodes at the base of the mesentery of the appendix, receiving lymph from the appendix and ileocaecal angle, drain into the lymph nodes about the superior mesenteric artery. Some of these lymph vessels are said, however, to pass in front of the pancreas and thence into pyloric nodes.

It is generally agreed that the lymph drainage of the anal canal below the pectinate line is perineal, while above this it is pelvic. There is a submucosal and an intermuscular lymphatic plexus. The lymph vessels from the anal canal below the pectinate line and from the perianal skin drain into the **superficial inguinal lymph nodes,** particularly into the upper and medial group, and hence into the **external iliac nodes.** A middle group of lymph vessels draining the upper anal canal and lower rectum follows the middle rectal vessels and drains into the **internal iliac nodes.** The superior group of lymph vessels passes through the muscular coat of the upper part of the rectum and ends in lymph nodes associated with the superior rectal artery, and hence drains to the nodes around the origin of the inferior mesenteric artery (**inferior mesenteric nodes**).

Nerves of the Colon

The nerve supply of the large intestine—sympathetic and parasympathetic—reaches it in the plexuses around the arteries. The ascending and traverse colons are supplied from the superior mesenteric plexus whose parasympathetic fibres are from the **vagus nerves.** The inferior mesenteric plexus probably receives all its parasympathetic fibres through the **pelvic splanchnic nerves,** although there is still some doubt about the lower limit of the vagal nerve supply. It has been shown that the emptying reflex of the large intestine, like that of the bladder, has both its afferent and efferent fibres in the pelvic splanchnic nerves.

Nerves of the Rectum

The sympathetic nerves reach the rectum along the superior rectal vessels, and in part from the **inferior hypogastric plexus.** The parasympathetic nerve supply is from the third, fourth, and sometimes the second **sacral nerves,** and runs in the **pelvic splanchnic nerves** to join the inferior hypogastric plexus which gives its contribution to the wall of the rectum via the lateral ligaments surrounding the middle rectal arteries. The sensory nerves follow the course of the parasympathetic nerves, except the afferent fibres to the anal canal and the perianal skin, which are sensory fibres of the inferior rectal branches of the pudendal nerves and are concerned with the cutaneous sensation of the anal skin.

It is generally stated that rectal continence depends on the integrity of the deep muscular ring round the anal canal, i.e., the ring formed by the deep part of the external sphincter and the puborectalis muscle, and that the internal sphincter is not an essential part of the mechanism, and can in fact be divided surgically without producing incontinence. There is, however, no evidence of sustained voluntary contraction of the external sphincter, and one suggestion is that there are two mechanisms involving rectal continence: one reflex, involving the rectum and the external anal sphincter; the other concerning the descending colon as a whole. According to this explanation the external sphincter, usually relaxed, contracts reflexly when contraction exerts pressure in the rectum, and remains contracted for the few seconds during which pressure is maintained, relaxing as rectal activity ceases. This would explain the disappointing results when, for example, the whole rectum is excised, leaving the sigmoid colon united to the anal canal. In practice it has been found that at least one quarter of the rectum is necessary to keep the afferent side of the reflex intact. The second mechanism has been called 'reservoir continence' and is based on the theory that the colon acts as a reservoir quite independent of sphincteric action, and this would account for the reasonably satisfactory continence which can be attained, given careful attention to regulation of diet, following colostomy.

THE DEVELOPMENT OF THE DIGESTIVE SYSTEM

During the formation of the head and tail folds, part of the yolk sac is taken into the embryonic body to form the **foregut** cranially and the **hindgut** caudally. These are connected by the **midgut** which is at first widely continuous with the yolk sac. With the rapid growth and further folding of the embryo, the elongation of the body stalk, and the extension of the amniotic cavity, the connexion with the yolk sac becomes narrowed to the **vitello-intestinal duct** which finally runs from the midgut through the umbilical cord to a diminutive yolk sac on the foetal surface of the placenta. That part of the foregut cranial to the septum transversum runs forward in the midline mesoderm dorsal to the developing heart as far as the **stomodaeum** or primitive oronasal cavity, through which it communicates with the exterior when the **buccopharyngeal membrane** disappears.

Caudal to the septum transversum, the splitting of the lateral mesoderm and consequent formation of the body cavity frees the gut from the body wall so that it comes to be in the coelom, broadly attached to the midline posteriorly by the **dorsal mesentery.** The abdominal part of the foregut, the future stomach and cranial half of the duodenum, is in contact ventrally with the mesodermal mass of the septum transversum

from which it draws out a **ventral mesentery.** The arteries of the foregut, midgut, and hindgut, namely the coeliac axis, the superior and inferior mesenteric arteries, run in the dorsal mesentery.

As the midgut grows it is pushed out of the shallow abdominal cavity into an extension of the coelom in the base of the umbilical cord which is still relatively large. Here it develops until the second month. To begin with, the midgut is a median loop which forms the gut tube from the caudal part of the duodenum to the transverse colon—developed from its cranial and caudal ends respectively. The superior mesenteric artery extends into the mesentery between the two limbs of the loop, giving branches to both, and continuing along the vitello-intestinal duct from the apex of the loop to the yolk sac. The duct gradually becomes obliterated to a fibrous cord which finally disappears together with that part of the artery [FIGS. 45 and 550].

Coils of small intestine develop from the cranial and adjacent part of the caudal limb of the loop within the coelom of the umbilical cord, and the caecum can be recognized at an early stage as a bulbous outgrowth on the caudal limb of the loop. The further development of the midgut involves an anticlockwise rotation through 180 degrees round the axis of the superior mesenteric artery, as seen from in front, and this is completed as the gut re-enters the abdominal cavity from the coelom in the umbilical cord. The small intestine returns first, passing into the centre of the abdomen and to the right of the mesentery of the hindgut, while the part of the colon derived from the midgut follows and takes up a transverse position superior to the main mass of the small intestine and in front of the duodenum. The superior mesenteric artery now gives off branches to the small intestine from its left (inferior) side and colic branches from its right (superior) side. The last part of the gut to leave the umbilical cord is the caecum, which at first lies below the liver and later is said to 'descend' to its adult position in the right iliac fossa.

Caecum and Appendix Vermiformis

The caecum appears in the sixth week of development as a small outgrowth of the antimesenteric wall of the primitive gut (midgut) not yet differentiated into small and large intestines. The outgrowth is of the same size throughout and is practically equal to the intestine in diameter. About the eleventh week it increases very considerably in length (its length being equal to about five times the diameter of the small intestine, and thus being relatively as long as in the adult). The base, about one-fifth of its length is as wide as the intestine, whilst the remaining four-fifths of the outgrowth—the future vermiform appendix—is only about one-half or one-third the diameter of the gut.

The base continues to expand with the intestine; the distal part grows mainly in length, so that towards the end of foetal life the caecum has attained a conical shape, the wider end joining the ascending colon, the narrow end tapering gradually and passing into the appendix. That form, known as the infantile type of caecum, is retained for some time after birth, and it may (in 2 or 3 per cent. of cases) persist even throughout life.

About the sixth or seventh month of intra-uterine life the wall of the terminal portion of the small intestine adheres to the medial side of the caecum for some distance below the

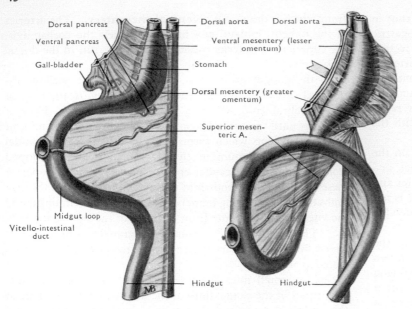

FIG. 550 A diagram to show two early stages in the development of the abdominal part of the gut tube and its mesenteries. Note the rotation of the midgut loop and the ballooning of the dorsal mesentery of the stomach to the left by the extension of the omental bursa (indicated by the arrow) into that mesentery.

ileocolic orifice. The connexion, which is made more intimate by the passage of two folds of peritoneum, one anterior, the other posterior, between the two parts, profoundly modifies the subsequent growth of the caecum and determines very largely its adult form. For, when the caecum begins to expand, the medial aspect lags behind, perhaps as a result of its connexion with the termination of the ileum. In consequence of this, the antimesenteric wall grows and expands more rapidly, making the caecum lopsided, and causing this wall to form the greater part of the caecum, including its distal wall. The original apex, with the vermiform appendix springing from it (anchored, as it were, to the end of the ileum), is relegated to

the medial and posterior aspect of the caecum, a little way inferior and posterior to the end of the ileum.

The position of the caecum varies at different periods of intra-uterine life. About the eleventh or twelfth week it lies immediately beneath the liver crossing the descending part of the duodenum. From there it then gradually travels to the right, crossing the second part of the duodenum, and is found lying on the right side, just beneath the liver, at the fourth month. From there, due to the growth in the length of the trunk, it separates from the liver and descends slowly to its adult position, which it usually approaches towards the end of foetal life, but it may not actually reach it until some time after birth.

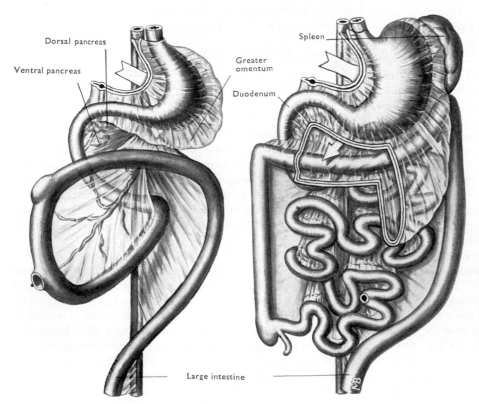

FIG. 551 A diagram of two later stages in the development of the abdominal part of the gut tube and its mesenteries. Note the changes in position of the various parts of the gut tube and the extension of the omental bursa with consequent formation of the greater omentum and the ligaments of the spleen from the dorsal mesentery of the stomach. The duodenum and the ascending and descending parts of the colon become fused to the posterior abdominal wall while the other parts of the gut tube retain their mesenteries.

Hindgut and Rectum

The hindgut develops into the left flexure of the transverse colon, the descending and sigmoid parts of the colon, and the rectum, but at first terminates in the wide cloaca completely surrounded by embryonic mesoderm.

The rectum and anal canal are formed from the tailward portion of the hindgut and from the **anal pit.** The primitive, closed, cloacal portion of the hindgut becomes divided by a septum into ventral and dorsal portions. The ventral portion, with the allantois growing from it, forms the **urogenital sinus,** and the dorsal portion forms the rectum and the upper part of the anal canal. The anal pit is separated from the primitive gut by the **anal membrane,** but that membrane disappears, and thus the rectum comes to open on the surface, through the anal canal.

A spindle-shaped enlargement of the terminal portion of the rectum is formed early. It persists until birth, and a smaller temporary enlargement is formed proximal to it. The wall of the distal swelling is thrown into numerous longitudinal ridges and folds from which the anal columns are formed.

The rectum and anal canal at first form a single, continuous, straight tube which passes downwards in front of the comparatively straight pelvic surface of the sacrum to the anal orifice. That is the condition which the parts present at birth. After birth, the bony pelvis undergoes great enlargement. The sacrum and coccyx become curved, and the anteroposterior diameter of the true pelvis increases very considerably. The urinary bladder and, in the female, the uterus—both of which are mainly in the abdomen at birth—descend into the true pelvis. The anal orifice appears to be moved further forwards in the perineum owing to the bending of the sacrum and coccyx, and the rectum is pushed back into the hollow of the sacrum, whose curvature it follows. Another curvature is formed by the junction of the curved rectum with the straight anal canal as it passes downwards and backwards through the tissues of the pelvic floor. The increase in the thickness of the pelvic floor gives to the anal canal the length which it attains in the adult.

Stomach

The abdominal part of the foregut, attached to the septum transversum widens with an increasing dorsal convexity. This is the primitive stomach, at this stage lying with its long axis in the median plane. Its convex, posterior border has the broad dorsal mesentery—the **dorsal mesogastrium**—attached to it, and a concave anterior border continuous with the mesoderm of the septum transversum, part of which becomes the **ventral mesentery,** or ventral mesogastrium [FIG. 550]. The stomach later rotates into the horizontal plane in such a way that its right surface is now posterior, the left surface anterior, the convex dorsal border passing to the left and forming the **greater curvature,** and the concave ventral margin becoming the right border, the **lesser curvature.**

Wall of the Gut

To begin with, the gut is a thin tube of entoderm and its muscular and serous coats are formed later from the inner (visceral) layer of the split lateral mesoderm which clothes the entodermal tube. This visceral mesoderm forms all the elements of the gut tube other than its epithelial lining and

that of the glands derived from it, and includes the visceral peritoneum and hence the mesenteries. The early gut thus lies within the broad dorsal mesentery by which it is attached to the midline of the dorsal wall of the coelom. The foregut has, in addition, its ventral mesodermal attachment to the septum transversum.

There is a proliferation of the epithelial lining of the embryonic gut, particularly of the duodenum, which may cause complete occlusion of its lumen at one stage, and this is later restored by a process of vacuolation. The villi appear as independent elevations of the epithelium and the intestinal glands—the crypts of Lieberkühn—as tubular outgrowths of the epithelium about the bases of the villi. Both villi and glands increase in number in early childhood.

Peritoneum

Three developmental processes are responsible for the complicated relations of the peritoneum in the adult:

1. The rotation of the gut.
2. The rotation of the stomach with the concomitant development of the omental bursa.
3. The absorption and fusion of layers of peritoneum, e.g., in the duodenum and the ascending and descending parts of the colon, whereby they adhere to the posterior abdominal wall and, losing their mesentery, are covered by peritoneum only on their anterior aspect.

Omental Bursa. The omental bursa commences as a shallow recess or hollowing out of the right side of the dorsal mesogastrium which at this stage is short and broad. The recess extends cranially, (1) to the right of the oesophagus as the pneumo-enteric recess, and (2) into the ventral mesentery to abut on the liver as the hepatico-enteric recess. A further enlargement—the pancreatico-enteric recess—extends caudally and to the left by the hollowing out of the dorsal mesogastrium, carrying the stomach within the two layers of its anterior wall, while the pancreas is enclosed between the two layers of the posterior wall. The site of the original invagination represents the **epiploic foramen,** the entrance to the common cavity or **vestibule.** The caudal end of the pneumo-enteric recess, with the hepato-enteric extension, forms the

FIG. 552 Diagrams to illustrate development of greater omentum. (After Hertwig, 1892.)

(1) shows the beginning of the greater omentum and its independence of the transverse mesocolon; in (2) the two come in contact; and in (3) they have fused along the line of contact.

A, stomach; B, transverse colon; C, small intestine; D, duodenum; E, pancreas; F, greater omentum; G, placed in peritoneal cavity; H, in omental bursa.

superior recess of the adult omental bursa and opens widely into the vestibule. The remainder of the pneumo-enteric recess is generally obliterated, but a small cranial remnant may remain as the infracardiac bursa to the right of the oesophagus, superior to the diaphragm. The pancreatico-enteric recess develops into the inferior recess of the omental bursa, communicating with the superior recess through a narrow neck between the left gastric and common hepatic arteries which form the gastropancreatic folds. To the left of these the bursa expands into the dorsal mesentery as the main extension of the bursa behind the stomach. The shifting and growing stomach draws out the mesoderm of the ventral mesentery from the region of the septum transversum, and the part of this ventral mesentery which becomes the adult lesser omentum forms part of the anterior wall of the superior recess and the vestibule of the omental bursa. The formation of the omental bursa is probably in part responsible for the rotation of the stomach into its transverse position, and thereafter the inferior recess of the omental bursa extends caudally anterior to the intestine, carrying the greater omentum (apron), which encloses this recess, downwards between the abdominal contents and the anterior abdominal wall.

Following the rotation of the gut and the formation of the

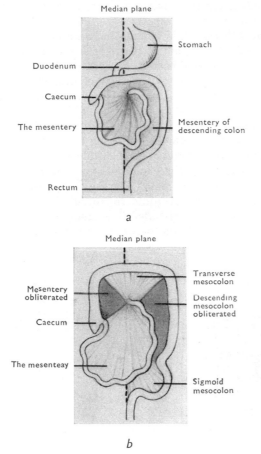

FIG. 553 Two diagrams to illustrate development of the mesenteries.

In (a) the rotation of the intestinal loop and the continuous primitive mesentery are shown. In (b) which shows a more advanced stage, the portions of the primitive mesentery (going to the ascending and descending colons) which disappear, through their adhesion to the posterior abdominal wall, are shaded dark; the portions which persist are lightly shaded.

omental bursa, fusion and absorption of opposed peritoneal layers occur. The posterior layer of the peritoneum of the omental bursa which encloses the pancreas fuses with the peritoneum of the posterior abdominal wall and the fused layers are then absorbed, leaving the pancreas fixed behind the omental bursa [FIG. 552]. Similarly, the posterior layers of the greater omentum forming the posterior wall of the omental bursa caudal to the stomach fuse with the transverse mesocolon to form the gastrocolic ligament. The adult omental bursa seldom extends lower than the attachment of the transverse mesocolon, and beyond this the four layers of the greater omentum are fused. The mass of small intestine filling up the central abdomen pushes the ascending and descending parts of the colon to the right and left respectively so that their mesenteries, applied to the posterior abdominal wall, fuse with it and are lost [FIG. 553].

ABNORMALITIES OF THE GUT

Diverticulum Ilei (Meckel's)

The commonest abnormality is a Meckel's diverticulum. This is an unobliterated remnant of the vitello-intestinal duct which originally extended from the apex of the midgut loop of the embryo along the body stalk to a degenerated yolk sac. A Meckel's diverticulum occurs in about 2 per cent. of subjects, and commonly it is connected to the ileum about 1 metre proximal to the ileocaecal valve, is 5–7·5 cm. in length, and has the same calibre as the ileum. Its blind extremity may be free or attached to the umbilicus by a fibrous band which may contain a continuation of the cavity, though this is rare. The mucous membrane of a Meckel's diverticulum usually resembles that of the small intestine, but occasionally shows patches of gastric mucosa, or even small pieces of pancreatic tissue.

Atresia

Complete congenital stenosis or atresia of the gut most often affects the duodenum or the rectum, but may occasionally be found in the jejunum or the ileum. Congenital cysts and duplications of different parts of the intestine occur, particularly in association with the lower end of the ileum.

Anomalies of Rotation

This results in some unusual arrangement of otherwise normal viscera within the abdominal cavity.

Situs Inversus. The most drastic alteration is found in the condition of situs inversus which may involve the thoracic as well as the abdominal viscera. The disposition of the viscera is a mirror image of the normal arrangement, the stomach passing from the right to the left side, the liver and gall-bladder on the left side, and the caecum and appendix in the lower left quadrant.

Reversed Rotation. In this anomaly the midgut loop has rotated in a clockwise instead of an anticlockwise direction, and as a result the small intestine crosses in front of the superior mesenteric artery, while the colon lies behind the artery, and there is a situs inversus involving the midgut loop.

Non-Rotation. Here the midgut loop is brought into the abdominal cavity without any essential change in position except, of course, that of increased growth. The small intestine lies superiorly and to the right side of the body, the colon inferiorly and to the left side, and both the small intestine and the ascending colon remain suspended by their original mesenteries. The caecum and appendix lie just above the pubic symphysis and close to the midline. There is no transverse colon, and the first part of the colon runs up towards the spleen, while the second part corresponds in position to the normal descending colon. An important anomaly, usually classed under this heading, is non-descent of the caecum, which is therefore situated immediately underneath the liver.

Congenital Umbilical Hernia

Congenital umbilical hernia is a form of incomplete rotation of the gut in which the midgut loop has not returned to the abdomen. The condition varies from one in which there is a small amount of gut, very often caecum, in a hernial sac, to the more severe omphalocele where most of the midgut loop has remained in the umbilical coelom.

Anomalies of the Rectum

Congenital anorectal anomalies are associated with an imperforate anus, and in the simplest type the anus and the rectum remain separated by a persistent anal membrane. The rectum, however, may end in the pelvis some distance above the anus, or the anus and a lower part of the rectum develop normally to form a blind pouch which is separated from the blind end of the upper portion of the rectum by a membrane or a considerable gap.

THE LIVER

The liver, an unpaired digestive gland, secretes bile which empties into the duodenum, and its so-called endocrine function is to deal with the products of digestion brought to it by the portal vein, e.g., the formation of glycogen and the controlled liberation of glucose into the blood stream. It is the largest and heaviest of the viscera, fills most of the abdominal cavity enclosed by the diaphragm and ribs [FIGS. 509 and 554], i.e., the right hypochondriac and epigastric regions, and extends into the left hypochondriac region as far as the left midclavicular line. In the foetus the liver constitutes as much as one-eighth of the total body weight, but the ratio is reduced to a thirty-sixth in the adult. In the living subject the liver is reddish brown in colour, firm and pliant to the touch, but friable, while the 'injected' liver of the dissecting room cadaver is hard and solid. When the abdomen is opened the liver is seen to be divided into a large right and a small left lobe by the attachment of the falciform ligament to its anterior surface. The great bulk of the liver is the right lobe with an anteroposterior diameter of 12·5–15 cm. reduced by half in the midline. The liver moves with the diaphragm in respiration but is kept in its upper position in the abdominal cavity by the intra-abdominal pressure, maintained largely by the tonus of the muscles of the anterior abdominal wall.

The liver has a continuous superior and anterior convex surface [FIG. 555] in contact above, at the sides, and in front with the diaphragm. In the midline it is related directly to the anterior abdominal wall in the infrasternal angle. The large right lobe is separated by the diaphragm from the base of the right lung, the pleura, and the sixth to eleventh right ribs and costal cartilages. To the left of the midline the diaphragm separates the liver from the ventricles of the heart and, to a small extent, the base of the left lung and pleura and the seventh and eighth costal cartilages. This convex surface is separated from the visceral surface by the sharp inferior border which normally is a little above the costal margin, except in the infrasternal angle where it lies in the transpyloric plane in the midline. On the visceral surface there are two small lobes—the **caudate** and the **quadrate**—marked out by one transverse and four vertical fissures [FIG. 556]. The transverse fissure, or **porta hepatis,** is the hilus of the gland for the transit of vessels and ducts. Inferior to the porta hepatis the quadrate lobe is demarcated by the fossa for the gall-bladder to the right and the fissure for the ligamentum teres to the left. The quadrate lobe overhangs the pylorus, the adjacent part of the duodenum, and the transverse colon. Above the porta hepatis is the caudate lobe, bounded by the groove for the inferior vena cava to the right and the fissure for the ligamentum venosum to the left. A bridge of liver tissue—the **caudate process**—connects the caudate and right lobes

FIG. 554 Right lateral aspect of trunk, showing surface topography of viscera.

L.   Liver.   RK.   Right kidney

PL.   Line of pleural reflexion.   RL.   Right lung

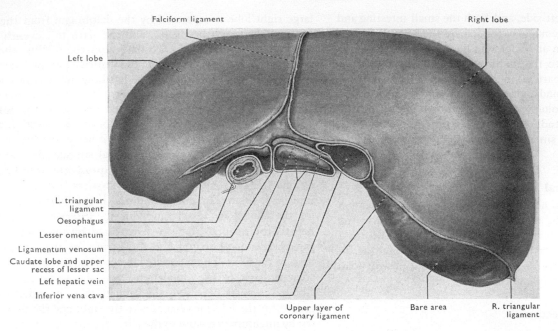

Falciform ligament

Right lobe

Left lobe

L. triangular
ligament

Oesophagus

Lesser omentum

Ligamentum venosum

Caudate lobe and upper
recess of lesser sac

Left hepatic vein

Inferior vena cava

Upper layer of
coronary ligament

Bare area

R. triangular
ligament

Fig. 555   Liver viewed from above.

between the inferior vena cava and the porta hepatis. The caudate lobe is related to the diaphragm anterior to the aorta, and forms the anterior wall of the superior recess of the omental bursa.

The visceral surface has the impressions of the viscera with which this aspect of the liver is in contact. The visceral surface of the left lobe has an elevated **omental tuber** related to the lesser omentum and lesser curvature of the stomach and the gastric impression continuous with an oesophageal impression near the upper end of the fissure for the ligamentum

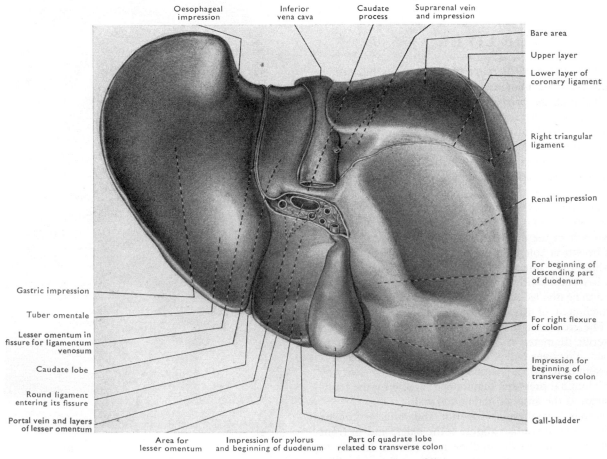

Oesophageal
impression

Inferior
vena cava

Caudate
process

Suprarenal vein
and impression

Bare area

Upper layer

Lower layer of
coronary ligament

Right triangular
ligament

Renal impression

For beginning of
descending part
of duodenum

For right flexure
of colon

Impression for
beginning of
transverse colon

Gall-bladder

Gastric impression

Tuber omentale

Lesser omentum in
fissure for ligamentum
venosum

Caudate lobe

Round ligament
entering its fissure

Portal vein and layers
of lesser omentum

Area for
lesser omentum

Impression for pylorus
and beginning of duodenum

Part of quadrate lobe
related to transverse colon

Fig. 556   Lower (visceral) and posterior surfaces of liver.

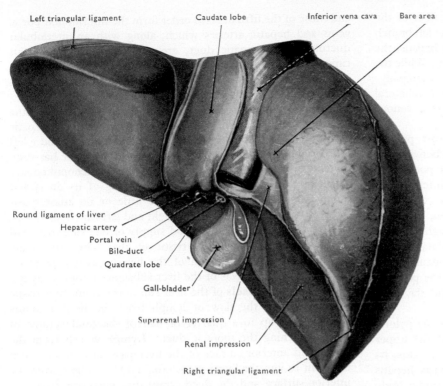

Left triangular ligament   Caudate lobe   Inferior vena cava   Bare area

Round ligament of liver
Hepatic artery
Portal vein
Bile-duct
Quadrate lobe

Gall-bladder

Suprarenal impression

Renal impression

Right triangular ligament

FIG. 557   Liver viewed from behind.

Comparison of the cut edges of peritoneum, as shown in this Figure, with the cut edges of peritoneum in Fig. 523, gives a very clear picture of the peritoneal reflexions from the surfaces of the liver.   The upper and lower layers of the coronary ligament diverge from the right triangular ligament to enclose the bare area.

gall-bladder is an impression made by the descending part of the duodenum [FIG. 556].

Peritoneum of the Liver

Except in the area of attachment of its ligaments, the liver is completely covered by the peritoneum.   The peritoneum is reflected in two layers from the liver to the diaphragm as the ligaments of the liver.   On the superior surface of the liver, the two layers forming the falciform ligament [FIG. 555] separate and pass to the right and to the left to form parts of the coronary and left triangular ligaments.   Posteriorly they converge on the lesser omentum forming the other layers of these ligaments, and enclosing between their layers [FIG. 557] an area of liver in direct contact with the diaphragm, the **bare area**.   At their extremities the two layers of the ligaments fuse to form the right and left triangular ligaments respectively [FIG. 557]. The **falciform ligament** extends from the midline of the supra-umbilical part of the anterior abdominal wall and the lower, anterior surface of the diaphragm, to the anterior and upper surface of the liver, where its right and left layers become continuous laterally with the peritoneum over the respective lobes, and superiorly with the anterior layers of the coronary and left triangular ligaments respectively.   Lying in the free edge of the falciform ligament is the **ligamentum teres** which is the obliterated remains of the left umbilical vein of the foetus, and extends from the umbilicus to the notch on the

venosum.   The right lobe has two large impressions, the upper for the anterior surface of the right kidney, and the lower for the right flexure of the colon.   Immediately to the right of the

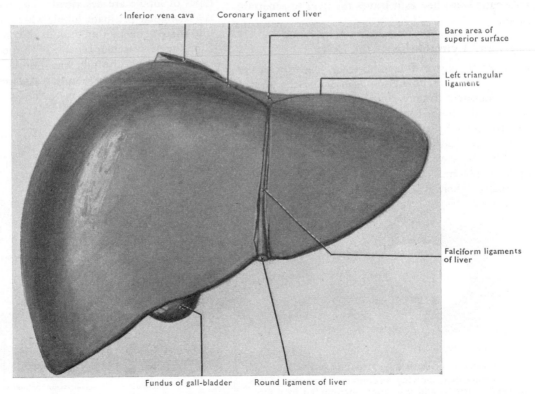

Inferior vena cava   Coronary ligament of liver

Bare area of
superior surface

Left triangular
ligament

Falciform ligaments
of liver

Fundus of gall-bladder   Round ligament of liver

FIG. 558   Liver viewed from the front.

inferior border of the liver. Here it runs in the floor of the fissure for the ligamentum teres [FIG. 556] to join the left branch of the portal vein at the left extremity of the porta hepatis, the lower end of the fissure for the ligamentum venosum. While the ligamentum teres is normally avascular, there are accompanying para-umbilical veins representing a potential collateral circulation between the portal vein and systemic venous system of the anterior abdominal wall.

The **ligamentum venosum** is a fibrous cord passing through the depths of the fissure for this ligament, between the caudate and left lobes, from the left branch of the portal vein to the left hepatic vein just before this enters the inferior vena cava. It represents a venous shunt—the **ductus venosus**—established in the foetus to allow blood returning from the placenta to reach the heart without passing through the liver sinusoids. It closes within the first 2 days of postnatal life. It is quite common to find a hepatocolic fold of peritoneum in the form of a ridge of parietal peritoneum between the postero-inferior surface of the right lobe of the liver and the right flexure.

The broader part of the lesser omentum from the pyloric half of the lesser curvature of the stomach and the upper border of the duodenum is attached to the porta hepatis, its free edge forming the right boundary of the porta hepatis [FIG. 556]. The narrower part of the lesser omentum (only 1–2·5 cm. broad) extends from the oesophageal half of the lesser curvature to the bottom of the fissure for the ligamentum venosum, its two layers separating superiorly to become continuous with the posterior layers of the coronary and left triangular ligaments respectively. The inferior vena cava is deeply embedded in its sulcus and is attached by dense connective tissue which may form a band known as the ligament of the vena cava, sometimes containing hepatic tissue. It receives the hepatic veins just as it leaves the liver to penetrate the diaphragm.

Blood Vessels and Lymphatics of the Liver

The blood from the gastro-intestinal tract, loaded with the products of digestion, is brought to the liver by the **portal vein,** while the **hepatic artery** supplies the liver substance. Both vessels enter the liver from the porta hepatis. Blood is drained from the liver by two or three hepatic veins which join the inferior vena cava just before it passes through the diaphragm. The portal vein and hepatic artery divide into a right and a left branch at or immediately within the porta hepatis. Thereafter, there is a regular sequence of branching,

and those of the fifth or sixth order form the interlobular portal veins and hepatic arteries which, along with an interlobular ductule of the hepatic duct, are enclosed within a fibrous capsule—the **perivascular fibrous capsule**—and known as the portal canals [FIG. 559]. Three or four such portal canals are grouped round the periphery of the liver lobule.

The functional division of the liver on the basis of the territory supplied by the right and left branches of the hepatic artery and portal veins respectively would be a right and a left lobe of approximately equal size. Indeed, the liver has been divided into segments comparable to the bronchopulmonary segments in accordance with the branching of its ducts and blood vessels. There is said to be little or no anastomosis between the blood vessels of the right and left functional lobes, so that interruption of the blood supply of one side may lead to destruction of the liver tissue of that lobe, and acute occlusion of a main branch of the hepatic artery generally produces rapid necrosis of the liver substance supplied by it.

The **lymph vessels** of the liver run in the connective tissue framework and the flow of lymph from the liver accounts for anything up to a quarter or half of the total volume of lymph reaching the thoracic duct. Lymph vessels from the upper and anterior surface of the liver pass into the falciform ligament to join the lower parasternal nodes. Those from the inferior surface and the deep lymphatics emerging from the porta hepatis drain into lymph nodes round the hepatic artery, or pass to the left in the lesser omentum to reach the gastric nodes. Other lymph vessels, especially from the bare area, accompany the inferior phrenic vessels to reach the coeliac nodes.

Structure of the Liver

The unit of liver structure is the lobule. Two different types of lobule are described: one, centred on a portal canal, is a triangular prismatic lobule which may be partially divided into three acini by thin strands of connective tissue radiating from the perivascular capsule of the portal canal. In the other, exemplified by the pig, the fibrous partitions of adjacent portal canals join, surrounding a hexagonal lobule [FIG. 560].

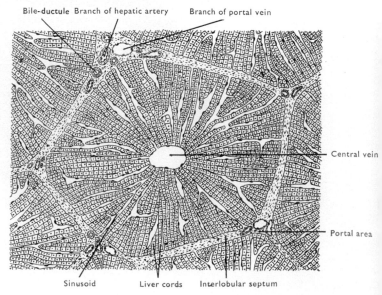

Bile-ductule   Branch of hepatic artery   Branch of portal vein

Central vein

Portal area

Sinusoid   Liver cords   Interlobular septum

FIG. 560   Diagram of the liver lobule (pig).

Hepatic cells

Hepatic artery

Fibrous capsule

Interlobular bile-ductule

Portal vein

FIG. 559   Diagram illustrating structure of liver.
Section of a portal canal, showing its contained branches of the portal vein, hepatic artery, and bile-duct, surrounded by a prolongation of the hepatobiliary capsule.

This lobule consists of the central vein with a radiate arrangement of sinusoids and liver cells. The liver cells are arranged in cords or plates extending radially from the **central vein** to the periphery, and separated from one another by adjacent sinusoids. It has been stated that the plates are one cell thick, but as the cells are secreting units it is more probable that the plates include at least two cells and secrete into the network of intercellular **bile canaliculi,** which are in fact spaces between thickened cell membranes of adjacent liver cells. At the periphery of the lobule the canaliculi drain into short branches which lead immediately into the **interlobular ductules** of the portal canals from the periphery of the lobules. **Sinusoids.** The sinusoids are irregular blood spaces which have a radial course in the lobule from the periphery to the central vein. They are separated from one another by the plates of liver cells, although it is said they communicate by perforating channels in the liver cell plates [FIG. 561]. The interlobular branches of the portal vein and hepatic artery open into the sinusoids at the periphery of the hexagonal lobule and the sinusoids drain into the central veins. Stellate reticulo-endothelial cells—the cells of von Kupffer—line the walls of the sinusoids separated from the liver plates by the space of Disse. One function of the liver is to deal with the products of destruction of red blood cells arriving from the spleen, and to form bile pigments from their haemoglobin. This process is started by the **stellate endothelial cells** and completed by the liver cells. The central veins of the lobules join to form **sublobular veins** which in turn unite to form hepatic veins. Because of the fibrous attachment of their walls the hepatic veins remain patent, and are obvious to naked-eye examination of the cut surface.

THE EXCRETORY BILIARY APPARATUS

The excretory apparatus consists of a right and a left **hepatic duct** which join to form the **common hepatic duct,** and the **gall-bladder** which through its duct—the **cystic duct**—unites with the common hepatic duct to form the **bile-duct.** After receiving the pancreatic duct the bile-duct opens into the duodenum.

GALL-BLADDER

[PLATE 33 A–C]

There is a continuous secretion of bile by the liver, but normally it enters the duodenum only during digestion. The bile is stored in the gall-bladder where there is some absorption of water and secretion of mucus by glands in the gall-bladder, producing in this way a more concentrated bile.

The gall-bladder is situated in the fossa on the visceral surface of the liver, to which it is attached by some areolar tissue, lymphatics, and small veins [FIG. 556]. The exposed or inferior surface of the gall-bladder is covered by peritoneum, but occasionally it may be completely covered with peritoneum and suspended from the liver by a short mesentery. It is about 7·5 cm. long, 2·5–3 cm. in diameter, and has a capacity of between 30 and 45 ml. It is a pear-shaped body, the wide end or **fundus** reaching the inferior border of the liver and touches the anterior abdominal wall behind the right ninth costal cartilage opposite the lateral border of the rectus abdominis. Superiorly, towards the porta hepatis, the body narrows to the neck which is curved medially, often with an S-shaped bend, as it becomes continuous with the cystic duct. The transverse colon and the first 2–3 cm. of the duodenum lie behind and below, and sometimes in actual contact with the gall-bladder, while the **neck** is a guide to the epiploic foramen which lies immediately to its left [FIG. 562]. Commonly there is a dilatation of the neck of the gall-bladder, the so-called Hartmann's pouch, close to the cystic duct, and if well developed the pouch may be adherent to the duct.

HEPATIC, CYSTIC AND BILE-DUCTS

[PLATE 33 D]

Normally the right and left hepatic ducts unite just outside the porta hepatis to form the common hepatic duct. They may, however, descend for a variable distance before uniting, so that the common hepatic duct may be from 2·5 to 7·5 cm. long. The **cystic duct** runs backwards, downwards, and to the left, from the neck of the gall-bladder to join the common hepatic duct and form with it the bile-duct. There is a great range of variation in the anatomy of this region in respect of the cystic duct, the manner of its junction with the common hepatic duct, and its relation to neighbouring arteries. On average it is up to 3·5 cm. long and runs parallel to the common hepatic duct before joining it just below the porta hepatis. It may, however, be as short as 1 cm. and join the right hepatic duct or the point of junction of the two hepatic ducts. On the other hand, a long cystic duct may join the common hepatic duct within the lesser omentum, or even behind the duodenum.

The **bile-duct,** about 5 mm. in diameter, can vary in length from 5 to 12 cm. according to the length of the common hepatic duct. Normally it begins just below

Branch of
hepatic artery   Bile-ductule   Bile canaliculus   Central vein

Branch of portal vein   Sinusoid   Liver cord

FIG. 561   Diagram of part of a liver lobule (human).

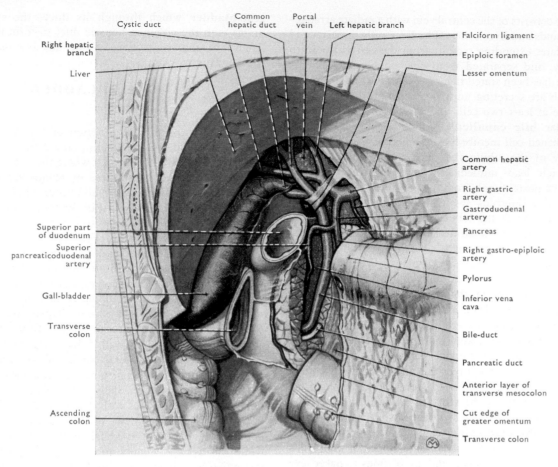

FIG. 562   Dissection showing relations of gall-bladder and bile passages.

the porta hepatis of the liver and in its course lies superior, posterior, and inferior to the superior part of the duodenum, before terminating in the greater duodenal papilla [FIG. 562]. The first part of the bile-duct is in the right free border of the lesser omentum as it forms the anterior boundary of the epiploic foramen, with the hepatic artery on its left side and the portal vein posterior.   The second part passes behind the superior part of the duodenum and has the gastroduodenal artery on the left side, with the portal vein still posterior.   Passing inferiorly, the bile-duct curves to the right, is usually embedded in the posterior part of the head of the pancreas, and is separated by pancreatic tissue from the front of the inferior vena cava.   As it reaches the descending part of the duodenum it lies posterosuperior to the pancreatic duct, and fuses with its external surface to run obliquely through the wall of the duodenum for about 2 cm.   At some point within the duodenal wall, the cavities of the two ducts unite and pierce the mucous membrane on the surface of the **greater duodenal papilla,** situated on the posteromedial wall of the descending part of the duodenum, about 10 cm. from the pylorus.

The bile-duct narrows abruptly where it enters the duodenal wall, but dilates after internal union with the pancreatic duct to form the **hepatopancreatic ampulla** (Vater) [FIG. 563]. The two ducts may, however, remain separate and open independently on the duodenal papilla.   Figures given for the length of the hepatopancreatic ampulla vary from 2 to 14 mm. Both the intramural part of the bile-duct and the ampulla have a sheath of smooth muscle fibres constituting the sphincters

of the bile-duct and of the hepatopancreatic ampulla, the latter consisting, it is said, of both longitudinal and circular muscle fibres.   There is some evidence of a reciprocal functioning of the musculature of the gall-bladder and the sphincter whereby, with the emptying of the gall-bladder, there is inhibition of the sphincter, but it has been shown that complete vagotomy has no appreciable effect on the rate of emptying of the gall-bladder,

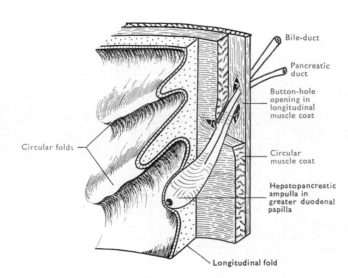

FIG. 563   Diagrammatic section of wall of duodenum to show passage of bile and pancreatic ducts and their junction to form the hepatopancreatic ampulla.

and it is generally accepted that there is no significant neural control of either the gall-bladder or the sphincter of the ampulla in Man. Certain substances, e.g., fats, induce the liberation of secretin from the wall of the small intestine, which in turn produces contraction of the gall-bladder and the progressive relaxation of the ampullary sphincter.

Blood Vessels and Nerves of the Gall-Bladder and Bile-Duct

The gall-bladder is supplied by the **cystic artery** which normally arises as a single stem from the right hepatic branch close to the porta hepatis. At the neck of the gall-bladder it divides into two branches supplying the two surfaces. The relations and course of the cystic artery are notoriously variable. It may be a branch of either the hepatic artery or its right branch; it may cross in front of the common hepatic duct and run parallel to the cystic duct on its left side. The blood supply of the bile-duct is by arterial twigs, mainly from the cystic artery and the superior pancreaticoduodenal artery. It should be noted that the supraduodenal portion of the bile-duct has, in a fair proportion of cases, an artery passing ventral to it, most commonly the right hepatic branch and occasionally the right gastric. The **veins** enter the portal vein, or pass directly from the gall-bladder into the hepatic sinusoids.

The **nerve supply** of the bile-ducts and gall-bladder is through the hepatic plexus associated with the hepatic artery and the portal vein, and joined at the porta hepatis by branches from the anterior vagal trunk. It is said that this plexus gives off a number of small descending branches which join the dorsal aspect of the bile-duct, some of which can be followed microscopically towards the duodenal papilla. Afferent fibres of the biliary tract obviously include a number of pain fibres which probably run by way of the sympathetic nerves, though fibres running with the phrenic nerve have been described. Pain is usually due to distension from gallstones.

Structure of the Gall-Bladder and Bile-Duct

The gall-bladder has an incomplete serous coat of peritoneum, a connective tissue coat, some plain muscle fibres, and a mucous membrane of tall columnar cells with a lamina propria. The mucous membrane is generally raised into a number of folds or ridges, and the epithelium has numerous mucous cells. A **spiral fold** in the mucous membrane of the neck of the gall-bladder, often prolonged into the first part of the cystic duct, is said to contain some smooth muscle fibres. There are a few tubular mucus-secreting glands in the lamina propria of the neck of the gall-bladder.

The **extrahepatic ducts** have a lining of tall columnar epithelial cells and the mucosa is often folded. There are scattered bundles of smooth muscle fibres which first become apparent in the bile-duct and form an incomplete layer round the wall of that duct.

DEVELOPMENT OF THE LIVER AND GALL-BLADDER

Both liver and pancreas develop as outgrowths of the primitive foregut lying caudal to the septum transversum. The liver diverticulum appears as early as the third week,

grows into the mesoderm of the caudal part of the septum transversum, and divides [FIG. 76]. The stem becomes the bile-duct, while the caudal limb of the first division forms the cystic duct and the gall-bladder, and the cranial limb is the common hepatic duct. From the branches of the latter, cell trabeculae grow into the mesoderm of the septum transversum. These proliferating and branching cell cords become associated with the vitelline veins bringing the blood from the yolk sac and embryonic digestive tube, and are in part responsible for breaking up these venous channels into sinusoids. The trabeculae acquire a central lumen—the bile canaliculi—flanked on each side by a single layer of cells. Sinusoids and trabeculae finally become arranged to form the liver lobule, each with a central vein (also derived from the vitelline veins) which becomes a tributary of an hepatic vein. Blood from the placenta reaches the liver by the left umbilical vein, but can bypass the sinusoids and reach the inferior vena cava through the ductus venosus and left hepatic vein.

The developing liver and stomach draw away from the septum transversum, bringing with them mesoderm which forms the ventral mesentery. The ventral mesentery between the liver and the stomach becomes the lesser omentum; between the liver and the diaphragm it is the tissue basis of the coronary ligaments above and of the falciform ligament below, and thus is continuous with the mesoderm of the supra-umbilical anterior abdominal wall. The liver enlarges rapidly and at one stage occupies almost the entire embryonic abdominal cavity, but its relative size and weight diminish, particularly as the gut enters the abdominal cavity. Although the secretion of bile and the glycogenic function of the liver are said to commence about the third foetal month, the embryonic liver is essentially a blood-forming organ.

THE PANCREAS

The pancreas, like the liver, is an unpaired digestive gland with an external secretion, the **pancreatic juice,** conveyed to the duodenum by the pancreatic duct, and internal secretions, insulin and glucagon, which pass directly into the blood stream. It is situated in the epigastrium and left hypochondrium about the level of the transpyloric plane, lying behind the omental bursa, in front of the bodies of the first and second lumbar vertebrae, and extending to the left to reach the spleen. The pancreas has a flattened head lodged in the duodenal curve, united to a body, triangular in section, and terminating in the tail which is in contact with the spleen. It has a total length of from 12·5 to 15 cm. and in the living subject is soft, lobulated, and yellowish in colour [FIG. 514].

Head

The head of the pancreas overlaps the descending and horizontal parts of the duodenum to the right and below, respectively, with the anterior and posterior pancreaticoduodenal arterial arcades closely related. Posterior to it are the medial border of the right kidney, the right renal vessels, the inferior vena cava, and the terminal portion of the left renal vein [FIG. 621]. The bile-duct is at first posterior to the head of the pancreas, then enters its substance. The immediate anterior relations are the transverse colon before it has acquired a mesentery, a variable upper part usually covered by the peritoneum of the posterior wall of the omental bursa, and a lower part covered by the peritoneum continuous with the inferior layer of the transverse mesocolon which separates it from the coils of jejunum. The common hepatic artery runs to the right along the upper border of the pancreas before turning upwards in the lesser omentum. The part of the head which extends to the left behind the superior mesenteric vessels is the **uncinate process.**

The junction of the head and body of the pancreas lies at the level of the pylorus, with the gastroduodenal artery on its right border, the beginning of the portal vein posteriorly,

and the superior mesenteric artery in a groove posterior and to the left. The uncinate process separates the mesenteric vessels from the aorta.

Body and Tail

The body of the pancreas has an anterior, a posterior, and a smaller inferior surface. It lies transversely across the vertebral column behind the omental bursa, passing upwards and to the left towards the spleen. Where the body joins the head there is an elevation of the anterior surface—the **omental tuber**—posterior to the lesser omentum and lesser curvature of the stomach. The posterior surface is in contact with the

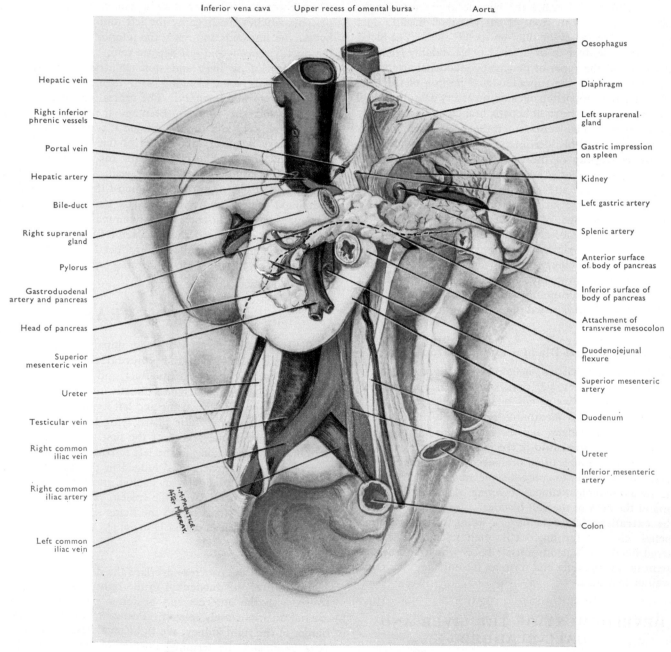

FIG. 564   Viscera and vessels on posterior abdominal wall.

The stomach, liver, and most of the intestines have been removed. The peritoneum has been preserved on the right kidney and the recess behind the caudate lobe. When the liver was taken out, the inferior vena cava was left *in situ*. The stomach bed is well shown.

aorta about the level of the origin of the superior mesenteric artery, the left crus of the diaphragm, the left suprarenal gland, the left kidney, and the left renal vessels, particularly the vein. Its most immediate posterior relation is the splenic vein with the splenic artery just above its superior border. The transverse mesocolon is attached along the anterior border separating the anterior and inferior surfaces, so that the anterior surface is covered by the continuation of the upper layer of the transverse mesocolon as the posterior wall of the omental bursa, and the inferior surface by the reflected inferior layer of the transverse mesocolon [FIG. 523]. The anterior surface faces anterosuperiorly and forms part of the stomach bed, while the inferior surface is related to the duodeno-jejunal flexure and the coils of jejunum. The tail of the pancreas lies within the lienorenal ligament along with the splenic vessels, and may reach the spleen directly anterior to its hilus.

In a dissection, the pancreas can be exposed by turning up the greater omentum and cutting the posterior parietal peritoneum above the horizontal part of the duodenum, or cutting through the greater omentum between the lower border of the stomach and the transverse colon, or through the peritoneum of the posterior wall of the omental bursa posterior to the stomach. All such incisions are liable to damage important blood vessels.

Structure of the Pancreas

The pancreas, like the salivary glands, is a compound acinar gland with lobules bound together by loose connective tissue, but it has no definite capsule. The **alveoli** which secrete the pancreatic juice have a single row of pyramidal cells, the bases resting on a basement membrane and the apices converging towards a central lumen which varies in size according to the functional condition of the organ [FIG. 565]. The inner part of the cell is generally filled with secretory granules and the outer part has a characteristic basophil staining. Between the alveoli are the **islets of Langerhans,** responsible for the internal secretion of insulin and glucagon. They vary in number from $\frac{1}{2}$ to $1\frac{1}{2}$ million, are highly vascular structures, and consist of cords of cells which are classified as A, B, or C, according to the specific staining of their contained granules. There are usually two pancreatic ducts. The **main duct** begins in the tail, runs through the substance of the gland nearer the posterior than the anterior aspect, receives tributaries from the neighbouring lobules, and joins the bile-duct at the duodenum. It may, however, open into the duodenum independently. An **accessory duct** which is usually present is about 6 mm. long, lies anterior and cranial to the main duct in the head of the pancreas, and opens into the duodenum about 2·5 cm. above and anterior to the opening of the bile-duct. There is considerable variation in the detailed anatomy of the pancreatic ducts.

FIG. 565 A section of the pancreas (human).

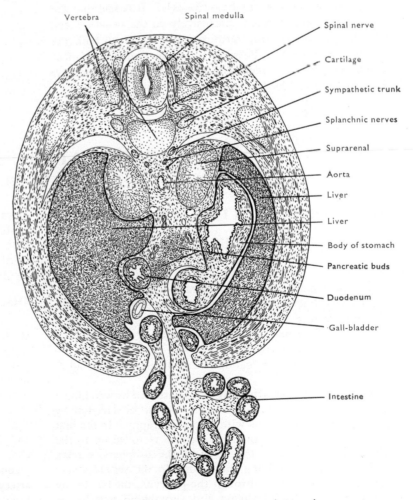

FIG. 566 A transverse section of a human embryo (26 mm.).

Blood Vessels, Lymphatics, and Nerves of the Pancreas

The arteries of the pancreas are branches from the **coeliac trunk** and the **superior mesenteric artery.** These include the two pancreaticoduodenal arteries forming the anterior and posterior pancreaticoduodenal arcades. A dorsal pancreatic branch from the splenic or the common hepatic artery, or sometimes direct from the coeliac axis, runs down behind the head of the pancreas and divides into a right branch supplying the uncinate process and a left branch which, as the transverse pancreatic artery, follows the lower border of the pancreas on its dorsal surface. In addition, the pancreas receives anything from two to ten branches from the splenic artery.

Most of the veins draining the pancreas empty into the **splenic vein,** but small pancreatic and pancreaticoduodenal veins drain into the superior mesenteric or portal veins.

The **lymph vessels** drain into the pancreaticosplenic nodes to the left, into the pyloric nodes to the right, and finally into the lumbar and coeliac nodes around the superior mesenteric and coeliac arteries.

Pacinian corpuscles [FIG. 974] are regularly associated with the pancreas and they are innervated by sensory fibres which travel with the sympathetic system. If there are any pain fibres in the pancreas they probably also run with the sympathetic. The motor innervation is by postganglionic sympathetic and preganglionic vagal fibres, passing to the pancreas in the plexuses on the blood vessels. It is said that the sympathetic fibres terminate exclusively on the blood vessels, while the parasympathetic terminate in relation to both the alveolar and islet cells. Possibly the splanchnic nerves affect the external pancreatic secretion only indirectly through their influence on the blood vessels.

Development of the Pancreas

The pancreas is developed from a dorsal and a ventral outgrowth of the primitive foregut, appearing about the same time as the liver bud. The dorsal outgrowth is just proximal to the liver bud, while the ventral outgrowth may be a primary branch of the liver bud. Following the rotation of the duodenum the origin of the liver outgrowth is shifted to the dorsomedial aspect of the gut so that the two pancreatic rudiments come into contact [FIG. 77]. The dorsal rudiment forms the body and the tail of the pancreas and the ventral rudiment most of the head. The duct systems of the two rudiments anastomose so that that of the body and the tail drains into the duct of the ventral rudiment which normally opens into the bile-duct. The proximal part of the duct of the dorsal rudiment remains in a considerable proportion of cases with an independent opening into the duodenum superior to the opening of the bile-duct. The islets of Langerhans differentiate from the pancreatic ducts simultaneously with the secretory alveoli, but later lose their connexion with the ducts. The pancreas starts to secrete about the third month of pregnancy. The body of the pancreas at first lies between the two layers of the dorsal mesoduodenum and dorsal mesogastrium, and, in the latter, lies in the posterior wall of the omental bursa. With fusion and absorption of peritoneal layers it finally becomes fixed behind the peritoneum.

Anomalies. The commonest anomaly is aberrant or accessory pancreatic tissue within the wall of the stomach, the duodenum, jejunum, or in a diverticulum ilei. Such accessory tissue may or may not contain islets. An anular pancreas forms a complete ring of pancreatic tissue round the descending part of the duodenum, and generally its duct runs to the right in front of the duodenum, turning round behind it to join the main pancreatic duct.

RADIOGRAPHIC ANATOMY

E. SAMUEL

THE GASTRO-INTESTINAL TRACT

Two of the **salivary glands,** the parotid and submandibular, can be demonstrated radiologically by cannulating their ducts with a fine cannula and introducing a radio-opaque oily medium (sialography). In a sialogram of the parotid gland, the accessory duct and the intraglandular branching of the duct can be seen easily [PLATE 28 A] whilst in the submandibular gland, the long course of the duct and the intraglandular branching are readily visible.

The sublingual glands cannot be demonstrated by radiological methods.

The mechanics of **swallowing** can be studied by watching a swallow of barium paste either on a fluorescent screen or by rapid filming of the process with cineradiography. In the first stage of swallowing, the bolus of barium is rolled on to the back of the tongue and the walls of the oropharynx relax in preparation for receiving the bolus. In the second stage, whilst the barium passes through the pharynx, the larynx, by the action from the infrahyoid and suprahyoid muscles, is

elevated beneath the hyoid bone and the epiglottis is tipped posteriorly over the laryngeal opening partly by the aryepiglottic muscle which also closes the inlet of the larynx. In the third stage, the bolus of barium passes down the oesophagus by the muscular action of the oesophageal wall.

Oesophagus

Anatomical impressions on the barium-filled oesophagus are made by some of the structures in the mediastinum [PLATE 28 D]. From above downwards they are:

1. The aortic arch which indents the left anterolateral surface of the oesophagus, the size of this indentation becoming progressively larger as the aortic arch widens with advancing years and loses its elasticity.

2. Immediately caudal to the aortic impression is a second and shallower impression caused by the left principal bronchus pressed posteriorly by the pulmonary artery. This second impression, frequently referred to as the pulmonary artery impression, should more correctly be termed the left bronchial impression.

3. The cardiac impression on the barium swallow is a smooth, anterior concavity and the actual chamber of the heart responsible for it is the left atrium.

4. Immediately above the diaphragm an impression is formed on the posterior wall of the oesophagus by the descending aorta as the oesophagus crosses to the left to enter the stomach.

Stomach

The mucosal pattern [PLATE 29 C] of the stomach as demonstrated by the first swallow of barium appears as linear folds extending along the lesser curvature of the stomach. The mucosal folds over the greater curvature are irregular in pattern and outline but, on the whole, have a more transverse course. As the stomach fills with barium the outlines of the mucosal folds disappear and the stomach fills out. The various shapes of the filled stomach have been designated as J-shaped (orthotonic), which was regarded as normal, steerhorn type (hypertonic) and hypotonic when the stomach is low in position. It is now accepted that all these changes in the shape of the stomach are normal variants and merely an index of body habitus, gastric muscular tone, and the position in which the X-ray is taken, i.e., whether erect or supine [PLATE 29A and B]. Certain areas which are well defined on X-ray examination, such as the pyloric antrum and pylorus [PLATE 29D], are less easy to define in anatomical specimens and the pyloric sphincter between the stomach and the duodenum is frequently difficult to define on the external aspect of the stomach. The pyloric sphincter can be more easily defined on the mucosal aspect by the alteration in pattern between the gastric and duodenal mucosae.

Small Intestine

The 'duodenal cap' [PLATE 29D], although an important site of peptic ulceration, is a radiological rather than a truly anatomical area and refers to the first 2.5 cm. or so of duodenum beyond the pylorus.

The pylorus exerts a dominant influence on the emptying of the stomach, and the barium meal becomes fragmented after it leaves the stomach. The delicate mucosal pattern of the duodenum, its flexures, and the remainder of the small intestine are seen in films taken at appropriate intervals after the barium meal and represent the mucosal folds [PLATE 30]. It is, however, difficult to appreciate the length of the small intestine from radiographs.

The mucosal pattern in the small bowel becomes progressively smoother from the upper jejunum to the ileum. In the terminal ileum the loops may appear quite smooth and devoid of any mucosal pattern.

The ileocaecal valve at the termination of the small intestine again forms a more distinct entity radiologically than anatomically. It is situated on the posterior surface of the caecum towards its medial border and appears as a horizontal opening with two overhanging lips. Radiologically, a well-defined upper lip overshadows a small and ill-defined lower lip.

The vermiform appendix fills with the barium solution during a barium meal or enema and its emptying is slower than the caecum even in normal subjects. The variability of position of the appendix is readily appreciated from barium enema examinations.

Large Intestine

The colon is customarily examined by a barium enema and the saccular appearances produced in the colon by the taeniae coli are reproduced on the radiograph. The sacculations in the colon (haustrations) [PLATE 32 A] fill with barium and the narrow areas between them are produced as a result of the contraction of the taeniae coli and localized contraction of the circular muscle. The transverse, ascending, and sigmoid colon show these sacculations more clearly than the relatively smooth descending colon.

The right and left flexures are relatively fixed points in the colon, but considerable overlapping of the colon frequently occurs as a normal feature at these sites. Likewise, the mobile transverse colon which courses between the flexures is subject to marked variations in its position [PLATE 31 A and D].

In the rectum the indentations due to rectal folds can be seen in the partly filled rectum, but when the rectum is distended with barium solution these folds disappear.

Some information with regard to the physiological and anatomical changes in the bowel can be obtained from a study of the normal gas distribution as seen on straight X-rays.

With the first breaths at birth, the infant inflates not only the lungs but also the stomach, and air so swallowed remains a constant feature of the gastro-intestinal anatomy and physiology throughout life. The most familiar evidence is the air 'cap' of the stomach as seen in examinations in the upright position but, though not so clearly evident, some air is at all times present in both stomach and colon. The small intestine seldom contains any quantity of air as swallowed air is rapidly passed to the colon.

In obstructive conditions of the intestine, air and fluid accumulate proximal to the site of obstruction and it is important to realize that in obstructions virtually all the gas is derived from swallowed air. The difference in X-ray density between the accumulating fluid and the contained air gives rise to fluid levels which are visible in the erect position and assist in the recognition of obstruction.

Gas in the bowel may obscure various organs such as the gall-bladder or the kidney and as such may play an important part in X-ray examination of these structures.

LIVER

The outlines of the soft tissue of the liver can be seen on a plain X-ray of the abdomen. Gross changes such as enlargements or shrinkage may be visualized in this way.

However, the size and shape of the liver can best be seen in life by the use of radioisotope methods. This allows substances excreted by the liver to be tagged with radioisotopes and by the use of digital scanners or scintillation counters the gross shape of the liver can be seen.

GALL-BLADDER [PLATE 33]

The gall-bladder cannot be seen by X-rays unless its contents are abnormal or contrast medium has been administered to fill its lumen. Very rarely, the calcium content of bile may increase to such an extent (limy bile) as to make the gall-bladder visible on the radiograph. Usually the outline of the gall-bladder is demonstrated after it has been filled with radio-opaque contrast medium. In the case of the gall-bladder, one of the iodine-containing cholecystographic agents is administered orally or intravenously and these substances are selectively excreted by the liver and concentrated in the gall-bladder along with the normal bile so that the shadow of the gall-bladder can be seen.

The administration of the oral preparations has been used as a test of gall-bladder function as the medium is concentrated in the gall-bladder. The newer intravenous preparations are concentrated less by the gall-bladder and are less useful in assessing gall-bladder function.

The assessment of function of any organ radiographically is beset with difficulties, as such an assessment is largely subjective, depending on the density of the shadow seen. This in itself is subject to such a large number of variables that great care must be taken in interpreting a poor shadow as an index of deficient function of the gall-bladder.

Radiologically, the gall-bladder appears as a pear-shaped shadow with the stalk directed upwards and medially towards the cystic duct.

With the modern contrast media, the detail is such that the spiral valve in the cystic duct can sometimes be shown. Gall-stones, which prior to the advent of cholecystography could be seen only when they contained sufficient calcium to be visualized, can be seen as negative shadows in the contrast-filled gall-bladder, and with the use of intravenous methods stones in the extrahepatic biliary passages can be seen. After the administration of a fatty meal, contraction of the gall-bladder occurs, and by the expulsion of contrast medium the bile-duct and lower part of the common hepatic duct may be seen.

By the newer intravenous methods the course of the intra- and extrahepatic ducts can be shown and the bile and cystic ducts can also be radiographed. Demonstration of the lower end of the bile-duct and its opening on the hepatopancreatic ampulla is, however, more difficult as the degree of contrast obtained by this method is frequently too low to show the detail necessary.

In operative and post-operative cholecystography the biliary ducts are filled by the introduction either of water-soluble or oily contrast media through a catheter in a surgical opening in the bile-duct or the cystic duct. Operative cholangiography is of utmost value in visualizing stones in the hepatic and bile-ducts at operation and in assessing anatomical variations which may be present. Post-operative cholangiography is usually performed to determine the patency of the bile-ducts and to exclude the presence of any stones remaining in them. The right and left hepatic ducts and the ventral and caudal tributaries of the right hepatic duct within the liver can be shown by these methods.

The common hepatic duct varies in width from 6 to 10 mm. Considerable variation in the site and position of junction of the cystic and hepatic ducts occurs. The cystic duct most frequently joins the posterolateral aspect of the common hepatic duct but it may unite with this on its posterior aspect, and still less frequently on its medial side. In each instance, the cystic and hepatic ducts run in a common sheath for a variable distance before actual union occurs.

Frequently there may be an accessory hepatic duct which represents the dorsal division of the right hepatic duct, and unites either with the common hepatic duct or the cystic duct. These variations in the anatomy of the extrahepatic ducts are important surgically as injuries to these structures during removal of the gall-bladder have serious results.

The bile-duct runs with the pancreatic duct in a common sheath to open into the duodenum. In approximately one-third of cases, the two structures have a common channel and during the performance of a cholangiogram slight reflux of contrast medium into the pancreatic duct may be seen.

PANCREAS

The pancreas cannot be seen on simple X-rays of the abdomen as it has no surrounding fat to allow contrast between it and the adjacent structures. Attempts to introduce air into the retroperitoneal tissues have been only partially successful but can be improved by the use of lateral tomography. In such cases the pancreatic shadow appears as a triangular soft tissue density between the gas-filled stomach and the gas in the retroperitoneal tissues.

The pancreas can also be demonstrated by the selective injection of contrast media into the superior pancreatico-duodenal artery or the superior mesenteric artery. In this way the arterial supply of the pancreas can be visualized and alterations in the shape of the pancreas noted.

The pancreatic duct system can be visualized by the introduction of X-ray-opaque material into it (pancreatogram). This can only be performed after the pancreatic duct has been catheterized at open operation. In a satisfactory pancreatogram, the whole length of the duct may be visualized and its branching seen.

Pancreatic stones can be clearly seen on plain films because of the high concentration of calcium which they contain. In a pancreatogram their relation to the pancreatic duct can be seen more clearly.

REFERENCES

ALEXANDER, W. F. (1940) The innervation of the biliary system, *J. comp. Neurol.*, **72**, 357.

ANSON, B. J., LYMAN, R. Y., and LANDER, H. H. (1936) The abdominal viscera in situ; a study of 125 consecutive cadavers *Anat. Rec.*, **67**, 17.

AREY, L. B., and TREMAINE, M. H. (1933) The muscle content of the lower oesophagus of man, *Anat. Rec.*, **56**, 315.

AREY, L. B., TREMAINE, M. J., and MONZINGO, F. L. (1935) The numerical and topographical relations of taste buds to human circumvallate papillae throughout the life span, *Anat. Rec.*, **64**, 9.

AREY, L. B. (1947) On the development, morphology, and interpretation of a system of cryptanalogues in the pharyngeal tonsil, *Amer. J. Anat.*, **80**, 203.

BORTONE, F. (1945) Esophageal diverticulum, *Amer. J. Surg.*, **70**, 64.

BOTHA, G. S. M., ASTLEY, R., and CARRE, I. J. (1957) A combined cineradiographic and manometric study of the gastro-oesophageal junction, *Lancet*, i, 659.

BROWNE, D. (1928) The surgical anatomy of the tonsil, *J. Anat.* (*Lond.*), **63**, 82.

BUCKSTEIN, J. (1953) *The Digestive Tract in Roentgenology*, Vols. I and II, 2nd ed., Philadelphia.

COCKER, R., and HATTON, J. M. (1955) A note on the innervation of dentine, *J. Anat. (Lond.)*, **89**, 189.

CORNES, J. S. (1965) Number, size and distribution of Peyer's patches in the human small intestine. I. The development of Peyer's patches, *Gut*, **6**, 225.

CORNES, J. S. (1965) Number, size and distribution of Peyer's patches in the human small intestine. II. The effect of age on Peyer's patches, *Gut*, **6**, 230.

COUPLAND, R. E. (1958) The innervation of pancreas of the rat, cat and rabbit as revealed by the cholinesterase technique, *J. Anat. (Lond.)*, **92**, 143.

DAVIES, F., and HARDING, H. E. (1942) Pouch of Hartmann, *Lancet*, i, 193.

DORRANCE, G. M. (1931) The so-called bursa pharyngea in man: Its origin, relationship with the adjoining nasopharyngeal structures and pathology, *Arch. Otolaryng.*, **13**, 187.

DOTT, N. M. (1923) Anomalies of intestinal rotation: Their embryology and surgical aspects; with report of five cases, *Brit. J. Surg.*, **11**, 251.

DOUBILET, H., SHAFIROFF, B. G. P., and MULHOLLAND, J. H. (1948) Anatomy of perieosphageal vagi, *Ann. Surg.*, **127**, 128.

ELIAS, H. (1955) Liver morphology, *Biol. Rev.*, **30**, 263.

EWING, M. R. (1954) Lingual thyroid, *Proc. roy. Soc. Med.*, **47**, 510.

EWING, M. R. (1954) White line of Hilton, *Proc. roy. Soc. Med.*, **47**, 525.

FALCONER, C. W. A., and GRIFFITHS, E. (1950) The anatomy of the blood-vessels in the region of the pancreas, *Brit. J. Surg.*, **37**, 334.

FULLER, A. P., FOZZARD, J. S., and WRIGHT, G. H. (1959) Sphincteric action of cricopharyngeus: radiographic demonstration, *Brit. J. Radiol.*, **32**, 32.

GOLIGHER, J. C., LEACOCK, A. G., and BROSSY, J. J. (1955) The surgical anatomy of the anal canal, *Brit. J. Surg.*, **43**, 51.

HALLY, A. D. (1959) The fine structure of the gastric parietal cell in the mouse, *J. Anat. (Lond.)*, **93**, 217.

HAMILTON, G. F. (1946) The longitudinal muscle coat of the human colon (Proc. Anat. Soc. June 1945), *J. Anat. (Lond.)*, **80**, 230.

HELLERSTROEM, C., HELLMAN, B., PETERSSON, B., and ALM, C. (1964) in *The Structure and Metabolism of the Pancreatic Islets*, Oxford.

HOLLINSHEAD, W. H. (1960) *Anatomy for Surgeons.* Volume 2. The Thorax, Abdomen, and Pelvis, New York.

JAMIESON, J. K., and DOBSON, J. F. (1920) The lymphatics of the tongue: with particular reference to the removal of lymphatic glands in cancer of the tongue, *Brit. J. Surg.*, **8**, 80.

JOHNS, B. A. E. (1952) Developmental changes in the oesophageal epithelium in man, *J. Anat. (Lond.)*, **86**, 431.

JONES, F. W. (1918) The sublingua and the plica fimbriata, *J. Anat. (Lond.)*, **52**, 345.

KIRK, J. (1944) Observations on the histology of the choledochoduodenal junction and papilla duodeni, with particular reference to the ampulla of Vater and sphincter of Oddi, *J. Anat. (Lond.)*, **78**, 118.

LEWINSKY, W., and STEWART, D. (1936) The innervation of the dentine, *J. Anat. (Lond.)*, **70**, 349.

LEWIS, F. T. (1912) The form of the stomach in human embryos with notes upon the nomenclature of the stomach, *Amer. J. Anat.*, **13**, 477.

LYTLE, F. T. (1912) The common bile-duct groove in the pancreas, *Brit. J. Surg.*, **47**, 209.

LEWINSKY, W., and STEWART, D. (1937) The innervation of the periodontal membrane of the cat, with some observations on the function of the end-organs found in that structure, *J. Anat. (Lond.)*, **71**, 232.

McKENZIE, J. (1948) The parotid gland in relation to the facial nerve, *J. Anat. (Lond.)*, **82**, 183.

McMINN, R. M. H. (1954) The rate of renewal of intestinal epithelium in the cat, *J. Anat. (Lond.)*, **88**, 527.

McMINN, R. M. H., and KUGLER, J. H. (1961) The glands of the bile and pancreatic ducts: autoradiographic and histochemical studies, *J. Anat. (Lond.)*, **95**, 1.

MARSHALL, S. F., and BECKER, W. F. (1949) Thyroglossal cysts and sinuses, *Ann. Surg.*, **129**, 642.

MAYO, C. W., and RICE, R. G. (1948) Situs inversus totalis: A statistical review of data on seventy-six cases with special reference to disease of the biliary tract, *Trans. west. surg. Ass.*, **56**, 188.

MENWISSEN, T. J. J. H. (1955) *X-ray Atlas and Manual of Esophagus, Stomach and Duodenum*, Amsterdam.

MICHELS, N. A. (1951) The hepatic, cystic and retroduodenal arteries and their relations to the biliary ducts: With samples of the entire celiacal blood supply, *Ann. Surg.*, **133**, 503.

MITCHELL, G. A. G. (1939) A specimen showing unusual arrangements of autonomic nerves, *J. Anat. (Lond.)*, **73**, 496.

MITCHELL, G. A. G. (1940) A macroscopic study of the nerve supply of the stomach, *J. Anat. (Lond.)*, **75**, 50.

MORLEY, J. (1945) Pharyngeal diverticula, *Brit. J. Surg.*, **33**, 101.

NEW, G. B. (1947) Congenital cysts of the tongue, the floor of the mouth, the pharynx, and the larynx, *Arch. Otolaryng.*, **45**, 145.

OLDFIELD, M. C. (1941) Cleft palate and the mechanism of speech (Arris and Gale Lecture), *Brit. J. Surg.*, **29**, 197.

PALAY, S. L., and KARIN, L. J. (1959) An electronmicroscopic study of the intestinal villus, *J. biophys. biochem. Cytol.*, **5**, 363.

RAPPAPORT, A. M. (1958) The structural and functional unit in the human liver (liver acinus), *Anat. Rec.*, **130**, 673.

SCHINZ, H. R., BAENSCH, W. E., FRIEDL, E., and VEHLINGER, L. (1971) *Roentgen Diagnostics, Gastrointestinal Tract, Gynecology, Urology*, Vol. IV, Eng. trans. Case, J. T., London.

SCHULZE, J. W., and BOYDEN, E. A. (1943) The blood supply and innervation of the choledochoduodenal junction in the cat, *Anat. Rec.*, **86**, 15.

SEDAR, A. W. (1962) The fine structure of the oxyntic cell in relation to functional activity of the stomach, *Ann. N.Y. Acad. Sci.*, **99**, 9.

SHAFIROFF, B. G. P., and HINTON, J. W. (1950) Surgical anatomy of the choledochal nerves, *Arch. Surg.*, **60**, 944.

SHANKS, S. C., and KERLEY, P. (1969) *A Text-book of X-Ray Diagnosis*, Vol. IV, 4th ed., London.

SHAPIRO, A. L., and ROBILLARD, G. L. (1948) The arterial blood supply of the common and hepatic bile ducts with reference to the problems of common duct injury and repair: Based on a series of twenty-three dissections, *Surgery*, **23**, 1.

SHEARMAN, D. J. C., and MUIR, A. R. (1960) Observations on the secretory cycle of goblet cells, *Quart. J. exp. Physiol.*, **45**, 337.

SHORTER, R. G., and CREAMER, B. (1962) Ribonucleic acid and protein metabolism in the gut. 1. Observations in gastro-intestinal cells with rapid turnover, *Gut*, **3**, 118.

SINGH, I. (1963) The terminal part of the accessory pancreatic duct and its musculature in the Rhesus monkey, *J. Anat. (Lond.)*, **97**, 107.

SINGH, I. (1964) On argyrophile and argentaffin reactions in individual granules of entero-chromaffin cells of the human gastro-intestinal tract, *J. Anat. (Lond.)* **98**, 497.

SPIRA, J. A. (1957) Comparison of cardiac and pyloric sphincters, *Lancet*, ii, 1008.

STERLING, J. A. (1955) *The Biliary Tract*, Baltimore.

STEWART, D. (1928) Further observations on the innervation of the teeth and the fibres of Mummery, *J. Anat. (Lond.)*, **63**, 1.

TESTUT, A., and LATARJET, A. (1949) *Traité d'anatomie humaine*, t. 4, 9th ed., Paris.

THIESSEN, N. W. (1941) The anatomy of the sphincter of Oddi, *Amer. J. Surg.*, **53**, 94.

WAKELEY, C. P. G. (1933) The position of the vermiform appendix as ascertained by an analysis of 10,000 cases, *J. Anat. (Lond.)*, **67**, 277.

WALLS, E. W. (1958) Observations on the microscopic anatomy of the human anal canal, *Brit. J. Surg.*, **45**, 504.

WHILLIS, J. (1930) A note on the muscles of the palate and the superior constrictor, *J. Anat. (Lond.)*, **65**, 92.

WHILLIS, J. (1946) Movements of the tongue in swallowing, *J. Anat. (Lond.)*, **80**, 115.

WIERNIK, G., SHORTER, R. G., and CREAMER, B. (1962) The arrest of intestinal epithelial 'turnover' by the use of X-irradiation, *Gut*, **3**, 26.

THE RESPIRATORY SYSTEM

G. M. WYBURN

THE parts of the respiratory system are the air passages, the nose, the pharynx, the larynx, the trachea, the bronchi, and their divisions, together with the two **lungs** covered by a serous membrane—the **pleura.** The lungs consist of the intrapulmonary bronchi with the respiratory tissue proper, the respiratory bronchioles, and air saccules [FIG. 567].

THE NOSE

EXTERNAL NOSE

The external nose has a **root** connected with the forehead, an **apex** situated inferiorly, and a **dorsum** extending from the root to the apex. The upper part of the dorsum, known as the

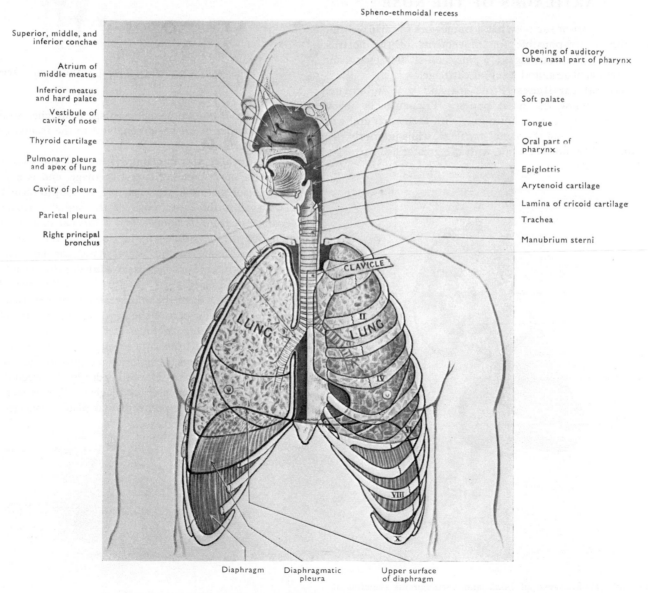

Spheno-ethmoidal recess

Superior, middle, and inferior conchae

Atrium of middle meatus

Inferior meatus and hard palate

Vestibule of cavity of nose

Thyroid cartilage

Pulmonary pleura and apex of lung

Cavity of pleura

Parietal pleura

Right principal bronchus

Opening of auditory tube, nasal part of pharynx

Soft palate

Tongue

Oral part of pharynx

Epiglottis

Arytenoid cartilage

Lamina of cricoid cartilage

Trachea

Manubrium sterni

CLAVICLE

LUNG

LUNG

II

IV

VI

VIII

X

Diaphragm   Diaphragmatic pleura   Upper surface of diaphragm

FIG. 567 Diagram of general arrangement of respiratory system. The lung, pleura, and ribs of the right side are represented as cut.

bridge of the nose, is formed by the nasal bones and the frontal processes of the maxilla. The lower part of the dorsum ends in the alae nasi forming the boundaries of the nostrils or nares. The nostrils are separated from each other by a median septum composed of fibrous tissue and skin. Within the circumference of each nostril there are several stout hairs or vibrissae.

Vessels and Nerves. The **arteries** are branches of the facial artery and of the ophthalmic artery. The **veins** drain to the facial vein and also communicate with the ophthalmic veins. The principal **lymph vessels** follow the course of the facial vein and open into the submandibular lymph nodes; but from the upper part of the nose a few vessels run laterally in the upper and lower eyelids and end in the superficial parotid lymph nodes. The **nerves** to the muscles of the nose come from the facial nerve, and to the skin from the infra-orbital branch of the maxillary nerve and from the infratrochlear and external nasal nerves, which are derived from the nasociliary branch of the ophthalmic nerve.

CARTILAGES OF THE NOSE

The lower part of the nose has a framework of cartilages and small muscles responsible for movements of the nostrils. There are five main cartilages; a pair each of lateral nasal and greater alar cartilages, and a septal cartilage.

The **lateral cartilages** [FIGS. 568 and 569] are situated immediately below the nasal bones. Triangular in shape, their anterior borders are continuous with each other and with the anterior margin of the septal cartilage, while inferiorly they are connected by fibrous tissue with the greater alar cartilages.

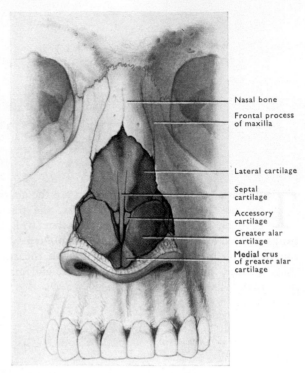

FIG. 569 Front view of bony and cartilaginous skeleton of external nose.

The posterior border of each cartilage is attached to the lower sloping border of the nasal bone and to the frontal process of the maxilla.

Each **greater alar cartilage** [FIGS. 568 and 569] is situated below the corresponding lateral cartilage, and is bent around the anterior margin of the nostril [FIG. 570] to form a **medial crus** against the lower edge of the septal cartilage (which separates it from its fellow), and a **lateral crus** which is the main body of the cartilage in the ala. This cartilage is attached by fibrous tissue to the lateral nasal cartilage and to the maxilla, with three or four **lesser alar cartilages** intervening.

The **septal cartilage** [FIG. 571] is usually situated slightly to one or other side of the midline, most frequently to the left. It forms the major part of the nasal septum anteriorly and has the form of an irregular four-sided plate. Its anterosuperior border is attached from above downwards to the back of the internasal suture and the anterior margins of the lateral nasal cartilages. The antero-inferior border lies between the medial crura of the greater alar cartilages. The posterosuperior border is joined to the perpendicular plate of the ethmoid and

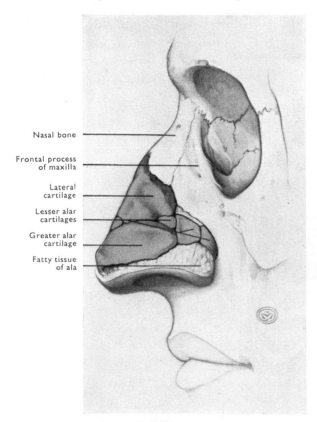

FIG. 568 Profile view of bony and cartilaginous skeleton of external nose.

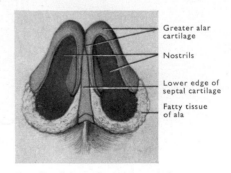

FIG. 570 Cartilages of nose from below.

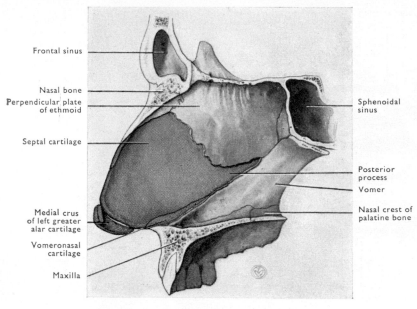

Frontal sinus

Nasal bone

Perpendicular plate of ethmoid

Septal cartilage

Medial crus of left greater alar cartilage

Vomeronasal cartilage

Maxilla

Sphenoidal sinus

Posterior process

Vomer

Nasal crest of palatine bone

FIG. 571   View of nasal septum from the left side.

groove on the vomer.   It may reach the body of the sphenoid.

Between the inferior border of the septal cartilage and the anterior border of the vomer, there is on each side a narrow strip of cartilage, the **vomeronasal cartilage** [FIG. 571].

CAVITY OF THE NOSE

A description of the bony boundaries of the nasal cavity and of its extensions, the air-filled paranasal sinuses and cells, is given on pages 119–123. Their general disposition is indicated in FIGURE 572.

The nasal cavity is divided into right and left compartments by the nasal septum, and they extend from the openings on the face (nostrils or **nares**) to the posterior nasal apertures or **choanae** which open in the nasal part of the pharynx.   Each cavity has lateral and medial walls, a roof and a floor, and tends to narrow from floor to roof. Three longitudinal, downturned laminae, the superior, middle, and inferior conchae project into it from the lateral wall, and each has a passage (meatus) below and lateral to it—respectively, the superior, middle, and inferior meatuses.   A fourth small concha suprema above the superior is not uncommon.   Above and behind the conchae there is a small space, the **spheno-ethmoidal recess.**   Three

the postero-inferior border is attached to the vomer and the anterior nasal spine.

A narrow **posterior process** of the septal cartilage is often prolonged backwards (especially in children) between the lower border of the perpendicular plate of the ethmoid and the

Crista galli

Uncinate process of ethmoid

Nasal septum

Maxillary sinus

Falx cerebri

Olfactory tract

Ethmoidal sinus

Ethmoidal sinus in bulla ethmoidalis

Middle concha

Inferior meatus

FIG. 572   Coronal section of head, viewed from the front. The contents of the right orbit have been removed.

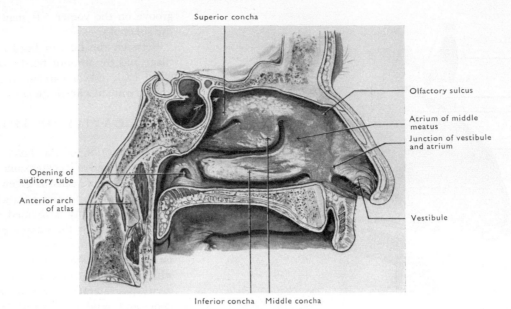

FIG. 573   Lateral wall of left nasal cavity.
The arrow passes from the sphenoidal sinus to the spheno-
ethmoidal recess.

regions are described in each nasal cavity: the vestibule, which forms the anterior and lower part immediately inside the nostril; the olfactory region, corresponding to about the upper third of the cavity; and below this and behind the vestibule, the respiratory region.

The Vestibule [FIGS. 573 and 574]. The vestibule is bounded laterally by the ala and the greater alar cartilage, medially by the medial crus of this cartilage. It extends forwards as a small recess towards the apex of the nose. The vestibule is limited above by a curved elevation (limen nasi) formed by the upper margin of the greater alar cartilage, and is lined with skin which has sebaceous glands, sweat glands, and vibrissae inferiorly.

The Lateral Wall [FIGS. 572–574]. The **superior meatus** is short, oblique, lies between the superior and middle conchae, and is confined to the posterior part of the lateral wall. The posterior ethmoidal cells open into it through one or more apertures. The sphenoidal sinus opens into the spheno-ethmoidal recess above and behind the superior concha [FIG. 573].

The **middle meatus** [FIG. 574], deeper in front than behind, is overhung by the middle concha, and is continued anteriorly into the **atrium,** a shallow depression above the vestibule, limited superiorly by a ridge known as the **agger nasi,** above which the olfactory sulcus leads to the olfactory part of the nasal cavity. The lateral wall of the middle meatus, displayed by removal of the middle concha, has a rounded prominence, the **bulla ethmoidalis,** caused by the bulging of the middle ethmoidal cells which open usually on or above the bulla. Immediately below the bulla, there is a deep curved groove, the **hiatus semilunaris** which may be continuous anteriorly with the lower end of the **ethmoidal infundibulum,** a funnel-shaped opening leading to the narrow passage or duct of the frontal sinus. Occasionally, however, the frontal sinus opens separately into the middle meatus. The opening

of the anterior ethmoidal cells is situated in the anterior, vertical part of the hiatus semilunaris [FIG. 574], and posteriorly there is a small opening for the maxillary sinus. This sinus may have an accessory opening below and behind the hiatus semilunaris.

The **inferior meatus** is situated below the inferior concha and has the opening of the nasolacrimal duct anteriorly.

Roof. The narrow roof has a central, horizontal, ethmoidal portion, while the frontonasal and sphenoidal parts slope downwards and forwards, and downwards and backwards respectively.

Floor. The broader floor is formed by the palatine process of the maxilla and the horizontal lamina of the palatine bone. In front is the incisive canal and related to this is a funnel-shaped impression known as the nasopalatine recess directed downwards and forwards, representing a wide communication which exists in early life between the nasal and buccal cavities.

The Medial Wall or Nasal Septum. The septum consists of the perpendicular plate of the ethmoid bone above, the vomer below, and the septal cartilage wedged between them in front.

Mucous Membrane of the Nose

The mucous membrane lines the nasal cavities with the exception of the vestibules, and its deeper layers are firmly adherent to the subjacent periosteum and perichondrium. It is continuous with the mucous membrane of the nasopharynx, of the paranasal sinuses, and through the nasolacrimal duct with that of the nasolacrimal sac. It is thickest over the conchae and the septum. On each side of the septum, just above the incisive canals, there is a minute orifice in the mucous membrane from which a blind pouch—the **vomeronasal organ**—extends upwards and backwards for a distance of 2–9 mm. In man, this vomeronasal organ is vestigial, but

Opening of middle ethmoidal cells

Openings of posterior ethmoidal cells

Frontal sinus, with a bristle passed from it into middle meatus

Agger nasi

Opening of anterior ethmoidal cells

Ethmoidal bulla

Hiatus semilunaris

Opening of maxillary sinus

Bristle in opening of nasolacrimal duct

Vestibule

Sphenoidal sinus

Spheno-ethmoidal recess

Cut edge of superior nasal concha

Cut edge of middle nasal concha

Opening of auditory tube

Cut edge of inferior nasal concha

FIG. 574   Lateral wall of right nasal cavity—the nasal conchae have been removed.

For the bones of the lateral wall, see Fig. 137.

in many of the lower animals, e.g. lizards and rabbits, it is well-developed, is lined with an epithelium similar to that of the olfactory region, and is supplied by branches of the olfactory nerve [FIG. 576].

Epithelium. The epithelium varies in different regions. In the vestibule it is stratified squamous epithelium. In the respiratory region, the mucous membrane is lined by a columnar ciliated epithelium interspersed among which are goblet cells and smaller cells. Beneath the epithelium there are regions with a diffuse lymphoid tissue and a layer of mucous and serous glands whose ducts open on the surface.

For structure of the mucous membrane in the olfactory region see FIGURES 1030 and 1031.

Blood Vessels of the Nose

The **sphenopalatine artery** gives the major supply to the walls of the nasal cavity. It reaches the nasal cavity from the maxillary artery through the sphenopalatine foramen, and it divides into (1) an artery to the **septum** and (2) branches (**posterior** and **lateral nasal**) which ramify over the meatuses and conchae and also supply the maxillary sinus and posterior ethmoidal cells. In the conchae the main arteries lie within the bone, and their branches then pass out to form a fine subepithelial plexus, which opens into wide venous 'blood-spaces'—an important constituent of the mucous membrane. In addition, there are direct arteriovenous anastomoses between the mucosal arteries and the 'blood-spaces' (Harper, 1947). These vascular arrangements permit rapid engorgement and swelling of the mucosa. The upper and anterior parts of the walls receive a supply from the **ethmoidal arteries,** and the posterior part some small branches from the greater palatine artery. The nostrils are supplied by a branch of the **facial artery** and by the septal branch of the **superior labial artery,** and all these vessels anastomose freely. The maxillary sinus is also supplied by the **infra-orbital** and **posterior superior alveolar arteries,** the frontal sinus by the anterior ethmoidal and supra-orbital

arteries, the ethmoidal cells by the ethmoidal arteries, and the sphenoidal sinus gets its chief supply from the sphenopalatine artery. The **veins** form a dense, cavernous plexus which is well seen in the respiratory region, and particularly over the inferior nasal concha and the lower and posterior parts of the septum. In the case of the conchae the veins drain first into a periosteal venous plexus and thence into a system of veins contained in bony grooves and canals in the conchae. The venous blood is carried forwards into the facial vein, backwards into the pterygoid plexus in the sphenopalatine vein, and upwards into the ethmoidal veins. The ethmoidal veins communicate with the ophthalmic veins and the veins of the dura mater; one ethmoidal vein passes up through the cribriform plate, and opens either into the venous plexus of the olfactory bulb or directly into one of the veins on the orbital surface of the brain.

Lymph Vessels

These form an irregular network in the superficial part of the mucous membrane; they are separated from prolongations of the subarachnoid space along the olfactory nerves by very thin walls which act as semipermeable membranes. The chief lymph vessels are directed backwards towards the choanae, and, on each side, are collected into two trunks, one passing to a retropharyngeal lymph node, the other to one or two upper deep cervical lymph nodes.

DEVELOPMENT OF THE NOSE

The first indication is the appearance of the **olfactory placode**—a proliferation of ectodermal cells on the ventro-lateral surface of the head—which can be recognized in an embryo of 4 mm. [FIG. 1025A]. The tissue on the medial and the lateral side of each placode becomes raised up into ridges which soon form rounded prominences called respectively the **globular** and the **lateral nasal processes.** The two globular processes, separated by a median groove, are parts of a **median**

nasal process, and the whole area on which the nasal processes and the olfactory placodes appear constitutes the **frontonasal process** [FIG. 67A].

As the nasal processes are developed, the olfactory placodes sink into the substance of the head so as to form the two **olfactory grooves.** The **maxillary process** grows from the first or mandibular pharyngeal arch and is at first separated from the lateral nasal process by a groove—the **nasolacrimal sulcus**—which leads to the region of the developing eye. When the maxillary process, by its growth in a ventral direction, joins the frontonasal process this ectodermal groove is buried, and may later give rise to the **nasolacrimal duct** [FIG. 575].

As each maxillary process grows ventromedially to form the upper lip, it fuses with the lateral nasal process and then with the globular process, and so forms a floor (primitive palate) for the original olfactory groove, which becomes converted to an **olfactory pit** that opens on to the surface through an almost circular aperture—the primitive nostril. The posterior end of the pit breaks through into the stomodaeum about the seventh week, and thus opens into it behind the maxillary process by the primitive posterior aperture of the nose.

Each olfactory pit enlarges, particularly in a superior direction, to form a **nasal cavity** by the extension of the surrounding mesoderm which, between the two nasal cavities, forms the **nasal septum,** and is continuous anteriorly with the

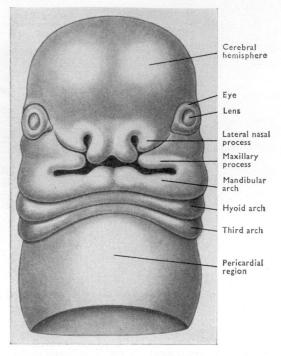

FIG. 575 Head of human embryo showing the relation of the nasal and maxillary processes to the olfactory pit and the nasolacrimal sulcus.

FIG. 576 Section through nose and mouth of 20 mm. human embryo to show position of vomeronasal organ.   × 28.

median nasal process. The original ectodermal cells of the olfactory placode now lie in the upper part of each cavity, and they differentiate into the olfactory epithelium. The right and left nasal cavities are still continuous with each other inferiorly through the future mouth cavity, below the nasal septum. Each maxillary process now develops a medially projecting shelf [FIG. 68]—the **palatine process**—which initially hangs down lateral to the developing tongue, the dorsum of which touches the inferior margin of the nasal septum. About the tenth week, the palatal processes begin to fuse with each other and with the nasal septum [FIG. 576]. This begins anteriorly, and extends backwards in the median plane until the entire palate and uvula are formed and the

stomodaeum is divided into oral and nasal cavities [FIG. 69]. This process may fail in part or in whole, thus giving rise to various degrees of cleft palate from a bifid uvula to complete absence of fusion. Failure of the maxillary process to fuse with the median nasal process anteriorly, leads to the production of a hare-lip—a split in the upper lip which may extend through the alveolar arch, lateral to the incisor teeth, to become continuous with a cleft palate.

The nasal conchae develop as a series of ridges on the lateral wall, known as the **maxilloturbinal,** which forms the inferior concha, and the **ethmoturbinals,** of which the first forms the middle concha and the second and third the superior concha; an additional ethmoturbinal may persist as the supreme concha.

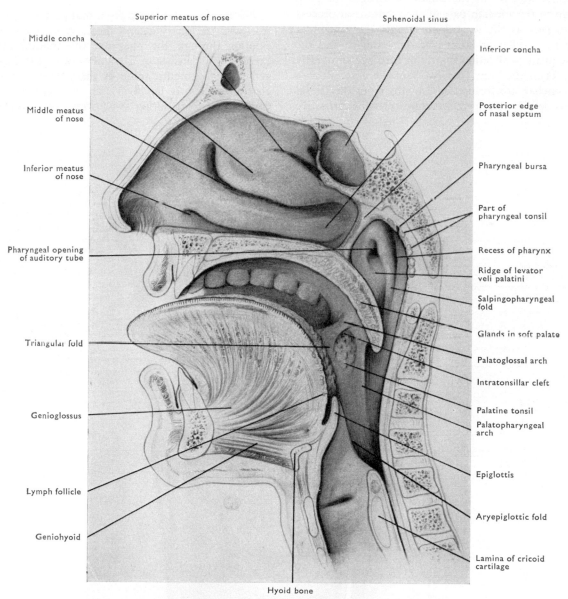

FIG. 577 Sagittal section through mouth, tongue, larynx, pharynx, and nasal cavity.

The section is slightly oblique, and the posterior edge of the nasal septum is preserved. The specimen is viewed slightly from below, hence the apparently low position of the inferior concha.

A more anterior elevation on the lateral wall—the **naso-turbinal**—is represented in the adult by the agger nasi.

The vomeronasal organ [FIG. 576] appears about the fifth or sixth week as an indentation of the nasal septum; it grows in a posterosuperior direction, and attains a depth of 4 mm. by about the sixth month; later, it atrophies.

With the continued growth of the maxillary processes, the frontonasal process, with its contained nares, is raised from the surface of the face so as to form the rudiment of an external nose [FIG. 1025 D] with the nostrils directed forwards, as they are even at birth.

The supporting skeleton of the nose constitutes what is called the **nasal capsule.** This arises by chondrification in both the medial and lateral walls of the developing nasal cavities and in the conchae; the process following the order of development of its parts. The chondrification in the lower part of the lateral wall forms the **nasal cartilages** [p. 470]. Chondrification in the median part of the frontonasal process also persists in part as the **septal cartilage,** the remainder becoming the perpendicular lamina of the ethmoid and the vomer. Subsequent ossification, partly in cartilage and partly in membrane, forms the bony framework of the nose.

When the conchae are formed as growths from the lateral wall, the areas left between them constitute the primitive meatuses, and outgrowths from these form the various **para-nasal sinuses.** The first of these to appear is the maxillary sinus, as an outgrowth from the middle meatus about the end of the third month; this is followed by the ethmoidal, frontal, and sphenoidal sinuses, but the sinuses undergo the major portion of their development after birth, though the ethmoidal cells are already formed but are not of adult size. Thus the frontal sinus does not reach that bone till the second year, the sphenoid sinus is not excavated until the fifth year, and the maxillary sinus, which is a mere groove at birth, does not acquire its full development till after puberty.

For **pharynx** see pages 415–420.

THE LARYNX
[PLATE 28 B and G]

The larynx is developed from that part of the foregut which will form the pharynx, and therefore communicates with the exterior through the pharynx, the mouth, and the nasal cavities. It consists of a framework of cartilages joined together by ligaments, some of them with a large amount of elastic tissue. It has a closing mechanism to prevent the entrance of food during swallowing, vocal cords responsible

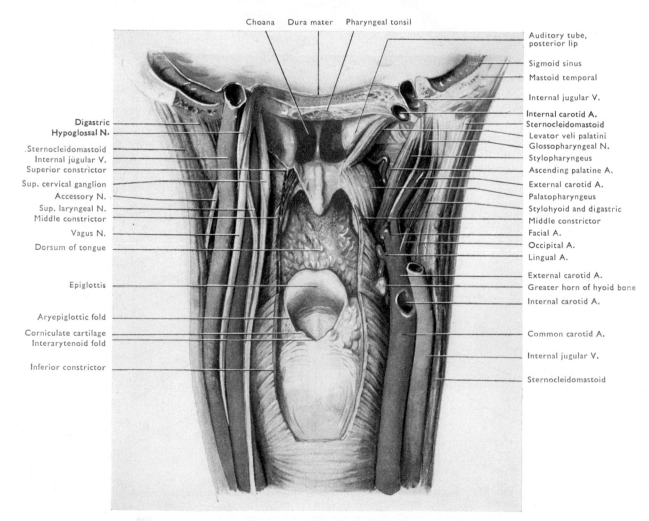

FIG. 578 Interior of pharynx and structures in relation to its side-walls viewed from behind.

for voice production and capable of approximating in the midline, thus shutting off the lower air passages. The larynx, about 4·5 cm. long in the adult male and 3·5 cm. in the adult female, extends from the tip of the epiglottis, at the level of the third cervical vertebra, to the lower border of the cricoid cartilage, where it becomes continuous with the trachea at the level of the sixth cervical vertebra [FIG. 577]. In children it lies slightly higher, and in infants its upper border is on the level with the axis vertebra (second cervical). The inferior part of the pharynx separates the larynx from the bodies of the vertebrae posteriorly, and anteriorly the laryngeal cartilages—the thyroid and the cricoid—can be readily palpated in the midline of the neck inferior to the hyoid bone, and more laterally are covered by the lobes of the thyroid gland and the infrahyoid muscles. On each side the larynx is closely related to the structures in the carotid sheath deep to the anterior border of the sternocleidomastoid muscle [FIG. 578]. The larynx is raised when the head is extended on the neck and depressed when the chin is carried downwards and forwards, and its upward movement is visible during swallowing. The relative positions of the laryngeal cartilages can be altered by the contraction of attached muscles, and these in turn can change the shape of the larynx as well as the size, position, and tension of the vocal cords during phonation.

CARTILAGES OF THE LARYNX

There are three single cartilages—the thyroid, the cricoid, and the epiglottis—and three paired cartilages—the arytenoid, the corniculate, and the cuneiform.

Thyroid Cartilage

The thyroid is the largest of the laryngeal cartilages and forms most of its anterior and lateral walls. It consists of two laminae fused in the midline anteriorly and diverging posteriorly [FIG. 579]. The angle of union of the thyroid laminae is about

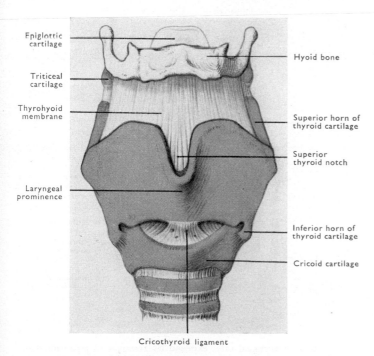

FIG. 579 Anterior aspect of cartilages and ligaments of larynx.

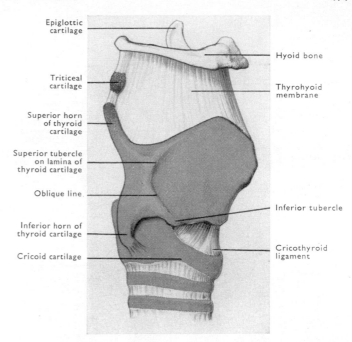

FIG. 580 Profile view of cartilages and ligaments of larynx.

90 degrees in men and about 120 degrees in women. The superior borders curve inferiorly in the midline forming a V-shaped thyroid notch, immediately below which is the **laryngeal prominence** (Adam's apple), more conspicuous in the male. The more pronounced laryngeal prominence, the longer vocal cords, and the lower pitch of the male voice are all associated with the smaller size of the thyroid angle. The thyrohyoid membrane joins the superior border of the thyroid cartilage to that of the hyoid bone above. The rounded posterior borders have long **superior horns** extending upwards, and shorter **inferior horns** curving inferomedially to articulate with the lateral surface of the cricoid cartilage [FIG. 580]. The outer aspect of the thyroid lamina is divided by an oblique line which gives attachment to the sternothyroid and thyrohyoid muscles, to the inferior constrictor of the pharynx, and to the pretracheal fascia. Laterally, the internal surface of the thyroid cartilage is covered by the mucous membrane of the lateral wall of the piriform recess [FIG. 583].

Cricoid Cartilage

The smaller cricoid cartilage lies immediately below the thyroid cartilage. It is shaped like a signet ring, with its band in front, and forms part of the anterior, lateral, and most of the posterior wall of the larynx. Its lower border is horizontal, but the upper border curves abruptly upwards and backwards so that the posterior lamina is 2·5 cm. in vertical extent, compared with about 0·75 cm. for the anterior arch. Laterally, there are articular surfaces for the inferior horns of the thyroid cartilage, while the arytenoid cartilages articulate with its upper borders posteriorly [FIG. 581]. Posteriorly is the **lamina** which has a median, vertical ridge on its posterior surface, and is covered on both sides of this by muscle and mucous membrane. This posterior surface forms the anterior wall of the lowest part of the pharynx, while the anterior surface forms the posterior wall of the larynx. The **cricothyroid** and **cricotracheal ligaments** connect the cricoid cartilage with the thyroid and the first tracheal ring

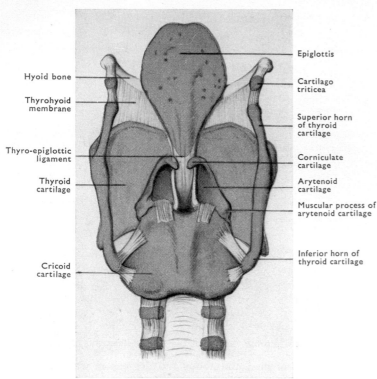

FIG. 581   Posterior aspect of cartilages and ligaments of larynx.

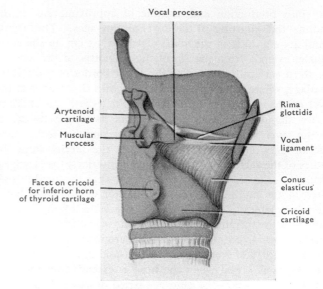

FIG. 582   Dissection to show laryngeal fibro-elastic membrane. The right lamina of the thyroid cartilage has been removed.

respectively.   The anterior **arch** of the cricoid is easily felt below the laryngeal prominence.

Epiglottis

The cartilage of the epiglottis is a pear-shaped lamina of elastic fibrocartilage lying behind the root of the tongue, the hyoid bone, and the thyrohyoid ligament [FIG. 585].   The narrow part of the cartilage is attached in front to the hyoid bone by the **hyo-epiglottic ligament** and to the inner aspect of the angle of the thyroid by the **thyro-epiglottic ligament.** The broad part is free and is covered on both surfaces by mucous membrane which is reflected from its anterior surface as the median and lateral **glosso-epiglottic folds** flanking the epiglottic valleculae [FIG. 583].   The posterior surface forms the anterior wall of the inlet and the vestibule of the larynx, and has a prominent bulge—the **tubercle.** The lateral margins of the epiglottis are attached to the arytenoid cartilages by the aryepiglottic folds of mucous membrane enclosing muscle fibres [FIG. 587].   The epiglottis has taste buds on its posterior surface and mucous glands which pierce the cartilage, and it helps to shut off the aperture of the larynx during swallowing. In some mammals it extends above the soft palate into the nasal part of the pharynx, making the latter directly continuous with the larynx.

Arytenoid Cartilages

The paired arytenoid cartilages are placed on the upper surface of the posterior lamina of the cricoid cartilage.   They are pyramidal in shape with three surfaces—posterior, medial, and anterolateral—and a base with projecting lateral, or **muscular** and anterior, or **vocal, processes.** The cartilage is covered by the mucous membrane, beneath which it gives attachment to laryngeal ligaments and muscles [FIG. 582].

Corniculate and Cuneiform Cartilages

The paired corniculate and cuneiform cartilages are small nodules in the posterior part of the aryepiglottic folds.   The corniculate cartilages may make contact with arytenoid cartilages.

JOINTS AND LIGAMENTS OF THE LARYNX

The joints of the larynx are the cricothyroid and crico-arytenoid joints, both with a number of small elastic ligaments. The movement at the cricothyroid joint consists of a rotation of the cricoid cartilage round an axis passing through the inferior horns of the thyroid to produce a backward tilting of

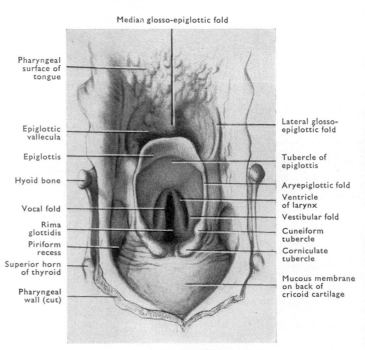

FIG. 583   Laryngeal inlet exposed by removal of posterior wall of pharynx and viewed from above and behind.

Epiglottis

Aryepiglottic fold

Tubercle of epiglottis

Vestibular fold

Vocal fold

Hyoid bone

Thyroid cartilage

Ventricle of larynx

Thyro-arytenoid muscle

Cricoid cartilage

FIG. 584   Coronal section through larynx to show its compartments.

the lamina of the cricoid, and a forward tilting of the thyroid cartilage, thus increasing the anteroposterior diameter of the larynx. Both a sliding and a rotatory movement occurs at the crico-arytenoid joint. The arytenoid cartilages can thus be approximated or separated, or have their vocal processes swung towards or away from the midline.

The ligaments of the larynx, particularly the intrinsic ligaments, have a large amount of elastic tissue.

One extrinsic ligament is the **thyrohyoid membrane.** It is attached below to the upper border of the thyroid cartilage, and passing up internal to the curve of the hyoid bone, is attached to its entire upper margin internally, but is separated from the posterior surface of the body of the hyoid by a bursa. The thick middle part is the **median thyrohyoid ligament,** and the posterior margins form rounded cords—the **thyrohyoid ligaments**—connecting the tips of the superior horns of the thyroid cartilage to the posterior ends of the greater horns of the hyoid. The **cricotracheal ligament** unites the lower border of the cricoid cartilage and the first ring of the trachea.

The intrinsic ligaments lie external to the mucous membrane of the larynx. A thin lamina of fibro-elastic tissue, sometimes known as the **quadrangular membrane,** extends from the sides of the epiglottic cartilage to the arytenoid cartilages. Covered by mucous membrane, it forms the boundary wall between the piriform recess and the larynx, with its upper border in the aryepiglottic fold, and its lower border in the vestibular fold (the **vestibular ligament**). A much thicker band of elastic tissue constitutes the remainder of the laryngeal fibro-elastic membrane connecting cricoid, thyroid, and arytenoid cartilages. The anterior part of this membrane is the **cricothyroid ligament** and the lateral part, lined with the mucous membrane of the larynx, is the **conus elasticus,**

directed upwards and medially, internal to the thyroid cartilage, from the sloping superior margin of the cricoid cartilage. On each side the conus elasticus has a thickened, free, upper margin—the **vocal ligament**—the basis of the corresponding vocal fold which is attached posteriorly to the vocal process of the arytenoid and anteriorly to the deep aspect of the thyroid angle [FIG. 582].

THE CAVITY OF THE LARYNX
[PLATES 28 B AND 34 A]

The oblique inlet to the larynx from the pharynx is bounded above and in front by the epiglottis, below and behind by the arytenoid cartilages, and laterally by the aryepiglottic folds with the corniculate and cuneiform cartilages [FIG. 583].

The cavity of the larynx has two pairs of projecting folds—the upper **vestibular** and the lower **vocal folds**—which divide it into three parts—the vestibule above the vestibular folds, a middle compartment between the two pairs of folds, and a lower compartment below the vocal folds, sometimes known as the **infraglottic larynx** [FIG. 584].

Vestibule

The vestibule of the larynx extends from the inlet to the vestibular folds, diminishes in width from above downwards, and has, of course, its maximum length in front where its wall is formed by the epiglottis with its tubercle. The lateral walls are the aryepiglottic folds with the quadrangular membranes covered by the mucous membrane, and the posterior wall is the mucous membrane stretching between the upper parts of the arytenoid cartilages.

Middle Compartment

The middle compartment of the larynx, bounded by the vestibular folds above and the vocal folds below, has a recess—the **ventricle** of the larynx—which undermines the vestibular folds. At the anterior end of the ventricle there is a narrow diverticulum—the **laryngeal saccule**—extending upwards between the vestibular folds and the inner surface of the thyroid cartilage, and containing a number of mucous glands whose secretion is said to lubricate the vocal folds. The size of the saccule varies, but it seldom extends beyond the upper border of the thyroid cartilage although it may be elongated, penetrate the thyrohyoid membrane, and terminate deep to the infrahyoid muscles. The laryngeal saccules are normally very much enlarged in certain apes where they form air sacs which may extend far into the tissues of the neck, but any similar enlargement in the human is an abnormal development and is known as a laryngocoele.

Vestibular Folds

The vestibular folds or false vocal cords are duplications of the mucous membrane of the side-wall of the larynx and include fibrous tissue, vestibular ligaments, and a few muscle fibres. The vestibular ligaments are attached to the thyroid angle in front and to the anterolateral surfaces of the arytenoid cartilages behind. The vestibular folds are movable and are approximated in forced closure of the glottis. The interval between the vestibular folds—the **rima vestibuli**—is wider than the corresponding interval (rima glottidis) between the

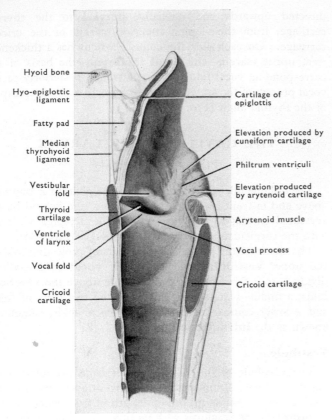

FIG. 585   Median section through larynx to show side-wall of its right half.

Labels (left side): Hyoid bone; Hyo-epiglottic ligament; Fatty pad; Median thyrohyoid ligament; Vestibular fold; Thyroid cartilage; Ventricle of larynx; Vocal fold; Cricoid cartilage

Labels (right side): Cartilage of epiglottis; Elevation produced by cuneiform cartilage; Philtrum ventriculi; Elevation produced by arytenoid cartilage; Arytenoid muscle; Vocal process; Cricoid cartilage

two vocal folds below; thus, when the larynx is examined from the mouth with the laryngoscope, all four folds are visible [FIG. 591].

Vocal Folds

The vocal folds, or true vocal cords, are prominent white folds of mucous membrane below the vestibular folds, attached in front to the angle of the thyroid cartilage and behind to the vocal processes of the arytenoid cartilages [FIG. 585]. They are 23–27 mm. long in the adult male and about five-sevenths of that length in the adult female. The substance of each vocal fold includes the **vocal ligament** (the thickened upper margin of the conus elasticus) and muscle fibres. The thin inner edges of the folds, consisting of fibrous tissue with closely adherent mucous membrane, are sometimes known as the **vocal bands**, in contrast to the thicker muscular lateral portions. In cross section the vocal folds are triangular with their free borders—the vocal bands—projecting upwards and medially. The **rima glottidis** is the interval between the vocal folds, and the term '**glottis**' refers to the narrow part of the larynx at the level of the vocal folds. The anterior three-fifths of the rima glottidis is described as the **intermembran-ous part**, and the posterior two-fifths between the vocal processes and bases of the arytenoid cartilages is the **inter-cartilaginous part**. The shape of the rima glottidis varies according to the position of the vocal folds which are concerned in voice production, while the vestibular folds are primarily part of the protective mechanism by which the airway is closed to prevent the entry of food or foreign bodies [FIG. 586].

Infraglottic Larynx

The lowest or infraglottic compartment of the larynx extends from the vocal cords above to the lower border of the cricoid cartilage below, and widens from above downwards towards the trachea.

MUCOUS MEMBRANE OF THE LARYNX

The epithelium of the laryngeal mucous membrane is mainly columnar ciliated like that of the trachea, but certain parts have the stratified squamous epithelium of the pharynx, e.g., the vocal folds, the upper half of the posterior surface of the epiglottis, and the margins of the aryepiglottic folds. Mucus-secreting glands are particularly numerous in front of the arytenoid cartilages, on the aryepiglottic folds, and in the epiglottis, but are absent from the free edges of the vocal folds. The mucous membrane of the vocal folds is firmly bound down at the margin and on the infero-medial surface, and is also closely adherent over the posterior surface of the epiglottis, but elsewhere is more loosely attached by submucous tissue. Inflammation of the larynx may give rise to swelling (oedema) of the tissues above the glottis, but because of the tight attachment of the mucous membrane at that level the oedema does not extend below the vocal cords. Supraglottic and infra-glottic oedema can thus occur independently. Above the level of the vocal folds the laryngeal mucous membrane is extremely sensitive, and contact with a foreign body immediately induces explosive coughing.

MUSCLES OF THE LARYNX
[FIGS. 587–590]

These are divided into the **extrinsic muscles** which move the larynx as a whole, and the **intrinsic muscles** concerned with modifying the size of the laryngeal inlet, the form of the rima glottidis, and with variations in the length and tension

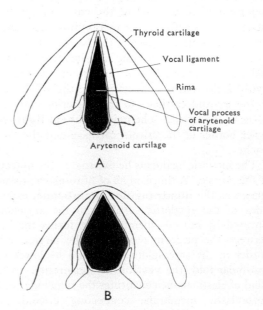

FIG. 586   Diagrams of rima glottidis.
A, during ordinary easy breathing.
B, widely opened.

Labels: Thyroid cartilage; Vocal ligament; Rima; Vocal process of arytenoid cartilage; Arytenoid cartilage

PLATE 33

A

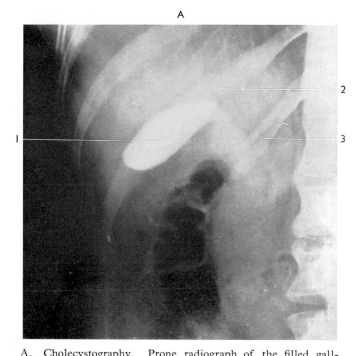

A. Cholecystography. Prone radiograph of the filled gall-bladder showing its relatively high position in this posture.

1. Gall-bladder.　　2. 11th rib.
3. 12th rib.

B

B. Cholecystography. The gall-bladder is contracting after a fatty meal.

1. Cystic duct.　　2. Gall-bladder.
3. Bile-duct.

C

C. Cholecystography. Erect radiograph of the filled gall-bladder, showing the fundus lying below the iliac crest in this posture.

1. Iliac crest.　　2. Gall-bladder.
3. 5th lumbar vertebra.

D

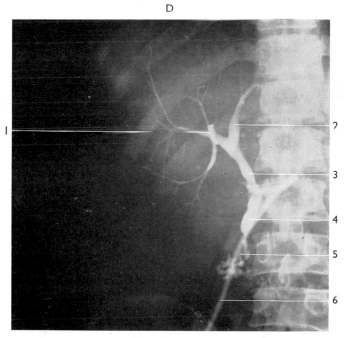

D. Post-operative cholangiogram to show the intra- and extra-hepatic ducts, the gall-bladder having been removed. The catheter which has been inserted at operation into the bile-duct is shown.

1. Right hepatic duct.
2. Left hepatic duct.
3. Common hepatic duct.
4. Bile-duct.
5. Contrast medium in duodenum.
6. Catheter.

PLATE 34

A

B

A. Anteroposterior tomograph of larynx. Only the structures lying in the plane of the optical section are sharp.

1. Piriform fossa.
2. Vestibular fold.
3. Ventricle of larynx.
4. Vocal fold separated from its
   fellow by a narrow rima glottidis.
5. Trachea.

B. Anteroposterior radiograph of male thorax of average physique.

1. Vascular shadows in lung.
2. Right atrium.
3. Left atrium.
4. Left ventricle.
5. Costodiaphragmatic recess.

C

D

C. Anteroposterior radiograph of female thorax of short broad physique.

1. Breast shadow.   2. Aortic arch.
3. Gas bubble in stomach.

D. Anteroposterior radiograph of thorax in which air has entered the right pleural cavity (pneumothorax). Note the shadow of the collapsed lung, and the extent to which all the vascular shadows of the lung have disappeared and only the ribs are visible.

1. Shadow of collapsed lung.

PLATE 35

A

B

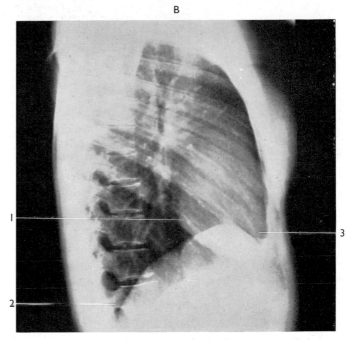

A. Anteroposterior radiograph of female chest. Average physique.

1. Superior vena cava. 4. Arch of aorta.
2. Right atrium. 5. Left ventricle.
3. Right diaphragm. 6. Breast shadow

B. Lateral radiograph of chest.

1. Base of ventricles.
2. Costodiaphragmatic recess.
3. Apex of heart.

C

D

C. Anteroposterior bronchogram. See Fig. 607.

1. Trachea.
2. Right superior lobe bronchus.
3. Lower branch of right bronchus.
4. Middle lobe bronchus.
5. Lower lobe bronchus.
6. Bifurcation of trachea.
7. Apicoposterior bronchus.
8. Left superior lobe bronchus.
9. Left inferior lobe bronchus.

D. Lateral bronchogram of right lung.

1. Right upper lobe bronchus.
2. Right lower lobe bronchus.
3. Branches of right posterior basal bronchus in the costodiaphragmatic recess.
4. Right middle lobe bronchus.

PLATE 36

A

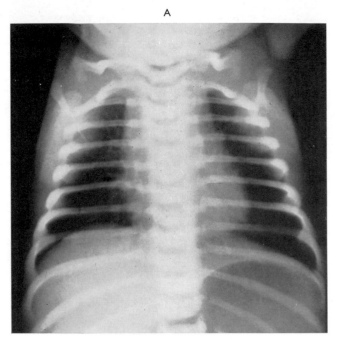

A. Radiograph of the chest of a new-born baby in inspiration.

B

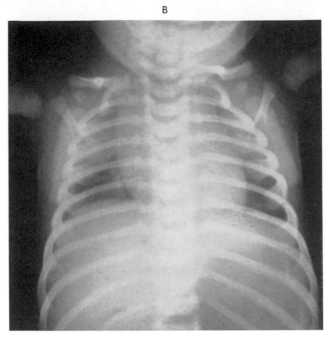

B. Radiograph of the chest of the same baby in expiration.

C

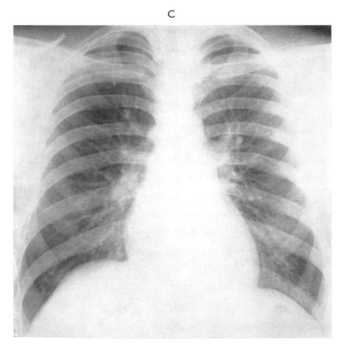

C. Radiograph of the chest of an adult in inspiration.

D

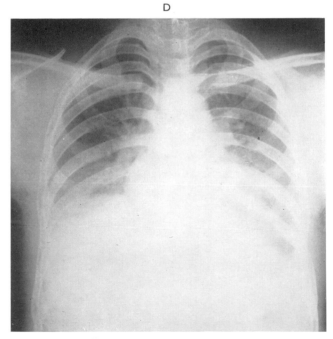

D. Radiograph of the chest of the same adult in expiration.

In the child, note the relatively large size of the heart, the horizontal arrangement of the ribs and the broad shadow of the superior mediastinum due to the thymus.

In the adult, note that in expiration the lung bases become dense due to loss of air, the heart shadow increases in transverse diameter, the diaphragm rises and the costodiaphragmatic recesses are obliterated.

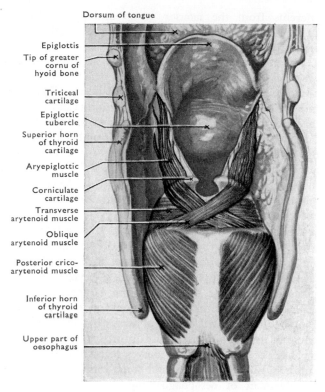

FIG. 587 Muscles on back of larynx.

larynx all lie between the mucous membrane and the cartilaginous framework, and their names indicate their attachments. They can be grouped functionally as the muscles of the laryngeal inlet, the abductors, the adductors, tensors, and relaxers of the vocal cords.

1. The muscles of the laryngeal inlet:

The **oblique arytenoids** and their extension into the aryepiglottic folds—the **aryepiglottic** muscles [FIG. 587]. Contraction of these muscles brings the aryepiglottic folds together and pulls the arytenoid cartilages towards the epiglottis, thus helping to close the laryngeal inlet. Muscle fibres running from the thyroid cartilage into the epiglottis—the **thyro-epiglotticus**—tend to widen the inlet of the larynx when they contract [FIG. 588].

2. The muscle groups which open and close the glottis:

i. ADDUCTORS. The **lateral crico-arytenoid muscles,** attached to the muscular processes of the arytenoid cartilages, rotate the arytenoid cartilages medially, swing the vocal processes inwards, and approximate the vocal folds. The transverse arytenoid muscle draws the arytenoid cartilages together.

ii. ABDUCTORS. The main abductors of the vocal cords are the **posterior crico-arytenoid muscles** attached to the muscular processes. They rotate the arytenoid cartilages laterally, swing the vocal processes outwards, and separate the vocal folds [FIG. 587].

iii. TENSORS. The chief tensors of the vocal cords are the **cricothyroid muscles.** They act on the cricothyroid joint so that the distance between the vocal processes behind and the angle of the thyroid in front is increased and the vocal folds are put on the stretch [FIG. 589].

iv. RELAXERS. The chief relaxers are the two **thyro-arytenoid muscles** which draw the arytenoid cartilages towards the thyroid, decreasing the anteroposterior diameter of the glottis,

of the vocal cords. Of the extrinsic muscles the infrahyoid muscles, omohyoid, sternohyoid, sternothyroid are depressors of the larynx, and the stylohyoid, digastric, mylohyoid, and stylopharyngeus muscles are elevators of the larynx, while the thyrohyoid draws the hyoid bone and thyroid cartilage together. With one exception the intrinsic muscles of the

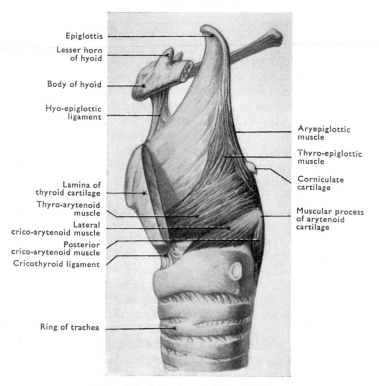

FIG. 588 Muscles of side-wall of larynx.

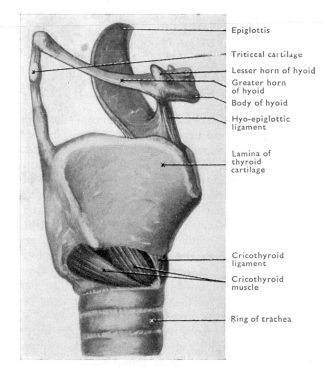

FIG. 589 Right cricothyroid muscle.

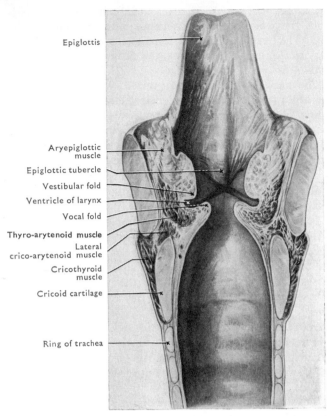

Epiglottis

Aryepiglottic muscle

Epiglottic tubercle

Vestibular fold

Ventricle of larynx

Vocal fold

Thyro-arytenoid muscle

Lateral crico-arytenoid muscle

Cricothyroid muscle

Cricoid cartilage

Ring of trachea

FIG. 590   Coronal section of larynx to show situation of muscles.

and relaxing the vocal folds. The superomedial fibres of this muscle, sometimes known as the **vocalis**, are within the substance of the vocal fold and are attached in such a way that they are spread along the length of each vocal ligament so that some fibres are attached anteriorly, others posteriorly, and the remainder at evenly distributed points along the ligament [FIG. 588]. This ensures that the vocal ligaments can be shortened without any folding or crinkling.

BLOOD VESSELS, LYMPHATICS, AND NERVES OF THE LARYNX

Two pairs of arteries supply the larynx—the **superior** and **inferior laryngeal arteries**. The superior laryngeal artery, a branch of the superior thyroid, enters the larynx by penetrating the thyrohyoid membrane along with the corresponding nerve. It descends beneath the mucosa of the lateral wall and the floor of the piriform recess to supply the mucous membrane and muscles of the larynx. The inferior laryngeal artery, a branch of the inferior thyroid, enters the larynx beneath the lower border of the inferior constrictor muscle along with the inferior laryngeal nerve. It also supplies the mucous membrane and muscles of the lower part of the larynx, and anastomoses with the superior laryngeal artery. The superior and inferior laryngeal veins accompany the arteries to join the superior and inferior thyroid veins respectively.

The **lymph vessels** are numerous except over the vocal folds, and are divided into a superior and an inferior group. The vessels of the superior group accompany the superior laryngeal artery and join the upper deep cervical nodes. The

efferents from the inferior group accompany the inferior laryngeal artery and join the lower deep cervical nodes, some of them going to the supraclavicular nodes.

The **nerves** of the larynx are the internal and external branches of the **superior laryngeal** nerve and the terminal branches of the **recurrent laryngeal** nerve. The **external branch** accompanies the superior thyroid artery and supplies the cricothyroid muscle. The **internal branch** pierces the thyrohyoid membrane, breaks into a number of branches in the lateral wall of the piriform recess, and is the sensory nerve supply of the mucous membrane of the epiglottis, the most posterior part of the tongue, and the larynx above the vocal folds. The terminal part of the recurrent laryngeal nerve—the **inferior laryngeal nerve**—enters the larynx just behind the corresponding cricothyroid articulation close to the inferior thyroid artery. It divides into an anterior and a posterior branch, communicates with the internal laryngeal nerve, and supplies the remaining laryngeal muscles and mucous membrane. Injury to the laryngeal nerves most commonly occurs during operations on the thyroid gland. Injury to the superior laryngeal nerves is seldom bilateral and does not usually give rise to any distinctive symptoms. Injury to both recurrent laryngeal nerves results in paralysis of the vocal folds, loss of voice, and some difficulty in swallowing (involvement of oesophagus and inferior pharyngeal constrictor). Paralysis of one recurrent laryngeal nerve is, however, partly compensated by movements of the normal vocal fold which may be abducted to widen the airway, or adducted across the midline for phonation. Nevertheless, the voice is hoarse and the cough abnormal. Temporary paralysis may occur due to post-operative oedema affecting the nerves. It is known that partial damage to the recurrent laryngeal nerve generally affects the abductor muscles, e.g., the posterior crico-arytenoid, more severely than the other muscles.

EXAMINATION OF THE LARYNX

The larynx can be examined in the living subject through the mouth using a laryngoscopic mirror. The vestibular and vocal folds are easily recognized and between them on each side the entrance to the laryngeal ventricle appears as a dark line on the side-wall of the larynx. The cavity seen in this way is much foreshortened. FIGURE 591A shows the shape of the rima glottidis during quiet respiration, and FIGURE 591B represents a condition of extreme abduction during forced inspiration when the arytenoid cartilages are rotated laterally and the vocal processes move widely apart. During certain phases of phonation the rima glottidis is reduced to a linear chink.

There is considerable variation in the size of the larynx which is quite independent of stature, and explains to a great extent the difference in pitch of voice in different people. The male larynx is not only absolutely but relatively larger than the female larynx, more particularly in the anteroposterior diameter, because of the more acute angulation of the thyroid laminae and the consequently greater laryngeal prominence in the male. In the new-born child the larynx is relatively large compared with the calibre of the trachea, but growth changes take place during the first 6 years, increasing the calibre of the trachea and the size of the epiglottis. At puberty in the male

A

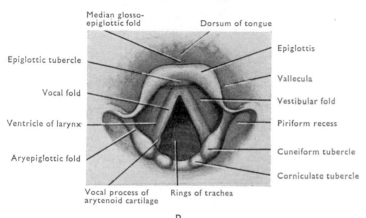

Median glosso-epiglottic fold

Dorsum of tongue

Epiglottic tubercle

Epiglottis

Vocal fold

Vallecula

Ventricle of larynx

Vestibular fold

Piriform recess

Aryepiglottic fold

Cuneiform tubercle

Corniculate tubercle

Vocal process of arytenoid cartilage

Rings of trachea

B

FIG. 591   Cavity of larynx seen with laryngoscope.
A, rima glottidis closed.
B, rima glottidis widely opened.

there is a growth spurt affecting the larynx which quite quickly increases in size. This growth spurt does not occur in males castrated before puberty.

THE VOCAL MECHANISM

The details of the mechanism of voice production, in particular the changes and movements of the vocal folds, are still far from clear. The parts of the body concerned with the production of speech sounds—the vocal tract—consist of the mouth with the tongue, teeth, palate, nasal cavities, the pharynx, the larynx, and the lungs. To produce a sound a stream of air is vibrated into sound waves mainly by the oscillation of the vocal folds. The voice has pitch, individual quality, and volume. High-speed motion pictures have shown that in phonation there is a rapid opening and closing of the vocal cords, i.e., vibration in the horizontal plane at a frequency appropriate to the pitch of the sound produced. During the closed phase pressure develops below the vocal folds, bursts them open while a momentary puff of air passes through, and this is the basis of the excitation of the air stream into sound waves.

The range and pitch of speech sounds is achieved by variation in the length, thickness, or tension of the vocal folds. Because, for example, their vocal folds are shorter and in particular thinner, women and children speak in higher pitched voices than men. The rhythmic interruption of the air stream by the vibration of the vocal cords establishes the sound waves

in the column of air which give the audible tone consisting not only of the fundamental tone or pitch but also upwards of fifty or more overtones. The emphasis of particular sets of overtones is responsible for the characteristic quality of the vowel sounds and the individual voice. This selection of overtones is made as the vibrating air stream passes up the vocal tract which acts as a resonator like the pipes of an organ, and an infinite range of overtone selection, which would otherwise require a great number of separate resonators, is made possible by adjustments and shifts in the position of the lips, the tongue, and the palate, etc. The nasal cavities also share in the resonating process, particularly for nasal consonants such as *m* and *n*. Thus, during speech the vocal tract is manipulated into a sequence of different shapes and configurations characteristic for any particular language, and each shape of the vocal tract forms a resonator with its own resonance frequencies, i.e., overtone selection. The volume of the voice is regulated by the amount of air set into vibration and that depends on the intensity of the glottal pulses. Alterations of the voice in old age are probably due to degenerative changes in the laryngeal muscles in connexion with poor vascularity.

THE TRACHEA

[PLATE 35 C and D]

The trachea is the continuation of the larynx. It is 4 cm. long at birth; 4·5 cm. between the first and second years; 5·5 cm. between the sixth and eighth years; 7 cm. between the fourteenth and sixteenth years; and from 9 to 15 cm. in the adult. The sagittal and coronal internal diameters of the trachea are: 5×6 mm. at birth; 9×9 mm. between the first and second years; 10×11 mm. between the sixth and eighth years; 10×13 mm. between the fourteenth and sixteenth years; and 16×14 mm. in the adult. Hence, in general, the coronal diameter is greater than the sagittal until adult life.

Until the third or fourth year the cross-sectional area of the trachea is equal to that of the two main bronchi; but from then until the tenth year the bronchi have slightly the larger area. After puberty the ratio continues to increase; and in the adult the area of the two main bronchi is 40 per cent. greater than that of the trachea.

The trachea commences at the lower border of the cricoid cartilage at the level of the sixth cervical vertebra and terminates at the upper border of the fifth thoracic vertebra where it bifurcates into a right and a left principal bronchus. In the living subject during deep inspiration in the erect position, the level of the bifurcation is said to descend to the sixth thoracic vertebra, in part due to a lengthening of the trachea itself to the extent of 2 cm. The external diameter of the trachea is about 2 cm. in the adult male and rather less in the adult female, and it is prevented from collapsing by fifteen to twenty incomplete rings of cartilage encircling the anterior and lateral parts of its circumference [FIG. 592].

Throughout its length the trachea is in the midline, inclines slightly to the right at the bifurcation, and posteriorly is separated from the vertebral column by the oesophagus. *In the neck* it is covered in front by the pretracheal part of the cervical deep fascia, the sternohyoid and sternothyroid muscles with the isthmus of the thyroid gland lying on its anterior surface at

FIG. 592   Trachea and bronchi.
The dotted area gives the outline of the thyroid gland.

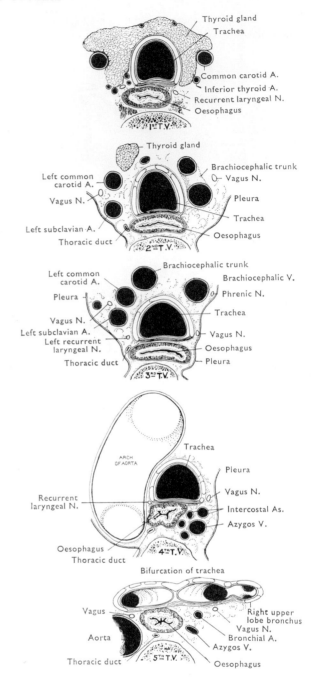

FIG. 593   Transverse sections through trachea and neighbouring structures at levels of upper five thoracic vertebrae.

the level of the second, third, and fourth tracheal cartilages, and below this the inferior thyroid veins, the tracheal lymph nodes, the thyroidea ima artery if present, and occasionally accessory thymic glandular tissue extending upwards as far as the lower border of the thyroid cartilage.

Lateral to the trachea are the lobes of the thyroid gland and the great vessels of the neck in the carotid sheath, both under cover of the anterior part of the sternocleidomastoid muscle. In the groove between the trachea and the oesophagus, on each side, is a recurrent laryngeal nerve (the left as far down as the arch of the aorta, the right to a level slightly above the subclavian artery) intermingled with branches of the inferior thyroid artery immediately below the larynx. The trachea may be opened surgically above the level of the isthmus of the thyroid gland—a high tracheostomy—or at the level of the sixth or seventh tracheal rings—a low tracheostomy—when, however, the inferior thyroid veins are encountered.

In the thorax, the jugular venous arch lies anterior to the trachea at the level of the jugular notch of the sternum, with the infrahyoid muscles and inferior thyroid veins intervening. The brachiocephalic trunk, the left common carotid artery, and below and to the left the arch of the aorta, all lie anterior to the trachea with the left brachiocephalic vein crossing obliquely from left to right in front of the arteries. The remains of the thymus gland are directly behind the sternum,

and the deep cardiac plexus of nerves lies on the trachea at its bifurcation behind the aortic arch [FIG. 592].

On the right side the trachea is in contact with the mediastinal pleura, with the right vagus passing downwards and backwards to reach the root of the lung and the azygos vein arching forwards to join the superior vena cava between them. On the left of the trachea are the arch of the aorta with the left recurrent laryngeal nerve looping round it, and posterolaterally the left subclavian artery. The relation of the trachea to the aortic arch makes it particularly liable to compression from dilatations of the aorta (aneurysms) or from vascular rings which occur in some forms of abnormal arterial development and partially encircle the oesophagus and trachea.

THE PRINCIPAL BRONCHI

The right and left principal bronchi differ in length, diameter, and the angle they make with the trachea [FIG. 592 and PLATE 35C and D]. The **left principal bronchus** is about 5 cm. long from its origin to its first branch (the bronchus of the superior lobe) is narrower than the right bronchus, and joins the trachea at an angle of 135 degrees. It passes downwards and laterally through the loop formed by the arch of the aorta, then in front of the descending aorta. The left pulmonary artery crosses in front of the bronchus to lie above it at the hilus of the lung with no branch of the bronchus superior to the artery. Below the level of the artery the upper left pulmonary vein lies in front of the bronchus which, at the hilus, runs downwards immediately postero-inferior to the oblique fissure. The left

bronchus is situated behind and a little above the transverse sinus of the pericardium and the left atrium of the heart which may be entered through the bronchial wall. The **right principal bronchus** is about 2·5 cm. in length from the tracheal bifurcation to its first branch which it gives off above the right pulmonary artery. It is wider than the left bronchus, forms an angle of 155 degrees with the trachea, and passes behind the ascending aorta and the superior vena cava at which point the terminal portion of the azygos vein arches above it. As it reaches the hilus of the lung, both the pulmonary artery and vein cross it anteriorly after it has given off its branch to the upper lobe. The stem branch then enters the lung medial to the oblique fissure. Because of the intimate relation of the left principal bronchus to the arch of the aorta the right principal bronchus is easier to approach surgically in spite of its shorter length.

STRUCTURE OF THE TRACHEA AND BRONCHI

This consists of three layers:

1. A mucous membrane with a pseudostratified columnar ciliated epithelium resting on a basement membrane. There are numerous goblet cells and the lamina propria contains a large number of elastic fibres and small glands opening on to the surface of the epithelium.

2. A muscular layer with external longitudinal fibres and deeper transverse fibres which form a more definite layer between the ends of the cartilages, sometimes known as the trachealis muscle.

3. An external fibrous membrane which encloses the cartilaginous rings and forms a single layer between them. This membrane has a large number of yellow elastic and white collagenous fibres so that the trachea as a whole is extensible and stretches longitudinally during respiration [FIG. 596].

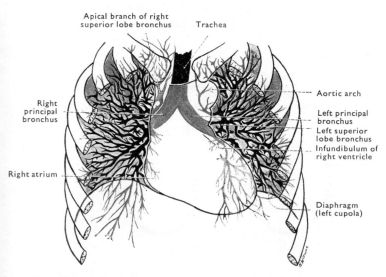

FIG. 594 Tracing from radiograph of chest in full inspiration. [See also Plate 36.]

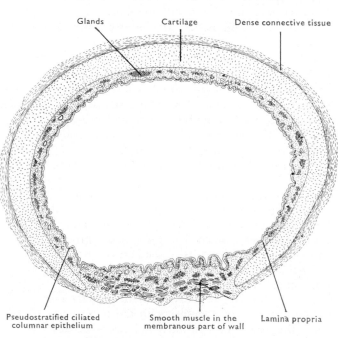

FIG. 595 Tracing from a radiograph of chest in full expiration. [See also Plate 36.]

FIG. 596 Section through wall of trachea.

The larger intrapulmonary bronchi have irregular plates of cartilage forming complete rings, while the smaller bronchioles have no cartilage.

When examined in the living subject through the broncho-scope, a projecting ridge between the bronchi, pointing up the trachea and known as the **carina** or keel, is a prominent land-mark at the bifurcation of the trachea. This is formed by a hook-like process of the lower border of the last cartilage of the trachea which curves between the two principal bronchi.

THE THORACIC CAVITY

[PLATES 34–36]

The thoracic cavity is enclosed by the ribs and costal cartilages, the vertebral column, and the sternum. The superior aperture slopes downwards and forwards, and is bounded by the first thoracic vertebra, the manubrium of the sternum, the first ribs and their costal cartilages. The inferior aperture slopes down-wards and backwards, is closed by the diaphragm and is bounded by the last thoracic vertebra, the twelfth ribs with the last six costal cartilages and the xiphoid process.

In the cavity of the thorax are the right and left lungs in the pleural sacs, separated in the midline by a thick mass of tissue called the **mediastinum.** This reaches from the sternum in front to the vertebral column behind, and extends throughout the length of the thorax. For descriptive purposes the mediastinum is divided into a superior and an inferior mediastinum by an anteroposterior plane through the disc between the fourth and fifth thoracic vertebrae and the manu-briosternal joint. The inferior mediastinum is further sub-divided into the anterior mediastinum in front of the peri-cardium, the middle mediastinum occupied by the heart and pericardium, and the posterior mediastinum behind the pericardium superiorly and the diaphram inferiorly.

Situated in the **superior mediastinum,** which is bridged by strands of fibrous tissue loosely connecting the manubrium of the sternum to the upper part of the pericardium and known as the **superior sternopericardial ligament,** are a number of structures: anteriorly, any thymic remains together with a number of lymph nodes, the **aortic arch** and its three branches—the brachiocephalic trunk, the left common carotid and the left subclavian arteries; the upper part of the superior vena cava, the arch of vena azygos, the brachiocephalic veins, and the left superior intercostal vein; the vagus, phrenic, cardiac, and left recurrent laryngeal nerves; the trachea, its bifurcation and principal bronchi, the oesophagus, and the thoracic and right lymph ducts behind [FIG. 593].

The **anterior mediastinum,** narrow above where it is encroached on by the pleural sacs, con-tains a few lymph nodes, small branches of the inter-nal thoracic (*mammary*) artery, and is also bridged by fibrous tissue—the **inferior sternoperi-cardial ligament**—attached to the xiphoid process and the inferior extremity of the pericardium.

The heart, with the ascending aorta and the pulmonary trunk within the pericardium, the lower part of the superior vena cava with the termination of the azygos vein, together with the pulmonary veins, the phrenic nerves, and lymph nodes, occupy the **middle mediastinum.**

The **posterior mediastinum** has the descending thoracic aorta and its branches, the oesophagus, the azygos and hemi-azygos veins, the vagus and splanchnic nerves, lymph nodes, and the main lymph channels—the thoracic and right lymph ducts.

The mediastinal viscera shift their position during respiration and with changes in the position of the body. The heart, the aortic arch, and the bifurcation of the trachea descend during inspiration or when changing from the supine to the erect posi-tion, and ascend during expiration or when there is distension of the abdominal cavity [FIGS. 594 and 595, and PLATE 36].

THE PLEURA

The developing lungs grow into and invaginate the serous lining of the body wall, in this way acquiring a closely investing visceral serous covering—the **pulmonary pleura**—while the **parietal pleura** lining the lateral thoracic wall, the upper sur-face of the diaphragm, and the mediastinum, completes the closed sac of the pleural cavity in which the lungs are situated. The parietal and pulmonary pleurae are continuous with one another at the root of the lungs [FIG. 597]. Because of the large right lobe of the liver which elevates the diaphragm on that side, and the position of the heart towards the left side, the left pleural cavity is narrower but slightly greater in vertical extent than the right pleural cavity [PLATE 34]. Normally, except for a small amount of lubricating fluid, the two layers of the pleura are in contact, and during respiration the pulmonary pleura slides smoothly on the parietal pleura, but they may become widely separated by fluid exudates caused by disease, or adherent as a late result of inflammation.

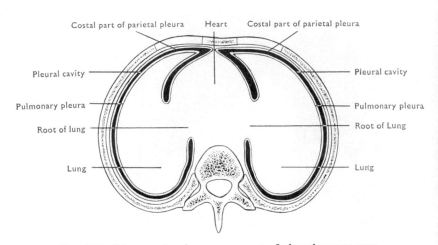

FIG. 597 Diagram showing arrangement of pleural sacs as seen in transverse section.

PARIETAL PLEURA

The parietal pleura, although a continuous sac, is described as the costal, diaphragmatic, cervical, and mediastinal pleura.

Cupula Pleurae (Cervical Pleura). The cervical pleura forms the dome-shaped roof of the pleural cavity, extending upwards within the concave internal margin of the first rib into the neck as far as the level of the neck of the first rib (level of the seventh cervical spine). The cervical pleura is strengthened by the **suprapleural membrane,** an aponeurotic expansion attached to the inner margin of the first rib, and derived from a small muscular slip (the scalenus minimus) that arises from the transverse process of the seventh cervical vertebra. It is covered laterally by the scalenus anterior and scalenus medius muscles, medially and superiorly it is grooved by the sub-clavian artery arching over it, and anteriorly by the brachio-cephalic and subclavian veins. The position of the cervical pleura can be marked out on the surface by a line, convex upwards, from the sternoclavicular joint to the junction of the medial and middle thirds of the clavicle, the highest point of which is 2–3 cm. above the medial third of the clavicle.

Costal Pleura. The costal pleura, which is the strongest part of the parietal pleura, lines the inner surface of the thoracic wall extending from the sternum in front, round to the sides of the bodies of the vertebrae behind, where it is firmly bound down. The medial extent of the pleural cavities anteriorly, i.e., the lines at which the costal and mediastinal pleurae meet on each side (the **costomediastinal recesses** of the pleura) converge from a point behind the sternoclavicular joints to the midline posterior to the sternal angle [FIG. 598]. Thereafter, the right and left pleural sacs descend in contact with one another as far as the level of the fourth costal cartilage where the reflexion line of the left pleura curves laterally to a point in the sixth left intercostal space, about 1 cm. from the border of the sternum, while the line of the right pleura continues down in or near the midline to the xiphisternal joint. Posteri-

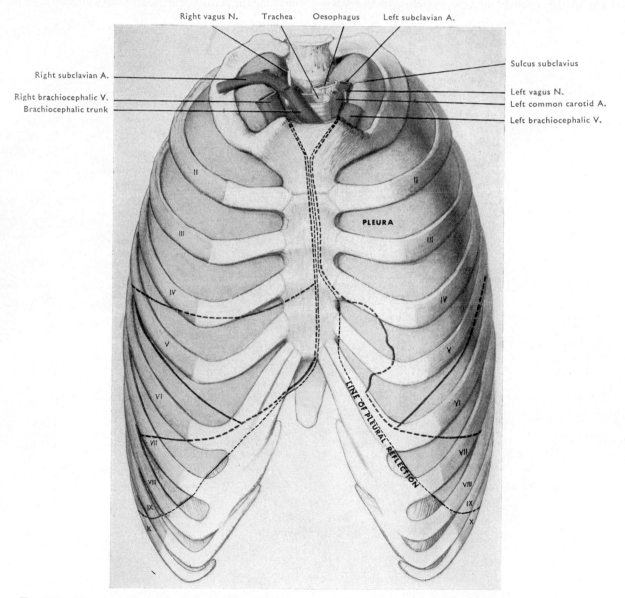

FIG. 598   Pleural sacs in subject hardened by formalin injection to show their relations to thoracic wall and inlet.

The anterior and diaphragmatic lines of pleural reflexion are exhibited by black dotted lines, whilst the outlines of the lungs and their fissures are indicated by the blue lines.

orly, the costal and mediastinal pleurae are continuous at the sides of the bodies of the vertebrae, and inferiorly the junction of the costal and diaphragmatic pleurae is indicated by a line commencing at the back of the xiphisternal joint, reaching the eighth costochondral junction in the midclavicular line, and inclining downwards and laterally to reach the tenth rib in the midaxillary line [FIG. 599]. Thence, it passes backwards and slightly upwards across the eleventh and twelfth ribs to reach the vertebral column midway between the head of the last rib and the transverse process of the first lumbar vertebra (spine of the twelfth thoracic vertebra). The lowest level of the pleural cavity is therefore the tenth rib in the midaxillary line. The pleura lines the inner surface of the medial part of the last rib and extends a short distance below it; and thus below the inferior border of the thoracic wall—a point of some importance in surgery of the kidney [FIG. 618]. In the living subject, according to radiographic evidence, this posterior line of pleural reflexion reaches the vertebral column at any level from the upper margin of the twelfth thoracic vertebra to the lower margin of the first lumbar vertebra [FIG. 600]. It is commonly stated that the left pleural sac reaches a lower level than the right, but this is by no means the rule.

Diaphragmatic Pleura. The thin diaphragmatic pleura covers the thoracic surface of the diaphragm lateral to the pericardium, but does not extend as far down as the costal attachment of the diaphragm (except posteromedially). The lower part of the diaphragm is therefore in direct contact with the costal cartilages and muscles of the intercostal spaces. From these adjacent surfaces the **phrenicopleural fascia** passes to the external surface of the costal and diaphragmatic pleurae at their line of junction and holds this reflexion, the **costo-diaphragmatic recess,** firmly in place.

Mediastinal Pleura. The mediastinal pleura forms the lateral boundary of the mediastinum with its contained structures and is continuous in front and behind with the costal pleura, below with the diaphragmatic pleura, and round the root of the lung with the pulmonary pleura [FIG. 601]. Above the root of the lung the mediastinal pleura passes directly from the sternum

FIG. 599 Lateral view of right pleural sac in subject hardened by formalin injection.

The blue lines indicate the outline of the right lung, and also the position of its fissures.

FIG. 600 Pleural sacs from behind in subject hardened by formalin injection. The blue lines indicate the outlines and the fissures of the lungs.

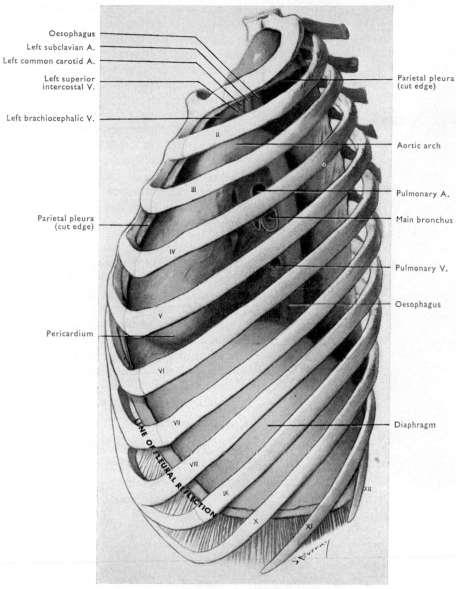

Oesophagus
Left subclavian A.
Left common carotid A.
Left superior intercostal V.
Left brachiocephalic V.
Parietal pleura (cut edge)
Pericardium

Parietal pleura (cut edge)
Aortic arch
Pulmonary A.
Main bronchus
Pulmonary V.
Oesophagus
Diaphragm

LINE OF PLEURAL REFLECTION

FIG. 601   Left pleural sac in subject hardened by formalin injection.

The sac has been opened by the removal of the costal part of the parietal pleura; and the lung also has been removed to display the mediastinal pleura.

the left side it lies on the descending thoracic aorta, and immediately above the diaphragm on the oesophagus (both sides) and the inferior vena cava (right side). At this level, then, the oesophagus is covered on both sides with a layer of pleura, and the two layers meet between the oesophagus and the aorta to form a thin partition between the right and left pleural cavities. This partition, formed by the meeting of the right and left mediastinal pleurae behind the oesophagus, commonly extends upwards to the level of the fourth thoracic vertebra. In consequence, there are right and left retro-oesophageal pleural recesses, occupied by lung. These are demonstrable by X-rays, for the recesses deepen during inspiration with the subject erect. The right recess may cross the median plane.

Along the sternal and diaphragmatic lines of reflexion the pleura bends acutely so that the two parts come into contact with one another to form recesses which are occupied by the borders of the lung in deep inspiration only. These are the **costomediastinal** and **costodiaphragmatic recesses.** The pleura, like other serous membranes, consists of a stratum of fibrous tissue with a sheet of mesothelial cells on its free surface. It is attached to the underlying tissues by a subserous layer of areolar tissue.

THE LUNGS

[PLATES 34 and 36]

The lungs, two in number, are the organs of respiration. They lie free in the pleural cavities except where they are attached by their roots to the trachea, to the heart, and to the pericardium, by the bronchi, the blood vessels, and the pulmonary ligaments respectively. The lungs are highly elastic organs, rose pink in colour in the infant but mottled dark grey in later life owing to the accumulation of atmospheric particles in subpleural tissue. On the surface of the upper lobes, which move less during inspiration than the lower lobes, the dark pigmentation tends to occur in bands corresponding to the intercostal spaces. As their substance is light and porous the lungs float in water and crepitate when handled owing to their content of air. The foetal lung, however, in which there is no air, is of a firm consistency like liver and sinks in water; it is only when air and an increased supply of blood are introduced into the lung with the first inspirations that it assumes its characteristic soft, spongy, and buoyant qualities.

The lungs are heavier in the male than in the female, the proportion being one thirty-seventh and one forty-third of the body weight respectively. The average weight of the right

to the vertebral column, and on the *left side* is in contact with the arch of the aorta, the left phrenic and vagus nerves, the left superior intercostal vein, the left common carotid and subclavian arteries, the oesophagus and thoracic duct. On the *right side* it is related to the right brachiocephalic vein, the superior vena cava, the phrenic nerve, the brachiocephalic trunk, the trachea, the right vagus nerve, and the oesophagus.

Below the root of the lung a fold of pleura—the **pulmonary ligament**—stretches from the mediastinal surface of the lung to the pericardium and ends below in a free border. In front of the root of the lung the mediastinal pleura covers and adheres to the pericardium, except where the phrenic nerve and pericardiacophrenic artery intervene. Behind the root of the lung on the right side, the mediastinal pleura passes over the oesophagus, azygos vein, and vertebral bodies, while on

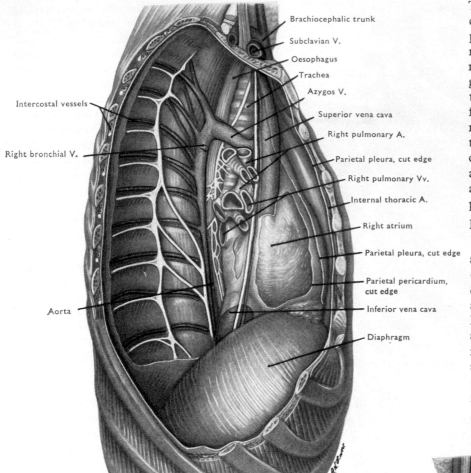

Brachiocephalic trunk

Subclavian V.

Oesophagus

Trachea

Azygos V.

Superior vena cava

Right pulmonary A.

Parietal pleura, cut edge

Right pulmonary Vv.

Internal thoracic A.

Right atrium

Parietal pleura, cut edge

Parietal pericardium, cut edge

Inferior vena cava

Diaphragm

Intercostal vessels

Right bronchial V.

Aorta

FIG. 602   The right side of the mediastinum and thoracic vertebral column.   The pleura has been removed together with part of the pericardium.

The apex of each lung is grooved by the subclavian artery arching over the pleura and passing posterior to the scalenus anterior muscle.   Behind is the anterior aspect of the neck of the first rib with the cervicothoracic ganglion of the sympathetic, the highest intercostal artery, and the ventral ramus of the first thoracic nerve.   Laterally is the scalenus medius muscle, and medially on the right side the brachiocephalic trunk, the right brachiocephalic vein, the trachea and oesophagus, and on the left side the common carotid artery, the left subclavian artery, the oesophagus, and the thoracic duct.

Base

The base of the lung is adapted to the thoracic surface of the diaphragm and is deeply concave.   The right base is separated by the diaphragm from the right lobe of the liver, and the left base from the left lobe of the liver, the fundus of the stomach, the spleen, and the adjacent left colic flexure.   The sharp inferior border of the lung between the costal surface and the base, which moves freely during respiration, fills the costodiaphragmatic recess only in deep inspiration.

Costal Surface

The costal surface is convex moulded to

lung is 620 g. and of the left lung 560 g. [FIG. 604].   They are conical in shape, and in the embalmed subject are moulded to the form of the thoracic cavity with an **apex**, a **base**, a costal and a medial surface. A sharp inferior border separates the base and costal surface, and a sharp anterior border separates the costal and medial surfaces.   The **right lung** is divided into three lobes, superior, middle, and inferior, by oblique and horizontal fissures.   The **left lung** has two lobes, superior and inferior, separated by an oblique fissure.

Apex

The apex of the lung fills the cervical pleura and is situated above the level of the superior aperture of the thorax, in front extending up to a point about 2·5 cm. above the medial third of the clavicle. Posteriorly, its highest point is at the side of the body of the first thoracic vertebra, at the level of the seventh cervical spine.

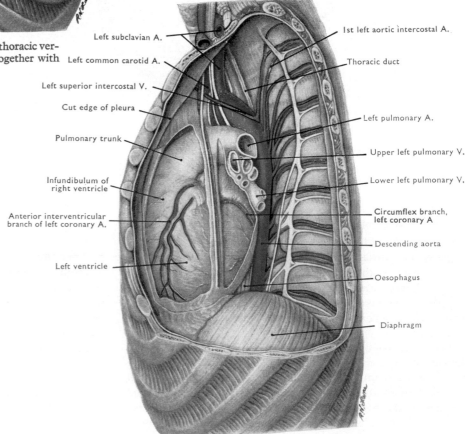

Left subclavian A.

Left common carotid A.

Left superior intercostal V.

Cut edge of pleura

Pulmonary trunk

Infundibulum of right ventricle

Anterior interventricular branch of left coronary A.

Left ventricle

1st left aortic intercostal A.

Thoracic duct

Left pulmonary A.

Upper left pulmonary V.

Lower left pulmonary V.

Circumflex branch, left coronary A

Descending aorta

Oesophagus

Diaphragm

FIG. 603   The left side of the mediastinum and thoracic vertebral column.   The pleura and part of the pericardium have been removed.

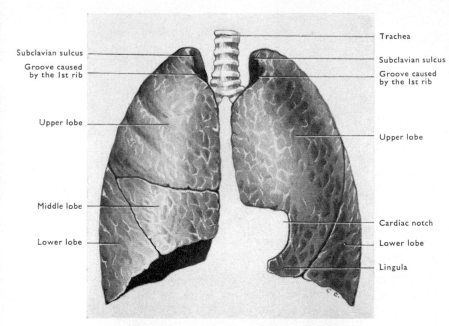

Subclavian sulcus

Groove caused
by the 1st rib

Upper lobe

Middle lobe

Lower lobe

Trachea

Subclavian sulcus

Groove caused
by the 1st rib

Upper lobe

Cardiac notch

Lower lobe

Lingula

FIG. 604   Trachea, bronchi, and lungs of a child, hardened
by formalin injection.

the wall of the pleural cavity, and may show shallow grooves
corresponding to the overlying ribs.

Medial Surface

The medial surface is divided into a posterior or vertebral
part and an anterior or mediastinal part by the **hilus,** where
the structures of the root enter or leave the lung.   The hilus
is enclosed by a short tube of pleura narrowing inferiorly to
become the pulmonary ligament.   The rounded **vertebral
part** of the medial surface is separated by the pleura from
the sides of the thoracic vertebrae, the posterior intercostal
vessels, sympathetic trunk, and the splanchnic nerves.   The
mediastinal part of the medial surface has in both lungs a
deep concavity—the **cardiac impression**—formed by the corresponding side of the heart, the more pronounced left cardiac
impression being due to the inclination of the heart to the left.
Lungs hardened by injection *in situ* are grooved by the mediastinal structures as shown in FIGURES 605 and 606, and it
should be noted that owing to its oblique course the oesophagus grooves both the right and the left lungs.

Anterior Border

The anterior border, separating the costal and medial surfaces, is thin and sharp.   It commences about 2·5 cm. below the
apex of the lung behind the sternoclavicular joint, and extends
to the base where it becomes continuous with the inferior
border [FIG. 598].   This part of the lung is carried forwards
into the costomediastinal recess of the pleural cavity.   It fills
this recess in deep inspiration, and in the upper part is separated
from the corresponding border of the left lung only by the
two layers of mediastinal pleura reflected from the sternum
to the pericardium.   The anterior border of the right lung,
which corresponds to the costomediastinal line of pleural reflexion, extends almost vertically to the xiphisternal joint.   The
anterior border of the left lung is interrupted by the **cardiac
notch,** and curves laterally at the level of the fourth costal

cartilage, to a point 3·5 cm. from the border of
the sternum and then medially to the sixth costal
cartilage 2–3 cm. from the side of the sternum.
This leaves an area in which the pericardium is
separated only by the pleura from the chest wall.
This is dull to percussion, and known as the area
of superficial cardiac dullness.   At the site of the
notch, the left costomediastinal recess of the
pleural cavity is, accordingly, occupied by lung
only in full inspiration.   The tongue-like portion
of the upper lobe of the left lung, situated between the cardiac notch and the oblique fissure,
is called the **lingula** [FIG. 604].

In childhood, when the thymus is active and
large, the anterior borders of the lungs, in expiration, are pressed apart by the thymus.   But
in inspiration the anterior borders of the lungs
approximate to each other anterior to the thymus.
After the age of 3 years, however, the thymic
shadow is no longer visible in radiographs of
healthy children.

Inferior Border [PLATES 35-36]

The inferior border, thin and sharp between
the diaphragmatic and costal surfaces, is obtuse and less well
defined on the mediastinal aspect.   During quiet respiration
the costodiaphragmatic margin is indicated on the surface of
the body by a line drawn from the sixth costal cartilage in
the midclavicular line to a point about 2 cm. lateral to the
tenth thoracic spine, and crossing the eighth rib in the midaxillary line.

Roots of the Lungs

The bronchi, the pulmonary arteries, veins, lymphatics, and
nerves, and the bronchial vessels constitute the roots of the
lungs, and towards the hilus are enclosed within the visceral
pleura where it becomes continuous with the parietal pleura.
The roots of the lungs lie opposite the bodies of the fifth, sixth,
and seventh thoracic vertebrae, with the phrenic nerve, the
pericardiophrenic artery and vein in front, and the vagus nerves
behind.   On the right, the superior vena cava lies in front of
the root of the lung and the azygos vein lies behind and arches
over its superior surface.   On the left side the aorta arches
over the lung root and the thoracic aorta descends behind it.

The relations of the main structures within the root of the
lung are seen in FIGURES 605 and 606.   The arrangement from
before backwards on both sides is : the pulmonary veins, the
pulmonary artery, and finally the principal bronchus.   From
above downwards on the right side there is the **superior
lobar bronchus** and pulmonary artery, the stem of the
middle and inferior lobar bronchi, and the lower
pulmonary vein, while on the left side the pulmonary artery is
superior, then the principal bronchus, then the lower pulmonary vein.

Fissures and Lobes in Life

The pleura-lined fissures vary in depth in different individuals, but never completely separate the lobes at the hilus
where there may be extensive fusion, particularly between the
middle and superior lobes of the right lung.   Here veins, arteries,

or bronchi may pass from one lobe into another.

The approximate position of the fissures and lobes in the fixed cadaver can be seen in FIGURES 598, 599, and 600.

The **oblique fissure** may be indicated posteriorly by the position of the medial border of the scapula when the hand is on top of the head. In the *right lung* it generally begins somewhat lower than that of the left lung. Its position is indicated approximately by a line beginning at the third to fifth rib in the paravertebral line, and following the curve of the sixth rib downwards and forwards across the midaxillary line. It meets the inferior border of the lung 7 to 8 cm. from the midline behind the sixth costal cartilage.

The oblique fissure of the *left lung* is similar to that of the right lung. It has a more vertical course, usually crossing either the fifth or sixth rib at the midaxillary line, and running forwards parallel to the sixth interspace or the upper border of the seventh rib, ends deep to the sixth or seventh costal cartilage.

The **horizontal fissure** of the right lung commonly begins at the oblique fissure in the midaxillary line at the level of the sixth rib, and runs almost horizontally forwards to end about the chondrosternal junction of the fourth rib. This fissure is rarely complete.

The **superior lobes** lie above and anterior to the oblique fissures and include the apex and the entire anterior border in the left lung. The posterior surface of the right superior lobe is adjacent to the upper three to five ribs posteriorly, but anteriorly descends only as far as the fourth costal cartilage. With the more vertical position of the left oblique fissure the superior lobe of the left lung is in contact with the upper three to five ribs posteriorly, but anteriorly it descends as far down as the sixth or seventh costal cartilage.

The **right middle lobe** is wedge-shaped, with its apex posteriorly deep to the sixth rib at the midaxillary line, and anteriorly it is in contact mainly with the fifth, sixth, and seventh costal cartilages. Each **inferior lobe** is in contact behind with the fifth to eleventh ribs and in front of the midaxillary line its highest point is at the sixth or seventh rib.

Variations in the anatomy of the lung take the form of partial fusion of adjacent lobes, or accessory fissures which tend to correspond to divisions between bronchopulmonary segments. A rare abnormality is the presence of an azygos lobe on the right side. The azygos lobe occurs as a consequence of an abnormal position of the arch of the azygos vein which lies in a fold of the parietal pleura—the mesazygos—and indents the superior lobe of the lung (often on the medial aspect), dividing it into lateral and medial portions.

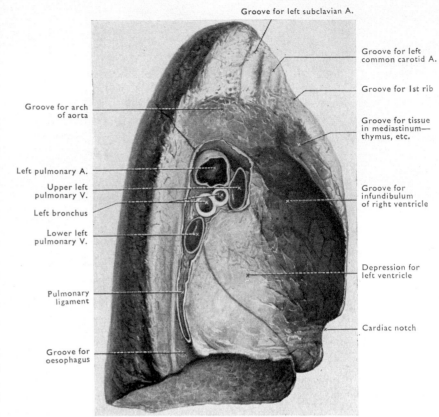

FIG. 605 Medial surface of left lung hardened *in situ*.

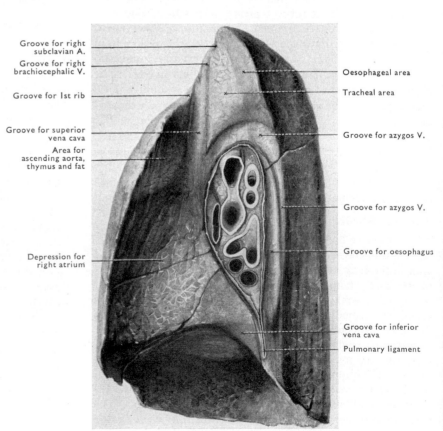

FIG. 606 Medial surface of right lung hardened *in situ*.

Structure of the Lungs

The structure of the lungs consists of the air ducts, the blood vessels, and the respiratory substance proper. There is a serous covering of visceral pleura and a framework, mainly of elastic fibres, forming a subserous layer and supporting septa between the lobes and the lobules of the lung.

Bronchopulmonary Segments [PLATE 35 C and D]

The right and left primary bronchi divide into secondary bronchi (three on the right and two on the left) each of which is distributed to a pulmonary lobe. Within the lobes the secondary bronchi in turn give off tertiary bronchi, and *the pyramid of lung supplied by each tertiary (segmental) bronchus constitutes a bronchopulmonary segment*. Within each bronchopulmonary segment there is further branching of the bronchi.

The short **superior bronchus** to the upper lobe of the right lung is given off before or just as the right principal bronchus enters the hilus, and after about a centimetre divides into three **segmental bronchi**, an **apical** directed superiorly towards the apex of the lung, a **posterior** directed backwards and laterally, and an **anterior** passing downwards and forwards. These are distributed to the corresponding bronchopulmonary segments [FIG. 607].

The **bronchus to the right middle lobe** arises from the stem bronchus very close to the origin of the superior segmental bronchus of the inferior lobe. It divides into a **medial** and a **lateral segmental** bronchus distributed to the corresponding bronchopulmonary segments. The first **branch to the right lower lobe** is the **superior segmental** bronchus. Thereafter, the stem bronchus gives off **medial** and **anterior basal segmental bronchi,** and then divides into **lateral** and **posterior basal segmental bronchi,** each distributed to the bronchopulmonary segments of the same name.

The **left principal bronchus** divides, inferior to the left pulmonary artery, into **superior** and **inferior lobar bronchi.** The superior lobar bronchus divides into **apicoposterior** and **anterior segmental bronchi.** In addition this bronchus gives off a stem which bifurcates into **superior** and **inferior lingular** bronchi distributed to the lingula.

The **inferior lobar bronchus** continues into the lower lobe where it has the five branches corresponding to those of the lower lobe of the right lung, though the anterior and medial basal may have a common stem.

A branch of the pulmonary artery accompanies the bronchus of each segment, generally lying posterolateral to it. The bronchopulmonary segment is not, however, a bronchovascular segment in the sense of having its individual bronchus, artery, and vein, but can in addition be supplied by small arteries from neighbouring segments. Tributaries of the pulmonary veins are situated at the periphery of the segments and, therefore, some lie adjacent to the fissures. From their intersegmental position they drain adjacent segments, and thus the connective tissue planes separating the bronchopulmonary segments are not necessarily avascular, and communications across the fissures may occur. These intersegmental planes may also be crossed by small bronchi from neighbouring segments. Although each pulmonary lobe has a recognized number of bronchopulmonary segments named according to their position, there is a considerable individual variation in their relation to the substance of the lung, the manner of branching of the bronchi, and in their blood supply. The upper lobe of the right lung has three segments including an apical one; the middle lobe two; and the right lower lobe five segments including a superior and four basal segments. The left upper lobe has four segments including an apicoposterior, an anterior, and a superior and an inferior lingular; and the left lower lobe has five segments including a superior and four basal.

While the shapes of the individual bronchopulmonary

FIG. 607   The bronchopulmonary segments.   (After Jackson and Huber, 1943.)
The asymmetry of the main bronchial pattern is imposed on the bronchial tree by the asymmetry of the heart.
Each bronchial branch is designated by the name of the subdivision of the lung supplied by it.

1. Apical.	9. Lateral basal.
2. Apicoposterior.	10. Medial basal.
3. Anterior.	11. Posterior basal.
4. Lateral.	12. Inferior lingular.
5. Medial.	13. Superior lingular.
6. Posterior.	14. Left principal bronchus.
7. Superior	15. Right principal bronchus.
8. Anterior basal.	16. Trachea.

segments are varied and complex they can be regarded in general as somewhat pyramidal, with their apex towards the hilus, but they vary considerably in the extent to which they present themselves on the surface of the lung from one individual to another. It is now recognized that segments rather than entire lobes may be localized sites of infection, e.g., in bronchiectasis, abscess, cystic disease, or tuberculosis, and in such conditions it may be possible for the surgeon to remove the affected bronchopulmonary segment. The practical importance of the bronchopulmonary segments is in connexion with the surgical resections of lung tissue in malignant disease, tuberculosis, or other forms of infection. A knowledge of the anatomy of the bronchopulmonary segments is also useful in postural drainage in such conditions as bronchiectasis.

Respiratory Substance [FIG. 608]

The bronchi of the bronchopulmonary segments divide into several orders of smaller tubules, the smallest—the **respiratory bronchioles**—having a diameter of about 0·5 mm. These are short tubes with no cartilage, and throughout their length the columnar epithelium loses its cilia and becomes low cuboidal. At intervals there are sac-like dilatations of the wall (**alveoli**) of the respiratory bronchiole. These alveoli are lined by thin, flattened epithelium, and are the first respiratory units of the lung where gaseous exchange takes

Bronchus

Bronchiole

Respiratory bronchiole

Alveoli

Alveolar saccule

Alveolar ductule

FIG. 608   Diagram to show terminal elements of the bronchial tree.

place. Each respiratory bronchiole divides into a number of tortuous branching passages (**alveolar ductules**) the walls of which are almost completely formed by openings into alveoli, though on the ends of the septa between these is some cuboidal epithelium. The alveolar ductules end in one or more dilated spaces (**alveolar saccules**) from which a continuous series of alveoli arise. The respiratory unit of the lung is the alveolus.

The walls of the **alveoli** have a thin epithelial lining with a basement membrane, a dense capillary plexus, and a network of supporting reticular and elastic fibres. Respiratory exchange takes place between the blood in the capillaries and the air in the alveoli. Free macrophages are found in the alveoli of most sections of the lung, and when they contain particles of dust are known as dust cells. They are probably derived from the mesodermal elements of the alveolar wall. The pulmonary lobules vary in size and shape, the larger superficial ones having basal portions towards the surface.

Expansion of the Lung in Respiration [PLATE 36]

Probably only about 3 mm. of the surface of the lung expands appreciably during quiet respiration, although pulmonary tissue to a depth of 30 mm. may be involved in forced breathing. The remainder of the lung tissue, which can be regarded as an extension of the root, is relatively fixed by the presence of large blood vessels and bronchi and the parts around the hilus. As the increased capacity of the thorax is made possible by the range of movement of its costosternal and diaphragmatic boundaries, the corresponding parts of the lung—costal and base—have the greatest degrees of freedom for expansion. During inspiration the root of the lung moves downwards and forwards, and it is this movement that makes it possible for the apical and posteromedial parts of the lung to expand.

Blood Vessels and Lymphatics of the Lung

Pulmonary Arteries. Venous blood is conveyed to the lung by the pulmonary artery. At the hilus of each lung the pulmonary artery divides into branches corresponding to the divisions of the bronchi, so that there is an artery accompanying the bronchus of each bronchopulmonary segment and generally lying posterolateral to it [FIG. 609]. The right pulmonary artery gives off its first branch to the upper lobe at the hilus, and thereafter the stem artery lies just below the medial part of the oblique fissure and in this situation it gives off segmental branches to the middle and lower lobes. The left pulmonary artery gives off its branches to the upper lobe as it lies in the hilus, medial to the oblique fissure. The bronchopulmonary segment is not, however, a bronchovascular segment in the sense of having its individual bronchus, artery, and vein, but in addition may be supplied by small arteries from neighbouring segments. The branches of the central artery of each segment generally accompany the bronchi as far as the respiratory bronchiole where they divide into smaller vessels which run to the walls of the air saccules, to be finally distributed as a capillary network around the alveoli.

Pulmonary Veins. The oxygenated blood is carried away

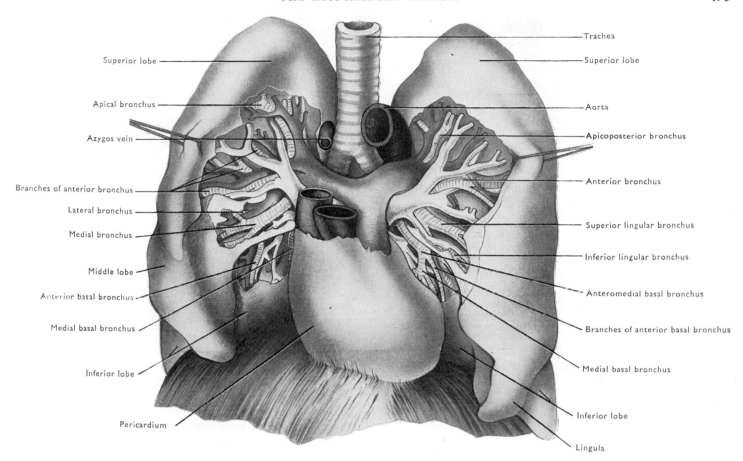

Fig. 609 Distribution of bronchi and pulmonary blood vessels in the lungs.

from the capillary plexuses by the pulmonary veins. A pulmonary vein emerges from each bronchopulmonary segment, and in addition there are peripheral intersegmental veins which drain adjacent bronchopulmonary segments, but towards the hilus the veins fuse to form about ten principal vessels running along with the bronchi. The veins lie below and in front of the bronchi, while the arteries are generally above and behind. The right superior pulmonary vein is formed close to the hilus by the union of the vein from the upper and middle lobes, and the inferior pulmonary vein drains the lower lobe. The left superior pulmonary vein drains only the upper lobe. The left inferior pulmonary vein can be located at the upper end of the pulmonary ligament. Some veins run superficially underneath the pleura, particularly on the mediastinal surface or within the fissures. When the fissures are incomplete it is common to find a vein passing from one lobe to another.

Bronchial Vessels. The walls of the air ducts within the lungs as far as the respiratory bronchioles are supplied by **bronchial arteries,** generally one for the right lung with a variable origin—from the aorta or posterior intercostal arteries —and two for the left lung, more constantly arising from the descending thoracic aorta. Branches of these arteries accompany the intrapulmonary bronchioles forming capillary plexuses in their walls and giving branches which communicate with those of the pulmonary artery. Much of the blood carried by the bronchial arteries is returned by the pulmonary veins. However, the larger **bronchial veins** lie on the posterior

aspect of the main bronchi and drain, on the right side, into the azygos and the superior intercostal veins, and on the left side, into the hemiazygos system.

Lymphatics. There is a superficial and a deep set of lymph vessels. The superficial vessels lie beneath the pulmonary pleura which they drain, and converge round the borders of the lung and the margins of the fissures, to end in the **bronchopulmonary lymph nodes** at the hilus. The deep lymph vessels commence as a plexus in the walls of the alveolar ductules, and accompany the bronchi and the pulmonary vessels to drain into the bronchopulmonary nodes, often through **pulmonary nodes.** Some anastomosis takes place between the deep and superficial lymph vessels.

From the bronchopulmonary lymph nodes the lymph flow drains to the **tracheobronchial nodes** and thence to the right and left **bronchomediastinal trunks** which frequently open independently into the junction of the internal jugular and subclavian veins of their own side.

One or two lymph vessels from the lower lobe of each lung pass through the pulmonary ligament to end in the posterior mediastinal nodes. Lymphoid tissue is scattered through the lung substance at the sites of division of the bronchi and bronchioles, on the alveolar ducts, and where the venous radicals arise beneath the pleura. This lymphoid tissue increases in old age and forms a sheath for the bronchial tree and pulmonary vessels, and also becomes diffused in the bronchial walls external to the epithelium and the plain muscle.

FIG. 610   Reconstruction of a bronchus, expanded in inspiration, with its muscles and nerves.   (After Miller, 1921, 1947.)

Innervation of the Lungs

Parasympathetic and sympathetic nerve fibres reach the lungs through the pulmonary plexuses which intermingle with the structures of the lung root, especially posteriorly. They are formed by branches of the vagus and the sympathetic trunk. Within the lung the nerves are grouped in peribronchial and periarterial plexuses, and it is said that the peribronchial plexuses contain parasympathetic ganglia and give off the post-ganglionic parasympathetic fibres to the smooth muscle of the bronchi, and that the periarterial plexus is composed of post-ganglionic sympathetic fibres supplying the musculature of the artery [FIG. 610]. It is generally supposed that the para-sympathetic efferent fibres are constrictors of the bronchi, while the sympathetic are vasoconstrictors. There are numerous afferent endings in the smooth musculature of the bronchi, particularly close to the hilus, and the afferent nerve fibres are concerned with the cough reflex. The evidence is that all the sensory fibres from the lung, including pain fibres, join the vagus nerve.

DEVELOPMENT OF THE RESPIRATORY SYSTEM

The larynx, trachea, bronchi, and lungs all arise from a median longitudinal groove, called the **laryngotracheal groove,** which makes its appearance in the ventral wall of the primitive pharynx between the sixth pharyngeal arches when the embryo is

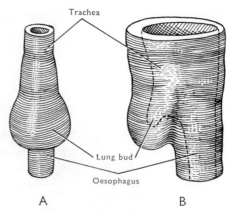

FIG. 611   Development of respiratory tube in 4·25 mm. human embryo.   (After Grosser, 1912.)

A, ventral view.   B, lateral view

about 3 mm. in length. The groove is soon pinched off from the foregut by lateral constrictions and is thus converted into the primitive respiratory tube [FIG. 611]. This separation begins at the caudal end and proceeds headwards, but it leaves the cephalic end of the tube in communication with the foregut.

The cephalic end of the primitive respiratory tube forms the larynx, the succeeding portion forms the trachea, whilst the caudal end bifurcates into two bulbous swellings called the **lung buds.** These grow laterally, surrounded by mesoderm of the mesentery, and expand with the pleural cavity into meso-dermal masses derived from the dorsal extensions (right and left) of the septum transversum [p. 498.] Thus the epithelium of the respiratory tube and of its branches is entoderm derived from the foregut. The remaining elements of the lungs are derived from the mesoderm of the mesentery.

Larynx. At first the cephalic end of the laryngotracheal groove lies caudal to the fourth pair of pharyngeal arches in a region corresponding to the sixth arches, but it soon moves headwards to a position between the ventral ends of the fourth pair of phar-yngeal arches [FIG. 612]. Most of the laryngeal muscles are derived from the sixth pair of arches, and hence are supplied by the nerves of these arches, namely, the recurrent laryngeal nerves. The slit-like opening of the primitive respiratory tube becomes the laryngeal orifice of the fully developed larynx [FIG. 612].

In embryos of about 5 mm. in length, a median elevation

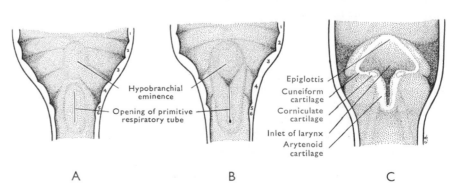

FIG. 612   Development of larynx.   (A and B after Frazer, 1940; C after Arey, 1954.)
A, 6 mm. human embryo;   B, 8·5 mm. human embryo;   C, 40 mm. human embryo.

(hypobranchial eminence) bulges into the pharynx from its ventral wall at the ventral ends of the third and fourth pharyngeal arches. The laryngotracheal groove is immediately caudal to this elevation. The caudal part of the median elevation gives rise to the **epiglottis,** and the lateral upgrowths become the aryepiglottic folds. About the fourth month, excavation of the medial side of each of these folds begins and gives rise to the corresponding **ventricle** of the larynx. The lower margin of each ventricle becomes a vocal fold.

At the end of the fourth week the cartilaginous framework of the larynx is indicated by mesodermal condensations, though actual cartilage does not appear until the seventh week. The arytenoid and corniculate cartilages which form in the upgrowths above the level of the glottis, the cricoid cartilage, and the cartilages of the trachea are all continuous at first. The epiglottic and cuneiform cartilages develop in the median elevation and likewise are originally continuous.

The **thyroid cartilage** first appears as a pair of separate mesodermal laminae, one on each side of the larynx, connected ventrally by dense mesoderm. Some time after chondrification has begun in the laminae, a cartilaginous element arises in this dense ventral mesoderm and above it two smaller nodules of cartilage appear. As chondrification proceeds, the thyroid cartilage is formed by the fusion of all these initially separate components. The thyroid laminae probably represent the cartilaginous elements of the fourth and fifth pharyngeal arches. A foramen in the fully developed thyroid lamina is the result of incomplete chondrification, probably attributable to the passage of a small blood vessel. The superior horn of the thyroid cartilage is at first continuous with the greater horn of the hyoid bone, and part of the connecting bar remains as the cartilago triticea [FIG. 581].

The precartilaginous rudiments of the **cricoid** and **arytenoid** cartilages, though at first continuous with each other, become differentiated by the appearance of a separate cartilaginous centre for each arytenoid and an incomplete ring (for a time deficient behind) for the cricoid. The cricoid thus resembles developmentally a tracheal ring, to which it probably corresponds morphologically. Chondrification proceeds in the cricoid from two centres, one on each side. These centres soon unite ventrally, but chondrification proceeds more slowly into the posterior lamina. The corniculate cartilages are merely detached portions of the arytenoid cartilages.

Trachea. The trachea is developed from the intermediate portion of the respiratory tube.

Lungs. The bronchi and lungs are developed from the two diverticula at the caudal end of the laryngotracheal groove and from the surrounding mesoderm. Both diverticula elongate, and almost immediately subdivide—the right into three and the left into two —thus early indicating the lobes of the respective lungs [FIG. 613]. As the primitive respiratory tube lies in the median mesentery posterior to the attachment of the septum transversum, each pulmonary diverticulum grows laterally into the corresponding pleural canal (pericardioperitoneal canal), i.e., the primitive pleural cavity [p. 66], carrying before it a covering of mesoderm from the mesentery. From this mesoderm are derived the blood vessels and other tissues which build up the lung, and from the entodermal lining of the primitive respiratory tube the

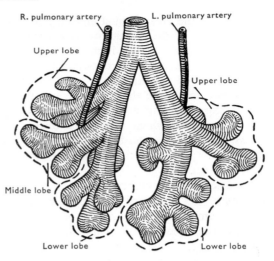

FIG. 613 Development of lobes of lungs in 14 mm. human embryo. (After Arey, 1954, after Ask.)

epithelial lining of the air passages is derived. The entodermal pulmonary diverticula branch and rebranch pushing their way into the pulmonary mesoderm until the complete bronchial tree is formed [FIG. 614]. When the ramification is complete, the bulbous ends of the various branches represent alveoli clustering around the alveolar saccules into which each alveolar ductule divides.

At first the lung is devoid of alveoli, but between the sixth month and full term they make their appearance on the alveolar saccules. By the close of the fourth month of intrauterine life the columnar cells which line the trachea and bronchi have become ciliated. Growth of the lung is not interrupted at birth, but since the terminal branches are now filled with air there must be a new form of budding. At any stage of postnatal growth the terminal branches bear evidence, in the alveoli projecting from them, that they may have been alveolar saccules. These alveoli become obliterated by the advance of the muscle sheet which extends over the length of the alveolar ductules [FIG. 610].

It was for long uncertain whether the **cuboidal cells,** which at one stage of foetal life form a continuous lining for the alveoli of the lungs, became flattened at birth but remained in continuity or were partially lost so that alveolar air and blood were separated in places only by vascular endothelium. Recent studies, however, indicate :

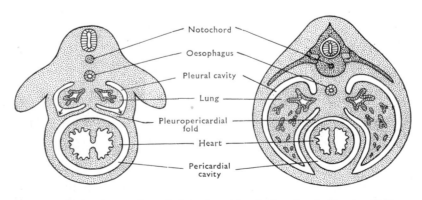

FIG. 614 Ventral extension of pleural cavities during development. (Schematic; after Patten, 1946.)

1. That towards the end of foetal life the cells that line the last two generations of the entodermal tree become flattened but continue to form a complete lining to the alveoli, so that alveolar epithelium, basement membrane, and vascular endothelium separate blood and air.

2. That in late foetal life the lungs are not gland-like and completely unexpanded but are partially opened up due to respiratory movements which take place during foetal life. Nevertheless, they sink in water because they contain amniotic fluid and not air. This fluid filling the lungs is very rapidly absorbed after birth.

Pleural Cavities. In the embryo there is a continuous body cavity or coelom. Its headward part, which becomes the pericardial cavity, communicates by means of a right and a left pericardioperitoneal opening, the primitive pleural canal, with the peritoneal cavity.

When the heart and pericardium are carried ventrocaudally the pleural passages are elongated, and the cephalic ends now open into the dorsal part of the pericardial cavity by the pleuropericardial openings, and their caudal ends open into the peritoneal cavity by the pleuroperitoneal openings.

As the common cardinal vein (duct of Cuvier) runs ventrally towards the septum transversum to enter the heart, it crosses lateral to the pleural canal, raising a ridge of body wall mesoderm, the dorsolateral part of the septum transversum. Each growing lung bud pushes the medial wall of the pleural canal laterally and expands into the ridge together with the pleural cavity, immediately caudal to this vein, thus splitting the ridge into cranial and caudal parts. After passing lateral to the vein, the bud expands within the body wall in three directions: headwards, on the lateral side of the common cardinal vein to form the apical part of the lung; ventrally, to form the anterior part which thus detaches or dissects the pericardium from the body wall [Fig. 614]; and tailwards, to form the base of the lung.

The pleural cavity enlarges within the body wall to keep pace with the ever-growing lung, and its openings into the peritoneal and pericardial cavities become closed. The pleuroperitoneal opening (in postnatal anatomy its site is the vertebrocostal trigone) is closed by the pleuroperitoneal membrane (the caudal part of the ridge) which grows across it from the septum transversum as a ledge excavated from the body wall by the growing base of the lung. Closure occurs earlier on the right side, and the liver covers the right opening. Hence, congenital diaphragmatic hernia, due to failure of the pleuroperitoneal channel to close, is rare on the right side.

The pleuropericardial opening is closed by the pleuropericardial membrane which grows medially across it as the cranial part of the ridge which contains the common cardinal vein (subsequently the superior vena cava on the right) and phrenic nerve. Closure is assisted by adhesion of the surfaces of the pleuropericardial passage. This opening is medial to the common cardinal vein and phrenic nerve, and is continuous with the definitive pleural cavity around the medial surface of the vein. With the descent of the heart and septum transversum and the consequent change in the position of the common cardinal vein and phrenic nerve, which come to run longitudinally anterior to the lung root, the continuity of the pleural cavity and the pleuropericardial passage now lies medial to the vein and nerve, and ventral to the lung root. In postnatal anatomy its site, on the right side, is between the root of the lung and the superior vena cava with the phrenic nerve. Failure of the right opening to close is unknown in Man, due presumably to the persistence of the right superior vena cava, but with the loss of the vessel on the left, the opening occasionally remains. In fact, it may be so dilated that the entire left side of the heart is in contact with the lung, posterior to the phrenic nerve.

Anomalies of the Respiratory System

These most commonly are fistulae between the upper trachea and oesophagus, due to incomplete separation of these elements derived from the supra-abdominal part of the foregut. Accessory fissures may occur subdividing otherwise normal lobes. These are usually incomplete, extending for a limited distance on the surface of the lung, or may be only subpleural furrows. Relatively common accessory fissures are: a partial fissure between the superior and basal bronchopulmonary segments of the lower lobes; a horizontal fissure in the left lung marking off a middle lobe, usually corresponding to the intersegmental plane between the lingula and the upper lobe.

Accessory and Aberrant Lobes. The distinction made between the two is that the bronchus of an accessory lobe is a branch of the normal bronchial tree but that of an aberrant lobe originates from some other source. The accessory **azygos lobe** has already been described [p. 492] and results from the lung bud splitting around the cardinal vein instead of lying entirely caudal to it. Examples of aberrant lobes are tracheal lobes typically arising from the wall of the trachea through a supernumerary branch. Lower accessory lungs are described attached to the pulmonary ligament and without any connexion with the bronchial tree.

RADIOGRAPHIC ANATOMY

J. B. KING

[Plates 34–36]

THE LARYNX

Lateral views of the larynx on X-ray films demonstrate the ventricle of the larynx, the vocal folds, the epiglottis, and the aryepiglottic folds.

Ossification in the thyroid and cricoid cartilages is also visible. The arytenoid cartilages ossify early in life and are seen through the unossified thyroid cartilage.

Tomographic sections [p. 12] in the anteroposterior plane allow the ventricle, and the ventricular and vocal folds to be shown. The piriform fossa can also be seen if it contains air. The outline of the mucosal surfaces of the larynx may be demonstrated by coating the larynx with oily contrast medium, and this method is a valuable means of detecting early tumors, and particularly in demonstrating the lower margin of a tumour as this is frequently invisible by laryngoscopy.

THE THORAX

The thoracic skeleton is dealt with in the sections on bones and joints, and the changes in the cardiac shadow due to respiration are covered in the section on the cardiovascular system.

That a radiograph of the thorax is a shadow picture of a three-dimensional object reduced to one plane is a difficult concept for many students, who see the shadows cast by the vertebral ends of the ribs as though they lay anteriorly, and find difficulty in understanding that the costal cartilages are not represented on a normal radiograph of the chest because they are surrounded by soft tissue of equal radiographic density, and that the sternum is almost entirely hidden by the overlapping shadows of the heart and vertebral column.

The fundamental feature of the normal radiographic anatomy of the lungs is that the degree of blackening of the radiograph is in proportion to the air content of the lung and is equal at equivalent points on both sides of the chest. That this does not always appear to be so is due to the predominantly left-sided position of the heart [PLATES 34 and 35]. Body weight, physical type, and age are the principal factors which determine the normal gross and detailed radiographic appearance of the lungs in any one individual, and no interpretation is possible without taking them into account. In taking a radiograph the intensity of the X-ray beam and the length of the exposure may be varied from one extreme where the lung structure is hardly penetrated, to another where it cannot be seen at all, and it is therefore essential that some standard of quality is agreed upon for the examination of any one structure. For the lungs in an anteroposterior position this is accepted as being achieved when the shadow of the thoracic spine can only just be seen through that of the heart in the lower half of the thorax [PLATE 35 A].

In the lungs no differentiation between vessel wall and blood is possible on radiographs because of their equal absorption of X-rays, and they appear as linear, oval, or circular shadows depending on the angle which they make with the beam of X-rays. By comparison, the walls of the larger bronchi may be seen because of their air content, and near the hila may appear as white shadows with black interiors which, like the blood vessels, vary in shape in proportion to their obliquity to the X-rays. It is known, however, that beyond the immediate root shadowing of the lungs, the walls of the bronchi and bronchioles play no significant part in forming the fine meshwork of lines which represent lung structure on the radiograph, and that this essentially represents the distribution of the pulmonary vessels. The actual degree of accuracy with which these vessels can be demonstrated in any one individual varies with the diameter of the thorax and with the amount of the overlying soft tissue which, when thick, obscures them by absorbing some of the rays and by preventing close apposition of the lungs to the film. Here, as elsewhere, the possibility of getting good photographic definition is largely dependent on obtaining close contact between the object and the photographic emulsion. By an anteroposterior radiograph is meant one in which the film has been placed against the back of the chest, the X-rays traversing the thorax from front to back. This is used if it is desired to examine the posterior thoracic wall, spine, or any abnormality lying well posterior in the lung. Similarly postero-anterior views are used in the radiography of anterior thoracic structures, including the heart.

The aeration of the lungs by the first breaths at birth enables them to be differentiated from the thorax and heart, and they remain so throughout life by virtue of the minimal absorption of X-rays by the air content, relative to these adjacent structures. At birth the heart is relatively large and the lungs are proportionally small, but the lungs grow rapidly, being said to increase three-fold in weight and six-fold in volume during the first year of life [PLATE 36 A and B]. In infancy the ribs are almost horizontal and therefore such costal movement as takes place has little effect on pulmonary ventilation which at this age is essentially due to diaphragmatic movement. The shadow of the thymus varies greatly in size, and in forced expiration, as at the end of a cry, may appear to fill almost the entire upper third of the thorax [PLATE 36 B]. These factors, together with the relatively large tidal respiratory volume and the practical difficulty of immobilization, make the establishment of a normal pattern for a chest radiograph in infancy a matter of much greater difficulty than in the adult.

Each lung field should show equal degrees of black and white shadowing at any two corresponding points on the radiograph, therefore some explanation is required for any deviation from this standard, but it by no means follows that the cause of the asymmetrical air content lies within the lung itself. Indeed, probably the most frequent single cause for asymmetrical lung density lies in a degree of postural lateral spinal curvature, which is frequently so slight as to evade observation on external examination. The right diaphragm is normally from 1·0 cm. to 2·5 cm. higher than the left but again quite trivial lateral spinal curvature will upset this relationship. Probably the next most frequent reason for a change in normal diaphragm levels is elevation on the left side due to an excess of swallowed air in the cardia of the stomach, frequently accompanied by air of similar origin at the left colic flexure [PLATE 34 C]. The musculature of one diaphragmatic cupola is occasionally under-developed and when this is the case on the left side there may be elevation of the diaphragm by as much as 10 cm. due to intra-abdominal pressure. On the right side, instead of forming one smooth curve convex upwards, the diaphragm quite frequently consists of a series of small overlapping arcs due to the uneven contraction of the diaphragm in full inspiration which is the phase in which radiographs are taken as a routine. The under surface of the diaphragm is of course only seen when there is free air in the abdomen as after an abdominal operation or in pathological perforation of the gastro-intestinal tract. The right and left costodiaphragmatic recesses are usually sharply defined acute angles, but sometimes on very full inspiration the smooth line of the diaphragm is interrupted by a series of sharp, pointed, soft tissue shadows running upwards and outwards. These are normal and represent the attachments of the diaphragm.

Lateral radiographs demonstrate the very great extent to which the costodiaphragmatic recesses extend below the cupola of the diaphragm as seen in anteroposterior views [PLATE 35 B], and also show the extent to which the posterior recess is greater than the anterior. Not only is the position of the dome of the diaphragm lower by 2·5–5 cm. in life than in death, but in the lateral recumbent position the weight of the abdominal viscera normally raises the lower half of the diaphragm by 2·5 cm. or more.

The cardiophrenic angles are described in the cardiovascular section. Diaphragm movement is equal on both sides and in quiet respiration varies from 1–2·5 cm., but forced respiration may increase this to as much as 10 cm. in the young and athletic. The extent of diaphragm movement may be observed by fluoroscopy but is more accurately recorded by comparison of films taken in inspiration and expiration.

Some of the shadows cast by the soft tissues of the thoracic wall are of importance. Those due to fully formed breasts are unmistakable when symmetrical [PLATE 35A], but deceptive asymmetry may be produced by uneven pressure. Adolescent breast shadows, lacking a clear outline, may closely simulate pulmonary abnormalities or mask such when present. The female nipples frequently cast a shadow which is usually unmistakable, but in the male it is rare for such a density to be seen and mistakes can arise. In oblique views of the thorax large breasts may cause real difficulty in interpretation. Well-developed pectoral muscles produce diffuse shadows over the upper lung fields which have a sharply-defined inferolateral margin which may be more marked on the right side in muscular males. Continuation of the soft tissue shadows of the neck across the lung apex can only be deceptive when asymmetry has been produced by rotation of the head and extension of the neck at the time of taking the radiograph.

The structures representing the medial borders of the lungs are given in the section on the radiographic anatomy of the heart.

The shadows forming the right hilus are mainly due to the main branches of the right pulmonary arteries and veins. On the left side these shadows represent the main stem of the pulmonary artery and its principal left branches, the main veins here being more obscured by the cardiac shadow than they are on the right side. It may be that when the mediastinal and cardiac shadows are narrow, as in persons of asthenic build, the tracheal and tracheobronchial lymph nodes form part of the complex of the upper portions of the hilar shadows, but the main aggregation of such nodes, being around the tracheal bifurcation, is overlapped by the cardiac shadow and takes no part in the formation of hilar shadows in normal anteroposterior chest radiographs. The pulmonary lymph nodes, being smaller and more discrete, are probably an insignificant factor in forming the peripheral hilar shadows, but enlargement of any or all of these nodes may cause them to cast shadows which are immediately evident. The great variability which occurs in the density of the hilar shadows is due firstly to a wide range in the normal diameter of the pulmonary arteries and veins, secondly to differences in the 'thoracic ratio', that is, the relative size of the anteroposterior and transverse diameters of the chest, and thirdly to variations in the transverse diameter of the heart. It follows that where so many variable factors arise the greatest care must be taken before shadows alone can be accepted as proof of abnormal conditions. This is one of the spheres where tomography and cardio-angiography are helpful.

The linear shadows throughout the lungs being caused by the vascular structures, it follows that these will have a symmetry in the peripheral lung affected only by differences in the arrangement of the vessels and at the hila by the orientation of the large pulmonary arteries and veins. It also follows that there is symmetry in their distribution, size, and density except at the hila, and that they will diminish in calibre as they pass towards the periphery of the lung. That all the peripheral shadows of radiographic lung structure are due to blood vessels has medical importance in that only those pathological conditions can be shown which produce some local alteration in the normal proportions of air and blood. Certain diseases do this to a very marked extent and can therefore be diagnosed by X-rays at an early stage, while others cause little such alteration and cannot be detected radiologically.

By having a density no different from that of the chest wall the pleura covering the lungs cannot be seen on a normal radiograph. But by lying 'edge on' and by having adjacent air content on both sides the pleura of the horizontal fissure on the right side can be seen in most good anteroposterior radiographs as a fine line lying at the level of the intercostal space between the third and fifth ribs in the anterior axillary line. The extent to which the pleura of the interlobar fissures on the right side is seen in right lateral radiographs, that is to say, when the film has been against the right side of the thorax, depends on the degree to which the pleura happens to lie 'end on' to the beam of X-rays. The trachea appears on anteroposterior radiographs as a dark column of relatively diminished density about 2 cm. wide, extending in the midline from the neck above to the level of the fifth thoracic vertebra inferiorly. The bifurcation and the two main bronchi can be demonstrated clearly in films taken as for the thoracic spine. The cartilaginous structures of the air passages in the neck are demonstrable in lateral radiographs of the neck to an extent partly dependent on the degree to which the pharynx is distended by air present in them, and partly on the amount of calcification. Moderate radiographic detail can be obtained by tomography.

The technique of bronchography has been developed in which a radio-opaque solution is introduced into the main bronchi and allowed to run into the smaller peripheral bronchi. It is important to appreciate that the opaque medium does not fill the lumina of the bronchi but merely coats the walls, and that the examination only extends as far as the smaller bronchioles [PLATE 35 C and D]. If the alveolar ductules are filled by coughing, loss of significant detail occurs.

A different pattern of lung structure must be accepted as 'normal' in old age from that of early life. The factors responsible for this are numerous and any one may predominate though all are usually present to some degree. The changes may be actually in the lung tissue and pulmonary vessels, or they may be consequent on an increased anteroposterior diameter of the thorax due to the stooping posture of increasing years which, with decreased elasticity of the chest wall, causes a diminution in air entry. Although strictly not an anatomical matter, the difficulty of getting an elderly person to hold his breath during full inspiration in an awkward posture is a real factor in determining what is acceptable as normal radiographic lung structure in the aged.

REFERENCES

APPLETON, A. B. (1944) Segments and blood-vessels of the lungs, *Lancet*, ii, 592.

APPLETON, A. B. (1945) The arteries and veins of the lungs : I. Right upper lobe, *J. Anat. (Lond.)*, **79,** 97.

AREY, L. B. (1965) *Developmental Anatomy*, 7th ed., Philadelphia.

BELLAIRS, A. D'A. (1942) Observations on Jacobson's organ and its innervation in Vipera Berus, *J. Anat. (Lond.)*, **76,** 167.

BELLAIRS, A. d'A., and BOYD, J. D. (1950) The lachrymal apparatus in lizards and snakes : II. The anterior part of the lachrymal duct and its relationship with the palate and with the nasal and vomeronasal organs, *Proc. zool. Soc., Lond.*, **120,** 269.

BERTALANFFY, F. D., and LEBLOND, C. P. (1953) The continuous renewal of the two types of alveolar cells in the lung of the rat, *Anat. Rec.*, **115,** 515.

BERTALANFFY, F. D., and LEBLOND, C. P. (1955) Structure of respiratory tissue, *Lancet*, ii, 1365.

BLOOMER, W. E., LIEBOW, A. E., and HALES, M. R. (1960) *Surgical Anatomy of the Bronchovascular Segments*, Oxford.

BOYDEN, E. A. (1949) A synthesis of prevailing patterns of the bronchopulmonary segments in the light of their variations, *Dis. Chest*, **15,** 657.

BOYDEN, E. A. (1955) *Segmental Anatomy of the Lungs*, New York.

BOYDEN, E. A., and SCANNELL, J. G. (1948) An analysis of variations in the bronchovascular pattern of the right upper lobe of fifty lungs, *Amer. J. Anat.*, **82,** 27.

BROCK, R. C. (1950) The nomenclature of bronchopulmonary anatomy : An international nomenclature accepted by the Thoracic Society, *Thorax*, **5,** 222.

BROCK, R. C. (1954) *The Anatomy of the Bronchial Tree*, 2nd ed., London.

FRABLE, M. A. (1961) Computation of motion at the cricoarytenoid joint, *Arch. Otolaryng.*, **73,** 551.

FRAZER, J. E. (1940) *A Manual of Embryology. The Development of the Human Body*, 2nd ed., London.

GROSSER, O. (1912) The development of the pharynx and of the organs of respiration, in *Kiebel and Mall's Manual of Human Embryology*, Vol. 2, London.

GRIESMAN, B. L. (1943) Mechanism of phonation demonstrated by planigraphy of the larynx, *Arch. Otolaryng.*, **38,** 17.

HUBERT, L. (1947) Thyroglossal cysts and sinuses : Analysis of 43 cases, *Arch. Otolaryng.*, **45,** 105.

JACKSON, C. L., and HUBER, J. F. (1943) Correlated applied anatomy of the bronchial tree and lungs with a system of nomenclature, *Dis. Chest*, **9,** 319.

KIRCHNER, J. A., and WYKE, B. (1965) Articular reflex mechanisms in the larynx, *Ann. Otol. (St. Louis)*, **74,** 749.

LEDEN, H. VON, and MOORE, P. (1960) Contact ulcer of the larynx. Experimental observations, *Arch. Otolaryng.*, **72,** 746.

LEMERE, F. (1934) Innervation of the larynx : IV. An analysis of Semon's law, *Ann. Otol. (St. Louis)*, **43,** 525.

LEWIS, D. D. (1905) The elastic tissue of the human larynx, *Amer. J. Anat.*, **4,** 175.

LUCAS KEENE, M. F. (1961) Muscle spindles in human laryngeal muscles, *J. Anat. (Lond.)*, **95,** 25.

LOW, F. N. (1953) The pulmonary alveolar epithelium of laboratory mammals and man, *Anat. Rec.*, **117,** 241.

MARCHAND, P., GILROY, J. C., and WILSON, V. H. (1950) An anatomical study of the bronchial vascular system and its variations in disease, *Thorax*, **5,** 207.

MILLER, W. SNOW (1921) The musculature of the finer divisions of the bronchial tree and its relation to certain pathological conditions, *Amer. Rev. Tuberc.*, **5,** 689.

MILLER, W. SNOW (1947) *The Lung*, 2nd ed., Springfield, Ill.

MITCHINSON, A. G., and YOFFEY, J. M. (1947) Respiratory displacement of larynx, hyoid bone and tongue, *J. Anat. (Lond.)*, **81,** 118.

MORTON, D. R., KLASSEN, K. P., and CURTIS, G. M. (1950) The clinical physiology of the human bronchi : I. Pain of tracheobronchial origin, *Surgery*, **28,** 699.

MORTON, D. R., KLASSEN, K. P., and CURTIS, G. M. (1951) The clinical physiology of the human bronchi : II. The effect of vagus section upon pain of tracheobronchial origin, *Surgery*, **30,** 800.

MURTAGH, J. A., and CAMPBELL, C. J. (1948) The respiratory function of the larynx : II. The elementary physiology of the recurrent laryngeal nerve, *Ann. Otol. (St. Louis)*, **57,** 465.

NEGUS, V. E. (1927) The function of the epiglottis, *J. Anat. (Lond.)*, **62,** 1.

NEGUS, V. E. (1947) Intrinsic cancer of larynx ; review of series of cases, *Proc. roy .Soc. Med.*, **40,** 515.

PATTEN, B. M. (1968) *Human Embryology*, 3rd ed., New York.

PIERCE, J. R., and DAVID, E. E. (1958) *Man's World of Sound*, New York.

PRESSMAN, J. J. (1941) Sphincter action of the larynx, *Arch. Otolaryng.*, **33,** 351.

PRESSMAN, J. J. (1942) Physiology of the vocal cords in phonation and respiration, *Arch. Otolaryng.*, **35,** 355.

SCHINZ, H. R., BAENSCH, W. E., FRIEDL, E., and UEHLINGER, E. (1971) *Roentgen Diagnostics*, 2nd ed., Vol. 3, Engl. trans. Case, J. T., London.

SHANKS, S. C., and KERLEY, P. (1969) *A Textbook of X-Ray Diagnosis*, 4th ed., Vol. 2, London.

SIMER, P. H. (1952) Drainage of pleural lymphatics, *Anat. Rec.*, **113,** 269.

STIBBE, E. P. (1919) The accessory pulmonary lobe of the vena azygos, *J. Anat. (Lond.)*, **53,** 305.

TAYLOR, H. M. (1944) Ventricular laryngocele, *Ann. Otol. (St. Louis)*, **53,** 536.

TOBIN, C. E. (1952) The bronchial arteries and their connections with other vessels in the human lung, *Surg. Gynec. Obstet.*, **95,** 741.

WARBRICK, J. G., McINTYRE, J. R., and FERGUSON, A. GIRDWOOD (1952) Remarks on the aetiology of congenital bilateral fistulae of the lower lip, *Brit. J. plast. Surg.*, Vol. IV, No. 4, 254.

WARBRICK, J. G. (1960) The early development of the nasal cavity and upper lip in the human embryo, *J. Anat. (Lond.)*, **94,** Part 3, 351.

WARBRICK, J. G. (1963) Aspects of facial and nasal development, in *The Scientific Basis of Medicine Annual Reviews*, 99.

THE UROGENITAL SYSTEM

R. G. HARRISON

THE urogenital system consists of: (1) the **urinary organs** proper, which are concerned with the formation, temporary storage, and discharge of the urine; and (2) the **reproductive** or **genital organs,** some of which arise from the same embryonic tissues as the urinary organs and retain an intimate relationship to them.

THE URINARY ORGANS

The urinary organs are: the **kidneys,** where the urine is elaborated from the blood; the **ureters,** which convey the urine from the kidneys to the urinary bladder; the **urinary bladder,** where the urine is temporarily stored, and the **urethra** through which it is voided.

THE KIDNEYS

[PLATES 37 and 38]

Form and Size. The kidneys are a pair of bean-shaped organs, brownish-red when fresh, and of a glistening appearance due to the fact that each of them is covered with a thin but tough **fibrous capsule** which is close-fitting though easily stripped off. They are approximately symmetrical in size and shape. Each kidney is about 11 cm. long, 5 cm. wide, and about 3 cm. thick; and it weighs about 130 g. It possesses

upper and **lower extremities** or poles (of which the upper is generally the more bulky), **anterior** and **posterior surfaces,** and **medial** and **lateral borders.** The lateral border is uniformly convex from the upper to the lower pole; the medial border has a deep concavity which leads into a hollow that extends almost half-way through the kidney. This hollow is the **sinus** of the kidney, and the entrance into it is called the **hilus** of the kidney. The renal artery enters the kidney through the hilus; the renal vein and the **pelvis of the kidney** (the upper, expanded part of the ureter) leave it by the same route.

Position. Each kidney lies in the upper, deep part of a gutter alongside the projecting vertebral bodies which are covered laterally by the psoas major muscle. Inferiorly, these gutters diverge and become shallower as the psoas major muscles, increasing in bulk, pass laterally and forwards. Thus each kidney, lying against the sloping lateral surface of the

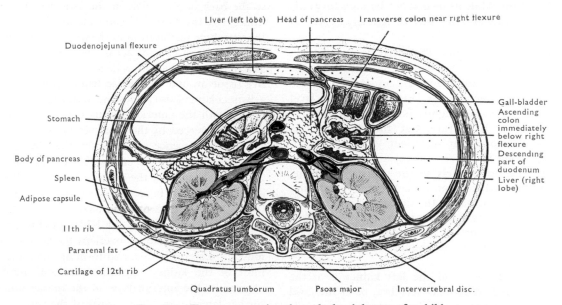

FIG. 615 Transverse section through the abdomen of a child.

The position and the general relations of the kidneys are well seen, and the arrangement of the renal fascia is indicated. The fascia is shown as a solid black line.

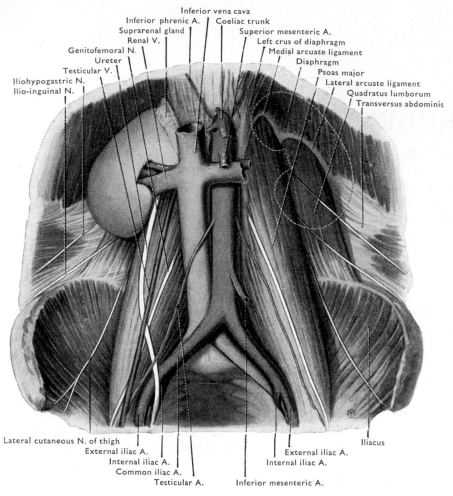

FIG. 616 Dissection to show the relation of the kidney and ureter to the posterior abdominal wall. The dotted outline indicates the position of the left kidney.

psoas major, has its long axis parallel to the lateral surface of that muscle [FIG. 616], while its medial border faces forwards as well as medially and its anterior surface laterally as well as forwards [FIG. 615].

The two kidneys are not quite symmetrical in position; almost always the left is about 1·5 cm. higher than the right and it is therefore slightly more medial, on account of the slope of the gutter in which it lies [FIG. 616]. The level of each kidney, however, varies slightly both with the respiratory movements and with posture, but the range of this normal physiological movement does not exceed 2·5 cm. During quiet respiration in the recumbent attitude, the hilus of the left kidney is approximately in the transpyloric plane about 3·5 cm. from the median plane of the body, or, anteriorly, about a finger's breadth medial to the tip of the ninth costal cartilage; the hilus of the right kidney is about 1·5 cm. below this point. The lower extremity of the kidney is about 7·5 cm. from the median plane—the left in the subcostal plane and the right 1·5 cm. lower.

Seen from behind, the hilus of the left kidney is at the level of the spine of the first lumbar vertebra, or at the level of the line where the lateral border of the erector spinae muscle meets the twelfth rib, and the lower extremity is 5 cm. or less above the highest point of the iliac crest [FIG. 618]. Corre-

sponding points on the right kidney are about 1·5 cm. lower.

Renal Fascia and Fat. Each kidney has an intimate relationship to the corresponding suprarenal gland and is enclosed with it in a common investment of weak fascia, known as the renal fascia, which is formed by the splitting of the transversalis fascia into anterior and posterior layers [FIG. 615]. Outside this investing fascia and between it and the kidney (surmounted by the suprarenal gland) there is a variable amount of fat called the renal fat. The fat outside the renal fascia is known as **pararenal fat**; it is most abundant behind the lower extremity of the kidney. The fat within the fascia is called the **adipose capsule** (perirenal fat) and is most abundant at the margins of the organ; it is traversed by weak fibrous strands which connect the renal fascia with the fibrous capsule of the kidney.

The two layers of the renal fascia fuse together a short distance above the suprarenal gland and become continuous with the fascia on the under surface of the diaphragm; lateral to the kidney also they fuse together and are continued as the transversalis fascia. Below the kidney the two layers remain separate, the anterior gradually fading into retroperitoneal areolar tissue and the posterior joining the fascia over the iliacus muscle. The posterior layer, passing over the quadratus lumborum, becomes continuous medially with the fascia of the psoas major at its posterior border. As it reaches the medial border of the kidney it gives off a lamina which turns laterally into the hilus of the kidney and is attached to the back of the pelvis of the kidney; the remainder continues over the psoas and gains an attachment to the bodies of the vertebrae just in front of the muscle. The anterior layer of the renal fascia extends across the aorta and inferior vena cava as a thin layer which loses its identity superiorly in the connective tissue around the coeliac trunk. From the anterior layer a distinct lamina also passes backwards medial to the kidney and blends with the anterior lamina of the posterior layer. This lamina is pierced by the renal vessels and the ureter (Martin, 1942); its presence accounts for the limitation of perirenal effusions to one side of the body and for the statement (Mitchell, 1939) that the anterior layer as well as the posterior is attached to the bodies of the vertebrae. See also Mitchell (1950), and FIGURES 615 and 682.

Posterior Relations of the Kidneys [FIGS. 615–619]. The medial part of the kidney overlaps the lateral side of the psoas muscle. Lateral to this, and forming the posterior relation of most of the kidney, lies the quadratus lumborum muscle covered by the anterior layer of the thoracolumbar fascia, and further laterally the aponeurosis of the transversus abdominis muscle, on to which the kidney extends for a short distance. Towards the upper end of the kidney, the posterior margin hypogastric and, below it, the ilio-inguinal nerve both emerge from the lateral side of the psoas major and run a parallel, inferolateral course. The subcostal and iliohypogastric nerves pierce the aponeurosis of the transversus abdominis a short distance lateral to the quadratus lumborum, and the ilio-inguinal nerve pierces it a little further forwards [FIG. 616].

In this region, some bony structures lie posterior to the diaphragm and quadratus lumborum, and are therefore indirect posterior relations of the kidneys. The twelfth rib runs downwards and laterally behind the diaphragm, and its lateral part passes for a variable distance into the abdominal wall just lateral to the upper end of quadratus lumborum. The transverse process of the first lumbar vertebra lies at the level of the junction of the medial and lateral arcuate ligaments, and those of the other vertebrae at correspondingly lower levels—their tips being behind the medial edge of the quadratus lumborum. If the kidneys have been hardened *in situ* shortly after death in a thin subject, their posterior surfaces are generally moulded by the muscles and bony structures against which they lie [FIG. 619], but this is merely a post-mortem phenomenon. Moreover, the part of the twelfth rib which lies behind the diaphragm is separated from the diaphragm by the costodiaphragmatic recess of the pleura, which extends down as far as a line drawn from the middle of the twelfth rib to a point midway between the head of the twelfth rib and the transverse process of the first lumbar vertebra [FIG. 618]. The pararenal fat behind the upper part of the kidney is therefore only separated from the pleura by a very thin sheet of diaphragm which is often partly deficient (vertebrocostal triangle, FIG. 416) especially on the left side of the body. Thus the pararenal fat and the pleura may be in direct contact.

More remote posterior relations of the kidneys are the middle layer of the thoracolumbar fascia, the upper three lumbar arteries, the erector spinae muscle, and the posterior layer of the thoracolumbar fascia giving origin to the serratus posterior inferior and the latissimus dorsi muscles.

FIG. 617 Posterior aspect of trunk, showing surface topography of viscera. Roman numbers, ribs; arabic numbers, vertebrae.

A.	Aorta, descending.	O.	Oesophagus.	S.	Spleen.
AC.	Ascending colon.	P.	Pancreas.	SC.	Sigmoid colon.
C.	Caecum.	PR.	Line of pleural	SR.	Suprarenal.
DC.	Descending colon.		reflexion.	St.	Stomach.
L.	Liver.	RK.	Right kidney.	T.	Trachea.
LL.	Left lung.	RL.	Right lung.		

Anterior Relations of the Kidneys [FIGS. 620 and 621]. The anterior aspect and medial border of the upper extremity of each kidney is covered by the corresponding suprarenal gland. The other anterior relationships of the two kidneys are quite different; and, as already pointed out, all other organs in front of the kidney have a less intimate relations hip to it, for they all lie anterior to the renal fascia, whereas

of the diaphragm (medial and lateral arcuate ligaments) crosses the psoas major and quadratus lumborum muscles, so that the superolateral third of the kidney lies in front of the diaphragm.

Emerging from behind the lower border of the lateral arcuate ligament and running downwards and laterally across the front of the quadratus lumborum muscle is the subcostal nerve accompanied by its vessels; and a little lower down the ilio-

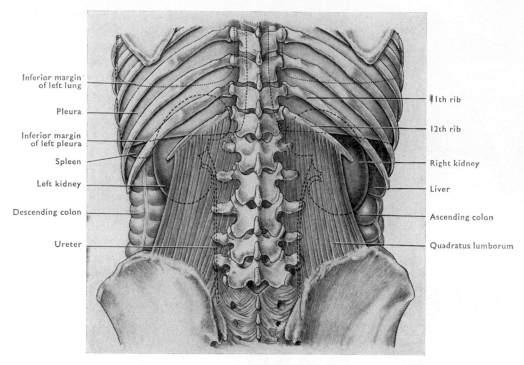

Inferior margin of left lung

Pleura

Inferior margin of left pleura

Spleen

Left kidney

Descending colon

Ureter

11th rib

12th rib

Right kidney

Liver

Ascending colon

Quadratus lumborum

FIG. 618   Dissection from behind to show the relation of the pleural sacs to the kidneys.

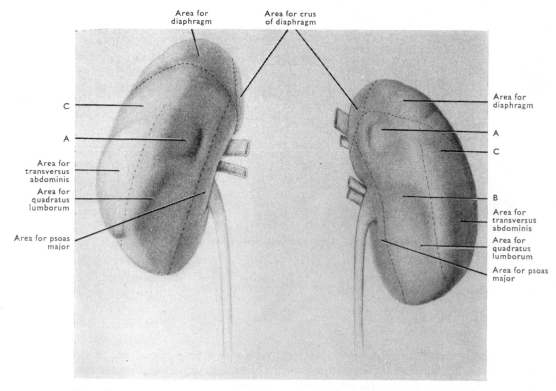

Area for diaphragm

Area for crus of diaphragm

C

A

Area for transversus abdominis

Area for quadratus lumborum

Area for psoas major

Area for diaphragm

A

C

B

Area for transversus abdominis

Area for quadratus lumborum

Area for psoas major

FIG. 619   Posterior surfaces of the kidneys.   The dotted lines mark out the areas in contact with muscles of the posterior abdominal wall.

A and B, depressions corresponding to transverse processes of first and second lumbar vertebrae.

C, depression corresponding to the twelfth rib.

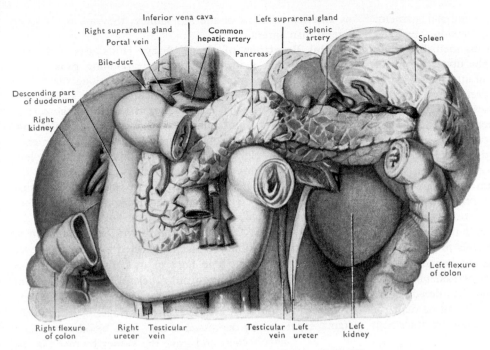

FIG. 620 Dissection to show relations of the kidneys.

The greater part of the stomach has been removed by an incision made close to the pylorus. The transverse colon has been taken away, and the small intestine has been cut across close to the duodenojejunal flexure. (From a model by Birmingham.)

covered by peritoneum which is reflected from the back of the liver on to the front of the kidney. Below the hepatic area, the region adjacent to the lower extremity is in contact with the right flexure of the colon, and the region adjoining the hilus is covered by the descending part of the duodenum. Below and medially, the peritoneum is reflected forwards from the hepatic area to cover the colon and the descending part of the duodenum. At the lower extremity of the kidney there is sometimes a small gap between the colic and duodenal areas, and, when this is present, it is covered with peritoneum that passes up from the posterior abdominal wall on to the front of the kidney and separates this area from coils of small intestine.

LEFT KIDNEY. Below and lateral to the area covered by the suprarenal gland there is a small triangular region covered with peritoneum of the omental bursa which is separated from the posterior surface of the stomach by the

the suprarenal gland is enclosed with the kidney inside this fascia.

RIGHT KIDNEY. Below the area covered by the suprarenal gland, about two-thirds of the anterior surface of the right kidney is in contact with the posterior surface of the right lobe of the liver. The uppermost part of this hepatic area may be related to the bare area of the liver, from which it is separated by some loose areolar tissue, but most of the hepatic area is

omental bursa. Below this gastric area the body of the pancreas lies across the middle of the kidney; and laterally the gastric area is bounded by the line of attachment of the lienorenal ligament, which runs approximately from the upper extremity of the kidney to the junction of its middle and lower thirds. The peritoneum which covers the gastric area passes downwards on to the front of the body of the pancreas, whence it continues as the anterior layer of the transverse mesocolon, while laterally it is

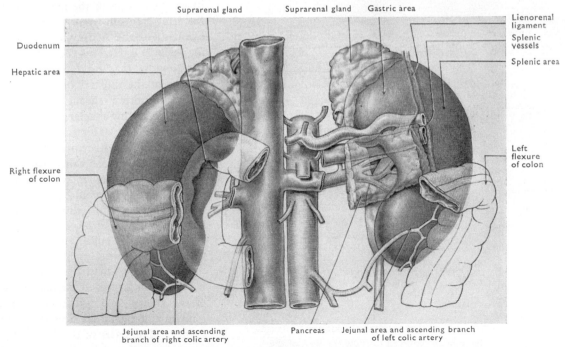

FIG. 621 Diagram of anterior relations of kidneys. The pancreas and other parts are shown as if transparent.

reflected forwards as the medial layer of the lienorenal ligament. The body of the pancreas is separated from the kidney by loose areolar tissue in which the splenic vein runs medially, while the artery runs its tortuous course along the upper border of the pancreas. The area lateral to the line of attachment of the lienorenal ligament is separated from the peritoneum covering the spleen by the peritoneum which passes on to the kidney from the posterior abdominal wall and is then reflected forwards as the lateral layer of the lienorenal ligament. Below the pancreatic area, the lateral part is covered by the left flexure of the colon and the upper part of the descending colon. The medial part is covered by the peritoneum which passes up from the posterior abdominal wall and is reflected forwards and downwards as the posterior layer of the transverse meso-colon and laterally on to the left flexure and the descending colon. This medial area is in relation to coils of jejunum in front of the peritoneum and to the ascending branch of the left colic artery, which crosses it behind the peritoneum. The obliquity of the anterior surface of the kidney has already been pointed out; and, if this is borne in mind, it will be realized that though the spleen and left flexure of the colon are described

as anterior relations of the left kidney, they are really more lateral than anterior.

Medial Relations of the Kidneys [FIGS. 615 and 616]. Medial to each kidney are the corresponding psoas major muscle and psoas minor (if present); and along the medial margin of the psoas lies the abdominal part of the sympathetic trunk. Anteromedially, each kidney is related to the corresponding suprarenal and renal vessels, ovarian or testicular vein, and upper end of the ureter. Further medially, is the inferior vena cava on the right side and the abdominal aorta on the left.

Sinus of the Kidney [FIGS. 622 and 623]. The sinus is the hollow within the kidney, and its long axis corresponds to that of the kidney itself. Its thick walls are formed by the substance of the kidney, and, except where they are covered by the minor calyces [p. 512], they are lined with a continuation of the capsule of the kidney, which is reflected into the sinus over the lips of the hilus. The floor of the sinus is not flat but presents a series of small conical elevations, called **renal papillae,** which vary from six to fifteen in number. Radiating from each papilla there are several small ridges of kidney substance separated by depressed areas. The blood vessels and nerves enter and leave the kidney by piercing the depressed areas. On the summit of each renal papilla there are many minute openings, which are the terminal apertures of the tubules, of which the kidney is mainly composed, and through which urine escapes into the minor calyces which are cupped over the papillae. The cavity of the sinus is filled with fat, embedded in which lie the major and minor calyces, part of the pelvis of the kidney, branches of the renal artery, and tributaries of the renal vein. The artery usually branches and the tributaries of the vein unite a short distance outside the hilus; and, within the sinus, the tributaries of the vein are most anterior, while further posteriorly are the calyces of the kidney, with branches of the artery both in front and behind.

Fixation of the Kidneys. The kidneys are held in position partly by the renal vessels and the surrounding fat and fascia, but mainly by the tension of the muscles of the anterior abdominal wall transmitted through the other abdominal viscera and acting against the inclined plane formed by the posterior abdominal wall.

The Kidney in Section. Sections through the kidney [FIGS. 622 and 623] show that it is composed to a large extent of a number of conical masses—known as **renal pyramids.** The pyramids constitute, collectively, the **medulla** of the kidney; their bases are directed towards the surface and their apices project into the minor calyces as the renal papillae. The pyramids are more numerous than the papillae, two or three usually ending in each papilla in the middle part of the kidney, and sometimes as many as six or more in a single papilla near the extremities of the kidney. The bases of the pyramids are separated from the surface by a layer called the **cortex** of the kidney. The cortex not only covers over the

FIG. 622 Longitudinal section through the kidney.

The vessels and fat have been removed to show the wall of the sinus of the kidney. The points where the vessels enter the substance of the kidney are seen as holes in the wall of the sinus.

bases of the pyramids, but also sends in prolongations, known as **renal columns,** between the pyramids, towards the sinus. In section, both parts have a striated appearance, but the cortex is less markedly striated and is usually darker in colour. The base of each pyramid appears to be composed of alternate dark and light streaks while the papillary part is often of a lighter colour, and is very faintly striated.

There are five main branches of the renal artery within the kidney ('segmental' arteries) which are visible between the pyramids; these divide into **interlobar arteries** and some of their main branches can be seen passing across the bases of the pyramids parallel to the surface of the kidney as the **arcuate arteries,** which do not anastomose with one another.

In the foetus and young child, and sometimes, though much less distinctly, in the adult, the surface of the kidney is marked by a number of grooves that divide it into polygonal areas [FIG. 680]. The areas represent the **lobes** or **reniculi** of which the kidney is originally composed, and each corresponds to one papilla with its pyramids and the surrounding cortex. There are fourteen such lobes, seven anterior and seven posterior (Löfgren, 1949). In addition there is a longitudinal sulcus situated along the lateral margin of the kidney which separates the anterior from the posterior lobes.

An examination with an ordinary pocket lens shows that it is the lighter striae of the bases of the pyramids which are continued into the cortex. As they pass through the cortex towards the surface of the kidney the striae become less distinct, and appear, when cut longitudinally, as separate ray-like prolongations carried outwards from the bases of the pyramids. The parts of the cortex which seem in this way to be continuations of the medulla are called medullary rays or the **radiate part** of the cortex. The part of the cortex which intervenes between them is darker in colour and forms what is known as the **convoluted part** or labyrinth. The appearance presented by the cortex in section varies much according to the plane in which the section has been taken. If the section is parallel to the axis of a pyramid, the medullary rays will appear as isolated streaks directed from the base of the pyramid towards the surface of the kidney, and separated from one another by narrow strips of the convoluted part. In sections made at right angles to the axis of a pyramid, the rays are circular in outline, and are surrounded by the convoluted part.

Kidney Tubules. The glandular substance of the kidney is composed of a vast number of minute **renal tubules,** each of which has a complicated course [FIG. 624]. The wall of each tubule consists throughout of a basement membrane and of an epithelial lining, but the lumen of the tubule and the character of the epithelium vary much in their different parts. Every tubule begins in a thin-walled spherical dilatation, known as a **glomerular capsule,** in which lies a complicated tuft of capillary blood vessels or **glomerulus** covered by a reflexion of the delicate wall of the capsule (visceral layer) [FIG. 625]. The capsules with their enclosed capillaries are called **renal corpuscles,** and they are all placed in the convoluted portion of the cortex, where they may be recognized as minute red points just visible to the unaided eye and best marked when the renal vessels are congested. The part of the tubule that leads from the capsule—the **proximal (first) convoluted tubule**—is very tortuous and lies within the convoluted part of the cortex. Passing from the convoluted part, the tubule enters a medullary ray, in which its course becomes less complicated, and here it receives the name of the **spiral tubule.** Leaving the medullary ray, the tubule enters the basal portion of the pyramid, and, diminishing in diameter, it pursues a straight course towards the apex of the pyramid, forming the so-called **descending limb of Henle's loop.** Within the apical portion of the pyramid the tubule suddenly bends upon itself, forming the loop of Henle, and, reversing its direction, passes back again through the base of the pyramid into a medullary ray of the cortex as the **ascending limb of Henle's loop.** Leaving the medullary ray again, the ascending limb enters the convoluted part of the cortex, where it returns to the glomerulus of its own renal tubule, and lies in between the afferent and efferent vessels

FIG. 623 Longitudinal section of the kidney and its pelvis to show the relations of the calyces to the renal papillae.

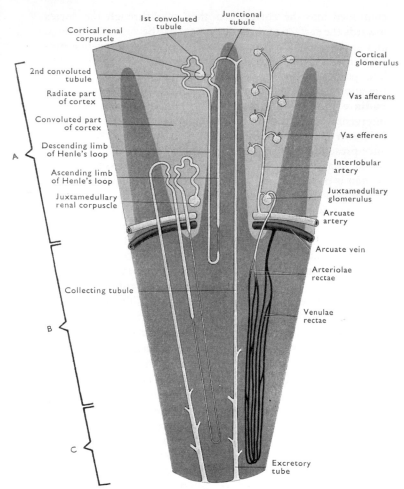

FIG. 624 Diagrammatic representation of the structures forming a kidney lobe. On the left of the Figure the course of a
renal tubule connected with a cortical renal corpuscle, and one
connected with a juxtamedullary renal corpuscle, are shown.
Note the differences in position and variations in calibre in
Henle's loop of cortical and juxtamedullary renal corpuscles.
On the right of the Figure are displayed the vascular arrangements within the kidney. The vasa efferentia of the cortical
renal corpuscles are shown cut across for purposes of clarity, but
they actually break up into capillaries around the renal tubules.

A, cortex; B, basal portion; C, papillary portion of pyramid.

of the glomerulus [FIG. 625]. The nuclei of the epithelial cells
of the tubule are concentrated in one wall of the tubule at this
point [FIG. 625] to form the **macula densa.** While still
within the convoluted part, the tubule again becomes tortuous
as the **distal (second) convoluted tubule**; this tubule
finally ends in a short **junctional tubule,** which passes back
into a medullary ray and joins a **collecting tubule.** Each
collecting tubule receives numerous renal tubules, and pursues
a straight course through a medullary ray of the cortex and
the pyramid. Finally, several collecting tubules, uniting together, form a **papillary duct,** which opens on the summit
of a renal papilla into a minor calyx by one of the minute
foramina already described. In microscopic sections the
various portions of the renal tubule may be distinguished by
the position which they occupy and by the character of the
lining epithelium. Each medullary ray with its surrounding
glomeruli and proximal and distal convoluted tubules which
empty into the collecting tubule forming the centre of the ray,

FIG. 625 Diagrammatic representation of a renal corpuscle.
(After Ham, 1957.)

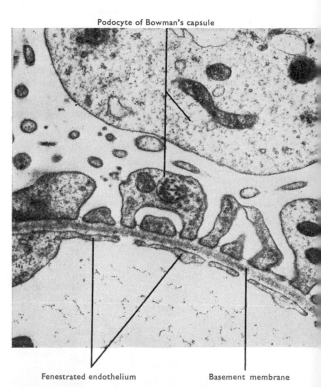

FIG. 626 An electron micrograph showing the filtration apparatus of the kidney. The section shows the structure separa-
ing the blood in the glomerular capillaries and the cavity
Bowman's capsule. It consists of a thick basement membra
partly covered on the vascular aspect by a layer of fenestrate
endothelium, and on the capsular side by the interdigitatin
foot processes of the podocytes which form the internal ep
thelium of Bowman's capsule. As can be seen, there are spac
between the foot processes where they expand on to the bas
ment membrane.

constitutes a **lobule** of kidney tissue. It is demarcated
each side by the **interlobular arteries,** fine vessels wh
arise from the arcuate arteries and mostly pass centrifuga
although a few course centripetally to supply the ju
medullary glomeruli [FIG. 624].

Supporting Tissue of the Kidney. The tubules and bl

vessels are all united together by a very small amount of areolar tissue which completely surrounds each tubule and blood vessel and binds it to its neighbours. The areolar tissue forms a continuous network, the spaces in which faithfully reproduce the outlines and the arrangement of the renal tubules. The network of the stroma is continuous with the capsule of the kidney.

Vessels and Nerves of the Kidney [PLATES 37B and 52C]

The **renal arteries,** one on each side, arise directly from the aorta. The left is usually slightly higher than the right and they are both large vessels relative to the size of the kidneys. Each artery lies behind the corresponding renal vein, and at or just outside the hilus of the kidney it divides typically into anterior and posterior divisions. The posterior division passes behind the **pelvis of the kidney** while the anterior division passes in front of it and divides into two to five branches. Within the sinus each branch divides further. The territories of distribution of the anterior and posterior divisions within the kidney meet at the junction of its anterior two-thirds and posterior one-third (Brödel's line) which does not correspond to the longitudinal sulcus (Sykes, 1964a). Having entered the substance of the kidney, the larger arteries lie in the intervals between the pyramids and are called **interlobar arteries.** These vessels divide and form a series of incomplete arterial arches—the **arcuate arteries**—which pass across the bases of the pyramids. No anastomosis between the branches of the interlobar arteries takes place, each artery which pierces the wall of the kidney sinus being an end-artery. The arcuate arteries give off a number of vessels which pass through the convoluted part of the cortex towards the surface of the kidney; they are known as the **interlobular arteries,** and they lie at very regular intervals. From them a number of short branches arise, termed **vasa afferentia,** each of which proceeds to a renal corpuscle. There, the vas afferens breaks up into the capillaries of the glomerulus, which are contained within the invaginated glomerular capsule. Just before breaking up into the glomerular capillaries, the cells of the vas afferens change in character, and are termed **juxtaglomerular cells** [FIG. 625], which are reputed to secrete a substance called renin. The small vessel—**vas efferens**—which issues from the glomerulus, instead of running directly into a larger vein, breaks up, after the manner of an artery, into capillaries which supply the convoluted tubules of the cortex and the longitudinal tubules of the medullary rays. At the bases of the pyramids there are some 'juxtamedullary' glomeruli whose efferent vessels pass as fine arterioles into the pyramids where they break up into a capillary network around the tubules. These clusters of fine arterioles, known as the **arteriolae rectae,** are just visible to the naked eye, and they give the bases of the pyramids their striated appearance [FIGS. 622–624]. The juxtamedullary glomeruli are larger than those in the convoluted part of the cortex and their afferent and efferent vessels are of equal calibre; they are apparently non-functional. In certain conditions the renal blood flow is said to be shunted through the vessels of these non-functional glomeruli and the arteriolae rectae directly from the renal artery into the renal vein without passing through the functional glomeruli of the cortex proper (Trueta *et al.,* 1947). Franklin (1949) gives an account of the history of research on the renal circulation.

A few interlobar arteries continue their radial course, perforate the fibrous capsule, and supply the adipose capsule.

Veins corresponding to the interlobular arteries and arteriolae rectae collect the blood from the capillaries around the tubules and unite to form a series of complete arcuate veins which anastomose freely across the bases of the pyramids. From the venous arcades, vessels arise which traverse the intervals between the pyramids and reach the sinus of the kidney, where they unite to form the tributaries of the renal vein. Some small veins in the superficial part of the cortex **(stellate venules)** communicate through the fibrous capsule with minute veins in the renal fat. Typically three main tributaries issue from the sinus and unite to form the **renal vein,** which runs a direct course to end in the inferior vena cava.

Variations in the renal vessels are common. Supernumerary arteries and veins or supernumerary branches of the renal artery and vein are frequent. The abnormal vessels may enter or leave the kidney within the sinus or at some other point. Small accessory renal veins from the lower extremity of the kidney are especially common.

The **lymph vessels** of the kidney end in the aortic lymph nodes.

The **nerves** of the kidney are derived mainly from the coeliac plexus (Mitchell, 1956) and accompany the branches of the artery, forming a renal plexus. The branches of this plexus form bundles of nerve fibres which travel along the branches of the renal artery into the kidney and also run between and supply the tubules.

From clinical evidence it would appear that the afferent nerve fibres which supply the kidney are connected with the tenth, eleventh, and twelfth thoracic nerves.

Variations

A marked difference in the size of the two kidneys is sometimes observed, a small kidney on one side of the body being usually compensated for by a large kidney on the opposite side. Congenital absence of one or other kidney is recorded. A few cases are on record in which an extra kidney was found on one or other side.

Traces of the superficial lobulation of the kidney, present in the foetus and young child, are often retained in the adult, representing remnants of the seven anterior and seven posterior lobes which, with their corresponding papillae and calyces, are present by the twenty-eighth week of development (Sykes, 1964b), but are subsequently reduced in number by fusion.

Horseshoe-kidney is a not infrequent abnormality. In such cases the two kidneys are united at their lower ends across the median plane by a connecting bar of kidney substance. The amount of fusion between the two kidneys varies considerably; it is sometimes very complete, while in other cases it is slight, the connexion being composed chiefly of fibrous tissue. The hilus of each kidney looks forwards, and the ureters always pass in front of the connecting piece which lies across the vertebral column caudal to the inferior mesenteric artery. Fusion between the two kidneys may also occur at other points on their medial borders.

Very rarely the kidney appears to be almost entirely surrounded by peritoneum and to be attached to the abdominal wall by a kind of mesentery which encloses the vessels and nerves passing to the hilus. The condition is believed to be congenital.

One or both kidneys may be at a much lower level than usual—in the iliac fossa or even in the pelvic cavity. This is associated with an arrest in the normal change in position, relative to surrounding structures, which the kidney experiences during development. In such cases the kidney does not receive its blood supply from the usual source but from vessels which arise from the lower end of the aorta, or from an iliac artery, or the median sacral artery. Such abnormally situated kidneys do not usually possess the typical outline of the normal organ but vary much in shape, and the hilus is often malformed and misplaced.

Methods of Clinical Anatomical Examination

A kidney of normal size and occupying its normal position cannot be palpated in the living subject, but, if the organ is much enlarged or abnormally mobile, it can usually be felt at the end of a deep inspiration. To search for a palpable kidney, place the patient on his back, pass one hand under him below the last rib, and with the other hand press the anterior abdominal wall opposite the fingers of the first hand. If the kidney is palpable it can be felt slipping between the opposed fingers when the patient takes a deep breath. The outline of the kidney cannot be defined by percussion as the organ is deeply placed and adjacent to other organs of similar consistency to itself.

In the angle between the lateral border of the erector spinae muscle and the lower border of the twelfth rib, the kidney lies fairly close to the posterior surface of the body, for there it is separated from the surface only by relatively thin structures— the quadratus lumborum, the layers of the lumbar fascia, the serratus posterior inferior, and the latissimus dorsi. Lesions, e.g., an abscess in the region of the kidney, can cause tenderness and swelling in this area. The close relationship of the psoas major muscle and the kidney explains why extension of the thigh may increase pain due to inflammation in the kidney region.

THE URETERS

[PLATES 37 and 38]

The ureters are the ducts that lead from the kidneys to the urinary bladder. Each ureter begins above in a thin-walled, funnel-shaped expansion, called the **pelvis of the kidney** which is placed partly inside and partly outside the sinus of the kidney. Towards the lower end of the kidney the part of the pelvis which lies outside the sinus diminishes in calibre, and forms a tube, the ureter, which conveys the urine to the bladder.

Pelvis of the Kidney [PLATE 38 c and d]

Within the sinus of the kidney the pelvis lies among the larger renal vessels [FIG. 623]. Its volume is about 8 ml. It is formed by the junction of three (more rarely, two or four) thin-walled tubes—the **major calyces**—each of which has a number of branches. The branches, called **minor calyces,** are short, and increase in diameter as they approach the wall of the sinus, to which they are attached. Their wide, funnel-like ends enclose the renal papillae and receive the urine. The

minor calyces vary in number from five to twenty, but there are usually eight or nine in the adult kidney (Sykes, 1964b) one calyx sometimes surrounding two or even four papillae, the result of a greater degree of fusion of the calyces than the papillae during development. The portion of the pelvis that lies outside the kidney has anterior to it, in addition to the renal vessels, on the right side, the descending part of the duodenum, and on the left side, a part of the body of the pancreas and the peritoneum of the posterior abdominal wall [FIGS. 620 and 621] or, sometimes, the duodenojejunal flexure.

Ureter [PLATES 37 and 38]

The ureter proper is a pale-coloured, thick-walled duct with a small lumen. When *in situ* it has a total length of about 25 cm., and lies throughout its whole course behind, and closely adherent to the peritoneum. The upper half of the ureter lies in the abdominal cavity, and the lower half in the lesser pelvis [FIGS. 616 and 627].

The **abdominal portion** of the ureter, about 12·5 cm. in length, is directed downwards and slightly medially, and lies on the psoas major muscle, crossing it obliquely from the lateral to the medial side [FIG. 616]. Anteriorly, the abdominal portions of both ureters are crossed obliquely by the testicular or ovarian vessels, while posteriorly, the genitofemoral nerve passes inferolaterally in almost the same line as the vessels. The right ureter has the descending part of the duodenum in front of its upper part, and below this is crossed anteriorly by the right colic and ileocolic vessels, and by the root of the mesentery shortly before the ureter enters the lesser pelvis. The left ureter is crossed anteriorly by the left colic vessels and by the root of the sigmoid mesocolon.

Crossing the common iliac or the external iliac artery, the ureter enters the lesser pelvis. Usually the left ureter crosses the common iliac artery, and the right ureter the external iliac; but that arrangement is by no means constant.

The **pelvic portion** of the ureter is about 12·5 cm. in length; it passes backwards and downwards on the side-wall of the lesser pelvis, describing a curve which is convex backwards and laterally [FIG. 627], the most convex point being close to the inferior extremity of the greater sciatic notch at the level of the ischial spine [FIG. 1097].

In its course within the pelvis, the ureter lies below and in front of the internal iliac artery, and crosses the medial side of the obturator nerve and vessels and of the umbilical artery.

IN THE MALE, about the level of the ischial spine, the ureter is crossed from before backwards by the ductus deferens, and from that point onwards is less intimately related to the peritoneum. It then bends forwards and slightly medially to reach the lateral angle of the bladder, where it is anterior to the upper extremity of the seminal vesicle [FIG. 631]. The ductus deferens having crossed the ureter also turns medially, and as it does so it lies on a plane posterior to the ureter. The lower end of the ureter is surrounded by a dense plexus of veins which brings the vesical plexus into communication with the internal iliac vein. The hypogastric nerve, which connects the superior hypogastric plexus with the inferior hypogastric plexus, divides to pass on each side of the ureter where it is crossed by the ductus deferens [FIGS. 627 and 965].

The right and left ureters reach the bladder 5 cm. apart, and

PLATE 37

A

B

A. Retroperitoneal air insufflation combined with an intravenous pyelogram. The outline of the kidney and renal pelvis are shown by this means.

1. Major calyx of right kidney.   4. Border of psoas.
2. Right renal pelvis.   5. Left iliac crest.
3. Left suprarenal gland.

B. Nephrographic phase of aortogram showing the increase in density of the renal tissue due to contrast medium in capillaries.

1. 12th rib.
2. Renal parenchyma.
3. Catheter lying in aorta.

C

D

C. Intravenous pyelogram, seven minutes after the intravenous injection of contrast medium. The renal pelvis and upper ends of the ureters are filled.

1. Right renal pelvis.   2. Upper end of right ureter.

D. Retrograde pyelogram. The renal pelvis and ureters have been filled by catheters inserted into the lower ends of both ureters.

1. Major calyx.   2. Minor calyx.   3. Renal pelvis.

PLATE 38

A

B

A. Tomogram of intravenous pyelogram. Note the obliteration of the overlying gas shadows.

B. Intravenous pyelogram showing a congenital abnormality (a bifid ureter and pelvis). The renal pelvis is in two parts and both parts drain by separate ureters. These fuse in their intra-abdominal course.

1. Upper ureter.
2. Lower ureter.
3. Point of junction of separate ureters.
4. Pelvis of kidney.

C

D

C and D. Detail of different types of renal pelvis visualized by intravenous pyelography.

1. Major calyx seen end on.
2. Transverse process of 1st lumbar vertebra
3. Pelvi-ureteric junction.
4. Minor calyces.

5. Major calyx.
6. Pelvis of kidney.
7. Transverse process of 2nd lumbar vertebra.

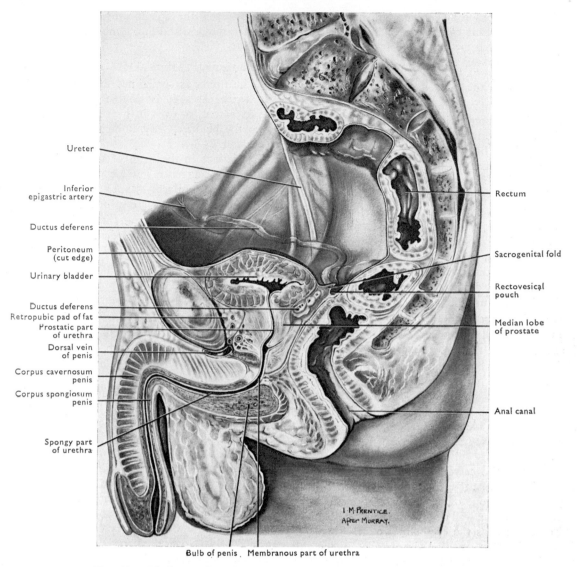

Ureter
Inferior epigastric artery
Ductus deferens
Peritoneum (cut edge)
Urinary bladder
Ductus deferens
Retropubic pad of fat
Prostatic part of urethra
Dorsal vein of penis
Corpus cavernosum penis
Corpus spongiosum penis
Spongy part of urethra

Rectum
Sacrogenital fold
Rectovesical pouch
Median lobe of prostate
Anal canal

I·M·PRENTICE.
After MURRAY.

Bulb of penis . Membranous part of urethra

FIG. 627 Median section of pelvis of adult male subject. The urinary bladder is empty and firmly contracted [cf. Fig. 633].

piercing the wall of the bladder very obliquely, are embedded within its muscular tissue for nearly 2 cm. Finally, they open into the bladder by a pair of minute, slit-like apertures which are of a valvular nature and prevent a backward passage of fluid from the bladder. It is probable, however, that an exaggerated idea of the valvular nature of the openings of the ureters into the bladder is obtained by an examination of the parts in the dead subject. When the bladder is empty the openings of the ureters are placed about 2·5 cm. apart, but when the bladder is distended they may be 5 cm. apart, or more. As the ureter pierces the wall of the bladder the muscular fibres of the bladder and ureter remain quite distinct, and so the ureter, remaining a thick-walled tubular structure, appears to pass through a gap in the muscular wall of the bladder. The mucous membrane of the ureter becomes continuous with that of the bladder.

The lumen of the ureter proper is not uniform throughout; it is slightly constricted near the renal pelvis and where it crosses the iliac artery; and it is narrowest where it passes through the bladder wall.

C.T.A.—S

IN THE FEMALE, the ureter descends in a peritoneal fold which forms the posterior boundary of the ovarian fossa [FIG. 658]. Near its termination it is accompanied by the uterine artery, and passing below the root of the broad ligament of the uterus and that artery, lies about 2 cm. lateral to the cervix uteri, above the lateral fornix of the vagina. Finally it inclines medially to lie anterior to the lateral margin of the vagina [FIG. 663], but since the vagina deviates to one side, usually the left, more of one ureter than the other lies anterior to the vagina (Brash, 1922). Its lower part is accompanied by the uterine artery, which crosses above and in front of it not far from its termination. Higher up it lies in the peritoneal fold which forms the posterior boundary of the ovarian fossa [FIG. 658].

Structure of the Ureter. The wall of the ureter, which is thick and of a whitish colour, is composed of mucous, muscular, and fibrous coats. The **mucous coat** is lined with a transitional epithelium composed of many layers of cells, those nearest the lumen being of large size. When the canal is empty the mucous coat is thrown into numerous longitudinal folds, and so its lumen has a stellate outline in transverse section. The **muscular coat** consists of plain muscle fibres collected into bundles which are separated by a considerable amount of fibro-areolar tissue, and are arranged some longitudinally, some circularly. In the upper part of the ureter a relatively large amount of fibro-areolar tissue is present deep to and among the bundles of muscle fibres, which are arranged in three strata—an inner longitudinal, middle circular, and an outer longitudinal. In the middle part of the tube the same layers may be recognized, but the circularly disposed bundles of fibres are more numerous than higher up; this may be due to the fibres of the middle stratum, which are helical, being more tightly coiled. In the lower part of the ureter the fibro-areolar tissue is relatively scanty and the inner longitudinal fibres lie close to the mucous coat; in this region also the longitudinal folds of the mucous coat become fewer and less marked. A short distance above the bladder, a number of coarse bundles of longitudinally arranged muscle fibres are applied to the outer surface of the muscular coat. These form the so-called sheath of the ureter, and are continuous with the superficial part of the muscular wall of the bladder. In the portion of the ureter which traverses the wall of the bladder, nearly all the fibres of the muscular coat are disposed longitudinally. The muscle fibres lie immediately outside the epithelium, and end just where the mucous coats of the bladder and ureter become continuous. The outer **adventitial coat** of fibrous tissue varies in thickness at different levels, and in its lower part blends with the fibro-areolar tissue which lies among the muscle fibres of the sheath of the ureter just mentioned.

The mucous coat of the calyces and of the pelvis of the kidney has an epithelium like that of the ureter. Where each renal papilla projects into one of the calyces a deep, circular recess is formed between the wall of the calyx and the papilla; at the bottom of this recess the epithelium of the calyx becomes continuous with that covering the papilla, and in this recess the epithelium is poorly supported and liable to rupture on sudden changes of pressure in the calyces. At the apex of the papilla the epithelium joins that of the renal tubules. The muscular fibres in the wall of the calyces and of the pelvis are collected into loosely arranged bundles separated by wide intervals occupied by fibrous tissue. As in the ureter proper, the outermost and innermost fibres are longitudinal, and the middle ones are circular. The circular fibres alone form a distinct layer.

Vessels and Nerves of the Ureter

The most constant arteries supplying the ureter are branches from the uterine artery in the floor of the lesser pelvis in the female. In the male, similar branches are derived from the inferior vesical artery. Branches from the renal arteries supply the upper end of the ureters. Usually intermediate arteries are derived either from the aorta, gonadal, common iliac, or internal iliac arteries. In the abdomen the arteries supplying the ureter approach it from the medial side; in the pelvis they approach it from the lateral side. In both abdomen and pelvis the supplying arteries are intimately connected with the peritoneum (Racker and Braithwaite, 1951).

The **nerves** of the ureter are derived from the renal, testicular or ovarian, and hypogastric plexuses. The afferent fibres reach the spinal medulla through the eleventh and twelfth thoracic and first lumbar nerves.

Variations [PLATE 38 B]

The ureter is sometimes duplicated in its upper portion. In rarer cases it is double throughout the greater part of its extent, or even in its whole length with two openings into the bladder. Asymmetry in such abnormalities is very common.

Variations in the form of the pelvis of the kidney are of frequent occurrence. Most usually the pelvis divides into three major calyces—upper, middle, and lower—though commonly only two are present. In some cases the two calyces spring directly from the ureter without the intervention of a pelvis, or a marked subdivision may lead to the formation of two pelves.

THE URINARY BLADDER

[PLATE 39 A and B]

Position. The urinary bladder is a hollow muscular organ situated below the peritoneum on the anterior part of the pelvic floor behind the pubic symphysis. The space in which it lies is three-sided. On each side, it is bounded above by the obturator internus and lower down by the levator ani muscles; and these two side-walls meet in front at the pubic symphysis. As the two levator ani muscles slope downwards, backwards, and medially they form the walls of a gully which is deeper behind than in front owing to the slope of its floor. This floor is defective in front because of the slight gap that exists between the two levator ani muscles, but the upper surface of these muscles is covered with a layer of fascia which bridges over the gap. Laterally where the fascia lies on the muscles it is sometimes known as the lateral puboprostatic (pubovesical in the female) ligaments; medially where it covers the gap between the two muscles the fascia is thickened and forms the **puboprostatic** (pubovesical in the female) **ligaments.** The gap is also partly closed at a lower level by the urogenital diaphragm. The space for the bladder is bounded behind in the male by the fascial tissue which stretches across the back of the bladder and is thickened and condensed as it passes from the bladder to be connected with the fascia of the levator ani on each side, the **rectovesical septum.**

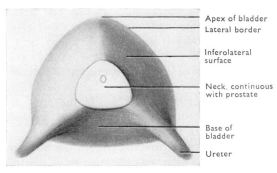

FIG. 628 Inferior aspect of empty male urinary bladder. From a subject in which the viscera had been hardened *in situ*.

Shape and Parts [FIGS. 628–630]. The empty adult bladder lies completely or almost completely within the above-mentioned space, and in dissecting-room fixed specimens it is usually moulded to a pyramidal shape by the walls of the space and by the pressure of overlying viscera. It is therefore usual to describe the empty adult bladder as possessing four surfaces: a **superior surface** looking straight upwards, a pair of **inferolateral surfaces** facing downwards, laterally, and forwards, and a **posterior surface** directed backwards and slightly downwards. The posterior surface is the **base**; the anterior extremity of the bladder, where the superior and the two inferolateral surfaces meet, is the **apex**; and the lowest part of the organ, where the base and the two inferolateral surfaces come together, is the **neck.** In life the bladder almost always contains some fluid so that it is more or less rounded, though its spherical shape may be slightly altered either by pressure of surrounding organs or by its attachment to surrounding structures. Hence, radiographs of the bladder in living subjects show it as a rounded organ, with its upper aspect frequently flattened by the weight of overlying viscera, or even slightly depressed by a heavy overlying viscus, e.g., a gravid uterus. It would therefore probably be more correct to speak of the various **aspects** of the bladder than of its **surfaces**; for its true shape appears to be that of a sphere or a spheroid, except when temporarily modified by pressure of surrounding structures [PLATE 39A].

Attachments. In both sexes the **neck** is the most firmly fixed part of the bladder. In the male it is held in place chiefly by the puboprostatic ligaments and by its firm connexion with the prostate, with which it is structurally continuous. Each side of the neck gives attachment to a lateral puboprostatic ligament which is merely the fascia on the anterior part of the levator ani muscle. The **puboprostatic ligaments** are a pair of thick, short, strong bands that lie side by side over the anterior part of the gap between the two

A

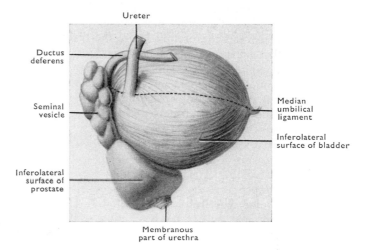

B

FIG. 630 Urinary bladder, prostate, and right seminal vesicle viewed from the right side.

Drawn from specimens in which the viscera were hardened *in situ*. In A the bladder contained a very small quantity of fluid; in B the quantity was a little greater. In A the peritoneum is shown covering the superior surface of the bladder, and its cut edge is seen where it is reflected along the lateral border. In B the level of the peritoneal reflexion is indicated by a dotted line.

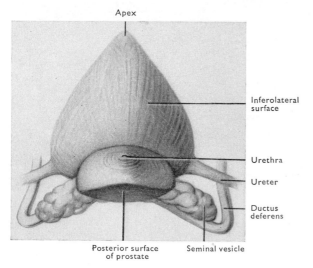

FIG. 629 Urinary bladder, prostate, and seminal vesicles, viewed from below.

Taken from a subject in which the viscera were hardened *in situ*. Same specimen as in Fig. 630A. The bladder contained a very small amount of fluid.

levatores ani in front of the neck of the bladder and attach it firmly to the back of the pubic bones; these ligaments contain bundles of plain muscle fibres which are continuous with the muscular coat of the bladder and are named, therefore, the **pubovesical muscle.** The prostate lies in the posterior part of the gap between the two levator ani muscles and is itself enclosed in a strong fibrous sheath which also is firmly connected with the puboprostatic ligaments anteriorly and with the superior fascia of the urogenital diaphragm inferiorly [p. 521]. In the female the neck of the bladder and upper part of the urethra are connected with the pubis and the levator ani by similar fascial bands which, in the absence of a prostate, are named **pubovesical** ligaments; and these parts are also directly connected with the superior fascia of the urogenital diaphragm.

The **apex** of the bladder is continuous with the median umbilical ligament, which extends up on the posterior surface of the anterior abdominal wall to the umbilicus. This ligament is the fibrous remnant of the urachus—the passage which, in the embryo, connects the developing bladder with the allantoic diverticulum.

The **base** of the bladder is attached by means of the condensation of fascia which, on each side, blends with the fascia on the levator ani muscle. The whole condensation is really a collection of fibrous tissue around the internal iliac artery and its branches, and it passes medially to be attached to the base of the bladder and seminal vesicles in the male, and the base of the bladder and the sides of the vagina and cervix uteri in the female.

In addition to these connexions, the base is loosely attached by areolar tissue to the anterior wall of the vagina in the female, but firmly to the seminal vesicles and ampullae of the ductus deferentes in the male.

The ureters enter the bladder at the upper and lateral aspects of the base, the points of entrance being about 5 cm. apart in the empty bladder; and the urethra leaves the bladder at its lowest point—that is the neck—by the internal urethral orifice.

In dissecting-room specimens, once the median umbilical ligament has been cut, the whole anterior part of the bladder is very mobile but the base and neck remain fixed.

Relations of the Male Bladder. The superior surface of the bladder [FIGS. 627 and 632] is covered with peritoneum, which separates it from coils of small intestine or from the pelvic colon. The peritoneum passes from the upper surface of the bladder forwards on to the posterior surface of the anterior abdominal wall at the level of the pubic crest; laterally, it passes slightly downwards and then upwards on to the side-wall of the pelvis, thus forming a shallow **paravesical fossa** on each side of the bladder. Posteriorly it passes backwards for about 1·5 cm., forming a prominent shelf-like fold—the **sacrogenital fold**—and then turns abruptly downwards for about 2·5 cm. to be reflected on to the front of the rectum, thus forming the **rectovesical pouch** [FIG. 632]. As the peritoneum turns downwards it may, in the median plane, cover a small portion of the base of the bladder but more laterally it is separated from the base by the ductus deferentes and the upper ends of the seminal vesicles. When the bladder is completely empty, the peritoneum on its upper aspect may be thrown into a transverse fold —the **transverse vesical fold** [FIG. 632].

The two inferolateral surfaces and the border between them are separated from the surrounding firm structures by a space—the **retropubic space**—which is filled with loose areolar tissue, fat, and a plexus of veins [FIG. 627]. This mass of tissue acts as a packing to fill the gaps between the bladder and the neighbouring, rather rigid, pelvic walls, and is known as the **retropubic pad**

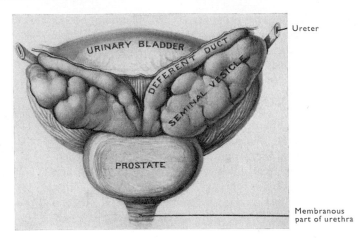

FIG. 631   Urinary bladder, seminal vesicles, and prostate, viewed from behind.

of fat. The retropubic space is bounded antero-inferiorly by the back of the pubis, at each side by the levator ani muscle below and the obturator internus muscle above, inferiorly by the puboprostatic ligaments, and posteriorly by the fascial condensation which passes from the base of the bladder to the side-walls of the pelvis. Superiorly, it is bounded by the peritoneum which passes from the superior surface of the bladder

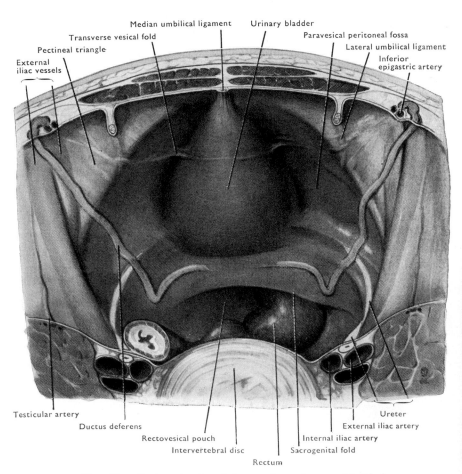

FIG. 632   Cavity of male pelvis seen from above and behind.

From a specimen in which the bladder was firmly contracted and contained only a small amount of fluid. The peritoneal pouch in front of the rectum is bounded on each side by the sacrogenital folds, which meet together in the median plane some distance behind the posterior border of the bladder.

to the anterior wall of the abdomen and the side-walls of the pelvis. As the bladder fills and expands, it lifts the peritoneum off the anterior abdominal wall and occupies the area between the lateral umbilical ligaments (obliterated umbilical arteries). Thus the retropubic space potentially extends upwards on to the back of the anterior abdominal wall and as far laterally as these ligaments. The obturator nerve and vessels extend into the posterior and upper part of the side-wall of the retropubic space even when the space is unexpanded, but as the bladder fills and the space enlarges the ductus deferens also may be included. The plexus of veins in the retropubic pad of fat is especially rich in the neighbour-hood of the neck of the bladder.

The **base** of the bladder faces back-wards and slightly downwards [Figs. 627 and 628]. Its inferolateral parts are in contact with the seminal vesicles [Fig. 631], which diverge from each other at rather more than a right angle as they pass laterally and upwards from the back of the prostate. Medial to each seminal vesicle, the base is in contact with the ampulla of the ductus deferens of the same side; between these two ducts it is separated by some loose areolar tissue from the front of the rectum, but the uppermost part of that interval may be covered with peri-toneum.

The **neck** is firmly attached to the prostate, and is structurally continuous with it, though a well-marked groove is present where the surfaces of the two organs meet.

Relations of the Female Bladder [Figs. 634 and 662]. The bladder occupies a slightly lower position than in the male. The superior surface is covered with peritoneum, which, as in the male, passes forwards on to the back of the anterior ab-dominal wall, and also laterally to the side-wall of the pelvis, forming, as it does so, a shallow **paravesical fossa.** Posteri-orly, the peritoneum passes from the superior surface of the bladder upwards to be attached to the uterus at the junction of its isthmus and body, and then passes forwards on the antero-inferior surface of the body of the uterus. In the female, therefore, the body of the uterus overlies the bladder and the peritoneal recess between the two organs is known as the **uterovesical pouch.** This pouch is usually slit-like, but in some cases coils of small intestine pass into it and partly separate the bladder and uterus.

The base of the bladder in the female is only loosely at-tached by areolar tissue to the anterior wall of the vagina. The upper part of the base, just as it passes into the superior surface of the organ, is separated from the anterior surface of the supravaginal part of the cervix uteri by loose areolar tissue; and laterally on each side the uterine vessels are about 2·5 cm. distant [Fig. 661].

Intestine

Urinary bladder

Reflexion of peritoneum onto anterior abdominal wall

Rectum

Rectovesical pouch

Terminal part of ductus deferens

Prostate

Ejaculatory duct

Corpus spongiosum penis

Corpus cavernosum penis

Sphincter urethrae

Anal canal

Bulb of penis   Bulbospongiosus muscle   Sphincter urethrae

Fig. 633   Median section of pelvis of adult male subject. The urinary bladder and rectum are both distended [cf. Fig. 627].

The inferolateral surfaces and the apex have the same general relations in the female as in the male.

The neck, in the absence of the prostate, is continuous only with the urethra, and the pubovesical ligaments are connected with it through the considerable amount of smooth muscle, glands, and fibrous tissue that surrounds it.

Bladder in the New-Born Child and Infant. At birth the empty bladder is spindle-shaped; its long axis extends from the apex to the external urethral orifice, and is directed down-wards and backwards [Fig. 635]. The lateral and posterior borders seen in the adult organ cannot be recognized at birth, nor is there any part of the bladder wall directed backwards and downwards, as in the base of the adult organ. In the foetus and young child the bladder occupies a much higher level than it does in the adult, and, even when empty, it ex-tends upwards into the abdominal cavity. Its anterior surface is in contact with the back of the anterior abdominal wall. At birth the peritoneum forming the rectovesical pouch covers the whole of the posterior surface of the bladder, and reaches as low as the upper limit of the prostate. The internal urethral orifice is placed at a high level, and sinks gradually after birth [Fig. 636]. In the new-born child the opening is on a level with the upper margin of the pubic symphysis and the openings of the ureters are almost in the plane of the

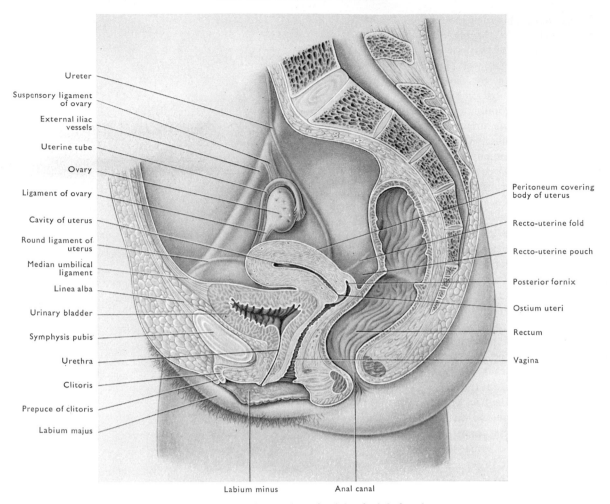

Ureter

Suspensory ligament
of ovary

External iliac
vessels

Uterine tube

Ovary

Ligament of ovary

Cavity of uterus

Round ligament of
uterus

Median umbilical
ligament

Linea alba

Urinary bladder

Symphysis pubis

Urethra

Clitoris

Prepuce of clitoris

Labium majus

Peritoneum covering
body of uterus

Recto-uterine fold

Recto-uterine pouch

Posterior fornix

Ostium uteri

Rectum

Vagina

Labium minus   Anal canal

FIG. 634   Median section of pelvis of adult female.

superior aperture of the pelvis. The obliterated umbilical arteries are more intimately related to the bladder in the child than in the adult, and lie close against its sides as they pass upwards on the back of the anterior abdominal wall towards the umbilicus [FIG. 645].

Interior of the Bladder [FIGS. 634, 638, and 639]. The mucous membrane which lines the bladder is loosely con-

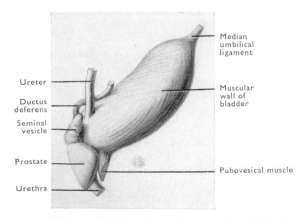

Ureter

Ductus
deferens

Seminal
vesicle

Prostate

Urethra

Median
umbilical
ligament

Muscular
wall of
bladder

Pubovesical muscle

FIG. 635   Urinary bladder of new-born male child, viewed from right side.

The drawing is from a specimen which had been hardened *in situ* [cf. FIG. 630A].

nected to the muscular coat, and when the bladder is contracted it is thrown into a number of prominent folds, except in a triangular area of the base (**trigone of the bladder**) where the mucous membrane is firmly attached to the muscular coat and remains smooth. The apex of the triangle is at the internal urethral orifice, and the base is a line drawn between the openings of the ureters. Immediately above and behind the internal urethral orifice the bladder wall in the male sometimes bulges slightly into the cavity owing to the presence of the median lobe of the prostate, which lies outside the mucous coat in that position. When well marked, as it often is in old men, the bulging is termed the **uvula of the bladder.** Stretching between the openings of the ureters there is usually a smooth **interureteric fold,** due to the presence of a transverse bundle of muscle fibres. It may be deficient near the median plane, and it is curved so as to be convex downwards and forwards. Lateral to the opening of each ureter there is a ridge, called the **ureteric fold,** produced by the terminal parts of the ureters as they traverse the bladder wall and lie outside the mucous coat of the bladder. In old people the region above and behind the trigone often bulges backwards and forms a shallow **retro-ureteric fossa.** A less distinct depression may sometimes be observed on each side of the trigone. Around the urethral orifice there are several minute, radially disposed folds which, disappearing into the urethra, become continuous

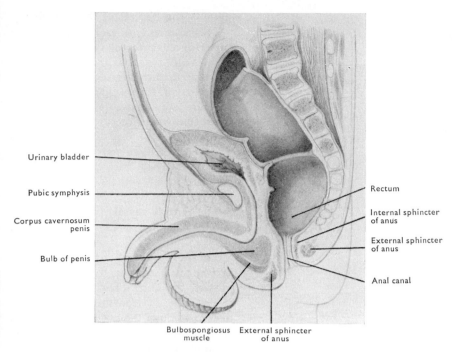

Urinary bladder

Pubic symphysis

Corpus cavernosum penis

Bulb of penis

Rectum

Internal sphincter of anus

External sphincter of anus

Anal canal

Bulbospongiosus muscle   External sphincter of anus

FIG. 636   Median section of new-born male pelvis.

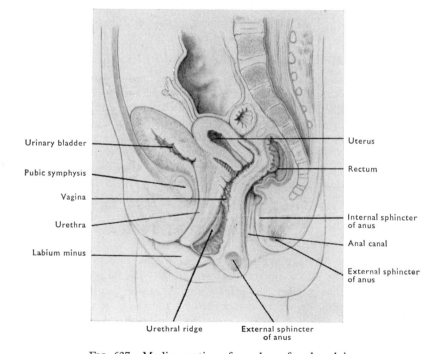

Urinary bladder

Pubic symphysis

Vagina

Urethra

Labium minus

Uterus

Rectum

Internal sphincter of anus

Anal canal

External sphincter of anus

Urethral ridge   External sphincter of anus

FIG. 637   Median section of new-born female pelvis.

distended, the distance between the openings may be increased to 5 cm. or more.

Position of the Internal Urethral Orifice. During the various changes in shape and size which the bladder undergoes, the region of the internal urethral orifice remains almost fixed in position. In the male the internal urethral orifice is immediately above the prostate, and behind and below the level of the upper margin of the pubic symphysis, from which it is about 5–6·5 cm. distant. It can be reached by a finger introduced into the bladder through the abdominal wall above the pubic symphysis. It is usually 1·5–2·5 cm. above the level of a plane which passes through the lower margin of the symphysis and the lower end of the sacrum, but in some cases it is slightly lower. In the female the internal urethral orifice is normally at a lower level than in the male.

Capacity of the Bladder. The distended bladder may contain 550 ml. or more; but in most cases the organ is emptied when it contains 160–300 ml. of fluid (Thompson, 1919). When the bladder contains 450 ml. it is in contact with the back of the anterior abdominal wall for a distance of about 7–8 cm. above the pubic crest and the peritoneal reflexion from its upper surface on to the abdominal wall has been raised above this level.

Structure of the Bladder. The wall of the bladder from without inwards is composed of a serous, a muscular, a submucous, and a mucous coat. The **serous coat** is the peritoneum, and covers only the upper surface of the bladder [FIGS. 627 and 634].

A considerable amount of loose fibroareolar tissue surrounds the **muscular coat,** and, penetrating it, divides it into numerous coarse bundles of muscle fibres. All the muscle fibres are of the plain variety, and the bundles which they form are arranged in three very imperfectly separated strata called external, middle, and internal.

1. The **external stratum** is for the most part made up of fibres which are directed longitudinally, and it is most marked near the median plane on the superior and inferolateral surfaces of the bladder. Further from the median plane, on the sides of the bladder, the fibres composing the external stratum run more obliquely, and frequently cross one another. In the male many of the fibres of the external stratum pass into the prostate in front of and behind the urethral opening, and in the female the corresponding fibres join the dense tissue which, in this sex, forms the upper part of the wall of the urethra. Other fibres of this stratum on each side of the median plane join the lower part of the pubic symphysis and constitute the

with the longitudinal folds of the mucous membrane of the upper part of that canal. The ureter pierces the bladder wall very obliquely, and so the ureteral orifice has an elliptical outline. The lateral boundary of each opening is a thin, crescentic fold which, when the bladder is artificially distended in the dead subject, acts as a valve in preventing water or air from entering the ureter. In the empty bladder the internal urethral orifice and the openings of the two ureters lie at the angles of an approximately equilateral triangle the sides of which are about 2·5 cm. in length. When the bladder is

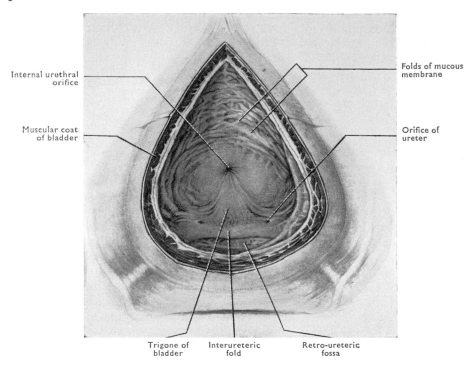

Internal urethral
orifice

Muscular coat
of bladder

Folds of mucous
membrane

Orifice of
ureter

Trigone of   Interureteric   Retro-ureteric
bladder   fold   fossa

FIG. 638 Empty and contracted urinary bladder, opened by
removal of its superior wall.

The peritoneum is seen spreading out from the lateral and
posterior borders of the organ [cf. Fig. 632].

Vessels and Nerves of the Bladder

The arteries on each side are the **superior** and **inferior vesical arteries.** Braithwaite (1952) has given details of the distribution of these arteries to the different parts of the bladder. The largest veins are found above the prostate and in the region where the ureter reaches the bladder. They form a dense plexus which pours its blood into tributaries of the internal iliac vein, and communicates below with the prostatic venous plexus.

The **lymph vessels** from the bladder join the iliac groups of lymph nodes [FIG. 1155].

The **nerve supply** of the bladder is derived on each side from the **vesical plexus,** the fibres of which come from two sources : (1) from the **upper lumbar nerves** through the hypogastric plexuses (sympathetic); and (2) from the **pelvic splanchnic nerves** (parasympathetic), which spring from the second and third, or the third and fourth, sacral ventral rami. These splanchnic nerves join the vesical plexus directly.

pubovesical muscle, which follows the course of the pubo-prostatic ligaments. Lastly, some fibres of the external stratum blend posteriorly with the front of the rectum and receive the name of the **rectovesical muscle.** In both sexes many muscular fibres of the bladder are continued along the course of the ureter to form the sheath of the ureter already mentioned [p. 514].

2. The **middle stratum** is composed of irregular interlacing bundles of muscle fibres which for the most part run circularly, and form the greater part of the thickness of the muscular coat. At the internal urethral orifice, the bundles of fibres are finer and more densely arranged **(sphincter vesicae),** and surround the opening in a plane which is directed obliquely downwards and forwards. Inferiorly the fibres of the sphincter vesicae are continuous with the muscular tissue of the prostate in the male, and with the muscular wall of the urethra in the female. In other parts of the bladder the bundles of the middle stratum are coarser and separated by intervals filled with fibro-areolar tissue.

3. The **internal stratum** is composed of a thin layer of muscle fibres directed for the most part longitudinally.

The **submucous coat** is composed of areolar tissue which contains numerous fine elastic fibres.

The **mucous coat** is loosely attached by the submucous layer to the muscular coat except in the region of the trigone. There, the muscular fibres are firmly adherent to the mucous coat. The mucous coat of the bladder is continuous with that of the ureters and urethra. Its epithelial covering is of the variety known as transitional epithelium; it varies much in appearance as the organ expands and contracts, being essentially an elastic epithelium, impervious to urine.

Methods of Clinical Anatomical Examination

The normal empty bladder cannot be palpated, and its outlines cannot be determined by percussion, but, as the organ fills and rises into contact with the anterior abdominal wall, its upper boundary can be mapped out by percussion. The base of the female bladder can be palpated by a vaginal examination.

The interior of the bladder can be inspected by means of a cystoscope ; and the outline of its cavity may be seen radiographically [PLATE 39A and B].

THE URETHRA

The urethra is the channel which conveys the urine from the bladder to the exterior. In the male its proximal part, less than 2·5 cm. in length, extends from the bladder into the prostate where it is joined by the ducts of the reproductive glands. A much longer distal portion serves as a common passage for the urine and the semen. The female urethra represents only the proximal part of the male canal. It is a short passage that leads from the bladder to the external urethral orifice—an aperture placed within the pudendal cleft immediately in front of the opening of the vagina.

THE MALE URETHRA

[PLATE 39B]

The male urethra [FIGS. 627 and 639] is a channel, about 20 cm. in length, which leads from the bladder to the external urethral orifice at the end of the glans penis. The canal not only serves for the passage of urine ; it also affords an exit for the seminal products, which enter it by the ejaculatory ducts, and for the secretion of the prostatic and bulbo-urethral glands; in addition, numerous minute **urethral glands** pour their mucous secretion into the urethra.

As it passes from the internal urethral orifice to its external opening the urethra describes an ∞-shaped course, and it is customary to divide it into three sections. The first part is within the lesser pelvis, and has a nearly vertical course as it traverses the prostate. Turning more forwards, the urethra leaves the pelvis by piercing the urogenital diaphragm 2·5 cm. below and behind the pubic symphysis, and enters the bulb of the penis. Throughout the rest of its course it lies in the erectile tissue of the corpus spongiosum and of the glans penis. The part of the urethra which is embedded in the prostate is called the prostatic part; the short part which pierces the urogenital diaphragm is called the membranous part; and the part surrounded by the corpus spongiosum is the spongy part. Of the three sections the spongy portion is much the longest, and the membranous is the shortest (slightly less than 1·5 cm.).

1. The **prostatic part** descends through the prostate from the base towards the apex, describing a slight curve which is concave forwards. It is about 2·5 cm. in length, and is narrower above and below than in the middle portion, which is, indeed, the widest part of the whole urethral canal. Except while fluid is passing, the canal is contracted, and the mucous membrane of the anterior and posterior walls is in contact and thrown into a series of longitudinal folds. When dis-

tended, the middle and widest part of the canal may normally have a diameter of about 0·8 cm. The posterior wall of the prostatic urethra presents a distinct median ridge called the **urethral crest** [FIG. 639]. The crest projects forwards into the urethra to such an extent that the canal in transverse section presents a crescentic outline [FIG. 657]. The groove on each side of the crest is known as the **prostatic sinus,** and into it the numerous ducts of the prostatic glands open by minute apertures. A few ducts from the middle lobe open nearer the median plane, on the sides of the urethral crest. The urethral crest is highest half-way down, forming an eminence called the **seminal colliculus,** on which is a small, slit-like opening which leads backwards and upwards for a distance of about 0·5 cm. into the substance of the prostate as a blind pouch. This diverticulum is known as the **prostatic utricle,** and it represents the fused caudal ends of the paramesonephric ducts, from which the uterus and vagina are developed. On each side of the mouth of the utricle there is the much more minute opening of the ejaculatory duct [p. 529]. When traced upwards, the urethral crest diminishes in height and becomes indistinct, but it can often be traced as a slight median ridge as far as the uvula of the bladder. In the opposite direction the ridge becomes less marked, but it can be followed into the membranous portion of the canal, where it divides into a pair of inconspicuous elevations which gradually fade out [FIG. 639]. The curvature and, to a less degree, the length of the prostatic urethra depend upon the amount of distension of the bladder and of the rectum [compare FIGS. 627 and 633].

2. The **membranous part** of the urethra, with the exception of the external orifice, is the least dilatable part of the urethra. It runs downwards and forwards, passing through the urogenital diaphragm. This diaphragm is a double layer of fascia; its upper layer, the **superior fascia of the urogenital diaphragm,** is thin and consists of the fascia covering the deep aspect of the sphincter urethrae and deep transverse perineal muscles. Its lower layer, strong in the male in which it affords a firm basis for the attachment of the muscles of the penis, but weak in the female, is known as the **inferior fascia of the uro-genital diaphragm,** or **perineal membrane.** It covers the superficial aspect of the two muscles mentioned above and is continuous with the upper layer around the posterior border of the deep transverse perineal muscle. The membranous urethra begins just in front of the apex of the prostate, and immediately pierces the superior fascia of the diaphragm. Within the diaphragm it is surrounded by the sphincter urethrae with a bulbo-urethral gland and a deep artery of the penis lying about 0·5 cm. laterally on each side. It then pierces the perineal membrane and ends about 0·5 cm. below the fascia by entering the bulb.

In addition to the folds continued down from the urethral crest there are other longitudinal folds of mucous membrane seen when the canal is empty, and the lumen of the empty tube is therefore stellate in transverse section.

FIG. 639 Dissection showing trigone of bladder and floor of the urethra in its prostatic, membranous, and proximal spongy parts.

Labels:
- Ureter
- Orifice of ureter
- Trigone of bladder
- Uvula of bladder
- Prostate
- Urethral crest
- Colliculus seminalis
- Mouth of prostatic utricle
- Bulbo-urethral gland
- Membranous part of urethra
- Opening of duct of bulbo-urethral gland
- Crus penis
- Opening of ejaculatory duct
- Bulb of penis
- Corpus cavernosum penis (cut)

Although the membranous part of the urethra appears to be firmly held, it is not strongly attached to the surrounding fascia, and in the living body is so mobile that it can be drawn upwards for a distance of about 4 cm. above the perineal membrane (Souttar, 1947). This is of some importance in the operation of retropubic prostatectomy.

It is important to note also that the lowest part of the membranous portion of the urethra, where it is overlapped by the bulb, lies below the perineal membrane.

Immediately after the urethra has pierced the perineal membrane its under surface is overlapped by the erectile tissue of the bulb, but the upper wall remains uncovered for a further distance of about 0·5 cm. [FIG. 627]. Here the wall of the urethra is very thin, and the passage is more readily dilatable than in other parts. In this region the urethral wall may readily be torn if undue force is used in passing an instrument or if the handle of the instrument is depressed too soon, when an attempt is being made to pass it into the narrower, more fixed part of the membranous urethra.

3. The **spongy part** of the urethra is about 15 cm. in length and is much the longest of the three divisions. It begins at the point where the urethra passes into the substance of the corpus spongiosum, and it ends at the external urethral orifice. Its proximal or perineal portion has a fixed position and direction, while its distal part varies with the position of the penis. It traverses the corpus spongiosum, including the bulb and the glans, and is therefore entirely surrounded by erectile tissue in the whole of its length. Directed at first forwards through the bulb, in the flaccid penis it turns downwards and forwards at the point where it comes to lie beneath the anterior part of the pubic symphysis [FIG. 627]. The bend in the canal corresponds approximately to the place of attachment of the suspensory ligament to the dorsum of the penis. When the penis is erect, the direction of the terminal half of the canal is changed, and at the same time the whole spongy urethra becomes more uniformly curved.

The urethra lies at first in the upper part of the erectile tissue, but as it passes forwards it sinks deeper and comes to occupy the middle part of the corpus spongiosum. In the glans, on the other hand, the erectile tissue lies mainly on the dorsal and lateral aspects of the urethra [FIG. 655]. Like the other parts of the urethra, the spongy portion is closed except during the passage of fluid, the closure being effected by the apposition of the dorsal and ventral walls except in the glans penis, where the side-walls of the canal are in contact. Thus the lumen of the first part of the canal, when empty, is represented in cross-section by a transverse slit [FIG. 654], and that of the terminal part by a vertical slit. The spongy part of the urethra does not present a uniform calibre throughout; it is wider in the bulb and glans than in the corpus spongiosum. In the glans the urethra expands to form a terminal dilated part named the **fossa navicularis** and it opens on the surface by the slit-like **external urethral orifice,** which is the narrowest and least dilatable point of the whole urethra.

The **ducts of the bulbo-urethral glands** [p. 534] open by very minute apertures in the inferior wall of the proximal part of the spongy urethra [FIG. 639]. Before opening into the canal, they lie for some distance immediately outside its mucous membrane. A number of small pit-like recesses, called the **urethral lacunae,** also open into the spongy part of the urethra, and their openings pass obliquely into the canal in the direction of its external orifice.

A valve-like fold of the mucous membrane, the valve of the fossa navicularis, is sometimes found in the upper wall of the fossa. Its free edge is directed towards the external orifice, and it may engage the point of a fine instrument introduced into the urethra.

Structure. The mucous membrane of the urethra contains numerous elastic fibres and varies in thickness in different parts of the canal. The lining epithelium is stratified columnar in type in the membranous and spongy parts of the canal; in the prostatic part of the urethra it changes to transitional epithelium which is continuous through the internal urethral orifice with the transitional epithelium of the bladder. In the region of the fossa navicularis the lining cells become stratified squamous.

Numerous minute mucous glands, the **urethral glands,** open into the urethra. They are most plentiful in the upper wall, but they occur in smaller numbers also in the floor and the side-walls. They are most numerous in the membranous part and in the anterior half of the spongy part.

There are also larger mucous glands deeply placed outside the mucous coat, which communicate with the urethra by long, slender, obliquely placed, branching ducts. The smaller glands lie in the mucous coat and form flask-like depressions with very short ducts. The ducts of some of the glands open into the lacunae, but many lacunae have no connexion with the urethral glands.

Frequently, two or more elongated ducts belonging to some of the larger glands open into the urethra quite close to its termination. They are termed **para-urethral ducts,** and may be traced backwards for some distance outside the mucous membrane of the roof of the urethra. Morphologically they do not correspond to the ducts in the female which have received the same name.

The muscular wall in the upper part of the urethra consists of plain muscle fibres directed for the most part longitudinally, but some circular fibres also are present. Throughout the greater part of the spongy urethra there is an obliquely placed band of circularly arranged plain muscle fibres which is continuous above with the sphincter vesicae [p. 520] whose fibres sweep downwards and forwards from below the trigone of the bladder. At a lower level, in front of the prostatic urethra, there is a band of striated muscular fibres which is continuous inferiorly with the inner circularly disposed part of the voluntary sphincter urethrae.

THE FEMALE URETHRA

The female urethra [FIG. 634] is a canal, about 4 cm. in length, which follows a slightly curved direction downwards and forwards, behind and below the lower border of the pubic symphysis. Except during the passage of urine the canal is closed by the apposition of its anterior and posterior walls. The **external urethral orifice** is placed between the labia minora, immediately in front of the opening of the vagina, about 2·5 cm. below and behind the clitoris [FIGS. 634 and 665]. The opening is slit-like, and is bounded by ill-defined lips. The mucous coat of the canal is raised into a number of low longitudinal folds, one of which, more distinct than the others and placed on the posterior wall of the passage, receives the name of **urethral crest.**

The upper part of the urethra is separated from the front of the vagina by a **vesicovaginal space** filled with loose connective tissue. Lower down, however, it becomes very intimately connected with the vagina, so that, as it approaches the external urethral orifice, it appears to be embedded in the anterior vaginal wall. This is due to the fact that the lower part of the urethra is bound to the front of the vagina by the fusion of their fasciae into a single dense layer.

Structure. The muscular coat of the female urethra is continuous above with that of the bladder, and is composed of layers of longitudinally and obliquely disposed plain muscle fibres. In the lower part, some of the fibres do not completely surround the urethra but are attached to the anterior wall of the vagina. Within the muscular coat the wall of the urethra is very vascular, and the canal itself is lined with a pale mucous membrane. The epithelium of the canal, in its upper part, is of the transitional variety, like that of the bladder; it then becomes stratified columnar in type, and finally, stratified squamous near the external urethral orifice. Numerous minute **urethral glands** and pit-like **urethral lacunae** open into the urethra. One group of these glands on each side possesses a minute common duct, known as the **para-urethral duct,** which opens into the pudendal cleft by the side of the external urethral orifice. These glands represent the prostatic glands of the male. The vascular layer which lies between the muscular coat and the mucous membrane contains elastic fibres, and in appearance resembles erectile tissue. Striated muscle fibres are present on the outer surface of the muscular coat of the urethra. In the upper part of the canal these fibres form a complete ring-like sphincter, so that the urethra is encircled by both plain and striated fibres. Lower down, some of the striated fibres are attached to the anterior vaginal wall, and in the lower third, where they are specially developed, they pass backwards on the outer surface of the vagina to enclose that passage together with the urethra in a single loop of muscle tissue, to form a sphincter.

THE GENITAL ORGANS

THE MALE GENITAL ORGANS

The male reproductive organs are: (1) the testes together with (2) their coverings and (3) their ducts and the seminal vesicles, (4) the prostate, (5) the bulbo-urethral glands, (6) the external genital organs, and (7) the male urethra.

The **testes** are the essential reproductive glands (gonads) of the male, and are a pair of nearly symmetrical, oval bodies situated in the scrotum. The duct of each testis is at first much twisted and convoluted to form a structure known as the **epididymis,** which is applied to the back of the testis. This duct emerges from the epididymis as the **ductus deferens,** which passes upwards towards the lower part of the anterior abdominal wall and pierces it obliquely to enter the abdominal cavity. There each ductus deferens is covered with peritoneum, and, crossing the linea terminalis of the pelvis runs on the side-wall of the lesser pelvis towards the spine of the ischium. On reaching the sacrogenital fold of peritoneum, the ductus deferens turns abruptly towards the base of the bladder, where it comes into relation with a branched tubular structure termed the **seminal vesicle.** Joined by the duct of the seminal vesicle, the ductus deferens forms a short canal, called the **ejaculatory duct,** which pierces the prostate and opens into the prostatic part of the **urethra.** The **prostate** and the **bulbo-urethral glands** are accessory organs connected with the male reproductive system. The ducts of the bulbo-urethral glands and those of the prostate, like the ejaculatory ducts, open into the urethra, which thus serves not only as a passage for urine, but also for the generative products. The external genital organs are the **penis** and **scrotum.**

THE TESTES

The testes are a pair of oval, slightly flattened bodies of whitish colour, measuring about 4 cm. in length, 2·5 cm. from before backwards, and rather less in thickness. Each testis projects into the cavity of the scrotum from its posterior wall, the left usually lying a little lower than the right. The long axis of the testis is nearly vertical, while the lateral and medial surfaces are somewhat flattened, and the anterior and posterior margins are rounded. The posterior margin is attached to the wall of the scrotum, while the remainder is covered by the visceral (serous) layer of the tunica vaginalis, and lies free in the cavity of that tunic. The **epididymis** and the lowest part of the **spermatic cord** [p. 501] are attached to the posterior margin of the testis [FIGS. 641 and 644].

Epididymis. The epididymis is a comma-shaped structure

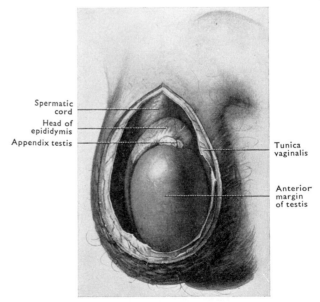

Spermatic cord
Head of epididymis
Appendix testis
Tunica vaginalis
Anterior margin of testis

FIG. 640 Right testis and epididymis, exposed by removal of anterior wall of scrotum.

that clasps the posterior margin of the testis and to some extent overlaps the posterior part of its lateral surface [FIG. 640]. The upper, larger part is called the **head** of the epididymis and overhangs the upper extremity of the testis, to which it is directly connected by numerous **efferent ductules,** by fibro-areolar tissue, and by the serous covering of the organ. The inferior and smaller part is termed the **tail** of the epididymis; it is attached by loose areolar tissue and by the serous covering to the lower extremity of the testis. The intervening part, or **body** of the epididymis, is applied against the posterior part of the lateral surface of the testis but is separated from it by a slit-like recess of their serous covering termed the **sinus of the epididymis.**

The epididymis is composed of an irregularly twisted tube, called the **duct of the epididymis** [FIGS. 643 and 644].

Minute sessile or pedunculated bodies are often found attached to the head of the epididymis and testis, and have a developmental interest. The **appendix of the testis** [FIG. 641] is a minute body which lies on the upper end of the testis and develops from the free end of the paramesonephric duct of the embryo; it is usually sessile. The **appendix of the epididymis** is attached to the head of the epididymis and is believed to represent a remnant of the cephalic end of the mesonephric duct. Other vestigial remains lying along the posterior margin of the testis, the **ductuli aberrantes** and the **paradidymis,** are believed to be derived from some of the mesonephric tubules.

Tunica Vaginalis. The walls of the cavity within which the testis and epididymis are placed are lined with a serous membrane—the tunica vaginalis—which resembles in appearance

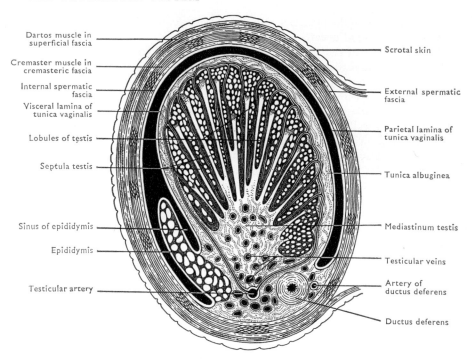

FIG. 642　Diagrammatic transverse section through the testis and scrotum. (After Harrison, 1963.)

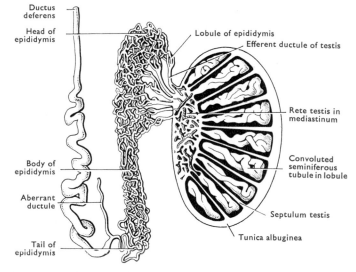

FIG. 643　Diagram to illustrate structure of testis and epididymis.

and structure the peritoneum, from which it was originally derived, but later separated. The cavity tapers to a point or ends blindly as it is traced upwards. The tunica vaginalis is an invaginated membrane doubled in on itself from behind, for, after lining the scrotal chamber, it is reflected forwards from the posterior wall to cover the testis, epididymis, and lower part of the spermatic cord immediately above the testis. The inner layer is known as the visceral layer and is closely applied to the enclosed organs; the outer or parietal layer is fairly strongly attached to the wall of the scrotum; and between the two layers there is a closed cavity that contains a little serous fluid. The visceral layer dips into the narrow interval between the body of the epididymis and the lateral

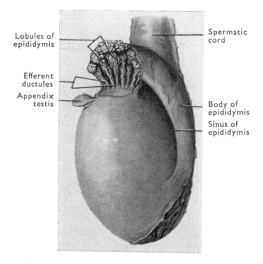

FIG. 641　Left testis and epididymis. Part of the tunica vaginalis has been removed in order to show the efferent ducts and the lobules of the epididymis.

surface of the testis to form the **sinus of the epididymis** [FIGS. 641 and 642]. The posterior margin of the testis is not in contact with the tunica vaginalis, being covered above by the head and below by the tail of the epididymis, and posteriorly by the spermatic cord from which the blood vessels and nerves enter the testis.

Structure of the Testis. Under cover of the tunica vaginalis the testis is invested by an external coat of dense, white, inelastic fibrous tissue called the **tunica albuginea,** from the deep surface of which a number of thin fibrous **septula** dip into the gland. These septula imperfectly divide the organ into a number of wedge-shaped **lobules** [FIG. 642]. All the septula end posteriorly in a mass of fibrous tissue called the **mediastinum testis,** which is directly continuous with the tunica albuginea and projects forwards into the testis along its posterior border. The mediastinum is traversed by an exceedingly complicated network of fine canals called the **rete testis;** the minute tubules that compose the substance proper of the testis open into the rete. The arteries, veins, and lymph vessels enter the posterior border of the testis, traverse the mediastinum, and spread out on the deep surface of the tunica albuginea to form the tunica vasculosa.

The mediastinum, the septula, and the tunica albuginea form a framework enclosing a number of imperfectly isolated spaces which are filled by a substance of a light brown colour called the **parenchyma testis.** The parenchyma is composed of large numbers of **convoluted seminiferous tubules** which look like fine threads to the unaided eye. Each tubule is a continuous loop with its convexity anterior. Its markedly convoluted limbs pass towards the mediastinum, and becoming less convoluted, unite with adjacent tubules to form a smaller number of slender, **straight seminiferous tubules,** which open into the rete testis [FIG. 643]. Usually two to four tubules are found in each lobule of the gland where they are loosely held together by delicate areolar tissue. The total number of tubules in the testis has been estimated at more than 800.

Histological sections show that the walls of the seminiferous tubules are composed of a basement membrane and of an epithelial lining formed of several layers of cells. Outside the basement membrane is a layer of smooth muscle-like cells which cause contractions of the seminiferous tubules; the contractions are thought to promote the transport of spermatozoa and fluid within the tubules towards the rete testis (see Suvanto and Kormano, 1970). In the adult, certain cells of the epithelium, called spermatogonia [p. 20], divide to form cells which mature into **spermatozoa,** and the histological appearance and spermatogenic activity of the tubules vary greatly with the age of the individual.

Structure of the Epididymis. The spermatozoa formed in the convoluted seminiferous tubules are carried through the straight tubules into the rete testis, and leave the rete, to reach the duct of the epididymis, through fifteen to twenty minute tubules called the **efferent ductules of the testis.** These efferent ductules pierce the tunica albuginea and enter the head of the epididymis. Each ductule is at first straight, but soon becomes much convoluted, and forms a little conical mass called a **lobule of the epididymis.** Within the head of the epididymis the duct of each lobule opens into the single, much-convoluted duct which constitutes the chief bulk of the

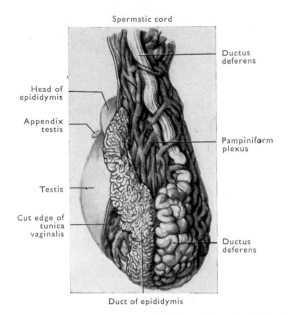

FIG. 644 Left testis and epididymis viewed from behind, showing duct of epididymis and first part of ductus deferens.

epididymis. This **duct of the epididymis,** which is about 600 cm. in length, begins in the head of the epididymis, and ends, after an extraordinarily tortuous course, at the tail by becoming the **ductus deferens** [FIGS. 643 and 644].

The duct has a muscular coat composed of an inner stratum of circular fibres and an outer stratum of longitudinally directed fibres. This muscular coat and its peristaltic activity (Macmillan and Aukland, 1960) are responsible for the passage of spermatozoa along the duct. The wall, at first thin, becomes much thicker as the canal approaches the ductus deferens.

Vessels and Nerves of the Testis

The testis is supplied by the **testicular artery,** a branch of the aorta. It is a slender vessel which, after a long course, reaches the posterior border of the testis, where it breaks up into branches. Harrison and Barclay (1948) found that there are commonly two main branches which pass forwards, one on each side of the organ, to ramify on the deep surface of the tunica albuginea in the **tunica vasculosa.** From this vascular plexus the small terminal arteries pass backwards into the substance of the testis along the septula and converge on the mediastinum. The **artery of the ductus deferens** and the **cremasteric artery** anastomose with the testicular artery (Harrison, 1949).

The **veins** running along the septula, and those of the tunica vasculosa, converge on the posterior border of the testis where they form a dense plexus, called the **pampiniform plexus,** which finally pours its blood through the **testicular vein,** on the right side, into the inferior vena cava, and, on the left side, into the left renal vein. The pampiniform plexus is also drained by the cremasteric vein which forms a venous plexus within and external to the cremasteric fascia (Harrison, 1966).

The **lymph vessels** of the testis pass upwards in the spermatic cord and end in the lymph nodes at the sides of the aorta and inferior vena cava below the renal veins [FIG. 1155].

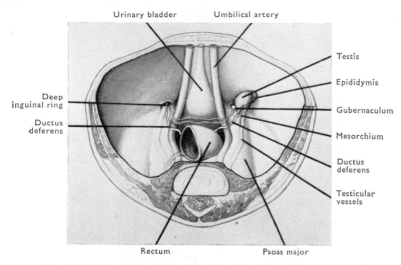

FIG. 645   View from above of cavity of pelvis and lower part of abdomen in male foetus about the 7th month.

On the left side, which represents a slightly more advanced condition than the right, the testis has entered the inguinal canal; on the right side the testis is still within the abdominal cavity.

The **nerves** for the testis and epididymis accompany the artery, and are derived through the **aortic** and **renal plexuses** from the tenth thoracic spinal nerve.

Descent of the Testis

The peculiar course pursued by the ductus deferens in the adult, and the manner in which it is related to the anterior abdominal wall, are explained by the development of the testis. Until nearly the end of intra-uterine life the testes are in the abdominal cavity.   The testis at first lies on the posterior abdominal wall at the level of the upper lumbar vertebrae.   To its lower pole is attached a ridge of tissue called the **gubernaculum testis** (Hunter, 1786) which extends down to the inguinal region and, passing through the abdominal wall, is attached to the skin.   Both testis and gubernaculum lie behind the primitive peritoneum; the gubernaculum thus forms a ridge covered with peritoneum, and the testis—almost completely surrounded by peritoneum—is attached to the posterior abdominal wall by a peritoneal fold called the mesorchium. As the foetus grows and the gubernaculum becomes relatively shorter, it carries the peritoneum on its anterior surface down through the anterior abdominal wall as a blind tube, the processus vaginalis, which thus traverses the inguinal region and reaches the genital swelling (rudimentary scrotum) (Backhouse and Butler, 1960).   As the processus vaginalis descends, the testis is guided by the gubernaculum down the posterior abdominal wall and the back of the processus vaginalis (Wells,

FIG. 646   Diagram to illustrate descent of the testis and origin of tunica vaginalis.

a.c., abdominal cavity; *p.v.*, processus vaginalis; *s.*, scrotum; *t.*, testis; *t.v.*, tunica vaginalis; *x.*, vestige of processus vaginalis.

1943) into the scrotum.   By the third month of foetal life the testis lies in the iliac fossa, and by the seventh it is near the deep inguinal ring [FIGS. 645 and 646]. Normally the communication between the peritoneal cavity and that part of the processus vaginalis surrounding the testis is obliterated, leaving the testis isolated in the lower part of the processus, which is known as the **tunica vaginalis.**

Often a small fibrous band may be found in the adult passing through the inguinal canal and joining the peritoneum superiorly in the region of the deep inguinal ring.   Sometimes the band is connected below with the tunica vaginalis, but more often it cannot be traced so far downwards.   When present it represents the obliterated portion of the processus vaginalis, and is therefore known as the **vestige of the processus vaginalis.**

The processus vaginalis normally closes shortly after birth but occasionally persists into adult life as a channel freely open to the abdominal cavity above; or the passage, becoming closed at intervals, may give rise to one or more cysts within the coats of the spermatic cord.

It sometimes happens that the descent of the testis is arrested, and then it either remains on the posterior abdominal wall or in a persisting upper part of the processus vaginalis in the inguinal canal.   The term **cryptorchism** is applied to such cases of undescended testis; such testes may be incapable of normal spermatogenesis.   On the other hand the testis may be **ectopic,** through irregular descent into the perineum or even into the upper part of the thigh.   In its descent the testis takes with it its ducts and vessels and nerves, which together form the **spermatic cord.**

In some mammals, such as the elephant, the testes remain permanently within the abdominal cavity; in others, such as the hedgehog, the peritoneal pouches remain widely open throughout life, and the testes are periodically withdrawn into the abdomen.   The descent of the mammalian testes ensures an intratesticular temperature lower than that in the abdomen (Harrison and Weiner, 1949).   The testes of normal children may frequently be retracted into the inguinal canal by the action of the cremaster muscle, particularly in cold weather, or after stroking the skin on the medial surface of the thigh (cremasteric reflex).

THE DUCTUS DEFERENS AND SPERMATIC CORD

The ductus (*vas*) deferens is the direct continuation of the duct of the epididymis.   It begins at the lower end of the epididymis and ends, after a course of nearly 45 cm., by joining the duct of the seminal vesicle to form the ejaculatory duct, which opens into the prostatic part of the urethra; in parts of its course it is slightly convoluted, and the actual distance traversed is not more than 30 cm.

In the scrotum, the ductus deferens ascends along the back of the testis on the medial side of the epididymis, and it is here that it is most convoluted [FIG. 644].   At the upper extremity of the testis it meets the vessels and nerves of the

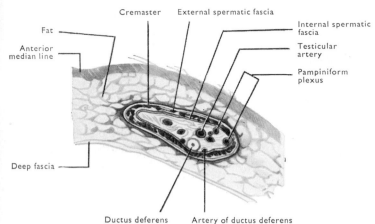

FIG. 647 Transverse section of spermatic cord immediately below superficial inguinal ring.

testis and epididymis, and is bound together with them by areolar tissue in a loose bundle called the spermatic cord.

The **spermatic cord** is composed of: (1) the ductus deferens and its own artery and vein; (2) the testicular artery and the pampiniform plexus of veins; (3) the lymph vessels and nerves of the testis and epididymis; and (4) remnants of the processus vaginalis—the plexus of veins being by far the bulkiest of these constituents. The cord extends from the testis to the deep inguinal ring, and it is enclosed in three tubular sheaths or **coats**—the external spermatic fascia, the cremasteric muscle and fascia, and the internal spermatic fascia

[FIG. 647]. When these coats reach the scrotum they expand and take a share in the formation of the wall of the scrotum; and the cremasteric coat carries with it its blood supply and nerve supply—the **cremasteric artery, vein,** and the **genital branch** of the **genitofemoral nerve.**

Between the scrotum and the superficial inguinal ring, the cord lies on the deep fascia of the muscles that spring from the pubis, and it is crossed by the superficial external pudendal vessels. In this part of the course the ductus is in the posterior part of the cord, and is easily distinguished from the other constituents by its hard, firm feel when the cord is gripped between the finger and thumb. When the cord enters the inguinal canal it lies on the lateral crus of the superficial ring immediately lateral to the pubic tubercle; and here it loses its external spermatic coat, which blends with the external oblique aponeurosis at the margins of the ring [FIG. 648].

During its passage through the inguinal canal, the ductus lies in the lower part of the cord on the grooved, upper surfaces of the lacunar and inguinal ligaments. The external oblique aponeurosis is anterior throughout and the lower fleshy fibres of the internal oblique are also in front in the lateral half of the canal. The conjoint tendon and the transversalis fascia are behind. Half-way along the canal the cord loses its cremasteric coat, since this is derived from the lower border of the internal oblique muscle, which arches over the cord to join transversus abdominis and form the conjoint tendon.

1·3 cm. above the midinguinal point, at the lateral side of the inferior epigastric artery, the cord reaches the deep inguinal ring. Here, the cord loses its internal spermatic coat which blends with the transversalis fascia at the margins of

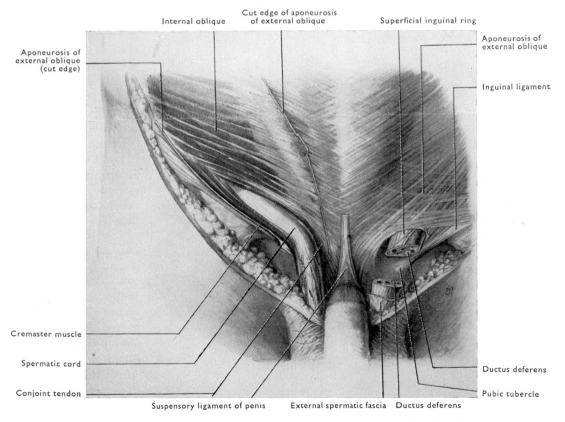

FIG. 648 Dissection to show the spermatic cord as it issues from the superficial inguinal ring. On the right side the external oblique muscle has been removed.

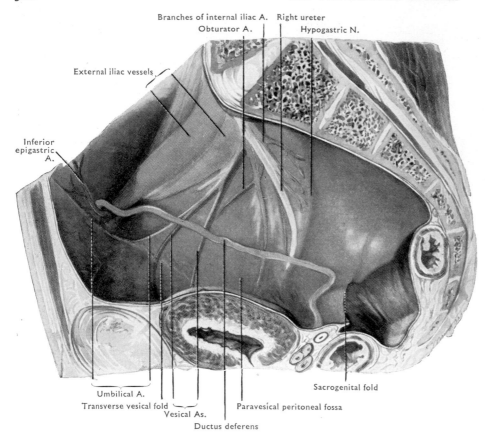

FIG. 649   Median section of adult male pelvis to show course of ductus deferens on side-wall of pelvic cavity.

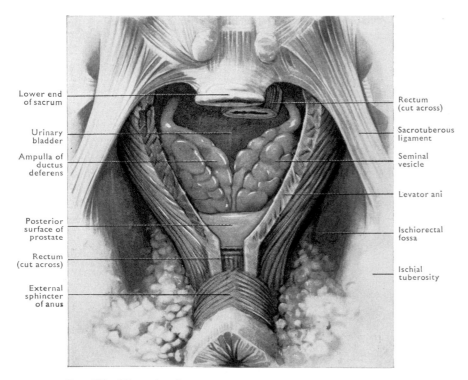

FIG. 650   Dissection from behind to display the seminal vesicles, the ampullae of the ductus deferentes, and the prostate.

the ring. Here also the cord ends, for at this point the ductus deferens parts company with the vessels and nerves.

The ductus deferens now curves medially across the external iliac artery behind the root of the inferior epigastric artery, and then, changing its direction, it runs for a short distance backwards, medially, and upwards, immediately outside the peritoneum, to a point 4–5 cm. from the pubic tubercle, where it crosses the linea terminalis and enters the lesser pelvis. In that part of its course the ductus usually lies at first in front of the external iliac vessels, and then in the floor of a small triangular fossa—the **pectineal triangle**—which lies between the external iliac vessels and the linea terminalis [FIG. 632]. On the side-wall of the pelvis the ductus continues backwards, and a little downwards and medially, towards the ischial spine, and lies immediately external to the peritoneum, through which it can usually be seen. In the pelvic part of its course the ductus crosses the medial side of: (1) the umbilical artery; (2) the obturator nerve and vessels; (3) the vesical vessels; and (4) the ureter [FIG. 649].

Crossing the ureter the ductus deferens hooks round it and passes downwards and medially outside the peritoneum. There, the ductus lies a short distance behind the terminal part of the ureter and immediately in front of the sacrogenital fold of peritoneum and upper end of the corresponding seminal vesicle [FIGS. 632 and 650]. Reaching the interval between the base of the bladder in front and the rectum behind, the two ductus deferentes occupy the angle between the right and left seminal vesicles [FIG. 650]. As they approach one another each ductus becomes slightly tortuous, sacculated, and dilated, and assumes a general resemblance in structure to a portion of the seminal vesicle. The dilated part of the ductus is termed its **ampulla.** Immediately above the base of the prostate the ductus deferens becomes once more a slender tube and is joined by the duct of the corresponding seminal vesicle to form the **ejaculatory duct** [FIG. 651].

In some cases the ductus deferens crosses the obliterated umbilical artery

PLATE 39

A

B

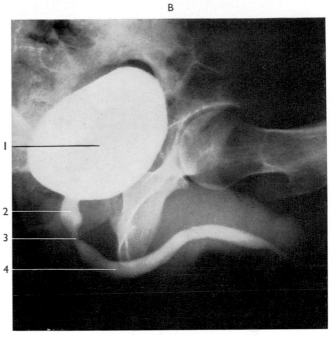

A. Urinary bladder containing contrast medium. The pyramidal shape of the bladder in the partly distended state is seen.

B. Cysto-urethrogram.

1. Bladder.
2. Prostatic part of urethra.
3. Membranous part of urethra.
4. Spongy part of urethra.

D
(i)

(ii)

C

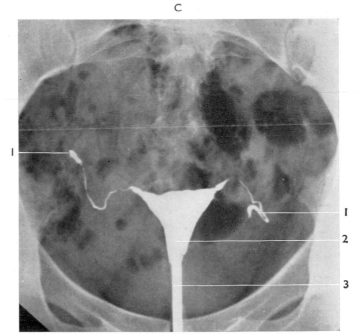

D. (i) X-ray of the uterus and uterine tubes after filling with a water-soluble contrast medium, part of which has escaped into the peritoneal cavity. The angle between the uterus and vagina has been straightened by traction on the cervix and the uterus lies vertically.

    1. Cavity of body of uterus.
    2. Recto-uterine pouch of peritoneum, lateral part.
    3. Uterine tube.
    4. Cervical canal of uterus.

   (ii) Late appearances of a spill of oily contrast medium from the uterine tubes. The peritoneal surfaces of the pelvis are outlined.

    1. Para-uterine pouch of peritoneum.
    2. Uterovesical pouch seen end on.
    3. Peritoneum on superior surface of uterus.

C. Cavity of uterus and uterine tubes demonstrated by injection of contrast medium. The uterus has been turned backwards.

    1. Uterine tubes.
    2. Uterine cavity.
    3. Cannula in vagina for introduction of X-ray opaque material.

Seminal
vesicle

Ampulla
of ductus
deferens

Ejaculatory
duct

Ampulla of
ductus deferens

Seminal vesicle   Ejaculatory duct

FIG. 651 Drawings of seminal vesicle and ampulla of ductus deferens taken from different subjects.

before it enters the cavity of the lesser pelvis ; it normally does so in the foetus [FIG. 645].

Ejaculatory Duct. This is the very slender tube formed by the union of the ductus deferens with the duct of the corresponding seminal vesicle [FIG. 651]. It is less than 2·5 cm. in length, and lies very close to its fellow of the opposite side as it passes downwards and forwards through the prostate behind its median lobe. The ducts open by slit-like apertures into the first part of the urethra, one on each side of the mouth of the **prostatic utricle.** They are well seen in sections through the upper part of the prostate [FIG. 657].

THE SEMINAL VESICLES

The seminal vesicles are a pair of hollow sacculated structures placed on the base of the bladder in front of the rectum [FIGS. 630 and 650]. Each vesicle is about 5 cm. in length, and has its long axis directed downwards, medially, and

slightly forwards. The upper end of the vesicle, which is partly covered with peritoneum, is large and rounded, and lies at a considerable distance from the median plane, behind the lower end of the ureter, and is separated from the rectum by the peritoneum of the rectovesical pouch. Below the level of the peritoneal cavity the seminal vesicle and rectum are more intimately related, but are still separated by a partition of smooth muscle fibres and fascia which is thicker than that separating the seminal vesicles from the bladder. The vesicle tapers towards its lower end, which is placed not far from the median plane immediately above the prostate. In this position the vesicle becomes constricted to form a short duct which joins the lateral side of the corresponding ductus deferens at an acute angle. The common duct thus formed is the **ejaculatory duct.** The medial side of each vesicle is related to the ductus deferens, and the lateral side, when the bladder is empty, lies close to the levator ani. The seminal vesicles are more intimately related to the bladder than to the rectum.

Their mode of development shows that the seminal vesicles are diverticula of the ductus deferentes, from which they originally arise as small, pouched outgrowths. The dense tissue in which the seminal vesicles are embedded contains much plain muscle tissue which sweeps round in the side-wall of the rectovesical pouch. Inferiorly this tissue is attached to the capsule of the prostate. The large veins draining the prostatic and vesical plexuses are closely related to the seminal vesicles.

Structure of the Ductus Deferens and Seminal Vesicle

Except at the ampulla, the ductus deferens is a thick-walled tube with a relatively very small lumen. The hard cord-like sensation which it conveys to the touch is due to the thickness and denseness of its wall. The wall of the ductus is composed of three layers—an outer **adventitious coat** of fibrous tissue, a middle **muscular coat,** and an inner **mucous coat.** The thickness of the wall is due to the great development of its coat of plain muscle, which is arranged in three layers, an outer and an inner of longitudinal fibres, and a middle layer—by far the thickest—of circular fibres. The mucous membrane of the ductus exhibits a number of slight longitudinal folds and is lined with a ciliated epithelium. The ampulla has a much thinner wall, and its mucous membrane is thrown into ridges so that it presents a honeycomb appearance. The wall of the seminal vesicle resembles that of the ampulla in being thin, and in having a mucous lining with uneven, honeycomb-like ridges.

Vessels and Nerves of the Ductus Deferens and Seminal Vesicle

The ductus receives its main **artery to the ductus** from the inferior vesical artery. This artery supplies the ductus, the seminal vesicle, the lower ureter, and the bladder, and could be called the vesiculodeferential artery (Braithwaite, 1952). The artery to the ductus accompanies it as far as the testis, where it ends by anastomosing with branches of the testicular artery. The **nerves** to the ductus seem to come mainly from the sympathetic through the hypogastric plexuses, and each of its muscle fibres appears to have a separate innervation (Richardson, 1962)—a unique arrangement in smooth muscle. The preganglionic sympathetic nerve fibres to the seminal vesicles emerge in the upper lumbar nerves, the parasympathetic in the pelvic splanchnics (S.2, 3, 4), and both traverse the inferior hypogastric plexus.

THE SCROTUM

The scrotum varies much in appearance in different subjects, and even in the same person at different times. As the result of cold or of exercise, the wall of the scrotum becomes contracted and firm, and its skin becomes wrinkled; at other times the wall may be relaxed and flaccid, the scrotum then assuming the appearance of a pendulous bag. The left side of the scrotum reaches to a lower level than the right, in correspondence with the lower level of the left testis. The skin of the scrotum is of a darker colour than the general skin of the body and is covered sparsely with hair; it is marked in the median plane by a ridge called the **raphe** of the scrotum, which is continued backwards towards the anus and forwards on the urethral surface of the penis. The difference in the appearance of the scrotum at different times is due to the degree of contraction or relaxation of a layer of plain muscular fibres, the **dartos,** situated in the superficial fascia. The contraction of this muscular layer is responsible for the scrotum becoming smaller, and the wrinkling of its skin. The layer of fascia containing the dartos is continuous superiorly with the superficial fascia of the penis, and with the deep layer of the superficial fascia of the abdomen, and is attached laterally to the sides of the pubic arch. The muscle fibres are arranged in a thick layer of interlacing bundles, and many of the deeper fibres are continued into the **septum** of the scrotum, which divides it into a pair of chambers, one for each testis. The wall of each chamber is formed by the fusion of the corresponding tunica vaginalis, internal spermatic fascia, cremasteric muscle and fascia and external spermatic fascia, while the skin, the superficial fascia, and the dartos muscle form coverings which are common to the whole scrotum, and they enclose both chambers. The layer of tissue immediately internal to the dartos is made up of exceedingly loose and easily stretched areolar tissue; throughout it and the superficial fascia of the scrotum there is no fat.

The scrotum in the foetus has no cavity until the processus vaginalis enters it, but, like the labia majora which correspond to it in the female, it is composed entirely of vascular and fatty areolar tissue.

Vessels and Nerves of the Scrotum

On each side there are **posterior scrotal branches** from the internal pudendal artery, which reach it from behind, and **anterior scrotal branches** from the external pudendal arteries, which supply its upper and anterior part.

The **nerves** of the scrotum are derived on each side from the posterior scrotal branches of the pudendal nerve, from the perineal branch of the posterior cutaneous nerve of the thigh, and from the ilio-inguinal nerve. The branches from the pudendal and posterior cutaneous nerves reach the scrotum from behind, while the ilio-inguinal supplies its upper and anterior part.

THE PENIS

The penis [Figs. 627 and 652] is composed chiefly of cavernous (erectile) tissue, and is traversed by the urethra. Engorge-ment of the cavernous tissue with blood produces considerable enlargement of the penis and its erection. The posterior surface of the flaccid penis is nearest the urethra and is called the **urethral surface;** the opposite and more extensive aspect is the **dorsum.** The erectile tissue is for the most part disposed in three longitudinal columns, which in the body of the organ are closely united, while at the root of the penis they separate from one another and become attached to the perineal membrane and the sides of the pubic arch. Two of the columns of erectile tissue are placed alongside the median plane; they form the dorsum and sides of the penis, and are called the **corpora cavernosa** of the penis. The third column is situated in the median plane near the urethral surface, and is called the **corpus spongiosum** of the penis. The corpus spongiosum is the part of the penis traversed by the urethra, and it is considerably smaller than the corpora cavernosa, which form the chief bulk of the organ [Figs. 653 and 654].

In the **body** of the penis each corpus cavernosum presents a rounded surface, except where it is flattened by contact with its fellow of the opposite side. They are separated on the dorsal surface by a shallow groove, and on the urethral aspect by a deeper and wider furrow in which the corpus spongiosum lies [Fig. 653]. Towards the distal end of the penis the corpus spongiosum expands, and, spreading towards the dorsal surface, forms a cap—the **glans penis**—which covers the conical end of the united corpora cavernosa [Fig. 655]. The prominent margin of the glans, called the **corona,** projects backwards and laterally over the ends of the corpora cavernosa. The groove so formed between the corona and the ends of the corpora cavernosa is the neck of the glans. The glans is traversed by the terminal part of the urethra, which ends near the summit of the glans in a vertical slit-like opening—the **external urethral orifice.** The united corpora cavernosa end in a blunt conical extremity, the apex of which fits into a hollow in the base of the glans. The skin of the body of the penis is thin, delicate, and freely movable, and, except near the pubis, is free from hairs; on the urethral aspect the skin is marked by a median raphe, continuous with the raphe of the scrotum. Traced towards the base of the glans, the skin forms a free fold—the **prepuce** or foreskin—which overlaps the glans to a variable extent. From the deep surface of the prepuce the skin is reflected on to the neck of the glans, and is continued over the entire glans to the external urethral orifice [Fig. 627]. A small median fold—the **frenulum of the prepuce**—passes to the deep surface of the prepuce from a point immediately below the external urethral orifice. The skin of the glans is firmly attached to the underlying erectile tissue, and here, as well as on the deep surface of the prepuce, it has some resemblance to mucous membrane.

The secretion which tends to collect beneath the prepuce is known as the **smegma,** and has its source in the desquamated and broken-down epithelial cells derived from the surface of the glans and prepuce, and the secretion of preputial glands. A long, narrow prepuce, which cannot be retracted from the glans and may be adherent to it, may have to be removed by the operation of **circumcision.**

At the **root of the penis** its three corpora separate from one another [Fig. 653]. The corpora cavernosa, diverging sidewards, at first become slightly swollen, and then, gradually tapering, gain a firm, fibrous attachment to the periosteum

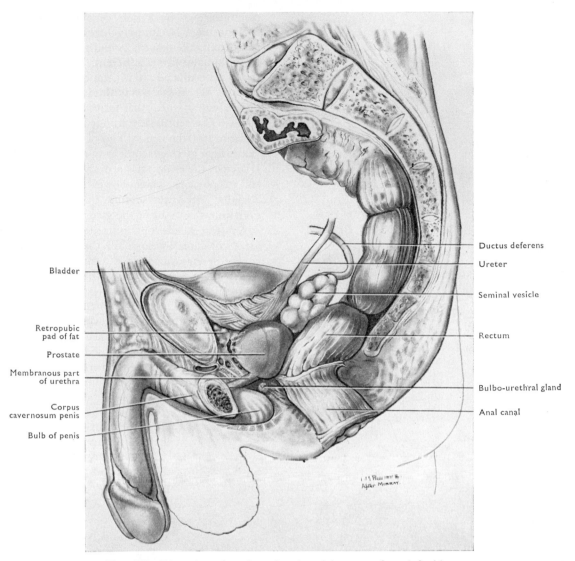

Bladder

Retropubic
pad of fat

Prostate

Membranous part
of urethra

Corpus
cavernosum penis

Bulb of penis

Ductus deferens

Ureter

Seminal vesicle

Rectum

Bulbo-urethral gland

Anal canal

FIG. 652 Dissection of penis and male pelvic organs from left side.

on the medial surface of the sides of the pubic arch. These diverging parts of the corpora cavernosa are the **crura** of the penis and each crus is covered by the corresponding ischiocavernosus muscle. The corpus spongiosum, lying between the crura, is enlarged to form a globular mass, the **bulb** of the penis. The bulb is variable in size and is attached to the inferior surface of the perineal membrane. Postero-inferiorly the bulb has a median notch, and deep to this a poorly developed, median fibrous septum—both features which indicate its development by the fusion of symmetrical halves; the cutaneous fusion being marked by the overlying raphe. The urethra, having pierced the perineal membrane, enters the bulb obliquely a short distance in front of its posterior end [FIGS. 430 and 652]. The superficial surface of the bulb is covered by the bulbospongiosus muscles.

A triangular band of strong fibrous tissue, the **suspensory ligament of the penis,** is attached to the front of the pubic symphysis and fuses with the fascial sheath of the penis [FIG. 648]. In addition the penis is suspended by the **fundi-form ligament** which extends downwards from the rectus sheath and linea alba to loop around the penis.

Structure of the Penis. Each **corpus cavernosum** penis is enclosed in a dense, white fibrous capsule (the tunica albuginea) which fuses with that of the opposite side to form the median **septum** of the penis. The septum is very incomplete, especially near the end of the penis, where it is interrupted by a number of nearly parallel slits [FIGS. 654 and 655]. Through these slits the erectile tissue of the two corpora cavernosa is continuous.

The **tunica albuginea** contains some elastic fibres; numerous fibrous strands, the **trabeculae,** proceed from its deep surface and, stretching across the interior of the corpus cavernosum, form a fine sponge-like framework whose interstices, the cavernous venous spaces, communicate freely with one another and are filled with blood. These spaces lead directly into the veins of the penis, and, like the veins, have a lining of flat endothelial cells. The size of the penis varies with the amount of blood in the erectile tissue. The structure of the corpus spongiosum resembles that of the corpora cavernosa, but its tunica albuginea is much thinner and more elastic, and the spongework is finer [FIG. 654].

The **glans penis** is also composed of erectile tissue which

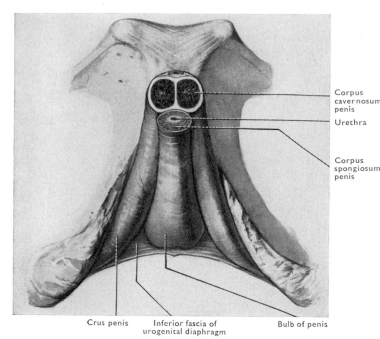

FIG. 653   Root of penis. The body of the penis is seen in section.

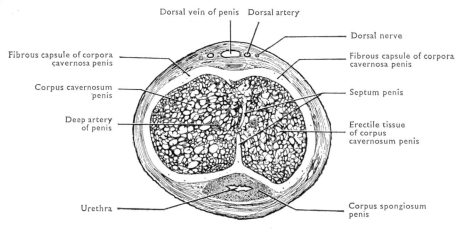

FIG. 654   Transverse section through anterior part of body of penis to show its structure.

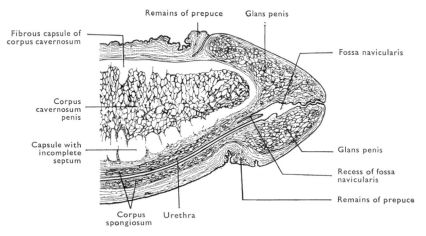

FIG. 655   Median section through terminal part of circumcised penis to show its structure.

communicates by a rich venous plexus with the corpus spongiosum. No strongly marked fibrous capsule is present, and the erectile tissue is bound to the skin, which is exceedingly thin and firmly adherent. The urethra in this part of the penis is dilated and is compressed laterally into a slit-like passage, the **fossa navicularis,** surrounded by a mass of fibro-elastic tissue which forms a median septum within the glans. This septum is continued backwards to join the tunica albuginea of the conical end of the corpora cavernosa, and ventrally it gives attachment to the frenulum of the prepuce. It divides the erectile tissue of the glans imperfectly into right and left portions, which, however, are in free communication dorsally. From the septum, trabeculae pass out in all directions into the tissue of the glans.

A fascial sheath, containing numerous elastic fibres, forms a loose common envelope for the corpora cavernosa and the corpus spongiosum. It is termed the **deep fascia of the penis,** and reaches as far as the base of the glans, where it becomes fixed to the neck of the glans. In its proximal part the sheath gives insertion to many of the fibres of the bulbo-spongiosus and ischiocavernosus muscles.

Superficial to the deep fascia of the penis is the **superficial fascia of the penis** composed of a layer of extremely lax areolar tissue, and more superficial still a prolongation of the dartos tunic of the scrotum, which lies underneath the delicate skin of the penis. This fascia contains no fat. Numerous sebaceous glands are present in the skin, especially on the urethral surface of the penis.

In some mammals, such as the walrus, dog, bear, and baboon, a bone called the **os penis** is developed in the septum between the corpora cavernosa penis.

Vessels and Nerves of the Penis

The arteries are derived from the internal pudendal artery. The erectile tissue of the corpora cavernosa is supplied chiefly by the **deep arteries** of the penis, and that of the corpus spongiosum by the **arteries to the bulb.** Branches of the **dorsal arteries** of the penis pierce the tunica albuginea of the corpora cavernosa and furnish additional twigs to the erectile tissue. The glans receives its chief blood supply from branches of the dorsal arteries. The small branches of the arteries run in the trabeculae of the erectile tissue, and the capillaries into which they lead, open directly into the cavernous spaces. As they lie in the finer trabeculae, the smaller branches often become tortuous, and the name **helicine arteries** is therefore applied to them.

The **veins** with which the cavernous spaces communicate carry the blood either directly into the **prostatic plexus,** or into the **dorsal vein** and so to that plexus. The dorsal vein of the penis begins in tributaries from the glans and prepuce, ascends in the groove between the corpora cavernosa, and

FIG. 656 Dorsal surface of penis showing the main blood vessels and nerves.

Glans penis

Dorsal vein

Dorsal artery

Dorsal nerve

Deep artery entering corpus cavernosum

Urethra

Crus penis

Internal pudendal artery

Artery of bulb   Bulb of penis

passes beneath the arcuate pubic ligament to join the prostatic plexus. On each side of it lies a dorsal artery, and, still further from the median plane, a dorsal nerve [FIGS. 654 and 656]. The **superficial dorsal vein** of the penis is either unpaired or double and runs on the deep fascia of the penis towards the symphysis pubis where it bends laterally to drain into the external pudendal vein.

The **lymph vessels** of the penis are arranged in a deep and superficial series, and they end in the medial groups of the **superficial inguinal lymph nodes.**

The **nerve supply** of the penis is derived from the pudendal nerve (second, third, and fourth sacral nerves), and from the pelvic sympathetic plexuses. The branches of the pudendal are the **dorsal nerve of the penis,** and branches from the perineal nerves. They supply the cutaneous structures of the penis, while the filaments from the **hypogastric plexuses,** which reach the penis through the prostatic nerve plexus, end in the erectile tissue.

THE PROSTATE

The prostate is a partly glandular, partly muscular organ surrounding the beginning of the urethra in the male. It lies within the pelvis behind the pubic symphysis and is enclosed by a dense fascial sheath. Through the various connexions of this sheath the prostate is firmly fixed within the pelvic cavity. The ejaculatory ducts traverse the upper, posterior part of the prostate in their course to join the urethra. The size of the prostate varies considerably, but usually its greatest transverse diameter is 4 cm., its anteroposterior diameter is 2 cm., and its vertical diameter 3 cm. Superficially, the prostate is separated from the bladder by deep, wide lateral grooves and by a narrow posterior groove [FIGS. 630 and 652].

The prostate has an apex which is directed downwards, a base superiorly, a posterior surface, and a pair of inferolateral surfaces. The **base** adjoins the inferior aspect of the bladder. in the neighbourhood of its urethral opening. The greater part of the base is structurally continuous with the bladder wall; only a narrow portion remains free on each side and forms the lower limit of the deep groove which marks the separation of the bladder and prostate [FIG. 630]. The **inferolateral surfaces** of the prostate are convex and prominent, and rest against the fascia covering the levator ani muscles. They also face slightly forwards, and meet in front in a rounded anterior surface. The **posterior surface** is flat and triangular, and faces backwards and very slightly downwards and is in contact with the urogenital diaphragm. From it may be felt in the living subject. The apex points downwards and is in contact with the urogenital diaphragm. From the apex, the anterior surface passes upwards in the median plane behind the pubic symphysis and retropubic pad of fat.

The urethra enters the prostate at a point near the middle of its base, and leaves it at a point on its anterior surface immediately above and in front of the apex.

The **ejaculatory ducts** enter a slit immediately in front of the posterior border of the base, in the groove between the bladder and prostate, and run downwards and forwards to open into the prostatic portion of the urethra on each side of the mouth of the prostatic utricle.

The wedge-shaped portion of the prostate which separates these ducts from the urethra is called the **median lobe** [FIG. 627]. It projects upwards against the bladder, and is continuous with the bladder wall immediately behind the internal urethral orifice. When hypertrophied, as it often is in old men, the median lobe of the prostate may cause an elevation in the cavity of the bladder, which may obstruct the flow of urine, and to which the term **uvula of the bladder** is applied. The remaining part of the prostate is described as a pair of large **lateral lobes,** which are, however, not marked off from one another superficially.

In front of the prostate, between it and the pubis, there is a close venous network called the **prostatic plexus,** which receives the dorsal vein of the penis. This plexus is continuous round the sides of the prostate with the large, thin-walled veins which lie principally in the deep sulcus between the bladder and the prostate, and communicate with the **vesical plexus.** Most of the veins of the plexus lie embedded in the fascial sheath of the prostate [FIG. 657].

Fascial Sheath of the Prostate. This sheath is a dense fibrous portion of the pelvic fascia, and closely invests the prostate. Inferiorly, the sheath becomes continuous with the superior fascia of the urogenital diaphragm and, through it, gains attachment to the sides of the pubic arch. In front and at the sides, it is fused with the puboprostatic ligaments, by which it is connected with the pubic bones and the fascia on the levatores ani. Between the puboprostatic ligaments of the two sides there is a shallow depression, the floor of which is formed by a thin layer of fascia which connects the anterior part of the sheath of the prostate with the back of the pubic symphysis. The medial edges of the levator ani muscles are immediately below the puboprostatic ligaments, and, when followed backwards, they are seen to embrace the lower part of the prostate to form the **levator prostatae.** The lower fibres of the pubovesical muscle [p. 520], which follow the

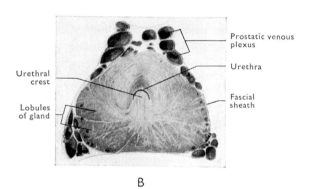

FIG. 657   Horizontal sections through prostate.   Section A is at a higher level than B.

course of the puboprostatic ligament, constitute the pubo-prostatic muscle.

Posteriorly the upward prolongation of the sheath is continuous with the fascial layers which enclose the ampullae of the ductus deferentes and the seminal vesicles, and it is adherent to the peritoneum of the rectovesical pouch. In this position it is spoken of as the **rectovesical septum.**

Structure of the Prostate.   Inside the fascial sheath the superficial part of the prostate is largely composed of interlacing bundles of plain muscle fibres and fibrous tissue which form the capsule of the organ.   The **capsule** is not sharply defined, for from its deep surface fibrous and muscular strands pass inwards, converging towards the posterior wall of the urethra, to become continuous with the mass of plain muscular tissue which surrounds this canal.   These radially arranged strands divide the prostate into a number of incompletely defined lobules, of which there are about fifty.   The yellowish-coloured glandular tissue of the lobules is composed of minute, slightly branched tubules, the walls of which in certain places show numerous saccular dilatations.   In the upper portion of the prostate the tubules are more convoluted, slightly dilated, and shorter than in the lower part.   The glandular tubules lead into twenty or thirty minute **prostatic ducts** which open for the most part into the prostatic sinuses [p. 521] on each side of the urethral crest in the posterior wall of the urethra.

The bulk of the glandular tissue is situated at the sides of the urethra and behind it.   In front of its upper part there is a mass of plain muscular fibres which is continued upwards and backwards on its sides to form a part of the sphincter vesicae.   Below the level of the opening of the ejaculatory ducts and prostatic utricle striated muscular tissue, which is continuous with the deep part of the sphincter urethrae, occupies a position in front of the urethra.   Behind the urethra is a layer of transversely arranged smooth muscle which is thickest above and in front of the ejaculatory ducts (Clegg, 1957).

The muscular tissue of the prostate is to be regarded as the thickened muscular layer of the wall of the urethra broken up and invaded by the prostatic glands, which arise and are developed from the lining layer of that canal during foetal life.   The large amount of plain muscle in the ductus deferens, prostate, prostatic urethra, and seminal vesicles is concerned with ejaculation.

Small at birth, the prostate enlarges rapidly at puberty.   After the fourth decade the glandular tissue may atrophy, and the organ becomes more fibrous and diminishes in size.   In many individuals, however, the prostate progressively enlarges (Swyer, 1944).

Vessels and Nerves of the Prostate

The **arteries** are constant branches of a trunk of variable origin.   This trunk, the prostatovesical artery (Clegg, 1955), divides into two terminal branches, the prostatic and inferior vesical arteries.   It arises most usually from the common origin of the internal pudendal and inferior gluteal arteries from the internal iliac artery, but may take origin either from the umbilical artery or the vesiculodeferential artery.   Within the prostate the intrinsic vessels form an outer (capsular) plexus, an intermediate zone of vessels, and a urethral plexus (Clegg, 1956), and are tortuous.   The **veins** are wide and thin-walled; they form the prostatic plexus, which communicates with the vesical plexus and is drained into the internal iliac veins.   The prostatic plexus communicates with the vertebral venous plexuses; a tumour in the prostate may therefore give rise to a secondary growth in the vertebral column.   In old men the veins of the prostate usually become enlarged.   The **lymph vessels** [FIG. 1155] are associated with those of the seminal vesicle and the neck of the bladder.   The **nerves** of the prostate are derived from the inferior hypogastric plexuses.

THE BULBO-URETHRAL GLANDS

The bulbo-urethral glands are a pair of small bodies placed behind the membranous part of the urethra among the fibres of the sphincter urethrae.   In old age they are often difficult to find without microscopical examination; in young adults they are each about the size of a pea and are of a yellowish-brown colour.   Placed within the urogenital diaphragm, deep to the perineal membrane, the glands lie below the level of the apex of the prostate and above that of the bulb of the penis [FIGS. 639 and 652].   Each gland is made up of a number of closely applied lobules, and is of the compound racemose type.   The ductules of the gland unite to form a single **duct** which pierces the perineal membrane and enters the bulb of the penis.   After a course of about 2·5 cm., it ends by opening into the inferior surface of the spongy portion of the urethra by a minute aperture. The secreting alveoli are lined with columnar epithelium, whose secretion is mostly mucous in character.

The glands receive their arterial supply from the arteries to the bulb.

METHODS OF CLINICAL ANATOMICAL EXAMINATION

The testis, epididymis, and lower end of the spermatic cord can be palpated easily in the living subject; and the back of the prostate and lower ends of the seminal vesicles and ductus deferentes can be palpated by rectal examination.

THE FEMALE GENITAL ORGANS

The female genital organs are: (1) the ovaries; (2) the uterine tubes; (3) the uterus; (4) the vagina; (5) the external genital organs; (6) the greater vestibular glands; and (7) the mammary glands.

The reproductive glands (gonads) in the female [FIG. 634] are a pair of **ovaries** placed one on each side in the cavity of the lesser pelvis. In connexion with each ovary there is an elongated tube—the **uterine tube**—which leads to the uterus and opens into its cavity. There is no direct continuity between the ovary and the uterine tube, such as exists between the other glands of the body and their ducts, but the ova, when shed from the ovary, pass into the peritoneal cavity and thence into the pelvic opening of the tube, whence they are transported to the uterine cavity. The **uterus** is a hollow muscular organ which occupies a nearly median position in the lesser pelvis. It is joined by the uterine tubes anterosuperiorly, and it communicates with the vagina postero-inferiorly. The ovum, having passed through the tube, reaches the cavity of the uterus, and undergoes development there if it has been fertilized. The **vagina** is the passage which leads from the uterus to the exterior, and has its external opening behind that of the urethra, within the **pudendal cleft**. In relation to the cleft there are a number of structures, included under the term **external genital organs**. They are the labia majora and the mons pubis, the labia minora, the clitoris, and the bulbs of the vestibule. The greater vestibular glands, placed one on each side of the lower part of the vagina, are accessory organs of the female reproductive system.

THE OVARIES

The ovaries are a pair of solid, flattened ovoids, 2·5–3·5 cm. long, 1·5 cm. broad, and 1 cm. thick in the multiparous

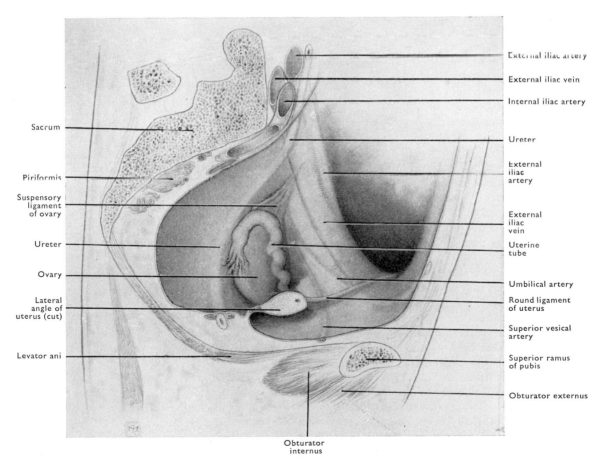

FIG. 658 Side-wall of female pelvis, showing position of the ovary and its relation to the uterine tube.

The pelvis was cut parallel to the median plane, but at some distance from it.

woman. In the adult the ovary is placed against the side-wall of the pelvic cavity, and is connected by a peritoneal fold with the broad ligament of the uterus [p. 539].

The ovary has two extremities—upper and lower. The upper extremity is the larger and more rounded; it is termed the **tubal extremity,** as it is most intimately connected with the uterine tube. The lower end is more pointed; it is called the **uterine extremity,** since it is connected with the uterus by a fibrous cord termed the **ligament of the ovary.** The surfaces of the ovary are called medial and lateral, and the borders separating them are the anterior or **mesovarian border** and the posterior or **free border.** The free border is convex; the mesovarian border is straighter and narrower, and it is connected by two very short peritoneal layers, called the **mesovarium,** with the upper or posterior layer of the broad ligament of the uterus. The vessels and nerves enter the ovary at the mesovarian border, which is therefore termed its **hilus.**

Position and Relations of the Ovary [FIGS. 634 and 658]. In the nulliparous woman the long axis of the ovary is vertical. Its lateral surface lies against the wall of the lesser pelvis, and its medial surface faces the pelvic cavity. The peritoneum of the pelvic wall, where the ovary lies against it, is depressed to form a shallow fossa between the ureter and the obliterated umbilical artery. Outside the peritoneum forming the floor of this fossa are the obturator nerve and vessels. The tubal extremity of the ovary lies a little below the level of the external iliac vessels, and its uterine extremity is immediately above the peritoneum of the pelvic floor. The mesovarian border is behind the line of the obliterated umbilical artery; the free border is immediately in front of the ureter [FIG. 658]. The medial surface of the ovary is largely covered by the uterine tube, which, passing upwards on it near its mesovarian border, arches over the tubal extremity and then turns downwards in relation to the free border and posterior part of the medial surface.

In women who have borne children the position of the ovary is variable, and its long axis may be horizontal.

Connexions of the Ovary. Anteriorly the ovary is connected with the broad ligament by a very short mesentery, or **mesovarium,** along the whole length of its mesovarian border. A small peritoneal fold passes upwards from the tubal end of the ovary to the peritoneum over the external iliac vessels and the psoas major muscle [FIGS. 634 and 658]. It is called the **suspensory ligament** and is continuous below with the mesovarium and, in front of it, with the broad ligament of the uterus. Between its two layers it contains the ovarian vessels and nerves as they pass down into the lesser pelvis to enter the mesovarium and thus reach the hilus of the ovary. The uterine extremity of the ovary is connected with the lateral border of the uterus by a round cord called the **ligament of the ovary;** this band lies in the free border of the medial

part of the mesovarium, and is attached to the lateral angle of the uterus posterosuperior to the point of entrance of the uterine tube. It is composed of fibrous tissue containing some plain muscle fibres continuous with those of the uterus. The tubal end of the ovary is directly connected with the **ovarian fimbria**—one of the largest of the fimbriae of the uterine tube [FIG. 659].

Descent of the Ovary. Like the testes, the ovaries at first lie in the abdominal cavity and only later descend to the lesser pelvis. At birth the ovary lies partly in the abdomen and partly in the lesser pelvis; soon, however, it takes up a position entirely within the lesser pelvis. As in the male, a **gubernaculum** is present in the early stages of development and is connected inferiorly with the tissues that become the labium majus; but in the female the gubernacula become attached to the paramesonephric ducts, from which the uterus is developed. Consequently, the gubernaculum becomes the round ligament of the uterus in front and the ligament of the ovary behind.

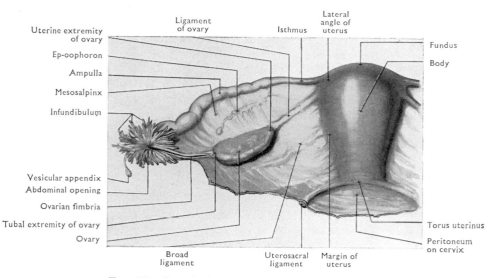

FIG. 659 Intestinal surface of uterus and broad ligament. The broad ligament has been spread out.

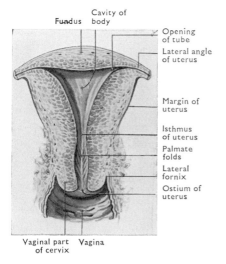

FIG. 660 Diagrammatic representation of uterine cavity. The tissue at the side of the cervix represents the lateral cervical ligament.

The ovary descends only as far as the lesser pelvis : it is a rare abnormality for the ovary, instead of entering the pelvis, to take a course similar to that of the testis, and pass through the inguinal canal into the labium majus.

Structure of the Ovary. The ovary is, for the most part, composed of a fibro-areolar tissue, called the **stroma of the ovary.** Superficially the stroma is dense and consists of fibroblasts and reticular fibres ; centrally it is looser in texture, richly supplied by blood vessels and nerves, and contains many elastic fibres. The surface of the ovary is covered with an epithelium composed of columnar or cubical cells, and is continuous with the flat-celled mesothelium of the peritoneum which covers the mesovarium near the hilus of the ovary. Shining through the epithelium of the fresh adult ovary (except in old age) a variable number of small vesicles can usually be seen—the **vesicular ovarian follicles**—in each of which is a developing egg-cell or oocyte. The number of follicles visible, and also the size which each follicle reaches before it ruptures and sheds it contents, are by no means constant. When a follicle ruptures and discharges the oocyte, its walls at first collapse, but later the cavity becomes filled with extravasated blood and cellular tissue of a yellowish colour. The resulting structure, called a **corpus luteum,** slowly degenerates unless pregnancy occurs, when it develops and becomes larger during the first 3 months. As it atrophies, the cells of the corpus luteum disappear, and the structure, losing its yellow colour as it is replaced by collagen, receives the name of **corpus albicans.** After a time the corpus albicans contracts to a mere scar. Owing to the periodic rupture of the ovarian follicles, the surface of the ovary, which is at first smooth and even, becomes dimpled and puckered in old age.

A histological section through the ovary, especially in the young child, shows that its superficial (cortical) part possesses large numbers of small **primary** (primordial) **ovarian follicles,** embedded in the fibro-areolar tissue of the stroma. Each primary follicle consists of a primary oocyte surrounded by a layer of investing follicular cells. Certain authorities once believed that the primary follicles are formed anew at each menstrual cycle from the epithelium covering the ovary, which is therefore termed the germinal epithelium. But these primary oocytes are present from the time of birth, and develop from the primordial germ cells which migrate into the developing ovary from the wall of the yolk sac (Brambell, 1956). After puberty, during each menstrual cycle, some of the primary follicles grow to form **vesicular ovarian follicles ;** one of these usually undergoes further maturation and ruptures at the time of ovulation to liberate a secondary oocyte in the process of undergoing the second maturation division [p. 19]. As a result there is a progressive decline in the total number of oocytes in the ovary with increasing age (Zuckerman, 1956, 1962). In the deeper part (medulla) of the ovary the blood vessels are most numerous, and some plain muscle fibres are found.

The ripe follicle contains a relatively large amount of fluid, and the surrounding stroma becomes differentiated to form a capsule or **theca folliculi.** The capsule is composed of an inner vascular coat, the **theca interna,** and an outer coat, the **theca externa,** composed of fibrous tissue.

Vessels and Nerves of the Ovary. The **ovarian arteries,** which correspond to the testicular arteries of the male, are a pair of long, slender vessels which spring from the front of the aorta below the level of origin of the renal vessels. Each artery runs in the suspensory ligament of the ovary, and enters the ovary at its mesovarian border. Close to the ovary the ovarian artery anastomoses with branches of the **uterine artery.** The smaller arteries within the adult ovary are spiral (Reynolds, 1948). The blood is returned by a series of communicating **veins,** which eventually form one vein draining into the inferior vena cava on the right side, and the renal vein on the left side.

The **lymph vessels** of the ovary follow the blood vessels and run with those from the uterine tube and upper part of the uterus to end in the lymph nodes beside the aorta and inferior vena cava [FIG. 1156].

The **nerves** of the ovary are derived chiefly from a plexus which accompanies the ovarian artery and is continuous above with the renal plexus. Other fibres are derived from the lower part of the aortic plexus and join the plexus on the ovarian artery. The afferent impulses from the ovary reach the central nervous system through the fibres of the posterior root of the tenth thoracic nerve.

THE UTERINE TUBES

[PLATE 39 C]

The uterine tubes (Fallopian tubes) are a pair of ducts which convey the oocytes, discharged from the follicles of the ovaries, to the cavity of the uterus [FIG. 660]. Each tube is about 10 cm. in length, and opens at one end into the peritoneal cavity near the ovary, and at the other end by a smaller opening into the anterolateral part of the uterine cavity. The tube lies in the margin of a fold of peritoneum called the **mesosalpinx,** which is a part of the broad ligament of the uterus.

The **abdominal opening** of the tube is only about 2 mm. in diameter when its walls are relaxed, and much narrower when the muscular coat of the tube is contracted. It is placed at the bottom of a funnel-like expansion of the ovarian extremity of the tube called the **infundibulum,** the margins of which are continued as a number of irregular processes, called **fimbriae,** many of which are branched or fringed. The surfaces of the fimbriae which look into the cavity of the infundibulum are covered with a mucous membrane continuous with that of the tube, while the outer surfaces are clothed with peritoneum. The mucous surfaces of the larger fimbriae present ridges and grooves which are continued into the folds and furrows of the mucous coat of the tube. One of the fimbriae, usually much larger than the rest, is connected either directly or indirectly with the tubal end of the ovary, and to it the name **ovarian fimbria** is applied. The part of the tube continuous with the infundibulum is called the **ampulla.** It is the widest and longest portion of the tube, and is usually slightly tortuous and of varying diameter, being slightly constricted in some places. The ampulla ends in the narrower, thicker-walled, and much shorter **isthmus** of the tube, which joins the lateral angle of the uterus. The **uterine part** of the tube is embedded in the uterine wall, which it traverses to reach the cavity of the uterus [FIG. 660]. The **uterine opening** is smaller than the pelvic opening, being about 1 mm. in diameter. The lumen of the canal gradually increases in width as it is traced from the uterus towards the ovary.

Course of the Uterine Tube. From the uterus, the uterine tube is directed at first horizontally towards the uterine extremity of the ovary, and then covers most of the medial surface of the ovary [p. 510]. The fimbriated extremity of the uterine tube lies in the abdominal cavity until the ovary in its descent has entered the lesser pelvis.

Structure of the Uterine Tubes. The wall of each tube is composed of a number of concentric layers. First there is a **serous coat** of peritoneum, under which lies a thin **subserous coat** of loose areolar tissue containing many vessels and nerves. Internal to the subserous coat there is the **muscular coat** composed of two strata of plain muscle fibres—a more superficial, thin stratum of longitudinal fibres and a deeper, thicker layer of circular fibres. Inside the circular muscular coat is a **submucous layer,** and then the lining membrane or **mucous coat.** In the part of the tube near the uterus the muscular layer is thicker than towards the ovarian end, and in the isthmus it forms the chief part of the wall. The mucous membrane, on the contrary, is thickest towards the fimbriated extremity, and there it forms the chief part of the wall. The mucous membrane is thrown into numerous longitudinal folds, which in the ampulla are exceedingly complex, the larger folds bearing on their surface numerous smaller ones which project into the lumen of the tube

and almost completely fill it. The mucous membrane is covered with an epithelium which is partly ciliated, the cilia of which tend to drive the fluid contents of the tube towards the uterus. **Vessels and Nerves of the Uterine Tube.** The **arteries** supplying the uterine tube are derived from the anastomosis between the uterine and ovarian arteries which course along the uterine tube in the mesosalpinx. The **veins** of the tube pour their blood partly into the **uterine** and partly into the **ovarian** veins. The **lymph vessels** run with those of the ovary and the upper part of the uterus to the lymph nodes beside the aorta. The **nerves** are derived from the plexus that supplies the ovary, and also from the plexus in connexion with the uterus. The afferent fibres are believed to belong to the eleventh and twelfth thoracic and first lumbar nerves.

THE UTERUS
[PLATE 39 C and D]

The uterus or womb [FIGS. 634 and 661] is a hollow, thick-walled muscular organ that projects upwards and forwards above the bladder from the upper part of the vagina [FIGS. 662 and 663].

Shape and Parts. The upper two-thirds of the uterus, known as the **body,** is pear-shaped with the blunt end anteriorly, and slightly flattened from above downwards. The lower third, called the **cervix,** is cylindrical and tapers slightly to its blunt lower end, which projects through the anterior vaginal wall into the upper part of the cavity of the vagina. The cervix is therefore divisible into a **vaginal** and a **supravaginal part;** the latter communicates with the body at the **isthmus** of the uterus which is indicated by a slight constriction.

In the centre of the extremity of the vaginal part of the cervix is a small opening—the **ostium uteri**—by which the cavity of the uterus communicates with that of the vagina. This part of the cervix is slightly flattened from before backwards, and the ostium of a uterus from a woman who has not been pregnant has the appearance of a short transverse slit bounded above and below by thick **anterior** and **posterior lips;** in women who have borne children the slit is larger and has an irregular outline. The posterior lip is thicker and more rounded than the anterior, and both lips are usually in contact with the posterior wall of the vagina.

The uterine tubes enter the organ at its lateral angles, where it attains its maximum breadth; and the rounded part above a line joining the points of

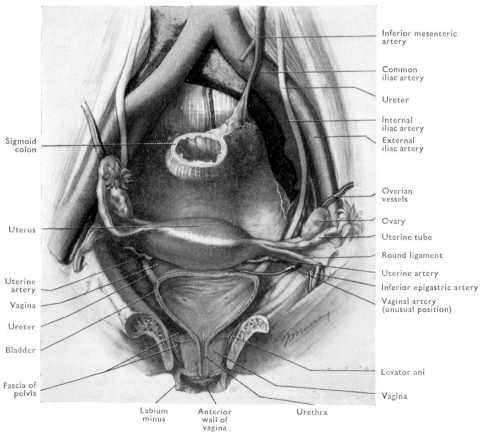

Sigmoid colon

Uterus

Uterine artery

Vagina

Ureter

Bladder

Fascia of pelvis

Labium minus

Anterior wall of vagina

Urethra

Inferior mesenteric artery

Common iliac artery

Ureter

Internal iliac artery

External iliac artery

Ovarian vessels

Ovary

Uterine tube

Round ligament

Uterine artery

Inferior epigastric artery

Vaginal artery (unusual position)

Levator ani

Vagina

FIG. 661 Dissection of pelvis of multiparous female showing some of the relations of the internal genital organs.

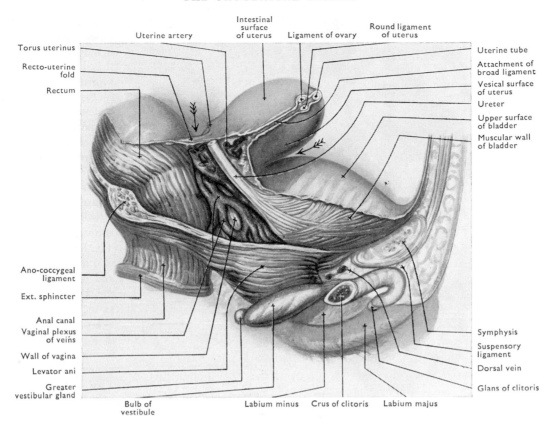

Intestinal
surface
of uterus

Uterine artery — Ligament of ovary — Round ligament of uterus

Torus uterinus

Recto-uterine fold

Rectum

Uterine tube

Attachment of broad ligament

Vesical surface of uterus

Ureter

Upper surface of bladder

Muscular wall of bladder

Ano-coccygeal ligament

Ext. sphincter

Anal canal
Vaginal plexus of veins

Wall of vagina

Levator ani

Greater vestibular gland

Bulb of vestibule

Labium minus — Crus of clitoris — Labium majus

Symphysis

Suspensory ligament

Dorsal vein

Glans of clitoris

FIG. 662   Model of dissection of female pelvic organs.   The arrows point to the vesico-uterine and recto-uterine pouches.

entrance of the two tubes is the **fundus.** The uterus is described as possessing an antero-inferior or **vesical surface** and a posterosuperior or **intestinal surface,** and these two surfaces are separated by right and left borders. Both surfaces are convex, but the intestinal is much more rounded. The length of the body is about 5 cm.; its maximum breadth also is 5 cm.; and its thickness or depth is 2·5 cm. The cervix is 2·5 cm. long and about 2·5 cm. in diameter. During pregnancy the whole organ is greatly enlarged.

Position and Attachments. The non-pregnant uterus lies completely within the cavity of the pelvis. It rarely lies exactly in the median plane, but is usually inclined slightly to one or other side, more frequently to the left. The cervix is fixed to, and structurally continuous with, the wall of the vagina where it pierces that organ. From the vagina the cervix passes upwards and forwards to become continuous with the isthmus and body which continue in a similar but more horizontal direction so that the body is flexed downwards on the cervix. In its usual position, therefore, the uterus is said to be **anteverted** and **anteflexed.** Although anteversion and anteflexion of the uterus are regarded as the normal condition, it should be realized that in many women the uterus is retroflexed or retroverted without producing any symptoms.

The peritoneum which covers the upper surface of the urinary bladder is reflected upwards near the posterior edge of that organ on to the uterus at the junction of the isthmus with the body and then passes upwards and forwards, covering the inferior or vesical surface of the body. Thus, a rather

deep **vesico-uterine pouch** is formed [FIG. 662]. From the vesical surface of the uterus the peritoneum passes over the fundus and then downwards and backwards, covering the intestinal surface of the body, the isthmus and supravaginal part of the cervix. It then passes on to the upper part of the vagina and is reflected from it upwards and backwards on the anterior wall of the rectum [FIGS. 634 and 664]. The peritoneal pouch between the back of the uterus and the front of the rectum is known as the **recto-uterine pouch,** the bottom of which is about 7·5 cm. or rather less above the level of the anus. The two layers of peritoneum which cover the vesical and intestinal surfaces of the uterus come together at its borders and are then carried laterally as extensive peritoneal folds to the side-walls of the pelvis. These folds are known as the **broad ligaments of the uterus** [FIGS. 659 and 663], and they enclose between their two layers various important structures associated with the uterus.

The two layers of each broad ligament are continuous with each other at a free edge which is directed forwards and upwards and surrounds the uterine tube; they also enclose some loose areolar tissue and a little plain muscle. To the contained tissue the name **parametrium** is given. In the upper part of the broad ligament the parametrium is small in quantity and the two layers of the ligament are close together, but towards its root the parametrium increases considerably in amount as the two peritoneal layers diverge from each other. The anterior layer passes forwards into the floor of the paravesical fossa; the posterior layer forms a prominent, shelf-like

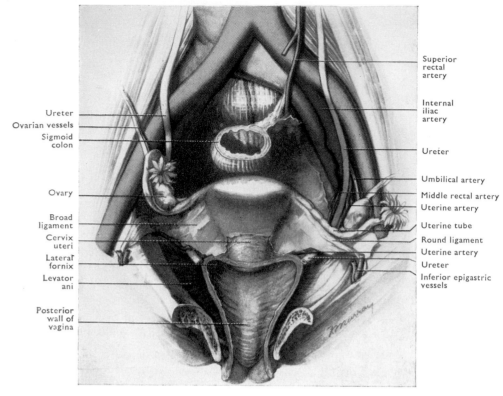

Ureter
Ovarian vessels
Sigmoid colon
Ovary
Broad ligament
Cervix uteri
Lateral fornix
Levator ani
Posterior wall of vagina

Superior rectal artery
Internal iliac artery
Ureter
Umbilical artery
Middle rectal artery
Uterine artery
Uterine tube
Round ligament
Uterine artery
Ureter
Inferior epigastric vessels

FIG. 663   Uterus and vagina displayed by further dissection of pelvis of multiparous female shown in FIG. 661.

recto-uterine fold, which corresponds to the sacrogenital fold in the male and curves backwards from the cervix uteri to the posterior wall of the pelvis a little lateral to the rectum. The two recto-uterine folds sometimes meet across the back of the cervix, forming a ridge known as the **torus uterinus** [FIG. 659].

Enclosed in the free edge of the broad ligament is the **uterine tube;** and just below the entrance of the tube into the uterine wall two other structures are attached to the uterus, namely, the **round ligament of the uterus** antero-inferiorly and the **ligament of the ovary** posterosuperiorly [FIG. 662]. These three structures are all enclosed in the broad ligament, and while they are very close together at their attachments to the uterus, they diverge widely as they approach the pelvic wall. Thus the attachment of the broad ligament to the uterus is almost a straight line but its attachment to the pelvic wall is drawn out into a sort of tripod [FIGS. 661 and 663]. The inferior part of the broad ligament passes laterally from the side of the cervix to the side of the pelvis, and as the whole uterus slopes from above downwards and backwards the lower part of the broad ligament slopes in the same general direction. The anterior part of the ligament is drawn forwards by the round ligament of the uterus, which is a narrow flat band of fibrous tissue containing some plain muscle at its uterine end. It passes from the upper part of the border of the uterus anterolaterally, and very slightly upwards to the deep inguinal ring, crossing the obliterated umbilical artery and external iliac vessels. After hooking round the lateral side of the inferior epigastric artery and traversing the inguinal canal, it ends in the skin and subcutaneous tissue of the labium majus. In some cases

a minute diverticulum of the peritoneal cavity accompanies the round ligament through the inguinal canal; it is the vestige of the **processus vaginalis,** which is formed in the foetus in both sexes.

The part of the broad ligament between the ligament of the ovary, the ovary, and the uterine tube is known as the **mesosalpinx.**

Laterally the mesosalpinx is continuous with the **suspensory ligament of the ovary,** which crosses the external iliac vessels about 2·5 cm. below their upper end. The lateral part of the mesosalpinx is more free than the medial part, and this permits the lateral part of the uterine tube to curve down along the posterior border of the ovary and almost enclose that organ in a fold of the mesosalpinx, thus forming a peritoneal compartment known as the **bursa ovarica.** The mesosalpinx contains the epoophoron and the paroophoron and the anastomosis between the ovarian and uterine arteries.

The part of the broad ligament that covers the ligament of the ovary is really the medial continuation of the mesovarium [FIG. 659], and, as the mesovarium is continuous laterally with the suspensory ligament of the ovary also, the complete fold forms the base of the mesosalpinx. The **ligament of the ovary** is a round fibrous band, about 2·5 cm. long, which contains a considerable amount of plain muscle, and is attached laterally to the uterine extremity of the ovary.

At the root of the broad ligament, where the two layers of peritoneum diverge to pass into the pelvic floor, a considerable amount of fibrous tissue and plain muscle occurs. This mass is the upper end of the fascial condensation which passes from the side of the vagina and the base of the urinary bladder to the fascia on the levator ani muscles. Its upper and more bulky part is attached medially to the side of the supravaginal part of the cervix and to the side of the upper third of the vagina, while laterally it fans out. Its posterior part passes backwards in the recto-uterine fold of peritoneum and gains an attachment to the front of the sacrum; this part is known as the **uterosacral ligament** and contains nerves supplying the uterus from the inferior hypogastric plexuses [FIG. 965]. The anterior part, passing directly laterally to blend with the fascia on the levator ani muscle, is called the **lateral cervical ligament** (Mackenrodt, 1895). The ureter and uterine artery lie in its upper part (Power, 1944). As already mentioned, it is really a condensation of fascia around the internal iliac artery and its branches.

Epoophoron and Paroophoron. The epoophoron and paroophoron are two vestigial structures sometimes found between the layers of the broad ligament.

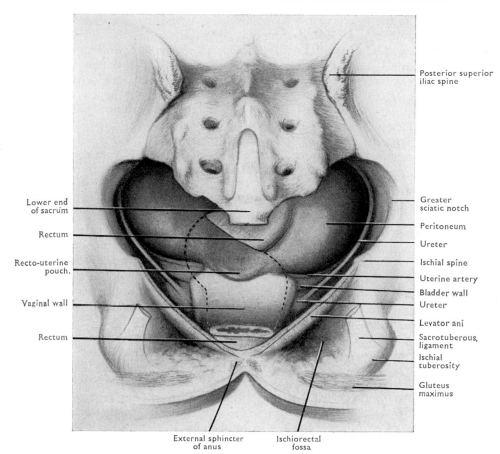

FIG. 664 Vagina, base of bladder, and recto-uterine pouch from behind.

The coccyx and sacrotuberous and sacrospinous ligaments have been removed. The levatores ani have been separated and drawn laterally. A considerable portion of the rectum has been removed, but its position is indicated by the dotted lines. The recto-uterine pouch is probably not quite so deep as usual. The triangular interval between the ureter and uterine artery was filled by a mass of fibromuscular tissue.

uterus is separated from the bladder by the uterovesical pouch of peritoneum; the front of the supravaginal part of the cervix is separated from the bladder by loose areolar tissue. Posteriorly the body and supravaginal part of the cervix are separated by a layer of peritoneum from the sigmoid colon or coils of small intestine. On each side, there is the broad ligament of the uterus with its contained structures. Of particular importance is the close relationship of the ureter which is crossed superiorly by the uterine artery at the side of the cervix. The uterus rises and falls as the urinary bladder fills and empties. The vaginal part of the cervix pierces the anterior wall of the vagina and projects into its cavity, where it is covered with vaginal mucous membrane.

Cavity of the Uterus. In comparison with the size of the organ, the cavity of the uterus is small, owing to the great thickness of the uterine wall. The **cavity of the body** of the uterus when viewed from the side, is a mere slit [FIG. 634]; from behind it is triangular in outline, the base in the fundus and the apex continuous with the canal of the cervix, with the sides of the triangle convex inwards. The lateral angles of the triangle mark the entrance of the two uterine tubes. The **canal of the cervix** extends from the isthmus to the **ostium,** where it opens into the vagina. It is a spindle-shaped passage which is narrower at the ends than in the middle, and is flattened anteroposteriorly. In the body of the uterus the walls of the cavity are smooth and even, but the mucous membrane of the nulliparous cervix forms a remarkable series of folds called the **palmate folds.** They consist of an anterior and a posterior longitudinal fold or ridge from which a large number of secondary folds or rugae branch off obliquely upwards and laterally [FIG. 660].

Isthmus of the Uterus. The lowest part of the cavity of the body, about 1 cm. in length, is constricted and is commonly known as the isthmus of the uterus [FIG. 660]. The separate identity of this part of the uterus has a functional basis, and the structure of its mucous coat is less complex than in the main part of the body (Frankl, 1933). Not only is it much less affected by the periodic changes of the menstrual cycle, but in the last months of pregnancy it becomes more distinct and during labour is transformed into the 'lower uterine segment' of the obstetrician.

Supports of the Uterus. The uterine supports are mainly the bony pelvis and the pelvic floor, for although the uterus itself does not rest directly on these structures, yet it lies on the urinary bladder and rectum which are supported by them.

The **epoophoron** [FIG. 659] lies in the mesosalpinx between the uterine tube and the ovary. In the adult it consists of a number of small rudimentary blind tubules lined with epithelium. One of the tubules—the **longitudinal duct of the epoophoron**—lies close to the uterine tube and runs nearly parallel with it. It is joined by a number of **transverse ductules** which enter it at right angles from the neighbourhood of the ovary. The longitudinal duct is a persistent portion of the mesonephric duct, and represents the duct of the epididymis; the transverse ductules are derived from the mesonephric tubules and represent the efferent ductules of the testis (and probably also the ductuli aberrantes).

One or more small pedunculated cystic structures, called **vesicular appendices,** are often seen near the infundibulum of the uterine tube. They are supposed to represent portions of the cephalic end of the mesonephric duct.

The **paroophoron** is a collection of very minute tubules that lie in the mesosalpinx nearer the uterus than the epoophoron. They represent the paradidymis in the male, and are derived from the part of the mesonephros which lies nearer the caudal end of the embryo. Though sometimes visible in the child at birth, the paroophoron in the adult can be recognized only by microscopic examination.

Relations of the Uterus. Anteriorly the body of the

The strength of the pelvic floor is therefore essential to the support of the uterus, and the strength of the floor is dependent on an intact **central perineal tendon** [p. 358]. This fibromuscular mass is liable to injury and laceration during parturition; and it is of such importance in obstetrical practice that some obstetricians refer to it simply as 'the perineum'. The lateral cervical and uterosacral ligaments are simple condensations of the pelvic fascia passing to the cervix (which is the part of the uterus most firmly held in position), but they are unable to resist sustained stress if the pelvic floor has given way (Harrison, 1963). The fundus of the uterus is mobile, and though the round ligaments play some part in maintaining the anteverted position, they do not prevent the uterus from enlarging into the abdomen in pregnancy.

Structure of the Uterus. The uterine wall is composed of three chief layers—the serous, the muscular, and the mucous coats.

The **serous coat,** or perimetrium, is the peritoneal covering. Over the fundus and body it is very firmly adherent to the deeper layers, and cannot be easily peeled off without tearing either it or the underlying muscular tissue. Near the borders of the uterus the peritoneum is less firmly attached, and over the back of the cervix it may be stripped easily without injury to the underlying structures.

The **muscular coat,** or myometrium, is composed of interlacing bundles of plain fibres, and forms the chief part of the uterine wall. Inferiorly the muscular coat of the uterus becomes continuous with that of the vagina. The more superficial layer of the muscular coat sends prolongations into the recto-uterine folds, into the round ligaments of the uterus, and into the ovarian ligaments. Other fibres join the walls of the uterine tubes. The main branches of the blood vessels and nerves of the uterus lie among the muscle fibres. In the internal layers of the muscular coat a considerable amount of fibrous tissue and some elastic fibres are to be found. At the isthmus the muscular fibres decrease and the fibrous tissue increases. The middle coat of the cervix contains mostly fibrous tissue with a small amount of muscular and elastic tissues; hence its greater firmness and rigidity.

The **mucous coat,** or endometrium, in the body of the uterus is smooth and soft and covered with columnar, partially ciliated epithelium. Simple tubular glands, also lined with a columnar epithelium, are present in the mucous membrane and extend to the muscle, as there is no submucosa. In the isthmus the glands are fewer, less deep, stain less intensely and are less affected by the menstrual cycle. In the cervix the mucous coat is firmer and more fibrous than in the body, and its surface is not smooth but presents a number of peculiarly disposed ridges which have been mentioned already. Like the mucous membrane of the body of the uterus, that of the cervix is covered with a columnar epithelium showing occasional ciliated cells, and this changes to squamous epithelium just inside the ostium of the uterus. In addition to unbranched tubular glands, numerous slightly branched glands are found in the cervix uteri. Both kinds of glands are lined with the same epithelium as the cervical canal. In many cases small, clear retention cysts are to be seen in the cervical mucous membrane. They arise as a result of obstruction at the mouths of the glands.

Differences in the Uterus at Different Ages. At birth the cervix uteri is relatively large, and its cavity is not distinctly marked off from that of the body by an isthmus. At that time also, the uterus is mainly an abdominal organ and the palmate folds extend throughout the whole of its length. The organ grows slowly until just before puberty, when its growth is rapid for a time. As the body increases in size the mucous membrane becomes smooth and the palmate folds become restricted to the cervix. In women who have borne children the cavity remains permanently a little wider and larger than in those who have never been pregnant. In old age the uterine wall becomes harder and has a paler colour than in the young subject.

Variations. In rare cases the uterus may be divided by a septum into two distinct cavities, or the upper aspect of one or both of its borders may be prolonged into straight or curved processes (uterine cornua) continuous with the uterine tubes. The latter abnormality recalls the appearance of the bicornuate uterus of some animals. Both the above conditions arise from an arrest in the fusion of the two separate tubes—the paramesonephric ducts—parts of which normally unite in the embryo to form the uterus.

Periodic Changes in the Uterine Wall. At each menstrual period (approximately every 28 days) there is a shedding of the superficial layers (stratum compactum and stratum spongiosum) of the endometrium. In the early part of the menstrual cycle the mucous membrane gradually thickens; after ovulation it becomes more vascular and its glands tortuous. Soon the superficial parts of the mucous membrane disintegrate and haemorrhage takes place from the endometrial blood vessels. When menstruation is over the mucous membrane is rapidly regenerated from the basal layer (stratum basalis) of the endometrium which remains unchanged throughout the menstrual cycle.

The Pregnant Uterus. The pregnant uterus increases rapidly in size and weight, so that from being a pelvic organ 7·5 cm. in length and about 40 g. in weight, it rises into the abdomen and becomes by the eighth month of pregnancy about 20 cm. in length and sometimes as much as 1 kg. in weight. In shape the uterus is then oval or rounded, with a thick wall composed chiefly of greatly enlarged, smooth muscle fibres arranged in distinct layers. The fundus is very round and prominent. The round ligaments are longer, and the layers of the medial part of the broad ligaments become separated by the growth of the uterus between them. The blood vessels, especially the arteries, are very large and tortuous. The changes which occur in the mucous membrane of the pregnant uterus are described on pages 35–6. For changes in the pelvic floor during parturition, see Power (1946).

Involution of Uterus. Immediately after delivery the uterus contracts, but it is still many times the size of the normal resting organ. It weighs about 1 kg. and lies against the anterior abdominal wall almost as high as the umbilicus. The peritoneum on its surface is thrown into wrinkles, and owing to the slackness of its ligaments the whole organ is extremely mobile. The cervix is lengthened and is flaccid, with thin walls and a widely open ostium. From the second day after delivery onwards the uterus rapidly shrinks. By the end of the first week it has diminished in weight to about 500 g., the wrinkles in its peritoneal covering have disappeared, and the cervix has shortened and become firmer. The ligaments of the uterus shorten as the uterus shrinks, and the pelvic floor also regains tone, so that the organ is less mobile. By the end of the second week it weighs 350 g. and by the end of the eighth week it should be reduced to its resting size and should weigh 40 g. or less. This process of gradual reduction of the uterus to its normal size is known as involution.

Vessels and Nerves of the Uterus

The uterus receives its arterial supply mainly from the **uterine arteries,** which are branches of the internal iliac arteries, but also from the **ovarian arteries,** derived from the aorta. The vessels from these two sources communicate freely with one another.

Numerous thin-walled **veins** form a plexus at the side of the cervix and pour their blood into the tributaries of the internal iliac vein.

The **lymph vessels** from the fundus of the uterus join those from the ovary and end in aortic lymph nodes. One or two run along the round ligament to the superficial inguinal lymph nodes. The lymph vessels from the cervix run to the sacral nodes and all the iliac groups ; those from the body end chiefly in the external iliac nodes [FIG. 1157].

The **nerves** of the uterus are derived chiefly from the **uterovaginal plexus** which is placed in the neighbourhood of the cervix uteri. Superiorly this plexus is continuous with the pelvic sympathetic plexuses, but it receives parasympathetic fibres also from the second and third or the third and fourth sacral nerves through the pelvic splanchnic nerves.

Clinical observations indicate that afferent impulses reach the central nervous system from the uterus through the posterior roots of the tenth, eleventh, and twelfth thoracic and the first lumbar nerves (Whitehouse and Featherstone, 1923).

THE VAGINA

The vagina [FIGS. 634 and 663] is a passage of variable size, but usually about 9 cm. in length, open at its lower end, and communicating above with the cavity of the uterus. The vagina is directed upwards and backwards, describing a slight curve which is convex forwards. When the urinary bladder is empty, the axis of the vagina makes with the axis of the uterus an angle of slightly more than 90 degrees, but this angle increases as the bladder fills and raises the fundus of the uterus. Its anterior and posterior walls are in contact except at its upper end, where the cervix is inserted ; and that is its widest part also. In transverse section the lower part is usually an H-shaped cleft, the middle part a simple transverse slit, while the lumen of the upper portion is more open. The cervix uteri enters the vagina through the upper portion of its anterior wall [FIG. 634]. As more of the posterior part of the cervix than of the anterior part projects into the vagina, the recess between the vaginal wall and the cervix is deeper behind than in front or laterally. The term **anterior fornix** is applied to the recess in front, **posterior fornix** to the recess behind, and **lateral fornix** to the recess on each side, though the four fornices are parts of a continuous recess surrounding the cervix. The **anterior wall** of the vagina is about 7·5 cm. in length ; the **posterior wall** is a little longer—about 9 cm. At its lower end the vagina opens into the pudendal cleft between the labia minora. The opening is partly closed in the virgin by a thin, crescentic or annular fold, called the **hymen,** torn fragments of which persist around the opening, as the **carunculae hymenales,** after the fold itself has been ruptured.

Relations of the Vagina. The vagina is in contact **anteriorly** from above downwards with the cervix uteri, the base of the bladder and ends of the ureters, from which it is separated by areolar tissue, and the urethra, to which it is intimately connected. Its upper end usually deviates a little to one side, with consequent asymmetrical relation to the ureters [p. 513]. **Posteriorly** it is in contact from above downwards with the recto-uterine peritoneal pouch for 1·3 cm. or more, then with the ampulla of the rectum, from which it is separated by a thin layer of very loose areolar tissue, and then with the central perineal tendon. **Laterally** on each side is the root of the broad ligament, and the ureter crossed by the uterine vessels runs just above the lateral fornix. Lower down is the levator ani muscle, a plexus of veins intervening ; and lower still is the greater vestibular gland and the bulb of the vestibule covered by the bulbospongiosus muscle. As the perineal membrane approaches the median plane, it turns downwards on each side of the vagina between it and the bulb of the vestibule and fuses below with the root of the labium minus.

Structure of the Vagina. The vaginal wall has a **muscular coat** composed of plain muscle fibres, most of which are longitudinally disposed. Towards the lower end there are circular bundles of striated muscle fibres, some of which are continuous with those which form a part of the urethral wall. The **mucous coat** is thick and is corrugated by a number of transverse ridges or elevations known as **vaginal rugae** covered with stratified squamous epithelium. In addition to the transverse rugae, longitudinal ridges are present on the anterior and posterior walls. These receive the name **columns of the rugae,** and, like the transverse rugae, they are seen best in young subjects and in the lower part of the vagina. The urethra lies in close relationship to the anterior column in its lower part, and hence that portion of the column is called the **urethral ridge.** Between the muscular and mucous coats there is a thin layer of vascular tissue which resembles erectile tissue ; and in the mucous coat there are small nodules of lymphoid tissue.

The vaginal wall is surrounded by a layer of loose, vascular areolar tissue which contains numerous large communicating veins.

Vessels and Nerves of the Vagina

The arteries on each side are the **vaginal artery,** the vaginal branch of the **uterine artery,** the vaginal branches of the **middle rectal artery,** and branches of the **internal pudendal artery.** The **veins** form a plexus around the vaginal wall, and drain their blood into the tributaries of the internal iliac veins.

The **lymph vessels** from the upper part of the vagina join the external and internal iliac nodes, and those from the lower part end in the superficial inguinal nodes [FIG. 1156].

The **nerves** of the vagina are derived from the uterovaginal plexus and from the vesical plexus. Parasympathetic fibres are derived from the second to fourth sacral nerves.

Methods of Clinical Anatomical Examination

The interior of the vagina and the vaginal part of the cervix can be inspected through a vaginal speculum. By a bimanual examination—two fingers of one hand in the vagina and the other hand on the anterior abdominal wall above the pubic crests—the fundus of the uterus can be felt when the fingers in the vagina press forwards and upwards. Normal uterine tubes cannot usually be felt, but if thickened or inflamed they can often be palpated in a bimanual examination. The ovaries may be palpated by placing the examining fingers in the appropriate lateral fornix.

The ostium of the uterus is usually at the level of the spines of the ischium, which can be felt from the vagina. In parous women the uterus may prolapse so that the ostium lies lower, and may even protrude through the vaginal orifice. The vaginal part of the cervix and the ostium can be felt by a finger in the rectum. The pregnant uterus can be palpated as soon as it has risen above the pelvic cavity and lies against the anterior abdominal wall.

THE FEMALE EXTERNAL GENITAL ORGANS

The collective term **pudendum femininum** or **vulva** is applied to the female external genital organs, i.e., to the mons pubis, the labia majora and the structures which lie between the labia [FIG. 665].

Labia Majora. The labia majora represent the scrotum in the male, and they form the largest part of the female external genital organs. They are the boundaries of the **pudendal**

cleft, into which the urethra and vagina open. Each labium is a prominent, rounded fold of skin, narrow behind where it approaches the anus, but increasing in size as it passes forwards and upwards to meet the other labium at the **anterior commissure** and end in a median elevation—the **mons pubis.** The mons pubis lies over the pubic symphysis, and, like the labia majora, it is composed chiefly of fatty and areolar tissue, and is covered with hair. The skin of the lateral, convex surface of each labium majus contains numerous sebaceous glands and corresponds to that of the scrotum in the male, but the medial surface is flatter and smooth and has a more delicate cutaneous covering. The posterior ends of the labia majora are connected across the middle line, in front of the anus, by the **posterior commissure.**

Usually the mons and the labia majora are the only visible parts of the external genital organs, since the labia are in contact with each other and completely conceal the structures within the pudendal cleft.

The round ligament of the uterus ends in the skin and fibro-fatty tissue of the labium majus. The superficial layer of the subcutaneous tissue resembles that of the scrotum, but contains no muscular fibres.

The **nerve supply** corresponds with that of the scrotum—the anterior part of each labium being supplied by branches of the ilio-inguinal nerve, and the posterior part by branches from the pudendal nerve and by the perineal branch of the posterior cutaneous nerve of the thigh. The **blood vessels** are derived from the external pudendal branches of the femoral and from the labial branches of the internal pudendal vessels.

Labia Minora. The labia minora are a pair of much smaller and narrower longitudinal folds of skin usually completely hidden in the cleft between the labia majora. Their posterior parts diminish in size and end by gradually joining the medial surfaces of the labia majora. In the young subject a slightly raised transverse fold, the **frenulum of the labia,** is usually seen connecting the posterior ends of the labia minora. Traced forwards, each labium minus divides into two portions—a lateral and a medial. The lateral portions of the two labia unite over the glans of the clitoris and form for it a fold or covering called the **prepuce of the clitoris.** The medial portions, uniting at an acute angle, join the ventral surface of the glans and form the **frenulum of the clitoris.** The skin of the labia minora resembles that on the medial surface of the labia majora, being smooth, moist and pink in colour. The medial surfaces of the labia minora are in contact with each other, and their lateral surfaces with the labia majora. The labia minora and the frenulum are devoid of fat.

The openings of the urethra and vagina are in the median plane, between the labia minora, which must be separated to bring them into view.

Vestibule of the Vagina. The vestibule is the name applied to the space between the labia minora. In its roof are the openings of the urethra, the vagina, and the ducts of the greater vestibular glands.

The **vestibular fossa** is the part of the

Prepuce of clitoris

Frenulum of clitoris

Vestibule of vagina (anterior part)

Labium minus

Vestibular fossa

Glans of clitoris

Labium majus

External urethral orifice

Orifice of vagina

Posterior commissure

Anus

FIG. 665   Female external genital organs.
The frenulum labiorum is seen stretching across behind the vestibular fossa

vestibule between the vaginal orifice and the frenulum of the labia.

The **external urethral orifice** is immediately in front of that of the vagina, and is about 2·5 cm. behind the glans clitoridis. The opening has the appearance of a vertical slit or of an inverted V-shaped cleft whose margins are prominent and in contact with each other. Sometimes on each side of the urethral orifice there may be seen the minute opening of the para-urethral duct [p. 523].

The **orifice of the vagina** varies in appearance with the condition of the **hymen.** When the hymen is intact the opening is small, and is seen only when that membrane is put on the stretch. When the hymen has been ruptured the opening is much larger, and round its margins are often seen small projections—**carunculae hymenales**—which are fragments of the hymen.

The **hymen** is a thin membranous fold which partially closes the lower end of the vagina, and is perforated usually in front of its middle point, though rarely it may be complete. The position of the opening gives the fold, when stretched, a crescentic appearance. The opening in the hymen is sometimes cleanly cut, sometimes fringed. The membrane is not stretched tightly across the lower end of the vagina, but is so ample that it projects downwards into the pudendal cleft. The opening is thus a median slit whose margins are normally in contact.

On each side of the vaginal opening, in the angle between the labium minus and the inferior surface of the carunculae hymenales, is the opening of the duct of the **greater vestibular gland** [see below]. It is usually just large enough to be visible to the unaided eye [FIG. 666].

Numerous minute mucous glands—the **lesser vestibular glands**—open into the vestibule between the urethral and vaginal orifices.

Clitoris. The clitoris is the morphological equivalent of the penis, and it has a body, a pair of crura, and a minute glans at the distal end of the body. Unlike the penis, the clitoris is not traversed by the urethra.

The **body** is composed for the most part of erectile tissue resembling that of the penis. It is about 2·5 cm. in length, and is bent upon itself [FIG. 666]. It tapers towards its distal end, which is covered by the glans. It is enclosed in a dense fibrous fascia and is divided by an incomplete **septum** into a pair of cylindrical **corpora cavernosa,** which diverge from each other at the root of the clitoris to form the **crura.** A fibrous **suspensory ligament** passes from the body of the clitoris to the pubic symphysis [FIG. 666].

The **glans** of the clitoris is a small mass of erectile tissue fitted over the pointed end of the body, and, like the glans penis, it is covered with a very sensitive epithelium. The prepuce, which hoods over the glans, and the frenulum, which is attached to it inferiorly, are continuous with the labia minora [FIG. 665].

The **crura** of the clitoris diverge from the body posteriorly, and are attached to the sides of the pubic arch. Each is continuous with one of the corpora cavernosa, and has a firm fibrous fascia which is covered by the corresponding ischio-cavernosus muscle. In structure the crura and body of the clitoris resemble the corpora cavernosa of the penis, while the glans more closely resembles the bulbs of the vestibule, with which it is connected by a slender band of erectile tissue [FIG. 666].

In the seal and some other animals a bone—the **os clitoridis,** which represents the os penis of the male—is developed in the septum of the clitoris.

The **arterial supply** of each crus is the deep artery of the clitoris, a branch of the internal pudendal. The glans is supplied by the dorsal arteries of the clitoris.

The **nerve supply** of the clitoris is derived partly from the inferior hypogastric plexuses and partly from the dorsal nerves of the clitoris, which are branches of the pudendal nerves.

Bulbs of the Vestibule. These are a pair of masses of erectile tissue which correspond developmentally to the two halves of the bulb of the penis, but are almost completely separated from each other by the vagina and the urethra—being only slightly connected in front by a narrow median part called the **commissure.** Each bulb is thick posteriorly, and more pointed in front, where it joins the commissure. It rests against the lateral wall of the vagina and the lower surface of the perineal membrane. Superficially it is covered by the bulbospongiosus muscle. The commissure lies in front of the opening of the urethra, and is connected by a slender band of erectile tissue with the glans clitoridis. The bulb is mostly composed of minute convoluted blood vessels held together by a very small amount of areolar tissue; they frequently anastomose with one another, and with the vessels of the commissure and the glans clitoridis.

The **blood supply** of the bulb is derived from a special branch of the internal pudendal artery.

Greater Vestibular Glands. The greater vestibular glands are placed one on each side of the lower part of the vagina, below the perineal membrane and concealed by the posterior parts of the bulbs of the vestibule. Each is about the size and shape of a small bean; and it has a long slender duct which pierces the perineal membrane and opens into the vestibule in the angle between the labium minus and the hymen [FIGS. 662 and 666]. They are tubulo-alveolar mucous glands, similar to the bulbo-urethral glands [p. 534].

Suspensory ligament

Clitoris

Commissure of the bulbs

Crus clitoridis

Urethral orifice

Bulb of vestibule

Greater vestibular gland

Inferior fascia of urogenital diaphragm

Vagina

Central tendon of perineum

FIG. 666 Dissection of female perineum to show clitoris, bulb of the vestibule, and the greater vestibular glands.

C.T.A.—T

THE DEVELOPMENT OF THE UROGENITAL ORGANS

General Account. The urogenital system develops from intermediate mesoderm (intermediate cell mass), situated originally between the paraxial mesoderm medially and the lateral plate mesoderm, with the coelomic epithelium (splanchnopleuric intra-embryonic mesoderm) covering its surface.

In an embryo which is 2·5 mm. in length, the intermediate mesoderm in the cervical region has differentiated into a mass of cells from which a duct is beginning to extend in a caudal direction. This mass of cells is the **pronephros.** It corresponds to the functional kidney of some chordates and forms a simple tubular system which connects the coelom with the

pronephric duct in the embryos of some mammals; in the human embryo, however, it is probably functionless and soon becomes vestigial. The duct which grows caudally from it is the **pronephric duct,** and this is then utilized by the intermediate mesoderm in the thoracic and upper lumbar segments of the embryo which forms an embryologically functional kidney known as the **mesonephros.** The part of the pronephric duct utilized by this structure then becomes the **mesonephric** (Wolffian) **duct.** With the development of the mesonephros the pronephros atrophies.

Mesonephric Duct and Embryonic Excretory Organ

The mesonephric duct lies in the lateral aspect of the mesonephros. At the end of the fourth week this duct in its growth reaches the cloaca. As soon as the cloaca is divided into dorsal and ventral parts by the urorectal septum of intermediate mesoderm, the mesonephric ducts end in the ventral division, thereby dividing it into two parts [FIG. 667 B]. The cephalic

FIG. 667   Diagram to illustrate manner in which ureter, ductus deferens, and urinary bladder arise in the embryo.

The structures developed from the cloaca are indicated in red, those from the mesonephric duct in yellow, and the ectoderm in black.

The manner in which the rectum and bladder become separated and acquire openings to the exterior is shown in B and C.

A.	Allantois.	R.	Rectum.
B.	Bladder.	Ur.	Urogenital canal.
C.	Cloaca.	U.	Ureter.
CM.	Ectoderm of cloacal membrane.	VD.	Ductus deferens.
		VS.	Seminal vesicle.
K.	Pelvis of kidney.	WD.	Mesonephric duct.

FIG. 668   Tail-end of 3-mm. human embryo, before the time at which the mesonephric ducts reach the cloaca. (Keibel, 1896.)

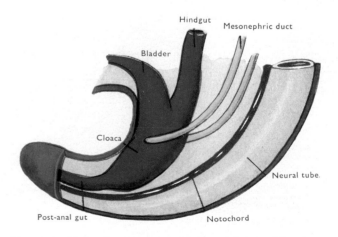

FIG. 669   Tail-end of 4·2-mm. human embryo. The mesonephric ducts open into the urogenital (anterior) part of the cloaca. (Keibel, 1896.)

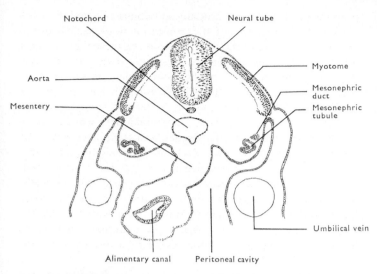

FIG. 670   Transverse section of 5-mm. human embryo.

vestigial structures, the epoophoron and paroophoron in the broad ligament of the uterus [p. 541].

Ureter and Permanent Kidney

The permanent kidney or **metanephros** has a double origin. The renal tubules are developed from a part of the intermediate cell mass caudal to that from which the mesonephros is derived; the ureter, the pelvis, calyces, and the collecting tubules of the kidney are formed from the ureteric bud of the mesonephric duct.

The **ureteric bud** arises as a tubular diverticulum from the mesonephric duct close to the point where it joins the cloaca [FIGS. 671 and 672]. The diverticulum appears during the fifth week, and grows headwards, dorsal to the body cavity. Even in its very early condition the portion of the outgrowth

part forms most of the urinary bladder, and all (in the female) or the cephalic part (in the male) of the urethra, while the caudal portion forms the urogenital sinus [FIGS. 667–673].

The **mesonephros** consists of a number of minute, transversely arranged tubules [FIG. 670], each of which opens laterally into the mesonephric duct, while its medial end is blind. The transverse tubules, like the canal into which they open, are at first solid cellular structures, and only later acquire a lumen. Increasing rapidly in length and number, the tubules become twisted and tortuous, and the blind end of each dilates to form a capsule invaginated upon itself and containing a bunch of capillary blood vessels similar to the **glomeruli** of the adult kidney. It would appear that primitively, as in the pronephros, one tubule is developed in the portion of the intermediate cell mass (nephrotome) which corresponds to each mesodermal somite, but, in higher vertebrates at all events, the mesonephros shows no sign of segmentation and there are several tubules opposite each somite, particularly at its caudal end where the tubules are very numerous. The tubules of the mesonephros arise in all segments from the sixth cervical to the third lumbar. The tubules in the headward part atrophy and disappear very early in development, even while others are being formed towards the caudal end of the embryo. When at its greatest development (fifth to eighth week) the mesonephros is a relatively large structure composed of tubules which resemble, but are shorter than, those of the adult kidney. At that time it bulges into the dorsal part of the body cavity, and extends from the region of the liver to the caudal end of the cavity.

In fishes and amphibia the mesonephros persists as the excretory organ of the adult, but when the metanephros develops in reptiles, birds, and mammals the mesonephros atrophies; yet some of its tubules persist in the male as the **efferent ductules** of the testis, the **ductuli aberrantes,** and the **paradidymis.**

The mesonephric duct becomes the duct of the epididymis, the ductus deferens, and the ejaculatory duct of the adult male [see p. 551]. In the female, when the permanent kidney is formed, the mesonephros and its duct atrophy to a greater extent than in the male, and they are represented only by

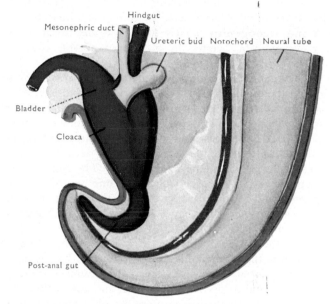

FIG. 671   Tail-end of 6·5-mm. human embryo.

The cloaca is dividing into rectal and urogenital parts. The ureter is arising as a bud from the mesonephric duct. (Keibel, 1896.)

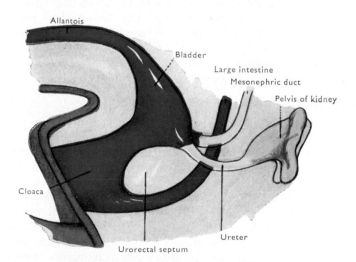

FIG. 672   Tail-end of 11·5-mm. human embryo.

The cloaca is becoming separated into rectal and urogenital portions by the formation of the urorectal septum. The ureter has acquired a separate opening into the urinary bladder. (Keibel, 1896.)

which lies nearest its origin from the mesonephric duct, and from which the adult ureter is developed, is more slender than the expanded distal part, which forms the pelvis of the kidney and gives rise to branches which become the calyces of the kidney. If the ureteric bud commences to branch too early, a partial or complete double ureter may result. From the calyces numerous collecting tubules grow out and acquire connexions with the renal tubules, which arise independently of the ureter in the **metanephric cap**—a condensation of the intermediate mesoderm covering the expanded ureteric bud. The blind, proximal end of each renal tubule soon dilates to form a capsule invaginated on itself to enclose a tuft of capillary blood vessels. The renal corpuscles arising in this manner are found in the human kidney as early as the eighth week. The renal tubules join the collecting tubules at their distal ends, at the site of the junctional tubule [p. 510] in the adult. If the tubules fail to join, the secreting tubules have no outlet and a **congenital cystic kidney** is the result; there are, however, other views on the origin of this condition (Willis, 1958).

The metanephric cap, in which the renal tubules arise, lies at first on the medial side of the ureteric bud and later on its dorsal aspect. As the ureter grows in a cephalic direction the metanephric cap follows it; hence, the metanephric tissue ceases to lie at the caudal end of the mesonephros, but extends cranially dorsal to it. If this process fails to occur, the kidney remains in the pelvic region in the adult. The right and left metanephric caps are closely related to one another at their caudal ends and fusion between them in this situation results in the formation of a 'horseshoe' kidney [p. 511]. Variations in the origin of the renal arteries are due to the change in position of the kidney during development; the blood supply is at first from the internal iliac artery, but as the kidney moves headwards it is supplied successively by branches from the common iliac and finally from the aorta, and any of these may persist, the most common in a normally placed kidney being an accessory artery to the lower pole.

As the ureteric bud divides to form the calyces, the metanephric cap breaks up into numerous cell masses—one for each calyx, and later one for each of the collecting tubules which grow out from the calyces. The formation of renal tubules within the cell masses is continued until shortly after birth. The kidney is at first distinctly lobulated; at birth, and sometimes even in the adult, distinct traces of its original division into lobules may be seen [FIG. 680].

Urinary Bladder

As the ureter increases in length, the part of the mesonephric duct uniting it to the ventral division of the cloaca is absorbed into that structure. Thus the mesonephric duct and ureter acquire independent openings into that part of the cloaca; the ureter entering nearer the head of the embryo into the part which will become the urinary bladder [FIG. 667].

The main portion of the urinary bladder is formed from the cephalic part of the ventral division of the cloaca. Early in development this becomes flattened laterally and has the two mesonephric ducts opening on either side of its narrow dorsal aspect [FIG. 672]. The tailward part of the ventral division of the cloaca, caudal to the openings of the mesonephric ducts, becomes constricted to form the **urogenital sinus.** The caudal ends of the mesonephric ducts, now entering separately

from the ureters, migrate caudally and obtain a secondary opening into the upper part of the urogenital sinus, caudal to the ureters, as a result of the absorption of the mesonephric ducts into the wall of the bladder and urogenital sinus. In the adult they therefore open close to the median plane in the prostatic part of the urethra. The urinary bladder has therefore a double origin. Its main portion is derived from the cloaca and is therefore entodermal; the remainder, which corresponds approximately to the trigone and the posterior wall of the urethra down to the entry of the mesonephric (here ejaculatory) ducts is mesodermal in origin for it arises from the parts of the mesonephric ducts absorbed into its walls. The extreme cephalic end of the ventral part of the cloaca tapers gradually, and is continuous with the allantoic diverticulum. This part of the cloaca is the **urachus**; it loses its lumen and in the adult is represented by a fibrous cord—the median umbilical ligament.

The cavity of the urachus is sometimes not lost so early, and in rare cases it persists in the child or adult as a patent channel that extends from the apex of the bladder to the umbilicus. The entoderm of the cloaca forms the mucous coat of the bladder except in the trigone. The muscular coat is derived from mesoderm which surrounds the bladder during development and which also contributes to the formation of the infra-umbilical region of the body wall (Wyburn, 1937; Glenister, 1958).

Malformations. Simple malformations of the bladder are uncommon. The gross malformation called **ectopia vesicae** or extroversion of the bladder is probably due to failure of fusion of the bilateral contributions of mesoderm which form the muscular coat of the bladder and the infra-umbilical region of the body wall. The lower part of the anterior abdominal wall is therefore deficient (in complete extroversion the pubic symphysis is absent) and the bladder is represented by an exposed, irregular area of mucous membrane—including the trigone—continuous with the surrounding skin. The ureters open on the exposed surface, and, in the male, this malformation is commonly associated with epispadias [p. 549].

Male Urethra

The posterior wall of the first part of the male urethra, down to the openings of the ejaculatory ducts, has an origin similar to that of the trigone of the bladder. The remaining portion—inferior to the openings of the ejaculatory ducts of the adult—is derived from the **urogenital sinus** or caudal division of the ventral part of the cloaca. The urogenital sinus is separated from the rectal part of the cloaca by a mesodermal septum, the **urorectal septum,** formed by fusion of the caudal ends of the columns of intermediate mesoderm, which also divides the original cloacal membrane into urogenital and anal membranes. Both of these membranes rapidly break down. The urogenital sinus itself is early divided into a pelvic part which lies in the future lesser pelvis and forms the prostatic urethra inferior to the entry of the ejaculatory ducts and the membranous urethra, and a phallic part which occupies the region where the corpus spongiosum is developed. The latter part of the urogenital sinus is flanked on each side by a genital fold and outside that by a **genital swelling.** Anteriorly lies the **genital tubercle** which, growing forwards, carries part of the sinus wall on its inferior surface (the **urethral plate**)

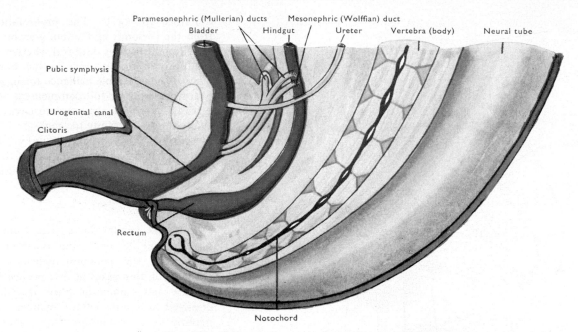

Paramesonephric (Mullerian) ducts   Mesonephric (Wolffian) duct
Bladder   Hindgut   Ureter   Vertebra (body)   Neural tube

Pubic symphysis

Urogenital canal

Clitoris

Rectum

Notochord

FIG. 673   Tail-end of 25-mm. female human embryo.
The rectum has acquired an external opening and the entoderm of the urogenital sinus is continued into the genital tubercle (clitoris).   (Keibel, 1896.)

flanked by the anterior extremities of the genital folds [FIG. 678]. The genital tubercle is formed by bilateral proliferation of the mesoderm of the infra-umbilical part of the anterior abdominal wall which forms the corpora cavernosa, while the apex develops into the glans penis. The **genital folds** fuse together beneath the opening of the urogenital sinus, and this fusion extends forwards on the inferior surface of the penis, closing off the urethral plate to form the spongy part of the urethra, and forming corpus spongiosum from its mesoderm (Paul and Kanagasuntheram, 1956; Glenister, 1958). This closure extends to the proximal margin of the glans penis where it is joined by a column of ectodermal cells which grows through the glans from its apex. This canalizes to form the terminal part of the urethra (including the fossa navicularis) and unites proximally with the spongy part, the opening of which on the inferior surface of the penis is then closed. The genital swellings, enlarging to accommodate the testes, extend beneath the posterior part of the fused genital folds, and meeting in the midline, fuse to form the scrotum; the line of fusion being indicated by the raphe.

Malformations. Malformations of the male urethra are not uncommon. They are due mainly to failure of closure of the genital folds at the two critical points—the junction of the membranous and spongy parts, and at the neck of the glans. Such abnormal openings of the urethra on the urethral surface of the penis constitute the malformation known as **hypospadias,** which varies in extent; abnormal formation of the fossa navicularis with hypospadias at the neck of the glans is the simplest and commonest variety. Complete hypospadias, associated with cryptorchism and failure of formation of the scrotum, occurs in male pseudohermaphroditism.

The cause of **epispadias,** in which the urethra opens on the dorsal surface of the root of the penis, is related to failure of fusion of the bilateral contributions of mesoderm which form the corpora cavernosa of the penis as well as the infra-

umbilical region of the body wall; it is therefore often associated with extroversion of the bladder.

Female Urethra

In the female the urogenital sinus opens out on to the surface and is continued forwards as a sulcus between the genital folds. The genital folds do not unite, but form the labia minora of the adult, the genital tubercle forms the clitoris, and the genital swellings develop into the labia majora. Later, a shortening and spreading out of the lower portion of the urogenital sinus, to form the vestibule of the adult, is responsible for bringing the opening of the urethra to the surface. The female urethra corresponds to the part of the male passage which lies above the opening of the prostatic utricle and ejaculatory ducts, since it is derived from the cephalic part of the ventral division of the cloaca caudal to the portion forming the bladder, and cranial to the opening of the mesonephric ducts.

Gonads

In the development of the gonads, male and female, the coelomic epithelium (splanchnopleuric intra-embryonic mesoderm) on the medial side of the mesonephros thickens to form a genital (gonadal) ridge. The **genital ridge** is soon found to have numerous **rete cells** embedded in its stroma of embryonic connective tissue, and some of them appear to originate, in both sexes, by a proliferation from the deep surface of the epithelium that covers the ridge.

The tissue which gives rise to the genital ridge occurs in all the body segments from the sixth thoracic to the second sacral, but the cephalic end of the ridge atrophies before the germinal epithelium can be recognized in the more caudal segments, and only about one-fourth to one-half of the ridge gives origin to the permanent gonad. As the gonad develops it becomes suspended by a mesentery (mesorchium in the case

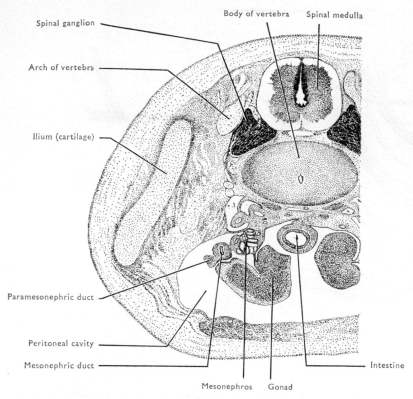

FIG. 674　Transverse section through lower part of trunk of human embryo of about 7 weeks. (Symington.)

of the testis, mesovarium for the ovary) from the medial aspect of the mesonephros [FIG. 674].

IN THE MALE, as early as the thirty-third day, the rete cells embedded in the stroma of the developing testis have become arranged into a network of **rete cords.** The rete cords undergo direct transformation into the **convoluted** and **straight seminiferous tubules,** and the **rete testis.** At a very early stage the superficial part of the stroma of the developing testis becomes denser, and gives origin to the **tunica albuginea.** The tissue surrounding the rete cords becomes converted into the septula of the testis and the mediastinum. A lumen can first be recognized in the seminiferous tubules in the seventh month. The rete testis becomes connected secondarily with the efferent ductules of the testis, which are derived from those tubules of the mesonephros which lie opposite the definitive gonad; these are connected with the mesonephric duct which becomes the epididymis and ductus deferens.

IN THE FEMALE, numerous large **primordial oocytes** are found in the stroma of the developing ovary beneath the germinal epithelium as early as the thirty-third day. They are formed outside the developing ovary, from the entoderm of the yolk sac, and later migrate through the root of the mesentery to occupy their permanent position within the stroma of the ovary (Witschi, 1948). The primordial spermatogenic cells, which become located in the rete cords, may have a similar origin. At first the primordial oocytes are isolated, but later— about the fifth week—they become surrounded by smaller cells which arise from the coelomic epithelium. Each primordial oocyte, surrounded by its cells, becomes a **primordial follicle,** the further development of which has already been

described [p. 537]. The proliferation of cells from the coelomic epithelium goes on until birth, but after birth it is doubtful whether oocytes are produced in this manner in the human ovary. There is considerable evidence to suggest that the ovary possesses its full complement of oocytes at birth and that these mature in successive menstrual cycles during adult life [p. 537].

Descent of the Testis and Ovary. The descent of the testis into the scrotum and of the ovary into the pelvis have already been considered [pp. 526 and 536].

Genital Ducts

The male genital ducts arise from the mesonephros and its duct; but those of the female arise from the paramesonephric ducts. The embryos of both sexes at first possess both mesonephric and paramesonephric ducts, which are arranged in a very definite manner. The mesonephric ducts, communicating directly with the tubules of the mesonephros, lie at first parallel to each other, and at a considerable distance apart. As they pass towards the caudal end of the embryo they approach each other, and each becomes enclosed in a fold of peritoneum called the **plica urogenitalis.** More caudally the ducts become closely approximated, and finally open into the dorsal aspect of the ventral division of the cloaca.

The **paramesonephric** (Müllerian) **ducts** are formed by an evagination of the coelomic epithelium at the cephalic end of the mesonephros (Faulconer, 1951), and therefore open freely into the body cavity. Each duct extends caudally on the lateral side of the corresponding mesonephric duct, and crossing ventral to it at the caudal extremity of the mesonephros, passes medially in the upper part of the urorectal septum to meet and fuse with its fellow. The median canal so formed descends in the septum to end blindly against the dorsal wall of the urogenital sinus between the points of entry of the two mesonephric ducts [FIG. 673]. Here it forms a solid tubercle of cells which invaginates the posterior wall of the urogenital sinus (Müllerian tubercle).

Ducts in the Male. By the third month the seminiferous tubules of the testis become connected with the mesonephric

FIG. 675　Urogenital passages at the indifferent stage of development.

Ureter: solid yellow. Mesonephric duct: dotted yellow; the origin of the seminal vesicle is indicated. Paramesonephric ducts: orange. Rectum, bladder, and urogenital sinus: red.

FIG. 676   Male urogenital passages.

Ureter, epididymis, and ductus deferens: yellow.   Prostatic utricle: orange.   Bladder, pelvic part of urethra and rectum: red.   Penile part of urethra: black.

FIG. 677   Female urogenital passages.

Derivatives of paramesonephric duct: orange.   Ureter: solid and dotted yellow.   The epoophoron is indicated in yellow between the uterine tube and the ovary.   Bladder, urethra, and rectum: red.

duct through a fusion of the tubules of the mesonephros adjacent to the gonad with the rete testis. The number of tubules that take part varies considerably, but each forms one of the efferent ductules found in the adult. The efferent ductules, becoming more convoluted where they join the mesonephric duct, form the **lobules of the epididymis,** while the **duct of the epididymis** is formed from the cephalic part of the mesonephric duct, and the **ductus deferens** from the more caudal portion. The ductuli aberrantes and the paradidymis are to be looked upon as persistent tubules of a more caudal portion of the mesonephros which have failed to become connected with the tubules of the testis, while the appendix of the epididymis represents the more cephalic remnants of the mesonephros and its duct.

The **seminal vesicles** are developed in the third month as evaginations from the mesonephric ducts near their caudal ends. The duct of the seminal vesicle therefore joins the ductus deferens to form the **ejaculatory duct.**

The **paramesonephric ducts** atrophy in the male embryo, but the appendices of the testes are vestigial remains of their cephalic portions and the prostatic utricle represents the fused caudal portions [FIG. 676].

Ducts in the Female. The paramesonephric ducts in the female retain their openings into the body cavity, and their cephalic portions become the uterine tubes. Their fused caudal parts, which at first abut on the urogenital sinus, give rise to the **uterus** and at least the greater part of the **vagina.** The mechanism whereby the definitive female urethra comes to open into the vestibule, involving shortening and opening out of the urogenital sinus to form the major part of the vestibule [p. 549], is also responsible for the lower end of the fused paramesonephric ducts being carried caudally till it lies above the vestibule and therefore forms the eventual opening of the vagina into the vestibule, the remnants of the intervening tissue forming the hymen. The epithelium lining the vagina may be derived from entodermal cells of the pelvic part of the urogenital sinus (Bulmer, 1959).

The mesonephric ducts and the mesonephros atrophy in the female, but traces of them are found as the epoophoron and paroophoron [FIG. 677]. In the foetus the mesonephric duct can be traced along the side of the uterus as far as the upper end of the vagina, where it is known as the duct of Gärtner.

Malformations. Variations of the uterus have been mentioned on page 542 as due to the arrest of the fusion of the paramesonephric ducts. Such failure of union may be so complete that the uterus and vagina appear to be double, or a dividing septum may persist in one or other of these organs or in both. The vagina may fail to reach the surface, or the simple malformation of 'imperforate hymen' may result from failure of the fused paramesonephric ducts to open into the vestibule.

Accessory Glands

The glandular portion of the **prostate** arises as a series of solid outgrowths from the epithelium of the urogenital sinus during the third month. The outgrowths are simple at first, but become branched and finally acquire a lumen. They are arranged in three groups—an upper and lower dorsal, and a ventral group. The glands of the ventral group soon become reduced in number and often completely disappear; those of the upper dorsal group form the chief part of the prostate. The prostatic glands arise in both sexes; but in the female, where they are known as **para-urethral glands,** they are few in number and not densely packed as in the male. The muscular tissue of the prostate is derived from the mesoderm surrounding it during development.

The **bulbo-urethral glands** arise in the third month as outgrowths from the epithelium of the urogenital sinus. The **greater vestibular glands** arise in the same manner.

External Genital Organs [FIG. 678]

The external genital organs are developed in the region of the cloacal membrane, and the male and female are alike in the earlier stages. Part of this membrane at first extends on the ventral aspect of the body almost from the tail to the umbilical cord. At its cephalic end there is a tubercle known as the

A

B   C   D

FIG. 678   External genital organs of human embryo.

A, embryo of 20 mm.   B, slightly larger.
Indifferent stage.   The genital folds and genital swellings are evident.
C, male embryo.
Formation of scrotum.   The genital swellings have grown backwards and united behind the primitive urogenital opening to form the scrotal raphe.   The genital folds are fusing below the base of the genital tubercle to form the body of the penis.   The glans, formed from the genital tubercle, is very prominent.
D, male embryo at later stage.
Behind the glans penis the urethra opens in a diamond-shaped fossa at the proximal end of which the median raphe ends.   The prepuce is forming behind the constriction which marks off the glans, and later grows forwards to cover it.

genital tubercle, and at its caudal end there is a **coccygeal tubercle.**   The cloacal membrane becomes divided by the urorectal septum into an **anal membrane,** which lies immediately in front of the coccygeal tubercle, and a **urogenital membrane** situated between the urorectal septum and the genital tubercle.   Both membranes rapidly break down: the anal orifice thus formed becomes surrounded completely by a circular ridge and therefore comes to lie at the base of an ectodermal pit, the **proctodaeum,** which forms the lower one-third of the anal canal.   Anteriorly, through the urogenital membrane, the urogenital sinus comes to open on the surface by a slit-like aperture, the **primitive urogenital opening,** the margins of which are raised up to form the **genital folds.**   A further pair of prominent folds, the **genital swellings** (labioscrotal folds), appear lateral to the genital folds.

IN THE FEMALE, the genital swellings give rise to the labia majora.   The genital folds give origin to the labia minora, and the genital tubercle becomes the clitoris.   On the clitoris at a very early date the glans is marked off by a surrounding sulcus.

IN THE MALE, the genital swellings grow backwards and, meeting behind the urogenital opening, fuse together to form the scrotum.   The genital tubercle develops into the glans penis, and the genital folds fuse together to complete the ventral wall of the penis, and the penile urethra.

Hermaphroditism.   True hermaphroditism—the presence of both testis and ovary—is exceedingly rare in the human subject; the production of an ovary and testis in the same individual, or the two constituents of a mixed ovotestis, is probably determined genetically.

Pseudohermaphroditism is defined as the association of the gonads of one sex with external genital organs that resemble those of the other sex.   Complete hypospadias in the male [p. 549] gives rise to the commonest variety; the external

genital organs, through failure of the testes to descend and of the genital swellings to unite, may closely resemble those of the female.   The opposite condition, with enlargement of the clitoris to simulate a penis and possible descent of the ovaries, is much less common.   It is probable that these conditions, in which secondary sex characters also are affected, are due to disturbance of normal regulation of development by sex hormones.   Such a disturbance is usually caused by abnormal activity of the suprarenal cortex.

THE MAMMARY GLANDS

The mammary glands are accessory organs of the female reproductive system.   Each gland is situated on the front of the thorax in the superficial fascia of the hemispherical elevation known as the mamma or **breast.**   It usually extends from the second rib to the sixth rib and lies on the pectoralis major and, to a lesser extent, on the obliquus externus abdominis and the serratus anterior.   A part of the gland extends towards the axilla partly under cover of the lateral border of the pectoralis major, and is known as the **axillary tail.**   The **nipple** is situated near the summit of the breast, and tends to lie at the level of the fourth intercostal space in the midclavicular line, though this is very variable; the lactiferous ducts open on it by minute apertures, and it is surrounded by a coloured, circular area of skin called the **areola.**   The skin of the nipple is thrown into numerous wrinkles, and on the areola it exhibits many minute, rounded projections due to the presence of underlying **areolar glands.**   The colour of the nipple and areola varies with the complexion, but in young subjects it is usually a rosy-pink, and it changes to a deep brown during the second and third months of the first pregnancy.   Also, during pregnancy, the areola increases in size and its glands become more marked.   The nipple contains a considerable number of plain muscle fibres, and it becomes firmer and more prominent as a result of mechanical stimulation.

The size and appearance of the breasts vary much, not only in different races of mankind, but also in the same person under different conditions.   In the young child they are small, and there is little difference between those of the male and female.   Their growth is slow until the approach of puberty, and then the female mammary glands increase rapidly in size.   At each pregnancy the breasts enlarge, and attain their greatest development during lactation.   The size of the breasts depends partly on the amount of superficial fat and partly on the amount of glandular tissue present, though the amount of glandular tissue is very small in the non-lactating breast.

Structure of Breast.   The mamma is composed of a mass of glandular tissue traversed and supported by strands of fibrous tissue, and covered with skin and a thick layer of fat.   The glandular tissue forms a conical mass, the apex of which corresponds to the position of the nipple, while its base is loosely connected to the deep fascia on which the gland lies.   In section it is readily distinguished from the surrounding fat by its firmer consistence and by its pinkish-white colour.   It is composed of from fifteen to twenty lobes

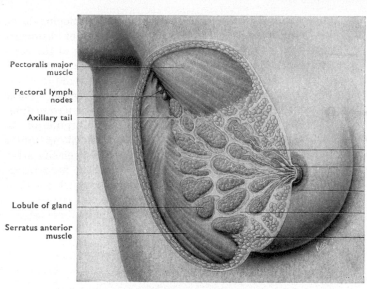

Pectoralis major muscle

Pectoral lymph nodes

Axillary tail

Lobule of gland

Serratus anterior muscle

Lactiferous duct

Lactiferous sinus

Areolar gland

Stroma

Subcutaneous fatty tissue

FIG. 679   Dissection of the right mammary gland.

divided into lobules, which make its superficial surface and edges very uneven—the inequalities of its surface being filled up by processes of the fatty tissue which covers the gland. The fatty covering is incomplete in the region of the areola, and here the lactiferous ducts pass into the nipple. The **lobes** radiate from the nipple, each lobe being quite distinct from the others and possessing its own duct; the **lobules** are bound together and supported by a considerable amount of fibro-areolar tissue, which forms the **stroma** of the gland [FIG. 679].

The alveoli of the gland and the secretory epithelium which lines them vary much under different conditions. At puberty the glandular tissue is composed chiefly of the ducts; at that time the alveoli are small and few in number. During lactation, when the gland is fully functional, the alveoli are enlarged, distended with fluid, and much more numerous. The epithelial cells are cubical and filled with fat globules. When the gland is not secreting, the alveoli become small and reduced in number, and the cells of the lining epithelium, which are then small, do not contain fat globules.

The **lactiferous ducts,** passing towards the nipple, enlarge to form small, spindle-shaped dilatations called **lactiferous sinuses;** then, becoming once more constricted, each duct passes, without communicating with its neighbours, to the summit of the nipple, where it opens on the surface.

In the male the various parts of the breast are represented in a rudimentary condition.

Variations. Asymmetry in the development of the breasts is very common—the left mamma being often larger than the right. Absence of one or both mammae is a very rare abnormality which may or may not be associated with absence of the nipples. When one nipple only is present it is usually the left. The presence of supernumerary glands or nipples is uncommon. The term **polymastia** has been applied to cases in which more than the normal number of mammae are present, and **polythelia** to those in which accessory nipples occur. Usually the accessory glands, or nipples, are present on the front of the thorax, and in most instances they occur below and a little to the medial side of the normal site. When the

abnormal glands are found above the normal site they generally lie further from the median plane. Much more rarely, accessory glands have been found on the abdomen, in the axilla, or in some other situation, such as the medial aspect of the thigh. Asymmetry is very common in such abnormal structures. Examples of polymastia and polythelia occur in the male as well as in the female.

Vessels and Nerves of the Breast

The mamma receives its arterial supply from the perforating branches of the **internal thoracic artery** and from the lateral mammary branches of the **lateral thoracic.** Additional supply is derived from some of the intercostal vessels. The veins from the gland pour their blood into the **axillary** and **internal thoracic veins.** Some small superficial veins from the breast join tributaries of the external jugular.

The **lymph vessels** of the breast are very numerous, and for the most part join the lymph nodes of the axilla. They take origin from an extensive **perilobular plexus.** The majority follow the lactiferous ducts and thus converge towards the nipple, and there join a plexus situated beneath the areola (subareolar plexus). From this plexus the main efferent vessels pass to the **pectoral axillary nodes,** but it is important to remember that free communication exists between all the subgroups of axillary nodes. In addition, vessels pass directly from the deep surface of the lateral part of the breast either through, above, or behind the pectoral muscles to the **apical axillary nodes.** A few vessels from the medial part of the breast, following the course pursued by the perforating arteries, may join the **parasternal** lymph nodes situated along the course of the internal thoracic artery. It is also to be remembered that a few, probably irregular, communications exist across the median plane with the lymph vessels of the opposite breast; and, further, that under diseased conditions communications may exist with lymph vessels in the upper part of the sheath of the rectus abdominis, and so with the hepatic nodes, via the falciform ligament, and parasternal nodes. The surgical importance of the facts regarding the lymphatic drainage of the breast cannot be exaggerated.

The **nerve supply** of the gland is derived from the **intercostal nerves** of the fourth, fifth, and sixth intercostal spaces. Along the course of these nerves sympathetic filaments reach the breast from the thoracic part of the sympathetic trunk. The nipple has a particularly rich nerve supply (Cathcart, Gairns, and Garven, 1948).

Development of the Mammary Glands

The mammary glands are developed as ingrowths of the ectoderm into the underlying mesodermal tissue. In the human embryo a thickened raised area of the ectoderm can be recognized in the region of the future gland at the end of the fourth week. The thickened ectoderm becomes depressed into the underlying mesoderm, and thus the surface of the mammary area soon becomes flat, and finally sinks below the level of the surrounding epidermis. The mesoderm, where it is in contact with the ingrowth of the ectoderm, is compressed, and its elements become arranged in concentric layers which, at a later stage, give rise to the stroma of the gland. The ingrowing mass of ectoderm cells soon becomes flask-shaped and then grows out into the surrounding mesoderm as a number of solid processes which represent the future ducts of the gland. These processes, by dividing and branching, give rise to the future lobes and lobules, and, much later, to the alveoli. The mammary area becomes gradually raised again in its central part to form the nipples. A lumen is formed in each part of the branching system of cellular processes only at birth, and the secretion of a fluid resembling milk may take place at this time as a result of maternal hormones which have passed across the placenta into the child's circulation. The lactiferous

sinuses appear before birth as swellings of the developing ducts.

In those animals which possess a number of mammary glands—such as the cat and pig—the thickening of the ectoderm, which is the first indication of the development of these glands, takes the form of a pair of ridges that extend from the level of the fore limb towards the inguinal region. These mammary ridges converge caudally, and at their terminations lie not far from the median line. By the absorption of the intervening portions, the ridges become divided up into a number of isolated areas in which the future glands arise. Similar linear thickenings of the ectoderm have been recognized in the human embryo also, and the usual positions assumed by the accessory glands when present suggest that polymastia and polythelia are caused by abnormal persistence of portions of the mammary ridges.

RADIOGRAPHIC ANATOMY

E. SAMUEL

THE RENAL TRACT

The kidney outlines usually can be clearly seen on a plain film of the abdomen as the surrounding fat in the perinephric tissues is less opaque to X-rays than kidney tissue. Likewise, the soft tissue shadow of the urinary bladder can frequently be seen against the gas-filled intestinal coils as they lie in the pelvis. The ureters, on the other hand, having the same density as the surrounding soft tissue structures, are not visible on the plain radiograph. Owing to the same relative absence of contrast, the renal pelvis, the calyces, and the internal structure of the kidney are not seen.

The outlines of the renal pelvis and calyces are demonstrated by the intravenous injection (intravenous pyelogram) [PLATE 37 C] of a contrast medium which is selectively excreted by the kidneys and in this way the lumen of the renal pelvis and of the major and minor calyces are seen. Another method of demonstrating these structures is by the injection of contrast medium through a catheter introduced into the ureter through a cystoscope (retrograde pyelogram) [PLATE 37 D]. This method has the merit of producing a shadow of greater density than that obtained by an intravenous pyelogram but has the disadvantage of requiring the introduction of a cystoscope.

Three major calyces (upper, middle, and lower) are seen to arise from the lateral aspect of the renal pelvis and from each of these major calyces, one to three minor calyces arise. The ends of the calyces are cup-shaped where they abut on the renal pyramids. In aortograms [PLATE 37 B] the pyramids of the kidneys may be distinguished from the remainder of the kidney substance first by their diminished and later by their increased, density, when the highest concentration of contrast medium is in the kidney substance. In an intravenous pyelogram, differentiation of the pyramids is seldom possible because of the low concentration of the contrast medium.

The **renal pelvis** is trumpet-shaped with the ureter arising from the caudal end. The ureter itself runs downwards and medially, the latter accentuated where the ureter crosses the

psoas muscle. Except in cases of obstruction, the ureter is not filled with contrast medium as the flow along it is intermittently produced by peristaltic waves.

In the lesser pelvis, the ureter turns medially to enter the posterolateral aspect of the bladder and a faint linear translucence joining the ureteric openings, the interureteric fold, can be seen.

Congenital anomalies in the site and shape of the kidneys can be recognized by pyelography and indeed even on a plain X-ray film the absence of a renal shadow from its normal position may suggest an ectopic kidney. Similarly, fusion of the two kidneys to form a horseshoe kidney can be seen on the pyelogram. Deformities of the renal pelvis and calyces are many and the number and pattern of calyces described above are subject to numerous variations, the commonest being an incomplete split in the renal pelvis, the bifid pelvis, which in varying degrees may be associated with a double ureter, the two fusing during their abdominal or pelvic course, or, less frequently, opening by separate orifices into the bladder.

Abnormalities in the position of the kidneys are frequently associated with abnormalities in the origin of the renal vessels. Thus, in a kidney situated in the pelvis the renal artery may arise from the adjoining iliac vessel, a fact that can be verified by aortography.

The radiographic appearances of the bladder shadow are naturally dependent on the amount of fluid present, being pyramidal in the partly empty phase and ovoid in the distended state.

The urethra is demonstrated radiographically by injecting water-soluble opaque media into the external urethral opening and taking films whilst the urethra is filled (urethrogram). The same procedure with the injection of more fluid may be used to outline the bladder (cystourethrogram) [PLATE 39 B]. In young children, the density of dye excreted into the bladder during an intravenous pyelogram may be such that the urethra may be seen during micturition.

To demonstrate the kidney outlines, oxygen, air, or carbon dioxide can be introduced into the retroperitoneal tissues, either by direct puncture of the perinephric space (perirenal pneumography) or by injection into the presacral tissues (presacral perirenal pneumography) and allowing the air to infiltrate gradually into the perirenal tissues [PLATE 37A].

THE GENITAL SYSTEM

Injection of the ductus deferens in the spermatic cord (after surgical exposure) to demonstrate the seminal vesicles (vesiculogram) is occasionally used to show the anatomy of these structures and to detect disease processes. This investigation is performed in cases of male infertility to detect obstructions in the ductus deferens and in the seminal vesicles.

Radiographic investigation of the female genital tract is performed almost as a routine in cases of infertility. Contrast medium is injected through the cervix, usually under fluoroscopic control, and the uterine cavity outlined [PLATE 39 C]. The uterine tubes can be seen as thin, thread-like structures extending towards the lateral walls of the pelvis. Obstructions and narrowings can thus be shown. If the contrast medium is allowed to flow out into the pelvis, it can be made to outline the ovaries and peritoneal pouches by appropriate positioning of the patient [PLATE 39 D]. Similar results can be obtained by the introduction of relatively large volumes of carbon dioxide or air, but in this case, the pelvic viscera appear as positive shadows surrounded by the gas (pelvigraphy).

Another useful method of demonstrating the pelvic structures on X-ray examination is by the injection of contrast material through a catheter introduced into the internal iliac artery; thereby outlining the organs by their blood vessels and showing displacements or change in shape of them.

REFERENCES

BACKHOUSE, K. M., and BUTLER, H. (1960) The gubernaculum testis of the pig (Sus scropha), J. Anat. (Lond.), **94,** 107.

BRAASCH, W. F., and EMMETT, J. L. (1951) *Clinical Urography,* Philadelphia.

BRAITHWAITE, J. L. (1952) The arterial supply of the male urinary bladder, *Brit. J. Urol.,* **24,** 64.

BRAMBELL, F. W. R. (1956) Ovarian changes, in *Marshall's Physiology of Reproduction,* Vol. I, Part I, ed. Parkes, A. S., London.

BRASH, J. C. (1922) The relation of the ureters to the vagina : with a note on the asymmetrical position of the uterus, *Brit. med. J.,* **2,** 790.

BULMER, D. (1959) The epithelium of the urogenital sinus in female human foetuses, *J. Anat. (Lond.),* **93,** 491.

CATHCART, E. P., GAIRNS, F. W., and GARVEN, H. S. D. (1948) The innervation of the human quiescent nipple, with notes on pigmentation, erection and hyperneury, *Trans. roy. Soc. Edinb.,* **61,** 699.

CLEGG, E. J. (1955) The arterial supply of the human prostate and seminal vesicles, *J. Anat. (Lond.),* **89,** 209.

CLEGG, E. J. (1956) The vascular arrangements within the human prostate gland, *Brit. J. Urol.,* **28,** 428.

CLEGG, E. J. (1957) The musculature of the human prostatic urethra, *J. Anat. (Lond.),* **91,** 345.

FAULCONER, R. J. (1951) Observations on the origin of the Müllerian groove in human embryos, *Contr. Embryol. Carneg. Inst.,* **34,** 159.

FRANKL, O. (1933) On the physiology and pathology of the isthmus uteri, *J. Obstet. Gynaec. Brit. Emp.,* **40,** 397.

FRANKLIN, K. J. (1949) The history of research upon the renal circulation, *Proc. roy. Soc. Med.,* **42,** 721.

GLENISTER, T. W. (1958) A correlation of the normal and abnormal development of the penile urethra and of the infra-umbilical abdominal wall, *Brit. J. Urol.,* **30,** 117.

HARRISON, R. G. (1949) The distribution of the vasal and cremasteric arteries to the testis and their functional importance, *J. Anat. (Lond.),* **83,** 267.

HARRISON, R. G. (1963) *A Textbook of Human Embryology,* 2nd ed., Oxford.

HARRISON, R. G. (1966) The anatomy of varicocele, *Proc. roy. Soc. Med.,* **59,** 763.

HARRISON, R. G., and BARCLAY, A. E. (1948) The distribution of the testicular artery (internal spermatic artery) to the human testis, *Brit. J. Urol.,* **20,** 57.

HARRISON, R. G., and WEINER, J. S. (1949) Vascular patterns of the mammalian testis and their functional significance, *J. exp. Biol.,* **26,** 304.

HUNTER, J. (1786) A description of the situation of the testis in the foetus, with its descent into the scrotum, Palmer's (1837) edition of *Works of John Hunter,* Vol. IV, p. 1, London.

LÖFGREN, F. (1949) *Das topographische System der Malpighischen Pyramiden der Menschenniere,* Lund.

MACKENRODT, A. (1895) Über die Ursachen der normalen und pathologischen Lagen des Uterus, *Arch. Gynäk.,* **48,** 393.

MACMILLAN, E. W., and AUKLAND, J. (1960) The transport of radiopaque medium through the initial segment of the rat epididymis, *J. Reprod. Fertil.,* **1,** 139.

MARTIN, C. P. (1942) A note on the renal fascia, *J. Anat. (Lond.),* **77,** 101.

MITCHELL, G. A. G. (1939) The spread of retroperitoneal effusions arising in the renal regions, *Brit. med J.,* **2,** 1134.

MITCHELL, G. A. G. (1950) The renal fascia, *Brit. J. Surg.,* **37,** 257.

MITCHELL, G. A. G. (1956) *Cardiovascular Innervation,* Edinburgh.

PAUL, M., and KANAGASUNTHERAM, R. (1956) The congenital anomalies of the lower urinary tract, *Brit. J. Urol.,* **28,** 118.

POWER, R. M. H. (1944) The exact anatomy and development of the ligaments attached to the cervix uteri, *Surg. Gynec. Obstet.,* **79,** 390.

POWER, R. M. H. (1946) The pelvic floor in parturition, *Surg. Gynec. Obstet.,* **83,** 296.

RACKER, D. C., and BRAITHWAITE, J. L. (1951) The blood supply to the lower end of the ureter and its relation to Wertheim's hysterectomy, *J. Obstet. Gynaec. Brit. Emp.,* **53,** 609.

REYNOLDS, S. R. M. (1948) Morphological determinants of the flow-characteristics between an artery and its branch, with special reference to the ovarian spiral artery in the rabbit, *Acta anat. (Basel),* **5,** 1.

RICHARDSON, K. C. (1962) The fine structure of autonomic nerve endings in smooth muscle of the rat vas deferens, *J. Anat. (Lond.),* **96,** 427.

SCHINZ, H. R., BAENSCH, W. E., FRIEDL, E., and UEHLINGER, E. (1954) *Roentgen-Diagnostics—Gastrointestinal Tract, Gynaecology, Urology,* Vol. IV, Eng. trans., Case, J. T., London.

SHANKS, S. C., and KERLEY, P. (1970) *A Textbook of X-Ray Diagnosis,* Vol. V, 3rd ed., London.

SOUTTAR, H. S. (1947) On complete removal of the prostate : a preliminary communication, *Brit. med. J.,* **1,** 917.

SUVANTO, O., and KORMANO, M. (1970) The relation between *in vitro* contractions of the rat seminiferous tubules and the cyclic stage of the seminiferous epithelium, *J. Reprod. Fertil.,* **21,** 227.

SWYER, G. I. M. (1944) Postnatal growth changes in the human prostate, *J. Anat. (Lond.),* **78,** 130.

SYKES, D. (1964a) Some aspects of the blood supply of the human kidney, *Symp. Zool. Soc. Lond.,* **11,** 49.

SYKES, D. (1964b) The morphology of renal lobulations and calyces, and their relationship to partial nephrectomy, *Brit. J. Surg.,* **51,** 294.

THOMPSON, R. (1919) The capacity of, and the pressure of fluid in, the urinary bladder, *J. Anat. (Lond.),* **53,** 241.

TRUETA, J., BARCLAY, A.E., DANIEL, P. M., FRANKLIN, K. J., and PRITCHARD, M. M. L. (1947) *Studies of the Renal Circulation*, Oxford.

WELLS, L. J. (1943) Descent of the testis : anatomical and hormonal considerations, *Surgery*, **14**, 436.

WHITEHOUSE, B., and FEATHERSTONE, H. (1923) Certain observations on the innervation of the uterus, *Brit. med. J.*, **2**, 406.

WILLIS, R. A. (1958) *The Borderland of Embryology and Pathology*, London.

WITSCHI, E. (1948) Migration of the germ cells of human embryos from the yolk sac to the primitive gonadal folds, *Contr. Embryol. Carneg. Instn.*, (No. 209) **32**, 67.

WYBURN, G. M. (1937) The development of the infra-umbilical portion of the abdominal wall, with remarks on the aetiology of ectopia vesicae, *J. Anat. (Lond.)*, **71**, 201.

ZUCKERMAN, S. (1956) The regenerative capacity of ovarian tissue, *Ciba Colloquia on Ageing*, **2**, 31.

ZUCKERMAN, S. (1962) *The Ovary*, London.

THE DUCTLESS GLANDS

R. G. HARRISON

THE term ductless is applied to certain glands whose products reach the circulation without passing through any special channels or ducts. They differ widely in structure, function and development, exhibiting a diversity similar to that of glands in general.

GLANDS

From the functional point of view glands may belong to one of two main categories:

1. **Exocrine,** in which the secretion leaves the gland by way of a duct (e.g., salivary glands);

2. **Endocrine** or ductless glands, in which the secretion is poured into the circulation without passing through any special channels or ducts (e.g., thyroid and suprarenal glands).

The **exocrine glands** may be further subdivided into: (i) **holocrine,** in which the cells composing the gland break down completely in order to liberate their secretion (e.g., sebaceous glands); (ii) **apocrine,** in which some of the cytoplasm at the free surface of the cells composing the gland is discharged with the secretion (e.g., axillary sweat glands); (iii) **merocrine,** which discharge their secretion without any fragmentation of the cytoplasm of their cells. Endocrine glands are merocrine in character. Exocrine glands may also be described as **simple** or **compound** depending on the complexity of their ducts, and **tubular** or **alveolar** according to the shape of their secretory units [FIG. 490].

Glands may be classified also with reference to their development. Thus, they may arise from one or other of the primary germ layers of the embryo—ectoderm, mesoderm or entoderm, or from certain regions, e.g., the thyroid and parathyroid glands which are **pharyngeal derivatives.** In a similar manner the hypophysis cerebri and the pineal body have been associated in a cerebroglandular group.

Endocrine Glands

These glands differ from exocrine glands in not retaining as a duct any connexion with the epithelium from which they develop, but they come into intimate contact with the intraglandular blood vessels. Their cells are arranged in columns, vesicles, or clusters, and they have a more profuse blood supply, often with specialized vascular arrangements such as a fenestrated capillary endothelium and arteriovenous anastomoses [FIG. 1063].

A single gland may contain both exocrine and endocrine parts (e.g., pancreas), and the gonads (testis and ovary), whilst having as a primary function the production of germ cells, are also concerned in the secretion of hormones; thus the testis manufactures **androgenic hormones** and the ovary **oestrogens.**

Endocrine glands produce chemical substances, commonly known as hormones, which control and modify the functional activity of the cells of other tissues in specific ways, and whose effects may be either excitatory or inhibitory. They are the basis of a great chemical system which co-ordinates the activities of the various tissues; and their physiological and clinical importance is such that special textbooks are devoted to them (Selye, 1947; Pincus, 1947–68; Pincus and Thimann, 1948–64). The study of the various aspects of these organs is known as endocrinology.

Histologically the endocrine glands exhibit various grades of specialization (Cowdry, 1932; 1944). These grades probably illustrate the steps by which the structural and functional differentiation of endocrine organs took place in the course of evolution. The production of a hormone has, indeed, fundamental resemblances to an activity displayed by all living cells, for these all give off substances into the circulation which affect the activities of other cells. Hormones are distinguished by the specific nature of the effects they produce; in some instances they are recognized only by these effects. For example, the adrenocorticotrophic hormone of the hypophysis exerts a direct effect only on the cortex of the suprarenal gland, and can only be recognized by virtue of this effect.

The first stage in the differentiation of a tissue as an endocrine organ is illustrated by certain tissues (e.g., gastric and duodenal mucosa) which, though primarily adapted by their structure to some other function, also produce a hormone as a secondary function without any evident cytological differentiation of the cells concerned. The placenta may be mentioned as an example of this stage.

A second stage in differentiation is illustrated in organs such as the pancreas, in which there is a recognizable specialization of structure as well as of function. In this organ some of the cells remain epithelial in structure and act as an exocrine secretory organ, whilst other cells are budded off from the epithelium in the course of development to form dissociated clusters, the islets of Langerhans, which act purely as an endocrine organ.

A further stage of differentiation is seen in the thyroid and parathyroid glands, for in the development of these glands there comes about a complete separation from the pharyngeal epithelium of masses of cells destined solely for an endocrine function.

Lastly, two masses of cells, each of specialized structure and function, may come into such intimate relation in the course of development as to constitute a single organ from the topographical point of view. Examples are furnished by the stomodaeal and neural parts of the hypophysis, and by the cortex and medulla of the suprarenal gland. It has been suggested that such close approximation may perhaps allow important interactions between their parts, thus influencing the respective functions.

In this section only the more specialized endocrine glands are described, namely those which belong to the third and fourth grades. The glomus caroticum, the corpus coccygeum, the pineal body, and the thymus are also included as a matter of convenience, although there is less evidence that these structures serve as endocrine glands.

The organs described in this section are taken in the following order:

1. The **Suprarenal Glands,** which are compound organs including chromaffin and cortical tissue.

   i. The **Chromaffin System,** of ectodermal origin, includes the medulla of the suprarenal gland and various 'paraganglionic' masses of chromaffin tissue.

   ii. The **Cortical System,** of mesodermal origin, is represented by the cortex of the suprarenal gland and by the occasional 'accessory cortical bodies'.

2. The **Organs of the Pharyngeal Pouches** (pharyngeal derivatives), of entodermal origin and developed from the walls of the pharynx, include the thyroid and parathyroid glands and the thymus.

3. The **Cerebroglandular Organs,** of ectodermal origin, consisting of the hypophysis cerebri and the pineal body.

Shape and Development of Glands

In the living body, glands are soft, plastic structures, and their shape is continuously subject to modification by the pressure of adjacent structures.

The special secreting cells of an exocrine gland are developed from the walls of a tubular structure which may be of simple form [FIG. 490] or may exhibit an elaborate system of branches such as is seen in compound racemose glands. The tubular structure with its branches eventually forms the gland duct and its tributaries. In some instances the tubular form is not seen while the secreting cells are developing, but instead elongated columns of cells are first formed which only later become canalized (e.g., in the development of the liver, page 461). The secreting cells of some of the endocrine glands likewise arise from the walls of tubular structures which later disappear—a mode of development which results from a progressive specialization of function (e.g., the organs of the pharyngeal pouches).

As glands grow, the glandular tissue accommodates itself between adjacent structures, and may become moulded to reflect their contours, e.g., the thyroid [FIG. 688]. A fixed gland from the cadaver thus frequently shows impressions made by adjacent blood vessels or bones; or it may even sur-

round structures during development, e.g., the parotid gland.

The majority of glands are extremely vascular, especially endocrine glands. The glandular cells are thus provided with the raw materials of secretion by the blood stream which also carries away the secretions of endocrine glands.

The nerves which supply glands terminate in two ways: (1) they may end in the walls of blood vessels and influence the functions of the gland through their control over the blood supply (e.g., the thyroid and the suprarenal cortex); (2) they may end in direct relation with the actual secreting cells (e.g., the suprarenal medulla).

THE SUPRARENAL GLANDS

[PLATE 37A]

The paired suprarenal (adrenal) glands lie on each side of the vertebral column in intimate relation with the superomedial aspects of the kidneys [FIGS. 680–686].

Each gland consists of a relatively thick layer of cortex enclosing a medulla, and is extremely vascular. Arterial blood enters the gland from a plexus on the surface of the cortex and passes through the cortical tissue to the medulla, whence it is drained usually by a single vein [FIG. 685].

The cortex and medulla are of different developmental origin and function; they are in effect two distinct endocrine organs. The **cortex** plays an indispensable role in the body. Interference with its function affects carbohydrate metabolism and the distribution of sodium, potassium, and water in the body. It secretes hormones which are closely related to the

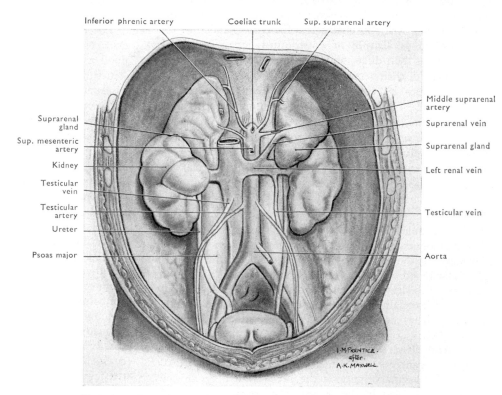

FIG. 680 Posterior abdominal wall of full-term foetus, illustrating the relatively large size of the suprarenal glands and the lobulation of the kidneys.

sex hormones. At least twenty-eight steroids have been isolated from suprarenal glands processed after death (Reichstein and Shoppee, 1943) but not all of them may be secreted by the suprarenal cortex in life. Of the steroids known to be present in the cortex, the mineralocorticoids, **aldosterone** and **deoxycortone,** control the metabolism of sodium and potassium. The glucocorticoid, **cortisone,** affects carbohydrate metabolism by increasing the deposition of glycogen in the liver. Oestrogenic, progestational, and androgenic hormones are also secreted by the cortex. The injection of cortisone causes a reduction in the number of eosinophils in the blood and involution of lymphatic tissue, particularly of the thymus. The cortex is also concerned in the reaction of the body to stress (Selye, 1950), and stores vitamin C and lipoids. The **medulla** secretes two hormones, **adrenaline** and **noradrenaline,** which have dissimilar actions, and are secreted separately from the medulla under different conditions (von Euler, 1955). Noradrenaline, also a regular constituent of postganglionic sympathetic nerves, causes general vasoconstriction and a rise in blood pressure. Adrenaline facilitates carbohydrate metabolism, and although it constricts the blood vessels in skin, it produces vasodilatation of the blood vessels of skeletal muscle. Both adrenaline and noradrenaline are liberated from the suprarenal gland when the splanchnic nerves of the cat are stimulated. Adrenaline may be a later product of evolution, for noradrenaline appears to be the hormone secreted in foetal and neonatal life and later replaced by adrenaline. Histochemical observations demonstrate that some cells of the suprarenal medulla contain much adrenaline and little noradrenaline, whilst others contain noradrenaline but little or no adrenaline.

Unlike the cortex, the medulla is not essential to life. The medulla does not display the functional changes of size which are seen in the cortex under various circumstances (e.g., during menstruation, after extirpation of the contralateral gland, or the influence of stress (Selye, 1950), and under the influence of the **adrenocorticotrophic hormone (ACTH)** of the hypophysis).

In colour the cortex is yellow, owing to the contained lipoid substances. The medulla is seen in a section of the fresh, healthy gland as a dark streak within the yellow cortex; it becomes brown on treatment with a dilute solution of chromic acid or potassium dichromate (chromaffin reaction).

The post-mortem size of the gland varies within wide limits, mainly on account of the great modifications produced by certain pathological conditions.

The average dimensions of the suprarenal glands are as follows: height, 3–5 cm.; breadth, 2·5–3 cm.; thickness, slightly under 1 cm.; and the weight is 7–12 g. The medulla forms only about one-tenth of the whole gland. The glands are relatively much larger in the foetus than in the adult. Even at birth they are still relatively large [FIG. 680], being little smaller than in the adult.

As a rule the glands are of unequal size (the left is more frequently the larger), occasionally the difference is extreme, and rarely one gland is absent. Sometimes the two glands are fused (cf. horseshoe-kidney).

Frequently there are accessory glands. These develop in the neighbourhood of the main gland and usually remain there, but may become attached early in embryonic life to adjacent organs which subsequently change their position. As a result, they may be found not only beside the main gland but also in the broad ligament of the uterus, in the spermatic cord, or even attached to the epididymis. Like the main glands, true accessory suprarenals are compounded of cortex and medulla, and require to be distinguished from the purely chromaffin bodies and accessory cortical bodies which may be found in any of the positions in which accessory suprarenal glands occur.

Shape and Relations. The right gland is pyramidal; the left is semilunar and extends further down the medial side of the kidney than the right one, and both are moulded by adjacent structures. Each presents posterior, anterior, and renal surfaces, and the right has an anteromedial surface. The renal surface is moulded on the medial aspect of the upper pole of the kidney, while the posterior surface lies against the

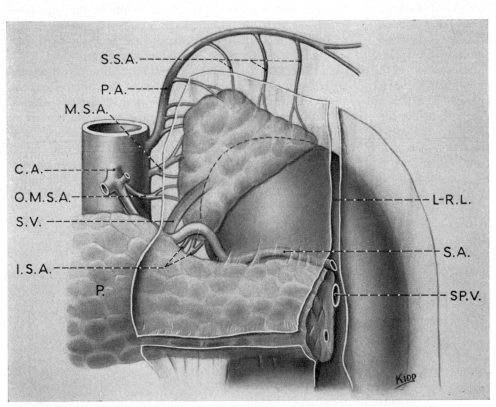

FIG. 681 Relations and blood supply of the left suprarenal gland. (After Harrison, 1963.)

C.A., coeliac trunk; I.S.A., inferior suprarenal arteries; L-R.L., lienorenal ligament; M.S.A., middle suprarenal artery; O.M.S.A., occasional middle suprarenal artery arising from coeliac trunk; P., pancreas; P.A., inferior phrenic artery; S.A., splenic artery; SP.V., splenic vein; S.S.A., superior suprarenal arteries; S.V., suprarenal vein.

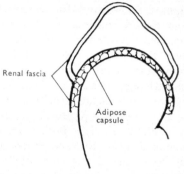

FIG. 682A Relations and blood supply of the right suprarenal gland, seen from the anterior aspect. (After Harrison, 1963.)

B.L.S., part of anterior surface related to bare area of liver; D., descending part of duodenum; I.S.A., inferior suprarenal artery; I.V.C.S., surface related to inferior vena cava (I.V.C.); L.R.V., left renal vein; M.S.A., middle suprarenal artery; P.A., inferior phrenic artery; P.L.S., peritoneal covered portion of anterior surface related to liver; S.S.A., superior suprarenal arteries; S.V., suprarenal vein.

FIG. 682B Schematic vertical section illustrating relation of the suprarenal gland to the kidney and the renal fascia.

diaphragm. The anterior surface of the left gland [FIGS. 681 and 683] is situated behind the omental bursa, which separates it from the stomach; the inferomedial extremity is, however, sometimes separated from the omental bursa by the splenic artery and the body of the pancreas. The anterior surface of the **right gland** is applied to the liver, being in immediate contact with the bare area in its upper part but separated from it in its lower part by peritoneum. The anteromedial surface is in contact with the inferior vena cava [FIGS. 680 and 682]. The coeliac plexus and ganglia are situated between the two glands.

The suprarenal gland is separated from the kidney by a small amount of the adipose capsule surrounding the kidney, and is enclosed with the kidney in the renal fascia [FIG. 682].

A cleft is found in each gland where the suprarenal vein leaves it. The cleft or **hilus** is situated in the right gland on the anteromedial surface, and near the lower end of the medial margin in the left gland.

The level of the gland varies in relation to the adjacent structures. Thus, the inferomedial part of the right gland is sometimes overlapped by the duodenum, and the area covered with peritoneum is correspondingly reduced. On the left side the pancreas and splenic artery usually lie at a lower level than the gland. In the infant the spleen makes contact with the superolateral part of the left suprarenal and sometimes it may do so in the adult. The suprarenal is displaced with the kidney by respiratory movements of the diaphragm.

Vessels and Nerves. The blood supply of the suprarenal glands is very abundant. Each gland receives a variable number of separate **arteries** from three sources—0–10 from the aorta, 0–27 from the inferior phrenic artery, and 10–30 from the renal artery [FIGS. 681 and 682]. Individual arteries are end-arteries to the zona fasciculata of the cortex (Harrison, 1951). Each gland may thus receive as many as sixty arteries, but is drained by a single large central **vein** which emerges through the hilus; the right, after a very short course, joins the inferior vena cava, the left enters the left renal vein behind the body of the pancreas after receiving the inferior phrenic vein, and anastomosing with the first lumbar vein. The suprarenal veins have little or no circular muscle in their walls. The

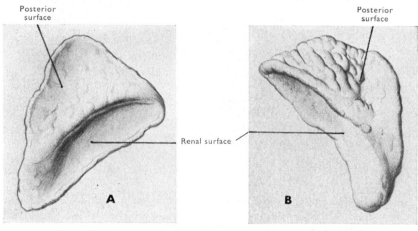

FIG. 683 Suprarenal glands isolated and viewed from behind
A, right. B left.

arrangement of the blood vessels within the gland is described with its structure. Numerous **lymph vessels** pass from the suprarenal glands to the aortic lymph nodes.

The **medulla** of the gland is more richly supplied with **nerves** than any other organ; they are derived from the sixth thoracic to the first lumbar ventral rami (Crowder, 1957) and reach the gland through the **splanchnic nerves** and the **coeliac plexus.** The nerve filaments form a plexus of medullated and non-medullated nerve fibres in the fibrous capsule before they pass into the gland. Some of the fibres are postganglionic fibres from nerve cells scattered along the sympathetic nerves or in the gland itself (Swinyard, 1937), but the majority are preganglionic fibres which end in synaptic relation with the secretory cells of the medulla (Hollinshead, 1936), which can therefore be considered as postganglionic sympathetic neurons.

There is no evidence that cells in the **cortex** receive any nerve supply; nerve fibres may end in relation to the arteriae medullae.

Structure. The suprarenal gland consists of a highly vascular central mass of chromaffin tissue—the **medulla**—enclosed within a thick parenchymatous layer of **cortex,** which in turn is enveloped in a capsule of fibrous tissue.

The **cortex** of the human suprarenal [FIGS. 684 and 685] is so folded or convoluted as to increase its surface of contact with the medulla. From the deep aspect of the capsule, trabeculae of fibrous tissue pass inwards to support the glandular parenchyma. In the superficial part of the cortex the trabeculae interlace freely so as to enclose a series of small rounded clusters of cortical cells, the **zona glomerulosa.** In the intermediate region of the cortex, elongated cell columns, usually formed of two or three rows of cortical cells, lie at right angles to the surface, the **zona fasciculata.** In the deepest part of the cortex the cell columns are more irregularly arranged and form a reticulum—the **zona reticularis.** There is some evidence for functional as well as morphological independence of the cortical zones; thus the mineralocorticoids are formed in the zona glomerulosa, the glucocorticoids in the zona fasciculata, and sex hormones in the zona reticularis.

The cortical parenchyma consists of large polyhedral cells arranged in the interstices of the fibrous trabeculae. The cells contain lipoid material, and are richer in cholesterol than any other tissue in the body. Ascorbic acid (vitamin C) is stored in the fasciculate and reticular zones. As the blood passes between the columns of the fasciculate zone it gains cortical hormones; the regulation of this process involves several mechanisms (Harrison and Hoey, 1960).

The **medulla** [FIG. 685] is formed of a spongework of cell columns

FIG. 684 Histological section through a human suprarenal gland. × 36.

FIG. 685 Stereogram of the suprarenal gland, showing the medulla (M.) with its sinusoids (M.S.) and central vein (C.V.), and the cortex with its three zones, the zona glomerulosa (Z.G.), zona fasciculata (Z.F.) and zona reticularis (Z.R.), enclosed by the capsule (C.). Two arteriae medullae (A.M.), and a recurrent cortical artery (R.A.C.) are shown. The columns of cells are mostly separated by cortical capillary sinusoids (C.C.). (After Harrison, 1963.)

separated by anastomosing venous sinusoids. The cells are large and granular and exhibit the characteristic chromaffin reaction. In a fresh gland the medulla is of a dark-red colour owing to the presence of blood in its sinusoidal spaces.

Vascular Pattern. From the main blood vessels, smaller vessels enter at numerous points in the fibrous capsule and run in the trabeculae, forming a close network around and between the cell masses and columns of the zona glomerulosa and zona fasciculata [FIG. 685]. In the zona reticularis the blood vessels open up to form a venous plexus which is continuous with the sinusoidal plexus in the medulla, and thus with the central efferent vein of the medulla which emerges at the hilus of the organ as the suprarenal vein. Some of the arterioles that enter the cortex pass directly through into the medulla, the **medullary arteries.** These act as channels to provide a blood supply to the medulla independent of that through the cortex. Constriction of these medullary arteries diverts blood into the capillaries of the cortex (Harrison and Hoey, 1960). The **cortical arteries** pass into the cortex to varying depths, and there end in capillaries, or form a loop which recurves to the capsule. Blood which reaches the medulla may therefore do so directly or first pass through capillaries in the cortex. Endothelial cells which line these capillaries show phagocytic properties. The medulla is, therefore, the first tissue in the body to receive suprarenal cortical hormones or vitamin C from the cortical tissue.

Development of the Suprarenal Glands. The cortical system is a derivative of the coelomic epithelium (splanchnic mesoderm), which proliferates in the medial coelomic bay between the root of the mesentery and the mesonephros at the 7-mm. stage, when numerous buds of cells develop from the deep surface of the mesothelium. These cells rapidly form a mass of cortical cells separate from the mesothelium. In Man the greater part of that tissue is ultimately included in the suprarenal cortex, but small masses may separate off to form either independent cortical bodies or portions of accessory suprarenals [p. 559].

Meanwhile the sympathochromaffin primordium of the medulla, derived mainly from the neural crest, has appeared (5-mm. stage). The cells of that tissue early make contact with the cortex primordium and commence to invade it at the 14-mm. stage. Not until the 19-cm. stage does the immigrant chromaffin tissue reach the neighbourhood of the central vein and form a true medulla (Zuckerkandl, 1912). Envelopment

of the medulla by the cortex is incomplete for a long time; this is probably due as much to progressive cortical overgrowth as to chromaffin invasion.

The relatively bulky **foetal cortex** degenerates progressively during the last 10 weeks of intra-uterine life. At birth two parts of the cortex are distinguished, a still bulky foetal cortex and a thin, overlying, true cortex [FIG. 686]. The characteristic cortical tissue of the foetus disappears in the course of the first year and the volume of the entire cortex thus diminishes rapidly for a time. The final specialization of the cortex is not complete until much later. The definitive zones of the permanent cortex are probably formed from the thin rim of adult cortex surrounding the foetal cortex at birth (Barr, 1954). In the prepubertal period there is accelerated growth of the cortex; and in middle age its involution begins.

THE CHROMAFFIN SYSTEM

This system comprises a large number of masses of tissue similar in development, structure, and histochemical reactions to the medulla of the suprarenal gland.

The tissue is called chromaffin on account of its supposed specific affinity for chromium salts, which give a brownish reaction with it. Chromaffin tissue, however, contains substances such as the catechol amines and indole amines which form coloured compounds following treatment with oxidants other than bichromate or chromic acid (Boyd, 1960). The term chromaffin is therefore inappropriate, but still used. Although the reaction is given by adrenaline and noradrenaline, it is not specific for these substances. The masses of tissue which form the system all originate in intimate association with the sympathetic nervous system. Indeed, the characteristic chromaffin cells and the neurons of the sympathetic ganglia are derived mainly from a common mother cell, the cell of the neural crest, though cells of the neural tube may also contribute.

The chromaffin system includes the medulla of the suprarenal gland and the extra-suprarenal chromaffin bodies which may occur in any part of the sympathetic system, though they are most numerous in the abdomen. The majority of the extra-suprarenal chromaffin bodies are adjacent to the aorta, the **para-aortic (chromaffin) bodies.** Some cells in the wall of the intestine contain granules giving the chromaffin reaction and are therefore termed **enterochromaffin cells.**

The extra-suprarenal chromaffin bodies are encapsulated structures, but non-encapsulated collections of chromaffin cells may be associated with any of the sympathetic ganglia. The abdominal para-aortic chromaffin bodies are dispersed during prepubertal growth due to relative growth changes. Some collections of chromaffin cells also remain in the abdominal prevertebral sympathetic plexuses in the adult, the largest collections being found in the coeliac and superior mesenteric plexuses. This dispersion also affects the largest discrete para-aortic bodies which lie adjacent to the aorta in the region of the origin of the inferior mesenteric artery and are sometimes known as the organs of Zuckerkandl (Coupland, 1952; 1954). In the new-born child they are paired, elongated structures, usually about a centimetre in length. They begin to disperse soon after birth, and by the period of puberty they have practically ceased to be visible to the naked

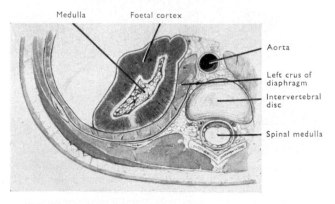

FIG. 686 Transverse section through suprarenal gland of a new-born child *in situ.*

eye, though vestiges may be recognized, at least histologically, to a much later period in life. Their regression is roughly paralleled by the maturation of the suprarenal medulla. In foetal life they contain from two to three times the quantity of noradrenaline present in the suprarenal medulla, but this decreases in postnatal life as the quantity of adrenaline and noradrenaline in the suprarenal medulla steadily increases.

Thus the function of the para-aortic bodies is taken over by the suprarenal medulla in postnatal life, and the character of the secretion changes, adrenaline rather than noradrenaline becoming the predominant hormone.

Glomus Caroticum

The carotid body (glomus caroticum) is situated close to the bifurcation of the common carotid artery [FIG. 687], adjacent to the carotid sinus. Frequently the glomus lies deep to the bifurcation; sometimes it is wedged in between the roots of the internal and external carotids, or it may be placed between them at a slightly higher level.

It is a small, neurovascular structure of a shape that varies with its position. When free from pressure from its surroundings it is oval; when situated between the internal and external carotids, it is wedge-shaped. On average its height is about 7 mm., its breadth 1·5–5 mm. It sometimes comprises two or more separate nodules. Its colour is yellowish-grey to brownish-red.

Structure. The carotid body consists of groups of 'epithelioid' cells, some of which are argyrophil, surrounded by and interspersed with fibro-areolar tissue containing sinusoids. The cells and the sinusoids are richly innervated by sensory nerve endings. The carotid body was formerly regarded as a part of the chromaffin system, but its cells show only a slight chromaffin reaction (Boyd, 1960). Its structure and the abundant sensory nerves which pass into the sinocarotid nerve are in keeping with its functions as a chemoreceptor, responsive to oxygen lack and carbon dioxide excess, although it is less sensitive to the latter.

Nerve Supply. Numerous afferent nerves pass from the carotid sinus and carotid body to a sinocarotid plexus. The nerve to the carotid sinus [FIG. 687] passes upwards with the pharyngeal branch of the glossopharyngeal nerve and divides into two branches, one of which joins the inferior ganglion of the vagus, the other the glossopharyngeal nerve. Other nerve filaments from the plexus pursue an independent but similar course, and join the glossopharyngeal and vagus nerves and the cervical sympathetic trunk. The afferent fibres pass mainly to the glossopharyngeal nerve; but some of those from the **carotid sinus** join the vagus, and it is probable that some afferent fibres from the carotid body do so also.

Development. The carotid body is developed as a condensation of mesenchyme around the third aortic arch (Boyd, 1937). Nerve fibres of the glossopharyngeal nerve enter this condensation, and neuroblasts derived from the ganglion of the nerve subsequently migrate along the fibres into the developing organ.

Similar structures are developed in relation to the fourth and sixth aortic arches and form smaller bodies which, on the left side, are adjacent to the aorta and the ductus arteriosus.

Corpus Coccygeum

This is a small vascular body measuring 2–3 mm. in diameter. It lies immediately anterior to the tip of the coccyx in intimate relation with a branch of the median sacral artery and with the ganglion impar of the sympathetic trunks. Usually it is associated with a group of smaller bodies of similar structure.

Structure. The corpus and its satellite bodies are enclosed in a fibrous capsule which ensheaths them individually. The characteristic feature of the corpus is the presence of masses of polyhedral 'epithelioid' cells, with large nuclei, surrounding the lumina of tortuous sinusoidal blood spaces. From these spaces capillary channels extend among the masses of closely packed cells. They resemble the cells of the carotid body and do not give the characteristic chromaffin reaction; hence the corpus cannot be included with the chromaffin system.

Development of the Chromaffin System

All chromaffin tissue develops in intimate relation with the sympathetic nervous system. The cells of both appear to be derivatives of a common mother-tissue whose elements originate from cells which migrate at an early period from the neural crest and neural tube.

In the 16-mm. embryo, sympathochromaffin tissue is met with in situations which more or less correspond to the distribution of the sympathetic nervous system. Differentiation of chromaffin cells from sympathetic neuroblasts begins about the 18-mm. stage, but is not completed till late in the period of gestation, if then. The process is characterized by increase in the size of the chromaffin cells and a diminution of the intensity of their reaction to ordinary stains; the chromaffin reaction does not develop until later.

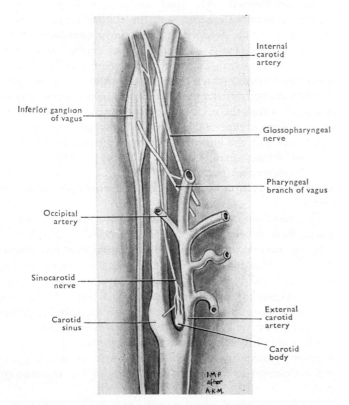

FIG. 687 Carotid body and the sinocarotid nerve with its principal connexions.

Internal carotid artery

Inferior ganglion of vagus

Glossopharyngeal nerve

Pharyngeal branch of vagus

Occipital artery

Sinocarotid nerve

External carotid artery

Carotid sinus

Carotid body

J.M.P. after A.K.M.

The aortic bodies develop as the chief masses of a paired discontinuous series associated with the prevertebral plexuses. Cranially this series is originally continuous with the primordium of the medulla of the suprarenal gland on each side. The aortic bodies are prominent structures in the 2-month embryo. The paraganglia associated with the ganglia of the sympathetic trunk appear a little later.

THE CORTICAL SYSTEM

The single constant representative of this system in higher vertebrates is the mass of tissue that forms the **cortex of the suprarenal gland.** Masses of similar tissue, however, are not infrequently met with in various situations, forming **accessory cortical bodies.**

Accessory cortical bodies may be associated with chromaffin medullary tissue, as is the case in the suprarenal itself, but the majority, including the smaller bodies, are purely cortical. The justification for regarding the suprarenal cortex, with the occasional accessory cortical masses, as representing a distinct system, is based upon phylogenetic and ontogenetic considerations. These warrant the concept of a distribution of cortical tissue which was originally more extensive and distinct from the chromaffin system. In the higher vertebrates the cortical system would appear to have undergone concentration, thus forming mainly the cortex of the suprarenal gland. In the cartilaginous fishes, such as the dogfish, the chromaffin bodies are arranged segmentally, while the cortical tissue is represented by a pair of interrenal bodies between the caudal ends of the kidneys.

It is only in the Tetrapoda (vertebrates possessing paired limbs) that cortical tissue and chromaffin tissue come into intimate topographical relation. In mammals the cortical tissue encloses the chromaffin tissue.

THE ORGANS OF THE PHARYNGEAL POUCHES

This group of organs is developed from epithelium derived from the embryonic pharynx, and they illustrate an advanced stage in the histological differentiation of endocrine glands [p. 557].

In an early phase of vertebrate evolution, entodermal epithelium lines the recesses of the pharyngeal cavity, and in one instance—the thyroid—it is sufficiently elaborated to justify the appellation of an exocrine gland (as still seen in ammocoetes, the larva of the lamprey, *Petromyzon*). Subsequently the epithelium of the recesses acquires specific endocrine functions and is isolated during embryonic development from the general pharyngeal epithelium to form separate extrapharyngeal organs.

Early Development of Pharynx. In the section on embryology, a description is given of the processes which lead to the formation of the embryonic mouth and pharynx [p. 53]. The primitive pharynx is formed from the cephalic part of the foregut and is brought into continuity with the stomodaeum by the breaking down of the **buccopharyngeal membrane** [FIG. 66]. The position of that membrane is such that the greater part of the floor of the mouth is derived from the foregut and is therefore lined with entoderm.

The part of the foregut which forms the primitive pharynx is flattened dorsoventrally and has narrow side-walls that contain thick, vertical bars of mesoderm [FIG. 71]. The thickenings are termed **pharyngeal arches,** and they contain the arterial **aortic arches** [FIG. 81].

At a stage when the cephalic part of the foregut is still cut off from the exterior by the buccopharyngeal membrane, there is a minute diverticulum in its floor known as the **median thyroid diverticulum** [FIG. 75]. This lies immediately posterior to a swelling called the **tuberculum impar** [p. 58]—a mass of tissue which is eventually incorporated in the tongue, in which the original site of formation of the median thyroid diverticulum is marked permanently by the foramen caecum.

In each side-wall of the pharynx, between the arches, are five entodermal depressions, the **pharyngeal pouches.** Corresponding ectodermal **pharyngeal grooves** are found on the external aspect of the pharynx. Between each groove and pouch is a **separating membrane.**

Only the first four **separating membranes** become so thinned that entoderm and ectoderm make contact, and in the case of the fourth membrane this phase is of short duration. The fifth pharyngeal pouch is vestigial and becomes associated with the fourth pharyngeal pouch as the **ultimobranchial (telobranchial) body** [p. 58]. The fourth pouch and ultimobranchial body are together termed the caudal pharyngeal complex.

Owing to the greater prominence of the first and second pharyngeal arches externally, the other arches are more deeply set, so that a **cervical sinus,** and later a **cervical vesicle** [p. 56], are formed. The epithelial cells of the vesicle later acquire an intimate relation with the epithelium of the third pharyngeal pouch, forming with it the primordium of the thymus (Norris, 1938). Portions of the cervical sinus may occasionally persist in the form of a slender tubular channel (**branchial sinus**) which opens to the exterior on the lower part of the anterior triangle of the neck, or a portion may be closed off within the neck to form a **branchial cyst.** If the separating membrane breaks down, a branchial sinus becomes continuous with the pharynx, and forms a **branchial fistula.** **Fate of Pharyngeal Pouches** [FIG. 73]. The first and second pharyngeal pouches are concerned in the development of the **tympanic cavity** and the **auditory tube** [pp. 56-57].

The third and fourth pouches, in embryos 10 mm. in length, have become small cavities connected to the pharynx only by narrow canals (the **pharyngobranchial ducts**) [FIG. 73]; shortly afterwards these slender connexions become solid and break, leaving small nests of cells (of entodermal origin) embedded in the pharyngeal wall. At the time of their isolation from the pharyngeal wall the **third pouches** each consist of: (1) a small, hollow, dorsal diverticulum, from which a **parathyroid gland** develops; and (2) a caudally directed cylinder of cells, the ventral side of which grows and is destined to form the lateral lobes of the **thymus** [p. 571].

The **fourth pharyngeal pouch** (with the ultimobranchial body) of each side forms a small, hollow mass at the time of its separation from the pharynx. This becomes differentiated into three parts—a dorsal portion from which a **parathyroid gland** is developed [p. 568], a ventral diverticulum which corresponds to the thymic part of the third pouch, and the ultimobranchial body.

THE THYROID GLAND

The thyroid gland is situated in the lower part of the front of the neck, and is enclosed in a fascial compartment formed by the sheath of pretracheal fascia which fixes it firmly to the trachea and larynx by its attachment to the oblique line on the thyroid cartilage. The thyroid is an endocrine gland which produces a hormone of the greatest importance for the proper growth and function of most of the tissues in the body. This hormone is known as **thyroxine**; its metabolically active form is probably **triiodothyronine.** Characteristic of this organ is the storage of its secretion within small closed cavities—the vesicles of the thyroid. It is an organ of very ancient history, yet one that exhibits remarkably little evolutionary change, for its histology and endocrine action are similar throughout the whole series of vertebrates. The gland is yellowish-red, soft and extremely vascular. It varies in size with age, sex, and general nutrition, being relatively large in youth, in women, and in the well nourished. In women it increases temporarily with menstruation and pregnancy. Its average dimensions are: height 5 cm., breadth 6 cm., thickness of each lobe 1–2 cm.; and its weight is 18–31 g. (Nolan, 1938). The maximum weight of the normal gland is attained in the young adult.

Usually the thyroid gland consists of a pair of conical **lobes** united across the median plane by a short band of gland tissue called the **isthmus.** But to many thyroid glands this description is inapplicable. In men and thin, elderly women the gland is not uncommonly horseshoe-shaped; in young, well-nourished women and during pregnancy its general contour is more rounded, deeply notched above to accommodate the larynx and deeply grooved behind for the trachea and oesophagus. Rarely, the gland is in two separate parts. Not infrequently, it is asymmetrical. In about 40 per cent. of specimens, a process of gland tissue called the **pyramidal lobe** extends upwards from the upper border of the isthmus, in front of the cricoid and thyroid cartilages, towards the hyoid bone. This process is seldom median, lying more often on the left than on the right; in rare cases, it is double; less rarely, it is double below and single above; sometimes it is represented by a strip of fibrous tissue or a narrow muscle (levator glandulae thyroideae). The muscle may, however, be present independently of the pyramidal lobe [p. 278].

Small, oval, accessory thyroid glands are common in the region of the hyoid bone, and are occasionally met with in relation to the right and left lobes. They may occur also in the superior mediastinum of the thorax.

Relations. The gland itself has an external layer of fibrous tissue —the fibrous capsule—which is loosely connected by areolar tissue to the fascial sheath; the gland with its capsule is thus readily taken out of its sheath of pretracheal fascia. It is under cover of the sternothyroid and sternohyoid muscles, and the superior bellies of the omohyoid muscles, and, more laterally, it is overlapped by the sternocleidomastoids [FIG. 688]. The isthmus is, however, comparatively superficial between the margins of the sternothyroid muscles a short distance above the sternum. The isthmus usually lies on the front of the second, third, and fourth rings of the trachea; but occasionally it lies as high as the cricoid cartilage or as low as the fourth, fifth, and sixth rings of the trachea. The lobes extend down to the level of the sixth ring of the trachea or even lower, and up as far as the oblique line of the thyroid cartilage. Superficially they are covered by the above muscles with twigs from the ansa cervicalis, and by the anterior jugular vein. The upper part of each lobe is moulded medially on the thyroid and cricoid cartilages, and the cricothyroid and inferior constrictor muscles, the latter crossed by the external branch of the superior laryngeal nerve. The lower part is moulded to the sides of the trachea and oesophagus and thus comes into close relation with the recurrent laryngeal nerve and inferior laryngeal artery.

The posterior surface of each lobe is variable in width. It is in contact with the prevertebral fascia covering the longus cervicis muscle, and more laterally with the carotid sheath, which becomes displaced in a posterolateral direction by simple

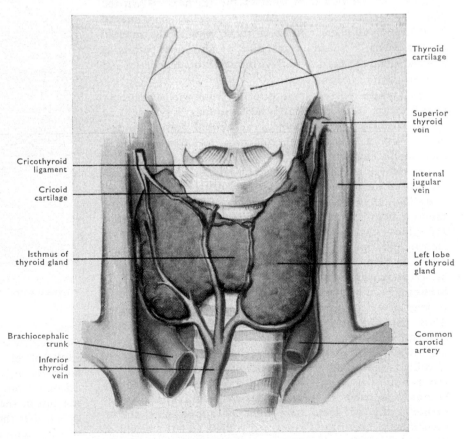

FIG. 688 Dissection of thyroid gland and of structures in immediate relation to it.

enlargements of the gland. The inferior end of the left lobe approaches or even makes contact with the thoracic duct. Closely applied to the posterior surface of each lobe, and within the fibrous capsule, is a pair of parathyroid glands; they may even be embedded within the substance of the thyroid.

Vessels and Nerves. The extraordinarily rich arterial supply is effected through the **superior** and **inferior thyroid arteries.** Occasionally (10 per cent. of cases) a fifth artery— the **thyroidea ima,** normally present in the embryo—persists in the adult as a branch of the brachiocephalic trunk. The pyramidal lobe, if well developed, receives a special branch from one of the superior thyroid arteries, usually the left. The arteries are remarkable for their large size and for the frequency of their anastomoses, which may be arterial or capillary in nature. An anastomosing trunk courses up over the back of each lobe within the fascial sheath and unites the inferior and superior thyroid arteries; it is a landmark for the identification of the parathyroid glands. The anterior branches of the superior thyroid arteries anastomose on the upper border of the isthmus. The anastomoses between the tracheal and oesophageal branches of the inferior thyroid artery and bronchial arteries are probably of importance in maintaining the arterial supply to thyroid tissue following partial thyroidectomy. Typically, three pairs of veins drain the gland. The upper two pairs—the **superior** and **middle thyroid veins**—join the internal jugular veins; the lower pair —the **inferior thyroid veins**—join the brachiocephalic veins. The veins take origin from the venous plexus on the surface of the gland or, in the case of the inferior, from a downward extension of the plexus in front of the trachea. When the gland is very large, accessory veins are present, sometimes in considerable numbers. Most of them pass to the internal jugular veins. A free, transverse, venous anastomosis is effected along the borders of the isthmus through superior and inferior communicating veins.

The **lymph vessels** pass directly from the subcapsular plexus to the **deep cervical lymph nodes**; a few descend in front of the trachea to the **pretracheal lymph nodes**, through which they are connected with sternal nodes behind the manubrium sterni.

The **nerves** are non-medullated postganglionic fibres which come from the **superior** and **middle cervical sympathetic ganglia.** They reach the gland by way of the cardiac and the superior and recurrent laryngeal nerves, and along the superior and inferior thyroid arteries. All of the nerve fibres are vaso-motor, and influence thyroid secretion indirectly by their action on blood vessels, particularly the arteriovenous anastomoses present in the thyroid.

Structure. The gland consists of a mass of minute, rounded vesicles of various sizes, commonly about 300μm. in diameter [FIGS. 689 and 690], groups of vesicles being enclosed in a connective tissue capsule to form lobules. Each vesicle consists of a layer of epithelial cells enclosing a variable quantity of colloid, depending on the activity of the gland: the colloid acts as a store of thyroid hormone in a resting vesicle, and this becomes depleted in an active vesicle. Individual vesicles within a thyroid gland have different appearances due to variations in quantity of contained colloid and the height of the epithelial cells. In addition to the epithelial cells, 'light' or 'C' cells are to be found between the thin basement mem-

FIG. 689   Injected specimen of dog's thyroid gland.

Enlarged drawing demonstrating the relative size of the lymph-plexus and blood-capillary plexus and their relation to the individual vesicles. Note that the lymphatic plexus (black) lies external and in less intimate contact with the individual vesicles than the blood-capillary plexus, which is specific for each vesicle. (Rienhoff, 1931.)

FIG. 690   Histological section through the relatively inactive thyroid gland of a man aged 62.   × 275.

brane of the vesicle and the epithelial cells. They may be found either singly or in groups, and are usually larger than the epithelial cells. They are responsible for the secretion of the hormone *thyrocalcitonin*, which inhibits the mobilization of calcium from bone and thereby depresses the amount of calcium in the blood. Microscopic study of living cells has shown that the epithelial cells produce cytoplasmic processes which extend into the lumen of the vesicle, a process which is associated with the addition of colloid to the lumen (Williams, 1941; de Robertis, 1942). The presence of a fenestrated vascular endothelium in this and other ductless glands suggests that the control of secretion into the blood stream is a function of the gland cells and not of the endothelium. The vesicles are embedded in a framework or stroma of fibrous tissue which is continuous externally with the capsule. Numerous lymph vessels, arteries, veins, and nerves course in the stroma. The lymph vessels appear to be mainly concerned in the drainage of the intervesicular tissue [FIG. 689]; the larger trunks pass to a dense plexus on the surface of the gland.

Development. The thyroid gland is developed from a median, ventral diverticulum of the floor of the pharynx [p. 58; FIG. 75]. It appears in embryos less than 2 mm. in length (possessing about six mesodermal somites) as a small outpocketing of the pharyngeal floor, with the tuberculum impar [p. 58] immediately anterior to it. The diverticulum by rapid downgrowth soon forms a plate-like structure ventral to the trachea, connected with the pharyngeal floor by a narrow tube called the **thyroglossal duct.**

This duct disappears normally in embryos of 6 weeks, but the site of its original opening into the pharynx is marked in the adult by the foramen caecum of the tongue [FIG. 485]. In the sixth week the fourth pharyngeal pouch, with its ventral diverticulum, is still connected by the fourth pharyngobranchial duct with the pharynx. At this stage the caudal pharyngeal complex [p. 564] on each side becomes closely applied to the lateral portion of the thyroid gland; the cells of the ultimobranchial body proliferate and develop into cells within the thyroid gland (Bejdl and Politzer, 1953) which form the 'light' or 'C' cells. These cells therefore become associated with the thyroid vesicles secondarily (Hirsch and Munson, 1969). It has been stated that the pharyngeal region as a whole has the potentiality for developing thymus and parathyroid tissue, so that parathyroid may develop from thymus, and thymus from parathyroid or thyroid tissue. Thymic tissue may develop from the ventral diverticulum of the fourth pouch, and there is also a theoretical possibility of thyroid tissue developing from this pouch (Kingsbury, 1939).

Histological differentiation of the thyroid gland proceeds by the arrangement of the embryonic epithelial cells in two sheets, with disappearance of the original lumen of the diverticulum, and, in the 50-mm. foetus, definite follicles are formed which later become vesicular. Colloid material is present in the vesicles, and the lining cells are apparently active, in the foetus of 11 weeks. Additional 'secondary' follicles are added throughout foetal life but the process slows down before birth and probably ceases about puberty. Postnatal growth of the gland consists mainly in an increase in size of the vesicles.

The development of the gland affords a ready explanation of its variations in the adult. Thus the pyramidal lobe, with partial or complete duplication, is due to the development of gland tissue from the inferior part of the thyroglossal duct, which may bifurcate at a higher or a lower level, with more or less complete fusion or separation of the masses thus formed. **Accessory thyroid glands** near the hyoid bone, or within the tongue, are the result of development of gland tissue from isolated remnants of the duct. Their occurrence behind the sternum above the arch of the aorta is probably due to a downward displacement of thyroid tissue along with the arteries (common carotid) of the third pharyngeal arch which, at their origin from the truncus arteriosus, have the median thyroid diverticulum in close relation at an early stage.

The occurrence in the adult of a duct that leads from the foramen caecum to, or towards, the hyoid bone (**lingual duct**) is due to a persistence of the upper part of the thyroglossal duct. Similarly, **thyroglossal cysts** are due to the persistence of short segments of the duct. The thyroglossal duct may be continuous with the pyramidal lobe [p. 565], and it usually passes in front of the hyoid bone since it is initially anterior to the second pharyngeal arch of the embryo. Any change in this relation is due to extension and growth of the primordium of the hyoid which may lead to the duct becoming embedded within the body of the hyoid or even lying posterior to it. Cyst-like structures, derived from the ultimobranchial body, may sometimes be found in the adult in close relation with the lower pole of the lateral lobe of the gland.

Methods of Clinical Anatomical Examination. The lateral lobe of the thyroid gland is felt indistinctly as a soft mass at the anterior border of the sternocleidomastoid muscle on the side of the lamina of the cricoid cartilage and the adjacent part of the thyroid cartilage. The isthmus also can be felt as a soft swelling extending over a small part of the front of the trachea a short distance below the cricoid cartilage. When the larynx and trachea are pulled up in the movement of swallowing, the thyroid gland accompanies them. The gland is often large enough to produce an evident fullness of the neck; the sternocleidomastoid can then be felt to lie superficial to the postero-inferior part of the lateral lobe.

THE PARATHYROID GLANDS

The parathyroid glands are two pairs of small glands closely applied to the back of the thyroid gland within its fibrous capsule [FIG. 691]. They are distinguished as the superior and inferior parathyroid glands; but with reference to their

FIG. 691   Dissection showing thyroid and parathyroid glands of adult from behind.

developmental origin [see below], the upper pair are known also as the parathyroids IV and the lower pair as the parathyroids III.

The parathyroid glands produce a hormone which mobilizes calcium from bone and therefore elevates the level of calcium in the blood. Extirpation of all four of the glands in a mammal causes death within a few days, but such is the margin of safety that removal of two of them produces no obvious effect.

The parathyroid glands are yellowish-brown, ovoid or lentiform structures, varying in size between extremes of 1 and 20 mm. in their long (generally vertical) axis. Most commonly they are from 3 to 6 mm. in length, from 2 to 4 mm. in width, and from 0·5 to 2 mm. in thickness, and their individual weights vary within wide limits. The total amount of parathyroid gland-tissue has been stated to show a mean variation from 0·10 to 0·14 g.

The normal number of parathyroids may be diminished to three, two, or even one. In some such cases, however, it is not easy to be certain that a minute ectopic gland may not have escaped notice. On the other hand, the number may be increased, probably as a result of division of the original primordia. Thus from five to eight—even in an extreme case as many as twelve—have been recorded. Six per cent. of subjects have only three parathyroid glands, and in 6 per cent. also there are five (Gilmour, 1938).

The **superior parathyroid gland** is usually embedded in the capsule of the thyroid gland at the back of the corresponding lobe, about its middle. The **inferior parathyroid gland** is similarly embedded in the back of the inferior end of the lobe. As a rule the anastomosing arterial channel which connects the inferior and superior thyroid arteries passes near both parathyroids and furnishes the best guide to them; but the range of exceptional positions which the glands may occupy is wide.

The superior parathyroid may be found: (1) behind the pharynx or oesophagus; (2) in the fibrous tissue at the side of the larynx, above the level of the thyroid gland; (3) behind any part of the corresponding lobe of the thyroid gland or even embedded in the thyroid substance (internal parathyroid).

The inferior parathyroid may be found: (1) near the bifurcation of the common carotid artery; (2) behind any part of the corresponding lobe of the thyroid gland; (3) on the side of the trachea; or (4) in the thorax close to the thymus. This wide range of variation in position is explicable by the close association of the inferior parathyroid with the main thymus element in development.

Vessels and Nerves. The artery to each parathyroid may spring from any branch of the **inferior** or **superior thyroid arteries,** but most commonly is a branch of the large anastomosing vessels between them. The lymph vessels drain with those of the thyroid gland. The **nerve supply** is abundant and comes from the plexus surrounding the thyroid vessels [see above]. The parathyroids produce their hormone after transplantation; the nerves are apparently vasomotor and not secretory.

Structure. The parathyroids are built up of interconnected trabeculae of epithelial cells with only a little vascular tissue between them. Sometimes the cellular arrangement is of a more compact character, or again, some of the cells may be arranged in follicle-like clumps enclosing a colloid material (devoid of iodine). The 'principal' cells, which contribute the majority, are large and clear; but, after the age of 8 or 9 years, some of the cells contain acidophil granules and are believed to be degenerating principal cells (Morgan, 1936).

Development. The parathyroid glands develop from the dorsal diverticula of the third and fourth pharyngeal pouches [FIG. 73]. The first indication of their development is a proliferation and thickening of the epithelium on the cephalic and lateral aspects of the diverticula, which may be seen in a 10-mm. embryo. Cords of cells grow out from the thickening, and fibrous tissue penetrates between the outgrowing cords, which soon lose their connexion with the pharynx.

Parathyroid III—the inferior parathyroid of the adult—being developed from the same pouch as the thymus, is normally drawn caudal to parathyroid IV by the migrating thymus. As a rule it halts at the caudal end of the lobe of the thyroid gland, but may continue with the thymus into the thorax. On the other hand it may not descend at all. In the latter case it remains near the bifurcation of the common carotid artery, where it is liable to be mistaken for the carotid body. (For details of development, consult Weller, 1933 and Boyd, 1950.)

THE THYMUS

[PLATE 36A and B]

The thymus is situated in the superior mediastinum between the sternum and the great vessels; and it usually extends down for a short distance into the anterior mediastinum in front of the pericardium. In the new-born child (in which it is relatively much larger than in the adolescent or adult) it commonly extends laterally between the thoracic wall and the anterior borders of the pleurae and lungs [FIG. 692].

The thymus is essentially an organ of the growth period of life. In the course of a few years after birth it shows a rapid diminution in the rate of its growth and hence of its relative size; but it continues to grow until puberty, and it is only after that period of life that it undergoes a gradual diminution in absolute size (Young and Turnbull, 1931).

During its period of growth the thymus appears as a pinkish mass (yellower in later life), consisting of a pair of laterally compressed, more or less pyramidal, asymmetrical lobes. The lobes are connected with each other, not by any bridge of glandular tissue but by areolar tissue. The surface of the organ, in the young, more actively glandular condition, is finely lobulated.

The thymus is soft in consistence, and the details of its shape are determined by its size and by the structures upon which it is moulded, namely, the pericardium and the great vessels of the superior mediastinum and the root of the neck. Its shape varies with its size and the age of the individual. In infants (with a short thorax) it is broad and squat; in adults (with a long thorax) it is drawn out into two irregular but more or less flattened bands [FIG. 693].

Posteriorly it is in relation from below upwards with the pericardium, ascending aorta, left brachiocephalic vein, the brachiocephalic trunk, the trachea, and inferior thyroid veins. Anteriorly it is related to the sternum and the lower ends of the sternothyroid muscles. In the young child the lateral margin is insinuated between the pleura and the upper costal cartilages, intercostal spaces, and internal thoracic artery.

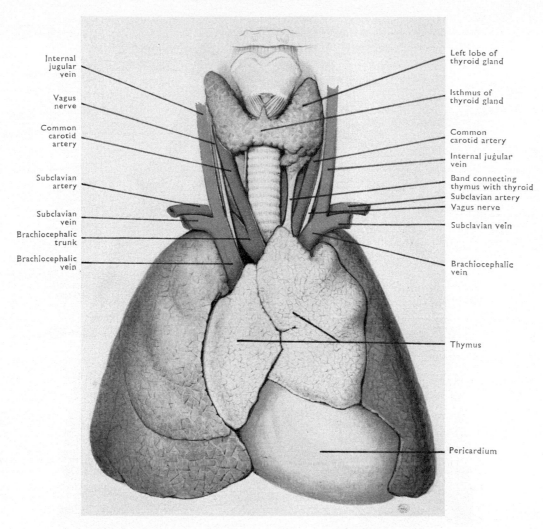

Internal jugular vein
Vagus nerve
Common carotid artery
Subclavian artery
Subclavian vein
Brachiocephalic trunk
Brachiocephalic vein

Left lobe of thyroid gland
Isthmus of thyroid gland
Common carotid artery
Internal jugular vein
Band connecting thymus with thyroid
Subclavian artery
Vagus nerve
Subclavian vein
Brachiocephalic vein
Thymus
Pericardium

FIG. 692 Thymus and thyroid gland in a full-term foetus hardened by formalin injection.

There is great individual variation in its size at any given age. Thus, at birth it ranges in weight from 2 to 17 g. with an average of about 13 g. At puberty the thymus is usually at the height of its development with an average weight of 37 g.; but sometimes it has already undergone considerable retrogression. In the young adult the average weight is reduced to about 25 g., but the organ is occasionally still quite large.

In old age it commonly weighs only about 10 g. These age changes may be secondary to the hormonal changes occurring in other endocrine glands.

Variations of the shape as well as in the size of the thymus sometimes occur, attributable to a partial persistence of the slender stalks by which at first the developing thymus remains attached to the third pharyngeal pouches. Thus, the thymus may exhibit slender prolongations into the neck on each side, anterolateral to the trachea. These processes may be connected to the lower parathyroid glands by strands of fibrous tissue [FIG. 696]; or again, the whole of the cervical processes may be represented by fibrous strands (atrophic cervical thymus). An isolated portion of thymus tissue may persist in close relation with the lower parathyroid (accessory cervical thymus III).

Vessels and Nerves. The arterial supply of the thymus is effected through inconstant branches, chiefly of the **internal thoracic arteries** and their branches. Its **veins** are irregular and mostly join the internal thoracic and left brachiocephalic veins. Its **lymph vessels** are abundant. Although they are not related to the organ in the same manner as in a lymph node, they have been found filled with lymphocytes presumably derived from the gland, though it is generally agreed that the great majority of the thymic lymphocytes do not leave the thymus. The lymph vessels enter the **anterior mediastinal lymph nodes.**

Its **nerves** are minute and are derived from the vagus and the sympathetic. The branches of the **vagus** descend directly to the thymus from about the level of the thyroid cartilage; the **sympathetic** fibres run with the blood vessels. The fibrous capsule of the thymus receives small irregular branches from the phrenic nerves, but they do not supply the gland tissue.

Structure. The thymus is invested by a fibrous capsule which sends septa into its substance between its constituent lobules. The two 'lobes', though really independent paired glands, are also bound together more or less intimately by their capsular investments. If the young gland is dissected

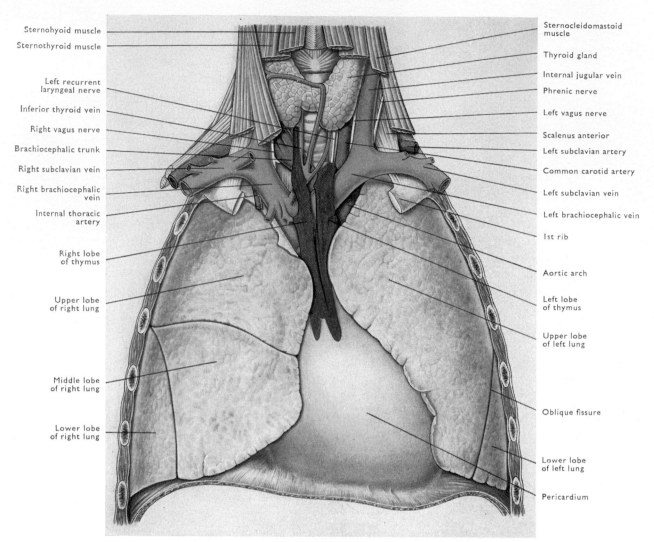

Sternohyoid muscle
Sternothyroid muscle
Left recurrent laryngeal nerve
Inferior thyroid vein
Right vagus nerve
Brachiocephalic trunk
Right subclavian vein
Right brachiocephalic vein
Internal thoracic artery
Right lobe of thymus
Upper lobe of right lung
Middle lobe of right lung
Lower lobe of right lung

Sternocleidomastoid muscle
Thyroid gland
Internal jugular vein
Phrenic nerve
Left vagus nerve
Scalenus anterior
Left subclavian artery
Common carotid artery
Left subclavian vein
Left brachiocephalic vein
1st rib
Aortic arch
Left lobe of thymus
Upper lobe of left lung
Oblique fissure
Lower lobe of left lung
Pericardium

FIG. 693   Dissection to show the thymus in an adult female.

so as to liberate its **lobules** from the fibro-areolar tissue between them, it will be found to consist of an elongated series of lobules, each connected with a central parenchymatous cord along which they are arranged in irregular necklace fashion. The cord, mainly of medullary substance, represents the original thymic diverticulum of which the lobular masses are secondary derivatives.

The lobules of the thymus are further subdivided into **secondary lobules** which average a little over a millimetre in diameter. They are imperfectly separated from one another by delicate areolar tissue. These are the units of the thymic structure : each shows a cortex and a medulla. The **cortex** consists of closely packed lymphocytes embedded in a delicate fibrous reticulum. It contains no lymph vessels. The cortex completely surrounds the medulla, which is continuous with that of adjacent lobules and, through the lobular stem, with the central cord of the gland [FIG. 694].

In its intimate structure the **medulla** differs from the cortex by the presence of many reticular cells and fibres but only few lymphocytes. The medulla, further, contains the so-called **corpuscles of Hassall,** which are concentrically arranged nests of epithelioid cells varying in size from about 25 to 75μm. in diameter. The central cells often show granular degeneration. They vary in number not only in different individuals but also at different periods of life. In a general way the structure of the thymus follicle resembles that of a lymph node, but there are neither lymph sinuses nor germinal centres. The vascular tissue elements and some associated areolar tissue are derived from invading mesenchyme during the course of development.

The thymus appears, then, to have some special relation with the cellular constituents of the vascular system. It has an endocrine function in that extracts of the medulla stimulate lymphopoiesis. Its presence is necessary in the new-born for the development of lymphoid tissue and therefore of some aspects of immunological competence. In the adult it is concerned with the production of lymphocytes which are immunologically uncommitted, though only a very small proportion of the lymphocytes it produces are released into the circulation, the remainder being destroyed in the thymus. The process of involution, by which there comes about a great reduction in the size of the gland after puberty, may be influenced by various endocrine glands, but is probably intrinsic to the thymus. A persistently enlarged thymus is

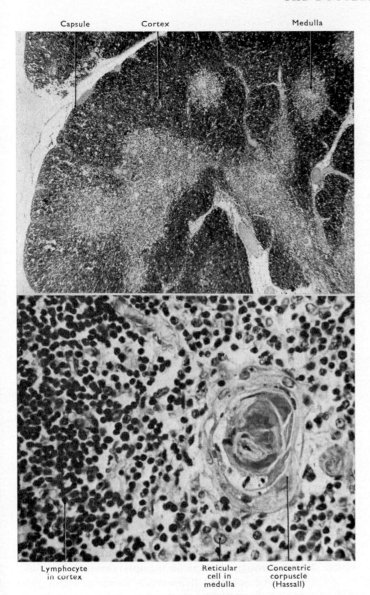

Capsule   Cortex   Medulla

Lymphocyte   Reticular   Concentric
in cortex   cell in   corpuscle
                    medulla   (Hassall)

associated with delayed sex development and obesity. For recent views on thymic function see Miller (1963) and Metcalf (1966).

Development. The paired thymus glands, or 'lobes', right and left, arise from the ventral diverticula of the third pharyngeal pouches and, in the pig, from ectoderm of the third pharyngeal cleft. The first indication of the thymus is seen in the 9-mm. embryo, when the third pouch has developed a caudally directed cylindrical elongation which migrates tailwards along the third pharyngeal arch arteries to reach the pericardium at the 15-mm. stage. As a result of the migration, the cephalic part, to which parathyroid III is attached, becomes drawn out and normally disappears [FIG. 73]. The relative time of the disappearance determines the permanent level of parathyroid III, for until it happens that gland is dragged in the wake of the thymus [p. 568]. Sometimes a small detached mass of formative tissue of the thymus may persist beside parathyroid III, and may differentiate to form an accessory cervical thymus III [FIG. 696].

The thymus soon becomes isolated from the parathyroid (22-mm. embryo) and forms a closely packed, cellular mass. The mass of cells is penetrated shortly afterwards by mesenchyme and blood vessels; and reticular fibres, apparently derived from the blood vessels, make their appearance between entodermal cells (Norris, 1938). Thymocytes appear in the eighth week, probably derived from the invading mesenchyme. The differentiation of cortex and medulla is visible soon afterwards. The concentric corpuscles are derived from the cellular mass which developed from the entoderm of the third pharyngeal pouch.

The growth of the thymus and its involution after puberty have been described above.

Thymus IV Vestiges [FIG. 696]. Although extremely rare

FIG. 694 Photomicrographs of the thymus of a young adult man (above, ×40; below, ×750). (By courtesy of Professor A. R. Muir.)

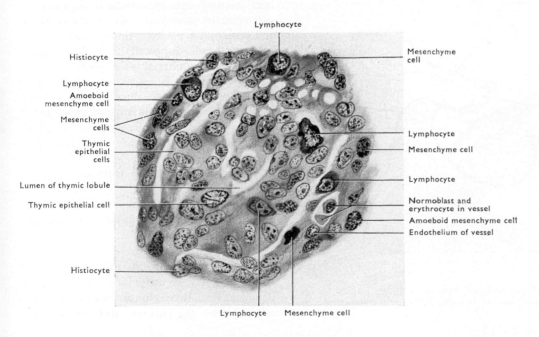

Lymphocyte

Histiocyte   Mesenchyme
                                                        cell

Lymphocyte
Amoeboid
mesenchyme cell

Mesenchyme
cells

Thymic   Lymphocyte
epithelial
cells   Mesenchyme cell

Lumen of thymic lobule   Lymphocyte

Thymic epithelial cell
                                                        Normoblast and
                                                        erythrocyte in vessel
                                                        Amoeboid mesenchyme cell
                                                        Endothelium of vessel

Histiocyte

Lymphocyte   Mesenchyme cell

FIG. 695 Section of embryonic thymus of rabbit. (Maximow and Bloom, *Textbook of Histology.*)

A hollow lobule is surrounded by mesenchyme in which some histiocytes of the reticulo-endothelial system are seen. Lymphocytes are present both around and among the thymic epithelial cells.

in Man, accessory thymic derivatives from the fourth pharyngeal pouch are not uncommonly met with in other mammals. They are found in close relation to parathyroids IV, and must not be confused with the accessory thymic lobules referred to above, which are to be regarded as detached portions of the cervical part of the main thymus.

Methods of Clinical Anatomical Examination. In children the apex of the thymus extends above the level of the manubrium sterni; but it is only occasionally recognizable on palpation as a soft mass situated in front of the trachea slightly to the left of the median line and behind the infrahyoid muscles.

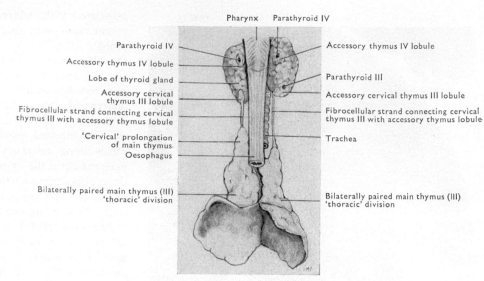

FIG. 696   Illustrating accessory thymus lobules described by Groschuff (1900).

THE HYPOPHYSIS

The hypophysis is a small gland of dual origin which forms a median basal appendage of the hypothalamus [FIGS. 698 and 699], and is known as the pituitary gland from an old notion that it secreted the nasal mucus.

Relations. It occupies the hypophysial fossa [PLATE 8D] between the dorsum sellae behind, and the tuberculum sellae in front. This cavity is lined with dura mater and roofed over by the **diaphragma sellae.** According to Wislocki (1937), no separate arachnoid or pial covering is formed around the hypophysis [FIG. 697] below the level of the diaphragma sellae Through a small aperture in the diaphragma sellae, the **infundibulum** connects the hypophysis with the tuber cinereum of the floor of the third ventricle. Lateral to the hypophysis are the **cavernous sinuses** which are united by the intercavernous sinuses passing below, behind, and in front of the

hypophysis. The internal carotid artery and abducent nerve in each cavernous sinus, and the oculomotor, trochlear, and ophthalmic nerves in its lateral wall are also close to the hypophysis. The **diaphragma sellae** intervenes between the hypophysis and the **optic chiasma,** which lies about 8 mm. above the diaphragma. The **interpeduncular cistern,** with the contained circulus arteriosus, surrounds the infundibulum above the diaphragma sellae. The **sphenoidal sinus** is situated below and in front of the hypophysial fossa, separated merely by a thin wall of bone; if small, it is an antero-inferior relation only. The close proximity of important structures to the hypophysis accounts for various symptoms associated with pathological enlargement of the gland (as in acromegaly). Enlargement of the anterior part of the hypophysis tends to thrust the forepart of the diaphragma sellae upwards and to cause visual defects through pressure on the optic chiasma [FIG. 698].

An enlarged hypophysis sometimes also gives rise to symptoms which are due to encroachment upon the roof of the sphenoidal sinus or to pressure upon the cavernous sinus and the abducent and oculomotor nerves. Enlargement may be recognized radiographically through the expansion of the hypophysial fossa produced by it.

The hypophysis is a rounded structure with the following average dimensions: 14 mm. transversely, 9 mm. anteroposteriorly, and 6 mm. vertically. It usually weighs a little over half a gramme. It undergoes some enlargement during pregnancy.

The following parts are distinguished [FIGS. 697 and 698]:

1. The **anterior lobe,** including: (i) the anterior part; (ii) the tuberal part; and (iii) the intermediate part.

2. The **posterior lobe.**

The anterior lobe forms the greater part of the entire gland; its **anterior part** [FIG. 698] is separated by an **intraglandular cleft** or a series of cysts from the **intermediate part,** which

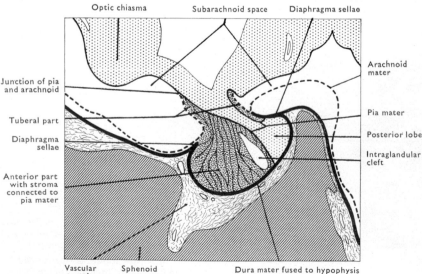

FIG. 697   Diagram of relation of meninges to hypophysis in 160-mm. human foetus. (After Wislocki, 1937.) The relative positions of the structures are not the same as in the adult.

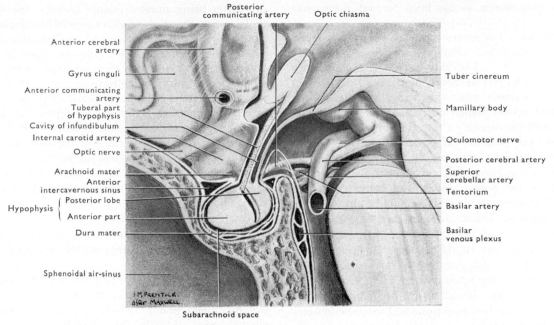

FIG. 698 Diagrammatic median section of hypophysis *in situ*, showing the parts of the organ and its relations to adjacent structures. (The existence of a subarachnoid space around the gland in the hypophysial fossa is doubtful [cf. Fig. 697].)

is a very thin layer of tissue applied to the surface of the **posterior lobe** [FIG. 703]. The **tuberal part** is a thin layer of tissue which encircles the front of the infundibulum, and extends as far as the tuber cinereum. When the infundibulum is cut across in the removal of the brain, most of the tuberal part usually remains adherent to the tuber cinereum. The anterior lobe weighs about half a gramme, or slightly more in women; it is five times as large as the posterior lobe, which contributes most of the remainder of the entire organ.

The two lobes of the hypophysis are of different developmental origin [FIG. 702]. The anterior lobe is derived from Rathke's pouch, a diverticulum of the stomodaeum [p. 53], the posterior lobe arises from a diverticulum of the floor of the diencephalon. The two lobes are different in structure and function, but the line of demarcation between them becomes obscured by their very intimate union [FIG. 704]. The hypophysis occupies a position of pre-eminence in the endocrine system by virtue of the control it exerts over most of the other endocrine glands by means of the various hormones it

FIG. 699 Camera lucida drawing of a coronal section through the cavernous sinus showing the relations of the hypophysis.

produces. For this reason it has been termed the master gland.

Vessels and Nerves. The anterior and posterior lobes of the gland have an abundant, but relatively independent, **arterial supply** (Stanfield, 1960). The anterior lobe is supplied by the **superior hypophysial arteries,** branches of the internal carotid artery immediately after this vessel has pierced the dura mater; these vessels mainly enter the infundibulum and there break up into capillaries. From this first capillary bed **hypophysial portal vessels** run down to the sinusoids of the anterior lobe. A large part of the blood supply of the anterior lobe therefore passes first through capillaries before being collected by the portal vessels and conveyed to the anterior lobe. The posterior lobe is supplied by the **inferior hypophysial arteries,** which arise on each side from the internal carotid artery as it lies within the cavernous sinus. They divide into medial and lateral branches, and anastomose with the branch of the superior hypophysial artery which supplies the lower part of the infundibulum and gives branches direct to the anterior part. Blood leaves the gland by means of **veins** which pass into the **cavernous** and **intercavernous sinuses. Lymph vessels** have not been demonstrated in the gland.

The **nerves** of the anterior lobe are a rich supply of non-medullated fibres from the carotid plexus. While many of these are vasomotor in function, there are some nerve fibres which have terminals in intimate relation with the epithelial cells and may be secretory in function. That the nervous connexions are not essential to all its endocrine functions is shown by the fact that grafts of the gland produce active secretions. The posterior lobe receives nerve fibres from the supra-optic and paraventricular hypothalamic nuclei by a well-marked tract which runs along the infundibulum. The fibres end by forming a dense network around the small cells of the posterior lobe. Functionally there is a close connexion between the hypophysis and the hypothalamus.

Structure and Function. The **anterior lobe** is composed in the adult mainly of columns of epithelial cells supported by two pockets of areolar tissue which contain vessels of considerable size. Capillaries pass from these vessels into intimate relation with the epithelial cells, and they eventually reach a vascular plexus on the periphery of the anterior part. The tubular arrangement of the epithelial cells seen in foetal life disappears before birth. Vesicles of colloid material are to be seen at all ages from the fourth month of intra-uterine life; colloid also appears in the intraglandular cleft.

Six principal kinds of cells are recognized: **chromophobe, acidophil, basophil, amphophil** (containing acidophil, basophil, or both types of granule), **violet** and **indifferent,** depending on the reaction of their cytoplasmic granules to acid Schiff-orange G stain (Sheehan, 1959) [FIG. 700]. Acidophil cells comprise about one-third of the total number of cells, basophils only about one-tenth of the total.

The anterior lobe is known to produce six distinct hormones, several of which influence the activity of other endocrine glands. The growth hormone appears to be derived from the acidophil cells; thyrotrophic hormone, and the gonadotrophic hormones (which influence the gonads), from the basophils. The precise source of the others is not known. The unequal distribution of basophils and acidophils in the anterior lobe

(the latter being situated mainly laterally, the basophils centrally) assists the study of alterations in histological appearance of the hypophysis in disease and experimental investigations. Tumours consisting of the different types of cell produce different effects—basophil tumours cause obesity and various other changes, known collectively as Cushing's syndrome, acidophil tumours cause acromegaly. Various conditions have been recognized which appear to depend on hyposecretion, e.g., the types of infantilism described by Simmonds and Lorain. A relation to periods of the oestrous cycle is indicated by the cyclic variations of cell types in guinea-pigs. The condition of diabetes insipidus is thought to be due to a lesion of the posterior lobe or of the hypothalamus or of both.

Absence of the acidophils is found to be related in mice to a defect in respect of a certain gene; the effect on the gonads is then found to be present, but not the growth-stimulating effect.

The **intermediate part** appears to be the least specialized portion of the hypophysis; it grows less, and shows cysts which contain colloid. These cysts are occasionally lined by ciliated cells. Mostly chromophobes and amphophils are present, but there are also a few basophils and acidophils. In the foetus, gland tubules are present and open into the cleft, which is a vestige of the lumen of the stomodaeal diverticulum (Rathke's pouch). These gland tubules subsequently extend into the posterior lobe [FIG. 704] and basophil cells also migrate from the intermediate part into the posterior lobe (Lewis and Lee, 1927). The **tuberal part** consists mainly of amphophils and chromophobe cells.

The **posterior lobe,** sometimes called the nervous part, consists of a mass of neuroglia, and of small cells that resemble neuroglial cells. The invasion of this part by basophil cells from the intermediate part of the anterior lobe has been mentioned above. Small hyaline bodies, described by Herring (1908), are found among these cells [p. 577]. Potent hormones are present in the posterior lobe; the **pressor** hormone **(vasopressin)** raises the blood pressure, the **antidiuretic** hormone diminishes urinary excretion of water and increases the excretion of chlorides, and **oxytocin** causes uterine muscle to contract. There is evidence suggesting that the pressor and antidiuretic effects are manifestations of the action of only one hormone. It is not known whether the immigrant basophil cells have the same function as those of the anterior lobe. The posterior lobe hormones are produced by the supra-optic and other hypothalamic nuclei, and pass along the axons of the hypothalamo-hypophysial tracts (such as the supra-optico-hypophysial, p. 679) into the posterior lobe, which serves as a storage-release centre rather than a gland of internal secretion (Scharrer and Scharrer, 1954; Stutinsky, 1967). Damage of the supra-opticohypophysial tract certainly causes **diabetes insipidus,** but this has always been considered previously as being due to release of the control over the secretions of the posterior lobe by the supra-optic nucleus.

Development. As already mentioned, the hypophysis originates from two entirely distinct rudiments. Both of these are hollow ectodermal diverticula, one neuro-ectodermal, derived from the floor of the primary forebrain, the other from the ectodermal lining of the stomodaeum. The stomodaeal diverticulum, known as Rathke's pouch, gives origin to the

FIG. 700 Photomicrographs demonstrating the cell types to be found in the anterior lobe of the human hypophysis, coloured with acid Schiff-orange G.

1. Low power view showing eosinophils (yellow) and basophils (brown). ×170.

2–5. High power views showing (2) eosinophils (yellow) with five basophils; (3) basophils flanked by a few yellow eosinophils; (4) chromophobe cells; (5) undifferentiated (indifferent) cells. ×780.

(Preparations and photomicrographs 2–5 by courtesy of Professor H. L. Sheehan.)

whole of the anterior lobe. The other diverticulum gives origin to the posterior lobe and the infundibulum.

The first indication of the appearance of the hypophysis is met with in embryos of 2–3 mm. in the shape of an angular depression from the stomodaeum [p. 53] immediately in front of the dorsal attachment of the buccopharyngeal membrane. It deepens progressively, and in the 7-mm. embryo it has become a deep and wide saccular diverticulum, compressed anteroposteriorly, and opening by an aperture of equal width into the primitive mouth cavity. The earliest indication of the appearance of the posterior lobe is a slight funnel-like depression in the floor of the forebrain vesicle [FIG. 75] in the 4–5-mm. embryo. At the 10–12-mm. stage the recess has become deeper and more sharply marked, and its anterior wall is in close apposition with the fundus of the saccular anterior lobe. After the 12-mm. stage the portion of the stomodaeal diverticulum nearest the mouth rapidly narrows

and elongates to form a slender tubular stalk. By the stage of the 20-mm. embryo this stalk has become interrupted, separating the embryonic anterior lobe from its original connexion with the epithelium of the mouth. Remnants of the obliterated stalk persist, not only at that stage but up to much later periods of intra-uterine life, and may sometimes be detected in the postnatal period, particularly in the pharyngeal roof at the posterior margin of the nasal septum, which marks the site of origin of Rathke's pouch. Very rarely the base of the skull remains incomplete immediately around the stem of the hypophysis, so that a **craniopharyngeal canal** may be found later in the osseous cranial base and sometimes even containing vestiges of the epithelial tissue of the stalk.

At the end of the second month the anterior lobe is a broad compressed sac, deeply notched for the reception of the posterior lobe. On each side of that notch the fundus of the sac is prolonged backwards in the form of paired hollow

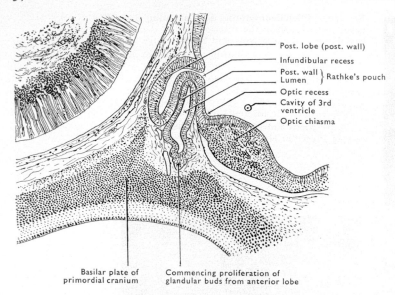

FIG. 701 Paramedian sagittal section of hypophysial region of a human embryo (20 mm. C.R. length).

cornual extensions of the anterior lobe at the sides of the neck of the posterior lobe (they are already conspicuous in that situation in transverse sections as early as the 14-mm. stage). At the 2-month period the original anterior surface of the sac has become secondarily cupped, its concavity now looking towards the floor of the brain in front of the infundibular region. The concavity is partially divided by a prominence in the middle into two fossae which are occupied by tissue

continuous with the surrounding mesoderm. These fossae are presently invaded by the proliferating glandular cell cords of the anterior wall of the sac during the process of formation of the anterior part of the organ. The bilateral pockets of areolar tissue, representing these mesodermal accumulations, were noted on page 574 in the course of the description of the anterior part of the adult gland.

The recurved margin of the anterior lobe extends forwards and upwards towards the brain and eventually reaches the tuberal region of the floor of the diencephalon, where it spreads out to a greater or lesser extent as the tuberal part.

Up to about the 20-mm. stage the wall of the anterior lobe preserves its simple epithelial character. Already, however, it has begun to show indications of a process of proliferative budding [FIG. 701] and in this manner the cell cords become intermingled with highly vascular mesenchyme which will form the sinusoidal stroma of the fully developed anterior part. In a foetus of the third month [FIG. 703] the epithelial posterior wall of the original Rathke's pouch is closely applied to the anterior surface of the posterior lobe and shows very little, if any, proliferative increase in thickness. It is this epithelial lamina which forms the intermediate part of the anterior lobe. In front of it the lumen of the anterior lobe of the hypophysis is still quite roomy. Later, it undergoes reduction and at least partial obliteration, the intraglandular cleft or clefts which represent it becoming occupied by globules and irregular masses of colloid. The lumen shown in FIGURE 703 does not, however, represent the whole of the earlier cavity of the anterior lobe, for portions of the sac become involved in the proliferative activity that produces the solid tissue of the anterior lobe.

During the course of development from the 20-mm. stage onwards the cavity of the posterior lobe and infundibulum tends to be obliterated in the human hypophysis, and they

FIG. 702 Diagrammatic median section of hypophysis in early stages of development, showing A, approximation of the infundibular recess to Rathke's pouch, and B, the origin of the various parts of the organ. The posterior wall of Rathke's pouch develops into the intermediate part of the hypophysis, its anterior wall into the anterior and tuberal parts.

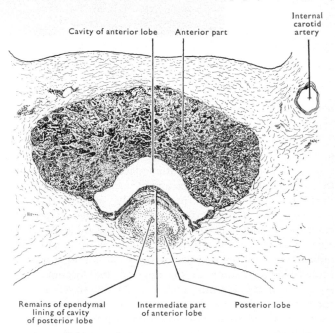

FIG. 703 Horizontal section of hypophysial region of a human foetus (71 mm. C.R. length).

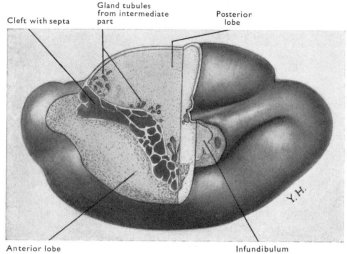

FIG. 704 Hypophysis of a child, aged 1 year, with a portion removed to show the stalk and glandular tubules projecting from intermediate part of anterior lobe into posterior lobe. (Lewis and Lee, 1927.)

become solid except in the root of the stalk which is occupied by the infundibular recess of the third ventricle.

In the later stages of antenatal growth there is a progressive invasion of the posterior lobe by groups of cells and by individual basophil cells that belong to the intermediate part of the anterior lobe, i.e., to the posterior wall of the original buccal diverticulum. These cells tend to undergo atrophy in the adult, but may persist throughout life. The basophil cells give rise to colloid and the hyaline bodies of Herring (1908) may represent the products of secretion of the supra-optic nucleus. Replacement of worn-out cells is apparently not a feature of the adult hypophysis, for cell division and degenerating cells are seldom seen. The intraglandular cleft, commonly filled with colloid in the child, becomes obliterated in many adults, though in others it persists as a space filled with colloid.

In the 20-mm. embryo the two lobes of the hypophysis are surrounded by mesenchyme, except where the infundibulum remains connected to the brain. This mesenchyme eventually differentiates into a lamina of dura mater which is firmly united both to the hypophysis and to the hypophysial fossa [FIG. 697].

THE PINEAL BODY

The pineal body is a small oval or cone-shaped structure, attached by a short, hollow stalk to the hinder end of the roof of the diencephalon, between the habenular and posterior commissures [FIGS. 705 and 773]. It is moderately firm in consistence, reddish-brown in colour, and measures 5–10 mm. anteroposteriorly, 3–7 mm. in its other diameters. It weighs about 0·2 g. Much discussion has centred around its endocrine function, and it contains several pharmacologically active indole hormones. One of these hormones has an inhibitory effect on the gonads. It commonly becomes calcified in middle age [see below].

Relations. It is covered with pia mater, and it lies in the sulcus between the superior colliculi. It is below the splenium of the corpus callosum, from which it is separated by the great cerebral vein. Its posterior end is free, and is a short

C.T.A.—U

distance in front of the vermis of the cerebellum and the free edge of the tentorium [FIG. 773].

The cavity of the stalk is the pineal recess of the third ventricle. On the dorsal lip of this recess is the habenular commissure. Immediately above the habenular commissure and the stalk of the pineal body is another and larger recess of the third ventricle called the suprapineal recess.

Structure. The pineal body consists mainly of a mass of distinctive cells (the pineal parenchyma) interspersed with neuroglial cells. This tissue becomes divided into lobules by the ingrowth of septa of fibrous tissue from the pia mater—a process which takes place mostly during childhood. Vascularized fibrous tissue invades the organ during early intra-uterine life.

The characteristic parenchyma cells have long processes which end in bulbous extremities, some of which are on the surfaces of the lobules next to the intervening fibrous septa, and others within the lobule on the fibrous tissue which encloses the blood vessels. The body of each cell is rounded and has a large round nucleus and granular cytoplasm. They offer a marked contrast with nerve cells—which are wanting in the pineal body except close to the habenular commissure. The granules of the cytoplasm have been claimed as being secretory in character, but they show no variation in number at different ages, nor does the pineal body display any differences in appearance under different physiological conditions. There is no evidence of histological change at puberty. The numerous, non-myelinated nerves which are found in the pineal body arise in the superior cervical sympathetic ganglion and appear to be concerned with the activity of its cells as well as being distributed to its blood vessels. Calcareous deposits usually make their appearance in the organ from adolescence onwards, and the pineal body thus becomes visible in radiographs [PLATE 46C].

Development. The pineal body develops as a pouch-like evagination of the hinder part of the roof of the diencephalon, separated from the midbrain by the posterior commissure only. Its epithelial wall becomes thickened, with restriction of the lumen to the stalk of attachment. Later, this epithelial mass is invaded by cellular sprouts from the ependyma around the cavity; these form the characteristic pineal parenchyma. Blood vessels also invade the organ. The development of the septa takes place actively in early postnatal life.

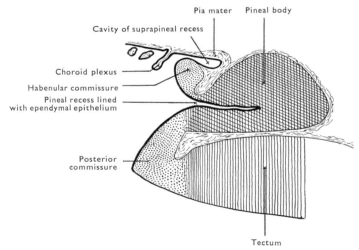

FIG. 705 Diagram of median sagittal section through pineal body and its immediate attachments.

RADIOGRAPHIC ANATOMY

E. SAMUEL

Hypophysis (Pituitary Gland)

Whilst gross endocrine disturbances involving the hypophysis may be present without any changes in the sella turcica, enlargement of the hypophysis, as in tumour formation, is readily shown by erosion and enlargement of the sella turcica. The introduction of air into the subarachnoid space or into the third ventricle may help to assess the size of the tumour mass.

Thyroid

The normal thyroid tissue is not recognizable on a plain radiograph from other soft tissues in the neck. Enlargement of one or both lobes may be recognizable by the compression, displacement, or indentation of the trachea resulting from pathological enlargement of the thyroid.

Parathyroid Glands

Normal parathyroid glands cannot be demonstrated radiographically.

Thymus

The thymic shadow is visible in the anterior mediastinum in many new-born infants and especially so in premature infants. The gland usually projects as a sail-like shadow to the right of the superior mediastinum. Enlargement of the gland may give rise to compression effects on the trachea and the lung.

Suprarenal Glands [PLATE 37A]

The suprarenal glands cannot be definitely separated from the soft tissue shadows produced by the fibro-fatty tissue surrounding the kidney. By the introduction of air, oxygen, or carbon dioxide into the tissues surrounding the kidney, the outline of the suprarenal glands may be shown. Tomographic films further assist by blurring overlying gas shadows. The investigation may be combined with an intravenous pyelogram to enable the renal outline to be seen more clearly. The right suprarenal gland lies on the medial aspect of the upper pole of the kidney, but in the radiographs a layer of air frequently separates the gland from the kidney substance. On the left side, there is an even wider air space between the gland and the upper pole of the left kidney. Even so, it must be appreciated that the normal suprarenal gland casts a relatively indefinite shadow, even with perirenal air insufflation.

Pineal Body [p. 703]

Pancreas [p. 466]

Ovary and Testes

The investigation of the endocrine functions of these structures is more a subject for biological and biochemical assay than for radiological methods.

REFERENCES

BARR, H. S. (1954) Foetal cortex of the adrenal glands, *Lancet* i, 670.

BEJDL, W., and POLITZER, G. (1953) Über die Frühentwicklung des telobranchialen Körpers beim Menschen, *Z. Anat. Entwickl.-Gesch.*, 117, 136.

BOYD, J. D. (1937) The development of the human carotid body, *Contr. Embryol. Carneg. Instn*, (No. 152), 26, 1.

BOYD, J. D. (1950) Development of the thyroid and parathyroid glands and the thymus, *Ann. roy. Coll. Surg. Engl.*, 7, 455.

BOYD, J. D. (1960) Origin, development and distribution of chromaffin cells, in Ciba Foundation Symposium on *Adrenergic Mechanisms*, London.

COWDRY, E. V. (1932) *Special Cytology. The Form and Functions of the Cell in Health and Disease*, 2nd ed., Vol. 2, New York.

COWDRY, E. V. (1944) *A Textbook of Histology. Functional Significance of Cells and Intercellular Substances*, 3rd ed., London.

COUPLAND, R. E. (1952) The prenatal development of the abdominal para-aortic bodies in man, *J. Anat. (Lond.)*, 86, 357.

COUPLAND, R. E. (1954) Post-natal fate of the abdominal para-aortic bodies in man, *J. Anat. (Lond.)*, 88, 455.

CROWDER, R. E. (1957) The development of the adrenal gland in man, with special reference to origin and ultimate location of cell types and evidence in favour of the 'cell migration' theory, *Contr. Embryol. Carneg. Instn*, 36, 195.

DE ROBERTIS, E. (1942) Intracellular colloid in the initial stages of thyroid activation, *Anat. Rec.*, 84, 125

EULER, U. S. VON (1955) Noradrenaline in hypotensive states and shock, *Lancet*, ii, 151.

GILMOUR, J. R. (1938) The gross anatomy of the parathyroid glands, *J. Path. Bact.*, 46, 133.

HARRISON, R. G. (1951) A comparative study of the vascularization of the adrenal gland in the rabbit, rat and cat, *J. Anat. (Lond.)*, 85, 12.

HARRISON, R. G. (1963) *A Textbook of Human Embryology*, Oxford.

HARRISON, R. G., and HOEY, M. J. (1960) *The Adrenal Circulation*, Oxford.

HERRING, P. T. (1908) The histological appearances of the mammalian pituitary body, *Quart. J. exp. Physiol.*, 1, 121.

HIRSCH, P. F., and MUNSON, P. L. (1969) Thyrocalcitonin, *Physiol. Rev.*, 49, 548.

HOLLINSHEAD, W. H. (1936) The innervation of the adrenal glands, *J. comp. Neurol.*, 64, 449.

KINGSBURY, B. F. (1939) The question of a lateral thyroid in mammals with special reference to man, *Amer. J. Anat.*, 65, 333.

LEWIS, D., and LEE, F. C. (1927) On the glandular elements in the posterior lobe of the human hypophysis, *Bull. Johns Hopk. Hosp.*, 41, 241.

METCALF, D. (1966) *The Thymus*, Berlin.

MILLER, J. F. A. P. (1963) Role of the thymus in immunity, *Brit. med. J.*, 2, 459.

MORGAN, J. R. E. (1936) The parathyroid glands. I. A study of the normal gland, *Arch. Path.*, 21, 10.

NOLAN, L. E. (1938) Variations in the size, weight and histologic structure of the thyroid gland, *Arch. Path.*, **25**, 1.

NORRIS, E. H. (1937) The parathyroid glands and the lateral thyroid in Man: their morphogenesis, histogenesis, topographic anatomy and prenatal growth, *Contr. Embryol. Carneg. Instn*, (No. 159), **26**, 247.

NORRIS, E. H. (1938) The morphogenesis and histogenesis of the thymus gland in Man: in which the origin of the Hassall's corpuscles of the human thymus is discovered, *Contr. Embryol. Carneg. Instn*, (No. 166), **27**, 191.

PINCUS, G. (1947–68) *Recent Progr. Hormone Res.*, **1–24**.

PINCUS, G., and THIMANN, K. V. (1948–64) *The Hormones*, **1–5**.

REICHSTEIN, T., and SHOPPEE, C. W. (1943) The hormones of the adrenal cortex, *Vitam. and Horm.*, **1**, 345.

SCHARRER, E., and SCHARRER, B. (1954) Hormones produced by neurosecretory cells, *Recent Progr. Hormone Res.*, **10**, 183.

SELYE, H. (1947) *Text-book of Endocrinology*, Montreal.

SELYE, H. (1950) *Stress*, Montreal.

SHEEHAN, H. L. (1959) La glande pituitaire dans la cirrhose du foie (modifications histologiques), *C.R. Soc. franç. Gynéc.*, **29**, 211.

STANFIELD, J. P. (1960) The blood supply of the human pituitary gland, *J. Anat. (Lond.)*, **94**, 257.

STUTINSKY, F. (1967) *Neurosecretion*, Berlin.

SWINYARD, C. A. (1937) The innervation of the suprarenal glands, *Anat. Rec.*, **68**, 417.

WELLER, G. L. (1933) Development of the thyroid, parathyroid and thymus glands in Man, *Contr. Embryol. Carneg. Instn*, (No. 141), **24**, 93.

WILLIAMS, R. G. (1941) Studies of vacuoles in the colloid of thyroid follicles in living mice, *Anat. Rec.*, **79**, 263.

WISLOCKI, G. B. (1937) The meningeal relations of the hypophysis cerebri. I. The relations in adult mammals, *Anat. Rec.*, **67**, 273.

WISLOCKI, G. B., and KING, L. S. (1936) The permeability of the hypophysis and hypothalamus to vital dyes, with a study of the hypophyseal vascular supply, *Amer. J. Anat.*, **58**, 421.

YOUNG, M., and TURNBULL, H. M. (1931) An analysis of the data collected by the status lymphaticus investigation committee, *J. Path. Bact.*, **34**, 213.

ZUCKERKANDL, E. (1912) The development of the chromaffin organs and of the suprarenal glands, in Keibel and Mall's *Manual of Human Embryology*, Vol. II, Chap. 15, Philadelphia.

THE CENTRAL NERVOUS SYSTEM

G. J. ROMANES

INTRODUCTION

THE nervous system is concerned with controlling and integrating the activity of the various parts of the body, and provides a mechanism by which the individual can react to a changing external environment, while maintaining an internal environment delicately balanced within the narrow limits which are necessary for survival.

The mechanisms by which the nervous system controls the tissues of the body are at first sight manifold, but in essence consist of the release of substances which alter the activity of the target organs.

This is achieved:

1. By the release of minute quantities of such substances from the terminations of nerve cells on the tissues in the presence of a mechanism (e.g., enzyme) which rapidly inactivates the substance, so that the tissue response is localized and transitory. This is the mechanism by which muscle and some gland cells are stimulated and by which many nerve cells transmit their activity to each other at synapses [p. 586].

2. By the release of substances into the blood stream, either from the nervous system itself, or from ductless glands directly under its control. These substances, known as hormones, are widely disseminated throughout the body and have more general effects which persist often for long periods. They may produce their effects directly on the tissues or through the intermediary of other glands which, in their turn, secrete varying concentrations of their hormones into the blood stream.

Nerve cells are therefore basically secretory in character, and though different substances are secreted by different nerve cells the mechanism is the same throughout.

The ability of an organism to react to change presupposes the presence of a mechanism which can detect such change and set in motion the appropriate response. This requires a sensory system which is able to record not only the wide range of changes imposed on the animal from without, but also the most minute changes in its internal processes, whether these are alterations of its chemical or physical structure. Thus control of gases and ions in the blood is dependent on a method for measuring them, as indeed the control of movement is dependent on a mechanism for measuring the progress of the movement from moment to moment. Such a sensory or receptive system must be able to differentiate between an enormous number of different circumstances and be capable of activating a mechanism which can integrate all the information showered upon it by the sensory system. Only in this way can the nervous system produce a response which is appropriate to all the conditions prevailing at the time. Thus manual dexterity in Man is not simply the result of the development of a hand with a wide range of movement, but

depends on the fact that it is also an important sensory organ. Loss of this sensation removes the ability to use the hand properly, even though the muscles, tendons, and joints which are essential to the production of the fine movements are in no way impaired. Similarly the full utilization of the hand in highly skilled activities requires that there should be adequate integration of the sensory information which records the position, speed, force, and direction of movements of the fingers with the visual or auditory impressions which these movements produce. Loss of the necessary information or failure to integrate it can disrupt the whole pattern of activity.

In every activity involving the nervous system, however complex or simple, there are therefore three elements involved —the sensory or receptive, the integrative, and the effector or motor. Of these three it is the integrative element which undergoes the greatest degree of development in higher animals to become the major part of the nervous system.

It is usual, for descriptive purposes, to divide the nervous system into central and peripheral parts. Nevertheless such a division is quite arbitrary, since the nerves which comprise the peripheral part, and the brain and spinal medulla which make up the central nervous system, are both essential parts of a functioning unit and cannot be clearly separated from each other on anatomical grounds.

A detailed knowledge of the structural organization of the nervous system is essential to an understanding of its function; yet the information which we possess is too scanty to allow of a full interpretation of the significance even of those structural details which are well known.

PHYLOGENY

The central nervous system in Man, and indeed in vertebrates in general, is so complex that a brief survey of the early stages of its evolutionary development may help to clarify the basis on which it is constructed.

In its simplest form, the nervous system consists of a network of interconnected, branching cells which lies between the external and internal epithelia. Such a network [FIG. 706 B] is in contact with sensory cells which form part of the epithelial surface, and is capable of receiving nervous impulses from them and transmitting these to activate the muscles. Such a diffuse nervous system is admirably suited to the protective closure of a sea-anemone and to the rhythmical activity by which a jelly-fish progresses, but it can also produce localized reactions such as the movement of a single tentacle. Indeed, as in all animals including the protozoa, the response obtained is related to the nature, intensity, and duration of the sensory stimulus. Thus even in this simple nervous system the sensory cells are able to discriminate

FIG. 706 A schematic diagram to indicate some of the features of the phylogeny of the nervous system.

A, hydra; B, coelenterate; C, flatworm; D, higher invertebrate, e.g. earthworm; E, vertebrate.

In A, the solid black cells are myoepithelial cells. In the remainder: blue, sensory cells; black, connector or internuncial cells; red, motor cells.

between stimuli, and some degree of interaction (integration) of different and coincident stimuli can take place.

This diffuse nervous system is found throughout the coelenterates and its first significant modification appears in the flatworms [FIG. 706 C]. Here directional progression, with the consequent development of a leading end, is associated with the orientation of the net into a series of longitudinal strands streaming tailwards from a collection of cells at the head end. This **head ganglion** is no doubt associated with the greater number of sensory impressions entering the animal at its leading end, and its formation represents the beginning of centralization in the nervous system, even though it is still just one part of an orientated nerve net. Further, while the sensory cells are still buried in the surface epithelium and have processes linking them to the underlying nerve net, certain special sensory cells have developed. These include taste buds, which are present over the whole surface of the body, together with rudimentary eye spots and gravity-measuring organs or statocysts, the latter two closely related to the head ganglion. These sensory organs contribute to the importance of the head ganglion and it seems not unlikely that this collection of cells, which now receives many different types of information, must be able to carry out an integrative function.

In the further development of the nervous system, the diffuse nerve net of the coelenterates is lost and the nerve cells become concentrated into a longitudinal cord [FIG. 706 D] extending tailwards from the head ganglion. The sensory cells which are situated in the epidermis of coelenterates, worms, and many other invertebrates, now send their processes into the centralized nerve cord. As a result of this centralization and the consequent separation of the nerve cells from the structures which they innervate, it is possible to distinguish two types of nerve cell within the cord.

1. Motor cells which innervate the muscles by sending relatively long processes to them from the cord.

2. Cells which lie entirely within the head ganglion and nerve cord, conducting nerve impulses through these structures, often in a longitudinal direction. Through these latter cells, which are intercalated (internuncial) between the sensory and motor cells, the various parts of the nervous system are linked together and impulses are relayed to the motor cells.

In higher animals [FIG. 706 E] the sensory cells migrate from their position in the epidermis to form groups of cells closely associated with, but still outside, the nerve cord, retaining their connexion with the surface through greatly elongated peripheral processes.

There is thus established a prototype of the vertebrate nervous system which extends throughout the length of the animal and contains the great majority of the nerve cells. This concentrated nervous system, which replaces the nerve net, allows of more rapid and effective integration of sensory stimuli from widely separated parts of the body, before these have each evoked local responses which may be opposed to each other.

The process of centralization outlined above is never complete, for even in the higher vertebrates the motor cells of that part of the nervous system (autonomic) which controls the involuntary organs of the body, e.g., the heart and alimentary canal, lie outside the central nervous system and tend to form plexuses or networks of interconnected nerve cells.

The sensory cells of the olfactory system also retain a primitive position in the surface epithelium even in Man.

While it is true that the gross appearances and position of the vertebrate and invertebrate nervous systems are very different, nevertheless the basic structural arrangements are similar in both. In the vertebrates, the fishes show a great increase in the number of nerve cells at all parts of the nervous system compared with the invertebrates. This increase is, however, primarily at the cephalic extremity which is influenced by a mass of sensory information entering it from the snout, eye, ear, and olfactory apparatus as well as from the heart, great vessels, and gills. In this way the cephalic part of the vertebrate nervous system becomes the largest correlating centre, or **brain,** whose activity is integrated with that of more caudal levels by the longitudinally directed intercalated cells. The processes of some of these intercalated cells carry information from the body to the brain while others give it a measure of control over local mechanisms in the caudal parts of the nervous system.

Once such a pattern has been established, the increased sensory experience which arises from the assumption of a terrestrial existence is associated with further increases in the size and complexity of the brain so that more complex behavioural patterns are possible. Thus in Man the assumption of the erect posture and the freeing of the upper limb from its supporting functions, with the consequent development of the hand as a potent new sensory organ, may well have played a significant part in the growth of the human nervous system to its present complexity.

A

B

FIG. 707 Photomicrographs showing A, a normal motor nerve cell of the spinal medulla, and B, a motor nerve cell which has undergone chromatolysis two weeks after its axon has been cut.

Note that the granules of Nissl substance which occupy the cell body and extend into the dendrites in A, have disappeared or dispersed as fine particles in B, and that the latter cell is swollen.

THE CELLULAR ELEMENTS OF THE NERVOUS SYSTEM

The central nervous system consists of nerve cells, or **neurons,** and **neuroglia.** The former are the active elements of the nervous system: the latter form a connective tissue which is peculiar to the central nervous system and is intimately related to the nervous elements both anatomically and functionally (Hyden, 1967). The neuroglia forms the coverings of the nerve cells and isolates them from the surrounding tissues, including the blood vessels which enter the nervous system, as well as forming the framework of the nervous system and carrying out phagocytic activities within it.

Neurons and neuroglial cells are both characterized by having **cytoplasmic processes** in the form of fine filaments which extend from the **cell body** with its contained nucleus [FIGS. 708 and 718]. In neurons these processes often reach a considerable length which allows them to propagate nervous impulses, often at great speed over relatively great distances.

The prodigious length of the processes of some neurons, which may exceed 1 metre, together with their invariably close association with neuroglial cells in the central nervous system and another type of sheath cell, **cells of Schwann,** in the peripheral nerves, led some investigators to assume that each axon was formed not by one cell but by many. Thus these investigators considered that the nervous system was a true syncytium whose cells were in continuity with each other without intervening cell membranes. In 1891 Waldeyer enunciated the **neuron theory,** which stated that the nervous system consisted of separate cells which, though they came into contact with each other, were never in continuity. This theory stimulated one of the most bitter controversies of neurohistology, which has continued almost to the present day despite the support it received from Cajal, the greatest of all neurohistologists, and the demonstration

in 1907 by R. G. Harrison that nerve cells could be grown
outside the body in clotted plasma in which their processes
extended into areas devoid of cells from which they could
have been formed. The solution of this controversy has had
to await the high resolution of the electron microscope. This
has made it possible to confirm the presence of spaces separating nerve cells from each other and from non-nervous tissues,
and to see the minutest processes of nerve cells [p. 585] as
separate entities.

Neurons (Nerve Cells)

Neurons, like other cells, consist of a nucleus and its surrounding cytoplasm. The **nucleus** appears almost empty in
stained sections. It has a large, single, central nucleolus and
close to it a small satellite present only in the cells of female
animals, the 'sex chromatin' (Barr, 1960). A distinct nuclear
membrane separates the nucleus from the cytoplasm which
contains well-defined bodies, stainable with basic aniline
dyes, and known as Nissl substance after their discoverer.
The **Nissl substance** is arranged in blocks which vary in size
from cell to cell, but tend to have a similar appearance in
cells of like function. Thus in the large motor neurons
which supply striated muscle [FIG. 707A] the blocks tend to be
of considerable size, while in the equally large sensory cells
they are smaller and appear as a fine dust when stained for
light microscopy. They may be virtually absent in some
neurons or present only in the peripheral cytoplasm. **Nissl
substance** [FIG. 2] consists of particles of ribose nucleic acid
arranged in a manner similar to that in secretory cells and
seems to be concerned with protein synthesis in the cell.
It tends to disappear from cells which have been kept in a
state of activity for a prolonged period, or subjected to injury
[FIG. 707].

Another characteristic of nerve cells is the fibrillary structure of the cytoplasm which can be seen in cells stained with
silver. Very fine fibres, **neurofibrillae,** appear to course
through the cytoplasm and extend into the processes. Such
fibrils have been described in living cells, and though fine,
apparently tubular, filaments are visible under the electron
microscope, it is not possible to be certain that they form
continuous structures. Their function is unknown, but they
indicate that there is some orientation of the elements of the
cytoplasm and that this is parallel to the long axis of the
processes. These neurotubules are finer in axons than they
are in dendrites.

The **cytoplasmic processes** of nerve cells are normally
described as being of two types called **axons** and **dendrites.**
In many cells it is possible to differentiate these from each
other on histological grounds [FIGS. 708 and 710], though
this is not invariably so. The terms indicate the direction in
which these processes normally conduct nervous impulses.
The dendrites are afferent or receptive and normally conduct
towards the cell body, while the axon is efferent and conducts
away from it.

The **axon** is usually a thin, single process of uniform calibre
throughout its length, and differs from the rest of the cytoplasm in not containing Nissl substance.

The **dendrites** are much more variable in structure, being
usually thicker and shorter than the axon. In cells with many
dendrites [**multipolar cells,** FIG. 709] they are of irregular

FIG. 708   Three nerve cells from anterior grey column of
human spinal medulla, stained to show the Nissl substance. The dendrites have many more branches.

outline, branch frequently and gradually taper to their termination, often forming a very complex arrangement in the
vicinity of the cell body [FIG. 710]. Some cells, especially the
sensory cells in the spinal and cranial nerve ganglia, have only
two processes which may either arise separately from the cell

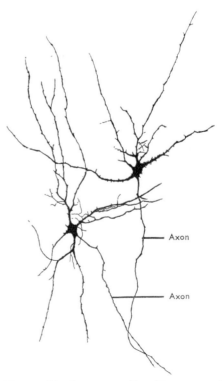

FIG. 709   Two multipolar nerve cells. (From a specimen prepared by the Golgi method.)

(vestibulocochlear nerve) making it **bipolar** in type, or as a single process which subsequently divides into two. These are **unipolar cells** which are normally derived from bipolar cells [FIG. 711]; in this case both processes have the same structure as an axon, though that which conducts towards the cell body is essentially a dendrite. The presence of many dendrites indicates that the cell receives impulses from several different sources, while single dendrites are found in those which are activated from a single source.

The **axis cylinders** or **axons** are, without exception, enclosed in a **cellular sheath**. This complex of axis cylinder and sheath is known as a **nerve fibre.** In peripheral nerves the sheath is composed of elongated cells each of which encloses a certain length of one or more axons, the complete sheath of any one axon being formed by a chain of these **Schwann cells,** arranged end to end along it. An almost identical arrangement exists within the central nervous system, but here the axons are ensheathed by neuroglial cells [p. 590].

All nerve fibres fall into two groups on account of the arrangement of their sheaths, which are either simple or complex. Axons vary in diameter from 0·0001 mm. to 0·01 mm., and of these the smaller, less than 0·001 mm. in diameter, are simply enveloped by the sheath cell. In this case there may be several such axons surrounded by a single sheath cell either as one bundle (e.g., olfactory nerves), or more usually each lying separately surrounded by the sheath cell [FIG. 712 A]. Axons of larger diameter are enclosed singly within their chain of sheath cells, and the ensheathing membrane of each is wound spirally around the axon in such a way that the number

FIG. 711 Three stages in the development of a unipolar cell from a bipolar cell (spinal ganglion).

of turns of the spiral increases with the diameter of the axon [FIGS. 712B and C; 713]. The membrane of the sheath cell is formed of alternate layers of lipoid and protein, the former gives it a white glistening appearance, unlike the grey colour of the nerve cell body, the axon, and the sheath cell. This complicated spiral is known as the **myelin sheath** ; nerve

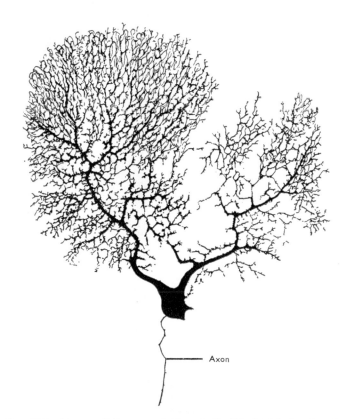

FIG. 710 Nerve cell from cerebellum (cell of Purkinje) showing the branching of the dendrites (Symington).

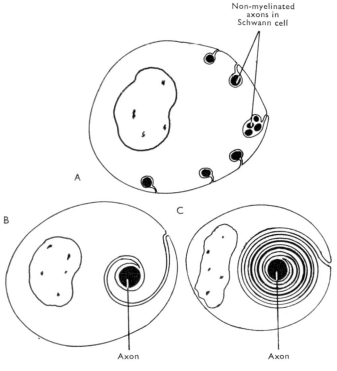

FIG. 712 Diagram to show the relation between peripheral axons and their covering Schwann cells.

A, non-myelinated fibres ; B, developing myelin sheath ; C, final stage of myelination.

Axolemma   Axoplasm

External tongue of   Myelin   Internal
oligodendroglial cell   mesaxon

FIG. 713   An electron micrograph of a myelinated axon in the central nervous system.   (By courtesy of Professor A. Peters.)

nervous system, or a motor cell to innervate 200 or more muscle fibres.

Axons and their collaterals terminate within the central nervous system by coming into contact with other nerve cells. Such **contacts,** or **synapses,** are found on the dendrites or body of the cell but only rarely on the axon of the receiving cell. They take the form either of expanded knobs **(boutons)** or the afferent process may touch the surface of the receiving cell as it passes, or be coiled around the cell body, sometimes forming a basket-like structure. In every case the bare surface of the axon comes directly into contact with the receiving cell without any intervening sheath, and has within it a large number of minute **vesicles** which are believed to contain the substance responsible for transmitting the nervous impulse across the synapse [FIGS. 715 and 716]. Some of these vesicles are thought to release their contents against the surface of the nerve cell when an impulse reaches the synapse. It is the presence of the synapses on the dendrites and cell body which determines that these are the receptive parts of the neuron. An impulse generated in this region can spread over the surface of the neuron and be carried away by the axon. Thus directional conduction within the nervous system is established as a result of the anatomical organization of its cells, even though the individual cells will conduct in either direction. Similar vesicles are found in the terminal parts of axons which are innervating muscle or gland cells

fibres which possess it are said to be **myelinated,** as distinct from the finer nerve fibres which are **non-myelinated** and grey in colour.

The great majority of peripheral nerves contain a mixture of myelinated and non-myelinated fibres and are consequently white though some in the autonomic or involuntary nervous system, are composed entirely of non-myelinated fibres and are grey. In the central nervous system the same mixture of fibres exists, forming the **white matter,** but there are also regions formed predominantly of nerve cell bodies and dendrites comprising the **grey matter.** This is pink in colour by transmitted light owing to the high concentration of cytochrome.

Myelinated and non-myelinated nerve fibres vary in diameter but, in the former, both the thickness of the myelin sheath and the length of the axis cylinder enclosed by each sheath cell are approximately in direct proportion to the diameter of the axis cylinder. Thus, as the diameter increases, the number of sheath cells per unit length of a myelinated nerve fibre decreases, hence the number of junctional regions (**nodes of Ranvier**) between adjacent sheath cells also decreases. At the nodes, there is a break in the continuity of the myelin; in peripheral nerve fibres the intervening portion of the axon is covered by interdigitating processes of the adjacent Schwann cells, while in the central nervous system it is covered by surrounding neuroglial elements other than those concerned with the formation of the myelin [FIG. 714]. The amount of cytoplasm of the myelin-forming cell external to the myelin is much less in the central than in the peripheral nervous system.

At the nodes, axis cylinders may give rise to branches or **collaterals** [FIG. 715]. It is this feature which allows a sensory nerve cell to activate many nerve cells in the central

AXON

MYELIN

NODE

A   B

FIG. 714   Diagrams to show the relation between the sheath cell and axon in myelinated fibres at nodal regions

A, in the central nervous system; B, in the peripheral nervous system.

Solid black, sheath cell cytoplasm.

and the same mechanism of chemical transmission is responsible for activating these structures [FIG. 715], e.g., at motor nerve endings on a muscle.

There is considerable evidence for the presence of a continuous flow of material in normal axons, particularly in

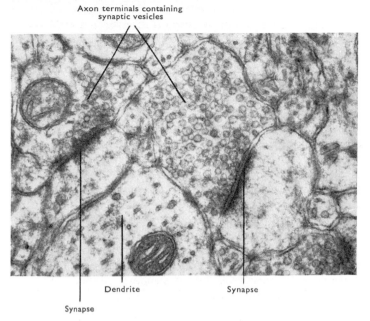

FIG. 716 An electron micrograph to show a synapse on each of two dendritic spines. Cerebral cortex. (By courtesy of Professor A. Peters.)

FIG. 715 Diagram to show the parts of a spinal motor neuron and its connexions. ×, recurrent collateral.

A, enlarged part of the cell body to show the terminations of other nerve fibres on it forming boutons.

B, drawing of electron micrograph of a bouton. Note the expanded axon termination containing mitochondria and vesicles, and the dense postsynaptic zone in the motor cell.

C, to indicate axon and myelin sheath arrangements at a branch.

D, diagram of motor ending on striated muscle. Note vesicles in the expanded axonal termination, the sheath of the axon and the complicated, folded, subneural apparatus which contains cholinesterase.

development or regeneration. This movement is presumably responsible for carrying those substances which maintain every branch of the cell, and for bringing to the terminations the materials for chemical transmission of nerve impulses. Thus the nerve cell can be looked upon as a type of gland cell in which the duct is the axon, and whose secretions are released in minute quantities in contact with the tissues to be affected. A special example of this type of activity is found in some nerve cells, e.g., in the hypothalamus. In these cells, droplets akin to those found in gland cells are present in the cell bodies and extend down the axons to accumulate within the posterior lobe of the hypophysis. This forms a storehouse from which these secretory products (hormones) can be released into the blood stream to produce their effects at a distance, thus involving sensitive tissues throughout the body. In keeping with the view that nerve cells are akin to gland cells is the large concentration of Nissl substance (with its contained ribose nucleic acid) which they have in their cytoplasm, indicating a considerable synthetic ability such as can be seen in the cells of the pancreas and other glands.

Degeneration and Regeneration in Nerve Cells

In all cells every part of the cytoplasm is dependent for its survival on being connected, however indirectly, with the nucleus, in which certain essential substances are produced. Thus if the axon of a nerve cell is divided, the part separated from the cell body undergoes a process of degeneration which ends in complete fragmentation and loss. If the axon is myelinated, the myelin sheath also degenerates although the myelin-producing sheath cells survive, proliferate, and may once again form myelin sheaths for the new processes which grow out from the central end of the divided axon. The process of **degeneration**, first described by Waller in 1850, has allowed anatomists to follow nerve fibres through the complexities of the central nervous system, and even to demonstrate their minute terminations when they have been damaged by pathological processes or by surgical intervention in Man and in experimental animals. This has been made possible by the development of staining methods which specifically demonstrate the degenerating fibres, their myelin sheaths, or their terminals.

In addition to the degeneration of the distal part of a divided axon, certain changes may take place in the cell body from which it arises (retrograde changes). Thus the blocks of Nissl substance may fragment and disappear from the central cytoplasm, the cell body may swell, and the nucleus on occasion is displaced to the periphery of the cell. In extreme cases there may be complete loss of Nissl substance and even cell death. This process, known as chromatolysis [FIG. 707B], occurs in many pathological conditions, but its main significance for anatomy is that it allows of the identification of the cell bodies from which divided axons have arisen. Unfortunately this reaction of the cell to injury is not invariable and, since many normal cells show a structure similar to that described above, the experimental results are liable to misinterpretation. The presence of chromatolysis in pathological conditions and the death of certain nerve cells following injury to their processes, has led to the assumption that chromatolysis invariably indicates an injured cell. However, the chromatolytic appearance of some normal nerve cells, and the fact that all nerve cells go through a phase of profound chromatolysis during their development, makes this unlikely. It has been shown, too, that motor cells of the spinal medulla which have undergone chromatolysis following injury to their axons do not succumb to the virus of poliomyelitis, as do their unchanged neighbours (Howe and Bodian, 1942).

Theoretically the study of axonal degeneration and chromatolysis should give a clear picture of the origin, course, and termination of nerve fibres which have been divided for this purpose. The methods are, however, fraught with difficulties because the silver stains, which demonstrate degenerating axons and boutons, frequently stain normal structures, and the classical Marchi method for staining degenerating myelin not only produces many artefacts but also is incapable of showing the exact point of termination of the fibres, since all axons lose their myelin sheath some distance before ending. In addition, some nerve cells, notably in the forebrain of adult animals and almost every cell in young animals, react to injury by complete degeneration not only of the peripheral part of the axon but also of its central part and the cell body. In such cases the direction in which the fibres are passing cannot be determined by simple injury methods and this causes particular difficulties when lesions are placed in regions where fibres are passing in many different directions.

In some situations, notably in the lateral geniculate body and the auditory and olfactory systems, nerve cells undergo extensive atrophy, with loss of cytoplasm and processes, when the nerve fibres leading to and synapsing with them are severed. This transneuronal atrophy may be much more common than has been supposed, and it cannot be assumed that because a nucleus or group of cells has decreased in size that the cells within it have sustained direct injury. The time taken for transneuronal atrophy to appear is much longer than the primary degeneration of severed axons, but in the human nervous system studies have often to be made on brains in which lesions have persisted for long periods of time. It is generally assumed that the degeneration of severed axons is complete in a few weeks at most, but recently evidence has been produced that it may take much longer than this (Van Crevel and Verhaart, 1963).

When a peripheral nerve fibre is divided the central stump usually degenerates at least to the nearest node but subsequently begins to grow out a new process which frequently divides to give rise to a number of minute fibres which grow rapidly into the surrounding tissue. Some of these may eventually reach and re-innervate the structure which was supplied by the parent fibre, though this is unlikely unless the original lesion is close to the organ supplied or the outgrowing processes enter the peripheral part of the divided nerve and are led to their destination by the surviving Schwann cells and the connective tissue sheaths of the nerves. Thus in crushed nerves, where the axons are divided but the connective tissue sheaths may still be in continuity, the regeneration may well be very satisfactory. Where all the elements are divided the chance of recovery is much less, either because the outgrowing fibres are directed by the peripheral part of the nerve to an incorrect destination (e.g., motor nerve fibres entering a cutaneous branch) or because the growing nerve sprouts escape into the surrounding tissue, where they form a complex tangle or neuroma and may even turn back and grow towards the central nervous system, reaching and invading it. Thus regeneration in a divided nerve is likely to be less satisfactory when the peripheral part contains many different kinds of branches than when it is either purely muscular or cutaneous.

Regeneration does not occur in the central nervous system, owing, it was believed, to the absence of Schwann cells. It now seems likely that the real reason is the rapid proliferation of neuroglial cells which blocks the outgrowth of axons. Certainly it is possible to grow cells of the central nervous system in tissue culture and some regeneration can be promoted in experimental animals which have been subjected to pyrogens. These raise the temperature of the animal and delay the proliferation of neuroglia.

Techniques in the Study of the Nervous System

In spite of the difficulties described above, it is possible to obtain much information on the central nervous system by the application of a number of staining methods to the normal nervous system. A very large number of such methods exist but they fall into four main complementary groups:

1. The staining of Nissl substance and nuclei with basic aniline dyes. This gives a clear picture of the cell bodies and the larger dendrites but no indication of axons or the finer ramifications of the dendrites.

2. Normal myelin may be stained with fat stains in frozen sections, e.g., with osmic acid which is reduced to a lower, black oxide of osmium by the unsaturated fats. It may also be stained by a number of methods which show the benzene insoluble fractions left behind after the preparation of sections in paraffin or celloidin. These methods show only the myelin sheath, not the axon, the cell body or its dendrites. Degenerating myelin is demonstrated by the Marchi method which depends on the fact that, in this state, myelin will reduce osmic acid in the presence of a strong oxidizing agent, e.g., potassium dichromate, while normal myelin will not. This

method has been extensively used to follow degenerating fibres, but is subject to artefact production, precludes the use of other staining methods on the same tissue, is valueless with non-myelinated fibres, and does not stain the terminal non-myelinated parts of myelinated fibres.

3. **Axons, dendrites, cell bodies, boutons,** etc., may be stained with silver. These methods show most features of the nerve cells but are difficult to interpret because they show all the elements in the complex tangle of nerve cells and fibres which forms the central nervous system. Modifications of silver methods are used to stain degenerating axons, boutons, etc., and though they are often unreliable they allow a more detailed analysis of the central nervous system than is possible with the Marchi method.

4. The **Golgi method** is a particular silver technique which picks out and stains in their entirety a few nerve cells only. These show up as opaque structures on a clear background and demonstrate the immense complexity of the branching patterns of different nerve cells [FIG. 710]. Although this is a method of great importance, it is not possible to be certain which, if any, nerve cells will be stained and, like most other methods used on the nervous system, lacks a rationale which would allow of its use with experimental material.

These methods are mentioned here to make it clear that anatomical analysis of the nervous system is a task of great difficulty, and that those details which are available give only a sketchy indication of the complexity of this system. In fact there are very few parts of the nervous system in which the anatomy is sufficiently well known to make it possible to link structure and function beyond reasonable doubt.

The Neuroglia

This is the supporting tissue of the central nervous system and consists of the cells which line the cavities of the brain and spinal medulla (the **ependyma**) and the neuroglia proper which pervade the central nervous system and form its **external limiting membrane** separating it from surrounding tissues and the blood vessels which enter it.

The **ependyma** is formed of columnar cells, some of which are ciliated. Processes of these cells pass through the central nervous system and reach the external surface at least in the early stages of development [FIG. 717]. This connexion appears to be broken in the adult except where the ependyma comes close to the surface as in the anterior median fissure of the spinal medulla. In each ventricle of the brain, one wall is formed only of ependyma which here separates the cavity from the pia mater [p. 698] investing the brain. In these regions tufts of vascular pia mater invaginate the ventricles carrying the ependyma before them to form the **choroid plexuses.** These are secretory structures responsible for forming a large part of the cerebrospinal fluid which fills the cavities of the brain and bathes its external surface. Since the capillaries in the choroid plexuses are separated from the ventricular cavities by ependyma, this epithelium probably plays a part in the formation of the fluid.

The **neuroglia proper** consists of astrocytes, oligodendroglia, and microglia. Of these the first two are derived from the ectoderm which forms the nervous system, while the latter is a mesodermal cell with active phagocytic properties. It becomes swollen and filled with the debris of

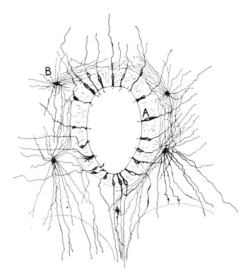

FIG. 717 Section through central canal of spinal medulla of human embryo, showing ependymal (A) and neuroglial (B) cells. (After v. Lenhossek.)

(Note that the dorsal (posterior) aspect is *below*.)

degenerating nerve cells and myelin sheaths when this tissue is destroyed.

Astrocytes are found throughout the central nervous system and consist of a cell body from which processes radiate in every direction. In the grey matter the processes are relatively thick and covered with branches giving the appearance of a dense bush (**protoplasmic astrocytes**), while in the white matter the processes are thinner, longer, contain

Oligodendroglial cell — Capillaries

Vascular end feet of fibrillary astrocytes

FIG. 718 Photomicrograph of fibrillary astrocytes. Astrocytes, capillaries, and a few oligodendroglial cells are stained.

minute fibrillae, and branch much less frequently, tending to run parallel to the nerve fibres among which they lie (**fibrillary astrocytes**). Both these types have at least one larger process which extends to an adjacent blood vessel over which it expands as an end foot [FIG. 718]. Such end feet form a complete layer outside the endothelium of capillaries in the central nervous system. Thus the astrocytes may be responsible for transporting substances to the neurons, or, together with the endothelium, may be responsible for the '**blood-brain barrier**' which prevents certain substances from entering nervous tissue which pass freely into other tissues from the circulation. Recent evidence suggests that the endothelium is the essential element in the barrier to certain substances (Reese and Karnovsky, 1967). On the surface of the central nervous system is a layer of neuroglia

FIG. 719   Photomicrograph of an oligodendroglial cell.

intimately adherent to the overlying pia mater and consisting of small astrocytes and the processes of other astrocytes lying in the subjacent nervous tissue. This external membrane is continued as a sleeve around the blood vessels which pierce the surface of the brain and becomes continuous with the layer of end feet around the intraneural blood vessels.

In spite of their complicated form, astrocytes are capable of

rapid proliferation in regions of the nervous system which have been injured and they form the scar tissue in it.

The **oligodendroglia** [FIG. 719] are smaller cells with fewer processes which branch infrequently. They lie in rows among the nerve fibres and are responsible for the formation of myelin sheaths in the central nervous system (Bunge, Bunge, and Pappas, 1962). They are therefore the counterparts of the Schwann cells of the peripheral nerves, but each oligodendroglial cell forms myelin segments on several different axons.

FIG. 720   Photomicrograph of a microglial cell.

The **microglia** [FIG. 720], the smallest of the neuroglia, are found throughout the central nervous system as slender, rod-like cells with a moderate number of branched processes. They are highly mobile and are easily demonstrated only during pathological processes when their macrophage activity is brought into play. It is not known if they have any function in the normal, uninjured nervous system.

THE DEVELOPMENT AND GROSS ANATOMY
OF THE NERVOUS SYSTEM

The following section deals with the major features of the development of this system to clarify the general arrangement in the adult. Many of the details have been omitted for the sake of clarity and no attempt has been made to discuss the

mechanisms responsible for the complicated morphogenetic processes which are involved, even in those regions where they are slightly understood.

The early stages of the formation of the neural tube and its differentiation into forebrain, midbrain, hindbrain, and spinal medulla have already been described [p. 50]. This structure is essentially a dorsal, longitudinal tube which shows no significant evidence of the segmentation that affects the surrounding mesoderm.

Lying along the dorsal aspect of the recently closed neural tube is a ridge of cells, the **neural crest**. These cells migrate from their original site and pass around each side of the neural tube towards the alimentary canal. They give rise to a large number of different structures (Horstadius, 1950), including the sheath cells of the peripheral nerves and all the nerve cells whose cell bodies lie outside the central nervous system, with the exception of some which migrate from the neural tube. This migration of the neural crest cells [FIG. 721] occurs throughout the length of the neural tube and is responsible for forming all the collections of nerve cells which lie beside the hindbrain and spinal medulla and produce the sensory nerve fibres in the cranial and spinal nerves. These and other peripheral clumps of nerve cells are termed **ganglia**, in contrast to similar aggregations within the central nervous system which are called **nuclei**. Other cells of the neural crest which migrate further are associated with the innervation of involuntary structures and form ganglia which are the **motor cells of the autonomic nervous system.** The latter cells are not derived uniformly from the whole length of the neural crest, but arise from those parts related to the hindbrain and spinal medulla which are not opposite the regions innervating the limbs. Thus they are formed in relation to:

1. The cranial nerves which arise from the midbrain and hindbrain and supply the eye, foregut, and midgut.

2. The spinal nerves which innervate the thoracic and upper two or three lumbar segments of the body.

3. The second, third, and fourth sacral segments concerned with the hindgut.

Of these 1 and 3 give rise to cells which pass almost exclusively to the gut tube, to form ganglia which are either closely associated with it or actually within its wall. In each case the structures innervated by the processes of the cells forming these ganglia include not only the gut tube, but also the structures which develop from it (e.g., respiratory system, digestive glands) and from its mesoderm (e.g., the heart and spleen).

The cells which arise from the cranial and sacral regions from the **parasympathetic part** of the autonomic nervous system, while the cells derived from the thoracolumbar region form the **sympathetic part**. The latter, in addition to forming a second supply to the gut tube, also supplies involuntary structures (glands, blood vessels, etc.) in the body wall.

From their formation the ganglia of the sympathetic part are closely associated with the arterial system and lie at some distance from the organs they supply. They are roughly divisible into groups which are laid down in relation to the three primary sets of branches of the dorsal aorta and, like these, supply the three primary divisions of the mesoderm: the mesoderm of the body wall, of the intermediate cell mass or urogenital mesoderm, and of the gut tube [FIG. 57B]. Since these and other autonomic ganglia are formed out of a migrating mass of cells, they are never entirely discrete but rather appear as local concentrations within a scattering of cells.

The ganglia concerned with the supply of involuntary structures in the body wall form the **sympathetic trunk**. At first this is limited to the thoracolumbar region (T.1–L.2), but later extends headwards and tailwards throughout the length of the body. Thus the cells forming it are able to send processes into every spinal (**grey rami communicantes**) or cranial nerve for distribution through them. The remaining ganglia of the sympathetic part supply the viscera and are therefore the visceral or **splanchnic ganglia**.

The **neural tube** consists initially of a thick layer of ectodermal cells or ependyma and is divided into quadrants by the thin, median **roof** and **floor plates** and a groove on each lateral wall, the **sulcus limitans**, which separates the **alar lamina** dorsally from the **basal lamina** ventrally. The ependyma of the basal lamina is concerned with the formation of the efferent or motor cells, while the alar lamina gives rise to the receptive cells. Shortly after the closure of the neural tube, the ependyma of the basal lamina proliferates and produces, on its anterolateral aspect, a mass of daughter cells which forms a second or mantle layer covering it. This **mantle layer,** which is subsequently formed from the alar lamina also, becomes the part of the neural tube composed mostly of cell bodies, the **grey matter,** as distinct from the nerve fibres which form on its external surface and later, developing myelin sheaths, become white in colour and constitute the **white matter.**

FIG. 721 Diagrammatic transverse section through the trunk of an embryo to show, on the right, the route of migration of neural crest cells (heavy arrows), the peripheral nerve cells derived from it (solid black bodies), the direction of growth of their postganglionic fibres (thin arrows), and their relation to the primary branches of the aorta (solid black lines).

On the left, the distribution of pre- and postganglionic fibres is indicated

THE SPINAL MEDULLA
(SPINAL CORD)

In that part of the neural tube which forms the spinal medulla, many of the cells in the mantle layer of the basal lamina begin to spin out processes, or axons, which pierce the surface of the neural tube over a narrow longitudinal strip and pass towards the surrounding tissues [FIG. 722]. These **ventral rootlets** gather together in bundles and pass out between the developing vertebrae, thus forming the ventral roots of individual spinal nerves. The ventral rootlets, which combine to form the efferent fibres of a single spinal nerve, arise from a length of the spinal medulla which also gives attachment to the corresponding dorsal root fibres which form the sensory component [FIG. 915]. This is the part of the spinal medulla supplying one segment of the body and is known as a **segment of the spinal medulla,** even though the nervous system shows no definite segmentation.

The **ventral root** fibres are of two distinct types and are formed separately from neurons in the dorsal and ventral parts of the basal lamina. The latter consist of fibres which pass out into the body wall or limb and supply striated muscles in these situations **(somatic motor),** while the former follow the general line of migration of the neural crest cells to end in one or more of the autonomic ganglia derived from the corresponding part of the neural crest. Thus they form a connexion between the central nervous system and the peripherally situated autonomic motor ganglia, and constitute the **preganglionic** or **visceral efferent fibres.** The cells giving origin to these fibres are found only in those parts of the neural tube where the neural crest contributes significantly to the autonomic nervous system, and it is therefore only through the nerves in these regions that preganglionic nerve fibres emerge from the neural tube; that is, through certain cranial nerves, the thoracic and upper two lumbar nerves and the second, third, and fourth sacral nerves. In the fully developed nervous system the preganglionic fibres are of smaller calibre (not more than 3 μm.) than most of the somatic motor fibres but are also myelinated and therefore white in colour. In the thoracolumbar region they form the **white rami communicantes** of the sympathetic trunk.

In following the line of migration of the neural crest cells, some of the preganglionic fibres of the sympathetic part end in the sympathetic trunk, spreading headwards and tailwards within that trunk so as to reach its extremities. Other fibres pass through the trunk to the splanchnic ganglia, thus forming the **splanchnic nerves.** In the parasympathetic part there is nothing equivalent to the sympathetic trunk and no supply to the body wall, only the splanchnic type of preganglionic nerve bundles being present.

While the motor cells are developing, each of the cells in the ganglia lying adjacent to the neural tube (sensory, spinal, and cranial ganglia) grows out two processes in opposite directions. One of these, directed towards the periphery, is destined to reach a sensory ending either in the body wall, by passing with the somatic motor cells, or in relation to the viscera, by coursing with the preganglionic fibres through the ganglia of the autonomic system. The opposite process is directed towards the alar lamina of the neural tube to form part of the **dorsal root of a spinal nerve** or part of a cranial nerve

[FIG. 750]. All the sensory cells of the body, with the exception of the olfactory and visual systems and some of the sensory cells of the trigeminal nerve [p. 621], lie in ganglia of the above type and all send nerve fibres into the alar lamina of

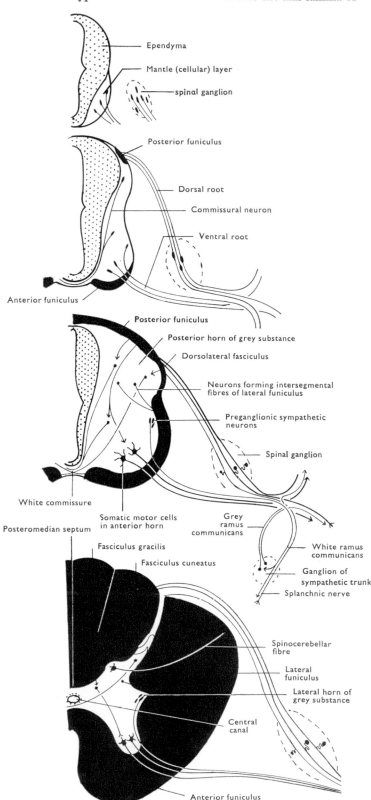

FIG. 722   Diagram to show the formation of the spinal medulla, its white (solid black) and grey (white) matter, and the general arrangement of the processes of the sensory, motor (somatic and visceral), and internuncial cells.

the neural tube. Originally bipolar, these cells become unipolar in type [FIG. 711] with the outstanding exception of the ganglion cells of the vestibulocochlear nerve which retain their bipolar form.

The ependyma of the alar lamina proliferates to form a mantle layer of receptive cells, as the central processes of the sensory cells run on to its surface in the form of a linear row of rootlets. Without entering the mantle layer these axons divide and run longitudinally headwards and tailwards forming, in aggregate, a third or marginal layer, composed of nerve fibres, on the dorsal surface of the tube. These fibres form a bundle on each side, extending from the midline to the point of entry of the dorsal rootlets, and constitute the **posterior funiculi** of the adult spinal medulla [FIG. 722]. The greater part of this mass of fibres extends upwards throughout the length of the spinal medulla to enter the caudal part of the brain, the medulla oblongata, and is so arranged that those fibres entering over the lowest dorsal roots lie most medially, while fibres from successively higher roots are laid down progressively in a more lateral position [FIG. 723]. In the midline the fibres of the two sides are separated by a **posterior median septum** of glial tissue which meets the surface at the **posterior median sulcus**. In the cephalic half of the spinal medulla, each is further subdivided by a similar but incomplete **posterior intermediate septum** into a thin, medial, **fasciculus gracilis** carrying dorsal root fibres from the lower half of the body, and a thicker, lateral, wedge-shaped **fasciculus cuneatus**, containing fibres which enter the spinal medulla through the upper thoracic and cervical dorsal roots [FIG. 726].

At the lateral edge of each posterior funiculus, pierced by the entering dorsal rootlets, is a smaller bundle in which the dorsal root fibres have a short course of a few segments only. This is the **dorsolateral fasciculus** which remains of approximately the same size throughout its length, since the fibres within it end as rapidly as others enter it, unlike the remainder of the posterior funiculus which steadily increases in size as it is followed headwards, owing to the continued addition of fibres to its lateral side.

These groups of longitudinal fibres subsequently give off large numbers of branches or collaterals into the mantle layer [FIG. 724] to make contact with its cells, the longitudinal course of the fibres allowing each of them to make a variety of different connexions throughout a variable length of the spinal medulla and even to send branches as far ventrally as the motor cells [FIG. 722].

The **receptive cells of the alar lamina,** together with similar cells of the basal lamina which take no part in the formation of the ventral root, therefore form a link between the incoming sensory fibres on the dorsal surface and the cells giving rise to the efferent fibres of the ventral roots. They are therefore known as **connector cells** or **interneurons** or **internuncial cells,** and the processes which they develop may link them either with other interneurons, or with efferent cells at the same level or, by extension along the spinal medulla, with similar cells at other levels of the nervous system, including the brain. The connexions which these interneurons produce are by no means simple but usually consist of complex chains of cells, often involving several pathways through which different sensory stimuli entering over several dorsal roots may interact with each other and play their part in producing the final reaction. To some extent the complexity of the response to any stimulus is related to the number of interneurons involved and the complexity of their connexions, and hence, in part, to the volume of active nervous tissue.

The **interneurons** [FIG. 722] which develop first in the alar and basal laminae send their processes through the mantle layer (grey matter) to pass on to the surface, particularly of its anterior and lateral parts, either on the same side, or on the opposite side after crossing ventral to the cavity of the neural tube, thereby forming a crossing or **white commissure.** They turn longitudinally and form a thin **marginal layer** covering the remainder of the mantle layer and, by their collaterals, connect the various segments of the spinal medulla, forming its **intersegmental tracts** or **fasciculi proprii** [FIG. 725].

At a later stage in development further fibres are added to the surface of the intersegmental tracts and these consist of the process of interneurons of the spinal medulla coursing to the brain—**long ascending tracts**—and similar fibres descending into the spinal medulla from the brain—the **long descending**

FIG. 723 Diagram to show the manner in which fibres of dorsal nerve roots enter and ascend in the posterior funiculus.

FIG. 724 Drawing, after Cajal, of a few of the collaterals of longitudinal fibres in the posterior funiculi of the spinal medulla.

tracts. Thus the lateral and anterior parts of the marginal layer (**funiculi** of white matter) of the spinal medulla, separated from each other only by the emergent ventral rootlets, consist of a mixture of long ascending and long descending tracts overlying a narrow zone of intersegmental tracts, while the posterior funiculi, lying between the dorsal rootlets, consist of longitudinally-running dorsal root fibres with a few intersegmental fibres in their anterior parts [FIG. 725].

FIGURE 722 gives an indication of the processes involved in the formation of the spinal medulla as seen in transverse section. It shows how the cavity of this part of the neural tube becomes relatively smaller at each stage, until it is represented by the small **central canal** lying in the narrow strip of grey matter connecting the two sides. In the spinal medulla of the adult, the central canal is frequently missing, being represented by a scattered group of ependymal cells without any cavity. This figure also indicates the formation of the **horns of the grey matter** (anterior, lateral, and posterior) and the arrangement of the surrounding white matter. It does not show that virtually all the structures seen in transverse section extend throughout the length of the spinal medulla, and that they are therefore arranged in columns of grey matter and funiculi of white matter.

The **grey columns** and **funiculi** (white columns) of the spinal medulla are not of the same size throughout their length [FIG. 726]. The funiculi increase steadily in size on ascending the spinal medulla owing to the greater number of fibres in each of the long ascending and descending tracts in its cephalic parts, while the amount of grey matter in each segment is proportional to the amount of tissue to be supplied by that segment. Thus, opposite the limbs, where the volume of muscle and the area of skin are greatly increased, the anterior and posterior horns of grey matter are correspondingly increased in size in comparison with similar structures in the thoracic and upper cervical regions. Similarly the lateral horn, containing the cells of origin of the preganglionic autonomic fibres, is formed only where such fibres arise, that is in the thoracolumbar and midsacral regions.

The increase in size of the anterior and posterior horns of the grey matter is greatest in the lumbosacral region and is responsible for the **enlargements** of the spinal medulla opposite the origin of the great limb plexuses, a feature missing from the spinal medulla in cases of congenital absence of the limbs, in animals such as snakes, and in the lumbosacral region of whales. The cervical enlargement has greater over-all dimensions than the lumbosacral, although the volume of the cervical grey matter is less; the ratio of grey to white matter being 1 : 2 in the sacral region and 1 : 5 in the cervical region. It is this preponderance of fibres in the cervical region which makes the superior limit of that enlargement less well defined, so that it is described as beginning at the first cervical segment though the highest limb nerve arises from the fifth.

The **posterior horn** varies markedly in shape in the different regions of the spinal medulla, but is capped throughout by a layer of cells, difficult to stain, which has a gelatinous appearance, the **substantia gelatinosa.** This lies immediately deep to the dorsolateral fasciculus whose fibres send branches into it and through it to the body of the posterior horn which contains scattered cells of variable size. Towards the junction of the posterior horn with the anterior and lateral horns, there are certain well-defined nuclei such as the **thoracic nucleus** which indents the posterior funiculi and is found between the first thoracic and second lumbar segments. It is composed of large cells which give rise to the posterior spinocerebellar tract. Between the anterior and posterior horns lies the **intermediate substance** (grey matter) which is divided into central and lateral parts. It contains many large internuncial neurons and its lateral portion lies against the medial edge of the lateral funiculus and extends into it to a greater or lesser degree, giving a mixture of nerve cells and fibres known as reticular formation.

The **reticular formation** is best developed in the upper cervical part of the spinal medulla where it gradually increases in amount and becomes continuous with the much more voluminous reticular formation of the medulla oblongata. The **lateral horn,** where present (T.1 – L.2, S.2 – S.4), is also an extension of the lateral intermediate substance, and contains the cells of origin of the preganglionic autonomic fibres [FIGS. 722 and 750].

The **anterior horn** is small in the thoracic and upper cervical regions but shows a marked lateral expansion in the enlargements. It is this lateral portion which contains the motor cells which supply the muscles of the limbs [FIG. 727]. At any level of the developing spinal medulla the ventromedial cells are at a more advanced stage of development than those situated dorsolaterally, while the corresponding groups at more caudal regions are at a less advanced stage. Thus, of the motor cells which are arranged around the anterior horn, the anteromedial are the first to spin out their axons and these pass into the most proximal or trunk muscles. The remaining motor cells, which develop in sequence around the periphery of the anterior horn in the enlargements, supply successively more peripheral muscles (Romanes, 1951 ; Sharrard, 1955). Thus the girdle muscles are supplied by cells in a anterolateral position while the muscles of the proximal (arm or thigh), intermediate (forearm or leg), and distal (hand or foot) segments of the limb are innervated by groups of cells which lie in sequence posterior and caudal to this. Each of these

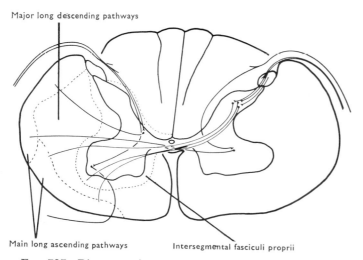

Major long descending pathways

Main long ascending pathways · Intersegmental fasciculi proprii

FIG. 725 Diagrammatic transverse section through the spinal medulla to show the position of the fasciculi proprii, and the general arrangement of the long ascending and descending pathways, in the anterior and lateral funiculi, formed by internuncial neurons.

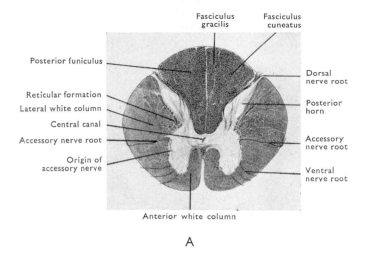

A

Fasciculus gracilis
Fasciculus cuneatus
Posterior funiculus
Reticular formation
Lateral white column
Central canal
Accessory nerve root
Origin of accessory nerve
Anterior white column
Dorsal nerve root
Posterior horn
Accessory nerve root
Ventral nerve root

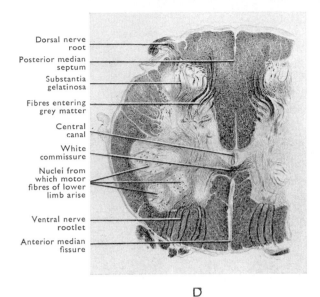

D

Dorsal nerve root
Posterior median septum
Substantia gelatinosa
Fibres entering grey matter
Central canal
White commissure
Nuclei from which motor fibres of lower limb arise
Ventral nerve rootlet
Anterior median fissure

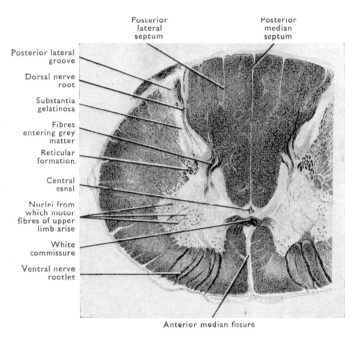

B

Posterior lateral septum
Posterior median septum
Posterior lateral groove
Dorsal nerve root
Substantia gelatinosa
Fibres entering grey matter
Reticular formation
Central canal
Nuclei from which motor fibres of upper limb arise
White commissure
Ventral nerve rootlet
Anterior median fissure

Dorsal nerve root
Posterior median septum
Substantia gelatinosa
Grey commissure
White commissure
Anterior median fissure

E

FIG. 726 Sections through spinal medulla, not to scale. (From specimens prepared by the Weigert-Pal method, by which the white matter is made dark and the grey matter bleached.)

A, transverse section of the upper part of the cervical region.
B, at the level of the fifth cervical nerve.
C, through the midthoracic region.
D, at the level of the fourth lumbar nerve.
E, at the level of the third sacral nerve.

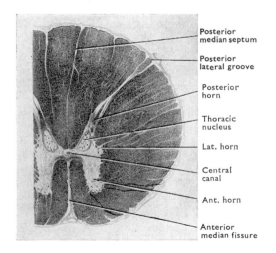

C

Posterior median septum
Posterior lateral groove
Posterior horn
Thoracic nucleus
Lat. horn
Central canal
Ant. horn
Anterior median fissure

groups, except the last, is split into lateral and medial parts which supply respectively the extensor and flexor muscles of the segment. Thus the small muscles of the hand and foot are supplied by cells lying in the extreme dorsolateral part of the anterior horn in the first thoracic and second and third sacral segments respectively. Each group of motor cells, seen in transverse section, forms a column extending for a variable length of the anterior horn, and is formed by a mixture of large, multipolar motor cells and smaller cells some of which are the source of the fine motor fibres (γ-efferents). These accompany the large motor fibres to somatic muscles and are responsible for the motor supply to the small muscle fibres which form the core of the sensory **muscle spindles.** The presence of such a motor supply to the muscle spindles allows these sensory organs to be adjustable in length and tension and, since they are tension recorders, the activation of their muscle fibres increases their sensitivity, and heightens, or may

FIG. 727　Sections through segments of spinal medulla.

A, fifth cervical segment.　　B, eighth thoracic segment.

C, third lumbar segment.　　D, second sacral segment.

(Founded on Plates in Bruce's *Atlas of the Spinal Cord*.)

initiate a discharge of, sensory impulses from them. This is one proven example of control by the central nervous system over the sensitivity of a sensory ending, though others exist [p. 665].

The spinal medulla extends throughout the length of the vertebral column up to the third month of intra-uterine life [FIG. 728], but subsequently grows less rapidly than that structure, so that a disproportion in length between the two

steadily develops. In normal development the junction of the brain and spinal medulla remains fixed at the foramen magnum of the skull, and the relative shortening of the spinal medulla causes it to rise in the vertebral canal so that its caudal extremity lies at the level of the third lumbar vertebra in the new-born, and in the adult has risen to the disc between the first and second lumbar vertebrae, the whole spinal medulla measuring from 43 to 45 cm. in length. This relative shortening of the spinal medulla is not uniform throughout its length, but affects the lumbosacral segments most and the thoracic segments least. As a result the lumbosacral roots are closely packed together at their attachment to the spinal medulla and the corresponding segments are extremely short; all the sacral segments taking up little more than the vertical extent of the first lumbar vertebra in the adult. The shortening also

produces a considerable lengthening and obliquity of the spinal roots which is in proportion to the extent to which any particular segment has risen within the vertebral canal. In the second sacral and more caudal spinal nerves, the spinal ganglia leave the corresponding intervertebral foramina and come to lie within the sacral canal, so reducing the lengthening of these roots which, in common with all spinal roots, join to form the spinal nerves at the spinal ganglia. [FIG. 729]. Thus the length of the roots varies from 8 mm. in the second cervical nerve to 260 mm. in the fifth sacral nerve, and the length of the spinal segments from 25 mm. in the thoracic region, through 15 mm. in the cervical region, to as little as 4 mm. in the sacral region. As the spinal medulla rises within the vertebral canal, it leaves behind it a thin filament attached to the dorsal surface of the coccyx, the **filum terminale** [FIG. 730], and a great leash of dorsal and ventral roots

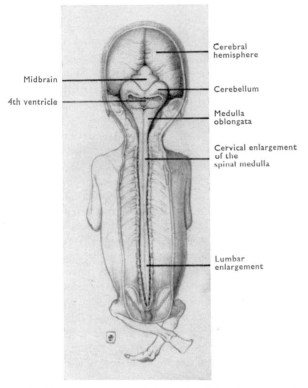

FIG. 728 Human foetus in third month of development, with brain and spinal medulla exposed from behind.

FIG. 730 Conus medullaris and the filum terminale.

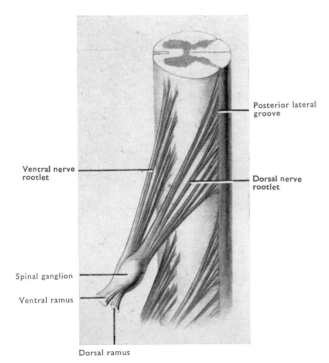

FIG. 729 Roots of the seventh thoracic nerve (semidiagrammatic).

FIG. 731 Section through conus medullaris and cauda equina as they lie in the vertebral canal.

descending to their point of exit from the vertebral canal, the **cauda equina** [FIG. 731], which surrounds the filum terminale. The filum terminale is 18 cm. in length and consists of the continuation of the immediate investment (the pia mater) of the spinal medulla [p. 693] from its tapering caudal extremity, the **conus medullaris.** This encloses a few nerve fibres and cells which may represent the remains of coccygeal nerves and, for a variable distance, the central canal which often expands into a small sac or **terminal ventricle** at its caudal end. Since the spinal medulla rises relative to the vertebral column, the segments of the spinal medulla no longer lie opposite the corresponding vertebrae. The relative levels in the adult are indicated in FIGURE 732, but it should be remem-

bered that in flexion and extension of the vertebral column there is a change in the position of the spinal medulla, the conus rising in flexion and descending in extension as the length of the anterior surface of the vertebral canal, composed of vertebral bodies and intervertebral discs, increases and decreases.

THE BLOOD SUPPLY OF THE SPINAL MEDULLA [FIG. 733]

The arteries of the spinal medulla reach it along the roots of the spinal nerves and are therefore divisible into ventral and dorsal root arteries arising from the intersegmental vessels in the different regions of the body, vertebral, intercostal, lumbar and lateral sacral arteries. The **dorsal radicular arteries** supply the spinal ganglia and continue on the dorsal roots to the spinal medulla where they anastomose along the line of entry of the dorsal roots, the **posterior spinal arteries.** They also form an irregular, circular, surface network of vessels with the ventral radicular arteries, from which radial branches pierce the spinal medulla and supply the white matter and the grey matter of the posterior horns near their tips. The **ventral radicular arteries** in the embryo not only assist in the formation of the surface anastomosis but also send a large branch ventrally to link up with similar branches from all the other ventral root arteries in a median, ventral, longitudinal vessel, the **anterior spinal artery.** Branches of this vessel pierce the spinal medulla through the anterior median fissure and, passing either to the right or the left, supply the greater part of the grey matter, the deeper part of the lateral and posterior funiculi, and not infrequently form a further longitudinal anastomosis just lateral to the central canal. Initially every ventral root artery takes part in this, but as development proceeds many lose their connexion with the anterior spinal artery and persist as smaller vessels ending in the surface anastomosis. Thus the meeting of branches of opposite ventral root arteries in the anterior spinal is lost,

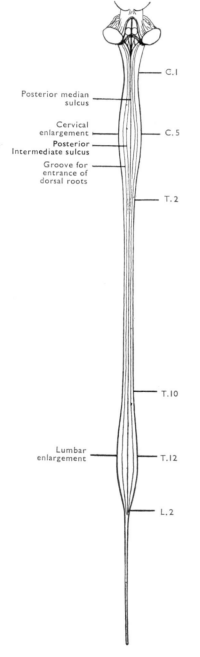

Posterior median sulcus

Cervical enlargement

Posterior Intermediate sulcus

Groove for entrance of dorsal roots

Lumbar enlargement

C.1

C.5

T.2

T.10

T.12

L.2

FIG. 732 Diagram of the spinal medulla as seen from behind. C.1, C.5, etc., indicate levels in relation to the vertebrae.

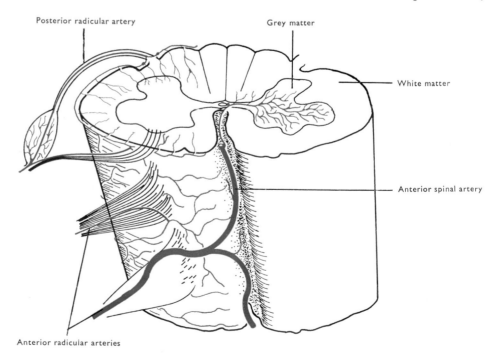

Posterior radicular artery

Grey matter

White matter

Anterior spinal artery

Anterior radicular arteries

FIG. 733 Semidiagrammatic drawing to show the blood supply of the spinal medulla.

except in the case of those on the first cervical nerve which are enlarged to form the upper parts of the **vertebral arteries.** These form the most cephalic part of the anterior spinal artery running caudally and its equivalent on the ventral surface of the hindbrain, the **basilar artery,** passing headwards [FIG. 1080]. This reduction of feeding vessels to the anterior spinal artery proceeds until only three to ten persist, most commonly on the left side. Thus the greater part of the grey matter of

the lumbosacral enlargement may be supplied by a single vessel entering on a ventral root between the ninth thoracic and the second lumbar.

The veins of the spinal medulla for the most part drain into a median posterior channel which connects with the intersegmental veins at various levels by running out along the dorsal roots. A smaller, more variable, anterior spinal vein is also present.

THE BRAIN (ENCEPHALON)

The part of the neural tube which is enclosed in the developing skull is the brain. Even before the closure of the neural tube is complete the brain shows three swellings, directly continuous one with another and only indistinctly separable. These are the hindbrain, midbrain, and forebrain. The **hindbrain** is continuous with the spinal medulla caudally and is separated from the midbrain cranially by a constricted zone or **isthmus.** At first a relatively large part of the brain, the **midbrain** becomes relatively smaller and narrower than the hindbrain or the forebrain with which it is continuous anteriorly. The midbrain and hindbrain together form the **brain stem.**

The cephalic extremity of the **forebrain** is closed by a thin membrane, the **lamina terminalis,** which represents the most cephalic part of the neural tube prior to the formation of the cerebral hemispheres, which later grow out from each side of the original forebrain and extend beyond it, hiding it between them.

Basically the structure of the brain is the same as that of the spinal medulla but it is greatly modified, firstly by the development of folds or flexures and secondly by the formation of three additional masses of nervous tissue which overshadow the original neural tube and come to make up the major part of the brain. These are:

1. The **two cerebral hemispheres** (telencephalon) one on each side, which develop from the forebrain as hollow extensions of the full thickness of the neural tube, so that their cavities, the **lateral ventricles,** are continuous with the remainder of the cavity of the neural tube.

2. The **cerebellum** forms as a solid organ on the dorsal aspect of the hindbrain. It expands to fill the greater part of the posterior cranial fossa of the skull, and completely hides the original hindbrain tube from the dorsal aspect.

The cerebral hemispheres and the cerebellum expanding independently fill the greater part of the cranial cavity and come to be wedged together within it, separated only by two thin but tough sheets of connective tissue, the meninges, left between them as they expand, and

lying in planes at right angles to each other : the one, between the two hemispheres, the **falx cerebri,** and the other separating both hemispheres from the cerebellum lying behind and below them, the **tentorium cerebelli** [p. 694].

The initial or **cephalic flexure** [FIG. 734] appears at the midbrain and allows the forebrain to turn ventrally so that it lies parallel to the hindbrain. At the same time it greatly elongates the dorsal surface of the midbrain while restricting its ventral surface. Although in later development this flexure is partly straightened, there remains a marked angulation in the midbrain region, such that the long axes of the fore- and hindbrains lie at an angle of approximately 140 degrees to each other and the dorsal surface of the midbrain is much more extensive than the ventral. The second or **pontine flexure** [FIG. 735] is in the opposite direction at the middle of the hindbrain which bulges ventrally towards the forebrain, increasing the cephalic flexure and forming an acute ventral

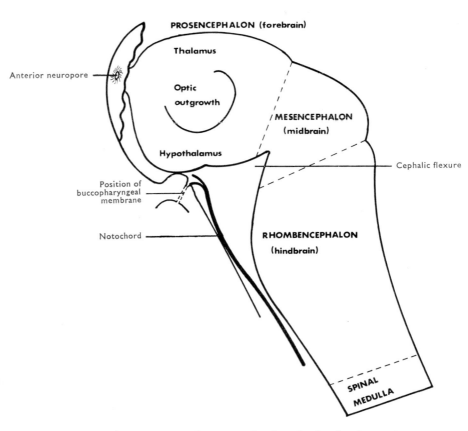

FIG. 734 Drawing to show an early phase in the development of the brain. The anterior neuropore is not yet closed.

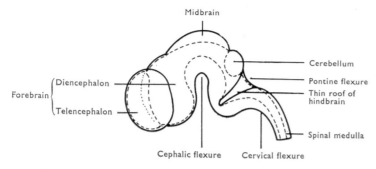

FIG. 735 Diagram to show the parts of the embryonic brain after the formation of the telencephalon. The optic outgrowth is not shown. Broken lines indicate the ventricular system; the dotted line, the roof of the diencephalon.

flexion between the hindbrain and the spinal medulla, the **cervical flexure,** where previously only the gentle curve due to the flexion of the embryo as a whole was present.

It should be appreciated from the outset that though three separate parts of the brain are described, these are developed from a single longitudinal tube and the line of demarcation between the parts is not sharp. Many structures pass longitudinally through the tube without interruption of any kind in the same way as the alar and basal laminae extend from the spinal medulla to the forebrain.

THE BLOOD SUPPLY OF THE BRAIN

[PLATE 49 and FIG. 1080]

The blood vessels which supply the brain are limited in number by comparison with the multiple vessels of the spinal medulla. The general arrangement is [FIG. 737] of four main arteries supplying the longitudinal anastomosis continued from the spinal medulla. The two vertebral arteries join to form the basilar artery, which supplies the hindbrain, midbrain, and the posterior part of the forebrain. It anastomoses with the cephalic ends of the dorsal aortae, the future internal carotids,

which enter the cranial cavity to supply the remainder of the forebrain. Here, as in the spinal medulla, the vessels are divisible into those forming a surface anastomosis, and perforating branches which pierce the ventral aspect of the neural tube to supply its deeper parts. However, the absence of the dorsal and ventral radicular arteries in this region causes the surface anastomosis to be fed by encircling branches of the ventral longitudinal vessels, so that the lateral and even the dorsal surfaces of the original tube are supplied by them. Such branches become:

1. The **posterior inferior cerebellar,** the **anterior inferior cerebellar,** and the **superior cerebellar.** Though the main area which these arteries supply is the cerebellum which develops on the dorsal surface of the hindbrain, they still retain their supply to the lateral and dorsal surfaces of the parent neural tube in the regions where they arise.

2. Three **cerebral arteries** supply each cerebral hemisphere when these develop from the forebrain. Thus the posterior cerebral comes to supply the occipital lobe of the cerebral hemisphere, but retains an extensive supply to the upper midbrain and to the dorsal wall and roof of the original forebrain tube, which become the thalamus and choroid plexus respectively [FIG. 737].

In the early stages of development the forebrain and hindbrain have many features in common. In particular the roof plate, which forms along the line of closure of the neural tube, remains thin, consisting of a single layer of ependymal cells from the greater part of which no neural elements are developed. This is later tucked into the cavity of the tube, on both sides, by the vascular connective tissue (pia mater) which overlies it, to form the **choroid plexus**—the source of much of the cerebrospinal fluid filling the cavities of the brain.

THE HINDBRAIN

With the development of the pontine flexure the shape of the hindbrain is markedly altered, in much the same manner

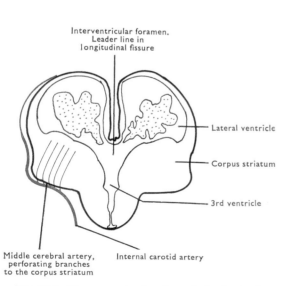

FIG. 736 Transverse section through the forebrain at the level of the interventricular foramen of the brain shown in Fig. 737.

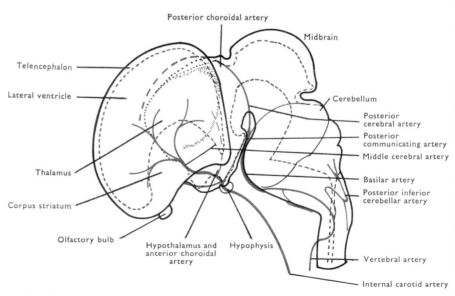

FIG. 737 Drawing of reconstruction of early human foetal brain derived from serial sections. The arteries are shown diagrammatically.

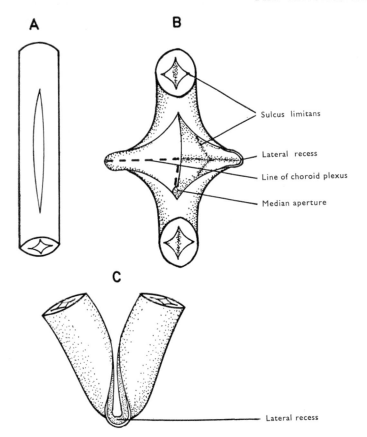

FIG. 738   Diagram to illustrate the probable mechanical effects of the pontine flexure in the formation of the fourth ventricle.

A, before folding. B, posterior view after folding, the thin roof has been removed on the right side. C, lateral view after folding with the thin roof removed.

the cavity of the nervous system and the surrounding tissues. These allow the cerebrospinal fluid to percolate into the loose connective tissue immediately surrounding the brain [p. 693]. The choroid plexus on each side forms an L-shaped invagination of the thin roof extending from the median aperture headwards to the line of the flexure. Here each turns laterally and passes into the lateral recess to reach the lateral aperture through which it may present on the surface of the brain. The evagination of the thin roof at the lateral recess forms a sleeve-like process extending ventrally around the side of the hindbrain with the aperture at its termination facing ventrally and the choroid plexus bulging through it [FIG. 753]. With the development of the cerebellum on its cephalic margin, the lateral recess is restricted and forms a funnel-shaped extension of the fourth ventricle passing over the posterolateral aspect of the upper medulla oblongata and separating it from the cerebellum. In the adult the cephalic margin of the median aperture is drawn posteriorly on the cerebellum, the aperture thus forming a triangular opening which faces inferiorly between the cerebellum and the medulla oblongata [FIG. 741]. On the ventricular aspect of this portion of the thin roof lies the choroid plexus which curves posteriorly on the anterior surface of the cerebellum.

THE CEREBELLUM

The cerebellum develops in the roof of the fourth ventricle as a result of proliferation of cells at the margins of the alar laminae of the metencephalon. The masses so formed extend from both sides into the thin roof covering this part of the hindbrain and meet in the midline. In this fashion, portions of the thin roof are left cephalic and caudal to the cerebellar rudiment, the latter being continuous with the thin roof which covers the myelencephalic portion of the fourth ventricle and carries the choroid plexus. The remaining parts of the thin roof in the metencephalon are invaded by nerve fibres connected with the cerebellum and consequently, becoming white in colour, are known respectively as the **superior** and **inferior medullary vela.**

At first there is a great proliferation of cells on the ventricular aspect of the cerebellar rudiment which causes it to bulge into the ventricle [FIG. 739, I]. Subsequently many of these cells, formed on its ependymal aspect, migrate to the posterior surface of the cerebellum and form a superficial layer (or cortex) of nerve cells which is never covered with white matter and hence remains grey in colour, the **cortex of the cerebellum.** Subsequently this aspect of the cerebellum enlarges rapidly, increasing its surface area partly by the development of transversely placed folds [FIGS. 739 and 740], but also by a general increase in size which eventually leads to the cerebellum overlying the posterior surface of the whole extent of the hindbrain and midbrain. Despite this enormous increase in size, and the complex folding of its surface to form the many deep and subsidiary transverse fissures which, on section, give the cerebellum the appearance of a tree (arbor vitae) with many leaves (**folia** [FIGS. 739, 741]) and confer on the cerebellum an enormous surface area, the extent of its attachment to the metencephalon is not increased. Thus the mass of the cerebellum as it expands posteriorly becomes

as a soft rubber tube is distorted when it is flexed after being cut along one surface [FIG. 738]. The neural tube is flattened opposite the line of flexure, and the thin roof plate drawn out into an extensive sheet stretching from side to side and even slightly everted at the lateral extremities of the flexure. Headwards and tailwards of the flexure, the amount of flattening progressively decreases until the margins of the alar laminae once again fuse with each other at the caudal margin of the midbrain and at the middle of the lower half of the hindbrain. In this fashion, the slit-like cavity of the hindbrain is converted into a diamond-shaped or rhomboid **fourth ventricle,** and the hindbrain is known as the **rhombencephalon.** It is divisible along the line of the pontine flexure into two triangular halves, of which the caudal is the **myelencephalon** or **medulla oblongata,** tapering towards the spinal medulla, and the cranial the **metencephalon** or **pons.** The acute angles of the fourth ventricle lie at the cephalic and caudal extremities where it is continuous, respectively, with the aqueduct of the cerebrum in the midbrain and the central canal of the lower medulla oblongata and spinal medulla. The obtuse angles lie at the widest part and constitute the **lateral recesses of the fourth ventricle.** Here, and at its caudal extremity, the thin roof subsequently breaks down, forming the two **lateral apertures** and the single **median aperture** which constitute the only communications between

FIG. 739 Six stages in the development of the cerebellum as shown in median sagittal section. Note the formation of cells on the ventricular aspect, their later migration to form the cortex and the folding of the cerebellum to form the fastigial recess of the fourth ventricle.

folded on itself with its caudal part rolled round against the thin roof of the inferior part of the fourth ventricle, and its cranial part against the dorsal surface of the midbrain. This type of growth leads to the fourth ventricle being carried up towards the centre of the cerebellum in the tent-like **fastigial recess** [FIG. 739], superior to which lies the narrow, transverse attachment of the cerebellum to the roof of the fourth ventricle. The nerve cells which were initially produced on the ventricular aspect of the cerebellar rudiment, and which have not migrated to form the cortex, remain close to the ventricle and adjacent to the fastigial recess, the presence of which makes this the deepest part of the fourth ventricle. These cells develop into four groups on each side extending laterally from the midline and forming the nuclei of the cerebellum, the most lateral and largest of which, the **dentate**

nucleus, so called because it is formed from a folded lamina of grey matter in the shape of a wrinkled bag [FIG. 742], has a ridged or toothed appearance on surface view [FIG. 744], and lies over the lateral extremity of the fastigial recess. The dentate and the other nuclei, which are much less well defined and are named from lateral to medial the **emboliform, globose,** and **fastigial** [FIG. 743], give rise to the nerve fibres which leave the cerebellum for other regions of the brain.

The cerebellum receives and emits large numbers of nerve fibres which are gathered together in a considerable bundle at the lateral margin of the metencephalon, emerging from the anterior parts of the **horizontal fissure** of the cerebellum [FIG. 748]. Leaving the cerebellum, this mass of fibres divides into three separate bundles of which the most lateral (the **middle cerebellar peduncle** [FIG. 749]) runs antero-superiorly to arch across the anterior surface of the metencephalon and fuse with its fellow, thus forming the **ventral part of the pons**—the name given to this part of the hindbrain. The other two bundles diverge, the most medial, or **superior peduncle,** passes in a cephalic direction along the margin of the superior medullary velum and sinks into the substance of the midbrain [FIG. 744]. The **inferior peduncle,** lying between the other two, runs anteriorly then turns caudally, anterior to the lateral recess of the fourth ventricle, to pass along the dorsolateral aspect of the medulla oblongata and gradually disappear as its constituent fibres diverge into the medulla oblongata.

The size and intricacy of the fissural pattern of the human cerebellum, as well as the number of fanciful descriptive terms applied to its several parts, all serve to emphasize the complexity of this organ. In addition, its mode of development and consequent curved nature make it difficult to describe. For this reason it is common practice to unfold the cerebellum diagrammatically, so that the superior and inferior medullary vela lie at the extremities of the diagram, instead of close together in the roof of the fourth ventricle [FIG. 745].

FIG. 740 Dorsal view of cerebellum and medulla oblongata from human foetus of 4 months. The ependymal roof of the fourth ventricle has been removed. At this stage the cerebellum is little more than a simple band or plate which arches over the back of the superior part of the cavity of the hindbrain.

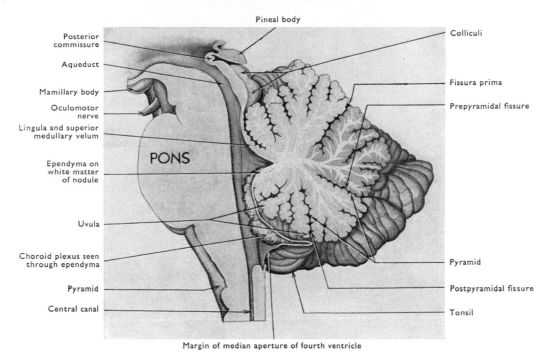

FIG. 741 Median section through brain stem and cerebellum, showing arbor vitae of vermis and the fourth ventricle.

Phylogenetically, three arbitrary stages may be recognized in the formation of the cerebellum and though these are not independent parts functionally, consideration of them does help to clarify the arrangement and connexions of this organ. Initially [FIG. 746] the cerebellum is an extension of the vestibular nuclei of the hindbrain and is concerned with them in the muscular adjustments necessary for the control of balance and posture resulting from information conveyed to it from the internal ear, the principal organ for recording changes in position in aquatic vertebrates. Control of muscular activity requires information regarding the state of the muscles, and from an early phase the cerebellum receives this from the spinal medulla by ascending nervous pathways, the **spinocerebellar tracts,** which also send fibres to the vestibular nuclei. With the increasing im-

portance of muscle, joint, and other senses for recording position, especially in terrestrial vertebrates, the spinocerebellar system and that part of the cerebellum to which it passes becomes much enlarged. This 'spinal part' is formed in the centre of the original cerebellum [FIG. 746 B] splitting the 'vestibular part' into a cephalic **lingula,** attached to the superior medullary velum, and a caudal **nodule** which, with its lateral extension the **flocculus,** are continuous inferiorly with the inferior medullary velum.

The third phase in the phylogenetic development of the cerebellum is represented by a growth *pari passu* with that of the cerebral hemispheres and with the development of increasingly voluminous connexions between them [see middle cerebellar peduncle, p. 611]. This third or 'cerebral' part divides the 'spinal' part into cephalic and caudal regions,

FIG. 742 Sagittal section through left hemisphere of cerebellum, showing the arbor vitae and the dentate nucleus.

Globose nucleus

Dentate nucleus

Emboliform nucleus

Fastigial nucleus

Inferior cerebellar peduncle

Nodule

Superior cerebellar peduncle

Middle cerebellar peduncle

Motor and superior sensory nuclei of trigeminal nerve

Fastigial recess of 4th ventricle

FIG. 743   Oblique transverse section, parallel to the fastigial recess of the fourth ventricle, to show the cerebellar nuclei.

and becomes the largest part of the cerebellum in Man. In other mammals where cerebral development is not so advanced, the lateral parts of the cerebellum are much smaller and the median portion stands out as a raised, sinuous ridge which, on account of the transverse fissures, has the appearance of a worm and is known as the **vermis.** The growth of the lateral lobes, or hemispheres, in Man makes the vermis much less obvious except in the caudal part.

This is displaced to the inferior surface, and lying in a notch between the hemispheres, is known as the **inferior vermis** [tuber, pyramid, uvula, and nodule, FIG. 748]. The more cephalic part is continuous with the hemispheres without interruption and is known as the **superior vermis** [FIG. 747].

FIGURE 745 shows the parts of the cerebellum belonging to each of the above divisions, the main fissures and the lobes. The general arrangement of these can readily be understood by reference to FIGURES 741, 747, and 748.   FIGURE 745 shows two of the methods of division of the cerebellum into lobes, that on the left being less open to criticism.

The Lateral Walls and Floor of the Fourth Ventricle

The lateral walls consist mainly of the cerebellar peduncles.   The inferior of these, together with the gracile and cuneate tubercles [FIG. 749], form the inferolateral wall, while the superolateral wall is composed of the superior cerebellar peduncle and the nervous tissue anterior to it, between the upper end of the inferior peduncle and the ventricle.   This is mainly the superior vestibular nucleus [p. 672] and the juxtarestiform body— scattered nerve fibres passing to and from the cerebellum on the medial side of the inferior cerebellar peduncle (restiform body).

The **floor of the fourth ventricle,** like the roof, is rhomboid in shape, the lateral recess extends over the posterior surface of the inferior cere-

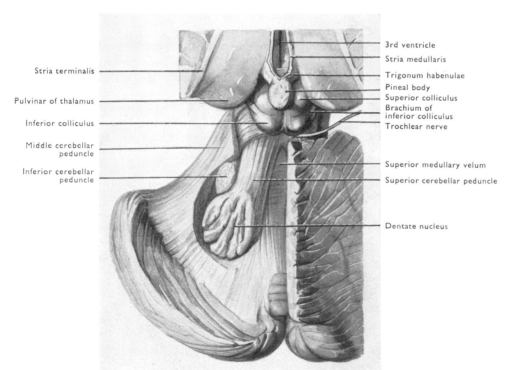

Stria terminalis

Pulvinar of thalamus

Inferior colliculus

Middle cerebellar peduncle

Inferior cerebellar peduncle

3rd ventricle

Stria medullaris

Trigonum habenulae

Pineal body

Superior colliculus

Brachium of inferior colliculus

Trochlear nerve

Superior medullary velum

Superior cerebellar peduncle

Dentate nucleus

FIG. 744   Dissection from above to show dentate nucleus, cerebellar peduncles and lateral lemniscus (between inferior colliculus and middle cerebellar peduncle).

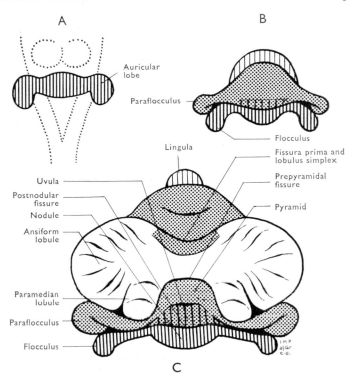

FIG. 745 Diagram of cerebellum. This has been unfolded so that the lingula (L) and the nodule (Nod) lie at opposite ends. The attachment of the superior and inferior medullary vela to these two parts respectively, and the latter to the flocculus and its peduncle is shown. The thin roof of the fourth ventricle and its apertures (arrows) are indicated cf. FIGURE 741.

C, lobulus centralis; Cul., culmen; D, declive; F, folium; T, tuber; Pyr., pyramid; U, uvula.

FIG. 746 Diagrams illustrating evolutionary development of cerebellum and the constitution of its three lobes.

A, hypothetical primitive cerebellum with vestibular connexions only.

B, cerebellum of reptile showing addition of part that receives spinocerebellar tracts.

C, mammalian cerebellum showing addition of part (lobus medius) that receives cortical impulses by way of the pontine nuclei.

Vestibular cerebellum, vertical hatching; spinal cerebellum, stipple; neocerebellum (lobus medius), white.

bellar peduncle. The general form of the floor can be seen in FIGURE 749, but particular note should be taken of the **median sulcus** and lateral to it the **medial eminence.** The latter, essentially representing the basal lamina, is a ridge extending throughout the length of the ventricle on each side of the median sulcus and showing in its lower half a series of ridges and shallow grooves dividing it into:

1. The **facial colliculus** (just cephalic to the level of the lateral recess) so named because the facial nerve passing to its exit partly encircles the abducent nucleus (VIth cranial nerve) which forms the colliculus.

2. The **hypoglossal triangle,** covering the hypoglossal nucleus.

3. The **vagal triangle,** covering the dorsal nucleus of the vagus.

4. The **area postrema,** just cephalic to the opening of the central canal of the lower half of the medulla oblongata. This consists of a mass of blood vessels and cells covered only by the ependyma of the floor of the ventricle.

5. **Foveae.** Lateral to the medial eminence lies an ill-defined groove, the remains of the sulcus limitans, often represented only by the superior and inferior foveae.

6. Lateral to the sulcus limitans and filling out the floor in the lateral angle is the **vestibular area.** This represents the position of the groups of cells, vestibular nuclei, which receive sensory fibres from the vestibular part of the internal ear via the vestibular part of the vestibulocochlear nerve.

THE BRAIN STEM (MEDULLA OBLONGATA, PONS, AND MIDBRAIN) [FIGS. 757–761]

The Medulla Oblongata

The pontine flexure produces the fourth ventricle and causes marked changes in the shape and general position of the parts

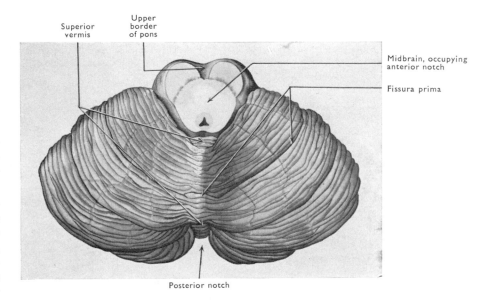

FIG. 747 Superior surface of the cerebellum.

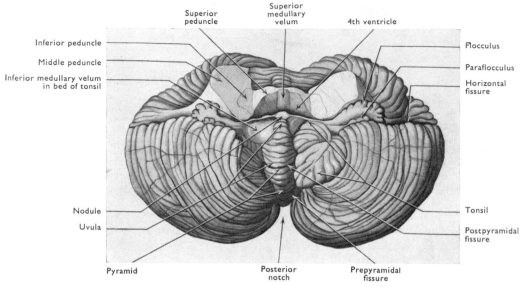

FIG. 748   Inferior surface of the cerebellum.   The right tonsil has been removed so as to display more fully the inferior medullary velum.

of the neural tube.   This is further altered by: (1) a considerable proliferation of the internuncial nerve cells, many of which are associated with the cerebellum; and (2) a change in the position of two of the major longitudinal pathways of the spinal medulla, the posterior funiculi, and the lateral corticospinal tract [FIGS. 755 F and G].   Traced headwards, both groups of fibres decussate and continue in a ventral position, the posterior funiculi as the medial lemniscus, after synapsing in the corresponding nuclei.

At the pontine flexure (junction of the pons and medulla

oblongata) the alar lamina lies lateral to the basal lamina and consequently the groups of sensory neurons which lie lateral to the hindbrain, and form the cephalic part of the series of sensory ganglia, send their central processes on to the lateral surface of this part of the brain as the equivalent of the dorsal roots of the spinal nerves.   Ventral root fibres grow outwards from the basal lamina, as in the spinal medulla, but fail to join the corresponding dorsal roots [FIG. 750].   Thus, excluding the olfactory and optic nerves which are formed in an entirely different manner and belong to the forebrain, two distinct

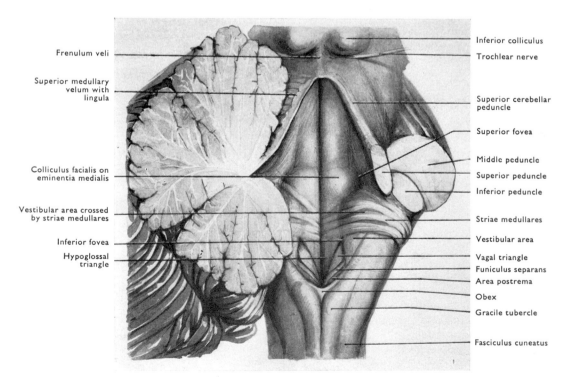

FIG. 749   Floor of the fourth ventricle.   The right half of the cerebellum has been removed.   The left half is drawn over to the left so as to expose the floor of the ventricle fully.

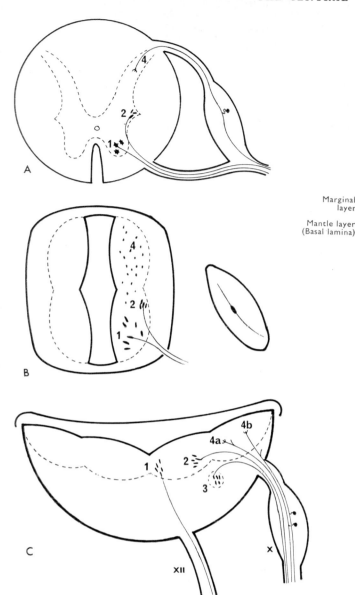

FIG. 750 Diagram to illustrate the difference in development between the spinal medulla and the hindbrain.

B, indicates the basic structure from which both A, thoracic spinal medulla, and C, medulla oblongata, arise.

1. Somatic motor supply to striated, somite muscle.
2. Preganglionic autonomic connector cells. } Basal lamina
3. Supply to striated visceral, or branchial muscle.
4. Sensory terminations :
    (a) Visceral sensory, tractus solitarius ;
    (b) Somatic sensory, trigeminal system.

groups of nerves are attached to the brain stem and constitute the cranial nerves. These two groups are :

1. Nerves attached to the ventral aspect of the brain stem consist of motor fibres supplying the head somites, which form the extrinsic ocular muscles and the muscles of the tongue except palatoglossus. These nerves lie in a row in series with the ventral spinal roots and are the **oculo-motor** (III), the **abducent** (VI), the **hypoglossal** (XII), and, belonging to the same group but arising from the dorsal surface of the brain after decussation, the **trochlear** (IV) nerve [FIGS. 753, 754].

2. Nerves attached in a row to the lateral aspect of the

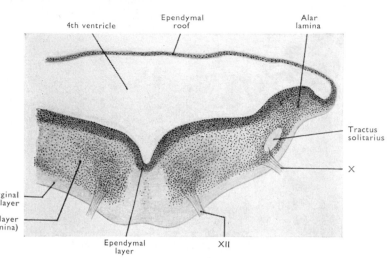

FIG. 751 Transverse section across the medulla oblongata of a human embryo. (From His, slightly modified.)

hindbrain. They are ganglionated and contain sensory nerve fibres and are the **trigeminal** (V), the **facial** (VII), the **vesti-bulocochlear** (VIII), the **glossopharyngeal** (IX), the **vagus** (X), and the **accessory** (XI).

FIGURES 752, 753, and 754 show the points of attachment of the cranial nerves and demonstrate that the largest, the trige-minal, is the only nerve attached to the pons. Since the nerves are numbered more or less serially from above downwards, it follows that the oculomotor and trochlear are attached to the midbrain, and those with numbers greater than V to the medulla oblongata. The deep connexions of these nerves are dealt with on pages 618-623.

In its caudal half, the medulla oblongata retains many of the features of the spinal medulla, including the presence of a **central canal** which opens out into the fourth ventricle about the middle of the medulla oblongata, having moved poster-iorly in its lower half. This posterior movement of certain structures within the medulla oblongata is the result of the de-cussations of the corticospinal tracts and medial lemnisci [p. 606 and FIG. 755]. In the first cervical segment a round bundle of fibres bulges medially into the intermediate grey matter from each lateral funiculus [FIG. 755 G], and swings into the position of the white commissure of the spinal medulla in the lowest part of the medulla oblongata [FIG. 755 F]. Here the bundles from the two sides interlace in alternate fascicles of considerable size and cross to the ventral surface of the medulla oblongata on which they can be followed upwards until they plunge into the inferior border of the pons. Since they cross in this manner they are at first narrow bundles but rapidly increase in size as each fascicle enters from the commissure so that they form the narrow elongated **pyramids** which are a characteristic feature of the ventral aspect of the medulla oblongata. This massive **decussation of the pyramids** [FIGS. 753, 756, and 761] vir-tually fills the lower third of the anterior median fissure in the medulla oblongata and, cutting through the grey matter, separ-ates the anterior horn of the first cervical segment of the spinal medulla from its equivalent in the medulla oblongata, the hypoglossal nucleus. It also cuts off the spinal nucleus of the accessory nerve from the nucleus ambiguus [FIG. 755 F, G], both of which belong to a single elongated group of cells

Optic chiasma

Infundibulum

Posterior
perforated substance

Crus cerebri

Abducent nerve

Hypoglossal nerve
(between olive and pyramid)

Medulla oblongata

FRONTAL LOBE

LONGITUDINAL FISSURE

TEMPORAL LOBE

PONS

CEREBELLUM

Olfactory bulb

Olfactory tract

Optic nerve

Anterior
perforated substance

Optic tract

Tuber cinereum

Oculomotor nerve

Left mamillary body

Trochlear nerve

Trigeminal nerve

Facial nerve

Vestibulocochlear nerve

Nervus intermedius

Glossopharyngeal nerve

Vagus nerve

Accessory nerve

Spinal medulla (cut)

Hypoglossal nerve

FIG. 752   Base of the brain with
the cranial nerves attached.

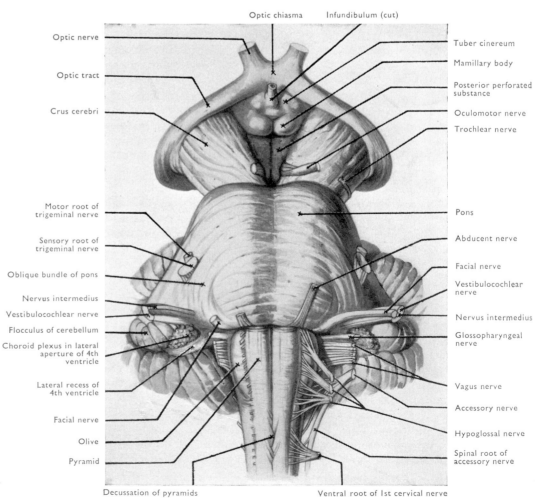

Optic chiasma

Infundibulum (cut)

Optic nerve

Optic tract

Crus cerebri

Motor root of
trigeminal nerve

Sensory root of
trigeminal nerve

Oblique bundle of pons

Nervus intermedius

Vestibulocochlear nerve

Flocculus of cerebellum

Choroid plexus in lateral
aperture of 4th
ventricle

Lateral recess of
4th ventricle

Facial nerve

Olive

Pyramid

Tuber cinereum

Mamillary body

Posterior perforated
substance

Oculomotor nerve

Trochlear nerve

Pons

Abducent nerve

Facial nerve

Vestibulocochlear
nerve

Nervus intermedius

Glossopharyngeal
nerve

Vagus nerve

Accessory nerve

Hypoglossal nerve

Spinal root of
accessory nerve

FIG. 753   Anterior view of medulla
oblongata, pons and midbrain.

Decussation of pyramids

Ventral root of 1st cervical nerve

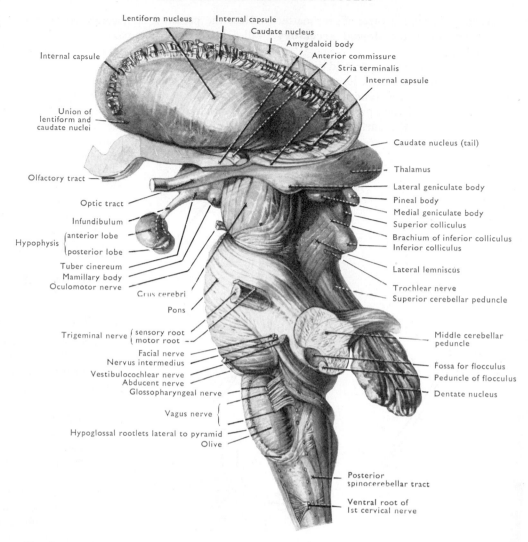

Lentiform nucleus
Internal capsule
Caudate nucleus
Amygdaloid body
Anterior commissure
Stria terminalis
Internal capsule
Internal capsule
Caudate nucleus (tail)
Union of lentiform and caudate nuclei
Thalamus
Olfactory tract
Lateral geniculate body
Optic tract
Pineal body
Infundibulum
Medial geniculate body
Hypophysis { anterior lobe / posterior lobe }
Superior colliculus
Brachium of inferior colliculus
Inferior colliculus
Tuber cinereum
Mamillary body
Oculomotor nerve
Lateral lemniscus
Crus cerebri
Trochlear nerve
Superior cerebellar peduncle
Pons
Trigeminal nerve { sensory root / motor root }
Middle cerebellar peduncle
Facial nerve
Nervus intermedius
Fossa for flocculus
Vestibulocochlear nerve
Peduncle of flocculus
Abducent nerve
Dentate nucleus
Glossopharyngeal nerve
Vagus nerve {
Hypoglossal rootlets lateral to pyramid
Olive
Posterior spinocerebellar tract
Ventral root of 1st cervical nerve

FIG. 754   Left lateral aspect of brain after removal of the cerebral hemisphere (except corpus striatum) and the cerebellum (except dentate nucleus).

supplying motor fibres to the glossopharyngeal, vagus, and accessory nerves [FIG. 762].

Also in the inferior part of the medulla oblongata, but cephalic to the decussation of the pyramids, the fasciculi gracilis and cuneatus enlarge owing to the presence of nerve cells (the **nuclei cuneatus** and **gracilis**) buried among their fibres forming the tubercles of the nuclei gracilis and cuneatus on the surface.   The former, smaller and more protuberant, extends to the obex [FIG. 749], while the latter, lying further headwards, is lateral to the most caudal part of the fourth ventricle.   The fibres of the fasciculi gracilis and cuneatus end on the cells of the corresponding nuclei.   These give rise to axons that sweep ventrally to cross in the white commissure, intermingling with the same fibres from the opposite side and, turning headwards, form a vertical sheet of fibres, the **medial lemniscus,** abutting on the midline just dorsal to the pyramid and ventral to the headward continuation of the anterior funiculus [FIG. 761].

The change in position of these two major bundles of the spinal medulla to the ventral aspect of the medulla oblongata not only displaces the central canal and hypoglossal nucleus dorsally but also separates most of the anterior funiculus from

the lateral funiculus, displacing the former dorsally and the latter dorsolaterally towards the dorsolateral fasciculus [FIG. 755 F] which is here enlarged by the addition of descending fibres of the trigeminal nerve and is known as its spinal tract.   On the lateral aspect of the medulla oblongata this **spinal tract of the trigeminal nerve** forms a faint ridge through the ventral part of which emerge the roots of the cranial nerves attached to the lateral surface.   Immediately ventral to the spinal tract of the trigeminal nerve a shallow groove, the posterior lateral sulcus, separates it from an oval swelling, the **olive,** which lies in the upper two-thirds of the medulla oblongata.   The olive is separated from the pyramid ventrally by the **anterior lateral sulcus** through which there emerge the rootlets of the hypoglossal nerve inferiorly and the abducent nerve at its pontine end [FIG. 753].   In section the olive is a crinkled bag of grey matter (olivary nucleus) through the open, medial, hilus of which emerges a mass of fibres which decussate, then sweep posterolaterally to enter the opposite inferior cerebellar peduncle on the posterolateral aspect of the medulla oblongata.   A thin plaque of grey matter medial to the **olivary nucleus,** but separated from it by the emergent rootlets of the hypoglossal nerve, is the **medial**

C.T.A.—X

accessory olivary nucleus, and a similar piece of grey matter dorsal to the olivary nucleus forms the **dorsal accessory olivary nucleus** [FIG. 765].

The appearance of the olive lateral to the pyramid displaces the lateral funiculus further posteriorly so that its most dorsal fibres, the **posterior spinocerebellar tract** [FIG. 754], escape posteriorly over the surface of the spinal tract of the trigeminal nerve, separating it from the surface [FIG. 836].

Here the posterior spinocerebellar tract may be visible as an ill-defined ridge as it passes posteriorly, to enter the inferior cerebellar peduncle just cranial to the tubercle of the nucleus cuneatus.

The displacement of tracts to the ventral surface and the formation here of the massive olive, together with the outfolding which produces the fourth ventricle, leads to the grey matter equivalent to that in the spinal medulla lying as a transverse strip in the floor of the ventricle where it forms the nuclei of the cranial nerves [FIG. 750]. In addition to this, many of the fibres which form the compact funiculi of the spinal medulla fan out in the intermediate region of the medulla oblongata, dorsal to the olive and ventral to the cranial nerve nuclei, and are here mixed with scattered groups of nerve cells to produce the reticular formation of the medulla oblongata.

The Reticular Formation

This apparently confused jumble of nerve cells and fibres is roughly divisible into lateral and medial parts, the former lying dorsolateral to the olive, the latter dorsomedial to it. The dispersal of the fibres of the anterior and lateral funiculi of the spinal medulla into the reticular formation obscures the presence of many nuclei in the medulla oblongata and makes the differentiation of grey and white matter less obvious except in certain well-defined nuclei, e.g., the olivary nucleus. The arrangement of the nerve cells and fibres which constitute the reticular formation suggests a different function from that of the spinal medulla, where a stricter separation of the grey and white matter exists, even though the spinal motor cells send dendrites into the anterior and lateral funiculi. Since the bundles of fibres running through the reticular formation are predominantly longitudinal, it seems likely that this portion of the nervous system allows both ascending and descending impulses to have an effect on the internuncial neurons which lie in it, with the possible exception of those tracts which remain discrete entities, e.g., the pyramid and the medial lemniscus. It would therefore seem to be a region concerned with the reception and integration of a large number of different nervous impulses. Certainly on physiological grounds it seems to be concerned with a large number of different functions.

FIG. 755 Diagram showing the variations in shape of the parts of the brain stem including its cavities (blue dots), the change in position of the long ascending (F, blue, medial lemniscus) and descending (G, red, corticospinal) pathways in the lower medulla oblongata, and the positions of the cranial nerves and their nuclei. Cranial nerve nuclei indicated by the same red numeral are of the same functional type and form parts of one column of cells, all lying in the same relative position: 1, somatic motor; 2, visceral motor (preganglionic); 3, branchial motor; 4, visceral sensory (tractus solitarius); 5, somatic sensory (parts of trigeminal).

ICP, inferior cerebellar peduncle; LP, longitudinal fibres of pons; MCP, middle cerebellar peduncle; ML, medial lemniscus; NC, nucleus cuneatus; NG, nucleus gracilis; SCP, superior cerebellar peduncle. Roman numerals indicate cranial nerves.

A, midbrain. B and C, mid and low pons. D, E, F, and G, medulla oblongata H, upper cervical spinal medulla.

FIG. 756 Diagram of the decussation of the pyramids (modified from van Gehuchten).

N.H., hypoglossal nucleus; N.V., vagoglossopharyngeal nucleus; F.S., tractus solitarius; N.A., nucleus ambiguus.

The Pons

The pons is the part of the hindbrain lying ventral to the upper half of the fourth ventricle. Externally it forms a rounded ridge which bulges ventrally separating the medulla oblongata below from the midbrain above [FIGS. 754 and 759]. Its anterior surface is grooved longitudinally by the basilar artery, and is continuous on each side of the hindbrain with a smooth, rounded **middle cerebellar peduncle**, which gradually sinks into the ventral extremity of the horizontal fissure of the cerebellum. The pons and middle cerebellar peduncles are directly continuous, but are arbitrarily separated by the trigeminal nerve which pierces them at their junction.

The pons consists of two parts:

1. The above ridge of tissue, or **ventral part of the pons**, appears to pass uninterruptedly from one cerebellar hemisphere to the other, but consists of a mass of nerve cells (**pontine nuclei**) which are intermingled with and covered by the fibres which arise from them. These fibres cross the midline ventrally and pass backwards into the opposite half of the cerebellum. The pontine nuclei are further split up by two very large bundles of fibres, the **crura cerebri**, which enter this ventral part of the pons, one on each side from the ventral surface of the midbrain, and split up into longitudinal bundles. The majority of the fibres in these bundles end on the cells of the pontine nuclei, the remainder emerge at the lower border of the pons as the pyramids of the medulla oblongata.

2. The **dorsal part of the pons**, which lies between the ventral part and the fourth ventricle, consists of those structures in the medulla oblongata which pass headwards into this region, and through it to the midbrain, or are passing in the opposite direction. It also contains the superior part of the vestibular area, nuclei of the trigeminal, abducent and facial cranial nerves, and the pontine reticular formation which is directly continuous above with the reticular formation in the midbrain and below with that in the medulla oblongata.

The Midbrain

The midbrain forms the narrow neck which connects the hindbrain (including the cerebellum) to the forebrain and passes through the aperture in the tentorium cerebelli [**tentorial notch,** p. 695]. It is the site of the cephalic flexure of the embryonic brain and hence its dorsal aspect is the more extensive; the forebrain, being bent ventrally on the hindbrain at this region, lies anterior as well as superior to it.

In the midbrain the roof plate remains thick and the proliferation of cells of the alar and basal laminae leads to a steady reduction in the relative size of its cavity so that it comes to form the narrow, tubular **aqueduct of the cerebrum** which traverses the midbrain from end to end and connects the median cavity of the forebrain (third ventricle) to the fourth ventricle. It thus forms a vital link through which the cerebrospinal fluid formed in the ventricles of the forebrain can reach the fourth ventricle and escape into the subarachnoid space [p. 693] through the apertures in its roof.

The aqueduct is surrounded by a thick cylindrical sheath of grey matter, the **central grey substance** which contains the nuclei of the cranial nerves lodged in the midbrain, the oculomotor, trochlear, and the mesencephalic nucleus of the trigeminal [pp. 618–622].

The part of the midbrain dorsal to the aqueduct, the **tectum,** is derived from the alar lamina and consists of the dorsal part of the central grey substance and four small swellings, two on each side, the **superior** and **inferior colliculi** which protrude from the dorsal surface and are composed of grey matter. The superior pair are broader and flatter than the inferior pair which rise abruptly immediately above the superior medullary velum and overhang the trochlear nerves as they emerge and pass laterally across the superior cerebellar peduncles which sink into the midbrain anterior to the inferior colliculi [FIG. 754]. As it enters the midbrain each superior cerebellar peduncle is covered by a thin strip of white matter which arches over it to reach the inferior colliculus; this is the **lateral lemniscus,** a portion of the auditory pathway. Arising from the cephalic margin of the inferior colliculus is a similar but narrower ribbon of white matter, the **brachium of the inferior colliculus,** which runs forwards and upwards over the lateral aspect of the midbrain to enter the deep surface of a rounded swelling lodged on the dorsal edge of the crus cerebri at its superior limit. This is the **medial geniculate body** [FIGS. 754 and 757], a portion of the thalamus the posterior end of which lies above the medial geniculate body.

On the same surface of the thalamus above the medial geniculate body is a bundle of nerve fibres, often invisible on the surface, which sweeps backwards and medially to the dorsal surface of the upper midbrain reaching the superior colliculus and the **pretectal region** immediately anterior to

FIGS. 757–761 Photographs of transverse sections, posterior (left) and lateral (right) views of the human brain stem, to show the surface features and internal structure of its various regions. The posterior and lateral views show the position of the individual sections.

FIG. 757A has most of the surface features labelled, but in the subsequent figures only those surface features at the level of the section are lettered. [See also FIG. 756.]

a. Aqueduct [FIG. 757]; accessory cuneate nucleus [FIG. 761].

an. Arcuate nucleus.

ap. Area postrema.

asc. Anterior spinocerebellar tract.

bic. Brachium of inferior colliculus.

bsc. Brachium of superior colliculus.

c. Central canal [FIG. 761]; central nucleus of thalamus [FIG. 757].

cc. Crus cerebri.

cgs. Central grey substance.

ch. Optic chiasma.

ci. Internal capsule.

cn. Cochlear nuclei.

dn. Dentate nucleus.

do. Dorsal accessory olivary nucleus.

fc. Facial colliculus.

gt. Gracile tubercle.

h. Habenular triangle.

ic. Inferior colliculus.

icc. Intercollicular commissure.

icp. Inferior cerebellar peduncle.

ipf. Interpeduncular fossa.

ivn. Inferior vestibular nucleus.

l. Lateral nucleus of thalamus.

lfp. Longitudinal fibres of pons.

lg. Lateral geniculate body.

ll. Lateral lemniscus.

m. Mamillary body.

mcp. Middle cerebellar peduncle.

me. Medial eminence.

mg. Medial geniculate body.

ml. Medial lemniscus.

mlf. Medial longitudinal fasciculus.

mo. Medial accessory olivary nucleus.

nc. Nucleus cuneatus.

ng. Nucleus gracilis.

nt. Dorsal nucleus of trapezoid body.

o. Optic nerve (FIGS. 757 and 758); obex [FIG. 760].

ol. Olive.

ot. Optic tract.

p. Pineal [FIG. 757]; pons [FIGS. 758 and 759].

pc. Posterior commissure.

pd. Pyramidal decussation.

po. Central tegmental fasciculus.

psc. Posterior spinocerebellar tract.

pyr. Pyramid.

r. Reticular nucleus of thalamus.

rf. Reticular formation.

rn. Red nucleus.

rs. Rubrospinal tract.

s. Subthalamic nucleus.

sc. Superior colliculus.

scp. Superior cerebellar peduncle.

sm. Striae medullares of fourth ventricle.

smt. Stria medullaris of thalamus.

smv. Superior medullary velum

sn. Substantia nigra.

st. Spinothalamic tract.

tb. Decussating fibres of trapezoid body.

tfp. Transverse fibres of pons.

th. Hypoglossal triangle.

ts. Tractus solitarius.

tsv. Thalamostriate vein

tv. Vagal triangle.

va. Vestibular area.

vm. Vestibulomesencephalic tract.

vn. Vestibular nuclei.

vpl. Nucleus ventralis posterolateralis of thalamus.

vpm. Nucleus ventralis posteromedialis of thalamus.

vs. Superior vestibular nucleus.

z. Zona incerta.

III. Oculomotor nerve.

IV. Trochlear nerve.

V. Trigeminal nerve.

VI. Abducent nerve.

VII. Facial nerve.

VIIn. Nucleus of facial nerve.

VIII. Vestibulocochlear nerve.

IX. Glossopharyngeal nerve.

X. Vagus.

XIc. Accessory nerve, cranial root.

XII. Hypoglossal nerve.

Fig. 757

FIG. 758

Fig. 759

Fig. 760

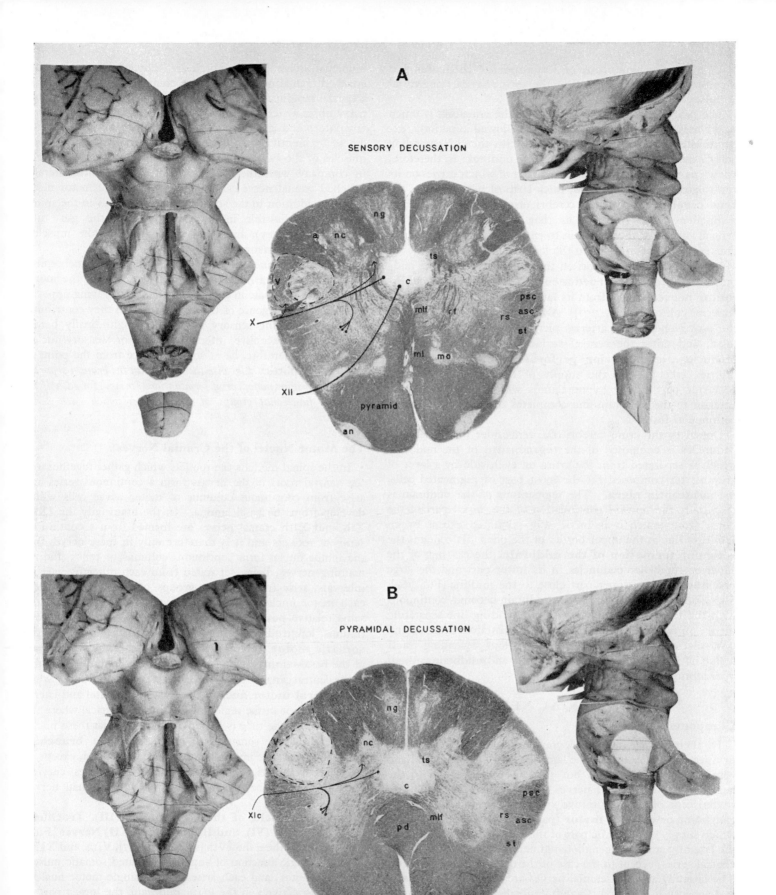

Fig. 761

it. This is the **brachium of the superior colliculus** and is composed of nerve fibres from the optic tract carrying visual impulses to the midbrain.

The part of the midbrain ventral to the aqueduct is much larger than the tectum and each half, followed superiorly, expands laterally to become continuous with the corresponding half of the cerebrum. This part of the midbrain is, therefore, known as the **cerebral peduncles** each of which carries on its ventrolateral surface a broad, thick strip of white matter, the **crus cerebri.** The crura cerebri originate in the cerebral hemispheres and converge as they pass inferiorly on the ventral surface of the midbrain to enter the ventral part of the pons where their fibres form the longitudinal bundles.

Between the crura cerebri on the ventral aspect of the midbrain is the deep **interpeduncular fossa** with the **oculo-motor nerves** arising from its lateral walls at the level of the superior colliculus [FIG. 757]. A leash of small branches from the posterior cerebral arteries pierce the median part of the fossa, and enter the ventral surface of the midbrain, thus producing the **posterior perforated substance.** These arteries take part in the supply of the midbrain and the posterior part of the hypothalamus which lies immediately anterior to the midbrain and completes the roof of the interpeduncular fossa.

Dorsal to the crura cerebri, the remainder of the cerebral peduncles is composed of the **tegmentum** of the midbrain which is separated from the crura on each side by a layer of grey matter composed for the most part of pigmented cells, the **substantia nigra.** The tegmentum of the midbrain is essentially the upward continuation of the dorsal part of the pons, from which it is in no way separated except by an arbitrary line at the upper border of the pons. It contains the **reticular formation of the midbrain,** the crossing of the superior cerebellar peduncles in its lower part and the large **red nuclei** in its upper part close to the midline [FIG. 764]. Superiorly the tegmentum of the midbrain becomes continuous with the forebrain without any clear dividing line separating them; in fact the red nucleus and the substantia nigra extend upwards into the forebrain as do the long ascending tracts which have traversed the medulla, pons, and midbrain without interruption.

THE CRANIAL NERVES

Components

In their arrangement, the cranial nerves are akin to the primitive spinal nerves of the lamprey in which the dorsal and ventral roots do not join but run a separate course to the periphery. The cranial nerves of all higher forms are also reminiscent of these primitive spinal nerves in that the pre-ganglionic or **visceral motor** (parasympathetic nerve) fibres, which supply the cephalic part of the gut tube, emerge through the laterally attached, ganglionated nerves and not through the ventral series, except in the case of the oculomotor (III) nerve. The laterally attached cranial nerves are therefore both motor and sensory in function, though equivalent in position to the dorsal roots of spinal nerves [FIG. 750], while the ventral series, including the trochlear, are almost exclusively **somatic motor.**

In addition, striated muscle is developed in the cephalic extremity of the gut tube and particularly in relation to the branchial arches [p. 56] which are supplied by the laterally attached cranial nerves. In these cranial nerves, therefore, as a special modification of the visceral efferent group, there are nerve fibres which pass directly to this striated muscle without any interposed ganglion cells, the **branchial motor** cells. These are identical with the nerve fibres supplying the striated muscles of the body wall and limbs, except that they emerge in company with the sensory fibres. Thus each laterally attached cranial nerve can contain two types of motor nerve fibres, in addition to the sensory fibres which, as in the spinal nerves, are divisible into those supplying the gut tube **(visceral sensory),** and those supplying the skin, muscles, etc. **(somatic sensory).**

The central connexions of the cranial nerves, like their spinal counterparts, are arranged in such a manner that the motor nerve fibres originate in groups of cells (nuclei) lying opposite the point of emergence of the nerve to which they contribute, while the entering sensory fibres run longitudinally before ending on the receptive, internuncial cells or 'sensory nuclei' which may, therefore, lie at some distance from the point of entry of the fibres. *Likewise there is within the brain a separate nucleus of origin (motor) or of termination (sensory) for each of the different functional groups of nerve fibres which each nerve contains.*

The Motor Nuclei of the Cranial Nerves

In the spinal medulla the rootlets which gather together into the ventral roots of the nerves form a continuous series and arise from continuous columns of motor nerve cells which develop from the basal lamina. In the brain only the IXth, Xth, and XIth cranial nerves are formed from a continuous series of rootlets and it is therefore only in these nerves that the motor nuclei form continuous columns of cells; the remaining nerves, being separated from each other by varying intervals, arise from separate groups of cells. Nevertheless each motor nucleus of the same functional group lies in the same relative position, as though it formed a part of a continuous longitudinal column or rod of cells. Thus the **somatic motor nuclei,** being derived from the ventral part of the basal lamina, always lie adjacent to the midline ventral to the central canal, the fourth ventricle or the aqueduct; while the **visceral motor nuclei** take up a more dorsal and lateral position except in the region of the fourth ventricle where the lateral wall of the tube is folded outwards so that these nuclei lie lateral to the somatic motor nuclei. The **branchial motor nuclei** take up a more ventral position and, except in the case of the trigeminal nerve, send their fibres curving dorsally to join one of the laterally attached cranial nerves [FIG. 755, 3].

The Motor Nuclei of the Oculomotor (III), Trochlear (IV), Abducent (VI), and Hypoglossal (XII) Nerves [FIGS. 755–762]. Of these the IVth [see also p. 607], VIth, and XIIth have the single function of supplying striated somatic muscle (somatic motor) and each arises from a single motor nucleus located respectively in the lower midbrain, the lowest part of the pons, and throughout the medulla oblongata [FIG. 762]. The IIIrd cranial nerve belongs to this same series, and the major portion of its nucleus lies in line with, and almost connected to, that of the IVth. It differs from the others in two respects:

1. The right and left IIIrd nerve nuclei are fused ventral to the aqueduct and interchange somatic motor fibres in the supply of the extrinsic ocular muscles [FIGS. 757 and 763].

2. The IIIrd cranial nerve also contains visceral motor fibres concerned with the supply of the intrinsic ocular muscles. These arise in an incompletely fused mass, the **accessory** or Edinger-Westphal (autonomic) **nucleus,** lying dorsal to the main nuclear complex.

The emergent fibres from these nuclei pass ventrally, except in the case of the trochlear nerve where the fibres pass laterally and wind spirally in a caudal direction around the margin of the central grey substance to reach the superior part of the superior medullary velum. Here the two nerves meet and decussate, deep to the frenulum of the superior medullary velum, each emerging on the opposite side dorsal to the superior cerebellar peduncle. As the fibres of the trochlear nerve wind round the central grey substance they pass close to the **mesencephalic nucleus** and **tract** of the trigeminal nerve and send some fibres into it. These are believed to be proprioceptive fibres from the superior oblique muscle which terminate in the mesencephalic nucleus (Tarkhan,

1934). The fibres of the IIIrd nerve pass through the tegmentum of the midbrain and pierce the medial part of the red nucleus as a series of separate bundles which collect in the medial part of the substantia nigra before emerging through the lateral wall of the interpeduncular fossa. The fibres of the abducent pass slightly caudally, as they run ventrally, and thus emerge, below their level of origin, at the lower border of the pons. The hypoglossal nerve passes ventrally through the medulla as a long row of fine rootlets lateral to the medial lemniscus, medial accessory olivary nucleus and pyramid, and medial to the olivary nucleus. The positions of the individual nuclei are shown in FIGURES 755–762.

The Motor Nuclei of the Trigeminal (V), Facial (VII), Glossopharyngeal (IX), Vagus (X), and Accessory (XI) Nerves. The positions of the motor nuclei of the laterally attached cranial nerves are shown in FIGURES 755–762. In FIGURE 755 the visceral motor nuclei are shown numbered 2 and the branchial nuclei 3. Of the laterally attached cranial nerves only the VIIth, IXth, Xth, and cranial portion of the XIth contain visceral motor or preganglionic fibres, and of these the latter three receive such fibres from a single nucleus

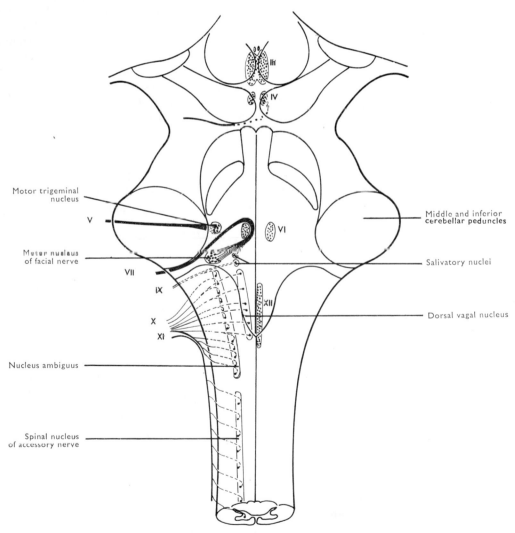

FIG. 762 Diagrammatic posterior view of the brain stem and upper spinal medulla, to show the positions of the cranial nerve motor nuclei and the course of their fibres to the surface.

A

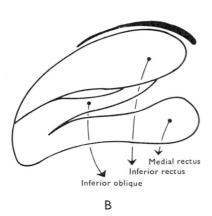

B

FIG. 763   Diagrams of the oculomotor nucleus showing its subdivisions and their functional significance. After Warwick (1953).

A, dorsal view. B, lateral view.
Accessory (autonomic) nucleus, solid black.

lying lateral to the hypoglossal nucleus and known as the **dorsal nucleus of the vagus.** At its extreme cephalic end is the **inferior salivatory nucleus** which supplies fibres to the IXth nerve, predominantly concerned with the innervation of the parotid and posterior intralingual salivary glands. Just cephalic to this lies the **superior salivatory nucleus** which contributes similar fibres to the facial (VII) nerve.

The **branchial motor** system of nuclei contributes to all the laterally attached cranial nerves except the purely sensory vestibulocochlear nerve. At the cephalic end of this column, in the middle of the pons, lies the motor trigeminal nucleus; at the lower border of the pons the motor nucleus of the facial nerve, and caudal to this is a continuous column of cells, broken into two parts by the decussation of the pyramidal tract. A cephalic part, the **nucleus ambiguus** [FIG. 765], sends a few fibres into the glosso-pharyngeal, but the majority into the vagus and cranial part of the accessory for the supply of the

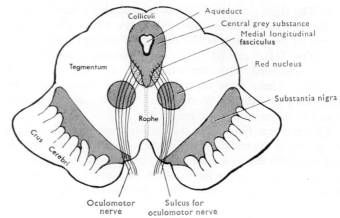

FIG. 764   Diagrammatic transverse section through upper part of midbrain.

pharyngeal and laryngeal muscles. The caudal part, the **spinal nucleus of the accessory nerve,** extends to the fifth cervical segment and gives rise to the spinal roots of the accessory nerve, which supply the sternocleidomastoid and trapezius muscles.

The fibres arising in the nucleus ambiguus and in the spinal nucleus of the accessory nerve, which lies in the posterior part of the cervical anterior horn [FIG. 766], pass dorsally and then curve laterally. The former join fibres from the dorsal nucleus of the vagus and emerge through the ventral part of the spinal tract of the trigeminal nerve. The latter fibres take a more ventral course and, emerging between the ligamentum denticulatum ventrally and the dorsal roots, turn cranially and join to form the spinal part of the accessory nerve [FIG. 753].

The motor nucleus of the facial nerve [FIGS. 759, 762 and 767] lies at the level of the lower border of the pons medial to the emergent root of the facial nerve. From the nucleus fibres converge on the caudal part of the abducent nucleus [FIGS. 759 and 762] then turn superiorly as a compact bundle between that nucleus and the midline, in the floor of the fourth

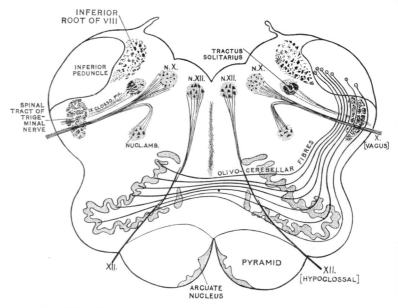

FIG. 765   Diagram of section of medulla oblongata to show the cranial nerve nuclei.

N.X., dorsal nucleus of vagus; N.XII, hypoglossal nucleus.

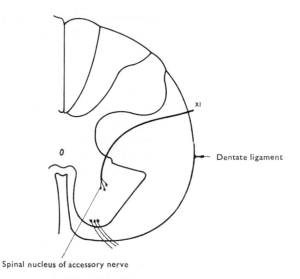

Dentate ligament

Spinal nucleus of accessory nerve

FIG. 766 Transverse section through the first cervical segment of the spinal medulla to show the nucleus of origin, and the mode of emergence of the spinal roots of the accessory nerve, XI.

ventricle. The facial nerve then curves laterally, cranial to the abducent nucleus, and turns ventrolaterally to emerge just lateral to the nucleus of origin, and ventral to the entering vestibulocochlear nerve. In the lower pons, the facial nerve is close to the abducent nucleus **(genu of the facial nerve)**, and to the opposite facial nerve. At its point of emergence it is separated from the vestibulocochlear nerve by the small **nervus intermedius** (*sensory root of the facial nerve*) which contains at least some of the sensory and parasympathetic fibres of the facial nerve.

Sensory Fibres of Cranial Nerves [FIG. 768]

These behave in a fashion very similar to those in the dorsal spinal roots with three major exceptions:

1. The majority of the **somatic sensory fibres** for the cranial region pass with the trigeminal nerve [FIG. 770]. For this reason it gives its name to the longitudinal bundles of somatic sensory fibres which course throughout the length of the brain stem from the upper midbrain to the upper cervical spinal medulla, even though similar fibres enter the same system in smaller numbers through the other laterally attached cranial nerves, just as the several dorsal roots of the spinal medulla send their fibres into the same longitudinal system, the dorsolateral fasciculus.

2. The midbrain or **mesencephalic portion** of this system differs from all other sensory systems in that the primary sensory cell bodies lie within the midbrain and upper pons, forming a long **mesencephalic nucleus,** instead of lying in the trigeminal ganglion.

3. The **visceral sensory fibres,** including those subserving taste, also form a well-defined longitudinal tract which, being surrounded by its nucleus of termination, forms an isolated bundle, the **tractus solitarius** [FIG. 765].

The **sensory fibres of the trigeminal nerve** enter the brain stem immediately caudal to the motor root. Having pierced the superficial part of the pons

at its junction with the middle cerebellar peduncle and entered the dorsal part of the pons, many of the fibres end immediately on a large **(superior) principal sensory nucleus** lying just lateral to the motor nucleus of the trigeminal nerve [FIG. 770]. The fibres which end here are believed to be concerned with tactile sensations. Of the remainder, the majority turn caudally to form the **spinal tract of the trigeminal nerve** which descends through the lateral part of the pons and medulla oblongata, where it lies immediately ventral to the inferior cerebellar peduncle and nucleus cuneatus and may form an ill-defined surface ridge which is obliterated where the posterior spinocerebellar tract crosses it [FIG. 760B]. Moving dorsally as the fourth ventricle closes inferiorly, it becomes continuous with the **dorsolateral fasciculus** in the cervical spinal medulla. Here the fibres of the spinal tract of the nerve intermingle with similar fibres from the cervical dorsal roots. Throughout its length the spinal tract is accompanied on its medial side by the nucleus of the spinal tract of the trigeminal nerve into which collaterals of the descending fibres pass and end, and which is continuous inferiorly with the substantia gelatinosa of the cervical spinal medulla.

The fibres of the spinal tract which descend below the inferior limit of the fourth ventricle (obex) seem to be concerned with relaying impulses which elsewhere are interpreted as painful and thermal sensations, since division of the spinal tract at the level of the obex abolishes these sensations in the ipsilateral trigeminal area. It is, therefore, functionally equivalent to the dorsolateral fasciculus [p. 658] with which it is continuous. The function of the fibres which terminate in the more cranial part of the nucleus of the spinal tract is not known, but there are differences between the cells in the cranial and caudal parts of the nucleus. The spinal tract of the trigeminal nerve is pierced by the facial, glossopharyngeal, and upper vagal

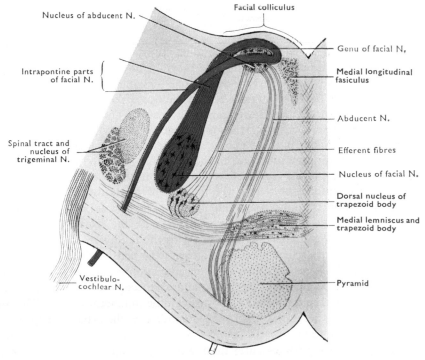

Nucleus of abducent N.

Facial colliculus

Intrapontine parts of facial N.

Genu of facial N.

Medial longitudinal fasiculus

Abducent N.

Spinal tract and nucleus of trigeminal N.

Efferent fibres

Nucleus of facial N.

Dorsal nucleus of trapezoid body

Medial lemniscus and trapezoid body

Vestibulo-cochlear N.

Pyramid

FIG. 767 Diagram of intrapontine course of facial nerve, seen from above.

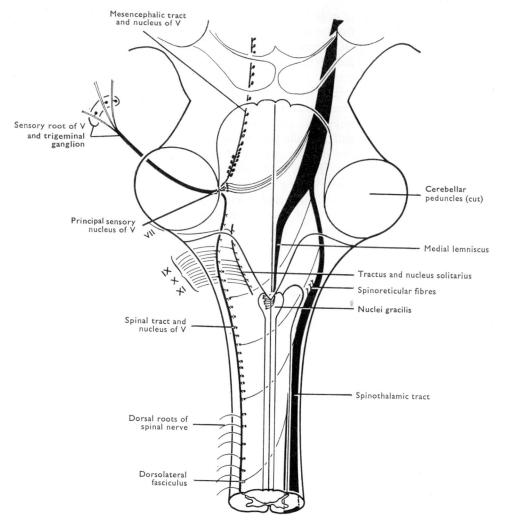

Mesencephalic tract
and nucleus of V

Sensory root of V
and trigeminal
ganglion

Principal sensory
nucleus of V

Spinal tract and
nucleus of V

Dorsal roots of
spinal nerve

Dorsolateral
fasciculus

Cerebellar
peduncles (cut)

Medial lemniscus

Tractus and nucleus solitarius

Spinoreticular fibres

Nuclei gracilis

Spinothalamic tract

FIG. 768 Diagrammatic posterior view of the brain stem and upper spinal medulla, to show the positions of the sensory nuclei of termination of the cranial nerves, except the eighth. The secondary ascending pathways are also shown.

rootlets, just as its caudal continuation, the dorsolateral fasciculus, is pierced by the dorsal spinal roots. The accessory rootlets emerge ventral to it [FIG. 766]. The remaining fibres of the trigeminal nerve pass cranially into the upper pons and midbrain constituting the **mesencephalic tract.** This lies at the anterolateral corner of the upper part of the fourth ventricle and lateral to the cerebral aqueduct at the outer edge of the central grey substance in the midbrain [FIG. 769]. This portion of the nerve is believed to be concerned with proprioceptive information from the muscles of mastication, temporomandibular joint, the tooth sockets, and the facial muscles, the latter through communications with the facial nerve in the face, and by fibres entering with the facial nerve which have been described ascending through the pons to the mesencephalic nucleus. It may also receive proprioceptive fibres from the oculomotor, trochlear, and abducent nerves, either directly in the midbrain (IV), or through the communications which these nerves have with the trigeminal in the orbit or middle cranial fossa.

Visceral sensory fibres enter the brain stem through the facial, glossopharyngeal, vagus, and cranial parts of the

accessory nerves [FIG. 768]. The complete picture of the termination of these fibres is not known, but the majority, including those concerned with taste (VII, IX, X), enter the tractus solitarius and, running caudally, end in the surrounding nucleus of the tractus solitarius. Some afferent fibres in the vagus have been described as ending on the dorsal nucleus of that nerve which has therefore been claimed to be a mixed nucleus, partially sensory, partially motor. The evidence that afferent fibres synapse with this nucleus cannot be taken to indicate that is has a sensory function. If it were, then many of the anterior horn cells of the spinal medulla which innervate the striated muscles of the body would have to be classified as sensory, since collaterals of the dorsal root fibres synapse directly with them. In any case all the cells of the dorsal nucleus of the vagus undergo retrograde change following section of that nerve at the base of the skull. It may be that all visceral sensory fibres do not run in the tractus solitarius since this bundle decreases in size opposite the entry of the vagus nerve, which must carry a large proportion of the total number of such fibres in view of its extensive distribution to the viscera. However, many of these fibres may be small, non-myelinated, and invisible in the standard preparations. Whatever their course it seems likely that the majority of the visceral afferent fibres end in the nucleus of the tractus solitarius.

The positioning of the nuclei of the vagus and hypoglossal nerves is well shown in FIGURE 765. Apart from the special development of the nucleus ambiguus, solitary tract, and vestibulocochlear nuclei, the arrangement is identical with that in the grey matter of the thoracic spinal medulla, except that the grey matter lies transversely after the eversion of the lateral wall of the neural tube in the formation of the fourth ventricle [FIG. 750].

The table in the Peripheral Nervous System section gives the functions of the cranial nerves [p. 710]. From this and from an understanding of the general plan of the cranial nerve nuclei, the immediate deep connexions of the cranial nerves may be plotted with some accuracy.

The **nuclei of termination of the vestibulocochlear nerve** lie at the level of entry of the nerve and are divided into cochlear and vestibular nuclei. The cochlear fibres, lying in the superior part of the nerve, end in a curved mass of cells over the lateral and posterior surfaces of the inferior

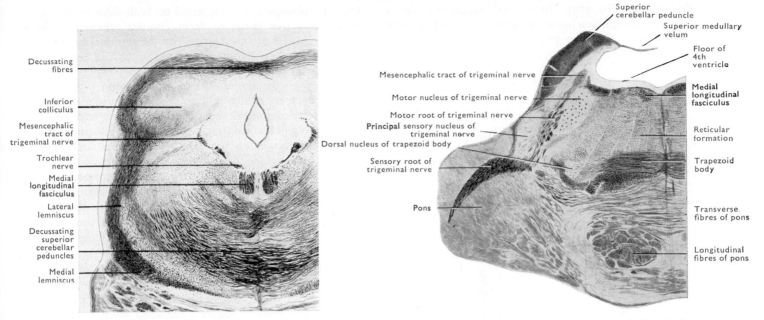

FIG. 769 Section through inferior colliculus and tegmentum below the level of nucleus of trochlear nerve.

FIG. 770 Transverse section through pons at level of trigeminal nerve.

FIG. 771 Section through adult medulla oblongata immediately above the decussation of the pyramids (Weigert-Pal specimen).

cerebellar peduncle. The vestibular fibres pass anterior to the inferior cerebellar peduncle and are distributed to the vestibular nuclei [FIG. 760] which form the raised vestibular area. This is a roughly triangular mass extending from the mid-pons almost to the obex between the sulcus limitans and the lateral angle of the floor of the fourth ventricle [FIG. 749 and p. 672].

THE FOREBRAIN [PLATES 40–44]

Initially the forebrain or prosencephalon consists simply of the anterior extremity of the neural tube. It has a deep, slit-like, median cavity closed in front by the thin **lamina terminalis** and continuous caudally with the aqueduct from which the sulcus limitans passes into its wall as the hypothalmic sulcus [FIG. 773]. In the early stages of its development the prosencephalon has many features in common with the hindbrain, including the presence of a thin roof plate

FIG. 772   Profile view of brain of human embryo of ten weeks. The cranial nerves are indicated by numerals. (After His.)

A, cerebral diverticulum (posterior lobe) of hypophysis cerebri; B, oral diverticulum, i.e., Rathke's pouch (anterior lobe) of hypophysis.

which is subsequently invaginated on both sides by the overlying connective tissue to form choroid plexuses. These choroid plexuses are vascularized by the terminal branches of the basilar artery which pass dorsally at the junction of the midbrain and forebrain, supplying this region and continuing on to the dorsal surface of the latter. The above vessels, which are similar in their arrangement to the posterior inferior cerebellar arteries in the hindbrain, are later involved in the supply of a considerable part of the expanding cerebral hemisphere, becoming the posterior cerebral arteries, the original branches to the choroid plexus remaining as the posterior choroidal branches of the posterior cerebral arteries [FIG. 784]. The posterior communicatin garteries connect the posterior cerebral arteries near their origins to the corresponding internal carotid arteries which reach the forebrain and extend on to its surface just dorsal to the buccopharyngeal membrane.

A series of outgrowths of the whole thickness of the forebrain wall forms six subsidiary structures; the two eyes, the paired telencephalic vesicles forming the greater part of the cerebral hemispheres, and two unpaired median structures, the hypophysis and the pineal body. The remaining, paramedian parts of the forebrain surround the third ventricle and form the diencephalon.

The Diencephalon. The first outgrowths from the forebrain are the optic vesicles [FIG. 734], followed shortly by the appearance of the **telencephalic vesicles** (cerebral hemispheres) from the upper anterior part of the forebrain [FIG.

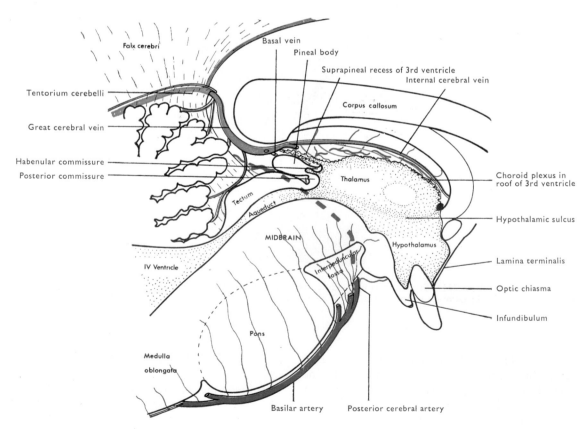

FIG. 773   Median sagittal section of human brain to show the relations of the pineal body and associated structures.

KEY TO PLATES 40–44

a. Adhesio interthalamica.

c. Caudate nucleus.

cc2. Genu of corpus callosum.

cc3. Trunk of corpus callosum.

cc4. Splenium of corpus callosum.

cplr. Tuft of choroid plexus appearing through lateral aperture of 4th ventricle.

cr. Corona radiata, visual radiation.

d. Dentate nucleus of cerebellum.

f. Fornix.

fl. Flocculus.

gc. Gyrus cinguli.

gr. Gyrus rectus.

h. Hippocampus.

ht. Hypothalamus.

ic. Internal capsule.

l. Lateral nucleus of thalamus.

lv. Lateral ventricle.

m. Medial nucleus of thalamus.

mb. Midbrain.

mo. Medulla oblongata.

o. Olfactory bulb.

ot. Olfactory tract.

p. Putamen of lentiform nucleus.

py. Pyramid.

sfg. Medial surface of superior frontal gyrus, medial frontal gyrus.

sp. Septum pellucidum.

t. Thalamus.

tp. Pulvinar of thalamus.

v. Ventral nucleus of thalamus.

II. Optic nerve.

III. Oculomotor nerve.

V. Trigeminal nerve.

VII. Facial nerve.

VIII. Vestibulocochlear nerve.

IX Glossopharyngeal nerve.

PLATE 40

PLATES 40 and 41. A series of photomontages showing stained horizontal sections of the brain in their correct relation to the hemispheres.

1. Through the central part of the lateral ventricle.
2. Through the superior part of the thalamus.
3. Through the superior margin of the lentiform nucleus.
4. Through the splenium of the corpus callosum.

PLATE 41

5. Through the posterior commissure.
6. Through the lateral geniculate body.
7. Through the lower midbrain.
8. Through the amygdaloid body.

PLATE 42

PLATES 42 to 44. A series of photomontages showing stained coronal sections of the brain in their correct relation to the hemispheres.

1. Through the genu of the corpus callosum.
2. Through the optic chiasma.
3. Through the tip of the inferior horn of the lateral ventricle.
4. Through the mamillary body.

PLATE 43

5. Showing the continuity of the corona radiata, the internal capsule, the crus cerebri, the longitudinal fibres of the pons and the pyramid.
6. Through the anterior end of the red nucleus.
7. Through the posterior extremity of the lentiform nucleus.
8. Through the posterior commissure and the 4th ventricle.

PLATE 44

9. Through the pulvinar of the thalamus.
10. Through the junction of the inferior and posterior horns with the central part of the lateral ventricle.

PLATE 45

A and B are ventriculograms taken after the introduction of air into the ventricular cavities by a needle passed through a burr hole in the skull.

1. Central part of lateral ventricle.   4. Posterior horn of lateral ventricle.
2. 3rd ventricle.   5. Anterior horn of lateral ventricle.
3. Inferior horn of lateral ventricle.   6. Interventricular foramen.

C and D are encephalograms taken after the introduction of air into the lumbar subarachnoid space. Some of the air has risen into the ventricular system, the remainder outlines the cerebral subarachnoid space.

1. Central part of lateral ventricle.
2. Anterior horn of lateral ventricle.
3. 3rd ventricle.
4. Air trapped beneath tentorium cerebelli.
5. Air in the subarachnoid space outlining the cerebral sulci.
6. Interpeduncular cistern.
7. Aqueduct of cerebrum.
8. 4th ventricle.

PLATE 46

A

(i)   (ii)

B

(i)   (ii)

A and B.   Myelograms.   These are produced by injecting X-ray opaque material into the spinal subarachnoid space either by cisternal or lumbar puncture.

A.   Cervical myelograms: (i) lateral and (ii) anteroposterior.

1. Basal cistern.
2. 5th cervical vertebra.
3. Foramen magnum.
4. Spinal subarachnoid space.
5. Central shadow due to the spinal medulla surrounded by a sheath of X-ray opaque material. Compare with the lumbar myelogram (B) where no such shadow is present because of the absence of the spinal medulla in this region.
6. Extension of subarachnoid space around nerve root.

B.   Lumbar myelograms: (i) lateral and (ii) anteroposterior.

1. 3rd lumbar vertebra.
2. Needle through which material is introduced into subarachnoid space.
3. Extension of subarachnoid space around nerve roots.

C

D

C.   Lateral radiograph of skull.

1. Anterior temporal diploic vein.
2. Shadow of the auricle.
3. Sphenoidal air sinus.
4. Posterior temporal diploic vein.
5. Calcification in the pineal body.

D. Low lumbar myelogram.   Anteroposterior, taken some days after injection, showing extension of contrast medium along sacral nerve roots.

1. 1st sacral vertebra.
2. Subarachnoid space surrounding the cauda equina.

785]. The optic outgrowths [FIG. 997], although hollow at first, rapidly become solid, but the telencephalic vesicles remain hollow and therefore contain lateral extensions of the cavity of the forebrain which constitute the **lateral ventricles,** and are continuous with each other through the part of the original median cavity immediately adjacent to the upper part of the lamina terminalis. This region of communication between the three cavities of the forebrain is known as the **interventricular foramen.** With the appearance of the lateral ventricles the median cavity is known as the **third ventricle** and its walls (the diencephalon) form the thalamus and hypothalamus, separated on the ventricular aspect by the **hypothalmic sulcus** which extends forwards to the interventricular foramen [FIGS. 773, 775].

The Hypophysis. The closure of the extreme anterior end of the neural tube is not as simple as elsewhere since this process, spreading headwards from the cervical region, fails to reach the extreme anterior end of the neural plate before that part begins to close from the buccopharyngeal membrane caudally [FIG. 734]. The two processes of closure meet at the **anterior neuropore** in the region of the lamina terminalis; the roof plate between this and the buccopharyngeal membrane is thus formed separately. In the roof plate adjacent to the buccopharyngeal membrane, the skin and neural ectoderm are in contact at the time of closure of the neural tube. This contact is maintained as mesoderm, proliferating to form the base of the skull, lifts the neural tube away from the roof of the mouth, so that a finger-like pouch of ectoderm is carried upwards from the stomodaeal roof. This is **Rathke's pouch,** which, together with the part of the floor of the neural tube with which it is in contact, forms the hypophysis [p. 574]. The hypophysis lies at the inferior end of the lamina terminalis immediately behind the roots of the optic outgrowths into which nerve fibres grow from the eyes to form the optic chiasma [p. 668]. This is the meeting-point of the stalks of the optic outgrowths which, filled with nerve fibres, form the optic nerves [FIG. 772].

The Pineal Body. This median diverticulum arises from the roof of the forebrain caudal to the thin roof of the third ventricle, at the junction of the midbrain with the forebrain. The cells of the ependymal diverticulum proliferate and form a small, reddish body in the shape of a flattened ovoid approximately 7–10 mm. long and 5–7 mm. wide, but very variable in size. The cells of which it is composed are secretory in type, and it contains many blood vessels but no duct system, thus suggesting that it produces a hormone or hormones. Until recently very little has been known about its function, though it is now known to contain several substances which are pharmacologically active, including noradrenaline, 5-hydroxytryptamine (serotonin) and melatonin. The latter produces marked contraction of melanophores when injected

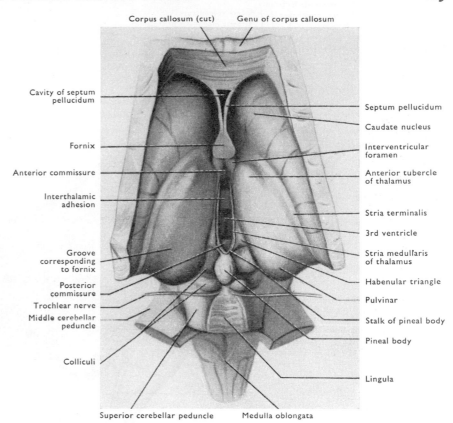

FIG. 774 Colliculi and neighbouring parts.

into frogs, and has many physiological effects in mammals (including inhibition of ovarian growth) produced mainly by direct action on the brain. The concentrations of serotonin and its derivative melatonin vary cyclically and reciprocally in the pineal body; serotonin increasing on exposure to light, melatonin in darkness. Surprisingly these effects are produced by the sympathetic supply to the pineal body and not by the optic tectum on which it lies. Thus the structure which is the pineal eye of lower forms has an important role in the hormonal reactions to light though no longer directly exposed to it in mammals (Wurtman, Axelrod, and Kelly, 1968). After middle age the pineal body is frequently calcified, and it may therefore be possible to demonstrate a shift of the brain to one side in a simple anteroposterior radiograph of the skull [PLATE 46]. The pineal body extends caudally to rest between the superior colliculi beneath the thick posterior end of the corpus callosum [p. 650] from which it is separated by an extension of the third ventricle, the **suprapineal recess,** and the junction of the internal cerebral veins to form the great cerebral vein [FIG. 773]. On either side of its base lie the posterior choroidal arteries passing forwards in company with the internal cerebral veins above the third ventricle and thalamus. A small pineal recess of the third ventricle extends into its base between two commissures which cross close to the attachment of the pineal, the **posterior commissure** below and the **habenular commissure** above. The latter is continuous with a bundle of fibres which passes forwards as a ridge, the **habenula,** on either side of the thin roof of the third ventricle to reach a triangular area on the posteromedial aspect of the thalamus, the **habenular triangle** [FIG. 774]. The two habenulae, the

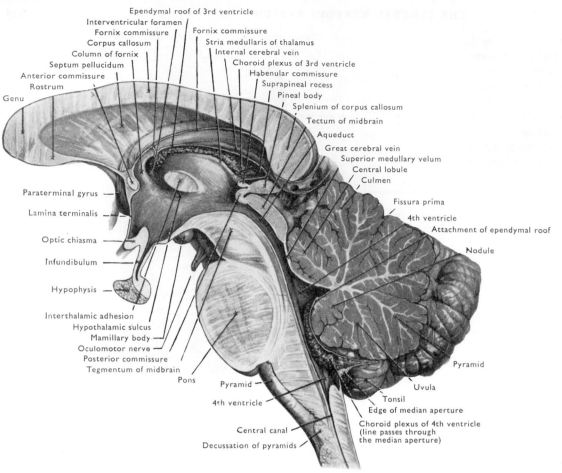

Ependymal roof of 3rd ventricle
Interventricular foramen
Fornix commissure
Corpus callosum
Column of fornix
Septum pellucidum
Anterior commissure
Rostrum
Genu
Fornix commissure
Stria medullaris of thalamus
Internal cerebral vein
Choroid plexus of 3rd ventricle
Habenular commissure
Suprapineal recess
Pineal body
Splenium of corpus callosum
Tectum of midbrain
Aqueduct
Great cerebral vein
Superior medullary velum
Central lobule
Culmen
Fissura prima
4th ventricle
Attachment of ependymal roof
Nodule
Paraterminal gyrus
Lamina terminalis
Optic chiasma
Infundibulum
Hypophysis
Interthalamic adhesion
Hypothalamic sulcus
Mamillary body
Oculomotor nerve
Posterior commissure
Tegmentum of midbrain
Pons
Pyramid
4th ventricle
Central canal
Decussation of pyramids
Pyramid
Uvula
Tonsil
Edge of median aperture
Choroid plexus of 4th ventricle
(line passes through
the median aperture)

FIG. 775 The parts of the brain in median section. The side-walls of the 3rd and 4th ventricles are shown.

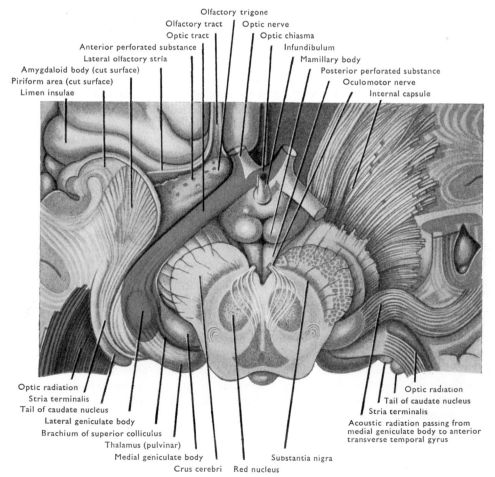

Olfactory trigone
Olfactory tract
Optic tract
Anterior perforated substance
Lateral olfactory stria
Amygdaloid body (cut surface)
Piriform area (cut surface)
Limen insulae
Optic nerve
Optic chiasma
Infundibulum
Mamillary body
Posterior perforated substance
Oculomotor nerve
Internal capsule

Optic radiation
Stria terminalis
Tail of caudate nucleus
Lateral geniculate body
Brachium of superior colliculus
Thalamus (pulvinar)
Medial geniculate body
Crus cerebri
Red nucleus
Substantia nigra
Optic radiation
Tail of caudate nucleus
Stria terminalis
Acoustic radiation passing from medial geniculate body to anterior transverse temporal gyrus

FIG. 776 Part of lower surface of forebrain, showing olfactory and optic tracts. The midbrain has been cut across.

habenular triangles and commissure, and the pineal body, constitute the **epithalamus.**

The **posterior commissure,** larger than the habenular commissure, belongs to the midbrain and allows passage of nerve fibres between the two sides of its cephalic extremity.

The Thalamus

The thalamus grows rapidly, forming a considerable mass of cells which bulges into the third ventricle to meet and fuse with the opposite thalamus, partially dividing the ventricle into dorsal and ventral parts by this **interthalamic adhesion** [FIG. 775, PLATES 43 and 44]. This growth is most marked at the posterior end so that the thalamus becomes triangular in shape (3 cm. long by 1·5 cm. wide in the adult) when seen from above [FIG. 774] with the apex at the interventricular foramen and the broad, rounded base **(pulvinar)** posteriorly, slightly overhanging the upper midbrain. The thalamus has therefore a medial surface covered by the ependyma of the third ventricle, and a dorsal surface, the lateral part of which is covered by the floor of the lateral ventricle [p. 652], while the medial part is covered with pia mater [FIG. 820]. The line of junction of these two areas is marked in the adult by a shallow groove which lodges the medial edge of the cerebral hemisphere, the **fornix** [FIG. 819], and marks the line of reflection of the ependyma of the lateral ventricle from the dorsal surface of the thalamus on to the choroid plexus of the lateral ventricle. In the same fashion, the ependyma of the third ventricle covering the medial wall of the thalamus is reflected at the junction of the medial and dorsal surfaces on to the thin roof and choroid plexus of the third ventricle. These two reflections are known as the **taeniae thalami,** and are continuous at the interventricular foramen through which the choroid plexuses of the two ventricles are continuous. Along the line of attachment of the roof of the third ventricle to each thalamus lies a small ridge of white matter, the **stria medullaris of the thalamus,** which extends posteriorly to the habenular triangle from the region of the interventricular foramen [FIG. 775].

On the dorsal surface of the thalamus at its apex is a small rounded swelling, the **anterior tubercle,** formed by the parts of the **anterior nucleus,** while two small rounded **geniculate bodies** [FIGS. 757 B and 776] protrude from its postero-inferior surface and comprise the **metathalamus.** Of the two geniculate bodies the **medial** is the more obvious from the surface. It lies wedged between the lateral surface of the midbrain, the dorsal margin of the crus cerebri, and the inferior surface of the pulvinar, and receives the brachium of the inferior colliculus [p. 611]. The **lateral geniculate body** forms a less well-defined swelling on the inferior surface of the pulvinar. This body is continuous anteriorly with the optic tract which enters it by sweeping posteriorly around the superior limit of the midbrain to reach the inferior surface of the thalamus. As it reaches the lateral

geniculate body the optic tract splits into medial and lateral roots which appear to become continuous with the corresponding geniculate bodies, though both roots enter the lateral geniculate body [p. 668]. The optic tract also gives rise to a slender bundle of fibres which sweeps dorsomedially over

FIG. 777 Diagram to show the main subdivisions of the thalamus. A, B, and C are coronal sections indicated by lines A, B, and C on the horizontal section shown above.

In all the figures the lateral medullary lamina is shown as a broken line, the medial medullary lamina by a solid line which splits to enclose the central nucleus postero-inferiorly and the anterior nuclei anterosuperiorly. In A, the medial medullary lamina is thickened by the presence of the mamillothalamic tract. The arrow indicates the brachium of the superior colliculus.

Lateral thalamic nucleus   Choroid plexus of lateral ventricle

Corpus callosum

Fornix

Internal cerebral veins in transverse fissure

Habenular triangle

Central thalamic nucleus

3rd ventricle

Prerubral field

Mamillothalamic tract

Fornix

Hypothalamus

Optic chiasma

Caudate nucleus

Choroid fissure
Lateral medullary lamina

Reticular thalamic nucleus

Ventral thalamic nucleus

Internal capsule

Zona incerta

Subthalamic nucleus

Optic tract

Amygdaloid body

Internal carotid artery

FIG. 778   Low power photomicrograph of horizontal section through the diencephalon.   See also Fig. 861, page 683.

the posterior surface of the thalamus, in a groove separating the pulvinar from the medial geniculate body, to reach the anterolateral part of the superior colliculus.   This is the **brachium of the superior colliculus** and its fibres convey nervous impulses from the optic tract to the most cephalic parts of the tectum of the midbrain [FIG. 757 B, 776].

In addition to the nuclei, described above, which form well-defined protrusions from the surface of the thalamus, the main mass consists of cells divided more or less clearly into a series of subsidiary nuclei.   This division is partially produced by two layers of white matter, the **medullary laminae,** which run approximately in a sagittal plane and are best visualized in sections [FIGS. 777 and 778].   The medial medullary lamina separates the **medial nucleus** from a lateral nuclear mass which is less clearly divisible into a dorsal **lateral nucleus,** whose caudal end forms the rounded posterior projection of the thalamus or **pulvinar,** and a **ventral nucleus** which can be subdivided into **anterior-ventral, intermediate-ventral,** and **posterior-ventral.**   The latter is divisible into lateral and medial parts, mainly on account of their different connexions.   The medial medullary lamina splits to enclose the

anterior nuclei anterosuperiorly, and the **central (centro-median) nucleus** posteroventrally, and has in association with it some small groups of cells, the **intralaminar nuclei.** It is thickened anteriorly by the mamillothalamic tract [p. 647] running in it.   At the external margin of the thalamus is a thin curved sheet of cells lateral to the lateral and ventral nuclei and separated from them by the **lateral medullary lamina.** This rim is traversed by many groups of fibres, is therefore known as the **reticular nucleus** of the thalamus, and lies against the posterior limb of the internal capsule [FIG. 778 and p. 638].   See pages 639–640 for the general arrangement of the thalamic projections.

Inferiorly the thalamus lies on the hypothalamus partially separated from it by a horizontal layer of nerve fibres, the thalamic fasciculus, through which certain ascending fibre bundles reach the ventral nucleus.

The Hypothalamus

This consists of two parts : (1) The medioventral part, which forms the roof of the interpeduncular fossa and corresponds with the hypothalamus of the physiologist. (2) The

FIG. 779 Median section through the hypophysial region.

superiorly by the anterior cerebral arteries. Lateral to the chiasma lie the internal carotid arteries where they emerge from the cavernous sinus, and on a slightly superior level the anterior perforated substance. From the posterolateral margins of the chiasma arise the optic tracts [FIG. 776] which sweep posterolaterally, forming the margin of the exposed part of the hypothalamus. The tracts are covered inferiorly by the medial margin of the temporal lobe of the cerebral hemisphere and by the posterior communicating artery, and are accompanied by the anterior choroidal branch of the internal carotid artery which turns laterally into the antero-inferior extremity of the choroid plexus near the tip of the inferior horn of the lateral ventricle [p. 635]. Reaching the anterior margin of the crus cerebri, the optic tract sweeps around the junction of the midbrain and forebrain to reach the postero-inferior surface of the thalamus where it appears to divide into its medial and lateral roots entering the lateral geniculate body.

Internal Structure of the Hypothalamus

The ventromedial portion consists of a medial part containing several nuclei, separated by the fornix [p. 647] (which arches through the hypothalamus) from a lateral part consisting of longitudinally directed fibres running through scattered nerve cells. The fibres constitute the **medial forebrain bundle,** a complicated collection of fibres originating from the olfactory cells in the region of the anterior perforated substance and connecting it to the hypothalamus. It extends caudally to the midbrain, carrying with it other fibres which originate in the hypothalamus. In addition the hypothalamus may be divided from before backwards into supra-optic, tuberal, and mamillary regions, corresponding approximately to the optic chiasma, tuber cinereum, and mamillary bodies.

The nuclei of the hypothalamus [FIG. 780] are not easily separated from each other, except in the case of the **supra-optic.** This lies over the beginning of the optic tract, lateral to the pre-optic nucleus which extends dorsally from the chiasma close to the midline. It is also possible to outline the

lateral or subthalamic part, which is nowhere visible on the surface of the brain. Both parts are directly continuous caudally with the tegmentum of the midbrain.

The medial part forms the floor and lateral wall of the third ventricle inferior to the hypothalamic sulcus, extending from the lamina terminalis and optic chiasma anteriorly to the tegmentum of the midbrain posteriorly. On the inferior aspect of the brain it consists of the optic chiasma and tracts anterolaterally, the tuber cinereum, the mamillary bodies, and the posterior perforated substance. It is limited posteriorly by the crura cerebri and the superior border of the pons, and forms the roof of the interpeduncular fossa. The **tuber cinereum** is a rounded swelling which projects inferiorly from the floor of the third ventricle immediately posterior to the optic chiasma. On the lateral aspect of the tuber cinereum are a series of small protuberances caused by groups of cells, the **tuberal nuclei,** lying immediately under the surface. Inferiorly the tuber cinereum tapers to a narrow stalk, the **infundibulum** [FIG. 776], by which it is continuous with the posterior (neural) lobe of the hypophysis [FIG. 779] and through which a considerable number of nerve fibres pass from the hypothalamus to the hypophysis. Posterior to this are the two spherical masses of grey matter, the **mamillary bodies,** lying side by side, and still further posteriorly is a triangular area extending caudally on the inferior surface of the midbrain, recessed between the crura cerebri, and reaching the pons at its apex. Into this area run a large number of small branches from the posterior cerebral arteries which pierce it, conferring on it the name of **posterior perforated substance.** These vessels supply the midbrain and posterior hypothalamus, and lateral to them emerge the oculomotor cranial nerves [FIG. 776].

The Optic Chiasma and Tracts

In the adult brain the optic chiasma or crossing lies at the inferior end of the lamina terminalis and receives the two optic nerves at its anterolateral margins just where they are crossed

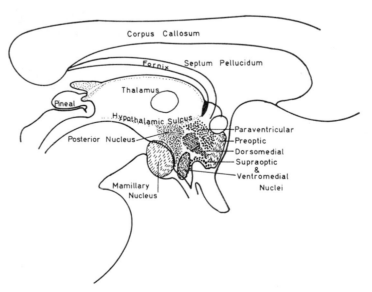

FIG. 780 Diagram to show the main hypothalamic nuclei and their position. After Le Gros Clark (1938).

paraventricular nucleus because of its rich supply of capillaries. From this nucleus, and from the tuberal and supra-optic nuclei, nerve fibres pass to every part of the neurohypophysis, that is, to the median eminence (in contact with the pars tuberalis), infundibulum, and posterior lobe.

The lateral or subthalamic part of the hypothalamus also lies inferior to the thalamus separating it from the inferior part of the internal capsule [p. 638] and the superior part of the crus cerebri. It consists of the **subthalamic nucleus** which is oval in section and lies against the internal capsule [Fig. 778], separated by it from the medial segment of the globus pallidus of the lentiform nucleus [p. 642]. Numerous fibres pass through the internal capsule or around its ventral margin between these two nuclei and constitute a part of the **ansa lenticularis** [p. 682]. Dorsal to the subthalamic nucleus lies the **zona incerta**, a nucleus which is separated from the thalamus by a layer of fibres, the **thalamic fasciculus**, consisting of the long ascending tracts passing to the thalamus. Medially this contains fibres from the capsule of the red nucleus [p. 675], which extends into the subthalamic region. Here fibres from the capsule of the red nucleus mingle with those of the thalamic fasciculus and fibres from the ansa lenticularis which pass medially, separating the subthalamic nucleus and zona incerta, to enter the hypothalamus, the red nucleus, and the substantia nigra. Medially this fused mass of fibres abuts on the mamillothalamic tract [p. 647].

The walls of the third ventricle, consisting of the thalamus dorsally and the hypothalamus ventrally, later come to lie between the two telencephalic vesicles (cerebral hemispheres) and then constitute the diencephalon or 'tweenbrain.

The Third Ventricle [Figs. 774, 781]

This is the cephalic extremity of the original median cavity of the neural tube from which the two lateral ventricles grow out. It is a deep slit-like cavity, lined with ependyma and all but separating the two halves of the diencephalon. It is limited *anteriorly* by the thin **lamina terminalis**, stretching between the **anterior commissure** superiorly and the optic chiasma inferiorly [Fig. 781]. This lamina lies at the anterior limit of the hypothalamus and not only joins the two hypothalami, but also the two cerebral hemispheres. *Inferiorly* its **floor** forms the roof of the interpeduncular fossa, lying superior to the optic chiasma (**optic recess**) anteriorly. Posterior to this it dips down for a variable distance into the infundibulum (**infundibular recess**), and then the floor passes posterosuperiorly towards the aqueduct, rapidly decreasing the vertical extent of the ventricle. Throughout most of its length, the **roof** of the third ventricle is composed of ependyma only, and is invaginated on both sides of the midline by minute tufts of choroid plexus. The thin roof begins anteriorly at the interventricular foramen, separated

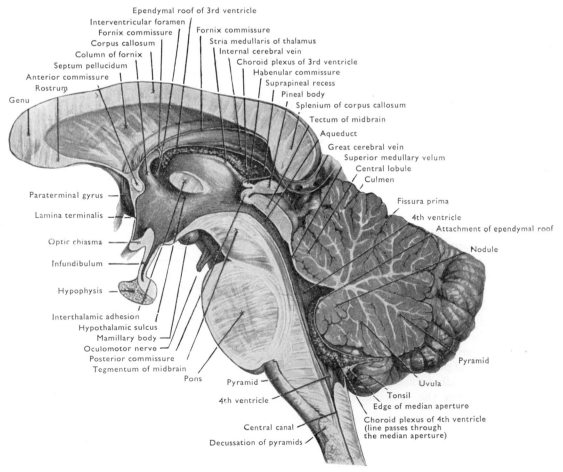

FIG. 781　The parts of the brain in median section. The side-walls of the third and fourth ventricles are shown.

from the anterior commissure by the columns of the fornix [FIG. 781] descending into the hypothalamus. Here the choroid plexus is tucked into the foramen, each half passing laterally through it into the corresponding lateral ventricle. Posteriorly, the thin roof bulges backwards over the superior surface of the pineal body (**suprapineal recess**), and passing to the **habenular commissure** becomes continuous with the ependyma on its anterior surface and that of the **posterior commissure** [FIG. 781]. Between these the ependyma enters a shallow recess in the base of the pineal body, the **pineal recess.** Laterally the thin roof is attached to the superomedial margins of the thalami, along the **striae medullares of the thalamus.** The lateral walls are formed by the thalamus and hypothalamus, separated by the hypothalamic sulcus, the former commonly, but not invariably, being joined across the midline by the **interthalamic adhesion.** The anterosuperior part of the third ventricle communicates on each side with a lateral ventricle and forms the **interventricular foramen** where the three ventricles of the forebrain communicate with each other.

THE TELENCEPHALON [PLATES 40–44]

The two telencephalic vesicles arise as outgrowths of the whole thickness of the walls of the forebrain at its antero-superior part [FIG. 782], and at first form hollow vesicles whose inferior wall is thick while the remainder is thin. The thick, cellular, inferior part of each telencephalic vesicle constitutes the rudiment of the corpus striatum of the telencephalon, and inferior to it lies the cephalic extremity of the dorsal aorta (internal carotid artery). This is continued as the middle cerebral artery, which sends several minute branches into the corpus striatum [FIG. 783], the future striate arteries. These pierce its inferior surface, forming the **anterior perforated substance** [FIG. 776], while the parent vessel runs laterally to supply that surface of the telencephalon.

The growth of the telencephalon, or endbrain, consists of a great expansion of the thin upper part of the vesicle, which thereby overgrows the thick basal part, rolling around it, so that before long it has virtually encircled the corpus striatum and all but hidden it from view on the ventral aspect. FIGURE 786 shows how the posterior end of the vesicle gradually becomes inferior in position and forms the future temporal lobe, lying below and behind the anterior end of the vesicle which forms the frontal lobe. Because of this method of growth, each cerebral hemisphere and its contained ventricle become C-shaped with the open part of the C facing downwards and forwards. The middle cerebral artery, as it passes to reach the lateral surface of the telencephalic vesicle, is caught between the two limbs of the C in the fissure thus formed between the frontal and temporal lobes, **the lateral sulcus** [FIG. 788]. While coursing laterally in this sulcus the artery

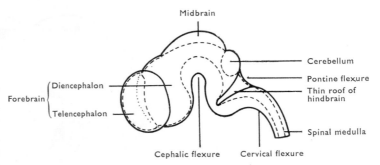

FIG. 782 Diagram to show the parts of the embryonic brain after the formation of the telencephalon. The optic outgrowth is not shown. Broken lines indicate the ventricular system; the dotted line, the roof of the diencephalon.

FIG. 783 Transverse section through the forebrain at the level of the interventricular foramen of the brain shown in Fig. 782.

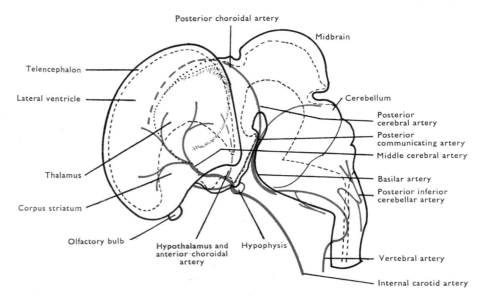

FIG. 784 Drawing of reconstruction of early human foetal brain derived from serial sections. The arteries are shown diagrammatically

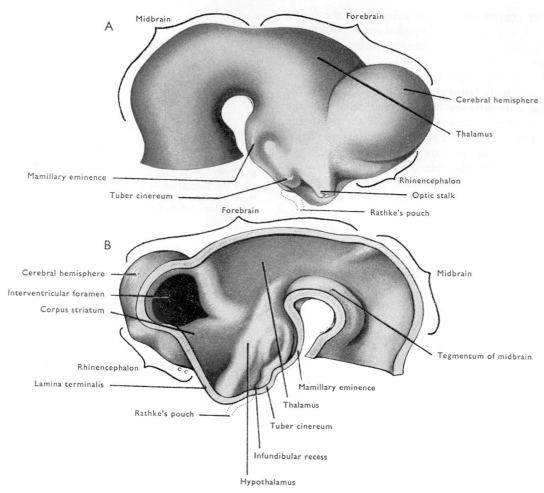

FIG. 785   Brain of human embryo in fifth week. (His.)
A, brain from right side.   B, median section of same brain.

gives off its branches [striate, FIG. 787] to the corpus striatum. Further growth of the surface of the hemisphere in the sagittal plane leads to the protrusion of the intermediate part posteriorly [FIG. 786, 5], to form the occipital lobe containing an extension of the ventricle. In this fashion the general shape of the hemisphere is produced, and the contained ventricle comes to have anterior, posterior, and inferior horns, occupying the frontal, occipital, and temporal lobes respectively and a central part between the first two.

Sulci and Gyri

The telencephalic vesicle also expands transversely so that the thin wall of the vesicle bulges laterally over the surface of the corpus striatum, which grows less rapidly in surface area. Since the corpus striatum is now virtually surrounded, the overgrowth occurs from in front, above, and below, and folds of the surface encroach on the part of the hemisphere overlying the centrally placed, approximately spherical, corpus striatum [FIGS. 786–792]. The covering folds, or **opercula,** gradually approach each other, and when they meet they hide from lateral view the cortex [p. 634] covering the corpus striatum, which thus forms a buried island or **insula,** and also the main part of the middle cerebral artery running over it [FIG. 789]. As a result the branches of the middle cerebral artery now appear

between the lips of the opercula which meet but do not fuse, thereby forming a deep triradiate sulcus on the lateral aspect of the hemisphere between the four opercula : orbital, frontal, frontoparietal, and temporal. The longest (posterior) ramus of the sulcus extends towards the occipital lobe which forms no operculum because of its freedom to expand posteriorly. At its anterior end this ramus is continuous with the lateral sulcus, which contains the first part of the middle cerebral artery. At their roots the deep surfaces of the opercula become continuous with the surface of the insula and the fold at which this occurs (**circular sulcus**) surrounds the insula, except at the lateral sulcus where a narrow tongue of insular cortex (the **limen insulae**) extends ventromedially towards the anterior perforated substance.

When the opercula meet, further expansion of the surface of the telencephalon can take place only by folding, and the pattern of this folding is partly the result of local growth patterns and partly the result of continued transverse expansion. This leads to the formation of a complicated series of grooves or **sulci,** with intervening ridges or **gyri.** Two major groups of these are formed on the superolateral surface in addition to the lateral sulcus and a large number of subsidiary ones.

1. Those running parallel to the posterior ramus of the

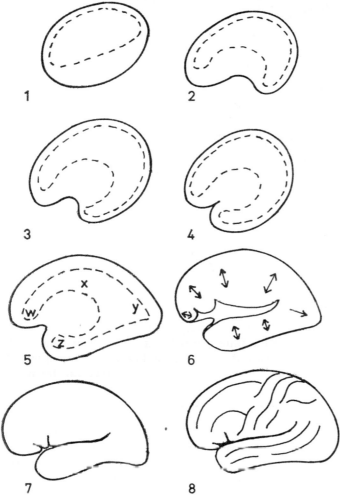

FIG. 786 Diagrams, not to scale, to show the growth of the cerebral hemisphere from the lateral view. The broken line indicates the position of the ventricle.

w, anterior horn; x, central part; y, posterior horn; z, inferior horn of lateral ventricle.

2. Between the frontal and parietal lobes a series of sulci, usually three, are formed running nearly transversely. These are the **central, precentral, and postcentral sulci.** The number of sulci formed in any region of the brain is a measure of the extent of the growth processes in that region, and these are never the same in any two brains. Thus, though the basic plan is similar in all brains [FIGS. 793–795], the details of the folding are so different that two or three major horizontal, frontal sulci may be formed, and there may be two, three, or four coronal sulci [FIG. 786, 8] lying transversely between the frontal and parietal lobes. Only if the normal pattern of three sulci is present and well defined can the **central sulcus** (between the frontal and parietal lobes) be readily determined.

The pattern and names of the sulci and gyri on the medial aspect of the hemisphere are shown in FIGURE 794. In the occipital region two deep, complicated sulci with subsidiary sulci and gyri in their depths are formed, the parieto-occipital and the calcarine, which are separated by a wedge-shaped area of cerebral cortex, the **cuneus.** The **parieto-occipital sulcus** separates the parietal and occipital lobes on the medial aspect, cutting the superomedial margin, while the **calcarine sulcus** is of such depth that it bulges into the medial wall of the posterior horn of the lateral ventricle, forming the **calcar avis** [FIG. 796]. Further anteriorly [FIG. 794 and p. 650] the **cingulate sulcus** runs parallel to the corpus callosum. In congenital absence of the corpus callosum, the medial aspect of the hemisphere is divided by a number of radially arranged sulci into a series of wedge-shaped gyri over the whole medial surface, suggesting that the gyrus and sulcus cinguli are formed as a result of the growth of the hemisphere against the corpus callosum.

Lobes of Cerebrum

Some of the sulci [FIG. 793] are used to divide the various lobes of the hemisphere from each other; thus the **frontal** and **parietal lobes** are separated by the central sulcus, the **frontal** and **temporal lobes** by the lateral sulcus, the **parietal** and **occipital lobes** by an imaginary line joining the superior part of the parieto-occipital sulcus to the **pre-occipital notch.** This line also separates the temporal and occipital lobes. The pre-occipital notch is formed where the

lateral sulcus in the temporal and frontal lobes, and a single similar sulcus in the parietal lobe, the **intraparietal sulcus** [FIGS. 786, 8; 793].

FIG. 787 Diagrammatic transverse section of developing cerebral hemisphere. The right side is further developed than the left, and the stipple indicates the corpus striatum.

FIG. 788   Diagram of the lateral view of the developing cerebral hemisphere, to show the formation of the insula and the course of the middle
cerebral artery.   The positions of the lateral ventricle (broken line) and
the corpus striatum (stipple) are shown.

inferolateral margin crosses the petrous temporal bone.   It
is perhaps unfortunate that the central sulcus also separates
two areas of the cerebral cortex, 'motor' and 'sensory', which
are significantly different in their structure and function,
since this tends to suggest that the division of the hemisphere
into lobes has a functional significance when, in fact, it is
purely for descriptive purposes.

Initially the two telencephalic vesicles lie some distance apart
and, as they expand, are held apart posteriorly by the width of
the thalamus and the brain stem.   Elsewhere they expand
medially over the thalamus towards each other, in such a manner
that the diencephalon is overgrown and the adult thalamus
appears to be inserted into the medial aspect of the telencephalon.   This expansion flattens the medial surfaces of the two
hemispheres against each other except where they are separated
by the diencephalon, midbrain, and the median structures of the
base of the skull.   It also allows the hemispheres to fuse above
and in front of the third ventricle, where the falx cerebri does
not intervene [p. 694].   This fusion permits the growth of
nerve fibres from one hemisphere to the other, thus forming
the great commissure [p. 650] of the hemispheres, the **corpus
callosum** [FIG. 789].

The convex superolateral surfaces of the hemispheres are
fitted to the growing skull, and the occipital lobes extend
backwards above the cerebellum which, bulging the tentorium
cerebelli [p. 694] upwards, separates the inferior parts of these
lobes.   Thus each **occipital lobe** is a three-sided pyramid
with a convex superolateral surface, a vertical medial surface
applied to the falx cerebri, and a sloping inferomedial or
tentorial surface [FIG. 796].

The inferior surfaces of the **frontal lobes** lie adjacent to
each other [FIG. 572] on the floor of the anterior cranial
fossa, superior to the nasal and orbital cavities.   The **temporal
lobes** fill the lateral parts of the middle cranial fossa, posterior
to the orbits.   They are widely separated from each other by
the hypophysial fossa and the cavernous sinuses [FIG. 893]
and above these by the interpeduncular fossa and the diencephalon [FIG. 698].

When the telencephalic vesicles first grow out from the

original forebrain, they carry with them an extension of the
thin roof of the third ventricle into their medial walls.   As
each vesicle expands backwards and downwards from the site
of the original outgrowth, the **interventricular foramen,**
this thin medial wall extends around the corpus striatum
into the medial wall of the inferior horn of the ventricle, one
edge of it attached to the convexity of the corpus striatum,
and the other to the medial margin of the hemisphere.   When
the hemisphere expands medially over the diencephalon, the
thin medial wall is carried over the thalamus and secondarily
becomes attached to it so that the lateral part of the thalamus
comes to lie in the floor of the central part of the lateral
ventricle, and in the roof of its inferior horn [p. 653], and the
thin medial wall of the ventricle lies against it [FIG. 789].
Here, as in the roofs of the third and fourth ventricles, the thin
medial wall is invaginated by the vascular connective tissue
lying outside, to form the **choroid plexus of the lateral
ventricle.**   It is continuous with the choroid plexus of the
third ventricle at the interventricular foramen [FIGS. 818 and
825], and passes from this point to the tip of the inferior horn
of the lateral ventricle, forming a C-shaped curve following the line of the ventricle.   It is invaginated between the
medial edge of the hemisphere and the thalamus through a
fissure known as the **choroid fissure** [FIG. 807].

Thus each cerebral hemisphere, its contained ventricle and
choroid plexus form a C-shaped structure overlapping the
thalamus medially and meeting with its fellow dorsally, but
diverging posteriorly around the lateral aspect of the brain
stem.   Laterally, by the formation of the insula, the corpus
striatum is deeply hidden, except in the lateral sulcus where
it still approaches the surface between the open limbs of the
C-shaped hemisphere at the anterior perforated substance.

The White Matter of the Cerebrum

At an early stage in the development of the hemisphere, cells
derived from the ependymal lining of the vesicle migrate
towards the surface and form the outer rind or **cortex** which
eventually covers the entire surface.   This process does not
occur uniformly, but appears first in the margin of the

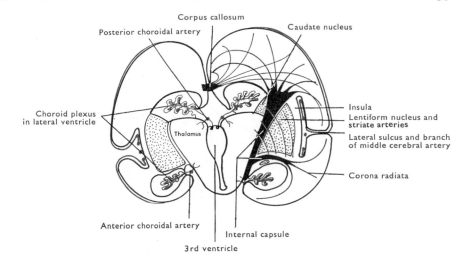

FIG. 789 Diagrammatic transverse section through the developing cerebral hemisphere at a later stage than that shown in Fig. 787. The right side is shown at a later stage of development than the left, and indicates the effects on the structure of the hemisphere of the development of the corpus callosum and internal capsule. Stipple, corpus striatum.

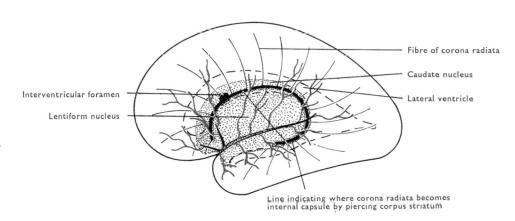

FIG. 790 Diagram of the lateral view of the developing cerebral hemisphere at a later stage than Fig. 788. The insula is entirely covered by the growth of the surrounding cerebral cortex. Internal structures shown as in Fig. 788.

FIG. 791 Right hemispheres of human foetuses showing three stages in development of insula and insular opercula.

A, right cerebral hemisphere from a foetus in latter part of fourth month; B, right hemisphere in fifth month; C, right hemisphere in latter part of eighth month.

In C the temporal operculum has been removed and a large part of the insula is thus exposed. The outline of the temporal operculum is indicated by a dotted line.

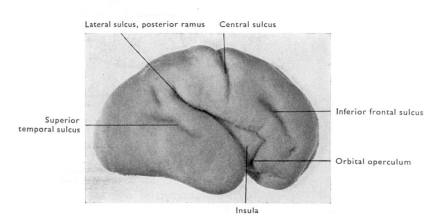

FIG. 792 Right cerebral hemisphere from early seventh-month foetus.

FIG. 793 Diagram of sulci and gyri on superolateral surface of hemisphere. The middle frontal sulcus, sometimes found between the superior and inferior frontal sulci, is not shown.

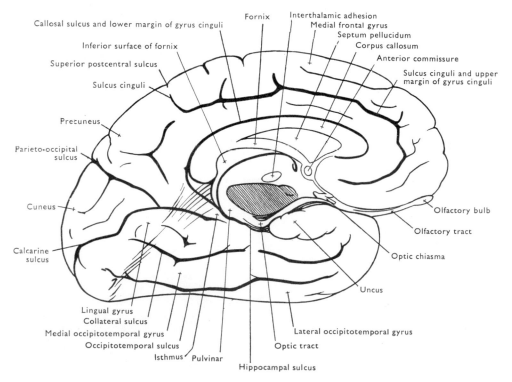

FIG. 794 Diagram of sulci and gyri on medial and tentorial surfaces of hemisphere.

Olfactory sulcus   Longitudinal fissure   Olfactory bulb

Olfactory tract

Optic nerve

Optic chiasma

Orbital sulcus

Lateral sulcus

Anterior perforated substance

Tuber cinereum

Mamillary body

Optic tract

Parahippocampal gyrus

Crus cerebri

Hippocampal sulcus

Aqueduct

Collateral sulcus

Superior colliculus

Pineal body

Occipitotemporal sulcus

Splenium of corpus callosum

Lateral occipitotemporal gyrus

Medial occipitotemporal gyrus

Calcarine sulcus

Lingual gyrus

Calcarine sulcus   Occipital pole

Fig. 795   Diagram of sulci and gyri on inferior surface of the hemispheres.

Splenium of corpus callosum

Bulb of the posterior horn

Bulb of the posterior horn

Calcar avis   Calcar avis

Tapetum of corpus callosum

Optic radiation

Inferior longitudinal fasciculus

Fig. 796   Coronal section through posterior horns of lateral ventricles, viewed from the front.

hemisphere (limbic cortex) bounding the choroid fissure, somewhat later in the olfactory cortex of the temporal lobe (piriform area) and finally in the remainder of the cerebral vesicle; the polar region of the frontal and occipital lobes being the last parts to develop a cortex.

With the appearance of the cerebral cortex and its gradual differentiation into a series of layers, the walls of the cerebral vesicle increase rapidly in thickness so that the ventricle and choroid plexus which originally filled the greater part of the vesicle become relatively reduced in size. This thickening is hastened by the outgrowth from the cortical cells of processes, some of which remain within the cortex while others pass beneath it to form a layer of nerve fibres which separates the cortex from the ependyma of the ventricle and from the corpus striatum in those parts which overlie it. The processes of the cortical cells are very numerous and can be divided into three groups on the basis of their destination.

1. Processes which pass from one part of a cerebral vesicle to another, the **association fibres** of the adult brain. These may be very short, looping between adjacent areas of the cerebral cortex; or of considerable length, even stretching throughout the length of the hemisphere [p. 643].

2. Processes which pass from one cerebral vesicle to the corresponding parts of the other. These are **commissural fibres** which cross the midline. The majority of these fibres run in the **corpus callosum** [p. 650], though a smaller number pass by other routes, especially the **anterior commissure** lying in the dorsal part of the lamina terminalis which stretches between the two hemispheres.

3. Processes which leave the cerebral cortex and pass to other parts of the central nervous system, the **projection fibres.**

Projection Fibres. These fibres arise from all areas of the cerebral cortex and, passing deeply, take part in the formation of the white matter of the cerebral hemisphere. They converge radially on the corpus striatum forming the **corona radiata,** which also contains fibres passing in the opposite direction, i.e., from the subjacent parts of the nervous system to the cerebral cortex. The fibres of the corona radiata reach the corpus striatum as a compact sheet which pierces it on

a C-shaped curve [FIGS. 790 and 798], passing inferomedially through it to reach the lateral aspect of the thalamus, on which it lies medial to the inferior part of the corpus striatum. It thus splits the corpus striatum into an outer rim, the comma-shaped **caudate nucleus,** and a central part, the lens-shaped or **lentiform nucleus.** The mass of fibres therefore covers the convex internal surface of the lentiform nucleus forming its **internal capsule,** which separates the anterior part and periphery of the lentiform nucleus from the caudate nucleus and the remainder of it from the thalamus. In horizontal sections, taken through the level of the interventricular foramen, the internal capsule presents two limbs sharply angled (the **genu**) on each other around the apex of the lentiform nucleus which lies opposite the interventricular foramen [FIG. 799]. The anterior limb lies between the caudate and lentiform nuclei and the posterior limb between the thalamus and the lentiform nucleus. The genu is only clearly visible in this plane of section and at this level, being absent both superiorly and inferiorly [PLATES 40 and 41].

Inferiorly, the fan-shaped internal capsule, stripped of the fibres which join it to the thalamus and corpus striatum of the telencephalon, is concentrated into a thick bundle, the crus cerebri. This emerges from the inferior surface of the forebrain between the thalamus medially and the medial edge of the hemisphere, thus separating this part of the thalamus from the choroid fissure at the inferior horn of the lateral ventricle. Since the thalamus and the midbrain are directly continuous, the internal capsule emerges from the forebrain on to the anterolateral surface of the midbrain forming the **crus cerebri** which pierces the ventral part of the pons and is split into a number of **longitudinal bundles** within its substance. The majority of these fibres end in the pons, the remainder emerging from its caudal margin to run on the ventral surface of the medulla oblongata, abutting on the corresponding fibres of the opposite side. Followed caudally, these two bundles, one on each side of the midline, taper gradually in the lower medulla oblongata as fascicles from each interdigitate and cross the midline, obliterating the anterior median fissure between them, and passing posterolaterally to enter the lateral funiculus of the spinal medulla. Here they form the **lateral**

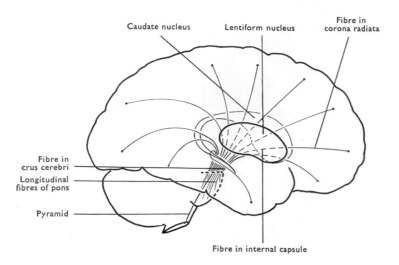

FIG. 779 Diagram to show arrangement of fibres in corona radiata, internal capsule, crus cerebri, pons and pyramid.

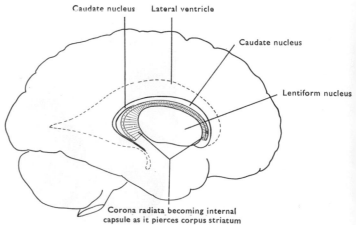

FIG. 798 Diagram to show the position of the corona radiata piercing the corpus striatum as the internal capsule.

corticospinal tract, part of which reaches the caudal extremity of the spinal medulla. A variable proportion of fibres does not cross the midline in this region but continues down to the thoracic region, as the **anterior corticospinal tract,** in the anterior funiculus of the spinal medulla. Because of their shape, the bundles on the anterior aspect of the medulla oblongata are known as the **pyramids** and their decussation as the **pyramidal decussation** [FIG. 756].

Thus the corona radiata, the internal capsule, the crus cerebri, the longitudinal fibres in the ventral part of the pons, the pyramid, and the corticospinal tracts represent the parts of a single efferent pathway from the cerebral cortex. The fibres forming this system converge radially through the corona radiata into the internal capsule and are concentrated into the crus cerebri. Thus the relative positions of fibres arising in the different parts of the cerebral cortex within this system remain unchanged [FIG. 797]. The fibres from the frontal pole run almost horizontally backwards and have a long course through the anterior part of the internal capsule before reaching the anteromedial part of the crus cerebri. The fibres from the occipital and temporal lobes, which hook round the posterior and inferior parts of the caudate nucleus, have only a transitory contact with the lentiform nucleus and thalamus in the **retrolentiform** and **sublentiform parts of the internal capsule,** almost immediately entering the posterolateral part of the crus cerebri. The intermediate fibres occupy intermediate positions, and those arising from the cerebral cortex in the region of the central sulcus occupy a central position [FIG. 800].

It should be appreciated that the corona radiata is a three-dimensional fountain of fibres converging on the internal capsule from all parts of the cortex. FIGURES 789, 797 and 800 show the general arrangement and indicate that fibres arising from the precentral gyrus or passing to the postcentral gyrus do not form a compact group in the internal capsule but are spread out in proportion to the obliquity of the central sulcus. The fibres which arise from, or end in the inferior extremity of each gyrus tend to lie anterior to those from the superior end [FIG. 800]. Thalamocortical fibres, which form a considerable portion of the internal capsule, enter it on its medial side from the thalamus and are at first medial to those descending to regions caudal to the thalamus, though superiorly they become mixed.

Thalamocortical Connexions. The fibres which enter the internal capsule from the thalamus on their way to the cerebral

FIG. 799 Horizontal section through the right cerebral hemisphere at the level of the interventricular foramen.

cortex, and similar fibres passing in the opposite direction (**corticothalamic fibres**) have the same radial arrangement and follow the other fibres of the internal capsule in their distribution. Thus [FIG. 801], in a horizontal section through the thalamus, each part sends fibres to the adjacent part of the cerebral cortex, so that beginning in the medial nucleus a line drawn around the thalamus (clockwise on the right, anticlockwise on the left) to the posteromedial part or medial geniculate body, passes successively through groups of cells

FIG. 800 Diagram to show the course of fibres arising in the precentral gyrus and passing through corona radiata, internal capsule, crus cerebri, pons and pyramid.

which project to the frontal lobe, medial, anterior- and intermediate-ventral nuclei; to the parietal lobe, intermediate- and posterior-ventral nuclei; to the occipital lobe, pulvinar and lateral geniculate body; and to the temporal lobe, the medial geniculate body. In this fashion the most posterior part of the thalamus projects to the part of the cerebral cortex which was originally the posterior end of the cerebral vesicle. Similarly in the coronal plane [FIG. 803] the thalamus projects to the cortex in an orderly fashion; its ventromedial cells to the cortex of the upper wall of the posterior ramus of the lateral sulcus, its dorsomedial cells to the cortex on the medial surface of the hemisphere, and its intermediate cells to the intermediate regions. Thus the anterior nuclei project to the cingulate gyrus while the medial part of the posterior-ventral nucleus (arcuate nucleus [FIG. 778]) projects to the inferior part of the postcentral gyrus. In addition to these specific thalamocortical connexions, which are exactly mirrored by corresponding corticothalamic connexions, the intralaminar and central nuclei are believed to project to the caudate and lentiform nuclei. The reticular nucleus may project to all areas of the cortex and has been said to be responsible for non-specific diffuse thalamocortical projections. Such a diffuse thalamic projection has not been demonstrated by anatomical methods (Chow, 1952), but see also page 679.

The Caudate and Lentiform Nuclei (Corpus Striatum), the Amygdaloid Body, and the Claustrum

[PLATES 40–44]

As a result of the growth of the internal capsule through the corpus striatum the main part of it is divided into the caudate and the lentiform nuclei.

The **caudate nucleus** lies between the internal capsule and the lateral ventricle and follows the concavity of the ventricle from the anterior horn to the tip of the inferior horn. Anteriorly the head of the caudate nucleus is large and extends almost to the inferior surface of the brain, being separated from it only by a thin sheet of nerve fibres from the rostrum of the corpus callosum [p. 651] and the anterior perforated substance. Here it forms the lateral wall of the anterior horn of the ventricle and is directly continuous on its lateral side with the anterior part of the putamen of the lentiform nucleus,

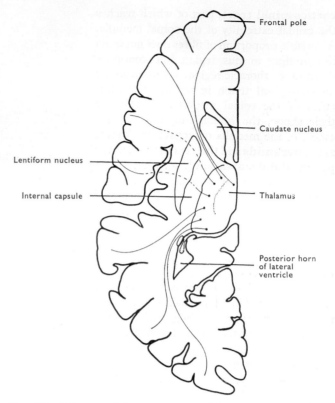

FIG. 801 Diagram of the arrangement of the thalamocortical radiations to the upper parts of the cerebral hemisphere. The broken lines indicate fibres arching superiorly out of the section over the lentiform nucleus.

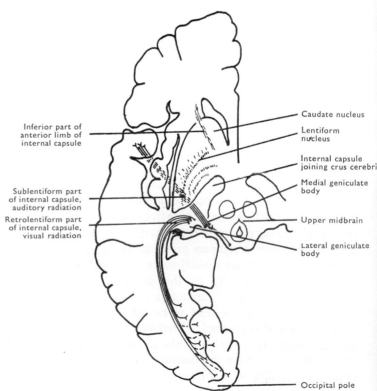

FIG. 802 Diagram of the thalamocortical radiations from the geniculate bodies. The broken lines indicate the fibres passing below the lentiform nucleus to the temporal lobe (auditory radiation).

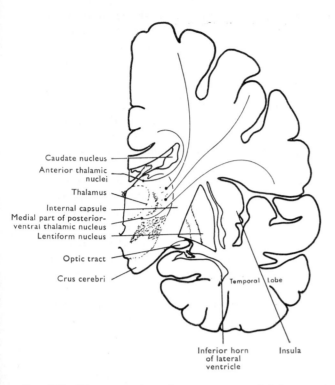

FIG. 803   Diagram to show the arrangement of thalamocortical radiations in a coronal section of the human brain.

FIG. 804   Coronal section through the cerebral hemispheres. The section (viewed from the front) cuts the putamen of the lentiform nucleus in front of the globus pallidus.

inferior to the anterior limb of the internal capsule which separates them superiorly [FIG. 804]. In this situation their fused masses are pierced horizontally from medial to lateral by the anterior commissure [p. 649] and vertically by the striate branches of the middle cerebral artery (reinforced medially by the recurrent branch of the anterior cerebral artery) and the corresponding veins running in the opposite direction. Many of these arteries course posterosuperiorly to supply the more posterior parts of both nuclei and the internal capsule between them, passing through the latter to reach and supply the anterolateral part of the thalamus [FIG. 789]. From its head the caudate nucleus rapidly tapers to a narrow body which forms the lateral wall of the central part of the lateral ventricle, and, becoming the tail, sweeps inferiorly, turning forwards into the roof of the inferior horn of the lateral ventricle as a very slender strip of grey matter. At the tip of the inferior horn it becomes continuous with the **amygdaloid body** which overlies this part of the ventricle. Throughout its length the caudate nucleus lies immediately external to the ventricular ependyma from which it is separated by a number of veins which issue from the hemisphere, cross the ventricular surface of the caudate nucleus, and gather together into a single **thalamostriate vein** on its medial aspect [FIG. 818]. This vein, accompanied by a bundle of nerve fibres from the amygdaloid body, the **stria terminalis,** courses towards the interventricular foramen where, turning posteriorly, the vein escapes from beneath the ventricular ependyma to form the anterior extremity of an **internal cerebral vein.** The thalamostriate vein and the stria terminalis separate the body of the caudate nucleus from the superior and posterior surfaces of the thalamus, but in the roof of the inferior horn of the

C.T.A.—Y

lateral ventricle the tail of the caudate with the vein and stria are separated from the thalamus and the lentiform nucleus by the sublentiform part of the internal capsule and the fibres of the anterior commissure spreading posteriorly into the temporal lobe.

The **amygdaloid body** [FIG. 806, PLATE 42, 2 and 3], an ovoid mass of nerve cells, is anterosuperior to the tip of the inferior horn of the lateral ventricle. It forms a solid mass of grey matter, continuous posteriorly with the tail of the caudate nucleus, superiorly with the lentiform nucleus, and medially with the cortex of the temporal lobe (piriform) which overlies it. It receives olfactory fibres and gives rise to the fibres of the stria terminalis. These pass with the caudate nucleus and thalamostriate vein, to end in the septal region [FIG. 816] and the anterior perforated substance. The amygdaloid body has several subdivisions.

The **lentiform nucleus,** approximately triangular in shape, underlies the cortex of the insula and is separated from it by two layers of nerve fibres; its external capsule, and the white matter of the insula, with a thin sheet of grey matter, the **claustrum,** lying between. The **lateral surface** of the lentiform nucleus [FIG. 805], smooth and rounded, is easily separated from the external capsule except where it is pierced by striate vessels, thus suggesting that few, if any, fibres enter or leave this surface. The **superomedial surface,** angled to fit the genu of the internal capsule, is firmly adherent to, and grooved by, the bundles of fibres of the internal capsule except antero-inferiorly where it fuses with the head of the caudate nucleus. Where it is separated from the caudate nucleus by the internal capsule, strands of grey matter pass between the bundles of the internal capsule joining the lateral part of the lentiform nucleus (putamen) to the caudate nucleus and giving them a striated appearance from which the name **corpus striatum** is derived. The inferior surface is horizontal and is separated from the inferior horn of the lateral ventricle by white matter composed of fibres passing to and from the temporal lobe of the hemisphere. This contains the sublentiform part of the internal capsule, fibres of the anterior commissure, and a part of the optic radiation [FIGS. 846, 850], passing forwards over the inferior horn of the ventricle before sweeping backwards to the occipital lobe.

On section the lentiform nucleus, unlike the caudate, is not of uniform structure but is composed of three (or occasionally four) parts separated by two (or three) layers of white matter which extend from the internal capsule to its inferior surface [FIG. 806]. The lateral of these **medullary laminae** separates the larger, external, darker **putamen** from the paler, wedge-shaped, medial **globus pallidus,** while the medial medullary lamina divides the globus pallidus into internal and external segments. The globus pallidus is almost entirely in contact with the posterior limb of the internal capsule. The putamen, which has colour and structure similar to the caudate nucleus, is the portion of the lentiform which fuses and

communicates with the caudate. If the lentiform nucleus is stripped from the internal capsule, branches of the striate vessels can be seen passing from it through the internal capsule, and the medullary laminae are seen to be continuous with the internal capsule. Even when it is entirely separated from the internal capsule the lentiform nucleus is still attached to the brain by a bundle of fibres which issues from its inferomedial angle (globus pallidus) and curves round the antero-inferior border of the internal capsule to enter the hypothalamic region, the **ansa lenticularis,** the main bundle of fibres leaving the lentiform nucleus [FIG. 861].

The **external capsule** lies against the lateral surface of the lentiform nucleus. It is composed of fibres passing over the nucleus to reach the claustrum; also association fibres which join the frontal and parietal lobes with the temporal lobe. It is a thin sheet except antero-inferiorly where it becomes continuous with the deep surface of the fasciculus uncinatus [p. 644] and with fibres of the corpus callosum over the anterior surface of the caudate nucleus. Superiorly it fuses with the corona radiata at the upper margin of the lentiform nucleus and inferiorly it is continuous with the white matter of the temporal lobe.

The **claustrum** is a thin sheet of grey matter of unknown

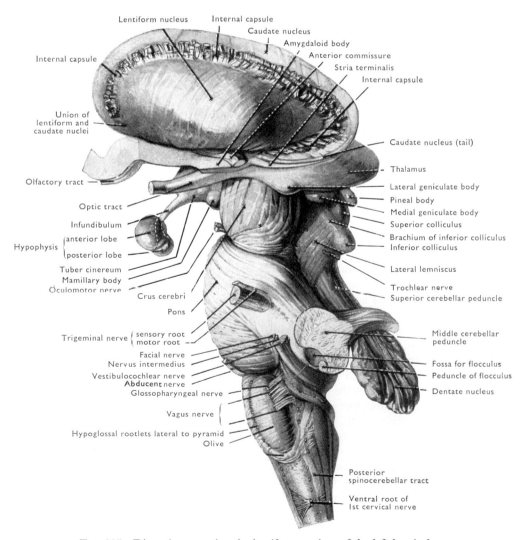

FIG. 805 Dissection exposing the lentiform nucleus of the left hemisphere.

function which lies between the external capsule and the white matter of the insula. It is smooth medially, scalloped on its lateral surface, and fades away at the superior margins of the insula, though inferiorly it is thickened and extends between the external capsule and those fibres of the fasciculus uncinatus [FIG. 809] which sweep posteriorly with the anterior commissure into the temporal and occipital lobes.

From the above description and the figures, it can be seen that the general structure of the telencephalon arises directly from its mode of development, and that in its final form it consists of a centrally placed lentiform nucleus almost completely surrounded by a series of concentric structures; the internal capsule, the caudate nucleus, the lateral ventricle and its choroid plexus, and the cerebral cortex.

The Association Fibres of the Cerebral Hemisphere

These fibres arise in the cells of the cerebral cortex and pass from one part of the hemisphere to another. Some are very short, looping from one gyrus to the next (**arcuate fibres**) while others are longer and may stretch from end to end of the hemisphere. The longer association fibres are for the most part gathered together into certain bundles related to the corona radiata as follows:

1. On the sides of the corona radiata, covered by its fibres which arch over to the medial and lateral sides of the hemisphere—cingulum and superior longitudinal (fasciculus) bundle [FIG. 808].

2. Along its inferior margin—inferior longitudinal (fasciculus) bundle.

3. Beneath its anterior part—uncinate fasciculus.

The first group consists of two bundles, the cingulum on the medial side and the superior longitudinal bundle on the lateral side. Each of these is C-shaped and curves from the anterior part of the frontal lobe through the parietal lobe to turn downwards and forwards into the temporal lobe, thus following the general configuration of the hemisphere. Each receives fibres from and gives off fibres to the cortex from its convex surface, while the concave surface is relatively free of such connexions and is capable of being dissected from its immediate surroundings.

The **cingulum** [FIGS. 807 and 822] starts in the medial part of the frontal lobe below the rostrum of the corpus callosum [p. 651] and sweeps around the superior surface of the length of the corpus callosum buried in the gyrus cinguli [FIG. 808]. Hooking round the splenium of the corpus callosum, where a large number of fibres leave its convex surface, it runs antero-inferiorly in the medial margin of the temporal lobe (parahippocampal gyrus) to end, inferior to the uncus, by spreading into the tip of the temporal lobe. In this latter part of its course it has a rope-like appearance, the bundles of fibres running spirally.

The **superior longitudinal (fasciculus) bundle** [FIGS. 808 and 809] lying lateral to the corona radiata has a shape and distribution very similar to the cingulum but it encircles the insula on the lateral aspect of the hemisphere. Beginning in the frontal lobe it runs posteriorly in the root of the fronto-

Choroid plexus
Lateral ventricle
Claustrum
Insula
Putamen
Globus pallidus
Amygdaloid body
Mamillothalamic tract
Anterior perforated substance
Corpus callosum
Fornix
Tela choroidea
Caudate nucleus
Thalamostriate vein
Thalamus
3rd ventricle
Choroid plexus
Internal capsule
Interventricular foramen
Column of fornix
Optic tract
Anterior commissure
Optic nerve
Olfactory tract
Optic chiasma
Infundibulum
Column of fornix

FIG. 806 Coronal section through the cerebrum (viewed from behind) to show the parts of the lentiform nucleus.

parietal operculum, and turning antero-inferiorly, fans out posteriorly into the occipital lobe and anteriorly into the temporal lobe. In the parietal part of the operculum it forms a compact bundle which is easily exposed by dissection, but in the frontal lobe it is traversed by many other fibres of the cerebral cortex, including fibres to and from the internal capsule and corpus callosum.

The **inferior longitudinal (fasciculus) bundle** runs between the temporal and occipital lobes along the inferior margin of the corona radiata. Its fibres blend with the corona radiata and extend anteriorly into the temporal lobe medial to the inferior part of the superior longitudinal bundle and inferior to the temporal part of the corona radiata.

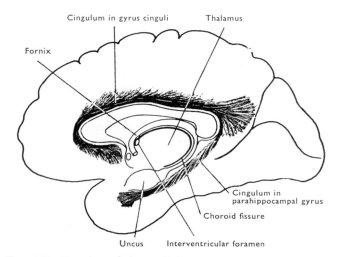

Cingulum in gyrus cinguli
Thalamus
Fornix
Cingulum in parahippocampal gyrus
Choroid fissure
Uncus
Interventricular foramen

FIG. 807 Drawing of the medial aspect of the hemisphere to show the position of the cingulum (schematic) and the choroid fissure.

The **uncinate fasciculus** hooks around the lateral sulcus from the frontal to the temporal and occipital lobes. It mingles in the frontal lobe with the anterior part of the superior longitudinal bundle and curves posterolaterally on the inferior surfaces of the anterior horn of the lateral ventricle and the conjoined parts of the caudate and lentiform nuclei. Posteriorly its fibres converge to form a narrow, thick bundle between the lateral sulcus and the anterior commissure where

FIG. 808 Diagram of a coronal section to show the position of the main association bundles, the corona radiata, and the corpus callosum.

FIG. 809 Dissection of left cerebral hemisphere to display the optic and auditory radiations and the general direction of the principal association bundles.

that commissure escapes from the lateral surface of the lentiform nucleus. The anterior fibres hook forwards beneath the lateral sulcus and pass into the temporal pole. The posterior fibres, running towards the occipital lobe, cover the antero-inferior part of the claustrum, and fusing with the expanded extremity of the anterior commissure, spread laterally into the temporal and occipital lobes [FIG. 809].

Each of these bundles consists of a mixture of fibres of longer and shorter trajectory, which enter and leave them throughout their length. The majority of the fibres in the cingulum and the superior longitudinal bundle seem, however, to run between the frontal and the occipital and posterior parietal lobes.

The Medial Margin of the Cerebral Cortex and the Commissures

The expansion of the telencephalic vesicle medially [p. 634] over the thalamus, results in that structure being inserted into the medial aspect of the telencephalon, in such a manner that a free margin of the telencephalon is formed which surrounds the thalamus and is attached to it only by the choroid plexus of the lateral ventricle invaginated between them. This circular rim of the cortex, which constitutes the 'rhinencephalon', is further complicated by the growth of the temporal lobe and the commissures of the telencephalon, and is continuous with the olfactory tract antero-inferiorly at the anterior perforated substance. This marginal cortex of the cerebral hemisphere consists of the olfactory bulb, tract and striae, the piriform area, the uncus, the hippocampus, the dentate

gyrus, the gyrus fasciolaris, the indusium griseum, the septum pellucidum, the subcallosal area, and the parateminal gyrus [FIG. 810].

The **olfactory bulb** and tract are formed as an outgrowth from the ventral aspect of the developing frontal part of the cerebral vesicle and at first contain an extension of the lateral ventricle which is later obliterated. The olfactory bulb receives bundles of non-myelinated olfactory nerve fibres which grow through the cribriform lamina of the ethmoid bone from the sensory cells buried in the olfactory mucous membrane into the grey matter covering the bulb. These fibres make contact with the numerous dendrites of the large mitral cells of the bulb, forming little knots of interlacing processes known as glomeruli, which lie close to the inferior surface of the bulb [FIG. 811]. The mitral cells give rise to the fibres in the olfactory tract which runs posteriorly on the inferior surface of the frontal lobe, wedged into the olfactory sulcus, and splits into medial and lateral **striae** [FIG. 817] which enclose between them the olfactory trigone [FIGS. 776, 812]. The **olfactory trigone** which is not usually visible in Man as a separate entity but is incorporated in the anterior perforated substance, is large in mammals with a well-developed sense of smell. Many of the olfactory tract fibres end in the olfactory trigone, the remainder skirt the anterolateral aspect of the anterior perforated substance in the lateral olfactory stria.

In most mammals the lateral olfactory stria is continuous posteriorly with a pear-shaped **piriform area** extending far posteriorly on the inferior surface of the hemisphere and

FIG. 810 Medial aspect of right cerebral hemisphere, with the 'rhinencephalon' coloured.

FIG. 811 Diagram of minute structure of olfactory bulb.

limited laterally by the curved rhinal sulcus [FIG. 812]. In the poorly developed olfactory system of Man, the anterior part of the piriform area is reduced to a slight ridge, the lateral olfactory stria, which is markedly bent by the forward growth of the temporal lobe. As a result it curves laterally towards the antero-inferior part of the insula, on the **limen insulae**, and then inferomedially behind the lateral sulcus to become continuous with the uncus [FIG. 810] and the anterior extremity of the parahippocampal gyrus, separated from the remainder of the temporal lobe by the **rhinal sulcus** [FIG. 810]. Fibres from the olfactory bulb extend into the anterior piriform area, but do not reach the posterior part.

The **uncus** is essentially a recurved portion of the anterior extremity of the parahippocampal gyrus and is continuous posteriorly with the inferior part of the marginal cortex of the hemisphere (the hippocampus and dentate gyrus) which curves around the inferior surface of the thalamus but is separated from it by the sublentiform part of the internal capsule and

the inferior horn of the lateral ventricle.

FIGURE 813 shows the mode of development of the hippocampus and dentate gyrus on the margin of the hemisphere and the formation of the fimbria.

In the adult, the **dentate gyrus** begins as a narrow ribbon of grey matter on the posteromedial surface of the uncus [FIG. 810]. It runs ventrally around the uncus (tail of the dentate gyrus), and then passes posteriorly, under cover of the fimbria in the hippocampal sulcus, as a crinkled strip of cortex lying above the superior margin of the parahippocampal gyrus (subiculum). Turning superiorly it reaches the splenium of the corpus callosum around which it curves as the **gyrus fasciolaris** to become continuous with a thin sheet of grey matter on the superior surface of the corpus callosum, the **indusium griseum.**

The **hippocampus** is a portion of the marginal cortex invaginated as a rounded ridge into the medial wall of the inferior horn of the lateral ventricle by the hippocampal sulcus. The hippocampus begins in the floor of the inferior horn of the lateral ventricle near its tip, where it forms the **pes hippocampi** [FIG. 826]. This extends laterally from the uncus as a ridged structure owing to the folded nature of the hippocampal cortex in this region. Turning posteriorly it runs parallel to the dentate gyrus to reach the splenium of the corpus callosum where it also becomes continuous with the indusium griseum and probably also with the septum pellucidum.

The hippocampus and dentate gyrus [FIG. 814] differ from

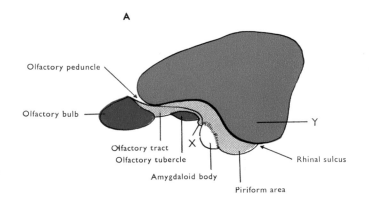

FIG. 812 Relation of olfactory part of hemisphere to neopallium.

A, lateral aspect of left cerebral hemisphere; B, inferior aspect of right half of rabbit's brain; C, corresponding view of human foetal brain at fifth month.

In Man, the great growth of the neopallium (phylogenetically the most recent part of the cerebral cortex which receives afferent fibres from the thalamus rather than from the olfactory system) produces a well-defined temporal lobe (Y) and not only flexes the more ancient piriform area at X but also relegates it to the medial surface of the hemisphere.

Olfactory areas, red; neopallium, blue.

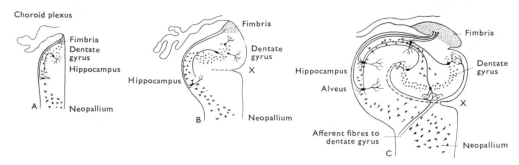

FIG. 813   Three stages in the development of the right hippocampal formation.
X   Hippocampal sulcus.

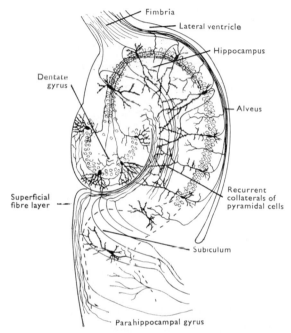

FIG. 814   Drawing of Golgi preparation of the left hippocampus, dentate gyrus, and subiculum.   Redrawn from Cajal (1955).

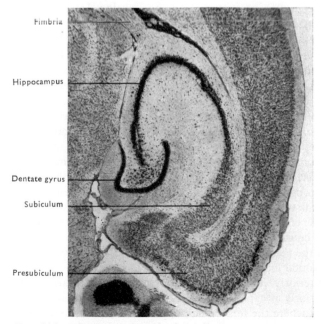

FIG. 815   Horizontal section of rat brain to show transition from hippocampus to neopallium. Cell bodies only are stained.

the remainder of the cerebral cortex in having a simpler structure, consisting essentially of three layers. The superficial layer contains few cells but a considerable number of tangentially running fibres which give this cortex its characteristic superficial white layer. In the dentate gyrus the middle layer consists of small granule cells which send their axons to the large pyramidal cells in the middle layer of the hippocampus. The deepest layer of cells in both parts of this cortex is formed of scattered cells of irregular shape.

The Fornix. The efferent fibres of the pyramidal cells of the hippocampus pass to its ventricular surface forming a thin sheet of white matter, the **alveus,** and these fibres pass medially to gather together in a bundle on the dorsomedial aspect of the hippocampus immediately inferior to the choroid fissure. This is the **fimbria** which sweeps posteriorly between the hippocampus and the choroid fissure, and turning superiorly as the **crus of the fornix** reaches the splenium of the corpus callosum beneath which it passes [FIGS. 807, 810, and 826]. Thence it follows the choroid fissure over the dorsal surface of the thalamus fused medially with the opposite fornix, **body of the fornix,** as far as the interventricular foramen. Throughout this part of their course the fimbria and fornix form the rim of the marginal cortex and have the ependyma of the choroid plexus of the lateral ventricle attached to them, the **taenia of the fornix.** At the interventricular foramen the fornix turns inferiorly **(column of the fornix)** between that foramen and the anterior commissure to curve posteroinferiorly from the superior part of the lamina terminalis into the hypothalamus. Here it ends in the mamillary body, separated from the opposite column of the fornix by the third ventricle. The fornix thus forms one complete turn of a spiral, starting in the medial part of the temporal lobe and ending close to the midline in the hypothalamus, having encircled the thalamus [FIG. 817].

Posteriorly the crura of the two fornices approach each other beneath the corpus callosum and exchange fibres forming the **commissure of the fornix.** At the interventricular foramen a number of fibres leave the fornix to pass ventrally in front of the anterior commissure into the septal region [FIG. 816]. These extend to the anterior perforated substance from which other fibres pass upwards to enter the fornix. Other fibres which leave it in this region pass posteriorly to the anterior thalamic nuclei and within the anterior hypothalamus the column of the fornix gives off many fibres.

From the mamillary body a bundle almost equal in size to the fornix, the **mamillothalamic tract,** passes laterally, its fibres giving rise to branches, the **mamillotegmental tract,** which run posteriorly into the midbrain. The mamillothalamic tract then turns superiorly in the medial medullary lamina of the thalamus to reach the parts of the anterior nuclei of the thalamus. The anterior thalamic nuclei in their turn have connexions with the cingulate gyrus through the internal capsule. Thus the hippocampus has connexions with other

3rd ventricle   Cingulum

Cingulate gyrus

corpus callosum

Indusium griseum

Anterior nuclei
of thalamus

Mamillothalamic tract
arising from
mamillary body

Column of fornix

Septal region

Anterior commissure

Diagonal band

Ansa lenticularis

Piriform cortex

Lateral ventricle

Stria terminalis

Crus of fornix

Dentate gyrus

Parahippocampal gyrus

Hippocampus

Optic nerve   Amygdaloid body   Tegmentum of midbrain

FIG. 816 Diagram to show the marginal structures of the hemisphere based on a dissection. The left thalamus and hypothalamus (except the mamillary body) have been removed exposing the third ventricle, interthalamic adhesion, and the upper part of the midbrain. The left half of the midbrain has been transected. Arrows in fornix indicate the passage of fibres in both directions within it.

parts of the marginal cortex, the hypothalamus, thalamus, midbrain, and cerebral cortex.

The hippocampus has long been considered an important part of the olfactory apparatus and, while this may be true in lower vertebrates, it seems certain that in Man the secondary olfactory fibres from the olfactory bulb do not reach the hippocampus, but end in the olfactory trigone and the anterior part of the piriform area. The hippocampus is also disproportionately large in Man in comparison with the size of the olfactory system and is well developed in anosmic aquatic mammals such as the whale. It seems certain, therefore, that it is concerned with other functions, and its massive hypothalamic and septal connexions should not be forgotten. Physiological and clinical investigations seem to suggest that it is an important part of the neural mechanism concerned with emotion and recent memory (Scoville and Milner, 1957). Certainly it has many connexions with the remainder of the temporal lobe which has also been implicated in the latter function.

In marsupials, where there is no corpus callosum, the hippocampus and dentate gyrus encircle the thalamus and can be traced forwards above it and the commissure of the fornix to become almost continuous with the medial olfactory stria anterior to the lamina terminalis. In eutherian mammals, including Man, the corpus callosum [FIG. 817] grows through the dorsal part of the medial margin of the hemisphere, obliterating the hippocampus and dentate gyrus in this region, but leaving a portion of the marginal cortex together

with the fornix inferior to the corpus callosum. This part forms the septum pellucidum, joining the corpus callosum and fornix. The part of the marginal cortex remaining superior to the corpus callosum persists as a thin layer of grey matter, the indusium griseum, together with two longitudinal bundles of nerve fibres, the **striae longitudinales** [FIG. 822].

The **septum pellucidum** in the fully developed brain appears to be a single membrane which connects each fornix to the corpus callosum and separates the anterior parts of the two lateral ventricles. It is, however, formed from right and left parts, one from each hemisphere, and though these are usually adherent over a large part of their area, a cavity (**cavum septi pellucidi**) is frequently present anteriorly; this represents the part of the longitudinal fissure lying inferior to the corpus callosum [FIG. 818]. The septum pellucidum extends anteriorly beyond the fornix into the concavity of the genu of the corpus callosum and therefore joins the anterior part of its trunk to the rostrum. It is covered on its ventricular surface with ependyma and is separated from this by a number of veins coursing posteriorly to join the internal cerebral veins from the genu of the corpus callosum and the white matter of the frontal lobe.

Anteriorly the septum pellucidum is extensive, while posteriorly it tapers rapidly towards the commissure of the fornix where the fornices are applied to the corpus callosum. The size of the septum is directly related to the size of the lateral ventricles and is usually enlarged in compensatory

FIG. 817 Drawings of the medial surface of the hemisphere of a kangaroo (left) and of Man (right).

hydrocephalus, where the ventricle expands with the loss of neural tissue in the aged.

The **indusium griseum** probably represents a remnant of the hippocampus [FIG. 810] and in the depths of the callosal sulcus lateral to the indusium griseum there is a small strip of cortex covered by a layer of white matter. This is continuous round the splenium of the corpus callosum with the medial part of the parahippocampal gyrus, the subiculum [FIG. 815], which also has a superficial fibrous layer. At the rostrum of the corpus callosum the indusium griseum and septum pellucidum become continuous with the subcallosal area and paraterminal gyrus which lie anterior to the lamina terminalis. It is into this area that the medial olfactory stria runs, thus completing a continuous ring of grey matter (limbic cortex) and nerve fibres forming the medial margin of the hemisphere and having the olfactory tract entering its antero-inferior part.

The Anterior Commissure

This small, rounded bundle of variable diameter (2–4 mm.) crosses the midline in the superior part of the lamina terminalis anterior to the columns of the fornices and the interventricular foramen [FIG. 781]. Traced laterally it pierces the head of the caudate nucleus and the anterior part of the putamen of the lentiform nucleus, passing inferior to the anterior limb of the internal capsule, above the anterior perforated substance. In this part of its course it may be joined by fibres from the olfactory tract which turn medially and, crossing to the opposite side, act as a commissure for the olfactory bulbs. This **anterior part** of the commissure is the major part in macrosmic mammals, but is small in Man where the majority of the fibres incline posteriorly (**posterior part**) while still in the putamen of the lentiform nucleus, and emerge from its lateral surface [FIG. 805] passing posterolaterally into the white matter of the temporal lobe in which its fibres fan out, some reaching as far posteriorly as the inferior part of the occipital lobe.

The Transverse Fissure and Corpus Callosum [FIG. 810]

The corpus callosum, found only in eutherian mammals, is formed of fibres which join corresponding parts of the cerebral cortex of the two hemispheres. It crosses the midline where the two hemispheres fuse superior to the roof of the third ventricle. It thus overlies the roof of the third ventricle and cuts off the inferior portion of the **longitudinal fissure** (which separates the two hemispheres) to form the **transverse fissure** and cave of the septum pellucidum, between it and the diencephalon. The transverse fissure

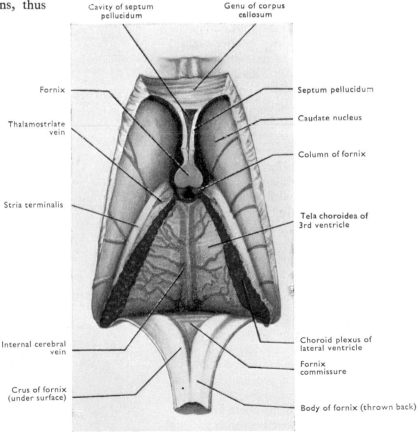

FIG. 818 Dissection to show tela choroidea of third and lateral ventricles, and the parts near it.

contains the connective tissue (pia mater) lying immediately superior to the roof of the third ventricle, with the posterior choroidal arteries and the internal cerebral veins [FIG. 778]. The pia mater in the transverse fissure invaginates the ependymal wall of the lateral ventricles through the choroid fissure at its lateral margins, and inferiorly, on either side of the midline, bulges into the third ventricle to form the vascular connective tissue of the third ventricular choroid plexuses; thus it is the **tela choroidea of the lateral and third ventricles.** The transverse fissure is closed above and anteriorly by the fusion of the two fornices; the enclosed sheet of pia mater forming the tela choroidea, narrows anteriorly as the lateral ventricles approach each other, and ends in an apex immediately posterior to the interventricular foramen [FIG. 818]. Here the **choroid plexus** of each lateral ventricle on the lateral margin of the tela choroidea becomes continuous, through the interventricular foramen, with the corresponding choroid plexus of the third ventricle on the inferior surface of the tela. Posteriorly the tela choroidea widens as the ventricles diverge, covers that part of the superior surface of the thalamus not lying in the floor of the lateral ventricle [FIG. 778], and becomes continuous with the pia mater covering the remainder of the brain by emerging between the splenium of the corpus callosum [p. 651] and the pineal body on the upper midbrain [FIG. 781]. In this region the two internal cerebral veins unite to form the **great cerebral vein** which turns superiorly around the splenium of the corpus callosum, joins the inferior sagittal sinus in the free edge of the falx cerebri to form the straight sinus. This runs posteriorly in the midline between the layers of dura mater at the junction of the falx cerebri with the tentorium cerebelli [p. 694] and reaches the internal occipital protuberance.

The fornix and the cortex immediately adjacent to it (septum pellucidum), which together form the medial part of the floor and medial wall of the lateral ventricle in this region, are separated from the remainder of the medial surface of the hemisphere by the corpus callosum [FIG. 825].

Nerve fibres from most parts of the cerebral cortex converge radially on the developing **corpus callosum** in a manner similar to the fibres running to the internal capsule, except that they are at right angles to the latter, and pass over the convex peripheral surface of the lateral ventricle [FIG. 821], while those entering the internal capsule run within the concavity of the ventricle [FIG. 798]. Thus the fibres from the superolateral surface of the hemisphere above the lateral sulcus, gather together, then passing horizontally through the fibres of the corona radiata, course superior to the caudate

nucleus and the lateral ventricle to reach the medial aspect of the hemisphere [FIG. 808]. Within this mass of fibres, those which arise immediately superior to the lateral sulcus lie most inferiorly and, crossing in the corpus callosum without mingling with adjacent fibres, run through the opposite hemisphere to end in a situation similar to that from which they arose, as do all the fibres in the corpus callosum.

In view of this arrangement of the callosal fibres it might be expected that the corpus callosum would develop the same C-shaped structure as the other parts of the hemisphere, but this is prevented by the intervention of the brain stem, which holds the hemispheres apart postero-inferiorly and interferes with the formation of any part of the corpus callosum inferior to the dorsal surface of the upper midbrain. Thus the corpus

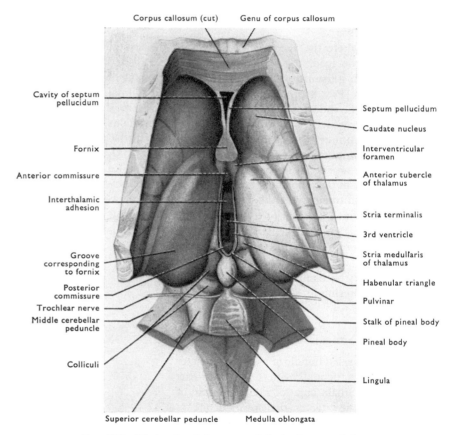

FIG. 819   Thalami and the parts of the brain around them.

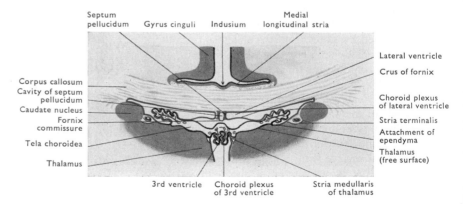

FIG. 820   Diagram of coronal section through central parts of lateral ventricles.

callosum is limited in extent from the superior part of the lamina terminalis anteriorly to the superior surface of the mid-brain posteriorly, and the callosal fibres from the temporal lobes and the inferior parts of the occipital lobes sweep superiorly, forming a compact layer immediately lateral to the posterior and inferior horns of the lateral ventricle and medial to the corona radiata [FIG. 821]. Superiorly this compact layer of fibres, the **tapetum,** hooks over the upper surface of the lateral ventricle to enter the posterior extremity of the corpus callosum, which is thereby increased in size and forms the rounded **splenium.** Fibres from the superior part of the occipital lobe also enter the splenium by coursing anteriorly over the superior surface of the posterior horn of the lateral ventricle, while those from its medial aspect are divided by the deep calcarine sulcus, which bulges into the posterior horn of the lateral ventricle forming the **calcar avis** [FIG. 796]. Callosal fibres arising inferior to the calcarine sulcus loop beneath the posterior horn of the lateral ventricle to enter the tapetum, while those arising superior to the calcarine sulcus pass anteriorly on the medial side of the ventricle and, curving far laterally around the deep parieto-occipital sulcus, bulge into

the medial aspect of the ventricle to form the **bulb of the posterior horn** [FIG. 825] immediately superior to the calcar avis, before turning medially into the splenium of the corpus callosum. Thus the posterior horn of the lateral ventricle is virtually surrounded by fibres of the corpus callosum passing to the splenium. When the fibres producing the bulb of the posterior horn of the lateral ventricle are traced from the splenium, they are seen to curve from the splenium laterally around the parieto-occipital sulcus to the medial side of the occipital lobe, thus forming the **forceps major** [FIG. 822] with the corresponding fibres of the opposite side. An exactly similar arrangement of fibres is found in the medial sides of the parietal and frontal lobes, except that in the latter the fibres do not sweep so far laterally and thus form a smaller curve or **forceps minor.**

Fibres passing to the anterior part of the corpus callosum arise from all surfaces of the frontal lobe and, converging on the anterior horn of the lateral ventricle and head of the caudate nucleus, surround them on the inferior, anterior, and superior surfaces, and thus form a sharply curved sheet of fibres which passes medially to become the **genu** and **rostrum**

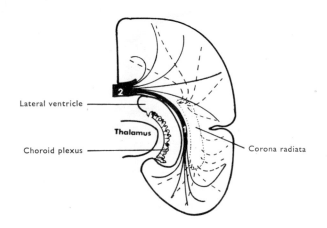

FIG. 821 Diagram to illustrate the course and position of the major commissural fibres of the hemisphere. In the lower figure, which is a coronal section along the line in the upper figure, (1) is the tapetum and (2) the splenium of the corpus callosum.

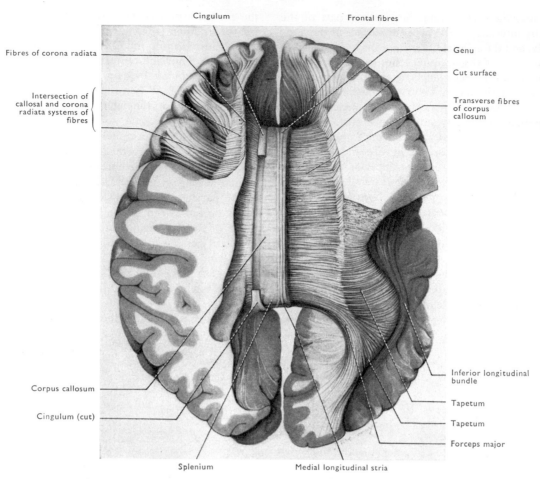

FIG. 822 **Corpus callosum**, exposed from **above and right half** dissected.

of the corpus callosum [FIG. 821]. The fibres from the in-feromedial part of the frontal lobe lie inferomedial to the ventricle and enter the rostrum of the corpus callosum. The remainder of the corpus callosum is the **trunk** or body.

The Lateral Ventricle [FIGS. 823 and 824]

The general shape of this ventricle has already been described [p. 631], but its relations can now be understood. The **anterior horns** of the two ventricles lie antero-inferior to the interventricular foramen and are separated from each other only by the **septum pellucidum** and its cavity which form their medial walls, while the lateral walls bulge inwards, being formed by the large convex heads of the **caudate nuclei.** The remaining walls of the anterior horns are formed by the corpus callosum [FIGS. 799, 821]. Posterior to the interventricular foramen the septum pellucidum continues to form the medial wall of the ventricle but rapidly decreases in height. The fornix, now on the inferior margin of the septum pellucidum, extends laterally into the floor of the **central part of the lateral ventricle** to the line of the choroid plexus, which is invaginated into the lateral ventricle between the fornix and the dorsal surface of the thalamus. The thalamus and the narrowing caudate nucleus, with the stria terminalis and thalamostriate vein between them, form the remainder of the floor of the ventricle [FIG. 825]. Laterally, the floor slopes superiorly to meet the roof, which is formed by the trunk of the corpus callosum and posteriorly by its splenium.

Posteriorly the two ventricles diverge because of the intervening brain stem, and the central part of the ventricle turns inferiorly around the posterior end of the thalamus to become continuous with the inferior horn anteriorly, and the posterior

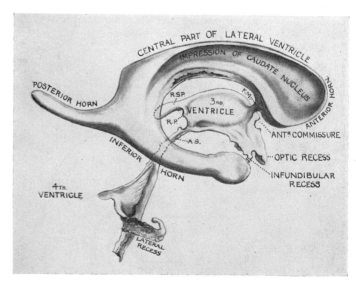

FIG. 823 Profile view of a cast of the ventricles of the brain. (From G. Retzius (1900) *Biol. Untersuch.*, **9**, 45.)

R.SP., suprapineal recess; A.S., aqueduct; R.P., pineal recess; F.M., interventricular foramen.

FIG. 824 Drawing of cast of ventricular system of the brain, seen from above. (After Retzius.)

R.SP., suprapineal recess.

FIG. 825 Dissection to show the fornix and lateral ventricles.

horn posteriorly. The **inferior horn** [FIG. 826] with the choroid fissure in its medial wall curves anteriorly around the crus cerebri. The choroid fissure follows the line of the optic tract antero-inferiorly from the lateral geniculate body to end against the uncus; the cavity of the ventricle, continuing anteriorly for a further 1 cm., turns inferiorly over the pes hippocampi, posterior to the amygdaloid body. In addition to the choroid fissure and fimbria, the rounded ridge of the **hippocampus** covered by the alveus lies in the medial wall, while the narrow floor is formed by the inferior longitudinal bundle and fibres of the corona radiata and tapetum passing to the medial side of the temporal lobe. Posteriorly, where the three parts of the ventricle meet, the floor widens into a triangular area slightly raised from below by the collateral sulcus, the **collateral trigone**. The roof of the inferior horn contains the stria terminalis and the tail of the caudate nucleus. The latter, thinned to a narrow ribbon of grey matter, is often broken in places, but is usually continuous anteriorly with the amygdaloid body. Superior to the tail of the caudate nucleus lie the most anterior fibres of the optic radiation in the

sublentiform part of the internal capsule, and the posteriorly directed fibres of the anterior commissure and fasciculus uncinatus, all spreading laterally into the temporal lobe [FIG. 805].

The **posterior horn** varies greatly in size, and is indented medially by the calcarine sulcus (calcar avis) and forceps major. It is virtually surrounded by fibres of the corpus callosum [p. 651].

The Choroid Plexuses of the Third and the Lateral Ventricles

The choroid plexuses of the third ventricle, consisting of a core of vascular pia mater covered on the ventricular aspect by ependyma, are little more than a minute fringe lying on each side of the midline, though their size increases slightly in the posterior part where they enter the suprapineal recess of the ventricle. Anteriorly each choroid plexus of the third ventricle passes laterally in the posterior wall of the interventricular foramen to become continuous with the corresponding choroid plexus of the lateral ventricle. In the lateral ventricle the choroid plexus sweeps posterolaterally between the dorsal surface of the thalamus and the lateral edge of the fornix [FIG. 825] as a small fringe which gradually increases in amount to reach its maximum size where it turns inferiorly over the posterior surface of the thalamus and expands into the meeting of the central part with the posterior and inferior horns of the ventricle. Here it usually expands posteriorly into the posterior horn, while a thin ribbon of choroid plexus continues antero-inferiorly to end at the notch between the fimbria and the uncus [FIG. 826]. Here the anterior choroidal artery, passing posteriorly from the internal carotid, enters the choroid fissure to supply the choroid plexus of the inferior horn. Further posteriorly where the choroid fissure sweeps round the lateral side of the upper midbrain parallel to the posterior cerebral artery, several branches of the artery pass laterally into it. In the central part of the ventricle the choroid plexus is supplied by a branch or branches passing in the tela choroidea dorsal to the thalamus. All these are the posterior choroidal branches of the posterior cerebral artery.

THE MICROSCOPIC ANATOMY OF THE CENTRAL NERVOUS SYSTEM

The previous section has shown the arrangement of the main masses of nerve cells and fibres which comprise the central nervous system and of the nerves which are attached to it. The gross anatomy gives little indication of the functions of this system beyond those which can be inferred from the peripheral connexions of the various nerves to be described in the next section. Knowledge of the functional significance of the various parts of this system, based on their structure, can only be obtained with any certainty from a study of the connexions of the groups of cells which comprise the central nervous system. However, no complete picture of the connexions of the cells in the nervous system is available, and the anatomical approach to the study of its function cannot yet produce more than a very sketchy outline.

The divisions of the central nervous system which are based on gross anatomy, i.e., forebrain, midbrain, hindbrain, and spinal medulla, do not necessarily have any functional significance, but rather represent parts of a unit which is concerned with integrating the activity of every part of the body. Such a system requires a high degree of integration of its parts, and this is attained by nerve fibres which connect the various regions of the central nervous system reciprocally.

Basically the central nervous system consists of three types of cells: sensory, connector or internuncial, and motor.

1. **Sensory neurons** have two processes, one of which passes to a sensory ending, while the cell body lies outside the central nervous system, either in the spinal ganglion of a dorsal root, or in the corresponding ganglion of a cranial nerve adjacent to the brain, or in the retina, or in the olfactory

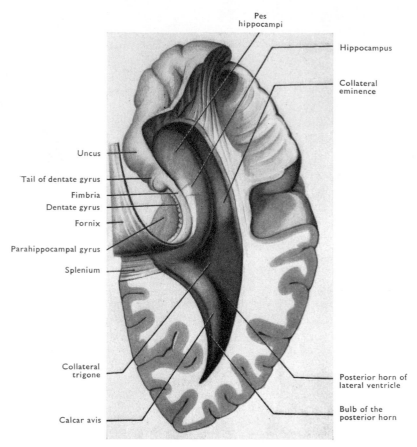

FIG. 826 Dissection from above to show posterior and inferior horns of lateral ventricle.

mucous membrane. The other process enters the central nervous system and, extending for a variable distance, gives rise to a number of branches through which it can activate several internuncial neurons.

2. **Internuncial neurons** are interposed between the sensory and motor cells and are not directly connected either with a sensory ending or with an effector organ. They comprise almost all the grey matter of the central nervous system, including such massive structures as the cerebrum and cerebellum, the thalamus, the corpus striatum, and a multitude of smaller nuclei and more scattered groups of nerve cells. Internuncial cells are of every shape and size, and lie entirely within the central nervous system, with the single exception of the preganglionic cells of the autonomic nervous system which send their axons to the peripherally situated autonomic ganglia which consist of 'motor' cells. Internuncial neurons form the pathways through which impulses pass for the production of local reflexes and their nerve fibres link together the various parts of the central nervous system. Such interregional links are of very variable length, ranging from fibres joining adjacent segments of the spinal medulla (intersegmental fibres of the fasciculi proprii), or adjacent gyri of the cerebral cortex (arcuate fibres), to nerve fibres which extend from the spinal medulla to the thalamus (spinothalamic tracts), or from the cerebral cortex even to the most caudal part of the spinal medulla (corticospinal tracts). Long tracts of the latter type may form short-distance con-

nexions by means of branches (collaterals), while conversely, short neurons may form long pathways when several such neurons are linked in series.

At all levels of the central nervous system and in their local and distant connexions, the internuncial neurons act together in constantly changing patterns which depend mainly on the varying sensory impulses entering the nervous system. Their interactions result in the free passage of impulses through certain nervous pathways, while other pathways are blocked or inhibited, so that inappropriate reactions do not take place. This continuous interplay of facilitation and inhibition within the complex network of internuncial neurons eventually has its effect on the motor cells to produce overt reactions, such as muscular contraction or inhibition, vasoconstriction or dilatation, glandular secretion or inhibition, etc.

Internuncial neurons interrelate the vast numbers of sensory impulses pouring into the nervous system over a wide area. They transmit this sensory information to every part of the nervous system (an essential feature of any integrating system), but only certain 'sensory' regions seem to be capable of evoking subjective awareness of sensation—in Man, the cerebral cortex. Thus the axons of internuncial neurons which form bundles of fibres (tracts) conveying impulses headwards to the 'sensory' regions from the spinal medulla and brain stem, are frequently called 'sensory' tracts, though they only differ from similar fibres carrying the same information to other regions, e.g., to the cerebellum, in that their destruction interferes with the conscious awareness of certain sensations. All the various regions of the central nervous system are also connected with those lying caudal to them, and such descending connexions produce effects on the caudal parts of the nervous system which are very similar to the effects of the ascending connexions on the more cranial parts. However, since all the motor cells lie in the brain stem and spinal medulla, activity in the descending pathways, e.g., from the cerebral cortex, often results in the production of a movement or other activity which can readily be observed. For this reason the descending pathways and the cells which give rise to them are frequently called 'motor'. It should be appreciated, however, that ascending pathways connecting the various segments of the spinal medulla with each other, or with the brain stem, may equally produce motor effects through the motor cells at these levels. Also, that any movement produced by activity in the internuncial neurons at the same or a different level of the central nervous system, calls for the inhibition of local activity which is at variance with the movement which the 'motor' pathway is producing, in addition to the inhibition of the antagonist muscles. The 'motor' pathways composed of internuncial neurons are therefore, of necessity, also inhibitory in their actions. The terms 'sensory' and 'motor' must therefore be applied with caution to the pathways composed of the axons of internuncial neurons, since they are not equivalent to the same terms applied to nerve cells attached either to sensory endings (true sensory neurons) or to effector organs (true motor neurons).

3. **Motor neurons** are connected with effector organs such as muscle, glands, etc., and have their cell bodies either in the motor nuclei of the cranial nerves, or in the anterior grey column of the spinal medulla, or in the peripheral autonomic ganglia. With the exception of certain autonomic ganglion cells which produce inhibition on stimulation (e.g., the action of the vagus on the heart), these cells invariably produce a response of a motor nature in the muscles which they innervate.

The evolution of the nervous system has been the result mainly of the elaboration of the parts at its cephalic end. It might therefore be expected that each level would receive sensory information from, and have descending connexions with, the region immediately caudal to it, thus having only the most indirect connexions with the caudal levels through chains of short ascending and descending neurons. However, in the nervous system of all mammals, the delays inherent in such multisynaptic pathways have been partly overcome by the development of long ascending and descending pathways in parallel with such multisynaptic pathways. These long pathways allow of more specific point-to-point connexions within the central nervous system, and are responsible for the steady increase in volume of nerve fibres at each more cephalic level of the spinal medulla in Man, a feature virtually absent in birds, reptiles, and lower forms, and particularly well illustrated in the posterior funiculi of the spinal medulla and in the corticospinal system. In the latter system there is wide variation even within the mammals, the corticospinal tracts reaching the most caudal part of the spinal medulla in some (e.g., primates) but entering only its most cephalic parts in others (e.g., sheep).

The presence of the long ascending and descending pathways, which are relatively simple to investigate anatomically, tends to overshadow the multisynaptic pathways, which are not readily amenable to anatomical techniques. However, the latter form a large and important part of the conducting routes even in the central nervous system of Man, and are not replaced by the longer and more direct tracts.

The remarkable variability in the structure of the nervous system, even within the mammals, makes it clear that experimental findings obtained in one animal cannot readily be transferred to Man, yet the great majority of the anatomical and physiological findings concerning the central nervous system have been obtained from mammals other than Man. It is therefore essential to view with caution some of the inferences drawn from animal sources which are inevitably included in the following account.

All pathways within the central nervous system are of the nature of intercommunications between its parts, but it is convenient to consider them in three groups :

1. The long ascending pathways, including those responsible for evoking sensation.

2. The connexions between the various parts of the brain.

3. The long descending pathways.

THE LONG ASCENDING PATHWAYS

Dorsal Roots

The dorsal root fibres of the spinal nerves and the fibres of the laterally attached cranial nerves have already been shown to form a series of rootlets entering the hindbrain and spinal medulla, and the distribution of the sensory fibres which they contain has been shown to follow the same general plan throughout [p. 621]. The **dorsal roots** of the spinal nerves form a linear series of rootlets attached to the posterior lateral sulcus of the spinal medulla. The fibres in these rootlets arise in the spinal ganglia from cells which vary markedly in size. The largest cells give rise to processes up to 10 μm. or more in diameter, with a correspondingly thick myelin sheath, while the smallest cells give rise to non-myelinated fibres many of which are 0·1 μm. or less in diameter, and between these two there is a range of intermediate sizes. The dorsal roots of spinal nerves transmit all types of sensory fibres from the peripheral nerves with which they are connected, including visceral sensory fibres. The area of skin supplied by each dorsal root, a **dermatome,** overlaps the similar areas supplied by adjacent dorsal roots to such an extent that the division of a single dorsal root does not usually produce an area of absolute sensory loss, though the quality of the sensation aroused by stimulation of the partially denervated skin is altered, because of the smaller number of sensory endings remaining in connexion with the central nervous system.

Efferent fibres have been described in the dorsal roots, and although these occur in birds and some lower vertebrates, there is no anatomical or physiological evidence for their presence in mammals. Vasodilatation which results from stimulation of the distal end of a cut dorsal root is attributed, without anatomical proof, to the distribution of cutaneous sensory fibres also to blood vessels : a feature which would explain the localized vasodilatation following cutaneous stimulation, even after recent complete denervation—the

axon reflex. However, the presence of large numbers of extremely small fibres in the dorsal roots has not been satisfactorily explained.

Within the dorsal rootlets a segregation of the fibres occurs, such that the coarser fibres tend to collect in the lateral part, while the finer ones lie in the medial portion. As the rootlets curve over into the dorsal surface of the spinal medulla the position is reversed and the coarser fibres enter the **posterior funiculus** of the same side while the finer fibres enter a smaller bundle lying lateral to this, the **dorsolateral fasciculus** [FIG. 827]. Each of the fibres on entering the spinal medulla divides into an ascending and a descending branch and from these longitudinally-running processes numerous collaterals are given off into the grey matter of the posterior horn, thus allowing each fibre to have several different connexions within the spinal medulla. Some of these collaterals can activate neurons responsible for producing local spinal reflexes, or more distant spinal activity through the intersegmental tracts ; others are responsible for the transmission of impulses headwards to the brain.

The Posterior Funiculi of the Spinal Medulla

The **coarser fibres of the dorsal root,** on entering the posterior funiculus divide into long ascending processes, many of which reach headwards to the medulla oblongata, and

Dorsolateral fasciculus

Fasciculus gracilis

Fasciculus cuneatus

Fibre in dorsolateral fasciculus

Fibre crossing to enter spinothalamic tract

Fibres in posterior funiculus

Fibre in dorsolateral fasciculus

Spinothalamic system

Substantia gelatinosa

Polysynaptic connexion

Motor cells

Extensor muscle innervation

Monosynaptic connexion

Flexor muscle innervation

FIG. 827 Diagram to show the course of dorsal root fibres of spinal nerves entering the cervical spinal medulla, and some of their collateral connexions.

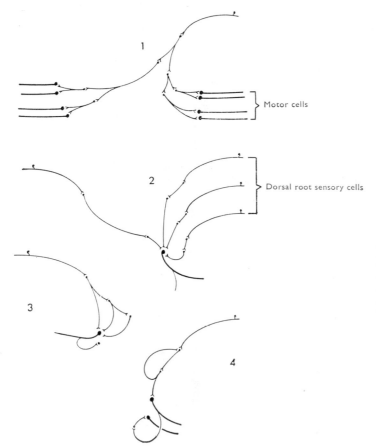

FIG. 828 Four diagrams to indicate some of the types of pathways which impulses, entering the spinal medulla over the dorsal roots, may follow through the internuncial neurons.

1. Divergent effect of internuncial neurons.

2. Convergent effect of internuncial neurons.

3. Arrangement of internuncial neurons capable of producing repetitive stimulation of a motor neuron from a single sensory impulse.

4. Shows recurrent collaterals of internuncial and motor neurons. The former can produce reverberating circuits, the latter are believed to stimulate internuncial neurons, Renshaw (1941) cells, which inhibit adjacent motor neurons.

These diagrams represent concepts derived from electrophysiological experiments, and bear no relation to the actual arrangement or appearance of these neurons under the microscope.

shorter descending processes which may terminate almost at once or extend for a few segments. Both processes rapidly decrease in diameter.

The long ascending processes of these fibres entering the spinal medulla through successively higher dorsal roots, are laid down on the lateral side of those which enter at lower levels. Hence, in its upper parts, the posterior funiculus on each side is greatly enlarged and composed of fibres, the most medial of which are processes of spinal ganglion cells in the sacral region, while the most lateral fibres come from the cervical region, the intermediate fibres lying between [FIG. 829].

Throughout its length, the posterior funiculus on each side is separated from its neighbour by the **posterior median septum,** but above the midthoracic region of the spinal medulla each posterior funiculus is further subdivided by a

posterior intermediate septum into a medial **fasciculus gracilis** and a lateral **fasciculus cuneatus** [FIG. 827]. The former contains fibres from the dorsal roots entering the caudal half of the spinal medulla, the latter those entering its cephalic half. In Man, unlike lower mammals, it seems likely that the majority of the long ascending branches reach the medulla oblongata, though probably fewer from the caudal than from the cranial half of the body, as evidenced by the small size of the fasciculus gracilis in the cervical region of the spinal medulla as compared with the fasciculus cuneatus. In the medulla oblongata, the fibres of the fasciculi gracilis and cuneatus end on the cells of the **nuclei gracilis** and **cuneatus** which are visible on the dorsal surface of the medulla oblongata as small swellings, the **gracile** and **cuneate tubercles** [FIG. 749].

The extent of the fibres which form the posterior funiculi allows them to give off collaterals to every level cranial to their point of entry, so that rapid reflex adjustments in the spinal medulla can be produced as a result of impulses reaching it over the large, rapidly conducting sensory fibres of the peripheral nerves. The impulses which these fibres carry originate

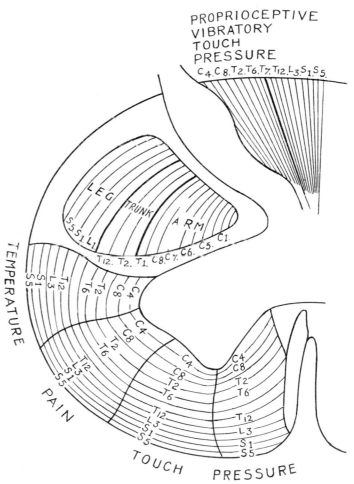

FIG. 829 Diagram to show: (1) the peripheral position in the anterior and lateral funiculi of fibres ascending from the lowest segments of the spinal medulla and the position progressively nearer the grey matter of those from higher segments; (2) the medial position in the posterior funiculi of fibres ascending from the lowest segments and the progressively lateral position of those from higher segments; (3) lamination of fibres within the lateral corticospinal tract. (After Foerster, 1936.)

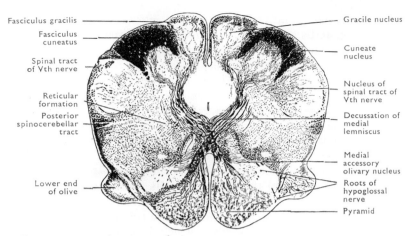

FIG. 830 Transverse section through foetal medulla oblongata immediately above the decussation of the pyramids (Weigert-Pal specimen).

in some hair follicles, the complex encapsulated endings of the skin, Pacinian corpuscles, muscle spindles, tendon organs, and the many complicated endings which are found in association with the joint capsules. Muscle and tendon sense fibres do not carry impulses which are consciously appreciated. On anatomical grounds, therefore, it might be expected that such impulses, evoked by tactile, pressure, and proprioceptive stimulation, would play a dominant part in spinal reflexes, since their impulses must reach all levels of the spinal medulla before those carried by the smaller, more slowly conducting fibres which have a much less extensive distribution within the spinal medulla.

The descending branches of fibres entering the posterior funiculi tend to be gathered together into two bundles : one, derived from fibres entering the fasciculus gracilis, lies adjacent to the posteromedian septum, the **septomarginal fasciculus** ; the other lies between the fasciculi gracilis and cuneatus and is the **fasciculus interfascicularis,** formed primarily by fibres entering the fasciculus cuneatus [FIG. 832]. Both bundles end as collaterals which enter the grey matter of the spinal medulla.

The **finer fibres of the dorsal roots** enter the **dorsolateral fasciculus** and again divide into ascending and descending branches which extend only for three or four segments in the fasciculus before terminating as collaterals entering the posterior horn. Because of the short course of its fibres, the dorsolateral fasciculus does not increase in size as it is traced cranially, though it is largest opposite the enlargements of the spinal medulla, where the number of fibres entering it is increased, and in its cephalic extension through the medulla oblongata and the lower half of the pons where it is known as the **spinal tract of the trigeminal nerve** [FIGS. 768 and 830], and receives a large number of sensory fibres which enter it principally from the trigeminal nerve.

Collaterals [FIG. 831]. The collaterals of fibres in the **posterior funiculi** curve ventromedially to enter the basal part of the posterior horn, the intermediate region, and the anterior horn where they may have direct connexions with the motor cells. Some of these collaterals may pass to the opposite side but the great majority of primary sensory neurons are distributed in the side of the spinal medulla which they enter. In the thoracic and upper lumbar regions, collaterals of the fasciculus gracilis enter a well-defined nucleus

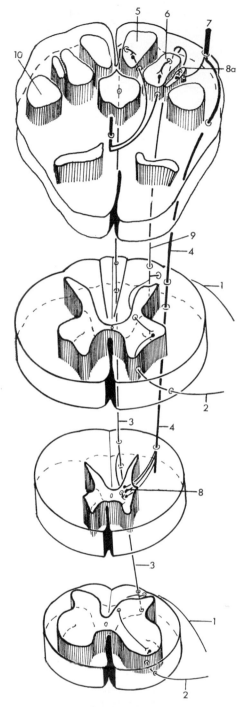

FIG. 831 Diagram, not to scale, composed of transparent slices of the lower medulla oblongata, cervical, thoracic, and sacral spinal medulla, to show the collaterals of the fibres in the fasciculi gracilis and cuneatus.

1. Dorsal root fibres.
2. Ventral root fibres.
3. Fibre in fasciculus gracilis.
4. Posterior spinocerebellar tract.
5. Nucleus gracilis.
6. Nucleus cuneatus.
7. Inferior cerebellar peduncle receiving posterior spinocerebellar tract and posterior external arcuate fibres.
8. Thoracic nucleus.
8a. Accessory cuneate nucleus.
9. Fibres in fasciculus cuneatus.
10. Nucleus of spinal tract of trigeminal nerve.

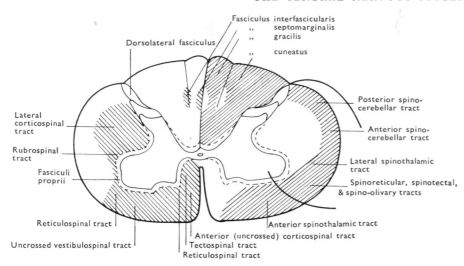

FIG. 832 Diagram to show the main ascending (right) and descending (left) pathways in the white matter of the spinal medulla. These have not been outlined individually because of the mixture of different pathways in the anterior and lateral funiculi.

situated in the medial part of the base of the posterior horn, the **thoracic nucleus.** This is composed of large cells giving rise to axons of considerable diameter which sweep laterally to the surface of the lateral funiculus of the same side and turning headwards form an ascending bundle immediately ventral to the dorsolateral fasciculus, the **posterior spinocerebellar tract** [FIG. 832]—a rapidly conducting tract destined for the cerebellum which carries proprioceptive, tactile, and pressure information from the lower half of the body. Corresponding information from the upper half of the body is relayed to the cerebellum by collaterals of the fasciculus cuneatus which enter the **accessory cuneate nucleus,** situated lateral to the nucleus cuneatus. This consists of cells, similar to those in the thoracic nucleus, which give rise to the **posterior external arcuate fibres,** coursing along the dorsolateral part of the medulla oblongata to join the posterior spinocerebellar tract and enter the cerebellum by the inferior cerebellar peduncle [FIG. 831].

Immediately ventral to the posterior spinocerebellar tract lies the **anterior spinocerebellar tract** [FIG. 832]. The exact origin and hence the significance of the fibres in this tract is unknown, but some of them appear to arise from cells in the dorsolateral part of the anterior horn, principally on the opposite side, and may therefore be activated by collaterals from the posterior funiculi. These two spinocerebellar tracts form the surface of the greater part of the white matter of the lateral funiculus, and though their fibres convey impulses originating in sensory fibres of the peripheral nerves, division of these pathways does not interfere with conscious appreciation of these sensations.

The Spinothalamic System

Situated over the tip of the posterior horn, the **fibres of the dorsolateral fasciculus** send their collaterals into this part of the horn, which contains the small cells of the **substantia gelatinosa** and the mixed cells of the body of the posterior horn [FIG. 827]. In addition to connexions which are made for local spinal reflexes and with cells which give rise to intersegmental fibres running in the **fasciculi proprii**

(including the dorsolateral fasciculus) [FIG. 832] of the spinal medulla, other cells, probably situated in the body of the posterior horn, give rise to fibres which pass obliquely headwards and sweep across the midline of the spinal medulla in the white commissure ventral to the central canal. They enter the anterolateral region of the white matter on the opposite side of the spinal medulla and ascend in this region forming the so-called **spinothalamic tracts** [FIG. 832]. Unlike the posterior spinocerebellar tract and the posterior funiculi which form well-defined pathways, the fibres forming the spinothalamic tracts are mixed with other axons including some descending fibres, and are widely scattered around the region of the emerging ventral rootlets which divide them into **lateral** and **anterior parts.**

The exact extent of the fibres forming this system is not clear and it is certain that many of its fibres do not reach the thalamus directly since, on division of this tract in Man, degeneration studies with the Marchi technique seem to indicate many fewer fibres in the pontine and mesencephalic parts of this pathway than are present in the spinal medulla. This may be due either to the presence of non-myelinated fibres in this system, or to the large proportion of the fibres which end in the grey matter at lower levels, e.g., in the reticular formation of the brain stem. Certainly there is good evidence that the fibres ascending in this part of the white matter of the spinal medulla, or their collaterals, end at various levels of the spinal medulla and brain stem.

Interruption of the fine fibres in the lateral parts of the dorsal roots at their entry to the spinal medulla, while leaving the larger medial fibres entering the posterior funiculi intact, abolishes reflexes produced by painful and thermal stimuli in some animals; interruption of the anterolateral region of the white matter of the spinal medulla in Man interferes with similar sensations from the opposite side of the body below the level of injury. These two facts, combined with the fact that at least some of the pain spots in the skin are innervated by fibres of the smallest calibre, indicate that these fibres are responsible for the transmission of impulses which evoke thermal and painful subjective sensations at a higher level. Experience with division of the spinothalamic system in Man seems to indicate that the lateral part is particularly concerned with the above sensory mechanisms and it is believed that the ventral part is concerned with tactile and proprioceptive sensations, since division or pathological degeneration of the posterior funiculi, though it interferes profoundly with the conscious appreciation of the latter sensations, does not abolish them.

Unilateral division of the spinothalamic system in Man does not always abolish painful sensations from the opposite side of the body, but it has been reported to produce bilateral retrograde changes in the cells of the posterior horn, and to be associated with a raised pain threshold on the side of the division. These facts seem to suggest that this pathway is composed partly of uncrossed fibres, which could account for failure to obtain relief after unilateral surgical division of the spinothalamic system, but this failure may sometimes be due to the fact that all the scattered fibres of this system have not been divided. Because of this failure other pathways have been postulated, including the passage of fibres headwards through the sympathetic trunk to enter the spinal medulla at higher levels. Evidence in favour of this has been produced (Nathan, 1956). It should be remembered, however, that this is a scattered system of fibres which gives off many collaterals throughout its length and there are therefore alternative pathways which would not be demonstrated readily by anatomical methods.

Since most of the fibres forming the spinothalamic system are crossed, and more fibres are added to it at each level, including the region of the brain stem where fibres from the nucleus of the spinal tract of the trigeminal nerve pass into it, it might be expected that the group of fibres nearest to the surface of the spinal medulla would be those derived from the sacral region, while the deepest fibres would be those which had just entered it. In keeping with this is the experience of surgical division (chordotomy) of this system in Man under local anaesthesia, when it can be shown that, as the cut is deepened into the anterolateral region, the loss of pain and temperature sensation on the opposite side rises towards the level of the section. The sensory loss never reaches that level since the fibres cross obliquely to enter the anterolateral region. This obliquity of crossing is believed to be greater in tactile than in pain and temperature fibres.

The division of a particular group of fibres in the main ascending pathways of the spinal medulla, either of the posterior funiculi or of the spinothalamic system, produces the same topographical distribution of sensory loss whatever the level of division. Thus superficial involvement of the spinothalamic system or injury to the midline fibres of the posterior funiculi, even in the cervical region, produces sensory loss limited to the lowest part of the body, ipsilateral in the latter case and contralateral in the former. A midline injury of fibres crossing to the spinothalamic system produces a bilateral belt of loss of pain and temperature sensation corresponding to the longitudinal extent of the lesion, but sparing this sensation both above and below the level of the injury; such a condition is found in lesions in the region of the central canal, e.g., syringomyelia.

The ascending pathways of the spinal medulla together form a mass of fibres which virtually covers its entire surface except in the midline anteriorly [FIG. 832]. These pathways retain this arrangement throughout the length of the spinal medulla except in its uppermost segments where the lateral corticospinal tract, which elsewhere lies deep to the spinocerebellar tracts, comes to the surface and temporarily displaces the spinocerebellar tracts ventrally [FIG. 836].

Ascending Pathways in the Medulla Oblongata. In the medulla oblongata [FIGS. 760 and 761] a redistribution of structures occurs [p. 607] as a result of the change in position of the lateral corticospinal tract from the deeper parts of the lateral funiculus of the spinal medulla. It passes through the massive pyramidal decussation in the white commissure, to the ventral surface of the opposite side of the medulla oblongata,

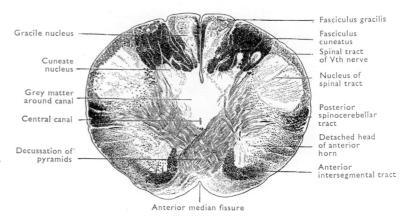

FIG. 833 Transverse section of lower end of medulla oblongata of full-term foetus. Treated by the Weigert-Pal method. The grey matter is bleached white, and the tracts of myelinated fibres are black.

where it forms the major part of the pyramid [FIG. 833]. This, combined with the appearance of the olive [FIGS. 755 and 834], on the dorsolateral aspect of the pyramid, displaces the medial part of the anterior funiculus of the spinal medulla dorsally (to become continuous with the medial longitudinal fasciculus of the medulla oblongata) and the remainder of the anterior and lateral funiculi dorsolaterally towards the spinal tract of the trigeminal nerve. Immediately superior to the pyramidal decussation, the posterior funiculi expand into their corresponding nuclei in which their fibres end, and from these nuclei (gracilis and cuneatus on both sides) secondary fibres arise which sweep ventrally (**internal arcuate fibres**) to cross the midline in the white commissure, forming the **decussation of the lemnisci** [FIGS. 831 and 834]. Leaving the decussation, the fibres on each side form a flat vertical sheet or fillet, the **medial lemniscus,** situated between the pyramid ventrally and the medial longitudinal fasciculus dor-

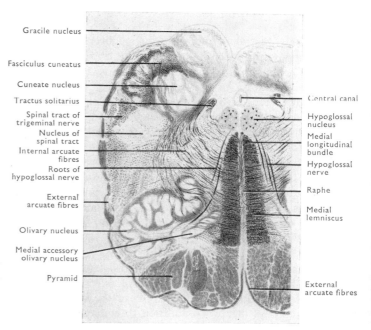

FIG. 834 Transverse section of medulla oblongata at the level of the olive.

sally. The median **raphe** separates the two medial lemnisci, and lateral to each the rootlets of the hypoglossal nerves pass ventrally.

The main bundles of white matter are redistributed in the lower medulla oblongata in such a fashion that much of the white matter of the lateral funiculus becomes continuous with the pyramid, and the pathway formed by the posterior funiculi in the spinal medulla is continued headwards as the medial lemniscus in a ventral paramedian position. Both these tracts lie ventral to the central canal, which is thereby displaced further dorsally and, cranial to the nucleus gracilis, opens out into the thin-roofed fourth ventricle. Like the secondary neurons which form the spinothalamic system, the secondary neurons which receive impulses from the posterior funiculi send their axons to the opposite side, so that in the medulla oblongata both of these ascending systems are crossed, but are still widely separated from each other. The medial lemniscus lies between the midline and the rootlets of the hypoglossal nerve, while the spinothalamic system lies at the lateral surface of the medulla oblongata between the olive and the spinal tract of the trigeminal nerve, close to the rootlets of the laterally attached cranial nerves, the accessory, the vagus, and the glossopharyngeal [FIG. 760]. Thus in the medulla oblongata, as in the spinal medulla, either pathway may be injured separately, but because the two medial lemnisci lie side by side they are likely to be injured together, as are the posterior funiculi in the spinal medulla, and injury either to the medial lemnisci or to the spinothalamic system tends to be associated with damage to adjacent structures. Thus the hypoglossal nerve may be involved with the medial lemniscus, and the VII, IX, X, XIth nerves, the spinal tract and nucleus of the trigeminal nerve and the spinocerebellar tracts with the spinothalamic system. The latter complex produces the bizarre loss of pain and temperature senses on the opposite side of the body (spinothalamic), and on the same side of the face (spinal root of the trigeminal nerve), combined with symptoms and signs referable to destruction of one or more of the above nerves, together with cerebellar signs consequent on the destruction of the spinocerebellar tracts.

Dorsal to the spinal tract and nucleus of the trigeminal nerve, opposite the inferior extremity of the fourth ventricle, lie the **cuneate** and **accessory cuneate nuclei.** Cranial to these nuclei is the inferior cerebellar peduncle, formed by the posterior external arcuate fibres from the accessory cuneate nucleus, joined by the **posterior spinocerebellar tract** [FIG. 835], numerous olivocerebellar fibres from the opposite olivary nucleus, and several fibres connecting the reticular formation and vestibular nuclei with the cerebellum. The posterior spinocerebellar tract joins the inferior cerebellar peduncle by passing dorsally across the spinal tract of the trigeminal nerve, where it can frequently be seen with the naked eye as an ill-defined oblique ridge [FIG. 805].

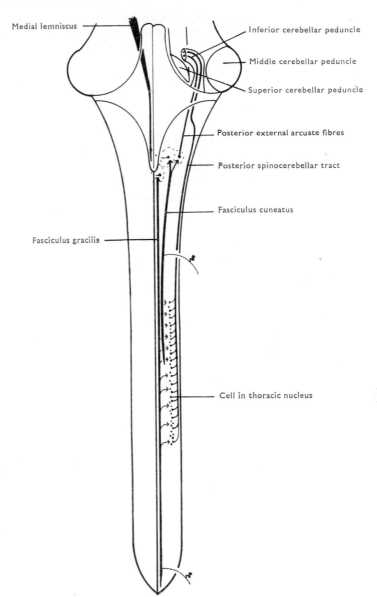

FIG. 835 A diagram of the origin and course of the spinocerebellar fibres entering by the inferior cerebellar peduncle.

FIG. 836 Diagram of the course of the main ascending tracts in the brain stem. The tracts have been separated to show each group individually.

FIGURE 836 shows the course of the main ascending tracts in the brain stem (except the medial lemniscus which is not shown in the medulla oblongata) and indicates how each in turn inclines dorsally to enter a derivative of the alar lamina, the first being the most dorsal of the laterally situated tracts [FIG. 832], the posterior spinocerebellar tract.

The passage of the posterior spinocerebellar tract into the inferior cerebellar peduncle leaves the anterior spinocerebellar tract and the **spinothalamic system** ventral to the spinal tract of the trigeminal nerve. In this position, the latter system gives off many fibres into the lateral reticular formation and the olive, thus forming what are sometimes described as separate bundles, the **spinoreticular** and **spino-olivary tracts**. Above this level the spinothalamic system is much reduced in size so that the number of direct fibres to the thalamus through this system is much smaller than might be expected from its size in the spinal medulla. This does not mean, however, that the impulses carried by fibres of the spinothalamic system ending in the brain stem do not reach the thalamus. There is good evidence that cells in the reticular formation give rise to ascending fibres which reach the thalamus and that these ascend on both sides of the brain stem so that this information is carried bilaterally to the thalamus.

Ascending Pathways in the Pons. [FIGS. 758 and 759.] The **medial lemniscus, spinothalamic,** and **anterior spinocerebellar tracts** enter the dorsal part of the pons. Here the medial lemniscus twists through a right angle from its sagittal position to form a coronal sheet [FIG. 837] with fibres from the nucleus gracilis in the lateral part close to the spinothalamic tract, while those from the nucleus cuneatus lie in the medial part and are at first close to the corresponding fibres in the opposite medial lemniscus. In this situation

the medial lemniscus is pierced by the transversely running fibres of the **trapezoid body** [FIG. 842] passing laterally to form the **lateral lemniscus** (the auditory path) which runs headwards between the spinothalamic and anterior spinocerebellar tracts. Thus, lying ventrally in the dorsal part of the pons there is a horizontal sheet of fibres, **the lemniscus,** formed from medial to lateral by the medial lemniscus, spinothalamic tract (**spinal lemniscus**), lateral lemniscus, and the anterior spinocerebellar tract, extending from the midline to the spinal tract of the trigeminal nerve laterally. The two main ascending pathways from the spinal medulla to the thalamus thus come together for the first time in the pons and headwards of this point remain in close association. This entire sheet of fibres which can readily be dissected out in the fixed brain, slides laterally and posteriorly as it is traced headwards. The **anterior spinocerebellar tract** turning posteriorly [FIG. 836], cephalic to the entry of the trigeminal nerve, runs into the cerebellum across the lateral surface of the superior cerebellar peduncle and on to the superior medullary velum, where many of its fibres cross to enter the opposite half of the cerebellum.

Ascending Pathways in the Midbrain and Diencephalon. In the midbrain [FIGS. 757 and 758], the **lateral lemniscus,** now forming the lateral edge of the lemniscus, curves dorsally as a well-defined ridge which emerges from the upper border of the pons immediately dorsal to the crus cerebri and curves posterosuperiorly over the lateral surface of the superior cerebellar peduncle, to expand into the **inferior colliculus** [FIGS. 805 and 838]. Here the majority of its fibres end, though a few pass on into the **brachium of the inferior colliculus.** This is a bundle of fibres carrying auditory impulses. It originates principally in the inferior colliculus [FIG. 844] and

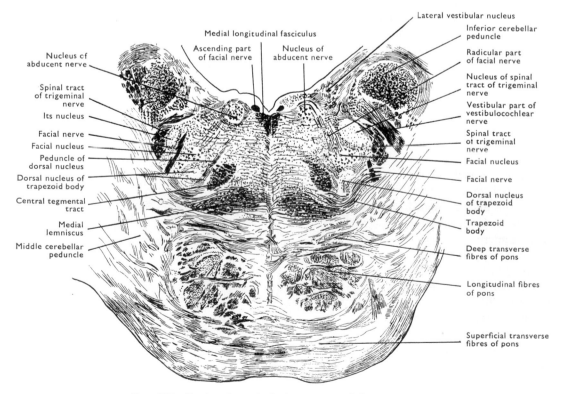

FIG. 837 Section through the lowest part of the pons.

Mesencephalic tract of trigeminal nerve
Anterior spinocerebellar tract
Superior cerebellar peduncle
Medial longitudinal fasciculus
Lateral lemniscus
Reticular formation
Medial lemniscus

Trochlear nerve
Upper part of 4th ventricle

A

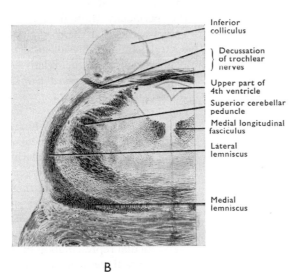

Inferior colliculus
} Decussation of trochlear nerves
Upper part of 4th ventricle
Superior cerebellar peduncle
Medial longitudinal fasciculus
Lateral lemniscus
Medial lemniscus

B

FIG. 838 Sections through dorsal part of pons close to the midbrain.

A is at a slightly lower level than B.

forms a ridge passing anterosuperiorly on the lateral surface of the superior part of the midbrain towards the angle between the crus cerebri and the posterior end of the thalamus beneath which the crus cerebri disappears. In this angle lies a small rounded swelling, the **medial geniculate body** [FIGS. 836 and 840], and the brachium of the inferior colliculus passes into its medial aspect to end among its cells, which thus form the thalamic relay station for auditory impulses.

In the lower part of the midbrain the **medial** and **spinal lemnisci** move further laterally and dorsally so that they come to lie deep to the lateral surface of the midbrain between the crus cerebri ventrally and the colliculi dorsally [FIGS. 839 and 840], covered by the brachium of the inferior colliculus. In this situation the spinal lemniscus gives off fibres dorsally into the superior colliculus (the **spinotectal tract**) and medially into the tegmentum of the midbrain. Both tracts then enter the posteroventral part of the thalamus where they fan out to enter and end in the medial and lateral parts of the **posterior-ventral nucleus of the thalamus** [FIGS. 757 and 778]. Because of the loss of fibres from the spinothalamic system to several levels of the brain stem the majority of the fibres ending

in the posterior-ventral nucleus of the thalamus are derived from the medial lemniscus.

It has already been noted that the spinal lemniscus receives fibres from the caudal part of the spinal nucleus of the trigeminal nerve. Similar fibres from the **principal sensory nucleus** of the trigeminal nerve are added to the medial portion of the medial lemniscus in the upper pons [FIG. 768] and it is presumed that impulses reaching the mesencephalic nucleus are similarly transmitted. Together these fibres form the **trigeminal lemniscus** which passes into the medial part of the posterior-ventral nucleus of the thalamus, while fibres carrying impulses from the lower parts of the body terminate in the lateral parts of this nucleus and those from the intermediate parts end in the intervening regions. These terminations have been determined by anatomical means but more specifically by recording the site of arrival of impulses at the thalamus in various mammals when different parts of the body are stimulated. In the cat, Poggio and Mountcastle (1959), on physiological grounds, differentiate between :

1. The above terminations, which are arranged in strict relation to the parts of the opposite half of the body and receive impulses carried principally by the medial lemniscus, but also through the spinothalamic tract.

2. Endings in the more posterior part of the thalamus (ventral pulvinar) where the same topographical localization does not exist, and individual neurons may be activated by stimulating widely separated parts of the body, even on either side. The pathways responsible for this can have little to do with localizing the source of the stimulus by comparison with the detailed point to point information transmitted by the medial lemniscus. Because of the long delay in the arrival of the former impulses at the thalamus, compared with those in the medial lemniscus, it seems that such impulses may come from neurons in the brain stem (e.g., reticular formation) which transmit a more general picture of the sensory input to the body as a whole, having been activated by spinoreticular fibres. The relatively small number of spinothalamic fibres which reach the thalamus directly (probably greater in Man than in other mammals) compared with the number of sensory endings at the periphery, makes it likely that each such fibre may transmit impulses from a number of endings, and so give less precise information on the point stimulated, possibly even leading to confusion between pain arising in superficial or deep tissues when other detailed information is lacking. These suggestions appear to have some validity in Man, since the destruction of the posterior funiculi or medial lemnisci with the spinothalamic tracts intact, profoundly affects the ability to localize the remaining sensations and to discriminate between different stimuli.

Ascending Pathways to the Cerebral Cortex. Thalamo-cortical fibres arise from almost every part of the thalamus including the posterior-ventral nucleus, which sends fibres through the posterior limb of the internal capsule to the posterior wall of the central sulcus and the adjacent surface of the **postcentral gyrus.** Within this area there is a marked difference between the arrangement of the cells in the cerebral cortex of the posterior wall of the central sulcus and the exposed part of the postcentral gyrus. Powell and Mountcastle (1959 a, b) indicate that this structural difference

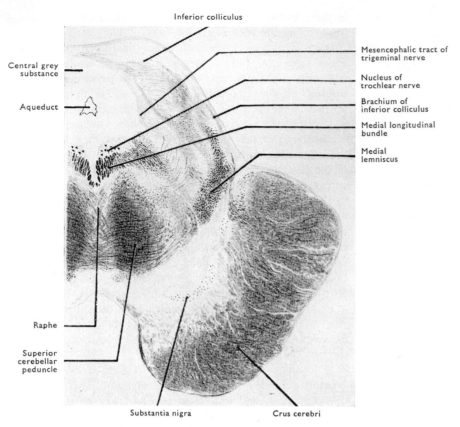

Inferior colliculus

Central grey substance

Aqueduct

Mesencephalic tract of trigeminal nerve

Nucleus of trochlear nerve

Brachium of inferior colliculus

Medial longitudinal bundle

Medial lemniscus

Raphe

Superior cerebellar peduncle

Substantia nigra

Crus cerebri

FIG. 839   Transverse section of midbrain at level of inferior colliculus.

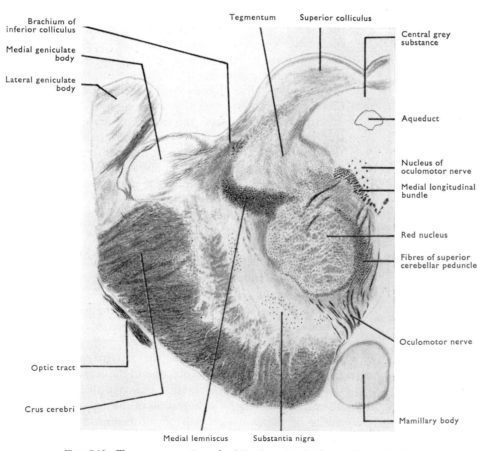

Brachium of inferior colliculus

Medial geniculate body

Lateral geniculate body

Tegmentum

Superior colliculus

Central grey substance

Aqueduct

Nucleus of oculomotor nerve

Medial longitudinal bundle

Red nucleus

Fibres of superior cerebellar peduncle

Optic tract

Crus cerebri

Oculomotor nerve

Mamillary body

Medial lemniscus

Substantia nigra

FIG. 840   Transverse section of midbrain at level of superior colliculus.

corresponds to a functional difference, since the fibres ending on the posterior wall of the sulcus convey impulses from the superficial tissues, while the exposed part of the gyrus receives impulses from the deep tissues.

Anatomical and physiological techniques demonstrate that the projections from the posterior-ventral nucleus of the thalamus to the postcentral gyrus follow a strictly defined pattern. Hence the pattern transmitted to the thalamus by the specific ascending pathways is passed on to the postcentral gyrus. Thus impulses from the trigeminal area reach the lowest part of the gyrus, followed superiorly by the cervical segments in reversed order, the thoracic segments, and the lumbosacral segments; impulses from the perineum through the lowest sacral nerves arriving at the medial surface (paracentral lobule). Physiological methods also demonstrate that the relative *area of the cerebral cortex concerned with each part of the body is not equivalent to the size of the part, but to its importance as a sensory organ.* Thus in Man the areas which receive impulses from the face (particularly the lips), the hand and foot (particularly the thumb and great toe) are much larger than would be expected on the basis of their size alone, just as the snout takes up the major part of the cortical sensory area in the pig. The reversal of the cervical segmental representation brings the lower cervical segmental area into association with the trigeminal area; hence the two main human sensory organs, the hand and the oral region, are topographically associated in the postcentral cerebral cortex.

The postcentral cerebral cortex is the area concerned with general sensations from the body and is therefore known as the **somatic sensory area.** It seems likely that other sensations are relayed to this area, e.g., taste to the inferior part of the gyrus, and possibly some visceral sensations, such as the sensations of fullness from bladder and rectum, known to be transmitted through the spinothalamic system, because of their loss when it is divided (Nathan and Smith, 1951, 1953).

Lesions of the postcentral gyrus do not produce significant alterations in the subjective appreciation of painful and thermal sensations, though they do interfere markedly with position, tactile and vibration sensations from the opposite side of the body. For this reason, and because stimulation of

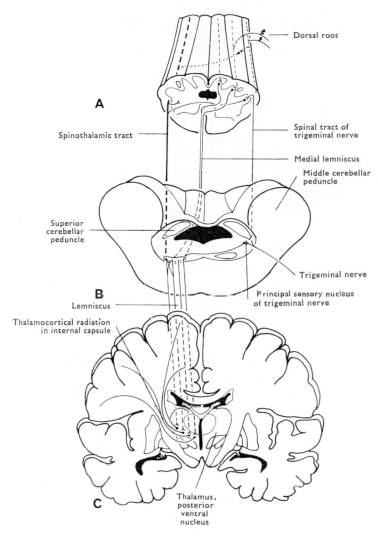

Spinothalamic tract

Superior cerebellar peduncle

Lemniscus

Thalamocortical radiation in internal capsule

Dorsal root

Spinal tract of trigeminal nerve

Medial lemniscus

Middle cerebellar peduncle

Trigeminal nerve

Principal sensory nucleus of trigeminal nerve

Thalamus, posterior ventral nucleus

FIG. 841   Diagram to show the course of the medial lemniscus and spinothalamic tracts through the brain, seen from above.

A, low medulla oblongata; B, pons; C, thalamocortical projections.

the corresponding primary areas is obscure, but the reversal of representation relative to that in the primary area seems to be common to all of them. The secondary somaesthetic area appears to be predominantly concerned with impulses carried by the spinothalamic system.

THE AUDITORY SYSTEM

The vestibulocochlear nerve enters the lateral aspect of the medulla oblongata at its junction with the pons. The cochlear fibres arch over the lateral and posterior surfaces of the inferior cerebellar peduncle to end in the cochlear nuclei which form a ridge of grey matter in this position, anterior to the lateral recess of the fourth ventricle.

The **cochlear nuclei** [FIG. 760], on which the fibres from the cochlea end in a well-defined pattern, are usually divided into the dorsal and ventral parts according to their relation to the inferior cerebellar peduncle. Each gives rise to fibres which pass superomedially, dorsal and ventral to the inferior cerebellar peduncle, to enter the dorsal part of the pons where they incline medially towards the midline forming a mass of crossing fibres, the **trapezoid body** [FIGS. 842 and 843], in association with which are the **nuclei of the trapezoid body** on which some of the fibres end by forming complex synapses. The fibres which cross in the trapezoid body pass through the substance of the medial and spinal lemnisci to turn headwards between the latter and the anterior spinocerebellar tract as the **lateral lemniscus** [FIG 844], the ascending auditory pathway [p. 662]. The latter is always recognizable in sections of the superior part of the pons by the presence of an elongated nucleus, the **nucleus of the lateral lemniscus,** within its substance [FIG. 759A].

The auditory pathway in the brain stem is characterized by the large number of cells associated with it throughout its course. Immediately anterior to the motor nucleus of the facial nerve at the lateral edge of the trapezoid body lies the **dorsal nucleus of the trapezoid body,** forming a slightly coiled mass of cells on which end fibres of the trapezoid body originating on both sides. This nucleus gives rise to fibres in the lateral lemniscus and to a bundle of fibres passing posteriorly towards the floor of the fourth ventricle. The latter bundle (peduncle of the dorsal nucleus of the trapezoid body) is said to make connexions with other motor cranial nerve nuclei, e.g., the facial, through which blink reflexes in relation to loud sounds may be effected, and has been suggested as the source of the efferent cochlear fibres (Rasmussen, 1953), which have been traced in the vestibulocochlear nerve to the cochlea. When these fibres are stimulated, they block the impulses in the cochlear nerve which would normally arise in response to noise (Galambos, 1956). Groups of cells are also present among the fibres of the trapezoid body, the **ventral nucleus of the trapezoid body.** In their turn the fibres of the lateral lemniscus surround the nucleus of that name and most of them ascending over the superior cerebellar peduncle [FIG. 805] end in the **inferior colliculus** [p. 662]. A few continue to the **medial geniculate body,** through the **brachium of the**

the postcentral gyrus in the unanaesthetized patient produces numbness, or sensations of light touch, or tingling in the appropriate part of the body, but does not evoke the sensation of pain, it has been assumed that the former sensations are appreciated here but that pain is appreciated in other parts of the brain. There is, however, evidence of alteration of painful sensations following lesions of the postcentral gyrus, and the failure to demonstrate it convincingly may well result from the fact that the spinothalamic system is bilaterally arranged. Also, because of the intermingling of impulses from different parts of the body at the thalamic level, pain and temperature sensations may well be diffusely represented in the postcentral cortex, so that localized lesions do not produce localized loss of pain sensation.

In addition to the main somaesthetic area, a secondary sensory area has been demonstrated by physiological means. This appears to lie at the lowest extremity of the postcentral gyrus, extending on to the inferior surface of the parietal operculum and insula. There is some evidence of a reversed but bilateral topographical representation of the parts of the body in this area in animals. The significance of this and similar secondary auditory and visual areas associated with

FIG. 842   Two sections through the pons. A is taken at a level slightly inferior to B.

inferior colliculus [FIGS. 758 and 805] which is mainly formed by fibres arising in the inferior colliculus.

The numerous nuclei on the course of the auditory pathway, cause the impulses that reach the medial geniculate body for transmission to the cerebral cortex (for conscious appreciation) to have passed through several synapses, as compared with the single synapse involved in other ascending pathways to the thalamus, e.g., the medial lemniscus. In keeping with the important reflex function of the various nuclei in the auditory pathway is their extreme development in the bat, which uses auditory impulses, generated by the reflection of its high-frequency cries, to locate and catch insects in flight and to avoid obstacles. This is a kind of echo-location which requires very rapid reflex adjustments to the flight of

the bat, and indicates the potential complexity of the connexions of the auditory pathway for reflexes.

The fibres in each lateral lemniscus arise from both right and left cochlear nuclei, some reaching it via the nuclei of the trapezoid body, and thus each lateral lemniscus contains fibres carrying impulses from both sides. Naturally this is also true of all parts of the auditory pathway cranial to the formation of the lateral lemniscus in the pons; hence unilateral lesions of this pathway or the cerebral cortex to which it finally passes, **superior temporal gyrus,** produce little or no loss of hearing. Unilateral deafness of nervous origin arises as a result of injury to the cochlear nerve, the cochlear nuclei, or the fibres passing from these nuclei to the trapezoid body.

The **medial geniculate body** [FIGS. 757 and 776] gives rise to fibres which leave its anterolateral aspect and passing laterally enter the sublentiform part of the internal capsule. In this they pass, superior to the inferior horn of the lateral ventricle, to mingle with fibres of the superior longitudinal bundle curving anteriorly into the temporal lobe and the posterior fibres of the fasciculus uncinatus and anterior commissure. The fibres curve superiorly to enter the superior temporal gyrus, and end mainly on its superior surface in the transverse temporal gyrus which abuts on the fronto-parietal operculum, but this receptive region also extends on to a small area of the lateral surface of the superior temporal gyrus [FIG. 845]. Thus, much of the auditory receptive area is hidden from view in the intact brain, as indeed are the major parts of the somatic sensory and visual areas. Within this auditory area of the superior temporal gyrus, fibres carrying impulses from the various parts of the cochlea end in a definite topographical manner in

FIG. 843   Section through the pons at level of entry of the trigeminal nerve.

FIG. 844 Diagram of the main connexions of the vestibulocochlear nerve and nuclei.

(1) Superior, (2) lateral, (3) medial, and (4) inferior vestibular nuclei.

animals, those from the apical parts, concerned with low frequencies, posterior to those from the basal parts ; the arrangement in Man is unknown. Further auditory areas have been described in animals but their presence in Man is unconfirmed.

If, as seems likely, they exist, they are probably in close association with the primary auditory area (Whitfield, 1967). The adjacent temporal cortex is believed to be responsible for interpreting the impulses which arrive at the primary receptive area.

THE VISUAL SYSTEM

The structure of the retina is illustrated in FIGURES 987 and 988. The sensory cells, the rods and cones, lie in the external layer of the retina and activate bipolar cells of the middle layer, from which impulses are passed to the innermost layer, consisting of ganglion cells. The density of packing of the sensory cells, and the number of bipolar and ganglion cells per unit area, increase markedly towards the posterior part of the retina, reaching its maximum in the region of the fovea centralis. In this region the number of fibres per unit area arising from ganglion cells and entering the optic nerve is greatest, there being virtually a 1 : 1 relationship between the tightly packed, slender cones of the fovea and the ganglion cells. Hence the area of the retina covered by a single optic nerve fibre in this region is extremely small, and the visual definition is correspondingly good. By comparison, in the periphery of the retina a single ganglion cell receives impulses from a large number

FIG. 845 Part of left cerebral hemisphere with opercula widely separated to expose insula and superior surface of temporal operculum.

Auditory area, blue.

of rods spread over a not inconsiderable area, and the definition is correspondingly reduced, even though the sensitivity to light is increased.

The fibres of the ganglion cells converge radially on the optic nerve, with the exception of the fibres from the temporal parts of the retina which lie adjacent to a horizontal line passing through the fovea; these curve superiorly or inferiorly so as to pass round the fovea. Thus fibres from the foveal region pass directly medially and enter the optic nerve on its lateral side. In this region the axons of the ganglion cells develop a myelin sheath and course posteriorly in the optic nerve through the optic canal towards the **optic chiasma.** This is situated at the inferior end of the lamina terminalis, above the sella turcica and some distance posterior to the sulcus chiasmatis on the sphenoid bone [FIG. 874]. Within the optic nerve the fibres from the central region of the retina take up a central position, with the fibres from the more peripheral fields arranged around them more or less in the same topographical position as the area of the retina from which they originate.

At the chiasma the two optic nerves meet and an interchange of fibres takes place so that the **optic tracts** which arise from the posterolateral angles of the chiasma contain some fibres from both eyes. The arrangement of fibres within the chiasma is not simple, but in general the fibres from the lower parts of the retina enter the inferior part of the chiasma, and the upper fibres the superior part, while the fibres from the foveal region lie between. Fibres which originate in the temporal half of each retina, including those from the temporal half of the foveal region, do not cross in the chiasma but pass back into the optic tract of the same side, while the fibres from the nasal halves of the retinae intermingle and cross to the opposite optic tract. In this manner each optic tract carries fibres from the parts of the two retinae which are stimulated by light coming from the opposite half of the field of vision, in fact from the corresponding parts of the two retinae [FIG. 846].

The **optic tracts,** emerging from the chiasma, have the fibres from the central regions of the corresponding halves of the two retinae in their intermediate part, the fibres from the lower quadrant of each retina lateral, and those from the superior quadrant medial. Each optic tract passes posterolaterally, medial to the anterior perforated substance and the internal carotid artery, and lateral to the hypothalamus. The anterior choroidal artery and the basal vein, which subsequently crosses the optic tract, lie immediately lateral to it in the groove which separates the tract from the uncus. Having passed lateral to the mamillary body the optic tract runs on to the crus cerebri and, curving around its lateral side at the junction of the crus cerebri and the internal capsule, continues in a horizontal plane [FIG. 846]. It appears partially to split into two roots as it leaves the crus cerebri to lie on the posteroinferior part of the thalamus, immediately lateral to the medial geniculate body. Here the lateral root of the tract sinks into a poorly defined swelling, the **lateral geniculate body** [FIG. 757], which is the thalamic relay station for the visual fibres in the optic tract. The medial root sweeps medially and appears to enter the anterolateral part of the medial geniculate body. This root has long been considered to be one of a number of fibres using the optic chiasma as a convenient commissure; the apparent division of the tract is, however, due to the

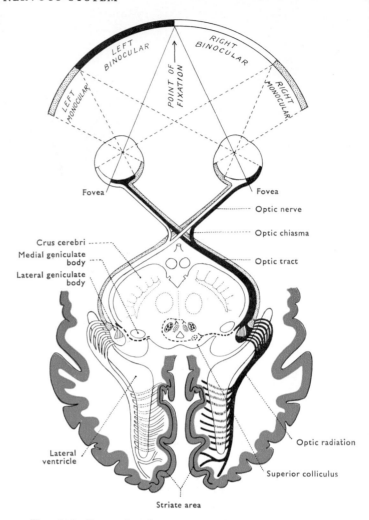

FIG. 846 Composite diagram to show the projection of the visual field upon the retinae, and the relationship of retinae to lateral geniculate bodies and striate area of cerebral hemispheres.

The fibres from the maculae, some crossed and some uncrossed, are not shown separately in the optic nerves and optic tracts, but their place of delivery in the lateral geniculate body—the intermediate part of its posterior two-thirds—is shown by hatching. The remainder of the posterior two-thirds receives the fibres from non-macular regions of the binocular retinal field. The monocular retinal field is shown projecting into the anterior third of the lateral geniculate body. No attempt has been made in this Figure to localize the delivery within the striate area of fibres from these different regions of the lateral geniculate body.

(After I. Mackenzie (1934) *J. Path. Bact.*, **39**, 113, and F. R. Winton and L. E. Bayliss (1948) *Human Physiology*, 3rd ed., London.)

inferior concavity of the hilus of the lateral geniculate body, and the medial root is the part of the optic tract passing to the medial portion of the lateral geniculate body (Wolff, 1953). In addition to these two major parts, a smaller bundle, which is not always visible on the surface of the brain, separates from the optic tract immediately anterior to the lateral geniculate body and, turning medially, passes over the postero-inferior aspect of the thalamus, superior to the medial geniculate body, to enter the anterolateral part of the superior colliculus where it frequently forms an obvious ridge. This bundle, the **brachium of the superior colliculus** [FIGS. 757 and 847], consists of collaterals of the optic tract fibres which pass to the optic stratum of the superior colliculus and are concerned with the

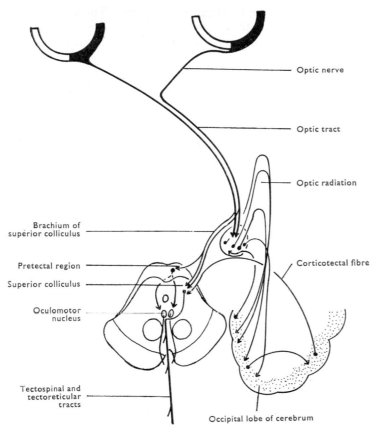

Optic nerve

Optic tract

Optic radiation

Brachium of
superior colliculus

Corticotectal fibre

Pretectal region

Superior colliculus

Oculomotor
nucleus

Tectospinal and
tectoreticular
tracts

Occipital lobe of cerebrum

FIG. 847 Composite diagram of the origin and distribution of fibres in one optic tract. The midbrain has been turned so as to be seen from below.

production of reflexes such as the involuntary turning of the head and eyes towards a sudden flash of light. From the complex laminated cortex of the superior colliculus, fibres arise which sweep ventrally skirting the **central grey substance** which surrounds the cerebral aqueduct and, crossing the midline ventral to the oculomotor nucleus (**dorsal tegmental decussation**), turn caudally. These descend immediately anterior to the medial longitudinal fasciculus [p. 672] and many are distributed to the reticular formation of the brain stem (**tectoreticular tract**); the remainder pass into the anterior funiculus of the spinal medulla (**tectospinal tract**) and end in its grey matter. Fibres from a similar source are believed to enter the cerebellum through the superior cerebellar peduncle and are distributed to part of the middle lobe of the cerebellar cortex [FIGS. 745 and 868].

Fibres of the brachium of the superior colliculus also pass to the small area immediately anterior to the superior colliculus, the **pretectal nucleus** [FIG. 848]. From the pretectal nucleus fibres arise which pass through the central grey substance of the midbrain to both oculomotor nuclei. These fibres terminate on the visceral efferent cells (the **accessory nucleus**) of the oculomotor nucleus which send preganglionic fibres through the oculomotor nerve to the **ciliary ganglion.** Axons of the large cells in this ganglion innervate the circular muscle of the iris and cause constriction of the pupil, in this case as a result of light falling on the retina.

Other fibres (the accessory optic tract) from the optic tract or from the brachium of the superior colliculus pass ventrally

around the crus cerebri to enter the interpeduncular region.

The fibres of the brachium of the superior colliculus are concerned with visual reflexes, while the fibres to the lateral geniculate body are concerned with transmission of retinal impulses to the cerebral cortex and their conscious appreciation.

The **lateral geniculate body** lies immediately medial to the choroid fissure as it sweeps inferiorly from the central part of the ventricle to pass anteriorly into the temporal lobe. In coronal section the body is a sinuous plate of grey matter, concave inferiorly in Man with the edges, especially the lateral, thinner and slightly everted. It consists essentially of six layers of cells separated by laminae of fibres consisting partly of fibres from the optic tract but mainly from the cells of the lateral geniculate body passing to the cerebral cortex. The six layers which are most pronounced in the intermediate region of the posterior part, tend to fuse together and become less distinct at the edges and are numbered serially from the inferior surface. The functions of the various layers are probably different, firstly because layers 1 and 2 are composed of larger cells than the others and are not concerned with colour vision; and secondly because layers 1, 4, and 6 receive fibres from the nasal half of the contralateral retina, while layers 2, 3, and 5 receive fibres from the temporal half of the ipsilateral retina [FIG. 849]. These connexions have been demonstrated by making restricted lesions in the retina and following the degenerating optic fibres to their termination. Such experiments have also shown that each spot on the retina is connected to the corresponding parts of all three layers. In Man each optic tract fibre may end predominantly on a single geniculate neuron, at least in the postero-intermediate part of the geniculate body which receives fibres from the central parts of the retina. The more peripheral parts of the retina send their fibres to the anteromedial (superior retina) and anterolateral (inferior retina) parts. The presence of three layers concerned with each spot on the retina, and the well-developed nature of these layers in the part of the geniculate body to which the fibres from the highly

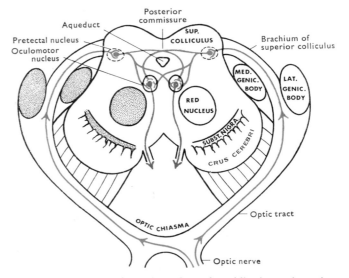

FIG. 848 Diagram of section through midbrain and optic chiasma showing the path of pupillary contrictor fibres. (After S. W. Ranson and H. W. Magoun (1933) *Arch. Neurol. Psychiat. (Chic.)*, **30**, 1193.)

FIG. 849   Diagram to show relationship of crossed and uncrossed fibres from retinae to cell laminae of lateral geniculate bodies.

The macular fibres, some of which are crossed and some uncrossed, are not shown separately. They are delivered to the cells that lie between the medial and lateral edges of the body in its posterior two-thirds [cf. Fig. 846].

(From I. Mackenzie (1934) *J. Path. Bact.*, **39**, 113, after Le Gros Clark.)

colour-sensitive central region of the retina pass, has led to the suggestion that each layer may be concerned with one primary colour. However, recent physiological experiments seem to show that the different layers are concerned with other retinal functions (De Valois, 1960).

The **efferent fibres** of the lateral geniculate body leave it as a compact bundle and run anteriorly, to pass within the concavity of the lateral ventricle [FIG. 850] and turn backwards towards the occipital lobe, separated from the ventricle by the tapetum. In general, the central fibres of this expanding sheet are those from the postero-intermediate part of the lateral geniculate body and are concerned with central vision, while the superior fibres carry impulses which originate in the upper peripheral parts of the retinae and the inferior fibres convey

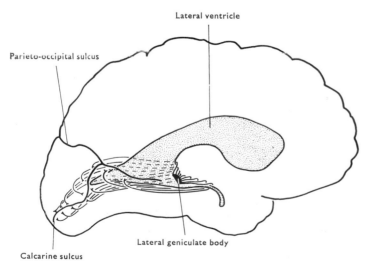

FIG. 850   Diagram of the optic radiation seen from the medial side. Note that the course of the lowest fibres allows them to reach the cortex anterior to the uppermost fibres.

impulses from the lower peripheral parts. The lowest fibres run forwards into the temporal lobe and, turning posteriorly, sweep beneath the posterior horn of the ventricle to curve superiorly on its medial side. Thus they reach the inferior aspect of the calcarine sulcus, which cuts deeply into this aspect of the hemisphere. The superior fibres similarly pass over the superior aspect of the ventricle and turn inferiorly towards the superior surface of the calcarine sulcus. The fibres at the margins of the radiation escape to the medial aspect of the occipital lobe anterior to the central fibres which pass towards the occipital pole, and may even extend with the sulcus to the cortex on the lateral surface of the pole. Thus the upper half of the radiation passes to the cerebral cortex on the superior surface of the calcarine fissure while the lower half is distributed to its lower surface [FIG. 867, 17]. The fibres carrying impulses from the central regions of the retina occupy a disproportionately large part of the cortex per unit area of the retina. Thus a large part of the polar cortex to which these fibres pass is taken up with macular vision which subtends only 1 degree of arc in the field of vision.

The visual system is of great *clinical importance* because defects are readily noticed by the patient, and since its intracranial extent is from the chiasma to the occipital pole, lying across the general direction of most of the long pathways in the brain stem, it is subject to damage throughout this extent. The eye is a simple lens system, hence the image on the retina is inverted and reversed. Any lesion of the visual system on the pathway to the occipital cerebral cortex produces defects in the field of vision which are the reverse of the parts of the retina thereby disconnected from the cortex. Thus, damage to fibres carrying impulses from the lower nasal quadrant of one retina causes loss of vision in the upper temporal quadrant of the field of vision of that eye. The effects of lesions in the various parts of the pathway can be understood readily provided it is remembered that all visual defects are described in terms of their effects on the field of vision :

1. Lesions of the retina cause blind spots involving the field of vision of that eye. These are frequently unnoticed if they involve the binocular field since the corresponding pathway from the opposite eye is intact.

2. Lesions of the optic nerve, if complete, cause blindness of that eye.

3. A midline section of the chiasma, by cutting the crossing fibres from the nasal halves of both retinae, causes loss of vision in both temporal fields—bitemporal hemianopia. If the lower fibres are affected first, as by external pressure, the loss of vision will begin in the upper temporal quadrants of the field of vision. Conversely, if the pressure involves the upper fibres, the lower temporal quadrants are affected.

4. Injury to an optic tract, by interfering with fibres coming from the corresponding parts of the two retinae, causes blindness involving the opposite half of the field of vision, homonymous hemianopia. Such an effect is also produced either by destruction of the lateral geniculate body, or by complete division of one optic radiation, or by destruction of the visual cortex on one side.

5. Because of the spread-out nature of the fibres comprising the optic radiation and the extent of the calcarine cortex, it is possible to damage these partially, thus producing blind spots

(scotomata) in the opposite half of the field of vision. Such scotomata involve the corresponding parts of both eyes and are therefore more obvious to the patient than a small, unilateral retinal lesion. They are present in the upper field of vision if the lower radiation or cortical area is involved, and in the lower field if the upper regions are affected.

Of the reflex pathways, the connexion with the pretectal region is important as an index of the integrity of the visual pathway. Since both pretectal regions, right and left, activate both third nerve nuclei, impulses passing through either optic tract produce an equal (consensual) contraction of both pupils. Thus with simple methods of testing, such as exposure of one eye to a bright light, a bilateral response will always occur even if one optic tract or one brachium of the superior colliculus is damaged, but no response in either eye will be elicited if the optic nerve of the exposed eye is destroyed. In the latter case a response will be elicited in the pupil of both eyes if the other eye is exposed to light. Unilateral loss of pupillary response to light is probably only found in lesions involving the efferent pathways, i.e., the oculomotor nerve, the ciliary ganglion or its branches.

THE RETICULAR FORMATION OF THE BRAIN STEM [FIGS. 757–761]

It is difficult to define exactly what is meant by the above term which is used in various ways by anatomists and physiologists. It is an anatomical concept and is properly applied to those parts of the brain stem (midbrain, pons, and medulla oblongata) composed of scattered cells of various sizes intermingled with nerve fibres running in all directions, but principally longitudinally. In the upper medulla oblongata it lies dorsal to the olive, extending from the region of the spinocerebellar tracts laterally to the medial lemniscus medially; further caudally it tapers to become continuous with the reticular formation of the upper cervical spinal medulla which lies at the medial edge of the lateral funiculus. In the pons it forms a large portion of the dorsal part, between the lemniscus and the floor of the fourth ventricle; while in the midbrain it forms the central part of each half of the tegmentum, between the central grey substance and the substantia nigra, and from the red nucleus medially to the lateral aspect of the mesencephalon.

The term 'reticular formation' is not usually applied to well-defined nuclei such as the red nucleus, the olivary nucleus or the vestibular nuclei which are closely associated with it. It is a purely descriptive term and, to some extent, a cloak for ignorance, indicating, as it does, the inability to analyse this type of neural complex under the microscope. It is obvious, however, that the close intermingling of nerve cells and fibres in the reticular formation permits complex intercommunications which would allow its cells to be used in different combinations for several functions. In aggregate it represents a collection of internuncial cells, the internal arrangements of which are not readily susceptible to anatomical analysis. Though it may be thought of as an anatomical entity, it is functionally very closely associated with the well-defined nuclei in close topographical relation to it.

Superficially, the reticular formation forms a continuous mass of similar tissue throughout the brain stem, but an analysis of its cyto-architecture and fibre connexions shows that its various parts are not as similar as would appear at first sight. Though the cells which lie in it are not collected into compact nuclei, certain groups can be recognized which have been shown to have different connexions. In general, it appears possible to divide the reticular formation into lateral and medial parts, and those parts which send fibres to the cerebellum. The lateral parts receive most, but not all, of the afferent fibres, and the medial parts give rise to the long ascending and descending pathways arising from the reticular formation.

Afferent fibres reach the reticular formation from a number of sources:

1. **Spinoreticular fibres,** passing with the spinothalamic system, are widely distributed in the reticular formation, but particularly to the medullary and pontine parts, from which ascending fibres arise and pass ipsilaterally and contralaterally through the pons and midbrain, probably to the thalamus. Some spinoreticular fibres end on the lateral reticular nucleus of the medulla oblongata which seems to act as a relay station for the impulses to the anterior lobe of the cerebellum [FIG. 852].

2. **Cerebelloreticular fibres** arise principally in the fastigial nucleus and are distributed to the medulla oblongata through the inferior cerebellar peduncle. Other fibres in the superior cerebellar peduncle give off branches which descend from the lower midbrain into the pontine and medullary reticular formation, as the descending limb of the superior cerebellar peduncle [FIG. 856].

3. **Cerebroreticular fibres** arise principally from the region of the central sulcus, but to a lesser extent from other parts of the cerebral cortex. They descend with the fibres of the corticospinal system and end principally in the cranial parts of the pontine and medullary reticular formation (the regions which give rise to most of the reticulospinal tracts) and also in the reticular nuclei which project to the cerebellum.

4. Some fibres which end in the reticular formation arise in the vestibular nuclei, the superior colliculi, and the red nucleus; others from the subthalamus and from the globus pallidus [p. 642] of the lentiform nucleus end in its cephalic part.

Efferent fibres. According to Brodal (1957) **long ascending fibres** arise mainly from the caudal parts of the pontine and medullary reticular formation and ascend bilaterally. Many of these fibres reach the thalamus (intralaminar, and reticular nuclei) and impulses are probably distributed from here to the cerebral cortex, the corpus striatum, and the hypothalamus. It has been claimed on physiological grounds which lack anatomical confirmation, that the impulses passing in these ascending fibres are distributed throughout the cerebral cortex by a diffuse thalamic projection system, thought to originate in the intralaminar or reticular thalamic nuclei [FIG. 777].

The **reticulospinal tracts** arise in the pons and medulla oblongata, particularly their superior parts. Bilateral pathways arise from each half of the medulla oblongata, the crossed pathway descending with the medial longitudinal fasciculus and the uncrossed pontine reticulospinal fibres into the anterior funiculus of the spinal medulla. The uncrossed medullary pathway, or pathways, descend in the anterolateral region of the white matter as scattered fibres in the cat; some lie as far posteriorly as the rubrospinal tract. The reticulospinal fibres end principally in the region of the anterior horn of the grey matter of the spinal medulla, and many are believed to terminate on the cells which give rise to the motor fibres to muscle spindles (γ-efferents).

Thus the reticular formation is: (1) a link in ascending pathways to the cerebellum, thalamus, corpus striatum, and cerebral cortex; and (2) a station on the descending pathways for most of the major masses of grey matter in the brain. In those mammals which have a rudimentary corticospinal system, e.g., the sheep, the reticulospinal tracts form a major part of the connexion between brain and spinal medulla, being reinforced by the rubrospinal [p. 676], tectospinal, vestibulo-spinal [FIG. 844], and olivospinal tracts.

In addition to its functions of controlling muscle tone, in conjunction with the parts of the brain acting through it, it is intimately concerned with the control of respiration and the cardiovascular system—a feature which is not unexpected in view of the fact that in lower forms (e.g., the fishes) the gills, heart, and great vessels lie opposite the hindbrain and are innervated by the nerves arising from it. The neural mechanisms developed for the control of these structures are retained in the brain stem of Man, even though gills are superseded by lungs and the heart is displaced caudally relative to the central nervous system.

THE VESTIBULAR SYSTEM

Nerve fibres of the vestibular part of the vestibulocochlear nerve are distributed to the vestibular nuclei. These lie under the raised, triangular vestibular area which fills out the lateral part of the floor of the fourth ventricle, medial to the lateral recess, and extends from the midpons almost to the obex. Other vestibular fibres pass directly to the cerebellum by coursing dorsally, medial to the inferior cerebellar peduncle.

The Vestibular Nuclei [FIG. 844].

All four vestibular nuclei, lateral (Deiters'), medial, superior, and inferior receive branches of the vestibular fibres of the vestibulocochlear nerve, though these are not distributed to all parts of the nuclei, but rather to those parts which do not give rise to efferent fibres. Other afferent fibres reach these nuclei from the posterior spinocerebellar tract as it passes in the inferior cerebellar peduncle, also from the cerebellum (mainly its fastigial nuclei and vermal cortex), and from the reticular formation of the midbrain and medulla oblongata. Efferent fibres from the vestibular nuclei pass throughout the length of the midbrain, hindbrain, and spinal medulla and are distributed to both sides. They fall into three groups:

1. Direct connexions with motor cells of the brain stem and spinal medulla. The lateral and superior nuclei give rise mainly to uncrossed fibres which respectively pass inferiorly and superiorly. The former arise from the large cells of the **lateral nucleus** and, passing inferomedially through the medulla oblongata, enter the anterolateral region of the white matter of the spinal medulla, sending fibres into the anterior horn. Fibres from the **superior nucleus** pass superomedially

under the floor of the fourth ventricle [FIG. 759], to reach the medial longitudinal fasciculus and run headwards in it through the midbrain to the level of the posterior commissure. Fibres are given off to the oculomotor, trochlear, and abducent nerve nuclei and to the tegmentum of the midbrain.

The **medial** and **inferior nuclei** give rise to fibres which pass into the medial longitudinal fasciculus of the same and of the opposite sides and extend longitudinally to the midbrain and spinal medulla.

The **medial longitudinal fasciculus** lies close to the midline and immediately beneath the floor of the fourth ventricle and the aqueduct in the midbrain. Caudally it is continuous with the fasciculi proprii of the anterior funiculus of the spinal medulla. Essentially, the medial longitudinal fasciculus is the brain stem continuation of that part of the fasciculi proprii of the spinal medulla closely associated with, and connecting the various levels of the anterior grey column. Within the brain stem the fasciculus forms a pathway for communications between the cranial nerve nuclei; fibres from the vestibular nuclei form the major part of this fasciculus.

Through their connexions, which are bilateral for each group of vestibular nuclei, right and left, vestibular impulses can effect changes in the tone of the extrinsic ocular muscles and the muscles of the neck, trunk, and limbs. Such arrangements allow of compensatory movements of the eyes and trunk in association with movements of the head, including nystagmus, the oscillatory eye movements which occur on rotation of the head and which persist after rotation so long as the endolymph is in motion in the appropriate semicircular ducts. There is evidence of a specific relationship between the parts of the vestibular apparatus of the internal ear, the vestibular nuclei, and the motor nuclei of the oculomotor, trochlear and abducent nerves. Thus, the major connexions of the ampullae of the lateral semicircular ducts are with the contralateral abducent and the ipsilateral part of the third nerve nucleus supplying the medial rectus [FIG. 851]. Such

FIG. 851 Diagram of the vestibulo-ocular connexions concerned in horizontal eye movements.

connexions would account for the slow component of the nystagmus, a co-ordinated movement of both eyes towards the opposite side. It should be appreciated that the semi-circular ducts in the right and left vestibular apparatus are so arranged as to give complementary information since they form corresponding pairs lying in the same plane, e.g., right and left lateral, and right anterior with left posterior semi-circular ducts, only one of each pair being maximally stimulated by rotation of the head in one direction parallel to a pair of canals, the other being activated by rotation in the opposite direction in the same plane.

2. The vestibular nuclei have reciprocal connexions with the **flocculonodular lobe** and **fastigial nuclei** of the cerebellum, as well as many connexions with the reticular formation of the brain stem. It seems likely that these connexions are concerned with the postural control of muscle tone in relation to gravitational forces acting on the maculae of the utricle and saccule, and with similar reactions to movement (e.g., the maintenance of balance) mediated through the reticular formation; also the rapid eye adjustments necessary during head movements in running and in walking, in order to stabilize the visual image on the retina. The anatomical pathways subserving these reactions are obscure.

3. The pathway by which vestibular impulses are transmitted to the cerebral cortex is unknown, though a group of fibres has been described which follows the path of the auditory fibres and presumably is relayed in the posterior part of the thalamus to the temporal lobe. It is, however, extremely difficult to analyse the efferent connexions of the vestibular nuclei as a result of lesions placed in them, because of the coincident injury to auditory and other fibres passing through them.

THE CEREBELLUM

The development and general arrangement of the cerebellum has already been considered [p. 601]. It has been noted that it consists of a highly convoluted cortex overlying a central mass of white matter in the depths of which are the nuclei of the cerebellum, closely related to the fastigial recess of the fourth ventricle [FIGS. 739 and 743]. The cortex is divisble into a median part, or **vermis**, and the large lateral lobes, **hemispheres**, which are directly continuous with the vermis and of exactly the same histological structure. On the inferior surface, the vermis (pyramid, uvula, and nodule) is separated by a deep groove from the corresponding parts of the hemispheres, the biventral lobule, the tonsil, and the flocculus. On the superior surface no clear line of demarcation exists between vermis and hemispheres, the parts of which are indicated in FIGURES 745 and 747.

Within each cerebellar hemisphere, the largest nucleus is the laterally placed **dentate nucleus** formed from a crinkled layer of grey matter, arranged in the shape of an open bag with the mouth, the **hilus**, facing anterosuperiorly, and the superior cerebellar peduncle emerging from it. Medial to the dentate lie three smaller, more nearly spherical nuclei, the **emboliform**, **globose**, and **fastigial**, the latter lying close to the midline [FIG. 743].

Afferent Cerebellar Connexions. These are numerous, and are mostly distributed to the cerebellar cortex, though a

C.T.A.—Z

few pass directly to the fastigial nuclei from the vestibular nerves and nuclei.

1. Fibres pass to the fastigial nuclei, flocculonodular lobe, and lingula from the **vestibular nerve** and some parts of the vestibular nuclei, especially the inferior. These fibres run dorsally, medial to the inferior cerebellar peduncle, and turn across the roof of the fourth ventricle, between it and the beginning of the superior cerebellar peduncle, to reach the fastigial nuclei and the nodule. As they lie in the roof of the fourth ventricle, the fibres are in the medial part of the attachment of the inferior medullary velum to the cerebellum, and some of them enter the **peduncles of flocculi,** to which the inferior medullary velum is attached [FIG. 745], to reach the flocculi [FIG. 748].

2. Fibres from the **spinal medulla** reach the cerebellum by way of the spinocerebellar tracts [FIG. 836] and the **posterior external arcuate fibres** [p. 659]. The latter, with the **posterior spinocerebellar tract,** enter the cerebellum through the inferior cerebellar peduncle [FIG. 852], and are distributed to the central lobule, the culmen, the declive, the pyramid, and the uvula, i.e., to the anterior and posterior parts of the vermis. The **anterior spinocerebellar tract** enters the cerebellum over the lateral surface of the superior cerebellar peduncle, and passing on to the superior medullary velum, enters the white matter of the cerebellum to be distributed to the same parts of the cortex cerebelli as the posterior spinocerebellar tract but in a more medial position. Anatomically there is no evidence of any localization in the distribution of these fibres to the cerebellar cortex, since the distribution of degeneration to the cerebellar cortex is the same whether the posterior spinocerebellar tract is divided at a lower or a higher level. Nevertheless, physiological experiments in animals have shown that there is a definite localization of the various parts of the body at least in the anterior lobe of the cerebellum [FIG. 853]. Thus, when the lower limb is stimulated, potentials due to impulses arriving at the cerebellar cortex are found in the lobulus centralis and adjacent parts of its ala. When, however, the upper limb is stimulated, the culmen and medial parts of the quadrangular lobule are

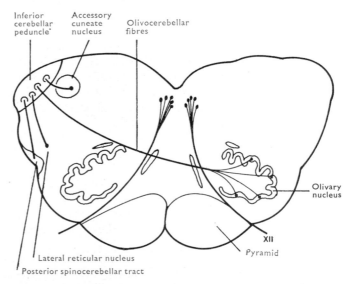

FIG. 852 Diagram of a transverse section of the medulla oblongata to show the fibres entering the inferior cerebellar peduncle.

FIG. 853 Diagram of projection areas of afferent cerebellar impulses from limbs and face on the cerebellar cortex of the monkey. (E. D. Adrian (1943) *Brain*, **66**, 289.)

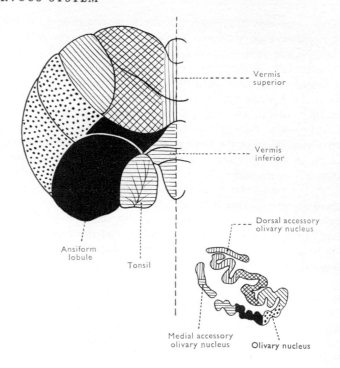

FIG. 854 Diagram to indicate the parts, respectively, of the right olivary nuclei in the medulla oblongata and of the left half of the cerebellum, which are linked together by olivocerebellar fibres. (After G. Holmes and T. G. Stewart (1908) *Brain*, **31**, 125.)

affected, and impulses from the head arrive at the anterior part of the declive and adjacent simple lobule. Fibres enter these parts of the cerebellum also from the lateral reticular nucleus [FIG. 852], which receives many spinoreticular fibres. There is evidence that spinoreticular fibres from different parts of the body end in separate parts of the lateral reticular nucleus (Brodal, 1957) and that these parts of this nucleus project in a similar manner to the vermis of the cerebellum, an arrangement which could account for the discrepancy between the anatomical and physiological findings.

3. In addition to the lateral reticular nucleus, fibres from certain other nuclei of the **reticular formation** also project to the cerebellum through the inferior cerebellar peduncle. These nuclei have connexions with the vermis and hemispheres of the cerebellum and receive spinoreticular and some cerebroreticular fibres.

4. A large number of afferent fibres reach the cerebellum from the opposite **olivary complex** in the medulla oblongata [FIG. 854]. These are distributed to all parts of the cerebellar cortex and are arranged in a definite point-to-point manner. The fibres arising in the medial accessory olive (which is situated medial to the hypoglossal rootlets) and the dorsal accessory olive pass to the vermis. These two nuclei form a much larger part of the olivary complex in lower mammals, e.g., the cat, than in Man, and it is of some interest that they are the site of termination of the spino-olivary fibres (Brodal, 1957). The main mass of the olivary nucleus, which has increased in size *pari passu* with the increase in size of the cerebral and cerebellar hemispheres, sends fibres to the lateral parts of the cerebellum and receives fibres from headward regions of the central nervous system. According to Walberg (1954), these fibres arise in the frontal cerebral cortex, the globus pallidus, the red nucleus and the peri-aqueductal, central grey substance of the midbrain. Thus, the olivary nucleus links the cerebral cortex, the corpus striatum, and the tegmentum of the midbrain to the cerebellum; the fact that its fibres are crossed is in keeping with the crossed efferent cerebellar pathway to these regions.

5. The largest single mass of afferents reaches the cerebellum from the opposite **pontine nuclei**, fibres from which form the massive middle cerebellar peduncle. Fibres from the pontine nuclei pass to all parts of the cerebellar cortex

except the flocculonodular lobe, and form a link between the cerebral cortex and the cerebellar cortex. Fibres reach the pons from all the lobes of the cerebrum and constitute the greater part of the crus cerebri. A brief inspection of the size of the crus cerebri and the pyramid gives some indication of the number of corticopontine fibres which the former transmits, since, apart from the fibres which pass through the pons to emerge as the pyramid, virtually all the fibres of the crus cerebri are corticopontine fibres.

On the ventral surface of the medulla oblongata, applied to the pyramids, are the **arcuate nuclei** [FIG. 869]. These give rise to fibres which pass dorsally in the raphe and laterally on the surface of the medulla, over or around the olive. Those which enter the raphe pass posteriorly to the floor of the fourth ventricle and then turn laterally across it as the **striae medullares** [FIG. 749], to enter the cerebellum by the inferior cerebellar peduncle. The other fibres, passing on the surface of the medulla, **anterior external arcuate fibres,** also reach the cerebellum. Both sets of fibres represent aberrant pontocerebellar fibres, and the arcuate nuclei are caudal extensions of the pontine nuclei.

6. **Auditory** and **visual fibres** also reach the cerebellar cortex. These have been demonstrated by recording impulses in that cortex following exposures of the eye to flashes of light, or the ear to controlled noise. The intermediate parts, the declive, folium, and tuber, and the corresponding parts of the hemisphere receive these impulses (Snider, 1950). The visual impulses from the superior colliculus probably enter through the superior cerebellar peduncle; the auditory impulses enter from the inferior colliculus by the same route, or through the inferior cerebellar peduncle from the cochlear nuclei (Fadiga and Pupilli, 1964).

It will be clear from the foregoing brief account that :

1. The cerebellum is not simply concerned with proprioceptive information but with all types of afferent information.

2. So far as its afferent connexions are concerned, each half of the cerebellum tends to be related to the same side of the spinal medulla, but to the opposite side of : (1) the cerebrum through the pons ; (2) the corpus striatum and the midbrain through the olive.

3. The afferent fibres are not uniformly distributed throughout the cortex, but tend to be relayed to specific regions, e.g., the vestibular and spinal afferent fibres, but several different afferents project to the same areas of the cerebellum.

The Structure of the Cerebellar Cortex

Most of the afferents to the cerebellum pass to the cortex and enter the deepest of its three layers [FIG. 855], the **granular layer.** This is composed of small granule cells which have complicated synaptic relations with the entering fibres which form **mossy endings,** and hence are called moss fibres. The granule cells send their axons peripherally through the stratum of large **Purkinje cell** bodies which forms the middle layer of the cortex into the outer or **molecular layer.** Here they turn through a right angle and run parallel to the surface of the cerebellum along the length of the folium. The molecular layer contains the dendrites of the Purkinje cells which form an extremely complicated branching mass spread out in one plane at right angles to the length of the folium, so that the axons of the granule cells pass among them. The Purkinje cells are therefore brought into synaptic contact with many thousands of granule cell axons, which pack the molecular layer, and synapse with the numerous spines which can be seen on Purkinje cell dendrites in Golgi preparations of the cerebellar cortex [FIG. 710].

Lying in the deepest parts of the molecular layer are other cells which send their axons parallel to the Purkinje cell layer to give off branches which encircle the bodies of several Purkinje cells in a complicated basketwork of synaptic contacts, the **basket cells.**

The **climbing fibres** enter the cortex, and running on to the Purkinje cell dendrites, extend over these dendrites forming many synapses with them. The exact origin of these fibres is not known, but it seems likely that most arise in the olivary nuclei (Szentagothai and Rajkovits, 1959).

The Purkinje cells give rise to axons which pierce the granular layer to enter the white matter and converge on the nuclei of the cerebellum. Most of the axons of the Purkinje cells give rise to branches which curve back into the cerebellar cortex and end, either on the cell of origin, or the adjacent Purkinje cells. Such recurrent collaterals may function to maintain or inhibit activity in the cerebellar cortex.

The cerebellar cortex is of uniform structure throughout, and is organized to bring together a great number of different impulses, which presumably interact at the surface of the Purkinje cells through their enormous synaptic field.

Efferent Cerebellar Connexions. The Purkinje cell axons (corticonuclear fibres) converge radially on the nuclei of the cerebellum, the vermis projecting to the fastigial nuclei, while the hemispheres send fibres principally to the dentate nuclei [FIG. 856]. That these fibres are arranged radially is indicated by the fact that the cephalic part of the fastigial nucleus receives fibres from the anterior part of the superior vermis, while its caudal part receives fibres from the inferior vermis. The same radial arrangement holds for the dentate nucleus, and the emboliform and globose (intermediate) nuclei receive fibres from the intermediate, **paravermal region** of the cerebellar cortex [FIG. 745]. No Purkinje cell axons leave the cerebellum except from the flocculonodular lobe directly to the vestibular nuclei.

The **fastigial nuclei** give rise to fibres to the vestibular and reticular nuclei of the medulla oblongata. These pass either directly through the ipsilateral inferior cerebellar peduncle, or by hooking over the contralateral superior cerebellar peduncle as the **hook bundle.**

The other cerebellar nuclei send efferent fibres into the **superior cerebellar peduncles,** passing anterosuperiorly into the midbrain beneath the inferior colliculi. In the lower midbrain these bundles sweep anteromedially and, intermingling in the midline [FIG. 839], cross to the opposite side ventral to the aqueduct, medial longitudinal bundle and trochlear nucleus. Having crossed, they ascend into the tegmentum of the upper midbrain where their fibres enter and surround the **red nuclei** [FIG. 840]. As the superior cerebellar peduncles are crossing, each gives rise to a descending bundle of fibres, the descending limb of the superior cerebellar peduncle, which passes to the pontine reticular

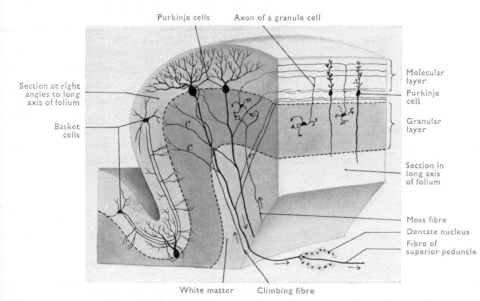

Section at right angles to long axis of folium

Basket cells

Purkinje cells   Axon of a granule cell

Molecular layer
Purkinje cell
Granular layer

Section in long axis of folium

Moss fibre
Dentate nucleus
Fibre of superior peduncle

White matter   Climbing fibre

FIG. 855 Diagram to show minute structure of a cerebellar folium. The surface to the right shows a section in the long axis of a folium and that to the left is a section at right angles to the long axis of a folium.

c, recurrent collaterals from the axons of Purkinje cells.

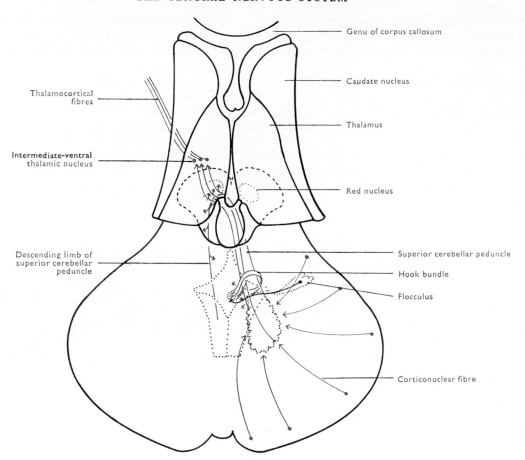

FIG. 856   Diagram of the main efferent connexions of the cerebellum, seen in a dorsal view of the brain stem.

formation and has been said to descend as far as the cervical spinal medulla.

The fibres of the superior cerebellar peduncle may end on the red nucleus, or pass beyond it into the **intermediate-ventral** and **intralaminar nuclei of the thalamus.** In the former they end around cells which send their axons to the precentral region of the frontal cerebral cortex, while the latter connexions form a route through which cerebellar impulses may reach the **corpus striatum** and other diencephalic structures.

The **red nucleus** consists of two groups of cells, a small, caudal, magnocellular part which gives rise to the **rubrospinal** and **rubroreticular tracts,** and a large parvocellular part which gives rise to ascending fibres. In mammals other than primates, the magnocellular part of the red nucleus is of considerable size and gives rise to fibres which immediately cross the midline in the ventral part of the tegmentum of the midbrain, the **ventral tegmental decussation.** These fibres turn caudally, pass through the dorsal parts of the pons and medulla oblongata to reach the lateral funiculus of the spinal medulla, and end in its grey matter. This rubrospinal tract gives off many fibres to the reticular formation of the pons and medulla oblongata, the rubroreticular tract. This part may form the major part of this pathway in Man, and constitutes most of the descending fibres from the red nucleus. The parvocellular part of the red nucleus gives rise to fibres

which are distributed to the thalamus, the midbrain reticular formation, and the cerebellum.

The cerebellum thus has extensive afferent and efferent fibres by which it is connected reciprocally with the vestibular nuclei, the reticular nuclei of the brain stem, the red nucleus, the thalamus, and through it with the cerebral cortex and corpus striatum. Functionally it has nothing to do with the conscious appreciation of sensation and has no direct mechanism for acting on the motor cells of the spinal medulla. All its activity must therefore be mediated through its action on the vestibular, reticular, and red nuclei, as well as on the cerebral cortex and corpus striatum. It is well known that the vestibulospinal, reticulospinal, and rubrospinal systems have marked effects on muscle tone in the trunk and limbs and are, presumably, the efferent pathways responsible for many of the righting reflexes known to be mediated in the brain stem. At least some of these pathways (e.g., reticulospinal) seem to act on muscle tone through the γ-efferent fibres, causing contraction or relaxation of muscle spindles and a consequent increase or decrease in the muscular contraction mediated by monosynaptic reflexes. There is evidence that the cerebellum plays a large part in controlling this activity, and it seems to be an essential element in the mechanism concerned with the correct distribution of tone in the muscles, necessary for the proper execution of voluntary and other activities. Cerebellar lesions in Man are usually associated

Labels on figure: Genu of corpus callosum; Caudate nucleus; Thalamus; Red nucleus; Superior cerebellar peduncle; Hook bundle; Flocculus; Corticonuclear fibre; Thalamocortical fibres; Intermediate-ventral thalamic nucleus; Descending limb of superior cerebellar peduncle with a marked decrease in muscle tone (atonia), and stimulation of the anterior lobe in animals decreases the decerebrate rigidity resulting from transection of the midbrain. Sprague and Chambers (1959) produce evidence, in the cat, of a functional difference between the vermis, the paravermal, and the lateral zones of the cortex, which project respectively by the fastigial, intermediate (globose and emboliform), and lateral (dentate) nuclei. Thus the vermis and the fastigial nuclei play a part in regulating bilaterally 'tone, posture, equilibrium, and locomotion of the whole body'. The paravermal and the lateral regions have their effects principally on the ipsilateral limbs, the activity of the two sides apparently being integrated in the brain stem.

Cerebellar lesions are characterized by the transitory nature of the signs which they produce, and most of these signs are potentiated by a simultaneous lesion of the sensorimotor cerebral cortex, or made to reappear by such a lesion if already 'compensated' for. The tremor during voluntary movement, which results from cerebellar cortical lesions, is dependent on the presence of the sensorimotor cerebral cortex, since destruction of this region abolishes it. It seems likely, in view of the massive connexions between the cerebellum and cerebrum, and the fact that both send fibres to the same brain stem nuclei, that their interaction is necessary for the smooth execution of normal movements which, in its turn, is dependent on the correct balance of tone in the opposing muscles.

The connexions of the flocculonodular lobe with the vestibular apparatus suggest a degree of control by the cerebellum on this system. This is confirmed in injuries of the flocculonodular lobe, when disturbances of balance are common without other cerebellar signs. Damage to the cerebellum in Man frequently produces nystagmus, an oscillating movement of the eyes most usually in the horizontal plane but possible in any direction. If horizontal, it is made most obvious by looking to one side. Nystagmus is common following lesions of the vestibular nuclei, and it may be that cerebellar nystagmus, which is similar in type, may represent a disturbance of the interaction of these two parts. On the other hand it seems likely that the cerebellum plays the same part in eye movements that it plays in other movements (Whitteridge, 1960), and if this is the case, the tremor of voluntary movement and of cerebellar nystagmus may have the same origin.

In general, the cerebellum is concerned in controlling, through the brain stem mechanisms, the adjustments in muscle tone which are necessary for posture and movement control (static and phasic tone). Each half of the cerebellum is predominantly concerned with the same side of the body and receives the sensory information necessary to carry out these adjustments through rapidly conducting, spinocerebellar pathways. In carrying out these actions, the cerebellum is intimately integrated with the other parts of the brain, which seem to be able to compensate for injuries to it.

THE DIENCEPHALON

The diencephalon consists of the **thalamus** dorsally and the **hypothalamus** ventrally, the latter being subdivided into a medial portion, the hypothalamus of the physiologist, and a lateral portion, the **subthalamus.** These parts have already

been described [pp. 627–30], including their positions, major nuclear groups, and connexions. The arrangement of the thalamic nuclei has been given in general [p. 627], and in more detail with each of the ascending pathways [pp. 662–9, Figs. 777 and 778].

The Thalamus

The thalamus is the part of the diencephalon which forms the great relay station for pathways leading to the telencephalon. Virtually all ascending fibres relay in this structure, and from its cells fibres arise which pass to the cerebral cortex and to the corpus striatum and hypothalamus; the only exception to this rule being a group of ascending fibres which pass directly to the cerebral cortex through the pyramid, crus cerebri, and internal capsule. These form a very small proportion of the afferent fibres to the cerebral cortex and their origin and destination are uncertain, though it seems that they arise in the cervical spinal medulla in Man (Nathan and Smith, 1955 b), and possibly throughout the spinal medulla and from the nuclei cuneatus and gracilis in the cat (Brodal and Walberg, 1952), the contribution from the cervical spinal medulla being larger than that from more caudal levels.

Each thalamic nucleus which projects to the cerebral cortex does so to a restricted area from which it receives corticothalamic fibres [FIG. 857]. Thus, the cerebral cortex can have direct effects on the thalamus, and it is possible to envisage

FIG. 857 Inferomedial and lateral views of the cerebral hemisphere, showing the projection areas of the thalamic nuclei.

these effects facilitating or inhibiting the passage of impulses through certain parts of the thalamus, or even being responsible for setting up continuous thalamocortical and cortico-thalamic activity.

The **thalamic nuclei** can be divided into :

1. **Nuclei transmitting specific ascending pathways to limited areas of the cerebral cortex.** These include :

(i) The **medial** (auditory and ? vestibular) and **lateral** (visual) **geniculate bodies** which project to the superior temporal and calcarine (striate) cortex respectively [pp. 666 and 669].

(ii) The **posterior-ventral nucleus** which relays the fibres of the spinal, medial, and trigeminal lemnisci to the postcentral gyrus of the parietal lobe, and is presumably also concerned with gustatory sensations which find their way to the same postcentral region [p. 663].

(iii) The **intermediate-ventral nucleus** which receives fibres from the superior cerebellar peduncle and red nucleus, and projects to the posterior part of the frontal lobe, including the precentral region.

2. **Nuclei which project to the cerebral cortex but do not form the termination of definite ascending pathways.** The reticular nucleus may belong to this group but is considered later [p. 679].

(i) The **medial nucleus** which has connexions with the hypothalamus, other thalamic nuclei, and the anterior part of the frontal lobe of the cerebrum through the anterior limb of the internal capsule. These thalamofrontal and fronto-thalamic fibres are divided in the operation of prefrontal leucotomy and may be responsible, at least in part, for the changes in emotional reactivity which this operation causes. Possibly because they form a route of communication between the frontal lobes and the hypothalamus, this system may be responsible for the visceral responses so often connected with emotional states, e.g., sweating, blushing, vomiting, increased pulse rate, etc.

(ii) The **dorsal-lateral nucleus** receives fibres from other thalamic nuclei and projects to the parietal lobe (except for the post-central region to which the posterior-ventral nucleus sends fibres) and probably also to the temporal lobe. The posterior extension of the lateral nucleus, the **pulvinar,** has similar connexions with the posterior parietal, occipital, and temporal lobes and seems to be concerned with the integration of auditory and visual impulses.

(iii) The **anterior nuclei** receive fibres from the fornix which leave it in the region of the anterior commissure [FIG. 816], and from the hypothalamus in the **mamillothalamic tract.** The various parts of the anterior nuclei project to the **gyrus cinguli** ; the significance of this system remains obscure. It forms a part of the so-called rhinencephalon, but it is very doubtful if this has more than the most remote connexion with the olfactory apparatus. The afferent fibres to the mamillary body come mostly through the fornix from the hippocampus [p. 646] ; the gyrus cinguli sends efferent fibres to the parahippocampal gyrus (entorhinal area), with which it is directly continuous, and the parahippocampal gyrus projects on to the hippocampus. Thus, a complex series of intercommunications is formed between the parts of the marginal cortex of the cerebral hemisphere, the so-called limbic system. It should be remembered that in marsupials [FIG. 817], where there is no corpus callosum and the hippocampus and dentate gyrus extend to the anterior commissure, the cerebral cortex corresponding to the gyrus cinguli bears the same relation to the hippocampus as the parahippocampal gyrus does in Man, both lying immediately adjacent to it, separated only by the narrow strip of cortex known as the subiculum. There are therefore intimate connexions between the parts of this marginal system [p. 645], which is connected with the hypothalamus by way of the fimbria and fornix, and with the brain stem, both indirectly through the mamillary body, and directly through the fornix.

3. **Thalamic nuclei with subcortical connexions.**

(i) The **anterior-ventral nucleus** receives fibres from the globus pallidus of the lentiform nucleus and sends efferent fibres to the corpus striatum as well as the cortex.

(ii) The **habenular nucleus** lies in the habenular triangle at the root of the pineal body, anterosuperior to the tectum of the midbrain and medial to the posterior part of the thalamus. Fibres enter the habenular triangle through the **medullary stria of the thalamus** which begins anteriorly in the anterior perforated substance (olfactory trigone) and amygdaloid body, and is reinforced by fibres from the septal region which join it near the interventricular foramen. From this point the bundle forms a raised ridge separating the medial and superior surfaces of the thalamus and, with the thin roof of the third ventricle attached to its medial margin, passes posteriorly to enter the habenular triangle. Fibres from the habenular nucleus cross to the opposite side in the **habenular commissure,** which lies in the root of the pineal body above the pineal recess of the third ventricle. A large bundle of fibres, the **fasciculus retroflexus,** passes antero-inferiorly from each habenular triangle, through the midbrain-diencephalic junction and the medial part of the red nucleus, to the **interpeduncular nucleus** on the ventral surface of the midbrain. The functions of this system are unknown but the interpeduncular nucleus probably forms part of the efferent pathway for this system, since it has connexions with the tegmentum of the lower midbrain.

(iii) The **central nucleus,** the largest nucleus lying in the medial medullary lamina, separates the medial part of the posterior-ventral nucleus from the medial nucleus [FIG. 778]. In common with the other intralaminar nuclei, it does not undergo degeneration following injury to the cerebral cortex (other than the rhinencephalon), as the nuclei with cortical projections do. Together with the other intralaminar nuclei, it has been found to undergo changes following injury to the corpus striatum and parts of the rhinencephalon (Cowan and Powell, 1955), and it seems that the main efferent fibres from these nuclei pass to the caudate nucleus, the putamen of the lentiform nucleus, and the anterior-ventral nucleus of the thalamus. The origin of its afferent fibres is not certain, but connexions have been described with the spinal medulla (collaterals of specific ascending pathways, e.g., spinothalamic tract (Getz, 1952)), the cerebral cortex, the striatum through the ansa lenticularis, and from the reticular formation of the brain stem (reticulothalamic fibres). Functionally this nucleus has been associated with the 'diffuse thalamic projections' which produce marked changes in the electrical activity of the cerebral cortex when they are stimulated. The efferent fibres of the central nucleus pass by way of the inferior thalamic radiation to the striatum and possibly to other thalamic nuclei, including the reticular nucleus.

The Reticular Nucleus of the Thalamus. This nucleus forms a thin shell external to the lateral medullary lamina and covers virtually the whole lateral surface of the thalamus. It has been claimed that this nucleus projects to a large area of the cerebral cortex (Chow, 1952). Though it receives fibres from every part of that structure (Carman, Cowan, and Powell, 1964) it appears to project mainly in a caudal direction (Schibel and Schibel, 1966) to other thalamic nuclei and to the midbrain. Since it forms a capsule for the thalamus, it is in a position to sample the thalamocortical and corticothalamic impulses traversing it. Thus it may form a feed-back mechanism, but does not seem to be part of the direct pathway for the so-called diffuse thalamic projection through which the reticular formation of the brain stem is known to alter the activity of the entire cerebral cortex. The anatomical pathways for this diffuse system are not known, but the intralaminar and anterior-ventral nuclei of the thalamus have been considered to be important elements.

The Subthalamus

This lateral part of the hypothalamus is essentially the upward continuation of the tegmentum of the midbrain. Laterally it consists of two nuclei, the **subthalamic** and the nucleus of the **zona incerta,** separated from each other by the **ansa lenticularis** [p. 682], with the zona incerta separated from the thalamus by the thalamic fasciculus [FIG. 778]. Medially it is composed of the subthalamic tegmental area, or prerubral field, in which the fibres from the capsule of the red nucleus, the ansa lenticularis, the trigeminothalamic tract, and medial lemniscus intermingle [FIG. 861].

The subthalamic nucleus is pierced by many fibres of the ansa lenticularis, some of which end on its cells, and it also receives fibres from the cerebral cortex. Its **efferents** pass towards the subthalamic tegmental area and red nuclei.

The nucleus of the zona incerta receives fibres from the ansa lenticularis and the cerebral cortex, and sends efferent fibres into the tegmentum, medial and caudal to it.

The fibres of the **ansa lenticularis,** in addition to ending in the two nuclei mentioned above, also pass directly to the red nucleus, possibly to the substantia nigra which abuts on the ventral surface of the caudal part of the subthalamic nucleus, to the medial part of the hypothalamus and dorsally into the thalamus. The latter connexion allows impulses originating in the lentiform nucleus to pass via the ansa lenticularis and thalamus to the cerebral cortex and, through the anterior-ventral nucleus of the thalamus, back to the corpus striatum.

The subthalamus is developed from the basal lamina and is intimately connected with the mechanisms of the midbrain tegmentum concerned with muscle tone. Its exact function is unknown but thrombosis in the subthalamic nucleus is associated with the development of spontaneous, writhing or sometimes violently jerky movements of the opposite upper limb, **hemiballismus.**

The Medial Part of the Hypothalamus

The general arrangement and relations of this region have already been described [pp. 628–30].

Afferent Fibres. These enter the hypothalamus from a large number of sources and the following account is by no means exhaustive :

1. **Connexions with the cerebral cortex.** Direct connexions with the cerebral cortex come from the rhinencephalon. The largest single bundle is the **fornix** [p. 647], which originates in the fimbria and consists of fibres from the hippocampus. It enters the anterosuperior part of the hypothalamus and curves inferiorly to the mamillary bodies, losing more than half its fibres to the anterior hypothalamus on the way. A large number of fibres from the fornix enter the septal region which, in Man, is grey matter ventral to, but continuous with the anterior part of the septum pellucidum and extending ventrally around the anterior commissure to become continuous with similar grey matter, the preoptic region, joining it with the hypothalamus. Many more fibres leave the fornix in the anterior hypothalamic area so that only a proportion of the fibres of the fornix actually reach the mamillary body.

2. Afferent fibres enter the hypothalamus from the **medial thalamic nucleus** by running as periventricular fibres in the wall of the third ventricle. These would seem to form a pathway by which impulses from the frontal lobe of the cerebral cortex can reach the hypothalamus, and also a route by which ascending fibres in the brain stem which pass to the thalamus can activate the hypothalamus, since collaterals of these fibres are said to pass to the medial nucleus.

3. The **medial forebrain bundle** arises in the region of the anterior perforated substance, septum, and preoptic region, and passes posteriorly through the lateral part of the hypothalamus conveying olfactory and rhinencephalic impulses. This pathway may also transmit impulses which originate in the frontal lobe and relay in the septal region.

4. Fibres reach the ventromedial hypothalamic nucleus from the **globus pallidus** of the lentiform nucleus by way of the ansa lenticularis, and may be accompanied by fibres from the subthalamic nucleus.

5. Fibres pass into the hypothalamus from the **amygdaloid body** by way of the stria terminalis.

Efferent Pathways. These fall into two groups :

1. Connexions with the posterior lobe of the hypophysis.

2. Connexions with other parts of the brain.

The **connexions with the posterior lobe of the hypophysis** arise in the supra-optic, paraventricular, and tuberal nuclei [FIG. 780]. Axons from all three situations descend through the hypothalamus, some ending in the **median eminence** (a small median protuberance immediately posterior to the root of the hypophysial stalk) and the hypophysial stalk. The majority pass into the posterior lobe which is a heavily innervated organ despite the paucity of structures in it which seem to require innervation. The cells of the three nuclei, and the fibres which arise from them, contain a substance which is stainable with chrome-haematoxylin and seems to accumulate in the posterior lobe. There is good evidence that the glial cells of the posterior lobe are not concerned with the secretion of its hormones, and that they remain intact after interruption of the stalk of the hypophysis, even though the pharmacological activity of the posterior lobe disappears. Again, degeneration of the supra-opticohypophysial system is associated with the excessive secretion of dilute urine, the converse of the activity of the antidiuretic hormone (vaso-pressin) of the posterior lobe which acts on the distal con-voluted and collecting tubules of the nephrons to cause reabsorption of water. Dehydration (Bargmann, 1966) and substances which cause antidiuresis by acting through the hypophysis (Bodian, 1951) cause a loss of stainable substance in the hypophysis. There is some evidence, too, that the paraventricular nuclei may be responsible, in the same way, for the production of oxytocin, a substance which causes uterine contractions (Olivecrona, 1954). Thus the hypothal-amic cells connected to the posterior lobe are secretory in type and their axons act as ducts, passing materials formed in the individual cells to the posterior lobe for storage. This is in keeping with the known function of nerve cells which elaborate substances in the cell body for release at the nerve endings.

The **connexions with other parts of the brain** are as follows:

(i) From the **mamillary body** a large bundle of fibres arises and, passing at first laterally, turns dorsally in the medial medullary lamina of the thalamus to pass to the anterior thalamic nuclei, the **mamillothalamic tract.** From the anterior nuclei, fibres pass to the gyrus cinguli, particularly its anterior part. In its lateral course the mamillo-thalamic tract fibres bifurcate and a bundle of **mamillo-tegmental fibres** passes caudally into the tegmentum of the midbrain.

(ii) By way of the periventricular fibres the hypothalamus is connected with the **medial nucleus of the thalamus** and through this nucleus with the frontal lobe of the cerebral cortex, particularly the polar and orbital regions.

(iii) Other fibres of the periventricular system pass to the midbrain and run through the central grey substance as the **dorsal longitudinal fasciculus,** connecting with autonomic centres in the peri-aqueductal grey matter, other than those that the mamillotegmental fibres make contact with in the tegmentum.

(iv) The **medial forebrain bundle,** in the lateral part of the medial hypothalamus, is continued into the midbrain mainly by fibres which arise in the posterior hypothalamus. These form a descending pathway from this part of the brain produced mainly by cells with short axons.

(v) The posterior hypothalamic region is said to be able to activate the 'diffuse thalamic projection system', and can thus play a part in modifying the activity of much of the cerebral cortex.

Thus the hypothalamus receives fibres from the ascending systems, has communications with other parts of the brain, and gives rise to fibres which pass caudally to act through the mechanisms of the brain stem: a pattern common to every part of the nervous system.

Physiological stimulation of the hypothalamus can elicit many autonomic activities, e.g., changes in heart rate and blood pressure, hyperglycaemia, bladder contraction, etc., and for this reason the medial hypothalamus has been considered the part of the forebrain concerned with visceral activity. It is possible, however, to produce similar effects from other parts of the forebrain, and the multiple connexions which the hypothalamus has with other regions of the brain, highlights the fact that it is an integral part of the brain mechanism working in association with the other parts. The function of

the nervous system is to integrate the activity of all the organs of the body, including the viscera whose functions are in-extricably bound up with somatic activity. Thus increased heart and respiratory rates, vasodilatation, vasoconstriction, etc., are the natural accompaniments of physical exercise.

THE NUCLEI OF THE TELENCEPHALON [PLATES 40–44]

These consist of:
The claustrum.
The amygdaloid body.
The corpus striatum—the caudate and lentiform nuclei.

The Claustrum

This is a thin sheet of grey matter between the external capsule and the white matter of the insula. Its connexions and function are relatively unknown, but it receives fibres from the cerebral cortex through the external capsule (Carman, Cowan, and Powell, 1964) in a manner similar to the corpus striatum.

The Amygdaloid Body

This lies near the pole of the temporal lobe, over the anterior extremity of the inferior horn of the lateral ventricle. Medially it is continuous with the cortex of the temporal lobe where it abuts on the anterior perforated substance [FIG. 778], but laterally it lies deep to the cortex. Superiorly its medial part is continuous with the inferior surface of the lentiform nucleus near the anterior extremity of that nucleus, with the anterior commissure passing between them [FIG. 806]. The amygdaloid body is divisible into a considerable number of subnuclei which seem to have different connexions and func-tions.

The connexions and functions of the amygdaloid body are very incompletely known, and it is not possible to make any clear estimate of its function from those connexions which are known. The **afferent fibres** which have been determined come mainly from the olfactory system. Direct fibres from the olfactory bulbs reach the dorsomedial portion through the lateral olfactory stria. Other afferent fibres have been de-scribed from the corpus striatum, the intralaminar nuclei of the thalamus, the olfactory trigone, the anterior piriform cortex, and from various regions of the cerebral cortex, in-cluding the temporal lobe. Electrophysiological studies indi-cate that ascending impulses from the spinal medulla reach the amygdaloid body, probably through multisynaptic path-ways such as those of the reticular formation.

The **efferent connexions** consist mainly of the **stria terminalis** which arises in the amygdaloid body and, running in the concavity of the lateral ventricle medial to the caudate nucleus, is distributed to the septal region and the anterior hypothalamus. Other fibres pass in the diagonal band to the olfactory trigone, septum, and hypothalamus through the medial forebrain bundle. The **diagonal band** [FIG. 816] originates in the amygdaloid body and, passing anteromedially across the posteromedial part of the anterior perforated sub-stance, immediately above the optic tract, turns superiorly on the medial surface of the hemisphere just anterior to the lamina terminalis in association with the medial olfactory stria. Here it reaches the septal region.

Thus both routes connect the amygdaloid body with similar regions, bringing it into relation with parts of the brain (e.g., hypothalamus and medial forebrain bundle) through which descending pathways may run to the brain stem, or ascending pathways pass to the thalamus and cerebral cortex. Fibres also pass directly to the piriform lobe; thus the amygdaloid body may have connexions with the hippocampus through this lobe.

From the above connexions it might be assumed that the amygdaloid body was primarily a part of the olfactory system, but it is well developed in the anosmic Cetacea (Breathnach and Goldby, 1954) and in this respect is similar to the hippocampus and dentate gyrus. Bilateral lesions of the amygdaloid in monkeys produce disturbances of emotional behaviour, increased motor activity associated with an apparent inability to ignore any visual stimulus, altered sexual behaviour, etc. Similar effects occur in Man in cases of temporal lobe, or psychomotor, epilepsy. There is loss of memory of recent events, confusion, illusions related to one or other sense, sometimes marked alteration of personality and automatic actions, often of a complex type, during the attack. Many of the changes seen during the attacks can be reproduced at operation by stimulation of the amygdaloid.

The Corpus Striatum [Figs. 858 and 859]

The corpus striatum consists of the **caudate** and **lentiform nuclei,** the latter having an external segment, the **putamen,**

FIG. 859   Frontal section through cerebrum.

FIG. 858 Coronal section through the cerebral hemisphere. The section (viewed from the front) cuts the anterior horns of the lateral ventricles, through which the central part of the ventricles, the columns of the fornix, and the openings of the interventricular foramen can be seen.

darker in colour than its inner part, the **globus pallidus** [PLATE 40, 4], which is separated from the putamen by a thin layer of white matter, the **lateral medullary lamina,** and divided into internal and external segments by the **medial medullary lamina.** The putamen of the lentiform nucleus and the caudate nucleus have the same structure (being composed of small and medium-sized cells) and communicate with each other by strands of grey matter which pass between the bundles of the internal capsule separating them. The globus pallidus on the other hand contains many myelinated fibres and a considerable number of large cells. For this reason the caudate and putamen are sometimes combined under the term **striatum,** while the globus pallidus is called the **pallidum.**

It seems likely that the division of the caudate from the putamen of the lentiform nucleus by the internal capsule is quite fortuitous, and that the use of the term striatum is advisable since they really form a single unit.

The general disposition of the parts of the corpus striatum has already been described [p. 640], the following section deals with the nervous connexions of its parts [FIG. 860].

Afferent Connexions of the Striatum. Afferent fibres have already been described passing to the striatum from the intralaminar nuclei of the thalamus, especially from the central nucleus. The fibres passing to this nucleus come from the midbrain reticular formation, the superior cerebellar peduncle (Jung and Hassler, 1960), and possibly as collaterals of ascending sensory pathways. The striatum thus receives information from the spinal medulla, the reticular formation, and the cerebellum. See also substantia nigra.

Other afferent fibres reach the striatum from the **cerebral cortex.** Such afferents have long been postulated and denied because of the absence of Marchi degeneration in the striatum following injury to the cerebral cortex. It is clear, however, that the fibres which enter the striatum from the internal capsule are mostly non-myelinated, and possibly collaterals of fibres passing to other destinations. Thus, the striatum probably samples all the impulses leaving the cerebral cortex.

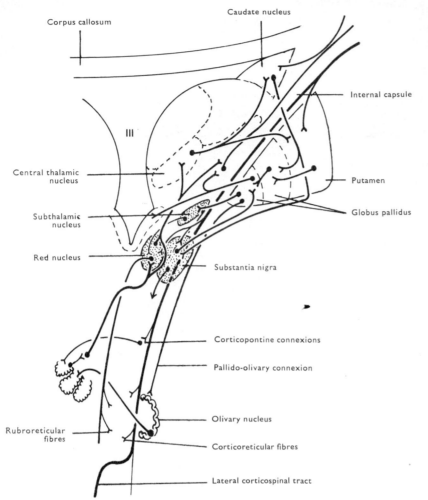

FIG. 860 Diagram to show some of the principal connexions of the corpus striatum.

efferent fibres to the caudate nucleus. These minute fibres, which have only been seen by fluorescence microscopy, and the dopamine in the caudate nucleus disappear when the substantia nigra is destroyed. The melanin contained in the cells of the substantia nigra may well be explained by their activity in producing dopa, and intermediary between tyrosine and melanin.

Afferent fibres to the pallidum arise mainly from the caudate nucleus and putamen and pass to the external segment, the fibres from the caudate passing through the internal capsule to reach it. Other afferent fibres reach the pallidum from the medial lemniscus, possibly the spinothalamic system, the substantia nigra, and from the intralaminar nuclei of the thalamus. The pallidum forms the major efferent pathway for the striatum (apart from its connexions with the substantia nigra), and the striatal efferents are associated with spinal and reticular impulses in the pallidum. The medial segment of the pallidum receives fibres from the lateral segment.

The **efferent fibres** of the pallidum form the **ansa lenticularis.** This is a mass of fibres which leaves the inferior and medial surfaces of the pallidum. Those leaving the inferior surface tend to hook around the ventral margin of the internal capsule, running parallel with the optic tract [FIGS. 816 and 861], and turn dorsally into the subthalamus, while those leaving the medial surface of the pallidum pierce the internal capsule to reach the subthalamus. The latter part is sometimes called the **fasciculus lenticularis.** The fibres gather together into a well-defined bundle on the ventral aspect of the zona incerta and either pass medially towards the **prerubral field,** or turn dorsally to enter the **thalamus,** where they end in the anterior-ventral nucleus. The latter fibres seem to come principally from the medial segment of the globus pallidus, while those from the lateral segment pass mainly to the subthalamic nucleus, the ventromedial hypothalamic nucleus, the subthalamic tegmentum, and the red nucleus. Through the latter connexion the corpus striatum can produce effects on the rubrospinal and rubro-olivary (Jung and Hassler, 1960) pathways, thus having potential connexions with the reticular formation of the brain stem and, through the olive, with the opposite cerebellar hemisphere. Through the thalamic connexions the corpus striatum can affect the activity of the cerebral cortex.

In general, there are many features which the corpus striatum and cerebellum have in common in their connexions. Both have extensive connexions with the cerebral cortex and brain stem, both receive spinal impulses, while neither has a direct pathway to the motor cells of the brain stem or spinal medulla; moreover the two systems are closely interconnected. The functions of the corpus striatum are not at all clear, but, so far as the anatomical information can allow of inferences, it seems likely that it acts with the cerebellum and cerebral cortex and must have some effect on the brain stem mechan-

Though the details of corticostriate fibres are not known in Man, it is likely that they converge radially on the striatum. Thus the head of the caudate would receive fibres from the medial, orbital, and anterior part of the superolateral surface of the frontal lobe; the body and tail of the caudate would receive similar fibres from the rest of the medial surface of the hemisphere, and the putamen from the remainder of the hemisphere (Carman, Cowan, and Powell, 1963). Many of the fibres to the caudate nucleus run in the **subcallosal fasciculus,** a longitudinal bundle of fibres in the angle between the corpus callosum and the caudate nucleus.

Efferent fibres from the striatum pass to the external segment of the globus pallidus and through the internal capsule to the substantia nigra.

The **substantia nigra** forms a slightly curved plate of cells lying between the crus cerebri and the tegmentum of the midbrain and extending from the caudal midbrain into the subthalamus. To the naked eye the substantia nigra is dark grey in colour, because of its melanin containing cells. The substantia nigra may receive afferents also from the medial lemniscus and sends efferent fibres to the thalamus (ventral nucleus), striatum, pallidum, and the tegmentum of the midbrain. Recent investigations seem to show that the large quantities of dopamine which are present in the caudate nucleus are derived from the substantia nigra, and are transmitted thence by

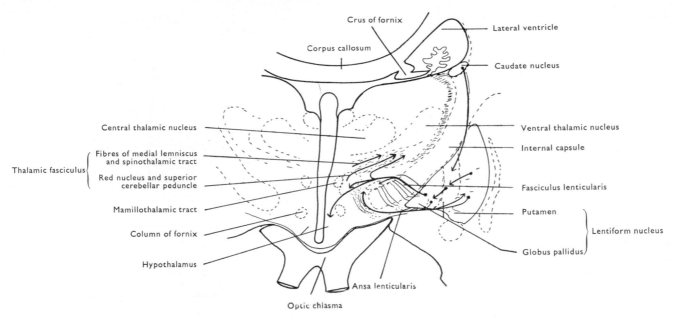

FIG. 861 Semidiagrammatic drawing of a horizontal section through the diencephalon, to show the efferent fibres of the globus pallidus, the ansa lenticularis and fasciculus lenticularis. [See also Fig. 778, p. 628.]

isms. Thus it presumably plays some part in controlling the tonic mechanisms of the brain stem, through which it acts in conjunction with the cerebellum and the cerebral cortex.

Clinically, lesions of the striatal mechanism are associated with alterations in muscle tone which tend to be the reverse of those caused by cerebellar lesions. There is a great increase in muscle tone with rigidity of the limbs and the same inco-ordination of antagonist muscles, so that the patient has to work against the unreleased tension of the antagonists. There is reduced activity with loss of associated movement (e.g., the swinging of the arms in walking, the smile of pleasure, the grimace of pain), and a tremor, particularly involving the limbs and neck, which is present at rest but tends to disappear during voluntary movement. To this extent it is the reverse of the cerebellar tremor, but it should be appreciated that the signs of disease of the striatal system have much in common with the signs of cerebellar disease, particularly when the midbrain-diencephalic region is involved since this ground is common to both.

The tremor of striatal disease seems to be primarily the result of damage to the substantia nigra, and it, like that of cerebellar disease, is abolished by damage to the efferent pathway of the motor cortex suggesting that an imbalance of the interaction of these systems is involved. It has been shown in Man that surgically produced lesions of the pallidum, the ansa lenticularis, or the anterior-ventral nucleus of the thalamus markedly reduce or even abolish the tremor and rigidity in striatal disease (Parkinson's disease). This is not understood, but it seems to suggest that the clinical condition represents an imbalance of the controlling mechanisms and that this is improved by lesions involving another element of a system normally in balance. Any explanation must remain speculative at this time, but it is highly probable that, in a delicately balanced system represented by the interacting masses of grey matter

(corpus striatum, thalamus, cerebral cortex, cerebellum, etc.) which comprise the brain, lesions in different situations may produce similar objective effects. *It is always tempting to ascribe particular functions to particular parts of the central nervous system, but it must be remembered that the results of a lesion do not indicate the function of the injured part, only what the remaining nervous tissue is capable of performing.*

THE CEREBRAL CORTEX

The cerebral cortex is the convoluted layer of grey matter which covers the entire surface of the cerebral hemispheres and is believed to be the site of conscious appreciation of sensations and the origin of 'voluntary' activity, as well as many other functions, at least a proportion of which do not seem to be under voluntary control. The cerebral cortex has a considerable area, approximately $2 - 2 \cdot 5 \times 10^3$ cm.^{2}, of which only about a third is visible on the surface, the remainder forming the walls of the sulci. This area contains a prodigious number of cells which has been estimated at approximately 5×10^9. This figure is only a very rough estimate, but it gives some indication of the possible number of neuronal complexes which can exist in this structure.

The cerebral cortex varies considerably in thickness in its different parts, being thickest (>4 mm.) in the superior part of the precentral gyrus, and thinnest (<2 mm.) in the regions receiving the ascending sensory pathways. The posterior part of the frontal lobe, the inferior parietal lobule, and the temporal pole also have a relatively thick cortex ($>3 \cdot 5$ mm.), while in the occipital and frontal poles the cortex is thin ($<2 \cdot 5$ mm.). The above measurements give only an approximate measure of thickness, since the cortex on the crown of any gyrus is approximately twice as thick as that in the depths of adjacent sulci.

Unlike the cerebellar cortex, the cerebral cortex varies considerably in structure from place to place, but it is possible to recognize two basically different types :

1. The **isocortex** which covers the greater part of the surface of the cerebral hemispheres and consists of nerve cells arranged in six layers. These layers lie parallel to the surface of the cortex and abut on each other, but are capable of differentiation because of the different types and densities of cells which each contains.

2. The **allocortex** forms a very small part of the surface of the hemisphere, limited to the rhinencephalon, is very poorly laminated and has always less than six layers. The allocortex forms the margin of the hemisphere adjacent to the choroid fissure, and is continuous with the isocortex through transitional zones. Thus the allocortex of the dentate gyrus and hippocampus is continuous with the isocortex of the periphery of the parahippocampal gyrus, through the adjacent **subiculum** and **presubiculum** [FIG. 815]. The allocortex is found in the above structures, in the gyrus fasciolaris with which the hippocampus and dentate gyrus are continuous, the indusium griseum and intralimbic gyrus (in the depths of the callosal sulcus), the subcallosal gyrus, the lateral and medial olfactory gyri which accompany the corresponding striae, the anterior perforated substance, and the uncus. Characteristic of the allocortex is the considerable amount of white matter on its external surface and this is most marked in the intralimbic gyrus and the subiculum, through which fibres pass into the hippocampus [PLATE 41, FIGS. 814 and 826]. In the uncus, the subiculum, and the presubiculum the cells are mostly small, and in the most superficial layer are arranged in nests or islands often formed of alternate groups of small and medium sized cells. The general arrangement of cells and fibres in the dentate gyrus and hippocampus is shown in FIGURE 814, while the transition from these to the isocortex in the rat is is shown in FIGURE 815.

The Isocortex

This cortex typically consists of six layers of cells which are most readily visible under the microscope when the cell bodies only are stained, as with a basic aniline dye. There are, however, several features readily visible to the naked eye when sections are made through a fresh brain. Even without magnification it is possible under these circumstances to recognize that the cortex is laminated because of a narrow white layer passing through certain parts of it [FIG. 862]. This is the **outer band of Baillarger,** which separates an outer greyish layer, surmounted by a thin whitish lamina on the surface, from a deeper yellowish-red stratum extending to the underlying white matter. Occasionally it is possible to see a second thin layer of white matter, the **inner band of Baillarger,** passing through the inner part of the grey matter. The outer band of Baillarger is most obvious in the regions of the cortex which receive ascending sensory fibres, i.e., the postcentral, superior temporal, and occipital regions. It is most marked in the visual area (which receives fibres from the lateral geniculate body) where the band was first noticed, and where it is called the **visual stria,** giving to this part of the cerebral cortex the name **striate area.**

The Histological Structure of the Isocortex. The following description is applicable to a large part of the cerebral

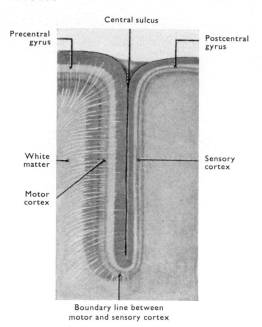

FIG. 862 Section across superior part of central sulcus in fresh brain.

cortex, but there are local differences in structure, and the architecture of the cells differs to a greater or lesser degree from that described below. The study of this cellular architecture (cyto-architectonics) has led some authors to describe a very large number of different areas, but minor variations, possibly of no great significance, are to be expected in a system of layers formed by scattered cells. Also, it is not possible to compare one area with another unless the sections used are exactly at right angles to the surface of the cortex : slight obliquity of a section increases the apparent thickness of the laminae and alters their appearance. In the convoluted cortex of Man and many other mammals, this is extremely difficult to achieve, and great care has to be exercised before differences of a minor nature can be accepted as significant. In some regions there is a sharp transition from one type of cortex to another, e.g., at the margins of the striate area and in the depths of the central sulcus [FIGS. 862 and 863]. Elsewhere the transitions are gradual (e.g., between the granular and agranular frontal cortex), and several intermediate types of cortex can be described. It remains to be seen whether some of these transitional areas are functionally different from the areas on either side of them, before too much significance is attached to these structural differences. There is agreement, however, concerning the position and extent of the main structural areas, and many of these have been shown to correspond with specific functional areas.

In a 'typical' part of the cortex the following laminae can be defined [FIGS. 864 and 865] :

1. The **lamina zonalis** is a thin, surface layer composed of non-myelinated and finely myelinated nerve fibres which form a closely meshed feltwork. They are derived from ascending axons of subjacent cortical cells, terminal branches of the apical dendrites of pyramidal cells, and processes of horizontally placed spindle-shaped cells, a few of which lie in this layer. The zonal lamina is probably of considerable importance in

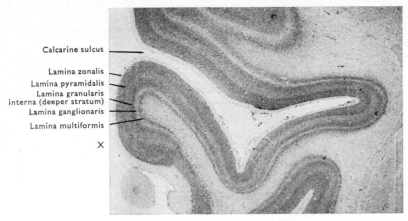

Calcarine sulcus

Lamina zonalis
Lamina pyramidalis
Lamina granularis
interna (deeper stratum)
Lamina ganglionaris
Lamina multiformis

X

FIG. 863 Low power photomicrograph of section of visual (striate) area of occipital cortex. × 6.

The visual stria is situated in the lamina granularis interna, i.e., superficial to the deep stratum of that lamina indicated in the FIGURE. The junction with the surrounding (peristriate) cortex is clearly seen opposite X.

Lamina zonalis 1

Lamina granularis externa 2

Lamina pyramidalis 3

Lamina granularis interna 4

Lamina ganglionaris 5

Lamina multiformis 6

FIG. 865 Photomicrograph of section of peristriate area of occipital cortex, showing the laminar differentiation. The differentiation between the lamina ganglionaris and the lamina multiformis is much less distinct than it is in some other cortical areas. × 45.

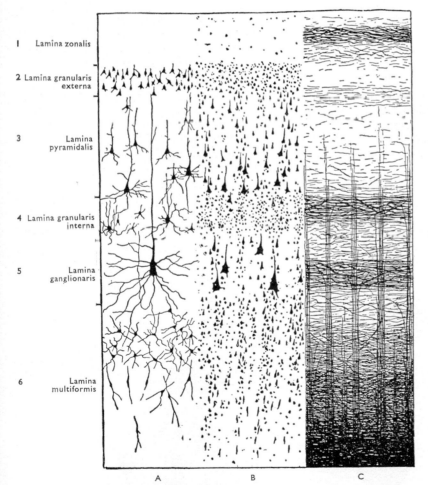

1 Lamina zonalis

2 Lamina granularis externa

3 Lamina pyramidalis

4 Lamina granularis interna

5 Lamina ganglionaris

6 Lamina multiformis

A   B   C

FIG. 864 Diagram of arrangement of cells and of myelinated nerve fibres in the cerebral cortex (cf. Fig. 865). (From C. J. Herrick (1938) An Introduction to Neurology, 5th ed., Philadelphia.)

A, cells shown by Golgi method; B, cells shown by Nissl method; C, nerve fibres shown by Weigert method.

the diffusion of impulses in all directions over the surface of the cortex.

2. The **lamina granularis externa** is a thin and usually ill-defined layer of small cells. It is composed of the same morphological types of cells as those in layer 3.

3. The **lamina pyramidalis** contains many small, stellate cells, but is characterized by the presence of pyramidal-shaped cells, which become progressively larger in its deeper parts. Each pyramidal cell gives rise to an apical dendrite which ascends towards the surface, usually reaching the lamina zonalis, and to basal dendrites which spread out from the cell body parallel to the surface. The axon emerges from the base of the cell and descending through the layers, usually reaches the white matter.

4. The **lamina granularis interna** is composed of numerous small stellate cells, each with several short, branching, dendritic processes, spreading in all directions. They also have a short axon which terminates in the vicinity of the cell. In addition there are minute pyramidal cells akin to those in layer 2.

5. The **lamina ganglionaris** contains cells of many types, but receives its name because of the very large pyramidal cells which are present in it in some areas. Throughout most of the cortex, the cells in layer 5 are smaller than those in the deeper part of layer 3.

6. The **lamina multiformis** lies adjacent to the white matter and is composed of cells of many types,

though the majority are spindle-shaped or pyramidal. It also contains many cells with axons which run through the more superficial layers towards the surface of the cortex. These are known as inverted pyramids and a few of them are to be found in the other layers of the cortex.

The bands of Baillarger have been described, but in many parts of the cerebral cortex, bundles of nerve fibres can be seen, in suitable preparations, running towards the surface from the white matter. The presence of such bundles at right angles to the cellular lamination, although to some extent to be expected, tends to suggest that the cortex may be organized in vertical columns, and there has been much physiological evidence that this may be so under certain experimental conditions. These gross bundles of fibres give very little indication of the complexity of the connexions between the cells of the cortex and the fibres which enter it from many sources. In fact, when sections are stained to show the nerve fibres in addition to the cell bodies, the cellular lamination seems to disappear, and there is excellent evidence of the most intimate contact between the cells of different layers. This is already clear from the fact that the apical dendrites of the pyramidal cells pass through all the more superficial layers, and since they are covered with synaptic spines throughout their length, they must be subjected to activity going on in these layers.

In spite of this, the horizontal lamination of the cortex is a fact, and its significance is heightened because certain layers seem to be concerned with particular functions. Thus the lamina granularis interna (layer 4) is predominantly receptive in function. Not only is it particularly well developed in the sensory projection areas and absent from the cortical areas which are mainly efferent in function (e.g., precentral gyrus), but afferent fibres have been found, by direct histological examination, to terminate to a great extent among its cells. It is their presence in this layer which accounts for the well-developed outer layer of Baillarger in the sensory projection areas. Layers 5 and 6 (infragranular layers) seem to be concerned primarily with the production of commissural and projection fibres, and layer 3 seems to receive a larger proportion of association fibres than the other layers. There is evidence then of some functional difference between the layers, though it must be understood that fibres of all types may pass to, or arise from, every layer of the cortex, and the functional differences between layers are by no means clear cut.

Whatever the significance of the lamination, there is no doubt that its pattern changes and that such changes seem to be associated with different functions. Thus, for example, the 'sensory' areas of the cortex all have a very similar structure, and this is very different from the pattern in the 'motor area' of the precentral gyrus. Many areas of the cortex differ from each other in some subtle way, but because the significance of most of these differences is in doubt, and most of them have not been related to function [see, however p. 663], only the major variations from the 'typical' (eulaminate) cortex described above will be considered. For further details see Bailey and Bonin (1951).

Variations in the Structure of the Cerebral Cortex.
1. The **agranular cortex** [horizontal hatching, FIG. 867], shows a virtual loss of layer 4 as a separate entity with an increase both in thickness of the cortex and in the number and size of the pyramidal cells, together with poorly differentiated

laminae. It is found in the posterior part of the frontal lobe. In layer 5 the agranular frontal cortex contains very large pyramidal cells which may be up to 120 μm. $\times$ 25 μm. in size. The largest of these, **giant cells of Betz,** are found in the posterosuperior part of this area, but they decrease in size in its more anterior and inferior parts, there being a complete gradation between the giant cells and the large pyramidal cells in the anterior part of the agranular cortex.

Some giant pyramidal cells are also found in the fifth layer of the cortex on the posterior wall of the central sulcus.

The large pyramidal cells give rise to a small proportion (approximately 3 per cent., Lassek, 1954) of the 1,200,000 fibres in the pyramid of the medulla (corticospinal tract), the remainder arising from other cells in the agranular frontal cortex (van Crevel, 1958).

A further variety of cortex, which is virtually agranular, with extremely ill-defined lamination, is found in the anterior part of the cingulate gyrus [vertical hatching, FIG. 867].

2. **Koniocortex** [solid black, FIG. 867]. At the other extreme there is a general reduction in the thickness of the cortex and of the size of its contained cells, with a considerable increase in the thickness of layer 4. The uniformly small size of the cells in this type of cortex gives the appearance of fine dust under the low power of the microscope and hence it is called dust cortex or koniocortex. This type of cortex is found in the areas to which the sensory, thalamocortical projection fibres pass, i.e., the posterior wall of the central sulcus, the visual area, and the auditory area in the superior temporal gyrus. In the visual area [FIG. 863] the fourth layer is subdivided by the presence in it of the visual stria (outer band of Baillarger). Also there are some large cells (cells of Meynert) in the infragranular layers, which are believed to give rise to efferent fibres which pass to the tectum of the midbrain and are concerned with direct cortical control over the visual tectal reflexes. In the walls of the central sulcus the two extreme cortical modifications, agranular and koniocortex, lie adjacent to each other, and it is possible with the naked eye to recognize the increase in thickness and the loss of the outer layer of Baillarger, in passing from the postcentral to the precentral cortex [FIGS. 862 and 866].

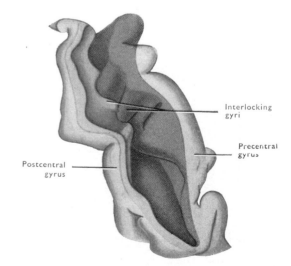

FIG. 866   Right central sulcus fully opened up, to exhibit interlocking and deep transitional gyri.

Motor cortex, red; sensory cortex, blue.

The eulaminate structure is found throughout the greater part of the remaining cerebral cortex, though it shows certain recognizable variations, e.g., the presence of particularly large cells in layers 3 and 5 of the inferior frontal gyrus. There are transitional zones between the eulaminate cortex and the agranular and koniocortex areas. In the former case there is a broad strip of so-called **dysgranular cortex** at the anterior margin of the agranular area; in this the laminae are poorly differentiated and the granular layer is often difficult to identify. Similar transitional zones (parakoniocortex) surround the koniocortices [large dots, Fig. 867]. The situations of the major types of cortex are shown in Figure 867. The above description is a much simplified version of that usually given and follows the general subdivision described by Bailey and Bonin (1951). While it is almost certain that further subdivision will become necessary as more information concerning the structure and functions of the various areas become available, it should be recognized that the map produced by Brodmann for the human brain has never been substantiated by histological evidence. As Bailey and Bonin (1951) and Walshe (1961) point out, it is surprising that this map has been so widely accepted in the

FIG. 867 Inferomedial and lateral views of the cerebral hemisphere, showing the distribution of the major types of cerebral cortex, based on histological structure. The numbers indicate some of those given by Brodmann, though the areas to which they correspond are not given.

▦ Agranular frontal cortex; ■ Koniocortex of the primary receptive areas. Of 3 and 41 only a narrow strip is visible; ▨ frontal dysgranular cortex; ⬚ parakoniocortex; ▥ cingulate agranular cortex; ▧ allocortex.

absence of the necessary proof that it represents a true histological picture of the human cerebral cortex.

The Connexions of the Cerebral Cortex. The majority of the **afferent connexions** have already been described with the ascending sensory tracts and the thalamus [pp. 663–680, FIG. 857]. Virtually all regions of the cortex receive fibres from the thalamus, except the anterior part of the temporal lobe, and each part of the cerebral cortex sends corticothalamic fibres back to the corresponding thalamic nucleus. In summary:

1. The **frontal lobe** receives fibres from the medial nucleus which projects to the granular frontal cortex, and from the anterior- and intermediate-ventral nuclei to the agranular area. The latter carry impulses from the pallidum and cerebellum respectively, while the connexions of the medial nucleus are principally with the hypothalamus.

2. The koniocortex of the **postcentral gyrus,** and the cortex adjacent to it, receive fibres from the medial and lateral parts of the posterior-ventral nucleus of the thalamus, thus relaying impulses from the medial and trigeminal lemnisci, gustatory impressions which seem to pass with the trigeminal lemniscus, and impulses transmitted by the spinothalamic system.

3. The remainder of the **parietal lobe** receives fibres from the lateral nucleus and pulvinar.

4. The **visual area** surrounding the calcarine sulcus receives fibres from the lateral geniculate body.

5. The remainder of the occipital lobe, and at least part of the temporal lobe receive fibres from the pulvinar.

6. The **superior temporal gyrus** (auditory area) receives fibres from the medial geniculate body.

7. On the medial surface of the hemisphere the anterior nuclear complex projects to the **cingulate gyrus** connecting this limbic cortex with the hypothalamus.

8. The exact distribution of the reticular thalamic nucleus is not known but it probably projects to the thalamus, midbrain, and other nuclei caudal to it. Other afferent fibres to the cerebral cortex are the commissural and association fibres.

The **efferent fibres of the cerebral cortex** can be divided into three separate groups:

1. **Association fibres** passing to other parts of the same hemisphere.

2. **Commissural fibres** passing to the opposite hemisphere.

3. **Projection fibres** passing to other parts of the central nervous system.

Association Fibres

The main bundles (fasciculi) of long association fibres have already been described [p. 643]. Detailed information regarding the organization of this system is lacking but in general it may be said that the primary receptive areas of the cerebral cortex, visual, auditory, and somaesthetic, send short association **(arcuate)** fibres into the adjacent cortex. From these areas, or from those adjacent to them, longer association fibres arise which pass to more distant regions of the cortex through one or other of the long association bundles. Thus the visual cortex (area 17, Brodmann) projects to the adjacent area 18; area 18 to area 19 external to it [FIG. 867], and the latter has long association fibres which pass as far afield as the agranular frontal cortex. There seems to be some difference in the functions of these three areas : thus stimulation of the primary visual area (17) produces flashes of light in the corresponding part of the opposite half of the field of vision [p. 668], while stimulation of the surrounding areas seems to give rise to more organized visual images. Recent investigations in cats in which all the commissures of the forebrain are divided, then one eye (i.e., the hemisphere of the same side) trained to discriminate various objects (Sperry, 1958), show that subsequent ablations of various parts of the trained hemisphere result in a decreased ability to discriminate various shapes, even though the ablations involve the temporal or frontal lobes. Initial ablation of the temporal lobe followed by ablation of the frontal lobe, leads to a deficit equal to the simultaneous ablation of both. This suggests that the complete hemisphere is necessary for full visual discrimination, though this does not seem to be true for the somaesthetic region. When the ablation is such as to leave only the visual area intact, learned discriminations are lost, but training can still be effective, though the level of performance is much lower. It is of course not clear what part the association fibres play in this mechanism, but if they are responsible for implicating the whole cortex in a particular function, it is clear that the cortex, like other regions of the central nervous system (e.g., the cerebellum), has a marked ability to compensate for the loss of neural tissue.

If the gross anatomy is any index of the connexions, then the association bundles of the superolateral surface connect the regions of that surface with each other through the superior longitudinal bundle, the lateral and posterosuperior fibres of the fasciculus uncinatus, and the external capsule [FIG. 809]. On the inferior surface the fibres of the inferior longitudinal bundle (temporo-occipital) and the medial fibres of the fasciculus uncinatus (frontotemporal) connect the various parts of this surface, while on the medial aspect the fibres of the cingulum connect the medial areas of the hemisphere [FIG. 807].

If the topographical position of the primary receptive areas is significant, it is clear that the somaesthetic area can have immediate connexions with the 'motor' area (precentral gyrus) and such connexions are known to exist. On the other hand, the three sensory areas are arrayed around the inferior parietal lobule, and it seems not impossible that such a region might be concerned with the association of the various incoming sensory impulses. The **inferior parietal lobule** lies inferior to the intraparietal sulcus and has the upturned ends of the posterior ramus of the lateral sulcus, and those of the superior and inferior temporal sulci entering it. In this fashion it is divided into three parts, each lying around the end of one sulcus ; of these the most anterior is the **supramarginal gyrus** and the middle, the **angular gyrus** [FIG. 793].

Damage to the inferior parietal lobule seems to interfere with the formation of the body image which is presumably the result of interaction of the various senses, and when the non-dominant hemisphere (usually the right in right-handed individuals) is injured the individual seems to be unable to recognize the presence of objects in the opposite half of the field of vision, being quite content to draw only one-half of objects (e.g., a bicycle wheel as a half-circle) and may not recognize his hand, on the side opposite to the lesion, as belonging to him.

Commissural Fibres

Commissural fibres pass between the corresponding parts of the two hemispheres by way of the corpus callosum [p. 650] and the anterior commissure [p. 649]. The **anterior commissure** connects the temporal lobes, particularly their anterior and inferior parts, the olfactory bulbs, and the anterior perforated substance. The **corpus callosum** transmits fibres from all lobes of the cerebrum, including the posterior and superior parts of the temporal lobes. It contains relatively few fibres from the anterior parts of the frontal lobes, and none either from the striate cortex [area 17, FIG. 867] or from the parts of the sensory areas which receive impulses from the distal parts of the limbs (Meyers, 1965 ; Jones and Powell, 1968). The auditory area also has relatively few callosal fibres.

From the size of the corpus callosum in the human brain it is reasonable to assume that this is a structure of great importance to the function of the brain, yet division of the corpus callosum or its congenital absence does not seem to be associated with commensurate signs or symptoms. It might be expected that there would be some loss of co-ordination of the two sides, but such defects are surprisingly slight and require careful testing to demonstrate any abnormality even in known cases—congenital absence of the corpus callosum usually being an incidental finding during air encephalography for other reasons. Experience with division of the commissures, including the optic chiasma, in animals demonstrates that the corpus callosum plays a part in the transfer of learned activities from one hemisphere to the other.

Projection Fibres

Many of the projection fibres have been mentioned already in their course from the cerebral cortex through the subjacent white matter into the internal capsule, crus cerebri, pyramid and corticospinal tracts. As indicated above, these can be divided into two main groups, and even though this division is artificial it is made here to stress the significance of the pathways by which the various parts of the brain are co-ordinated.

1. Projection fibres which pass to other grey masses of the central nervous system, presumably to interrelate their functions with those of the cerebral cortex. The cerebral cortex receives reciprocal connexions from these areas through the thalamus. Such pathways are the corticostriate, corticothalamic, corticopontocerebellar, corticorubral, and corticoreticular [see (i) below].

2. Projection fibres which pass to the brain stem and spinal medulla and have their effects on the internuncial and motor cells of those levels, the corticonuclear and corticospinal tracts [see (ii) below].

(i) The largest single group of projection fibres about which there is evidence in Man is the **corticopontocerebellar system**. Fibres enter this system from all the lobes of the brain, though not necessarily from all parts of each lobe. They pass through the internal capsule and enter the corresponding crus cerebri, where they form at least two-thirds of the total fibre complement, and surrounding the fibres from the posterior part of the frontal lobe which are destined to pass through the pons to the pyramid, are relayed in the pontine nuclei principally to the opposite cerebellar hemisphere.

There is very little evidence concerning the **corticostriate** fibres in Man, though it seems likely that they are arranged on the same principle as those in laboratory animals. They may be separate fibres or collaterals of other fibres passing through the internal capsule. There is evidence that some at least are collaterals, and though their origin is not known it may well be that they are branches of the corticopontine, corticothalamic, or even corticospinal fibres. Although the origin of the corticostriate fibres is not known in Man, the position of the striatum relative to the internal capsule makes it not unlikely that the striatum receives fibres from the greater part of the cerebral cortex, though the large size of the head of the caudate nucleus and the anterior part of the putamen of the lentiform nucleus makes it likely that the majority of the fibres entering the striatum from the cortex come from the frontal lobe.

Cortical Projections to Brain Stem Nuclei. Fibres have been described passing from the cerebral cortex to the subthalamic nucleus, the red nucleus, the substantia nigra, and to the reticular nuclei of the pons and medulla oblongata. The latter end principally on those reticular nuclei which give rise to the reticulospinal tracts (Brodal, 1957). It is not clear from which parts of the cerebral cortex these fibres arise, though it is likely that most of them originate in the posterior part of the frontal lobe. These connexions form an important part of the efferent cortical pathway through which it can affect the brain stem mechanisms responsible for muscle tone. It is interesting that these projections correspond closely with those already described for the cerebellum and the corpus striatum, and it is probable that the impulses from these masses of grey matter can interact in the brain stem nuclei, in addition to the direct pathways which exist for interaction between the masses of grey matter themselves. FIGURE 868 summarizes the distribution of the ascending 'sensory' systems (other than the olfactory) to the major masses of grey matter in the brain. A similar figure could show that these masses are reciprocally interconnected and that each of them has the common feature of projecting to the reticular formation of the brain stem. The positions of the major descending pathways are indicated in FIGURES 832 and 869.

Recently, attention has been focused on the descending fibres from the cerebral cortex which end in relation to many primary **sensory nuclei** in the brain stem and spinal medulla (Brodal, Szabo and Torvik, 1956; Walberg, 1957a; Kuypers and Lawrence, 1967). Following injury to the cerebral cortex in the region of the central sulcus, degenerating nerve terminals have been described in the sensory nuclei of the trigeminal nerve, the nuclei cuneatus and gracilis, the nucleus of the tractus solitarius, and the posterior horns of the spinal medulla. Like the corticothalamic fibres these may be concerned with facilitating or inhibiting the passage of impulses through the sensory nuclei, thus affecting the information reaching the cerebral cortex, and the passage of impulses through the local internuncial neurons. This could form a mechanism whereby the cortex could facilitate or inhibit brain stem or spinal reflexes almost at their source, thus preventing inappropriate activity during voluntary movement.

(ii) The **corticospinal** and **corticonuclear tracts** [FIG. 870] enter the **internal capsule** and lie in the anterior part of its posterior limb, with the fibres from the lower part of the

FIG. 868 Diagram of the 'sensory' pathways (other than olfactory) in the brain stem and cerebrum, seen from the dorsal aspect. The cerebellum has been split sagittally and each half moved to the side.

(1) Posterior root fibres (a) passing to nucleus cuneatus (2) in the fasciculus cuneatus, and (b) relaying in the posterior horn from which a fibre passes into the anterolateral ascending system (spinothalamic and spinocerebellar) of the spinal medulla (3). This gives off:

(4) Spinoreticular fibres.

(5) Posterior spinocerebellar tract fibres, some of which enter the vestibular nuclei (5a, broken outline) as spinovestibular fibres. Fibres from the vestibular nerve (22) and nuclei enter the cerebellum.

(6) Anterior spinocerebellar tract.

(7) Spinotectal fibres to superior colliculus.

(2) Nucleus cuneatus gives rise to posterior external arcuate fibres (8) and to the medial lemniscus (9). The medial lemniscus joins the other spinothalamic fibres and is distributed to the thalamus (10), from which thalamic radiations pass to many regions including the cerebral cortex (11) and the striatum.

The cochlear nerve (12) ends on the cochlear nuclei, from which fibres pass to the cerebellum, and through the trapezoid body to the lateral lemniscus (13). The fibres in the lateral lemniscus reach the thalamus (medial geniculate body) via the brachium of the inferior colliculus (14) after relay in the inferior colliculus. The medial geniculate body gives rise to the auditory radiation (15) passing beneath the lentiform nucleus (in the sub-lentiform part of the internal capsule) to the superior temporal gyrus.

Fibres in the optic nerves (16) enter the optic tract (17) from the corresponding halves of the two retinae. The optic tract is distributed to the lateral geniculate body (18) and to the superior colliculus (20). The former sends fibres through the optic radiation (19) to the occipital lobe, the latter to the cerebellum (21) and brain stem. The dotted line indicates a possible pathway for vestibulo-thalamo-cerebral fibres with the lateral lemniscus and auditory radiation.

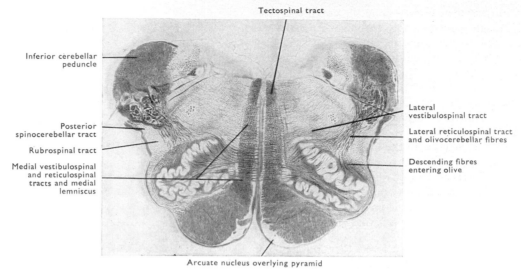

FIG. 869   To show the approximate areas occupied by the various long descending tracts in the medulla oblongata.

precentral gyrus anterior to those from its upper part. As they descend into the **crus cerebri,** the bundle turns almost through a right angle so that the anterior fibres take up a medial position in the central part of the crus with the remainder in a more lateral position. Entering the **ventral part of the pons** these fibres are split up into the longitudinal bundles which gather together again at the lower border of the pons and emerge on to the ventral aspect of the medulla oblongata as the **pyramid.**   Throughout this part of their course, bundles are given off which pass across the midline into the region of the motor nuclei of the cranial nerves (corticonuclear fibres).   The available experimental evidence shows that most

FIG. 870 Coronal sections through two levels of one brain. Left half shows corticonuclear and corticospinal fibres and is anterior to right half showing main ascending 'sensory' systems.

of these do not end on the motor cells, but pass to the internuncial neurons adjacent to them (Walberg, 1957b; Kuypers, 1958). Thus the fibres which pass to activate the oculomotor nucleus end in the tegmental reticular formation of the midbrain, in common with fibres from the superior colliculus, (tectoreticular tract), and probably other systems acting on the oculomotor nuclei.

In the lower part of the medulla oblongata the fibres of both pyramids pass in large bundles into the white commissure and, interdigitating, cross to the opposite side, entering the lateral funiculi of the spinal medulla, to form the **lateral corticospinal tracts.** This **pyramidal decussation** is rarely complete and may be asymmetrical, a variable number of fibres passing caudally into the anterior funiculus of the spinal medulla. In most cases this **uncrossed, anterior corticospinal tract** does not reach further caudally than the mid-thoracic region, most of its fibres crossing in the white commissure of the spinal medulla to enter the grey matter of the opposite side. Cases have been described where there was only a small pyramidal decussation and most of the pyramid passed into the anterior funiculus (Marie, 1895). Whatever the arrangement at the pyramidal decussation, the majority of the fibres in this system end on the opposite side of the brain stem and spinal medulla. Thus the cerebral cortex produces its effects predominantly on the opposite side of the body, and lesions of the cerebral cortex similarly produce contralateral disturbances.

The **lateral corticospinal tract** at first bulges medially into the grey matter of the first cervical spinal segment then sinks laterally into the lateral funiculus, so that in the second and third cervical segments it reaches the surface of the spinal medulla and displaces the spinocerebellar tracts ventrally [FIG. 836] (Smith, 1957). Caudal to this it is more medially placed in the posterior part of the lateral funiculus, and gives off fibres to every level of the spinal medulla as it passes caudally. The medial fibres terminate in the superior levels of the spinal medulla [FIG. 829]. The fibres pass medially from the lateral corticospinal tract and enter the intermediate part of the grey matter of the spinal medulla to end on the internuncial cells in the lateral part of the base of the posterior horn and the lateral intermediate substance [FIG. 871]. In

Dorsal root

Corticospinal

Ventral root

Vestibulospinal and reticulospinal

FIG. 871 Diagram to show the sites of termination of some of the major descending pathways in the spinal medulla. The corticospinal fibres to the motor cells are not shown.

Note that the corticospinal system ends predominantly in the posterior horn and intermediate substance. The single fibre shown representing the vestibulospinal and reticulospinal tracts, is indicated ending on a small motor cell, which is intended to indicate the source of a γ-efferent fibre.

most experimental animals, only a few of the fibres end directly on the motor cells of the anterior horn (Hoff and Hoff, 1934), but these are in the minority. Thus the corticospinal system produces its effects principally through the same internuncial cells that are in use by the local spinal reflexes. It could, therefore, be said to activate or inhibit the spinal reflex mechanisms rather than the motor cells. Such connexions persist in higher mammals, but in the primates an increasing number of corticospinal fibres end directly on the motor cells of the spinal medulla, particularly on those with axons passing to the distal muscles of the limbs (Kuypers, 1960). Though the extent of these in Man is not known, it is likely that they are more numerous than in the apes: a feature in keeping with the greater involvement of the muscles of the peripheral parts of the limbs in the paralysis following a stroke.

The origin of the corticonuclear and corticospinal tracts (Walshe, 1961) has been a source of controversy ever since Holmes and May (1909) demonstrated retrograde changes in the giant cells of Betz following injury to this descending system. It is natural that these cells should have been considered for many years to be the source of this system, since they had a structure and size akin to the motor cells of the spinal medulla, and lay in the excitable 'motor' cortex of the precentral gyrus. However, certain facts made it clear that there must be other cells of origin :

(a) Giant cells of Betz are not present in the inferior part of the precentral gyrus, though stimulation of this area causes movements of the head, neck, and upper limb which are abolished by section of the pyramids.

(b) Counts of fibres in the pyramids, and of Betz cells in the 'motor' cortex (Lassek, 1954) show that the latter form only 3 per cent. of the former, and approximately 3 per cent. of the fibres in the pyramid are large, suggesting that these alone may originate in Betz cells.

In view of this it seems reasonable to assume that small cells in the precentral region may give rise to the remaining fibres. During the last decade, however, doubts have been cast on the generally accepted idea that the fibres carried by the pyramid all originate in the precentral gyrus, and descend to the lower brain stem and spinal medulla. These doubts arise from the following facts :

1. Extensive lesions of the precentral gyrus do not lead to complete degeneration of the fibres of the pyramid, though unilateral decortication does so. Even when the electrically excitable cortex anterior to the precentral gyrus, the **premotor area,** is removed in addition to the precentral gyrus, complete degeneration of the pyramid is said not to occur. Some investigations (van Crevel, 1958) indicate that the entire pyramid degenerates following removal of the agranular frontal cortex, provided the survival time after operation is prolonged considerably beyond that normally used for such degeneration experiments.

2. Injuries to the temporal and occipital lobes of the cerebral cortex also lead to degeneration of fibres in the pyramid (Walberg and Brodal, 1953). Though the number of these fibres is small, it is difficult to see how their presence can be denied, even if there is apparently total degeneration of the pyramid, following removal of the posterior part of the frontal lobe (van Crevel and Verhaart, 1963). Fibres from all parts of the cerebral cortex have been described entering the

pyramid. Some arising in the precentral region pass through the corpus callosum and descend in the opposite pyramid (Walberg and Brodal, 1953). Thus the cerebral cortex giving rise to the fibres in the pyramid may send them into all four corticospinal tracts.

3. Ascending fibres have been described passing to the cerebral cortex through the lateral corticospinal tract, the pyramid, crus cerebri, and internal capsule, without relay in the thalamus (Brodal and Walberg, 1952, in the cat; Nathan and Smith, 1955b, in Man). These fibres arise in the contralateral brachial and lumbosacral regions, and the ipsilateral brachial region of the spinal medulla, and are distributed widely in the cerebral cortex, to agranular frontal cortex, and more posterior parts [FIG. 867]. The presence of these fibres has been confirmed by physiological methods in the cat (Brodal and Kaada, 1953), and it has been shown that they transmit all varieties of sensation from the spinal medulla.

Clearly, then, the pyramid is not the simple efferent pathway from the posterior part of the frontal lobe that it was previously considered to be. The significance of the different groups of fibres which it contains has still to be assessed, and the presence of all of them in Man requires confirmation. Until this is done, the findings of van Crevel and Verhaart (1963) should be remembered, as should the fact that most of these discoveries depend on staining methods which are capricious.

Stimulation of the 'motor' cortex in Man has shown that there is a well-defined relationship between the parts of this cortex, and the site of movements produced by threshold stimuli applied to it. The arrangement is similar to that in the 'sensory', postcentral gyrus. Thus, the inferior part of the precentral gyrus is concerned with the face, tongue, chewing, and swallowing movements, while the neck, upper limb, trunk, lower limb, and perineal areas follow in sequence, reaching the medially placed paracentral lobule superiorly. This has led to the belief that there is a mosaic of points in the precentral gyrus, each connected with a particular group of motor cells in the brain stem or spinal medulla. The anatomical evidence is quite contrary to this, since it was shown by Sherrington (1889) that localized lesions of the motor cortex led to degeneration of fibres to all parts of the spinal medulla, though predominantly to one particular region. This is hardly surprising, if the cortex is concerned with movements, since almost every movement involves muscles distant from the moving part, e.g., stabilization of the trunk and legs, in arm movements, particularly when these are vigorous and against resistance. The significance of localization in the precentral gyrus is discussed critically by Walshe (1948) and Clark (1948).

Stimulation of the cerebral cortex has uncovered the presence of two regions, adjacent to, but outside the classical motor area, from which motor responses can be elicited in a point-to-point fashion. The supplementary 'motor' area lies on the medial aspect of the cortex, antero-inferior to the leg region of the 'motor' area, while the second 'motor' area lies anterior to the second sensory area [p. 665] extending towards the insula from the motor area. It is difficult to estimate the significance of these and other 'motor' areas, though injury of the supplementary motor area in the monkey causes hypertonia (Travis, 1955). The term 'motor' must be used with caution in this situation since the production of movement by stimulation of a particular area does not give anyone the right to call that area 'motor', any more than it is legitimate to call a dorsal root of a spinal nerve (sensory root) 'motor' because stimulation of it causes movement. The cells being stimulated are internuncial neurons, and the results obtained depend on the nature and timing of the stimulus, and the existing activity of the cortex. In keeping with this is the fact that subliminal stimulation of the 'motor' cortex, in unanaesthetized monkeys, produces relaxation when applied to the same point, from which a stronger stimulus would produce motor effects (Clark and Ward, 1948).

Damage to the corticospinal system in Man commonly occurs as a result of injuries to the internal capsule, when a number of different pathways are involved, often including the sensory radiations. Such a lesion produces a contralateral paralysis of voluntary movement with increase in muscle tone, **spastic paralysis** (cf. the flaccid paralysis following the destruction of motor cells innervating muscle), which involves the peripheral parts of the limbs more than their proximal parts, and has little or no effect on the trunk. Voluntary movements normally produced by the activation of certain cranial nerves are affected, e.g., the hypoglossal, producing tongue paralysis on the opposite side; the nucleus ambiguus producing contralateral laryngeal and pharyngeal paralysis; the facial nucleus, resulting in paralysis of the contralateral facial muscles from the orbicularis oculi inferiorly, but not of the occipitofrontalis. The paralysis of voluntary movement affecting some muscle groups more than others, has been ascribed to a bilateral supply of cortical fibres to some motor nuclei, and not to others. The corticospinal system, however, has direct connexions mainly with the motor cells innervating the distal muscles of the limbs, while other descending tracts (reticulospinal, vestibulospinal, tectospinal) end principally on those innervating the trunk and proximal limb muscles. The sparing of the latter in injury to the corticospinal system may be a measure of the extent to which their control is by these other systems. Nevertheless, the ending of the corticospinal system also on internuncial cells would allow it to affect motor cells on both sides of the brain stem and spinal medulla—a feature which may well be necessary for those movements in which the corresponding muscle groups of the opposite side are also involved. Thus movements of the trunk invariably use the muscles of both sides, while the movements of one hand may be entirely independent of the other. The absence of any commissural fibres between the parts of the 'sensory' cerebral cortex which receive impulses from the distal parts of the limbs [p. 688] may have a bearing on this. However, the same absence of commissural fibres is seen in the visual cortex which receives impulses from the highly co-ordinated eyes. It is interesting that animals with little or no independent movement of the distal parts of the limbs tend to have a very restricted corticospinal system, e.g., the sheep.

In addition to the paralysis, lesions of the corticospinal system are associated with loss of the abdominal reflexes, and an alteration of the plantar reflexes. The former consist of contraction of the underlying abdominal muscles when the abdominal skin is lightly scratched. In the latter, the flexion of all the toes which normally follows scratching the outer side of the sole of the foot is reversed—the big toe being extended and the other toes flexed and spread, the **Babinski response.**

It is possible that the ascending spinocortical fibres in the pyramid form the sensory pathway for these reflexes.

Efferent fibres from the cerebral cortex have already been described passing to other parts of the brain, e.g., corpus striatum, tectum of the midbrain, red nucleus, cerebellum, and reticular formation, through which the cerebral cortex can have indirect effects on the motor cells of the brain stem, and spinal medulla. These pathways form part of an extensive system, known as the **extrapyramidal system,** because the pathways involved do not pass through the pyramid. It is concerned with complex movements of a more automatic nature, such as the maintenance of balance during movements, the correct distribution of muscle tone necessary for normal movements, etc. The long descending pathways of the extrapyramidal system are the reticulospinal, the vestibulospinal, the rubrospinal, and the tectospinal tracts. These are almost certainly reinforced by many other paths consisting of chains of short neurons, such as those which connect the hypothalamus and subthalamus with the reticular formation. The extrapyramidal system is, anatomically, extremely complicated and the method by which its numerous parts interact is not clear. It is characterized by multiple feedback systems linking its several parts. The major parts of this system have already been described, but it should be realized that the separation of the pyramidal and extrapyramidal systems is completely artificial, and that the two systems are complementary parts of the normal nervous system.

THE MENINGES

The meninges form the protective coverings of the central nervous system and separate it from the bony elements which surround it, the skull and vertebral column. They consist of an external, thick, fibrous **dura mater**, or pachymeninx, and a mass of delicate tissue, the leptomeninges, believed to be developed from the neural crest (Horstadius, 1950). The latter is partly split into two layers (arachnoid mater and pia mater) by the presence within it of the **subarachnoid space,** filled with the clear, watery **cerebrospinal fluid.** The external of these two layers, the **arachnoid mater,** is applied to the internal surface of the dura mater, but separated from it by a capillary interval, the **subdural space.** The adjacent surfaces of the arachnoid and dura are covered with a flattened layer of mesothelial cells so that the subdural space forms a large bursa which surrounds the central nervous system and allows of movement of the nervous system and leptomeninges on the dura. Every structure, blood vessels, nerves, etc., passing to or from the brain and spinal medulla must cross the subdural space, and it is only where such structures cross it that the continuity of the subdural space is interrupted.

The external surface of the arachnoid is everywhere applied to the deep surface of the dura and its shape is therefore the same as that of the dura. On the other hand, the **pia mater** immediately invests the central nervous system, following its every contour. The blood vessels running on the nervous system are adherent to the external surface of the pia mater. Thus the branches of these vessels which pierce the nervous system carry a sleeve of pia mater with an extension of the subarachnoid space passing for a short distance into the nervous tissue, the **perivascular space.** Where the shape of the dural sac differs from that of the underlying nervous system, the distance between the pia and arachnoid mater is altered and hence the depth of the subarachnoid space, which forms deep spaces or **cisterns** in some situations and is relatively shallow in others. Apart from variations in its depth, the subarachnoid space also contains a variable number of delicate strands of tissue lying in the cerebrospinal fluid. These join the arachnoid to the pia and form a delicate web, from which the arachnoid obtains its name.

The meninges, as protective coverings of the central nervous system, have two entirely different mechanical problems to overcome:

1. The protection of the relatively firm spinal medulla within the moving vertebral column, by slinging it in a fluid-filled tube, protected externally by fat and venous spaces.

2. The support and protection of the voluminous, soft brain within the skull. This is achieved by the formation of rigid compartments within the skull and a special fluid-filled 'sponge' immediately surrounding the brain.

The Spinal Meninges

The spinal dura mater forms a long tubular sheath extending from the margin of the foramen magnum to the upper part of the sacrum. It is not of uniform width throughout, being widest in the cervical region and narrowest in the thoracic region where the sac is closest to the spinal medulla. Throughout the length of the vertebral canal it is separated from the periosteum lining the bony elements of the vertebral canal by a variable amount of soft fat and a considerable number of venous spaces, the **internal vertebral venous plexus.** The fat is found mostly over its dorsal surface forming a buffer between it and the moving laminae, and, though a few venous channels are present on this surface, the largest extradural veins form channels on either side extending the length of the vertebral canal. These communicate with the intersegmental veins through the intervertebral foramina, and with the basivertebral veins anterior to the posterior longitudinal ligament. Principally in the thoracic region, the dural sac is lightly attached to the posterior longitudinal ligament by fine collagenous strands which hold the sac close to the posterior surface of the vertebral bodies and thus to the axes of vertebral movements.

Throughout its length the spinal dura mater is lined by arachnoid while the pia mater forms a thick, tough membrane on the surface of the spinal medulla. It is thickened into a glistening ribbon, the **linea splendens,** which runs anterior to the anterior median fissure and the anterior spinal artery. The pia is also markedly thickened to form a linear fold extending laterally from the surface of the spinal medulla between the ventral and dorsal roots. This flange of the pia mater has a thick, rounded, lateral border which forms the free edge of a scalloped or toothed ligament, the **ligamentum denticulatum** [FIG. 872], which passes laterally between the points of emergence of the spinal nerve roots, forming a sharp point attached to the dura mater through the arachnoid. The ligamentum denticulatum thus suspends the spinal medulla within the spinal subarachnoid space at some twenty to twenty-two points on each side, and allows for a

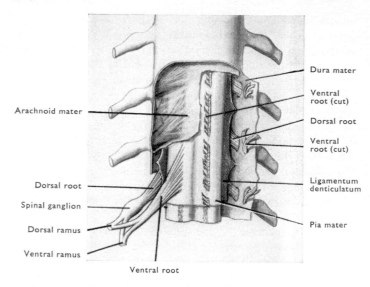

Arachnoid mater

Dorsal root

Spinal ganglion

Dorsal ramus

Ventral ramus

Ventral root

Dura mater

Ventral root (cut)

Dorsal root

Ventral root (cut)

Ligamentum denticulatum

Pia mater

FIG. 872 Membranes of spinal medulla and mode of origin of spinal nerves. The lowest spinal nerve on the left is shown semi-diagrammatically dissected, to indicate the participation of both roots in the formation of each ramus.

considerable range of movement of the dura without a corresponding movement of the spinal medulla. The spinal subarachnoid space is relatively wide with only a few delicate strands connecting the arachnoid to the pia on the posterior surface of the spinal medulla.

The spinal medulla ends immediately below the first lumbar vertebra, but the dural and arachnoid sheaths extend inferiorly covering the cauda equina and carrying the subarachnoid space to the second part of the sacrum. Here the **filum terminale**, the thread-like continuation of the spinal medulla and its covering of pia mater, pierce the arachnoid and dura and, carrying off a sheath from them, pass to be attached to the dorsal surface of the coccyx, thus holding down the caudal part of the dural sheath [FIGS. 730 and 731].

At its cranial end, the dural sheath tends to be more firmly attached to the second and third cervical vertebrae and at the foramen magnum becomes continuous with the cranial dura mater which, unlike the spinal part, is fused with the periosteum lining the skull. Also at its cranial end the ligamentum denticulatum is continued into the cranial cavity just anterior to the spinal accessory nerve and is attached to the dura posterior to the hypoglossal canal.

The Cranial Meninges

The Cranial Dura Mater. This is closely applied to the interior of the skull without any intervening fat, and is fused with the periosteum except where the venous sinuses of the dura mater intervene or the dura is folded inwards between the parts of the brain. The combined periosteum and dura mater, sometimes erroneously called the outer and inner layers of the dura mater, are less firmly adherent over the cranial vault than they are at the base of the brain, where both periosteum and dura mater are continued through the various foramina, the one to become continuous with the periosteum on the external surface of the skull, the other with the fibrous sheaths of the nerves. The **venous sinuses of the dura mater** are endothelial tubes lying in the same layer as the

internal vertebral venous plexus (between the dura and the periosteum) and are continuous with it through the foramen magnum. Like the internal vertebral venous plexus, the venous sinuses of the dura mater communicate:

1. With veins outside the skull:
    (i) Through the various foramina for the cranial nerves.
    (ii) Through emissary foramina.
2. With the diploic veins in the cranial bones.

The method of growth of the brain, whereby its three main masses, two cerebral hemispheres and the cerebellum, expand separately into the surrounding tissues, results in the formation of three separate pockets in that tissue, pockets which communicate with each other in the plane of the original brain tube. In this fashion an incomplete septum of tissue, the **falx cerebri,** is left between the two hemispheres and a similar septum, the **tentorium cerebelli,** lies between the cerebellum below and the occipital lobes of the hemispheres above. These two septa meet in a λ-shaped arrangement in the midline posteriorly where the two occipital lobes and the cerebellum come closest together. A similar but much smaller fold **(falx cerebelli)** extends anteriorly from the internal occipital crest between the cerebellar hemispheres; another, the **diaphragma sellae,** roofs over the hypophysis in the sella turcica. Each of these septa is in the form of a fold of dura mater with the arachnoid applied to its cerebral or cerebellar surface. *In situ*, the falx cerebri and tentorium cerebelli form rigid structures which act as baffles within the cranial cavity, the former preventing, to some extent, side-to-side movement of the brain, while the tentorium tends to prevent downward displacement of the occipital lobes and holds the cerebellum in position.

The **falx cerebri** [FIGS. 873 and 874] is a sickle-shaped structure lying in the median plane, attached to the cerebral surface of the cranial vault from the crista galli of the ethmoid to the internal occipital protuberance. It passes into the longitudinal fissure between the hemispheres, increasing in depth from before backwards so that its free, inferior margin comes close to the splenium of the corpus callosum posteriorly. Here its inferior edge joins the anterior extremity (apex) of the tentorium cerebelli, while anteriorly its free edge lies some distance from the genu of the corpus callosum [FIG. 874]. Within its fixed border, enclosed between its diverging surfaces and the periosteum, lies the superior sagittal sinus, the lateral lacunae of which separate the dura and periosteum further laterally. Between the layers of its free, concave border lies the inferior sagittal sinus which is joined by the great cerebral vein where that border meets the tentorium. The straight sinus, so formed, passes postero-inferiorly in the line of junction of the tentorium and the falx to reach the internal occipital protuberance. The anterior part of the falx, and the dura mater of the superior sagittal sinus and its lacunae are perforated or cribriform. The cribriform parts of the superior sagittal sinus allow the superior cerebral veins and extensions of the underlying arachnoid (**arachnoid villi** and **granulations**) to pass through the dura and come into association with the endothelial lining of the sinus and its lacunae.

The **tentorium cerebelli** [FIGS. 873 and 874] is a large, crescentic, transversely placed fold of dura mater which roofs

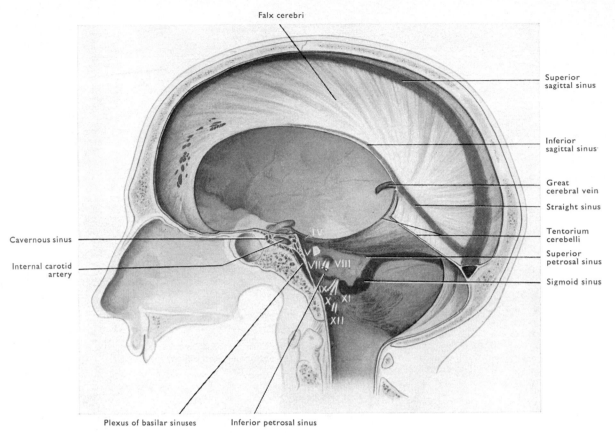

Falx cerebri

Superior sagittal sinus

Inferior sagittal sinus

Great cerebral vein

Straight sinus

Tentorium cerebelli

Superior petrosal sinus

Sigmoid sinus

Cavernous sinus

Internal carotid artery

Plexus of basilar sinuses

Inferior petrosal sinus

FIG. 873 Sagittal section of skull, a little to the left of median plane, to show arrangement of dura mater. Cranial nerves are indicated by numerals.

in the posterior cranial fossa and is shaped like a tent, the apex of which lies at its meeting with the free edge of the falx cerebri, posterosuperior to the splenium of the corpus callosum.

It is accurately fitted to the superior surface of the cerebellum and every part of it slopes inferiorly from its apex towards its fixed margin. The **fixed margin** passes from the posterior clinoid process of the sphenoid bone, posterolaterally along the margin of the petrous temporal bone which separates the middle and posterior cranial fossae and carries the superior petrosal sinus. Passing over the mastoid part of the temporal bone to the postero-inferior angle of the parietal bone, it straddles a broad groove formed by the transverse sinus, and is pierced from above by inferior anastomotic and other inferior cerebral veins passing to the transverse sinus. The attached margin then sweeps medially along the transverse sinus to the internal occipital protuberance where it meets the fixed margin of the falx cerebri. The **free margin** forms the anterosuperior border of the tentorium and clasps the midbrain, thus forming the **tentorial notch.** The free margin is attached antero-inferiorly to the anterior clinoid processes from which it passes posteriorly as a slightly raised ridge, lateral to the oculomotor nerve, to cross the fixed margin of the tentorium lateral

Falx Cerebri

Thalamus

Tentorium cerebelli

Transverse venous sinus

Cerebellum

Medulla oblongata

Posterior arch of atlas

Transverse ligament of atlas

Corpus callosum

Septum pellucidum

Pons

Lamina terminalis

Frontal sinus

Anterior cerebral artery

Optic nerve

Hypophysis

Basilar artery

FIG. 874 Drawing prepared from a photograph of a section (supplied by Professor R. Walmsley) to show the relation between the skull, meninges, and brain. The brain stem is slightly displaced posteriorly, the pons normally lies closer to the clivus.

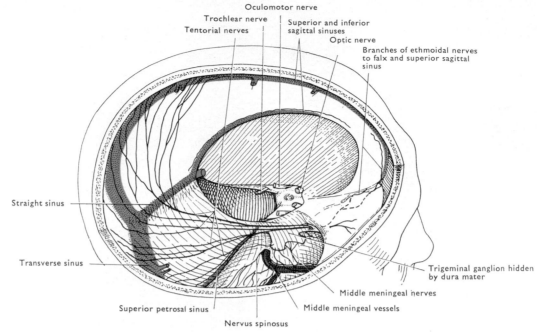

FIG. 875 Distribution of sensory nerves to dura mater of anterior and middle cranial fossae, falx cerebri, and tentorium cerebelli. (After Penfield and McNaughton, 1940.)

to the posterior clinoid process. Above the anterior part of the free edge lies the medial side of the temporal lobe, and further posteriorly the posterior cerebral artery, while below this margin lies the trochlear nerve and the superior cerebellar artery. This relation of the temporal lobe to the tentorium results in the uncus being displaced downwards medial to it when there is any space-occupying lesion in the supratentorial compartment of the skull.

The Innervation of the Dura Mater. The cranial dura mater is innervated by branches of the trigeminal, vagus, and by sympathetic nerves on the meningeal arteries which supply the dura mater and skull but play no part in the vascular supply of the brain. The anterior cranial fossa and the anterior part of the falx cerebri are supplied by the anterior (and posterior) ethmoidal nerves, passing from the cribriform plates of the ethmoid. The middle cranial fossa is supplied by branches of the maxillary and mandibular divisions of the trigeminal nerves, one of which (the nervus spinosus) is a recurrent branch of the mandibular nerve entering by the foramen spinosum. The tentorium and the posterior part of the falx cerebri are supplied by recurrent tentorial branches of the ophthalmic divisions of the trigeminal nerves, which run backwards near the free margin of the tentorium [FIG. 875]. The posterior cranial fossa receives a recurrent meningeal branch of the vagus which enters the skull through the jugular foramen.

The spinal dura mater is supplied by the recurrent meningeal branches of the spinal nerves.

The Cranial Arachnoid Mater. This differs from the spinal arachnoid in two ways:

1. On the cerebral and cerebellar hemispheres it is connected by a dense meshwork of processes to the pia mater. The interstices of this mesh are filled with cerebrospinal fluid which thus forms a sort of fluid-filled sponge which may act as a cushion preventing, in part, injury to the brain when the head is suddenly moved.

2. **Arachnoid villi and granulations** [FIGS. 876 and 877]. Along the line of the superior sagittal sinus, where the venous pressure is at its lowest, finger-like prolongations of the arachnoid pass through the fenestrations in the dura and project into the sinus and its lateral lacunae. These arachnoid villi are present but very small in the new-born but enlarge progressively throughout life until in elderly individuals they may be so large as to indent the overlying bone (foveolae granulares). The villi and granulations are covered on their vascular surfaces by the endothelium of the sinus which becomes continuous with capillary-like tubules leading through these structures to become continuous with the subarachnoid space. In the villi and granulations the tubules are surrounded by the subarachnoid meshwork which becomes distended with

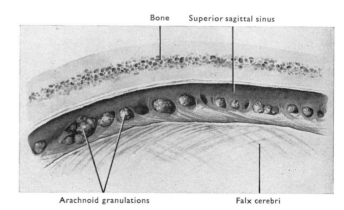

FIG. 876 Median section through cranial vault in parietal region (enlarged). Displays a portion of the superior sagittal sinus and the arachnoid granulations protruding into it.

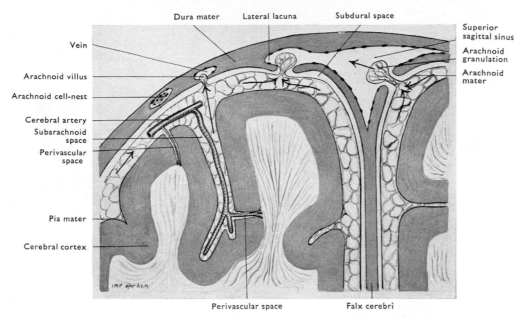

FIG. 877 Diagram to show relation of meninges to brain and of subarachnoid space to arteries, neuronal elements, and venous channels of dura mater. The necks of the villi are fused with the dura mater, and there is no continuity between the venous sinus and the subdural space.

cerebrospinal fluid when the pressure of that fluid is greater than the pressure in the sinus. The strands of the meshwork hold open the tubules which discharge cerebrospinal fluid directly into the sinus or its lacunae. When the venous pressure exceeds that of the cerebrospinal fluid, the villi or granulations collapse, the tubules are closed, and blood cannot pass into the subarachnoid space (Jayatilaka, 1965). Thus these structures are valvular openings from the subarachnoid space into the venous system, and they prevent increases in intracranial pressure. This arrangement explains the great increase in intracranial pressure which follows clotting (thrombosis) of the blood in the superior sagittal sinus.

The Cranial Subarachnoid Space. Because the brain and skull show marked differences in shape, there are several situations where the subarachnoid space is deep, forming a series of subarachnoid cisterns within the skull. These are found particularly on the base of the brain, and in them the trabeculae joining arachnoid to pia mater are very much reduced so that the arachnoid can easily be lifted away from the pia.

The largest of the **cisterns** is the **cerebellomedullary** which lies between and behind the cerebellum and medulla oblongata. It can readily be entered by a needle introduced immediately below the posterior margin of the foramen magnum and passed upwards **(cisternal puncture).** In its depth lies the median aperture of the fourth ventricle. It is continuous inferiorly with the spinal subarachnoid space, and around the sides of the medulla oblongata with a shallower but extensive cistern surrounding the pons and medulla oblongata. This pontine cistern is continuous anteriorly on the base of the brain with a deep **interpeduncular cistern,** formed by the arachnoid mater bridging across between the temporal lobes. The interpeduncular cistern contains the arterial circle of the base of the brain, and extensions of this cistern follow the main branches of this circle :

1. With the posterior cerebral arteries around the superior part of the midbrain at the margin of the tentorium, cisterna ambiens.

2. With the middle cerebral artery into the lateral sulcus and its branches, the **cistern of the lateral fossa.**

3. With the anterior cerebral arteries, surrounding the superior and inferior surfaces of the optic chiasma, the **cisterna chiasmatis.** Within the lateral sulcus the middle cerebral artery is so closely surrounded by brain tissue that, it is believed, the pulsations of the artery drive the cerebrospinal fluid laterally from the interpeduncular fossa over the lateral surface of the hemisphere. Possibly the other major arteries have similar effects in causing directional flow of cerebrospinal fluid in the subarachnoid space.

The Formation and Circulation of Cerebrospinal Fluid. The bulk of the cerebrospinal fluid is formed in the choroid plexuses of the lateral ventricles and in lesser amounts in the third and fourth ventricles. It is probably also produced by ependyma throughout the ventricular system and some may even be derived from the capillaries on the surface of the brain and spinal medulla. It is a clear, protein-free fluid which contains two or three lymphocytes per mm.[3] and is in osmotic equilibrium with blood plasma, principally because of a higher concentration of chlorides. The normal course of the fluid is from the lateral ventricles through the interventricular foramen to the third ventricle, thence through the midbrain in the **cerebral aqueduct** to the fourth ventricle. Here the cerebrospinal fluid escapes through the median and the two lateral apertures [p. 601] into the subarachnoid space, entering the cerebello-medullary and pontine cisterns respectively. From this site the fluid may pass downwards around the spinal medulla, but most of it flows slowly through the tentorial notch and spreads over the cerebral hemispheres, partly aided by the active pulsations of the cerebral arteries, to reach the arachnoid granulations where it can pass into the

FIG. 878 Diagram to show the main points relating to the production, course, and absorption of the cerebrospinal fluid. (After N. M. Dott (1928) *Edinb. med. J.*, **35**, 182.)

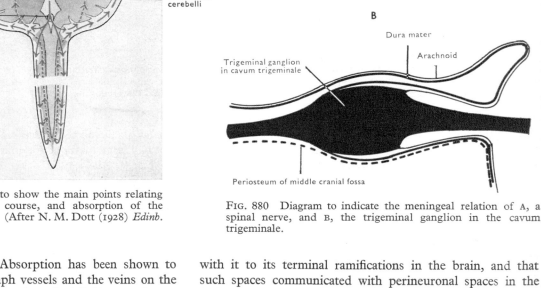

FIG. 880 Diagram to indicate the meningeal relation of A, a spinal nerve, and B, the trigeminal ganglion in the cavum trigeminale.

venous system [FIG. 878]. Absorption has been shown to occur into the perineural lymph vessels and the veins on the spinal nerves. In particular, cerebrospinal fluid passes into :

1. The orbital lymph vessels, by passing along the sub-arachnoid sheath which surrounds the optic nerve to the lamina cribrosa of the sclera.

2. The lymph vessels of the nose from the region of the olfactory bulb, also along the vestibulocochlear and facial nerves. At one time it was thought that every branch of the arteries in the subarachnoid space carried a sleeve of the space

with it to its terminal ramifications in the brain, and that such spaces communicated with perineuronal spaces in the brain. This is now known not to be the case, since no such spaces are visible under the electron microscope and the peri-vascular spaces only accompany the larger vessels for a short distance as they pierce the nervous system.

The Pia Mater of the Brain. In most situations this is a much more delicate membrane than its spinal counterpart, though over the medulla oblongata it is thick and firmly adherent. Elsewhere it forms the intimate covering of the brain, dipping into every sulcus. In certain places it lies in apposition with the ependymal lining of the brain, i.e., in the roofs of the fourth and third ventricles and in the medial walls of the lateral ventricles. Here it is known as the **tela choroidea,** and forms the vascular connective tissue core of the choroid plexuses, being invaginated into the ventricle with a covering of ependyma in these situations [pp. 600 and 653].

The Relation of the Meninges to the Nerves. [FIG. 880] In the spinal nerves the dura mater is prolonged separately over the ventral and dorsal roots to fuse with the epineurium at the distal part of the spinal ganglion. The arach-noid also forms a separate sheath for both roots, the subarachnoid space extending to the

FIG. 879 Diagram of coronal section through tela choroidea of third ventricle.

proxima part of the spinal ganglion. It is in this region that reabsorption of cerebrospinal fluid is believed to occur. Exactly the same arrangement is present in the cranial nerves, the dura fusing with the distal part of the ganglia and the arachnoid with their proximal parts. The arrangement around the trigeminal ganglion is the same as that of spinal ganglia, only in this case the dura mater is folded over it, forming the floor of the middle cranial fossa, so that the cavum trigeminale has two layers of dura mater superior to it, but is otherwise the same as the space surrounding any spinal ganglion. The pia mater is continuous with the perineurium.

CRANIOCEREBRAL TOPOGRAPHY

The relationship between the brain and the internal cranial base is readily visualized. In the midline [FIG. 874] the medulla and pons lie on the clivus with the upper border of the pons reaching almost to the dorsum sellae and its anterior surface separated from the bone by the basilar venous plexus [FIG. 1119], the meninges, and the basilar artery in the subarachnoid space. The optic chiasma does not lie in the sulcus chiasmatis of the sphenoid bone but considerably posterior to this, almost vertically above the hypophysis. From the

chiasma the optic nerves pass forwards, above the hypophysis in the sella turcica and below the corresponding anterior cerebral artery, to reach the optic canal immediately lateral to the sphenoidal air sinus. In the anterior cranial fossa lies the crista galli and the attachment of the falx cerebri to the ethmoid bone. Immediately on either side of this, the olfactory bulb lies on the cribriform lamina of the ethmoid which, together with the dura mater and nasal epithelium, separate it from the nasal cavity. Further laterally the orbital surface of the frontal lobe lies in contact with the orbital part of the frontal bone and the lesser wing of the sphenoid. The orbital part of the frontal bone separates the brain from the ethmoidal air sinuses medially and the orbit laterally. It is a thin plate of bone which is in close apposition to the surface of the frontal lobe and is marked with ridges and impressions corresponding to the sulci and gyri overlying it. Medially it slopes inferiorly to overlie the ethmoid as far as the lateral edge of the cribriform lamina, while posteriorly it is continuous with the lesser wing of the sphenoid, the sharp concave posterior edge of which is inserted into the lateral sulcus and overlies the extreme anterior part of the temporal lobe. The floor of the anterior cranial fossa is very thin throughout, except for the occasional extension of the frontal air sinus posteriorly between the inner and outer tables of the orbital part of the frontal

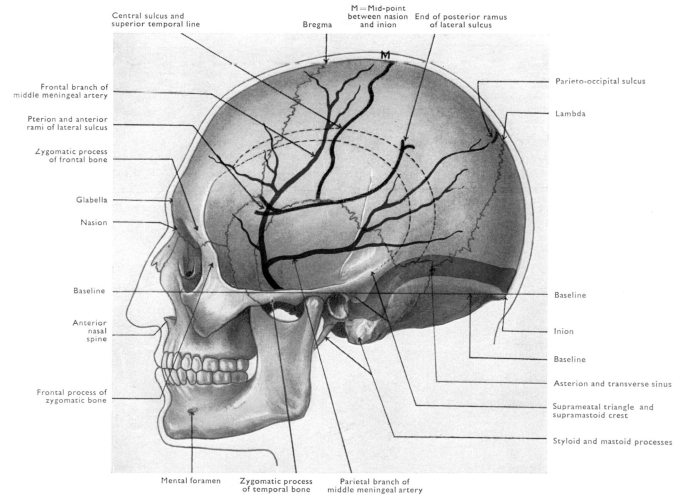

FIG. 881 Craniocerebral topography : landmarks of skull ; chief sulci of cerebrum ; transverse and sigmoid venous sinuses ; middle meningeal artery.

bone. Because it is thin, the floor of the anterior cranial fossa is not uncommonly fractured in injuries to the skull; blood and cerebrospinal fluid may then escape into the nose or orbit, or communication may be established between the cranial cavity and the ethmoidal or frontal air sinuses.

The inferior surfaces of the temporal and occipital lobes, which cannot be separated from each other on this surface, rest on the greater wing of the sphenoid [FIG. 134] (separated from it by the middle meningeal artery and veins curving anterolaterally towards the pterion), the inferomedial part of the squamous temporal bone and the petrous temporal bone. On the latter it is separated from the middle ear by the thin tegmen tympani which is cartilaginous in early childhood. Posterior to the temporal bone the inferior surface is separated from the cerebellum by the tentorium cerebelli.

The superolateral surface of the brain [FIGS. 881 and 882] is related to the medial wall of the temporal fossa and the cranial vault and, over a large part of this surface, to the branches of the middle meningeal artery [FIG. 883]. The relationship of individual parts of the superolateral surface to this region of the skull is remarkably variable in different skull types, and exact positioning of any given sulcus from the surface is not possible. It is of some value, however, to know the approxi-

mate position of the main sulci and gyri, so that an estimate may be given of the part of the brain likely to be injured in direct blows producing depressed fractures, or to assist in localization of a subdural haemorrhage if signs of local brain damage are present. Much more detailed information is required in modern neurosurgery when it may be necessary to injure particular parts of the thalamus or corpus striatum by the introduction of a fine electrode through a trephine hole in the skull. For this purpose, the surface configuration of the brain does not give sufficiently accurate information, and the estimation of the position of deep structures in the brain depends on outlining the third and lateral ventricles by radiography after filling them with air [FIG. 883]. Then midline structures, e.g., the anterior and posterior commissures, may be visualized and used as a baseline from which to estimate the position of other structures. The relation of the lateral ventricle to the brain is shown in FIGURE 884.

On the surface of the head a baseline may be drawn from the lower margin of the orbit backwards through the upper margin of the external acoustic meatus. This line cuts the occiput close to the external occipital protuberance and has the cerebrum lying entirely above it, while that part of the cere-

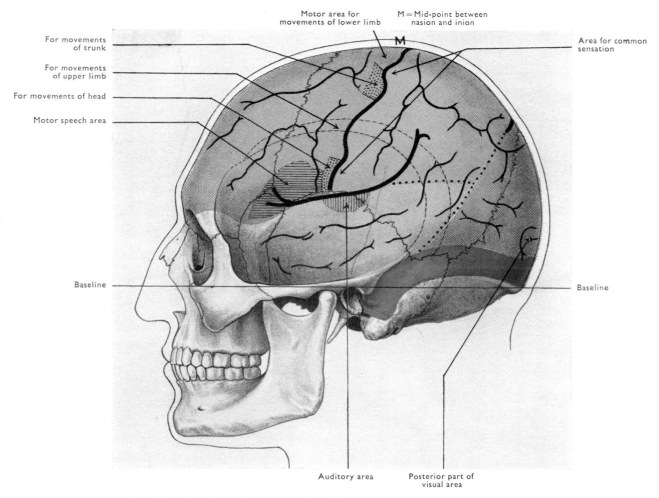

FIG. 882 Craniocerebral topography : lobes of cerebrum ; motor and sensory areas.
Frontal and occipital lobes, red ; parietal lobe, blue ; temporal lobe, pale blue.

FIG. 883 Topography of normal lateral ventricle and middle meningeal artery. (By permission of the editor of *The British Journal of Surgery*.) [See also PLATE 45.]

FIG. 884 Drawing to show the position of the lateral ventricle in the brain. Prepared from photographs taken during dissection, this ventricle is slightly enlarged and has a very small posterior horn.

bellum below its horizontal fissure lies in the posterior cranial fossa inferior to the posterior third of the line.

A point, two finger-breadths above the zygomatic arch and one finger-breadth posterior to the frontozygomatic suture (which can be felt as a notch) marks the inferior extremity of the coronal suture and the pterion. From this point the coronal suture slopes posterosuperiorly to the bregma, which may be felt as a slight depression and lies immediately anterior to the point at which a vertical line through the external acoustic meatus cuts the median plane. The central sulcus lies approximately parallel to the coronal suture, approaching it slightly in its inferior part, and at its superior extremity is approximately 1 cm. posterior to the midpoint [M, FIG. 882] of a line from nasion to inion. The parieto-occipital sulcus is marked by a line extending 2 cm. transversely from the midline at the lambda. The lambda can either be felt, or taken as 8–10 cm. posterior to the superior end of the central sulcus, or 6–7 cm. above the inion. A point 3·5 cm. behind and 1 cm. above the frontozygomatic suture marks the anterior end of the posterior ramus of

the lateral sulcus. From this point the posterior ramus may be marked out as a posterior extension of the line joining it to the frontozygomatic suture and curving superiorly to the parietal eminence at its posterior end. The anterior ramus of the lateral sulcus passes forwards from the point approximately to the lower end of the coronal suture, while the ascending ramus passes superiorly at right angles to the posterior ramus for 2 cm. 1 cm. above a point 2·5 cm. posterior to the origin of the posterior ramus from the lateral sulcus, lies the inferior end of the central sulcus.

The frontal branch of the middle meningeal artery, often in the form of several parallel vessels, lies between the central sulcus and the coronal suture.

The lateral ventricle [PLATE 45 B and D] lies within the area of the temporal fossa except for its posterior horn. The interventricular foramen lies on the level of the posterior ramus of the lateral sulcus, 1·5 cm. behind its anterior extremity. The anterior horn reaches the inferior part of the coronal suture, while the central part extends horizontally backwards, turns inferiorly with its dorsal surface parallel to the inferior temporal line, and divides into inferior and posterior horns. The inferior horn passes forwards in the line of the middle temporal gyrus to end about 3 cm. posterior to the temporal pole, just posterior to the surface marking of the hypophysis and postero-inferior to the optic chiasma.

RADIOGRAPHIC ANATOMY OF THE
CENTRAL NERVOUS SYSTEM

E. SAMUEL

Nerve cells, glial tissue, and the vascular components of the brain are not seen on the plain radiograph, but, using various contrast media to replace the cerebrospinal fluid, it is possible to outline certain parts of the brain.

The main methods employed to produce this contrast are ventriculography and encephalography in which air is used to

replace some of the cerebrospinal fluid. In this way the ventricular system, the subarachnoid spaces, and the convolutions of the cerebral hemisphere can be seen.

Ventriculography is performed by making a hole in each parietal bone and injecting air through a needle or catheter into one or both lateral ventricles. By suitably positioning the subject's head, the air can be floated into the various recesses of the lateral ventricles, into the third ventricle, and finally into the subarachnoid spaces.

The ventricle thus filled shows a characteristic shape on the radiograph [PLATE 45A and B]. In the frontal projection the X-ray beam is directed in the long axis of the lateral ventricle, thus the shadows cast by the anterior horn, the central part, and the posterior horn are superimposed. As air is less absorbent of X-rays than is brain tissue, this shadow is black in contrast to the white shadow of the skull outline. In this view, the composite shadow of both lateral ventricles has been called the butterfly shadow and each 'wing' is composed of a less dense oval shadow, the anterior horn, superimposed on a darker triangular shadow corresponding to the central part, and a third, less dense, intermediate shadow caused by the posterior horn. The shadow representing the central part of the lateral ventricle has a greater density owing to its length and the fact that the anterior and posterior horns overlie it. The third ventricle may be seen as a narrow, vertical, median shadow inferior to the butterfly shadow.

In the lateral view, the whole of the lateral ventricle can be seen in profile without any superimposition except of its fellow on the opposite side. The rounded anterior horn is marked posteriorly by a dense shadow corresponding to the inter-ventricular foramen, whilst lying posteriorly is the central part of the lateral ventricle with a gentle elevation of its floor caused by the thalamus. The posterior horn, often long and pointed, varies considerably in size but never attains the pro-portions of the anterior horn. The inferior horn arises at the junction of the posterior horn with the central part and sweeps forwards into the temporal lobe in a crescent-like curve.

The third ventricle is less well defined owing to its narrow slit-like shape. Projecting from the posterior end of the third ventricle is the pineal body outlined by contrast with the air in the suprapineal recess. The cerebral aqueduct connects the posterior aspect of the third ventricle with the fourth ventricle [PLATE 45]. Special views must be taken to show the relatively small quantity of air in this structure.

The fourth ventricle in the lateral view appears as a tri-angular shadow with its base directed anteriorly and its apex (fastigial recess) posteriorly. In the frontal projection the fourth ventricle appears to have approximately the same shape.

To produce the contrast in **encephalography,** a quantity of the cerebrospinal fluid is replaced by air. A needle is intro-duced into the subarachnoid space, usually between the third and fourth lumbar vertebrae, and after withdrawal of 5–30 ml. of cerebrospinal fluid, a slightly smaller quantity of air is injected. The injected air ascends in the spinal subarachnoid space to the cerebellomedullary cistern in the skull, thence through the median aperture into the fourth ventricle and aqueduct, rising to the third ventricle. By positioning the head it may be made to enter the lateral ventricles.

Ventriculography and encephalography enable intracranial space-occupying lesions, such as tumours and blood clots, to be shown by the displacement and deformity of the ventricular system. Similarly, from the deformity pattern of the ven-tricular system it may be possible to locate accurately the site of the intracranial lesion. Both these methods may also be used to determine sites of obstruction to the flow of cerebro-

spinal fluid which may be responsible for the development of a hydrocephalus.

When a block is suspected in the spinal subarachnoid space, the needle may be inserted between the skull and the posterior arch of the atlas into the cerebellomedullary cistern [PLATE 46A], and heavy contrast medium allowed to gravitate to the site of the obstruction.

Under selected conditions, a positive contrast medium may be introduced into the ventricle and this is especially useful when the patency of the cerebral aqueduct is being investi-gated. Because of its small size the aqueduct is poorly demonstrated by air contrast studies.

The second method for outlining the brain structure relies on information gained by filling the intracerebral vessels—angiography. This method attempts to assess changes in the brain from alterations in the anatomy and distribution of these vessels.

The method employed is **cerebral angiography** either after puncture of the common carotid, or if the vessels in the posterior cranial fossa are under review, by puncture of the vertebral artery. Only 5–10 ml. of contrast medium is em-ployed, and by taking rapid serial films, demonstration of filling of the intracranial vessels in arterial, capillary, and venous phases is obtained.

By means of compression of the opposite common carotid vessels, cross-circulation films are obtained showing the extent of communication to the opposite hemisphere via the arterial circle of the brain. These are essential before certain surgical procedures are carried out.

The cerebral angiogram [PLATE 49] shows the whole course of the internal carotid artery within the skull and its circuitous course along the side of the pituitary fossa before its division into the anterior and middle cerebral vessels.

The capillary phase of filling of the vessels follows immedi-ately on the arterial phase, a fine network of slender vessels being visible.

The venous phase is indicated by the filling of the super-ficial veins as they cross the cerebral hemispheres and increase in density as they reach the midline.

The deep cerebral veins, the internal cerebral vein, the great cerebral vein, and the straight sinus do not show the same variability as the superficial veins, and displacements of these veins can be used as an index of intracranial shift.

Sinuses of the Dura Mater. The great venous sinuses of the skull fill in the last phase of the angiogram. The large superior sagittal sinus extends in the midline to reach the occipital bone where it turns acutely to the right to fill the transverse sinus, the left transverse sinus being the continua-tion of the straight sinus. The actual junction of these sinuses is subject to considerable variation, and fusion of one or both transverse sinuses may occur, forming the confluence of the sinuses.

Anteriorly, both transverse sinuses bend acutely downwards to form the sigmoid sinus, and as the contrast medium becomes increasingly dilute and the sigmoid sinus leaves the skull only a faint trace can be seen in the internal jugular vein.

The vascular grooves on the inner table of the skull also form a definite pattern which can be seen in plain radiographs. They can be grouped into arterial, diploic, and venous mark-ings. The most readily visible marking in the lateral projec-tion is that of the middle meningeal vessels which are etched on the inner table of the sphenoid, temporal, and parietal bones. The markings of some of the other vessels, such as the supra-orbital and the superficial temporal, are less clearly defined than those of the middle meningeal but should be recognized as normal markings in the appropriate projections.

The **diploic vessels** [PLATE 46C] have a more variable pattern than the arterial markings but occur essentially in three groups—those in the frontal, occipital, and parietal regions of the skull. The diploic vessels in the parietal region of the skull form the most constant group of these vessels and have a radiating appearance fancifully compared to a spider's web, 'the parietal spider'.

Emissary veins are a specialized group of veins connecting the extracranial veins with the venous sinuses of the dura mater. At their passage through the skull, certain foramina can be seen. The emissary vein in the mastoid region is most commonly seen. Less frequently, the foramina caused by the frontal and parietal emissary veins can be seen.

Certain intracranial structures are so frequently the sites of calcification that they can almost be regarded as physiological although, in fact, they represent, in most instances, degenerative changes.

Calcification in the pineal body [PLATE 46C] occurs in approximately 40 per cent. of people over the age of 40 years, and its displacement from the normal median position has proved of great value in indicating displacements of the brain. Calcification in the habenular commissure anterior to the gland is so intimately associated with it that it is frequently regarded as pineal calcification.

Less frequently, calcification is seen in the choroid plexuses of the lateral ventricle. Calcification may also be seen in the arachnoid granulations along the superior sagittal sinus.

GENERAL REFERENCES

BAILEY, P., and BONIN, G. (1951) *The Isocortex of Man*, Urbana.

BRODAL, A. (1959) *The Cranial Nerves*, Oxford.

BRODAL, A. (1957) *The Reticular Formation of the Brain Stem: Anatomical Aspects and Functional Correlations*, Edinburgh.

BRODAL, A. (1969) *Neurological Anatomy in Relation to Clinical Medicine*, 2nd ed., New York.

BRODAL, A., WALBERG, F., and POMPEIANO, O. (1961) *The Vestibular Nuclei and their Connections. Anatomy and Functional Correlations*, Edinburgh.

BUCY, P. C. (1949) *The Precentral Motor Cortex*, Chicago.

CAJAL, S. RAMÓN Y (1955) *Histologie du Système Nerveux de L'Homme et les Vertébrés*, Madrid.

CLARK, W. E. LE GROS (1942) The visual centres of the brain and their connexions, *Physiol. Rev.*, **22**, 205.

CLARK, W. E. LE GROS (1958) *The Tissues of the Body; An Introduction to the Study of Anatomy*, 4th ed., Chapter 13, p. 326.

CLARK, W. E. LE GROS, BEATTIE, J., RIDDOCH, G., and DOTT, N. M. (1938) *The Hypothalamus: Morphological, Functional, Clinical and Surgical Aspects*, Edinburgh.

DOW, R. S. (1942) The evolution and anatomy of the cerebellum, *Biol. Rev.*, **17**, 179.

ECONOMO, C. VON (1929) *The Cytoarchitectonics of the Human Cerebral Cortex*, trans. Parker, S., London.

FULTON, J. F. (1949) *Functional Localization in the Frontal Lobes and Cerebellum*, Oxford.

FULTON, J. F., ARING, C. D., and WORTIS, S. B. (Eds.) (1948) The frontal lobes, *Res. Publ. Ass. nerv. ment. Dis.*, **27**.

FULTON, J. F., RAWSON, S. W., and FRANTZ, A. M. (Eds.) (1939) The hypothalamus, *Res. Publ. Ass. nerv. ment. Dis.*, **20**.

GLEES, P. (1955) *Neuroglia, Morphology and Function*, Oxford.

HERRICK, C. J. (1938) *An Introduction to Neurology*, 5th ed., Philadelphia.

HORSTADIUS, S. (1950) *The Neural Crest, its Properties and Derivatives in the Light of Experimental Research*, London.

JANSEN, J., and BRODAL, A. (1954) *Aspects of Cerebellar Anatomy*, Oslo.

JASPER, H. H., PROCTOR, L. D., KNIGHTON, R. S., NOSHAY, W. C., and COSTELLO, R. T. (Eds.) (1958) *Reticular Formation of the Brain*, London.

KAPPERS, C. U. A., HUBER, G. C., and CROSBY, E. C. (1936) *The Comparative Anatomy of the Nervous System of Vertebrates*, New York.

LASSEK, A. M. (1954) *The Pyramidal Tract*, Springfield.

OLSZEWSKI, J., and BAXTER, D. (1954) *Cytoarchitecture of the Human Brain Stem*, New York.

PENFIELD, W., and JASPER, H. H. (1954) *Epilepsy and the Functional Anatomy of the Human Brain*, Boston.

PENFIELD, W., and RASMUSSEN, T. (1950) *The Cerebral Cortex of Man*, New York.

PUTNAM, T. J. (Ed.) (1942) The diseases of the basal ganglia, *Res. Publ. Ass. nerv. ment. Dis.*, **21**.

RILEY, H. A. (1943) *An Atlas of the Basal Ganglia, Brain Stem and Spinal Cord*, Baltimore.

WALKER, A. E. (1938) *The Primate Thalamus*, Chicago.

WALSHE, F. M. R. (1948) *Critical Studies in Neurology*, Edinburgh.

WHITFIELD, I. C. (1967) *The Auditory Pathway*, London.

WOOLLAM, D. H. M., and MILLEN, J. W. (1954) Perivascular spaces of the mammalian central nervous system, *Biol. Rev.*, **29**, 251.

WURTMAN, R. J., AXELROD, J., and KELLY, D. E. (1968) *The Pineal*, New York.

YOUNG, J. Z. (1942) The functional repair of nervous tissue, *Physiol. Rev.*, **22**, 318.

SPECIFIC REFERENCES

BARGMANN, W. (1966) Neurosecretion, *Int. Rec. Cytol.*, **19**, 183.

BARR, M. L. (1960) Sexual dimorphism in interphase nuclei, *Amer. J. hum. Genet.*, **12**, 118.

BODIAN, D. (1951) Nerve endings, neurosecretory substance and lobular organisation of the neurohypophysis, *Bull. Johns Hopk. Hosp.*, **89**, 354.

BREATHNACH, A. S., and GOLDBY, F. (1954) The amygdaloid nuclei, hippocampus and other parts of the rhinencephalon in the porpoise, *J. Anat. (Lond.)*, **88**, 267.

BRODAL, A., and KAADA, B. R. (1953) Exteroceptive and proprioceptive ascending impulses in the pyramidal tract of cat, *J. Neurophysiol.*, **16**, 567.

BRODAL, A., SZABO, T., and TORVIK, A. (1956) Corticofugal fibres to sensory trigeminal nuclei and nucleus of solitary tract, *J. comp. Neurol.*, **106**, 527.

BRODAL, A., and WALBERG, F. (1952) Ascending fibres in the pyramidal tract of cat, *Arch. Neurol. Psychiat. (Chic.)*, **68**, 1.

BUNGE, M. B., BUNGE, R. P., and PAPPAS, G. D. (1962) Electron microscopic demonstration of connections between glia and myelin sheaths in the developing mammalian central nervous system, *J. Cell. Biol.*, **12**, 448.

CARMAN, J. B., COWAN, W. M., and POWELL, T. P. S. (1963) The organization of cortico-striate connexions in the rabbit, *Brain*, **86**, 525.

CARMAN, J. B., COWAN, W. M., and POWELL, T. P. S. (1964a) The cortical projection upon the claustrum, *J. Neurol. Neurosurg. Psychiat.*, **27**, 46.

CARMAN, J. B., COWAN, W. M., and POWELL, T. P. S. (1964b) Cortical connexions of the thalamic reticular nucleus, *J. Anat. (Lond.)*, **98**, 587.

CHOW, K. L. (1952) Regional degeneration of the thalamic reticular nucleus following cortical ablations in monkeys, *J. comp. Neurol.*, **97**, 37.

CLARK, G. (1948) The mode of representation in the motor cortex, *Brain*, **71**, 320.

CLARK, G., and WARD, J. W. (1948) Responses elicited from the cortex of monkeys by electrical stimulation through fixed electrodes, *Brain*, **71**, 332.

CLARK, W. E. LE GROS (1952) A note on cortical cytoarchitectonics *Brain*, **75**, 96.

CLARK, W. E. LE GROS (1957) Enquiries into the anatomical basis of olfactory discrimination, *Proc. roy. Soc. B*, **146**, 299.

COWAN, W. M., and POWELL, T. P. S. (1955) The projection of the midline and intralaminar nuclei of the thalamus of the rabbit, *J. Neurol. Neurosurg. Psychiat.*, **18**, 266.

CREVEL, H. VAN (1958) *The Rate of Secondary Degeneration in the Central Nervous System*, Leiden.

CREVEL, H. VAN, and VERHAART, W. J. C. (1963) The rate of secondary degeneration in the central nervous system. 1. The pyramidal tract of the cat, *J. Anat. (Lond.)*, **97**, 429.

DE VALOIS, R. L. (1960) Colour vision mechanisms in the monkey, *J. gen. Physiol.*, **43**, No. 6, pt. II, 115.

FADIGA, E., and PUPILLI, G. C. (1964) Teleceptive components of the cerebellar function, *Physiol. Rev.*, **44**, 432.

GALAMBOS, R. (1956) Suppression of auditory nerve activity by stimulation of efferent fibres to the cochlea, *J. Neurophysiol.*, **19**, 424.

GETZ, B. (1952) The terminations of spino-thalamic fibres in the cat as studied by the method of terminal degeneration, *Acta anat. (Basel)*, **16**, 271.

HARRISON, R. G. (1907) Observations on the living developing nerve fibre, *Anat. Rec.*, **1**, 116.

HOFF, E. C., and HOFF, H. E. (1934) Spinal terminations of the projection fibres from the motor cortex of primates, *Brain*, **57**, 454.

HOLMES, G., and MAY, W. P. (1909) On the exact origin of the pyramidal tracts in Man and other mammals, *Brain*, **32**, 1.

HOWE, H. A., and BODIAN, D. (1942) *Neural Mechanisms in Poliomyelitis*, The Commonwealth Fund, New York.

HYDEN, H. (1967) Dynamic aspects of the neuron-glia relationship, a study with microchemical methods, in *The Neuron*, ed. Hyden, H., Amsterdam.

JAYATILAKA, A. D. P. (1965) An electron microscopic study of sheep arachnoid granulations. *J. Anat. (Lond.)*, **99**, 635.

JONES, E. G., and POWELL, T. P. S. (1968) The commissural connexions of the somatic sensory cortex in the cat, *J. Anat. (Lond.)*, **103**, 433.

JUNG, R., and HASSLER, R. (1960) The extrapyramidal motor system, in *Handbook of Physiology*, American Physiological Society, Section 1, **2**, 863.

KUYPERS, H. G. J. M. (1958) Corticobulbar connexions to the pons and lower brainstem in Man, *Brain*, **81**, 364.

KUYPERS, H. G. J. M. (1960) Central cortical projections to motor and somatic sensory groups, *Brain*, **83**, 161.

KUYPERS, H. G. J. M., and LAWRENCE, D. G. (1967) Cortical projections to the red nucleus and the brainstem in the rhesus monkey, *Brain Research*, **4**, 151.

MARIE, P. (1895) *Lectures on Diseases of the Spinal Cord*, trans. Lubbock, for New Sydenham Soc., London.

McLARDY, T. (1950) Thalamic projection to the frontal cortex in Man, *J. Neurol. Neurosurg. Psychiat.*, **13**, 198.

MYERS, R. E. (1965) in *Functions of the Corpus Callosum*, London.

NATHAN, P. W. (1956) Awareness of bladder filling with divided sensory tract, *J. Neurol. Neurosurg. Psychiat.*, **19**, 101.

NATHAN, P. W., and SMITH, M. C. (1951) The centripetal pathway from the bladder and urethra within the spinal cord, *J. Neurol. Neurosurg. Psychiat.*, **14**, 262.

NATHAN, P. W., and SMITH, M. C. (1953) Spinal pathways subserving defaecation and sensation from the lower bowel, *J. Neurol. Neurosurg. Psychiat.*, **16**, 245.

NATHAN, P. W., and SMITH, M. C. (1955a) Long descending tracts in Man, *Brain*, **78**, 248.

NATHAN, P. W., and SMITH, M. C. (1955b) Spino-cortical fibres in Man, *J. Neurol. Neurosurg. Psychiat.*, **18**, 181.

NATHAN, P. W., and SMITH, M. C. (1959) Fasciculi proprii of the spinal cord in Man, *Brain*, **82**, 610.

OLIVECRONA, H. (1954) Relation of the paraventricular nucleus to the pituitary gland, *Nature (Lond.)*, **173**, 1001.

PENFIELD, W., and FAULK, M. E. (1955) The insula. Further observations on its function, *Brain*, **78**, 445.

POGGIO, G. F., and MOUNTCASTLE, V. B. (1959) A study of the functional contributions of the lemniscal and spinothalamic systems to somatic sensibility, *Bull. Johns Hopk. Hosp.*, **105**, 201.

POWELL, T. P. S. (1958) The organization and connexions of the hippocampal and intralaminar systems, *Recent Progr. Psychiat.*, **3**, 54.

POWELL, T. P. S., and MOUNTCASTLE, V. B. (1959a) The cytoarchitecture of the postcentral gyrus of the monkey, Macaca mulatta, *Bull. Johns Hopk. Hosp.*, **105**, 108.

POWELL, T. P. S., and MOUNTCASTLE, V. B. (1959b) Some aspects of the functional organisation of the cortex of the postcentral gyrus of the monkey: a correlation of findings obtained in a single unit analysis with cytoarchitecture, *Bull. Johns Hopk. Hosp.*, **105**, 133.

RASMUSSEN, G. L. (1953) Further observations on the efferent cochlear bundle, *J. comp. Neurol.*, **99**, 61.

REESE, T. S., and KARNOVSKY, M. J. (1967) Fine structural localization of a blood-brain barrier to exogenous peroxidase, *J. Cell. Biol.*, **34**, 207.

RENSHAW, B. (1941) Influence of the discharge of motoneurons upon excitation of neighbouring motoneurons, *J. Neurophysiol.*, **4**, 167.

ROMANES, G. J. (1951) The motor cell columns of the lumbosacral spinal cord of the cat, *J. comp. Neurol.*, **94**, 313.

SCHIBEL, M. A., and SCHIBEL, A. B. (1966) The organization of the nucleus reticularis thalami: A Golgi study, *Brain Research*, **1**, 43.

SCOVILLE, W. B., and MILNER, B. (1957) Loss of recent memory after bilateral hippocampal lesions, *J. Neurol. Neurosurg. Psychiat.* **20**, 11.

SHARRARD, W. J. W. (1955) The distribution of the permanent paralysis in the lower limb in poliomyelitis. A clinical and pathological study, *J. Bone Jt Surg.*, **37-B**, 540.

SHERRINGTON, C. S. (1889) On nerve tracts degenerating secondarily to lesions of the cortex cerebri, *J. Physiol. (Lond.)*, **10**, 429.

SMITH, M. C. (1957) Observations on the topography of the lateral column of the human cervical spinal cord, *Brain*, **80**, 263.

SNIDER, R. S. (1950) Recent contributions to the anatomy and physiology of the cerebellum, *Arch. Neurol. Psychiat. (Chic.)*, **64**, 196.

SPALDING, J. M. K. (1952) Wounds of the visual pathway. Part I. The visual radiation, *J. Neurol. Neurosurg. Psychiat.*, **15**, 99.

SPERRY, R. W. (1958) Physiological plasticity and brain circuit theory, in *Biological and Biochemical Bases of Behaviour*, ed. Harlow, H. F., and Woolsey, C. N., p. 401.

SPRAGUE, J. M., and CHAMBERS, W. W. (1959) An analysis of cerebellar function in the cat, as revealed by its partial and complete destruction, and its interaction with the cerebral cortex, *Arch. ital. Biol.*, **97**, 68.

SZENTAGOTHAI, J., and RAJKOVITS, K. (1959) Über den Ursprung der Kletterfasern des Kleinhirns, *Z. Anat. Entwickl.-Gesch.*, **121**, 130.

TARKHAN, A. A. (1934) The innervation of the extrinsic ocular muscles, *J. Anat. (Lond.)*, **68**, 293.

TRAVIS, A. M. (1955) Neurological deficiencies following supplementary motor area lesions in Macaca mulatta, *Brain*, **78**, 174.

WALBERG, F. (1954) Descending connections to the inferior olive, in *Aspects of Cerebellar Anatomy*, ed. Jansen, J., and Brodal A., Oslo.

WALBERG, F. (1957a) Corticofugal fibres to the nuclei of the dorsal columns, *Brain*, **80**, 273.

WALBERG, F. (1957b) Do the motor nuclei of the cranial nerves receive corticofugal fibres?, *Brain*, **80**, 597.

WALBERG, F., and BRODAL, A. (1953) Pyramidal tract fibres from temporal and occipital lobes. An experimental study in the cat, *Brain*, **76**, 491.

WALSHE, F. M. R. (1961) The problem of the origin of the pyramidal tract, in *Scientific Aspects of Neurology*, ed. Garland, H., Edinburgh

WARWICK, R. (1953) Representation of the extra-ocular muscles in the oculomotor nuclei of the monkey, *J. comp. Neurol.*, **98**, 449.

WHITTERIDGE, D. (1960) Central control of eye movements, in *Handbook of Physiology*, American Physiological Society, Section 1, **2**, 1089.

WOLFF, E. (1953) The so-called medial root of the optic tract is essentially a visual commissure, *Brain*, **76**, 455.

THE PERIPHERAL NERVOUS SYSTEM

G. J. ROMANES

THE peripheral nervous system is that part of the nervous system which lies external to the brain and spinal medulla. It consists of nerve fibres [p. 585] and their connective tissue sheaths which extend to all parts of the body. These nerve fibres are the processes of nerve cells the majority of which lie outside the brain and spinal medulla, though a proportion lie inside these organs and send their processes into the peripheral nervous system.

Peripheral nerves consist of bundles of nerve fibres, both myelinated and non-myelinated, enclosed in three layers of connective tissue. The outermost layer, **epineurium,** is formed of collagen (and fibrocytes) which gives much of the strength to the peripheral nerve, and binds together the various bundles of nerve fibres which comprise it. The middle layer, **perineurium,** encloses each separate bundle of nerve fibres. It consists of a variable number of layers of squamous, epithelial cells which form continuous sheets separated by minute amounts of collagen (Gamble and Eames, 1964; Shanthaveerappa and Bourne, 1966). The innermost layer, **endoneurium,** is a delicate connective tissue which separates the individual nerve fibres of the nerve, inside the perineurium. Traced towards the central nervous system, the epineurium becomes continuous with the dura mater [p. 698] and the perineurium with the pia-arachnoid; also the number of layers of the perineurium increases as does the size of the individual nerve fibre bundles. Traced towards the periphery, the number of layers of the perineurium decreases, and it becomes continuous with the capsules of the various nerve endings (Shanthaveerappa and Bourne, 1966). The perineurium thus isolates the peripheral nerve fibres from the surrounding tissues, and seems to form a layer which is impervious to many substances.

The peripheral nervous system forms the pathways through which the central nervous system receives sensory information from the periphery and produces changes in the activity of muscles, glands, etc. On topographical grounds it is broadly divisible into three parts.

1. **The spinal nerves** are attached to the spinal medulla by two separate rows of rootlets, dorsal and ventral. These rootlets unite in groups to form respectively the dorsal and ventral roots of the spinal nerves; one of each of these roots combine to form a spinal nerve. Each spinal nerve emerges through a separate intervertebral foramen to play its part in the innervation of the trunk and limbs.

2. **The cranial nerves** are attached to the brain by a single root or by a single row of rootlets, and they leave the skull through its various foramina to be distributed predominantly to the head region.

Some of the nerves in both of the above groups also contain nerve fibres **(preganglionic)** which pass from cells in the central nervous system to control the activity of the nerve cells in 3.

3. **The autonomic nervous system** is divided into sympathetic and parasympathetic parts, and is concerned in the main with the non-volitional activities of the body. Both parts consist of groups of nerve cells and their processes which lie entirely outside the central nervous system. They are often arranged in complicated, loose meshworks (plexuses) unlike the compact bundles of nerve fibres which constitute the cranial and spinal nerves.

The **sympathetic part** receives preganglionic nerve fibres from the central nervous system through the thoracic and upper lumbar spinal nerves; it innervates involuntary structures throughout the body.

The **parasympathetic part** receives preganglionic nerve fibres from the central nervous system through certain cranial and sacral spinal nerves; it innervates only the viscera and some structures in the orbit.

The peripheral nervous system is also divisible on functional grounds into motor (efferent) and sensory (afferent) parts.

The Efferent Part

(a) The nerve **supply to striated, voluntary muscles** which are developed from the somites is by nerve cells **(somatic motor)** which are situated within the central nervous system and send their thick myelinated processes (axons) to these muscles through certain cranial nerves and the ventral roots of all spinal nerves. These fibres are accompanied by smaller motor fibres which also arise from cells in the central nervous system but supply the muscle spindles [p. 976] in the voluntary muscles. The latter fibres form one example of efferent fibres which control the sensitivity of peripheral sensory organs. It is not known if this type of arrangement is common to all sensory endings, but efferent fibres, which may have the same function, are found in the optic and vestibulocochlear nerves which are commonly thought to be purely sensory.

(b) The **supply to involuntary (non-striated, or smooth) muscles and glands (visceral motor)** is by cells of the autonomic nervous system which lie entirely outside the central nervous system either in discrete clumps **(ganglia)** or scattered throughout the plexuses. These cells **(autonomic ganglion cells)** send their axons (thin, non-myelinated, **postganglionic nerve fibres)** to the viscera through the plexuses, and to similar structures in the territories of the spinal and cranial nerves by entering and running in these nerves. Thin myelinated nerve fibres pass from nerve cells in the central nervous system via certain cranial and spinal nerves. These control the activity of the autonomic nerve cells and are the **preganglionic nerve fibres.**

The Afferent Part

This part consists of the peripheral (dendrites) and central (axons) processes of sensory nerve cells which, with three exceptions [p. 592], lie close to the central nervous system in

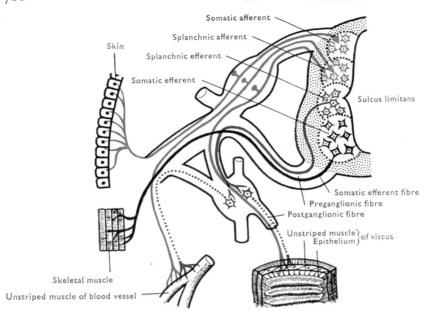

FIG. 885　Diagram of components of a spinal nerve and their relation to nuclei of origin or termination in section of embryonic spinal medulla.　Compare with components of cranial nerves in Fig. 886 and also with Fig. 962.

sensory ganglia are attached.　Those that pass to visceral structures course with the preganglionic and postganglionic fibres of the autonomic nervous system but have no direct functional connexion with them.　The central processes of all the sensory cells pass directly into the central nervous system ; hence reactions to sensory stimuli are the result of the activity of the central nervous system and are not due to peripheral connexions between the sensory and motor nerve cells.

Thus several different types of fibres are to be found in peripheral nerves.　**Cutaneous nerves** are not only sensory to the skin and subcutaneous tissues but contain postganglionic sympathetic fibres to blood vessels, sweat glands, and the muscles of the hairs (arrectores pilorum).　**Nerves to muscles** contain sensory fibres to muscles, tendons, bones, and joints, as well as motor fibres to the muscles and sympathetic fibres to their blood vessels, and the **autonomic nerve plexuses** transmit sensory fibres in addition to autonomic nerve fibres. *Thus almost every nerve, whatever its destination, consists of a mixture of efferent and afferent fibres.*

The artificial subdivisions given above and the separate sections dealing with the central and peripheral nervous systems are convenient for the purposes of study and description, but it is important to realize that the peripheral nervous system is an extension of the central nervous system, that neither system can function in the absence of the other, and that together they are responsible for integrating all the activities of the body.　It is incorrect, therefore, to think of either system or their parts as though they acted independently.

compact ganglia on certain of the cranial nerves and on the dorsal roots of the spinal nerves (spinal ganglia).

The diameters of these cells and their processes (0·1–20 μm.) are very variable, and the smaller, slowly conducting processes (< 2 μm. in diameter) have no myelin sheath.　The peripheral processes of the cells are distributed to the sensory endings of the body through the spinal and cranial nerves to which the

THE CRANIAL NERVES

There are twelve pairs of symmetrically arranged cranial nerves attached to the brain, and all are distributed in the head and neck with the exception of the tenth pair (the vagus nerves) which also supply structures in the thorax and abdomen.　Each cranial nerve is attached to a specific region on the surface of the brain (superficial origin) and though some of the nerves consist almost entirely of one type of nerve fibre at this point, the majority contain a mixture of (1) **afferent** (incoming or sensory) **fibres** which enter the brain and end on various groups of nerve cells (nuclei) therein, and (2) **efferent** (outgoing or motor) **fibres** which arise from one or more nuclei in the brain.　These afferent and efferent nuclei constitute respectively the deep terminations and origins of the nerves and are dealt with in the account of the brain.　In this section the course of each cranial nerve will be described from the superficial origin to the peripheral distribution irrespective of the type or types of nerve fibres which it contains.

The table on page 710 shows the various types of nerve fibres in the cranial nerves, their distribution and function.　It indicates that certain cranial nerves contain **branchial efferent nerve fibres** in addition to the types already described.　These

fibres, found only in some cranial nerves, are structurally identical with the somatic motor fibres of other cranial and spinal nerves which supply the striated muscles developed from the somites of the embryo.　Unlike these, they supply the striated muscles of voluntary type which lie in the anterior extremity of the gut tube, and are used for the rapid movements of swallowing (pharynx and oesophagus), speaking (larynx), mastication, expression, etc.　Not all these muscles are immediately under voluntary control.

At their **superficial attachments** the cranial nerves are arranged in two rows.　(1) A ventral row consisting of the first, second, third, sixth, and twelfth which lie in line with the ventral roots of the spinal nerves.　(2) A lateral row comprising the fifth and seventh to eleventh inclusive.　These lie approximately in line with the dorsal roots of the spinal nerves and, like them, have sensory ganglia.　The fourth nerve arises from the dorsal surface of the brain and does not lie in either row.

The first two cranial nerves (olfactory and optic) differ from the others.　The **olfactory nerves** consist of bundles of fine, non-myelinated fibres which arise in sensory cells situated in the olfactory epithelium, and enter the olfactory bulb through

FIG. 886 Diagram of components of cranial nerves and their relation to nuclei of origin or termination in transverse section of embryonic hindbrain. Compare with components of spinal nerve in Fig. 885 and also with Fig. 750.

they supply the muscles derived from the head somites: the extrinsic ocular muscles and the muscles of the tongue. In addition they carry from these muscles a few sensory fibres, though the majority enter the brain through the lateral row of cranial nerves. The third nerve also transmits preganglionic fibres to the parasympathetic cells (ciliary ganglion) innervating certain structures in the eyeball.

The cranial nerves of the lateral row contain both afferent and efferent fibres (the eighth or vestibulocochlear is the only nerve with a mainly sensory function). They transmit the great majority of the sensory fibres from the territories which they innervate, but they also contain branchial motor fibres and some (VII, IX, X) transmit preganglionic parasympathetic fibres.

OLFACTORY NERVE

The olfactory nerve consists of fine, non-myelinated fibres which arise as the central processes (axons) of bipolar nerve cells lying between the columnar

the cribriform plate of the ethmoid. The **optic nerves,** and the retinae from which their nerve fibres arise, are formed as an outgrowth of the entire thickness of the embryonic brain tube. Hence the nerves correspond to a brain tract rather than a peripheral nerve which is formed by the outgrowth of nerve fibres into the surrounding tissues from nuclei within the brain or from ganglia external to it.

The remaining members of the ventral row and the fourth nerve resemble the ventral roots of the spinal nerves in that

epithelial cells of the olfactory mucous membrane over the upper third of the nasal septum and most of the superior concha [FIGS. 898 and 899]. The peripheral processes of these nerve cells [FIG. 1031] are short and each ends in a swelling which lies in the mucus covering the free surface of the epithelial cells, and has olfactory hairs projecting from its margin (Clark and Warwick, 1946).

The olfactory nerve fibres gather into approximately twenty filaments which pass through the cribriform plate of the

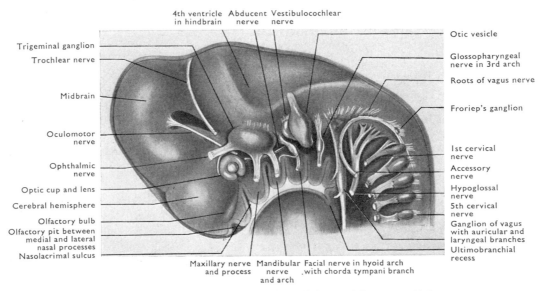

FIG. 887 Developmental arrangement of the cranial nerves. (After Streeter, 1908.)

COMPONENT AND FUNCTIONAL ANALYSIS
OF CRANIAL NERVES

No.	Name	Components	Distribution	Functions
I.	Olfactory	Afferent	Olfactory mucous membrane	Smell
II.	Optic	Afferent	Retina	Vision
		Efferent	To retina	
III.	Oculomotor	Efferent:		
		i. Somatic	Sup., med., and inf. recti and inf. oblique muscles of eyeball; levator palp. sup. muscle	Movements of eyeball and elevation of upper eyelid
		ii. Parasympathetic preganglionic	Ciliary ganglion; ciliary and sphincter pupillae muscles of eyeball	Focussing; control of pupillary size
		Afferent	The above muscles	Proprioceptive (? via ophthalmic part of trigeminal)
IV.	Trochlear	Efferent somatic	Sup. oblique muscle of eyeball	Depression of the medially rotated cornea.
		Afferent	The above muscle	Proprioceptive (? via ophthalmic part of trigeminal)
V.	Trigeminal	Efferent branchial	Muscles of mastication	Mastication and swallowing
			Digastric (ant. belly) and mylohyoid muscles	
			Tensor veli palatini	Tenses anterior part of soft palate
			Tensor tympani	Modifies movements of tympanic membrane and malleus
		Afferent	Skin of face and anterior scalp	General sensibility
			Mucous membrane of mouth including teeth, gums, and tongue (ant. $\frac{2}{3}$)	
			Mucous membrane of nasal cavities and paranasal sinuses	
			Meninges (dura mater) and skull	
			Sensory to muscles of mastication and facial expression, and of tongue and eyeball	Proprioceptive
VI.	Abducent	Efferent somatic	Lateral rectus muscle of eyeball	Turns cornea laterally
		Afferent	The above muscle	Proprioceptive
VII.	Facial	Efferent:		
		i. Branchial	Muscles of facial expression and scalp	Facial expression, closure of eye, movements of lips
			Digastric (post. belly), stylohyoid and stapedius muscles	Elevation of hyoid bone in swallowing. Control of stapes
		ii. Parasympathetic preganglionic	Pterygopalatine ganglion; glands of nasal cavity, of hard and soft palates, and lacrimal gland	Secretomotor
			Submandibular ganglion; submandibular and sublingual glands and glands of tongue (ant. $\frac{2}{3}$)	
		Afferent	Mucous membrane of tongue (ant. $\frac{2}{3}$) (excluding vallate papillae)	Taste
VIII.	Vestibulocochlear	Afferent	From duct of cochlea	Hearing
			From maculae of utricle and saccule	Equilibration (measurement of position of head relative to gravity and to changes in angular velocity)
			From ampullae of semicircular ducts	
		Efferent	To organ of Corti and vestibular apparatus	

COMPONENT AND FUNCTIONAL ANALYSIS

OF CRANIAL NERVES (contd.)

No.	Name	Components	Distribution	Functions
IX.	Glosso-pharyngeal	Efferent: i. Branchial	Stylopharyngeus muscle Middle constrictor	Pharyngeal movements and elevation of larynx
		ii. Parasympathetic preganglionic	Otic ganglion Lingual ganglia	Secretomotor to parotid gland, and glands in posterior ⅓ of tongue
		Afferent	Mucous membrane of pharynx and tongue (post. ⅓) including vallate papillae, and tonsillar region	General sensibility and taste
			Mucous membrane of auditory tube, tympanic cavity and antrum, and mastoid air cells	General sensibility
			Carotid body and carotid sinus	Vasosensory (chemoreceptors and pressure receptors)
X. (XI)	Vagus (and Cranial Root of Accessory)	Efferent: i. Branchial	Levator veli palatini muscle, palatoglossus	Lifts soft palate
			Pharyngeal muscles	Swallowing and other pharyngeal movements
			Laryngeal muscles	Control of larynx in respiration and phonation
		ii. Parasympathetic preganglionic	Plain muscle of oesophagus, stomach, intestine (as far as splenic flexure), and gall-bladder	Movements of these viscera
			Pancreas and gastric glands	Secretomotor
			Heart (nodal tissue and cardiac muscle)	Cardiac depressor
			Lungs (plain muscle and glands of bronchi and bronchioles)	Bronchoconstriction
		Afferent	Mucous membrane of alimentary and respiratory passages from epiglottis to transverse colon	General and visceral sensibility, and some taste
			Part of auricle and ext. acoustic meatus	General sensibility
			Lungs	Respiratory reflexes
			Glomus aorticum and wall of aorta	Vasosensory
XI.	Accessory (Spinal Root)	Efferent branchial	Trapezius and sternocleidomastoid muscles	Movements of head and shoulder
XII.	Hypoglossal	Efferent somatic	Extrinsic and intrinsic muscles of tongue, except palatoglossus	Movements of tongue

ethmoid and the meninges to enter the olfactory bulb. Here they form complicated synapses with the processes of the mitral cells of the bulb which send their axons to form the olfactory tract [pp. 645–646].

Nervus Terminalis

Lying medial and parallel to each olfactory tract is a fine, plexiform bundle of nerve fibres interspersed with nerve cells. These terminal nerves are attached superficially to the anterior perforated substance and pass into the nose with the olfactory nerves [FIG. 889]. Their function is unknown, but they may be partly autonomic.

OPTIC NERVE

The fibres of the optic nerve are the axons of the ganglion cells on the internal surface of the retina. The axons course over this surface and converge on the circular optic disc which lies 3 mm. medial to the sagittal meridian of the eyeball and slightly above the horizontal meridian (Traquair, 1949). At the optic disc each fibre acquires a myelin sheath; a feature which accounts for the white colour of the optic disc and occasionally for white streaks which extend outwards from the disc when some fibres myelinate before they reach it.

The fibres at the optic disc gather into bundles which pierce the sclera and divide it into a mesh (**lamina cribrosa**). They then combine to form the compact optic nerve which is immediately enclosed in the three layers of the meninges and the spaces which separate them [p. 693]. It passes backwards and slightly medially to enter the optic canal within the tendinous ring from which the four rectus muscles of the eye arise. Approximately 1 cm. posterior to the eyeball, the central artery and vein of the retina pierce the inferior surface of the optic

Olfactory bulb
Olfactory tract
Olfactory trigone
Optic nerve
Optic chiasma
Oculomotor nerve
Trochlear nerve
Trigeminal nerve
Abducent nerve
Vestibulocochlear nerve
Nervus intermedius
Facial nerve (motor root)
Glossopharyngeal nerve
Vagus nerve
Accessory nerve (cranial roots)
Accessory nerve (spinal roots)

Olfactory bulb
Olfactory tract
Area subcallosa
Olfactory trigone
Medial olfactory stria
Lateral olfactory stria
Optic chiasma
Anterior perforated substance
Temporal lobe (cut)
Optic tract
Oculomotor nerve
Trochlear nerve
Stria terminalis
Trigeminal nerve
Lateral geniculate body
Abducent nerve
Medial geniculate body
Pulvinar
Facial nerve
Nervus intermedius
Vestibulocochlear nerve
Middle cerebellar peduncle
Glossopharyngeal nerve
Vagus nerve
Accessory nerve (cranial roots)
Accessory nerve (spinal roots)
Occipital lobe (cut)

Hypoglossal nerve   Spinal medulla   Hypoglossal nerve

FIG. 888   Inferior surface of brain to show superficial attachments of cranial nerves.   The lower portion of the left temporal and occipital lobes and the left half of the cerebellum have been removed.

Olfactory tract
Olfactory bulb   Nervi terminales
Olfactory tract
Lamina terminalis
Optic chiasma

FIG. 889   Diagram of part of lower surface of adult human brain to show position of nervi terminales and their relations to olfactory tracts.   (After Brookover, 1914.)

The olfactory bulb and part of the tract have been removed on the left side, and the right optic nerve and the greater part of the optic chiasma have been cut away to expose the lamina terminalis.

nerve and its coverings and run in its substance to the optic disc as the only vascular supply to the retina.

In the optic canal, immediately lateral to the sphenoidal sinus, the nerve lies with the ophthalmic artery, and here may receive the central artery of the retina. It emerges from the canal into the middle cranial fossa, and ends by joining its fellow in the optic chiasma.

The **optic chiasma** lies in the anterior part of the floor of the third ventricle of the brain, immediately anterior to the tuber cinereum and infundibulum [FIG. 698], a short distance superior to the diaphragma sellae and hypophysis [FIG. 699], and between the two internal carotid arteries. Thus pathological changes either in the carotid arteries, or in the hypophysis, or in the floor of the third ventricle may affect the chiasma.

In the chiasma the fibres from the nasal halves of the two retinae decussate, and joining the uncrossed fibres from the temporal half of the opposite retina, form an optic tract at each posterolateral angle of the chiasma. Each **optic tract** then sweeps posterolaterally around the hypothalamus and crus cerebri [FIG. 890] to the **lateral geniculate body.** Here the fibres concerned with vision end, but many of them give off branches concerned with visual reflexes. The latter continue between the lateral and medial geniculate bodies to reach the **superior colliculus** and **pretectal nucleus** on the dorsal aspect of the midbrain [p. 668 and FIG. 848].

The effect which the course of these fibres has on the symptoms following lesions of the different parts of the optic pathways is dealt with on page 670.

There is no evidence that fibres cross in the chiasma to unite the two retinae (Woollard and Harpman, 1939; Magoun and Ranson, 1942; Wolff, 1953) though extensive injuries to one eye may be followed by degenerative changes in the other. In many animals and probably also in Man there are considerable numbers of **efferent fibres** to the retinae running in the optic tracts and nerves, but their function is unknown (Cowan and Powell, 1963). Branching fibres in the optic chiasma have been described, but they seem to be too few to account for a significant bilateral pathway through both optic tracts from any part of the retina.

OCULOMOTOR NERVE

This is predominantly a **somatic motor nerve** which supplies the striated muscle in levator palpebrae superioris and all the extrinsic ocular muscles, except the lateral rectus and the superior oblique. It also carries **preganglionic parasympathetic fibres** to the ciliary ganglion which supplies the sphincter of the pupil and the ciliary muscle within the eyeball. Some **afferent nerve fibres** from the extrinsic ocular muscles may pass to the brain through the oculomotor

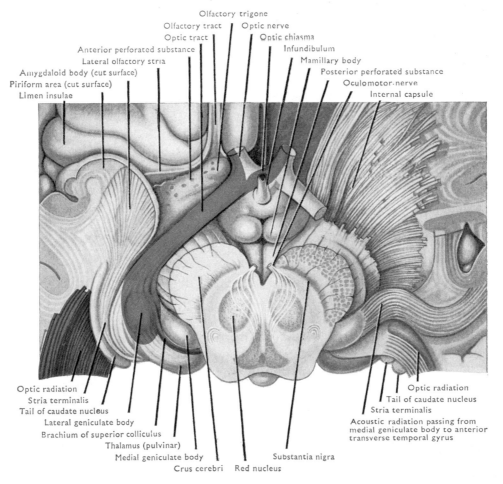

FIG. 890 Part of lower surface of forebrain, showing olfactory and optic tracts. The midbrain has been cut across.

nerve [p. 709], but the majority are probably transmitted through the trigeminal nerve via communications with the oculomotor nerve in the wall of the cavernous sinus. **Sympathetic postganglionic fibres** also join the nerve from the plexus on the internal carotid artery in the same region, and are distributed through the nerve.

Superficial Origin and Course

The oculomotor nerve arises from the ventral surface of the midbrain by several closely packed rootlets attached to the medial sulcus of the crus cerebri, immediately lateral to the posterior perforated substance and just superior to the pons [FIGS. 888 and 753. Deep origin, p. 618]. The nerve passes horizontally forwards between the superior cerebellar and posterior cerebral arteries, then lateral to the posterior communicating artey in the interpeduncular cistern. At a point between the free and attached margins of the tentorium cerebelli the nerve pierces the small triangular area of dura mater which forms the roof of the **cavernous sinus**. It then runs forwards in the lateral wall of the cavernous sinus, superior to the trochlear nerve, and dividing into **superior and inferior branches,** enters the orbit through the superior orbital fissure within the tendinous ring from which the rectus muscles arise [FIG. 896]. The **preganglionic parasympathetic fibres** which enter the inferior branch lie in the superior part of the nerve at its superficial origin but are irregular in position after the nerve pierces the dura mater (Sunderland and Hughes, 1946).

Branches

The **superior branch** supplies the superior rectus and the levator palpebrae superioris.

The **inferior branch** passes forwards, and giving branches to the medial and inferior recti and to the ciliary ganglion (preganglionic parasympathetic fibres) continues to the inferior oblique muscle.

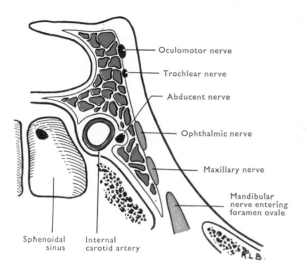

FIG. 892 Diagrammatic oblique coronal section through the right cavernous sinus, viewed from behind

Ciliary Ganglion

This small reddish body about 1 mm. in diameter lies in the posterior part of the orbit between the lateral rectus muscle and the optic nerve, anterior to the ophthalmic artery. It is one of the cranial parasympathetic ganglia, but does not contain all the ganglion cells of the orbit as some scattered cells are also found, particularly superior to the eyeball [p. 778]. The ganglion receives (1) **preganglionic parasympathetic fibres** from the nerve to the inferior oblique muscle (**oculomotor root**) and these are the only fibres which have functional, synaptic connexions with the ganglion cells. (2) **Postganglionic sympathetic fibres** from the **carotid plexus** pass through the ganglion on their way to supply the vessels and dilator pupillae of the eyeball. They may either reach it as branches of the nasociliary nerve or directly from the extension of the plexus on the ophthalmic artery (sympathetic branch to ciliary ganglion). (3) The **communicating branch from the naso-ciliary nerve** consists of sensory fibres of the trigeminal nerve which pass through the ganglion to the coats of the eyeball.

The ciliary ganglion sends twelve to fifteen **short ciliary nerves** to pierce the sclera above and below the optic nerve. These nerves contain the fibres from the sympathetic branch to the ganglion and the communicating branch from the nasociliary nerve which have passed through the ganglion unchanged. In addition they carry the myelinated **postganglionic processes** of the large cells of the ciliary ganglion which pass to supply the sphincter pupillae muscle of the iris and the ciliary muscle. These fibres, together with the preganglionic fibres in the oculomotor nerve comprise the efferent pathway through which constriction of the pupil is achieved (e.g. in the light reflex) and

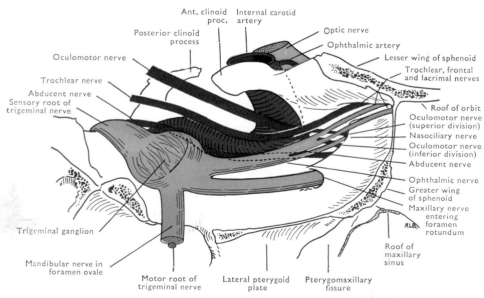

FIG. 891 Relations of structures in cavernous sinus and superior orbital fissure, viewed from the right side.

The section of the skull passes sagittally through the foramen ovale.

changes in the focal length of the lens are produced, e.g., in focussing on a near object as in the convergence-accommodation reflex. Damage to the oculomotor nerve paralyses most of the extrinsic ocular muscles and levator palpebrae superioris, and disconnects the cells of the ciliary ganglion from the central nervous system. The unopposed action of the lateral rectus and superior oblique muscles turns the cornea laterally and downwards, yet the patient does not complain of diplopia (double vision) because the eye is shut owing to the paralysis of the levator. The dilated pupil on the injured side is not constricted when either eye is exposed to light, or in the convergence-accommodation reflex.

TROCHLEAR NERVE

Superficial Origin and Course

This somatic motor nerve supplies the superior oblique muscle of the eyeball. The slender nerve arises from the dorsal surface of the brain immediately below the inferior colliculus and lateral to the frenulum veli [FIG. 774]. For deep origin, see page 618. It runs a long intracranial course forwards on the lateral aspect of the midbrain parallel to and immediately below the free edge of the tentorium, between the posterior cerebral and superior cerebellar arteries. It pierces the inferior surface of the tentorium close to the free edge and superior to the trigeminal nerve. It continues forwards to the superior orbital fissure in the lateral wall of the cavernous sinus [FIG. 892] between the oculomotor and ophthalmic nerves. Here it receives **sympathetic filaments** from the plexus on the internal carotid artery, and it may receive **sensory fibres** from the ophthalmic nerve. In the orbit it lies lateral to the fibrous ring from which the rectus muscles arise and passes medially between levator palpebrae superioris and the orbital periosteum near the posterior extremity of the orbit [FIG. 896]. It then enters the orbital surface of the superior oblique muscle, the only structure this nerve supplies. This nerve differs from the

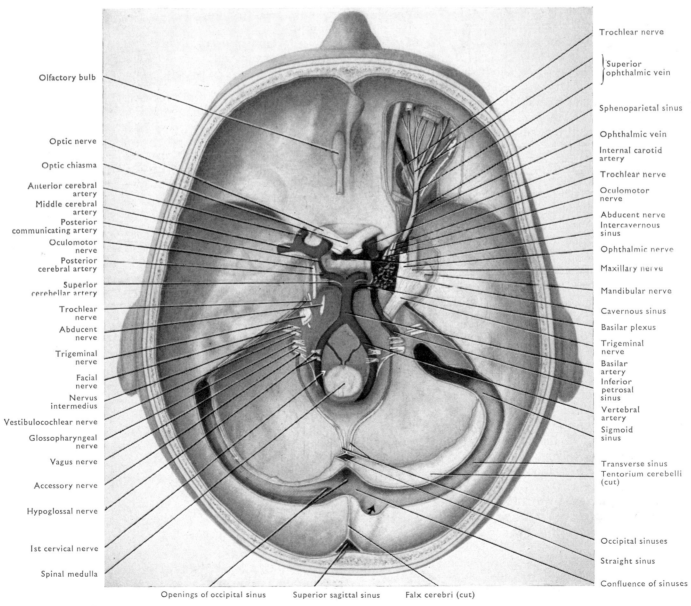

FIG. 893   Base of skull, to show dura mater, venous sinuses, arteries, and nerves.

other somatic motor nerves (III, VI, XII) in that its fibres emerge dorsally, having decussated within the brain substance.

Damage to the trochlear nerve causes paralysis of the superior oblique muscle of the eye with consequent inability to turn the cornea inferiorly when it is already turned medially by the medial rectus. Thus in right superior oblique paralysis the patient complains of double vision when looking downwards and to the left. Failure of the right eye to follow the left results in the image of an object failing to fall on the corresponding parts of the two retinae, thereby producing double vision.

TRIGEMINAL NERVE

This is the major sensory nerve of the head but it is also motor to the muscles of mastication.

It is **sensory** to the face (except over the angle of the jaw), the temple, the anterior scalp as far as the vertex, and parts of the auricle and external acoustic meatus. Also to the contents of the orbit (except the retina of the eye), the walls of the nasal cavity and paranasal air sinuses, the mouth (except the posterior one-third of the tongue), the temporomandibular joint, and to portions of the nasal part of the pharynx, auditory tube, and the cranial dura mater and periosteum.

Each of the three main divisions of the nerve (ophthalmic, maxillary, and mandibular) has a distribution which approximately corresponds to the structures that develop from one of the three processes (frontonasal, maxillary, and mandibular) which combine to produce the face in the embryo [p. 53]. Thus the divisions supply three clearly defined cutaneous areas [FIG. 900] and the structures which lie deep to them.

Pain arising in the deep structures (e.g. teeth or air sinuses) may be referred incorrectly to other parts of the distribution and felt there. This may be because of the widespread sensory distribution, or possibly because either some nerve fibres innervate more then one structure, or different fibres have a common pathway in the central nervous system.

The **motor** fibres which emerge with the trigeminal nerve from the brain pass entirely into its mandibular division. They supply the muscles of the first pharyngeal (branchial) arch of the embryo, i.e. the muscles that move the mandible, the anterior belly of digastric and mylohyoid, tensor veli palatini, and tensor tympani.

Some of the peripheral branches of all three divisions of the trigeminal nerve are joined by preganglionic or postganglionic fibres of the parasympathetic nervous system and by postganglionic fibres of the sympathetic nervous system. These autonomic motor nerve fibres course with the fibres of the trigeminal nerve to supply salivary, sweat (Wilson, 1936) and other glands, in the areas to which that nerve is sensory, but none of them emerges from the brain with that nerve.

The sympathetic fibres arise either in the cells of the superior cervical ganglion or in the cells of the internal carotid plexus. The preganglionic parasympathetic fibres leave the brain in the oculomotor, facial, and glossopharyngeal nerves, and passing to the ciliary,

pterygopalatine, otic, and submandibular ganglia, end on their cells. The postganglionic parasympathetic nerve fibres arise from these cells and pass with the sensory (trigeminal) and sympathetic fibres which often traverse these ganglia but have no functional connexion with them.

Superficial Origin

The trigeminal nerve arises from the lateral surface of the pons in the posterior cranial fossa [for deep connexions, see pp. 619, 621]. It consists of a large, inferior, sensory root and a small, superior, motor root which emerge together. The roots twist through 180 degrees (Davies and Haven, 1933) as they pass forward to run beneath the dural floor of the middle cranial fossa [FIG. 880] carrying with them a loose sleeve of dura and arachnoid mater which surrounds the **cavum trigeminale** in which they lie. The dural sleeve begins where the roots leave the posterior cranial fossa grooving the superior border of the petrous temporal bone [FIG. 134] inferior to the attached edge of the tentorium cerebelli [FIG. 1119] and the superior petrosal sinus. It ends anteriorly by enlarging to enclose and fuse with the trigeminal ganglion into which the sensory root expands.

The **trigeminal ganglion** is a large, flattened, semilunar, sensory ganglion which extends obliquely from the lateral wall of the cavernous sinus to a shallow fossa on the anterior surface

FIG. 894 Diagram of ciliary ganglion and connexions of autonomic nervous system with the eye.

Sympathetic, red ; parasympathetic, green ; sensory, blue.

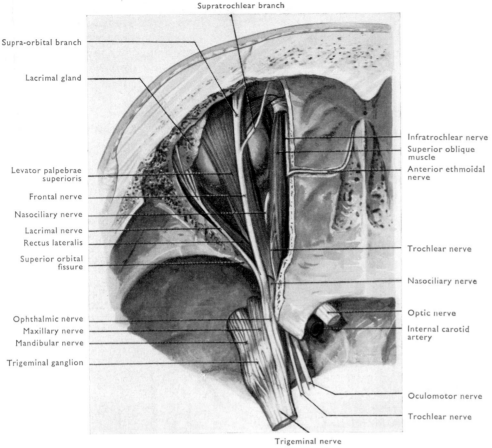

FIG. 895   Nerves of orbit from above.

OPHTHALMIC NERVE

The ophthalmic nerve passes forwards to the orbit in the lateral wall of the cavernous sinus inferior to the trochlear nerve [FIG. 892]. Here it receives, and transmits through its branches, postganglionic sympathetic filaments from the plexus on the internal carotid artery; and gives off a small, recurrent branch to the dura mater of the tentorium and posterior parts of the falx cerebri [FIG. 875]. Fine filaments which pass between the ophthalmic nerve and the oculomotor, trochlear, and abducent nerves may be purely temporary communications (Sunderland and Hughes, 1946), but could transfer sensory fibres from the trigeminal nerve for distribution to the extrinsic ocular muscles. The ophthalmic nerve divides into lacrimal, frontal, and nasociliary branches [FIG. 891] which enter the orbit through the superior orbital fissure.

Lacrimal Nerve

This nerve traverses the lateral angle of the superior orbital fissure and passes above and parallel to the lateral rectus muscle to the supero-

of the petrous temporal bone near its apex. Three large nerves (ophthalmic, maxillary, and mandibular [FIG. 891]) arise from the convex, distal margin of the ganglion, while the motor root passes deep to its inferolateral part to join the mandibular nerve.

lateral part of the orbital margin [FIG. 895]. It receives parasympathetic postganglionic nerve fibres for the lacrimal gland through a branch from the zygomatic nerve [FIG. 901] and divides into branches to (1) the lacrimal gland; (2) the lateral part of the upper eyelid.

FIG. 896   Schematic representation of nerves which traverse the cavity of the orbit.

Note that the lacrimal, frontal, and trochlear nerves, after entering the orbit through the superior orbital fissure, pass outside the common tendinous ring from which the recti muscles arise; whilst the oculomotor, abducent, and nasociliary nerves, together with the optic nerve and ophthalmic artery, enter the orbit within the ring, and therefore lie at first within the cone formed by the ocular muscles as they diverge forwards from it.

Frontal Nerve

This branch passes forwards above the levator palpebrae superioris and divides at a variable point into a larger supra-orbital and a smaller supratrochlear branch [FIGS. 895 and 897].

The **supra-orbital nerve** passes directly forwards through the supra-orbital groove or foramen, sends twigs to the frontal sinus, the upper eyelid, and turning upwards, supplies the forehead and scalp as far as the vertex.

The **supratrochlear nerve** passes anteromedially above the tendon of the superior oblique muscle. It turns superiorly on the medial part of the supra-orbital margin and supplies the medial parts of the upper eyelid and forehead.

Nasociliary Nerve

This nerve passes with the two divisions of the oculomotor nerve and the abducent nerve through the superior orbital fissure within the tendinous ring from which the rectus muscles arise [FIG. 896], to lie within the cone formed by the extrinsic ocular muscles. The nasociliary nerve then curves anteromedially, superior to the ophthalmic artery and the optic nerve, to leave the cone between the superior oblique and medial rectus muscles, close to the anterior ethmoidal foramen. Here it gives off the small **infratrochlear nerve,** and continues as the **anterior ethmoidal nerve** through that foramen between the frontal and ethmoid bones to supply the ethmoidal sinuses and pass into the anterior cranial fossa on the cribriform plate of the ethmoid, supplying the adjacent dura mater. The anterior ethmoidal nerve then descends into the roof of the nasal cavity through the **nasal slit** [FIG. 165] and supplies the anterosuperior parts of the medial (septal) and lateral walls of that cavity by **medial and lateral nasal branches** [FIGS. 898 and 899]. The lateral branch sends an **external nasal branch** between the nasal bone and the upper nasal cartilage to supply the lower part of the dorsum and side of the external nose [FIG. 901].

The **infratrochlear nerve** arises close to the anterior ethmoidal foramen. It passes forwards inferior to the pulley (trochlea) of the superior oblique muscle to supply skin and fasciae of the most medial parts of the eyelids and adjacent region of the external nose [FIG. 901]. It *communicates* with the supratrochlear nerve in the orbit or on the face.

Within the cone of the extrinsic ocular muscles, the nasociliary nerve sends sensory and sympathetic postganglionic fibres to the coats of the eyeball and their vessels, including the dilator pupillae muscle of the iris. (1) The communicating branch to the ciliary ganglion arises while the nerve lies lateral to the optic nerve; its fibres pass directly through the ganglion into the **short ciliary nerves** [p. 714]. (2) The **long ciliary nerves** arise superior to the optic nerve and pass along it to pierce the sclera beside that nerve [FIG. 986].

The nasociliary nerve occasionally sends a minute posterior ethmoidal nerve through the foramen of the same name to supply the posterior ethmoidal and sphenoidal air sinuses.

MAXILLARY NERVE

The maxillary nerve passes forwards in the lower part of the lateral wall of the cavernous sinus [FIG. 892], then through the foramen rotundum [FIG. 163] into the upper part of the pterygopalatine fossa. Here it turns anterolaterally in the pterygomaxillary fissure and passing along the inferior orbital fissure, curves anteromedially through that fissure to enter and traverse the infra-orbital groove (**infra-orbital nerve**), canal, and foramen and emerge on the face [FIG. 902].

Branches

1. In the cranial cavity it gives a minute **meningeal** branch to the dura mater of the middle cranial fossa.

2. As the nerve enters the pterygopalatine fossa, two large **pterygopalatine nerves** pass inferomedially to the pterygopalatine ganglion [p. 726]. These branches consist of sensory nerve fibres which, together with postganglionic sympathetic nerve fibres from the **deep petrosal nerve** [p. 784], pass through or around the ganglion and form most of the nerve fibres in its so-called branches, without being functionally related to its cells. The cells of the ganglion contribute postganglionic parasympathetic fibres to the branches which are, therefore, both sensory and secretomotor to the territories they supply.

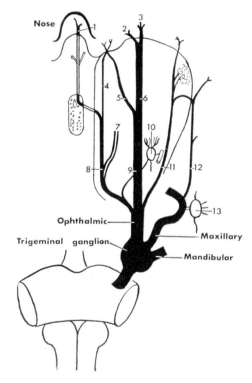

FIG. 897 Scheme to show the distribution of the ophthalmic nerve.

9, frontal nerve, giving off supratrochlear branch (5) and proceeding as supra orbital (6), which may divide into medial (2) and lateral (3) branches before leaving the orbit.

8, nasociliary nerve ; gives off afferent root to ciliary ganglion (10), two long ciliary nerves (7) to eyeball, and, just before leaving the orbit through the anterior ethmoidal foramen, the infratrochlear nerve (4) ; the rest of its course up to its termination on the nose as the external nasal nerve (1) illustrates the description given in the text.

11, lacrimal nerve, communicates with zygomatic nerve (12).

13, pterygopalatine ganglion.

Right nasopalatine nerve

Branches of the olfactory nerves

Medial nasal branch

Left nasopalatine nerve

Lesser palatine nerve

Lesser palatine nerve

Greater palatine nerve

FIG. 898   Innervation of nasal septum and palate.

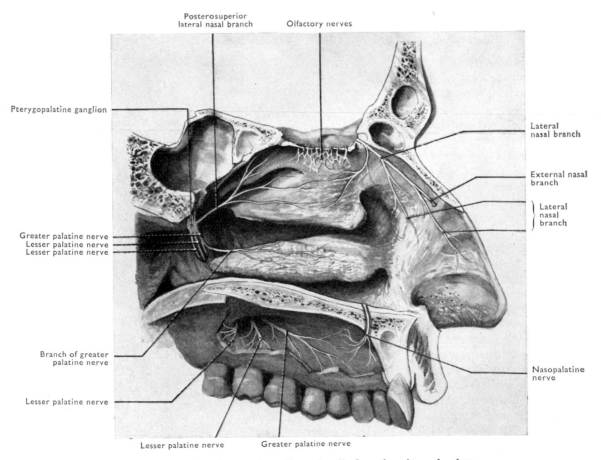

Posterosuperior lateral nasal branch

Olfactory nerves

Pterygopalatine ganglion

Lateral nasal branch

External nasal branch

Lateral nasal branch

Greater palatine nerve
Lesser palatine nerve
Lesser palatine nerve

Branch of greater palatine nerve

Nasopalatine nerve

Lesser palatine nerve

Lesser palatine nerve

Greater palatine nerve

FIG. 899   Innervation of lateral wall of nasal cavity and palate.

The **branches of the pterygopalatine ganglion** are:

(i) **The pharyngeal branch** passes posteriorly through the palatovaginal canal to supply the roof of the pharynx, the sphenoidal air sinus, and the pharyngeal opening of the auditory tube.

(ii) **The three palatine nerves** descend vertically through the pterygopalatine fossa and the palatine canals.

(a) The **greater palatine nerve** emerges from the greater palatine foramen [FIGS. 476, 477] and passing forwards near the lateral margin of the hard palate, breaks into branches which supply the structures of the hard palate and adjacent gum as far as the incisor teeth, where it communicates with branches of the nasopalatine nerves. In the palatine canal the main nerve gives off small **posterior inferior nasal branches** which pierce the vertical plate of the palatine bone and supply the mucous membrane of the postero-inferior part of the nasal cavity including the inferior concha [FIG. 899]. Other small branches pierce the maxilla and supply the posterior wall of the **maxillary sinus.**

(b) Two **lesser palatine nerves** pierce the palatine bone behind the greater palatine foramen [FIG. 477]. The more medial supplies the anteromedial part of the soft palate, the more lateral innervates the posterolateral part, the adjacent part of the tonsil, and the posterior part of the gum.

(iii) **The nasopalatine nerve** passes medially through the sphenopalatine foramen into the roof of the posterior part of the nasal cavity. It supplies this part of the roof and runs antero-inferiorly over the nasal septum (vomer) giving branches to the posterior part of that septum and floor of the nasal cavity. It then traverses the incisive canal with its fellow, communicates with the branches of the greater palatine nerve, and supplies the anterior part of the hard palate and gum behind the incisor teeth [FIG. 898].

(iv) Small **posterior superior nasal branches** pass through the sphenopalatine foramen to supply the posterosuperior part of the lateral wall of the nasal cavity including the superior and middle conchae.

(v) Small **orbital branches** pass superiorly to supply the orbital periosteum, and may contain secretomotor nerve fibres for the lacrimal gland.

3. The posterior superior alveolar nerves descend through the pterygomaxillary fissure to the posterior surface of the maxilla [FIG. 902]. One branch passes to the gum and adjacent part of the cheek, while the other two enter canaliculi in the bone which course round the maxillary sinus between the zygomatic process and the alveolar margin. They supply the molar teeth and mingle **(dental plexus)** in the canaliculi with fibres of the **anterior** (and sometimes middle) **superior alveolar nerve** both of them supplying the first molar tooth (Westwater, 1960), the wall of the maxillary sinus, the gum, and the alveolar periosteum.

4. The small **zygomatic nerve** arises in the pterygomaxillary fissure, passes through the inferior orbital fissure, and runs on the inferolateral wall of the orbit to enter the zygomatic bone and divide into: (i) a **zygomaticofacial branch** which emerges from the facial aspect of the zygomatic bone, and is sensory to the structures overlying it; (ii) a **zygomaticotemporal branch** which emerges from the temporal surface of the bone and pierces the temporal fascia to supply the anterior part of the temple [FIGS. 900, 907]. In the orbit the zygomatic nerve **communicates with the lacrimal nerve** [FIG. 901] and may transmit postganglionic parasympathetic nerve fibres to the lacrimal gland from the pterygopalatine ganglion.

5. The **infra-orbital nerve** is the continuation of the maxillary through the infra-orbital groove and canal to the face.

The **anterior superior alveolar branch** arises from the infra-orbital nerve in the canal and enters a sinuous canaliculus (Jones, 1939) in the maxilla through which it passes forwards in the roof of the maxillary sinus, then medially in its anterior wall below the infra-orbital foramen to reach the lateral wall of the nasal cavity near the anterior end of the inferior concha. Here it turns inferiorly and then medially [FIG. 902] to enter the palatine process of the maxilla and emerge at the side of

FIG. 900   Distribution of cutaneous nerves to head and neck. The ophthalmic, maxillary, and mandibular divisions of the trigeminal nerve are indicated by different shading.

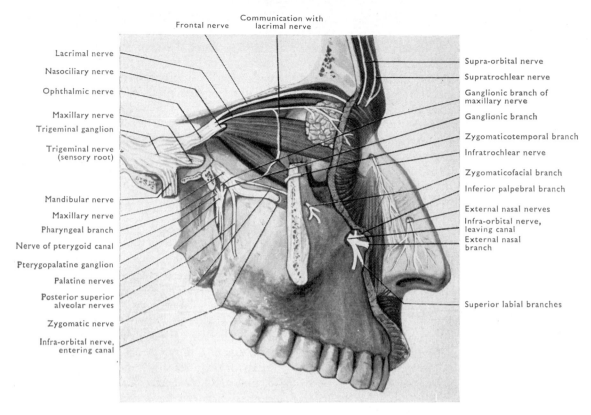

Communication with
Frontal nerve   lacrimal nerve

Lacrimal nerve

Nasociliary nerve

Ophthalmic nerve

Maxillary nerve

Trigeminal ganglion

Trigeminal nerve
(sensory root)

Mandibular nerve

Maxillary nerve

Pharyngeal branch

Nerve of pterygoid canal

Pterygopalatine ganglion

Palatine nerves

Posterior superior
alveolar nerves

Zygomatic nerve

Infra-orbital nerve,
entering canal

Supra-orbital nerve

Supratrochlear nerve

Ganglionic branch of
maxillary nerve

Ganglionic branch

Zygomaticotemporal branch

Infratrochlear nerve

Zygomaticofacial branch

Inferior palpebral branch

External nasal nerves

Infra-orbital nerve,
leaving canal

External nasal
branch

Superior labial branches

FIG. 901   Course of the ophthalmic and maxillary nerves.

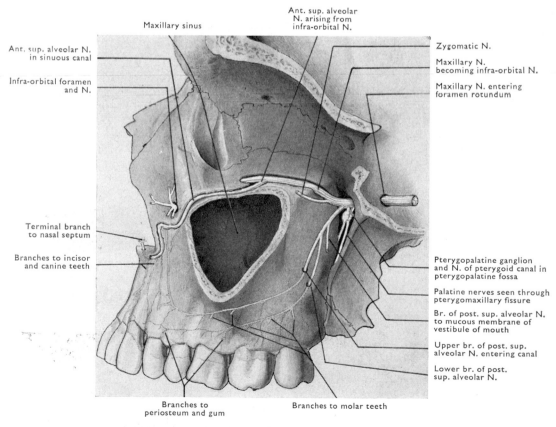

Ant. sup. alveolar
N. arising from
Maxillary sinus   infra-orbital N.

Ant. sup. alveolar N.
in sinuous canal

Infra-orbital foramen
and N.

Terminal branch
to nasal septum

Branches to incisor
and canine teeth

Zygomatic N.

Maxillary N.
becoming infra-orbital N.

Maxillary N. entering
foramen rotundum

Pterygopalatine ganglion
and N. of pterygoid canal in
pterygopalatine fossa

Palatine nerves seen through
pterygomaxillary fissure

Br. of post. sup. alveolar N.
to mucous membrane of
vestibule of mouth

Upper br. of post. sup.
alveolar N. entering canal

Lower br. of post.
sup. alveolar N.

Branches to
periosteum and gum

Branches to molar teeth

FIG. 902   Course and branches of maxillary nerve.

the nasal septum close to the anterior nasal spine. In the palate and anterior part of the maxilla the nerve gives **branches** to the incisor, canine, premolar, and first molar teeth, to the adjacent gum and periosteum, to the maxillary sinus, and to the lateral wall, floor, and septum of the nasal cavity anteriorly. The dental branches form a **plexus** with those of the posterior superior alveolar nerves [FIG. 902].

In 30 per cent. of cases a **middle superior alveolar branch** arises from the infra-orbital nerve posterior to the anterior superior alveolar nerve. It passes laterally in the maxilla and turns inferiorly in the lateral wall of the maxillary sinus to join the branches of the other alveolar nerves in the canaliculi.

In the face, the infra-orbital nerve divides into a series of radiating *branches:* (i) **inferior palpebral**, to the lower eyelids; (ii) **external nasal**, to the side of the nose; (iii) **superior labial**, to the upper lip and anterior cheek. These branches form an infra-orbital plexus with the zygomatic branches of the facial nerve [FIG. 907].

MANDIBULAR NERVE

The mandibular nerve consists of a large sensory part from the trigeminal ganglion and the entire motor root of the trigeminal nerve. The two parts emerge through the foramen ovale combining into a short, thick nerve situated between the lateral pterygoid and tensor veli palatini muscles, immediately anterior to the middle meningeal artery. Two small branches arise from the nerve: (1) a small **meningeal branch** passes through the foramen spinosum with the middle meningeal artery to supply the periosteum, dura mater, and mastoid air cells; (2) a branch which passes medially (through or beside the otic ganglion [p. 729]), **the nerve to the medial pterygoid,** tensor tympani, and tensor veli palatini muscles.

The main nerve then divides into a small, mainly motor, anterior trunk and a large, mainly sensory, posterior trunk.

The Anterior Trunk

This passes antero-inferiorly medial to the lateral pterygoid muscle.
Branches. 1. The **nerve to lateral pterygoid** passes into the deep surface of that muscle.

2. The **masseteric nerve** passes laterally superior to the lateral pterygoid muscle and in contact with the base of the skull immediately anterior to the capsule of the temporomandibular joint, to which it sends a filament. The nerve enters the deep surface of masseter by passing posterior to the tendon of temporalis in the mandibular notch.

3. Two or three **deep temporal nerves** in contact with the skull run laterally above the lateral pterygoid muscle and curve upwards to enter the deep

surface of temporalis. The anterior nerve often passes with the buccal nerve.

4. The **buccal nerve,** the only purely sensory branch of the trunk, passes between the two heads of the lateral pterygoid muscle and descends towards the cheek. It grooves or pierces the lowest anterior fibres of the tendon of temporalis close to the posterior wall of the vestibule of the mouth. The branches of the nerve are sensory to the entire thickness of the cheek and to the gums. They may partly supply the premolar and first molar teeth of the lower jaw. The motor fibres to the buccinator muscle reach it via the buccal branch of the facial nerve which communicates with the buccal nerve in the cheek.

The Posterior Trunk

This trunk gives off the auriculotemporal nerve and then divides into the inferior alveolar and lingual nerves while still medial to the lateral pterygoid muscle.
Branches. 1. The **auriculotemporal nerve** arises as two purely sensory roots which unite after embracing the middle meningeal artery. Each of these roots receives a small bundle of postganglionic parasympathetic fibres from the otic ganglion (secretomotor to the parotid gland, p. 779) and postganglionic sympathetic fibres which traverse the ganglion from the plexus on the middle meningeal artery. The nerve passes posterolaterally between the lateral pterygoid muscle and the spine of the sphenoid, then curves laterally on the posterior surface of the neck of the mandible. Here it sends branches to the

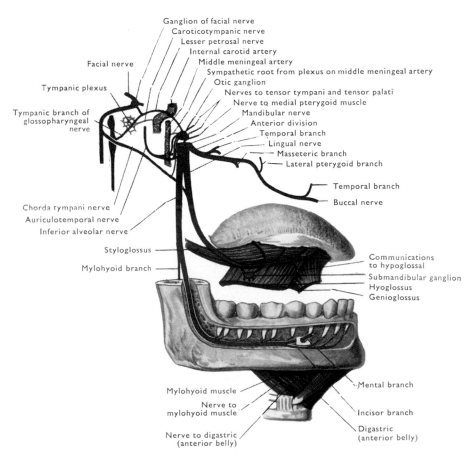

FIG. 903   Scheme of distribution and connexions of mandibular nerve.

temporomandibular joint and to the parotid gland (sensory and secreto-motor). It then runs through the upper part of the parotid gland or its fascia, sends branches to the tympanic membrane and external acoustic meatus, and to the lateral surface of the upper half of the auricle. It crosses the zygomatic process of the temporal bone, and ends by supplying the skin and fascia of the temple and scalp almost to the vertex [FIG. 900].

2. The **lingual nerve** is the anterior and smaller of the descending branches of the posterior trunk, and is *entirely sensory*. While deep to the lateral pterygoid muscle, it is joined by the **chorda tympani** branch of the facial nerve [FIG. 903] which adds taste and preganglionic parasympathetic nerve fibres to it. The lingual nerve curves anteriorly between the ramus of the mandible and the medial pterygoid muscle, and passes below the inferior margin of the superior constrictor muscle to lie between the mucous membrane of the mouth and the body of the mandible with the last molar tooth superiorly and the posterior end of the mylohyoid line inferiorly [FIGS. 345, 492]. Continuing antero-inferiorly between the lateral surface of the hyoglossus and the mylohyoid muscle, and separating the sublingual gland anteriorly from the deep part of the submandibular gland postero-inferiorly [FIG. 492], the nerve curves below the alveololingual sulcus and the submandibular duct to turn superiorly on the lateral surface of genioglossus and supply the anterior two-thirds of the tongue.

FIG. 905 Dissection of the mouth and pharynx from the interior. The tongue has been removed to show the lingual nerve.

The lingual nerve gives off a series of **branches** on the lateral surface of hyoglossus.

(i) To the submandibular ganglion. This conveys sensory trigeminal fibres and preganglionic parasympathetic fibres from the chorda tympani.

(ii) To the floor and lateral wall of the alveololingual sulcus, the lingual aspect of the lower gum, and the premolar and first molar teeth. These branches may pass through the submandibular ganglion and receive postganglionic parasympathetic fibres from it.

(iii) A communicating branch with the hypoglossal nerve through which sensory fibres may be distributed with that nerve [FIG. 904].

The lingual nerve receives branches from the submandibular ganglion. These contain postganglionic parasympathetic fibres which pass to the glands in the territory of the lingual nerve.

FIG. 904 Lingual and hypoglossal nerves in the submandibular region.

3. The **inferior alveolar nerve** contains both motor and sensory fibres. It descends to the mandibular foramen posterior to the lingual nerve from which it receives a communication. At the foramen it lies between the spheno-mandibular ligament and the bone, and gives rise to its only motor branch, the **mylohyoid nerve.** This pierces the ligament and descends with the corresponding vessels in the mylohyoid groove [FIG. 141] to pass over the inferior surface of the mylohyoid muscle deep to the submandibular gland. It supplies the mylohyoid muscle and the anterior belly of the digastric muscle.

The inferior alveolar nerve enters the mandibular canal with the inferior alveolar vessels. Here it gives branches which form a fine **inferior dental plexus** from which the mandibular teeth, alveolar bone, and gums are supplied. The branches that supply the molar, premolar, and canine teeth arise before the mental branch leaves the nerve, while the branches that form the plexus supplying the incisor (and canine) teeth arise distal to this (Starkie and Stewart, 1931). The premolar and first molar teeth are also partly supplied by the lingual and buccal nerves (Stewart and Wilson, 1928).

The **mental nerve** arises from the inferior alveolar nerve in the mandibular canal. It passes through the mental foramen, communicates with the marginal mandibular branch of the facial nerve [FIG. 907] and immediately divides into numerous branches deep to the facial muscles. It supplies the skin of the lower lip and chin and the mucous membrane and glands of the alveololabial sulcus and adjacent gum.

Injury to the trigeminal nerve mainly produces sensory loss in the territory of the nerve, or of the part involved. If the mandibular nerve is destroyed, the muscles of mastication are paralysed. Thus the jaw cannot be protruded or retracted symmetrically; on protrusion the chin is swung towards the paralysed side (lateral pterygoid paralysis), on retraction towards the uninjured side (paralysis of temporalis).

ABDUCENT NERVE

The sixth cranial or abducent nerve supplies only the lateral rectus muscle of the eyeball. It emerges from the brain in the pontomedullary sulcus [FIG. 888] immediately superior to the pyramid [FIG. 753. For deep origin, see p. 618] and curves superiorly, anterior to the pons, and usually posterior to the anterior inferior cerebellar artery. At the apex of the petrous temporal bone it turns abruptly forwards, pierces the arachnoid and dura mater of the posterior cranial fossa to enter or pass above, the inferior petrosal sinus and below the petroclinoid ligament [FIG. 1076]. The nerve then enters the cavernous sinus and runs forwards immediately lateral to the internal carotid artery, receiving a **sympathetic** filament from the internal carotid plexus. The nerve leaves the sinus anteriorly, communicates with the ophthalmic nerve, and enters the orbit through the superior orbital fissure within the tendinous ring from which the rectus muscles arise, lying inferior to the oculomotor and nasociliary nerves. It enters the medial surface of the lateral rectus muscle.

The position of the nerve in the cavernous sinus makes it susceptible to pathological changes in the sinus, while its vertical course in the subarachnoid space results in increased tension on the nerve whenever the brain stem is displaced inferiorly. Both conditions lead to weakness or paralysis of the lateral rectus muscle with a consequent internal squint.

FACIAL NERVE

The seventh cranial or facial nerve supplies the structures derived from the second pharyngeal (branchial) arch of the embryo. It is predominantly a **motor** nerve (1) to the muscles of facial expression, also to the posterior belly of digastric, the stylohyoid, and the stapedius muscles, and (2) to many of the glands of the head by preganglionic parasympathetic nerve fibres that pass to the cells of the pterygopalatine and sub-mandibular ganglia. It also contains a few **sensory** fibres which originate in the cells of its geniculate ganglion and are predominantly concerned with taste sensations from the anterior two-thirds of the tongue and the palate.

The facial nerve arises by two roots from the lateral part of the pontomedullary sulcus [FIG. 754. For deep connexions see p. 620] immediately anterior to the vestibulocochlear (eighth cranial) nerve. The **roots** are a large, motor, facial nerve proper which lies anterior to the small **nervus intermedius** which transmits sensory and preganglionic parasympathetic nerve fibres [FIG. 888]. They pass laterally with, and superior to, the eighth nerve into the internal acoustic meatus, surrounded by a sheath of the meninges. Here the branches of the nervus intermedius join the seventh and eighth nerves, though all its fibres probably enter the seventh nerve distally [FIG. 909]. The facial nerve pierces the meninges at the lateral end of the internal acoustic meatus [FIG. 1016] and continues laterally in the bony facial canal lying above and between the cochlea and vestibule [FIGS. 135, 1013]. At the hiatus for the greater petrosal nerve, the facial nerve is enlarged by the sensory **genicular ganglion** and gives off (1) the **greater petrosal nerve,** (2) a **branch to the tympanic plexus,** and (3) a branch to the sympathetic plexus on the middle meningeal artery. The facial nerve now turns abruptly backwards **(geniculum of the facial nerve)** in the bone of the upper part of the medial (labyrinthine) wall of the middle ear cavity, superior to the fenestra vestibuli and then inferior to the prominence caused by the lateral semicircular canal in the aditus to the mastoid antrum [FIG. 1005]. Medial to the aditus, the nerve turns vertically downwards in the bony septum which separates the middle ear from the mastoid antrum and air cells, and gives off first the **nerve to stapedius,** then the **chorda tympani,** and finally a **communicating branch to the auricular branch of the vagus.** The latter arises immediately before the nerve emerges from the stylo-mastoid foramen [FIG. 906].

The facial nerve emerges from the stylomastoid foramen under cover of the mastoid process; here it is more liable to injury in the child because that process is incompletely developed. The nerve passes anterolaterally between the styloid process and the posterior belly of digastric [FIG. 1075], and gives off: (1) one or two descending **branches** which supply the posterior belly of the **digastric** and the **stylohyoid** muscle; and (2) the **posterior auricular nerve.** The nerve then enters the posteromedial surface of the parotid gland, and crossing superficial to the external carotid artery and the retromandibular vein, breaks into a number of branches which

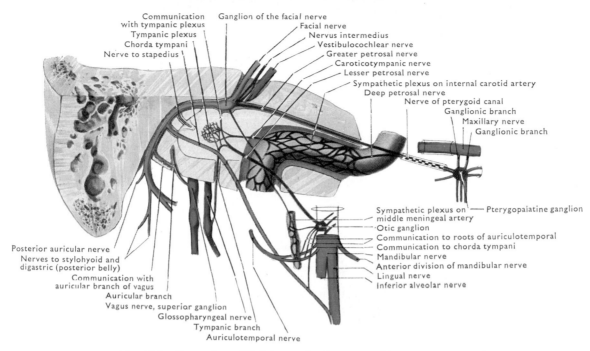

FIG. 906   Connexions of facial nerve in the temporal bone.

emerge separately from the gland [FIG. 907] and pass to supply the muscles of facial expression on their deep surfaces.

Branches

1. *From the geniculum of the facial nerve:*

i. The **greater petrosal nerve** (sensory, taste to the palate, and preganglionic parasympathetic to pterygopalatine ganglion) passes forwards in its canal and then in the dural floor of the middle cranial fossa and crosses the upper part of the foramen lacerum.   Here it is joined by a branch (**deep petrosal nerve**) from the internal carotid plexus (sympathetic) and passes through the pterygoid canal to the pterygopalatine ganglion as the **nerve of the pterygoid canal** [FIG. 163].

ii. A small branch passes through the temporal bone to join **the tympanic plexus** of the glossopharyngeal nerve (q.v.).

iii. A minute, inconstant branch to the sympathetic plexus on the middle meningeal artery.

2. *In the descending part of the canal:*

i. The small **stapedial nerve** passes forwards to supply the stapedius muscle.   If the facial nerve is destroyed proximal to this branch, stapedius is paralysed and sounds are heard louder (hyperacousia) in the affected ear in addition to the effects of destruction of the facial nerve distal to this branch.

ii. The **chorda tympani** passes forwards into the middle ear through the posterior canaliculus, and crossing the medial aspect of the fibrous tympanic membrane and handle of the malleus, leaves it through the anterior canaliculus and the petrotympanic fissure [FIG. 157] medial to the temporo-mandibular joint.   It then passes antero-inferiorly in the infratemporal fossa (medial to the spine of the sphenoid and the lateral pterygoid muscle), receives a filament from the otic ganglion, and joins the lingual nerve [FIG. 906].   The sensory (taste) fibres are distributed with that nerve to the anterior two-thirds of the tongue (excluding the vallate papillae).   The preganglionic parasympathetic fibres pass to

the cells of the submandibular ganglion [see below] in a branch from the lingual nerve.

iii. A fine branch passes to the **auricular branch of the vagus** just superior to the stylomastoid foramen.   It presumably transmits sensory fibres to the external acoustic meatus.

3. *In the neck.*

i. A descending **digastric branch** (or branches) passes to supply the posterior belly of digastric and the stylohoid muscle.

ii. The **posterior auricular nerve** bends backwards lateral to the mastoid process with the posterior auricular artery.   It supplies the intrinsic and posterior auricular muscles, and sends an **occipital branch** along the superior nuchal line closely applied to the skull to the occipital belly of epicranius.

4. *In the parotid gland:*

i. **Temporal branches** pass anterosuperiorly over the zygomatic arch to supply the facial muscles above the level of the palpebral fissure.   A small branch passes posterosuperiorly to the anterior and superior auricular muscles.

ii. The **zygomatic branches** supply the facial muscles between the levels of the palpebral and oral fissures.   The smaller, upper branches cross the zygomatic arch towards the orbicularis oculi and may be indistinguishable from the temporal branches.   The larger, lower branches pass with the parotid duct over masseter, and run medially to form the infra-orbital plexus with the **infra-orbital nerve.**

iii. The **buccal branches** run towards the angle of the mouth, and forming a plexus with the **buccal nerve,** supply buccinator and the other muscles in this region.

iv. The **marginal mandibular branch** passes along the inferior border of the mandible, often looping down into the neck.   It ends after forming a plexus with the mental nerve by supplying the muscles in the lower lip and the chin.

v. The **cervical branch** emerges from the lower end of

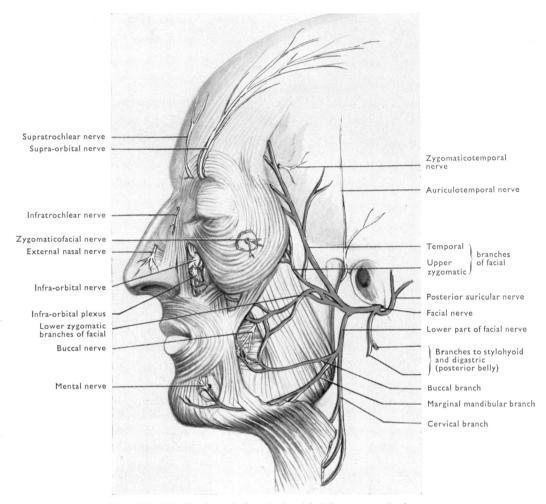

Supratrochlear nerve
Supra-orbital nerve

Infratrochlear nerve

Zygomaticofacial nerve
External nasal nerve

Infra-orbital nerve

Infra-orbital plexus
Lower zygomatic
branches of facial
Buccal nerve

Mental nerve

Zygomaticotemporal
nerve

Auriculotemporal nerve

Temporal ⎫ branches
         ⎬ of facial
Upper   ⎪
zygomatic ⎭

Posterior auricular nerve
Facial nerve
Lower part of facial nerve

⎫ Branches to stylohyoid
⎬ and digastric
⎭ (posterior belly)

Buccal branch
Marginal mandibular branch
Cervical branch

FIG. 907   Distribution of trigeminal and facial nerves on the face.

the parotid gland and passes antero-inferiorly, below the angle of the mandible, to supply platysma from its deep surface.

These branches are the only motor supply to the facial muscles, but the exact distribution of each cannot be given because they vary in size from individual to individual and communicate with each other on the face. Though these branches may transmit some proprioceptive fibres from the facial muscles, it is probable that the majority enter the brain through the trigeminal nerve via the communications on the face.

Injury to the facial nerve in the parotid gland causes paralysis of these muscles, and, amongst other things, inability to close the eye (orbicularis oculi) or mouth (orbicularis oris), or to use the lips or cheeks (buccinator) properly in eating or speaking. Thus the conjunctiva is not kept moist by blinking, while the drooping lower eyelid fills with lacrimal fluid which cannot reach the lacrimal puncta and consequently overflows on to the face. Speech is blurred, fluid and food collects in the vestibule of the mouth and escapes from between the lips. The mouth is pulled towards the normal side by the unopposed action of that half of the orbicularis oris. If the nerve is damaged in the bony canal there may also be loss of taste on the anterior two-thirds of the tongue and hyperacousia on the affected side. Damage in the internal acoustic meatus is

usually associated with injury to the eighth nerve and thus deafness, not hyperacousia, results.

Pterygopalatine Ganglion [FIG. 899]

This parasympathetic ganglion is a small reddish body lying in the upper part of the pterygopalatine fossa [FIG. 902] close to the sphenopalatine foramen. Of the different types of nerve fibres which are associated with the ganglion the only fibres that synapse with the cells are the **preganglionic parasympathetic nerve fibres** which reach it from the **greater petrosal** branch of the facial nerve via the **nerve of the pterygoid canal** [FIG. 908]. The **postganglionic sympathetic fibres** of the deep petrosal nerve which reach the ganglion through the nerve of the pterygoid canal, and the **sensory fibres** which join the maxillary nerve to the ganglion pass directly into its various branches [p. 720]. In addition to sending postganglionic fibres into these branches, the cells of the ganglion also send fibres into the maxillary nerve (q.v.) for distribution through some of its branches. Thus **postganglionic parasympathetic nerve fibres** from the ganglion reach gland cells of the nasal cavity, some of the paranasal sinuses, the palate, the roof of the pharynx, the auditory tube, and the lacrimal gland. In addition to the sensory fibres of the maxillary nerve which traverse the ganglion,

FIG. 908 Diagram of connexions of pterygopalatine and submandibular ganglia with distribution of parasympathetic components of seventh cranial nerve.

Sympathetic, red; parasympathetic, **green**; sensory, blue.

taste fibres from the palate (especially in children) presumably pass through the ganglion to their cell bodies in the genicular ganglion via the nerve of the pterygoid canal and then the greater petrosal nerve.

Submandibular Ganglion

This parasympathetic ganglion lies on the hyoglossus muscle between the lingual nerve and the submandibular duct [FIG. 904]. Its connexions are similar to those of the pterygopalatine ganglion. The **preganglionic parasympathetic fibres** come from the chorda tympani via the lingual nerve, while **sensory fibres** from the lingual nerve and **postganglionic sympathetic fibres** from the plexus on the facial artery traverse it. **Postganglionic parasympathetic fibres** from the cells of the ganglion pass with the sensory and sympathetic fibres to the submandibular and sublingual glands, to the lingual gum, and to the alveololingual sulcus, while others pass directly into the lingual nerve for distribution to the glands in the tongue. Possibly such fibres also pass through the communication which the lingual nerve has with the inferior alveolar nerve and thus pass to supply glands in the lower lip, etc.

VESTIBULOCOCHLEAR NERVE

The eighth cranial or vestibulocochlear nerve is purely sensory and supplies the organs of hearing (cochlea) and equilibration (utricle, saccule, and semicircular ducts) within the petrous temporal bone. The nerve consists of a **superior** (vestibular) **root** and an **inferior** (cochlear) **root** which arise together from the pontomedullary sulcus of the brain stem immediately posterior to the facial nerve [FIGS. 754, 909. For deep connexions, see p. 622]. The two roots unite to form the vestibulocochlear nerve and pass into the internal acoustic meatus inferior to the facial nerve and nervus intermedius with which they have temporary communications. Near the fundus of the meatus, two or more swellings appear on the eighth nerve. These consist of the bipolar cells of the **vestibular ganglion** the peripheral process of which innervate

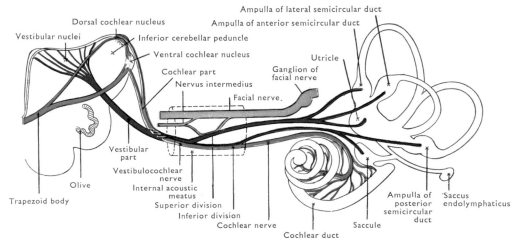

FIG. 909 Scheme of origin and distribution of vestibulocochlear nerve [cf. Fig. 1024].

The bundle of fibres shown coursing across the floor of the fourth ventricle are not auditory; they are the striae medullares of fourth ventricle [p. 674] and are probably aberrant pontocerebellar fibres.

the organs of equilibrium, while the central processes form the **vestibular part of the nerve.** Distal to the ganglion, the vestibular part divides into superior and inferior parts, the inferior part lying with the cochlear part.

Vestibular Part

The branches of the superior vestibular part pierce the superior vestibular area in the fundus of the meatus [FIG. 1016] and supply: (1) the macula of the utricle (**utricular nerve**), (2) and (3) supply the ampullae of the anterior and lateral semicircular ducts (**anterior and lateral ampullary nerves**), and (4) a small branch to the macula of the saccule.

The branches of the inferior vestibular part supply: (1) the macula of the saccule through the inferior vestibular area of the meatal fundus [FIG. 1016]; and (2) the ampulla of the posterior semicircular duct through the foramen singulare (**posterior ampullary nerve**).

Cochlear Part

The cochlear part of the eighth nerve passes through the tractus spiralis foraminosus [FIG. 1016] as a series of fine bundles which enter the modiolus [FIG. 1015] and turning outwards towards the osseous spiral lamina expand into the spiral ganglion which fills the spiral canal medial to the base of the lamina [FIG. 1020]. The spiral ganglion, the counterpart of the vestibular ganglion on the vestibular fibres, contains bipolar cell bodies, the peripheral processes of which innervate the hair cells of the cochlea.

The vestibulocochlear nerve may be injured in a number of pathological conditions, and the cells of its ganglia are particularly susceptible to certain drugs, e.g., streptomycin and salicylates. Tumours may form from the sheath cells of this nerve and first irritate and then destroy it. The irritation causes noises in the head (cochlear part) and giddiness which is observed as a rapid nystagmus (oscillating movement) of the eyes, and a tendency to stagger and fall (vestibular part). Unilateral total deafness and loss of reaction from the vestibular part of the internal ear supervene when the nerve is destroyed. The facial nerve (q.v.) is also likely to be involved by such a tumour, especially if it is in the internal acoustic meatus.

GLOSSOPHARYNGEAL NERVE

The ninth cranial or glossopharyngeal nerve supplies the structures that develop from the third pharyngeal (branchial) arch of the embryo. It is a mixed nerve with a large **sensory** element which supplies part of the pharynx and the posterior one-third of the tongue, and a small **motor** part which supplies the stylopharyngeus and middle constrictor muscles (branchial motor) and also sends preganglionic **parasympathetic** nerve fibres predominantly to the otic ganglion.

Superficial Origin and Course

The glossopharyngeal nerve arises by a few fine rootlets from the lateral surface of the upper medulla oblongata between the seventh and tenth nerves [FIG. 753. For deep connexions see pp. 620, 622[. The rootlets unite as they enter the jugular foramen between the sigmoid and inferior petrosal sinuses, close to the vagus and accessory nerves, but in a separate dural sheath [FIG. 913]. Within the foramen there are usually two enlargements of the nerve, the **superior and inferior sensory ganglia.** The nerve descends into the neck between the internal carotid artery and the internal jugular vein and, passing between the internal and external carotid arteries, curves behind the stylopharyngeus to reach its lateral aspect. It then runs forwards deep to the stylohyoid ligament and hyoglossus muscle, and enters the base of the tongue, close to the lower pole of the palatine tonsil, by passing below the inferior border of the superior constrictor muscle with the stylopharyngeus.

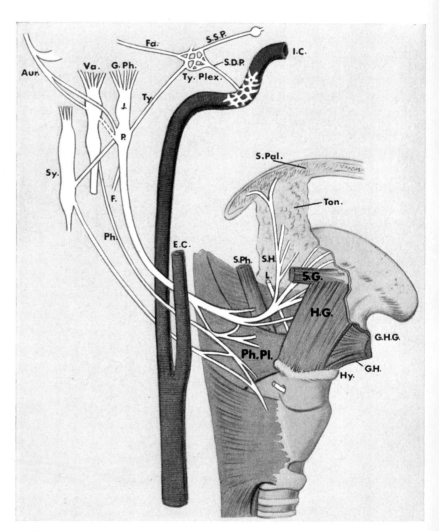

FIG. 910   Scheme of connexions and distribution of glossopharyngeal nerve.

G.Ph., glossopharyngeal nerve; J., superior, and P., inferior ganglia; Ty., tympanic nerve; Ty. Plex., tympanic plexus; Fa., root from ganglion of facial nerve; S.S.P., lesser petrosal nerve to the otic ganglion; S.D.P., caroticotympanic nerve; I.C., internal carotid artery; Va., vagus nerve; Aur., auricular branch of vagus; Sy., superior cervical sympathetic ganglion; F., communicating branch to facial nerve; Ph., pharyngeal branch of vagus; E.C., external carotid artery; Ph.Pl., pharyngeal plexus; S.Ph., stylopharyngeus muscle; S.H.L., stylohyoid ligament; H.G., hyoglossus; S.G., styloglossus; S.Pal., soft palate; G.H.G., genioglossus; G.H., geniohyoid; Hy., hyoid bone; Ton., palatine tonsil.

Branches

1. The **tympanic nerve** arises from the inferior ganglion and passes superiorly in the bone between the jugular foramen and the carotid canal into the middle ear. Here it forms the **tympanic plexus** [FIG. 906] on the medial wall of the middle ear with sympathetic fibres from the internal carotid plexus (**caroticotympanic nerves**) and a twig from the genicular ganglion of the facial nerve. Branches from the plexus supply the walls of the middle ear, the mastoid air cells, and the auditory tube (**tubal branch**). Some of the glosso-pharyngeal (parasympathetic) and facial fibres unite to form the **lesser petrosal nerve**. This passes into the dural floor of the middle cranial fossa, and piercing the base of the skull close to the foramen ovale, enters the **otic ganglion.**

The **inferior ganglion** also receives sympathetic fibres from the superior cervical ganglion and has branches joining it to the auricular branch and superior ganglion of the vagus and to the facial nerve.

2. In the neck: (i) A **branch to stylopharyngeus** which also sends fibres through that muscle to supply pharyngeal mucous membrane (**tonsillar branches**); (ii) **pharyngeal branches** to the middle constrictor and the mucous membrane of the pharynx, which either pass directly through the superior constrictor or unite with pharyngeal plexus [FIG. 910]: (iii) the **carotid sinus branch** runs downwards anterior to the internal carotid artery [FIG. 687], communicates with the vagus and sympathetic (Sheehan, Mulholland and Shafiroff, 1941) then divides in the angle of bifurcation of the

common carotid artery to supply the carotid body and carotid sinus. It carries impulses from the baroreceptors in the carotid sinus, and from chemoreceptors in the carotid body.

3. The **lingual branches** supply the mucous membrane of the posterior one-third of the tongue (including the vallate papillae) with taste and general sensory fibres. They also contain preganglionic parasympathetic fibres which may relay in small ganglia scattered in the substance of the tongue to supply the posterior lingual glands (Fitzgerald and Alexander, 1969). **Tonsillar branches** supply the mucous membrane covering the tonsil, the adjacent part of the soft palate, and the palatine arches.

Otic Ganglion

This small, parasympathetic ganglion lies immediately inferior to the foramen ovale, between the mandibular nerve and the tensor veli palatini muscle. **Preganglionic parasympathetic fibres** reach it from the glossopharyngeal nerve (and possibly also from the facial nerve) via the lesser petrosal nerve (q.v.). As in the other parasympathetic ganglia, these are the only fibres which synapse with the cells of the ganglion; the postganglionic sympathetic fibres from the plexus on the middle meningeal artery and the sensory and motor fibres (supply to the medial pterygoid, tensor veli palatini, and tensor tympani muscles) from the mandibular nerve merely pass through the ganglion.

The **postganglionic parasympathetic fibres** from the ganglion pass with the sympathetic fibres mainly into the auriculotemporal nerve to supply the parotid gland. Other small twigs join the chorda tympani, the nerve of the pterygoid canal and branches of the mandibular nerve.

VAGUS NERVE

The tenth cranial or vagus nerve is mixed. It supplies (i) the structures developed from the fourth and sixth pharyngeal (branchial) arches of the embryo, and (ii) the foregut and midgut and the structures developed from the entoderm or mesoderm of these parts of the gut tube. It, therefore, carries **afferent nerve fibres** from the pharynx, oesophagus, stomach, small and part of the large intestine, larynx, trachea, lungs, heart and great vessels, pancreas and liver, though some of the sensory fibres from most of these organs enter the central nervous system by running through the sympathetic system into the spinal nerves. The **motor fibres** of the vagus are: (1) preganglionic parasympathetic fibres distributed to the parasympathetic ganglia that innervate smooth or cardiac muscle and gland cells in the above organs; (2) fibres (branchial) direct to the striated muscles of the larynx and pharynx which also emerge from the brain in the cranial part of the accessory nerve.

Superficial Origin and Course

The vagus arises from the lateral surface of the medulla oblongata by numerous rootlets in series with the glosso-pharyngeal above and the accessory nerve below [for deep connexions, pp. 620, 622] The rootlets unite and pass through the jugular foramen immediately posterior to the glosso-pharyngeal nerve, but in a separate dural sheath shared with

FIG. 911 Diagram of connexions of otic ganglion with distribution of parasympathetic components of ninth cranial nerve.

Sympathetic, red; parasympathetic, green; sensory, blue.

the accessory nerve, a feature which makes the accessory nerve essentially a part of the vagus nerve. The vagus has a small, **superior ganglion** in the jugular foramen and a larger, **inferior ganglion** immediately below the foramen. Both ganglia are fusiform in shape and consist of sensory nerve cells.

In the neck, the vagus descends vertically in the carotid sheath, posterior to and between the internal jugular vein and first the internal and then the common carotid artery. It enters the superior mediastinum of the thorax posterior to the formation of the brachiocephalic vein: on the right side, after crossing anterior to the first part of the subclavian artery; on the left side, in the interval between the common carotid and subclavian arteries.

In the mediastinum both nerves pass postero-inferiorly to reach the posterior surface of the corresponding lung root, the right nerve on the trachea; the left at first between the two arteries and then on the left side of the arch of the aorta, posterior to the phrenic nerve. At the root of the lung each nerve gives branches to the **pulmonary plexuses** [q.v. FIG. 912] and appears inferior to the lung root as two or more branches. These pass inferomedially to unite with the corresponding branches of the opposite vagus and form a **plexus surrounding the oesophagus** which receives one or more branches from each greater splanchnic nerve (sympathetic). Near the oesophageal opening in the diaphragm, the plexus reforms into **anterior** and **posterior vagal trunks** each of which passes as one or more bundles on to the corresponding surface of the stomach **(gastric branches).** Each trunk contains fibres from both vagus nerves and from the sympathetic.

COMMUNICATIONS AND BRANCHES

Superior Ganglion

In the jugular foramen this ganglion communicates with the superior cervical sympathetic ganglion, the accessory nerve, and (sometimes) the inferior glossopharyngeal ganglion. It gives off two branches.

1. A **meningeal branch** passes posteriorly to supply the dura mater of the posterior cranial fossa.

2. The **auricular branch** receives a twig from the tympanic branch of the glossopharyngeal nerve, and entering the mastoid canaliculus on the lateral wall of the jugular fossa, is joined by a branch from the facial nerve. This complex emerges from the tympanomastoid fissure and supplies the inferior part of the tympanic membrane and the floor of the external acoustic meatus. The importance of this branch lies in the fact that irritation of the skin in this region, e.g. a foreign body in the external acoustic meatus in children, may mimic irritation of other parts of the sensory distribution of the vagus and give rise to coughing or vomiting.

Inferior Ganglion

This ganglion communicates with the superior cervical sympathetic ganglion, the hypoglossal nerve, and the loop between the first and second cervical ventral rami. The **accessory nerve** is closely applied to the ganglion and sends its cranial fibres to it for distribution through the vagus to the muscles of the larynx. It gives off two branches.

1. The **pharyngeal branch** (often double, and said to contain fibres from the accessory nerve) passes inferomedially between the internal and external carotid arteries to form the pharyngeal plexus on the middle constrictor muscle with branches from the glossopharyngeal nerve and the superior cervical sympathetic ganglion [FIG. 910]. This plexus supplies most of the muscles of the pharynx and soft palate (except stylopharyngeus and tensor veli palatini) and sends a lingual branch to join the hypoglossal nerve in the anterior triangle of the neck.

2. The **superior laryngeal nerve** passes inferomedially, deep to both carotid arteries, towards the thyroid cartilage. It receives twigs from the sympathetic and pharyngeal plexus, and is said to send a branch to the internal carotid artery. It has two branches. (i) A larger, sensory, **internal branch** which passes anteromedially, deep to the thyrohyoid muscle, with the corresponding branch of the superior thyroid artery, and piercing the thyrohyoid membrane, divides into three branches immediately external to the mucous lining of the pharynx. It supplies the mucous membrane of the pharynx and larynx from the level of the vocal folds to the posterior part of the dorsum of the tongue (including the epiglottis), and communicates with a branch **(inferior laryngeal)** of the recurrent laryngeal nerve deep to the lamina of the thyroid cartilage. (ii) The **external branch** descends medial to the carotid sheath, on the inferior constrictor muscle of the pharynx. It supplies branches to that muscle, and ends in the cricothyroid muscle.

Branches of the Vagus in the Neck

1. The **superior** and **inferior cervical cardiac branches** descend into the thorax. *On the right side,* they pass posterior to the subclavian artery on the side of the trachea to enter the deep cardiac plexus [p. 788]. *On the left,* the superior branch passes between the arch of the aorta and the trachea to the deep cardiac plexus; the inferior branch descends with the vagus and the superior cervical cardiac nerve (sympathetic) lateral to the arch of the aorta, to end in the superficial cardiac plexus [p. 788].

2. The **right recurrent laryngeal nerve** arises from the vagus anterior to the first part of the subclavian artery, hooks under it, and ascends superomedially, passing posterior to the common carotid artery and either posterior or anterior to the inferior thyroid artery. It then lies in the groove between the oesophagus and trachea, medial to the right lobe of the thyroid gland, and passes deep to the lower border of the inferior constrictor muscle as the **inferior laryngeal nerve.**

(i) **Cardiac branches** arise from the nerve on the subclavian artery. They descend beside the trachea to the deep cardiac plexus.

(ii) Communicating branches to the **cervicothoracic ganglion** arise from the nerve posterior to the subclavian artery.

(iii) Branches supply the **trachea, oesophagus,** and the **inferior constrictor** muscle of the pharynx.

(iv) The **inferior laryngeal nerve** supplies the intrinsic muscles of the larynx, except cricothyroid, and the mucous membrane of the larynx and pharynx inferior to the vocal folds. It communicates with the internal branch of the superior laryngeal nerve medial to the lamina of the thyroid cartilage.

Branches of the Vagus in the Thorax

1. The **left recurrent laryngeal nerve** arises from the vagus on the aortic arch. It hooks round the inferior surface of the arch, posterior to the ligamentum arteriosum, and passes upwards through the superior mediastinum and lower part of the neck in the groove between the oesophagus and trachea, supplying both. Its branches are the same as those of the right nerve, but larger **cardiac branches** arise inferior to the aortic arch and pass to the deep cardiac plexus.

2. The **right vagus** on the trachea sends thoracic cardiac branches to the deep cardiac plexus [p. 788], and **tracheal** and **oesophageal branches** medially to these structures.

3. As each vagus approaches the lung root it sends **bronchial** and pulmonary branches to form a complex **pulmonary plexus** [p. 788] mainly on the posterior surface of the lung root, though small branches pass to the anterior surface from the upper border of the root. The bronchial branches of the two sides communicate, and the plexus behind the lung root is joined by fine branches from the second, third, and fourth thoracic ganglia of the sympathetic trunk. The pulmonary plexus extends into the lung surrounding the bronchi and vessels and passing to the pulmonary pleura.

4. Each vagus emerges from behind the corresponding lung root as two or more trunks which pass inferomedially to break up and communicate on the oesophagus as the **oesophageal plexus**. This plexus receives filaments from the greater splanchnic nerve and its ganglion (sympathetic). It supplies the oesophagus, sends small filaments to the aorta (sensory), and reconstitutes into the anterior and posterior **vagal trunks** towards the inferior end of the thoracic oesophagus.

The cardiac, pulmonary, and oesophageal **plexuses** are predominantly parasympathetic. The fibres that reach them from the vagus are preganglionic in nature. They end on the ganglionic cells (parasympathetic) which are scattered throughout these plexuses and send their axons to the heart, lungs, and oesophagus. The vagal branches also contain many sensory fibres whose cell bodies lie in the superior and inferior ganglia of the vagus. These fibres convey impulses from the heart (but not pain), great vessels (including the glomus aorticum), and lung, and are of importance in vascular and respiratory reflexes.

Branches of the Vagus in the Abdomen

The anterior and posterior **vagal trunks** pass with the oesophagus through the diaphragm into the abdomen. Both trunks supply the corresponding surfaces of the stomach through several anterior and posterior **gastric branches.** In addition the posterior trunk sends fibres to the coeliac and renal plexuses of the autonomic nervous system [p. 789] from which they are distributed along the visceral branches of the aorta to the abdominal structures, including the gut tube as far as the transverse colon, beyond which the parasympathetic innervation comes from the sacral region. The anterior gastric branches also send filaments between the layers of the lesser omentum to the liver and biliary system **(hepatic branches),** and to the pylorus and superior part of the duodenum.

All the **efferent vagal fibres** that enter the abdomen are preganglionic in type and they end on scattered ganglionic neurons, many of which lie in the intramural plexuses of the walls of the gut tube (submucous and myenteric plexuses and

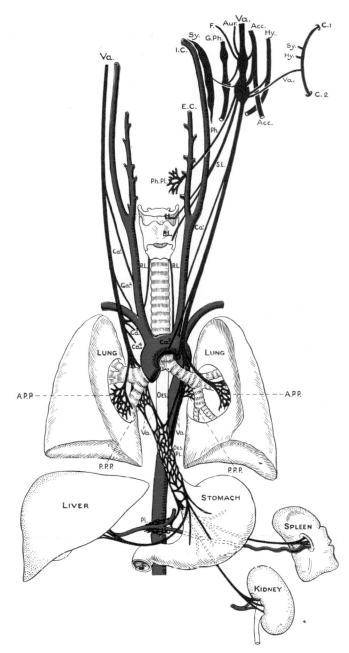

FIG. 912 Scheme of connexions and distribution of vagus nerves.

Acc., accessory nerve; A.P.P., pulmonary plexus; Aur., auricular branch; C.1 C.2, loop between first and second cervical ventral rami; Ca.¹, Ca.², upper and lower cervical cardiac branches; Ca.³, cardiac branches of recurrent laryngeal; Ca.⁴, thoracic cardiac branch (right vagus); E.C., external carotid artery; E.L., external laryngeal branch; F., meningeal branch; G.Ph., glossopharyngeal nerve; Hy., hypoglossal nerve; Hy., cervical communication to hypoglossal nerve; I.C., internal carotid artery; I.L., internal laryngeal branch; Oes., oesophagus; Oes.Pl., oesophageal plexus; Ph., pharyngeal branch; Ph.Pl., pharyngeal plexus; Pl., coeliac plexus; P.P.P., pulmonary plexus; R.L., recurrent laryngeal nerve; S.L., superior laryngeal nerve; Sy., superior cervical ganglion of sympathetic; Sy., grey ramus communicans; Va., right and left vagi; Va., cervical communication with inferior ganglion of vagus.

in other viscera. These neurons innervate the gland cells and smooth muscle of these viscera with postganglionic parasympathetic nerve fibres. The sympathetic fibres which also innervate these viscera for the most part have their ganglia either at the origin from the aorta of the arteries that supply the viscera or scattered along these vessels. Thus the nerve fibres entering a particular viscus consist mainly of postganglionic sympathetic and preganglionic parasympathetic fibres.

The *asymmetry* of certain parts of the vagus nerves is due to the developmental asymmetry of the pharyngeal (branchial) arch arteries between which the pharyngeal and laryngeal branches of the vagus nerves pass to innervate the cranial part of the foregut tube. On both sides the **recurrent laryngeal nerve** passes caudal to the dorsal part of the corresponding sixth arch artery, and thus forms a recurrent loop around the artery as it slides caudally with the heart relative to the foregut tube. *On the right*, the dorsal part of the sixth arch artery disappears allowing the nerve to loop round the fourth arch (subclavian) artery at a higher level. *On the left*, the dorsal part of the sixth arch artery persists (ligamentum arteriosum) and the nerve continues to pass caudal to it [FIG. 1133]. The more caudal position of the left recurrent laryngeal nerve also explains the absence of thoracic cardiac branches of the left vagus, for these are incorporated in the branches of the recurrent nerve. The positions of the anterior and posterior **vagal trunks** and their gastric branches is explained by the rotation of the stomach which causes its left side to become anterior. The mixing of the fibres of both vagus nerves in the **oesophageal plexus** and vagal trunks allows both these nerves to send fibres to the abdominal viscera through the communication of the posterior vagal trunk with the abdominal sympathetic plexuses [p. 789].

Destruction of the vagus near its emergence from the brain produces defects which are visible mainly because of the paralysis of the pharyngeal and laryngeal muscles on the affected side. Thus there is difficulty in swallowing and in speech but relatively little to show for the loss of the preganglionic innervation of the heart, lungs, and abdominal viscera, though there may be a decrease in gastric and intestinal secretions. Vagotomy has occasionally been used to reduce the acid secretion from the stomach in cases of duodenal ulceration, but the bilateral distribution of the vagus in the abdomen probably reduces the effect of unilateral injury to the nerve. Involvement of the laryngeal nerves in any pathological process may irritate the sensory fibres from the larynx and lead to protracted coughing of an ineffective, brassy type because there is partial or complete paralysis of the laryngeal muscles which prevents effective closure of the glottis.

ACCESSORY NERVE

The eleventh cranial or accessory nerve is essentially the caudal part of the vagus nerve. It therefore supplies the most caudal pharyngeal (branchial) arch, the muscle of which becomes the muscles of the larynx and the sternocleidomastoid and trapezius muscles ; the former are supplied by the cranial roots, the latter two by the spinal roots.

The **cranial roots** arise from the lateral surface of the medulla oblongata caudal to and in series with the vagal

rootlets [FIG. 752. For deep origin see p. 620]. These rootlets join the spinal part and pass into the jugular foramen with it and the vagus. Here the cranial roots send a twig to the superior ganglion of the vagus, then join that nerve at the inferior ganglion, forming the **internal branch** of the accessory nerve. They are distributed through the laryngeal (and possibly pharyngeal) branches of the vagus (q.v.).

The **spinal roots** arise from the lateral aspect of the spinal medulla dorsal to the ligamentum denticulatum as far caudally as the fifth or sixth cervical segment [FIG. 913. For deep origin, see p. 620]. Turning cranially, these roots unite successively to form a trunk which ascends through the foramen magnum in the subarachnoid space, receives the cranial roots, and passes through the jugular foramen in the same dural sheath as the vagus [FIG. 893]. *In the foramen*, the cranial part joins the vagus, while the spinal part (**external branch**) emerging between the internal carotid artery and the internal jugular vein, passes postero-inferiorly either superficial or deep to the internal jugular vein, and crosses the transverse process of the atlas deep to the posterior belly of the digastric muscle. It then enters the deep surfaces of sternocleidomastoid, receives a sensory **communicating branch** from the second cervical ventral ramus, and supplies the muscle. It emerges from the posterior border of the sternocleidomastoid at or below the junction of the upper and middle thirds, and usually has the lesser occipital nerve turning round its inferior aspect to ascend along the posterior border of the muscle. The accessory nerve then runs postero-inferiorly in the fascial roof of the **posterior triangle**, receives **communicating branches** from the third and fourth cervical ventral rami (sensory fibres), and passes

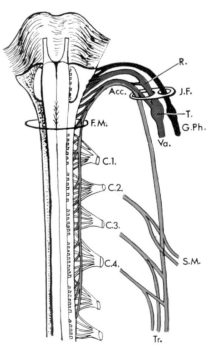

FIG. 913 Scheme of the origin, connexions and distribution of the accessory nerve.

Acc., accessory nerve ; C.1-4, first four cervical nerves (dorsal roots) ; Va., vagus nerve ; R., superior ganglion ; T., inferior ganglion ; G.Ph., glossopharyngeal nerve ; S.M., nerves to sternocleidomastoid ; Tr., nerves to trapezius ; F.M., foramen magnum ; J.F., jugular foramen.

on to the deep surface of the trapezius muscle supplying it and again receiving communications from the same cervical nerves.

The nerve has a number of deep cervical lymph nodes in contact with it adjacent to the sternocleidomastoid muscle and is prone to damage in operations which require mobilization of the muscle.

When the external branch of the accessory nerve is destroyed, the sternocleidomastoid and trapezius muscles are paralysed, for the communicating branches from the cervical plexus are sensory in nature. The shoulder droops and its elevation is weak. The unopposed pull of the opposite sternocleidomastoid and trapezius draws the occiput towards the shoulder of that side, turning the face upwards and towards the paralysed side (wryneck).

HYPOGLOSSAL NERVE

The twelfth cranial or hypoglossal nerve is a pure motor nerve to the extrinsic and intrinsic muscles of the tongue, except palatoglossus. It arises by a number of rootlets from the ventral aspect of the medulla oblongata between the pyramid and the olive in line with abducent and oculomotor nerves and the ventral root of the first cervical nerve [FIG. 753. For deep origin, see p. 618]. The rootlets pierce the dura mater separately in two bundles which unite in the hypoglossal canal or outside the skull. Here the nerve lies posteromedial to the nerves emerging from the jugular foramen, receives a bundle of fibres from the loop uniting the ventral rami of the first and second cervical nerves and from the superior cervical sympathetic ganglion. The hypoglossal nerve then passes inferolaterally and winds closely round the posterior surface of the inferior ganglion of the vagus (with which it communicates) to reach its lateral side deep to the posterior belly of digastric and the occipital artery. It then passes inferior to the digastric, and curving forwards, crosses the lateral surfaces of the internal carotid artery, the occipital and external carotid arteries near their point of separation, and the loop of the lingual artery on the middle constrictor muscle. It then runs deep to the posterior belly of the digastric muscle on the hyoglossus muscle, and passing superior to the free posterior border of the mylohyoid muscle, lies between it and hyoglossus. Here it communicates with the lingual nerve and divides into its lingual branches which pass forwards between mylohyoid and genioglossus.

Branches

1. The branches of the hypoglossal nerve proper are the **lingual** branches to the hyoglossus, styloglossus, genioglossus, and the intrinsic muscles of the tongue. The branches to **geniohyoid and to thyrohyoid** are formed by fibres which enter and run with the hypoglossal nerve from the loop between the first and second cervical ventral rami [FIG. 914].

2. The **meningeal** branch arises close to the skull, and passes to the dura mater and periosteum of the posterior cranial fossa.

3. The **superior root of the ansa cervicalis** arises where the nerve crosses the internal carotid artery. It descends anterior to the carotid sheath in the anterior triangle, and is joined, below the middle of the neck, either superficial or deep to the internal jugular vein, by the **inferior root** which consists of fibres from the ventral rami of the second and third cervical nerves. The loop which this junction forms is the **ansa cervicalis,** and it sends branches to sternohyoid, sternothyroid, and both bellies of omohyoid.

4. The **nerve to the thyrohyoid muscle** arises in the carotid triangle and descends on the middle constrictor muscle, posterior to the greater cornu of the hyoid bone [FIG. 914].

If the hypoglossal nerve is destroyed, the intrinsic and extrinsic muscles of the corresponding half of the tongue are paralysed: if the injury is not too high in the neck, the thyrohyoid and geniohyoid muscles are also paralysed. The paralysis of tongue muscles causes difficulty with the lingual part of speech and swallowing, and the protruded tongue is thrust towards the paralysed side by the unopposed action of the muscles on the normal side. The paralysis of the thyrohyoid muscle interferes with laryngeal movements in swallowing.

FIG. 914 Muscles of hyoid bone and styloid process, and extrinsic muscles of the tongue, showing connexions and distribution of hypoglossal nerve.

THE SPINAL NERVES

Thirty-one pairs of spinal nerves arise from the spinal medulla and leave the vertebral canal through the intervertebral foramina. Eight of these are cervical, the first emerging between the occiput and the first cervical vertebra **(suboccipital nerve),** and the eighth between the seventh cervical and first thoracic vertebra. There are twelve thoracic, five lumbar, five sacral, and one coccygeal nerve, and each of these emerges below the corresponding vertebra. The coccygeal nerve is occasionally absent, but there may be one or two additional filaments caudal to it which remain within the sacral canal.

The size of a spinal nerve is directly proportional to the volume of tissue which it supplies. Thus the limb nerves (fifth cervical to first thoracic, and second lumbar to third sacral) are the largest, and of these the nerves innervating the lower limbs are the larger. The coccygeal nerve is the smallest, and the upper cervical nerves decrease in size from below upwards.

ORIGIN

Each spinal nerve is attached to the spinal medulla by a **dorsal** (sensory or afferent) and a **ventral** (motor or efferent) **root,** except the first cervical which may have no dorsal root. Every root consists of a number of rootlets, each of which

contains many nerve fibres surrounded by a layer of pia mater [p. 698]. The **dorsal rootlets** form a continuous linear series attached to the posterior lateral sulcus of the spinal medulla [FIG. 729]. The **ventral rootlets** arise irregularly from a strip on the anterolateral surface of the spinal medulla, and are separated from the dorsal rootlets by the ligamentum denticulatum [FIG. 872, p. 693], and above the sixth cervical segment also by the spinal part of the accessory nerve.

The dorsal and ventral rootlets of a spinal nerve separately converge on the arachnoid and dura opposite the corresponding intervertebral foramen to form a dorsal and a ventral root. The roots passing separately through these membranes receive a sleeve from each of them [FIGS. 872, 880]. The dorsal root, which is larger and contains more nerve fibres than the corresponding ventral root, immediately expands into the spinal ganglion in the intervertebral foramen.

The **spinal ganglion** consists of many spherical, unipolar [FIG. 711] **sensory nerve cells,** each of which gives rise to a short process that divides into a central branch forming part of a dorsal rootlet, and a peripheral branch passing into the spinal nerve. Scattered ganglion cells may be found on the dorsal rootlets and even on the ventral roots (which have no ganglion) particularly in the lumbosacral region. The function of those on the ventral roots is unknown though they have the structure of sensory cells (Windle, 1931). The ventral root

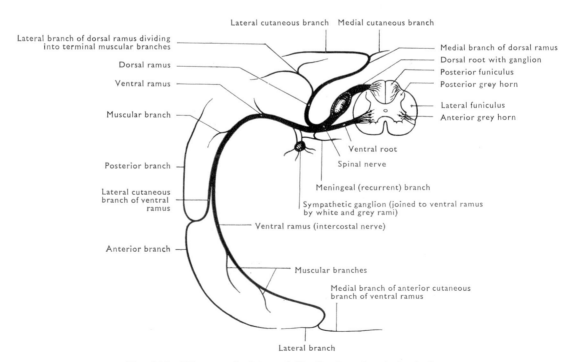

FIG. 915 Diagram of origin and distribution of typical spinal nerve.
Note that the medial branch of the dorsal ramus is represented as distributed to skin, whilst the lateral branch terminates at a deeper level in muscle. Both branches, however, supply muscles; and in the lower half of the body it is the lateral branch that supplies the skin.

grooves the anterior surface of the distal part of the ganglion, and beyond this *the two roots are fused into the spinal nerve.* The arachnoid sleeve of the dorsal root fuses with the proximal part of the ganglion, while the dural sleeves of the roots become continuous with the epineurium of the nerve.

The spinal medulla is shorter than the vertebral canal, and ends between the first and second lumbar vertebrae. Thus the segment of the spinal medulla from which each pair of roots arises does not lie opposite the corresponding vertebra [FIG. 732] and the rootlets run with increasing obliquity from above downwards to reach the appropriate intervertebral foramen. Those roots which emerge from the vertebral column below the termination of the spinal medulla, run longitudinally in the vertebral canal with the filum terminale [p. 597; enclosed together in the dura and arachnoid with the cerebrospinal fluid of the subarachnoid space, and thus form the **cauda equina** [FIG. 731]. The arachnoid and dura which surround the spinal medulla and cauda equina end at the level of the second sacral vertebra. Thus the nerves caudal to the first sacral pierce these meninges before reaching the appropriate intervertebral foramina. They also have their ganglia within the sacral part of the vertebral canal and not in the intervertebral foramina. The ganglion of the coccygeal nerve may be very small or consist of scattered cells on the dorsal root.

The Ventral Roots

These consist of coarse and fine myelinated nerve fibres. The **coarse fibres** are motor to the striated muscles of the body-wall and limbs. The **fine fibres** are of two types: (1) Motor fibres distributed to the intrafusal muscle fibres of the **muscle spindles**; (2) **Preganglionic autonomic nerve fibres.** (i) The **sympathetic** preganglionic fibres emerge from the spinal medulla with the first thoracic to second or third lumbar nerves inclusive, and pass from the ventral ramus of each of these nerves as a **white ramus communicans** to the corresponding ganglion of the sympathetic trunk. (ii) The preganglionic **parasympathetic** fibres emerge with the second and third, or third and fourth, sacral nerves, and leave their ventral rami to form the **pelvic splanchnic nerves** (q.v.).

The Dorsal Roots

These consist of nerve fibres of all diameters, both myelinated and non-myelinated. It seems certain that the nerve fibres in the dorsal roots are all processes of cells in the spinal ganglia; there is no conclusive evidence for efferent fibres within these roots, despite the fact that stimulation of the distal end of a divided dorsal root produces vasodilatation in its territory of distribution (for discussion, see Mitchell, 1953).

In the **intervertebral foramen,** the spinal nerve and ganglion in their fibrous sheath are surrounded by veins which unite the **internal vertebral venous plexus** in the vertebral canal with the veins external to the vertebral column at every vertebral level. These structures in the intervertebral foramen lie posterior to the posterolateral part of the intervertebral disc and vertebral body, and anterior to the joint between the vertebral articular processes [FIG. 110]. They are liable to compression from a herniated nucleus pulposus [p. 215] or from the bony walls of the intervertebral foramen if the adjacent vertebral bodies are approximated, after rupture of an

FIG. 916 Diagram to show level of origin of spinal nerve roots from spinal medulla relative to point of emergence from the vertebral column.

The nerves are shown as thick black lines on the left side.

intervertebral disc. Pathological changes in the intervertebral joints may also involve the spinal nerves or their dorsal rami (see below) in this site. Neither the first nor the second cervical nerves are related to an intervertebral disc, but lie posterior to the superior and inferior parts of the lateral mass of the atlas, and may be involved in pathological changes in the joints which these form with the skull and axis. The sacral nerves pass between the fused sacral vertebra and are not liable to injury from the above causes.

Rami of the Spinal Nerves

As the spinal nerve emerges from the intervertebral foramen it gives off a **meningeal branch** and divides into dorsal and ventral rami, both of which contain motor and sensory fibres. The meningeal branch receives a branch from the sympathetic trunk, and re-enters the vertebral canal to supply the meninges and blood vessels therein.

The dorsal rami pass posteriorly, immediately lateral to the joint between the corresponding vertebral articular processes (except in the case of the first and second cervical nerves) and are distributed to the axial muscles, fascia, ligaments, and skin of the back. They play no part in the supply of limb muscles even though their cutaneous branches may pierce some of them, e.g. the rhomboids and trapezius [FIG. 373].

The ventral rami are larger than the dorsal rami. They supply the lateral and anterior parts of the body, the limbs, and the perineum, and they tend to unite with each other to form complicated plexuses particularly for the innervation of the limbs.

Distribution of Spinal Nerves

The distribution of the spinal nerves is, like their origin, essentially segmental, though the pattern is modified and almost obliterated in some regions. This is because (1) most of the muscles are formed by the fusion of a number of segmental myotomes (e.g., the long muscles of the back) and consequently are innervated by more than one spinal nerve; (2) the area of skin supplied by each spinal nerve (known as a **dermatome**) overlaps the areas of adjacent nerves. This overlap is sufficient to prevent total cutaneous sensory loss when a single spinal nerve is destroyed, though there is altered sensation in its area of distribution. When three adjacent spinal nerves are destroyed there is a total cutaneous sensory loss corresponding to the area supplied by the intermediate nerve (Head, 1893; Foerster, 1933; Keegan and Garrett,

1948). Some indication of the cutaneous distribution of nerves may be obtained from the vesicular eruption of the skin which occurs in the virus infection of shingles (herpes zoster) which affects part or all of the cutaneous distribution of one nerve and its sensory ganglion.

The outgrowth of the limbs, which can be looked upon as parts of the body-wall, carries certain dermatomes entirely into the limb so that they are missing from the expected sequence in the trunk. Yet the sequence remains unaltered if the limbs and trunk are looked on as a whole [FIGS. 917, 919, 930, 943, and 954].

In the description of nerves, the terms 'muscular' and 'cutaneous' are applied to the branches passing to muscles and skin. It is important to realize that **nerves to muscles** contain sensory nerve fibres (from muscle, tendon, joints, bone, and intermuscular fascial planes) and postganglionic sympathetic fibres (to blood vessels) as well as motor fibres to the muscle. Also that **cutaneous nerves** invariably supply subcutaneous tissue and sometimes joints, ligaments, and bones (e.g., the digital nerves) and that they always contain postganglionic sympathetic fibres to blood vessels and sweat glands. Thus an area of skin which is anaesthetic due to a peripheral nerve injury does not sweat.

DORSAL RAMI OF SPINAL NERVES

These branches supply the erector spinae muscles, and the skin fasciae, ligaments, bones, and joints of the posterior part of the trunk and neck. Their cutaneous area also extends on to the back of the scalp and shoulder, and the gluteal region [FIG. 917].

The majority of the dorsal rami divide into medial and lateral branches [FIG. 915]. Both of these supply the erector spinae; also the medial are cutaneous in the upper half of the trunk, while the lateral have this function in the lower half. In the upper half the cutaneous branches pass posteromedially through the muscle to enter the skin close to the spines of the vertebrae, while in the lower half they extend inferolaterally and enter the skin further from the median plane and at a considerable distance below their level of origin.

Cervical Dorsal Rami

First Cervical (Suboccipital) Dorsal Ramus. The dorsal root of this nerve may be small or absent. The dorsal ramus is larger than the ventral, does not divide into medial and lateral branches, and has no direct cutaneous branch.

The dorsal ramus arises from the nerve on the posterior arch of the atlas, inferior to the vertebral artery [FIG. 918]. It passes posteriorly into the suboccipital triangle deep to semispinalis capitis, gives muscular branches to the muscles of the triangle (obliquus capitis superior and inferior, rectus capitis posterior major and minor) and to semispinalis capitis, and communicates with the second cervical dorsal ramus over the posterior surface of obliquus capitis inferior. This communication carries sensory fibres to the muscles supplied by the first cervical dorsal ramus when the first cervical nerve has no dorsal root.

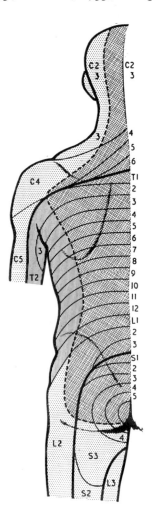

FIG. 917 Distribution of cutaneous nerves on the back of the trunk to illustrate the dermatomes.

Second Cervical Dorsal Ramus. This is the largest of the dorsal rami. It passes posteriorly between obliquus capitis inferior and the semispinalis cervicis muscles, gives communicating branches to the first and third cervical dorsal rami, and sends branches to semispinalis and splenius capitis, semispinalis cervicis, and multifidus. It pierces semispinalis capitis close to the midline, and running superolaterally between it and trapezius, pierces the tendinous origin of trapezius and the deep fascia approximately 2·5 cm. lateral to the external occipital protuberance. It then runs with the occipital artery (as the **greater occipital nerve**) to supply the posterior part of the scalp as far as the vertex. It communicates in the scalp with the great auricular, lesser occipital, posterior auricular, supra-orbital, and third occipital nerves.

Third Cervical Dorsal Ramus. This dorsal ramus is much smaller than the second cervical. It communicates with the second and fourth dorsal rami, and divides into medial and lateral branches. The lateral branch enters erector spinae; the medial branch passes posteromedially, pierces semispinalis capitis and trapezius close to the median plane as the **third occipital nerve.** It supplies the skin and fasciae of the upper part of the back of the neck and a small adjacent area of the scalp.

The dorsal rami of the **fourth, fifth, and sixth cervical nerves** are still smaller. Deep to semispinalis capitis, they divide into medial and lateral branches to erector spinae, and the medial branches reach the skin close to the median plane. The sixth is the smallest, and the fifth and sixth may have no cutaneous branches.

Seventh and Eighth Cervical Dorsal Rami. These are the smallest of the cervical dorsal rami. They may have no cutaneous branches, though the eighth commonly has.

Thoracic Dorsal Rami

All the thoracic dorsal rami divide into medial and lateral branches. In the upper six or seven the medial branch is mainly cutaneous and innervates the scapular region. The medial emergence of these branches prevents them from interfering with medial movements of the scapula. The **first** is small, the **second** is very large and reaches the acromion. The rest diminish in size from above downwards and become progressively more oblique. In the lower five or six the lateral branches give rise to the cutaneous nerves. These branches curve downwards around or through the lateral part of erector spinae and pierce latissimus dorsi, emerging further laterally than the upper dorsal rami. They supply the skin over the back of the lower chest and loin, the **eleventh and twelfth** passing over the iliac crest into the gluteal skin. Because of the descent of the lower dorsal rami, the area of skin supplied lies below the point of emergence of the nerve from the vertebral column. This descent is commensurate with that of the corresponding ventral rami in the intercostal spaces, and thus the lower border of the dermatomes remains nearly horizontal despite the oblique course of the nerves [FIG. 917].

Lumbar Dorsal Rami

The **upper three** lumbar dorsal rami divide into medial and lateral branches which innervate the erector spinae. The lateral branches are mainly cutaneous. They pass inferolaterally lateral to or through erector spinae, and piercing the lumbar fascia immediately superior to the iliac crest, a short distance lateral to the posterior superior iliac spine, supply a strip of skin in the gluteal region. This strip extends from the posterior median line above the iliac crest to a point distal and posterior to the greater trochanter of the femur [FIG. 917].

Almost all the dorsal rami communicate in erector spinae, and the third lumbar nerve which may not send a direct branch to the skin, can transmit fibres to it through the second.

The dorsal rami of the **fourth and fifth** lumbar nerves usually supply only muscular branches to the erector spinae in the same manner as the lower cervical dorsal rami.

Sacral Dorsal Rami

The sacral dorsal rami emerge through the dorsal sacral foramina. The first three supply medial branches to multifidus, and lateral branches which pierce the sacrotuberous ligament and gluteus maximus to supply skin and fascia over the dorsal surface of the sacrum and the medial part of the gluteal region.

The dorsal rami of the fourth and fifth unite and are joined by the minute dorsal ramus of the coccygeal nerve. The combined nerve pierces the sacrotuberous ligament and supplies skin and fascia in the neighbourhood of the coccyx.

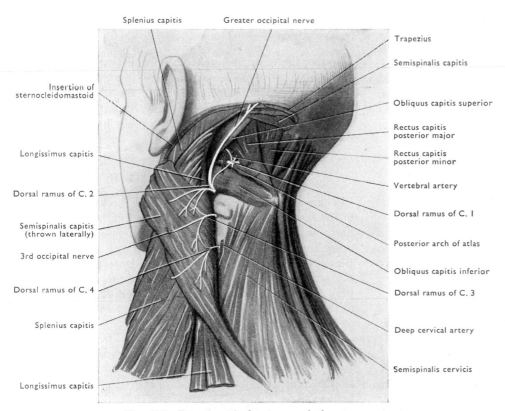

Splenius capitis   Greater occipital nerve

Trapezius

Semispinalis capitis

Insertion of sternocleidomastoid

Obliquus capitis superior

Rectus capitis posterior major

Rectus capitis posterior minor

Longissimus capitis

Vertebral artery

Dorsal ramus of C. 2

Dorsal ramus of C. I

Semispinalis capitis (thrown laterally)

Posterior arch of atlas

3rd occipital nerve

Obliquus capitis inferior

Dorsal ramus of C. 4

Dorsal ramus of C. 3

Splenius capitis

Deep cervical artery

Longissimus capitis

Semispinalis cervicis

FIG. 918 Dorsal rami of upper cervical nerves.

C.T.A.—2 B

General Arrangement of the Dorsal Rami

The **muscular distribution** is confined to erector spinae.

The **cutaneous distribution** for the greater part covers the area of the erectores spinae, but on the head and at the roots of the limbs the cutaneous branches extend beyond the regions supplied by their motor fibres as though drawn out by the growth of these regions. Thus the second and third cervical extend on to the scalp; the upper thoracic run laterally over the scapular region, and the upper lumbar and sacral dorsal rami extend their supply into the gluteal region.

At the level of each limb, where the ventral rami are particularly large, a variable number of the dorsal rami fail to innervate the skin (usually the fifth, sixth, and seventh cervical and the fourth and fifth lumbar); thus the regular sequence of dermatomes is interrupted dorsally at the level of the vertebra prominens and of the posterior superior iliac spines. At each of these interruptions a hypothetical **dorsal axial line** may be drawn [FIG. 917]. This line can be continued into the limb where a similar interruption exists between the areas supplied by the ventral rami [p. 771]. Injuries to the nerves whose dorsal rami do not reach the skin will produce no alteration of cutaneous sensation in the areas of the dorsal rami, hence their integrity cannot be assured by testing cutaneous sensation in these areas.

VENTRAL RAMI OF SPINAL NERVES

The ventral rami are distributed to the lateral and anterior aspects of the body, including the limbs. Each separates from the dorsal ramus on emerging from the intervertebral foramen, and consists of the same mixture of dorsal and ventral root fibres except that the sympathetic preganglionic fibres in the thoracic and upper three lumbar nerves and the preganglionic parasympathetic fibres in the third and fourth sacral nerves pass only into the ventral rami. These fibres leave the ventral rami almost at once, the sympathetic passing to the sympathetic trunk as **white rami communicantes,** while the parasympathetic pass to the autonomic plexuses in the pelvis as **pelvic splanchnic nerves.**

Every ventral ramus is joined near its origin by a delicate bundle of non-myelinated, postganglionic fibres (**grey ramus communicans**) from the sympathetic trunk. These fibres are distributed to blood vessels and sweat glands in the territory of both rami of the nerve they enter.

In the thoracic region the ventral rami are distributed in a regular segmental manner, but elsewhere they combine to form plexuses and consequently their arrangement is more complex.

A **typical thoracic ventral ramus,** e.g. the fifth intercostal nerve [FIG. 915], is first connected to the sympathetic trunk by a grey and a white ramus communicans, and sweeps forwards in the upper part of an intercostal space, frequently sending a collateral branch along the lower part of the same space. This branch either rejoins the main trunk or becomes cutaneous anteriorly. The ventral ramus supplies the muscles of the intercostal space and sends a lateral and an anterior cutaneous branch through the overlying muscles [FIG. 915].

FIG. 919   Distribution of cutaneous nerves on the front of the trunk to illustrate the dermatomes.

Together these supply a strip of skin which extends posteriorly from the anterior median line to become continuous with the area supplied by the dorsal ramus of the same spinal nerve [FIGS. 917, 919]. The entire strip of skin so innervated between the anterior and posterior median lines, constitutes a **dermatome.**

CERVICAL VENTRAL RAMI

The ventral rami of the cervical nerves, together with parts of the first and second thoracic nerves, are distributed to the head, neck, and upper limb. The first four cervical nerves innervate the neck and part of the head through the **cervical plexus;** the last four cervical and a large part of the first thoracic nerve supply the upper limb through the **brachial plexus.** The second thoracic ventral ramus may contribute to this plexus, and usually helps to innervate the arm through its lateral cutaneous branch.

The **first cervical ventral ramus** emerges from the vertebral canal inferomedial to the vertebral artery on the posterior arch of the atlas. It curves round the posterior and lateral surfaces of the lateral mass of that vertebra, and emerges between rectus capitis anterior and rectus capitis lateralis. Here it descends anterior to the transverse process of the atlas to join the ascending branch of the second cervical ventral ramus and send a large part of its fibres to join the hypoglossal nerve (see below).

The **second cervical ventral ramus** emerges posterior to the **superior articular process** of the axis. It passes forwards, lateral to the vertebral artery and divides into: (1) an ascending branch which unites with the first cervical ventral ramus and contributes some fibres to the hypoglossal nerve; (2) a descending part which unites with the third cervical ventral ramus.

The remaining cervical ventral rami emerge through the intervertebral foramina, and pass between the anterior and posterior intertransverse muscles.

THE CERVICAL PLEXUS

This plexus is formed by an irregular series of loop communications between the ventral rami of the first four cervical nerves, each of which is joined by one or more **grey rami communicantes** from the superior cervical sympathetic ganglion. The loop communications lie close to the vertebral column between the prevertebral muscles and those arising from the posterior tubercles of the transverse processes, and they give rise to the branches of the plexus posteromedial to the internal jugular vein under cover of the sternocleidomastoid.

CUTANEOUS BRANCHES

(C. 2, 3)	(C. 3, 4)
Ascending to head	Descending to side of neck, shoulder and front of chest
Lesser occipital	Medial, intermediate, and lateral supraclavicular
Great auricular	
To neck	
Transverse nerve of neck	

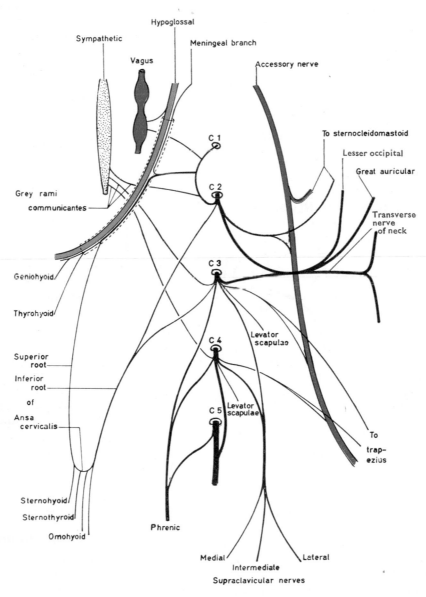

FIG. 920 Diagram of left cervical plexus.

MUSCULAR BRANCHES

LATERAL	MEDIAL
Sternocleidomastoid (C. 2) sensory	Prevertebral muscles (C. 1, 2, 3, 4)
Trapezius (C. 3, 4) sensory	Geniohyoid (C. 1)
Levator scapulae (C. 3, 4)	Infrahyoid muscles (C. 1, 2, 3; through ansa cervicalis)
Scalenus medius and posterior (C. 3, 4)	Diaphragm (C. 3, 4, 5; phrenic nerve)

COMMUNICATING BRANCHES

LATERAL	MEDIAL
With accessory nerve (C. 2, 3, 4)	Vagus and hypoglossal nerves (C. 1, or C. 1, 2)
	Sympathetic (grey rami to C. 1, 2, 3, 4).

Cutaneous Branches

These nerves appear in the posterior triangle a little above the midpoint of the posterior border of sternocleidomastoid [FIG. 922].

The **lesser occipital nerve** (C. 2, 3) is variable in size and is sometimes double. It passes posteriorly deep to sterno-cleidomastoid, and hooking below the accessory nerve at the posterior border of the muscle, ascends along that border to pierce the deep fascia at the apex of the posterior triangle.

Here it divides into branches which supply the skin and fascia: (1) on the upper lateral part of the neck; (2) on the cranial surface of the auricle and mastoid process; (3) on the adjacent part of the scalp. When double, it is usually only the smaller branch which is in contact with the accessory nerve.

The **great auricular nerve** (C. 2, 3, or rarely C. 3 alone) is usually the largest cutaneous branch of the plexus, but its size and territory varies reciprocally with the lesser occipital. It winds round the posterior border of sternocleidomastoid [FIG. 922] and passes anterosuperiorly across the superficial surface of that muscle, deep to platysma, towards the inferior part of the auricle. On the muscle it divides into a number of branches. (1) **Posterior branches** ascend over the mastoid process, communicate with the lesser occipital and posterior auricular nerves, and supply skin and fascia in this region. (2) **Intermediate branches** pass to supply the lower part of the auricle on both surfaces. (3) **Anterior branches** pass through the substance of the parotid gland and over the angle of the mandible, to supply skin and fascia over the postero-inferior part of the face [FIG. 921]. It communicates with the facial nerve in the parotid gland and may send branches almost to the zygomatic bone.

The **transverse nerve of the neck** (C. 2, 3) winds horizontally round the posterior and lateral aspects of sternocleidomastoid, deep to platysma and the external jugular vein. It divides into **superior** and **inferior branches** near the anterior margin of sternocleidomastoid and supplies skin and fascia over the anterior triangle of the neck. The superior branches meet the area supplied by the trigeminal nerve at the inferior border of the body of the mandible, and communicate with the cervical branch of the facial nerve.

The **supraclavicular nerves** (C. 3, 4) appear at the posterior border of sterno-cleidomastoid as a large trunk which descends through the lower part of the posterior triangle and divides into **medial, intermediate,** and **lateral supraclavicular nerves.** These nerves pierce the deep fascia above the clavicle, supply the skin and fascia of the inferior part of the side of the neck, and descend superficial to the corresponding thirds of the clavicle, deep to platysma, to supply skin and fascia to the level of the sternal angle. The medial nerve also supplies the **sternoclavicular joint,** and the lateral nerve the **acromioclavicular joint.** Branches of the intermediate and lateral nerves may groove or pierce the clavicle, and some of the branches of the lateral nerve usually pass deep to trapezius and pierce it to reach the skin.

Muscular Branches

These arise with the other branches of the cervical plexus deep to sternocleidomastoid, and pass either laterally into the posterior triangle or medially into the anterior triangle.

Labels (top to bottom, left side): Auriculotemporal, Supra-orbital, Zygomaticotemporal, Greater occipital, Supratrochlear, Lacrimal, Infratrochlear, Zygomaticofacial, Nasociliary, Infra-orbital, Great auricular, Buccal, Mental, Transverse nerve of neck, Supraclavicular nerves

In figure: TRIGEMINAL NERVE, SPINAL NERVES, C2.3, C2, DORSAL RAMI, C3, C4, C5, C6, VENTRAL RAMI, C2.3, C3.4.

FIG. 921 Distribution of cutaneous nerves to head and neck. The ophthalmic, maxillary, and mandibular divisions of the trigeminal nerve are indicated by different shading.

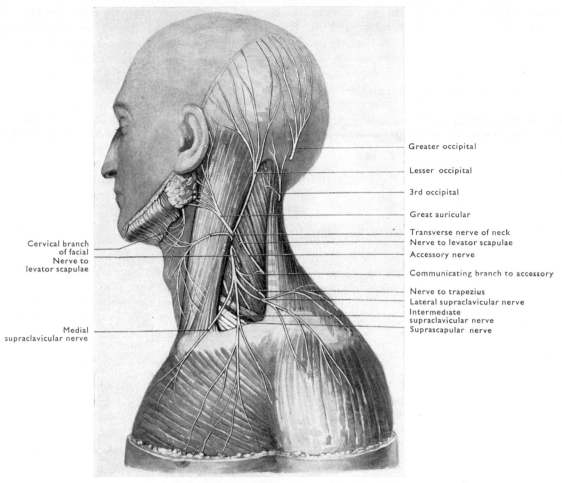

Greater occipital

Lesser occipital

3rd occipital

Great auricular

Transverse nerve of neck
Nerve to levator scapulae
Accessory nerve

Communicating branch to accessory

Nerve to trapezius
Lateral supraclavicular nerve
Intermediate
supraclavicular nerve
Suprascapular nerve

Cervical branch
of facial
Nerve to
levator scapulae

Medial
supraclavicular nerve

FIG. 922   Nerves of side of neck.

Lateral Branches. 1. From the second cervical ventral ramus a sensory branch enters the deep surface of **sternocleidomastoid** having communicated with the accessory nerve deep to that muscle.

2. From the third and fourth cervical ventral rami. (i) Branches cross the posterior triangle to enter the deep surface of **trapezius.** They supply sensory fibres to that muscle and communicate with the accessory nerve in the posterior triangle and deep to the trapezius. (ii) Separate branches supply the **levator scapulae** muscle by entering its lateral surface in the posterior triangle. (iii) Branches supply **scalenus medius** and **posterior.**

Medial Branches. 1. From the loop uniting the **first and second cervical ventral rami** anterior to the transverse process of the atlas. (i) A branch joins the **hypoglossal nerve** as it emerges from the skull [FIGS. 914, 920]. A few sensory fibres of this branch pass superiorly in the hypoglossal nerve to supply the skull and dura mater of the posterior cranial fossa (meningeal branch). Most of the fibres descend in the hypoglossal nerve and form three branches of that nerve which probably do not contain any nerve fibres of the hypoglossal nerve proper. (a) The nerve to **thyrohyoid** [p. 733]. (b) The nerve to **geniohyoid** [p. 733]. (c) The **superior root of the ansa cervicalis** arises from the hypoglossal nerve anterior to the internal carotid artery. It descends in front of

the internal and common carotid arteries, and joins the inferior root of ansa cervicalis (see below) to form this loop superficial to the carotid sheath. The **ansa** sends branches to the **sternohyoid, sternothyroid,** and **omohyoid muscles.** The superior root of the ansa may occasionally arise from the vagus. In this case the fibres from the first two cervical ventral rami pass through the communication to the vagus and not to the hypoglossal.

(ii) A small branch supplies the rectus capitis lateralis and anterior, and the longus capitis muscles.

2. From the **second and third cervical ventral rami.** Each of these nerves sends a slender branch downwards along the internal jugular vein to unite on its anterior surface **(inferior root of the ansa cervicalis).** This forms the ansa cervicalis with the superior root (see above) and supplies the infrahyoid muscles except thyrohyoid. Thus medial branches of the first three cervical ventral rami communicate to supply a paramedian strip of muscle from the chin to the sternum, and they also communicate with the hypoglossal nerve which supplies the tongue muscles immediately superior to this strip.

3. From the **second, third,** and **fourth cervical ventral rami,** small branches supply the intertransverse, longus colli, and longus capitis muscles.

4. From the **third, fourth,** and **fifth cervical ventral rami. The phrenic nerve** [FIGS. 920, 923] arises mainly

from the fourth, but receives branches from the third (either directly or through the nerve to sternohyoid) and fifth (either directly or from the nerve to subclavius as the **accessory phrenic nerve**). The phrenic nerve descends on scalenus anterior posterolateral to the internal jugular vein, and passing anterior to the cervical pleura between the subclavian artery and vein (anterior to the first part of the subclavian artery on the left and the second part on the right) deviates medially in front of the internal thoracic artery (occasionally posterior to it on the left).

The nerve descends through the thorax to the diaphragm, separated from the pleural cavity only by the mediastinal pleura. In the *superior mediastinum*, the left nerve lies between the common carotid and subclavian arteries and crosses the aortic arch anterior to the vagus nerve; the right nerve lies on the right brachiocephalic vein and the superior vena cava, and is not in contact with the vagus nerve. Both nerves pass down the *middle mediastinum* between the pleura and pericardium, anterior to the lung root, the right reaching the diaphragm with the inferior vena cava. Most of the phrenic nerve fibres pierce the diaphragm and supply it from the inferior surface, but some pass over its pleural surface.

Branches of the Phrenic Nerve. 1. Muscular to the diaphragm. 2. Sensory to mediastinal and diaphragmatic pleura and to pericardium **(pericardial branch)**. 3. Sensory to diaphragmatic peritoneum and probably to liver, gallbladder, and inferior vena cava by **phrenico-abdominal branches.**

The **accessory phrenic nerve** arises from the fifth, or fifth and sixth cervical ventral rami, and passes inferomedially, deep to sternocleidomastoid to join the main nerve in the lower part of the neck or in the thorax. It commonly arises from the nerve to the subclavius muscle, but may be absent. In addition to this nerve, the phrenic may receive nerve fibres from the ansa cervicalis and from the cervical sympathetic trunk. In the abdomen it communicates with the coeliac plexus.

Communicating Branches of the Cervical Plexus. In addition to the fibres which the plexus sends to the **accessory** and **hypoglossal** nerves, the loop between the first and second cervical ventral rami sends a branch to the **vagus,** and the **grey rami communicantes** pass from the superior cervical ganglion of the sympathetic trunk to the first four cervical ventral rami.

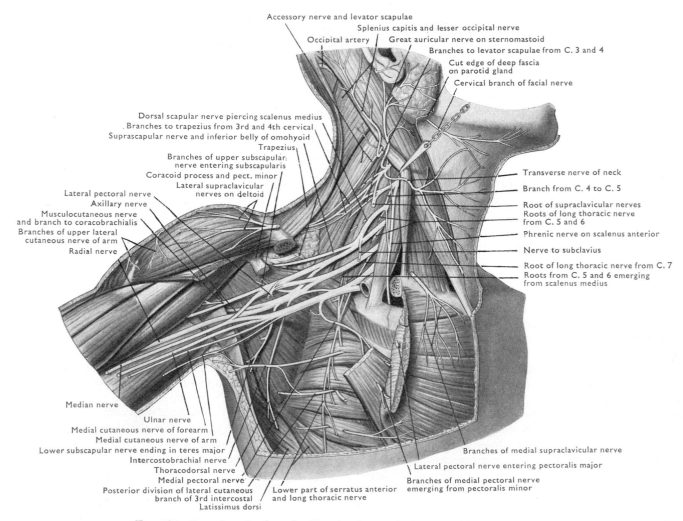

Fig. 923 Posterior triangle and axilla, showing cervical and brachial plexuses and origin of their principal branches.

Summary of Distribution of Cervical Plexus [Fig. 920].

The cutaneous distribution, in which the first cervical nerve plays no part, extends from the side of the head (lesser occipital nerve, C. 2, 3), the lower part and posterior surface of the auricle and the skin over the parotid gland and angle of the jaw (great auricular nerve, C. 2, 3), and the anterior triangle of the neck (transverse nerve of the neck, C. 2, 3), to the front of the chest and to the shoulder down to the level of the sternal angle (supraclavicular nerves, C. 3, 4).

The muscular distribution is to the prevertebral and intertransverse muscles, to the other muscles attached to the transverse processes of the vertebra as far posteriorly as the levator scapulae, behind which the muscles are supplied by the dorsal rami. The cervical plexus also supplies the anterior strip of muscle from the chin to the sternum consisting of geniohyoid, sternohyoid, omohyoid, thyrohyoid, and sternothyroid, and the diaphragm with the assistance of the fifth cervical nerve. It also sends sensory fibres to sternocleidomastoid and trapezius.

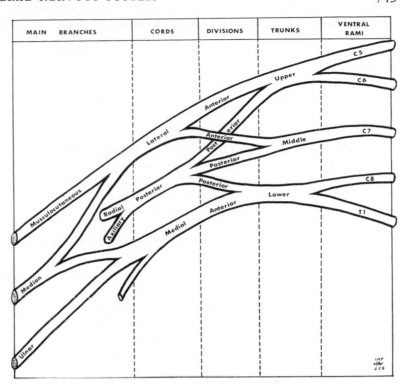

FIG. 924 Scheme to show stages of formation of brachial plexus.

THE BRACHIAL PLEXUS

[Figs. 923–925]

The brachial plexus extends from the neck to the axilla. It is formed by the ventral rami of the fifth, sixth, seventh, and eighth cervical nerves with the greater (ascending) part of that of the first thoracic nerve. Frequently a slender branch from the fourth cervical ventral ramus passes to the plexus, and the second thoracic ventral ramus not uncommonly sends a communication to the first thoracic ventral ramus. The lateral cutaneous branch of the second thoracic (**intercostobrachial nerve**) almost invariably innervates some skin on the medial side of the arm and the floor of the axilla.

Position of Plexus

The ventral rami which form the plexus enter the lower part of the posterior triangle of the neck in series with the ventral rami of the cervical plexus [Fig. 923], with the second part of the subclavian artery lying immediately anterior to its lowest rami. The rami, intermingling to form the plexus, and the artery pass inferolaterally towards the axilla enclosed in an extension of the prevertebral fascia. In the neck, the plexus is deep to platysma, the supraclavicular nerves, the inferior belly of omohyoid, and the transverse cervical artery. It then passes deep to the clavicle and the suprascapular vessels, to enter the axilla, and surround the second part of the axillary artery.

Constitution of Brachial Plexus

The plexus is subject to some variation, but four stages in the redistribution of

its contained nerve fibres can always be identified.

1. The ventral rami, the roots of the plexus, lie between the scalenus anterior and medius.

2. As they enter the posterior triangle, the upper two (C. 5, 6) and the lower two (C. 8, T. 1) roots of the plexus unite to form the **upper** and **lower trunks** respectively, while the middle (C. 7) continues as the **middle trunk.** The lower trunk may groove the superior surface of the first rib posterior to the subclavian artery, and the root from the first thoracic ventral ramus is always in contact with it.

3. Each trunk divides [Fig. 924] into ventral and dorsal

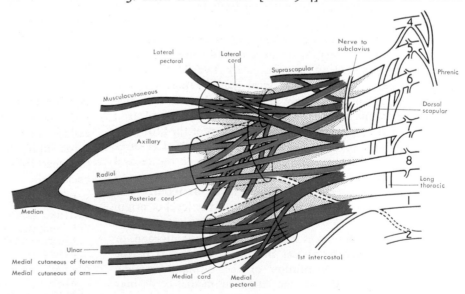

FIG.925 Diagram of brachial plexus, showing the origins of its branches. Ventral divisions, red; dorsal divisions, blue. Broken lines indicate variable connexions.

divisions which are destined to supply the anterior (flexor) and posterior (extensor) parts of the limb respectively. Occasionally the ventral rami divide before or during the formation of the trunks, it can then be seen that the dorsal division of the eighth is small and that of the first thoracic may be minute or absent, while the others divide equally.

4. The **cords (fasciculi)** of the plexus are formed in the axilla. The dorsal divisions unite to form the **posterior cord** (C. 5–9 ± T. 1). The ventral divisions of the upper and middle trunks unite to form the **lateral cord** (C. 5–7), while the ventral division of the lower trunk continues as the **medial cord** (C. 8 and T. 1). The various cords bear the relation suggested by their names to the second part of the axillary artery; each ends by dividing into two main branches at the beginning of the third part of the artery

Nerves arising from a cord do not necessarily contain fibres from all the ventral rami which send fibres into that cord; e.g., the musculocutaneous and axillary nerves, though arising from the lateral and posterior cords respectively, contain only fibres of the fifth and sixth cervical nerves. Each part of the brachial plexus, including its branches, shows an internal plexiform arrangement which prevents the dissection of individual fibres back through the plexus to determine their segmental origin.

Communications from Sympathetic. The fifth and sixth cervical ventral rami receive **grey rami communicantes** from the **middle cervical ganglion,** while two or more grey rami pass from the **inferior cervical** or **cervicothoracic ganglion** to the seventh and eighth cervical ventral rami. These grey rami either pierce the prevertebral muscles or pass posterior to the medial border of the scalenus anterior muscle to reach the ventral rami. The first thoracic ventral ramus receives its grey ramus from the cervicothoracic ganglion. These sympathetic fibres to the brachial plexus are supplemented by branches from the sympathetic plexus on the vertebral artery to the fourth to sixth cervical ventral rami, while other sympathetic fibres enter the limb on the subclavian artery.

Variations. The plexus as a whole may arise from a slightly higher or lower level relative to the vertebral column than the usual description: (1) in approximately two-thirds, of all plexuses, a part of the fourth cervical ventral ramus joins the plexus; (2) in approximately one-third, there is no contribution from the fourth but all the fifth joins the plexus; (3) in less than 10 per cent. of cases only part of the fifth joins the plexus (Kerr, 1918); (4) in more than one-third of cases there is an intrathoracic communication between the second and first thoracic ventral rami. In type (1) the plexus is said to be **prefixed**; in types (3) and (4) it is **postfixed.** The presence of a cervical rib is independent of the level of origin of the plexus.

The pattern of the plexus may vary: (1) trunk division, or cord formation may be absent in one or other part of the plexus, though the constitution of the ultimate terminal branches is unchanged; (2) the lateral or medial cords may receive fibres from ventral rami below or above the normal levels respectively (see also, Herringham, 1886; Harris, 1904, 1939; Jones, 1910).

BRANCHES OF THE BRACHIAL PLEXUS
Arising from Supraclavicular Part

These are nerves that arise from the plexus superior to the clavicle, either from the anterior or the posterior surface of the plexus.

From the Anterior Surface. 1. Muscular twigs to the scalenus anterior and longus colli muscles arise from the lower four cervical ventral rami as they emerge from the intervertebral foramina.

2. A branch to the **phrenic nerve** arises from the fifth cervical at the lateral border of scalenus anterior [see also accessory phrenic nerve, p. 742].

3. The **nerve to subclavius** arises from the upper trunk (C. (4), 5, 6). It descends in the posterior triangle of the neck anterior to the third part of the subclavian artery, and often communicates with the phrenic nerve [p. 742].

From the Posterior Surface. 1. Small branches to the scalenus medius and posterior arise from the lower four cervical ventral rami as they emerge from the intervertebral foramina.

2. The **dorsal scapular nerve** arises from the fifth cervical ventral ramus. It pierces the scalenus medius, and descending in the posterior triangle of the neck, passes deep to the levator scapulae and to the rhomboid muscles close to the medial border of the scapula. It supplies the rhomboid muscles and levator scapulae, and occasionally pierces the latter muscle.

3. The **long thoracic nerve** arises from the fifth, sixth, and seventh cervical ventral rami close to the intervertebral foramina. The upper two roots unite after piercing scalenus medius and are joined by the lowest root crossing the anterior surface of that muscle. The nerve descends on the muscle, posterior to the trunks of the brachial plexus, and passing between the first rib and the axillary artery on the first digitation of serratus anterior, runs on the axillary surface of that muscle and supplies it.

4. The **suprascapular nerve** arises from the upper trunk of the plexus (C. (4) 5, 6). It lies superior to the trunks of the brachial plexus, and passes inferolaterally parallel to them to pass through the scapular notch on the superior border of the scapula deep to trapezius. It supplies and passes deep to supraspinatus, sends a slender filament to the shoulder joint, and ends in infraspinatus by traversing the spinoglenoid notch with the suprascapular artery.

Arising from Infraclavicular Part

These branches arise from the cords of the plexus in the axilla. The anterior branches from the lateral and medial cords supply the front of the chest and limb; the posterior branches from the posterior cord supply the shoulder and back of the limb.

BRANCHES OF CORDS OF BRACHIAL PLEXUS
LATERAL CORD

Lateral pectoral (C. 5, 6, 7)
Musculocutaneous (C. 5, 6, 7)   } Main branches
Lateral root of median (C. 5, 6, 7)

MEDIAL CORD

Medial pectoral (C. 8, T. 1)
Medial cutaneous of arm (T. 1)
Medial cutaneous of forearm (C. 8, T. 1)
Ulnar (C. 8, T. 1)
Medial root of median (C. 8, T. 1)   } Main branches

BRANCHES OF THE MEDIAL AND LATERAL CORDS OF THE PLEXUS

PECTORAL NERVES

The **lateral pectoral nerve** arises from the ventral division of the upper and middle trunks (C. 5, 6, 7) just before they unite to form the lateral cord. It passes anteromedially, and giving a branch to the medial pectoral nerve, pierces the clavipectoral fascia and supplies pectoralis major.

The **medial pectoral nerve** arises from the medial cord (C. 8, T. 1), passes anteriorly between the axillary artery and vein, and receiving the communication from the lateral pectoral nerve anterior to the artery, supplies and pierces pectoralis minor to end in pectoralis major. Part of the nerve may pass below the inferior border of pectoralis minor.

MUSCULOCUTANEOUS NERVE

This nerve arises from the lateral cord of the plexus (C. 5, 6, and C. 4 in more than 50 per cent. of cases) and contains the nerve to the coracobrachialis muscle (C. 7, or 6 and 7) [FIG. 925]. The nerve descends between the axillary artery and coracobrachialis, sends a branch to that muscle and then pierces it to continue distally between the biceps and brachialis muscles, often communicating with the median nerve. At the level of the elbow, it pierces the deep fascia between biceps and brachioradialis as the lateral cutaneous nerve of the forearm.

Muscular branches pass to coracobrachialis and to biceps and brachialis.

The Lateral Cutaneous Nerve of the Forearm. The **anterior branch** descends along the front of the lateral aspect of the forearm. It supplies the skin and fascia of the lateral half of this surface, and extends to the ball of the thumb. It supplies branches to the radial artery and communicates with the superficial branch of the radial nerve proximal to the wrist. The **posterior branch** passes backwards to supply a variable

FIG. 926 Diagram of the origin and distribution of the nerves to the pectoral muscles.

L.P., lateral pectoral nerve; M.P., medial pectoral nerve; C.5, 6, 7, C.8, T.1, nerves of the brachial plexus; Art., axillary artery; Cl., clavicle; Scl., subclavius muscle; P.Mi., pectoralis minor, joined to subclavius by clavipectoral fascia; P.Ma., pectoralis major.

FIG. 927   Deeper nerves of upper arm.

The nerve to the brachialis usually enters the medial side of the muscle as in Fig. 928.

area of skin and fascia over the extensor muscles of the forearm, wrist, and sometimes the first metacarpal bone. It communicates with the lower lateral cutaneous nerve of the arm and the superficial branch of the radial nerve.

The musculocutaneous nerve also sends twigs to the marrow and to the periosteum of the distal anterior surface of the humerus, and to the brachial artery.

MEDIAN NERVE

This nerve is formed by the union of a lateral root from the lateral cord (C. 5, 6, 7) and a medial root from the medial cord (C. 8 and T. 1) which passes anterior to the third part of the axillary artery. After its formation, the median nerve descends in the upper part of the arm, lateral to the brachial artery, crosses to its medial side (usually anterior to the artery), and is deep to the bicipital aponeurosis and the median cubital vein at the elbow [FIGS. 928, 931]. It enters the forearm between the two heads of the pronator teres muscle, and descends closely bound to the deep surface of flexor digitorum superficialis by its fascial sheath. Near the wrist it becomes superficial by passing anteriorly between the tendons of flexor digitorum superficialis and flexor carpi radialis, deep to palmaris longus tendon. It enters the palm by passing deep to the flexor retinaculum anterior to the long flexor tendons. At the distal margin of the flexor retinaculum it separates into its six terminal branches deep to the palmar aponeurosis [FIG. 932]. In the forearm, the median nerve is accompanied by the small median artery.

Branches

There are no branches in the arm, though the nerve commonly communicates with the musculocutaneous nerve.

In the Forearm. 1. **Muscular branches** pass to pronator teres, flexor carpi radialis, palmaris longus, and flexor digitorum superficialis in the cubital fossa. Branches are usually present here to the upper fibres of both flexor pollicis longus and flexor digitorum profundus. The nerve to pronator teres is the first to be given off and often arises independently. The index belly of flexor digitorum superficialis often receives a separate branch in the middle of the forearm.

2. The **anterior interosseous nerve** arises from the posterior surface of the median nerve in the cubital fossa. It descends on the anterior surface of the interosseous membrane with the anterior interosseous artery, supplying the flexor pollicis longus and the lateral part of the flexor digitorum profundus. It then passes posterior to the pronator quadratus, and supplies that muscle and the radiocarpal and intercarpal joints. The nerve also sends medullary and periosteal twigs to the radius and ulna and to the interosseous membrane.

3. **Articular branches** to the anterior surface of the elbow joint arise from the muscular branches in the cubital fossa.

4. The **palmar cutaneous branch** arises in the distal one-third of the forearm, pierces the deep fascia, and enters the palm of the hand superficial to the flexor retinaculum. It

FIG. 928 Deeper nerves of front of upper arm and forearm.

supplies a small area of the skin and fascia of the palm and adjacent thenar eminence, and communicates with the corresponding branch of the ulnar nerve. It may be absent.

In the Hand. 1. The **main muscular branch** arises near the distal border of the flexor retinaculum and curves laterally and proximally into the thenar muscles. It supplies branches to the abductor pollicis brevis, opponens pollicis, and flexor pollicis brevis.

2. The **palmar digital branches** arise in a variable manner near the distal border of the flexor retinaculum. Frequently there is one **proper palmar digital nerve** to the radial side of the index finger (which also supplies the first lumbrical muscle), and three **common palmar digital nerves** each of which divides into two proper palmar digital nerves: (i) to both sides of the thumb; (ii) to the adjacent sides of the index and middle fingers (which supplies the second lumbrical muscle); (iii) to the adjacent sides of the middle and ring fingers. The latter common palmar digital nerve receives a communication from the common palmar digital branch of the ulnar nerve [FIG. 932].

The digital nerves lie deep to the palmar aponeurosis and the superficial palmar arch, but superficial to the flexor tendons in the palm. The nerves to the thumb and the radial side of the index finger emerge from under the lateral edge of the palmar aponeurosis, while the other two pass with the corresponding lumbrical muscles, between the slips of the aponeurosis to the fingers, and send branches to the dorsal digital nerves to supply the skin on the dorsal surfaces of the distal phalanx of the thumb and the distal two phalanges of the lateral two and a half fingers. They also send **articular branches** to the interphalangeal and metacarpophalangeal joints, and vascular branches to the adjacent arteries. The nerves are beset with numerous lamellated corpuscles [FIG. 974] some of which are very large.

Communications. The median nerve frequently communicates with the musculocutaneous nerve in the arm, and sometimes with the ulnar nerve posterior to the flexor muscles of the forearm. The latter communication may transfer to the ulnar nerve fibres destined for the thenar muscles (Hovelacque, 1927). It also communicates with the ulnar nerve in the palm of the hand.

Destruction of the median nerve above the elbow produces a cutaneous sensory loss on the palm of the hand, the palmar surfaces of the thumb, index, and middle fingers, and on the dorsal surfaces of the distal phalanx of the thumb, and the distal two phalanges of the index and middle fingers. There may also be some sensory loss on the radial surface of the ring finger, though the area of absolute sensory loss is always smaller than the anatomical distribution of the nerve because of the overlap from adjacent nerves (ulnar and radial). Pronation is severely affected (pronator teres, pronator quadratus, and flexor carpi radialis). Flexion of the wrist is associated with ulnar deviation (action of flexor carpi ulnaris). Flexion (flexor digitorum superficialis and lateral half of flexor digitorum profundus) is absent in the index and middle fingers and is weakened in the ring and little fingers. Flexion of the metacarpophalangeal joints with extension of the interphalangeal joints of index and middle fingers is weak (lumbricals) though some action may be present from the interossei. All movements of the thumb except adduction, extension, and abduction by abductor pollicis longus are lost; thus opposition and flexion are impossible.

Destruction of the median nerve at the wrist has similar sensory effects though the action of the flexors and pronators in the forearm is not affected. Opposition, true abduction, and flexion of the thumb at the metacarpophalangeal joint are severely affected (thenar muscles) as is lumbrical action on the index and middle fingers.

MEDIAL CUTANEOUS NERVE OF THE ARM

This nerve arises from the medial cord of the brachial plexus, principally from the first thoracic nerve [FIG. 925]. It descends between the axillary artery and vein, passes anterior or posterior, or even through the vein, to pierce the deep fascia and supply the skin and fascia on the medial side of the proximal half of the arm. The nerve usually communicates and varies inversely in size with the intercostobrachial nerve which may entirely replace it, together with the posterior cutaneous branch of the radial nerve to the arm.

MEDIAL CUTANEOUS NERVE OF THE FOREARM

This nerve arises from the medial cord of the brachial plexus (C. 8 and T. 1) and passes through the axilla and proximal half of the arm medial to the main artery. At the middle of the arm it pierces the deep fascia with the basilic vein, and accompanies it to the anterior surface of the elbow. Here it divides into anterior and ulnar branches.

Branches

As it pierces the deep fascia the nerve gives a small branch to supply the skin on the anteromedial surface of the distal half of the arm. The terminal branches pass superficial or deep to the median cubital vein to supply the medial surface of the forearm [FIG. 929]. The **anterior branch** supplies the medial part of the front of the forearm as far as the wrist. The smaller **ulnar** (posterior) **branch** passes postero-inferiorly over the common flexor origin to supply skin and fascia on the proximal three-quarters of the medial part of the back of the forearm. Both branches may communicate with the ulnar nerve in the distal part of the forearm.

ULNAR NERVE

The ulnar nerve arises from the medial cord of the brachial plexus (C. 8 and T. 1), and receives a communication (C. 7) either from the lateral cord of the plexus or from the lateral root of the median nerve in more than half the cases. The nerve descends between the axillary artery and vein, posterior to the medial cutaneous nerve of the forearm, and then lies anterior to triceps muscle on the medial side of the brachial artery. In the distal half of the arm it passes backwards with an ulnar collateral artery, through the medial intermuscular septum, and continues between this structure and the medial head of triceps muscle to enter the forearm between the medial epicondyle of the humerus and the olecranon, deep to the fibrous arch between the origins of flexor carpi ulnaris from these two bones. It descends, deep to the flexor carpi ulnaris, first on the ulnar collateral ligament and then on the flexor digitorum profundus, and joins the ulnar artery above the middle of the forearm. In the distal half of the forearm it becomes comparatively superficial, sends branches to the ulnar artery, and pierces the deep fascia with it just proximal to the flexor retinaculum, lateral to flexor carpi ulnaris. It passes

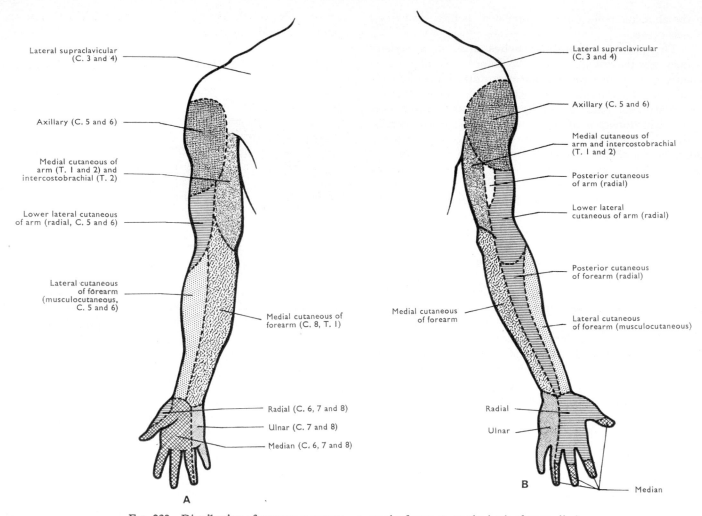

FIG. 929 Distribution of cutaneous nerves. A, on the front; B, on the back of upper limb.

on the anterior surface of the flexor retinaculum immediately lateral to the pisiform bone, and divides into superficial and deep branches under cover of the palmaris brevis muscle.

Branches

These arise in the forearm and hand.

An **articular branch** passes to the elbow joint as the nerve lies posterior to the medial epicondyle of the humerus.

Muscular branches in the forearm pass to flexor carpi ulnaris and the medial half of flexor digitorum profundus immediately distal to the elbow. The former muscle receives two branches, one each to its olecranon and epicondylar heads, and one or two additional branches at a lower level.

Cutaneous Branches in the Forearm. I. The **palmar cutaneous branch** is variable in size and position. It pierces the deep fascia in the distal third of the forearm and supplies the skin of the medial part of the palm. It supplies the ulnar artery, and often communicates with the anterior branch of the medial cutaneous nerve of the forearm and the palmar cutaneous branch of the median nerve.

2. The **dorsal branch** arises in the distal half of the forearm. It passes postero-inferiorly between the ulna and flexor

carpi ulnaris to become cutaneous on the medial side of the distal quarter of the forearm. At the medial side of the wrist it crosses the triquetrum, against which it can be palpated, and gives branches to the skin and fascia of the dorsal surfaces of the wrist and hand. These branches may communicate with branches of the posterior cutaneous nerve of the forearm and with the corresponding branches from the superficial branch of the radial nerve. The areas supplied by these three nerves on the dorsum of the wrist and hand is very variable, though commonly the radial and ulnar nerves are the sole supply to the dorsum of the hand and share it almost equally. The dorsal branch of the ulnar nerve passes on to the base of the fifth metacarpal and divides into three **dorsal digital nerves.** One passes along the ulnar side of the dorsum of the hand and little finger as far as the nail bed; another divides at the cleft between the little and ring fingers and supplies their adjacent sides to the same extent; a third dorsal digital nerve supplies the adjacent sides of the middle and ring fingers as far as the proximal interphalangeal joints, the more distal parts being supplied by the palmar digital branches of the median nerve. This last branch communicates with the most medial dorsal digital branch of the radial nerve which may supply part of the cleft between the ring and middle fingers.

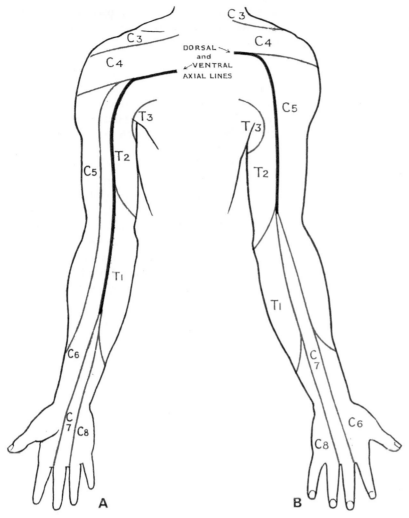

FIG. 930 Dermatomes of upper limb, showing segmental cutaneous distribution of spinal nerves (C-T) on A, the front, and B, the back of the limb. (After Head, 1893, and Foerster, 1933.)

See page 736 for definition of 'dermatome' and page 771 for more particular reference to the segmental innervation of the upper limb.

Branches in the Wrist and Hand. 1. The nerve sends a small branch to palmaris brevis and then divides into superficial and deep branches.

2. The **superficial branch** lies deep to the palmaris brevis. It divides into one **proper** (medial) and one **common palmar digital nerve.** The former passes distally on the flexor digiti minimi brevis and the medial aspect of the little finger, supplying its palmar surface. The common palmar digital nerve passes between the palmar aponeurosis and the long flexor tendons to the little finger. It sends a communicating branch to the medial common palmar digital nerve from the median, and reaches the loose tissue in the web between the ring and little fingers. Here, anterior to the fourth lumbrical muscle, it divides into two **proper palmar digital nerves** which pass to the adjacent sides, palmar surfaces, nail beds and the metacarpophalangeal and interphalangeal joints of these fingers [FIG. 932].

The ulnar nerve thus supplies the skin and fascia of the medial one-third of the palm and the palmar surfaces of one and a half fingers. On the dorsal surface it supplies skin of

half of the hand and wrist, of the same one and a half fingers together with the cleft between the ring and middle fingers, and thus most of the skin over the proximal phalanx of the ring finger. This common arrangement (Stopford, 1918) is subject to considerable variation; rarely the radial nerve may supply almost all the skin on the dorsum usually innervated by the ulnar nerve, or the ulnar may supply the greater part of the dorsum of the hand and the dorsal surfaces of the fingers not supplied by the median nerve. In all cases it is probable that these variations represent an abnormal distribution of fibres through the peripheral nerves rather than an abnormal innervation.

3. The **deep branch** arises from the nerve on the flexor retinaculum lateral to the pisiform bone [FIG. 932]. It passes posteriorly between the abductor and short flexor of the little finger, supplying them, and piercing the opponens digiti minimi (which it also supplies) near its origin from the flexor retinaculum, turns laterally over the distal surface of the hook of the hamate bone. It then runs laterally between the proximal parts of the metacarpal bones and the long flexor tendons, to pass between the two heads of adductor pollicis with the deep palmar arch, and end in the first dorsal interosseous muscle. In the palm the deep branch supplies the interossei, the third and fourth lumbrical muscles, the adductor pollicis, and the deep vessels of the palm. The first dorsal interosseous muscle may receive a partial or complete nerve supply from the median nerve (Sunderland, 1946). Very rarely the ulnar nerve may supply the thenar muscles, presumably by fibres which reach it from the median nerve through the communication in the forearm.

Destruction of the ulnar nerve at or above the elbow produces a sensory alteration in the medial part of the hand, and in the little and parts of the ring and middle (dorsal) fingers. The absolute sensory loss is usually less extensive than the anatomical distribution, indicating an overlap from adjacent nerves. Flexion of the wrist occurs with radial deviation (paralysis of flexor carpi ulnaris); the distal phalanges of the little and ring fingers cannot be flexed (medial half of flexor digitorum profundus); abduction and adduction of the fingers is missing (the interossei), though some adduction can still occur on flexion of the fingers (long flexor tendons). Adduction of the thumb is absent though a spurious adduction can be produced by flexor pollicis longus. Other movements of the thumb are normal. Paralysis of the third and fourth lumbricals results in extension of the metacarpophalangeal joints of the ring and little fingers (occasionally the middle finger also) with flexion of the interphalangeal joints (claw hand).

BRANCHES OF THE POSTERIOR CORD OF THE BRACHIAL PLEXUS

SUBSCAPULAR NERVES [FIGS. 923, 927]

The **upper subscapular nerve** (C. (4), 5, 6, (7)) arises high in the axilla often as two or three branches. It passes

Brachioradialis
Radial nerve
Lateral cutaneous nerve of forearm
Deep branch of radial nerve
Extensor carpi radialis longus
Extensor carpi radialis brevis
Branch to supinator
Pronator teres
Superficial branch of radial nerve
Anterior interosseous nerve
Radial head of flexor superficialis
Extensor carpi radialis brevis
Flexor pollicis longus
Brachioradialis
Abductor pollicis longus
Superficial branch of radial nerve
Pronator quadratus
Flexor carpi radialis
Abductor pollicis longus and extensor pollicis brevis
A branch of radial nerve
Abductor pollicis brevis and flexor pollicis brevis

Biceps
Ulnar nerve
Brachialis
Median nerve
Bicipital aponeurosis
Pronator teres
Flexor carpi radialis
Palmaris longus
Humeral head of flexor superficialis
Ulnar nerve
Flexor carpi ulnaris
Flexor digitorum profundus
Dorsal branch of ulnar nerve
Tendons of flexor superficialis
Palmar branch of median nerve
Palmar branch of ulnar nerve
Pisiform bone
Terminal branches of ulnar nerve
Abductor digiti minimi and palmaris brevis
Flexor digiti minimi brevis
Palmar aponeurosis

FIG. 931 Median, ulnar, and radial nerves in forearm.

posteromedially into subscapularis, supplying it and the shoulder joint.

The **lower subscapular nerve** (C. 5, 6) arises close to the axillary nerve, and coursing inferiorly behind the subscapular vessels, enters the teres major muscle. It supplies that muscle and the inferolateral part of the subscapularis.

THORACODORSAL NERVE
[FIGS. 923, 927]

This nerve (C. (6), 7, 8) usually arises between the two subscapular nerves. It passes inferomedially over subscapularis, and accompanies the subscapular vessels along the anterolateral surface of latissimus dorsi, sending branches over the deep surface of that muscle to supply it.

AXILLARY NERVE

This terminal branch (C. 5, 6) of the posterior cord descends posterior to the axillary artery to the lower border of sub-

scapularis. Here it passes posteriorly through the quadrangular space [FIGS. 386, 933] with the posterior humeral circumflex vessels, gives one or two branches to the shoulder joint, and divides into anterior and posterior branches.

The **anterior branch** winds round the posterior surface of the surgical neck of the humerus, deep to deltoid. It sends branches into deltoid, a few of which descend through the muscle to supply skin over its lower part.

The **posterior branch** sends a nerve with a small swelling on it into the inferolateral surface of teres minor, and after giving a few branches to the posterior part of deltoid, descends round the posterior border of that muscle as the **upper lateral cutaneous nerve of the arm**. This nerve supplies the skin and fascia over the lower part of deltoid and the lateral head of triceps as far as the middle of the arm [FIGS. 929, 933].

RADIAL NERVE

The radial nerve is the direct continuation of the posterior cord of the brachial plexus. It arises from the fifth to eighth cervical and the first thoracic ventral rami, with the contributions from the first and last of these being small and inconstant [FIG. 925].

In the axilla it lies posterior to the axillary artery on the subscapularis, latissimus dorsi, and teres major, and descends into the arm between the brachial artery and the long head of triceps, medial to the humerus. It then inclines inferolaterally with the profunda brachii artery, and passing between the long head and the upper part of the origin of the medial head of triceps from the humerus, curves round the posterior surface of that bone in the groove for the radial nerve, deep to the lateral head of triceps. Reaching the distal third of the arm at the lateral border of the humerus, the radial nerve pierces the lateral intermuscular septum and descends to the front of the lateral epicondyle, lying deeply between the brachialis and brachioradialis muscles. Here it divides into its terminal branches; deep and superficial.

Branches

A. Medial to the Humerus. 1. The **posterior cutaneous nerve of the arm** arises in the axilla and pierces the deep fascia on the medial side of the arm near the posterior axillary fold. It supplies skin and fascia on the posterior surface of the proximal third of the arm [FIG. 929].

2. Four branches pass **to the triceps** in the following order: (i) to the long head, arising from the nerve in the axilla; (ii) to the distal part of the medial head, arising near the lower border of teres major it accompanies the ulnar nerve to the middle of the arm (ulnar collateral nerve); (iii) to the lateral head, arising just distal to the last-mentioned branch; (iv) the main nerve to the medial head, arising at the entry to

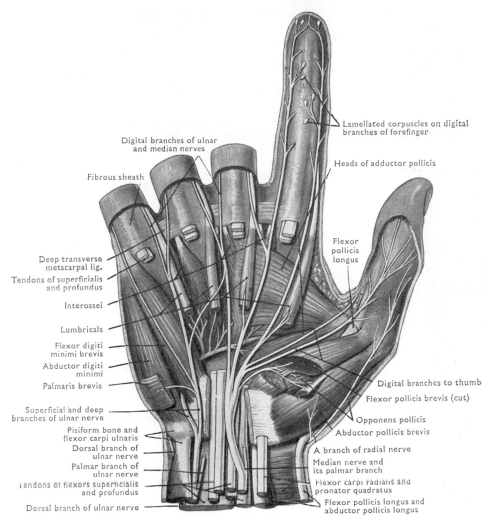

Digital branches of ulnar
and median nerves

Lamellated corpuscles on digital
branches of forefinger

Fibrous sheath

Heads of adductor pollicis

Deep transverse
metacarpal lig.

Tendons of superficialis
and profundus

Interossei

Lumbricals

Flexor digiti
minimi brevis

Flexor
pollicis
longus

Abductor digiti
minimi

Palmaris brevis

Superficial and deep
branches of ulnar nerve

Digital branches to thumb

Flexor pollicis brevis (cut)

Pisiform bone and
flexor carpi ulnaris

Opponens pollicis

Dorsal branch of
ulnar nerve

Abductor pollicis brevis

Palmar branch of
ulnar nerve

A branch of radial nerve

Tendons of flexors superficialis
and profundus

Median nerve and
its palmar branch

Flexor carpi radialis and
pronator quadratus

Dorsal branch of ulnar nerve

Flexor pollicis longus and
abductor pollicis longus

FIG. 932   Median and ulnar nerves in hand.

(occasionally to extensor carpi radialis brevis also), and an inconstant branch to brachialis. The latter appears to be entirely sensory to that muscle.

D. Terminal Branches. 1. The **superficial branch** is the direct continuation of the radial nerve along the anterolateral side of the forearm and is entirely sensory. It lies successively on supinator, pronator teres, flexor digitorum superficialis, and flexor pollicis longus, deep to brachioradialis; with the radial artery on its medial side. It gives a branch to the artery, and passing obliquely backwards in the distal third of the forearm deep to the tendon of brachioradialis, pierces the deep fascia to supply skin and fascia of the dorsum of the wrist, the lateral side of the dorsum of the hand, and the dorsum of the thumb and lateral one (or two) and a half fingers [FIG. 929]. Of the four or five small **dorsal digital nerves,** two pass on the dorsum of the thumb as far as the interphalangeal joint, supplying also that joint, the metacarpophalangeal joint, and skin on the lateral part of the thenar eminence. One passes to the lateral side of the index finger as far as the middle phalanx. The remaining one or two divide at the clefts between the second and third and the third and fourth fingers, and innervate the adjacent sides of these fingers over the proximal phalanges. The area supplied by the superficial branch of the radial nerve is subject to wide variation [p. 748], but it usually does not supply skin medial to the second metacarpal bone (Stopford, 1918).

2. The **deep branch** is entirely muscular and articular in its distribution. It passes postero-inferiorly, and giving branches to extensor carpi radialis brevis and supinator, enters the latter muscle [FIG. 935] and curves round the lateral and posterior surfaces of the radius either in the muscle or between it and the radius (Davies and Laird, 1948). It emerges from supinator as the **posterior interosseous nerve** between the superficial extensor muscles and the lowest fibres of supinator. Here it gives recurrent branches which enter extensor digitorum, extensor digiti minimi, and extensor carpi ulnaris close to their origins, and occasionally to anconeus. It then passes with the posterior interosseous artery superficial to the abductor pollicis longus supplying it. It supplies branches to extensor pollicis longus and brevis and extensor indicis, and ends in a small gangliform enlargement on the back of the carpus from which the intercarpal joints are supplied.

Radial Nerve Injury

Damage to the radial nerve above the origin of the branches to triceps leads to paralysis of triceps, brachioradialis,

the groove on the humerus for the radial nerve. This nerve supplies the medial head, and descending through the muscle, passes posterior to the lateral epicondyle of the humerus and ends by supplying the anconeus muscle.

B. Posterior to the Humerus. 1. The **lower lateral cutaneous nerve of the arm** [FIGS. 929, 933] arises before the radial nerve pierces the lateral intermuscular septum. It emerges a little inferior to deltoid, and supplies the skin and fascia on the posterior surface of the distal third of the arm and a small area on the back of the forearm.

2. The **posterior cutaneous nerve of the forearm** arises with (or immediately distal to) the previous nerve, and pierces the deep fascia 2 cm. lower down. It descends into the back of the forearm, and supplies a variable area of skin and fascia between the areas innervated by the musculocutaneous nerve and the ulnar branch of the medial cutaneous nerve of the forearm. It may reach the level of the wrist or extend on to the dorsum of the hand, supplementing the area supplied by the radial nerve there.

C. Lateral to the Humerus. As it lies between brachialis and brachioradialis, the radial nerve gives one or two branches to brachioradialis, a branch to extensor carpi radialis longus

FIG. 933   Axillary and radial nerves.

of the wrist renders the flexors incapable of acting on the fingers to produce a fist, so that objects cannot be firmly held in the hand (active insufficiency p. 264). The paralysis of supinator and brachioradialis is not significant because biceps and brachialis are adequate to produce, respectively, supination and flexion at the elbow. If, however, triceps is paralysed, supination cannot occur without flexion at the elbow.

THORACIC VENTRAL RAMI

Twelve thoracic ventral rami emerge inferior to the corresponding vertebra and rib; all of these are intercostal in position (**intercostal nerves**) except the twelfth which lies below the twelfth rib and is called, therefore, the **subcostal nerve.** The first, second, third, and twelfth thoracic ventral rami show variations in their course and distribution from the typical arrangement of the remainder.

First Thoracic Ventral Ramus

This, the largest of the series, enters the first intercostal space below the neck of the first rib, and divides into (1) a smaller, inferior, **intercostal nerve,** and (2) a larger, **ascending branch** to the

FIG. 934   Diagram of branches of radial nerve.

supinator, and the extensors of the wrist, thumb, and fingers. Extension of the interphalangeal joints can still be produced by the lumbricals and interossei. The injury also produces impairment of sensation in the territories of distribution of the lower lateral cutaneous nerve of the arm, the posterior cutaneous nerves of the arm and forearm, and the superficial branch of the radial nerve [p. 751]. When the nerve is injured in the groove for the radial nerve on the humerus, the triceps is not completely paralysed and the posterior cutaneous nerve of the arm is unaffected, but the other effects are similar.

The outstanding defect in all injuries to the radial nerve is the inability to extend, or even straighten the wrist, which is held flexed by the flexor muscles (wrist drop). This position

Radial nerve

Superficial branch of radial nerve
Deep branch of
radial nerve

Deep branch of radial nerve
passing through supinator

Muscular branches to superficial muscles

Posterior interosseous artery

Posterior interosseous nerve

Muscular branch to abductor pollicis longus
Muscular branch to extensor pollicis longus
Muscular branch to extensor indicis
Muscular branch to extensor
pollicis brevis

Terminal branch to carpal joints

FIG. 935   Posterior interosseous nerve.

brachial plexus. The latter passes superolaterally across the internal border of the first rib to join the eighth cervical ventral ramus between the scalenus medius muscle and the subclavian artery. The intercostal nerve runs forwards on the pleural surface of the first rib, and supplying the inter-costal muscles of the first space enters that space near the costal cartilage. It then passes anterior to the internal thoracic artery and usually turns forwards to end in the superficial fascia as a small **anterior branch** which does not usually supply skin. A **lateral branch** is frequently absent; if present it communicates with the intercosto-brachial nerve or, rarely, with the medial cutaneous nerve of the arm.

Communications. In addition to the above, the first thoracic ventral ramus is connected to the cervicothoracic ganglion (or first thoracic ganglion if separate) by grey and white **rami communicantes.** The first thoracic ventral ramus also receives a branch from the second thoracic ventral ramus. This branch varies in size and distribution, and may pass into either or both of the branches of the first thoracic ventral ramus; thus postganglionic sympathetic fibres from the second thoracic ganglion in addition to fibres of the second thoracic ventral ramus may reach the brachial plexus by this route.

Second Thoracic Ventral Ramus

This ventral ramus runs forwards between the second rib and the parietal pleura and passes between the inner-most and internal intercostal muscles far forwards. Near the midaxillary line it gives off its large lateral cutaneous

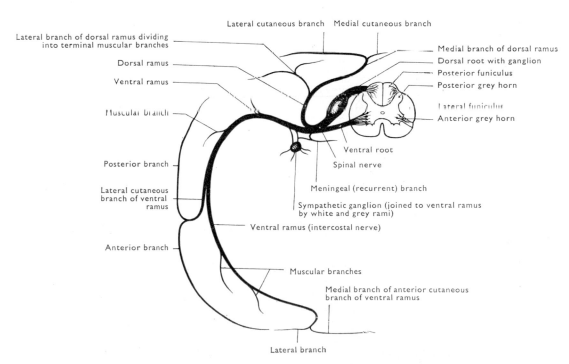

Lateral branch of dorsal ramus dividing
into terminal muscular branches

Dorsal ramus

Ventral ramus

Muscular branch

Posterior branch

Lateral cutaneous
branch of ventral
ramus

Anterior branch

Lateral cutaneous branch   Medial cutaneous branch

Medial branch of dorsal ramus
Dorsal root with ganglion
Posterior funiculus
Posterior grey horn
Lateral funiculus
Anterior grey horn

Ventral root

Spinal nerve

Meningeal (recurrent) branch

Sympathetic ganglion (joined to ventral ramus
by white and grey rami)

Ventral ramus (intercostal nerve)

Muscular branches

Medial branch of anterior cutaneous
branch of ventral ramus

Lateral branch

FIG. 936   Diagram of **origin** and distribution of typical spinal nerve.

Note that the medial branch of the dorsal ramus is represented as distributed to skin, whilst the lateral branch terminates at a deeper level in muscle. Both branches, however, supply muscles; and in the lower half of the body it is the lateral branch that supplies the skin.

branch **(intercostobrachial nerve),** and then continues towards the sternum either as a single trunk or with a small **collateral branch** below [p. 738]. Anterior to the internal thoracic artery, it passes forwards **(anterior cutaneous branch** [FIG. 936]) through the internal intercostal muscle, the external intercostal membrane, and pectoralis major to supply skin and fascia in the region of the anterior part of the second interspace and the adjacent part of the sternum.

Branches. 1. It supplies the muscles of the second intercostal space.

2. The **intercostobrachial nerve** pierces the intercostal and serratus anterior muscles, and passes postero-inferiorly in the axilla to pierce the deep fascia near the posterior axillary fold. It supplies the skin and fascia of the floor of the axilla and the posteromedial surface of the arm as far as the elbow [FIG. 929]. The nerve varies in size inversely with the **medial cutaneous nerve of the arm** with which it communicates either before or after piercing the deep fascia. It also communicates with the posterior branch of the lateral cutaneous branch of the third intercostal nerve in the floor of the axilla, and communicates with the lateral cutaneous branch of the first intercostal nerve when that branch is present.

3. A branch to the first thoracic ventral ramus ascends anterior to the neck of the second rib [p. 753].

4. A grey and a white **ramus communicans** connects it with the second thoracic ganglion of the sympathetic trunk.

5. Anterior cutaneous branch, see above.

Third Thoracic Ventral Ramus

This nerve differs from the typical thoracic ventral ramus in that the posterior part of its lateral cutaneous branch communicates with the **intercostobrachial nerve** and supplies skin and fascia on the medial side of the root of the limb and the floor of the axilla.

Fourth, Fifth, and Sixth Thoracic Ventral Rami

These ventral rami have a typical course and distribution as indeed have the second and third apart from the peculiarities mentioned above [FIG. 936].

At first lying in the posterior wall of the thorax between the ribs, each enters the costal groove of the upper rib, and passes forwards between the innermost and internal intercostal muscles inferior to the corresponding vessels. Anterior to the innermost intercostal muscle, each nerve is immediately external to the pleura, but then passes anterior to the transversus thoracis and the internal thoracic artery. Close to the sternum, it pierces the internal intercostal muscle, the external intercostal membrane, and the pectoralis major as the anterior cutaneous branch.

Branches. Each of these nerves supplies a segment of the thoracic wall from the skin to the parietal pleura by direct **muscular branches** and, usually, by a **collateral branch** [p. 738] to the intrinsic thoracic muscles. The skin and fascia are supplied by the **anterior and lateral cutaneous branches;** the latter piercing the intercostal and serratus anterior muscles and dividing into anterior and posterior branches, supplies an area which is continuous with that innervated by the anterior cutaneous branch and the dorsal ramus of the same spinal nerve. This whole area constitutes a **dermatome** which does not necessarily overlie the muscle strip (myotome) supplied by the nerve [FIGS. 917, 919]. The second intercostal nerve reaches the skin mainly inferior to the sternal angle, the sixth at the base of the xiphoid process, while the fourth innervates the nipple.

Each ventral ramus receives a grey **ramus communicans** from the corresponding ganglion of the sympathetic trunk and sends a white ramus to it.

Seventh to Eleventh Thoracic Ventral Rami

These nerves differ from the preceding nerves in that their anterior parts run in the abdominal wall and supply its skin and fascia, muscles, and peritoneum. When these nerves reach the anterior end of their intercostal spaces they pass deep to the upturned part of the costal cartilage between the attachments of the transversus abdominis and diaphragm to these costal cartilages, and run antero-inferiorly in the abdominal wall between the transversus and the internal oblique muscles. Each nerve then pierces the rectus sheath, and divides into a smaller branch which enters the lateral border of the rectus abdominis and a larger branch which passes deep to that muscle. The latter sends a branch into the medial part of the muscle, and piercing it and the anterior wall of its sheath forms an anterior cutaneous branch.

Branches. Muscular branches pass to the intercostal muscles and to the internal and external oblique, transverse, and rectus muscles of the abdomen. These branches may arise from the cutaneous branches. Branches are also given to the diaphragm, but these are sensory in nature.

The lateral and anterior cutaneous branches are similar

FIG. 937 Diagram to show relations of typical intercostal nerve to the intercostal muscles and membranes.

- Anterior cutaneous branch
- Sternum
- External intercostal membrane
- Internal thoracic vessels
- External intercostal muscle
- Transversus thoracis muscle
- Internal intercostal muscle
- Pleura
- Lateral cutaneous branch
- Ventral root of spinal nerve
- Innermost intercostal muscle
- Ganglion of sympathetic trunk
- Ramus communicans
- Superior costotransverse lig.
- Internal intercostal membrane
- Intercostal nerve
- Dorsal root and spinal ganglion
- Spinal nerve trunk
- Medial branch of dorsal ramus (cutaneous)
- Lateral branch of dorsal ramus (muscular)

to those of the other intercostal nerves. The **lateral cutaneous branches** emerge where the external oblique interdigitates with the serratus anterior or latissimus dorsi muscles. They descend further than the lateral branches of the higher nerves, the tenth extending as far inferiorly as the iliac crest. The lateral branch of the eleventh crosses the iliac crest into the gluteal region. The **anterior cutaneous branches** are small. The seventh to ninth supply the skin between the xiphoid process and the umbilicus, the tenth the umbilicus, and the eleventh immediately below the umbilicus.

Communications with the sympathetic trunk are by grey and white rami communicantes [p. 738].

Subcostal Nerve

This ventral ramus of the twelfth thoracic nerve emerges below the twelfth rib, posterior to the upper part of the psoas muscle and the lowest part of the pleura. It passes inferolaterally into the abdomen posterior to the lateral arcuate ligament, and lies on quadratus lumborum posterior to the kidney. It then pierces the transversus abdominis to pass between it and the internal oblique muscle in the abdominal wall. It passes anterior to rectus abdominis, and its anterior cutaneous branch pierces the rectus sheath midway between the umbilicus and the pubis.
Branches. Muscular branches pass to quadratus lumborum, to the four muscles of the abdominal wall, and to pyramidalis.

A large **lateral cutaneous branch** descends through the muscles of abdominal wall to become superficial immediately above the iliac crest, 5 cm. behind the anterior superior iliac spine. It descends over the iliac crest to supply skin and fascia in the gluteal region down to the level of the greater trochanter of the femur [FIG. 941]. The **anterior cutaneous branch** supplies skin and fascia midway between the umbilicus and the pubis.

The subcostal nerve may receive a communicating branch from the eleventh intercostal near its origin, more frequently it sends a fine branch through psoas to the first lumbar ventral ramus. It may communicate with the iliohypogastric nerve in the abdominal wall. It communicates with the sympathetic trunk by grey and white **rami communicantes** [p. 738].

LUMBAR, SACRAL AND COCCYGEAL PLEXUSES

These plexuses are formed by the ventral rami of the lumbar (5), sacral (5), and coccygeal (1) nerves. The lower limb is supplied by the lumbar and sacral plexuses; the upper lumbar nerves also supply part of the trunk, and the lower sacral and coccygeal nerves also supply the perineum.

Lumbar Plexus

This plexus is formed by the first four lumbar ventral rami in the substance of the psoas muscle and

supplies that muscle, the quadratus lumborum, and the lowest parts of the muscles of the abdominal wall as well as the limb.

Sacral Plexus

This plexus is formed by the ventral rami of the fourth and fifth lumbar and the upper four sacral nerves. The fourth lumbar splits to pass into both the lumbar and the sacral plexuses (nervus furcalis); the fourth sacral divides to enter the sacral and coccygeal plexuses. The sacral plexus lies on the dorsal wall of the pelvis and supplies the lower limb and perineum.

Coccygeal Plexus

This minute plexus is formed from the ventral rami of the fourth and fifth sacral and coccygeal nerves. It lies in the

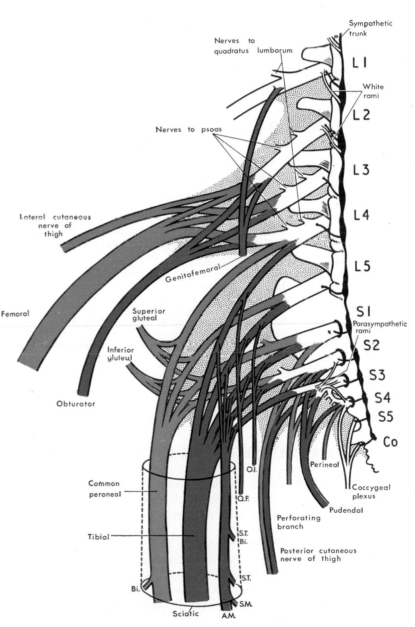

FIG. 938 Nerves of the lumbar, sacral, and coccygeal plexuses. Ventral divisions, red; dorsal, blue.

A.M. Adductor magnus.	S.M. Semimembranosus.	
O.I. Obturator internus.	Bi. Biceps.	
Q.F. Quadratus femoris.	S.T. Semitendinosus.	

dorsal wall of the pelvis on coccygeus, and it supplies that muscle, the levator ani, and the skin over the coccyx.

Communications with Autonomic System

Each of the nerves receives a **grey ramus communicans** from the sympathetic trunk. Because of the anterior position of the sympathetic trunk in the lumbar region, these rami are long and pass posteriorly between the body of the corresponding vertebra and the psoas muscle. One ganglion of the sympathetic trunk may send grey rami to two adjacent ventral rami, or one ventral ramus may receive two rami from adjacent ganglia.

The first two or three lumbar ventral rami each send a **white ramus communicans** to the sympathetic trunk. These either accompany the corresponding grey ramus communicans or are incorporated with it. The remaining ventral rami do not give rise to white rami communicantes, but the second and third, or third and fourth, sacral ventral rami send fine branches **(pelvic splanchnic nerves)** anteromedially to join the hypogastric plexuses of the autonomic system. These consist of **preganglionic parasympathetic fibres** which have no connexion with the sympathetic trunk.

Variation in Origin of Lumbar and Sacral Plexuses

The lumbar plexus may begin as high as the eleventh thoracic ventral ramus or as low as the first lumbar (Eisler, 1891). Similarly the lowest component to join the sciatic may be from the second, third, or fourth sacral ventral ramus. The table shows some of the extreme cases.

	PREFIXED VARIETY	NORMAL	POSTFIXED VARIETY
Nervus furcalis	L. 3, 4	L. 4	L. 5
Obturator	L. 1, 2, 3	L. 2, 3, 4	L. 2, 3, 4, 5
Femoral	T. 12, L. 1, 2	L. 2, 3, 4	L. 2, 3, 4, 5
Tibial	L. 3, 4, 5, S. 1, 2	L. 4, 5, S. 1, 2, 3	L. 5, S. 1, 2, 3, 4
Common peroneal	L. 3, 4, 5, S. 1	L. 4, 5, S. 1, 2	L. 5, S. 1, 2, 3

The commonest variation is towards the postfixed variety, and some evidence exists that vertebral anomalies are often associated.

THE LUMBAR PLEXUS

This plexus is formed by the ventral rami of the first three lumbar spinal nerves and part of the fourth, with a small branch from the subcostal nerve in half the cases.

Position and Constitution [FIGS. 938, 939]

The nerves emerge from the intervertebral foramina and their ventral rami communicate with the sympathetic trunk (see above). They then divide and recombine in the substance of the psoas muscle anterior to the plane of the transverse processes of the lumbar vertebrae.

The first two lumbar ventral rami divide into superior and inferior branches. The superior branch of the first, with or without a branch from the subcostal nerve, divides into the **iliohypogastric** and **ilio-inguinal nerves**. The inferior branch unites with the superior branch of the second to form the **genitofemoral nerve.** The inferior branch of the second, the third, and the upper part of the fourth all divide into smaller ventral and larger dorsal divisions. The ventral divisions combine to form the **obturator nerve** (sometimes without the branch from the second nerve). The dorsal divisions unite to form the **femoral nerve** and the **lateral cutaneous nerve of the thigh** which arises from the second and third only. Irregular muscular branches pass to the psoas and quadratus lumborum muscles.

NERVES ARISING FROM THE LUMBAR PLEXUS
[FIG. 939]

1. Muscular branches to the quadratus lumborum (T. 12, L. 1-4) and psoas (L. (1), 2, 3, (4))
2. Iliohypogastric (L. 1, (T. 12))
3. Ilio-inguinal (L. 1, (T. 12))
4. Genitofemoral (L. 1, 2)
5. Lateral cutaneous of thigh (L. 2, 3)
6. Obturator (L. 2, 3, 4)
7. Femoral (L. 2, 3, 4)

Direct Muscular Branches

The nerves to quadratus lumborum and psoas arise separately from the ventral rami, though those to psoas are often associated at their origin with the nerve to iliacus from the femoral nerve. Psoas minor, when present, is supplied by the first or the second lumbar ventral ramus.

ILIOHYPOGASTRIC NERVE

This nerve and the ilio-inguinal are in series with the subcostal nerve. They have a similar course and distribution.

The iliohypogastric nerve arises from the first lumbar ventral ramus with a small contribution from the twelfth thoracic. It passes inferolaterally through the substance of psoas major and over the anterior surface of quadratus lumborum posterior to the kidney. It then pierces transversus abdominis, and running forwards in the abdominal wall, superior to the iliac crest, between that muscle and the internal oblique, pierces the latter muscle 2–3 cm. medial to the anterior superior iliac spine [FIG. 941]. It then passes medially, deep to the aponeurosis of the external oblique, and becomes cutaneous by piercing that aponeurosis 4 cm. superior to the superficial inguinal ring.

Branches

1. To the muscles of the abdominal wall except rectus abdominis.

2. A small **lateral cutaneous branch** is usually present. It pierces obliquus internus and externus just superior to the iliac crest and posterior to the corresponding branch of the subcostal nerve. It supplies skin and fascia in the upper lateral part of the gluteal region adjacent to the area supplied by the dorsal ramus of the first lumbar nerve.

3. The **anterior cutaneous branch** supplies the skin and fascia of the anterior abdominal wall immediately superior to the pubis.

ILIO-INGUINAL NERVE

This nerve arises from the first lumbar ventral ramus with the iliohypogastric nerve, and like it, may contain some fibres from the twelfth thoracic ventral ramus. It may be combined

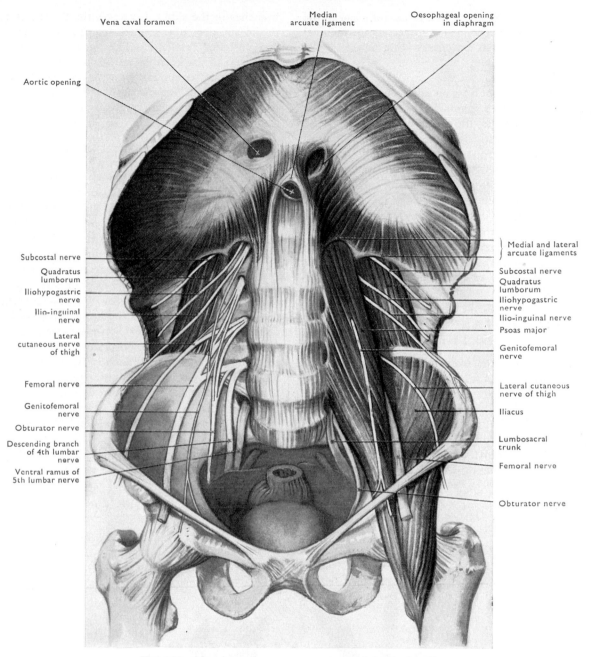

FIG. 939 Muscles and nerves on posterior abdominal wall.

with the iliohypogastric nerve for a variable distance, but when they separate it follows the same course on a lower level, usually communicating with the iliohypogastric nerve. It pierces the internal oblique muscle further forward and lower down than the iliohypogastric [FIG. 941] and passing deep to the aponeurosis of the external oblique muscle close to the inguinal ligament, enters the inguinal canal, and reaches the skin through the superficial inguinal ring and the external spermatic fascia [FIG. 941].

Branches

1. It supplies the muscles of the abdominal wall between which it passes.

2. Its terminal branches [FIG. 942] innervate the skin and

fascia over (i) the pubic symphysis, (ii) the superomedial part of the femoral triangle, (iii) the anterior surface of the scrotum. and root and dorsum of the penis (**anterior scrotal nerves**), or the mons pubis and labium majus in the female (**anterior labial nerves**). The latter nerves supply skin adjacent to that supplied by branches of the pudendal nerve. There is no lateral cutaneous branch.

GENITOFEMORAL NERVE

This nerve is formed in the substance of psoas major by the union of two roots which pass forwards from the first and second lumbar ventral rami. The nerve emerges through the anterior surface of psoas major medial to psoas minor, pierces the psoas fascia, and descending on that fascia posterior to the ureter, divides into two branches as it approaches the inguinal ligament lateral to the external iliac artery [FIG. 939].

Branches

1. The small **genital branch** passes anterior to the external iliac artery through the deep inguinal ring into the inguinal canal. It gives branches to the external iliac artery, the cremaster muscle, the testicular autonomic plexus, and emerging with the spermatic cord (or the round ligament) supplies the skin and fascia of the scrotum (or labium majus) and the adjacent part of the thigh. It communicates with the ilio-inguinal nerve in the inguinal canal.

2. The **femoral branch** enters the thigh posterior to the inguinal ligament, lateral to the femoral artery to which it gives a small branch. It then passes through the saphenous opening of the fascia lata and supplies an area of skin and fascia over the femoral triangle lateral to that innervated by the ilio-inguinal nerve [FIG. 942].

LATERAL CUTANEOUS NERVE OF THE THIGH

This cutaneous nerve arises from the posterior surface of the second and third lumbar ventral rami. It emerges from the lateral border of psoas major anterior to the iliac crest, and passing between iliacus and the iliac fascia, enters the thigh posterior to the lateral end of the inguinal ligament, medial to the anterior superior iliac spine. It then pierces (or passes superficial or deep to) the proximal part of sartorius [FIG. 941] and descends deep to the fascia lata giving off a few small branches to the skin. It pierces the fascia lata 10 cm. distal to the anterior superior iliac spine, and divides into anterior and posterior branches. The anterior branch supplies skin and fascia of the anterolateral surface of the thigh to the knee, where it may communicate with the patellar plexus. The smaller posterior branch supplies the skin and fascia on the lateral surface between the greater trochanter and distal third of the thigh [FIG. 942].

OBTURATOR NERVE

This nerve supplies the obturator externus and the adductor muscles of the thigh, gives articular branches to the hip and knee joints, and occasionally has a cutaneous branch. It arises by a branch from the ventral divisions of each of the second, third, and fourth lumbar ventral rami [FIG. 939]. The contribution from the second lumbar is occasionally small or absent. These branches unite in the substance of psoas and descend vertically through its posterior part to emerge from its medial border on the lateral part of the sacrum. It then crosses the sacro-iliac joint, enters the lesser pelvis, and descends on obturator internus a short distance posterior to the linea terminalis, to enter the obturator groove and divide into anterior and posterior branches, anterior to the obturator internus [FIG. 940]. In the lesser pelvis, the nerve is lateral to the internal iliac vessels and ureter, and is joined by the obturator vessels lateral to the ovary or the ductus deferens.

Branches

1. The **anterior branch** descends into the thigh anterior to obturator externus and adductor brevis, posterior to pectineus

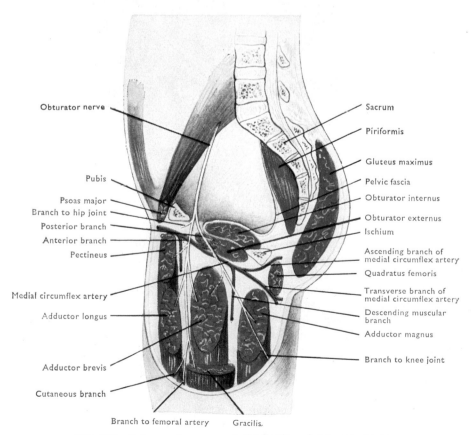

FIG. 940 Scheme of course and distribution of obturator nerve.

and adductor longus [FIG. 941]. As it enters the thigh, it gives a lateral branch to the hip joint through the acetabular notch, and may send a communicating branch to the femoral nerve. As it lies between the muscles, it sends branches into adductor longus, gracilis, adductor brevis (usually), and pectineus (occasionally), and it divides into terminal cutaneous and vascular branches.

The **cutaneous branch** is frequently absent. When present, it passes between gracilis and adductor longus near the middle of the thigh, and supplies the skin and fascia of the distal two-thirds of its medial side. In the distal third of the thigh the nerve assists in the formation of a **plexus** with branches of the saphenous nerve and the anterior cutaneous nerve of the thigh, deep to sartorius. The **vascular branch**

FIG. 941 Nerves of front of thigh.

reaches the femoral artery in the adductor canal by passing along the medial border of adductor longus.

2. The **posterior branch** of the obturator nerve supplies and pierces obturator externus, and descends between adductor brevis and magnus. It runs obliquely through adductor magnus supplying its upper (adductor) part, and entering the popliteal fossa posterior to the popliteal artery, ends by piercing the oblique popliteal ligament to supply the posterior part of the knee joint, including the cruciate ligaments.

The **accessory obturator nerve** is only occasionally present (29 per cent., Eisler, 1891; 8 per cent., Sunderland, 1969). It arises from the third, or third and fourth lumbar ventral rami between the obturator and femoral nerves. It runs with the obturator nerve till that nerve emerges from psoas, and descending between the linea terminalis and the external iliac vessels, enters the thigh between the pubic bone and the femoral vessels.

The nerve commonly replaces the femoral branch to pectineus, sends a branch to the hip joint, and communicates with the anterior branch of the obturator nerve. It may only form the nerve to pectineus, or make a significant contribution to the adductor muscles through the obturator nerve.

FEMORAL NERVE

This nerve supplies the muscles of the front of the thigh, gives articular nerves to the hip and knee joints, and has an extensive cutaneous distribution to the anteromedial surface of the limb as far distally as the medial side of the foot.

It arises from the dorsal divisions of the second, third, and fourth lumbar ventral rami behind the obturator nerve in the substance of psoas major. It passes inferolaterally through the muscle and lies in the groove between psoas and iliacus immediately below the iliac crest [FIG. 939]. It descends in this groove, posterior to the fascia iliaca, and enters the thigh posterior to the inguinal ligament and lateral to the femoral sheath. In the femoral triangle it ends by dividing into a number of branches [FIG. 941].

Branches

In the abdomen a lateral branch passes into iliacus, and a descending branch passes to the femoral artery, which receives further branches in the thigh.

In the femoral triangle

Muscular Branches. 1. The nerve **to pectineus,** frequently double, arises close to the inguinal ligament, and passing

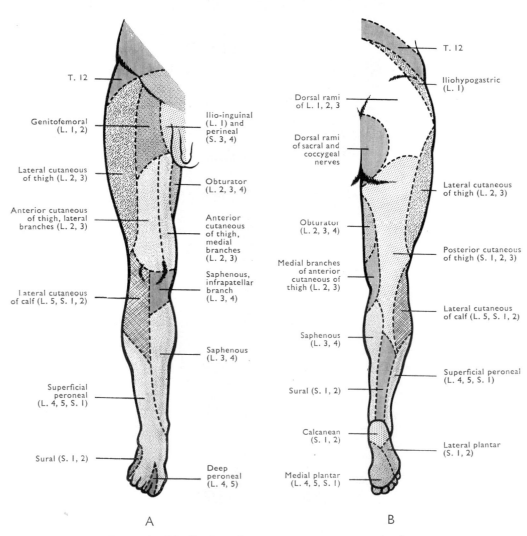

FIG. 942 Distribution of cutaneous nerves. A, on the front; B, on the back of lower limb.

inferomedially posterior to the femoral vessels, enters the lateral border of the muscle. It may give off a small branch to the anterior branch of the obturator nerve. (See accessory obturator nerve.)

2. The nerves **to sartorius** arise from the anterior cutaneous branches of the femoral nerve. They consist of a lateral group of short branches and a medial group of longer branches which enter the muscle about its middle.

3. The nerves **to the quadriceps.** The vastus lateralis and rectus femoris are supplied on their deep surfaces by separate branches which accompany branches of the lateral femoral circumflex artery. The vastus intermedius is supplied by a branch which runs on its superficial surface and pierces it to supply the articularis genus muscle. The vastus medialis is supplied by a proximal branch which enters the superior part of the muscle and sends branches to the vastus intermedius. A distal branch to the same muscle descends on the lateral side of the femoral artery with the saphenous nerve and enters the medial surface of the muscle either from the adductor canal or after piercing its fascial roof. The nerve sends a small branch through the nutrient canal of the femur, and to the articularis genus muscle.

Articular Branches. A branch to the hip joint arises from the nerve to rectus femoris, and passes with branches from the lateral circumflex femoral artery. Four branches pass to the knee joint. One arises from each of the nerves to the three vastus muscles which send descending branches anterior to the femur. The fourth branch may arise from the saphenous nerve medial to the knee.

Cutaneous Branches [FIGS. 941, 942]. 1. The **anterior cutaneous nerves** of the thigh arise as lateral and medial branches in the proximal part of the femoral triangle. Two lateral branches descend vertically on the anterior surface of the thigh, and become cutaneous by piercing the fascia lata over the proximal third of sartorius. They supply muscular branches to sartorius and the more lateral nerve may pierce the muscle. They supply the anterior surface of the thigh, communicate with the femoral branch of the genitofemoral nerve, and take part in the formation of the **patellar plexus,** anterior to the patella. In the femoral triangle, lateral to the femoral vessels, the medial cutaneous branch gives off two or more twigs to the skin of the proximal part of the thigh in the region of the saphenous vein. It then runs inferomedially, and passing superficial to the femoral vessels at the apex of the femoral triangle, divides into anterior and posterior branches. The **anterior branch** passes downwards, superficial to and supplying sartorius, to end medial to the patella in the **patellar plexus** [FIG. 941]. It supplies the skin and fascia of the distal half of the medial side of the thigh. The **posterior branch** passes superficially parallel to the sartorius muscle, and communicates with the saphenous and obturator nerves in the middle of the thigh to form a **subsartorial plexus.** It pierces the deep fascia on the medial side of the distal third of the thigh, and sends branches over the medial side of the knee to form part of the **patellar plexus.** The size of the medial branch varies inversely with that of the cutaneous branch of the obturator nerve and the infrapatellar branch of the saphenous nerve.

2. The **saphenous nerve** descends through the femoral triangle into the adductor canal on the lateral side of the femoral vessels. Here it gives a branch to a **subsartorial plexus** formed by the obturator and anterior cutaneous nerves, and passing obliquely across the anterior surface of the femoral artery, lies anterior to adductor magnus with the saphenous branch of the descending genicular artery. Here it gives off an **infrapatellar branch** which becomes superficial by piercing sartorius, then passes antero-inferiorly below the patella to supply skin and fascia over the front and medial side of the knee and proximal part of the leg. It communicates with the patellar plexus.

The saphenous nerve becomes superficial in the leg by passing between the sartorius and gracilis muscles, and may give an articular branch to the medial side of the knee joint. It then descends with the great saphenous vein on the medial side of the leg, and passing anterior to the medial malleolus, ends by

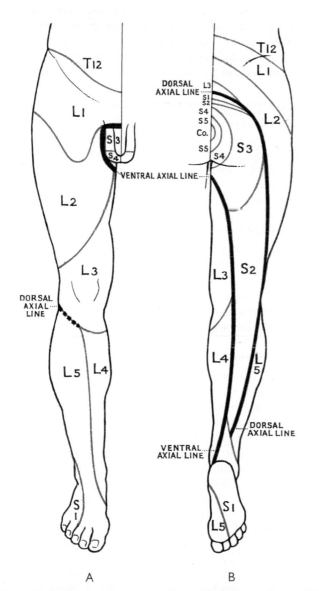

FIG. 943 Dermatomes of lower limb, showing the segmental cutaneous distribution of spinal nerves (T. 12, L. 1-5, S. 1-5, Co.) on A, the front and B, the back of the limb and lower part of trunk. (After Head, 1893, and Foerster, 1933.)

See page 736 for definition of 'dermatome' and page 771 for more particular reference to the segmental innervation of the lower limb.

supplying skin and fascia on the proximal half of the medial side and adjacent dorsum of the foot. Branches of the saphenous nerve supply skin and fascia on the front and medial side of the leg, some passing over the anterior border of the tibia.

The **patellar plexus** consists of fine intercommunicating branches of the nerves which supply the skin of the front of the knee, i.e. the infrapatellar branch of the saphenous nerve, the anterior cutaneous branches of the femoral nerve, and sometimes the lateral cutaneous nerve of the thigh [FIG. 941].

THE SACRAL PLEXUS

[FIG. 938]

This plexus is formed from the ventral rami of the fourth lumbar to the fourth sacral spinal nerves [p. 755].

Position

The plexus lies on the dorsal wall of the pelvis, mainly between the piriformis muscle and its fascia. Anterior to that fascia lie the internal iliac vessels and the ureter, with the sigmoid colon on the left and coils of ileum on the right, further anteriorly.

The nerves entering the plexus converge to form a broad triangular band [FIG. 945] which passes through the inferior part of the greater sciatic foramen, below piriformis, and enters the gluteal region as the **sciatic nerve** [FIG. 948]. Numerous small branches arise from the anterior and posterior surfaces of the triangular band, and from its lower margin the pudendal nerve passes to the perineum.

The sciatic nerve divides into the **tibial nerve** and the **common peroneal nerve** at a variable point in the thigh. These nerves are often distinct at their origin (in which case the latter nerve may pierce the piriformis muscle), and it is always possible to separate them to the plexus within the sheath of the sciatic nerve. If this is done, the origin of these nerves from ventral (tibial nerve) and dorsal divisions (common peroneal nerve) can be confirmed.

Formation

The descending part of the fourth lumbar ventral ramus emerges from the medial border of the psoas muscle, medial to the obturator nerve, and posterior to the common iliac vessels. The nerve descends into the lesser pelvis, and joins the ventral ramus of the fifth lumbar nerve which has reached the pelvic surface of the sacro-iliac joint by passing over the lateral part of the sacrum. The combined nerves constitute the **lumbo-sacral trunk**. The first, second, and third sacral ventral rami pass laterally on piriformis from the pelvic sacral foramina. The fourth contributes to the sacral and coccygeal plexuses.

These nerves (of which the fifth lumbar and first sacral are the largest) are divisible into dorsal and ventral divisions. They converge on the lower part of the greater sciatic foramen, the dorsal divisions (L. 4, 5, S. 1, 2) form the common peroneal nerve, while the ventral divisions (L. 4, 5, S. 1, 2, 3) unite to form the tibial nerve [FIG. 938].

BRANCHES OF THE SACRAL PLEXUS

ANTERIOR BRANCHES	POSTERIOR BRANCHES
Nerve to quadratus femoris and inferior gemellus (L. 4, 5, S. 1)	Muscular twigs to piriformis (S. 1, 2) and to coccygeus and levator ani (S. 3, 4)
Nerve to obturator internus and superior gemellus (L. 5, S. 1, 2)	Superior gluteal nerve (L. 4, 5, S. 1)
Pelvic splanchnic nerves (S. (2), 3, (4))	Inferior gluteal nerve (L. 5, S. 1, 2)
Posterior femoral cutaneous nerve (S. 2, 3)	Posterior femoral cutaneous nerve (S. 1, 2)
Pudendal nerve (S. (1), 2, 3, 4, (5))	Perforating cutaneous nerve (S. 2, 3)
	Perineal branch of S. 4

SCIATIC NERVE (L. 4, 5, S. 1, 2, 3)

Nerve to Quadratus Femoris

This nerve arises from the ventral divisions (L. 4, 5, S. 1), descends through the lower part of the greater sciatic foramen, and passing on to the posterior surface of the ischium anterior to the sciatic nerve, obturator internus, and the gemelli, sends a fine branch to the adjacent hip joint. It supplies gemellus inferior and ends in the deep surface of quadratus femoris.

Nerve to Obturator Internus [FIG. 945]

This nerve arises from the ventral divisions (L. 5, S. 1, 2) and descends into the gluteal region between the sciatic nerve (laterally) and the internal pudendal vessels. It sends a branch to the superior gemellus, and turns forwards below the ischial spine to enter the ischiorectal fossa on the medial surface of obturator internus.

Pelvic Splanchnic Nerves

These consist of preganglionic parasympathetic nerve fibres. They arise as two slender filaments from the second and third, or third and fourth (or rarely fourth and fifth) sacral ventral rami, and pass forwards to join the autonomic plexuses of the pelvis, through which they supply the urogenital organs and the lower part of the large intestine [see p. 780].

Pudendal Nerve

This principal nerve of the perineum arises in the pelvis usually by three roots (S. (1) 2, 3, 4) [FIG. 938]. It enters the gluteal region through the inferomedial part of the greater sciatic foramen, to lie on the sacrospinous ligament medial to the internal pudendal artery. It passes through the lesser sciatic foramen into the perineum with the artery, entering the **pudendal canal** on the obturator fascia of the lateral wall of the ischiorectal fossa. Here it gives off the inferior rectal nerve, and shortly afterwards divides into the perineal nerve and the dorsal nerve of the penis or clitoris.

Inferior Rectal Nerve. This nerve usually arises as the pudendal nerve enters the pudendal canal, though it may arise separately from the third and fourth sacral ventral rami. It crosses the ischiorectal fossa inferomedially with the inferior rectal vessels, supplies levator ani, and reaching the anus, innervates the skin and fascia around it and the sphincter ani externus. The cutaneous branches communicate with the perineal branches of the (1) posterior femoral cutaneous, (2) perineal, and (3) fourth sacral nerves.

Perineal Nerve. This nerve passes anteromedially and

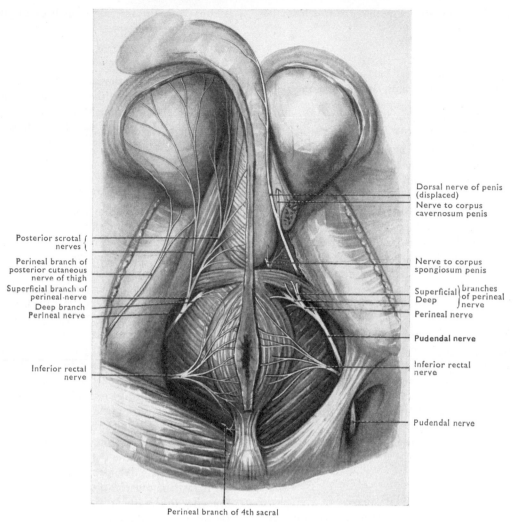

Posterior scrotal
nerves

Perineal branch of
posterior cutaneous
nerve of thigh

Superficial branch of
perineal nerve

Deep branch
Perineal nerve

Inferior rectal
nerve

Dorsal nerve of penis
(displaced)
Nerve to corpus
cavernosum penis

Nerve to corpus
spongiosum penis

Superficial } branches
Deep } of perineal
nerve

Perineal nerve

Pudendal nerve

Inferior rectal
nerve

Pudendal nerve

Perineal branch of 4th sacral

FIG. 944   Distribution of the pudendal nerve.

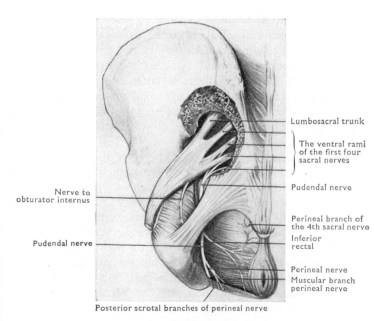

Nerve to
obturator internus

Pudendal nerve

Lumbosacral trunk

The ventral rami
of the first four
sacral nerves

Pudendal nerve

Perineal branch of
the 4th sacral nerve

Inferior
rectal

Perineal nerve
Muscular branch
perineal nerve

Posterior scrotal branches of perineal nerve

FIG. 945   Origin and course of the pudendal nerve.

divides into posterior scrotal or posterior labial nerves and deep and superficial branches.

The **posterior scrotal** or **posterior labial nerves** pass superficially to supply the skin and fascia of the perineum and the scrotum or the labium majus by medial and lateral branches. The lateral may send branches to the skin of the medial aspect of the thigh, and it communicates with the inferior rectal and posterior femoral cutaneous nerves. The superficial and deep branches pass into the corresponding perineal spaces to supply the muscles and the erectile tissue and mucous membrane of the bulb and corpus spongiosum penis, extending as far as the glans penis.

Dorsal Nerve of the Penis or Clitoris. This nerve passes forwards through the deep perineal space with the internal pudendal artery. It lies close to the medial surface of the inferior pubic ramus, superior to the perineal membrane and the crus of the penis or clitoris, and inferior to the sphincter urethrae muscle. It sends a twig through the perineal membrane to the crus and corpus cavernosum of the penis or clitoris; then piercing the perineal membrane near its anterior border, passes on to the dorsal surface of the penis or clitoris, lateral to the dorsal artery. It is distributed to the distal two-thirds of the penis, sending branches round the sides to reach the inferior surface of that organ. The nerve is much smaller in the female than in the male.

Nerve to Piriformis

This nerve arises from the dorsal surface of the first and second, or second sacral ventral rami, and enters the pelvic surface of the muscle, often as two separate nerves.

Nerves to Coccygeus and Levator Ani

In addition to branches from the inferior rectal and the perineal nerves, nerves arise from a loop between the third and fourth sacral ventral rami, and descend to these muscles.

The **perineal branch of the fourth sacral nerve** arises from the lower part of the same loop. It supplies and pierces coccygeus, and entering the posterior angle of the ischiorectal fossa, passes anteriorly supplying levator ani, the posterior part of the external anal sphincter, and the overlying skin and fascia.

Perforating Cutaneous Nerve

This nerve arises (S. 2, 3) close to the lower roots of the posterior femoral cutaneous nerve [Fig. 938] and may be incorporated in the gluteal branches of that nerve or in the pudendal nerve. The nerve descends, pierces the sacro-tuberous ligament, and becomes cutaneous close to the coccyx by turning round the inferior border or piercing the lower fibres of gluteus maximus. It supplies the skin and fascia of the lower part of the buttock and the medial part of the gluteal fold.

Superior Gluteal Nerve

This nerve arises from the dorsal surface of the plexus (L. 4, 5, S. 1) and passes posterolaterally into the gluteal region with the superior gluteal vessels, superior to piriformis. It passes forwards as an upper and a lower branch between gluteus medius and minimus. The upper branch supplies gluteus medius. The lower branch crosses the middle of gluteus minimus, sends branches to that muscle and to gluteus medius, and ends in the deep surface of tensor fasciae latae [Fig. 948].

It runs with the inferior branch of the deep division of the superior gluteal artery.

Inferior Gluteal Nerve

It arises from the dorsal surface of the plexus (L. 5, S. 1, 2) and running medial to the posterior femoral cutaneous nerve, enters the gluteal region at the inferior border of piriformis, superficial to the sciatic nerve. It divides into a number of branches which pass posteriorly into the deep surface of gluteus maximus.

Posterior Femoral Cutaneous Nerve [Fig. 938]

This nerve arises from the upper three sacral ventral rami, or from the second and third. It is distributed to the skin and fascia of the posterior surface of the thigh and upper part of the leg, the perineum, and the inferior half of the buttock. At its origin the upper roots of the nerve are associated with the origin of the inferior gluteal nerve, and the lowest root with the pudendal nerve. It enters the gluteal region on the posterior surface of the sciatic nerve with the inferior gluteal vessels and nerve. It descends to the inferior border of gluteus maximus, near which it gives off its **gluteal** and **perineal** branches, then passes down the back of the thigh deep to the fascia lata covering the hamstring muscles. Over the lower part of the popliteal fossa it pierces the deep fascia as two or more cutaneous branches which supply the skin and fascia over the upper half of the back of the leg, though they may reach the ankle, or stop at the lower border of the popliteal fossa [Fig. 942].

Branches. 1. The **perineal branch** curves medially over the hamstring muscles [Fig. 948] and the pubic arch, and sends branches to the upper medial part of the thigh and perineum. Its branches [Fig. 944] pass forwards to supply the skin and fascia of the scrotum (labium majus) and root of the penis (clitoris), communicating with branches of the ilio-inguinal nerve. Other branches pass medially to the skin and fascia anterior to the anus, communicating with the inferior rectal and perineal branches of the pudendal nerve.

2. The **gluteal branches** arise deep to the inferior border of gluteus maximus, and pierce the deep fascia along that border. They supply the skin and fascia of the inferior half of the buttock, the lateral branches overlapping the areas supplied by the lateral femoral cutaneous nerve and the dorsal rami of the upper three lumbar nerves. The medial branches may pierce the sacrotuberous ligament and replace the perforating cutaneous nerve, or supply the area adjacent to it.

3. The **branches to the thigh** pierce the fascia lata at intervals, and supply the skin and fascia of its posterior surface. For the branches to the leg, see above.

When the two parts of the sciatic nerve are separated by piriformis, the posterior femoral cutaneous nerve is also in two parts. (i) A dorsal part (S. 1, 2), arising with the common peroneal nerve in association with the lower roots of the inferior gluteal nerve, supplies the buttock and lateral part of the back of the thigh. (iii) A ventral part (S. 2, 3), arising with the tibial nerve in association with the perforating cutaneous and pudendal nerves, supplies the perineum and medial part of the cutaneous area in the limb.

Fig. 946 Perineal dermatomes, showing segmental cutaneous distribution of spinal nerves (L. 1-3, S. 2-5, Co) to perineum and adjacent parts of trunk and lower limb (After Head, 1893, and Foerster, 1933.)

Gluteus medius
Posterior superior iliac spine
Sacrotuberous ligament
Pudendal nerve on sacrospinous ligament
Obturator internus and its nerve
Coccyx
Perforating cutaneous nerve
Inferior gemellus
Semimembranosus
Perineal branch of posterior cutaneous nerve of thigh
Semitendinosus
Branch to adductor magnus disappearing in front of semimembranosus
Long head of biceps
Tendinous intersection of semitendinosus
Semimembranosus
Gracilis
Tibial nerve
Gracilis
Semimembranosus
Medial head of gastrocnemius
Sural nerve
Popliteus and its nerve
Soleus
Branch to flexor digitorum longus
Tibial nerve
Flexor digitorum longus
Flexor retinaculum
Medial calcanean nerve
Tendo calcaneus

Superior gluteal nerve and gluteus minimus
Piriformis and inferior gluteal nerve
Sciatic nerve and superior gemellus
Gluteus medius
Gluteus maximus
Quadratus femoris and its nerve
Adductor magnus
Branch to semitendinosus
Iliotibial tract on vastus lateralis
Linea aspera
Branch to semimembranosus
Heads of biceps
Iliotibial tract
Common peroneal nerve
Peroneal communicating branch
Plantaris
Lateral cutaneous nerve of calf
Lateral head of gastrocnemius
Head of fibula
Branch to tibialis posterior
Posterior surface of fibula
Peroneus longus
Tibialis posterior
Flexor hallucis longus and its nerve
Peroneus brevis
Sural nerve
Peroneus longus
Lateral calcanean nerve

FIG. 947   Nerves of back of lower limb.

SCIATIC NERVE

This is the thickest nerve in the body. It consists of a medially placed tibial nerve and its branches to the hamstring muscles, and a laterally placed common peroneal nerve and its

branch to the short head of biceps, bound together in a common sheath. It is formed by nerve fibres from the ventral rami of the fourth lumbar to third sacral spinal nerves, and emerges from the pelvis into the gluteal region through the greater sciatic foramen. At first, deep to piriformis, it runs inferolaterally on the nerve to quadratus femoris and the ischium with the inferior gluteal and posterior femoral cutaneous nerves [FIG. 948]. Inferior to piriformis, it lies deep to gluteus maximus, and descends more vertically posterior to the tendon of obturator internus and the gemelli. It then passes on to quadratus femoris in the hollow between the ischial tuberosity and the greater trochanter of the femur, and enters the thigh by reaching the inferior border of gluteus maximus on the posterior surface of adductor magnus. Here the nerves to the hamstrings separate from the sciatic nerve, which descending, is crossed superficially from medial to lateral by the long head of biceps femoris.

The sciatic nerve usually ends at the proximal angle of the popliteal fossa by dividing into the tibial and common peroneal nerves, but this division may occur at any higher level [FIGS. 947, 948].

Branches

The nerve **to the hamstring muscles** (L. 4, 5, S. 1, 2, 3) lies in the medial part of the sciatic nerve in the gluteal region, and leaves it near the lower border of quadratus femoris. It is distributed to the hamstring muscles (except the short head of biceps) by a series of branches, which are usually given off separately by the sciatic nerve [FIG. 948]. The nerve to semitendinosus is double; the branch to the part superior to the tendinous intersection may be associated with the nerve to the long head of biceps. The nerve to the ischial part of adductor magnus arises in common with the nerve to semimembranosus, and descends anterior to the lateral margin of that muscle supplying both.

The nerve **to the short head of biceps** (L. 5, S. 1, 2) is formed at the plexus in common with the inferior gluteal nerve, and leaves the lateral side of the common peroneal nerve above the middle of the thigh, usually with the branches of that nerve to the knee joint. It may have an independent course in the thigh, or be associated with the inferior gluteal nerve in the buttock.

An **articular branch** (L. 4, 5, S. 1) to the lateral and anterior parts of the knee joint usually arises with the nerve to the short head of biceps. It descends through the superior part of the popliteal fossa anterior to biceps femoris, and divides into proximal and distal branches, which pass with the superior and inferior lateral genicular arteries respectively.

FIG. 948   Nerves of gluteal region and back of thigh.

COMMON PERONEAL NERVE

This nerve supplies the skin and fascia of the anterolateral surface of the leg and the dorsum of the foot, the muscles of the anterior and peroneal compartments of the leg, extensor digitorum brevis, and the knee, ankle, and joints of the foot.

The nerve arises from the dorsal divisions of the sacral plexus (L. 4, 5, S. 1, 2) and is laterally placed in the sciatic nerve. From the bifurcation of that nerve it passes inferolaterally in the proximal and lateral part of the popliteal fossa, under cover of the medial edge of the biceps femoris and its tendon, to reach the back of the head of the fibula. It ends by dividing into the deep and superficial peroneal nerves where it winds round the lateral surface of the neck of the fibula in the substance of peroneus longus, approximately 2 cm. distal to the apex of the head of the fibula. It is palpable posterior to the

head of the fibula and posterolateral to its neck [FIGS. 947, 949, 951].

Branches

In the thigh. The nerve to the short head of biceps, and the articular branch to knee joint are described above.

In the popliteal fossa. 1. The **lateral cutaneous nerve of the calf** often arises in common with the peroneal communicating nerve, pierces the deep fascia over the lateral head of gastrocnemius, and is distributed to the skin and fascia in the proximal two-thirds of the posterolateral surface of the leg. Its size and distribution varies inversely with that of the posterior femoral cutaneous and sural nerves [FIG. 942].

2. The **peroneal communicating branch** passes between the deep fascia and the lateral head of gastrocnemius to join

Vastus lateralis
Iliotibial tract
Patella
Ligamentum patellae
Head of fibula
Peroneus longus
Extensor digitorum longus
Recurrent branch of common peroneal entering tibialis anterior
Deep peroneal nerve
Superficial peroneal nerve
Soleus
Branch to extensor digitorum longus (cut)
Branches to peroneus longus
Branches to extensor hallucis longus
Branches to peroneus brevis
Extensor hallucis longus
Peroneus longus
Extensor digitorum longus
Gastrocnemius
Soleus
Peroneus brevis
Tibialis anterior
Deep fascia
Extensor hallucis longus
Medial branch of superficial peroneal
Superior extensor retinaculum
Lateral branch of superficial peroneal
Lateral malleolus
Upper band of inferior extensor retinaculum
Extensor digitorum brevis
Extensor digitorum longus
Tibialis anterior
Lateral terminal branch of deep peroneal nerve
Extensor hallucis longus
Medial terminal branch of deep peroneal nerve
Extensor hallucis brevis

Biceps
Common peroneal nerve
Lateral cutaneous nerve of calf
Lateral head of gastrocnemius

Sural nerve
Peroneus longus
Peroneus brevis
Peroneus tertius

Lateral tendon of extensor digitorum brevis

FIG. 949 Nerves of front and lateral side of leg and dorsum of foot.

Deep Peroneal Nerve [FIG. 949]. This nerve begins between the neck of the fibula and peroneus longus. It runs inferomedially on the fibula, deep to extensor digitorum longus, to join the anterior tibial vessels on the anterior surface of the interosseous membrane, lateral to tibialis anterior. It then descends on the interosseous membrane medial to the origin of extensor hallucis longus but overlapped by that muscle, and passes on to the distal part of the tibia deep to the muscle and the superior extensor retinaculum. At the ankle joint, the nerve is covered only by the tendon of extensor hallucis longus (which passes obliquely across it) and the inferior extensor retinaculum. On the proximal part of the dorsum of the foot, the nerve lies superficially between the tendons of extensor digitorum longus and extensor hallucis longus.

In the leg it supplies extensor digitorum longus, tibialis anterior, extensor hallucis longus, and peroneus tertius, and sends a filament to the anterior surface of the ankle joint.

It then gives a lateral, **muscular branch** obliquely over the tarsus, deep to extensor digitorum brevis, ending in a gangliform enlargement. Branches pass from this enlargement to extensor digitorum brevis, to the tarsal, tarsometatarsal, and metatarsophalangeal joints. Of these articular branches, the lateral may only reach the tarsometatarsal joints, while the medial may give sensory branches also to the second and third dorsal interosseous muscles.

The nerve continues along the lateral side of the dorsalis pedis artery and sends twigs to the medial tarsal and tarsometatarsal joints. Proximal to the first interdigital cleft, it divides into two **dorsal digital nerves,** one to the lateral side of the hallux, the other to the medial side of the second toe. Each of these nerves communicates with the terminal branches of the superficial peroneal nerve, supplies the metatarsophalangeal and interphalangeal joints, and the skin as far as the distal interphalangeal joint. A sensory branch enters the first dorsal interosseous muscle.

Superficial Peroneal Nerve. This nerve descends in the intermuscular septum between the peronei and extensor digitorum longus, anterior to the fibula, and sends muscular branches to peroneus longus and brevis. As it enters the lower third of the leg, the nerve pierces the deep fascia on the anterior surface [FIG. 949], and divides into **medial** and **intermediate dorsal cutaneous nerves.** The nerve may divide into these branches before piercing the deep fascia. These nerves descend in the subcutaneous tissue to the dorsum of the foot, giving branches to the skin and fascia of the distal third of the anterior surface of the leg and to the dorsum of the foot.

the sural nerve in the middle third of the leg. Sometimes this union does not take place; then the peroneal communicating nerve may supply skin and fascia on the lateral side of the leg, or pass to the area usually supplied by the sural nerve.

3. A small **recurrent branch** arises immediately above the termination of the common peroneal nerve. It passes forwards through the origin of peroneus longus and extensor digitorum longus, and divides, inferior to the lateral condyle of the tibia, into branches to the proximal part of the tibialis anterior muscle and the superior tibiofibular and knee joints [FIG. 949].

FIG. 950   Distribution of cutaneous nerves on dorsum of the foot.
I.S., saphenous nerve ; S.P., superficial peroneal nerve ; D.P.,
deep peroneal nerve ; E.S., sural nerve. The extremities of the toes
are supplied by the medial and lateral plantar nerves (I.P., E.P.).

The **medial dorsal cutaneous nerve** divides into three branches. 1. The most medial branch supplies the skin and fascia of the medial side of the dorsum of the foot and of the hallux [FIG. 950] and communicates with the saphenous nerve. 2. The intermediate branch passes towards the cleft between the hallux and the second toe, and divides into two branches which communicate with the branches of the deep peroneal nerve to that cleft. 3. The lateral branch passes to the cleft between the second and third toes, and divides into dorsal digital nerves to their adjacent sides [FIG. 950].

The **intermediate dorsal cutaneous nerve** divides into two branches on the dorsum of the foot. They pass to the lateral two interdigital clefts to form dorsal digital nerves to the adjacent side of the toes. The lateral branch communicates with the **lateral dorsal cutaneous nerve** from the sural nerve [FIG. 950].

The area supplied by the intermediate dorsal cutaneous nerve is variable. It may additionally supply the adjacent sides of the second and third toes, or the lateral dorsal cutaneous nerve may extend into its territory and supply as many as two and a half toes (Kosinski, 1926).

The dorsal digital nerves are much smaller than the plantar digital nerves which supply all the skin and fascia over the distal phalanges.

TIBIAL NERVE

This nerve (L. 4, 5, S. 1, 2, 3) arises from the ventral surface of the plexus. It forms the medial part of the sciatic nerve, and continues the line of that nerve from its bifurcation, through the popliteal fossa to lie deep in the posterior part of the leg.

In the popliteal fossa it is at first concealed by the semi-membranosus, but then comes to lie immediately deep to the popliteal fascia, and passes obliquely across the posterior surface of the popliteal vessels to lie medial to them on popliteus, deep to gastrocnemius and plantaris [FIG. 951]. It then passes deep to the tendinous arch of soleus, and lies medial to the posterior tibial vessels in the intermuscular septum superficial to tibialis posterior. The nerve descends in this plane, between flexor digitorum longus and flexor hallucis longus, and crossing the posterior surface of the ankle joint between the tendons of these muscles, curves antero-inferiorly into the foot with them. Here, postero-inferior to the medial malleolus it is covered superficially only by the flexor retinaculum and skin and divides into lateral and medial plantar nerves [FIG. 953].

Branches

1. While it is incorporated in the sciatic nerve it gives off its **nerves to the hamstrings** and an occasional **nerve to the hip joint,** see above.

2. In the popliteal fossa. (i) One slender **articular branch** pierces the oblique popliteal ligament, the other passes with the inferior medial genicular vessels to the knee joint.

(ii) **Muscular branches** enter the two heads of gastrocnemius and plantaris at the borders of the popliteal fossa [FIG. 951], and the nerve to soleus enters the superficial surface of that muscle. The nerve to popliteus descends over the posterior surface of that muscle, turns round its inferior border, to enter its anterior surface. At the inferior border of popliteus this nerve gives branches (a) to tibialis posterior, (b) to the whole length of the interosseous membrane, (c) to the tibio-fibular joints, and (d) a medullary branch to the tibia.

(iii) The cutaneous branch is the **sural nerve.** It arises in the lower part of the popliteal fossa, and descending between the heads of gastrocnemius, pierces the deep fascia in the middle third of the posterior surface of the leg. It is then usually joined by the peroneal communicating branch of the common peroneal nerve, and passes on to the tendo calcaneus with the small saphenous vein, giving cutaneous branches to the back and lateral side of the lower third of the leg. At the ankle it lies posterior to the tendons of the peroneal muscles, and giving branches to the ankle and heel (**lateral calcanean branches),** passes forwards into the lateral border of the foot (**lateral dorsal cutaneous nerve)** to supply it and send **articular branches** to the ankle and tarsal joints. In the foot it communicates with the intermediate cutaneous branch of the superficial peroneal nerve [FIG. 950], and may either reinforce or replace that branch. The peroneal communicating branch of the common peroneal nerve is distributed within the territory of the sural nerve if they do not unite.

3. Branches in the Leg. (i) **Muscular branches** (often by a common trunk) to the deep surface of soleus and to tibialis posterior, the latter sending branches to the posterior tibial artery. Separate branches pass to flexor digitorum longus and flexor hallucis longus ; the latter accompanies the peroneal artery for some distance and supplies it.

(ii) Cutaneous, **medial calcanean branches** pierce the flexor retinaculum, and supply the skin and fascia of the heel and posterior part of the foot. (iii) A long, slender branch arises with the nerves to the deep muscles. It runs with the

Gracilis

Semimembranosus

Medial head of gastrocnemius

Sural nerve

Popliteus and its nerve

Soleus

Branch to flexor digitorum longus

Tibial nerve

Flexor digitorum longus

Flexor retinaculum

Medial calcanean nerve

Tendo calcaneus

Peroneal communicating branch

Plantaris

Lateral cutaneous nerve of calf

Lateral head of gastrocnemius

Head of fibula

Branch to tibialis posterior

Posterior surface of fibula

Peroneus longus

Tibialis posterior

Flexor hallucis longus and its nerve

Peroneus brevis

Sural nerve

Peroneus longus

Lateral calcanean nerve

FIG. 951   Nerves of popliteal fossa and back of leg.

branches to tarsal and tarsometatarsal joints. (iv) Vascular nerves to the adjacent arteries from the nerve and its branches. (v) One proper and three common plantar digital nerves.

The **proper plantar digital nerve** [FIG. 953] separates from the nerve before the others, and passing forwards, pierces the deep fascia posterior to the ball of the big toe. It supplies a muscular branch to flexor hallucis brevis, and cutaneous branches to the medial side of the foot, the ball and medial aspect of the great toe [FIG. 950].

The three **common plantar digital nerves** arise together. The first (medial) supplies the first lumbrical muscle, and becoming superficial just proximal to the first interdigital cleft, gives a proper plantar digital nerve to the adjacent sides of the first and the second toes. The second and third common plantar digital nerves have no muscular branches, but otherwise behave as the first [FIG. 953]. The most lateral proper plantar digital nerve communicates with the most medial proper digital branch of the lateral plantar nerve.

Lateral Plantar Nerve. This nerve, homologous with the ulnar nerve in the hand, passes anterolaterally towards the base of the fifth metatarsal between flexor digitorum brevis and flexor accessorius, on the medial side of the lateral plantar artery. It gives branches to flexor accessorius and abductor digiti minimi, and divides into superficial and deep branches medial to the base of the fifth metatarsal [FIG. 953].

Branches. (i) **Muscular,** see above. (ii) Small **cutaneous branches** pierce the plantar fascia along the line between flexor

peroneal vessels as far as the ankle, gives a medullary branch to the fibula which also supplies its periosteum. One or two articular branches to the ankle arise from the tibial nerve at the ankle.

4. Terminal Branches.

Medial Plantar Nerve. This nerve, homologous with the median nerve in the hand, is rather larger than the lateral plantar [FIGS. 952, 953]. It passes forwards with the medial plantar artery under cover of the flexor retinaculum and abductor hallucis, to the interval between that muscle and the flexor digitorum brevis in the sole of the foot.

Branches. (i) **Muscular branches** to abductor hallucis and flexor digitorum brevis. (ii) Small **cutaneous branches** pierce the plantar fascia between the above muscles to supply the skin of the medial part of the sole of the foot. (iii) **Articular**

C.T.A.—2C

FIG. 952   Scheme of distribution of the plantar nerves.

On the medial side, medial plantar nerve, and its cutaneous and muscular branches; F.D.B., flexor digitorum brevis; A.H., abductor hallucis. On the lateral side, lateral plantar nerve, and its cutaneous and muscular branches; F.D.A., flexor digitorum accessorius; A.D.M., abductor digiti minimi; F.D.M.B., flexor digiti minimi brevis.

Transverse head of adductor hallucis

Fibrous flexor sheath

Tendons of short flexor (cut) and long flexor

Flexor digiti minimi

Abductor digiti minimi

4th dorsal interosseous

3rd plantar interosseous

Common digital branches of superficial branch of lateral plantar nerve

A branch to skin of sole

Branch to flexor digiti minimi

Deep branch of lateral plantar nerve

Base of 5th metatarsal bone

Oblique head of adductor hallucis

Sheath of peroneus longus

Branches to skin of sole

Branch to flexor digitorum brevis (cut)

Branch to abductor digiti minimi

Flexor digitorum brevis

Plantar aponeurosis

Abductor digiti minimi

Lateral process of tuber calcanei

Branches of lateral calcanean nerve

Proper and common digital branches of medial plantar nerve

Lumbricals

Branches to the two parts of flexor hallucis brevis

A branch to skin of sole

Flexor hallucis longus

Branch to flexor accessorius between its two heads

A branch to skin of sole

Flexor digitorum longus

Medial plantar nerve

Abductor hallucis

Lateral plantar nerve crossing flexor hallucis longus

Branches of medial calcanean nerve

Medial process of tuber calcanei

FIG. 953   Plantar nerves.

digitorum brevis and abductor digiti minimi to supply the lateral part of the sole. (iii) Vascular branches to adjacent arteries.

The superficial branch passes forwards between flexor digitorum brevis and abductor digiti minimi, gives cutaneous branches to the sole, and divides into a proper (lateral) and a common (medial) **plantar digital nerve.** The former supplies flexor digiti minimi brevis and sometimes the fourth dorsal and third plantar interosseous muscles. It becomes superficial posterior to the ball of the little toe, supplies the skin of that area and on the plantar surface, lateral side, and distal part of the dorsum of the little toe. The **common plantar digital nerve** divides proximal to the fourth interdigital cleft into **proper plantar digital nerves** to the adjacent sides of the fourth and fifth toes. The more medial of these nerves communicates with the adjacent branch of the medial plantar nerve.

Proper plantar digital nerves supply the skin and fascia on the plantar, tip, and distal part of the dorsal surfaces of the

toes. They also supply the metatarsophalangeal and interphalangeal joints.

The deep branch of the lateral plantar nerve runs medially with the plantar arch on the inferior surface of the metatarsal bases, superior to flexor digitorum accessorius and the oblique head of adductor hallucis. It supplies the tarsal and tarsometatarsal joints, the interossei (except any supplied by the superficial branch), the adductor hallucis, and the lateral three lumbricals. The muscular branches enter the superior surfaces of the muscles (except the interossei), that to the second lumbrical passing anteriorly above the transverse head of adductor hallucis.

THE COCCYGEAL PLEXUS

The fourth sacral ventral ramus descends to join the fifth sacral, which passes caudally beside the coccyx and is joined by the coccygeal nerve to form a plexiform cord, the coccygeal

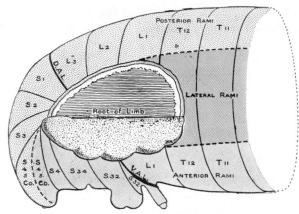

FIG. 954   Scheme of innervation of hinder portion of trunk and perineum.   The scheme shows the interruption of the segmental arrangement of the nerves due to the formation of the limb. V.A.L. and D.A.L., ventral and dorsal axial lines [See also Figs. 943 and 946.]

plexus.   Fine twigs arise from the plexus, and some of these enter and supply coccygeus and the adjacent part of levator ani, while others pierce the coccygeus and the overlying ligaments (sacrospinous and sacrotuberous) to supply skin adjacent to the coccyx and posterior to the anus.

The sacral and coccygeal ventral rami receive **grey rami communicantes** from the sacral part of the sympathetic trunk which lies medial to their point of emergence from the vertebral column.

GENERAL ARRANGEMENT OF THE PERINEAL NERVES

The perineum is supplied by the ventral rami of the spinal nerves which emerge from the vertebral column caudal to the lumbosacral plexus.   It represents, therefore, the ventral part of the trunk which lies caudal to the lower limb and tapers to the coccyx.   It is equivalent to that ventral part of the trunk caudal to the upper limb which is supplied by the ventral rami of the thoracic nerves inferior to the brachial plexus.   Just as the latter nerves are serially arranged from the second thoracic caudally, so the perineal nerves are serially arranged so that the highest (S. 2, 3) innervates skin near the groin (adjacent to the area supplied by the first lumbar nerve) while the lowest (coccygeal) supplies skin in the region of the coccyx [FIG. 946]. Because of the overlap of all adjacent dermatomes [p. 736], no area of skin in the perineum is supplied by a single spinal nerve, but by two adjacent nerves at least.   Between the areas of skin supplied by the ilio-inguinal nerve (L. 1) and the dorsal nerve of the penis (S. (2), 3) is the beginning of the ventral axial line (q.v.) of the lower limb [FIG. 943] and the beginning of the dorsal axial line lies between the areas of skin supplied by the dorsal rami of the third lumbar and first sacral nerves [FIG. 943, p. 737].

DISTRIBUTION OF SPINAL NERVES TO THE SKIN AND MUSCLES OF THE LIMBS

The evidence of dissection, experiment, and clinical observation indicate that virtually every peripheral nerve to the limbs contains nerve fibres derived from more than one ventral ramus, and that the cutaneous distribution of individual peripheral nerves is such that the area of skin supplied overlaps the areas of adjacent cutaneous nerves.   The ventral rami of the spinal nerves are, nevertheless, distributed through these peripheral nerves in a standard manner such that certain general rules can be laid down regarding their cutaneous and muscular distribution (Herringham, 1886).   All ventral rami entering a limb plexus eventually divide into dorsal and ventral divisions which supply respectively the sheets of muscle and the overlying skin of the dorsal (extensor) and ventral (flexor) surfaces of the limbs prior to their rotation into the adult position [p. 50].

INNERVATION OF THE SKIN OF THE LIMBS

[FIGS. 930, 943]

The growth of the embryonic limb causes the segmentally innervated skin which overlies it to be drawn out like a stretched sheet, in such a manner that the dermatome supplied by the middle ventral ramus of the plexus is carried away from the trunk with the distal part of the limb.   The dermatomes supplied by the more cephalic ventral rami are drawn out to a correspondingly lesser extent along the superior (preaxial; thumb or great toe) border of the limb, while those supplied by the more caudal ventral rami lie along the inferior (postaxial; little finger or toe) border of the limb.   Consequently: (1) the dermatome supplied by the middle ventral ramus of the plexus and a number of those superior and inferior to it disappear from the trunk; (2) the dermatomes lie serially along the superior and inferior borders of each limb, but those of the inferior border abutting on those of the superior border are not necessarily supplied by adjacent ventral rami [FIGS. 930, 943], being separated by the imaginary dorsal and ventral **axial lines.**   At the root of the limb, the axial lines separate areas of skin which are supplied by the ventral rami from the superior and inferior nerves of the plexus, and it is only at the periphery of the limbs that the dermatomes lie in numerical sequence from preaxial to postaxial borders.

Thus of two spots on the skin of the limb: (1) that in the preaxial part will be supplied by a higher ventral ramus than that in the postaxial part; (2) that more proximally placed on the preaxial border, or more distally placed on the postaxial border will also be supplied by a higher ventral ramus.

Axial Lines

In the upper limb the dorsal axial line extends from the vertebra prominens down the back of the limb to the elbow. The ventral axial line extends from the manubriosternal joint along the ventral surface of the arm and forearm [FIG. 930].

In the lower limb the dorsal axial line extends across the posterior superior iliac spine, the gluteal region, and the back of the thigh to the head of the fibula and the lateral side of the leg.   The ventral axial line extends from the penis along the medial side of the thigh and knee and the posteromedial aspect of the leg to the heel [FIG. 943].

When the developing limbs rotate into the adult position, the flexor surfaces of the upper limbs and the extensor surfaces of the lower limbs become anterior, and the nerves of these respective surfaces form the preponderant cutaneous supply for each limb. Thus nerves arising from dorsal divisions of the ventral rami in the lower limb and from ventral divisions of the ventral rami in the upper limb supply a disproportionately a rge part of the skin of the limb [FIGS. 929, 942].

INNERVATION OF THE MUSCLES OF THE LIMBS

1. The limb muscles are entirely supplied by the ventral rami, and each muscle is supplied by more than one ventral ramus.

2. The dorsal (extensor) and ventral (flexor) strata of muscles are always supplied by the dorsal and ventral divisions of the ventral rami respectively. The ventral rami that supply the flexor muscles are more numerous, the extra ventral rami being inferior. Thus in the upper limb dorsal divisions supplying muscles arise from C. 5, 6, 7, 8, and ventral divisions from

C. 5, 6, 7, 8, T. 1. In the lower limb the corresponding figures are L. 2, 3, 4, 5, S. 1, 2 (dorsal) and L. 2, 3, 4, 5, S. 1, 2, 3 (ventral).

3. The muscles of the preaxial part of the limb are supplied by nerves higher than those of the postaxial part of the limb.

4. The most distal muscles of the limb are supplied by the lowest nerve or nerves of the plexus; small muscles of the hand by the first thoracic nerve, and those of the foot by the second and third sacral nerves.

Muscles with a Double Nerve Supply

Virtually all muscles of the limbs receive nerve fibres from more than one ventral ramus; in some cases (e.g., pectoralis major) from all the ventral rami of the plexus. In such cases, no mixing of fibres from dorsal or ventral divisions occurs. It is assumed in these cases that the muscles are formed from the fusion of muscle primordia from several segments. In a very few cases (e.g., biceps femoris and sometimes pectineus), the

THE SEGMENTAL INNERVATION OF THE MUSCLES OF THE UPPER LIMB

Muscle	C.5	C.6	C.7	C.8	T.1
Supraspinatus	■	■			
Infraspinatus	■	■			
Teres minor	■				
Deltoid	■	■			
Biceps		■			
Brachialis		■			
Brachioradialis		■			
Serratus anterior	■	■	■		
Subscapularis	■	■	■		
Pectoralis major	■	■	■	■	■
Coracobrachialis		■	■		
Teres major		■	■		
Pectoralis minor		■	■	■	
Latissimus dorsi		■	■	■	
Pronator teres		■	■		
Ext. carpi radialis longus and brevis		■	■		
Triceps		■	■	■	
Flexor carpi radialis		■	■		
Extensor digitorum			■	■	
Flexor carpi ulnaris			■	■	
Extensor digiti minimi			■	■	
Extensor carpi ulnaris			■	■	
Extensor pollicis longus			■	■	
Abductor pollicis longus			■	■	
Extensor pollicis brevis			■	■	
Extensor indicis			■	■	
Flexor digitorum superficialis			■	■	■
Flexor digitorum profundus			■	■	■
Pronator quadratus				■	■
Flexor pollicis longus				■	■
Adductor pollicis				■	■
Flexor pollicis brevis				■	■
Opponens pollicis				■	■
Interossei				■	■
Abductor pollicis brevis				■	■
Lumbricals				■	■

THE SEGMENTAL INNERVATION OF THE MUSCLES OF THE LOWER LIMB

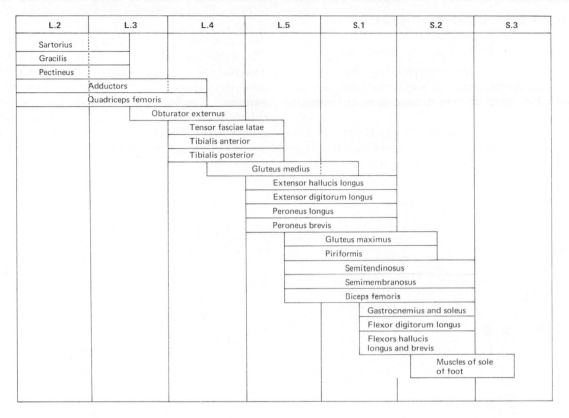

dual nerve supply is from different divisions of the ventral rami and this is thought to represent fusions between muscle rudiments of the flexor and extensor sheets. The terms 'flexor' and 'extensor' used in connexion with the muscle sheets in the ventral and dorsal parts of the developing limb do not apply to the final functions of the muscles. Thus though most of the muscles of the dorsal sheet become extensors of joints, many may act as flexors and vice versa; e.g., brachioradialis is a flexor of the elbow, and rectus femoris and sartorius are flexors of the hip, though all these muscles belong to the 'extensor' muscle sheet and are supplied by dorsal divisions of the ventral rami.

Intraneural Plexuses

Within peripheral nerves, the bundles of nerve fibres are constantly changing by branching and recombining to form different fasciculi surrounded by perineurium. The significance of these intraneural plexuses is obscure (Sunderland, 1968) since the amount of change is greater than necessary to rearrange the individual nerve fibres into the final peripheral branches. The variations which occur in the peripheral distribution of some nerves, and the communications between various nerves after they have left a common stem (e.g., the median and musculocutaneous nerves) show that the presence of these intraneural plexuses is not always associated with the standard distribution of individual fibres into specific branches. Whatever the reasons for these variable communications within nerves, their presence complicates the process of regeneration in a divided nerve where the outgrowths from the central stump have to reach the correct peripheral branches if regeneration is to be successful.

BLOOD SUPPLY OF PERIPHERAL NERVES

All peripheral nerves are supplied with blood vessels which are essential to their proper function. If the blood supply is interrupted, failure of conduction results; the large diameter fibres being more sensitive to anoxia than the small diameter fibres.

Every peripheral nerve receives a number of small arterial branches from adjacent vessels along its length. These are irregular in their arrangement, and they divide into longitudinal branches on the epineurium which anastomose with similar branches of adjacent nutrient vessels. The epineurial vessels send branches on to the perineurium of the fascicles in which they form a rich, longitudinal plexus of capillaries which passes along the whole length of the nerve in unbroken continuity, reinforced at intervals by the various nutrient arteries reaching the epineurium in such a manner that there is no one special nutrient artery (Durward, 1948).

Special arteries associated with certain nerves, e.g., the sciatic and median, do not form a preferential supply, but are embryonic remnants. The arteries to nerves are accompanied by postganglionic sympathetic nerve fibres which supply their muscle coats (Adams, 1942, 1943; Sunderland, 1945 a, b, c).

Veins drain into adjacent vessels, but do not unite to form longitudinal channels of any size.

DEVELOPMENT OF SPINAL NERVES AND GANGLIA

[FIG. 955]

The peripheral nervous system is derived from two sources, the cells of the neural crest [FIG. 955 and p. 591] and those of the neural tube, both of which are derived from the original neural plate.

The neural crest forms a considerable number of different varieties of cells (Horstadius, 1950) and all of these are migratory in character. The crest extends from the level of the hindbrain to the caudal extremity of the neural tube, and from it cells migrate outwards with the developing nerve fibres of the peripheral nerves. These form: (1) spongioblasts (Schwann cells), the cellular sheaths of the nerve fibres (Harrison, 1924); (2) neuroblasts, which form the **spinal ganglia,** and a special variety of these which migrate further from the central nervous system to form the **autonomic ganglia.** The latter are sometimes known as sympathoblasts, although they form both sympathetic and parasympathetic ganglia from different regions of the neural crest. (3) A number of other types of cells including the pigment-forming **dendritic cells** of the skin and phaeochromoblasts.

1. Spinal ganglion,
2. Ganglion of sympathetic trunk.
3. Sympathetic cells innervating kidneys and gonads (x) or forming suprarenal medulla.
4. Sympathetic ganglia of alimentary canal and structures developed from it.
5. Spinal medulla.
6. Alimentary canal.
7. Dorsal ramus,
8. Ventral ramus.
9.}
10.} Lateral and anterior cutaneous branches.
11.}
12.} Dorsal and ventral divisions of ventral ramus to plexus.
13. Intercostal nerve.
14. Cutaneous branch of dorsal ramus.
15. Thoracic splanchnic nerve.
16. Anterior wall of trunk.
17. Cervical splanchnic nerve.
18. Ventral root.
19. Dorsal root.
20. Grey ramus communicans.
21. White ramus communicans.

FIG. 955 Diagrams of the development and arrangement of certain spinal nerves. A. Arrangement of nerve cells derived from the neural crest. B. The elements of a typical spinal nerve. C. The elements of a spinal nerve adjacent to a limb plexus, e.g., 1st thoracic. D. The elements of a spinal nerve in a limb plexus. Black, somatic and sympathetic (postganglionic) motor cells. Red, preganglionic sympathetic cells. Blue, sensory cells

The **spinal ganglia** appear as laterally directed, segmental swellings of the neural crest which extend ventrally to lie lateral to the neural tube. The neuroblasts which form these ganglia each send a process (axon) centrally to enter the dorso-lateral aspect of the neural tube and together these form the **dorsal roots** of the spinal nerves. They also give rise to peripheral processes (dendrites) which immediately join the corresponding ventral root to form a **spinal nerve.** By the continued outgrowth of their peripheral processes the spinal ganglion cells make contact with the various sensory areas of the body, and thus link them to the spinal medulla. The original bipolar cells of the spinal ganglia become unipolar by excentric growth of the cells [FIG. 711].

The **sympathoblasts** migrate ventrally from the spinal ganglia to form separate groups of sympathetic nerve cells, which are primarily associated with the three basic groups of arteries arising from the aorta: to the body wall mesoderm, to the intermediate mesoderm, and to the visceral mesoderm [FIG. 721]. By the outgrowth of axons, each of these groups of sympathetic neurons innervates the vessel adjacent to which it lies and the tissues supplied by that vessel [see also p. 591 and p. 791].

The **phaeochromoblasts** are intimately associated with the sympathoblasts in their migration, but form isolated clumps of cells which stain dark brown with chromium salts, **chromaffin tissue.** This forms the medulla of the supra-renal glands, and the para-aortic bodies which are temporary structures especially well developed in the foetus anterior to the abdominal aorta.

The **ventral roots** of the spinal nerves arise from neuro-blasts situated in the mantle layer of the basal lamina [p. 592] of the neural tube. These neuroblasts send their axons out in small bundles from the neural tube to unite segmentally into the ventral roots, which join the dorsal roots at the spinal ganglia between the myotome and the neural tube. The ventral roots contain two types of nerve fibres. (1) The **somatic motor nerve fibres** which are distributed to the striated muscles of the body wall and limbs. (2) **Preganglionic**
autonomic nerve fibres which grow out into the ventral rami of the spinal nerves, and then pass along the line of migration of the sympathoblasts to reach, and make synaptic connexion, with these cells and with the phaeochromoblasts. In the first part of this course, but only in the first thoracic to third lumbar nerves, these fibres pass from the ventral ramus to the sympathetic trunk as the **white ramus communicans.** Those fibres which continue to the visceral ganglia of the sympathetic system (accompanied by fibres from the spinal ganglia which also follow this route) pass through the sympathetic trunk and become the **splanchnic nerves.** In the third, and second or fourth sacral spinal nerves, equivalent fibres leave the ventral rami and pass directly to the corresponding visceral ganglia (parasympathetic) as **pelvic splanchnic nerves** [p. 780]. They have no connexion with the sympathetic trunk.

The body wall (somatic) part of each spinal nerve is formed by the lateral growth of its nerve fibres, which pass from the spinal ganglia and ventral roots into the corresponding dorsal and ventral rami. The dorsal ramus enters the myotome, supplies it, and branching, continues through it to the overlying skin. The ventral ramus grows ventrolaterally, anterior to the myotome, gives off its smaller visceral branch (white ramus communicans), and continues into the body wall giving rise to branches to muscles and skin [FIG. 955].

Formation of the Limb Plexuses

The growing nerves in the regions of the limbs behave in the same manner, but the ventral rami give no visceral branches to the autonomic nervous system. They pass directly into the limb bud ventral to the myotomes, and divide into dorsal and ventral divisions which unite with the corresponding branches of adjacent ventral rami to form the nerves of the dorsal and ventral aspects of the limb bud. The dorsal and ventral divisions of the ventral rami that enter the limb plexuses may correspond to the lateral and anterior cutaneous branches of the typical thoracic ventral ramus.

THE AUTONOMIC NERVOUS SYSTEM

The autonomic nervous system may be defined as that part of the nervous system which innervates glands, smooth muscle, cardiac muscle, and muscle which is histologically of the 'voluntary' type in the oesophagus. It has sometimes been called the 'involuntary' nervous system because the structures which it supplies do not appear to be immediately under voluntary control. Yet this is not a satisfactory name since there are smooth muscles which, though innervated by the autonomic nervous system, are obviously used as part of 'voluntary' acts, e.g., the urinary bladder in micturition, the ciliary muscle in focussing the lens of the eye, etc. Also many muscles which are histologically of the 'voluntary' type cannot usually be contracted at will, e.g., the pharyngeal constrictors, and any voluntary muscle may be used in 'involuntary' acts, e.g., shivering. In fact the functional arrangements which are dictated by the central nervous system
are such that the structures innervated by the autonomic nervous system play their part in every kind of activity whether this activity is apparently initiated voluntarily or not. Thus changes occur in blood pressure and in the calibre of blood vessels so that extra blood is carried to active tissues, e.g., the voluntary muscles in muscular exercise, and in states of fear.

The only satisfactory anatomical definition of the autonomic nervous system, which allows it to be differentiated from the remainder of the nervous system, is that of Langley (1921). This regards it as a peripheral efferent system consisting of groups of nerve cells (**ganglia**) situated outside the central nervous system and interposed between it and the organs which are supplied by the axons (**postganglionic nerve fibres**) of the cells in these ganglia. It is in no sense functionally separate from the central nervous system, but receiving axons (**preganglionic nerve fibres**) from cells within that system, forms one of the routes by which the central nervous system controls the tissues of the body. The significant difference between this efferent route **(visceral efferent)** and that which supplies the muscles of the body wall and limbs **(somatic efferent)** is that the cells of the visceral efferent route which actually innervate the tissues lie outside the central nervous system, while those of the somatic efferent route lie inside the central nervous system.

This definition excludes all parts of the sensory system from the autonomic system. Yet sensory nerve fibres in large numbers run in the same fibrous sheaths as the nerve fibres of the autonomic nervous system, and are distributed to the same territories. However, they make no direct connexions with the cells of the autonomic nervous system, but discharge directly into the central nervous system through the dorsal roots of the spinal nerves, or through the cranial nerves, and have their

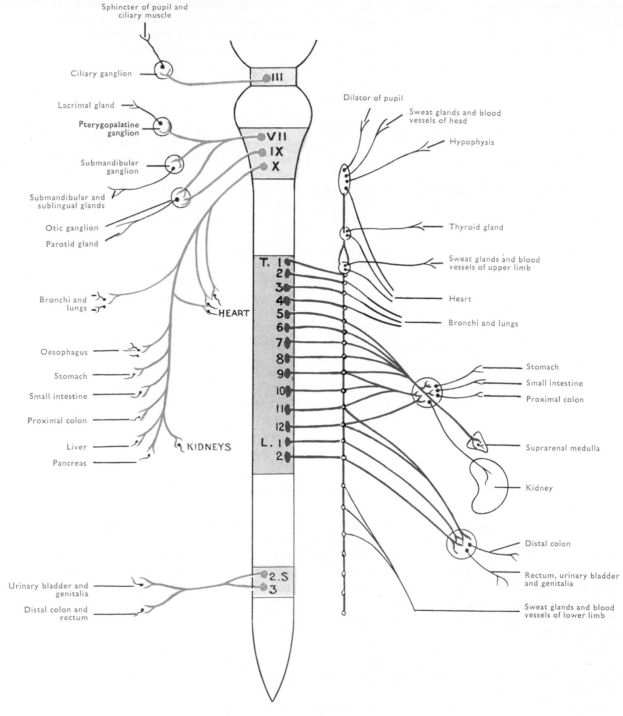

Fig. 956   Scheme of general arrangement of autonomic nervous system.

Preganglionic fibres of the parasympathetic system, blue, and of the sympathetic system, red. Postganglionic fibres of both systems, black. Afferent fibres not included.

cell bodies in the ganglia of these nerves. They are, therefore, known as **visceral afferent,** or visceral sensory fibres. Since the reactions their impulses produce through the central nervous system may be somatic or visceral or both combined, they cannot be considered as specifically autonomic. The information which the visceral afferent fibres transmit is, however, different from that transmitted by the somatic afferent fibres. Most of it does not reach a conscious level (though a sense of fullness and occasionally pain may be appreciated), but it is of profound importance for the maintenance of the internal environment, especially the information from pressure and chemical receptors related to the blood vessels.

The autonomic nervous system is distributed throughout the body. Though many of its cells are arranged in ganglia which can readily be dissected, in other situations, particularly in association with the gut tube and blood vessels, the cells are scattered or in minute groups in a tangle of preganglionic, postganglionic, and sensory nerve fibres. Such arrangements constitute the **autonomic plexuses.** The ganglion cells greatly exceed in number the preganglionic fibres which pass to synapse with them. Thus each preganglionic fibre tends to synapse with many ganglion cells, hence widespread effects are likely to arise from stimulation of a few preganglionic nerve fibres. There must, however, be a topographical organization within the autonomic nervous system which allows for localized responses for particular functions.

The autonomic nervous system is divisible into **sympathetic** and **parasympathetic parts** since the preganglionic fibres to the sympathetic part emerge in the thoracic and upper two or three lumbar spinal nerves (**thoracolumbar outflow**), while those of the parasympathetic part emerge with certain cranial and sacral nerves (**craniosacral outflow**).

The two parts can also be differentiated, for the sympathetic is distributed to all parts of the body, more especially to the blood vessels; while the parasympathetic has a distribution limited to the alimentary tract and structures developed in or from its walls, to the respiratory and urogenital systems, and the eye, all of which are therefore innervated by both sympathetic and parasympathetic parts. *The parasympathetic plays no part in the innervation of the body wall or limbs.* It has been claimed that there are parasympathetic fibres in the dorsal roots of the spinal nerves (Kure, 1931) because stimulation of the distal end of a cut dorsal root produces vasodilatation in its territory of distribution. There is, however, no evidence of efferent fibres in the dorsal roots, and the vasodilatation is believed to be due to stimulation of the sensory fibres.

Where sympathetic and parasympathetic parts supply the same structure, their effects are frequently opposite. Thus the heart is accelerated by the sympathetic, and slowed or arrested by the parasympathetic; the pupil is dilated by the sympathetic, and constricted by the parasympathetic; the parasympathetic increases peristalsis in the alimentary canal, while the sympathetic reduces it and causes contraction of the sphincters. This antagonism is not real for the two parts co-operate to produce a balance within which it is possible to obtain the multiple gradations necessary for normal functioning in a wide variety of circumstances.

In the somatic tissues, the sympathetic part controls the calibre of blood vessels, causes sweat glands to secrete, and the arrectores pilorum to contract. In a general way the sympathetic part can be looked upon as preparing the body for reaction to some external crisis, while the parasympathetic part is concerned with conservation and the building up of energy.

Chemical transmission of nerve impulses has mainly been studied in the autonomic nervous system because of the ease of access to the synapses between preganglionic axons and ganglion cells (Elliott, 1905; Loewi, 1935; Dale, 1934, 1938). Acetylcholine has long been known to play an important part in the transmission at the above synapses, at some synapses in the central nervous system, and at the junctions both of postganglionic parasympathetic and of somatic motor axons with their effector organs. On the other hand noradrenaline is the chemical concerned with transmission of impulses at most of the endings of the sympathetic postganglionic fibres, though in the case of sweat glands the transmitter substance is acetylcholine. Cholinergic fibres cannot, therefore, be equated with postganglionic parasympathetic fibres, nor adrenergic fibres with postganglionic sympathetic fibres, though this basic chemical difference exists between most of them.

In general, the ganglion cells of the parasympathetic part differ from those of the sympathetic part in that they tend to lie closer to the organs which they supply. Again, the sympathetic ganglia are commonly associated with blood vessels, and their postganglionic fibres are frequently transmitted along blood vessels [FIG. 721]—features which are less common with the parasympathetic cells and fibres.

THE PARASYMPATHETIC PART

The **preganglionic fibres** of this part emerge from two widely separated regions of the central nervous system. The cranial fibres leave the brain through the oculomotor, facial, glossopharyngeal, and vagus nerves, while the preganglionic fibres of the sacral parasympathetic leave the spinal medulla in the third, and second or fourth sacral spinal nerves.

The preganglionic fibres of the cranial and sacral parasympathetic are relatively long and pass directly to parasympathetic ganglia. They have no connexions with the sympathetic trunk or ganglia, and though they may be intermingled with sympathetic postganglionic fibres as they approach their termination, they are merely running together without any direct connexions between them. In the case of the oculomotor, facial, and glossopharyngeal nerves, the preganglionic parasympathetic fibres end on discrete ganglia which have a circumscribed distribution. Hence the effects of stimulation of these fibres are localized, unlike the widespread effects of preganglionic sympathetic stimulation. On the other hand preganglionic parasympathetic fibres in the vagus have a very widespread distribution in the head, neck, thorax, and abdomen, extending as far inferiorly as the left flexure of the colon where their territory of distribution meets that of the preganglionic fibres of the sacral outflow. The sacral parasympathetic preganglionic fibres are distributed to ganglionic neurons on or near the terminal parts of the alimentary and urogenital systems.

CRANIAL PARASYMPATHETIC

Oculomotor Nerve

The preganglionic parasympathetic nerve fibres in this nerve probably arise in the accessory nucleus of the oculomotor nerve in the midbrain [p. 619]. They pass in the oculomotor nerve [p. 713], and leaving the branch to the inferior oblique muscle in the orbit, run as the parasympathetic, **oculomotor root** to the ciliary ganglion [FIG. 957]. In the ganglion, they synapse with the large cells, from which myelinated **postganglionic fibres** enter the **short ciliary nerves.** These nerves pierce the sclera, and running forwards between it and the choroid, supply the ciliary muscle and the circular muscle of the iris (sphincter pupillae), and thus decrease the focal length of the lens and the diameter of the pupil. When looking from a distant to a near object (convergence-accommodation) the eyes are converged on it (medial rectus muscles), the lens focussed, and the pupil contracted ; all these are activities produced by nerve fibres in the oculomotor nerve. Other ganglion cells have been described in the orbit (Nathan and Turner, 1942), but their function, if different from those of the ciliary ganglion, is not known.

Facial Nerve

The preganglionic fibres in this nerve arise in the superior salivatory nucleus [p. 620] at the ponto-medullary junction. They emerge in the nervus intermedius, and joining the facial nerve in the internal acoustic meatus, leave it in the greater petrosal and chorda tympani branches [FIG. 958].

1. The **greater petrosal nerve** [p. 725] is joined by

FIG. 957 Diagram of ciliary ganglion and connexions of autonomic nervous system with the eye.

Sympathetic, red ; parasympathetic, green ; sensory, blue.

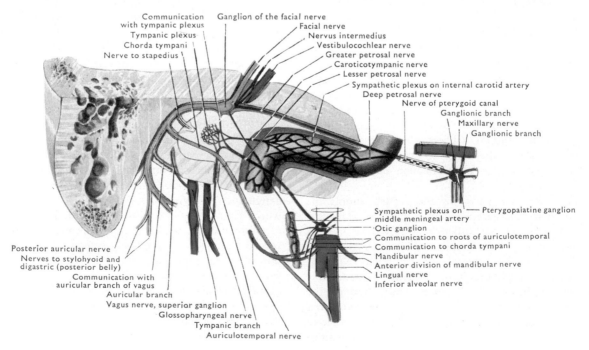

FIG. 958 Branches of facial and glossopharyngeal nerves in the temporal bone, and connexions of pterygopalatine and otic ganglia.

the deep petrosal nerve, and passes to the **pterygopalatine ganglion.** Here its parasympathetic fibres synapse on the cells of the ganglion whose **postganglionic fibres** are distributed with the branches of the maxillary nerve to the palate, to the walls of the cavity and air sinuses of the nose, and to the lacrimal gland [FIG. 959], to increase the secretory activity of glands in these regions.

2. The **chorda tympani** [p. 725] joins the lingual nerve in the infratemporal fossa, the **preganglionic fibres** leave that nerve on the hyoglossus to pass to the **submandibular ganglion** and synapse with its cells. The **postganglionic fibres** from the cells of the ganglion pass: (i) directly to the submandibular gland [FIG. 904] and sublingual gland; and (ii) into the lingual nerve for distribution to the glands in the anterior part of the tongue and to the sublingual gland [FIG. 959]. Some preganglionic fibres may pass directly to scattered ganglion cells adjacent to these glands (Fitzgerald and Alexander, 1969). Stimulation of the chorda tympani causes vasodilatation in the glands it supplies, though their vasomotor supply is through sympathetic fibres from the plexuses on the arteries that supply them (Kunz and Richins, 1946).

Preganglionic fibres of the facial nerve are also described as passing to the **tympanic plexus,** and thence through the lesser petrosal nerve to the otic ganglion. If this is their course,

then the parasympathetic fibres in the facial nerve may assist those of the glossopharyngeal nerve in supplying the parotid gland.

Glossopharyngeal Nerve

The preganglionic parasympathetic nerve fibres pass into this nerve from the inferior salivatory nucleus [p. 620]. They are distributed through: (1) its **tympanic branch** to the tympanic plexus, and thence via the lesser petrosal nerve to the **otic ganglion** [FIG. 960]; and (2) to scattered ganglion cells in the posterior third of the tongue. The cells of the otic ganglion send their **postganglionic fibres** through the auriculotemporal nerve to the parotid gland, while those in the posterior third of the tongue supply the numerous glands in that part of the organ; both increase the secretory activity of the gland cells.

Vagus Nerve

This nerve contains the largest group of preganglionic parasympathetic fibres in any nerve, together with numerous visceral afferent fibres. The preganglionic fibres arise from the cells of the dorsal vagal nucleus [p. 620]. The ganglion cells to which the preganglionic fibres pass are all either scattered close to the organs they supply, or included in the

FIG. 959 Diagram of connexions of pterygopalatine and submandibular ganglia with distribution of parasympathetic components of seventh cranial nerve.

Sympathetic, red; parasmypathetic, green; sensory, blue.

FIG. 960 Diagram of connexions of otic ganglion with distribution of parasympathetic components of ninth cranial nerve.

Sympathetic, red; parasympathetic, green; sensory, blue.

autonomic plexuses in which the sympathetic fibres to the organs also run. The latter remain functionally separate though impossible to disentangle by dissection. All branches of the vagus, except possibly the meningeal and auricular, contain preganglionic parasympathetic fibres. Through these they are distributed to: (1) scattered ganglion cells associated with the glands in the pharynx and larynx, and the glands and smooth muscle in the trachea and bronchi; (2) the cardiac, pulmonary, and oesophageal plexuses; (3) the stomach via the vagal trunks and gastric nerves; and (4) from the posterior vagal trunk, fibres pass through the great abdominal plexuses, associated with the coeliac and superior mesenteric arteries and their branches [pp. 789, 790] to the myenteric and submucous plexuses of the gut tube as far as the left flexure of the colon, and to the liver and pancreas (Schofield, 1956).

In general these fibres are concerned with: (1) decreasing the rate of cardiac contraction; (2) bronchoconstriction; (3) increasing peristalsis of the gut tube and relaxing the sphincters; and (4) causing increased secretion by the glands in the whole territory of the vagus nerve. Section of the vagus nerves decreases, for example, the gastric secretion of hydrochloric acid.

SACRAL PARASYMPATHETIC

The sacral parasympathetic nerves (**pelvic splanchnic nerves**) arise as delicate branches from the third and fourth sacral ventral rami, less commonly also from either the second or the fifth (Sheehan, 1941). Their cells of origin lie in the sacral part of the spinal medulla at the level of the first lumbar vertebra, and thus their nerve fibres course through the cauda equina to reach the sacral foramina. The two

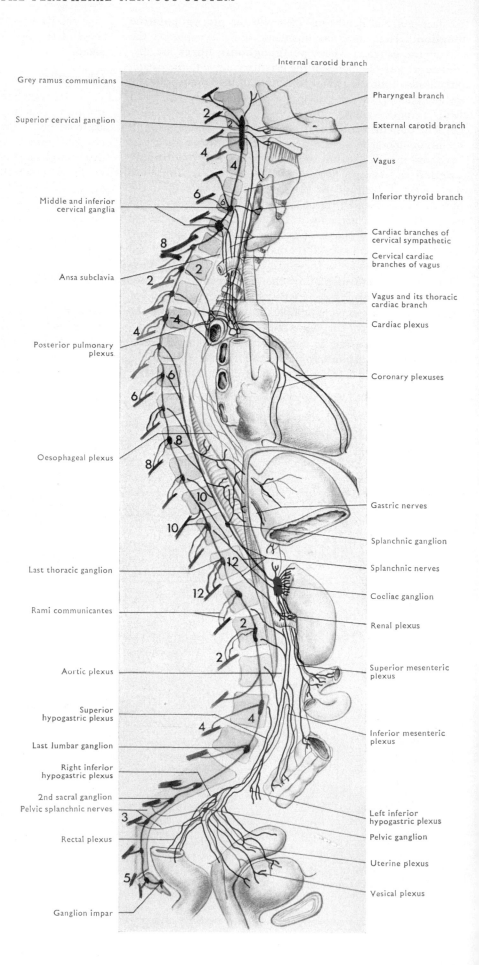

FIG. 961 Topographical plan of main parts of autonomic nervous system.

The spinal nerves, sympathetic trunk, and rami communicantes, black; sympathetic branches of distribution and plexuses, red; parasympathetic nerves (vagus and sacral), green. The vertebrae and spinal nerves are numbered.

(or three) pelvic splanchnic nerves consist of fine myelinated, preganglionic axons, and pass directly forwards to the inferior hypogastric (pelvic) plexuses [p. 790]. Here they intermingle with sympathetic fibres, and synapse with the parasympathetic ganglionic neurons either in these plexuses, or in the walls of the large intestine from the left colic flexure caudally, or in the pelvic parts of the urogenital system. Thus they replace the vagus from the region of the left colic flexure distally, and co-operate with the sympathetic in the control of the above organs. They are particularly concerned with the emptying mechanisms of the pelvic viscera, and with erection of the genital organs (vasodilation).

THE SYMPATHETIC PART

Preganglionic Fibres

The preganglionic fibres of this part arise from cells which lie in the lateral grey column of the spinal medulla between the two limb plexuses, and emerge only in those spinal nerves which arise from this region, that is, only in the first thoracic to second or third lumbar spinal nerves (**thoracolumbar out-flow**). The preganglionic fibres are slender and thinly myelinated, and traverse the ventral roots of the spinal nerves to enter the ventral rami. They leave the latter almost immediately to run to an adjacent part of the sympathetic trunk as the **white rami com-municantes.** Each sympathetic trunk consists of a chain of more or less segmentally arranged ganglia (groups of sympathetic nerve cells) which extend along the whole length of the vertebral column. The ganglia of each trunk are linked by a bundle, or bundles, of nerve fibres (**interganglionic rami),** and these contain longitudinally directed branches of the pregang-lionic fibres which enter the trunk through the white rami communi-cantes. In this way the ganglia of the sympathetic trunk which lie cranial to the first thoracic nerve and caudal to the second or third lumbar nerve receive preganglionic fibres, even though these fibres only emerge from the central nervous system through the restricted thora-columbar outflow.

Each **white ramus communi-cans** usually enters the sympathetic trunk at a ganglion, not infrequently at a level inferior to the ventral ramus from which it emerges. Here the individual preganglionic fibres may give direct branches to the ganglion cells in addition to sending branches longitudinally in the trunk

to end by synapsing with ganglion cells at higher and/or lower levels. In the upper thoracic region the latter bran-ches are predominantly ascending, in the lower thoracic and upper lumbar region they are predominantly descending. Other preganglionic fibres, sometimes accompanied by post-ganglionic fibres arising in the cells of the sympathetic trunk, pierce that trunk either directly or after running longitudinally in it for a variable distance, and pass anteromedially as visceral (**splanchnic**) nerves to reach the sympathetic ganglia (visceral or prevertebral) which do not lie in the sympathetic trunk.

The arrangement of the branches of the preganglionic fibres and of the ganglion cells (see below) is such that each pre-ganglionic fibre can make synapses with a large number of ganglion cells at various levels in the sympathetic part of the autonomic nervous system.

Ganglia

The cells of the sympathetic ganglia are all efferent, and there is no adequate evidence of afferent neurons in these ganglia. The nerve cells of the sympathetic ganglia, each surrounded by a cellular capsule, mostly have many dendrites (multipolar) though a few are bipolar and some are pseudo-unipolar. Some of the dendrites ramify within the capsule : others pierce it and

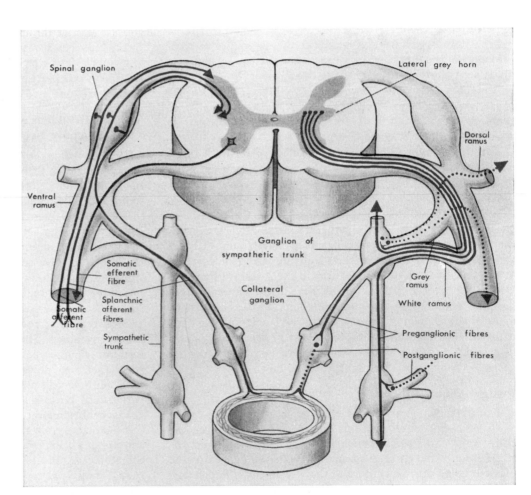

FIG. 962 Scheme showing relation of sympathetic system to spinal nerves and spinal medulla.

Sympathetic efferent fibres (preganglionic and postganglionic red), are shown on the right side; a somatic efferent fibre (black) and somatic and visceral afferent fibres (blue) on the left. For relative position of white a grey rami, see page 782. [Cf. Figs. 750 and 955.]

branching in the ganglion make contact with the capsules or extracapsular dendrites of its cells, and are exposed to the preganglionic fibres entering the ganglion. The preganglionic fibres also pierce the capsules of the ganglion cells and make contact with the intracapsular dendrites. Thus a few preganglionic fibres entering a ganglion can synapse with a large number of the cells in the ganglion.

There is some evidence (Matthews and Raisman, 1969) of interneurons in the ganglia, but it appears that most of the cells give rise to axons (postganglionic) which pass to effector organs. There is no evidence that the visceral sensory fibres which traverse all parts of the sympathetic system have any synapses with the sympathetic cells of the ganglia or plexuses, but merely transmit their impulses direct to the spinal medulla through the dorsal roots of the spinal nerves.

There are twenty-one or twenty-two ganglia of the sympathetic trunk; three cervical, ten or eleven thoracic, four lumbar, and four sacral. At their caudal ends the two sympathetic trunks unite either by communicating strands or in a median, **ganglion impar** on the front of the coccyx. At their cranial ends each sympathetic trunk is prolonged into the cranial cavity as a plexus around the internal carotid artery. This plexus contains some ganglion cells, and sends postganglionic fibres into a number of the cranial nerves. It unites with its fellow in a small ganglion on the anterior communicating artery (Mitchell, 1953).

Prevertebral ganglia are found in the large prevertebral **plexuses** of the thorax and abdomen, and their postganglionic fibres are mainly distributed along branches of these plexuses on the main arteries to the visceral structures. The plexuses consist of nerve fibres in which are scattered groups of ganglion cells only a few of which are sufficiently large to be dissected and named.

Terminal ganglia are tiny groups of nerve cells in the distal ramifications of the prevertebral plexuses buried in the walls of the viscera. They belong mainly to the parasympathetic part.

While it is possible to differentiate these three types of ganglia, only those of the sympathetic trunk are clearly defined, the others are larger aggregations of ganglion cells within a gangliated plexus which extends from the splanchnic nerves to the terminations of the plexuses in the viscera. Even the ganglia of the sympathetic trunk have small **intermediate ganglia** scattered on the interganglionic rami, and two or more of the ganglia of the trunk may be fused. Every sympathetic ganglion cell, wherever situated, receives synapses from preganglionic fibres which emerge from the central nervous system in the thoracolumbar outflow.

Postganglionic Fibres

The fibres which arise from the ganglia of the sympathetic trunk are distributed predominantly to structures in the body wall, either in somatic nerves or as plexuses on the arteries supplying these structures. From the ganglia of the plexuses they are distributed to the viscera through plexuses on the arteries which supply these structures. From the terminal ganglia they are distributed within the organs themselves.

Postganglionic fibres pass from the cephalic part of the sympathetic trunk (superior cervical ganglion) and its con-

tinuation, the internal carotid plexus, into most of the cranial nerves, and from all parts of the sympathetic trunk into the ventral rami of the corresponding spinal nerves (grey rami communicantes). In both cases these fibres are distributed through every branch of the nerves they enter, and pass to blood vessels (**vasomotor fibres**), sweat (**sudomotor fibres**) and other glands, and arrectores pilorum (**pilomotor fibres**).

1. **Grey Rami Communicantes.** These bundles of postganglionic fibres pass from the cells of the sympathetic trunk into the ventral ramus of every spinal nerve. They arise directly from the ganglia or from the interganglionic rami, and usually pass transversely to enter the ventral ramus proximal to the point where the white ramus, if present (T. 1–L. 3), leaves it (Pick and Sheehan, 1946; Botar, 1932). Where present, the white ramus may be fused with the corresponding grey ramus, and both also transmit visceral sensory fibres. The grey ramus contains non-myelinated and a few finely myelinated postganglionic fibres. On entering the ventral ramus, some of the fibres pass into the dorsal ramus, but the distribution of a grey ramus is limited to the territory of a single spinal nerve, unlike the widespread distribution of the fibres of a white ramus communicans within the sympathetic trunk and through its visceral branches.

2. **Vascular Branches.** The majority of these pass to the somatic arteries through the grey rami and spinal nerves, but direct branches from the sympathetic trunk pass to arteries (e.g. subclavian) which lie in close association with it. Most of the visceral postganglionic fibres pass with and supply the visceral arteries.

Visceral Branches of the Sympathetic Trunk (Splanchnic Nerves)

Such branches consist mainly of preganglionic fibres which have entered the sympathetic trunk in the white rami communicantes, and either pass directly through, or run longitudinally in the trunk for a variable distance before leaving it in the visceral branches. These branches arise from every level of the sympathetic trunk, and all course caudally towards the viscus they supply because of the caudal displacement of the viscera relative to the vertebral column and sympathetic trunks during development. In addition to the preganglionic fibres, some postganglionic fibres in the visceral branches arise from cells in the trunk, from ganglion cells scattered along the visceral branches, and in the prevertebral plexuses to which they are passing. The preganglionic fibres synapse with these scattered ganglion cells and with the macroscopically visible ganglia of the prevertebral plexuses. The postganglionic fibres of these cells are distributed to the viscera principally as perivascular plexuses of nerves.

Visceral sensory fibres traverse all the ramifications of these branches from their sensory endings at the periphery, to enter a spinal nerve either through the grey or white ramus communicans, and pass into the spinal medulla in the dorsal root. These sensory fibres have no synapses with the ganglion cells of the autonomic nervous system, but their presence in its branches means that all contain a mixture of myelinated (preganglionic and sensory) and non-myelinated (postganglionic and sensory) fibres. It is impossible, therefore, to identify particular fibres by virtue of the presence or absence of a myelin sheath, and this complexity is further increased by the admixture of parasympathetic fibres in the peripheral parts of the distribution.

THE SYMPATHETIC TRUNK

Cervical Part

This part consists of three (occasionally two or rarely four) cervical sympathetic ganglia, united by long, slender, interganglionic rami. It has no white rami communicantes, but it gives grey rami communicantes to all the cervical ventral rami. This part of the trunk lies on the prevertebral muscles, posterior to the carotid sheath. It is continued superiorly into the cranium as the internal carotid nerve which ascends with the internal carotid artery and breaks up into the internal carotid plexus on it. Inferiorly, the cervical part is continuous with the thoracic part anterior to the neck of the first rib, and here the trunk turns backwards to follow the vertebral curvature.

The **superior cervical ganglion** is a fusiform structure 2·5 cm. or more in length. It lies posterior to the internal carotid artery, internal jugular vein, and the last four cranial nerves, on the fascia covering the longus capitis muscle. It extends from a point 2 cm. inferior to the entrance of the carotid canal, to the level of the angle of the mandible, but its lower limit varies with the level of bifurcation of the common carotid artery. It represents the fused upper four cervical segmental ganglia, and hence gives grey rami to the first four cervical nerves.

The **middle cervical ganglion** is the smallest and most variable of the cervical ganglia. It usually lies on the inferior thyroid artery as it passes between the common carotid artery and the sixth cervical transverse process. It may lie further inferiorly on the vertebral artery (Axford, 1928) or even be partly or completely fused with the inferior cervical ganglion. The ganglion represents the fifth and sixth cervical ganglia and sends grey rami to the corresponding nerves.

The **cervicothoracic (stellate) ganglion** is large and irregular in shape. It is formed by the fusion of the inferior cervical ganglion (and sometimes the middle) with the first thoracic ganglion (and sometimes the second). The inferior cervical ganglion may form a separate element, in which case it represents the fused lower two cervical ganglia, and therefore sends grey rami communicantes to the seventh and eighth cervical nerves. The cervicothoracic ganglion lies at the junction of the cervical and thoracic parts of the sympathetic trunk, anterior to the transverse process of the seventh cervical vertebra and the neck of the first rib. It is posterior to the vertebral artery and veins, and the dome of the pleura; its inferior pole lies close to the termination of the costocervical trunk [FIG. 963]. The anterosuperior part of the ganglion (often fused with the middle cervical ganglion) may lie anterior to the vertebral artery (ganglion vertebrale) and be joined to the remainder by strands encircling the artery. The middle cervical and cervicothoracic ganglia are united by an interganglionic ramus (commonly multiple) and by a connexion which loops around the subclavian artery (**ansa subclavia**) [FIG. 963]. When the two ganglia are fused, the ansa unites the upper and lower parts of the common ganglionic mass.

Thoracic Part

The thoracic part of the sympathetic trunk descends posterior to the pleura on the intercostal vessels. Superiorly it lies anterior to the necks of the ribs, but as the thoracic vertebrae widen inferiorly, the trunk passes on to the costovertebral joints (midthoracic region) and then on to the sides of the vertebral bodies (lower thoracic region), to become continuous with the lumbar part by passing behind the medial arcuate ligament.

There are usually ten or eleven ganglia united by thick interganglionic rami which are often duplicated.

The thoracic part receives most of the preganglionic fibres through the white ramus communicans which each thoracic ventral ramus sends to it. Of the fibres in these rami a large number ascend to provide the preganglionic supply to the cervical part of the sympathetic trunk, while others descend to aid those from the first and second lumbar nerves to form the preganglionic supply for the lower lumbar and sacral parts of the sympathetic trunk. Clinical evidence suggests that the preganglionic fibres which are concerned with particular peripheral structures emerge at constant levels from the spinal medulla. Thus (1) preganglionic fibres for the head and neck leave the spinal medulla in the upper two thoracic nerves, and passing in the white rami to the sympathetic trunk, ascend in it to the superior cervical sympathetic ganglion. (2) Preganglionic fibres concerned with the heart emerge with the upper four thoracic nerves, and ascend to all three cervical ganglia, whence they pass in **cervical cardiac nerves** to the cardiac plexus [p. 788]. Direct cardiac branches from the upper thoracic part of the trunk also pass to the cardiac plexus [p. 785]. **Visceral sensory fibres** from the heart pass in the latter branches and in the middle and inferior cervical cardiac nerves, to enter the spinal medulla in the first to fifth thoracic spinal nerves. (3) Preganglionic fibres for the upper limbs emerge in the third to seventh thoracic nerves, and pass in the trunk to the middle cervical and cervicothoracic ganglia before they synapse. (4) The preganglionic outflow for abdominal structures is in the lower seven thoracic spinal nerves via the splanchnic nerves. (5) Preganglionic fibres for the lower limbs emerge in the tenth thoracic to second or third lumbar spinal nerves, and synapse in the second lumbar to third sacral ganglia. (6) Preganglionic fibres for the pelvic viscera emerge in the upper lumbar nerves.

Lumbar Part

This part of the trunk descends on to the anterior surfaces of the lumbar vertebrae [FIG. 965], medial to the psoas muscle. The right trunk is overlapped by the inferior vena cava, the left by para-aortic, lumbar lymph nodes. The trunks pass posterior to the suprarenal, renal, gonadal, and common iliac vessels, and anterior to the lumbar vessels, but may be crossed anteriorly by some of the lumbar veins, thus complicating the removal of this part of the trunk in the operation of 'lumbar sympathectomy'. There are usually four lumbar ganglia, but a greater or lesser number is common (Pick and Sheehan, 1946; Cowley and Yeager, 1949).

The anterior position of the lumbar part of the trunk separates it from the emerging ventral rami of the lumbar nerves and increases the length of the **rami communicantes.** These pass with the lumbar vessels medial to psoas on the lumbar vertebral bodies. The upper two (or three) lumbar ventral rami send white rami communicantes to the trunk, and ganglion cells of the trunk send grey rami communicantes to all the lumbar ventral rami. The white and grey rami are frequently fused in this region, and commonly carry small, intermediate ganglia on them (Wrete, 1935); a feature which is commonest in this and the cervical regions. Such ganglia may be responsible for continued sympathetic activity in the territory of the lumbar nerves after lumbar sympathectomy (Pick and Sheehan, 1946; Boyd and Monro, 1949).

Pelvic Part

The lumbar part of the sympathetic trunk becomes continuous with the pelvic part posterior to the common iliac vessels, and extends inferiorly and a little medially on the pelvic surface of the sacrum and coccyx. It lies medial to the first three pelvic sacral foramina, but crosses the fourth on the fourth sacral ventral ramus. Below this the two trunks end by irregular communications on the terminal pieces of the sacrum and the coccyx, or in a small **ganglion impar** on the front of the coccyx, or they may end independently. There are usually four small sacral ganglia, the main branches of which are the grey rami to the sacral and coccygeal nerves.

BRANCHES OF THE SYMPATHETIC TRUNK

Cervical Part

FROM THE SUPERIOR CERVICAL GANGLION. **Branches accompanying Arteries.** (1) From the apex of the ganglion arise one or more **internal carotid nerves** which form the **internal carotid plexus** around the internal carotid artery in the carotid canal. The nerves consist of preganglionic and postganglionic fibres, and the plexus contains some small ganglia on which the preganglionic fibres end. The **postganglionic fibres** are distributed: (i) as **plexuses** on the branches of the internal carotid artery to brain, orbit, and forehead, and hypophysis; (ii) to the third, fourth, fifth, and sixth **cranial nerves** at variable points, commonly between the cavernous sinus and orbit; (iii) from the internal carotid plexus in the upper part of the foramen lacerum, the **deep petrosal nerve** passes to join the greater petrosal branch of the facial nerve and form the nerve of the pterygoid canal, which runs to the pterygopalatine ganglion; (iv) the **caroticotympanic nerves** enter the middle ear cavity and run with fibres of the tympanic branch of the glossopharyngeal nerve; (v) the **sympathetic root of the ciliary ganglion** reaches that ganglion either from the nasociliary nerve or from the plexus on the ophthalmic artery. The last group of postganglionic fibres passes straight through the ciliary ganglion to enter the eye in the short ciliary nerves. They supply the vessels of the choroid and iris, and the dilator of the pupil.

Thus the internal carotid plexus has a widespread distribution in the cranial cavity, orbit, forehead, nose, and palate (deep petrosal nerve); its interruption not only interferes with vasomotor control and sweating (forehead) in that area, but results in constriction of the pupil (paralysis of the dilator), drooping (ptosis) of the upper eyelid (paralysis of the smooth muscle in levator palpebrae superioris), and some sinking in of the eye (enophthalmos), said to be due to paralysis of smooth muscle in the orbit. The same effect on the eye (Horner's syndrome) can be produced by damage to the preganglionic fibres in any part of the cervical sympathetic trunk, but in addition there will be effects involving the other branches of the superior, middle, and cervicothoracic ganglia depending on the level of injury.

(2) Branches also pass to the external carotid artery from the superior ganglion to form the **external carotid plexus.** This supplies glands (sweat and salivary) in the territory of that artery, many of its postganglionic fibres passing with branches of the trigeminal nerve near their termination, e.g. sudomotor fibres to the face (Wilson, 1936).

Branches accompanying nerves. In addition to the **grey rami communicantes** to the upper four cervical ventral rami and the branches of the internal and external carotid plexuses to cranial nerves, the superior cervical ganglion gives a **jugular branch** to the glossopharyngeal and vagus nerves, and sends other postganglionic fibres to the **hypoglossal** and **accessory nerves.**

Visceral branches. These are the **laryngopharyngeal branch** and **superior cervical cardiac nerve.** The former passes anteromedially, deep to the carotid sheath to join the pharyngeal plexus of the glossopharyngeal and vagus nerves, and innervate glands and blood vessels in the pharynx and larynx. Cardiac nerve; see below.

FROM THE MIDDLE CERVICAL GANGLION. **Grey rami communicantes** to the fifth and sixth cervical ventral rami.

Branches accompanying arteries. Medial branches of the ganglion form a plexus around the **inferior thyroid artery.** This supplies that artery and the gland, gives branches to the recurrent and external laryngeal nerves, and communicates with the superior cervical cardiac nerve. The ansa subclavia, which joins the ganglion to the cervicothoracic ganglion, gives branches to the **subclavian artery** (subclavian plexus), but these are probably derived mainly from the latter ganglion.

Visceral branch. The **middle cervical cardiac nerve;** see below.

FROM THE CERVICOTHORACIC GANGLION. **Grey rami communicantes** to the last two cervical and the first thoracic ventral rami.

Branches accompanying arteries. (1) Several large branches **(vertebral nerves)** from the superior part of the ganglion (and from the ganglion vertebrale) pass with the vertebral artery **(vertebral plexus).** This plexus supplies the intracranial and extracranial branches of the vertebral artery, and gives postganglionic fibres to some of the cervical spinal nerves (usually 4th–6th). (2) Postganglionic fibres in the **ansa subclavia** pass on to the proximal part of the subclavian artery **(subclavian plexus).** This is reinforced in the axilla by postganglionic fibres which reach the artery through adjacent peripheral nerves.

Visceral branch. The **inferior cervical cardiac nerve;** see below.

Cardiac Nerves

There are six **cervical cardiac nerves,** one arising from each of the three cervical sympathetic ganglia (or the interganglionic rami) on both sides. All of these nerves end in the deep part of the **cardiac plexus** except the left superior which

Pharyngeal branch of
vagus nerve   Glossopharyngeal nerve

Internal carotid artery
and plexus
Ventral ramus, C. 1
Transverse process of atlas
Grey ramus to C. 1
Grey ramus to C. 2
Transverse process of axis
Ventral ramus, C. 2
Grey ramus to C. 3
Superior cervical ganglion

Ventral ramus, C. 4

Phrenic nerve
Superior cardiac branch
Ventral ramus, C. 5
Grey rami to C. 5 and 6

Ventral ramus, C. 6
Scalenus anterior
Ventral ramus, C. 7
Costocervical trunk
Cervical pleura
Thyrocervical trunk
Vertebral artery
Scalenus medius

Subclavian artery and lower
trunk brachial plexus

Pharyngeal branch of
superior cervical
ganglion
Stylopharyngeus

Pharyngeal plexus
External carotid plexus
Internal laryngeal nerve
External laryngeal nerve
Thyroid cartilage
Superior thyroid artery and plexus
Middle cervical ganglion
Cricoid cartilage
Inferior thyroid artery and plexus
Vertebral artery and plexus
Thyroid gland
Inferior cervical ganglion and
grey rami 7 and 8
1st thoracic ganglion
on neck of 1st rib
Rami communicantes of 1st thoracic nerve
Middle and inferior cardiac branches
Vagus nerve
Common carotid artery

Recurrent laryngeal nerve
Ansa subclavia

Scalenus anterior   Internal thoracic artery
and phrenic nerve

FIG. 963   Distribution of branches of sympathetic trunk and ganglia in the neck.

enters the superficial part of the plexus after passing lateral to the arch of the aorta.

The **superior cervical cardiac nerve** descends from the superior ganglion in the posterior wall of the carotid sheath, and usually passes anterior to the inferior thyroid artery. *On the right*, in common with the middle nerve of that side, it lies either anterior or posterior to the subclavian artery, and descends behind the brachiocephalic artery to reach the posterior surface of the arch of the aorta, anterior to the tracheal bifurcation. In the neck, both have connexions with the external and recurrent laryngeal nerves, the vagus (superior cardiac branches), and the plexus on the inferior thyroid artery. *On the left*, the superior nerve descends with the common carotid artery, and usually passing lateral to the arch of the aorta, enters the superficial part of the cardiac plexus inferior to the arch.

The **middle cervical cardiac nerve** passes inferomedially. On the right, it runs with the superior cervical cardiac nerve; on the left, it unites with the inferior nerve, and passes medial to the aortic arch, close to the recurrent laryngeal nerve.

The **inferior cervical cardiac nerve** passes antero-inferiorly; on the right, posterior to the subclavian and brachiocephalic arteries; on the left it joins the middle nerve. Both communicate with the corresponding recurrent laryngeal nerve.

With the possible exception of the superior cervical cardiac nerves, all the cardiac nerves contain visceral sensory fibres in addition to preganglionic and postganglionic sympathetic fibres. For the arrangement of the cardiac plexus, see page 788.

Thoracic Part

Grey rami communicantes pass to every thoracic ventral ramus. Not infrequently a single ventral ramus may receive a grey ramus from more than one ganglion, and those that pass to the first (and second) thoracic ventral rami send postganglionic fibres into the brachial plexus. These are important in surgical operations designed to remove the sympathetic supply from the upper limb.

Branches accompanying blood vessels. Fine branches pass anteriorly from the upper five thoracic ganglia to form a plexus around the aorta [FIG. 964] and pulmonary trunk. These consist of postganglionic and visceral sensory fibres.

Visceral branches. 1. Several fine, **pulmonary** and **cardiac branches** pass anteromedially from the second to fourth thoracic ganglia to join the corresponding plexuses. Those passing to the heart carry both efferent and afferent (pain) fibres (Mitchell, 1956).

2. **The splanchnic nerves.** These arise from the fifth and subsequent thoracic ganglia. They pass antero-inferiorly

Rami communicantes

Rami communicantes

Rami communicantes

} Left vagus

} Right vagus

Thoracic sympathetic trunk

Oesophageal plexus

Oesophageal branches from greater splanchnic nerve

Aortic branches

Splanchnic ganglion

Lesser splanchnic nerve

Lowest splanchnic nerve

Coeliac plexus
Suprarenal plexus
Lowest splanchnic nerve
Lesser splanchnic nerve
Renal plexus

Greater splanchnic nerve
Lesser splanchnic nerve
Coeliac ganglion
Lowest splanchnic nerve
Aorticorenal ganglion
Superior mesenteric plexus

Aortic plexus

FIG. 964 Sympathetic trunk and parasympathetic fibres from the vagi in the thorax and upper abdomen.

ganglion, and the nerve sends branches to join the **oesophageal** and **aortic plexuses** in the thorax [FIG. 964].

(ii) The **lesser splanchnic nerve** arises in the region of the ninth and tenth ganglia, and descends between the trunk and the greater splanchnic nerve. It pierces the diaphragm close to the latter nerve, and enters in the **coeliac ganglion.**

(iii) The **lowest splanchnic nerve,** if present, arises from the last thoracic ganglion, or from the lesser splanchnic nerve. It pierces the diaphragm lateral to the lesser splanchnic nerve, and enters the **renal plexus.**

Lumbar Part

Grey rami communicantes pass to every lumbar ventral ramus. They are long, pass backwards medial to psoas, and may divide to pass to adjacent ventral rami, or one ventral ramus may receive more than one grey ramus from different levels of the trunk. They also transmit visceral sensory fibres.

Branches accompanying blood vessels. Small branches pass on the lumbar arteries to the lumbar part of the **abdominal aortic plexus.** This is continuous with the thoracic aortic plexus, and is continued as plexuses around the common iliac and median sacral arteries and their branches.

Visceral branches. These **lumbar splanchnic nerves** arise from all the lumbar ganglia of the sympathetic trunk [FIG. 965]. Those from the upper two ganglia frequently unite on the aorta near the origin of the inferior mesenteric artery and send branches on to that artery, (inferior mesenteric plexus). They descend, anterior to the common iliac vessels, and are joined by the lumbar splanchnic nerve from the third ganglion, which may pass between the common iliac artery and vein. That from the fourth ganglion, passes deep to both vessels. The four nerves of both sides form a loose-meshed plexus (superior hypogastric plexus) in the interval between the right and left common iliac vessels, anterior to the fifth lumbar vertebra.

The **superior hypogastric plexus** (presacral nerve) descends into the pelvis dividing into right and left **hypogastric nerves,** through which it is distributed to the **inferior hypogastric plexuses.** Thus the lumbar splanchnic nerves are predominantly distributed to the pelvic viscera, the only exceptions being their fibres which pass with the abdominal branches

on the thoracic vertebral bodies and consist of a mixture of preganglionic, postganglionic, and visceral sensory fibres, and also scattered ganglion cells. The preganglionic fibres end partly on these cells, but principally on the ganglion cells in the prevertebral plexuses of the abdomen to which the splanchnic nerves are passing.

(i) The **greater splanchnic nerve** arises by branches from the fifth to ninth thoracic ganglia of the sympathetic trunk [FIGS. 961, 964]. These unite as the nerve descends to pierce the crus of the diaphragm and enter the **coeliac ganglion** in the abdomen. It frequently carries a small, **splanchnic ganglion,** approximately at the level of the eleventh thoracic of the inferior mesenteric artery, and any filaments which they supply to the abdominal aortic plexus. Structurally the lumbar splanchnic nerves are similar to the splanchnic nerves arising in the thorax. Indeed they form a continuous series with them, and branches from the coeliac and renal ganglia, which receive their preganglionic fibres from the thoracic splanchnic nerves, pass to join the lumbar splanchnic nerves on the aorta [FIG. 966].

Pelvic Part

Grey rami communicantes pass irregularly from the pelvic part of the trunk or its ganglia to the ventral rami of the sacral nerves and to the coccygeal nerve.

Branches accompanying blood vessels. Small branches pass to the median and lateral sacral arteries.

Visceral branches. Two or three small **sacral splanchnic nerves** pass from the second and third ganglia to the inferior hypogastric plexus. Direct branches to the rectum and ureter have been described.

THE SYMPATHETIC PLEXUSES

There are two types of sympathetic plexus which differ from each other only in distribution, size, and complexity. (1) The **perivascular plexuses** consist principally of postganglionic and sensory fibres which are mainly distributed to the smooth muscle of the arterial tree, but also to glands and other smooth muscles in the territory of distribution of the arteries. Such plexuses are the carotid, inferior thyroid, vertebral, subclavian,

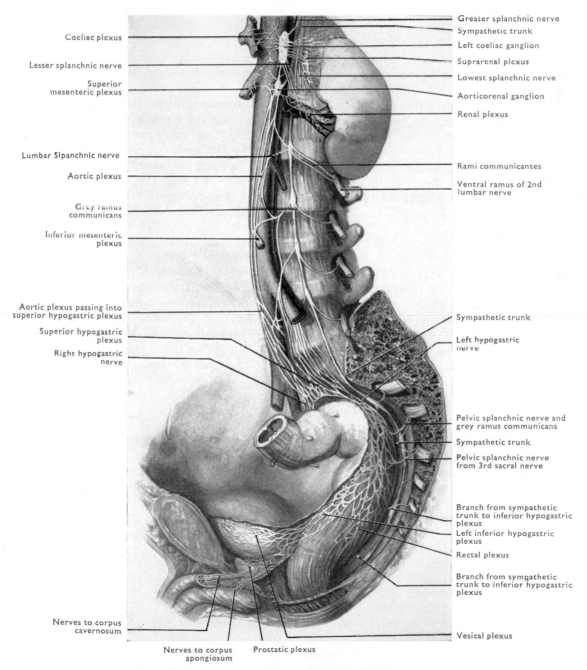

FIG. 965 Plexuses of autonomic system in abdomen and pelvis.

aortic, and iliac plexuses, but a similar mesh of postganglionic sympathetic fibres is found on all arteries and to a lesser extent on the veins. Together with the peripheral nerves, these plexuses form a major route of distribution for the postganglionic sympathetic nerve fibres. Few of these plexuses contain a significant number of ganglion cells, and none of those on vessels supplying the body wall or limbs contains any parasympathetic fibres.

(2) The **prevertebral plexuses** are concerned with the innervation of the thoracic, abdominal, and pelvic viscera. They are also closely associated with the blood vessels supplying these viscera, but they consist of a mixture of preganglionic, postganglionic, and sensory nerve fibres with many ganglion cells scattered among them, some of which are aggregated into large, readily visible ganglia. In addition, they contain parasympathetic fibres and cells, and it is through these plexuses that both parts of the autonomic nervous system are distributed to the viscera.

A number of separate plexuses are described, but they form what is virtually a continuous mass from the cardiac plexus at the tracheal bifurcation to the inferior hypogastric plexuses in the pelvis. The parts of this mass are differentiated by the territories to which their postganglionic fibres pass, but their diffuse nature prevents accurate description of their individual extent. Indeed the most peripheral part of this system, the gangliated plexuses on the wall of the gut tube, are continuous from the upper part of the oesophagus to the anal canal.

THORACIC PLEXUSES

Cardiac Plexus

The various cardiac branches of the vagus and sympathetic enter a large, gangliated cardiac plexus, which lies on the anterior surface of the lower part of the trachea and its bifurcation, behind the great vessels emerging from the base of the heart. Inferolaterally it is continuous along the principal bronchi with the **pulmonary plexuses,** and it extends antero-inferiorly around the aorta and pulmonary trunk to the coronary vessels of the heart. To the left, it extends into the concavity of the arch of the aorta, anterior to the ligamentum arteriosum. This part, frequently called the superficial cardiac plexus, receives the **left superior cervical cardiac nerve** (sympathetic) and the **inferior cervical cardiac branch** of the left vagus (parasympathetic). It is principally distributed along the pulmonary trunk to the right coronary plexus, and to the left pulmonary plexus. The deeper part, between the tracheal bifurcation and the aortic arch, receives all the other cardiac branches of the vagus and sympathetic of both sides, and is distributed to the heart partly along the aorta and pulmonary trunk, and partly along the superior vena cava (sinuatrial node, etc.) and the pulmonary veins.

Early in development, the arterial and venous ends of the heart are widely separated, and each has its own mesocardium through which autonomic nerve fibres and cells reach these two ends of the heart separately. As the heart folds, the venous and arterial ends are approximated, thus bringing together both parts of the autonomic supply into a single mass (the cardiac plexus). These two parts cannot be differentiated in the adult heart, though the atrial part of the plexus passes predominantly with the great veins, the ventricular part with the great arteries. The folding of the heart [FIG. 1129] and the asymmetrical development of the great veins [FIG. 1142] and hence of the sinus venosus into which they discharge, leads to the left horn of that organ (with its part of the nodal tissue of the heart—atrioventricular node) becoming the coronary sinus, while the right horn (with its nodal tissue, sinu-atrial node) forms the right atrium adjacent to the superior vena cava. Thus the corresponding parts of the nodal tissue lie adjacent to the entry of these two veins, and while the atrioventricular node is supplied with nerves predominantly from the left side, the sinu-atrial node is supplied predominantly from the right.

The cardiac plexus, like the remainder of the prevertebral plexuses, is a mixture of sympathetic, parasympathetic, and sensory fibres. The ganglion cells in the plexus are predominantly parasympathetic (inhibitor) as are other ganglion cells which are found in the subepicardial tissues of the atria and ventricles and in the myocardium, principally with the conducting tissue. These ganglion cells are mixed in type (multipolar, bipolar, and unipolar) and it has been suggested that there may be sensory cells among them (Davies, Francis, and King, 1952) though a similar variability of cells is found in the ganglia of the sympathetic trunk.

The **coronary plexuses** follow the right and left coronary arteries and are extensions of the cardiac plexus. Both sympathetic and parasympathetic fibres and ganglion cells are present; the former appear to dilate the coronary vessels, while the latter may constrict them. Sensory fibres are present with these and with the nerve fibres which innervate the subendocardial tissue (Stotler and McMahon, 1947), though those which pass to the nodal tissue and follow the conducting tissue may be purely efferent.

Pulmonary Plexus

This plexus surrounds the root of each lung but is principally posterior to it. The plexus consists mainly of parasympathetic fibres derived from the vagus nerve [p. 731], but it also receives branches from the second, third, and fourth thoracic ganglia of the sympathetic trunk, which pass mainly to the posterior part. On each side, the plexus is continuous with the corresponding part of the cardiac plexus, and communicates with its fellow. The pulmonary plexuses are distributed to the lung (e.g., smooth muscle of the bronchial tree), its blood vessels, and the pulmonary pleura (Larsell, 1922).

Oesophageal Plexus

This plexus is formed principally by the vagus nerves [p. 731] and extends inferiorly from the pulmonary plexus. It is mainly a parasympathetic plexus, but it receives sympathetic fibres from the greater splanchnic nerves.

ABDOMINOPELVIC PLEXUSES

This great mass of autonomic plexus lies principally in association with the abdominal aorta and the internal iliac arteries, and extends along their branches to the abdominal and pelvic viscera. Three major parts are recognized, the **coeliac, superior hypogastric,** and **inferior hypogastric plexuses,** but they are continuous with each other and with the splanchnic nerves and the oesophageal plexus in the thorax, and with the extensions of this complex along the main branches of the arteries. Major ganglia tend to be associated with the aorta and the internal iliac arteries, but smaller groups and scattered cells are found throughout the system.

In general the abdominal plexuses receive **the splanchnic nerves** from the thoracic part of the sympathetic trunk and **vagal branches** which enter the abdomen with the oesophagus from the oesophageal plexus. The hypogastric plexuses receive fibres from the sympathetic trunk through the **lumbar splanchnic nerves,** and parasympathetic fibres through the **pelvic splanchnic nerves.** The sacral splanchnic nerves are small and unimportant, but the pelvic splanchnic nerves (parasympathetic) also extend into the abdomen to innervate the descending and sigmoid parts of the colon.

Coeliac Plexus

This is the largest of the prevertebral plexuses and lies on the aorta at the level of the twelfth thoracic and first lumbar vertebrae, in close association with the origins of the coeliac, superior mesenteric, and renal arteries from the aorta. Minute ganglia are scattered throughout this plexus and its extensions along the above aortic branches, but certain large ganglia are also present.

The large **coeliac ganglia** flank the aorta at the level of origin of the coeliac artery. Each lies on the corresponding crus of the diaphragm; the right overlapped by the inferior vena cava, while the left lies posterior to the omental bursa superior to the pancreas.

The **aorticorenal ganglia** lie at the level of the renal arteries and may be partly posterior to them. They are often continuous with the coeliac ganglia, and are always linked by many nerve strands.

The **superior mesenteric ganglia** lie on the aorta around the origin of the superior mesenteric artery. They may form a single, lobulated mass, or a number of smaller ganglia.

All these ganglia are united by numerous nerve strands many of which pass from side to side across the aorta.

Nerves to the Coeliac Plexus. The **greater splanchnic nerve** pierces the crus of the diaphragm and enters the superior pole of the coeliac ganglion. Many of its fibres end here, but others pass through the ganglion into the branches of the plexus and to the medulla of the suprarenal gland. The **lesser splanchnic nerve** pierces the diaphragm lateral to the greater, and is distributed to all the named ganglia of the coeliac plexus. The **lowest splanchnic nerve,** if present, joins the aorticorenal ganglion or the adjacent plexus.

The **posterior vagal trunk** enters the abdomen with the oesophagus, gives off posterior gastric and hepatic branches, and branches which pass to the coeliac plexus for distribution through the parts of that plexus, including the territory of the superior mesenteric artery.

Branches of the Coeliac Plexus. These consist of the same mixture of preganglionic parasympathetic, preganglionic and postganglionic sympathetic, and sensory nerve fibres, and ganglion cells, except in the case of the suprarenal plexus where the fibres are principally preganglionic. The majority of these branches follow the branches of the aorta and receive their names from them. They have the function of vasomotor control

and also of control over the glands and smooth muscle of the viscera to which the blood vessels are distributed; and they transmit sensory fibres. The arrangement of these plexuses is sufficiently irregular to make a detailed gross description valueless, but the general arrangement is indicated below.

1. Extensions of the plexus along the coeliac artery and its branches form; the left **gastric plexus** to stomach and oesophagus; the **hepatic plexus** to liver, gall-bladder, stomach, duodenum, and pancreas; the **splenic plexus** supplies the pancreas, spleen, and stomach.

2. The **phrenic plexus** passes with the inferior phrenic artery to the diaphragm, suprarenal plexus, inferior vena cava, and oesophagus. Some of its fibres unite with the phrenic nerve on the inferior surface of the diaphragm, and there may be a **phrenic ganglion** where it joins that nerve on the right side.

3. The **suprarenal plexus** arises from the coeliac plexus and from direct branches of the splanchnic nerves. The fibres are mostly preganglionic to the cells of the suprarenal medulla, which are developmentally equivalent to ganglionic neurons of the sympathetic system, and like them, secrete biologically active amines, mostly adrenaline (Elliott, 1913; Young, 1939; MacFarland and Davenport, 1941).

4. The **renal plexus** is derived from the aorticorenal ganglion, the lesser and lowest splanchnic nerves, and the

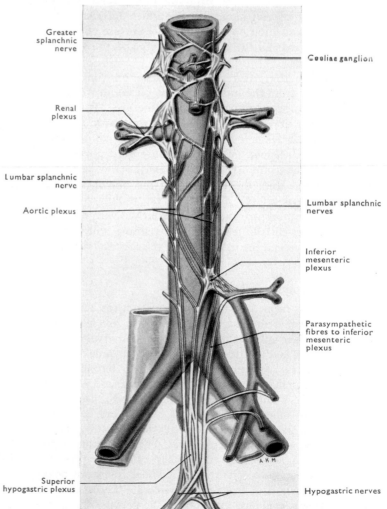

Greater splanchnic nerve

Coeliac ganglion

Renal plexus

Lumbar splanchnic nerve

Aortic plexus

Lumbar splanchnic nerves

Inferior mesenteric plexus

Parasympathetic fibres to inferior mesenteric plexus

Superior hypogastric plexus

Hypogastric nerves

FIG. 966 The aortic and superior hypogastric plexuses.

aortic plexus (Mitchell, 1935a). It extends along the renal arteries to the hilus of the kidney, communicates with the suprarenal plexus along the inferior suprarenal artery, and forms the upper part of the **ureteric plexus,** which contains few if any ganglion cells (Notley, 1969), but has both cholinergic and adrenergic fibres in it. The latter plexus extends along the ureter, and receives further fibres along the various arteries to the ureter.

5. The **testicular** or **ovarian plexuses** are extensions from the beginning of the renal plexus. They accompany the corresponding arteries and supply respectively the spermatic cord, epididymis, and testis, or ovary, broad ligament, and uterine tube. The latter communicates with the **uterine plexus** from the inferior hypogastric plexus.

6. The **superior mesenteric plexus** is an extension of the coeliac plexus along the superior mesenteric artery and its branches. It carries sympathetic and parasympathetic fibres to the small intestine, pancreas, caecum, appendix vermiformis, ascending and transverse parts of the colon. For the most part the parasympathetic fibres synapse with the intramural plexuses (myenteric and submucous) in the gut tube.

Inferior to the renal and superior mesenteric plexuses, sympathetic strands extend from these plexuses along the aorta (**abdominal aortic** and **intermesenteric plexuses**). This part is joined by the upper two lumbar splanchnic nerves [FIG. 965] and gives off the **inferior mesenteric plexus** along the artery of that name. It continues anterior to the aorta and its bifurcation on to the anterior surface of the left common iliac vein and fifth lumbar vertebra as the superior hypogastric plexus.

The **inferior mesenteric plexus** is composed of sympathetic strands from the intermesenteric plexus in which are fibres from the upper lumbar splanchnic nerves. It receives parasympathetic fibres from the **pelvic splanchnic nerves,** though the route taken by these fibres is not known with certainty. It is claimed (Telford and Stopford, 1934) that they ascend through the hypogastric plexuses [FIG. 966], but they may ascend from the pelvis along the branches of the inferior mesenteric artery. Ganglia are not usually visible on the inferior mesenteric artery; its plexus, mixed with scattered ganglion cells, is distributed to the descending and sigmoid parts of the colon, and the upper part of the rectum (**superior rectal plexus**).

Superior Hypogastric Plexus

This plexus is formed by the continuation of the abdominal aortic and intermesenteric plexuses to which are added the lower two **lumbar splanchnic nerves** in the interval between the two common iliac arteries. It consists of the same elements as the other parts of the autonomic plexuses, but the nerve fibres are predominantly sympathetic in origin, and join their parasympathetic counterparts in the lesser pelvis. It consists of a number of strands of gangliated plexus, and unites the abdominal and pelvic (inferior hypogastric plexuses) parts of the autonomic prevertebral plexuses. The superior hypogastric plexus is sometimes erroneously called the presacral nerve, and is occasionally surgically divided (presacral neurectomy) to relieve pain from pelvic viscera. The success of this procedure is evidence in favour of the passage of sensory fibres by this route.

Inferior Hypogastric (Pelvic) Plexuses

In the lesser pelvis the superior hypogastric plexus splits right and left into hypogastric nerves. Each joins the corresponding inferior hypogastric plexus in the areolar tissue between the lateral wall of the pelvis and the rectum and adjacent parts of the bladder and seminal vesicle (male), or the rectum and vagina (female), and also in the sacrogenital fold.

The inferior hypogastric plexus receives fibres from: (1) the **superior hypogastric plexus** (sympathetic and sensory fibres); (2) the **pelvic splanchnic nerves** (sacral parasympathetic and sensory) from the third and second or fourth sacral ventral rami; (3) the small **sacral splanchnic nerves** from the second and third sacral ganglia of the sympathetic trunk.

Some of the parasympathetic fibres may run a recurrent course through the inferior and superior hypogastric plexuses to take part in the formation of the inferior mesenteric plexus (Telford and Stopford, 1934; Mitchell, 1935b).

Branches. As with the other plexuses, the branches of the inferior hypogastric plexus pass with branches of the internal iliac artery to the pelvic viscera.

1. The **rectal plexus** is completed by plexuses along the middle and inferior rectal arteries (middle and inferior rectal plexuses) which mingle with the superior rectal plexus from the inferior mesenteric plexus. They supply rectum and anal canal and meet in the rectal wall.

2. The **vesical plexus** arises as a number of nerves from the anterior part of the inferior hypogastric plexus. They pass medially to surround the terminal part of the ureter and run on the walls of the bladder, the seminal vesicle, and the ductus deferens. The latter part may extend as far as the epididymis. In the female, they pass medially in the base of the broad ligament, and giving branches to the uterus, pass forwards inferolateral to the cervix, below the uterine artery. They send branches on the upper vagina, in the vesicovaginal septum to the base of the bladder (Ballantyne and Smith, 1968).

3. The **prostatic plexus** lies inferior to, and communicates with, the vesical plexus. It supplies the prostate, the structures that pierce it, and the seminal vesicle and membranous urethra. Sensory fibres from the pudendal nerve may enter this plexus.

4. The **uterine plexus** passes into the broad ligament with the uterine artery. It supplies all parts of the uterus and communicates with the ovarian plexus in supplying the uterine tube.

5. The **vaginal plexus** receives fibres which come mainly from the pelvic splanchnic nerves (parasympathetic). It supplies vagina and urethra, and together with the uterine plexus corresponds with the prostatic plexus in the male.

6. The **cavernous nerves.** Extensions of the prostatic plexus pass to the erectile tissue of the penis, and similar extensions of the vaginal plexus pass to the clitoris. Autonomic nerve fibres also reach the external genitalia through the pudendal nerve and along the internal pudendal artery.

Innervation of the Urinary Bladder and Rectum

1. The **pudendal nerve** (S. (1), 2, 3, 4) supplies the voluntary muscle of the external anal and urethral sphincters which give voluntary control over these orifices. It gives sensory fibres to the mucosa of the anal canal and also to the

urethra, probably including the prostatic part in the male which is sensitive to distention by urine entering it from the bladder.

2. **Sympathetic fibres** reach the rectum both by the inferior mesenteric and the inferior hypogastric plexuses, and the urinary bladder through the inferior hypogastric plexus. The preganglionic fibres emerge from the spinal medulla in the upper lumbar ventral rami.

3. **Parasympathetic fibres** reach the rectum and urinary bladder from the inferior hypogastric plexus. The preganglionic fibres enter this plexus by the pelvic splanchnic nerves.

Both sympathetic and parasympathetic pathways transmit sensory fibres. The latter seem to be associated with the normal sensations of bladder and rectal distention, the former with pain from these pelvic organs, e.g. on over-distention.

The efferent parasympathetic fibres seem to be concerned with the contraction of urinary bladder and rectal muscle associated with micturition and defaecation, arising as a result of afferent impulses entering with the pelvic splanchnic nerves. The sympathetic efferents cause contraction of the internal anal and vesical sphincters, with relaxation of the muscular walls of these organs. They play an important part in ejaculation for they cause contraction of the ductus deferentes, seminal vesicles, prostatic musculature, and internal vesical sphincter. The latter prevents the entry of urine into the prostatic urethra and the reflux of seminal fluid into the urinary bladder.

INNERVATION OF BLOOD VESSELS

All arteries, though not all veins (Lever, Spriggs and Graham, 1968) are accompanied by nerve fibres which form a plexus of varying density on them. These are postganglionic sympathetic and sensory fibres. Occasionally parasympathetic fibres appear to supply blood vessels; e.g., stimulation of the chorda tympani produces vasodilatation in salivary glands and the pelvic splanchnics in the erectile tissue of the external genitalia. For the most part the nervous control of vascular diameter depends on the sympathetic alone. Vasodilatation in the skin can be produced by stimulation of the distal end of a cut dorsal root, but it is not known to play any part in vasomotor control in normal circumstances.

The most extensive perivascular plexuses are the prevertebral plexuses, but only the minority of the fibres in these are concerned with control of the blood vessels. In the body wall and limbs, the perivascular plexuses are most extensive in the peripheral part of the limbs and exposed parts of the skin, particularly where there are **arteriovenous anastomoses** [FIG. 1063] concerned with temperature control of such skin.

The vessels of the body wall and limbs receive a proximal plexus direct from the sympathetic trunk. This plexus is reinforced peripherally by branches from the nerves in the territory of each artery. These branches contain: (1) postganglionic sympathetic fibres which have entered the peripheral nerves through the grey rami communicantes; and (2) sensory nerve fibres.

Sensory fibres to blood vessels are widely distributed and are frequently myelinated. They do not usually give rise to conscious sensations, though pain may be appreciated from arteries and veins. The function of these nerve fibres is improperly understood except in certain regions, e.g., the pressure receptors (baroreceptors) in the aorta, carotid sinus, and great veins (glossopharyngeal and vagus nerves), and the chemoreceptors of the carotid body (glossopharyngeal nerve). The carotid body contains innervated nests of epithelioid cells in the walls of blood vessels. These cells seem to be sensitive to oxygen-lack and possibly also to excess of carbon dioxide in the blood. Similar structures found in relation to the heart and great vessels are innervated by the vagus nerve, and may also have a chemoreceptor function.

DEVELOPMENT OF THE AUTONOMIC NERVOUS SYSTEM

The cells of the autonomic ganglia and plexuses (sympathoblasts) migrate to their final position (1) from the neural crest and (2) from the neural tube (Raven, 1937). The neural crest extends over most of the length of the neural tube and is concerned throughout with the formation of the sensory ganglia of the spinal and cranial nerves, while the formation of sympathoblasts is limited mainly to the thoracolumbar (sympathetic) and craniosacral (parasympathetic) regions (Kunz, 1953).

In the thoracolumbar region the sympathoblasts which arise from the neural crest, separate from the ventral border of the primitive spinal ganglia and meet the cells which migrate outwards from the neural tube along the ventral roots. This mixture of cells, in which it is not possible to distinguish those of different origin, leave the spinal nerves and migrate towards the aorta. Each group proliferates and fuses with adjacent groups to form an irregular, longitudinal rod of cells associated with the intersegmental body wall branches of the aorta. This primitive, thoracolumbar sympathetic trunk extends cranially and caudally by migration of cells from its two extremities, thus forming the definitive **sympathetic trunk.** With the growth of the embryo and the development of longitudinal fibres within the trunk, the rod breaks up into a number of separate ganglia united by interganglionic rami. This method of development accounts for the variable arrangement of the ganglia and of the grey rami communicantes which grow out from these cells to enter the ventral rami of all the spinal nerves.

Some cells continue to migrate ventromedially beyond the position of the sympathetic trunk [FIG. 955]. These come into relation first with the lateral, visceral branches of the aorta (suprarenal, renal, and gonadal) and then with the ventral, visceral branches of the aorta (coeliac, superior mesenteric, inferior mesenteric), and those of the internal iliac arteries. In association with all of these they form groups of cells of various sizes (ganglia), while other cells migrate along the arteries forming scattered cells and minor ganglia, and even reach the organs supplied by the arteries.

In this manner there is formed a scattering of cells between the ventral rami of the spinal nerves and the visceral organs, and within this are a certain number of larger ganglia, with smaller ganglia between them; e.g., the intermediate ganglia on rami communicantes, sympathetic trunk, and splanchnic nerves. The axon of each sympathoblast grows out along the line of the artery with which it is primarily associated, and is thus distributed to that artery and the other structures innervated by the sympathetic system within the territory of the vessel [FIG. 955].

Subsequently nerve cells in the grey matter of the spinal medulla (the future lateral grey horn) send processes through the ventral roots into the ventral rami (T. 1–L. 3). These processes (preganglionic axons) leave the ventral rami, and follow all the migrating groups of sympathetic neuroblasts, thus reaching and synapsing with all of them. Thus some of the preganglionic axons end in the sympathetic trunk by synapsing with its cells at the level of entry, or at higher or lower levels by extending longitudinally within the trunk. Other preganglionic fibres pass straight through the trunk, or run longitudinally in it for some distance before leaving it to follow the further line of migration of sympathetic cells and form the **splanchnic nerves** of the sympathetic trunk. These axons, predominantly preganglionic, synapse with the scattered and larger groups of ganglion cells along their route (coeliac, aorticorenal, etc.) and pass through the prevertebral plexuses, which they form together with the ganglion cells and their postganglionic axons. Thus the number of preganglionic axons steadily decreases from the sympathetic trunk to the peripheral distribution of the prevertebral plexuses, while the number of postganglionic axons steadily increases, only to diminish as they end on the organs which they supply.

Processes of the **spinal ganglion cells** grow outwards with the preganglionic fibres, and following the same routes, are distributed to the visceral organs which are innervated by the postganglionic fibres. They do not appear to develop any synaptic connexions with the autonomic ganglion cells.

In the head region the autonomic **(parasympathetic)** cells are derived in a similar manner from the neural crest and neural tube. These cells migrate to their final destination along the pathway of one or other of the cranial nerves, including those through which the preganglionic fibres later pass to synapse with these cells. The ganglia formed by these cells lie close to the organs which they eventually innervate, and many of the cells are not aggregated into ganglia but lie in plexuses, similar to those of the sympathetic part, but are often buried in the organs which they supply.

Many of the cells of the cranial outflow migrate over considerable distances and are responsible for the formation of most of the neurons in the **vagal plexuses** (cardiac, pulmonary, oesophageal, and the submucous and myenteric plexuses of the abdomen as far as the left colic flexure). Within the head, some of the parasympathetic ganglion cells arise in association with the trigeminal ganglion, others migrate along the oculomotor, facial or glossopharyngeal nerves, which later transmit the preganglionic fibres to these ganglion cells.

GANGLION	ORIGIN OF GANGLION CELLS	ORIGIN AND COURSE OF PREGANGLIONIC FIBRES
Ciliary	From trigeminal via ophthalmic nerve; from oculomotor nerve	From accessory oculomotor nucleus via oculomotor nerve
Pterygopalatine	From trigeminal via maxillary nerve; from facial via greater petrosal nerve	From superior salivatory nucleus via facial nerve (greater petrosal nerve)
Submandibular	From trigeminal via mandibular nerve; from facial via chorda tympani	From superior salivatory nucleus via facial nerve (chorda tympani)
Otic	From trigeminal via mandibular nerve; from glossopharyngeal via lesser petrosal nerve	From inferior salivatory nucleus via glossopharyngeal (lesser petrosal nerve)

The origin of the sacral parasympathetic cells is not certain. It seems probable that they arise, in part at least, from the sacral parts of the neural crest and neural tube. They are mixed almost immediately with the thoracolumbar cells forming the inferior hypogastric plexus, and their identity is lost.

REFERENCES

ADAMS, W. E. (1942) The blood supply of nerves, I. Historical review, *J. Anat. (Lond.)*, **76**, 323.

ADAMS, W. E. (1943) The blood supply of nerves, II. The effects of exclusion of its regional sources of supply on the sciatic nerve of the rabbit, *J. Anat. (Lond.)*, **77**, 243.

AXFORD, M. (1928) Some observations on the cervical sympathetic in Man, *J. Anat. (Lond.)*, **62**, 301.

BALLANTYNE, B., and SMITH, P. H. (1968) Surgical anatomy of the extrinsic innervation of the human female urinary bladder, *J. Anat. (Lond.)*, **103**, 199.

BOTAR, J. (1932) Études sur les rapports des rameaux communicants thoraco-lumbaires avec les nerfs viscéraux chez l'homme et chez l'animal, *Ann. Anat. path. méd.-chir.*, **9**, 88.

BOYD, J. D., and MONRO, P. A. G. (1949) Partial retention of autonomic function after paravertebral sympathectomy, *Lancet*, ii, 892.

CLARK, W. E. LE GROS, and WARWICK, R. T. T. (1946) The pattern of olfactory innervation, *J. Neurol. Neurosurg. Psychiat.*, **9**, 101.

COWAN, W. M., and POWELL, T. P. S. (1963) Centrifugal fibres in the avian visual system, *Proc. roy. Soc. B*, **158**, 232.

COWLEY, R. A., and YEAGER, G. H. (1949) Anatomic observations on the lumbar sympathetic system, *Surgery*, **25**, 880.

DALE, H. (1934) Pharmacology and nerve-endings (Dixon Memorial Lecture), *Proc. roy. Soc. Med.*, **28**, 319.

DALE, H. (1938) Chemical agents transmitting nervous excitation, *Irish J. med. Sci.*, **150**, 245.

DAVIES, F., FRANCIS, E. T. B., and KING, T. S. (1952) Neurological studies of the cardiac ventricles of mammals, *J. Anat. (Lond.)*, **86**, 130.

DAVIES, F., and LAIRD, M. (1948) The supinator muscle and the deep radial (posterior interosseous) nerve, *Anat. Rec.*, **101**, 243.

DAVIES, L., and HAVEN, H. A. (1933) Surgical anatomy of the sensory root of the trigeminal nerve, *Arch. Neurol. Psychiat. (Chic.)*, **29**, I.

DURWARD, A. (1948) The blood supply of nerves, *Postgrad. med. J.*, **24**, 11.

EISLER, P. (1891) Der Plexus lumbosacralis des Menschen, *Anat. Anz.*, **6**, 274.

ELLIOTT, T. R. (1905) The action of adrenalin, *J. Physiol. (Lond.)*, **32**, 401.

ELLIOTT, T. R. (1913) The innervation of the adrenal glands, *J. Physiol. (Lond.)*, **46**, 285.

FITZGERALD, M. T. J., and ALEXANDER, R. W. (1969) The intramuscular ganglia of the cat's tongue, *J. Anat. (Lond.)*, **105**, 27.

FOERSTER, O. (1933) The dermatomes in Man, *Brain*, **56**, I.

GAMBLE, H. J., and EAMES, R. A. (1964) An electron microscope study of the connective tissues of human peripheral nerve, *J. Anat. (Lond.)*, **98**, 655.

HARRIS, W. (1904) The true form of the brachial plexus and its motor distribution, *J. Anat. Physiol.*, **38**, 399.

HARRIS, W. (1939) *The Morphology of the Brachial Plexus*, London.

HARRISON, R. G. (1924) Neuroblast versus sheath cell in the development of peripheral nerves, *J. comp. Neurol.*, **37**, 123.

HEAD, H. (1893) On disturbances of sensation with especial reference to the pain of visceral disease, *Brain*, **16**, I.

HERRINGHAM, W. P. (1886) The minute anatomy of the brachial plexus, *Proc. roy. Soc. B*, **41**, 423.

HORSTADIUS, S. (1950) *The Neural Crest*, London.

HOVELACQUE, A. (1927) *Anatomie des nerfs craniens et rachidiens et du système grand sympathique chez l'homme*, Paris.

JONES, F. W. (1910) On the relation of the limb plexuses to the ribs and vertebral column, *J. Anat. Physiol.*, **44**, 377.

JONES, F. W. (1939) The anterior superior alveolar nerve and vessels, *J. Anat. (Lond.)*, **73**, 583.

KEEGAN, J. J., and GARRETT, F. D. (1948) The segmental distribution of the cutaneous nerves in the limbs of Man, *Anat. Rec.*, **102**, 409.

KERR, A. T. (1918) The brachial plexus of nerves in Man. The variations in its formation and branches, *Amer. J. Anat.*, **23**, 285.

KOSINSKI, C. (1926) The course, mutual relations and distribution of the cutaneous nerves of the metazonal region of leg and foot, *J. Anat. (Lond.)*, **60**, 274.

KUNTZ, A. (1953) *The Autonomic Nervous System*, 4th ed., Philadelphia.

KUNTZ, A., and RICHINS, C. A. (1946) Components and distribution of the nerves of the parotid and submandibular glands, *J. comp. Neurol.*, **85**, 21.

KURE, K. (1931) *Über den Spinal-parasympathikus*, Basel.

LANGLEY, J. N. (1921) *The Autonomic Nervous System*, Cambridge.

LARSELL, O. (1922) The ganglia, plexuses, and nerve-terminations of the mammalian lung and pleura pulmonalis, *J. comp. Neurol.*, **35**, 97.

LEVER, J. D., SPRIGGS, T. L. B., and GRAHAM, J. D. P. (1968) A formol-fluorescence, fine-structure and autoradiographic study of the adrenergic innervation of the vascular tree in the intact and sympathectomised pancreas of the cat, *J. Anat. (Lond.)*, **103**, 15.

LOEWI, O. (1935) Problems connected with the principle of humoral transmission of nervous impulses (Ferrier Lecture), *Proc. roy. Soc. B*, **118**, 299.

MACFARLAND, W. E., and DAVENPORT, H. A. (1941) Adrenal innervation, *J. comp. Neurol.*, **75**, 219.

MAGOUN, H. W., and RANSON, M. (1942) The supraoptic decussations in the cat and monkey, *J. comp. Neurol.*, **76**, 435.

MATTHEWS, M. R., and RAISMAN, G. (1969) The ultrastructure and somatic efferent synapses of small granule-containing cells in the superior cervical ganglion, *J. Anat. (Lond.)*, **105**, 255.

MITCHELL, G. A. G. (1935a) The innervation of the kidney, ureter, testicle and epididymis, *J. Anat. (Lond.)*, **70**, 10.

MITCHELL, G. A. G. (1935b) The innervation of the distal colon, *Edinb. med. J.*, **42**, 11.

MITCHELL, G. A. G. (1953) *Anatomy of the Autonomic Nervous System*, Edinburgh.

MITCHELL, G. A. G. (1956) *Cardiovascular Innervation*, Edinburgh.

NATHAN, P. W., and TURNER, J. W. A. (1942) The efferent pathway for pupillary contraction, *Brain*, **65**, 343.

NOTLEY, R. G. (1969) The innervation of the upper ureter in man and in the rat: an ultrastructural study. Ref. to ureteric plexus, *J. Anat. (Lond.)*, **105**, 393.

NYBERG-HANSEN, R. (1956) Anatomical demonstration of gamma motoneurons in the cat's spinal cord, *Exp. Neurol.*, **13**, 71.

PICK, J., and SHEEHAN, D. (1946) Sympathetic rami in Man, *J. Anat. (Lond.)*, **80**, 12.

RAVEN, C. P. (1937) Experiments on the origin of the sheath cells and sympathetic neuroblasts in amphibia, *J. comp. Neurol.*, **67**, 221.

SCHOFIELD, G. (1956) Observations on the distribution of extrinsic nerve fibres in the gut of the rat, *J. Anat. (Lond.)*, **90**, 592.

SHANTHAVEERAPPA, T. R., and BOURNE, G. H. (1966) Perineurial epithelium: a new concept of its role in the integrity of the peripheral nervous system, *Science*, **154**, 1464.

SHEEHAN, D. (1941) Spinal autonomic outflows in Man and monkey, *J. comp. Neurol.*, **75**, 341.

SHEEHAN, D., MULHOLLAND, J. H., and SHAFIROFF, B. (1941) Surgical anatomy of the carotid sinus nerve, *Anat. Rec.*, **80**, 431.

STARKIE, C., and STEWART, D. (1931) The intra-mandibular course of the inferior dental nerve, *J. Anat. (Lond.)*, **65**, 319.

STEWART, D., and WILSON, S. L. (1928) Regional anaesthesia and innervation of teeth, *Lancet*, ii, 809.

STOPFORD, J. S. B. (1918) The variation in the distribution of the cutaneous nerves of the hand and digits, *J. Anat. (Lond.)*, **53**, 14.

STOPFORD, J. S. B. (1930) *Sensation and the Sensory Pathway*, London.

STOTLER, W. A., and McMAHON, R. A. (1947) The innervation and structure of the conducting system of the human heart, *J. comp. Neurol.*, **87**, 57.

SUNDERLAND, S. (1945a) Blood supply of the nerves of the upper limb in Man, *Arch. Neurol. Psychiat. (Chic.)*, **53**, 91.

SUNDERLAND, S. (1945b) Blood supply of peripheral nerves, *Arch. Neurol. Psychiat. (Chic.)*, **54**, 280.

SUNDERLAND, S. (1945c) Blood supply of the sciatic nerve and its popliteal divisions in Man, *Arch. Neurol. Psychiat. (Chic.)*, **54**, 283.

SUNDERLAND, S. (1946) The innervation of the first dorsal interosseous muscle of the hand, *Anat. Rec.*, **95**, 7.

SUNDERLAND, S. (1968) *Nerves and Nerve Injuries*, Edinburgh.

SUNDERLAND, S., and HUGHES, E. S. R. (1946) The pupilloconstrictor pathway and the nerves to the ocular muscles in Man, *Brain*, **69**, 301.

TELFORD, E. D., and STOPFORD, J. S. B. (1934) The autonomic nerve supply of the distal colon, *Brit. med. J.*, **1**, 572.

TRAQUAIR, H. M. (1949) *An Introduction to Clinical Perimetry*, 6th ed., London.

WESTWATER, L. A. (1960) The innervation of the pulps of the teeth, *Brit. dent. J.* **109**, 407.

WILSON, W. C. (1936) Observations relating to the innervation of the sweat glands of the face, *Clin. Sci.*, **2**, 273.

WINDLE, W. F. (1931) Neurons of the sensory type in the ventral roots of Man and of other mammals, *Arch. Neurol. Psychiat. (Chic.)*, **26**, 791.

WOLFF, E. (1953) The so-called medial root of the optic tract is essentially a visual commissure, *Brain*, **76**, 455.

WOOLLARD, H. H., and HARPMAN, A. (1939) The cortical projection of the medial geniculate body, *J. Neurol. Psychiat.*, **2**, 35.

WRETE, M. (1935) Die Entwicklung der intermediären Ganglien beim Menschen, *Morph. Jb.*, **75**, 229.

YOUNG, J. Z. (1939) Partial degeneration of the nerve supply of the adrenal. A study in autonomic innervation, *J. Anat. (Lond.)*, **73**, 540.

THE SKIN AND THE SENSORY ORGANS

G. J. ROMANES

THE sense organs consist of the apparatus by which the individual is made aware of changes in the internal and external environments. These changes are monitored by special receptive tissues which may be highly developed, localized organs such as the eye and the internal ear, or may consist of structures of varying complexity scattered over the surface and in the deeper parts of the body. All of them receive the terminal parts of sensory nerve fibres through which impulses are transmitted to the central nervous system as a result of the application of appropriate stimuli to the end-organ. The skin, because of its function as the largest single sensory organ, is included in this section.

The functions of many of the receptive end-organs is not known, but they are divided into the organs of the **special senses,** sight, hearing, equilibration, smell, and taste, and those concerned with **general sensations,** e.g., touch, pressure, pain, temperature, position (joint sense), and others arising in muscles and tendons which do not appear to evoke conscious sensation.

THE SKIN

The skin is the largest single organ in the body. Apart from its important sensory function, it is a waterproof layer concerned with the prevention of fluid loss; a temperature regulating organ, both because of the variability of its vascular supply and the presence of sweat glands; an excretory organ for certain crystalloids in the sweat; and a secretory organ by means of its sebaceous glands whose fatty secretion not only helps to maintain the waterproofing but also may be acted on by sunlight to produce vitamin D. The skin can react to continued friction by increasing the thickness of its superficial

Duct of sweat gland
Hair
Papillae of corium
Sebaceous glands
Arrector pili muscle
Hair follicle
Body of sweat gland

Epidermis
Corium
Subcutaneous fatty tissue

Papilla of hair
Oblique section through a lamellated corpuscle

FIG. 967   Schematic vertical section of skin.

layers, to sunlight and heat by increased pigmentation, and to wounds by increased growth and repair. It contains hairs over most of the surface, though these are more concerned with sensation than heat control in Man, and it produces the nails. Despite its waterproof nature, certain substances can be absorbed through the skin, particularly those which are fat soluble, and the amounts absorbed may be sufficient to cause acute poisoning particularly if a large area of skin is involved.

The skin is very **elastic,** though decreasingly so with age. Unless firmly adherent to the underlying tissues, it stretches readily, but most freely in one direction because of the tendency of the **collagen** in its deeper layers to run predominantly at right angles to that direction, i.e., parallel to the minute inter-communicating grooves which are present on most of the skin surface. When pierced by a cylindrical spike, it splits along the major direction of the collagen thus giving rise to a linear wound (Jones, 1941). The colour of the skin depends on the presence of **pigment** (melanin) particularly in the deeper layers of the epidermis, and especially in certain regions, e.g., exposed skin, skin of the external genitalia, perianal region, axillae, and mammary areolae. There are also carotenoids in the horny layer, and in the fat of the skin and subcutaneous tissues. Blood, containing either reduced or oxyhaemoglobin depending on the state of the circulation, also colours the skin. In darker skinned races, the melanin is distributed throughout the layers of the epidermis.

The surface of the skin is pierced by the orifices of the sweat glands and hair follicles. Sebaceous glands discharge into the superficial parts of the hair follicles, though some sebaceous glands are found in the absence of hairs on the mammary areolae, on the glans and coronal sulcus of the penis, the labia minora, and the lips. On the palmar surface of the hand and the plantar surface of the foot are numerous, permanent, parallel **epidermal ridges** the arrangement of which is characteristic for each individual, is even different in detail in identical twins (Penrose, 1968), and is responsible for the finger prints.

In most parts of the body the skin is freely mobile on the subjacent tissues, especially where there are cutaneous muscles or subcutaneous bursae. It is bound firmly to the deep fascia on the palms, soles, and lateral aspects of the fingers, at the skin creases over joints, and on the subcutaneous parts of the tibia.

STRUCTURE

[FIGS. 967, 968]

The skin consists of a superficial, cellular layer of ectodermal origin, the **epidermis,** and a deeper layer derived from the mesoderm, the **dermis** or corium.

Epidermis

This layer of stratified squamous epithelium varies in thickness from 0·3 mm. to 1 mm. or more. It is thickest on the palms and soles, and thinnest on the eyelids and penis. In thick skin it may be divided into a number of layers from within outwards; stratum basale, stratum spinosum, stratum granulo-sum, stratum lucidum and the stratum corneum.

The **stratum basale** consists of a single layer of columnar cells separated from the dermis only by a basement membrane. It is in this layer and the stratum spinosum that new cells are produced to replace the cells constantly being worn away from the surface.

The **stratum spinosum** consists of several layers of irregularly shaped cells which tend to become more flattened as the stratum granulosum is approached. These cells are readily separated by shrinkage, except where they are attached to each other by adhesion plaques or desmosomes; hence they have a spiny appearance in many histological preparations.

In the **stratum granulosum** the cells are further flattened, the cytoplasm contains many keratohyalin granules which are believed to be implicated in the formation of keratin, and the cells are in the process of dying.

Superficial to the stratum granulosum is the **stratum lucidum,** a transparent layer in which boundaries of the elongated flattened cells can only be seen with the electron microscope. Nuclei and intracellular organelles of the deeper layers have disappeared.

The **stratum corneum** consists of many layers of flattened, keratinized squames which are continually shed from the surface. This is the layer which is most responsible for the variations in thickness of the different regions of the skin.

In the deepest layers of the epidermis are **dendritic cells,** the branching processes of which pass superficially between the adjacent epithelial cells. These are the epidermal **melano-cytes** which are responsible for the production of melanin and its transfer to the other cells of the epidermis.

The deep surface of the epidermis always interdigitates with irregularities of the surface of the dermis [FIG. 968]. The depth of the interdigitations varies with the degree of shearing force to which the skin is subjected, being deepest in the palms

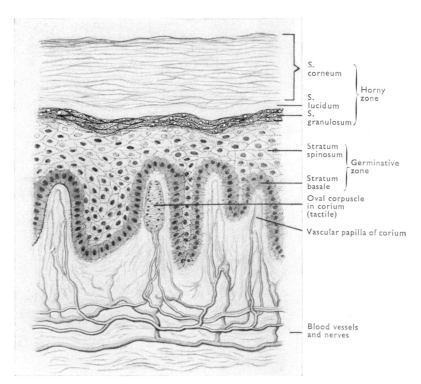

S. corneum

S. lucidum
S. granulosum

Horny zone

Stratum spinosum

Stratum basale

Germinative zone

Oval corpuscle in corium (tactile)

Vascular papilla of corium

Blood vessels and nerves

FIG. 968 Vertical section of epidermis and papillae of corium.

and soles. Here there are usually subsidiary epidermal ridges between the main extensions into the dermis, and it is through these smaller ridges that the ducts of the sweat glands pass to open on the surface epidermal ridges.

Dermis

The basis of this is a felt of collagen and elastic fibres. The superficial, or **papillary layer,** interdigitates with the epidermis and is clearly demarcated from it. The deeper, **reticular layer** fades into the subcutaneous tissue without a clear line of demarcation. The dermis contains blood and lymphatic vessels, nerves, sebaceous glands, the superficial parts of the sweat glands and of the hair follicles, and the muscles of the hair follicles, arrectores pilorum. Most of the cutaneous nerve endings lie in the dermis, especially its papillary layer, though some nerve fibres pass between the cells of the epidermis.

Vessels and Nerves. From the arterial plexus in the subcutaneous tissue, branches pass into the reticular layer of the dermis to form a subsidiary plexus from which capillary loops pass into the papillary layer. Veins and lymphatics begin in the papillae of the dermis, pass to plexuses in the reticular layer, and then join the corresponding subcutaneous vessels. Most of the lymph vessels in the dermis are valveless. Within the dermis, especially in the exposed parts of the body, there are many **arteriovenous anastomoses** [FIG. 1063] which can allow a considerable flow of blood direct from the arterioles to the venules, thus increasing the heat supply to the superficial tissues (Lewis, 1927; Popoff, 1934).

The skin is heavily innervated by **sensory nerve fibres** but it also receives many **postganglionic sympathetic fibres** for the supply of the blood vessels, the sweat glands, the arrectores pilorum, and the dartos muscle of the scrotum. The fibres that supply the arrectores pilorum and sweat glands are cholingeric and do not release noradrenaline as do the fibres ending on the blood vessels.

APPENDAGES OF THE SKIN

These are nails, hairs, sebaceous and sweat glands, and all are derived from the epidermis.

Nails [FIGS. 969, 970]

These consist of tightly packed, cornified cells which form a translucent plate, similar to but harder than the stratum lucidum. They are a special modification of the superficial

FIG. 969 Transverse section of nail.

FIG. 970 Longitudinal section through root of nail.

epidermal layers, particularly of the stratum lucidum, but without a stratum granulosum. The visible part of the nail is its **body;** the hidden part, or **root,** from which growth of the nail principally occurs, is deeply buried in an invagination of the epidermis. The body and root both lie on the **nail bed** [FIG. 970], the epidermal surface of which is heavily cornified to form the nail.

The fold of skin which overlies the root of the nail is the **eponychium,** and its free, cornified margin forms the cuticle. A cornified layer under the free margin of the nail is the **hyponychium.** The downturned sides of the nail are partly buried by the **nail walls.** The epithelium of the nail bed consists only of stratum basale and spinosum, and is thrown into longitudinal folds corresponding to the longitudinal striations on the nail surface. The proximal part of this epithelium proliferates to form the nail over a region extending distally as far as the **lunule,** the white semilunar region which may extend beyond the eponychium.

Hairs

These are present over the entire surface of the body except on the palms and soles, the sides of the fingers, feet and toes, the distal phalanges, the knuckles, the red parts of the lips, the glans and prepuce of the penis, the areolae and nipples, and the skin between the labia majora. Most of the hairs are very fine (vellus hairs) so that the skin may appear hairless, but their numbers may be greater in some regions than in hairy mammals. There is a marked sexual differentiation in the distribution of coarse hairs, particularly on the face and trunk (Dupertuis et al., 1945), and in its loss from the scalp. All hairs, except the eyelashes, emerge obliquely from the surface of the skin; the hairs in any one site point in the same direction forming a hair tract (Kidd, 1903; Jones, 1941; Clark, 1971).

The root of the hair is ensheathed in a sleeve of epidermis called the **hair follicle** [FIG. 967] which extends into the subcutaneous tissue. The **shaft** is the free portion of the hair. Throughout most of its length the hair consists of the keratinized remains of cells. The outermost layer **(cuticle)** is formed of overlapping squames with their free edges pointing towards the tip of the hair. Internal to this is the **cortex** which forms the greater part of the hair. It consists of tightly packed hard keratin which contains pigment and a few small air spaces.

The **medulla** forms the core of the hair. It consists of soft keratin, may be broken into discontinuous pieces, and only forms a significant part in coarse hairs, though it is said to be present in all but the finest hairs (Montagna, 1956). It also

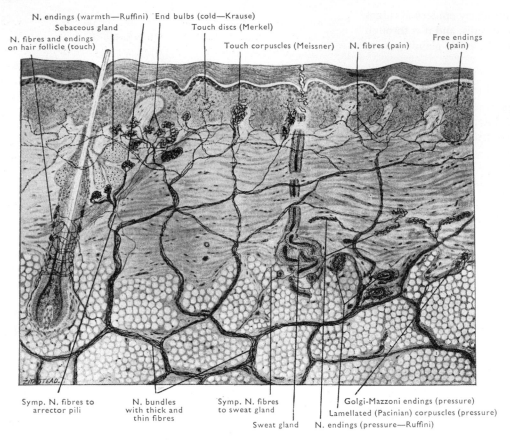

N. endings (warmth—Ruffini)   End bulbs (cold—Krause)
Sebaceous gland   Touch discs (Merkel)
N. fibres and endings
on hair follicle (touch)   Touch corpuscles (Meissner)   N. fibres (pain)   Free endings (pain)

Symp. N. fibres to   N. bundles   Symp. N. fibres   Golgi-Mazzoni endings (pressure)
arrector pili   with thick and   to sweat gland   Lamellated (Pacinian) corpuscles (pressure)
    thin fibres   Sweat gland   N. endings (pressure—Ruffini)

FIG. 971   Composite diagram showing the innervation of the human skin. (Woollard, Weddell, and Harpman, 1940.)

contains larger air spaces, fat globules and pigment, even in apparently white hairs.

In the growing hair, the deepest part of the hair follicle expands to form an onion-shaped cap (**bulb of the hair**) which almost encloses some loose, vascular connective tissue, the **papilla** [FIG. 967]. The cells of the follicle which surmount the papilla proliferate to form all three layers of the hair; the most central cells, which form the medulla, keratinize less rapidly than those forming the cortex, and hence retain their nuclei for a greater distance up the hair.

In the resting hair follicle, the bulb and papilla shrink and the deepest part of the follicle becomes irregular in shape.

The hair follicle extends obliquely through the skin into the superficial part of the subcutaneous tissue, and is spirally arranged in curly hair. It has a fibrous covering, consisting of outer longitudinal and inner circular layers; the latter separated from the epidermal part of the follicle by a translucent layer which is continuous with the basement membrane of the epidermis. The epidermal part consists of **inner** and **outer root sheaths,** and expands laterally to form the **sebaceous glands** immediately below the papillary layer of the dermis, usually in the angle between the follicle and arrector pili muscle. Deep to the sebaceous

glands only the two deeper layers of the epidermis can be identified, though all layers are visible superficial to them.

The **outer root sheath** is the thickest part of the follicle. It is continuous above with the stratum basale and spinosum of the epidermis, but is thinner in the depths of the follicle. The **inner root sheath,** which disappears in the region of the sebaceous gland consists of: (1) an inner thin **cuticle** composed of flattened, overlapping cells whose free edges face inwards and interlock with the squames of the hair cuticle, thus holding the hair in the follicle; (2) one or two layers of polyhedral, nucleated cells (Huxley's layer); (3) an outer (Henle's) single layer of nucleated, cubical cells.

Arrectores Pilorum Muscles [FIG. 971]

Bundles of smooth muscle fibres are attached to the fibrous sheath of the hair follicle deep to the sebaceous glands, and pass to the papillary layer of the dermis on the side towards which the hair slopes. These smooth muscle fibres are supplied by sympathetic postganglionic fibres in the cutaneous nerves. When they contract they erect the hairs, and pulling them slightly towards the surface of the skin, elevate the skin around the mouths of the hair follicles to produce the so-called gooseflesh. They also compress the sebaceous gland.

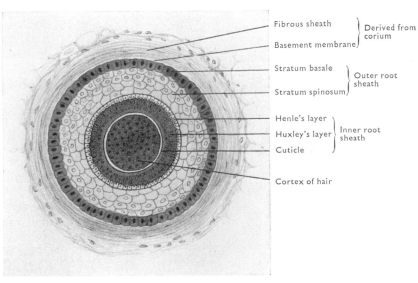

Fibrous sheath   } Derived from corium
Basement membrane   }

Stratum basale   } Outer root sheath
Stratum spinosum   }

Henle's layer   }
Huxley's layer   } Inner root sheath
Cuticle   }

Cortex of hair

FIG. 972   Transverse section of hair follicle and hair (semidiagrammatic).

Sebaceous Glands

These glands are formed as diverticula of the superficial part of every hair follicle, but they also occur on the labia minora and majora, the glans penis, and the mammary areolae in the absence of hairs. Those in the mammary areolae enlarge considerably in pregnancy. Each of the glands (1–4) associated with a hair follicle expands into a cluster of flask-shaped alveoli which lie in the angle between the shaft of the hair and the arrector pili muscle. The alveoli are filled with tightly packed cells containing oil droplets, and these disintegrate (holocrine secretion) to discharge their contents (sebum) into the hair follicle, or directly on to the surface where they are not associated with hairs. Disintegrating cells are replaced by proliferation of the neck cells of the alveoli, the daughter cells descending into the alveoli and gradually developing into mature sebaceous cells. The secretion waterproofs the hairs and skin, and helps to account for the relative ease of absorption of fat-soluble substances through the skin. Secretion of these glands is not affected by sympathectomy.

Sweat Glands

These glands are found throughout the skin, but tend to be more numerous on its exposed parts, especially the palms and soles where the ducts open on the summits of its epidermal ridges. Each gland consists of a long tube (duct) which may extend into the subcutaneous tissue, where it is coiled to form the secretory body of the gland, 0·1–0·5 mm. in diameter [FIG. 967]. In thick skin, the duct takes a spiral course through the epidermis to open through a funnel-shaped sweat pore. Sweat is produced by the cells of the body of the gland without any loss of cytoplasm (eccrine secretion), except in the case of some large, apocrine sweat glands (1–4 mm. in diameter) which tend to open into hair follicles in the axilla, groin, and perianal regions. In these glands the secretory cells have bulbous processes projecting into the lumen; these may be lost into the sweat (apocrine secretion).

The body of the gland is a tightly coiled tube with a single layer of pale-staining pyramidal, secretory cells which are partly separated from a thick basement membrane by flattened, branched cells which contain myofilaments similar to those in smooth muscle cells. These myo-epithelial cells are believed to assist with the expulsion of the secretion, but they are not found in the narrower duct which has two layers of darkly staining cells with many fine filaments in the internal cytoplasmic layer. Sweat glands are supplied by post-ganglionic, cholinergic, sympathetic nerve fibres. No sweating occurs after sympathetic denervation.

The ciliary glands which open at the margins of the eyelids are modified, uncoiled sweat glands, as also are the branched ceruminous glands in the external acoustic meatus. The ducts of the latter glands open into hair follicles, and their cells contain a yellowish pigment which colours the wax secretion.

DEVELOPMENT OF THE SKIN AND ITS APPENDAGES

The dermis is developed from the mesoderm, the epidermis, nails, hair, sweat, sebaceous, and mammary glands from the ectoderm. By the first month of intra-uterine life the meso-derm forms a subectodermal layer which differentiates into two layers (dermis and subcutaneous tissue), by the third month. Dermal papillae appear in the fourth month.

The epidermis is a single layer of cells at first, but it differentiates into deep, cubical and superficial, flattened layers of cells, by the second intra-uterine month. Three layers are visible in the third month, the deeper polyhedral cells giving rise to the stratum basale and spinosum, while the flattened, superficial layer is known as the epitrichium because the growing hairs lie deep to it and push it off to form part of the greasy covering of the child at birth, the vernix caseosa. Formation of daughter cells by the deeper layers leads eventually to the production of the definitive epidermis of the skin (Montagna 1956).

Nails. These structures begin to develop in the third foetal month as a thickening of the epidermis on the dorsal surfaces of the terminal phalanges. This thickening is overgrown proximally and on each side by the surrounding epidermis to form the nail fold and walls. The nail thickening keratinizes without the appearance of a stratum granulosum to form the hard keratin of the nail, which grows in length by the proliferation of the cells of the stratum basale and spinosum of the nail bed under the nail fold and lunule (germinal matrix). Distal to this, the corresponding layers of the nail bed (sterile matrix) make no contribution to the nail. Nail growth is most rapid in the hand (3 mm. per month) especially in well nourished boys, and particularly in the third digit (Clark and Buxton, 1938; Gilchrist and Buxton, 1939).

Hair. Solid downgrowths of the stratum basale of the epidermis begin to pass obliquely into the dermis in the third to fourth month of pregnancy. The deep extremity of each downgrowth expands to form the bulb, producing the germinal matrix over the dermal papilla. Growth of the cells of the matrix forms the hair and its inner root sheath; the outer root sheath is derived from the original downgrowth. Condensation of the surrounding mesoderm forms the fibrous sheath and vascular plexus of the follicle. The growing hairs appear at the surface about the fifth month, and pushing off the epitrichium, form the first crop of hairs, the lanugo. These very delicate hairs are fully formed by the seventh month, but are shed at birth or shortly afterwards; the eyelashes and scalp hairs, being the last to be shed, are replaced by stronger hairs.

The shedding and renewal of hairs continues throughout life, and periods of growth and quiescence alternate (Trotter, 1924). In quiescence, proliferation of the germinal matrix ceases, the bulb and papilla shrink, the base of the hair becomes club-shaped, and eventually the deepest part of the follicle degenerates. Despite the absence of growth from below, the club hair is shed, and subsequently the outer root sheath cells proliferate to produce downgrowth of the follicle, reformation of the bulb and papilla, and regrowth of the hair.

Hair growth rates differ considerably in different parts of the body, but an average figure is probably 1·5–2 mm. per week.

Sebaceous Glands. Solid outgrowths of the outer root sheath cells of the hair follicles appear in the fifth foetal month. These enlarge distally forming a number of lobules. The sebaceous secretion plays a major part in the formation of vernix caseosa.

Sweat Glands. These solid, yellowish downgrowths of the

stratum basale of the epidermis pass through the dermis into the subcutaneous tissue. Here they become coiled to form the secretory part. They appear first on the palms and soles in the fourth month of pregnancy, but much later in the hairy regions. The ducts open on the surface during the seventh month.

ORGANS OF GENERAL SENSATION

This group of sensory nerve endings is widely distributed throughout the body and is concerned with sensations such as pain, heat, cold, pressure, touch, and the afferent impulses which arise in muscles, tendons, and joints. These nerve endings may either be surrounded by capsules or form fine meshes of bare axons in the tissues (Cauna, 1959).

Although many different varieties of encapsulated nerve endings have been described, most of them do not fall into clearly defined histological groups; attempts to correlate particular microscopic appearances with sensitivity to certain stimuli have not been outstandingly successful. Indeed the structure of nerve endings may alter with use, and some have been shown to increase in size and complexity with age while decreasing in number (Cauna, 1964).

Encapsulated endings are found in many tissues, usually arranged in small groups. In the skin they are commonest in the hairless regions, and their relative absence in certain other regions is not associated with the inability to appreciate certain stimuli, though the threshold may be higher. Thus it appears that highly differentiated sensory endings are not essential for the appreciation of most types of sensory stimuli, and though certain histological types of endings may be sensitive to particular stimuli, it does not follow that they are always concerned with producing the same kind of subjective sensation. Thus the known sensitivity of Pacinian corpuscles to deformation (Sinclair, 1967) means that they could record any kind of tissue deformation, from pressure applied to the skin, to distension of a blood vessel under pressure.

Again, the majority of encapsulated sensory nerve endings are supplied by several nerve fibres, often both myelinated and non-myelinated; a feature which does not suggest a high degree of specificity between the nerve ending and the central nervous system which interprets its impulses. Very little is known about the chemical specificity of nerve endings; thus endings which appear similar when stained with silver, may be chemically different and have a different stimulus specificity.

Under normal circumstances, stimuli applied to the body give rise to nerve impulses in a wide variety of nerve fibres from both superficial and deep tissues, and normal subjective sensation is not the result of stimulation of a few highly specific nerve endings, but rather results from the analysis of a shower of impulses delivered to the central nervous system in a particular pattern. Many problems remain to be solved before it is possible to interpret the significance of the structure of peripheral nerve endings in the genesis of subjective sensations (for review, see Sinclair, 1967).

FREE NERVE ENDINGS
[FIG. 973]

Such networks are found throughout the body, in skin, mucous membranes, and deep tissues. They form the only type of innervation in the cornea, teeth, tympanic membrane, and the walls of blood vessels, and they cannot be distinguished from sympathetic plexuses. They contain both myelinated and non-myelinated nerve fibres and tend to form coarser, deep plexuses, the meshes of which become finer as they approach the epidermis of the skin. While most of the nerve fibres in the skin end in the dermis, they also enter the epidermis and pass between its basal cells. Individual nerve fibres may form both deep and superficial endings, and though their over-all distribution is not known, it is clear that many fibres supply each area of skin.

Hairs

All hair follicles are innervated [FIG. 971], and these probably form the main tactile organs in hairy skin. Each hair receives branches from several nerve fibres, and these form an inner longitudinal and outer circular arrangement around the follicle. The manner of ending is not known, but there is evidence that single nerve fibres may innervate more than one hundred hairs (Weddell and Pallie, 1955).

SPECIAL END-ORGANS

Lamellated (Pacinian) Corpuscles [FIG. 974B]

These corpuscles consist of many concentric layers of flattened epithelial-type cells which are continuous with the perineurium (Shanthaveerappa and Bourne, 1963, 1966), but separated by layers of collagen. A single, thickly myelinated, nerve fibre, enters and extends in the central core, unlike the multiple innervation of most sensory corpuscles. Lamellated corpuscles vary greatly in size (1 mm.–4 mm. or more), are often oval in shape, but may be lobulated or flattened. They have a very wide distribution, in the dermis (especially of hairless skin and more especially in the fingers), in fascial planes, in

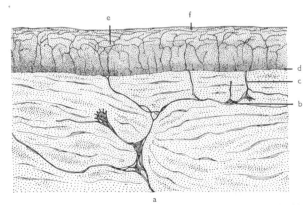

FIG. 973   Transverse section of cornea stained with gold chloride.
a, b, plexus in substantia propria of cornea; c, branch passing to subepithelial plexus, d; e, intra-epithelial plexus; f, terminations of fibrils.

joint capsules, close to blood vessels, in the pleura and extra-peritoneal tissues, in the external genitalia and the mammary gland, and in many other situations. These corpuscles are sensitive to deformation, and may, therefore, record a wide variety of different changes depending on the tissue in which they are placed.

Tactile (Meissner) Corpuscles [FIG. 974 C]

These oval, circular, or lobulated corpuscles lie in small groups in the dermal papillae of the palmar surface of the fingers, the palm, and the sole of the foot. They consist of a stack of flattened cells enclosed in a capsule continuous with the perineurium of the nerves which enter it. Several medium-sized nerve fibres enter each corpuscle, and each of these fibres supplies several corpuscles. The nerve fibres lose their myelin sheath as they enter the corpuscle, and branching, pass

between the cells of the corpuscle to form a complex mesh of fine fibres which do not unite. These corpuscles decrease in number, but increase in size and complexity with age. They are believed to be sensitive to touch, but absolute proof of this is lacking, though it is clear that several corpuscles are concerned with activating a single nerve fibre, and each corpuscle has many routes along which it can send impulses to the central nervous system.

Tactile Discs (Merkel) [FIG. 975]

These disc-like expansions of branches of heavily myelinated fibres are usually found apparently surrounded by epithelial cells in the basal layers of the epidermis with the same distribution as the tactile corpuscles. They are also found in other regions, especially in other mammals, but though their functions are unknown, their close relation to the epidermis and the fact that each unit has an appearance like one element of a tactile corpuscle, suggests that their name may be appropriate.

Bulbous Corpuscles (Krause) [FIG. 974 A]

These are small tufts formed by the non-myelinated branches of myelinated nerve fibres looped and coiled together in a cylindrical or spherical bundle. There is no capsule but often a condensation of the surrounding connective tissue, and individual fibres may leave the corpuscle and pass to take part with others in the formation of further bulbous corpuscles. They are found principally in mucocutaneous regions, e.g., lip, gum, external genitalia, perianal region, etc. Similar types of nerve endings are also found in other regions of the skin, but these are of such variable structure that it is often impossible to determine to which group of sensory endings they belong (Weddell and Miller, 1962). It seems likely that the morphology of these and of the other special cutaneous endings is so variable that attempts to divide them into distinct groups is

FIG. 974 Four types of special end-organs.

A, bulbous corpuscle.
B, lamellated corpuscle.
C, oval corpuscle.

Nucleus of stratum germinativum

Basement membrane of epidermis

Myelinated nerve fibre

Tactile cell surrounding tactile disc

FIG. 975 Semidiagrammatic representation of a cutaneous tactile organ found in hairy skin and supplied by a single thick nerve fibre ending in a series of tactile discs.

unrealistic, and merely leads to the assumption that such distinct morphological types have separate functions; an assumption which is not justifiable on the evidence (Quilliam, 1966).

SENSORY ENDINGS IN MUSCLE AND TENDON

Nerve endings of various kinds are found in intermuscular fascial planes, and in the capsules of joints especially near their attachments to the periosteum. The main sensory endings in muscle and tendon are the neuromuscular and neuro-tendinous spindles.

Neuromuscular Spindles

These complicated sensory organs are present in most of the somatic, striated muscles of the body [FIG. 976], though simpler varieties are present in the extrinsic ocular muscles (Cooper and Daniel, 1949) [FIG. 977]. The fusiform spindles consist of a group (3–10) of thin intrafusal muscle fibres lying parallel to the ordinary muscle fibres, and enclosed together in a connective tissue **sheath** from which they are separated in their central region by a fluid-filled space. Within the space, each small muscle fibre contains a number of nuclei which are either arranged in a mass (**nuclear bag**) or in a row (**nuclear chain**) filling the muscle fibres (Boyd, 1962). Two types of myelin-ated sensory fibres enter the spindle: (1) heavily myelinated fibres pass to the central region and form spiral coiled (**anulo-spiral**) endings around the nuclear region of the muscle fibres; (2) medium sized myelinated fibres pass to regions further from the centre and break up into a complicated series of branches (**flower-spray endings**) on the contractile part of the intrafusal muscle fibres. The anulospiral endings are sensitive to stretch, and it is probable that the flower-spray endings are sensitive to increased tension. However, the exact sensitivity of the parts of the spindle, if different, is not known, though they react to increasing length or tension.

In addition to the sensory fibres, fine motor fibres (**γ-efferent fibres**) innervate the intrafusal muscle fibres near their ends. These motor fibres arise independently from small cells in the anterior horn of the spinal medulla (Nyberg-

FIG. 977   Sensory nerve ending enveloping a fibre of an ocular muscle. (After Dogiel.)

Hansen, 1965), and are used to increase the tension or decrease the length of the intrafusal muscle fibres. Thus they can adjust the length of the spindles to that of the surrounding muscle fibres so that the spindles are available to record stretching of the muscle in any position, and can increase the spindle tension so that it generates nerve impulses even when the muscle is not being stretched.

Neurotendinous Spindles

There are many types of endings on tendons, including simple branched nerve fibres, while complex and sometimes encapsulated endings are also present especially near the myo-tendinous junction. The latter have a resemblance to neuro-muscular spindles, but are arranged in series with the muscle fibres. Presumably they record increasing tension in the tendon whether this is due to stretch or muscle contraction. It is of some interest that most of the tendon organs are arranged in the form of flower-spray endings, but their structure is variable.

The afferent impulses from muscle and tendon spindles do not appear to evoke any conscious sensation; position sense seems to depend on the sensory endings from joints.

DEVELOPMENT OF NERVE ENDINGS

Most sensory endings develop during foetal life (Jalowy, 1939), and though they are recognizable at birth, some of them continue to develop throughout postnatal life (Cauna, 1964), but there is a general decrease in the number of cutaneous sensory endings with age.

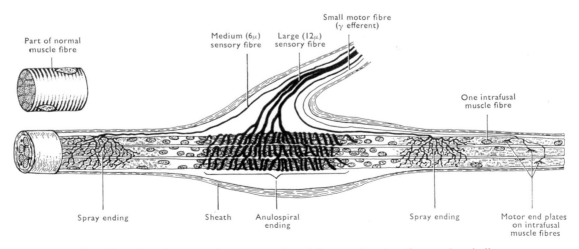

FIG. 976   Semidiagrammatic representation of the central region of a muscle spindle.

ORGANS OF THE SPECIAL SENSES

ORGAN OF VISION—THE EYEBALL

The eyeball is composed of segments of two unequal spheres; the transparent, anterior, **corneal segment** has a smaller radius of curvature than the opaque, white, posterior, **scleral segment** [FIG. 978], and projects forwards from it. The two parts join at a slight, external groove, the **sulcus sclerae.** The central points of the anterior and posterior surfaces of the eyeball are known as the anterior and posterior **poles;** the line joining the poles is the **optic axis;** an imaginary line encircling the eyeball, midway between the poles, is the **equator.** These are purely descriptive terms and have little significance since the posterior pole cannot be determined in life. The **visual axis** is of more importance, and joins the centre of the cornea to the fovea centralis of the retina [FIG. 978]. It represents the course taken through the eye by a ray of light from the centre point of vision. When looking at a distant object, the visual axes of the two eyes are parallel, though the optic axes are slightly, and the optic nerves markedly convergent posteriorly.

The eyeball measures approximately 2·5 cm. in all directions, and has a volume just over 8 cm³. The eyeball completes the greater part of its growth in the antenatal period. Hence relative to body size it is much larger in infants and children, and slightly larger in females than in males (Schultz, 1940).

FASCIAL SHEATH OF THE EYEBALL

This is a thin fibrous sheath which surrounds the sclera, forms a socket for the eyeball, and separates it from the other contents of the orbit. The sheath fuses with the sclera both immediately posterior to the corneoscleral junction, and where the optic nerve, ciliary nerves, and posterior ciliary arteries pierce it. Elsewhere the sheath is very loosely united to the sclera by a few minute trabeculae of connective tissue which cross this potential, **episcleral (intervaginal) space.** The sheath is pierced by the venae vorticosae close to the equator [FIG. 986], and sends tubular expansions outwards over each of the extrinsic ocular muscles as they pass through it to the sclera [FIG. 370]. These **sheaths** become continuous with the epimysium of the muscles (in the case of the superior oblique with its trochlea), but each has an extension **(check ligament)** to the periosteum of the orbit close to its origin from the fascial sheath of the eyeball. The check ligament of the medial and lateral rectus muscles are well developed and form triangular masses of fibrous tissue, while the remainder are delicate structures. The check ligament of the lateral rectus is attached to a tubercle on the orbital surface of the zygomatic bone and to the lateral palpebral ligament, while that of the medial rectus is attached to the lacrimal bone posterior to its crest. Between these two check ligaments, the fascial sheath inferior to the eyeball is thicker than elsewhere, is pierced by the inferior rectus and inferior oblique muscles, and forms the **suspensory ligament** (Lockwood, 1885) which thus passes like a hammock under the eyeball from the medial to the lateral orbital walls.

The presence of the episcleral space between the eyeball and its fascial sheath allows some movement of the eyeball within the sheath, but in most movements of the eye, sheath and eyeball move together, the check ligaments limiting the range of movement.

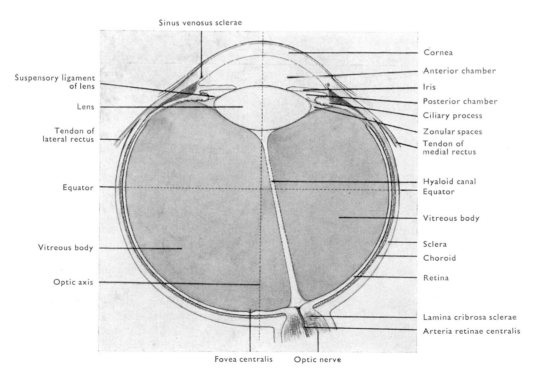

Sinus venosus sclerae

Suspensory ligament of lens
Lens
Tendon of lateral rectus
Equator
Vitreous body
Optic axis

Cornea
Anterior chamber
Iris
Posterior chamber
Ciliary process
Zonular spaces
Tendon of medial rectus
Hyaloid canal
Equator
Vitreous body
Sclera
Choroid
Retina
Lamina cribrosa sclerae
Arteria retinae centralis

Fovea centralis   Optic nerve

FIG. 978 Diagram of horizontal section through left eyeball and optic nerve. × 4.

At the margins of the orbit, the orbital periosteum (**peri-orbita**) is continued into the eyelids as the **palpebral fascia** which extends into the tarsal plates. This sheet, split by the palpebral fissure, constitutes the **orbital septum**. The part in the inferior eyelid receives an extension from the sheaths of the inferior rectus and inferior oblique muscles, while the sheath of the superior rectus, fused with that of levator palpebrae superioris, is inserted into the superior conjunctival fornix deep to the orbital septum.

COATS OF THE EYEBALL

There are three concentric coats forming the wall of the eyeball. (1) The outer **fibrous coat**, consisting of the sclera and cornea. (2) A middle, **vascular, pigmented coat** (or uvea), the choroid, ciliary body and iris. (3) An inner, **nervous coat**, the retina, which is continued forwards in a modified form on the ciliary body and iris [Fig. 983]. These coats surround and partly subdivide the contents of the eyeball, which consist of the **vitreous body** between the lens and the retina, and a space between the lens and the cornea which contains the **aqueous humour** and is partly subdivided by the iris into the **anterior** and **posterior chambers**.

FIBROUS COAT

Sclera

This is the firm, opaque, white coat of the eye. It has a bluish tinge in young children due to the pigment of the choroid shining through, and is yellowish in old age. It is thickest posteriorly (1 mm.), thinnest at the equator (0·5 mm.), and intermediate anteriorly (0·6 mm.). The optic nerve pierces the sclera as a considerable number of separate bundles slightly more than 3 mm. medial to the fovea centralis, and just above the horizontal meridian (Traquair, 1949). This part of

the sclera is therefore known as the **cribriform part**, and externally, the dural sheath of the optic nerve fuses with the sclera at the margin of that part. Around this fusion there are fifteen to twenty small apertures in the sclera for the **short ciliary nerves** and **short posterior ciliary arteries**, while a **long posterior ciliary artery** and a **long ciliary nerve** pierce it a little further distant on each side [Fig. 986]. Their blood is drained by the **venae vorticosae**, two superior and two inferior, which pierce the sclera at equidistant points near the equator. There are openings for the **anterior ciliary arteries and veins** near the sulcus sclerae [Fig. 986].

The deep surface of the sclera is attached to the choroid by a delicate mesh of pigmented areolar tissue, the **suprachoroid lamina**. The ciliary arteries and nerves run in this layer which is often looked on as a space (**perichoroidal space**) because of the delicate nature of its areolar tissue. When choroid and sclera are separated, some of the lamina remains adherent to the sclera, and is known as the **lamina fusca** of the sclera.

At the corneoscleral junction, the sclera is continuous with the cornea, the deepest part projecting inwards as a shelf (the **scleral spur**) posterior to a canal (**sinus venosus sclerae**) which encircles the eye in the corneoscleral junction [Fig. 979]. The scleral spur is continuous with the deepest part of the cornea which forms the fenestrated **pectinate ligament of the iridocorneal angle**, medial to the sinus venosus sclerae. The sinus, *which does not contain blood*, receives fluid from the aqueous humour through the fenestrations of the pectinate ligament (**spaces of the iridocorneal angle**), and communicates with the anterior ciliary veins.

Structure. The sclera consists of bundles of white fibrous tissue arranged in interlacing equatorial and meridional layers. There are some fine elastic fibres and many fibrous tissue cells, with some pigment cells between the bundles of the inner layers.

Vessels and Nerves. The **arteries** are from the anterior and short posterior ciliary arteries; the **veins** enter the venae vorticosae and anterior ciliary veins. The **nerves** are branches of the ciliary nerves, mainly consisting of afferent fibres from the trigeminal nerve, but some are sympathetic fibres to the blood vessels.

Cornea

This avascular, transparent, anterior part of the eye forms one sixth of the total eyeball surface, and is slightly thicker than the sclera, 0·9 mm. at the centre and 1·2 mm. at the periphery. Its anterior surface is covered with stratified squamous epithelium continuous with the **conjunctiva**; its posterior surface forms the anterior wall of the anterior chamber of the eye and is lined by the **endothelium** of that chamber. The curvature of the cornea differs in different individuals, and is greater in the young than in the aged; it is slightly greater in the vertical than in the horizontal plane, and diminishes from the centre to the periphery. As there is a considerable change in refractive index from air to cornea, this junction forms an important part of the focusing mechanism of the eye. Thus irregularities of its surface interfere with the ability to form a sharp image on the retina, and inequalities of curvature in the vertical or horizontal planes are associated with an inability to focus either vertical or horizontal lines sharply on the retina, a condition known as *astigmatism*.

Sclera — Sinus venosus sclerae — Scleral spur — Pectinate ligament of iridocorneal angle — Dilator pupillae — Iris

Meridional fibres of ciliary muscle — Iridocorneal angle — Parts of ciliary processes — Circular fibres of ciliary muscle

Fig. 979 Section of iridocorneal angle. (Thomson, 1912.)

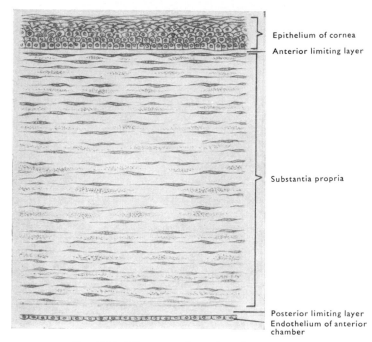

FIG. 980 Diagrammatic section of cornea.

Epithelium of cornea

Anterior limiting layer

Substantia propria

Posterior limiting layer
Endothelium of anterior chamber

The cornea fits inside the anterior margin of the sclera, the junction being oblique. The obliquity is not uniform around the entire junction, so that the cornea appears slightly elliptical from the front, the horizontal diameter greater than the vertical. The iris is attached immediately posterior to the internal junction of the cornea and sclera, and the angle formed here between the iris and the cornea is the **iridocorneal angle** [FIG. 979].

Structure [FIG. 980]. 1. The **anterior epithelium of the cornea** consists of six to eight layers of cells continuous with the epithelium of the conjunctiva at the corneal margins. The deepest layer consists of columnar cells, the intermediate layer of polygonal cells with interdigitating processes attached by desmosomes, the superficial layer of nucleated squames. This is a highly sensitive layer [FIG. 973] which regenerates rapidly after injury.

2. The **anterior limiting layer,** approximately 8 μm. thick, is a part of the substantia propria of the cornea which appears different because of the irregular arrangement of its collagen fibres, and the absence of elastic fibres.

3. The **substantia propria** of the cornea consists of bundles of collagen fibres arranged in many layers in each of which the fibres run in different directions. The layers are woven together by the passage of fibres from one to another. There are also fine elastic meshes, fibrous tissue cells, and a polysaccharide containing ground substance which may be responsible for the transparency of the tissue. It is avascular, but lymphocytes can be found between the layers, and appear in large numbers in infections. In persons over middle age there may be a deposition of fat between the layers at the periphery, particularly inferiorly, and this forms an opaque, greyish ring, the *arcus senilis.*

4. The **posterior limiting layer** appears to be a thick, transparent basement membrane, probably formed from the endothelium of the anterior chamber. It separates easily from

the remainder of the cornea, and contains an atypical collagen mesh. At the periphery of the cornea **(limbus),** this layer becomes continuous with the **pectinate ligament** whose fenestrations are lined by the endothelium of the anterior chamber and lead from that chamber to the sinus venosus sclerae. The deepest part of the pectinate ligament continues round the iridocorneal angle into the iris in close association with the myoepithelial fibres of dilator pupillae (Thomson, 1912).

5. The **endothelium of the anterior chamber** consists of a single layer of large squamous cells which lines the entire anterior chamber, including the posterior surface of the cornea.

At the margin of the cornea, the anterior limiting layer is replaced by a layer of loose connective tissue containing many capillaries. These do not extend into the cornea, but are the blood vessels of the conjunctiva from which the cornea is partly nourished by diffusion.

Vessels and Nerves. *The cornea does not contain blood vessels.* Spaces may be seen in it in microscopic preparations, and these have been thought to represent pathways for the passage of fluid, but they are probably artefacts.

The **nerves** are derived from the ciliary nerves which form a plexus round the limbus. Fibres enter the substantia propria, and forming a subepithelial plexus, send fine filaments through the anterior epithelium into its surface layers [FIG. 973].

VASCULAR COAT

The Choroid

This vascular layer contains stellate pigment cells and phagocytes in a fibro-elastic mesh. It is loosely attached to the sclera by the **suprachoroid lamina,** is missing where the optic nerve pierces it, and is firmly bound to the sclera where the ciliary arteries enter it. The choroid extends forwards to the ora serrata of the retina [FIG. 983]. where it becomes continuous with the ciliary body, and through that with the iris.

Structure. 1. The **suprachoroid lamina** transmits the ciliary nerves, and long posterior ciliary arteries. It consists of the above structures.

2. The **vascular layer** of the choroid consists of an outer zone which contains large blood vessels (branches of the short posterior ciliary arteries) and superficial to these the whorled tributaries of the venae vorticosae [FIG. 986]. These are

Basal lamina

Choriocapillary lamina
Intermediate stratum

Vascular lamina

Suprachoroid lamina

Sclera

FIG. 981 Transverse section of choroid and inner part of sclera.

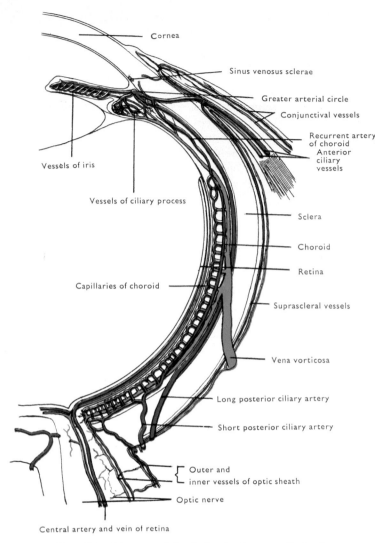

FIG. 982   Diagram of circulation in the eye.   (Leber.)

thelium (the outer of which is pigmented) representing the two layers of the optic cup of the embryo. Both ciliary ring and processes have a similar structure to the choroid, except that the suprachoroid lamina is poorly developed, and the choriocapillary lamina is absent, in keeping with the absence of a functional retinal layer.

The **ciliary ring** is a slightly ridged zone, about 4 mm. wide, which extends from the ora serrata of the retina [FIG. 984] to the external extremities of the ciliary processes.

The **ciliary processes** are about seventy pale, radially arranged folds which begin abruptly at the base of the iris (with their free ends facing the periphery of the lens), and extend outwards to merge into the ciliary ring. Each fold is approximately 2 mm. long, and its surface is thrown into a number of complicated folds each of which is covered by the ciliary part of the retina. Close to the lens, the ciliary processes give attachment to some of the fibrils of the **suspensory ligament of the lens.** These fibrils appear to arise from the cells of the ciliary part of the retina, and passing to the lens, fuse into thicker fibres which adhere to the ciliary processes on their way.

The **ciliary muscle** lies in the angle between the sclera and the ciliary processes and ring, external to the attached margin of the iris [FIG. 983]. This smooth muscle consists of meridional fibres which radiate backwards from the scleral spur to the ciliary processes and ring. In addition, a small bundle of circular muscle fibres lies between the iridocorneal angle and the ciliary processes, medial to the meridional fibres. **Action.** When the ciliary muscle contracts, it pulls the ciliary ring and processes antero-internally, and thus relaxing the fibres of the suspensory ligament of the lens, allows the elasticity of that organ to increase its convexity, thereby decreasing its focal length. The degree of development of the parts of the ciliary muscle varies ; the smaller bundle of circular fibres is absent or rudimentary in myopic eyes, but is usually well developed in hypermetropic eyes.

Iris

The iris is the coloured part of the eye. It is a contractile, bilaminar diaphragm which extends inwards in front of the

connected through an intermediate stratum of fibro-elastic tissue [FIG. 981] to the thin layer of wide bore capillaries which lie in the **choriocapillary lamina** and are separated from the pigmented cells of the retina by the basal lamina only.

The **basal lamina** is a thin, glassy membrane containing collagen and fine elastic fibres in its outer part. The inner, homogeneous layer is the basement membrane of the pigment cells of the retina.

Tapetum. In many mammals, but not in Man, the choroid contains a brilliant, iridescent layer by which light passing through the retina is reflected back into it, thus increasing its sensitivity in low intensity illumination. This may be a thin fibrous layer (horse, sheep) or several layers of flat, iridescent cells (seal) in the intermediate stratum.

Ciliary Body

This body is the anterior extension of the choroid to the iris, and consists of the ciliary ring, the ciliary processes, and the ciliary muscle [FIG. 984]. It is lined by the **ciliary part of the retina** which consists of a double layer of columnar epi-

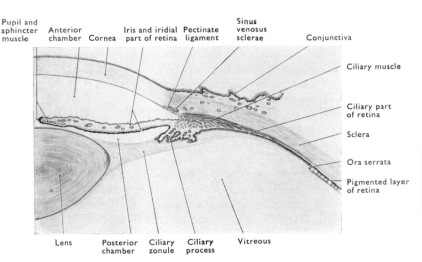

FIG. 983   Meridional section through ciliary region of human eyeball.   (After Schaffer.)

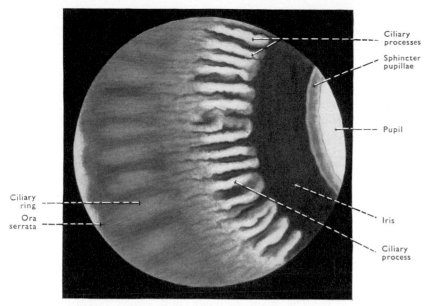

FIG. 984 A segment of the anterior part of eyeball seen from within.

same nerve supply) in the **accommodation-convergence reaction.** The dilator pupillae is formed from the heavily pigmented cells of the anterior layer of the iridial part of the retina which develops myo-epithelial elements. It is supplied by **postganglionic sympathetic fibres** from the internal carotid plexus.

The iris tends to be slightly thicker in its intermediate part than at either margin, and its internal part lies on the anterior surface of the lens with only a narrow space between them through which the anterior and posterior chambers of the eye communicate [FIG. 983].

The anterior surface of the iris is said to be covered with the same endothelium of the anterior chamber which covers the posterior surface of the cornea. This surface is highly irregular with a number of jagged folds and crypts, and when the iris is fully dilated tends to bunch against the pectinate ligament, and may interfere with the drainage of aqueous humour through that ligament into the sinus venosus sclerae.

lens and consists anteriorly of a layer of vascular connective tissue **(stroma)** containing some circular and radial smooth muscle fibres and usually some pigment cells, and posteriorly of a double layer of pigmented epithelial cells, the **iridial part of the retina.** These layers are continuous respectively: (1) with the ciliary body and the pectinate ligament; and (2) with the ciliary part of the retina. Slightly to the nasal side of its centre, the iris is perforated by an almost circular aperture, **the pupil,** through which a variable amount of light can enter the eye depending on the size of the pupil which is altered by contraction of the circular **(sphincter pupillae)** or radial **(dilator pupillae)** muscle fibres of the iris. The sphincter pupillae is said to be derived from the iridial part of the retina. It is a thin ring of smooth muscle in the pupillary margin of the iris which is innervated by **postganglionic parasympathetic fibres** from the ciliary ganglion, and tends to contract in unison with the ciliary muscle (which has the

Vessels and Nerves. The **arteries** of the ciliary body and iris are long posterior ciliary and anterior ciliary branches of the ophthalmic artery. The two former pierce the sclera medial and lateral to the optic nerve, and running forwards deep to the sclera, anastomose at the outer margin of the iris with four to six anterior ciliary arteries to form the **greater arterial circle** of the iris [FIG. 985]. Branches from this circle pass radially inwards towards the pupil in the stroma of the iris. They anastomose close to the pupillary margin to form the **lesser arterial circle** of the iris [FIG. 985]. Before birth the pupil is covered by a thin, vascular, **pupillary membrane** which extends inwards from the anterior surface of the iris initially supplied from the hyaloid artery, but later from the vessels of the iris. About the seventh month of pregnancy the central vessels begin to atrophy. This process extends outwards to the lesser arterial circle, the pupillary membrane

FIG. 985 Blood vessels of the iris and anterior part of the choroid.

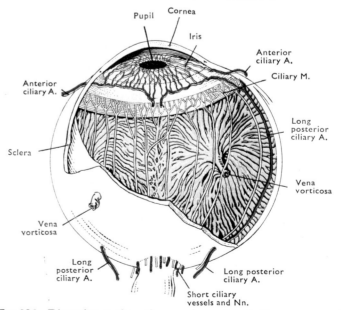

FIG 986 Dissection to show the vascular coat and ciliary nerves and vessels.

breaks down and is lost. Fragments of the membrane (or even a complete membrane) may be present at birth and may persist into adult life. The **veins** of the iris pass radially outwards and communicate with the veins of the ciliary processes and the sinus venosus sclerae [p. 804].

The **nerves** of the choroid and iris are derived from the **long** and **short ciliary nerves** [FIG. 986]. The former (2–3) arise from the nasociliary nerve, the latter (8–14) from the ciliary ganglion. Both groups contain sensory (trigeminal) and sympathetic fibres, only the short ciliary nerves contain postganglionic parasympathetic fibres and occasional ganglion cells (Wolff, 1948). The nerves pierce the sclera around the optic nerve, and passing forwards deep to the sclera, form a plexus on the ciliary muscle from which fibres are distributed to the ciliary body, iris, and cornea. Preganglionic parasympathetic fibres reach the ciliary ganglion through the **oculomotor nerve.**

INNER COAT

Retina

The retina is a double layer of epithelium derived from the optic cup of the embryo. It lines the internal surfaces of the choroid **(optic part),** the ciliary body and processes **(ciliary part),** and the iris **(iridial part).** Posteriorly its inner layer is greatly thickened to form the **cerebral layer** which contains the light-sensitive elements of the retina, and from which the optic nerve fibres arise. The cerebral layer extends peripherally from the optic nerve and is gradually reduced in thickness from 0·4 mm. adjacent to that nerve, to 0·1 mm. at the **ora serrata**—its irregular edge at the posterior border of the ciliary body [FIG. 983]. Here the cerebral layer loses its sensory elements, contains a number of spaces, and is then suddenly reduced to a single stratum of non-pigmented epithelial cells which passes over the ciliary body with the outer, pigmented epithelial layer of the retina. As these two layers pass on to the posterior surface of the iris, the inner layer also becomes pigmented.

The outer or **pigmented layer of the retina** is a single layer of heavily pigmented cells which have thick, internal processes which interdigitate with the photosensitive cells (rods and cones) of the cerebral layer. Pigment accumulates in the processes with increasing intensity of illumination of the retina. The pigment layer is sometimes considered to be part of the choroid to which its cells are more closely attached by a thick basement membrane than to the cerebral layer of the retina, from which it is sometimes separated in life (detachment of the retina). The pigment layer tends to absorb any light passing through the retina, and thus prevents it from being reflected back into the cerebral layer at various angles; a feature which would reduce the resolving power of that layer.

The cerebral layer has a small, oval, yellowish spot in the visual axis. This is the **macula,** and its central part is excavated to form the **fovea centralis.** About 3 mm. to the nasal side and slightly above the macula (Fison, 1920; Polyak, 1957) there is the temporal edge of a circular white disc surrounded by a narrow ring of pigment. This **disc of the optic nerve** is formed by nerve fibres passing out of the eye and all obtaining their myelin sheaths at the disc. It has a slightly raised margin and a central excavation, is approxim-

ately 1·5 mm. in diameter, and is non-pigmented and insensitive to light **(blind spot),** since both retinal layers are absent; the two layers fusing around the periphery of the disc at the pigment ring.

The cerebral layer is transparent during life, but in an animal kept in the dark, a purple tinge develops due to the accumulation of visual purple or rhodopsin in the rod cells. The purple is absent from the macula and in a narrow zone, 3–4 mm. wide, close to the ora serrata.

Structure [FIGS. 987, 988]. The **cerebral layer** of the retina consists of a number of layers of neural elements enclosed in the processes of slender neuroglial cells which extend radially between the external and internal limiting membranes [FIG. 988], forming these two 'membranes' by their extremities.

The nervous elements consist of three basic layers of cells with intermediate zones in which lie the synapses between the processes of these cells, which conduct impulses from the rods and cones to the nerve fibres in the optic nerve.

1a. The **layer of rods and cones (stratum neuro-epitheliale).** This is the outermost, light sensitive layer of cells which interdigitates with the processes of the cells of the pigment layer. Each cell consists of a body containing the nucleus with peripheral and central processes. The body lies internal to the external limiting membrane, while the peripheral process extends external to it. Under the electron microscope the peripheral process of each **rod cell** is seen to consist of an **outer segment**—a slender cylinder containing a stack of flattened membrane sacs joined to the thicker, mitochondria-filled, inner segment by a narrow, cilium-like neck. The **inner segment** is continued as a slender fibre (dendrite) into the **outer nuclear layer** [FIG. 987] to the **body of the rod cell,** whence an axon passes inwards to synapse with elements of the ganglionic layer of the retina. The outer segment of the

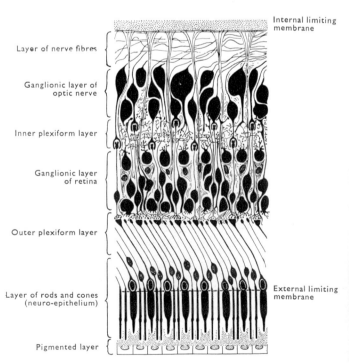

FIG. 987 Schematic reconstruction of human retina. (After Polyak, 1941.)

Internal limiting membrane

Layer of nerve fibres

Ganglionic layer of optic nerve

Inner plexiform layer

Ganglionic layer of retina

Outer plexiform layer

Layer of rods and cones (neuro-epithelium)

External limiting membrane

Pigmented layer

Nucleus of sustentacular fibre   Centrifugal fibre   Horizontal cell

Cell of ganglionic layer of optic nerve

Amacrine cells
Bipolar cell
Horizontal cell

Rod nucleus
Cone nucleus
Rod
Cone

Internal limiting membrane
Layer of nerve fibres
Ganglionic layer of optic nerve

Inner plexiform layer

Ganglionic layer of retina

Outer plexiform layer

Layer of rod and cone nuclei

External limiting membrane

Layer of rods and cones

Pigmented layer

Neuro-epithelium

FIG. 988   Plan of the retinal neurones.   (After Cajal.)

cone processes are conical in shape, particularly towards the periphery of the retina where they are short and fat. Towards the fovea they increase in height and become thinner and more cylindrical; the fovea consisting solely of these tall, thin cones tightly packed together. The **outer segment** does not contain rhodopsin (hence the fovea is never purple in colour, but yellow), and is united to an inner segment similar to that in the rod cell except that it is fused to the body and its contained nucleus. The **body** therefore lies in the outer nuclear layer immediately internal to the external limiting membrane, and forms a single row of nuclei with those of the other cone cells, while the rod cell nuclei lie further internally and are more scattered [FIG. 988]. The axon of each cone cell passes through the entire thickness of the outer nuclear layer to synapse with a cell or cells of the ganglionic layer of the retina.

In the **periphery of the retina** the cones are few and squat. They increase gradually in number and density towards the fovea centralis, becoming taller and thinner, while the relative number of rods decreases, though they are more densely packed. Over the retina as a whole there are approximately twenty times as many rods as cones. The rods have a lower threshold to light than the cones, but are not sensitive to colour. Hence the monochrome appearance of objects seen in low intensity illumination and the inability to see objects with the rod-free fovea in these conditions.

1b. The **outer nuclear layer** consists of the cell bodies of the rod and cone cells (the latter externally), and the processes of both.

2. The axons of the rod and cone cells enter the **outer plexiform** (synaptic) **layer,** and here they form synapses with the dendrites of the intermediate layer of cells; the bipolar cells of the **ganglionic layer of the retina.** The rod axons end in small spherules, the cone axons expand in the plane of the retina, and both synapse with a variety of different bipolar cells, some of which have contact with both rod and cone terminals by branching dendrites, while a few make contact with a single cone process or a group of rod spherules only. In addition there are **horizontal cells** which link various parts of this layer, and the processes of some bipolar cells which are believed to conduct peripherally.

3. The great majority of the **bipolar cells** conduct impulses through their axons to the inner plexiform layer. Here they synapse with the ganglion cells, from which the axons in the optic nerve arise. Cell bodies of the bipolar, horizontal, and amacrine cells form the **ganglionic layer of the retina** [FIG. 988]. The number of bipolar cells gradually increases from the periphery of the retina towards the fovea centralis; the number of those with a simple cone connexion increasing most noticeably. The total increase is greater than the increase in number of light-sensitive elements, hence the size of the bipolar cell unit (the number of light-sensitive cells discharging on a single bipolar cell) decreases proportionately.

Horizontal cells are of several types. The small variety send dendrites on to cone axons, and an axon which extends horizontally through the outer plexiform layer to end on other cone processes at some distance. The larger variety seem to have the same type of connexion predominantly with the endings of rod cells. Similar cells are found in the inner plexiform layer. Here they unite the dendrites of the ganglion cells and the axons of the bipolar cells. Such cells could allow for considerable integration of impulses within adjacent parts of the retina.

Amacrine cells appear to have no axon. They are found adjacent to the inner plexiform layer in which they send their dendrites either in a single stratum or in a diffuse manner. Their function is unknown.

4. The **inner plexiform layer** consists of the branched, expanded axonal endings of the bipolar cells on the extensive dendritic trees and on the cell bodies of the ganglion cells of the innermost cellular layer of the retina, and on the processes of horizontal and amacrine cells of the ganglionic layer of the retina. The bipolar cells which make a simple connexion with a cone axon peripherally have a similar relation to a single ganglion cell centrally, their axons each being enclosed in the expanded dendrite of the ganglion cell [FIG. 987], while giving branches also to other ganglion, horizontal and amacrine cells.

5. The cells of the **ganglionic layer of the optic nerve** each give rise to one axon which passes over the internal surface of the retina to reach the optic disc and enter the optic nerve. These cells vary in size but are larger than the bipolar cells.

At the periphery of the retina they are very sparse and widely separated, so that the area served by each is considerable. Further posteriorly they increase in number to form a continuous layer of cells, and near the fovea centralis they form several layers. Thus the average size of the retinal unit (the area stimulating a single ganglion cell) decreases considerably towards the fovea, and the number of single cone units increases markedly to reach its maximum in the fovea centralis which consists almost exclusively of such units.

Structure of the Macula and Fovea Centralis. In this region the light-sensitive elements are purely cones, but their outer segments are greatly elongated, thin and tightly packed. At the fovea centralis the inner layers of the retina are swept aside so that the cones are virtually on the internal surface, and the light reaching them does not have to pass through the internal layers as in other parts of the retina. Hence the image thrown on the retina by the refracting tissues of the eye is least distorted when it reaches the light sensitive elements in this part. This feature combined with the tightly packed single cone elements gives this part of the retina the highest resolving power of any part, and it corresponds to the sharp area in the field of vision which covers only one degree of arc. At the fovea the axons of the cones pass obliquely outwards to the bipolar and ganglion cells which are heaped up around it. The large number of single cone units in the fovea results in a disproportionately large number of optic nerve fibres arising from this region and streaming away from its perimeter to the optic disc without crossing the fovea. Nerve fibres approaching the optic disc from the temporal parts of the retina sweep superior and inferior to the fovea, so that the otherwise radial arrangement of these fibres is distorted in this region.

Vessels [FIG. 989]. *The sole arterial supply to the retina is the* **central artery of the retina;** a branch of the ophthalmic artery. At a variable point posterior to the eyeball it pierces the sheaths of the optic nerve to lie centrally in it. Passing distally in this position it appears in the centre of the optic disc and divides into superior and inferior branches. Each of these gives off a medial arteriole to the nasal retina and a lateral branch (**superior and inferior macular arterioles**) and then divides into **temporal** and **nasal retinal arterioles** which supply the corresponding quadrants of the retina. The superior and inferior temporal arterioles arch laterally above

and below the macula sending small branches towards it. But neither these nor the macular arterioles reach the fovea centralis which is devoid of blood vessels. These arteries, and the corresponding veins which lie close to them, are in the nerve fibre layer of the retina, the veins external to the arteries. The veins consist of a layer of endothelium surrounded by delicate connective tissue, and are therefore easily distended. They unite to form the **central vein of the retina.** This lies with the central artery and either enters the superior ophthalmic vein or runs directly to the cavernous sinus.

REFRACTING MEDIA

Light on its way to the retina passes through a number of structures with different refractive indices. These are cornea, aqueous humour, lens, and vitreous body [FIG. 978]. The change in refractive index from air to cornea is the greatest, and thus minor imperfections in the curvature of the cornea alter the course of light into the lens in such a manner that the lens may be unable to focus that light on the retina.

Aqueous Humour and Chambers of the Eye

The aqueous humour is a watery fluid which contains 1·4 per cent. sodium chloride and has a refractive index of 1·336. It fills the space between the cornea and the lens, and bathes both surfaces of the iris which incompletely divides that space into the anterior and posterior chambers of the eye. At the lateral margins of these chambers the fluid extends into the spaces of the iridocorneal angle and the zonular spaces [FIG. 991] respectively.

Lens

The lens is a biconvex, transparent body [FIG. 978] which lies between the iris and the vitreous body. The central points of its anterior and posterior surfaces are the anterior and posterior **poles,** and the line joining these is the **axis;** the peripheral circumference is the **equator.** The axis measures approximately 4 mm., the transverse diameter is 9–10 mm. The anterior surface lies immediately behind the posterior chamber. The posterior surface of the lens is lodged in the **hyaloid fossa** of the vitreous body. The anterior surface is less convex than the posterior surface, and its shape is maintained by the constant tension of the ciliary zonule (q.v.) which is released by contraction of the ciliary muscle allowing the elastic lens to increase its convexity and shorten its focal length to bring near objects into sharp focus on the retina. In the foetus the lens is soft, pink, and elastic, but with age it becomes increasingly harder, yellowish, and fixed in the flattened shape. Thus focussing on near objects becomes progressively more difficult; a condition known as presbyopia. The lens may also become opaque in the aged; a condition known as cataract. **Structure.** The lens is formed from an ectodermal vesicle the posterior cells of which elongate greatly to fill the cav-

Superior nasal branch —

Optic disc —

Inferior nasal branch —

Superior temporal branch

Superior and inferior macular arterioles

Fovea centralis in macula

Inferior temporal branch

FIG. 989   Blood vessels of left retina.

FIG. 990 Diagrammatic section through equator of lens, showing the gradual transition of the epithelium into lens fibres. (After Babuchin.)

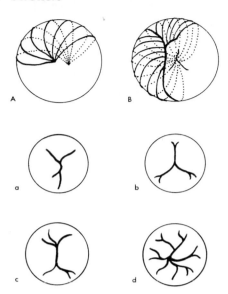

FIG. 992 Diagram showing formation of radii of lens. (After Ida Mann, 1928.)

A, the simplest arrangement of lens fibres all of the same length and all extending from pole to pole, with resulting absence of radii. B, the manner in which lens radii develop when all the fibres are not of the same length and do not reach from pole to pole. a, b, c, d, varieties of radii.

ity. Thus they form the **lens fibres** which are arranged more or less concentrically around the axis of the lens, the outer fibres near the equator becoming continuous with the layer of cubical cells (**epithelium of the lens**) which covers the anterior surfaces of the lens fibres [FIG. 990] and is derived from the anterior wall of the embryonic lens vesicle. New lens fibres are formed throughout life from the epithelium at the equator, but since they fail to grow to the poles, they meet at lines on the surfaces of the lens. These **radii of the lens** become more complex with continuing growth and age [FIG. 992], the older lens fibres at the centre being harder than those at the periphery. The epithelium anteriorly, and the lens fibres posteriorly, are covered by a thin (15 µm.), homogeneous, transparent capsule of highly refractile material to the outer layer of which the fibres of the ciliary zonule are attached.

Vitreous Body

This transparent, jelly-like substance contains hyaluronic acid and a mesh of fine fibres. It fills the space between the lens and the retina (**vitreous chamber**), and is especially adherent over the ora serrata and ciliary body. Here its fibres are more obvious and pass towards the equator of the lens with fibres which arise from the cells of the ciliary part of the retina. Together these form a ridged, fibrous (**zonular fibres**) sheet (**ciliary zonule** [FIG. 983]) which is intimately adherent to the surfaces of the ciliary processes but not to the grooves between them, aqueous humour of the posterior chamber passing between the grooves and the zonule. As the zonule approaches the equator of the lens, it splits into two layers, the coarse fibres pass to the anterior surface of the lens capsule while the fine fibres pass to its posterior surface; and between these groups of fibres lie the **zonular spaces.** The thicker, anterior fibres are sometimes known as the suspensory ligament of the lens, and together with the posterior fibres of the zonule maintain tension on the lens. The vitreous body is traversed by the sinuous **hyaloid canal** [FIG. 978] from the middle of the optic disc to the posterior surface of the lens. This contains the remains of the **hyaloid artery** which vascularized the lens in the earlier stages of its development and an aqueous fluid which is of sufficiently different refractive index from the vitreous body to be seen in the living eye with a slit lamp microscope.

FIG. 991 The zonular spaces distended (viewed from the front).

Ciliary zonule

Zonular spaces

Lens

Retina

EYEBROWS

Each eyebrow is a fold of skin which is thickened by the presence of fatty areolar tissue between it and the periosteum of the skull. Above the margin of the fold is a zone covered with short, stout hairs, which extends below the supra-orbital ridges of the skull medially. Between the skin and fatty areolar tissue are the interlacing fibres of the frontalis and orbicularis oculi muscles.

FIG. 993   Eye with lids slightly everted to show structures at the medial angle.

EYELIDS

The eyelids are two thin, surface folds which are strengthened by the presence in each of a firm plate of condensed fibrous tissue, **the tarsus.** Each is covered by thin skin on its external surface and by conjunctiva on its deep surface. In the fat-free subcutaneous tissue both eyelids contain the palpebral fibres of the orbicularis oculi muscle which approximates their flat, free margins **(limbi)** from the outer edges of which the eyelashes **(cilia)** arise, while the ducts of 20–30 modified sebaceous **tarsal glands** (which lie on the conjunctiva) open

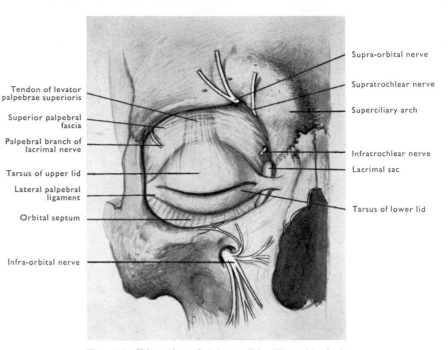

FIG 995   Dissection of right eyelid. The orbicularis oculi has been removed.

along the inner edge of each lid [FIG. 994]. Between the free margins is the **palpebral fissure** which forms a transverse slit opposite the lower margin of the cornea when the eye is gently closed; when open, the edge of the lower eyelid lies immediately below the cornea, while the upper covers its superior quarter. The eyelids meet **(commissures of the eyelids)** at the lateral and medial **angles of the eye;** the eyelashes and tarsal glands cease some 6 mm. from the medial angle where there is a small posteriorly directed elevation **(lacrimal papilla)** on the posterior part of the free margin of each lid. A minute opening **(lacrimal punctum)** on the apex of the papilla marks the beginning of the **lacrimal canaliculus** through which lacrimal fluid is transported to the lacrimal sac. Medial to the papilla the margins of the lids are rounded and enclose a triangular area **(lacus lacrimalis)** within which is a small raised island of modified skin, the **lacrimal caruncle.** This is surmounted by tiny hairs with minute sebaceous glands. Lateral to the caruncle, the conjunctiva passing towards the sclera forms a vertical fold, the **plica semilunaris.** This corresponds to the nictitating membrane of lower vertebrates which may contain cartilage and smooth muscle fibres.

The medial angle of the eye (medial canthus) is partly hidden in some races, notably Mongolians, by a fold of skin which extends from the area between the eyebrow and the upper eyelid to the side of the nose. This **epicanthic fold** gives an oblique appear-

FIG. 994   Diagrammatic sagittal section through upper eyelid.

ance to the palpebral fissure, and may be well formed in so-called mongoloid idiots.

The Tarsus

The superior tarsus is larger and firmer than the inferior tarsus and has a straight lower margin and a thin, curved upper edge into which is inserted the deep lamella of levator palpebrae superioris. The inferior tarsus is a strip about half the height of the superior tarsus [FIG. 995], and both have the tarsal glands embedded in their deep surface. Laterally and medially the tarsi unite to form the lateral and medial **palpebral ligaments** which are attached respectively to the orbital surface of the zygomatic bone and the frontal process of the maxilla anterior to the lacrimal fossa. The periosteum of the orbital margin is continued as a fibrous sheet (**orbital septum**) into each eyelid. In the upper lid it fuses with the superficial layer of levator palpebrae superioris, anterior to the tarsus, but in the lower lid it is thin and fuses with the inferior tarsus.

Among the follicles of the eyelashes are two or three rows of modified sweat glands (**ciliary glands**) with relatively straight ducts which open into the follicles. Immediately posterior to these is a small, separate slip of orbicularis oculi in the lid margin, together with some smooth muscle fibres (Whitnall, 1932).

Conjunctiva

This highly sensitive membrane covers the deep surfaces of the eyelids and is reflected from them on to the anterior surface of the eyeball at the superior and inferior **conjunctival fornices.** It is continuous with the skin at the margins of the eyelids, and is closely adherent to the deep surfaces of the tarsi and tarsal glands. Here it is highly vascular and thicker than the transparent part which covers the sclera and cornea. It is separated from the sclera by the fascial sheath of the eyeball, but is firmly adherent to the cornea where it forms the stratified squamous **anterior corneal epithelium.** Elsewhere its epithelium is stratified columnar in nature and of variable thickness.

Vessels and Nerves

A **medial palpebral artery** enters each eyelid from the ophthalmic artery by piercing the orbital septum superior and inferior to the medial palpebral ligament. Each forms a palpebral arch near the free margin of the eyelid, the superior reinforced from the supra-orbital and supratrochlear arteries, the inferior from the facial artery.

Conjunctival **veins** drain into the muscular tributaries of the ophthalmic veins, and pretarsal, palpebral veins enter the facial and superficial temporal veins. **Lymph** plexuses are present on both surfaces of the tarsus; some drain medially to join the lymph vessels that accompany the facial vein to the submandibular lymph nodes, but the main drainage is laterally to the superficial **parotid lymph nodes**

(Burch, 1939). The **sensory nerve** supply is from the trigeminal nerve via the supratrochlear and supra-orbital to the upper lid and the infra-orbital to the lower lid, while infratrochlear and lacrimal supply both at the corresponding angles [FIG. 995]. The **motor** supply to orbicularis oculi is from the facial nerve, that to levator palpebrae superioris is from the oculomotor (striated muscle fibres) and the sympathetic (smooth muscle fibres in that muscle and in the eyelids).

LACRIMAL APPARATUS

Lacrimal Gland

The greater, **orbital part** of the lacrimal gland [FIG. 996] lies in the lacrimal fossa on the medial surface of the zygomatic process of the frontal bone, extending down almost to the lateral angle of the eye. Here, lateral to the aponeurosis of levator palpebrae superioris, it becomes continuous with the **palpebral part** which arches upwards and medially, deep to the aponeurosis, in the root of the upper eyelid, close to the superior conjunctival fornix, and may extend downwards beyond the lateral angle of the eye. Three to nine **excretory ductules** open into the superolateral part of the superior conjunctival fornix. Numerous, small, **accessory lacrimal glands** open close to both conjunctival fornices (especially the superior) and there are mucous cells in the conjunctival epithelium.

Structure. The lacrimal gland is a tubulo-alveolar, serous gland which moistens the conjunctival sac with a watery fluid containing lysozyme which destroys bacteria. It receives sympathetic, parasympathetic (facial nerve [FIG. 959]) and sensory (lacrimal nerve) **nerves,** and branches from the **lacrimal artery.** Its venous drainage is to the **ophthalmic vein.** The lacrimal fluid helps to remove particulate material and flows towards the medial angle of the eye, assisted by contraction of the orbicularis oculi.

Temporal fascia
Lacrimal gland
Temporal muscle
Ductules of lacrimal gland
Palpebral part
Infra-orbital nerve
Maxillary sinus
Buccinator

Superior fornix of conjunctiva
Puncta lacrimalia
Lacrimal canaliculi
Lacrimal sac
Medial palpebral ligament
Nasolacrimal duct
Middle nasal concha
Mucoperiosteum
Lacrimal fold
Inferior meatus of nose
Inferior nasal concha

FIG. 996   Dissection to show the lacrimal apparatus (semidiagrammatic).

Lacrimal Canaliculi

Just lateral to the lacrimal caruncle the fluid enters the **puncta lacrimalia** [FIG. 996] to reach the lacrimal canaliculi. Each canaliculus (approximately 10 mm. long) passes upwards or downwards into the corresponding eyelid, and then turning medially is distended to form an **ampulla** before coursing respectively superior or inferior to the medial palpebral ligament, to open into the lacrimal sac a little above its mid point, close to its fellow. The canaliculi are lined with stratified squamous epithelium surrounded by connective tissue and the lacrimal part of the orbicularis oculi muscle. If the lower eyelid droops away from the eyeball either as a result of paralysis of orbicularis oculi or because of scarring contracture (e.g. after burns), the lacrimal fluid collects in the dependent eyelid and spills over its free margin without even reaching the level of the inferior punctum.

Lacrimal Sac

The lacrimal sac [FIGS. 995 and 996] is the blind, upper part of the nasolacrimal duct through which lacrimal fluid is conveyed to the nasal cavity. The sac lies in its fossa formed by the lacrimal and maxillary bones, and is covered laterally by a sheet of fascia which joins the anterior and posterior edges of the fossa and gives origin to the lacrimal part of orbicularis oculi. The medial palpebral ligament and palpebral part of orbicularis oculi lie anterior to the upper half of the sac.

Nasolacrimal Duct

The nasolacrimal duct [FIG. 996] is approximately 18 mm. long and 3–4 mm. wide. It passes downwards with a slight inclination backwards and laterally through a canal formed by the maxilla, lacrimal bone, and inferior concha, to open into the anterior part of the inferior meatus of the nose, approximately 30 mm. behind the nostril. The opening, which is variable in size and position, is guarded by a flap (**lacrimal fold**) of mucous membrane which prevents the entry of air into the duct when the nose is blown [FIG. 996]. The lining of the sac and duct is stratified columnar epithelium which is folded in the duct.

Vessels and Nerves. The arteries of the lacrimal ducts are derived from the palpebral, facial, and infra-orbital arteries. The veins form a plexus which drains to face, orbit, and nose. The canaliculi and sac are supplied by the infratrochlear nerve, the nasolacrimal duct receives fibres from the anterior superior alveolar nerve.

DEVELOPMENT OF THE EYE

The retina and optic nerve are formed by an outgrowth of the full thickness of the antero-inferior part of the diencephalon [p. 624]. The distal part is distended to form the optic vesicle; the proximal part, elongated and narrow, forms the optic stalk, the rudiment of the optic nerve. The ectoderm overlying the optic vesicle thickens, invaginates, and finally separates from the surrounding ectoderm as the **lens vesicle** [FIG. 998]. As the lens vesicle approaches the optic vesicle, the inferolateral wall of that vesicle is invaginated, virtually obliterating the cavity of the vesicle and turning it into a two-layered **optic**

FIG. 997 Sections of developing eye of human embryo. × 20. A, 5 mm.; B, 10 mm.

cup with a deep **choroidal fissure** [FIG. 998] in its inferolateral wall. The fissure extends for some distance into the optic stalk, the surrounding mesoderm passes through it to fill the cup carrying with it the rudiments of the central vessels of the retina and assisting in the formation of the vitreous body. At first the artery is continued forwards as a number of branches which form a plexus on the surface of the lens, but by the fifth to sixth month only one remains (the **hyaloid artery**), and this, together with the remains of the lens plexus, atrophies during the last intra-uterine month, leaving only the **hyaloid canal** to mark its continuity with the central artery of the retina.

The edges of the choroidal fissure grow together and fuse, thus completing the optic cup and enclosing the future central artery and vein of the retina in the cup and stalk. The distal margin of the cup grows over the superficial surface of the lens to form the epithelium of the iris (iridial part of the retina), while posterior to this the two layers of the cup form the ciliary part of the retina, and further posteriorly the cerebral and pigment layers of the retina.

The **vitreous body** appears to be formed partly from the inner layer of the cup by fine protoplasmic fibrils which project from its cells. Some of these later become condensed to form the **ciliary zonule** and the walls of the hyaloid canal, though others remain throughout the vitreous with interfibrillar material probably derived from the mesoderm.

The **lens** [FIG. 990] is formed by the ectoderm in response to the presence of the optic vesicle. At first a somewhat flattened, spherical vesicle, the cells of its deeper wall elongate into and fill the cavity of the vesicle to form the lens fibres, while the superficial cells remain a simple cubical layer, the anterior **epithelium of the lens.** The lens fibres at the centre of the lens are the earliest formed and the longest, while growth in diameter is achieved as new lens fibres are continu-

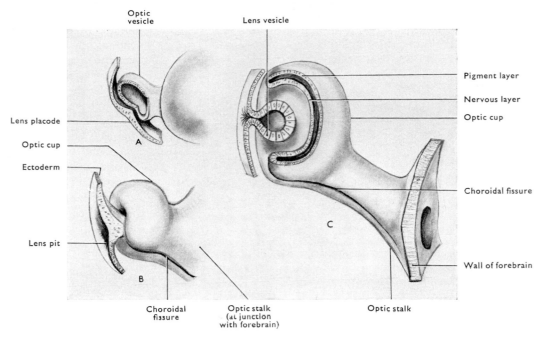

FIG. 998   Models of developing human eye.   × 100.   (From Arey's *Developmental Anatomy*, after Ida Mann, 1928.)

The lens has been sectioned, and the optic vesicle has been partly cut away in A and C.

A, 4·5 mm. embryo; B, 5·5 mm.; C, 7·5 mm.

ally formed at the equator by transformation of the cells continuous with the anterior epithelium [FIG. 990]. During the greater part of intra-uterine development the lens is surrounded by a vascular plexus fed by the hyaloid and iridial arteries. The part of this vascular coat superficial to the lens forms the **pupillary membrane** which normally disappears before birth [p. 807].

The walls of the hollow optic stalk are thickened by the ingrowth of nerve fibres from the cerebral layer of the retina, so that the original cavity is obliterated and the optic nerve formed.

The **cerebral layer of the retina** consists initially of a neuro-epithelial (ependymal) layer adjacent to the outer, pigment layer. From the former, cells migrate towards the internal aspect of the optic cup (original external surface of optic vesicle) to form a second (inner) neuroblastic layer. This gives rise to the ganglion cells, amacrine cells, and glial cells, while the outer layer gives rise to the rods and cones and the bipolar and horizontal cells of the inner nuclear layer. Thus the cells of the inner nuclear layer are derived from both of the primitive neuroblastic layers (Mann, 1928).

Mesoderm surrounding the optic cup condenses to form the sclera, cornea, and choroid, while anterior to the lens a cleft (anterior chamber of the eye) appears in it between the cornea and the layer which forms the mesodermal stroma of the iris and the interior vascular coat of the lens. The deep surface of the iris is the double layer of epithelium derived from the anterior part of the optic cup which also gives rise to the radial myoepithelial elements of the iris.

The **eyelids** arise as surface, ectoderm-covered folds above and below the cornea. These folds meet anterior to the cornea in the third intra-uterine month, and only separate shortly before birth; though they remain fused until after birth in many mammals. The ectoderm forms the epithelium of the conjunctiva, the anterior epithelium of the cornea, hair follicles of the eyelashes, and the tarsal, ciliary, and lacrimal glands.

The **lacrimal sac and nasolacrimal duct** are formed as a solid rod of ectodermal cells between the fusing maxillary and frontonasal processes [FIG. 1025], which are separated laterally by the eye. The central cells of the rod disintegrate to form a tube, the lower end of which remains closed until shortly before birth. From the upper end, two small rods of cells which subsequently canalize, grow out to form the lacrimal canaliculi. The lacrimal caruncle is said to be a separated portion of the lower eyelid (Mann, 1928).

THE ORGANS OF HEARING AND EQUILIBRATION

The ear [FIG. 999] consists of three parts. (1) The **external ear** which is a sound collecting apparatus, the auricle, continuous with a tube (external acoustic meatus) which conveys air vibrations to the tympanic membrane which forms the wall between the external and middle ear cavities. (2) The **middle ear (tympanic cavity)** is an extension of the nasal part of the pharynx. Vibrations of the tympanic membrane are transmitted across the cavity by a chain of three ossicles;

FIG. 999   Diagrammatic section of organ of hearing.

movements of the first and last of which can be controlled by the actions of minute muscles, thus preventing extreme movements being transmitted to the delicate internal ear which is buried in the petrous temporal bone, medial to the middle ear. (3) The **internal ear** consists of a number of parts, but it is divisible functionally into (i) the part concerned with hearing **(cochlea),** and (ii) the parts concerned with assessing (a) the position of the head relative to gravity, and (b) movements of the head (the **vestibular apparatus).** The sensory endings in both these parts are supplied by the eighth cranial or vestibulocochlear nerve.

THE EXTERNAL EAR

AURICLE

The auricle projects from the side of the head at a variable angle (circa 30 degrees). Its skeleton is a folded plate of elastic fibrocartilage, and its lateral surface [FIG. 1000] shows two irregular, concentric, C-shaped grooves above and behind

the opening of the external acoustic meatus which lies at the antero-inferior end of the deeper, inner groove (the **concha),** partly overlapped from in front by the tragus. The two grooves are separated by the **anthelix,** which frequently splits above to enclose the **triangular fossa,** while the outer groove **(scaphoid fossa)** is limited by the rolled, outer margin of the auricle, the **helix.** The helix begins **(crus of the helix)** in the concavity of the concha above the external acoustic meatus. It ends postero-inferiorly by entering the lobule of the ear close to the anthelix which runs along the superior margin of the lobule and ends in a small protrusion. This **antitragus** usually points laterally, and partly overlapping the postero-inferior part of the concha, is separated from the tragus by the **incisura intertragica.**

The medial surface of the auricle shows a series of elevations which correspond to the depressions on the lateral surface and carry the same names, e.g., eminence of the concha.

The shape of the auricle is extremely variable. In the newborn its length is approximately one-third of that in the adult, and it increases in size and thickness in old age.

Structure of the Auricle

The skin of the auricle is thin, smooth, and adherent to the perichondrium on the lateral aspect, but mobile on the medial surface. Fine hairs are found over the greater part of the auricle, and coarse hairs (tragi) are present on the tragus, antitragus, and in the incisura intertragica in men. **Sebaceous glands** are present on both surfaces, particularly in the concha and triangular fossa, while sweat glands are virtually limited to the medial surface.

The elastic fibrocartilage is absent from the lobule which is composed of fat and fibro-areolar tissue. The cartilage has the same contours as the auricle and is continuous with that of the external acoustic meatus by a narrow strip (8–9 mm. wide) medial to the deepest part of the incisura intertragica [FIG.

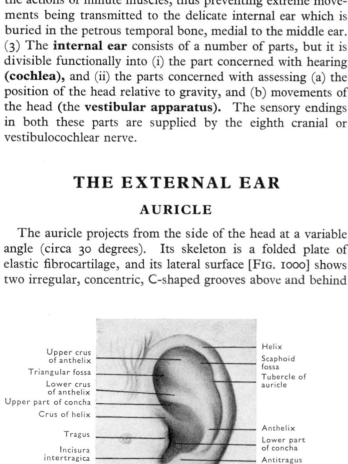

FIG. 1000   Lateral surface of left auricle.

FIG. 1001   Lateral surface of cartilage of left auricle, with muscles.

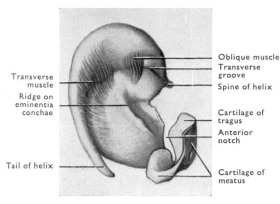

FIG. 1002 Medial surface of cartilage of left auricle, with muscles.

1002], where the skin of auricle and meatus also become continuous. The cartilage of the helix ends anteriorly in the spine [FIGS. 1001, 1002] and postero-inferiorly in the tail.

Ligaments. An **anterior ligament** extends from the spine of the helix and tragus to the zygomatic process, and a **posterior ligament** from the eminence of the concha to the mastoid part of the temporal bone. Small ligaments unite the various parts of the cartilage.

Muscles. The extrinsic muscles which attach the auricle to the skull and scalp are described elsewhere [p. 280]. In addition there are six rudimentary intrinsic muscles, four on the lateral surface and two on the medial surface [FIGS. 1001, 1002].

Vessels and Nerves. The **arteries** to the lateral surface come principally from the superficial temporal artery; those to the medial surface from the posterior auricular artery. The latter sends three or four branches which also give twigs to the lateral surface through the cartilage and round the margin of the helix. The **veins** drain to the corresponding veins, but some communicate with the mastoid emissary vein. Lymph vessels pass: (1) forwards to the parotid lymph nodes, especially to a node immediately in front of the tragus; (2) downwards to lymph nodes along the external and internal jugular veins; (3) backwards to retro-auricular nodes on the mastoid process. The **nerve supply** to all the muscles is from the facial nerve. The sensory supply is from: (1) the **great auricular nerve** to most of the medial surface and the lower part of the lateral surface; (2) the **auriculotemporal nerve** to the tragus and above principally on the lateral surface; and (3) to the upper part of the medial surface from the **lesser occipital nerve.** The latter is very variable in its extent and may supply most of the medial surface.

EXTERNAL ACOUSTIC MEATUS

This passage leads from the lower part of the concha to the tympanic membrane. In the adult it is 35 mm. long from the tragus, and 24 mm. from the medial part of the concha; of this the lateral cartilaginous part is about 8 mm. long, and the medial bony part about 16 mm. but is relatively much shorter in the newborn where the bony part has not formed. The tympanic membrane lies obliquely across the end of the meatus, thus the anterior and inferior walls are longer than the posterior and superior walls [FIGS. 1003, 1004]. The meatus passes medially: the lateral part also inclines forwards and upwards; the intermediate part backwards; and the medial part (the longest) forwards and slightly downwards. Throughout most of its course it is elliptical in transverse section with the greatest diameter directed downwards and backwards, but is more nearly circular close to the tympanic membrane. It is widest at the concha, becomes slightly narrower towards the medial end of the cartilaginous part, and is again constricted to its narrowest **(isthmus)** near the medial end, 20 mm. from the concha. The bony part of the meatus lies between the head of the mandible and the mastoid process, the former also abutting slightly on to the cartilaginous part, but separated from it by a slip of the parotid gland. When the head of the mandible moves forwards, e.g., on opening the mouth, the cartilaginous part is widened.

Structure

The **cartilage of the meatus** is a curved plate which forms only the anterior and inferior walls [FIG. 1002], the remainder being fibrous tissue. Medially the cartilage and fibrous tissue are fused to the bony part, laterally the cartilage is continuous with that of the tragus. The cartilage of the anterior wall contains two fissures filled with fibrous tissue.

The bony part is described on page 107. In the new-born it consists of the incomplete **tympanic ring** [FIG. 161] which is closed superiorly by part of the squamous temporal bone. The margin of the fibrous layer of the tympanic membrane is attached to the tympanic groove on the internal surface of the tympanic ring. A little inferior to its anterosuperior extremity, the medial surface of the tympanic ring has a groove which passes antero-inferiorly and transmits the **anterior ligament of the malleus,** the **chorda tympani nerve,** and the **anterior tympanic artery.** The tympanic ring is joined to the cartilage of the meatus by a fibrous tympanic plate (Symington, 1885) which ossifies from the ring to form the bony tympanic part of the adult meatus.

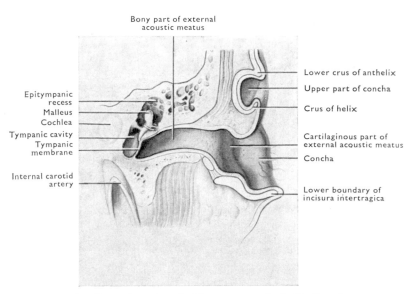

FIG. 1003 Coronal section of right ear; anterior half of section, viewed from behind.

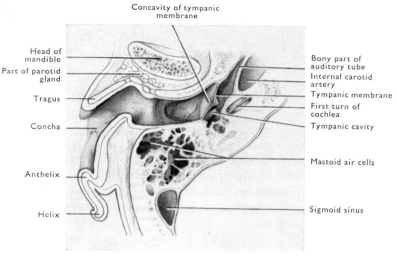

FIG. 1004 Horizontal section through right ear; upper half of section, viewed from below.

In the new-born the lumen of the meatus is small. A lateral funnel-shaped part is continuous with the slit-like medial part which lies between the fibrous tympanic plate inferiorly and the tympanic membrane superiorly. The **tympanic membrane** is close to the surface, is more obliquely set than in the adult, and lies almost parallel to the inferior aspect of the skull.

The skin of the external acoustic meatus covers the external surface of the tympanic membrane and is continuous with the skin of the auricle laterally. It is thick in the cartilaginous part where it carries fine hairs and sebaceous glands, the latter continuing along the posterosuperior wall of the bony part. Numerous enlarged 'sweat' glands are found in the cartilaginous part. These secrete wax or cerumen and are known as **ceruminous glands** [p. 799].

Vessels and Nerves. The **arteries** are branches of the superficial temporal (anterior auricular), posterior auricular, and maxillary (deep auricular) arteries; the last also partly supplies the tympanic membrane. The **veins** join the maxillary, external jugular, and pterygoid veins. Lymph drainage is in common with the auricle. Sensory **nerves** are supplied by the auriculotemporal, vagus (auricular branch), and facial nerves [p. 725].

THE MIDDLE EAR

TYMPANIC CAVITY

The tympanic cavity is an air-filled space between the tympanic membrane and the internal ear [FIGS. 999, 1004]. It is lined with **mucous membrane** the epithelium of which varies between simple squamous and ciliated columnar in its various parts. It is contin-

uous with that of the nasal part of the pharynx through the auditory tube, and covers: (1) the chain of ossicles (malleus, incus, and stapes) which extends across the cavity from the tympanic membrane to the wall of the inner ear; and (2) the ligaments and tendons passing to the ossicles from the walls of the cavity [FIG. 1012]. With the exception of a few goblet cells, no glands are found in the tympanic cavity.

The tympanic cavity is little more than a vertical fissure most of which lies medial and parallel to the tympanic membrane [FIG. 1004]. Its upper part **(epitympanic recess)** extends above the membrane and contains the greater part of the incus and the superior part of the malleus. The vertical and longitudinal measurements are each approximately 15 mm., while its width is 6mm. in the epitympanic recess, 4 mm. near its floor, and 1·5 or 2 mm. in its central part where the lateral and medial walls bulge into the cavity [FIG. 1003].

Posteriorly the epitympanic recess is continuous with air-filled spaces in the mastoid process [FIG. 1005].

Walls of the Tympanic Cavity

The **roof** or **tegmental wall** is formed by a thin plate of bone (tegmen tympani) which separates it from the dura mater of the floor of the middle cranial fossa. Posteriorly, it covers the mastoid antrum, and anteromedially the semicanal for tensor tympani. It may contain air cells, or be deficient in part, and is formed of cartilage in the child. Anterolaterally it curves downwards to appear in the **tympanosquamous fissure** on the base of the skull.

The **floor** or **jugular wall** is narrower than the roof. It consists of a plate of bone which posteriorly separates the tympanic cavity from the jugular fossa, and anteriorly slopes upwards as the posterior wall of the carotid canal [FIG. 1004]. The **canaliculus for the tympanic nerve** (branch of the

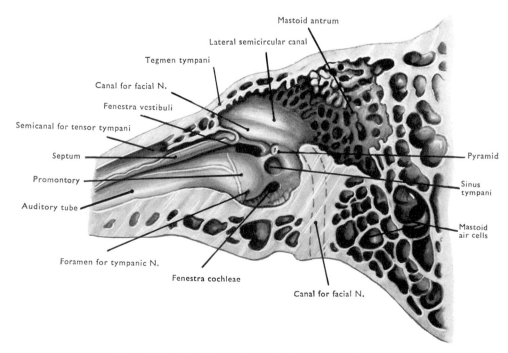

FIG. 1005 The labyrinthine wall of the tympanic cavity.

glossopharyngeal nerve) pierces the floor close to the medial wall.

The **medial** or **labyrinthine wall** [FIG. 1005] is the lateral wall of the internal ear. The bone covering the lateral surface of the base of the cochlea bulges into the middle ear cavity (**the promontory**) and is grooved by the tympanic plexus of nerves. There are two gaps in the medial wall. (1) Postero-superior to the promontory is the **fenestra vestibuli** (3 × 1·5 mm.) filled by the base of the stapes which is attached to the margin of the fenestra by the elastic **anular ligament** and has the vestibule of the bony labyrinth medial to it. (2) The **fenestra cochleae** lying postero-inferior to, and almost covered by the promontory, is filled by the **secondary tympanic membrane**—a layer of fibrous tissue covered laterally by the mucosa of the middle ear and medially by the epithelium of the cochlear part of the labyrinth. In the epitympanic recess, above the fenestra vestibuli, is an ill-defined, horizontal ridge produced by the bone covering the **canal for the facial nerve.** This prominence of the canal for the facial nerve passes back into the medial wall of the opening (**aditus**) to the mastoid antrum, and there lies inferior to the prominence of the **lateral semicircular canal.** Superior to the anterior part of the fenestra vestibuli lies the **processus cochleariformis**—the posterior end of a thin shelf of bone which forms the floor of the semicanal for tensor tympani, and here bends laterally to form a pulley for the tendon of that muscle, passing to the malleus. Immediately posterior to the promontory and between the two fenestrae is a small depression in the medial wall, the **sinus tympani.** It marks the position of the ampulla of the posterior semicircular canal.

The **posterior** or **mastoid wall** [FIGS. 1005, 1011] is pierced superiorly by the round or triangular opening (**aditus**) into the mastoid antrum. Immediately below this is the **fossa for the incus** which receives the short crus and posterior ligament of that bone. Further inferiorly lies a small, conical projection (the **pyramid**) through the apex of which the tendon of **stapedius** passes anteriorly to the stapes. The **facial nerve** descends vertically within the posterior bony wall [FIG. 906] and sends a fine branch arching upwards and forwards to supply stapedius. This branch may arise in the bony canal for the facial nerve or immediately after that nerve emerges from the base of the skull. Lateral to the pyramid, the small **posterior canaliculus for the chorda tympani** opens close to the posterior edge of the tympanic membrane [FIG. 1007].

The **lateral** or **membranous wall** [FIG. 1007] is almost entirely formed by the tympanic membrane. This is inserted into the tympanic groove except superiorly where the groove is missing and the flaccid part of the membrane is attached to the margin of the **tympanic notch.** Superior to this is the lateral wall of the epitympanic recess formed by the squamous temporal bone. Immediately anterior to the upper part of the tympanic membrane, the tegmen tympani turns downwards in the tympanosquamous fissure, and beneath the tegmen, through the petrotympanic fissure, the anterior ligament of

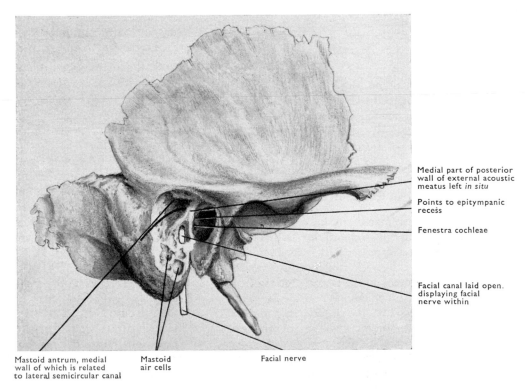

Medial part of posterior wall of external acoustic meatus left *in situ*

Points to epitympanic recess

Fenestra cochleae

Facial canal laid open, displaying facial nerve within

Mastoid antrum, medial wall of which is related to lateral semicircular canal

Mastoid air cells

Facial nerve

FIG. 1006 Preparation displaying position and relations of right mastoid antrum.

The greater part of the posterior wall of the external meatus has been removed, leaving only a bridge of bone at its medial end; under this a bristle passes from the mastoid antrum to the tympanic cavity.

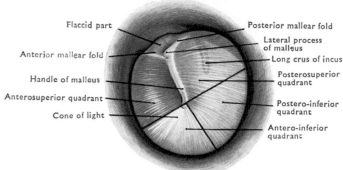

FIG. 1007 View of left tympanic membrane and epitympanic recess from medial side.

FIG. 1008 Left tympanic membrane seen from lateral side. The four arbitrary quadrants are indicated by solid lines and by the handle of the malleus.

the malleus and the chorda tympani (**anterior canaliculus**) leave the middle ear, while the anterior tympanic artery enters it.

The carotid wall. Anteromedially the walls of the tympanic cavity converge on the openings of two canals which continue in this direction lateral to the carotid canal, to open on the base of the skull in the angle between the greater wing of the sphenoid and the petrous part of the temporal bone. The upper, smaller, **semicanal for tensor tympani** (2 mm. in diameter) lies immediately below the tegmen tympani, and extends posteriorly to the processus cochleariformis which is produced by the thin plate of bone separating it from the lower, larger, **semicanal for** (i.e., bony part of) **the auditory tube.** The latter is funnel-shaped, increasing in diameter posteriorly. The floor of the tympanic cavity sloping upwards to the auditory tube is a thin plate of bone immediately above the ascending part of the carotid canal. It is sometimes deficient, and is always perforated by the **caroticotympanic nerves** passing to the tympanic plexus from the internal carotid plexus (sympathetic).

TYMPANIC MEMBRANE

This oblique, almost circular membrane lies at an angle of almost 55 degrees to the anterior and inferior walls of the external acoustic meatus [FIGS. 1003, 1004]. Its thickened margin, the fibrocartilaginous ring, is inserted into the tympanic groove, but leaves the bone to pass to the lateral process of the malleus from the ends of the tympanic notch as the **anterior** and **posterior mallear folds** [FIG. 1007]. Superior to these is the thin, **flaccid part of the tympanic membrane,** while the remainder (**tense part**) is tightly stretched. The handle of the malleus is firmly attached to the tympanic membrane, and its lower end lies opposite the point of maximum indrawing of the membrane, the **umbo** of the tympanic membrane.

The **chorda tympani** crosses the deep surfaces of the fibrous layer of the tympanic membrane and the neck of the malleus, raising a fold of the tympanic mucosa.

Structure

The basis of the tympanic membrane is a fibrous sheet consisting of connective tissue bundles radiating outwards from the handle of the malleus (**stratum radiatum**) with an increasing number of circumferential fibres (**stratum circulare**) towards the periphery. Both groups stop at the fibrocartilaginous ring, and hence are absent from the pars flaccida.

The fibrous sheet is covered on its external surface by skin of the external acoustic meatus (**cutaneous layer**) and on its internal surface by the mucous membrane of the tympanic cavity (**mucosal layer**). The latter is thickest on the superior part of the membrane and on the bone immediately surrounding it.

Otoscopic Examination of Tympanic Membrane
[FIG. 1008]

In life the tympanic membrane is pearl-grey in colour, sometimes with a yellowish tinge, and the posterior segment is more transparent than the anterior. The lateral process of the malleus appears as a whitish spot projecting into the meatus close to the anterosuperior edge of the membrane with the mallear folds passing from it to the ends of the tympanic notch. Below this the **handle of the malleus** forms a ridge extending down to a rounded extremity at the umbo. Posterior to the handle of the malleus, the long crus of the incus may be seen through the membrane descending parallel to the handle of the malleus as far as its mid-point. From the umbo a bright area, the cone of light, radiates antero-inferiorly.

Vessels and Nerves

Separate **arteries** supply the two surfaces of the membrane, but they anastomose by small branches which pass mainly through the periphery. The inner surface is supplied mainly by the **auricular branch of the occipital artery** and the **anterior tympanic artery,** the outer surface by the **deep auricular artery.** The latter two arise from the maxillary artery. The **veins** of the outer surface drain to the external jugular vein, those from the inner surface drain with the rest of the tympanic cavity partly to the veins of the auditory tube, and partly to the sigmoid sinus, and to the veins of the dura mater through the tegmen tympani. The **lymph vessels** drain in a comparable manner to the veins, those from both sides of the membrane communicating freely. The **auriculo-temporal nerve** and the auricular branch of the **vagus** supply the lateral aspect, while the medial aspect is supplied via the **tympanic plexus** by the glossopharyngeal nerve. When a foreign body is lodged in the meatus, or it is subject to inflammation, the supply by the vagus accounts for the tendency to develop an intractable cough or even vomiting.

MASTOID ANTRUM AND MASTOID CELLS

[FIG. 1005 and PLATE 9D]

The **mastoid antrum** is an air-filled space continuous anteriorly with the epitympanic recess through the aditus. The antrum is approximately 12–15 mm. in length, 8–10 mm. high, and 6–8 mm. wide. The **roof** is the tegmen tympani, the **floor** the mastoid process, the **medial wall** the petrous temporal bone, and the **lateral wall** the squamous temporal bone posterosuperior to the meatus. The latter wall is only 1–2 mm. thick at birth, but increases to 10 mm. by the ninth year, and 15 mm. in the adult.

The **mastoid cells** form as air-filled extensions of the antrum as the mastoid process grows during childhood. They are very variable in size and extent. They may extend to the tip of the mastoid process and even upwards into the squamous part, passing into the roof of the external acoustic meatus, or even medially into the jugular process of the occipital bone. When the cells are very extensive they may only be separated by a thin lamina of bone from the posterior cranial fossa and the sigmoid sinus. The lamina may even be perforated, allowing infection to spread readily from the mastoid air cells. In other cases the mastoid cells are small and fill only a fraction of the mastoid process. The antrum and cells are lined by an extension of the mucous membrane of the tympanic cavity with a lining of flattened cells firmly bound by connective tissue to the bone.

AUDITORY TUBE

The auditory tube connects the tympanic cavity to the nasal part of the pharynx. By allowing air to enter or leave the cavity it balances the pressure on both sides of the tympanic membrane and so allows it to vibrate freely. If the tube is blocked, the air in the tympanic cavity is absorbed into the mucosal vessels with loss of pressure, increasing concavity of the tympanic membrane, and progressive deafness. If associated with infection, there may be a collection of fluid under increasing pressure in the tympanic cavity with bulging of the tympanic membrane into the meatus, and increasing deafness due to the failure of the tympanic membrane to vibrate freely.

The tube is approximately 3·5 cm. long. The cartilaginous, anteromedial two thirds passes backwards, laterally (45 degrees to the saggital plane) and upwards (40 degrees to the horizontal) to join the more horizontal bony third at an angle of 160 degrees. The **cartilage** is a triangular plate with its base in the pharyngeal wall and its apex at the bony part, is folded downwards over the lining of the tube to cover the medial and

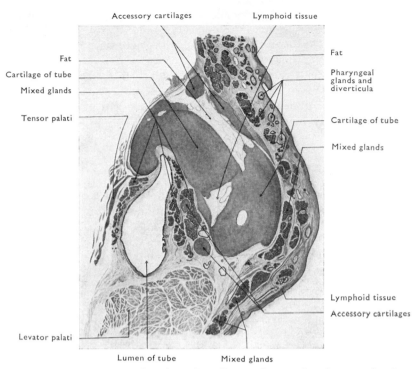

Accessory cartilages   Lymphoid tissue

Fat

Cartilage of tube

Mixed glands

Tensor palati

Fat

Pharyngeal glands and diverticula

Cartilage of tube

Mixed glands

Lymphoid tissue

Accessory cartilages

Levator palati

Lumen of tube   Mixed glands

FIG. 1009   Transverse-section through auditory tube near its pharyngeal end.

superolateral parts of its wall; the gap between the edges of the plate being filled with fibrous tissue to complete the supporting wall of the tube. The base of the cartilage produces a hook-like ridge (**tubal elevation** [FIG. 496]) on the lateral wall of the nasal part of the pharynx, with a fold of mucous membrane (**salpingopharyngeal fold,** covering the salpingopharyngeus muscle) descending in the pharyngeal wall from its posterior edge. The size of tube and cartilage gradually decreases towards the bony part, and its folded edge is firmly fixed to the base of the skull in the groove between the greater wing of the sphenoid and the petrous temporal bone. At its pharyngeal orifice, the tube is attached to the posterior edge of the medial pterygoid lamina and then lies medial to the tensor veli palatini muscle which partly arises from it and also separates it from the mandibular nerve and the middle meningeal artery. The levator veli palatini lies between the tube and the pharyngeal mucosa behind the tubal elevation. The tube is opened, as in swallowing and yawning, by the simultaneous contraction of tensor veli palatini and salpingopharyngeus which are attached to opposite sides of the cartilage plate. The **bony part** of the tube lies between the petrous and tympanic parts of the temporal bone, below the semicanal for tensor tympani muscle, lateral to the carotid canal. It increases in diameter as it approaches the tympanic cavity.

Mucous Membrane. In the bony part this is thin, tightly bound to the periosteum, and lined with a single layer of columnar ciliated cells. In the cartilaginous part it is thick, thrown into folds, and lined with pseudostratified columnar ciliated epithelium containing goblet cells. There are also some mucus-secreting tubulo-alveolar **glands** near the pharyngeal end, and scattered lymphocytes which may be aggregated to form a small **tubal tonsil** in the same region.

Vessels and Nerves. The ascending pharyngeal, middle meningeal, and artery of the pterygoid canal supply the tube.

Its **veins** drain into the pterygoid and pharyngeal plexuses. Sensory **nerves** are derived from the tympanic plexus and the pharyngeal branch of the pterygopalatine ganglion.

In the foetus the pharyngeal opening of the tube is below the level of the hard palate, at birth level with it, at the fourth year 3–4 mm. above it, and in the adult 10 mm. above it. In the child the lumen of the tube is wider than that of the adult, it is more horizontal, and the bony part is relatively shorter.

AUDITORY OSSICLES

The three ossicles, malleus, incus, and stapes—form a jointed arch between the tympanic membrane and the fenestra vestibuli.

The Malleus [Fig. 1010]

This is the largest of the three (8–9 mm. long) with a head, a neck, a handle, and anterior and lateral processes.

The **handle** is fused to the tympanic membrane by its periosteum and a thin layer of cartilage. It begins at the umbo, where its inferior end is curved anterolaterally, and ascends with a forward inclination (125–150 degrees to the horizontal) to become continuous with the conical **lateral process** immediately inferior to the pars flaccida. This process has the mallear folds attached to it, is the upper limit of the attachment of the malleus to the tympanic membrane, and protrudes the membrane laterally. Above this the narrow **neck** and rounded **head** rise vertically into the epitympanic recess at a slight angle to the handle. On the medial surface of the upper part of the handle is a slight **tubercle** for the attachment of the tendon of tensor tympani. From the anterior surface of the neck, on the level of the pars flaccida, the slender **anterior process** passes towards the petrotympanic fissure, with the **anterior ligament** continuing from it through that fissure. The rounded **head** has an oval facet for articulation with the incus on the posteromedial surface below the summit. The facet is divided by an oblique ridge into a larger, upper, posterior surface, and a smaller, lower, medial surface. The inferior margin of the facet is prominent and forms a marked spur where it meets the oblique ridge. Immediately below the spur is the point of attachment of the **lateral ligament** of the malleus.

The Incus [Fig. 1010]

This is named from its supposed likeness to an anvil on which the hammer (malleus) strikes. The anterior surface of the body of the incus has a saddle-shaped articulation for the head of the malleus, and a recess for its spur. The pyramidal **short crus** [Fig. 1011] extends horizontally backwards from the body to the **fossa for the incus,** to which its cartilage tip is attached by the **posterior ligament of the incus.** The **long crus** extends vertically downwards from the body (posteromedial and parallel to the handle of the malleus) and in the same sagittal plane but approximately at right angles to the short crus. Inferiorly the long crus bends medially to end in the **lenticular process,** a small knob of bone which articulates with the stapes.

The Stapes [Fig. 1010]

This stirrup-shaped bone has a small, laterally directed head which articulates with the lenticular process of the incus, and medial to this a slightly constricted neck gives rise to the two crura which pass to the anterior and posterior ends of the base. The anterior crus is shorter and more nearly in line with the neck than the posterior crus which arches backwards before sweeping medially. In the fresh state the two crura are united by a membrane. The base is oval or kidney-shaped and fits into the fenestra vestibuli.

Joints of the Auditory Ossicles

The **incudomallear** and **incudostapedial** joints are both synovial; the former is saddle-shaped, the latter of the ball-and-socket type, and both have considerable amounts of elastic tissue in their capsules.

Ligaments of the Auditory Ossicles. The **malleus** has anterior, superior, and lateral ligaments [Fig. 1011]. The **anterior ligament** consists of (1) a part which passes forwards from the anterior process to run through the petrotympanic fissure to the spine of the sphenoid.

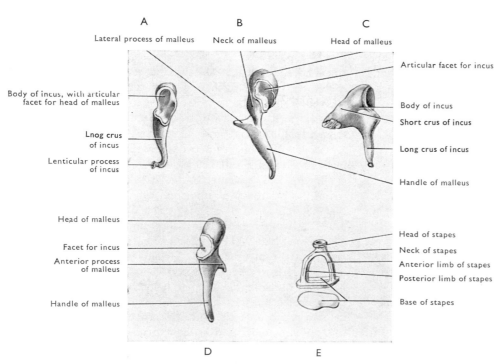

FIG. 1010 Auditory ossicles of left ear (enlarged about three times).

A, incus, seen from the front; B, malleus, seen from behind; C, incus, and D, malleus, seen from the medial side; E, stapes, seen from above and from the medial side.

Thence it is continued as the sphenomandibular ligament, the whole representing parts of the first pharyngeal arch (Meckel's) cartilage. It may contain some muscle fibres (Soemmerring, 1806; Walls, 1946). (2) The other part of the ligament passes to the anterior end of the tympanic notch. The **superior ligament** passes upwards from the head to the roof of the epitympanic recess. The **lateral ligament** is short and fan-shaped. Its fibres pass from the malleus to the posterior edge of the tympanic notch. With the anterior ligament it forms the axis around which the malleus moves (Helmholtz, 1868).

The **incus** is held in position by the capsule of the incudomallear joint and its **posterior ligament** on which it can swing medially and laterally with the head of the malleus. The so-called superior ligament is a mucosal fold.

The base of the **stapes** is covered with hyaline cartilage on its rim and medial surface. That on the rim is attached to the margin of the fenestra vestibuli by the elastic **anular ligament** of the stapes.

Muscles of the Auditory Ossicles [FIG. 1011]

The **tensor tympani** arises from the upper surface of the cartilage of the auditory tube, the adjacent greater wing of the sphenoid, and the walls of its semicanal. In the semicanal it runs posterolaterally to end in a tendon which turns laterally around the processus cochleariformis, and crossing the tympanic cavity, is inserted into the medial and anterior surfaces of the handle of the malleus near its upper end. When the muscle contracts it pulls the handle of the malleus medially, tenses the tympanic membrane, and thus reduces the amplitude of its oscillations. **Nerve supply:** mandibular nerve.

The **stapedius** arises from the walls of a canal which leads postero-inferiorly from the pyramid. Its tendon passes forwards through the apex of the pyramid to the neck of the stapes. On contraction it pulls the neck of the stapes posteriorly, tilts the base in the fenestra vestibuli tightening the anular ligament and reducing the oscillatory range. **Nerve supply:** the facial nerve.

Movements of the Auditory Ossicles

The handle of the malleus moves with the tympanic membrane. When this moves inwards, the head of the malleus moves outwards around the axis of its anterior and lateral ligaments. This movement rotates the incus around its short crus, swinging the long crus inwards, and pressing the stapes into the fenestra vestibuli. This movement causes a wave of compression to pass through the perilymph of the internal ear, the displacement of this fluid by the stapes being allowed for by a corresponding bulging of the **secondary tympanic membrane.** The spur on the head of the malleus locks the incudomallear joint when the tympanic membrane is forcibly pushed inwards. In the opposite movement, excessive movement of the malleus is prevented by sliding of the head of the malleus on the incus, and thus the stapes is not torn from the fenestra vestibuli.

Vessels and Nerves of the Tympanic Cavity

The **arteries** are: (1) the **anterior tympanic,** which enters through the petrotympanic fissure; (2) from the stylomastoid which enters by that foramen, the **posterior tympanic;** (3) **inferior tympanic** [p. 867]; (4) **superior tympanic** which accompanies the lesser petrosal nerve. The **veins** drain to the pterygoid plexus and the superior petrosal sinus. A plexus of **lymph vessels** in the mucous membrane drains to parotid and retropharyngeal lymph nodes. The sensory supply to the mucous membrane is through the tympanic plexus [p. 729]; the fibres in the chorda tympani nerve pass through this region to reach the lingual nerve [p. 725].

In foetal life much of the tympanic cavity is filled with loose mesenchymatous tissue, the epithelial-lined space being relatively small. By birth the epithelial cavity has its definitive arrangement but is filled with fluid which is absorbed shortly after birth. The middle and internal parts of the ear are relatively large at birth, and the amount of postnatal growth is small.

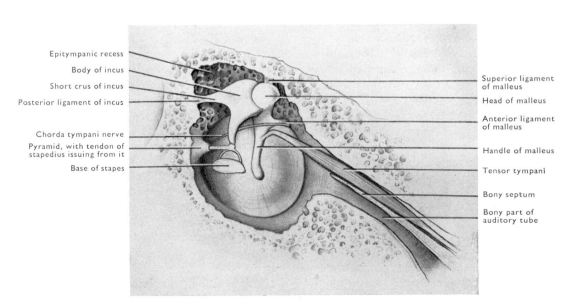

Epitympanic recess
Body of incus
Short crus of incus
Posterior ligament of incus
Chorda tympani nerve
Pyramid, with tendon of stapedius issuing from it
Base of stapes

Superior ligament of malleus
Head of malleus
Anterior ligament of malleus
Handle of malleus
Tensor tympani
Bony septum
Bony part of auditory tube

FIG. 1011 Left tympanic membrane and chain of auditory ossicles (seen from medial side).

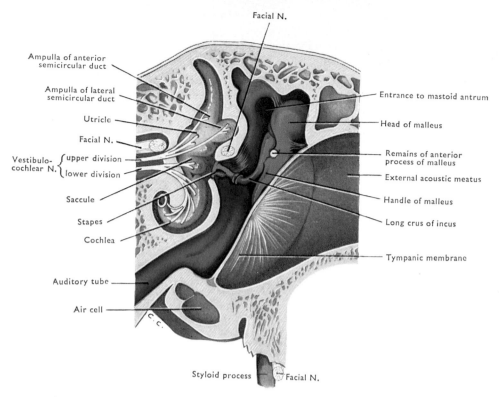

FIG. 1012 Semidiagrammatic coronal section through the left tympanic cavity viewed from the front; mucous membrane, blue.

THE INTERNAL EAR

The internal ear lies buried in the petrous temporal bone. It consists of a complex of membranous tubes and spaces filled with **endolymph**, which are continuous with each other, the **membranous labyrinth**. This fits, for the most part very loosely inside a similarly shaped bony cavity (the **bony labyrinth**) and the space between the membrane and bone is filled with fluid (**perilymph**) and loose connective tissue. The parts of the internal ear from anterior to posterior are: (1) the spiral duct of the **cochlea** in the bony cochlea; (2) the **saccule** and **utricle** in the bony **vestibule,** the first continuous with the duct of the cochlea; (3) three **semi-circular ducts** which arise from the utricle and are enclosed in bony **semicircular canals.**

THE BONY LABYRINTH

Vestibule

This ovoid cavity is the middle part of the bony labyrinth communicating anteriorly with the bony cochlea and posteriorly with the semicircular canals [FIG. 1014] by five circular openings. Its long axis (6 mm.) lies anteroposteriorly, vertically it measures 4–5 mm., and transversely 3 mm. Its **lateral wall,** pierced by the fenestra vestibuli [FIG. 1013], is part of the medial wall of the tympanic cavity. Its **medial wall** corresponds to the fundus of the internal acoustic meatus through which the various peripheral branches of the vestibulo-cochlear nerve pass to supply the sensory endings in the membranous labyrinth. On the medial wall of the vestibule anteriorly is a circular depression (**spherical recess**) pierced by 12–15 tiny foramina (**macula cribrosa media**) for the nerves to the saccule which lies in the depression. Above and behind the spherical recess is a curved ridge, the **vestibular crest,** which is triangular in shape above (**pyramid of the vestibule**) but inferiorly splits to enclose a small depression (**cochlear recess**) pierced by nerves to the vestibular extremity of the cochlear duct. In the roof of the vestibule, postero-superior to the vestibular crest, is the **elliptical recess** containing the utricle. The inferomedial part of the recess and the adjacent pyramid are pierced by 25–30 nerves to the utricle and the ampullae of the anterior and lateral semi-circular ducts. Posterior to the cochlear recess is a furrow which deepens to a canal posteriorly. This **aqueduct of the vestibule** pierces the petrous temporal bone and opens through a slit-like fissure on its posterior surface midway between the internal acoustic meatus and the groove for the sigmoid sinus. It is 8–10 mm. long, and transmits the **endolymphatic duct** and a small vein. The cochlea communicates with the vestibule through an elliptical opening which leads anteriorly from the spherical recess into the scala vestibuli of the cochlea [FIG. 1014]. This opening is bounded inferolaterally by a thin plate of bone (the **osseous spiral lamina**) which begins immediately anterior to the cochlear recess, passes forwards into the cochlea, and partly separates the scala vestibuli from the scala tympani inferolateral to it.

Semicircular Canals

The three semicircular canals (anterior, posterior, and lateral) lie posterosuperior to the vestibule [FIGS. 1013, 1014] into which they open. Each forms approximately two-thirds of a circle, and has one dilated end, the **ampulla.** They are oval

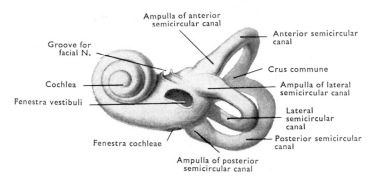

FIG. 1013 Lateral view of left bony labyrinth

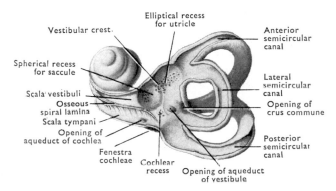

FIG. 1014 Lateral view of interior of left bony labyrinth.

in transverse section, measuring 1–1·5 mm. in diameter, while the ampullae are approximately 2 mm. across. The anterior and posterior canals together open into the vestibule by three apertures, having one stem, **crus commune,** common to both, while the lateral canal has two separate openings.

The **anterior semicircular canal** is 15–20 mm. long, and lies in a vertical plane across the long axis of the petrous temporal bone deep to the arcuate eminence [FIGS. 135, 158], with its convexity upwards. The ampulla lies at the antero-lateral end, and opens into the vestibule immediately superior to the opening of the ampulla of the lateral canal. The posteromedial end joins the superior end of the posterior semi-circular canal to form the **crus commune** (4 mm. long) which opens into the superomedial part of the vestibule. The **posterior semicircular canal,** 18–20 mm. long, lies in a vertical plane parallel to the long axis of the petrous temporal bone, with its convexity directed poster-iorly. The ampulla is on the inferior extremity and opens into the postero-inferior part of the vestibule close to six to eight small foramina (**macula crib-rosa inferior**) for the transmission of nerves to that ampulla. The **lateral semicircular canal,** 12–15 mm. long, is nearly horizontal, but slopes upwards and forwards at approximately 30 degrees. Its convexity faces posterolaterally, and the ampulla at the anterolateral end opens into the vestibule immediately above the fenestra vestibuli. This end of the canal produces an elevation of the medial wall of the aditus to the mastoid antrum above

that for the facial nerve [FIG. 1012]. The semicircular canals of the two internal ears are so arranged that the anterior canal of one side is parallel to the posterior canal of the other, and the lateral canals are in the same plane.

Cochlea

The bony cochlea surrounds and forms the central pillar (**modiolus**) for a spiral tube of $2\frac{1}{2}$–$2\frac{3}{4}$ turns arranged in the form of a short cone. The **base** of the cochlea lies at the fundus of the internal acoustic meatus ; the apex or **cupula,** close to the semicanal for tensor tympani, points antero-laterally ; thus the axis of the cochlea is at right angles to the long axis of the petrous temporal bone. The diameter of the base is approximately 9 mm. ; from base to apex is 5 mm. The tube is 32 mm. long and 2 mm. in diameter at its widest in the basal turn, the first part of which bulges towards the tympanic cavity forming the **promontory.**

The **modiolus** is about 3 mm. long, conical in shape, and has a thin shelf of bone, the **osseus spiral lamina,** projecting from it into the spiral canal like the thread of a screw. This partly divides the canal into scala vestibuli and scala tympani [FIG. 1015] ; a division which is completed by the **basilar lamina,** which membrane passes from the free edge of the lamina to the outer wall of the canal. The osseous spiral lamina begins in the floor of the vestibule superomedial to the fenestra cochleae and scala tympani ; it is separated from a smaller ridge of bone on the outer wall of the basal turn of the spiral canal (the **secondary spiral lamina**) by the fissure which is closed by the basilar lamina [FIG. 1015]. At first at right angles to the modiolus, the osseous spiral lamina becomes progressively angled towards the apex, and ends there in a sickle-shaped margin, **hamulus** of the spiral lamina. The hamulus forms part of the boundary of an aperture, the **helicotrema,** through which the scalae vestibuli and tympani are continuous with each other in the fresh state.

The base of the modiolus forms the cochlear area of the fundus of the internal acoustic meatus [FIG. 1016]. It is pierced by a large number of foramina arranged in a spiral form, **tractus spiralis foraminosus.** Each foramen transmits fibres of the cochlear part of the vestibulocochlear nerve longitudinally through the modiolus [FIG. 1015] together with a branch of the labyrinthine artery. Where each foramen approaches the base of the corresponding part of the osseous

FIG. 1015 Section of bony cochlea.

FIG. 1016   Fundus of left internal acoustic meatus divided into upper and lower areas by transverse crest.

spiral lamina it turns outwards to enter a small **spiral canal of the modiolus** which contains the cell bodies **(spiral ganglion)** of the sensory neurons of the cochlear part. These cells send peripheral processes outwards through a number of minute canals in the osseous spiral lamina to reach the basilar lamina.

The **scala tympani** begins at the fenestra cochleae infero-lateral to the smaller **scala vestibuli** which begins in the vestibule. They take a spiral course around each other into the cochlea so that the scala vestibuli rapidly becomes lateral to the scala tympani (i.e., nearer the apex of the cochlea), and retains this position to the helicotrema, becoming larger than the scala tympani in the apical coils. Near the beginning of the scala tympani, a small orifice on its medial wall leads into a narrow canal, **aqueduct of the cochlea,** which passes infero-medially through the temporal bone to open on the upper border of the jugular fossa close to the posterior surface of that bone. This extension of the scala tympani has been said to communicate with the subarachnoid space.

The lateral end (fundus) of the **internal acoustic meatus** abuts on the medial wall of the vestibule and the base of the modiolus. It is divided into upper and lower areas by a **transverse crest** [FIG. 1016]. The large single foramen for the facial nerve pierces the anterior part of the upper area, anterior to the **superior vestibular area** which transmits nerve fibres and minute arteries to the utricle and the ampullae of the anterior and lateral semicircular ducts (q.v.). The cochlear area lies in the anterior part of the lower area, while the posterior part **(inferior vestibular area)** transmits nerves to the saccule. Behind this, the **foramen singulare** carries vestibular nerve fibres to the ampulla of the posterior semicircular duct.

MEMBRANOUS LABYRINTH

The membranous labyrinth [FIGS. 1017, 1024] is an ectodermal cavity loosely fitted within the bony laby-rinth and filled with endolymph, while the spaces be-tween it and the bony walls are filled with loose connec-tive tissue and **perilymph**—a fluid akin to cerebrospinal fluid but containing more protein and much less chloride, and different in composition from **endolymph** which has a high potassium and low sodium content. The

membranous labyrinth in the vestibule consists of two com-municating sacs, the utricle and saccule. The semicircular ducts arise from the utricle and traverse the semi circular canals, while the saccule gives rise to a blind cochlear duct which extends through the spiral cochlea sandwiched between the scalae vestibuli and tympani.

Utricle and Saccule

The **utricle** is an elongated sac in the posterosuperior part of the vestibule. The anterosuperior part of the utricle lies in the elliptical recess and receives the ampullae of the anterior and lateral semicircular ducts [FIG. 1017], while the central part receives the crus commune and the non-ampullated end of the lateral semicircular duct. The inferomedial part receives the ampulla of the posterior semicircular duct. The **macula of the utricle** is a pale, oval (3 mm. × 2·3 mm.) thickening of the floor and adjacent anterior wall of the anterior part of the utricle. This sensory area is innervated by utricular nerve fibres of the vestibular part of the vestibulocochlear nerve which pierce the pyramid of the vestibule.

The **saccule** is an oval vesicle (3 mm. × 2 mm.), smaller than the utricle. It lies in the antero-inferior part of the vestibule, and has a thickening of its anterior wall, **macula of the saccule** (circa 1·5 mm. in diameter) innervated by the saccular fibres of the vestibulocochlear nerve, and similar but at right angles to the macula of the utricle. The posterosuperior part of the saccule is fused with the utricle (Dickie, 1920), and the endolymphatic duct arises from the saccule just inferior to this. A small **utriculosaccular duct** passes from the utricle to the **endolymphatic duct,** which then traverses the aqueduct of the vestibule and ends as a blind dilatation, the **endolymph-atic sac,** external to the dura mater of the posterior cranial fossa on the posterior surface of the petrous temporal bone. **Structure.** The walls of the utricle and saccule consist of epithelium and a thin layer of connective tissue which is continuous through the connective tissue of the perilymph

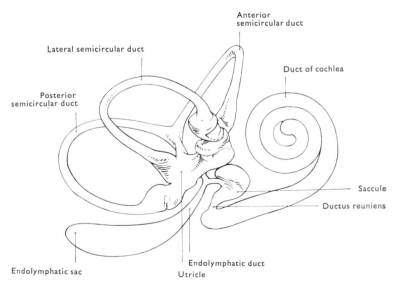

FIG.1017   Diagrammatic representation of model of right membranous labyrinth. (From a model by J. K. Milne Dickie.)

FIG. 1018 Vertical section of wall of highest part of utricle with the macula and bundles of nerve fibres.

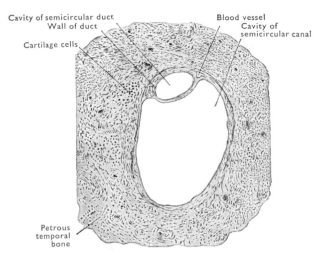

FIG. 1019 Section of a semicircular canal and duct.

space with the endosteum of the bony vestibule. The lining epithelium is separated from the fibrous layer by a basement membrane, and is simple squamous in type except at the maculae where the epithelium (**neuroepithelium**) is of a columnar type and the connective tissue is greatly thickened and contains many nerve fibres. The epithelium of the maculae [FIG. 1018] consists of supporting and **sensory cells,** the latter with hair-like processes projecting into the cavity and buried in a jelly-like mass (**membrana statoconiorum**) of polysaccharide-protein complex containing numerous, minute, crystalline bodies (the otoliths or **statoconia**) consisting of a calcium carbonate-protein complex. The free surfaces of the flask-shaped sensory cells have a mass of long microvilli projecting from them. These narrow markedly towards their bases, and increase progressively in length towards the side of the cell where a cilium is present. The deep parts of these cells contact the ends of the nerve fibres which enter the neuroepithelium. One type of sensory cell is virtually enclosed in a cup-shaped nerve ending, while a second type appears to have a ribbon-like sensory ending together with another type of ending which has all the appearances of being efferent (Wersall, 1956). The significance of these two types of sensory cell, one apparently capable of being modified by an efferent fibre, or fibres, is not known, but it is believed that the action of gravity on the statoconia is an adequate stimulus to the hair cells. The maculae of the utricle and saccule being at right angles to each other are differently affected by the pull of gravity whatever the position of the head relative to that force, and they may also be stimulated by inertial movements of the statoconia on acceleration or deceleration of the head.

Semicircular Ducts

The semicircular ducts are oval in transverse section, and are attached to one wall of the much larger semicircular canal in which they lie [FIG. 1019]. In the bony ampulla, each duct is dilated and almost fills the space available: the wall of this **membranous ampulla** is folded transversely from one side into its lumen to form a ridge or crest (**crista ampullaris** [FIG. 1023]). On the sides and apex of the crista the epithelial lining is thickened to form an innervated structure identical with that of the maculae but for the absence of statoconia; the processes of the hair cells also being embedded in a jelly-like mass, the **cupula.** Movement of the endolymph which fills the semicircular ducts appears to be the stimulus

necessary to excite the sensory cells, and this occurs when the head is rotated in the plane of any pair of semicircular ducts causing inertial movement of the endolymph. The broad base of the crista and the absence of sensory cells in this region, combined with the narrow bases of the long microvilli, makes it likely that it is movement of the hairs and not of the entire crista which causes stimulation of the innervated hair cells.

Nerve impulses generated in the hair cells enter the brain through the vestibular part of the vestibulocochlear nerve, and these evoke reflex movements of the eye, trunk, and limbs, and the conscious sensation of movement of the head.

Duct of the Cochlea

This spiral, blind, tube [FIG. 1020] lies in the bony cochlea between the external part of the osseous spiral lamina and the outer wall of the cochlea. It is compressed between the scala vestibuli (which is separated from it by the delicate **spiral membrane—vestibular wall),** and the scala tympani which is separated from it by the basilar lamina and osseous spiral lamina (**tympanic wall**). These two scalae unite around the

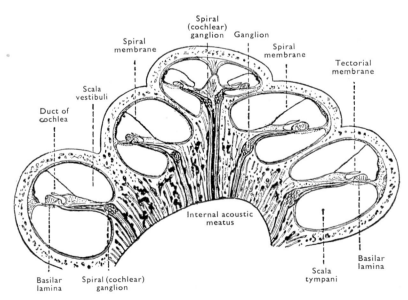

FIG. 1020 Axial section of human cochlea. (Schaefer's *Essentials of Histology.*)

blind end of the cochlear duct (**caecum cupulare**) at the heli-cotrema. The basal part of the cochlear duct lies in the cochlear recess of the vestibule, and is connected to the lower part of the saccule by the short, conical **ductus reuniens** [FIG. 1017]. The **basilar lamina** extends from the free edge of the osseous spiral lamina to the **basilar crest** on the outer wall of the cochlea [FIG. 1021] where the endosteum is thickened to form the **spiral ligament** of the cochlea. The **spiral membrane** lies between the apical surface of the osseous spiral lamina and the outer wall of the cochlea ; the duct of the cochlea enclosed between it and the basilar lamina being triangular in transverse section. The spiral membrane consists of little more than a layer of flattened mesothelium of the scala vestibuli applied to the somewhat thicker epithelial cells of the cochlear duct which bear short microvilli projecting into the cochlear duct.

On the outer wall of the cochlear duct, the epithelium attached to the endosteum of the bony cochlea, is thickened, stratified, and contains capillaries between its mitochondria rich cells. This **stria vascularis** is probably concerned with the secretion or maintenance of the endolymph. It bulges into the cochlear duct to form the **spiral prominence,** which contains a larger vessel (the **vas prominens**) close to the attachment of the basilar lamina, from which it is separated by the **external spiral sulcus** [FIG. 1021].

The **epithelium of the cochlear duct** is modified to form the complex spiral organ on the internal part of the basilar lamina, while internal to this the endosteum on the apical surface of the osseous spiral lamina is considerably thickened (**limbus of the spiral lamina**) and grooved externally (**internal spiral sulcus** [FIG. 1021]). The **tympanic and vestibular lips** of this groove pass respectively to the basilar lamina and the tectorial membrane, the former lip being perforated by the dendrites of the spiral ganglion which pass outwards through the osseous spiral lamina on their way to the spiral organ. The surface of the limbus of the spiral lamina is markedly ridged towards the vestibular lip where the ridges project like a row of teeth (**acoustic teeth**). The ridges and grooves between them are covered by the epithelium of the cochlear duct, and these cells form a thin cuticle on their free surface, the **tectorial membrane.** This membrane thickens markedly as it projects outwards from the free edge of the vestibular lip to lie over the spiral organ [FIG. 1022].

The **basilar lamina** lies between the tympanic lip of the limbus and the basilar crest of the spiral ligament, and carries the spiral organ on its thin internal part, while its outer part is thicker and striated. On the tympanic surface is a layer of vascular connective tissue with one larger vessel, the **spiral vessel,** coursing along the basilar lamina under the spiral organ. The width of the basilar lamina increases from the base (0·2 mm.) of the cochlea to the apex (0·35 mm.), and its over-all length is 32 mm. (Hardy, 1938 ; Keen, 1940).

Spiral Organ

This complex sensory organ consists of several elements formed from the thickened epithelium of the cochlear duct on the basilar lamina [FIG. 1022]. Each of these elements seen on radial section forms one part of a long row of cells spiralling on the basilar lamina from the base to the apex of the cochlea.

1. Towards the internal edge of the spiral organ, two rows (internal and external) of tall pillar cells (rods of Corti) surround the innermost of three spaces in the spiral organ, the triangular **internal tunnel** (of Corti), which communicates between the external pillar cells with the paracuniculus (space of Nuel) immediately lateral to them. These supporting, pillar cells are filled with microtubules and have their bases widely

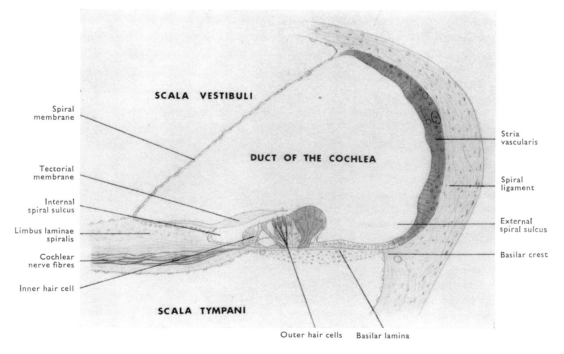

FIG. 1021   Section across duct of cochlea. (Retzius, 1881.)

spread on the basilar lamina, but lean towards each other so that they fuse at the free surface between the inner and outer hair and phalangeal cell groups. At the fusion of the pillar cells, the head of the inner cell splits into medial and lateral beak-like (coracoid) processes, the latter overlapping the laterally curved, finger-like (phalangeal) process of the corresponding lateral pillar cell, which separates the paracuniculus from the cochlear duct. The outer pillar cells lie more obliquely, are longer, and less numerous than the inner pillar cells, and groups of nerve fibres from the spiral ganglion pass between the intermediate portions of both rows of pillar cells, crossing the inner tunnel between them.

Hair Cells and Phalangeal (Deiters') Cells. A smaller group of these cell complexes lies internal to the pillar cells, while a larger group lies external to them. The phalangeal cells have their bases on the basilar lamina and extend towards the free surface **(reticular membrane)** of the spiral organ. Approximately half way they suddenly narrow, thus forming a recess in which the base of a hair cell is situated; the narrowed part of the phalangeal cell passes between the surrounding hair cells to reach the reticular membrane. Each phalangeal cell contains a bundle of fibrils, some of which pass from the basilar lamina to the base of the hair cell, while the remainder pass through the narrow process of the cell to expand on the reticular membrane and separate the free surfaces of adjacent hair cells. The inner hair cell-phalangeal cell complex (approximately 3,500 hair cells) forms a single row, while the outer complex (approximately 12,000 hair cells) forms four rows in the apical region and three such rows in the basal region the two groups being separated by the expanded laminae of the heads of the inner pillar cells. The hair cells are cylindirical in shape and their free ends protrude between the expanded ends of the phalangeal cells forming the reticular membrane. Forty or more microvilli project from the free ends and are arranged as in the hair cells of the vestibular apparatus. In life they lie against or may even be embedded in the tectorial membrane, and are arranged in parallel rows along the long axis of the basilar lamina. Nerve fibres reach the hair cells by piercing the tympanic lip below the inner hair cells, and supplying them, pass outwards between the pillar cells and across the inner tunnel and the paracuniculus to reach the outer hair cells.

External to the outer hair cells is the **outer tunnel,** a space which communicates with the paracuniculus and is limited laterally by a mound of external limiting cells (of Hensen)

which are larger and more numerous than the internal limiting cells internal to the inner hair cells [FIG. 1022]. The former are attached to the basilar lamina by thin processes, but expand towards the free surface to form a prominent ridge. On the basilar membrane external to the limiting cells is a layer (or layers) of large cuboidal supporting cells (of Claudius).

The **reticular membrane** consists of the heads of the outer pillar cells, the processes of the external phalangeal cells, and the surfaces of the external limiting cells. The free ends of the hair cells project through it in rows.

The Tectorial Membrane. This shelf of jelly-like material containing a mixture of delicate fibrils projects outwards from the vestibular lip, and thickening over the internal part of the spiral organ, lies on the processes of the hair cells. Externally it rapidly thins, and ends in an irregular margin close to the external limiting cells. Internally it lies on the limbus between the vestibular lip and the attachment of the spiral membrane.

Sound waves in the air cause movements of the tympanic membrane and ear ossicles so that similar waves are transmitted to the perilymph at the fenestra vestibuli, and so to the basilar membrane. It was long thought that this was a tuned membrane, the short basal parts of which would respond to high frequencies while the long apical parts would respond to low frequencies. However, the entire basilar membrane responds to low frequency vibrations; moreover the basilar membrane is lax and not taut. It has been shown that the basal part responds with nerve impulses to high and low frequencies, while the apical part responds to low frequencies only. Thus it is not clear exactly how individual frequencies are recognized by the cochlea and transmitted to the brain. There are efferent fibres in the vestibulocochlear nerve which pass to the cochlea, and these fibres can reduce the nerve impulses arising from stimulation of the cochlea by sound (Rasmussen, 1953; Galambos, 1956).

VESTIBULOCOCHLEAR NERVE
[FIG. 1024]

This nerve divides in the internal acoustic meatus into cochlear and vestibular parts.

The **cochlear part** consists of nerve fibres which are the central processes of cells in the spiral ganglion in the modiolus of the cochlea. From the ganglion they pierce the modiolus in

FIG. 1022 Transverse section of spiral organ from central coil of duct of cochlea. (Retzius, 1881.)

Cavity of
auditory tube

Middle
Ear

Perilymph

Malleus

Foot plate
of stapes

Incus

Facial nerve

Spiral ganglion

Cochlea and cochlear nerve

Macula of saccule

Bony labyrinth

Utricle and macula of utricle

Crista ampullaris

Semicircular duct

FIG. 1023   Horizontal section through the inner and middle ear of a 10-day
old kitten.   At this age the middle ear is not completely pneumatized,
and is still filled with loose connective tissue.

bundles, those from the apical coil of the ganglion in the central foramen, the more basal parts sending fibres through the foramina of the tractus spiralis foraminosus and the cochlear recess.   The peripheral processes of the ganglion cells pierce the osseous spiral lamina and the tympanic lip of the limbus to reach the inner hair cells or to pass between the pillar cells and across the inner tunnel of the spiral organ to the outer hair cells.   Other coarser fibres accompany the former to the internal edge of the spiral organ, and then turn to take a spiral course on the internal part of the basilar membrane.

The **vestibular part** consists of the processes of the bipolar cells of the vestibular ganglion in the internal acoustic meatus. The ganglion is usually in two (or three) parts, the superior of which sends fibres through the superior vestibular area to the macula of the utricle, the ampullae of the anterior and lateral semicircular ducts, and the anterior part of the macula of the saccule.   The inferior part sends fibres to the macula of the saccule, and through the foramen singulare to the ampulla of the posterior semicircular duct.   A branch from the vestibular

to the cochlear part of the nerve appears to contain some of the efferent cochlear fibres.   A branch has been described passing to the macula of the saccule from the basal part of the spiral ganglion (Hardy, 1934), but there is no evidence that fibres from that macula join the cochlear part of the nerve (Stein and Carpenter, 1967).

VESSELS OF INTERNAL EAR

The **arteries** are the **labyrinthine,** a branch of the basilar or anterior inferior cerebellar artery, and the **stylomastoid,** a branch of the posterior auricular or occipital artery.   The former passes with the vestibulocochlear nerve and is distributed with its branches.   The **veins** drain through the various apertures at the lateral extremity of the internal acoustic meatus, and unite to form a **labyrinthine vein** which passes to the inferior petrosal sinus or the sigmoid sinus.   Small veins traverse the aqueducts of the vestibule and cochlea to enter the superior and inferior petrosal sinuses respectively.

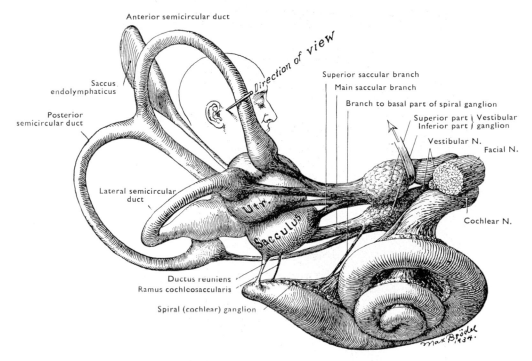

FIG. 1024   Distribution of vestibulocochlear nerve to the membranous labyrinth.
(M. Hardy, *Anatomical Record*, 1934.)

DEVELOPMENT OF THE EAR

External Ear

The development of the external and middle ears is intimately connected with the pharyngeal arches and their intervening grooves and pouches [pp. 56–57]. The **auricle** is developed from the dorsal ends of the first and second pharyngeal arches each of which forms three temporary tubercles by proliferation of its mesoderm [FIG. 1025]. These surround the upper part of the first ectodermal groove which forms the external acoustic meatus, and though the part played by each arch in the formation of the auricle is not clear, it is likely that the first (mandibular) only forms the tragus, while the second forms the remainder of the auricle (Jones and I-Chuan, 1934). The ectoderm of the dorsal part of the first groove initially is in contact with the entoderm of the first pouch, though later separated from it by the ingrowth of a considerable thickness of mesoderm. Subsequent proliferation of the ectoderm to form a solid **meatal plate** carries it medially towards the entoderm of the tympanic cavity, thinning the intervening mesoderm to form the fibrous part of the tympanic membrane and the cartilage of the handle of the malleus. The central cells of the meatal plate then break down, extending the meatus medially to the remaining layer of ectoderm which forms the cutaneous layer of the tympanic membrane. Extension of the entodermal lining of the tympanic cavity to cover the superomedial surface of the tympanic membrane completes that structure.

Middle Ear and Auditory Tube

This cavity is formed from the dorsal wings of the first and second pharyngeal pouches. The caudal growth of the second pharyngeal arch, combined with a forward growth of the third arch medial to it, draws the dorsal wings of the first and second pharyngeal pouches posterolaterally from the pharyngeal wall (lateral to the third arch), and thus compounds them into a single cavity with the second arch and its contained facial nerve in the posterolateral wall, while the first and third arches (with the corresponding nerves) form the anterolateral and posteromedial walls respectively. This complex extension of the pharyngeal cavity, the **tubotympanic recess,** forms the auditory tube and the middle ear cavity, and its subsequent extension in the sixth month of pregnancy produces the **mastoid antrum** and after birth the **mastoid cells.** At this stage the tubotympanic recess is very oblique with its future lateral and medial walls lying inferolateral and superomedial respectively, with the rudiment of the membranous labyrinth, the otic vesicle, on the superomedial wall. The later dorsal extension of the part forming the middle ear cavity brings the labyrinth into the medial wall and makes that wall more vertical. The dorsal ends of the cartilages of the first and second pharyngeal arches articulate superior to the tubotympanic recess. These cartilages form the **malleus, incus,** and **stapes.** At least part of the malleus is formed from the first arch cartilage (Lightoller, 1939), and the stapes (perforated by the transitory artery of the second arch) is formed from the second, but the origin of the incus at least is in doubt (Hanson and Anson, 1962). Subsequently the mesoderm surrounding these ear ossicles is absorbed and the tympanic cavity extends dorsolaterally to surround and cover them with the mucous lining of the middle ear. The ear ossicles begin to ossify in the third month of pregnancy; the malleus from one centre for the head and handle with another for the anterior process; the incus from a single centre in the upper part of the long crus; the stapes from a single centre in its base.

The tensor tympani and stapedius muscles are developed in

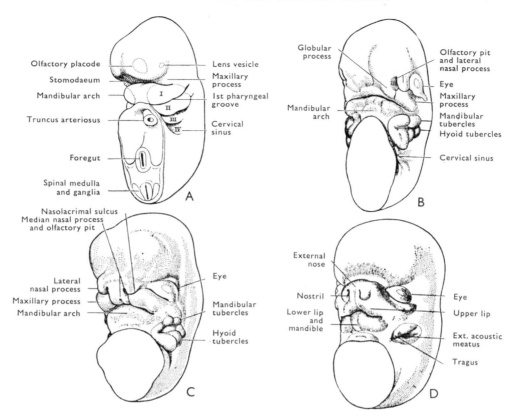

FIG. 1025 Ventrolateral view of head in series of human embryos, showing development of face and external ear. Note the lateral and dorsal migration of the auricle coincident with the formation of the lower lip and mandible.

A, 6 mm. embryo (× 11); B, 12 mm. (× 7·5); C, 14 mm. (× 7·5); D, 18 mm. (× 6).

I, II, III, IV, pharyngeal arches.

the dorsal ends of the first and second pharyngeal arches respectively, and are supplied by the corresponding fifth and seventh cranial nerves.

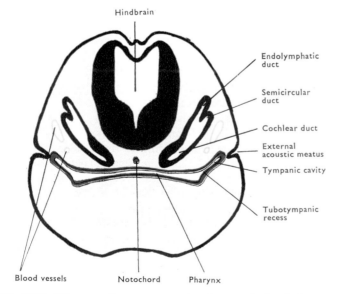

FIG. 1026 Diagram of transverse section through the head of an embryo, showing the rudiments of the three parts of the ear and their relation to the tubotympanic recess and the first pharyngeal groove.

Internal Ear

The epithelium of the membranous labyrinth is derived from the ectoderm by a thickening, **auditory placode,** which appears at the side of the hindbrain dorsal to the second pharyngeal arch. This placode sinks into the underlying mesoderm to form an auditory pit, which finally separates from the surface ectoderm as the **otic vesicle.** As it sinks into the mesoderm beside the hindbrain at the origin of the seventh and eighth cranial nerves, it elongates vertically and gives off from its medial side a dorsally directed diverticulum, the **endolymphatic duct** [FIG. 1027]. Between the fifth and sixth weeks of embryonic life the ventral part of the vesicle extends anteromedially as a blind process, the beginning of the **cochlear duct.** This slowly increases in length, and coiling on itself, forms all the turns of the cochlear duct by the twelfth week. From the posterosuperior part of the vesicle three disc-like evaginations arise lateral to the endolymphatic duct. The central parts of each disc coalesce and disappear leaving only the peripheral **semicircular ducts.** Of these the first to develop is the anterior, and the last the lateral, though all three are formed by the beginning of the second month. At the attachment of the endolymphatic duct the wall of the vesicle is constricted, thus separating the utricle postero-superiorly from the saccule antero-inferiorly. This constriction extends into the base of the endolymphatic duct dividing it into a Y-shaped tube connecting the utricle and saccule. A

FIG. 1027 Sections through hindbrain of foetal rabbits to illustrate development of membranous labyrinth.

A, B, and C, successive phases in the development of the auditory pit and the otic vesicle.

further constriction occurs between the saccule and the cochlear duct to form the **ductus reuniens.**

The **epithelial lining,** at first columnar, becomes cubical except opposite the terminations of the vestibulocochlear nerve fibres. Here it forms the complex, columnar epithelium of the maculae and cristae, and of the spiral organ. Later much of the epithelium becomes more flattened, e.g., on the spiral membrane, in the semicircular ducts, and in the non-macular areas of the utricle and saccule. Two ridges appear on the basal surface of the cochlear duct. The inner, formed by connective tissue, is the limbus laminae spiralis, while the outer is epithelial and forms the spiral organ by differentiation of its cells (Bast *et al.*, 1947).

The connective tissue surrounding the otic vesicle differentiates into an inner, loose reticular layer, a middle precartilaginous layer, and an outer zone of cartilage, the **otic** (auditory) **capsule** [p. 131]. With the growth of the membranous labyrinth, the surrounding cartilage is eroded, and the loose reticular layer progressively absorbed to produce the **perilymphatic spaces** surrounding the membranous labyrinth. Initially these are formed as a number of separate spaces which subsequently coalesce and expand at the expense of the precartilage and cartilaginous zones from which the loose connective tissue of the perilymphatic spaces is formed. The perilymphatic space first appears as a cistern around the utricle and saccule [FIG. 1029 A], and from this the **scala vestibuli** extends along the cochlear duct to meet the **scala tympani** (which develops independently and slightly earlier) at the helicotrema about the sixteenth intra-uterine week. The perilymphatic spaces in the semicircular canals contain more connective tissue than the scalae of the cochlea, but all communicate with the space surrounding the utricle and saccule.

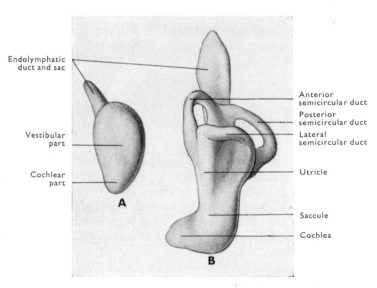

FIG. 1028 The left membranous labyrinth seen from the lateral side. A, at about the 4th week; B, at about the 5th week. (W. His, Jr.)

C.T.A.—2 E

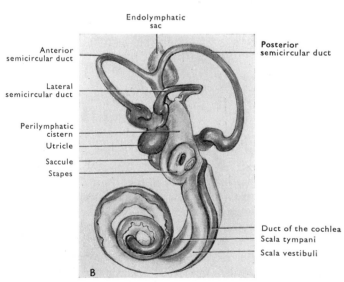

FIG. 1029 Membranous labyrinths of human embryos (Streeter, 1918). A, 50 mm.; B, 85 mm.

OLFACTORY ORGAN

The general anatomy of the nasal cavity and paranasal sinuses is dealt with in the respiratory system. This section is concerned only with the olfactory apparatus.

The olfactory epithelium lies in the posterosuperior part of the nasal cavity. This brownish-yellow mucous membrane covers the superior conchae and the upper 1 cm. of the sides of the nasal septum, the total area being about 10 cm². but the area decreases greatly with age (Naessen, 1970). The epithelial covering is tall pseudostratified columnar, and it contains three types of cells. (1) The cell bodies and dendrites of the bipolar **olfactory sensory cells.** These cells have their round nuclei in the middle of the thickness of the epithelium [FIG. 1030] and conspicuous neurofibrils in their cytoplasm. The distal process (dendrite) passes superficially between the **supporting cells** (oval nuclei), and expanding into a rounded olfactory vesicle among the long microvilli on their surface, gives rise to 6–8 extremely long, non-motile cilia which run parallel to the surface of the epithelium buried in the surface mucus, and contain a considerable amount of lipid. (2) The supporting cells contain axial filaments and pigment granules and have a number of long microvilli projecting from their

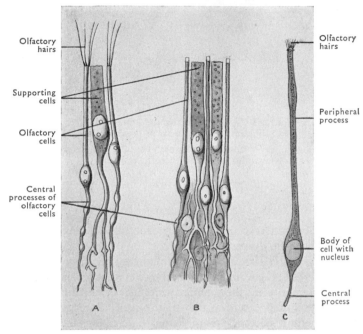

FIG. 1031   Olfactory and supporting cells.
A, frog; B, human; C, human olfactory cell.

FIG. 1030   Section through olfactory mucous membrane.

surface. They are packed around the projecting dendrites and are fused to them at the surface. (3) A thin layer of basal cells surrounds the axons of the sensory cells, which leave the epithelium surrounded by sheath cells, and form bundles that pierce the cribriform plate of the ethmoid to enter the olfactory bulb.

The olfactory mucous membrane contains many serous, tubulo-alveolar **glands** whose secretion contains many enzymes. It appears that the secretions of these glands wash away substances dissolved in the surface mucus and make it sensitive to new substances passing into solution on the surface of the mucous membrane.

THE ORGAN OF TASTE

The peripheral organs which are believed to be mainly responsible for the sensation of taste are oval or flask-shaped collections of cells, **taste buds,** in the epithelium of the oral and pharyngeal cavities. They are most numerous on the tongue, especially on the sides of the vallate papillae and the opposed walls of the vallum [FIG. 1032], and on the foliate papillae anterior to the palatoglossal arch. They are also present: (1) on the fungiform papillae of the back and sides of the tongue, though less numerous towards the tip; (2) on the anterior part of the soft palate; (3) on the posterior surface of the epiglottis (especially its lateral parts); (4) on the superior

surfaces of the arytenoid cartilages; and (5) on the posterior wall of the oral part of the pharynx. The number of taste buds is maximal in children, but decreases rapidly with age (Arey et al., 1935). Many salivary glands in the tongue open into the sulci of the vallate papillae. Their secretions first dissolve and subsequently wash away the substances which stimulate the taste buds.

Structure [FIG. 1034]

The taste buds are pale-staining structures buried in the darker epithelium. They taper from expanded bases to end

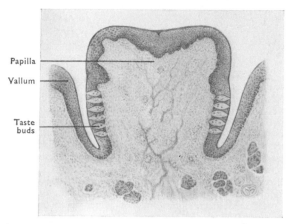

Papilla

Vallum

Taste
buds

FIG. 1032   Section through vallate papilla of human tongue.

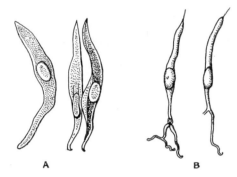

FIG. 1033   Isolated cells from taste bud of rabbit (Engelmann).
A, supporting cells; B, gustatory cells.

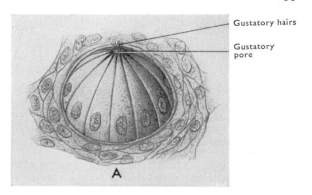

Gustatory hairs

Gustatory
pore

A

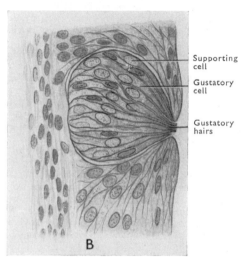

Supporting
cell

Gustatory
cell

Gustatory
hairs

B

FIG. 1034   Taste buds from folia linguae of rabbit.
A, three-quarter surface view; B, vertical section.

in a few fine, hair-like processes which project through a minute **gustatory pore** in the superficial surface of the epithelium. Most of the cells of the taste buds [FIG. 1033] are elongated, slightly curved spindles (supporting cells) which enclose a group (5–15) of centrally placed gustatory cells. Each of these slender, rod-like cells has a nucleus near its middle, a peripheral process which ends in a gustatory hair at the gustatory pore, and a central process which branches at the base of the taste bud.

Innervation

Nerve fibres approaching taste buds in the epithelium branch frequently to form a subepithelial plexus from which some branches pass into the bases of the taste buds, while others closely surround them or pass into the epithelium between them. On the anterior two-thirds of the tongue they are innervated by fibres from the chorda tympani branch of the facial nerve (through the lingual nerve), while those on the posterior third, including the vallate and foliate papillae are supplied by the glossopharyngeal nerve. On the palate they

are again supplied by the facial nerve through its greater petrosal branch which sends fibres into the palatine nerves, while the vagus (internal branch of the superior laryngeal nerve) supplies those on the epiglottis and arytenoid cartilages. Thus the nerves which supply taste buds also carry parasympathetic fibres to the associated glands.

It is not known if the taste buds are the sole sources of appreciation of the four taste sensations (salt, bitter, sweet, acid) for it seems that such sensations may be appreciated from areas where taste buds are not present. Whether the nerve fibres which end in the epithelium around the taste buds are also concerned with taste sensation is not known, but subjective testing is remarkably unreliable as the sensation of 'taste' is usually a combination of taste, texture (ordinary tactile sensation), and smell. Thus injuries to the trigeminal nerve, which is not thought to transmit any fibres concerned with the innervation of taste buds, may cause a subjective alteration in the sensation of taste, and interference with the sensation of smell markedly alters the subjective sense of taste.

REFERENCES

AREY, L. B., TREMAINE, M. J., and MONZINGO, F. L. (1935) The numerical and topographical relations of taste buds to human circumvallate papillae throughout the life span, *Anat. Rec.*, **64**, 9.

BAST, T. H., ANSON, B. J., and GARDNER, W. D. (1947) The developmental course of the human auditory vesicle, *Anat. Rec.*, **99**, 55.

BOYD, I. A. (1962) The structure and innervation of the nuclear bag muscle fibre system and the nuclear chain muscle fibre system in mammalian muscle spindles, *Phil. Trans. B*, **245**, 81.

BURCH, G. E. (1939) Superficial lymphatics of the human eyelids observed by injection *in vivo*, *Anat. Rec.*, **73**, 443.

CAUNA, N. (1959) The mode of termination of the sensory nerves and its significance, *J. comp. Neurol.*, **113**, 169.

CAUNA, N. (1964) in *Advances in Biology of the Skin*, Vol. VI, Aging, New York.

CLARK, W. E. LE GROS (1971) *The Tissues of the Body*, 6th ed., Oxford.

CLARK, W. E. LE GROS, and BUXTON, L. H. D. (1938) Studies in nail growth, *Brit. J. Derm.*, **50**, 221.

COOPER, S., and DANIEL, P. M. (1949) Muscle spindles in human extrinsic eye muscles, *Brain*, **72**, 1.

DICKIE, J. K. M. (1920) Note on the anatomy of the membranous labyrinth, *J. Laryng.*, **35**, 76.

DUPERTUIS, C. W., ATKINSON, W. B., and ELFTMAN, H. (1945) Sex differences in pubic hair distribution, *Hum. Biol.*, **17**, 137.

FISON, J. (1920) The relative positions of the optic disc and macula latea to the posterior pole of the eye, *J. Anat. Physiol.*, **54**, 184–8.

GALAMBOS, R. (1956) Suppression of auditory nerve activity by stimulation of efferent fibres to the cochlea, *J. Neurophysiol.*, **19**, 424.

GILCHRIST, M. L., and BUXTON, L. H. D. (1939) The relation of finger-nail growth to nutritional status, *J. Anat. (Lond.)*, **73**, 575.

HANSON, J. R., and ANSON, B. J., and STRICKLAND, E. M. (1962) Branchial sources of the auditory ossicles in Man, *Arch. Otolaryng.*, **76**, 200.

HARDY, M. (1934) Observations on the innervation of the macula sacculi in Man, *Anat. Rec.*, **59**, 403.

HARDY, M. (1938) The length of the organ of Corti in Man, *Amer. J. Anat.*, **62**, 291.

HELMHOLTZ, H. VON (1868) *The Mechanism of the Ossicles of the Ear and Membrana Tympani*, trans. Buck, A. H., and Smith, N., 1873, New York.

JALOWY, B. (1939) Über die Entwicklung der Nervenendigungen in der Haut des Menschen, *Z. ges. Anat. I. Z. Anat. Entwickl.-Gesch.*, **109**, 344.

JONES, F. W. (1941) Tension lines, cleavage lines and hair tracts in Man, *J. Anat. (Lond.)*, **75**, 248.

JONES, F. W., and I-CHUAN, W. (1934) The development of the external ear, *J. Anat. (Lond.)*, **68**, 525.

KEEN, J. A. (1940) A note on the length of the basilar membrane in Man and in various mammals, *J. Anat. (Lond.)*, **74**, 524.

KIDD, W. (1903) *The Direction of Hair in Animals and Man*, London.

LEWIS, T. (1927) *The Blood Vessels of the Human Skin and their Responses*, London.

LIGHTOLLER, G. H. S. (1939) On the comparative anatomy of the mandibular and hyoid arches and their musculature, *Trans. zool. Soc. Lond.*, **24**, 349.

LOCKWOOD, C. B. (1885) The anatomy of the muscles, ligaments and fasciae of the orbit, including an account of the capsule of Tenon, the check ligaments of the recti, and the suspensory ligament of the eye, *J. Anat. Physiol.*, **20**, 1.

MANN, IDA C. (1928) *The Development of the Human Eye*, Cambridge.

MONTAGNA, W. (1956) *The Structure and Function of Skin*, New York.

NAESSEN, R. (1971) An enquiry on the morphological characteristics and possible changes with age in the olfactory region of man, *Acta Otolaryng.*, **71**, 49.

NYBERG-HANSEN, R. (1965) Anatomical demonstration of gamma motoneurons in the cat's spinal cord, *Exp. Neurol.*, **13**, 71.

PENROSE (1968) Medical significance of finger-prints and related phenomena, *Brit. med. J.*, **2**, 321.

POLYAK, S. (1957) *The Vertebrate Visual System*, Chicago.

POPOFF, N. W. (1934) The digital vascular system, *Arch. Path.*, **18**, 295.

QUILLIAM, T. A. (1966) Unit design and array patterns in receptor organs, in *Touch, Heat and Pain*, ed., de REUCK, A. V. S., and KNIGHT, J., Ciba Symposium, London.

RASMUSSEN, G. L. (1953) Further observations on the efferent cochlear bundle, *J. comp. Neurol.*, **99**, 61.

RETZIUS, G. (1881–4) *Das Gehörorgan der Wirbelthiere, morphologisch-histologische Studien*, Stockholm.

SCHULTZ, A. H. (1940) The size of the orbit and of the eye in primates, *Amer. J. phys. Anthrop.*, **26**, 389.

SHANTHAVEERAPPA, T. R., and BOURNE, G. H. (1963) New observations on the structure of the Pacinian corpuscle, and its relation to the perineurial epithelium of peripheral nerves, *Amer. J. Anat.*, **112**, 97.

SHANTHAVEERAPPA, T. R., and BOURNE, G. H. (1966) Perineurial epithelium: a new concept of its role in the integrity of the nervous system, *Science*, **154**, 1464.

SINCLAIR, D. (1967) *Cutaneous Sensation*, London.

SOEMMERRING, S. T. (1806) *Icones Organi Auditus Humani*, table 2, fig. 10, Frankfurt.

STEIN, B. M., and CARPENTER, N. B. (1967) Central projections of portions of the vestibular ganglia innervating specific parts of the labyrinth in the rhesus monkey, *Amer. J. Anat.*, **120**, 281.

SYMINGTON, J. (1885) The external auditory meatus in the child, *J. Anat. Physiol.*, **19**, 280.

THOMSON, A. (1912) *Atlas of the Eye*, Oxford.

TRAQUAIR, H. M. (1949) *An Introduction to Clinical Perimetry*, 6th ed., London.

TROTTER, M. (1924) The life cycles of hair in selected regions of the body, *Amer. J. phys. Anthrop.*, **7**, 427.

WALLS, E. W. (1946) The laxator tympani muscle, *J. Anat. (Lond.)*, **80**, 210.

WEDDELL, G., and MILLER, S. (1962) Cutaneous sensibility, *Ann. Rev. Physiol.*, **24**, 199.

WEDDELL, G., and PALLIE, W. (1955) Studies on the innervation of skin. The number, size and distribution of hairs, hair follicles and orifices from which hairs emerge in the rabbit ear, *J. Anat. (Lond.)*, **89**, 175.

WERSALL, T. (1956) Studies on the structure and innervation of the sensory epithelium of the cristae ampullares in the guinea pig. A light and electron microscopic investigation, *Acta Otolaryng. (Stockh.)* Suppl., **126**, 1.

WHITNALL, S. E. (1932) *The Anatomy of the Human Orbit and Accessory Organs of Vision*, 2nd ed., London.

WOLFF, E. (1948) *The Anatomy of the Eye and Orbit*, 3rd ed., London.

THE BLOOD VASCULAR AND LYMPHATIC SYSTEMS

E. W. WALLS

THE BLOOD VASCULAR SYSTEM

THE movement of the blood is constantly in a circle, and is brought about by the beat of the heart.' When in 1628 William Harvey reached this conclusion the proof he offered was complete, and clearly demonstrated that the same blood which leaves the heart through **arteries** returns to it along **veins.** His reasoning had been masterly, and had forged a chain of evidence only one link of which awaited visual confirmation. This came when the introduction of the microscope enabled Malpighi (1661) to observe **capillaries** in the lung of the frog; tiny vessels which Harvey knew must exist between the smallest arteries, now called **arterioles,** and the smallest veins or **venules.**

The heart is a rhythmically contractile muscular pump divided by septa into two thin-walled receiving chambers, or **atria,** and two thicker-walled ejecting chambers or **ventricles.** During **diastole,** the interval between two contractions, blood carried by afferent vessels called veins enters the atria and continues on into the ventricles; and then by atrial contraction **(systole)** the onward flow is completed and the ventricles distended. Ventricular contraction then follows and the blood is projected into efferent vessels called arteries.

Two great arteries leave the heart, the **aorta** and **pulmonary trunk,** and from them are given off branches which themselves divide, and so ever smaller and more numerous vessels are formed. As noted above, the arterioles end in plexuses of vessels known as capillaries [FIG. 1035].

Capillary networks are found in all tissues except cartilage, epidermis, the cornea and most epithelia. They are drained by the veins which unite to form progressively larger trunks, the largest of which return the blood to the atria.

As blood flows along the capillaries some of its fluid and other substances pass through their thin walls. In this way the interstitial fluid, which is the medium of metabolic exchange, is replenished. Means of draining this fluid away are provided in two ways. Part re-enters the blood capillaries towards their venous ends, and so is returned to the heart; and part is collected by quite another set of vessels, the **lymph capillaries,** which form a meshwork of blindly ending tubes of fine calibre. **Lymph vessels** drain this meshwork and eventually discharge into the great veins at the root of the neck. Thus, all the fluid which irrigates the tissues is eventually returned to the heart.

The food and water required by the tissues is taken by mouth and then absorbed in suitable form by the blood capillaries and lymph capillaries in the wall of the abdominal part of the alimentary canal. Elimination of fluid and of waste metabolic products is shared by the skin, lungs, and kidneys.

A constant supply of **oxygen** is indispensable for the continued activity, and indeed survival, of the tissues; equally necessary is the efficient discharge of **carbon dioxide.** In the higher vertebrates these needs are met by the lungs, through which all blood first passes before returning to the heart for distribution throughout the rest of the body. Such a system requires a double pump, one to receive venous blood from the body **(right atrium)** and pump it to the lungs **(right ventricle)** while another receives blood from the lungs **(left atrium)** and pumps it to the body **(left ventricle).** This completely separate, double system is only achieved in mammals and birds, and is quite different from the arrangement in fishes where the heart pumps blood through the gill capillaries to the body—a mechanism which precludes the circulation of blood to the rest of the body at a high pressure because of the capillary resistance in the gills. Thus in the fish all blood passes through two sets of capillaries before returning to the heart, while it only passes through one set in the higher forms, except in the portal circulations (see below). The arteries, capillaries, and veins of the lungs form the

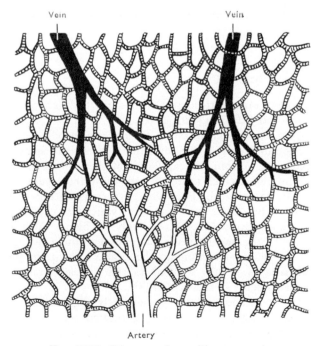

FIG. 1035   Diagram of a capillary network.

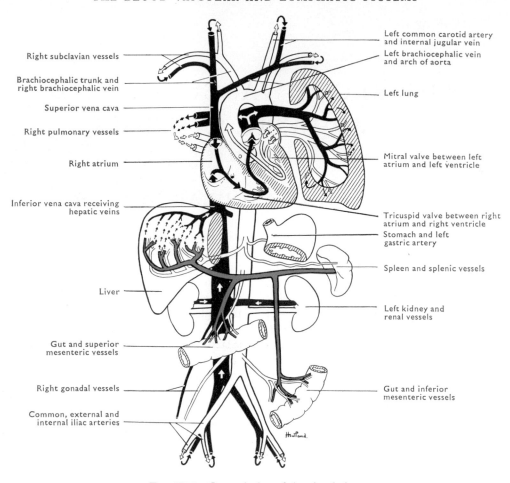

Right subclavian vessels

Brachiocephalic trunk and
right brachiocephalic vein

Superior vena cava

Right pulmonary vessels

Right atrium

Inferior vena cava receiving
hepatic veins

Liver

Gut and superior
mesenteric vessels

Right gonadal vessels

Common, external and
internal iliac arteries

Left common carotid artery
and internal jugular vein

Left brachiocephalic vein
and arch of aorta

Left lung

Mitral valve between left
atrium and left ventricle

Tricuspid valve between right
atrium and right ventricle

Stomach and left
gastric artery

Spleen and splenic vessels

Left kidney and
renal vessels

Gut and inferior
mesenteric vessels

FIG. 1036   General plan of the circulation.

The division of the pulmonary trunk is indicated by a large white bifurcating
arrow, and only the left pulmonary veins are shown entering the left atrium; the
liver is shown receiving the portal vein and the hepatic artery; the three large
arteries which spring from the abdominal aorta to supply the alimentary canal,
i.e. the coeliac trunk, and the superior and inferior mesenteric arteries are shown
arising in succession from the front of the aorta; the gonadal vessels are indicated
only on the right side.

pulmonary circulation, and those of the body generally, the **systemic circulation** [FIG. 1036].

A special arrangement exists for the return of blood from the abdominal part of the alimentary canal, the spleen and the pancreas. From the capillaries of these parts the blood flows into veins which unite to form the **portal vein.** This large vein passes to the liver where it breaks up into increasingly small branches, the smallest of which are capillary-like vessels called **sinusoids.** Accordingly the blood which leaves the liver by the hepatic veins has passed through two sets of capillaries before returning to the heart. The term **portal circulation** is usually applied to this special arrangement which also occurs in the hypophysis.

There is, in addition, the system of lymph vessels already mentioned. Lymph nodes are associated with the lymph vessels, and together they constitute the lymphatic system [p. 941].

TISSUES OF THE VASCULAR SYSTEM

Of the tissues which compose the vascular system only endothelium is present throughout. The other constituents of the system are areolar tissue, white fibrous tissue, elastic tissue, plain muscle, and cardiac muscle.

The **heart** consists mainly of cardiac muscle, the special features of which are considered on page 850.

ARTERIES

The arteries are all lined with endothelium, but the other tissues are present in different amounts according to functional needs. Although the diminution in arterial size from the aorta to the periphery is gradual, and the change in the structure of the wall usually correspondingly gradual, it is customary to recognize three orders of vessel, large arteries, medium arteries, and small arteries or arterioles. In all three categories the tissues are arranged in three coats, the **tunica intima, tunica media,** and **tunica adventitia,** from within outwards.

When the heart contracts, blood is suddenly forced into the aorta and its primary large branches such as the brachiocephalic, common carotid, subclavian, and common iliac. These vessels have a predominance of **elastic tissue** in their walls and are considerably distended with each systole. This distension protects the smaller vessels beyond from the sudden strain of systole, whilst the elastic recoil of the walls during diastole continues to drive the blood on. In this way the blood flow is converted to a steady stream in the smaller vessels.

The need to regulate the supply of blood to the various tissues and organs, according to their state of activity, is met by the **visceral muscle** which predominates in arteries of medium and small size. This muscle consists of spindle-shaped cells in the tunica media, whose degree of contraction is under involuntary nervous control. Thus, should a given area need more blood, the muscle cells of the arteries concerned can be relaxed and the lumina of the vessels increased; should less blood be required, the muscle cells contract and the lumina are correspondingly diminished. The two processes, acting in different areas at the same time, allow of a more effective distribution of the limited output of the heart without causing a fall in the pressure of the circulating blood.

It is convenient to describe first the structure of the wall of medium-sized arteries, and thereafter to indicate the distinguishing features of the walls of large and small arteries.

Medium-sized Arteries

Known also as distributing, or muscular, arteries these vessels all possess a tunica media which is predominantly muscular. Commencing as branches or continuations of large elastic vessels, they range down to something less than half a millimetre in diameter.

Tunica Intima. The constituent parts of this, the **inner coat,** are the endothelial lining, a thin subendothelial layer of fine areolar tissue, and the **internal elastic lamina.** This last consists of elastic fibres fused together in the form of a fenestrated membrane which, owing to the contracted state of

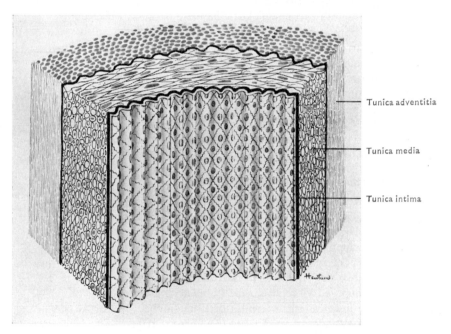

Tunica adventitia

Tunica media

Tunica intima

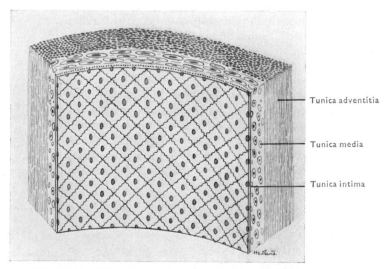

Tunica adventitia

Tunica media

Tunica intima

FIG. 1037 Transverse section of wall of large artery.

1. Tunica intima.   2. Tunica media.
3. Tunica adventitia.

FIG. 1038 Medium-sized artery and vein as seen in section.
The distinction between the tunica media and tunica adventitia of the vein is in fact less clear than shown, the diagram being designed to emphasize the relative poverty of muscular tissue in the media of the venous wall.

the usual type of preparation, appears wavy in transverse section [FIG. 1038].

Tunica Media. This, the **middle coat,** consists mainly of visceral muscle fibres bound together in more or less spiral strata by reticular fibres, a few collagen fibres, and some delicate elastic fibres.

Tunica Adventitia. This, the **outer coat,** although thinner than the tunica media, is a strong, thick layer formed mainly of collagen fibres. It blends gradually with the surrounding fibro-areolar tissue which forms a sheath for the vessel. Longitudinal or tangential bundles of elastic fibres are also present, and at the outer border of the tunica media they form the **external elastic lamina.**

Large Arteries

Known also as elastic or conducting arteries, these vessels all possess a very thick **tunica media** which consists of fifty or more concentrically arranged elastic laminae between which are collagen fibres, fine elastic fibres, and fibroblasts, all embedded in an amorphous ground substance. Small visceral muscle cells lie on the surface of the laminae [FIG. 1037]. The **tunica intima** is noticeably thicker than in medium-sized arteries. It contains more elastic fibres, and so the boundary between the subendothelial tissue and the internal elastic lamina is ill defined. Thus the tunica intima is not sharply delimited from the tunica media. The **tunica adventitia** is relatively thin.

Small Arteries

Known also as **arterioles,** the smallest have a diameter only slightly greater than that of capillaries. With regard to the size of the largest arterioles there is no universal agreement, figures given ranging from 0·1 to 0·5 mm. The **tunica intima** consists simply of endothelium with a thin internal elastic lamina in all but the smallest. The **tunica media** consists of circularly disposed visceral muscle fibres arranged in layers which become fewer as the arterioles become smaller; ultimately only a single layer of small scattered fibres remains. The **tunica adventitia** in the larger arterioles may equal the tunica media in thickness, but in the smallest it becomes very thin and the external elastic lamina disappears.

CAPILLARIES

Capillaries are on average 8–10 μm in diameter and about 0·75 mm. in length. Their walls consist essentially of **endothelium,** i.e., pavement cells with sinuous, tapering edges which overlap and adhere to those of adjacent cells, and oval central nuclei [FIG. 1039]. With the light microscope the cells appear cemented to one another along their margins by inter-

FIG. 1040   Capillary loops of the nail fold photographed in life.

FIG. 1041   Electron micrograph of a transverse section through a myocardial capillary (mouse). × 7,000.

Portions of striated muscle fibres occupy the upper left and lower right edges of the field and contain numerous mitochondria (m′).

The capillary, whose lumen is partly filled by an erythrocyte (rbc), is lined by a single endothelial cell with margins joined at j. The pericapillary space (s) contains a pericapillary cell at c, in which a mitochondrion is well seen. The cytoplasm of the endothelial cell also contains a mitochondrion (m″) as well as numerous pinocytosis vesicles. Some of these are separated from the cell membrane (V1), some are contiguous with it (V2), whilst others are open at the cell membranes (V3) communicating with both the lumen and the pericapillary space.

The layer of amorphous dense material (bm) covering the outside of the endothelium represents the basement membrane of the capillary. (By courtesy of Dr. M. A. Epstein.)

cellular substance which stains with silver nitrate, but electron microscopy reveals no evidence of such material (Florey, Poole, and Meek, 1959). Investing the capillary endothelium there is a layer of fine reticular fibres, with collagen fibres in places. Several types of cells occur in a pericapillary position, but there is not complete agreement as to their nature. Some are probably undifferentiated mesenchymal cells, while others seem to be fibroblasts and fixed macrophages. However, cells of a different character, with long branching processes, have been observed closely applied to capillary endothelium, and to them a contractile function has been ascribed.

Rouget (1879) described such cells in the frog, and in Man somewhat similar cells are present in certain sites; but whether these cells, sometimes called pericytes, are responsible for capillary contractility is not yet agreed. Physiological evidence strongly suggests that capillaries possess inherent powers of contraction and dilatation, and it may be that swelling of their endothelial cells obliterates the vessel lumen which is restored when the cells become flattened again. The whole question is a difficult one and Clark and Clark (1943) should be consulted.

FIG. 1039   Capillary blood vessel in surface view.

Capillaries are arranged in networks [FIG. 1035] the meshes of which show great variations in size and form in different tissues, and also with functional activity. In the skin of the nail fold the capillaries form characteristic loops which may be examined with the microscope in the living [FIG. 1040].

The structure of capillaries as seen by the electron microscope is shown in FIGURE 1041.

In some organs (e.g., liver, suprarenal glands, bone marrow, *inter alia*) the finest blood channels are not true capillaries but **sinusoids** (Minot, 1900). These are formed by the growth of columns of cells into blood spaces which are thus broken up into narrow channels of somewhat irregular calibre. Wider than capillaries, they differ also in having phagocytic cells of the reticulo-endothelial system present amongst their endothelial cells.

In erectile tissue also there are no true capillaries. Small arteries open directly into cavernous spaces, lined with endothelium, from which blood is drained by small veins.

VEINS

In veins the blood flows more slowly than in arteries and at a considerably lower pressure. Accordingly, their lumina are larger and their walls thinner than those of their accompanying arteries; and, being thin-walled, veins collapse when empty.

The walls of veins resemble those of arteries, but in general they are poorer in muscular and elastic tissue and richer in white fibrous tissue.

The structure of the three coats varies in different veins and it is not always possible to distinguish the boundaries between them [FIG. 1038].

Tunica Intima. This consists of the endothelial lining, a layer of subendothelial areolar tissue, less abundant than in arteries and often absent, and an outer layer of longitudinally arranged elastic fibres which seldom gives the appearance of a fenestrated membrane.

A feature of the inner coat in most veins is the presence of **valves** [FIG. 1042] formed by folds of endothelium strengthened by a little fibrous and elastic tissue, and so disposed that they prevent flow of blood towards the periphery, and help to support a column of blood in veins in which there is an upward flow. The valve cusps are semilunar in shape, and usually are arranged in pairs. On the cardiac side of each valve the venous wall is, as a rule, dilated into a shallow pouch so that when the veins are distended they appear nodulated.

Competent valves do not occur in the large veins of the trunk or in those of the adult portal system [p. 922]. They are more numerous in the deep veins than in the superficial veins of the limbs, and in the veins of children than in those of adults. (See also Franklin, 1937.)

The presence of valves in the superficial veins of the forearm is readily demonstrated by Harvey's simple, classical experiment [PLATE 47]. The venous return being obstructed by pressure on the upper arm, a finger is placed on one of the distended veins and held firmly in position while the vein is emptied of its blood as far as the next valve by stroking upwards along it. The column of blood will be sustained by that valve while the part of the vein below it remains empty. If the pressure of the finger is now removed, the vein fills from below, so that the flow of blood towards the heart is also demonstrated.

Tunica Media. This is much thinner than the corresponding coat in an artery, and contains less muscle and more collagen fibres. In some veins, e.g., the majority of the veins of the brain, muscular tissue is absent.

Tunica Adventitia. This consists of white fibrous and elastic tissue. In medium-sized veins it is thicker than the tunica media, and in large veins considerably so. Longitudinally disposed muscle fibres in the outer coat of some of the large veins doubtless allow of change of length besides affording extra protection against external pressure (Franklin, 1937).

VASCULAR AND NERVOUS SUPPLY OF ARTERIES AND VEINS

The walls of large and medium-sized arteries and veins are supplied by small vessels, called **vasa vasorum,** which are distributed to the outer and middle coats. They may arise from the vessels they supply or from their branches, or from adjacent arteries, and after a short course enter the walls of the vessels in which they end in a capillary plexus. Draining the plexus are small veins which run to adjacent veins of larger size.

Lymph vessels, which drain into the perivascular lymph plexuses, are present in the outer coat of all the larger arteries and veins.

Arteries are well supplied with **nerves** which form a plexus in the outer coat. The non-medullated fibres, of sympathetic origin, are efferent and supply the muscular tissue of the middle coat. The myelinated fibres, which are considered to be afferent, end mainly in the outer coat but some may reach to the subendothelial tissue. A few lamellated (Pacinian) corpuscles occur in the outer coat of large vessels. The veins have a similar, but less rich, nerve supply. Sites of specially rich sensory innervation which are concerned with the reflex adjustment of heart rate and blood pressure are the carotid sinus [p. 563], the arch of the aorta, and the roots of the caval and pulmonary veins [p. 791].

FIG. 1042 Diagrams showing the valves of veins (W. Sharpey).

A, vein laid open, showing the folds of the inner coat forming a valve. B, longititudinal section of a vein through a valve. C, distended vein, showing the swellings opposite a valve. (Schäfer's *Essentials of Histology*.)

THE HEART

The heart is a hollow, muscular organ, enclosed in a fibro-serous sac called the pericardium, and situated in the middle mediastinum. The cavity of the fully developed heart is separated into right and left halves by an obliquely placed longitudinal septum, and each half consists of a posterior, receiving chamber—the **atrium**—and an anterior, ejecting chamber—the **ventricle.** Externally the coronary sulcus indicates the separation of the atria from the ventricles. Internally each atrium opens into the corresponding ventricle by a wide **atrioventricular orifice** provided with a valve which allows the passage of blood from atrium to ventricle, but effectually prevents its return.

In shape the heart is bluntly conical and somewhat flattened from front to back; a base (posterior surface), an apex, and three surfaces (sternocostal, diaphragmatic, and left or pulmonary) are described [FIG. 1043]. The axis of the heart runs forwards, to the left and slightly downwards, from base to apex.

The **coronary sulcus** runs obliquely round the heart and separates a posterosuperior, atrial portion from an antero-inferior ventricular portion. Anterosuperiorly the sulcus is interrupted by the roots of the aorta and pulmonary trunk. At the base of the heart a faint vertical groove indicates the separation into right and left atria. Anterior and posterior interventricular grooves, which meet at the lower border of the sternocostal surface to the right of the apex, mark the division into right and left ventricles.

The **base of the heart** is formed by the atria—almost entirely the left atrium—and is directed mainly backwards, but also slightly upwards and to the right. It is separated from the middle four thoracic vertebrae by the descending thoracic aorta, the oesophagus and the vagi, the thoracic duct, and the azygos and hemiazygos veins [FIG. 1051].

The base is more or less flat and somewhat quadrilateral in outline. It extends from the postero-inferior part of the coronary sulcus, in which the principal vein of the heart, the coronary sinus, lies [FIG. 1044], to the pulmonary arteries above.

Six large veins enter the base of the heart. The four pulmonary veins pierce the left atrium, the left pair near the left border of the base, and the right pair just to the left of the interatrial sulcus. The superior and inferior venae cavae enter the right atrium, the superior at the right upper angle of the base, and the inferior at its right lower angle. The part of the surface between the right and left pulmonary veins forms the anterior boundary of a cul-de-sac of the pericardial cavity called the oblique sinus [FIG. 1051].

The **apex of the heart,** bluntly rounded, is formed by the left ventricle and is directed forwards, down-wards, and to the left. It lies under cover of the left lung and pleura, behind the fifth left intercostal space within the midclavicular line (7·5–9 cm. from the anterior median line in the average adult). The apex-beat may be seen and felt in that situation, where also the stethoscope is placed to hear the sound of the mitral valve. Age and bodily build influence the position of the apex beat; in children it is usually as high as the fourth interspace, while in adults with long narrow chests it may be as low as the sixth.

The **sternocostal surface** faces forwards, upwards, and to the left. It lies behind the body of the sternum, the medial ends of the third to the sixth right costal cartilages, and the greater parts of the corresponding cartilages of the left side. Separating it from these structures, and from the transversus thoracis muscles is the fibrous pericardium, which is itself overlaid by the lungs and pleurae save for a small area where it is in direct contact with the chest wall [p. 855]. The sterno-costal surface is formed mainly by the right ventricle but partly by the other chambers. The ventricular por-tion of this surface, which is divided by the **anterior interventricular groove** into a smaller left and a larger right part, lies below and to the left of the antero-superior part of the coronary sulcus [FIG. 1045]. The atrial part of the surface is formed by the right atrium and by the auricles of both atria which embrace the roots of the ascending aorta and pulmonary trunk. These great vessels ascend in front of the left atrium so that

FIG. 1043 Diagram of a median section to show the general distribution of the structures in the mediastinum.

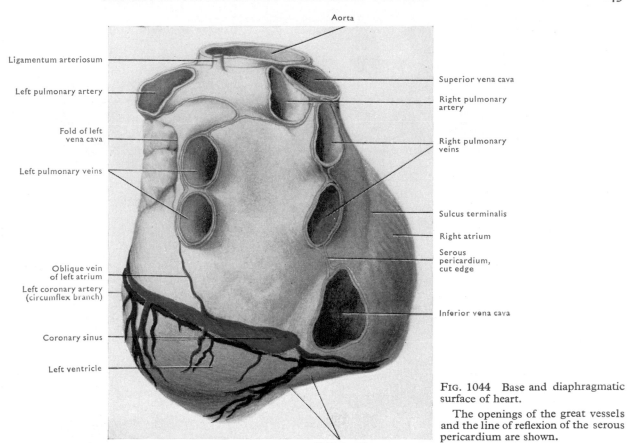

Aorta

Ligamentum arteriosum

Left pulmonary artery

Fold of left vena cava

Left pulmonary veins

Oblique vein of left atrium

Left coronary artery (circumflex branch)

Coronary sinus

Left ventricle

Superior vena cava

Right pulmonary artery

Right pulmonary veins

Sulcus terminalis

Right atrium

Serous pericardium, cut edge

Inferior vena cava

Right ventricle

FIG. 1044 Base and diaphragmatic surface of heart.

The openings of the great vessels and the line of reflexion of the serous pericardium are shown.

Superior vena cava

Aorta

Right coronary artery

Right auricle

Right coronary artery

Right ventricle

Right marginal artery

Left atrium

Left auricle

Pulmonary trunk

Left coronary artery (circumflex branch)

Anterior interventricular branch of left coronary artery with great cardiac vein

Left ventricle

FIG. 1045 Sternocostal surface of heart.

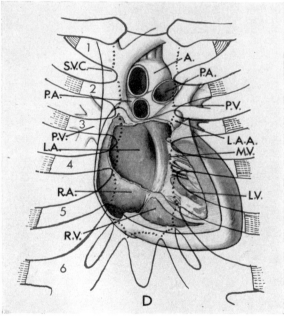

FIG. 1046 Relations of cavities and valves of heart to anterior wall of thorax. (From photographs of a formalin-hardened subject, with the heart dissected *in situ*.)

In A the anterior wall of the right ventricle has been removed and the pulmonary trunk opened.

In B the anterior walls of the ascending aorta and of the right atrium have been removed; also the anterior cusp of the right atrioventricular (tricuspid) valve.

In C the greater part of the interventricular septum has been removed, and the anterior cusp of the left atrioventricular (mitral) valve exposed.

In D the ascending aorta, anterior cusp of left atrioventricular (mitral) valve, pulmonary trunk, and interatrial septum have been removed; the cavities of the left atrium and left ventricle are exposed.

A.	Aortic arch.	P.	Pulmonary valve.
A.V.	Aortic valve.	P.V.	Pulmonary vein.
L.A.	Left atrium.	R.A.	Right atrium.
L.A.A.	Left auricle.	R.V.	Right ventricle.
L.V.	Left ventricle.	S.T.	Septomarginal trabecula.
M.V.	Left atrioventricular	S.V.	Interventricular septum.
	(mitral) valve.	S.V.C.	Superior vena cava.
P.A.	Pulmonary trunk and	T.V.	Right atrioventricular (tricuspid) valve.
	arteries.		

only its auricle contributes to the sternocostal surface. If the ascending aorta and pulmonary trunk are removed, the left atrium is exposed on the sternocostal surface, and its upper margin (which is the **upper margin** of the heart) lies immediately inferior to the bifurcation of the pulmonary trunk and its two branches. The superior vena cava enters the heart at the right end of this margin.

The **right margin,** gently convex to the right, is formed by the right atrium. It is sometimes marked by a shallow groove, the **sulcus terminalis,** which runs from the front of the superior vena cava to the right side of the inferior vena cava.

The **lower margin** of the sternocostal surface, where it meets the diaphragmatic surface, is sharp and, in the cadaver, usually concave corresponding with the curvature of the anterior part of the diaphragm. It is formed mainly by the right ventricle, but near the apex by the left ventricle. The anterior and posterior interventricular grooves meet on this border in a slight depression, called the **apical incisure** or notch, a little to the right of the apex. The lower margin lies in the angle between the diaphragm and the anterior wall of the thorax, and is commonly described as passing from the lower end of the right border behind the xiphisternal joint to the apex [but see p. 854]. To the left, the sternocostal surface becomes continuous with the left or pulmonary surface. The transition is gradual, and cannot be said to be marked by a left margin.

The **left surface,** convex from above downwards and from before backwards, is formed by the left ventricle and only to a small extent by the left atrium and its auricle. It is crossed at its widest part (i.e., above) by the left part of the coronary sulcus, and narrows towards the apex. It faces upwards and backwards, and is separated from the left lung and pleura by the pericardium and phrenic nerve.

The **diaphragmatic surface** is formed by the ventricles and rests upon the diaphragm, chiefly on the central tendon. It is divided into a smaller right portion, and a larger left, by the **posterior interventricular groove** which runs obliquely across it from behind forwards. The postero-inferior part of the coronary sulcus separates it from the base.

The surface anatomy of the heart is described on page 854.

THE CHAMBERS OF THE HEART

ATRIA

The atrial portion of the heart is cuboidal in shape, but is bent round the lateral and posterior surfaces of the ascending aorta and pulmonary trunk. Its cavity is divided by the inter-atrial septum which runs from the anterior wall backwards and to the right.

The long axis of each atrium is vertical, and each possesses an ear-shaped prolongation, known as the auricle, which projects forwards from its anterior and upper angle. The margins of the auricles are crenated.

The Right Atrium [PLATE 48A]

The upper part of this chamber is in contact with the right side of the ascending aorta and its auricle overlaps the anterior surface of the root of that vessel. Its right side forms the right

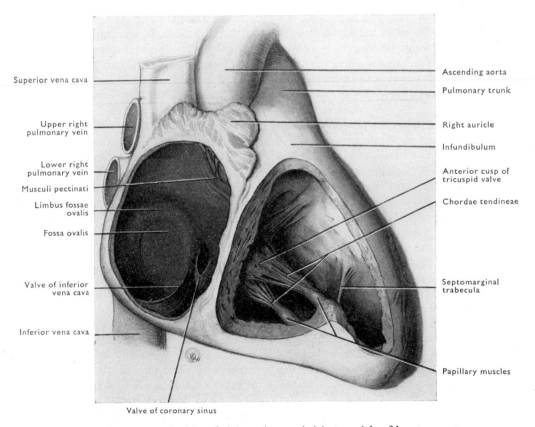

Superior vena cava

Upper right pulmonary vein

Lower right pulmonary vein

Musculi pectinati

Limbus fossae ovalis

Fossa ovalis

Valve of inferior vena cava

Inferior vena cava

Valve of coronary sinus

Ascending aorta

Pulmonary trunk

Right auricle

Infundibulum

Anterior cusp of tricuspid valve

Chordae tendineae

Septomarginal trabecula

Papillary muscles

FIG. 1047   Cavities of right atrium and right ventricle of heart.

border of the heart, and is separated by the pericardium from the right pleura and lung (the right phrenic nerve and pericardiacophrenic vessels intervening) as is the part of the atrium which contributes to the sternocostal surface of the heart. Posteriorly and on the left the interatrial septum separates it from the left atrium. Joining the right atrium posteriorly there are the superior vena cava above and the inferior vena cava below; and a little above its middle, it is crossed behind by the lower right pulmonary vein.

Interior of the Right Atrium. The walls are lined with a glistening membrane called **endocardium,** and are smooth except anteriorly and in the auricle, where muscular bundles, the **musculi pectinati,** form a series of small, oblique columns [FIG. 1047]. They spring from a network of ridges in the auricle and run downwards and backwards to end behind in a crest, the **crista terminalis.** The crista terminalis corresponds in position with the sulcus terminalis, and indicates the junction of the right part of the primitive sinus venosus—the smooth-walled portion of the right atrium—with the atrium proper [p. 926].

In the posterior part of the cavity there is, above, the orifice of the superior vena cava, and below, that of the inferior vena cava. The former has no valve, but the latter is bounded anteriorly by the vestigial valve of the inferior vena cava. Immediately in front and to the left of that valve, between it and the atrioventricular orifice, is the opening of the coronary sinus, guarded by the unicuspid valve of the coronary sinus. The **right atrioventricular orifice** is guarded on the ventricular side by a tricuspid valve, and so is known also as the tricuspid orifice. It is situated in the antero-inferior part of the left wall of the atrium, and admits the tips of three fingers. Scattered over the walls are the orifices of the **venae cordis minimae,** tiny veins which return some of the blood from the substance of the heart. Two or three small vessels, the anterior cardiac veins, also open into the right atrium.

In the septal wall is an oval depression called the **fossa ovalis.** Its floor, which is formed from the septum primum of the embryonic heart, is bounded above and in front by a raised margin, the **limbus fossae ovalis.** The limbus represents the free edge of the septum secundum which faces the orifice of the inferior vena cava and partly overlaps the septum primum in the foetus [p. 925]. Before birth the two atria are in communication through the **foramen ovale;** this is the name given to the gap then present in the septum primum which is overlapped on the right by the thick, but incomplete, septum secundum [FIG. 1048]. The pressure changes consequent upon birth bring the septa into apposition and fusion, but even in the adult, part of the foramen ovale sometimes persists at the upper part of the fossa ovalis as a small slit. On the posterior atrial wall a slight bulge may be seen, between the openings of the venae cavae, just behind the upper part of the fossa ovalis. It is called the **intervenous tubercle,** and it has been supposed to help in directing the blood from the superior vena cava to the tricuspid orifice during foetal life [p. 937].

The **valve of the inferior vena cava** is a thin, sometimes fenestrated, sickle-shaped fold of endocardium and subendo-

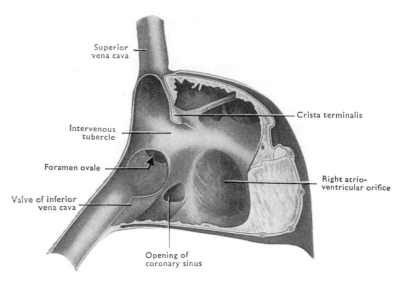

FIG. 1048   Sagittal section of heart of full-term foetus.

cardial tissue which is attached by its convex border to the anterior margin of the caval orifice. Its free concave margin is continuous with the anterior part of the limbus fossae ovalis; below and to the right some of the fibres of the crista terminalis can be traced into it. In the foetus the valve directs the greater part of the blood from the inferior vena cava through the foramen ovale into the left atrium; a measure of its functional importance is that, unlike the adult valve, it shows little variation. Variations in the form of the adult valve are described by Powell and Mullaney (1960).

The **valve of the coronary sinus** is a semilunar fold of endocardium at the right margin of the orifice of the sinus. It is very variable in size and form, and the role commonly ascribed to it of preventing reflux of blood into the sinus during atrial contraction is seldom possible.

The Left Atrium [PLATE 48 D]

This chamber is irregularly cuboidal in shape with its long axis running vertically. It forms most of the base of the heart and, posteriorly, is separated by the pericardium and its oblique sinus from the descending thoracic aorta and the oesophagus [FIG. 1051]. Its concave sternocostal surface is hidden by the ascending aorta and the pulmonary trunk. The interatrial septum, which faces forwards and to the right, forms its right side. Its left side forms a very small portion of the left surface of the heart, and its long, narrow left auricle overlaps the coronary sulcus and the left side of the root of the pulmonary trunk. Four pulmonary veins, two on each side, enter the upper, posterior part, but it is not rare to find the two veins of one or both sides united before their entry into the atrium.

Interior of the Left Atrium. The walls are smooth except in the auricle where musculi pectinati are present. All four orifices of the pulmonary veins are devoid of valves. On the septal wall there may be observed an oval area which corresponds to the fossa ovalis of the right atrium; it is more easily seen when this wall is held to the light. Foramina for venae cordis minimae are scattered over the atrial wall, and below and in front there is the **left atrioventricular** or mitral orifice which opens forwards, and slightly downwards and to the left. It admits the tips of two fingers, and is guarded on the ventricular side by a valve formed of two cusps arranged like a bishop's mitre, known as the **left atrioventricular** or **mitral valve** [FIG. 1054].

VENTRICLES

The ventricular part of the heart forms a slightly flattened cone. Its base is directed backwards and upwards, and is pierced by the four orifices shown in FIGURE 1049. In this region of junction there is dense fibrous and fibrocartilaginous tissue, sometimes called the heart skeleton [p. 85].

The **interventricular septum** is placed obliquely, with one surface facing forwards and to the right and the other backwards and to the left. Its margins are indicated by the anterior and posterior interventricular grooves.

The Right Ventricle [PLATE 48A]

This chamber forms the greater part of the sternocostal surface of the heart and of its lower border, but the smaller part of the diaphragmatic surface. Above and to its right lies the right atrium, and leading out of the upper part of the ventricle is the pulmonary trunk. The left or septal wall bulges into its interior.

Interior of the Right Ventricle. Superiorly, between the pulmonary and atrioventricular orifices there is a thick arch of muscle called the supraventricular crest [FIG. 1052]. This crest separates the lower or inflow part of the cavity from the anterosuperior part or **infundibulum**—so called because of its funnel shape—which carries the outflowing blood to the pulmonary trunk.

The **right atrioventricular** (or tricuspid) **orifice** is guarded by the **right atrioventricular** (or **tricuspid**) **valve** whose three cusps, of roughly triangular shape, are named **posterior** (inferior), **septal** and **anterior** (superior). Small

FIG. 1050   Relation of heart and great vessels to anterior wall of thorax.

1–7. Ribs and costal cartilages.	M. Mitral orifice.
A. Aortic orifice.	P. Pulmonary orifice.
Ao. Ascending aorta.	R.V. Right ventricle.
C. Clavicle.	S.V.C. Superior vena cava.
L.V. Left ventricle.	T. Tricuspid orifice.

accessory cusps are frequently present in the angles between the named cusps, and like them consist of a fold of endocardium strengthened by fibrous tissue. The bases of the cusps are attached to a fibrous ring at the atrioventricular orifice where they may be continuous with one another or separated by one or more of the accessory cusps; the cusps project into the ventricle. The cusp margins are irregularly notched and thinner than the central portions. The atrial surfaces over which the entering blood flows, are smooth, whereas the ventricular surfaces are roughened and give insertion to the **chordae tendineae**. These last are fine tendinous cords which pass to the margins and ventricular surfaces of the cusps from conical muscular projections of the ventricle wall called **papillary muscles** [FIG. 1047].

The **pulmonary orifice** is in front and to the left of the tricuspid orifice. It is guarded by a **pulmonary valve** composed of three semilunar, pocket-like valvules, named, from their positions, anterior, right, and left. The convex outer border of each valvule is attached to the root of the pulmonary trunk. The free inner border is thickened at its middle to form the **nodule** on each side of which there is a small, thin, crescentic area called the **lunule**. Each valvule of

FIG. 1049   Base of the ventricular part of the heart with the atria and great vessels removed.

the valve is clad with endothelium on its arterial surface and endocardium on its ventricular surface, between which is a stratum of fibrous tissue. When the valve closes, the nodules are closely opposed, the lunules are pressed together, and both nodules and lunules project upwards into the lumen of the pulmonary trunk.

The wall of the ventricle is lined with endocardium. In the infundibulum, which is formed from part of the embryonic bulbus cordis [p. 925], the wall is smooth, but elsewhere it is ridged by many inwardly projecting muscular columns called the **trabeculae carneae.** Some are simply ridges, while others are rounded bundles attached at each end to the wall of the ventricle but free in the middle. Many of this second group cross the cavity especially towards its apical part where they form a meshwork. One member of this class runs from the septum to the sternocostal wall, where it joins the base of the anterior papillary muscle [FIG. 1052]. It is called the **septomarginal trabecula** and serves to carry the right limb of the atrioventricular bundle to the anterior wall [p. 852]. The papillary muscles form a third type of trabecula. They

are attached by their bases to the ventricular wall; from their apices numerous chordae tendineae run to their insertion on the tricuspid valve cusps.

The **papillary muscles** of the right ventricle are: (1) a large **anterior** muscle from which the chordae pass to the anterior and posterior cusps of the valve; (2) a **posterior** muscle, often represented by two or more parts, from which chordae pass to the posterior and septal cusps; and (3) a variable group of small **septal** papillary muscles whose chordae run to the anterior and septal cusps. When the ventricle contracts, the papillary muscles do so also, and their distribution ensures that the cusp margins are held together and that the cusps themselves are not driven into the atrium.

The wall of the right ventricle is one-third the thickness of the left—in harmony with their respective tasks.

The Left Ventricle [PLATE 48C]

This ventricle is conical, and its apex forms the apex of the heart. At its base, the cavity communicates postero-inferiorly with the left atrium through the mitral valve, and with the

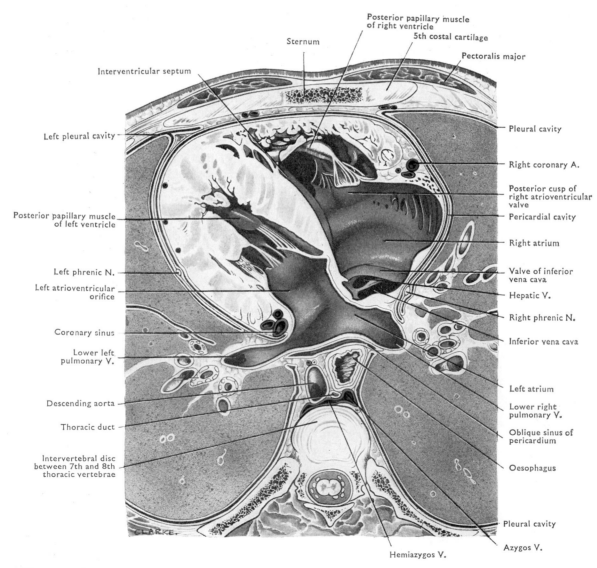

FIG. 1051   Horizontal section through the thorax at the level of the intervertebral disc between the seventh and eighth thoracic vertebrae.

PLATE 47

Plate from William Harvey's *Exercitatio Anatomica de Motu Cordis et Sanguinis in Animalibus* (1628).

These figures illustrate Harvey's simple, classical experiments that demonstrate the presence and function of the valves of the veins and the movement of the blood in the veins towards the heart (see p. 841).

PLATE 48

Cardio-angiography. Four X-rays showing two phases of the passage of a contrast medium through the heart after its introduction through the right basilic vein. (Venous angiocardiography.)

Aand B are anteroposterior and lateral views with right heart outlined.

C and D are also anteroposterior and lateral views with left heart outlined.

1. Right subclavian vein.	8. Site of right atrioventricular orifice.
2. Superior vena cava.	9. Aortic arch.
3. Right pulmonary artery.	10. Ascending aorta.
4. Right atrium.	11. Descending aorta.
5. Right ventricle.	12. Left ventricle.
6. Left pulmonary artery.	13. Left atrium.
7. Pulmonary trunk.	

PLATE 49

A

B

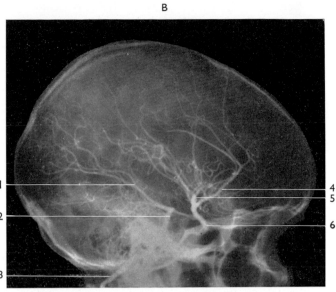

B. Lateral cerebral angiogram with filling of vertebral and carotid arteries.

1. Posterior cerebral artery.
2. Basilar artery.
3. Vertebral artery.
4. Anterior cerebral artery.
5. Middle cerebral artery.
6. Internal carotid artery ('carotid siphon').

A. Cerebral angiogram produced by injection of the left internal carotid artery. Filling of the contralateral arteries has occurred through the anterior communicating artery, the right internal carotid artery being occluded.

1. Both anterior cerebral arteries.
2. Middle cerebral artery branching on the insula.
3. Anterior cerebral artery.
4. Middle cerebral artery.
5. Bend of internal carotid artery ('carotid siphon').

C

D

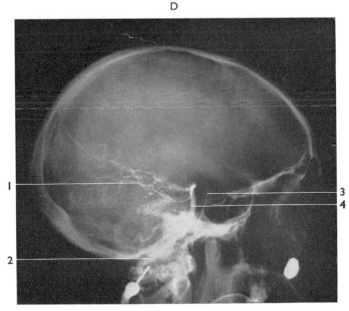

C. Lateral view of cerebral angiogram showing later phase than B with venous filling.

1. Superior sagittal sinus.
2. Superior anastomotic vein.
3. Great cerebral vein.
4. Superior cerebral vein.
5. Internal cerebral vein.

D. Lateral view of cerebral angiogram following injection of the vertebral artery. Note the needle entering the artery on the posterior arch of the atlas.

1. Posterior cerebral artery.
2. Vertebral artery.
3. Dorsum sellae.
4. Basilar artery.

PLATE 50

A

A. Arteriogram of the aorta and its branches.

1. Vertebral artery.
2. Common carotid artery.
3. Vertebral artery on the 7th cervical transverse process.
4. Brachiocephalic trunk.
5. Air in trachea.
6. Left subclavian artery.
7. Internal thoracic artery.
8. Arch of aorta.

B

B. Arteriogram to show abdominal aorta and its branches.

1. Hepatic artery.
2. Right renal artery.
3. 3rd lumbar artery.
4. Splenic artery.
5. Left renal artery.
6. Superior mesenteric artery with ileal and jejunal branches.

C

C. Abdominal venogram. Injection of the uterine veins with a contrast medium in a pregnant woman. The X-ray is taken with the patient lying supine and the inferior vena cava is obliterated by the weight of the uterus. Note the filling of the ascending veins which bypass this obstruction.

1. Internal vertebral venous plexus.
2. Common iliac vein.
3. Anastomosing sacral veins.
4. Internal iliac vein.
5. Uterine vein.
6. Ascending lumbar vein.

D

D. X-ray of injected specimen to show the arteries of the wrist and hand.

1. Ulnar artery.
2. Superficial palmar arterial arch.
3. Deep palmar arterial arch.
4. Radial component of superficial palmar arterial arch.
5. Radial artery.

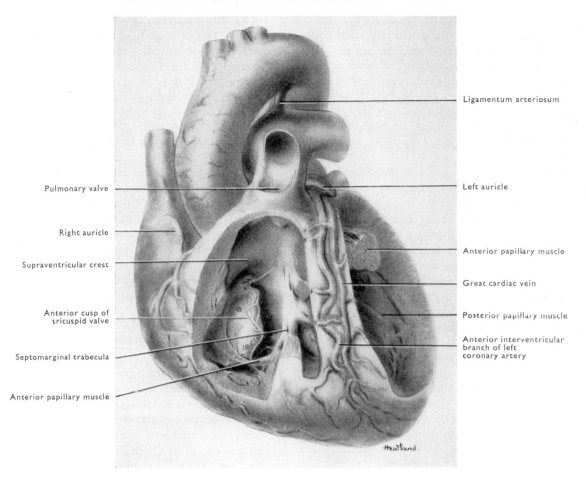

Pulmonary valve

Right auricle

Supraventricular crest

Anterior cusp of
tricuspid valve

Septomarginal trabecula

Anterior papillary muscle

Ligamentum arteriosum

Left auricle

Anterior papillary muscle

Great cardiac vein

Posterior papillary muscle

Anterior interventricular
branch of left
coronary artery

FIG. 1052 The ventricles opened from in front.

aorta anterosuperiorly and to the right. Inferiorly its surface is flattened and forms two-thirds of the diaphragmatic surface of the heart, but it is convex in front and on the left where it forms nearly one-third of the ventricular part of the sterno-costal surface and almost all of the left surface. The right or septal wall is concave towards the cavity [FIG. 1053]. The muscular wall is thickest around the widest part of the cavity, which is about one-quarter of its length from the base, and thinnest at the apex. But the upper, posterior part of the septum, which is membranous, is the thinnest region of the whole wall. **Interior of the Left Ventricle.** Whereas in the right ventricle the 'inflow' and 'outflow' orifices—tricuspid and pulmonary—are 2·5 cm. apart, the corresponding orifices in the left ventricle—mitral and aortic—are only separated by the anterior cusp of the mitral valve [FIGS. 1049 and 1054]. The part of the cavity which leads into the aorta is called the aortic vestibule; its tough fibrous wall is non-contractile.

The **left atrioventricular** (or mitral) **orifice** is oval, with its long axis running downwards, backwards, and to the right. It is guarded by the **left atrioventricular** or **mitral valve**, a bicuspid valve with triangular unequal cusps. The smaller **posterior cusp** is postero-inferior and to the left of the orifice; the larger **anterior cusp** is anterosuperior and to the right, between the mitral and aortic ori-

fices. The bases of the cusps are attached to the fibrous ring round the mitral orifice where they are either continuous with one another or separated by accessory cusps. According to Rusted, Scheifley, and Edwards (1952) accessory cusps are

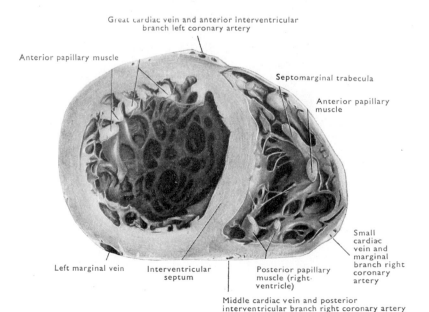

Great cardiac vein and anterior interventricular
branch left coronary artery

Anterior papillary muscle

Septomarginal trabecula

Anterior papillary muscle

Left marginal vein

Interventricular septum

Posterior papillary muscle (right ventricle)

Small cardiac vein and marginal branch right coronary artery

Middle cardiac vein and posterior interventricular branch right coronary artery

FIG. 1053 Transverse section of ventricles of heart.

seldom seen; these authors point out, however, that there is always junctional tissue between the two cusps of the valve over a depth of from 0·5 to 1 cm. The cusps are smooth on their atrial surfaces and their structure is the same as that of the tricuspid valve cusps; and like them their margins receive the insertions of chordae tendineae. The chordae do not extend far beyond the margin of the anterior cusp on its ventricular surface. This cusp is therefore smooth on both its surfaces over which blood flows into and out of the ventricle. Consult also Brock (1952, 1955).

The **aortic orifice** is circular and lies immediately above and in front of the mitral orifice from which it is separated by the anterior cusp of the mitral valve. It is guarded by the **aortic valve,** formed of three semilunar valvules, named, from their position, right, left, and posterior. The structure and attachments of the valvules of the aortic valve are similar to those of the valvules of the pulmonary valve [p. 847].

The walls of the left ventricle are lined with endocardium. Many fine **trabeculae carneae** form a meshwork which is especially rich at the apex and on the inferior wall, but the septum and upper part of the sternocostal wall are relatively smooth. There are two **papillary muscles,** an **anterior** arising from the sternocostal wall and a **posterior** from the inferior wall, and each is connected by chordae tendineae with both cusps of the mitral valve. Fine strands, which may be formed entirely of branches of the left limb of the atrioventricular bundle, are often found crossing the cavity from the septum to the papillary muscles [Fig. 1054].

The **interventricular septum** extends from the right of the apex of the heart to the interval that separates the pulmonary and tricuspid orifices from the aortic and mitral orifices. Its borders correspond with the anterior and posterior interventricular grooves. It is so placed that one surface looks forwards and to the right and bulges into the right ventricle, whilst the other looks backwards and to the left and is concave towards the left ventricle. Almost the whole of the septum is thick and muscular, but its upper posterior part is fibrous and not more than 1 mm. in thickness; it is called the **membranous part** of the interventricular septum and shows as a small oval transparency when held to the light. From the left side it is seen to lie just below the angle between the right and posterior valvules of the aortic valve; its right side is divided by the upper part of the attached border of the septal cusp of the tricuspid valve into anterior and posterior parts. The anterior part separates the right ventricle from the aortic vestibule, whereas the posterior part separates the lower part of the right atrium from the aortic vestibule.

THE STRUCTURE OF THE HEART

The heart consists mainly of muscle, called the myocardium, which is enclosed between the epicardium, or visceral layer of the pericardium, and the endocardium.

The **epicardium** consists of white fibrous and elastic tissue which is covered with flat polygonal mesothelial cells on its free surface.

The **endocardium** is continuous with the inner coats of the vessels. Much thinner than the epicardium, it consists of a lining of endothelial cells, a layer of fine areolar tissue, and an external elastic layer usually in the form of a fenestrated membrane.

The fibres of the **myocardium** differ from those of voluntary striated muscle in several ways: they are less distinctly striated; they are made up of short, cylindrical segments joined end to end by transverse discs; the nuclei lie centrally, one (seldom two) in each segment or cell; the sarcolemma is finer than in voluntary muscle fibres; finally, the cells of heart muscle give off branches which, uniting with branches of neighbouring cells, convert the whole into a network [Fig. 329].

Immediately subjacent to the subendocardial tissue are found the muscle **fibres of Purkinje.** First observed by Purkinje (1845) these fibres are part of the conducting system of the heart and in many animals are of considerably larger diameter than ordinary myocardial fibres; the central portion of each consists of granular protoplasm in which are embedded two or more nuclei, and the peripheral portion is transversely striated. Purkinje fibres of much smaller calibre

Fig. 1054 Cavity of left ventricle of heart.
The greater part of the pulmonary trunk has been removed, and the aorta opened to show the aortic valves.

also occur (Kugler and Parkin, 1956). In the sheep and ox heart, typical large Purkinje fibres are readily observed, but in the human heart they are much less plentiful.

The network of cardiac muscle fibres forms sheets and strands which have a more or less characteristic and definite arrangement in different parts of the heart.

In the **atria** the muscular fasciculi fall naturally into two groups : (1) superficial fibres common to both atria ; (2) deep fibres peculiar to each atrium.

The superficial fibres run transversely across the atria and a few of them dip into the atrial septum.

The deep fibres are : (1) looped fibres ; the ends of these are attached to the fibrous rings around the atrioventricular orifices, and the fibres pass anteroposteriorly over the atria ; (2) anular fibres which surround (i) the extremities of the large vessels which open into the atria, (ii) the auricles, and (iii) the fossa ovalis. Cardiac muscle fibres extend for variable distances along the great veins entering the atria.

In the **ventricles** the muscular fasciculi form more or less definite V-shaped loops which begin and end at the fibrous rings at the bases of the ventricles. These loops embrace the cavities of either one or both ventricles, one limb of each loop lying on the outer surface of the heart and the other in the interior.

The superficial fibres on the sternocostal surface pass towards the left, those on the inferior surface towards the right. At the apex all are coiled into a whorl or vortex through which they pass into the interior of the ventricular walls and run towards the base, some in the septum and others to end in the papillary muscles. The arrangement is exceedingly complex and for further details reference should be made to Mall (1911), and Thomas (1957).

The heart also possesses a fibrous skeleton which strengthens its orifices. It consists of:

1. **Fibrous rings** which surround the atrioventricular orifices.

2. Short tubular fibrous zones at the roots of the aorta and pulmonary trunk to which are attached, proximally, some of the ventricular muscle fibres, and distally, the arterial walls.

3. A mass of tough fibrous tissue, in which there may be some cartilage, which lies between the aortic root in front and the two fibrous rings : it is called the **right fibrous trigone.**

4. The **left fibrous trigone** which is a smaller mass joining the left side of the aortic root to the front of the left atrioventricular ring [FIG. 1049].

5. The membranous part of the septum which is continuous above and behind with the right fibrous trigone.

The valves of the heart consist of a thin layer of fibrous tissue, continuous with the fibrous rings that surround the orifices and covered by a reduplication of the lining membrane.

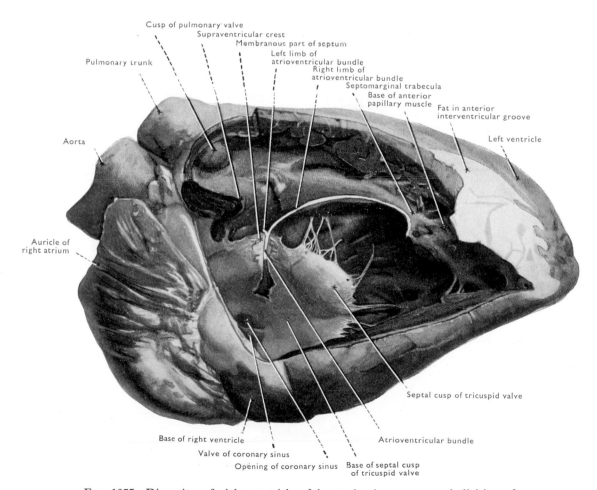

FIG. 1055 Dissection of right ventricle of heart showing course and division of atrioventricular bundle.

Atrial muscle fibres extend into the base of the atrioventricular valves, but the associated capillary vessels pass into the valves for a distance of not more than 3 mm. in health (Dow and Harper, 1932). Otherwise, these valves, like the pulmonary and aortic, are avascular.

CONDUCTING SYSTEM OF HEART

From the account, given above, of the fibrous skeleton of the heart, it would appear that the atrial and ventricular muscle fibres are everywhere effectively separated. This, however, is not the case, for there is connecting them a bundle of muscle fibres, called the **atrioventricular bundle,** which traverses the right fibrous trigone. The bundle forms part of the conducting system of the heart which is composed of muscle fibres specially differentiated for their task of initiating the sequence of events in the cardiac cycle, of controlling its regularity, and of transmitting the impulses from atria to ventricles.

The parts of which the conducting system is composed are these :

The Sinu-atrial Node. This structure, first described by Keith and Flack (1907), is commonly called the pacemaker of the heart. It is situated in the wall of the right atrium in the upper part of the sulcus terminalis from which it extends over the front of the opening of the superior vena cava. The nodal fibres, of smaller calibre than the neighbouring atrial fibres with which they are continuous, are arranged in a meshwork intermingled with which there is a considerable quantity of connective tissue.

The Atrioventricular Node. First described by Tawara (1906), this structure lies in the atrial septum just above the opening of the coronary sinus. Its fibres form a dense meshwork which establishes continuity with the surrounding atrial muscle fibres. There is no special connexion between the two nodes, the impulse spreading from the pacemaker through the atrial musculature.

The Atrioventricular Bundle. Known also as the bundle of His (1893), this slender strand springs from the atrioventricular node and runs forwards through the right fibrous trigone to reach the posterior border of the membranous part of the interventricular septum. Continuing forwards under cover of the septal cusp of the tricuspid valve, it runs along the lower border of the membranous septum before dividing into right and left limbs which straddle the upper border of the muscular septum [FIGS. 1055 and 1056]. The bundle is about 1 cm. long and 2 mm. in diameter.

The **right limb** continues on the right side of the muscular part of the septum, where it is covered by the endocardium and a thin layer of myocardium which, however, may be absent. From the septum the right limb passes in the septomarginal trabecula to the base of the anterior papillary muscle.

The **left limb** of the bundle descends immediately beneath the endocardium on the left side of the muscular interventricular septum, as a broad thin band which usually divides into two branches about half-way down the septum. These soon divide into lesser branches which pass in the trabeculae carneae to the papillary muscles. Strands of the left limb or its branches sometimes cross the cavity of the ventricle to reach the papillary muscles, forming as they do the so-called 'false tendons' [FIG. 1054].

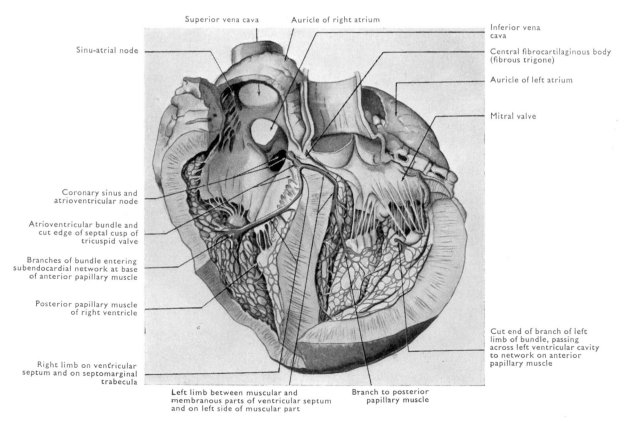

FIG. 1056 Conducting system of sheep's heart; from dissected and injected specimens.

In many animals the bundle and its divisions are composed of large Purkinje fibres, but in the human heart such fibres do not appear until some way down the limbs; indeed, the fibres which compose the bundle and the upper parts of its limbs in Man are no larger than those of the ordinary ventricular myocardium.

In the heart of an ungulate the bundle and its branches are readily dissected, but in the human heart they are less easily displayed (Walls, 1945).

The **terminal subendocardial network** (Purkinje) is continuous with the branches of the bundle at the bases of the papillary muscles from which it spreads out beneath the endocardium of the greater part of both ventricles. The subendocardial network and the course of the bundle and its branches may be shown by injecting suitable coloured liquid into the sheath which surrounds the whole system. This method gives good results on the ox or sheep heart [FIG. 1056], but it is unsuccessful in the human heart because of the extreme thinness of the sheath.

When the atrioventricular bundle is severed, experimentally or by disease, the ventricles contract independently of the atria, and with a much slower rhythm. This condition is known as 'heart block'. However, since the bundle contains nerves, it has been argued that the impulse for cardiac contraction might well be carried by these nerves rather than by the bundle itself. Evidence that conduction throughout the bundle and its limbs is indeed myogenic has been advanced by Davies *et al.* (1956).

THE SIZE OF THE HEART

The heart in health is roughly the size of the individual's closed fist. It is about 12·5 cm. long, 8·5 cm. broad, and its greatest depth from sternocostal to diaphragmatic surfaces is 6 cm.

Weight. At birth the heart weighs 20–25 g., in the adult male the average is 310 g. and in the adult female 255 g.

The increase in weight with age is rapid up to the seventh year, then more gradual to the age of puberty when a second acceleration sets in. The gradual increase in adult heart weight commonly observed with increasing age is probably more to be linked with increasing body weight than advancing years (Gray and Mahan, 1943).

The several parts of the heart also vary in their relation to one another at different periods of life. This is most marked in the ventricles which, at one stage of foetal development, are of equal bulk. After birth the left ventricle grows rapidly, while the right, after an initial reduction in absolute weight, grows more slowly. After the age of 6 months the proportion of the right ventricle to the left remains unaltered throughout infancy, adolescence, and normal adult life (Keen, 1955).

Capacity. Although after death the cavity of the right ventricle appears larger than that of the left, during life their capacities are probably the same— 90–120 ml. The atria are slightly less capacious.

THE BLOOD SUPPLY OF THE HEART

The walls of the heart are supplied by the coronary arteries [p. 861], the capillaries forming a close-meshed network around the muscular fibres.

In most cases the sinu-atrial and atrioventricular nodes, and the bundle, are supplied by the right coronary artery. The limbs of the bundle derive their supply from both coronary arteries.

The veins of the heart are described on page 903.

Lymph Vessels of Heart. These are described on page 955.

NERVES OF HEART

The heart receives its nerves from the superficial and deep cardiac plexuses [p. 788]. These are formed by branches from the vagi and sympathetic trunks, all of which—with one exception—contain both efferent and afferent fibres; the exception is the superior cervical cardiac branch of the sympathetic which, on both sides, consists solely of efferent (postganglionic) fibres.

The cardiac plexuses send small branches to the heart along the great arteries and veins, and continue as the right and left coronary plexuses which innervate these arteries and their territories of supply.

The vagal efferent fibres (parasympathetic) exert restraint

FIG. 1057   The nerve supply of the heart (diagrammatic).

Afferent fibres are represented by interrupted lines. The superior, middle, and inferior cervical ganglia of the sympathetic trunk are marked S, M, and I, the first five thoracic ganglia i–v. The third thoracic segment of the spinal medulla is shown in section. V.G., inferior vagal ganglion.

on the heart rate. They emerge from the brain stem in the roots of the vagus and accessory nerves, and run as preganglionic fibres to the ganglia present in the cardiac plexuses, and in the heart itself [see below] where they synapse.

The vagal afferent fibres have their cell bodies in the inferior vagal ganglion; they carry impulses from the heart, **arch of the aorta,** and the roots of the great veins, which bring about reflex adjustment of the heart rate.

The sympathetic (preganglionic) fibres emerge from the spinal medulla in the upper four or five thoracic nerves and pass in the white rami to the corresponding ganglia of the sympathetic trunk [FIG. 1057]. Some of the fibres synapse in these ganglia, from which postganglionic fibres then proceed as small cardiac branches to the cardiac plexuses; most of the preganglionic fibres, however, ascend in the sympathetic trunk to relay in the cervical sympathetic ganglia, from which postganglionic branches run to the cardiac plexuses [p. 788].

Afferent fibres, whose cell bodies are in the spinal ganglia of the upper four or five thoracic spinal nerves, travel in all the sympathetic cardiac nerves except the superior cervical. They subserve pain sensibility, and the severe pain of angina pectoris may be greatly relieved by surgical removal of the related sympathetic ganglia.

Ganglion cells are widely distributed in the atria, including the regions close to the sinu-atrial and atrioventricular nodes, but appear to be absent from the ventricles (King and Coakley, 1958). Reports on the types of nerve ending found in the heart are contradictory, and Davies, Francis, and King (1952) should be consulted. These authors found no endings resembling the motor endings or muscle spindles of skeletal muscle. However, they did find non-medullated fibres ending in spiral formations around individual muscle fibres; medullated fibres were present in the epicardium, endocardium, and myocardium, but all lost their myelin sheaths some distance from their terminations.

THE SURFACE ANATOMY OF THE HEART
[PLATES 34 and 36]

Several factors affect the position of the heart in health. The first is general body build, or habitus, with which the level of the diaphragm is commonly associated. In thick-set subjects the diaphragm is frequently at a higher level than in those of slender build, and in consequence the heart lies more horizontally in the first-mentioned group and more vertically in the second. Bodily build apart, there are factors which modify the position of the heart in any individual. Thus the beating of the heart itself, respiratory movements, and the position of the body all play their part; moreover, as the barrel-shaped, high-set thorax of infancy gradually assumes adult proportions, the heart comes to lie less horizontally, while undergoing a relative reduction in size.

It is therefore not possible to give a surface marking for the heart which is of general application.

In the *adult cadaver* the outline of the heart is shown in FIGURES 1046, 1050, and 1058C, and may be drawn on the surface by connecting the following points: (1) at the lower border of the second left costal cartilage 3·5 cm. from the median plane; (2) at the upper border of the third right cartilage 2·5 cm. from the median plane; (3) on the sixth right

cartilage 2·5 cm. from the median plane; (4) over the fifth left intercostal space about 9 cm. from the median plane, medial to the midclavicular line. The upper border is straight, and corresponds approximately to the upper level of the atria; the right border is gently convex to the right; the lower border is almost horizontal, and passes behind the xiphisternal joint; the left border is convex to the left and upwards.

The orifices of the valves may be represented on the surface by lines corresponding to those shown in FIGURE 1050. Those for the pulmonary and aortic valves should be 2·5 cm. in length, that for the mitral valve 3 cm., and for the tricuspid valve 4 cm.

The Position of the Heart during Life

While traditional methods of examination such as inspection, palpation, and percussion unquestionably yield useful information, it cannot be disputed that radiology has provided the most exact knowledge of the position of the heart in the living subject. Using radiographs corrected for dispersion of X-rays, Mainland and Gordon (1941) investigated heart position in young adult males, standing and in the midphase of respiration. Their findings of the average relationships of the heart outline to the anterior thoracic wall are shown in FIGURE 1058. Certain points call for special comment.

The lower border of the heart is just over 5 cm. below the level of the xiphisternal joint, so that more than a third of the vertical cardiac height lies below the joint. The upper border of the heart lies a little above the middle of the body of the sternum, and slopes slightly downwards from left to right.

When lying supine the upper border is about a rib's width

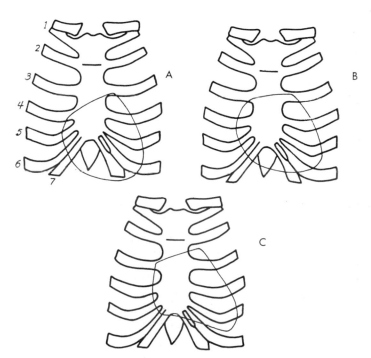

FIG. 1058 Average relationships of heart outline to anterior thoracic wall.

The upper two diagrams (A, living erect; B, living recumbent) are derived from radiographs corrected for dispersion of X-rays. Diagram C represents a common textbook statement. (After Mainland (1946) with kind permission.)

higher, and rather less than a third of the vertical cardiac height lies below the xiphisternal joint.

The range of variation between individuals is considerable. **Heart Sounds.** The sounds produced by the individual valves are heard best, not directly over their anatomical situation, but over the area where the cavity in which the valve lies is nearest the surface. Hence the mitral sound is best heard over the apex (**mitral area**) which, though most often in the fifth left intercostal space about 9 cm. from the midline, ranges from the fourth space to the sixth costal cartilage and from 6·5 to 10 cm. from the midline; the **tricuspid area** is over the lower part of the body of the sternum or the xiphoid process; the **aortic area** is close to the sternum over the second right costal cartilage or second interspace; and the **pulmonary area** is also near the sternum over the third left costal cartilage or second left interspace.

THE PERICARDIUM

The pericardium is the fibroserous sac which surrounds the heart in the middle mediastinum. It consists of a strong, conical, fibrous sac called the fibrous pericardium, within which is the serous pericardium arranged as shown schematically in FIGURE 1059.

The **fibrous pericardium:** inferiorly it is attached to the central tendon and to the adjacent muscular substance of the diaphragm, and is pierced by the inferior vena cava. Above, and behind, it becomes continuous with the adventitia of the great vessels which pierce it on their way to and from the heart. Thus it sheaths the aorta, the two branches of the pulmonary trunk, the superior vena cava, the four pulmonary veins, and the ligamentum arteriosum.

Its anterior surface forms the posterior boundary of the anterior mediastinum, and it is attached to the sternum, above and below, by the **sternopericardial ligaments.** In most of its extent it is separated from the thoracic wall by the anterior parts of the lungs and pleural sacs, but it is in contact with the left half of the lower portion of the body of the sternum and

often with the medial ends of the left fourth, fifth, and sixth costal cartilages and the left transversus thoracis muscle. The extent of its contact with the thymus decreases with age [FIGS. 692 and 693].

Its posterior surface forms the anterior boundary of the upper part of the posterior mediastinum, and each lateral surface is in contact with the mediastinal pleura, the phrenic nerve and its accompanying vessels intervening [FIG. 1051].

The inner surface of the fibrous sac is closely adherent to the parietal part of the serous pericardium.

The **serous pericardium** is a transparent membrane which forms a closed sac containing a little fluid, and essentially forms a large bursa which facilitates cardiac movement. It is invested by the fibrous pericardium and is itself invaginated by the heart in the manner of a finger pushed into a balloon. It is therefore described as having two parts—the parietal, which lines the fibrous pericardium, and the visceral (or epicardium), which partially ensheathes the heart and great vessels; but the two parts are continuous where the great vessels pierce the fibrous pericardium. By reference to FIGURE 1060 it will be seen that there are two such sites of continuity: (1) at the arterial end of the heart where the aorta and pulmonary trunk are enclosed in one tube of the visceral layer; and (2) at the venous end where the two caval and four pulmonary veins are enclosed in a second tube; but whereas

FIG. 1059 Diagram of the pericardial sac. The serous pericardium (interrupted line) covers the heart and lines the fibrous pericardium. P.C., pericardial cavity; D, diaphragm.

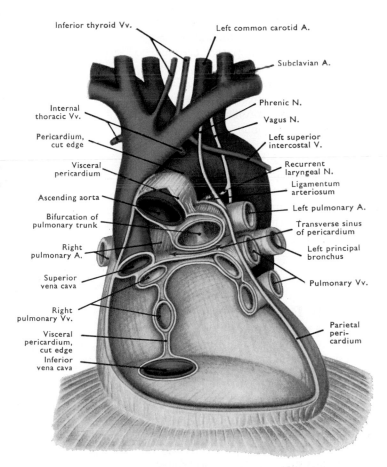

Inferior thyroid Vv.
Left common carotid A.
Subclavian A.
Internal thoracic Vv.
Phrenic N.
Vagus N.
Left superior intercostal V.
Pericardium, cut edge
Recurrent laryngeal N.
Visceral pericardium
Ligamentum arteriosum
Ascending aorta
Left pulmonary A.
Bifurcation of pulmonary trunk
Transverse sinus of pericardium
Right pulmonary A.
Left principal bronchus
Superior vena cava
Pulmonary Vv.
Right pulmonary Vv.
Visceral pericardium, cut edge
Inferior vena cava
Parietal pericardium

FIG. 1060 The pericardium and great vessels after removal of the heart. The arrow lies in the transverse sinus of the pericardium, and the posterior wall of the oblique sinus lies between the right and left pulmonary veins.

the investment of the visceral layer around the great arteries is of a simple tubular form, that round the six veins is shaped like an inverted U with unequal limbs.

When the pericardial sac is opened from the front it is possible to pass a finger behind the aorta and pulmonary trunk, and in front of the atria, through the **transverse sinus** of the pericardium. This passage appears early in development as a result of the breakdown of the dorsal mesocardium which suspends the simple tubular heart within the sac of the serous pericardium: this breakdown affords greater freedom for growth and movement, and in the fully developed heart the sinus is said to serve as a bursa, containing pericardial fluid, between the pulsating arteries in front and the contracting atria behind.

The **oblique sinus** of the pericardium [FIG. 1060] forms a cul-de-sac bounded by the serous pericardium surrounding the pulmonary and caval veins. It owes its shape to the changes which attend the absorption of the sinus venosus and the pulmonary veins into the atria. The oblique sinus lies behind the left atrium, and in front of the oesophagus and the descending thoracic aorta.

A small fold of the serous pericardium, called the **fold of the left vena cava,** passes from the left pulmonary artery to the upper left pulmonary vein, behind the left end of the transverse sinus. It encloses a fibrous strand, called the ligament of the left vena cava, a remnant of the left superior vena cava—or common cardinal vein—which atrophies early in foetal life. From the lower end of the ligament the oblique vein of the left atrium descends to the coronary sinus; it represents the continuation of the left common cardinal vein [p. 935].

Structure. The fibrous pericardium is a dense unyielding fibrous membrane. The serous pericardium is covered on its inner surface by a layer of flat mesothelial cells which rests upon a basis of mixed white and elastic fibres in which run numerous blood vessels, lymph vessels, and nerves.

THE PULMONARY CIRCULATION

THE PULMONARY TRUNK

[PLATE 48A and B]

This vessel arises from the infundibulum of the right ventricle, and passes upwards and backwards from the front of the ascending aorta to its left side. It is slightly dilated, immediately above the valvules of the pulmonary valve, into three shallow **sinuses.** After a course of just over 5 cm. it ends,

within the concavity of the aortic arch, by dividing into right and left pulmonary arteries at the level of the sternal end of the second left costal cartilage.

Within the fibrous pericardium it is enveloped, along with the ascending aorta, in a common sheath of the visceral layer of the serous pericardium [FIG. 1060]. It lies behind the sternal end of the second left intercostal space, from which it is separated by the anterior margins of the left lung and pleural sac, and by the pericardium. Posterior to it are first the ascending aorta, and then the left atrium [FIGS. 1045 and 1049]. To its right are the auricle of the right atrium and the ascending aorta, and to its left the auricle of the left atrium. Between its bifurcation and the aortic arch, is the superficial part of the cardiac plexus of nerves. Both coronary arteries are related to its root—the right artery on its right, the left at first behind and then on its left.

The **right pulmonary artery,** longer and wider than the left, commences in front of the oesophagus and below the right principal bronchus [FIG. 1062]; it passes behind the ascending aorta, superior vena cava, and upper right pulmonary

Internal thoracic vessels

Anterior papillary M.

Infundibulum

Left pulmonary semilunar valvule

Left coronary A.

Left phrenic N.

Left auricle

Left principal bronchus

Left vagus N.

Descending aorta

Accessory hemiazygos V.

Right auricle

Orifice of right coronary A.

Ascending aorta

Superior vena cava

Right phrenic N.

Right pulmonary A.

Right principal bronchus

Right vagus N.

Tracheobronchial lymph nodes

Oesophagus

Vena azygos

Thoracic duct

6th rib

FIG. 1061 Horizontal section between the fifth and sixth thoracic vertebrae.

vein, and in front of the right bronchus immediately below its superior lobar branch to reach the hilus of the lung.

The **left pulmonary artery** passes laterally and backwards in front of the left principal bronchus and descending aorta [FIG. 1061]. Above are, the aortic arch—to which it is connected by the ligamentum arteriosum [p. 937]—and the left recurrent laryngeal nerve. Before entering the lung it is crossed anteriorly by the upper left pulmonary vein.

THE PULMONARY VEINS

The tributaries of the pulmonary veins arise in the capillary plexuses of the alveoli, and by the union of the smaller veins, eventually one vein, formed by tributaries which correspond more or less to the bronchopulmonary segments (though many are intersegmental), passes from each lobe into the root of the lung. Thus are formed five pulmonary veins, but those from the upper and middle lobes of the right lung usually unite on entering the lung root, and so only four pulmonary veins open into the left atrium [FIGS. 1044, 1051, and 1060]. Neither the main stems nor their tributaries possess valves.

In the root of each lung the upper pulmonary vein lies below and in front of the pulmonary artery [FIGS. 605, 606]. The lower pulmonary vein is below and posterior to the upper vein.

On the right side the upper pulmonary vein passes behind the superior vena cava, and the lower behind the right atrium ; both end in the upper and posterior part of the left atrium close to the atrial septum. On the left side both veins cross in front of the descending aorta, to the upper, posterior part of the left atrium near its left border.

All four veins pierce the fibrous layer of the pericardium, and then receive partial coverings of the serous layer.

Fusion of the left veins into a single trunk before entry is not uncommon ; on the other hand, there may be separate openings for the veins from all three lobes of the right lung.

The Intrapulmonary Course of the Pulmonary Vessels. In general, the branches of the pulmonary arteries accompany the bronchi, divide with them, and lie on their posterosuperior aspect. The veins behave differently : peripherally, they run apart from the arteries, but as larger veins are formed they join the arteries and bronchi and run with them to the lung root. They lie on the antero-inferior aspects of the corresponding bronchi.

Modern thoracic surgery demands a detailed knowledge of intrapulmonary vascular anatomy which is, in fact, variable (Boyden, 1955).

FIG. 1062   Dissection of the bronchi and pulmonary vessels.

THE SYSTEMIC CIRCULATION

ARTERIES

The systemic arteries all stem from the aorta, and by branching and rebranching form what is sometimes spoken of as the 'arterial tree'.

The situation of the larger arteries in the trunk and limbs affords them protection from injury. Thus the main artery of each limb runs distally on the flexor aspect; accordingly, as the limb is bent, the vessel is increasingly safeguarded. Further, its position lessens the tension exerted upon it during movements of the limb joints. Most main arteries pursue a more or less straight or evenly curved course, but some, for functional reasons, are very tortuous; for example, the facial artery which has to adapt itself to the movements of the mandible and of the facial muscles, and the uterine artery—probably in relation to necessary enlargement during pregnancy. The splenic artery, which runs behind the stomach, is another vessel whose tortuosity is linked with mechanical considerations, but such a direct association of tortuosity and adaptation to changing position is not always apparent.

The cross-sectional area of the 'arterial tree' increases steadily towards the periphery, and the combined sectional area of all the arterioles greatly exceeds that of the aorta—a fact of profound haemodynamic importance. The manner in which arteries divide varies. The usual arrangement is illustrated by the arteries of the limbs, which give off successive branches yet retain their individuality as main vessels. Some arteries—the aorta is the most notable—end by dividing into two more or less equal branches; in some cases several arteries arise together from a short parent branch, e.g., the coeliac trunk.

Functional Distribution. Some arteries, such as the pulmonary and renal, clearly have a functional distribution, but Hilton (1863) considered the distribution of other arteries, not confined to single organs, to have a general functional significance. For example, he named the maxillary artery the 'masticatory artery' because of its distribution to the bones, teeth, and muscles of mastication. The subclavian, apart from its continuation to the upper limb, he considered on an analysis of the distribution of its branches to be essentially a 'respiratory artery', with the vertebral artery giving supply to the respiratory nerve centres, the internal thoracic to the diaphragm and phrenic nerve, etc. Hilton's views are attractive, but it has to be recognized that structures which are linked functionally do tend to be in natural proximity.

Anastomoses. The great majority of arteries communicate with others through branches that unite to form anastomoses. In some regions anastomosis occurs directly between arteries of considerable size, so that there is—potentially—an even distribution of blood to the parts supplied, e.g., the formation of the 'arterial arcades' in the supply of the intestines [FIG. 1093]. At the base of the brain the circulus arteriosus is a remarkable example of such an anastomosis [p. 875]. But the most widespread anastomoses take place between the smaller arteries, and as a rule those between arterioles form extensive

networks from which the capillaries arise. Around the limb joints the anastomoses between medium-sized arteries are of clinical importance as the means by which the blood supply is maintained distally by the opening up of a **collateral circulation** should the main artery be blocked by ligature or disease. Quiring (1949) provides a survey of the chief anastomoses, both arterial and venous, in the body.

Arteriovenous Anastomoses. In certain regions the small arteries and veins are connected directly, as well as through the capillary bed. Such direct connexions are called arteriovenous anastomoses and they consist of thick-walled, heavily innervated, contractile vessels. When they are fully closed all the blood reaching the site passes through the capillaries, but as they become open they will by-pass the capillary circulation partially or completely. The best-known are the complex anastomoses which form the basis of the 'glomera' found in the skin of the palmar surface of the digits [FIG. 1063], but they have also been found in the skin of the lips, ears, and nose, and in the nasal mucous membrane [p. 473]. All these sites are subjected to temperature changes, and there is evidence that arteriovenous anastomoses are concerned *inter alia* with the regulation of body temperature. Their occurrence in the mucous membrane of the stomach and intestine is related to the intermittent activity of these parts, periods of digestion and absorption being accompanied by circulation through the capillary bed (when the anastomoses are closed), and periods of fasting by circulation through the anastomoses which are then open. For a general review of the subject Clark (1938) should be consulted; the part played by the anastomoses in temperature regulation is discussed by Prichard and Daniel (1956).

End Arteries. Small arteries which do not anastomose with others and therefore (apart from capillary communications) constitute the sole source of blood to the areas they supply are called end arteries; they are of pathological importance since

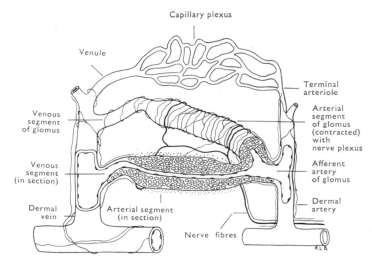

FIG. 1063 Arteriovenous anastomosis in the skin of the hand. (After Masson, 1937.)

occlusion of such an artery causes death of the tissues it supplies. End arteries are found in the kidney [p. 511] and spleen [p. 966]; and in the case of the blood supply of the spinal medulla and brain, although communications do exist, the smaller vessels probably act as 'functional end arteries' owing to the difficulty of establishing an effective collateral circulation. One very important end artery is the central artery of the retina [p. 871], since its occlusion causes blindness.

THE AORTA

[PLATE 50]

The aorta is the main arterial trunk of the systemic circulation, and for descriptive purposes it is divided into the **ascending aorta,** the **arch of the aorta,** and the **descending aorta,** the last being further divided into **thoracic** and **abdominal** parts.

THE ASCENDING AORTA

The ascending aorta, about 5 cm. long and 2·5 cm. in diameter, springs from the base of the left ventricle, behind the left half of the sternum, at the level of the lower border of the third costal cartilage. Passing upwards, forwards, and to the right, it becomes the arch of the aorta behind the sternal angle and part of the second right costal cartilage. Its root shows three bulgings, called the **aortic sinuses,** each opposite a

valvule of the aortic valve; and frequently the right side of the ascending aorta and the beginning of the arch show a slight dilatation which is a common site of aneurysm (an aneurysm is a pathological dilatation of an artery, of limited extent).

Relations. The ascending aorta is enclosed in the fibrous pericardium, and is enveloped, together with the pulmonary trunk, in a tube of serous pericardium. It lies at first behind the infundibulum and pulmonary trunk, but its upper part is nearer the sternum from which it is separated by the pericardium, the anterior margins of the right lung and pleura, and the remains of the thymus. Posteriorly it rests successively on the transverse sinus of the pericardium and left atrium, the right pulmonary artery, and the right bronchus. On its left side are the left atrium below, and the pulmonary trunk above. On its right side are first the right atrium whose auricle overlaps it, and higher up the superior vena cava which it overlaps to a varying degree.

Branches. The right coronary artery springs from the right sinus of the aorta, and the left coronary artery from the left sinus [FIG. 1049 and p. 861].

THE ARCH OF THE AORTA

The arch of the aorta lies in the superior mediastinum opposite the lower half of the manubrium sterni. It first runs upwards, backwards, and to the left; then backwards towards the left side of the body of the fourth thoracic vertebra; finally, turning downwards, it becomes continuous with the descending aorta at the lower border of that vertebra [FIG. 1064].

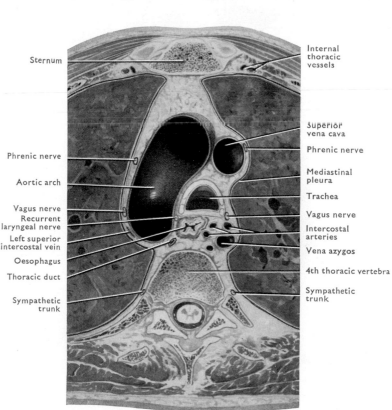

Sternum

Internal
thoracic
vessels

Phrenic nerve

Aortic arch

Vagus nerve
Recurrent
laryngeal nerve
Left superior
intercostal vein

Oesophagus

Thoracic duct

Sympathetic
trunk

Superior
vena cava

Phrenic nerve

Mediastinal
pleura

Trachea

Vagus nerve

Intercostal
arteries

Vena azygos

4th thoracic vertebra

Sympathetic
trunk

FIG. 1064 Transverse section through superior mediastinum at level of fourth thoracic vertebra.

Over-all the arch forms an upward convexity [FIG. 1066], and lies more in a sagittal than a coronal plane.

Relations. Anterior to the arch is the remains of the thymus gland, overlapped by the lungs and pleurae. On its left the following structures lie between it and the pleura: (1) four nerves, arranged from before backwards in this order: the left phrenic, the inferior cervical cardiac branch of the left vagus, the superior cervical cardiac branch of the left sympathetic, and the trunk of the left vagus; (2) the left superior intercostal vein which crosses it obliquely, superficial to the left vagus nerve and deep to the left phrenic nerve. Posteriorly and on the right of the arch [FIG. 1064] there are, from before backwards, the deep cardiac plexus, the trachea, the left recurrent laryngeal nerve, the left border of the oesophagus, the thoracic duct, and the vertebral column. Above are its three large branches—the brachiocephalic trunk, the left common carotid, and the left subclavian arteries; and crossing anterior to them is the left brachiocephalic vein. Below the arch lie the bifurcation of the pulmonary trunk, the ligamentum arteriosum [p. 937], the root of the left lung, and the left recurrent laryngeal nerve. The latter curves medially behind the ligament from the left vagus, and ascends medial to the arch. The superficial cardiac plexus lies in front and to the right of the ligament.

Branches. The branches of the arch of the aorta are described on page 863.

Surface Anatomy. The ascending aorta, the arch of the aorta, and the pulmonary trunk may be indicated on the surface as shown in FIGURE 1050.

Variations. As a rule, the summit of the arch of the aorta reaches half-way up the manubrium, but it may reach the level of the upper border, or only as high as the sternal angle.

THE DESCENDING AORTA

THORACIC AORTA

This part of the vessel lies in the posterior mediastinum [FIGS. 1061 and 1066]; it extends to the aortic opening in

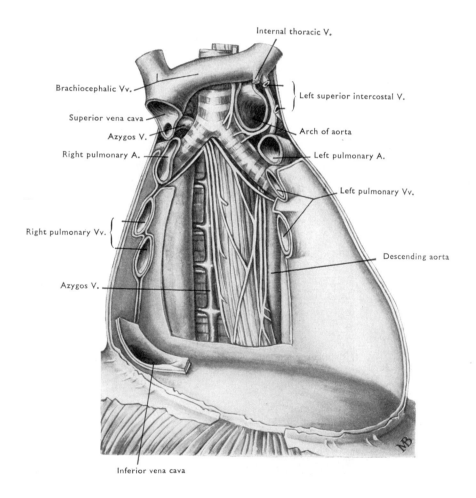

Internal thoracic V.

Brachiocephalic Vv.

Left superior intercostal V.

Superior vena cava

Azygos V.

Arch of aorta

Right pulmonary A.

Left pulmonary A.

Left pulmonary Vv.

Right pulmonary Vv.

Descending aorta

Azygos V.

Inferior vena cava

FIG. 1065   Dissection of the upper part of the posterior mediastinum after removal of the heart and posterior wall of the pericardium.

the diaphragm, where, opposite the middle of the lower border of the twelfth thoracic vertebra, it become continuous with the abdominal aorta.

Relations. Posteriorly it lies on the vertebral column and on the hemiazygos and accessory hemiazygos veins as they pass to join the azygos vein. Anteriorly, from above downwards, are the root of the left lung, the pericardium over the left atrium, the oesophagus (which with its plexus of nerves crosses it from right to left) and the diaphragm. Along the whole length of the right side of the thoracic aorta are the thoracic duct and the azygos vein, and in its lower part the right lung and pleura; while the left lung and pleura are on its left side.

Branches. The branches of the thoracic aorta are described on page 884.

Surface Anatomy. The thoracic aorta may be indicated by a band, 2·5 cm. wide, centred above on the second left chondrosternal junction and below on the midline one finger's-breadth above the transpyloric plane. For variations see page 885.

THE ABDOMINAL AORTA

[PLATE 50B]

This vessel extends to the body of the fourth lumbar vertebra, where, to the left of the median plane, it bifurcates into the two common iliac arteries [FIG. 1067]. The point of division is on a level with the highest points of the iliac crests.

Relations. Posteriorly it lies on the anterior longitudinal

ligament covering the twelfth thoracic to fourth lumbar vertebrae, with the third and fourth left lumbar veins intervening. Anterior to it lies the aortic plexus of nerves [p. 788], and from above downwards, the coeliac plexus, the omental bursa (which lies between it and the lower part of the caudate lobe of the liver and the lesser omentum), the body of the pancreas and the splenic vein, the left renal vein, the horizontal part of the duodenum, the root of the mesentery, the posterior parietal peritoneum and coils of small intestine. It lies at first between the crura of the diaphragm. The thoracic duct and cisterna chyli, and the vena azygos lie between it and the right crus which separates it from the inferior vena cava. Below, it is in direct contact with the inferior vena cava. Lumbar lymph nodes lie in front and on both sides of it.

Branches. The branches of the abdominal aorta are described on page 885.

Surface Anatomy. The abdominal aorta may be represented by a band, about 2 cm. wide, centred above on the midline one finger's-breadth above the transpyloric plane, and below on a point 1·5 cm. below and to the left of the umbilicus.

BRANCHES OF THE ASCENDING AORTA

CORONARY ARTERIES

There are two coronary arteries, right and left; they are distributed to the heart [FIGS. 1044, 1045, and 1049].

The **right coronary artery** arises from the right aortic sinus, and passes forwards, between the pulmonary trunk and the right auricle, to the coronary sulcus. It follows the sulcus to the inferior extremity of the right border of the heart, and then, turning backwards and to the left, continues in the coronary sulcus to the posterior interventricular groove where it divides into two branches. The larger is called the **posterior interventricular branch,** and it runs in the interventricular groove towards the apex of the heart, supplying branches to the ventricles and the interventricular septum. The smaller terminal branch continues in the coronary sulcus towards the termination of the circumflex branch of the left coronary artery, and ends in small branches to the left ventricle. The right coronary artery gives branches to the right atrium, and a larger **right marginal** branch which runs along the lower border of the heart. The right coronary artery is accompanied by branches from the cardiac plexus and by lymph vessels; and in the second part of its course by the small cardiac vein and then the coronary sinus.

FIG. 1066 Left side of the mediastinum and thoracic vertebral column.

Left subclavian A.
Left common carotid A.
Left superior intercostal V.
Cut edge of pleura
Pulmonary trunk
Infundibulum of right ventricle
Anterior interventricular branch of left coronary A.
Left ventricle

1st left aortic intercostal A.
Thoracic duct
Left pulmonary A.
Upper left pulmonary V.
Lower left pulmonary V.
Circumflex branch, left coronary A.
Descending aorta
Oesophagus
Diaphragm

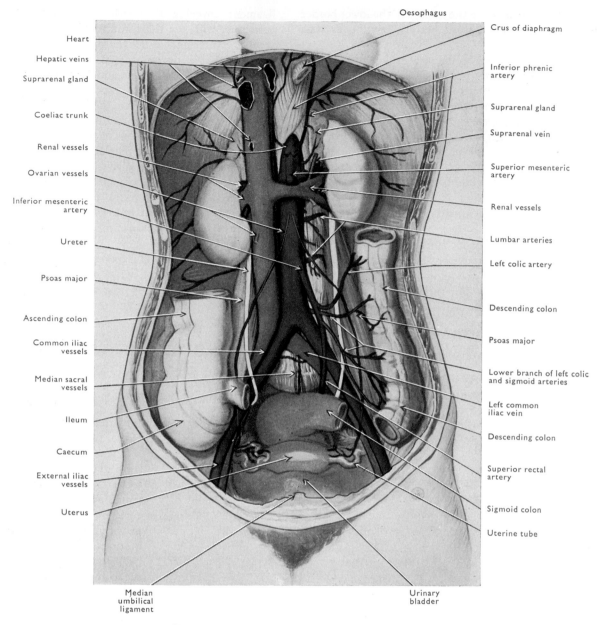

Oesophagus

Heart

Hepatic veins

Suprarenal gland

Coeliac trunk

Renal vessels

Ovarian vessels

Inferior mesenteric
artery

Ureter

Psoas major

Ascending colon

Common iliac
vessels

Median sacral
vessels

Ileum

Caecum

External iliac
vessels

Uterus

Crus of diaphragm

Inferior phrenic
artery

Suprarenal gland

Suprarenal vein

Superior mesenteric
artery

Renal vessels

Lumbar arteries

Left colic artery

Descending colon

Psoas major

Lower branch of left colic
and sigmoid arteries

Left common
iliac vein

Descending colon

Superior rectal
artery

Sigmoid colon

Uterine tube

Median
umbilical
ligament

Urinary
bladder

Fig. 1067   Abdominal aorta and its branches.

The **left coronary artery** arises from the left aortic sinus. It runs to the left behind the pulmonary trunk and then forwards to the upper end of the anterior interventricular groove. Here it divides into a larger **anterior interventricular branch,** and a smaller **circumflex branch.** The latter runs in the left part of the coronary sulcus, giving branches to the left atrium, the left (pulmonary) surface of the heart **(left marginal artery),** and the base of the left ventricle inferiorly. The anterior interventricular branch is an important vessel which passes down the anterior interventricular groove, and turns on to the posterior interventricular groove close to the apex. It supplies both ventricles and the interventricular septum, and is accompanied by cardiac nerves, lymph vessels, and the great cardiac vein [PLATE 52A].

Variations. Rarely the coronary arteries arise by a single stem, or both may spring from the same aortic sinus. Such variability is not remarkable, as the arteries are essentially enlarged vasa vasorum.

Anastomoses. Obstruction of a coronary artery by the formation of blood clot in its lumen (thrombosis) is a common and very serious disease. Accordingly the question of the extent to which the branches of the coronary arteries anastomose with one another has received much attention. It may be said that many anastomoses exist between arterioles of up to 200 μm. in diameter, but that they probably carry little blood in health. Gradual onset of obstruction to the coronary flow will allow the collateral circulation provided by these small vessels to enlarge ; but a sudden blockage will not be supported by prior adaptation, and so the extent of cardiac damage will be greater.

Extracardiac Anastomoses. Small twigs from the coronary arteries pass along the great vessels and form their vasa vasorum. They also pass to the pericardium where they anastomose, *inter alia*, with branches of the internal thoracic artery and of its pericardiacophrenic and musculophrenic branches.

For more detailed information on the anatomy of the coronary arteries Fulton (1965) should be consulted.

BRANCHES OF THE ARCH OF THE AORTA

The branches which arise from the arch of the aorta supply the head and neck, the upper limbs, and part of the body wall.

They are three in number—the brachiocephalic trunk, the left common carotid, and the left subclavian arteries.

Variations. Six branches have been recorded—right and left subclavian, common carotid, and vertebral; more commonly, the left vertebral artery forms an additional branch between the left common carotid and the left subclavian.

An artery called the **thyroidea ima** occasionally springs from the arch or from the brachiocephalic trunk.

An uncommon, but not unimportant, variation is that in which the right subclavian springs from the descending aorta and passes upwards and to the right behind the oesophagus.

BRACHIOCEPHALIC TRUNK

Course and Surface Anatomy. The brachiocephalic trunk [FIG. 1060] arises from the arch of the aorta behind the middle of the manubrium sterni. It is 3·5 to 5 cm. long, and runs upwards, backwards, and to the right from the superior mediastinum into the root of the neck. It ends at the level of the upper part of the right sternoclavicular joint by dividing into the right subclavian and common carotid arteries. Its surface marking may be gauged by reference to FIGURE 1050.

Relations. Anteriorly the left brachiocephalic vein crosses between it and the thymus; at a higher level, the sternothyroid muscle separates it from the sternohyoid and the sternoclavicular joint. Posteriorly it lies on the trachea and then on the right pleura. On its right side are the right brachiocephalic vein and the upper part of the superior vena cava. On its left side is the origin of the left common carotid artery, and at a higher level the trachea is in contact with it.

Branches. As well as its two terminal branches the brachiocephalic trunk may give off the thyroidea ima which runs on the front of the trachea to the thyroid gland.

ARTERIES OF THE HEAD AND NECK

The vessels to the head and neck are derived chiefly from the carotid trunks, but others arise from the main arterial stems of the upper limbs. The most important of those, the vertebral arteries, will be described with the carotid system since they supply structures in the neck and much of the brain.

COMMON CAROTID ARTERIES

The **right** common carotid artery arises at the bifurcation of the brachiocephalic trunk, the **left** common carotid from the arch of the aorta; but each terminates at the level of the upper border of the thyroid cartilage.

LEFT COMMON CAROTID ARTERY

This vessel arises immediately to the left and slightly behind the origin of the brachiocephalic trunk. Its **thoracic portion,** 2·5–3·5 cm. in length, runs upwards and laterally to the left sternoclavicular joint.

Surface Anatomy [FIG. 1068]. A line from a point just below and to the left of the centre of the manubrium sterni to the sternoclavicular joint represents the thoracic portion of the left common carotid artery. The cervical portion of the left artery, and all of the right, is indicated by a line drawn from the appropriate sternoclavicular joint to a point 1 cm. behind the superior horn of the thyroid cartilage.

Relations. *In the thorax*, the vessel is in contact, from below upwards, with the trachea, the left recurrent laryngeal nerve, and the left margin of the oesophagus [FIG. 1081]. Anteriorly the left brachiocephalic vein runs across the artery, and cardiac branches from the left vagus and sympathetic

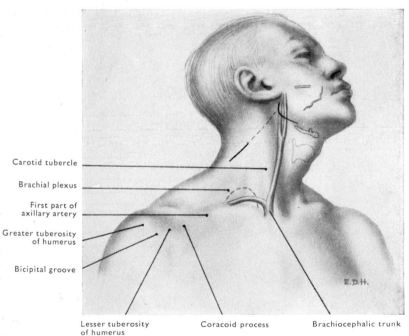

Carotid tubercle

Brachial plexus

First part of axillary artery

Greater tuberosity of humerus

Bicipital groove

E.D.H.

Lesser tuberosity of humerus   Coracoid process   Brachiocephalic trunk

FIG. 1068   Side of the neck.

Of the structures shown, only the brachiocephalic trunk is labelled. The spinal portion of the accessory nerve is shown as an interrupted line between the transverse process of the atlas and the posterior border of the sternocleidomastoid, and as a continuous line between that muscle and the trapezius. The hypoglossal nerve is shown, above the hyoid bone, crossing the internal carotid and then the external carotid arteries. On the face the parotid duct is indicated above and the facial artery below. The apex of the lung is represented as an interrupted line arching above the clavicle.

descend in front of it. Those structures, together with the thymus and the anterior margins of the left lung and pleura, separate the artery from the manubrium sterni, and from the origins of the sternohyoid and sternothyroid muscles. Medially there are the brachiocephalic trunk below, and the trachea above; laterally is the left pleura; and posteriorly, the left phrenic and vagus nerves and the left subclavian artery.

In the neck, the left common carotid artery continues to the level of the upper border of the thyroid cartilage at the lower border of the third cervical vertebra, where it divides into external and internal carotid arteries.

It runs upwards at first deep to the sternocleidomastoid, and then in the anterior triangle of the neck, enclosed with the internal jugular vein (lateral) and the vagus nerve (posterolateral), in the carotid sheath of cervical fascia [FIG. 1070].

Posteriorly the sympathetic trunk intervenes between the artery and the prevertebral muscles and fascia. At the level of the seventh cervical vertebra the vertebral artery and the thoracic duct pass behind it, as does the inferior thyroid artery at the level of the cricoid cartilage.

The anterior tubercle of the sixth cervical transverse process is prominent and, since the common carotid artery can be compressed against it, is known as the **carotid tubercle**. It corresponds in level with the cricoid cartilage.

The superior root of the ansa cervicalis lies on the carotid sheath [FIG. 1069], and opposite the sixth cervical vertebra the omohyoid muscle crosses superficial to the artery. Above that muscle it is overlapped by the anterior border of the sterno-cleidomastoid and by deep cervical lymph nodes; and it is crossed by the superior thyroid vein. Below the omohyoid it is covered by the sternothyroid, the sternohyoid, and the sternocleidomastoid muscles, and may be overlapped by the thyroid gland; it is also crossed by the middle thyroid vein. Just above the sternum the anterior jugular vein is separated from it by the sternohyoid and sternothyroid muscles.

The trachea and oesophagus, with the recurrent laryngeal nerve between them, and the larynx and pharynx, are medial to the artery; and the carotid body lies just superior to its termination.

The internal jugular vein occupies the lateral part of the carotid sheath.

Branches. These are the external and internal carotid arteries and some minute twigs to the carotid body.

FIG. 1069　Dissection of head and neck showing the carotid arteries.

RIGHT COMMON CAROTID ARTERY

The right common carotid artery corresponds with the cervical portion of the left common carotid, and its relations are similar. Such differences as exist are as follows : there is no thoracic duct on the right ; the left recurrent laryngeal nerve is posterior to the mediastinal part of the left artery, and medial to its cervical part, whereas the right nerve passes posterior to the lower part of the corresponding artery to reach its medial side. The oesophagus has a less intimate relation with the right artery than with the left.

The terminal portion of each common carotid artery and the root of its internal carotid branch are dilated to form the **carotid sinus.** This sinus is part of the mechanism that regulates blood pressure ; its walls are specially innervated by the glossopharyngeal and vagus nerves, which also supply the carotid body [FIG. 687].

Variations. To what has been said earlier [p. 863], it is only necessary to add that the common carotid arteries may divide at a higher or lower level than usual—more often higher.

EXTERNAL CAROTID ARTERY

Course and Surface Anatomy. The external carotid artery [FIGS. 1069 and 1074] extends from the upper border of the thyroid cartilage to the back of the neck of the mandible, where it ends by dividing into the superficial temporal and maxillary arteries.

It begins in the carotid triangle, and is at first anterior and medial to the internal carotid artery ; it inclines backwards as it ascends, and thus becomes lateral to the internal carotid. Its course is indicated by a line drawn from a point just behind the tip of the greater horn of the hyoid bone to the lobule of the ear.

Deep Relations. At first the inferior constrictor muscle is close to its medial side, but at a higher level the structures which intervene between it and the internal carotid—the stylopharyngeus muscle, the styloid process (or the styloglossus muscle), the glossopharyngeal nerve, the pharyngeal branch of the vagus and a portion of the parotid gland—separate it from the wall of the pharynx.

Superficial Relations. In the carotid triangle it is overlapped by the anterior border of the sternocleidomastoid, and is crossed by the superior thyroid (sometimes), lingual, and facial veins, and by the hypoglossal nerve. At the level of the angle of the mandible it passes under cover of the posterior belly of the digastric and stylohyoid muscles, which separate it from the inferior part of the parotid gland. It then passes laterally into the gland, within which it lies deep to the retromandibular vein and branches of the facial nerve.

Branches. Eight branches arise from the external carotid artery ; three—the superior thyroid, the lingual, and the facial —from its anterior surface ; two from its posterior surface, namely, the occipital and the posterior auricular ; one from its medial side, namely, the ascending pharyngeal ; its terminal branches are the superficial temporal and maxillary arteries.

BRANCHES OF THE EXTERNAL CAROTID ARTERY

The Superior Thyroid Artery [FIGS. 1069 and 1074]

This vessel arises just below the tip of the greater horn of the hyoid bone, and runs downwards and forwards in contact medially with the inferior constrictor and the external laryngeal branch of the superior laryngeal nerve to the upper pole of the corresponding lobe of the thyroid gland. Its upper part is deep to the anterior border of the sternocleidomastoid, and its lower part to the infrahyoid muscles.

Branches in the Carotid Triangle. The small **infrahyoid branch** runs along the lower border of the hyoid bone, deep to the thyrohyoid muscle.

The **superior laryngeal artery** pierces the thyrohyoid membrane and enters the lateral wall of the piriform fossa, in company with the internal laryngeal nerve.

The **sternocleidomastoid branch** passes across the common carotid artery to the deep surface of the muscle.

Branches in the Muscular Triangle. The **cricothyroid branch** anastomoses in front of the cricothyroid ligament with its fellow of the opposite side.

The terminal **glandular branches** are anterior and posterior. (1) The **anterior branch** descends along the anterior border of the gland to the upper border of the isthmus where it anastomoses with its fellow. (2) The **posterior branch** is distributed to the deep surface of the lobe. (3) A lateral branch is inconstant. The branches anastomose with one another and with branches from the inferior thyroid artery, especially the posterior branch.

The Lingual Artery [FIGS. 904, 1069, and 1074]

This vessel arises opposite the tip of the greater horn of the hyoid bone, and ends beneath the tip of the tongue.

The first part of the artery is deep to the platysma, and rests medially on the middle constrictor. It forms an upward loop which is crossed by the hypoglossal nerve, and then passes forwards on the middle constrictor, deep to the hyoglossus which separates it from the hypoglossal nerve and its vena comitans, and from the lower part of the submandibular gland. It then ascends between the genioglossus and the anterior border of the hyoglossus, and having given off the sublingual artery, ends as the deep artery of the tongue. The deep artery is covered, on its lower surface, by the mucous membrane, and may be felt pulsating when the tongue is held between finger and thumb. At

FIG. 1070 Diagram of relation of cervical fascia to thyroid gland.

Deep cervical fascia (sternocleidomastoid layer), blue ; sheath of thyroid gland (pretracheal fascia), pale red, and capsule of thyroid gland, deep to it, white.

(Modified from De Quervain, 1924.)

C.T.A.—2 F

the tip of the tongue it anastomoses with its fellow of the opposite side.

Branches. The **suprahyoid branch** is very small. It runs along the upper border of the hyoid bone.

The **dorsal branches of the tongue**—usually two—arise deep to the hyoglossus, and ascend to the dorsum of the tongue. They supply the posterior part of the tongue, and send branches to the palatine tonsil.

The **sublingual artery** runs between the mylohyoid and the genioglossus to supply the sublingual gland.

The Facial Artery [Figs. 1069 and 1071]

This vessel arises immediately above the lingual and ends at the medial angle of the eye after a tortuous course. It first ascends on the lateral surface of the pharynx, and comes to lie deep to the posterior belly of the digastric and the stylohyoid, separated from the palatine tonsil only by the superior constrictor muscle. Just above the stylohyoid it enters a groove in the submandibular gland, and runs downwards between the lateral surface of the gland and the medial pterygoid muscle to the lower border of the mandible. Turning round this border of the bone it enters the face at the anterior border of the masseter where its pulsations may be felt. Thence it passes sinuously towards the angle of the mouth, giving off its labial branches 1·0–1·5 cm. from the angle, and ascends more vertically to the medial angle of the eye (**angular artery**).

The facial vein is posterior to the artery in the face, and runs a straighter course.

Branches in the Neck. The **ascending palatine artery** [Fig. 1079] ascends between the styloglossus and stylopharyngeus muscles and turns downwards over the upper border of the superior constrictor, accompanying the levator veli palatini [Fig. 1075]. It supplies the pharynx, soft palate, tonsil, and auditory tube.

The **tonsillar branch** pierces the superior constrictor, and ends in the palatine tonsil.

The **glandular branches** pass into the submandibular gland.

The **submental artery** runs forwards on the mylohyoid muscle, medial to the submandibular gland, to the chin.

Branches in the Face. The **inferior labial artery** runs medially, under cover of the muscles of the lower lip, and anastomoses with the mental artery, and with its fellow of the opposite side.

The **superior labial artery** runs medially, between the orbicularis oris and the mucous membrane. It supplies the upper lip, and, by a **septal branch,** the lower and anterior part of the nasal septum.

The **lateral nasal** branch ramifies on the side of the nose [Fig. 1071].

The labial arteries are readily palpable when the lips are held between finger and thumb.

The facial artery may arise from a stem common to it and the lingual artery—the linguofacial trunk.

The Occipital Artery [Figs. 1069 and 1074]

This vessel arises opposite the facial artery, and ends near the medial end of the superior nuchal line of the occipital bone, by dividing into medial and lateral branches. It runs deep and parallel to the lower fibres of the posterior belly of the digastric, and at the base of the skull deep to the attachment of that muscle, enters a groove on the temporal bone from which it passes on to the superior oblique muscle. It enters the superficial fascia at the junction of the medial and intermediate thirds of the superior nuchal line.

In its course it crosses the carotid sheath, the hypoglossal nerve (which hooks round it), and the accessory nerve. Near its termination it is crossed by the greater occipital nerve, and it passes either through the trapezius or between the trapezius and the sternocleidomastoid.

Branches. Muscular branches include two **sternocleidomastoid branches.** One is looped downwards across the hypoglossal nerve; the other accompanies the accessory nerve into the muscle.

The **descending branch** is given off on the surface of the superior oblique; it runs down among the muscles of the back of the neck, anastomosing with the deep cervical artery.

The **meningeal** branches enter the skull through the hypoglossal canal and the jugular foramen.

The **auricular branch** aids in the supply of the auricle.

The **mastoid branch** passes through the mastoid foramen; it supplies the mastoid air cells on its way to the dura mater.

The terminal **occipital branches,** medial and lateral,

Superficial temporal

Supratrochlear and
supra-orbital branches
of ophthalmic artery

Zygomatico-orbital

Transverse facial

Terminal part
of facial

Lateral nasal

Infra-orbital

Superior labial

Inferior labial

Facial   Buccal, from maxillary
FIG. 1071   Arteries of the face.

ramify tortuously in the scalp, where they anastomose with the posterior auricular and superficial temporal arteries.

The Posterior Auricular Artery [FIGS. 1074 and 1079]

From its origin just above the posterior belly of the digastric it runs to the interval between the mastoid process and the auricle where it divides.

Branches. There are three named branches.

The **stylomastoid artery** enters the stylomastoid foramen and accompanies the facial nerve. It supplies the tympanic cavity and mastoid antrum, the vestibule and semicircular canals, the mastoid air cells (**mastoid branches**), and the stapedius muscle through the **posterior tympanic artery,** which also forms a vascular circle around the tympanic membrane with the anterior tympanic branch of the maxillary artery.

The **auricular branch** supplies the scalp and both surfaces of the auricle.

The **occipital branch** supplies the occipital belly of the occipitofrontalis and the adjacent skin.

The Ascending Pharyngeal Artery [FIG. 1079]

This long, slender vessel, usually the first or second branch of the external carotid, ascends on the pharynx, medial to the internal carotid artery.

Branches. Small **pharyngeal branches** supply the pharynx, the palatine tonsil, and the auditory tube.

Small **meningeal** branches enter the cranium by the hypoglossal canal, the jugular foramen (**posterior meningeal artery),** and the foramen lacerum.

The **inferior tympanic artery** accompanies the tympanic branch of the glossopharyngeal nerve.

The Superficial Temporal Artery [FIG. 1071]

This terminal branch of the external carotid begins between the parotid gland and the neck of the mandible, and ends in the scalp, from 2·5 to 5 cm. above the zygomatic arch, by dividing into frontal and parietal branches. As it crosses the posterior root of the zygoma it may easily be compressed. The auriculotemporal nerve runs behind it, and the superficial temporal vein is usually superficial to it.

Branches. Small **parotid branches** supply the gland and the temporomandibular joint.

The **transverse facial artery** runs across the masseter, below the zygomatic arch and above the parotid duct.

Small **anterior auricular branches** supply the auricle and the external acoustic meatus.

The **middle temporal artery** crosses the zygomatic arch, pierces the temporal fascia, and ascends in the temporal fossa grooving the skull wall.

The **zygomatico-orbital artery** runs forwards above the zygomatic arch.

The **frontal branch** runs sinuously towards the frontal tuber. It is frequently visible through the skin in the living person.

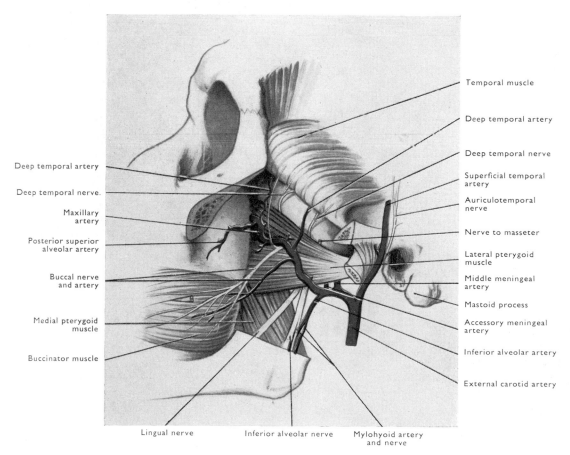

FIG 1072 Dissection showing termination of external carotid artery and first and second parts of maxillary artery.

The **parietal branch,** less tortuous, runs towards the parietal tuber.

Wounds of the scalp bleed freely owing to the fixation of its vessels' walls to the dense subcutaneous tissue. Moreover, so freely do the arteries of the two sides anastomose, it is not possible to control haemorrhage from one side by ligature of the corresponding external carotid artery. The power of healing possessed by the scalp is remarkable—an index of its very rich vascularity.

The Maxillary Artery

This, the larger terminal branch of the external carotid, begins between the parotid gland and the back of the neck of the mandible and ends in the pterygopalatine fossa [FIGS. 1072, 1074, and 1075]. It is described in three parts. The **first part** extends forwards with the maxillary vein, between the sphenomandibular ligament and the neck of the mandible, below the auriculotemporal nerve, to the middle of the lower border of the lateral pterygoid muscle. The **second part** is in the infratemporal fossa, and lies superficial (or deep) to the lateral pterygoid muscle. The **third part** passes between the heads of the lateral pterygoid, and through the pterygo-maxillary fissure into the pterygopalatine fossa.

Branches from the First Part. The deep auricular artery pierces the wall of the external acoustic meatus and supplies its lining and the tympanic membrane.

The anterior tympanic artery traverses the petro-tympanic fissure and enters the middle ear.

The middle meningeal artery [FIGS. 1073 and 1076], the largest of the meningeal arteries, ascends through the auriculo-temporal nerve, and enters the skull through the foramen spinosum. In the middle cranial fossa it passes for a short distance forwards and laterally and divides into frontal and parietal branches at a point 1–2 cm. above the level of the centre of the zygomatic arch.

The middle meningeal artery sends a small **petrosal branch** through the hiatus for the greater petrosal nerve, and a small **superior tympanic artery** through either the canal for the tensor tympani or the petrosquamous suture. It also sends an **anastomotic branch to the lacrimal artery** through the lateral part of the superior orbital fissure, and may form that artery.

The **frontal branch,** the larger and more important of the two, passes upwards to the antero-inferior angle of the parietal bone, where it is often enclosed in a bony canal. It continues upwards and backwards, parallel to and a variable distance

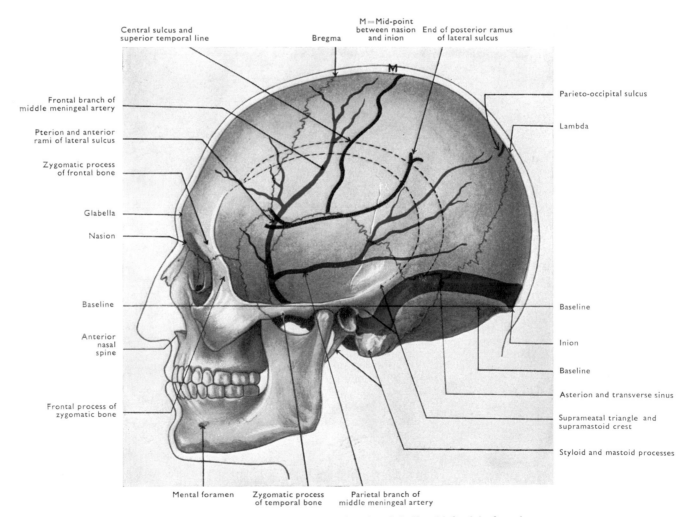

FIG. 1073 Craniocerebral topography: landmarks of skull; chief sulci of cerebrum; transverse and sigmoid venous sinuses; middle meningeal artery. M = midpoint between nasion and inion.

(0·5–1·5 cm.) behind the coronal suture, almost to the vertex of the skull, often as two branches. Here it lies in the periosteum lining the skull and is separated by the corresponding veins from the bone and by the meninges from the precentral gyrus [FIG. 1073]. It sends branches forwards and backwards.

The **parietal branch** passes backwards on to the squamous part of the temporal bone.

Because of its liability to rupture from blows on the side of the head, with resultant haemorrhage between the skull and the dura mater, the surface marking of the middle meningeal artery [FIG. 1073] is of surgical importance.

An accessory meningeal branch may arise from the first part of the maxillary or from the middle meningeal artery. It enters the skull through the foramen ovale.

The inferior alveolar artery passes downwards to enter the mandibular foramen with the inferior alveolar nerve. It descends in the mandibular canal and, after giving off the mental artery, is continued in the bone to the median plane. The **mylohyoid branch,** given off immediately above the foramen, descends with the mylohyoid nerve, in the mylohyoid groove to the superficial surface of the muscle. In the mandibular canal branches are given off to the teeth. The **mental artery** passes through the mental foramen, supplies the tissues of the chin, and anastomises with the inferior labial artery.

Branches from the Second Part. Branches pass to the muscles of mastication. **The masseteric artery** passes through the mandibular notch to the deep surface of the

muscle. **The deep temporal arteries,** anterior and posterior, ascend between the temporalis muscle and the skull. Small **pterygoid branches** supply the pterygoid muscles. **The buccal artery** [FIG. 1072] supplies the buccinator muscle, and the skin and mucous membrane of the cheek.

Branches from the Third Part. One or more **posterior superior alveolar arteries** descend on the posterior surface of the maxilla to supply the molar and premolar teeth and the lining of the maxillary sinus.

The infra-orbital artery enters the orbit through the inferior orbital fissure and runs forwards, in the infra-orbital groove and canal, to the infra-orbital foramen, through which it emerges on the face. In the infra-orbital canal it gives twigs to the canine and incisor teeth (**anterior superior alveolar arteries**) and to the walls of the maxillary sinus. In the face it sends branches to the lower eyelid, upper lip, and cheek.

The descending palatine artery descends through the greater palatine canal giving off several **lesser palatine arteries** which pass through the lesser palatine canals to the soft palate. Continuing as the **greater palatine artery** it passes through the greater palatine foramen and runs forwards on the bony palate, medial to the alveolar process. Its terminal portion ascends through the incisive canal to the nasal septum.

The artery of the pterygoid canal runs backwards through the canal with the corresponding nerve and supplies branches to the pharynx, to the auditory tube, and soft palate.

The pharyngeal branch runs through the palatovaginal canal to the roof of the pharynx.

FIG. 1074   External carotid, maxillary, and meningeal arteries.

The sphenopalatine artery enters the nose through the sphenopalatine foramen, and divides into the **posterior lateral nasal** and **posterior septal nasal arteries.** The latter accompanies the nasopalatine nerve across the roof of the cavity and downwards and forwards in the groove on the vomer.

INTERNAL CAROTID ARTERY
[PLATE 49]

The internal carotid artery [FIGS. 1075, 1076, and 1079] springs from the common carotid opposite the upper border of the thyroid cartilage, and ascends to the base of the skull where it enters the carotid canal. From the canal it passes into the cavernous sinus above which it divides into the anterior and middle cerebral arteries [FIG. 1080].

Relations. It is convenient to describe the vessel in three parts.

In the Neck. Posterior to the artery are the longus capitis, the prevertebral fascia, and the sympathetic trunk, which separate it from the cervical transverse processes. The superior laryngeal nerve runs obliquely behind the vessel. Medial to the internal carotid is the external carotid artery for

a short distance below, and above this is the wall of the pharynx with the ascending pharyngeal artery intervening throughout, and the superior laryngeal nerve dividing into its two branches inferiorly. Immediately lateral to the internal carotid artery are the internal jugular vein and, more posteriorly, the vagus nerve; but the vein becomes progressively more posterior until, just below the skull, it lies behind the artery, separated from it by the last four cranial nerves. Superficially, the artery is overlapped by the sternocleidomastoid, and deep to this muscle is crossed by the lingual and facial veins, and by the hypoglossal nerve from which the superior root of the ansa cervicalis descends upon it. The posterior belly of the digastric, the occipital and posterior auricular arteries, and the stylohyoid muscle separate it from the parotid gland, and above this level it is separated from the external carotid artery by the deeper part of the gland and by the styloid process, styloglossus, stylopharyngeus, the glossopharyngeal nerve, and the pharyngeal branch of the vagus.

In the Carotid Canal. The artery ascends into the canal with the internal carotid plexus, and bending anteromedially is antero-inferior to the cochlea and the middle ear cavity, medial and then superior to the auditory tube, and then inferior

1. Posterior cerebral A.
2. Basal vein.
3. Lentiform nucleus.
4. Cerebral peduncle.
5. Anterior commissure.
6. Middle cerebral A.
7. Anterior cerebral A.
8. Oculomotor nerve.
9. Internal carotid A.
10. Interclinoid ligament.
11. Ophthalmic A.
12. Inferior rectus M.
13. Optic nerve.
14. Nasociliary nerve (cut).
15. Nerve to inf. oblique M.
16. Tendon of sup. oblique M.
17. Superior rectus M.
18. Levator palpebrae M.
19. Lateral rectus M.
20. Inferior oblique M.
21. Maxillary nerve.
22. Infra-orbital nerve and A.
23. Ostium of maxillary sinus.
24. Maxillary sinus.
25. Posterior superior alveolar N.
26. Maxillary artery.
27. Medial pterygoid lamina.
28. Tensor veli palatini M.
29. Auditory tube.
30. Superior constrictor M.
31. Lingual nerve.
32. Ascending palatine A.
33. Styloglossus M.
34. Stylopharyngeus M.
35. Stylohyoid M.
36. External carotid A.
37. Levator veli palatini M.
38. Internal jugular vein.
39. Longissimus capitis M.
40. Posterior belly of digastric M.
41. Occipital artery.
42. Sternocleidomastoid M.
43. Internal carotid A.
44. Facial nerve.
45. Sigmoid sinus.
46. Tympanic membrane.
47. Bony auditory tube.
48. Head of malleus.
49. Mastoid antrum.
50. Trigeminal ganglion.
51. Abducent nerve.
52. Trochlear nerve.
53. Tentorium cerebelli

FIG. 1075 Dissection showing course and relations of upper part of internal carotid artery.

to the trigeminal ganglion. Leaving the canal it comes to lie above the fibrocartilage which fills the lower part of the foramen lacerum during life. It then turns upwards into the cavernous sinus.

In the Cranial Cavity. The artery bends sharply forwards in the cavernous sinus [FIGS. 891, 892] with the abducent nerve close to its inferolateral aspect. The oculomotor, trochlear, the ophthalmic division of the trigeminal, and, usually, the maxillary division, lie in the lateral wall of the sinus. Near the superior orbital fissure, it turns back on itself, and passing posteromedially, pierces the roof of the cavernous sinus inferior to the optic nerve. It then passes between the optic and oculomotor nerves to end below the anterior perforated substance by dividing into the anterior and middle cerebral arteries.

The **carotid sinus** [p. 865] includes the root of the internal carotid artery, and may be limited to it.

Variations. The internal carotid artery occasionally springs from the arch of the aorta. Its course in the neck may be tortuous and brings it into close relation with the tonsil.

BRANCHES OF INTERNAL CAROTID ARTERY

No branches are given off in the neck.

Very small **caroticotympanic branches** perforate the posterior wall of the carotid canal and enter the tympanum.

In the cranium twigs are given to the trigeminal ganglion.

Branches to the hypophysis are constant vessels which have been fully described by Stanfield (1960); see page 574.

The Ophthalmic Artery [FIGS. 1076 and 1079]

This vessel springs from the internal carotid immediately after it has pierced the dura and arachnoid. It runs through the optic canal below the optic nerve and within its arachnoid and dural sheaths. Piercing the sheaths, it runs forwards for a short distance on the lateral side of the nerve; it then crosses, between the optic nerve and the superior rectus, to the medial wall of the orbit, where it turns forwards to end by dividing into the supratrochlear and dorsal nasal arteries.

Branches. The **central artery of the retina** arises in, or

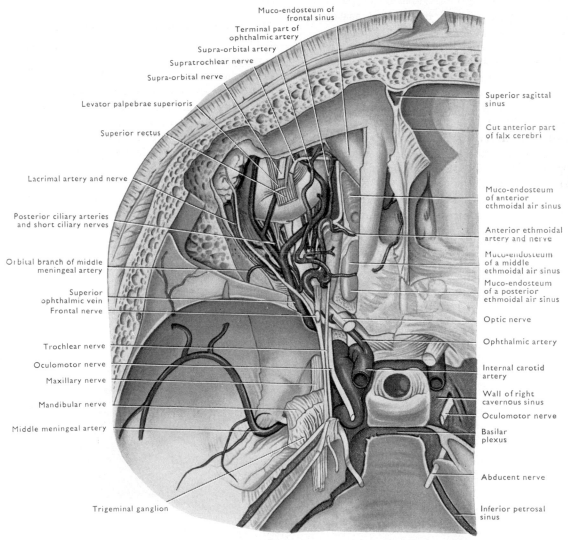

FIG. 1076   Dissection of orbit and middle cranial fossa.

On the right side the trochlear nerve has been removed, and in the left orbit portions of the structures above the ophthalmic artery have been taken away.

close to, the optic canal and runs within the dural sheath of the optic nerve. It pierces the nerve about half-way along its intra-orbital course, and runs in the centre of the nerve with its companion vein to the retina [p. 810]. It is an 'end artery' and the most important branch of the ophthalmic.

The **short posterior ciliary arteries,** usually six to eight in number, run forwards beside the optic nerve; they divide into numerous branches which pierce the sclera near the nerve and end in the choroid coat. The **long posterior ciliary arteries,** one on each side, pierce the sclera further laterally in the horizontal meridian, and pass forwards between it and the choroid to the periphery of the iris. Here their branches anastomose with anterior ciliary arteries to form the **greater arterial circle** at the periphery of the iris, from which secondary branches run to a **lesser arterial circle** near the pupillary border of the iris.

The **lacrimal artery** runs to the upper lateral corner of the orbit. It gives branches to the lacrimal gland and ends in the lateral palpebral arteries which supply the conjunctiva and the eyelids.

Small **meningeal** branches, of which one (the **anterior meningeal artery**) arises from the lacrimal artery, pass backwards through the superior orbital fissure and anastomose with the middle meningeal artery.

The **muscular** branches supply orbital muscles, and they also give off anterior ciliary arteries.

The **anterior ciliary arteries** pierce the sclera behind the corneoscleral junction, and join the greater arterial circle of the iris. [For the circulation in the eye and its accessory organs, see pages 810, 813, 814.]

The **supra-orbital artery** runs forwards, between the levator palpebrae superioris and the periosteum, to the supra-orbital notch or foramen. It supplies the skin of the forehead and anterior part of the scalp.

The **anterior** and **posterior ethmoidal arteries.** The posterior, much the smaller, traverses the posterior ethmoidal

foramen and supplies the posterior ethmoidal cells. The anterior ethmoidal artery passes through the anterior ethmoidal foramen, crosses the cribriform plate to the nasal slit, and descends in a groove on the inner surface of the nasal bone. It passes between the lateral cartilage and the lower border of the bone to reach the tip of the nose. It supplies the anterior and middle ethmoidal cells, the meninges, the frontal sinus, the anterosuperior part of the nasal mucoperiosteum, and the skin of the nose.

Medial palpebral arteries, upper and lower, are distributed to the eyelids. With the **lateral palpebral arteries** they form the **superior** and **inferior palpebral arches.** Together with the anterior ciliary arteries, they send branches to the conjunctiva.

The **dorsal nasal artery** ends on the side of the nose.

The **supratrochlear artery** pierces the palpebral fascia and ascends in the superficial fascia of the scalp.

The Posterior Communicating Artery

This branch of the internal carotid forms part of the circulus arteriosus cerebri [p. 875]. It joins the posterior cerebral artery.

The Anterior Choroidal Artery

This small branch passes backwards and laterally, and ends in the choroid plexus of the inferior horn of the lateral ventricle. It supplies the optic tract, the lateral geniculate body (Abbie, 1938), the central part of the crus cerebri, the greater part of the posterior limb and sublentiform parts of the internal capsule, and the adjacent parts of the globus pallidus and thalamus (Abbie, 1937). For a full account of the blood supply of the visual pathway, see Abbie (1938).

The Anterior Cerebral Artery [PLATE 49B]

This vessel passes forwards and medially above the optic nerve, and anterosuperior to the optic chiasma where it is

FIG. 1077 Arteries of the superolateral surface of the cerebral hemisphere. Stipple, anterior cerebral artery; cross-hatching, posterior cerebral artery; no markings, middle cerebral artery.

joined to its fellow by the **anterior communicating artery.** It then ascends anterior to the lamina terminalis, and continues either on the external surface of the corpus callosum or in the sulcus cinguli. It ends by passing superiorly, anterior to the parieto-occipital sulcus.

Branches. Branches of all the cerebral arteries form two distinct groups which do not communicate with one another: (1) **central,** (2) **cortical.** The central vessels pierce the brain at once, but the cortical branches ramify and anastomose on the piamater [p. 600].

The Branches of the Anterior Cerebral Artery

Central Branches. These slender arteries pierce the lamina terminalis and rostrum of the corpus callosum, and supply the anterior parts of the caudate nucleus, lentiform nucleus, and internal capsule. Small branches from the region of the anterior communicating enter the optic chiasma and nerve. A larger branch arising lateral to the anterior communicating artery, runs back along the main vessel and enters the medial part of the anterior perforated substance, to supply the lower, anterior part of the corpus striatum.

Cortical Branches. The named cortical branches are **orbital, frontal,** and **parietal,** and their area of distribution is shown in FIGURES 1078 and 1080. It should be noted that this includes the superomedial parts of the motor and sensory areas of the cortex.

The Middle Cerebral Artery [PLATE 49 A and B]

This, the larger terminal branch of the internal carotid artery, is in more direct continuation with it. It passes in the lateral sulcus to the surface of the insula, where it divides into numerous cortical branches.

Central branches pierce the anterior perforated substance. They may be divided into **medial** and **lateral striate branches.** The former traverse and supply the globus pallidus and internal capsule, reaching the lateral part of the thalamus. The latter pass over and through the putamen of the lentiform nucleus, reach the internal capsule further superiorly, and supply the upper part of the head and the body of the caudate nucleus. These vessels are commonly involved in cerebral thrombosis or haemorrhage.

The cortical branches are named **orbital, frontal, parietal,** and **temporal,** and their distribution is shown in FIGURES 1077 and 1078.

VERTEBRAL ARTERY
[PLATE 49]

The vertebral artery [FIGS. 1079 and 1080] is the first branch of the subclavian trunk, and is divisible into four parts.

The **first part** runs to the foramen in the sixth cervical transverse process. Posterior to it are the ventral rami of the seventh and eighth cervical nerves, and also, somewhat medial to its origin, the inferior cervical sympathetic ganglion. Anterior to it are the vertebral and internal jugular veins, and, on the left side, the terminal part of the thoracic duct.

The **second part** runs upwards through the foramina in the upper six cervical transverse processes. On passing through the transverse process of the axis, it turns laterally and then upwards to the atlas. It is accompanied by branches of the inferior cervical sympathetic ganglion, and by a plexus of veins.

The **third part** winds backwards and medially on the lateral mass of the atlas with the first cervical ventral ramus, and enters the suboccipital triangle in the groove on the upper surface of the posterior arch of the atlas. It then passes anterior to the edge of the posterior atlanto-occipital membrane and enters the vertebral canal.

The **fourth part** pierces the spinal dura and arachnoid and runs upwards and forwards on the side of the medulla oblongata, passing among the rootlets of the hypoglossal nerve. Reaching the lower border of the pons, it unites with its fellow to form the basilar artery [FIG. 1080].

Branches

Muscular branches from the vertebral artery supply the neighbouring muscles.

Spinal branches pass through the intervertebral foramina and reinforce the anterior and posterior spinal arteries.

Small **meningeal** branches ascend into the posterior fossa of the skull.

The **posterior spinal artery** usually arises from the posterior inferior cerebellar branch of the vertebral artery, and seldom from the vertebral itself. It descends along the line of attachment of the dorsal nerve rootlets, being reinforced by spinal twigs of the vertebral, ascending cervical, posterior intercostal, lumbar, and lateral sacral arteries. Its branches are similar to the branches of the anterior spinal artery on the pia mater [see below].

FIG. 1078 Arteries of the medial and tentorial surfaces of the cerebral hemisphere. Markings as in Fig. 1077.

Posterior communicating arteries

Posterior cerebral arteries
Superior cerebellar artery
Basilar artery

Vertebral arteries

Internal carotid artery

Ascending pharyngeal artery

Styloglossus muscle
Stylopharyngeus muscle
Posterior auricular artery
Occipital artery
Facial artery
Lingual artery
External carotid artery

Common carotid artery

Deep cervical artery

Highest intercostal artery
Anastomosis with (aortic) posterior intercostal artery III.

Stalk of hypophysis
Middle cerebral artery
Anterior cerebral artery
Optic nerve

Ophthalmic artery
Ethmoidal arteries
Supra-orbital artery
Supratrochlear artery
Dorsal nasal artery
Ciliary arteries

Facial artery

Ascending palatine artery

Superior thyroid artery
Sternocleidomastoid branch

Vertebral artery
Cricothyroid artery

Thyrocervical trunk
Subclavian artery
Internal thoracic artery
Brachiocephalic trunk

FIG. 1079   Carotid, subclavian, and vertebral arteries and their main branches.

The **anterior spinal artery** [FIG. 1080] runs downwards and medially, and unites with its fellow in front of the decussation of the pyramids to form a single median vessel. A variable number (2–10) of radicular arteries (Romanes, 1965) pass along the ventral roots and reinforce it.

The anterior spinal artery supplies all the grey matter of the spinal medulla save part of the posterior columns which are supplied by the posterior spinal arteries, and the deepest parts of the lateral and posterior columns of white matter. It anastomoses on the pia mater with the posterior spinal arteries. The central branches enter the anterior median fissure, and though they do not anastomose with adjacent central vessels, their capillary territories overlap. The white matter receives most of its supply from the surface anastomotic vessels between the anterior and posterior spinal arteries. (See also Woollam and Millen, 1955, 1958.)

The anterior spinal arteries also supply the paramedian parts of the medulla oblongata, including the hypoglossal nucleus.

The **posterior inferior cerebellar artery** pursues a tortuous course on the side of the medulla oblongata, and, entering the vallecula of the cerebellum, divides into lateral and medial terminal branches. The trunk of the artery gives branches to the posterolateral part of the medulla oblongata and to the choroid plexus of the fourth ventricle. For its detailed distribution consult Bury and Stopford (1913), and Shellshear (1922).

The clinical picture of thrombosis of the posterior inferior cerebellar artery is very typical. The following are a few of its features with, in parentheses, the structure involved: loss of pain and temperature sensation on the same side of the face (spinal tract of the fifth nerve), and on the opposite side of the body (lateral spinothalamic tract); muscular inco-ordination of the limbs on the same side (afferent fibres entering inferior cerebellar peduncle); difficulty in swallowing (nucleus ambiguus).

Basilar Artery

The basilar artery is formed by the junction of the two vertebral arteries at the lower border of the anteromedian surface of the pons. It ends at the upper border of the pons by bifurcating into the two posterior cerebral arteries.

Branches. Small branches supply the pons.

The **labyrinthine arteries** arise either from the basilar or from the anterior inferior cerebellar artery of the same side. Each enters the internal acoustic meatus, and is distributed to the internal ear.

The **anterior inferior cerebellar arteries,** one on each side, pass backwards to the anterior surfaces of the cerebellar hemispheres.

The **superior cerebellar arteries** arise near the termination of the basilar. Each passes backwards on the lateral surface of the inferior part of the midbrain to the upper surface of the cerebellum.

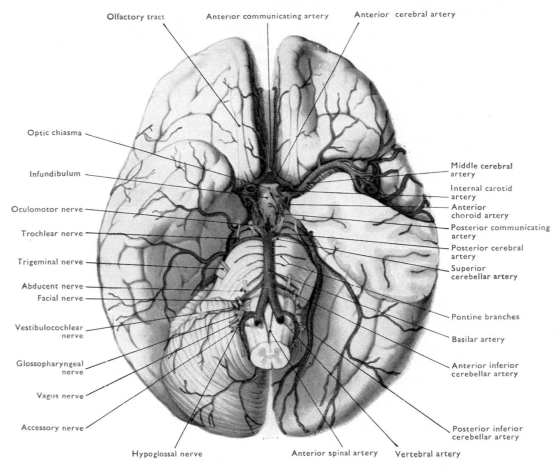

Olfactory tract
Anterior communicating artery
Anterior cerebral artery
Optic chiasma
Infundibulum
Oculomotor nerve
Trochlear nerve
Trigeminal nerve
Abducent nerve
Facial nerve
Vestibulocochlear nerve
Glossopharyngeal nerve
Vagus nerve
Accessory nerve
Hypoglossal nerve
Anterior spinal artery
Vertebral artery
Middle cerebral artery
Internal carotid artery
Anterior choroid artery
Posterior communicating artery
Posterior cerebral artery
Superior cerebellar artery
Pontine branches
Basilar artery
Anterior inferior cerebellar artery
Posterior inferior cerebellar artery

FIG. 1080   Arteries of base of brain showing the circulus arteriosus.

The **posterior cerebral arteries** [FIGS. 1075 and 1080] are the terminal branches of the basilar. Each runs round the cerebral peduncle, parallel to the superior cerebellar artery from which it is separated by the oculomotor and trochlear nerves. Reaching the calcarine sulcus, it divides into occipital and parieto-occipital branches. It is connected with the internal carotid by the posterior communicating artery.

The branches of the posterior cerebral artery are of three kinds: (1) central; (2) choroidal; and (3) cortical.

1. A set of small **posteromedial central branches** pierces the posterior perforated substance; with similar branches of the posterior communicating artery they supply the midbrain and part of the floor and walls of the third ventricle.

A set of small **posterolateral central branches** supply the colliculi, the medial geniculate bodies, the pineal body, the peduncle, and the posterior part of the thalamus.

2. Small **posterior choroidal branches** enter the choroid fissure, and others run forwards in the tela choroidea of the third ventricle to supply the choroid plexuses of the third and lateral ventricles.

3. The named cortical branches are **temporal, occipital, and parieto-occipital;** their positions and the areas which they supply are shown in FIGURES 1078 and 1080. The occipital branches are especially associated with the supply of the visual area of the cortex of the brain.

Circulus Arteriosus [FIG. 1080]. The cerebral arteries of opposite sides are intimately connected together at the base of the brain by anastomosing channels, an arrangement which provides for the continuation of a regular blood supply if one or more of the main trunks should be obstructed.

These vessels form the so-called **circulus arteriosus,** described by Willis (1664) and long known as the 'circle of Willis'. It is situated at the base of the brain in the interpeduncular cistern and is irregularly polygonal in outline. It is formed by the termination of the basilar and the two posterior cerebral arteries, the posterior communicating arteries and the internal carotids, the anterior cerebral arteries, and the anterior communicating artery. A recent comprehensive account of the circle is that by Alpers, Berry, and Paddison (1959). It is subject to many variations, but the commonest is an enlargement of one (or both) posterior communicating artery so that a greater proportion (or all) of the blood in the posterior cerebral artery is derived from the internal carotid artery.

ARTERIES OF THE UPPER LIMB

The main arterial stem of the upper limb passes through the root of the neck, the axilla, and the upper arm to the forearm, where, opposite the neck of the radius, it divides into the radial and ulnar arteries. In the root of the neck it is named the subclavian artery, in the axilla the axillary artery, and thereafter the brachial artery.

SUBCLAVIAN ARTERIES

The right subclavian artery [FIGS. 1060, 1069, and 1079] arises from the brachiocephalic trunk behind the right sterno-clavicular joint, and lies wholly in the root of the neck.

The left artery arises from the aortic arch in the superior mediastinum. In the root of the neck each artery arches over the cervical pleura behind the scalenus anterior muscle, and is described in three parts which lie respectively medial to, behind, and lateral to the muscle. The extent to which the artery rises above the clavicle varies from 1·2 to 2·5 cm. The first parts of the subclavian arteries differ both in extent and position, but the second and third parts are similar on the two sides.

First Part of the Right Subclavian Artery [FIG. 963]

Relations. Anteriorly the vessel is crossed by the right vagus, by cardiac branches of the vagus and sympathetic, and by the internal jugular and vertebral veins; the ansa subclavia curves beneath it. More superficially lie the sternohyoid and sterno-thyroid muscles, the anterior jugular vein, the sternocleido-mastoid, the medial supraclavicular nerves, and the platysma. Posteriorly and inferiorly it is separated from the pleura over the apex of the lung by the suprapleural membrane and by the recurrent laryngeal nerve which winds round its inferior and posterior surfaces.

First Part of the Left Subclavian Artery [FIG. 1066]

Relations. 1. In its thoracic course the artery has the relations shown in FIGURE 1081. All that need be added is that, medially, from below up, it is related to the trachea, the left recurrent laryngeal nerve, the oesophagus, and the thoracic duct.

2. In its cervical course the relations of the first part of the left subclavian artery differ from those of the corres-ponding part of the right artery in that the phrenic nerve and thoracic duct cross it anteriorly, and the left recurrent laryngeal nerve does not curve below it.

Second Part of the Subclavian Artery

Relations. Postero-inferiorly it rests on the pleural sac and suprapleural membrane. Anteriorly it is covered by the scalenus anterior, behind the sternocleidomastoid muscle. The scalenus anterior separates it from the subclavian vein, which lies at a slightly lower level, from the transverse cervical and suprascapular arteries, from the anterior jugular vein, and, on the right side, from the phrenic nerve.

Third Part of the Subclavian Artery

This portion extends from the lateral border of the scalenus anterior to the outer border of the first rib, lying partly in the posterior triangle of the neck and partly behind the clavicle, the subclavius muscle and the supra-scapular vessels.

Relations. It rests upon the upper surface of the first rib. Immediately posterior to it the lower trunk of the brachial plexus separates it from the scalenus medius muscle. Anterior to it, and at a slightly lower level, lies the subclavian vein. The external jugular vein crosses this por-tion of the artery, as does the nerve to the subclavius muscle.

Surface Anatomy [FIG. 1068]

To map out the course of the subclavian artery in the neck, a line is drawn, convex upwards, from the upper border of the sternoclavicular joint to the middle of the clavicle, the highest part of the convexity to reach about 2 cm. above the clavicle. The third part of the artery can be compressed against the first rib by pressure directed downwards and backwards, just behind the clavicle, a little lateral to the posterior border of the sternocleidomastoid.

BRANCHES OF SUBCLAVIAN ARTERY

Four branches arise from the subclavian artery—the vertebral artery, the thyrocervical trunk, the internal thoracic artery, and the costocervical trunk.

The Vertebral Artery [p. 873]

The Thyrocervical Trunk [FIGS. 1079 and 1113]

This short vessel arises close to the medial border of the scalenus anterior. It ends by dividing into three branches: (1) the inferior thyroid; (2) the transverse cervical; and (3) the suprascapular.

The Inferior Thyroid Artery [FIG. 963]. This sinuous artery first ascends behind the carotid sheath. It then turns medially, opposite the cricoid cartilage, to the middle of the

Fig. 1081   Horizontal section through the third thoracic vertebra.

posterior border of the thyroid gland, and then descends to the lower end of the lobe, where it terminates in ascending and inferior branches. The recurrent laryngeal nerve passes either in front of or behind the vessel, a variable relationship of great surgical importance (Bowden, 1955); and, on the left side, the thoracic duct passes in front of it near its origin.

Branches. The **ascending cervical artery** ascends medial to the phrenic nerve, and sends spinal branches through the intervertebral foramina.

Small **pharyngeal** branches supply the pharynx, and twigs pass to the oesophagus and trachea.

An **inferior laryngeal artery** accompanies the recurrent laryngeal nerve to the larynx.

Inferior and ascending terminal **glandular branches** supply the posterior and lower parts of the thyroid gland, and give small branches to the parathyroid glands.

The Transverse Cervical Artery [FIGS. 1069 and 1150]. This vessel crosses the scalenus anterior, and runs through the posterior triangle lying on the brachial plexus, scalenus medius, and levator scapulae. At the anterior border of the levator scapulae it divides into a **superficial branch** which supplies the trapezius, and a **deep branch** [FIG. 1082] which descends close to the medial border of the scapula. The **deep branch** frequently arises from the third part of the subclavian artery, and is then called the descending scapular artery.

The Suprascapular Artery. This vessel crosses the scalenus anterior, and continuing laterally behind the clavicle, anterior to the brachial plexus, reaches the scapular notch and passes over the superior transverse scapular ligament. It then descends through the supraspinous fossa deep to supraspinatus, and enters the infraspinous fossa lateral to the spine of the scapula [FIG. 1082].

Branches. The suprascapular artery gives an acromial branch which joins the **acromial rete** [p. 879], and nutrient arteries to the clavicle and scapula.

It sends a branch to the subscapular fossa, and this, together

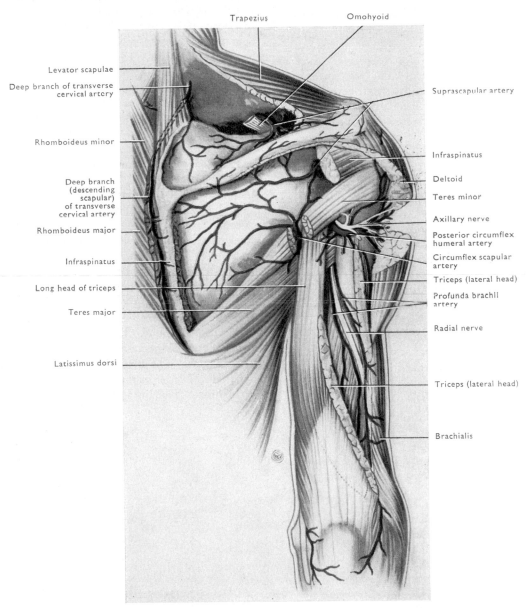

FIG. 1082 Anastomosing arteries on dorsum of scapula and posterior humeral circumflex and profunda brachii arteries.

with those given off in the supraspinous and infraspinous fossae, anastomose with branches of the subscapular and transverse cervical arteries.

The Internal Thoracic Artery [FIGS. 415, 1079, 1081]

This vessel arises immediately below the thyrocervical trunk, and ends behind the sternal end of the sixth intercostal space by dividing into the musculophrenic and superior epigastric arteries.

It lies at first on the pleura and then on the transversus thoracis, and runs behind the brachiocephalic vein, the sternal end of the clavicle, the first six costal cartilages and the intervening spaces. On the right side the phrenic nerve crosses in front of it, from lateral to medial. The artery is accompanied by lymph vessels and nodes, and by two veins which unite behind the third cartilage to form a single vessel that ends in the brachiocephalic vein. Below the first cartilage the **surface marking** of the artery is a vertical line drawn 1 cm. from the border of the sternum.

Branches. The **pericardiacophrenic artery** accompanies the phrenic nerve to the diaphragm.

Mediastinal branches supply the back of the sternum, the thymus, and the pericardium.

The **anterior intercostal arteries,** two in number in each of the upper six spaces, pass laterally, first between the transversus thoracis and the internal intercostal muscles, and then between these muscles and the intercostales intimi; they end by anastomosing with the posterior intercostal arteries.

The **perforating branches,** one in each of the upper six intercostal spaces, pass forwards to the skin. Those in the second, third, and fourth spaces give off **mammary branches** which are of special importance in the female.

The **musculophrenic artery** runs downwards and laterally on the thoracic surface of the diaphragm, but it pierces that muscle and ends on its abdominal surface. It gives branches to the diaphragm, and two **anterior intercostal branches** to each of the seventh, eighth, and ninth intercostal spaces.

The **superior epigastric artery** leaves the thorax between the sternal and costal origins of the diaphragm, and enters the sheath of the rectus abdominis muscle. It ends by anastomosing with branches of the inferior epigastric artery. It gives branches to the rectus, to the other muscles of the abdominal wall, to the diaphragm, and to the skin.

The Costocervical Trunk [FIG. 1079]

This vessel springs from the back of the second part of the subclavian artery on the right side and from the first part on the left side. It runs over the pleura to the neck of the first rib, where it divides into the deep cervical and highest intercostal arteries.

The **deep cervical artery** runs backwards between the seventh cervical transverse process and the neck of the first rib. It then ascends on the semispinalis cervicis muscle, and ends by anastomosing with the descending branch of the occipital artery.

The **highest intercostal artery** descends in front of the neck of the first rib. It gives off the posterior intercostal artery of the first

space, then, crossing in front of the neck of the second rib, becomes the posterior intercostal artery of the second space. These intercostal arteries are distributed in the same manner as the other posterior intercostal arteries [p. 884].

AXILLARY ARTERY

The axillary artery extends from the outer border of the first rib to the lower border of the teres major, where it becomes the brachial artery [FIG. 1084].

For descriptive purposes the artery is divided into three parts: the first part lies above the pectoralis minor, the second behind, and the third part below it.

Relations of the First Part. The first part of the artery is enclosed, together with the vein and the cords of the brachial plexus, in a prolongation of the prevertebral fascia known as the **axillary sheath.** Posteriorly the sheath lies on the superior slip of the serratus anterior, the first intercostal space, and the nerve to serratus anterior. Within the sheath, the medial cord of the brachial plexus lies behind the artery. Anteriorly it is covered by the clavipectoral fascia, which separates it from the cephalic vein and the clavicular part of the pectoralis major. Laterally and above are the lateral and posterior cords of the brachial plexus. Medially and below is the axillary vein.

Relations of the Second Part. Posteriorly are the posterior cord of the plexus and the subscapularis muscle. Anteriorly are the pectoralis minor and the pectoralis major. Laterally lies the lateral cord of the brachial plexus. Medially the medial cord of the plexus lies close to the artery, between it and the axillary vein.

Relations of the Third Part. Posteriorly the artery rests upon the subscapularis, the latissimus dorsi, and the teres major, separated from the subscapularis by the axillary and radial nerves, and from the lower two muscles by the radial nerve alone. Anteriorly it is crossed by the medial root of the median nerve. Its upper half lies under cover of the lower part of the pectoralis major, but its lower part is covered by skin and fasciae only. Laterally lie the median and musculocutaneous nerves and the coracobrachialis muscle. Medially is the axillary vein. The two vessels are, however, separated by the medial cutaneous nerve of the forearm in front and the ulnar nerve behind. The **brachial veins** end in the axillary vein at the lower border of the subscapularis muscle.

Variations. The distal part of the artery may be crossed

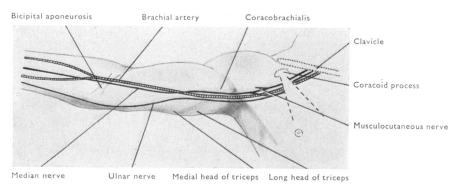

FIG. 1083 Axilla and medial side of arm and elbow. The pectoralis minor is indicated by interrupted lines.

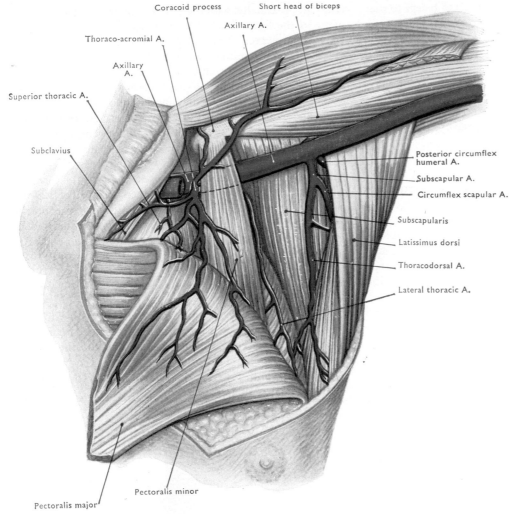

FIG. 1084   The axillary artery and its branches.

pectoralis major, by dividing into four terminal branches. The acromial branch runs to the acromion where it joins the **acromial** network or **rete**. The clavicular branch runs to the sternoclavicular joint. The deltoid branch descends alongside the cephalic vein, as far as the insertion of the deltoid. The pectoral branch descends between and supplies the two pectoral muscles.

Lateral Thoracic Artery

The lateral thoracic artery arises from the second part of the axillary, and descends along the lateral border of the pectoralis minor. It supplies the adjacent muscles, and sends **lateral mammary branches** to the mammary gland.

Subscapular Artery

This, the largest branch of the axillary, arises from the third part of the artery, and runs along the lower border of the subscapularis. It gives off the circumflex scapular artery, and then, as the **thoracodorsal artery** accompanied by the thoracodorsal nerve, continues to the wall of the thorax. A large branch enters the latissimus dorsi with its nerve at a well-defined neurovascular hilus [p. 261].

The **circumflex scapular artery** passes backwards into the triangular space [p. 310]. Turning round the lateral border of the scapula, it enters the infraspinous fossa, where it anastomoses with the deep branch of the transverse cervical artery and the suprascapular artery. While in the triangular space it sends a branch into the subscapular fossa.

Posterior Circumflex Humeral Artery

This vessel arises from the third part of the axillary artery and passes backwards, with the axillary nerve, through the quadrangular space [FIG. 1082]. It turns round the surgical neck of the humerus, and ends in branches which supply the deltoid and anastomose with the anterior circumflex humeral artery.

It supplies the shoulder joint, and sends a branch to the acromial rete. A **descending branch** anastomoses with the ascending branch of the profunda brachii artery.

Anterior Circumflex Humeral Artery

This branch of the third part of the axillary passes deep to the coracobrachialis and the two heads of the biceps, around the front of the surgical neck of the humerus, and ends by anastomosing with the posterior circumflex humeral. It sends

superficially by a muscular or tendinous 'axillary arch' passing between the latissimus dorsi and the pectoralis major ; and sometimes the axillary artery divides into the radial and ulnar arteries.

Surface Anatomy. With the arm abducted to a right angle, a line drawn from the middle of the clavicle to the medial border of the prominence of the coracobrachialis indicates the position and direction of the artery which, in this position of the limb, is almost straight. With the arm hanging by the side the axillary artery describes a curve with the concavity directed downwards and medially, and the vein lies along its medial side.

BRANCHES OF AXILLARY ARTERY

Highest Thoracic Artery

The highest thoracic artery arises from the first part of the axillary and runs across the first intercostal space.

Thoraco-acromial Artery

The thoraco-acromial artery [FIG. 1084] arises from the second part of the axillary artery. It is a short, wide trunk which pierces the clavipectoral fascia, and ends, deep to the a branch along the tendon of the long head of the biceps to the shoulder joint.

BRACHIAL ARTERY

The brachial artery begins at the lower border of the teres major, and ends, in the cubital fossa, opposite the neck of the radius, by dividing into the radial and ulnar arteries [FIG. 1086]. It lies at first on the medial side of the humerus and then in front of it.

Relations. Posteriorly it lies, successively, on the long head of the triceps with the radial nerve and the profunda vessels intervening; the medial head of the triceps; the insertion of the coracobrachialis; and the brachialis. Anteriorly it is overlapped by the medial border of the biceps, and it is crossed, at the middle of the arm, by the median nerve. In the cubital fossa the bicipital aponeurosis separates it from the median cubital vein. Laterally it is in relation, proximally, with the median nerve, and, distally, with the biceps. Medially it is in relation, proximally, with the basilic vein, the medial cutaneous nerve of the forearm, and the ulnar nerve, and, distally, with the median nerve. Two brachial veins accompany the artery, and freely communicate across it.

Variations. The brachial artery sometimes divides at a higher level than usual. In such cases the ulnar artery may cross superficial to the flexor muscles, and may even be subcutaneous; and the radial artery may descend in the superficial fascia of the forearm. In performing venesection at the elbow such variations have to be borne in mind.

Sometimes the brachial artery accompanies the median nerve behind a supracondylar process [p. 153], or ligament; and it may pass in front of the median nerve instead of behind it.

Surface Anatomy. To mark the position of the brachial artery, the line of the axillary artery [p. 879] should be continued to a point 1 cm. beyond the centre of the bend of the elbow [FIG. 1083].

BRANCHES OF BRACHIAL ARTERY

1. The **profunda brachii artery** runs with the radial nerve, and divides at the back of the humerus into two descending branches, the **radial collateral artery** and the **median collateral artery**. The former accompanies the radial nerve to the elbow, and in its course sends a branch to the back of the lateral epicondyle [FIG. 1086]; the latter descends in the

FIG. 1085   Brachial artery and its branches.

Deltoid

Pectoralis major

Latissimus dorsi
Teres major

Triceps

Profunda brachii artery

Ulnar nerve
Superior ulnar collateral artery

Median nerve

Biceps

Brachial artery

Inferior ulnar collateral artery

Brachialis

Pronator teres

Bicipital aponeurosis

Brachioradialis

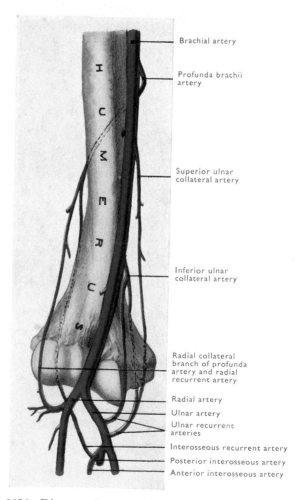

FIG. 1086   Diagram of anastomosis around elbow joint.

Brachial artery

Profunda brachii artery

Superior ulnar collateral artery

Inferior ulnar collateral artery

Radial collateral branch of profunda artery and radial recurrent artery

Radial artery

Ulnar artery

Ulnar recurrent arteries

Interosseous recurrent artery

Posterior interosseous artery

Anterior interosseous artery

substance of the medial head of the triceps. One of the descending branches gives off a **nutrient branch** (occasional) to the humerus; and a **deltoid branch** which anastomoses with the descending branch of the posterior circumflex humeral artery.

2. **Muscular branches** are given to the adjacent muscles.

3. A small **nutrient branch** enters the nutrient canal on the anteromedial surface of the humerus.

4. The **superior ulnar collateral artery** runs with the ulnar nerve to the back of the medial epicondyle of the humerus, where it ends by anastomosing with the posterior ulnar recurrent and inferior ulnar collateral arteries.

5. The **inferior ulnar collateral artery** arises about 5 cm. above the elbow. It takes part in the anastomoses in front and behind the medial epicondyle, and behind the lateral epicondyle.

RADIAL ARTERY

The radial artery [FIGS. 1088 to 1090] begins in the cubital fossa opposite the neck of the radius, and ends by completing the deep palmar arch.

It is described in three parts. The **first part** runs down the forearm to the apex of the styloid process of the radius; the **second part** curves round the lateral side of the wrist, to the proximal end of the first interosseous space; the **third part** passes through that space to the palm.

Relations of First Part. Posteriorly, it rests on the tendon of the biceps, then on the muscles covering the front of the radius, and finally on the lower end of the radius. To its radial side are the brachioradialis which overlaps it proximally, and the superficial branch of the radial nerve which lies near its middle third. To its ulnar side are the pronator teres

proximally, and the flexor carpi radialis, distally. Two veins accompany the artery throughout.

Branches of First Part. The **radial recurrent artery** arises in the cubital fossa. It runs between the superficial and deep branches of the radial nerve and then turns upwards to anastomose with the radial collateral artery.

Muscular branches supply the muscles on the radial side of the forearm.

The **superficial palmar branch** [FIG. 1088] arises just above the wrist. It usually pierces the muscles of the thenar

FIG. 1088 Superficial dissection of forearm and hand, showing radial and ulnar arteries and superficial palmar arch with its branches.

FIG. 1087 Front of elbow, forearm, and hand.

eminence, and ends either in their substance or by completing the superficial palmar arch.

The **palmar carpal branch** passes deep to the flexor tendons, and with the palmar carpal branch of the ulnar artery forms the palmar carpal arch.

Relations of Second Part. As it curves over the wrist, the radial artery lies upon the radial collateral carpal ligament and upon the scaphoid and the trapezium. It is crossed by the tendons of the abductor pollicis longus, the extensor pollicis brevis, and the extensor pollicis longus; and more superficially by the commencement of the cephalic vein and some filaments of the superficial branch of the radial nerve.

Branches of Second Part [FIG. 1090]. The dorsal carpal branch, together with the corresponding branch of the ulnar artery, forms the **dorsal carpal rete** in which the anterior interosseous artery ends. From the rete three **dorsal meta-carpal arteries** run distally: each divides into two **dorsal digital arteries** at the webs between the fingers.

The **dorsal digital arteries of the thumb** and of the **radial side of the index finger** take independent origin from the radial artery.

Each dorsal metacarpal artery is connected with the deep palmar arch by a **proximal perforating branch,** and with a digital branch from the superficial palmar arch by a **distal perforating branch.**

Relations of Third Part. The radial artery passes into the palm between the heads of the first dorsal interosseous muscle, and, after passing between the two heads of the adductor pollicis, completes the **deep palmar arch** [FIG. 1089].

Branches of Third Part. The **princeps pollicis artery** runs distally under cover of the long flexor tendon. It divides into the two **palmar digital arteries** of the thumb [PLATE 50D].

The **radialis indicis artery** descends between the first dorsal interosseous muscle and the adductor pollicis, and along the radial side of the index finger.

Variations. The radial artery may pass to the back of the wrist superficial to the extensor tendons. [See also p. 880.]

Surface Anatomy. To mark the course of the radial artery draw a line from the bifurcation of the brachial to the tubercle of the scaphoid; above the tubercle and below the tip of the styloid process the artery winds dorsally [FIG. 1087]. Its pulsations can readily be felt at the wrist, lateral to the tendon of flexor carpi radialis.

ULNAR ARTERY

The ulnar artery [FIGS. 1088 and 1089] begins in the cubital fossa, opposite the neck of the radius, and ends at the radial side of the pisiform bone, by dividing into its **deep palmar branch** and the **superficial palmar arch** [PLATE 50D].

Relations. Posteriorly, it lies on the brachialis, the flexor digitorum profundus, and the flexor retinaculum. Anteriorly, it is crossed by the pronator teres, the median nerve (separated from it by the ulnar head of the pronator teres), the flexor carpi radialis, the palmaris longus, and the flexor digitorum superficialis. In the middle third of the forearm it is overlapped by the flexor carpi ulnaris, and on the flexor retinaculum it is covered by the superficial part of that structure [p. 338]. In its distal two thirds the flexor digitorum superficialis lies

FIG. 1089 Deep dissection of forearm and hand, showing radial and ulnar arteries and deep palmar arch with its branches.

to its radial side, and the ulnar nerve is on its ulnar side. Two veins accompany the artery.

Branches. The **anterior ulnar recurrent artery** and the **posterior ulnar recurrent artery** arise in the cubital fossa and ascend to the front and back of the medial epicondyle respectively. Both take part in the anastomosis round the joint.

The **common interosseous artery** arises in the distal part of the cubital fossa. It passes backwards, and divides into anterior and posterior interosseous branches.

The **anterior interosseous artery** descends on the anterior surface of the membrane to the proximal border of the pronator quadratus. Here it pierces the membrane and continues downwards to join the dorsal carpal rete. It gives off **muscular branches, nutrient arteries** to the radius and ulna, and a **communicating** branch to the palmar carpal arch. The **median artery,** a slender branch from the proximal part of the anterior interosseous, runs with the median nerve to the palm.

The **posterior interosseous artery** passes backwards above the proximal border of the interosseous membrane; thereafter it descends between, and supplies, the superficial and deep muscles on the back of the forearm. It ends by anastomosing with the anterior interosseous artery and so

with the dorsal carpal rete. It gives off an **interosseous recurrent artery** which runs to the back of the lateral epicondyle [FIG. 1086].

The **palmar carpal branch** of the ulnar artery completes the palmar carpal arch.

The **dorsal carpal branch** passes to the dorsal carpal rete.

The **deep palmar branch** of the ulnar artery descends between the abductor and flexor digiti minimi, and, turning laterally, deep to the opponens and the flexor tendons, it joins the radial artery to complete the **deep palmar arch.**

Variations. The ulnar artery may arise higher than usual, and then commonly passes superficial to the muscles which spring from the medial epicondyle [p. 880]; but even when it begins at the usual level it may pass superficial to these muscles.

The median artery may be quite large, and may end either as digital arteries or by joining the superficial palmar arch.

Surface Anatomy. The upper third of the ulnar artery takes a curved course towards the medial part of the front of the forearm; its lower two-thirds correspond to the lower two-thirds of a line drawn from the front of the medial epicondyle to the radial border of the pisiform [FIG. 1087].

ARTERIAL ARCHES OF HAND

The Superficial Palmar Arch [FIG. 1088]. This is the main terminal branch of the ulnar artery. It begins on the flexor retinaculum immediately distal to the pisiform, and descends across the hook of the hamate deep to palmaris brevis. It then turns laterally and is completed by the superficial palmar branch of the radial, or by the radialis indicis or the princeps pollicis artery. It is accompanied by two veins and lies just beneath the palmar aponeurosis. The arch is often irregular [PLATE 50D], and indeed may be absent.

Four arteries arise from the convexity of the arch, the **palmar digital artery** of the ulnar border of the little finger and three **common palmar digital arteries.** Each of the latter divides into two proper digital branches which supply the contiguous sides of the related fingers. On the sides of the fingers they lie behind the corresponding digital nerves, and give off dorsal branches which anastomose with the dorsal digital arteries.

Each common palmar digital artery is joined, just before it divides, by a palmar metacarpal artery from the deep palmar arch and a distal perforating branch from a dorsal metacarpal artery. The digital artery to the ulnar side of the little finger is joined by a branch which arises either from the medial palmar metacarpal artery or from the deep palmar arch.

The Deep Palmar Arch [FIG. 1089]. This is formed by the radial artery and its anastomosis

Triceps

Anconeus

Supinator

Interosseous recurrent artery

Posterior interosseous artery

Flexor carpi ulnaris

Extensor carpi ulnaris

Anterior interosseous artery

Posterior interosseous artery

Dorsal carpal rete

Perforating branch

Dorsal metacarpal arteries

Brachioradialis

Extensor carpi radialis longus

Extensor carpi ulnaris

Extensor digitorum

Extensor carpi radialis brevis

Posterior interosseous nerve

Pronator teres

Abductor pollicis longus

Extensor pollicis longus

Extensor pollicis brevis

Extensor indicis

Radial artery

Dorsal metacarpal arteries of thumb

Radial dorsal metacarpal artery of index

FIG. 1090 Posterior interosseous artery and second part of radial artery, with their branches.

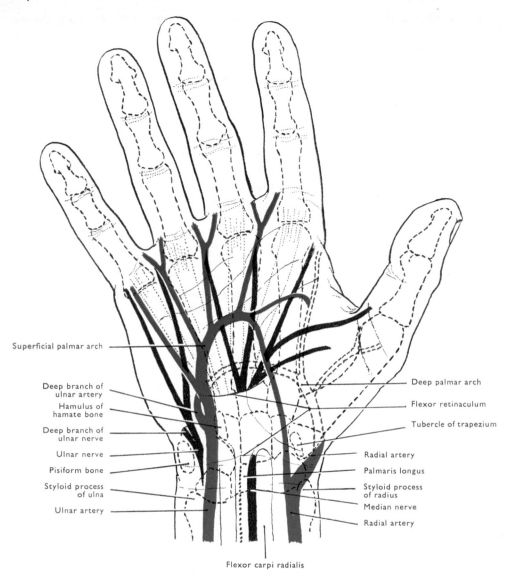

Superficial palmar arch

Deep branch of
ulnar artery

Hamulus of
hamate bone

Deep branch of
ulnar nerve

Ulnar nerve

Pisiform bone

Styloid process
of ulna

Ulnar artery

Deep palmar arch

Flexor retinaculum

Tubercle of trapezium

Radial artery

Palmaris longus

Styloid process
of radius

Median nerve

Radial artery

Flexor carpi radialis

FIG. 1091   Palm of hand. [Cf. Plate 50 D.]

the spinal medulla, the vertebral column, and the upper part of the abdominal wall.

VISCERAL BRANCHES OF THE THORACIC AORTA

The **upper left bronchial artery** arises from the front of the aorta opposite the fifth thoracic vertebra, and the **lower left bronchial artery** at a slightly lower level. Both vessels run to the back of the bronchus, and, dividing similarly, follow its ramifications [p. 495]. They supply the walls of the bronchial tubes, the connective tissue of the lungs, and the visceral pleura.

As a rule there is only one **right bronchial artery**. It arises from the third right posterior intercostal artery, and in its course and distribution corresponds to the bronchial arteries of the left side. It may arise from the upper left bronchial artery or, more rarely, from the aorta.

The anastomoses between the branches of the bronchial and pulmonary arteries enlarge when one or other system is blocked (Tobin, 1952).

Variations. Amongst the anomalous sites of origin of the bronchial arteries are the subclavian, internal thoracic, and inferior thyroid arteries. Consult also Nakamura (1924).

The **oesophageal branches** are small twigs that spring from the front of the aorta.

The **pericardial branches** are small, as are the **mediastinal branches** which pass to the lymph nodes in the posterior mediastinum.

with the deep palmar branch of the ulnar. It lies deep to the long flexor tendons and their synovial sheaths.

From the arch three **palmar metacarpal arteries** pass distally to join the common palmar digital arteries.

Three small **perforating branches** join the dorsal metacarpal arteries, and small **recurrent branches** run to the palmar carpal arch.

Surface Anatomy. The most distal part of the superficial palmar arch reaches to a point level with the distal border of the fully extended thumb. The deep arch lies almost transversely about 1–2 cm. proximal to the superficial arch [FIG. 1091].

BRANCHES OF THE THORACIC AORTA

The branches of the thoracic aorta are distributed chiefly to the thoracic walls and viscera. They also help to supply

PARIETAL BRANCHES OF DESCENDING THORACIC AORTA

Posterior Intercostal Arteries. Nine pairs of posterior intercostal arteries (III–XI) arise from the back of the aorta. The arteries of opposite sides correspond, but, since the upper part of the descending thoracic aorta lies on the left of the vertebral column, most of the right posterior intercostal arteries cross the front of the column, behind the oesophagus, the thoracic duct, and the vena azygos. As each artery runs round the side of the vertebral column, to an intercostal space, it passes behind the pleura and the sympathetic trunk. The lower arteries are crossed by the splanchnic nerves also, and those on the left side by the hemiazygos or accessory hemi-azygos vein.

Each artery passes laterally, and ascends to the upper border of the space to which it belongs; at the angle of the upper rib it enters the **costal groove** and continues forwards between the innermost intercostal and the internal intercostal muscles [p. 340]. In the groove the artery lies between the corresponding vein above and the intercostal nerve below, and it ends by anastomosing with an anterior intercostal artery. The lower two posterior intercostal arteries, on each side, continue into the abdominal wall.

Several branches arise from each posterior intercostal artery.

The **dorsal branch** passes backwards, between the necks of the ribs, and then between the transverse processes, to the vertebral groove; here it divides into medial and lateral cutaneous branches which supply the muscles and skin of the back. A **spinal branch** passes through the corresponding intervertebral foramen; twigs run medially, on the roots of the spinal nerve, and may reinforce the spinal arteries. Others anastomose with adjacent spinal branches and supply the vertebrae [FIG. 1136].

A **collateral branch** arises near the angle of the rib and runs along the lower border of the intercostal space.

Several **nutrient** branches supply the rib above, and **muscular branches** arise from the main trunk and its collateral branch.

A **lateral cutaneous branch** accompanies the lateral cutaneous branch of the intercostal nerve. Those of the third, fourth, and fifth spaces give a **mammary branch** to the mammary gland.

The position of the intercostal arteries determines that in aspirating the chest the needle should pass immediately above a rib and should not be introduced medial to the rib angle.

Subcostal Arteries. Each of this pair of vessels, which has the same branches as the intercostal arteries, runs with the subcostal nerve along the lower border of the twelfth rib. Passing behind the lateral arcuate ligament, it crosses in front of the quadratus lumborum, posterior to the kidney, and piercing the aponeurosis of origin of transversus abdominis, runs between that muscle and the internal oblique.

Superior Phrenic Arteries. These small vessels arise from the lower part of the thoracic aorta and ramify on the surface of the diaphragm.

The **vas aberrans** is a remnant of the right dorsal aorta of the embryo. When present it arises near the upper left bronchial artery, and passes upwards and to the right behind the oesophagus. It may be enlarged and form the first part of the right subclavian artery.

Variations. Coarctation of the aorta, i.e., partial or complete obliteration of the lumen, occurs most commonly just beyond the origin of the left subclavian artery and proximal to the attachment of the ligamentum arteriosum. This part of the aorta is constricted in the foetus and is known as the aortic isthmus; it is succeeded by a fusiform dilatation called the aortic spindle. Both these features persist, but much less markedly, in the adult. Coarctation leads in time to a well-marked collateral circulation. For example, the posterior intercostal arteries (which then transmit blood from the internal thoracic artery to the aorta) may become so enlarged and tortuous as to cause notching of the lower borders of the ribs; this is detectable on radiological examination.

BRANCHES OF THE ABDOMINAL AORTA

[PLATES 50 and 52]

The branches of the abdominal portion of the aorta may be classified conveniently as follows:

$$
\text{Visceral} \begin{cases} \text{Paired} \\ \text{(lateral)} \\ \\ \text{Single} \\ \text{(median} \\ \text{ventral)} \end{cases} \begin{cases} \text{Middle suprarenal} \\ \text{Renal} \\ \text{Testicular or ovarian} \\ \text{Coeliac} \\ \text{Superior mesenteric} \\ \text{Inferior mesenteric} \end{cases}
$$

$$
\text{Parietal} \begin{cases} \text{Paired} \\ \\ \text{Single} \end{cases} \begin{cases} \text{Inferior phrenic} \\ \text{Lumbar} \\ \text{Common iliac} \\ \text{Median sacral} \end{cases}
$$

PAIRED VISCERAL BRANCHES OF THE ABDOMINAL AORTA

There are three sets of **suprarenal arteries** [p. 560]—the superior, middle, and inferior. Of these only the middle arise directly from the aorta; the superior spring from the inferior phrenic arteries, and the inferior from the renal arteries.

The **middle suprarenal arteries** arise close to the origin of the superior mesenteric artery. They run, one or more on each side, upon the crura of the diaphragm just above the renal arteries, to the suprarenal glands.

Renal Arteries

The renal arteries [FIG. 621] arise about 1·5 cm. below the origin of the superior mesenteric artery and opposite the first lumbar vertebra.

Both are large, and the right is longer, and frequently slightly lower than the left. Each runs to the hilus of the corresponding kidney, passing anterior to the crus of the diaphragm and the psoas muscle. The left artery lies posterior to the pancreas; the right vessel passes behind the inferior vena cava, the head of the pancreas, and the descending part of the duodenum. The renal vein lies anterior to the artery and, on the left side, at a slightly lower level.

Approaching the hilus of the kidney, each artery divides into three or four branches of which one passes behind the pelvis of the kidney and the rest in front of it. From these branches five segments of the kidney are supplied, and it is of surgical importance that anastomosis between segments does not occur (Graves, 1956).

Each renal artery gives off small **ureteric branches** to the upper part of the ureter, and one or more **inferior suprarenal arteries,** which pass upwards to the lower part of the suprarenal gland.

Variations. Accessory renal arteries, more common on the left side, may be derived from the aorta, or from the common iliac or internal iliac arteries. They usually enter towards one pole of the kidney, and are 'end arteries' to a portion of the kidney substance (Graves, 1956). As the kidney ascends during development it is not surprising that it may retain one or more of its primitive vessels of supply.

Testicular or Ovarian Arteries

The testicular or ovarian arteries are long, slender vessels which spring from the front of the aorta a short distance below the renal arteries.

Each **testicular artery** runs on the psoas major to the deep inguinal ring, and then accompanies the ductus deferens through the inguinal canal to the testis. Posteriorly, each artery passes in front of the genitofemoral nerve, the ureter, and the lower part of the external iliac artery. The right artery also crosses the inferior vena cava. Anterior to the right artery are the horizontal part of the duodenum, the right colic and ileocolic vessels, and the root of the mesentery with the distal parts of the superior mesenteric vessels. Crossing anterior to the left artery are the left colic and sigmoid vessels and the terminal part of the descending colon.

In its lower abdominal course each testicular artery is accompanied by two veins which issue from the pampiniform plexus and enter the abdomen through the deep inguinal ring; at a higher level they fuse into a single stem which is usually lateral to the artery. Beyond the superficial inguinal ring the artery lies anterolateral to the ductus deferens and in close association with the anterior group of testicular veins. At the posterior border of the testis it breaks up into branches, some of which pass to the testis and others to the epididymis. Further details are given on page 525.

Each testicular artery gives off **ureteric branches** to the abdominal part of the ureter.

The course and the relations of each **ovarian artery,** as far as the external iliac artery, are the same as those of the corresponding testicular artery; it then enters the lesser pelvis by crossing the external iliac vessels, and passes between the layers of the suspensory ligament of the ovary to reach the broad ligament of the uterus. In the broad ligament it runs below the uterine tube, and turns backwards into the mesovarium, where it breaks up into branches which enter the ovary at its hilus. In the broad ligament each ovarian artery is accompanied by the pampiniform plexus of ovarian veins. In its lower abdominal course it is accompanied by two veins which issue from the plexus at the brim of the lesser pelvis and unite at a higher level into a single trunk.

In the abdomen it gives branches to the ureter and, in the pelvis, branches which supply the uterine tube, the round ligament of the uterus, and anastomose with the uterine artery.

SINGLE VISCERAL BRANCHES OF THE ABDOMINAL AORTA

THE COELIAC TRUNK

This short, wide vessel arises immediately below the aortic orifice of the diaphragm and between its crura. It runs forwards for about 1 cm., and ends by dividing into three branches—the left gastric, the common hepatic, and the splenic [FIGS. 1092 and 1906; PLATE 52 B]. Developmentally it is the artery of the abdominal portion of the foregut.

Relations. The trunk lies behind the omental bursa below the caudate lobe of the liver and above the upper border of the pancreas and the splenic vein. It has a coeliac ganglion on each side and is surrounded by the coeliac plexus of nerves.

The Left Gastric Artery

This is the smallest branch of the coeliac trunk. It runs upwards, forwards, and to the left, behind the omental bursa, and reaches the lesser curvature of the stomach close to the oesophagus. It then turns sharply forwards, downwards, and

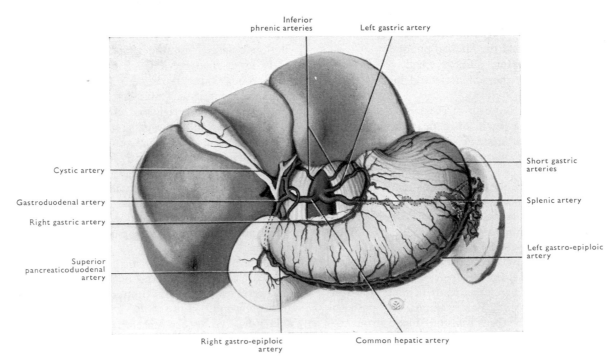

FIG. 1092   Coeliac trunk and its branches.

to the right, and runs within the lesser omentum to anastomose with the right gastric branch of the common hepatic artery.

The left gastric artery sends branches to both surfaces of the stomach, and also gives off **oesophageal branches** which anastomose with oesophageal branches of the thoracic aorta.

The Splenic Artery [Fig. 1092]

This is the largest branch of the coeliac trunk. It runs a tortuous course along the upper border of the body and tail of the pancreas, behind the omental bursa, and superior to the splenic vein. It lies in front of the left suprarenal gland and left kidney, and passes forwards in the lienorenal ligament, with the splenic vein, dividing into five to eight terminal **splenic branches** which enter the hilus of the spleen. For their detailed distribution see pages 965, 966.

Branches. Numerous small **pancreatic branches** are given off along its length.

Four or five **short gastric arteries** pass between the layers of the gastrosplenic ligament to the upper part of the greater curvature and adjacent surfaces of the stomach.

The **left gastro-epiploic artery** passes downwards and forwards in the gastrosplenic ligament, to run parallel to and a short distance from the greater curvature of the stomach, between the anterior two layers of the greater omentum. It anastomoses with the right gastro-epiploic artery, gives numerous branches to both surfaces of the stomach, and long, slender **omental** (epiploic) **branches** to the greater omentum.

The Common Hepatic Artery [Fig. 1092]

This vessel runs forwards and to the right, below the epiploic foramen, to the upper border of the superior part of the duodenum. Here it enters the lesser omentum, and having given off the gastroduodenal artery, becomes the **hepatic artery** which ascends near the free margin of the omentum, in front of the foramen. In the omentum it is in front of the portal vein and to the left of the bile-duct; and near the porta hepatis it divides into right and left terminal branches. It is accompanied by the hepatic plexus of nerves, and by lymph vessels.

Branches. The **gastroduodenal artery** arises at the superior part of the duodenum, descends behind it, and ends opposite its lower border. It runs between the pancreas and the duodenum, anterior to the portal vein, and has the bile-duct on its right side; it ends by dividing into the right gastro-epiploic and superior pancreaticoduodenal arteries. The **right gastro-epiploic artery** passes to the left between the layers of the greater omentum to anastomose with the left gastro-epiploic artery. The **superior pancreaticoduodenal artery** passes to the right, and divides into anterior and posterior branches which descend between the duodenum and the head of the pancreas to anastomose with branches of the inferior pancreaticoduodenal artery. They supply the head of the pancreas, the descending part of the duodenum, and the bile-duct.

Branches of Hepatic Artery. The **right gastric artery** is small and arises as the hepatic enters the lesser omentum. It runs within the omentum to the pylorus, and then along the lesser curvature. It supplies the stomach, and also gives a branch to the superior part of the duodenum.

The **right branch** passes behind the common hepatic duct and the cystic duct, giving off the cystic artery, and reaching the right end of the porta hepatis, divides into two or more branches which enter the liver and accompany the branches of the portal vein and the hepatic duct. The **cystic artery** runs along the cystic duct to the gall-bladder, where it divides into anterior and posterior branches; the anterior passes between the gall-bladder and the visceral surface of the liver, to both of which it gives offsets; the posterior branch is distributed on the posterior surface of the gall-bladder, and lies between it and the peritoneum. The **left branch** runs to the left end of the porta hepatis, gives branches to the caudate and quadrate lobes, crosses the fissure for the ligamentum venosum, and breaks up into branches which enter the left lobe of the liver.

Variations

The coeliac trunk may be absent, its branches arising separately from the aorta; the left gastric artery may give off the left branch of the hepatic, or an accessory hepatic artery. The accessory hepatic or the entire common hepatic artery may arise from the superior mesenteric artery. For further details Michels (1949) should be consulted.

THE SUPERIOR MESENTERIC ARTERY

This, the artery of the embryonic midgut, arises about 1 cm. below the coeliac trunk and opposite the first lumbar vertebra [Figs. 1093, 1095, and 1096] behind the pancreas and splenic vein.

It passes downwards and forwards, anterior to the left renal vein, the uncinate process of the head of the pancreas, and the horizontal part of the duodenum. Here it enters the root of the mesentery, behind the transverse colon or its mesentery, and curves downwards to the right iliac fossa behind the coils of small intestine; crossing the aorta, the inferior vena cava, the right ureter, and the right psoas major muscle. The superior mesenteric vein runs along its right side. It ends, much reduced in size, by anastomosing with the ileocolic artery.

Branches

The branches of the superior mesenteric artery supply part of the duodenum and pancreas, the rest of the small intestine, and the large intestine approximately to the left colic flexure.

The **inferior pancreaticoduodenal artery** arises either from the superior mesenteric or from the first jejunal branch. It runs to the right, and ends in two branches, anterior and posterior, which anastomose with the corresponding branches of the superior pancreaticoduodenal artery.

The branches to the small intestine, from ten to sixteen in number, are separable into two groups—**jejunal** and **ileal.** They enter the mesentery, the branches of adjacent vessels uniting to form arcades, from which secondary branches arise and may also unite in a similar manner. The upper jejunal branches form only one or two tiers of arcades, but the ileal branches form four or five in the longer, lower part of the mesentery. The branches from the terminal arcades pass to alternate sides of the wall of the intestine, and anastomose in it, but these intramural anastomoses are insufficient to maintain

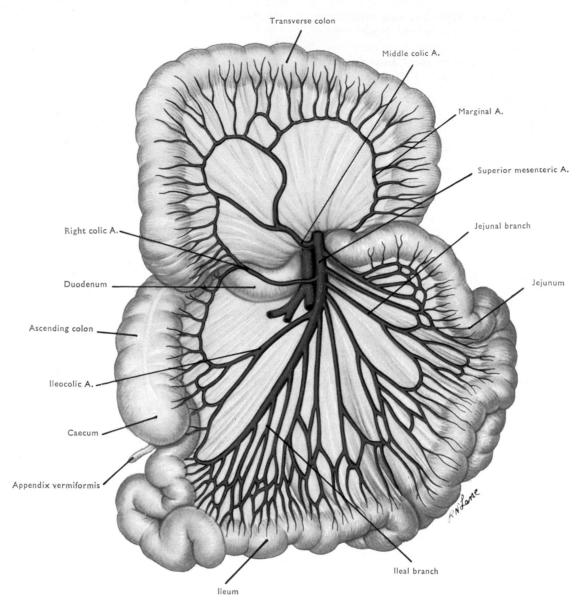

FIG. 1093   The superior mesenteric artery and its branches.

a segment of the jejunum divorced from the direct supply of an arcade (Barlow, 1952).

The remaining branches of the superior mesenteric artery supply the caecum, appendix, ascending colon, and much of the transverse colon. The rest of the colon is supplied by branches of the inferior mesenteric artery [see below], and the major branches of both arteries to the colon unite to form a continuous vessel (**marginal artery**) which runs along the length of the colon and gives branches to it [FIGS. 1093, 1095].

The **ileocolic artery** may arise with the right colic, or separately lower down. It passes downwards and to the right, and ends by dividing into an ascending branch which unites with the lower branch of the right colic, and a descending branch which supplies part of the ascending colon, the caecum, the appendix, the terminal ileum, and ends by uniting with the continuation of the superior mesenteric trunk. The **anterior caecal branch** crosses the ileocaecal junction in the fold of peritoneum called the vascular fold of the caecum; the **posterior caecal branch** crosses posterior to the junction. The **appendicular artery** passes behind the terminal part of the ileum into the meso-appendix [FIG. 1094].

The **right colic artery** arises either from the superior mesenteric as it crosses the duodenum [FIGS. 1093 and 1095] or below the duodenum from a common trunk with the ileo-colic artery. It runs in front of the right ureter and gonadal vessels, towards the ascending colon, near which it divides into ascending and descending branches. The former unites with the middle colic artery; the latter descends to unite with the upper branch of the ileocolic. From this part of the marginal artery branches pass to the ascending colon and the beginning of the transverse colon.

The **middle colic artery** arises from the superior mesenteric as it escapes from behind the pancreas. It enters the transverse mesocolon, and ends by dividing into two branches which unite with the right and left colic arteries, continuing the marginal

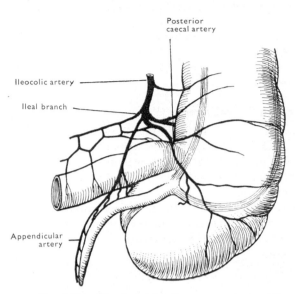

FIG. 1094 Blood supply of caecum and appendix.

The caecum from behind. The artery of the vermiform appendix, and the three taeniae coli springing from its root, should be specially noted. (Modified from Jonnesco, 1895.)

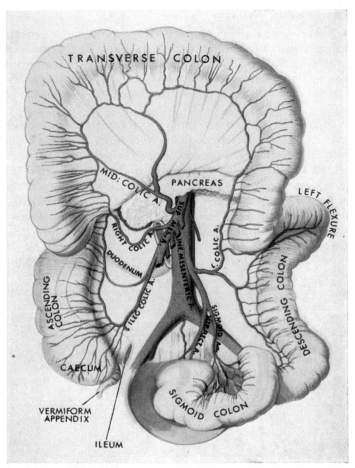

FIG. 1095 Distribution of superior and inferior mesenteric arteries.

There are usually two or more sigmoid arteries.

artery from which branches are distributed to the transverse colon.

THE INFERIOR MESENTERIC ARTERY

[FIGS. 1095 and 1096]

This, the hindgut artery of the embryo, arises about 3–4 cm. above the aortic bifurcation; it passes downwards, at first in front of the aorta, and then on its left side, and at the upper border of the left common iliac artery it becomes the superior rectal artery. Its vein, close to its left side below, separates from it above.

Branches

The **left colic artery** arises from the inferior mesenteric near its origin, and, crossing anterior to the ureter and the gonadal vessels, divides into ascending and descending branches. The **ascending branch** runs across the lower part of the left kidney, and divides into two branches; one enters the transverse mesocolon to join the left branch of the middle colic artery, and the other joins the upper division of the descending branch. The **descending branch** also divides into two branches; one unites with the lower division of the ascending branch and the other with a branch of the first sigmoid artery.

The **sigmoid arteries,** usually three in number, pass downwards and to the left in front of the psoas major, the ureter, and the upper part of the iliacus; the lowest entering the left limb of the sigmoid mesocolon. They end in branches which anastomose with those of the descending branch of the left colic above and of the superior rectal below, thus continuing the marginal artery from which branches pass to the colon.

The **superior rectal artery** is the direct continuation of the inferior mesenteric. It crosses the front of the left common iliac artery, enters the sigmoid mesocolon, descends as far as the third piece of the sacrum, and divides into two branches which run on the sides of the rectum. Half-way down the rectum each terminal branch divides into two or more branches which pass through the muscular coats into the submucous tissue; from them numerous small vessels pass vertically downwards, anastomosing with one another, and with offsets from the middle rectal and inferior rectal arteries [p. 450].

The anastomosis between the lowest sigmoid artery and the superior rectal artery has long been surgically suspect; however, Griffiths (1956) states that this anastomosis is always adequate, but that the truly critical point is at the left colic flexure.

PARIETAL BRANCHES OF ABDOMINAL AORTA

The Inferior Phrenic Arteries [FIG. 1096]

These small arteries, right and left, arise immediately below the diaphragm to which they are distributed. Each gives several **superior suprarenal arteries** to the gland of its own side.

The Lumbar Arteries

These vessels are in series with the posterior intercostal arteries, and their distribution is similar. The upper four pairs arise from the back of the aorta, the fifth pair from the median sacral artery, and all pass round the bodies of the lumbar vertebrae to reach the lateral part of the abdominal wall.

Each artery, except the fifth, is protected in its backward course by one of the fibrous arches from which the psoas major arises. It runs behind the muscle and the lumbar plexus, then behind the quadratus lumborum, and finally pierces the aponeurosis of origin of the transversus abdominis to gain the interval between the transversus and internal oblique muscles. On the right side the arteries pass behind the inferior vena cava, but the upper two are separated from it by the right crus of the diaphragm.

The lumbar arteries anastomose with one another, with the lower posterior intercostal and the subcostal arteries, and with branches of the superior and inferior epigastric and of the deep circumflex iliac and iliolumbar arteries.

Branches. Each lumbar artery gives off muscular branches, and a dorsal branch which is distributed like that of a posterior intercostal artery [p. 885].

The Median Sacral Artery [FIG. 1096]

This small vessel, originally the dorsal aorta in the sacral region, arises from the back of the aorta about 1 cm. above its bifurcation, and ends in the coccygeal body. Its companion veins unite to form a single median sacral vein which joins the left common iliac.

It gives off on each side the **arteria lumbalis ima** (fifth lumbar artery), which is distributed like an ordinary lumbar artery. Small branches of the median sacral artery also pass to the rectum.

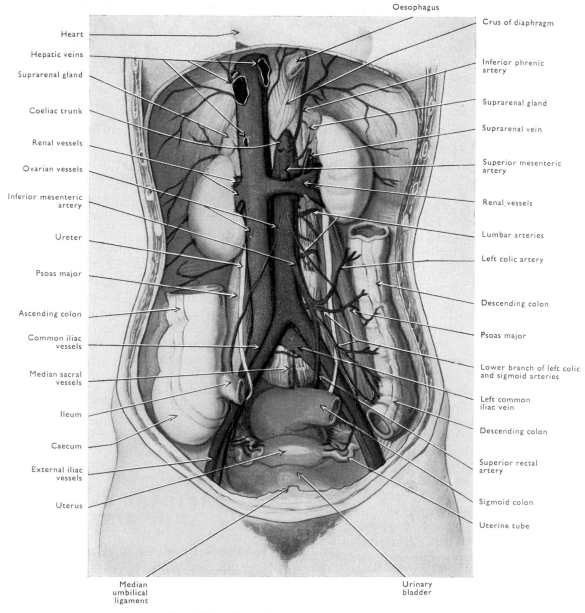

Oesophagus

Heart
Hepatic veins
Suprarenal gland
Coeliac trunk
Renal vessels
Ovarian vessels
Inferior mesenteric artery
Ureter
Psoas major
Ascending colon
Common iliac vessels
Median sacral vessels
Ileum
Caecum
External iliac vessels
Uterus

Crus of diaphragm
Inferior phrenic artery
Suprarenal gland
Suprarenal vein
Superior mesenteric artery
Renal vessels
Lumbar arteries
Left colic artery
Descending colon
Psoas major
Lower branch of left colic and sigmoid arteries
Left common iliac vein
Descending colon
Superior rectal artery
Sigmoid colon
Uterine tube

Median umbilical ligament

Urinary bladder

FIG. 1096   Abdominal aorta and its branches.

THE COMMON ILIAC ARTERIES

The common iliac arteries [Fig. 1096 and Plate 51 a] are the terminal branches of the abdominal aorta. They begin opposite the middle of the body of the fourth lumbar vertebra, a little to the left of the median plane, and passing infero-laterally, end anterior to the sacro-iliac joint at the level of the lumbosacral disc, by dividing into external and internal iliac branches.

Relations. Both arteries are covered by peritoneum in front and medially, and overlaid by small intestine. The superior hypogastric plexus [Fig. 966] lies anterior to their proximal ends, and the left artery is crossed by the superior rectal vessels. On each side, near its termination, the artery is often crossed by the ureter. The last two lumbar vertebral bodies and intervening disc, the sympathetic trunk and the psoas major lie behind each artery, but are separated from the right artery by the beginning of the inferior vena cava. Deep to the psoas are the obturator nerve, the lumbosacral trunk and the iliolumbar artery. Lateral to each artery are the ureter, gonadal vessels, and the genitofemoral nerve. The changing relation of each common iliac vein to its companion artery can be followed in Figure 1096.

Branches. The external and internal iliac arteries are the only branches.

Surface Anatomy. The common and external iliac arteries may be indicated by a line drawn with a slight lateral convexity, from a point opposite the bifurcation of the aorta [p. 861] to the midinguinal point. The upper third of the line corresponds to the common iliac artery, the lower two-thirds to the external iliac.

THE INTERNAL ILIAC ARTERY

[Figs. 1097 and 1100]

In the foetus this vessel is the direct continuation of the common iliac trunk to the umbilicus, and is prolonged through the umbilical cord as the **umbilical artery** [p. 893].

When the umbilical cord is severed at birth, the part of the internal iliac trunk between the pelvis and the umbilicus atrophies to a fibrous cord known as the **lateral umbilical ligament.** However, the proximal part of the artery remains patent to convey blood to its superior vesical branch. The permanent internal iliac artery is short, and appears to end in an anterior and posterior division; the anterior is the original continuation of the vessel, and the posterior is simply a common stem of origin for some of the branches.

The artery arises from the common iliac anteromedial to the sacro-iliac joint at the level of the lumbosacral disc, and descends into the lesser pelvis, to end near the upper border of the greater sciatic notch in its two divisions.

Relations. Anteromedially, each internal iliac artery is covered by peritoneum, beneath which the ureter descends along the anterior surface of the artery. The sigmoid colon usually overlaps the left vessel, and the terminal ileum the

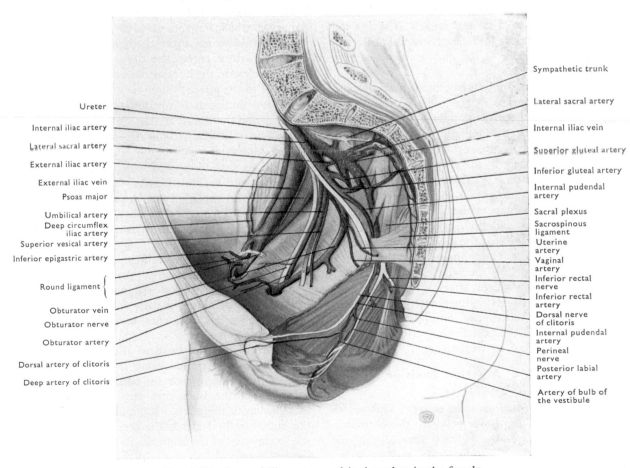

Fig. 1097 Internal iliac artery and its branches in the female.

right. In the female, the ovary and the infundibulum of the uterine tube usually lie close in front of it, separated by the peritoneum and the ureter [FIG. 1100]. Posterior to it are the internal iliac vein and the commencement of the common iliac vein; behind these are the lumbosacral trunk and the sacro-iliac joint. Laterally, the external iliac vein separates it from the psoas major muscle, and at a lower level the obturator nerve lies between the artery and the side-wall of the pelvis.

Branches. The posterior division of the internal iliac artery gives rise to parietal branches, the anterior division to parietal and visceral branches.

Branches of the Posterior Division of the Internal Iliac Artery

The Iliolumbar Artery. This vessel runs to the iliac fossa, passing over the lateral part of the sacrum between the lumbosacral trunk and the obturator nerve, and divides into an iliac and a lumbar branch. The **iliac branch** gives offsets to the iliacus, and a large nutrient branch to the ilium. The **lumbar branch** ascends behind the psoas major, anastomoses with the fifth lumbar artery, and ends in the quadratus lumborum. Its **spinal branch** enters the intervertebral foramen between the fifth lumbar vertebra and the sacrum.

The Lateral Sacral Arteries. Two in number, superior and inferior, they descend in front of the sacral ventral rami, giving off spinal branches which pass through the anterior sacral foramina.

The Superior Gluteal Artery [FIGS. 1097 and 1099]. This large vessel, the continuation of the posterior division of the internal iliac artery, passes backwards between the lumbosacral trunk and the first sacral ventral ramus. Leaving the pelvis through the greater sciatic foramen, above the piriformis muscle, it divides under cover of the gluteus maximus into superficial and deep branches.

The **superficial branch** supplies the gluteus maximus and the skin over that muscle's origin.

The **deep branch** lies under the gluteus medius and very soon divides into superior and inferior branches. The **superior branch** follows the upper border of the gluteus minimus and ends under cover of the tensor fasciae latae, by anastomosing with the ascending branch of the lateral circumflex femoral artery. It also anastomoses with the deep circumflex iliac artery, and gives branches to the adjacent muscles. The **inferior branch** runs with the superior gluteal nerve across the gluteus minimus towards the greater trochanter. It supplies the gluteal muscles, and anastomoses with the ascending branches of the lateral and medial circumflex femoral arteries, and with the inferior gluteal artery.

Surface Anatomy. The superior gluteal artery reaches the gluteal region opposite the junction of the upper and middle thirds of a line drawn from the posterior superior iliac spine to the upper border of the greater trochanter.

Parietal Branches of the Anterior Division of the Internal Iliac Artery

The Obturator Artery [FIGS. 426 and 1097]. This vessel runs along the side-wall of the lesser pelvis to the obturator canal through which it leaves the pelvis. Immediately on entering the thigh it divides into anterior and posterior terminal branches which surround the obturator foramen deep to the obturator externus muscle. In the pelvis the obturator nerve lies above the artery and the vein below it; the ureter, behind, and the ductus deferens, in front, lie between the artery and the peritoneum. In the female, the ovary and broad ligament are medial.

Branches. Within the pelvis the obturator artery gives off a **pubic branch** which anastomoses with the pubic branch of the inferior epigastric. The latter may pass either lateral or medial to the femoral ring. In the medial position it lies on the lacunar ligament medial to the site of a femoral hernia, and is especially important when the obturator artery is formed by this enlarged branch of the inferior epigastric artery [p. 894]. The **anterior** and **posterior terminal branches** supply the adjacent muscles, and the posterior branch also supplies the hip joint and the ligament of the head of the femur by a branch which passes through the acetabular notch. According to Harty (1953) the head and neck of the femur derive their significant blood supply from vessels on the neck of the femur.

The Internal Pudendal Artery [FIGS. 1097 and 1098]. This vessel descends anterior to the sacral plexus to the lower part of the greater sciatic foramen, and passes between the piriformis and coccygeus to enter the gluteal region. It is accompanied by two veins. From the gluteal region, where it lies on the ischial spine between the pudendal nerve medially and the nerve to the obturator internus laterally, it passes through the lesser sciatic foramen and enters the perineum.

In the perineum it lies at first in the lateral wall of the ischiorectal fossa, where it is enclosed in the pudendal canal in the obturator fascia. This canal, which is 3–4 cm. above the lower margin of the ischial tuberosity, contains also the pudendal nerve, which lies above the artery, and the perineal nerve, which lies below it. From the ischiorectal fossa, the internal pudendal artery enters the deep perineal space, and runs forwards close to the inferior ramus of the pubis. About 1 cm. posterior to the pubic symphysis it pierces the perineal membrane and immediately divides into its terminal branches—the deep and the dorsal arteries of the penis (or clitoris).

Branches. In the ischiorectal fossa the **inferior rectal artery** pierces the wall of the pudendal canal, and divides into two or three branches which pass across the fossa to levator ani and the anal canal. There they anastomose with the middle and superior rectal arteries.

Two **posterior scrotal** or **posterior labial** branches arise in the anterior part of the ischiorectal fossa, and pierce the posterior border of the perineal fascia to enter the superficial perineal space in which they pass forwards to the scrotum or labium majus. They supply the muscles and subcutaneous structures of the urogenital triangle.

The **perineal artery** usually arises from one of the scrotal or labial branches, and runs to the central tendon of the perineum.

The **artery of the bulb of the penis** arises in the deep perineal space. It runs along the posterior border of the sphincter urethrae, pierces the perineal membrane to enter the substance of the bulb, and passes onwards in the corpus spongiosum to the glans. It supplies the sphincter urethrae, the bulbo-urethral gland, the corpus spongiosum penis, and the spongy part of the urethra. In the female, the corresponding artery supplies the bulb of the vestibule and the erectile tissue of the lower part of the vagina.

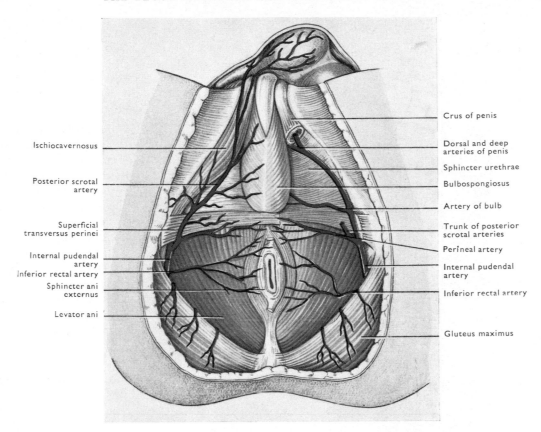

FIG. 1098 Perineal distribution of internal pudendal artery in the male. The perineal membrane has been removed on the right.

The **deep artery of the penis** (or **clitoris**) immediately enters the crus, and runs forwards in the corpus cavernosum of the penis (or clitoris) which it supplies.

The **dorsal artery of the penis** (or **clitoris**) ascends between the crus and the bone, passes forwards between the layers of the suspensory ligament of the penis (or clitoris), and runs along the dorsal surface of the organ, with the dorsal nerve immediately lateral to it, and separated from its fellow by the dorsal vein in the median plane. It supplies the superficial tissues on the dorsum of the penis, sends branches into the corpus cavernosum, and terminates in the glans and prepuce.

The Inferior Gluteal Artery [FIGS. 1097 and 1099]. This vessel runs backwards and laterally between the first and second, or second and third, sacral ventral rami, and traverses the greater sciatic foramen to enter the gluteal region just below the piriformis. It then descends along the posteromedial side of the sciatic nerve to reach the proximal part of the thigh.

Branches. Muscular branches pass to the gluteus maximus, and one or two **coccygeal branches** pierce the sacrotuberous ligament to reach the skin over the coccyx. An **anastomosing** branch joins with branches of the medial and lateral femoral circumflex, and the first perforating arteries, in the so-called 'cruciate anastomosis' of the thigh. The **companion artery of the sciatic nerve** is a long slender branch which runs on the surface of the sciatic nerve or in its substance.

Surface Anatomy. The inferior gluteal artery enters the gluteal region just above the midpoint of a line joining the posterior superior iliac spine to the ischial tuberosity.

Visceral Branches of the Anterior Division of the Internal Iliac Artery

Umbilical Artery. The lumen of this vessel persists, though much diminished, as far as the origin of the superior vesical artery. Thereafter the umbilical artery is reduced to a fibrous cord which runs along the side of the bladder and then, as the **lateral umbilical ligament,** on the deep surface of the anterior abdominal wall to the umbilicus. In the pelvis it is crossed by the ductus deferens or by the round ligament of the uterus.

The **superior vesical artery** arises from the umbilical artery, and passes to the upper part of the urinary bladder.

Inferior Vesical Artery. This runs on the levator ani to supply the lower part of the bladder, the seminal vesicle, the ductus deferens, the lower part of the ureter, and the prostate.

Artery of the Ductus Deferens. This may arise from the umbilical, the superior vesical, or the inferior vesical artery. It is a slender vessel which accompanies the ductus deferens to the testis.

Middle Rectal Artery. This runs medially to the rectum, where it anastomoses with its fellow, and with the other rectal arteries.

Vaginal Artery. This usually replaces the inferior vesical artery. It runs to the side of the vagina. The branches of opposite sides anastomose and form anterior and posterior longitudinal vessels—the so-called azygos arteries which anastomose above with the vaginal branches of the uterine arteries.

Uterine Artery. This runs medially and slightly forwards

Gluteus medius

Gluteus Maximus

Quadratus femoris

2nd perforating A.

Vastus lateralis

Biceps (short head)

Common peroneal N.

Popliteal A.

Peroneal communicating branch

Superior lateral genicular A.

Piriformis

Inferior gluteal A.

Medial femoral circumflex A.

Sciatic N.

Biceps (long head)

3rd perforating A.

Tibial N.

Popliteal A.

Genicular branch of obturator N.

Superior medial genicular A.

Gastrocnemius

FIG. 1099   The arteries of the gluteal region and back of the thigh.

on the levator ani to the lower border of the broad ligament, and passes medially between its layers. About 3 cm. from the uterus it crosses above and in front of the ureter [FIGS. 661 and 1097], and passes above the lateral fornix of the vagina to the side of the cervix. It then ascends in a tortuous manner alongside the uterus, and, turning laterally below the uterine tube, ends by anastomosing with the ovarian artery. The uterine artery sends many branches to the uterus, and aids in the supply of the vagina, uterine tube, and ovary.

Variations

The inferior gluteal artery may, as in the foetus, constitute the main artery of the lower limb, and become continuous with

the popliteal artery; probably the normal companion artery of the sciatic nerve represents this original vessel [FIG. 1099].

The obturator artery may arise from the inferior epigastric artery [p. 895]. The course of such an **accessory obturator artery** is of surgical importance. It descends on the medial side of the external iliac vein, and usually on the lateral side of the femoral ring; but in 30 per cent. of cases, and more often in males, it descends on the medial side of the ring and may be injured in the operation for the relief of a strangulated femoral hernia [FIG. 426].

For further details of variations in the branches of the internal iliac artery Braithwaite (1952) should be consulted.

EXTERNAL ILIAC ARTERY

The external iliac artery [FIGS. 1096 and 1100] extends from a point opposite the sacro-iliac joint, at the level of the lumbosacral disc, to a point behind the inguinal ligament, midway between the anterior superior iliac spine and the pubic symphysis, where it becomes the femoral artery.

It runs along the linea terminalis of the pelvis, and is enclosed, with its accompanying vein, in a thin fascial sheath.

Relations. Anteriorly, peritoneum separates it on the right side from the terminal ileum, and sometimes from the vermiform appendix, and on the left side from the small intestine and the termination of the descending colon. Several branches of the lower sigmoid arteries also cross it on the left side. The artery is crossed at its origin by the ureter and, 2·5 cm. lower down, by the ovarian vessels. Near its lower end the testicular vessels and the genital branch of the genitofemoral nerve lie on the artery, and all are crossed by the deep circumflex iliac vein. That part of the artery is crossed also by the ductus deferens or the round ligament of the uterus. External iliac lymph nodes lie in front and at the sides of the artery. Posteriorly, it lies on the fascia iliaca, at first medial and then anterior to the psoas major muscle. To its lateral side is the genitofemoral nerve, and to its medial side peritoneum. Below, the external iliac vein is medial to the artery, but, above, the vein comes to lie partly behind the artery.

Psoas major muscle

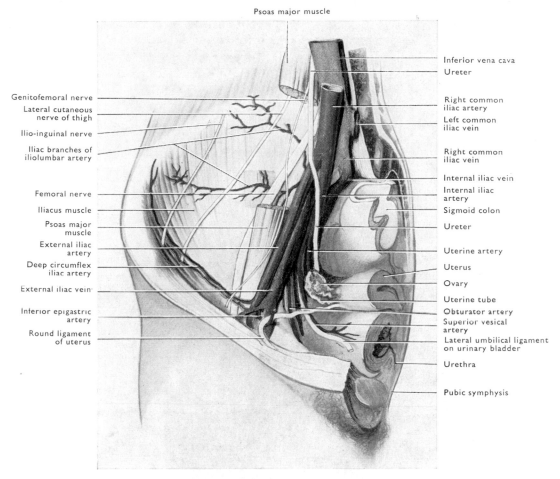

Genitofemoral nerve

Lateral cutaneous
nerve of thigh

Ilio-inguinal nerve

Iliac branches of
iliolumbar artery

Femoral nerve

Iliacus muscle

Psoas major
muscle

External iliac
artery

Deep circumflex
iliac artery

External iliac vein

Inferior epigastric
artery

Round ligament
of uterus

Inferior vena cava

Ureter

Right common
iliac artery

Left common
iliac vein

Right common
iliac vein

Internal iliac vein

Internal iliac
artery

Sigmoid colon

Ureter

Uterine artery

Uterus

Ovary

Uterine tube

Obturator artery

Superior vesical
artery

Lateral umbilical ligament
on urinary bladder

Urethra

Pubic symphysis

FIG. 1100   Iliac arteries and veins in the female.

Branches

Two large branches spring from the external iliac artery : (1) the inferior epigastric ; and (2) the deep circumflex iliac.
The Inferior Epigastric Artery [FIGS. 426, 1097, and 1100]. This arises immediately above the inguinal ligament and ascends towards the umbilicus on the medial side of the deep inguinal ring to the deep surface of the rectus abdominis. In its course it raises a fold of peritoneum called the **lateral umbilical fold**—not to be confused with the **medial umbilical fold** which is raised by the **lateral umbilical ligament**. It then pierces the transversalis fascia, and passes anterior to the arcuate line into the rectus sheath. It penetrates the muscle, and its branches anastomose with those of the superior epigastric artery and lower posterior intercostal arteries. At the deep inguinal ring, the ductus deferens (or the round ligament of the uterus) curves round the lateral and anterior surfaces of the artery.
Branches.   Muscular and cutaneous branches supply the anterior abdominal wall. The **cremasteric artery** in the male (**artery of the round ligament of the uterus** in the female) is small. It accompanies the spermatic cord, supplying its coverings ; in the female it runs with the round ligament. The **pubic branch** anastomoses with the pubic branch of the obturator artery [p. 892].
The Deep Circumflex Iliac Artery [FIGS. 1100 and 1101].

This arises immediately above the inguinal ligament, and runs to the anterior superior iliac spine. A little beyond the spine it pierces the transversus abdominis, and is continued between the transversus and the internal oblique.
Branches.   Of its muscular branches one is frequently large and is known as the **ascending branch;** it pierces the transversus muscle a short distance in front of the anterior superior spine, and ascends vertically on the internal oblique.

ARTERIES OF THE LOWER LIMB

The main artery of the lower limb is the continuation of the external iliac artery. It descends through the thigh (femoral artery) and popliteal fossa (popliteal artery), and divides into the anterior and posterior tibial arteries. In the proximal two-thirds of the thigh it is named the femoral artery ; its distal part, which is situated in the popliteal fossa, is named the popliteal artery.

FEMORAL ARTERY
[PLATE 51A]

The femoral artery [FIGS. 1101 and 1102] begins behind the inguinal ligament, and becomes continuous with the popliteal artery at the opening in the adductor magnus.

FIG. 1101   Femoral artery and its branches.
The outlines of the sartorius, the upper part of the rectus femoris, and the
adductor longus are indicated by broken black lines.

Its proximal half lies in the femoral triangle and is comparatively superficial; at the apex of the triangle it passes deep to the sartorius, enters the **adductor canal,** and is then more deeply placed.

At their entry into the femoral triangle both the artery and its vein are enclosed for a distance of 3 cm. in a funnel-shaped fascial sheath formed of the fascia transversalis in front and the fascia iliaca behind. It is called the **femoral sheath** and is divided by septa into three compartments. The lateral compartment is occupied by the femoral artery and the femoral

branch of the genitofemoral nerve; the intermediate compartment contains the femoral vein; and the medial compartment is the **femoral canal** [p. 393].

Relations in the Femoral Triangle

Anteriorly, the femoral artery is covered by skin and fasciae, and by superficial inguinal lymph nodes. The femoral sheath and the cribriform fascia cover its proximal part, and the fascia lata its distal part; and near the apex of the triangle the artery is crossed by an anterior cutaneous

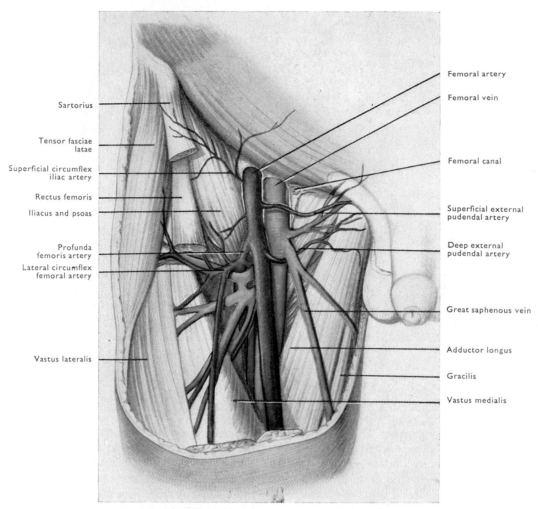

FIG. 1102   Femoral vessels in the femoral triangle.

nerve.   Posteriorly and proximodistally it lies on the femoral sheath and the psoas major, the pectineus, and the adductor longus.   The nerve to the pectineus passes between the artery and the psoas major; the femoral vein and the profunda artery intervene between it and the pectineus, and the femoral vein separates it from the adductor longus also.

The femoral vein lies behind the artery in the lower part of the triangle, but passes to its medial side above.   On the lateral side of the artery, proximally, are the femoral branch of the genitofemoral nerve, the femoral sheath, and the femoral nerve; more distally the saphenous nerve and the nerve to the vastus medialis are continued on its lateral side.

Relations in the Adductor Canal

The adductors longus and magnus are posterior to the artery, but are separated from it by the femoral vein, which is posterior to the artery proximally, and posterolateral distally. The vastus medialis is anterolateral.   The fascial roof of the canal, the subsartorial plexus of nerves, and the sartorius are anteromedial.   The saphenous nerve enters the canal lateral to the artery, and then crosses in front of it to become medial.

Surface Anatomy

To map out the course of the femoral artery the thigh should be slightly flexed, abducted, and rotated laterally [FIG. 1103];

C.T.A.—2 G

a line is drawn from the **midinguinal point,** i.e., midway between the anterior superior iliac spine and the symphysis pubis, to the adductor tubercle.   Rather less than the upper third of the line corresponds to the artery in the femoral triangle and rather more than its middle third corresponds to the artery as it lies in the adductor canal.   The femoral pulse is easily felt in the upper part of the vessel.

Branches of the Femoral Artery

The **superficial circumflex iliac artery** arises just below the inguinal ligament.   It enters the superficial fascia lateral to the saphenous opening, and runs as far as the anterior superior iliac spine.   It supplies lymph nodes and the skin of the groin.

The **superficial epigastric artery** pierces the cribriform fascia, and passes upwards towards the umbilicus supplying the skin.

The **superficial external pudendal artery** also pierces the cribriform fascia, and runs towards the pubic tubercle.   It supplies the skin of the lower abdomen and pubis.

The **deep external pudendal artery** runs medially, pierces the deep fascia, and ends in the scrotum or the labium majus.

The **profunda femoris artery** [FIG. 1102], the largest branch of the femoral artery, arises from its lateral side usually

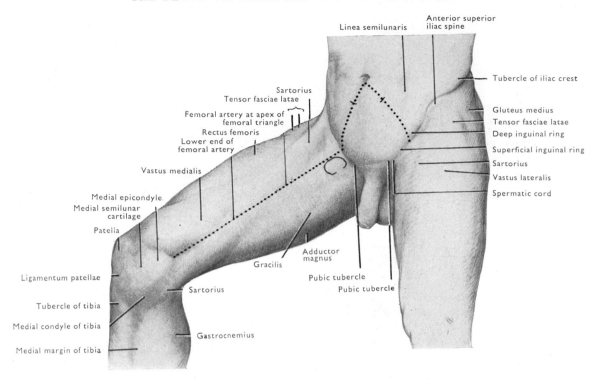

FIG. 1103   The thigh and groin.

about 4 cm. distal to the inguinal ligament. It curves behind the femoral artery, and leaving the femoral triangle between pectineus and adductor longus, descends posterior to the latter, close to the medial aspect of the femur, anterior to the adductor brevis and magnus muscles. In the distal third of the thigh it pierces the latter muscle as the fourth perforating artery. Initially the profunda and femoral veins separate it from the femoral artery, while the adductor longus separates them inferiorly.

Branches from the profunda supply the adductor muscles; some of them pass through the adductor magnus to the hamstrings and vastus lateralis.

1. The **lateral circumflex femoral artery** [FIGS. 1101 and 1102] springs from the profunda near its origin. It runs laterally among the branches of the femoral nerve, then, passing posterior to the sartorius and the rectus femoris, it ends by dividing into three branches. The **ascending branch** runs upwards, posterior to the tensor fasciae latae, along the trochanteric line. It then passes between gluteus medius and gluteus minimus, and anastomoses with branches of the superior gluteal artery. It supplies the neighbouring muscles, and gives a branch to the hip joint. The **transverse branch** passes into the vastus lateralis, winds round the femur and joins the cruciate anastomosis [p. 893]. The **descending branch** runs along the anterior border of the vastus lateralis, accompanied by the nerve to that muscle. It gives branches to the quadriceps femoris, and takes part in the anastomosis round the knee.

2. The **medial circumflex femoral artery** arises from the profunda at the same level as the lateral circumflex, and runs backwards, out of the femoral triangle, between the psoas major and the pectineus. It continues between the obturator externus superiorly and adductor brevis inferiorly, to the

upper border of the adductor magnus, where it divides into a transverse and an ascending branch. Before dividing it sends a branch through the acetabular notch to the hip joint. The **ascending branch** passes between the obturator externus and the quadratus femoris to the trochanteric fossa of the femur, where it anastomoses with branches of both gluteal arteries. The **transverse branch** passes backwards between the quadratus femoris and the adductor magnus. It takes part in the cruciate anastomosis [p. 893].

3. The **perforating arteries** [FIG. 1104], including the terminal branch of the profunda, are usually four in number. They curve backwards and laterally, lying between the femur and tendinous arches which interrupt the insertion of the adductor magnus to the linea aspera, and communicate freely with each other. The **first** perforating artery pierces the insertions of the adductors brevis and magnus, and usually that of gluteus maximus also. It takes part in the cruciate anastomosis. The **second** perforating artery pierces the adductors brevis and magnus, and then passes between the gluteus maximus and the short head of the biceps femoris into the vastus lateralis. The **third** and **fourth** perforating arteries pass through the adductor magnus and the short head of the biceps femoris into the vastus lateralis.

A **nutrient branch** to the femur is given off from either the second or third perforating artery.

It should be noted that there is a chain of *anastomoses*, extending from the gluteal arteries to the muscular branches of the popliteal artery, by which a collateral circulation can be provided in cases of occlusion of the femoral artery [FIG. 1104].

The **descending genicular artery** arises low in the adductor canal, and divides into a superficial and a deep branch. The superficial or **saphenous branch** continues through the

Gluteus maximus

Sacrotuberous ligament

Internal pudendal artery

Inferior gluteal artery

Companion artery of sciatic nerve

Biceps and semitendinosus

Semimembranosus

Adductor magnus

Muscular branch of profunda artery

Gracilis

Popliteal artery

Medial superior genicular artery

Semitendinosus

Gastrocnemius

Muscular artery

Gluteus medius

Gluteus minimus

Deep branch of superior gluteal artery

Piriformis

Obturator internus and gemelli

Ascending branch of medial circumflex femoral artery

Quadratus femoris

Transverse branch of medial circumflex femoral artery

Perforating artery

Perforating artery

Perforating artery

Termination of profunda artery (perforating)

Short head of biceps

Long head of biceps

Popliteal vein

Lateral superior genicular artery

Gastrocnemius

FIG. 1104 Arteries of gluteal region and posterior surface of thigh and knee.

canal and appears superficially between the gracilis and the sartorius. It gives twigs to the skin of the proximal part of the leg. The deep branch runs in the substance of the vastus medialis, and gives off **articular branches** which take part in the anastomosis around the knee.

Variations

In rare cases the femoral artery is small and ends in the profunda and circumflex branches, and the inferior gluteal artery forms the principal vessel of the lower limb.

POPLITEAL ARTERY

The popliteal artery is the direct continuation of the femoral. It begins at the medial and proximal side of the popliteal fossa, and ends by dividing into the anterior and posterior tibial arteries at the distal border of the popliteus muscle, on a level with the distal part of the tuberosity of the tibia.

Relations

Anteriorly, from above downwards, it rests on fat covering the popliteal surface of the femur, the back of the knee joint, and the fascia covering the popliteus. Posteriorly, it is overlapped proximally by the lateral border of the semimembranosus; it is crossed, about its middle, by the popliteal vein and the tibial nerve, the vein intervening between the artery and the nerve. Distally, it is overlapped by the adjacent borders of the heads of the gastrocnemius, and is crossed by the nerves to the soleus and popliteus and by the plantaris muscle. Laterally, it is in contact with the tibial nerve and the popliteal vein proximally, then with the lateral condyle of the femur and the plantaris, and, distally, with the lateral head of the gastrocnemius. Medially, it is in contact with the semimembranosus, proximally then with the medial condyle of the femur, and distally with the tibial nerve, the popliteal vein, and the medial head of the gastrocnemius. Popliteal lymph nodes are arranged irregularly around the artery.

Surface Anatomy

To map out the popliteal artery a line is drawn from a point just medial to the upper angle of the fossa to a point midway between the condyles of the femur, and thence along the middle of the fossa to the level of the lower part of the tuberosity of the tibia. The pulsations of the vessel may be felt when the limb is flexed.

Branches

Branches are given off to the hamstring muscles and to the muscles of the calf, those to the latter being called the **sural arteries.** Each head of the gastrocnemius receives a large branch which enters at a definite neurovascular hilus; these branches are virtually end arteries.

There are five **genicular arteries**—namely, lateral superior and inferior, medial superior and inferior, and a middle [FIGS 1104 and 1105].

The **lateral superior genicular artery** passes anterior to the biceps tendon, into the vastus lateralis.

The **medial superior genicular artery** passes anterior to the tendon of the adductor magnus, into the vastus medialis.

The **lateral inferior genicular artery** runs across the

popliteus muscle, and passes deep to the fibular collateral ligament.

The **medial inferior genicular artery** passes along the proximal border of the popliteus, and then turns forwards between the tibia and the tibial collateral ligament.

The **middle genicular artery** pierces the oblique popliteal ligament of the knee joint, and supplies the cruciate ligaments and the synovial membrane.

Of the **cutaneous branches,** one, the **superficial sural artery,** accompanies the small saphenous vein.

Variations

The popliteal artery may, very rarely, form the direct continuation of the inferior gluteal artery; and it sometimes divides at a more proximal or more distal level than usual.

Popliteal vessels
Medial superior genicular artery
Gastrocnemius (medial head)
Semimembranosus
Oblique popliteal ligament of knee joint
Medial inferior genicular artery
Popliteus
Posterior tibial artery
Soleus
Flexor digitorum longus
Posterior tibial artery
Tibialis posterior
Flexor digitorum longus
Communicating branch
Tibia
Tendon of tibialis posterior
Medial plantar nerve
Medial plantar artery
Lateral plantar artery
Lateral plantar nerve
Flexor retinaculum
Calcaneus

Biceps
Tibial nerve
Lateral superior genicular artery
Gastrocnemius
Plantaris
Lateral inferior genicular artery
Common peroneal nerve
Soleus
Soleus
Peroneal artery
Peroneus longus
Flexor hallucis longus
Tendon of peroneus longus
Peroneus brevis
Peroneal artery
Flexor hallucis longus
Posterior talofibular ligament
Superior peroneal retinaculum
Bursa of tendo calcaneus
Tendo calcaneus

FIG. 1105 Popliteal and posterior tibial arteries and their branches. [Cf. Fig. 458.]

The anastomosis round the knee joint is rich and is formed by the following vessels : the medial and lateral genicular, the descending genicular, the descending branch of the lateral circumflex femoral, the anterior tibial recurrent, and the circumflex fibular [see below].

POSTERIOR TIBIAL ARTERY

The posterior tibial artery, the larger terminal branch of the popliteal, begins at the distal border of the popliteus and ends, midway between the tip of the medial malleolus and the most prominent part of the heel, by dividing into the medial and the lateral plantar arteries [FIG. 1106; PLATE 51 B]. Two veins accompany it throughout.

Relations

Anteriorly, it is in contact, proximodistally, with the tibialis posterior, the flexor digitorum longus, and the posterior surfaces of the tibia and the ankle joint. Posteriorly, it is crossed by the tibial nerve about an inch distal to its origin, and elsewhere it is in contact with the fascia over the deep layer of muscles. The proximal half of the artery is also covered by the soleus and gastrocnemius muscles, but its distal half only by skin and fasciae. At its termination it lies deep to the flexor retinaculum. The tibial nerve lies at first on the medial side of the vessel, but soon crosses to its lateral side. The artery is separated from the medial malleolus by the tendons of the tibialis posterior and the flexor digitorum longus, whilst the tendon of the flexor hallucis longus lies lateral to it.

Surface Anatomy

The position of the posterior tibial artery corresponds with a line drawn from the lower angle of the popliteal fossa, at the level of the neck of the fibula, to a point midway between the medial malleolus and the most prominent part of the heel.

Branches

The posterior tibial gives branches to the adjacent muscles and to the skin of the medial part of the leg.

A **circumflex fibular branch** passes over the neck of the fibula to the anastomosis round the knee. It may arise from the anterior tibial.

The **nutrient artery to the tibia,** the largest of the nutrient arteries to long bones, springs from the proximal part of the posterior tibial and enters the nutrient foramen on the posterior surface of the bone.

The **communicating branch** unites the posterior tibial to the peroneal artery about 2·5 cm. above the tibiofibular syndesmosis.

Calcanean branches pierce the flexor retinaculum and supply the tissues of the heel. Over the tuber calcanei they form a rete with branches of the peroneal artery.

A **malleolar branch** joins the network on the medial malleolus.

The Peroneal Artery [FIG. 1105]

This is the largest branch of the posterior tibial [PLATE 51 B]. Arising about 2·5 cm. below the distal border of the popliteus, it descends along the medial crest of the fibula between the tibialis posterior in front and the flexor hallucis longus behind. About 2·5 cm. proximal to the ankle joint it gives off a perforating branch, and then passes behind the lateral malleolus to join the **rete calcaneum.** Its two companion veins anastomose around it.

Branches. The peroneal artery gives off muscular branches, the nutrient artery to the fibula, and a communicating branch which joins that of the posterior tibial.

The **perforating branch** passes through the interosseous membrane to the dorsum of the foot, where it anastomoses with branches of the anterior tibial and dorsalis pedis arteries.

A **malleolar branch** and **calcanean branches** take part in the anastomosis on the lateral side of the ankle and around the heel.

Variations. The perforating branch of the peroneal is almost invariably large when the anterior tibial artery is small; in some cases it replaces the whole of the dorsalis pedis continuation of the latter vessel.

PLANTAR ARTERIES

The medial and lateral plantar arteries arise, under cover of the flexor retinaculum, midway between the tip of the medial malleolus and the most prominent part of the medial side of the heel [FIGS. 1105 and 1106; PLATE 51B].

Medial Plantar Artery

The medial plantar artery, the smaller of the two, passes forwards, medial to the medial plantar nerve, in the interval between the abductor hallucis and the flexor digitorum brevis; it ends by uniting with the branch of the first plantar metatarsal artery to the medial side of the big toe. It gives off a superficial branch which ramifies on the abductor hallucis, and branches to the adjacent muscles, articulations, and skin; it also gives **three digital branches** which join the medial plantar metatarsal arteries.

Surface Anatomy. To mark the position of the medial plantar artery a line is drawn from the point of bifurcation of the posterior tibial artery to the first interdigital cleft. Just distal to the midpoint of this line the artery breaks up into its digital branches.

Lateral Plantar Artery

The lateral plantar artery runs obliquely across the sole on the lateral side of the lateral plantar nerve, first between the flexor digitorum brevis superficially and the flexor accessorius deeply, and then, in the interval between the flexor digitorum brevis and the abductor digiti minimi. At the base of the fifth metatarsal bone it arches across the foot, on the metatarsals, to the lateral side of the first metatarsal bone. The last part of the artery forms the **plantar arch,** which is completed by its junction with the dorsalis pedis.

Surface Anatomy. To mark the position of the lateral plantar artery a line is drawn from the point of bifurcation of the posterior tibial artery towards the fourth toe. At the level

FIG. 1106   Plantar arteries and their branches.

of the tuberosity of the fifth metatarsal it is continued across the sole, with a distal convexity, to the proximal end of the first intermetatarsal space.

Branches. The lateral plantar artery gives off: a **calcanean** branch to the skin and subcutaneous tissue of the heel; **muscular** branches; and **cutaneous** branches to the lateral side of the foot.

The **plantar arch** gives off four plantar metatarsal arteries, and twigs to the tarsal joints and muscles.

The **first plantar metatarsal artery** [p. 903] arises where the plantar arch is joined by the deep plantar branch of the dorsalis pedis artery. The **second, third,** and **fourth plantar metatarsal arteries** run on the plantar surfaces of the interossei, and each gives off a **perforating branch** to the corresponding dorsal metatarsal artery before continuing as a **common plantar digital artery.** At the interdigital clefts, each of these vessels divides into two **proper plantar digital arteries** for the adjacent sides of the toes. The proper plantar digital artery to the lateral border of the little toe arises from the lateral plantar artery where it becomes the plantar arch.

Perforating branches also arise from the plantar metatarsal arteries just before they divide.

ANTERIOR TIBIAL ARTERY

The anterior tibial artery begins opposite the distal border of the popliteus muscle, and ends in front of the ankle, where it is continued as the dorsalis pedis artery [FIG. 1107].

Course. The artery passes forwards through an aperture in the interosseous membrane, and descends on the anterior surface of the membrane, the distal part of the tibia, and the front of the ankle joint. It lies lateral to tibialis anterior, and is first medial to extensor digitorum longus and then extensor hallucis longus, the tendon of which crosses anterior to it in its distal third. It then continues between that tendon and the medial tendon of extensor digitorum longus. Two veins accompany the artery.

The deep peroneal nerve approaches the vessel from the lateral side and comes to lie in front of its middle third; more distally the nerve again becomes lateral.

Branches

The **circumflex fibular branch** may arise from the posterior tibial artery [p. 900].

The **posterior tibial recurrent branch,** small and inconstant, runs to the back of the knee joint.

The **anterior tibial recurrent artery** arises in front of the interosseous membrane. It runs upwards to the anastomosis round the knee joint [p. 900].

Muscular branches pass to the adjacent muscles, and **cutaneous** branches supply the skin of the front of the leg.

The **medial anterior malleolar artery** ramifies over the medial malleolus.

The **lateral anterior malleolar artery** anastomoses with the perforating branch of the peroneal artery and with the lateral tarsal artery.

On the surface of each malleolus the anastomosing vessels form a well-marked **malleolar rete.**

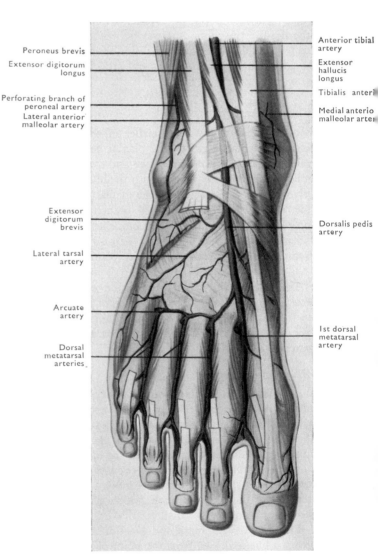

FIG. 1107 Anterior tibial artery and its branches.

FIG. 1108 Dorsalis pedis artery and its branches.

Dorsalis Pedis Artery [FIG. 1108]

This vessel is the direct continuation of the anterior tibial. It begins on the front of the ankle joint and runs to the proximal end of the first interosseous space, where it gives off the arcuate artery and continues as the first dorsal metatarsal artery which sends its deep plantar branch between the two heads of the first dorsal interosseous muscle to unite with the plantar arch [PLATE 51B]. Two veins accompany it throughout.

It is covered by skin and fasciae, including the inferior extensor retinaculum, and is crossed by the extensor hallucis brevis. It rests upon the ankle joint, the head of the talus, the navicular and intermediate cuneiform bones, and the connecting ligaments. On its lateral side is the medial terminal branch of the deep peroneal nerve, between it and the extensor digitorum brevis and the most medial tendon of the extensor digitorum longus. On its medial side is the tendon of the extensor hallucis longus.

Surface Anatomy. The line of the dorsalis pedis artery is from a point opposite the ankle joint, midway between the malleoli, to the proximal end of the first intermetatarsal space. Because of its superficial position and its close contact with the tarsal bones (against which it can be felt pulsating) the artery is frequently divided in wounds of this region.

Branches. Cutaneous branches pass to the skin on the dorsum and medial side of the foot.

The **lateral tarsal artery** runs laterally, deep to the extensor digitorum brevis.

The **medial tarsal arteries** are small branches which run to the medial border of the foot.

The **arcuate artery** runs laterally, on the bases of the metatarsals, deep to the long and short extensor tendons. It gives off **three dorsal metatarsal arteries,** second, third, and fourth, which run to the clefts of the toes, where each divides into two **dorsal digital arteries.** The lateral side of the little toe receives a branch from the most lateral artery. Each dorsal metatarsal artery gives off a **deep plantar branch** to the corresponding plantar metatarsal artery, and an anterior perforating branch to the end of the corresponding common plantar digital artery.

The **first dorsal metatarsal artery** runs forwards from the dorsalis pedis, and ends by dividing into dorsal digital branches for the adjacent sides of the big toe and second toe, and for the medial side of the big toe.

As the deep plantar branch of the first space unites with the plantar arch, the **first plantar metatarsal artery** is given off. It passes forwards, and giving off the **plantar digital artery** to the medial side of the first (big) toe, becomes the **first common plantar digital artery.** This divides into **plantar digital arteries** to the adjacent sides of the first and second toes.

VEINS

Veins commence in networks of capillaries; but whereas those of the **pulmonary circulation,** and those of the **systemic circulation,** progressively unite to form large trunks which open into the heart, the veins of the **portal system** both begin and end in capillaries [p. 922].

The four pulmonary veins return oxygenated blood from the lungs, and open into the left atrium. They are described on page 857.

The veins of the systemic circulation, and of the portal system, will now be considered.

SYSTEMIC VEINS

The systemic veins return blood to the right atrium of the heart through the **superior vena cava,** the **inferior vena cava,** and the **coronary sinus.** The coronary sinus drains blood only from the walls of the heart, while the venae cavae drain the rest of the body except the lungs.

The veins in general are much more numerous than the arteries and of greater capacity; also they anastomose more freely and in many regions form **venous plexuses** composed of networks of thin-walled, intercommunicating vessels, e.g., the pterygoid plexus [p. 909]. The presence of **valves** in the veins [p. 841] is of importance in the maintenance of the circulation.

General Arrangement

The veins of the head and neck, the body wall, and limbs form two groups: superficial and deep.

The **superficial veins** lie in the superficial fascia; they begin in the capillaries of the skin and subcutaneous tissues, and are very numerous. They freely anastomose with one another, and they also communicate with the deep veins, in which they end after piercing the deep fascia. They may or may not accompany superficial arteries.

The **deep veins** accompany arteries, except in the cranium. The large arteries have only one companion vein, but with the medium-sized and small arteries there are usually two (**venae comitantes**) which anastomose freely with each other by short transverse channels which surround the artery.

Visceral veins usually accompany the arteries which supply viscera in the head, neck, thorax, and abdomen. As a rule one vein runs with each visceral artery, and, with the exception of those which form the portal system, they end in the deep systemic veins.

The **venous sinuses** which are found inside the cranium [p. 909] differ from ordinary veins in that they have no muscular coat and are simply endothelial channels usually between the dura mater and the endocranium.

CORONARY SINUS AND VEINS OF THE HEART

The coronary sinus [FIGS. 1044 and 1109] is a short, but relatively wide, venous trunk which receives the majority of the veins of the heart. It lies in the posterior part of the coronary sulcus, between the left atrium and the left ventricle, and it is covered superficially by some of the muscular fibres

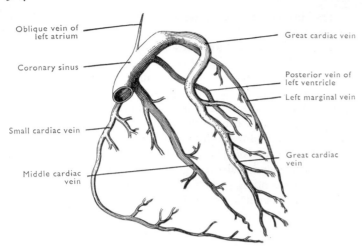

FIG. 1109   Diagram of veins on surface of heart.

of the atrium. Its opening into the right atrium has been described [p. 846].

The apertures of its tributaries are not provided with valves, except those of the great and small cardiac veins, and their valves are often incompetent.

Tributaries

The **great cardiac vein** [FIG. 1045] begins at the apex of the heart. It ascends in the anterior interventricular groove to the coronary sulcus, then passing round the left surface of the heart, ends in the left extremity of the coronary sinus. It receives the **left marginal vein** which ascends along the left (pulmonary) surface of the heart.

The **oblique vein of the left atrium** [FIG. 1044] is a slender vessel which descends on the posterior wall of the left atrium, and ends in the coronary sinus. It is a remnant of the left common cardinal vein, as is the **fold of the left vena cava** [p. 856] with which it is continuous above.

The **posterior vein of the left ventricle** lies to the left of the posterior interventricular sulcus. It usually ends in the coronary sinus, but it may end in the great cardiac vein.

The **middle cardiac vein** begins at the apex of the heart, and runs in the posterior interventricular groove to end in the coronary sinus near its right extremity.

The **small cardiac vein** begins near the apex, and follows the inferior margin of the sternocostal surface to reach the antero-inferior part of the coronary sulcus. It then turns to the left in the posterior part of the sulcus, and ends in the right extremity of the coronary sinus.

Veins of the Heart which do not end in the Coronary Sinus

The **anterior cardiac veins** are small vessels which ascend on the anterior wall of the right ventricle. Having crossed the coronary sulcus, they end directly in the right atrium [FIG. 1045]. The **venae cordis minimae** are small veins which begin in the walls of the heart and end directly in its cavities, principally in the atria. Direct observation of the interior of the left atrium during life has given proof of their functional importance (Butterworth, 1954).

SUPERIOR VENA CAVA

The superior vena cava [FIGS. 1060, 1110, and 1112] returns the blood from the upper half of the body. It is formed behind the lower border of the first right costal cartilage by the union of the two brachiocephalic veins, and it opens into the right atrium at the level of the third right cartilage. Its lower half is within the fibrous pericardium, and is covered in front and on the sides by serous pericardium.

Relations. Anteriorly, the margins of the right lung and pleura separate it from the chest wall and from the internal thoracic vessels. Posteriorly it is related to the right margin of the trachea, and then to the root of the right lung : and halfway along its length the vena azygos opens into it from behind. The ascending aorta and the brachiocephalic trunk lie to its left side (the aorta partly overlapping it), and on its right side the phrenic nerve and the pericardiacophrenic vessels lie between it and the pleura.

Surface Anatomy. The superior vena cava lies immediately to the right of the sternal margin, opposite the first and second interspaces and the second and third costal cartilages.

Tributaries. In addition to the brachiocephalic veins, the superior vena cava receives the vena azygos, and several small pericardial and mediastinal veins.

Variations. Persistence of both common cardinal veins results in the presence of two superior venae cavae, the transverse anastomosis which usually forms the left brachiocephalic vein being small or absent. In such cases the left brachiocephalic vein descends across the aortic arch, and being joined by the left superior intercostal vein, becomes the left superior vena cava which traverses the enlarged oblique vein of the left atrium and the coronary sinus to the postero-inferior part of the right atrium. More usually only the upper part of the left vein persists, particularly when it drains the corresponding pulmonary veins upwards to the left brachiocephalic vein. The right pulmonary veins may also drain into the corresponding vena cava.

THE AZYGOS VEIN AND ITS TRIBUTARIES

The vena azygos usually springs from the back of the inferior vena cava at the level of the renal veins ; but it may begin as the continuation of the right subcostal vein, or from the junction of that vein and the right ascending lumbar vein. It enters the thorax through the aortic opening of the diaphragm, and having ascended through the posterior mediastinum, arches forwards above the root of the right lung, and enters the back of the superior vena cava at the level of the second costal cartilage [FIG. 1111].

Relations

Posteriorly, t rests upon the lower eight thoracic vertebrae, and on the right posterior intercostal arteries. To its left side are the thoracic duct and the descending aorta, and, as it arches forwards above the root of the right lung, the oesophagus, trachea, and right vagus. To its right are the right lung and pleura and the greater splanchnic nerve. The oesophagus, diaphragm, right pleura and lung, and right lung root are anterior relations from below upwards in the posterior mediastinum.

Fig. 1110—Sagittal section through the superior and inferior venae cavae.

Tributaries

1. The **accessory hemiazygos vein** begins as the continuation of the fourth left posterior intercostal vein. It descends alongside the descending aorta as far as the seventh thoracic vertebra, where, turning sharply to the right, it crosses behind the aorta to end in the azygos vein. It receives the left bronchial veins, and the left posterior intercostal veins of the fourth, fifth, sixth, and seventh spaces. It communicates above with the left superior intercostal vein.

2. The **hemiazygos vein** usually springs from the back of the left renal vein and enters the thorax by piercing the left crus of the diaphragm : but it may begin as the continuation of the left subcostal vein or from the junction of that vein and the left ascending lumbar vein. It runs upwards to the eighth thoracic vertebra, where it turns to the right behind the aorta to end in the vena azygos. Its tributaries are the left ascending lumbar vein, the left subcostal vein, and the left posterior intercostal veins of the lower four spaces.

3. The **bronchial veins** are usually two on each side ; those of the right side open into the azygos vein, and those of the left into the accessory hemiazygos vein, or into the left superior intercostal vein. Many veins from the smaller bronchi open into the pulmonary veins.

4. The **ascending lumbar vein** [p. 917] is a longitudinal channel that connects the lateral sacral, iliolumbar, and lumbar veins. It ascends posterior to the psoas major, and ends in the azygos vein on the right side and the hemiazygos vein on the left side.

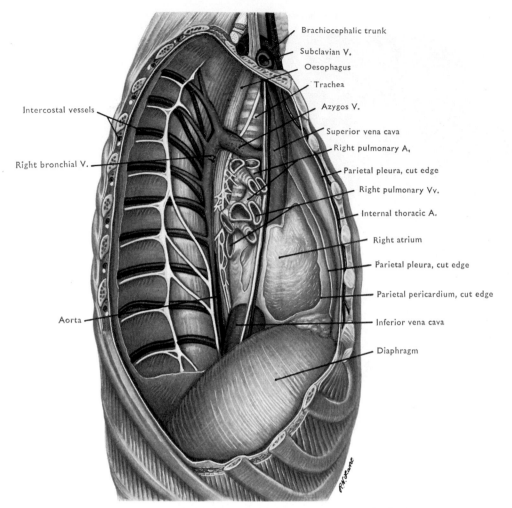

Brachiocephalic trunk

Subclavian V.

Oesophagus

Trachea

Azygos V.

Superior vena cava

Right pulmonary A.

Parietal pleura, cut edge

Right pulmonary Vv.

Internal thoracic A.

Right atrium

Parietal pleura, cut edge

Parietal pericardium, cut edge

Inferior vena cava

Diaphragm

Intercostal vessels

Right bronchial V.

Aorta

FIG. 1111   The right side of the mediastinum and vertebral column with pleura and part of pericardium removed.

For further information on the azygos and hemiazygos veins consult Gladstone (1929).

BRACHIOCEPHALIC VEINS

The brachiocephalic veins [FIGS. 1060 and 1081] are two in number, right and left. Each is formed behind the medial end of the clavicle by the union of the internal jugular and subclavian veins; they terminate, at the lower border of the right first costal cartilage, by forming the superior vena cava. They do not possess valves.

The Right Brachiocephalic Vein

This vessel is just over 2·5 cm. in length and runs almost vertically on the right side of the brachiocephalic trunk. Above, it rests on the pleura, and is anterior to the right phrenic nerve and internal thoracic artery; but lower down these structures become lateral relations, the phrenic nerve lying along its right side between it and the pleura. The right vagus nerve passes behind the upper part of the vein to reach the side of the trachea, and so becomes posteromedial.

Its **tributaries** are the right vertebral and internal thoracic veins, the highest right intercostal vein, and sometimes the right inferior thyroid vein or a common trunk of the two veins. The **right lymph duct** (or separate lymph trunks) also opens into it.

The Left Brachiocephalic Vein

This vessel, about 6·5 cm. long, runs downwards and to the right behind the upper half of the manubrium sterni, and immediately above the arch of the aorta whose branches it crosses superficially. In the root of the neck it lies in front of the left pleura, the internal thoracic artery, and the phrenic and vagus nerves. Behind the manubrium it is covered by the thymus, and is separated from the trachea by the left common carotid artery and brachiocephalic trunk [FIG. 693.]

Its **tributaries** are the **thoracic duct** (which opens into it at the angle of junction of the internal jugular and subclavian veins), the vertebral, internal thoracic, inferior thyroid, and superior intercostal veins of its own side, the highest left intercostal vein, and some pericardial and thymic veins. Sometimes the right inferior thyroid vein joins it.

Surface Anatomy. The left brachiocephalic vein lies behind the left sternoclavicular joint and the upper part of the manubrium sterni; the right vein lies behind the right sternoclavicular joint and the first right costal cartilage. The veins are about 1·5 cm. wide.

5. The **posterior intercostal veins** are eleven on each side. Each lies in the uppermost part of its intercostal space, and receives a lateral cutaneous tributary, and a dorsal tributary which passes forwards, between the transverse processes, from the vertebral plexuses.

The **highest posterior intercostal vein** ascends in front of the neck of the first rib, and arches over the pleura to end in either the brachiocephalic or the vertebral vein. The second and third (and sometimes the fourth) unite to form the **superior intercostal vein**. The **right** superior intercostal vein joins the azygos vein; the **left** crosses the aortic arch, superficial to the left vagus and deep to the left phrenic nerve, and joins the left brachiocephalic vein. The remaining posterior intercostal veins and the subcostal vein of the right side end in the azygos vein; those of the left side end in the hemiazygos and accessory hemiazygos veins.

The azygos vein also receives small twigs from the oesophagus, pericardium and diaphragm.

Variations

The arch of the azygos vein is sometimes enclosed in a fold of pleura and sunk deeply into the right lung, cutting off an accessory lobe—the lobe of the azygos vein [p. 492] medial to it.

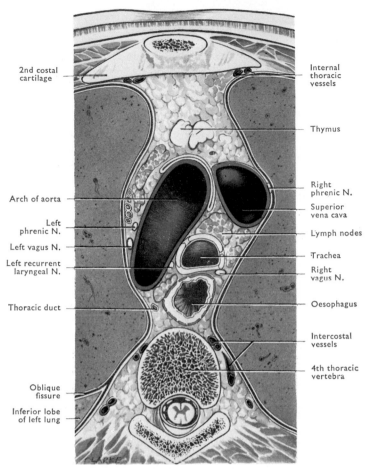

2nd costal
cartilage

Internal
thoracic
vessels

Thymus

Right
phrenic N.

Arch of aorta

Superior
vena cava

Left
phrenic N.

Lymph nodes

Left vagus N.

Left recurrent
laryngeal N.

Trachea

Right
vagus N.

Thoracic duct

Oesophagus

Intercostal
vessels

4th thoracic
vertebra

Oblique
fissure

Inferior lobe
of left lung

FIG. 1112 A horizontal section at the level of
the fourth thoracic vertebra.

Internal Thoracic Veins. Each internal thoracic artery
is accompanied by two veins which unite behind the third
costal cartilage; the single vessel then runs on the medial
side of the artery to the brachiocephalic vein of the same
side.

The tributaries of the internal thoracic veins correspond
with the branches of the internal thoracic arteries [p. 878].

Vertebral Veins. These vessels correspond only to the
extracranial parts of the vertebral arteries. Each consists of a
plexus of veins which surrounds the artery in the foramina of
the cervical transverse processes. Inferiorly, the plexus forms
a single trunk which emerges from the foramen in the transverse process of the sixth cervical vertebra, and descends to the
brachiocephalic vein.

Tributaries. The vertebral vein receives tributaries from the
external and internal vertebral plexuses, and also communicates with the suboccipital venous plexus. The **anterior
vertebral vein,** which lies alongside the ascending cervical
artery, ends in the vertebral vein immediately below the
foramen in the sixth cervical transverse process. The **deep
cervical vein** begins in the suboccipital venous plexus and
descends beside the deep cervical artery. At the root of the
neck it passes forwards into the vertebral vein.

The venous plexus around the vertebral artery may end
below in two terminal trunks. The second vessel, the
accessory vertebral vein, passes through the foramen in the

transverse process of the seventh cervical vertebra (which
otherwise transmits a small tributary), and joins the brachiocephalic vein.

Inferior Thyroid Veins [FIG. 1113].   These two vessels,
formed by tributaries from the isthmus and lobes of the gland,
descend in front of the trachea.   The right vein ends either
in the right brachiocephalic vein or in the junction of the two
brachiocephalic veins, and the left in the left brachiocephalic
vein.   The two veins may unite to form a single trunk, which
ends usually in the left brachiocephalic vein.   The inferior
thyroid veins are connected by anastomosing channels, and
sometimes form a plexus in front of the trachea.

VEINS OF THE HEAD AND NECK

Internal Jugular Vein [FIGS. 1069, 1113, 1117, and 1172]

The internal jugular vein returns the blood from the brain,
face, and much of the neck.   Each begins in the jugular
foramen as the continuation of the sigmoid sinus, and ends
behind the medial end of the clavicle by uniting with the subclavian to form the brachiocephalic vein.   Its commencement
is dilated and forms the **superior bulb.** A second dilatation,
called the **inferior bulb,** marks its lower end; it is bounded
above by a valve of two cusps.

Surface Anatomy. The position of the vein corresponds
to a line drawn from the medial end of the clavicle to the
interval between the ramus of the mandible and the mastoid
process.

Relations. Throughout its course the vein is enclosed within
the carotid sheath.   Accordingly, its relations correspond
closely with those of its companion artery, and only certain
points need be mentioned.

Immediately below the skull the internal carotid artery and
the last four cranial nerves are anteromedial to the vein; but
thereafter, it is in contact, medially, first with the internal and
then with the common carotid artery, the vagus nerve lying
posteromedially, between it and the large arteries.

The sternocleidomastoid overlaps the vein above, and covers
it below; numerous deep cervical lymph nodes lie along its
course, mostly on its superficial surface.

Just below the transverse process of the atlas it is crossed,
on its lateral side, by the accessory nerve; about its middle it
is crossed by the inferior root of the ansa cervicalis; and near
its lower end by the anterior jugular vein which runs between
it and the infrahyoid muscles.

Posterior to the vein are the cervical transverse processes
and their attached muscles, the phrenic nerve as it descends
on the scalenus anterior, and the first part of the subclavian
artery; and, on the left side, the terminal part of the thoracic
duct.

Tributaries. The **inferior petrosal sinus** joins it at its
commencement.   Two or three **pharyngeal veins** drain the
pharyngeal venous plexus. The common terminal trunk
of the **lingual veins** returns part of the blood from the tongue,
and may be joined by the vein which accompanies the hypoglossal nerve and usually ends in the facial vein [p. 908].   The
superior thyroid vein accompanies the corresponding artery;
it receives the **superior laryngeal vein.** The **middle
thyroid vein** passes backwards across the common carotid
artery.

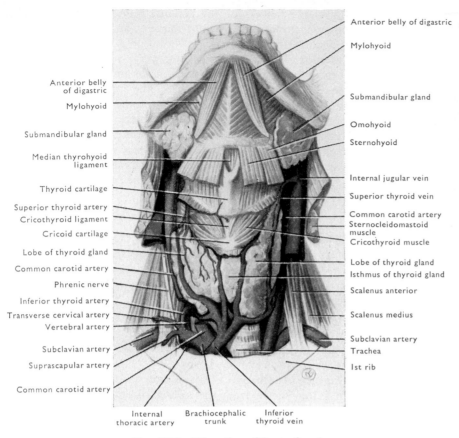

Anterior belly of digastric
Mylohyoid
Submandibular gland
Omohyoid
Sternohyoid
Internal jugular vein
Superior thyroid vein
Common carotid artery
Sternocleidomastoid muscle
Cricothyroid muscle
Lobe of thyroid gland
Isthmus of thyroid gland
Scalenus anterior
Scalenus medius
Subclavian artery
Trachea
1st rib

Anterior belly of digastric
Mylohyoid
Submandibular gland
Median thyrohyoid ligament
Thyroid cartilage
Superior thyroid artery
Cricothyroid ligament
Cricoid cartilage
Lobe of thyroid gland
Common carotid artery
Phrenic nerve
Inferior thyroid artery
Transverse cervical artery
Vertebral artery
Subclavian artery
Suprascapular artery
Common carotid artery

Internal thoracic artery
Brachiocephalic trunk
Inferior thyroid vein

FIG. 1113   Dissection of front of neck.
The lower portions of the sternocleidomastoid muscles and right common carotid artery have been removed to show the deeper parts.

Subclavian Vein

The subclavian vein is the continuation of the main vein of the upper limb, i.e., the axillary vein.

It commences at the outer border of the first rib, and ends behind the medial end of the clavicle by joining the internal jugular vein to form the brachiocephalic vein. Its surface marking may be gauged by reference to FIGURE 1050.

Each subclavian vein possesses a bicuspid valve immediately lateral to the opening of the external jugular vein.

Relations. The subclavian vein lies behind the subclavius muscle and the medial part of the clavicle. It rests on the first rib and pleura, below and in front of the third part of the subclavian artery, and then on the scalenus anterior which separates it from the second part of the artery. As soon as it reaches the medial border of the muscle it unites with the internal jugular vein.

Tributaries. The subclavian vein usually has only one named tributary—the external jugular vein.

The **external jugular vein** [FIG. 1114] is formed by the union of the posterior auricular vein with the posterior division of the retromandibular vein. It descends, deep to the platysma, across the sternocleidomastoid on which it lies in front of and parallel with the great auricular nerve. Reaching

The **facial vein** [FIG. 1114] begins at the medial angle of the eye by the union of the supra-orbital and **supratrochlear veins**. It runs behind the facial artery, and with a much straighter course. After crossing the body of the mandible at the anterior border of masseter, it leaves the facial artery and passes superficial to the submandibular gland. A little below the angle of the jaw it is joined by the anterior division of the **retromandibular vein** [p. 909], and continues across the external and internal carotid arteries to end in the internal jugular vein.

The facial vein receives tributaries corresponding with all the branches of the facial artery, except the ascending palatine and the tonsillar which have no accompanying veins. It communicates with the cavernous sinus [p. 912] through the superior ophthalmic vein, and also through the pterygoid plexus by means of the deep facial vein. Thrombosis of the facial vein —consequent upon sepsis—may thus spread to the cavernous sinus.

Supra-orbital vein
Facial vein
External nasal vein
Superior labial vein
Inferior labial vein
Facial vein
Inferior labial vein
Anastomosis between facial and anterior jugular veins
Anterior jugular vein
External jugular vein

Superficial temporal vein
Occipital vein
Maxillary veins
Retromandibular vein
Posterior auricular vein
Anterior branch of retromandibular vein
Posterior external jugular vein
Transverse cervical vein

FIG. 1114   Superficial veins of head and neck.

the subclavian triangle it pierces the deep fascia and ends in the subclavian vein.

It usually has two valves, one at its termination, and a second 4 cm. higher.

Surface Anatomy. The external jugular vein—usually visible through the skin, or easily made so by compressing it just above the clavicle, or by raising the intrathoracic pressure —runs in a line from the angle of the jaw to the junction of the medial and middle thirds of the clavicle.

Tributaries. The **posterior auricular vein** and the **posterior external jugular vein** drain the areas shown in Figure 1114.

The **transverse cervical** and **suprascapular veins** accompany the corresponding arteries.

The **anterior jugular vein** is formed in the submental region, and descends, in the superficial fascia, at a variable distance from the median plane. Inferiorly, it enters the suprasternal space [p. 301] where it is connected with its fellow of the opposite side by a transverse channel, the **jugular venous arch.** It then turns laterally, deep to the sternocleidomastoid muscle, and ends in the external jugular vein.

The external jugular vein sometimes receives the cephalic vein or a communication from it.

Variations. The external jugular vein may be absent or smaller than usual, and if so the anterior or internal jugular vein is enlarged.

Veins of the Scalp

The veins which drain the superficial parts of the scalp are the supratrochlear, the supra-orbital, the superficial temporal, the posterior auricular, and the occipital. The deeper part of the scalp, in the region of the temporal fossa, drains into the **deep temporal veins,** which pass medial to the temporalis muscle into the pterygoid plexus.

The **supratrochlear** and **supra-orbital** veins unite to form the facial vein [p. 908]. A branch of the supra-orbital vein passes through the supra-orbital notch (where it receives the frontal diploic vein—page 910) and joins the superior ophthalmic vein.

The **superficial temporal vein** [Figs. 1069 and 1114] descends to the upper border of the zygomatic arch, immediately anterior to the auricle. Here it receives the **middle temporal vein** and crosses the arch to enter the parotid gland. Within the gland it unites with the transverse facial and maxillary veins to form the retromandibular vein.

The **occipital vein** [Figs. 1069 and 1114] pierces the trapezius, and passes into the suboccipital plexus which is drained by the vertebral and deep cervical veins. Occasionally it continues along the occipital artery to end in the internal jugular vein.

Veins of the Orbit, Nose, and Infratemporal Region

The veins of these regions are closely associated, for the ophthalmic veins, which drain the orbit, are connected with the pterygoid plexus.

Veins of Orbit. With the exception of the supra-orbital and supratrochlear, the veins of the orbit correspond with the branches of the ophthalmic artery. They converge to form two main trunks—a **superior ophthalmic vein** and an **inferior ophthalmic vein**—which pass to the cavernous sinus through the superior orbital fissure. The **central vein of the retina** joins the superior ophthalmic or runs direct to the sinus.

Veins of Nose. The majority of the **internal nasal veins** join to form the **sphenopalatine vein** which passes via the sphenopalatine foramen and pterygopalatine fossa to the pterygoid plexus; some end in the ethmoidal tributaries of the superior ophthalmic vein, and others in the septal tributaries of the superior labial and in the **external nasal veins.**

Pterygoid Plexus and Maxillary Vein. The pterygoid plexus of veins lies in the infratemporal fossa on the lateral surface of the medial pterygoid muscle and around the lateral pterygoid. Its tributaries correspond with many of the branches of the maxillary artery. It communicates with the cavernous sinus by emissary veins through the foramen ovale; with the inferior ophthalmic vein through the inferior orbital fissure; and with the facial vein by the deep facial branch. Posteriorly it forms and drains into the maxillary vein.

The **maxillary vein** accompanies the first part of the maxillary artery, and unites with the superficial temporal vein to form the retromandibular vein.

The **retromandibular vein,** so formed, descends through the parotid gland within which it bifurcates. The anterior division joins the facial vein, and the posterior division forms one of the vessels of origin of the external jugular vein.

VENOUS SINUSES AND VEINS OF THE CRANIUM AND CONTENTS

The venous channels met with in the cranial walls and cavity are: the diploic veins, the meningeal veins, the veins of the brain, and the cranial venous sinuses.

FIG. 1115 Veins of the diploë.

Diploic and Meningeal Veins [PLATE 46 C]

Diploic veins are anastomosing spaces, lined with endothelium, in the marrow cavity or diploë of the flat bones of the skull [FIG. 1115] and may be seen especially in radiographs of older individuals. There are usually at least four efferent vessels on each side (Jefferson and Stewart, 1928).

The **frontal diploic vein** passes through a small aperture in the upper margin of the supra-orbital notch and ends in the supra-orbital vein.

The **anterior temporal diploic vein** ends either in the sphenoparietal sinus or in the anterior deep temporal vein.

The **posterior temporal diploic vein** ends either in the transverse sinus, or, more commonly, in the mastoid emissary vein.

The **occipital diploic vein** is usually the largest of the series, and ends either in the occipital emissary vein or in the transverse sinus.

Small unnamed diploic veins pierce the inner table of the skull and enter the venous sinuses and meningeal veins.
Meningeal Veins. These vessels arise from the diploë and from plexuses in the fused dura mater and periosteum. Some terminate in cranial venous sinuses, while others run with the meningeal arteries and end in extracranial veins. Those which accompany the middle meningeal artery separate that vessel from the bone, and are readily torn when the bone is fractured; they leave the skull through the foramen ovale or foramen spinosum and end in the pterygoid plexus.

VEINS OF THE BRAIN

The veins of the brain are thin-walled, valveless vessels which open into the cranial venous sinuses.

Veins of the Cerebrum

The cerebral veins are arranged in two groups—deep and superficial. The deep veins issue from the interior of the brain. The superficial veins lie upon its surface in the pia mater and the subarachnoid space. The terminal trunks of both sets pierce the arachnoid and the dura mater, and open into the venous sinuses.
The Deep Cerebral Veins. These are the choroid veins, the thalamostriate veins, the internal cerebral veins, the great cerebral vein, and the striate veins.

Each **choroid vein** drains the choroid plexus of a lateral ventricle. It begins in the inferior horn and curves round with the plexus to the interventricular foramen; it there receives efferents from the choroid plexus of the third ventricle, and unites with the thalamostriate vein to form the internal cerebral vein.

The **thalamostriate vein,** on each side, is formed by tributaries from the corpus striatum, the cerebral white matter, and thalamus. It runs with the stria terminalis, immediately deep to the ependyma of the lateral ventricle, in the groove between the thalamus and caudate nucleus. It ends by joining the choroid vein.

Each **internal cerebral vein** runs backwards between the layers of the tela choroidea of the third ventricle. Beneath the splenium of the corpus callosum it unites with its fellow of the opposite side to form the great cerebral vein.

The **great cerebral vein** (Galen) curves upwards round the splenium and ends in the anterior extremity of the straight sinus. It receives the basal, midbrain, and superior cerebellar veins of each side [see below].
The Superficial Cerebral Veins. These vessels lie on the surface of the cerebrum, and are divisible into two sets—the superior and the inferior [FIG. 1116].

The **superior cerebral veins,** from six to twelve in number on each side, lie on the upper and lateral aspect of the cerebral hemispheres. They converge on and enter the superior sagittal sinus, and those that encounter the lacunae laterales [p. 912] pass beneath them to reach the sinus.

The **inferior cerebral veins** lie on the tentorial and inferolateral aspects of the hemispheres, and they end in the cavernous, the superior petrosal, and the transverse sinuses. The **superficial middle cerebral vein** runs superficially along the posterior ramus of the lateral sulcus and the sulcus itself to the cavernous sinus; frequently it is united either by a **superior anastomotic vein** with the superior sagittal sinus, or by an **inferior anastomotic vein** with the transverse sinus [FIG. 1116].

The **basal vein** is formed at the anterior perforated substance by the union of: (1) the **anterior cerebral vein** which accompanies the corresponding artery; (2) the **deep middle cerebral vein** which lies deeply in the lateral sulcus and drains the insula; and (3) the **striate veins** which emerge through the anterior perforated substance. So formed, the basal vein passes backwards round the cerebral peduncle and ends in the great cerebral vein.
Veins of Midbrain. These end either in the great cerebral vein or in the basal veins.
Cerebellar Veins. Those veins form two sets, **superior**

FIG. 1116 Veins of superolateral surface of right cerebral hemisphere seen through arachnoid mater.

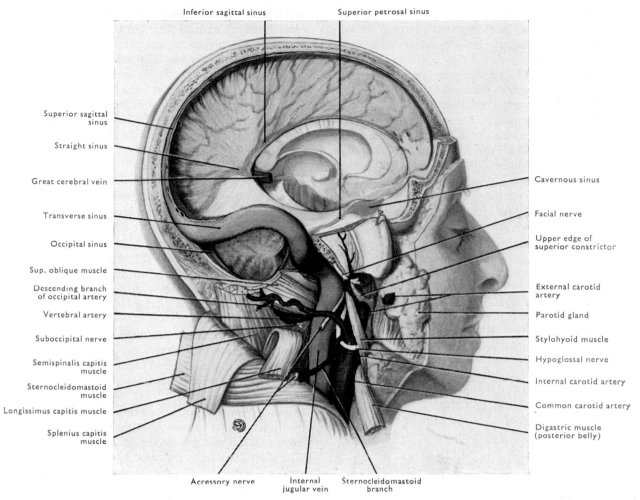

Inferior sagittal sinus   Superior petrosal sinus

Superior sagittal sinus

Straight sinus

Great cerebral vein

Transverse sinus

Occipital sinus

Sup. oblique muscle

Descending branch of occipital artery

Vertebral artery

Suboccipital nerve

Semispinalis capitis muscle

Sternocleidomastoid muscle

Longissimus capitis muscle

Splenius capitis muscle

Cavernous sinus

Facial nerve

Upper edge of superior constrictor

External carotid artery

Parotid gland

Stylohyoid muscle

Hypoglossal nerve

Internal carotid artery

Common carotid artery

Digastric muscle (posterior belly)

Accessory nerve   Internal jugular vein   Sternocleidomastoid branch

FIG. 1117 Dissection of head and neck, showing venous sinuses of dura mater and superior part of internal jugular vein.

Artery in superficial fascia

Vein in superficial fascia communicating through skull with lateral lacuna

Lax areolar tissue

Outer table of cranial wall

Diplöe

Inner table of cranial wall

Dura mater

Cerebral vein in subarachnoid space opening into superior sagittal sinus

Cortex cerebri

Skin

Loculated fat

Epicranial aponeurosis

Pericranium

Vein of diplöe connecting veins of scalp with those of dura mater

Arachnoid granulation protruding into superior sagittal sinus

Arachnoid mater

Pia mater

FIG. 1118 Diagrammatic representation of coronal section through scalp, cranium, meninges, and cerebral cortex.

and **inferior.** The former end in the great cerebral vein, and in the transverse, straight, and superior petrosal sinuses; and the latter in the straight, inferior petrosal, transverse, and occipital sinuses.

Veins of Pons. The veins of the pons pass to a ventral plexus which drains superiorly to the basal veins, and inferiorly to the cerebellar veins or petrosal sinuses.

Veins of Medulla Oblongata. The veins of the medulla end in a superficial plexus which is drained by an anterior and a posterior median vein and by radicular veins. The median veins are continuous with the corresponding veins of the spinal medulla; above, the anterior vein communicates with the plexus on the pons, and the posterior divides into two branches which join the inferior petrosal sinuses or the basilar plexus. The radicular veins run with the roots of the last four cranial nerves, and end in the inferior petrosal and occipital sinuses or in the upper part of the internal jugular vein.

VENOUS SINUSES OF THE DURA MATER

The cranial venous sinuses are spaces between the dura mater and the periosteum lining the skull, and they are lined with an endothelium continuous with that of the veins. They receive the veins of the brain, communicate with the meningeal and diploic veins and with veins external to the cranium by emissary veins, and end directly or indirectly in the internal jugular vein. Some of the sinuses are unpaired, others are paired.

Unpaired Sinuses

The **superior sagittal sinus** begins at the crista galli, where it may communicate, through the foramen caecum, with the veins of the frontal sinus, and sometimes with those of the nasal cavity. It arches backwards in the fixed margin of the falx cerebri, grooving the cranial vault [FIG. 1117], and ends at the internal occipital protuberance by becoming the right transverse sinus. Occasionally it ends in the left transverse sinus, and sometimes in both. Its termination shows a dilatation, called the **confluence of the sinuses,** which is connected by an anastomosing channel with a similar dilatation which marks the junction of the straight sinus with the transverse sinus of the opposite side. Opening into the superior sagittal sinus are the superior cerebral veins; and it communicates on each side by slit-like openings with a series of spaces in the same plane, called the **lacunae laterales,** into which arachnoid granulations project [p. 693; FIG. 876]. The parietal emissary veins [p. 913] link it with the veins on the exterior of the cranium. Its cavity is triangular in transverse section, and is crossed by several fibrous strands.

Surface Anatomy. The superior sagittal sinus occupies the median plane of the vertex from the glabella to the inion. Narrow in front, it broadens to a width of 1 cm. behind.

The **inferior sagittal sinus** runs in the posterior two-thirds of the free edge of the falx cerebri. It receives tributaries from the falx and from the medial surface of each hemisphere, and ends behind by joining with the great cerebral vein to form the straight sinus.

The **straight sinus** runs downwards and backwards in the line of union of the falx cerebri with the tentorium cerebelli,

and usually becomes continuous with the left transverse sinus. Occasionally it ends in the right transverse sinus, and in that case the superior sagittal sinus ends in the left.

Two **intercavernous sinuses** lie in the anterior and posterior margins of the diaphragma sellae; others pass across the floor of the hypophysial fossa [FIG. 698].

The **basilar plexus** lies on the clivus of the skull. It connects the cavernous and the inferior petrosal sinuses of the two sides, and communicates with the internal vertebral venous plexus.

Paired Sinuses

Each **transverse sinus,** beginning at the internal occipital protuberance, passes laterally in the attached border of the tentorium cerebelli, in a groove which runs from the occipital bone to the posterior inferior angle of the parietal bone; it then leaves the tentorium and becomes the sigmoid sinus. Its tributaries are some inferior cerebral and cerebellar veins, the posterior temporal diploic vein, the inferior anastomotic vein, and, as it becomes the sigmoid sinus, the superior petrosal sinus.

The **sigmoid sinus** first runs downwards and medially, then turning forwards enters the jugular foramen to become continuous with the internal jugular vein [FIG. 1117]. It gives rise to emissary veins which pass through the mastoid foramen and the condylar canal.

Surface Anatomy. The transverse sinus may be marked by a line, slightly convex upwards, drawn from just above the inion to the upper part of the back of the root of the auricle. The sigmoid sinus is indicated by a continuation of this line along the back of the root of the auricle to the level of the lower margin of the meatus, and then forwards to meet that part of the margin, opposite the jugular foramen [FIG. 1073].

The **occipital sinuses** lie in the attached border of the falx cerebelli—in which they may be fused—and in the dura along the posterolateral boundaries of the foramen magnum. Each links the transverse and sigmoid sinuses of its own side, and both communicate with the posterior internal vertebral plexus.

The **cavernous sinuses** [FIG. 699] lie at the sides of the body of the sphenoid bone. Each sinus begins at the superior orbital fissure, and ends at the apex of the petrous temporal by dividing into the superior and inferior petrosal sinuses. Its cavity is so divided by fibrous strands as to give the appearance of cavernous tissue. In its lateral wall are embedded the oculomotor, the trochlear, the ophthalmic, and maxillary nerves, and passing through the sinus are the internal carotid artery with its sympathetic plexus, and the abducent nerve. The ophthalmic, and inferior cerebral veins (including the superficial middle cerebral vein) and the sphenoparietal sinus enter the cavernous sinus. It communicates with its fellow through the intercavernous sinuses; with the pterygoid plexus by emissary veins [see below]; with the internal jugular vein by the inferior petrosal sinus; and through the superior ophthalmic vein with the facial vein.

The **sphenoparietal sinuses** run along the lesser wings of the sphenoid bone close to their posterior borders. Each ends in the anterior part of the cavernous sinus.

Each **superior petrosal sinus** runs in the attached margin of the tentorium cerebelli to the junction of the transverse and sigmoid sinuses. It lies on the superior margin of the petrous

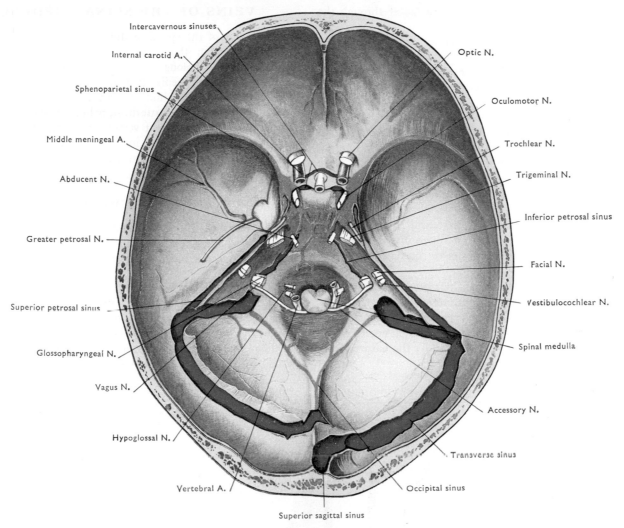

Intercavernous sinuses

Internal carotid A.

Sphenoparietal sinus

Middle meningeal A.

Abducent N.

Greater petrosal N.

Superior petrosal sinus

Glossopharyngeal N.

Vagus N.

Hypoglossal N.

Vertebral A.

Superior sagittal sinus

Optic N.

Oculomotor N.

Trochlear N.

Trigeminal N.

Inferior petrosal sinus

Facial N.

Vestibulocochlear N.

Spinal medulla

Accessory N.

Transverse sinus

Occipital sinus

FIG. 1119   The venous sinuses of the dura mater.

temporal bone, and receives inferior cerebral, superior cerebellar, and small tympanic veins.

Each **inferior petrosal sinus** begins at the posterior end of the cavernous sinus; it runs backwards, laterally, and downwards, in the groove between the temporal and occipital bones, to the anterior compartment of the jugular foramen, through which it passes, to end in the internal jugular vein. Its tributaries include cerebellar veins and the **labyrinthine vein.**

Emissary Veins

Emissary veins connect the venous sinuses with the veins on the exterior of the skull. They are of clinical importance, for they may serve as pathways for the spread of infection from outside the skull to within.

The **parietal emissary veins,** one on each side, pass through the parietal foramina and connect the superior sagittal sinus with the occipital veins. A **mastoid emissary vein** connects the sigmoid sinus, through the mastoid foramen, with the occipital or the posterior auricular vein. A **condylar emissary vein** passes through the condylar canal when present, and connects the sigmoid sinus with the suboccipital plexus of veins.

The cavernous sinus is connected to the pterygoid and pharyngeal plexuses by veins which traverse the foramen ovale, the foramen lacerum, and the sphenoidal emissary foramen, when present.

VEINS OF THE VERTEBRAL COLUMN

The veins of the vertebral column form freely communicating plexuses along its length, both inside and outside the vertebral canal. Accordingly, internal and external plexuses are described, in each of which anterior and posterior plexuses are recognized.

Internal Vertebral Plexuses. These form a continuous network, between the dura mater and the walls of the vertebral canal.

In the **anterior internal vertebral plexus** there are two longitudinal channels which communicate with each other by transverse channels that lie deep to the posterior longitudinal ligament. Two less obvious longitudinal channels can sometimes be distinguished as the **posterior internal vertebral plexus** on the internal surfaces of the vertebral arches and the ligamenta flava. Both of these plexuses communicate with the corresponding venous sinuses of the dura mater through the foramen magnum.

External Vertebral Plexuses. The anterior external vertebral plexus lies on the anterior surfaces of the vertebral bodies; the **posterior external vertebral plexus** lies on the posterolateral surfaces of the vertebrae, and around the spines and transverse processes.

The **basivertebral veins** are wide, thin-walled channels that arise within the haemopoietic tissue of the vertebral bodies. In front they communicate with the anterior external venous plexus, but their major drainage is through a large foramen in the back of the vertebral body into the transverse anastomoses of the anterior internal vertebral plexus.

Intervertebral Veins. The intervertebral veins, which convey blood from the internal vertebral plexuses, pass through the intervertebral foramina and open into the vertebral veins, the dorsal branches of the posterior intercostal, lumbar, and the lateral sacral veins.

The veins of the vertebral plexuses possess few if any valves and are in widespread connexion with veins of many parts of the body. The work of Batson (1957) strongly indicates that, *inter alia*, the mode of spread of cancer from such organs as the prostate and breast may well be explained by the nature of the vertebral plexuses and their connexions.

VEINS OF THE SPINAL MEDULLA

The veins of the spinal medulla end on the pia mater in a plexus in which there are six longitudinal channels—one **anteromedian,** along the anterior median fissure; two **anterolateral,** immediately behind the ventral nerve roots; two **posterolateral,** immediately behind the dorsal nerve roots; and one **posteromedian,** behind the posterior median sulcus. Radicular efferent vessels pass along the nerve roots to communicate with the internal vertebral venous plexus. For further details Woollam and Millen (1958) should be consulted.

VEINS OF UPPER LIMB

The veins of the upper limb are divisible into two sets, **superficial** and **deep,** both sets of which open eventually into a common trunk known as the **axillary vein.**

Deep Veins of Upper Limb

The deep veins, with the exception of the axillary vein, are arranged in pairs which accompany the arteries and are similarly named. They possess valves, and lie one on each side of their companion artery across which they are usually united by numerous transverse anastomoses.

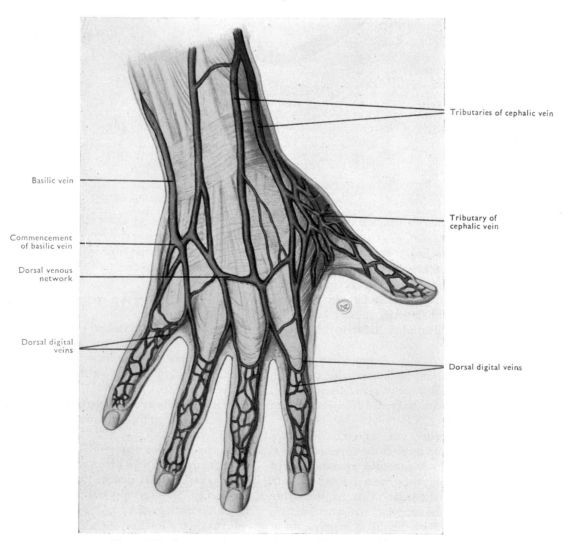

FIG. 1120   Superficial veins on dorsum of hand and digits.

The **axillary vein** [FIGS. 1069 and 1166] begins, as the continuation of the basilic vein, opposite the lower border of the teres major muscle, and ends at the outer border of the first rib by becoming the subclavian vein.

It runs on the medial side of the axillary artery, overlapping it slightly, and has similar relations.

Tributaries. Its tributaries correspond with the branches of the axillary artery, except the thoraco-acromial which joins the cephalic vein; it also receives the brachial veins at the lower border of the subscapularis, and the cephalic vein above the upper border of the pectoralis minor muscle.

Superficial Veins of Upper Limb

Veins of Digits and Hand. The **palmar digital veins** [FIG. 1121] drain to a fine plexus in the superficial fascia of the palm, and also, by means of **intercapitular veins** which pass between the heads of the metacarpal bones, to the dorsal digital veins.

The **dorsal digital veins,** two in each digit, anastomose freely and unite to form a number of **dorsal metacarpal veins** which end in a **dorsal venous network** of variable shape [FIG. 1120].

Veins of Forearm and Upper Arm. The veins of the forearm emerge from the dorsal venous network and from the palmar venous plexus. They vary considerably, but as a rule there are two main longitudinal channels—the **cephalic vein** on the radial side and the **basilic vein** on the ulnar side. Sometimes there is a **median vein** on the front of the forearm.

The **cephalic vein** begins in the dorsal venous network, receives the dorsal veins of the thumb, and ascends in the anterolateral part of the forearm to the elbow. It then runs along the lateral border of the biceps to the interval between the deltoid and pectoralis major, along which it ascends to the infraclavicular fossa. Turning medially, it pierces the clavipectoral fascia and ends in the axillary vein.

A number of tributaries join it in the forearm; at the elbow

FIG. 1121 Superficial veins on flexor aspect of right upper limb.

The median cubital vein connects the cephalic and basilic veins.

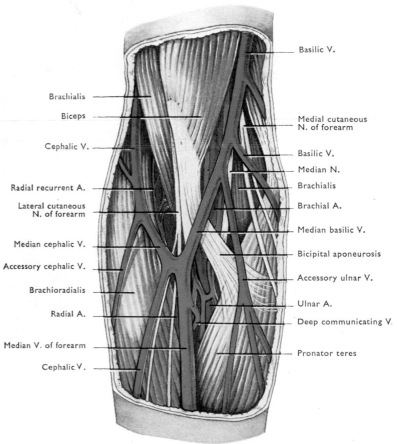

FIG. 1122 Superficial veins at the elbow.
The median vein is large.

Diaphragm

Ligamentum flavum

Portal V.

1st lumbar N.

Right renal A.

Left renal V.

Lumbar vessels

Intervertebral joint

5th lumbar N.

Rectum

Gluteus maximus

Ductus deferens

Seminal vesicle

Levator ani M

T 12

S 1

Liver

Xiphoid process

Lesser omentum

Hepatic A.

Superior part of duodenum

Omental bursa

Greater omentum

Pancreas

Transverse mesocolon

Horizontal part of duodenum

Transverse colon

Greater omentum

Right common iliac A.

The mesentery

Sigmoid colon

Urinary bladder

Pubic bone

Prostate

Scrotum

Crus of penis

FIG. 1123   A sagittal section through the abdomen and pelvis along the inferior vena cava.

it is connected with the basilic vein by the **median cubital vein;** and in the infraclavicular fossa it is joined by tributaries which correspond with the branches of the thoraco-acromial artery.

The **median cubital vein** slants upwards and medially superficial to the bicipital aponeurosis. It receives a tributary

from the deep veins of the forearm, and one or more superficial veins from the front of the forearm.

When the median cubital vein is large, the upper part of the cephalic vein is correspondingly small.

The **basilic vein** begins in the dorsal venous network of the hand. It ascends along the ulnar border of the back of the forearm for about two-thirds of its extent, and then inclines forwards to the medial bicipital groove. At the middle of the arm it pierces the deep fascia, and ascends on the medial side of the brachial artery to the axilla where it becomes the axillary vein.

It is joined by tributaries from the forearm, and in front of the elbow by the median cubital vein.

The **median vein of the forearm,** when present, begins in the palmar venous plexus and runs to the elbow. It may join either the median cubital vein or the basilic, or it may divide into the **median cephalic** and the **median basilic** veins which end in the cephalic and basilic veins respectively. The median basilic vein then replaces the median cubital vein [FIG. 1122].

Variations. The superficial veins of the forearm are extremely variable, and any of them may be absent. The cephalic vein may end in the external jugular vein, or it may be connected with it by a vessel which passes over or through the clavicle.

INFERIOR VENA CAVA

The inferior vena cava [FIGS. 1123 and 1124] returns the blood from the lower limbs, and from the walls and viscera of the abdominopelvic cavity—most of the visceral return joining it after passing through the portal system.

It begins on the body of the fifth lumbar vertebra, immediately to the right of the median plane, and ends in the right atrium very shortly after passing through the diaphragm.

Relations. It first ascends on the vertebral column, and on the anterior edge of the right psoas major, and is in contact with the right side of the aorta; higher up, however, it becomes separated from the aorta by the right crus on which it continues forwards and upwards to the caval opening in the diaphragm at the level of the eighth thoracic vertebra. It covers the right lumbar sympathetic trunk, and the right renal and middle suprarenal arteries cross behind it. Anterior to it are the right common iliac artery, the mesentery and the superior mesenteric vessels, the right gonadal artery, and the horizontal part of the duodenum. It then passes behind the head of the pancreas, the bile duct, the portal vein, the superior part of the duodenum, and the epiploic foramen. Finally it occupies a deep groove on the back of the liver, between the caudate and right lobes. To its right side there are the right ureter and kidney, and the right suprarenal gland which also lies partly behind it.

Obstruction of the inferior vena cava enlarges the collateral circulation whereby blood may pass to the heart via the superior vena cava. Thus the epigastric, circumflex iliac, lateral thoracic, thoraco-epigastric, and internal thoracic veins all become enlarged and may become exceedingly prominent; and a deep by-pass is provided by the azygos and hemiazygos veins and their tributaries, principally the vertebral plexuses.

Variations and Abnormalities. Occasionally the inferior vena cava is continuous with the vena azygos, which is much enlarged, and all the inferior caval blood is then carried to the superior vena cava. In these cases the hepatic veins open into the right atrium by a channel which represents the upper end of the inferior vena cava.

Such a variation—and there are many more, both of the inferior vena cava and of its tributaries—find an explanation in the complex mode of development of the inferior vena caval system [p. 935].

Tributaries. In addition to the common iliac veins, the inferior vena cava receives the third and fourth lumbar veins of both sides, the right gonadal vein, the right and left renal veins, the right suprarenal vein, the right inferior phrenic vein, and the hepatic veins.

There are usually five **lumbar veins** on each side, one with each aortic lumbar artery and one with the lumbar branch of the median sacral artery. Their mode of termination varies. The upper two may end in the ascending lumbar vein, or in the azygos or the hemiazygos vein; usually the third and fourth end in the inferior vena cava; the fifth ends in the iliolumbar vein. And they are all united by a longitudinal anastomosing vessel, the ascending lumbar vein, that lies deep to the psoas major.

Each **ascending lumbar vein** begins in the lateral sacral vein of the same side, anastomoses with the iliolumbar vein, connects the lumbar veins together, and anastomoses with the inferior vena cava and the renal vein. The right ascending lumbar vein ends in the azygos vein and the left in the hemiazygos vein.

The **testicular veins,** on each side, issue from the testis and epididymis and form the **pampiniform plexus** in the spermatic cord. It consists of from eight to ten veins, most of which lie in front of the ductus deferens, and ends near the deep inguinal ring in two main trunks which ascend with the corresponding testicular artery; the two veins soon unite to form a single vein which on the right side opens into the inferior vena cava, and on the left side into the left renal vein.

The much greater frequency of **varicocele,** that is, a varicose condition of the pampiniform plexus, on the left side, is said to be associated with the fact that the left testicular vein joins the left renal at right angles; moreover, it is overlaid by the descending colon.

The **ovarian veins,** on each side, issue from the hilus of the ovary. Between the layers of the broad ligament they form the **pampiniform plexus,** from which two veins accompany the ovarian artery. They soon unite to form a single vein which ends, on the right side, in the inferior vena cava, and on the left side in the left renal vein.

The **renal veins** are formed by the union of five or six tributaries which issue from the hilus of each kidney.

The **right renal vein,** about 2·5 cm. long, lies behind the descending part of the duodenum, and ends in the right side of the inferior vena cava.

The **left renal vein,** about 7·5 cm. long, crosses in front of the aorta immediately behind the superior mesenteric artery. It runs behind the body of the pancreas and above the horizontal part of the duodenum to end in the left side of the inferior vena cava. The left gonadal vein and the left suprarenal vein open into it.

A single **suprarenal vein** issues from the hilus of each gland; the right vein is very short and ends in the back of the inferior vena cava; the left descends to join the left renal vein but may cross directly to the inferior vena cava.

The **inferior phrenic veins** correspond to the inferior

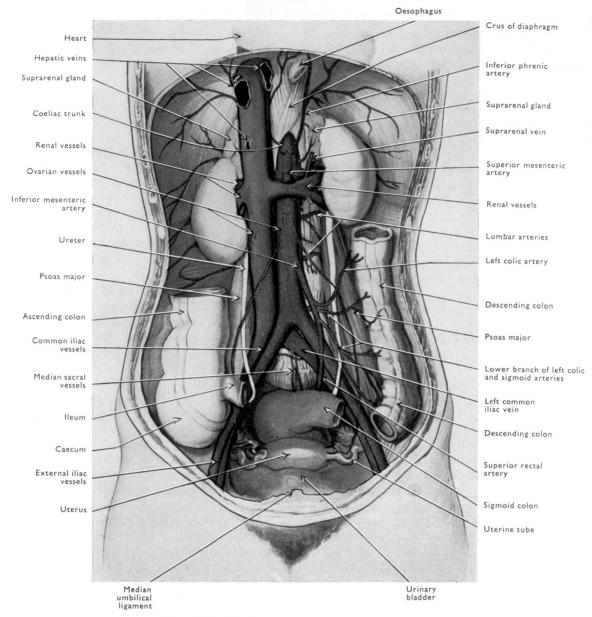

Oesophagus

Heart

Hepatic veins

Suprarenal gland

Coeliac trunk

Renal vessels

Ovarian vessels

Inferior mesenteric artery

Ureter

Psoas major

Ascending colon

Common iliac vessels

Median sacral vessels

Ileum

Caecum

External iliac vessels

Uterus

Crus of diaphragm

Inferior phrenic artery

Suprarenal gland

Suprarenal vein

Superior mesenteric artery

Renal vessels

Lumbar arteries

Left colic artery

Descending colon

Psoas major

Lower branch of left colic and sigmoid arteries

Left common iliac vein

Descending colon

Superior rectal artery

Sigmoid colon

Uterine tube

Median umbilical ligament

Urinary bladder

FIG. 1124   Inferior vena cava and its tributaries.

phrenic arteries; the right vein ends in the inferior vena cava, the left usually in the left suprarenal vein.

The **hepatic veins** [FIG. 1124] open into the inferior vena cava immediately below the diaphragm as it lies in the back of the liver. They form two groups, upper and lower. The upper group may consist of only two veins, a right and a left; more frequently there is also a middle vein which issues from the caudate lobe. All three are large trunks. The veins of the lower group vary in number from six to twenty and are smaller; they return blood from the right and caudate lobes.

Detailed accounts of the hepatic veins are given by Elias and Petty (1952), and Gibson (1959) deals with their sphincter mechanisms.

COMMON ILIAC VEINS

The common iliac veins [FIGS. 1100 and 1124], right and left, are formed by the union of the external and internal iliac

veins. Each begins at the brim of the lesser pelvis immediately behind the upper part of the internal iliac artery, and both veins pass superomedially to the front of the body of the fifth lumbar vertebra where they unite to form the inferior vena cava to the right of the midline.

The **right common iliac vein** is much shorter than the left; it lies anterior to the obturator nerve and the iliolumbar artery, and posterior to the right common iliac artery.

The **left common iliac vein** runs first on the medial side of its own artery, and then behind the right common iliac artery where it ends. Near its origin it is crossed by the sigmoid mesocolon and the superior rectal vessels.

Tributaries. Each common iliac vein receives the corresponding external iliac, internal iliac, and iliolumbar veins. The left common iliac vein receives, in addition, the **median sacral vein,** the single vessel formed by the fusion of the companion veins of the median sacral artery.

The **iliolumbar vein** receives the fifth lumbar vein, and ends in the back of the common iliac vein.

Internal Iliac Vein [FIG. 1097]

This short trunk is formed by the union of tributaries which correspond to all the branches of the internal iliac artery except the iliolumbar and umbilical arteries.

It begins at the upper border of the greater sciatic notch and ends at the brim of the lesser pelvis by uniting with the external iliac to form the common iliac vein. It lies immediately posterior to the internal iliac artery, and is crossed laterally by the obturator nerve. The left vein is in contact medially with the sigmoid colon, the right with the lower part of the ileum.

Tributaries. These form extrapelvic and intrapelvic groups.

The **extrapelvic tributaries** are all parietal; they are:

1. The **superior gluteal veins** and the **inferior gluteal veins.** They accompany the corresponding arteries through the greater sciatic foramen.

2. The **obturator vein** enters the lesser pelvis through the obturator canal, and runs along the side-wall of the pelvis immediately below the obturator artery.

3. The **internal pudendal veins** accompany the internal pudendal artery, and usually unite on the ischial spine to form a single vessel that joins the internal iliac vein. They arise in the deep perineal space by the union of the deep vein of the penis, or clitoris, with the vein from the bulb, and at their origin are in communication with the prostatic or vesical venous plexus. They receive tributaries corresponding to the branches of the artery with the exception of the dorsal vein of the penis or clitoris.

4. The **inferior rectal veins** begin in the substance of the external anal sphincter and in the walls of the anal canal; they anastomose through the submucous rectal plexus with the middle and superior rectal veins, and so connect the portal and systemic veins.

The **intrapelvic tributaries** of the internal iliac vein are **parietal** and **visceral:** the former are the lateral sacral veins; the latter are the efferent vessels from the plexuses around the pelvic viscera.

1. The **lateral sacral veins** accompany the lateral sacral arteries and end in the corresponding internal iliac vein.

2. The **pelvic venous plexuses** form dense networks of thin-walled veins associated with the rectum, bladder, prostate, uterus, and vagina; they communicate freely with one another.

The **rectal plexuses** consist of an **internal plexus** in the submucosa of the rectum and anal canal, and an **external plexus** on the outer surface of their muscular coats; free communication exists between the two. The rectal plexuses are drained by the superior, middle, and inferior rectal veins; the superior rectal vein joins the portal system, whereas the middle and inferior are tributaries of the systemic veins. The main drainage of the internal plexus is upwards by a series of large vessels which ascend in the anal columns, and then, after a further upward course in the submucosa, pierce the muscle layers and unite to form the superior rectal vein. Varicosity of the internal rectal plexus results in haemorrhoids or piles.

3. The **middle rectal veins** are inconstant. When present they begin in the submucosa of the rectum, where they communicate with the superior and inferior rectal veins in the rectal plexus; they pass through the muscular coat, and fuse to form two middle rectal veins, right and left, which end in the corresponding internal iliac vein.

The **prostatic plexus** is situated mainly within the fascial sheath of the prostate [FIG. 657]. In front it receives the dorsal vein of the penis, and behind it communicates with the vesical plexus. Efferent vessels pass to the internal iliac and internal pudendal veins, but it drains mainly through the vesical plexus.

The **vesical plexus** lies on the outer surface of the muscular coat of the lower part of the bladder, and communicates either with the prostatic plexus or with the vaginal plexus; it receives the dorsal vein of the clitoris. 4. Several **vesical veins** pass from it on each side to the internal iliac vein.

The **uterine plexuses** lie within the broad ligament, and they communicate with the ovarian and vaginal plexuses.

5. The **uterine veins,** usually two on each side, issue from the uterine plexuses and end in the internal iliac vein.

The **vaginal plexuses** lie at the sides of the vagina. They communicate with the uterine, vesical, and rectal plexuses. 6. A **vaginal vein** issues from the vaginal plexus on each side, and ends in the internal iliac vein.

Dorsal Veins of the Penis. There are two dorsal veins of the penis.

The **superficial dorsal vein** runs immediately beneath the skin, to the pubic symphysis, where it divides into right and left branches which end in the superficial external pudendal veins.

The **dorsal vein** lies beneath the deep fascia. It begins in the sulcus behind the glans, and runs backwards in the mid-dorsal line in the sulcus between the corpora cavernosa. At the root of the penis it passes below the arcuate pubic ligament, and ends by dividing into two branches which join the prostatic plexus [FIG. 652].

The **dorsal vein of the clitoris** has a similar course to that of the dorsal vein of the penis. It ends in the vesical plexus.

External Iliac Vein [FIGS. 1097, 1100, and 1124]

This vessel is the continuation of the femoral vein, and so begins immediately behind the inguinal ligament. It ascends along the brim of the lesser pelvis, and, opposite the sacro-iliac joint, it joins the internal iliac vein to form the common iliac vein. The right vein, at first medial to its artery, becomes posterior to it above; the left vein is medial to its artery throughout. In its whole course, on each side, the vein lies anterior to the obturator nerve. Its tributaries, the deep circumflex iliac and inferior epigastric veins, open into it close to its commencement. In addition, it frequently receives the pubic vein which links the obturator and external iliac veins, and may form the main termination of the obturator vein.

VEINS OF LOWER LIMB

The veins of the lower limb, like those of the upper limb, are arranged in two groups, **superficial** and **deep;** but there are important differences in detail, for in the lower limb each main artery has two companion veins only as far as the knee, where a single trunk, the **popliteal vein,** is formed. It is continued through the thigh as the **femoral vein.**

Moreover, in the upper limb the superficial veins drain into the axillary vein at the root of the limb; but in the lower limb the superficial veins of the lateral parts of the leg and foot join the popliteal vein, whilst those of the medial part of the limb join the femoral vein near the root of the limb.

The differences noted above should not obscure certain basic similarities in the superficial venous drainage of the upper and lower limbs, for the basilic and cephalic veins have their counterparts, known respectively as the small and great saphenous veins.

Deep Veins of Lower Limb

Two veins accompany each of the arteries of the limb, except the popliteal and femoral trunks and the profunda femoris, and are similarly named. They possess numerous valves.

The **popliteal vein** [FIG. 1104] is formed, at the distal border of the popliteus muscle, by the union of the anterior and posterior tibial veins. As it ascends through the popliteal fossa it crosses from medial to lateral behind the popliteal artery—to which it is bound by a dense fascial sheath. It then passes through the opening in the adductor magnus muscle and becomes the femoral vein.

Its tributaries correspond with the branches of the popliteal artery, and it receives also the small saphenous vein.

The **femoral vein** [FIG. 1125] begins at the opening in the adductor magnus muscle. It ascends through the adductor canal and the femoral triangle, and ends a little medial to the midinguinal point by becoming the external iliac vein.

In the adductor canal it lies at first posterolateral to the femoral artery and then posterior to it. As it enters the femoral triangle it is still behind the artery, but in the proximal part of the triangle it is on the medial side of the femoral artery. About 3 cm. below the inguinal ligament it enters the middle compartment of the femoral sheath.

Its **tributaries** are: (1) the profunda femoris vein; (2) the great saphenous vein; (3) several muscular tributaries, the descending genicular, the lateral and medial circumflex femoral, and the deep external pudendal veins corresponding with the arteries.

The **profunda femoris vein** lies in front of its artery and separates it from the adductor longus and the femoral vein. It receives tributaries corresponding to the branches of its artery, but only occasionally the circumflex femoral veins.

Superficial Veins of Lower Limb

The superficial veins of the lower limb terminate in two trunks, one of which, the **small saphenous vein,** passes from the foot to the back of the knee; the other, the **great saphenous vein,** extends from the foot to the groin. These veins often become varicose (i.e., dilated and tortuous).

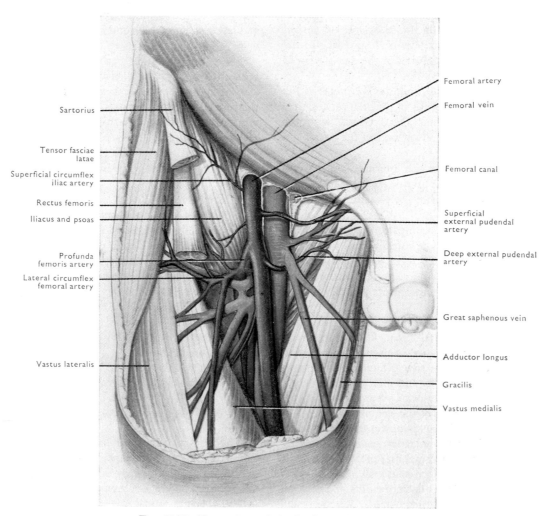

FIG. 1125 Femoral vessels in the femoral triangle.

The superficial veins of the sole form a fine plexus **(plantar rete),** immediately under the skin, from which efferents pass medially to the great saphenous vein, laterally to the small saphenous vein, and anteriorly to the **plantar venous arch.** The arch, which lies at the root of the toes, also receives small **plantar digital veins,** and it communicates between the heads of the metatarsals with the veins on the dorsum.

The superficial veins on the dorsal aspect of each toe form two **dorsal digital veins.** Those of the adjacent borders of the interdigital clefts unite to form four **dorsal metatarsal veins** which end in the dorsal venous arch of the foot.

The **dorsal venous arch** lies opposite the anterior parts of the shafts of the metatarsal bones. It unites medially with the medial dorsal digital vein of the big toe to form the **great saphenous vein,** and laterally with the lateral dorsal digital vein of the little toe to form the **small saphenous vein.** The arch receives the dorsal metatarsal veins and efferents from the plantar venous arch; and numerous tributaries from the wide-meshed **dorsal venous rete** are continuous with it posteriorly.

The **great saphenous vein** [FIG. 1126], formed as stated above, passes in front of the medial malleolus, ascends obliquely across the medial surface of the distal third of the shaft of the tibia, and reaching the knee, lies medial to the posterior part of the medial condyle of the femur. Continuing upwards, forwards, and laterally in the thigh, it perforates the cribriform fascia and the femoral sheath and ends in the femoral vein. In the foot and leg it is accompanied by the saphenous nerve. It has from eight to twenty bicuspid valves, but these become incompetent if the vein is varicose.

Tributaries. It communicates through the deep fascia with the deep, intermuscular veins at several points, particularly near the ankle and knee joints. Tributaries join it from the medial part of the sole and from the dorsal venous network, from the medial and posterior parts of the heel, the front of the leg, and the calf; and it anastomoses freely with the small saphenous vein. Cockett and Jones (1953) point out that the subcutaneous veins of the lower half of the leg on the inner

FIG. 1126   Great saphenous vein and its tributaries.

FIG. 1127   Small saphenous vein and its tributaries.

side—the area prone to ulceration—normally drain not into the great saphenous vein but directly into the posterior tibial veins through three easily definable perforating veins. In the thigh it receives numerous unnamed tributaries, and also the **superficial circumflex iliac, superficial epigastric,** and **superficial external pudendal veins** which join it just before it perforates the cribriform fascia [FIG. 1126].

The tributaries of the superficial epigastric vein communicate above with those of the lateral thoracic veins which join the axillary vein; e.g., the **thoraco-epigastric vein** on the anterolateral surface of the trunk. These communications unite vessels which pass to the inferior vena cava and the superior vena cava respectively; and they may enlarge if there is any obstruction in either of the venae cavae [p. 917].

The **small saphenous vein** [FIG. 1127] passes with the sural nerve, along the lateral side of the foot and below the lateral malleolus. It then runs behind the lateral malleolus, and along the lateral border of the tendo calcaneus, to the middle of the calf, whence it reaches the inferior angle of the popliteal fossa, pierces the deep fascia and ends in the popliteal vein. It communicates with the great saphenous vein, and through the deep fascia with the deep veins. It contains from six to twelve bicuspid valves.

Tributaries. It receives tributaries from the lateral side of the foot, the lateral side and back of the heel, and the back of the leg. Just before it pierces the popliteal fascia it frequently gives off a branch which on the medial side of the thigh unites with a vein of some size to form the **accessory saphenous vein,** a channel through which a communication is established between the great and small saphenous veins.

THE PORTAL SYSTEM OF VEINS

The veins which, with their tributaries, form the portal system are the portal, the superior and inferior mesenteric, and the splenic. They convey blood to the liver : (1) from the whole of the abdominal and pelvic parts of the alimentary canal, except the terminal part of the anal canal ; (2) from the pancreas ; and (3) from the spleen. All the larger vessels of this system are devoid of valves in the adult.

Unlike other veins, the portal vein ends by breaking up into branches which ultimately terminate in the sinusoidal capillaries in the substance of the liver. From these capillaries, which receive also the blood conveyed to the liver by the hepatic artery, the hepatic veins arise and open into the inferior vena cava. Thus the portal blood ultimately reaches the general systemic circulation.

The portal vein conveys products of digestion from the intestine, and of red blood cell destruction from the spleen to the liver ; in addition, the whole portal system also acts as a reservoir of blood for the needs of the general circulation.

THE PORTAL VEIN

This wide channel, which is about 7·5 cm. long, is formed by the union of the superior mesenteric and the splenic veins behind and to the left of the head of the pancreas. It ascends in front of the inferior vena cava and behind the pancreas and the superior part of the duodenum, until, just above the duodenum, it passes forwards and enters the lesser omentum. Continuing upwards, it lies behind the bile-duct and hepatic artery, and in front of the epiploic foramen. At the right end of the porta hepatis it ends by dividing into a short and wide right branch and a longer and narrower left branch.

The **right branch** receives the cystic vein, and then enters the right lobe of the liver in which it breaks up into numerous branches which end as interlobular veins ; these are connected to the tributaries of the hepatic veins by the sinusoids [p. 459].

The **left branch** runs from right to left, in the porta hepatis, giving off branches to the caudate and quadrate lobes ; it crosses the fissure for the round ligament of the liver, and ends in the left lobe of the liver in the same manner as the right branch. As it crosses the fissure it is joined in front by the **round ligament of the liver** and some small accompanying veins, and behind by the **ligamentum venosum.** The round ligament of the liver is a fibrous cord which passes from the umbilicus to the left branch of the portal vein ; it is the remains of the umbilical vein of the foetus. The small **para-umbilical veins** which accompany the round ligament connect the left branch of the portal vein with the superficial veins round the umbilicus. The ligamentum venosum connects the left branch of the portal vein with the upper part of the inferior vena cava. It is the remains of a foetal blood vessel—the **ductus venosus**—through which blood, carried from the placenta by the umbilical vein, passed to the inferior vena cava without going through the liver sinusoids. The portal vein is accompanied by numerous lymph vessels, and it is surrounded, in the lesser omentum, by filaments of the hepatic plexus of nerves.

For further details on the distribution of the portal vein within the liver Elias and Popper (1955) should be consulted. **Tributaries.** The **left gastric vein** runs with its artery as far as the coeliac trunk, and then with the common hepatic artery to the superior part of the duodenum where it joins the portal vein.

The **right gastric vein** runs with its artery along the lesser curvature, and ends in the portal vein as that vessel enters the lesser omentum. It receives the **prepyloric vein** which marks the position of the pylorus [p. 428].

The **cystic vein** ascends along the cystic duct and ends as a rule in the right branch of the portal vein. Some small veins pass directly from the gall-bladder into the substance of the liver.

The Superior Mesenteric Vein. This large vessel ascends along the right side of the superior mesenteric artery, and has similar relations. Behind the pancreas it unites with the splenic vein to form the portal vein.

Its tributaries are the veins that accompany the branches of the superior mesenteric artery, and also the right gastro-epiploic vein and pancreaticoduodenal veins, which enter it near its termination.

The **right gastro-epiploic vein** runs from left to right in the greater omentum, near the greater curvature of the stomach. Near the pylorus it turns downwards and backwards, and ends in the superior mesenteric vein.

The **pancreaticoduodenal veins** run with their corresponding arteries ; the lower vein ends either in the superior mesenteric or in the right gastro-epiploic, and the upper vein either in the superior mesenteric or in the portal vein.

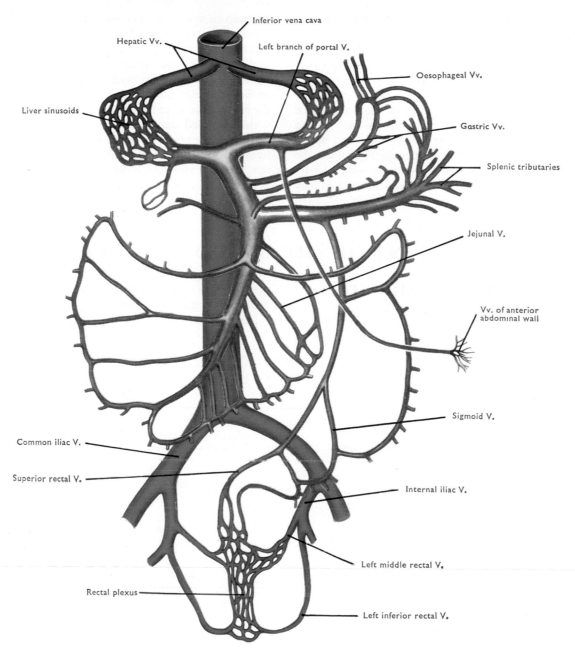

FIG. 1128   Diagram of the portal venous system (grey) to show its anastomoses with the systemic venous system (blue).

The Splenic Vein. This vessel is formed by the union of five or six tributaries which issue from the hilus of the spleen. It passes in the lienorenal ligament [p. 427] to the kidney, turns to the right and runs behind the body of the pancreas and below the splenic artery; it is separated from the aorta by the superior mesenteric artery and the left renal vein, and ends by joining the superior mesenteric vein to form the portal vein.

Its tributaries correspond to the branches of the splenic artery, and in addition it receives the inferior mesenteric vein.

The **inferior mesenteric vein** is continuous with the **superior rectal vein** on the left common iliac artery, and it receives the **sigmoid veins** from the sigmoid colon and lower portion of the descending colon, and the **left colic vein** from

the upper part of the descending colon and the left flexure. It begins on the middle of the left common iliac artery, and ascends to the left of its own artery and to the right of the ureter. It lies on the left psoas major, and crosses superficial to the left gonadal vessels. Near its upper end it passes behind, or to the left of the duodenojejunal flexure, and then, crossing in front of the left renal vein, it ends in the splenic vein behind the pancreas. Occasionally it ends in the junction of the splenic and superior mesenteric veins. Its relation to the duodenal peritoneal recesses has been mentioned on page 435.

The **superior rectal vein** is formed by tributaries of the rectal venous plexuses, through which it communicates with the middle and inferior rectal veins. It ascends with the superior rectal artery between the layers of the sigmoid mesocolon, and in front of the left common iliac artery it becomes the inferior mesenteric vein.

COMMUNICATIONS BETWEEN PORTAL AND SYSTEMIC VEINS

The most notable of these venous anastomoses occur at the lower end of the oesophagus (between the left gastric vein and oesophageal veins which join the azygos system) and in the rectal plexus of veins (between the superior rectal vein and the middle and inferior rectal veins which pass to the internal iliac) [FIG. 1128]. In both situations the communicating veins may enlarge; in the anal region such varicose enlargements of the

submucous vessels are known as haemorrhoids or piles and in the oesophagus are called oesophageal varices.

The communication with superficial veins around the umbilicus by para-umbilical veins which pass along the round ligament of the liver has been mentioned already. Obstruction to the passage of blood through the liver is often indicated by the appearance of enlarged veins radiating from the umbilicus —the so-called caput medusae. Enlarged veins may also appear in the falciform ligament or on the 'bare area' of the liver; they communicate with the veins of the diaphragm and the internal thoracic veins. Other communications occur between intestinal tributaries of the portal vein and the renal or lumbar veins.

THE DEVELOPMENT OF THE BLOOD VASCULAR SYSTEM

THE DEVELOPMENT OF THE HEART AND ARTERIES

A general account of the development of the primitive vascular system has been given on pages 62–65; before considering the further development of the heart and blood vessels certain points will be amplified.

The first evidence of the heart is found between the primitive pericardial cavity and the yolk sac as a network of cells from which two endothelial tubes are formed. Later, the tubes fuse into a single heart tube, which, following the formation of the head fold, comes to lie dorsal to the pericardial cavity. Meanwhile, the pericardial wall next to the primitive heart tube becomes thickened to form the **myo-epicardial mantle,** and, for a time, it is separated from the endothelial heart tube by a loose mesh of endocardial tissue. Later, as the heart tube sinks into the pericardial cavity, the myo-epicardial mantle forms the myocardium and epicardium.

A series of dilatations now appear on the surface of the median tubular heart, and the following parts can be recognized from the caudal end forwards: the sinus venosus in the septum transversum; the atrium, the narrow atrioventricular canal, the ventricles, and the bulbus cordis in the pericardial cavity; and the truncus arteriosus which lies beyond the pericardium.

It has been seen [pp. 63-64] that the truncus is linked through the short ventral aortae and first pair of aortic arches to the dorsal aortae. In the human embryo the ventral aortae soon fuse to form a dilated vessel, called the **aortic sac,** which is in continuity with the truncus arteriosus. When, in succession, the remaining five pairs of aortic arches are formed behind the first pair, they pass between the dorsal aortae and the aortic sac.

After its formation the heart elongates more rapidly than the pericardium and so assumes the form of a loop, the apex of which is formed by the ventricle [FIG. 1129]. The interval between the limbs of the loop is occupied by the mesocardium—a fold of the lining pericardium.

As the heart folds longitudinally it is also bent in the transverse plane, and its parts take up the positions shown in FIGURE 1129, 3A and 4A. Later, as the ventricle enlarges

ventrally and tailwards, the right segment of the sinus venosus is absorbed into the atrium, and the atrium, thus reinforced, expands round the sides of the bulbus cordis [FIG. 1129].

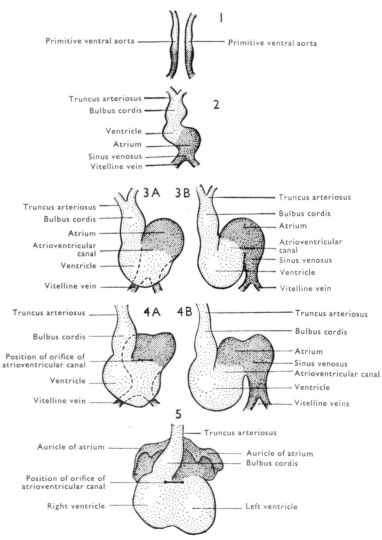

FIG. 1129 Development of the heart.
Diagrams showing the changes in form and relation of parts at different stages. Modified from His's models. 3B and 4B are side views; the other figures represent the heart as seen from the front.

Subsequently the bulbus cordis is absorbed, partly into the ventricle and partly into the truncus arteriosus. Thereafter the aortic sac and truncus divide into the pulmonary trunk and the ascending part of the aorta, the gap left by the earlier disappearance of the mesocardium now forming the transverse sinus of the pericardium between these great vessels ventrally and the atrium dorsally.

While these changes are taking place, the cavities of the atrium, atrioventricular canal, and ventricle are all being divided into right and left halves by means of septa, which it is convenient to consider separately although they are formed simultaneously and fuse with one another in parts of their extent.

Division of the Atrioventricular Canal

After the longitudinal folding of the primitive cardiac tube has occurred, the atrioventricular canal runs dorsoventrally from the atrium to the ventricle. Two thickenings, called endocardial cushions, gradually project into the canal from its cephalic and caudal walls, and by the fusion of their intermediate parts a septum of the atrioventricular canal is formed, dividing the canal into the right and left atrioventricular orifices [FIG. 1130].

Later, by excavation of the ventricular aspect of the cushions the atrioventricular valve cusps are formed. Since the cushions are continuous with the loose meshwork of ventricular muscle, this surface of the cusps remains attached to the papillary muscles by chordae tendineae.

Division of the Atrium

The primitive atrium is divided into right and left atria by the formation and later fusion of two septa—the septum primum and the septum secundum, which appear in that order —and the fusion of the septum primum with the septum of the atrioventricular canal.

The **septum primum** is a thin sheet which grows ventrally from the dorsal and cephalic walls of the atrium. As it grows,

the area of communication between the right and left parts of the atrium is gradually reduced to an aperture, called the **ostium primum** [FIG. 1130]. It is obliterated when the septum primum fuses with the septum of the atrioventricular canal, but, before that occurs, an aperture called the **ostium secundum** appears in the dorsal part of the septum primum [FIG. 1130].

The **septum secundum** appears immediately to the right of the septum primum, and is a thick, solid, incomplete septum, with its free edge facing towards the inferior vena cava.

The growth of the septum secundum ceases when its free margin overlaps the ostium secundum, and then the septum primum acts as a flap-valve against the septum secundum. This permits blood to pass through the foramen ovale—the name given to the interatrial communication after the formation of the septum secundum—from the right to the left atrium, but prevents its return [FIG. 1048]. After birth, pulmonary venous blood returning to the left atrium raises the pressure in that chamber, the septum primum is held against the septum secundum and subsequently fuses with it, so that the orifice is permanently closed. As was pointed out in the description of the adult right atrium [p. 846], the floor of the fossa ovalis is formed from the septum primum, while the limbus fossae ovalis represents the free edge of the septum secundum.

Division of the Truncus Arteriosus and the Cranial Part of the Bulbus Cordis

While the bulbus cordis is being absorbed—partly into the truncus arteriosus, and partly into the ventricular portion of the heart—spiral ridges, similar in structure to the endocardial cushions, develop in it and in the truncus arteriosus and aortic sac. There are two such ridges, and they arise on opposite walls; later, two very short ridges appear between them in the ventricular end of the part of the bulbus which will be absorbed into the truncus.

Within the aortic sac, truncus, and upper bulbus, the main ridges fuse to form a spiral septum which is so disposed that when the vessel splits longitudinally, the aorta separating from the pulmonary trunk, the latter lies on a plane posterior to the aorta above, then on its left side, and is anterior to it below. At the level of the aortic and pulmonary valves, subendothelial swellings beneath the main and accessory ridges form the valvules of the corresponding valves.

Thus far the lower (ventricular) part of the bulbus and the ventricle are still undivided, and, clearly, the process of separation must be such as to bring the right ventricle into communication with the pulmonary trunk (and so with the sixth aortic arterial arches which become parts of the pulmonary arteries—see later), and the left ventricle into communication with the aorta (and so with the remaining aortic arches which help to form the main arterial trunks of the head, neck, and upper limbs) [FIG. 1132].

Division of the lower part of the bulbus is effected by the continuous descent of the spiral septum, which separates a pulmonary channel (infundibulum) anterior and to the right, from an aortic channel, posterior and to the left. Until this septum is united to the developing interventricular septum, a communication between the two ventricles persists.

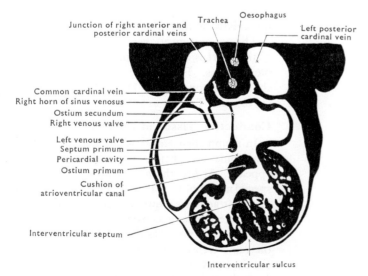

Junction of right anterior and posterior cardinal veins
Trachea
Oesophagus
Left posterior cardinal vein

Common cardinal vein
Right horn of sinus venosus
Ostium secundum
Right venous valve
Left venous valve
Septum primum
Pericardial cavity
Ostium primum
Cushion of atrioventricular canal
Interventricular septum
Interventricular sulcus

FIG. 1130   Section of heart of human embryo, showing development of septa. (Edinburgh University collection.)

Division of the Ventricle

The **interventricular septum** begins as a fold of the ventricle wall, immediately to the right of its most ventral point, and its position is indicated on the surface by a sulcus [FIG. 1130] which sometimes remains to produce a bifid apex of the heart. At first the septum is formed mainly as a result of active growth of the right and left ventricles on either side of it, and is semilunar in shape. As the septum grows, its extremities, or horns, meet the atrioventricular cushions and bulbar septum, the posterior horn joining the septum of the atrioventricular canal to the right of the attachment of the septum primum superiorly, while the anterior horn joins the anterior part of the lower margin of the bulbar septum in front. The gap which still exists above the free edge of the interventricular septum (i.e., the **interventricular foramen**) and inferior to the posteriorly placed aortic part of the bulbus, is now closed by tissue which grows forwards from the antero-inferior part of the septum of the atrioventricular canal [FIG. 1131]. This not only unites the lower end of the bulbar septum to the interventricular septum, but also forms the right wall of the aortic part of the bulbus (aortic vestibule). The fully formed interventricular septum therefore consists of two parts —a ventral and larger muscular part formed from the ventricular wall, and a dorsal and smaller fibrous part, the **pars membranacea,** formed from extensions of the atrioventricular cushions, no part of which forms muscle, but only connective tissue.

Fate of the Sinus Venosus

The sinus venosus was formed, in the septum transversum, by the fusion of the terminal parts of the common vitello-umbilical venous trunks. It consists of a single chamber possessing a right and left horn and a middle section.

Each horn receives a common cardinal vein, a vitelline vein, and an umbilical vein. Later, as the longitudinal folding of the cardiac tube occurs, the sinus venosus emerges from the septum transversum and appears in the caudal part of the pericardium. There it lies dorsal to the atrium into which its right part opens through an orifice guarded by right and left **venous valves** which fuse with each other at their cephalic and caudal ends.

As development proceeds, the right part of the sinus venosus is absorbed into the right atrium, the right common cardinal vein becoming the superior vena cava. With atrophy of the left common cardinal and right umbilical veins, and changes in the vitelline and left umbilical veins [p. 931], the left horn remains as the coronary sinus of the fully developed heart. The left venous valve disappears; the lower portion of the right valve persists in part to form the valve of the inferior vena cava and the valve of the coronary sinus, and its upper portion contributes to the formation of the **crista terminalis.**

The account given here of the development of the heart is a general outline of a very complex process. For further details the following should be consulted: Davis (1927), Odgers (1938), and Tandler (1912, 1913).

Conducting Tissue of the Heart

In mammalian embryos the heart begins to beat some time before the conducting tissue appears. At first the heart rate is slow, and no special means of propagating the impulse to cardiac contraction is required. Later, however, the heart rate increases considerably, and the conducting tissue provides the mechanism both for the rapid transmission of the impulse throughout the heart, and for the delay in transmission from the atria to the ventricles whose muscular continuity is broken by the fibrous ring.

FIG. 1131 Diagrams to show how the lower bulbar region is divided and the interventricular septum completed. (After Frazer.)

In the bottom Figure the contribution to the septum from the endocardial cushions is shown by an interrupted line.

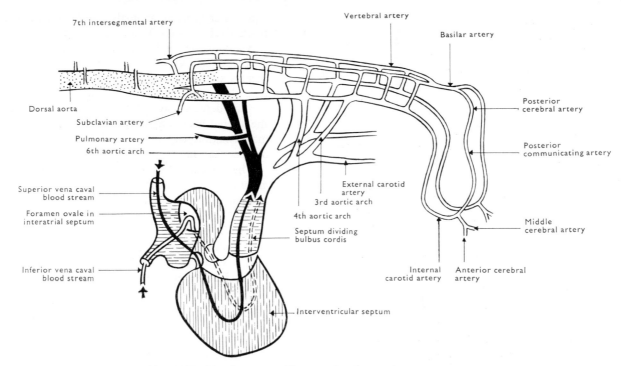

FIG. 1132 Development of heart and main arteries.
Diagram of the heart, showing the formation of its septa, and of the cephalic portion of the arterial system.

The **atrioventricular node** and bundle can be recognized when the embryo is about 10 mm. in length—by the sixth week or so. Opinions differ as to whether the node or the bundle appears first, for there are those (Shaner, 1929; Walls, 1947) who consider that the bundle grows from the node situated on the posterior wall of the atrioventricular canal, and others (Field, 1951; Muir, 1954) who are of the view that the bundle is the first structure to differentiate and that the node appears somewhat later.

The **sinu-atrial node** is the last part of the system to appear. As befits its role of pacemaker it is, from an early period, closely associated with nervous elements, but the special histological features which characterize the adult node do not appear until after birth (Duckworth, 1952).

Fate of the Aortic Arches and Unfused Parts of the Dorsal Aortae

Little need be added to the account already given [p. 65] of the formation of the aortic arches. The first arch of each side connects the corresponding dorsal aorta with the primitive ventral aorta which soon fuses with its partner to form the aortic sac. Additional pairs of arches then appear in this order, the second, the third, the fourth, and the sixth—the fifth pair appear briefly and later disappear completely—and connect the aortic sac with the dorsal aortae. Not all the arches are present at the same time, for the first two pairs have disappeared before the fifth and sixth have formed.

By reference to FIGURES 1133 and 1134 the fate of the aortic arches and of the unfused parts of the dorsal aortae may be seen. Stated briefly, the first

and second arch arteries disappear in great part. Their ventral roots contribute to the external carotid artery, which, however, is formed mainly as a new outgrowth from the third arch. In addition, the first arch artery probably contributes to the maxillary artery, and the second certainly forms the **stapedial artery**—a transient vessel which traverses the primordium of the stapes, and so helps to separate its crura.

The third aortic arches form the common carotid arteries and the proximal parts of the internal carotids; the distal parts of the latter vessels are formed by the extension headwards of the dorsal aortae. On the left side the fourth aortic arch artery becomes the part of the arch of the aorta between the left

FIG. 1133 Scheme of part of arterial system of foetus seen from left side.

Parts of the first and second arches, the ductus caroticus on both sides, the dorsal part of the right sixth arch, and the dorsal aorta caudal to the right fourth and fifth arches have atrophied. The position of the fifth arch is not indicated.

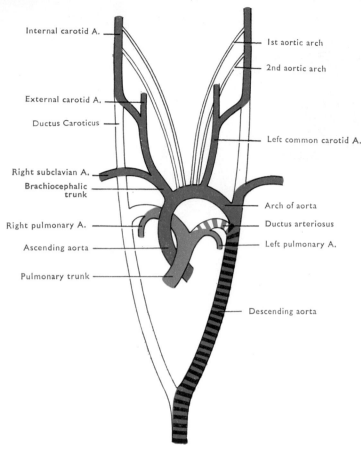

FIG. 1134   Scheme of part of the foetal arterial system. Persisting vessels are shown in solid colour.

common carotid and the left subclavian arteries, the rest of the arch of the aorta being formed from the aortic sac and from the left of two upward extensions of the sac, called horns, which result from lengthening of the neck. The right horn of the sac extends upwards as the brachiocephalic trunk and joins with the right third and fourth arch arteries, the latter of which becomes the root of the right subclavian artery. From the sixth arch artery of each side the pulmonary artery arises and thereafter the dorsal part of the right sixth artery disappears. On the left side, however, this part persists as the **ductus arteriosus** through which blood passes from the pulmonary trunk into the aortic arch until birth. The ductus becomes obliterated after birth and is converted into the ligamentum arteriosum. The ventral part of the right sixth arch artery becomes the extrapulmonary portion of the right pulmonary artery, but on the left side this part is absorbed into the pulmonary trunk.

The fates of the dorsal aortae are indicated in FIGURE 1134. For further details consult Congdon (1922) and Padget (1948).

CONGENITAL MALFORMATION OF THE HEART AND GREAT VESSELS

Although such malformations have long been of academic interest, their recognition during life is now—thanks to advances in surgery—of great practical importance. Only the commoner types of anomaly will be mentioned here, and for further information Abbott (1936) should be consulted.

Many of the conditions listed are due to arrest or failure of development. On the other hand, parts which should normally regress sometimes persist. Again, a process of rotation may take place in a reverse direction.

Such defects may occur either singly or in combination, for the altered haemodynamics which result from certain types of malformation will be themselves causative of compensatory defects elsewhere.

The system of classification which follows is broadly topographical, but it should be evident which developmental error is responsible for the defects mentioned.

1. MALPOSITION
   (i) *Dextrocardia*. The heart is transposed and is simply a mirror image of the normal. It is usually associated with transposition of the other viscera (situs inversus). Life expectancy is not affected. This anomaly is very rare.
   (ii) *Ectopia cordis*. The heart projects outside the body wall. As yet surgical treatment is of no avail and the infant survives only a matter of weeks. Very rare.
2. ATRIAL SEPTAL DEFECTS
   (i) *Patent foramen ovale*. In such cases, which are common, the opening is small and of no clinical importance. In a very few cases a wider opening is present, when, as a rule, it is associated with and probably due to other anomalies such as pulmonary stenosis (narrowing) or transposition of the aorta and pulmonary trunk.
   (ii) *Persistence of the ostium primum*. This results in a fairly large opening in the lower part of the interatrial septum.
   (iii) *Persistence of the primitive ostium secundum*. This results in a fairly large opening in the upper part of the septum.
   (iv) *Absence of the interatrial septum*. This results either in a three-chambered heart if there are two ventricles (cor triloculare biventriculare), or in a two-chambered heart if there is no interventricular septum (cor biloculare).
3. VENTRICULAR SEPTAL DEFECTS
   (i) *Absence, or near absence, of the septum*. If two atria are present, the condition is called cor triloculare biatriatum. See also above.
   (ii) *Localized defects*. These occur just below the aortic valve and are most often associated with other abnormalities such as occur in the tetralogy of Fallot: in this condition the four essentials of the tetralogy are a high ventricular septal defect, apparent dextroposition of the aorta which overrides the defect, due to the absence of the membranous septum, pulmonary stenosis, and consequent hypertrophy of the right ventricle.
4. VALVULAR ATRESIA (IMPERFORATION) AND STENOSIS
   (i) *Pulmonary*. Most often pulmonary valve atresia or stenosis is associated with septal defects. Its contribution to Fallot's tetralogy has been noted. The cause of pulmonary atresia and stenosis is failure of the bulbus cordis to develop normally.
   (ii) *Aortic*. Congenital stenosis of this valve is rare; atresia allows survival for only a few months after birth.

(iii) *Mitral and Tricuspid.* Stenosis and atresia of these valves are rare congenital anomalies, usually associated with compensatory septal defects.

5. ANOMALIES OF THE AORTA

(i) *Coarctation.* This has been mentioned on page 885.

(ii) *Transposition of the aorta and pulmonary trunk, with the aorta arising from the right ventricle and the pulmonary trunk from the left.* The condition is due to a reversal of the normal direction of rotation of the spiral, bulbar septum.

(iii) *Double Aortic Arch.* This is due to the persistence of both aortic arches and of the associated parts of the dorsal aortae.

(iv) *Right Aortic Arch.* This is due to the persistence of the fourth right aortic arch, and the associated portion of the right dorsal aorta, with disappearance of the corresponding vessels on the left.

(v) *Communications between Aorta and Pulmonary Artery.*
  a. Complete, i.e., absence of the spiral, bulbar septum with a resultant common arterial trunk.
  b. Partial, due to a localized defect in the septum.

6. PATENT DUCTUS ARTERIOSUS

This may be compensatory to some other defect, but in many cases it seems to result from some arrest of development still to be explained. Its presence results in an abnormally high pulmonary arterial pressure and a consequent hypertrophy of the right ventricular wall which causes the heart to have a globular shape—a condition common to all abnormal communications between the right and left ventricles or the great arteries.

DESCENDING AORTA

The greater part of the descending aorta is formed by the progressive fusion of the primitive dorsal aortae down to the twenty-third body segment—the level of the fourth lumbar vertebra—where the common iliac arteries arise. Later the small terminal portions of the dorsal aortae fuse to form the median sacral artery.

Branches of the Dorsal Aortae

The branches of the primitive dorsal aortae are lateral splanchnic, ventral splanchnic, and dorsolateral or somatic intersegmental.

The lateral splanchnic arteries are distributed segmentally to organs developed in or near the intermediate cell masses. They become the suprarenal, renal, and gonadal arteries.

The ventral splanchnic arteries pass to the primitive alimentary canal and its diverticula, and, in the early stages, to the walls of the yolk sac. As the yolk sac atrophies, the

C.T.A.—2 H

branches to it disappear, and simultaneously the corresponding vessels of opposite sides fuse to form unpaired stem trunks from which the three great vessels of the abdominal part of the alimentary canal are derived, namely, the coeliac, the superior mesenteric, and the inferior mesenteric arteries. But all three vessels eventually arise from the abdominal part of the aorta at levels far below those at which they first took origin.

The superior mesenteric artery retains longest its connexion with the extra-embryonic part of the yolk sac, and occasionally a fibrous strand, representing the original channel, extends from one of its ileal branches to the umbilicus.

The somatic intersegmental arteries early form a series of paired vessels from the cervical to the sacral regions. However, only in the thoracic and lumbar regions are their original characters retained. The vessels pass backwards, and divide into dorsal and ventral branches which accompany the corresponding rami of the spinal nerves [FIGS. 1135, 1136, and 1137].

The **ventral branches,** together with the stems, form the main trunks of the intercostal and lumbar arteries in the adult. They are connected together by precostal and ventral anastomoses [see below] and each gives off a branch which runs with the lateral cutaneous branch of a spinal nerve.

The **dorsal branches** form the posterior branches of the intercostal and lumbar arteries of the adult; they are connected by postcostal and by post-transverse anastomoses [see below]. Each dorsal branch gives off a spinal offset which divides into a dorsal, a ventral, and a neural branch which are linked by longitudinal anastomoses as shown in FIGURES 1135–1137.

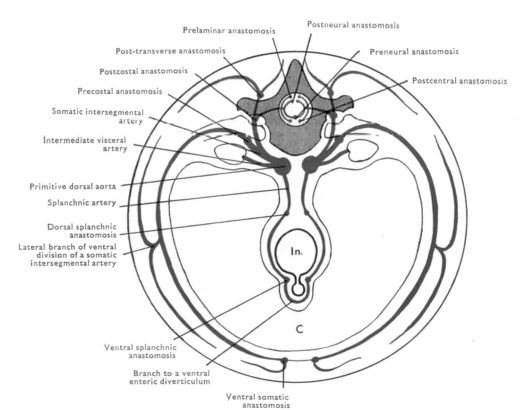

FIG. 1135 Diagram of arrangement and communications of segmental and intersegmental arteries at early stage of development.

C, coelom; In, intestine.

FIG. 1136   Diagram of segmental and intersegmental arteries at later period of development than in Fig. 1135.

C, coelom; DA, dorsal aorta; DSp, dorsal splanchnic anastomosis; In, intestine; VED, branch to ventral enteric diverticulum; VSp, ventral splanchnic anastomosis.

The precostal anastomoses persist in the thoracic region as the highest intercostal arteries; in the lumbar region they disappear. The ventral somatic anastomoses persist as the internal thoracic and superior and inferior epigastric arteries.

The postcostal and post-transverse anastomoses disappear in the thoracic and lumbar regions.

The anastomoses within the vertebral canal persist, the pre- and postneural anastomoses aiding in the formation of the anterior and posterior spinal arteries respectively.

In the cervical region, the first six pairs of somatic intersegmental arteries lose their connexions with the dorsal

aortae. The seventh pair, however, persist in their entirety; and from them are formed part of the right subclavian and the whole of the left subclavian artery. The continuation of the subclavian artery, beyond the inner margin of the first rib, is the persistent and enlarged lateral offset of the ventral branch of the seventh somatic intersegmental artery, which is continued into the upper limb. The thyrocervical trunk is a persistent precostal anastomosis, as is the ascending cervical artery. The vertebral artery is, morphologically, somewhat complex. The first part represents the dorsal branch of the seventh somatic intersegmental artery; the second part consists of persistent postcostal anastomoses; the part on the arch of the atlas is the spinal branch of the first somatic intersegmental artery and its neural continuation; whilst, finally, the part of the vertebral artery in the cranial cavity appears to represent a prolongation of the preneural anastomoses, which further upwards are probably represented by the basilar artery.

The deep cervical artery is a remnant of the post-transverse longitudinal anastomoses.

The origin of the seventh somatic intersegmental artery is, at first, caudal to the sixth aortic arch, but with the elongation of the neck and the descent of the heart, it comes to lie opposite the dorsal end of the fourth aortic arch.

UMBILICAL AND ILIAC ARTERIES

The umbilical arteries first arise cranial to their final origin from the ventral wall of the primitive dorsal aorta at the fourth lumbar segment. As each umbilical artery passes to the body stalk it lies medial to the pronephric duct; this, however, is temporary, for a new vessel arises on each side, from the lateral wall of the aorta. This new vessel passes lateral to the mesonephric duct, and then unites with the primitive umbilical artery. Thereafter the ventral origin of the umbilical artery disappears [FIG. 1138]. From the newly formed vessel there arises the **inferior gluteal artery,** the first-formed main artery of the lower limb. Later, a second branch becomes the **external iliac** and the **femoral arteries** of the adult. The portion of the umbilical stem dorsal to the external iliac becomes the **common iliac artery,** and the ventral part becomes the **internal iliac artery.** But the part of the original umbilical artery that runs to the umbilicus, and on to the placenta, is still called the umbilical artery. After birth its intra-abdominal portion atrophies and becomes the **lateral umbilical ligament;** but a portion in the pelvis remains patent, and gives origin to the **superior vesical artery.**

ARTERIES OF THE UPPER LIMB

At first a number of branches pass from the dorsal aortae to the upper limb buds where they end in vascular plexuses.

Later, the only connexion with the aorta, on each side, is the ventral branch of the seventh intersegmental artery; and a single **axial artery,** divisible into subclavian, axillary, brachial, and interosseous segments, serves the limb. Secon-

Prelaminar anastomosis
Postcentral anastomosis
Postneural anastomosis
Post-transverse anastomosis
Preneural anastomosis
Postcostal anastomosis
Precostal anastomosis
Somatic intersegmental artery
Primitive dorsal aorta
Lateral branch of a somatic intersegmental artery
Cephalic aortic arch
Branch to a lateral enteric diverticulum
Primitive ventral aorta
Ventral somatic anastomosis
Branch to a ventral enteric diverticulum

In

FIG. 1137   Diagram of arrangement and communications of segmental and intersegmental arteries in region of cephalic aortic arches.

In, Intestine.

FIG. 1138 Diagrams showing formation of secondary part of primitive umbilical artery.

dary vessels then appear, the first of which, the **median artery,** is destined to be superseded by the **radial** and **ulnar arteries.** The sequence of events can be followed very clearly in FIGURE 1139.

ARTERIES OF THE LOWER LIMB

A primary **axial artery** springs from the secondary umbilical artery, and leaves the pelvis through the greater sciatic foramen. It descends between the hamstring and adductor muscles, passes through the popliteal fossa in front of the popliteus muscle, and descends along the interosseous membrane to the foot, where it ends in a capillary plexus [FIG. 1140].

The main remnants of the primitive axial artery in the adult are—the inferior gluteal artery and its sciatic branch (the companion artery of the sciatic nerve), the proximal part of the popliteal artery, and the distal part of the peroneal artery.

A new branch, the external iliac artery, arises from the umbilical artery above the origin of the axial artery, and enters the front of the thigh. It becomes the **femoral artery** and is joined by a recurrent branch of the axial artery (**ramus com-**

municans superior) which pierces the adductor magnus so that the continuity between femoral and popliteal arteries is established.

The distal part of the **popliteal artery** and the proximal part of the **posterior tibial artery** are produced by the fusion of two branches of the axial artery. The **anterior tibial artery** arises from the 'perforating' branch of the axial artery and is joined to the popliteal by the **ramus communicans medius.** The **peroneal artery** obtains its connexion with the posterior tibial through the **ramus communicans inferior.** For further details Senior (1919, 1920) should be consulted.

THE DEVELOPMENT OF VEINS

The veins which appear during embryonic life may be classified in two groups: (1) those which are partly extra-embryonic and partly intra-embryonic; (2) those which are entirely intra-embryonic.

Veins Partly Intra-embryonic and Partly Extra-embryonic

The earliest veins to appear are the paired vitelline veins, which return blood from the yolk sac, and the paired umbilical veins, which return purified blood from the placenta [p. 341]. Their further history may now be considered.

Vitelline and Umbilical Veins, Portal Vein, Hepatic Veins, and Cardiac End of the Inferior Vena Cava. After the entodermal vesicle is separated into the embryonic alimentary canal, the yolk sac, and the vitello-intestinal duct, each vitelline vein passes along the duct to the umbilical orifice, where it passes at once into the septum transversum; there it unites with the cardiac end of the corresponding umbilical vein to form a common vitello-umbilical trunk, which terminates in the sinus venosus [FIGS. 78-79]. Subsequently, as the sinus venosus passes into the pericardial cavity, the common trunk is absorbed into it, and each vein thus acquires its own opening into the sinus [FIG. 80].

The liver is developed in the septum transversum from the branchings of a diverticulum of the duodenum. As a result, the portions of the vitelline and umbilical veins which lie in the septum are expanded as the liver grows, and at the same time broken up into many anastomosing channels called **sinusoids.** Strands of liver cells multiply and invade the sinusoidal blood spaces until they are, in the main, reduced to the size of capillaries, and the liver becomes a comparatively solid organ.

As the duodenum is delimited, the vitelline veins pass along its

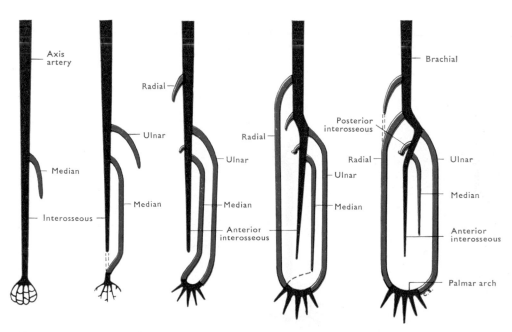

FIG. 1139 Diagrams of stages in development of arteries of upper limb. (After Arey, 1946.)

The original axial artery is black and the later-formed arteries red.

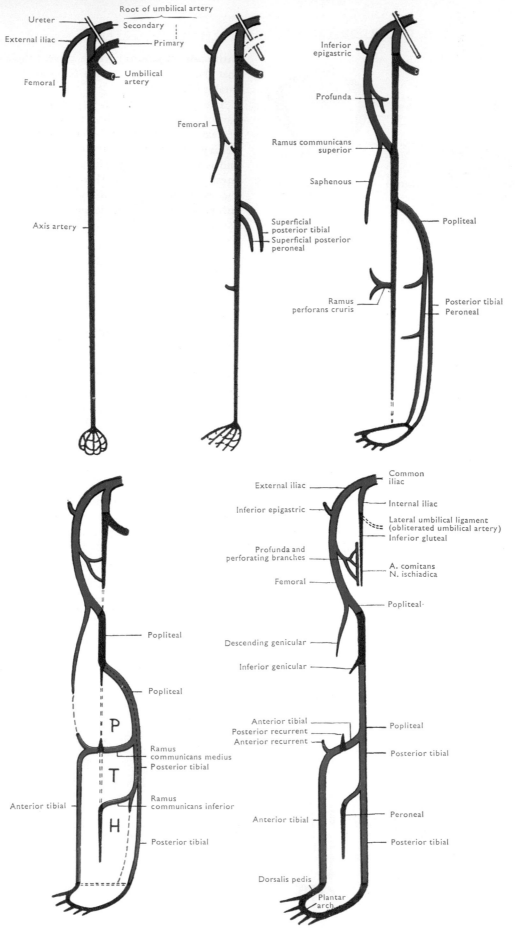

Fig. 1140 Diagrams of stages in development of arteries of lower limb. (After Senior, 1919, and Arey, 1946.)

The original umbilical artery and the axial artery are black, the later-formed arteries red. The development of the arteries of the foot is schematic and details are not shown. P, T, and H: position of popliteus, tibialis posterior and flexor hallucis longus muscles.

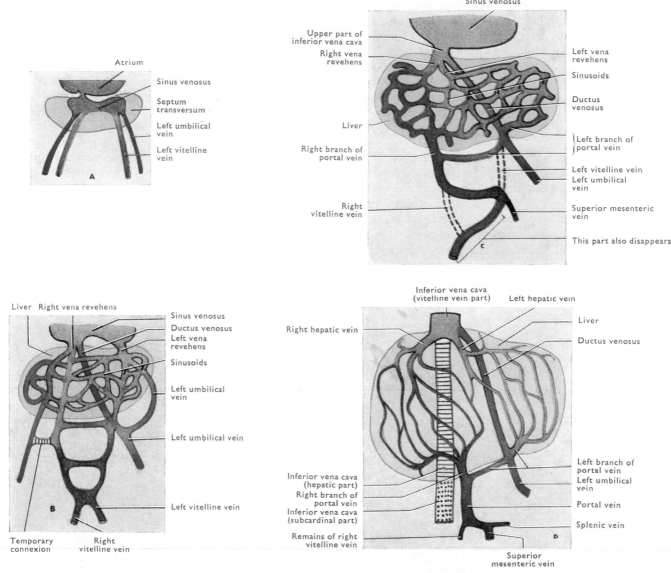

FIG. 1141   Diagram showing four stages of development of portal system and parts of inferior vena cava.

sides and are connected around it by transverse anastomoses, two of which lie ventral and one dorsal to the duodenum. Passing through the septum transversum each vitelline vein is broken up into : a caudal part, the vena advehens, which enters the liver ; an area of sinusoidal channels in the liver substance ; and a cephalic part, the vena revehens, which passes from the liver to the heart. After a time the **left vena revehens** no longer opens directly into the sinus venosus, but into the cephalic end of the **right vena revehens** [FIG. 1141]. This part of the right vena revehens becomes the terminal part of the inferior vena cava, and thus the left vena revehens forms the left hepatic vein ; the right hepatic vein is formed from the intrahepatic part of the right vena revehens. Therefore all the blood passing to the liver by the vitelline veins eventually reaches the heart by the inferior vena cava, which also receives the ductus venosus [see below].

Caudal to the duodenum the vitelline veins are replaced by the superior mesenteric vein which opens into the left vitelline vein where, a little later, the splenic vein also enters. The final result is the formation of the portal vein from parts of both vitelline veins and of their transverse anastomoses [FIG. 1141]. The right branch of the portal vein is the right vena advehens. The left branch of the portal vein is formed from the left vena advehens, and the more cephalic of the two ventral anastomoses between the vitelline veins.

Umbilical Veins. A single umbilical vein passes from the placenta to the umbilical orifice, where it divides into the left and right umbilical veins. Each of the latter becomes directly connected with the sinus venosus, and later with sinusoidal spaces of the liver and the vena advehens. The right umbilical vein undergoes early atrophy and completely disappears. The left umbilical vein persists until birth. After the disappearance of the right umbilical vein, the left conveys blood from the placenta to the liver, where part of the placental blood passes into the left vena advehens and so through the sinusoids and left vena revehens to the inferior vena cava, and part passes into the ductus venosus.

Ductus Venosus. The ductus venosus is developed as the left umbilical vein becomes united to the left vena advehens. It is formed from the sinusoidal spaces of the liver, and connects the commencement of the left vena advehens with the cephalic part of the right vena revehens. It forms the channel by which the greater part of the blood from the placenta is passed to the heart through the upper end of the inferior vena cava without passing through the liver sinusoids. After birth, the left umbilical vein becomes the ligamentum teres of the liver and the ductus venosus is converted into the ligamentum venosum [p. 922]. For further details consult Dickson (1957).

Intra-embryonic Veins

In the subsequent account the embryo and foetus are considered as being in the quadrupedal position. Therefore the terms anterior and posterior are equivalent to cephalic and caudal, and to superior and inferior in the erect posture; and the prefixes sub- and supra- are equivalent to ventral and dorsal, and to anterior and posterior in the erect posture.

Anterior Cardinal Veins. These extend, one on each side, from the region of the eye to the sinus venosus, and each is separable into cephalic, nuchal, and thoracic parts. The cephalic part, or primary head vein, runs medial to the tri-

geminal ganglion, but lateral to the otic vesicle and the seventh to the eleventh cranial nerves. Its anterior portion becomes the cavernous sinus, and its posterior portion disappears [see below].

The anterior tributaries of the primary head vein become the ophthalmic vein, whereas its dorsal tributaries form three plexuses, anterior, middle, and posterior, associated respectively with the forebrain and midbrain, with the cerebellar region, and with the region of the medulla oblongata [FIG. 1142A]. A stem vessel drains from each plexus to the primary head vein; later, when an anastomosis forms between the middle and posterior stems, the part of the primary head vein lateral to the otic vesicle disappears [FIG. 1142B].

The main parts of the plexuses are carried away from the brain with the membrane that will form the dura mater, and they later establish connexions with the veins which appear on the surfaces of the cerebral hemispheres as these develop from the forebrain.

When the cerebral hemispheres increase in size, the dural tissue is compressed between them and between the hemi-

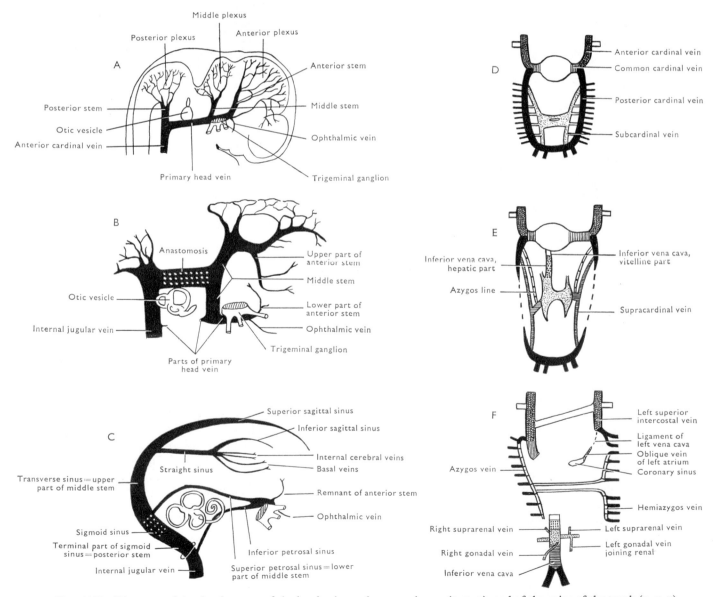

FIG. 1142 Diagrams of the development of the head veins and venous sinuses (A, B, C), and of the veins of the trunk (D, E, F).

spheres above and the midbrain and hindbrain below, in the form of folds. At the same time the anterior and middle plexuses (now conjoined) are carried towards those of the opposite side, fuse with them, and so form the superior and inferior sagittal sinuses and the straight sinus [FIG. 1142C].

Meantime, the growth of the hemispheres forces the upper part of the middle stem tributary backwards and downwards until it becomes the transverse sinus, and the anastomosis above the otic region and the posterior stem tributary are converted into the sigmoid sinus; the anterior portion of the primary head vein which lies medial to the trigeminal ganglion becomes the cavernous sinus, and the lower part of the middle stem tributary is converted into the superior petrosal sinus.

The inferior petrosal sinus appears to be an independently formed anastomosis [FIG. 1142C]. For a full account of the development of the cranial venous sinuses Streeter (1915) should be consulted.

The history of the nuchal and thoracic parts of the anterior cardinal veins is different on the two sides. On both sides the anterior cardinal vein is joined, at the root of the neck, by the chief vein of the upper limb. The part cephalic to the junction becomes the internal jugular vein. Caudal to the junction a transverse anastomosis forms between the two anterior cardinal veins. The part between the anastomosis and the junction with the limb vein on the right side, becomes the right brachiocephalic vein; on the left side it becomes part of the left brachiocephalic vein, the remainder of which is formed by the transverse anastomosis. The right anterior cardinal, caudal to the transverse anastomosis, is joined by the right posterior cardinal vein. The part cephalic to that junction becomes the extrapericardial part of the superior vena cava. The part caudal to the junction is the right **common cardinal vein** and it becomes the intrapericardial part of the superior vena cava. On the left side, the anterior and common cardinal veins virtually disappear caudal to the transverse anastomosis, losing the connexion with the left horn of the sinus venosus. The caudal part, the common cardinal vein, is represented by the oblique vein of the left atrium and the fold of the left vena cava, whilst the cephalic part becomes part of the left superior intercostal vein.

The **external jugular vein** is a new formation which receives for a time the cephalic vein of the upper limb.

Anomalies. The left superior vena cava may persist *in toto*, and then it drains into the coronary sinus (left horn of the sinus venosus). However, its upper part alone most commonly persists, particularly when the left pulmonary veins (or vein) enter it. In this case it drains upwards into the left brachiocephalic vein, and is joined by the left superior intercostal vein. Such partial or complete anomalous pulmonary venous drainage (which may occur from either or both lungs into the corresponding superior vena cava) leads to mixing of oxygenated and deoxygenated blood. Also, because of the greater volume of blood reaching the right atrium, there is a higher pressure in the right than the left atrium, and thus the foramen ovale remains open after birth, allowing venous blood to enter the left atrium and pass to the systemic circulation.

Development of the Thoraco-abdominal Veins caudal to the Heart

The veins caudal to the heart form a complicated plexus from which three main longitudinal systems develop on the posterior thoracic and abdominal walls. The definitive veins of the adult arise from parts of these systems, which because of their free anastomosis can take over each other's functions as development proceeds. The three systems are: (1) The **posterior cardinal veins** which lie lateral to the intermediate mesoderm and primarily drain the intersegmental veins of the body wall. (2) The **subcardinal veins** which lie ventromedial to the intermediate mesoderm and drain its derivatives—the kidneys, gonads, and suprarenals. The two subcardinal veins, right and left, anastomose ventral to the aorta. (3) The **supracardinal system** of veins which surrounds the sympathetic trunk and forms medial (azygos) and lateral (thoracolumbar) longitudinal channels. The former anastomoses with its fellow across the midline posterior to the aorta, and with the subcardinal vein of the same side lateral to the aorta. At their cranial and caudal ends the subcardinal and supracardinal systems anastomose with the posterior cardinal vein of the same side, and also have intermediate communications with it.

The **posterior cardinal veins** extend from the pelvis to unite with the anterior cardinal veins at the level of the septum transversum and form the common cardinal veins. As the supracardinal systems of veins take over the intersegmental veins of the trunk, the greater part of both posterior cardinal veins disappears, their most cranial parts (which still drain the other systems) remain as the upper parts of the azygos vein (on the right) and the left superior intercostal vein (on the left), while their caudal parts remain as the internal iliac veins.

The **common iliac veins,** which now drain the pelvic veins and the developing veins of the lower limbs, represent a communication between the posterior cardinal veins and the medial supracardinal veins; the left common iliac in the midline representing a communication between the two medial supracardinal veins, posterior to the aorta.

The **subcardinal veins** originally extend throughout the thorax and abdomen. As the thoracic part of the mesonephros disappears, the subcardinal veins become limited to the abdomen, where they not only anastomose with the medial supracardinal veins, but the right subcardinal also develops a communication ventrally with the hepatic venous system, thus forming an alternative route to the heart for the veins of the posterior abdominal wall. This rapidly enlarges to become the **prerenal part of the inferior vena cava** which drains the right subcardinal vein, the left subcardinal vein through an enlarged communication between these veins (the part of the left renal vein anterior to the aorta), and a large part of the supracardinal system of veins caudal to the renal veins. Probably as a result of the asymmetrical development of the prerenal part of the inferior vena cava, the right medial supracardinal vein and its communication with the right subcardinal enlarge to form the caudal, or **postrenal part of the inferior vena cava,** while its communications with the shrinking left medial supracardinal become the parts of the lumbar veins posterior to the aorta and the terminal part of the left common iliac vein.

As the mesonephros disappears and the more restricted metanephros reaches its definitive position, the subcardinal veins are limited to this level. On the left, the subcardinal vein forms a short length of the renal vein into which both gonadal and suprarenal veins drain, while on the right it forms a short segment of the inferior vena cava which receives the same tributaries from the right side. The lateral parts of the left renal vein and all of the right renal vein are formed by a communication from the posteriorly placed metanephros through the lateral and medial parts of the supracardinal system to the subcardinal. Hence the cranial parts of the supracardinal system, the **hemiazygos** and **azygos veins,** begin inferiorly in the left renal vein and the postrenal part of the inferior vena cava respectively, both of which are formed from the supracardinal veins.

Thus the **inferior vena cava** consists from below upwards of parts of : (1) the right medial supracardinal vein; (2) the right subcardinal vein; (3) the subcardinal-hepatic anastomosis; (4) hepatic sinusoids (vitelline veins); and (5) the right vitelline vein [FIG. 1142F].

For further information on the development of the veins of the trunk, Gladstone (1929) and Franklin (1948) should be consulted.

VEINS OF LIMBS

In each limb the earliest veins are a superficial distal arch and a postaxial trunk; later, digital veins connect with the distal arch, and a pre-axial trunk is formed. In the upper limb the arch and its tributaries remain as the dorsal venous network and the digital veins, and the postaxial vein becomes the basilic, axillary, and subclavian veins. The pre-axial vein of the upper limb becomes the cephalic vein, which originally terminated in the external jugular vein, above the clavicle; its union with the axillary vein is secondary.

The distal arch and its tributaries in the lower limb persist as the dorsal venous arch of the foot and the digital veins. The postaxial vein becomes the small saphenous vein, which originally continued proximally as the popliteal and inferior gluteal veins to the internal iliac portion of the posterior cardinal vein.

The pre-axial vein of the lower limb becomes the great saphenous vein, which is continued as the proximal part of the femoral and the external iliac veins.

The companion veins of the arteries in the limbs are secondarily developed vessels.

THE FOETAL CIRCULATION

During foetal life metabolic exchange occurs between the maternal and foetal blood in the placenta; accordingly, the umbilical arteries and the umbilical vein persist till birth.

The foetal lungs do not function as respiratory organs, but are bypassed by the foramen ovale and the ductus arteriosus, so that most of the blood may be transmitted to the aorta without passing through the lungs.

As a consequence, the course of the circulation is different before and after birth.

In the foetus, blood laden with food and oxygen is carried from the placenta by the umbilical vein, and transmitted to the inferior vena cava, partly directly through the ductus venosus, and partly indirectly

through the liver by branches which are ultimately taken over by the left branch of the portal vein [FIG. 1141]. In the inferior vena cava it joins venous blood returning from lower levels, and so becomes mixed blood. The mixed stream then passes by the inferior vena cava into the right atrium, and the greater part of it traverses the foramen ovale into the left atrium, whence it passes to the left ventricle, which ejects it into the ascending aorta [FIG. 1143].

However, the whole stream is not directed by the valve of the inferior vena cava through the foramen ovale, for it impinges on the limbus fossae ovalis so that a small part of it (probably less than one quarter) passes forwards and is carried into the right ventricle with the blood from the superior vena cava.

On the other hand, the blood returning by the superior vena cava to the right atrium passes wholly into the right ventricle and so to the pulmonary trunk. Only a small quantity of this stream enters the pulmonary arteries; the greater part of it

FIG. 1143   Diagram of foetal circulation.

The inferior vena caval blood is represented as passing wholly through the foramen ovale into the left atrium, but recent work [see p. 937] has confirmed the older view that part of it is directed by the limbus fossae ovalis through the right atrioventricular orifice into the right ventricle.

passes through the ductus arteriosus—which is in direct line with the pulmonary trunk [FIG. 1144]—into the descending aorta, beyond the origin of the left subclavian artery, where it mixes with the blood from the left ventricle which has not entered the branches of the arch of the aorta.

The part played in the human heart by the 'intervenous tubercle' [p. 846] in directing the stream from the superior vena cava through the right atrioventricular orifice has been exaggerated. The factors that prevent complete mixture of the blood entering the foetal right atrium through the venae cavae are the different directions of the two streams (the inferior vena cava lies on a plane posterior to the superior vena cava) so that the inferior caval blood impinges directly on the fossa ovalis. Doubts about the separation of the two streams of blood have been set at rest by the researches of Barclay and his colleagues on the foetal circulation as demonstrated by X-ray cinematography. Their method of injecting radio-opaque media into the blood stream of foetal sheep has given a clear picture of the routes taken by the blood through the living heart; see Barclay, Barcroft, Barron, and Franklin (1939), and Barclay, Franklin, and Prichard (1944).

In the foetus, therefore, pure oxygenated blood is found only in the left umbilical vein and the ductus venosus, and that becomes mixed with venous blood in the inferior vena cava. This mixed stream is delivered, via the ascending aorta to the head and neck and the upper limbs; the remainder passing into the descending aorta is joined, beyond the origin of the left subclavian artery, by the predominantly venous stream which has passed from the superior vena cava through the right side of the heart, the pulmonary trunk, and the ductus arteriosus. The latter blood, containing relatively less oxygen and food material and more waste products than the blood which is distributed from the arch of the aorta, passes to the walls of the thorax, to the abdomen and pelvis, to the lower limbs, and through the umbilical branches of the internal iliac arteries to the placenta.

TRANSITION FROM FOETAL TO ADULT CIRCULATION

The changes that occur in the vascular system at birth depend on changes in the lungs with the onset of breathing and on the cessation of the placental circulation.

At birth the ductus arteriosus begins to contract, and the **pulmonary arteries** dilate so that the pulmonary vascular resistance and pressure fall. Thus, for a while, blood passes through the ductus arteriosus from the aorta to the pulmonary artery. The consequent increase in blood flow through the lungs and the low pulmonary resistance, raise the pressure in the left atrium so that it exceeds that in the right. Thus the foramen ovale ceases to function as a passage for the blood in the inferior vena cava, and the opposing surfaces of the valve-like aperture come together and begin to unite. Eventually the foramen is structurally as well as functionally closed in most cases, though the presence of an inadequate valve (failure of the septum secundum to grow sufficiently, or persistence of an ostium primum [p. 925] will now result in blood flowing from the left atrium to the right.

The functional closure of the **ductus arteriosus** occurs within a day or two, apparently by sphincter-like action of its heavily innervated but cholinesterase-free muscular wall. Thereafter, the ductus is gradually transformed into the **ligamentum arteriosum.**

The cessation of the placental circulation, completed by ligature of the umbilical cord, leads to the obliteration of the **umbilical arteries** and **vein** which are transformed into the lateral umbilical ligaments [p. 891] and the round ligament of the liver respectively. The closure of the **ductus venosus,** like that of the ductus arteriosus, seems in the first instance to be due to active sphincteric contraction, and it is gradually reduced to a fibrous cord—the ligamentum venosum.

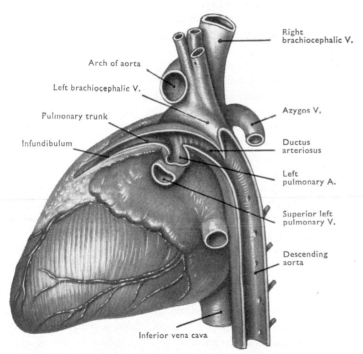

Right brachiocephalic V.

Arch of aorta

Left brachiocephalic V.

Pulmonary trunk

Infundibulum

Azygos V.

Ductus arteriosus

Left pulmonary A.

Superior left pulmonary V.

Descending aorta

Inferior vena cava

FIG. 1144 The heart and great vessels of a foetus in which the pulmonary trunk, ductus arteriosus, and descending aorta have been opened.

RADIOGRAPHIC ANATOMY

E. SAMUEL

By virtue of the differential absorption of X-rays between the heart and the adjoining lung fields, the outline of the heart can be studied easily on a radiograph or X-ray screen. Films taken during full inspiration and expiration show clearly that the transverse cardiac diameter may apparently alter by 5 per cent. according to the position of the diaphram [PLATE 36]. Studies of cardiac motion are made by observing the movement on a fluorescent screen (fluoroscopy) or by the more complicated methods of cineradiography which allow detailed analysis of the films.

Permanent records of the heart's outline are usually made by taking films in the postero-anterior, right and left anterior oblique, and lateral positions, all films being taken at 2 m. to minimize enlargement of the heart's outline by the divergent X-ray beam. This distortion effect is especially pronounced in the oblique positions where the distance between the heart and the film is increased. The right anterior oblique position, i.e., with the right shoulder against the film or screen and the incident ray directed through the chest from behind forwards, shows the left and right auricular shadows forming the posterior border of the heart shadow and the right ventricle forming it anteriorly. The arch of the aorta and the descending aorta can be seen clearly. In the left oblique position (the left shoulder against the film) the left ventricle and left auricle form the inferior and superior parts of the posterior edge of the cardiac shadow. The ascending aorta and its arch are readily seen in the left oblique or left lateral positions.

The outlines of the heart can be recognized by virtue of the surrounding translucent lung, but the individual chambers cannot be seen either on the fluorescent screen or on the radiograph as the density is uniform throughout. Likewise, the aorta and the pulmonary artery can be seen only where they abut on the lung outside the heart shadow.

The individual chambers of the heart may be visualized by the introduction of a bolus of contrast medium which is followed through the heart by taking rapid, serial radiographs. The contrast medium may be introduced either directly into an antecubital vein (angiocardiography), or through a catheter passed along it into the right atrium or ventricle (selective angiocardiography), or through a catheter passed towards the heart through one of the peripheral arteries (retrograde arterial catheterization), or even by direct ventricular puncture. These methods of study of the cardiac chambers are of the greatest importance in the analysis of congenital cardiac abnormalities.

Enlargements of the various chambers produce significant changes in the shape of the heart's outline and these appearances may be of immense value in assessing cardiac disease. It is, therefore, of fundamental importance to appreciate the part that each chamber plays in the formation of the cardiac shadow. In the postero-anterior view [PLATES 34–36] the right border of the cardiovascular shadow is formed, in its upper third, by the superior vena cava above and by the overlapping shadows of the superior vena cava and the ascending aorta below. A distinct notch on the right border indicates the junction of this vascular shadow with the bulge of the cardiac shadow proper. The remainder of the right border of the shadow is formed by the smooth, rounded curve of the right atrium. Sometimes, and especially in thin subjects, a small triangular shadow where this border reaches the diaphragm indicates the end the inferior vena cava entering the right atrium.

The left border of the cardiovascular shadow is formed craniocaudally by the rounded, knob-like prominence of the aortic arch. This tends to be more pronounced with advancing age due to degenerative changes which occur in the aortic arch. Immediately below this is the prominence caused by the pulmonary trunk, and adjoining this, a relatively insignificant flattening of the left border indicates the position of the left auricle on this border. The fourth and main prominence on the left border of the heart is the left ventricle, which also forms the apex of the heart.

The diaphragmatic surface of the heart is formed by the right and left ventricles ; the latter being seen through the stomach gas bubble in the postero-anterior projection.

The relative positions of the various chambers of the heart and of the great vessels in the lateral position are illustrated in PLATE 35 B.

Angiocardiography provides a clearer picture of the topography of the cardiac chambers than the examination of postmortem specimens. PLATE 48 A and B show that the right atrium and ventricle form an almost continuous U-shaped structure and that the separation of atrium and ventricle is less well defined than in the cadaver. The right ventricle, lying immediately posterior to the sternum, forms the main bulk of the U, the right limb of the U is formed by the right atrium, and the left limb by the infundibulum of the right ventricle and the pulmonary trunk. The left atrium is an ovoid structure lying symmetrically across the midline in the posterosuperior part of the cardiac shadow [PLATE 48 D]. Antero-inferiorly, the pyramidial cavity of the left ventricle is directed towards the apex and the relatively small size of the cavity in relation to the thickness of the ventricular wall can be appreciated readily.

In the lateral projections of the angiocardiogram, the relative positions of the pulmonary and aortic valves and the anterior position of the pulmonary trunk relative to the aorta at their emergence from the heart can be seen.

The pulmonary trunk and pulmonary arteries can be seen in the postero-anterior view [PLATE 48 A], the right pulmonary artery passing behind the ascending aorta and the superior vena cava, while the left pulmonary artery lies in the window beneath the aortic arch [PLATE 48 B].

The pulmonary veins show as faint shadows running transversely into the left artium. They are only readily visible in the right lung as the cardiac shadow obscures them on the left [PLATE 48 C].

The vena azygos can be seen immediately above the root of the right lung and may occasionally be situated within the lung substance at the inferior end of a pleural fissure which separates an accessory, azygos lobe from the right upper lobe of the lung.

Individual peripheral arteries may be demonstrated by injecting X-ray opaque medium directly into those which are accessible, or by passing a catheter from a peripheral artery (e.g., femoral) and threading it into the mouth of the appropriate branch of the aorta after visualizing it by injecting a test dose into the aorta (aortogram, PLATE 50 B). The contrast medium is then injected under pressure to form a bolus which can be followed through the branches of the vessel by taking rapid serial films. Thus the pattern and displacements of the vessels may be seen and alterations in the rate of flow through them estimated [PLATE 52]. Veins may be demonstrated by direct injection of a contrast medium (venography). This is especially valuable when pathological dilatations of veins (varicosities) are being investigated.

For further information on the radiographic anatomy of the heart and blood vessels Cooley and Schreiber (1967), Schinz and Baensch (1970), and Shanks and Kerley (1969) should be consulted.

REFERENCES

ABBIE, A. A. (1937) The anatomy of capsular vascular disease, *Med. J. Aust.*, ii, 564.

ABBIE, A. A. (1938) The blood supply of the visual pathways, *Med. J. Aust.*, ii, 199.

ABBOTT, MAUDE E. (1936) *Atlas of Congenital Cardiac Disease*, New York, Amer. Heart Assoc.

ALPERS, B. J., BERRY, R. G., and PADDISON, R. M. (1959) Anatomical studies of the circle of Willis in normal brain, *Arch. Neurol. Psychiat.* (*Chic.*), **81**, 409.

BARCLAY, A. E., BARCROFT, J., BARRON, D. H., and FRANKLIN, K. J. (1939) A radiographic demonstration of the circulation through the heart in the adult and in the foetus, and the identification of the ductus arteriosus, *Brit. J. Radiol.*, **12**, 505; *Amer. J. Roentgenol.* (1942) **47**, 678.

BARCLAY, A. E., FRANKLIN, K. J., and PRICHARD, M. M. L. (1944) *The Foetal Circulation*, Oxford.

BARLOW, T. E. (1952) Vascular patterns in the alimentary canal, in *Visceral Circulation*, Ciba Foundation Symposium, **21**, London.

BATSON, O. V. (1957) The vertebral venous system, *Amer. J. Roentgenol.*, **78**, 195.

BOWDEN, R. E. M. (1955) The surgical anatomy of the recurrent laryngeal nerve, *Brit. J. Surg.*, **43**, 153.

BOYDEN, E. A. (1955) *Segmental Anatomy of the Lungs*, New York.

BRAITHWAITE, J. L. (1952) Variation in origin of the parietal branches of the internal iliac artery, *J. Anat.* (*Lond.*), **86**, 423.

BROCK, R. C. (1952) The surgical and pathological anatomy of the mitral valve, *Brit. Heart J.*, **14**, 489.

BROCK, R. C. (1955) Control mechanisms in the outflow tract of the right ventricle in health and disease, *Guy's Hosp. Rep.*, **104**, 356.

BURY, J. S., and STOPFORD, J. S. B. (1913) On a case of occlusion of the posterior inferior cerebellar artery, *Med. Chron.*, **58**, 200.

BUTTERWORTH, R. F. (1954) The venous drainage of the left atrium, *J. Anat.* (*Lond.*), **88**, 131.

CLARK, E. R. (1938) Arterio-venous anastomoses, *Physiol. Rev.*, **18**, 229.

CLARK, E. R., and CLARK, E. L. (1943) Calibre changes in minute blood vessels observed in the living mammal, *Amer. J. Anat.*, **73**, 215.

COCKETT, F. B., and JONES, D. E. E. (1953) The ankle blow-out syndrome. A new approach to the varicose ulcer problem, *Lancet*, i, 17.

CONGDON, E. D. (1922) Transformation of the aortic-arch system during the development of the human embryo, *Contr. Embryol. Carneg. Instn*, **14**, 47.

COOLEY, R. N., and SCHREIBER, M. H. (1967) *Radiology of the Heart and Great Vessels*, 2nd ed., Baltimore.

DAVIES, F., FRANCIS, E. T. B., and KING, T. S. (1952) Neurological studies of the cardiac ventricles of mammals, *J. Anat.* (*Lond.*), **86**, 130.

DAVIES, F., FRANCIS, E. T. B., WOOD, D. R., and JOHNSON, E. A. (1956) The atrioventricular pathway for conduction of the impulse for cardiac contraction in the dog, *Trans. roy. Soc. Edinb.*, **63**, 71.

DAVIS, C. L. (1927) Development of the human heart from its first appearance to the stage found in embryos of twenty paired somites, *Contr. Embryol. Carneg. Instn*, **19**, 245.

DICKSON, A. D. (1957) The development of the ductus venosus in man and the goat, *J. Anat.* (*Lond.*), **91**, 358.

DOW, D. R., and HARPER, W. F. (1932) The vascularity of the valves of the human heart, *J. Anat.* (*Lond.*), **66**, 610.

DUCKWORTH, J. W. A. (1952) *The Development of the Sinu-atrial and Atrioventricular Nodes of the Human Heart*, M.D. Thesis, University of Edinburgh.

ELIAS, H., and PETTY, D. (1952) The gross anatomy of the blood vessels and ducts within the human liver, *Amer. J. Anat.*, **90**, 59.

ELIAS, H., and POPPER, H. (1955) Venous distribution in livers, *Arch. Path. Lab. Med.*, **59**, 332.

FIELD, E. J. (1951) The development of the conducting system in the heart of the sheep, *Brit. Heart J.*, **13**, 129.

FLOREY, H. W., POOLE, J. C. F., and MEEK, G. A. (1959) Endothelial cells and cement lines, *J. Path. Bact.*, **77**, 625.

FRANKLIN, K. J. (1937) *A Monograph on Veins*, Springfield, Ill.

FRANKLIN, K. J. (1948) *Cardiovascular Studies*, Oxford.

FULTON, W. F. M. (1965) *The Coronary Arteries*, Springfield, Ill.

GIBSON, J. B. (1959) The hepatic veins in man and their sphincteric mechanisms, *J. Anat.* (*Lond.*), **93**, 368.

GLADSTONE, R. J. (1929) Development of the inferior vena cava in the light of recent research, with special reference to certain abnormalities, and current descriptions of the ascending lumbar and azygos veins, *J. Anat.* (*Lond.*), **64**, 70.

GRAVES, F. T. (1956) The anatomy of the intrarenal arteries in health and disease, *Brit. J. Surg.*, **43**, 605.

GRAY, H., and MAHAN, E. (1943) Prediction of heart weight in man, *Amer. J. phys. Anthrop.*, **1**, 271.

GRIFFITHS, J. D. (1956) Surgical anatomy of the blood supply of the distal colon, *Ann. roy. Coll. Surg. Engl.*, **19**, 241.

HARTY, M. (1953) Blood supply of the femoral head, *Brit. med. J.*, **2**, 1236.

HARVEY, W. (1628) *Exercitatio Anatomica de Motu Cordis et Sanguinis in Animalibus*, Gulielmi Fitzeri, Frankfurt.

HILTON, J. (1863) *Lectures on Rest and Pain*, 6th ed. (1950), ed. Philipp, E. E., and Walls, E. W., London.

HIS, W. (1893) Die Thätigkeit des embryonalen Herzens, *Arb. med. Klin. Lpz.*, cited by Mall, F. P., in *Amer. J. Anat.*, **13**, 278.

JEFFERSON, G., and STEWART, D. (1928) On the veins of the diploë, *Brit. J. Surg.*, **16**, 70.

KEEN, E. N. (1955) The postnatal development of the human cardiac ventricles, *J. Anat.* (*Lond.*), **89**, 484.

KEITH, A., and FLACK, M. (1907) The form and nature of the muscular connexion between the primary divisions of the vertebrate heart, *J. Anat.* (*Lond.*), **41**, 172.

KING, T. S., and COAKLEY, J. B. (1958) The intrinsic nerve cells of the cardiac atria of mammals and man, *J. Anat.* (*Lond.*), **92**, 353.

KUGLER, J. H., and PARKIN, J. B. (1956) Continuity of Purkinje fibres with cardiac muscle, *Anat. Rec.*, **126**, 335.

MAINLAND, D., and GORDON, E. J. (1941) The position of organs determined from thoracic radiographs of young adult males, with a study of the cardiac apex beat, *Amer. J. Anat.*, **68**, 457.

MALL, F. P. (1911) On the muscular architecture of the ventricles of the human heart, *Amer. J. Anat.*, **11**, 211.

MALPIGHI, M. (1661) *De Pulmonibus*, Bononiae, quoted by Cole, F. J. (1944), in *A History of Comparative Anatomy*, p. 180, London.

MASSON, P. (1937) *Les Glomus neuro-vasculaires*, Paris.

MICHELS, N. A. (1949) Variations in the blood supply of the supra-mesocolonic organs, *J. int. Coll. Surg.*, **12**, 625.

MINOT, C. S. (1900) On a hitherto unrecognized form of blood circulation without capillaries in the organs of the vertebrata, *Proc. Boston Soc. Nat. Hist.*, **29**, 185.

MUIR, A. (1954) The development of the ventricular part of the conducting tissue in the heart of the sheep, *J. Anat.* (*Lond.*), **88**, 381.

NAKAMURA, N. (1924) Zur Anatomie der Bronchialarterien, *Anat. Anz.*, **58**, 508.

ODGERS, P. N. B. (1938) The development of the pars membranacea septi in the human heart, *J. Anat.* (*Lond.*), **72**, 247.

PADGET, D. H. (1948) The development of the cranial arteries in the human embryo, *Contr. Embryol. Carneg. Instn*, **32**, 205.

POWELL, E. D. U., and MULLANEY, J. M. (1960) The Chiari network and the valve of the inferior vena cava, *Brit. Heart J.*, **22**, 579.

PRICHARD, M. M. L., and DANIEL, P. M. (1956) Arteriovenous anastomoses in the human external ear, *J. Anat.* (*Lond.*), **90**, 309.

PURKINJE, J. E. VON (1845) Mikroskopisch-neurologische Beobachtungen, *Arch. Anat. Physiol.*, **12**, 281.

QUIRING, D. P. (1949) *Collateral Circulation (Anatomical Aspects)*, London.

ROMANES, G. J. (1965) The arterial blood supply of the human spinal cord, *Paraplegia*, **2**, 199.

ROUGET, C. (1879) Sur la contractilité des capillaires sanguins, *C. R. Acad. Sci.* (*Paris*), **88**, 916.

RUSTED, I. E., SCHEIFLEY, C. H., and EDWARDS, J. E. (1952) Studies of the mitral valve. I. Anatomic features of the normal mitral valve and associated structures, *Circulation*, **6**, 825.

SCHINZ, H. R., and BAENSCH, W. S. (1970) *Roentgen-Diagnostics*, Vol. 4, Pt. 1 The Heart and Great Vessels, New York.

SENIOR, H. D. (1919a) The development of the arteries of the human lower extremity, *Amer. J. Anat.*, **25**, 55.

SENIOR, H. D. (1919b) An interpretation of the recorded arterial anomalies of the human leg and foot, *J. Anat. (Lond.)*, **53**, 130.

SENIOR, H. D. (1920) The development of the human femoral artery, a correction, *Anat. Rec.*, **17**, 271.

SHANER, R. F. (1929) The development of the atrio-ventricular node, bundle of His, and sino-atrial node in the calf: with a description of a third embryonic node-like structure, *Anat. Rec.*, **44**, 85.

SHANKS, S. C., and KERLEY, P. (1969) *A Text-book of X-ray Diagnosis*, Vol. 2, 4th ed., London.

SHELLSHEAR, J. L. (1922) Blood supply of the dentate nucleus of the cerebellum, *Lancet*, i, 1046.

STANFIELD, J. P. (1960) The blood supply of the human pituitary gland, *J. Anat. (Lond.)*, **94**, 257.

STREETER, G. L. (1915) The development of the venous sinuses of the dura mater in the human embryo, *Amer. J. Anat.*, **18**, 145.

TANDLER, J. (1912) The development of the heart, in *Manual of Human Embryology*, ed. Keibel, F., and Mall, F. P., Vol. II, p. 534, Philadelphia.

TANDLER, J. (1913) Anatomie des Herzens, *Bardeleben's Handbuch der Anatomie des Menschen*, Bd. III, Abth. I, Jena.

TAWARA, S. (1906) *Das Reizleitungssystem des Saugetierherzens*, Jena.

THOMAS, C. E. (1957) The muscular architecture of the ventricles of hog and dog hearts, *Amer. J. Anat.*, **101**, 17.

TOBIN, C. E. (1952) The bronchial arteries and their connections with other vessels in the human lung, *Surg. Gynec. Obstet.*, **95**, 741.

WALLS, E. W. (1945) Dissection of the atrio-ventricular bundle in the human heart, *J. Anat. (Lond.)*, **79**, 45.

WALLS, E. W. (1947) The development of the specialized conducting tissue of the human heart, *J. Anat. (Lond.)*, **81**, 93.

WALMSLEY, T. (1929) The heart, in *Quain's Elements of Anatomy*, 11th ed., Vol. IV, pt. III, London.

WILLIS, T. (1664) *Cerebri anatome: cui accessit nervorum descriptio et usus*, London.

WOOLLAM, D. H. M., and MILLEN, J. W. (1955) The arterial supply of the spinal cord and its significance, *J. Neurol. Neurosurg. Psychiat.*, **18**, 97.

WOOLLAM, D. H. M., and MILLEN, J. W. (1958) The anatomical background to vascular disease of the spinal cord, *Proc. roy. Soc. Med.*, **51**, 540.

THE LYMPHATIC SYSTEM

The role of the lymphatic system in absorption is mentioned on page 837. The fluid absorbed by its vessels is called lymph, and is colourless, except that from the intestine (chyle), which appears milky during digestion because of its fatty content. Not all parts of the body have lymph vessels, for they are absent from avascular tissues and from the central nervous system; and possibly from muscle and bone marrow.

In their course lymph vessels are interrupted by lymph nodes which serve in part as filters (see below), and in part as sources of lymphocytes. Most of the lymph passes through at least one lymph node—generally more—before reaching the blood stream, but in a proportion of the animals studied by Engeset (1959a) lymph travelled from the testis to the thoracic duct without passing through any nodes.

The lymph capillaries from which lymph vessels arise are abundant in the skin, mucous membranes, glands, serous membranes and synovial membranes and are essentially sub-epithelial in position.

The **superficial lymph vessels** lie in the skin and sub-cutaneous tissues; they frequently accompany the superficial veins and, in the limbs, they join the deep vessels at constant sites.

On each side of the body the cutaneous lymph vessels converge from three large areas upon three groups of lymph nodes [FIGS. 1146, 1147, and 1173]: (1) from the skin of the lower limb, perineum, external genital organs, and the trunk below the level of the umbilicus—to the superficial inguinal lymph nodes in the groin; (2) from the skin of the upper limb and the trunk above the umbilicus to the level of the clavicle in front and halfway up the back of the neck behind—to the lymph nodes in the axilla; (3) from the scalp, face, and the rest of the neck—to the cervical nodes.

The **deep lymph vessels** drain the lymph from parts deep to the deep fascia, and they tend to accompany the blood vessels of the region.

In the limbs and trunk they are relatively scanty and arise primarily in the synovial membranes of the joints. In relation to muscle they are probably mainly passengers although lymph capillary networks have been described on tendons; and some observers believe that they occur also in the fibro-areolar tissues of muscle.

The **lymph nodes** also are divided into superficial and deep groups. The former lie in the superficial fascia and are comparatively few; they are associated more particularly with the superficial lymph vessels of the limbs and the trunk. The deep nodes of the limbs also are comparatively few, but those of the head, neck, and trunk are numerous.

GENERAL PLAN OF THE LYMPHATIC SYSTEM

From the capillary plexuses, collecting vessels run towards the lymph nodes. On reaching a node the vessels, called **afferent vessels,** penetrate the capsule at numerous points, and the lymph percolates slowly through a meshwork of lymph sinuses lined by phagocytic cells [p. 943]. These cells take up from the lymph particulate material, such as bacteria which have entered the tissues, or inhaled carbon particles, and are an important part of the body's defence. **Efferent vessels** from the nodes either run with afferent vessels into another node of the same group or pass on to another group; and from the most proximal group of each chain of nodes the efferent vessels unite to form **lymph trunks** which are named: (1) lumbar; (2) intestinal; (3) bronchomediastinal; (4) subclavian; (5) jugular. Each trunk drains a definite territory of the body, and they all empty into great terminal vessels—the thoracic duct and right lymph duct [pp. 943-4]—which open into the great veins in the root of the neck [FIG. 1145], forming the only lymphaticovenous communications, apart from the occasional direct entry into the great veins of the subclavian, jugular, or bronchomediastinal trunks (Engeset, 1959b).

For a more detailed account of the lymphatic system than that which follows, Rouvière (1932) should be consulted. The

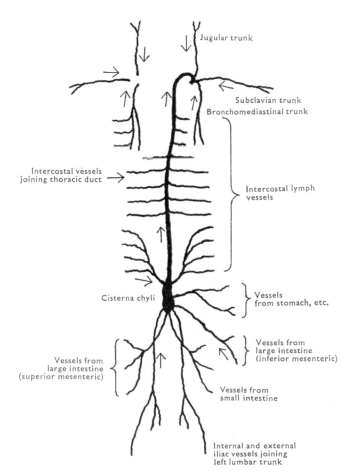

Jugular trunk

Subclavian trunk
Bronchomediastinal trunk

Intercostal vessels
joining thoracic duct

Intercostal lymph
vessels

Cisterna chyli

Vessels
from stomach, etc.

Vessels from
large intestine
(inferior mesenteric)

Vessels from
large intestine
(superior mesenteric)

Vessels from
small intestine

Internal and external
iliac vessels joining
left lumbar trunk

FIG. 1145 Diagram of main lymph vessels.

FIG. 1146   Cutaneous lymph vessels of anterior surface of trunk. (After Sappey, 1874.)

Note the 'lymphshed' between the vessels that pass to the axillary and to the inguinal lymph nodes.

The **collecting vessels** are small, and are distinguished by their beaded appearance due to valves placed at close and regular intervals; but, unlike veins, they run in streams of individual vessels rather than by confluence into larger trunks [FIG. 1167]. On reaching a node each afferent vessel usually breaks up into smaller branches which penetrate the capsule. The efferent vessels are only slightly larger than the afferents and have the same structure. Owing to the valves, the collecting lymph vessels can be injected only through a hollow needle introduced into the plexus of origin (Jamieson and Dobson, 1910).

In recent years techniques have been developed whereby lymph vessels can be rendered visible in patients by the injection of a harmless dye or X-ray opaque material. Information obtained in this way is of value in assessing cases where there is faulty lymph flow (Kinmonth, 1954) [PLATE 51 C and D].

Structure. The walls of the largest lymph vessels resemble those of veins. The **inner coat** is a layer of endothelium covered by elastic fibres or a fenestrated elastic membrane. The **middle coat** is formed of plain muscle fibres, intermingled with elastic fibres. The **outer coat** consists of fibrous tissue amongst which are plain muscle fibres.

The bicuspid **valves** are semilunar folds of the inner coat similar to those of veins. They are numerous in the collecting vessels and are present also near the entrances of the great lymph channels into the venous system.

Lymph Nodes

These are nodules which vary in size and shape; they are usually ovoid or reniform, and from 1 to 25 mm. in length. The larger nodes show a depressed area, the hilus, through which the blood vessels pass and the efferent lymph vessel emerges. Their colour is usually greyish pink, but those of the lungs are commonly blackened by carbon particles, and those of the liver and spleen are often brownish. Lymph nodes tend to arrest for a while the central spread of cancerous tumours—one reason why exact knowledge of the lymphatic system is essential for surgical practice.

Structure of Lymph Nodes. Lymph nodes consist essentially of: (1) masses of lymphoid tissue, supported by (2) a fibrous framework, which includes an external capsule and internal trabeculae, separated from the lymphoid tissue by (3) spaces called lymph sinuses. Each node is separable into cortex and medulla. The **cortex** lies immediately internal to the capsule, except at the hilus, where it is absent. The **medulla** forms the internal part of the node and reaches the capsule at the hilus.

The **capsule** is formed mainly of white fibrous tissue and from its deep surface **trabeculae** pass towards the medulla.

physiological and clinical significance of the lymphatic system is fully discussed by Yoffey and Courtice (1956).

Lymph Vessels

The **capillaries** of origin are closed, and have no communication with blood vessels or with tissue spaces. They are wider than blood capillaries, more irregular in calibre, and have a similar structure except for the absence of pericytes. They are able to absorb particulate matter and larger molecules (e.g., protein) than can pass through blood capillary walls. The capillary plexuses have no valves, and are drained by vessels of the venule type; they form wider-meshed networks from which the collecting vessels arise.

FIG. 1147 Cutaneous lymph vessels of posterior surface of trunk. (After Sappey, 1874.)

Note the 'lymphsheds' between the vessels that pass to the axillary and to the inguinal lymph nodes, and on the buttock between vessels that pass to the inguinal nodes by lateral and medial routes

forms lymph sinuses which afford passage to the lymph. The sinuses are known as subcapsular (or marginal), trabecular, and medullary, and their walls are formed by the reticular cells which clothe the reticular fibres. As noted earlier some of these cells are phagocytic.

The lymphocytes produced in the germinal centres of the follicles are added to the lymph flowing through the node, and eventually enter the blood through the thoracic duct; and, as Gowans and Knight (1964) have shown, there is a continuous recirculation of lymphocytes from the blood back into the lymph. For an account of modern views on the role of lymphocytes in the immunological responses of the body, Roitt *et al.* (1969) should be consulted.

Solitary Lymph Follicles. Tiny collections of lymphoid tissue are found in the mucous coat of the alimentary canal. In the pharynx there are great aggregations of units which form tonsils; in the ileum many of the follicles coalesce to form elongated raised patches (aggregated lymph follicles); and in the vermiform appendix they form masses which constitute the most striking feature of its structure.

TERMINAL LYMPH VESSELS

The terminal lymph vessels are the thoracic duct and the right lymph duct (or the vessels that represent it).

Thoracic Duct

The thoracic duct begins in the **cisterna chyli,** a dilatation some 6 mm. wide and 6 cm. long, which lies between the aorta and vena azygos on the bodies of the first two lumbar vertebrae [FIG. 1149]. From the cisterna the duct passes through the aortic opening of the diaphragm, and enters the posterior mediastinum in which it lies on the vertebral column to the right of the median plane. At the level of the fifth thoracic vertebra it crosses behind the oesophagus, and continues along its left side through the superior mediastinum. At the root of the neck [FIG. 1150] the duct arches laterally over the apex of the pleura, and between the left carotid sheath in front and the vertebral vessels behind. It then descends across the first part of the left subclavian artery, and enters the angle of junction of the left internal jugular and subclavian veins.

The thoracic duct is about 2 mm. in diameter, and has many paired valves—one near its termination.

Tributaries. The cisterna chyli commonly receives: (1) the intestinal trunk [p. 951]; (2) a pair of lumbar trunks [p. 944]; and (3) a pair of descending intercostal lymph trunks [p. 954].

In the posterior mediastinum the thoracic duct receives

Within this collagenous framework there is a fine meshwork, composed of reticular fibres and their associated cells, which pervades the node and supports the lymphoid tissue [FIG. 1148].

The lymphoid tissue in the medulla forms branching and anastomosing cords, but in the cortex it is in the form of rounded **lymph follicles** which consist of masses of densely packed lymphocytes supported by reticular tissue. The centre of each follicle consists of somewhat larger cells, called lymphoblasts, which by their division form lymphocytes; the name **germinal centre** is given to this part of the follicle.

Beneath the capsule, alongside the trabeculae, and in the medulla, the reticular tissue is more loosely arranged and

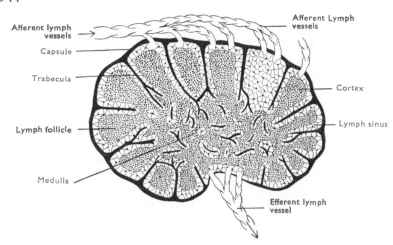

FIG. 1148   Diagram of section of lymph node.

efferents from some of the intercostal and posterior mediastinal nodes; and in the superior mediastinum the efferents of the upper intercostal nodes of both sides open into it.

Immediately before its termination it may receive: (1) the efferents from the nodes of the left upper limb, which frequently unite to form a **subclavian trunk**; and (2) the **left jugular trunk,** which conveys the lymph from the left side of the head and neck; but those vessels may end separately in the brachiocephalic or the subclavian vein. The **left bronchomediastinal trunk** may join the thoracic duct in the neck, but it usually opens independently into the left brachiocephalic vein. It collects lymph from the left side of the deeper parts of the anterior thoracic wall, from the upper part of the anterior abdominal wall, and from the anterior part of the diaphragm, the left lung and the left side of the heart.

Right Lymph Duct [FIG. 1151]

This duct rarely exists as such, since the three vessels which may unite to form it usually open separately into the right internal jugular, subclavian, and brachiocephalic veins. These vessels are: (1) the **right jugular trunk**; (2) the **right subclavian trunk**; and (3) the **right bronchomediastinal trunk.** The jugular and subclavian trunks convey lymph from the right side of the head and neck and from the right upper limb respectively; the right bronchomediastinal trunk collects lymph from an area corresponding to that drained by the left vessel, but including also the upper part of the right lobe of the liver.

Variations

The cisterna chyli may be replaced by a plexus of vessels from which the thoracic duct takes origin by several roots.

The thoracic duct sometimes ends in a number of branches which enter the great veins separately; or it may divide in the upper part of the thorax into two stems, of which the right joins the right jugular trunk to form one variety of the right lymph duct [FIG. 1151].

The arrangement and mode of termination of the three lymph trunks—subclavian, jugular, bronchomediastinal—are very variable, but the commonest arrangements are as follows.

1. On the left side the jugular trunk ends in the thoracic duct, and the other two trunks may do so also; but the subclavian trunk often ends in the subclavian vein, and the bronchomediastinal trunk usually ends in the brachiocephalic vein.

2. On the right side the trunks usually have separate entrances into the subclavian, internal jugular, and brachiocephalic veins; but it is not uncommon for the subclavian and jugular trunks to unite to form the right lymph duct.

LUMBAR LYMPH TRUNKS

These para-aortic lymph trunks receive the lymph from the lower limbs, the abdominal wall below the level of the umbilicus, the urogenital system, and the portion of the digestive tract supplied by the inferior mesenteric artery [FIG. 1157].

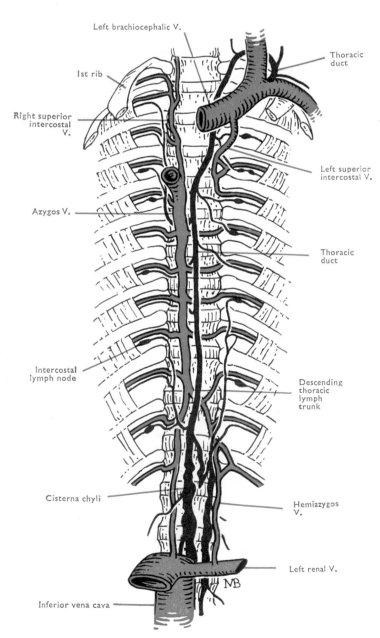

FIG. 1149   The thoracic duct and cisterna chyli.

PLATE 51

A

A. Aortogram to show aortic bifurcation and main vessels.

1. Lumbar artery.
2. Aorta.
3. Common iliac artery.
4. Internal iliac artery.
5. External iliac artery.

B

B. Arteries of the ankle and foot injected after death.

1. Peroneal artery.
2. Anterior tibial artery.
3. Posterior tibial artery.
4. Dense arterial plexus in the sole of the foot.
5. Dorsalis pedis artery.
6. Plantar arch.

C

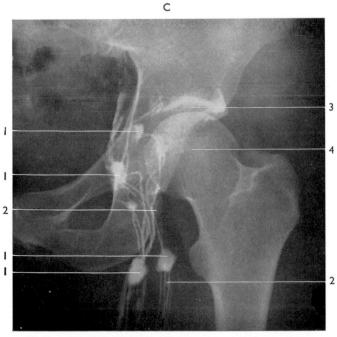

C. Lymphangiogram of inguinal region.

1. Lymph nodes.
2. Lymph vessels.
3. Acetabulum.
4. Head of femur.

D

D. Lymphangiograms to show some of the main superficial lymph channels of the leg and thigh.

PLATE 52

A

B

A. Left coronary artery catheterized through the aorta, oblique view.

1. Catheter in superior vena cava.
2. Catheter in arch of aorta.
3. Circumflex branch.
4. Left marginal artery.
5. Apex of heart.
6. Anterior interventricular artery.
7. Left coronary artery.

B. Coeliac trunk catheterized through the aorta, antero-posterior view.

1. Right branch of hepatic artery.
2. Common hepatic artery.
3. Coeliac trunk.
4. Spleen.
5. Splenic artery.
6. Pelvis of left kidney outlined by excreted test dose.
7. Catheter entering aorta from right common iliac artery.
8. Right gastro-epiploic artery.
9. Superior pancreaticoduodenal artery.
10. Hepatic artery.

C

C. Serial angiograms of left kidney. (i) At injection. (ii) One second later. (iii) Three seconds after injection.

1. Renal artery. 2. Ureter outlined by excreted test dose. 3. Left renal vein. 4. Edge of psoas major muscle.

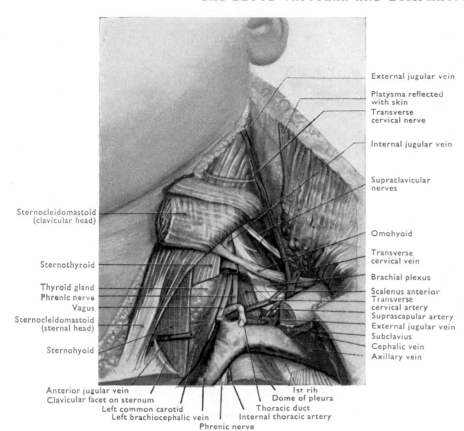

External jugular vein

Platysma reflected with skin

Transverse cervical nerve

Internal jugular vein

Supraclavicular nerves

Omohyoid

Transverse cervical vein

Brachial plexus

Scalenus anterior
Transverse cervical artery
Suprascapular artery
External jugular vein
Subclavius
Cephalic vein
Axillary vein

Sternocleidomastoid (clavicular head)

Sternothyroid

Thyroid gland
Phrenic nerve
Vagus
Sternocleidomastoid (sternal head)

Sternohyoid

Anterior jugular vein
Clavicular facet on sternum
Left common carotid
Left brachiocephalic vein
Phrenic nerve

1st rib
Dome of pleura
Thoracic duct
Internal thoracic artery

FIG. 1150 Deep dissection of root of neck on left side to show dome of pleura and relations of thoracic duct.

Parts of the sternocleidomastoid and the sternothyroid have been removed.

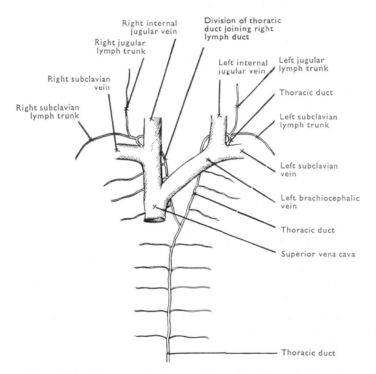

Right internal jugular vein

Right jugular lymph trunk

Right subclavian vein

Right subclavian lymph trunk

Division of thoracic duct joining right lymph duct

Left internal jugular vein

Left jugular lymph trunk

Thoracic duct

Left subclavian lymph trunk

Left subclavian vein

Left brachiocephalic vein

Thoracic duct

Superior vena cava

Thoracic duct

FIG. 1151 Diagram showing variation in ending of thoracic duct.

The bronchomediastinal lymph trunks are not represented.

Superficial Lymph Nodes of Lower Limb

The **superficial inguinal lymph nodes** [FIGS. 1146 and 1152] lie in the subcutaneous fat and are normally palpable. They form a **proximal** set (five or six) below and parallel with the inguinal ligament and a **distal** set (four or five) associated with the great saphenous vein. Their **afferents** are derived from all the skin below the level of the umbilicus, except the area drained by the popliteal nodes; and from the lining membranes of the anal canal and of the penile urethra, or of the vulva and lower end of the vagina. They receive also a few vessels from the uterus which run along the round ligament through the inguinal canal. The **efferents** pass through the cribriform fascia; some enter the deep inguinal nodes, but most pass through the femoral canal to end in the external iliac nodes.

Superficial Lymph Vessels of the Lower Limb and Trunk [PLATE 51 C and D]

These vessels are the afferents of the superficial inguinal nodes upon which they converge in a great whorl [FIGS. 1146, 1147, 1152, and 1153].

Certain points must be noted: (1) lymph vessels from the lateral part of the plantar surface of the foot, from its lateral border, and from the heel, accompany the small saphenous vein and end in the popliteal nodes; (2) the lymph vessels from the glans penis end mainly in the deep inguinal nodes, but a few run with the dorsal vein into the pelvis; (3) *the testis does not send lymph vessels to the inguinal nodes.*

The arrangement of the superficial lymph vessels of the toes and the foot is similar to that of the fingers and hand [p. 957]. From plexuses on the plantar surface, vessels pass to a dorsal network from which most of the collecting vessels ascend parallel to the great saphenous vein and end in the distal group of superficial inguinal nodes.

Deep Lymph Nodes and Lymph Vessels of the Lower Limb

The deep nodes are few and small. The **anterior tibial lymph node** is a tiny nodule on the upper part of the interosseous membrane.

The **popliteal lymph nodes** (six or seven) are small nodules in the popliteal fat alongside the great blood vessels. They receive the anterior and posterior tibial lymph vessels and others which issue from the knee joint. The lymph vessels running with the small saphenous vein end in a node which lies on the upper end of that vein under the deep fascia. The **efferents** of the popliteal nodes end in the deep inguinal nodes.

The **deep inguinal lymph nodes** [FIGS. 1155, 1157, and 1158] lie on the medial side of the femoral vein. They are

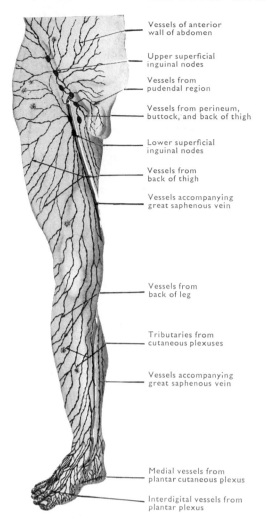

Vessels of anterior
wall of abdomen

Upper superficial
inguinal nodes

Vessels from
pudendal region

Vessels from perineum,
buttock, and back of thigh

Lower superficial
inguinal nodes

Vessels from
back of thigh

Vessels accompanying
great saphenous vein

Vessels from
back of leg

Tributaries from
cutaneous plexuses

Vessels accompanying
great saphenous vein

Medial vessels from
plantar cutaneous plexus

Interdigital vessels from
plantar plexus

FIG. 1152   Superficial lymph vessels of anterior surface of lower limb.

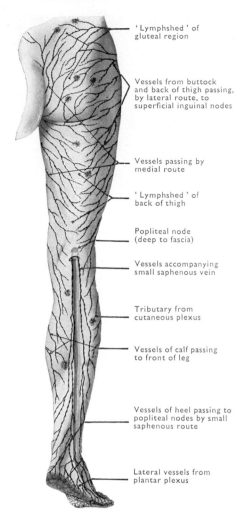

'Lymphshed' of
gluteal region

Vessels from buttock
and back of thigh passing,
by lateral route, to
superficial inguinal nodes

Vessels passing by
medial route

'Lymphshed' of
back of thigh

Popliteal node
(deep to fascia)

Vessels accompanying
small saphenous vein

Tributary from
cutaneous plexus

Vessels of calf passing
to front of leg

Vessels of heel passing to
popliteal nodes by small
saphenous route

Lateral vessels from
plantar plexus

FIG. 1153   Superficial lymph vessels of posterior surface of lower limb.

small and few. They receive some of the efferents from the superficial inguinal nodes, and all the deep vessels in the region of distribution of the femoral artery. Their **efferents** pass to the external iliac nodes.

The **deep lymph vessels** of the lower limb follow the main blood vessels. Some pass directly to the deep inguinal nodes, but others end in the popliteal nodes. The deep lymph vessels of the gluteal region and perineum accompany branches of the internal iliac artery into the pelvis and will be noted later.

Lymph Vessels of the Anterior Wall of the Abdomen

The *superficial lymph vessels* of the upper part of the anterior wall of the abdomen go mainly to the **pectoral** group of axillary nodes; but some pierce the wall of the thorax and end in the parasternal nodes. Those of the lower part of the abdominal wall end in the superficial inguinal nodes [FIG. 1146].

The *deep lymph vessels* of the upper part of the anterior abdominal wall accompany the superior epigastric blood vessels and end in the parasternal nodes. Those of the lower part accompany the inferior epigastric and deep circumflex iliac vessels, and end in the external iliac nodes.

Lymph Vessels of the External Genital Organs

The lymph vessels of the scrotum or of the vulva pass to the medial, proximal superficial inguinal nodes. The superficial lymph vessels of the penis go to the medial, proximal superficial inguinal group, but the deep vessels, including those of the penile urethra, end either in the medial, proximal superficial inguinal group or in the deep inguinal nodes. A few vessels follow the dorsal vein and join the internal iliac nodes. The lymph vessels of the clitoris end like those of the penis.

LYMPH NODES OF THE PELVIS

The lymph nodes of the pelvis may be grouped with the blood vessels [FIGS. 1154–1157].

The **external iliac lymph nodes** (from eight to ten in number) lie along the external iliac vessels. They receive the efferents of the inguinal nodes, the deep lymph vessels of the iliac fossa and of the lower part of the anterior abdominal wall, and a number of afferents **direct** from the pelvic viscera (q.v.). Their **efferents** pass to the common iliac nodes.

The **internal iliac lymph nodes** lie along the trunk and branches of the internal iliac vessels. They receive **afferents** from all the pelvic viscera except the ovary and adjacent part of the uterine tube, from the deep structures of the perineum, and from the gluteal region including most of the hip joint (the anterior part of the capsule drains to the deep inguinal nodes). Their **efferents** pass to the common iliac nodes. Only one subgroup of the internal iliac nodes need be named—the **sacral lymph nodes,** which lie in the hollow of the sacrum.

The **common iliac lymph nodes** (from four to six in number) extend along the common iliac vessels to the aortic bifurcation where they form the median common iliac nodes. Their **efferents** pass to the lumbar nodes.

The nodes of the lesser pelvis thus form two major subdivisions: those of the **cavity**—the internal iliac nodes, and those of the **brim**—the external and common iliac nodes.

Both receive direct afferents from the pelvic viscera, and in addition those of the brim receive the efferent vessels from the nodes of the cavity. *Nevertheless, vessels from the upper part of the rectum, from the upper part of the uterus, and from the uterine tube and ovary, escape even the nodes of the brim and pass to the lumbar nodes in the abdomen.*

LYMPH VESSELS OF THE PELVIC VISCERA

The **lymph vessels of the urinary bladder** pass mainly to the external iliac lymph nodes, but some from the base end in internal iliac nodes. Those from the neck of the bladder run to the sacral and the median common iliac nodes.

The **lymph vessels of the prostate** mainly end in the internal iliac nodes; some pass to the sacral nodes, and others follow the ductus deferens to external iliac nodes.

Lymph Vessels of the Urethra. In the male, the vessels of the spongy part of the urethra pass to the superficial and deep inguinal nodes. Those of the membranous and prostatic parts of the urethra mainly end in the internal iliac nodes.

The **lymph vessels of the female urethra** correspond with those of the membranous and prostatic parts of the male urethra and end mainly in the internal iliac nodes, though its inferior extremity drains with the skin of the vulva.

The lymph vessels of the **ductus deferens** and of the **seminal vesicle** pass to the external and internal iliac nodes.

Lymph Vessels of the Vagina. Below the hymen, the vessels pass with those of the vulva to the superficial inguinal

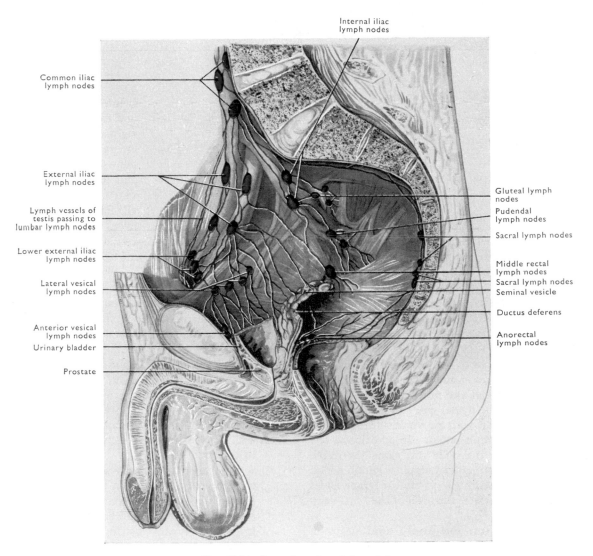

FIG. 1154 Lymph nodes of the pelvis.

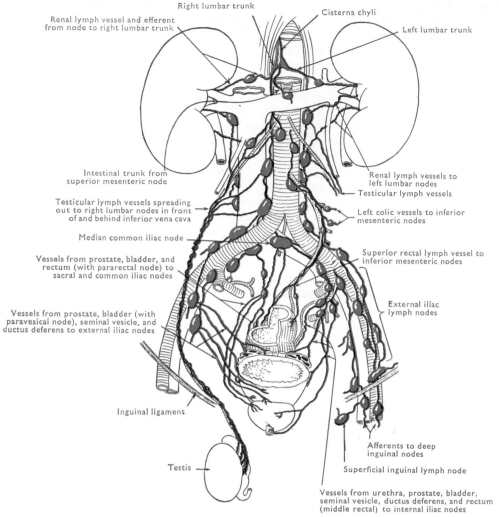

Right lumbar trunk

Renal lymph vessel and efferent
from node to right lumbar trunk

Cisterna chyli

Left lumbar trunk

Intestinal trunk from
superior mesenteric node

Renal lymph vessels to
left lumbar nodes

Testicular lymph vessels

Testicular lymph vessels spreading
out to right lumbar nodes in front
of and behind inferior vena cava

Left colic vessels to inferior
mesenteric nodes

Median common iliac node

Superior rectal lymph vessel to
inferior mesenteric nodes

Vessels from prostate, bladder, and
rectum (with pararectal node) to
sacral and common iliac nodes

External iliac
lymph nodes

Vessels from prostate, bladder (with
paravesical node), seminal vesicle, and
ductus deferens to external iliac nodes

Inguinal ligament

Afferents to deep
inguinal nodes

Testis

Superficial inguinal lymph node

Vessels from urethra, prostate, bladder,
seminal vesicle, ductus deferens, and rectum
(middle rectal) to internal iliac nodes

FIG. 1155 Diagram of lymph vessels and lymph nodes of male pelvis
and abdomen.

nodes. From the middle part they accompany the vaginal artery to the internal iliac nodes; and from the upper part they accompany those of the cervix of the uterus to external and internal iliac nodes [FIGS. 1156 and 1157].

Lymph Vessels of the Uterus. The capillary plexus in the mucous coat communicates with a subserous plexus from which the collecting vessels pass to widely separated groups of lymph nodes [FIGS. 1156 and 1157].

From the **cervix,** vessels pass to both external and internal iliac nodes, and also to the sacral and median common iliac nodes. From the **lower part of the body** most vessels go to external iliac nodes. From the **upper part of the body** and the **fundus,** the main outflow is with the vessels of the uterine tube and ovary to lumbar lymph nodes. In addition, a few vessels from the fundus and the body run along the round ligament of the uterus to the superficial inguinal lymph nodes.

The lymph vessels of the **ovary** pass to lumbar lymph nodes from the bifurcation of the aorta to the level of the renal veins [FIG. 1157].

Lymph Vessels of the Rectum and Anal Canal. Most of the vessels end in pelvic nodes, but those from the anal canal below the pectinate line pass to superficial inguinal nodes, and

those from the upper part of the rectum pass upwards to the inferior mesenteric nodes.

From the greater part of the anal canal the lymph vessels run either across the ischiorectal fossa, or upwards with those of the lower part of the rectum along the middle rectal blood vessels, to internal iliac nodes. Other vessels from the rectum end in the sacral and median common iliac nodes; and, as noted, an important group ascends along the superior rectal blood vessels to the inferior mesenteric nodes. Vessels from the anal canal and lower rectum may be interrupted in small anorectal lymph nodes; and others from the middle and upper parts of the rectum pass through pararectal nodes which lie on the wall of the rectum.

LUMBAR AND INFERIOR MESENTERIC LYMPH NODES

The lumbar lymph nodes [FIGS. 1155 and 1158] are in continuity below with the common iliac nodes, and from their upper members the right and left lumbar lymph trunks emerge. Their relation to the aorta allows of a subdivision into **right** and **left lateral, pre-aortic** and **retro-aortic** groups, but they are all freely interconnected; moreover, they are not easily distinguished from the nodes around the stems of origin of the inferior and superior mesenteric arteries, and of the coeliac trunk.

The territory drained by the lumbar nodes corresponds to the distribution of the paired branches of the aorta, plus that of the inferior mesenteric artery; for although the nodes associated with the unpaired visceral branches of the abdominal aorta can be regarded as offshoots of the lumbar nodes, the inferior mesenteric group drains into the lumbar lymph trunk while the coeliac and superior mesenteric groups reach the cisterna chyli separately through the intestinal trunk.

The lumbar nodes thus receive the following lymph vessels:

1. The efferents of the common iliac nodes.
2. The efferents of the inferior mesenteric nodes [see below].
3. Lymph vessels of the **testis and epididymis.** Some six to eight vessels ascend in the spermatic cord, and then follow the testicular vessels upwards. They end in lumbar nodes, from the bifurcation of the aorta to the level of the renal veins [FIG. 1155]. The corresponding vessels from the **ovary,** the **uterine tube,** and upper part of the uterus have the same termination.

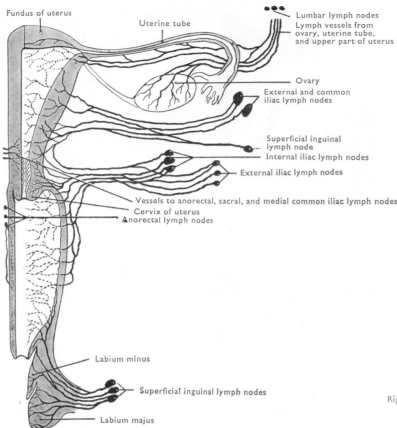

FIG. 1156 Diagram of lymph vessels of female genital organs.

4. Lymph vessels of the **ureter.** From the part of the ureter in the pelvis vessels pass to the nodes of the cavity and brim of the lesser pelvis; from the middle part of the ureter vessels pass to the lumbar nodes; and those from its upper part accompany the renal lymph vessels.

5. Lymph vessels of the **kidney.** These emerge from the hilus, and pass medially to lumbar nodes.

6. Lymph vessels of the **suprarenal gland.** These end mainly in upper lumbar nodes, but some pass through the diaphragm to the posterior mediastinal nodes.

7. Lymph vessels of the lumbar synovial joints.

8. Lymph vessels of the **peritoneum.** The lymph vessels of the visceral peritoneum run with those of the organs it covers. Those from the parietal peritoneum tend to converge on the neighbouring blood vessels. In the lesser pelvis they pass to the nearest nodes of the cavity or brim. On the anterior abdominal wall and in the iliac fossa they accompany the epigastric and the deep circumflex iliac blood vessels. On the posterior wall of the abdomen, in its lower part, they run to the nodes on the stems of the superior and inferior mesenteric arteries; laterally they accompany the lumbar blood vessels to the lateral lumbar nodes; higher up

they join the vessels from the kidneys. The lymph plexus on the abdominal surface of the diaphragm drains, partly, to the uppermost lumbar nodes; but it also communicates freely with the rich subpleural plexus on the upper surface of the muscle.

The **inferior mesenteric lymph nodes** lie along the stem of the artery of that name. They receive lymph from nodes scattered along the branches of the artery (including **left colic nodes**) as far as the wall of the intestine. These outlying nodes are lymph nodes of the colon, and are arranged similarly to those along the colic branches of the superior mesenteric artery.

The **lymph nodes of the colon** [FIG. 1161] form epicolic, paracolic, intermediate, and main colic groups. The **epicolic nodes** are tiny nodules in the appendices epiploicae and the wall of the gut. The **paracolic nodes** lie along the medial borders of the ascending and descending colon, and along the mesenteric borders of the transverse and sigmoid colon. The **intermediate colic nodes** lie along

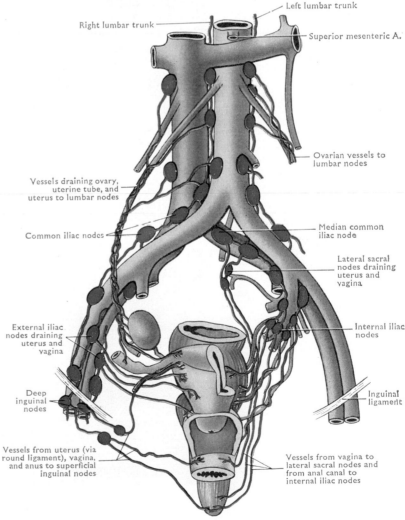

FIG. 1157 Diagram of lymph vessels and lymph nodes of female pelvis and abdomen.

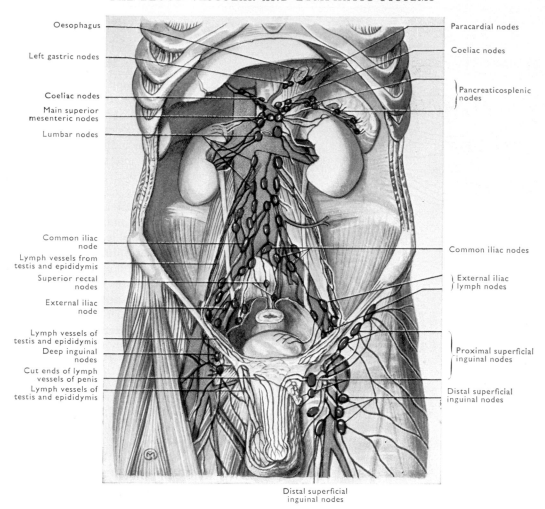

Oesophagus

Left gastric nodes

Coeliac nodes

Main superior
mesenteric nodes

Lumbar nodes

Common iliac
node

Lymph vessels from
testis and epididymis

Superior rectal
nodes

External iliac
node

Lymph vessels of
testis and epididymis

Deep inguinal
nodes

Cut ends of lymph
vessels of penis

Lymph vessels of
testis and epididymis

Paracardial nodes

Coeliac nodes

Pancreaticosplenic
nodes

Common iliac nodes

External iliac
lymph nodes

Proximal superficial
inguinal nodes

Distal superficial
inguinal nodes

Distal superficial
inguinal nodes

FIG. 1158   Semidiagrammatic view of lymph nodes and lymph
vessels of inguinal region, pelvis, and posterior wall of abdomen.

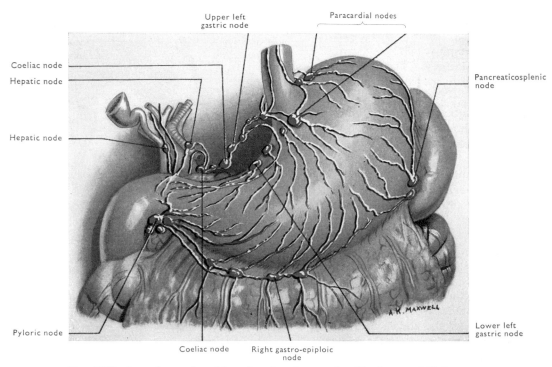

Upper left
gastric node

Paracardial nodes

Coeliac node

Hepatic node

Hepatic node

Pancreaticosplenic
node

Pyloric node

Coeliac node

Right gastro-epiploic
node

Lower left
gastric node

FIG. 1159   Lymph vessels and lymph nodes of stomach.   (Jamieson and Dobson,
1907a, redrawn.)

the branches of the colic arteries. The **main colic nodes** are situated on the stems of the colic arteries. In addition to the rectal lymphatic vessels that run with the superior rectal artery, the main inferior mesenteric nodes receive lymph from the descending and sigmoid colon; their **efferents** pass to the lumbar lymph nodes which drain mainly into the left lumbar lymph trunk.

INTESTINAL LYMPH TRUNK

The intestinal lymph trunk enters the cisterna chyli and is formed by the union of the efferents of the proximal members of the groups of lymph nodes situated on the coeliac and superior mesenteric arteries and their branches.

COELIAC AND SUPERIOR MESENTERIC LYMPH NODES

The coeliac lymph nodes surround the stem of the coeliac trunk, and drain the abdominal part of the foregut; but some lymph from the liver, duodenum, and pancreas passes to other nodes.

In association with the branches of the coeliac trunk there are the following groups of lymph nodes: left gastric, right gastric, hepatic, pancreaticosplenic, and pyloric.

The **left gastric lymph nodes** form three subgroups. The **lower** nodes lie on the lesser curvature to the left of the angular notch. Around the cardiac end of the stomach **paracardial** nodes are disposed like a string of beads; the **upper** nodes lie on the trunk of the left gastric artery. Finally the chain merges with the coeliac group. The left gastric nodes drain the abdominal part of the oesophagus, and the right two-thirds of the surfaces of the stomach from the fundus as far as the pyloric canal.

The **right gastric lymph nodes** lie in the lesser omentum along the right gastric artery; their efferents pass to the hepatic group.

The **hepatic lymph nodes** lie along the hepatic artery as far as the porta hepatis; one of the group is the **cystic lymph node** which lies in the curve of the neck of the gall-bladder. Some nodes descend along the bile-duct and may be called **biliary lymph nodes**. The hepatic nodes receive afferents from the liver and gall-bladder, stomach, duodenum and pancreas; their efferents pass to the coeliac group.

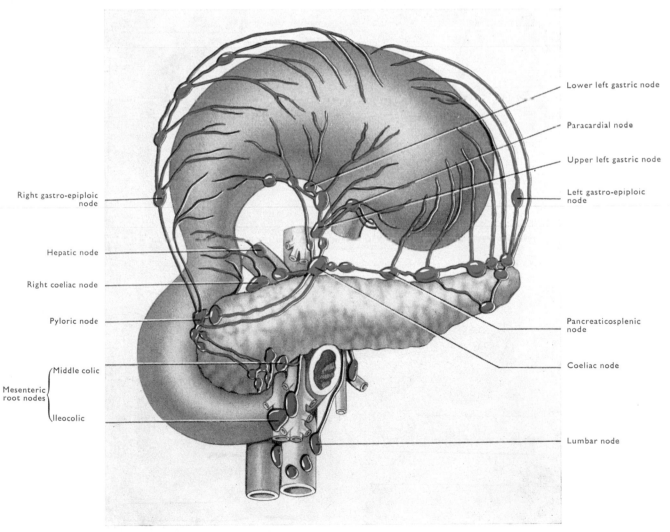

FIG. 1160 Lymph vessels and lymph nodes of posterior surface of stomach. The stomach has been turned upwards to show the pancreaticosplenic and other groups of nodes. (Jamieson and Dobson, 1907a, redrawn.)

The **pancreaticosplenic lymph nodes** lie along the splenic artery. The nodes in the hilus of the spleen receive the lymph vessels from the capsule of that organ, and also vessels which accompany the left gastro-epiploic artery from the stomach; small left gastro-epiploic nodes may be found in the gastrosplenic ligament. The efferents from the pancreaticosplenic nodes pass to the coeliac group.

The **pyloric lymph nodes** lie on the head of the pancreas below the duodenum, and receive efferent vessels from the nodes between the layers of the greater omentum—the **right gastro-epiploic lymph nodes**. These nodes together drain the adjacent parts of the stomach and duodenum; from the pyloric nodes some efferents run to nodes on the stem of the superior mesenteric, some to the coeliac group, and some to the hepatic nodes [FIG. 1160].

The efferent vessels of the coeliac and superior mesenteric nodes combine to form the (single or multiple) **intestinal trunk**.

Superior Mesenteric Lymph Nodes

1. *On the stem of the artery* these are large and numerous. They receive the efferents of all the subsidiary groups, some efferents from the pyloric nodes, direct afferents from the head of the pancreas and the duodenum, and some efferents from the upper left colic group [FIG. 1161].

2. Subsidiary groups related to the branches of the artery.

(i) The **lymph nodes of the mesentery** (100–200 in number) are scattered throughout the mesentery: they drain the jejunum and the ileum (excepting the lowest few inches).

(ii) The epicolic and paracolic members of the **ileocolic lymph nodes** form a cluster in the ileocolic angle, below which a few nodes descend with the caecal vessels in front of and behind the ileocolic junction—**caecal lymph nodes**. One **appendicular node** may be found in the mesentery of the appendix. The intermediate group lies half-way up the stem of the ileocolic artery, and the main group lies on the origin of that artery. The ileocolic nodes drain the lower end of the ileum, the appendix, caecum, and lower end of the ascending colon.

(iii) The **right colic lymph nodes** lie along the right colic artery and receive the lymph from the ascending colon.

(iv) The **middle colic lymph nodes** accompany the branches and stem of the middle colic artery. They drain the right flexure and the right two-thirds of the transverse colon.

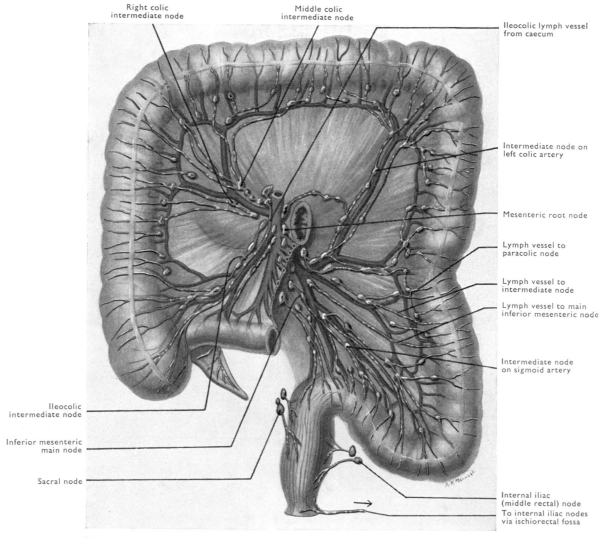

FIG. 1161 Lymph vessels and lymph nodes of large intestine. (Jamieson and Dobson, 1909, redrawn.)

LYMPH VESSELS OF THE DIGESTIVE SYSTEM IN THE ABDOMEN

The lymph vessels of the abdominal part of the alimentary canal arise from a capillary plexus under the lining epithelium and around the gastric and intestinal glands. In the small intestine they also have an important origin in the lacteals [p. 440]. The lymph vessels perforate the muscular coats, and are joined by tributaries from the subserous network; they then follow the blood vessels.

Lymph Vessels of the Large Intestine. The blood supply of the large intestine is a guide to its lymph drainage; in general, the lymph vessels accompany the arteries, and, having traversed or bypassed intervening nodes, eventually end in the nodes round the stems of the inferior and superior mesenteric arteries. However, the lymph drainage from the **left third** of the **transverse colon** and the **left flexure** must be specially mentioned because the stream divides at the intermediate nodes [FIG. 1161] to follow two paths—one leading to the inferior mesenteric nodes, the other to the superior mesenteric nodes. The course of the lymph vessels from the caecum and appendix is shown in FIGURES 1161 and 1162. Those from the appendix unite in its mesentery to form four or five trunks which follow the appendicular blood vessels behind the ileum and end in nodes along the whole length of the ileocolic chain.

Lymph Vessels of the Small Intestine. From the **ileum** and **jejunum** the vessels run to the nodes of the mesentery. The lower half of the **duodenum** drains to the superior mesenteric nodes, the upper half to the pyloric nodes in front and to the biliary nodes behind.

Lymph Vessels of the Stomach. These run in three streams:

1. From the front and back of the fundus and body to the right of a line (nearer the greater than the lesser curvature) from the fundus as far as the pyloric canal, to the left gastric nodes.

2. From the area to the left of the line as far as the angular notch, to the pancreaticosplenic nodes.

3. From the pyloric canal, downwards to the right gastro-epiploic and pyloric nodes, and upwards to the right gastric nodes, to the biliary chain and to the coeliac nodes [FIGS. 1159 and 1160].

Lymph Vessels of the Pancreas. The pancreas drains to the pancreaticosplenic nodes above, to the superior mesenteric nodes below, and to the lumbar nodes behind.

Lymph Vessels of the Liver. Superficial and deep collecting vessels pass from the capillaries of the liver lobules. The superficial vessels run beneath the peritoneum, and many traverse the falciform ligament and follow the superior epigastric blood vessels to the lower parasternal nodes [p. 954]. Others run round the lower edge of the liver towards the porta hepatis, where they pick up the vessels of the visceral surface and the **gall-bladder,** and a deep stream of vessels from the interior of the liver. Most of these vessels end in the hepatic nodes, but some in the left gastric nodes. From the posterior surface of the liver a few vessels run to the coeliac nodes. Deep vessels also emerge with the hepatic veins and pass with the inferior vena cava to middle diaphragmatic nodes in the thorax [see below]. There are therefore two main lymph streams from the liver—one via the coeliac nodes to the cisterna chyli, the other via the thorax to the mediastinal trunks of both sides [FIG. 1163]. Even so, figures for the hepatic contribution to thoracic duct lymph flow indicate that it contributes nearly half of the total (Yoffey and Courtice, 1956).

Lymph Vessels of the Spleen. The capsule drains to nodes of the pancreaticosplenic group which lie near its hilus. The pulp drains via the splenic vein and not by lymph vessels.

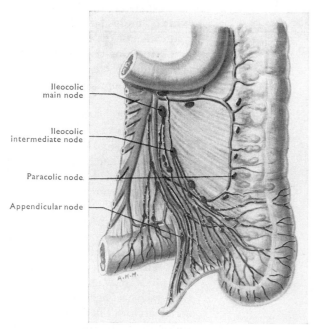

FIG. 1162 Lymph vessels and lymph nodes of caecum and appendix, posterior aspect. (Jamieson and Dobson, 1907b, redrawn.)

FIG. 1163 Diagrammatic sagittal section showing lymph drainage of liver.

LYMPH DRAINAGE OF THE THORAX

The chief pathway of the lymph vessels of the thorax is the **bronchomediastinal lymph trunk** [p. 944]. In addition, there are two **descending intercostal lymph trunks** [see below], and also direct tributaries of the thoracic duct from intercostal and posterior mediastinal lymph nodes.

The drainage of the superficial tissues of the thoracic wall is mainly to the axillary lymph nodes [FIGS. 1146 and 1147, and p. 956]. The deeper tissues drain mainly to the intercostal, parasternal, and phrenic lymph nodes.

Lymph Nodes of the Thorax

The **intercostal lymph nodes** are one or two small nodules in the vertebral end of each space. They receive afferents from the parietal pleura and deep tissues of the posterior thoracic wall. The efferents of the upper nodes run to the thoracic duct. Those of the lower four or five spaces join on each side to form a **descending intercostal trunk** which passes through the aortic opening to the cisterna chyli.

The **posterior mediastinal lymph nodes** lie around the lower part of the oesophagus. They receive afferents from the oesophagus, pericardium, and diaphragm; their efferents pass to the thoracic duct and to the descending intercostal lymph trunks.

The **phrenic lymph nodes** [FIGS. 1163 and 1164] form: (1) a posterior group, better considered as the lowest members of the posterior mediastinal group; (2) an anterior group better considered as the lowest members of the parasternal chain; and (3) a middle group on each half of the diaphragm which receives afferents from the diaphragm and from the liver, and sends efferents to the parasternal and posterior mediastinal nodes.

The **parasternal lymph nodes** [FIG. 1164] lie along the internal thoracic artery, in the intercostal spaces and behind the xiphoid process. The lowest nodes receive vessels from the pericardium, the diaphragm, and the liver. Their efferents pass to those alongside the sternum which receive the lymph vessels running with the intercostal and perforating branches of the internal thoracic artery. The intercostal afferents drain the parietal pleura, and with the perforating arteries there travel lymph vessels from the mammary gland. The importance of the parasternal nodes in the surgery of malignant disease of the breast has been stressed by Handley and Thackray (1954).

The **efferents** of the parasternal nodes unite into a single vessel which joins with the efferents of the tracheobronchial and anterior mediastinal nodes to form the bronchomediastinal lymph trunk.

In association with the pulmonary system and the heart there are the following groups of nodes [FIG. 1165].

1. **Bronchopulmonary lymph nodes** in the hilus of the lung. Small **pulmonary nodes** occur on the larger bronchi within the lung.

2. **Inferior tracheobronchial lymph nodes** in the angle of bifurcation of the trachea.

3. **Superior tracheobronchial lymph nodes** continuous with the upper bronchopulmonary nodes on each side.

These three groups receive the lymph vessels of the lung, visceral pleura, bronchi, and lower end of the trachea; and vessels from the left side of the heart end in the superior tracheobronchial group.

4. **Tracheal lymph nodes** extend along each side of the trachea into the neck. They drain the trachea and part of the oesophagus.

5. **Anterior mediastinal lymph nodes** in the vicinity of the brachiocephalic veins and the aortic arch. They receive lymph vessels from the thymus, the pericardium, and the right side of the heart.

Parasternal node

Lower parasternal (retrosternal) node

Middle phrenic node near right phrenic nerve and inferior vena cava

Retrosternal (anterior phrenic) node

Middle phrenic node near left phrenic nerve

Posterior mediastinal (posterior phrenic) node near aorta and oesophagus

FIG. 1164 Parasternal and phrenic lymph nodes, posterior aspect, and lymph drainage of thoracic surface of diaphragm.

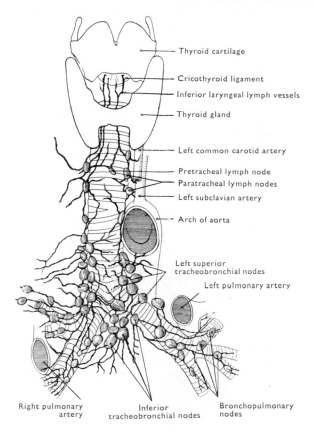

Thyroid cartilage

Cricothyroid ligament

Inferior laryngeal lymph vessels

Thyroid gland

Left common carotid artery

Pretracheal lymph node

Paratracheal lymph nodes

Left subclavian artery

Arch of aorta

Left superior tracheobronchial nodes

Left pulmonary artery

Right pulmonary artery

Inferior tracheobronchial nodes

Bronchopulmonary nodes

FIG. 1165   Lymph nodes in relation to trachea and main bronchi.

Lymph Vessels of the Lungs and Pulmonary Pleura

A plexus of vessels on the bronchial tree reaches the smallest bronchioles; but the alveoli have no lymph vessels. Beneath the pulmonary pleura there is also a plexus of vessels which extends over the lung surface to the hilus. Between the two plexuses there are communicating vessels which run in the interlobular septa, and in which lymph may run in either direction. All emerge at the hilus and end in the broncho-pulmonary and tracheobronchial nodes.

Lymph Vessels of the Bronchi and Trachea. The bronchi and the intrathoracic part of the trachea drain into the bronchopulmonary, tracheobronchial, and tracheal nodes.

Lymph Vessels of the Pericardium. The pericardium drains into the nearest nodes—anterior and posterior mediastinal, and the parasternal.

Lymph Vessels of the Heart. From a plexus under the endocardium efferents pierce the myocardium to join others from the richer plexus under the epicardium. The drainage of the atria is not well known, but seems to be chiefly from the upper parts of the atria to tracheobronchial nodes. The collecting vessels from the ventricles pass into the coronary sulcus and follow the coronary arteries towards their origin. Those with the right coronary artery usually form a single trunk which ascends in front of the ascending aorta to an anterior mediastinal node; those with the left coronary artery also usually end in a single vessel which passes behind the pulmonary trunk to a node of the right superior tracheo-bronchial group (Shore, 1929). (Consult also Patek (1939).)

Lymph Vessels of the Oesophagus. The abdominal part of the oesophagus drains to the left gastric nodes, the lower thoracic part to the posterior mediastinal nodes, and the upper thoracic and cervical parts to the tracheal nodes.

Lymph Vessels of the Thymus. From the interlobular connective tissue vessels pass to the anterior mediastinal and to the parasternal nodes of both sides.

The ultimate **efferents** of all these nodes unite into a single vessel on each side which usually joins the single efferent of the parasternal nodes to form the **bronchomediastinal lymph trunk** [p. 944]. After a short course upwards, the trunks usually end in the corresponding brachiocephalic vein. The right trunk may ascend deep to that vein, and passing behind the internal jugular vein, join the right lymph duct, while the left may join the thoracic duct.

Lymph Vessels of the Diaphragm and Thorax

Lymph Vessels of the Diaphragm [FIG. 1164]. An extensive lymph plexus is present on each surface of the diaphragm, and the two plexuses are united by numerous vessels that pierce the diaphragm.

Lymph vessels from both surfaces of the diaphragm run to the parasternal, middle phrenic, and posterior mediastinal nodes: only a few vessels from the abdominal surface run to lumbar nodes.

Lymph Vessels of the Parietal Pleura. The vessels from the named parts of the parietal pleura drain to the following nodes: from the **costal pleura**—mainly to the intercostal and parasternal nodes, but some vessels from the pleura lining the upper spaces, except the first, pass to the axillary nodes; from the costal pleura of the first space and from the **cervical pleura**—to the lower deep cervical nodes; from the **mediastinal pleura**—to the tracheobronchial and mediastinal nodes; and from the **diaphragmatic pleura**—to the nodes draining the corresponding parts of the diaphragm.

SUBCLAVIAN LYMPH TRUNKS

The subclavian lymph trunk [p. 944] drains the upper limb, a cutaneous area of the trunk from the umbilical plane to the clavicle in front and half-way up the neck behind, and most of the mammary gland.

The nodes in the territory of the subclavian trunk comprise a few superficial nodes and a greater number of deep nodes including the important axillary group.

Superficial Lymph Nodes of Upper Limb. One or two **cubital nodes** lie medial to the basilic vein a little above the medial epicondyle of the humerus: they receive afferents from the medial fingers and from the ulnar side of the hand and forearm. Their efferents run with the basilic vein and join the deep lymph vessels.

The **infraclavicular lymph nodes,** one or two in number, lie in the infraclavicular fossa on the cephalic vein and receive vessels from the skin of the 'vaccination area' and of the shoulder, and from the upper part of the mammary gland: their efferents perforate the clavipectoral fascia to join the apical axillary nodes. Outlying members occur in two situations.

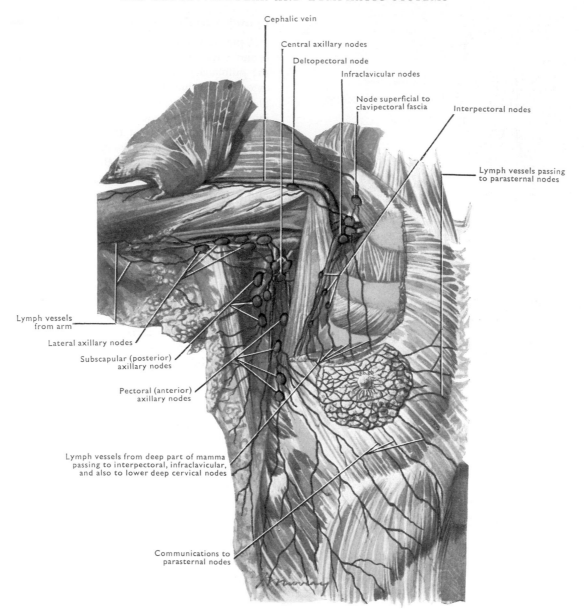

Cephalic vein

Central axillary nodes

Deltopectoral node

Infraclavicular nodes

Node superficial to
clavipectoral fascia

Interpectoral nodes

Lymph vessels passing
to parasternal nodes

Lymph vessels
from arm

Lateral axillary nodes

Subscapular (posterior)
axillary nodes

Pectoral (anterior)
axillary nodes

Lymph vessels from deep part of mamma
passing to interpectoral, infraclavicular,
and also to lower deep cervical nodes

Communications to
parasternal nodes

FIG. 1166   Dissection of axilla and anterior part of thoracic wall, showing lymph nodes and lymph vessels. (Semidiagrammatic, with only a portion of the mamma represented.)

A **deltopectoral lymph node** may be found in the groove between the deltoid and the pectoralis major, and small **interpectoral lymph nodes** are sometimes present between the pectoral muscles [FIG. 1166].

Deep Lymph Nodes of the Upper Limb. Small nodules may occur along the radial, ulnar, and interosseous arteries, or as deep cubital nodes at the bifurcation of the brachial artery. The brachial lymph nodes are inconstant, outlying axillary nodes along the upper part of the brachial artery. The efferents of all these nodes pass to the lateral axillary group.

The **axillary lymph nodes** are widely distributed in the axillary fat, and may be divided into five groups; four of the groups lie below the pectoralis minor tendon, and one (the apical group) above.

The **lateral** axillary nodes lie alongside the axillary vein, and from them the **pectoral** and the **subscapular** nodes radiate along the lateral thoracic and subscapular arteries respectively. The **central group** is composed of members of the former three groups which lie superior to the fascia of the axillary floor. The afferents of these four groups are: (1) the deep vessels of the whole limb; (2) the cutaneous vessels of the limb and trunk above the umbilical plane; and (3) vessels of the mammary gland. Though all the groups receive vessels from all these areas, the main drainage of the limb is to the lateral nodes, of the mammary gland and the anterior thoracic wall to the pectoral nodes, and of the scapular region and the back of the chest to the subscapular nodes. The **apical** axillary nodes receive the efferents of the lower groups and of the infraclavicular nodes. They are also in continuity with the lower deep cervical nodes [p. 958], but their efferents go mainly to form the **subclavian lymph trunk.**

Superficial Lymph Vessels of the Upper Limb and Trunk

The superficial lymph vessels of the upper limb run upwards on the front of the forearm with an inclination towards the axilla, and are reinforced by vessels that slope upwards from the back of the limb round its margins [FIGS. 1167 and 1168]. Some vessels are interrupted by the superficial cubital nodes, and some pass through the deep fascia with the basilic vein to join the deep stream; but nearly all run to the axilla before they pierce the deep fascia. On the back of the elbow and of the arm there is a 'lymphshed' from which vessels pass round the lateral and medial margins to the front. A few vessels from the skin over the upper part of the cephalic vein pass to the infraclavicular nodes, and may be interrupted in deltopectoral nodes.

The cutaneous plexuses are finest and most dense on the palmar surfaces of the fingers and hand. The efferents from the palmar plexus of each finger pass to the dorsum where they form dorsal digital vessels which run to the dorsum of the hand [FIG. 1169].

The efferents from the plexus on the palm of the hand run in various directions. Those passing laterally and medially join the efferents of the thumb and little finger respectively; a few pass superiorly along the median vein of the forearm; and those that pass inferiorly turn backwards at the interdigital clefts to join the vessels on the dorsum of the hand [FIGS. 1167 and 1168].

The **superficial lymph vessels of the trunk** form a great whorl as they converge on the lower axillary nodes [FIGS. 1146 and 1147]. The plexus of origin is continuous over the median line, and many vessels draining to one side begin on the other. Several vessels penetrate deeply, and thus reach intercostal and parasternal nodes. A few ascend over the clavicle to reach the lower deep cervical nodes.

Lymph Vessels of the Mammary Gland

The mammary gland receives its blood supply from three main sources—the axillary artery, the internal thoracic artery, and the lateral cutaneous branches of the posterior intercostals—and the lymph drainage is approximately in proportion. Thus, most of the lymph from the breast drains to the axillary nodes, a considerable proportion to the parasternal nodes, and an inconstant, small amount to the intercostal nodes (Turner-Warwick, 1959).

The lymph vessels of the main axillary pathway arise in the lobules of the mammary gland and most of them drain directly to the pectoral nodes. They run within the substance of the gland and pass through the axillary fascia within the axillary tail of the mammary gland; at no time do they run in the deep fascia.

Drainage to the parasternal nodes proceeds from both the lateral and medial halves of the gland, the lymph vessels running with the perforating branches of the internal thoracic

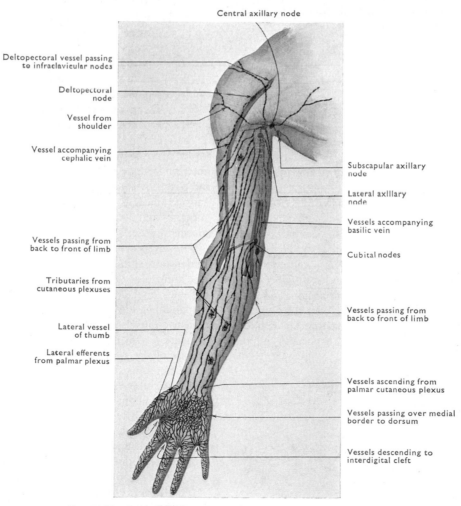

FIG. 1167 Superficial lymph vessels and lymph nodes of anterior surface of upper limb.

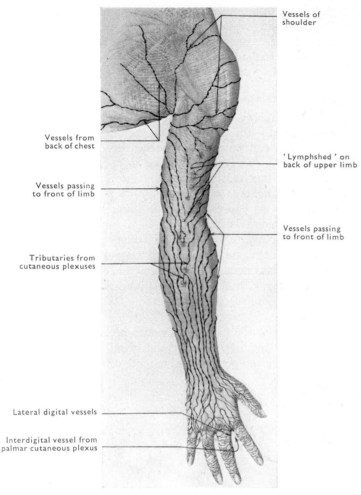

Vessels of shoulder

Vessels from back of chest

Vessels passing to front of limb

'Lymphshed' on back of upper limb

Vessels passing to front of limb

Tributaries from cutaneous plexuses

Lateral digital vessels

Interdigital vessel from palmar cutaneous plexus

FIG. 1168 Superficial lymph vessels of posterior surface of upper limb.

artery. In a small proportion of cases some lymph vessels run with the lateral cutaneous branches of the posterior intercostal arteries to reach the intercostal nodes.

Because the various groups of axillary lymph nodes are interconnected, lymph vessels from the mammary gland may pass directly to any group. From the upper part of the mammary gland lymph vessels run to the apical axillary nodes, either directly or via the infraclavicular group; some of these are interrupted in the small interpectoral nodes.

The cutaneous lymph vessels of the breast are in continuity with neighbouring vessels across the midline, over the clavicles, and on the abdominal wall; and they anastomose with the deeper vessels of the mammary gland, especially in the

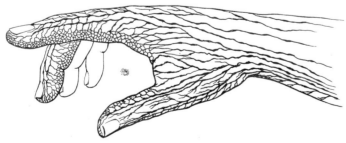

FIG. 1169 Superficial lymph vessels of thumb and forefinger.

nipple region. They drain to the axillary, infraclavicular, and parasternal nodes, and a few from the upper part of the breast pass to the lower deep cervical nodes.

Deep Lymph Vessels of the Upper Limb

These accompany the deeper blood vessels. Some pass to deep cubital or brachial nodes, but most go directly to the lateral group of axillary nodes.

JUGULAR LYMPH TRUNKS

The area of drainage of the jugular lymph trunk [p. 944] comprises the head and neck, except the skin of the lower part of the back of the neck. The trunk is formed by the efferents of the deep cervical lymph nodes.

Lymph Nodes of the Head and Neck

The lymph nodes of the head and neck form one continuous group with numerous offshoots, but it is convenient to describe a main chain of nodes along the great vessels of the neck, and subsidiary groups along the course of the branches of the external carotid artery and on the inferior thyroid artery.

The nodes thus arranged are:

1. Deep cervical nodes (jugular chain).
2. Anterior cervical nodes: (i) infrahyoid; (ii) prelaryngeal; (iii) tracheal—pretracheal and paratracheal.
3. Nodes associated with the scalp and face, forming a 'collar-chain' between the head and the neck: (i) occipital; (ii) retro-auricular; (iii) parotid and superficial cervical; (iv) submandibular, with facial nodes; (v) submental.

1. **Deep cervical lymph nodes.** The deep cervical lymph nodes form a chain along the carotid sheath from the base of the skull to the root of the neck. Most of them lie under cover of the sternocleidomastoid, but some appear in the anterior triangle of the neck, and quite a number extend into the posterior triangle, particularly in its lower part [FIGS. 1171 and 1172]. It is customary to divide the continuous chain into superior and inferior parts, above and below the omohyoid muscle.

The highest members of the **superior deep cervical nodes** lie behind the lateral border of the nasopharynx, and are known as **retropharyngeal nodes**. In the anterior triangle, one of the nodes (**jugulodigastric**) is frequently palpable below the angle of the jaw, in the angle between the digastric muscle and the internal jugular vein. It is especially associated with the lymph drainage of the tongue and tonsil. Others of the superior group which escape from beneath the sternocleidomastoid appear in the posterior triangle: some along the accessory nerve, and one, the **jugulo-omohyoid node** just above the tendon of the omohyoid. It is said to receive lymph vessels direct from the tongue.

The parts which drain, directly or indirectly, to the upper deep cervical nodes will be mentioned below. The efferents from these nodes pass to the lower deep cervical nodes and to the jugular lymph trunk.

The **inferior deep cervical nodes** [FIGS. 1171 and 1172] lie on the internal jugular vein deep to the heads of the sternocleidomastoid, and also in front of the brachial plexus and the

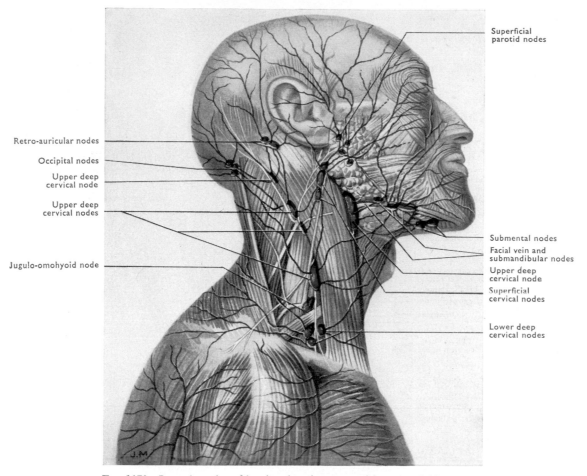

FIG. 1170   Lymph nodes of head and neck as seen with sternocleidomastoid in position.

The occipital, retro-auricular, and superficial parotid nodes are inserted in accordance with descriptions.   The other nodes were present in one or other of the two bodies from which the drawing was made.   [Cf. FIG. 1172.]

third part of the subclavian artery.   The latter nodes are in communication with the apical axillary nodes, and, as noted, receive vessels from the skin of the breast.   The efferents of all the inferior nodes pass to the jugular trunk.

2. **Anterior cervical lymph nodes.**   These are small groups all of which send their efferents to the deep cervical nodes.

(a) With the superior thyroid artery : (i) **infrahyoid nodes** on the thyrohyoid membrane ; (ii) **prelaryngeal nodes** on the cricothyroid ligament.

(b) With the inferior thyroid artery : (iii, a) **pretracheal nodes** at the lower end of the thyroid gland ; (iii, b) the **para-tracheal nodes** in the groove between trachea and oeso-phagus.

They receive lymph from the larynx, trachea, cervical oesophagus, and thyroid gland.

3. The **occipital lymph nodes** (i) lie on the occipital vessels as they pierce the trapezius ; their afferents come from the scalp.

(ii) The **retro-auricular lymph nodes** (two or three) lie on the mastoid process, and the largest is frequently palpable. Their afferents drain the scalp and the back of the auricle.

(iii) The **parotid lymph nodes** lie on the superficial sur-

face of the salivary gland both superficial and deep to its fascial sheath, and also in its substance.   The **superficial** nodes receive afferents from the front of the auricle and from the scalp, forehead, eyebrow, eyelids, and cheek ; and their efferents pass to the deep parotid nodes and to the upper deep cervical nodes.   The **deep** parotid lymph nodes receive afferents from the external acoustic meatus, the auditory tube and the middle ear, the soft palate, the posterior part of the nasal cavity, and the deeper portions of the cheek.   Their efferents pass to the upper deep cervical nodes.

The **superficial cervical lymph nodes** [FIG. 1170] are merely superficial parotid nodes scattered along the external jugular vein.

(iv) The **submandibular lymph nodes,** from four to six in number, lie between the jaw and the submandibular salivary gland ; they receive the vessels of the face below the eye, and many from the tongue ; the efferents run to the deep cervical nodes, notably the jugulodigastric and jugulo-omohyoid.

Minute nodes may be found on the course of the vessels of the face [FIG. 1173].

(v) The **submental lymph nodes** form one median group on the mylohyoid muscles ; their afferents are derived from

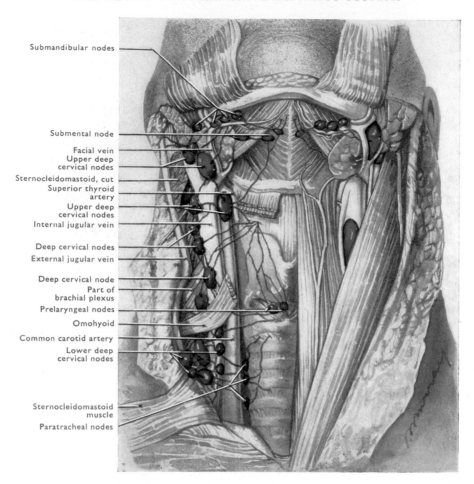

Submandibular nodes

Submental node

Facial vein
Upper deep
cervical nodes

Sternocleidomastoid, cut
Superior thyroid
artery

Upper deep
cervical nodes

Internal jugular vein

Deep cervical nodes

External jugular vein

Deep cervical node
Part of
brachial plexus

Prelaryngeal nodes

Omohyoid

Common carotid artery

Lower deep
cervical nodes

Sternocleidomastoid
muscle

Paratracheal nodes

FIG. 1171 Lymph nodes of the neck, **anterior** aspect. Infrahyoid
nodes and pretracheal nodes were not present.

the floor of the mouth, mandible, and possibly also the lower lip and chin; their efferents run to the deep cervical chain as far down as the jugulo-omohyoid node.

Lymph Vessels of the Head and Neck

The lymph vessels of the **scalp** and **forehead,** and of the various parts of the **face,** pass to the nodes of the 'collar-chain' [FIG. 1172]; but many vessels proceed directly to the deep cervical nodes [FIG. 1173]. The superficial vessels of the neck perforate the deep fascia at numerous points to enter the deep cervical nodes; those from the lower part of the back of the neck pass with the vessels of the scapular region to the axillary nodes.

The lymph vessels of the **eyelids** form two groups.

1. The **medial vessels** drain a small part of the skin only of the upper eyelid, and about half the skin and conjunctiva of the lower eyelid; they pass to the submandibular lymph nodes.

2. The **lateral vessels** drain most of the skin of the upper eyelid and all its conjunctiva, and about half the skin and conjunctiva of the lower eyelid. They pass to the parotid and superficial cervical nodes [see also p. 813, and Burch, 1939].

The lymph vessels of the **lacrimal gland** run to the parotid lymph nodes but from the **lacrimal sac** mainly to the submandibular nodes.

The lymph vessels from the **auricle** and the **external acoustic meatus** end in the parotid and retro-auricular nodes.

The lymph vessels of the lateral wall of the **middle ear** join those of the **tympanic membrane** and external acoustic meatus and end in the parotid nodes. The vessels of the medial wall and **auditory tube** end in the retropharyngeal and upper deep cervical nodes.

The lymph vessels from the **nose** and from the lining of the anterior part of the **nasal cavity** end in the submandibular nodes; those from the posterior part pass backwards through the pharyngeal wall to the upper deep cervical (including the retropharyngeal) nodes.

The lymph vessels of the **paranasal sinuses** follow those from the walls of the nasal cavities. Channels along the course of the olfactory nerves allow the passage of cerebrospinal fluid to the lymph vessels of the nasal cavity, but these are common to many nerves.

The lymph drainage of the skin and mucous membrane of the median part of the lower lip may be to the submental nodes; but, with that exception, the lymph vessels of the skin and mucous membrane of both **lips** end mainly in the submandibular nodes—as do those of the **cheeks.**

The vessels of the **teeth,** and of the outer parts of both **gums,** pass to the submandibular nodes. Some of the vessels

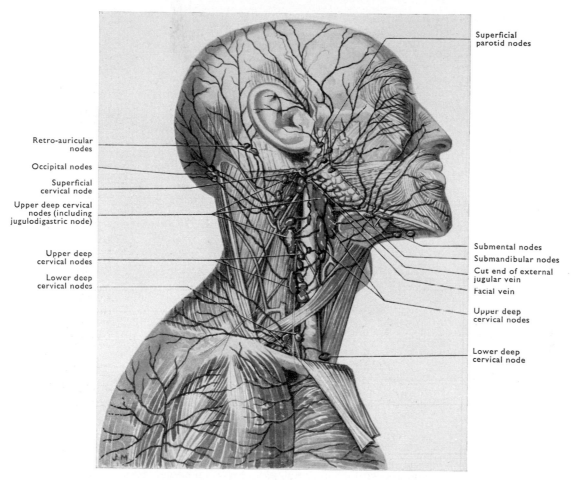

FIG. 1172 Lymph nodes of head and neck as seen after removal of sterno-cleidomastoid muscle.

The occipital, retro-auricular, and superficial parotid nodes are inserted in accordance with descriptions. The other nodes were present in one or both of the bodies from which the Figure was made. [Cf. Fig. 1170.]

from the inner part of the mandibular gum also end in these nodes; but others run to the submental nodes from the incisor region, and from the more posterior part to the upper deep cervical nodes which also receive the vessels from the inner part of the maxillary gum and from the **palate.**

The lymph vessels of the **tongue** pass from a rich capillary plexus in the mucous membrane. They are arranged in three groups.

1. The **anterior vessels** drain the anterior two-thirds of the dorsum and margins and pass to the submandibular and upper deep cervical nodes [FIG. 1174], though some vessels from the tip may pass to the submental nodes or even the jugulo-omohyoid node [FIGS. 1174 and 1175].

2. The **posterior vessels** drain the posterior third of the dorsum, and also end in the deep cervical chain—principally in the jugulodigastric node.

3. The **central vessels** arise in the dorsal plexus near the median plane. They descend between the genioglossi, and with vessels from the deep parts of the tongue pass to the upper deep cervical nodes below the digastric muscle [FIG. 1177].

It is important to note that any of the upper deep cervical nodes below the digastric may receive vessels direct from the

tongue; and, in general, the further forward the origin of the vessel the lower down is the node in which it ends. Note also that some anterior, posterior, and central vessels cross over to nodes of the opposite side [FIG. 1177].

The lymph vessels from the mucous membrane of the anterior part of the floor of the mouth drain to the submental nodes, while more posteriorly they accompany the vessels of the tongue.

The lymph vessels of the **parotid gland** end in the parotid lymph nodes; those of the **submandibular** and **sublingual glands** end in the submandibular and upper deep cervical nodes.

From the **pharyngeal walls,** lymph vessels pass to the retropharyngeal and deep cervical nodes at the corresponding level. The nasal part and **pharyngeal tonsil** draining particularly to the posterior upper, deep cervical nodes.

Lymph vessels of the **palatine tonsil** and the adjacent parts of the palatoglossal and palatopharyngeal arches end in upper deep cervical nodes, many in the jugulodigastric node.

From the **piriform fossae** and the posterior surface of the larynx, the lymph vessels pierce the thyrohyoid membrane, and end in the upper deep cervical nodes.

The lymph plexus of the **larynx** is divisible into upper and

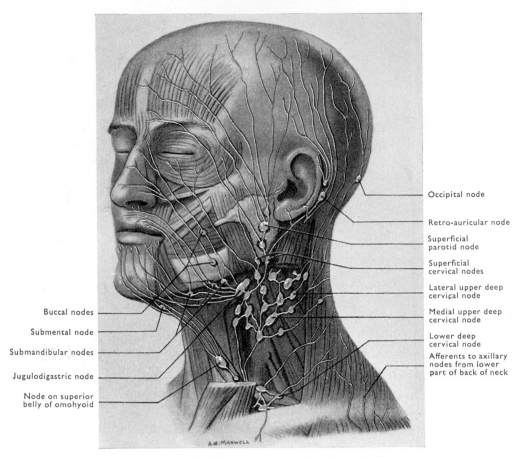

Occipital node

Retro-auricular node

Superficial parotid node

Superficial cervical nodes

Lateral upper deep cervical node

Medial upper deep cervical node

Lower deep cervical node

Afferents to axillary nodes from lower part of back of neck

Buccal nodes

Submental node

Submandibular nodes

Jugulodigastric node

Node on superior belly of omohyoid

FIG. 1173 Superficial lymph vessels of scalp and face, and superficial ('collar chain') lymph nodes between head and neck.

lower parts which are separated, laterally and anteriorly, by the vocal folds. The lymph vessels of the upper part pierce the thyrohyoid membrane and end in the upper deep cervical nodes. From the lower part of the larynx some vessels pierce the cricothyroid ligament and end in the prelaryngeal, the pretracheal, and the lower deep cervical nodes; others pierce the cricotracheal membrane and end in the paratracheal and lower deep cervical nodes [FIG. 1171].

Vallate papillae

Superficial lymph vessels of side and dorsum of tongue

Lymph vessels of apex of tongue

Afferents to submandibular nodes

Sublingual salivary gland

Submental nodes

Mylohyoid (cut)

Afferents to deep cervical nodes

Anterior belly of digastric (cut)

Styloglossus

Stylohyoid

Digastric

Afferents to deep cervical nodes from posterior third of tongue

Facial vein

Deep cervical lymph nodes

Omohyoid

FIG. 1174 Lymph vessels of tongue.

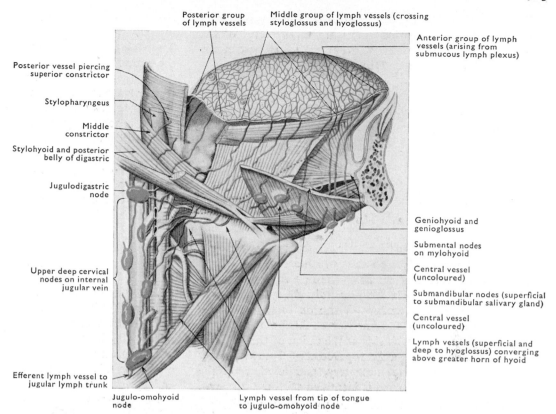

Posterior group of lymph vessels

Middle group of lymph vessels (crossing styloglossus and hyoglossus)

Anterior group of lymph vessels (arising from submucous lymph plexus)

Posterior vessel piercing superior constrictor

Stylopharyngeus

Middle constrictor

Stylohyoid and posterior belly of digastric

Jugulodigastric node

Upper deep cervical nodes on internal jugular vein

Geniohyoid and genioglossus

Submental nodes on mylohyoid

Central vessel (uncoloured)

Submandibular nodes (superficial to submandibular salivary gland)

Central vessel (uncoloured)

Lymph vessels (superficial and deep to hyoglossus) converging above greater horn of hyoid

Efferent lymph vessel to jugular lymph trunk

Jugulo-omohyoid node

Lymph vessel from tip of tongue to jugulo-omohyoid node

Fig. 1175 Diagram of course of lymph vessels of tongue.

Only one central vessel is shown (uncoloured). Central vessels and one or more of the posterior and anterior groups may cross the median plane and end in deep cervical nodes of the opposite side.

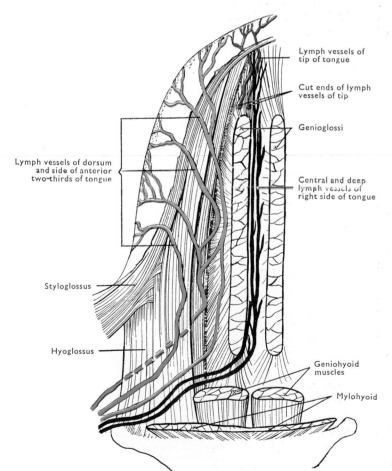

Lymph vessels of tip of tongue

Cut ends of lymph vessels of tip

Genioglossi

Lymph vessels of dorsum and side of anterior two-thirds of tongue

Central and deep lymph vessels of right side of tongue

Styloglossus

Hyoglossus

Geniohyoid muscles

Mylohyoid

FIG. 1176 Diagram of lymph vessels of anterior two-thirds of tongue, inferior aspect (After Poirier and Cunéo, 1902, modified.)
Anterior vessels, red ; middle vessels, blue ; central vessels, black.

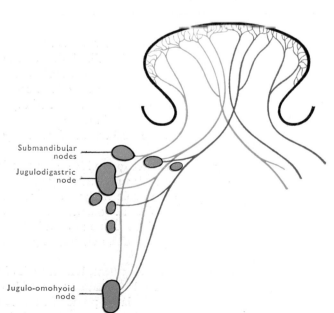

Submandibular nodes

Jugulodigastric node

Jugulo-omohyoid node

FIG. 1177 Diagram of course of central lymph vessels of tongue to show crossing of median plane. (Jamieson and Dobson, 1920.)

From much of the **thyroid gland** the lymph vessels pass directly to the deep cervical nodes up to the level of the superior thyroid artery. Those from the lower part of the isthmus and each lobe end in the pretracheal and paratracheal nodes.

The lymph vessels of the cervical part of the **trachea** end in the prelaryngeal, paratracheal, and inferior deep cervical nodes.

THE DEVELOPMENT OF THE LYMPHATIC SYSTEM

The first evidence of the lymphatic system is the appearance of structures known as **lymph sacs** which are closely related to the veins in certain situations. The first to appear is the jugular lymph sac, and six are present in the 3-cm. embryo—an **anterior** or **jugular lymph sac** on each side at the union of the subclavian and anterior cardinal veins, a **posterior lymph sac** on each side at the union of the iliac and posterior cardinal veins, a **retroperitoneal lymph sac** at the root of the primitive mesentery, and another on the posterior wall of the abdomen is identified with the **cisterna chyli.**

With the exception of the anterior part of the sac from which the cisterna chyli is developed all become broken up into anastomosing channels and are later converted into groups of lymph nodes; and the jugular sacs acquire secondary connexions with the veins.

According to Sabin (1916), the lymph sacs are developed as outgrowths of the endothelium of the veins near by, and from the sacs all the lymph vessels of the body sprout in a radiating manner. In this view, the lymph sacs lose their primary connexions with the veins, and only later do the jugular sacs acquire connexions again.

According to Huntington (1911) and McClure (1915), on the other hand, all lymph vessels are originally formed as clefts in the mesenchyme exactly as blood vessels are developed [p. 63]. These clefts acquire an endothelial lining and unite into plexuses of capillary vessels. In this view, the lymph sacs are formed by the running together of parts of the capillary plexuses along the primitive venous trunks, the main lymph trunks by the development of channels in these plexuses, and all peripheral lymph vessels by the local fusion of mesenchymal spaces.

Lymph Nodes. Lymph nodes are formed from aggregations of cells in the mesenchymal strands surrounded by plexuses of lymph vessels. Around each nodule of lymphoid tissue so formed, the vessels are transformed into the peripheral portion of the lymph sinus. The lymphoid tissue is then broken up into cords by anastomosing lymph channels which grow into it from the sinus.

In the early stages of their development, lymph nodes have a more extensive system of blood capillaries than of lymph vessels, though in postnatal life the latter predominate. If the blood vessels predominate in the later stages, the node has a reddish colour and is known as a haemal lymph node. Some such nodes are found in the neck, thorax, and retroperitoneal

regions in Man, but are much more common in the sheep and the rat.

Thoracic Duct. A rich plexus of lymph vessels is formed in the thorax around the aorta; it communicates below with the lymph sac that forms the cisterna chyli and above with both jugular lymph sacs. From the plexus two longitudinal vessels, with numerous transverse anastomoses, are developed, but as a rule the continuity of each vessel is broken at the level of the fifth thoracic vertebra, where one of the transverse anastomoses enlarges. The lower part of the left vessel and the upper part of the right disappear. From the primitive condition many varieties of the thoracic duct might be developed, and the majority of the possible variations have been found in adult bodies (Davis, 1915).

THE SPLEEN

The spleen (lien) [FIG. 1178] is a soft, freely movable organ of purplish colour, situated far back in the upper left part of the abdomen, behind the stomach. Its size varies greatly during life, depending on the amount of its contained blood, and this factor also affects its weight at death. As a rule the adult spleen is about 12·5 cm. in length, 7·5 cm. in breadth, and 3·5 cm. in thickness; and its weight is about 150 g. with a range of 50–250 g.

The **position** of the spleen varies with the excursions of the diaphragm, with the state of adjacent viscera, and with posture. It lies obliquely with its long axis parallel with the posterior parts of the ninth, tenth, and eleventh ribs, from which it is separated by the diaphragm and pleural cavity. Its posterior end is about 4 cm. from the median plane at the level of the tenth thoracic spine, and its anterior end lies just behind the midaxillary line. The normal spleen is not palpable through the abdominal wall, but its **surface projection** can be assessed

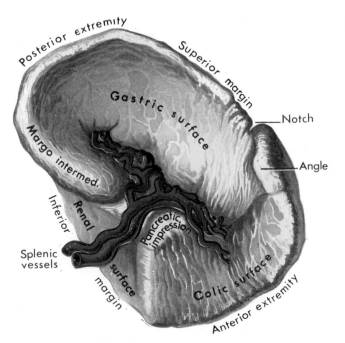

FIG. 1178   The visceral surface of the spleen.

either from the above data, or by percussion which reveals an area of impaired resonance, the splenic dullness; this reaches almost to the midaxillary line where it gives way to the area of stomach resonance.

The **shape** of the spleen is influenced largely by the stomach and left colic flexure. When the stomach is distended the spleen resembles a segment of an orange, while distension of the colon causes it to become of irregular tetrahedral form. The spleen is described as having two **ends,** anterior and posterior; two **borders,** superior and inferior; and two principal **surfaces,** diaphragmatic and visceral.

Relations

The **diaphragmatic surface** is smooth, convex, and lies against the diaphragm which intervenes between it and the pleura and lower border of the lung. It is separated from the visceral surface by the superior and inferior borders. The superior border is notched near the anterior end, and when the spleen is sufficiently enlarged to be palpable, the notches can be felt easily.

The **visceral surface** is related to the stomach, left kidney, left colic flexure, and pancreas; and when the spleen has been well fixed *in situ* the visceral surface shows four corresponding surfaces. The **gastric surface** lies against the upper part of the posterior surface of the stomach; it is separated by a raised, blunt margin from the **renal surface** which is in contact with the upper lateral part of the left kidney. In the young child the spleen touches the left suprarenal gland also. The **colic surface** rests on the left colic flexure and the phrenicocolic ligament.

The **hilus** is on the lower part of the gastric surface, and here the splenic vessels enter and leave. The peritoneum which surrounds the spleen is firmly adherent to its capsule, and is continuous with the peritoneal folds known as the **lienorenal ligament** and the **gastrosplenic ligament.** These connect the hilus of the spleen to the left kidney and stomach respectively, and, originally part of the dorsal mesogastrium, form the left wall of the omental bursa in the adult. The tail of the pancreas lies in the lienorenal ligament, and may touch the spleen immediately below the anterior part of the hilus.

Small rounded **accessory spleens** are often present, usually on the gastrosplenic ligament.

Radiographic Examination of Spleen

The normal spleen may be visible radiographically if there are large amounts of gas in both the stomach and the left colic flexure. The spleen, like the liver, may also be made semi-opaque to X-rays by the injection of an X-ray opaque medium into the splenic artery [PLATE 52].

Vessels and Nerves

The splenic artery runs with the vein in the lienorenal ligament, and at the hilus breaks up into six or more branches which enter independently. The splenic vein, which passes to the portal vein, is formed by several tributaries which emerge at the hilus. The lymph vessels, which arise only from the capsule and trabeculae, pass to the pancreaticosplenic lymph nodes.

The nerves are vasomotor in function, and come through the coeliac plexus from the greater splanchnic nerve.

When the spleen is distended it acts as a reservoir from which blood can be returned to the general circulation in harmony with the body's needs; but the absence of smooth muscle in the capsule of the human spleen precludes active splenic contraction such as occurs in the dog.

Structure

Beneath its peritoneal coat the spleen is invested by a fibro-elastic capsule. **Trabeculae,** of similar structure, pass into the organ both from the hilus (carrying the larger blood vessels with them) and from the capsule [FIG. 1179]. The branches of the trabeculae ultimately become continuous with the reticular meshwork which pervades the main substance or **pulp** of the spleen. Associated with the reticular fibres are reticular cells of the reticulo-endothelial system [see below].

The pulp is composed of **white pulp** and **red pulp.** The white pulp consists of ovoid masses of lymphoid tissue called Malpighian corpuscles, or **lymph follicles,** within which

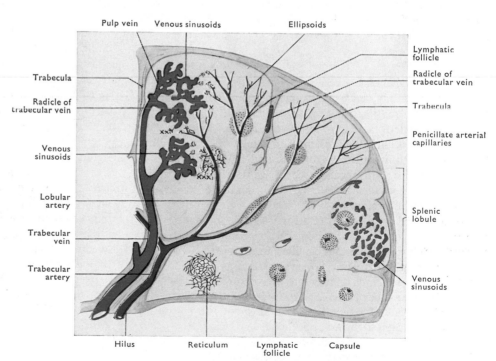

FIG. 1179 Schematic representation of splenic structure, much simplified by the suppression of ramifications of the trabeculae and vascular tree. (After Stöhr, *Lehrbuch der Histologie*, Fischer.)

Note. Both the pulp reticulum and the system of venous sinusoids supported by it must be understood as pervading the entire extent of the area figured except the trabeculae.

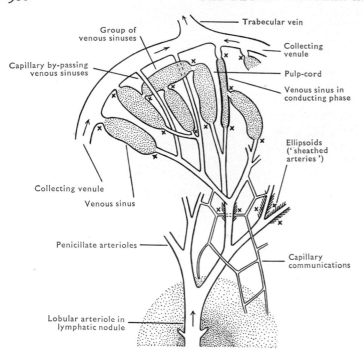

FIG. 1180 Diagram of the types of vascular connexions that have been observed in the spleen. (After Knisely, 1936.)

x, sphincters.

may be seen germinal centres. The red pulp forms the greater part of the splenic substance. It consists of the following: the reticular meshwork; **venous sinuses** whose endothelial cells can become phagocytic; and, between the sinuses, cords of cells known as splenic cords. The cords are in fact sinusoids, but this may be disguised by collapse of the vessels and by their endothelial cells becoming phagocytic with loss of their endothelial form.

Vascular Pattern

From the trabeculae arteries pass into the red pulp and, almost immediately, each is invested by a lymph follicle (white pulp) through which it runs eccentrically. Having given capillaries to the follicle it re-enters the red pulp and divides into several parallel '**penicillate**' **vessels** of fine calibre. These soon acquire thickened walls and have been termed sheathed arteries; the sheaths, which are poorly developed in Man, are composed of reticular cells and are known as **ellipsoids** or Schweigger-Seidel sheaths.

The sheathed arteries end in arterial capillaries, but beyond this point the question of whether the splenic circulation is 'open' or 'closed' has long remained unresolved. By 'open' is meant that the capillaries open into the pulp whence the extravasated blood finds its way into the venous sinuses, which, in this view, must have openings in their walls. According to the 'closed' view the blood passing through the spleen is, throughout, confined within endothelial-lined channels.

Knisely (1936a, b), who studied the circulation in the living spleen, states that the capillaries which leave the sheathed arteries end in the trabecular veins either: (1) directly, after

a long course, or (2) through bulbous venous sinuses; those which take the first course provide a bypass or shunt mechanism. Knisely also describes sphincters at the entrances and exits of the sinuses which can therefore expand to serve as reservoirs, or become narrowed to act as conductors [FIG. 1180]. With such changes the spleen as a whole would vary in size.

According to the same author there is a complete endothelial lining to the vascular system of the spleen, but Whipple, Parpart, and Chang (1954) claim that some capillaries may open directly into pulp spaces, and that some arteriovenous anastomoses exist. Moreover, Lewis (1957) supports the view of an 'open' circulation.

Weiss (1966) reviews the whole question very thoroughly. That author, while agreeing that the flow in the living spleen may appear to be through continuous vessels, believes that blood may pass between the lining cells of the sinuses and so leave their lumen.

Development

The spleen is formed by the fusion of several condensations of mesoderm which appear in the dorsal mesogastrium about the sixth week (10-mm. stage). Failure of fusion of one or more of the condensations accounts for the presence of accessory spleens [p. 965]; moreover, the adult spleen still retains evidence of its multiple origin in possessing compartments each with a separate vascular supply from its own artery and vein (Braithwaite and Adams, 1957). During intra-uterine life the spleen is actively engaged in haemopoiesis, but by birth this function has been taken over entirely by the bone marrow; nevertheless, should an increased demand for blood formation arise after birth, the spleen may resume its former role.

The Reticulo-endothelial or Macrophage System

The cells of this system are widely distributed in the body, and serve as a protective and scavenging mechanism. They are actively phagocytic, and all have the property of taking up vital dyes, such as trypan blue, which have been injected into the body, and storing them as coarse granules in their cytoplasm. It is, indeed, this quality which distinguishes the cells of the system, rather than special morphological characters. Should circumstances demand, they are capable of detaching themselves from their usual sites and of acquiring amoeboid movement. They are engaged in the destruction of red blood corpuscles, in bile pigment production, in the metabolism of iron, fats, and proteins, and in clearing the body of bacteria, protozoa, and non-living particles. They occur: (1) in the blood where they are represented by the monocytes; (2) lining the walls of the **venous sinuses** of the spleen, of the bone marrow, of the liver (where they are known as the **stellate endothelial cells** of von Kupffer), of the suprarenal, and of the anterior lobe of the hypophysis; (3) in the reticular tissue of the spleen and lymph nodes; (4) in connective tissue where they are present as histiocytes or 'resting wandering cells'; (5) in the central nervous system where they are represented by the microglial cells.

RADIOGRAPHIC ANATOMY

E. SAMUEL

The lymph vessels and nodes, having the same X-ray density as the surrounding structures are not visible on plain films. They can, however, be demonstrated by the injection of a special contrast medium after identification of the vessels by the injection of a dye. This lymphangiography is most commonly performed in the lower limb as it also demonstrates the lymph vessels and nodes in the abdominal cavity. Less frequently a lymph vessel on the dorsum of the hand is used to show some of the axillary lymph nodes.

The lymph vessels may be X-rayed immediately after injection but the fine detail of the lymph nodes and its alteration in disease are best seen in films taken twenty–four hours later [PLATE 51].

The Spleen

The air in the lung fields and the gas bubble in the stomach and colon may allow the normal soft tissue outline of the spleen to be seen in plain films of the abdomen. Enlargement of the spleen may be detected by displacement of the surrounding structures—stomach, left colic flexure, or kidney.

The splenic and portal veins can be shown by direct puncture of the spleen through the skin and the injection of a suitable contrast medium. The medium is rapidly carried into the splenic vein and can demonstrate its patency and that of the portal vein in addition to changes in the intrahepatic veins in liver disease.

REFERENCES

BRAITHWAITE, J. L., and ADAMS, D. J. (1957) The venous drainage of the rat spleen, *J. Anat. (Lond.)*, **91**, 352.

BURCH, G. E. (1939) Superficial lymphatics of human eyelids observed by injection *in vivo*, *Anat. Rec.*, **73**, 443.

DAVIS, H. K. (1915) A statistical study of the thoracic duct in man, *Amer. J. Anat.*, **17**, 211.

ENGESET, A. (1959a) The route of peripheral lymph to the blood stream An X-ray study of the barrier theory, *J. Anat. (Lond.)*, **93**, 96.

ENGESET, A. (1959b) Lymphatico-venous communications in the albino rat, *J. Anat. (Lond.)*, **93**, 380.

GOWANS, J. L., and KNIGHT, E. J. (1964) The route of recirculation of lymphocytes in the rat, *Proc. roy. Soc. B*, **159**, 257.

HANDLEY, R. S., and THACKRAY, A. C. (1954) Invasion of internal mammary lymph nodes in carcinoma of breast, *Brit. med. J.*, **1**, 61.

HUNTINGTON, G. S. (1911) The anatomy and development of the systemic lymphatic vessels in the domestic cat, *Amer. Anat. Mem.*, **I**, Philadelphia, Wistar Institute.

JAMIESON, J. K., and DOBSON, J. F. (1907a) The lymphatic system of the stomach, *Lancet*, i, 1061.

JAMIESON, J. K., and DOBSON, J. F. (1907b) The lymphatic system of the caecum and appendix, *Lancet*, i, 1137.

JAMIESON, J. K., and DOBSON, J. F. (1909) The lymphatics of the colon, *Proc. roy. Soc. Med.*, **2**, 149.

JAMIESON, J. K., and DOBSON, J. F. (1910) On the injection of lymphatics by Prussian blue, *J. Anat. Physiol.*, **45**, 7.

KINMONTH, J. B. (1954) Lymphangiography in clinical surgery and particularly in treatment of lymphoedema, *Ann. roy. Coll. Surg. Engl.*, **15**, 300.

KNISELY, M. H. (1936a) Spleen studies. I. Microscopic observations of the circulatory system of living unstimulated mammalian spleens, *Anat. Rec.*, **65**, 23.

KNISELY, M. H. (1936b) Spleen studies. II. Microscopic observations of the circulatory system of living traumatized spleens, and of dying spleens, *Anat. Rec.*, **65**, 131.

LEWIS, O. J. (1957) The blood vessels of the adult mammalian spleen, *J. Anat. (Lond.)*, **91**, 245.

McCLURE, C. F. W. (1915) The development of the lymphatic system in the light of the more recent investigations in the field of vasculogenesis, *Anat. Rec.*, **9**, 563.

PATEK, P. R. (1939) The morphology of the lymphatics of the mammalian heart, *Amer. J. Anat.*, **64**, 203.

POIRIER, P., and CUNEO, B. (1902) Les lymphatiques, in Poirier and Charpy's *Traité d'anatomie humaine*, t. 11, fasc. 4, Paris. Engl. ed. trans. and ed. Leaf, C. H. (1903), London.

ROITT, I. M., GREAVES, M. F., TORRIGIANI, G., BROSTOFF, J., and PLAYFAIR, J. H. L. (1969) The cellular basis of immunological responses, *Lancet*, ii, 367.

ROUVIÈRE, H. (1932) *Anatomie des lymphatiques de l'homme*, Paris.

SABIN, F. R. (1916) The origin and development of the lymphatic system, *Johns Hopk. Hosp. Rep.*, **17**, 347.

SAPPEY, P. C. (1874) *Traité d'anatomie, physiologie et pathologie des vaisseaux lymphatiques considérés chez l'homme et les vertébrés*, Paris.

SHORE, L. R. (1929) The lymphatic drainage of the human heart, *J. Anat. (Lond.)*, **63**, 291.

TURNER-WARWICK, R. T. (1959) The lymphatics of the breast, *Brit. J. Surg.*, **46**, 574.

WEISS, L. (1966) Spleen, in *Histology*, ed. Greep, R. O., New York.

WHIPPLE, A. O., PARPART, A. K., and CHANG, J. J. (1954) Study of circulation of blood in spleen of living mouse, *Ann. Surg.*, **140**, 266.

YOFFEY, J. M., and COURTICE, F. C. (1956) *Lymphatics, Lymph and Lymphoid Tissue*, 2nd ed., London.

INDEX

Page numbers appearing in bold type indicate major descriptions.